CALCULUS
A COMPLETE COURSE

SECOND EDITION

Ross L. Finney

Franklin D. Demana
The Ohio State University

Bert K. Waits
The Ohio State University

Daniel Kennedy
Baylor School

With

Marianne H. Lepp

GTE Laboratories

 ADDISON-WESLEY

An imprint of Addison Wesley Longman, Inc.

Reading, Massachusetts • Menlo Park, California • New York • Harlow, England
Don Mills, Ontario • Sydney • Mexico City • Madrid • Amsterdam

Acknowledgments

Photographs

Cover Boris Starosta/Stock Connection

Chapter 1 **xxii** Sylvain Coffie/Tony Stone Images; **29** Culver Pictures; **43** Hewlett-Packard Company

Chapter 2 **54** Ron Stanford/Tony Stone Images; **74** John Elk III/Bruce Coleman Inc.; **89** Jet Propulsion Labs

Chapter 3 **94** Michael W. Davidson/Photo Researchers; **132** From *PSSC Physics,* 2/e, 1965; D.C. Heath & Company, with Education Development Center, Inc., Newton, MA. Reprinted with permission; **148** Jenny Thomas*; **171** Steve Wewerka*

Chapter 4 **176** Peter Poulides/Tony Stone Images; **193** Focus on Sports; **231** NASA

Chapter 5 **246** Ralph Mercer/Tony Stone Images; **260** German Information Center; **277** Corbis-Bettmann; **285** Archive Photos; **297** Marshall Henrichs*; **301** Jenny Thomas*

Chapter 6 **302** Ralph Oberlander/Stock, Boston; **341** Norman Kent Productions; **357** Jenny Thomas*

Chapter 7 **362** Douglas Pulsipher/Third Coast Stock; **400** Jenny Thomas*; **403** Corbis-Bettmann; **412** Keith Burke*

Chapter 8 **416** NASA; **430** Hank Morgan/Rainbow

Chapter 9 **456** Dan McCoy/Rainbow; **467** From *PSSC Physics,* 2/e, 1965; D.C. Heath & Co. with Education Development Center, Inc., Newton, MA. Reprinted with permission; **499** Frank Siteman/Stock, Boston; **511** Peter Yates*

Chapter 10 **512** Roger Tully/Tony Stone Images; **543** Agence France Presse/Corbis-Bettmann; **550** From *PSSC Physics,* 2/e, 1965; D.C. Heath & Co. with Education Development Center, Inc., Newton, MA. Reprinted with permission; **572** Corbis-Bettmann

Chapter 11 **574** © 1999 PhotoDisc

Chapter 12 **630** © 1998 Tony Stone Images; **667** Figure 12.37 from *PSSC Physics,* 2/e, 1965; D.C. Heath & Co. with Education Development Center, Inc., Newton, MA. Reprinted with permission.

Chapter 13 **674** © 1999 PhotoDisc; **708** AIT 13.1 Roger-Viollet

Chapter 14 **760** © 1999 PhotoDisc

Chapter 15 **828** © 1999 PhotoDisc; **836** Figure 15.16 InterNetwork, Inc., and NASA/Jet Propulsion Laboratory.

*Photographed expressly for Addison Wesley Longman

Illustrations

Technical illustrative and technical artwork for this edition was produced by Tech-Graphics.

Text

Chapter 2 **75** Figure 2.20 from A. J. Lotka, *Elements of Mathematical Biology,* p. 69. (New York: Dover, 1956). Reprinted by permission.

Chapter 12 **657** Figure 12.29 © 1994 Nelson L. Max, University of California/Biological Photo Service.

Chapter 13 **680** Figure 13.7 Appalachian Mountain Club.

Chapter 15 **836** Figure 15.11 Adapted from *NCFMF Book of Film Notes,* 1974, MIT Press with Education Development Center, Inc.; **836** Figure 15.12 Adapted from *NCFMF Book of Film Notes,* 1974, MIT Press with Education Development Center, Inc.

Library of Congress Cataloging-in-Publication Data

Calculus: a complete course / Ross L. Finney ... [et al.].—2nd ed.
  p.   cm.
  Includes index.
  ISBN 0-201-44140-3
  1. Calculus   I. Finney, Ross L.
QA303.C158 2000   99-39840
515—dc21   CIP

Printed in the United States of America

ISBN 0-201-44140-3

2 3 4 5 6 7 8 9 10–VH–020100

Preface

Calculus: A Complete Course is the lean and lively technology-based calculus book college instructors have been waiting for. Without distraction from inessential material, students get right into the calculus to see how it works from the very beginning. This text combines an appropriate use of technology with standard analytic techniques to provide a balanced approach to the study and implementation of calculus. Technology is fully integrated, rather than just added. The text encourages group work, mathematical modeling (problem solving), conceptual understanding, and facility with technology. It fosters an appreciation of calculus both as a coherent body of knowledge and as a human accomplishment.

Audience and Prerequisites

This book is a complete revision of the first edition and explores all concepts necessary for the standard calculus sequence. Its purpose, in addition to making it possible to learn calculus, is to teach students how to use calculus effectively; it also shows how knowing calculus can pay off in any profession they decide to enter. The applications described within the text are real, and their presentations are self-contained; students will not need any experience in the fields from which the applications are drawn. The prerequisites are exposure to algebra, geometry, and trigonometry. Chapter 1 reviews this material.

Philosophy

We believe in using the full array of calculator functions from the beginning, and Chapter 1 introduces them with appropriate examples and applications. In addition to the usual linear, polynomial, power, and rational functions, there are exponential functions, parametrizations, logarithmic functions, inverses of familiar functions, trigonometric functions, and inverse trigonometric functions. Chapter 1 also introduces the capability that most graphing calculators now have for regression analysis and shows how to use this capability to summarize and visualize numerical data and to evaluate the predictions of analytic models.

We explore calculus through the interpretation of graphs and tables as well as the application of analytic methods. Derivatives are interpreted both as rates of change and as local linear approximations. Local linearity is a recurring theme. The definite integral is introduced both as a description of the net effect of rate of change over an interval of time and as a limit of Riemann sums. There are applications to such diverse topics as pollution control, medicine, diesel oil consumption, household electricity, economics, projectile motion, population dynamics, and work done pumping. Please see the Index of Applications on page 1139 for a complete listing of all applications by subject.

Chapter 6 focuses on the use of differential equations to model real-world problems. We interpret differential equations using slope fields and solve initial value problems analytically and numerically. In presenting infinite series, the emphasis (as it was historically and still is in applied fields) is on the utility of power series and the calculus connection, not on sequences and series of constants. Convergence and divergence are interpreted graphically, numerically, and analytically, and Lagrange error bounds are used to confirm the accuracies of Taylor-polynomial approximations.

Technology is integrated throughout the book to provide a balanced approach to teaching and learning calculus. The presentation uses graphical, numerical, and algebraic/analytic techniques and encourages written and oral communication. Students are expected to use a multi-representational approach to investigating problems, to write about their conclusions, and often to work in groups to communicate mathematics orally. This book reflects what we have learned about the appropriate use of technology in the classroom.

Rich and varied problems have been a hallmark of Finney/Thomas books, and to these we have added a wide variety of new graphical and data-based problems. Enhanced versions of the visualizations and technological explorations pioneered by Demana and Waits have been incorporated throughout.

The Rule of Four: A Balanced Approach

A principal feature of this edition is the balance attained among graphical, numerical, and algebraic/analytic techniques and communication (the rule of four). We believe that students must value all these methods of representation, understand how they are connected in a given problem, and learn how to choose the one or ones most appropriate for solving a particular problem. Suggestions from numerous instructors have helped us shape this modern, balanced, technological approach to the discipline of calculus.

In support of the rule of four, we use a variety of techniques to solve problems and to communicate their solutions. For instance, we obtain solutions algebraically or analytically, support our results graphically or numerically with technology, and then interpret the results in the original problem context. We have written exercises that ask students to solve problems by one method and then support or confirm their solutions by using another method. We want students to understand that technology can be used to support, but not prove, results and that algebraic or analytic techniques are needed to prove results. We want students to understand that it is *mathematics* that provides the foundation that allows us to use technology to solve problems.

Applications

The text includes a rich array of interesting applications in biology, business, chemistry, economics, engineering, finance, physics, the social sciences, and statistics. For a complete listing of all applications organized by subject, please see the Index of Applications on page 1139. Many applications are based on real data from cited sources. Students are first exposed to functions as mechanisms for modeling data and then learn how various functions can model real-life problems. They learn to analyze and model data, represent data graphically, interpret from graphs, and fit curves. Additionally, the tabular representations of data presented in the text highlight the concept that a function is a correspondence between numerical variables, helping students to build the connection between the numbers and the graphs.

Explorations

We want students to be actively involved in understanding calculus concepts and solving problems. Stepped Explorations in the text provide guided investigations of many key concepts. These Explorations help build problem-solving ability by guiding students to develop a mathematical model of a problem, solve the mathematical model, support or confirm the solution, and interpret the solution. The ability to communicate their understanding is just as important to the students' learning process as reading and studying, not only in mathematics but in every academic pursuit. Students gain an entirely new perspective on their knowledge when they explain what they know in writing.

Graphing Utilities

This book assumes familiarity with a graphing utility that will produce the graph of a function within an arbitrary viewing window, find the zeros of a function, compute the derivative of a function numerically, and compute definite integrals numerically. Students are expected to recognize that a graph is reasonable, identify all the important characteristics of a graph, interpret those characteristics, and confirm them using analytic methods. Toward that end, most of the two-dimensional graphs appearing in this book resemble students' actual grapher output or suggest hand-drawn sketches. Many of the three-dimensional graphics have been generated by computer. This is one of the first calculus textbooks to take full advantage of graphing calculators, philosophically restructuring the course to teach new things in new ways to achieve new understanding, while abandoning some old things and old ways that are no longer serving a purpose.

Exercise Sets and Key Terms

The exercise sets were revised extensively for this edition, with many new exercises added. There are more than 6300 exercises, incuding more than 890 Quick Review Exercises. There are an additional 300 exercises in the apendices. Two sets of Cumulative Review Exercises are also included, one covering Chapters 1 through 10 and another covering Chapters 11 through 15. Exercises involved, as appropriate,

- algebraic and analytic manipulation
- data analysis
- exploration
- extending ideas
- graphical and numerical representations of function
- group activities
- interpretations of graphs
- titled application
- writing to learn

Each section's exercise set begins with Quick Review Exercises, which can be used to introduce lessons, support examples, and review prerequisite skills. The exercises that follow are graded from routine to challenging. Additional blocks of exercises, entitled Explorations and Extending the Ideas, may be used in a variety of ways, including group work. Each chapter concludes with a list of Key Terms with appropriate page references, followed by a set of Review Exercises. A closer look at the features of this revision can be found on page xvi.

Acknowledgments

The authors would like to express their gratitude to all who have used the first edition of the text, with special thanks to the professors who have commented on this edition. In particular, we thank Carl W. Anderson, Johnson County Community College; Marie Aratari, Oakland Community College; Bill Ardis, Collin County Community College; Jon Beal, Clarion University; Dan Burkett, Indiana University of Pennsylvania; Mark Farris, Midwestern State University; Stephen Merrill, Marquette University; Cheryl Slayden, Pellissippi State Technical Community College; and Randall Westhoff, Bemidji State University. We are particularly indebted to Judith Broadwin, who consulted with us throughout the development of this revision, for her dedication and excellent advice.

In Conclusion

We are excited by the new directions our text has taken, and hope you will contact us with your comments.

Ross L. Finney
Franklin D. Demana
Bert K. Waits
Daniel Kennedy
Math@awl.com

Supplements

The following products are designed to accompany *Calculus: A Complete Course*, Second Edition:

For the Student

Student's Solutions Manual, Volume I
Solutions to all odd-numbered exercises in Chapters 1 through 10.

Student's Solutions Manual, Volume II
Solutions to all odd-numbered exercises in Chapters 11 through 15.

Technology Resource Manual, Volume I
For TI Calculators

Technology Resource Manual, Volume II
For Casio and Hewlett-Packard Calculators

For the Instructor

Teacher's Guide
Bibliography • Teaching Notes • Exploration Extensions • Assignment Guides • Alternate Assessment Ideas • Answer Key

Assessment
Quizzes • Tests • Semester and Final Tests • Alternate Assessment

Transparencies
50 four-color transparencies

Instructor's Solutions Manual, Volume I
Complete solutions for all problems in Chapters 1 through 10.

Instructor's Solutions Manual, Volume II
Complete solutions for all problems in Chapters 11 through 15.

Technology Resources

Teacher Resource Planner CD-ROM
All Supplements • Calculator Programs

Calculus Testworks
Electronic Test Bank

Contents

CHAPTER

1

Prerequisites for Calculus

$x = 3 \cos t,\ y = 4 \sin t$

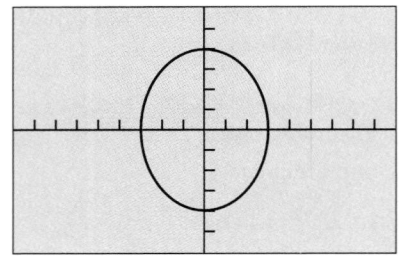

[−9, 9] by [−6, 6]

CHAPTER

2

Limits and Continuity

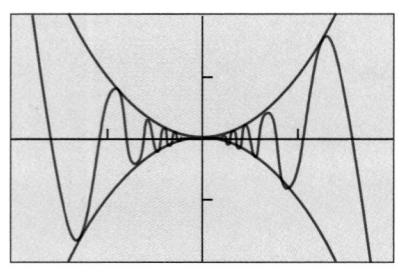

[−0.2, 0.2] by [−0.02, 0.02]

CHAPTER

3
Derivatives

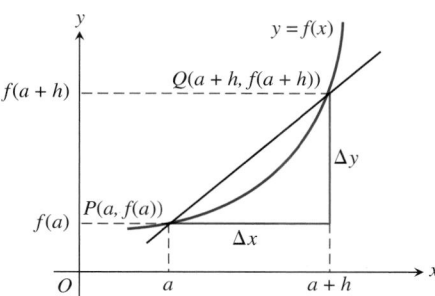

CHAPTER

4
Applications of Derivatives

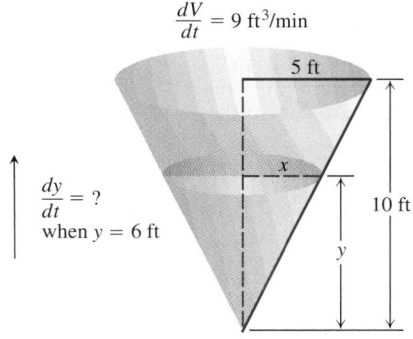

CHAPTER

5

The Definite Integral

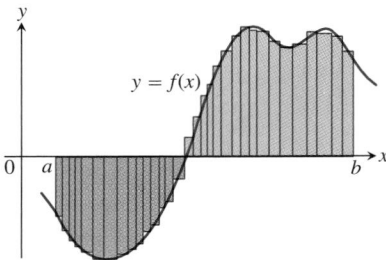

CHAPTER

6

Differential Equations and Mathematical Modeling

[0, 20] by [0, 300]

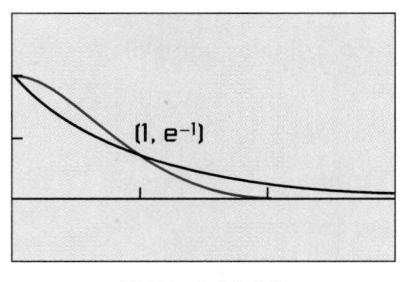

CHAPTER
9
Infinite Series

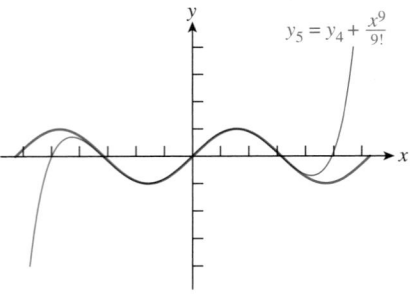

CHAPTER
10
Parametric, Vector, and Polar Functions

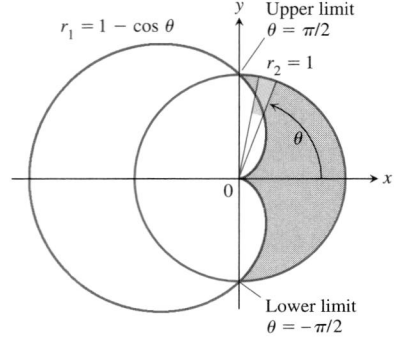

CHAPTER

11

Vectors and Analytic Geometry in Space

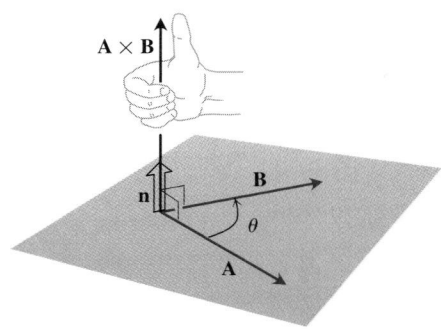

CHAPTER

12

Vector-Valued Functions and Motion in Space

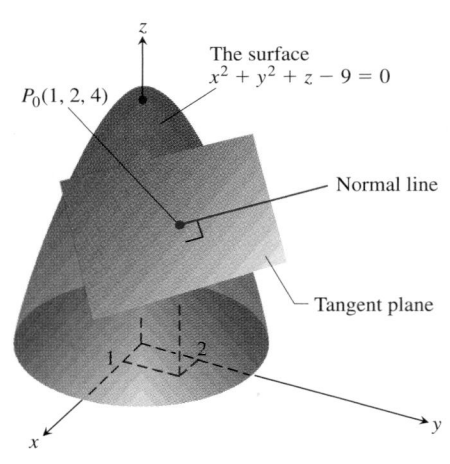

$P_0(1, 2, 4)$

The surface
$x^2 + y^2 + z - 9 = 0$

Normal line

Tangent plane

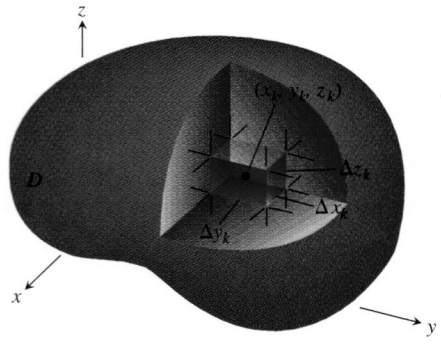

CHAPTER
15
Integration in Vector Fields

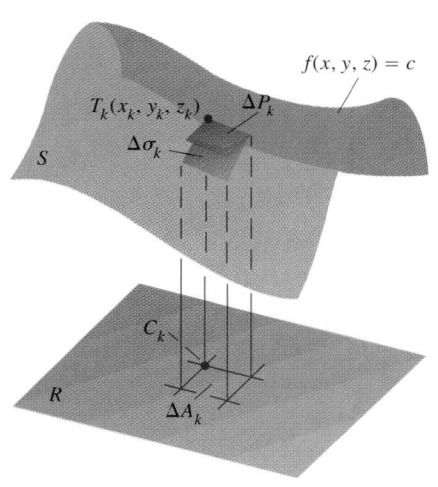

$f(x, y, z) = c$

$T_k(x_k, y_k, z_k)$

ΔP_k

$\Delta \sigma_k$

S

C_k

R

ΔA_k

To the Student

Working with graphical, numerical, analytical, and verbal forms of functions

You are about to embark on what we hope will be the beginning of your adventures into higher level mathematics. All of your previous math courses will serve as the foundation for calculus. Your study of algebra has taught you how to work with functions, equations, and graphs. Your study of geometry has taught you how to work with figures, areas, volumes, and coordinatization.

Theorems and definitions continue to be important. Some of the proofs will be provided and some will be developed graphically. You will be writing proofs in some of the exercises.

Theorems, Definitions, and Rules will always be boxed and shaded in yellow.

(a)

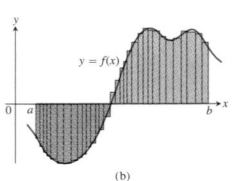

(b)

Figure 5.15 The curve of Figure 5.12 with rectangles from finer partitions of $[a, b]$. Finer partitions create more rectangles, with shorter bases.

Definition The Definite Integral as a Limit of Riemann Sums

Let f be a function defined on a closed interval $[a, b]$. For any partition P of $[a, b]$, let the numbers c_k be chosen arbitrarily in the subintervals $[x_{k-1}, x_k]$.

If there exists a number I such that

$$\lim_{\|P\| \to 0} \sum_{k=1}^{n} f(c_k) \Delta x_k = I$$

no matter how P and the c_k's are chosen, then f is **integrable** on $[a, b]$ and I is the **definite integral** of f over $[a, b]$.

Theorem 4 (continued) The Fundamental Theorem of Calculus, Part 2

If f is continuous at every point of $[a, b]$, and if F is any antiderivative of f on $[a, b]$, then

$$\int_a^b f(x)\, dx = F(b) - F(a).$$

This part of the Fundamental Theorem is also called the **Integral Evaluation Theorem.**

Proof Part 1 of the Fundamental Theorem tells us that an antiderivative of f exists, namely

$$G(x) = \int_a^x f(t)\, dt.$$

Thus, if F is *any* antiderivative of f, then $F(x) = G(x) + C$ for some constant C (by Corollary 3 of the Mean Value Theorem for Derivatives, Section 4.2). Evaluating $F(b) - F(a)$, we have

$$F(b) - F(a) = [G(b) + C] - [G(a) + C]$$
$$= G(b) - G(a)$$
$$= \int_a^b f(t)\, dt - \int_a^a f(t)\, dt$$
$$= \int_a^b f(t)\, dt - 0$$
$$= \int_a^b f(t)\, dt. \qquad \blacksquare$$

Numerous **Examples** show how the theorems and definitions are applied.

Justification for each step in an Example often appears in red type to the right of the step.

Example 1 APPLYING THE FUNDAMENTAL THEOREM

Find

$$\frac{d}{dx} \int_{-\pi}^{x} \cos t\, dt \quad \text{and} \quad \frac{d}{dx} \int_{0}^{x} \frac{1}{1 + t^2}\, dt$$

by using the Fundamental Theorem.

Solution

$$\frac{d}{dx} \int_{-\pi}^{x} \cos t\, dt = \cos x \qquad \text{Eq. 1 with } f(t) = \cos t$$

$$\frac{d}{dx} \int_{0}^{x} \frac{1}{1 + t^2}\, dt = \frac{1}{1 + x^2}. \qquad \text{Eq. 1 with } f(t) = \frac{1}{1 + t^2}$$

Green boxes help summarize important skills and strategies.

Graph Viewing Skills

1. Recognize that the graph is reasonable.
2. See all the important characteristics of the graph.
3. Interpret those characteristics.
4. Recognize grapher failure.

Linking derivatives and integrals with the Fundamental Theorem of Calculus

In calculus, you will study limits, derivatives, and integrals, and applications of all of these ideas. Your study will be based on a balanced approach. You will be asked to solve graphically, support numerically, confirm analytically, and solve algebraically, all while applying calculus to problem situations.

Solving graphically gives insight into the behavior of functions.

Often you will **confirm** the graphical results by using the algebraic techniques based on the theorems, definitions, and rules you have acquired throughout your math courses.

You will **analyze** graphs and use **numerical** tables to **support** your conjectures.

You will participate actively in your learning. Working through the **Explorations** will give you insights into calculus.

Many **Explorations** are designed for you to work in groups. Working together will give you the opportunity to reinforce each other's ideas and verify your work.

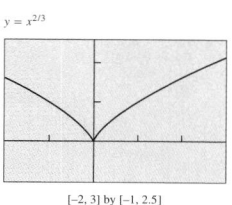

$y = x^{2/3}$

[−2, 3] by [−1, 2.5]

Figure 4.5 (Example 3)

[−2, 4] by [−1, 3]
(a)

X	Y1
.1	10
.2	5
.3	3.3333
.4	2.5
.5	2
.6	1.6667
.7	1.4286

X = .1
(b)

Figure 3.17 (a) The graph of NDER ln (x) and (b) a table of values. What graph could this be? (Example 4)

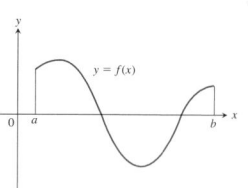

$y = f(x)$

Figure 5.27 The graph of the function in Exploration 2.

Example 3 FINDING ABSOLUTE EXTREMA

Find the absolute maximum and minimum values of $f(x) = x^{2/3}$ on the interval $[-2, 3]$.

Solution
Solve Graphically
Figure 4.5 suggests that f has an absolute maximum value of about 2 at $x = 3$ and an absolute minimum value of 0 at $x = 0$.

Confirm Analytically
We evaluate the function at the critical points and endpoints and take the largest and smallest of the resulting values.

The first derivative

$$f'(x) = \frac{2}{3}x^{-1/3} = \frac{2}{3\sqrt[3]{x}}$$

has no zeros but is undefined at $x = 0$. The values of f at this one critical point and at the endpoints are

Critical point value: $f(0) = 0$;

Endpoint values: $f(-2) = (-2)^{2/3} = \sqrt[3]{4}$;
$f(3) = (3)^{2/3} = \sqrt[3]{9}$.

We can see from this list that the function's absolute maximum value is $\sqrt[3]{9} \approx 2.08$, and occurs at the right endpoint $x = 3$. The absolute minimum value is 0, and occurs at the interior point $x = 0$.

Example 4 GRAPHING A DERIVATIVE USING NDER

Let $f(x) = \ln x$. Use NDER to graph $y = f'(x)$. Can you guess what function $f'(x)$ is by analyzing its graph?

Solution The graph is shown in Figure 3.17a. The shape of the graph suggests, and the table of values in Figure 3.17b supports, the conjecture that this is the graph of $y = 1/x$. We will prove in Section 3.9 (using analytic methods) that this is indeed the case.

Differentiability Implies Continuity
We began this section with a look at the typical ways that a function could fail to have a derivative at a point. As one example, we indicated graphically that a discontinuity in the graph of f would cause one or both of the one-sided derivatives to be nonexistent. It is actually not difficult to give an analytic proof that continuity is an essential condition for the derivative to exist, so we include that as a theorem here.

Exploration 2 Finding the Derivative of an Integral

Suppose we are given the graph of a continuous function f, as in Figure 5.27. *Work in groups of two or three.*

1. Copy the graph of f onto your own paper. Choose any x greater than a in the interval $[a, b]$ and mark it on the x-axis.
2. Using only *vertical line segments*, shade in the region between the graph of f and the x-axis from a to x. (Some shading might be below the x-axis.)
3. Your shaded region represents a definite integral. Explain why this integral can be written as $\int_a^x f(t)\, dt$. (Why don't we write it as $\int_a^x f(x)\, dx$?)
4. Compare your picture with others produced by your group. Notice how your integral (a real number) depends on which x you chose in the interval $[a, b]$. The integral is therefore a *function of x* on $[a, b]$. Call it F.
5. Recall that $F'(x)$ is the limit of $\Delta F/\Delta x$ as Δx gets smaller and smaller. Represent ΔF in your picture by drawing *one more vertical shading segment* to the right of the last one you drew in step 2. ΔF is the (signed) *area* of your vertical segment.
6. Represent Δx in your picture by moving x to beneath your newly-drawn segment. That small change in Δx is the *thickness* of your vertical segment.
7. What is now the *height* of your vertical segment?
8. Can you see why Newton and Leibniz concluded that $F'(x) = f(x)$?

Creating a mathematical model for physical situations

Each chapter begins with a problem that can be solved using the mathematics in the chapter. These problems are intended to pique your interest. Problem solving is continued throughout the chapter to develop the concepts in application contexts.

Practical applications and situations (many with sources cited) are included. You will learn how to evaluate a problem critically, formulate a strategy, solve the problem, and analyze and interpret the results.

You will model problems using various methods, including logistical regression, linearizations, and solving differential equations.

One of the main topics in calculus is differential calculus, which includes the study of rates of change of functions and approximations.

We model rates of change using differential equations, and use slope fields to interpret differential equations. We also solve differential equations using numerical methods such as Euler's method.

Another important calculus topic is integral calculus.

Integrals are used to model physical, economic, and social situations. We use integrals of rates of change to get accumulated change, and often use Riemann sums to approximate accumulated change.

Shown here is the pain reliever acetaminophen in crystalline form, photographed under a transmitted light microscope. While acetaminophen relieves pain with few side effects, it is toxic in large doses. One study found that only 30% of parents who gave acetaminophen to their children could accurately calculate and measure the correct dose.

One rule for calculating the dosage (mg) of acetaminophen for children ages 1 to 12 years old is $D(t) = 750t/(t + 12)$, where t is age in years. What is an expression for the rate of change of a child's dosage with respect to the child's age? How does the rate of change of the dosage relate to the growth rate of children? This problem can be solved with the information covered in Section 3.4.

Example 2 MODELING BEAR POPULATION

A national park is known to be capable of supporting no more than 100 grizzly bears. Ten bears are in the park at present. We model the population with a logistic differential equation with $k = 0.1$.

(a) Draw and describe a slope field for the differential equation.

(b) Find a logistic growth model $P(t)$ for the population and draw its graph.

(c) When will the bear population reach 50?

Solution **(a)** The carrying capacity is 100, so $M = 100$. The model we seek is a solution to the following differential equation.

$$\frac{dP}{dt} = \frac{0.1}{100}P(100 - P) \quad \text{Eq. 1 with } k = 0.1, M = 100$$

$$= 0.001P(100 - P)$$

Figure 6.12 shows a slope field for this differential equation. There appears to be a horizontal asymptote at $P = 100$. The solution curves fall toward this level from above, and rise toward it from below.

[0, 150] by [0, 150]

Figure 6.12 A slope field for the logistic differential equation

$$\frac{dP}{dt} = 0.001P(100 - P).$$

(Example 2)

Example 5 POTATO CONSUMPTION

From 1970 to 1980, the rate of potato consumption in a particular country was $C(t) = 2.2 + 1.1^t$ millions of bushels per year, with t being years since the beginning of 1970. How many bushels were consumed from the beginning of 1972 to the end of 1973?

Solution We seek the cumulative effect of the consumption rate for $2 \le t \le 4$.

Step 1:
Riemann sum. We partition $[2, 4]$ into subintervals of length Δt and let t_k be a time in the kth subinterval. The amount consumed during this interval is approximately

$$C(t_k) \, \Delta t \text{ million bushels.}$$

The consumption for $2 \le t \le 4$ is approximately

$$\sum C(t_k) \, \Delta t \text{ million bushels.}$$

Step 2:
Definite integral. The amount consumed from $t = 2$ to $t = 4$ is the limit of these sums as the norms of the partitions go to zero.

$$\int_2^4 C(t) \, dt = \int_2^4 (2.2 + 1.1^t) \, dt \text{ million bushels}$$

Step 3:
Evaluate. Evaluating numerically, we obtain

$$\text{NINT} (2.2 + 1.1^t, t, 2, 4) \approx 7.066 \text{ million bushels.}$$

Interpreting data

We hope that you will find the applications in this text relevant and interesting. Examples provide problems from a wide variety of subjects, including medicine, health, nature, economics, and physics. Exercises also include numerous opportunities to apply calculus to many situations.

Applications come from many sources, including newspapers, government data, and experimental data obtained with CBL™ software.

Angiography

An opaque dye is injected into a partially blocked artery to make the inside visible under X-rays. This reveals the location and severity of the blockage.

 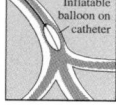

Angioplasty

A balloon-tipped catheter is inflated inside the artery to widen it at the blockage site.

Example 12 UNCLOGGING ARTERIES

In the late 1830s, the French physiologist Jean Poiseuille ("pwa-ZOY") discovered the formula we use today to predict how much the radius of a partially clogged artery has to be expanded to restore normal flow. His formula,

$$V = kr^4,$$

says that the volume V of fluid flowing through a small pipe or tube in a unit of time at a fixed pressure is a constant times the fourth power of the tube's radius r. How will a 10% increase in r affect V?

Solution The differentials of r and V are related by the equation

$$dV = \frac{dV}{dr}\,dr = 4kr^3\,dr.$$

The relative change in V is

$$\frac{dV}{V} = \frac{4kr^3\,dr}{kr^4} = 4\,\frac{dr}{r}.$$

The relative change in V is 4 times the relative change in r, so a 10% increase in r will produce a 40% increase in the flow.

You will be given data from up-to-date sources and asked to interpret this data mathematically. You will be asked to find regression equations that, in some cases, provide a good approximation for the mathematical model of the data. From these equations you will be able to predict the behavior for new values of the function.

Graphing calculator art shows what will appear as you work with your grapher. The window is provided to help you duplicate the graph.

Table 6.3 Experimental Data

Time (sec)	T (°C)	$T - T_s$ (°C)
2	64.8	60.3
5	49.0	44.5
10	31.4	26.9
15	22.0	17.5
20	16.5	12.0
25	14.2	9.7
30	12.0	7.5

[0, 35] by [0, 70]

(a)

Intersection
X = 38.227514   Y = 8

[0, 60] by [−20, 70]

(b)

Figure 6.9 (Example 5)

Example 5 USING NEWTON'S LAW OF COOLING

A temperature probe (thermometer) is removed from a cup of coffee and placed in water that has a temperature of $T_s = 4.5°C$. Temperature readings T, as recorded in Table 6.3, are taken after 2 sec, 5 sec, and every 5 sec thereafter. Estimate

(a) the coffee's temperature at the time the temperature probe was removed.

(b) the time when the temperature probe reading will be 8°C.

Solution

Model

According to Newton's Law of Cooling, $T - T_s = (T_0 - T_s)e^{-kt}$, where $T_s = 4.5$ and T_0 is the temperature of the coffee (probe reading) at $t = 0$.

We use exponential regression to find that

$$T - 4.5 = 61.66(0.9277^t)$$

is a model for the $(t, T - T_s) = (t, T - 4.5)$ data.

Thus,

$$T = 4.5 + 61.66(0.9277^t)$$

is a model for the (t, T) data.

Figure 6.9a shows the graph of the model superimposed on a scatter plot of the (t, T) data.

(a) At time $t = 0$, when the probe was removed, the temperature was

$$T = 4.5 + 61.66(0.9277^0) \approx 66.16°C.$$

(b) Solve Graphically

Figure 6.9b shows that the graphs of

$$y = 8 \quad \text{and} \quad y = T = 4.5 + 61.66(0.9277^t)$$

intersect at about $t = 38$.

Interpret

The temperature of the coffee was about 66.2°C when the temperature probe was removed. The temperature probe will reach 8°C about 38 sec after it is removed from the coffee and placed in the water.

Explorations using the graphing calculator provide simulations of motion, an important calculus application.

> **Exploration 3 Seeing Motion on a Graphing Calculator**
>
> The graphs in Figure 3.26b give us plenty of information about the flight of the rock in Example 4, but neither graph shows the path of the rock in flight. We can simulate the moving rock by graphing the parametric equations
>
> $$x_1(t) = 3(t < 5) + 3.1(t \geq 5), \quad y_1(t) = 160t - 16t^2$$
>
> in dot mode.
>
> This will show the upward flight of the rock along the vertical line $x = 3$, and the downward flight of the rock along the line $x = 3.1$.
>
> **1.** To see the flight of the rock from beginning to end, what should we use for tMin and tMax in our graphing window?
>
> **2.** Set xMin = 0, xMax = 6, and yMin = −10. Use the results from Example 4 to determine an appropriate value for yMax.

Studying a human accomplishment enhanced by technology

Calculus has been developed through the centuries by individuals such as Leibniz and Agnesi, and it has evolved over time. This evolution has been forwarded dramatically by advances in technology. The graphing calculator will play an important part in your study by enabling you to explore functions in ways that were not available to the individuals who first discovered calculus.

This book assumes you are familiar with a graphing utility that will produce the graph of a function within an arbitrary viewing window, find the zeros of a function, compute the derivative of a function numerically, and compute definite integrals numerically.

Graphing technology manuals that give more specific information about each specific calculator are available. The textbook uses special programs such as SLOPEFLD. These programs are given in the *Technology Resource Manuals*.

$y = 100e^{0.056x}$

[0, 20] by [0, 300]

Figure 6.2 The solution of the initial value problem in Example 1.

Exploration 1 **Constructing a Slope Field**

1. Let $h = (\text{Xmax} - \text{Xmin})/10$ and $k = (\text{Ymax} - \text{Ymin})/10$. Explain why

$$\left(\text{Xmin} + (2i - 1)\frac{h}{2}, \ \text{Ymin} + (2j - 1)\frac{k}{2}\right)$$

for $i, j = 1, 2, ..., 10$, represents 100 points in the viewing window [Xmin, Xmax] by [Ymin, Ymax].

2. Show that the horizontal distance between adjacent points with the same y-coordinate in part 1 is h.

3. Show that the vertical distance between adjacent points with the same x-coordinate in part 1 is k.

4. Consider the differential equation $dy/dx = 2/(x + 1)$ and the viewing window [0, 10] by [0, 10]. Find the slope (given by the differential equation) and construct a short line segment at each point in the third column ($i = 3$) of the points given by part 1. What do the line segments have in common?

5. Use the same differential equation and viewing window as in part 4. Find the slope and construct a short line segment at each point in the fourth row ($j = 4$) of the points given by part 1.

6. Explain how you could complete the slope field using the information in parts 4 and 5.

7. Let L and W be positive integers, $h = (\text{Xmax} - \text{Xmin})/W$, and $k = (\text{Ymax} - \text{Ymin})/L$. If $i = 1, 2, ..., W$ and $j = 1, 2, ..., L$, how many points are produced in part 1?

Gottfried Wilhelm Leibniz (1646–1716)

The method of limits used in this book was not discovered until nearly a century after Newton and Leibniz, the discoverers of calculus, had died.

To Leibniz, the key idea was the *differential*, an infinitely small quantity that was almost like zero, but which—unlike zero—could be used in the denominator of a fraction. Thus, Leibniz thought of the derivative dy/dx as the quotient of two differentials, dy and dx.

Maria Agnesi (1718–1799)

The first text to include differential and integral calculus along with analytic geometry, infinite series, and differential equations was written in the 1740s by the Italian mathematician Maria Gaetana Agnesi. Agnesi, a gifted scholar and linguist whose Latin essay defending higher education for women was published when she was only nine years old, was a well-published scientist by age 20, and an honorary faculty member of the University of Bologna by age 30.

Calculus at Work features introduce you to individuals who are using calculus in their jobs. Some of the applications of calculus they encounter are mentioned throughout the text.

Calculus at Work

I have a Bachelor's and Master's degree in Aerospace Engineering from the University of California at Davis. I started my professional career as a Facility Engineer managing productivity and maintenance projects in the Unitary Project Wind Tunnel facility at NASA Ames Research Center. I used calculus and differential equations in fluid mechanic analyses of the tunnels. I then moved to the position of Test Manager, still using some fluid mechanics and other mechanical engineering analysis tools to solve problems. For example, the lift and drag forces acting on an airplane wing can be determined by integrating the known pressure distribution on the wing.

I am currently a NASA On-Site Systems Engineer for the Lunar Prospector spacecraft project, at Lockheed Martin Missiles and Space in Sunnyvale, California. Differential equations and integration are used to design some of the flight hardware for the spacecraft. I work on ensuring that the different systems of the spacecraft are adequately integrated together to meet the specified design requirements. This often means doing some analysis to determine if the systems will function properly and within the constraints of the space environment. Some of these analyses require use of differential equations and integration to determine the most exact results, within some margin of error.

Ross Shaw
NASA Ames Research Center
Sunnyvale, CA

Exercises provide review, build skill and conceptual understanding, and promote exploration and discovery

Each set of exercises is carefully graded to run from routine at the beginning to more challenging at the end. Answers for most odd-numbered problems appear in the back of the text. Complete solutions are included for representative problems.

Quick Review exercises help you review skills from previous courses or chapters that you will need in the rest of the exercise set.

Many of the exercises are similar to Examples worked out in the lesson, but you will also be expected to apply the principles to new problems.

Explorations are opportunities to discover exciting mathematics on your own or in *groups of two or three.* These exercises will often ask you to use critical thinking to discover ideas.

Many of the exercises ask you to practice your writing skills by **Writing to Learn.** Communicating your ideas can give you a new perspective on your own knowledge.

Extending the Ideas exercises go beyond what is presented in the textbook. These exercises are challenging extensions of the book's material. You may be asked to work on these problems in groups or to solve them as projects.

Review Exercises at the end of each chapter prepare you for Chapter Tests. These exercises cover the entire chapter and include many additional applications.

Quick Review 4.2

In Exercises 1 and 2, find exact solutions to the inequality.

1. $2x^2 - 6 < 0$ **2.** $3x^2 - 6 > 0$

In Exercises 3–5, let $f(x) = \sqrt{8 - 2x^2}$.

3. Find the domain of f. **4.** Where is f continuous?

5. Where is f differentiable?

In Exercises 6–8, let $f(x) = \dfrac{x}{x^2 - 1}$.

6. Find the domain of f. **7.** Where is f continuous?

8. Where is f differentiable?

In Exercises 9 and 10, find C so that the graph of the function f passes through the specified point.

9. $f(x) = -2x + C,\quad (-2, 7)$

10. $g(x) = x^2 + 2x + C,\quad (1, -1)$

In Exercises 35–38, *work in groups of two or three* and sketch a graph of a differentiable function $y = f(x)$ that has the given properties.

35. (a) local minimum at $(1, 1)$, local maximum at $(3, 3)$

(b) local minima at $(1, 1)$ and $(3, 3)$

(c) local maxima at $(1, 1)$ and $(3, 3)$

36. $f(2) = 3,\quad f'(2) = 0,\quad$ and

(a) $f'(x) > 0$ for $x < 2,\quad f'(x) < 0$ for $x > 2$.

(b) $f'(x) < 0$ for $x < 2,\quad f'(x) > 0$ for $x > 2$.

(c) $f'(x) < 0$ for $x \neq 2$.

(d) $f'(x) > 0$ for $x \neq 2$.

37. $f'(-1) = f'(1) = 0,\quad f'(x) > 0$ on $(-1, 1)$, $f'(x) < 0$ for $x < -1,\quad f'(x) > 0$ for $x > 1$.

38. A local minimum value that is greater than one of its local maximum values

39. *Speeding* A trucker handed in a ticket at a toll booth showing that in 2 h she had covered 159 mi on a toll road with speed limit 65 mph. The trucker was cited for speeding. Why?

40. *Temperature Change* It took 20 sec for the temperature to rise from 0°F to 212°F when a thermometer was taken from a freezer and placed in boiling water. Explain why at some moment in that interval the mercury was rising at exactly 10.1°F/sec.

44. *Diving* (a) With what velocity will you hit the water if you step off from a 10-m diving platform?

(b) With what velocity will you hit the water if you dive off the platform with an upward velocity of 2 m/sec?

45. Writing to Learn The function

$$f(x) = \begin{cases} x, & 0 \leq x < 1 \\ 0, & x = 1 \end{cases}$$

is zero at $x = 0$ and at $x = 1$. Its derivative is equal to 1 at every point between 0 and 1, so f' is never zero between 0 and 1, and the graph of f has no tangent parallel to the chord from $(0, 0)$ to $(1, 0)$. Explain why this does not contradict the Mean Value Theorem.

46. Writing to Learn Explain why there is a zero of $y = \cos x$ between every two zeros of $y = \sin x$.

24. *Oil Flow* Oil flows through a cylindrical pipe of radius 3 inches, but friction from the pipe slows the flow toward the outer edge. The speed at which the oil flows at a distance r inches from the center is $8(10 - r^2)$ inches per second.

(a) In a plane cross section of the pipe, a thin ring with thickness Δr at a distance r inches from the center approximates a rectangular strip when you straighten it out. What is the area of the strip (and hence the approximate area of the ring)?

(b) Explain why we know that oil passes through this ring at approximately $8(10 - r^2)(2\pi r)\,\Delta r$ cubic inches per second.

(c) Set up and evaluate a definite integral that will give the rate (in cubic inches per second) at which oil is flowing through the pipe.

Explorations

51. *Analyzing Derivative Data* Assume that f is continuous on $[-2, 2]$ and differentiable on $(-2, 2)$. The table gives some values of $f'(x)$.

x	$f'(x)$	x	$f'(x)$
-2	7	0.25	-4.81
-1.75	4.19	0.5	-4.25
-1.5	1.75	0.75	-3.31
-1.25	-0.31	1	-2
-1	-2	1.25	-0.31
-0.75	-3.31	1.5	1.75
-0.5	-4.25	1.75	4.19
-0.25	-4.81	2	7
0	-5		

(a) Estimate where f is increasing, decreasing, and has local extrema.

(b) Find a quadratic regression equation for the data in the table and superimpose its graph on a scatter plot of the data.

(c) Use the model in (b) for f' and find a formula for f that satisfies $f(0) = 0$. ∎

Extending the Ideas

53. *Geometric Mean* The **geometric mean** of two positive numbers a and b is $\sqrt{ab}$. Show that for $f(x) = 1/x$ on any interval $[a, b]$ of positive numbers, the value of c in the conclusion of the Mean Value Theorem is $c = \sqrt{ab}$.

54. *Arithmetic Mean* The **arithmetic mean** of two numbers a and b is $(a + b)/2$. Show that for $f(x) = x^2$ on any interval $[a, b]$, the value of c in the conclusion of the Mean Value Theorem is $c = (a + b)/2$.

1

Prerequisites for Calculus

Exponential functions are used to model situations in which growth or decay change dramatically. Such situations are found in nuclear power plants, which contain rods of plutonium-239, an extremely toxic radioactive isotope.

Operating at full capacity for one year, a 1,000 megawatt power plant discharges about 435 lb of plutonium-239. With a half-life of 24,400 years, how much of the isotope will remain after 1,000 years? This question can be answered with the mathematics covered in Section 1.3.

Chapter 1 Overview

This chapter reviews the most important things you need to know to start learning calculus. It also introduces the use of a graphing utility as a tool to investigate mathematical ideas, to support analytic work, and to solve problems with numerical and graphical methods. The emphasis is on functions and graphs, the main building blocks of calculus.

Functions and parametric equations are the major tools for describing the real world in mathematical terms, from temperature variations to planetary motions, from brain waves to business cycles, and from heartbeat patterns to population growth. Many functions have particular importance because of the behavior they describe. Trigonometric functions describe cyclic, repetitive activity; exponential, logarithmic, and logistic functions describe growth and decay; and polynomial functions can approximate these and most other functions.

1.1 Lines

Increments • Slope of a Line • Parallel and Perpendicular Lines • Equations of Lines • Applications

Increments

One reason calculus has proved to be so useful is that it is the right mathematics for relating the rate of change of a quantity to the graph of the quantity. Explaining that relationship is one goal of this book. It all begins with the slopes of lines.

When a particle in the plane moves from one point to another, the net changes or *increments* in its coordinates are found by subtracting the coordinates of its starting point from the coordinates of its stopping point.

Definition Increments

If a particle moves from the point (x_1, y_1) to the point (x_2, y_2), the **increments** in its coordinates are

$$\Delta x = x_2 - x_1 \quad \text{and} \quad \Delta y = y_2 - y_1.$$

The symbols Δx and Δy are read "delta x" and "delta y." The letter Δ is a Greek capital d for "difference." Neither Δx nor Δy denotes multiplication; Δx is not "delta times x" nor is Δy "delta times y."

Increments can be positive, negative, or zero, as shown in Example 1.

Example 1 FINDING INCREMENTS

The coordinate increments from $(4, -3)$ to $(2, 5)$ are

$$\Delta x = 2 - 4 = -2, \quad \Delta y = 5 - (-3) = 8.$$

From $(5, 6)$ to $(5, 1)$, the increments are

$$\Delta x = 5 - 5 = 0, \quad \Delta y = 1 - 6 = -5.$$

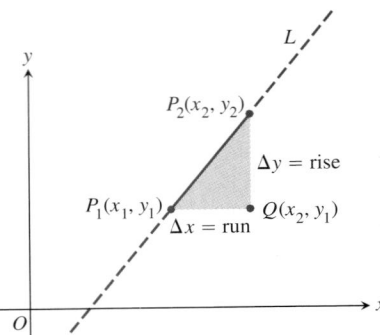

Figure 1.1 The slope of line L is
$$m = \frac{\text{rise}}{\text{run}} = \frac{\Delta y}{\Delta x}.$$

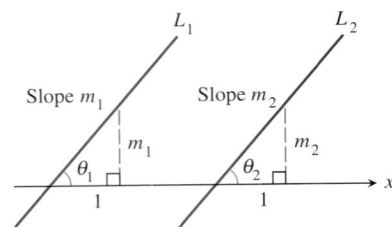

Figure 1.2 If $L_1 \parallel L_2$, then $\theta_1 = \theta_2$ and $m_1 = m_2$. Conversely, if $m_1 = m_2$, then $\theta_1 = \theta_2$ and $L_1 \parallel L_2$.

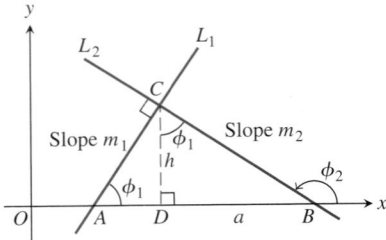

Figure 1.3 $\triangle ADC$ is similar to $\triangle CDB$. Hence ϕ_1 is also the upper angle in $\triangle CDB$, where $\tan \phi_1 = a/h$.

Slope of a Line

Each nonvertical line has a *slope*, which we can calculate from increments in coordinates.

Let L be a nonvertical line in the plane and $P_1(x_1, y_1)$ and $P_2(x_2, y_2)$ two points on L (Figure 1.1). We call $\Delta y = y_2 - y_1$ the **rise** from P_1 to P_2 and $\Delta x = x_2 - x_1$ the **run** from P_1 to P_2. Since L is not vertical, $\Delta x \neq 0$ and we define the slope of L to be the amount of rise per unit of run. It is conventional to denote the slope by the letter m.

Definition Slope

Let $P_1(x_1, y_1)$ and $P_2(x_2, y_2)$ be points on a nonvertical line, L. The **slope** of L is
$$m = \frac{\text{rise}}{\text{run}} = \frac{\Delta y}{\Delta x} = \frac{y_2 - y_1}{x_2 - x_1}.$$

A line that goes uphill as x increases has a positive slope. A line that goes downhill as x increases has a negative slope. A horizontal line has slope zero since all of its points have the same y-coordinate, making $\Delta y = 0$. For vertical lines, $\Delta x = 0$ and the ratio $\Delta y/\Delta x$ is undefined. We express this by saying that vertical lines *have no slope*.

Parallel and Perpendicular Lines

Parallel lines form equal angles with the x-axis (Figure 1.2). Hence, nonvertical parallel lines have the same slope. Conversely, lines with equal slopes form equal angles with the x-axis and are therefore parallel.

If two nonvertical lines L_1 and L_2 are perpendicular, their slopes m_1 and m_2 satisfy $m_1 m_2 = -1$, so each slope is the *negative reciprocal* of the other:
$$m_1 = -\frac{1}{m_2}, \qquad m_2 = -\frac{1}{m_1}.$$

The argument goes like this: In the notation of Figure 1.3, $m_1 = \tan \phi_1 = a/h$, while $m_2 = \tan \phi_2 = -h/a$. Hence, $m_1 m_2 = (a/h)(-h/a) = -1$.

Example 2 DETERMINING PERPENDICULARITY FROM SLOPE

Let L be a line with slope 3/4. Then any line with slope $-4/3$ will be perpendicular to L.

Equations of Lines

The vertical line through the point (a, b) has equation $x = a$ since every x-coordinate on the line has the value a. Similarly, the horizontal line through (a, b) has equation $y = b$.

Example 3 FINDING EQUATIONS OF VERTICAL AND HORIZONTAL LINES

The vertical and horizontal lines through the point $(2, 3)$ have equations $x = 2$ and $y = 3$, respectively (Figure 1.4).

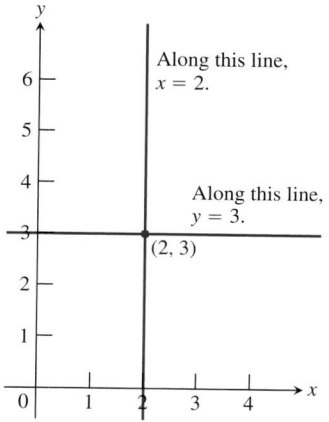

Figure 1.4 The standard equations for the vertical and horizontal lines through the point $(2, 3)$ are $x = 2$ and $y = 3$. (Example 3)

We can write an equation for any nonvertical line L if we know its slope m and the coordinates of one point $P_1(x_1, y_1)$ on it. If $P(x, y)$ is *any* other point on L, then

$$\frac{y - y_1}{x - x_1} = m,$$

so that

$$y - y_1 = m(x - x_1) \quad \text{or} \quad y = m(x - x_1) + y_1.$$

Definition Point-slope Equation

The equation

$$y = m(x - x_1) + y_1$$

is the **point-slope equation** of the line through the point (x_1, y_1) with slope m.

Example 4 USING THE POINT-SLOPE EQUATION

Write an equation for the line through the point $(2, 3)$ with slope $-3/2$.

Solution We substitute $x_1 = 2$, $y_1 = 3$, and $m = -3/2$ into the point-slope equation and obtain

$$y = -\frac{3}{2}(x - 2) + 3 \quad \text{or} \quad y = -\frac{3}{2}x + 6.$$

Example 5 USING THE POINT-SLOPE EQUATION

Write an equation for the line through $(-2, -1)$ and $(3, 4)$.

Solution The line's slope is

$$m = \frac{4 - (-1)}{3 - (-2)} = \frac{5}{5} = 1.$$

We can use this slope with either of the two given points in the point-slope equation. For $(x_1, y_1) = (-2, -1)$, we obtain

$$y = 1 \cdot (x - (-2)) + (-1)$$
$$y = x + 2 + (-1)$$
$$y = x + 1.$$

The y-coordinate of the point where a nonvertical line intersects the y-axis is the **y-intercept** of the line. Similarly, the x-coordinate of the point where a nonhorizontal line intersects the x-axis is the **x-intercept** of the line. A line with slope m and y-intercept b passes through $(0, b)$ (Figure 1.5), so

$$y = m(x - 0) + b, \quad \text{or, more simply,} \quad y = mx + b.$$

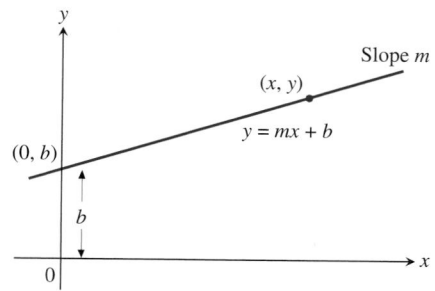

Figure 1.5 A line with slope m and y-intercept b.

Definition Slope-intercept Equation

The equation

$$y = mx + b$$

is the **slope-intercept equation** of the line with slope m and y-intercept b.

Example 6 WRITING EQUATIONS FOR LINES

Write an equation for the line through the point $(-1, 2)$ that is **(a)** parallel, and **(b)** perpendicular to the line L: $y = 3x - 4$.

Solution The line L, $y = 3x - 4$, has slope 3.

(a) The line $y = 3(x + 1) + 2$, or $y = 3x + 5$, passes through the point $(-1, 2)$, and is parallel to L because it has slope 3.

(b) The line $y = (-1/3)(x + 1) + 2$, or $y = (-1/3)x + 5/3$, passes through the point $(-1, 2)$, and is perpendicular to L because it has slope $-1/3$.

If A and B are not both zero, the graph of the equation $Ax + By = C$ is a line. Every line has an equation in this form, even lines with undefined slopes.

Definition General Linear Equation

The equation

$$Ax + By = C \quad (A \text{ and } B \text{ not both } 0)$$

is a **general linear equation** in x and y.

Although the general linear form helps in the quick identification of lines, the slope-intercept form is the one to enter into a calculator for graphing.

Example 7 ANALYZING AND GRAPHING A GENERAL LINEAR EQUATION

Find the slope and y-intercept of the line $8x + 5y = 20$. Graph the line.

Solution Solve the equation for y to put the equation in slope-intercept form:

$$8x + 5y = 20$$
$$5y = -8x + 20$$
$$y = -\frac{8}{5}x + 4$$

This form reveals the slope ($m = -8/5$) and y-intercept ($b = 4$), and puts the equation in a form suitable for graphing (Figure 1.6).

$y = -\dfrac{8}{5}x + 4$

[–5, 7] by [–2, 6]

Figure 1.6 The line $8x + 5y = 20$. (Example 7)

Applications

Many important variables are related by linear equations. For example, the relationship between Fahrenheit temperature and Celsius temperature is linear, a fact we use to advantage in the next example.

Example 8   TEMPERATURE CONVERSION

Find the relationship between Fahrenheit and Celsius temperature. Then find the Celsius equivalent of 90°F and the Fahrenheit equivalent of −5°C.

Solution Because the relationship between the two temperature scales is linear, it has the form $F = mC + b$. The freezing point of water is $F = 32°$ or $C = 0°$, while the boiling point is $F = 212°$ or $C = 100°$. Thus,

$$32 = m \cdot 0 + b \quad \text{and} \quad 212 = m \cdot 100 + b,$$

so $b = 32$ and $m = (212 - 32)/100 = 9/5$. Therefore,

$$F = \frac{9}{5}C + 32, \quad \text{or} \quad C = \frac{5}{9}(F - 32).$$

These relationships let us find equivalent temperatures. The Celsius equivalent of 90°F is

$$C = \frac{5}{9}(90 - 32) \approx 32.2°.$$

The Fahrenheit equivalent of −5°C is

$$F = \frac{9}{5}(-5) + 32 = 23°.$$

It can be difficult to see patterns or trends in lists of paired numbers. For this reason, we sometimes begin by plotting the pairs (such a plot is called a **scatter plot**) to see whether the corresponding points lie close to a curve of some kind. If they do, and if we can find an equation $y = f(x)$ for the curve, then we have a formula that

1. summarizes the data with a simple expression, and

2. lets us predict values of y for other values of x.

The process of finding a curve to fit data is called **regression analysis** and the curve is called a **regression curve.**

There are many useful types of regression curves—power, exponential, logarithmic, sinusoidal, and so on. In the next example, we use the calculator's linear regression feature to fit the data in Table 1.1 with a line.

Example 9   REGRESSION ANALYSIS – PREDICTING WORLD POPULATION

Starting with the data in Table 1.1, build a linear model for the growth of the world population. Use the model to predict the world population in the year 2010.

Some graphing utilities have a feature that enables them to approximate the relationship between variables with a linear equation. We use this feature in Example 9.

Table 1.1 World Population

Year	Population (millions)
1986	4936
1987	5023
1988	5111
1989	5201
1990	5329
1991	5422

Source: Statistical Office of the United Nations, *Monthly Bulletin of Statistics,* 1991.

[1985, 1992] by [4800, 5500]

(a)

[1985, 2015] by [4400, 8000]

(b)

Figure 1.7 (Example 9)

Why Not Round the Decimals in Equation 1 Even More?

If we do, our final calculation will be way off. Using $y = 98x - 190{,}157$, for instance, gives $y = 6823$ when $x = 2010$, as compared to $y = 7282$, a shortfall of 459 million. The rule is: *Retain all decimal places while working a problem. Round only at the end.* We rounded the co-efficients in Equation 1 enough to make it readable, but not enough to hurt the outcome. However, we knew how much we could safely round *only from first having done the entire calculation with numbers unrounded.*

Rounding Rule

Round your answer as appropriate, but do not round the numbers in the calculations that lead to it.

Solution

Model

Upon entering the data into the grapher, we find the regression equation to be approximately

$$y = 98.2286x - 190{,}157.181, \tag{1}$$

where x represents the year and y the population *in millions.*

Figure 1.7a shows the scatter plot for Table 1.1 together with a graph of the regression line just found. You can see how well the line fits the data.

Solve Graphically

Our goal is to predict the population in the year 2010. Reading from the graph in Figure 1.7b, we conclude that when x is 2010, y is approximately 7282.

Confirm Algebraically

Evaluating Equation 1 for $x = 2010$ gives

$$y = 98.2286(2010) - 190{,}157.181$$

$$\approx 7282.$$

Interpret

The linear regression equation suggests that the world population in the year 2010 will be about 7282 million, or approximately 7.3 billion.

Regression Analysis

Regression analysis has four steps:

1. Plot the data (scatter plot).
2. Find the regression equation. For a line, it has the form $y = mx + b$.
3. Superimpose the graph of the regression equation on the scatter plot to see the fit.
4. Use the regression equation to predict y-values for particular values of x.

Quick Review 1.1

1. Find the value of y that corresponds to $x = 3$ in $y = -2 + 4(x - 3)$.

2. Find the value of x that corresponds to $y = 3$ in $y = 3 - 2(x + 1)$.

In Exercises 3 and 4, find the value of m that corresponds to the values of x and y.

3. $x = 5$, $y = 2$, $m = \dfrac{y - 3}{x - 4}$

4. $x = -1$, $y = -3$, $m = \dfrac{2 - y}{3 - x}$

In Exercises 5 and 6, determine whether the ordered pair is a solution to the equation.

5. $3x - 4y = 5$
   (a) $(2, 1/4)$ (b) $(3, -1)$

6. $y = -2x + 5$
   (a) $(-1, 7)$ (b) $(-2, 1)$

In Exercises 7 and 8, find the distance between the points.

7. $(1, 0)$, $(0, 1)$

8. $(2, 1)$, $(1, -1/3)$

In Exercises 9 and 10, solve for y in terms of x.

9. $4x - 3y = 7$

10. $-2x + 5y = -3$

Section 1.1 Exercises

In Exercises 1–4, find the coordinate increments from A to B.

1. $A(1, 2)$, $B(-1, -1)$
2. $A(-3, 2)$, $B(-1, -2)$
3. $A(-3, 1)$, $B(-8, 1)$
4. $A(0, 4)$, $B(0, -2)$

In Exercises 5–8, let L be the line determined by points A and B.

(a) Plot A and B.   (b) Find the slope of L.

(c) Draw the graph of L.

5. $A(1, -2)$, $B(2, 1)$
6. $A(-2, -1)$, $B(1, -2)$
7. $A(2, 3)$, $B(-1, 3)$
8. $A(1, 2)$, $B(1, -3)$

In Exercise 9–12, write an equation for (a) the vertical line and (b) the horizontal line through the point P.

9. $P(2, 3)$
10. $P(-1, 4/3)$
11. $P(0, -\sqrt{2})$
12. $P(-\pi, 0)$

In Exercises 13–16, write the point-slope equation for the line through the point P with slope m.

13. $P(1, 1)$, $m = 1$
14. $P(-1, 1)$, $m = -1$
15. $P(0, 3)$, $m = 2$
16. $P(-4, 0)$, $m = -2$

In Exercises 17–20, write a general linear equation for the line through the two points.

17. $(0, 0)$, $(2, 3)$
18. $(1, 1)$, $(2, 1)$
19. $(-2, 0)$, $(-2, -2)$
20. $(-2, 1)$, $(2, -2)$

In Exercises 21–24, write the slope-intercept equation for the line with slope m and y-intercept b.

21. $m = 3$, $b = -2$
22. $m = -1$, $b = 2$
23. $m = -1/2$, $b = -3$
24. $m = 1/3$, $b = -1$

In Exercises 25 and 26, the line contains the origin and the point in the upper right corner of the grapher screen. Write an equation for the line.

25.

26.

$[-10, 10]$ by $[-25, 25]$ $[-5, 5]$ by $[-2, 2]$

In Exercises 27–30, find the (a) slope and (b) y-intercept, and (c) graph the line.

27. $3x + 4y = 12$
28. $x + y = 2$
29. $\dfrac{x}{3} + \dfrac{y}{4} = 1$
30. $y = 2x + 4$

In Exercises 31–34, write an equation for the line through P that is (a) parallel to L, (b) perpendicular to L.

31. $P(0, 0)$, $L: y = -x + 2$
32. $P(-2, 2)$, $L: 2x + y = 4$
33. $P(-2, 4)$, $L: x = 5$
34. $P(-1, 1/2)$, $L: y = 3$

In Exercises 35 and 36, a table of values is given for the linear function $f(x) = mx + b$. Determine m and b.

35.

x	$f(x)$
1	2
3	9
5	16

36.

x	$f(x)$
2	-1
4	-4
6	-7

In Exercises 37 and 38, find the value of x or y for which the line through A and B has the given slope m.

37. $A(-2, 3)$, $B(4, y)$, $m = -2/3$

38. $A(-8, -2)$, $B(x, 2)$, $m = 2$

In Exercises 39 and 40, use linear regression analysis.

39. Table 1.2 lists the ages and weights of nine girls.

Table 1.2 Girls' Ages and Weights

Age (months)	Weight (pounds)
19	22
21	23
24	25
27	28
29	31
31	28
34	32
38	34
43	39

(a) Find the linear regression equation for the data.

(b) Find the slope of the regression line. What does the slope represent?

(c) Superimpose the graph of the linear regression equation on a scatter plot of the data.

(d) Use the regression equation to predict the approximate weight of a 30-month-old girl.

40. Table 1.3 shows the mean annual compensation of construction workers.

Table 1.3 Construction Workers' Average Annual Compensation

Year	Annual Compensation (dollars)
1980	22,033
1985	27,581
1988	30,466
1989	31,465
1990	32,836

Source: U.S. Bureau of Economic Analysis.

(a) Find the linear regression equation for the data.

(b) Find the slope of the regression line. What does the slope represent?

(c) Superimpose the graph of the linear regression equation on a scatter plot of the data.

(d) Use the regression equation to predict the construction workers' average annual compensation in the year 2000.

41. *Revisiting Example 5* Show that you get the same equation in Example 5 if you use the point $(3, 4)$ to write the equation.

42. Writing to Learn *x- and y-intercepts*

(a) Explain why c and d are the x-intercept and y-intercept, respectively, of the line
$$\frac{x}{c} + \frac{y}{d} = 1.$$

(b) How are the x-intercept and y-intercept related to c and d in the line
$$\frac{x}{c} + \frac{y}{d} = 2?$$

43. *Parallel and Perpendicular Lines* For what value of k are the two lines $2x + ky = 3$ and $x + y = 1$ **(a)** parallel? **(b)** perpendicular?

In Exercises 44–46, *work in groups of two or three to solve the problem.*

44. *Insulation* By measuring slopes in the figure below, find the temperature change in degrees per inch for the following materials.

(a) gypsum wallboard

(b) fiberglass insulation

(c) wood sheathing

(d) Writing to Learn Which of the materials in (a)–(c) is the best insulator? the poorest? Explain.

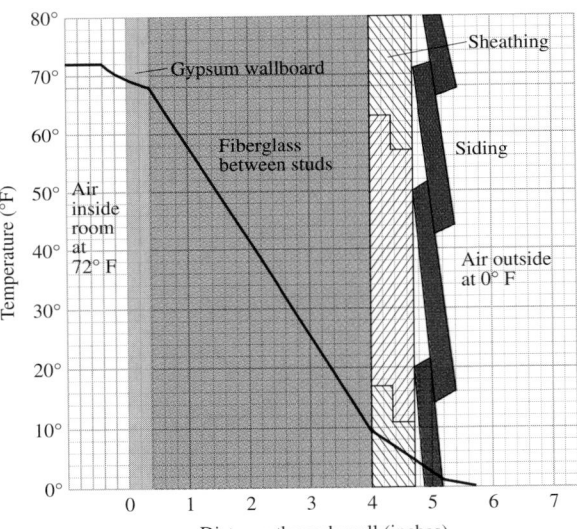

45. *Pressure under Water* The pressure p experienced by a diver under water is related to the diver's depth d by an equation of the form $p = kd + 1$ (k a constant). When $d = 0$ meters, the pressure is 1 atmosphere. The pressure at 100 meters is 10.94 atmospheres. Find the pressure at 50 meters.

46. *Modeling Distance Traveled* A car starts from point P at time $t = 0$ and travels at 45 mph.

(a) Write an expression $d(t)$ for the distance the car travels from P.

(b) Graph $y = d(t)$.

(c) What is the slope of the graph in (b)? What does it have to do with the car?

(d) Writing to Learn Create a scenario in which t could have negative values.

(e) Writing to Learn Create a scenario in which the y-intercept of $y = d(t)$ could be 30.

Extending the Ideas

47. The median price of existing single-family homes has increased consistently during the past two decades. However, the data in Table 1.4 show that there have been differences in various parts of the country.

Table 1.4 Median Price of Single-Family Homes

Year	Northeast (dollars)	Midwest (dollars)
1970	25,200	20,100
1975	39,300	30,100
1980	60,800	51,900
1985	88,900	58,900
1990	141,200	74,000

Source: National Association of Realtors®, *Home Sales Yearbook* (Washington, DC, 1990).

(a) Find the linear regression equation for home cost in the Northeast.

(b) What does the slope of the regression line represent?

(c) Find the linear regression equation for home cost in the Midwest.

(d) Where is the median price increasing more rapidly, in the Northeast or the Midwest?

48. *Fahrenheit versus Celsius* We found a relationship between Fahrenheit temperature and Celsius temperature in Example 8.

(a) Is there a temperature at which a Fahrenheit thermometer and a Celsius thermometer give the same reading? If so, what is it?

(b) Writing to Learn Graph $y_1 = (9/5)x + 32$, $y_2 = (5/9)(x - 32)$, and $y_3 = x$ in the same viewing window. Explain how this figure is related to the question in (a).

49. *Parallelogram* Three different parallelograms have vertices at $(-1, 1)$, $(2, 0)$, and $(2, 3)$. Draw the three and give the coordinates of the missing vertices.

50. *Parallelogram* Show that if the midpoints of consecutive sides of any quadrilateral are connected, the result is a parallelogram.

51. *Tangent Line* Consider the circle of radius 5 centered at $(0, 0)$. Find an equation of the line tangent to the circle at the point $(3, 4)$.

52. *Distance From a Point to a Line* This activity investigates how to find the distance from a point $P(a, b)$ to a line $L: Ax + By = C$.

We suggest that students *work in groups of two or three.*

(a) Write an equation for the line M through P perpendicular to L.

(b) Find the coordinates of the point Q in which M and L intersect.

(c) Find the distance from P to Q.

1.2 Functions and Graphs

Functions • Domains and Ranges • Viewing and Interpreting Graphs • Even Functions and Odd Functions—Symmetry • Functions Defined in Pieces • Absolute Value Function • Composite Functions

Functions

The values of one variable often depend on the values for another:

- The temperature at which water boils depends on elevation (the boiling point drops as you go up).

- The amount by which your savings will grow in a year depends on the interest rate offered by the bank.

- The area of a circle depends on the circle's radius.

In each of these examples, the value of one variable quantity depends on the value of another. For example, the boiling temperature of water, b, depends on the elevation, e; the amount of interest, I, depends on the interest rate, r. We call b and I **dependent variables** because they are determined by the values of the variables e and r on which they depend. The variables e and r are **independent variables.**

A rule that assigns to each element in one set a unique element in another set is called a *function*. The sets may be sets of any kind and do not have to be the same. A function is like a machine that assigns a unique output to every allowable input. The inputs make up the **domain** of the function; the outputs make up the **range** (Figure 1.8).

Figure 1.8 A "machine" diagram for a function.

Definition Function

A **function** from a set D to a set R is a rule that assigns a unique element in R to each element in D.

In this definition, D is the domain of the function and R is a set *containing* the range (Figure 1.9).

Euler invented a symbolic way to say "y is a function of x":

$$y = f(x),$$

which we read as "y equals f of x." This notation enables us to give different functions different names by changing the letters we use. To say that the boiling point of water is a function of elevation, we can write $b = f(e)$. To say that the area of a circle is a function of the circle's radius, we can write $A = A(r)$, giving the function the same name as the dependent variable.

The notation $y = f(x)$ gives a way to denote specific values of a function. The value of f at a can be written as $f(a)$, read "f of a."

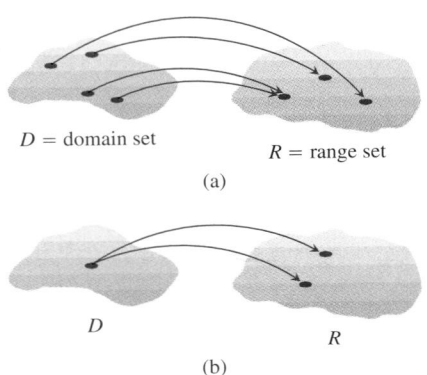

Figure 1.9 (a) A function from a set D to a set R. (b) *Not* a function. The assignment is not unique.

Example 1   THE CIRCLE-AREA FUNCTION

The domain of the circle-area function $A(r) = \pi r^2$ is the set of all possible radii—the set of all positive real numbers. The range is also the set of all positive real numbers.

The value of the function at $r = 2$ is

$$A(2) = \pi(2)^2 = 4\pi.$$

The area of a circle of radius 2 is 4π.

Domains and Ranges

In Example 1, the domain of the function is restricted by context: the independent variable is a radius and must be positive. When we define a function $y = f(x)$ with a formula and the domain is not stated explicitly or restricted by context, the domain is assumed to be the largest set of x-values for which the formula gives real y-values—the so-called **natural domain.** If we want to restrict the domain, we must say so. The domain of $y = x^2$ is understood to be the entire set of real numbers. We must write "$y = x^2, \; x > 0$" if we want to restrict the function to positive values of x.

Leonhard Euler (1707–1783)

Leonhard Euler, the dominant mathematical figure of his century and the most prolific mathematician ever, was also an astronomer, physicist, botanist, and chemist, and an expert in oriental languages. His work was the first to give the function concept the prominence that it has in mathematics today. Euler's collected books and papers fill 72 volumes. This does not count his enormous correspondence to approximately 300 addresses. His introductory algebra text, written originally in German (Euler was Swiss), is still available in English translation.

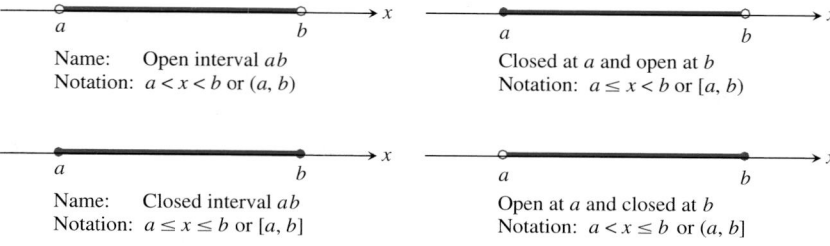

Name: Open interval ab
Notation: $a < x < b$ or (a, b)

Closed at a and open at b
Notation: $a \leq x < b$ or $[a, b)$

Name: Closed interval ab
Notation: $a \leq x \leq b$ or $[a, b]$

Open at a and closed at b
Notation: $a < x \leq b$ or $(a, b]$

Figure 1.10 Open and closed finite intervals.

Figure 1.11 Half-open finite intervals.

0

Name: The set of all real numbers
Notation: $-\infty < x < \infty$ or $(-\infty, \infty)$

a

Name: The set of numbers greater than a
Notation: $a < x$ or (a, ∞)

a

Name: The set of numbers greater than or equal to a
Notation: $a \leq x$ or $[a, \infty)$

b

Name: The set of numbers less than b
Notation: $x < b$ or $(-\infty, b)$

b

Name: The set of numbers less than or equal to b
Notation: $x \leq b$ or $(-\infty, b]$

Figure 1.12 Infinite intervals—rays on the number line and the number line itself. The symbol ∞ (infinity) is used merely for convenience; it does not mean there is a number ∞.

The domains and ranges of many real-valued functions of a real variable are intervals or combinations of intervals. The intervals may be open, closed, or half-open (Figures 1.10 and 1.11) and finite or infinite (Figure 1.12).

The endpoints of an interval make up the interval's **boundary** and are called **boundary points.** The remaining points make up the interval's **interior** and are called **interior points. Closed intervals** contain their boundary points. **Open intervals** contain no boundary points. Every point of an open interval is an interior point of the interval.

Example 2 IDENTIFYING DOMAIN AND RANGE OF A FUNCTION

Verify the domains of these functions.

Function	Domain (x)	Range (y)
$y = x^2$	$(-\infty, \infty)$	$[0, \infty)$
$y = 1/x$	$(-\infty, 0) \cup (0, \infty)$	$(-\infty, 0) \cup (0, \infty)$
$y = \sqrt{x}$	$[0, \infty)$	$[0, \infty)$
$y = \sqrt{4 - x}$	$(-\infty, 4]$	$[0, \infty)$
$y = \sqrt{1 - x^2}$	$[-1, 1]$	$[0, 1]$

Solution The formula $y = x^2$ gives a real y-value for any real number x, so the domain is $(-\infty, \infty)$.

The formula $y = 1/x$ gives a real y-value for every real x-value except $x = 0$. *We cannot divide any number by 0.*

The formula $y = \sqrt{x}$ gives a real y-value only when x is positive or zero.

The formula $y = \sqrt{4 - x}$ gives a real y-value only when $4 - x$ is greater than or equal to zero. So, $0 \leq 4 - x$, or $x \leq 4$.

The formula $y = \sqrt{1 - x^2}$ gives a real y-value for every value of x in the closed interval from -1 to 1. Outside this interval $1 - x^2$ is negative and its square root is not a real number. The domain is $[-1, 1]$.

Viewing and Interpreting Graphs

The points (x, y) in the plane whose coordinates are the input-output pairs of a function $y = f(x)$ make up the function's **graph.** The graph of the function $y = x + 2$, for example, is the set of points with coordinates (x, y) for which y equals $x + 2$.

$y = \dfrac{1}{\sqrt{4-x^2}}$

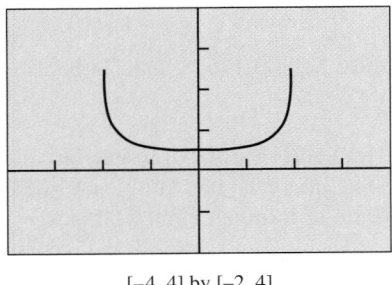

[-4, 4] by [-2, 4]

(a)

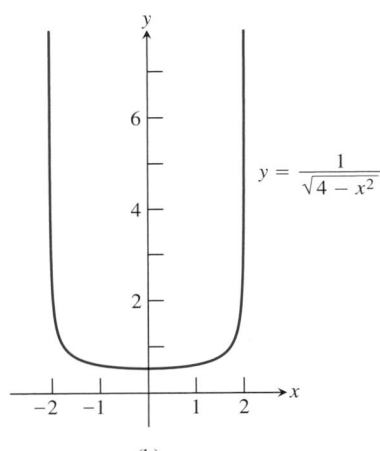

(b)

Figure 1.13 (a) Grapher failure.
(b) A more accurate graph of
$y = 1/\sqrt{4-x^2}$. (Example 3)

Graphing $y = x^{2/3}$ — Possible Grapher Failure

On some graphing calculators you need to enter this function as $y = (x^2)^{1/3}$ or $y = (x^{1/3})^2$ to obtain a correct graph. Try graphing this function on your grapher.

Graphing with pencil and paper requires that you develop graph *drawing* skills. Graphing with a grapher requires that you develop graph *viewing* skills.

Graph Viewing Skills

1. Recognize that the graph is reasonable.

2. See all the important characteristics of the graph.

3. Interpret those characteristics.

4. Recognize grapher failure.

Being able to recognize that a graph is reasonable comes with experience. You need to know the basic functions, their graphs, and how changes in their equations affect the graphs.

Grapher failure occurs when the graph produced by a grapher is less than precise—or even incorrect—usually due to the limitations of the screen resolution of the grapher.

Example 3   RECOGNIZING GRAPHER FAILURE

Find the domain and range of $y = f(x) = 1/\sqrt{4 - x^2}$.

Solution The graph of f in Figure 1.13a seems to suggest that the domain of f is an interval between -2 and 2, and that the range is also a finite interval. The latter observation is the result of grapher failure; in this case, we can recognize the failure using algebra.

Solve Algebraically

The expression $4 - x^2$ must be greater than zero.

$$4 - x^2 > 0$$
$$x^2 < 4$$

Thus, $-2 < x < 2$, and the domain is $(-2, 2)$.

The smallest value of f is $1/2$ and occurs when $x = 0$. The values of f get very large as x approaches 2 from the left or -2 from the right, as suggested by the following table. (The values of f are rounded to three decimal places.)

x	± 1.99	± 1.999	± 1.9999	± 1.99999
$f(x)$	5.006	15.813	50.001	158.114

The range of f is $[0.5, \infty)$.

Figure 1.14 shows graphs of *power functions* that arise frequently in calculus. Knowing the general shapes of these graphs will help you recognize grapher failure. We will review other functions as the chapter continues.

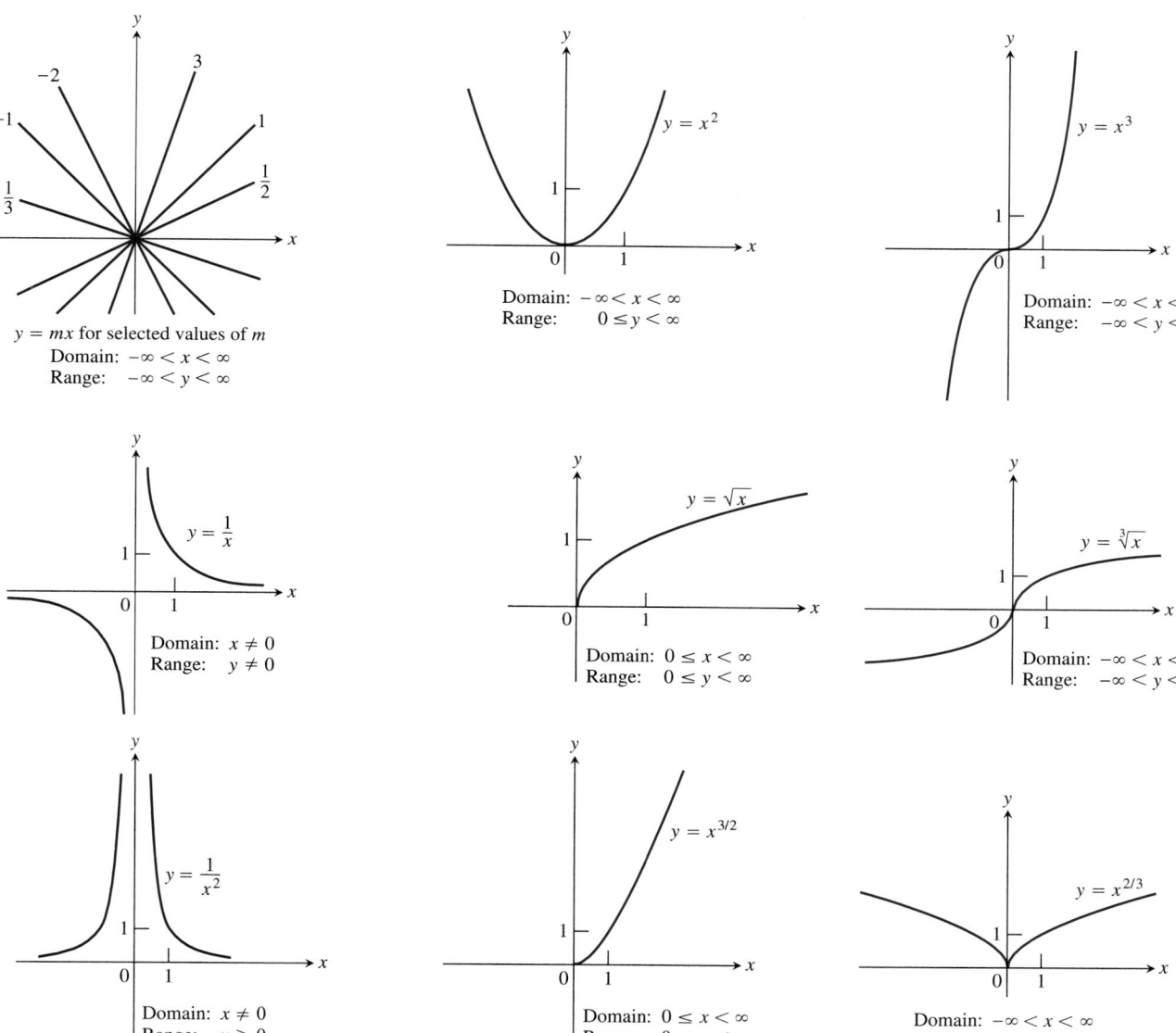

Figure 1.14 Useful power functions.

Even Functions and Odd Functions—Symmetry

The graphs of *even* and *odd* functions have important symmetry properties.

Definitions Even Function, Odd Function

A function $y = f(x)$ is an

$$\textbf{even function of } x \text{ if } f(-x) = f(x),$$

$$\textbf{odd function of } x \text{ if } f(-x) = -f(x),$$

for every x in the function's domain.

(a)

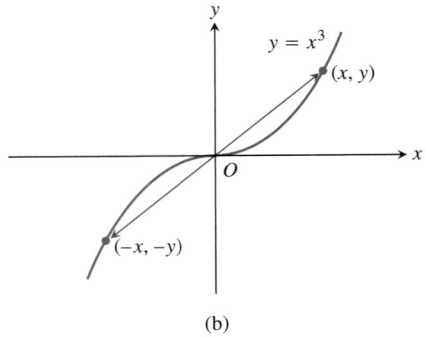

(b)

Figure 1.15 (a) The graph of $y = x^2$ (an even function) is symmetric about the y-axis. (b) The graph of $y = x^3$ (an odd function) is symmetric about the origin.

The names even and odd come from powers of x. If y is an even power of x, as in $y = x^2$ or $y = x^4$, it is an even function of x (because $(-x)^2 = x^2$ and $(-x)^4 = x^4$). If y is an odd power of x, as in $y = x$ or $y = x^3$, it is an odd function of x (because $(-x)^1 = -x$ and $(-x)^3 = -x^3$).

The graph of an even function is **symmetric about the y-axis.** Since $f(-x) = f(x)$, a point (x, y) lies on the graph if and only if the point $(-x, y)$ lies on the graph (Figure 1.15a).

The graph of an odd function is **symmetric about the origin.** Since $f(-x) = -f(x)$, a point (x, y) lies on the graph if and only if the point $(-x, -y)$ lies on the graph (Figure 1.15b). Equivalently, a graph is symmetric about the origin if a rotation of 180° about the origin leaves the graph unchanged.

Example 4 RECOGNIZING EVEN AND ODD FUNCTIONS

$f(x) = x^2$ Even function: $(-x)^2 = x^2$ for all x; symmetry about y-axis.

$f(x) = x^2 + 1$ Even function: $(-x)^2 + 1 = x^2 + 1$ for all x; symmetry about y-axis (Figure 1.16a).

$f(x) = x$ Odd function: $(-x) = -x$ for all x; symmetry about the origin.

$f(x) = x + 1$ Not odd: $f(-x) = -x + 1$, but $-f(x) = -x - 1$. The two are not equal. Not even: $(-x) + 1 \neq x + 1$ for all $x \neq 0$ (Figure 1.16b).

It is useful in graphing to recognize even and odd functions. Once we know the graph of either type of function on one side of the y-axis, we know its graph on both sides.

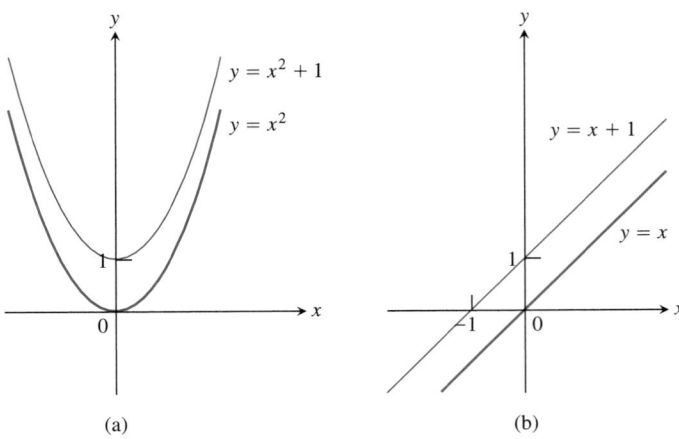

(a)   (b)

Figure 1.16 (a) When we add the constant term 1 to the function $y = x^2$, the resulting function $y = x^2 + 1$ is still even and its graph is still symmetric about the y-axis. (b) When we add the constant term 1 to the function $y = x$, the resulting function $y = x + 1$ is no longer odd. The symmetry about the origin is lost. (Example 4)

$$y = \begin{cases} -x, & x < 0 \\ x^2, & 0 \le x \le 1 \\ 1, & x > 1 \end{cases}$$

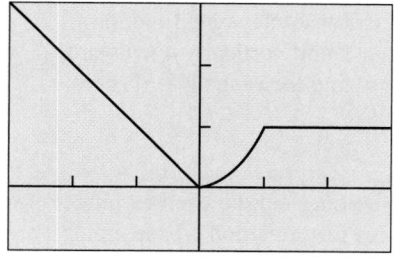

[−3, 3] by [−1, 3]

Figure 1.17 The graph of a piecewise defined function. (Example 5)

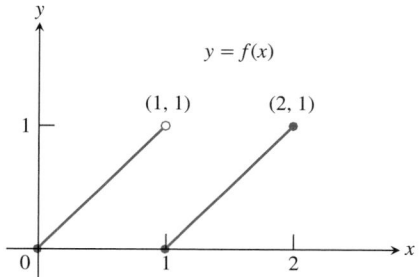

Figure 1.18 The segment on the left contains (0, 0) but not (1, 1). The segment on the right contains both of its endpoints. (Example 6)

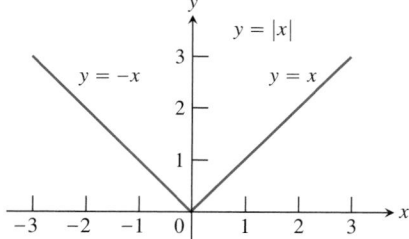

Figure 1.19 The absolute value function has domain $(-\infty, \infty)$ and range $[0, \infty)$.

Functions Defined in Pieces

While some functions are defined by single formulas, others are defined by applying different formulas to different parts of their domains.

Example 5 GRAPHING PIECEWISE DEFINED FUNCTIONS

Graph $y = f(x) = \begin{cases} -x, & x < 0 \\ x^2, & 0 \le x \le 1 \\ 1, & x > 1. \end{cases}$

Solution The values of f are given by three separate formulas: $y = -x$ when $x < 0$, $y = x^2$ when $0 \le x \le 1$, and $y = 1$ when $x > 1$. However, the function is *just one function,* whose domain is the entire set of real numbers (Figure 1.17).

Example 6 WRITING FORMULAS FOR PIECEWISE FUNCTIONS

Write a formula for the function $y = f(x)$ whose graph consists of the two line segments in Figure 1.18.

Solution We find formulas for the segments from (0, 0) to (1, 1) and from (1, 0) to (2, 1) and piece them together in the manner of Example 5.

Segment from (0, 0) to (1, 1) The line through (0, 0) and (1, 1) has slope $m = (1 - 0)/(1 - 0) = 1$ and y-intercept $b = 0$. Its slope-intercept equation is $y = x$. The segment from (0, 0) to (1, 1) that includes the point (0, 0) but not the point (1, 1) is the graph of the function $y = x$ restricted to the half-open interval $0 \le x < 1$, namely,

$$y = x, \quad 0 \le x < 1.$$

Segment from (1, 0) to (2, 1) The line through (1, 0) and (2, 1) has slope $m = (1 - 0)/(2 - 1) = 1$ and passes through the point (1, 0). The corresponding point-slope equation for the line is

$$y = 1(x - 1) + 0, \quad \text{or} \quad y = x - 1.$$

The segment from (1, 0) to (2, 1) that includes both endpoints is the graph of $y = x - 1$ restricted to the closed interval $1 \le x \le 2$, namely,

$$y = x - 1, \quad 1 \le x \le 2.$$

Piecewise formula Combining the formulas for the two pieces of the graph, we obtain

$$f(x) = \begin{cases} x, & 0 \le x < 1 \\ x - 1, & 1 \le x \le 2. \end{cases}$$

Absolute Value Function

The **absolute value function** $y = |x|$ is defined piecewise by the formula

$$|x| = \begin{cases} -x, & x < 0 \\ x, & x \ge 0. \end{cases}$$

The function is even, and its graph (Figure 1.19) is symmetric about the y-axis.

$y = |x - 2| - 1$

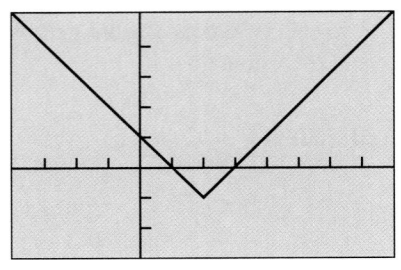

[–4, 8] by [–3, 5]

Figure 1.20 The lowest point of the graph of $f(x) = |x - 2| - 1$ is $(2, -1)$. (Example 7)

Figure 1.21 Two functions can be composed when a portion of the range of the first lies in the domain of the second.

Example 7 USING TRANSFORMATIONS

Find the domain, range, and draw the graph of $f(x) = |x - 2| - 1$.

Solution The graph of f is the graph of the absolute value function shifted 2 units horizontally to the right and 1 unit vertically downward (Figure 1.20). The domain of f is $(-\infty, \infty)$ and the range is $[-1, \infty)$.

Composite Functions

Suppose that some of the outputs of a function g can be used as inputs of a function f. We can then link g and f to form a new function whose inputs x are inputs of g and whose outputs are the numbers $f(g(x))$, as in Figure 1.21. We say that the function $f(g(x))$ (read "f of g of x") is the **composite of g and f.** It is made by *composing g and f* in the order of first g, then f. The usual "standalone" notation for this composite is $f \circ g$, which is read as "f of g." Thus, the value of $f \circ g$ at x is $(f \circ g)(x) = f(g(x))$.

Example 8 COMPOSING FUNCTIONS

Find a formula for $f(g(x))$ if $g(x) = x^2$ and $f(x) = x - 7$. Then find $f(g(2))$.

Solution To find $f(g(x))$, we replace x in the formula $f(x) = x - 7$ by the expression given for $g(x)$.

$$f(x) = x - 7$$

$$f(g(x)) = g(x) - 7 = x^2 - 7$$

We then find the value of $f(g(2))$ by substituting 2 for x.

$$f(g(2)) = (2)^2 - 7 = -3$$

Exploration 1   Composing Functions

Some graphers allow a function such as y_1 to be used as the independent variable of another function. With such a grapher, we can compose functions.

1. Enter the functions $y_1 = f(x) = 4 - x^2$, $y_2 = g(x) = \sqrt{x}$, $y_3 = y_2(y_1(x))$, and $y_4 = y_1(y_2(x))$. Which of y_3 and y_4 corresponds to $f \circ g$? to $g \circ f$?

2. Graph y_1, y_2, and y_3 and make conjectures about the domain and range of y_3.

3. Graph y_1, y_2, and y_4 and make conjectures about the domain and range of y_4.

4. Confirm your conjectures algebraically by finding formulas for y_3 and y_4.

Quick Review 1.2

In Exercises 1–6, solve for x.

1. $3x - 1 \leq 5x + 3$

2. $x(x - 2) > 0$

3. $|x - 3| \leq 4$

4. $|x - 2| \geq 5$

5. $x^2 < 16$

6. $9 - x^2 \geq 0$

In Exercises 7 and 8, describe how the graph of f can be transformed to the graph of g.

7. $f(x) = x^2$, $g(x) = (x + 2)^2 - 3$

8. $f(x) = |x|$, $g(x) = |x - 5| + 2$

In Exercises 9–12, find all real solutions to the equations.

9. $f(x) = x^2 - 5$

    (a) $f(x) = 4$ **(b)** $f(x) = -6$

10. $f(x) = 1/x$

    (a) $f(x) = -5$ **(b)** $f(x) = 0$

11. $f(x) = \sqrt{x + 7}$

    (a) $f(x) = 4$ **(b)** $f(x) = 1$

12. $f(x) = \sqrt[3]{x - 1}$

    (a) $f(x) = -2$ **(b)** $f(x) = 3$

Section 1.2 Exercises

In Exercises 1–4, write a formula that expresses the first variable as a function of the second.

1. the area of a circle as a function of its diameter

2. the height of an equilateral triangle as a function of its side length

3. the surface area of a cube as a function of the length of the cube's edges

4. the volume of a sphere as a function of the sphere's radius

In Exercises 5–18, do the following for the function.

    (a) Find the domain.   **(b)** Find the range.

    (c) Draw its graph.

    (d) Determine any symmetries discussed in this section that are characteristics of the graph.

5. $y = 4 - x^2$

6. $y = x^2 - 9$

7. $y = 2 + \sqrt{x - 1}$

8. $y = -\sqrt{-x}$

9. $y = 2\sqrt{3 - x}$

10. $y = \dfrac{1}{x - 2}$

11. $y = \sqrt[3]{x - 3}$

12. $y = \sqrt[3]{1 - x^2}$

13. $y = \sqrt[4]{-x}$

14. $y = 1 + \dfrac{1}{x}$

15. $y = \sqrt{4 - x^2}$

16. $y = x^{2/3}$

17. $y = 1 + \dfrac{1}{x^2}$

18. $y = x^{3/2}$

In Exercises 19–28, determine whether the function is even, odd, or neither. Try to answer without writing anything (except the answer).

19. $y = x^4$

20. $y = x + x^2$

21. $y = x + 2$

22. $y = x^2 - 3$

23. $y = \sqrt{x^2 + 2}$

24. $y = x + x^3$

25. $y = \dfrac{x^3}{x^2 - 1}$

26. $y = \sqrt[3]{2 - x}$

27. $y = \dfrac{1}{x - 1}$

28. $y = \dfrac{1}{x^2 - 1}$

In Exercises 29–34, **(a)** draw the graph of the function. Then find its **(b)** domain and **(c)** range.

29. $f(x) = -|3 - x| + 2$

30. $f(x) = 2|x + 4| - 3$

31. $f(x) = \begin{cases} 3 - x, & x \leq 1 \\ 2x, & 1 < x \end{cases}$

32. $f(x) = \begin{cases} 1, & x < 0 \\ \sqrt{x}, & x \geq 0 \end{cases}$

33. $f(x) = \begin{cases} 4 - x^2, & x < 1 \\ (3/2)x + 3/2, & 1 \leq x \leq 3 \\ x + 3, & x > 3 \end{cases}$

34. $f(x) = \begin{cases} x^2, & x < 0 \\ x^3, & 0 \leq x \leq 1 \\ 2x - 1, & x > 1 \end{cases}$

35. Writing to Learn The *vertical line test* to determine whether a curve is the graph of a function states: If every vertical line in the xy-plane intersects a given curve in at most one point, then the curve is the graph of a function. Explain why this is true.

36. Writing to Learn For a curve to be *symmetric about the x-axis*, the point (x, y) must lie on the curve if and only if the point $(x, -y)$ lies on the curve. Explain why a curve that is symmetric about the x-axis is not the graph of a function, unless the function is $y = 0$.

In Exercises 37–40, use the vertical line test (see Exercise 35) to determine whether the curve is the graph of a function.

37.

38.

39.

40.

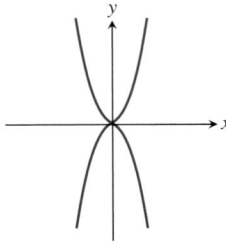

In Exercises 41–48, write a piecewise formula for the function.

41.

42.

43.

44.

45.

46.

47.

48.

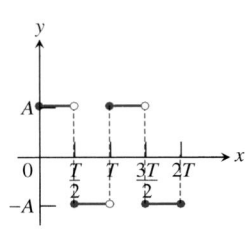

In Exercises 49 and 50, find

(a) $f(g(x))$ (b) $g(f(x))$ (c) $f(g(0))$

(d) $g(f(0))$ (e) $g(g(-2))$ (f) $f(f(x))$

49. $f(x) = x + 5$, $g(x) = x^2 - 3$

50. $f(x) = x + 1$, $g(x) = x - 1$

Explorations

In Exercises **51–54**, (a) graph $f \circ g$ and $g \circ f$ and make a conjecture about the domain and range of each function. (b) Then confirm your conjectures by finding formulas for $f \circ g$ and $g \circ f$.

51. $f(x) = x - 7$, $g(x) = \sqrt{x}$

52. $f(x) = 1 - x^2$, $g(x) = \sqrt{x}$

53. $f(x) = x^2 - 3$, $g(x) = \sqrt{x + 2}$

54. $f(x) = \dfrac{2x - 1}{x + 3}$, $g(x) = \dfrac{3x + 1}{2 - x}$

In Exercises 55–58, use a graph to determine the domain and range of the function. Confirm your answer algebraically.

55. $f(x) = \dfrac{1}{\sqrt{x^2 - 4}}$ **56.** $f(x) = \dfrac{2}{\sqrt[4]{9 - x^2}}$

57. $f(x) = \dfrac{2}{\sqrt[3]{9 - x^2}}$ **58.** $f(x) = \dfrac{1}{\sqrt[3]{x^2 - 1}}$

In Exercises 59–62, a portion of the graph of a function defined on $[-2, 2]$ is shown. *Work in groups of two or three* to complete each graph assuming that the graph is (a) even, (b) odd.

59.

60.

61.

62.

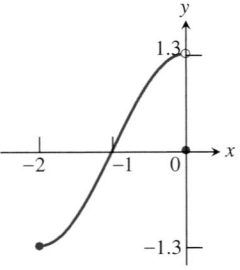

63. Copy and complete the following table.

	$g(x)$	$f(x)$	$(f \circ g)(x)$
(a)	?	$\sqrt{x-5}$	$\sqrt{x^2-5}$
(b)	?	$1 + 1/x$	x
(c)	$1/x$	?	x
(d)	$\sqrt{x}$	?	$|x|$

64. *Mutual Fund* Table 1.5 shows how an initial investment of $10,000 on December 31, 1977, grew over time.

Table 1.5 Merrill Lynch Capital Fund

Year	Amount (dollars)
1990	74,240
1991	92,570
1992	97,225
1993	110,551
1994	111,561
1995	148,226

Source: Data provided by Morningstar, Inc., as reported in *USA Today*, February 1, 1996.

(a) Find the power regression equation for the data in Table 1.5. Let $x = 0$ represent 1970, $x = 1$ represent 1971, and so forth.

(b) Superimpose the graph of the power regression equation on a scatter plot of the data.

(c) Use the graph of the power regression equation to predict the amount to which the initial $10,000 investment will grow in the year 2000. Confirm algebraically.

(d) Now use linear regression to predict the amount to which the initial $10,000 investment will grow in the year 2000.

65. *The Cone Problem* Begin with a circular piece of paper with a 4-in. radius as shown in (a). Cut out a sector with an arc length of x. Join the two edges of the remaining portion to form a cone with radius r and height h, as shown in (b).

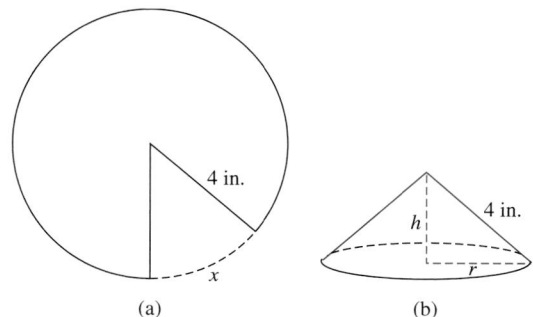

(a)   (b)

(a) Explain why the circumference of the base of the cone is $8\pi - x$.

(b) Express the radius r as a function of x.

(c) Express the height h as a function of x.

(d) Express the volume V of the cone as a function of x.

66. *Industrial Costs* Dayton Power and Light, Inc., has a power plant on the Miami River where the river is 800 ft wide. To lay a new cable from the plant to a location in the city 2 mi downstream on the opposite side costs $180 per foot across the river and $100 per foot along the land.

(a) Suppose that the cable goes from the plant to a point Q on the opposite side that is x ft from the point P directly opposite the plant. Write a function $C(x)$ that gives the cost of laying the cable in terms of the distance x.

(b) Generate a table of values to determine if the least expensive location for point Q is less than 2000 ft or greater than 2000 ft from point P.

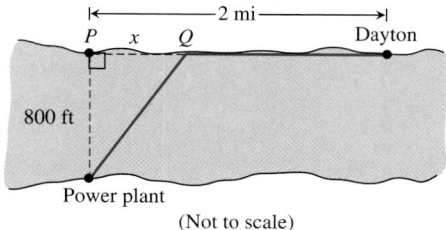

(Not to scale)

Extending the Ideas

67. Enter $y_1 = \sqrt{x}$, $y_2 = \sqrt{1-x}$ and $y_3 = y_1 + y_2$ on your grapher.

(a) Graph y_3 in $[-3, 3]$ by $[-1, 3]$.

(b) Compare the domain of the graph of y_3 with the domains of the graphs of y_1 and y_2.

(c) Replace y_3 by

$$y_1 - y_2, \quad y_2 - y_1, \quad y_1 \cdot y_2, \quad y_1/y_2, \quad \text{and} \quad y_2/y_1,$$

in turn, and repeat the comparison of part (b).

(d) Based on your observations in (b) and (c), what would you conjecture about the domains of sums, differences, products, and quotients of functions?

68. *Even and Odd Functions*

(a) Must the product of two even functions always be even? Give reasons for your answer.

(b) Can anything be said about the product of two odd functions? Give reasons for your answer.

1.3 Exponential Functions

Exponential Growth • Exponential Decay • Applications •
The Number *e*

Exponential Growth

Table 1.6 shows the growth of $100 invested in 1996 at an interest rate of 5.5%, compounded annually.

Table 1.6 Savings Account Growth		
Year	Amount (dollars)	Increase (dollars)
1996	100	
		5.50
1997	$100(1.055) = 105.50$	
		5.80
1998	$100(1.055)^2 = 111.30$	
		6.12
1999	$100(1.055)^3 = 117.42$	
		6.46
2000	$100(1.055)^4 = 123.88$	

After the first year, the value of the account is always 1.055 times its value in the previous year. After *n* years, the value is $y = 100 \cdot (1.055)^n$.

Compound interest provides an example of *exponential growth* and is modeled by a function of the form $y = P \cdot a^x$, where P is the initial investment and a is equal to 1 plus the interest rate expressed as a decimal.

The equation $y = P \cdot a^x$, $a > 0$, $a \neq 1$, identifies a family of functions called *exponential functions*.

Exploration 1 **Exponential Functions**

1. Graph the function $y = a^x$ for $a = 2, 3, 5$, in a $[-5, 5]$ by $[-2, 5]$ viewing window.

2. For what values of x is it true that $2^x < 3^x < 5^x$?

3. For what values of x is it true that $2^x > 3^x > 5^x$?

4. For what values of x is it true that $2^x = 3^x = 5^x$?

5. Graph the function $y = (1/a)^x = a^{-x}$ for $a = 2, 3, 5$.

6. Repeat parts 2–4 for the functions in part 5.

Definition **Exponential Function**

Let a be a positive real number other than 1. The function

$$f(x) = a^x$$

is the **exponential function with base *a*.**

$y = 2^x$

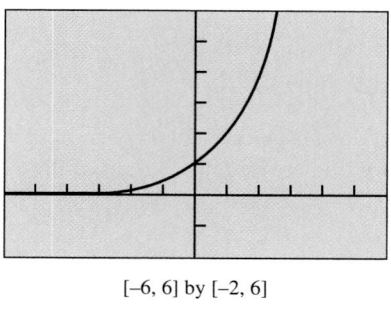

[–6, 6] by [–2, 6]

(a)

$y = 2^{-x}$

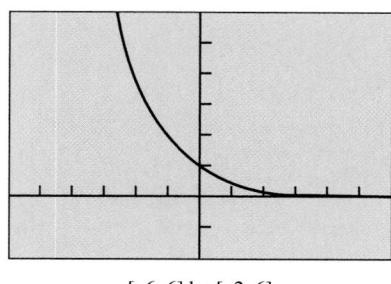

[–6, 6] by [–2, 6]

(b)

Figure 1.22 A graph of (a) $y = 2^x$ and (b) $y = 2^{-x}$.

The domain of $f(x) = a^x$ is $(-\infty, \infty)$ and the range is $(0, \infty)$. If $a > 1$, the graph of f looks like the graph of $y = 2^x$ in Figure 1.22a. If $0 < a < 1$, the graph of f looks like the graph of $y = 2^{-x}$ in Figure 1.22b.

Exponential functions obey the rules for exponents.

Rules for Exponents

If $a > 0$ and $b > 0$, the following hold for all real numbers x and y.

1. $a^x \cdot a^y = a^{x+y}$

2. $\dfrac{a^x}{a^y} = a^{x-y}$

3. $(a^x)^y = (a^y)^x = a^{xy}$

4. $a^x \cdot b^x = (ab)^x$

5. $\left(\dfrac{a}{b}\right)^x = \dfrac{a^x}{b^x}$

Population growth can sometimes be modeled with an exponential function. In Section 1.1 we gave some values for the population of the world in Table 1.1, which we repeat here as Table 1.7. We have also divided the population in one year by the population in the previous year to get an idea of how the population is growing. These ratios are in the third column.

Table 1.7 World Population

Year	Population (millions)	Ratio
1986	4936	
1987	5023	$5023/4936 \approx 1.0176$
1988	5111	$5111/5023 \approx 1.0175$
1989	5201	$5201/5111 \approx 1.0176$
1990	5329	$5329/5201 \approx 1.0246$
1991	5422	$5422/5329 \approx 1.0175$

Source: Statistical Office of the United Nations, *Monthly Bulletin of Statistics,* 1991.

Example 1   PREDICTING WORLD POPULATION

Use the data in Table 1.7 and an exponential model to predict the population of the world in the year 2010.

Solution   Based on the third column of Table 1.7, we might be willing to conjecture that the population of the world in any year is about 1.018 times the population in the previous year.

If we start with the population in 1986, then according to the model the population (in millions) in 2010 would be about

$$4936(1.018)^{24} \approx 7573.9,$$

or about 7.6 billion people.

Exponential Decay

Exponential functions can also model phenomena that produce a decrease over time, such as happens with radioactive decay. The **half-life** of a radioactive substance is the amount of time it takes for half of the substance to change from its original radioactive state to a nonradioactive state by emitting energy in the form of radiation.

Example 2   MODELING RADIOACTIVE DECAY

Suppose the half-life of a certain radioactive substance is 20 days and that there are 5 grams present initially. When will there be only 1 gram of the substance remaining?

Solution

Model

The number of grams remaining after 20 days is

$$5\left(\frac{1}{2}\right) = \frac{5}{2}.$$

The number of grams remaining after 40 days is

$$5\left(\frac{1}{2}\right)\left(\frac{1}{2}\right) = 5\left(\frac{1}{2}\right)^2 = \frac{5}{4}.$$

The function $y = 5(1/2)^{t/20}$ models the mass in grams of the radioactive substance after t days.

Solve Graphically

Figure 1.23 shows that the graphs of $y_1 = 5(1/2)^{t/20}$ and $y_2 = 1$ (for 1 gram) intersect when t is approximately 46.44.

Interpret

There will be 1 gram of the radioactive substance left after approximately 46.44 days, or about 46 days 10.5 hours.

$y = 5\left(\frac{1}{2}\right)^{t/20}, y = 1$

Intersection
X=46.438562   Y=1

[0, 80] by [−3, 5]

Figure 1.23 (Example 2)

Compound interest investments, population growth, and radioactive decay are all examples of *exponential growth and decay.*

> **Definitions Exponential Growth, Exponential Decay**
>
> The function $y = k \cdot a^x$, $k > 0$ is a model for **exponential growth** if $a > 1$, and a model for **exponential decay** if $0 < a < 1$.

Applications

Most graphers have the exponential growth and decay model $y = k \cdot a^x$ built in as an exponential regression equation. We use this feature in Example 3 to analyze the U.S. population from the data in Table 1.8.

Table 1.8 U.S. Population

Year	Population (millions)
1880	50.2
1890	63.0
1900	76.0
1910	92.0
1920	105.7
1930	122.8
1940	131.7
1950	151.3
1960	179.3
1970	203.3

Source: The Statesman's Yearbook, 129th ed. (London: The Macmillan Press, Ltd., 1992).

Example 3 PREDICTING THE U.S. POPULATION

Use the population data in Table 1.8 to estimate the population for the year 1990. Compare the result with the actual 1990 population of approximately 250 million.

Solution

Model

Let $x = 0$ represent 1880, $x = 1$ represent 1890, and so on. We enter the data into the grapher and find the exponential regression equation to be

$$f(x) = 55.05(1.16063^x).$$

Figure 1.24 shows the graph of f superimposed on the scatter plot of the data.

Solve Graphically

The year 1990 is represented by $x = 11$. Reading from the curve, we find

$$f(11) \approx 283.4.$$

The exponential model estimates the 1990 population to be 283.4 million, an overestimate of approximately 33 million, or about 13%.

$y = 55.05(1.16063^x)$

X=11   Y=283.40396

[−1, 15] by [−50, 300]

Figure 1.24 (Example 3)

Example 4 INTERPRETING EXPONENTIAL REGRESSION

What *annual* rate of growth can we infer from the exponential regression equation in Example 3?

Solution Let r be the annual rate of growth of the U.S. population, expressed as a decimal. Because the time increments we used were 10-year intervals, we have

$$(1 + r)^{10} \approx 1.16063$$
$$r \approx \sqrt[10]{1.16063} - 1$$
$$r \approx 0.015.$$

The annual rate of growth is about 1.5%. This is surprisingly close to the 1.8% rate in Example 1.

The Number e

Many natural, physical, and economic phenomena are best modeled by an exponential function whose base is the famous number e, which is 2.718281828 to nine decimal places. We can define e to be the number that the function $f(x) = (1 + 1/x)^x$ approaches as x approaches infinity. The graph and table in Figure 1.25 strongly suggest that such a number exists.

The exponential functions $y = e^x$ and $y = e^{-x}$ are frequently used as models of exponential growth or decay. For example, interest **compounded continuously** uses the model $y = P \cdot e^{rt}$, where P is the initial investment, r is the interest rate as a decimal, and t is time in years.

$y = (1 + 1/x)^x$

[−10, 10] by [−5, 10]

X	Y₁
1000	2.7169
2000	2.7176
3000	2.7178
4000	2.7179
5000	2.718
6000	2.7181
7000	2.7181

Y₁ = (1+1/X)^X

Figure 1.25 A graph and table of values for $f(x) = (1 + 1/x)^x$ both suggest that as $x \to \infty$, $f(x) \to e \approx 2.718$.

Quick Review 1.3

In Exercises 1–3, evaluate the expression. Round your answers to 3 decimal places.

1. $5^{2/3}$

2. $3^{\sqrt{2}}$

3. $3^{-1.5}$

In Exercises 4–6, solve the equation. Round your answers to 4 decimal places.

4. $x^3 = 17$

5. $x^5 = 24$

6. $x^{10} = 1.4567$

In Exercises 7 and 8, find the value of investing P dollars for n years with the interest rate r compounded annually.

7. $P = \$500, \quad r = 4.75\%, \quad n = 5$ years

8. $P = \$1000, \quad r = 6.3\%, \quad n = 3$ years

In Exercises 9 and 10, simplify the exponential expression.

9. $\dfrac{(x^{-3}y^2)^2}{(x^4y^3)^3}$

10. $\left(\dfrac{a^3b^{-2}}{c^4}\right)^2 \left(\dfrac{a^4c^{-2}}{b^3}\right)^{-1}$

Section 1.3 Exercises

In Exercises 1–6 , match the function with its graph. Try to do it without using your grapher.

1. $y = 2^x$

2. $y = 3^{-x}$

3. $y = -3^{-x}$

4. $y = -0.5^{-x}$

5. $y = 2^{-x} - 2$

6. $y = 1.5^x - 2$

(a)

(b)

(c)

(d)

(e)

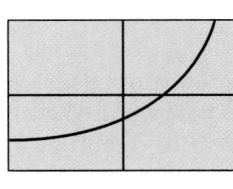

(f)

In Exercises 7–10, graph the function. State its domain, range, and intercepts.

7. $y = -2^x + 3$

8. $y = e^x + 3$

9. $y = 3 \cdot e^{-x} - 2$

10. $y = -2^{-x} - 1$

In Exercises 11–14, rewrite the exponential expression to have the indicated base.

11. 9^{2x}, base 3

12. 16^{3x}, base 2

13. $(1/8)^{2x}$, base 2

14. $(1/27)^x$, base 3

In Exercises 15–18, use graphs to solve the equations.

15. $2^x = 5$

16. $e^x = 4$

17. $3^x - 0.5 = 0$

18. $3 - 2^{-x} = 0$

In Exercises 19–22, copy and *work in groups of two or three* to complete the table for the function.

19. $y = 2x - 3$

x	y	Change (Δy)
1	?	
2	?	?
3	?	?
4	?	?

20. $y = -3x + 4$

x	y	Change (Δy)
1	?	
2	?	?
3	?	?
4	?	?

21. $y = x^2$

x	y	Change (Δy)
1	?	
2	?	?
3	?	?
4	?	?

22. $y = 3e^x$

x	y	Ratio (y_i/y_{i-1})
1	?	
2	?	?
3	?	?
4	?	?

In Exercises 23–34, use an exponential model to solve the problem.

23. *Population Growth* The population of Knoxville is 500,000 and is increasing at the rate of 3.75% each year. Approximately when will the population reach 1 million?

24. *Population Growth* The population of Silver Run in the year 1890 was 6250. Assume the population increased at a rate of 2.75% per year.

(a) Estimate the population in 1915 and 1940.

(b) Approximately when did the population reach 50,000?

25. *Radioactive Decay* The half-life of phosphorus-32 is about 14 days. There are 6.6 grams present initially.

(a) Express the amount of phosphorus-32 remaining as a function of time t.

(b) When will there be 1 gram remaining?

26. *Finding Time* If John invests $2300 in a savings account with a 6% interest rate compounded annually, how long will it take until John's account has a balance of $4150?

27. *Doubling Your Money* Determine how much time is required for an investment to double in value if interest is earned at the rate of 6.25% compounded annually.

28. *Doubling Your Money* Determine how much time is required for an investment to double in value if interest is earned at the rate of 6.25% compounded monthly.

29. *Doubling Your Money* Determine how much time is required for an investment to double in value if interest is earned at the rate of 6.25% compounded continuously.

30. *Tripling Your Money* Determine how much time is required for an investment to triple in value if interest is earned at the rate of 5.75% compounded annually.

31. *Tripling Your Money* Determine how much time is required for an investment to triple in value if interest is earned at the rate of 5.75% compounded daily.

32. *Tripling Your Money* Determine how much time is required for an investment to triple in value if interest is earned at the rate of 5.75% compounded continuously.

33. *Cholera Bacteria* Suppose that a colony of bacteria starts with 1 bacterium and doubles in number every half hour. How many bacteria will the colony contain at the end of 24 h?

34. *Eliminating a Disease* Suppose that in any given year, the number of cases of a disease is reduced by 20%. If there are 10,000 cases today, how many years will it take

(a) to reduce the number of cases to 1000?

(b) to eliminate the disease; that is, to reduce the number of cases to less than 1?

35. Writing to Learn Explain how the change Δy is related to the slopes of the lines in Exercises 19 and 20. If the changes in x are constant for a linear function, what would you conclude about the corresponding changes in y?

36. *Bacteria Growth* The number of bacteria in a petri dish culture after t hours is

$$B = 100e^{0.693t}.$$

(a) What was the initial number of bacteria present?

(b) How many bacteria are present after 6 hours?

(c) Approximately when will the number of bacteria be 200? Estimate the doubling time of the bacteria.

37. Table 1.9 gives some data about the population of Mexico.

Table 1.9 Population of Mexico

Year	Population (millions)
1950	25.8
1960	34.9
1970	48.2
1980	66.8
1990	81.1

Source: The Statesman's Yearbook, 129th ed. (London: The Macmillan Press, Ltd., 1992).

(a) Let $x = 0$ represent 1900, $x = 1$ represent 1901, and so forth. Find an exponential regression equation for the data and superimpose its graph on a scatter plot of the data.

(b) Use the exponential regression equation to estimate the population of Mexico in 1900. How close is the estimate to the actual population in 1900 of 13,607,272?

(c) Use the exponential regression equation to estimate the annual rate of growth of the population of Mexico.

38. Table 1.10 gives population data for South Africa.

Table 1.10 Population of South Africa

Year	Population (millions)
1904	5.2
1911	6.0
1921	6.9
1936	9.6
1946	11.4
1951	12.7
1960	16.0
1970	18.3
1980	20.6

Source: The Statesman's Yearbook, 129th ed. (London: The Macmillan Press, Ltd., 1992).

(a) Let $x = 0$ represent 1900, $x = 1$ represent 1901, and so forth. Find an exponential regression equation for the data and superimpose its graph on a scatter plot of the data.

(b) Use the exponential regression equation to estimate the population of South Africa in 1990.

(c) Use the exponential regression equation to estimate the annual rate of growth of the population of South Africa.

39. World Population *(continuation of Example 1)* Use 1.018 and the population in 1991 to estimate the population of the world in the year 2010.

Exploration

40. Let $y_1 = x^2$ and $y_2 = 2^x$.

(a) Graph y_1 and y_2 in $[-5, 5]$ by $[-2, 10]$. How many times do you think the two graphs cross?

(b) Compare the corresponding changes in y_1 and y_2 as x changes from 1 to 2, 2 to 3, and so on. How large must x be for the changes in y_2 to overtake the changes in y_1?

(c) Solve for x: $x^2 = 2^x$.

(d) Solve for x: $x^2 < 2^x$.

Extending the Ideas

In Exercises 41 and 42, assume that the graph of the exponential function $f(x) = k \cdot a^x$ passes through the two points. Find the values of a and k.

41. $(1, 4.5), (-1, 0.5)$ **42.** $(1, 1.5), (-1, 6)$

1.4 Parametric Equations

Relations • Circles • Ellipses • Lines and Other Curves

Relations

A **relation** is a set of ordered pairs (x, y) of real numbers. The **graph of a relation** is the set of points in the plane that correspond to the ordered pairs of the relation. If x and y are *functions* of a third variable t, called a *parameter*, then we can use the *parametric mode* of a grapher to obtain a graph of the relation.

$x = \sqrt{t}, y = t$

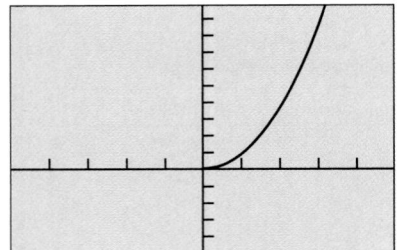

[-5, 5] by [-5, 10]

Figure 1.26 You must choose a *smallest* and *largest* value for t in parametric mode. Here we used 0 and 10, respectively. (Example 1)

Example 1   GRAPHING HALF A PARABOLA

Describe the graph of the relation determined by

$$x = \sqrt{t}, \quad y = t, \quad t \geq 0.$$

Solution Set $x_1 = \sqrt{t}$, $y_1 = t$, and use the parametric mode of the grapher to draw the graph in Figure 1.26. The graph appears to be the right half of the parabola $y = x^2$. Notice that there is no information about t on the graph itself.

Confirm Algebraically
Both x and y will be greater than or equal to zero because $t \geq 0$. Eliminating t we find that for every value of t,

$$y = t = (\sqrt{t})^2 = x^2.$$

Thus, the relation is the function $y = x^2$, $x \geq 0$.

Definitions **Parametric Curve, Parametric Equations**

If x and y are given as functions

$$x = f(t), \quad y = g(t)$$

over an interval of t-values, then the set of points $(x, y) = (f(t), g(t))$ defined by these equations is a **parametric curve**. The equations are **parametric equations** for the curve.

The variable t is a **parameter** for the curve and its domain I is the **parameter interval.** If I is a closed interval, $a \leq t \leq b$, the point $(f(a), g(a))$ is the **initial point** of the curve and the point $(f(b), g(b))$ is the **terminal point** of the curve. When we give parametric equations and a parameter interval for a curve, we say that we have **parametrized** the curve. The equations and interval constitute a **parametrization** of the curve.

In Example 1, the parameter interval is $[0, \infty)$, so $(0, 0)$ is the initial point and there is no terminal point.

A grapher can draw a parametrized curve only over a closed interval, so the portion it draws has endpoints even when the curve being graphed does not. Keep this in mind when you graph.

Circles

In applications, t often denotes time, an angle, or the distance a particle has traveled along its path from its starting point. In fact, parametric graphing can be used to simulate the motion of the particle.

Exploration 1 **Parametrizing Circles**

Let $x = a \cos t$ and $y = a \sin t$.

1. Let $a = 1$, 2, or 3 and graph the parametric equations in a *square viewing window* using the parameter interval $[0, 2\pi]$. How does changing a affect this graph?

2. Let $a = 2$ and graph the parametric equations using the following parameter intervals: $[0, \pi/2]$, $[0, \pi]$, $[0, 3\pi/2]$, $[2\pi, 4\pi]$, and $[0, 4\pi]$. Describe the role of the length of the parameter interval.

3. Let $a = 3$ and graph the parametric equations using the intervals $[\pi/2, 3\pi/2]$, $[\pi, 2\pi]$, $[3\pi/2, 3\pi]$, and $[\pi, 5\pi]$. What are the initial point and terminal point in each case?

4. Graph $x = 2 \cos (-t)$ and $y = 2 \sin (-t)$ using the parameter intervals $[0, 2\pi]$, $[\pi, 3\pi]$, and $[\pi/2, 3\pi/2]$. In each case, describe how the graph is traced.

For $x = a \cos t$ and $y = a \sin t$, we have

$$x^2 + y^2 = a^2 \cos^2 t + a^2 \sin^2 t = a^2(\cos^2 t + \sin^2 t) = a^2(1) = a^2,$$

using the identity $\cos^2 t + \sin^2 t = 1$. Thus, the curves in Exploration 1 were either circles or portions of circles, each with center at the origin.

Ellipses

Parametrizations of ellipses are similar to parametrizations of circles. Recall that the standard form of an ellipse centered at $(0, 0)$ is

$$\frac{x^2}{a^2} + \frac{y^2}{b^2} = 1.$$

Example 2 GRAPHING AN ELLIPSE

Graph the parametric curve $x = 3 \cos t, \quad y = 4 \sin t, \quad 0 \le t \le 2\pi$.

Find a Cartesian equation for a curve that contains the parametric curve. What portion of the graph of the Cartesian equation is traced by the parametric curve?

Solution Figure 1.27 suggests that the curve is an ellipse. The Cartesian equation is

$$\left(\frac{x}{3}\right)^2 + \left(\frac{y}{4}\right)^2 = \cos^2 t + \sin^2 t = 1,$$

so the parametrized curve lies along an ellipse with major axis endpoints $(0, \pm 4)$ and minor axis endpoints $(\pm 3, 0)$. As t increases from 0 to 2π, the point $(x, y) = (3 \cos t, 4 \sin t)$ starts at $(3, 0)$ and traces the entire ellipse once counterclockwise.

$x = 3 \cos t, y = 4 \sin t$

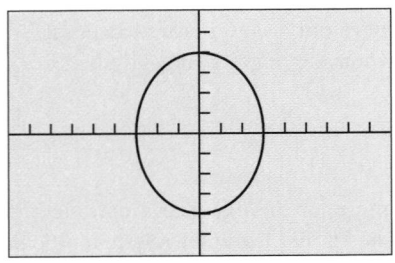

[–9, 9] by [–6, 6]

Figure 1.27 A graph of the parametric equations $x = 3 \cos t, y = 4 \sin t$ for $0 \le t \le 2\pi$. (Example 2)

Exploration 2   Parametrizing Ellipses

Let $x = a \cos t$ and $y = b \sin t$.

1. Let $a = 2$ and $b = 3$. Then graph using the parameter interval $[0, 2\pi]$. Repeat, changing b to 4, 5, and 6.

2. Let $a = 3$ and $b = 4$. Then graph using the parameter interval $[0, 2\pi]$. Repeat, changing a to 5, 6, and 7.

3. Based on parts 1 and 2, how do you identify the axis that contains the major axis of the ellipse? the minor axis?

4. Let $a = 4$ and $b = 3$. Then graph using the parameter intervals $[0, \pi/2]$, $[0, \pi]$, $[0, 3\pi/2]$, and $[0, 4\pi]$. Describe the role of the length of the parameter interval.

5. Graph $x = 5 \cos(-t)$ and $y = 2 \sin(-t)$ using the parameter intervals $[0, 2\pi]$, $[\pi, 3\pi]$, and $[\pi/2, 3\pi/2]$. Describe how the graph is traced. What are the initial point and terminal point in each case?

For $x = a \cos t$ and $y = b \sin t$, we have $(x/a)^2 + (y/b)^2 = \cos^2 t + \sin^2 t = 1$. Thus, the curves in Exploration 2 were either ellipses or portions of ellipses, each with center at the origin.

In the exercises you will see how to graph hyperbolas parametrically.

Lines and Other Curves

Lines, line segments, and many other curves can be defined parametrically.

Example 3   GRAPHING A LINE SEGMENT

Draw and identify the graph of the parametric curve determined by

$$x = 3t, \quad y = 2 - 2t, \quad 0 \le t \le 1.$$

Solution The graph (Figure 1.28) appears to be a line segment with endpoints $(0, 2)$ and $(3, 0)$.

$x = 3t, y = 2 - 2t$

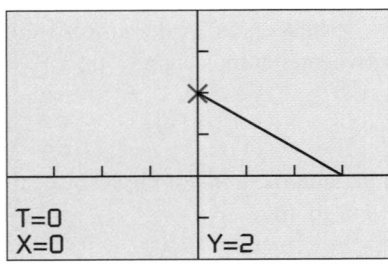

T=0
X=0   Y=2

[–4, 4] by [–2, 4]

Figure 1.28 The graph of the line segment $x = 3t, y = 2 - 2t, 0 \le t \le 1$, with trace on the initial point $(0, 2)$. (Example 3)

$x = 2 \cot t,\ y = 2 \sin^2 t$

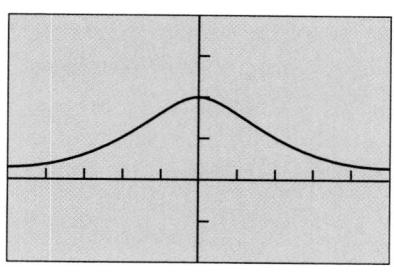

[−5, 5] by [−2, 4]

Figure 1.29 The witch of Agnesi. (Exploration 3)

Maria Agnesi (1718–1799)

The first text to include differential and integral calculus along with analytic geometry, infinite series, and differential equations was written in the 1740s by the Italian mathematician Maria Gaetana Agnesi. Agnesi, a gifted scholar and linguist whose Latin essay defending higher education for women was published when she was only nine years old, was a well-published scientist by age 20, and an honorary faculty member of the University of Bologna by age 30.

Today, Agnesi is remembered chiefly for a bell-shaped curve called *the witch of Agnesi*. This name, found only in English texts, is the result of a mistranslation. Agnesi's own name for the curve was *versiera* or "turning curve." John Colson, a noted Cambridge mathematician, probably confused versiera with *avversiera,* which means "wife of the devil" and translated it into "witch."

Confirm Algebraically

When $t = 0$, the equations give $x = 0$ and $y = 2$. When $t = 1$, they give $x = 3$ and $y = 0$. When we substitute $t = x/3$ into the y equation, we obtain

$$y = 2 - 2\left(\frac{x}{3}\right) = -\frac{2}{3}x + 2.$$

Thus, the parametric curve traces the segment of the line $y = -(2/3)x + 2$ from the point $(0, 2)$ to $(3, 0)$.

If we change the parameter interval $[0, 1]$ in Example 3 to $(-\infty, \infty)$, the parametrization will trace the entire line $y = -(2/3)x + 2$.

The bell-shaped curve in Exploration 3 is the famous witch of Agnesi. You will find more information about this curve in Exercise 41.

Exploration 3 **Graphing the Witch of Agnesi**

The witch of Agnesi is the curve

$$x = 2 \cot t, \quad y = 2 \sin^2 t, \quad 0 < t < \pi.$$

1. Draw the curve using the window in Figure 1.29. What did you choose as a closed parameter interval for your grapher? In what direction is the curve traced? How far to the left and right of the origin do you think the curve extends?

2. Graph the same parametric equations using the parameter intervals $(-\pi/2, \pi/2)$, $(0, \pi/2)$, and $(\pi/2, \pi)$. In each case, describe the curve you see and the direction in which it is traced by your grapher.

3. What happens if you replace $x = 2 \cot t$ by $x = -2 \cot t$ in the original parametrization? What happens if you use $x = 2 \cot (\pi - t)$?

Example 4 PARAMETRIZING A LINE SEGMENT

Find a parametrization for the line segment with endpoints $(-2, 1)$ and $(3, 5)$.

Solution Using $(-2, 1)$ we create the parametric equations

$$x = -2 + at, \quad y = 1 + bt.$$

These represent a line, as we can see by solving each equation for t and equating to obtain

$$\frac{x + 2}{a} = \frac{y - 1}{b}.$$

This line goes through the point $(-2, 1)$ when $t = 0$. We determine a and b so that the line goes through $(3, 5)$ when $t = 1$.

$$3 = -2 + a \quad \Rightarrow \quad a = 5 \quad \text{$x = 3$ when $t = 1$.}$$

$$5 = 1 + b \quad \Rightarrow \quad b = 4 \quad \text{$y = 5$ when $t = 1$.}$$

Therefore,

$$x = -2 + 5t, \quad y = 1 + 4t, \quad 0 \le t \le 1$$

is a parametrization of the line segment with initial point $(-2, 1)$ and terminal point $(3, 5)$.

Quick Review 1.4

In Exercises 1–3, write an equation for the line.

1. the line through the points $(1, 8)$ and $(4, 3)$

2. the horizontal line through the point $(3, -4)$

3. the vertical line through the point $(2, -3)$

In Exercises 4–6, find the x- and y-intercepts of the graph of the relation.

4. $\dfrac{x^2}{9} + \dfrac{y^2}{16} = 1$

5. $\dfrac{x^2}{16} - \dfrac{y^2}{9} = 1$

6. $2y^2 = x + 1$

In Exercises 7 and 8, determine whether the given points lie on the graph of the relation.

7. $2x^2 y + y^2 = 3$

  (a) $(1, 1)$ (b) $(-1, -1)$ (c) $(1/2, -2)$

8. $9x^2 - 18x + 4y^2 = 27$

  (a) $(1, 3)$ (b) $(1, -3)$ (c) $(-1, 3)$

9. Solve for t.

  (a) $2x + 3t = -5$ (b) $3y - 2t = -1$

10. For what values of a is each equation true?

  (a) $\sqrt{a^2} = a$ (b) $\sqrt{a^2} = \pm a$ (c) $\sqrt{4a^2} = 2|a|$

Section 1.4 Exercises

In Exercises 1–4, match the parametric equations with their graph. State the approximate dimensions of the viewing window. Give a parameter interval that traces the curve exactly once.

1. $x = 3 \sin (2t), \quad y = 1.5 \cos t$

2. $x = \sin^3 t, \quad y = \cos^3 t$

3. $x = 7 \sin t - \sin (7t), \quad y = 7 \cos t - \cos (7t)$

4. $x = 12 \sin t - 3 \sin (6t), \quad y = 12 \cos t + 3 \cos (6t)$

(a)

(b)

(c)

(d)

Explorations

5. *Hyperbolas* Let $x = a \sec t$ and $y = b \tan t$.

  (a) **Writing to Learn** Let $a = 1, 2,$ or 3, $b = 1, 2,$ or 3, and graph using the parameter interval $(-\pi/2, \pi/2)$. Explain what you see, and describe the role of a and b in these parametric equations. (Caution: If you get what appear to be asymptotes, try using the approximation $[-1.57, 1.57]$ for the parameter interval.)

  (b) Let $a = 2$, $b = 3$, and graph in the parameter interval $(\pi/2, 3\pi/2)$. Explain what you see.

  (c) **Writing to Learn** Let $a = 2$, $b = 3$, and graph using the parameter interval $(-\pi/2, 3\pi/2)$. Explain why you must be careful about graphing in this interval or any interval that contains $\pm\pi/2$.

  (d) Use algebra to explain why

$$\left(\frac{x}{a}\right)^2 - \left(\frac{y}{b}\right)^2 = 1.$$

  (e) Let $x = a \tan t$ and $y = b \sec t$. Repeat (a), (b), and (d) using an appropriate version of (d).

6. *Transformations* Let $x = (2 \cos t) + h$ and $y = (2 \sin t) + k$.

  (a) **Writing to Learn** Let $k = 0$ and $h = -2, -1, 1,$ and 2, in turn. Graph using the parameter interval $[0, 2\pi]$. Describe the role of h.

  (b) **Writing to Learn** Let $h = 0$ and $k = -2, -1, 1,$ and 2, in turn. Graph using the parameter interval $[0, 2\pi]$. Describe the role of k.

  (c) Find a parametrization for the circle with radius 5 and center at $(2, -3)$.

  (d) Find a parametrization for the ellipse centered at $(-3, 4)$ with semimajor axis of length 5 parallel to the x-axis and semiminor axis of length 2 parallel to the y-axis. ∎

In Exercises 7–26, a parametrization is given for a curve.

  (a) Graph the curve. What are the initial and terminal points, if any? Indicate the direction in which the curve is traced.

  (b) Find a Cartesian equation for a curve that contains the parametrized curve. What portion of the graph of the Cartesian equation is traced by the parametrized curve?

7. $x = \cos t, \quad y = \sin t, \quad 0 \le t \le \pi$

8. $x = \sin (2\pi t), \quad y = \cos (2\pi t), \quad 0 \le t \le 1$

9. $x = \cos (\pi - t), \quad y = \sin (\pi - t), \quad 0 \le t \le \pi$

10. $x = 4 \cos t, \quad y = 2 \sin t, \quad 0 \le t \le 2\pi$

11. $x = 4 \sin t, \quad y = 2 \cos t, \quad 0 \le t \le \pi$

12. $x = 4 \sin t, \quad y = 5 \cos t, \quad 0 \le t \le 2\pi$

13. $x = 3t, \quad y = 9t^2, \quad -\infty < t < \infty$

14. $x = -\sqrt{t}, \quad y = t, \quad t \ge 0$

15. $x = t, \quad y = \sqrt{t}, \quad t \ge 0$

16. $x = (\sec^2 t) - 1, \quad y = \tan t, \quad -\pi/2 < t < \pi/2$

17. $x = -\sec t, \quad y = \tan t, \quad -\pi/2 < t < \pi/2$

18. $x = \tan t, \quad y = -2 \sec t, \quad -\pi/2 < t < \pi/2$

19. $x = 2t - 5, \quad y = 4t - 7, \quad -\infty < t < \infty$

20. $x = 1 - t, \quad y = 1 + t, \quad -\infty < t < \infty$

21. $x = t, \quad y = 1 - t, \quad 0 \le t \le 1$

22. $x = 3 - 3t, \quad y = 2t, \quad 0 \le t \le 1$

23. $x = 4 - \sqrt{t}, \quad y = \sqrt{t}, \quad 0 \le t$

24. $x = t^2, \quad y = \sqrt{4 - t^2}, \quad 0 \le t \le 2$

25. $x = \sin t, \quad y = \cos 2t, \quad -\infty < t < \infty$

26. $x = t^2 - 3, \quad y = t, \quad t \le 0$

In Exercises 27–32, find a parametrization for the curve.

27. the line segment with endpoints $(-1, -3)$ and $(4, 1)$

28. the line segment with endpoints $(-1, 3)$ and $(3, -2)$

29. the lower half of the parabola $x - 1 = y^2$

30. the left half of the parabola $y = x^2 + 2x$

31. the ray (half line) with initial point $(2, 3)$ that passes through the point $(-1, -1)$

32. the ray (half line) with initial point $(-1, 2)$ that passes through the point $(0, 0)$

In Exercises 33–36, refer to the graph of

$$x = 3 - |t|, \quad y = t - 1, \quad -5 \le t \le 5,$$

shown in the figure. *Work in groups of two or three* to find the values of t that produce the graph in the given quadrant.

33. Quadrant I   **34.** Quadrant II

35. Quadrant III   **36.** Quadrant IV

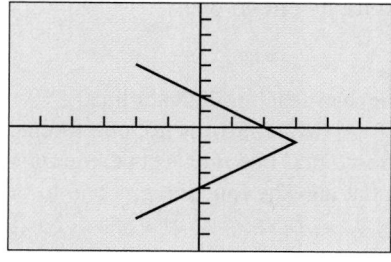

[−6, 6] by [−8, 8]

In Exercises 37 and 38, find a parametrization for the part of the graph that lies in Quadrant I.

37. $y = x^2 + 2x + 2$ **38.** $y = \sqrt{x + 3}$

39. *Circles* Find parametrizations to model the motion of a particle that starts at $(a, 0)$ and traces the circle $x^2 + y^2 = a^2$, $a > 0$, as indicated.

(a) once clockwise   **(b)** once counterclockwise

(c) twice clockwise   **(d)** twice counterclockwise

40. *Ellipses* Find parametrizations to model the motion of a particle that starts at $(-a, 0)$ and traces the ellipse

$$\left(\frac{x}{a}\right)^2 + \left(\frac{y}{b}\right)^2 = 1, \quad a > 0, b > 0,$$

as indicated.

(a) once clockwise   **(b)** once counterclockwise

(c) twice clockwise   **(d)** twice counterclockwise

Extending the Ideas

41. *The Witch of Agnesi* The bell-shaped witch of Agnesi can be constructed as follows. Start with the circle of radius 1, centered at the point $(0, 1)$ as shown in the figure.

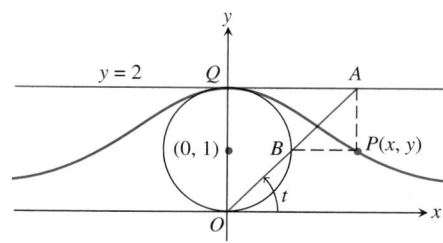

Choose a point A on the line $y = 2$, and connect it to the origin with a line segment. Call the point where the segment crosses the circle B. Let P be the point where the vertical line through A crosses the horizontal line through B. The witch is the curve traced by P as A moves along the line $y = 2$.

Find a parametrization for the witch by expressing the coordinates of P in terms of t, the radian measure of the angle that segment OA makes with the positive x-axis. The following equalities (which you may assume) will help:

(i) $x = AQ$

(ii) $y = 2 - AB \sin t$

(iii) $AB \cdot AO = (AQ)^2$

42. *Parametrizing Lines and Segments*

(a) Show that $x = x_1 + (x_2 - x_1)t, \quad y = y_1 + (y_2 - y_1)t,$ $-\infty < t < \infty$ is a parametrization for the line through the points (x_1, y_1) and (x_2, y_2).

(b) Find a parametrization for the line segment with endpoints (x_1, y_1) and (x_2, y_2).

1.5 Functions and Logarithms

One-to-One Functions • Inverses • Finding Inverses •
Logarithmic Functions • Properties of Logarithms • Applications

One-to-One Functions

As you know, a function is a rule that assigns a single value in its range to each point in its domain. Some functions assign the same output to more than one input. For example, $f(x) = x^2$ assigns the output 4 to both 2 and -2. Other functions never output a given value more than once. For example, the cubes of different numbers are always different.

If each output value of a function is associated with exactly one input value, the function is *one-to-one*.

Definition One-to-One Function

A function $f(x)$ is **one-to-one** on a domain D if $f(a) \neq f(b)$ whenever $a \neq b$.

The graph of a one-to-one function $y = f(x)$ can intersect any horizontal line at most once (the *horizontal line test*). If it intersects such a line more than once it assumes the same y-value more than once, and is therefore not one-to-one (Figure 1.30).

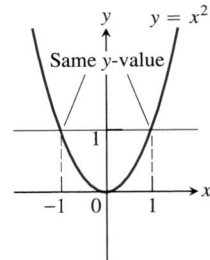

One-to-one: Graph meets each
horizontal line once.

Not one-to-one: Graph meets some
horizontal lines more than once.

Figure 1.30 Using the horizontal line test, we see that $y = x^3$ is one-to-one and $y = x^2$ is not.

$y = |x|$

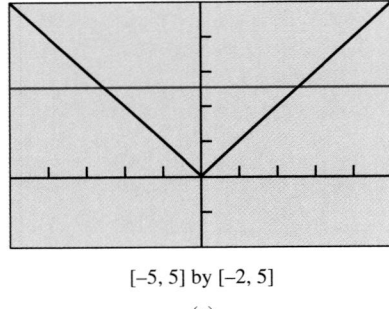

[−5, 5] by [−2, 5]

(a)

$y = \sqrt{x}$

[−5, 5] by [−2, 3]

(b)

Figure 1.31 (a) The graph of $f(x) = |x|$ and a horizontal line. (b) The graph of $g(x) = \sqrt{x}$ and a horizontal line. (Example 1)

Example 1 USING THE HORIZONTAL LINE TEST

Determine whether the functions are one-to-one.

$$\textbf{(a)}\, f(x) = |x| \qquad \textbf{(b)}\, g(x) = \sqrt{x}$$

Solution As Figure 1.31a suggests, each horizontal line $y = c,\, c > 0$, intersects the graph of $f(x) = |x|$ twice. So f is not one-to-one. As Figure 1.31b suggests, each horizontal line intersects the graph of $g(x) = \sqrt{x}$ either once or not at all. The function g is one-to-one.

Inverses

Since each output of a one-to-one function comes from just one input, a one-to-one function can be reversed to send outputs back to the inputs from which they came. The function defined by reversing a one-to-one function f is the **inverse of f.** The functions in Tables 1.11 and 1.12 are inverses of one another. The symbol for the inverse of f is f^{-1}, read "f inverse." The -1 in f^{-1} is not an exponent; $f^{-1}(x)$ does not mean $1/f(x)$.

Table 1.11 Rental Charge versus Time		Table 1.12 Time versus Rental Charge	
Time x (hours)	Charge y (dollars)	Charge x (dollars)	Time y (hours)
1	5.00	5.00	1
2	7.50	7.50	2
3	10.00	10.00	3
4	12.50	12.50	4
5	15.00	15.00	5
6	17.50	17.50	6

As Tables 1.11 and 1.12 suggest, composing a function with its inverse in either order sends each output back to the input from which it came. In other words, the result of composing a function and its inverse in either order is the **identity function,** the function that assigns each number to itself. This gives a way to test whether two functions f and g are inverses of one another. Compute $f \circ g$ and $g \circ f$. If $(f \circ g)(x) = (g \circ f)(x) = x$, then f and g are inverses of one another; otherwise they are not. The functions $f(x) = x^3$ and $g(x) = x^{1/3}$ are inverses of one another because $(x^3)^{1/3} = x$ and $(x^{1/3})^3 = x$ for every number x.

Exploration 1 **Testing For Inverses Graphically**

For each of the function pairs below,

(a) Graph f and g together in a square window.
(b) Graph $f \circ g$.
(c) Graph $g \circ f$.

What can you conclude from the graphs?

1. $f(x) = x^3$, $g(x) = x^{1/3}$
2. $f(x) = x$, $g(x) = 1/x$
3. $f(x) = 3x$, $g(x) = x/3$
4. $f(x) = e^x$, $g(x) = \ln x$

Finding Inverses

How do we find the graph of the inverse of a function? Suppose, for example, that the function is the one pictured in Figure 1.32a. To read the graph, we start at the point x on the x-axis, go up to the graph, and then move over to the *y*-axis to read the value of *y*. If we start with *y* and want to find the *x* from which it came, we reverse the process (Figure 1.32b).

The graph of *f* is already the graph of f^{-1}, although the latter graph is not drawn in the usual way with the domain axis horizontal and the range axis vertical. For f^{-1}, the input-output pairs are reversed. To display the graph of f^{-1} in the usual way, we have to reverse the pairs by reflecting the graph across the 45° line $y = x$ (Figure 1.32c) and interchanging the letters *x* and *y* (Figure 1.32d). This puts the independent variable, now called *x*, on the horizontal axis and the dependent variable, now called *y*, on the vertical axis.

The fact that the graphs of *f* and f^{-1} are reflections of each other across the line $y = x$ is to be expected because the input-output pairs (a, b) of *f* have been reversed to produce the input-output pairs (b, a) of f^{-1}.

The pictures in Figure 1.32 tell us how to express f^{-1} as a function of *x* algebraically.

Writing f^{-1} as a Function of x

1. Solve the equation $y = f(x)$ for *x* in terms of *y*.

2. Interchange *x* and *y*. The resulting formula will be $y = f^{-1}(x)$.

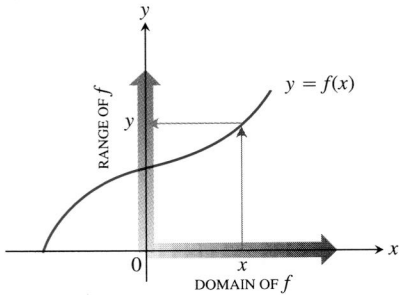

(a) To find the value of *f* at *x*, we start at *x*, go up to the curve, and then over to the *y*-axis.

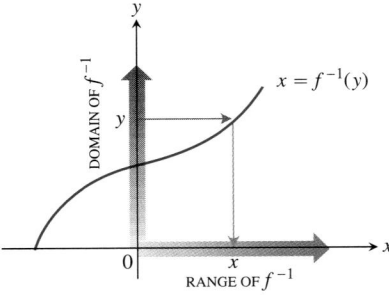

(b) The graph of *f* is also the graph of f^{-1}. To find the *x* that gave *y*, we start at *y* and go over to the curve and down to the *x*-axis. The domain of f^{-1} is the range of *f*. The range of f^{-1} is the domain of *f*.

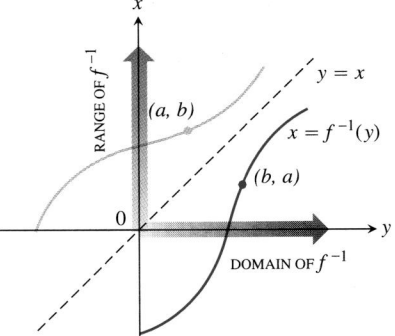

(c) To draw the graph of f^{-1} in the usual way, we reflect the system across the line $y = x$.

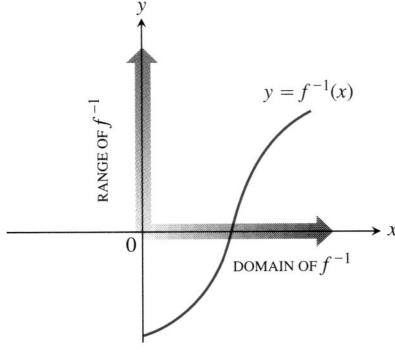

(d) Then we interchange the letters *x* and *y*. We now have a normal-looking graph of f^{-1} as a function of *x*.

Figure 1.32 The graph of $y = f^{-1}(x)$.

Example 2 FINDING THE INVERSE FUNCTION

Show that the function $y = f(x) = -2x + 4$ is one-to-one and find its inverse function.

Solution Every horizontal line intersects the graph of f exactly once, so f is one-to-one and has an inverse.

Step 1:

Solve for x in terms of y:
$$y = -2x + 4$$
$$x = -\frac{1}{2}y + 2$$

Step 2:

Interchange x and y:
$$y = -\frac{1}{2}x + 2$$

The inverse of the function $f(x) = -2x + 4$ is the function $f^{-1}(x) = -(1/2)x + 2$. We can verify that both composites are the identity function.

$$f^{-1}(f(x)) = -\frac{1}{2}(-2x + 4) + 2 = x - 2 + 2 = x$$
$$f(f^{-1}(x)) = -2(-\frac{1}{2}x + 2) + 4 = x - 4 + 4 = x$$

We can use parametric graphing to graph the inverse of a function without finding an explicit rule for the inverse, as illustrated in Example 3.

Graphing $y = f(x)$ and $y = f^{-1}(x)$ Parametrically

We can graph any function $y = f(x)$ as
$$x_1 = t, \quad y_1 = f(t).$$
Interchanging t and $f(t)$ produces parametric equations for the inverse:
$$x_2 = f(t), \quad y_2 = t.$$

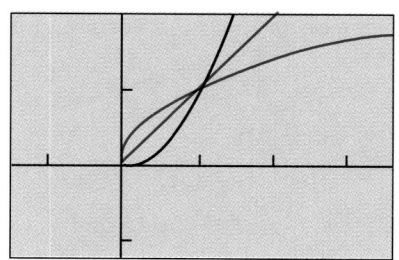

[–1.5, 3.5] by [–1, 2]

Figure 1.33 The graphs of f and f^{-1} are reflections of each other across the line $y = x$. (Example 3)

Example 3 GRAPHING THE INVERSE PARAMETRICALLY

(a) Graph the one-to-one function $f(x) = x^2$, $x \geq 0$, together with its inverse and the line $y = x$, $x \geq 0$.

(b) Express the inverse of f as a function of x.

Solution

(a) We can graph the three functions parametrically as follows:

Graph of f: $x_1 = t, \quad y_1 = t^2, \quad t \geq 0$
Graph of f^{-1}: $x_2 = t^2, \quad y_2 = t$
Graph of $y = x$: $x_3 = t, \quad y_3 = t$

Figure 1.33 shows the three graphs.

(b) Next we find a formula for $f^{-1}(x)$.

Step 1:

Solve for x in terms of y.
$$y = x^2$$
$$\sqrt{y} = \sqrt{x^2}$$
$$\sqrt{y} = x \qquad \text{Because } x \geq 0$$

Step 2:

Interchange x and y.
$$\sqrt{x} = y$$

Thus, $f^{-1}(x) = \sqrt{x}$.

Logarithmic Functions

If a is any positive real number other than 1, the base a exponential function $f(x) = a^x$ is one-to-one. It therefore has an inverse. Its inverse is called the *base a logarithm function.*

Definition Base *a* Logarithm Function

The **base *a* logarithm function** $y = \log_a x$ is the inverse of the base *a* exponential function $y = a^x$ $(a > 0, a \neq 1)$.

The domain of $\log_a x$ is $(0, \infty)$, the range of a^x. The range of $\log_a x$ is $(-\infty, \infty)$, the domain of a^x.

Because we have no technique for solving for x in terms of y in the equation $y = a^x$, we do not have an explicit formula for the logarithm function as a function of x. However, the graph of $y = \log_a x$ can be obtained by reflecting the graph of $y = a^x$ across the line $y = x$, or by using parametric graphing (Figure 1.34).

Logarithms with base e and base 10 are so important in applications that calculators have special keys for them. They also have their own special notation and names:

$$\log_e x = \ln x,$$

$$\log_{10} x = \log x.$$

The function $y = \ln x$ is called the **natural logarithm function** and $y = \log x$ is often called the **common logarithm function.**

[–6, 6] by [–4, 4]

Figure 1.34 The graphs of $y = 2^x$ $(x_1 = t, y_1 = 2^t)$, its inverse $y = \log_2 x$ $(x_2 = 2^t, y_2 = t)$, and $y = x$ $(x_3 = t, y_3 = t)$.

Properties of Logarithms

Because a^x and $\log_a x$ are inverses of each other, composing them in either order gives the identity function. This gives two useful properties.

Inverse Properties for a^x and $\log_a x$

1. **Base *a*:** $a^{\log_a x} = x$, $\log_a a^x = x$, $a > 1, x > 0$
2. **Base *e*:** $e^{\ln x} = x$, $\ln e^x = x$, $x > 0$

These properties help us with the solution of equations that contain logarithms and exponential functions.

Example 4   USING THE INVERSE PROPERTIES

Solve for x: **(a)** $\ln x = 3t + 5$ **(b)** $e^{2x} = 10$

Solution

(a) $\ln x = 3t + 5$

$\quad e^{\ln x} = e^{3t+5}$ Exponentiate both sides.

$\quad\quad x = e^{3t+5}$ Inverse Property

(b) $e^{2x} = 10$

$\ln e^{2x} = \ln 10$ Take logarithms of both sides.

$2x = \ln 10$ Inverse Property

$x = \dfrac{1}{2}\ln 10 \approx 1.15$

The logarithm function has the following useful arithmetic properties.

Properties of Logarithms

For any real numbers $x > 0$ and $y > 0$,

1. *Product Rule:* $\log_a xy = \log_a x + \log_a y$

2. *Quotient Rule:* $\log_a \dfrac{x}{y} = \log_a x - \log_a y$

3. *Power Rule:* $\log_a x^y = y \log_a x$

Exploration 2   **Supporting the Product Rule**

Let $y_1 = \ln(ax)$, $y_2 = \ln x$, and $y_3 = y_1 - y_2$.

1. Graph y_1 and y_2 for $a = 2, 3, 4,$ and 5. How do the graphs of y_1 and y_2 appear to be related?

2. Support your finding by graphing y_3.

3. Confirm your finding algebraically.

The following formula allows us to evaluate $\log_a x$ for any base $a > 0$, $a \neq 1$, and to obtain its graph using the natural logarithm function on our grapher.

Change of Base Formula

$$\log_a x = \frac{\ln x}{\ln a}$$

$y = \dfrac{\ln x}{\ln 2}$

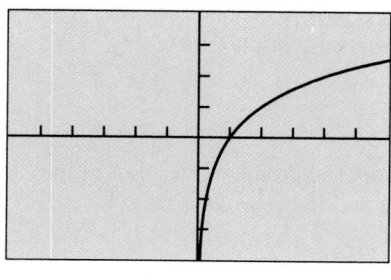

[–6, 6] by [–4, 4]

Figure 1.35 The graph of $f(x) = \log_2 x$ using $f(x) = (\ln x)/(\ln 2)$. (Example 5)

Example 5   **GRAPHING A BASE *a* LOGARITHM FUNCTION**

Graph $f(x) = \log_2 x$.

Solution   We use the change of base formula to rewrite $f(x)$.

$$f(x) = \log_2 x = \frac{\ln x}{\ln 2}.$$

Figure 1.35 gives the graph of f.

Applications

In Section 1.3 we used graphical methods to solve exponential growth and decay problems. Now we can use the properties of logarithms to solve the same problems algebraically.

Example 6   FINDING TIME

Sarah invests $1000 in an account that earns 5.25% interest compounded annually. How long will it take the account to reach $2500?

Solution

Model

The amount in the account at any time t in years is $1000(1.0525)^t$, so we need to solve the equation

$$1000(1.0525)^t = 2500.$$

Solve Algebraically

$$(1.0525)^t = 2.5 \qquad \text{Divide by 1000.}$$

$$\ln (1.0525)^t = \ln 2.5 \qquad \text{Take logarithms of both sides.}$$

$$t \ln 1.0525 = \ln 2.5 \qquad \text{Power Rule}$$

$$t = \frac{\ln 2.5}{\ln 1.0525} \approx 17.9$$

Interpret

The amount in Sarah's account will be $2500 in about 17.9 years, or about 17 years and 11 months.

Example 7   ESTIMATING OIL PRODUCTION

Table 1.13 shows the number of metric tons of oil produced by Indonesia for three different years.

Find the natural logarithm regression equation for the data in Table 1.13 and use it to estimate the number of metric tons of oil produced by Indonesia in 1982 and 2000.

Solution

Model

We let $x = 60$ represent 1960, $x = 70$ represent 1970, and so forth. We compute the natural logarithm regression equation to be

$$f(x) = -474.31 + 121.1346 \ln x.$$

Solve Graphically

Figure 1.36 shows the graph of f superimposed on the scatter plot of the data. The year 1982 is represented by 82 and the year 2000 by 100. Reading from the graph, we find

$$f(82) \approx 59.5 \quad \text{and} \quad f(100) \approx 83.5.$$

Interpret

The natural logarithm regression model estimates the 1982 oil production to have been 59.5 million metric tons, and the 2000 oil production to be 83.5 million metric tons.

Table 1.13 Indonesia's Oil Production

Year	Metric Tons (millions)
1960	20.56
1970	42.10
1990	70.10

Source: The Statesman's Yearbook, 129th ed. (London: The Macmillan Press, Ltd., 1992).

$y = -474.31 + 121.1346 \ln x$

X=82   Y=59.496438

[50, 110] by [−20, 100]

Figure 1.36 The value of f at $x = 82$ is about 59.5. (Example 7)

Quick Review 1.5

In Exercises 1–4, let $f(x) = \sqrt[3]{x - 1}$, $g(x) = x^2 + 1$, and evaluate the expression.

1. $(f \circ g)(1)$

2. $(g \circ f)(-7)$

3. $(f \circ g)(x)$

4. $(g \circ f)(x)$

In Exercises 5 and 6, choose parametric equations and a parameter interval to represent the function on the interval specified.

5. $y = \dfrac{1}{x - 1}$, $x \geq 2$

6. $y = x$, $x < -3$

In Exercises 7–10, find the points of intersection of the two curves. Round your answers to 2 decimal places.

7. $y = 2x - 3$, $y = 5$

8. $y = -3x + 5$, $y = -3$

9. (a) $y = 2^x$, $y = 3$ (b) $y = 2^x$, $y = -1$

10. (a) $y = e^{-x}$, $y = 4$ (b) $y = e^{-x}$, $y = -1$

Section 1.5 Exercises

In Exercises 1–6, determine whether the function is one-to-one.

1.

$y = 2|x|$

2.

$y = x + 1$

3.

$y = \dfrac{1}{x}$

4.
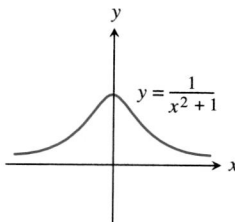
$y = \dfrac{1}{x^2 + 1}$

5.

$y = -3x^3$

6.
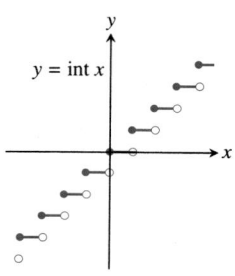
$y = \text{int } x$

In Exercises 7–12, determine whether the function has an inverse function.

7. $y = \dfrac{3}{x - 2} - 1$

8. $y = x^2 + 5x$

9. $y = x^3 - 4x + 6$

10. $y = x^3 + x$

11. $y = \ln x^2$

12. $y = 2^{3-x}$

In Exercises 13–24, find f^{-1} and verify that
$$(f \circ f^{-1})(x) = (f^{-1} \circ f)(x) = x.$$

13. $f(x) = 2x + 3$

14. $f(x) = 5 - 4x$

15. $f(x) = x^3 - 1$

16. $f(x) = x^2 + 1$, $x \geq 0$

17. $f(x) = x^2$, $x \leq 0$

18. $f(x) = x^{2/3}$, $x \geq 0$

19. $f(x) = -(x - 2)^2$, $x \leq 2$

20. $f(x) = x^2 + 2x + 1$, $x \geq -1$

21. $f(x) = \dfrac{1}{x^2}$, $x > 0$

22. $f(x) = \dfrac{1}{x^3}$

23. $f(x) = \dfrac{2x + 1}{x + 3}$

24. $f(x) = \dfrac{x + 3}{x - 2}$

In Exercises 25–32, use parametric graphing to graph f, f^{-1}, and $y = x$.

25. $f(x) = e^x$

26. $f(x) = 3^x$

27. $f(x) = 2^{-x}$

28. $f(x) = 3^{-x}$

29. $f(x) = \ln x$

30. $f(x) = \log x$

31. $f(x) = \sin^{-1} x$

32. $f(x) = \tan^{-1} x$

In Exercises 33–36, draw the graph and determine the domain and range of the function.

33. $y = 2 \ln (3 - x) - 4$

34. $y = -3 \log (x + 2) + 1$

35. $y = \log_2 (x + 1)$

36. $y = \log_3 (x - 4)$

In Exercises 37–40, solve the equation algebraically. Support your solution graphically.

37. $(1.045)^t = 2$

38. $e^{0.05t} = 3$

39. $e^x + e^{-x} = 3$

40. $2^x + 2^{-x} = 5$

In Exercises 41 and 42, solve for y.

41. $\ln y = 2t + 4$

42. $\ln (y - 1) - \ln 2 = x + \ln x$

In Exercises 43 and 44, find a formula for f^{-1} and verify that $(f \circ f^{-1})(x) = (f^{-1} \circ f)(x) = x$.

43. $f(x) = \dfrac{100}{1 + 2^{-x}}$ **44.** $f(x) = \dfrac{50}{1 + 1.1^{-x}}$

45. *Self-inverse* Prove that the function f is its own inverse.

(a) $f(x) = \sqrt{1 - x^2}, \quad x \geq 0$ (b) $f(x) = 1/x$

46. *Radioactive Decay* The half-life of a certain radioactive substance is 12 hours. There are 8 grams present initially.

(a) Express the amount of substance remaining as a function of time t.

(b) When will there be 1 gram remaining?

47. *Doubling Your Money* Determine how much time is required for a \$500 investment to double in value if interest is earned at the rate of 4.75% compounded annually.

48. *Population Growth* The population of Glenbrook is 375,000 and is increasing at the rate of 2.25% per year. Predict when the population will be 1 million.

In Exercises 49 and 50, let $x = 60$ represent 1960, $x = 70$ represent 1970, and so forth.

49. *Oil Production*

(a) Find a natural logarithm regression equation for the data in Table 1.14.

(b) Estimate the number of metric tons of oil produced by Saudi Arabia in 1975.

(c) Predict when Saudi Arabian oil production will reach 400 million metric tons.

Table 1.14 Saudi Arabian Oil Production

Year	Metric Tons (millions)
1960	61.09
1970	176.85
1990	321.93

Source: The Statesman's Yearbook, 129th ed. (London: The Macmillan Press, Ltd., 1992).

50. *Oil Production*

(a) Find a natural logarithm regression equation for the data in Table 1.15.

(b) Estimate the number of metric tons of oil produced by Canada in 1985.

(c) Predict when Canadian oil production will reach 120 metric tons.

Table 1.15 Canadian Oil Production

Year	Metric Tons (millions)
1960	27.48
1970	69.95
1990	92.24

Source: The Statesman's Yearbook, 129th ed. (London: The Macmillan Press, Ltd., 1992).

51. *Inverse Functions* We suggest that students *work in groups of two or three.*

Let $y = f(x) = mx + b, \; m \neq 0$.

(a) **Writing to Learn** Give a convincing argument that f is a one-to-one function.

(b) Find a formula for the inverse of f. How are the slopes of f and f^{-1} related?

(c) If the graphs of two functions are parallel lines with a nonzero slope, what can you say about the graphs of the inverses of the functions?

(d) If the graphs of two functions are perpendicular lines with a nonzero slope, what can you say about the graphs of the inverses of the functions?

Exploration

52. *Supporting the Quotient Rule* Let $y_1 = \ln (x/a)$, $y_2 = \ln x$, $y_3 = y_2 - y_1$, and $y_4 = e^{y_3}$.

(a) Graph y_1 and y_2 for $a = 2, 3, 4,$ and 5. How are the graphs of y_1 and y_2 related?

(b) Graph y_3 for $a = 2, 3, 4,$ and 5. Describe the graphs.

(c) Graph y_4 for $a = 2, 3, 4,$ and 5. Compare the graphs to the graph of $y = a$.

(d) Use $e^{y_3} = e^{y_2 - y_1} = a$ to solve for y_1.   ■

Extending the Ideas

53. *One-to-One Functions* If f is a one-to-one function, prove that $g(x) = -f(x)$ is also one-to-one.

54. *One-to-One Functions* If f is a one-to-one function and $f(x)$ is never zero, prove that $g(x) = 1/f(x)$ is also one-to-one.

55. *Domain and Range* Suppose that $a \neq 0$, $b \neq 1$, and $b > 0$. Determine the domain and range of the function.

(a) $y = a(b^{c-x}) + d$ (b) $y = a \log_b (x - c) + d$

56. *Inverse Functions* We suggest that students *work in groups of two or three.*

Let $f(x) = \dfrac{ax + b}{cx + d}, \quad c \neq 0, \quad ad - bc \neq 0$.

(a) **Writing to Learn** Give a convincing argument that f is one-to-one.

(b) Find a formula for the inverse of f.

(c) Find the horizontal and vertical asymptotes of f.

(d) Find the horizontal and vertical asymptotes of f^{-1}. How are they related to those of f?

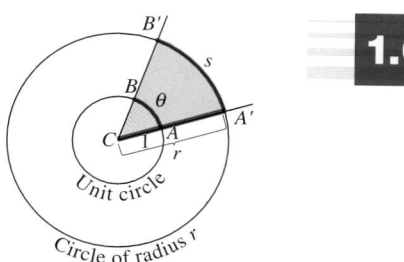

Figure 1.37 The radian measure of angle *ACB* is the length θ of arc *AB* on the unit circle centered at *C*. The value of θ can be found from any other circle, however, as the ratio s/r.

1.6 Trigonometric Functions

Radian Measure • Graphs of Trigonometric Functions • Periodicity • Even and Odd Trigonometric Functions • Transformations of Trigonometric Graphs • Applications • Inverse Trigonometric Functions

Radian Measure

The **radian measure** of the angle *ACB* at the center of the unit circle (Figure 1.37) equals the length of the arc that *ACB* cuts from the unit circle.

When an angle of measure θ is placed in *standard position* at the center of a circle of radius *r* (Figure 1.38), the six basic trigonometric functions of θ are defined as follows:

$$\textbf{sine: } \sin \theta = \frac{y}{r} \qquad \textbf{cosecant: } \csc \theta = \frac{r}{y}$$

$$\textbf{cosine: } \cos \theta = \frac{x}{r} \qquad \textbf{secant: } \sec \theta = \frac{r}{x}$$

$$\textbf{tangent: } \tan \theta = \frac{y}{x} \qquad \textbf{cotangent: } \cot \theta = \frac{x}{y}$$

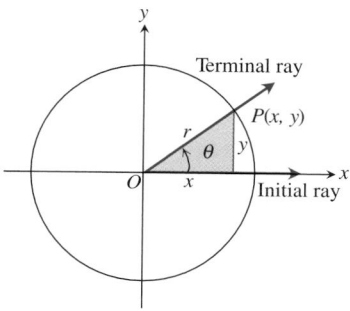

Figure 1.38 An angle θ in standard position.

Graphs of Trigonometric Functions

When we graph trigonometric functions in the coordinate plane, we usually denote the independent variable (radians) by *x* instead of θ. Figure 1.39 shows sketches of the six trigonometric functions. It is a good exercise for you to compare these with what you see in a grapher viewing window. (Some graphers have a "trig viewing window.")

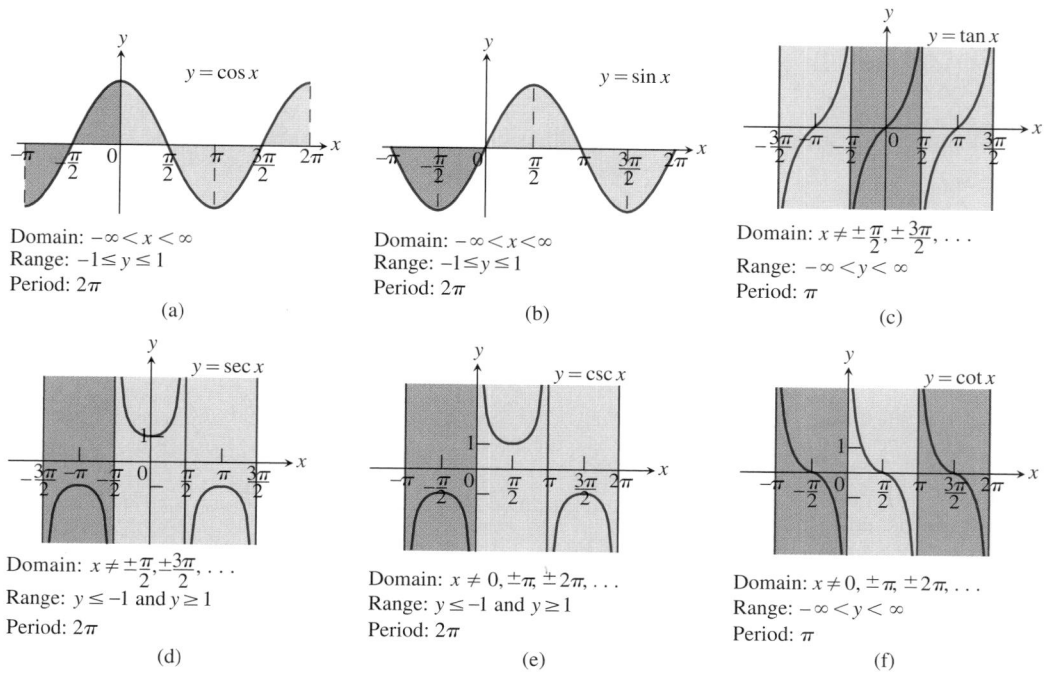

Figure 1.39 Graphs of the (a) cosine, (b) sine, (c) tangent, (d) secant, (e) cosecant, and (f) cotangent functions using radian measure.

Exploration 1 **Unwrapping Trigonometric Functions**

Set your grapher in *radian mode,* parametric mode, and *simultaneous mode* (all three). Enter the parametric equations

$$x_1 = \cos t, \quad y_1 = \sin t \quad \text{and} \quad x_2 = t, \quad y_2 = \sin t.$$

1. Graph for $0 \le t \le 2\pi$ in the window $[-1.5, 2\pi]$ by $[-2.5, 2.5]$. Describe the two curves. (You may wish to make the viewing window square.)
2. Use trace to compare the y-values of the two curves.
3. Repeat part 2 in the window $[-1.5, 4\pi]$ by $[-5, 5]$, using the parameter interval $0 \le t \le 4\pi$.
4. Let $y_2 = \cos t$. Use trace to compare the x-values of curve 1 (the unit circle) with the y-values of curve 2 using the parameter intervals $[0, 2\pi]$ and $[0, 4\pi]$.
5. Set $y_2 = \tan t$, $\csc t$, $\sec t$, and $\cot t$. Graph each in the window $[-1.5, 2\pi]$ by $[-2.5, 2.5]$ using the interval $0 \le t \le 2\pi$. How is a y-value of curve 2 related to the corresponding point on curve 1? (Use trace to explore the curves.)

A grapher can find trigonometric values, as the next exploration demonstrates.

$x = \cos t, y = \sin t$

T=1
X=.54030231   Y=.84147098

[−3, 3] by [−2, 2]

Figure 1.40 A graph of the unit circle $x^2 + y^2 = 1$ obtained in parametric mode. The values of $x = \cos 1$ and $y = \sin 1$ are shown using trace. (Exploration 2)

Exploration 2 **Finding Sines and Cosines**

Set your grapher in radian mode and enter the parametric equations

$$x_1 = \cos t, \quad y_1 = \sin t.$$

1. Graph in a square viewing window using the parameter interval $0 \le t \le 2\pi$ with t-step $= 0.1$. Use trace to read x and y for different values of t as suggested by Figure 1.40.
2. Try to match the entries in Table 1.16.

Table 1.16 Some Values of Sine and Cosine

t (radians)	$\cos t$	$\sin t$
0.5	0.87758256	0.47942554
1	0.54030231	0.84147098
2	−0.41614684	0.90929743
3.5	−0.93645669	−0.35078323
6.2	0.99654210	−0.08308940

3. Set your grapher in degree mode and use the parameter interval $0 \le t \le 360$ with t-step $= 15$. Graph and trace. For what values of t are $\cos t$ and $\sin t$ being computed?

Angle Convention: Use Radians

From now on in this book it is assumed that all angles are measured in radians unless degrees or some other unit is stated explicitly. When we talk about the angle $\pi/3$, we mean $\pi/3$ radians (which is 60°), not $\pi/3$ degrees. When you do calculus, keep your calculator in radian mode.

Periodicity

When an angle of measure θ and an angle of measure $\theta + 2\pi$ are in standard position, their terminal rays coincide. The two angles therefore have the same trigonometric function values:

$$\cos(\theta + 2\pi) = \cos\theta \quad \sin(\theta + 2\pi) = \sin\theta \quad \tan(\theta + 2\pi) = \tan\theta$$
$$\sec(\theta + 2\pi) = \sec\theta \quad \csc(\theta + 2\pi) = \csc\theta \quad \cot(\theta + 2\pi) = \cot\theta \tag{1}$$

Similarly, $\cos(\theta - 2\pi) = \cos\theta$, $\sin(\theta - 2\pi) = \sin\theta$, and so on.

We see the values of the trigonometric functions repeat at regular intervals. We describe this behavior by saying that the six basic trigonometric functions are *periodic*.

Periods of Trigonometric Functions

Period π: $\tan(x + \pi) = \tan x$
$\cot(x + \pi) = \cot x$

Period 2π: $\sin(x + 2\pi) = \sin x$
$\cos(x + 2\pi) = \cos x$
$\sec(x + 2\pi) = \sec x$
$\csc(x + 2\pi) = \csc x$

> **Definition Periodic Function, Period**
>
> A function $f(x)$ is **periodic** if there is a positive number p such that $f(x + p) = f(x)$ for every value of x. The smallest such value of p is the **period** of f.

As we can see in Figure 1.39, the functions $\cos x$, $\sin x$, $\sec x$, and $\csc x$ are periodic with period 2π. The functions $\tan x$ and $\cot x$ are periodic with period π.

The importance of periodic functions stems from the fact that much of the behavior we study in science is periodic (Figure 1.41). Brain waves and heartbeats are periodic, as are household voltage and electric current. The electromagnetic field that heats food in a microwave oven is periodic, as are cash flows in seasonal businesses and the behavior of rotational machinery. The seasons are periodic; so is the weather. The phases of the moon are periodic, as are the motions of the planets. There is strong evidence that the ice ages are periodic, with a period of 90,000–100,000 years.

Why are trigonometric functions so important in the study of things periodic? The answer lies in a surprising and beautiful theorem from advanced calculus that says that every periodic function we want to use in mathematical modeling can be written as an algebraic combination of sines and cosines. Thus, once we learn the calculus of sines and cosines, we will know everything we need to know to model the mathematical behavior of most periodic phenomena.

Figure 1.41 This compact patient monitor shows several periodic functions associated with the human body. This device dynamically monitors electrocardiogram (ECG) and respiration, and blood pressure.

Even and Odd Trigonometric Functions

The graphs in Figure 1.39 suggest that $\cos x$ and $\sec x$ are even functions because their graphs are symmetric about the y-axis. The other four basic trigonometric functions are odd.

Example 1 CONFIRMING EVEN AND ODD

Show that cosine is an even function and sine is odd.

Solution From Figure 1.42 it follows that

$$\cos(-\theta) = \frac{x}{r} = \cos\theta, \quad \sin(-\theta) = \frac{-y}{r} = -\sin\theta,$$

so cosine is an even function and sine is odd.

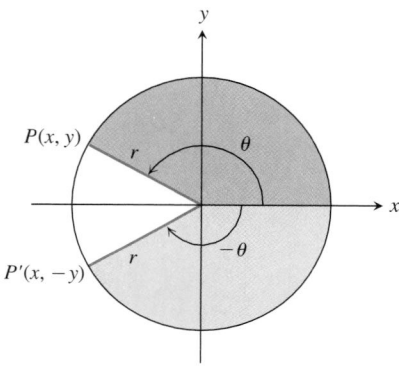

Figure 1.42 Angles of opposite sign. (Example 1)

We can use the results of Example 1 to establish the *parity* of the other four basic trigonometric functions. For example,

$$\sec(-\theta) = \frac{1}{\cos(-\theta)} = \frac{1}{\cos\theta} = \sec\theta,$$

$$\tan(-\theta) = \frac{\sin(-\theta)}{\cos(-\theta)} = \frac{-\sin\theta}{\cos\theta} = -\tan\theta,$$

so secant is an even function and tangent is odd. Similar steps will show that cosecant and cotangent are odd functions.

Transformations of Trigonometric Graphs

The rules for shifting, stretching, shrinking, and reflecting the graph of a function apply to the trigonometric functions. The following diagram will remind you of the controlling parameters.

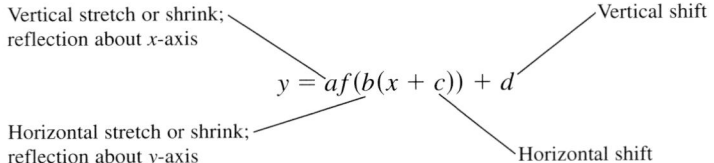

Vertical stretch or shrink; reflection about *x*-axis

Vertical shift

$$y = af(b(x + c)) + d$$

Horizontal stretch or shrink; reflection about *y*-axis

Horizontal shift

Applications

Example 2   FINDING A TRIGONOMETRIC MODEL

The builders of the Trans-Alaska Pipeline used insulated pads to keep the pipeline heat from melting the permanently frozen soil beneath. To design the pads, it was necessary to take into account the variation in air temperature throughout the year. The variation was represented in the calculations by a general sine function or **sinusoid** of the form

$$f(x) = A \sin\left[\frac{2\pi}{B}(x - C)\right] + D,$$

where $|A|$ is the *amplitude*, $|B|$ is the *period*, C is the *horizontal shift*, and D is the *vertical shift* (Figure 1.43).

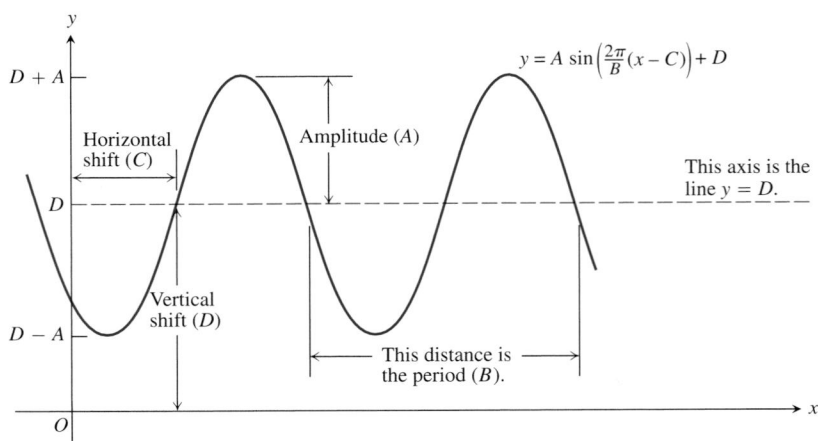

Figure 1.43 The general sine curve $y = A \sin\left[(2\pi/B)(x - C)\right] + D$, shown for A, B, C, and D positive. (Example 2)

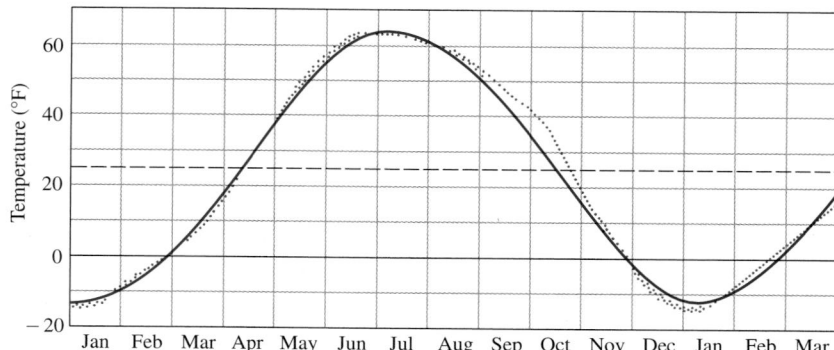

Figure 1.44 Normal mean air temperatures for Fairbanks, Alaska, plotted as data points (red). The approximating sine function is $f(x) = 37 \sin[(2\pi/365)(x - 101)] + 25$. *Source:* "Is the Curve of Temperature Variation a Sine Curve?" by B. M. Lando and C. A. Lando, *The Mathematics Teacher,* 7:6, Fig. 2, p. 535 (September 1977).

Figure 1.44 shows how we can use such a function to represent temperature data. The data points in the figure are plots of the mean daily air temperatures for Fairbanks, Alaska, based on records of the National Weather Service from 1941 to 1970. The sine function used to fit the data is

$$f(x) = 37 \sin\left[\frac{2\pi}{365}(x - 101)\right] + 25,$$

where f is temperature in degrees Fahrenheit and x is the number of the day counting from the beginning of the year. The fit is remarkably good.

Musical notes are pressure waves in the air. The wave behavior can be modeled with great accuracy by general sine curves. New devices called Calculator Based Laboratory™ (CBL) systems can record these waves with the aid of a microphone. The data in Table 1.17 give pressure displacement versus time in seconds of a musical note produced by a tuning fork and recorded with a CBL system.

Table 1.17 Tuning Fork Data

Time	Pressure	Time	Pressure	Time	Pressure
0.00091	−0.080	0.00271	−0.141	0.00453	0.749
0.00108	0.200	0.00289	−0.309	0.00471	0.581
0.00125	0.480	0.00307	−0.348	0.00489	0.346
0.00144	0.693	0.00325	−0.248	0.00507	0.077
0.00162	0.816	0.00344	−0.041	0.00525	−0.164
0.00180	0.844	0.00362	0.217	0.00543	−0.320
0.00198	0.771	0.00379	0.480	0.00562	−0.354
0.00216	0.603	0.00398	0.681	0.00579	−0.248
0.00234	0.368	0.00416	0.810	0.00598	−0.035
0.00253	0.099	0.00435	0.827		

$y = 0.6 \sin(2488.6x - 2.832) + 0.266$

[0, 0.0062] by [−0.5, 1]

Figure 1.45 A sinusoidal regression model for the tuning fork data in Table 1.17. (Example 3)

$x = t, \ y = \sin t, \ -\dfrac{\pi}{2} \le t \le \dfrac{\pi}{2}$

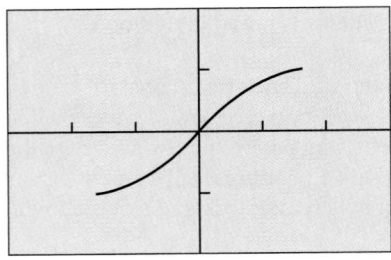

[−3, 3] by [−2, 2]

(a)

$x = \sin t, \ y = t, \ -\dfrac{\pi}{2} \le t \le \dfrac{\pi}{2}$

[−3, 3] by [−2, 2]

(b)

Figure 1.46 (a) A restricted sine function and (b) its inverse. (Example 4)

Example 3 FINDING THE FREQUENCY OF A MUSICAL NOTE

Consider the tuning fork data in Table 1.17 on the previous page.

(a) Find a sinusoidal regression equation (general sine curve) for the data and superimpose its graph on a scatter plot of the data.

(b) The *frequency* of a musical note, or wave, is measured in cycles per second, or hertz (1 Hz = 1 cycle per second). The frequency is the reciprocal of the *period* of the wave, which is measured in seconds per cycle. Estimate the frequency of the note produced by the tuning fork.

Solution

(a) The sinusoidal regression equation produced by our calculator is approximately

$$y = 0.6 \sin(2488.6x - 2.832) + 0.266.$$

Figure 1.45 shows its graph together with a scatter plot of the tuning fork data.

(b) The period is $\dfrac{2\pi}{2488.6}$ sec, so the frequency is $\dfrac{2488.6}{2\pi} \approx 396$ Hz.

Interpretation

The tuning fork is vibrating at a frequency of about 396 Hz. On the pure tone scale, this is the note G above middle C. It is a few cycles per second different from the frequency of the G we hear on a piano's tempered scale, 392 Hz.

Inverse Trigonometric Functions

None of the six basic trigonometric functions graphed in Figure 1.39 is one-to-one. These functions do not have inverses. However, in each case the domain can be restricted to produce a new function that does have an inverse, as illustrated in Example 4.

Example 4 RESTRICTING THE DOMAIN OF THE SINE

Show that the function $y = \sin x, \ -\pi/2 \le x \le \pi/2,$ is one-to-one, and graph its inverse.

Solution Figure 1.46a shows the graph of this restricted sine function using the parametric equations

$$x_1 = t, \qquad y_1 = \sin t, \qquad -\frac{\pi}{2} \le t \le \frac{\pi}{2}.$$

This restricted sine function is one-to-one because it does not repeat any output values. It therefore has an inverse, which we graph in Figure 1.46b by interchanging the ordered pairs using the parametric equations

$$x_2 = \sin t, \qquad y_2 = t, \qquad -\frac{\pi}{2} \le t \le \frac{\pi}{2}.$$

The inverse of the restricted sine function of Example 4 is called the *inverse sine function*. The inverse sine of x is the angle whose sine is x. It is denoted by $\sin^{-1} x$ or arcsin x. Either notation is read "arcsine of x" or "the inverse sine of x."

The domains of the other basic trigonometric functions can also be restricted to produce a function with an inverse. The domains and ranges of the resulting inverse functions become parts of their definitions.

Definitions Inverse Trigonometric Functions		
Function	**Domain**	**Range**
$y = \cos^{-1} x$	$-1 \le x \le 1$	$0 \le y \le \pi$
$y = \sin^{-1} x$	$-1 \le x \le 1$	$-\dfrac{\pi}{2} \le y \le \dfrac{\pi}{2}$
$y = \tan^{-1} x$	$-\infty < x < \infty$	$-\dfrac{\pi}{2} < y < \dfrac{\pi}{2}$
$y = \sec^{-1} x$	$|x| \ge 1$	$0 \le y \le \pi,\, y \ne \dfrac{\pi}{2}$
$y = \csc^{-1} x$	$|x| \ge 1$	$-\dfrac{\pi}{2} \le y \le \dfrac{\pi}{2},\, y \ne 0$
$y = \cot^{-1} x$	$-\infty < x < \infty$	$0 < y < \pi$

The graphs of the six inverse trigonometric functions are shown in Figure 1.47.

Domain: $-1 \le x \le 1$
Range: $0 \le y \le \pi$

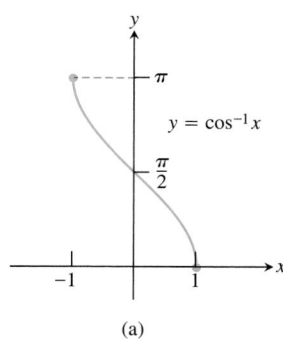

(a)

Domain: $-1 \le x \le 1$
Range: $-\dfrac{\pi}{2} \le y \le \dfrac{\pi}{2}$

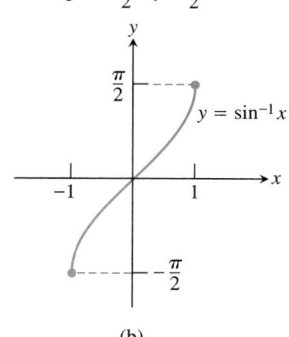

(b)

Domain: $-\infty < x < \infty$
Range: $-\dfrac{\pi}{2} < y < \dfrac{\pi}{2}$

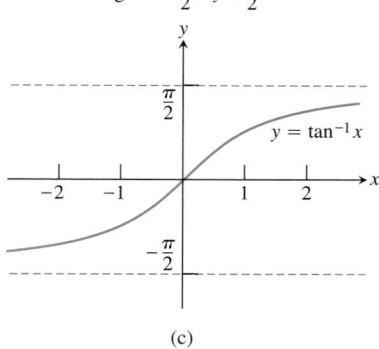

(c)

Domain: $x \le -1$ or $x \ge 1$
Range: $0 \le y \le \pi,\, y \ne \dfrac{\pi}{2}$

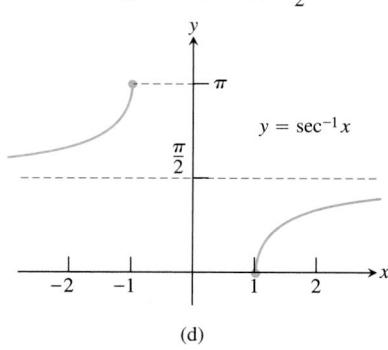

(d)

Domain: $x \le -1$ or $x \ge 1$
Range: $-\dfrac{\pi}{2} \le y \le \dfrac{\pi}{2},\, y \ne 0$

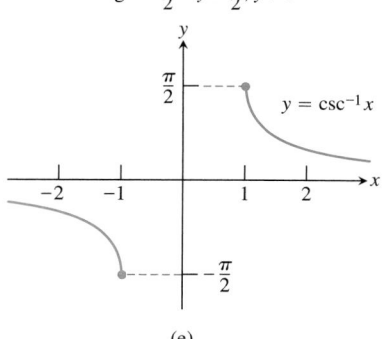

(e)

Domain: $-\infty < x < \infty$
Range: $0 < y < \pi$

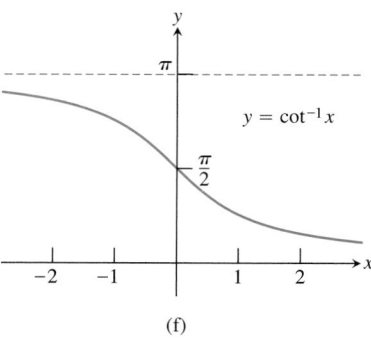

(f)

Figure 1.47 Graphs of (a) $y = \cos^{-1} x$, (b) $y = \sin^{-1} x$, (c) $y = \tan^{-1} x$, (d) $y = \sec^{-1} x$, (e) $y = \csc^{-1} x$, and (f) $y = \cot^{-1} x$.

Example 5 USING THE INVERSE TRIGONOMETRIC FUNCTIONS

Solve for x.

(a) $\sin x = 0.7$ in $0 \leq x < 2\pi$

(b) $\tan x = -2$ in $-\infty < x < \infty$

Solution

(a) Notice that $x = \sin^{-1}(0.7) \approx 0.775$ is in the first quadrant, so 0.775 is one solution of this equation. The angle $\pi - x$ is in the second quadrant and has sine equal to 0.7. Thus two solutions in this interval are

$$\sin^{-1}(0.7) \approx 0.775 \quad \text{and} \quad \pi - \sin^{-1}(0.7) \approx 2.366.$$

(b) The angle $x = \tan^{-1}(-2) \approx -1.107$ is in the fourth quadrant and is the only solution to this equation in the interval $-\pi/2 < x < \pi/2$ where $\tan x$ is one-to-one. Since $\tan x$ is periodic with period π, the solutions to this equation are of the form

$$\tan^{-1}(-2) + k\pi \approx -1.107 + k\pi$$

where k is any integer.

Quick Review 1.6

In Exercises 1–4, convert from radians to degrees or degrees to radians.

1. $\pi/3$

2. -2.5

3. $-40°$

4. $45°$

In Exercises 5–7, solve the equation graphically in the given interval.

5. $\sin x = 0.6$, $\quad 0 \leq x < 2\pi$

6. $\cos x = -0.4$, $\quad 0 \leq x < 2\pi$

7. $\tan x = 1$, $\quad -\dfrac{\pi}{2} \leq x < \dfrac{3\pi}{2}$

8. Show that $f(x) = 2x^2 - 3$ is an even function. Explain why its graph is symmetric about the y-axis.

9. Show that $f(x) = x^3 - 3x$ is an odd function. Explain why its graph is symmetric about the origin.

10. Give one way to restrict the domain of the function $f(x) = x^4 - 2$ to make the resulting function one-to-one.

Section 1.6 Exercises

In Exercises 1–4, the angle lies at the center of a circle and subtends an arc of the circle. Find the missing angle measure, circle radius, or arc length.

Angle	Radius	Arc Length
1. $5\pi/8$	2	?
2. 175°	?	10
3. ?	14	7
4. ?	6	$3\pi/2$

In Exercises 5 and 6, choose an appropriate viewing window to display two complete periods of each trigonometric function in radian mode.

5. **(a)** $y = \sec x$ **(b)** $y = \csc x$ **(c)** $y = \cot x$

6. **(a)** $y = \sin x$ **(b)** $y = \cos x$ **(c)** $y = \tan x$

In Exercises 7–10, give the measure of the angle in radians and degrees. Give exact answers whenever possible.

7. $\sin^{-1}(0.5)$

8. $\sin^{-1}\left(-\dfrac{\sqrt{2}}{2}\right)$

9. $\tan^{-1}(-5)$

10. $\cos^{-1}(0.7)$

In Exercises 11–16, specify **(a)** the period, **(b)** the amplitude, and **(c)** identify the viewing window that is shown.

11. $y = 1.5 \sin 2x$

12. $y = 2 \cos 3x$

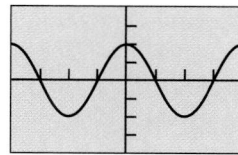

13. $y = -3 \cos 2x$

14. $y = 5 \sin \dfrac{x}{2}$

15. $y = -4 \sin \dfrac{\pi}{3} x$

16. $y = \cos \pi x$

In Exercises 17–20, determine **(a)** the period, **(b)** the domain, and **(c)** the range, and **(d)** draw the graph of the function.

17. $y = 3 \csc (3x + \pi) - 2$

18. $y = 2 \sin (4x + \pi) + 3$

19. $y = -3 \tan (3x + \pi) + 2$

20. $y = 2 \sin \left(2x + \dfrac{\pi}{3}\right)$

In Exercises 21–24, use the given information to find the values of the six trigonometric functions at the angle θ. Give exact answers.

21. $\theta = \sin^{-1}\left(\dfrac{8}{17}\right)$

22. $\theta = \tan^{-1}\left(-\dfrac{5}{12}\right)$

23. The point $P(-3, 4)$ is on the terminal side of θ.

24. The point $P(-2, 2)$ is on the terminal side of θ.

In Exercises 25–30, solve the equation in the specified interval.

25. $\tan x = 2.5, \quad 0 \le x < 2\pi$

26. $\cos x = -0.7, \quad 2\pi \le x < 4\pi$

27. $\csc x = 2, \quad 0 < x < 2\pi$

28. $\sec x = -3, \quad -\pi \le x < \pi$

29. $\sin x = -0.5, \quad -\infty < x < \infty$

30. $\cot x = -1, \quad -\infty < x < \infty$

In Exercises 31 and 32, evaluate the expression.

31. $\sin\left(\cos^{-1}\left(\dfrac{7}{11}\right)\right)$

32. $\tan\left(\sin^{-1}\left(\dfrac{9}{13}\right)\right)$

33. We suggest that students *work in groups of two or three.* A musical note like that produced with a tuning fork or pitch meter is a pressure wave. Table 1.18 gives frequencies (in Hz) of musical notes on the tempered scale. The pressure versus time tuning fork data in Table 1.19 were collected using a CBL™ and a microphone.

Table 1.18 Frequencies of Notes

Note	Frequency (Hz)
C	262
C$^{\sharp}$ or D$^{\flat}$	277
D	294
D$^{\sharp}$ or E$^{\flat}$	311
E	330
F	349
F$^{\sharp}$ or G$^{\flat}$	370
G	392
G$^{\sharp}$ or A$^{\flat}$	415
A	440
A$^{\sharp}$ or B$^{\flat}$	466
B	494
C (next octave)	524

Source: CBL™ System Experimental Workbook, Texas Instruments, Inc., 1994.

Table 1.19 Tuning Fork Data

Time (s)	Pressure	Time (s)	Pressure
0.0002368	1.29021	0.0049024	−1.06632
0.0005664	1.50851	0.0051520	0.09235
0.0008256	1.51971	0.0054112	1.44694
0.0010752	1.51411	0.0056608	1.51411
0.0013344	1.47493	0.0059200	1.51971
0.0015840	0.45619	0.0061696	1.51411
0.0018432	−0.89280	0.0064288	1.43015
0.0020928	−1.51412	0.0066784	0.19871
0.0023520	−1.15588	0.0069408	−1.06072
0.0026016	−0.04758	0.0071904	−1.51412
0.0028640	1.36858	0.0074496	−0.97116
0.0031136	1.50851	0.0076992	0.23229
0.0033728	1.51971	0.0079584	1.46933
0.0036224	1.51411	0.0082080	1.51411
0.0038816	1.45813	0.0084672	1.51971
0.0041312	0.32185	0.0087168	1.50851
0.0043904	−0.97676	0.0089792	1.36298
0.0046400	−1.51971		

(a) Find a sinusoidal regression equation for the data in Table 1.19 and superimpose its graph on a scatter plot of the data.

(b) Determine the frequency of and identify the musical note produced by the tuning fork.

34. *Temperature Data*   Table 1.20 gives the average monthly temperatures for St. Louis for a 12-month period starting with January. Model the monthly temperature with an equation of the form

$$y = a \sin [b(t - h)] + k,$$

y in degrees Fahrenheit, *t* in months, as follows:

(a) Find the value of *b* assuming that the period is 12 months.

(b) How is the amplitude *a* related to the difference $80° - 30°$?

(c) Use the information in (b) to find *k*.

(d) Find *h*, and write an equation for *y*.

(e) Superimpose a graph of *y* on a scatter plot of the data.

Table 1.20 Temperature Data for St. Louis

Time (months)	Temperature (°F)
1	34
2	30
3	39
4	44
5	58
6	67
7	78
8	80
9	72
10	63
11	51
12	40

35. *Temperatures in Fairbanks, Alaska*   Find the **(a)** amplitude, **(b)** period, **(c)** horizontal shift, and **(d)** vertical shift of the model

$$f(x) = 37 \sin \left[\frac{2\pi}{365}(x - 101) \right] + 25$$

used in Example 2.

36. *Temperatures in Fairbanks, Alaska*   Use the equation of Exercise 35 to approximate the answers to the following questions about the temperatures in Fairbanks, Alaska, shown in Figure 1.44 on page 45. Assume that the year has 365 days.

(a) What are the highest and lowest mean daily temperatures?

(b) What is the average of the highest and lowest mean daily temperatures? Why is this average the vertical shift of the function?

37. *Even-Odd*

(a) Show that cot *x* is an odd function of *x*.

(b) Show that the quotient of an even function and an odd function is an odd function.

38. *Even-Odd*

(a) Show that csc *x* is an odd function of *x*.

(b) Show that the reciprocal of an odd function is odd.

39. *Even-Odd*   Show that the product of an even function and an odd function is an odd function.

40. *Finding the Period*   Give a convincing argument that the period of tan *x* is π.

41. *Sinusoidal Regression*   Table 1.21 gives the values of the function

$$f(x) = a \sin (bx + c) + d$$

accurate to two decimals.

(a) Find a sinusoidal regression equation for the data.

(b) Rewrite the equation with *a*, *b*, *c*, and *d* rounded to the nearest integer.

Table 1.21 Values of a Function

x	*f(x)*
1	3.42
2	0.73
3	0.12
4	2.16
5	4.97
6	5.97

Exploration

42. *Trigonometric Identities*   Let $f(x) = \sin x + \cos x$.

(a) Graph $y = f(x)$. Describe the graph.

(b) Use the graph to identify the amplitude, period, horizontal shift, and vertical shift.

(c) Use the formula

$$\sin \alpha \cos \beta + \cos \alpha \sin \beta = \sin (\alpha + \beta)$$

for the sine of the sum of two angles to confirm your answers.   ∎

Extending the Ideas

43. **Exploration**   Let $y = \sin (ax) + \cos (ax)$.

Use the symbolic manipulator of a computer algebra system (CAS) to help you with the following:

(a) Express *y* as a sinusoid for $a = 2, 3, 4,$ and 5.

(b) Conjecture another formula for *y* for *a* equal to any positive integer *n*.

(c) Check your conjecture with a CAS.

(d) Use the formula for the sine of the sum of two angles (see Exercise 42c) to confirm your conjecture.

44. Exploration Let $y = a \sin x + b \cos x$.

Use the symbolic manipulator of a computer algebra system (CAS) to help you with the following:

(a) Express y as a sinusoid for the following pairs of values:
$a = 2, b = 1$; $a = 1, b = 2$; $a = 5, b = 2$;
$a = 2, b = 5$; $a = 3, b = 4$.

(b) Conjecture another formula for y for any pair of positive integers. Try other values if necessary.

(c) Check your conjecture with a CAS.

(d) Use the following formulas for the sine or cosine of a sum or difference of two angles to confirm your conjecture.

$$\sin \alpha \cos \beta \pm \cos \alpha \sin \beta = \sin (\alpha \pm \beta)$$
$$\cos \alpha \cos \beta \pm \sin \alpha \sin \beta = \cos (\alpha \pm \beta)$$

In Exercises 45 and 46, show that the function is periodic and find its period.

45. $y = \sin^3 x$ **46.** $y = |\tan x|$

In Exercises 47 and 48, graph one period of the function.

47. $f(x) = \sin (60x)$ **48.** $f(x) = \cos (60\pi x)$

Chapter 1 Key Terms

absolute value function (p. 15)
base a logarithm function (p. 36)
boundary of an interval (p. 11)
boundary points (p. 11)
change of base formula (p. 37)
closed interval (p. 11)
common logarithm function (p. 36)
composing (p. 16)
composite function (p. 16)
compounded continuously (p. 23)
cosecant function (p. 41)
cosine function (p. 41)
cotangent function (p. 41)
dependent variable (p. 10)
domain (p. 10)
even function (p. 13)
exponential decay (p. 22)
exponential function base a (p. 20)
exponential growth (p. 22)
function (p. 10)
general linear equation (p. 4)
graph of a function (p. 11)
graph of a relation (p. 26)
grapher failure (p. 12)
half-life (p. 21)
half-open interval (p. 11)
identity function (p. 33)
increments (p. 1)
independent variable (p. 10)

initial point of parametrized curve (p. 27)
interior of an interval (p. 11)
interior points of an interval (p. 11)
inverse cosecant function (p. 47)
inverse cosine function (p. 47)
inverse cotangent function (p. 47)
inverse function (p. 33)
inverse properties for a^x and $\log_a x$ (p. 36)
inverse secant function (p. 47)
inverse sine function (p. 47)
inverse tangent function (p. 47)
linear regression (p. 5)
natural domain (p. 10)
natural logarithm function (p. 36)
odd function (p. 13)
one-to-one function (p. 32)
open interval (p. 11)
parallel lines (p. 2)
parameter (p. 27)
parameter interval (p. 27)
parametric curve (p. 26)
parametric equations (p. 26)
parametrization of a curve (p. 27)
parametrize (p. 27)
period of a function (p. 43)
periodic function (p. 43)
perpendicular lines (p. 2)
piecewise defined function (p. 15)
point-slope equation (p. 3)

power function (p. 12)
power rule for logarithms (p. 37)
product rule for logarithms (p. 37)
quotient rule for logarithms (p. 37)
radian measure (p. 41)
range (p. 10)
regression analysis (p. 5)
regression curve (p. 5)
relation (p. 26)
rise (p. 2)
rules for exponents (p. 21)
run (p. 2)
scatter plot (p. 5)
secant function (p. 41)
sine function (p. 41)
sinusoid (p. 44)
sinusoidal regression (p. 46)
slope (p. 2)
slope-intercept equation (p. 4)
symmetry about the origin (p. 14)
symmetry about the y-axis (p. 14)
tangent function (p. 41)
terminal point of parametrized curve (p. 27)
witch of Agnesi (p. 29)
x-intercept (p. 3)
y-intercept (p. 3)

Chapter 1 Review Exercises

In Exercises 1–14, write an equation for the specified line.

1. through $(1, -6)$ with slope 3

2. through $(-1, 2)$ with slope $-1/2$

3. the vertical line through $(0, -3)$

4. through $(-3, 6)$ and $(1, -2)$

5. the horizontal line through $(0, 2)$

6. through $(3, 3)$ and $(-2, 5)$

7. with slope -3 and y-intercept 3

8. through $(3, 1)$ and parallel to $2x - y = -2$

9. through $(4, -12)$ and parallel to $4x + 3y = 12$

10. through $(-2, -3)$ and perpendicular to $3x - 5y = 1$

11. through $(-1, 2)$ and perpendicular to $\dfrac{1}{2}x + \dfrac{1}{3}y = 1$

12. with x-intercept 3 and y-intercept -5

13. the line $y = f(x)$, where f has the following values:

x	-2	2	4
$f(x)$	4	2	1

14. through $(4, -2)$ with x-intercept -3

In Exercises 15–18, determine whether the graph of the function is symmetric about the y-axis, the origin, or neither.

15. $y = x^{1/5}$

16. $y = x^{2/5}$

17. $y = x^2 - 2x - 1$

18. $y = e^{-x^2}$

In Exercises 19–26, determine whether the function is even, odd, or neither.

19. $y = x^2 + 1$

20. $y = x^5 - x^3 - x$

21. $y = 1 - \cos x$

22. $y = \sec x \tan x$

23. $y = \dfrac{x^4 + 1}{x^3 - 2x}$

24. $y = 1 - \sin x$

25. $y = x + \cos x$

26. $y = \sqrt{x^4 - 1}$

In Exercises 27–38, find the **(a)** domain and **(b)** range, and **(c)** graph the function.

27. $y = |x| - 2$

28. $y = -2 + \sqrt{1 - x}$

29. $y = \sqrt{16 - x^2}$

30. $y = 3^{2-x} + 1$

31. $y = 2e^{-x} - 3$

32. $y = \tan (2x - \pi)$

33. $y = 2 \sin (3x + \pi) - 1$

34. $y = x^{2/5}$

35. $y = \ln (x - 3) + 1$

36. $y = -1 + \sqrt[3]{2 - x}$

37. $y = \begin{cases} \sqrt{-x}, & -4 \le x \le 0 \\ \sqrt{x}, & 0 < x \le 4 \end{cases}$

38. $y = \begin{cases} -x - 2, & -2 \le x \le -1 \\ x, & -1 < x \le 1 \\ -x + 2, & 1 < x \le 2 \end{cases}$

In Exercises 39 and 40, write a piecewise formula for the function.

39.

40.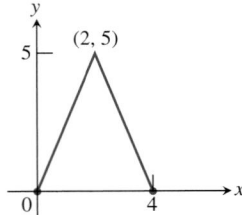

In Exercises 41 and 42, find

(a) $(f \circ g)(-1)$ **(b)** $(g \circ f)(2)$

(c) $(f \circ f)(x)$ **(d)** $(g \circ g)(x)$

41. $f(x) = \dfrac{1}{x}$, $g(x) = \dfrac{1}{\sqrt{x + 2}}$

42. $f(x) = 2 - x$, $g(x) = \sqrt[3]{x + 1}$

In Exercises 43 and 44, **(a)** write a formula for $f \circ g$ and $g \circ f$ and find the **(b)** domain and **(c)** range of each.

43. $f(x) = 2 - x^2$, $g(x) = \sqrt{x + 2}$

44. $f(x) = \sqrt{x}$, $g(x) = \sqrt{1 - x}$

In Exercises 45–48, a parametrization is given for a curve.

(a) Graph the curve. Identify the initial and terminal points, if any. Indicate the direction in which the curve is traced.

(b) Find a Cartesian equation for a curve that contains the parametrized curve. What portion of the graph of the Cartesian equation is traced by the parametrized curve?

45. $x = 5 \cos t$, $y = 2 \sin t$, $0 \le t \le 2\pi$

46. $x = 4 \cos t$, $y = 4 \sin t$, $\pi/2 \le t < 3\pi/2$

47. $x = 2 - t$, $y = 11 - 2t$, $-2 \le t \le 4$

48. $x = 1 + t$, $y = \sqrt{4 - 2t}$, $t \le 2$

In Exercises 49–52, give a parametrization for the curve.

49. the line segment with endpoints $(-2, 5)$ and $(4, 3)$

50. the line through $(-3, -2)$ and $(4, -1)$

51. the ray with initial point $(2, 5)$ that passes through $(-1, 0)$

52. $y = x (x - 4)$, $x \le 2$

In Exercises 53 and 54, *work in groups of two or three* to
(a) find f^{-1} and show that $(f \circ f^{-1})(x) = (f^{-1} \circ f)(x) = x$.
(b) graph f and f^{-1} in the same viewing window.

53. $f(x) = 2 - 3x$

54. $f(x) = (x + 2)^2$, $x \ge -2$

In Exercises 55 and 56, find the measure of the angle in radians and degrees.

55. $\sin^{-1}(0.6)$ **56.** $\tan^{-1}(-2.3)$

57. Find the six trigonometric values of $\theta = \cos^{-1}(3/7)$. Give exact answers.

58. Solve the equation $\sin x = -0.2$ in the following intervals.

    (a) $0 \le x < 2\pi$ **(b)** $-\infty < x < \infty$

59. Solve for x: $e^{-0.2x} = 4$

60. The graph of f is shown. Draw the graph of each function.

    (a) $y = f(-x)$ **(b)** $y = -f(x)$

    (c) $y = -2f(x + 1) + 1$ **(d)** $y = 3f(x - 2) - 2$

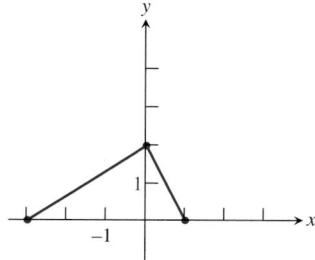

61. A portion of the graph of a function defined on $[-3, 3]$ is shown. Complete the graph assuming that the function is

    (a) even.   **(b)** odd.

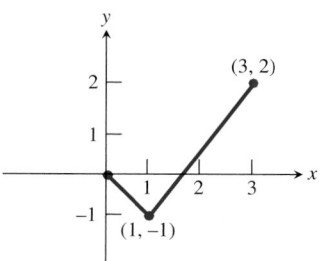

62. *Depreciation* Smith Hauling purchased an 18-wheel truck for $100,000. The truck depreciates at the constant rate of $10,000 per year for 10 years.

    (a) Write an expression that gives the value y after x years.

    (b) When is the value of the truck $55,000?

63. *Drug Absorption* A drug is administered intravenously for pain. The function

$$f(t) = 90 - 52 \ln(1 + t), \quad 0 \le t \le 4$$

gives the number of units of the drug in the body after t hours.

    (a) What was the initial number of units of the drug administered?

    (b) How much is present after 2 hours?

    (c) Draw the graph of f.

64. *Finding Time* If Joenita invests $1500 in a retirement account that earns 8% compounded annually, how long will it take this single payment to grow to $5000?

65. *Guppy Population* The number of guppies in Susan's aquarium doubles every day. There are four guppies initially.

    (a) Write the number of guppies as a function of time t.

    (b) How many guppies were present after 4 days? after 1 week?

    (c) When will there be 2000 guppies?

    (d) **Writing to Learn** Give reasons why this might not be a good model for the growth of Susan's guppy population.

66. *Doctoral Degrees* Table 1.22 shows the number of doctoral degrees earned in the given academic year by Hispanic students. Let $x = 0$ represent 1970–71, $x = 1$ represent 1971–72, and so forth.

Table 1.22 Doctorates Earned by Hispanic Americans

Year	Number of Degrees
1976–77	520
1980–81	460
1984–85	680
1988–89	630
1990–91	730
1991–92	810
1992–93	830

Source: U.S. Department of Education, as reported in the *Chronicle of Higher Education,* April 28, 1995.

    (a) Find the linear regression equation for the data and superimpose its graph on a scatter plot of the data.

    (b) Use the regression equation to predict the number of doctoral degrees that will be earned by Hispanic Americans in the academic year 2000–01.

    (c) **Writing to Learn** Find the slope of the regression line. What does the slope represent?

67. *Estimating Population Growth* Use the data in Table 1.23 about the population of New York State. Let $x = 60$ represent 1960, $x = 70$ represent 1970, and so forth.

Table 1.23 Population of New York State

Year	Population (millions)
1960	16.78
1980	17.56
1990	17.99

Source: The Statesman's Yearbook, 129th ed. (London: The Macmillan Press, Ltd., 1992).

    (a) Find the exponential regression equation for the data.

    (b) Use the regression equation to predict when the population will be 25 million.

    (c) What annual rate of growth can we infer from the regression equation?

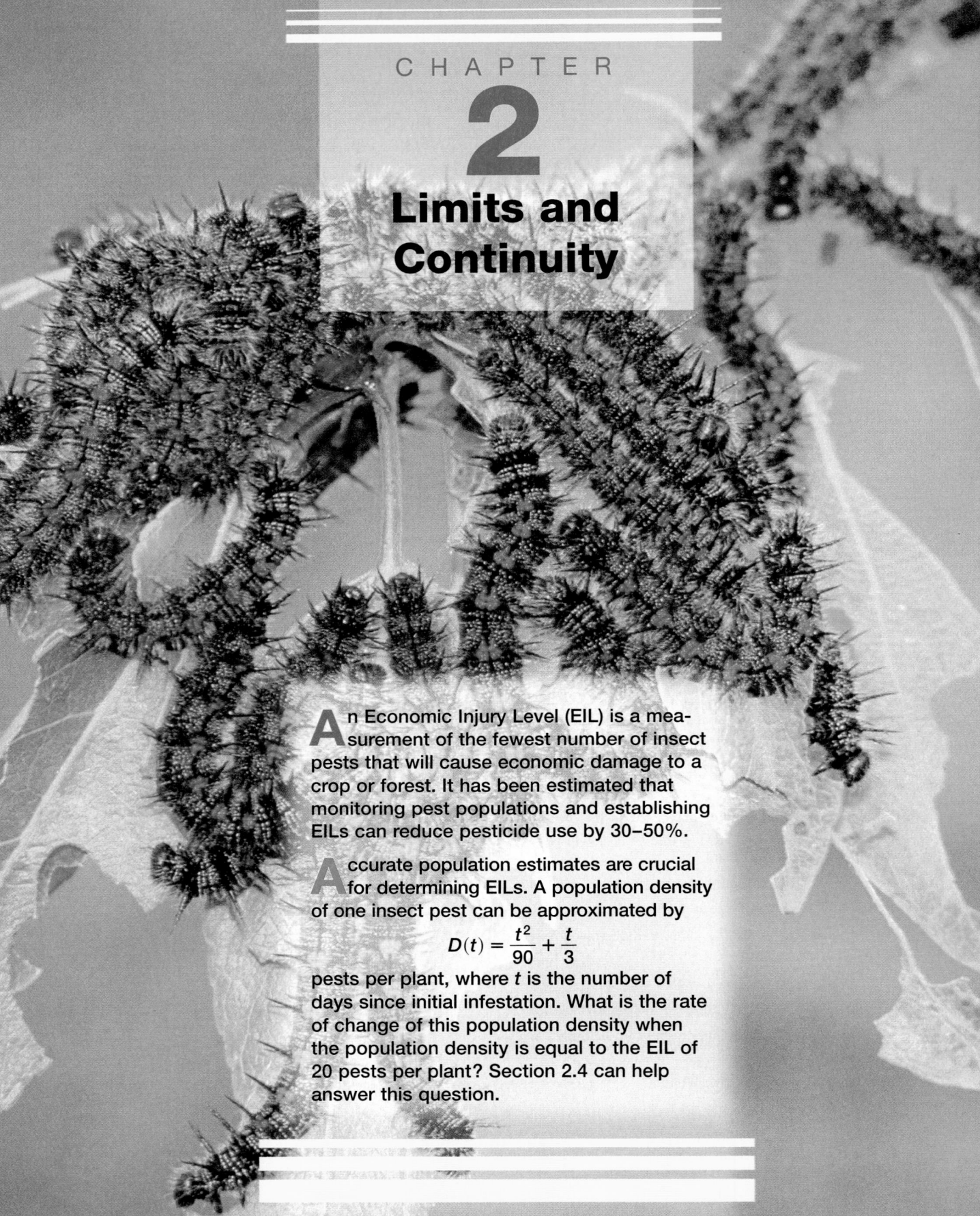

CHAPTER
2
Limits and Continuity

An Economic Injury Level (EIL) is a measurement of the fewest number of insect pests that will cause economic damage to a crop or forest. It has been estimated that monitoring pest populations and establishing EILs can reduce pesticide use by 30–50%.

Accurate population estimates are crucial for determining EILs. A population density of one insect pest can be approximated by

$$D(t) = \frac{t^2}{90} + \frac{t}{3}$$

pests per plant, where t is the number of days since initial infestation. What is the rate of change of this population density when the population density is equal to the EIL of 20 pests per plant? Section 2.4 can help answer this question.

Chapter 2 Overview

The concept of limit is one of the ideas that distinguish calculus from algebra and trigonometry.

In this chapter, we show how to define and calculate limits of function values. The calculation rules are straightforward and most of the limits we need can be found by substitution, graphical investigation, numerical approximation, algebra, or some combination of these.

One of the uses of limits is to test functions for continuity. Continuous functions arise frequently in scientific work because they model such an enormous range of natural behavior. They also have special mathematical properties, not otherwise guaranteed.

2.1 Rates of Change and Limits

Average and Instantaneous Speed • Definition of Limit • Properties of Limits • One-sided and Two-sided Limits • Sandwich Theorem

Average and Instantaneous Speed

A moving body's **average speed** during an interval of time is found by dividing the distance covered by the elapsed time. The unit of measure is length per unit time—kilometers per hour, feet per second, or whatever is appropriate to the problem at hand.

Free Fall

Near the surface of the earth, all bodies fall with the same constant acceleration. The distance a body falls after it is released from rest is a constant multiple of the square of the time fallen. At least, that is what happens when a body falls in a vacuum, where there is no air to slow it down. The square-of-time rule also holds for dense, heavy objects like rocks, ball bearings, and steel tools during the first few seconds of fall through air, before the velocity builds up to where air resistance begins to matter. When air resistance is absent or insignificant and the only force acting on a falling body is the force of gravity, we call the way the body falls *free fall*.

Example 1 FINDING AN AVERAGE SPEED

A rock breaks loose from the top of a tall cliff. What is its average speed during the first 2 seconds of fall?

Solution Experiments show that a dense solid object dropped from rest to fall freely near the surface of the earth will fall

$$y = 16t^2$$

feet in the first t seconds. The average speed of the rock over any given time interval is the distance traveled, Δy, divided by the length of the interval Δt. For the first 2 seconds of fall, from $t = 0$ to $t = 2$, we have

$$\frac{\Delta y}{\Delta t} = \frac{16(2)^2 - 16(0)^2}{2 - 0} = 32\frac{\text{ft}}{\text{sec}}.$$

Example 2 FINDING AN INSTANTANEOUS SPEED

Find the speed of the rock in Example 1 at the instant $t = 2$.

Solution

Solve Numerically

We can calculate the average speed of the rock over the interval from time $t = 2$ to any slightly later time $t = 2 + h$ as

$$\frac{\Delta y}{\Delta t} = \frac{16(2 + h)^2 - 16(2)^2}{h}. \tag{1}$$

Table 2.1 Average Speeds over Short Time Intervals Starting at $t = 2$

$$\frac{\Delta y}{\Delta t} = \frac{16(2 + h)^2 - 16(2)^2}{h}$$

Length of Time Interval, h (sec)	Average Speed for Interval $\Delta y/\Delta t$ (ft/sec)
1	80
0.1	65.6
0.01	64.16
0.001	64.016
0.0001	64.0016
0.00001	64.00016

We cannot use this formula to calculate the speed at the exact instant $t = 2$ because that would require taking $h = 0$, and $0/0$ is undefined. However, we can get a good idea of what is happening at $t = 2$ by evaluating the formula at values of h *close* to 0. When we do, we see a clear pattern (Table 2.1). As h approaches 0, the average speed approaches the limiting value 64 ft/sec.

Confirm Algebraically

If we expand the numerator of Equation 1 and simplify, we find that

$$\frac{\Delta y}{\Delta t} = \frac{16(2 + h)^2 - 16(2)^2}{h} = \frac{16(4 + 4h + h^2) - 64}{h}$$

$$= \frac{64h + 16h^2}{h} = 64 + 16h.$$

For values of h different from 0, the expressions on the right and left are equivalent and the average speed is $64 + 16h$ ft/sec. We can now see why the average speed has the limiting value $64 + 16(0) = 64$ ft/sec as h approaches 0.

Definition of Limit

As in the preceding example, most limits of interest in the real world can be viewed as numerical limits of values of functions. And this is where a graphing utility and calculus come in. A calculator can suggest the limits, and calculus can give the mathematics for confirming the limits analytically.

Limits give us a language for describing how the outputs of a function behave as the inputs approach some particular value. In Example 2, the average speed was not defined at $h = 0$ but approached the limit 64 as h approached 0. We were able to see this numerically and to confirm it algebraically by eliminating h from the denominator. But we cannot always do that. For instance, we can see both graphically and numerically (Figure 2.1) that the values of $f(x) = (\sin x)/x$ approach 1 as x approaches 0.

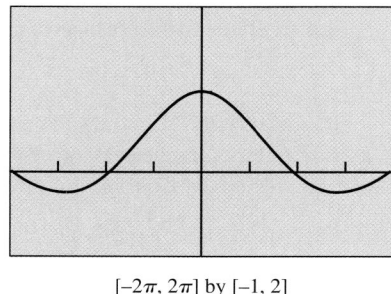

X	Y1	
-.3	.98507	
-.2	.99335	
-.1	.99833	
0	ERROR	
.1	.99833	
.2	.99335	
.3	.98507	

Y1 ◻ sin(X)/X

$[-2\pi, 2\pi]$ by $[-1, 2]$

(a)

(b)

Figure 2.1 (a) A graph and (b) table of values for $f(x) = (\sin x)/x$ that suggest the limit of f as x approaches 0 is 1.

We cannot eliminate the x from the denominator of $(\sin x)/x$ to confirm the observation algebraically. We need to use a theorem about limits to make that confirmation, as you will see in Exercise 63.

Definition Limit

Let c and L be real numbers. The function f **has limit L as x approaches c** if, given any positive number ε, there is a positive number δ such that for all x,

$$0 < |x - c| < \delta \Rightarrow |f(x) - L| < \varepsilon.$$

We write

$$\lim_{x \to c} f(x) = L.$$

The sentence $\lim_{x \to c} f(x) = L$ is read, "The limit of f of x as x approaches c equals L." The notation means that the values $f(x)$ of the function f approach or equal L as the values of x approach *(but do not equal)* c. Appendix A3 provides practice applying the definition of limit.

We saw in Example 2 that $\lim_{h \to 0} (64 + 16h) = 64$.

As suggested in Figure 2.1,

$$\lim_{x \to 0} \frac{\sin x}{x} = 1.$$

Figure 2.2 illustrates the fact that the existence of a limit as $x \to c$ never depends on how the function may or may not be defined at c. The function f has limit 2 as $x \to 1$ even though f is not defined at 1. The function g has limit 2 as $x \to 1$ even though $g(1) \neq 2$. The function h is the only one whose limit as $x \to 1$ equals its value at $x = 1$.

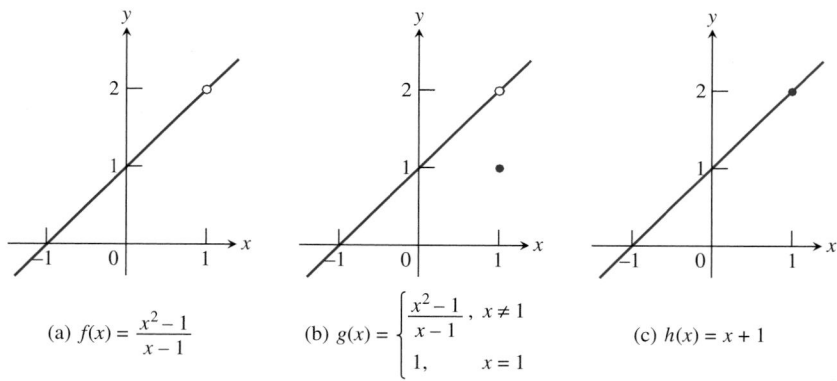

(a) $f(x) = \dfrac{x^2 - 1}{x - 1}$

(b) $g(x) = \begin{cases} \dfrac{x^2 - 1}{x - 1}, & x \neq 1 \\ 1, & x = 1 \end{cases}$

(c) $h(x) = x + 1$

Figure 2.2 $\lim_{x \to 1} f(x) = \lim_{x \to 1} g(x) = \lim_{x \to 1} h(x) = 2$

Properties of Limits

By applying six basic facts about limits, we can calculate many unfamiliar limits from limits we already know. For instance, from knowing that

$$\lim_{x \to c} (k) = k \qquad \text{Limit of the function with constant value } k$$

and

$$\lim_{x \to c} (x) = c, \qquad \text{Limit of the identity function at } x = c$$

we can calculate the limits of all polynomial and rational functions. The facts are listed in Theorem 1.

Theorem 1 Properties of Limits

If L, M, c, and k are real numbers and

$$\lim_{x\to c} f(x) = L \quad \text{and} \quad \lim_{x\to c} g(x) = M, \text{ then}$$

1. *Sum Rule:* $\quad\lim_{x\to c} (f(x) + g(x)) = L + M$

The limit of the sum of two functions is the sum of their limits.

2. *Difference Rule:* $\quad\lim_{x\to c} (f(x) - g(x)) = L - M$

The limit of the difference of two functions is the difference of their limits.

3. *Product Rule:* $\quad\lim_{x\to c} (f(x) \cdot g(x)) = L \cdot M$

The limit of a product of two functions is the product of their limits.

4. *Constant Multiple Rule:* $\quad\lim_{x\to c} (k \cdot f(x)) = k \cdot L$

The limit of a constant times a function is the constant times the limit of the function.

5. *Quotient Rule:* $\quad\lim_{x\to c} \dfrac{f(x)}{g(x)} = \dfrac{L}{M}, \; M \neq 0$

The limit of a quotient of two functions is the quotient of their limits, provided the limit of the denominator is not zero.

6. *Power Rule:* If r and s are integers, $s \neq 0$, then

$$\lim_{x\to c} (f(x))^{r/s} = L^{r/s}$$

provided that $L^{r/s}$ is a real number.

The limit of a rational power of a function is that power of the limit of the function, provided the latter is a real number.

Here are some examples of how Theorem 1 can be used to find limits of polynomial and rational functions.

Example 3   USING PROPERTIES OF LIMITS

Use the observations $\lim_{x\to c} k = k$ and $\lim_{x\to c} x = c$, and the properties of limits to find the following limits.

(a) $\lim_{x\to c} (x^3 + 4x^2 - 3)$ **(b)** $\lim_{x\to c} \dfrac{x^4 + x^2 - 1}{x^2 + 5}$

Solution

(a) $\lim_{x\to c} (x^3 + 4x^2 - 3) = \lim_{x\to c} x^3 + \lim_{x\to c} 4x^2 - \lim_{x\to c} 3$ Sum & Difference Rules

$$= c^3 + 4c^2 - 3 \qquad \text{Product \& Multiple Rules}$$

(b) $\lim_{x\to c} \dfrac{x^4 + x^2 - 1}{x^2 + 5} = \dfrac{\lim_{x\to c} (x^4 + x^2 - 1)}{\lim_{x\to c} (x^2 + 5)}$ Quotient Rule

$$= \frac{\lim_{x\to c} x^4 + \lim_{x\to c} x^2 - \lim_{x\to c} 1}{\lim_{x\to c} x^2 + \lim_{x\to c} 5} \qquad \text{Sum \& Difference Rules}$$

$$= \frac{c^4 + c^2 - 1}{c^2 + 5} \qquad \text{Product Rule}$$

Example 3 shows the remarkable strength of Theorem 1. From the two simple observations that $\lim_{x \to c} k = k$ and $\lim_{x \to c} x = c$, we can immediately work our way to limits of polynomial functions and most rational functions using substitution.

Theorem 2 Polynomial and Rational Functions

1. If $f(x) = a_n x^n + a_{n-1} x^{n-1} + \cdots + a_0$ is any polynomial function and c is any real number, then

$$\lim_{x \to c} f(x) = f(c) = a_n c^n + a_{n-1} c^{n-1} + \cdots + a_0.$$

2. If $f(x)$ and $g(x)$ are polynomials and c is any real number, then

$$\lim_{x \to c} \frac{f(x)}{g(x)} = \frac{f(c)}{g(c)}, \qquad \text{provided that } g(c) \neq 0.$$

Example 4 USING THEOREM 2

(a) $\lim_{x \to 3} [x^2(2 - x)] = (3)^2(2 - 3) = -9$

(b) $\lim_{x \to 2} \dfrac{x^2 + 2x + 4}{x + 2} = \dfrac{(2)^2 + 2(2) + 4}{2 + 2} = \dfrac{12}{4} = 3$

As with polynomials, limits of many familiar functions can be found by substitution at points where they are defined. This includes trigonometric functions, exponential and logarithmic functions, and composites of these functions. Feel free to use these properties.

Example 5 USING THE PRODUCT RULE

Determine $\lim_{x \to 0} \dfrac{\tan x}{x}$.

Solution

Solve Graphically

The graph of $f(x) = (\tan x)/x$ in Figure 2.3 suggests that the limit exists and is about 1.

Confirm Analytically

Using the analytic result of Exercise 63, we have

$$\lim_{x \to 0} \frac{\tan x}{x} = \lim_{x \to 0} \left(\frac{\sin x}{x} \cdot \frac{1}{\cos x} \right) \qquad \tan x = \frac{\sin x}{\cos x}$$

$$= \lim_{x \to 0} \frac{\sin x}{x} \cdot \lim_{x \to 0} \frac{1}{\cos x} \qquad \text{Product Rule}$$

$$= 1 \cdot \frac{1}{\cos 0} = 1 \cdot \frac{1}{1} = 1.$$

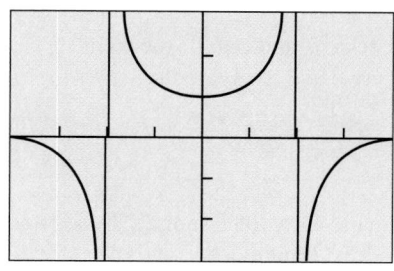

$[-\pi, \pi]$ by $[-3, 3]$

Figure 2.3 The graph of $f(x) = (\tan x)/x$ suggests that $f(x) \to 1$ as $x \to 0$. (Example 5)

Sometimes we can use a graph to discover that limits do not exist, as illustrated by Example 6.

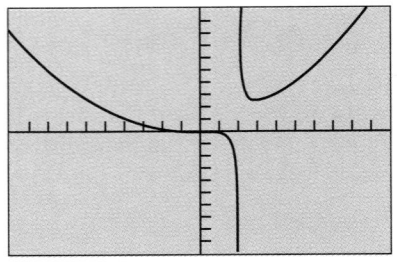

[-10, 10] by [-10, 10]

Figure 2.4 The graph of
$$f(x) = (x^3 - 1)/(x - 2)$$
obtained using parametric graphing to produce a more accurate graph. (Example 6)

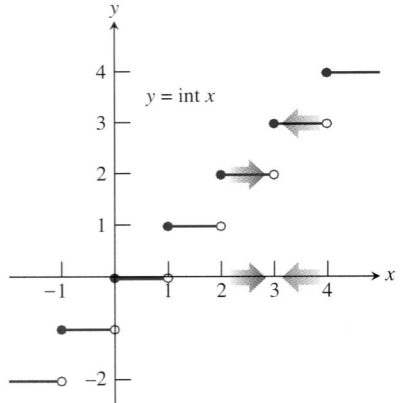

Figure 2.5 At each integer, the greatest integer function $y = \text{int } x$ has different right-hand and left-hand limits. (Example 7)

On the Far Side

If f is not defined to the left of $x = c$, then f does not have a left-hand limit at c. Similarly, if f is not defined to the right of $x = c$, then f does not have a right-hand limit at c.

Example 6   EXPLORING A NONEXISTENT LIMIT

Use a graph to show that
$$\lim_{x \to 2} \frac{x^3 - 1}{x - 2}$$
does not exist.

Solution Notice that the denominator is 0 when x is replaced by 2, so we cannot use substitution to determine the limit. The graph in Figure 2.4 of $f(x) = (x^3 - 1)/(x - 2)$ strongly suggests that as $x \to 2$ from either side, the absolute values of the function values get very large. This, in turn, suggests that the limit does not exist.

One-sided and Two-sided Limits

Sometimes the values of a function f tend to different limits as x approaches a number c from opposite sides. When this happens, we call the limit of f as x approaches c from the right the **right-hand limit** of f at c and the limit as x approaches c from the left the **left-hand limit** of f at c. Here is the notation we use:

right-hand: $\lim_{x \to c^+} f(x)$ *The limit of f as x approaches c from the right.*

left-hand: $\lim_{x \to c^-} f(x)$ *The limit of f as x approaches c from the left.*

Example 7   FUNCTION VALUES APPROACH TWO NUMBERS

The greatest integer function $f(x) = \text{int } x$ has different right-hand and left-hand limits at each integer, as we can see in Figure 2.5. For example,
$$\lim_{x \to 3^+} \text{int } x = 3 \quad \text{and} \quad \lim_{x \to 3^-} \text{int } x = 2.$$
The limit of int x as x approaches an integer n from the right is n, while the limit as x approaches n from the left is $n - 1$.

We sometimes call $\lim_{x \to c} f(x)$ the **two-sided limit** of f at c to distinguish it from the *one-sided* right-hand and left-hand limits of f at c. Theorem 3 shows how these limits are related.

Theorem 3   One-sided and Two-sided Limits

A function $f(x)$ has a limit as x approaches c if and only if the right-hand and left-hand limits at c exist and are equal. In symbols,
$$\lim_{x \to c} f(x) = L \Leftrightarrow \lim_{x \to c^+} f(x) = L \quad \text{and} \quad \lim_{x \to c^-} f(x) = L.$$

Thus, the greatest integer function $f(x) = \text{int } x$ of Example 7 does not have a limit as $x \to 3$ even though each one-sided limit exists.

Example 8   EXPLORING RIGHT- AND LEFT-HAND LIMITS

All the following statements about the function $y = f(x)$ graphed in Figure 2.6 are true.

At $x = 0$: $\lim_{x \to 0^+} f(x) = 1.$

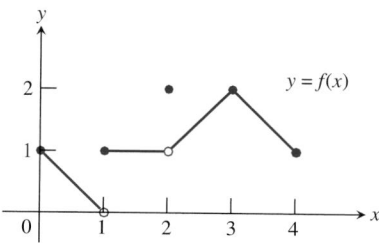

Figure 2.6 The graph of the function

$$f(x) = \begin{cases} -x + 1, & 0 \le x < 1 \\ 1, & 1 \le x < 2 \\ 2, & x = 2 \\ x - 1, & 2 < x \le 3 \\ -x + 5, & 3 < x \le 4. \end{cases}$$

(Example 8)

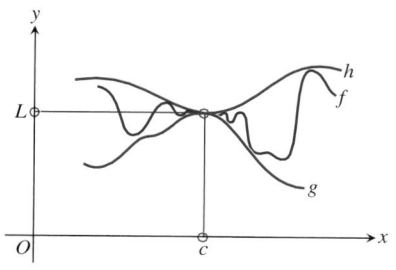

Figure 2.7 Sandwiching f between g and h forces the limiting value of f to be between the limiting values of g and h.

At $x = 1$: $\lim\limits_{x \to 1^-} f(x) = 0$ even though $f(1) = 1$,

$\lim\limits_{x \to 1^+} f(x) = 1$,

f has no limit as $x \to 1$. (The right- and left-hand limits at 1 are not equal, so $\lim_{x \to 1} f(x)$ does not exist.)

At $x = 2$: $\lim\limits_{x \to 2^-} f(x) = 1$,

$\lim\limits_{x \to 2^+} f(x) = 1$,

$\lim\limits_{x \to 2} f(x) = 1$ even though $f(2) = 2$.

At $x = 3$: $\lim\limits_{x \to 3^-} f(x) = \lim\limits_{x \to 3^+} f(x) = 2 = f(3) = \lim\limits_{x \to 3} f(x)$.

At $x = 4$: $\lim\limits_{x \to 4^-} f(x) = 1$.

At noninteger values of c between 0 and 4, f has a limit as $x \to c$.

Sandwich Theorem

If we cannot find a limit directly, we may be able to find it indirectly with the Sandwich Theorem. The theorem refers to a function f whose values are sandwiched between the values of two other functions, g and h. If g and h have the same limit as $x \to c$, then f has that limit too, as suggested by Figure 2.7.

Theorem 4   The Sandwich Theorem

If $g(x) \le f(x) \le h(x)$ for all $x \ne c$ in some interval about c, and

$$\lim_{x \to c} g(x) = \lim_{x \to c} h(x) = L,$$

then

$$\lim_{x \to c} f(x) = L.$$

Example 9   USING THE SANDWICH THEOREM

Show that $\lim\limits_{x \to 0} [x^2 \sin (1/x)] = 0$.

Solution We know that the values of the sine function lie between -1 and 1. So, it follows that

$$\left| x^2 \sin \frac{1}{x} \right| = |x^2| \cdot \left| \sin \frac{1}{x} \right| \le |x^2| \cdot 1 = x^2$$

and

$$-x^2 \le x^2 \sin \frac{1}{x} \le x^2.$$

Because $\lim\limits_{x \to 0} (-x^2) = \lim\limits_{x \to 0} x^2 = 0$, the Sandwich Theorem gives

$$\lim_{x \to 0} \left(x^2 \sin \frac{1}{x} \right) = 0.$$

The graphs in Figure 2.8 support this result.

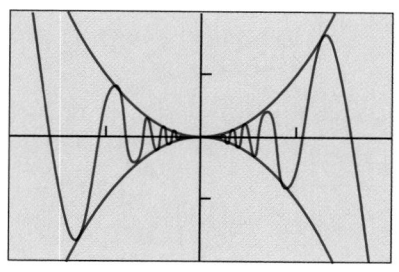

[−0.2, 0.2] by [−0.02, 0.02]

Figure 2.8 The graphs of $y_1 = x^2$, $y_2 = x^2 \sin (1/x)$, and $y_3 = -x^2$. Notice that $y_3 \le y_2 \le y_1$. (Example 9)

Quick Review 2.1

In Exercises 1–4, find $f(2)$.

1. $f(x) = 2x^3 - 5x^2 + 4$ **2.** $f(x) = \dfrac{4x^2 - 5}{x^3 + 4}$

3. $f(x) = \sin\left(\pi\dfrac{x}{2}\right)$

4. $f(x) = \begin{cases} 3x - 1, & x < 2 \\ \dfrac{1}{x^2 - 1}, & x \geq 2 \end{cases}$

In Exercises 5–8, write the inequality in the form $a < x < b$.

5. $|x| < 4$ **6.** $|x| < c^2$

7. $|x - 2| < 3$ **8.** $|x - c| < d^2$

In Exercises 9 and 10, write the fraction in reduced form.

9. $\dfrac{x^2 - 3x - 18}{x + 3}$ **10.** $\dfrac{2x^2 - x}{2x^2 + x - 1}$

Section 2.1 Exercises

In Exercises 1–6, use the graph to estimate the limits and value of the function, or explain why the limits do not exist.

1.

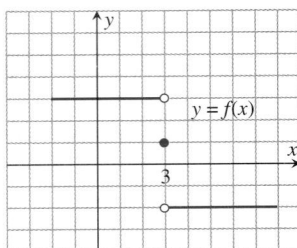

(a) $\lim\limits_{x \to 3^-} f(x)$ **(b)** $\lim\limits_{x \to 3^+} f(x)$ **(c)** $\lim\limits_{x \to 3} f(x)$ **(d)** $f(3)$

2.

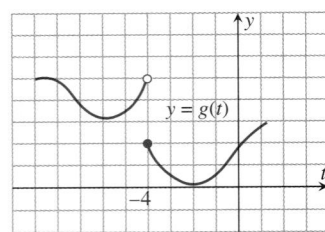

(a) $\lim\limits_{t \to -4^-} g(t)$ **(b)** $\lim\limits_{t \to -4^+} g(t)$ **(c)** $\lim\limits_{t \to -4} g(t)$ **(d)** $g(-4)$

3.

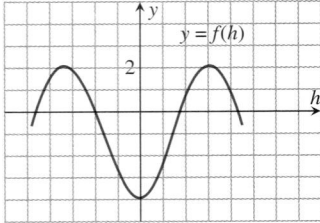

(a) $\lim\limits_{h \to 0^-} f(h)$ **(b)** $\lim\limits_{h \to 0^+} f(h)$ **(c)** $\lim\limits_{h \to 0} f(h)$ **(d)** $f(0)$

4.

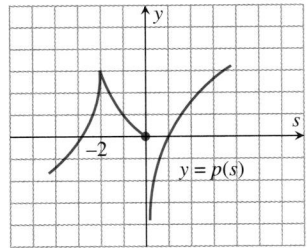

(a) $\lim\limits_{s \to -2^-} p(s)$ **(b)** $\lim\limits_{s \to -2^+} p(s)$ **(c)** $\lim\limits_{s \to -2} p(s)$ **(d)** $p(-2)$

5.

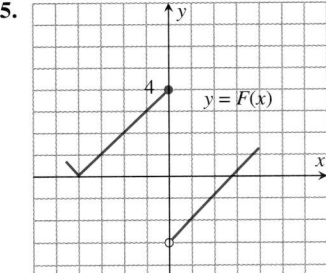

(a) $\lim\limits_{x \to 0^-} F(x)$ **(b)** $\lim\limits_{x \to 0^+} F(x)$ **(c)** $\lim\limits_{x \to 0} F(x)$ **(d)** $F(0)$

6.

(a) $\lim\limits_{x \to 2^-} G(x)$ **(b)** $\lim\limits_{x \to 2^+} G(x)$ **(c)** $\lim\limits_{x \to 2} G(x)$ **(d)** $G(2)$

In Exercises 7–16, determine the limit by substitution. Support graphically.

7. $\lim\limits_{x \to -1/2} 3x^2(2x - 1)$ **8.** $\lim\limits_{x \to -4} (x + 3)^{1998}$

9. $\lim\limits_{x \to 1} (x^3 + 3x^2 - 2x - 17)$

10. $\lim\limits_{y \to 2} \dfrac{y^2 + 5y + 6}{y + 2}$

11. $\lim\limits_{y \to -3} \dfrac{y^2 + 4y + 3}{y^2 - 3}$

12. $\lim\limits_{x \to 1/2}$ int x

13. $\lim\limits_{x \to -2} (x - 6)^{2/3}$

14. $\lim\limits_{x \to 2} \sqrt{x + 3}$

15. $\lim\limits_{x \to 0} e^x \cos x$

16. $\lim\limits_{x \to \pi/2} \ln (\sin x)$

In Exercises 17–20, explain why you cannot use substitution to determine the limit. Find the limit if it exists.

17. $\lim\limits_{x \to -2} \sqrt{x - 2}$

18. $\lim\limits_{x \to 0} \dfrac{1}{x^2}$

19. $\lim\limits_{x \to 0} \dfrac{|x|}{x}$

20. $\lim\limits_{x \to 0} \dfrac{(4 + x)^2 - 16}{x}$

In Exercises 21–30, determine the limit graphically. Confirm algebraically.

21. $\lim\limits_{x \to 1} \dfrac{x - 1}{x^2 - 1}$

22. $\lim\limits_{t \to 2} \dfrac{t^2 - 3t + 2}{t^2 - 4}$

23. $\lim\limits_{x \to 0} \dfrac{5x^3 + 8x^2}{3x^4 - 16x^2}$

24. $\lim\limits_{x \to 0} \dfrac{\dfrac{1}{2 + x} - \dfrac{1}{2}}{x}$

25. $\lim\limits_{x \to 0} \dfrac{(2 + x)^3 - 8}{x}$

26. $\lim\limits_{x \to 0} \dfrac{\sin 2x}{x}$

27. $\lim\limits_{x \to 0} \dfrac{\sin x}{2x^2 - x}$

28. $\lim\limits_{x \to 0} \dfrac{x + \sin x}{x}$

29. $\lim\limits_{x \to 0} \dfrac{\sin^2 x}{x}$

30. $\lim\limits_{x \to 0} \dfrac{3 \sin 4x}{\sin 3x}$

In Exercises 31 and 32, which of the statements are true about the function $y = f(x)$ graphed there, and which are false?

31.

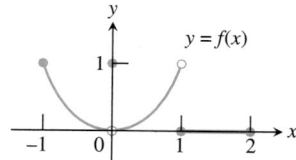

(a) $\lim\limits_{x \to -1^+} f(x) = 1$

(b) $\lim\limits_{x \to 0^-} f(x) = 0$

(c) $\lim\limits_{x \to 0^-} f(x) = 1$

(d) $\lim\limits_{x \to 0^-} f(x) = \lim\limits_{x \to 0^+} f(x)$

(e) $\lim\limits_{x \to 0} f(x)$ exists

(f) $\lim\limits_{x \to 0} f(x) = 0$

(g) $\lim\limits_{x \to 0} f(x) = 1$

(h) $\lim\limits_{x \to 1} f(x) = 1$

(i) $\lim\limits_{x \to 1} f(x) = 0$

(j) $\lim\limits_{x \to 2^-} f(x) = 2$

32.

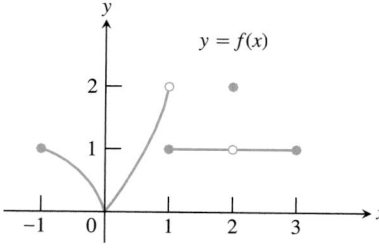

(a) $\lim\limits_{x \to -1^+} f(x) = 1$

(b) $\lim\limits_{x \to 2} f(x)$ does not exist.

(c) $\lim\limits_{x \to 2} f(x) = 2$

(d) $\lim\limits_{x \to 1^-} f(x) = 2$

(e) $\lim\limits_{x \to 1^+} f(x) = 1$

(f) $\lim\limits_{x \to 1} f(x)$ does not exist.

(g) $\lim\limits_{x \to 0^+} f(x) = \lim\limits_{x \to 0^-} f(x)$

(h) $\lim\limits_{x \to c} f(x)$ exists at every c in $(-1, 1)$.

(i) $\lim\limits_{x \to c} f(x)$ exists at every c in $(1, 3)$.

In Exercises 33–36, match the function with the table.

33. $y_1 = \dfrac{x^2 + x - 2}{x - 1}$

34. $y_1 = \dfrac{x^2 - x - 2}{x - 1}$

35. $y_1 = \dfrac{x^2 - 2x + 1}{x - 1}$

36. $y_1 = \dfrac{x^2 + x - 2}{x + 1}$

X	Y₁	
.7	-.4765	
.8	-.3111	
.9	-.1526	
1	0	
1.1	.14762	
1.2	.29091	
1.3	.43043	
X = .7		

(a)

X	Y₁	
.7	7.3667	
.8	10.8	
.9	20.9	
1	ERROR	
1.1	-18.9	
1.2	-8.8	
1.3	-5.367	
X = .7		

(b)

X	Y₁	
.7	2.7	
.8	2.8	
.9	2.9	
1	ERROR	
1.1	3.1	
1.2	3.2	
1.3	3.3	
X = .7		

(c)

X	Y₁	
.7	-.3	
.8	-.2	
.9	-.1	
1	ERROR	
1.1	.1	
1.2	.2	
1.3	.3	
X = .7		

(d)

In Exercises 37–44, determine the limit.

37. $\lim\limits_{x \to 0^+}$ int x

38. $\lim\limits_{x \to 0^-}$ int x

39. $\lim\limits_{x \to 0.01}$ int x

40. $\lim\limits_{x \to 2^-}$ int x

41. $\lim\limits_{x \to 0^+} \dfrac{x}{|x|}$

42. $\lim\limits_{x \to 0^-} \dfrac{x}{|x|}$

43. Assume that $\lim\limits_{x \to 4} f(x) = 0$ and $\lim\limits_{x \to 4} g(x) = 3$.

(a) $\lim\limits_{x \to 4} (g(x) + 3)$

(b) $\lim\limits_{x \to 4} x f(x)$

(c) $\lim\limits_{x \to 4} g^2(x)$

(d) $\lim\limits_{x \to 4} \dfrac{g(x)}{f(x) - 1}$

44. Assume that $\lim\limits_{x \to b} f(x) = 7$ and $\lim\limits_{x \to b} g(x) = -3$.

(a) $\lim\limits_{x \to b} (f(x) + g(x))$

(b) $\lim\limits_{x \to b} (f(x) \cdot g(x))$

(c) $\lim\limits_{x \to b} 4 g(x)$

(d) $\lim\limits_{x \to b} \dfrac{f(x)}{g(x)}$

In Exercises 45–48, complete parts (a), (b), and (c) for the piecewise-defined function.

(a) Draw the graph of f.

(b) Determine $\lim_{x \to c^+} f(x)$ and $\lim_{x \to c^-} f(x)$.

(c) Writing to Learn Does $\lim_{x \to c} f(x)$ exist? If so, what is it? If not, explain.

45. $c = 2, f(x) = \begin{cases} 3 - x, & x < 2 \\ \dfrac{x}{2} + 1, & x > 2 \end{cases}$

46. $c = 2, f(x) = \begin{cases} 3 - x, & x < 2 \\ 2, & x = 2 \\ x/2, & x > 2 \end{cases}$

47. $c = 1, f(x) = \begin{cases} \dfrac{1}{x - 1}, & x < 1 \\ x^3 - 2x + 5, & x \geq 1 \end{cases}$

48. $c = -1, f(x) = \begin{cases} 1 - x^2, & x \neq -1 \\ 2, & x = -1 \end{cases}$

In Exercises 49–52, complete parts (a)–(d) for the piecewise-defined function.

(a) Draw the graph of f.

(b) At what points c in the domain of f does $\lim_{x \to c} f(x)$ exist?

(c) At what points c does only the left-hand limit exist?

(d) At what points c does only the right-hand limit exist?

49. $f(x) = \begin{cases} \sin x, & -2\pi \leq x < 0 \\ \cos x, & 0 \leq x \leq 2\pi \end{cases}$

50. $f(x) = \begin{cases} \cos x, & -\pi \leq x < 0 \\ \sec x, & 0 \leq x \leq \pi \end{cases}$

51. $f(x) = \begin{cases} \sqrt{1 - x^2}, & 0 \leq x < 1 \\ 1, & 1 \leq x < 2 \\ 2, & x = 2 \end{cases}$

52. $f(x) = \begin{cases} x, & -1 \leq x < 0, \text{ or } 0 < x \leq 1 \\ 1, & x = 0 \\ 0, & x < -1, \text{ or } x > 1 \end{cases}$

In Exercises 53–56, find the limit graphically. Use the Sandwich Theorem to confirm your answer.

53. $\lim_{x \to 0} x \sin x$

54. $\lim_{x \to 0} x^2 \sin x$

55. $\lim_{x \to 0} x^2 \sin \dfrac{1}{x^2}$

56. $\lim_{x \to 0} x^2 \cos \dfrac{1}{x^2}$

57. *Free Fall* A water balloon dropped from a window high above the ground falls $y = 4.9t^2$ m in t sec. Find the balloon's

(a) average speed during the first 3 sec of fall.

(b) speed at the instant $t = 3$.

58. *Free Fall on a Small Airless Planet* A rock released from rest to fall on a small airless planet falls $y = gt^2$ m in t sec, g a constant. Suppose that the rock falls to the bottom of a crevasse 20 m below and reaches the bottom in 4 sec.

(a) Find the value of g.

(b) Find the average speed for the fall.

(c) With what speed did the rock hit the bottom?

Explorations

In Exercises 59–62, complete the following tables and state what you believe $\lim_{x \to 0} f(x)$ to be.

(a)

x	-0.1	-0.01	-0.001	-0.0001	...
$f(x)$	?	?	?	?	

(b)

x	0.1	0.01	0.001	0.0001	...
$f(x)$	?	?	?	?	

59. $f(x) = x \sin \dfrac{1}{x}$

60. $f(x) = \sin \dfrac{1}{x}$

61. $f(x) = \dfrac{10^x - 1}{x}$

62. $f(x) = x \sin (\ln |x|)$ ■

63. *Proof that* $\lim_{\theta \to 0} (\sin \theta)/\theta = 1$ *when θ is Measured in Radians* *Work in groups of two or three.* The plan is to show that the right- and left-hand limits are both 1.

(a) To show that the right-hand limit is 1, explain why we can restrict our attention to $0 < \theta < \pi/2$.

(b) Use the figure to show that

$$\text{area of } \triangle OAP = \frac{1}{2} \sin \theta,$$

$$\text{area of sector } OAP = \frac{\theta}{2},$$

$$\text{area of } \triangle OAT = \frac{1}{2} \tan \theta.$$

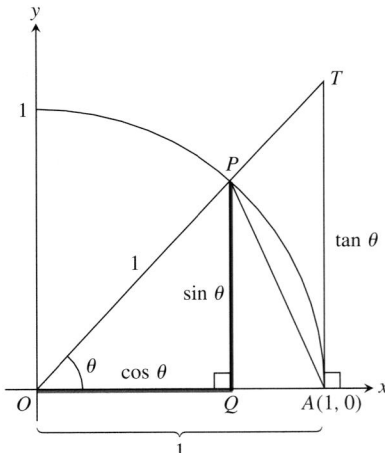

(c) Use (b) and the figure to show that for $0 < \theta < \pi/2$,

$$\frac{1}{2} \sin \theta < \frac{1}{2}\theta < \frac{1}{2} \tan \theta.$$

(d) Show that for $0 < \theta < \pi/2$ the inequality of (c) can be written in the form

$$1 < \frac{\theta}{\sin \theta} < \frac{1}{\cos \theta}.$$

(e) Show that for $0 < \theta < \pi/2$ the inequality of (d) can be written in the form

$$\cos \theta < \frac{\sin \theta}{\theta} < 1.$$

(f) Use the Sandwich Theorem to show that

$$\lim_{\theta \to 0^+} \frac{\sin \theta}{\theta} = 1.$$

(g) Show that $(\sin \theta)/\theta$ is an even function.

(h) Use (g) to show that

$$\lim_{\theta \to 0^-} \frac{\sin \theta}{\theta} = 1.$$

(i) Finally, show that

$$\lim_{\theta \to 0} \frac{\sin \theta}{\theta} = 1.$$

Extending the Ideas

64. *Controlling Outputs* Let $f(x) = \sqrt{3x - 2}$.

(a) Show that $\lim_{x \to 2} f(x) = 2 = f(2)$.

(b) Use a graph to estimate values for a and b so that $1.8 < f(x) < 2.2$ provided $a < x < b$.

(c) Use a graph to estimate values for a and b so that $1.99 < f(x) < 2.01$ provided $a < x < b$.

65. *Controlling Outputs* Let $f(x) = \sin x$.

(a) Find $f(\pi/6)$.

(b) Use a graph to estimate an interval (a, b) about $x = \pi/6$ so that $0.3 < f(x) < 0.7$ provided $a < x < b$.

(c) Use a graph to estimate an interval (a, b) about $x = \pi/6$ so that $0.49 < f(x) < 0.51$ provided $a < x < b$.

66. *Limits and Geometry* Let $P(a, a^2)$ be a point on the parabola $y = x^2$, $a > 0$. Let O be the origin and $(0, b)$ the y-intercept of the perpendicular bisector of line segment OP. Find $\lim_{P \to O} b$.

Limits Involving Infinity

Finite Limits as $x \to \pm\infty$ • Sandwich Theorem Revisited • Infinite Limits as $x \to a$ • End Behavior Models • "Seeing" Limits as $x \to \pm\infty$

Finite Limits as $x \to \pm\infty$

The symbol for infinity (∞) does not represent a real number. We use ∞ to describe the behavior of a function when the values in its domain or range out-grow all finite bounds. For example, when we say "the limit of f as x approaches infinity" we mean the limit of f as x moves increasingly far to the right on the number line. When we say "the limit of f as x approaches negative infinity $(-\infty)$" we mean the limit of f as x moves increasingly far to the left. (The limit in each case may or may not exist.)

Looking at $f(x) = 1/x$ (Figure 2.9), we observe

(a) as $x \to \infty$, $(1/x) \to 0$ and we write $\lim\limits_{x \to \infty} (1/x) = 0$,

(b) as $x \to -\infty$, $(1/x) \to 0$ and we write $\lim\limits_{x \to -\infty} (1/x) = 0$.

We say that the line $y = 0$ is a *horizontal asymptote* of the graph of f.

[−6, 6] by [−4, 4]

Figure 2.9 The graph of $f(x) = 1/x$.

Definition Horizontal Asymptote

The line $y = b$ is a **horizontal asymptote** of the graph of a function $y = f(x)$ if either

$$\lim_{x \to \infty} f(x) = b \quad \text{or} \quad \lim_{x \to -\infty} f(x) = b.$$

The graph of $f(x) = 2 + (1/x)$ has the single horizontal asymptote $y = 2$ because

$$\lim_{x \to \infty} \left(2 + \frac{1}{x} \right) = 2 \quad \text{and} \quad \lim_{x \to -\infty} \left(2 + \frac{1}{x} \right) = 2.$$

A function can have more than one horizontal asymptote, as Example 1 demonstrates.

Example 1 LOOKING FOR HORIZONTAL ASYMPTOTES

Investigate the graph of $f(x) = x/\sqrt{x^2 + 1}$.

Solution

Solve Graphically

Figure 2.10a shows the graph for $-10 \le x \le 10$. The graph climbs rapidly toward the line $y = 1$ as x moves away from the origin to the right. On our calculator screen, the graph soon becomes indistinguishable from the line. Similarly, as x moves away from the origin to the left, the graph drops rapidly toward the line $y = -1$ and soon appears to overlap the line.

Confirm Numerically

The table in Figure 2.10b confirms the rapid approach of $f(x)$ toward 1 as $x \to \infty$. Since f is an odd function of x, we can expect its values to approach -1 in a similar way as $x \to -\infty$.

Sandwich Theorem Revisited

The Sandwich Theorem also holds for limits as $x \to \pm\infty$.

Example 2 FINDING A LIMIT AS x APPROACHES ∞

Find $\lim_{x \to \infty} f(x)$ for $f(x) = \dfrac{\sin x}{x}$.

Solution

Solve Graphically and Numerically

The graph and table of values in Figure 2.11 suggest that $y = 0$ is the horizontal asymptote of f.

[–10, 10] by [–1.5, 1.5]

(a)

X	Y1	
0	0	
1	.70711	
2	.89443	
3	.94868	
4	.97014	
5	.98058	
6	.98639	

Y1 ▊ X/√(X² + 1)

(b)

Figure 2.10 (a) The graph of $f(x) = x/\sqrt{x^2 + 1}$ has two horizontal asymptotes, $y = -1$ and $y = 1$. (b) Selected values of f. (Example 1)

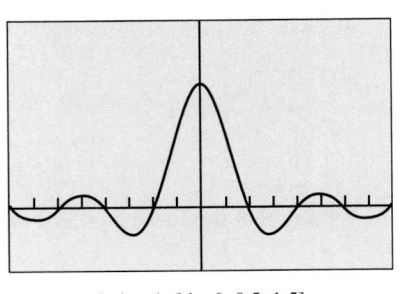

[–4π, 4π] by [–0.5, 1.5]

(a)

X	Y1	
100	–.0051	
200	–.0044	
300	–.0033	
400	–.0021	
500	–9E–4	
600	7.4E–5	
700	7.8E–4	

Y1 ▊ sin(X)/X

(b)

Figure 2.11 (a) The graph of $f(x) = (\sin x)/x$ oscillates about the x-axis. The amplitude of the oscillations decreases toward zero as $x \to \pm\infty$. (b) A table of values for f that suggests $f(x) \to 0$ as $x \to \infty$. (Example 2)

Confirm Analytically

We know that $-1 \le \sin x \le 1$. So, for $x > 0$ we have

$$-\frac{1}{x} \le \frac{\sin x}{x} \le \frac{1}{x}.$$

Therefore, by the Sandwich Theorem,

$$0 = \lim_{x \to \infty}\left(-\frac{1}{x}\right) = \lim_{x \to \infty} \frac{\sin x}{x} = \lim_{x \to \infty} \frac{1}{x} = 0.$$

Since $(\sin x)/x$ is an even function of x, we can also conclude that

$$\lim_{x \to -\infty} \frac{\sin x}{x} = 0.$$

Limits at infinity have properties similar to those of finite limits.

Theorem 5   Properties of Limits as $x \to \pm\infty$

If L, M, and k are real numbers and

$$\lim_{x \to \pm\infty} f(x) = L \quad \text{and} \quad \lim_{x \to \pm\infty} g(x) = M, \text{ then}$$

1. *Sum Rule:* $\displaystyle \lim_{x \to \pm\infty}(f(x) + g(x)) = L + M$

2. *Difference Rule:* $\displaystyle \lim_{x \to \pm\infty}(f(x) - g(x)) = L - M$

3. *Product Rule:* $\displaystyle \lim_{x \to \pm\infty}(f(x) \cdot g(x)) = L \cdot M$

4. *Constant Multiple Rule:* $\displaystyle \lim_{x \to \pm\infty}(k \cdot f(x)) = k \cdot L$

5. *Quotient Rule:* $\displaystyle \lim_{x \to \pm\infty} \frac{f(x)}{g(x)} = \frac{L}{M}, M \ne 0$

6. *Power Rule:* If r and s are integers, $s \ne 0$, then

$$\lim_{x \to \pm\infty}(f(x))^{r/s} = L^{r/s}$$

provided that $L^{r/s}$ is a real number.

We can use Theorem 5 to find limits at infinity of functions with complicated expressions, as illustrated in Example 3.

Example 3   USING THEOREM 5

Find $\displaystyle \lim_{x \to \infty} \frac{5x + \sin x}{x}$.

Solution Notice that

$$\frac{5x + \sin x}{x} = \frac{5x}{x} + \frac{\sin x}{x} = 5 + \frac{\sin x}{x}.$$

So,

$$\lim_{x \to \infty} \frac{5x + \sin x}{x} = \lim_{x \to \infty} 5 + \lim_{x \to \infty} \frac{\sin x}{x} \qquad \text{Sum Rule}$$

$$= 5 + 0 = 5. \qquad \text{Known values}$$

Exploration 1   Exploring Theorem 5

We must be careful how we apply Theorem 5.

1. (Example 3 again) Let $f(x) = 5x + \sin x$ and $g(x) = x$. Do the limits as $x \to \infty$ of f and g exist? Can we apply the Quotient Rule to $\lim_{x \to \infty} f(x)/g(x)$? Explain. Does the limit of the quotient exist?

2. Let $f(x) = \sin^2 x$ and $g(x) = \cos^2 x$. Describe the behavior of f and g as $x \to \infty$. Can we apply the Sum Rule to $\lim_{x \to \infty} (f(x) + g(x))$? Explain. Does the limit of the sum exist?

3. Let $f(x) = \ln (2x)$ and $g(x) = \ln (x + 1)$. Find the limits as $x \to \infty$ of f and g. Can we apply the Difference Rule to $\lim_{x \to \infty} (f(x) - g(x))$? Explain. Does the limit of the difference exist?

4. Based on parts 1–3, what advice might you give about applying Theorem 5?

Infinite Limits as $x \to a$

If the values of a function $f(x)$ outgrow all positive bounds as x approaches a finite number a, we say that $\lim_{x \to a} f(x) = \infty$. If the values of f become large and negative, exceeding all negative bounds as $x \to a$, we say that $\lim_{x \to a} f(x) = -\infty$.

Looking at $f(x) = 1/x$ (Figure 2.9, page 65), we observe that

$$\lim_{x \to 0^+} 1/x = \infty \quad \text{and} \quad \lim_{x \to 0^-} 1/x = -\infty.$$

We say that the line $x = 0$ is a *vertical asymptote* of the graph of f.

Definition   Vertical Asymptote

The line $x = a$ is a **vertical asymptote** of the graph of a function $y = f(x)$ if either

$$\lim_{x \to a^+} f(x) = \pm\infty \quad \text{or} \quad \lim_{x \to a^-} f(x) = \pm\infty.$$

Example 4   FINDING VERTICAL ASYMPTOTES

Find the vertical asymptotes of $f(x) = \dfrac{1}{x^2}$.

Solution The values of the function approach ∞ on either side of $x = 0$.

$$\lim_{x \to 0^+} \frac{1}{x^2} = \infty \quad \text{and} \quad \lim_{x \to 0^-} \frac{1}{x^2} = \infty.$$

The line $x = 0$ is the only vertical asymptote.

We can also say that $\lim_{x \to 0} (1/x^2) = \infty$. We can make no such statement about $1/x$.

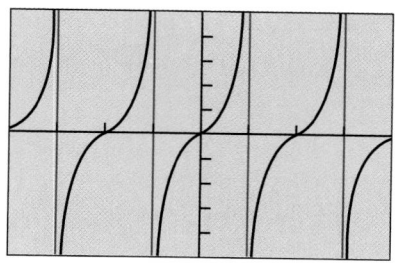

$[-2\pi, 2\pi]$ by $[-5, 5]$

Figure 2.12 The graph of $f(x) = \tan x$ has a vertical asymptote at
$$\dots, -\frac{3\pi}{2}, -\frac{\pi}{2}, \frac{\pi}{2}, \frac{3\pi}{2}, \dots . \text{(Example 5)}$$

$y = 3x^4 - 2x^3 + 3x^2 - 5x + 6$

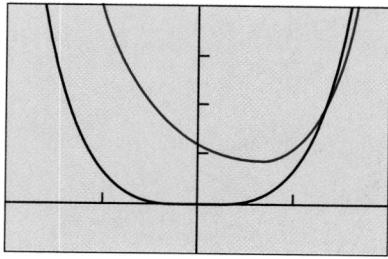

$[-2, 2]$ by $[-5, 20]$

(a)

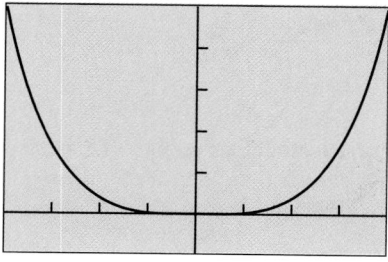

$[-20, 20]$ by $[-100000, 500000]$

(b)

Figure 2.13 The graphs of f and g, (a) distinct for $|x|$ small, are (b) nearly identical for $|x|$ large. (Example 6)

Example 5 FINDING VERTICAL ASYMPTOTES

The graph of $f(x) = \tan x = (\sin x)/(\cos x)$ has infinitely many vertical asymptotes, one at each point where the cosine is zero. If a is an odd multiple of $\pi/2$, then

$$\lim_{x \to a^+} \tan x = -\infty \quad \text{and} \quad \lim_{x \to a^-} \tan x = \infty,$$

as suggested by Figure 2.12.

You might think that the graph of a quotient always has a vertical asymptote where the denominator is zero, but that need not be the case. For example, we observed in Section 2.1 that $\lim_{x \to 0} (\sin x)/x = 1$.

End Behavior Models

For numerically large values of x, we can sometimes model the behavior of a complicated function by a simpler one that acts virtually in the same way.

Example 6 MODELING FUNCTIONS FOR $|x|$ LARGE

Let $f(x) = 3x^4 - 2x^3 + 3x^2 - 5x + 6$ and $g(x) = 3x^4$. Show that while f and g are quite different for numerically small values of x, they are virtually identical for $|x|$ large.

Solution

Solve Graphically

The graphs of f and g (Figure 2.13a), quite different near the origin, are virtually identical on a larger scale (Figure 2.13b).

Confirm Analytically

We can test the claim that g models f for numerically large values of x by examining the ratio of the two functions as $x \to \pm\infty$. We find that

$$\lim_{x \to \pm\infty} \frac{f(x)}{g(x)} = \lim_{x \to \pm\infty} \frac{3x^4 - 2x^3 + 3x^2 - 5x + 6}{3x^4}$$

$$= \lim_{x \to \pm\infty} \left(1 - \frac{2}{3x} + \frac{1}{x^2} - \frac{5}{3x^3} + \frac{2}{x^4}\right)$$

$$= 1,$$

convincing evidence that f and g behave alike for $|x|$ large.

Definition End Behavior Model

The function g is

(a) a **right end behavior model** for f if and only if $\lim_{x \to \infty} \dfrac{f(x)}{g(x)} = 1$.

(b) a **left end behavior model** for f if and only if $\lim_{x \to -\infty} \dfrac{f(x)}{g(x)} = 1$.

A function's right and left end behavior models need not be the same function.

Example 7 FINDING END BEHAVIOR MODELS

Let $f(x) = x + e^{-x}$. Show that $g(x) = x$ is a right end behavior model for f while $h(x) = e^{-x}$ is a left end behavior model for f.

Solution On the right,

$$\lim_{x \to \infty} \frac{f(x)}{g(x)} = \lim_{x \to \infty} \frac{x + e^{-x}}{x} = \lim_{x \to \infty} \left(1 + \frac{e^{-x}}{x}\right) = 1 \text{ because } \lim_{x \to \infty} \frac{e^{-x}}{x} = 0.$$

On the left,

$$\lim_{x \to -\infty} \frac{f(x)}{h(x)} = \lim_{x \to -\infty} \frac{x + e^{-x}}{e^{-x}} = \lim_{x \to -\infty} \left(\frac{x}{e^{-x}} + 1\right) = 1 \text{ because } \lim_{x \to -\infty} \frac{x}{e^{-x}} = 0.$$

The graph of f in Figure 2.14 supports these end behavior conclusions.

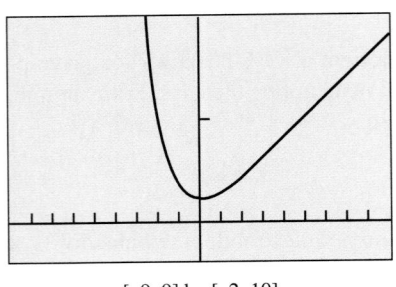

[–9, 9] by [–2, 10]

Figure 2.14 The graph of $f(x) = x + e^{-x}$ looks like the graph of $g(x) = x$ to the right of the y-axis, and like the graph of $h(x) = e^{-x}$ to the left of the y-axis. (Example 7)

If one function provides both a left and right end behavior model, it is simply called an **end behavior model.** Thus, $g(x) = 3x^4$ is an end behavior model for $f(x) = 3x^4 - 2x^3 + 3x^2 - 5x + 6$ (Example 6).

In general, $g(x) = a_n x^n$ is an end behavior model for the polynomial function $f(x) = a_n x^n + a_{n-1} x^{n-1} + \cdots + a_0, a_n \neq 0$. In the large, all polynomials behave like monomials. This is the key to the end behavior of rational functions, as illustrated in Example 8.

Example 8 FINDING END BEHAVIOR MODELS

Find an end behavior model for

(a) $f(x) = \dfrac{2x^5 + x^4 - x^2 + 1}{3x^2 - 5x + 7}$. **(b)** $g(x) = \dfrac{2x^3 - x^2 + x - 1}{5x^3 + x^2 + x - 5}$.

Solution

(a) Notice that $2x^5$ is an end behavior model for the numerator of f, and $3x^2$ is one for the denominator. This makes

$$\frac{2x^5}{3x^2} = \frac{2}{3}x^3$$

an end behavior model for f.

(b) Similarly, $2x^3$ is an end behavior model for the numerator of g, and $5x^3$ is one for the denominator of g. This makes

$$\frac{2x^3}{5x^3} = \frac{2}{5}$$

an end behavior model for g.

Notice in Example 8b that the end behavior model for g, $y = 2/5$, is also a horizontal asymptote of the graph of g, while in 8a, the graph of f does not have a horizontal asymptote. We can use the end behavior model of a rational function to identify any horizontal asymptote.

Example 9 FINDING A HORIZONTAL ASYMPTOTE

Does the graph of

$$f(x) = \frac{4x^2 - 3x + 5}{2x^3 + x - 1}$$

have a horizontal asymptote? If so, what is it?

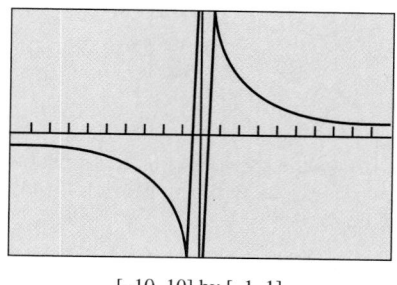

[-10, 10] by [-1, 1]

(a)

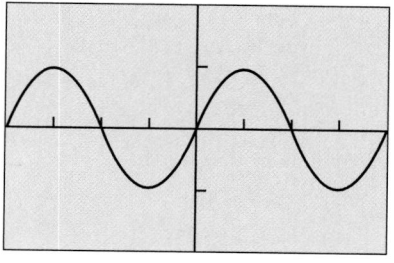

[-2π, 2π] by [-2, 2]

(b)

Figure 2.15 The graphs of (a) $f(x) =$ sin $(1/x)$ and (b) $g(x) = f(1/x) = \sin x$. (Example 10)

Solution An end behavior model for f is

$$\frac{4x^2}{2x^3} = \frac{2}{x}.$$

So, $\lim_{x \to \pm\infty} f(x) = \lim_{x \to \pm\infty} 2/x = 0$ and $y = 0$ is the horizontal asymptote of f.

We can see from Examples 8 and 9 that a rational function always has a simple power function as an end behavior model.

"Seeing" Limits as $x \to \pm\infty$

We can investigate the graph of $y = f(x)$ as $x \to \pm\infty$ by investigating the graph of $y = f(1/x)$ as $x \to 0$.

Example 10   USING SUBSTITUTION

Find $\lim\limits_{x \to \infty} \sin (1/x)$.

Solution Figure 2.15a suggests that the limit is 0. Indeed, replacing $\lim_{x \to \infty} \sin (1/x)$ by the equivalent $\lim_{x \to 0^+} \sin x = 0$ (Figure 2.15b), we find

$$\lim_{x \to \infty} \sin 1/x = \lim_{x \to 0^+} \sin x = 0.$$

Quick Review 2.2

In Exercises 1–4, find f^{-1} and graph f, f^{-1}, and $y = x$ in the same square viewing window.

1. $f(x) = 2x - 3$

2. $f(x) = e^x$

3. $f(x) = \tan^{-1} x$

4. $f(x) = \cot^{-1} x$

In Exercises 5 and 6, find the quotient $q(x)$ and remainder $r(x)$ when $f(x)$ is divided by $g(x)$.

5. $f(x) = 2x^3 - 3x^2 + x - 1$, $g(x) = 3x^3 + 4x - 5$

6. $f(x) = 2x^5 - x^3 + x - 1$, $g(x) = x^3 - x^2 + 1$

In Exercises 7–10, write a formula for **(a)** $f(-x)$ and **(b)** $f(1/x)$. Simplify where possible.

7. $f(x) = \cos x$

8. $f(x) = e^{-x}$

9. $f(x) = \dfrac{\ln x}{x}$

10. $f(x) = \left(x + \dfrac{1}{x}\right) \sin x$

Section 2.2 Exercises

In Exercises 1–8, use graphs and tables to find **(a)** $\lim_{x \to \infty} f(x)$ and **(b)** $\lim_{x \to -\infty} f(x)$. **(c)** Identify all horizontal asymptotes.

1. $f(x) = \cos\left(\dfrac{1}{x}\right)$

2. $f(x) = \dfrac{\sin 2x}{x}$

3. $f(x) = \dfrac{e^{-x}}{x}$

4. $f(x) = \dfrac{3x^3 - x + 1}{x + 3}$

5. $f(x) = \dfrac{3x + 1}{|x| + 2}$

6. $f(x) = \dfrac{2x - 1}{|x| - 3}$

7. $f(x) = \dfrac{x}{|x|}$

8. $f(x) = \dfrac{|x|}{|x| + 1}$

In Exercises 9–16, use graphs and tables to find the limits.

9. $\lim\limits_{x \to 2^+} \dfrac{1}{x - 2}$

10. $\lim\limits_{x \to 2^-} \dfrac{x}{x - 2}$

11. $\lim\limits_{x \to -3^-} \dfrac{1}{x + 3}$

12. $\lim\limits_{x \to -3^+} \dfrac{x}{x + 3}$

13. $\lim_{x\to 0^+} \dfrac{\text{int } x}{x}$

14. $\lim_{x\to 0^-} \dfrac{\text{int } x}{x}$

15. $\lim_{x\to 0^+} \csc x$

16. $\lim_{x\to (\pi/2)^+} \sec x$

In Exercises 17–22, **(a)** find the vertical asymptotes of the graph of $f(x)$. **(b)** Describe the behavior of $f(x)$ to the left and right of each vertical asymptote.

17. $f(x) = \dfrac{1}{x^2 - 4}$

18. $f(x) = \dfrac{x^2 - 1}{2x + 4}$

19. $f(x) = \dfrac{x^2 - 2x}{x + 1}$

20. $f(x) = \dfrac{1 - x}{2x^2 - 5x - 3}$

21. $f(x) = \cot x$

22. $f(x) = \sec x$

In Exercises 23–28, find $\lim_{x\to\infty} y$ and $\lim_{x\to -\infty} y$.

23. $y = \left(2 - \dfrac{x}{x+1}\right)\left(\dfrac{x^2}{5 + x^2}\right)$

24. $y = \left(\dfrac{2}{x} + 1\right)\left(\dfrac{5x^2 - 1}{x^2}\right)$

25. $y = \dfrac{\cos (1/x)}{1 + (1/x)}$

26. $y = \dfrac{2x + \sin x}{x}$

27. $y = \dfrac{\sin x}{2x^2 + x}$

28. $y = \dfrac{x \sin x + 2 \sin x}{2x^2}$

In Exercises 29–32, match the function with the graph of its end behavior model.

29. $y = \dfrac{2x^3 - 3x^2 + 1}{x + 3}$

30. $y = \dfrac{x^5 - x^4 + x + 1}{2x^2 + x - 3}$

31. $y = \dfrac{2x^4 - x^3 + x^2 - 1}{2 - x}$

32. $y = \dfrac{x^4 - 3x^3 + x^2 - 1}{1 - x^2}$

(a)

(b)

(c)

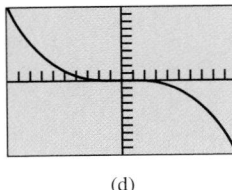

(d)

In Exercises 33–38, **(a)** find a power function end behavior model for f. **(b)** Identify any horizontal asymptotes.

33. $f(x) = 3x^2 - 2x + 1$

34. $f(x) = -4x^3 + x^2 - 2x - 1$

35. $f(x) = \dfrac{x - 2}{2x^2 + 3x - 5}$

36. $f(x) = \dfrac{3x^2 - x + 5}{x^2 - 4}$

37. $f(x) = \dfrac{4x^3 - 2x + 1}{x - 2}$

38. $f(x) = \dfrac{-x^4 + 2x^2 + x - 3}{x^2 - 4}$

In Exercises 39–42, find **(a)** a simple basic function as a right end behavior model and **(b)** a simple basic function as a left end behavior model for the function.

39. $y = e^x - 2x$

40. $y = x^2 + e^{-x}$

41. $y = x + \ln |x|$

42. $y = x^2 + \sin x$

In Exercises 43–46, use the graph of $y = f(1/x)$ to find $\lim_{x\to\infty} f(x)$ and $\lim_{x\to -\infty} f(x)$.

43. $f(x) = xe^x$

44. $f(x) = x^2 e^{-x}$

45. $f(x) = \dfrac{\ln |x|}{x}$

46. $f(x) = x \sin \dfrac{1}{x}$

In Exercises 47 and 48, find the limit of $f(x)$ as **(a)** $x\to -\infty$, **(b)** $x\to\infty$, **(c)** $x\to 0^-$, and **(d)** $x\to 0^+$.

47. $f(x) = \begin{cases} 1/x, & x < 0 \\ -1, & x \ge 0 \end{cases}$

48. $f(x) = \begin{cases} \dfrac{x - 2}{x - 1}, & x \le 0 \\ 1/x^2, & x > 0 \end{cases}$

In Exercises 49 and 50, *work in groups of two or three* and sketch a graph of a function $y = f(x)$ that satisfies the stated conditions. Include any asymptotes.

49. $\lim_{x\to 1} f(x) = 2$, $\lim_{x\to 5^-} f(x) = \infty$, $\lim_{x\to 5^+} f(x) = \infty$,

$\lim_{x\to\infty} f(x) = -1$, $\lim_{x\to -2^+} f(x) = -\infty$,

$\lim_{x\to -2^-} f(x) = \infty$, $\lim_{x\to -\infty} f(x) = 0$

50. $\lim_{x\to 2} f(x) = -1$, $\lim_{x\to 4^+} f(x) = -\infty$, $\lim_{x\to 4^-} f(x) = \infty$,

$\lim_{x\to\infty} f(x) = \infty$, $\lim_{x\to -\infty} f(x) = 2$

51. End Behavior Models *Work in groups of two or three.* Suppose that $g_1(x)$ is a right end behavior model for $f_1(x)$ and that $g_2(x)$ is a right end behavior model for $f_2(x)$. Explain why this makes $g_1(x)/g_2(x)$ a right end behavior model for $f_1(x)/f_2(x)$.

52. Writing to Learn Let L be a real number, $\lim_{x\to c} f(x) = L$, and $\lim_{x\to c} g(x) = \infty$ or $-\infty$. Can $\lim_{x\to c} (f(x) + g(x))$ be determined? Explain.

53. Table 2.2 gives the average amount per grant of emergency financial assistance for veterans in Franklin County, Ohio. Let $x = 0$ represent 1980, $x = 1$ represent 1981, and so forth.

Table 2.2 Emergency Financial Assistance

Year	Average Amount (dollars)
1986	145.38
1987	155.00
1988	230.43
1989	420.70
1990	494.55
1991	555.00
1992	508.77
1993	460.92
1994	453.77

Source: Franklin County Veterans Service Commission as reported by Mary Stephens in *The Columbus [Ohio] Dispatch*, May 31, 1995.

(a) Find a cubic regression equation and superimpose its graph on a scatter plot of the data.

(b) Find a quartic regression equation and superimpose its graph on a scatter plot of the data.

(c) Use each regression equation to predict the average amount per grant of financial assistance for veterans in the year 2000.

(d) Writing to Learn Find an end behavior model for each regression equation. What does each predict about the future sizes of grants? Does this seem reasonable? Explain.

Exploration

54. *Exploring Properties of Limits* Find the limits of f, g, and fg as $x \to c$.

(a) $f(x) = \dfrac{1}{x}$, $\quad g(x) = x$, $\quad c = 0$

(b) $f(x) = -\dfrac{2}{x^3}$, $\quad g(x) = 4x^3$, $\quad c = 0$

(c) $f(x) = \dfrac{3}{x - 2}$, $\quad g(x) = (x - 2)^3$, $\quad c = 2$

(d) $f(x) = \dfrac{5}{(3 - x)^4}$, $\quad g(x) = (x - 3)^2$, $\quad c = 3$

(e) Writing to Learn Suppose that $\lim_{x \to c} f(x) = 0$ and $\lim_{x \to c} g(x) = \infty$. Based on your observations in (a)–(d), what can you say about $\lim_{x \to c} (f(x) \cdot g(x))$? ∎

Extending the Ideas

55. *The Greatest Integer Function*

(a) Show that

$$\frac{x - 1}{x} < \frac{\text{int } x}{x} \le 1 \ (x > 0) \text{ and } \frac{x - 1}{x} > \frac{\text{int } x}{x} \ge 1 \ (x < 0).$$

(b) Determine $\lim_{x \to \infty} \dfrac{\text{int } x}{x}$.   **(c)** Determine $\lim_{x \to -\infty} \dfrac{\text{int } x}{x}$.

56. *Sandwich Theorem* Use the Sandwich Theorem to confirm the limit as $x \to \infty$ found in Exercise 3.

57. Writing to Learn Explain why there is no value L for which $\lim_{x \to \infty} \sin x = L$.

In Exercises 58–60, find the limit. Give a convincing argument that the value is correct.

58. $\lim_{x \to \infty} \dfrac{\ln x^2}{\ln x}$ **59.** $\lim_{x \to \infty} \dfrac{\ln x}{\log x}$

60. $\lim_{x \to \infty} \dfrac{\ln (x + 1)}{\ln x}$

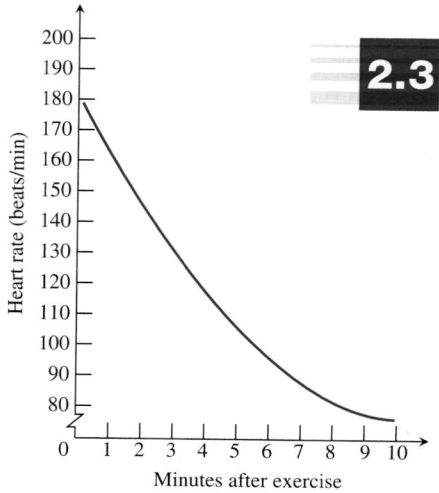

Figure 2.16 How the heartbeat returns to a normal rate after running.

2.3 Continuity

Continuity at a Point • Continuous Functions • Algebraic Combinations • Composites • Intermediate Value Theorem for Continuous Functions

Continuity at a Point

When we plot function values generated in the laboratory or collected in the field, we often connect the plotted points with an unbroken curve to show what the function's values are likely to have been at the times we did not measure (Figure 2.16). In doing so, we are assuming that we are working with a *continuous function,* a function whose outputs vary continuously with the inputs and do not jump from one value to another without taking on the values in between. Any function $y = f(x)$ whose graph can be sketched in one continuous motion without lifting the pencil is an example of a continuous function.

Figure 2.17 The laser was developed as a result of an understanding of the nature of the atom.

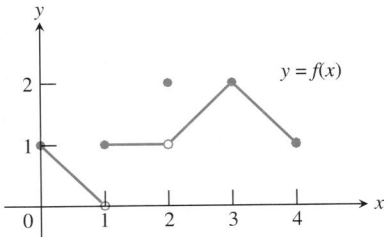

Figure 2.18 The function is continuous on $[0, 4]$ except at $x = 1$ and $x = 2$. (Example 1)

Continuous functions are the functions we use to find a planet's closest point of approach to the sun or the peak concentration of antibodies in blood plasma. They are also the functions we use to describe how a body moves through space or how the speed of a chemical reaction changes with time. In fact, so many physical processes proceed continuously that throughout the eighteenth and nineteenth centuries it rarely occurred to anyone to look for any other kind of behavior. It came as a surprise when the physicists of the 1920s discovered that light comes in particles and that heated atoms emit light at discrete frequencies (Figure 2.17). As a result of these and other discoveries, and because of the heavy use of discontinuous functions in computer science, statistics, and mathematical modeling, the issue of continuity has become one of practical as well as theoretical importance.

To understand continuity, we need to consider a function like the one in Figure 2.18, whose limits we investigated in Example 8, Section 2.1.

Example 1 INVESTIGATING CONTINUITY

Find the points at which the function f in Figure 2.18 is continuous, and the points at which f is discontinuous.

Solution The function f is continuous at every point in its domain $[0, 4]$ except at $x = 1$ and $x = 2$. At these points there are breaks in the graph. Note the relationship between the limit of f and the value of f at each point of the function's domain.

Points at which f is continuous:

At $x = 0$, $\lim\limits_{x \to 0^+} f(x) = f(0)$.

At $x = 4$, $\lim\limits_{x \to 4^-} f(x) = f(4)$.

At $0 < c < 4$, $c \neq 1, 2$, $\lim\limits_{x \to c} f(x) = f(c)$.

Points at which f is discontinuous:

At $x = 1$, $\lim\limits_{x \to 1} f(x)$ does not exist.

At $x = 2$, $\lim\limits_{x \to 2} f(x) = 1$, but $1 \neq f(2)$.

At $c < 0$, $c > 4$,   these points are not in the domain of f.

To define continuity at a point in a function's domain, we need to define continuity at an interior point (which involves a two-sided limit) and continuity at an endpoint (which involves a one-sided limit). (Figure 2.19)

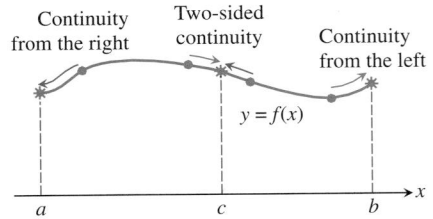

Figure 2.19 Continuity at points a, b, and c for a function $y = f(x)$ that is continuous on the interval $[a, b]$.

> **Definition** **Continuity at a Point**
>
> *Interior Point:* A function $y = f(x)$ is **continuous at an interior point c** of its domain if
>
> $$\lim_{x \to c} f(x) = f(c).$$
>
> *Endpoint:* A function $y = f(x)$ is **continuous at a left endpoint a** or is **continuous at a right endpoint b** of its domain if
>
> $$\lim_{x \to a^+} f(x) = f(a) \quad \text{or} \quad \lim_{x \to b^-} f(x) = f(b), \quad \text{respectively.}$$

If a function f is not continuous at a point c, we say that f is **discontinuous** at c and c is a **point of discontinuity** of f. Note that c need not be in the domain of f.

Example 2 FINDING POINTS OF CONTINUITY & DISCONTINUITY

Find the points of continuity and the points of discontinuity of the greatest integer function (Figure 2.20).

Solution For the function to be continuous at $x = c$, the limit as $x \to c$ must exist and must equal the value of the function at $x = c$. The greatest integer function is discontinuous at every integer. For example,

$$\lim_{x \to 3^-} \text{int } x = 2 \quad \text{and} \quad \lim_{x \to 3^+} \text{int } x = 3$$

so the limit as $x \to 3$ does not exist. Notice that int $3 = 3$. In general, if n is any integer,

$$\lim_{x \to n^-} \text{int } x = n - 1 \quad \text{and} \quad \lim_{x \to n^+} \text{int } x = n,$$

so the limit as $x \to n$ does not exist.

The greatest integer function is continuous at every other real number. For example,

$$\lim_{x \to 1.5} \text{int } x = 1 = \text{int } 1.5.$$

In general, if $n - 1 < c < n$, n an integer, then

$$\lim_{x \to c} \text{int } x = n - 1 = \text{int } c.$$

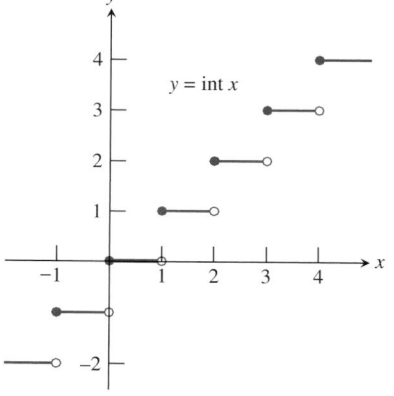

Figure 2.20 The function int x is continuous at every noninteger point. (Example 2)

Shirley Ann Jackson
1946–

Distinguished scientist, Shirley Jackson credits her interest in science to her parents and excellent mathematics and science teachers in high school. She studied physics, and in 1973, became the first African American woman to earn a Ph.D. at the Massachusetts Institute of Technology. Since then, Dr. Jackson has done research on topics relating to theoretical material sciences, has received numerous scholarships and honors, and has published more than one hundred scientific articles.

Figure 2.21 is a catalog of discontinuity types. The function in (a) is continuous at $x = 0$. The function in (b) would be continuous if it had $f(0) = 1$. The function in (c) would be continuous if $f(0)$ were 1 instead of 2. The discontinuities in (b) and (c) are **removable.** Each function has a limit as $x \to 0$, and we can remove the discontinuity by setting $f(0)$ equal to this limit.

The discontinuities in (d)–(f) of Figure 2.21 are more serious: $\lim_{x \to 0} f(x)$ does not exist and there is no way to improve the situation by changing f at 0. The step function in (d) has a **jump discontinuity:** the one-sided limits exist but have different values. The function $f(x) = 1/x^2$ in (e) has an **infinite discontinuity.** The function in (f) has an **oscillating discontinuity:** it oscillates too much to have a limit as $x \to 0$.

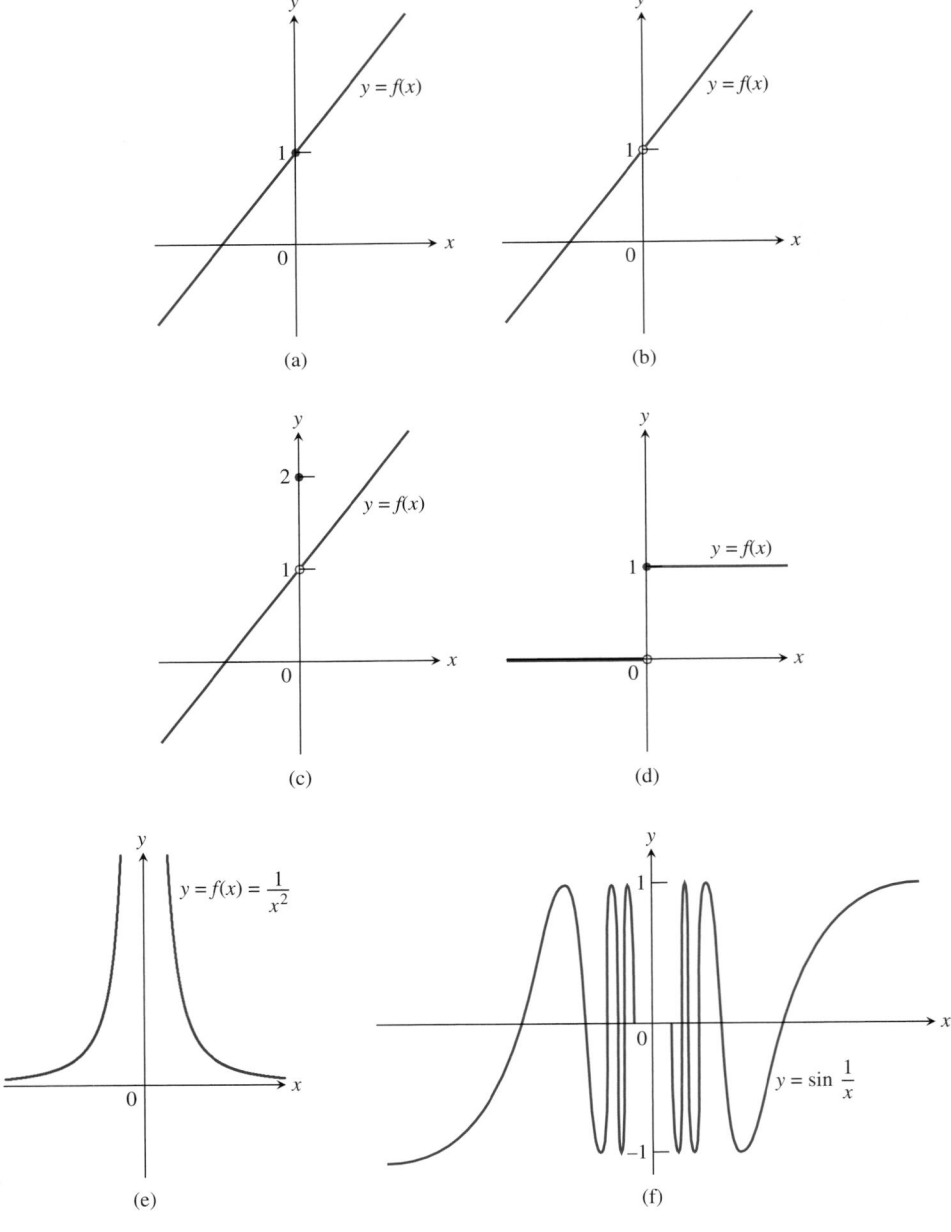

Figure 2.21 The function in (a) is continuous at $x = 0$. The functions in (b)–(f) are not.

Exploration 1 **Removing a Discontinuity**

Let $f(x) = \dfrac{x^3 - 7x - 6}{x^2 - 9}$.

1. Factor the denominator. What is the domain of f?

2. Investigate the graph of f around $x = 3$ to see that f has a removable discontinuity at $x = 3$.

3. How should f be defined at $x = 3$ to remove the discontinuity? Use zoom-in and tables as necessary.

4. Show that $(x - 3)$ is a factor of the numerator of f, and remove all common factors. Now compute the limit as $x \to 3$ of the reduced form for f.

5. Show that the *extended function*

$$g(x) = \begin{cases} \dfrac{x^3 - 7x - 6}{x^2 - 9}, & x \neq 3 \\ 10/3, & x = 3 \end{cases}$$

is continuous at $x = 3$. The function g is the **continuous extension** of the original function f to include $x = 3$.

Continuous Functions

A function is **continuous on an interval** if and only if it is continuous at every point of the interval. A **continuous function** is one that is continuous at every point of its domain. A continuous function need not be continuous on every interval. For example, $y = 1/x$ is not continuous on $[-1, 1]$.

Example 3   IDENTIFYING CONTINUOUS FUNCTIONS

The reciprocal function $y = 1/x$ (Figure 2.22) is a continuous function because it is continuous at every point of its domain. However, it has a point of discontinuity at $x = 0$ because it is not defined there.

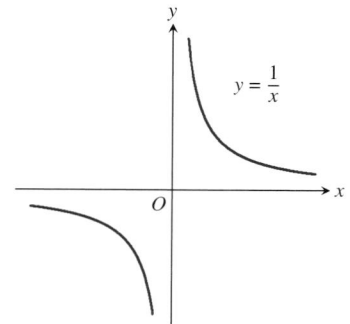

Figure 2.22 The function $y = 1/x$ is continuous at every value of x except $x = 0$. It has a point of discontinuity at $x = 0$. (Example 3)

Polynomial functions f are continuous at every real number c because $\lim_{x \to c} f(x) = f(c)$. Rational functions are continuous at every point of their domains. They have points of discontinuity at the zeros of their denominators. The absolute value function $y = |x|$ is continuous at every real number. The exponential functions, logarithmic functions, trigonometric functions, and radical functions like $y = \sqrt[n]{x}$ (n a positive integer greater than 1) are continuous at every point of their domains. All of these functions are continuous functions.

Algebraic Combinations

As you may have guessed, algebraic combinations of continuous functions are continuous wherever they are defined.

> **Theorem 6 Properties of Continuous Functions**
>
> If the functions f and g are continuous at $x = c$, then the following combinations are continuous at $x = c$.
>
> 1. *Sums:* $f + g$
> 2. *Differences:* $f - g$
> 3. *Products:* $f \cdot g$
> 4. *Constant multiples:* $k \cdot f,$ for any number k
> 5. *Quotients:* $f/g,$ provided $g(c) \neq 0$

Composites

All composites of continuous functions are continuous. This means composites like

$$y = \sin(x^2) \quad \text{and} \quad y = |\cos x|$$

are continuous at every point at which they are defined. The idea is that if $f(x)$ is continuous at $x = c$ and $g(x)$ is continuous at $x = f(c)$, then $g \circ f$ is continuous at $x = c$ (Figure 2.23). In this case, the limit as $x \to c$ is $g(f(c))$.

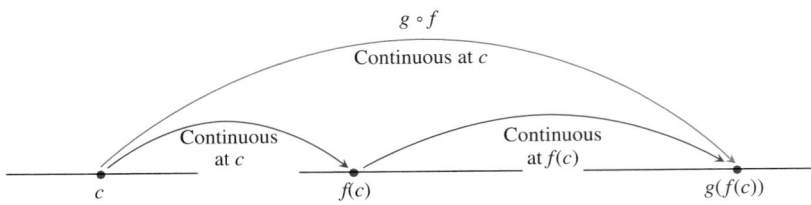

Figure 2.23 Composites of continuous functions are continuous.

> **Theorem 7 Composite of Continuous Functions**
>
> If f is continuous at c and g is continuous at $f(c)$, then the composite $g \circ f$ is continuous at c.

Example 4   USING THEOREM 7

Show that $y = \left| \dfrac{x \sin x}{x^2 + 2} \right|$ is continuous.

Solution The graph (Figure 2.24) of $y = |(x \sin x)/(x^2 + 2)|$ suggests that the function is continuous at every value of x. By letting

$$g(x) = |x| \quad \text{and} \quad f(x) = \frac{x \sin x}{x^2 + 2},$$

we see that y is the composite $g \circ f$.

We know that the absolute value function g is continuous. The function f is continuous by Theorem 6. Their composite is continuous by Theorem 7.

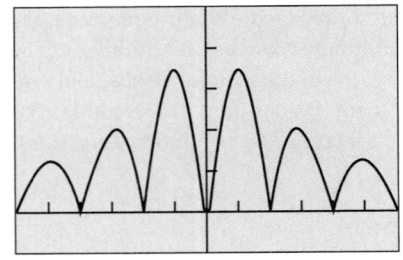

$[-3\pi, 3\pi]$ by $[-0.1, 0.5]$

Figure 2.24 The graph suggests that $y = |(x \sin x)/(x^2 + 2)|$ is continuous. (Example 4)

Intermediate Value Theorem for Continuous Functions

Functions that are continuous on intervals have properties that make them particularly useful in mathematics and its applications. One of these is the *intermediate value property*. A function is said to have the **intermediate value property** if it never takes on two values without taking on all the values in between.

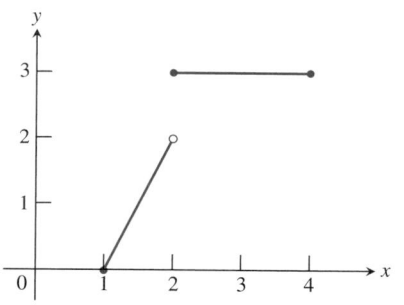

Figure 2.25 The function

$$f(x) = \begin{cases} 2x - 2, & 1 \le x < 2 \\ 3, & 2 \le x \le 4 \end{cases}$$

does not take on all values between $f(1) = 0$ and $f(4) = 3$; it misses all the values between 2 and 3.

Grapher Failure

In connected mode, a grapher may conceal a function's discontinuities by portraying the graph as a connected curve when it is not. To see what we mean, graph $y = $ int (x) in a $[-10, 10]$ by $[-10, 10]$ window in both connected and dot modes. A knowledge of where to expect discontinuities will help you recognize this form of grapher failure.

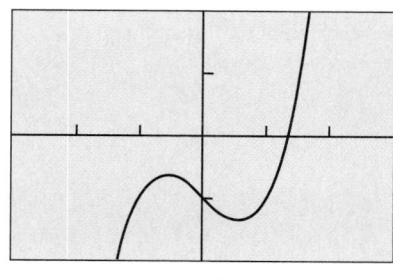

$[-3, 3]$ by $[-2, 2]$

Figure 2.26 The graph of $f(x) = x^3 - x - 1$. (Example 5)

> **Theorem 8** **The Intermediate Value Theorem for Continuous Functions**
>
> A function $y = f(x)$ that is continuous on a closed interval $[a, b]$ takes on every value between $f(a)$ and $f(b)$. In other words, if y_0 is between $f(a)$ and $f(b)$, then $y_0 = f(c)$ for some c in $[a, b]$.
>
>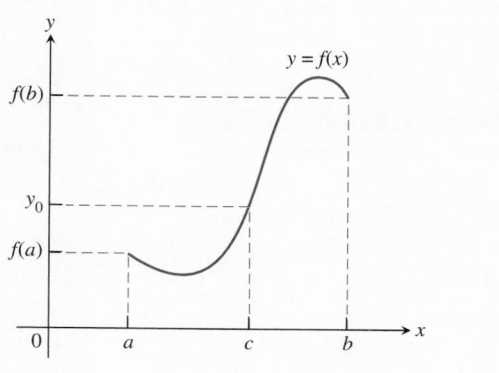

The continuity of f on the interval is essential to Theorem 8. If f is discontinuous at even one point of the interval, the theorem's conclusion may fail, as it does for the function graphed in Figure 2.25.

A Consequence for Graphing: Connectivity Theorem 8 is the reason why the graph of a function continuous on an interval cannot have any breaks. The graph will be **connected**, a single, unbroken curve, like the graph of sin x. It will not have jumps like those in the graph of the greatest integer function int x, or separate branches like we see in the graph of $1/x$.

Most graphers can plot points *(dot mode)*. Some can turn on pixels between plotted points to suggest an unbroken curve *(connected mode)*. For functions, the connected format basically assumes that outputs *vary continuously* with inputs and do not jump from one value to another without taking on all values in between.

Example 5 USING THEOREM 8

Is any real number exactly 1 less than its cube?

Solution We answer this question by applying the Intermediate Value Theorem in the following way. Any such number must satisfy the equation $x = x^3 - 1$ or, equivalently, $x^3 - x - 1 = 0$. Hence, we are looking for a zero value of the continuous function $f(x) = x^3 - x - 1$ (Figure 2.26). The function changes sign between 1 and 2, so there must be a point c between 1 and 2 where $f(c) = 0$.

Quick Review 2.3

1. Find $\lim\limits_{x \to -1} \dfrac{3x^2 - 2x + 1}{x^3 + 4}$.

2. Let $f(x) = \text{int } x$. Find each limit.

(a) $\lim\limits_{x \to -1^-} f(x)$ (b) $\lim\limits_{x \to -1^+} f(x)$ (c) $\lim\limits_{x \to -1} f(x)$ (d) $f(-1)$

3. Let $f(x) = \begin{cases} x^2 - 4x + 5, & x < 2 \\ 4 - x, & x \geq 2. \end{cases}$

Find each limit.

(a) $\lim\limits_{x \to 2^-} f(x)$ (b) $\lim\limits_{x \to 2^+} f(x)$ (c) $\lim\limits_{x \to 2} f(x)$ (d) $f(2)$

In Exercises 4–6, find the remaining functions in the list of functions: f, g, $f \circ g$, $g \circ f$.

4. $f(x) = \dfrac{2x - 1}{x + 5}$, $g(x) = \dfrac{1}{x} + 1$

5. $f(x) = x^2$, $(g \circ f)(x) = \sin x^2$,   domain of $g = [0, \infty)$

6. $g(x) = \sqrt{x - 1}$, $(g \circ f)(x) = 1/x$, $x > 0$

7. Use factoring to solve $2x^2 + 9x - 5 = 0$.

8. Use graphing to solve $x^3 + 2x - 1 = 0$.

In Exercises 9 and 10, let

$$f(x) = \begin{cases} 5 - x, & x \leq 3 \\ -x^2 + 6x - 8, & x > 3. \end{cases}$$

9. Solve the equation $f(x) = 4$.

10. Find a value of c for which the equation $f(x) = c$ has no solution.

Section 2.3 Exercises

In Exercises 1–10, find the points of discontinuity of the function. Identify each type of discontinuity.

1. $y = \dfrac{1}{(x + 2)^2}$

2. $y = \dfrac{x + 1}{x^2 - 4x + 3}$

3. $y = \dfrac{1}{x^2 + 1}$

4. $y = |x - 1|$

5. $y = \sqrt{2x + 3}$

6. $y = \sqrt[3]{2x - 1}$

7. $y = |x|/x$

8. $y = \cot x$

9. $y = e^{1/x}$

10. $y = \ln(x + 1)$

In Exercises 11–18, use the function f defined and graphed below to answer the questions.

$$f(x) = \begin{cases} x^2 - 1, & -1 \leq x < 0 \\ 2x, & 0 < x < 1 \\ 1, & x = 1 \\ -2x + 4, & 1 < x < 2 \\ 0, & 2 < x < 3 \end{cases}$$

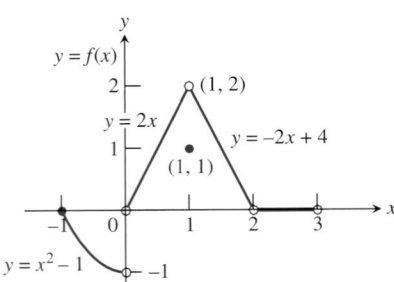

11. (a) Does $f(-1)$ exist?

(b) Does $\lim\limits_{x \to -1^+} f(x)$ exist?

(c) Does $\lim\limits_{x \to -1^+} f(x) = f(-1)$?

(d) Is f continuous at $x = -1$?

12. (a) Does $f(1)$ exist?

(b) Does $\lim\limits_{x \to 1} f(x)$ exist?

(c) Does $\lim\limits_{x \to 1} f(x) = f(1)$?

(d) Is f continuous at $x = 1$?

13. (a) Is f defined at $x = 2$? (Look at the definition of f.)

(b) Is f continuous at $x = 2$?

14. At what values of x is f continuous?

15. What value should be assigned to $f(2)$ to make the extended function continuous at $x = 2$?

16. What new value should be assigned to $f(1)$ to make the new function continuous at $x = 1$?

17. Writing to Learn Is it possible to extend f to be continuous at $x = 0$? If so, what value should the extended function have there? If not, why not?

18. Writing to Learn Is it possible to extend f to be continuous at $x = 3$? If so, what value should the extended function have there? If not, why not?

In Exercises 19–24, (a) find each point of discontinuity.
(b) Which of the discontinuities are removable? not removable? Give reasons for your answers.

19. $f(x) = \begin{cases} 3 - x, & x < 2 \\ \dfrac{x}{2} + 1, & x > 2 \end{cases}$

20. $f(x) = \begin{cases} 3 - x, & x < 2 \\ 2, & x = 2 \\ x/2, & x > 2 \end{cases}$

21. $f(x) = \begin{cases} \dfrac{1}{x - 1}, & x < 1 \\ x^3 - 2x + 5, & x \geq 1 \end{cases}$

22. $f(x) = \begin{cases} 1 - x^2, & x \neq -1 \\ 2, & x = -1 \end{cases}$

23.

24.

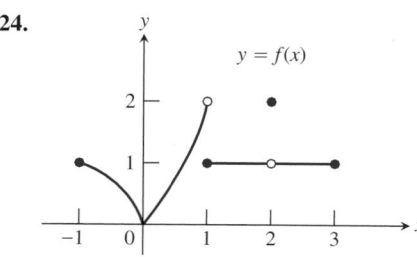

In Exercises 25–30, give a formula for the extended function that is continuous at the indicated point.

25. $f(x) = \dfrac{x^2 - 9}{x + 3}, \quad x = -3$ **26.** $f(x) = \dfrac{x^3 - 1}{x^2 - 1}, \quad x = 1$

27. $f(x) = \dfrac{\sin x}{x}, \quad x = 0$ **28.** $f(x) = \dfrac{\sin 4x}{x}, \quad x = 0$

29. $f(x) = \dfrac{x - 4}{\sqrt{x} - 2}, \quad x = 4$

30. $f(x) = \dfrac{x^3 - 4x^2 - 11x + 30}{x^2 - 4}, \quad x = 2$

In Exercises 31–34, *work in groups of two or three.* Verify that the function is continuous and state its domain. Indicate which theorems you are using, and which functions you are assuming to be continuous.

31. $y = \dfrac{1}{\sqrt{x + 2}}$ **32.** $y = x^2 + \sqrt[4]{4 - x}$

33. $y = |x^2 - 4x|$ **34.** $y = \begin{cases} \dfrac{x^2 - 1}{x - 1}, & x \neq 1 \\ 2, & x = 1 \end{cases}$

In Exercises 35–38, sketch a possible graph for a function f that has the stated properties.

35. $f(3)$ exists but $\lim_{x \to 3} f(x)$ does not.

36. $f(-2)$ exists, $\lim_{x \to -2^+} f(x) = f(-2)$, but $\lim_{x \to -2} f(x)$ does not exist.

37. $f(4)$ exists, $\lim_{x \to 4} f(x)$ exists, but f is not continuous at $x = 4$.

38. $f(x)$ is continuous for all x except $x = 1$, where f has a nonremovable discontinuity.

39. *Solving Equations* Is any real number exactly 1 less than its fourth power? Give any such values accurate to 3 decimal places.

40. *Solving Equations* Is any real number exactly 2 more than its cube? Give any such values accurate to 3 decimal places.

41. *Continuous Function* Find a value for a so that the function

$$f(x) = \begin{cases} x^2 - 1, & x < 3 \\ 2ax, & x \geq 3 \end{cases}$$

is continuous.

42. **Writing to Learn** Explain why the equation $e^{-x} = x$ has at least one solution.

43. *Salary Negotiation* A welder's contract promises a 3.5% salary increase each year for 4 years and Luisa has an initial salary of $36,500.

(a) Show that Luisa's salary is given by

$$y = 36{,}500(1.035)^{\mathrm{int}\ t},$$

where t is the time, measured in years, since Luisa signed the contract.

(b) Graph Luisa's salary function. At what values of t is it continuous?

44. *Airport Parking* Valuepark charge $1.10 per hour or fraction of an hour for airport parking. The maximum charge per day is $7.25.

(a) Write a formula that gives the charge for x hours with $0 \leq x \leq 24$. (*Hint:* See Exercise 43.)

(b) Graph the function in (a). At what values of x is it continuous?

Exploration

45. Let $f(x) = \left(1 + \dfrac{1}{x}\right)^x$.

(a) Find the domain of f.

(b) Draw the graph of f.

(c) **Writing to Learn** Explain why $x = -1$ and $x = 0$ are points of discontinuity of f.

(d) **Writing to Learn** Are either of the discontinuities in (c) removable? Explain.

(e) Use graphs and tables to estimate $\lim_{x \to \infty} f(x)$.   ■

Extending the Ideas

46. *Continuity at a Point* Show that $f(x)$ is continuous at $x = a$ if and only if

$$\lim_{h \to 0} f(a + h) = f(a).$$

47. *Continuity on Closed Intervals* Let f be continuous and never zero on $[a, b]$. Show that either $f(x) > 0$ for all x in $[a, b]$ or $f(x) < 0$ for all x in $[a, b]$.

48. *Properties of Continuity* Prove that if f is continuous on an interval, then so is $|f|$.

49. *Everywhere Discontinuous* Give a convincing argument that the following function is not continuous at any real number.

$$f(x) = \begin{cases} 1, & \text{if } x \text{ is rational} \\ 0, & \text{if } x \text{ is irrational} \end{cases}$$

Rates of Change and Tangent Lines

Average Rates of Change   •   Tangent to a Curve   •   Slope of a
Curve   •   Normal to a Curve   •   Speed Revisited

Average Rates of Change

We encounter average rates of change in such forms as average speed (in miles
per hour), growth rates of populations (in percent per year), and average
monthly rainfall (in inches per month). The **average rate of change** of a quantity over a period of time is the amount of change divided by the time it takes.
In general, the *average rate of change* of a function over an interval is the
amount of change divided by the length of the interval.

Example 1   FINDING AVERAGE RATE OF CHANGE

Find the average rate of change of $f(x) = x^3 - x$ over the interval $[1, 3]$.

Solution   Since $f(1) = 0$ and $f(3) = 24$, the average rate of change over
the interval $[1, 3]$ is

$$\frac{f(3) - f(1)}{3 - 1} = \frac{24 - 0}{2} = 12.$$

Experimental biologists often want to know the rates at which populations
grow under controlled laboratory conditions. Figure 2.27 shows how the number of fruit flies *(Drosophila)* grew in a controlled 50-day experiment. The
graph was made by counting flies at regular intervals, plotting a point for each
count, and drawing a smooth curve through the plotted points.

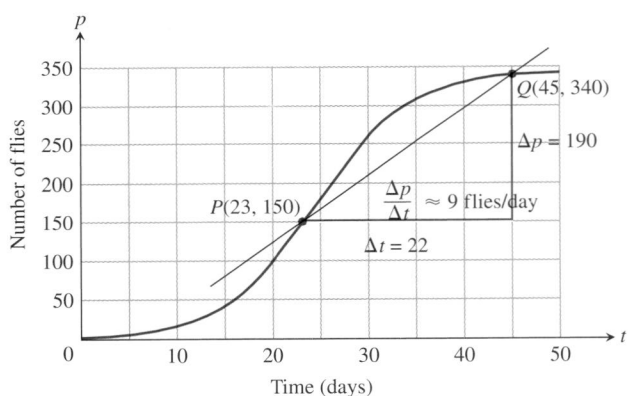

Figure 2.27   Growth of a fruit fly population in a controlled experiment.
Source: Elements of Mathematical Biology. (Example 2)

Secant to a Curve

A line through two points on a curve is a
secant to the curve.

Example 2   GROWING *DROSOPHILA* IN A LABORATORY

Use the points $P(23, 150)$ and $Q(45, 340)$ in Figure 2.27 to compute the
average rate of change and the slope of the secant line *PQ*.

Solution   There were 150 flies on day 23 and 340 flies on day 45. This
gives an increase of $340 - 150 = 190$ flies in $45 - 23 = 22$ days.

Marjorie Lee Browne

1914–1979

When Marjorie Browne graduated from the University of Michigan in 1949, she was one of the first two African American women to be awarded a Ph.D. in Mathematics. Browne went on to become chairperson of the mathematics department at North Carolina Central University, and succeeded in obtaining grants for retraining high school mathematics teachers. Problem 34 in this section deals with the rate of change of congressional funding for collegiate academic programs.

The average rate of change in the population p from day 23 to day 45 was

$$\textit{Average rate of change: } \frac{\Delta p}{\Delta t} = \frac{340 - 150}{45 - 23} = \frac{190}{22} \approx 8.6 \text{ flies/day,}$$

or about 9 flies per day.

This average rate of change is also the slope of the secant line through the two points P and Q on the population curve. We can calculate the slope of the secant PQ from the coordinates of P and Q.

$$\textit{Secant slope: } \frac{\Delta p}{\Delta t} = \frac{340 - 150}{45 - 23} = \frac{190}{22} \approx 8.6 \text{ flies/day}$$

As suggested by Example 2, *we can always think of an average rate of change as the slope of a secant line.*

In addition to knowing the average rate at which the population grew from day 23 to day 45, we may also want to know how fast the population was growing on day 23 itself. To find out, we can watch the slope of the secant PQ change as we back Q along the curve toward P. The results for four positions of Q are shown in Figure 2.28.

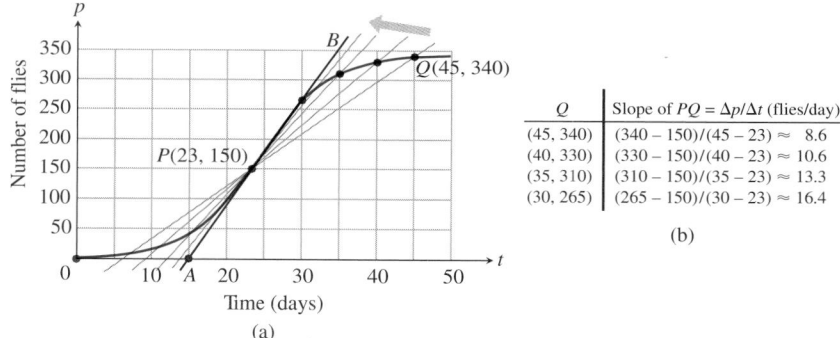

Figure 2.28 (a) Four secants to the fruit fly graph of Figure 2.27, through the point $P(23, 150)$. (b) The slopes of the four secants.

In terms of geometry, what we see as Q approaches P along the curve is this: The secant PQ approaches the tangent line AB that we drew by eye at P. This means that within the limitations of our drawing, the slopes of the secants approach the slope of the tangent, which we calculate from the coordinates of A and B to be

$$\frac{350 - 0}{35 - 15} = 17.5 \text{ flies/day.}$$

In terms of population, what we see as Q approaches P is this: The average growth rates for increasingly smaller time intervals approach the slope of the tangent to the curve at P (17.5 flies per day). The slope of the tangent line is therefore the number we take as the rate at which the fly population was growing on day $t = 23$.

Tangent to a Curve

The moral of the fruit fly story would seem to be that we should define the rate at which the value of the function $y = f(x)$ is changing with respect to x at any particular value $x = a$ to be the slope of the tangent to the curve

Why Find Tangents to Curves?

In mechanics, the tangent determines the direction of a body's motion at every point along its path.

In geometry, the tangents to two curves at a point of intersection determine the angle at which the curves intersect.

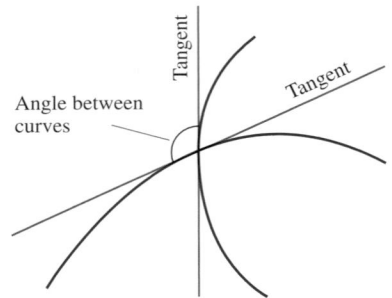

In optics, the tangent determines the angle at which a ray of light enters a curved lens (more about this in Section 3.7). The problem of how to find a tangent to a curve became the dominant mathematical problem of the early seventeenth century and it is hard to overestimate how badly the scientists of the day wanted to know the answer. Descartes went so far as to say that the problem was the most useful and most general problem not only that he knew but that he had any desire to know.

$y = f(x)$ at $x = a$. But how are we to define the tangent line at an arbitrary point P on the curve and find its slope from the formula $y = f(x)$? The problem here is that we know only one point. Our usual definition of slope requires two points.

The solution that mathematician Pierre Fermat found in 1629 proved to be one of that century's major contributions to calculus. We still use his method of defining tangents to produce formulas for slopes of curves and rates of change:

1. We start with what we *can* calculate, namely, the slope of a secant through P and a point Q nearby on the curve.

2. We find the limiting value of the secant slope (if it exists) as Q approaches P along the curve.

3. We define the *slope of the curve* at P to be this number and define the *tangent to the curve* at P to be the line through P with this slope.

Example 3   FINDING SLOPE AND TANGENT LINE

Find the slope of the parabola $y = x^2$ at the point $P(2, 4)$. Write an equation for the tangent to the parabola at this point.

Solution We begin with a secant line through $P(2, 4)$ and a nearby point $Q(2 + h, (2 + h)^2)$ on the curve (Figure 2.29).

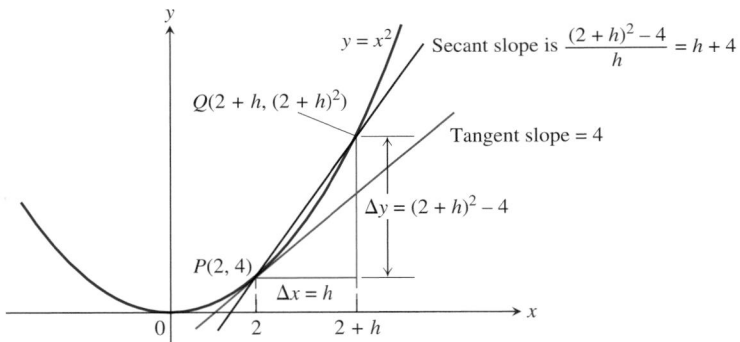

Figure 2.29 The slope of the tangent to the parabola $y = x^2$ at $P(2, 4)$ is 4.

We then write an expression for the slope of the secant line and find the limiting value of this slope as Q approaches P along the curve.

$$\text{Secant slope} = \frac{\Delta y}{\Delta x} = \frac{(2 + h)^2 - 4}{h}$$

$$= \frac{h^2 + 4h + 4 - 4}{h}$$

$$= \frac{h^2 + 4h}{h} = h + 4$$

The limit of the secant slope as Q approaches P along the curve is

$$\lim_{Q \to P} (\text{secant slope}) = \lim_{h \to 0} (h + 4) = 4.$$

Thus, the slope of the parabola at P is 4.

The tangent to the parabola at P is the line through $P(2, 4)$ with slope $m = 4$.

$$y - 4 = 4(x - 2)$$
$$y = 4x - 8 + 4$$
$$y = 4x - 4$$

Slope of a Curve

To find the tangent to a curve $y = f(x)$ at a point $P(a, f(a))$ we use the same dynamic procedure. We calculate the slope of the secant line through P and a point $Q(a + h, f(a + h))$. We then investigate the limit of the slope as $h \to 0$ (Figure 2.30). If the limit exists, it is the slope of the curve at P and we define the tangent at P to be the line through P having this slope.

Figure 2.30 at left:

y
$y = f(x)$
$Q(a + h, f(a + h))$
$f(a + h) - f(a)$
$P(a, f(a))$
h
0 a $a + h$ x

Figure 2.30 The tangent slope is
$$\lim_{h \to 0} \frac{f(a + h) - f(a)}{h}.$$

Definition Slope of a Curve at a Point

The **slope of the curve** $y = f(x)$ at the point $P(a, f(a))$ is the number
$$m = \lim_{h \to 0} \frac{f(a + h) - f(a)}{h},$$
provided the limit exists.

The **tangent line to the curve** at P is the line through P with this slope.

Example 4 EXPLORING SLOPE AND TANGENT

Let $f(x) = 1/x$.

(a) Find the slope of the curve at $x = a$.

(b) Where does the slope equal $-1/4$?

(c) What happens to the tangent to the curve at the point $(a, 1/a)$ for different values of a?

Solution

(a) The slope at $x = a$ is
$$\lim_{h \to 0} \frac{f(a + h) - f(a)}{h} = \lim_{h \to 0} \frac{\frac{1}{a + h} - \frac{1}{a}}{h}$$
$$= \lim_{h \to 0} \frac{1}{h} \cdot \frac{a - (a + h)}{a(a + h)}$$
$$= \lim_{h \to 0} \frac{-h}{ha(a + h)}$$
$$= \lim_{h \to 0} \frac{-1}{a(a + h)} = -\frac{1}{a^2}.$$

(b) The slope will be $-1/4$ if
$$-\frac{1}{a^2} = -\frac{1}{4}$$
$$a^2 = 4 \qquad \text{Multiply by } -4a^2.$$
$$a = \pm 2.$$

Pierre de Fermat (1601–1665)

The dynamic approach to tangency, invented by Fermat in 1629, proved to be one of the seventeenth century's major contributions to calculus.

Fermat, a skilled linguist and one of his century's greatest mathematicians, tended to confine his writing to professional correspondence and to papers written for personal friends. He rarely wrote completed descriptions of his work, even for his personal use. His name slipped into relative obscurity until the late 1800s, and it was only from a four-volume edition of his works published at the beginning of this century that the true importance of his many achievements became clear.

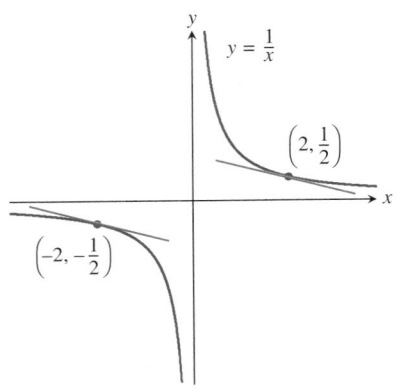

Figure 2.31 The two tangent lines to $y = 1/x$ having slope $-1/4$. (Example 4)

All of these are the same:

1. the slope of $y = f(x)$ at $x = a$

2. the slope of the tangent to $y = f(x)$ at $x = a$

3. the (instantaneous) rate of change of $f(x)$ with respect to x at $x = a$

4. $\lim\limits_{h \to 0} \dfrac{f(a + h) - f(a)}{h}$

About the Word *Normal*

When analytic geometry was developed in the seventeenth century, European scientists still wrote about their work and ideas in Latin, the one language that all educated Europeans could read and understand. The Latin word *normalis,* which scholars used for *perpendicular,* became *normal* when they discussed geometry in English.

The curve has the slope $-1/4$ at the two points $(2, 1/2)$ and $(-2, -1/2)$ (Figure 2.31).

(c) The slope $-1/a^2$ is always negative. As $a \to 0^+$, the slope approaches $-\infty$ and the tangent becomes increasingly steep. We see this again as $a \to 0^-$. As a moves away from the origin in either direction, the slope approaches 0 and the tangent becomes increasingly horizontal.

The expression

$$\frac{f(a + h) - f(a)}{h}$$

is the **difference quotient of f at a.** Suppose the difference quotient has a limit as h approaches zero. If we interpret the difference quotient as a secant slope, the limit is the slope of both the curve and the tangent to the curve at the point $x = a$. If we interpret the difference quotient as an average rate of change, the limit is the function's rate of change with respect to x at the point $x = a$. This limit is one of the two most important mathematical objects considered in calculus. We will begin a thorough study of it in Chapter 3.

Normal to a Curve

The **normal line** to a curve at a point is the line perpendicular to the tangent at that point.

Example 5   FINDING A NORMAL LINE

Write an equation for the normal to the curve $f(x) = 4 - x^2$ at $x = 1$.

Solution   The slope of the tangent to the curve at $x = 1$ is

$$\lim_{h \to 0} \frac{f(1 + h) - f(1)}{h} = \lim_{h \to 0} \frac{4 - (1 + h)^2 - 3}{h}$$

$$= \lim_{h \to 0} \frac{4 - 1 - 2h - h^2 - 3}{h}$$

$$= \lim_{h \to 0} \frac{-h(2 + h)}{h} = -2.$$

Thus, the slope of the normal is $1/2$, the negative reciprocal of -2. The normal to the curve at $(1, f(1)) = (1, 3)$ is the line through $(1, 3)$ with slope $m = 1/2$.

$$y - 3 = \frac{1}{2}(x - 1)$$

$$y = \frac{1}{2}x - \frac{1}{2} + 3$$

$$y = \frac{1}{2}x + \frac{5}{2}$$

You can support this result by drawing the graphs in a square viewing window.

Speed Revisited

The function $y = 16t^2$ that gave the distance fallen by the rock in Example 1, Section 2.1, was the rock's *position function*. A body's average speed along a coordinate axis (here, the y-axis) for a given period of time is the average rate

Particle Motion

We only have considered objects moving in one direction in this chapter. In Chapter 3, we will deal with more complicated motion.

of change of its *position* $y = f(t)$. Its **instantaneous speed** at any time t is the **instantaneous rate of change** of position with respect to time at time t, or

$$\lim_{h \to 0} \frac{f(t + h) - f(t)}{h}.$$

We saw in Example 1, Section 2.1, that the rock's instantaneous speed at $t = 2$ sec was 64 ft/sec.

Example 6 INVESTIGATING FREE FALL

Find the speed of the falling rock in Example 1, Section 2.1, at $t = 1$ sec.

Solution The position function of the rock is $f(t) = 16t^2$. The average speed of the rock over the interval between $t = 1$ and $t = 1 + h$ sec was

$$\frac{f(1 + h) - f(1)}{h} = \frac{16(1 + h)^2 - 16(1)^2}{h} = \frac{16(h^2 + 2h)}{h} = 16(h + 2).$$

The rock's speed at the instant $t = 1$ was

$$\lim_{h \to 0} 16(h + 2) = 32 \text{ ft/sec.}$$

Quick Review 2.4

In Exercises 1 and 2, find the increments Δx and Δy from point A to point B.

1. $A(-5, 2)$, $B(3, 5)$ **2.** $A(1, 3)$, $B(a, b)$

In Exercises 3 and 4, find the slope of the line determined by the points.

3. $(-2, 3)$, $(5, -1)$ **4.** $(-3, -1)$, $(3, 3)$

In Exercises 5–9, write an equation for the specified line.

5. through $(-2, 3)$ with slope $= 3/2$

6. through $(1, 6)$ and $(4, -1)$

7. through $(1, 4)$ and parallel to $y = -\dfrac{3}{4}x + 2$

8. through $(1, 4)$ and perpendicular to $y = -\dfrac{3}{4}x + 2$

9. through $(-1, 3)$ and parallel to $2x + 3y = 5$

10. For what value of b will the slope of the line through $(2, 3)$ and $(4, b)$ be 5/3?

Section 2.4 Exercises

In Exercises 1–6, find the average rate of change of the function over each interval.

1. $f(x) = x^3 + 1$
     (a) $[2, 3]$ **(b)** $[-1, 1]$

2. $f(x) = \sqrt{4x + 1}$
     (a) $[0, 2]$ **(b)** $[10, 12]$

3. $f(x) = e^x$
     (a) $[-2, 0]$ **(b)** $[1, 3]$

4. $f(x) = \ln x$
     (a) $[1, 4]$ **(b)** $[100, 103]$

5. $f(x) = \cot t$
     (a) $[\pi/4, 3\pi/4]$ **(b)** $[\pi/6, \pi/2]$

6. $f(x) = 2 + \cos t$
     (a) $[0, \pi]$ **(b)** $[-\pi, \pi]$

In Exercises 7 and 8, a distance-time graph is shown.

(a) Estimate the slopes of the secants PQ_1, PQ_2, PQ_3, and PQ_4, arranging them in order in a table. What is the appropriate unit for these slopes?

(b) Estimate the speed at point P.

7. *Accelerating from a Standstill* The figure shows the distance-time graph for a 1994 Ford® Mustang Cobra™ accelerating from a standstill.

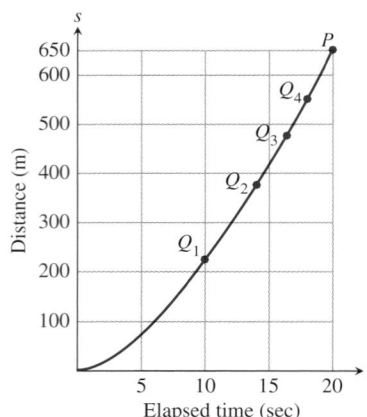

8. *Lunar Data* The figure shows a distance-time graph for a wrench that fell from the top platform of a communication mast on the moon to the station roof 80 m below.

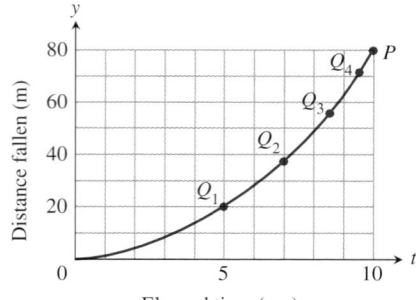

In Exercises 9–12, at the indicated point find

(a) the slope of the curve,

(b) an equation of the tangent, and

(c) an equation of the normal.

(d) Then draw a graph of the curve, tangent line, and normal line in the same square viewing window.

9. $y = x^2$ at $x = -2$

10. $y = x^2 - 4x$ at $x = 1$

11. $y = \dfrac{1}{x - 1}$ at $x = 2$

12. $y = x^2 - 3x - 1$ at $x = 0$

In Exercises 13 and 14, find the slope of the curve at the indicated point.

13. $f(x) = |x|$ at   **(a)** $x = 2$ **(b)** $x = -3$

14. $f(x) = |x - 2|$ at $x = 1$

In Exercises 15–18, determine whether the curve has a tangent at the indicated point. If it does, give its slope. If not, explain why not.

15. $f(x) = \begin{cases} 2 - 2x - x^2, & x < 0 \\ 2x + 2, & x \geq 0 \end{cases}$ at $x = 0$

16. $f(x) = \begin{cases} -x, & x < 0 \\ x^2 - x, & x \geq 0 \end{cases}$ at $x = 0$

17. $f(x) = \begin{cases} 1/x, & x \leq 2 \\ \dfrac{4 - x}{4}, & x > 2 \end{cases}$ at $x = 2$

18. $f(x) = \begin{cases} \sin x, & 0 \leq x < 3\pi/4 \\ \cos x, & 3\pi/4 \leq x \leq 2\pi \end{cases}$ at $x = 3\pi/4$

In Exercises 19–22, **(a)** find the slope of the curve at $x = a$ and **(b)** **Writing to Learn** describe what happens to the tangent at $x = a$ as a changes.

19. $y = x^2 + 2$ **20.** $y = 2/x$

21. $y = \dfrac{1}{x - 1}$ **22.** $y = 9 - x^2$

23. *Free Fall* An object is dropped from the top of a 100-m tower. Its height above ground after t sec is $100 - 4.9t^2$ m. How fast is it falling 2 sec after it is dropped?

24. *Rocket Launch* At t sec after lift-off, the height of a rocket is $3t^2$ ft. How fast is the rocket climbing after 10 sec?

25. *Area of Circle* What is the rate of change of the area of a circle with respect to the radius when the radius is $r = 3$ in.?

26. *Volume of Sphere* What is the rate of change of the volume of a sphere with respect to the radius when the radius is $r = 2$ in.?

27. *Free Fall on Mars*   The equation for free fall at the surface of Mars is $s = 1.86t^2$ m with t in seconds. Assume a rock is dropped from the top of a 200-m cliff. Find the speed of the rock at $t = 1$ sec.

28. *Free Fall on Jupiter*   The equation for free fall at the surface of Jupiter is $s = 11.44t^2$ m with t in seconds. Assume a rock is dropped from the top of a 500-m cliff. Find the speed of the rock at $t = 2$ sec.

29. *Horizontal Tangent*   At what point is the tangent to $f(x) = x^2 + 4x - 1$ horizontal?

30. *Horizontal Tangent*   At what point is the tangent to $f(x) = 3 - 4x - x^2$ horizontal?

31. *Finding Tangents and Normals*

(a) Find an equation for each tangent to the curve $y = 1/(x - 1)$ that has slope -1. (See Exercise 21.)

(b) Find an equation for each normal to the curve $y = 1/(x - 1)$ that has slope 1.

32. *Finding Tangents*   Find the equations of all lines tangent to $y = 9 - x^2$ that pass through the point (1, 12).

33. Table 2.3 gives the amount of federal funding for the Immigration and Naturalization Service (INS) for several years.

(a) Find the average rate of change in funding from 1993 to 1995.

(b) Find the average rate of change from 1995 to 1997.

(c) Let $x = 0$ represent 1990, $x = 1$ represent 1991, and so forth. Find the quadratic regression equation for the data and superimpose its graph on a scatter plot of the data.

(d) Compute the average rates of change in (a) and (b) using the regression equation.

(e) Use the regression equation to find how fast the funding was growing in 1997.

Table 2.3 Federal Funding

Year	INS Funding ($ billions)
1993	1.5
1994	1.6
1995	2.1
1996	2.6
1997	3.1

Source: Immigration and Naturalization Service as reported by Bob Laird in *USA Today,* February 18, 1997.

34. Table 2.4 gives the amounts of money earmarked by the U.S. Congress for collegiate academic programs for several years.

(a) Let $x = 0$ represent 1980, $x = 1$ represent 1981, and so forth. Make a scatter plot of the data.

(b) Let P represent the point corresponding to 1997, and Q the point for any one of the previous years. Make a table of the slopes possible for the secant line PQ.

(c) **Writing to Learn**   Based on the computations, explain why someone might be hesitant to make a prediction about the rate of change of congressional funding in 1997.

Table 2.4 Congressional Academic Funding

Year	Funding ($ millions)
1988	225
1989	289
1990	270
1991	493
1992	684
1993	763
1994	651
1995	600
1996	296
1997	440

Source: The Chronicle of Higher Education, March 28, 1997.

Explorations

In Exercises **35** and **36,** complete the following for the function.

(a) Compute the difference quotient
$$\frac{f(1 + h) - f(1)}{h}.$$

(b) Use graphs and tables to estimate the limit of the difference quotient in (a) as $h \to 0$.

(c) Compare your estimate in (b) with the given number.

(d) **Writing to Learn**   Based on your computations, do you think the graph of f has a tangent at $x = 1$? If so, estimate its slope. If not, explain why not.

35. $f(x) = e^x,\ e$ **36.** $f(x) = 2^x,\ \ln 4$ ∎

In Exercises 37–40, *work in groups of two or three.*

The curve $y = f(x)$ has a **vertical tangent** at $x = a$ if

$$\lim_{h \to 0} \frac{f(a + h) - f(a)}{h} = \infty$$

or if

$$\lim_{h \to 0} \frac{f(a + h) - f(a)}{h} = -\infty.$$

In each case, the right- and left-hand limits are required to be the same: both $+\infty$ or both $-\infty$.

Use graphs to investigate whether the curve has a vertical tangent at $x = 0$.

37. $y = x^{2/5}$ **38.** $y = x^{3/5}$

39. $y = x^{1/3}$ **40.** $y = x^{2/3}$

Extending the Ideas

In Exercises 41 and 42, determine whether the graph of the function has a tangent at the origin. Explain your answer.

41. $f(x) = \begin{cases} x^2 \sin \dfrac{1}{x}, & x \ne 0 \\ 0, & x = 0 \end{cases}$

42. $f(x) = \begin{cases} x \sin \dfrac{1}{x}, & x \ne 0 \\ 0, & x = 0 \end{cases}$

43. *Sine Function* Estimate the slope of the curve $y = \sin x$ at $x = 1$. (*Hint:* See Exercises 35 and 36.)

Chapter 2   Key Terms

average rate of change (p. 82)
average speed (p. 55)
connected graph (p. 79)
Constant Multiple Rule for Limits (p. 67)
continuity at a point (p. 75)
continuous at an endpoint (p. 75)
continuous at an interior point (p. 75)
continuous extension (p. 77)
continuous function (p. 77)
continuous on an interval (p. 77)
difference quotient (p. 86)
Difference Rule for Limits (p. 67)
discontinuous (p. 75)
end behavior model (p. 70)
free fall (p. 55)
horizontal asymptote (p. 65)
infinite discontinuity (p. 76)
instantaneous rate of change (p. 87)
instantaneous speed (p. 87)
intermediate value property (p. 79)
Intermediate Value Theorem for
     Continuous Functions (p. 79)

jump discontinuity (p. 76)
left end behavior model (p. 69)
left-hand limit (p. 60)
limit of a function (p. 57)
normal to a curve (p. 86)
oscillating discontinuity (p. 76)
point of discontinuity (p. 75)
Power Rule for Limits (p. 67)
Product Rule for Limits (p. 67)
Properties of Continuous Functions (p. 78)
Quotient Rule for Limits (p. 67)
removable discontinuity (p. 76)
right end behavior model (p. 69)
right-hand limit (p. 60)
Sandwich Theorem (p. 61)
secant to a curve (p. 82)
slope of a curve (p. 85)
Sum Rule for Limits (p. 67)
tangent line to a curve (p. 85)
two-sided limit (p. 60)
vertical asymptote (p. 68)
vertical tangent (p. 90)

Chapter 2 Review Exercises

In Exercises 1–14, find the limits.

1. $\lim_{x \to -2} (x^3 - 2x^2 + 1)$

2. $\lim_{x \to -2} \dfrac{x^2 + 1}{3x^2 - 2x + 5}$

3. $\lim_{x \to 4} \sqrt{1 - 2x}$

4. $\lim_{x \to 5} \sqrt[4]{9 - x^2}$

5. $\lim_{x \to 0} \dfrac{\dfrac{1}{2 + x} - \dfrac{1}{2}}{x}$

6. $\lim_{x \to \pm\infty} \dfrac{2x^2 + 3}{5x^2 + 7}$

7. $\lim_{x \to \pm\infty} \dfrac{x^4 + x^3}{12x^3 + 128}$

8. $\lim_{x \to 0} \dfrac{\sin 2x}{4x}$

9. $\lim_{x \to 0} \dfrac{x \csc x + 1}{x \csc x}$

10. $\lim_{x \to 0} e^x \sin x$

11. $\lim_{x \to 7/2^+} \text{int}\,(2x - 1)$

12. $\lim_{x \to 7/2^-} \text{int}\,(2x - 1)$

13. $\lim_{x \to \infty} e^{-x} \cos x$

14. $\lim_{x \to \infty} \dfrac{x + \sin x}{x + \cos x}$

In Exercises 15–20, determine whether the limit exists on the basis of the graph of $y = f(x)$. The domain of f is the set of real numbers.

15. $\lim_{x \to d} f(x)$

16. $\lim_{x \to c^+} f(x)$

17. $\lim_{x \to c^-} f(x)$

18. $\lim_{x \to c} f(x)$

19. $\lim_{x \to b} f(x)$

20. $\lim_{x \to a} f(x)$

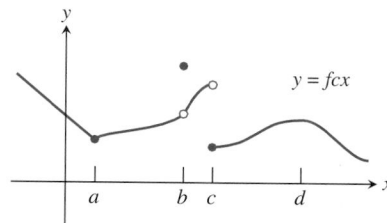

In Exercises 21–24, determine whether the function f used in Exercises 15–20 is continuous at the indicated point.

21. $x = a$

22. $x = b$

23. $x = c$

24. $x = d$

In Exercises 25 and 26, use the graph of the function with domain $-1 \le x \le 3$.

25. Determine

(a) $\lim_{x \to 3^-} g(x)$.

(b) $g(3)$.

(c) whether $g(x)$ is continuous at $x = 3$.

(d) the points of discontinuity of $g(x)$.

(e) **Writing to Learn** whether any points of discontinuity are removable. If so, describe the extended function. If not, explain why not.

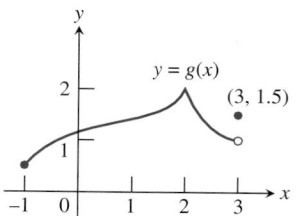

26. Determine

(a) $\lim_{x \to 1^-} k(x)$.

(b) $\lim_{x \to 1^+} k(x)$.

(c) $k(1)$.

(d) whether $k(x)$ is continuous at $x = 1$.

(e) the points of discontinuity of $k(x)$.

(f) **Writing to Learn** whether any points of discontinuity are removable. If so, describe the extended function. If not, explain why not.

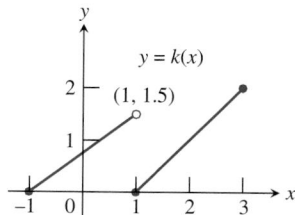

In Exercises 27 and 28, (a) find the vertical asymptotes of the graph of $y = f(x)$, and (b) describe the behavior of $f(x)$ to the left and right of any vertical asymptote.

27. $f(x) = \dfrac{x + 3}{x + 2}$

28. $f(x) = \dfrac{x - 1}{x^2\,(x + 2)}$

In Exercises 29 and 30, answer the questions for the piecewise-defined function.

29. $f(x) = \begin{cases} 1, & x \le -1 \\ -x, & -1 < x < 0 \\ 1, & x = 0 \\ -x, & 0 < x < 1 \\ 1, & x \ge 1 \end{cases}$

(a) Find the right-hand and left-hand limits of f at $x = -1$, 0, and 1.

(b) Does f have a limit as x approaches -1? 0? 1? If so, what is it? If not, why not?

(c) Is f continuous at $x = -1$? 0? 1? Explain.

30. $f(x) = \begin{cases} |x^3 - 4x|, & x < 1 \\ x^2 - 2x - 2, & x \ge 1 \end{cases}$

(a) Find the right-hand and left-hand limits of f at $x = 1$.

(b) Does f have a limit as $x \to 1$? If so, what is it? If not, why not?

(c) At what points is f continuous?

(d) At what points is f discontinuous?

In Exercises 31 and 32, find all points of discontinuity of the function.

31. $f(x) = \dfrac{x + 1}{4 - x^2}$

32. $g(x) = \sqrt[3]{3x + 2}$

In Exercises 33–36, find (a) a power function end behavior model and (b) any horizontal asymptotes.

33. $f(x) = \dfrac{2x + 1}{x^2 - 2x + 1}$

34. $f(x) = \dfrac{2x^2 + 5x - 1}{x^2 + 2x}$

35. $f(x) = \dfrac{x^3 - 4x^2 + 3x + 3}{x - 3}$

36. $f(x) = \dfrac{x^4 - 3x^2 + x - 1}{x^3 - x + 1}$

In Exercises 37 and 38, find (a) a right end behavior model and (b) a left end behavior model for the function.

37. $f(x) = x + e^x$

38. $f(x) = \ln |x| + \sin x$

In Exercises 39 and 40, *work in groups of two or three.* What value should be assigned to k to make f a continuous function?

39. $f(x) = \begin{cases} \dfrac{x^2 + 2x - 15}{x - 3}, & x \ne 3 \\ k, & x = 3 \end{cases}$

40. $f(x) = \begin{cases} \dfrac{\sin x}{2x}, & x \ne 0 \\ k, & x = 0 \end{cases}$

In Exercises 41 and 42, *work in groups of two or three.* Sketch a graph of a function f that satisfies the given conditions.

41. $\lim_{x \to \infty} f(x) = 3$, $\lim_{x \to -\infty} f(x) = \infty$,

$\lim_{x \to 3^+} f(x) = \infty$, $\lim_{x \to 3^-} f(x) = -\infty$

42. $\lim_{x \to 2} f(x)$ does not exist, $\lim_{x \to 2^+} f(x) = f(2) = 3$

43. *Average Rate of Change* Find the average rate of change of $f(x) = 1 + \sin x$ over the interval $[0, \pi/2]$.

44. *Rate of Change* Find the instantaneous rate of change of the volume $V = (1/3)\pi r^2 H$ of a cone with respect to the radius r at $r = a$ if the height H does not change.

45. *Rate of Change* Find the instantaneous rate of change of the surface area $S = 6x^2$ of a cube with respect to the edge length x at $x = a$.

46. *Slope of a Curve* Find the slope of the curve $y = x^2 - x - 2$ at $x = a$.

47. *Tangent and Normal* Let $f(x) = x^2 - 3x$ and $P = (1, f(1))$. Find (a) the slope of the curve $y = f(x)$ at P, (b) an equation of the tangent at P, and (c) an equation of the normal at P.

48. *Horizontal Tangents* At what points, if any, are the tangents to the graph of $f(x) = x^2 - 3x$ horizontal? (See Exercise 47)

49. *Bear Population* The number of bears in a federal wildlife reserve is given by the population equation

$$p(t) = \dfrac{200}{1 + 7e^{-0.1t}},$$

where t is in years.

(a) **Writing to Learn** Find $p(0)$. Give a possible interpretation of this number.

(b) Find $\lim_{t \to \infty} p(t)$.

(c) **Writing to Learn** Give a possible interpretation of the result in (b).

50. *Taxi Fares* Bluetop Cab charges \$3.20 for the first mile and \$1.35 for each additional mile or part of a mile.

(a) Write a formula that gives the charge for x miles with $0 \le x \le 20$.

(b) Graph the function in (a). At what values of x is it discontinuous?

51. *Congressional Academic Funding* Consider Table 2.4 in Exercise 34 of Section 2.4.

(a) Let $x = 0$ represent 1980, $x = 1$ represent 1981, and so forth. Find a cubic and a quartic regression equation for the data.

(b) Find an end behavior model for each regression equation. What does each predict about future funding?

52. *Limit Properties* Assume that

$$\lim_{x \to c} [f(x) + g(x)] = 2,$$

$$\lim_{x \to c} [f(x) - g(x)] = 1,$$

and that $\lim_{x \to c} f(x)$ and $\lim_{x \to c} g(x)$ exist. Find $\lim_{x \to c} f(x)$ and $\lim_{x \to c} g(x)$.

53. Table 2.5 gives the spending for software, equipment, and services to create and run *intranets*—private, Internet-like networks that link a company's operations.

(a) Let $x = 0$ represent 1990, $x = 1$ represent 1991, and so forth. Make a scatter plot of the data.

(b) Let P represent the point corresponding to 2000, and Q the point for any one of the previous years. Make a table of the slopes possible for the secant line PQ.

(c) Predict the rate of change of spending in 2000.

(d) Find a quadratic regression equation for the data, and use it to calculate the rate of change of spending in 2000.

Table 2.5 Intranet Spending

Year	Spending (\$ billions)
1995	2.7
1996	4.8
1997	7.8
1998	11.2
1999	15.2
2000	20.1

Source: Killen & Associates as reported by Anne R. Carey and Elys A. McLean in *USA Today,* April 14, 1997.

CHAPTER
3
Derivatives

Shown here is the pain reliever acetaminophen in crystalline form, photographed under a transmitted light microscope. While acetaminophen relieves pain with few side effects, it is toxic in large doses. One study found that only 30% of parents who gave acetaminophen to their children could accurately calculate and measure the correct dose.

One rule for calculating the dosage (mg) of acetaminophen for children ages 1 to 12 years old is $D(t) = 750t/(t + 12)$, where t is age in years. What is an expression for the rate of change of a child's dosage with respect to the child's age? How does the rate of change of the dosage relate to the growth rate of children? This problem can be solved with the information covered in Section 3.4.

Chapter 3 Overview

In Chapter 2, we learned how to find the slope of a tangent to a curve as the limit of the slopes of secant lines. In Example 4 of Section 2.4, we derived a formula for the slope of the tangent at an arbitrary point $(a, 1/a)$ on the graph of the function $f(x) = 1/x$ and showed that it was $-1/a^2$.

This seemingly unimportant result is more powerful than it might appear at first glance, as it gives us a simple way to calculate the instantaneous rate of change of f at any point. The study of rates of change of functions is called *differential calculus*, and the formula $-1/a^2$ was our first look at a *derivative*. The derivative was the 17th-century breakthrough that enabled mathematicians to unlock the secrets of planetary motion and gravitational attraction—of objects changing position over time. We will learn many uses for derivatives in Chapter 4, but first we will concentrate in this chapter on understanding what derivatives are and how they work.

3.1 Derivative of a Function

Definition of Derivative • Notation • Relationships between the Graphs of f and f' • Graphing the Derivative from Data • One-sided Derivatives

Definition of Derivative

In Section 2.4, we defined the slope of a curve $y = f(x)$ at the point where $x = a$ to be

$$m = \lim_{h \to 0} \frac{f(a + h) - f(a)}{h}.$$

When it exists, this limit is called the **derivative of f at a.** In this section, we investigate the derivative as a *function* derived from f by considering the limit at each point of the domain of f.

Definition Derivative

The **derivative** of the function f with respect to the variable x is the function f' whose value at x is

$$f'(x) = \lim_{h \to 0} \frac{f(x + h) - f(x)}{h}, \tag{1}$$

provided the limit exists.

The domain of f', the set of points in the domain of f for which the limit exists, may be smaller than the domain of f. If $f'(x)$ exists, we say that f **has a derivative (is differentiable)** at x. A function that is differentiable at every point of its domain is a **differentiable function.**

Example 1 APPLYING THE DEFINITION

Differentiate (that is, find the derivative of) $f(x) = x^3$.

Solution Applying the definition, we have

$$f'(x) = \lim_{h \to 0} \frac{f(x+h) - f(x)}{h}$$

$$= \lim_{h \to 0} \frac{(x+h)^3 - x^3}{h} \qquad \text{Eq. 1 with } f(x) = x^3, \\ f(x+h) = (x+h)^3$$

$$= \lim_{h \to 0} \frac{(x^3 + 3x^2 h + 3xh^2 + h^3) - x^3}{h} \qquad (x+h)^3 \text{ expanded}$$

$$= \lim_{h \to 0} \frac{(3x^2 + 3xh + h^2)h}{h} \qquad x^3\text{s cancelled, } h \text{ factored out}$$

$$= \lim_{h \to 0} (3x^2 + 3xh + h^2)$$

$$= 3x^2.$$

The derivative of $f(x)$ at a point where $x = a$ is found by taking the limit as $h \to 0$ of slopes of secant lines, as shown in Figure 3.1.

By relabeling the picture as in Figure 3.2, we arrive at a useful alternate formula for calculating the derivative. This time, the limit is taken as x approaches a.

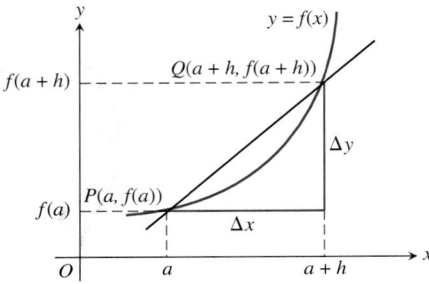

Figure 3.1 The slope of the secant line PQ is
$$\frac{\Delta y}{\Delta x} = \frac{f(a+h) - f(a)}{(a+h) - a} = \frac{f(a+h) - f(a)}{h}.$$

Definition (Alternate) Derivative at a Point

The **derivative** of the function f **at the point** $x = a$ is the limit

$$f'(a) = \lim_{x \to a} \frac{f(x) - f(a)}{x - a}, \qquad (2)$$

provided the limit exists.

After we find the derivative of f at a point $x = a$ using the alternate form, we can find the derivative of f as a function by applying the resulting formula to an arbitrary x in the domain of f.

Example 2 APPLYING THE ALTERNATE DEFINITION

Differentiate $f(x) = \sqrt{x}$ using the alternate definition.

Solution At the point $x = a$,

$$f'(a) = \lim_{x \to a} \frac{f(x) - f(a)}{x - a}$$

$$= \lim_{x \to a} \frac{\sqrt{x} - \sqrt{a}}{x - a} \qquad \text{Eq. 2 with } f(x) = \sqrt{x}$$

$$= \lim_{x \to a} \frac{\sqrt{x} - \sqrt{a}}{x - a} \cdot \frac{\sqrt{x} + \sqrt{a}}{\sqrt{x} + \sqrt{a}} \qquad \text{Rationalize...}$$

$$= \lim_{x \to a} \frac{x - a}{(x - a)(\sqrt{x} + \sqrt{a})} \qquad \text{...the numerator.}$$

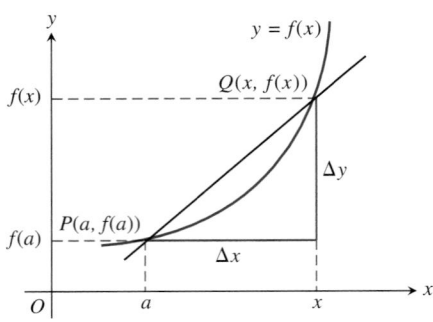

Figure 3.2 The slope of the secant line PQ is
$$\frac{\Delta y}{\Delta x} = \frac{f(x) - f(a)}{x - a}.$$

$$= \lim_{x \to a} \frac{1}{\sqrt{x} + \sqrt{a}}$$

We can now take the limit.

$$= \frac{1}{2\sqrt{a}}.$$

Applying this formula to an arbitrary $x > 0$ in the domain of f identifies the derivative as the function $f'(x) = 1/(2\sqrt{x})$ with domain $(0, \infty)$.

Notation

Why all the notation?

The "prime" notations y' and f' come from notations that Newton used for derivatives. The d/dx notations are similar to those used by Leibniz. Each has its advantages and disadvantages.

There are many ways to denote the derivative of a function $y = f(x)$. Besides $f'(x)$, the most common notations are these:

y'	"y prime"	Nice and brief, but does not name the independent variable.
$\dfrac{dy}{dx}$	"dy dx" or "the derivative of y with respect to x"	Names both variables and uses d for derivative.
$\dfrac{df}{dx}$	"df dx" or "the derivative of f with respect to x"	Emphasizes the function's name.
$\dfrac{d}{dx}f(x)$	"d dx of f at x" or "the derivative of f at x"	Emphasizes the idea that differentiation is an operation performed on f.

Relationships between the Graphs of *f* and *f′*

When we have the explicit formula for $f(x)$, we can derive a formula for $f'(x)$ using methods like those in Examples 1 and 2. We have already seen, however, that functions are encountered in other ways: graphically, for example, or in tables of data.

Because we can think of the derivative at a point in graphical terms as *slope,* we can get a good idea of what the graph of the function f' looks like by *estimating the slopes* at various points along the graph of f.

(a)

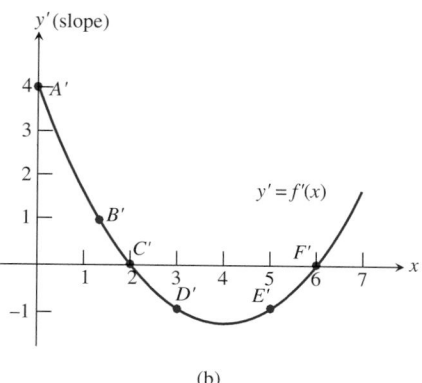

(b)

Figure 3.3 By plotting the slopes at points on the graph of $y = f(x)$, we obtain a graph of $y' = f'(x)$. The slope at point A of the graph of f in part (a) is the y-coordinate of point A' on the graph of f' in part (b), and so on. (Example 3)

Example 3   GRAPHING *f′* FROM *f*

Graph the derivative of the function f whose graph is shown in Figure 3.3a. Discuss the behavior of f in terms of the signs and values of f'.

Solution First, we draw a pair of coordinate axes, marking the horizontal axis in x-units and the vertical axis in slope units (Figure 3.3b). Next, we estimate the slope of the graph of f at various points, plotting the corresponding slope values using the new axes. At $A(0, f(0))$, the graph of f has slope 4, so $f'(0) = 4$. At B, the graph of f has slope 1, so $f' = 1$ at B', and so on.

We complete our estimate of the graph of f' by connecting the plotted points with a smooth curve.

Although we do not have a formula for either f or f', the graph of each reveals important information about the behavior of the other. In particular, notice that f is decreasing where f' is negative and increasing where f' is positive. Where f' is zero, the graph of f has a horizontal tangent, changing from increasing to decreasing (point C) or from decreasing to increasing (point F).

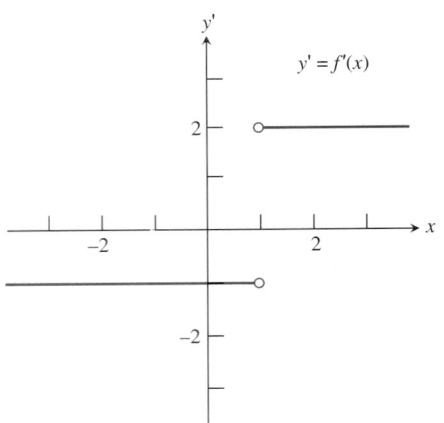

Figure 3.4 The graph of the derivative. (Example 4)

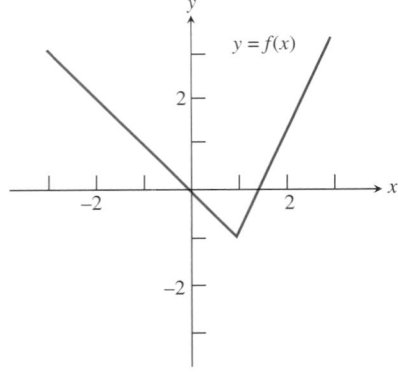

Figure 3.5 The graph of *f*, constructed from the graph of *f'* and two other conditions. (Example 4)

What's happening at x = 1?

Notice that *f* is defined at *x* = 1, while *f'* is not. It is the continuity of *f* that enables us to conclude that *f*(1) = −1. Looking at the graph of *f*, can you see why *f'* could not possibly be defined at *x* = 1? We will explore the reason for this in Example 6.

Exploration 1 **Reading the Graphs**

Suppose that the function *f* in Figure 3.3a on the previous page represents the depth *y* (in inches) of water in a ditch alongside a dirt road as a function of time *x* (in days). How would you answer the following questions?

1. What does the graph in Figure 3.3b represent? What units would you use along the *y'*-axis?

2. Describe as carefully as you can what happened to the water in the ditch over the course of the 7-day period.

3. Can you describe the weather during the 7 days? When was it the wettest? When was it the driest?

4. How does the graph of the derivative help in finding when the weather was wettest or driest?

5. Interpret the significance of point *C* in terms of the water in the ditch. How does the significance of point *C'* reflect that in terms of rate of change?

6. It is tempting to say that it rains right up until the beginning of the second day, but that overlooks a fact about rainwater that is important in flood control. Explain.

Construct your own "real-world" scenario for the function in Example 3, and pose a similar set of questions that could be answered by considering the two graphs in Figure 3.3.

Example 4 **GRAPHING f FROM f'**

Sketch the graph of a function *f* that has the following properties:

   i. $f(0) = 0$;

   ii. the graph of *f'*, the derivative of *f*, is as shown in Figure 3.4;

   iii. *f* is continuous for all *x*.

Solution To satisfy property (i), we begin with a point at the origin.

To satisfy property (ii), we consider what the graph of the derivative tells us about slopes. To the left of *x* = 1, the graph of *f* has a constant slope of −1; therefore we draw a line with slope −1 to the left of *x* = 1, making sure that it goes through the origin.

To the right of *x* = 1, the graph of *f* has a constant slope of 2, so it must be a line with slope 2. There are infinitely many such lines but only one— the one that meets the left side of the graph at (1, −1)—will satisfy the continuity requirement. The resulting graph is shown in Figure 3.5.

Graphing the Derivative from Data

Discrete points plotted from sets of data do not yield a continuous curve, but we have seen that the shape and pattern of the graphed points (called a scatter plot) can be meaningful nonetheless. It is often possible to fit a curve to the points using regression techniques. If the fit is good, we could use the curve to get a graph of the derivative visually, as in Example 3. However, it is also possible to get a scatter plot of the derivative numerically, directly from the data, by computing the slopes between successive points, as in Example 5.

Example 5 ESTIMATING THE PROBABILITY OF SHARED BIRTHDAYS

Suppose 30 people are in a room. What is the probability that two of them share the same birthday? Ignore the year of birth.

Solution It may surprise you to learn that the probability of a shared birthday among 30 people is at least 0.706, well above two-thirds! In fact, if we assume that no one day is more likely to be a birthday than any other day, the probabilities shown in Table 3.1 are not hard to determine (see Exercise 29).

Table 3.1 Probabilities of Shared Birthdays

People in Room (x)	Probability (y)
0	0
5	0.027
10	0.117
15	0.253
20	0.411
25	0.569
30	0.706
35	0.814
40	0.891
45	0.941
50	0.970
55	0.986
60	0.994
65	0.998
70	0.999

Table 3.2 Estimates of Slopes on the Probability Curve

Midpoint of Interval (x)	Change (slope $\Delta y / \Delta x$)
2.5	0.0054
7.5	0.0180
12.5	0.0272
17.5	0.0316
22.5	0.0316
27.5	0.0274
32.5	0.0216
37.5	0.0154
42.5	0.0100
47.5	0.0058
52.5	0.0032
57.5	0.0016
62.5	0.0008
67.5	0.0002

A scatter plot of the data in Table 3.1 is shown in Figure 3.6.

Notice that the probabilities grow slowly at first, then faster, then much more slowly past $x = 45$. At which x are they growing the fastest? To answer the question, we need the graph of the derivative.

Using the data in Table 3.1, we compute the slopes between successive points on the probability plot. For example, from $x = 0$ to $x = 5$ the slope is

$$\frac{0.027 - 0}{5 - 0} = 0.0054.$$

We make a new table showing the slopes, beginning with slope 0.0054 on the interval $[0, 5]$ (Table 3.2). A logical x value to use to represent the interval is its midpoint 2.5.

David H. Blackwell

By the age of 22, David Blackwell had earned a Ph.D. in Mathematics from the University of Illinois. He taught at Howard University, where his research included statistics, Markov chains, and sequential analysis. He then went on to teach and continue his research at the University of California at Berkeley. Dr. Blackwell served as president of the American Statistical Association and was the first African American mathematician of the National Academy of Sciences.

[−5, 75] by [−0.2, 1.1]

Figure 3.6 Scatter plot of the probabilities (y) of shared birthdays among x people, for $x = 0, 5, 10, \ldots, 70$. (Example 5)

[−5, 75] by [−0.01, 0.04]

Figure 3.7 A scatter plot of the derivative data in Table 3.2. (Example 5)

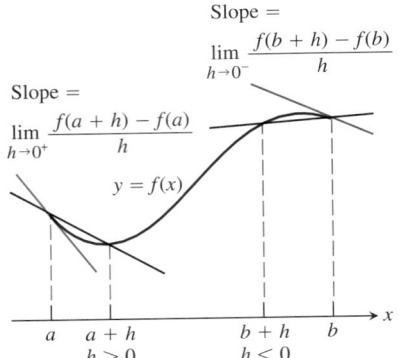

Figure 3.8 Derivatives at endpoints are one-sided limits.

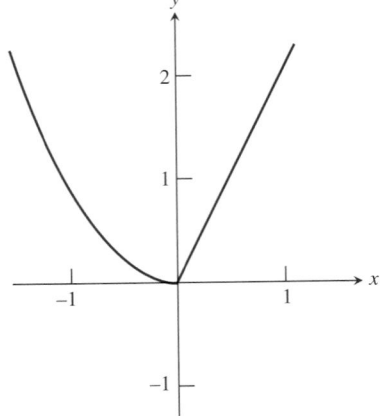

Figure 3.9 A function with different one-sided derivatives at $x = 0$. (Example 6)

A scatter plot of the derivative data in Table 3.2 is shown in Figure 3.7.

From the derivative plot, we can see that the rate of change peaks near $x = 20$. You can impress your friends with your "psychic powers" by predicting a shared birthday in a room of just 25 people (since you will be right about 57% of the time), but the derivative warns you to be cautious: a few less people can make quite a difference. On the other hand, going from 40 people to 100 people will not improve your chances much at all.

Generating shared birthday probabilities: If you know a little about probability, you might try generating the probabilities in Table 3.1. Exploration Exercise 29 at the end of this section shows how to generate them on a calculator.

One-sided Derivatives

A function $y = f(x)$ is **differentiable on a closed interval** $[a, b]$ if it has a derivative at every interior point of the interval, and if the limits

$$\lim_{h \to 0^+} \frac{f(a + h) - f(a)}{h} \quad \text{[the \textbf{right-hand derivative at} } a\text{]}$$

$$\lim_{h \to 0^-} \frac{f(b + h) - f(b)}{h} \quad \text{[the \textbf{left-hand derivative at} } b\text{]}$$

exist at the endpoints. In the right-hand derivative, h is positive and $a + h$ approaches a from the right. In the left-hand derivative, h is negative and $b + h$ approaches b from the left (Figure 3.8).

Right-hand and left-hand derivatives may be defined at any point of a function's domain.

The usual relationship between one-sided and two-sided limits holds for derivatives. Theorem 3, Section 2.1, allows us to conclude that a function has a (two-sided) derivative at a point if and only if the function's right-hand and left-hand derivatives are defined and equal at that point.

Example 6 ONE-SIDED DERIVATIVES CAN DIFFER AT A POINT

Show that the following function has left-hand and right-hand derivatives at $x = 0$, but no derivative there (Figure 3.9).

$$y = \begin{cases} x^2, & x \le 0 \\ 2x, & x > 0 \end{cases}$$

Solution We verify the existence of the left-hand derivative:

$$\lim_{h \to 0^-} \frac{(0 + h)^2 - 0^2}{h} = \lim_{h \to 0^-} \frac{h^2}{h} = 0.$$

We verify the existence of the right-hand derivative:

$$\lim_{h \to 0^+} \frac{2(0 + h) - 0^2}{h} = \lim_{h \to 0^+} \frac{2h}{h} = 2.$$

Since the left-hand derivative equals zero and the right-hand derivative equals 2, the derivatives are not equal at $x = 0$. The function does not have a derivative at 0.

Quick Review 3.1

In Exercises 1–4, evaluate the indicated limit algebraically.

1. $\lim\limits_{h\to 0} \dfrac{(2+h)^2 - 4}{h}$

2. $\lim\limits_{x\to 2^+} \dfrac{x+3}{2}$

3. $\lim\limits_{y\to 0^-} \dfrac{|y|}{y}$

4. $\lim\limits_{x\to 4} \dfrac{2x - 8}{\sqrt{x} - 2}$

5. Find the slope of the line tangent to the parabola $y = x^2 + 1$ at its vertex.

6. By considering the graph of $f(x) = x^3 - 3x^2 + 2$, find the intervals on which f is increasing.

In Exercises 7–10, let

$$f(x) = \begin{cases} x + 2, & x \le 1 \\ (x - 1)^2, & x > 1. \end{cases}$$

7. Find $\lim\limits_{x\to 1^+} f(x)$ and $\lim\limits_{x\to 1^-} f(x)$.

8. Find $\lim\limits_{h\to 0^+} f(1 + h)$.

9. Does $\lim\limits_{x\to 1} f(x)$ exist? Explain.

10. Is f continuous? Explain.

Section 3.1 Exercises

1. If $f(2) = 3$ and $f'(2) = 5$, find an equation of **(a)** the *tangent* line, and **(b)** the *normal* line to the graph of $y = f(x)$ at the point where $x = 2$.

2. Use the definition

$$f'(a) = \lim_{h\to 0} \frac{f(a + h) - f(a)}{h}$$

to find the derivative of $f(x) = (1/x)$ at $x = 3$.

3. Use the definition

$$f'(a) = \lim_{x\to a} \frac{f(x) - f(a)}{x - a}$$

to find the derivative of $f(x) = (1/x)$ at $x = 3$.

4. Find $f'(x)$ if $f(x) = 3x - 12$.

5. Find dy/dx if $y = 7x$.

6. Find $\dfrac{d}{dx}(x^2)$.

In Exercises 7–10, match the graph of the function with the graph of the derivative shown here:

(a)

(b)

(c)

(d)

7.

$y = f_1(x)$

8.

$y = f_2(x)$

9.

$y = f_3(x)$

10.

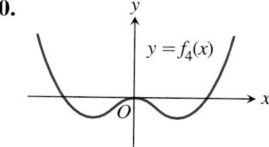

$y = f_4(x)$

11. Find the derivative of the function $y = 2x^2 - 13x + 5$ and use it to find the equation of the line tangent to the curve at $x = 3$.

12. Find the lines that are **(a)** tangent and **(b)** normal to the curve $y = x^3$ at the point $(1, 1)$.

13. Shown below is the graph of $f(x) = x \ln x - x$. From what you know about the graphs of functions (i) through (v), pick out the one that is the *derivative* of f for $x > 0$.

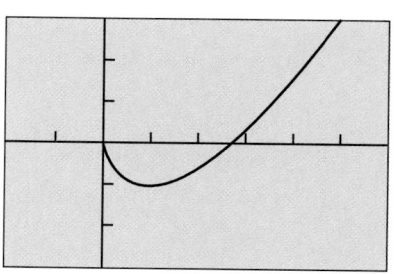

[–2, 6] by [–3, 3]

i. $y = \sin x$ **ii.** $y = \ln x$ **iii.** $y = \sqrt{x}$

iv. $y = x^2$ **v.** $y = 3x - 1$

14. From what you know about the graphs of functions (i) through (v), pick out the one that is *its own derivative*.

i. $y = \sin x$ **ii.** $y = x$ **iii.** $y = \sqrt{x}$

iv. $y = e^x$ **v.** $y = x^2$

15. *Daylight in Fairbanks* The viewing window below shows the number of hours of daylight in Fairbanks, Alaska, on each day for a typical 365-day period from January 1 to December 31. Answer the following questions by estimating slopes on the graph in hours per day. For the purposes of estimation, assume that each month has 30 days.

[0, 365] by [0, 24]

(a) On about what date is the amount of daylight increasing at the fastest rate? What is that rate?

(b) Do there appear to be days on which the rate of change in the amount of daylight is zero? If so, which ones?

(c) On what dates is the rate of change in the number of daylight hours positive? negative?

16. *Graphing f′ from f* Given the graph of the function f below, sketch a graph of the *derivative* of f.

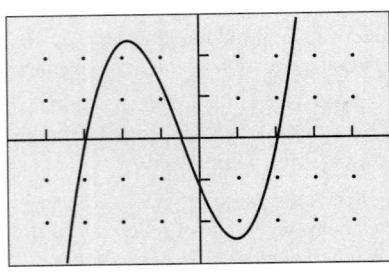

[−5, 5] by [−3, 3]

17. The graphs in Figure 3.10a show the numbers of rabbits and foxes in a small arctic population. They are plotted as functions of time for 200 days. The number of rabbits increases at first, as the rabbits reproduce. But the foxes prey on the rabbits and, as the number of foxes increases, the rabbit population levels off and then drops. Figure 3.10b shows the graph of the derivative of the rabbit population. We made it by plotting slopes, as in Example 3.

(a) What is the value of the derivative of the rabbit population in Figure 3.10 when the number of rabbits is largest? smallest?

(b) What is the size of the rabbit population in Figure 3.10 when its derivative is largest? smallest?

(a)

Time (days)

Derivative of the rabbit population

(b)

Figure 3.10 Rabbits and foxes in an arctic predator-prey food chain. *Source: Differentiation* by W. U. Walton et al., Project CALC, Education Development Center, Inc., Newton, MA, 1975, p. 86.

18. The graph of the function $y = f(x)$ shown here is made of line segments joined end to end.

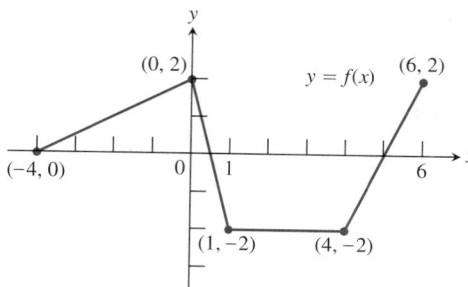

(a) Graph the function's derivative.

(b) At what values of x between $x = -4$ and $x = 6$ is the function not differentiable?

In Exercises 19 and 20, use the data to answer the questions.

19. *A Downhill Skier*   Table 3.3 gives the approximate distance traveled by a downhill skier after t seconds for $0 \le t \le 10$. Use the method of Example 5 to sketch a graph of the derivative; then answer the following questions:

(a) What does the derivative represent?

(b) In what units would the derivative be measured?

(c) Can you guess an equation of the derivative by considering its graph?

Table 3.3 Skiing Distances

Time t (seconds)	Distance Traveled (feet)
0	0
1	3.3
2	13.3
3	29.9
4	53.2
5	83.2
6	119.8
7	163.0
8	212.9
9	269.5
10	332.7

20. *A Whitewater River*   Bear Creek, a Georgia river known to kayaking enthusiasts, drops more than 770 feet over one stretch of 3.24 miles. By reading a contour map, one can estimate the elevations (y) at various distances (x) downriver from the start of the kayaking route (Table 3.4).

Table 3.4 Elevations along Bear Creek

Distance Downriver (miles)	River Elevation (feet)
0.00	1577
0.56	1512
0.92	1448
1.19	1384
1.30	1319
1.39	1255
1.57	1191
1.74	1126
1.98	1062
2.18	998
2.41	933
2.64	869
3.24	805

(a) Sketch a graph of elevation (y) as a function of distance downriver (x).

(b) Use the technique of Example 5 to get an approximate graph of the derivative, dy/dx.

(c) The average change in elevation over a given distance is called a *gradient*. In this problem, what units of measure would be appropriate for a gradient?

(d) In this problem, what units of measure would be appropriate for the derivative?

(e) How would you identify the most dangerous section of the river (ignoring rocks) by analyzing the graph in (a)? Explain.

(f) How would you identify the most dangerous section of the river by analyzing the graph in (b)? Explain.

21. Writing to Learn   Graph $y = \sin x$ and $y = \cos x$ in the same viewing window. Which function could be the derivative of the other? Defend your answer in terms of the behavior of the graphs.

22. Using one-sided derivatives, show that the function

$$f(x) = \begin{cases} x^3, & x \le 1 \\ 3x, & x > 1 \end{cases}$$

does not have a derivative at $x = 1$.

23. In Example 2 of this section we showed that the derivative of $y = \sqrt{x}$ is a function with domain $(0, \infty)$. However, the function $y = \sqrt{x}$ itself has domain $[0, \infty)$, so it could have a right-hand derivative at $x = 0$. Prove that it does not.

24. Writing to Learn   Use the concept of the derivative to define what it might mean for two parabolas to be parallel. Construct equations for two such parallel parabolas and graph them. Are the parabolas "everywhere equidistant," and if so, in what sense?

25. *Graphing f from f′* Sketch the graph of a continuous function f with $f(0) = -1$ and

$$f'(x) = \begin{cases} 1, & x < -1 \\ -2, & x > -1. \end{cases}$$

Explorations

26. Let $f(x) = \begin{cases} x^2, & x \le 1 \\ 2x, & x > 1. \end{cases}$

(a) Find $f'(x)$ for $x < 1$.

(b) Find $f'(x)$ for $x > 1$.

(c) Find $\lim_{x \to 1^-} f'(x)$.

(d) Find $\lim_{x \to 1^+} f'(x)$.

(e) Does $\lim_{x \to 1} f'(x)$ exist? Explain.

(f) Use the definition to find the left-hand derivative of f at $x = 1$ if it exists.

(g) Use the definition to find the right-hand derivative of f at $x = 1$ if it exists.

(h) Does $f'(1)$ exist? Explain.

27. *Work in groups of two or three.* Using graphing calculators, have each person in your group do the following:

(a) pick two numbers a and b between 1 and 10;

(b) graph the function $y = (x - a)(x + b)$;

(c) graph the *derivative* of your function (it will be a line with slope 2);

(d) find the y-intercept of your derivative graph.

(e) Compare your answers and determine a simple way to predict the y-intercept, given the values of a and b. Test your result. ■

Extending the Ideas

28. Find the unique value of k that makes the function

$$f(x) = \begin{cases} x^3, & x \le 1 \\ 3x + k, & x > 1 \end{cases}$$

differentiable at $x = 1$.

29. *Generating the Birthday Probabilities* Example 5 of this section concerns the probability that, in a group of n people, at least two people will share a common birthday. You can generate these probabilities on your calculator for values of n from 1 to 365.

Step 1: Set the values of N and P to zero:

Step 2: Type in this single, multi-step command:

Now each time you press the ENTER key, the command will print a new value of N (the number of people in the room) alongside P (the probability that at least two of them share a common birthday):

```
                           {1  0}
           {2  .002739726}
          {3  .0082041659}
          {4  .0163559125}
          {5  .0271355737}
          {6  .0404624836}
          {7  .0562357031}
```

If you have some experience with probability, try to answer the following questions without looking at the table:

(a) If there are three people in the room, what is the probability that they all have *different* birthdays? (Assume that there are 365 possible birthdays, all of them equally likely.)

(b) If there are three people in the room, what is the probability that at least two of them share a common birthday?

(c) Explain how you can use the answer in (b) to find the probability of a shared birthday when there are *four* people in the room. (This is how the calculator statement in Step 2 generates the probabilities.)

(d) Is it reasonable to assume that all calendar dates are equally likely birthdays? Explain your answer.

3.2 Differentiability

How *f'(a)* Might Fail to Exist • Differentiability Implies Local Linearity • Derivatives on a Calculator • Differentiability Implies Continuity • Intermediate Value Theorem for Derivatives

How *f'(a)* Might Fail to Exist

A function will not have a derivative at a point $P(a, f(a))$ where the slopes of the secant lines,

$$\frac{f(x) - f(a)}{x - a},$$

fail to approach a limit as x approaches a. The next figures illustrate four different instances where this occurs. For example, a function whose graph is otherwise smooth will fail to have a derivative at a point where the graph has

1. a *corner*, where the one-sided derivatives differ;

   Example: $f(x) = |x|$

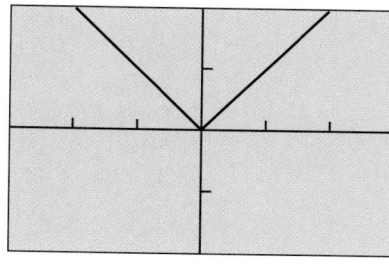

[–3, 3] by [–2, 2]

Figure 3.11 There is a "corner" at $x = 0$.

2. a *cusp*, where the slopes of the secant lines approach ∞ from one side and $-\infty$ from the other (an extreme case of a corner);

   Example: $f(x) = x^{2/3}$

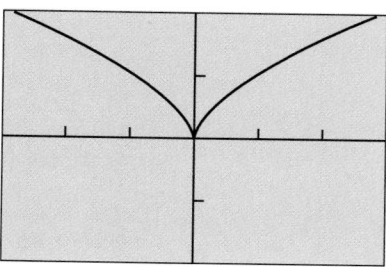

[–3, 3] by [–2, 2]

Figure 3.12 There is a "cusp" at $x = 0$.

3. a *vertical tangent*, where the slopes of the secant lines approach either ∞ or $-\infty$ from both sides (in this example, ∞);

Example: $f(x) = \sqrt[3]{x}$

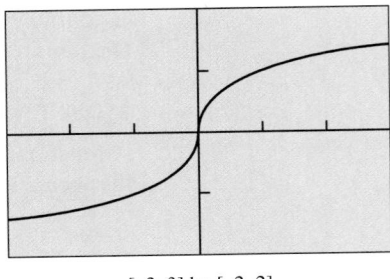

[–3, 3] by [–2, 2]

Figure 3.13 There is a vertical tangent line at $x = 0$.

4. a *discontinuity* (which will cause one or both of the one-sided derivatives to be nonexistent).

Example: The *Unit Step Function*

$$U(x) = \begin{cases} -1, & x < 0 \\ 1, & x \geq 0 \end{cases}$$

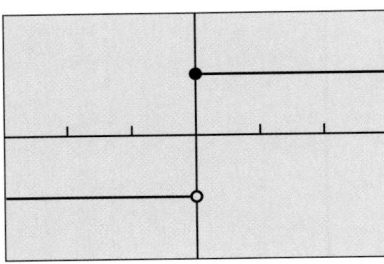

[–3, 3] by [–2, 2]

Figure 3.14 There is a discontinuity at $x = 0$.

In this example, the left-hand derivative fails to exist:

$$\lim_{h \to 0^-} \frac{(-1) - (1)}{h} = \lim_{h \to 0^-} \frac{-2}{h} = \infty.$$

Later in this section we will prove a theorem that states that a function *must* be continuous at a to be differentiable at a. This theorem would provide a quick and easy verification that U is not differentiable at $x = 0$.

Example 1   FINDING WHERE A FUNCTION IS NOT DIFFERENTIABLE

Find all points in the domain of $f(x) = |x - 2| + 3$ where f is not differentiable.

Solution   Think graphically! The graph of this function is the same as that of $y = |x|$, translated 2 units to the right and 3 units up. This puts the corner at the point $(2, 3)$, so this function is not differentiable at $x = 2$.

At every other point, the graph is (locally) a straight line and f has derivative $+1$ or -1 (again, just like $y = |x|$).

How rough can the graph of a continuous function be?

The graph of the absolute value function fails to be differentiable at a single point. If you graph $y = \sin^{-1}(\sin(x))$ on your calculator, you will see a continuous function with an *infinite* number of points of non-differentiability. But can a continuous function fail to be differentiable at *every* point?

The answer, surprisingly enough, is yes, as Karl Weierstrass showed in 1872. One of his formulas (there are many like it) was

$$f(x) = \sum_{n=0}^{\infty} \left(\frac{2}{3}\right)^n \cos(9^n \pi x),$$

a formula that expresses f as an infinite (but converging) sum of cosines with increasingly higher frequencies. By adding wiggles to wiggles infinitely many times, so to speak, the formula produces a function whose graph is too bumpy in the limit to have a tangent anywhere!

Most of the functions we encounter in calculus are differentiable wherever they are defined, which means that they will *not* have corners, cusps, vertical tangent lines, or points of discontinuity within their domains. Their graphs will be unbroken and smooth, with a well-defined slope at each point. Polynomials are differentiable, as are rational functions, trigonometric functions, exponential functions, and logarithmic functions. Composites of differentiable functions are differentiable, and so are sums, products, integer powers, and quotients of differentiable functions, where defined. We will see why all of this is true as the chapter continues.

Differentiability Implies Local Linearity

A good way to think of differentiable functions is that they are **locally linear;** that is, a function that is differentiable at a closely resembles its own tangent line very close to a. In the jargon of graphing calculators, differentiable curves will "straighten out" when we zoom in on them at a point of differentiability. (See Figure 3.15.)

Exploration 1   Zooming in to "See" Differentiability

Is either of these functions differentiable at $x = 0$?

$$\textbf{(a) } f(x) = |x| + 1 \qquad \textbf{(b) } g(x) = \sqrt{x^2 + 0.0001} + 0.99$$

1. We already know that f is not differentiable at $x = 0$; its graph has a corner there. Graph f and zoom in at the point $(0, 1)$ several times. Does the corner show signs of straightening out?

2. Now do the same thing with g. Does the graph of g show signs of straightening out? We will learn a quick way to differentiate g in Section 3.6, but for now suffice it to say that it *is* differentiable at $x = 0$, and in fact has a horizontal tangent there.

3. How many zooms does it take before the graph of g looks exactly like a horizontal line?

4. Now graph f and g *together* in a standard square viewing window. They appear to be identical until you start zooming in. The differentiable function eventually straightens out, while the nondifferentiable function remains impressively unchanged.

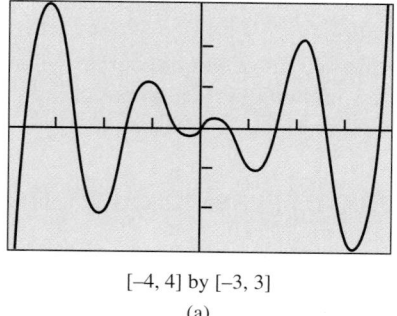

[−4, 4] by [−3, 3]

(a)

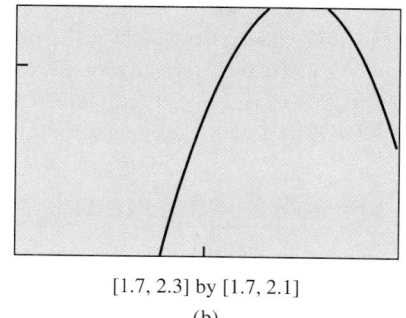

[1.7, 2.3] by [1.7, 2.1]

(b)

[1.93, 2.07] by [1.85, 1.95]

(c)

Figure 3.15 Three different views of the differentiable function $f(x) = x \cos(3x)$. We have zoomed in here at the point $(2, 1.9)$.

Derivatives on a Calculator

Many graphing utilities can approximate derivatives numerically with good accuracy at most points of their domains.

For small values of h, the difference quotient

$$\frac{f(a + h) - f(a)}{h}$$

is often a good numerical approximation of $f'(a)$. However, as suggested by Figure 3.16, the same value of h will usually yield a *better* approximation if we use the **symmetric difference quotient**

$$\frac{f(a + h) - f(a - h)}{2h},$$

which is what our graphing calculator uses to calculate NDER $f(a)$, the **numerical derivative of f at a point a**. The **numerical derivative of f** as a function is denoted by NDER $f(x)$. Sometimes we will use NDER $(f(x), a)$ for NDER $f(a)$ when we want to emphasize both the function *and* the point.

Although the symmetric difference quotient is not the quotient used in the definition of $f'(a)$, it can be proven that

$$\lim_{h \to 0} \frac{f(a + h) - f(a - h)}{2h}$$

equals $f'(a)$ wherever $f'(a)$ exists.

You might think that an extremely small value of h would be required to give an accurate approximation of $f'(a)$, but in most cases $h = 0.001$ is more than adequate. In fact, your calculator probably assumes such a value for h unless you choose to specify otherwise (consult your *Owner's Manual*). The numerical derivatives we compute in this book will use $h = 0.001$; that is,

$$\text{NDER } f(a) = \frac{f(a + 0.001) - f(a - 0.001)}{0.002}.$$

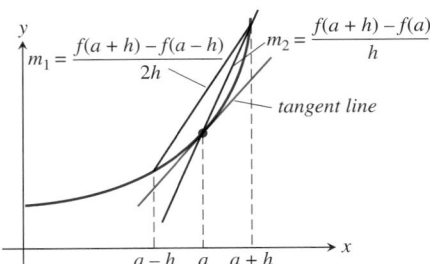

Figure 3.16 The symmetric difference quotient (slope m_1) usually gives a better approximation of the derivative for a given value of h than does the regular difference quotient (slope m_2), which is why the symmetric difference quotient is used in the numerical derivative.

Example 2   COMPUTING A NUMERICAL DERIVATIVE

Compute NDER $(x^3, 2)$, the numerical derivative of x^3 at $x = 2$.

Solution   Using $h = 0.001$,

$$\text{NDER } (x^3, 2) = \frac{(2.001)^3 - (1.999)^3}{0.002} = 12.000001.$$

In Example 1 of Section 3.1, we found the derivative of x^3 to be $3x^2$, whose value at $x = 2$ is $3(2)^2 = 12$. The numerical derivative is accurate to 5 decimal places. Not bad for the push of a button.

Example 2 gives dramatic evidence that NDER is very accurate when $h = 0.001$. Such accuracy is usually the case, although it is also possible for NDER to produce some surprisingly inaccurate results, as in Example 3.

Example 3   FOOLING THE SYMMETRIC DIFFERENCE QUOTIENT

Compute NDER $(|x|, 0)$, the numerical derivative of $|x|$ at $x = 0$.

Solution   We saw at the start of this section that $|x|$ is not differentiable at $x = 0$ since its right-hand and left-hand derivatives at $x = 0$ are not the same. Nonetheless,

$$\text{NDER}\,(|x|, 0) = \lim_{h \to 0} \frac{|0 + h| - |0 - h|}{2h}$$

$$= \lim_{h \to 0} \frac{|h| - |h|}{2h}$$

$$= \lim_{h \to 0} \frac{0}{2h}$$

$$= 0.$$

The symmetric difference quotient, which works symmetrically on either side of 0, never detects the corner! Consequently, most graphing utilities will indicate (wrongly) that $y = |x|$ is differentiable at $x = 0$, with derivative 0.

In light of Example 3, it is worth repeating here that NDER $f(a)$ actually does approach $f'(a)$ *when $f'(a)$ exists,* and in fact approximates it quite well (as in Example 2).

An Alternative to NDER

Graphing

$$y = \frac{f(x + 0.001) - f(x - 0.001)}{0.002}$$

is equivalent to graphing $y = $ NDER $f(x)$ (useful if NDER is not readily available on your calculator).

Exploration 2 Looking at the Symmetric Difference Quotient Analytically

Let $f(x) = x^2$ and let $h = 0.01$.

1. Find

$$\frac{f(10 + h) - f(10)}{h}.$$

   How close is it to $f'(10)$?

2. Find

$$\frac{f(10 + h) - f(10 - h)}{2h}.$$

   How close is it to $f'(10)$?

3. Repeat this comparison for $f(x) = x^3$.

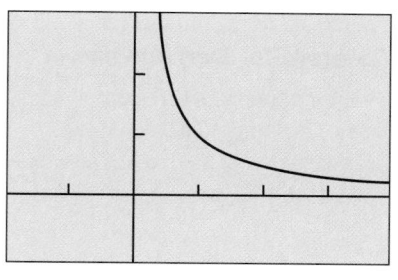

[–2, 4] by [–1, 3]

(a)

X	Y₁	
.1	10	
.2	5	
.3	3.3333	
.4	2.5	
.5	2	
.6	1.6667	
.7	1.4286	
X = .1		

(b)

Figure 3.17 (a) The graph of NDER ln (x) and (b) a table of values. What graph could this be? (Example 4)

Example 4 GRAPHING A DERIVATIVE USING NDER

Let $f(x) = \ln x$. Use NDER to graph $y = f'(x)$. Can you guess what function $f'(x)$ is by analyzing its graph?

Solution The graph is shown in Figure 3.17a. The shape of the graph suggests, and the table of values in Figure 3.17b supports, the conjecture that this is the graph of $y = 1/x$. We will prove in Section 3.9 (using analytic methods) that this is indeed the case.

Differentiability Implies Continuity

We began this section with a look at the typical ways that a function could fail to have a derivative at a point. As one example, we indicated graphically that a discontinuity in the graph of f would cause one or both of the one-sided derivatives to be nonexistent. It is actually not difficult to give an analytic proof that continuity is an essential condition for the derivative to exist, so we include that as a theorem here.

> **Theorem 1   Differentiability Implies Continuity**
>
> If f has a derivative at $x = a$, then f is continuous at $x = a$.

Proof   Our task is to show that $\lim_{x \to a} f(x) = f(a)$, or, equivalently, that

$$\lim_{x \to a} [f(x) - f(a)] = 0.$$

Using the Limit Product Rule (and noting that $x - a$ is not zero), we can write

$$\lim_{x \to a} [f(x) - f(a)] = \lim_{x \to a} \left[(x - a) \frac{f(x) - f(a)}{x - a} \right]$$

$$= \lim_{x \to a} (x - a) \cdot \lim_{x \to a} \frac{f(x) - f(a)}{x - a}$$

$$= 0 \cdot f'(a)$$

$$= 0. \qquad \blacksquare$$

The converse of Theorem 1 is false, as we have already seen. A continuous function might have a corner, a cusp, or a vertical tangent line, and hence not be differentiable at a given point.

Intermediate Value Theorem for Derivatives

Not every function can be a derivative. A derivative must have the intermediate value property, as stated in the following theorem (the proof of which can be found in advanced texts).

> **Theorem 2   Intermediate Value Theorem for Derivatives**
>
> If a and b are any two points in an interval on which f is differentiable, then f' takes on every value between $f'(a)$ and $f'(b)$.

Example 5   APPLYING THEOREM 2

Does any function have the Unit Step Function (Figure 3.14) as its derivative?

Solution   No. Choose some $a < 0$ and some $b > 0$. Then $U(a) = -1$ and $U(b) = 1$, but U does not take on any value between -1 and 1.

The question of when a function is a derivative of some function is one of the central questions in all of calculus. The answer, found by Newton and Leibniz, would revolutionize the world of mathematics. We will see what that answer is when we reach Chapter 5.

Quick Review 3.2

In Exercises 1–5, tell whether the limit could be used to define $f'(a)$ (assuming that f is differentiable at a).

1. $\lim\limits_{h \to 0} \dfrac{f(a + h) - f(a)}{h}$

2. $\lim\limits_{h \to 0} \dfrac{f(a + h) - f(h)}{h}$

3. $\lim\limits_{x \to a} \dfrac{f(x) - f(a)}{x - a}$

4. $\lim\limits_{x \to a} \dfrac{f(a) - f(x)}{a - x}$

5. $\lim\limits_{h \to 0} \dfrac{f(a + h) + f(a - h)}{h}$

6. Find the domain of the function $y = x^{4/3}$.

7. Find the domain of the function $y = x^{3/4}$.

8. Find the range of the function $y = |x - 2| + 3$.

9. Find the slope of the line $y - 5 = 3.2(x + \pi)$.

10. If $f(x) = 5x$, find

$$\frac{f(3 + 0.001) - f(3 - 0.001)}{0.002}.$$

Section 3.2 Exercises

In Exercises 1–4, compare the right-hand and left-hand derivatives to show that the function is not differentiable at the point *P.*

1.

2.

3.

4.
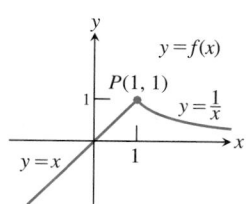

In Exercises 5–10, the graph of a function over a closed interval *D* is given. At what domain points does the function appear to be

(a) differentiable?

(b) continuous but not differentiable?

(c) neither continuous nor differentiable?

5.

6.

7.

8.

9.

10.
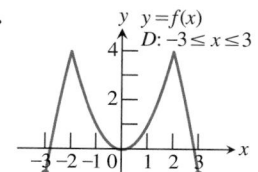

In Exercises 11–16, the function fails to be differentiable at $x = 0$. Tell whether the problem is a corner, a cusp, a vertical tangent, or a discontinuity.

11. $y = \begin{cases} \tan^{-1} x, & x \neq 0 \\ 1, & x = 0 \end{cases}$

12. $y = x^{4/5}$

13. $y = x + \sqrt{x^2} + 2$

14. $y = 3 - \sqrt[3]{x}$

15. $y = 3x - 2|x| - 1$

16. $y = \sqrt[3]{|x|}$

In Exercises 17–22, find all values of *x* for which the function is differentiable.

17. $f(x) = \dfrac{x^3 - 8}{x^2 - 4x - 5}$

18. $h(x) = \sqrt[3]{3x - 6} + 5$

19. $P(x) = \sin(|x|) - 1$

20. $Q(x) = 3\cos(|x|)$

21. $g(x) = \begin{cases} (x + 1)^2, & x \leq 0 \\ 2x + 1, & 0 < x < 3 \\ (4 - x)^2, & x \geq 3 \end{cases}$

22. $C(x) = x|x|$

23. Writing to Learn Recall that the numerical derivative (NDER) can give meaningless values at points where a function is not differentiable. In this exercise, we consider the numerical derivatives of the functions $1/x$ and $1/x^2$ at $x = 0$.

(a) Explain why neither function is differentiable at $x = 0$.

(b) Find NDER at $x = 0$ for each function.

(c) By analyzing the definition of the symmetric difference quotient, explain why NDER returns wrong responses that are so different from each other for these two functions.

In Exercises 24–28, *work in groups of two or three.* Use NDER to graph the derivative of the function. If possible, identify the derivative function by looking at the graph.

24. $y = 0.25x^4$

25. $y = -\cos x$

26. $y = \dfrac{x|x|}{2}$

27. $y = -\ln|\cos x|$

28. $y = 2x \cos(2x) + 4x \sin x$

29. Let f be the function defined as

$$f(x) = \begin{cases} 3 - x, & x < 1 \\ ax^2 + bx, & x \geq 1 \end{cases}$$

where a and b are constants.

(a) If the function is continuous for all x, what is the relationship between a and b?

(b) Find the unique values for a and b that will make f both continuous and differentiable.

30. Show that the function

$$f(x) = \begin{cases} 0, & -1 \leq x < 0 \\ 1, & 0 \leq x \leq 1 \end{cases}$$

is not the derivative of any function on the interval $-1 \leq x \leq 1$.

Extending the Ideas

31. *Oscillation* There is another way that a function might fail to be differentiable, and that is by *oscillation*. Let

$$f(x) = \begin{cases} x \sin \dfrac{1}{x}, & x \neq 0 \\ 0, & x = 0. \end{cases}$$

(a) Show that f is continuous at $x = 0$.

(b) Show that

$$\frac{f(0 + h) - f(0)}{h} = \sin \frac{1}{h}.$$

(c) Explain why

$$\lim_{h \to 0} \frac{f(0 + h) - f(0)}{h}$$

does not exist.

(d) Does f have either a left-hand or right-hand derivative at $x = 0$?

(e) Now consider the function

$$g(x) = \begin{cases} x^2 \sin \dfrac{1}{x}, & x \neq 0 \\ 0, & x = 0. \end{cases}$$

Use the definition of the derivative to show that g is differentiable at $x = 0$ and that $g'(0) = 0$.

3.3 Rules for Differentiation

Positive Integer Powers, Multiples, Sums, and Differences • Products and Quotients • Negative Integer Powers of x • Second and Higher Order Derivatives

Positive Integer Powers, Multiples, Sums, and Differences

The first rule of differentiation is that the derivative of every constant function is the zero function.

Rule 1 Derivative of a Constant Function

If f is the function with the constant value c, then

$$\frac{df}{dx} = \frac{d}{dx}(c) = 0.$$

Proof of Rule 1 If $f(x) = c$ is a function with a constant value c, then

$$\lim_{h \to 0} \frac{f(x + h) - f(x)}{h} = \lim_{h \to 0} \frac{c - c}{h} = \lim_{h \to 0} 0 = 0.$$ ∎

The next rule is a first step toward a rule for differentiating any polynomial.

Rule 2 Power Rule for Positive Integer Powers of x

If n is a positive integer, then

$$\frac{d}{dx}(x^n) = nx^{n-1}.$$

Proof of Rule 2 If $f(x) = x^n$, then $f(x + h) = (x + h)^n$ and the difference quotient for f is

$$\frac{(x + h)^n - x^n}{h}.$$

We can readily find the limit of this quotient as $h \to 0$ if we apply the algebraic identity

$$a^n - b^n = (a - b)(a^{n-1} + a^{n-2}b + \cdots + ab^{n-2} + b^{n-1}) \quad \text{n a positive integer}$$

with $a = x + h$ and $b = x$. For then $(a - b) = h$ and the h's in the numerator and denominator of the quotient cancel, giving

$$\frac{f(x + h) - f(x)}{h} = \frac{(x + h)^n - x^n}{h}$$

$$= \frac{h[(x + h)^{n-1} + (x + h)^{n-2}x + \cdots + (x + h)x^{n-2} + x^{n-1}]}{h}$$

$$= \underbrace{(x + h)^{n-1} + (x + h)^{n-2}x + \cdots + (x + h)x^{n-2} + x^{n-1}}_{\text{n terms, each with limit x^{n-1} as $h \to 0$}}.$$

Hence,

$$\frac{d}{dx}(x^n) = \lim_{h \to 0} \frac{f(x + h) - f(x)}{h} = nx^{n-1}.$$ ∎

The Power Rule says: To differentiate x^n, multiply by n and subtract 1 from the exponent. For example, the derivatives of x^2, x^3, and x^4 are $2x^1$, $3x^2$, and $4x^3$, respectively.

Rule 3 The Constant Multiple Rule

If u is a differentiable function of x and c is a constant, then

$$\frac{d}{dx}(cu) = c\frac{du}{dx}.$$

Proof of Rule 3

$$\frac{d}{dx}(cu) = \lim_{h \to 0} \frac{cu(x + h) - cu(x)}{h}$$

$$= c \lim_{h \to 0} \frac{u(x + h) - u(x)}{h} = c\frac{du}{dx}$$ ∎

Denoting Functions by *u* and *v*

The functions we work with when we need a differentiation formula are likely to be denoted by letters like *f* and *g*. When we apply the formula, we do not want to find the formula using these same letters in some other way. To guard against this, we denote the functions in differentiation rules by letters like *u* and *v* that are not likely to be already in use.

Rule 3 says that if a differentiable function is multiplied by a constant, then its derivative is multiplied by the same constant. Combined with Rule 2, it enables us to find the derivative of any monomial quickly; for example, the derivative of $7x^4$ is $7(4x^3) = 28x^3$.

To find the derivatives of polynomials, we need to be able to differentiate sums and differences of monomials. We can accomplish this by applying the Sum and Difference Rule.

Rule 4 The Sum and Difference Rule

If *u* and *v* are differentiable functions of *x*, then their sum and difference are differentiable at every point where *u* and *v* are differentiable. At such points,

$$\frac{d}{dx}(u \pm v) = \frac{du}{dx} \pm \frac{dv}{dx}.$$

Proof of Rule 4

We use the difference quotient for $f(x) = u(x) + v(x)$.

$$\frac{d}{dx}[u(x) + v(x)] = \lim_{h \to 0} \frac{[u(x + h) + v(x + h)] - [u(x) + v(x)]}{h}$$

$$= \lim_{h \to 0} \left[\frac{u(x + h) - u(x)}{h} + \frac{v(x + h) - v(x)}{h} \right]$$

$$= \lim_{h \to 0} \frac{u(x + h) - u(x)}{h} + \lim_{h \to 0} \frac{v(x + h) - v(x)}{h}$$

$$= \frac{du}{dx} + \frac{dv}{dx}$$

The proof of the rule for the difference of two functions is similar. ∎

Example 1   DIFFERENTIATING A POLYNOMIAL

Find $\dfrac{dp}{dt}$ if $p = t^3 + 6t^2 - \dfrac{5}{3}t + 16$.

Solution By Rule 4 we can differentiate the polynomial term-by-term, applying Rules 1 through 3 as we go.

$$\frac{dp}{dt} = \frac{d}{dt}(t^3) + \frac{d}{dt}(6t^2) - \frac{d}{dt}\left(\frac{5}{3}t\right) + \frac{d}{dt}(16) \qquad \text{Sum and Difference Rule}$$

$$= 3t^2 + 6 \cdot 2t - \frac{5}{3} + 0 \qquad \text{Constant and Power Rules}$$

$$= 3t^2 + 12t - \frac{5}{3}$$

Example 2   FINDING HORIZONTAL TANGENTS

Does the curve $y = x^4 - 2x^2 + 2$ have any horizontal tangents? If so, where?

Solution The horizontal tangents, if any, occur where the slope dy/dx is zero. To find these points, we

(a) calculate dy/dx:

$$\frac{dy}{dx} = \frac{d}{dx}(x^4 - 2x^2 + 2) = 4x^3 - 4x.$$

(b) solve the equation $dy/dx = 0$ for x:

$$4x^3 - 4x = 0$$

$$4x(x^2 - 1) = 0$$

$$x = 0, 1, -1.$$

The curve has horizontal tangents at $x = 0, 1,$ and -1. The corresponding points on the curve (found from the equation $y = x^4 - 2x^2 + 2$) are $(0, 2)$, $(1, 1)$, and $(-1, 1)$. You might wish to graph the curve to see where the horizontal tangents go.

The derivative in Example 2 was easily factored, making an algebraic solution of the equation $dy/dx = 0$ correspondingly simple. When a simple algebraic solution is not possible, the solutions to $dy/dx = 0$ can still be found to a high degree of accuracy by using the SOLVE capability of your calculator.

Example 3 USING CALCULUS AND CALCULATOR

As can be seen in the viewing window $[-10, 10]$ by $[-10, 10]$, the graph of $y = 0.2x^4 - 0.7x^3 - 2x^2 + 5x + 4$ has three horizontal tangents (Figure 3.18). At what points do these horizontal tangents occur?

Solution First we find the derivative

$$\frac{dy}{dx} = 0.8x^3 - 2.1x^2 - 4x + 5.$$

Using the calculator solver, we find that $0.8x^3 - 2.1x^2 - 4x + 5 = 0$ when $x \approx -1.862, 0.9484,$ and 3.539. We use the calculator again to evaluate the original function at these x-values and find the corresponding points to be approximately $(-1.862, -5.321)$, $(0.9484, 6.508)$, and $(3.539, -3.008)$.

Products and Quotients

While the derivative of the sum of two functions is the sum of their derivatives and the derivative of the difference of two functions is the difference of their derivatives, the derivative of the product of two functions is *not* the product of their derivatives.

For instance,

$$\frac{d}{dx}(x \cdot x) = \frac{d}{dx}(x^2) = 2x, \quad \text{while} \quad \frac{d}{dx}(x) \cdot \frac{d}{dx}(x) = 1 \cdot 1 = 1.$$

The derivative of a product is actually the sum of *two* products, as we now explain.

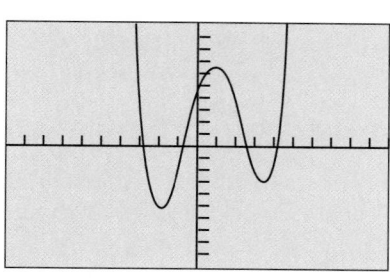

[–10, 10] by [–10, 10]

Figure 3.18 The graph of

$$y = 0.2x^4 - 0.7x^3 - 2x^2 + 5x + 4$$

has three horizontal tangents. (Example 3)

On Rounding Calculator Values

Notice in Example 3 that we rounded the x-values to four significant digits when we presented the answers. The calculator actually presented many more digits, but there was no practical reason for writing all of them. When we used the calculator to compute the corresponding y-values, however, we *used the x-values stored in the calculator,* not the rounded values. We then rounded the y-values to four significant digits when we presented the ordered pairs. Significant "round-off errors" can accumulate in a problem if you use rounded intermediate values for doing additional computations, so avoid rounding until the final answer.

You can remember the Product Rule with the phrase "the first times the derivative of the second plus the second times the derivative of the first."

Rule 5 The Product Rule

The product of two differentiable functions u and v is differentiable, and

$$\frac{d}{dx}(uv) = u\frac{dv}{dx} + v\frac{du}{dx}.$$

Gottfried Wilhelm Leibniz (1646–1716)

The method of limits used in this book was not discovered until nearly a century after Newton and Leibniz, the discoverers of calculus, had died.

To Leibniz, the key idea was the *differential,* an infinitely small quantity that was almost like zero, but which—unlike zero—could be used in the denominator of a fraction. Thus, Leibniz thought of the derivative *dy/dx* as the quotient of two differentials, *dy* and *dx*.

The problem was explaining why these differentials sometimes became zero and sometimes did not! See Exercise 40.

Some 17th-century mathematicians were confident that the calculus of Newton and Leibniz would eventually be found to be fatally flawed because of these mysterious quantities. It was only after later generations of mathematicians had found better ways to prove their results that the calculus of Newton and Leibniz was accepted by the entire scientific community.

Proof of Rule 5 We begin, as usual, by applying the definition.

$$\frac{d}{dx}(uv) = \lim_{h \to 0} \frac{u(x+h)v(x+h) - u(x)v(x)}{h}$$

To change the fraction into an equivalent one that contains difference quotients for the derivatives of u and v, we subtract and add $u(x+h)v(x)$ in the numerator. Then,

$$\frac{d}{dx}(uv) = \lim_{h \to 0} \frac{u(x+h)v(x+h) - u(x+h)v(x) + u(x+h)v(x) - u(x)v(x)}{h}$$

$$= \lim_{h \to 0} \left[u(x+h)\frac{v(x+h) - v(x)}{h} + v(x)\frac{u(x+h) - u(x)}{h} \right] \quad \text{Factor and separate.}$$

$$= \lim_{h \to 0} u(x+h) \cdot \lim_{h \to 0} \frac{v(x+h) - v(x)}{h} + v(x) \cdot \lim_{h \to 0} \frac{u(x+h) - u(x)}{h}.$$

As h approaches 0, $u(x+h)$ approaches $u(x)$ because u, being differentiable at x, is continuous at x. The two fractions approach the values of dv/dx and du/dx, respectively, at x. Therefore

$$\frac{d}{dx}(uv) = u\frac{dv}{dx} + v\frac{du}{dx}. \qquad \blacksquare$$

Example 4   DIFFERENTIATING A PRODUCT

Find $f'(x)$ if $f(x) = (x^2 + 1)(x^3 + 3)$.

Solution From the Product Rule with $u = x^2 + 1$ and $v = x^3 + 3$, we find

$$f'(x) = \frac{d}{dx}[(x^2 + 1)(x^3 + 3)] = (x^2 + 1)(3x^2) + (x^3 + 3)(2x)$$

$$= 3x^4 + 3x^2 + 2x^4 + 6x$$

$$= 5x^4 + 3x^2 + 6x.$$

We could also have done Example 4 by multiplying out the original expression and then differentiating the resulting polynomial. That alternate strategy will not work, however, on a product like $x^2 \sin x$, nor in a problem like Example 5.

Example 5   WORKING WITH NUMERICAL VALUES

Let $y = uv$ be the product of the functions u and v. Find $y'(2)$ if

$$u(2) = 3, \quad u'(2) = -4, \quad v(2) = 1, \quad \text{and} \quad v'(2) = 2.$$

Solution From the Product Rule, $y' = (uv)' = uv' + vu'$. In particular,

$$y'(2) = u(2)v'(2) + v(2)u'(2)$$

$$= (3)(2) + (1)(-4)$$

$$= 2.$$

Just as the derivative of the product of two differentiable functions is not the product of their derivatives, the derivative of a quotient of two functions is not the quotient of their derivatives. What happens instead is this:

Using the Quotient Rule

Since order is important in subtraction, be sure to set up the numerator of the Quotient Rule correctly:

> *v* times the derivative of *u*

minus

> *u* times the derivative of *v*.

You can remember the Quotient Rule with the phrase "bottom times the derivative of the top minus the top times the derivative of the bottom, all over the bottom squared."

Rule 6 The Quotient Rule

At a point where $v \neq 0$, the quotient $y = u/v$ of two differentiable functions is differentiable, and

$$\frac{d}{dx}\left(\frac{u}{v}\right) = \frac{v\dfrac{du}{dx} - u\dfrac{dv}{dx}}{v^2}.$$

Proof of Rule 6

$$\frac{d}{dx}\left(\frac{u}{v}\right) = \lim_{h \to 0} \frac{\dfrac{u(x+h)}{v(x+h)} - \dfrac{u(x)}{v(x)}}{h}$$

$$= \lim_{h \to 0} \frac{v(x)u(x+h) - u(x)v(x+h)}{hv(x+h)v(x)}$$

To change the last fraction into an equivalent one that contains the difference quotients for the derivatives of *u* and *v*, we subtract and add $v(x)u(x)$ in the numerator. This allows us to continue with

$$\frac{d}{dx}\left(\frac{u}{v}\right) = \lim_{h \to 0} \frac{v(x)u(x+h) - v(x)u(x) + v(x)u(x) - u(x)v(x+h)}{hv(x+h)v(x)}$$

$$= \lim_{h \to 0} \frac{v(x)\dfrac{u(x+h) - u(x)}{h} - u(x)\dfrac{v(x+h) - v(x)}{h}}{v(x+h)v(x)}.$$

Taking the limits in both the numerator and denominator now gives us the Quotient Rule. ∎

Example 6 SUPPORTING COMPUTATIONS GRAPHICALLY

Differentiate $f(x) = \dfrac{x^2 - 1}{x^2 + 1}$. Support graphically.

Solution We apply the Quotient Rule with $u = x^2 - 1$ and $v = x^2 + 1$:

$$f'(x) = \frac{(x^2 + 1) \cdot 2x - (x^2 - 1) \cdot 2x}{(x^2 + 1)^2} \qquad \frac{v(du/dx) - u(dv/dx)}{v^2}$$

$$= \frac{2x^3 + 2x - 2x^3 + 2x}{(x^2 + 1)^2}$$

$$= \frac{4x}{(x^2 + 1)^2}.$$

The graphs of $y_1 = f'(x)$ calculated above and of $y_2 = \text{NDER } f(x)$ are shown in Figure 3.19. The fact that they appear to be identical provides strong graphical support that our calculations are indeed correct.

[–3, 3] by [–2, 2]

Figure 3.19 The graph of

$$y = \frac{4x}{(x^2 + 1)^2}$$

and the graph of

$$y = \text{NDER}\left(\frac{x^2 - 1}{x^2 + 1}\right)$$

appear to be the same. (Example 6)

Negative Integer Powers of *x*

The rule for differentiating negative powers of *x* is the same as Rule 2 for differentiating positive powers of *x*, although our proof of Rule 2 does not work for negative values of *n*. We can now extend the Power Rule to negative integer powers by a clever use of the Quotient Rule.

Rule 7 Power Rule for Negative Integer Powers of *x*

If *n* is a negative integer and $x \neq 0$, then

$$\frac{d}{dx}(x^n) = nx^{n-1}.$$

Proof of Rule 7 If *n* is a negative integer, then $n = -m$, where *m* is a positive integer. Hence, $x^n = x^{-m} = 1/x^m$, and

$$\frac{d}{dx}(x^n) = \frac{d}{dx}\left(\frac{1}{x^m}\right)$$

$$= \frac{x^m \cdot \dfrac{d}{dx}(1) - 1 \cdot \dfrac{d}{dx}(x^m)}{(x^m)^2}$$

$$= \frac{0 - mx^{m-1}}{x^{2m}}$$

$$= -mx^{-m-1}$$

$$= nx^{n-1}. \qquad \blacksquare$$

Example 7 USING THE POWER RULE

Find an equation for the line tangent to the curve

$$y = \frac{x^2 + 3}{2x}$$

at the point $(1, 2)$. Support your answer graphically.

Solution We could find the derivative by the Quotient Rule, but it is easier to first simplify the function as a sum of two powers of *x*.

$$\frac{dy}{dx} = \frac{d}{dx}\left(\frac{x^2}{2x} + \frac{3}{2x}\right)$$

$$= \frac{d}{dx}\left(\frac{1}{2}x + \frac{3}{2}x^{-1}\right)$$

$$= \frac{1}{2} - \frac{3}{2}x^{-2}$$

The slope at $x = 1$ is

$$\left.\frac{dy}{dx}\right|_{x=1} = \left[\frac{1}{2} - \frac{3}{2}x^{-2}\right]_{x=1} = \frac{1}{2} - \frac{3}{2} = -1.$$

The line through $(1, 2)$ with slope $m = -1$ is

$$y - 2 = (-1)(x - 1)$$

$$y = -x + 1 + 2$$

$$y = -x + 3.$$

We graph $y = (x^2 + 3)/2x$ and $y = -x + 3$ (Figure 3.20), observing that the line appears to be tangent to the curve at $(1, 2)$. Thus, we have graphical support that our computations are correct.

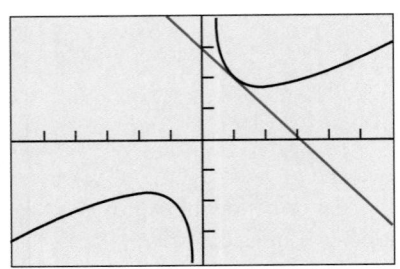

[−6, 6] by [−4, 4]

Figure 3.20 The line $y = -x + 3$ appears to be tangent to the graph of

$$y = \frac{x^2 + 3}{2x}$$

at the point $(1, 2)$. (Example 7)

Second and Higher Order Derivatives

The derivative $y' = dy/dx$ is called the *first derivative* of y with respect to x. The first derivative may itself be a differentiable function of x. If so, its derivative,

$$y'' = \frac{dy'}{dx} = \frac{d}{dx}\left(\frac{dy}{dx}\right) = \frac{d^2y}{dx^2},$$

is called the *second derivative* of y with respect to x. If y'' ("y double-prime") is differentiable, its derivative,

$$y''' = \frac{dy''}{dx} = \frac{d^3y}{dx^3},$$

is called the *third derivative* of y with respect to x. The names continue as you might expect they would, except that the multiple-prime notation begins to lose its usefulness after about three primes. We use

$$y^{(n)} = \frac{d}{dx}y^{(n-1)} \quad \text{"}y \text{ super } n\text{"}$$

to denote the **nth derivative** of y with respect to x. (We also use d^ny/dx^n.) Do not confuse $y^{(n)}$ with the nth power of y, which is y^n.

Technology Tip

HIGHER ORDER DERIVATIVES WITH NDER

Some graphers will allow the *nesting* of the NDER function,

NDER2 f = NDER(NDER f),

but such nesting, in general, is safe only to the second derivative. Beyond that, the error buildup in the algorithm makes the results unreliable.

Example 8 FINDING HIGHER ORDER DERIVATIVES

Find the first four derivatives of $y = x^3 - 5x^2 + 2$.

Solution The first four derivatives are:

First derivative: $\quad y' = 3x^2 - 10x;$

Second derivative: $\quad y'' = 6x - 10;$

Third derivative: $\quad y''' = 6;$

Fourth derivative: $\quad y^{(4)} = 0.$

This function has derivatives of all orders, the fourth and higher order derivatives all being zero.

Quick Review 3.3

In Exercises 1–6, write the expression as a sum of powers of x.

1. $(x^2 - 2)(x^{-1} + 1)$

2. $\left(\dfrac{x}{x^2+1}\right)^{-1}$

3. $3x^2 - \dfrac{2}{x} + \dfrac{5}{x^2}$

4. $\dfrac{3x^4 - 2x^3 + 4}{2x^2}$

5. $(x^{-1} + 2)(x^{-2} + 1)$

6. $\dfrac{x^{-1} + x^{-2}}{x^{-3}}$

7. Find the positive roots of the equation
$$2x^3 - 5x^2 - 2x + 6 = 0$$
and evaluate the function $y = 500x^6$ at each root. Round your answers to the nearest integer, but only in the final step.

8. If $f(x) = 7$ for all real numbers x, find
  (a) $f(10).$ **(b)** $f(0).$
  (c) $f(x + h).$ **(d)** $\lim\limits_{x \to a} \dfrac{f(x) - f(a)}{x - a}.$

9. Find the derivatives of these functions with respect to x.
  (a) $f(x) = \pi$ **(b)** $f(x) = \pi^2$ **(c)** $f(x) = \pi^{15}$

10. Find the derivatives of these functions with respect to x using the definition of the derivative.
  (a) $f(x) = \dfrac{x}{\pi}$ **(b)** $f(x) = \dfrac{\pi}{x}$

Section 3.3 Exercises

In Exercises 1–10, find dy/dx and d^2y/dx^2.

1. $y = -x^2 + 3$

2. $y = \dfrac{x^3}{3} - x$

3. $y = 2x + 1$

4. $y = x^2 + x + 1$

5. $y = \dfrac{x^3}{3} + \dfrac{x^2}{2} + x$

6. $y = 1 - x + x^2 - x^3$

7. $y = x^4 - 7x^3 + 2x^2 + 15$

8. $y = 5x^3 - 3x^5$

9. $y = 4x^{-2} - 8x + 1$

10. $y = \dfrac{x^{-4}}{4} - \dfrac{x^{-3}}{3} + \dfrac{x^{-2}}{2} - x^{-1} + 3$

11. Let $y = (x + 1)(x^2 + 1)$. Find dy/dx **(a)** by applying the Product Rule, and **(b)** by multiplying the factors first and then differentiating.

12. Let $y = (x^2 + 3)/x$. Find dy/dx **(a)** by using the Quotient Rule, and **(b)** by first dividing the terms in the numerator by the denominator and then differentiating.

In Exercises 13–19, find dy/dx.

13. $y = \dfrac{2x + 5}{3x - 2}$

14. $y = \dfrac{x^2 + 5x - 1}{x^2}$

15. $y = \dfrac{(x - 1)(x^2 + x + 1)}{x^3}$

16. $y = (1 - x)(1 + x^2)^{-1}$

17. $y = \dfrac{x^2}{1 - x^3}$

18. $y = \dfrac{\sqrt{x} - 1}{\sqrt{x} + 1}$

19. $y = \dfrac{(x + 1)(x + 2)}{(x - 1)(x - 2)}$

20. Use the definition of derivative (given in Section 3.1, Equation 1) to show that

(a) $\dfrac{d}{dx}(x) = 1$.

(b) $\dfrac{d}{dx}(-u) = -\dfrac{du}{dx}$.

21. Use the Product Rule to show that

$$\frac{d}{dx}(c \cdot f(x)) = c \cdot \frac{d}{dx}f(x)$$

for any constant c.

22. Devise a rule for $\dfrac{d}{dx}\left(\dfrac{1}{f(x)}\right)$.

23. Suppose u and v are functions of x that are differentiable at $x = 0$, and that $u(0) = 5$, $u'(0) = -3$, $v(0) = -1$, $v'(0) = 2$. Find the values of the following derivatives at $x = 0$.

(a) $\dfrac{d}{dx}(uv)$

(b) $\dfrac{d}{dx}\left(\dfrac{u}{v}\right)$

(c) $\dfrac{d}{dx}\left(\dfrac{v}{u}\right)$

(d) $\dfrac{d}{dx}(7v - 2u)$

24. Suppose u and v are functions of x that are differentiable at $x = 2$ and that $u(2) = 3$, $u'(2) = -4$, $v(2) = 1$, and $v'(2) = 2$. Find the values of the following derivatives at $x = 2$.

(a) $\dfrac{d}{dx}(uv)$

(b) $\dfrac{d}{dx}\left(\dfrac{u}{v}\right)$

(c) $\dfrac{d}{dx}\left(\dfrac{v}{u}\right)$

(d) $\dfrac{d}{dx}(3u - 2v + 2uv)$

25. Which of the following numbers is the slope of the line tangent to the curve $y = x^2 + 5x$ at $x = 3$?

i. 24 **ii.** $-5/2$ **iii.** 11 **iv.** 8

26. Which of the following numbers is the slope of the line $3x - 2y + 12 = 0$?

i. 6 **ii.** 3 **iii.** 3/2 **iv.** 2/3

In Exercises 27–32, support your answer graphically.

27. Find an equation of the line perpendicular to the tangent to the curve $y = x^3 - 3x + 1$ at the point $(2, 3)$.

28. Find the tangents to the curve $y = x^3 + x$ at the points where the slope is 4. What is the smallest slope of the curve? At what value of x does the curve have this slope?

29. Find the points on the curve $y = 2x^3 - 3x^2 - 12x + 20$ where the tangent is parallel to the x-axis.

30. Find the x- and y-intercepts of the line that is tangent to the curve $y = x^3$ at the point $(-2, -8)$.

31. Find the tangents to *Newton's serpentine,*

$$y = \frac{4x}{x^2 + 1},$$

at the origin and the point $(1, 2)$.

32. Find the tangent to the *witch of Agnesi,*

$$y = \frac{8}{4 + x^2},$$

at the point $(2, 1)$.

When we work with functions of a single variable in mathematics, we often call the independent variable x and the dependent variable y. Applied fields use many different letters, however. Here are some examples.

33. *Cylinder Pressure* If gas in a cylinder is maintained at a constant temperature T, the pressure P is related to the volume V by a formula of the form

$$P = \frac{nRT}{V - nb} - \frac{an^2}{V^2},$$

in which a, b, n, and R are constants. Find dP/dV.

34. *Free Fall* When a rock falls from rest near the surface of the earth, the distance it covers during the first few seconds is given by the equation

$$s = 4.9t^2.$$

In this equation, s is the distance in meters and t is the elapsed time in seconds. Find ds/dt and d^2s/dt^2.

In Exercises 35–39, *work in groups of two or three* to solve the problem.

35. *The Body's Reaction to Medicine* The reaction of the body to a dose of medicine can often be represented by an equation of the form

$$R = M^2\left(\frac{C}{2} - \frac{M}{3}\right),$$

where C is a positive constant and M is the amount of medicine absorbed in the blood. If the reaction is a change in blood pressure, R is measured in millimeters of mercury. If the reaction is a change in temperature, R is measured in degrees, and so on.

Find dR/dM. This derivative, as a function of M, is called the sensitivity of the body to medicine. In Chapter 4, we shall see how to find the amount of medicine to which the body is most sensitive. *Source: Some Mathematical Models in Biology*, Revised Edition, December 1967, PB-202 364, p. 221; distributed by N.T.I.S., U.S. Department of Commerce.

36. Writing to Learn Recall that the area A of a circle with radius r is πr^2 and that the circumference C is $2\pi r$. Notice that $dA/dr = C$. Explain in terms of geometry why the instantaneous rate of change of the area with respect to the radius should equal the circumference.

37. Writing to Learn Recall that the volume V of a sphere of radius r is $(4/3)\pi r^3$ and that the surface area A is $4\pi r^2$. Notice that $dV/dr = A$. Explain in terms of geometry why the instantaneous rate of change of the volume with respect to the radius should equal the surface area.

38. *Orchard Farming* An apple farmer currently has 156 trees yielding an average of 12 bushels of apples per tree. He is expanding his farm at a rate of 13 trees per year, while improved husbandry is improving his average annual yield by 1.5 bushels per tree. What is the current (instantaneous) rate of increase of his total annual production of apples? Answer in appropriate units of measure.

39. *Picnic Pavilion Rental* The members of the Blue Boar society always divide the pavilion rental fee for their picnics equally among the members. Currently there are 65 members and the pavilion rents for $250. The pavilion cost is increasing at a rate of $10 per year, while the Blue Boar membership is increasing at a rate of 6 members per year. What is the current (instantaneous) rate of change in each member's share of the pavilion rental fee? Answer in appropriate units of measure.

Extending the Ideas

40. *Leibniz's Proof of the Product Rule* Here's how Leibniz explained the Product Rule in a letter to his colleague John Wallis:

It is useful to consider quantities infinitely small such that when their ratio is sought, they may not be considered zero, but which are rejected as often as they occur with quantities incomparably greater. Thus if we have $x + dx$, dx is rejected. Similarly we cannot have xdx and $dxdx$ standing together, as xdx is incomparably greater than $dxdx$. Hence if we are to differentiate uv, we write

$$d(uv) = (u + du)(v + dv) - uv$$
$$= uv + vdu + udv + dudv - uv$$
$$= vdu + udv.$$

Answer the following questions about Leibniz's proof.

(a) What does Leibniz mean by a quantity being "rejected"?

(b) What happened to $dudv$ in the last step of Leibniz's proof?

(c) Divide both sides of Leibniz's formula

$$[d(uv) = vdu + udv]$$

by the differential dx. What formula results?

(d) Why would the critics of Leibniz's time have objected to dividing both sides of the equation by dx?

(e) Leibniz had a similar simple (but not-so-clean) proof of the Quotient Rule. Can you reconstruct it?

3.4 Velocity and Other Rates of Change

Instantaneous Rates of Change • Motion along a Line •
Sensitivity to Change • Derivatives in Economics

Instantaneous Rates of Change

In this section we examine some applications in which derivatives as functions are used to represent the rates at which things change in the world around us. It is natural to think of change as change with respect to time, but other variables can be treated in the same way. For example, a physician may want to know how change in dosage affects the body's response to a drug. An economist may want to study how the cost of producing steel varies with the number of tons produced.

If we interpret the difference quotient

$$\frac{f(x + h) - f(x)}{h}$$

as the average rate of change of the function f over the interval from x to $x + h$, we can interpret its limit as h approaches 0 to be the rate at which f is changing at the point x.

Definition Instantaneous Rate of Change

The (**instantaneous**) **rate of change** of f with respect to x at a is the derivative

$$f'(a) = \lim_{h \to 0} \frac{f(a + h) - f(a)}{h},$$

provided the limit exists.

It is conventional to use the word *instantaneous* even when x does not represent time. The word, however, is frequently omitted in practice. When we say *rate of change,* we mean *instantaneous rate of change.*

Example 1 ENLARGING CIRCLES

(**a**) Find the rate of change of the area A of a circle with respect to its radius r.

(**b**) Evaluate the rate of change of A at $r = 5$ and at $r = 10$.

(**c**) If r is measured in inches and A is measured in square inches, what units would be appropriate for dA/dr?

Solution The area of a circle is related to its radius by the equation $A = \pi r^2$.

(**a**) The (instantaneous) rate of change of A with respect to r is

$$\frac{dA}{dr} = \frac{d}{dr}(\pi r^2) = \pi \cdot 2r = 2\pi r.$$

(**b**) At $r = 5$, the rate is 10π (about 31.4). At $r = 10$, the rate is 20π (about 62.8).

Notice that the rate of change gets bigger as r gets bigger. As can be seen in Figure 3.21, the same change in radius brings about a bigger change in area as the circles grow radially away from the center.

(c) The appropriate units for dA/dr are square inches (of area) per inch (of radius).

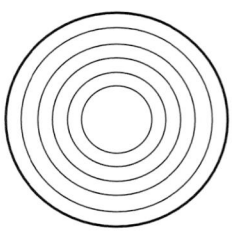

Figure 3.21 The same change in radius brings about a larger change in area as the circles grow radially away from the center. (Example 1, Exploration 1)

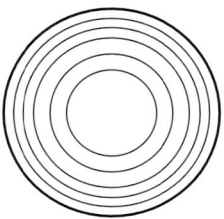

Figure 3.22 Which is the more appropriate model for the growth of rings in a tree—the circles here or those in Figure 3.21? (Exploration 1)

Exploration 1   Growth Rings on a Tree

The phenomenon observed in Example 1, that the rate of change in area of a circle with respect to its radius gets larger as the radius gets larger, is reflected in nature in many ways. When trees grow, they add layers of wood directly under the inner bark during the growing season, then form a darker, protective layer for protection during the winter. This results in concentric rings that can be seen in a cross-sectional slice of the trunk. The age of the tree can be determined by counting the rings.

1. Look at the concentric rings in Figure 3.21 and Figure 3.22. Which is a better model for the pattern of growth rings in a tree? Is it likely that a tree could find the nutrients and light necessary to increase its amount of growth every year?

2. Considering how trees grow, explain why the change in *area* of the rings remains relatively constant from year to year.

3. If the change in area is constant, and if

$$\frac{dA}{dr} = \frac{\text{change in area}}{\text{change in radius}} = 2\pi r,$$

explain why the change in radius must get smaller as r gets bigger.

Motion along a Line

Suppose that an object is moving along a coordinate line (say an s-axis) so that we know its position s on that line as a function of time t:

$$s = f(t).$$

The **displacement** of the object over the time interval from t to $t + \Delta t$ is

$$\Delta s = f(t + \Delta t) - f(t)$$

(Figure 3.23) and the **average velocity** of the object over that time interval is

$$v_{av} = \frac{\text{displacement}}{\text{travel time}} = \frac{\Delta s}{\Delta t} = \frac{f(t + \Delta t) - f(t)}{\Delta t}.$$

To find the object's velocity at the exact instant t, we take the limit of the average velocity over the interval from t to $t + \Delta t$ as Δt shrinks to zero. The limit is the derivative of f with respect to t.

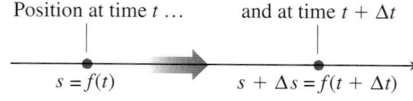

Position at time t ...   and at time $t + \Delta t$

$s = f(t)$ $s + \Delta s = f(t + \Delta t)$

Figure 3.23 The positions of an object moving along a coordinate line at time t and shortly later at time $t + \Delta t$.

Definition Instantaneous Velocity

The **(instantaneous) velocity** is the derivative of the position function $s = f(t)$ with respect to time. At time t the velocity is

$$v(t) = \frac{ds}{dt} = \lim_{\Delta t \to 0} \frac{f(t + \Delta t) - f(t)}{\Delta t}.$$

Example 2 FINDING THE VELOCITY OF A RACE CAR

Figure 3.24 shows the time-to-distance graph of a 1996 Riley & Scott Mk III-Olds WSC race car. The slope of the secant PQ is the average velocity for the 3-second interval from $t = 2$ to $t = 5$ sec, in this case, about 100 ft/sec or 68 mph. The slope of the tangent at P is the speedometer reading at $t = 2$ sec, about 57 ft/sec or 39 mph. The acceleration for the period shown is a nearly constant 28.5 ft/sec during each second, which is about $0.89g$ where g is the acceleration due to gravity. The race car's top speed is an estimated 190 mph. *Source: Road and Track*, March 1997.

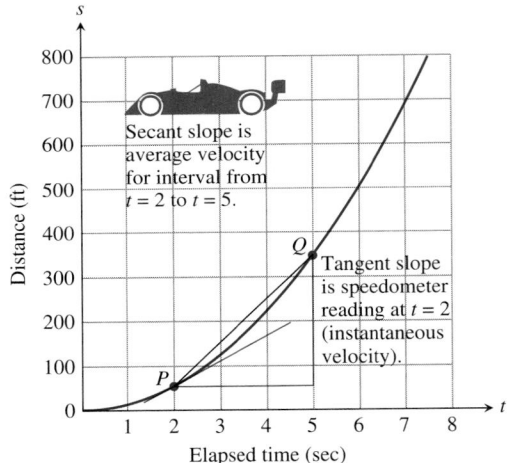

Figure 3.24 The time-to-distance graph for Example 2.

Besides telling how fast an object is moving, velocity tells the direction of motion. When the object is moving forward (when s is increasing), the velocity is positive; when the object is moving backward (when s is decreasing), the velocity is negative.

If we drive to a friend's house and back at 30 mph, the speedometer will show 30 on the way over but will not show -30 on the way back, even though our distance from home is decreasing. The speedometer always shows *speed*, which is the absolute value of velocity. Speed measures the rate of motion regardless of direction.

Definition Speed

Speed is the absolute value of velocity.

$$\text{Speed} = |v(t)| = \left| \frac{ds}{dt} \right|$$

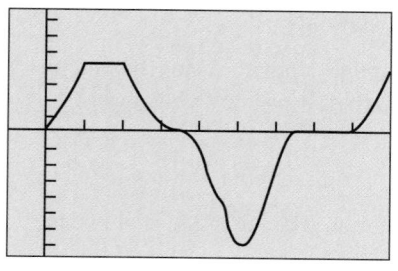

[–4, 36] by [–7.5, 7.5]

Figure 3.25 A student's velocity graph from data recorded by a motion detector. (Example 3)

Example 3 READING A VELOCITY GRAPH

A student walks around in front of a motion detector that records her velocity at 1-second intervals for 36 seconds. She stores the data in her graphing calculator and uses it to generate the time-velocity graph shown in Figure 3.25. Describe her motion as a function of time by reading the velocity graph. When is her *speed* a maximum?

Solution The student moves forward for the first 14 seconds, moves backward for the next 12 seconds, stands still for 6 seconds, and then moves forward again. She achieves her maximum speed at $t \approx 20$, while moving backward.

The rate at which a body's velocity changes is called the body's *acceleration*. The acceleration measures how quickly the body picks up or loses speed.

Definition Acceleration

Acceleration is the derivative of velocity with respect to time. If a body's velocity at time t is $v(t) = ds/dt$, then the body's acceleration at time t is

$$a(t) = \frac{dv}{dt} = \frac{d^2s}{dt^2}.$$

The earliest questions that motivated the discovery of calculus were concerned with velocity and acceleration, particularly the motion of freely falling bodies under the force of gravity. (See Examples 1 and 2 in Section 2.1.) The mathematical description of this type of motion captured the imagination of many great scientists, including Aristotle, Galileo, and Newton. Experimental and theoretical investigations revealed that the distance a body released from rest falls freely is proportional to the square of the amount of time it has fallen. We express this by saying that

$$s = \frac{1}{2}gt^2,$$

where s is distance, g is the acceleration due to Earth's gravity, and t is time. The value of g in the equation depends on the units used to measure s and t. With t in seconds (the usual unit), we have the following values:

Free-fall Constants (Earth)

English units: $g = 32\dfrac{\text{ft}}{\text{sec}^2},$ $s = \dfrac{1}{2}(32)t^2 = 16t^2$ (s in feet)

Metric units: $g = 9.8\dfrac{\text{m}}{\text{sec}^2},$ $s = \dfrac{1}{2}(9.8)t^2 = 4.9t^2$ (s in meters)

The abbreviation ft/sec² is read "feet per second squared" or "feet per second per second," and m/sec² is read "meters per second squared."

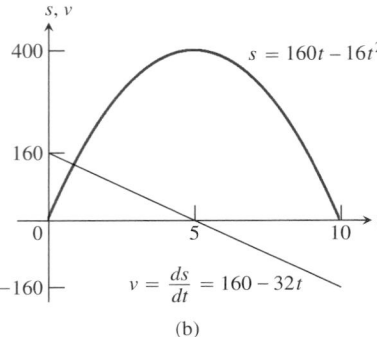

Figure 3.26 (a) The rock in Example 4. (b) The graphs of s and v as functions of time t, showing that s is largest when $v = ds/dt = 0$. (The graph of s is *not* the path of the rock; it is a plot of height as a function of time.) (Example 4)

Example 4   MODELING VERTICAL MOTION

A dynamite blast propels a heavy rock straight up with a launch velocity of 160 ft/sec (about 109 mph) (Figure 3.26a). It reaches a height of $s = 160t - 16t^2$ ft after t seconds.

(a) How high does the rock go?

(b) What is the velocity and speed of the rock when it is 256 ft above the ground on the way up? on the way down?

(c) What is the acceleration of the rock at any time t during its flight (after the blast)?

(d) When does the rock hit the ground?

Solution In the coordinate system we have chosen, s measures height from the ground up, so velocity is positive on the way up and negative on the way down.

(a) The instant when the rock is at its highest point is the one instant during the flight when the velocity is 0. At any time t, the velocity is

$$v = \frac{ds}{dt} = \frac{d}{dt}(160t - 16t^2) = 160 - 32t \text{ ft/sec}.$$

The velocity is zero when $160 - 32t = 0$, or at $t = 5$ sec.

The maximum height is the height of the rock at $t = 5$ sec. That is,

$$s_{max} = s(5) = 160(5) - 16(5)^2 = 400 \text{ ft}.$$

See Figure 3.26b.

(b) To find the velocity when the height is 256 ft, we determine the two values of t for which $s(t) = 256$ ft.

$$s(t) = 160t - 16t^2 = 256$$
$$16t^2 - 160t + 256 = 0$$
$$16(t^2 - 10t + 16) = 0$$
$$(t - 2)(t - 8) = 0$$
$$t = 2 \text{ sec} \quad \text{or} \quad t = 8 \text{ sec}$$

The velocity of the rock at each of these times is

$$v(2) = 160 - 32(2) = 96 \text{ ft/sec},$$
$$v(8) = 160 - 32(8) = -96 \text{ ft/sec}.$$

At both instants, the speed of the rock is 96 ft/sec.

(c) At any time during its flight after the explosion, the rock's acceleration is

$$a = \frac{dv}{dt} = \frac{d}{dt}(160 - 32t) = -32 \text{ ft/sec}^2.$$

The acceleration is always downward. When the rock is rising, it is slowing down; when it is falling, it is speeding up.

(d) The rock hits the ground at the positive time for which $s = 0$. The equation $160t - 16t^2 = 0$ has two solutions: $t = 0$ and $t = 10$. The blast initiated the flight of the rock from ground level at $t = 0$. The rock returned to the ground 10 seconds later.

Exploration 2   **Modeling Horizontal Motion**

The position (*x*-coordinate) of a particle moving on the horizontal line $y = 2$ is given by $x(t) = 4t^3 - 16t^2 + 15t$ for $t \geq 0$.

1. Graph the parametric equations $x_1(t) = 4t^3 - 16t^2 + 15t$, $y_1(t) = 2$ in $[-4, 6]$ by $[-3, 5]$. Use TRACE to support that the particle starts at the point $(0, 2)$, moves to the right, then to the left, and finally to the right. At what times does the particle reverse direction?

2. Graph the parametric equations $x_2(t) = x_1(t)$, $y_2(t) = t$ in the same viewing window. Explain how this graph shows the back and forth motion of the particle. Use this graph to find when the particle reverses direction.

3. Graph the parametric equations $x_3(t) = t$, $y_3(t) = x_1(t)$ in the same viewing window. Explain how this graph shows the back and forth motion of the particle. Use this graph to find when the particle reverses direction.

4. Use the methods in parts 1, 2, and 3 to represent and describe the *velocity* of the particle.

Exploration 3   **Seeing Motion on a Graphing Calculator**

The graphs in Figure 3.26b give us plenty of information about the flight of the rock in Example 4, but neither graph shows the path of the rock in flight. We can simulate the moving rock by graphing the parametric equations

$$x_1(t) = 3(t < 5) + 3.1(t \geq 5), \qquad y_1(t) = 160t - 16t^2$$

in dot mode.

This will show the upward flight of the rock along the vertical line $x = 3$, and the downward flight of the rock along the line $x = 3.1$.

1. To see the flight of the rock from beginning to end, what should we use for *t*Min and *t*Max in our graphing window?

2. Set $x\text{Min} = 0$, $x\text{Max} = 6$, and $y\text{Min} = -10$. Use the results from Example 4 to determine an appropriate value for $y\text{Max}$. (You will want the entire flight of the rock to fit within the vertical range of the screen.)

3. Set *t*Step initially at 0.1. (A higher number will make the simulation move faster. A lower number will slow it down.)

4. Can you explain why the grapher actually slows down when the rock would slow down, and speeds up when the rock would speed up?

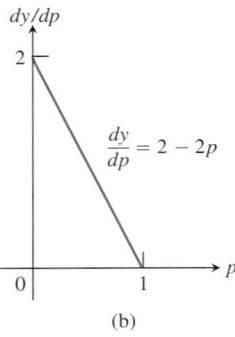

Figure 3.27 (a) The graph of $y = 2p - p^2$ describing the proportion of smooth-skinned peas. (b) The graph of dy/dp. (Example 5)

Sensitivity to Change

When a small change in x produces a large change in the value of a function $f(x)$, we say that the function is relatively **sensitive** to changes in x. The derivative $f'(x)$ is a measure of this sensitivity.

Example 5 GENETIC DATA AND SENSITIVITY TO CHANGE

The Austrian monk Gregor Johann Mendel (1822–1884), working with garden peas and other plants, provided the first scientific explanation of hybridization. His careful records showed that if p (a number between 0 and 1) is the relative frequency of the gene for smooth skin in peas (dominant) and $(1 - p)$ is the relative frequency of the gene for wrinkled skin in peas (recessive), then the proportion of smooth-skinned peas in the next generation will be

$$y = 2p(1 - p) + p^2 = 2p - p^2.$$

The graph of y versus p in Figure 3.27a suggests that the value of y is more sensitive to a change in p when p is small than it is to a change in p when p is large. Indeed, this is borne out by the derivative graph in Figure 3.27b, which shows that dy/dp is close to 2 when p is near 0 and close to 0 when p is near 1.

The implication for genetics is that introducing a few more dominant genes into a highly recessive population will have a more dramatic effect on later generations than will a similar increase in a highly dominant population.

Derivatives in Economics

Engineers use the terms *velocity* and *acceleration* to refer to the derivatives of functions describing motion. Economists, too, have a specialized vocabulary for rates of change and derivatives. They call them *marginals*.

In a manufacturing operation, *the cost of production* $c(x)$ is a function of x, the number of units produced. The *marginal cost of production* is the rate of change of cost with respect to the level of production, so it is dc/dx.

Suppose $c(x)$ represents the dollars needed to produce x tons of steel in one week. It costs more to produce $x + h$ tons per week, and the cost difference divided by h is the average cost of producing each additional ton.

$$\frac{c(x + h) - c(x)}{h} = \begin{cases} \text{the average cost of each of the} \\ \text{additional } h \text{ tons produced} \end{cases}$$

The limit of this ratio as $h \to 0$ is the **marginal cost** of producing more steel per week when the current production is x tons (Figure 3.28).

$$\frac{dc}{dx} = \lim_{h \to 0} \frac{c(x + h) - c(x)}{h} = \text{marginal cost of production}$$

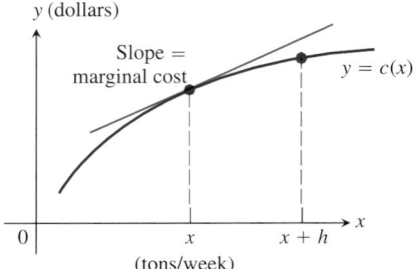

Figure 3.28 Weekly steel production: $c(x)$ is the cost of producing x tons per week. The cost of producing an additional h tons per week is $c(x + h) - c(x)$.

Sometimes the marginal cost of production is loosely defined to be the extra cost of producing one more unit,

$$\frac{\Delta c}{\Delta x} = \frac{c(x + 1) - c(x)}{1},$$

which is approximated by the value of dc/dx at x. This approximation is acceptable if the slope of c does not change quickly near x, for then the difference quotient is close to its limit dc/dx even if $\Delta x = 1$ (Figure 3.29). The approximation works best for large values of x.

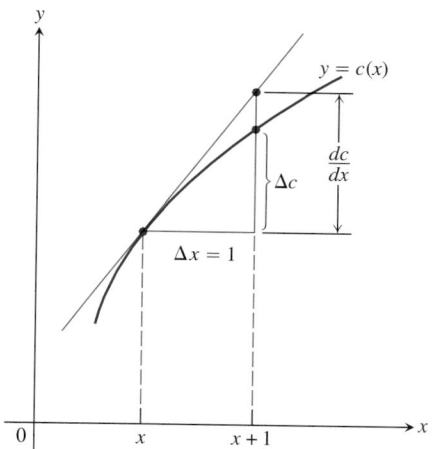

Figure 3.29 Because dc/dx is the slope of the tangent at x, the marginal cost dc/dx approximates the extra cost Δc of producing $\Delta x = 1$ more unit.

Example 6 MARGINAL COST AND MARGINAL REVENUE

Suppose it costs

$$c(x) = x^3 - 6x^2 + 15x$$

dollars to produce x radiators when 8 to 10 radiators are produced, and that

$$r(x) = x^3 - 3x^2 + 12x$$

gives the dollar revenue from selling x radiators. Your shop currently produces 10 radiators a day. Find the marginal cost and **marginal revenue.**

Solution The marginal cost of producing one more radiator a day when 10 are being produced is $c'(10)$.

$$c'(x) = \frac{d}{dx}(x^3 - 6x^2 + 15x) = 3x^2 - 12x + 15$$

$$c'(10) = 3(100) - 12(10) + 15 = 195 \text{ dollars}$$

The marginal revenue is

$$r'(x) = \frac{d}{dx}(x^3 - 3x^2 + 12x) = 3x^2 - 6x + 12,$$

so,

$$r'(10) = 3(100) - 6(10) + 12 = 252 \text{ dollars.}$$

Quick Review 3.4

In Exercises 1–10, answer the questions about the graph of the quadratic function $y = f(x) = -16x^2 + 160x - 256$ by analyzing the equation algebraically. Then support your answers graphically.

1. Does the graph open upward or downward?

2. What is the y-intercept?

3. What are the x-intercepts?

4. What is the range of the function?

5. What point is the vertex of the parabola?

6. At what x-values does $f(x) = 80$?

7. For what x-value does $dy/dx = 100$?

8. On what interval is $dy/dx > 0$?

9. Find $\lim\limits_{h \to 0} \dfrac{f(3 + h) - f(3)}{h}$.

10. Find d^2y/dx^2 at $x = 7$.

Section 3.4 Exercises

1. Find the instantaneous rate of change of the volume V of a cube with respect to a side s.

2. *Particle Motion* A particle moves along a line so that its position at any time $t \geq 0$ is given by the function

$$s(t) = t^2 - 3t + 2,$$

where s is measured in meters and t is measured in seconds.

(a) Find the displacement during the first 5 seconds.

(b) Find the average velocity during the first 5 seconds.

(c) Find the instantaneous velocity when $t = 4$.

(d) Find the acceleration of the particle when $t = 4$.

(e) At what values of t does the particle change direction?

(f) Where is the particle when s is a minimum?

3. *Lunar Projectile Motion* A rock thrown vertically upward from the surface of the moon at a velocity of 24 m/sec (about 86 km/h) reaches a height of $s = 24t - 0.8t^2$ meters in t seconds.

(a) Find the rock's velocity and acceleration as functions of time. (The acceleration in this case is the acceleration of gravity on the moon.)

(b) How long did it take the rock to reach its highest point?

(c) How high did the rock go?

(d) When did the rock reach half its maximum height?

(e) How long was the rock aloft?

4. *Free Fall* The equations for free fall near the surfaces of Mars and Jupiter (s in meters, t in seconds) are: Mars, $s = 1.86t^2$; Jupiter, $s = 11.44t^2$. How long would it take a rock falling from rest to reach a velocity of 16.6 m/sec (about 60 km/h) on each planet?

5. *Projectile Motion* On Earth, in the absence of air, the rock in Exercise 3 would reach a height of $s = 24t - 4.9t^2$ meters in t seconds. How high would the rock go?

6. *Speeding Bullet* A bullet fired straight up from the moon's surface would reach a height of $s = 832t - 2.6t^2$ ft after t sec. On Earth, in the absence of air, its height would be $s = 832t - 16t^2$ ft after t sec. How long would it take the bullet to get back down in each case?

7. *Parametric Graphing* Devise a grapher simulation of the problem situation in Exercise 6. Use it to support the answers obtained analytically.

8. *Bacterium Population* When a bactericide was added to a nutrient broth in which bacteria were growing, the bacterium population continued to grow for a while but then stopped growing and began to decline. The size of the population at time t (hours) was $b(t) = 10^6 + 10^4 t - 10^3 t^2$. Find the growth rates at $t = 0$, $t = 5$, and $t = 10$ hours.

9. *Draining a Tank* The number of gallons of water in a tank t minutes after the tank has started to drain is $Q(t) = 200(30 - t)^2$. How fast is the water running out at the end of 10 min? What is the average rate at which the water flows out during the first 10 min?

10. *Marginal Cost* Suppose that the dollar cost of producing x washing machines is $c(x) = 2000 + 100x - 0.1x^2$.

    (a) Find the average cost of producing 100 washing machines.

    (b) Find the marginal cost when 100 machines are produced.

    (c) Show that the marginal cost when 100 washing machines are produced is approximately the cost of producing one more washing machine after the first 100 have been made, by calculating the latter cost directly.

11. *Marginal Revenue* Suppose the weekly revenue in dollars from selling x custom-made office desks is

$$r(x) = 2000\left(1 - \frac{1}{x + 1}\right).$$

    (a) Draw the graph of r. What values of x make sense in this problem situation?

    (b) Find the marginal revenue when x desks are sold.

    (c) Use the function $r'(x)$ to estimate the increase in revenue that will result from increasing sales from 5 desks a week to 6 desks a week.

    (d) **Writing to Learn** Find the limit of $r'(x)$ as $x \to \infty$. How would you interpret this number?

12. *Particle Motion* The position of a body at time t sec is $s = t^3 - 6t^2 + 9t$ m. Find the body's acceleration each time the velocity is zero.

13. *Finding Speed* A body's velocity at time t sec is $v = 2t^3 - 9t^2 + 12t - 5$ m/sec. Find the body's speed each time the acceleration is zero.

14. *Finding f from f'* Let $f'(x) = 3x^2$.

    (a) Compute the derivatives of $g(x) = x^3$, $h(x) = x^3 - 2$, and $t(x) = x^3 + 3$.

    (b) Graph the numerical derivatives of g, h, and t.

    (c) Describe a *family* of functions, $f(x)$, that have the property that $f'(x) = 3x^2$.

    (d) Is there a function f such that $f'(x) = 3x^2$ *and* $f(0) = 0$? If so, what is it?

    (e) Is there a function f such that $f'(x) = 3x^2$ *and* $f(0) = 3$? If so, what is it?

15. *Finding Profit* The monthly profit (in thousands of dollars) of a software company is given by

$$P(x) = \frac{10}{1 + 50 \cdot 2^{5 - 0.1x}},$$

where x is the number of software packages sold.

    (a) Graph $P(x)$.

    (b) What values of x make sense in the problem situation?

    (c) Use NDER to graph $P'(x)$. For what values of x is P relatively sensitive to changes in x?

    (d) What is the profit when the marginal profit is greatest?

    (e) What is the marginal profit when 50 units are sold? 100 units, 125 units, 150 units, 175 units, and 300 units?

    (f) What is $\lim_{x \to \infty} P(x)$? What is the maximum profit possible?

    (g) **Writing to Learn** Is there a practical explanation to the maximum profit answer? Explain your reasoning.

Explorations

16. *Launching a Rocket* When a model rocket is launched, the propellant burns for a few seconds, accelerating the rocket upward. After burnout, the rocket coasts upward for a while and then begins to fall. A small explosive charge pops out a parachute shortly after the rocket starts downward. The parachute slows the rocket to keep it from breaking when it lands. This graph shows velocity data from the flight.

Use the graph to answer the following.

(a) How fast was the rocket climbing when the engine stopped?

(b) For how many seconds did the engine burn?

(c) When did the rocket reach its highest point? What was its velocity then?

(d) When did the parachute pop out? How fast was the rocket falling then?

(e) How long did the rocket fall before the parachute opened?

(f) When was the rocket's acceleration greatest? When was the acceleration constant?

17. *Pisa by Parachute* *(continuation of Exercise 16)* A few years ago, Mike McCarthy parachuted 179 ft from the top of the Tower of Pisa. Make a rough sketch to show the shape of the graph of his downward velocity during the jump.

18. *Fruit Flies* *(Example 2, Section 2.4 continued) Work in groups of two or three.* Populations starting out in closed environments grow slowly at first, when there are relatively few members, then more rapidly as the number of reproducing individuals increases and resources are still abundant, then slowly again as the population reaches the carrying capacity of the environment.

(a) Use the graphical technique of Section 3.1, Example 3, to graph the derivative of the fruit fly population introduced in Section 2.4. The graph of the population is reproduced below. What units should be used on the horizontal and vertical axes for the derivative's graph?

(b) During what days does the population seem to be increasing fastest? slowest?

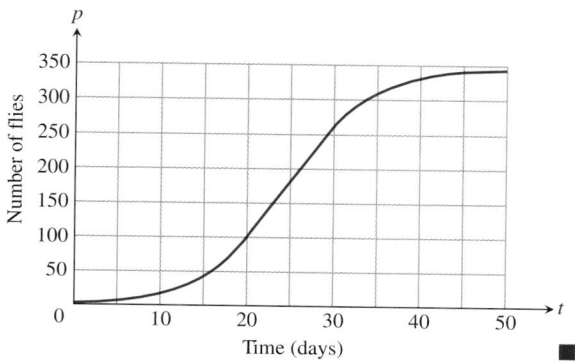

19. In step 1 of Exploration 2, at what time is the particle at the point (5, 2)?

20. *Particle Motion* *Work in groups of two or three.* The position (*x*-coordinate) of a particle moving on the line *y* = 2 is given by $x(t) = 2t^3 - 13t^2 + 22t - 5$ where *t* is time in seconds.

(a) Describe the motion of the particle for $t \geq 0$.

(b) When does the particle speed up? slow down?

(c) When does the particle change direction?

(d) When is the particle at rest?

(e) Describe the velocity and speed of the particle.

(f) When is the particle at the point (5, 2)?

In Exercises 21 and 22, the coordinates *s* of a moving body for various values of *t* are given. *Work in groups of two or three.* **(a)** Plot *s* versus *t* on coordinate paper, and sketch a smooth curve through the given points. **(b)** Assuming that this smooth curve represents the motion of the body, estimate the velocity at $t = 1.0$, $t = 2.5$, and $t = 3.5$.

21.

t (sec)	0	0.5	1.0	1.5	2.0	2.5	3.0	3.5	4.0
s (ft)	12.5	26	36.5	44	48.5	50	48.5	44	36.5

22.

t (sec)	0	0.5	1.0	1.5	2.0	2.5	3.0	3.5	4.0
s (ft)	3.5	−4	−8.5	−10	−8.5	−4	3.5	14	27.5

23. *Particle Motion* The accompanying figure shows the velocity $v = ds/dt = f(t)$ (m/sec) of a body moving along a coordinate line.

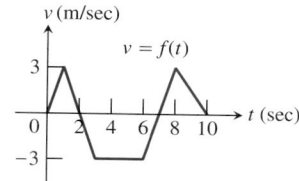

(a) When does the body reverse direction?

(b) When (approximately) is the body moving at a constant speed?

(c) Graph the body's speed for $0 \leq t \leq 10$.

(d) Graph the acceleration, where defined.

24. *Particle Motion* A particle *P* moves on the number line shown in part (a) of the accompanying figure. Part (b) shows the position of *P* as a function of time *t*.

(a)

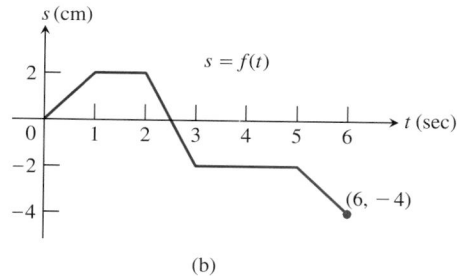

(b)

(a) When is *P* moving to the left? moving to the right? standing still?

(b) Graph the particle's velocity and speed (where defined).

25. *Particle Motion* The accompanying figure shows the velocity $v = f(t)$ of a particle moving on a coordinate line.

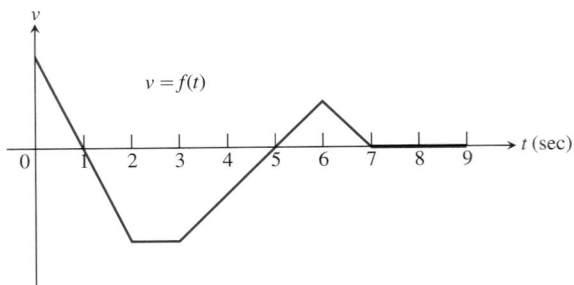

(a) When does the particle move forward? move backward? speed up? slow down?

(b) When is the particle's acceleration positive? negative? zero?

(c) When does the particle move at its greatest speed?

(d) When does the particle stand still for more than an instant?

26. *Moving Truck* The graph here shows the position s of a truck traveling on a highway. The truck starts at $t = 0$ and returns 15 hours later at $t = 15$.

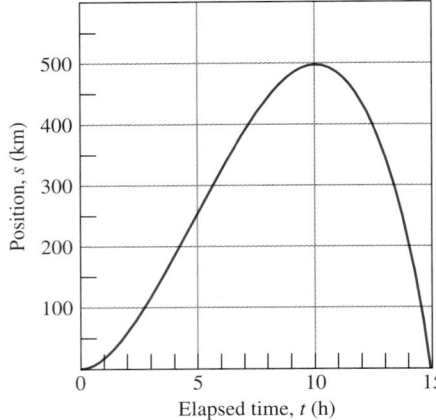

(a) Use the technique described in Section 3.1, Example 3, to graph the truck's velocity $v = ds/dt$ for $0 \leq t \leq 15$. Then repeat the process, with the velocity curve, to graph the truck's acceleration dv/dt.

(b) Suppose $s = 15t^2 - t^3$. Graph ds/dt and d^2s/dt^2, and compare your graphs with those in (a).

27. *Falling Objects* The multiflash photograph in Figure 3.30 shows two balls falling from rest. The vertical rulers are marked in centimeters. Use the equation $s = 490t^2$ (the free-fall equation for s in centimeters and t in seconds) to answer the following questions. *Work in groups of two or three.*

(a) How long did it take the balls to fall the first 160 cm? What was their average velocity for the period?

(b) How fast were the balls falling when they reached the 160-cm mark? What was their acceleration then?

Figure 3.30 Two balls falling from rest. (Exercise 27)

(c) About how fast was the light flashing (flashes per second)?

28. *Writing to Learn* The graphs in Figure 3.31 show as functions of time t the position s, velocity $v = ds/dt$, and acceleration $a = d^2s/dt^2$ of a body moving along a coordinate line. Which graph is which? Give reasons for your answers.

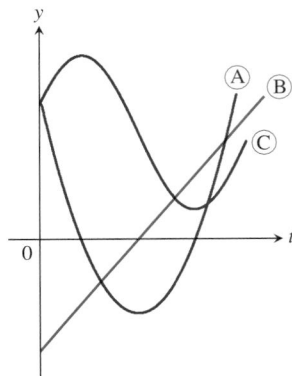

Figure 3.31 The graphs for Exercise 28.

29. Writing to Learn The graphs in Figure 3.32 show as functions of time t the position s, the velocity $v = ds/dt$, and the acceleration $a = d^2s/dt^2$ of a body moving along a coordinate line. Which graph is which? Give reasons for your answers.

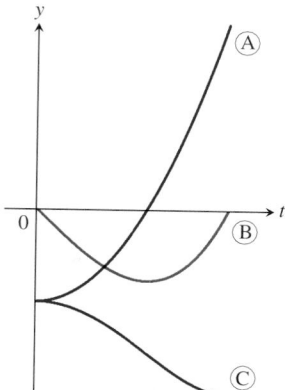

Figure 3.32 The graphs for Exercise 29.

30. *Draining a Tank* It takes 12 hours to drain a storage tank by opening the valve at the bottom. The depth y of fluid in the tank t hours after the valve is opened is given by the formula

$$y = 6\left(1 - \frac{t}{12}\right)^2 \text{ m.}$$

(a) Find the rate dy/dt (m/h) at which the water level is changing at time t.

(b) When is the fluid level in the tank falling fastest? slowest? What are the values of dy/dt at these times?

(c) Graph y and dy/dt together and discuss the behavior of y in relation to the signs and values of dy/dt.

31. *Inflating a Balloon* The volume $V = (4/3)\pi r^3$ of a spherical balloon changes with the radius.

(a) At what rate does the volume change with respect to the radius when $r = 2$ ft?

(b) By approximately how much does the volume increase when the radius changes from 2 to 2.2 ft?

32. *Airplane Takeoff* Suppose that the distance an aircraft travels along a runway before takeoff is given by $D = (10/9)t^2$, where D is measured in meters from the starting point and t is measured in seconds from the time the brakes are released. If the aircraft will become airborne when its speed reaches 200 km/h, how long will it take to become airborne, and what distance will it have traveled by that time?

33. *Volcanic Lava Fountains* Although the November 1959 Kilauea Iki eruption on the island of Hawaii began with a line of fountains along the wall of the crater, activity was later confined to a single vent in the crater's floor, which at one point shot lava 1900 ft straight into the air (a world record). What was the lava's exit velocity in feet per second? in miles per hour? (*Hint:* If v_0 is the exit velocity of a particle of lava, its height t seconds later will be $s = v_0 t - 16t^2$ feet. Begin by finding the time at which $ds/dt = 0$. Neglect air resistance.)

34. Writing to Learn Suppose you are looking at a graph of velocity as a function of time. How can you estimate the acceleration at a given point in time?

35. Writing to Learn Explain how the Sum and Difference Rule (Rule 4 in Section 3.3) can be used to derive a formula for *marginal profit* in terms of marginal revenue and marginal cost.

36. *Thoroughbred Racing* A racehorse is running a 10-furlong race. (A furlong is 220 yards, although we will use furlongs and seconds as our units in this exercise.) As the horse passes each furlong marker (F), a steward records the time elapsed (t) since the beginning of the race, as shown in the table below:

F	0	1	2	3	4	5	6	7	8	9	10
t	0	20	33	46	59	73	86	100	112	124	135

(a) How long does it take the horse to finish the race?

(b) What is the average speed of the horse over the first 5 furlongs?

(c) What is the approximate speed of the horse as it passes the 3-furlong marker?

(d) During which portion of the race is the horse running the fastest?

(e) During which portion of the race is the horse accelerating the fastest?

Extending the Ideas

37. *Even and Odd Functions*

(a) Show that if f is a differentiable even function, then f' is an odd function.

(b) Show that if f is a differentiable odd function, then f' is an even function.

38. *Extended Product Rule* Derive a formula for the derivative of the product fgh of three differentiable functions.

3.5   Derivatives of Trigonometric Functions

Derivative of the Sine Function • Derivative of the Cosine Function
• Simple Harmonic Motion • Jerk • Derivatives of the Other
Basic Trigonometric Functions

Derivative of the Sine Function

Trigonometric functions are important because so many of the phenomena we
want information about are periodic (heart rhythms, earthquakes, tides, weather).
It is known that continuous periodic functions can always be expressed in terms
of sines and cosines, so the derivatives of sines and cosines play a key role in
describing periodic change. This section introduces the derivatives of the six
basic trigonometric functions.

Exploration 1 **Making a Conjecture with NDER**

In the window $[-2\pi, 2\pi]$ by $[-4, 4]$, graph $y_1 = \sin x$ and
$y_2 = \text{NDER}\,(\sin x)$ (Figure 3.33).

1. When the graph of $y_1 = \sin x$ is increasing, what is true about the
   graph of $y_2 = \text{NDER}\,(\sin x)$?

2. When the graph of $y_1 = \sin x$ is decreasing, what is true about
   the graph of $y_2 = \text{NDER}\,(\sin x)$?

3. When the graph of $y_1 = \sin x$ stops increasing and starts decreasing, what is true about the graph of $y_2 = \text{NDER}\,(\sin x)$?

4. At the places where $\text{NDER}\,(\sin x) = \pm 1$, what appears to be the
   slope of the graph of $y_1 = \sin x$?

5. Make a conjecture about what function the derivative of sine
   might be. Test your conjecture by graphing your function and
   $\text{NDER}\,(\sin x)$ in the same viewing window.

6. Now let $y_1 = \cos x$ and $y_2 = \text{NDER}\,(\cos x)$. Answer questions
   (1) through (5) *without* looking at the graph of $\text{NDER}\,(\cos x)$
   until you are ready to test your conjecture about what function
   the derivative of cosine might be.

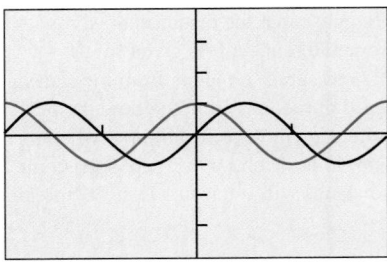

$[-2\pi, 2\pi]$ by $[-4, 4]$

Figure 3.33 Sine and its derivative. What is the derivative? (Exploration 1)

If you conjectured that the derivative of the sine function is the cosine function, then you are right. We will confirm this analytically, but first we appeal to technology one more time to evaluate two limits needed in the proof:

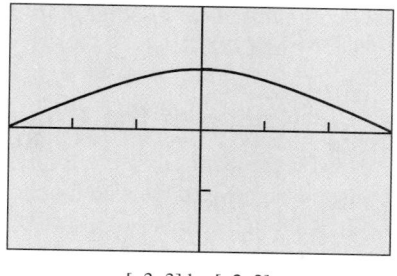

[−3, 3] by [−2, 2]

(a)

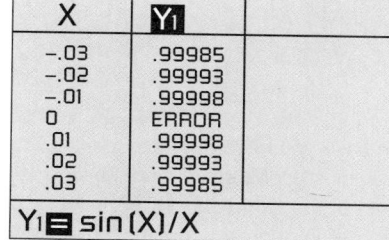

X	Y₁
−.03	.99985
−.02	.99993
−.01	.99998
0	ERROR
.01	.99998
.02	.99993
.03	.99985

Y₁ ☰ sin(X)/X

(b)

Figure 3.34 (a) Graphical and (b) tabular support that $\lim\limits_{h \to 0} \dfrac{\sin(h)}{h} = 1$.

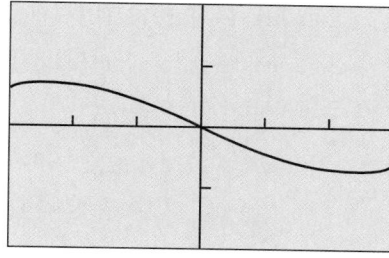

[−3, 3] by [−2, 2]

(a)

X	Y₁
−.03	.015
−.02	.01
−.01	.005
0	ERROR
.01	−.005
.02	−.01
.03	−.015

Y₁ ☰ (cos(X)−1)/X

(b)

Figure 3.35 (a) Graphical and (b) tabular support that $\lim\limits_{h \to 0} \dfrac{\cos(h) - 1}{h} = 0$.

Confirm Analytically

(Also, see Section 2.1, Exercise 63.) Now, let $y = \sin x$. Then

$$\frac{dy}{dx} = \lim_{h \to 0} \frac{\sin(x + h) - \sin x}{h}$$

$$= \lim_{h \to 0} \frac{\sin x \cos h + \cos x \sin h - \sin x}{h} \qquad \text{Angle sum identity}$$

$$= \lim_{h \to 0} \frac{(\sin x)(\cos h - 1) + \cos x \sin h}{h}$$

$$= \lim_{h \to 0} \sin x \cdot \lim_{h \to 0} \frac{(\cos h - 1)}{h} + \lim_{h \to 0} \cos x \cdot \lim_{h \to 0} \frac{\sin h}{h}$$

$$= \sin x \cdot 0 + \cos x \cdot 1$$

$$= \cos x.$$

In short, the derivative of the sine is the cosine.

$$\frac{d}{dx} \sin x = \cos x$$

Now that we know that the sine function is differentiable, we know that sine and its derivative obey all the rules for differentiation. We also know that $\sin x$ is continuous. The same holds for the other trigonometric functions in this section. Each one is differentiable at every point in its domain, so each one is continuous at every point in its domain, and the differentiation rules apply for each one.

Radian Measure in Calculus

In case you have been wondering why calculus uses radian measure when the rest of the world seems to measure angles in degrees, you are now ready to understand the answer. The derivative of $\sin x$ is $\cos x$ only if x is measured in radians! If you look at the analytic confirmation, you will note that the derivative comes down to

$$\cos x \text{ times } \lim_{h \to 0} \frac{\sin h}{h}.$$

We saw that

$$\lim_{h \to 0} \frac{\sin h}{h} = 1$$

in Figure 3.34, but only because the graph in Figure 3.34 is in *radian mode.* If you look at the limit of the same function in *degree* mode you will get a very different limit (and hence a different derivative for sine). See Exercise 25.

Derivative of the Cosine Function

If you conjectured in Exploration 1 that the derivative of the cosine function is the negative of the sine function, you were correct. You can confirm this analytically in Exercise 14.

$$\frac{d}{dx}\cos x = -\sin x$$

Example 1 REVISITING THE DIFFERENTIATION RULES

Find the derivatives of **(a)** $y = x^2 \sin x$ and **(b)** $u = \cos x/(1 - \sin x)$.

Solution

(a)
$$\frac{dy}{dx} = x^2 \cdot \frac{d}{dx}(\sin x) + \sin x \cdot \frac{d}{dx}(x^2) \quad \text{Product Rule}$$

$$= x^2 \cos x + 2x \sin x$$

(b)
$$\frac{du}{dx} = \frac{(1 - \sin x) \cdot \dfrac{d}{dx}(\cos x) - \cos x \cdot \dfrac{d}{dx}(1 - \sin x)}{(1 - \sin x)^2} \quad \text{Quotient Rule}$$

$$= \frac{(1 - \sin x)(-\sin x) - \cos x\,(0 - \cos x)}{(1 - \sin x)^2}$$

$$= \frac{-\sin x + \sin^2 x + \cos^2 x}{(1 - \sin x)^2}$$

$$= \frac{1 - \sin x}{(1 - \sin x)^2} \qquad \sin^2 x + \cos^2 x = 1$$

$$= \frac{1}{1 - \sin x}$$

Simple Harmonic Motion

The motion of a weight bobbing up and down on the end of a spring is an example of **simple harmonic motion.** Example 2 describes a case in which there are no opposing forces like friction or buoyancy to slow down the motion.

Example 2 THE MOTION OF A WEIGHT ON A SPRING

A weight hanging from a spring (Figure 3.36) is stretched 5 units beyond its rest position $(s = 0)$ and released at time $t = 0$ to bob up and down. Its position at any later time t is

$$s = 5 \cos t.$$

What are its velocity and acceleration at time t? Describe its motion.

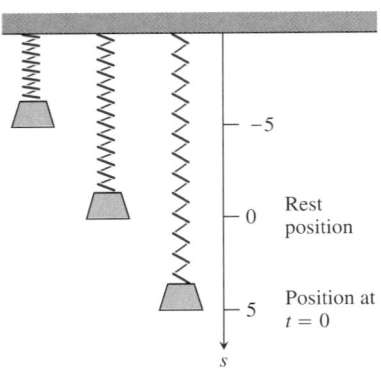

Figure 3.36 The weighted spring in Example 2.

Solution We have:

Position: $s = 5 \cos t;$

Velocity: $v = \dfrac{ds}{dt} = \dfrac{d}{dt}(5 \cos t) = -5 \sin t;$

Acceleration: $a = \dfrac{dv}{dt} = \dfrac{d}{dt}(-5 \sin t) = -5 \cos t.$

Notice how much we can learn from these equations:

1. As time passes, the weight moves down and up between $s = -5$ and $s = 5$ on the *s*-axis. The amplitude of the motion is 5. The period of the motion is 2π.

2. The velocity $v = -5 \sin t$ attains its greatest magnitude, 5, when $\cos t = 0$, as the graphs show in Figure 3.37. Hence the speed of the weight, $|v| = 5\,|\sin t|$, is greatest when $\cos t = 0$, that is, when $s = 0$ (the rest position). The speed of the weight is zero when $\sin t = 0$. This occurs when $s = 5 \cos t = \pm 5$, at the endpoints of the interval of motion.

3. The acceleration value is always the exact opposite of the position value. When the weight is above the rest position, gravity is pulling it back down; when the weight is below the rest position, the spring is pulling it back up.

4. The acceleration, $a = -5 \cos t$, is zero only at the rest position where $\cos t = 0$ and the force of gravity and the force from the spring offset each other. When the weight is anywhere else, the two forces are unequal and acceleration is nonzero. The acceleration is greatest in magnitude at the points farthest from the rest position, where $\cos t = \pm 1$.

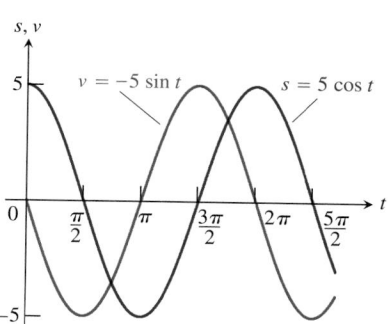

Figure 3.37 Graphs of the position and velocity of the weight in Example 2.

Jerk

A sudden change in acceleration is called a "jerk." When a ride in a car or a bus is jerky, it is not that the accelerations involved are necessarily large but that the changes in acceleration are abrupt. Jerk is what spills your soft drink. The derivative responsible for jerk is the *third* derivative of position.

Definition Jerk

Jerk is the derivative of acceleration. If a body's position at time t is $s(t)$, the body's jerk at time t is

$$j(t) = \frac{da}{dt} = \frac{d^3s}{dt^3}.$$

Recent tests have shown that motion sickness comes from accelerations whose changes in magnitude or direction take us by surprise. Keeping an eye on the road helps us to see the changes coming. A driver is less likely to become sick than a passenger who is reading in the back seat.

Example 3 A COUPLE OF JERKS

(a) The jerk caused by the constant acceleration of gravity ($g = -32$ ft/sec^2) is zero:

$$j = \frac{d}{dt}(g) = 0.$$

This explains why we don't experience motion sickness while just sitting around.

(b) The jerk of the simple harmonic motion in Example 2 is

$$j = \frac{da}{dt} = \frac{d}{dt}(-5 \cos t)$$

$$= 5 \sin t.$$

It has its greatest magnitude when $\sin t = \pm 1$. This does not occur at the extremes of the displacement, but at the rest position, where the acceleration changes direction and sign.

Derivatives of the Other Basic Trigonometric Functions

Because $\sin x$ and $\cos x$ are differentiable functions of x, the related functions

$$\tan x = \frac{\sin x}{\cos x}, \qquad \sec x = \frac{1}{\cos x},$$

$$\cot x = \frac{\cos x}{\sin x}, \qquad \csc x = \frac{1}{\sin x}$$

are differentiable at every value of x for which they are defined. Their derivatives (Exercises 15 and 16) are given by the following formulas.

$$\frac{d}{dx}\tan x = \sec^2 x \qquad \frac{d}{dx}\sec x = \sec x \tan x$$

$$\frac{d}{dx}\cot x = -\csc^2 x \qquad \frac{d}{dx}\csc x = -\csc x \cot x$$

Example 4 FINDING TANGENT AND NORMAL LINES

Find equations for the lines that are tangent and normal to the graph of

$$f(x) = \frac{\tan x}{x}$$

at $x = 2$. Support graphically.

Solution

Solve Numerically

Since we will be using a calculator approximation for $f(2)$ anyway, this is a good place to use NDER.

We compute $(\tan 2)/2$ on the calculator and store it as k. The slope of the tangent line at $(2, k)$ is

$$\text{NDER}\left(\frac{\tan x}{x}, 2\right),$$

which we compute and store as m. The equation of the tangent line is $y - k = m(x - 2)$, or

$$y = mx + k - 2m.$$

Only after we have found m and $k - 2m$ do we round the coefficients, giving the tangent line as

$$y = 3.43x - 7.96.$$

The equation of the normal line is

$$y - k = -\frac{1}{m}(x - 2), \text{ or}$$

$$y = -\frac{1}{m}x + k + \frac{2}{m}.$$

Again we wait until the end to round the coefficients, giving the normal line as

$$y = -0.291x - 0.51.$$

Confirm Graphically

Figure 3.38, showing the original function and the two lines, supports our computations.

$y_1 = \tan(x)/x$
$y_2 = 3.43x - 7.96$
$y_3 = -0.291x - 0.51$

X=2   Y=-1.09252

$[-3\pi/2, 3\pi/2]$ by $[-3, 3]$

Figure 3.38 Graphical support for Example 4.

Example 5 A TRIGONOMETRIC SECOND DERIVATIVE

Find y'' if $y = \sec x$.

Solution $y = \sec x$

$$y' = \sec x \tan x$$

$$y'' = \frac{d}{dx}(\sec x \tan x)$$

$$= \sec x \frac{d}{dx}(\tan x) + \tan x \frac{d}{dx}(\sec x)$$

$$= \sec x (\sec^2 x) + \tan x (\sec x \tan x)$$

$$= \sec^3 x + \sec x \tan^2 x$$

Quick Review 3.5

1. Convert 135 degrees to radians.

2. Convert 1.7 radians to degrees.

3. Find the exact value of $\sin(\pi/3)$ without a calculator.

4. State the domain and the range of the cosine function.

5. State the domain and the range of the tangent function.

6. If $\sin a = -1$, what is $\cos a$?

7. If $\tan a = -1$, what are two possible values of $\sin a$?

8. Verify the identity:

$$\frac{1 - \cos h}{h} = \frac{\sin^2 h}{h(1 + \cos h)}.$$

9. Find an equation of the line tangent to the curve $y = 2x^3 - 7x^2 + 10$ at the point $(3, 1)$.

10. A particle moves along a line with velocity $v = 2t^3 - 7t^2 + 10$ for time $t \geq 0$. Find the acceleration of the particle at $t = 3$.

Section 3.5 Exercises

In Exercises 1–10, find dy/dx. Use your grapher to support your analysis if you are unsure of your answer.

1. $y = 1 + x - \cos x$

2. $y = 2 \sin x - \tan x$

3. $y = \dfrac{1}{x} + 5 \sin x$

4. $y = x \sec x$

5. $y = 4 - x^2 \sin x$

6. $y = 3x + x \tan x$

7. $y = \dfrac{4}{\cos x}$

8. $y = \dfrac{x}{1 + \cos x}$

9. $y = \dfrac{\cot x}{1 + \cot x}$

10. $y = \dfrac{\cos x}{1 + \sin x}$

11. Find an equation of the line tangent to the graph of $y = \sin x + 3$ at $x = \pi$.

12. Find an equation of the line normal to the graph of $y = (\tan x)/x$ at $x = \pi/4$.

13. Find an equation of the line tangent to the graph of $y = x^2 \sin x$ at $x = 3$.

14. Use the definition of the derivative to prove that $(d/dx)(\cos x) = -\sin x$. (You will need the limits found at the beginning of this section.)

15. Assuming that $(d/dx)(\sin x) = \cos x$ and $(d/dx)(\cos x) = -\sin x$, prove each of the following.

(a) $\dfrac{d}{dx}\tan x = \sec^2 x$

(b) $\dfrac{d}{dx}\sec x = \sec x \tan x$

16. Assuming that $(d/dx)(\sin x) = \cos x$ and $(d/dx)(\cos x) = -\sin x$, prove each of the following.

(a) $\dfrac{d}{dx}\cot x = -\csc^2 x$

(b) $\dfrac{d}{dx}\csc x = -\csc x \cot x$

17. Show that the graphs of $y = \sec x$ and $y = \cos x$ have horizontal tangents at $x = 0$.

18. Show that the graphs of $y = \tan x$ and $y = \cot x$ have no horizontal tangents.

19. Find equations for the lines that are tangent and normal to the curve $y = \sqrt{2} \cos x$ at the point $(\pi/4, 1)$.

20. Find the points on the curve $y = \tan x$, $-\pi/2 < x < \pi/2$, where the tangent is parallel to the line $y = 2x$.

In Exercises 21 and 22, find an equation for (a) the tangent to the curve at P and (b) the horizontal tangent to the curve at Q.

21.
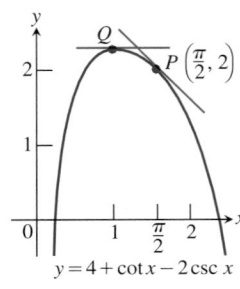
$y = 4 + \cot x - 2\csc x$

22.
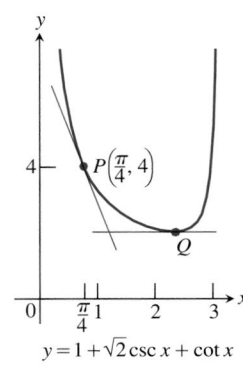
$y = 1 + \sqrt{2}\csc x + \cot x$

In Exercises 23 and 24, a body is moving in simple harmonic motion with position $s = f(t)$ (s in meters, t in seconds). *Work in groups of two or three.*

(a) Find the body's velocity, speed, acceleration, and jerk at time t.

(b) Find the body's velocity, speed, acceleration, and jerk at time $t = \pi/4$ sec.

(c) Describe the motion of the body.

23. $s = 2 - 2\sin t$

24. $s = \sin t + \cos t$

Exploration

25. *Radians vs. Degrees* What happens to the derivatives of $\sin x$ and $\cos x$ if x is measured in degrees instead of radians? To find out, take the following steps.

(a) With your grapher in degree mode, graph
$$f(h) = \frac{\sin h}{h}$$
and estimate $\lim_{h \to 0} f(h)$. Compare your estimate with $\pi/180$. Is there any reason to believe the limit should be $\pi/180$?

(b) With your grapher in degree mode, estimate
$$\lim_{h \to 0} \frac{\cos h - 1}{h}.$$

(c) Now go back to the derivation of the formula for the derivative of $\sin x$ in the text and carry out the steps of the derivation using degree-mode limits. What formula do you obtain for the derivative?

(d) Derive the formula for the derivative of $\cos x$ using degree-mode limits.

(e) The disadvantages of the degree-mode formulas become apparent as you start taking derivatives of higher order. What are the second and third degree-mode derivatives of $\sin x$ and $\cos x$?   ■

26. Find y'' if $y = \csc x$.

27. Find y'' if $y = \theta \tan \theta$.

28. **Writing to Learn** Is there a value of b that will make
$$g(x) = \begin{cases} x + b, & x < 0 \\ \cos x, & x \ge 0 \end{cases}$$
continuous at $x = 0$? differentiable at $x = 0$? Give reasons for your answers.

29. Find $\dfrac{d^{999}}{dx^{999}}(\cos x)$.

30. Find $\dfrac{d^{725}}{dx^{725}}(\sin x)$.

31. *Local Linearity* This is the graph of the function $y = \sin x$ close to the origin. Since $\sin x$ is differentiable, this graph resembles a line. Find an equation for this line.

32. *(Continuation of Exercise 31)* For values of x close to 0, the linear equation found in Exercise 31 gives a good approximation of $\sin x$.

(a) Use this fact to estimate $\sin (0.12)$.

(b) Find $\sin (0.12)$ with a calculator. How close is the approximation in (a)?

33. Use the identity $\sin 2x = 2 \sin x \cos x$ to find the derivative of $\sin 2x$. Then use the identity $\cos 2x = \cos^2 x - \sin^2 x$ to express that derivative in terms of $\cos 2x$.

34. Use the identity $\cos 2x = \cos x \cos x - \sin x \sin x$ to find the derivative of $\cos 2x$. Express the derivative in terms of $\sin 2x$.

Extending the Ideas

35. Use analytic methods to show that

$$\lim_{h \to 0} \frac{\cos h - 1}{h} = 0.$$

(Hint: Multiply numerator and denominator by $(\cos h + 1)$.)

36. Find A and B in $y = A \sin x + B \cos x$ so that $y'' - y = \sin x$.

3.6

Chain Rule

Derivative of a Composite Function • "Outside-Inside" Rule •
Repeated Use of the Chain Rule • Slopes of Parametrized Curves
• Power Chain Rule

Derivative of a Composite Function

We now know how to differentiate $\sin x$ and $x^2 - 4$, but how do we differentiate a composite like $\sin (x^2 - 4)$? The answer is with the Chain Rule, which is probably the most widely used differentiation rule in mathematics. This section describes the rule and how to use it.

Example 1   RELATING DERIVATIVES

The function $y = 6x - 10 = 2(3x - 5)$ is the composite of the functions $y = 2u$ and $u = 3x - 5$. How are the derivatives of these three functions related?

Solution We have

$$\frac{dy}{dx} = 6, \qquad \frac{dy}{du} = 2, \qquad \frac{du}{dx} = 3.$$

Since $6 = 2 \cdot 3$,

$$\frac{dy}{dx} = \frac{dy}{du} \cdot \frac{du}{dx}.$$

C: *y* turns   B: *u* turns   A: *x* turns

Figure 3.39 When gear A makes x turns, gear B makes u turns and gear C makes y turns. By comparing circumferences or counting teeth, we see that $y = u/2$ and $u = 3x$, so $y = 3x/2$. Thus $dy/du = 1/2$, $du/dx = 3$, and $dy/dx = 3/2 = (dy/du)(du/dx)$.

Is it an accident that $dy/dx = dy/du \cdot du/dx$?

If we think of the derivative as a rate of change, our intuition allows us to see that this relationship is reasonable. For $y = f(u)$ and $u = g(x)$, if y changes twice as fast as u and u changes three times as fast as x, then we expect y to change six times as fast as x. This is much like the effect of a multiple gear train (Figure 3.39).

Let us try again on another function.

Example 2 RELATING DERIVATIVES

The polynomial $y = 9x^4 + 6x^2 + 1 = (3x^2 + 1)^2$ is the composite of $y = u^2$ and $u = 3x^2 + 1$. Calculating derivatives, we see that

$$\frac{dy}{du} \cdot \frac{du}{dx} = 2u \cdot 6x$$

$$= 2(3x^2 + 1) \cdot 6x$$

$$= 36x^3 + 12x.$$

Also,

$$\frac{dy}{dx} = \frac{d}{dx}(9x^4 + 6x^2 + 1)$$

$$= 36x^3 + 12x.$$

Once again,

$$\frac{dy}{du} \cdot \frac{du}{dx} = \frac{dy}{dx}.$$

The derivative of the composite function $f(g(x))$ at x is the derivative of f at $g(x)$ times the derivative of g at x (Figure 3.40). This is known as the Chain Rule.

Figure 3.40 Rates of change multiply: the derivative of $f \circ g$ at x is the derivative of f at the point $g(x)$ times the derivative of g at x.

Rule 8 The Chain Rule

If f is differentiable at the point $u = g(x)$, and g is differentiable at x, then the composite function $(f \circ g)(x) = f(g(x))$ is differentiable at x, and

$$(f \circ g)'(x) = f'(g(x)) \cdot g'(x).$$

In Leibniz notation, if $y = f(u)$ and $u = g(x)$, then

$$\frac{dy}{dx} = \frac{dy}{du} \cdot \frac{du}{dx},$$

where dy/du is evaluated at $u = g(x)$.

It would be tempting to try to prove the Chain Rule by writing

$$\frac{\Delta y}{\Delta x} = \frac{\Delta y}{\Delta u} \cdot \frac{\Delta u}{\Delta x}$$

(a true statement about fractions with nonzero denominators) and taking the limit as $\Delta x \to 0$. This is essentially what is happening, and it would work as a proof if we knew that Δu, the change in u, was nonzero; but we do not know this. A small change in x could conceivably produce no change in u. An airtight proof of the Chain Rule can be constructed through a different approach, but we will omit it here.

Example 3   APPLYING THE CHAIN RULE

An object moves along the x-axis so that its position at any time $t \geq 0$ is given by $x(t) = \cos(t^2 + 1)$. Find the velocity of the object as a function of t.

Solution   We know that the velocity is dx/dt. In this instance, x is a composite function: $x = \cos(u)$ and $u = t^2 + 1$. We have

$$\frac{dx}{du} = -\sin(u) \quad x = \cos(u)$$

$$\frac{du}{dt} = 2t. \qquad u = t^2 + 1$$

By the Chain Rule,

$$\frac{dx}{dt} = \frac{dx}{du} \cdot \frac{du}{dt}$$

$$= -\sin(u) \cdot 2t$$

$$= -\sin(t^2 + 1) \cdot 2t$$

$$= -2t \sin(t^2 + 1).$$

"Outside-Inside" Rule

It sometimes helps to think about the Chain Rule this way: If $y = f(g(x))$, then

$$\frac{dy}{dx} = f'(g(x)) \cdot g'(x).$$

In words, differentiate the "outside" function f and evaluate it at the "inside" function $g(x)$ left alone; then multiply by the derivative of the "inside function."

Example 4   DIFFERENTIATING FROM THE OUTSIDE IN

Differentiate $\sin(x^2 + x)$ with respect to x.

Solution

$$\frac{d}{dx} \sin \underbrace{(x^2 + x)}_{\text{inside}} = \cos \underbrace{(x^2 + x)}_{\substack{\text{inside} \\ \text{left alone}}} \cdot \underbrace{(2x + 1)}_{\substack{\text{derivative of} \\ \text{the inside}}}$$

Repeated Use of the Chain Rule

We sometimes have to use the Chain Rule two or more times to find a derivative. Here is an example:

Example 5   A THREE-LINK "CHAIN"

Find the derivative of $g(t) = \tan(5 - \sin 2t)$.

Solution   Notice here that tan is a function of $5 - \sin 2t$, while sin is a function of $2t$, which is itself a function of t. Therefore, by the Chain Rule,

$$g'(t) = \frac{d}{dt}(\tan(5 - \sin 2t))$$

$$= \sec^2(5 - \sin 2t) \cdot \frac{d}{dt}(5 - \sin 2t)$$
Derivative of tan *u* with *u* = 5 − sin 2*t*

$$= \sec^2(5 - \sin 2t) \cdot (0 - \cos 2t \cdot \frac{d}{dt}(2t))$$
Derivative of 5 − sin *u* with *u* = 2*t*

$$= \sec^2(5 - \sin 2t) \cdot (-\cos 2t) \cdot 2$$

$$= -2(\cos 2t)\sec^2(5 - \sin 2t).$$

Slopes of Parametrized Curves

A parametrized curve $(x(t), y(t))$ is *differentiable at t* if x and y are differentiable at t. At a point on a differentiable parametrized curve where y is also a differentiable function of x, the derivatives dy/dt, dx/dt, and dy/dx are related by the Chain Rule:

$$\frac{dy}{dt} = \frac{dy}{dx} \cdot \frac{dx}{dt}.$$

If $dx/dt \neq 0$, we may divide both sides of this equation by dx/dt to solve for dy/dx.

Finding *dy/dx* Parametrically

If all three derivatives exist and $dx/dt \neq 0$,

$$\frac{dy}{dx} = \frac{dy/dt}{dx/dt}. \tag{3}$$

Example 6   DIFFERENTIATING WITH A PARAMETER

Find the line tangent to the right-hand hyperbola branch defined parametrically by

$$x = \sec t, \quad y = \tan t, \quad -\frac{\pi}{2} < t < \frac{\pi}{2}$$

at the point $(\sqrt{2}, 1)$, where $t = \pi/4$ (Figure 3.41).

Solution All three of the derivatives in Equation 3 exist and $dx/dt = \sec t \tan t \neq 0$ at the indicated point. Therefore, Equation 3 applies and

$$\frac{dy}{dx} = \frac{dy/dt}{dx/dt}$$

$$= \frac{\sec^2 t}{\sec t \tan t}$$

$$= \frac{\sec t}{\tan t}$$

$$= \csc t.$$

Setting $t = \pi/4$ gives

$$\left.\frac{dy}{dx}\right|_{t=\pi/4} = \csc(\pi/4) = \sqrt{2}.$$

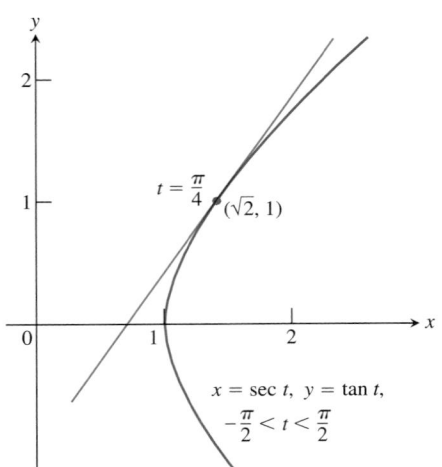

$t = \dfrac{\pi}{4}$ ($\sqrt{2}$, 1)

$x = \sec t, \; y = \tan t,$

$-\dfrac{\pi}{2} < t < \dfrac{\pi}{2}$

Figure 3.41 The hyperbola branch in Example 6. Eq. 3 applies for every point on the graph except (1, 0). Can you state why Eq. 3 fails at (1, 0)?

The equation of the tangent line is

$$y - 1 = \sqrt{2}(x - \sqrt{2})$$
$$y = \sqrt{2}x - 2 + 1$$
$$y = \sqrt{2}x - 1.$$

Power Chain Rule

If f is a differentiable function of u, and u is a differentiable function of x, then substituting $y = f(u)$ into the Chain Rule formula

$$\frac{dy}{dx} = \frac{dy}{du} \cdot \frac{du}{dx}$$

leads to the formula

$$\frac{d}{dx}f(u) = f'(u)\frac{du}{dx}.$$

Here's an example of how it works: If n is an integer and $f(u) = u^n$, the Power Rules (Rules 2 and 7) tell us that $f'(u) = nu^{n-1}$. If u is a differentiable function of x, then we can use the Chain Rule to extend this to the **Power Chain Rule:**

$$\frac{d}{dx}u^n = nu^{n-1}\frac{du}{dx}. \qquad \frac{d}{du}(u^n) = nu^{n-1}$$

Example 7   FINDING SLOPE

(a) Find the slope of the line tangent to the curve $y = \sin^5 x$ at the point where $x = \pi/3$.

(b) Show that the slope of every line tangent to the curve $y = 1/(1 - 2x)^3$ is positive.

Solution

(a) $\quad \dfrac{dy}{dx} = 5\sin^4 x \cdot \dfrac{d}{dx}\sin x \qquad$ Power Chain Rule with $u = \sin x$, $n = 5$

$$= 5\sin^4 x \cos x$$

The tangent line has slope

$$\frac{dy}{dx}\bigg|_{x=\pi/3} = 5\left(\frac{\sqrt{3}}{2}\right)^4\left(\frac{1}{2}\right) = \frac{45}{32}.$$

(b) $\quad \dfrac{dy}{dx} = \dfrac{d}{dx}(1 - 2x)^{-3}$

$$= -3(1-2x)^{-4}\cdot\frac{d}{dx}(1-2x) \qquad \begin{matrix}\text{Power Chain Rule with}\\ u = (1-2x),\ n = -3\end{matrix}$$

$$= -3(1-2x)^{-4}\cdot(-2)$$

$$= \frac{6}{(1-2x)^4}$$

At any point (x, y) on the curve, $x \neq 1/2$ and the slope of the tangent line is

$$\frac{dy}{dx} = \frac{6}{(1-2x)^4},$$

the quotient of two positive numbers.

Example 8 RADIANS VERSUS DEGREES

It is important to remember that the formulas for the derivatives of both sin *x* and cos *x* were obtained under the assumption that *x* is measured in radians, *not* degrees. The Chain Rule gives us new insight into the difference between the two. Since $180° = \pi$ radians, $x° = \pi x/180$ radians. By the Chain Rule,

$$\frac{d}{dx}\sin(x°) = \frac{d}{dx}\sin\left(\frac{\pi x}{180}\right) = \frac{\pi}{180}\cos\left(\frac{\pi x}{180}\right) = \frac{\pi}{180}\cos(x°).$$

See Figure 3.42. Similarly, the derivative of cos $(x°)$ is $-(\pi/180)\sin(x°)$.

The factor $\pi/180$, annoying in the first derivative, would compound with repeated differentiation. We see at a glance the compelling reason for the use of radian measure.

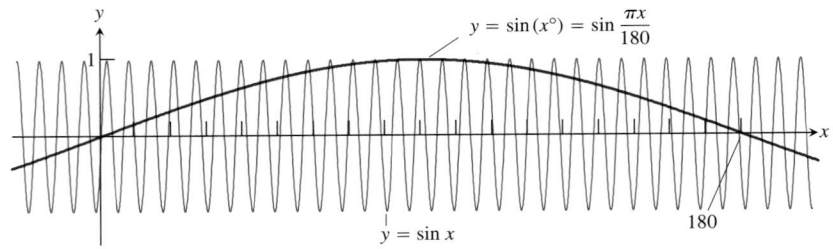

Figure 3.42 sin $(x°)$ oscillates only $\pi/180$ times as often as sin *x* oscillates. Its maximum slope is $\pi/180$. (Example 8)

Quick Review 3.6

In Exercises 1–5, let $f(x) = \sin x$, $g(x) = x^2 + 1$, and $h(x) = 7x$. Write a simplified expression for the composite function.

1. $f(g(x))$ **2.** $f(g(h(x)))$

3. $(g \circ h)(x)$ **4.** $(h \circ g)(x)$

5. $f\left(\dfrac{g(x)}{h(x)}\right)$

In Exercises 6–10, let $f(x) = \cos x$, $g(x) = \sqrt{x+2}$, and $h(x) = 3x^2$. Write the given function as a composite of two or more of *f*, *g*, and *h*. For example, cos $3x^2$ is $f(h(x))$.

6. $\sqrt{\cos x + 2}$ **7.** $\sqrt{3\cos^2 x + 2}$

8. $3\cos x + 6$ **9.** $\cos 27x^4$

10. $\cos\sqrt{2 + 3x^2}$

Section 3.6 Exercises

In Exercises 1–20, find dy/dx. If you are unsure of your answer, use NDER to support your computation.

1. $y = \sin(3x + 1)$ **2.** $y = \sin(7 - 5x)$

3. $y = \cos(\sqrt{3}x)$ **4.** $y = \tan(2x - x^3)$

5. $y = 5\cot\left(\dfrac{2}{x}\right)$ **6.** $y = \left(\dfrac{\sin x}{1 + \cos x}\right)^2$

7. $y = \cos(\sin x)$ **8.** $y = \sec(\tan x)$

9. $y = (x + \sqrt{x})^{-2}$ **10.** $y = (\csc x + \cot x)^{-1}$

11. $y = \sin^{-5} x - \cos^3 x$ **12.** $y = x^3(2x - 5)^4$

13. $y = \sin^3 x \tan 4x$

14. $y = 4\sqrt{\sec x + \tan x}$

15. $y = \dfrac{3}{\sqrt{2x + 1}}$

16. $y = \dfrac{x}{\sqrt{1 + x^2}}$

17. $y = \sin^2(3x - 2)$

18. $y = (1 + \cos 2x)^2$

19. $y = (1 + \cos^2 7x)^3$

20. $y = \sqrt{\tan 5x}$

In Exercises 21–24, find ds/dt.

21. $s = \cos\left(\dfrac{\pi}{2} - 3t\right)$

22. $s = t\cos(\pi - 4t)$

23. $s = \dfrac{4}{3\pi} \sin 3t + \dfrac{4}{5\pi} \cos 5t$

24. $s = \sin\left(\dfrac{3\pi}{2}t\right) + \cos\left(\dfrac{7\pi}{4}t\right)$

In Exercises 25–28 find $dr/d\theta$.

25. $r = \tan(2 - \theta)$

26. $r = \sec 2\theta \tan 2\theta$

27. $r = \sqrt{\theta} \sin \theta$

28. $r = 2\theta\sqrt{\sec \theta}$

In Exercises 29–32, find y''.

29. $y = \tan x$

30. $y = \cot x$

31. $y = \cot(3x - 1)$

32. $y = 9\tan(x/3)$

In Exercises 33–38, find the value of $(f \circ g)'$ at the given value of x.

33. $f(u) = u^5 + 1, \quad u = g(x) = \sqrt{x}, \quad x = 1$

34. $f(u) = 1 - \dfrac{1}{u}, \quad u = g(x) = \dfrac{1}{1 - x}, \quad x = -1$

35. $f(u) = \cot\dfrac{\pi u}{10}, \quad u = g(x) = 5\sqrt{x}, \quad x = 1$

36. $f(u) = u + \dfrac{1}{\cos^2 u}, \quad u = g(x) = \pi x, \quad x = \dfrac{1}{4}$

37. $f(u) = \dfrac{2u}{u^2 + 1}, \quad u = g(x) = 10x^2 + x + 1, \quad x = 0$

38. $f(u) = \left(\dfrac{u - 1}{u + 1}\right)^2, \quad u = g(x) = \dfrac{1}{x^2} - 1, \quad x = -1$

What happens if you can write a function as a composite in different ways? Do you get the same derivative each time? The Chain Rule says you should. Try it with the functions in Exercises 39 and 40.

39. Find dy/dx if $y = \cos(6x + 2)$ by writing y as a composite with

    (a) $y = \cos u$ and $u = 6x + 2$.

    (b) $y = \cos 2u$ and $u = 3x + 1$.

40. Find dy/dx if $y = \sin(x^2 + 1)$ by writing y as a composite with

    (a) $y = \sin(u + 1)$ and $u = x^2$.

    (b) $y = \sin u$ and $u = x^2 + 1$.

In Exercises 41–48, find the equation of the line tangent to the curve at the point defined by the given value of t.

41. $x = 2\cos t, \quad y = 2\sin t, \quad t = \pi/4$

42. $x = \sin 2\pi t, \quad y = \cos 2\pi t, \quad t = -1/6$

43. $x = \sec^2 t - 1, \quad y = \tan t, \quad t = -\pi/4$

44. $x = \sec t, \quad y = \tan t, \quad t = \pi/6$

45. $x = t, \quad y = \sqrt{t}, \quad t = 1/4$

46. $x = 2t^2 + 3, \quad y = t^4, \quad t = -1$

47. $x = t - \sin t, \quad y = 1 - \cos t, \quad t = \pi/3$

48. $x = \cos t, \quad y = 1 + \sin t, \quad t = \pi/2$

49. Let $x = t^2 + t$, and let $y = \sin t$.

    (a) Find dy/dx as a function of t.

    (b) Find $\dfrac{d}{dt}\left(\dfrac{dy}{dx}\right)$ as a function of t.

    (c) Find $\dfrac{d}{dx}\left(\dfrac{dy}{dx}\right)$ as a function of t.

    Use the Chain Rule and your answer from (b).

    (d) Which of the expressions in parts (b) and (c) is d^2y/dx^2?

50. A circle of radius 2 and center $(0, 0)$ can be parametrized by the equations $x = 2\cos t$ and $y = 2\sin t$. Show that for any value of t, the line tangent to the circle at $(2\cos t, 2\sin t)$ is perpendicular to the radius.

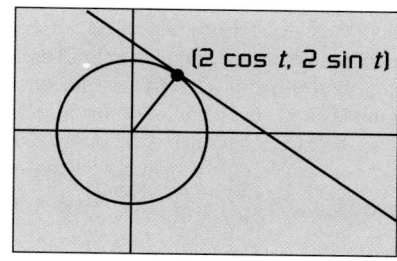

$(2\cos t, 2\sin t)$

51. Let $s = \cos\theta$. Evaluate ds/dt when $\theta = 3\pi/2$ and $d\theta/dt = 5$.

52. Let $y = x^2 + 7x - 5$. Evaluate dy/dt when $x = 1$ and $dx/dt = 1/3$.

53. What is the largest value possible for the slope of the curve $y = \sin(x/2)$?

54. Write an equation for the tangent to the curve $y = \sin mx$ at the origin.

55. Find the lines that are tangent and normal to the curve $y = 2\tan(\pi x/4)$ at $x = 1$. Support your answer graphically.

56. *Working with Numerical Values* Suppose that functions f and g and their derivatives have the following values at $x = 2$ and $x = 3$.

x	$f(x)$	$g(x)$	$f'(x)$	$g'(x)$
2	8	2	$1/3$	-3
3	3	-4	2π	5

Evaluate the derivatives with respect to x of the following combinations at the given value of x.

    (a) $2f(x)$ at $x = 2$

    (b) $f(x) + g(x)$ at $x = 3$

    (c) $f(x) \cdot g(x)$ at $x = 3$

    (d) $f(x)/g(x)$ at $x = 2$ **(e)** $f(g(x))$ at $x = 2$

    (f) $\sqrt{f(x)}$ at $x = 2$ **(g)** $1/g^2(x)$ at $x = 3$

    (h) $\sqrt{f^2(x) + g^2(x)}$ at $x = 2$

57. Working with Numerical Values Suppose that the functions f and g and their derivatives with respect to x have the following values at $x = 0$ and $x = 1$.

x	$f(x)$	$g(x)$	$f'(x)$	$g'(x)$
0	1	1	5	1/3
1	3	-4	$-1/3$	$-8/3$

Evaluate the derivatives with respect to x of the following combinations at the given value of x.

(a) $5f(x) - g(x)$, $x = 1$ **(b)** $f(x)g^3(x)$, $x = 0$

(c) $\dfrac{f(x)}{g(x) + 1}$, $x = 1$ **(d)** $f(g(x))$, $x = 0$

(e) $g(f(x))$, $x = 0$

(f) $(g(x) + f(x))^{-2}$, $x = 1$

(g) $f(x + g(x))$, $x = 0$

58. Orthogonal Curves Two curves are said to cross at right angles if their tangents are perpendicular at the crossing point. The technical word for "crossing at right angles" is **orthogonal.** Show that the curves $y = \sin 2x$ and $y = -\sin(x/2)$ are orthogonal at the origin. Draw both graphs and both tangents in a square viewing window.

59. Writing to Learn Explain why the Chain Rule formula

$$\frac{dy}{dx} = \frac{dy}{du} \cdot \frac{du}{dx}$$

is not simply the well-known rule for multiplying fractions.

60. Running Machinery Too Fast Suppose that a piston is moving straight up and down and that its position at time t seconds is

$$s = A \cos(2\pi bt),$$

with A and b positive. The value of A is the amplitude of the motion, and b is the frequency (number of times the piston moves up and down each second). What effect does doubling the frequency have on the piston's velocity, acceleration, and jerk? (Once you find out, you will know why machinery breaks when you run it too fast.)

Figure 3.43 The internal forces in the engine get so large that they tear the engine apart when the velocity is too great.

Figure 3.44 Normal mean air temperatures at Fairbanks, Alaska, plotted as data points, and the approximating sine function (Exercise 61).

61. Temperatures in Fairbanks, Alaska *Work in groups of two or three.* The graph in Figure 3.44 shows the average Fahrenheit temperature in Fairbanks, Alaska, during a typical 365-day year. The equation that approximates the temperature on day x is

$$y = 37 \sin\left[\frac{2\pi}{365}(x - 101)\right] + 25.$$

(a) On what day is the temperature increasing the fastest?

(b) About how many degrees per day is the temperature increasing when it is increasing at its fastest?

62. Particle Motion The position of a particle moving along a coordinate line is $s = \sqrt{1 + 4t}$, with s in meters and t in seconds. Find the particle's velocity and acceleration at $t = 6$ sec.

63. Constant Acceleration Suppose the velocity of a falling body is $v = k\sqrt{s}$ m/sec (k a constant) at the instant the body has fallen s meters from its starting point. Show that the body's acceleration is constant.

64. Falling Meteorite The velocity of a heavy meteorite entering the earth's atmosphere is inversely proportional to $\sqrt{s}$ when it is s kilometers from the earth's center. Show that the meteorite's acceleration is inversely proportional to s^2.

65. Particle Acceleration A particle moves along the x-axis with velocity $dx/dt = f(x)$. Show that the particle's acceleration is $f(x)f'(x)$.

66. Temperature and the Period of a Pendulum For oscillations of small amplitude (short swings), we may safely model the relationship between the period T and the length L of a simple pendulum with the equation

$$T = 2\pi\sqrt{\frac{L}{g}},$$

where g is the constant acceleration of gravity at the pendulum's location. If we measure g in centimeters per second squared, we measure L in centimeters and T in seconds. If the pendulum is made of metal, its length will vary with temperature, either increasing or decreasing at a rate that is roughly proportional to L. In symbols, with u being temperature and k the proportionality constant,

$$\frac{dL}{du} = kL.$$

Assuming this to be the case, show that the rate at which the period changes with respect to temperature is $kT/2$.

67. Writing to Learn *Chain Rule* Suppose that $f(x) = x^2$ and $g(x) = |x|$. Then the composites

$$(f \circ g)(x) = |x|^2 = x^2 \quad \text{and} \quad (g \circ f)(x) = |x^2| = x^2$$

are both differentiable at $x = 0$ even though g itself is not differentiable at $x = 0$. Does this contradict the Chain Rule? Explain.

68. *Tangents* Suppose that $u = g(x)$ is differentiable at $x = 1$ and that $y = f(u)$ is differentiable at $u = g(1)$. If the graph of $y = f(g(x))$ has a horizontal tangent at $x = 1$, can we conclude anything about the tangent to the graph of g at $x = 1$ or the tangent to the graph of f at $u = g(1)$? Give reasons for your answer.

Explorations

69. *The Derivative of sin 2x* Graph the function $y = 2 \cos 2x$ for $-2 \le x \le 3.5$. Then, on the same screen, graph

$$y = \frac{\sin 2(x + h) - \sin 2x}{h}$$

for $h = 1.0, 0.5,$ and 0.2. Experiment with other values of h, including negative values. What do you see happening as $h \to 0$? Explain this behavior.

70. *The Derivative of cos* (x^2) Graph $y = -2x \sin(x^2)$ for $-2 \le x \le 3$. Then, on screen, graph

$$y = \frac{\cos [(x + h)^2] - \cos (x^2)}{h}$$

for $h = 1.0, 0.7,$ and 0.3. Experiment with other values of h. What do you see happening as $h \to 0$? Explain this behavior. ∎

Extending the Ideas

71. *Absolute Value Functions* Let u be a differentiable function of x.

(a) Show that $\dfrac{d}{dx}|u| = u' \dfrac{u}{|u|}$.

(b) Use part (a) to find the derivatives of $f(x) = |x^2 - 9|$ and $g(x) = |x| \sin x$.

72. *Geometric and Arithmetic Mean* The geometric mean of u and v is $G = \sqrt{uv}$ and the arithmetic mean is $A = (u + v)/2$. Show that if $u = x$, $v = x + c$, c a real number, then

$$\frac{dG}{dx} = \frac{A}{G}.$$

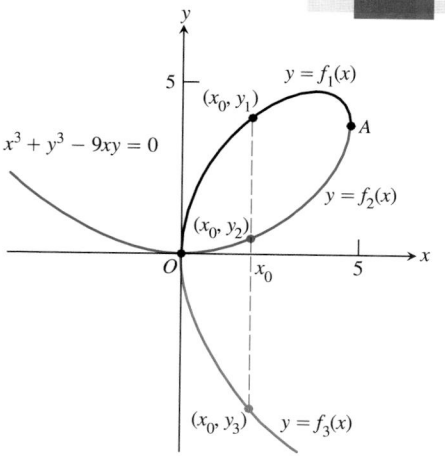

Figure 3.45 The graph of

$$x^3 + y^3 - 9xy = 0$$

(called a *folium*). Although not the graph of a function, it can be seen as the union of the graphs of three separate functions. This particular curve dates to Descartes in 1638.

3.7 Implicit Differentiation

Implicitly Defined Functions • Lenses, Tangents, and Normal Lines • Derivatives of Higher Order • Rational Powers of Differentiable Functions

Implicitly Defined Functions

The graph of the equation $x^3 + y^3 - 9xy = 0$ (Figure 3.45) has a well-defined slope at nearly every point because it is the union of the graphs of the functions $y = f_1(x)$, $y = f_2(x)$, and $y = f_3(x)$, which are differentiable except at O and A. But how do we find the slope when we cannot conveniently solve the equation to find the functions? The answer is to treat y as a differentiable function of x and differentiate both sides of the equation with respect to x, using the differentiation rules for sums, products, and quotients, and the Chain Rule. Then solve for dy/dx in terms of x and y *together* to obtain a formula that calculates the slope at any point (x, y) on the graph from the values of x and y.

The process by which we find dy/dx is called **implicit differentiation.** The phrase derives from the fact that the equation $x^3 + y^3 - 9xy = 0$ defines the functions $f_1, f_2,$ and f_3 implicitly (i.e., hidden inside the equation), without giving us *explicit* formulas to work with.

Example 1 DIFFERENTIATING IMPLICITLY

Find dy/dx if $y^2 = x$.

Solution To find dy/dx, we simply differentiate both sides of the equation $y^2 = x$ with respect to x, treating y as a differentiable function of x and applying the Chain Rule:

$$y^2 = x$$

$$2y\frac{dy}{dx} = 1 \qquad \frac{d}{dx}(y^2) = \frac{d}{dy}(y^2) \cdot \frac{dy}{dx}$$

$$\frac{dy}{dx} = \frac{1}{2y}.$$

In the previous example we differentiated with respect to x, and yet the derivative we obtained appeared as a function of y. Not only is this acceptable, it is actually quite useful. Figure 3.46, for example, shows that the curve has two different tangent lines when $x = 4$: one at the point $(4, 2)$ and the other at the point $(4, -2)$. Since the formula for dy/dx depends on y, our single formula gives the slope in both cases.

Implicit differentiation will frequently yield a derivative that is expressed in terms of both x and y, as in Example 2.

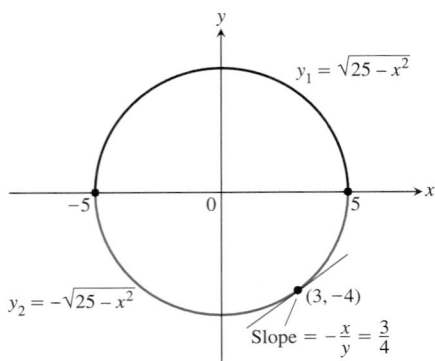

Figure 3.46 The derivative found in Example 1 gives the slope for the tangent lines at both P and Q, because it is a function of y.

Example 2 FINDING SLOPE ON A CIRCLE

Find the slope of the circle $x^2 + y^2 = 25$ at the point $(3, -4)$.

Solution The circle is not the graph of a single function of x, but it is the union of the graphs of two differentiable functions, $y_1 = \sqrt{25 - x^2}$ and $y_2 = -\sqrt{25 - x^2}$ (Figure 3.47). The point $(3, -4)$ lies on the graph of y_2, so it is possible to find the slope by calculating explicitly:

$$\frac{dy_2}{dx}\bigg|_{x=3} = -\frac{-2x}{2\sqrt{25 - x^2}}\bigg|_{x=3} = -\frac{-6}{2\sqrt{25 - 9}} = \frac{3}{4}.$$

But we can also find this slope more easily by differentiating both sides of the equation of the circle implicitly with respect to x:

$$\frac{d}{dx}(x^2 + y^2) = \frac{d}{dx}(25) \qquad \text{Differentiate both sides with respect to } x.$$

$$2x + 2y\frac{dy}{dx} = 0$$

$$\frac{dy}{dx} = -\frac{x}{y}.$$

The slope at $(3, -4)$ is

$$-\frac{x}{y}\bigg|_{(3, -4)} = -\frac{3}{-4} = \frac{3}{4}.$$

Figure 3.47 The circle combines the graphs of two functions. The graph of y_2 is the lower semicircle and passes through $(3, -4)$. (Example 2)

The implicit solution, besides being computationally easier, yields a formula for dy/dx that applies at any point on the circle (except, of course, $(\pm 5, 0)$, where slope is undefined). The explicit solution derived from the formula for y_2 applies only to the lower half of the circle.

Ellen Ochoa

1958–

After earning a doctorate degree in electrical engineering from Stanford University, Ellen Ochoa became a research engineer and, within a few years, received three patents in the field of optics. In 1990, Ochoa joined the NASA astronaut program, and, three years later, became the first Hispanic female to travel in space. Ochoa's message to young people is: "If you stay in school you have the potential to achieve what you want in the future."

To calculate the derivatives of other implicitly defined functions, we proceed as in Examples 1 and 2. We treat y as a differentiable function of x and apply the usual rules to differentiate both sides of the defining equation.

Example 3 SOLVING FOR dy/dx

Show that the slope dy/dx is defined at every point on the graph of $2y = x^2 + \sin y$.

Solution First we need to know dy/dx, which we find by implicit differentiation:

$$2y = x^2 + \sin y$$

$$\frac{d}{dx}(2y) = \frac{d}{dx}(x^2 + \sin y) \qquad \text{Differentiate both sides with respect to } x \ldots$$

$$= \frac{d}{dx}(x^2) + \frac{d}{dx}(\sin y)$$

$$2\frac{dy}{dx} = 2x + \cos y\,\frac{dy}{dx} \qquad \ldots \text{treating } y \text{ as a function of } x \text{ and using the Chain Rule.}$$

$$2\frac{dy}{dx} - (\cos y)\frac{dy}{dx} = 2x \qquad \text{Collect terms with } dy/dx \ldots$$

$$(2 - \cos y)\frac{dy}{dx} = 2x \qquad \ldots \text{and factor out } dy/dx.$$

$$\frac{dy}{dx} = \frac{2x}{2 - \cos y}. \qquad \text{Solve for } dy/dx \text{ by dividing.}$$

The formula for dy/dx is defined at every point (x, y), except for those points at which $\cos y = 2$. Since $\cos y$ cannot be greater than 1, this never happens.

Implicit Differentiation Process

1. Differentiate both sides of the equation with respect to x.

2. Collect the terms with dy/dx on one side of the equation.

3. Factor out dy/dx.

4. Solve for dy/dx.

Lenses, Tangents, and Normal Lines

In the law that describes how light changes direction as it enters a lens, the important angles are the angles the light makes with the line perpendicular to the surface of the lens at the point of entry (angles A and B in Figure 3.48). This line is called the *normal to the surface* at the point of entry. In a profile view of a lens like the one in Figure 3.48, the normal is a line perpendicular to the tangent to the profile curve at the point of entry.

The profiles of lenses are often described by quadratic curves like the one on the next page. When they are, we can use implicit differentiation to find the tangents and normals.

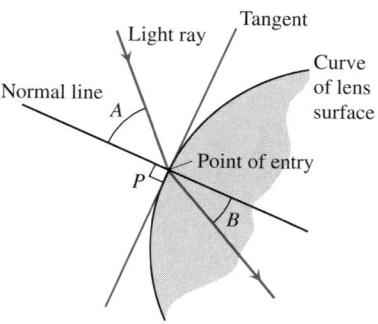

Figure 3.48 The profile of a lens, showing the bending (refraction) of a ray of light as it passes through the lens surface.

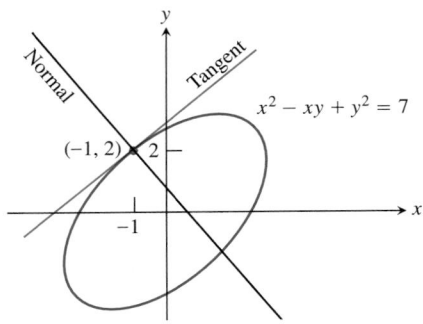

Figure 3.49 Tangent and normal lines to the ellipse $x^2 - xy + y^2 = 7$ at the point $(-1, 2)$. (Example 4)

Example 4   TANGENT AND NORMAL TO AN ELLIPSE

Find the tangent and normal to the ellipse $x^2 - xy + y^2 = 7$ at the point $(-1, 2)$. (See Figure 3.49.)

Solution We first use implicit differentiation to find dy/dx:

$$x^2 - xy + y^2 = 7$$

$$\frac{d}{dx}(x^2) - \frac{d}{dx}(xy) + \frac{d}{dx}(y^2) = \frac{d}{dx}(7) \qquad \text{Differentiate both sides with respect to } x \ldots$$

$$2x - \left(x\frac{dy}{dx} + y\frac{dx}{dx}\right) + 2y\frac{dy}{dx} = 0 \qquad \text{\ldots treating } xy \text{ as a product and } y \text{ as a function of } x.$$

$$(2y - x)\frac{dy}{dx} = y - 2x \qquad \text{Collect terms.}$$

$$\frac{dy}{dx} = \frac{y - 2x}{2y - x}. \qquad \text{Solve for } dy/dx.$$

We then evaluate the derivative at $x = -1$, $y = 2$ to obtain

$$\left.\frac{dy}{dx}\right|_{(-1, 2)} = \left.\frac{y - 2x}{2y - x}\right|_{(-1, 2)}$$

$$= \frac{2 - 2(-1)}{2(2) - (-1)}$$

$$= \frac{4}{5}.$$

The tangent to the curve at $(-1, 2)$ is

$$y - 2 = \frac{4}{5}(x - (-1))$$

$$y = \frac{4}{5}x + \frac{14}{5}.$$

The normal to the curve at $(-1, 2)$ is

$$y - 2 = -\frac{5}{4}(x + 1)$$

$$y = -\frac{5}{4}x + \frac{3}{4}.$$

Derivatives of Higher Order

Implicit differentiation can also be used to find derivatives of higher order. Here is an example.

Example 5   FINDING A SECOND DERIVATIVE IMPLICITLY

Find d^2y/dx^2 if $2x^3 - 3y^2 = 8$.

Solution To start, we differentiate both sides of the equation with respect to x in order to find $y' = dy/dx$.

$$\frac{d}{dx}(2x^3 - 3y^2) = \frac{d}{dx}(8)$$

$$6x^2 - 6yy' = 0$$

$$x^2 - yy' = 0$$

$$y' = \frac{x^2}{y}, \text{ when } y \neq 0$$

We now apply the Quotient Rule to find y''.

$$y'' = \frac{d}{dx}\left(\frac{x^2}{y}\right) = \frac{2xy - x^2 y'}{y^2} = \frac{2x}{y} - \frac{x^2}{y^2} \cdot y'$$

Finally, we substitute $y' = x^2/y$ to express y'' in terms of x and y.

$$y'' = \frac{2x}{y} - \frac{x^2}{y^2}\left(\frac{x^2}{y}\right) = \frac{2x}{y} - \frac{x^4}{y^3}, \text{ when } y \neq 0$$

Exploration 1 An Unexpected Derivative

Consider the set of all points (x, y) satisfying the equation $x^2 - 2xy + y^2 = 4$. What does the graph of the equation look like? You can find out in two ways in this Exploration.

1. Use implicit differentiation to find dy/dx. Are you surprised by this derivative?

2. Knowing the derivative, what do you conjecture about the graph?

3. What are the possible values of y when $x = 0$? Does this information enable you to refine your conjecture about the graph?

4. The original equation can be written as $(x - y)^2 - 4 = 0$. By factoring the expression on the left, write two equations whose graphs combine to give the graph of the original equation. Then sketch the graph.

5. Explain why your graph is consistent with the derivative found in part 1.

Rational Powers of Differentiable Functions

We know that the Power Rule

$$\frac{d}{dx}x^n = nx^{n-1}$$

holds for any integer n (Rules 2 and 7 of this chapter). We can now prove that it holds when n is any rational number.

Rule 9 Power Rule for Rational Powers of x

If n is any rational number, then

$$\frac{d}{dx}x^n = nx^{n-1}.$$

If $n < 1$, then the derivative does not exist at $x = 0$.

Proof   Let p and q be integers with $q > 0$ and suppose that $y = \sqrt[q]{x^p} = x^{p/q}$. Then

$$y^q = x^p.$$

Since p and q are integers (for which we already have the Power Rule), we can differentiate both sides of the equation with respect to x and obtain

$$qy^{q-1}\frac{dy}{dx} = px^{p-1}.$$

If $y \neq 0$, we can divide both sides of the equation by qy^{q-1} to solve for dy/dx, obtaining

$$\frac{dy}{dx} = \frac{px^{p-1}}{qy^{q-1}}$$

$$= \frac{p}{q} \cdot \frac{x^{p-1}}{(x^{p/q})^{q-1}} \qquad y = x^{p/q}$$

$$= \frac{p}{q} \cdot \frac{x^{p-1}}{x^{p-p/q}} \qquad \frac{p}{q}(q-1) = p - \frac{p}{q}$$

$$= \frac{p}{q} \cdot x^{(p-1)-(p-p/q)} \qquad \text{A law of exponents}$$

$$= \frac{p}{q} \cdot x^{(p/q)-1}.$$

This proves the rule.   ∎

By combining this result with the Chain Rule, we get an extension of the Power Chain Rule to rational powers of u:

If n is a rational number and u is a differentiable function of x, then u^n is a differentiable function of x and

$$\frac{d}{dx}u^n = nu^{n-1}\frac{du}{dx},$$

provided that $u \neq 0$ if $n < 1$.

The restriction that $u \neq 0$ when $n < 1$ is necessary because 0 might be in the domain of u^n but not in the domain of u^{n-1}, as we see in the first two parts of Example 6.

Example 6   USING THE RATIONAL POWER RULE

(a) $\dfrac{d}{dx}(\sqrt{x}) = \dfrac{d}{dx}(x^{1/2}) = \dfrac{1}{2}x^{-1/2} = \dfrac{1}{2\sqrt{x}}$

Notice that $\sqrt{x}$ is defined at $x = 0$, but $1/(2\sqrt{x})$ is not.

(b) $\dfrac{d}{dx}(x^{2/3}) = \dfrac{2}{3}(x^{-1/3}) = \dfrac{2}{3x^{1/3}}$

The original function is defined for all real numbers, but the derivative is undefined at $x = 0$. Recall Figure 3.12, which showed that this function's graph has a *cusp* at $x = 0$.

(c) $\dfrac{d}{dx}(\cos x)^{-1/5} = -\dfrac{1}{5}(\cos x)^{-6/5} \cdot \dfrac{d}{dx}(\cos x)$

$$= -\frac{1}{5}(\cos x)^{-6/5}(-\sin x)$$

$$= \frac{1}{5}\sin x(\cos x)^{-6/5}$$

Quick Review 3.7

In Exercises 1–5, sketch the curve defined by the equation and find two functions y_1 and y_2 whose graphs will combine to give the curve.

1. $x - y^2 = 0$

2. $4x^2 + 9y^2 = 36$

3. $x^2 - 4y^2 = 0$

4. $x^2 + y^2 = 9$

5. $x^2 + y^2 = 2x + 3$

In Exercises 6–8, solve for y' in terms of y and x.

6. $x^2 y' - 2xy = 4x - y$

7. $y' \sin x - x \cos x = xy' + y$

8. $x(y^2 - y') = y'(x^2 - y)$

In Exercises 9–10, find an expression for the function using rational powers rather than radicals.

9. $\sqrt{x}(x - \sqrt[3]{x})$

10. $\dfrac{x + \sqrt[3]{x^2}}{\sqrt{x^3}}$

Section 3.7 Exercises

In Exercises 1–20, find dy/dx.

1. $y = x^{9/4}$

2. $y = x^{-3/5}$

3. $y = \sqrt[3]{x}$

4. $y = \sqrt[4]{x}$

5. $y = (2x + 5)^{-1/2}$

6. $y = (1 - 6x)^{2/3}$

7. $y = x\sqrt{x^2 + 1}$

8. $y = \dfrac{x}{\sqrt{x^2 + 1}}$

9. $x^2 y + xy^2 = 6$

10. $x^3 + y^3 = 18xy$

11. $y^2 = \dfrac{x - 1}{x + 1}$

12. $x^2 = \dfrac{x - y}{x + y}$

13. $y = \sqrt{1 - \sqrt{x}}$

14. $y = 3(2x^{-1/2} + 1)^{-1/3}$

15. $y = 3(\csc x)^{3/2}$

16. $y = [\sin (x + 5)]^{5/4}$

17. $x = \tan y$

18. $x = \sin y$

19. $x + \tan (xy) = 0$

20. $x + \sin y = xy$

21. Which of the following could be true if $f''(x) = x^{-1/3}$?

(a) $f(x) = \dfrac{3}{2}x^{2/3} - 3$

(b) $f(x) = \dfrac{9}{10}x^{5/3} - 7$

(c) $f'''(x) = -\dfrac{1}{3}x^{-4/3}$

(d) $f'(x) = \dfrac{3}{2}x^{2/3} + 6$

22. Which of the following could be true if $g''(t) = 1/t^{3/4}$?

(a) $g'(t) = 4\sqrt[4]{t} - 4$

(b) $g'''(t) = -4/\sqrt[4]{t}$

(c) $g(t) = t - 7 + (16/5)t^{5/4}$

(d) $g'(t) = (1/4)t^{1/4}$

In Exercises 23–26, use implicit differentiation to find dy/dx and then d^2y/dx^2.

23. $x^2 + y^2 = 1$

24. $x^{2/3} + y^{2/3} = 1$

25. $y^2 = x^2 + 2x$

26. $y^2 + 2y = 2x + 1$

In Exercises 27–36, find the lines that are **(a)** tangent and **(b)** normal to the curve at the given point.

27. $x^2 + xy - y^2 = 1$, $(2, 3)$

28. $x^2 + y^2 = 25$, $(3, -4)$

29. $x^2 y^2 = 9$, $(-1, 3)$

30. $y^2 - 2x - 4y - 1 = 0$, $(-2, 1)$

31. $6x^2 + 3xy + 2y^2 + 17y - 6 = 0$, $(-1, 0)$

32. $x^2 - \sqrt{3}xy + 2y^2 = 5$, $(\sqrt{3}, 2)$

33. $2xy + \pi \sin y = 2\pi$, $(1, \pi/2)$

34. $x \sin 2y = y \cos 2x$, $(\pi/4, \pi/2)$

35. $y = 2 \sin (\pi x - y)$, $(1, 0)$

36. $x^2 \cos^2 y - \sin y = 0$, $(0, \pi)$

37. *The Eight Curve* **(a)** Find the slopes of the figure-eight-shaped curve
$$y^4 = y^2 - x^2$$
at the two points shown on the graph that follows.

(b) Use parametric mode and the two pairs of parametric equations
$$x_1(t) = \sqrt{t^2 - t^4}, \quad y_1(t) = t,$$
$$x_2(t) = -\sqrt{t^2 - t^4}, \quad y_2(t) = t,$$
to graph the curve. Specify a window and a parameter interval.

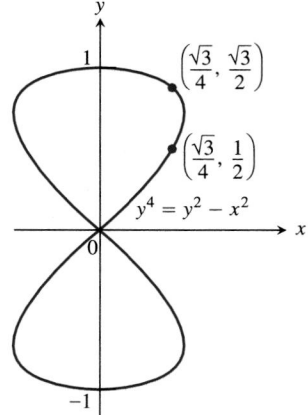

38. *The Cissoid of Diocles (dates from about 200 B.C.)*

(a) Find equations for the tangent and normal to the cissoid of Diocles,

$$y^2(2 - x) = x^3,$$

at the point $(1, 1)$ as pictured below.

(b) Explain how to reproduce the graph on a grapher.

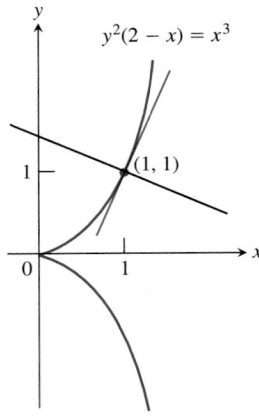

39. (a) Confirm that $(-1, 1)$ is on the curve defined by $x^3 y^2 = \cos(\pi y)$.

(b) Use part (a) to find the slope of the line tangent to the curve at $(-1, 1)$.

40. *Work in groups of two or three.*

(a) Show that the relation

$$y^3 - xy = -1$$

cannot be a function of x by showing that there is more than one possible y-value when $x = 2$.

(b) On a small enough square with center $(2, 1)$, the part of the graph of the relation within the square will define a function $y = f(x)$. For this function, find $f'(2)$ and $f''(2)$.

41. Find the two points where the curve $x^2 + xy + y^2 = 7$ crosses the x-axis, and show that the tangents to the curve at these points are parallel. What is the common slope of these tangents?

42. Find points on the curve $x^2 + xy + y^2 = 7$ **(a)** where the tangent is parallel to the x-axis and **(b)** where the tangent is parallel to the y-axis. (In the latter case, dy/dx is not defined, but dx/dy is. What value does dx/dy have at these points?)

43. *Orthogonal Curves* Two curves are *orthogonal* at a point of intersection if their tangents at that point cross at right angles. Show that the curves $2x^2 + 3y^2 = 5$ and $y^2 = x^3$ are orthogonal at $(1, 1)$ and $(1, -1)$. Use parametric mode to draw the curves and to show the tangent lines.

44. The position of a body moving along a coordinate line at time t is $s = (4 + 6t)^{3/2}$, with s in meters and t in seconds. Find the body's velocity and acceleration when $t = 2$ sec.

45. The velocity of a falling body is $v = 8\sqrt{s - t} + 1$ feet per second at the instant t (sec) the body has fallen s feet from its starting point. Show that the body's acceleration is 32 ft/sec^2.

46. *The Devil's Curve (Gabriel Cramer [the Cramer of Cramer's Rule], 1750)* Find the slopes of the devil's curve $y^4 - 4y^2 = x^4 - 9x^2$ at the four indicated points.

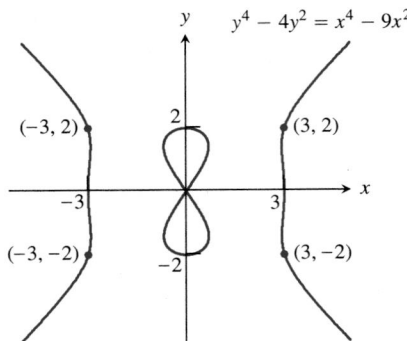

47. *The Folium of Descartes* (See Figure 3.45 on page 149)

(a) Find the slope of the folium of Descartes, $x^3 + y^3 - 9xy = 0$ at the points $(4, 2)$ and $(2, 4)$.

(b) At what point other than the origin does the folium have a horizontal tangent?

(c) Find the coordinates of the point A in Figure 3.45, where the folium has a vertical tangent.

48. The line that is normal to the curve $x^2 + 2xy - 3y^2 = 0$ at $(1, 1)$ intersects the curve at what other point?

49. Find the normals to the curve $xy + 2x - y = 0$ that are parallel to the line $2x + y = 0$.

50. Show that if it is possible to draw these three normals from the point $(a, 0)$ to the parabola $x = y^2$ shown here, then a must be greater than $1/2$. One of the normals is the x-axis. For what value of a are the other two normals perpendicular?

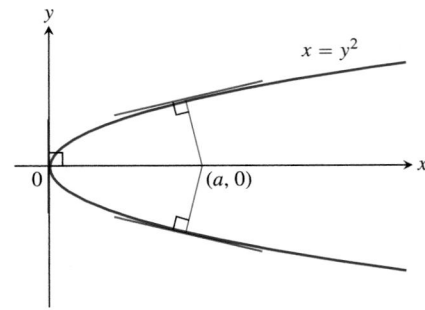

Extending the Ideas

51. *Finding Tangents*

(a) Show that the tangent to the ellipse

$$\frac{x^2}{a^2} + \frac{y^2}{b^2} = 1$$

at the point (x_1, y_1) has equation

$$\frac{x_1 x}{a^2} + \frac{y_1 y}{b^2} = 1.$$

(b) Find an equation for the tangent to the hyperbola

$$\frac{x^2}{a^2} - \frac{y^2}{b^2} = 1$$

at the point (x_1, y_1).

52. *End Behavior Model* Consider the hyperbola

$$\frac{x^2}{a^2} - \frac{y^2}{b^2} = 1.$$

Show that

(a) $y = \pm\dfrac{b}{a}\sqrt{x^2 - a^2}$.

(b) $g(x) = (b/a)|x|$ is an end behavior model for

$$f(x) = (b/a)\sqrt{x^2 - a^2}.$$

(c) $g(x) = -(b/a)|x|$ is an end behavior model for

$$f(x) = -(b/a)\sqrt{x^2 - a^2}.$$

3.8 Derivatives of Inverse Trigonometric Functions

Derivatives of Inverse Functions • Derivative of the Arcsine •
Derivative of the Arctangent • Derivative of the Arcsecant •
Derivatives of the Other Three

Derivatives of Inverse Functions

In Section 1.5 we learned that the graph of the inverse of a function f can be obtained by reflecting the graph of f across the line $y = x$. If we combine that with our understanding of what makes a function differentiable, we can gain some quick insights into the differentiability of inverse functions.

As Figure 3.50 suggests, the reflection of a continuous curve with no cusps or corners will be another continuous curve with no cusps or corners. Indeed, if there is a tangent line to the graph of f at the point $(a, f(a))$, then that line will reflect across $y = x$ to become a tangent line to the graph of f^{-1} at the point $(f(a), a)$. We can even see geometrically that the *slope* of the reflected tangent line (when it exists and is not zero) will be the *reciprocal* of the slope of the original tangent line, since a change in y becomes a change in x in the reflection, and a change in x becomes a change in y.

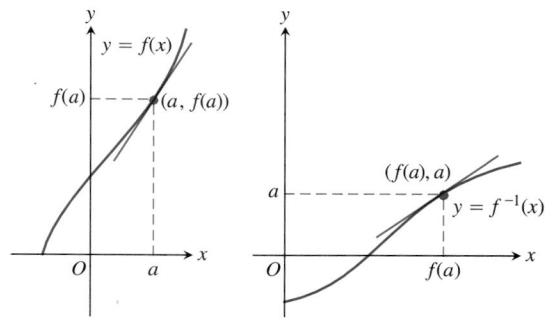

The slopes are reciprocal: $\left.\dfrac{df^{-1}}{dx}\right|_{f(a)} = \dfrac{1}{\left.\dfrac{df}{dx}\right|_a}$

Figure 3.50 The graphs of a function and its inverse. Notice that the tangent lines have reciprocal slopes.

All of this serves as an introduction to the following theorem, which we will assume as we proceed to find derivatives of inverse functions. Although the essentials of the proof are illustrated in the geometry of Figure 3.50, a careful analytic proof is more appropriate for an advanced calculus text and will be omitted here.

Theorem 3 Derivatives of Inverse Functions

If f is differentiable at every point of an interval I and df/dx is never zero on I, then f has an inverse and f^{-1} is differentiable at every point of the interval $f(I)$.

Exploration 1   **Finding a Derivative on an Inverse Graph Geometrically**

Let $f(x) = x^5 + 2x - 1$. Since the point $(1, 2)$ is on the graph of f, it follows that the point $(2, 1)$ is on the graph of f^{-1}. Can you find

$$\frac{df^{-1}}{dx}(2),$$

the value of df^{-1}/dx at 2, without knowing a formula for f^{-1}?

1. Graph $f(x) = x^5 + 2x - 1$. A function must be one-to-one to have an inverse function. Is this function one-to-one?

2. Find $f'(x)$. How could this derivative help you to conclude that f has an inverse?

3. Reflect the graph of f across the line $y = x$ to obtain a graph of f^{-1}.

4. Sketch the tangent line to the graph of f^{-1} at the point $(2, 1)$. Call it L.

5. Reflect the line L across the line $y = x$. At what point is the reflection of L tangent to the graph of f?

6. What is the slope of the reflection of L?

7. What is the slope of L?

8. What is $\dfrac{df^{-1}}{dx}(2)$?

Derivative of the Arcsine

We know that the function $x = \sin y$ is differentiable in the open interval $-\pi/2 < y < \pi/2$ and that its derivative, the cosine, is positive there. Theorem 3 therefore assures us that the inverse function $y = \sin^{-1}(x)$ (the *arcsine* of x) is differentiable throughout the interval $-1 < x < 1$. We cannot expect it to be differentiable at $x = -1$ or $x = 1$, however, because the tangents to the graph are vertical at these points (Figure 3.51).

We find the derivative of $y = \sin^{-1}(x)$ as follows:

$$y = \sin^{-1} x$$

$$\sin y = x \qquad \text{Inverse function relationship}$$

$$\frac{d}{dx}(\sin y) = \frac{d}{dx}x \qquad \text{Differentiate both sides.}$$

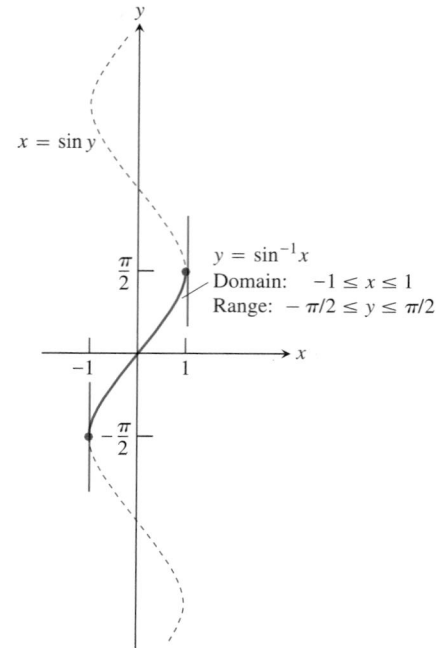

$x = \sin y$

$y = \sin^{-1} x$
Domain: $-1 \le x \le 1$
Range: $-\pi/2 \le y \le \pi/2$

Figure 3.51 The graph of $y = \sin^{-1} x$ has vertical tangents $x = -1$ and $x = 1$.

$$\cos y \frac{dy}{dx} = 1 \qquad \text{Implicit differentiation}$$

$$\frac{dy}{dx} = \frac{1}{\cos y}$$

The division in the last step is safe because $\cos y \neq 0$ for $-\pi/2 < y < \pi/2$. In fact, $\cos y$ is *positive* for $-\pi/2 < y < \pi/2$, so we can replace $\cos y$ with $\sqrt{1 - (\sin y)^2}$, which is $\sqrt{1 - x^2}$. Thus

$$\frac{d}{dx}(\sin^{-1} x) = \frac{1}{\sqrt{1 - x^2}}.$$

If u is a differentiable function of x with $|u| < 1$, we apply the Chain Rule to get

$$\frac{d}{dx}\sin^{-1} u = \frac{1}{\sqrt{1 - u^2}}\frac{du}{dx}, \qquad |u| < 1.$$

Example 1   APPLYING THE FORMULA

$$\frac{d}{dx}(\sin^{-1} x^2) = \frac{1}{\sqrt{1 - (x^2)^2}} \cdot \frac{d}{dx}(x^2) = \frac{2x}{\sqrt{1 - x^4}}$$

Derivative of the Arctangent

Although the function $y = \sin^{-1}(x)$ has a rather narrow domain of $[-1, 1]$, the function $y = \tan^{-1} x$ is defined for all real numbers. It is also differentiable for all real numbers, as we will now see. The differentiation proceeds exactly as with the arcsine function above.

$$y = \tan^{-1} x$$

$$\tan y = x \qquad \text{Inverse function relationship}$$

$$\frac{d}{dx}(\tan y) = \frac{d}{dx}x$$

$$\sec^2 y \frac{dy}{dx} = 1 \qquad \text{Implicit differentiation}$$

$$\frac{dy}{dx} = \frac{1}{\sec^2 y}$$

$$= \frac{1}{1 + (\tan y)^2} \qquad \text{Trig identity: } \sec^2 y = 1 + \tan^2 y$$

$$= \frac{1}{1 + x^2}$$

The derivative is defined for all real numbers. If u is a differentiable function of x, we get the Chain Rule form:

$$\frac{d}{dx}\tan^{-1} u = \frac{1}{1 + u^2}\frac{du}{dx}.$$

Example 2 A MOVING PARTICLE

A particle moves along the x-axis so that its position at any time $t \geq 0$ is $x(t) = \tan^{-1} \sqrt{t}$. What is the velocity of the particle when $t = 16$?

Solution

$$v(t) = \frac{d}{dt} \tan^{-1} \sqrt{t} = \frac{1}{1 + (\sqrt{t})^2} \cdot \frac{d}{dt} \sqrt{t} = \frac{1}{1 + t} \cdot \frac{1}{2\sqrt{t}}$$

When $t = 16$, the velocity is

$$v(16) = \frac{1}{1 + 16} \cdot \frac{1}{2\sqrt{16}} = \frac{1}{136}.$$

Derivative of the Arcsecant

We find the derivative of $y = \sec^{-1} x$, $|x| > 1$, beginning as we did with the other inverse trigonometric functions.

$$y = \sec^{-1} x$$

$$\sec y = x \qquad \text{Inverse function relationship}$$

$$\frac{d}{dx}(\sec y) = \frac{d}{dx} x$$

$$\sec y \tan y \frac{dy}{dx} = 1$$

$$\frac{dy}{dx} = \frac{1}{\sec y \tan y} \qquad \begin{array}{l} \text{Since } |x| > 1, \ y \text{ lies in } (0, \pi/2) \cup (\pi/2, \pi) \\ \text{and } \sec y \tan y \neq 0. \end{array}$$

To express the result in terms of x, we use the relationships

$$\sec y = x \quad \text{and} \quad \tan y = \pm\sqrt{\sec^2 y - 1} = \pm\sqrt{x^2 - 1}$$

to get

$$\frac{dy}{dx} = \pm \frac{1}{x\sqrt{x^2 - 1}}.$$

Can we do anything about the $\pm$ sign? A glance at Figure 3.52 shows that the slope of the graph $y = \sec^{-1} x$ is always positive. That must mean that

$$\frac{d}{dx} \sec^{-1} x = \begin{cases} + \dfrac{1}{x\sqrt{x^2 - 1}} & \text{if } x > 1 \\[3mm] - \dfrac{1}{x\sqrt{x^2 - 1}} & \text{if } x < -1. \end{cases}$$

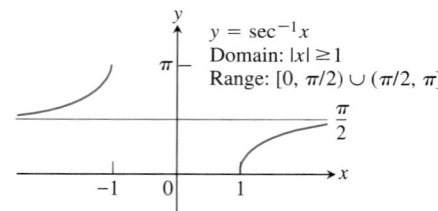

$y = \sec^{-1} x$
Domain: $|x| \geq 1$
Range: $[0, \pi/2) \cup (\pi/2, \pi]$

Figure 3.52 The slope of the curve $y = \sec^{-1} x$ is positive for both $x < -1$ and $x > 1$.

With the absolute value symbol we can write a single expression that eliminates the "$\pm$" ambiguity:

$$\frac{d}{dx} \sec^{-1} x = \frac{1}{|x|\sqrt{x^2 - 1}}.$$

If u is a differentiable function of x with $|u| > 1$, we have the formula

$$\frac{d}{dx} \sec^{-1} u = \frac{1}{|u|\sqrt{u^2 - 1}} \frac{du}{dx}, \qquad |u| > 1.$$

Example 3 USING THE FORMULA

$$\frac{d}{dx}\sec^{-1}(5x^4) = \frac{1}{|5x^4|\sqrt{(5x^4)^2 - 1}}\frac{d}{dx}(5x^4)$$

$$= \frac{1}{5x^4\sqrt{25x^8 - 1}}(20x^3)$$

$$= \frac{4}{x\sqrt{25x^8 - 1}}$$

Derivatives of the Other Three

We could use the same technique to find the derivatives of the other three inverse trigonometric functions: arccosine, arccotangent, and arccosecant, but there is a much easier way, thanks to the following identities.

Inverse Function – Inverse Cofunction Identities

$$\cos^{-1}x = \pi/2 - \sin^{-1}x$$

$$\cot^{-1}x = \pi/2 - \tan^{-1}x$$

$$\csc^{-1}x = \pi/2 - \sec^{-1}x$$

It follows easily that the derivatives of the inverse cofunctions are the negatives of the derivatives of the corresponding inverse functions (see Exercises 24–26).

You have probably noticed by now that most calculators do not have buttons for $\cot^{-1}$, $\sec^{-1}$, or $\csc^{-1}$. They are not needed because of the following identities:

Calculator Conversion Identities

$$\sec^{-1}x = \cos^{-1}(1/x)$$

$$\cot^{-1}x = \pi/2 - \tan^{-1}x$$

$$\csc^{-1}x = \sin^{-1}(1/x)$$

Notice that we do not use $\tan^{-1}(1/x)$ as an identity for $\cot^{-1}x$. A glance at the graphs of $y = \tan^{-1}(1/x)$ and $y = \pi/2 - \tan^{-1}x$ reveals the problem (Figure 3.53).

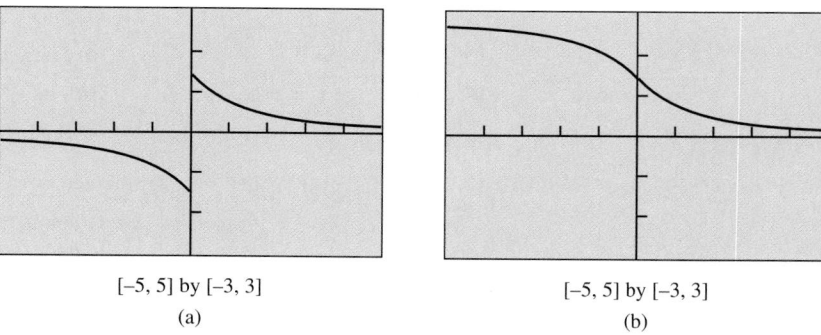

[–5, 5] by [–3, 3]

(a)

[–5, 5] by [–3, 3]

(b)

Figure 3.53 The graphs of (a) $y = \tan^{-1}(1/x)$ and (b) $y = \pi/2 - \tan^{-1}x$. The graph in (b) is the same as the graph of $y = \cot^{-1}x$.

We cannot replace $\cot^{-1} x$ by the function $y = \tan^{-1}(1/x)$ in the identity for the inverse functions and inverse cofunctions, and so it is not the function we want for $\cot^{-1} x$. The ranges of the inverse trigonometric functions have been chosen in part to make the two sets of identities above hold.

Example 4 A TANGENT LINE TO THE ARCCOTANGENT CURVE

Find an equation for the line tangent to the graph of $y = \cot^{-1} x$ at $x = -1$.

Solution First, we note that
$$\cot^{-1}(-1) = \pi/2 - \tan^{-1}(-1) = \pi/2 - (-\pi/4) = 3\pi/4.$$

The slope of the tangent line is
$$\left.\frac{dy}{dx}\right|_{x=-1} = -\left.\frac{1}{1+x^2}\right|_{x=-1} = -\frac{1}{1+(-1)^2} = -\frac{1}{2}.$$

So the tangent line has equation $y - 3\pi/4 = (-1/2)(x + 1)$.

Quick Review 3.8

In Exercises 1–5, give the *domain* and *range* of the function, and evaluate the function at $x = 1$.

1. $y = \sin^{-1} x$
2. $y = \cos^{-1} x$
3. $y = \tan^{-1} x$
4. $y = \sec^{-1} x$
5. $y = \tan(\tan^{-1} x)$

In Exercises 6–10, find the inverse of the given function.

6. $y = 3x - 8$
7. $y = \sqrt[3]{x + 5}$
8. $y = \dfrac{8}{x}$
9. $y = \dfrac{3x - 2}{x}$
10. $y = \arctan(x/3)$

Section 3.8 Exercises

In Exercises 1–18, find the derivative of y with respect to the appropriate variable.

1. $y = \cos^{-1}(x^2)$
2. $y = \cos^{-1}(1/x)$
3. $y = \sin^{-1}\sqrt{2t}$
4. $y = \sin^{-1}(1 - t)$
5. $y = \sec^{-1}(2s + 1)$
6. $y = \sec^{-1} 5s$
7. $y = \csc^{-1}(x^2 + 1), \quad x > 0$
8. $y = \csc^{-1} x/2$
9. $y = \sec^{-1}\dfrac{1}{t}, \quad 0 < t < 1$
10. $y = \sin^{-1}\dfrac{3}{t^2}$
11. $y = \cot^{-1}\sqrt{t}$
12. $y = \cot^{-1}\sqrt{t - 1}$
13. $y = s\sqrt{1 - s^2} + \cos^{-1} s$
14. $y = \sqrt{s^2 - 1} - \sec^{-1} s$
15. $y = \tan^{-1}\sqrt{x^2 - 1} + \csc^{-1} x, \quad x > 1$
16. $y = \cot^{-1}\dfrac{1}{x} - \tan^{-1} x$
17. $y = x \sin^{-1} x + \sqrt{1 - x^2}$
18. $y = \dfrac{1}{\sin^{-1}(2x)}$

19. (a) Find an equation for the line tangent to the graph of $y = \tan x$ at the point $(\pi/4, 1)$.

(b) Find an equation for the line tangent to the graph of $y = \tan^{-1} x$ at the point $(1, \pi/4)$.

20. Let $f(x) = x^5 + 2x^3 + x - 1$.

(a) Find $f(1)$ and $f'(1)$.

(b) Find $f^{-1}(3)$ and $(f^{-1})'(3)$.

21. Let $f(x) = \cos x + 3x$.

(a) Show that f has a differentiable inverse.

(b) Find $f(0)$ and $f'(0)$.

(c) Find $f^{-1}(1)$ and $(f^{-1})'(1)$.

In Exercises 22 and 23, *work in groups of two or three.*

22. Graph the function $f(x) = \sin^{-1}(\sin x)$ in the viewing window $[-2\pi, 2\pi]$ by $[-4, 4]$. Then answer the following questions:

(a) What is the domain of f?

(b) What is the range of f?

(c) At which points is f not differentiable?

(d) Sketch a graph of $y = f'(x)$ without using NDER or computing the derivative.

(e) Find $f'(x)$ algebraically. Can you reconcile your answer with the graph in (d)?

23. A particle moves along the x-axis so that its position at any time $t \geq 0$ is given by $x = \arctan t$.

(a) Prove that the particle is always moving to the right.

(b) Prove that the particle is always decelerating.

(c) What is the limiting position of the particle as t approaches infinity?

In Exercises 24–26, use the inverse function–inverse cofunction identities to derive the formula for the derivative of the function.

24. arccosine **25.** arccotangent

26. arccosecant

Explorations

In Exercises **27–32**, find **(a)** the right end behavior model, **(b)** the left end behavior model, and **(c)** any horizontal tangents for the function if they exist.

27. $y = \tan^{-1} x$ **28.** $y = \cot^{-1} x$

29. $y = \sec^{-1} x$ **30.** $y = \csc^{-1} x$

31. $y = \sin^{-1} x$ **32.** $y = \cos^{-1} x$ ∎

Extending the Ideas

33. *Identities* Confirm the following identities for $x > 0$.

(a) $\cos^{-1} x + \sin^{-1} x = \pi/2$

(b) $\tan^{-1} x + \cot^{-1} x = \pi/2$

(c) $\sec^{-1} x + \csc^{-1} x = \pi/2$

34. *Proof Without Words* The figure gives a proof without words that $\tan^{-1} 1 + \tan^{-1} 2 + \tan^{-1} 3 = \pi$. Explain what is going on.

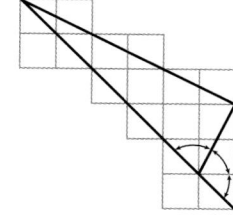

35. *(Continuation of Exercise 34)* Here is a way to construct $\tan^{-1} 1$, $\tan^{-1} 2$, and $\tan^{-1} 3$ by folding a square of paper. Try it and explain what is going on.

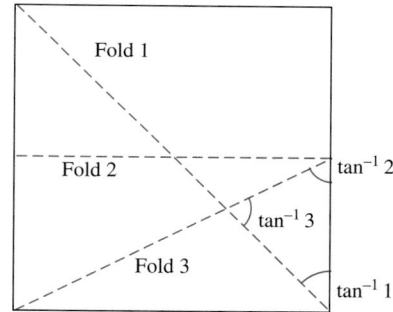

3.9 Derivatives of Exponential and Logarithmic Functions

Derivative of e^x • Derivative of a^x • Derivative of $\ln x$ • Derivative of $\log_a x$ • Power Rule for Arbitrary Real Powers

Derivative of e^x

At the end of the brief review of exponential functions in Section 1.3, we mentioned that the function $y = e^x$ was a particularly important function for modeling exponential growth. The number e was defined in that section to be the limit of $(1 + 1/x)^x$ as $x \to \infty$. This intriguing number shows up in other interesting limits as well, but the one with the most interesting implications for the *calculus* of exponential functions is this one:

$$\lim_{h \to 0} \frac{e^h - 1}{h} = 1.$$

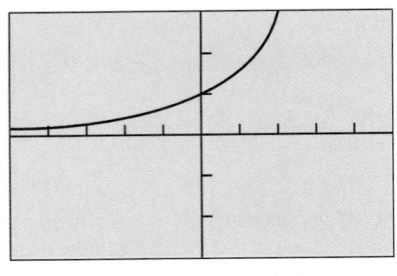

[−4.9, 4.9] by [−2.9, 2.9]

(a)

X	Y₁
−.03	.98515
−.02	.99007
−.01	.99502
0	ERROR
.01	1.005
.02	1.0101
.03	1.0152

X=0

(b)

Figure 3.54 (a) The graph and (b) the table support the conclusion that

$$\lim_{h \to 0} \frac{e^h - 1}{h} = 1.$$

Is any other function its own derivative?

The zero function is also its own derivative, but this hardly seems worth mentioning. (Its value is always 0 and its slope is always 0.) In addition to e^x, however, we can also say that any constant *multiple* of e^x is its own derivative:

$$\frac{d}{dx}(c \cdot e^x) = c \cdot e^x.$$

The next obvious question is whether there are still *other* functions that are their own derivatives, and this time the answer is no. The only functions that satisfy the condition $dy/dx = y$ are functions of the form $y = ke^x$ (and notice that the zero function can be included in this category). We will prove this significant fact in Chapter 6.

(The graph and the table in Figure 3.54 provide strong support for this limit being 1. A formal algebraic proof that begins with our limit definition of e would require some rather subtle limit arguments, so we will not include one here.)

The fact that the limit is 1 creates a remarkable relationship between the function e^x and its derivative, as we will now see.

$$\frac{d}{dx}(e^x) = \lim_{h \to 0} \frac{e^{x+h} - e^x}{h}$$

$$= \lim_{h \to 0} \frac{e^x \cdot e^h - e^x}{h}$$

$$= \lim_{h \to 0} \left(e^x \cdot \frac{e^h - 1}{h} \right)$$

$$= e^x \cdot \lim_{h \to 0} \left(\frac{e^h - 1}{h} \right)$$

$$= e^x \cdot 1$$

$$= e^x$$

In other words, the derivative of this particular function is itself!

$$\frac{d}{dx}(e^x) = e^x$$

If u is a differentiable function of x, then we have

$$\frac{d}{dx}e^u = e^u \frac{du}{dx}.$$

We will make extensive use of this formula when we study exponential growth and decay in Chapter 6.

Example 1   HOW FAST DOES A FLU SPREAD?

The spread of a flu in a certain school is modeled by the equation

$$P(t) = \frac{100}{1 + e^{3-t}},$$

where $P(t)$ is the total number of students infected t days after the flu was first noticed. Many of them may already be well again at time t.

(a) Estimate the initial number of students infected with the flu.

(b) How fast is the flu spreading after 3 days?

(c) When will the flu spread at its maximum rate? What is this rate?

Solution   The graph of P as a function of t is shown in Figure 3.55.

(a) $P(0) = 100/(1 + e^3) = 5$ students (to the nearest whole number).

(b) To find the rate at which the flu spreads, we find dP/dt. To find dP/dt, we need to invoke the Chain Rule twice:

$$\frac{dP}{dt} = \frac{d}{dt}(100(1 + e^{3-t})^{-1}) = 100 \cdot (-1)(1 + e^{3-t})^{-2} \cdot \frac{d}{dt}(1 + e^{3-t})$$

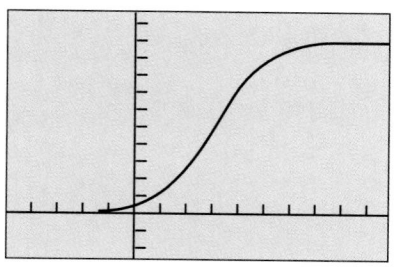

[−5, 10] by [−25, 120]

Figure 3.55 The graph of

$$P(t) = \frac{100}{1 + e^{3-t}},$$

modeling the spread of a flu. (Example 1)

[−5, 10] by [−10, 30]

Figure 3.56 The graph of dP/dt, the rate of spread of the flu in Example 1. The graph of P is shown in Figure 3.55

$$= -100(1 + e^{3-t})^{-2} \cdot (0 + e^{3-t} \cdot \frac{d}{dt}(3 - t))$$

$$= -100(1 + e^{3-t})^{-2}(e^{3-t} \cdot (-1))$$

$$= \frac{100e^{3-t}}{(1 + e^{3-t})^2}$$

At $t = 3$, then, $dP/dt = 100/4 = 25$. The flu is spreading to 25 students per day.

(c) We could estimate when the flu is spreading the fastest by seeing where the graph of $y = P(t)$ has the steepest upward slope, but we can answer both the "when" and the "what" parts of this question most easily by finding the maximum point on the graph of the derivative (Figure 3.56).

We see by tracing on the curve that the maximum rate occurs at about 3 days, when (as we have just calculated) the flu is spreading at a rate of 25 students per day.

Derivative of a^x

What about an exponential function with a base other than e? We will assume that the base is positive and different from 1, since negative numbers to arbitrary real powers are not always real numbers, and $y = 1^x$ is a constant function.

If $a > 0$ and $a \neq 1$, we can use the properties of logarithms to write a^x in terms of e^x. The formula for doing so is

$$a^x = e^{x \ln a}. \qquad e^{x \ln a} = e^{\ln(a^x)} = a^x$$

We can then find the derivative of a^x with the Chain Rule.

$$\frac{d}{dx}a^x = \frac{d}{dx}e^{x \ln a} = e^{x \ln a} \cdot \frac{d}{dx}(x \ln a) = e^{x \ln a} \cdot \ln a = a^x \ln a$$

Thus, if u is a differentiable function of x, we get the following rule.

For $a > 0$ and $a \neq 1$,

$$\frac{d}{dx}(a^u) = a^u \ln a \frac{du}{dx}.$$

Example 2   REVIEWING THE ALGEBRA OF LOGARITHMS

At what point on the graph of the function $y = 2^t - 3$ does the tangent line have slope 21?

Solution The slope is the derivative:

$$\frac{d}{dt}(2^t - 3) = 2^t \cdot \ln 2 - 0 = 2^t \ln 2.$$

We want the value of t for which $2^t \ln 2 = 21$. We could use the solver on the calculator, but we will use logarithms for the sake of review.

$$2^t \ln 2 = 21$$

$$2^t = \frac{21}{\ln 2}$$

$$\ln 2^t = \ln\left(\frac{21}{\ln 2}\right) \qquad \text{Logarithm of both sides}$$

$$t \cdot \ln 2 = \ln 21 - \ln(\ln 2) \qquad \text{Properties of logarithms}$$

$$t = \frac{\ln 21 - \ln(\ln 2)}{\ln 2}$$

$$t \approx 4.921$$

$$y = 2^t - 3 \approx 27.297 \qquad \text{Using the stored value of } t$$

The point is approximately (4.9, 27.3).

Exploration 1 Leaving Milk on the Counter

A glass of cold milk from the refrigerator is left on the counter on a warm summer day. Its temperature y (in degrees Fahrenheit) after sitting on the counter t minutes is

$$y = 72 - 30(0.98)^t.$$

Answer the following questions by interpreting y and dy/dt.

1. What is the temperature of the refrigerator? How can you tell?
2. What is the temperature of the room? How can you tell?
3. When is the milk warming up the fastest? How can you tell?
4. Determine algebraically when the temperature of the milk reaches 55°F.
5. At what rate is the milk warming when its temperature is 55°F? Answer with an appropriate unit of measure.

Derivative of ln x

Now that we know the derivative of e^x, it is relatively easy to find the derivative of its inverse function, $\ln x$.

$$y = \ln x$$
$$e^y = x \qquad \text{Inverse function relationship}$$
$$\frac{d}{dx}(e^y) = \frac{d}{dx}(x) \qquad \text{Differentiate implicitly.}$$
$$e^y \frac{dy}{dx} = 1$$
$$\frac{dy}{dx} = \frac{1}{e^y} = \frac{1}{x}$$

If u is a differentiable function of x and $u > 0$,

$$\frac{d}{dx}\ln u = \frac{1}{u}\frac{du}{dx}.$$

This equation answers what was once a perplexing problem: Is there a function with derivative x^{-1}? All of the other power functions follow the Power Rule,

$$\frac{d}{dx}x^n = nx^{n-1}.$$

However, this formula is not much help if one is looking for a function with x^{-1} as its derivative! Now we know why: The function we should be looking for is not a power function at all; it is the natural logarithm function.

Example 3 A TANGENT THROUGH THE ORIGIN

A line with slope m passes through the origin and is tangent to the graph of $y = \ln x$. What is the value of m?

Solution This problem is a little harder than it looks, since we do not know the point of tangency. However, we do know two important facts about that point:

1. it has coordinates $(a, \ln a)$ for some positive a, and

2. the tangent line there has slope $m = 1/a$ (Figure 3.57).

Since the tangent line passes through the origin, its slope is

$$m = \frac{\ln a - 0}{a - 0} = \frac{\ln a}{a}.$$

Setting these two formulas for m equal to each other, we have

$$\frac{\ln a}{a} = \frac{1}{a}$$

$$\ln a = 1$$

$$e^{\ln a} = e^1$$

$$a = e$$

$$m = \frac{1}{e}.$$

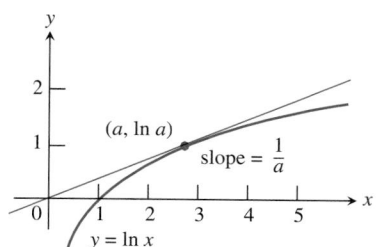

Figure 3.57 The tangent line intersects the curve at some point $(a, \ln a)$, where the slope of the curve is $1/a$. (Example 3)

Derivative of $\log_a x$

To find the derivative of $\log_a x$ for an arbitrary base ($a > 0$, $a \neq 1$), we use the change-of-base formula for logarithms to express $\log_a x$ in terms of natural logarithms, as follows:

$$\log_a x = \frac{\ln x}{\ln a}.$$

The rest is easy:

$$\frac{d}{dx}\log_a x = \frac{d}{dx}\left(\frac{\ln x}{\ln a}\right)$$

$$= \frac{1}{\ln a} \cdot \frac{d}{dx}\ln x \quad \text{Since ln } a \text{ is a constant}$$

$$= \frac{1}{\ln a} \cdot \frac{1}{x}$$

$$= \frac{1}{x \ln a}.$$

So, if u is a differentiable function of x and $u > 0$, the formula is as follows.

For $a > 0$ and $a \neq 1$,

$$\frac{d}{dx}\log_a u = \frac{1}{u \ln a}\frac{du}{dx}.$$

Example 4   GOING THE LONG WAY WITH THE CHAIN RULE

Find dy/dx if $y = \log_a a^{\sin x}$.

Solution Carefully working from the outside in, we apply the Chain Rule to get:

$$\frac{d}{dx}(\log_a a^{\sin x}) = \frac{1}{a^{\sin x}\ln a} \cdot \frac{d}{dx}(a^{\sin x}) \qquad \log_a u, \ u = a^{\sin x}$$

$$= \frac{1}{a^{\sin x}\ln a} \cdot a^{\sin x}\ln a \cdot \frac{d}{dx}(\sin x) \qquad a^u, \ u = \sin x$$

$$= \frac{a^{\sin x}\ln a}{a^{\sin x}\ln a} \cdot \cos x$$

$$= \cos x.$$

We could have saved ourselves a lot of work in Example 4 if we had noticed at the beginning that $\log_a a^{\sin x}$, being the composite of inverse functions, is equal to $\sin x$. It is always a good idea to simplify functions *before* differentiating, wherever possible. On the other hand, it is comforting to know that all these rules do work if applied correctly.

Power Rule for Arbitrary Real Powers

We are now ready to prove the Power Rule in its final form. As long as $x > 0$, we can write any real power of x as a power of e, specifically

$$x^n = e^{n \ln x}.$$

This enables us to differentiate x^n for any real power n, as follows:

$$\frac{d}{dx}(x^n) = \frac{d}{dx}(e^{n \ln x})$$

$$= e^{n \ln x} \cdot \frac{d}{dx}(n \ln x) \qquad e^u, \ u = n \ln x$$

$$= e^{n \ln x} \cdot \frac{n}{x}$$

$$= x^n \cdot \frac{n}{x}$$

$$= nx^{n-1}.$$

The Chain Rule extends this result to the Power Rule's final form.

Rule 10 Power Rule for Arbitrary Real Powers

If u is a positive differentiable function of x and n is any real number, then u^n is a differentiable function of x, and

$$\frac{d}{dx}u^n = nu^{n-1}\frac{du}{dx}.$$

Example 5 USING THE POWER RULE IN ALL ITS POWER

(a) If $y = x^{\sqrt{2}}$, then
$$\frac{dy}{dx} = \sqrt{2}\,x^{(\sqrt{2}-1)}.$$

(b) If $y = (2 + \sin 3x)^{\pi}$, then
$$\frac{d}{dx}(2 + \sin 3x)^{\pi} = \pi(2 + \sin 3x)^{\pi-1}(\cos 3x)\cdot 3$$
$$= 3\pi(2 + \sin 3x)^{\pi-1}(\cos 3x).$$

Example 6 FINDING DOMAIN

If $f(x) = \ln(x-3)$, find $f'(x)$. State the domain of f'.

Solution The domain of f is $(3, \infty)$ and
$$f'(x) = \frac{1}{x-3}.$$

The domain of f' appears to be all $x \neq 3$. However, since f is not defined for $x < 3$, neither is f'. Thus,
$$f'(x) = \frac{1}{x-3}, \qquad x > 3.$$

That is, the domain of f' is $(3, \infty)$.

Sometimes the properties of logarithms can be used to simplify the differentiation process, even if we must introduce the logarithms ourselves as a step in the process. Example 7 shows a clever way to differentiate $y = x^x$ for $x > 0$.

Example 7 LOGARITHMIC DIFFERENTIATION

Find dy/dx for $y = x^x$, $x > 0$.

Solution
$$y = x^x$$
$$\ln y = \ln x^x \qquad \text{Logs of both sides}$$
$$\ln y = x \ln x \qquad \text{Property of logs}$$
$$\frac{d}{dx}(\ln y) = \frac{d}{dx}(x \ln x) \qquad \text{Differentiate implicitly.}$$
$$\frac{1}{y}\frac{dy}{dx} = 1\cdot \ln x + x\cdot \frac{1}{x}$$
$$\frac{dy}{dx} = y(\ln x + 1)$$
$$\frac{dy}{dx} = x^x(\ln x + 1)$$

Quick Review 3.9

1. Write $\log_5 8$ in terms of natural logarithms.

2. Write 7^x as a power of e.

In Exercises 3–7, simplify the expression using properties of exponents and logarithms.

3. $\ln (e^{\tan x})$

4. $\ln (x^2 - 4) - \ln (x + 2)$

5. $\log_2 (8^{x-5})$

6. $(\log_4 x^{15})/(\log_4 x^{12})$

7. $3 \ln x - \ln 3x + \ln (12x^2)$

In Exercises 8–10, solve the equation algebraically using logarithms. Give an *exact* answer, such as $(\ln 2)/3$, and also an approximate answer to the nearest hundredth.

8. $3^x = 19$

9. $5^t \ln 5 = 18$

10. $3^{x+1} = 2^x$

Section 3.9 Exercises

In Exercises 1–40, find dy/dx. Remember that you can use NDER to support your computations.

1. $y = 2e^x$

2. $y = e^{2x}$

3. $y = e^{-x}$

4. $y = e^{-5x}$

5. $y = e^{2x/3}$

6. $y = e^{-x/4}$

7. $y = xe^2 - e^x$

8. $y = x^2 e^x - xe^x$

9. $y = e^{\sqrt{x}}$

10. $y = e^{(x^2)}$

11. $y = x^\pi$

12. $y = x^{1+\sqrt{2}}$

13. $y = x^{-\sqrt{2}}$

14. $y = x^{1-e}$

15. $y = 8^x$

16. $y = 9^{-x}$

17. $y = 3^{\csc x}$

18. $y = 3^{\cot x}$

19. $y = x^{\ln x}$

20. $y = x^{(1/\ln x)}$

21. $y = \ln (x^2)$

22. $y = (\ln x)^2$

23. $y = \ln (1/x)$

24. $y = \ln (10/x)$

25. $y = \ln (x + 2)$

26. $y = \ln (2x + 2)$

27. $y = \ln (2 - \cos x)$

28. $y = \ln (x^2 + 1)$

29. $y = \ln (\ln x)$

30. $y = x \ln x - x$

31. $y = \log_4 x^2$

32. $y = \log_5 \sqrt{x}$

33. $y = \log_2 (3x + 1)$

34. $y = \log_{10} \sqrt{x + 1}$

35. $y = \log_2 (1/x)$

36. $y = 1/\log_2 x$

37. $y = \ln 2 \cdot \log_2 x$

38. $y = \log_3 (1 + x \ln 3)$

39. $y = \log_{10} e^x$

40. $y = \ln 10^x$

41. Find an equation for a line that is tangent to the graph of $y = e^x$ and goes through the origin.

42. Find an equation for a line that is normal to the graph of $y = xe^x$ and goes through the origin.

In Exercises 43–46, use the technique of logarithmic differentiation to find dy/dx. *Work in groups of two or three.*

43. $y = (\sin x)^x, \quad 0 < x < \pi/2$

44. $y = x^{\tan x}, \quad x > 0$

45. $y = \sqrt[5]{\dfrac{(x - 3)^4(x^2 + 1)}{(2x + 5)^3}}$

46. $y = \dfrac{x\sqrt{x^2 + 1}}{(x + 1)^{2/3}}$

47. *Radioactive Decay* The amount A (in grams) of radioactive plutonium remaining in a 20-gram sample after t days is given by the formula

$$A = 20 \cdot (1/2)^{t/140}.$$

At what rate is the plutonium decaying when $t = 2$ days? Answer in appropriate units.

48. For any positive constant k, the derivative of $\ln (kx)$ is $1/x$. Prove this fact

(a) by using the Chain Rule.

(b) by using a property of logarithms and differentiating.

49. Let $f(x) = 2^x$.

(a) Find $f'(0)$.

(b) Use the definition of the derivative to write $f'(0)$ as a limit.

(c) Deduce the exact value of

$$\lim_{h \to 0} \frac{2^h - 1}{h}.$$

(d) What is the exact value of

$$\lim_{h \to 0} \frac{7^h - 1}{h}?$$

50. Writing to Learn The graph of $y = \ln x$ looks as though it might be approaching a horizontal asymptote. Write an argument based on the graph of $y = e^x$ to explain why it does not.

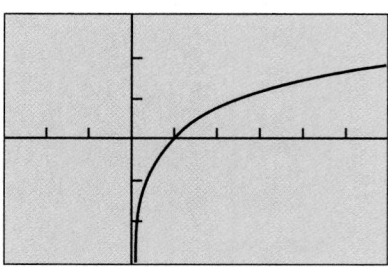

$[-3, 6]$ by $[-3, 3]$

Exploration

51. Let $y_1 = a^x$, $y_2 = \text{NDER } y_1$, $y_3 = y_2/y_1$, and $y_4 = e^{y_3}$.

(a) Describe the graph of y_4 for $a = 2, 3, 4, 5$. Generalize your description to an arbitrary $a > 1$.

(b) Describe the graph of y_3 for $a = 2, 3, 4, 5$. Compare a table of values for y_3 for $a = 2, 3, 4, 5$ with $\ln a$. Generalize your description to an arbitrary $a > 1$.

(c) Explain how parts (a) and (b) support the statement

$$\frac{d}{dx}\, a^x = a^x \quad \text{if and only if} \quad a = e.$$

(d) Show algebraically that $y_1 = y_2$ if and only if $a = e$. ∎

Extending the Ideas

52. *Orthogonal Families of Curves* Prove that all curves in the family

$$y = -\frac{1}{2}x^2 + k$$

(k any constant) are perpendicular to all curves in the family $y = \ln x + c$ (c any constant) at their points of intersection. (See figure below.)

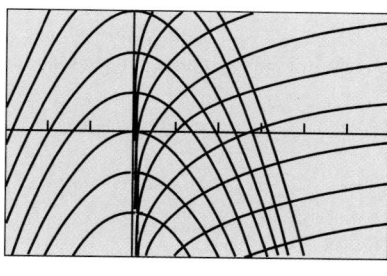

[–3, 6] by [–3, 3]

53. *Which is Bigger, π^e or e^π?* Calculators have taken some of the mystery out of this once-challenging question. (Go ahead and check; you will see that it is a surprisingly close call.) You can answer the question without a calculator, though, by using the result from Example 3 of this section.

Recall from that example that the line through the origin tangent to the graph of $y = \ln x$ has slope $1/e$.

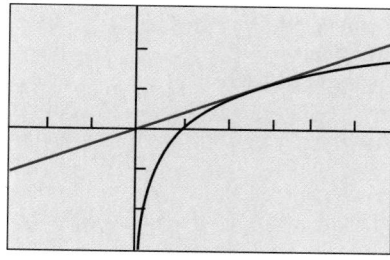

[–3, 6] by [–3, 3]

(a) Find an equation for this tangent line.

(b) Give an argument based on the graphs of $y = \ln x$ and the tangent line to explain why $\ln x < x/e$ for all positive $x \neq e$.

(c) Show that $\ln (x^e) < x$ for all positive $x \neq e$.

(d) Conclude that $x^e < e^x$ for all positive $x \neq e$.

(e) So which is bigger, π^e or e^π?

Calculus at Work

I work at Ramsey County Hospital and other community hospitals in the Minneapolis area, both with patients and in a laboratory. I have wanted to be a physician since I was about 12 years old, and I began attending medical school when I was 30 years old. I am now working in the field of internal medicine.

Cardiac patients are common in my field, especially in the diagnostic stages. One of the machines that is sometimes used in the emergency room to diagnose problems is called a Swan-Ganz catheter, named after its inventors Harold James Swan and William Ganz. The catheter is inserted into the pulmonary artery and then is hooked up to a cardiac monitor. A program measures cardiac output by looking at changes of slope in the curve. This information alerts me to left-sided heart failure.

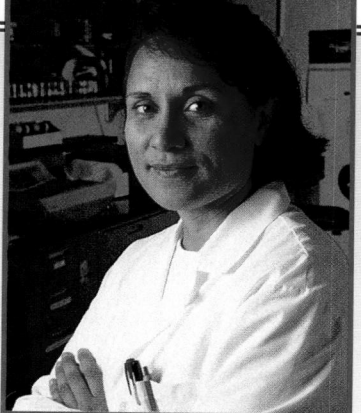

Lupe Bolding, M.D.
Ramsey County Hospital
Minneapolis, MN

Chapter 3 Key Terms

acceleration (p. 125)
average velocity (p. 123)
Chain Rule (p. 142)
Constant Multiple Rule (p. 113)
Derivative of a Constant Function (p. 112)
derivative of f at a (p. 95)
differentiable function (p. 95)
differentiable on a closed interval
　(p. 100)
displacement (p. 123)
free-fall constants (p. 125)
implicit differentiation (p. 149)
instantaneous rate of change (p. 122)
instantaneous velocity (p. 124)
Intermediate Value Theorem for
　Derivatives (p. 110)

inverse function–inverse
　cofunction identities (p. 161)
jerk (p. 137)
left-hand derivative (p. 100)
local linearity (p. 107)
logarithmic differentiation (p. 169)
marginal cost (p. 128)
marginal revenue (p. 129)
nth derivative (p. 119)
normal to the surface (p. 151)
numerical derivative (NDER) (p. 108)
orthogonal curves (p. 148)
orthogonal families (p. 171)
Power Chain Rule (p. 145)
Power Rule for Arbitrary Real
　Powers (p. 168)

Power Rule for Negative Integer
　Powers of x (p. 118)
Power Rule for Positive Integer
　Powers of x (p. 113)
Power Rule for Rational Powers
　of x (p. 153)
Product Rule (p. 115)
Quotient Rule (p. 117)
right-hand derivative (p. 100)
sensitivity to change (p. 128)
simple harmonic motion (p. 136)
speed (p. 124)
Sum and Difference Rule (p. 114)
symmetric difference quotient (p. 108)
velocity (p. 124)

Chapter 3 Review Exercises

In Exercises 1–30, find the derivative of the function.

1. $y = x^5 - \dfrac{1}{8}x^2 + \dfrac{1}{4}x$

2. $y = 3 - 7x^3 + 3x^7$

3. $y = 2 \sin x \cos x$

4. $y = \dfrac{2x + 1}{2x - 1}$

5. $s = \cos (1 - 2t)$

6. $s = \cot \dfrac{2}{t}$

7. $y = \sqrt{x} + 1 + \dfrac{1}{\sqrt{x}}$

8. $y = x\sqrt{2x + 1}$

9. $r = \sec (1 + 3\theta)$

10. $r = \tan^2 (3 - \theta^2)$

11. $y = x^2 \csc 5x$

12. $y = \ln \sqrt{x}$

13. $y = \ln (1 + e^x)$

14. $y = xe^{-x}$

15. $y = e^{(1 + \ln x)}$

16. $y = \ln (\sin x)$

17. $r = \ln (\cos^{-1} x)$

18. $r = \log_2 (\theta^2)$

19. $s = \log_5 (t - 7)$

20. $s = 8^{-t}$

21. $y = x^{\ln x}$

22. $y = \dfrac{(2x)2^x}{\sqrt{x^2 + 1}}$

23. $y = e^{\tan^{-1} x}$

24. $y = \sin^{-1}\sqrt{1 - u^2}$

25. $y = t \sec^{-1} t - \dfrac{1}{2}\ln t$

26. $y = (1 + t^2) \cot^{-1} 2t$

27. $y = z \cos^{-1} z - \sqrt{1 - z^2}$

28. $y = 2\sqrt{x - 1} \, \csc^{-1}\sqrt{x}$

29. $y = \csc^{-1} (\sec x), \ 0 \le x \le 2\pi$

30. $r = \left(\dfrac{1 + \sin \theta}{1 - \cos \theta} \right)^2$

In Exercises 31–34, find all values of x for which the function is differentiable.

31. $y = \ln x^2$

32. $y = \sin x - x \cos x$

33. $y = \sqrt{\dfrac{1 - x}{1 + x^2}}$

34. $y = (2x - 7)^{-1}(x + 5)$

In Exercises 35–38, find dy/dx.

35. $xy + 2x + 3y = 1$

36. $5x^{4/5} + 10y^{6/5} = 15$

37. $\sqrt{xy} = 1$

38. $y^2 = \dfrac{x}{x + 1}$

In Exercises 39–42, find d^2y/dx^2 by implicit differentiation.

39. $x^3 + y^3 = 1$

40. $y^2 = 1 - \dfrac{2}{x}$

41. $y^3 + y = 2 \cos x$

42. $x^{1/3} + y^{1/3} = 4$

In Exercises 43 and 44, find all derivatives of the function.

43. $y = \dfrac{x^4}{2} - \dfrac{3}{2}x^2 - x$

44. $y = \dfrac{x^5}{120}$

In Exercises 45–48, find an equation for the **(a)** tangent and **(b)** normal to the curve at the indicated point.

45. $y = \sqrt{x^2 - 2x}, \quad x = 3$

46. $y = 4 + \cot x - 2 \csc x, \quad x = \pi/2$

47. $x^2 + 2y^2 = 9, \quad (1, 2)$　　**48.** $x + \sqrt{xy} = 6, \quad (4, 1)$

In Exercises 49–52, find an equation for the line tangent to the curve at the point defined by the given value of t.

49. $x = 2 \sin t, \quad y = 2 \cos t, \quad t = 3\pi/4$

50. $x = 3 \cos t, \quad y = 4 \sin t, \quad t = 3\pi/4$

51. $x = 3 \sec t, \quad y = 5 \tan t, \quad t = \pi/6$

52. $x = \cos t, \quad y = t + \sin t, \quad t = -\pi/4$

53. Writing to Learn

(a) Graph the function

$$f(x) = \begin{cases} x, & 0 \le x \le 1 \\ 2 - x, & 1 < x \le 2. \end{cases}$$

(b) Is f continuous at $x = 1$? Explain.

(c) Is f differentiable at $x = 1$? Explain.

54. Writing to Learn For what values of the constant m is

$$f(x) = \begin{cases} \sin 2x, & x \le 0 \\ mx, & x > 0 \end{cases}$$

(a) continuous at $x = 0$? Explain.

(b) differentiable at $x = 0$? Explain.

In Exercises 55–58, determine where the function is
(a) differentiable, (b) continuous but not differentiable, and
(c) neither continuous nor differentiable.

55. $f(x) = x^{4/5}$ **56.** $g(x) = \sin(x^2 + 1)$

57. $f(x) = \begin{cases} 2x - 3, & -1 \le x < 0 \\ x - 3, & 0 \le x \le 4 \end{cases}$

58. $g(x) = \begin{cases} \dfrac{x - 1}{x}, & -2 \le x < 0 \\[2mm] \dfrac{x + 1}{x}, & 0 \le x \le 2 \end{cases}$

In Exercises 59 and 60, use the graph of f to sketch the graph of f'.

59. Sketching f' from f

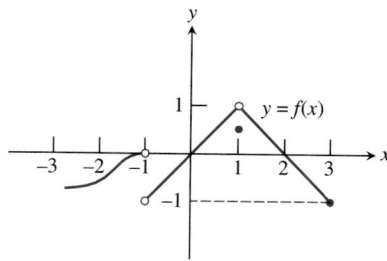

60. Sketching f' from f

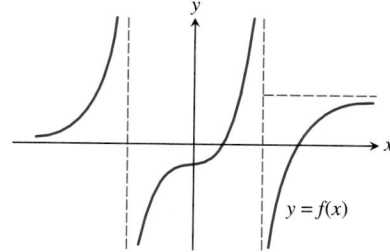

61. Recognizing Graphs The following graphs show the distance traveled, velocity, and acceleration for each second of a 2-minute automobile trip. Which graph shows

(a) distance?   (b) velocity?   (c) acceleration?

(i)

(ii)

(iii)
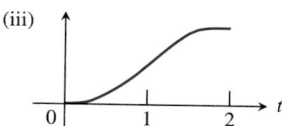

62. Sketching f from f' Sketch the graph of a continuous function f with $f(0) = 5$ and

$$f'(x) = \begin{cases} -2, & x < 2 \\ -0.5, & x > 2. \end{cases}$$

63. Sketching f from f' Sketch the graph of a continuous function f with $f(-1) = 2$ and

$$f'(x) = \begin{cases} -2, & x < 1 \\ 1, & 1 < x < 4 \\ -1, & 4 < x < 6. \end{cases}$$

64. Which of the following statements could be true if $f''(x) = x^{1/3}$?

i. $f(x) = \dfrac{9}{28} x^{7/3} + 9$ **ii.** $f'(x) = \dfrac{9}{28} x^{7/3} - 2$

iii. $f'(x) = \dfrac{3}{4} x^{4/3} + 6$ **iv.** $f(x) = \dfrac{3}{4} x^{4/3} - 4$

A. i only   **B.** iii only

C. ii and iv only   **D.** i and iii only

65. Derivative from Data The following data give the coordinates of a moving body for various values of t.

t (sec)	0	0.5	1	1.5	2	2.5	3	3.5	4
s (ft)	10	38	58	70	74	70	58	38	10

(a) Make a scatter plot of the (t, s) data and sketch a smooth curve through the points.

(b) Compute the average velocity between consecutive points of the table.

(c) Make a scatter plot of the data in (b) using the midpoints of the t values to represent the data. Then sketch a smooth curve through the points.

(d) **Writing to Learn** Why does the curve in (c) approximate the graph of ds/dt?

66. *Working with Numerical Values* Suppose that a function f and its first derivative have the following values at $x = 0$ and $x = 1$.

x	$f(x)$	$f'(x)$
0	9	-2
1	-3	1/5

Find the first derivative of the following combinations at the given value of x.

(a) $\sqrt{x}\, f(x)$, $x = 1$

(b) $\sqrt{f(x)}$, $x = 0$

(c) $f(\sqrt{x})$, $x = 1$

(d) $f(1 - 5 \tan x)$, $x = 0$

(e) $\dfrac{f(x)}{2 + \cos x}$, $x = 0$

(f) $10 \sin\left(\dfrac{\pi x}{2}\right) f^2(x)$, $x = 1$

67. *Working with Numerical Values* Suppose that functions f and g and their first derivatives have the following values at $x = -1$ and $x = 0$.

x	$f(x)$	$g(x)$	$f'(x)$	$g'(x)$
-1	0	-1	2	1
0	-1	-3	-2	4

Find the first derivative of the following combinations at the given value of x.

(a) $3f(x) - g(x)$, $x = -1$

(b) $f^2(x)g^3(x)$, $x = 0$

(c) $g(f(x))$, $x = -1$

(d) $f(g(x))$, $x = -1$

(e) $\dfrac{f(x)}{g(x) + 2}$, $x = 0$

(f) $g(x + f(x))$, $x = 0$

68. Find the value of dw/ds at $s = 0$ if $w = \sin(\sqrt{r} - 2)$ and $r = 8 \sin(s + \pi/6)$.

69. Find the value of dr/dt at $t = 0$ if $r = (\theta^2 + 7)^{1/3}$ and $\theta^2 t + \theta = 1$.

70. *Particle Motion* The position at time $t \geq 0$ of a particle moving along the s-axis is

$$s(t) = 10 \cos(t + \pi/4).$$

(a) Give parametric equations that can be used to simulate the motion of the particle.

(b) What is the particle's initial position $(t = 0)$?

(c) What points reached by the particle are farthest to the left and right of the origin?

(d) When does the particle first reach the origin? What are its velocity, speed, and acceleration then?

71. *Vertical Motion* On Earth, if you shoot a paper clip 64 ft straight up into the air with a rubber band, the paper clip will be $s(t) = 64t - 16t^2$ feet above your hand at t sec after firing.

(a) Find ds/dt and d^2s/dt^2.

(b) How long does it take the paper clip to reach its maximum height?

(c) With what velocity does it leave your hand?

(d) On the moon, the same force will send the paper clip to a height of $s(t) = 64t - 2.6t^2$ ft in t sec. About how long will it take the paper clip to reach its maximum height, and how high will it go?

72. *Free Fall* Suppose two balls are falling from rest at a certain height in centimeters above the ground. Use the equation $s = 490t^2$ to answer the following questions.

(a) How long does it take the balls to fall the first 160 cm? What is their average velocity for the period?

(b) How fast are the balls falling when they reach the 160-cm mark? What is their acceleration then?

73. *Filling a Bowl* If a hemispherical bowl of radius 10 in. is filled with water to a depth of x in., the volume of water is given by $V = \pi[10 - (x/3)]x^2$. Find the rate of increase of the volume per inch increase of depth.

74. *Marginal Revenue* A bus will hold 60 people. The fare charged (p dollars) is related to the number x of people who use the bus by the formula $p = [3 - (x/40)]^2$.

(a) Write a formula for the total revenue per trip received by the bus company.

(b) What number of people per trip will make the marginal revenue equal to zero? What is the corresponding fare?

(c) *Writing to Learn* Do you think the bus company's fare policy is good for its business?

75. *Searchlight* The figure shows a boat 1 km offshore sweeping the shore with a searchlight. The light turns at a constant rate, $d\theta/dt = -0.6$ rad/sec.

(a) How fast is the light moving along the shore when it reaches point A?

(b) How many revolutions per minute is 0.6 rad/sec?

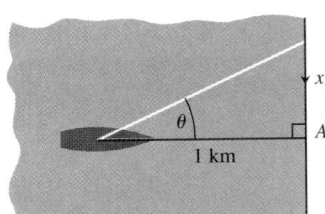

76. *Horizontal Tangents* The graph of $y = \sin(x - \sin x)$ appears to have horizontal tangents at the x-axis. Does it?

77. *Fundamental Frequency of a Vibrating Piano String* We measure the frequencies at which wires vibrate in cycles (trips back and forth) per sec. The unit of measure is a *hertz:* 1 cycle per sec. Middle A on a piano has a frequency 440 hertz. For any given wire, the fundamental frequency y is a function of four variables:

$r:$ the radius of the wire;

$l:$ the length;

$d:$ the density of the wire;

$T:$ the tension (force) holding the wire taut.

With r and l in centimeters, d in grams per cubic centimeter, and T in dynes (it takes about 100,000 dynes to lift an apple), the fundamental frequency of the wire is

$$y = \frac{1}{2rl}\sqrt{\frac{T}{\pi d}}.$$

If we keep all the variables fixed except one, then y can be alternatively thought of as four different functions of one variable, $y(r)$, $y(l)$, $y(d)$, and $y(T)$. How would changing each variable affect the string's fundamental frequency? To find out, calculate $y'(r)$, $y'(l)$, $y'(d)$, and $y'(T)$.

78. *Spread of Measles* The spread of measles in a certain school is given by

$$P(t) = \frac{200}{1 + e^{5-t}},$$

where t is the number of days since the measles first appeared, and $P(t)$ is the total number of students who have caught the measles to date.

(a) Estimate the initial number of students infected with measles.

(b) About how many students in all will get the measles?

(c) When will the rate of spread of measles be greatest? What is this rate?

79. Graph the function $f(x) = \tan^{-1}(\tan 2x)$ in the window $[-\pi, \pi]$ by $[-4, 4]$. Then answer the following questions.

(a) What is the domain of f?

(b) What is the range of f?

(c) At which points is f not differentiable?

(d) Describe the graph of f'.

80. If $x^2 - y^2 = 1$, find d^2y/dx^2 at the point $(2, \sqrt{3})$.

Applications of Derivatives

An automobile's gas mileage is a function of many variables, including road surface, tire type, velocity, fuel octane rating, road angle, and the speed and direction of the wind. If we look only at velocity's effect on gas mileage, the mileage of a certain car can be approximated by:

$$m(v) = 0.00015v^3 - 0.032v^2 + 1.8v + 1.7$$

(where v is velocity)

At what speed should you drive this car to obtain the best gas mileage? The ideas in Section 4.1 will help you find the answer.

Chapter 4 Overview

In the past, when virtually all graphing was done by hand—often laboriously—derivatives were the key tool used to sketch the graph of a function. Now we can graph a function quickly, and usually correctly, using a grapher. However, confirmation of much of what we see and conclude true from a grapher view must still come from calculus.

This chapter shows how to draw conclusions from derivatives about the extreme values of a function and about the general shape of a function's graph. We will also see how a tangent line captures the shape of a curve near the point of tangency, how to deduce rates of change we cannot measure from rates of change we already know, and how to find a function when we know only its first derivative and its value at a single point. The key to recovering functions from derivatives is the Mean Value Theorem, a theorem whose corollaries provide the gateway to *integral calculus,* which we begin in Chapter 5.

4.1 Extreme Values of Functions

Absolute (Global) Extreme Values • Local (Relative) Extreme Values • Finding Extreme Values

Absolute (Global) Extreme Values

One of the most useful things we can learn from a function's derivative is whether the function assumes any maximum or minimum values on a given interval and where these values are located if it does. Once we know how to find a function's extreme values, we will be able to answer such questions as "What is the most effective size for a dose of medicine?" and "What is the least expensive way to pipe oil from an offshore well to a refinery down the coast?" We will see how to answer questions like these in Section 4.4.

Definition Absolute Extreme Values

Let f be a function with domain D. Then $f(c)$ is the

(a) absolute maximum value on D if and only if $f(x) \leq f(c)$ for all x in D.

(b) absolute minimum value on D if and only if $f(x) \geq f(c)$ for all x in D.

Absolute (or **global**) maximum and minimum values are also called **absolute extrema** (plural of the Latin *extremum*). We often omit the term "absolute" or "global" and just say maximum and minimum.

Example 1 shows that extreme values can occur at interior points or endpoints of intervals.

Example 1 EXPLORING EXTREME VALUES

On $[-\pi/2, \pi/2]$, $f(x) = \cos x$ takes on a maximum value of 1 (once) and a minimum value of 0 (twice). The function $g(x) = \sin x$ takes on a maximum value of 1 and a minimum value of -1 (Figure 4.1).

Figure 4.1 (Example 1)

(a) abs min only

(b) abs max and min

(c) abs max only

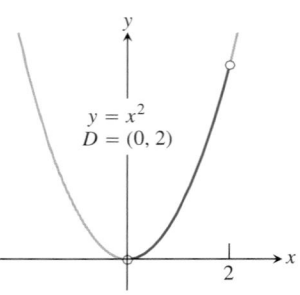

(d) no abs max or min

Figure 4.2 (Example 2)

Functions with the same defining rule can have different extrema, depending on the domain.

Example 2 EXPLORING ABSOLUTE EXTREMA

The absolute extrema of the following functions on their domains can be seen in Figure 4.2.

	Function Rule	Domain D	Absolute Extrema on D
(a)	$y = x^2$	$(-\infty, \infty)$	No absolute maximum. Absolute minimum of 0 at $x = 0$.
(b)	$y = x^2$	$[0, 2]$	Absolute maximum of 4 at $x = 2$. Absolute minimum of 0 at $x = 0$.
(c)	$y = x^2$	$(0, 2]$	Absolute maximum of 4 at $x = 2$. No absolute minimum.
(d)	$y = x^2$	$(0, 2)$	No absolute extrema.

Example 2 shows that a function may fail to have a maximum or minimum value. This cannot happen with a continuous function on a finite closed interval.

Theorem 1 The Extreme Value Theorem

If f is continuous on a closed interval $[a, b]$, then f has both a maximum value and a minimum value on the interval. (Figure 4.3)

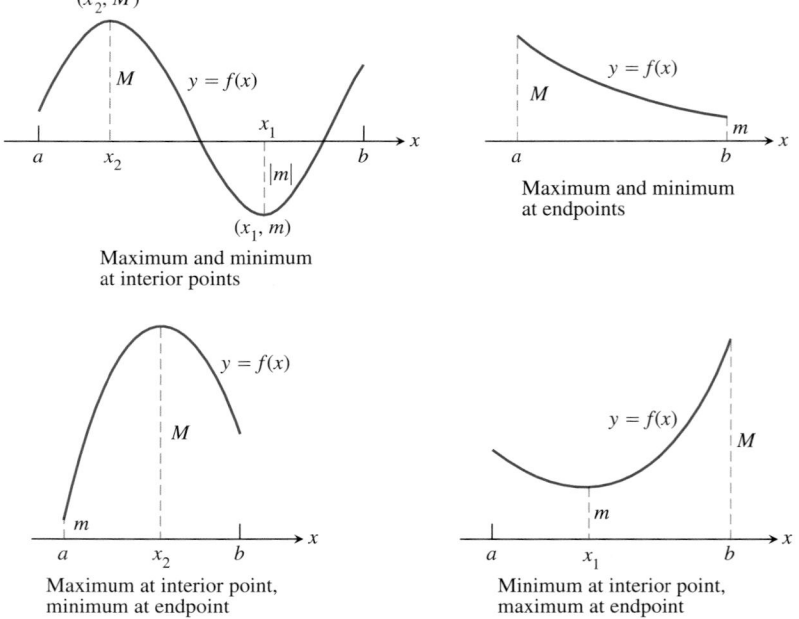

Figure 4.3 Some possibilities for a continuous function's maximum and minimum on a closed interval $[a, b]$.

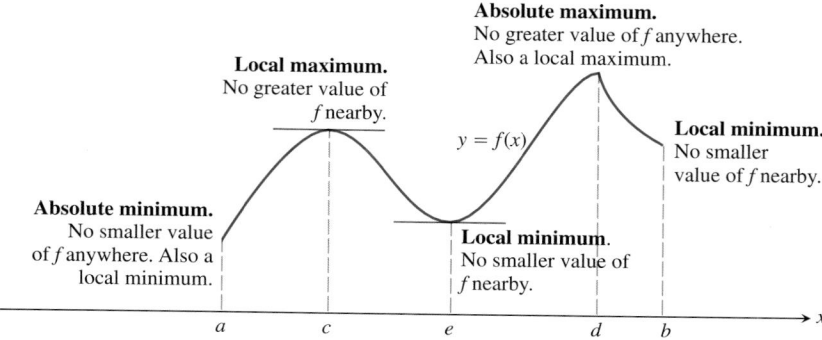

Figure 4.4 Classifying extreme values.

Local (Relative) Extreme Values

Figure 4.4 shows a graph with five points where a function has extreme values on its domain $[a, b]$. The function's absolute minimum occurs at a even though at e the function's value is smaller than at any other point *nearby*. The curve rises to the left and falls to the right around c, making $f(c)$ a maximum locally. The function attains its absolute maximum at d.

Definition   Local Extreme Values

Let c be an interior point of the domain of the function f. Then $f(c)$ is a

(a) local maximum value at c if and only if $f(x) \leq f(c)$ for all x in some open interval containing c.

(b) local minimum value at c if and only if $f(x) \geq f(c)$ for all x in some open interval containing c.

Local extrema are also called **relative extrema.** We can extend the definitions of local extrema to endpoints of intervals. A function f has a local maximum or local minimum *at an endpoint c* if the appropriate inequality holds for all x in some half-open domain interval containing c.

An absolute extremum is also a local extremum, because being an extreme value overall makes it an extreme value in its immediate neighborhood. Hence, *a list of local extrema will automatically include absolute extrema if there are any.*

Finding Extreme Values

The interior domain points where the function in Figure 4.4 has local extreme values are points where either f' is zero or f' does not exist. This is generally the case, as we see from the following theorem.

Theorem 2   Local Extreme Values

If a function f has a local maximum value or a local minimum value at an interior point c of its domain, and if f' exists at c, then

$$f'(c) = 0.$$

Because of Theorem 2, we usually need to look at only a few points to find a function's extrema. These consist of the interior domain points where $f' = 0$ or f' does not exist (the domain points covered by the theorem) and the domain endpoints (the domain points not covered by the theorem). At all other domain points, $f' > 0$ or $f' < 0$.

The following definition helps us summarize these findings.

Definition **Critical Point**

A point in the interior of the domain of a function f at which $f' = 0$ or f' does not exist is a **critical point** of f.

Thus, in summary, extreme values occur only at critical points and endpoints.

Example 3   FINDING ABSOLUTE EXTREMA

Find the absolute maximum and minimum values of $f(x) = x^{2/3}$ on the interval $[-2, 3]$.

Solution

Solve Graphically

Figure 4.5 suggests that f has an absolute maximum value of about 2 at $x = 3$ and an absolute minimum value of 0 at $x = 0$.

Confirm Analytically

We evaluate the function at the critical points and endpoints and take the largest and smallest of the resulting values.

The first derivative

$$f'(x) = \frac{2}{3}x^{-1/3} = \frac{2}{3\sqrt[3]{x}}$$

has no zeros but is undefined at $x = 0$. The values of f at this one critical point and at the endpoints are

Critical point value: $f(0) = 0;$

Endpoint values: $f(-2) = (-2)^{2/3} = \sqrt[3]{4};$

$f(3) = (3)^{2/3} = \sqrt[3]{9}.$

We can see from this list that the function's absolute maximum value is $\sqrt[3]{9} \approx 2.08$, and occurs at the right endpoint $x = 3$. The absolute minimum value is 0, and occurs at the interior point $x = 0$.

In Example 4, we investigate the function whose graph was drawn in Example 3 of Section 1.2.

Example 4   FINDING EXTREME VALUES

Find the extreme values of $f(x) = \dfrac{1}{\sqrt{4 - x^2}}$.

$y = x^{2/3}$

$[-2, 3]$ by $[-1, 2.5]$

Figure 4.5 (Example 3)

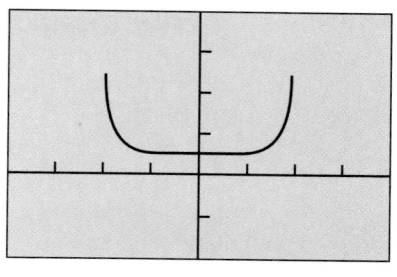

[–4, 4] by [–2, 4]

Figure 4.6 The graph of

$$f(x) = \frac{1}{\sqrt{4 - x^2}}.$$

(Example 4)

Solution

Solve Graphically

Figure 4.6 suggests that f has an absolute minimum of about 0.5 at $x = 0$. There also appear to be local maxima at $x = -2$ and $x = 2$. However, f is not defined at these points and there do not appear to be maxima anywhere else.

Confirm Analytically

The function f is defined only for $4 - x^2 > 0$, so its domain is the open interval $(-2, 2)$. The domain has no endpoints, so all the extreme values must occur at critical points. We rewrite the formula for f to find f':

$$f(x) = \frac{1}{\sqrt{4 - x^2}} = (4 - x^2)^{-1/2}.$$

Thus,

$$f'(x) = -\frac{1}{2}(4 - x^2)^{-3/2}(-2x) = \frac{x}{(4 - x^2)^{3/2}}.$$

The only critical point in the domain $(-2, 2)$ is $x = 0$. The value

$$f(0) = -\frac{1}{\sqrt{4 - 0^2}} = \frac{1}{2}$$

is therefore the sole candidate for an extreme value.

To determine whether $1/2$ is an extreme value of f, we examine the formula

$$f(x) = \frac{1}{\sqrt{4 - x^2}}.$$

As x moves away from 0 on either side, the denominator gets smaller, the values of f increase, and the graph rises. We have a minimum value at $x = 0$, and the minimum is absolute.

The function has no maxima, either local or absolute. This does not violate Theorem 1 (The Extreme Value Theorem) because here f is defined on an *open* interval. To invoke Theorem 1's guarantee of extreme points, the interval must be closed.

While a function's extrema can occur only at critical points and endpoints, not every critical point or endpoint signals the presence of an extreme value. Figure 4.7 illustrates this for interior points. Exercise 53 describes a function that fails to assume an extreme value at an endpoint of its domain.

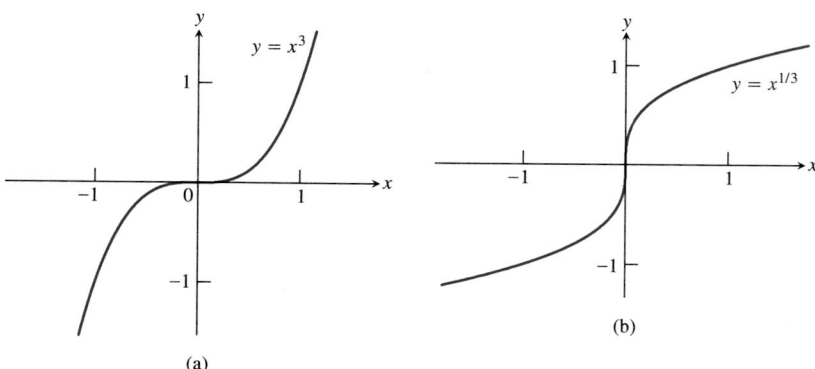

Figure 4.7 Critical points without extreme values. (a) $y' = 3x^2$ is 0 at $x = 0$, but $y = x^3$ has no extremum there. (b) $y' = (1/3)x^{-2/3}$ is undefined at $x = 0$, but $y = x^{1/3}$ has no extremum there.

Example 5 FINDING EXTREME VALUES

Find the extreme values of

$$f(x) = \begin{cases} 5 - 2x^2, & x \leq 1 \\ x + 2, & x > 1. \end{cases}$$

Solution

Solve Graphically

The graph in Figure 4.8 suggests that $f'(0) = 0$ and that $f'(1)$ does not exist. There appears to be a local maximum value of 5 at $x = 0$ and a local minimum value of 3 at $x = 1$.

Confirm Analytically

For $x \neq 1$, the derivative is

$$f'(x) = \begin{cases} \dfrac{d}{dx}(5 - 2x^2) = -4x, & x < 1 \\ \dfrac{d}{dx}(x + 2) = 1, & x > 1. \end{cases}$$

The only point where $f' = 0$ is $x = 0$. What happens at $x = 1$?

At $x = 1$, the right- and left-hand derivatives are respectively

$$\lim_{h \to 0^+} \frac{f(1 + h) - f(1)}{h} = \lim_{h \to 0^+} \frac{(1 + h) + 2 - 3}{h} = \lim_{h \to 0^+} \frac{h}{h} = 1,$$

$$\lim_{h \to 0^-} \frac{f(1 + h) - f(1)}{h} = \lim_{h \to 0^-} \frac{5 - 2(1 + h)^2 - 3}{h}$$

$$= \lim_{h \to 0^-} \frac{-2h(2 + h)}{h} = -4.$$

Since these one-sided derivatives differ, f has no derivative at $x = 1$, and 1 is a second critical point of f.

The domain $(-\infty, \infty)$ has no endpoints, so the only values of f that might be local extrema are those at the critical points:

$$f(0) = 5 \quad \text{and} \quad f(1) = 3.$$

From the formula for f, we see that the values of f immediately to either side of $x = 0$ are less than 5, so 5 is a local maximum. Similarly, the values of f immediately to either side of $x = 1$ are greater than 3, so 3 is a local minimum.

Most graphing calculators have built-in methods to find the coordinates of points where extreme values occur. We must, of course, be sure that we use correct graphs to find these values. The calculus that you learn in this chapter should make you feel more confident about working with graphs.

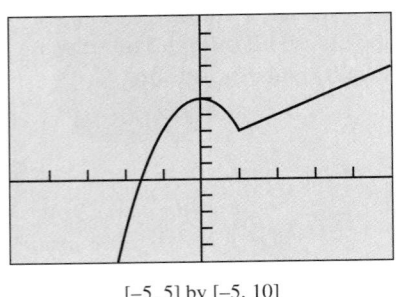

[−5, 5] by [−5, 10]

Figure 4.8 The function in Example 5.

Example 6 USING GRAPHICAL METHODS

Find the extreme values of $f(x) = \ln \left| \dfrac{x}{1 + x^2} \right|$.

Solution

Solve Graphically

The domain of f is the set of all nonzero real numbers. Figure 4.9 suggests that f is an even function with a maximum value at two points. The coordinates found in this window suggest an extreme value of about -0.69 at approximately $x = 1$. Because f is even, there is another extreme of the same value at approximately $x = -1$. The figure also suggests a minimum value at $x = 0$, but f is not defined there.

Confirm Analytically

The derivative

$$f'(x) = \frac{1 - x^2}{x(1 + x^2)}$$

is defined at every point of the function's domain. The critical points where $f'(x) = 0$ are $x = 1$ and $x = -1$. The corresponding values of f are both $\ln (1/2) = -\ln 2 \approx -0.69$.

Maximum
X = .9999988   Y = -.6931472

[-4.5, 4.5] by [-4, 2]

Figure 4.9 The function in Example 6.

Exploration 1 **Finding Extreme Values**

Let $f(x) = \left| \dfrac{x}{x^2 + 1} \right|$, $-2 \le x \le 2$.

1. Determine graphically the extreme values of f and where they occur. Find f' at these values of x.
2. Graph f and f' (or NDER $(f(x), x, x)$) in the same viewing window. Comment on the relationship between the graphs.
3. Find a formula for $f'(x)$.

Quick Review 4.1

In Exercises 1–8, find the first derivative of the function.

1. $f(x) = \sqrt{4 - x}$

2. $f(x) = x^{3/4}$

3. $f(x) = \dfrac{2}{\sqrt{9 - x^2}}$

4. $f(x) = \dfrac{1}{\sqrt[3]{x^2 - 1}}$

5. $g(x) = \ln (x^2 + 1)$

6. $g(x) = \cos (\ln x)$

7. $h(x) = e^{2x}$

8. $h(x) = e^{\ln x}$

In Exercises 9 and 10, find the limit for

$$f(x) = \frac{2}{\sqrt{9 - x^2}}.$$

9. $\lim\limits_{x \to 3^-} f(x)$

10. $\lim\limits_{x \to -3^+} f(x)$

In Exercises 11 and 12, let

$$f(x) = \begin{cases} x^3 - 2x, & x \le 2 \\ x + 2, & x > 2. \end{cases}$$

11. Find **(a)** $f'(1)$, **(b)** $f'(3)$, **(c)** $f'(2)$.

12. (a) Find the domain of f'.

  (b) Write a formula for $f'(x)$.

Section 4.1 Exercises

In Exercises 1–6, identify each x-value at which any absolute extreme value occurs. Explain how your answer is consistent with the Extreme Value Theorem.

1.

2.

3.

4.

5.

6.
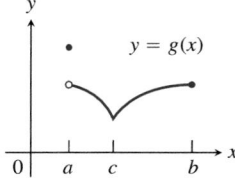

In Exercises 7–10, find the extreme values and where they occur.

7.

8.

9.

10.
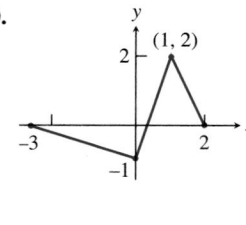

In Exercises 11–18, use analytic methods to find the extreme values of the function on the interval and where they occur.

11. $f(x) = \dfrac{1}{x} + \ln x, \quad 0.5 \le x \le 4$

12. $g(x) = e^{-x}, \quad -1 \le x \le 1$

13. $h(x) = \ln(x + 1), \quad 0 \le x \le 3$

14. $k(x) = e^{-x^2}, \quad -\infty < x < \infty$

15. $f(x) = \sin\left(x + \dfrac{\pi}{4}\right), \quad 0 \le x \le \dfrac{7\pi}{4}$

16. $g(x) = \sec x, \quad -\dfrac{\pi}{2} < x < \dfrac{3\pi}{2}$

17. $f(x) = x^{2/5}, \quad -3 \le x < 1$

18. $f(x) = x^{3/5}, \quad -2 < x \le 3$

In Exercises 19–30, find the extreme values of the function and where they occur.

19. $y = 2x^2 - 8x + 9$

20. $y = x^3 - 2x + 4$

21. $y = x^3 + x^2 - 8x + 5$

22. $y = x^3 - 3x^2 + 3x - 2$

23. $y = \sqrt{x^2 - 1}$

24. $y = \dfrac{1}{x^2 - 1}$

25. $y = \dfrac{1}{\sqrt{1 - x^2}}$

26. $y = \dfrac{1}{\sqrt[3]{1 - x^2}}$

27. $y = \sqrt{3 + 2x - x^2}$

28. $y = \dfrac{3}{2}x^4 + 4x^3 - 9x^2 + 10$

29. $y = \dfrac{x}{x^2 + 1}$

30. $y = \dfrac{x + 1}{x^2 + 2x + 2}$

In Exercises 31–34, *work in groups of two or three* to find the extreme values of the function on the interval and where they occur.

31. $f(x) = |x - 2| + |x + 3|, \quad -5 \le x \le 5$

32. $g(x) = |x - 1| - |x - 5|, \quad -2 \le x \le 7$

33. $h(x) = |x + 2| - |x - 3|, \quad -\infty < x < \infty$

34. $k(x) = |x + 1| + |x - 3|, \quad -\infty < x < \infty$

Explorations

In Exercises 35 and 36, give reasons for your answers.

35. Writing to Learn Let $f(x) = (x - 2)^{2/3}$.

(a) Does $f'(2)$ exist?

(b) Show that the only local extreme value of f occurs at $x = 2$.

(c) Does the result in (b) contradict the Extreme Value Theorem?

(d) Repeat (a) and (b) for $f(x) = (x - a)^{2/3}$, replacing 2 by a.

36. Writing to Learn Let $f(x) = |x^3 - 9x|$.

(a) Does $f'(0)$ exist?

(b) Does $f'(3)$ exist?

(c) Does $f'(-3)$ exist?

(d) Determine all extrema of f.   ∎

In Exercises 37–44, find the derivative at each critical point and determine the local extreme values.

37. $y = x^{2/3}(x + 2)$

38. $y = x^{2/3}(x^2 - 4)$

39. $y = x\sqrt{4 - x^2}$

40. $y = x^2\sqrt{3 - x}$

41. $y = \begin{cases} 4 - 2x, & x \le 1 \\ x + 1, & x > 1 \end{cases}$

42. $y = \begin{cases} 3 - x, & x < 0 \\ 3 + 2x - x^2, & x \ge 0 \end{cases}$

43. $y = \begin{cases} -x^2 - 2x + 4, & x \le 1 \\ -x^2 + 6x - 4, & x > 1 \end{cases}$

44. $y = \begin{cases} -\dfrac{1}{4}x^2 - \dfrac{1}{2}x + \dfrac{15}{4}, & x \le 1 \\ x^3 - 6x^2 + 8x, & x > 1 \end{cases}$

In Exercises 45–48, match the table with a graph of $f(x)$.

45.

x	$f'(x)$
a	0
b	0
c	5

46.

x	$f'(x)$
a	0
b	0
c	-5

47.

x	$f'(x)$
a	does not exist
b	0
c	-2

48.

x	$f'(x)$
a	does not exist
b	does not exist
c	-1.7

(a)

(b)

(c)

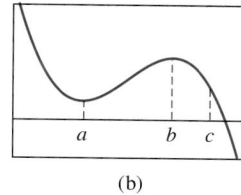

(d)

49. Writing to Learn The function

$$V(x) = x(10 - 2x)(16 - 2x), \quad 0 < x < 5,$$

models the volume of a box.

(a) Find the extreme values of V.

(b) Interpret any values found in (a) in terms of volume of the box.

50. Writing to Learn The function

$$P(x) = 2x + \frac{200}{x}, \quad 0 < x < \infty,$$

models the perimeter of a rectangle of dimensions x by $100/x$.

(a) Find any extreme values of P.

(b) Give an interpretation in terms of perimeter of the rectangle for any values found in (a).

Extending the Ideas

51. *Cubic Functions* Consider the cubic function

$$f(x) = ax^3 + bx^2 + cx + d.$$

(a) Show that f can have 0, 1, or 2 critical points. Give examples and graphs to support your argument.

(b) How many local extreme values can f have?

52. *Proving Theorem 2* Assume that the function f has a local maximum value at the interior point c of its domain.

(a) Show that there is an open interval containing c such that $f(x) - f(c) \le 0$ for all x in the open interval.

(b) Writing to Learn Now explain why we may say

$$\lim_{x \to c^+} \frac{f(x) - f(c)}{x - c} \le 0.$$

(c) Writing to Learn Now explain why we may say

$$\lim_{x \to c^-} \frac{f(x) - f(c)}{x - c} \ge 0.$$

(d) Writing to Learn Explain how (b) and (c) allow us to conclude $f'(c) = 0$.

(e) Writing to Learn Give a similar argument if f has a local minimum value at an interior point.

In Exercise 53, *work in groups of two or three.*

53. *Functions with No Extreme Values at Endpoints*

(a) Graph the function

$$f(x) = \begin{cases} \sin\dfrac{1}{x}, & x > 0 \\ 0, & x = 0. \end{cases}$$

Explain why $f(0) = 0$ is not a local extreme value of f.

(b) Construct a function of your own that fails to have an extreme value at a domain endpoint.

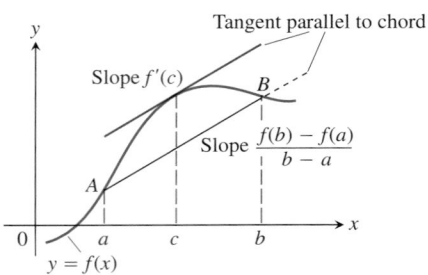

Figure 4.10 Figure for the Mean Value Theorem.

4.2 Mean Value Theorem

Mean Value Theorem • Physical Interpretation • Increasing and Decreasing Functions • Other Consequences

Mean Value Theorem

The Mean Value Theorem connects the average rate of change of a function over an interval with the instantaneous rate of change of the function at a point within the interval. Its powerful corollaries lie at the heart of some of the most important applications of the calculus.

The theorem says that somewhere between points A and B on a differentiable curve, there is at least one tangent line parallel to chord AB (Figure 4.10).

Theorem 3 Mean Value Theorem for Derivatives

If $y = f(x)$ is continuous at every point of the closed interval $[a, b]$ and differentiable at every point of its interior (a, b), then there is at least one point c in (a, b) at which

$$f'(c) = \frac{f(b) - f(a)}{b - a}.$$

Rolle's Theorem

The first version of the Mean Value Theorem was proved by French mathematician Michel Rolle (1652–1719). His version had $f(a) = f(b) = 0$ and was proved only for polynomials, using algebra and geometry.

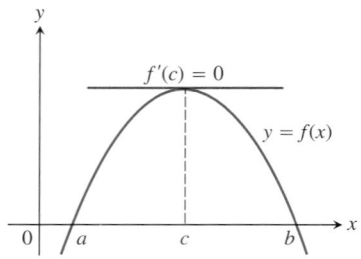

Rolle distrusted calculus and spent most of his life denouncing it. It is ironic that he is known today only for an unintended contribution to a field he tried to suppress.

The hypotheses of Theorem 3 cannot be relaxed. If they fail at even one point, the graph may fail to have a tangent parallel to the chord. For instance, the function $f(x) = |x|$ is continuous on $[-1, 1]$ and differentiable at every point of the interior $(-1, 1)$ except $x = 0$. The graph has no tangent parallel to chord AB (Figure 4.11a). The function $g(x) = \text{int}(x)$ is differentiable at every point of $(1, 2)$ and continuous at every point of $[1, 2]$ except $x = 2$. Again, the graph has no tangent parallel to chord AB (Figure 4.11b).

The Mean Value Theorem is an *existence theorem*. It tells us the number c exists without telling how to find it. We can sometimes satisfy our curiosity about the value of c but the real importance of the theorem lies in the surprising conclusions we can draw from it.

(a)

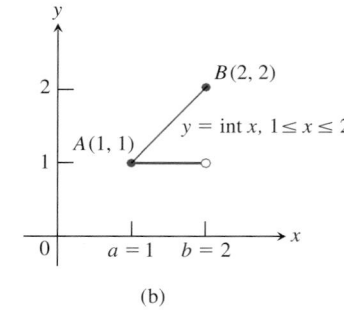

(b)

Figure 4.11 No tangent parallel to chord AB.

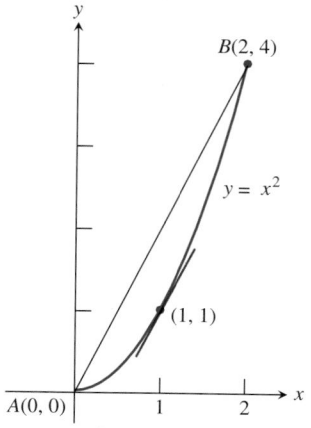

Figure 4.12 (Example 1)

Example 1   EXPLORING THE MEAN VALUE THEOREM

Show that the function $f(x) = x^2$ satisfies the hypotheses of the Mean Value Theorem on the interval $[0, 2]$. Then find a solution c to the equation

$$f'(c) = \frac{f(b) - f(a)}{b - a}$$

on this interval.

Solution The function $f(x) = x^2$ is continuous on $[0, 2]$ and differentiable on $(0, 2)$. Since $f(0) = 0$ and $f(2) = 4$, the Mean Value Theorem guarantees a point c in the interval $(0, 2)$ for which

$$f'(c) = \frac{f(b) - f(a)}{b - a}$$

$$2c = \frac{f(2) - f(0)}{2 - 0} = 2 \quad f'(x) = 2x$$

$$c = 1.$$

Interpret

The tangent line to $f(x) = x^2$ at $x = 1$ has slope 2 and is parallel to the chord joining $A(0, 0)$ and $B(2, 4)$ (Figure 4.12).

Example 2   APPLYING THE MEAN VALUE THEOREM

Let $f(x) = \sqrt{1 - x^2}$, $A = (-1, f(-1))$, and $B = (1, f(1))$. Find a tangent to f in the interval $(-1, 1)$ that is parallel to the secant AB.

Solution The function f (Figure 4.13) is continuous on the interval $[-1, 1]$ and

$$f'(x) = \frac{-x}{\sqrt{1 - x^2}}$$

is defined on the interval $(-1, 1)$. The function is not differentiable at $x = -1$ and $x = 1$, but it does not need to be for the theorem to apply. Since $f(-1) = f(1) = 0$, the tangent we are looking for is horizontal. We find that $f' = 0$ at $x = 0$, where the graph has the horizontal tangent $y = 1$.

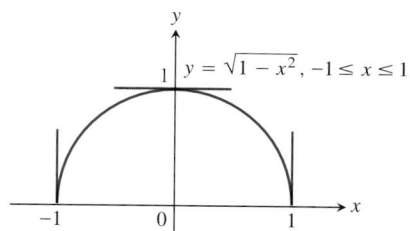

Figure 4.13 (Example 2)

Physical Interpretation

If we think of the difference quotient $(f(b) - f(a))/(b - a)$ as the average change in f over $[a, b]$ and $f'(c)$ as an instantaneous change, then the Mean Value Theorem says that the instantaneous change at some interior point must equal the average change over the entire interval.

Example 3   INTERPRETING THE MEAN VALUE THEOREM

If a car accelerating from zero takes 8 sec to go 352 ft, its average velocity for the 8-sec interval is $352/8 = 44$ ft/sec, or 30 mph. At some point during the acceleration, the theorem says, the speedometer must read exactly 30 mph (Figure 4.14).

Figure 4.14 (Example 3)

Increasing and Decreasing Functions

Our first use of the Mean Value Theorem will be its application to increasing and decreasing functions.

Monotonic Functions

A function that is always increasing on an interval or always decreasing on an interval is said to be **monotonic** there.

Definitions Increasing Function, Decreasing Function

Let f be a function defined on an interval I and let x_1 and x_2 be any two points in I.

1. f **increases** on I if $\quad x_1 < x_2 \quad \Rightarrow \quad f(x_1) < f(x_2)$.

2. f **decreases** on I if $\quad x_1 < x_2 \quad \Rightarrow \quad f(x_1) > f(x_2)$.

The Mean Value Theorem allows us to identify exactly where graphs rise and fall. Functions with positive derivatives are increasing functions; functions with negative derivatives are decreasing functions.

Corollary 1 Increasing and Decreasing Functions

Let f be continuous on $[a, b]$ and differentiable on (a, b).

1. If $f' > 0$ at each point of (a, b), then f increases on $[a, b]$.

2. If $f' < 0$ at each point of (a, b), then f decreases on $[a, b]$.

Proof Let x_1 and x_2 be any two points in $[a, b]$ with $x_1 < x_2$. The Mean Value Theorem applied to f on $[x_1, x_2]$ gives

$$f(x_2) - f(x_1) = f'(c)(x_2 - x_1)$$

for some c between x_1 and x_2. The sign of the right-hand side of this equation is the same as the sign of $f'(c)$ because $x_2 - x_1$ is positive. Therefore,

(a) $f(x_1) < f(x_2)$ if $f' > 0$ on (a, b) (f is increasing), or

(b) $f(x_1) > f(x_2)$ if $f' < 0$ on (a, b) (f is decreasing). ∎

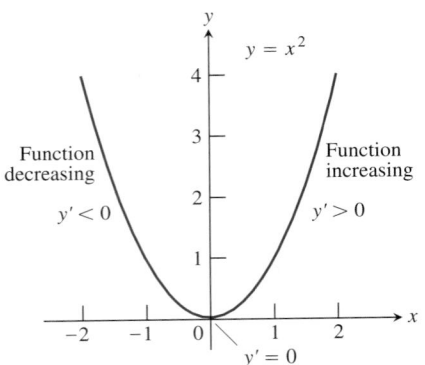

Figure 4.15 (Example 4)

Example 4 DETERMINING WHERE GRAPHS RISE OR FALL

The function $y = x^2$ (Figure 4.15) is

(a) decreasing on $(-\infty, 0]$ because $y' = 2x < 0$ on $(-\infty, 0)$.

(b) increasing on $[0, \infty)$ because $y' = 2x > 0$ on $(0, \infty)$.

Example 5 DETERMINING WHERE GRAPHS RISE OR FALL

Where is the function $f(x) = x^3 - 4x$ increasing and where is it decreasing?

Solution

Solve Graphically

The graph of f in Figure 4.16 suggests that f is increasing from $-\infty$ to the x-coordinate of the local maximum, decreasing between the two local extrema, and increasing again from the x-coordinate of the local minimum to ∞. This information is supported by the superimposed graph of $f'(x) = 3x^2 - 4$.

Confirm Analytically

The function is increasing where $f'(x) > 0$.

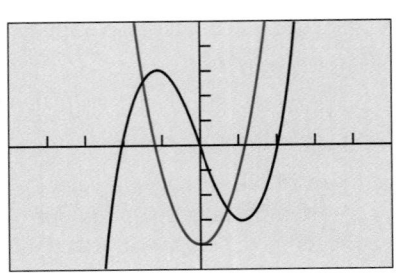

[–5, 5] by [–5, 5]

Figure 4.16 By comparing the graphs of $f(x) = x^3 - 4x$ and $f'(x) = 3x^2 - 4$ we can relate the increasing and decreasing behavior of f to the sign of f'. (Example 5)

$$3x^2 - 4 > 0$$

$$x^2 > \frac{4}{3}$$

$$x < -\sqrt{\frac{4}{3}} \quad \text{or} \quad x > \sqrt{\frac{4}{3}}$$

The function is decreasing where $f'(x) < 0$.

$$3x^2 - 4 < 0$$

$$x^2 < \frac{4}{3}$$

$$-\sqrt{\frac{4}{3}} < x < \sqrt{\frac{4}{3}}$$

Interpret

To describe the intervals we use the approximation $\sqrt{4/3} \approx 1.15$. The function increases on $(-\infty, -1.15]$ and $[1.15, \infty)$. It decreases on $[-1.15, 1.15]$.

In Example 5, the exact interval on which the function f decreases is $[-\sqrt{4/3}, \sqrt{4/3}]$. We reported it as $[-1.15, 1.15]$. As we have been doing all along, we continue to display results to a number of decimal places that seems appropriate to the problem.

Other Consequences

We know that constant functions have the zero function as their derivative. We can now use the Mean Value Theorem to show conversely that the only functions with the zero function as derivative are constant functions.

Corollary 2 Functions with $f' = 0$ are Constant

If $f'(x) = 0$ at each point of an interval I, then there is a constant C for which $f(x) = C$ for all x in I.

Proof Our plan is to show that $f(x_1) = f(x_2)$ for any two points x_1 and x_2 in I. We can assume the points are numbered so that $x_1 < x_2$. Since f is differentiable at every point of $[x_1, x_2]$, it is continuous at every point as well. Thus, f satisfies the hypotheses of the Mean Value Theorem on $[x_1, x_2]$. Therefore, there is a point c between x_1 and x_2 for which

$$f'(c) = \frac{f(x_2) - f(x_1)}{x_2 - x_1}.$$

Because $f'(c) = 0$, it follows that $f(x_1) = f(x_2)$. ∎

We can use Corollary 2 to show that if two functions have the same derivative, they differ by a constant.

Corollary 3 Functions with the Same Derivative Differ by a Constant

If $f'(x) = g'(x)$ at each point of an interval I, then there is a constant C such that $f(x) = g(x) + C$ for all x in I.

Proof Let $h = f - g$. Then for each point x in I,

$$h'(x) = f'(x) - g'(x) = 0.$$

It follows from Corollary 2 that there is a constant C such that $h(x) = C$ for all x in I. Thus, $h(x) = f(x) - g(x) = C$, or $f(x) = g(x) + C$. ∎

We know that the derivative of $f(x) = x^2$ is $2x$ on the interval $(-\infty, \infty)$. So, any other function $g(x)$ with derivative $2x$ on $(-\infty, \infty)$ must have the formula $g(x) = x^2 + C$ for some constant C.

Example 6 APPLYING COROLLARY 3

Find the function $f(x)$ whose derivative is $\sin x$ and whose graph passes through the point $(0, 2)$.

Solution Since f has the same derivative as $g(x) = -\cos x$, we know that $f(x) = -\cos x + C$ for some constant C. To identify C, we use the condition that the graph must pass through $(0, 2)$. This is equivalent to saying that

$$f(0) = 2$$
$$-\cos(0) + C = 2 \quad f(x) = -\cos x + C$$
$$-1 + C = 2$$
$$C = 3.$$

The formula for f is $f(x) = -\cos x + 3$.

In Example 6 we were given a derivative and asked to find a function with that derivative. This type of function is so important that it has a name.

> **Definition Antiderivative**
>
> A function $F(x)$ is an **antiderivative** of a function $f(x)$ if $F'(x) = f(x)$ for all x in the domain of f. The process of finding an antiderivative is **antidifferentiation.**

We know that if f has one antiderivative F then it has infinitely many antiderivatives, each differing from F by a constant. Corollary 3 says these are all there are. In Example 6, we found the particular antiderivative of $\sin x$ whose graph passed through the point $(0, 2)$.

Example 7 FINDING VELOCITY AND POSITION

Find the velocity and position functions of a freely falling body for each of the following sets of conditions:

(a) The acceleration is 9.8 m/sec^2 and the body falls from rest.

(b) The acceleration is 9.8 m/sec^2 and the body is propelled downward with an initial velocity of 1 m/sec.

Solution

(a) *Falling from rest.* We measure distance fallen in meters and time in seconds, and assume that the body is released from rest at time $t = 0$.

Velocity: We know that the velocity $v(t)$ is an antiderivative of the constant function 9.8. We also know that $g(t) = 9.8t$ is an antiderivative of 9.8. By Corollary 3,

$$v(t) = 9.8t + C$$

for some constant C. Since the body falls from rest, $v(0) = 0$. Thus,

$$9.8(0) + C = 0 \quad \text{and} \quad C = 0.$$

The body's velocity function is $v(t) = 9.8t$.

Position: We know that the position $s(t)$ is an antiderivative of $9.8t$. We also know that $h(t) = 4.9t^2$ is an antiderivative of $9.8t$. By Corollary 3,

$$s(t) = 4.9t^2 + C$$

for some constant C. Since $s(0) = 0$,

$$4.9(0)^2 + C = 0 \quad \text{and} \quad C = 0.$$

The body's position function is $s(t) = 4.9t^2$.

(b) *Propelled downward.* We measure distance fallen in meters and time in seconds, and assume that the body is propelled downward with velocity of 1 m/sec at time $t = 0$.

Velocity: The velocity function still has the form $9.8t + C$, but instead of being zero, the initial velocity (velocity at $t = 0$) is now 1 m/sec. Thus,

$$9.8(0) + C = 1 \quad \text{and} \quad C = 1.$$

The body's velocity function is $v(t) = 9.8t + 1$.

Position: We know that the position $s(t)$ is an antiderivative of $9.8t + 1$. We also know that $k(t) = 4.9t^2 + t$ is an antiderivative of $9.8t + 1$. By Corollary 3,

$$s(t) = 4.9t^2 + t + C$$

for some constant C. Since $s(0) = 0$,

$$4.9(0)^2 + 0 + C = 0 \quad \text{and} \quad C = 0.$$

The body's position function is $s(t) = 4.9t^2 + t$.

Quick Review 4.2

In Exercises 1 and 2, find exact solutions to the inequality.

1. $2x^2 - 6 < 0$

2. $3x^2 - 6 > 0$

In Exercises 3–5, let $f(x) = \sqrt{8 - 2x^2}$.

3. Find the domain of f.

4. Where is f continuous?

5. Where is f differentiable?

In Exercises 6–8, let $f(x) = \dfrac{x}{x^2 - 1}$.

6. Find the domain of f.

7. Where is f continuous?

8. Where is f differentiable?

In Exercises 9 and 10, find C so that the graph of the function f passes through the specified point.

9. $f(x) = -2x + C, \quad (-2, 7)$

10. $g(x) = x^2 + 2x + C, \quad (1, -1)$

Section 4.2 Exercises

In Exercises 1–8, use analytic methods to find **(a)** the local extrema, **(b)** the intervals on which the function is increasing, and **(c)** the intervals on which the function is decreasing.

1. $f(x) = 5x - x^2$

2. $g(x) = x^2 - x - 12$

3. $h(x) = \dfrac{2}{x}$

4. $k(x) = \dfrac{1}{x^2}$

5. $f(x) = e^{2x}$

6. $f(x) = e^{-0.5x}$

7. $y = 4 - \sqrt{x + 2}$

8. $y = x^4 - 10x^2 + 9$

In Exercises 9–14, find **(a)** the local extrema, **(b)** the intervals on which the function is increasing, and **(c)** the intervals on which the function is decreasing.

9. $f(x) = x \sqrt{4 - x}$

10. $g(x) = x^{1/3}(x + 8)$

11. $h(x) = \dfrac{-x}{x^2 + 4}$

12. $k(x) = \dfrac{x}{x^2 - 4}$

13. $f(x) = x^3 - 2x - 2 \cos x$

14. $g(x) = 2x + \cos x$

In Exercises 15–18, **(a)** show that the function f satisfies the hypotheses of the Mean Value Theorem on the given interval $[a, b]$. **(b)** Find each value of c in (a, b) that satisfies the equation

$$f'(c) = \frac{f(b) - f(a)}{b - a}.$$

15. $f(x) = x^2 + 2x - 1, \quad [0, 1]$

16. $f(x) = x^{2/3}, \quad [0, 1]$

17. $f(x) = \sin^{-1} x, \quad [-1, 1]$

18. $f(x) = \ln (x - 1), \quad [2, 4]$

In Exercises 19 and 20, the interval $a \le x \le b$ is given. Let $A = (a, f(a))$ and $B = (b, f(b))$. Write an equation for

(a) the secant line AB.

(b) a tangent line to f in the interval (a, b) that is parallel to AB.

19. $f(x) = x + \dfrac{1}{x}, \quad 0.5 \le x \le 2$

20. $f(x) = \sqrt{x - 1}, \quad 1 \le x \le 3$

In Exercises 21–24, **(a)** show that the function f does not satisfy the hypotheses of the Mean Value Theorem on the given interval $[a, b]$. **(b)** Graph f together with the line through the points $A(a, f(a))$ and $B(b, f(b))$. **(c)** Find any values of c in (a, b) that satisfy the equation

$$f'(c) = \frac{f(b) - f(a)}{b - a}.$$

21. $f(x) = x^{1/3}, \quad [-1, 1]$

22. $f(x) = |x - 1|, \quad [0, 3]$

23. $f(x) = 1 - |x|, \quad [-1, 1]$

24. $f(x) = \begin{cases} \cos x, & -\pi \le x < 0, \\ 1 + \sin x, & 0 \le x \le \pi, \end{cases}$ on $[-\pi, \pi]$

In Exercises 25–30, find all possible functions f with the given derivative.

25. $f'(x) = x$

26. $f'(x) = 2$

27. $f'(x) = 3x^2 - 2x + 1$

28. $f'(x) = \sin x$

29. $f'(x) = e^x$

30. $f'(x) = \dfrac{1}{x - 1}, \quad x > 1$

In Exercises 31–34, find the function with the given derivative whose graph passes through the point P.

31. $f'(x) = -\dfrac{1}{x^2}, \quad x > 0, \quad P(2, 1)$

32. $f'(x) = \dfrac{1}{4x^{3/4}}, \quad P(1, -2)$

33. $f'(x) = \dfrac{1}{x + 2}, \quad x > -2, \quad P(-1, 3)$

34. $f'(x) = 2x + 1 - \cos x, \quad P(0, 3)$

In Exercises 35–38, *work in groups of two or three* and sketch a graph of a differentiable function $y = f(x)$ that has the given properties.

35. **(a)** local minimum at $(1, 1)$, local maximum at $(3, 3)$

　　(b) local minima at $(1, 1)$ and $(3, 3)$

　　(c) local maxima at $(1, 1)$ and $(3, 3)$

36. $f(2) = 3, \quad f'(2) = 0, \quad$ and

　　(a) $f'(x) > 0$ for $x < 2, \quad f'(x) < 0$ for $x > 2$.

　　(b) $f'(x) < 0$ for $x < 2, \quad f'(x) > 0$ for $x > 2$.

　　(c) $f'(x) < 0$ for $x \ne 2$.

　　(d) $f'(x) > 0$ for $x \ne 2$.

37. $f'(-1) = f'(1) = 0, \quad f'(x) > 0$ on $(-1, 1)$, $f'(x) < 0$ for $x < -1, \quad f'(x) > 0$ for $x > 1$.

38. A local minimum value that is greater than one of its local maximum values.

39. *Speeding* A trucker handed in a ticket at a toll booth showing that in 2 h she had covered 159 mi on a toll road with speed limit 65 mph. The trucker was cited for speeding. Why?

40. *Temperature Change* It took 20 sec for the temperature to rise from 0°F to 212°F when a thermometer was taken from a freezer and placed in boiling water. Explain why at some moment in that interval the mercury was rising at exactly 10.1°F/sec.

41. *Triremes* Classical accounts tell us that a 170-oar trireme (ancient Greek or Roman warship) once covered 184 sea miles in 24 h. Explain why at some point during this feat the trireme's speed exceeded 7.5 knots (sea miles per hour).

42. *Running a Marathon* A marathoner ran the 26.2-mi New York City Marathon in 2.2 h. Show that at least twice, the marathoner was running at exactly 11 mph.

43. Free Fall On the moon, the acceleration due to gravity is 1.6 m/sec².

(a) If a rock is dropped into a crevasse, how fast will it be going just before it hits bottom 30 sec later?

(b) How far below the point of release is the bottom of the crevasse?

(c) If instead of being released from rest, the rock is thrown into the crevasse from the same point with a downward velocity of 4 m/sec, when will it hit the bottom and how fast will it be going when it does?

44. Diving (a) With what velocity will you hit the water if you step off from a 10-m diving platform?

(b) With what velocity will you hit the water if you dive off the platform with an upward velocity of 2 m/sec?

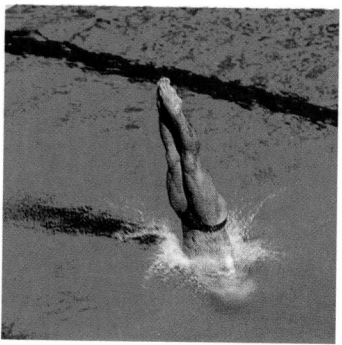

45. Writing to Learn The function

$$f(x) = \begin{cases} x, & 0 \le x < 1 \\ 0, & x = 1 \end{cases}$$

is zero at $x = 0$ and at $x = 1$. Its derivative is equal to 1 at every point between 0 and 1, so f' is never zero between 0 and 1, and the graph of f has no tangent parallel to the chord from $(0, 0)$ to $(1, 0)$. Explain why this does not contradict the Mean Value Theorem.

46. Writing to Learn Explain why there is a zero of $y = \cos x$ between every two zeros of $y = \sin x$.

47. Unique Solution Assume that f is continuous on $[a, b]$ and differentiable on (a, b). Also assume that $f(a)$ and $f(b)$ have opposite signs and $f' \ne 0$ between a and b. Show that $f(x) = 0$ exactly once between a and b.

In Exercises 48 and 49, show that the equation has exactly one solution in the interval. (*Hint:* See Exercise 47.)

48. $x^4 + 3x + 1 = 0, \quad -2 \le x \le -1$

49. $x + \ln(x + 1) = 0, \quad 0 \le x \le 3$

50. Parallel Tangents Assume that f and g are differentiable on $[a, b]$ and that $f(a) = g(a)$ and $f(b) = g(b)$. Show that there is at least one point between a and b where the tangents to the graphs of f and g are parallel or the same line. Illustrate with a sketch.

Explorations

51. Analyzing Derivative Data Assume that f is continuous on $[-2, 2]$ and differentiable on $(-2, 2)$. The table gives some values of $f'(x)$.

x	$f'(x)$	x	$f'(x)$
-2	7	0.25	-4.81
-1.75	4.19	0.5	-4.25
-1.5	1.75	0.75	-3.31
-1.25	-0.31	1	-2
-1	-2	1.25	-0.31
-0.75	-3.31	1.5	1.75
-0.5	-4.25	1.75	4.19
-0.25	-4.81	2	7
0	-5		

(a) Estimate where f is increasing, decreasing, and has local extrema.

(b) Find a quadratic regression equation for the data in the table and superimpose its graph on a scatter plot of the data.

(c) Use the model in (b) for f' and find a formula for f that satisfies $f(0) = 0$.

52. Analyzing Motion Data Priya's distance D in meters from a motion detector is given by the data in Table 4.1.

Table 4.1 Motion Detector Data

t (sec)	D (m)	t (sec)	D (m)
0.0	3.36	4.5	3.59
0.5	2.61	5.0	4.15
1.0	1.86	5.5	3.99
1.5	1.27	6.0	3.37
2.0	0.91	6.5	2.58
2.5	1.14	7.0	1.93
3.0	1.69	7.5	1.25
3.5	2.37	8.0	0.67
4.0	3.01		

(a) Estimate when Priya is moving toward the motion detector; away from the motion detector.

(b) Writing to Learn Give an interpretation of any local extreme values in terms of this problem situation.

(c) Find a cubic regression equation $D = f(t)$ for the data in Table 4.1 and superimpose its graph on a scatter plot of the data.

(d) Use the model in (c) for f to find a formula for f'. Use this formula to estimate the answers to (a). ∎

Extending the Ideas

53. *Geometric Mean* The **geometric mean** of two positive numbers a and b is $\sqrt{ab}$. Show that for $f(x) = 1/x$ on any interval $[a, b]$ of positive numbers, the value of c in the conclusion of the Mean Value Theorem is $c = \sqrt{ab}$.

54. *Arithmetic Mean* The **arithmetic mean** of two numbers a and b is $(a + b)/2$. Show that for $f(x) = x^2$ on any interval $[a, b]$, the value of c in the conclusion of the Mean Value Theorem is $c = (a + b)/2$.

55. *Upper Bounds* Show that for any numbers a and b, $|\sin b - \sin a| \le |b - a|$.

56. *Sign of f'* Assume that f is differentiable on $a \le x \le b$ and that $f(b) < f(a)$. Show that f' is negative at some point between a and b.

57. *Monotonic Functions* Show that monotonic increasing and decreasing functions are one-to-one.

4.3 Connecting f' and f'' with the Graph of f

First Derivative Test for Local Extrema • Concavity • Points of Inflection • Second Derivative Test for Local Extrema • Learning about Functions from Derivatives

First Derivative Test for Local Extrema

As we see once again in Figure 4.17, a function f may have local extrema at some critical points while failing to have local extrema at others. The key is the sign of f' in a critical point's immediate vicinity. As x moves from left to right, the values of f increase where $f' > 0$ and decrease where $f' < 0$.

At the points where f has a minimum value, we see that $f' < 0$ on the interval immediately to the left and $f' > 0$ on the interval immediately to the right. (If the point is an endpoint, there is only the interval on the appropriate side to consider.) This means that the curve is falling (values decreasing) on the left of the minimum value and rising (values increasing) on its right. Similarly, at the points where f has a maximum value, $f' > 0$ on the interval immediately to the left and $f' < 0$ on the interval immediately to the right. This means that the curve is rising (values increasing) on the left of the maximum value and falling (values decreasing) on its right.

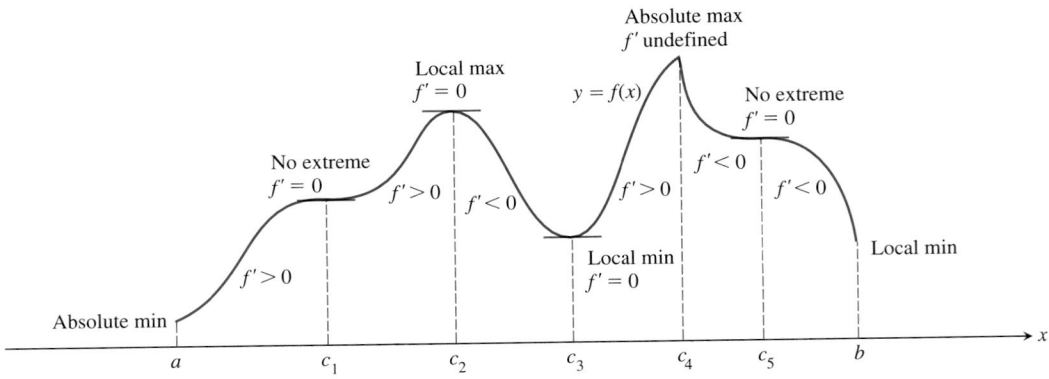

Figure 4.17 A function's first derivative tells how the graph rises and falls.

Theorem 4 **First Derivative Test for Local Extrema**

The following test applies to a continuous function $f(x)$.

At a critical point *c*:

1. If f' changes sign from positive to negative at c ($f' > 0$ for $x < c$ and $f' < 0$ for $x > c$), then f has a local maximum value at c.

2. If f' changes sign from negative to positive at c ($f' < 0$ for $x < c$ and $f' > 0$ for $x > c$), then f has a local minimum value at c.

3. If f' does not change sign at c (f' has the same sign on both sides of c), then f has no local extreme value at c.

At a left endpoint *a*:

If $f' < 0$ ($f' > 0$) for $x > a$, then f has a local maximum (minimum) value at a.

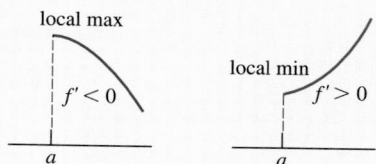

At a right endpoint *b*:

If $f' < 0$ ($f' > 0$) for $x < b$, then f has a local minimum (maximum) value at b.

Here is how we apply the first derivative test to find the local extrema of a function. The critical points of a function f partition the x-axis into intervals on which f' is either positive or negative. We determine the sign of f' in each interval by evaluating f' for one value of x in the interval. Then we apply Theorem 4 as shown in Examples 1 and 2.

Example 1 USING THE FIRST DERIVATIVE TEST

Find the critical points of $f(x) = x^3 - 12x - 5$. Find the function's local and absolute extreme values. Identify the intervals on which f is increasing and decreasing.

Solution Figure 4.18 suggests that f has two critical points.

Solve Analytically

Since f is continuous and differentiable for all real numbers, the critical points occur only at the zeros of f'.

$$f'(x) = 3x^2 - 12 = 0$$
$$x^2 = 4$$
$$x = -2, 2$$

The zeros of f' partition the x-axis into intervals as follows.

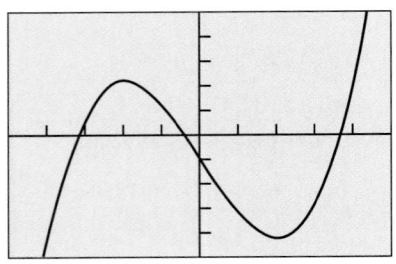

[−5, 5] by [−25, 25]

Figure 4.18 The graph of $f(x) = x^3 - 12x - 5$. (Example 1)

Intervals	$-\infty < x < -2$	$-2 < x < 2$	$2 < x < \infty$
Sign of f'	$+$	$-$	$+$
Behavior of f	increasing	decreasing	increasing

We can see from the table that there is a local maximum at $x = -2$ and a local minimum at $x = 2$. The local maximum value is $f(-2) = 11$, and the local minimum value is $f(2) = -21$. There are no absolute extrema. The function increases on the intervals $(-\infty, -2]$ and $[2, \infty)$, and decreases on the interval $[-2, 2]$.

If a function increases, then decreases, and finally increases again like the function in Example 1, you might conjecture that it is a polynomial function. Example 2 shows that this is not necessarily true.

Example 2 USING THE FIRST DERIVATIVE TEST

Find the critical points of $f(x) = (x^2 - 3)e^x$. Find the function's local and absolute extreme values. Identify the intervals on which f is increasing and decreasing.

Solution This time it is a little harder to see one of the extrema (Figure 4.19).

Solve Analytically

The function f is continuous and differentiable for all real numbers, so the critical points occur only at the zeros of f'.

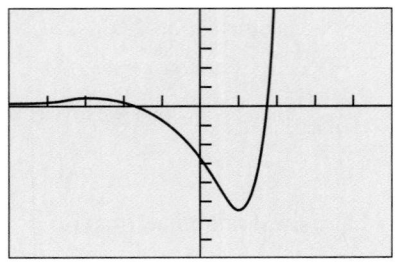

[−5, 5] by [−8, 5]

Figure 4.19 The graph of $f(x) = (x^2 - 3)e^x$. (Example 2)

Using the Product Rule we find

$$f'(x) = (x^2 - 3) \cdot \frac{d}{dx}e^x + \frac{d}{dx}(x^2 - 3) \cdot e^x$$

$$= (x^2 - 3) \cdot e^x + (2x) \cdot e^x$$

$$= (x^2 + 2x - 3)e^x.$$

Since e^x is never zero, the first derivative is zero if and only if

$$x^2 + 2x - 3 = 0$$

$$(x + 3)(x - 1) = 0.$$

The zeros $x = -3$ and $x = 1$ partition the x-axis into intervals as follows.

Intervals	$x < -3$	$-3 < x < 1$	$1 < x$
Sign of f'	$+$	$-$	$+$
Behavior of f	increasing	decreasing	increasing

We can see from the table that there is a local maximum (about 0.299) at $x = -3$, and a local minimum (about -5.437) at $x = 1$. The local minimum value is also an absolute minimum because $f(x) > 0$ for $|x| > \sqrt{3}$. There is no absolute maximum. The function increases on $(-\infty, -3]$ and $[1, \infty)$, and decreases on $[-3, 1]$.

Concavity

As you can see in Figure 4.20, the function $y = x^3$ rises as x increases, but the portions defined on the intervals $(-\infty, 0)$ and $(0, \infty)$ *turn* in different ways. Looking at tangents as we scan from left to right, we see that the slope y' of the curve decreases on the interval $(-\infty, 0)$ and then increases on the interval $(0, \infty)$. The curve $y = x^3$ is *concave down* on $(-\infty, 0)$ and *concave up* on $(0, \infty)$. The curve lies below the tangents where it is concave down, and above the tangents where it is concave up.

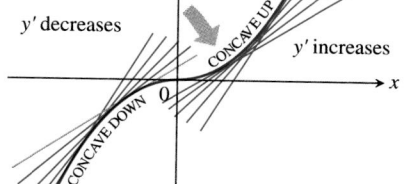

Figure 4.20 The graph of $y = x^3$ is concave down on $(-\infty, 0)$ and concave up on $(0, \infty)$.

Definition **Concavity**

The graph of a differentiable function $y = f(x)$ is

(a) concave up on an open interval *I* if y' is increasing on *I.*

(b) concave down on an open interval *I* if y' is decreasing on *I.*

If a function $y = f(x)$ has a second derivative, then we can conclude that y' increases if $y'' > 0$ and y' decreases if $y'' < 0$.

Concavity Test

The graph of a twice-differentiable function $y = f(x)$ is

(a) concave up on any interval where $y'' > 0$.

(b) concave down on any interval where $y'' < 0$.

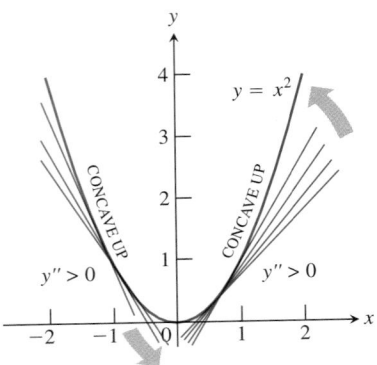

Figure 4.21 The graph of $y = x^2$ is concave up on any interval. (Example 3)

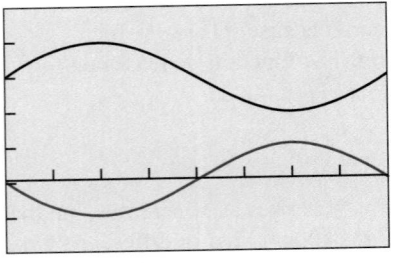

$[0, 2\pi]$ by $[-2, 5]$

Figure 4.22 Using the graph of y'' to determine the concavity of y. (Example 4)

Example 3 APPLYING THE CONCAVITY TEST

The curve $y = x^2$ (Figure 4.21) is concave up on $(-\infty, \infty)$ because its second derivative $y'' = 2$ is always positive.

Example 4 DETERMINING CONCAVITY

Determine the concavity of $y = 3 + \sin x$ on $[0, 2\pi]$.

Solution The graph of $y = 3 + \sin x$ is concave down on $(0, \pi)$, where $y'' = -\sin x$ is negative. It is concave up on $(\pi, 2\pi)$, where $y'' = -\sin x$ is positive (Figure 4.22).

Points of Inflection

The curve $y = 3 + \sin x$ in Example 4 changes concavity at the point $(\pi, 3)$. We call $(\pi, 3)$ a *point of inflection* of the curve.

Definition Point of Inflection

A point where the graph of a function has a tangent line and where the concavity changes is a **point of inflection.**

A point on a curve where y'' is positive on one side and negative on the other is a point of inflection. At such a point, y'' is either zero (because derivatives have the intermediate value property) or undefined. If y is a twice-differentiable function, $y'' = 0$ at a point of inflection and y' has a local maximum or minimum.

To study the motion of a body moving along a line, we often graph the body's position as a function of time. One reason for doing so is to reveal where the body's acceleration, given by the second derivative, changes sign. On the graph, these are the points of inflection.

Example 5 STUDYING MOTION ALONG A LINE

A particle is moving along a horizontal line with position function

$$s(t) = 2t^3 - 14t^2 + 22t - 5, \quad t \geq 0.$$

Find the velocity and acceleration, and describe the motion of the particle.

Solution

Solve Analytically

The velocity is

$$v(t) = s'(t) = 6t^2 - 28t + 22 = 2(t - 1)(3t - 11),$$

and the acceleration is

$$a(t) = v'(t) = s''(t) = 12t - 28 = 4(3t - 7).$$

When the function $s(t)$ is increasing, the particle is moving to the right; when $s(t)$ is decreasing, the particle is moving to the left. Figure 4.23 shows the graphs of the position, velocity, and acceleration of the particle.

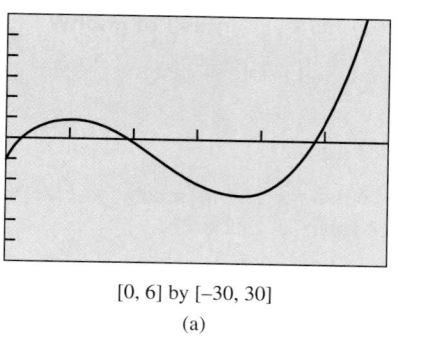

[0, 6] by [−30, 30]

(a)

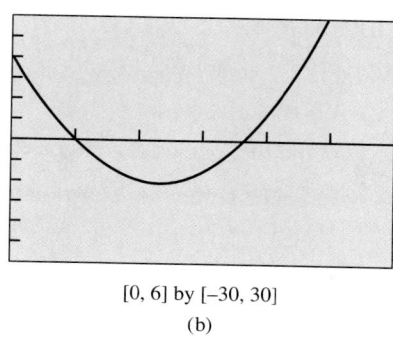

[0, 6] by [−30, 30]

(b)

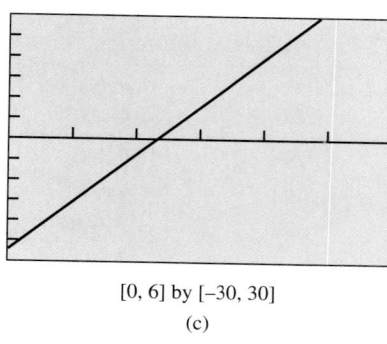

[0, 6] by [−30, 30]

(c)

Figure 4.23 The graph of (a) $s(t) = 2t^3 - 14t^2 + 22t - 5, t \geq 0$, (b) $s'(t) = 6t^2 - 28t + 22$, and (c) $s''(t) = 12t - 28$. (Example 5)

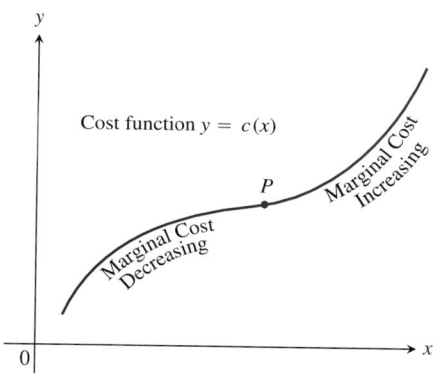

Figure 4.24 On a typical cost curve, a point of inflection separates an interval of decreasing marginal cost from an interval of increasing marginal cost.

Notice that the first derivative $(v = s')$ is zero when $t = 1$ and $t = 11/3$.

Intervals	$0 < t < 1$	$1 < t < 11/3$	$11/3 < t$
Sign of $v = s'$	+	−	+
Behavior of s	increasing	decreasing	increasing
Particle motion	right	left	right

The particle is moving to the right in the time intervals $[0, 1)$ and $(11/3, \infty)$, and moving to the left in $(1, 11/3)$.

The acceleration $a(t) = s''(t) = 12t - 28 = 4(3t - 7)$ is zero when $t = 7/3$.

Intervals	$0 < t < 7/3$	$7/3 < t$
Sign of $a = s''$	−	+
Graph of s	concave down	concave up

The accelerating force is directed toward the left during the time interval $[0, 7/3)$, is momentarily zero at $t = 7/3$, and is directed toward the right thereafter.

Inflection points have applications in some areas of economics. If $y = c(x)$ (Figure 4.24) is the total cost of producing x units of something, the point of inflection at P is then the point at which the marginal cost (the cost of producing one more unit) changes from decreasing to increasing; i.e., it is where the marginal cost reaches a minimum.

The growth of an individual company, of a population, in sales of a new product, or of salaries often follows a *logistic* or *life cycle curve* like the one shown in Figure 4.25. For example, sales of a new product will generally grow slowly at first, then experience a period of rapid growth. Eventually, sales growth slows down again. The function f in Figure 4.25 is increasing. Its rate of increase, f', is at first increasing $(f'' > 0)$ up to the point of inflection, and then its rate of increase, f', is decreasing $(f'' < 0)$. This is, in a sense, the opposite of what happens in Figure 4.20.

Some graphers have the logistic curve as a built-in regression model. We use this feature in Example 6.

Figure 4.25 A logistic curve

$$y = \frac{c}{1 + ae^{-bx}}.$$

Table 4.2 NFL Average Salaries	
Year	Average Salary (dollars)
1983	141,000
1984	206,000
1985	217,000
1986	220,000
1987	220,000
1988	250,000
1989	319,000
1990	365,000
1991	425,000
1992	492,000
1993	683,000
1994	636,000
1995	714,000
1996	795,000

Source: NFLPA as reported by Gordon Forbes in *USA Today,* May 7, 1997.

[0, 25] by [0, 1500]

(a)

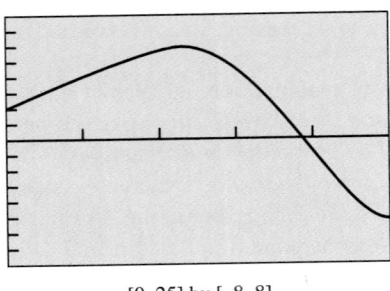

[0, 25] by [−8, 8]

(b)

Figure 4.26 The graph of (a)
$$y = \frac{2171}{1 + 24.72e^{-0.168x}}$$
and (b) y''. (Example 6)

Example 6   USING LOGISTIC REGRESSION

Table 4.2 shows the average salaries for National Football League (NFL) players.

(a) Find the logistic regression equation for the data.

(b) Use the regression equation to predict when the rate of salary increases (f') will start to decrease, and the average salary at that time.

(c) Is there a ceiling to the salaries? If so, what is it?

Solution

(a) We let $x = 0$ represent 1980, $x = 1$ represent 1981, and so forth. We enter the salaries in thousands, using 141 for the salary in 1983, and so on. The logistic regression equation is approximately

$$y = \frac{2171}{1 + 24.72e^{-0.168x}}.$$

Its graph is superimposed on a scatter plot of the data in Figure 4.26a.

(b) We need to find the point of inflection, so we need to solve the equation

$$y'' = 0.$$

Using the graph of y'' in Figure 4.26b we find that $y'' = 0$ when $x \approx 19.097$. The corresponding salary is $y(19.097) \approx 1086$. So, the rate of increase should start to decrease in 1999, and the average player salary at that time should be approximately $1,086,000.

(c) Notice that

$$\lim_{x \to \infty} \frac{2171}{1 + 24.72e^{-0.168x}} = 2171,$$

so the average salaries have a ceiling of about $2,171,000.

Second Derivative Test for Local Extrema

Instead of looking for sign changes in y' at critical points, we can sometimes use the following test to determine the presence of local extrema.

Theorem 5   Second Derivative Test for Local Extrema

1. If $f'(c) = 0$ and $f''(c) < 0$, then f has a local maximum at $x = c$.

2. If $f'(c) = 0$ and $f''(c) > 0$, then f has a local minimum at $x = c$.

This test requires us to know f'' *only at c itself* and not in an interval about c. This makes the test easy to apply. That's the good news. The bad news is that the test fails if $f''(c) = 0$ or if $f''(c)$ fails to exist. When this happens, go back to the first derivative test for local extreme values.

In Example 7, we apply the second derivative test to the function in Example 1.

Example 7   USING THE SECOND DERIVATIVE TEST

Find the extreme values of $f(x) = x^3 - 12x - 5$.

Solution We have

$$f'(x) = 3x^2 - 12 = 3(x^2 - 4)$$
$$f''(x) = 6x.$$

Testing the critical points $x = \pm 2$ (there are no endpoints), we find

$$f''(-2) = -12 < 0 \Rightarrow f \text{ has a local maximum at } x = -2 \text{ and}$$
$$f''(2) = 12 > 0 \Rightarrow f \text{ has a local minimum at } x = 2.$$

Example 8 USING f' AND f" TO GRAPH f

Let $f'(x) = 4x^3 - 12x^2$.

(a) Identify where the extrema of f occur.

(b) Find the intervals on which f is increasing and the intervals on which f is decreasing.

(c) Find where the graph of f is concave up and where it is concave down.

(d) Sketch a possible graph for f.

Solution f is continuous since f' exists. The domain of f' is $(-\infty, \infty)$, so the domain of f is also $(-\infty, \infty)$. Thus, the critical points of f occur only at the zeros of f'. Since

$$f'(x) = 4x^3 - 12x^2 = 4x^2(x - 3),$$

the first derivative is zero at $x = 0$ and $x = 3$.

Intervals	$x < 0$	$0 < x < 3$	$3 < x$
Sign of f'	$-$	$-$	$+$
Behavior of f	decreasing	decreasing	increasing

(a) Using the first derivative test and the table above we see that there is no extremum at $x = 0$ and a local minimum at $x = 3$.

(b) Using the table above we see that f is decreasing in $(-\infty, 0]$ and $[0, 3]$, and increasing in $[3, \infty)$.

(c) $f''(x) = 12x^2 - 24x = 12x(x - 2)$ is zero at $x = 0$ and $x = 2$.

Intervals	$x < 0$	$0 < x < 2$	$2 < x$
Sign of f''	$+$	$-$	$+$
Behavior of f	concave up	concave down	concave up

We see that f is concave up on the intervals $(-\infty, 0)$ and $(2, \infty)$, and concave down on $(0, 2)$.

(d) Summarizing the information in the two tables above we obtain

$x < 0$	$0 < x < 2$	$2 < x < 3$	$x < 3$
decreasing	decreasing	decreasing	increasing
concave up	concave down	concave up	concave up

Figure 4.27 shows one possibility for the graph of f.

Note

The *second derivative test* does not apply at $x = 0$ because $f''(0) = 0$. We need the first derivative test to see that there is no local extremum at $x = 0$.

Figure 4.27 The graph for f has no extremum but has points of inflection where $x = 0$ and $x = 2$, and a local minimum where $x = 3$. (Example 8)

Let $f'(x) = 4x^3 - 12x^2$.

1. Find three different functions with derivative equal to $f'(x)$. How are the graphs of the three functions related?

2. Compare their behavior with the behavior found in Example 8.

Learning about Functions from Derivatives

We have seen in Example 8 and Exploration 1 that we are able to recover almost everything we need to know about a differentiable function $y = f(x)$ by examining y'. We can find where the graph rises and falls and where any local extrema are assumed. We can differentiate y' to learn how the graph bends as it passes over the intervals of rise and fall. We can determine the shape of the function's graph. The only information we cannot get from the derivative is how to place the graph in the xy-plane. As we discovered in Section 4.2, the only additional information we need to position the graph is the value of f at one point.

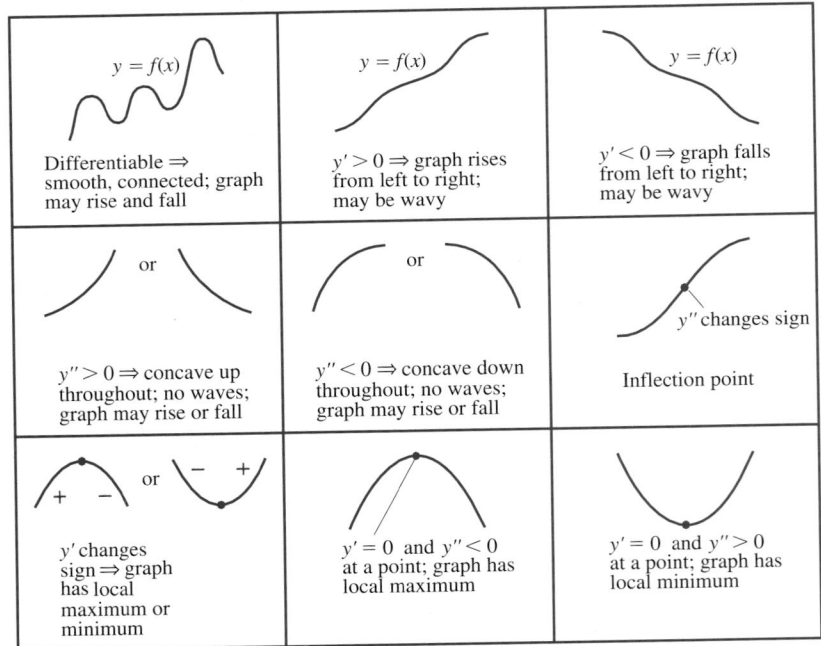

$y = f(x)$ Differentiable $\Rightarrow$ smooth, connected; graph may rise and fall	$y = f(x)$ $y' > 0 \Rightarrow$ graph rises from left to right; may be wavy	$y = f(x)$ $y' < 0 \Rightarrow$ graph falls from left to right; may be wavy
or $y'' > 0 \Rightarrow$ concave up throughout; no waves; graph may rise or fall	*or* $y'' < 0 \Rightarrow$ concave down throughout; no waves; graph may rise or fall	y'' changes sign Inflection point
or $+$ $-$ $-$ $+$ y' changes sign $\Rightarrow$ graph has local maximum or minimum	$y' = 0$ and $y'' < 0$ at a point; graph has local maximum	$y' = 0$ and $y'' > 0$ at a point; graph has local minimum

> **Exploration 2**   **Finding f from f' and f"**
>
> A function f is continuous on its domain $[-2, 4]$, $f(-2) = 5$, $f(4) = 1$, and f' and $f"$ have the following properties.
>
> | x | $-2 < x < 0$ | $x = 0$ | $0 < x < 2$ | $x = 2$ | $2 < x < 4$ |
> |---|---|---|---|---|---|
> | f' | $+$ | does not exist | $-$ | 0 | $-$ |
> | $f"$ | $+$ | does not exist | $+$ | 0 | $-$ |
>
> 1. Find where all absolute extrema of f occur.
> 2. Find where the points of inflection of f occur.
> 3. Sketch a possible graph of f.

Quick Review 4.3

In Exercises 1 and 2, factor the expression and use sign charts to solve the inequality.

1. $x^2 - 9 < 0$
2. $x^3 - 4x > 0$

In Exercises 3–6, find the domains of f and f'.

3. $f(x) = xe^x$
4. $f(x) = x^{3/5}$
5. $f(x) = \dfrac{x}{x - 2}$
6. $f(x) = x^{2/5}$

In Exercises 7–10, find the horizontal asymptotes of the function's graph.

7. $y = (4 - x^2)e^x$
8. $y = (x^2 - x)e^{-x}$
9. $y = \dfrac{200}{1 + 10e^{-0.5x}}$
10. $y = \dfrac{750}{2 + 5e^{-0.1x}}$

Section 4.3 Exercises

In Exercises 1 and 2, use the graph of the function f to estimate where (a) f' and (b) $f"$ are 0, positive, and negative.

1.

2.
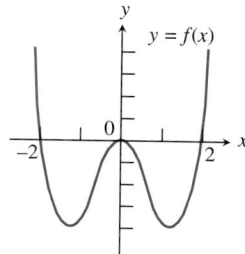

In Exercises 3–6, use the graph of f' to estimate the intervals on which the function f is (a) increasing or (b) decreasing. (c) Estimate where f has local extreme values.

3.

4.
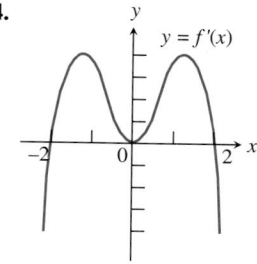

5. The domain of f' is $[0, 4) \cup (4, 6]$.

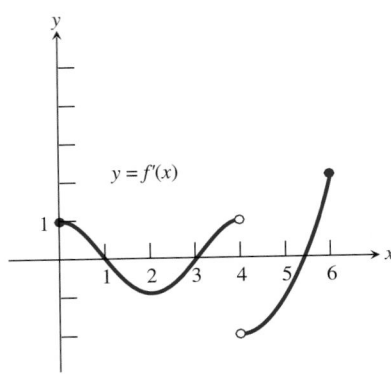

6. The domain of f' is $[0, 1) \cup (1, 2) \cup (2, 3]$.

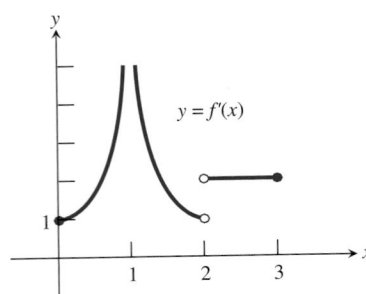

In Exercises 7–12, use analytic methods to find the intervals on which the function is

    (a) increasing,   **(b)** decreasing,

    (c) concave up,   **(d)** concave down.

Then find any

    (e) local extreme values,   **(f)** inflection points.

Support your answers graphically.

7. $y = x^2 - x - 1$ **8.** $y = -2x^3 + 6x^2 - 3$

9. $y = 2x^4 - 4x^2 + 1$ **10.** $y = xe^{1/x}$

11. $y = x\sqrt{8 - x^2}$ **12.** $y = \begin{cases} 3 - x^2, & x < 0 \\ x^2 + 1, & x \ge 0 \end{cases}$

In Exercises 13–28, find the intervals on which the function is

    (a) increasing,   **(b)** decreasing,

    (c) concave up,   **(d)** concave down.

Then find any

    (e) local extreme values,   **(f)** inflection points.

13. $y = 4x^3 + 21x^2 + 36x - 20$ **14.** $y = -x^4 + 4x^3 - 4x + 1$

15. $y = 2x^{1/5} + 3$ **16.** $y = 5 - x^{1/3}$

17. $y = \dfrac{5e^x}{e^x + 3e^{0.8x}}$ **18.** $y = \dfrac{8e^{-x}}{2e^{-x} + 5e^{-1.5x}}$

19. $y = \begin{cases} 2x, & x < 1 \\ 2 - x^2, & x \ge 1 \end{cases}$ **20.** $y = e^x, \quad 0 \le x \le 2\pi$

21. $y = xe^{1/x^2}$ **22.** $y = x^2\sqrt{9 - x^2}$

23. $y = \tan^{-1} x$ **24.** $y = x^{3/4}(5 - x)$

25. $y = x^{1/3}(x - 4)$ **26.** $y = x^{1/4}(x + 3)$

27. $y = \dfrac{x^3 - 2x^2 + x - 1}{x - 2}$ **28.** $y = \dfrac{x}{x^2 + 1}$

In Exercises 29 and 30, use the derivative of the function $y = f(x)$ to find the points at which f has a

    (a) local maximum,   **(b)** local minimum, or

    (c) point of inflection.

29. $y' = (x - 1)^2(x - 2)$

30. $y' = (x - 1)^2(x - 2)(x - 4)$

Exercises 31 and 32 show the graphs of the first and second derivatives of a function $y = f(x)$. *Work in groups of two or three.* Copy the figure and add a sketch of a possible graph of f that passes through the point P.

31.

32.

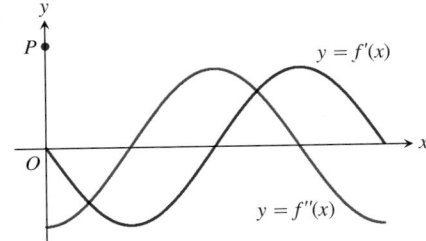

In Exercises 33 and 34, *work in groups of two or three.*

    (a) Find the absolute extrema of f and where they occur.

    (b) Find any points of inflection.

    (c) Sketch a possible graph of f.

33. f is continuous on $[0, 3]$ and satisfies the following.

x	0	1	2	3
f	0	2	0	-2
f'	3	0	does not exist	-3
f''	0	-1	does not exist	0

x	$0 < x < 1$	$1 < x < 2$	$2 < x < 3$
f	+	+	-
f'	+	-	-
f''	-	-	-

34. f is an even function, continuous on $[-3, 3]$, and satisfies the following.

x	0	1	2
f	2	0	-1
f'	does not exist	0	does not exist
f''	does not exist	0	does not exist

x	$0 < x < 1$	$1 < x < 2$	$2 < x < 3$
f	+	$-$	$-$
f'	$-$	$-$	+
f''	+	$-$	$-$

(d) What can you conclude about $f(3)$ and $f(-3)$?

In Exercises 35 and 36, *work in groups of two or three*. Sketch a possible graph of a continuous function f that has the given properties.

35. Domain $[0, 6]$, graph of f' given in Exercise 5, and $f(0) = 2$.

36. Domain $[0, 3]$, graph of f' given in Exercise 6, and $f(0) = -3$.

In Exercises 37–40, a particle is moving along a line with position function $s(t)$. Find the **(a)** velocity and **(b)** acceleration, and **(c)** describe the motion of the particle for $t \geq 0$.

37. $s(t) = t^2 - 4t + 3$

38. $s(t) = 6 - 2t - t^2$

39. $s(t) = t^3 - 3t + 3$

40. $s(t) = 3t^2 - 2t^3$

In Exercises 41 and 42, the graph of the position function $y = s(t)$ of a particle moving along a line is given. At approximately what times is the particle's **(a)** velocity equal to zero? **(b)** acceleration equal to zero?

41.

42.

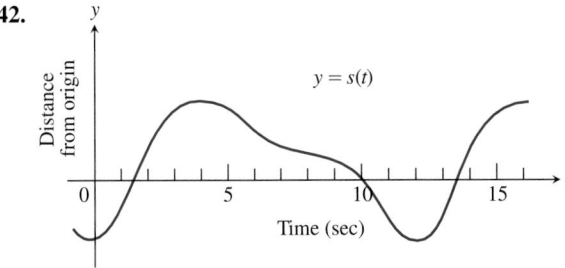

43. **Writing to Learn** If $f(x)$ is a differentiable function and $f'(c) = 0$ at an interior point c of f's domain, must f have a local maximum or minimum at $x = c$? Explain.

44. **Writing to Learn** If $f(x)$ is a twice-differentiable function and $f''(c) = 0$ at an interior point c of f's domain, must f have an inflection point at $x = c$? Explain.

45. *Connecting f and f′* Sketch a smooth curve $y = f(x)$ through the origin with the properties that $f'(x) < 0$ for $x < 0$ and $f'(x) > 0$ for $x > 0$.

46. *Connecting f and f″* Sketch a smooth curve $y = f(x)$ through the origin with the properties that $f''(x) < 0$ for $x < 0$ and $f''(x) > 0$ for $x > 0$.

47. *Connecting f, f′, and f″* Sketch a continuous curve $y = f(x)$ with the following properties. Label coordinates where possible.

$f(-2) = 8$	$f'(x) > 0$ for $|x| > 2$
$f(0) = 4$	$f'(x) < 0$ for $|x| < 2$
$f(2) = 0$	$f''(x) < 0$ for $x < 0$
$f'(2) = f'(-2) = 0$	$f''(x) > 0$ for $x > 0$

48. *Using Behavior to Sketch* Sketch a continuous curve $y = f(x)$ with the following properties. Label coordinates where possible.

x	y	Curve
$x < 2$		falling, concave up
2	1	horizontal tangent
$2 < x < 4$		rising, concave up
4	4	inflection point
$4 < x < 6$		rising, concave down
6	7	horizontal tangent
$x > 6$		falling, concave down

49. Table 4.3 gives the sales of prepaid phone cards for several years (the values for 1997 and 1998 were estimated). Let $x = 0$ represent 1992, $x = 1$ represent 1993, and so forth.

(a) Find the logistic regression equation and superimpose its graph on a scatter plot of the data.

(b) Use the regression equation to predict when the rate of increase in sales will start to decrease, and to predict the sales at that time.

(c) At what amount does the regression equation predict that sales will stabilize?

Table 4.3 Prepaid Phone Card Sales

Year	Sales (millions of dollars)
1992	12
1993	114
1994	348
1995	792
1996	1110
1997	1520
1998	1900

Source: Atlantic ACM as reported by Grant Jerding in *USA Today*, April 16, 1997.

Exploration

50. *Graphs of Cubics* There is almost no leeway in the locations of the inflection point and the extrema of $f(x) = ax^3 + bx^2 + cx + d$, $a \neq 0$, because the one inflection point occurs at $x = -b/(3a)$ and the extrema, if any, must be located symmetrically about this value of x. Check this out by examining **(a)** the cubic in Exercise 13 and **(b)** the cubic in Exercise 8. Then **(c)** prove the general case. ∎

Extending the Ideas

In Exercises 51 and 52, feel free to use a CAS (computer algebra system), if you have one, to solve the problem.

51. *Logistic Functions* Let $f(x) = c/(1 + ae^{-bx})$ with $a > 0$, $abc \neq 0$.

    (a) Show that f is increasing on the interval $(-\infty, \infty)$ if $abc > 0$, and decreasing if $abc < 0$.

    (b) Show that the point of inflection of f occurs at $x = (\ln |a|)/b$.

52. *Quartic Polynomial Functions* Let $f(x) = ax^4 + bx^3 + cx^2 + dx + e$ with $a \neq 0$.

    (a) Show that the graph of f has 0 or 2 points of inflection.

    (b) Write a condition that must be satisfied by the coefficients if the graph of f has 0 or 2 points of inflection.

4.4 Modeling and Optimization

Examples from Business and Industry • Examples from Mathematics • Examples from Economics • Modeling Discrete Phenomena with Differentiable Functions

Examples from Business and Industry

To *optimize* something means to maximize or minimize some aspect of it. What is the size of the most profitable production run? What is the least expensive shape for an oil can? What is the stiffest rectangular beam we can cut from a 12-inch log? We usually answer such questions by finding the greatest or smallest value of some function that we have used to model the situation.

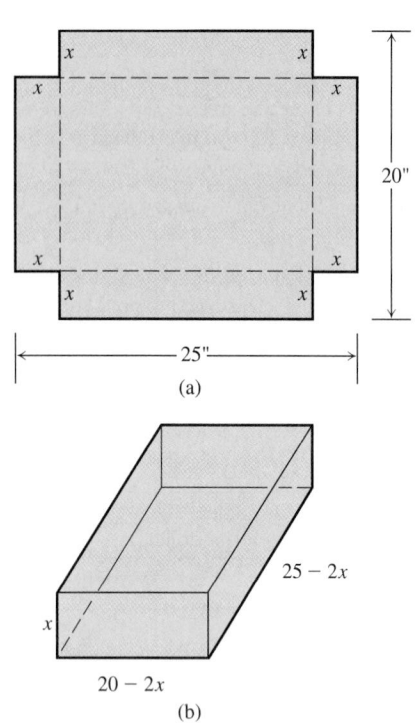

Figure 4.28 An open box made by cutting the corners from a piece of tin. (Example 1)

Example 1 FABRICATING A BOX

An open-top box is to be made by cutting congruent squares of side length x from the corners of a 20- by 25-inch sheet of tin and bending up the sides (Figure 4.28). How large should the squares be to make the box hold as much as possible? What is the resulting maximum volume?

Solution

Model

The height of the box is x, and the other two dimensions are $(20 - 2x)$ and $(25 - 2x)$. Thus, the volume of the box is

$$V(x) = x(20 - 2x)(25 - 2x).$$

Solve Graphically

Because $2x$ cannot exceed 20, we have $0 \leq x \leq 10$. Figure 4.29 suggests that the maximum value of V is about 820.53 and occurs at $x \approx 3.68$.

Confirm Analytically

Expanding, we obtain $V(x) = 4x^3 - 90x^2 + 500x$. The first derivative of V is

$$V'(x) = 12x^2 - 180x + 500.$$

$y = x(20 - 2x)(25 - 2x)$

Maximum
X = 3.6811856 Y = 820.52819

[0, 10] by [−300, 1000]

Figure 4.29 We chose the −300 in −300 ≤ y ≤ 1000 so that the coordinates of the local maximum at the bottom of the screen would not interfere with the graph. (Example 1)

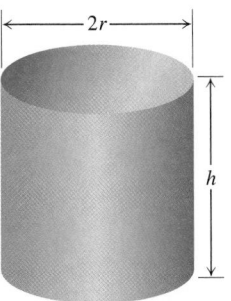

Figure 4.30 This one-liter can uses the least material when $h = 2r$. (Example 2)

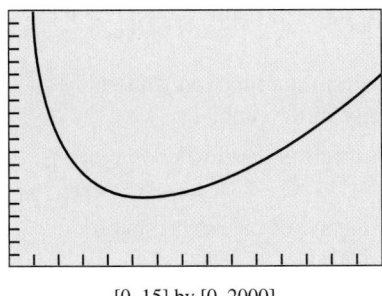

[0, 15] by [0, 2000]

Figure 4.31 The graph of $A = 2\pi r^2 + 2000/r$, $r > 0$. (Example 2)

The two solutions of the quadratic equation $V'(x) = 0$ are

$$c_1 = \frac{180 - \sqrt{180^2 - 48(500)}}{24} \approx 3.68 \text{ and}$$

$$c_2 = \frac{180 + \sqrt{180^2 - 48(500)}}{24} \approx 11.32.$$

Only c_1 is in the domain $[0, 10]$ of V. The values of V at this one critical point and the two endpoints are

Critical point value: $V(c_1) \approx 820.53$

Endpoint values: $V(0) = 0, \quad V(10) = 0.$

Interpret

Cutout squares that are about 3.68 in. on a side give the maximum volume, about 820.53 in³.

Example 2   DESIGNING A CAN

You have been asked to design a one-liter oil can shaped like a right circular cylinder (Figure 4.30). What dimensions will use the least material?

Solution

Volume of can: If r and h are measured in centimeters, then the volume of the can in cubic centimeters is

$$\pi r^2 h = 1000. \qquad \text{1 liter} = 1000 \text{ cm}^3$$

Surface area of can: $A = \underbrace{2\pi r^2}_{\substack{\text{circular} \\ \text{ends}}} + \underbrace{2\pi rh}_{\substack{\text{cylinder} \\ \text{wall}}}$

How can we interpret the phrase "least material"? One possibility is to ignore the thickness of the material and the waste in manufacturing. Then we ask for dimensions r and h that make the total surface area as small as possible while satisfying the constraint $\pi r^2 h = 1000$. (Exercise 17 describes one way to take waste into account.)

Model

To express the surface area as a function of one variable, we solve for one of the variables in $\pi r^2 h = 1000$ and substitute that expression into the surface area formula. Solving for h is easier,

$$h = \frac{1000}{\pi r^2}.$$

Thus,

$$A = 2\pi r^2 + 2\pi rh$$

$$= 2\pi r^2 + 2\pi r\left(\frac{1000}{\pi r^2}\right)$$

$$= 2\pi r^2 + \frac{2000}{r}.$$

Solve Analytically

Our goal is to find a value of $r > 0$ that minimizes the value of A. Figure 4.31 suggests that such a value exists.

Notice from the graph that for small r (a tall thin container, like a piece of pipe), the term $2000/r$ dominates and A is large. For large r (a short wide container, like a pizza pan), the term $2\pi r^2$ dominates and A again is large.

Since A is differentiable on $r > 0$, an interval with no endpoints, it can have a minimum value only where its first derivative is zero.

$$\frac{dA}{dr} = 4\pi r - \frac{2000}{r^2}$$

$$0 = 4\pi r - \frac{2000}{r^2} \qquad \text{Set } dA/dr = 0.$$

$$4\pi r^3 = 2000 \qquad \text{Multiply by } r^2.$$

$$r = \sqrt[3]{\frac{500}{\pi}} \approx 5.42 \qquad \text{Solve for } r.$$

Something happens at $r = \sqrt[3]{500/\pi}$, but what?

If the domain of A were a closed interval, we could find out by evaluating A at this critical point and the endpoints and comparing the results. But the domain is an open interval, so we must learn what is happening at $r = \sqrt[3]{500/\pi}$ by referring to the shape of A's graph. The second derivative

$$\frac{d^2A}{dr^2} = 4\pi + \frac{4000}{r^3}$$

is positive throughout the domain of A. The graph is therefore concave up and the value of A at $r = \sqrt[3]{500/\pi}$ an absolute minimum.

The corresponding value of h (after a little algebra) is

$$h = \frac{1000}{\pi r^2} = 2\sqrt[3]{\frac{500}{\pi}} = 2r.$$

Interpret

The one-liter can that uses the least material has height equal to the diameter, with $r \approx 5.42$ cm and $h \approx 10.84$ cm.

Strategy for Solving Max-Min Problems

1. **Understand the Problem** Read the problem carefully. Identify the information you need to solve the problem.

2. **Develop a Mathematical Model of the Problem** Draw pictures and label the parts that are important to the problem. Introduce a variable to represent the quantity to be maximized or minimized. Using that variable, write a function whose extreme value gives the information sought.

3. **Graph the Function** Find the domain of the function. Determine what values of the variable make sense in the problem.

4. **Identify the Critical Points and Endpoints** Find where the derivative is zero or fails to exist.

5. **Solve the Mathematical Model** If unsure of the result, support or confirm your solution with another method.

6. **Interpret the Solution** Translate your mathematical result into the problem setting and decide whether the result makes sense.

Examples from Mathematics

Example 3 USING THE STRATEGY

Find two numbers whose sum is 20 and whose product is as large as possible.

Solution

Model

If one number is x, the other is $(20 - x)$, and their product is
$f(x) = x(20 - x)$.

Solve Graphically

We can see from the graph of f in Figure 4.32 that there is a maximum. From what we know about parabolas, the maximum occurs at $x = 10$.

Interpret

The two numbers we seek are $x = 10$ and $20 - x = 10$.

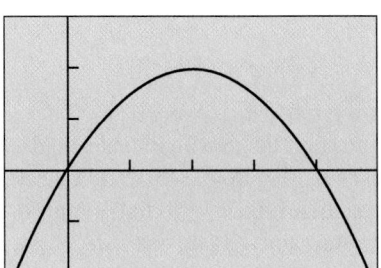

[–5, 25] by [–100, 150]

Figure 4.32 The graph of $f(x) = x(20 - x)$ with domain $(-\infty, \infty)$ has an absolute maximum of 100 at $x = 10$. (Example 3)

Sometimes we find it helpful to use both analytic and graphical methods together, as in Example 4.

Example 4 INSCRIBING RECTANGLES

A rectangle is to be inscribed under one arch of the sine curve (Figure 4.33). What is the largest area the rectangle can have, and what dimensions give that area?

Solution

Model

Let $(x, \sin x)$ be the coordinates of point P in Figure 4.33. From what we know about the sine function the x-coordinate of point Q is $(\pi - x)$. Thus,

$$\pi - 2x = \text{length of rectangle}$$

and

$$\sin x = \text{height of rectangle}.$$

The area of the rectangle is

$$A(x) = (\pi - 2x) \sin x.$$

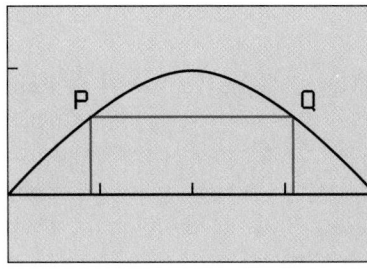

[0, π] by [–0.5, 1.5]

Figure 4.33 A rectangle inscribed under one arch of $y = \sin x$. (Example 4)

Maximum
X = .71046344 Y = 1.1221927

[0, π/2] by [–1, 2]

(a)

Zero
X = .71046274 Y = 0

[0, π/2] by [–4, 4]

(b)

Figure 4.34 The graph of (a) $A(x) = (\pi - 2x) \sin x$ and (b) A' in the interval $0 \leq x \leq \pi/2$. (Example 4)

Solve Analytically and Graphically

We can assume that $0 \leq x \leq \pi/2$. Notice that $A = 0$ at the endpoints $x = 0$ and $x = \pi/2$. Since A is differentiable, the only critical points occur at the zeros of the first derivative,

$$A'(x) = -2 \sin x + (\pi - 2x) \cos x.$$

It is not possible to solve the equation $A'(x) = 0$ using algebraic methods. We can use the graph of A (Figure 4.34a) to find the maximum value and where it occurs. Or, we can use the graph of A' (Figure 4.34b) to find where the derivative is zero, and then evaluate A at this value of x to find the maximum value. The two x-values appear to be the same, as they should.

Interpret

The rectangle has a maximum area of about 1.12 square units when $x \approx 0.71$. At this point, the rectangle is $\pi - 2x \approx 1.72$ units long by $\sin x \approx 0.65$ unit high.

Exploration 1 **Constructing Cones**

A cone of height h and radius r is constructed from a flat, circular disk of radius 4 in. by removing a sector AOC of arc length x in. and then connecting the edges OA and OC. What arc length x will produce the cone of maximum volume, and what is that volume?

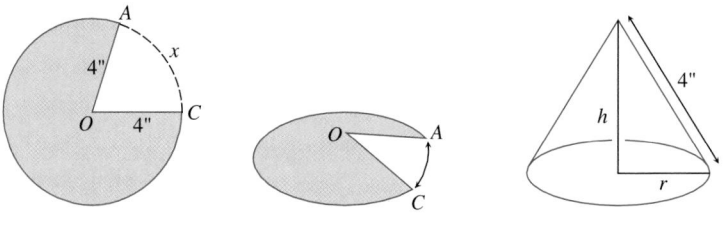

NOT TO SCALE

1. Show that
$$r = \frac{8\pi - x}{2\pi}, \quad h = \sqrt{16 - r^2}, \quad \text{and}$$
$$V(x) = \frac{\pi}{3}\left(\frac{8\pi - x}{2\pi}\right)^2 \sqrt{16 - \left(\frac{8\pi - x}{2\pi}\right)^2}.$$

2. Show that the natural domain of V is $0 \leq x \leq 16\pi$. Graph V over this domain.

3. Explain why the restriction $0 \leq x \leq 8\pi$ makes sense in the problem situation. Graph V over this domain.

4. Use graphical methods to find where the cone has its maximum volume, and what that volume is.

5. Confirm your findings in part 4 analytically. (*Hint:* Use $V(x) = (1/3)\pi r^2 h$, $h^2 + r^2 = 16$, and the Chain Rule.)

Examples from Economics

Here we want to point out two more places where calculus makes a contribution to economic theory. The first has to do with maximizing profit. The second has to do with minimizing average cost.

Suppose that

$$r(x) = \text{the revenue from selling } x \text{ items,}$$

$$c(x) = \text{the cost of producing the } x \text{ items,}$$

$$p(x) = r(x) - c(x) = \text{the profit from selling } x \text{ items.}$$

Marginal Analysis

Because differentiable functions are locally linear, we can use the marginals to approximate the extra revenue, cost, or profit resulting from selling or producing one more item. Using these approximations is referred to as *marginal analysis*.

The marginal revenue, marginal cost, and marginal profit at this production level (x items) are

$$\frac{dr}{dx} = \text{marginal revenue,}$$

$$\frac{dc}{dx} = \text{marginal cost,}$$

$$\frac{dp}{dx} = \text{marginal profit.}$$

The first observation is about the relationship of p to these derivatives.

Theorem 6 Maximum Profit

Maximum profit (if any) occurs at a production level at which marginal revenue equals marginal cost.

Proof We assume that $r(x)$ and $c(x)$ are differentiable for all $x > 0$, so if $p(x) = r(x) - c(x)$ has a maximum value, it occurs at a production level at which $p'(x) = 0$. Since $p'(x) = r'(x) - c'(x)$, $p'(x) = 0$ implies that

$$r'(x) - c'(x) = 0 \quad \text{or} \quad r'(x) = c'(x). \qquad \blacksquare$$

Figure 4.35 gives more information about this situation.

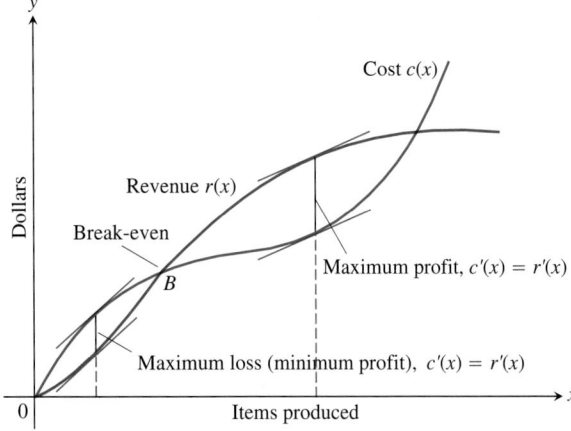

Figure 4.35 The graph of a typical cost function starts concave down and later turns concave up. It crosses the revenue curve at the break-even point *B*. To the left of *B*, the company operates at a loss. To the right, the company operates at a profit, the maximum profit occurring where $r'(x) = c'(x)$. Farther to the right, cost exceeds revenue (perhaps because of a combination of market saturation and rising labor and material costs) and production levels become unprofitable again.

What guidance do we get from this observation? We know that a production level at which $p'(x) = 0$ need not be a level of maximum profit. It might be a level of minimum profit, for example. But if we are making financial projections for our company, we should look for production levels at which marginal cost seems to equal marginal revenue. If there is a most profitable production level, it will be one of these.

Example 5 MAXIMIZING PROFIT

Suppose that $r(x) = 9x$ and $c(x) = x^3 - 6x^2 + 15x$, where x represents thousands of units. Is there a production level that maximizes profit? If so, what is it?

Solution Notice that $r'(x) = 9$ and $c'(x) = 3x^2 - 12x + 15$.

$$3x^2 - 12x + 15 = 9 \qquad \text{Set } c'(x) = r'(x).$$

$$3x^2 - 12x + 6 = 0$$

The two solutions of the quadratic equation are

$$x_1 = \frac{12 - \sqrt{72}}{6} = 2 - \sqrt{2} \approx 0.586 \qquad \text{and}$$

$$x_2 = \frac{12 + \sqrt{72}}{6} = 2 + \sqrt{2} \approx 3.414.$$

The possible production levels for maximum profit are $x \approx 0.586$ thousand units or $x \approx 3.414$ thousand units. The graphs in Figure 4.36 show that maximum profit occurs at about $x = 3.414$ and maximum loss occurs at about $x = 0.586$.

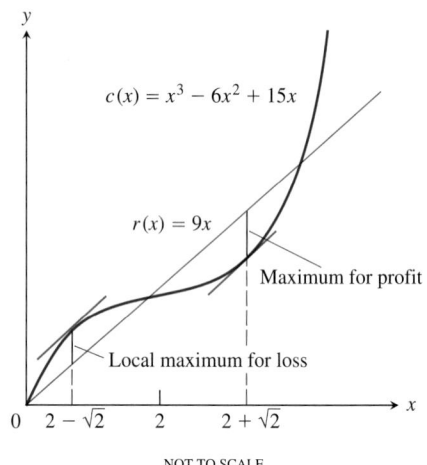

$c(x) = x^3 - 6x^2 + 15x$

$r(x) = 9x$

Maximum for profit

Local maximum for loss

$0 \quad 2 - \sqrt{2} \quad 2 \quad 2 + \sqrt{2}$

NOT TO SCALE

Figure 4.36 The cost and revenue curves for Example 5.

Another way to look for optimal production levels is to look for levels that minimize the average cost of the units produced. Theorem 7 helps us find them.

> ### Theorem 7 Minimizing Average Cost
>
> The production level (if any) at which average cost is smallest is a level at which the average cost equals the marginal cost.

Proof We assume that $c(x)$ is differentiable.

$$c(x) = \text{cost of producing } x \text{ items,} \quad x > 0$$

$$\frac{c(x)}{x} = \text{average cost of producing } x \text{ items}$$

If the average cost can be minimized, it will be a production level at which

$$\frac{d}{dx}\left(\frac{c(x)}{x}\right) = 0$$

$$\frac{xc'(x) - c(x)}{x^2} = 0 \qquad \text{Quotient Rule}$$

$$xc'(x) - c(x) = 0 \qquad \text{Multiply by } x^2.$$

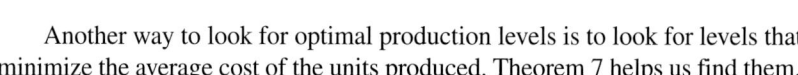

$$\underbrace{c'(x)}_{\text{marginal cost}} = \underbrace{\frac{c(x)}{x}}_{\text{average cost}}.$$

∎

Again we have to be careful about what Theorem 7 does and does not say. It does not say that there is a production level of minimum average cost—it says where to look to see if there is one. Look for production levels at which average cost and marginal cost are equal. Then check to see if any of them gives a minimum average cost.

Example 6 MINIMIZING AVERAGE COST

Suppose $c(x) = x^3 - 6x^2 + 15x,$ where x represents thousands of units. Is there a production level that minimizes average cost? If so, what is it?

Solution We look for levels at which average cost equals marginal cost.

Marginal cost: $c'(x) = 3x^2 - 12x + 15$

Average cost: $\dfrac{c(x)}{x} = x^2 - 6x + 15$

$$3x^2 - 12x + 15 = x^2 - 6x + 15 \qquad \text{Marginal cost = Average cost}$$
$$2x^2 - 6x = 0$$
$$2x(x - 3) = 0$$
$$x = 0 \quad \text{or} \quad x = 3$$

Since $x > 0,$ the only production level that might minimize average cost is $x = 3$ thousand units.

We use the second derivative test.

$$\frac{d}{dx}\left(\frac{c(x)}{x}\right) = 2x - 6$$
$$\frac{d^2}{dx^2}\left(\frac{c(x)}{x}\right) = 2 > 0$$

The second derivative is positive for all $x > 0,$ so $x = 3$ gives an absolute minimum.

Modeling Discrete Phenomena with Differentiable Functions

In case you are wondering how we can use differentiable functions $c(x)$ and $r(x)$ to describe the cost and revenue that comes from producing a number of items x that can only be an integer, here is the rationale.

When x is large, we can reasonably fit the cost and revenue data with smooth curves $c(x)$ and $r(x)$ that are defined not only at integer values of x but at the values in between just as we do when we use regression equations. Once we have these differentiable functions, which are supposed to behave like the real cost and revenue when x is an integer, we can apply calculus to draw conclusions about their values. We then translate these mathematical conclusions into inferences about the real world that we hope will have predictive value. When they do, as is the case with the economic theory here, we say that the functions give a good model of reality.

What do we do when our calculus tells us that the best production level is a value of x that isn't an integer, as it did in Example 5? We use the nearest convenient integer. For $x \approx 3.414$ thousand units in Example 5, we might use 3414, or perhaps 3410 or 3420 if we ship in boxes of 10.

Quick Review 4.4

1. Use the first derivative test to identify the local extrema of $y = x^3 - 6x^2 + 12x - 8$.

2. Use the second derivative test to identify the local extrema of $y = 2x^3 + 3x^2 - 12x - 3$.

3. Find the volume of a cone with radius 5 cm and height 8 cm.

4. Find the dimensions of a right circular cylinder with volume 1000 cm^3 and surface area 600 cm^2.

In Exercises 5–8, rewrite the expression as a trigonometric function of the angle α.

5. $\sin(-\alpha)$ **6.** $\cos(-\alpha)$

7. $\sin(\pi - \alpha)$ **8.** $\cos(\pi - \alpha)$

In Exercises 9 and 10, use substitution to find the exact solutions of the system of equations.

9. $\begin{cases} x^2 + y^2 = 4 \\ y = \sqrt{3}x \end{cases}$ **10.** $\begin{cases} \dfrac{x^2}{4} + \dfrac{y^2}{9} = 1 \\ y = x + 3 \end{cases}$

Section 4.4 Exercises

In Exercises 1–10, solve the problem analytically. Support your answer graphically.

1. *Finding Numbers* The sum of two nonnegative numbers is 20. Find the numbers if

(a) the sum of their squares is as large as possible; as small as possible.

(b) one number plus the square root of the other is as large as possible; as small as possible.

2. *Maximizing Area* What is the largest possible area for a right triangle whose hypotenuse is 5 cm long, and what are its dimensions?

3. *Maximizing Perimeter* What is the smallest perimeter possible for a rectangle whose area is 16 in^2, and what are its dimensions?

4. *Finding Area* Show that among all rectangles with an 8-m perimeter, the one with largest area is a square.

5. *Inscribing Rectangles* The figure shows a rectangle inscribed in an isosceles right triangle whose hypotenuse is 2 units long.

(a) Express the y-coordinate of P in terms of x. (*Hint:* Write an equation for the line *AB*.)

(b) Express the area of the rectangle in terms of x.

(c) What is the largest area the rectangle can have, and what are its dimensions?

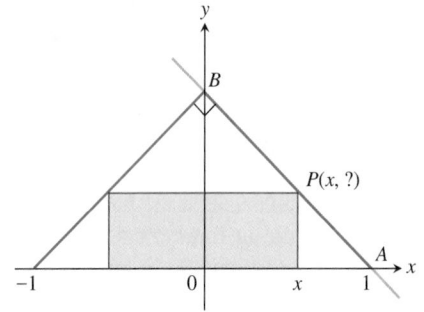

6. *Largest Rectangle* A rectangle has its base on the x-axis and its upper two vertices on the parabola $y = 12 - x^2$. What is the largest area the rectangle can have, and what are its dimensions?

7. *Optimal Dimensions* You are planning to make an open rectangular box from an 8- by 15-in. piece of cardboard by cutting congruent squares from the corners and folding up the sides. What are the dimensions of the box of largest volume you can make this way, and what is its volume?

8. *Closing Off the First Quadrant* You are planning to close off a corner of the first quadrant with a line segment 20 units long running from $(a, 0)$ to $(0, b)$. Show that the area of the triangle enclosed by the segment is largest when $a = b$.

9. *The Best Fencing Plan* A rectangular plot of farmland will be bounded on one side by a river and on the other three sides by a single-strand electric fence. With 800 m of wire at your disposal, what is the largest area you can enclose, and what are its dimensions?

10. *The Shortest Fence* A 216-m^2 rectangular pea patch is to be enclosed by a fence and divided into two equal parts by another fence parallel to one of the sides. What dimensions for the outer rectangle will require the smallest total length of fence? How much fence will be needed?

11. *Designing a Tank* Your iron works has contracted to design and build a 500-ft^3, square-based, open-top, rectangular steel holding tank for a paper company. The tank is to be made by welding thin stainless steel plates together along their edges. As the production engineer, your job is to find dimensions for the base and height that will make the tank weigh as little as possible.

(a) What dimensions do you tell the shop to use?

(b) **Writing to Learn** Briefly describe how you took weight into account.

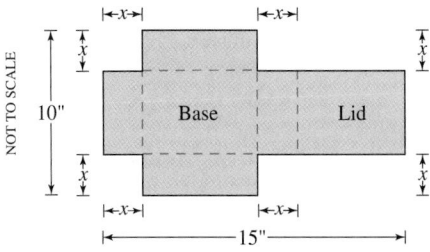

12. *Catching Rainwater* A 1125-ft^3 open-top rectangular tank with a square base x ft on a side and y ft deep is to be built with its top flush with the ground to catch runoff water. The costs associated with the tank involve not only the material from which the tank is made but also an excavation charge proportional to the product xy.

(a) If the total cost is

$$c = 5(x^2 + 4xy) + 10xy,$$

what values of x and y will minimize it?

(b) Writing to Learn Give a possible scenario for the cost function in (a).

13. *Designing a Poster* You are designing a rectangular poster to contain 50 in^2 of printing with a 4-in. margin at the top and bottom and a 2-in. margin at each side. What overall dimensions will minimize the amount of paper used?

14. *Vertical Motion* The height of an object moving vertically is given by

$$s = -16t^2 + 96t + 112,$$

with s in ft and t in sec. Find **(a)** the object's velocity when $t = 0$, **(b)** its maximum height and when it occurs, and **(c)** its velocity when $s = 0$.

15. *Finding an Angle* Two sides of a triangle have lengths a and b, and the angle between them is θ. What value of θ will maximize the triangle's area? (*Hint:* $A = (1/2)\,ab\sin\theta$.)

16. *Designing a Can* What are the dimensions of the lightest open-top right circular cylindrical can that will hold a volume of 1000 cm^3? Compare the result here with the result in Example 2.

17. *Designing a Can* You are designing a 1000-cm^3 right circular cylindrical can whose manufacture will take waste into account. There is no waste in cutting the aluminum for the side, but the top and bottom of radius r will be cut from squares that measure $2r$ units on a side. The total amount of aluminum used up by the can will therefore be

$$A = 8r^2 + 2\pi rh$$

rather than the $A = 2\pi r^2 + 2\pi rh$ in Example 2. In Example 2 the ratio of h to r for the most economical can was 2 to 1. What is the ratio now?

In Exercises 18 and 19, *work in groups of two or three*.

18. *Designing a Box with Lid* A piece of cardboard measures 10- by 15-in. Two equal squares are removed from the corners of a 10-in. side as shown in the figure. Two equal rectangles are removed from the other corners so that the tabs can be folded to form a rectangular box with lid.

(a) Write a formula $V(x)$ for the volume of the box.

(b) Find the domain of V for the problem situation and graph V over this domain.

(c) Use a graphical method to find the maximum volume and the value of x that gives it.

(d) Confirm your result in (c) analytically.

19. *Designing a Suitcase* A 24- by 36-in. sheet of cardboard is folded in half to form a 24- by 18-in. rectangle as shown in the figure. Then four congruent squares of side length x are cut from the corners of the folded rectangle. The sheet is unfolded, and the six tabs are folded up to form a box with sides and a lid.

(a) Write a formula $V(x)$ for the volume of the box.

(b) Find the domain of V for the problem situation and graph V over this domain.

(c) Use a graphical method to find the maximum volume and the value of x that gives it.

(d) Confirm your result in (c) analytically.

(e) Find a value of x that yields a volume of 1120 in^3.

(f) Writing to Learn Write a paragraph describing the issues that arise in part (b).

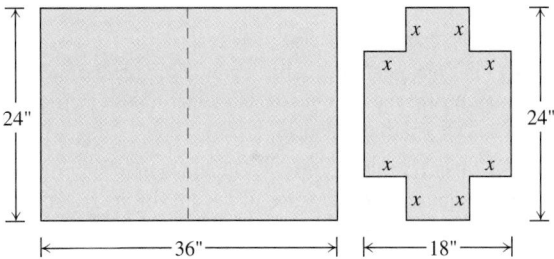

The sheet is then unfolded.

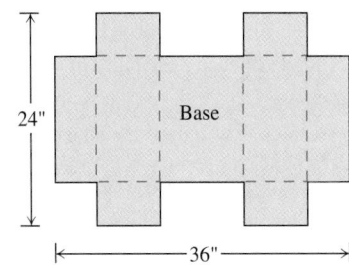

20. *Quickest Route* Jane is 2 mi offshore in a boat and wishes to reach a coastal village 6 mi down a straight shoreline from the point nearest the boat. She can row 2 mph and can walk 5 mph. Where should she land her boat to reach the village in the least amount of time?

21. *Inscribing Rectangles* A rectangle is to be inscribed under the arch of the curve $y = 4\cos(0.5x)$ from $x = -\pi$ to $x = \pi$. What are the dimensions of the rectangle with largest area, and what is the largest area?

22. *Maximizing Volume* Find the dimensions of a right circular cylinder of maximum volume that can be inscribed in a sphere of radius 10 cm. What is the maximum volume?

Exploration

23. The figure shows the graph of $f(x) = xe^{-x}$, $x \geq 0$.

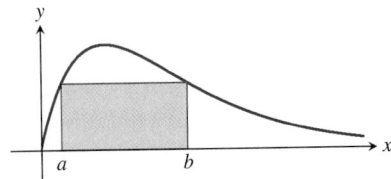

(a) Find where the absolute maximum of f occurs.

(b) Let $a > 0$ and $b > 0$ be given as shown in the figure. Complete the following table where A is the area of the rectangle in the figure.

a	b	A
0.1		
0.2		
0.3		
$\vdots$		
1		

(c) Draw a scatter plot of the data (a, A).

(d) Find the quadratic, cubic, and quartic regression equations for the data in (b), and superimpose their graphs on a scatter plot of the data.

(e) Use each of the regression equations in (d) to estimate the maximum possible value of the area of the rectangle. ■

24. *Cubic Polynomial Functions*
Let $f(x) = ax^3 + bx^2 + cx + d$, $a \neq 0$.

(a) Show that f has either 0 or 2 local extrema.

(b) Give an example of each possibility in (a).

25. *Shipping Packages* The U.S. Postal Service will accept a box for domestic shipment only if the sum of its length and girth (distance around), as shown in the figure, does not exceed 108 in. What dimensions will give a box with a square end the largest possible volume?

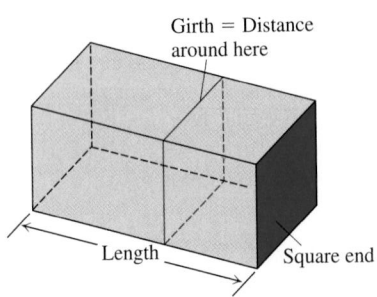

Girth = Distance around here

Length

Square end

26. *Constructing Cylinders* Compare the answers to the following two construction problems.

(a) A rectangular sheet of perimeter 36 cm and dimensions x cm by y cm is to be rolled into a cylinder as shown in part (a) of the figure. What values of x and y give the largest volume?

(b) The same sheet is to be revolved about one of the sides of length y to sweep out the cylinder as shown in part (b) of the figure. What values of x and y give the largest volume?

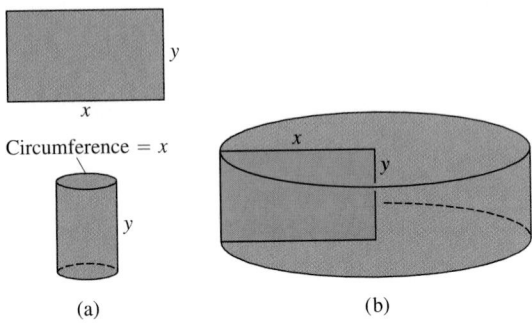

Circumference = x

(a)

(b)

27. *Constructing Cones* A right triangle whose hypotenuse is $\sqrt{3}$ m long is revolved about one of its legs to generate a right circular cone. Find the radius, height, and volume of the cone of greatest volume that can be made this way.

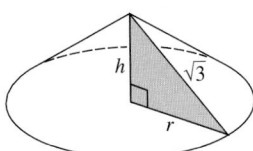

28. *Finding Parameter Values* What value of a makes $f(x) = x^2 + (a/x)$ have **(a)** a local minimum at $x = 2$?
(b) a point of inflection at $x = 1$?

29. *Finding Parameter Values* Show that $f(x) = x^2 + (a/x)$ cannot have a local maximum for any value of a.

30. *Finding Parameter Values* What values of a and b make $f(x) = x^3 + ax^2 + bx$ have **(a)** a local maximum at $x = -1$ and a local minimum at $x = 3$? **(b)** a local minimum at $x = 4$ and a point of inflection at $x = 1$?

31. *Inscribing a Cone* Find the volume of the largest right circular cone that can be inscribed in a sphere of radius 3.

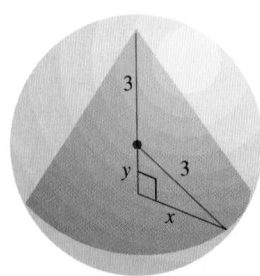

32. *Strength of a Beam* The strength S of a rectangular wooden beam is proportional to its width times the square of its depth.

(a) Find the dimensions of the strongest beam that can be cut from a 12-in. diameter cylindrical log.

(b) Writing to Learn Graph S as a function of the beam's width w, assuming the proportionality constant to be $k = 1$. Reconcile what you see with your answer in (a).

(c) Writing to Learn On the same screen, graph S as a function of the beam's depth d, again taking $k = 1$. Compare the graphs with one another and with your answer in (a). What would be the effect of changing to some other value of k? Try it.

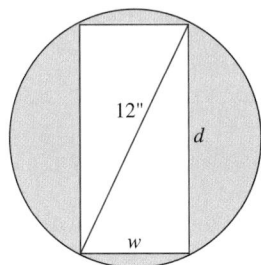

33. *Stiffness of a Beam* The stiffness S of a rectangular beam is proportional to its width times the cube of its depth.

(a) Find the dimensions of the stiffest beam that can be cut from a 12-in. diameter cylindrical log.

(b) Writing to Learn Graph S as a function of the beam's width w, assuming the proportionality constant to be $k = 1$. Reconcile what you see with your answer in (a).

(c) Writing to Learn On the same screen, graph S as a function of the beam's depth d, again taking $k = 1$. Compare the graphs with one another and with your answer in (a). What would be the effect of changing to some other value of k? Try it.

34. *Frictionless Cart* A small frictionless cart, attached to the wall by a spring, is pulled 10 cm from its rest position and released at time $t = 0$ to roll back and forth for 4 sec. Its position at time t is $s = 10 \cos \pi t$.

(a) What is the cart's maximum speed? When is the cart moving that fast? Where is it then? What is the magnitude of the acceleration then?

(b) Where is the cart when the magnitude of the acceleration is greatest? What is the cart's speed then?

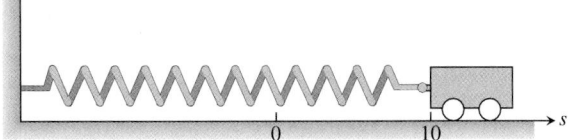

35. *Electrical Current* Suppose that at any time t (sec) the current i (amp) in an alternating current circuit is $i = 2 \cos t + 2 \sin t$. What is the peak (largest magnitude) current for this circuit?

36. *Calculus and Geometry* How close does the curve $y = \sqrt{x}$ come to the point $(3/2, 0)$? (*Hint:* If you minimize the *square* of the distance, you can avoid square roots.)

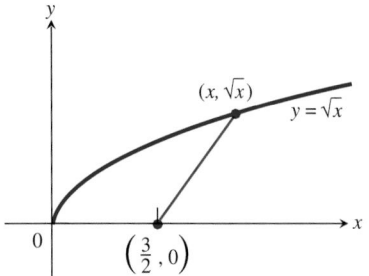

37. *Calculus and Geometry* How close does the semicircle $y = \sqrt{16 - x^2}$ come to the point $(1, \sqrt{3})$?

38. Writing to Learn Is the function $f(x) = x^2 - x + 1$ ever negative? Explain.

39. Writing to Learn You have been asked to determine whether the function $f(x) = 3 + 4 \cos x + \cos 2x$ is ever negative.

(a) Explain why you need to consider values of x only in the interval $[0, 2\pi]$.

(b) Is f ever negative? Explain.

40. *Vertical Motion* Two masses hanging side by side from springs have positions $s_1 = 2 \sin t$ and $s_2 = \sin 2t$, respectively, with s_1 and s_2 in meters and t in seconds.

(a) At what times in the interval $t > 0$ do the masses pass each other? (*Hint:* $\sin 2t = 2 \sin t \cos t$.)

(b) When in the interval $0 \le t \le 2\pi$ is the vertical distance between the masses the greatest? What is this distance? (*Hint:* $\cos 2t = 2 \cos^2 t - 1$.)

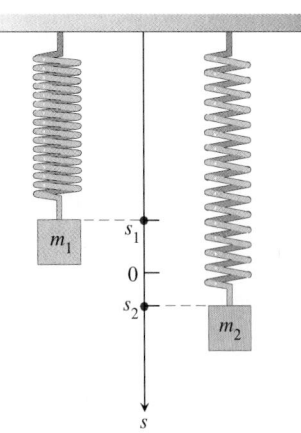

41. *Motion on a Line* The positions of two particles on the s-axis are $s_1 = \sin t$ and $s_2 = \sin (t + \pi/3)$, with s_1 and s_2 in meters and t in seconds.

(a) At what time(s) in the interval $0 \le t \le 2\pi$ do the particles meet?

(b) What is the farthest apart that the particles ever get?

(c) When in the interval $0 \le t \le 2\pi$ is the distance between the particles changing the fastest?

42. *Finding an Angle* The trough in the figure is to be made to the dimensions shown. Only the angle θ can be varied. What value of θ will maximize the trough's volume?

43. *Paper Folding* *Work in groups of two or three.* A rectangular sheet of 8 1/2- by 11-in. paper is placed on a flat surface. One of the corners is placed on the opposite longer edge, as shown in the figure, and held there as the paper is smoothed flat. The problem is to make the length of the crease as small as possible. Call the length L. Try it with paper.

(a) Show that $L^2 = 2x^3/(2x - 8.5)$.

(b) What value of x minimizes L^2?

(c) What is the minimum value of L?

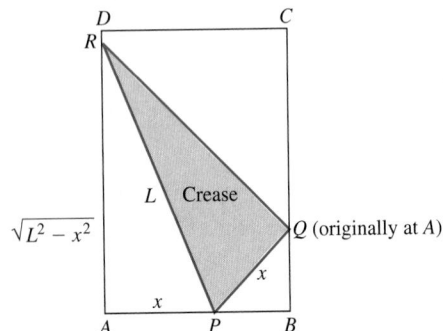

44. *Sensitivity to Medicine* *(continuation of Exercise 35, Section 3.3)* Find the amount of medicine to which the body is most sensitive by finding the value of M that maximizes the derivative dR/dM.

45. *Selling Backpacks* It costs you c dollars each to manufacture and distribute backpacks. If the backpacks sell at x dollars each, the number sold is given by

$$n = \frac{a}{x - c} + b(100 - x),$$

where a and b are certain positive constants. What selling price will bring a maximum profit?

46. *Fermat's Principle in Optics* Fermat's principle in optics states that light always travels from one point to another along a path that minimizes the travel time. Light from a source A is reflected by a plane mirror to a receiver at point B, as shown in the figure. Show that for the light to obey Fermat's principle, the angle of incidence must equal the angle of reflection, both measured from the line normal to the reflecting surface. (This result can also be derived without calculus. There is a purely geometric argument, which you may prefer.)

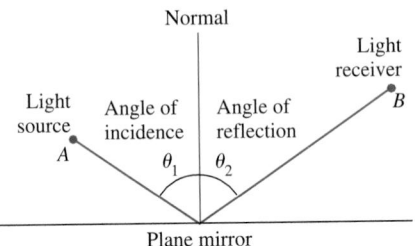

47. *Tin Pest* When metallic tin is kept below 13.2°C, it slowly becomes brittle and crumbles to a gray powder. Tin objects eventually crumble to this gray powder spontaneously if kept in a cold climate for years. The Europeans who saw tin organ pipes in their churches crumble away years ago called the change *tin pest* because it seemed to be contagious. And indeed it was, for the gray powder is a catalyst for its own formation.

A *catalyst* for a chemical reaction is a substance that controls the rate of reaction without undergoing any permanent change in itself. An *autocatalytic reaction* is one whose product is a catalyst for its own formation. Such a reaction may proceed slowly at first if the amount of catalyst present is small and slowly again at the end, when most of the original substance is used up. But in between, when both the substance and its catalyst product are abundant, the reaction proceeds at a faster pace.

In some cases it is reasonable to assume that the rate $v = dx/dt$ of the reaction is proportional both to the amount of the original substance present and to the amount of product. That is, v may be considered to be a function of x alone, and

$$v = kx(a - x) = kax - kx^2,$$

where

$x =$ the amount of product,

$a =$ the amount of substance at the beginning,

$k =$ a positive constant.

At what value of x does the rate v have a maximum? What is the maximum value of v?

48. *How We Cough* When we cough, the trachea (windpipe) contracts to increase the velocity of the air going out. This raises the question of how much it should contract to maximize the velocity and whether it really contracts that much when we cough.

Under reasonable assumptions about the elasticity of the tracheal wall and about how the air near the wall is slowed by friction, the average flow velocity v (in cm/sec) can be modeled by the equation

$$v = c(r_0 - r)r^2, \quad \frac{r_0}{2} \leq r \leq r_0,$$

where r_0 is the rest radius of the trachea in cm and c is a positive constant whose value depends in part on the length of the trachea.

(a) Show that v is greatest when $r = (2/3)r_0$, that is, when the trachea is about 33% contracted. The remarkable fact is that X-ray photographs confirm that the trachea contracts about this much during a cough.

(b) Take r_0 to be 0.5 and c to be 1, and graph v over the interval $0 \leq r \leq 0.5$. Compare what you see to the claim that v is a maximum when $r = (2/3)r_0$.

49. *Tour Service* You operate a tour service that offers the following rates:

• $200 per person if 50 people (the minimum number to book the tour) go on the tour.

• For each additional person, up to a maximum of 80 people total, the rate per person is reduced by $2.

It costs $6000 (a fixed cost) plus $32 per person to conduct the tour. How many people does it take to maximize your profit?

50. *Wilson Lot Size Formula* One of the formulas for inventory management says that the average weekly cost of ordering, paying for, and holding merchandise is

$$A(q) = \frac{km}{q} + cm + \frac{hq}{2},$$

where q is the quantity you order when things run low (shoes, radios, brooms, or whatever the item might be), k is the cost of placing an order (the same, no matter how often you order), c is the cost of one item (a constant), m is the number of items sold each week (a constant), and h is the weekly holding cost per item (a constant that takes into account things such as space, utilities, insurance, and security).

(a) Your job, as the inventory manager for your store, is to find the quantity that will minimize $A(q)$. What is it? (The formula you get for the answer is called the *Wilson lot size formula*.)

(b) Shipping costs sometimes depend on order size. When they do, it is more realistic to replace k by $k + bq$, the sum of k and a constant multiple of q. What is the most economical quantity to order now?

51. *Production Level* Show that if $r(x) = 6x$ and $c(x) = x^3 - 6x^2 + 15x$ are your revenue and cost functions, then the best you can do is break even (have revenue equal cost).

52. *Production Level* Suppose $c(x) = x^3 - 20x^2 + 20{,}000x$ is the cost of manufacturing x items. Find a production level that will minimize the average cost of making x items.

Extending the Ideas

53. *Airplane Landing Path* An airplane is flying at altitude H when it begins its descent to an airport runway that is at horizontal ground distance L from the airplane, as shown in the figure. Assume that the landing path of the airplane is the graph of a cubic polynomial function $y = ax^3 + bx^2 + cx + d$ where $y(-L) = H$ and $y(0) = 0$.

(a) What is dy/dx at $x = 0$?

(b) What is dy/dx at $x = -L$?

(c) Use the values for dy/dx at $x = 0$ and $x = -L$ together with $y(0) = 0$ and $y(-L) = H$ to show that

$$y(x) = H\left[2\left(\frac{x}{L}\right)^3 + 3\left(\frac{x}{L}\right)^2\right].$$

In Exercises 54 and 55, you might find it helpful to use a CAS and to *work in groups of two or three.*

54. *Generalized Cone Problem* A cone of height h and radius r is constructed from a flat, circular disk of radius a in. as described in Exploration 1.

(a) Find a formula for the volume V of the cone in terms of x and a.

(b) Find r and h in the cone of maximum volume for $a = 4$, 5, 6, 8.

(c) **Writing to Learn** Find a simple relationship between r and h that is independent of a for the cone of maximum volume. Explain how you arrived at your relationship.

55. *Circumscribing an Ellipse* Let $P(x, a)$ and $Q(-x, a)$ be two points on the upper half of the ellipse

$$\frac{x^2}{100} + \frac{(y-5)^2}{25} = 1$$

centered at $(0, 5)$. A triangle RST is formed by using the tangent lines to the ellipse at Q and P as shown in the figure.

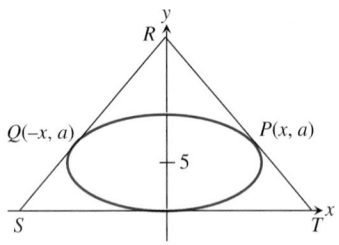

(a) Show that the area of the triangle is

$$A(x) = -f'(x)\left[x - \frac{f(x)}{f'(x)}\right]^2,$$

where $y = f(x)$ is the function representing the upper half of the ellipse.

(b) What is the domain of A? Draw the graph of A. How are the asymptotes of the graph related to the problem situation?

(c) Determine the height of the triangle with minimum area. How is it related to the y-coordinate of the center of the ellipse?

(d) Repeat parts (a)–(c) for the ellipse

$$\frac{x^2}{C^2} + \frac{(y-B)^2}{B^2} = 1$$

centered at $(0, B)$. Show that the triangle has minimum area when its height is $3B$.

4.5 **Linearization and Newton's Method**

Linear Approximation • Newton's Method • Differentials • Estimating Change with Differentials • Absolute, Relative, and Percentage Change • Sensitivity to Change

Linear Approximation

The line $y = mx + b$ tangent to a curve $y = f(x)$ lies close to the curve near the point of tangency. You can observe this phenomenon by zooming in on the two graphs at the point of tangency, or by looking at tables of values for the difference $y = f(x) - (mx + b)$ near the x-coordinate of the point of tangency. Exploration 1 gives more details.

Exploration 1 **Approximating with Tangent Lines**

Let $f(x) = x^2$.

1. Show that the line tangent to the graph of f at the point $(1, 1)$ is $y = 2x - 1$.

2. Set $y_1 = x^2$ and $y_2 = 2x - 1$. Zoom in on the two graphs at $(1, 1)$.

3. For a more dramatic view, set $y_3 = y_1 - y_2$. Then turn off y_1 and y_2, zoom in on the point $(1, 0)$, and watch the difference collapse to zero. Explain why this gives a graphical measure of how well the tangent line approximates the function near $x = 1$.

4. Look at tables of values for y_3 with Δ Table $= 0.1, 0.01, 0.001$, and 0.0001. Explain why this gives a numerical measure of how well the tangent line approximates the function near $x = 1$.

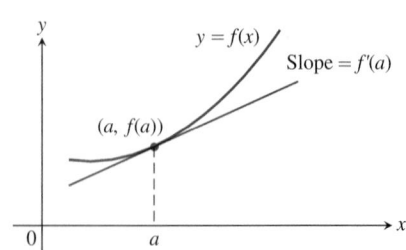

Figure 4.37 The tangent to the curve $y = f(x)$ at $x = a$ is the line $y = f(a) + f'(a)(x - a)$.

In general, the tangent to $y = f(x)$ at $x = a$ (Figure 4.37) passes through the point $(a, f(a))$, so its point-slope equation is

$$y = f(a) + f'(a)(x - a).$$

Thus, the tangent is the graph of the linear function

$$L(x) = f(a) + f'(a)(x - a).$$

For as long as the line remains close to the graph of f, $L(x)$ gives a good approximation to $f(x)$.

Definition Linearization

If f is differentiable at $x = a$, then the approximating function

$$L(x) = f(a) + f'(a)(x - a)$$

is the **linearization** of f at a.

The approximation $f(x) \approx L(x)$ is the **standard linear approximation** of f at a. The point $x = a$ is the **center** of the approximation.

Example 1   FINDING A LINEARIZATION

Find the linearization of $f(x) = \sqrt{1 + x}$ at $x = 0$ (Figure 4.38).

Solution Since

$$f'(x) = \frac{1}{2}(1 + x)^{-1/2},$$

we have $f(0) = 1$, $f'(0) = 1/2$, and

$$L(x) = f(a) + f'(a)(x - a) = 1 + \frac{1}{2}(x - 0) = 1 + \frac{x}{2}.$$

Look at how accurate the approximation $\sqrt{1 + x} \approx 1 + (x/2)$ is for values of x near 0.

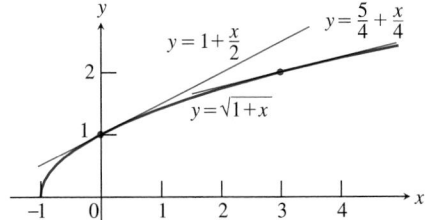

Figure 4.38 The graph of $f(x) = \sqrt{1 + x}$ and its linearization at $x = 0$ and $x = 3$. (Example 1)

Approximation	|True Value − Approximation|
$\sqrt{1.2} \approx 1 + \dfrac{0.2}{2} = 1.10$	$< 10^{-2}$
$\sqrt{1.05} \approx 1 + \dfrac{0.05}{2} = 1.025$	$< 10^{-3}$
$\sqrt{1.005} \approx 1 + \dfrac{0.005}{2} = 1.00250$	$< 10^{-5}$

As we move away from zero we lose accuracy. For example, for $x = 2$ the linearization gives 2 as the approximation for $\sqrt{3}$, which is not even accurate to one decimal place.

Do not be misled by the preceding calculations into thinking that whatever we do with a linearization is better done with a calculator. In practice, we would never use a linearization to find a particular square root. The utility of a linearization is its ability to replace a complicated formula by a simpler one over an entire interval of values. If we have to work with $\sqrt{1 + x}$ for x close to 0 and can tolerate the small amount of error involved, we can work with $1 + (x/2)$ instead. Of course, we then need to know how much error there is. We will have a full story on error in Chapter 9.

In Exercise 7 you will provide the details for the linearization in Example 2.

Example 2   FINDING A LINEARIZATION

The most important linear approximation for roots and powers is

$$(1 + x)^k \approx 1 + kx, \qquad (x \approx 0, \text{ any number } k).$$

Example 3   APPLYING EXAMPLE 2

The following approximations are consequences of Example 2.

$$\sqrt{1 + x} \approx 1 + \frac{1}{2}x \qquad\qquad\qquad k = 1/2$$

$$\frac{1}{1 - x} = (1 - x)^{-1} \approx 1 + (-1)(-x) = 1 + x \qquad \begin{array}{l} k = -1; \text{ replace} \\ x \text{ by } -x. \end{array}$$

$$\sqrt[3]{1 + 5x^4} = (1 + 5x^4)^{1/3} \approx 1 + \frac{1}{3}(5x^4) = 1 + \frac{5}{3}x^4 \qquad \begin{array}{l} k = 1/3; \text{ replace} \\ x \text{ by } 5x^4. \end{array}$$

$$\frac{1}{\sqrt{1 - x^2}} = (1 - x^2)^{-1/2} \approx 1 + \left(-\frac{1}{2}\right)(-x^2) = 1 + \frac{1}{2}x^2 \qquad \begin{array}{l} k = -1/2; \\ \text{replace } x \text{ by } -x^2. \end{array}$$

Example 4   FINDING A LINEARIZATION

Find the linearization of $f(x) = \cos x$ at $x = \pi/2$ (Figure 4.39).

Solution Since $f(\pi/2) = \cos(\pi/2) = 0$, $f'(x) = -\sin x$, and $f'(\pi/2) = -\sin(\pi/2) = -1$, we have

$$L(x) = f(a) + f'(a)(x - a)$$

$$= 0 + (-1)\left(x - \frac{\pi}{2}\right)$$

$$= -x + \frac{\pi}{2}.$$

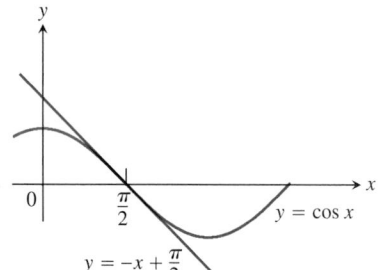

Figure 4.39 The graph of $f(x) = \cos x$ and its linearization at $x = \pi/2$. Near $x = \pi/2$, $\cos x \approx -x + (\pi/2)$. (Example 4)

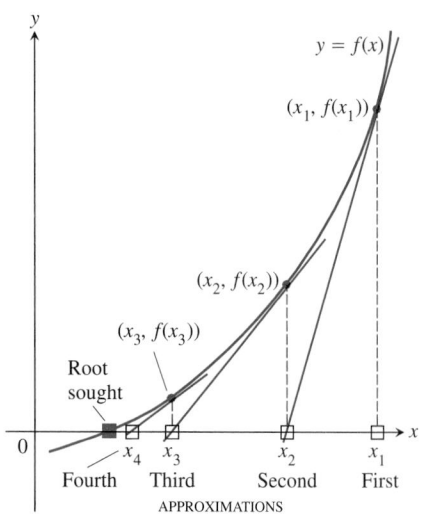

Figure 4.40 Usually the approximations rapidly approach an actual zero of $y = f(x)$.

Newton's Method

Newton's method is a numerical technique for approximating a zero of a function with zeros of its linearizations. Under favorable circumstances, the zeros of the linearizations *converge* rapidly to an accurate approximation. Many calculators use the method because it applies to a wide range of functions and usually gets results in only a few steps. Here is how it works.

To find a solution of an equation $f(x) = 0$, we begin with an initial estimate x_1, found either by looking at a graph or simply guessing. Then we use the tangent to the curve $y = f(x)$ at $(x_1, f(x_1))$ to approximate the curve (Figure 4.40). The point where the tangent crosses the x-axis is the next approximation x_2. The number x_2 is usually a better approximation to the solution than is x_1. The point where the tangent to the curve at $(x_2, f(x_2))$ crosses the x-axis is the next approximation x_3. We continue on, using each approximation to generate the next, until we are close enough to the zero to stop.

There is a formula for finding the $(n + 1)$st approximation x_{n+1} from the nth approximation x_n. The point-slope equation for the tangent to the curve at $(x_n, f(x_n))$ is

$$y - f(x_n) = f'(x_n)(x - x_n).$$

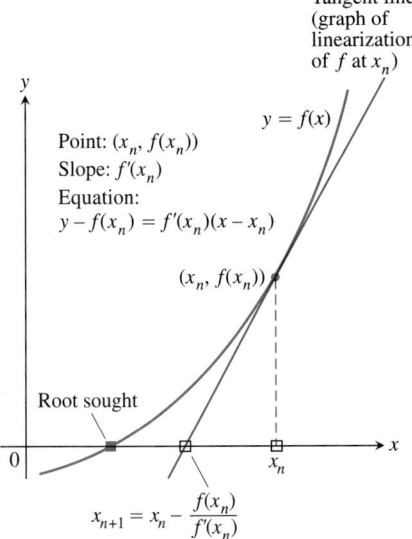

Tangent line (graph of linearization of f at x_n)

$y = f(x)$

Point: $(x_n, f(x_n))$
Slope: $f'(x_n)$
Equation:
$y - f(x_n) = f'(x_n)(x - x_n)$

$(x_n, f(x_n))$

Root sought

x_n

$x_{n+1} = x_n - \dfrac{f(x_n)}{f'(x_n)}$

Figure 4.41 From x_n we go up to the curve and follow the tangent line down to find x_{n+1}.

Fan Chung Graham

1949–

"Don't be intimidated!" is Dr. Fan Chung Graham's advice to young women considering careers in mathematics. Fan Chung Graham came to the U.S. from Taiwan to earn a Ph.D. in Mathematics from the University of Pennsylvania. She worked in the field of combinatorics at Bell Labs and Bellcore, and then, in 1994, returned to her alma mater as a Professor of Mathematics. Her research interests include spectral graph theory, discrete geometry, algorithms, and communication networks.

We can find where it crosses the x-axis by setting $y = 0$ (Figure 4.41).

$$0 - f(x_n) = f'(x_n)(x - x_n)$$
$$-f(x_n) = f'(x_n) \cdot x - f'(x_n) \cdot x_n$$
$$f'(x_n) \cdot x = f'(x_n) \cdot x_n - f(x_n)$$
$$x = x_n - \frac{f(x_n)}{f'(x_n)} \qquad \text{If } f'(x_n) \neq 0$$

This value of x is the next approximation x_{n+1}. Here is a summary of Newton's method.

Procedure for Newton's Method

1. Guess a first approximation to a solution of the equation $f(x) = 0$. A graph of $y = f(x)$ may help.

2. Use the first approximation to get a second, the second to get a third, and so on, using the formula

$$x_{n+1} = x_n - \frac{f(x_n)}{f'(x_n)}.$$

Example 5 USING NEWTON'S METHOD

Use Newton's method to solve $x^3 + 3x + 1 = 0$.

Solution Let $f(x) = x^3 + 3x + 1$, then $f'(x) = 3x^2 + 3$ and

$$x_{n+1} = x_n - \frac{f(x_n)}{f'(x_n)} = x_n - \frac{x_n^3 + 3x_n + 1}{3x_n^2 + 3}.$$

The graph of f in Figure 4.42 suggests that $x_1 = -0.3$ is a good first approximation to the zero of f in the interval $-1 \leq x \leq 0$. Then,

$$x_1 = -0.3,$$
$$x_2 = -0.322324159,$$
$$x_3 = -0.3221853603,$$
$$x_4 = -0.3221853546.$$

The x_n for $n \geq 5$ all appear to equal x_4 on the calculator we used to do the computations. We conclude that the solution to the equation $x^3 + 3x + 1 = 0$ is about -0.3221853546.

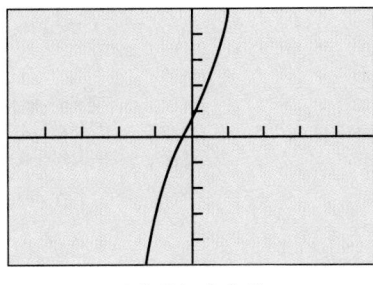

[–5, 5] by [–5, 5]

Figure 4.42 Our graphing calculator's root finder reports -0.3221853546 as the zero of $f(x) = x^3 + 3x + 1$. (Example 5)

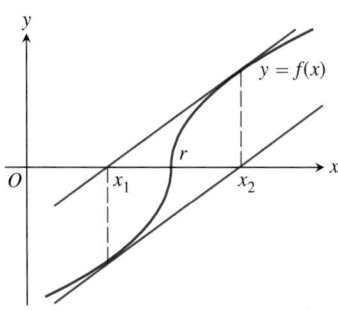

Figure 4.43 The graph of the function

$$f(x) = \begin{cases} -\sqrt{r-x}, & x < r \\ \sqrt{x-r}, & x \ge r. \end{cases}$$

If $x_1 = r - h$, then $x_2 = r + h$. Successive approximations go back and forth between these two values, and Newton's method fails to converge.

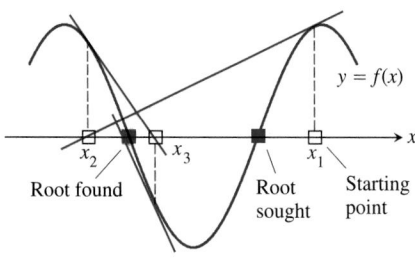

Figure 4.44 Newton's method may miss the zero you want if you start too far away.

Exploration 2 **Using Newton's Method on the Grapher**

Let $f(x) = x^3 + 3x + 1$. Here is a home screen program to perform the computations in Newton's method.

1. Let $y_1 = f(x)$ and $y_2 = \text{NDER } f(x)$.
2. Store $x_1 = -0.3$ into x.
3. Then store $x - (y_1/y_2)$ into x and press the "Enter" key over and over. Watch as the numbers converge to the zero of f.
4. Use different values for x_1 and repeat steps 2 and 3.
5. Write your own equation and use this approach to solve it using Newton's method. Compare your answer with the answer given by the built-in feature of your calculator that gives zeros of functions.

Newton's method does not work if $f'(x_1) = 0$. In that case, choose a new starting point.

Newton's method does not always converge. For instance (see Figure 4.43), successive approximations $r - h$ and $r + h$ can go back and forth between these two values, and no amount of iteration will bring us any closer to the zero r.

If Newton's method does converge, it converges to a zero of f. However, the method may converge to a zero that is different from the expected one if the starting value is not close enough to the zero sought. Figure 4.44 shows how this might happen.

Differentials

We sometimes use the notation dy/dx to represent the derivative y' of y with respect to x. Contrary to its appearance, it is not a ratio. We now introduce two new variables dx and dy with the property that if their ratio exists, it will be equal to the derivative.

Definition Differentials

Let $y = f(x)$ be a differentiable function. The **differential dx** is an independent variable. The **differential dy** is

$$dy = f'(x)\, dx.$$

Unlike the independent variable dx, the variable dy is always a dependent variable. It depends on both x and dx.

Example 6 FINDING THE DIFFERENTIAL dy

Find dy if

(a) $y = x^5 + 37x$.   **(b)** $y = \sin 3x$.

Solution

(a) $dy = (5x^4 + 37)\, dx$ **(b)** $dy = (3 \cos 3x)\, dx$

If $dx \neq 0$, then the quotient of the differential dy by the differential dx is equal to the derivative $f'(x)$ because

$$\frac{dy}{dx} = \frac{f'(x)\,dx}{dx} = f'(x).$$

We sometimes write

$$df = f'(x)\,dx$$

in place of $dy = f'(x)\,dx$, calling df the **differential of f.** For instance, if $f(x) = 3x^2 - 6$, then

$$df = d(3x^2 - 6) = 6x\,dx.$$

Every differentiation formula like

$$\frac{d(u + v)}{dx} = \frac{du}{dx} + \frac{dv}{dx} \qquad \text{or} \qquad \frac{d(\sin u)}{dx} = \cos u \frac{du}{dx}$$

has a corresponding differential form like

$$d(u + v) = du + dv \qquad \text{or} \qquad d(\sin u) = \cos u\,du.$$

Example 7   FINDING DIFFERENTIALS OF FUNCTIONS

(a) $d(\tan 2x) = \sec^2 (2x)\,d(2x) = 2 \sec^2 2x\,dx$

(b) $d\left(\dfrac{x}{x + 1}\right) = \dfrac{(x + 1)\,dx - x\,d(x + 1)}{(x + 1)^2} = \dfrac{x\,dx + dx - x\,dx}{(x + 1)^2} = \dfrac{dx}{(x + 1)^2}$

Estimating Change with Differentials

Suppose we know the value of a differentiable function $f(x)$ at a point a and we want to predict how much this value will change if we move to a nearby point $a + dx$. If dx is small, f and its linearization L at a will change by nearly the same amount (Figure 4.45). Since the values of L are simple to calculate, calculating the change in L offers a practical way to estimate the change in f.

In the notation of Figure 4.45, the change in f is

$$\Delta f = f(a + dx) - f(a).$$

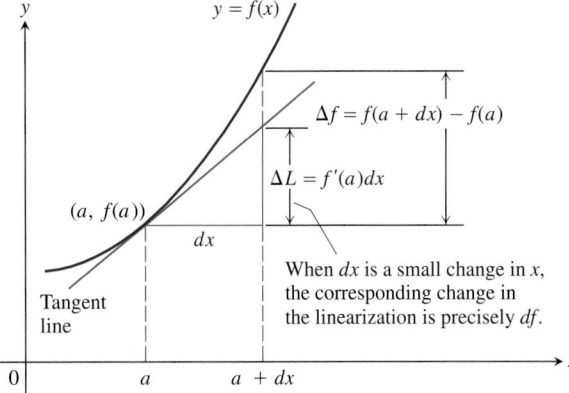

Figure 4.45 Approximating the change in the function f by the change in the linearization of f.

The corresponding change in L is

$$\Delta L = L(a + dx) - L(a)$$

$$= \underbrace{f(a) + f'(a)[(a + dx) - a]}_{L(a + dx)} - \underbrace{f(a)}_{L(a)}$$

$$= f'(a)\, dx.$$

Thus, the differential $df = f'(x)\, dx$ has a geometric interpretation: The value of df at $x = a$ is ΔL, the change in the linearization of f corresponding to the change dx.

Differential Estimate of Change

Let $f(x)$ be differentiable at $x = a$. The approximate change in the value of f when x changes from a to $a + dx$ is

$$df = f'(a)\, dx.$$

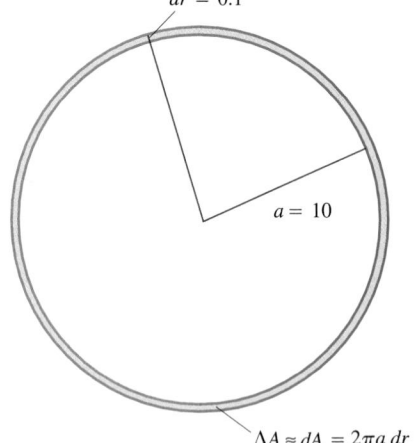

$dr = 0.1$

$a = 10$

$\Delta A \approx dA = 2\pi a\, dr$

Figure 4.46 When dr is small compared with a, as it is when $dr = 0.1$ and $a = 10$, the differential $dA = 2\pi a\, dr$ gives a good estimate of ΔA. (Example 8)

Example 8 ESTIMATING CHANGE WITH DIFFERENTIALS

The radius r of a circle increases from $a = 10$ m to 10.1 m (Figure 4.46). Use dA to estimate the increase in the circle's area A. Compare this estimate with the true change ΔA.

Solution Since $A = \pi r^2$, the estimated increase is

$$dA = A'(a)\, dr = 2\pi a\, dr = 2\pi(10)(0.1) = 2\pi \text{ m}^2.$$

The true change is

$$\Delta A = \pi(10.1)^2 - \pi(10)^2 = (102.01 - 100)\pi = (\underbrace{2\pi}_{dA} + \underbrace{0.01\pi}_{\text{error}}) \text{ m}^2.$$

Absolute, Relative, and Percentage Change

As we move from a to a nearby point $a + dx$, we can describe the change in f in three ways:

	True	**Estimated**
Absolute change	$\Delta f = f(a + dx) - f(a)$	$df = f'(a)\, dx$
Relative change	$\dfrac{\Delta f}{f(a)}$	$\dfrac{df}{f(a)}$
Percentage change	$\dfrac{\Delta f}{f(a)} \times 100$	$\dfrac{df}{f(a)} \times 100$

Example 9 COMPUTING PERCENTAGE CHANGE

The estimated percentage change in the area of the circle in Example 8 is

$$\frac{dA}{A(a)} \times 100 = \frac{2\pi}{100\pi} \times 100 = 2\%.$$

The true percentage change is

$$\frac{\Delta A}{A(a)} \times 100 = \frac{2.01\pi}{100\pi} \times 100 = 2.01\%.$$

Usually it is not possible or easy to compute the exact (true) change as we did in Example 9. This is sometimes due to uncertainty in measurement, as shown in Example 10.

Example 10 ESTIMATING THE EARTH'S SURFACE AREA

Suppose the earth were a perfect sphere and we determined its radius to be 3959 ± 0.1 miles. What effect would the tolerance of ± 0.1 mi have on our estimate of the earth's surface area?

Solution The surface area of a sphere of radius r is $S = 4\pi r^2$. The uncertainty in the calculation of S that arises from measuring r with a tolerance of dr miles is

$$dS = \left(\frac{dS}{dr}\right) dr = 8\pi r\, dr.$$

With $r = 3959$ and $dr = 0.1$, our estimate of S could be off by as much as

$$dS = 8\pi(3959)(0.1) \approx 9950 \text{ mi}^2,$$

to the nearest square mile, which is about the area of the state of Maryland.

Note

If we underestimated the radius of the earth by 528 ft during a calculation of the earth's surface area, we would leave out an area the size of the state of Maryland.

Example 11 DETERMINING TOLERANCE

About how accurately should we measure the radius r of a sphere to calculate the surface area $S = 4\pi r^2$ within 1% of its true value?

Solution We want any inaccuracy in our measurement to be small enough to make the corresponding increment ΔS in the surface area satisfy the inequality

$$|\Delta S| \le \frac{1}{100}S = \frac{4\pi r^2}{100}.$$

We replace ΔS in this inequality by its approximation

$$dS = \left(\frac{dS}{dr}\right) dr = 8\pi r\, dr.$$

This gives

$$|8\pi r\, dr| \le \frac{4\pi r^2}{100}, \quad \text{or} \quad |dr| \le \frac{1}{8\pi r} \cdot \frac{4\pi r^2}{100} = \frac{1}{2} \cdot \frac{r}{100}.$$

We should measure r with an error dr that is no more than 0.5% of the true value.

Angiography

An opaque dye is injected into a partially blocked artery to make the inside visible under X-rays. This reveals the location and severity of the blockage.

Angioplasty

A balloon-tipped catheter is inflated inside the artery to widen it at the blockage site.

Example 12   UNCLOGGING ARTERIES

In the late 1830s, the French physiologist Jean Poiseuille ("pwa-ZOY") discovered the formula we use today to predict how much the radius of a partially clogged artery has to be expanded to restore normal flow. His formula,

$$V = kr^4,$$

says that the volume V of fluid flowing through a small pipe or tube in a unit of time at a fixed pressure is a constant times the fourth power of the tube's radius r. How will a 10% increase in r affect V?

Solution The differentials of r and V are related by the equation

$$dV = \frac{dV}{dr}\, dr = 4kr^3\, dr.$$

The relative change in V is

$$\frac{dV}{V} = \frac{4kr^3 dr}{kr^4} = 4\frac{dr}{r}.$$

The relative change in V is 4 times the relative change in r, so a 10% increase in r will produce a 40% increase in the flow.

Sensitivity to Change

The equation $df = f'(x)\, dx$ tells how *sensitive* the output of f is to a change in input at different values of x. The larger the value of f' at x, the greater the effect of a given change dx.

Example 13   FINDING DEPTH OF A WELL

You want to calculate the depth of a well from the equation $s = 16t^2$ by timing how long it takes a heavy stone you drop to splash into the water below. How sensitive will your calculations be to a 0.1 sec error in measuring the time?

Solution The size of ds in the equation

$$ds = 32t\, dt$$

depends on how big t is. If $t = 2$ sec, the error caused by $dt = 0.1$ is only

$$ds = 32(2)(0.1) = 6.4 \text{ ft.}$$

Three seconds later at $t = 5$ sec, the error caused by the same dt is

$$ds = 32(5)(0.1) = 16 \text{ ft.}$$

Quick Review 4.5

In Exercises 1 and 2, find dy/dx.

1. $y = \sin(x^2 + 1)$

2. $y = \dfrac{x + \cos x}{x + 1}$

In Exercises 3 and 4, solve the equation graphically.

3. $xe^{-x} + 1 = 0$

4. $x^3 + 3x + 1 = 0$

In Exercises 5 and 6, let $f(x) = xe^{-x} + 1$. Write an equation for the line tangent to f at $x = c$.

5. $c = 0$

6. $c = -1$

7. Find where the tangent line in **(a)** Exercise 5 and **(b)** Exercise 6 crosses the x-axis.

8. Let $g(x)$ be the function whose graph is the tangent line to the graph of $f(x) = x^3 - 4x + 1$ at $x = 1$. Complete the table.

x	$f(x)$	$g(x)$
0.7		
0.8		
0.9		
1		
1.1		
1.2		
1.3		

In Exercises 9 and 10, graph $y = f(x)$ and its tangent line at $x = c$.

9. $c = 1.5$, $f(x) = \sin x$

10. $c = 4$, $f(x) = \begin{cases} -\sqrt{3 - x}, & x < 3 \\ \sqrt{x - 3}, & x \geq 3 \end{cases}$

Section 4.5 Exercises

In Exercises 1–6, **(a)** find the linearization $L(x)$ of $f(x)$ at $x = a$. **(b)** How accurate is the approximation $L(a + 0.1) \approx f(a + 0.1)$? See the comparisons following Example 1.

1. $f(x) = x^3 - 2x + 3$, $a = 2$

2. $f(x) = \sqrt{x^2 + 9}$, $a = -4$

3. $f(x) = x + \dfrac{1}{x}$, $a = 1$

4. $f(x) = \ln(x + 1)$, $a = 0$

5. $f(x) = \tan x$, $a = \pi$

6. $f(x) = \cos^{-1} x$, $a = 0$

7. Show that the linearization of $f(x) = (1 + x)^k$ at $x = 0$ is $L(x) = 1 + kx$.

8. Use the linear approximation $(1 + x)^k \approx 1 + kx$ to find an approximation for the function $f(x)$ for values of x near zero.

(a) $f(x) = (1 - x)^6$

(b) $f(x) = \dfrac{2}{1 - x}$

(c) $f(x) = \dfrac{1}{\sqrt{1 + x}}$

(d) $f(x) = \sqrt{2 + x^2}$

(e) $f(x) = (4 + 3x)^{1/3}$

(f) $f(x) = \sqrt[3]{\left(1 - \dfrac{1}{2 + x}\right)^2}$

9. Writing to Learn Find the linearization of $f(x) = \sqrt{x + 1} + \sin x$ at $x = 0$. How is it related to the individual linearizations for $\sqrt{x + 1}$ and $\sin x$?

10. Use the linearization $(1 + x)^k \approx 1 + kx$ to approximate the following. State how accurate your approximation is.

(a) $(1.002)^{100}$

(b) $\sqrt[3]{1.009}$

In Exercises 11–14, choose a linearization with center not at $x = a$ but at a nearby value at which the function and its derivative are easy to evaluate. State the linearization and the center.

11. $f(x) = 2x^2 + 4x - 3$, $a = -0.9$

12. $f(x) = \sqrt[3]{x}$, $a = 8.5$

13. $f(x) = \dfrac{x}{x + 1}$, $a = 1.3$

14. $f(x) = \cos x$, $a = 1.7$

In Exercises 15–18, use Newton's method to estimate all real solutions of the equation. Make your answers accurate to 6 decimal places.

15. $x^3 + x - 1 = 0$

16. $x^4 + x - 3 = 0$

17. $x^2 - 2x + 1 = \sin x$

18. $x^4 - 2 = 0$

In Exercises 19–26, **(a)** find dy, and **(b)** evaluate dy for the given value of x and dx.

19. $y = x^3 - 3x$, $x = 2$, $dx = 0.05$

20. $y = \dfrac{2x}{1 + x^2}$, $x = -2$, $dx = 0.1$

21. $y = x^2 \ln x$, $x = 1$, $dx = 0.01$

22. $y = x\sqrt{1 - x^2}$, $x = 0$, $dx = -0.2$

23. $y = e^{\sin x}, \quad x = \pi, \quad dx = -0.1$

24. $y = 3 \csc \left(1 - \dfrac{x}{3}\right), \quad x = 1, \quad dx = 0.1$

25. $y + xy - x = 0, \quad x = 0, \quad dx = 0.01$

26. $y = \sec(x^2 - 1), \quad x = 1.5, \quad dx = 0.05$

In Exercises 27–30, the function f changes value when x changes from a to $a + dx$. Find

    (a) the absolute change $\Delta f = f(a + dx) - f(a)$.

    (b) the estimated change $df = f'(a)\, dx$.

    (c) the approximation error $|\Delta f - df|$.

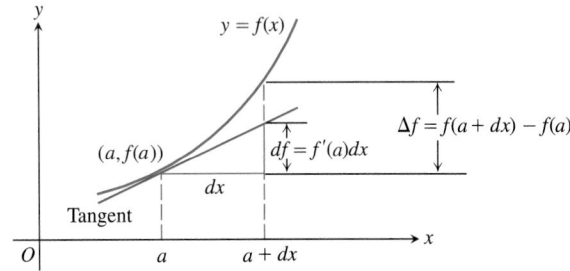

27. $f(x) = x^2 + 2x, \quad a = 0, \quad dx = 0.1$

28. $f(x) = x^3 - x, \quad a = 1, \quad dx = 0.1$

29. $f(x) = x^{-1}, \quad a = 0.5, \quad dx = 0.05$

30. $f(x) = x^4, \quad a = 1, \quad dx = 0.01$

In Exercises 31–36, write a differential formula that estimates the given change in volume or surface area.

31. *Volume* The change in the volume $V = (4/3)\pi r^3$ of a sphere when the radius changes from a to $a + dr$

32. *Surface Area* The change in the surface area $S = 4\pi r^2$ of a sphere when the radius changes from a to $a + dr$

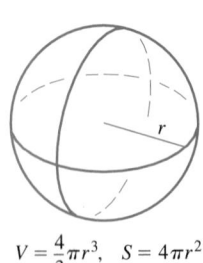

$$V = \frac{4}{3}\pi r^3, \quad S = 4\pi r^2$$

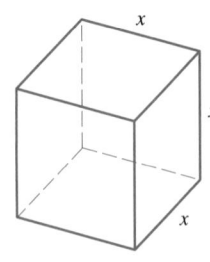

$$V = x^3, \quad S = 6x^2$$

33. *Volume* The change in the volume $V = x^3$ of a cube when the edge lengths change from a to $a + dx$

34. *Surface Area* The change in the surface area $S = 6x^2$ of a cube when the edge lengths change from a to $a + dx$

35. *Volume* The change in the volume $V = \pi r^2 h$ of a right circular cylinder when the radius changes from a to $a + dr$ and the height does not change

36. *Surface Area* The change in the lateral surface area $S = 2\pi rh$ of a right circular cylinder when the height changes from a to $a + dh$ and the radius does not change

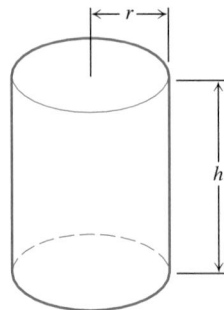

$$V = \pi r^2 h, \quad S = 2\pi rh$$

37. *Linear Approximation* Let f be a function with $f(0) = 1$ and $f'(x) = \cos(x^2)$.

    (a) Find the linearization of f at $x = 0$.

    (b) Estimate the value of f at $x = 0.1$.

    (c) **Writing to Learn** Do you think the actual value of f at $x = 0.1$ is greater than or less than the estimate in (b)? Explain.

38. *Expanding Circle* The radius of a circle is increased from 2.00 to 2.02 m.

    (a) Estimate the resulting change in area.

    (b) Estimate as a percentage of the circle's original area.

39. *Growing Tree* The diameter of a tree was 10 in. During the following year, the circumference increased 2 in. About how much did the tree's diameter increase? the tree's cross section area?

40. *Percentage Error* The edge of a cube is measured as 10 cm with an error of 1%. The cube's volume is to be calculated from this measurement. Estimate the percentage error in the volume calculation.

41. *Percentage Error* About how accurately should you measure the side of a square to be sure of calculating the area to within 2% of its true value?

42. *Estimating Volume* Estimate the volume of material in a cylindrical shell with height 30 in., radius 6 in., and shell thickness 0.5 in.

0.5 in.

30 in.

6 in.

43. *Estimating Volume* A surveyor is standing 30 ft from the base of a building. She measures the angle of elevation to the top of the building to be 75°. How accurately must the angle be measured for the percentage error in estimating the height of the building to be less than 4%?

44. *Tolerance* The height and radius of a right circular cylinder are equal, so the cylinder's volume is $V = \pi h^3$. The volume is to be calculated with an error of no more than 1% of the true value. Find approximately the greatest error that can be tolerated in the measurement of h, expressed as a percentage of h.

45. *Tolerance* **(a)** About how accurately must the interior diameter of a 10-m high cylindrical storage tank be measured to calculate the tank's volume to within 1% of its true value?

(b) About how accurately must the tank's exterior diameter be measured to calculate the amount of paint it will take to paint the side of the tank to within 5% of the true amount?

46. *Minting Coins* A manufacturer contracts to mint coins for the federal government. How much variation dr in the radius of the coins can be tolerated if the coins are to weigh within 1/1000 of their ideal weight? Assume the thickness does not vary.

47. *The Effect of Flight Maneuvers on the Heart* The amount of work done by the heart's main pumping chamber, the left ventricle, is given by the equation

$$W = PV + \frac{V\delta v^2}{2g},$$

where W is the work per unit time, P is the average blood pressure, V is the volume of blood pumped out during the unit of time, δ ("delta") is the density of the blood, v is the average velocity of the exiting blood, and g is the acceleration of gravity.

When P, V, δ, and v remain constant, W becomes a function of g, and the equation takes the simplified form

$$W = a + \frac{b}{g} \quad (a, b \text{ constant}).$$

As a member of NASA's medical team, you want to know how sensitive W is to apparent changes in g caused by flight maneuvers, and this depends on the initial value of g. As part of your investigation, you decide to compare the effect on W of a given change dg on the moon, where $g = 5.2$ ft/sec², with the effect the same change dg would have on Earth, where $g = 32$ ft/sec². Use the simplified equation above to find the ratio of dW_{moon} to dW_{Earth}.

48. *Measuring Acceleration of Gravity* When the length L of a clock pendulum is held constant by controlling its temperature, the pendulum's period T depends on the acceleration of gravity g. The period will therefore vary slightly as the clock is moved from place to place on the earth's surface, depending on the change in g. By keeping track of ΔT, we can estimate the variation in g from the equation $T = 2\pi(L/g)^{1/2}$ that relates T, g, and L.

(a) With L held constant and g as the independent variable, calculate dT and use it to answer (b) and (c).

(b) Writing to Learn If g increases, will T increase or decrease? Will a pendulum clock speed up or slow down? Explain.

(c) A clock with a 100-cm pendulum is moved from a location where $g = 980$ cm/sec² to a new location. This increases the period by $dT = 0.001$ sec. Find dg and estimate the value of g at the new location.

49. *Newton's Method* Suppose your first guess in using Newton's method is lucky in the sense that x_1 is a root of $f(x) = 0$. What happens to x_2 and later approximations?

50. *Oscillation* Show that if $h > 0$, applying Newton's method to

$$f(x) = \begin{cases} \sqrt{x}, & x \geq 0 \\ \sqrt{-x}, & x < 0 \end{cases}$$

leads to $x_2 = -h$ if $x_1 = h$, and to $x_2 = h$ if $x_1 = -h$. Draw a picture that shows what is going on.

51. *Approximations that Get Worse and Worse* Apply Newton's method to $f(x) = x^{1/3}$ with $x_1 = 1$, and calculate $x_2, x_3, x_4,$ and x_5. Find a formula for $|x_n|$. What happens to $|x_n|$ as $n \to \infty$? Draw a picture that shows what is going on.

Exploration

52. *Quadratic Approximations*

(a) Let $Q(x) = b_0 + b_1(x - a) + b_2(x - a)^2$ be a quadratic approximation to $f(x)$ at $x = a$ with the properties:

i. $Q(a) = f(a)$,

ii. $Q'(a) = f'(a)$,

iii. $Q''(a) = f''(a)$.

Determine the coefficients b_0, b_1, and b_2.

(b) Find the quadratic approximation to $f(x) = 1/(1 - x)$ at $x = 0$.

(c) Graph $f(x) = 1/(1 - x)$ and its quadratic approximation at $x = 0$. Then zoom in on the two graphs at the point $(0, 1)$. Comment on what you see.

(d) Find the quadratic approximation to $g(x) = 1/x$ at $x = 1$. Graph g and its quadratic approximation together. Comment on what you see.

(e) Find the quadratic approximation to $h(x) = \sqrt{1 + x}$ at $x = 0$. Graph h and its quadratic approximation together. Comment on what you see.

(f) What are the linearizations of f, g, and h at the respective points in (b), (d), and (e)? ∎

Extending the Ideas

53. *Formulas for Differentials* Verify the following formulas.

(a) $d(c) = 0$ (c a constant)

(b) $d(cu) = c\,du$ (c a constant)

(c) $d(u + v) = du + dv$

(d) $d(u \cdot v) = u\,dv + v\,du$

(e) $d\left(\dfrac{u}{v}\right) = \dfrac{v\,du - u\,dv}{v^2}$

(f) $d(u^n) = nu^{n-1}\,du$

54. *Linearization* Show that the approximation of $\tan x$ by its linearization at the origin must improve as $x \to 0$ by showing that

$$\lim_{x \to 0} \frac{\tan x}{x} = 1.$$

55. *The Linearization is the Best Linear Approximation*
Suppose that $y = f(x)$ is differentiable at $x = a$ and that $g(x) = m(x - a) + c$ (m and c constants). If the error $E(x) = f(x) - g(x)$ were small enough near $x = a$, we might think of using g as a linear approximation of f instead of the linearization $L(x) = f(a) + f'(a)(x - a)$. Show that if we impose on g the conditions

i. $E(a) = 0$,   The error is zero at $x = a$.

ii. $\displaystyle \lim_{x \to a} \frac{E(x)}{x - a} = 0$,   The error is negligible when compared with $(x - a)$.

then $g(x) = f(a) + f'(a)(x - a)$. Thus, the linearization gives the only linear approximation whose error is both zero at $x = a$ and negligible in comparison with $(x - a)$.

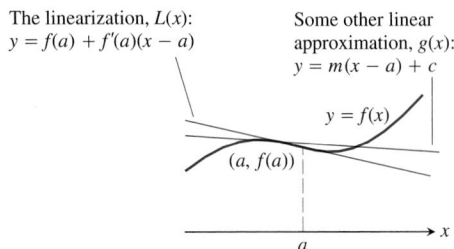

The linearization, $L(x)$:
$y = f(a) + f'(a)(x - a)$

Some other linear approximation, $g(x)$:
$y = m(x - a) + c$

$y = f(x)$

$(a, f(a))$

4.6   Related Rates

Related Rate Equations • Solution Strategy • Simulating Related Motion

Related Rate Equations

Suppose that a particle $P(x, y)$ is moving along a curve C in the plane so that its coordinates x and y are differentiable functions of time t. If D is the distance from the origin to P, then using the Chain Rule we can find an equation that relates dD/dt, dx/dt, and dy/dt.

$$D = \sqrt{x^2 + y^2}$$

$$\frac{dD}{dt} = \frac{1}{2}(x^2 + y^2)^{-1/2}\left(2x\frac{dx}{dt} + 2y\frac{dy}{dt}\right)$$

Any equation involving two or more variables that are differentiable functions of time t can be used to find an equation that relates their corresponding rates.

Example 1   FINDING RELATED RATE EQUATIONS

Assume that the radius r and height h of a cone are differentiable functions of t and let V be the volume of the cone. Find an equation that relates dV/dt, dr/dt, and dh/dt.

Solution $V = \dfrac{\pi}{3}r^2 h$ Cone volume formula

$$\frac{dV}{dt} = \frac{\pi}{3}\left(r^2 \cdot \frac{dh}{dt} + 2r\frac{dr}{dt} \cdot h\right) = \frac{\pi}{3}\left(r^2\frac{dh}{dt} + 2rh\frac{dr}{dt}\right)$$

Solution Strategy

How fast is a balloon rising at a given instant? How fast does the water level drop when a tank is drained at a certain rate? Questions like these ask us to calculate a rate that may be difficult to measure from a rate that we know, or is easy to measure. We begin by writing an equation that relates the variables involved and then differentiate it to get an equation that relates the rate we seek to the rates we know.

Example 2   A RISING BALLOON

A hot-air balloon rising straight up from a level field is tracked by a range finder 500 ft from the lift-off point. At the moment the range finder's elevation angle is $\pi/4$, the angle is increasing at the rate of 0.14 rad/min. How fast is the balloon rising at that moment?

Solution   We answer the question in six steps.

Step 1:
Draw a picture and name the variables and constants (Figure 4.47). The variables in the picture are

$$\theta = \text{the angle in radians the range finder makes with the ground,}$$

$$y = \text{the height in feet of the balloon.}$$

We let t represent time in minutes and assume θ and y are differentiable functions of t.

The one constant in the picture is the distance from the range finder to the lift-off point (500 ft). There is no need to give it a special symbol.

Step 2:
Write down the additional numerical information.

$$\frac{d\theta}{dt} = 0.14 \text{ rad/min} \quad \text{when} \quad \theta = \frac{\pi}{4}$$

Step 3:
Write down what we are to find. We want dy/dt when $\theta = \pi/4$.

Step 4:
Write an equation that relates the variables y and θ.

$$\frac{y}{500} = \tan \theta \quad \text{or} \quad y = 500 \tan \theta$$

Step 5:
Differentiate with respect to t using the Chain Rule. The result tells how dy/dt (which we want) is related to $d\theta/dt$ (which we know).

$$\frac{dy}{dt} = 500(\sec^2 \theta) \frac{d\theta}{dt}$$

Step 6:
Evaluate with $\theta = \pi/4$ and $d\theta/dt = 0.14$ to find dy/dt.

$$\frac{dy}{dt} = 500(\sqrt{2})^2(0.14) = 140 \quad \sec \frac{\pi}{4} = \sqrt{2}$$

Interpret

At the moment in question, the balloon is rising at the rate of 140 ft/min.

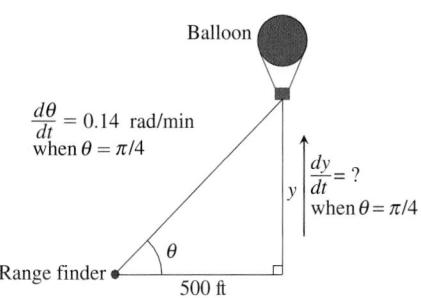

$$\frac{d\theta}{dt} = 0.14 \text{ rad/min}$$
when $\theta = \pi/4$

y $\left. \frac{dy}{dt} = ? \right.$ when $\theta = \pi/4$

θ

Range finder

500 ft

Figure 4.47 (Example 2)

The units in the formula for *dy/dt* (Step 6)

A radian is the ratio of two lengths, meaning its units are length/length. In Example 2, the units are ft/ft. Radians per minute are (ft/ft)/min = ft/(ft × min). The secant of an angle is also a ratio of lengths. The units of the secant in Example 2 are ft/ft. For the equation

$$\frac{dy}{dt} = 500 \times (\sec \theta)^2 \times \frac{d\theta}{dt} = 140,$$

we therefore have

$$\frac{dy}{dt} = \text{ft} \times \left(\frac{\text{ft}}{\text{ft}}\right)^2 \times \frac{\text{ft}}{\text{ft} \times \text{min}} = \frac{\text{ft}}{\text{min}}.$$

Because the units of radians (length/length) cancel, we usually regard radians as *dimensionless* and don't write anything in. The trigonometric ratios (length/length) are dimensionless, too. With this understanding, the units of radians/minute, instead of being (length/length)/minute, are 1/minute or "per minute," the secant's units do not appear at all, and the units of

$$500 \times (\sqrt{2})^2 \times (0.14) = 140$$

are calculated as

$$\text{ft} \times \frac{1}{\text{min}} = \frac{\text{ft}}{\text{min}}.$$

Related Rate Problem Strategy

1. *Draw a picture and name the variables and constants.* Use *t* for time. Assume all variables are differentiable functions of *t*.

2. *Write down the numerical information* (in terms of the symbols you have chosen).

3. *Write down what we are asked to find* (usually a rate, expressed as a derivative).

4. *Write an equation that relates the variables.* You may have to combine two or more equations to get a single equation that relates the variable whose rate you want to the variables whose rates you know.

5. *Differentiate with respect to t.* Then express the rate you want in terms of the rate and variables whose values you know.

6. *Evaluate.* Use known values to find the unknown rate.

Example 3 A HIGHWAY CHASE

A police cruiser, approaching a right-angled intersection from the north, is chasing a speeding car that has turned the corner and is now moving straight east. When the cruiser is 0.6 mi north of the intersection and the car is 0.8 mi to the east, the police determine with radar that the distance between them and the car is increasing at 20 mph. If the cruiser is moving at 60 mph at the instant of measurement, what is the speed of the car?

Solution We carry out the steps of the strategy.

Step 1:

Picture and variables. We picture the car and cruiser in the coordinate plane, using the positive *x*-axis as the eastbound highway and the positive *y*-axis as the southbound highway (Figure 4.48). We let *t* represent time and set

$$x = \text{position of car at time } t,$$

$$y = \text{position of cruiser at time } t, \text{ and}$$

$$s = \text{distance between car and cruiser at time } t.$$

We assume *x*, *y*, and *s* are differentiable functions of *t*.

Step 2:

Numerical information. At the instant in question,

$$x = 0.8 \text{ mi}, \quad y = 0.6 \text{ mi}, \quad \frac{dy}{dt} = -60 \text{ mph}, \quad \frac{ds}{dt} = 20 \text{ mph}.$$

Note *dy/dt* is negative because *y* is decreasing.

Step 3:

To find: dx/dt

Step 4:

How the variables are related: $s^2 = x^2 + y^2$
(We could also use $s = \sqrt{x^2 + y^2}$.)

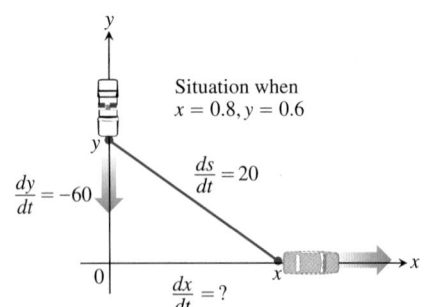

Situation when
$x = 0.8, y = 0.6$

$\frac{dy}{dt} = -60$

$\frac{ds}{dt} = 20$

$\frac{dx}{dt} = ?$

Figure 4.48 (Example 3)

Step 5:

Differentiate with respect to t.

$$2s\frac{ds}{dt} = 2x\frac{dx}{dt} + 2y\frac{dy}{dt}$$

$$\frac{ds}{dt} = \frac{1}{s}\left(x\frac{dx}{dt} + y\frac{dy}{dt}\right)$$

$$= \frac{1}{\sqrt{x^2 + y^2}}\left(x\frac{dx}{dt} + y\frac{dy}{dt}\right)$$

Step 6:

Evaluate. Use $x = 0.8$, $y = 0.6$, $dy/dt = -60$, $ds/dt = 20$, and solve for dx/dt.

$$20 = \frac{1}{\sqrt{(0.8)^2 + (0.6)^2}}\left(0.8\frac{dx}{dt} + (0.6)(-60)\right)$$

$$\frac{dx}{dt} = \frac{20\sqrt{(0.8)^2 + (0.6)^2} + (0.6)(60)}{0.8} = 70$$

Interpret

At the moment in question, the car's speed is 70 mph.

Example 4   FILLING A CONICAL TANK

Water runs into a conical tank at the rate of 9 ft³/min. The tank stands point down and has a height of 10 ft and a base radius of 5 ft. How fast is the water level rising when the water is 6 ft deep?

Solution   We carry out the steps of the strategy.

Step 1:

Picture and variables. Figure 4.49 on the next page shows a partially filled conical tank. The variables in the problem are

$V =$ volume (ft³) of the water in the tank at time t (min),

$x =$ radius (ft) of the surface of the water at time t, and

$y =$ depth (ft) of water in tank at time t.

We assume V, x, and y are differentiable functions of t. The constants are the dimensions of the tank.

Step 2:

Numerical information. At the time in question,

$$y = 6 \text{ ft} \quad \text{and} \quad \frac{dV}{dt} = 9 \text{ ft}^3/\text{min}.$$

Step 3:

To find: dy/dt

Step 4:

How the variables are related: The water forms a cone with volume

$$V = \frac{1}{3}\pi x^2 y.$$

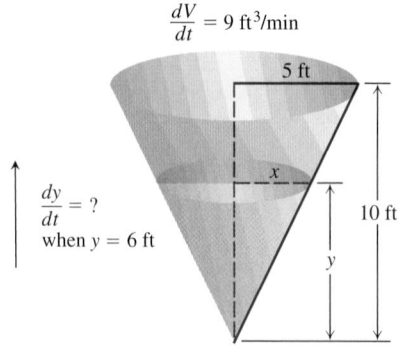

$\frac{dV}{dt} = 9 \text{ ft}^3/\text{min}$

5 ft

$\frac{dy}{dt} = ?$
when $y = 6$ ft

x

10 ft

y

Figure 4.49 (Example 4)

This equation involves x as well as V and y. Because no information is given about x and dx/dt at the time in question, we need to eliminate x. The similar triangles in Figure 4.49 give us a way to express x in terms of y:

$$\frac{x}{y} = \frac{5}{10} \quad \text{or} \quad x = \frac{y}{2}.$$

Therefore,

$$V = \frac{1}{3}\pi\left(\frac{y}{2}\right)^2 y = \frac{\pi}{12}y^3.$$

Step 5:

Differentiate with respect to t.

$$\frac{dV}{dt} = \frac{\pi}{12} \cdot 3y^2 \frac{dy}{dt} = \frac{\pi}{4}y^2\frac{dy}{dt}$$

Step 6:

Evaluate. Use $y = 6$ and $dV/dt = 9$ to solve for dy/dt.

$$9 = \frac{\pi}{4}(6)^2\frac{dy}{dt}$$

$$\frac{dy}{dt} = \frac{1}{\pi} \approx 0.32 \text{ ft/min}$$

Interpret

At the moment in question, the water level is rising at about 0.32 ft/min.

Simulating Related Motion

Parametric mode on a grapher can be used to simulate the motion of moving objects when the motion of each can be expressed as a function of time.

Exploration 1 **Sliding Ladder**

A 13-ft ladder is leaning against a wall. Suppose that the base of the ladder slides away from the wall at the constant rate of 3 ft/sec.

1. Explain why the motion of the two ends of the ladder can be represented by the parametric equations:

$$x_1(t) = 3t, \quad y_1(t) = 0$$
$$x_2(t) = 0, \quad y_2(t) = \sqrt{13^2 - (3t)^2}.$$

2. What values of t make sense in this problem situation?

3. Use simultaneous mode. Give a viewing window that shows the action. (*Hint:* It may be helpful to "hide" the coordinate axes if your grapher has this feature. If not, adjust the parametric equations to move the action away from the axes.)

4. Use analytic methods to find the rates at which the top of the ladder is moving down the wall at $t = 0.5$, 1, 1.5, and 2 sec. In theory, how fast is the top of the ladder moving as it hits the ground?

Quick Review 4.6

In Exercises 1 and 2, find the distance between the points *A* and *B*.

1. $A(0, 5)$, $B(7, 0)$

2. $A(0, a)$, $B(b, 0)$

In Exercises 3–6, find dy/dx.

3. $2xy + y^2 = x + y$

4. $x \sin y = 1 - xy$

5. $x^2 = \tan y$

6. $\ln (x + y) = 2x$

In Exercises 7 and 8, find a parametrization for the line segment with endpoints *A* and *B*.

7. $A(-2, 1)$, $B(4, -3)$

8. $A(0, -4)$, $B(5, 0)$

In Exercises 9 and 10, let $x = 2 \cos t$, $y = 2 \sin t$. Find a parameter interval that produces the indicated portion of the graph.

9. The portion in the second and third quadrants, including the points on the axes.

10. The portion in the fourth quadrant, including the points on the axes.

Section 4.6 Exercises

In Exercises 1–41, assume all variables are differentiable functions of *t*.

1. *Area* The radius *r* and area *A* of a circle are related by the equation $A = \pi r^2$. Write an equation that relates dA/dt to dr/dt.

2. *Surface Area* The radius *r* and surface area *S* of a sphere are related by the equation $S = 4\pi r^2$. Write an equation that relates dS/dt to dr/dt.

3. *Volume* The radius *r*, height *h*, and volume *V* of a right circular cylinder are related by the equation $V = \pi r^2 h$.

(a) How is dV/dt related to dh/dt if *r* is constant?

(b) How is dV/dt related to dr/dt if *h* is constant?

(c) How is dV/dt related to dr/dt and dh/dt if neither *r* nor *h* is constant?

4. *Electrical Power* The power *P* (watts) of an electric circuit is related to the circuit's resistance *R* (ohms) and current *I* (amperes) by the equation $P = RI^2$.

(a) How is dP/dt related to dR/dt and dI/dt?

(b) How is dR/dt related to dI/dt if *P* is constant?

5. *Diagonals* If *x, y,* and *z* are lengths of the edges of a rectangular box, the common length of the box's diagonals is $s = \sqrt{x^2 + y^2 + z^2}$. How is ds/dt related to dx/dt, dy/dt, and dz/dt?

6. *Area* If *a* and *b* are the lengths of two sides of a triangle, and θ the measure of the included angle, the area *A* of the triangle is $A = (1/2)\, ab \sin \theta$. How is dA/dt related to da/dt, db/dt, and $d\theta/dt$?

7. *Changing Voltage* The voltage *V* (volts), current *I* (amperes), and resistance *R* (ohms) of an electric circuit like the one shown here are related by the equation $V = IR$. Suppose that *V* is increasing at the rate of 1 volt/sec while *I* is decreasing at the rate of 1/3 amp/sec. Let *t* denote time in sec.

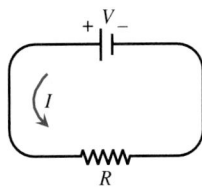

(a) What is the value of dV/dt?

(b) What is the value of dI/dt?

(c) Write an equation that relates dR/dt to dV/dt and dI/dt.

(d) **Writing to Learn** Find the rate at which *R* is changing when $V = 12$ volts and $I = 2$ amp. Is *R* increasing, or decreasing? Explain.

8. *Heating a Plate* When a circular plate of metal is heated in an oven, its radius increases at the rate of 0.01 cm/sec. At what rate is the plate's area increasing when the radius is 50 cm?

9. *Changing Dimensions in a Rectangle* The length ℓ of a rectangle is decreasing at the rate of 2 cm/sec while the width *w* is increasing at the rate of 2 cm/sec. When $\ell = 12$ cm and $w = 5$ cm, find the rates of change of

(a) the area,   (b) the perimeter, and

(c) the length of a diagonal of the rectangle.

(d) **Writing to Learn** Which of these quantities are decreasing, and which are increasing? Explain.

10. *Changing Dimensions in a Rectangular Box* Suppose that the edge lengths x, y, and z of a closed rectangular box are changing at the following rates:

$$\frac{dx}{dt} = 1 \text{ m/sec}, \quad \frac{dy}{dt} = -2 \text{ m/sec}, \quad \frac{dz}{dt} = 1 \text{ m/sec}.$$

Find the rates at which the box's **(a)** volume, **(b)** surface area, and **(c)** diagonal length $s = \sqrt{x^2 + y^2 + z^2}$ are changing at the instant when $x = 4$, $y = 3$, and $z = 2$.

11. *Air Traffic Control* An airplane is flying at an altitude of 7 mi and passes directly over a radar antenna as shown in the figure. When the plane is 10 mi from the antenna ($s = 10$), the radar detects that the distance s is changing at the rate of 300 mph. What is the speed of the airplane at that moment?

12. *Filling a Trough* A trough is 15 ft long and 4 ft across the top as shown in the figure. Its ends are isosceles triangles with height 3 ft. Water runs into the trough at the rate of 2.5 ft³/min. How fast is the water level rising when it is 2 ft deep?

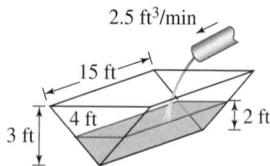

13. *Sliding Ladder* A 13-ft ladder is leaning against a house (see figure) when its base starts to slide away. By the time the base is 12 ft from the house, the base is moving at the rate of 5 ft/sec.

(a) How fast is the top of the ladder sliding down the wall at that moment?

(b) At what rate is the area of the triangle formed by the ladder, wall, and ground changing at that moment?

(c) At what rate is the angle θ between the ladder and the ground changing at that moment?

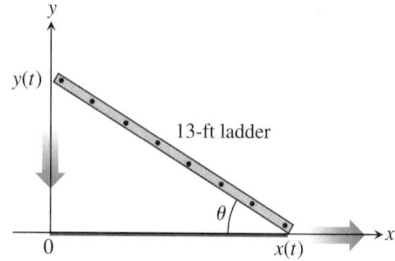

14. *Flying a Kite* Inge flies a kite at a height of 300 ft, the wind carrying the kite horizontally away at a rate of 25 ft/sec. How fast must she let out the string when the kite is 500 ft away from her?

15. *Boring a Cylinder* The mechanics at Lincoln Automotive are reboring a 6-in. deep cylinder to fit a new piston. The machine they are using increases the cylinder's radius one-thousandth of an inch every 3 min. How rapidly is the cylinder volume increasing when the bore (diameter) is 3.800 in.?

16. *Growing Sand Pile* Sand falls from a conveyor belt at the rate of 10 m³/min onto the top of a conical pile. The height of the pile is always three-eighths of the base diameter. How fast are the **(a)** height and **(b)** radius changing when the pile is 4 m high? Give your answer in cm/min.

17. *Draining Conical Reservoir* Water is flowing at the rate of 50 m³/min from a concrete conical reservoir (vertex down) of base radius 45 m and height 6 m. **(a)** How fast is the water level falling when the water is 5 m deep? **(b)** How fast is the radius of the water's surface changing at that moment? Give your answer in cm/min.

18. *Draining Hemispherical Reservoir* Water is flowing at the rate of 6 m³/min from a reservoir shaped like a hemispherical bowl of radius 13 m, shown here in profile. Answer the following questions given that the volume of water in a hemispherical bowl of radius R is $V = (\pi/3)y^2(3R - y)$ when the water is y units deep.

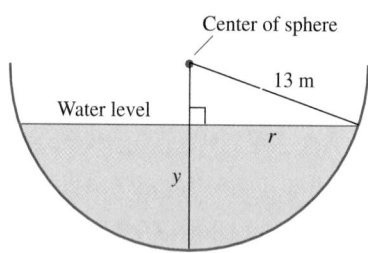

(a) At what rate is the water level changing when the water is 8 m deep?

(b) What is the radius r of the water's surface when the water is y m deep?

(c) At what rate is the radius r changing when the water is 8 m deep?

19. *Growing Raindrop* Suppose that a droplet of mist is a perfect sphere and that, through condensation, the droplet picks up moisture at a rate proportional to its surface area. Show that under these circumstances the droplet's radius increases at a constant rate.

20. *Inflating Balloon* A spherical balloon is inflated with helium at the rate of 100π ft³/min.

(a) How fast is the balloon's radius increasing at the instant the radius is 5 ft?

(b) How fast is the surface area increasing at that instant?

21. *Hauling in a Dinghy* A dinghy is pulled toward a dock by a rope from the bow through a ring on the dock 6 ft above the bow as shown in the figure. The rope is hauled in at the rate of 2 ft/sec.

(a) How fast is the boat approaching the dock when 10 ft of rope are out?

(b) At what rate is angle θ changing at that moment?

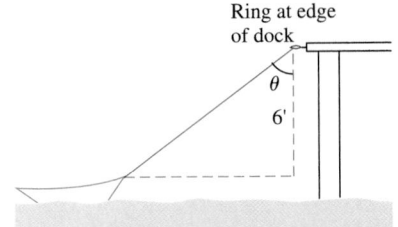

Ring at edge
of dock
θ
6'

22. *Rising Balloon* A balloon is rising vertically above a level, straight road at a constant rate of 1 ft/sec. Just when the balloon is 65 ft above the ground, a bicycle moving at a constant rate of 17 ft/sec passes under it. How fast is the distance between the bicycle and balloon increasing 3 sec later (see figure)?

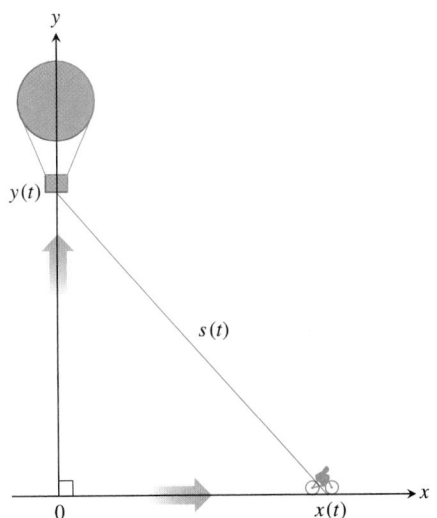

23. *Cost, Revenue, and Profit* A company can manufacture x items at a cost of $c(x)$ dollars, a sales revenue of $r(x)$ dollars, and a profit of $p(x) = r(x) - c(x)$ dollars (all amounts in thousands). Find dc/dt, dr/dt, and dp/dt for the following values of x and dx/dt.

(a) $r(x) = 9x$, $c(x) = x^3 - 6x^2 + 15x$,
and $dx/dt = 0.1$ when $x = 2$.

(b) $r(x) = 70x$, $c(x) = x^3 - 6x^2 + 45/x$,
and $dx/dt = 0.05$ when $x = 1.5$.

24. *Making Coffee* Coffee is draining from a conical filter into a cylindrical coffeepot at the rate of 10 in³/min.

(a) How fast is the level in the pot rising when the coffee in the cone is 5 in. deep?

(b) How fast is the level in the cone falling at that moment?

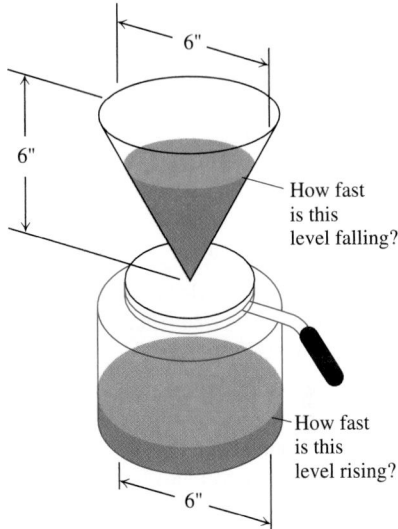

How fast
is this
level falling?

How fast
is this
level rising?

25. *Cardiac Output* Work in groups of two or three. In the late 1860s, Adolf Fick, a professor of physiology in the Faculty of Medicine in Würtzberg, Germany, developed one of the methods we use today for measuring how much blood your heart pumps in a minute. Your cardiac output as you read this sentence is probably about 7 liters a minute. At rest it is likely to be a bit under 6 L/min. If you are a trained marathon runner running a marathon, your cardiac output can be as high as 30 L/min.

Your cardiac output can be calculated with the formula

$$y = \frac{Q}{D},$$

where Q is the number of milliliters of CO_2 you exhale in a minute and D is the difference between the CO_2 concentration (mL/L) in the blood pumped to the lungs and the CO_2 concentration in the blood returning from the lungs. With $Q = 233$ mL/min and $D = 97 - 56 = 41$ mL/L,

$$y = \frac{233 \text{ mL/min}}{41 \text{ mL/L}} \approx 5.68 \text{ L/min},$$

fairly close to the 6 L/min that most people have at basal (resting) conditions. (Data courtesy of J. Kenneth Herd, M.D., Quillan College of Medicine, East Tennessee State University.)

Suppose that when $Q = 233$ and $D = 41$, we also know that D is decreasing at the rate of 2 units a minute but that Q remains unchanged. What is happening to the cardiac output?

26. *Particle Motion* A particle moves along the parabola $y = x^2$ in the first quadrant in such a way that its x-coordinate (in meters) increases at a constant rate of 10 m/sec. How fast is the angle of inclination θ of the line joining the particle to the origin changing when $x = 3$?

27. *Particle Motion* A particle moves from right to left along the parabolic curve $y = \sqrt{-x}$ in such a way that its x-coordinate (in meters) decreases at the rate of 8 m/sec. How fast is the angle of inclination θ of the line joining the particle to the origin changing when $x = -4$?

28. *Particle Motion* A particle $P(x, y)$ is moving in the coordinate plane in such a way that $dx/dt = -1$ m/sec and $dy/dt = -5$ m/sec. How fast is the particle's distance from the origin changing as it passes through the point $(5, 12)$?

29. *Moving Shadow* A man 6 ft tall walks at the rate of 5 ft/sec toward a streetlight that is 16 ft above the ground. At what rate is the length of his shadow changing when he is 10 ft from the base of the light?

30. *Moving Shadow* A light shines from the top of a pole 50 ft high. A ball is dropped from the same height from a point 30 ft away from the light as shown in the figure. How fast is the shadow of the ball moving along the ground 1/2 sec later? (Assume the ball falls a distance $s = 16t^2$ in t sec.)

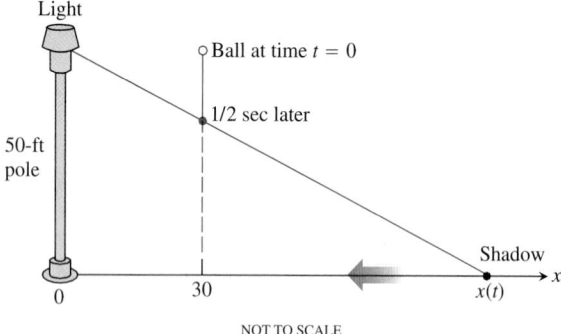

NOT TO SCALE

31. *Moving Race Car* You are videotaping a race from a stand 132 ft from the track, following a car that is moving at 180 mph (264 ft/sec) as shown in the figure. About how fast will your camera angle θ be changing when the car is right in front of you? a half second later?

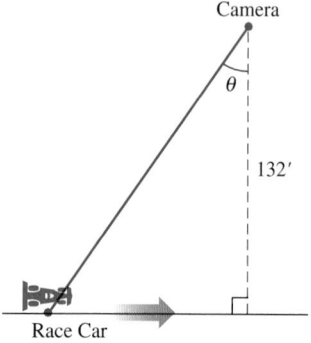

32. *Melting Ice* A spherical iron ball is coated with a layer of ice of uniform thickness. If the ice melts at the rate of 8 mL/min, how fast is the outer surface area of ice decreasing when the outer diameter (ball plus ice) is 20 cm?

33. *Speed Trap* A highway patrol airplane flies 3 mi above a level, straight road at a constant rate of 120 mph. The pilot sees an oncoming car and with radar determines that at the instant the line-of-sight distance from plane to car is 5 mi the line-of-sight distance is decreasing at the rate of 160 mph. Find the car's speed along the highway.

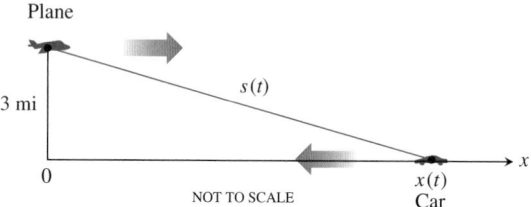

NOT TO SCALE

34. *Building's Shadow* On a morning of a day when the sun will pass directly overhead, the shadow of an 80-ft building on level ground is 60 ft long as shown in the figure. At the moment in question, the angle θ the sun makes with the ground is increasing at the rate of $0.27°/\text{min}$. At what rate is the shadow length decreasing? Express your answer in in./min, to the nearest tenth. (Remember to use radians.)

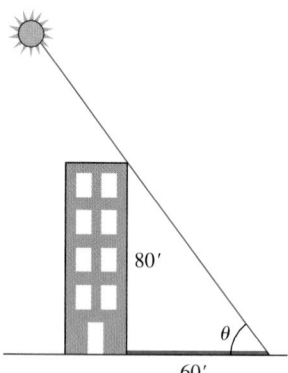

35. *Walkers* A and B are walking on straight streets that meet at right angles. A approaches the intersection at 2 m/sec and B moves away from the intersection at 1 m/sec as shown in the figure. At what rate is the angle θ changing when A is 10 m from the intersection and B is 20 m from the intersection? Express your answer in degrees per second to the nearest degree.

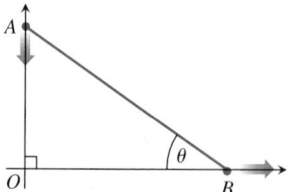

36. *Moving Ships* Two ships are steaming away from a point O along routes that make a 120° angle. Ship A moves at 14 knots (nautical miles per hour; a nautical mile is 2000 yards). Ship B moves at 21 knots. How fast are the ships moving apart when $OA = 5$ and $OB = 3$ nautical miles?

In Exercises 37 and 38, a particle is moving along the curve $y = f(x)$.

37. Let $y = f(x) = \dfrac{10}{1 + x^2}$.

If $dx/dt = 3$ cm/sec, find dy/dt at the point where

(a) $x = -2$.   **(b)** $x = 0$.   **(c)** $x = 20$.

38. Let $y = f(x) = x^3 - 4x$.

If $dx/dt = -2$ cm/sec, find dy/dt at the point where

(a) $x = -3$.   **(b)** $x = 1$.   **(c)** $x = 4$.

Extending the Ideas

39. *Motion along a Circle* A wheel of radius 2 ft makes 8 revolutions about its center every second.

(a) Explain how the parametric equations

$$x = 2 \cos \theta, \quad y = 2 \sin \theta$$

can be used to represent the motion of the wheel.

(b) Express θ as a function of time t.

(c) Find the rate of horizontal movement and the rate of vertical movement of a point on the edge of the wheel when it is at the position given by $\theta = \pi/4$, $\pi/2$, and π.

40. *Ferris Wheel* A Ferris wheel with radius 30 ft makes one revolution every 10 sec.

(a) Assume that the center of the Ferris wheel is located at the point (0, 40), and write parametric equations to model its motion. (*Hint:* See Exercise 39.)

(b) At $t = 0$ the point P on the Ferris wheel is located at (30, 40). Find the rate of horizontal movement, and the rate of vertical movement of the point P when $t = 5$ sec and $t = 8$ sec.

41. *Industrial Production* **(a)** Economists often use the expression "rate of growth" in relative rather than absolute terms. For example, let $u = f(t)$ be the number of people in the labor force at time t in a given industry. (We treat this function as though it were differentiable even though it is an integer-valued step function.)

Let $v = g(t)$ be the average production per person in the labor force at time t. The total production is then $y = uv$. If the labor force is growing at the rate of 4% per year ($du/dt = 0.04u$) and the production per worker is growing at the rate of 5% per year ($dv/dt = 0.05v$), find the rate of growth of the total production, y.

(b) Suppose that the labor force in (a) is decreasing at the rate of 2% per year while the production per person is increasing at the rate of 3% per year. Is the total production increasing, or is it decreasing, and at what rate?

Chapter 4   Key Terms

absolute change (p. 226)
absolute maximum value (p. 177)
absolute minimum value (p. 177)
antiderivative (p. 190)
antidifferentiation (p. 190)
arithmetic mean (p. 194)
average cost (p. 212)
center of linear approximation (p. 221)
concave down (p. 197)
concave up (p. 197)
concavity test (p. 197)
critical point (p. 180)
decreasing function (p. 188)
differential (p. 224)
differential estimate of change (p. 226)
differential of a function (p. 225)

extrema (p. 177)
Extreme Value Theorem (p. 178)
first derivative test (p. 194)
first derivative test for
   local extrema (p. 195)
geometric mean (p. 194)
global maximum value (p. 177)
global minimum value (p. 177)
increasing function (p. 188)
linear approximation (p. 220)
linearization (p. 221)
local maximum value (p. 179)
local minimum value (p. 179)
logistic curve (p. 199)
logistic regression (p. 200)
marginal analysis (p. 211)

marginal cost and revenue (p. 211)
Mean Value Theorem (p. 186)
monotonic function (p. 188)
Newton's method (p. 222)
optimization (p. 206)
percentage change (p. 226)
point of inflection (p. 198)
profit (p. 211)
quadratic approximation (p. 231)
related rates (p. 232)
relative change (p. 226)
relative extrema (p. 179)
Rolle's Theorem (p. 186)
second derivative test for
   local extrema (p. 200)
standard linear approximation (p. 221)

Chapter 4 Review Exercises

In Exercises 1 and 2, use analytic methods to find the global extreme values of the function on the interval and state where they occur.

1. $y = x\sqrt{2 - x}, \quad -2 \le x \le 2$

2. $y = x^3 - 9x^2 - 21x - 11, \quad -\infty < x < \infty$

In Exercises 3 and 4, use analytic methods. Find the intervals on which the function is

    (a) increasing,   **(b)** decreasing,

    (c) concave up,   **(d)** concave down.

Then find any

    (e) local extreme values,   **(f)** inflection points.

3. $y = x^2 e^{1/x^2}$ **4.** $y = x\sqrt{4 - x^2}$

In Exercises 5–16, find the intervals on which the function is

    (a) increasing,   **(b)** decreasing,

    (c) concave up,   **(d)** concave down.

Then find any

    (e) local extreme values,   **(f)** inflection points.

5. $y = 1 + x - x^2 - x^4$ **6.** $y = e^{x-1} - x$

7. $y = \dfrac{1}{\sqrt[4]{1 - x^2}}$ **8.** $y = \dfrac{x}{x^3 - 1}$

9. $y = \cos^{-1} x$ **10.** $y = \dfrac{x}{x^2 + 2x + 3}$

11. $y = \ln|x|, \quad -2 \le x \le 2, \quad x \ne 0$

12. $y = \sin 3x + \cos 4x, \quad 0 \le x \le 2\pi$

13. $y = \begin{cases} e^{-x}, & x \le 0 \\ 4x - x^3, & x > 0 \end{cases}$

14. $y = -x^5 + \dfrac{7}{3}x^3 + 5x^2 + 4x + 2$

15. $y = x^{4/5}(2 - x)$

16. $y = \dfrac{5 - 4x + 4x^2 - x^3}{x - 2}$

In Exercises 17 and 18, use the derivative of the function $y = f(x)$ to find the points at which f has a

    (a) local maximum,   **(b)** local minimum, or

    (c) point of inflection.

17. $y' = 6(x + 1)(x - 2)^2$ **18.** $y' = 6(x + 1)(x - 2)$

In Exercises 19–22, find all possible functions with the given derivative.

19. $f'(x) = x^{-5} + e^{-x}$ **20.** $f'(x) = \sec x \tan x$

21. $f'(x) = \dfrac{2}{x} + x^2 + 1, \; x > 0$ **22.** $f'(x) = \sqrt{x} + \dfrac{1}{\sqrt{x}}$

In Exercises 23 and 24, find the function with the given derivative whose graph passes through the point P.

23. $f'(x) = \sin x + \cos x, \quad P(\pi, 3)$

24. $f'(x) = x^{1/3} + x^2 + x + 1, \quad P(1, 0)$

In Exercises 25 and 26, the velocity v or acceleration a of a particle is given. Find the particle's position s at time t.

25. $v = 9.8t + 5, \quad s = 10 \quad$ when $\quad t = 0$

26. $a = 32, \quad v = 20 \quad$ and $\quad s = 5 \quad$ when $\quad t = 0$

In Exercises 27–30, find the linearization $L(x)$ of $f(x)$ at $x = a$.

27. $f(x) = \tan x, \quad a = -\pi/4$

28. $f(x) = \sec x, \quad a = \pi/4$

29. $f(x) = \dfrac{1}{1 + \tan x}, \quad a = 0$

30. $f(x) = e^x + \sin x, \quad a = 0$

In Exercises 31–34, use the graph to answer the questions.

31. Identify any global extreme values of f and the values of x at which they occur.

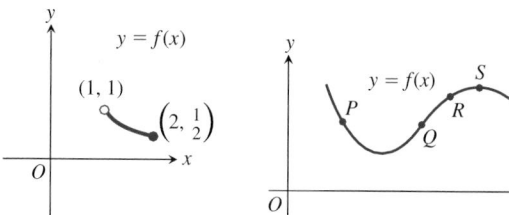

Figure for Exercise 31   Figure for Exercise 32

32. At which of the five points on the graph of $y = f(x)$ shown here

    (a) are y' and y'' both negative?

    (b) is y' negative and y'' positive?

33. Estimate the intervals on which the function $y = f(x)$ is **(a)** increasing; **(b)** decreasing. **(c)** Estimate any local extreme values of the function and where they occur.

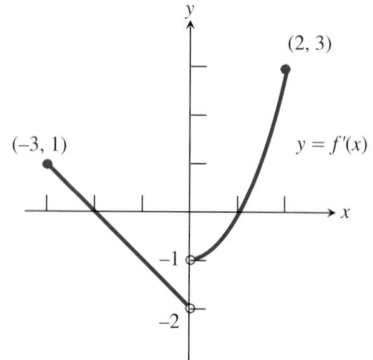

34. Here is the graph of the fruit fly population from Section 2.4, Example 1. On approximately what day did the population's growth rate change from increasing to decreasing?

Time (days)

35. *Connecting f and f′* The graph of f' is shown in Exercise 33. Sketch a possible graph of f given that it is continuous with domain $[-3, 2]$ and $f(-3) = 0$.

36. *Connecting f, f′, and f″* The function f is continuous on $[0, 3]$ and satisfies the following.

x	0	1	2	3
f	0	-2	0	3
f'	-3	0	does not exist	4
f''	0	1	does not exist	0

x	$0 < x < 1$	$1 < x < 2$	$2 < x < 3$
f	$-$	$-$	$+$
f'	$-$	$+$	$+$
f''	$+$	$+$	$+$

(a) Find the absolute extrema of f and where they occur.

(b) Find any points of inflection.

(c) Sketch a possible graph of f.

37. *Mean Value Theorem* Let $f(x) = x \ln x$.

(a) Writing to Learn Show that f satisfies the hypotheses of the Mean Value Theorem on the interval $[a, b] = [0.5, 3]$.

(b) Find the value(s) of c in (a, b) for which

$$f'(c) = \frac{f(b) - f(a)}{b - a}.$$

(c) Write an equation for the secant line AB where $A = (a, f(a))$ and $B = (b, f(b))$.

(d) Write an equation for the tangent line that is parallel to the secant line AB.

38. *Motion along a Line* A particle is moving along a line with position function $s(t) = 3 + 4t - 3t^2 - t^3$. Find the **(a)** velocity and **(b)** acceleration, and **(c)** describe the motion of the particle for $t \geq 0$.

39. *Approximating Functions* Let f be a function with $f'(x) = \sin x^2$ and $f(0) = -1$.

(a) Find the linearization of f at $x = 0$.

(b) Approximate the value of f at $x = 0.1$.

(c) Writing to Learn Is the actual value of f at $x = 0.1$ greater than or less than the approximation in (b)?

40. *Differentials* Let $y = x^2 e^{-x}$. Find **(a)** dy and **(b)** evaluate dy for $x = 1$ and $dx = 0.01$.

41. *Logistic Regression* Table 4.4 gives the numbers of on-line debit card transactions for several years. Let $x = 0$ represent 1990, $x = 1$ represent 1991, and so on.

(a) Find the logistic regression equation and superimpose its graph on a scatter plot of the data.

(b) Use the regression equation to predict when the rate of increase in debit card transactions will start to decrease, and the number of transactions at that time.

(c) What is the ceiling of the number of debit card transactions predicted by the regression equation?

Table 4.4 Debit Card Transactions

Year	Number (millions)
1991	211
1992	280
1993	386
1994	551
1995	684
1996	905

Source: "Debit Card News" as reported by Julie Stacey in *USA Today*, May 30, 1997.

42. *Newton's Method* Use Newton's method to estimate all real solutions to $2 \cos x - \sqrt{1 + x} = 0$. State your answers accurate to 6 decimal places.

43. *Rocket Launch* A rocket lifts off the surface of Earth with a constant acceleration of 20 m/sec². How fast will the rocket be going 1 min later?

44. *Launching on Mars* The acceleration of gravity near the surface of Mars is 3.72 m/sec². If a rock is blasted straight up from the surface with an initial velocity of 93 m/sec (about 208 mph), how high does it go?

45. *Area of Sector* If the perimeter of the circular sector shown here is fixed at 100 ft, what values of r and s will give the sector the greatest area?

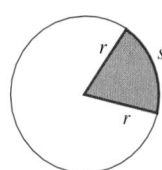

46. *Area of Triangle* An isosceles triangle has its vertex at the origin and its base parallel to the *x*-axis with the vertices above the axis on the curve $y = 27 - x^2$. Find the largest area the triangle can have.

47. *Storage Bin* Find the dimensions of the largest open-top storage bin with a square base and vertical sides that can be made from 108 ft² of sheet steel. (Neglect the thickness of the steel and assume that there is no waste.)

48. *Designing a Vat* You are to design an open-top rectangular stainless-steel vat. It is to have a square base and a volume of 32 ft³; to be welded from quarter-inch plate, and weigh no more than necessary. What dimensions do you recommend?

49. *Inscribing a Cylinder* Find the height and radius of the largest right circular cylinder that can be put into a sphere of radius $\sqrt{3}$ as described in the figure.

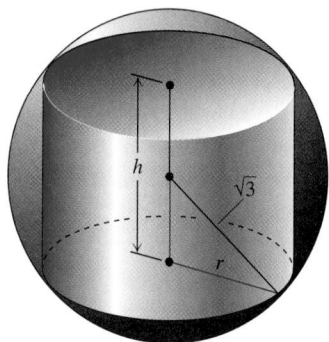

50. *Cone in a Cone* The figure shows two right circular cones, one upside down inside the other. The two bases are parallel, and the vertex of the smaller cone lies at the center of the larger cone's base. What values of *r* and *h* will give the smaller cone the largest possible volume?

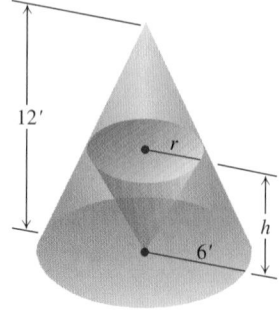

51. *Box with Lid* Repeat Exercise 18 of Section 4.4 but this time remove the two equal squares from the corners of a 15-in. side.

52. *Inscribing a Rectangle* A rectangle is inscribed under one arch of $y = 8 \cos{(0.3x)}$ with its base on the *x*-axis and its upper two vertices on the curve symmetric about the *y*-axis. What is the largest area the rectangle can have?

53. *Oil Refinery* A drilling rig 12 mi offshore is to be connected by a pipe to a refinery onshore, 20 mi down the coast from the rig as shown in the figure. If underwater pipe costs $40,000 per mile and land-based pipe costs $30,000 per mile, what values of *x* and *y* give the least expensive connection?

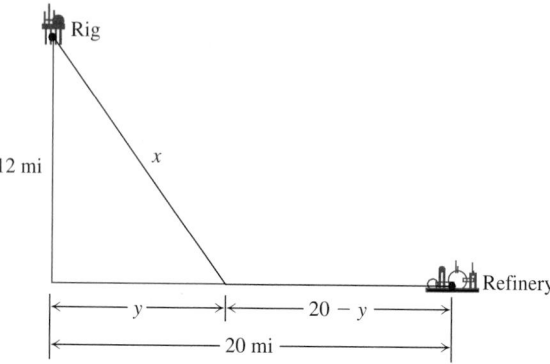

54. *Designing an Athletic Field* An athletic field is to be built in the shape of a rectangle *x* units long capped by semicircular regions of radius *r* at the two ends. The field is to be bounded by a 400-m running track. What values of *x* and *r* will give the rectangle the largest possible area?

55. *Manufacturing Tires* Your company can manufacture *x* hundred grade A tires and *y* hundred grade B tires a day, where $0 \le x \le 4$ and

$$y = \frac{40 - 10x}{5 - x}.$$

Your profit on a grade A tire is twice your profit on a grade B tire. What is the most profitable number of each kind to make?

56. *Particle Motion* The positions of two particles on the *s*-axis are $s_1 = \cos{t}$ and $s_2 = \cos{(t + \pi/4)}$.

(a) What is the farthest apart the particles ever get?

(b) When do the particles collide?

57. *Open-top Box* An open-top rectangular box is constructed from a 10- by 16-in. piece of cardboard by cutting squares of equal side length from the corners and folding up the sides. Find analytically the dimensions of the box of largest volume and the maximum volume. Support your answers graphically.

58. *Changing Area* The radius of a circle is changing at the rate of $-2/\pi$ m/sec. At what rate is the circle's area changing when $r = 10$ m?

59. *Particle Motion* The coordinates of a particle moving in the plane are differentiable functions of time *t* with $dx/dt = -1$ m/sec and $dy/dt = -5$ m/sec. How fast is the particle approaching the origin as it passes through the point (5, 12)?

60. *Changing Cube* The volume of a cube is increasing at the rate of 1200 cm³/min at the instant its edges are 20 cm long. At what rate are the edges changing at that instant?

61. *Particle Motion* A point moves smoothly along the curve $y = x^{3/2}$ in the first quadrant in such a way that its distance from the origin increases at the constant rate of 11 units per second. Find dx/dt when $x = 3$.

62. *Draining Water* Water drains from the conical tank shown in the figure at the rate of 5 ft³/min.

(a) What is the relation between the variables h and r?

(b) How fast is the water level dropping when $h = 6$ ft?

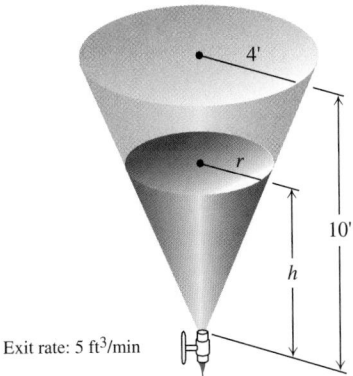

Exit rate: 5 ft³/min

63. *Stringing Telephone Cable* As telephone cable is pulled from a large spool to be strung from the telephone poles along a street, it unwinds from the spool in layers of constant radius as suggested in the figure. If the truck pulling the cable moves at a constant rate of 6 ft/sec, use the equation $s = r\theta$ to find how fast (in rad/sec) the spool is turning when the layer of radius 1.2 ft is being unwound.

64. *Throwing Dirt* You sling a shovelful of dirt up from the bottom of a 17-ft hole with an initial velocity of 32 ft/sec. Is that enough speed to get the dirt out of the hole, or had you better duck?

65. *Estimating Change* Write a formula that estimates the change that occurs in the volume of a right circular cone (see figure) when the radius changes from a to $a + dr$ and the height does not change.

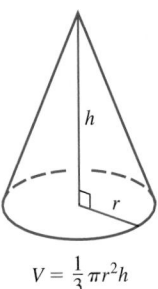

$$V = \frac{1}{3}\pi r^2 h$$

66. *Controlling Error*

(a) How accurately should you measure the edge of a cube to be reasonably sure of calculating the cube's surface area with an error of no more than 2%?

(b) Suppose the edge is measured with the accuracy required in part (a). About how accurately can the cube's volume be calculated from the edge measurement? To find out, estimate the percentage error in the volume calculation that might result from using the edge measurement.

67. *Compounding Error* The circumference of the equator of a sphere is measured as 10 cm with a possible error of 0.4 cm. This measurement is then used to calculate the radius. The radius is then used to calculate the surface area and volume of the sphere. Estimate the percentage errors in the calculated values of (a) the radius, (b) the surface area, and (c) the volume.

68. *Finding Height* To find the height of a lamppost (see figure), you stand a 6-ft pole 20 ft from the lamp and measure the length a of its shadow, finding it to be 15 ft, give or take an inch. Calculate the height of the lamppost using the value $a = 15$, and estimate the possible error in the result.

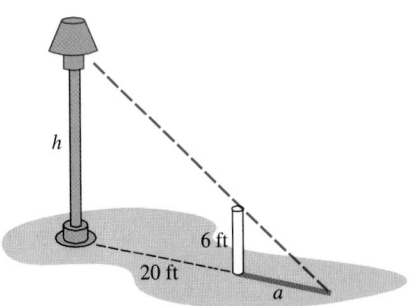

69. *Decreasing Function* Show that the function $y = \sin^2 x - 3x$ decreases on every interval in its domain.

The Definite Integral

The 1995 Reader's Digest Sweepstakes grand prize winner is being paid a total of $5,010,000 over 30 years. If invested, the winnings plus the interest earned generate an amount defined by:

$$A = e^{rT}\int_0^T mPe^{-rt}\,dt$$

(r = interest rate, P = size of payment, T = term in years, m = number of payments per year.)

Would the prize have a different value if it were paid in 15 annual installments of $334,000 instead of 30 annual installments of $167,000? Section 5.4 can help you compare the total amounts.

Chapter 5 Overview

We have seen how the need to calculate instantaneous rates of change led the discoverers of calculus to an investigation of the slopes of tangent lines and, ultimately, to the derivative—to what we call *differential* calculus. But they knew that derivatives revealed only half the story. In addition to a calculation method (a "calculus") to describe how functions were changing at a given instant, they also needed a method to describe how those instantaneous changes could accumulate over an interval to produce the function. That is why they were also investigating *areas under curves,* an investigation that ultimately led to the second main branch of calculus, called *integral* calculus.

Once they had the calculus for finding slopes of tangent lines and the calculus for finding areas under curves—two geometric operations that would seem to have nothing at all to do with each other—the challenge for Newton and Leibniz was to prove the connection that they knew intuitively had to be there. The discovery of this connection (called the Fundamental Theorem of Calculus) brought differential and integral calculus together to become the single most powerful insight mathematicians had ever acquired for understanding how the universe worked.

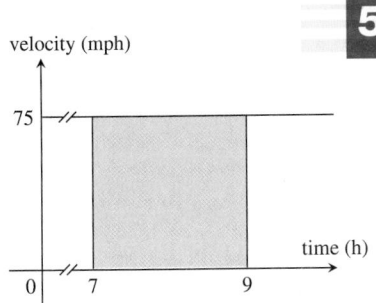

Figure 5.1 The distance traveled by a 75 mph train in 2 hours is 150 miles, which corresponds to the area of the shaded rectangle.

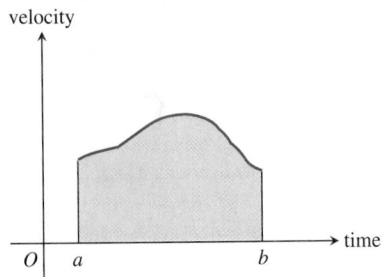

Figure 5.2 If the velocity varies over the time interval [a, b], does the shaded region give the distance traveled?

| 5.1 | Estimating with Finite Sums |

Distance Traveled • Rectangular Approximation Method (RAM) • Volume of a Sphere • Cardiac Output

Distance Traveled

We know why a mathematician pondering motion problems might have been led to consider slopes of curves, but what do those same motion problems have to do with areas under curves? Consider the following problem from a typical elementary school textbook:

> A train moves along a track at a steady rate of 75 miles per hour from 7:00 A.M. to 9:00 A.M. What is the total distance traveled by the train?

Applying the well-known formula *distance = rate × time,* we find that the answer is 150 miles. Simple. Now suppose that you are Isaac Newton trying to make a connection between this formula and the graph of the velocity function.

You might notice that the distance traveled by the train (150 miles) is exactly the *area* of the rectangle whose base is the time interval [7, 9] and whose height at each point is the value of the constant velocity function $v = 75$ (Figure 5.1). This is no accident, either, since *the distance traveled* and *the area* in this case are both found by multiplying the rate (75) by the change in time (2).

This same connection between distance traveled and rectangle area could be made no matter how fast the train was going or how long or short the time interval was. But what if the train had a velocity v that *varied* as a function of time? The graph (Figure 5.2) would no longer be a horizontal line, so the region under the graph would no longer be rectangular.

velocity

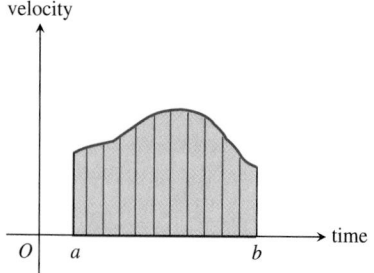

Figure 5.3 The region is partitioned into vertical strips. If the strips are narrow enough, they are almost indistinguishable from rectangles. The sum of the areas of these "rectangles" will give the total area and can be interpreted as distance traveled.

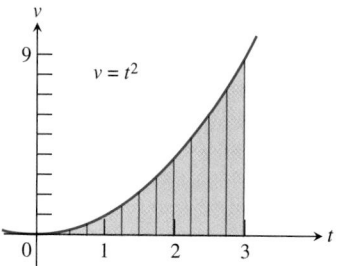

Figure 5.4 The region under the parabola $v = t^2$ from $t = 0$ to $t = 3$ is partitioned into 12 thin strips, each with base $\Delta t = 1/4$. The strips have curved tops. (Example 1)

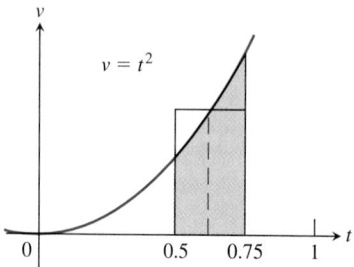

Figure 5.5 The area of the shaded region is approximated by the area of the rectangle whose height is the function value at the midpoint of the interval. (Example 1)

Would the area of this irregular region still give the total distance traveled over the time interval? Newton and Leibniz (and, actually, many others who had considered this question) thought that it obviously would, and that is why they were interested in a calculus for finding areas under curves. They imagined the time interval being partitioned into many tiny subintervals, each one so small that the velocity over it would essentially be constant. Geometrically, this was equivalent to slicing the irregular region into narrow strips, each of which would be nearly indistinguishable from a narrow rectangle (Figure 5.3).

They argued that, just as the total area could be found by summing the areas of the (essentially rectangular) strips, the total distance traveled could be found by summing the small distances traveled over the tiny time intervals.

Example 1 FINDING DISTANCE TRAVELED WHEN VELOCITY VARIES

A particle starts at $x = 0$ and moves along the x-axis with velocity $v(t) = t^2$ for time $t \geq 0$. Where is the particle at $t = 3$?

Solution We graph v and partition the time interval $[0, 3]$ into subintervals of length Δt. (Figure 5.4 shows twelve subintervals of length 3/12 each.)

Notice that the region under the curve is partitioned into thin strips with bases of length 1/4 and *curved* tops that slope upward from left to right. You might not know how to find the area of such a strip, but you can get a good approximation of it by finding the area of a suitable rectangle. In Figure 5.5, we use the rectangle whose height is the y-coordinate of the function at the midpoint of its base.

The area of this narrow rectangle approximates the distance traveled over the time subinterval. Adding all the areas (distances) gives an approximation of the total area under the curve (total distance traveled) from $t = 0$ to $t = 3$ (Figure 5.6).

Computing this sum of areas is straightforward. Each rectangle has a base of length $\Delta t = 1/4$, while the height of each rectangle can be found by evaluating the function at the midpoint of the subinterval. Table 5.1 shows the computations for the first four rectangles.

Table 5.1

Subinterval	$\left[0, \dfrac{1}{4}\right]$	$\left[\dfrac{1}{4}, \dfrac{1}{2}\right]$	$\left[\dfrac{1}{2}, \dfrac{3}{4}\right]$	$\left[\dfrac{3}{4}, 1\right]$
Midpoint m_i	$\dfrac{1}{8}$	$\dfrac{3}{8}$	$\dfrac{5}{8}$	$\dfrac{7}{8}$
Height = $(m_i)^2$	$\dfrac{1}{64}$	$\dfrac{9}{64}$	$\dfrac{25}{64}$	$\dfrac{49}{64}$
Area = $(1/4)(m_i)^2$	$\dfrac{1}{256}$	$\dfrac{9}{256}$	$\dfrac{25}{256}$	$\dfrac{49}{256}$

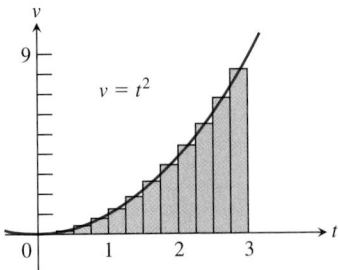

Figure 5.6 These rectangles have approximately the same areas as the strips in Figure 5.4. Each rectangle has height m_i^2, where m_i is the midpoint of its base. (Example 1)

Approximation by Rectangles

Approximating irregularly-shaped regions by regularly-shaped regions for the purpose of computing areas is not new. Archimedes used the idea more than 2200 years ago to find the area of a circle, demonstrating in the process that π was located between 3.140845 and 3.142857. He also used approximations to find the area under a parabolic arch, anticipating the answer to an important seventeenth-century question nearly 2000 years before anyone thought to ask it. The fact that we still measure the area of anything—even a circle—in "square units" is obvious testimony to the historical effectiveness of using rectangles for approximating areas.

Continuing in this manner, we derive the area $(1/4)(m_i)^2$ for each of the twelve subintervals and add them:

$$\frac{1}{256} + \frac{9}{256} + \frac{25}{256} + \frac{49}{256} + \frac{81}{256} + \frac{121}{256} + \frac{169}{256} + \frac{225}{256} +$$

$$\frac{289}{256} + \frac{361}{256} + \frac{441}{256} + \frac{529}{256} = \frac{2300}{256} \approx 8.98.$$

Since this number approximates the area and hence the total distance traveled by the particle, we conclude that the particle has moved approximately 9 units in 3 seconds. If it starts at $x = 0$, then it is very close to $x = 9$ when $t = 3$.

To make it easier to talk about approximations with rectangles, we now introduce some new terminology.

Rectangular Approximation Method (RAM)

In Example 1 we used the *Midpoint Rectangular Approximation Method (MRAM)* to approximate the area under the curve. The name suggests the choice we made when determining the heights of the approximating rectangles: We evaluated the function at the midpoint of each subinterval. If instead we had evaluated the function at the left-hand endpoint we would have obtained the *LRAM* approximation, and if we had used the right-hand endpoints we would have obtained the *RRAM* approximation. Figure 5.7 shows what the three approximations look like graphically when we approximate the area under the curve $y = x^2$ from $x = 0$ to $x = 3$ with six subintervals.

No matter which RAM approximation we compute, we are adding products of the form $f(x_i) \cdot \Delta x$, or, in this case, $(x_i)^2 \cdot (3/6)$.

LRAM:

$$\left(0\right)^2\left(\frac{1}{2}\right) + \left(\frac{1}{2}\right)^2\left(\frac{1}{2}\right) + \left(1\right)^2\left(\frac{1}{2}\right) + \left(\frac{3}{2}\right)^2\left(\frac{1}{2}\right) + \left(2\right)^2\left(\frac{1}{2}\right) + \left(\frac{5}{2}\right)^2\left(\frac{1}{2}\right) = 6.875$$

MRAM:

$$\left(\frac{1}{4}\right)^2\left(\frac{1}{2}\right) + \left(\frac{3}{4}\right)^2\left(\frac{1}{2}\right) + \left(\frac{5}{4}\right)^2\left(\frac{1}{2}\right) + \left(\frac{7}{4}\right)^2\left(\frac{1}{2}\right) + \left(\frac{9}{4}\right)^2\left(\frac{1}{2}\right) + \left(\frac{11}{4}\right)^2\left(\frac{1}{2}\right) = 8.9375$$

RRAM:

$$\left(\frac{1}{2}\right)^2\left(\frac{1}{2}\right) + \left(1\right)^2\left(\frac{1}{2}\right) + \left(\frac{3}{2}\right)^2\left(\frac{1}{2}\right) + \left(2\right)^2\left(\frac{1}{2}\right) + \left(\frac{5}{2}\right)^2\left(\frac{1}{2}\right) + \left(3\right)^2\left(\frac{1}{2}\right) = 11.375$$

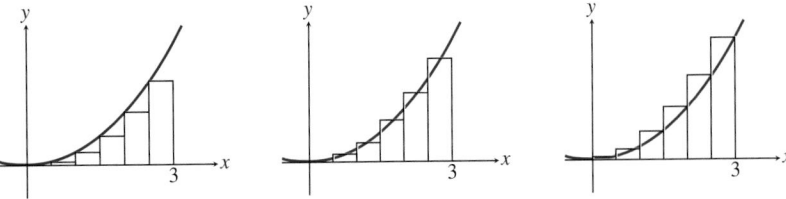

Figure 5.7 LRAM, MRAM, and RRAM approximations to the area under the graph of $y = x^2$ from $x = 0$ to $x = 3$.

As we can see from Figure 5.7, LRAM is smaller than the true area and RRAM is larger. MRAM appears to be the closest of the three approximations. However, observe what happens as the number n of subintervals increases:

n	$LRAM_n$	$MRAM_n$	$RRAM_n$
6	6.875	8.9375	11.375
12	7.90625	8.984375	10.15625
24	8.4453125	8.99609375	9.5703125
48	8.720703125	8.999023438	9.283203125
100	8.86545	8.999775	9.13545
1000	8.9865045	8.99999775	9.0135045

We computed the numbers in this table using a graphing calculator and a summing program called RAM. A version of this program for most graphing calculators can be found in the *Technology Resource Manual* that accompanies this textbook. *All three sums* approach the same number (in this case, 9).

Example 2 ESTIMATING AREA UNDER THE GRAPH OF A NONNEGATIVE FUNCTION

Figure 5.8 shows the graph of $f(x) = x^2 \sin x$ on the interval $[0, 3]$. Estimate the area under the curve from $x = 0$ to $x = 3$.

Solution We apply our RAM program to get the numbers in this table.

n	$LRAM_n$	$MRAM_n$	$RRAM_n$
5	5.15480	5.89668	5.91685
10	5.52574	5.80685	5.90677
25	5.69079	5.78150	5.84320
50	5.73615	5.77788	5.81235
100	5.75701	5.77697	5.79511
1000	5.77476	5.77667	5.77857

It is not necessary to compute all three sums each time just to approximate the area, but we wanted to show again how all three sums approach the same number. With 1000 subintervals, all three agree in the first three digits. (The *exact* area is $-7 \cos 3 + 6 \sin 3 - 2$, which is 5.77666752456 to twelve digits.)

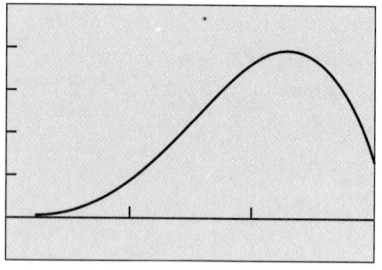

[0, 3] by [−1, 5]

Figure 5.8 The graph of $y = x^2 \sin x$ over the interval $[0, 3]$. (Example 2)

You might think from the previous two RAM tables that LRAM is always a little low and RRAM a little high, with MRAM somewhere in between. That, however, depends on n and on the shape of the curve.

1. Graph $y = 5 - 4 \sin (x/2)$ in the window $[0, 3]$ by $[0, 5]$. Copy the graph on paper and sketch the rectangles for the LRAM, MRAM, and RRAM sums with $n = 3$. Order the three approximations from greatest to smallest.

2. Graph $y = 2 \sin (5x) + 3$ in the same window. Copy the graph on paper and sketch the rectangles for the LRAM, MRAM, and RRAM sums with $n = 3$. Order the three approximations from greatest to smallest.

3. If a positive, continuous function is increasing on an interval, what can we say about the relative sizes of LRAM, MRAM, and RRAM? Explain.

4. If a positive, continuous function is decreasing on an interval, what can we say about the relative sizes of LRAM, MRAM, and RRAM? Explain.

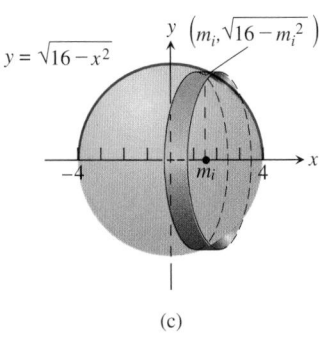

Figure 5.9 (a) The semicircle $y = \sqrt{16 - x^2}$ revolved about the x-axis to generate a sphere. (b) Slices of the solid sphere approximated with cylinders (drawn for $n = 8$). (c) The typical approximating cylinder has radius $f(m_i) = \sqrt{16 - m_i^2}$. (Example 3)

Volume of a Sphere

Although the visual representation of RAM approximation focuses on area, remember that our original motivation for looking at sums of this type was to find distance traveled by an object moving with a nonconstant velocity. The connection between Examples 1 and 2 is that in each case, we have a function f defined on a closed interval and estimate what we want to know with a sum of function values multiplied by interval lengths. Many other physical quantities can be estimated this way.

Example 3 ESTIMATING THE VOLUME OF A SPHERE

Estimate the volume of a solid sphere of radius 4.

Solution We picture the sphere as if its surface were generated by revolving the graph of the function $f(x) = \sqrt{16 - x^2}$ about the x-axis (Figure 5.9a). We partition the interval $-4 \leq x \leq 4$ into n subintervals of equal length $\Delta x = 8/n$. We then slice the sphere with planes perpendicular to the x-axis at the partition points, cutting it like a round loaf of bread into n parallel slices of width Δx. When n is large, each slice can be approximated by a cylinder, a familiar geometric shape of known volume, $\pi r^2 h$. In our case, the cylinders lie on their sides and h is Δx while r varies according to where we are on the x-axis (Figure 5.9b). A logical radius to choose for each cylinder is $f(m_i) = \sqrt{16 - m_i^2}$, where m_i is the midpoint of the interval where the ith slice intersects the x-axis (Figure 5.9c).

We can now approximate the volume of the sphere by using MRAM to sum the cylinder volumes,

$$\pi r^2 h = \pi(\sqrt{16 - m_i^2})^2 \Delta x.$$

Keeping Track of Units

Notice in Example 3 that we are summing products of the form $\pi(16 - x^2) \times$ (a cross section area, measured in square units) times Δx (a length, measured in units). The products are therefore measured in cubic units, which are the correct units for volume.

The function we use in the RAM program is $\pi(\sqrt{16 - x^2})^2 = \pi(16 - x^2)$. The interval is $[-4, 4]$.

Number of Slices (n)	MRAM$_n$
10	269.42299
25	268.29704
50	268.13619
100	268.09598
1000	268.08271

The value for $n = 1000$ compares *very* favorably with the true volume,

$$V = \frac{4}{3}\pi r^3 = \frac{4}{3}\pi(4)^3 = \frac{256\pi}{3} \approx 268.0825731.$$

Even for $n = 10$ the difference between the MRAM approximation and the true volume is a small percentage of V:

$$\frac{|\text{MRAM}_{10} - V|}{V} = \frac{\text{MRAM}_{10} - 256\pi/3}{256\pi/3} \le 0.005.$$

That is, the error percentage is about one half of one percent!

Cardiac Output

So far we have seen applications of the RAM process to finding distance traveled and volume. These applications hint at the usefulness of this technique. To suggest its versatility we will present an application from human physiology.

The number of liters of blood your heart pumps in a fixed time interval is called your *cardiac output*. For a person at rest, the rate might be 5 or 6 liters per minute. During strenuous exercise the rate might be as high as 30 liters per minute. It might also be altered significantly by disease. How can a physician measure a patient's cardiac output without interrupting the flow of blood?

One technique is to inject a dye into a main vein near the heart. The dye is drawn into the right side of the heart and pumped through the lungs and out the left side of the heart into the aorta, where its concentration can be measured every few seconds as the blood flows past. The data in Table 5.2 and the plot in Figure 5.10 (obtained from the data) show the response of a healthy, resting patient to an injection of 5.6 mg of dye.

Table 5.2 Dye Concentration Data

Seconds after Injection t	Dye Concentration (adjusted for recirculation) c
5	0
7	3.8
9	8.0
11	6.1
13	3.6
15	2.3
17	1.45
19	0.91
21	0.57
23	0.36
25	0.23
27	0.14
29	0.09
31	0

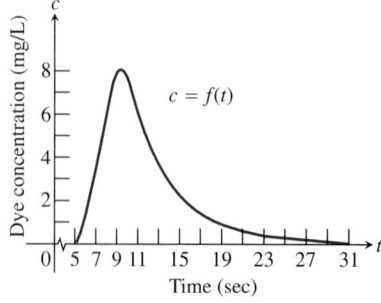

Figure 5.10 The dye concentration data from Table 5.2, plotted and fitted with a smooth curve. Time is measured with $t = 0$ at the time of injection. The dye concentration is zero at the beginning while the dye passes through the lungs. It then rises to a maximum at about $t = 9$ sec and tapers to zero by $t = 31$ sec.

Charles Richard Drew

1904–1950

Millions of people are alive today because of Charles Drew's pioneering work on blood plasma and the preservation of human blood for transfusion. After directing the Red Cross program that collected plasma for the Armed Forces in World War II, Dr. Drew went on to become Head of Surgery at Howard University and Chief of Staff at Freedmen's Hospital in Washington, D. C.

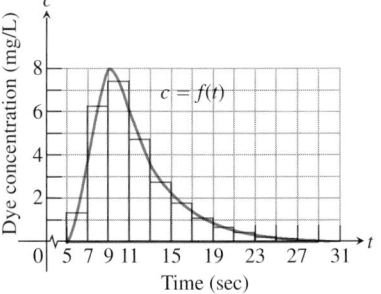

Figure 5.11 The region under the concentration curve of Figure 5.10 is approximated with rectangles. We ignore the portion from $t = 29$ to $t = 31$; its concentration is negligible. (Example 4)

The graph shows dye concentration (measured in milligrams of dye per liter of blood) as a function of time (in seconds). How can we use this graph to obtain the cardiac output (measured in liters of blood per second)? The trick is to divide the *number of mg of dye* by the *area under the dye concentration curve*. You can see why this works if you consider what happens to the units:

$$\frac{\text{mg of dye}}{\text{units of area under curve}} = \frac{\text{mg of dye}}{\dfrac{\text{mg of dye}}{\text{L of blood}} \cdot \text{sec}}$$

$$= \frac{\text{mg of dye}}{\text{sec}} \cdot \frac{\text{L of blood}}{\text{mg of dye}}$$

$$= \frac{\text{L of blood}}{\text{sec}}.$$

So you are now ready to compute like a cardiologist.

Example 4 COMPUTING CARDIAC OUTPUT FROM DYE CONCENTRATION

Estimate the cardiac output of the patient whose data appear in Table 5.2 and Figure 5.10. Give the estimate in liters per minute.

Solution We have seen that we can obtain the cardiac output by dividing the amount of dye (5.6 mg for our patient) by the area under the curve in Figure 5.10. Now we need to find the area. Our geometry formulas do not apply to this irregularly shaped region, and the RAM program is useless without a formula for the function. Nonetheless, we can draw the MRAM rectangles ourselves and estimate their heights from the graph. In Figure 5.11 each rectangle has a base 2 units long and a height $f(m_i)$ equal to the height of the curve above the midpoint of the base.

The area of each rectangle, then, is $f(m_i)$ times 2, and the sum of the rectangular areas is the MRAM estimate for the area under the curve:

$$\text{Area} \approx f(6) \cdot 2 + f(8) \cdot 2 + f(10) \cdot 2 + \cdots + f(28) \cdot 2$$

$$\approx 2 \cdot (1.4 + 6.3 + 7.5 + 4.8 + 2.8 + 1.9 + 1.1 + 0.7 + 0.5 + 0.3 + 0.2 + 0.1)$$

$$= 2 \cdot (27.6) = 55.2 \ (\text{mg/L}) \cdot \text{sec}.$$

Dividing 5.6 mg by this figure gives an estimate for cardiac output in liters per second. Multiplying by 60 converts the estimate to liters per minute:

$$\frac{5.6 \ \text{mg}}{55.2 \ \text{mg} \cdot \text{sec/L}} \cdot \frac{60 \ \text{sec}}{1 \ \text{min}} \approx 6.09 \ \text{L/min}.$$

Quick Review 5.1

As you answer the questions in Exercises 1–10, try to associate the answers with area, as in Figure 5.1.

1. A train travels at 80 mph for 5 hours. How far does it travel?

2. A truck travels at an average speed of 48 mph for 3 hours. How far does it travel?

3. Beginning at a standstill, a car maintains a constant acceleration of 10 ft/sec² for 10 seconds. What is its velocity after 10 seconds? Give your answer in ft/sec and then convert it to mi/h.

4. In a vacuum, light travels at a speed of 300,000 kilometers per second. How many kilometers does it travel in a year? (This distance equals one *light-year.*)

5. A long distance runner ran a race in 5 hours, averaging 6 mph for the first 3 hours and 5 mph for the last 2 hours. How far did she run?

6. A pump working at 20 gallons/minute pumps for an hour. How many gallons are pumped?

7. At 8:00 P.M. the temperature began dropping at a rate of 1 degree Celsius per hour. Twelve hours later it began rising at a rate of 1.5 degrees per hour for six hours. What was the net change in temperature over the 18-hour period?

8. Water flows over a spillway at a steady rate of 300 cubic feet per second. How many cubic feet of water pass over the spillway in one day?

9. A city has a population density of 350 people per square mile in an area of 50 square miles. What is the population of the city?

10. A hummingbird in flight beats its wings at a rate of 70 times per second. How many times does it beat its wings in an hour if it is in flight 70% of the time?

Section 5.1 Exercises

Exercises 1–4 refer to the region R enclosed between the graph of the function $y = 2x - x^2$ and the x-axis for $0 \le x \le 2$.

1. (a) Sketch the region R.

   (b) Partition $[0, 2]$ into 4 subintervals and show the four rectangles that LRAM uses to approximate the area of R. Compute the LRAM sum without a calculator.

2. Repeat Exercise 1(b) for RRAM and MRAM.

3. Using a calculator program, find the RAM sums that complete the following table.

n	LRAM_n	MRAM_n	RRAM_n
10			
50			
100			
500			

4. Make a conjecture about the area of the region R.

In Exercises 5–8, use RAM to estimate the area of the region enclosed between the graph of f and the x-axis for $a \le x \le b$.

5. $f(x) = x^2 - x + 3,\quad a = 0,\quad b = 3$

6. $f(x) = \dfrac{1}{x},\quad a = 1,\quad b = 3$

7. $f(x) = e^{-x^2},\quad a = 0,\quad b = 2$

8. $f(x) = \sin x,\quad a = 0,\quad b = \pi$

9. *Cardiac Output* The following table gives dye concentrations for a dye-concentration cardiac-output determination like the one in Example 4. The amount of dye injected in this patient was 5 mg instead of 5.6 mg. Use

rectangles to estimate the area under the dye concentration curve and then go on to estimate the patient's cardiac output.

Seconds after Injection t	Dye Concentration (adjusted for recirculation) c
2	0
4	0.6
6	1.4
8	2.7
10	3.7
12	4.1
14	3.8
16	2.9
18	1.7
20	1.0
22	0.5
24	0

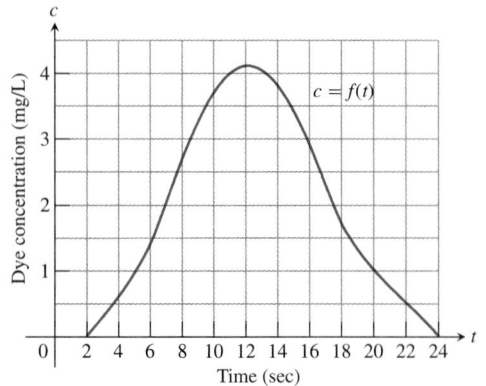

10. *Distance Traveled*   The table below shows the velocity of a model train engine moving along a track for 10 sec. Estimate the distance traveled by the engine, using 10 subintervals of length 1 with **(a)** left-endpoint values (LRAM) and **(b)** right-endpoint values (RRAM).

Time (sec)	Velocity (in./sec)	Time (sec)	Velocity (in./sec)
0	0	6	11
1	12	7	6
2	22	8	2
3	10	9	6
4	5	10	0
5	13		

11. *Distance Traveled Upstream*   You are walking along the bank of a tidal river watching the incoming tide carry a bottle upstream. You record the velocity of the flow every 5 minutes for an hour, with the results shown in the table below. About how far upstream does the bottle travel during that hour? Find the **(a)** LRAM and **(b)** RRAM estimates using 12 subintervals of length 5.

Time (min)	Velocity (m/sec)	Time (min)	Velocity (m/sec)
0	1	35	1.2
5	1.2	40	1.0
10	1.7	45	1.8
15	2.0	50	1.5
20	1.8	55	1.2
25	1.6	60	0
30	1.4		

12. *Length of a Road*   You and a companion are driving along a twisty stretch of dirt road in a car whose speedometer works but whose odometer (mileage counter) is broken. To find out how long this particular stretch of road is, you record the car's velocity at 10-sec intervals, with the results shown in the table below. (The velocity was converted from mi/h to ft/sec using 30 mi/h = 44 ft/sec.) Estimate the length of the road by averaging the LRAM and RRAM sums.

Time (sec)	Velocity (ft/sec)	Time (sec)	Velocity (ft/sec)
0	0	70	15
10	44	80	22
20	15	90	35
30	35	100	44
40	30	110	30
50	44	120	35
60	35		

13. *Distance from Velocity Data*   The table below gives data for the velocity of a vintage sports car accelerating from 0 to 142 mi/h in 36 sec (10 thousandths of an hour.)

Time (h)	Velocity (mi/h)	Time (h)	Velocity (mi/h)
0.0	0	0.006	116
0.001	40	0.007	125
0.002	62	0.008	132
0.003	82	0.009	137
0.004	96	0.010	142
0.005	108		

(a) Use rectangles to estimate how far the car traveled during the 36 sec it took to reach 142 mi/h.

(b) Roughly how many seconds did it take the car to reach the halfway point? About how fast was the car going then?

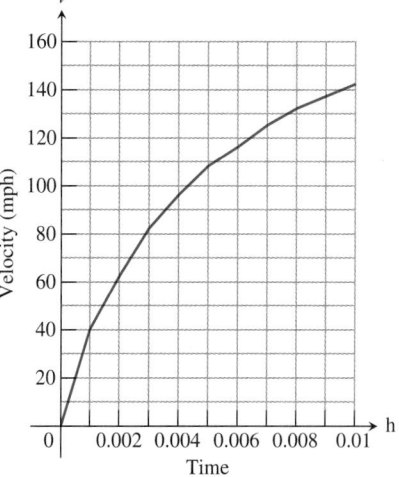

14. *(Continuation of Example 3)* Use the slicing technique of Example 3 to find the MRAM sums that approximate the volume of a sphere of radius 5. Use $n = 10, 20, 40, 80,$ and 160.

15. *(Continuation of Exercise 14)* Use a geometry formula to find the volume V of the sphere in Exercise 14 and find **(a)** the error and **(b)** the percentage error in the MRAM approximation for each value of n given.

16. *Volume of a Solid Hemisphere* To estimate the volume of a solid hemisphere of radius 4, imagine its axis of symmetry to be the interval $[0, 4]$ on the x-axis. Partition $[0, 4]$ into eight subintervals of equal length and approximate the solid with cylinders based on the circular cross sections of the hemisphere perpendicular to the x-axis at the subintervals' left endpoints. (See the accompanying profile view.)

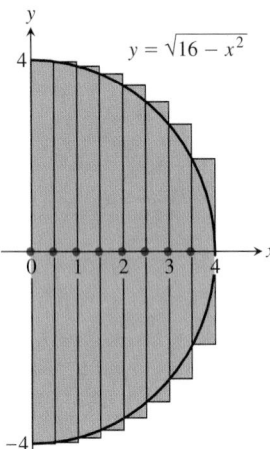

$y = \sqrt{16 - x^2}$

(a) **Writing to Learn** Find the sum S_8 of the volumes of the cylinders. Do you expect S_8 to overestimate V? Give reasons for your answer.

(b) Express $|V - S_8|$ as a percentage of V to the nearest percent.

17. Repeat Exercise 16 using cylinders based on cross sections at the right endpoints of the subintervals.

18. *Volume of Water in a Reservoir* A reservoir shaped like a hemispherical bowl of radius 8 m is filled with water to a depth of 4 m.

(a) Find an estimate S of the water's volume by approximating the water with eight circumscribed solid cylinders.

(b) It can be shown that the water's volume is $V = (320\pi)/3$ m³. Find the error $|V - S|$ as a percentage of V to the nearest percent.

19. *Volume of Water in a Swimming Pool* A rectangular swimming pool is 30 ft wide and 50 ft long. The table below shows the depth $h(x)$ of the water at 5-ft intervals from one end of the pool to the other. Estimate the volume of water in the pool using (a) left-endpoint values and (b) right-endpoint values.

Position (ft) x	Depth (ft) $h(x)$	Position (ft) x	Depth (ft) $h(x)$
0	6.0	30	11.5
5	8.2	35	11.9
10	9.1	40	12.3
15	9.9	45	12.7
20	10.5	50	13.0
25	11.0		

20. *Volume of a Nose Cone* The nose "cone" of a rocket is a *paraboloid* obtained by revolving the curve $y = \sqrt{x}$, $0 \le x \le 5$ about the x-axis, where x is measured in feet. Estimate the volume V of the nose cone by partitioning $[0, 5]$ into five subintervals of equal length, slicing the cone with planes perpendicular to the x-axis at the subintervals' left endpoints, constructing cylinders of height 1 based on cross sections at these points, and finding the volumes of these cylinders. (See the accompanying figure.)

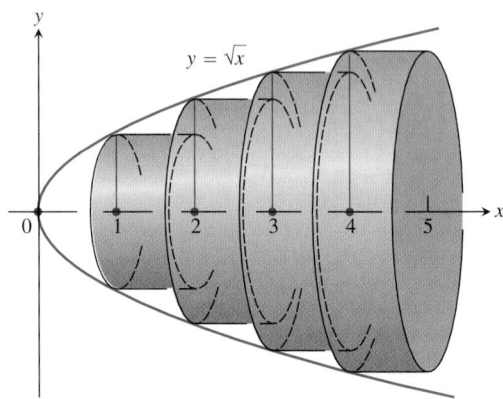

$y = \sqrt{x}$

21. *Volume of a Nose Cone* Repeat Exercise 20 using cylinders based on cross sections at the *midpoints* of the subintervals.

22. *Free Fall with Air Resistance* An object is dropped straight down from a helicopter. The object falls faster and faster but its acceleration (rate of change of its velocity) decreases over time because of air resistance. The acceleration is measured in ft/sec² and recorded every second after the drop for 5 sec, as shown in the table below.

t	0	1	2	3	4	5
a	32.00	19.41	11.77	7.14	4.33	2.63

(a) Use LRAM$_5$ to find an upper estimate for the speed when $t = 5$.

(b) Use RRAM$_5$ to find a lower estimate for the speed when $t = 5$.

(c) Use upper estimates for the speed during the first second, second second, and third second to find an upper estimate for the distance fallen when $t = 3$.

23. *Distance Traveled by a Projectile* An object is shot straight upward from sea level with an initial velocity of 400 ft/sec.

(a) Assuming gravity is the only force acting on the object, give an upper estimate for its velocity after 5 sec have elapsed. Use $g = 32$ ft/sec² for the gravitational constant.

(b) Find a lower estimate for the height attained after 5 sec.

24. *Water Pollution* Oil is leaking out of a tanker damaged at sea. The damage to the tanker is worsening as evidenced by the increased leakage each hour, recorded in the table below.

Time (h)	0	1	2	3	4
Leakage (gal/h)	50	70	97	136	190

Time (h)	5	6	7	8
Leakage (gal/h)	265	369	516	720

(a) Give an upper and lower estimate of the total quantity of oil that has escaped after 5 hours.

(b) Repeat (a) for the quantity of oil that has escaped after 8 hours.

(c) The tanker continues to leak 720 gal/h after the first 8 hours. If the tanker originally contained 25,000 gal of oil, approximately how many more hours will elapse in the worst case before all of the oil has leaked? in the best case?

25. *Air Pollution* A power plant generates electricity by burning oil. Pollutants produced by the burning process are removed by scrubbers in the smokestacks. Over time the scrubbers become less efficient and eventually must be replaced when the amount of pollutants released exceeds government standards. Measurements taken at the end of each month determine the rate at which pollutants are released into the atmosphere as recorded in the table below.

Month	Jan	Feb	Mar	Apr	May	Jun
Pollutant Release Rate (tons/day)	0.20	0.25	0.27	0.34	0.45	0.52

Month	Jul	Aug	Sep	Oct	Nov	Dec
Pollutant Release Rate (tons/day)	0.63	0.70	0.81	0.85	0.89	0.95

(a) Assuming a 30-day month and that new scrubbers allow only 0.05 ton/day released, give an upper estimate of the total tonnage of pollutants released by the end of June. What is a lower estimate?

(b) In the best case, approximately when will a total of 125 tons of pollutants have been released into the atmosphere?

26. Writing to Learn The graph shows the sales record for a company over a 10-year period. If sales are measured in millions of units per year, explain what information can be obtained from the area of the region, and why.

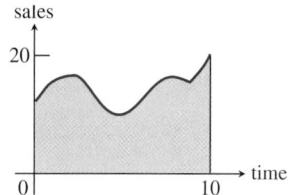

Exploration

27. *Area of a Circle* *Work in groups of two or three.* Inscribe a regular n-sided polygon inside a circle of radius 1 and compute the area of the polygon for the following values of n.

(a) 4 (square)   **(b)** 8 (octagon)   **(c)** 16

(d) Compare the areas in parts (a), (b), and (c) with the area of the circle. ∎

Extending the Ideas

28. *Rectangular Approximation Methods* Prove or disprove the following statement: $MRAM_n$ is always the average of $LRAM_n$ and $RRAM_n$.

29. *Rectangular Approximation Methods* Show that if f is a nonnegative function on the interval $[a, b]$ and the line $x = (a + b)/2$ is a line of symmetry of the graph of $y = f(x)$, then $LRAM_n f = RRAM_n f$ for every positive integer n.

30. *(Continuation of Exercise 27)*

(a) Inscribe a regular n-sided polygon inside a circle of radius 1 and compute the area of one of the n congruent triangles formed by drawing radii to the vertices of the polygon.

(b) Compute the limit of the area of the inscribed polygon as $n \to \infty$.

(c) Repeat the computations in (a) and (b) for a circle of radius r.

5.2 Definite Integrals

Riemann Sums • Terminology and Notation of Integration • Definite Integral and Area • Constant Functions • Integrals on a Calculator • Discontinuous Integrable Functions

Riemann Sums

In the preceding section, we estimated distances, areas, and volumes with finite sums. The terms in the sums were obtained by multiplying selected function values by the lengths of intervals. In this section we move beyond finite sums to see what happens in the limit, as the terms become infinitely small and their number infinitely large.

Sigma notation enables us to express a large sum in compact form:

$$\sum_{k=1}^{n} a_k = a_1 + a_2 + a_3 + \cdots + a_{n-1} + a_n.$$

The Greek capital letter $\sum$ (sigma) stands for "sum." The index k tells us where to begin the sum (at the number below the $\sum$) and where to end (at the number above). If the symbol ∞ appears above the $\sum$, it indicates that the terms go on indefinitely.

The sums in which we will be interested are called *Riemann* ("*ree*-mahn") *sums,* after Georg Friedrich Bernhard Riemann (1826–1866). LRAM, MRAM, and RRAM in the previous section are all examples of Riemann sums—not because they estimated area, but because they were constructed in a particular way. We now describe that construction formally, in a more general context that does not confine us to nonnegative functions.

We begin with an arbitrary continuous function $f(x)$ defined on a closed interval $[a, b]$. Like the function graphed in Figure 5.12, it may have negative values as well as positive values.

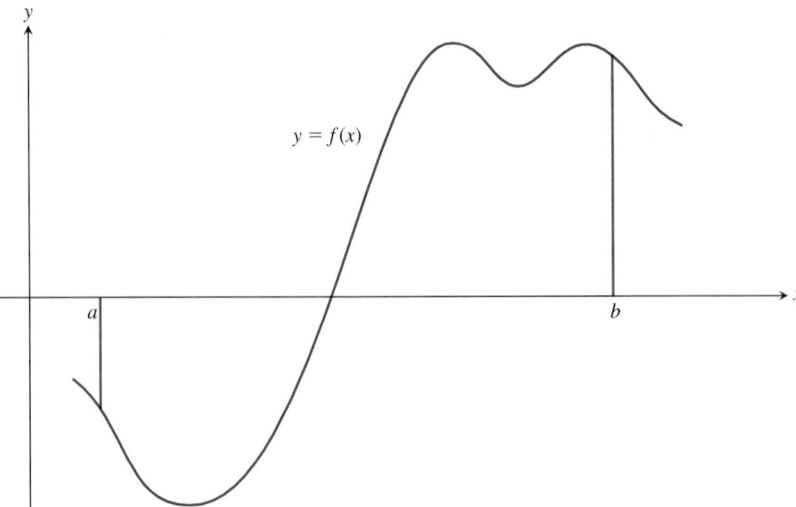

Figure 5.12 The graph of a typical function $y = f(x)$ over a closed interval $[a, b]$.

We then partition the interval $[a, b]$ into n subintervals by choosing $n - 1$ points, say $x_1, x_2, \cdots, x_{n-1}$, between a and b subject only to the condition that

$$a < x_1 < x_2 < \cdots < x_{n-1} < b.$$

To make the notation consistent, we denote a by x_0 and b by x_n. The set

$$P = \{x_0, x_1, x_2, \cdots, x_n\}$$

is called a **partition** of $[a, b]$.

The partition P determines n closed **subintervals,** as shown in Figure 5.13. The kth subinterval is $[x_{k-1}, x_k]$, which has length $\Delta x_k = x_k - x_{k-1}$.

Figure 5.13 The partition $P = \{a = x_0, x_1, x_2, \cdots, x_n = b\}$ divides $[a, b]$ into n subintervals of lengths $\Delta x_1, \Delta x_2, \cdots, \Delta x_n$. The kth subinterval has length Δx_k.

In each subinterval we select some number. Denote the number chosen from the kth subinterval by c_k.

Then, on each subinterval we stand a vertical rectangle that reaches from the x-axis to touch the curve at $(c_k, f(c_k))$. These rectangles could lie either above or below the x-axis (Figure 5.14).

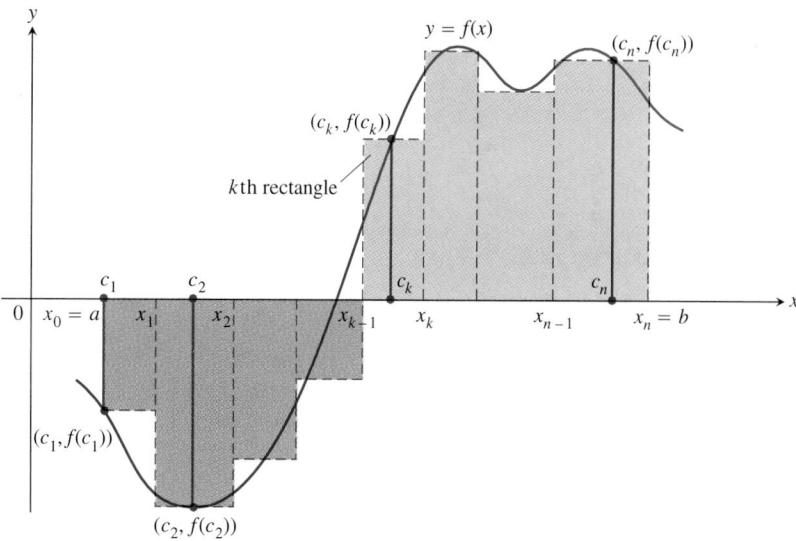

Figure 5.14 Rectangles extending from the x-axis to intersect the curve at the points $(c_k, f(c_k))$. The rectangles approximate the region between the x-axis and the graph of the function.

On each subinterval, we form the product $f(c_k) \cdot \Delta x_k$. This product can be positive, negative, or zero, depending on $f(c_k)$.

Finally, we take the sum of these products:

$$S_n = \sum_{k=1}^{n} f(c_k) \cdot \Delta x_k.$$

(a)

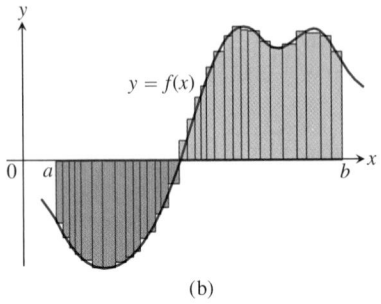

(b)

Figure 5.15 The curve of Figure 5.12 with rectangles from finer partitions of $[a, b]$. Finer partitions create more rectangles, with shorter bases.

Georg Riemann (1826–1866)

The mathematicians of the 17th and 18th centuries blithely assumed the existence of limits of Riemann sums (as we admittedly did in our RAM explorations of the last section), but the existence was not established mathematically until Georg Riemann proved Theorem 1 in 1854. You can find a current version of Riemann's proof in most advanced calculus books.

This sum, which depends on the partition P and the choice of the numbers c_k, is a **Riemann sum for f on the interval $[a, b]$**.

As the partitions of $[a, b]$ become finer and finer, we would expect the rectangles defined by the partitions to approximate the region between the x-axis and the graph of f with increasing accuracy (Figure 5.15).

Just as LRAM, MRAM, and RRAM in our earlier examples converged to a common value in the limit, *all* Riemann sums for a given function on $[a, b]$ converge to a common value, as long as the lengths of the subintervals all tend to zero. This latter condition is assured by requiring the longest subinterval length (called the **norm** of the partition and denoted by $\|P\|$) to tend to zero.

Definition **The Definite Integral as a Limit of Riemann Sums**

Let f be a function defined on a closed interval $[a, b]$. For any partition P of $[a, b]$, let the numbers c_k be chosen arbitrarily in the subintervals $[x_{k-1}, x_k]$.

If there exists a number I such that

$$\lim_{\|P\| \to 0} \sum_{k=1}^{n} f(c_k)\Delta x_k = I$$

no matter how P and the c_k's are chosen, then f is **integrable** on $[a, b]$ and I is the **definite integral** of f over $[a, b]$.

Despite the potential for variety in the sums $\sum f(c_k)\Delta x_k$ as the partitions change and as the c_k's are chosen arbitrarily in the intervals of each partition, the sums always have the same limit as $\|P\| \to 0$ as long as f is *continuous* on $[a, b]$.

Theorem 1 The Existence of Definite Integrals

All continuous functions are integrable. That is, if a function f is continuous on an interval $[a, b]$, then its definite integral over $[a, b]$ exists.

Because of Theorem 1, we can get by with a simpler construction for definite integrals of continuous functions. Since we know for these functions that the Riemann sums tend to the same limit for *all* partitions in which $\|P\| \to 0$, we need only to consider the limit of the so-called **regular partitions,** in which all the subintervals have the same length.

The Definite Integral of a Continuous Function on $[a, b]$

Let f be continuous on $[a, b]$, and let $[a, b]$ be partitioned into n subintervals of equal length $\Delta x = (b - a)/n$. Then the definite integral of f over $[a, b]$ is given by

$$\lim_{n \to \infty} \sum_{k=1}^{n} f(c_k)\Delta x,$$

where each c_k is chosen arbitrarily in the kth subinterval.

Terminology and Notation of Integration

Leibniz's clever choice of notation for the derivative, *dy/dx,* had the advantage of retaining an identity as a "fraction" even though both numerator and denominator had tended to zero. Although not really fractions, derivatives can *behave* like fractions, so the notation makes profound results like the Chain Rule

$$\frac{dy}{dx} = \frac{dy}{du} \cdot \frac{du}{dx}$$

seem almost simple.

The notation that Leibniz introduced for the definite integral was equally inspired. In his derivative notation, the Greek letters ("Δ" for "difference") switch to Roman letters ("d" for "differential") in the limit,

$$\lim_{\Delta x \to 0} \frac{\Delta y}{\Delta x} = \frac{dy}{dx}.$$

In his definite integral notation, the Greek letters again become Roman letters in the limit,

$$\lim_{n \to \infty} \sum_{k=1}^{n} f(c_k)\Delta x = \int_a^b f(x)\, dx.$$

Notice that the difference Δx has again tended to zero, becoming a differential dx. The Greek "Σ" has become an elongated Roman "S," so that the integral can retain its identity as a "sum." The c_k's have become so crowded together in the limit that we no longer think of a choppy selection of x values between a and b, but rather of a continuous, unbroken sampling of x values from a to b. It is as if we were summing *all* products of the form $f(x)\, dx$ as x goes from a to b, so we can abandon the k and the n used in the finite sum expression.

The symbol

$$\int_a^b f(x)\, dx$$

is read as "the integral from a to b of f of x dee x," or sometimes as "the integral from a to b of f of x with respect to x." The component parts also have names:

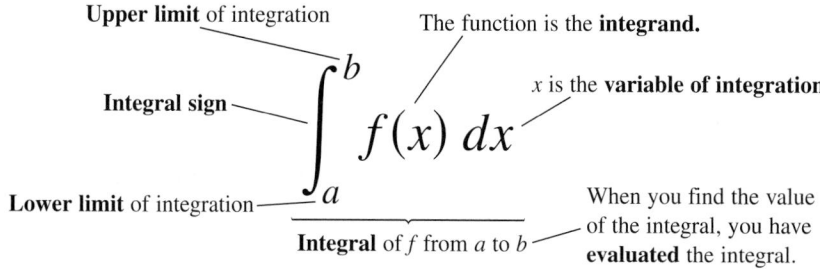

Upper limit of integration

The function is the **integrand.**

Integral sign

x is the **variable of integration.**

Lower limit of integration

$$\int_a^b f(x)\, dx$$

Integral of f from a to b

When you find the value of the integral, you have **evaluated** the integral.

The value of the definite integral of a function over any particular interval depends on the function and not on the letter we choose to represent its independent variable. If we decide to use t or u instead of x, we simply write the integral as

$$\int_a^b f(t)\, dt \quad \text{or} \quad \int_a^b f(u)\, du \quad \text{instead of} \quad \int_a^b f(x)\, dx.$$

No matter how we represent the integral, it is the same *number,* defined as a limit of Riemann sums. Since it does not matter what letter we use to run from a to b, the variable of integration is called a **dummy variable.**

Example 1   USING THE NOTATION

The interval $[-1, 3]$ is partitioned into n subintervals of equal length $\Delta x = 4/n$. Let m_k denote the midpoint of the kth subinterval. Express the limit

$$\lim_{n\to\infty} \sum_{k=1}^{n} (3(m_k)^2 - 2m_k + 5)\, \Delta x$$

as an integral.

Solution   Since the midpoints m_k have been chosen from the subintervals of the partition, this expression is indeed a limit of Riemann sums. (The points chosen did not have to be midpoints; they could have been chosen from the subintervals in any arbitrary fashion.) The function being integrated is $f(x) = 3x^2 - 2x + 5$ over the interval $[-1, 3]$. Therefore,

$$\lim_{n\to\infty} \sum_{k=1}^{n} (3(m_k)^2 - 2m_k + 5)\, \Delta x = \int_{-1}^{3} (3x^2 - 2x + 5)\, dx.$$

Definite Integral and Area

If an integrable function $y = f(x)$ is nonnegative throughout an interval $[a, b]$, each nonzero term $f(c_k)\Delta x_k$ is the area of a rectangle reaching from the x-axis up to the curve $y = f(x)$. (See Figure 5.16.)

The Riemann sum

$$\sum f(c_k)\, \Delta x_k,$$

which is the sum of the areas of these rectangles, gives an estimate of the area of the region between the curve and the x-axis from a to b. Since the rectangles give an increasingly good approximation of the region as we use partitions with smaller and smaller norms, we call the limiting value the area under the curve.

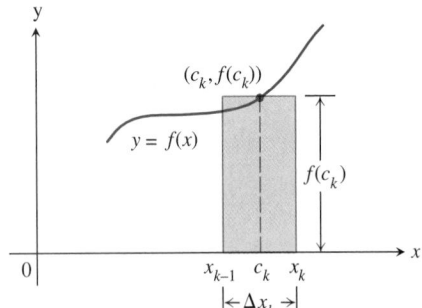

Figure 5.16 A term of a Riemann sum $\sum f(c_k)\Delta x_k$ for a nonnegative function f is either zero or the area of a rectangle such as the one shown.

Definition   Area Under a Curve (as a Definite Integral)

If $y = f(x)$ is nonnegative and integrable over a closed interval $[a, b]$, then the **area under the curve** $y = f(x)$ **from a to b** is the integral of f from a to b,

$$A = \int_{a}^{b} f(x)\, dx.$$

This definition works both ways: We can use integrals to calculate areas *and* we can use areas to calculate integrals.

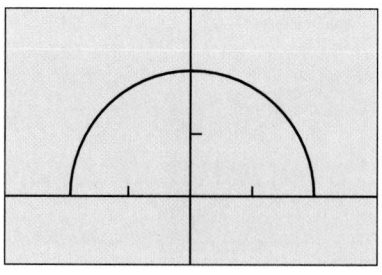

[-3, 3] by [-1, 3]

Figure 5.17 A square viewing window on $y = \sqrt{4 - x^2}$. The graph is a semicircle because $y = \sqrt{4 - x^2}$ is the same as $y^2 = 4 - x^2$, or $x^2 + y^2 = 4$, with $y \geq 0$. (Example 2)

Example 2   REVISITING AREA UNDER A CURVE

Evaluate the integral $\int_{-2}^{2} \sqrt{4 - x^2}\, dx$.

Solution We recognize $f(x) = \sqrt{4 - x^2}$ as a function whose graph is a semicircle of radius 2 centered at the origin (Figure 5.17).

The area between the semicircle and the x-axis from -2 to 2 can be computed using the geometry formula

$$\text{Area} = \frac{1}{2} \cdot \pi r^2 = \frac{1}{2} \cdot \pi (2)^2 = 2\pi.$$

Because the area is also the value of the integral of f from -2 to 2,

$$\int_{-2}^{2} \sqrt{4 - x^2}\, dx = 2\pi.$$

If an integrable function $y = f(x)$ is nonpositive, the nonzero terms $f(c_k)\Delta x_k$ in the Riemann sums for f over an interval $[a, b]$ are negatives of rectangle areas. The limit of the sums, the integral of f from a to b, is therefore the *negative* of the area of the region between the graph of f and the x-axis (Figure 5.18).

$$\int_{a}^{b} f(x)\, dx = -(\text{the area}) \quad \text{if} \quad f(x) \leq 0.$$

Or, turning this around,

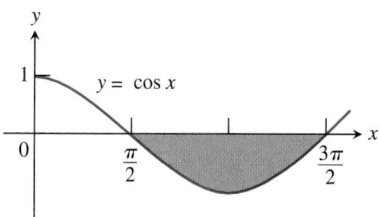

Figure 5.18 Because $f(x) = \cos x$ is nonpositive on $[\pi/2, 3\pi/2]$, the integral of f is a negative number. The area of the shaded region is the opposite of this integral,

$$\text{Area} = -\int_{\pi/2}^{3\pi/2} \cos x\, dx.$$

$$\text{Area} = -\int_{a}^{b} f(x)\, dx \quad \text{when} \quad f(x) \leq 0.$$

If an integrable function $y = f(x)$ has both positive and negative values on an interval $[a, b]$, then the Riemann sums for f on $[a, b]$ add areas of rectangles that lie above the x-axis to the negatives of areas of rectangles that lie below the x-axis, as in Figure 5.19. The resulting cancellations mean that the limiting value is a number whose magnitude is less than the total area between the curve and the x-axis. The value of the integral is the area above the x-axis minus the area below.

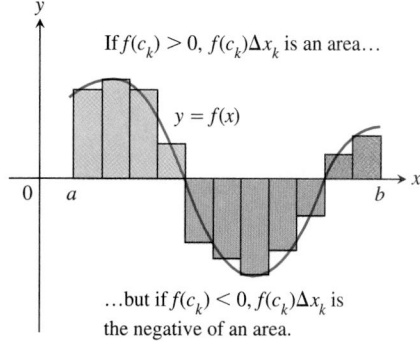

If $f(c_k) > 0$, $f(c_k)\Delta x_k$ is an area...

$y = f(x)$

...but if $f(c_k) < 0$, $f(c_k)\Delta x_k$ is the negative of an area.

Figure 5.19 An integrable function f with negative as well as positive values.

Net Area

Sometimes $\int_a^b f(x)\,dx$ is called the *net area* of the region determined by the curve $y = f(x)$ and the x-axis between $x = a$ and $x = b$.

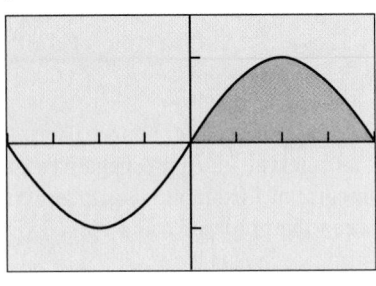

$y = \sin x$

$[-\pi, \pi]$ by $[-1.5, 1.5]$

Figure 5.20

$\int_0^\pi \sin x\,dx = 2$. (Exploration 1)

For any integrable function,

$$\int_a^b f(x)\,dx = (\text{area above the } x\text{-axis}) - (\text{area below the } x\text{-axis}).$$

Exploration 1 Finding Integrals by Signed Areas

It is a fact (which we will revisit) that $\int_0^\pi \sin x\,dx = 2$ (Figure 5.20). With that information, what you know about integrals and areas, what you know about graphing curves, and sometimes a bit of intuition, determine the values of the following integrals. Give as convincing an argument as you can for each value, based on the graph of the function.

1. $\int_\pi^{2\pi} \sin x\,dx$ **2.** $\int_0^{2\pi} \sin x\,dx$ **3.** $\int_0^{\pi/2} \sin x\,dx$

4. $\int_0^\pi (2 + \sin x)\,dx$ **5.** $\int_0^\pi 2\sin x\,dx$ **6.** $\int_2^{\pi+2} \sin(x - 2)\,dx$

7. $\int_{-\pi}^\pi \sin u\,du$ **8.** $\int_0^{2\pi} \sin(x/2)\,dx$ **9.** $\int_0^\pi \cos x\,dx$

10. Suppose k is *any* positive number. Make a conjecture about $\int_{-k}^k \sin x\,dx$ and support your conjecture.

Constant Functions

Integrals of constant functions are easy to evaluate. Over a closed interval, they are simply the constant times the length of the interval (Figure 5.21).

Theorem 2 The Integral of a Constant

If $f(x) = c$, where c is a constant, on the interval $[a, b]$, then

$$\int_a^b f(x)\,dx = \int_a^b c\,dx = c(b - a).$$

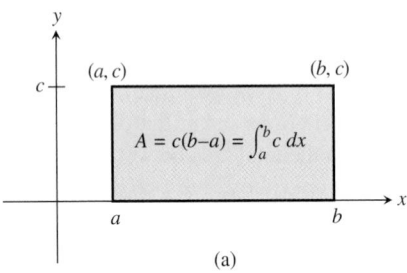

(a, c) (b, c)

$A = c(b-a) = \int_a^b c\,dx$

(a)

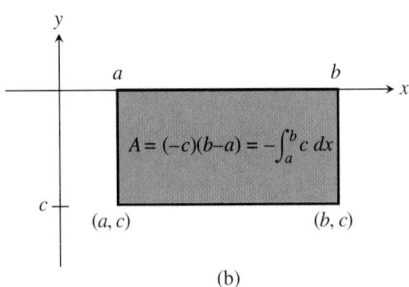

$A = (-c)(b-a) = -\int_a^b c\,dx$

(a, c) (b, c)

(b)

Figure 5.21 (a) If c is a positive constant, then $\int_a^b c\,dx$ is the area of the rectangle shown. (b) If c is negative, then $\int_a^b c\,dx$ is the opposite of the area of the rectangle.

Proof A constant function is continuous, so the integral exists, and we can evaluate it as a limit of Riemann sums with subintervals of equal length $(b - a)/n$. Any such sum looks like

$$\sum_{k=1}^n f(c_k) \cdot \Delta x, \quad \text{which is} \quad \sum_{k=1}^n c \cdot \frac{b - a}{n}.$$

Then

$$\sum_{k=1}^{n} c \cdot \frac{b-a}{n} = c \cdot (b-a) \sum_{k=1}^{n} \frac{1}{n}$$

$$= c(b-a) \cdot n\left(\frac{1}{n}\right)$$

$$= c(b-a).$$

Since the sum is *always* $c(b-a)$ for any value of n, it follows that the limit of the sums, the integral to which they converge, is also $c(b-a)$. ∎

Example 3   REVISITING THE TRAIN PROBLEM

A train moves along a track at a steady 75 miles per hour from 7:00 A.M. to 9:00 A.M. Express its total distance traveled as an integral. Evaluate the integral using Theorem 2.

Solution (See Figure 5.22.)

$$\text{Distance traveled} = \int_{7}^{9} 75 \, dt = 75 \cdot (9-7) = 150$$

Since the 75 is measured in miles/hour and the $(9-7)$ is measured in hours, the 150 is measured in miles. The train traveled 150 miles.

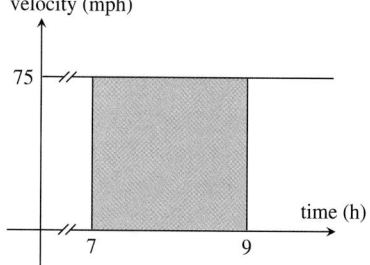

Figure 5.22 The area of the rectangle is a special case of Theorem 2. (Example 3)

Integrals on a Calculator

You do not have to know much about your calculator to realize that finding the limit of a Riemann sum is exactly the kind of thing that it does best. We have seen how effectively it can approximate areas using MRAM, but most modern calculators have sophisticated built-in programs that converge to integrals with much greater speed and precision than that. We will assume that your calculator has such a numerical integration capability, which we will denote as **NINT.** In particular, we will use NINT $(f(x), x, a, b)$ to denote a calculator (or computer) approximation of $\int_{a}^{b} f(x) \, dx$. When we write

$$\int_{a}^{b} f(x) \, dx = \text{NINT} \, (f(x), x, a, b),$$

we do so with the understanding that the right-hand side of the equation is an approximation of the left-hand side.

Example 4   USING NINT

Evaluate the following integrals numerically.

(a) $\int_{-1}^{2} x \sin x \, dx$ **(b)** $\int_{0}^{1} \frac{4}{1+x^2} \, dx$ **(c)** $\int_{0}^{5} e^{-x^2} \, dx$

Solution

(a) NINT $(x \sin x, x, -1, 2) \approx 2.04$

(b) NINT $(4/(1+x^2), x, 0, 1) \approx 3.14$

(c) NINT $(e^{-x^2}, x, 0, 5) \approx 0.89$

Bounded Functions

We say a function is *bounded* on a given domain if its range is confined between some minimum value m and some maximum value M. That is, given any x in the domain, $m \le f(x) \le M$. Equivalently, the graph of $y = f(x)$ lies between the horizontal lines $y = m$ and $y = M$.

$y = |x|/x$

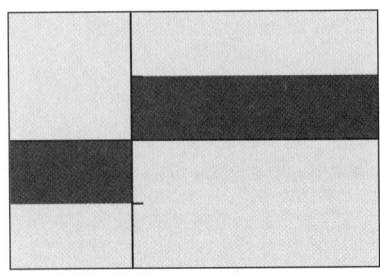

$[-1, 2]$ by $[-2, 2]$

Figure 5.23 A discontinuous integrable function:

$$\int_{-1}^{2} \frac{|x|}{x}\, dx =$$

$-$(area below x-axis) $+$ (area above x-axis). (Example 5)

A Nonintegrable Function

How "bad" does a function have to be before it is *not* integrable? One way to defeat integrability is to be unbounded (like $y = 1/x$ near $x = 0$), which can prevent the Riemann sums from tending to a finite limit. Another, more subtle, way is to be bounded but badly discontinuous, like the *characteristic function of the rationals*:

$$f(x) = \begin{cases} 1 & \text{if } x \text{ is rational} \\ 0 & \text{if } x \text{ is irrational.} \end{cases}$$

No matter what partition we take of the closed interval $[0, 1]$, every subinterval contains both rational and irrational numbers. That means that we can always form a Riemann sum with all rational c_k's (a Riemann sum of 1) or all irrational c_k's (a Riemann sum of 0). The sums can therefore never tend toward a unique limit.

We will eventually be able to confirm that the exact value for Example 4a is $-2\cos 2 + \sin 2 - \cos 1 + \sin 1$. You might want to conjecture for yourself what the exact answer to Example 4b might be. As for Example 4c, no explicit *exact* value has ever been found for this integral! The best we can do in this case (and in many like it) is to approximate the integral numerically. Here, technology is not only useful, it is essential.

Discontinuous Integrable Functions

Theorem 1 guarantees that all continuous functions are integrable. But some functions with discontinuities are also integrable. For example, a bounded function (see margin note) that has a finite number of points of discontinuity on an interval $[a, b]$ will still be integrable on the interval if it is continuous everywhere else.

Example 5   INTEGRATING A DISCONTINUOUS FUNCTION

Find $\displaystyle\int_{-1}^{2} \frac{|x|}{x}\, dx$.

Solution This function has a discontinuity at $x = 0$, where the graph jumps from $y = -1$ to $y = 1$. The graph, however, determines two rectangles, one below the x-axis and one above (Figure 5.23).

Using the idea of net area, we have

$$\int_{-1}^{2} \frac{|x|}{x}\, dx = -1 + 2 = 1.$$

Exploration 2 **More Discontinuous Integrands**

1. Explain why the function

$$f(x) = \frac{x^2 - 4}{x - 2}$$

   is not continuous on $[0, 3]$. What kind of discontinuity occurs?

2. Use areas to show that

$$\int_{0}^{3} \frac{x^2 - 4}{x - 2}\, dx = 10.5.$$

3. Use areas to show that

$$\int_{0}^{5} \text{int}\,(x)\, dx = 10.$$

Quick Review 5.2

In Exercises 1–3, evaluate the sum.

1. $\displaystyle\sum_{n=1}^{5} n^2$

2. $\displaystyle\sum_{k=0}^{4} (3k - 2)$

3. $\displaystyle\sum_{j=0}^{4} 100\,(j + 1)^2$

In Exercises 4–6, write the sum in sigma notation.

4. $1 + 2 + 3 + \cdots + 98 + 99$

5. $0 + 2 + 4 + \cdots + 48 + 50$

6. $3(1)^2 + 3(2)^2 + \cdots + 3(500)^2$

In Exercises 7 and 8, write the expression as a single sum in sigma notation.

7. $\displaystyle 2\sum_{x=1}^{50} x^2 + 3\sum_{x=1}^{50} x$

8. $\displaystyle\sum_{k=0}^{8} x^k + \sum_{k=9}^{20} x^k$

9. Find $\displaystyle\sum_{k=0}^{n} (-1)^k$ if n is odd.

10. Find $\displaystyle\sum_{k=0}^{n} (-1)^k$ if n is even.

Section 5.2 Exercises

In Exercises 1–6, express the limit as a definite integral.

1. $\displaystyle\lim_{\|P\|\to 0} \sum_{k=1}^{n} c_k^2 \Delta x_k$,   where P is any partition of $[0, 2]$

2. $\displaystyle\lim_{\|P\|\to 0} \sum_{k=1}^{n} (c_k^2 - 3c_k)\, \Delta x_k$, where P is any partition of $[-7, 5]$

3. $\displaystyle\lim_{\|P\|\to 0} \sum_{k=1}^{n} \frac{1}{c_k} \Delta x_k$,   where P is any partition of $[1, 4]$

4. $\displaystyle\lim_{\|P\|\to 0} \sum_{k=1}^{n} \frac{1}{1 - c_k} \Delta x_k$,   where P is any partition of $[2, 3]$

5. $\displaystyle\lim_{\|P\|\to 0} \sum_{k=1}^{n} \sqrt{4 - c_k^2}\, \Delta x_k$,   where P is any partition of $[0, 1]$

6. $\displaystyle\lim_{\|P\|\to 0} \sum_{k=1}^{n} (\sin^3 c_k)\, \Delta x_k$,   where P is any partition of $[-\pi, \pi]$

In Exercises 7–12, evaluate the integral.

7. $\displaystyle\int_{-2}^{1} 5\,dx$

8. $\displaystyle\int_{3}^{7} (-20)\,dx$

9. $\displaystyle\int_{0}^{3} (-160)\,dt$

10. $\displaystyle\int_{-4}^{-1} \frac{\pi}{2}\,d\theta$

11. $\displaystyle\int_{-2.1}^{3.4} 0.5\,ds$

12. $\displaystyle\int_{\sqrt{2}}^{\sqrt{18}} \sqrt{2}\,dr$

In Exercises 13–22, use the graph of the integrand and areas to evaluate the integral.

13. $\displaystyle\int_{-2}^{4} \left(\frac{x}{2} + 3\right) dx$

14. $\displaystyle\int_{1/2}^{3/2} (-2x + 4)\,dx$

15. $\displaystyle\int_{-3}^{3} \sqrt{9 - x^2}\,dx$

16. $\displaystyle\int_{-4}^{0} \sqrt{16 - x^2}\,dx$

17. $\displaystyle\int_{-2}^{1} |x|\,dx$

18. $\displaystyle\int_{-1}^{1} (1 - |x|)\,dx$

19. $\displaystyle\int_{-1}^{1} (2 - |x|)\,dx$

20. $\displaystyle\int_{-1}^{1} (1 + \sqrt{1 - x^2})\,dx$

21. $\displaystyle\int_{\pi}^{2\pi} \theta\,d\theta$

22. $\displaystyle\int_{\sqrt{2}}^{5\sqrt{2}} r\,dr$

In Exercises 23–28, use areas to evaluate the integral.

23. $\displaystyle\int_{0}^{b} x\,dx, \quad b > 0$

24. $\displaystyle\int_{0}^{b} 4x\,dx, \quad b > 0$

25. $\displaystyle\int_{a}^{b} 2s\,ds, \quad 0 < a < b$

26. $\displaystyle\int_{a}^{b} 3t\,dt, \quad 0 < a < b$

27. $\displaystyle\int_{a}^{2a} x\,dx, \quad a > 0$

28. $\displaystyle\int_{a}^{\sqrt{3}a} x\,dx, \quad a > 0$

In Exercises 29–38, *work in groups of two or three.* Use graphs, your knowledge of area, and the fact that

$$\int_{0}^{1} x^3\,dx = \frac{1}{4}$$

to evaluate the integral.

29. $\displaystyle\int_{-1}^{1} x^3\,dx$

30. $\displaystyle\int_{0}^{1} (x^3 + 3)\,dx$

31. $\displaystyle\int_{2}^{3} (x - 2)^3\,dx$

32. $\displaystyle\int_{-1}^{1} |x|^3\,dx$

33. $\displaystyle\int_{0}^{1} (1 - x^3)\,dx$

34. $\displaystyle\int_{-1}^{2} (|x| - 1)^3\,dx$

35. $\displaystyle\int_{0}^{2} \left(\frac{x}{2}\right)^3 dx$

36. $\displaystyle\int_{-8}^{8} x^3\,dx$

37. $\displaystyle\int_0^1 (x^3 - 1)\, dx$

38. $\displaystyle\int_0^1 \sqrt[3]{x}\, dx$

In Exercises 39–42, use NINT to evaluate the expression.

39. $\displaystyle\int_0^5 \frac{x}{x^2 + 4}\, dx$

40. $3 + 2\displaystyle\int_0^{\pi/3} \tan x\, dx$

41. Find the area enclosed between the x-axis and the graph of $y = 4 - x^2$ from $x = -2$ to $x = 2$.

42. Find the area enclosed between the x-axis and the graph of $y = x^2 e^{-x}$ from $x = -1$ to $x = 3$.

In Exercises 43–46, **(a)** find the points of discontinuity of the integrand on the interval of integration, and **(b)** use area to evaluate the integral.

43. $\displaystyle\int_{-2}^3 \frac{x}{|x|}\, dx$

44. $\displaystyle\int_{-6}^5 2\,\text{int}\,(x - 3)\, dx$

45. $\displaystyle\int_{-3}^4 \frac{x^2 - 1}{x + 1}\, dx$

46. $\displaystyle\int_{-5}^6 \frac{9 - x^2}{x - 3}\, dx$

Extending the Ideas

47. Writing to Learn The function

$$f(x) = \begin{cases} \dfrac{1}{x^2}, & 0 < x \le 1 \\ 0, & x = 0 \end{cases}$$

is defined on $[0, 1]$ and has a single point of discontinuity at $x = 0$.

(a) What happens to the graph of f as x approaches 0 from the right?

(b) The function f is not integrable on $[0, 1]$. Give a convincing argument based on Riemann sums to explain why it is not.

48. It can be shown by mathematical induction (see Appendix 2) that

$$\sum_{k=1}^n k^2 = \frac{n(n + 1)(2n + 1)}{6}.$$

Use this fact to give a formal proof that

$$\int_0^1 x^2\, dx = \frac{1}{3}$$

by following the steps given below.

(a) Partition $[0, 1]$ into n subintervals of length $1/n$. Show that the RRAM Riemann sum for the integral is

$$\sum_{k=1}^n \left(\left(\frac{k}{n}\right)^2 \cdot \frac{1}{n} \right).$$

(b) Show that this sum can be written as

$$\frac{1}{n^3} \cdot \sum_{k=1}^n k^2.$$

(c) Show that the sum can therefore be written as

$$\frac{(n + 1)(2n + 1)}{6n^2}.$$

(d) Show that

$$\lim_{n \to \infty} \sum_{k=1}^n \left(\left(\frac{k}{n}\right)^2 \cdot \frac{1}{n} \right) = \frac{1}{3}.$$

(e) Explain why the equation in (d) proves that

$$\int_0^1 x^2\, dx = \frac{1}{3}.$$

5.3 # Definite Integrals and Antiderivatives

Properties of Definite Integrals • Average Value of a Function • Mean Value Theorem for Definite Integrals • Connecting Differential and Integral Calculus

Properties of Definite Integrals

In defining $\int_a^b f(x)$ as a limit of sums $\sum c_k \Delta x_k$, we moved from left to right across the interval $[a, b]$. What would happen if we integrated in the *opposite direction*? The integral would become $\int_b^a f(x)\, dx$—again a limit of sums of the form $\sum f(c_k)\Delta x_k$—but this time each of the Δx_k's would be negative as the x-values *decreased* from b to a. This would change the signs of all the terms in each Riemann sum, and ultimately the sign of the definite integral. This suggests the rule

$$\int_b^a f(x)\,dx = -\int_a^b f(x)\,dx.$$

Since the original definition did not apply to integrating backwards over an interval, we can treat this rule as a logical extension of the definition.

Although $[a, a]$ is technically not an interval, another logical extension of the definition is that $\int_a^a f(x)\,dx = 0$.

These are the first two rules in Table 5.3. The others are inherited from rules that hold for Riemann sums. However, the limit step required to *prove* that these rules hold in the limit (as the norms of the partitions tend to zero) places their mathematical verification beyond the scope of this course. You can see intuitively why each rule makes sense, however.

Table 5.3 Rules for Definite Integrals

1. *Order of Integration:* $\displaystyle\int_b^a f(x)\,dx = -\int_a^b f(x)\,dx$ A definition

2. *Zero:* $\displaystyle\int_a^a f(x)\,dx = 0$ Also a definition

3. *Constant Multiple:* $\displaystyle\int_a^b kf(x)\,dx = k\int_a^b f(x)\,dx$ Any number k

$\displaystyle\int_a^b -f(x)\,dx = -\int_a^b f(x)\,dx$ $k = -1$

4. *Sum and Difference:* $\displaystyle\int_a^b (f(x) \pm g(x))\,dx = \int_a^b f(x)\,dx \pm \int_a^b g(x)\,dx$

5. *Additivity:* $\displaystyle\int_a^b f(x)\,dx + \int_b^c f(x)\,dx = \int_a^c f(x)\,dx$

6. *Max-Min Inequality:* If max f and min f are the maximum and minimum values of f on $[a, b]$, then

$$\min f \cdot (b - a) \le \int_a^b f(x)\,dx \le \max f \cdot (b - a).$$

7. *Domination:* $f(x) \ge g(x)$ on $[a, b] \Rightarrow \displaystyle\int_a^b f(x)\,dx \ge \int_a^b g(x)\,dx$

$f(x) \ge 0$ on $[a, b] \Rightarrow \displaystyle\int_a^b f(x)\,dx \ge 0$ $g = 0$

Example 1 USING THE RULES FOR DEFINITE INTEGRALS

Suppose

$$\int_{-1}^{1} f(x)\, dx = 5, \qquad \int_{1}^{4} f(x)\, dx = -2, \qquad \text{and} \qquad \int_{-1}^{1} h(x)\, dx = 7.$$

Find each of the following integrals, if possible.

(a) $\displaystyle\int_{4}^{1} f(x)\, dx$ **(b)** $\displaystyle\int_{-1}^{4} f(x)\, dx$ **(c)** $\displaystyle\int_{-1}^{1} [2f(x) + 3h(x)]\, dx$

(d) $\displaystyle\int_{0}^{1} f(x)\, dx$ **(e)** $\displaystyle\int_{-2}^{2} h(x)\, dx$ **(f)** $\displaystyle\int_{-1}^{4} [f(x) + h(x)]\, dx$

Solution

(a) $\displaystyle\int_{4}^{1} f(x)\, dx = -\int_{1}^{4} f(x)\, dx = -(-2) = 2$

(b) $\displaystyle\int_{-1}^{4} f(x)\, dx = \int_{-1}^{1} f(x)\, dx + \int_{1}^{4} f(x)\, dx = 5 + (-2) = 3$

(c) $\displaystyle\int_{-1}^{1} [2f(x) + 3h(x)]\, dx = 2\int_{-1}^{1} f(x)\, dx + 3\int_{-1}^{1} h(x)\, dx = 2(5) + 3(7) = 31$

(d) Not enough information given. (We cannot assume, for example, that integrating over half the interval would give half the integral!)

(e) Not enough information given. (We have no information about the function h outside the interval $[-1, 1]$.)

(f) Not enough information given (same reason as in (e)).

Example 2 FINDING BOUNDS FOR AN INTEGRAL

Show that the value of $\int_{0}^{1} \sqrt{1 + \cos x}\, dx$ is less than $3/2$.

Solution The Max-Min Inequality for definite integrals (Rule 6) says that $\min f \cdot (b - a)$ is a *lower bound* for the value of $\int_{a}^{b} f(x)\, dx$ and that $\max f \cdot (b - a)$ is an *upper bound*. The maximum value of $\sqrt{1 + \cos x}$ on $[0, 1]$ is $\sqrt{2}$, so

$$\int_{0}^{1} \sqrt{1 + \cos x}\, dx \le \sqrt{2} \cdot (1 - 0) = \sqrt{2}.$$

Since $\int_{0}^{1} \sqrt{1 + \cos x}\, dx$ is bounded above by $\sqrt{2}$ (which is $1.414...$), it is less than $3/2$.

Average Value of a Function

The *average* of n numbers is the sum of the numbers divided by n. How would we define the average value of an arbitrary function f over a closed interval $[a, b]$? As there are infinitely many values to consider, adding them and then dividing by infinity is not an option.

Consider, then, what happens if we take a large *sample* of n numbers from regular subintervals of the interval $[a, b]$. One way would be to take some number c_k from each of the n subintervals of length

$$\Delta x = \frac{b - a}{n}.$$

The average of the n sampled values is

$$\frac{f(c_1) + f(c_2) + \cdots + f(c_n)}{n} = \frac{1}{n} \cdot \sum_{k=1}^{n} f(c_k)$$

$$= \frac{\Delta x}{b - a} \sum_{k=1}^{n} f(c_k) \qquad \frac{1}{n} = \frac{\Delta x}{b - a}$$

$$= \frac{1}{b - a} \cdot \sum_{k=1}^{n} f(c_k) \Delta x.$$

Does this last sum look familiar? It is $1/(b - a)$ times a Riemann sum for f on $[a, b]$. That means that when we consider this averaging process as $n \to \infty$, we find it *has a limit*, namely $1/(b - a)$ times the integral of f over $[a, b]$. We are led by this remarkable fact to the following definition.

Definition **Average (Mean) Value**

If f is integrable on $[a, b]$, its **average (mean) value** on $[a, b]$ is

$$av(f) = \frac{1}{b - a} \int_a^b f(x)\, dx.$$

Example 3 APPLYING THE DEFINITION

Find the average value of $f(x) = 4 - x^2$ on $[0, 3]$. Does f actually take on this value at some point in the given interval?

Solution

$$av(f) = \frac{1}{b - a} \int_a^b f(x)\, dx$$

$$= \frac{1}{3 - 0} \int_0^3 (4 - x^2)\, dx$$

$$= \frac{1}{3 - 0} \cdot 3 \qquad \text{Using NINT}$$

$$= 1$$

The average value of $f(x) = 4 - x^2$ over the interval $[0, 3]$ is 1. The function assumes this value when $4 - x^2 = 1$ or $x = \pm\sqrt{3}$. Since $x = \sqrt{3}$ lies in the interval $[0, 3]$, the function does assume its average value in the given interval (Figure 5.24).

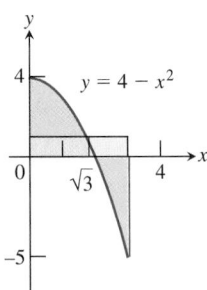

Figure 5.24 The rectangle with base $[0, 3]$ and with height equal to 1 (the average value of the function $f(x) = 4 - x^2$) has area equal to the net area between f and the x-axis from 0 to 3. (Example 3)

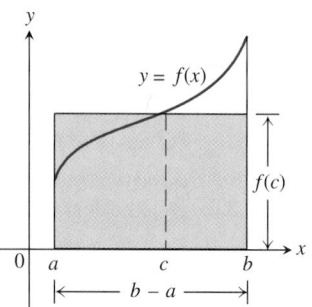

Figure 5.25 The value $f(c)$ in the Mean Value Theorem is, in a sense, the average (or *mean*) height of f on $[a, b]$. When $f \geq 0$, the area of the shaded rectangle

$$f(c)(b - a) = \int_a^b f(x)\, dx,$$

is the area under the graph of f from a to b.

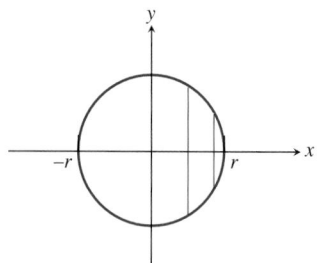

Figure 5.26 Chords perpendicular to the diameter $[-r, r]$ in a circle of radius r centered at the origin. (Exploration 1)

Mean Value Theorem for Definite Integrals

It was no mere coincidence that the function in Example 3 took on its average value at some point in the interval. Look at the graph in Figure 5.25 and imagine rectangles with base $(b - a)$ and heights ranging from the minimum of f (a rectangle too small to give the integral) to the maximum of f (a rectangle too large). Somewhere in between there is a "just right" rectangle, and its topside will intersect the graph of f if f is continuous. The statement that a continuous function on a closed interval *always* assumes its average value at least once in the interval is known as the Mean Value Theorem for Definite Integrals.

Theorem 3   The Mean Value Theorem for Definite Integrals

If f is continuous on $[a, b]$, then at some point c in $[a, b]$,

$$f(c) = \frac{1}{b - a} \int_a^b f(x)\, dx.$$

Exploration 1   How Long Is the Average Chord of a Circle?

Suppose we have a circle of radius r centered at the origin. We want to know the average length of the chords perpendicular to the diameter $[-r, r]$ on the x-axis.

1. Show that the length of the chord at x is $2\sqrt{r^2 - x^2}$ (Figure 5.26).

2. Set up an integral expression for the average value of $2\sqrt{r^2 - x^2}$ over the interval $[-r, r]$.

3. Evaluate the integral by identifying its value as an area.

4. So, what is the average length of a chord of a circle of radius r?

5. Explain how we can use the Mean Value Theorem for Definite Integrals (Theorem 3) to show that the function assumes the value in step 4.

Connecting Differential and Integral Calculus

Before we move on to the next section, let us pause for a moment of historical perspective that can help you to appreciate the power of the theorem that you are about to encounter. In Example 3 we used NINT to find the integral, and in Section 5.2, Example 2 we were fortunate that we could use our knowledge of the area of a circle. The area of a circle has been around for a long time, but NINT has not; so how did people evaluate definite integrals when they could not apply some known area formula? For example, in Exploration 1 of the previous section we used the fact that

$$\int_0^\pi \sin x\, dx = 2.$$

Would Newton and Leibniz have known this fact? How?

They did know that *quotients of infinitely small quantities,* as they put it, could be used to get velocity functions from position functions, and that *sums of infinitely thin "rectangle areas"* could be used to get position functions from velocity functions. In some way, then, there had to be a connection between these two seemingly different processes. Newton and Leibniz were able to picture that connection, and it led them to the Fundamental Theorem of Calculus. Can you picture it? Try Exploration 2.

Exploration 2 **Finding the Derivative of an Integral**

Suppose we are given the graph of a continuous function f, as in Figure 5.27. *Work in groups of two or three.*

1. Copy the graph of f onto your own paper. Choose any x greater than a in the interval $[a, b]$ and mark it on the x-axis.

2. Using only *vertical line segments*, shade in the region between the graph of f and the x-axis from a to x. (Some shading might be below the x-axis.)

3. Your shaded region represents a definite integral. Explain why this integral can be written as $\int_a^x f(t)\, dt$. (Why don't we write it as $\int_a^x f(x)\, dx$?)

4. Compare your picture with others produced by your group. Notice how your integral (a real number) depends on which x you chose in the interval $[a, b]$. The integral is therefore a *function of x* on $[a, b]$. Call it F.

5. Recall that $F'(x)$ is the limit of $\Delta F / \Delta x$ as Δx gets smaller and smaller. Represent ΔF in your picture by drawing *one more vertical shading segment* to the right of the last one you drew in step 2. ΔF is the (signed) *area* of your vertical segment.

6. Represent Δx in your picture by moving x to beneath your newly-drawn segment. That small change in Δx is the *thickness* of your vertical segment.

7. What is now the *height* of your vertical segment?

8. Can you see why Newton and Leibniz concluded that $F'(x) = f(x)$?

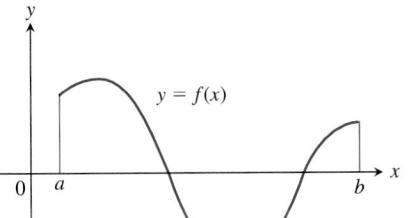

Figure 5.27 The graph of the function in Exploration 2.

If all went well in Exploration 2, you concluded that the derivative with respect to x of the integral of f from a to x is simply f. Specifically,

$$\frac{d}{dx} \int_a^x f(t)\, dt = f(x).$$

This means that the integral is an *antiderivative* of f, a fact we can exploit in the following way.

If F is any antiderivative of f, then

$$\int_a^x f(t)\, dt = F(x) + C$$

for some constant C. Setting x in this equation equal to a gives

$$\int_a^a f(t)\,dt = F(a) + C$$

$$0 = F(a) + C$$

$$C = -F(a).$$

Putting it all together,

$$\int_a^x f(t)\,dt = F(x) - F(a).$$

The implications of this last equation were enormous for the discoverers of calculus. It meant that they could evaluate the definite integral of f from a to any number x simply by computing $F(x) - F(a)$, where F is *any antiderivative of f*.

Example 4   FINDING AN INTEGRAL USING ANTIDERIVATIVES

Find $\int_0^\pi \sin x\,dx$ using the formula $\int_a^x f(t)\,dt = F(x) - F(a)$.

Solution   Since $F(x) = -\cos x$ is an antiderivative of $\sin x$, we have

$$\int_0^\pi \sin x\,dx = -\cos(\pi) - (-\cos(0))$$

$$= -(-1) - (-1)$$

$$= 2.$$

This explains how we obtained the value for Exploration 1 of the previous section.

Quick Review 5.3

In Exercises 1–10, find dy/dx.

1. $y = -\cos x$

2. $y = \sin x$

3. $y = \ln(\sec x)$

4. $y = \ln(\sin x)$

5. $y = \ln(\sec x + \tan x)$

6. $y = x\ln x - x$

7. $y = \dfrac{x^{n+1}}{n+1}$ $(n \ne -1)$

8. $y = \dfrac{1}{2^x + 1}$

9. $y = xe^x$

10. $y = \tan^{-1} x$

Section 5.3 Exercises

The exercises in this section are designed to reinforce your understanding of the definite integral from the algebraic and geometric points of view. For this reason, you should not use the numerical integration capability of your calculator (NINT) except perhaps to support an answer.

1. Suppose that f and g are continuous functions and that

$$\int_1^2 f(x)\,dx = -4, \quad \int_1^5 f(x)\,dx = 6, \quad \int_1^5 g(x)\,dx = 8.$$

Use the rules in Table 5.3 to find each integral.

(a) $\displaystyle\int_2^2 g(x)\,dx$

(b) $\displaystyle\int_5^1 g(x)\,dx$

(c) $\displaystyle\int_1^2 3f(x)\,dx$

(d) $\displaystyle\int_2^5 f(x)\,dx$

(e) $\displaystyle\int_1^5 [f(x) - g(x)]\,dx$

(f) $\displaystyle\int_1^5 [4f(x) - g(x)]\,dx$

2. Suppose that f and h are continuous functions and that

$$\int_1^9 f(x)\, dx = -1, \quad \int_7^9 f(x)\, dx = 5, \quad \int_7^9 h(x)\, dx = 4.$$

Use the rules in Table 5.3 to find each integral.

(a) $\displaystyle\int_1^9 -2f(x)\, dx$
(b) $\displaystyle\int_7^9 [f(x) + h(x)]\, dx$

(c) $\displaystyle\int_7^9 [2f(x) - 3h(x)]\, dx$
(d) $\displaystyle\int_9^1 f(x)\, dx$

(e) $\displaystyle\int_1^7 f(x)\, dx$
(f) $\displaystyle\int_9^7 [h(x) - f(x)]\, dx$

3. Suppose that $\int_1^2 f(x)\, dx = 5$. Find each integral.

(a) $\displaystyle\int_1^2 f(u)\, du$
(b) $\displaystyle\int_1^2 \sqrt{3}\, f(z)\, dz$

(c) $\displaystyle\int_2^1 f(t)\, dt$
(d) $\displaystyle\int_1^2 [-f(x)]\, dx$

4. Suppose that $\int_{-3}^0 g(t)\, dt = \sqrt{2}$. Find each integral.

(a) $\displaystyle\int_0^{-3} g(t)\, dt$
(b) $\displaystyle\int_{-3}^0 g(u)\, du$

(c) $\displaystyle\int_{-3}^0 [-g(x)]\, dx$
(d) $\displaystyle\int_{-3}^0 \frac{g(r)}{\sqrt{2}}\, dr$

5. Suppose that f is continuous and that

$$\int_0^3 f(z)\, dz = 3 \quad \text{and} \quad \int_0^4 f(z)\, dz = 7.$$

Find each integral.

(a) $\displaystyle\int_3^4 f(z)\, dz$
(b) $\displaystyle\int_4^3 f(t)\, dt$

6. Suppose that h is continuous and that

$$\int_{-1}^1 h(r)\, dr = 0 \quad \text{and} \quad \int_{-1}^3 h(r)\, dr = 6.$$

Find each integral.

(a) $\displaystyle\int_1^3 h(r)\, dr$
(b) $\displaystyle -\int_3^1 h(u)\, du$

In Exercises 7–16, evaluate the integral.

7. $\displaystyle\int_3^1 7\, dx$
8. $\displaystyle\int_0^2 5x\, dx$

9. $\displaystyle\int_3^5 \frac{x}{8}\, dx$
10. $\displaystyle\int_0^2 (2t - 3)\, dt$

11. $\displaystyle\int_0^{\sqrt{2}} (t - \sqrt{2})\, dt$
12. $\displaystyle\int_2^1 \left(1 + \frac{z}{2}\right) dz$

13. $\displaystyle\int_{-1}^1 \frac{dx}{1 + x^2}$
14. $\displaystyle\int_{-1/2}^{1/2} \frac{dx}{\sqrt{1 - x^2}}$

15. $\displaystyle\int_0^2 e^x\, dx$
16. $\displaystyle\int_0^3 \frac{3\, dx}{x + 1}$

In Exercises 17–20, find the total shaded area.

17.

18.

19.

20.

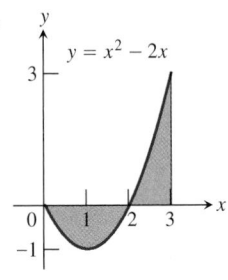

In Exercises 21–24, graph the function over the interval. Then **(a)** integrate the function over the interval and **(b)** find the area of the region between the graph and the x-axis.

21. $y = x^2 - 6x + 8, \quad [0, 3]$

22. $y = -x^2 + 5x - 4, \quad [0, 2]$

23. $y = 2x - x^2, \quad [0, 3]$
24. $y = x^2 - 4x, \quad [0, 5]$

In Exercises 25–28, find the average value of the function on the interval. At what point(s) in the interval does the function assume its average value?

25. $y = x^2 - 1, \quad [0, \sqrt{3}]$
26. $y = -\dfrac{x^2}{2}, \quad [0, 3]$

27. $y = -3x^2 - 1, \quad [0, 1]$
28. $y = (x - 1)^2, \quad [0, 3]$

In Exercises 29–32, find the average value of the function on the interval without integrating, by appealing to the geometry of the region between the graph and the x-axis.

29. $f(x) = \begin{cases} x + 4, & -4 \leq x \leq -1, \\ -x + 2, & -1 < x \leq 2, \end{cases}$ on $[-4, 2]$

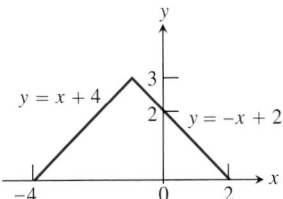

30. $f(t) = 1 - \sqrt{1 - t^2}, \quad [-1, 1]$

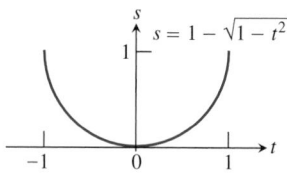

31. $f(t) = \sin t, \quad [0, 2\pi]$

32. $f(\theta) = \tan \theta, \quad \left[-\dfrac{\pi}{4}, \dfrac{\pi}{4}\right]$

In Exercises 33 and 34, *work in groups of two or three.*

33. Use the Max-Min Inequality to find upper and lower bounds for the value of
$$\int_0^1 \frac{1}{1 + x^4}\, dx.$$

34. *(Continuation of Exercise 33)* Use the Max-Min Inequality to find upper and lower bounds for the values of
$$\int_0^{0.5} \frac{1}{1 + x^4}\, dx \quad \text{and} \quad \int_{0.5}^1 \frac{1}{1 + x^4}\, dx.$$

Add these to arrive at an improved estimate for
$$\int_0^1 \frac{1}{1 + x^4}\, dx.$$

35. Show that the value of $\int_0^1 \sin(x^2)\, dx$ cannot possibly be 2.

36. Show that the value of $\int_0^1 \sqrt{x + 8}\, dx$ lies between $2\sqrt{2} \approx 2.8$ and 3.

37. *Integrals of Nonnegative Functions* Use the Max-Min Inequality to show that if f is integrable then
$$f(x) \geq 0 \text{ on } [a, b] \quad \Rightarrow \quad \int_a^b f(x)\, dx \geq 0.$$

38. *Integrals of Nonpositive Functions* Show that if f is integrable then
$$f(x) \leq 0 \text{ on } [a, b] \quad \Rightarrow \quad \int_a^b f(x)\, dx \leq 0.$$

39. Writing to Learn If $av(f)$ really is a typical value of the integrable function $f(x)$ on $[a, b]$, then the number $av(f)$ should have the same integral over $[a, b]$ that f does. Does it? That is, does
$$\int_a^b av(f)\, dx = \int_a^b f(x)\, dx?$$

Give reasons for your answer.

40. Writing to Learn A driver averaged 30 mph on a 150-mile trip and then returned over the same 150 miles at the rate of 50 mph. He figured that his average speed was 40 mph for the entire trip.

(a) What was his total distance traveled?

(b) What was his total time spent for the trip?

(c) What was his average speed for the trip?

(d) Explain the error in the driver's reasoning.

41. Writing to Learn A dam released 1000 m³ of water at 10 m³/min and then released another 1000 m³ at 20 m³/min. What was the average rate at which the water was released? Give reasons for your answer.

42. Use the inequality $\sin x \leq x$, which holds for $x \geq 0$, to find an upper bound for the value of $\int_0^1 \sin x\, dx$.

43. The inequality $\sec x \geq 1 + (x^2/2)$ holds on $(-\pi/2, \pi/2)$. Use it to find a lower bound for the value of $\int_0^1 \sec x\, dx$.

Exploration

44. *Comparing Area Formulas* Consider the region in the first quadrant under the curve $y = (h/b)\, x$ from $x = 0$ to $x = b$ (see figure).

(a) Use a geometry formula to calculate the area of the region.

(b) Find all antiderivatives of y.

(c) Use an antiderivative of y to evaluate $\int_0^b y(x)\, dx$.

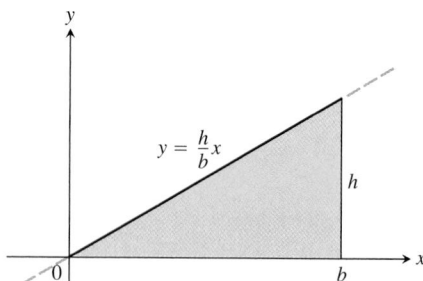

Extending the Ideas

45. *Graphing Calculator Challenge* If $k > 1$, and if the average value of x^k on $[0, k]$ is k, what is k? Check your result with a CAS if you have one available.

46. Show that if $F'(x) = G'(x)$ on $[a, b]$, then
$$F(b) - F(a) = G(b) - G(a).$$

5.4 Fundamental Theorem of Calculus

Fundamental Theorem, Part 1 • Graphing the Function $\int_a^x f(t)\,dt$ •
Fundamental Theorem, Part 2 • Area Connection • More
Applications

Fundamental Theorem, Part 1

Sir Isaac Newton (1642–1727)

Sir Isaac Newton is considered to be one of the most influential mathematicians of all time. Moreover, by the age of 25, he had also made revolutionary advances in optics, physics, and astronomy.

This section presents the discovery by Newton and Leibniz of the astonishing connection between integration and differentiation. This connection started the mathematical development that fueled the scientific revolution for the next 200 years, and is still regarded as the most important computational discovery in the history of mathematics: The Fundamental Theorem of Calculus.

The Fundamental Theorem comes in two parts, both of which were previewed in Exploration 2 of the previous section. The first part says that the definite integral of a continuous function is a differentiable function of its upper limit of integration. Moreover, it tells us what that derivative is. The second part says that the definite integral of a continuous function from a to b can be found from any one of the function's antiderivatives F as the number $F(b) - F(a)$.

Theorem 4 The Fundamental Theorem of Calculus, Part 1

If f is continuous on $[a, b]$, then the function

$$F(x) = \int_a^x f(t)\,dt$$

has a derivative at every point x in $[a, b]$, and

$$\frac{dF}{dx} = \frac{d}{dx}\int_a^x f(t)\,dt = f(x).$$

Proof The geometric exploration at the end of the previous section contained the idea of the proof, but it glossed over the necessary limit arguments. Here we will be more precise.

Apply the definition of the derivative directly to the function F. That is,

$$\frac{dF}{dx} = \lim_{h \to 0} \frac{F(x + h) - F(x)}{h}$$

$$= \lim_{h \to 0} \frac{\displaystyle\int_a^{x+h} f(t)\,dt - \int_a^x f(t)\,dt}{h}$$

$$= \lim_{h \to 0} \frac{\displaystyle\int_x^{x+h} f(t)\,dt}{h} \qquad \text{Rules for integrals, Section 5.3}$$

$$= \lim_{h \to 0} \left[\frac{1}{h} \int_x^{x+h} f(t)\,dt \right].$$

The expression in brackets in the last line is the average value of f from x to $x + h$. We know from the Mean Value Theorem for Definite Integrals (Theorem 3, Section 5.3) that f, being continuous, takes on its average value at least once in the interval; that is,

$$\frac{1}{h}\int_x^{x+h} f(t)\, dt = f(c) \quad \text{for some } c \text{ between } x \text{ and } x + h.$$

We can therefore continue our proof, letting $(1/h)\int_x^{x+h} f(t)\, dt = f(c)$,

$$\frac{dF}{dx} = \lim_{h \to 0} \frac{1}{h}\int_x^{x+h} f(t)\, dt$$

$$= \lim_{h \to 0} f(c), \quad \text{where } c \text{ lies between } x \text{ and } x + h.$$

What happens to c as h goes to zero? As $x + h$ gets closer to x, it carries c along with it like a bead on a wire, forcing c to approach x. Since f is continuous, this means that $f(c)$ approaches $f(x)$:

$$\lim_{h \to 0} f(c) = f(x).$$

Putting it all together,

$$\frac{dF}{dx} = \lim_{h \to 0} \frac{F(x + h) - F(x)}{h} \qquad \text{Definition of derivatives}$$

$$= \lim_{h \to 0} \frac{\displaystyle\int_x^{x+h} f(t)\, dt}{h} \qquad \text{Rules for integrals}$$

$$= \lim_{h \to 0} f(c) \quad \text{for some } c \text{ between } x \text{ and } x + h.$$

$$= f(x). \qquad \text{Because } f \text{ is continuous}$$

This concludes the proof. ∎

It is difficult to overestimate the power of the equation

$$\frac{d}{dx}\int_a^x f(t)\, dt = f(x). \tag{1}$$

It says that every continuous function f is the derivative of some other function, namely $\int_a^x f(t)\, dt$. It says that every continuous function has an antiderivative. And it says that the processes of integration and differentiation are inverses of one another. If any equation deserves to be called the Fundamental Theorem of Calculus, this equation is surely the one.

Example 1 APPLYING THE FUNDAMENTAL THEOREM

Find

$$\frac{d}{dx}\int_{-\pi}^x \cos t \, dt \quad \text{and} \quad \frac{d}{dx}\int_0^x \frac{1}{1 + t^2}\, dt$$

by using the Fundamental Theorem.

Solution

$$\frac{d}{dx}\int_{-\pi}^{x}\cos t\,dt = \cos x \qquad \text{Eq. 1 with } f(t)=\cos t$$

$$\frac{d}{dx}\int_{0}^{x}\frac{1}{1+t^2}\,dt = \frac{1}{1+x^2}. \qquad \text{Eq. 1 with } f(t)=\frac{1}{1+t^2}$$

Example 2 THE FUNDAMENTAL THEOREM WITH THE CHAIN RULE

Find dy/dx if $y=\int_{1}^{x^2}\cos t\,dt$.

Solution The upper limit of integration is not x but x^2. This makes y a composite of

$$y=\int_{1}^{u}\cos t\,dt \quad \text{and} \quad u=x^2.$$

We must therefore apply the Chain Rule when finding dy/dx.

$$\frac{dy}{dx}=\frac{dy}{du}\cdot\frac{du}{dx}$$

$$=\left(\frac{d}{du}\int_{1}^{u}\cos t\,dt\right)\cdot\frac{du}{dx}$$

$$=\cos u\cdot\frac{du}{dx}$$

$$=\cos(x^2)\cdot 2x$$

$$=2x\cos x^2$$

Example 3 VARIABLE LOWER LIMITS OF INTEGRATION

Find dy/dx.

(a) $y=\int_{x}^{5}3t\sin t\,dt$ **(b)** $y=\int_{2x}^{x^2}\frac{1}{2+e^t}\,dt$

Solution The rules for integrals set these up for the Fundamental Theorem.

(a) $\dfrac{d}{dx}\displaystyle\int_{x}^{5}3t\sin t\,dt=\dfrac{d}{dx}\left(-\int_{5}^{x}3t\sin t\,dt\right)$

$$=-\frac{d}{dx}\int_{5}^{x}3t\sin t\,dt$$

$$=-3x\sin x$$

(b) $\dfrac{d}{dx}\displaystyle\int_{2x}^{x^2}\frac{1}{2+e^t}\,dt=\dfrac{d}{dx}\left(\int_{0}^{x^2}\frac{1}{2+e^t}\,dt-\int_{0}^{2x}\frac{1}{2+e^t}\,dt\right)$

$$=\frac{1}{2+e^{x^2}}\frac{d}{dx}(x^2)-\frac{1}{2+e^{2x}}\frac{d}{dx}(2x) \quad \text{Chain Rule}$$

$$=\frac{1}{2+e^{x^2}}\cdot 2x-\frac{1}{2+e^{2x}}\cdot 2$$

$$=\frac{2x}{2+e^{x^2}}-\frac{2}{2+e^{2x}}$$

Example 4 CONSTRUCTING A FUNCTION WITH A GIVEN DERIVATIVE AND VALUE

Find a function $y = f(x)$ with derivative

$$\frac{dy}{dx} = \tan x$$

that satisfies the condition $f(3) = 5$.

Solution The Fundamental Theorem makes it easy to construct a function with derivative $\tan x$:

$$y = \int_3^x \tan t \, dt.$$

Since $y(3) = 0$, we have only to add 5 to this function to construct one with derivative $\tan x$ whose value at $x = 3$ is 5:

$$f(x) = \int_3^x \tan t \, dt + 5.$$

Although the solution to the problem in Example 4 satisfies the two required conditions, you might question whether it is in a useful form. Not many years ago, this form might have posed a computation problem. Indeed, for such problems much effort has been expended over the centuries trying to find solutions that do not involve integrals. We will see some in Chapter 6, where we will learn (for example) how to write the solution in Example 4 as

$$y = \ln \left| \frac{\cos 3}{\cos x} \right| + 5.$$

However, now that computers and calculators are capable of evaluating integrals, the form given in Example 4 is not only useful, but in some ways preferable. It is certainly easier to find and is always available.

Graphing the Function $\int_a^x f(t) \, dt$

Consider for a moment the two forms of the function we have just been discussing,

$$F(x) = \int_3^x \tan t \, dt + 5 \quad \text{and} \quad F(x) = \ln \left| \frac{\cos 3}{\cos x} \right| + 5.$$

With which expression is it easier to evaluate, say, $F(4)$? From the time of Newton almost to the present, there has been no contest: the expression on the right. At least it provides something to compute, and there have always been tables or slide rules or calculators to facilitate that computation. The expression on the left involved at best a tedious summing process and almost certainly an increased opportunity for error.

Today we can find $F(4)$ from either expression on the same machine. The choice is between NINT $(\tan x, x, 3, 4) + 5$ and $\ln(\text{abs}(\cos(3)/\cos(4))) + 5$. Both calculations give 5.415135083 in approximately the same amount of time.

We can even use NINT to graph the function. This modest technology feat would have absolutely dazzled the mathematicians of the 18th and 19th centuries, who knew how the solutions of differential equations, such as $dy/dx = \tan x$, could be written as integrals, but for whom integrals were of no practical use computationally unless they could be written in exact form. Since so few integrals could, in fact, be written in exact form, NINT would have spared generations of scientists much frustration.

Nevertheless, one must not proceed blindly into the world of calculator computation. Exploration 1 will demonstrate the need for caution.

Exploration 1 Graphing NINT *f*

Let us use NINT to attempt to graph the function we just discussed,

$$F(x) = \int_3^x \tan t \, dt + 5.$$

1. Graph the function $y = F(x)$ in the window $[-10, 10]$ by $[-10, 10]$. You will probably wait a long time and see no graph. Break out of the graphing program if necessary.

2. Recall that the graph of the function $y = \tan x$ has vertical asymptotes. Where do they occur on the interval $[-10, 10]$?

3. When attempting to graph the function $F(x) = \int_3^x \tan t \, dt + 5$ on the interval $[-10, 10]$, your grapher begins by trying to find $F(-10)$. Explain why this might cause a problem for your calculator.

4. Set your viewing window so that your calculator graphs only over the domain of the continuous branch of the tangent function that contains the point $(3, \tan 3)$.

5. What is the domain in step 4? Is it an open interval or a closed interval?

6. What is the domain of $F(x)$? Is it an open interval or a closed interval?

7. Your calculator graphs over the closed interval $[x_{min}, x_{max}]$. Find a viewing window that will give you a good look at the graph of F and produce the graph on your calculator.

8. Describe the graph of F.

Graphing NINT *f*

Some graphers can graph the numerical integral $y = \text{NINT}(f(x), x, a, x)$ directly as a function of x. Others will require a toolbox program such as the one called NINT-GRAF provided in the *Technology Resource Manual*.

You have probably noticed that your grapher moves slowly when graphing NINT. This is because it must compute each value as a limit of sums—comparatively slow work even for a microprocessor. Here are some ways to speed up the process:

1. Change the *tolerance* on your grapher. The smaller the tolerance, the more accurate the calculator will try to be when finding the limiting value of each sum (and the longer it will take to do so). The default value is usually quite small (like 0.00001), but a value as large as 1 can be used for graphing in a typical viewing window.

2. Change the *x-resolution*. The default resolution is 1, which means that the grapher will compute a function value for every vertical column of pixels. At resolution 2 it computes only every second value, and so on. With higher resolutions, some graph smoothness is sacrificed for speed.

3. Switch to parametric mode. To graph $y = \text{NINT}(f(x), x, a, x)$ in parametric mode, let $x(t) = t$ and let $y(t) = \text{NINT}(f(t), t, a, t)$. You can then control the speed of the grapher by changing the t-step. (Choosing a bigger t-step has the same effect as choosing a larger x-resolution.)

Exploration 2 **The Effect of Changing a in $\int_a^x f(t)\,dt$**

The first part of the Fundamental Theorem of Calculus asserts that the derivative of $\int_a^x f(t)\,dt$ is $f(x)$, regardless of the value of a.

1. Graph NDER (NINT $(x^2, x, 0, x)$).
2. Graph NDER (NINT $(x^2, x, 5, x)$).
3. Without graphing, tell what the x-intercept of NINT $(x^2, x, 0, x)$ is. Explain.
4. Without graphing, tell what the x-intercept of NINT $(x^2, x, 5, x)$ is. Explain.
5. How does changing a affect the graph of $y = (d/dx)\int_a^x f(t)\,dt$?
6. How does changing a affect the graph of $y = \int_a^x f(t)\,dt$?

Fundamental Theorem, Part 2

The second part of the Fundamental Theorem of Calculus shows how to evaluate definite integrals directly from antiderivatives.

Theorem 4 (continued) **The Fundamental Theorem of Calculus, Part 2**

If f is continuous at every point of $[a, b]$, and if F is any antiderivative of f on $[a, b]$, then

$$\int_a^b f(x)\,dx = F(b) - F(a).$$

This part of the Fundamental Theorem is also called the **Integral Evaluation Theorem.**

Proof Part 1 of the Fundamental Theorem tells us that an antiderivative of f exists, namely

$$G(x) = \int_a^x f(t)\,dt.$$

Thus, if F is *any* antiderivative of f, then $F(x) = G(x) + C$ for some constant C (by Corollary 3 of the Mean Value Theorem for Derivatives, Section 4.2). Evaluating $F(b) - F(a)$, we have

$$F(b) - F(a) = [G(b) + C] - [G(a) + C]$$

$$= G(b) - G(a)$$

$$= \int_a^b f(t)\,dt - \int_a^a f(t)\,dt$$

$$= \int_a^b f(t)\,dt - 0$$

$$= \int_a^b f(t)\,dt. \qquad \blacksquare$$

At the risk of repeating ourselves: It is difficult to overestimate the power of the simple equation

$$\int_a^b f(x)\, dx = F(b) - F(a).$$

It says that any definite integral of any continuous function f can be calculated without taking limits, without calculating Riemann sums, and often without effort—so long as an antiderivative of f can be found. If you can imagine what it was like before this theorem (and before computing machines), when approximations by tedious sums were the only alternative for solving many real-world problems, then you can imagine what a miracle calculus was thought to be. If any equation deserves to be called the Fundamental Theorem of Calculus, this equation is surely the (second) one.

Example 5   EVALUATING AN INTEGRAL

Evaluate $\int_{-1}^{3} (x^3 + 1)\, dx$ using an antiderivative.

Solution

Solve Analytically

A simple antiderivative of $x^3 + 1$ is $(x^4/4) + x$. Therefore,

$$\int_{-1}^{3}\left(x^3 + 1\right) dx = \left[\frac{x^4}{4} + x\right]_{-1}^{3}$$

$$= \left(\frac{81}{4} + 3\right) - \left(\frac{1}{4} - 1\right)$$

$$= 24.$$

Support Numerically

NINT $(x^3 + 1, x, -1, 3) = 24$.

<hr/>

Integral Evaluation Notation

The usual notation for $F(b) - F(a)$ is

$$F(x)\Big]_a^b \quad \text{or} \quad \left[F(x)\right]_a^b,$$

depending on whether F has one or more terms. This notation provides a compact "recipe" for the evaluation, allowing us to show the antiderivative in an intermediate step.

<hr/>

Area Connection

In Section 5.2 we saw that the definite integral could be interpreted as the net area between the graph of a function and the x-axis. We can therefore compute areas using antiderivatives, but we must again be careful to distinguish net area (in which area below the x-axis is counted as negative) from total area. The unmodified word "area" will be taken to mean *total area*.

Example 6   FINDING AREA USING ANTIDERIVATIVES

Find the area of the region between the curve $y = 4 - x^2$, $0 \le x \le 3$, and the x-axis.

Solution   The curve crosses the x-axis at $x = 2$, partitioning the interval $[0, 3]$ into two subintervals, on each of which $f(x) = 4 - x^2$ will not change sign.

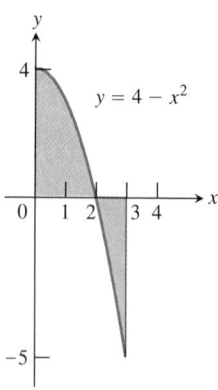

Figure 5.28 The function $f(x) = 4 - x^2$ changes sign only at $x = 2$ on the interval $[0, 3]$. (Example 6)

We can see from the graph (Figure 5.28) that $f(x) > 0$ on $[0, 2)$ and $f(x) < 0$ on $(2, 3]$.

Over $[0, 2]$: $\displaystyle\int_0^2 (4 - x^2)\, dx = \left[4x - \frac{x^3}{3}\right]_0^2 = \frac{16}{3}.$

Over $[2, 3]$: $\displaystyle\int_2^3 (4 - x^2)\, dx = \left[4x - \frac{x^3}{3}\right]_2^3 = -\frac{7}{3}.$

The area of the region is $\left|\dfrac{16}{3}\right| + \left|-\dfrac{7}{3}\right| = \dfrac{23}{3}.$

How to Find Total Area Analytically

To find the area between the graph of $y = f(x)$ and the x-axis over the interval $[a, b]$ analytically,

1. partition $[a, b]$ with the zeros of f,
2. integrate f over each subinterval,
3. add the absolute values of the integrals.

We can find area numerically by using NINT to integrate the *absolute value* of the function over the given interval. There is no need to partition. By taking absolute values, we automatically reflect the negative portions of the graph across the x-axis to count all area as positive (Figure 5.29).

Example 7 FINDING AREA USING NINT

Find the area of the region between the curve $y = x \cos 2x$ and the x-axis over the interval $-3 \le x \le 3$ (Figure 5.29).

Solution Rounded to two decimal places, we have

$$\text{NINT}\,(|x \cos 2x|, x, -3, 3) = 5.43.$$

How to Find Total Area Numerically

To find the area between the graph of $y = f(x)$ and the x-axis over the interval $[a, b]$ numerically, evaluate

$$\text{NINT}\,(|f(x)|, x, a, b).$$

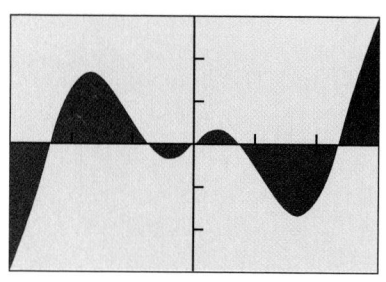

[−3, 3] by [−3, 3]

(a)

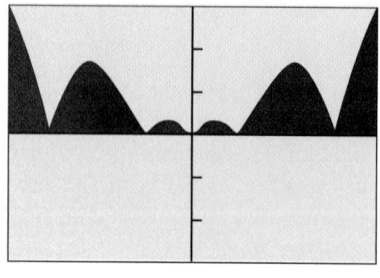

[−3, 3] by [−3, 3]

(b)

Figure 5.29 The graphs of (a) $y = x \cos 2x$ and (b) $y = |x \cos 2x|$ over $[-3, 3]$. The shaded regions have the same area.

More Applications

We close the section with two applications from the world of economics.

Example 8 DETERMINING COST FROM MARGINAL COST

The fixed cost of starting a manufacturing run and producing the first 10 units is $200. After that, the marginal cost at x units output is

$$\frac{dc}{dx} = \frac{1000}{x^2}.$$

Find the total cost of producing the first 100 units.

Section 5.4 Fundamental Theorem of Calculus **285**

Solution If $c(x)$ is the cost of x units, then

$$\underbrace{c(100)}_{\substack{\text{Cost of}\\100\text{ units}}} = \underbrace{200}_{\substack{\text{Startup and}\\\text{first 10 units}}} + \underbrace{c(100) - c(10)}_{\substack{\text{Cost of units}\\11\text{ through }100}}$$

$$= 200 + \int_{10}^{100} \frac{dc}{dx}\,dx \qquad \text{Since } \int_{10}^{100} \frac{dc}{dx}\,dx = c(100) - c(10)$$

$$= 200 + \int_{10}^{100} \frac{1000}{x^2}\,dx$$

$$= 200 + 1000\left[-\frac{1}{x}\right]_{10}^{100}$$

$$= 290.$$

The total cost of producing the first 100 units is $290.

The notion of a function's average value is used in economics to study things like average daily inventory. An **inventory function** $I(x)$ gives the number of radios, shoes, tires, or whatever product a firm has on hand on day x. The average value of I over a time period $a \le x \le b$ is the firm's average daily inventory for the period.

Definition Average Daily Inventory

If $I(x)$ is the number of items on hand on day x, the **average daily inventory** of the items for the period $a \le x \le b$ is

$$av(I) = \frac{1}{b-a}\int_a^b I(x)\,dx.$$

If h is the dollar cost of holding one item per day, the **average daily holding cost** for the period $a \le x \le b$ is $av(I)\cdot h$.

Example 9 FINDING AVERAGE DAILY INVENTORY

Suppose a wholesaler receives a shipment of 1200 cases of boxes of chocolates every 30 days. The chocolate is sold to retailers at a steady rate, and x days after the shipment arrives, the inventory of cases still on hand is $I(x) = 1200 - 40x$. Find the average daily inventory. Also, find the average daily holding cost if the cost of holding one case is 3 cents a day.

Solution The average daily inventory is

$$av(I) = \frac{1}{30-0}\int_0^{30}(1200 - 40x)\,dx$$

$$= \frac{1}{30}\left[1200x - 20x^2\right]_0^{30}$$

$$= 600.$$

The average daily holding cost, then, is $(600)(0.03) = 18$ dollars a day.

Figure 5.30 Maintaining a steady inventory to ship to the wholesaler. (Example 9)

Quick Review 5.4

In Exercises 1–10, find dy/dx.

1. $y = \sin(x^2)$

2. $y = (\sin x)^2$

3. $y = \sec^2 x - \tan^2 x$

4. $y = \ln(3x) - \ln(7x)$

5. $y = 2^x$

6. $y = \sqrt{x}$

7. $y = \dfrac{\cos x}{x}$

8. $y = \sin t$ and $x = \cos t$

9. $xy + x = y^2$

10. $dx/dy = 3x$

Section 5.4 Exercises

In Exercises 1–14, evaluate each integral using Part 2 of the Fundamental Theorem. Support your answer with NINT if you are unsure.

1. $\displaystyle\int_{1/2}^{3} \left(2 - \frac{1}{x}\right) dx$

2. $\displaystyle\int_{2}^{-1} 3^x \, dx$

3. $\displaystyle\int_{0}^{1} (x^2 + \sqrt{x}) \, dx$

4. $\displaystyle\int_{0}^{5} x^{3/2} \, dx$

5. $\displaystyle\int_{1}^{32} x^{-6/5} \, dx$

6. $\displaystyle\int_{-2}^{-1} \frac{2}{x^2} \, dx$

7. $\displaystyle\int_{0}^{\pi} \sin x \, dx$

8. $\displaystyle\int_{0}^{\pi} (1 + \cos x) \, dx$

9. $\displaystyle\int_{0}^{\pi/3} 2 \sec^2 \theta \, d\theta$

10. $\displaystyle\int_{\pi/6}^{5\pi/6} \csc^2 \theta \, d\theta$

11. $\displaystyle\int_{\pi/4}^{3\pi/4} \csc x \cot x \, dx$

12. $\displaystyle\int_{0}^{\pi/3} 4 \sec x \tan x \, dx$

13. $\displaystyle\int_{-1}^{1} (r + 1)^2 \, dr$

14. $\displaystyle\int_{0}^{4} \frac{1 - \sqrt{u}}{\sqrt{u}} \, du$

In Exercises 15–18, find the total area of the region between the curve and the x-axis.

15. $y = 2 - x, \quad 0 \le x \le 3$

16. $y = 3x^2 - 3, \quad -2 \le x \le 2$

17. $y = x^3 - 3x^2 + 2x, \quad 0 \le x \le 2$

18. $y = x^3 - 4x, \quad -2 \le x \le 2$

In Exercises 19–24, answer the following questions about the integral.

(a) **Writing to Learn** Can the Fundamental Theorem of Calculus, Part 2 (Theorem 4) be used to evaluate the integral? Explain.

(b) Does the integral have a value? If so, what is it? Explain.

19. $\displaystyle\int_{-2}^{3} \frac{x^2 - 1}{x + 1} \, dx$

20. $\displaystyle\int_{0}^{5} \frac{9 - x^2}{3x - 9} \, dx$

21. $\displaystyle\int_{0}^{2\pi} \tan x \, dx$

22. $\displaystyle\int_{0}^{2} \frac{x + 1}{x^2 - 1} \, dx$

23. $\displaystyle\int_{-1}^{2} \frac{\sin x}{x} \, dx$

24. $\displaystyle\int_{-2}^{3} \frac{1 - \cos x}{x^2} \, dx$

In Exercises 25–28, find the area of the shaded region.

25.

26.

27.

28.

In Exercises 29–34, use NINT to solve the problem.

29. Evaluate $\displaystyle\int_{0}^{10} \frac{1}{3 + 2 \sin x} \, dx.$

30. Evaluate $\displaystyle\int_{-0.8}^{0.8} \frac{2x^4 - 1}{x^4 - 1} \, dx.$

31. Find the average value of $\sqrt{\cos x}$ on the interval $[-1, 1]$.

32. Find the area of the semielliptical region between the x-axis and the graph of $y = \sqrt{8 - 2x^2}$.

33. For what value of x does $\displaystyle\int_{0}^{x} e^{-t^2} \, dt = 0.6$?

34. Find the area of the region in the first quadrant enclosed by the coordinate axes and the graph of $x^3 + y^3 = 1$.

In Exercises 35 and 36, find K so that

$$\int_a^x f(t)\, dt + K = \int_b^x f(t)\, dt.$$

35. $f(x) = x^2 - 3x + 1;\quad a = -1;\quad b = 2$

36. $f(x) = \sin^2 x;\quad a = 0;\quad b = 2$

In Exercises 37–42, find dy/dx.

37. $y = \displaystyle\int_0^x \sqrt{1 + t^2}\, dt$

38. $y = \displaystyle\int_x^1 \frac{1}{t}\, dt,\quad x > 0$

39. $y = \displaystyle\int_0^{\sqrt{x}} \sin(t^2)\, dt$

40. $y = \displaystyle\int_0^{2x} \cos t\, dt$

41. $y = \displaystyle\int_{x^2}^{x^3} \cos(2t)\, dt$

42. $y = \displaystyle\int_{\sin x}^{\cos x} t^2\, dt$

In Exercises 43–46, choose which of the following functions has the given derivative and numerical value. Confirm your answer.

(a) $y = \displaystyle\int_1^x e^{-t^2} dt - 3$

(b) $y = \displaystyle\int_0^x \sec t\, dt + 4$

(c) $y = \displaystyle\int_{-1}^x \sec t\, dt + 4$

(d) $y = \displaystyle\int_\pi^x e^{-t^2}\, dt - 3$

43. $\dfrac{dy}{dx} = e^{-x^2},\quad y(\pi) = -3$

44. $\dfrac{dy}{dx} = \sec x,\quad y(-1) = 4$

45. $\dfrac{dy}{dx} = \sec x,\quad y(0) = 4$

46. $\dfrac{dy}{dx} = e^{-x^2},\quad y(1) = -3$

47. *Identifying a Zero* For what value of x is $\int_a^x f(t)\, dt$ sure to be zero?

48. Suppose $\int_1^x f(t)\, dt = x^2 - 2x + 1$. Find $f(x)$.

49. *Linearization* Find the linearization of

$$f(x) = 2 + \int_0^x \frac{10}{1+t}\, dt \quad \text{at} \quad x = 0.$$

50. Find $f(4)$ if $\int_0^x f(t)\, dt = x \cos \pi x$.

51. *Finding Area* Show that if k is a positive constant, then the area between the x-axis and one arch of the curve $y = \sin kx$ is always $2/k$.

52. *Archimedes' Area Formula for Parabolas* Archimedes (287–212 B.C.), inventor, military engineer, physicist, and the greatest mathematician of classical times, discovered that the area under a parabolic arch like the one shown here is always two-thirds the base times the height.

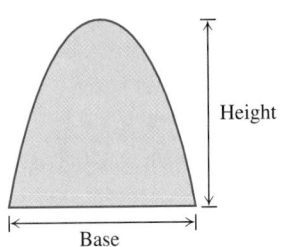

Height

Base

(a) Find the area under the parabolic arch

$$y = 6 - x - x^2, \quad -3 \le x \le 2.$$

(b) Find the height of the arch.

(c) Show that the area is two-thirds the base times the height.

In Exercises 53–55, *work in groups of two or three.*

53. Let

$$H(x) = \int_0^x f(t)\, dt,$$

where f is the continuous function with domain $[0, 12]$ graphed here.

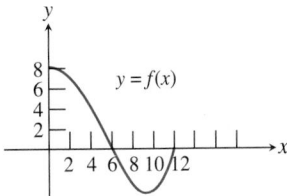

(a) Find $H(0)$.

(b) On what interval is H increasing? Explain.

(c) On what interval is the graph of H concave up? Explain.

(d) Is $H(12)$ positive or negative? Explain.

(e) Where does H achieve its maximum value? Explain.

(f) Where does H achieve its minimum value? Explain.

In Exercises 54 and 55, f is the differentiable function whose graph is shown in the figure. The position at time t (sec) of a particle moving along a coordinate axis is

$$s = \int_0^t f(x)\, dx$$

meters. Use the graph to answer the questions. Give reasons for your answers.

54.

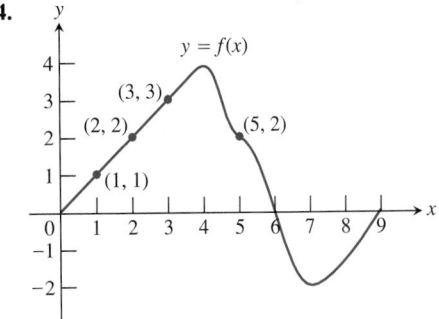

(a) What is the particle's velocity at time $t = 5$?

(b) Is the acceleration of the particle at time $t = 5$ positive or negative?

(c) What is the particle's position at time $t = 3$?

(d) At what time during the first 9 sec does s have its largest value?

(e) Approximately when is the acceleration zero?

(f) When is the particle moving toward the origin? away from the origin?

(g) On which side of the origin does the particle lie at time $t = 9$?

55.

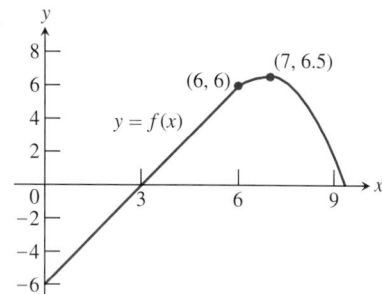

(a) What is the particle's velocity at time $t = 3$?

(b) Is the acceleration of the particle at time $t = 3$ positive or negative?

(c) What is the particle's position at time $t = 3$?

(d) When does the particle pass through the origin?

(e) Approximately when is the acceleration zero?

(f) When is the particle moving toward the origin? away from the origin?

(g) On which side of the origin does the particle lie at time $t = 9$?

Exploration

56. *The Sine Integral Function* The sine integral function

$$\text{Si}(x) = \int_0^x \frac{\sin t}{t}\, dt$$

is one of the many useful functions in engineering that are defined as integrals. Although the notation does not show it, the function being integrated is

$$f(t) = \begin{cases} \dfrac{\sin t}{t}, & t \neq 0 \\ 1, & t = 0, \end{cases}$$

the continuous extension of $(\sin t)/t$ to the origin.

(a) Show that $\text{Si}(x)$ is an odd function of x.

(b) What is the value of $\text{Si}(0)$?

(c) Find the values of x at which $\text{Si}(x)$ has a local extreme value.

(d) Use NINT to graph $\text{Si}(x)$. ∎

57. *Cost from Marginal Cost* The marginal cost of printing a poster when x posters have been printed is

$$\frac{dc}{dx} = \frac{1}{2\sqrt{x}}$$

dollars. Find

(a) $c(100) - c(1)$, the cost of printing posters 2 to 100.

(b) $c(400) - c(100)$, the cost of printing posters 101 to 400.

58. *Revenue from Marginal Revenue* Suppose that a company's marginal revenue from the manufacture and sale of eggbeaters is

$$\frac{dr}{dx} = 2 - \frac{2}{(x+1)^2},$$

where r is measured in thousands of dollars and x in thousands of units. How much money should the company expect from a production run of $x = 3$ thousand eggbeaters? To find out, integrate the marginal revenue from $x = 0$ to $x = 3$.

59. *Average Daily Holding Cost* Solon Container receives 450 drums of plastic pellets every 30 days. The inventory function (drums on hand as a function of days) is $I(x) = 450 - x^2/2$.

(a) Find the average daily inventory.

(b) If the holding cost for one drum is $0.02 per day, find the average daily holding cost.

60. Suppose that f has a negative derivative for all values of x and that $f(1) = 0$. Which of the following statements must be true of the function

$$h(x) = \int_0^x f(t)\, dt?$$

Give reasons for your answers.

(a) h is a twice-differentiable function of x.

(b) h and dh/dx are both continuous.

(c) The graph of h has a horizontal tangent at $x = 1$.

(d) h has a local maximum at $x = 1$.

(e) h has a local minimum at $x = 1$.

(f) The graph of h has an inflection point at $x = 1$.

(g) The graph of dh/dx crosses the x-axis at $x = 1$.

Extending the Ideas

61. Writing to Learn If f is an odd continuous function, give a graphical argument to explain why $\int_0^x f(t)\, dt$ is even.

62. Writing to Learn If f is an even continuous function, give a graphical argument to explain why $\int_0^x f(t)\, dt$ is odd.

63. Writing to Learn Explain why we can conclude from Exercises 61 and 62 that every even continuous function is the derivative of an odd continuous function and vice versa.

64. Give a convincing argument that the equation

$$\int_0^x \frac{\sin t}{t}\, dt = 1$$

has exactly one solution. Give its approximate value.

5.5 Trapezoidal Rule

Trapezoidal Approximations • Other Algorithms • Error Analysis

Trapezoidal Approximations

You probably noticed in Section 5.1 that MRAM was generally more efficient in approximating integrals than either LRAM or RRAM, even though all three RAM approximations approached the same limit. All three RAM approximations, however, depend on the areas of rectangles. Are there other geometric shapes with known areas that can do the job more efficiently? The answer is yes, and the most obvious one is the trapezoid.

As shown in Figure 5.31, if $[a, b]$ is partitioned into n subintervals of equal length $h = (b - a)/n$, the graph of f on $[a, b]$ can be approximated by a straight line segment over each subinterval.

The region between the curve and the x-axis is then approximated by the trapezoids, the area of each trapezoid being the length of its horizontal "altitude" times the average of its two vertical "bases." That is,

$$\int_a^b f(x)\, dx \approx h \cdot \frac{y_0 + y_1}{2} + h \cdot \frac{y_1 + y_2}{2} + \cdots + h \cdot \frac{y_{n-1} + y_n}{2}$$

$$= h\left(\frac{y_0}{2} + y_1 + y_2 + \cdots + y_{n-1} + \frac{y_n}{2}\right)$$

$$= \frac{h}{2}\left(y_0 + 2y_1 + 2y_2 + \cdots + 2y_{n-1} + y_n\right),$$

where

$$y_0 = f(a), \quad y_1 = f(x_1), \quad \cdots, \quad y_{n-1} = f(x_{n-1}), \quad y_n = f(b).$$

This is algebraically equivalent to finding the numerical average of LRAM and RRAM; indeed, that is how some texts define the Trapezoidal Rule.

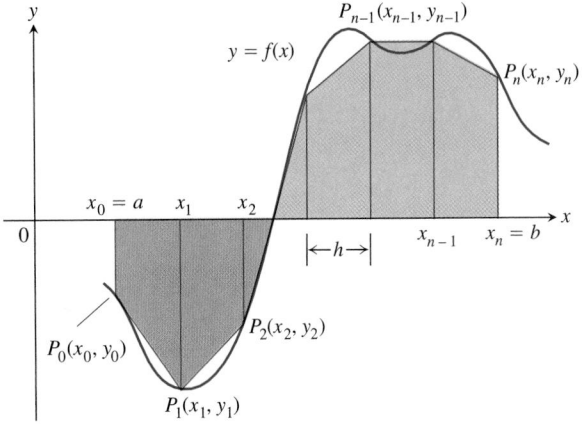

Figure 5.31 The trapezoidal rule approximates short stretches of the curve $y = f(x)$ with line segments. To approximate the integral of f from a to b, we add the "signed" areas of the trapezoids made by joining the ends of the segments to the x-axis.

Figure 5.32 The trapezoidal approximation of the area under the graph of $y = x^2$ from $x = 1$ to $x = 2$ is a slight overestimate. (Example 1)

Table 5.4

x	$y = x^2$
1	1
$\frac{5}{4}$	$\frac{25}{16}$
$\frac{6}{4}$	$\frac{36}{16}$
$\frac{7}{4}$	$\frac{49}{16}$
2	4

The Trapezoidal Rule

To approximate $\int_a^b f(x)\, dx$, use

$$T = \frac{h}{2}\left(y_0 + 2y_1 + 2y_2 + \cdots + 2y_{n-1} + y_n\right),$$

where $[a, b]$ is partitioned into n subintervals of equal length $h = (b - a)/n$.

Equivalently,

$$T = \frac{\text{LRAM}_n + \text{RRAM}_n}{2},$$

where LRAM_n and RRAM_n are the Riemann sums using the left and right endpoints, respectively, for f for the partition.

Example 1 APPLYING THE TRAPEZOIDAL RULE

Use the Trapezoidal Rule with $n = 4$ to estimate $\int_1^2 x^2\, dx$. Compare the estimate with the value of NINT $(x^2, x, 1, 2)$ and with the exact value.

Solution Partition $[1, 2]$ into four subintervals of equal length (Figure 5.32). Then evaluate $y = x^2$ at each partition point (Table 5.4).

Using these y values, $n = 4$, and $h = (2 - 1)/4 = 1/4$ in the Trapezoidal Rule, we have

$$T = \frac{h}{2}\left(y_0 + 2y_1 + 2y_2 + 2y_3 + y_4\right)$$

$$= \frac{1}{8}\left(1 + 2\left(\frac{25}{16}\right) + 2\left(\frac{36}{16}\right) + 2\left(\frac{49}{16}\right) + 4\right)$$

$$= \frac{75}{32} = 2.34375.$$

The value of NINT $(x^2, x, 1, 2)$ is 2.333333333.

The exact value of the integral is

$$\int_1^2 x^2\, dx = \frac{x^3}{3}\Bigg]_1^2 = \frac{8}{3} - \frac{1}{3} = \frac{7}{3}.$$

The T approximation overestimates the integral by about half a percent of its true value of 7/3. The percentage error is $(2.34375 - 7/3)/(7/3) \approx 0.00446$.

We could have predicted that the Trapezoidal Rule would overestimate the integral in Example 1 by considering the geometry of the graph in Figure 5.32. Since the parabola is concave *up*, the approximating segments lie above the curve, giving each trapezoid slightly more area than the corresponding strip under the curve. In Figure 5.31 we see that the straight segments lie *under* the curve on those intervals where the curve is concave *down*, causing the Trapezoidal Rule to *underestimate* the integral on those intervals. The interpretation of "area" changes where the curve lies below the x-axis but it is still the case that the higher y-values give the greater signed area. So we can always say that T overestimates the integral where the graph is concave up and underestimates the integral where the graph is concave down.

Example 2   AVERAGING TEMPERATURES

An observer measures the outside temperature every hour from noon until midnight, recording the temperatures in the following table.

Time	N	1	2	3	4	5	6	7	8	9	10	11	M
Temp	63	65	66	68	70	69	68	68	65	64	62	58	55

What was the average temperature for the 12-hour period?

Solution We are looking for the average value of a continuous function (temperature) for which we know values at discrete times that are one unit apart. We need to find

$$av(f) = \frac{1}{b-a}\int_a^b f(x)\,dx,$$

without having a formula for $f(x)$. The integral, however, can be approximated by the Trapezoidal Rule, using the temperatures in the table as function values at the points of a 12-subinterval partition of the 12-hour interval (making $h = 1$).

$$T = \frac{h}{2}\left(y_0 + 2y_1 + 2y_2 + \cdots + 2y_{11} + y_{12}\right)$$

$$= \frac{1}{2}\left(63 + 2\cdot 65 + 2\cdot 66 + \cdots + 2\cdot 58 + 55\right)$$

$$= 782$$

Using T to approximate $\int_a^b f(x)\,dx,$ we have

$$av(f) \approx \frac{1}{b-a}\cdot T = \frac{1}{12}\cdot 782 \approx 65.17.$$

Rounding to be consistent with the data given, we estimate the average temperature as 65 degrees.

Other Algorithms

LRAM, MRAM, RRAM, and the Trapezoidal Rule all give reasonable approximations to the integral of a continuous function over a closed interval. The Trapezoidal Rule is more efficient, giving a better approximation for small values of n, which makes it a faster algorithm for numerical integration.

Indeed, the only shortcoming of the Trapezoidal Rule seems to be that it depends on approximating curved arcs with straight segments. You might think that an algorithm that approximates the curve with *curved* pieces would be even more efficient (and hence faster for machines), and you would be right. All we need to do is find a geometric figure with a straight base, straight sides, and a curved top that has a known area. You might not know one, but the ancient Greeks did; it is one of the things they knew about parabolas.

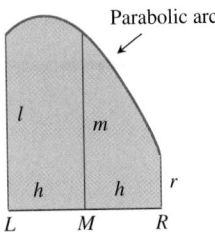

Parabolic arc

Figure 5.33 The area under the parabolic arc can be computed from the length of the base LR and the lengths of the altitudes constructed at L, R and midpoint M. (Exploration 1)

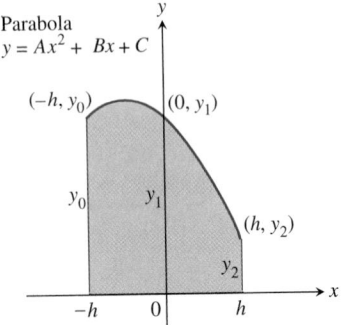

Figure 5.34 A convenient coordinatization of Figure 5.33. The parabola has equation $y = Ax^2 + Bx + C$, and the midpoint of the base is at the origin. (Exploration 1)

What's in a Name?

The formula that underlies Simpson's Rule (see Exploration 1) was discovered long before Thomas Simpson (1720–1761) was born. Just as Pythagoras did not discover the Pythagorean Theorem, Simpson did not discover Simpson's Rule. It is another of history's beautiful quirks that one of the ablest mathematicians of eighteenth-century England is remembered not for his successful textbooks and his contributions to mathematical analysis, but for a rule that was never his, that he never laid claim to, and that bears his name only because he happened to mention it in one of his books.

Exploration 1   **Area Under a Parabolic Arc**

The area A_P of a figure having a horizontal base, vertical sides, and a parabolic top (Figure 5.33) can be computed by the formula

$$A_P = \frac{h}{3}\left(l + 4m + r\right),$$

where h is half the length of the base, l and r are the lengths of the left and right sides, and m is the altitude at the midpoint of the base. This formula, once a profound discovery of ancient geometers, is readily verified today with calculus.

1. Coordinatize Figure 5.33 by centering the base at the origin, as shown in Figure 5.34. Let $y = Ax^2 + Bx + C$ be the equation of the parabola. Using this equation, show that $y_0 = Ah^2 - Bh + C$, $y_1 = C$, and $y_2 = Ah^2 + Bh + C$.

2. Show that $y_0 + 4y_1 + y_2 = 2Ah^2 + 6C$.

3. Integrate to show that the area A_P is

$$\frac{h}{3}(2Ah^2 + 6C).$$

4. Combine these results to derive the formula

$$A_P = \frac{h}{3}\left(y_0 + 4y_1 + y_2\right).$$

This last formula leads to an efficient rule for approximating integrals numerically. Partition the interval of integration into an even number of subintervals, apply the formula for A_P to successive interval pairs, and add the results. This algorithm is known as Simpson's Rule.

Simpson's Rule

To approximate $\int_a^b f(x)\,dx$, use

$$S = \frac{h}{3}\left(y_0 + 4y_1 + 2y_2 + 4y_3 + \cdots + 2y_{n-2} + 4y_{n-1} + y_n\right),$$

where $[a, b]$ is partitioned into an *even* number n of subintervals of equal length $h = (b - a)/n$.

Example 3   APPLYING SIMPSON'S RULE

Use Simpson's Rule with $n = 4$ to approximate $\int_0^2 5x^4\,dx$.

Solution   Partition $[0, 2]$ into four subintervals and evaluate $y = 5x^4$ at the partition points (Table 5.5).

Table 5.5

x	$y = 5x^4$
0	0
$\dfrac{1}{2}$	$\dfrac{5}{16}$
1	5
$\dfrac{3}{2}$	$\dfrac{405}{16}$
2	80

Then apply Simpson's Rule with $n = 4$ and $h = 1/2$:

$$S = \frac{h}{3}\left(y_0 + 4y_1 + 2y_2 + 4y_3 + y_4\right)$$

$$= \frac{1}{6}\left(0 + 4\left(\frac{5}{16}\right) + 2(5) + 4\left(\frac{405}{16}\right) + 80\right)$$

$$= \frac{385}{12}.$$

This estimate differs from the exact value (32) by only 1/12, a percentage error of less than three-tenths of one percent—and this was with just 4 subintervals.

There are still other algorithms for approximating definite integrals, most of them involving fancy numerical analysis designed to make the calculations more efficient for high-speed computers. Some are kept secret by the companies that design the machines. In any case, we will not deal with them here.

Error Analysis

After finding that the trapezoidal approximation in Example 1 overestimated the integral, we pointed out that this could have been predicted from the concavity of the curve we were approximating.

Knowing something about the error in an approximation is more than just an interesting sidelight. Despite what your years of classroom experience might have suggested, exact answers are not always easy to find in mathematics. It is fortunate that for all *practical* purposes exact answers are also rarely necessary. (For example, a carpenter who computes the need for a board of length $\sqrt{34}$ feet will happily settle for an approximation when cutting the board.)

Suppose that an exact answer really can *not* be found, but that we know that an approximation within 0.001 unit is good enough. How can we tell that our approximation is within 0.001 if we do not know the exact answer? This is where knowing something about the error is critical.

Since the Trapezoidal Rule approximates curves with straight lines, it seems reasonable that the error depends on how "curvy" the graph is. This suggests that the error depends on the second derivative. It is also apparent that the error depends on the length h of the subintervals. It can be shown that if f'' is continuous the error in the trapezoidal approximation, denoted E_T, satisfies the inequality

$$|E_T| \le \frac{b-a}{12}\, h^2 M_{f''},$$

where $[a, b]$ is the interval of integration, h is the length of each subinterval, and $M_{f''}$ is the maximum value of $|f''|$ on $[a, b]$.

It can also be shown that the error E_S in Simpson's Rule depends on h and the *fourth* derivative. It satisfies the inequality

$$|E_S| \le \frac{b-a}{180}\, h^4 M_{f^{(4)}},$$

where $[a, b]$ is the interval of integration, h is the length of each subinterval, and $M_{f^{(4)}}$ is the maximum value of $|f^{(4)}|$ on $[a, b]$, provided that $f^{(4)}$ is continuous.

For comparison's sake, if all the assumptions hold, we have the following *error bounds*.

Error Bounds

If T and S represent the approximations to $\int_a^b f(x)\,dx$ given by the Trapezoidal Rule and Simpson's Rule, respectively, then the errors E_T and E_S satisfy

$$|E_T| \le \frac{b-a}{12}\,h^2 M_{f''} \quad \text{and} \quad |E_S| \le \frac{b-a}{180}\,h^4 M_{f^{(4)}}.$$

If we disregard possible differences in magnitude between $M_{f''}$ and $M_{f^{(4)}}$, we notice immediately that $(b-a)/180$ is one-fifteenth the size of $(b-a)/12$, giving S an obvious advantage over T as an approximation. That, however, is almost insignificant when compared to the fact that the trapezoid error varies as the *square* of h, while Simpson's error varies as the *fourth power* of h. (Remember that h is already a small number in most partitions.)

Table 5.6 shows T and S values for approximations of $\int_1^2 1/x\,dx$ using various values of n. Notice how Simpson's Rule dramatically improves over the Trapezoidal Rule. In particular, notice that when we double the value of n (thereby halving the value of h), the T error is divided by 2 *squared*, while the S error is divided by 2 *to the fourth*.

Table 5.6 Trapezoidal Rule Approximations (T_n) and Simpson's Rule Approximations (S_n) of $\ln 2 = \int_1^2 (1/x)\,dx$

n	T_n	|Error| less than …	S_n	|Error| less than …
10	.6937714032	.0006242227	.6931502307	.0000030502
20	.6933033818	.0001562013	.6931473747	.0000001942
30	.6932166154	.0000694349	.6931472190	.0000000385
40	.6931862400	.0000390595	.6931471927	.0000000122
50	.6931721793	.0000249988	.6931471856	.0000000050
100	.6931534305	.0000062500	.6931471809	.0000000004

This has a dramatic effect as h gets very small. The Simpson approximation for $n = 50$ rounds accurately to seven places, and for $n = 100$ agrees to nine decimal places (billionths)!

We close by showing you the values (Table 5.7) we found for $\int_1^5 (\sin x)/x\,dx$ by six different calculator methods. The exact value of this integral to six decimal places is 0.603848, so both Simpson's method with 50 subintervals and NINT give results accurate to at least six places (millionths).

Table 5.7 Approximations of $\int_1^5 (\sin x)/x\ dx$

Method	Subintervals	Value
LRAM	50	0.6453898
RRAM	50	0.5627293
MRAM	50	0.6037425
TRAP	50	0.6040595
SIMP	50	0.6038481
NINT	Tol = 0.00001	0.6038482

Quick Review 5.5

In Exercises 1–10, tell whether the curve is concave up or concave down on the given interval.

1. $y = \cos x$ on $[-1, 1]$

2. $y = x^4 - 12x - 5$ on $[8, 17]$

3. $y = 4x^3 - 3x^2 + 6$ on $[-8, 0]$

4. $y = \sin (x/2)$ on $[48\pi, 50\pi]$

5. $y = e^{2x}$ on $[-5, 5]$

6. $y = \ln x$ on $[100, 200]$

7. $y = \dfrac{1}{x}$ on $[3, 6]$

8. $y = \csc x$ on $[0, \pi]$

9. $y = 10^{10} - 10x^{10}$ on $[10, 10^{10}]$

10. $y = \sin x - \cos x$ on $[1, 2]$

Section 5.5 Exercises

In Exercises 1–6, **(a)** use the Trapezoidal Rule with $n = 4$ to approximate the value of the integral. **(b)** Use the concavity of the function to predict whether the approximation is an overestimate or an underestimate. Finally, **(c)** find the integral's exact value to check your answer.

1. $\displaystyle\int_0^2 x\, dx$

2. $\displaystyle\int_0^2 x^2\, dx$

3. $\displaystyle\int_0^2 x^3\, dx$

4. $\displaystyle\int_1^2 \frac{1}{x}\, dx$

5. $\displaystyle\int_0^4 \sqrt{x}\, dx$

6. $\displaystyle\int_0^\pi \sin x\, dx$

7. *Volume of Water in a Swimming Pool* A rectangular swimming pool is 30 ft wide and 50 ft long. The table below shows the depth $h(x)$ of the water at 5-ft intervals from one end of the pool to the other. Estimate the volume of water in the pool using the Trapezoidal Rule with $n = 10$, applied to the integral

$$V = \int_0^{50} 30 \cdot h(x)\, dx.$$

Position (ft) x	Depth (ft) $h(x)$	Position (ft) x	Depth (ft) $h(x)$
0	6.0	30	11.5
5	8.2	35	11.9
10	9.1	40	12.3
15	9.9	45	12.7
20	10.5	50	13.0
25	11.0		

8. *Stocking a Fish Pond* As the fish and game warden of your township, you are responsible for stocking the town pond with fish before the fishing season. The average depth of the pond is 20 feet. Using a scaled map, you measure distances across the pond at 200-foot intervals, as shown in the diagram.

(a) Use the Trapezoidal Rule to estimate the volume of the pond.

(b) You plan to start the season with one fish per 1000 cubic feet. You intend to have at least 25% of the opening day's fish population left at the end of the season. What is the maximum number of licenses the town can sell if the average seasonal catch is 20 fish per license?

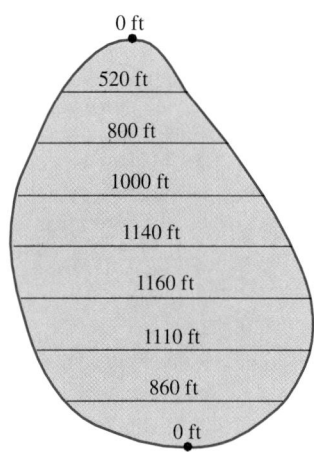

0 ft

520 ft

800 ft

1000 ft

1140 ft

1160 ft

1110 ft

860 ft

0 ft

Vertical spacing = 200 ft

9. *Ford® Mustang Cobra™* The accompanying table shows time-to-speed data for a 1994 Ford Mustang Cobra accelerating from rest to 130 mph. How far had the Mustang traveled by the time it reached this speed? (Use trapezoids to estimate the area under the velocity curve, but be careful: the time intervals vary in length.)

Speed Change	Time (sec)
Zero to 30 mph	2.2
40 mph	3.2
50 mph	4.5
60 mph	5.9
70 mph	7.8
80 mph	10.2
90 mph	12.7
100 mph	16.0
110 mph	20.6
120 mph	26.2
130 mph	37.1

Source: Car and Driver, April 1994.

10. Consider the integral $\int_{-1}^{3} (x^3 - 2x) \, dx$.

(a) Use Simpson's Rule with $n = 4$ to approximate its value.

(b) Find the exact value of the integral. What is the error, $|E_S|$?

(c) Explain how you could have predicted what you found in (b) from knowing the error-bound formula.

(d) **Writing to Learn** Is it possible to make a general statement about using Simpson's Rule to approximate integrals of cubic polynomials? Explain.

11. Writing to Learn In Example 2 (before rounding) we found the average temperature to be 65.17 degrees when we used the integral approximation, yet the average of the 13 discrete temperatures is only 64.69 degrees. Considering the shape of the temperature curve, explain why you would expect the average of the 13 discrete temperatures to be less than the average value of the temperature function on the entire interval.

12. *(Continuation of Exercise 11)*

(a) In the Trapezoidal Rule, every function value in the sum is doubled except for the two endpoint values. Show that if you double the endpoint values, you get 70.08 for the average temperature.

(b) Explain why it makes more sense to not double the endpoint values if we are interested in the average temperature over the entire 12-hour period.

In Exercises 13–16, use a calculator program to find the Simpson's Rule approximations with $n = 50$ and $n = 100$.

13. $\int_{-1}^{1} 2\sqrt{1 - x^2} \, dx$ The exact value is π.

14. $\int_{0}^{1} \sqrt{1 + x^4} \, dx$ An integral that came up in Newton's research

15. $\int_{0}^{\pi/2} \frac{\sin x}{x} \, dx$

16. $\int_{0}^{\pi/2} \sin (x^2) \, dx$ An integral associated with the diffraction of light

17. Consider the integral $\int_{0}^{\pi} \sin x \, dx$.

(a) Use a calculator program to find the Trapezoidal Rule approximations for $n = 10$, 100, and 1000.

(b) Record the errors with as many decimal places of accuracy as you can.

(c) What pattern do you see?

(d) **Writing to Learn** Explain how the error bound for E_T accounts for the pattern.

18. *(Continuation of Exercise 17)* Repeat Exercise 17 with Simpson's Rule and E_S.

Explorations

In Exercises 19 and 20, *work in groups of two or three.*

19. Consider the integral $\int_{-1}^{1} \sin (x^2) \, dx$.

(a) Find f'' for $f(x) = \sin (x^2)$.

(b) Graph $y = f''(x)$ in the viewing window $[-1, 1]$ by $[-3, 3]$.

(c) Explain why the graph in (b) suggests that $|f''(x)| \le 3$ for $-1 \le x \le 1$.

(d) Show that the error estimate for the Trapezoidal Rule in this case becomes

$$|E_T| \le \frac{h^2}{2}.$$

(e) Show that the Trapezoidal Rule error will be less than or equal to 0.01 if $h \le 0.1$.

(f) How large must n be for $h \le 0.1$?

20. Consider the integral $\int_{-1}^{1} \sin (x^2) \, dx$.

(a) Find $f^{(4)}$ for $f(x) = \sin (x^2)$. (You may want to check your work with a CAS if you have one available.)

(b) Graph $y = f^{(4)}(x)$ in the viewing window $[-1, 1]$ by $[-30, 10]$.

(c) Explain why the graph in (b) suggests that $|f^{(4)}(x)| \le 30$ for $-1 \le x \le 1$.

(d) Show that the error estimate for Simpson's Rule in this case becomes

$$|E_S| \le \frac{h^4}{3}.$$

(e) Show that the Simpson's Rule error will be less than or equal to 0.01 if $h \le 0.4$.

(f) How large must n be for $h \le 0.4$?

21. *Aerodynamic Drag* A vehicle's aerodynamic drag is determined in part by its cross section area, so, all other things being equal, engineers try to make this area as small as possible. Use Simpson's Rule to estimate the cross section area of the body of James Worden's solar-powered Solectria® automobile at M.I.T. from the diagram below.

22. *Wing Design* The design of a new airplane requires a gasoline tank of constant cross section area in each wing. A scale drawing of a cross section is shown here. The tank must hold 5000 lb of gasoline, which has a density of 42 lb/ft^3. Estimate the length of the tank.

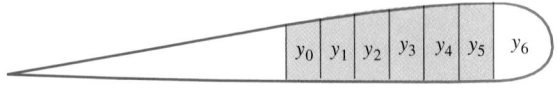

$y_0 = 1.5$ ft, $y_1 = 1.6$ ft, $y_2 = 1.8$ ft, $y_3 = 1.9$ ft, $y_4 = 2.0$ ft, $y_5 = y_6 = 2.1$ ft   Horizontal spacing = 1 ft

Extending the Ideas

23. Using the definitions, prove that, in general,

$$T_n = \frac{\text{LRAM}_n + \text{RRAM}_n}{2}.$$

24. Using the definitions, prove that, in general,

$$S_{2n} = \frac{\text{MRAM}_n + 2T_{2n}}{3}.$$

Chapter 5   Key Terms

Chapter 5 Review Exercises

Exercises 1–6 refer to the region R in the first quadrant enclosed by the x-axis and the graph of the function $y = 4x - x^3$.

1. Sketch R and partition it into four subregions, each with a base of length $\Delta x = 1/2$.

2. Sketch the rectangles and compute (by hand) the area for the LRAM$_4$ approximation.

3. Sketch the rectangles and compute (by hand) the area for the MRAM$_4$ approximation.

4. Sketch the rectangles and compute (by hand) the area for the RRAM$_4$ approximation.

5. Sketch the trapezoids and compute (by hand) the area for the T_4 approximation.

6. Find the exact area of R by using the Fundamental Theorem of Calculus.

7. Use a calculator program to compute the RAM approximations in the following table for the area under the graph of $y = 1/x$ from $x = 1$ to $x = 5$.

n	LRAM$_n$	MRAM$_n$	RRAM$_n$
10			
20			
30			
50			
100			
1000			

8. *(Continuation of Exercise 7)* Use the Fundamental Theorem of Calculus to determine the value to which the sums in the table are converging.

9. Suppose

$$\int_{-2}^{2} f(x)\, dx = 4, \quad \int_{2}^{5} f(x)\, dx = 3, \quad \int_{-2}^{5} g(x)\, dx = 2.$$

Which of the following statements are true, and which, if any, are false?

(a) $\int_{5}^{2} f(x)\, dx = -3$ (b) $\int_{-2}^{5} [f(x) + g(x)]\, dx = 9$

(c) $f(x) \le g(x)$ on the interval $-2 \le x \le 5$

10. The region under one arch of the curve $y = \sin x$ is revolved around the x-axis to form a solid. (a) Use the method of Example 3, Section 5.1, to set up a Riemann sum that approximates the volume of the solid. (b) Find the volume using NINT.

11. The accompanying graph shows the velocity (m/sec) of a body moving along the s-axis during the time interval from $t = 0$ to $t = 10$ sec. (a) About how far did the body travel during those 10 seconds?

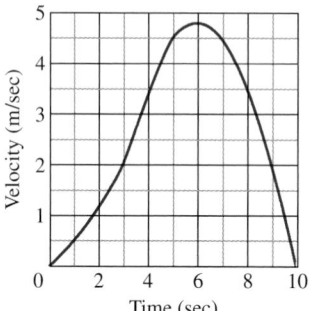

(b) Sketch a graph of position (s) as a function of time (t) for $0 \le t \le 10$, assuming $s(0) = 0$.

12. The interval $[0, 10]$ is partitioned into n subintervals of length $\Delta x = 10/n$. We form the following Riemann sums, choosing each c_k in the kth subinterval. Write the limit as $n \to \infty$ of each Riemann sum as a definite integral.

(a) $\sum_{k=1}^{n} (c_k)^3 \Delta x$ (b) $\sum_{k=1}^{n} c_k (\sin c_k) \Delta x$

(c) $\sum_{k=1}^{n} c_k (3c_k - 2)^2 \Delta x$ (d) $\sum_{k=1}^{n} (1 + c_k^2)^{-1} \Delta x$

(e) $\sum_{k=1}^{n} \pi (9 - \sin^2 (\pi c_k / 10)) \Delta x$

In Exercises 13 and 14, find the total area between the curve and the x-axis.

13. $y = 4 - x, \quad 0 \le x \le 6$ 14. $y = \cos x, \quad 0 \le x \le \pi$

In Exercises 15–24, evaluate the integral analytically by using the Integral Evaluation Theorem (Part 2 of the Fundamental Theorem, Theorem 4).

15. $\int_{-2}^{2} 5\, dx$ 16. $\int_{2}^{5} 4x\, dx$

17. $\int_{0}^{\pi/4} \cos x\, dx$ 18. $\int_{-1}^{1} (3x^2 - 4x + 7)\, dx$

19. $\int_{0}^{1} (8s^3 - 12s^2 + 5)\, ds$ 20. $\int_{1}^{2} \frac{4}{x^2}\, dx$

21. $\int_{1}^{27} y^{-4/3}\, dy$ 22. $\int_{1}^{4} \frac{dt}{t\sqrt{t}}$

23. $\int_{0}^{\pi/3} \sec^2 \theta\, d\theta$ 24. $\int_{1}^{e} (1/x)\, dx$

In Exercises 25–29, evaluate the integral.

25. $\int_0^1 \dfrac{36}{(2x+1)^3}\,dx$

26. $\int_1^2 \left(x + \dfrac{1}{x^2}\right) dx$

27. $\int_{-\pi/3}^0 \sec x \tan x\,dx$

28. $\int_{-1}^1 2x \sin(1 - x^2)\,dx$

29. $\int_0^2 \dfrac{2}{y+1}\,dy$

In Exercises 30–32, evaluate the integral by interpreting it as area and using formulas from geometry.

30. $\int_0^2 \sqrt{4 - x^2}\,dx$

31. $\int_{-4}^8 |x|\,dx$

32. $\int_{-8}^8 2\sqrt{64 - x^2}\,dx$

33. *Oil Consumption on Pathfinder Island* A diesel generator runs continuously, consuming oil at a gradually increasing rate until it must be temporarily shut down to have the filters replaced.

Day	Oil Consumption Rate (liters/hour)
Sun	0.019
Mon	0.020
Tue	0.021
Wed	0.023
Thu	0.025
Fri	0.028
Sat	0.031
Sun	0.035

(a) Give an upper estimate and a lower estimate for the amount of oil consumed by the generator during that week.

(b) Use the Trapezoidal Rule to estimate the amount of oil consumed by the generator during that week.

34. *Rubber-Band–Powered Sled* A sled powered by a wound rubber band moves along a track until friction and the unwinding of the rubber band gradually slow it to a stop. A speedometer in the sled monitors its speed, which is recorded at 3-second intervals during the 27-second run.

Time (sec)	Speed (ft/sec)
0	5.30
3	5.25
6	5.04
9	4.71
12	4.25
15	3.66
18	2.94
21	2.09
24	1.11
27	0

(a) Give an upper estimate and a lower estimate for the distance traveled by the sled.

(b) Use the Trapezoidal Rule to estimate the distance traveled by the sled.

35. **Writing to Learn** Your friend knows how to compute integrals but never could understand what difference the "*dx*" makes, claiming that it is irrelevant. How would you explain to your friend why it is necessary?

36. The function
$$f(x) = \begin{cases} x^2, & x \ge 0 \\ x - 2, & x < 0 \end{cases}$$
is discontinuous at 0, but integrable on $[-4, 4]$. Find $\int_{-4}^4 f(x)\,dx$.

37. Show that $0 \le \int_0^1 \sqrt{1 + \sin^2 x}\,dx \le \sqrt{2}$.

38. Find the average value of
(a) $y = \sqrt{x}$ over the interval $[0, 4]$.
(b) $y = a\sqrt{x}$ over the interval $[0, a]$.

In Exercises 39–42, find dy/dx.

39. $y = \int_2^x \sqrt{2 + \cos^3 t}\,dt$

40. $y = \int_2^{7x^2} \sqrt{2 + \cos^3 t}\,dt$

41. $y = \int_x^1 \dfrac{6}{3 + t^4}\,dt$

42. $y = \int_x^{2x} \dfrac{1}{t^2 + 1}\,dt$

43. *Printing Costs* Including start-up costs, it costs a printer $50 to print 25 copies of a newsletter, after which the marginal cost at x copies is
$$\dfrac{dc}{dx} = \dfrac{2}{\sqrt{x}} \text{ dollars per copy.}$$
Find the total cost of printing 2500 newsletters.

44. *Average Daily Inventory* Rich Wholesale Foods, a manufacturer of cookies, stores its cases of cookies in an air-conditioned warehouse for shipment every 14 days. Rich tries to keep 600 cases on reserve to meet occasional peaks in demand, so a typical 14-day inventory function is $I(t) = 600 + 600t$, $0 \le t \le 14$. The holding cost for each case is 4¢ per day. Find Rich's average daily inventory and average daily holding cost.

45. Solve for x: $\int_0^x (t^3 - 2t + 3)\,dt = 4$.

46. Suppose $f(x)$ has a positive derivative for all values of x and that $f(1) = 0$. Which of the following statements must be true of
$$g(x) = \int_0^x f(t)\,dt?$$

(a) g is a differentiable function of x.
(b) g is a continuous function of x.
(c) The graph of g has a horizontal tangent line at $x = 1$.
(d) g has a local maximum at $x = 1$.
(e) g has a local minimum at $x = 1$.
(f) The graph of g has an inflection point at $x = 1$.
(g) The graph of dg/dx crosses the x-axis at $x = 1$.

47. Suppose $F(x)$ is an antiderivative of $f(x) = \sqrt{1 + x^4}$. Express $\int_0^1 \sqrt{1 + x^4}\, dx$ in terms of F.

48. Express the function $y(x)$ with

$$\frac{dy}{dx} = \frac{\sin x}{x} \quad \text{and} \quad y(5) = 3$$

as a definite integral.

49. Show that $y = x^2 + \int_1^x 1/t\, dt + 1$ satisfies both of the following conditions:

i. $y'' = 2 - \dfrac{1}{x^2}$

ii. $y = 2$ and $y' = 3$ when $x = 1$.

50. Writing to Learn Which of the following is the graph of the function whose derivative is $dy/dx = 2x$ and whose value at $x = 1$ is 4? Explain your answer.

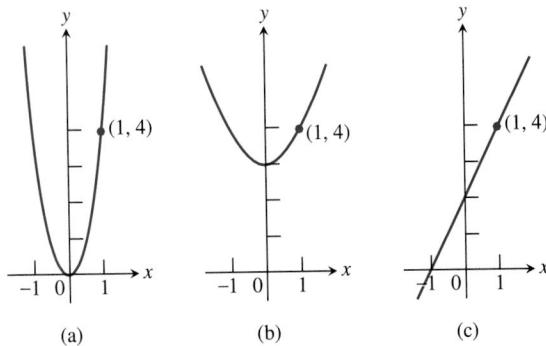

(a)   (b)   (c)

51. *Fuel Efficiency* An automobile computer gives a digital readout of fuel consumption in gallons per hour. During a trip, a passenger recorded the fuel consumption every 5 minutes for a full hour of travel.

time	gal/h	time	gal/h
0	2.5	35	2.5
5	2.4	40	2.4
10	2.3	45	2.3
15	2.4	50	2.4
20	2.4	55	2.4
25	2.5	60	2.3
30	2.6		

(a) Use the Trapezoidal Rule to approximate the total fuel consumption during the hour.

(b) If the automobile covered 60 miles in the hour, what was its fuel efficiency (in miles per gallon) for that portion of the trip?

52. *Skydiving* Skydivers A and B are in a helicopter hovering at 6400 feet. Skydiver A jumps and descends for 4 sec before opening her parachute. The helicopter then climbs to 7000 feet and hovers there. Forty-five seconds after A leaves the aircraft, B jumps and descends for 13 sec before opening her parachute. Both skydivers descend at 16 ft/sec with parachutes open. Assume that the skydivers fall freely (with acceleration -32 ft/sec^2) before their parachutes open.

(a) At what altitude does A's parachute open?

(b) At what altitude does B's parachute open?

(c) Which skydiver lands first?

53. *Relating Simpson's Rule, MRAM, and T* The figure below shows an interval of length $2h$ with a trapezoid, a midpoint rectangle, and a parabolic region on it.

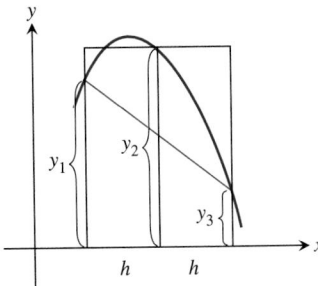

(a) Show that the area of the trapezoid plus twice the area of the rectangle equals

$$h(y_1 + 4y_2 + y_3).$$

(b) Use the result in (a) to prove that

$$S_{2n} = \frac{2 \cdot \text{MRAM}_n + T_n}{3}.$$

54. The graph of a function f consists of a semicircle and two line segments as shown below.

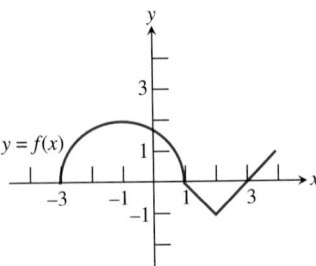

Let $g(x) = \int_1^x f(t)\, dt$.

(a) Find $g(1)$.   **(b)** Find $g(3)$.   **(c)** Find $g(-1)$.

(d) Find all values of x on the open interval $(-3, 4)$ at which g has a relative maximum.

(e) Write an equation for the line tangent to the graph of g at $x = -1$.

(f) Find the x-coordinate of each point of inflection of the graph of g on the open interval $(-3, 4)$.

(g) Find the range of g.

55. What is the total area under the curve $y = e^{-x^2/2}$?

The graph approaches the x-axis as an asymptote both to the left and the right, but quickly enough so that the total area is a finite number. In fact,

$$\text{NINT } (e^{-x^2/2}, x, -10, 10)$$

computes all but a negligible amount of the area.

(a) Find this number on your calculator. Verify that NINT $(e^{-x^2/2}, x, -20, 20)$ does not increase the number enough for the calculator to distinguish the difference.

(b) This area has an interesting relationship to π. Perform various (simple) algebraic operations on the number to discover what it is.

56. *Filling a Swamp* A town wants to drain and fill the small polluted swamp shown below. The swamp averages 5 ft deep. About how many cubic yards of dirt will it take to fill the area after the swamp is drained?

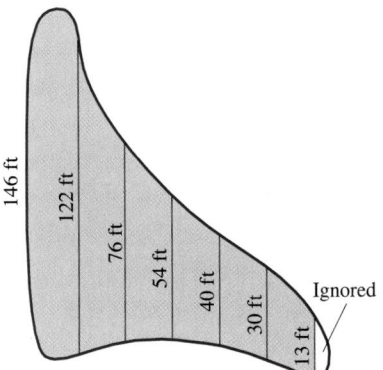

Horizontal spacing = 20 ft

57. *Household Electricity* We model the voltage V in our homes with the sine function

$$V = V_{\max} \sin (120 \pi t),$$

which expresses V in volts as a function of time t in seconds. The function runs through 60 cycles each second. The number $V_{\max}$ is the *peak voltage*.

To measure the voltage effectively, we use an instrument that measures the square root of the average value of the square of the voltage over a 1-second interval:

$$V_{\text{rms}} = \sqrt{(V^2)_{\text{av}}}.$$

The subscript "rms" stands for "root mean square." It turns out that

$$V_{\text{rms}} = \frac{V_{\max}}{\sqrt{2}}. \tag{1}$$

The familiar phrase "115 volts ac" means that the rms voltage is 115. The peak voltage, obtained from Equation 1 as $V_{\max} = 115\sqrt{2}$, is about 163 volts.

(a) Find the average value of V^2 over a 1-sec interval. Then find V_{rms}, and verify Equation 1.

(b) The circuit that runs your electric stove is rated 240 volts rms. What is the peak value of the allowable voltage?

Calculus at Work

I have a degree in Mechanical Engineering with a minor in Psychology. I am a Research and Development Engineer at Komag, which designs and manufactures hard disks in Santa Clara, California. My job is to test the durability and reliability of the disks, measuring the rest friction between the read/write heads and the disk surface, which is called "Contact-Start-Stop" testing.

I use calculus to evaluate the moment of inertia of different disk stacks, which consist of disks on a spindle, separated by spacer rings. Because the rings vary in size as well as material, the mass of each

ring must be determined. For such problems, I refer to my college calculus textbook and its tables of summations and integrals. For instance, I use:

Moment of Inertia =

$$\left[\sum_{i=1}^{n} M_i L_i^2 \right] + \frac{1}{3} M_{rod} L_{rod}^2$$

where i = components 1 to n;
M_i = mass of component i such as the disk and/or ring stack;
L_i = distance of component i from a reference point;
M_{rod} = mass of the spindle that rotates; L_{rod} = length of the spindle.

Andrea Woo
Komag
Santa Clara, CA

CHAPTER

6
Differential Equations and Mathematical Modeling

One way to measure how light in the ocean diminishes as water depth increases involves using a Secchi disk. This white disk is 30 centimeters in diameter, and is lowered into the ocean until it disappears from view. The depth of this point (in meters), divided into 1.7, yields the coefficient k used in the equation $I_x = I_0 e^{-kx}$. This equation estimates the intensity I_x of light at depth x using I_0, the intensity of light at the surface.

In an ocean experiment, if the Secchi disk disappears at 55 meters, at what depth will only 1% of surface radiation remain? Section 6.4 will help you answer this question.

Chapter 6 Overview

One of the early accomplishments of calculus was predicting the future position of a planet from its present position and velocity. Today this is just one of a number of occasions on which we deduce everything we need to know about a function from one of its known values and its rate of change. From this kind of information, we can tell how long a sample of radioactive polonium will last; whether, given current trends, a population will grow or become extinct; and how large major league baseball salaries are likely to be in the year 2010. In this chapter, we examine the analytic, graphical, and numerical techniques on which such predictions are based.

6.1 Antiderivatives and Slope Fields

Solving Initial Value Problems • Antiderivatives and Indefinite Integrals • Properties of Indefinite Integrals • Applications

Solving Initial Value Problems

An equation like

$$\frac{dy}{dx} = y \ln x$$

containing a derivative is a **differential equation.** The problem of finding a function y of x when we are given its derivative and its value at a particular point is called an **initial value problem.** The value of f for one value of x is the **initial condition** of the problem. When we find all the functions y that satisfy the differential equation we have **solved the differential equation.** When we then find the **particular solution** that fulfills the initial condition, we have **solved the initial value problem.**

Order

The **order** of a differential equation is the order of the highest derivative involved in the equation.

Table 6.1 Initial Value Problems from Earlier Chapters

Original Problem…	… as an Initial Value Problem
(a) Find the function $y = f(x)$ whose derivative is sin x and whose graph passes through the point $(0, 2)$. (Section 4.2, Example 6)	Solve the initial value problem: Differential equation: $\frac{dy}{dx} = \sin x,$ Initial condition: $y(0) = 2.$
(b) Find the function $y = f(x)$ with derivative $dy/dx = \tan x$ satisfying the condition $f(3) = 5$. (Section 5.4, Example 4)	Solve the initial value problem: Differential equation: $\frac{dy}{dx} = \tan x,$ Initial condition: $y(3) = 5.$

Suppose that y_0 dollars are invested in an account at an interest rate of 5.6% per year, compounded *continuously.* This means that at any given instant the amount y in the account is increasing at a rate that is 5.6% of the current balance. If t is the time in years since the account was opened, we can model the situation with the following initial value problem:

Differential equation: $\dfrac{dy}{dt} = 0.056y,$ At any instant the rate of change of y is 5.6% of y.

Initial condition: $y(0) = y_0.$ There are y_0 dollars in the account when $t = 0$.

Example 1   COMPOUNDING INTEREST CONTINUOUSLY

Suppose \$100 is invested in an account that pays 5.6% interest compounded continuously. Find a formula for the amount in the account at any time t.

Solution   We let $t = 0$ when the initial \$100 is deposited in the account.

Model

We can model this problem with the initial value problem

$$\frac{dy}{dt} = 0.056y \quad \text{and} \quad y(0) = 100.$$

Solve Analytically

We are looking for a function whose derivative is a constant multiple of itself. We know that exponential functions have this property. In fact, the function $y(t) = Ce^{0.056t}$ is a solution of the differential equation for any real number C because

$$\frac{dy}{dt} = C(0.056 \cdot e^{0.056t}) = 0.056(Ce^{0.056t}) = 0.056y(t).$$

In Section 6.4 we will see that there are no other solutions.

Applying the initial condition we find

$$y(t) = Ce^{0.056t}$$
$$y(0) = Ce^{(0.056)(0)}$$
$$100 = C.$$

Interpret

The amount in the account at any time t is $y(t) = 100e^{0.056t}$. At the end of the first year, for instance, the balance will be $y(1) = 100e^{0.056(1)} = \$105.76$.

Each member of the family of functions $y(t) = Ce^{0.056t}$ is a solution of the equation

$$\frac{dy}{dt} = 0.056y. \tag{1}$$

We can use our grapher to obtain a reasonably accurate visualization of these solutions by using a slope field program (see *Technology Resource Manual*) that plots short line segments of slope $0.056y$ at selected points (x, y) in a viewing window (Figure 6.1a). In the window of Figure 6.1a, line segments are constructed at 100 points in the plane. The line segments are arranged in 10 columns and 10 rows. (See Exploration 1 for more details.) This is equivalent to plotting short pieces of the linearizations of the solution curves that pass through these points. The plot is called a *slope field* or *direction field* for the differential equation. Figure 6.1b shows the graphs of four solutions of Equation 1.

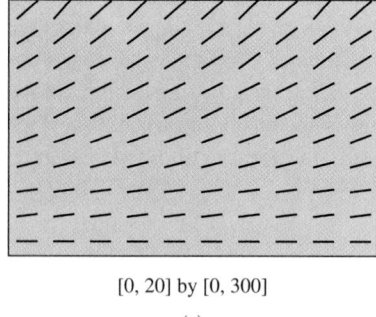

[0, 20] by [0, 300]

(a)

[0, 20] by [0, 300]

(b)

Figure 6.1 Notice how the slope field (a) gives a visualization of a solution curve (b) by indicating the direction of the curve as we proceed from left to right across the viewing window.

$y = 100e^{0.056x}$

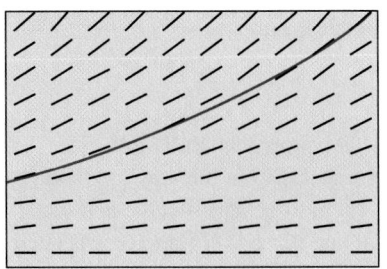

[0, 20] by [0, 300]

Figure 6.2 The solution of the initial value problem in Example 1.

The solution $y(t) = 100e^{0.056t}$ in Example 1 is the one passing through the point $(0, 100)$ (Figure 6.2).

Definition Slope Field or Direction Field

A **slope field** or **direction field** for the first order differential equation

$$\frac{dy}{dx} = f(x, y)$$

is a plot of short line segments with slopes $f(x, y)$ for a lattice of points (x, y) in the plane.

Here $f(x, y)$ is a function of the two variables x and y, for example, $f(x, y) = -2xy/(1 + x^2)$. Slope fields enable us to graph solution curves without solving the differential equation analytically (handy if it is an equation we cannot solve). This is especially useful when the formula given for dy/dx involves both x and y, as in the formulas we get from implicit differentiation.

Visual Support from a Slope Field

A slope field can be used to check that we have solved an initial value problem correctly.

Exploration 1 Constructing a Slope Field

1. Let $h = (\text{Xmax} - \text{Xmin})/10$ and $k = (\text{Ymax} - \text{Ymin})/10$. Explain why

$$\left(\text{Xmin} + (2i - 1)\frac{h}{2}, \text{Ymin} + (2j - 1)\frac{k}{2} \right)$$

for $i, j = 1, 2, \dots, 10,$ represents 100 points in the viewing window $[\text{Xmin}, \text{Xmax}]$ by $[\text{Ymin}, \text{Ymax}]$.

2. Show that the horizontal distance between adjacent points with the same y-coordinate in part 1 is h.

3. Show that the vertical distance between adjacent points with the same x-coordinate in part 1 is k.

4. Consider the differential equation $dy/dx = 2/(x + 1)$ and the viewing window $[0, 10]$ by $[0, 10]$. Find the slope (given by the differential equation) and construct a short line segment at each point in the third column ($i = 3$) of the points given by part 1. What do the line segments have in common?

5. Use the same differential equation and viewing window as in part 4. Find the slope and construct a short line segment at each point in the fourth row ($j = 4$) of the points given by part 1.

6. Explain how you could complete the slope field using the information in parts 4 and 5.

7. Let L and W be positive integers, $h = (\text{Xmax} - \text{Xmin})/W,$ and $k = (\text{Ymax} - \text{Ymin})/L$. If $i = 1, 2, \dots, W$ and $j = 1, 2, \dots, L$, how many points are produced in part 1?

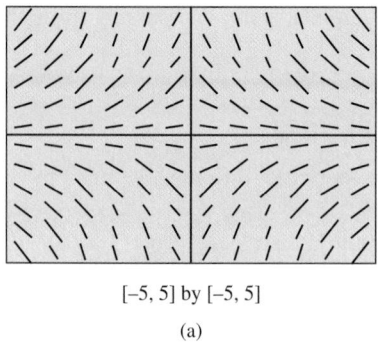

[–5, 5] by [–5, 5]

(a)

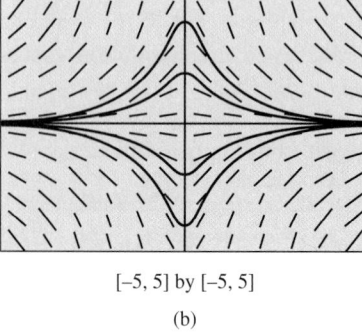

[–5, 5] by [–5, 5]

(b)

Figure 6.3 Notice that the slope $y' = -2xy/(1 + x^2)$ is zero when $x = 0$. (Example 2)

Example 2 PORTRAYING SOLUTIONS WITH A SLOPE FIELD

Plot the solution curves of the differential equation

$$y' = -\frac{2xy}{1 + x^2}.$$

Solution With our grapher in *differential equation graphing mode,* we generate the slope field (Figure 6.3a). We can see from the way the field "flows" how the solution curves will go. Then, to be specific, we add the solutions through the points $(0, 0)$, $(0, \pm2)$ and $(0, \pm4)$ (Figure 6.3b). These are the solutions satisfying the initial conditions $y(0) = 0$, $y(0) = \pm2$, and $y(0) = \pm4$.

We can tell from the concavity of the curves where the second derivatives of the solutions are positive, where they are negative, and approximately where they are zero, observations that can be significant in an application. We can also see that the solutions approach 0 as $x \to \pm\infty$. And we learn all this without touching the equation except to enter it into the machine.

Antiderivatives and Indefinite Integrals

Recall that a function $F(x)$ is an antiderivative of a function $f(x)$ if $F'(x) = f(x)$.

Definition Indefinite Integral

The set of all antiderivatives of a function $f(x)$ is the **indefinite integral** of f with respect to x and is denoted by

$$\int f(x)\, dx.$$

As in Chapter 5, the symbol $\int$ is an **integral sign,** the function f is the **integrand** of the integral, and x is the **variable of integration.**

According to Corollary 3 of the Mean Value Theorem for Derivatives in Section 4.2, once we have found one antiderivative F of a function f, all the other antiderivatives differ from F by constants. We indicate this in integral notation in the following way.

$$\int f(x)\, dx = F(x) + C$$

The equation is read, "The indefinite integral of f with respect to x is $F(x) + C$." The constant C is the **constant of integration** and is an **arbitrary constant.** When we find $F(x) + C$ we have **integrated** f or **evaluated** the integral.

Example 3 FINDING INDEFINITE INTEGRALS ANALYTICALLY

Evaluate $\int 2x \, dx$.

Solution

$$\int 2x \, dx = x^2 + C$$

an antiderivative of $2x$

the arbitrary constant

The formula $x^2 + C$ generates all the antiderivatives of the function $2x$. The functions $x^2 + 1$, $x^2 - \pi$, and $x^2 + \sqrt{2}$ are all antiderivatives of $2x$, as you can check by differentiation.

Many indefinite integrals needed in scientific work are found by reversing derivative formulas. You will see what we mean if you look at Table 6.2, which lists standard integral forms side by side with their derivative-formula sources.

Table 6.2 Integral Formulas

Indefinite Integral	Reversed Derivative Formula
1. (a) $\displaystyle\int x^n \, dx = \frac{x^{n+1}}{n+1} + C, \; n \neq -1$	$\dfrac{d}{dx}\left(\dfrac{x^{n+1}}{n+1}\right) = x^n, \; n \neq -1$
(b) $\displaystyle\int \frac{dx}{x} = \ln|x| + C$	$\dfrac{d}{dx}\ln|x| = \dfrac{1}{x}$
2. $\displaystyle\int e^{kx} \, dx = \frac{e^{kx}}{k} + C$	$\dfrac{d}{dx}\left(\dfrac{e^{kx}}{k}\right) = e^{kx}$
3. $\displaystyle\int \sin kx \, dx = -\frac{\cos kx}{k} + C$	$\dfrac{d}{dx}\left(-\dfrac{\cos kx}{k}\right) = \sin kx$
4. $\displaystyle\int \cos kx \, dx = \frac{\sin kx}{k} + C$	$\dfrac{d}{dx}\left(\dfrac{\sin kx}{k}\right) = \cos kx$
5. $\displaystyle\int \sec^2 x \, dx = \tan x + C$	$\dfrac{d}{dx}\tan x = \sec^2 x$
6. $\displaystyle\int \csc^2 x \, dx = -\cot x + C$	$\dfrac{d}{dx}(-\cot x) = \csc^2 x$
7. $\displaystyle\int \sec x \tan x \, dx = \sec x + C$	$\dfrac{d}{dx}\sec x = \sec x \tan x$
8. $\displaystyle\int \csc x \cot x \, dx = -\csc x + C$	$\dfrac{d}{dx}(-\csc x) = \csc x \cot x$

Example 4 USING INTEGRAL FORMULAS

(a) $\displaystyle\int x^5\,dx = \frac{x^6}{6} + C$ Formula 1a with $n = 5$

(b) $\displaystyle\int \frac{1}{\sqrt{x}}\,dx = \int x^{-1/2}\,dx = 2x^{1/2} + C = 2\sqrt{x} + C$ Formula 1a with $n = -1/2$

(c) $\displaystyle\int e^{-3x}\,dx = \frac{e^{-3x}}{-3} + C = -\frac{1}{3}e^{-3x} + C$ Formula 2 with $k = -3$

(d) $\displaystyle\int \cos\frac{x}{2}\,dx = \frac{\sin(1/2)x}{1/2} + C = 2\sin\frac{x}{2} + C$ Formula 4 with $k = 1/2$

When we cannot find a ready-made antiderivative like the ones in Table 6.2, we can construct one of our own using Part 1 of the Fundamental Theorem of Calculus from Section 5.4. This is the part that says that if $f(x)$ is continuous for $a \leq x \leq b$ then

$$\frac{d}{dx}\int_a^x f(t)\,dt = f(x)$$

at every point x in $[a, b]$. The integral is the antiderivative we seek.

$y = \text{NINT}\,(x \sin x, x, 0, x)$

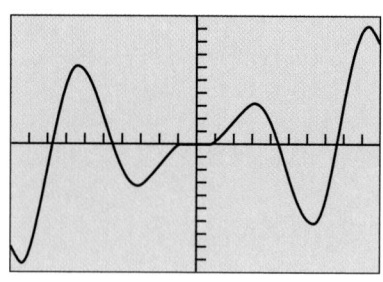

[–10, 10] by [–10, 10]

Figure 6.4 An antiderivative of $x \sin x$. (Example 5)

Example 5 USING THE FUNDAMENTAL THEOREM

Evaluate $\displaystyle\int x \sin x\,dx$.

Solution The function $f(x) = x \sin x$ is continuous for all x, so

$$F(x) = \int_0^x t \sin t\,dt$$

is an antiderivative of f. Therefore

$$\int x \sin x\,dx = F(x) + C = \int_0^x t \sin t\,dt + C.$$

Figure 6.4 shows $F(x) = \int_0^x t \sin t\,dt$ graphed as NINT $(x \sin x, x, 0, x)$. All other antiderivatives have graphs that are vertical shifts of this graph.

Properties of Indefinite Integrals

Here are the standard arithmetic rules for indefinite integration.

Checking Your Work

To confirm an indefinite integration analytically, differentiate the right-hand side. The derivative should be the integrand.

To support an indefinite integration graphically, draw a slope field for the integrand and overlay the graph of the right-hand side for a specific value of C. This graph should follow the direction of the slope field.

Properties of Indefinite Integrals

Let k be a real number.

1. *Constant Multiple Rule:* $\displaystyle\int k f(x)\,dx = k\int f(x)\,dx$

    If $k = -1$, *then:* $\displaystyle\int -f(x)\,dx = -\int f(x)\,dx$

2. *Sum and Difference Rule:* $\displaystyle\int (f(x) \pm g(x))\,dx = \int f(x)\,dx \pm \int g(x)\,dx$

Just as the Sum and Difference Rule for differentiation enables us to differentiate expressions term by term, the Sum and Difference Rule for integration enables us to integrate expressions term by term. When we do so, we combine the individual constants of integration into a single arbitrary constant at the end.

Example 6 INTEGRATING TERM BY TERM

Evaluate $\int (x^2 - 2x + 5)\, dx$.

Solution If we recognize that $(x^3/3) - x^2 + 5x$ is an antiderivative of $x^2 - 2x + 5$, we can evaluate the integral as

$$\int (x^2 - 2x + 5)\, dx = \overbrace{\frac{x^3}{3} - x^2 + 5x}^{\text{antiderivative}} + \underset{\text{arbitrary constant}}{C.}$$

If we do not recognize the antiderivative right away, we can generate it term by term with the Sum and Difference Rule.

$$\int (x^2 - 2x + 5)\, dx = \int x^2\, dx - \int 2x\, dx + \int 5\, dx$$

$$= \frac{x^3}{3} + C_1 - x^2 + C_2 + 5x + C_3$$

This formula is more complicated than it needs to be. If we combine C_1, C_2, and C_3 into a single constant $C = C_1 + C_2 + C_3$, the formula simplifies to

$$\frac{x^3}{3} - x^2 + 5x + C$$

and still gives *all* the antiderivatives. For this reason we recommend that you go right to the final form even if you elect to integrate term by term. Write

$$\int (x^2 - 2x + 5)\, dx = \int x^2\, dx - \int 2x\, dx + \int 5\, dx$$

$$= \frac{x^3}{3} - x^2 + 5x + C.$$

Find the simplest antiderivative you can for each part and add the constant at the end.

Applications

Torricelli's Law says that if you drain a tank like the one in Figure 6.5, the rate at which the water runs out is a constant times the square root of the water's depth x. The constant depends on the size of the exit valve. In Example 7 we assume the constant is 1/2.

Example 7 DRAINING A TANK

A right circular cylindrical tank with radius 5 ft and height 16 ft that was initially full of water is being drained at the rate of $0.5\sqrt{x}$ ft³/min. Find a formula for the depth and the amount of water in the tank at any time t. How long will it take the tank to empty?

Figure 6.5 The rate at which water runs out is $0.5\sqrt{x}$ ft³/min. (Example 7)

Solution The volume of a right circular cylinder with radius r and height h is $V = \pi r^2 h$.

Model

The volume of water in the tank (Figure 6.5) is

$$V = \pi r^2 h = \pi(5)^2 x = 25\pi x.$$

Differential equation: $\quad \dfrac{dV}{dt} = -25\pi\dfrac{dx}{dt} \qquad$ Negative because V is decreasing

$$0.5\sqrt{x} = -25\pi\dfrac{dx}{dt} \qquad \text{Torricelli's Law}$$

$$\dfrac{dx}{dt} = -\dfrac{\sqrt{x}}{50\pi}$$

Initial condition: $\qquad x(0) = 16 \qquad$ The water is 16 ft deep when $t = 0$.

Solve Analytically

We first solve the differential equation.

$$x^{-1/2}\dfrac{dx}{dt} = -\dfrac{1}{50\pi} \qquad \text{Terms rearranged}$$

$$\int x^{-1/2}\dfrac{dx}{dt}\,dt = -\int\dfrac{1}{50\pi}\,dt \qquad \begin{array}{l}\text{Integrate both sides with}\\\text{respect to } t.\end{array}$$

$$\int x^{-1/2}\,dx = -\int\dfrac{1}{50\pi}\,dt \qquad \begin{array}{l}dx = (dx/dt)\,dt, \text{ property}\\\text{of differentials}\end{array}$$

$$2x^{1/2} = -\dfrac{1}{50\pi}t + C \qquad \begin{array}{l}\text{Example 6 (constants of}\\\text{integration combined as one)}\end{array}$$

The initial condition $x(0) = 16$ determines the value of C.

$$2(16)^{1/2} = -\dfrac{1}{50\pi}(0) + C$$

$$C = 8$$

With $C = 8$, we have

$$2x^{1/2} = -\dfrac{1}{50\pi}t + 8 \qquad \text{or} \qquad x^{1/2} = 4 - \dfrac{t}{100\pi}.$$

The formulas we seek are

$$x = \left(4 - \dfrac{t}{100\pi}\right)^2 \qquad \text{and} \qquad V = 25\pi x = 25\pi\left(4 - \dfrac{t}{100\pi}\right)^2.$$

Interpret

At any time t, the water in the tank is $(4 - t/(100\pi))^2$ ft deep and the amount of water is $25\pi(4 - t/(100\pi))^2$ ft³. At $t = 0$, we have $x = 16$ ft and $V = 400\pi$ ft³, as required. The tank will empty ($V = 0$) in $t = 400\pi$ minutes, which is about 21 hours.

In Example 8 we have to integrate twice, starting with a second derivative, to find the function we seek.

Example 8 FINDING A PROJECTILE'S HEIGHT

A heavy projectile is fired straight up from a platform 3 m above the ground with an initial velocity of 160 m/sec. Assume that the only force affecting the projectile during its flight is from gravity, which produces a downward acceleration of 9.8 m/sec². If $t = 0$ when the projectile is fired, find a formula for the projectile's

(a) velocity as a function of time t.

(b) height above the ground as a function of time t.

Solution

Model

To model the problem we draw Figure 6.6 and let s represent the projectile's height above the ground at time t. We assume s to be a twice-differentiable function of t and represent the projectile's velocity and acceleration with the derivatives

$$v = \frac{ds}{dt} \quad \text{and} \quad a = \frac{dv}{dt} = \frac{d^2s}{dt^2}.$$

Since gravity acts in the direction of decreasing s in Figure 6.6, the initial value problem to solve is

Differential equation: $\quad \dfrac{d^2s}{dt^2} = -9.8,$

Initial conditions: $\quad \dfrac{ds}{dt}(0) = 160 \quad$ and $\quad s(0) = 3.$

Solve Analytically

(a) Integrate both sides of the differential equation with respect to t to find ds/dt.

$$\int \frac{d^2s}{dt^2}\, dt = \int (-9.8)\, dt$$

$$\frac{ds}{dt} = -9.8t + C_1$$

We apply the first initial condition to find C_1.

$$160 = -9.8(0) + C_1 \qquad \frac{ds}{dt}(0) = 160$$

$$C_1 = 160$$

Thus,

$$v = \frac{ds}{dt} = -9.8t + 160.$$

(b) We integrate with respect to t again to find s.

$$\int \frac{ds}{dt}\, dt = \int (-9.8t + 160)\, dt$$

$$s = -4.9t^2 + 160t + C_2$$

We apply the second initial condition to find C_2.

$$3 = -4.9(0)^2 + 160(0) + C_2 \qquad s(0) = 3$$

$$C_2 = 3$$

Therefore,

$$s = -4.9t^2 + 160t + 3.$$

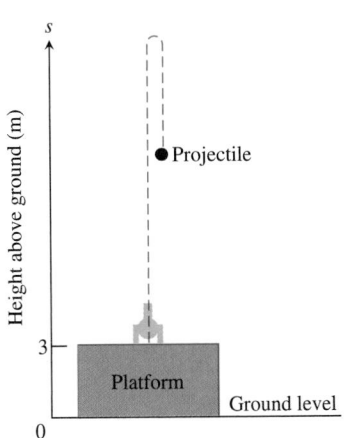

Figure 6.6 (Example 8)

When we solve an initial value problem involving a first derivative, we have one arbitrary constant, as in Examples 1 and 7. When we find a function from its second derivative, we have to deal with two constants, one from each antidifferentiation, as in Example 8. To find a function from its third derivative would require finding the values of three constants, and so on. In each case, the values of the constants are determined by the problem's initial conditions. Each time we find an antiderivative, we need an initial condition to tell us the value of C.

Quick Review 6.1

In Exercises 1–4, determine the amount in the account one year later if $100 is invested at 6% interest compounded k times per year.

1. $k = 1$ (annually)

2. $k = 4$ (quarterly)

3. $k = 12$ (monthly)

4. $k = 365$ (daily)

In Exercises 5–8, find dy/dx.

5. $y = \sin 3x$

6. $y = \tan \dfrac{5}{2} x$

7. $y = Ce^{2x}$, C a constant

8. $y = \ln (x + 2)$

In Exercises 9 and 10, use NINT to graph the function.

9. Graph $\int_1^x dt/t$ for $x > 0$. How does the graph compare with the graph of $y = \ln x$ for $x > 0$?

10. Graph $\int_{-1}^x dt/t$ for $x < 0$. How does the graph compare with the graph of $y = \ln (-x)$ for $x < 0$?

Section 6.1 Exercises

In Exercises 1–6, find an antiderivative for the function. Confirm your answer by differentiation.

1. $x^2 - 2x + 1$

2. $-3x^{-4}$

3. $x^2 - 4\sqrt{x}$

4. $8 + \csc x \cot x$

5. e^{4x}

6. $\dfrac{1}{x + 3}$

In Exercises 7–24, evaluate the integral.

7. $\displaystyle\int (x^5 - 6x + 3)\, dx$

8. $\displaystyle\int (-x^{-3} + x - 1)\, dx$

9. $\displaystyle\int \left(e^{t/2} - \dfrac{5}{t^2} \right) dt$

10. $\displaystyle\int \dfrac{4}{3} \sqrt[3]{t}\, dt$

11. $\displaystyle\int \left(x^3 - \dfrac{1}{x^3} \right) dx$

12. $\displaystyle\int \left(\sqrt[3]{x} + \dfrac{1}{\sqrt[3]{x}} \right) dx$

13. $\displaystyle\int \dfrac{1}{3} x^{-2/3}\, dx$

14. $\displaystyle\int (3 \sin x - \sin 3x)\, dx$

15. $\displaystyle\int \dfrac{\pi}{2} \cos \dfrac{\pi x}{2}\, dx$

16. $\displaystyle\int 2 \sec t \tan t\, dt$

17. $\displaystyle\int \left(\dfrac{2}{x - 1} + \dfrac{1}{x} \right) dx$

18. $\displaystyle\int \left(\dfrac{1}{x - 2} + \sin 5x - e^{-2x} \right) dx$

19. $\displaystyle\int 5 \sec^2 5r\, dr$

20. $\displaystyle\int \csc^2 7t\, dt$

21. $\displaystyle\int \cos^2 x\, dx$ (*Hint:* $\cos^2 x = \dfrac{1 + \cos 2x}{2}$)

22. $\displaystyle\int \sin^2 x\, dx$ (*Hint:* $\sin^2 x = \dfrac{1 - \cos 2x}{2}$)

23. $\displaystyle\int \tan^2 \theta\, d\theta$

24. $\displaystyle\int \cot^2 t\, dt$

In Exercises 25 and 26, **(a)** determine which graph shows the solution of the initial value problem without actually solving the problem. **(b) Writing to Learn** Explain why you eliminated two of the possibilities.

25. $\dfrac{dy}{dx} = \dfrac{1}{1 + x^2}$, $y(0) = \dfrac{\pi}{2}$

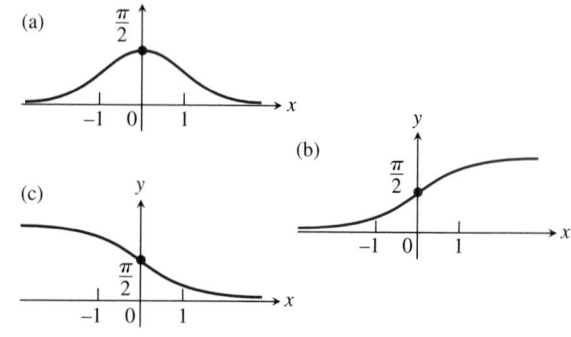

26. $\dfrac{dy}{dx} = -x, \quad y(-1) = 1$

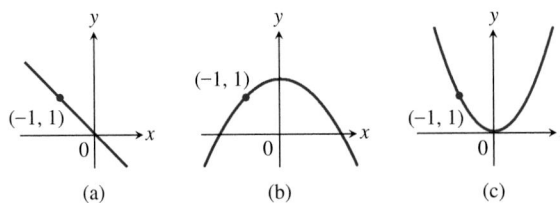

(a)　　　　　(b)　　　　　(c)

In Exercises 27–30, solve the initial value problem. Support your answer by overlaying your solution on a slope field for the differential equation.

27. $\dfrac{dy}{dx} = 2x - 1, \quad y(2) = 0$　　**28.** $\dfrac{dy}{dx} = \dfrac{1}{x^2} + x, \quad y(2) = 1$

29. $\dfrac{dy}{dx} = \sec^2 x, \quad y(\pi/4) = -1$　　**30.** $\dfrac{dy}{dx} = x^{-2/3}, \quad y(-1) = -5$

In Exercises 31–38, solve the initial value problem.

31. $\dfrac{dy}{dx} = 9x^2 - 4x + 5, \quad y(-1) = 0$

32. $\dfrac{dy}{dx} = \cos x + \sin x, \quad y(\pi) = 1$

33. $\dfrac{dy}{dt} = 2e^{-t}, \quad y(\ln 2) = 0$

34. $\dfrac{dy}{dx} = \dfrac{1}{x}, \quad y(e^3) = 0$

35. $\dfrac{d^2y}{d\theta^2} = \sin \theta, \quad y(0) = -3, \quad y'(0) = 0$

36. $\dfrac{d^2y}{dx^2} = 2 - 6x, \quad y(0) = 1, \quad y'(0) = 4$

37. $\dfrac{d^3y}{dt^3} = \dfrac{1}{t^3}, \quad y(1) = 1, \quad y'(1) = 3, \quad y''(1) = 2$

38. $\dfrac{d^4y}{d\theta^4} = \sin \theta + \cos \theta, \quad y(0) = -3, \quad y'(0) = -1,$
$y''(0) = -1, \quad y^{(3)}(0) = -3$

In Exercises 39–42, the velocity $v = ds/dt$ or acceleration $a = dv/dt$ of a body moving along a coordinate line is given. Find the body's position s at time t.

39. $v = 9.8t + 5, \quad s(0) = 10$　　**40.** $v = \sin \pi t, \quad s(1) = 0$

41. $a = 32, \quad s(0) = 0, \quad v(0) = 20$

42. $a = \cos t, \quad s(0) = 1, \quad v(0) = -1$

In Exercises 43 and 44, *work in groups of two or three.* Draw a possible graph for the function f with the given slope field that satisfies the stated condition.

43. $f(0) = 0$

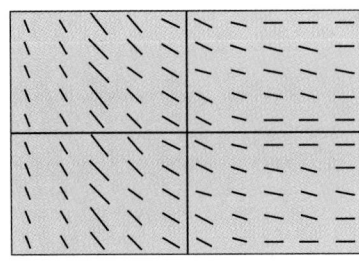

$[-2, 2]$ by $[-3, 3]$

44. $f(-1) = -2$

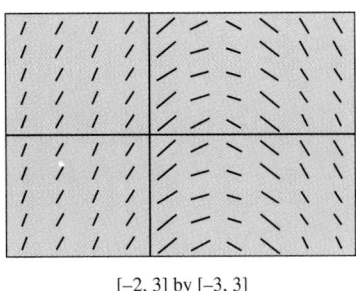

$[-2, 3]$ by $[-3, 3]$

In Exercises 45–48, confirm the integration formula by differentiation.

45. $\displaystyle\int \dfrac{dx}{1 + x^2} = \tan^{-1} x + C$　　**46.** $\displaystyle\int \dfrac{dx}{\sqrt{1 - x^2}} = \sin^{-1} x + C$

47. $\displaystyle\int \dfrac{dx}{|x|\sqrt{x^2 - 1}} = \sec^{-1} x + C$

48. $\displaystyle\int \dfrac{dx}{\sqrt{1 - x^2}} = -\cos^{-1} x + C$

Exploration

49. Let $\dfrac{dy}{dx} = x - \dfrac{1}{x^2}$.

(a) Find a solution to the differential equation in the interval $(0, \infty)$ that satisfies $y(1) = 2$.

(b) Find a solution to the differential equation in the interval $(-\infty, 0)$ that satisfies $y(-1) = 1$.

(c) Show that the following piecewise function is a solution to the differential equation for any values of C_1 and C_2.

$$y = \begin{cases} \dfrac{1}{x} + \dfrac{x^2}{2} + C_1, & x < 0 \\[2mm] \dfrac{1}{x} + \dfrac{x^2}{2} + C_2, & x > 0 \end{cases}$$

(d) Choose values for C_1 and C_2 so that the solution in part (c) agrees with the solutions in parts (a) and (b).

(e) Choose values for C_1 and C_2 so that the solution in part (c) satisfies $y(2) = -1$ and $y(-2) = 2$. ■

50. *Marginal Revenue* Suppose that the marginal revenue when x thousand units are sold is

$$\frac{dr}{dx} = 3x^2 - 6x + 12$$

dollars per unit. Find the revenue function $r(x)$ if $r(0) = 0$.

51. *Marginal Cost* Suppose that the marginal cost of manufacturing an item when x thousand units are produced is

$$\frac{dc}{dx} = 3x^2 - 12x + 15$$

dollars per item. Find the cost function $c(x)$ if $c(0) = 400$.

52. *Indefinite Integration* Suppose that

$$f(x) = \frac{d}{dx}(x^2 e^x) \quad \text{and} \quad g(x) = \frac{d}{dx}(x \sin x).$$

Evaluate each integral.

(a) $\displaystyle\int f(x)\, dx$

(b) $\displaystyle\int g(x)\, dx$

(c) $\displaystyle\int [-f(x)]\, dx$

(d) $\displaystyle\int [-g(x)]\, dx$

(e) $\displaystyle\int [f(x) + g(x)]\, dx$

(f) $\displaystyle\int [f(x) - g(x)]\, dx$

(g) $\displaystyle\int [x + f(x)]\, dx$

(h) $\displaystyle\int [g(x) - 4]\, dx$

53. *Stopping a Car in Time* You are driving along a highway at a steady 60 mph (88 ft/sec) when you see an accident ahead and slam on the brakes. What constant deceleration is required to stop your car in 242 ft? To find out, do the following.

(a) Solve the initial value problem

Differential equation: $\dfrac{d^2 s}{dt^2} = -k$ (k constant),

Initial conditions: $\dfrac{ds}{dt} = 88$ and $s = 0$ when $t = 0$,

measuring time and distance from when the brakes are applied.

(b) Find the value of t that makes $ds/dt = 0$. (The answer will involve k.)

(c) Find the value of k that makes $s = 242$ for the value of t you found in (b).

54. *Stopping a Motorcycle* The State of Illinois Cycle Rider Safety Program requires riders to be able to brake from 30 mph (44 ft/sec) to 0 in 45 ft. What constant deceleration does it take to do that? (*Hint:* See Exercise 53.)

55. *The Hammer and the Feather* When Apollo 15 astronaut David Scott dropped a hammer and a feather on the moon to demonstrate that in a vacuum all bodies fall with the same (constant) acceleration, he dropped them from about 4 ft above the ground. The television footage of the event shows the hammer and feather falling more slowly than on Earth, where, in a vacuum, they would have taken

only half a second to fall the 4 ft. How long did it take the hammer and feather to fall 4 ft on the moon? To find out, solve the following initial value problem for s as a function of t. Then find the value of t that makes s equal to 0.

Differential equation: $\dfrac{d^2 s}{dt^2} = -5.2$ ft/sec^2

Initial conditions: $\dfrac{ds}{dt} = 0$ and $s = 4$ when $t = 0$

56. *Motion with Constant Acceleration* The standard equation for the position s of a body moving with a constant acceleration a along a coordinate line is

$$s = \frac{a}{2}t^2 + v_0 t + s_0,$$

where v_0 and s_0 are the body's velocity and position at time $t = 0$. Derive this equation by setting up and solving an initial value problem.

57. *Draining a Tank* A conical tank with radius 4 ft and height 10 ft that was initially full of water is being drained at the rate of $(1/6)\sqrt{h}$ (see figure). Find a formula for the depth and the amount of water in the tank at any time t.

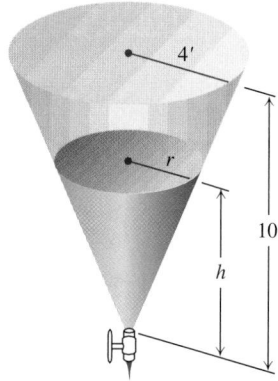

58. *Continuous Compound Interest* Suppose $500 is invested in an account that pays 4.75% interest compounded continuously.

(a) Find a formula for the amount in the account at any time t.

(b) When will the amount in the account be double the initial value?

59. *Continuous Compound Interest* Suppose $1200 is invested in an account that pays 6.25% interest compounded continuously.

(a) Find a formula for the amount in the account at any time t.

(b) When will the amount in the account be triple the initial value?

In Exercises 60 and 61, **(a)** use Part 1 of the Fundamental Theorem of Calculus to evaluate the indefinite integral. **(b)** Find the antiderivative whose graph passes through the point (0, 1).

60. $\displaystyle\int x^2 \cos x\, dx$

61. $\displaystyle\int x e^x\, dx$

Extending the Ideas

62. *Solving Differential Equations* Let

$$\frac{d^2y}{dx^2} = 6x.$$

(a) Find a solution to the differential equation that is continuous for $-\infty < x < \infty$ and whose graph passes through the point $(0, 1)$ and has a horizontal tangent there.

(b) Writing to Learn How many functions like this are there? How do you know?

In Exercises 63–66, *work in groups of two or three.* Generate a slope field for the differential equation in a viewing window large enough to display the listed points. Then add to your display the solution curves through these points. What can you learn from the concavity of the solution curves?

63. $y' = xy$; $(1, 1)$, $(-1, 2)$, $(0, -2)$, $(-2, -1)$

64. $y' = y$; $(0, 1)$, $(0, 2)$, $(0, -1)$

65. $y' = 2(y - 4)$; $(0, 1)$, $(0, 4)$, $(0, 5)$

66. $y' = y^2$; $(0, 1)$, $(0, 2)$, $(0, -1)$, $(0, 0)$

67. *Solving Differential Equations* Let

$$\frac{dy}{dx} = \frac{1}{x}.$$

(a) Show that $y = \ln x + C$ is a solution to the differential equation in the interval $(0, \infty)$.

(b) Show that $y = \ln(-x) + C$ is a solution to the differential equation in the interval $(-\infty, 0)$.

(c) Writing to Learn Explain why $y = \ln|x| + C$ is a solution to the differential equation in the domain $(-\infty, 0) \cup (0, \infty)$.

(d) Show that the function

$$y = \begin{cases} \ln(-x) + C_1, & x < 0 \\ \ln x + C_2, & x > 0 \end{cases}$$

is a solution to the differential equation for any values of C_1 and C_2.

6.2 Integration by Substitution

Power Rule in Integral Form • Trigonometric Integrands • Substitution in Indefinite Integrals • Substitution in Definite Integrals • Separable Differential Equations

Power Rule in Integral Form

When u is a differentiable function of x and n is a real number different from -1, the Chain Rule tells us that

$$\frac{d}{dx}\left(\frac{u^{n+1}}{n+1}\right) = u^n \frac{du}{dx}.$$

By reversing this derivative formula, we obtain the integral formula

$$\int \left(u^n \frac{du}{dx}\right) dx = \frac{u^{n+1}}{n+1} + C.$$

When the integrand of this integral is rewritten in the simpler *differential* form $u^n \, du$, we obtain the *Power Rule for Integration.*

Power Rule for Integration

If u is any differentiable function of x, then

$$\int u^n \, du = \frac{u^{n+1}}{n+1} + C, \quad n \neq -1.$$

A change of variables can often turn an unfamiliar integral into one that we can evaluate. The method for doing this is called the **substitution method of integration.** In Example 1 we use substitution to put the integral in the form

$$\int u^n \, du.$$

Example 1 USING SUBSTITUTION

Evaluate $\int (x+2)^5 \, dx$.

Solution

$$\int (x+2)^5 \, dx = \int u^5 du \qquad\qquad u = x+2,\ du = d(x+2) = dx$$

$$= \frac{u^6}{6} + C \qquad\qquad \text{Power Rule}$$

$$= \frac{(x+2)^6}{6} + C \quad u = x+2$$

Example 2 USING SUBSTITUTION

Evaluate $\int \sqrt{4x-1} \, dx$.

Solution

$$\int \sqrt{4x-1} \, dx = \int u^{1/2} \cdot \frac{1}{4} \, du \qquad \begin{array}{l} u = 4x-1,\ du = d(4x-1) = 4dx, \\ dx = du/4 \end{array}$$

$$= \frac{1}{4} \int u^{1/2} \, du \qquad\qquad \text{Constant Multiple Rule}$$

$$= \frac{1}{4} \cdot \frac{u^{3/2}}{3/2} + C \qquad\qquad \text{Power Rule}$$

$$= \frac{1}{6} u^{3/2} + C \qquad\qquad \text{Simplify.}$$

$$= \frac{1}{6} (4x-1)^{3/2} + C \quad u = 4x-1$$

Exploration 1 **Supporting Indefinite Integrals Graphically**

Consider the integral $\int \sqrt{1+x^2} \cdot 2x \, dx$.

1. Use the substitution $u = 1+x^2$ to show that
$$\int \sqrt{1+x^2} \cdot 2x \, dx = \frac{2}{3}\left(1+x^2\right)^{3/2} + C.$$

2. Explain why
$$y_1 = \frac{2}{3}\left(1+x^2\right)^{3/2} \qquad \text{and} \qquad y_2 = \int_0^x \sqrt{1+t^2} \cdot 2t \, dt$$
are both antiderivatives of $2x\sqrt{1+x^2}$.

3. Graph y_1 and y_2 in the same viewing window. Find a value of C for which $y_1 = y_2 + C$. Explain why this shift is possible.

4. If the factor $2x$ is removed from the integrand, the resulting integral has no easily obtained explicit form. You can understand why by trying to find a function whose derivative is $\sqrt{1+x^2}$. Graph $\int_a^x \sqrt{1+t^2} \, dt$ for $a = -4, -2, 0, 2, 4$ in $[-10, 10]$ by $[-30, 30]$ to see what five antiderivatives of $\sqrt{1+x^2}$ look like.

5. Explain why the five curves in part 4 are solution curves of the differential equation $dy/dx = \sqrt{1+x^2}$.

Trigonometric Integrands

When u is a differentiable function of x, the six trigonometric functions of u are also differentiable functions of x. Their derivative formulas, obtained using the Chain Rule, can be reversed to obtain integral formulas. For example,

$$\frac{d}{dx}\sin u = \cos u\,\frac{du}{dx} \quad \text{gives} \quad \int \cos u\,du = \sin u + C.$$

In this way we obtain the six formulas.

Trigonometric Integral Formulas

If u is any differentiable function of x, then

1. $\displaystyle\int \cos u\,du = \sin u + C.$ **2.** $\displaystyle\int \sin u\,du = -\cos u + C.$

3. $\displaystyle\int \sec^2 u\,du = \tan u + C.$ **4.** $\displaystyle\int \csc^2 u\,du = -\cot u + C.$

5. $\displaystyle\int \sec u \tan u\,du = \sec u + C.$ **6.** $\displaystyle\int \csc u \cot u\,du = -\csc u + C.$

Example 3   USING SUBSTITUTION

Evaluate $\int \cos (7x + 5)\,dx.$

Solution

$$\int \cos (7x + 5)\,dx = \int \cos u \cdot \frac{1}{7}\,du \qquad \begin{array}{l} u = 7x + 5,\ du = 7dx, \\ dx = du/7 \end{array}$$

$$= \frac{1}{7}\int \cos u\,du \qquad \text{Constant Multiple Rule}$$

$$= \frac{1}{7}\sin u + C \qquad \text{Formula 1}$$

$$= \frac{1}{7}\sin (7x + 5) + C \quad u = 7x + 5$$

Example 4   USING IDENTITIES AND SUBSTITUTION

Evaluate $\displaystyle\int \frac{1}{\cos^2 2x}\,dx.$

Solution

$$\int \frac{1}{\cos^2 2x}\,dx = \int \sec^2 2x\,dx \qquad \frac{1}{\cos 2x} = \sec 2x$$

$$= \int \sec^2 u \cdot \frac{1}{2}\,du \quad u = 2x,\ du = 2dx,\ dx = (1/2)\,du$$

$$= \frac{1}{2}\int \sec^2 u\,du$$

$$= \frac{1}{2} \tan u + C \qquad \text{Formula 3}$$

$$= \frac{1}{2} \tan 2x + C \qquad u = 2x$$

Confirm Analytically

$$\frac{d}{dx}\left(\frac{1}{2} \tan 2x + C\right) = \frac{1}{2}(\sec^2 2x)(2) = \sec^2 2x = \frac{1}{\cos^2 2x}$$

The integral of the tangent function involves the logarithm function as shown in Example 5. The same is true for the cotangent, secant, and cosecant functions, as you will see if you do Exercises 23, 29, and 30.

Example 5 USING SUBSTITUTION

Evaluate $\int \tan x \, dx$.

Solution

$$\int \tan x \, dx = \int \frac{\sin x}{\cos x} \, dx \qquad \tan x = \frac{\sin x}{\cos x}$$

$$= \int \frac{-du}{u} \qquad u = \cos x,\ du = -\sin x \, dx$$

$$= -\int \frac{du}{u}$$

$$= -\ln |u| + C \qquad \text{Integrate.}$$

$$= \ln \frac{1}{|u|} + C \qquad \ln \frac{1}{|u|} = -\ln |u|$$

$$= \ln \frac{1}{|\cos x|} + C \qquad u = \cos x$$

$$= \ln |\sec x| + C \qquad \frac{1}{\cos x} = \sec x$$

The Substitution Method

Take these steps to evaluate the integral

$$\int f(g(x)) \cdot g'(x) \, dx,$$

when f and g' are continuous functions.

1. Substitute $u = g(x)$ and $du = g'(x) \, dx$ to obtain the integral

$$\int f(u) \, du.$$

2. Integrate with respect to u.
3. Replace u by $g(x)$ in the result.

Substitution in Indefinite Integrals

The substitutions in the preceding Examples and Exploration are all instances of the following general method.

$$\int f(g(x)) \cdot g'(x) \, dx = \int f(u) \, du \qquad \text{1. Substitute } u = g(x),\ du = g'(x)\, dx.$$

$$= F(u) + C \qquad \text{2. Evaluate by finding an antiderivative } F(u) \text{ of } f(u).$$

$$= F(g(x)) + C \qquad \text{3. Replace } u \text{ by } g(x).$$

These three steps are the steps of the substitution method of integration. The method works because $F(g(x))$ is an antiderivative of $f(g(x)) \cdot g'(x)$ whenever F is an antiderivative of f.

Example 6 USING THE SUBSTITUTION METHOD

Evaluate $\int (x^2 + 2x - 3)^2(x + 1)\, dx$.

Solution Let $u = x^2 + 2x - 3$, then $du = (2x + 2)\, dx = 2(x + 1)\, dx$ and $(x + 1)\, dx = (1/2)\, du$.

$$\int (x^2 + 2x - 3)^2(x + 1)\, dx = \int u^2 \cdot \frac{1}{2}\, du = \frac{1}{2}\int u^2\, du \qquad \text{Step 1}$$

$$= \frac{1}{2} \cdot \frac{u^3}{3} + C = \frac{1}{6}u^3 + C \qquad \text{Step 2}$$

$$= \frac{1}{6}(x^2 + 2x - 3)^3 + C \qquad \text{Step 3}$$

Example 7 USING THE SUBSTITUTION METHOD

$$\int \sin^4 x \cos x\, dx = \int u^4\, du \qquad u = \sin x,\ du = \cos x\, dx$$

$$= \frac{u^5}{5} + C \qquad \text{Integrate.}$$

$$= \frac{\sin^5 x}{5} + C \qquad u = \sin x$$

Substitution in Definite Integrals

To evaluate a *definite* integral by substitution, we can avoid Step 3 of the substitution method (replacing u by $g(x)$). Here is the alternative procedure.

Substitution in Definite Integrals

Substitute $u = g(x)$, $du = g'(x)\, dx$, and integrate with respect to u from $u = g(a)$ to $u = g(b)$.

$$\int_a^b f(g(x)) \cdot g'(x)\, dx = \int_{g(a)}^{g(b)} f(u)\, du$$

Example 8 EVALUATING DEFINITE INTEGRALS

Evaluate $\int_0^{\pi/4} \tan x \sec^2 x\, dx$.

Solution Let $u = \tan x$, then $du = \sec^2 x\, dx$. Notice that

$$u(0) = \tan 0 = 0 \quad \text{and} \quad u\left(\frac{\pi}{4}\right) = \tan\frac{\pi}{4} = 1.$$

Thus,

$$\int_0^{\pi/4} \tan x \sec^2 x\, dx = \int_0^1 u\, du$$

$$= \frac{u^2}{2}\Bigg]_0^1$$

$$= \frac{1}{2} - 0 = \frac{1}{2}.$$

Exploration 2 compares two methods for evaluating a definite integral.

Evaluate $\int_{-1}^{1} 3x^2 \sqrt{x^3 + 1}\ dx$ by the two methods below.

1. Substitute $u = x^3 + 1$, $du = 3x^2 dx$ and integrate from $u(-1)$ to $u(1)$.

2. Substitute $u = x^3 + 1$, $du = 3x^2 dx$. Then find an antiderivative with respect to u. Replace u with $x^3 + 1$. Evaluate using the original x-limits of integration, $x = -1$ and 1.

Separable Differential Equations

A differential equation $y' = f(x, y)$ is **separable** if f can be expressed as a product of a function of x and a function of y. The differential equation then has the form

$$\frac{dy}{dx} = g(x)h(y). \tag{1}$$

If $h(y) \neq 0$, we can **separate the variables** by dividing both sides of the equation by $h(y)$, obtaining, in succession,

$$\frac{1}{h(y)} \frac{dy}{dx} = g(x)$$

$$\int \frac{1}{h(y)} \frac{dy}{dx} dx = \int g(x)\ dx \qquad \text{Integrate both sides with respect to } x.$$

$$\int \frac{1}{h(y)} dy = \int g(x)\ dx. \qquad dy = (dy/dx)\,dx \tag{2}$$

With x and y now separated, the integrated equation provides the solution we seek by expressing y either explicitly or implicitly as a function of x, up to an arbitrary constant.

In practice, we go directly from Equation 1 to Equation 2 without writing the intervening steps.

Example 9 SOLVING BY SEPARATION OF VARIABLES

Solve the differential equation

$$\frac{dy}{dx} = 2x(1 + y^2)e^{x^2}.$$

Solution Since $h(y) = 1 + y^2$ is never zero, we can solve the equation by separating the variables. Dividing both sides by $(1 + y^2)$ and integrating as above, we have

$$\frac{dy}{dx} = 2x(1 + y^2)e^{x^2}$$

$$\int \frac{1}{1 + y^2} dy = \int 2xe^{x^2}\ dx. \qquad \text{Separate the variables and integrate.}$$

On the left, $\displaystyle\int \frac{1}{1 + y^2}\, dy = \tan^{-1} y + C.$

On the right, $\displaystyle\int 2xe^{x^2}\, dx = \int e^u\, du$ $u = x^2,\ du = 2x\, dx$

$$= e^u + C$$
$$= e^{x^2} + C.$$

Combining the constants of integration, we have

$$\tan^{-1} y = e^{x^2} + C.$$ *C* now represents the com-bined constants of integration.

The equation $\tan^{-1} y = e^{x^2} + C$ gives *y* as an implicit function of *x*. In this case, we can solve for *y* as an explicit function of *x* by taking the tangent of both sides:

$$\tan\,(\tan^{-1} y) = \tan\,(e^{x^2} + C)$$
$$y = \tan\,(e^{x^2} + C).$$

Quick Review 6.2

In Exercises 1 and 2, evaluate the definite integral.

1. $\displaystyle\int_0^2 x^4\, dx$

2. $\displaystyle\int_1^5 \sqrt{x - 1}\, dx$

In Exercises 3–10, find dy/dx.

3. $\displaystyle y = \int_2^x 3^t\, dt$

4. $\displaystyle y = \int_0^x 3^t\, dt$

5. $y = (x^3 - 2x^2 + 3)^4$

6. $y = \sin^2 (4x - 5)$

7. $y = \ln \cos x$

8. $y = \ln \sin x$

9. $y = \ln (\sec x + \tan x)$

10. $y = \ln (\csc x + \cot x)$

Section 6.2 Exercises

In Exercises 1–8, use the indicated substitution to evaluate the integral. Confirm your answer by differentiation.

1. $\displaystyle\int \sin 3x\, dx, \quad u = 3x$

2. $\displaystyle\int x \cos (2x^2)\, dx, \quad u = 2x^2$

3. $\displaystyle\int \sec 2x \tan 2x\, dx, \quad u = 2x$

4. $\displaystyle\int 28(7x - 2)^3\, dx, \quad u = 7x - 2$

5. $\displaystyle\int \frac{dx}{x^2 + 9}, \quad u = \frac{x}{3}$

6. $\displaystyle\int \frac{9r^2\, dr}{\sqrt{1 - r^3}}, \quad u = 1 - r^3$

7. $\displaystyle\int \left(1 - \cos \frac{t}{2}\right)^2 \sin \frac{t}{2}\, dt, \quad u = 1 - \cos \frac{t}{2}$

8. $\displaystyle\int 8(y^4 + 4y^2 + 1)^2 (y^3 + 2y)\, dy, \quad u = y^4 + 4y^2 + 1$

In Exercises 9–30, use substitution to evaluate the integral.

9. $\displaystyle\int \frac{dx}{(1 - x)^2}$

10. $\displaystyle\int \sec^2 (x + 2)\, dx$

11. $\displaystyle\int \sqrt{\tan x}\ \sec^2 x\, dx$

12. $\int \sec\left(\theta + \dfrac{\pi}{2}\right) \tan\left(\theta + \dfrac{\pi}{2}\right) d\theta$

13. $\displaystyle\int_e^6 \dfrac{dx}{x \ln x}$

14. $\displaystyle\int_{-\pi/4}^{\pi/4} \tan^2 x \sec^2 x \, dx$

15. $\int \cos(3z + 4) \, dz$

16. $\int \sqrt{\cot x} \, \csc^2 x \, dx$

17. $\int \dfrac{\ln^6 x}{x} \, dx$

18. $\int \tan^7 \dfrac{x}{2} \sec^2 \dfrac{x}{2} \, dx$

19. $\int s^{1/3} \cos(s^{4/3} - 8) \, ds$

20. $\int \dfrac{dx}{\sin^2 3x}$

21. $\int \dfrac{\sin(2t + 1)}{\cos^2(2t + 1)} \, dt$

22. $\int \dfrac{6 \cos t}{(2 + \sin t)^2} \, dt$

23. $\displaystyle\int_{\pi/4}^{3\pi/4} \cot x \, dx$

24. $\displaystyle\int_0^7 \dfrac{dx}{x + 2}$

25. $\displaystyle\int_{-1}^3 \dfrac{x \, dx}{x^2 + 1}$

26. $\displaystyle\int_0^5 \dfrac{40 \, dx}{x^2 + 25}$

27. $\int \dfrac{dx}{\cot 3x}$

28. $\int \dfrac{dx}{\sqrt{5x + 8}}$

29. $\int \sec x \, dx$ (*Hint:* Multiply the integrand by
$$\dfrac{\sec x + \tan x}{\sec x + \tan x}$$
and then use a substitution to integrate the result.)

30. $\int \csc x \, dx$ (*Hint:* Multiply the integrand by
$$\dfrac{\csc x + \cot x}{\csc x + \cot x}$$
and then use a substitution to integrate the result.)

In Exercises 31–38, make a u-substitution and integrate from $u(a)$ to $u(b)$.

31. $\displaystyle\int_0^3 \sqrt{y + 1} \, dy$

32. $\displaystyle\int_0^1 r\sqrt{1 - r^2} \, dr$

33. $\displaystyle\int_{-\pi/4}^0 \tan x \sec^2 x \, dx$

34. $\displaystyle\int_{-1}^1 \dfrac{5r}{(4 + r^2)^2} \, dr$

35. $\displaystyle\int_0^1 \dfrac{10\sqrt{\theta}}{(1 + \theta^{3/2})^2} \, d\theta$

36. $\displaystyle\int_{-\pi}^{\pi} \dfrac{\cos x}{\sqrt{4 + 3 \sin x}} \, dx$

37. $\displaystyle\int_0^1 \sqrt{t^5 + 2t} \, (5t^4 + 2) \, dt$

38. $\displaystyle\int_0^{\pi/6} \cos^{-3} 2\theta \sin 2\theta \, d\theta$

In Exercises 39–42, solve the differential equation by separation of variables.

39. $\dfrac{dy}{dx} = (y + 5)(x + 2)$

40. $\dfrac{dy}{dx} = x\sqrt{y} \cos^2 \sqrt{y}$

41. $\dfrac{dy}{dx} = (\cos x)e^{y + \sin x}$

42. $\dfrac{dy}{dx} = e^{x - y}$

In Exercises 43 and 44, solve the initial value problem by separation of variables.

43. $\dfrac{dy}{dx} = -2xy^2$, $\quad y(1) = 0.25$

44. $\dfrac{dy}{dx} = \dfrac{4\sqrt{y} \ln x}{x}$, $\quad y(e) = 1$

Explorations

45. *Constant of Integration* Consider the integral
$$\int \sqrt{x + 1} \, dx.$$

(a) Show that $\displaystyle\int \sqrt{x + 1} \, dx = \dfrac{2}{3}(x + 1)^{3/2} + C.$

(b) Writing to Learn Explain why
$$y_1 = \int_0^x \sqrt{t + 1} \, dt \quad \text{and} \quad y_2 = \int_3^x \sqrt{t + 1} \, dt$$
are antiderivatives of $\sqrt{x + 1}$.

(c) Use a table of values for $y_1 - y_2$ to find the value of C for which $y_1 = y_2 + C$.

(d) Writing to Learn Give a convincing argument that
$$C = \int_0^3 \sqrt{x + 1} \, dx.$$

46. *Making Connections* Work in groups of two or three. Suppose that
$$\int f(x) \, dx = F(x) + C.$$

(a) Explain how you can use the derivative of $F(x) + C$ to confirm the integration is correct.

(b) Explain how you can use a slope field of f and the graph of $y = F(x)$ to support your evaluation of the integral.

(c) Explain how you can use the graphs of $y_1 = F(x)$ and $y_2 = \int_0^x f(t) \, dt$ to support your evaluation of the integral.

(d) Explain how you can use a table of values for $y_1 - y_2$, y_1 and y_2 defined as in (c), to support your evaluation of the integral.

(e) Explain how you can use graphs of f and NDER of $F(x)$ to support your evaluation of the integral.

(f) Illustrate (a)–(e) for $f(x) = \dfrac{x}{\sqrt{x^2 + 1}}$. ∎

Two Routes to the Integral In Exercises 47 and 48, make a substitution $u = \cdots$ (an expression in x), $du = \cdots$. Then

(a) integrate with respect to u from $u(a)$ to $u(b)$.

(b) find an antiderivative with respect to u, replace u by the expression in x, then evaluate from a to b.

47. $\displaystyle\int_0^1 \frac{x^3}{\sqrt{x^4 + 9}}\, dx$

48. $\displaystyle\int_{\pi/6}^{\pi/3} (1 - \cos 3x)\sin 3x \, dx$

49. Show that
$$y = \ln \left| \frac{\cos 3}{\cos x} \right| + 5$$

is the solution to the initial value problem
$$\frac{dy}{dx} = \tan x, \quad f(3) = 5.$$

(See the discussion following Example 4, Section 5.4.)

Extending the Ideas

50. *Connecting Antiderivatives* Consider the integral
$$\int \csc^2 2\theta \cot 2\theta \, d\theta.$$

(a) Use the substitution $u = \cot 2\theta$ to write the integral in the form $F_1(\theta) + C$. Identify F_1.

(b) Use the substitution $u = \csc 2\theta$ to write the integral in the form $F_2(\theta) + C$. Identify F_2.

(c) Confirm that each integration is correct.

(d) Determine a constant b so that $F_1 = F_2 + b$.

51. *Connecting Antiderivatives* Consider the integral
$$\int 2 \sin x \cos x \, dx.$$

(a) Use the substitution $u = \sin x$ to evaluate the integral.

(b) Use the substitution $u = \cos x$ to evaluate the integral.

(c) Rewrite the integrand as $\sin 2x$ and evaluate the integral.

(d) Confirm that each integration is correct.

6.3	**Integration by Parts**

Product Rule in Integral Form • Repeated Use • Solving for the Unknown Integral • Tabular Integration

Product Rule in Integral Form

When u and v are differentiable functions of x, the Product Rule for differentiation tells us that
$$\frac{d}{dx}(uv) = u\frac{dv}{dx} + v\frac{du}{dx}.$$

Integrating both sides with respect to x and rearranging leads to the integral equation
$$\int \left(u\frac{dv}{dx} \right) dx = \int \left(\frac{d}{dx}(uv) \right) dx - \int \left(v\frac{du}{dx} \right) dx$$
$$= uv - \int \left(v\frac{du}{dx} \right) dx.$$

When this equation is written in the simpler differential notation we obtain the following formula.

Integration by Parts Formula

$$\int u \, dv = uv - \int v \, du$$

This formula expresses one integral, $\int u \, dv$, in terms of a second integral, $\int v \, du$. With a proper choice of u and v, the second integral may be easier to evaluate than the first. This is the reason for the importance of the formula. When faced with an integral that we cannot handle analytically, we can replace it by one with which we might have more success.

Example 1 USING INTEGRATION BY PARTS

Evaluate $\int x \cos x \, dx$.

Solution We use the formula $\int u \, dv = uv - \int v \, du$ with

$$u = x, \qquad dv = \cos x \, dx.$$

To complete the formula, we take the differential of u and find the simplest antiderivative of $\cos x$.

$$du = dx \qquad v = \sin x$$

Then,

$$\int x \cos x \, dx = x \sin x - \int \sin x \, dx = x \sin x + \cos x + C.$$

Let's examine the choices available for u and v in Example 1.

Example 2 INVESTIGATING INTEGRATION BY PARTS

What are the choices for u and dv when we apply integration by parts to

$$\int x \cos x \, dx = \int u \, dv?$$

Which choices lead to a successful evaluation of the original integral?

Solution There are four possible choices.

1. $u = 1$ and $dv = x \cos x \, dx$ **2.** $u = x$ and $dv = \cos x \, dx$

3. $u = x \cos x$ and $dv = dx$ **4.** $u = \cos x$ and $dv = x \, dx$

Choice 1 won't do because we still don't know how to integrate $dv = x \cos x \, dx$ to get v.

Choice 2 works well as we saw in Example 1.

Choice 3 leads to $u = x \cos x, \qquad\qquad dv = dx,$

$$du = (\cos x - x \sin x) \, dx, \qquad v = x,$$

and the new integral

$$\int v \, du = \int (x \cos x - x^2 \sin x) \, dx.$$

This is worse than the integral we started with.

Choice 4 leads to $u = \cos x, \qquad dv = x \, dx,$

$$du = -\sin x \, dx, \qquad v = x^2/2,$$

and the new integral

$$\int v \, du = -\int \frac{x^2}{2} \sin x \, dx.$$

This, too, is worse.

LIPET

If you are wondering what to choose for u, here is what we usually do. Our first choice is a natural logarithm (L), if there is one. If there isn't, we look for an inverse trigonometric function (I). If there isn't one of these either, look for a polynomial (P). Barring that, look for an exponential (E) or a trigonometric function (T). That's the preference order: **L I P E T.**

In general, we want u to be something that simplifies when differentiated, and dv to be something that remains manageable when integrated.

The goal of integration by parts is to go from an integral $\int u\, dv$ that we don't see how to evaluate to an integral $\int v\, du$ that we can evaluate. Keep in mind that integration by parts does not always work.

Example 3 FINDING AREA

Find the area of the region bounded by the curve $y = xe^{-x}$ and the x-axis from $x = 0$ to $x = 3$.

Solution The region is shaded in Figure 6.7. Its area is

$$\int_0^3 xe^{-x}\, dx.$$

Solve Analytically

We use the formula $\int u\, dv = uv - \int v\, du$ with

$$u = x, \qquad dv = e^{-x}\, dx,$$
$$du = dx, \qquad v = -e^{-x}.$$

Then

$$\int xe^{-x}\, dx = -xe^{-x} - \int (-e^{-x})\, dx$$

$$= -xe^{-x} + \int e^{-x}\, dx$$

$$= -xe^{-x} - e^{-x} + C.$$

Thus,

$$\int_0^3 xe^{-x}\, dx = \left[-xe^{-x} - e^{-x} \right]_0^3$$

$$= (-3e^{-3} - e^{-3}) - (-e^{0}) = 1 - 4e^{-3} \approx 0.80.$$

Support Numerically

NINT $(xe^{-x}, x, 0, 3)$ gives the same value.

$y = xe^{-x}$

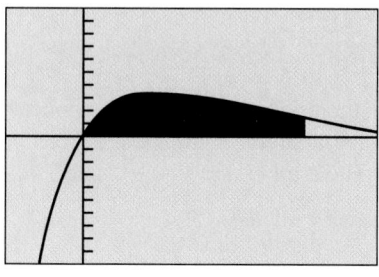

[–1, 4] by [–1, 1]

Figure 6.7 The region in Example 3.

Exploration 1 **Evaluating and Checking Integrals**

Consider $\int \ln x\, dx$.

1. Use integration by parts with $u = \ln x$ and $dv = dx$ to show that the integral is $x \ln x - x + C$.

2. Differentiate $x \ln x - x$ to confirm the result in part 1.

3. Use a slope field of the differential equation $dy/dx = \ln x$ and the graph of $y = x \ln x - x$ to support the result in part 1. Explain the connection.

4. Use the graphs of $y = x \ln x - x$ and $y = \int_1^x \ln t\, dt$ to support the result in part 1. Explain the connection.

Repeated Use

Sometimes we have to use integration by parts more than once to evaluate an integral.

Example 4 REPEATED USE OF INTEGRATION BY PARTS

Evaluate $\int x^2 e^x \, dx$.

Solution With $u = x^2$, $dv = e^x \, dx$, $du = 2x \, dx$, and $v = e^x$, we have

$$\int x^2 e^x \, dx = x^2 e^x - 2 \int x e^x \, dx.$$

The new integral is less complicated than the original because the exponent on x is reduced by one. To evaluate the integral on the right, we integrate by parts again with $u = x$, $dv = e^x \, dx$. Then $du = dx$, $v = e^x$, and

$$\int x e^x \, dx = x e^x - \int e^x \, dx = x e^x - e^x + C.$$

Hence,

$$\int x^2 e^x \, dx = x^2 e^x - 2 \int x e^x \, dx$$

$$= x^2 e^x - 2x e^x + 2e^x + C.$$

The technique of Example 4 works for any integral $\int x^n e^x \, dx$ in which n is a positive integer, because differentiating x^n will eventually lead to zero and integrating e^x is easy. We will say more on this later in this section when we discuss *tabular integration*.

Solving for the Unknown Integral

Integrals like the one in the next example occur in electrical engineering. Their evaluation requires two integrations by parts, followed by solving for the unknown integral.

Example 5 SOLVING FOR THE UNKNOWN INTEGRAL

Evaluate $\int e^x \cos x \, dx$.

Solution Let $u = e^x$, $dv = \cos x \, dx$. Then $du = e^x \, dx$, $v = \sin x$, and

$$\int e^x \cos x \, dx = e^x \sin x - \int e^x \sin x \, dx.$$

The second integral is like the first, except it has $\sin x$ in place of $\cos x$. To evaluate it, we use integration by parts with

$$u = e^x, \qquad dv = \sin x \, dx, \qquad v = -\cos x, \qquad du = e^x \, dx.$$

Then

$$\int e^x \cos x \, dx = e^x \sin x - \left(-e^x \cos x - \int (-\cos x)(e^x \, dx) \right)$$

$$= e^x \sin x + e^x \cos x - \int e^x \cos x \, dx.$$

The unknown integral now appears on both sides of the equation. Adding the integral to both sides gives

$$2 \int e^x \cos x \, dx = e^x \sin x + e^x \cos x + C.$$

Dividing by 2 and renaming the constant of integration gives

$$\int e^x \cos x \, dx = \frac{e^x \sin x + e^x \cos x}{2} + C.$$

When making repeated use of integration by parts in circumstances like Example 5, once a choice for u and dv is made, it is usually not a good idea to switch choices in the second stage of the problem. Doing so may result in undoing the work. For example, if we had switched to the substitution $u = \sin x$, $dv = e^x dx$ in the second integration, we would have obtained

$$\int e^x \cos x \, dx = e^x \sin x - \left(e^x \sin x - \int e^x \cos x \, dx \right)$$

$$= \int e^x \cos x \, dx,$$

undoing the first integration by parts.

Tabular Integration

We have seen that integrals of the form $\int f(x) g(x) dx$, in which f can be differentiated repeatedly to become zero and g can be integrated repeatedly without difficulty, are natural candidates for integration by parts. However, if many repetitions are required, the calculations can be cumbersome. In situations like this, there is a way to organize the calculations that saves a great deal of work. It is **tabular integration,** as shown in Examples 6 and 7.

Example 6 USING TABULAR INTEGRATION

Evaluate $\int x^2 e^x \, dx$.

Solution With $f(x) = x^2$ and $g(x) = e^x$, we list:

$f(x)$ and its derivatives		$g(x)$ and its integrals
x^2	(+)	e^x
$2x$	(−)	e^x
2	(+)	e^x
0		e^x

We combine the products of the functions connected by the arrows according to the operation signs above the arrows to obtain

$$\int x^2 e^x \, dx = x^2 e^x - 2x e^x + 2 e^x + C.$$

Compare this with the result in Example 4.

Example 7 USING TABULAR INTEGRATION

Evaluate $\int x^3 \sin x \, dx$.

Solution With $f(x) = x^3$ and $g(x) = \sin x$, we list:

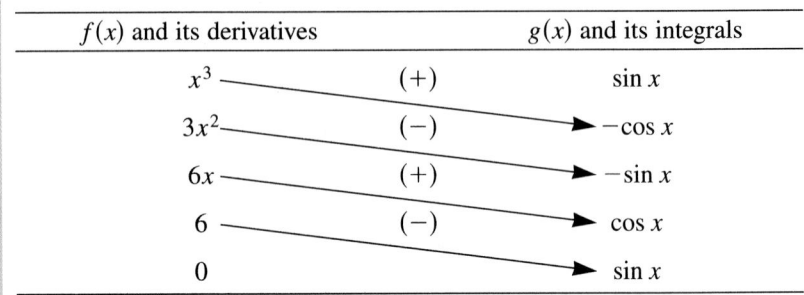

$f(x)$ and its derivatives		$g(x)$ and its integrals
x^3	$(+)$	$\sin x$
$3x^2$	$(-)$	$-\cos x$
$6x$	$(+)$	$-\sin x$
6	$(-)$	$\cos x$
0		$\sin x$

Again we combine the products of the functions connected by the arrows according to the operation signs above the arrows to obtain

$$\int x^3 \sin x \, dx = -x^3 \cos x + 3x^2 \sin x + 6x \cos x - 6 \sin x + C.$$

Quick Review 6.3

In Exercises 1–4, find dy/dx.

1. $y = x^3 \sin 2x$

2. $y = e^{2x} \ln (3x + 1)$

3. $y = \tan^{-1} 2x$

4. $y = \sin^{-1} (x + 3)$

In Exercises 5 and 6, solve for x in terms of y.

5. $y = \tan^{-1} 3x$

6. $y = \cos^{-1} (x + 1)$

7. Find the area under the arch of the curve $y = \sin \pi x$ from $x = 0$ to $x = 1$.

8. Solve the differential equation $\dfrac{dy}{dx} = e^{2x}$.

9. Solve the initial value problem $\dfrac{dy}{dx} = x + \sin x, \quad y(0) = 2$.

10. Use differentiation to confirm the integration formula

$$\int e^x \sin x \, dx = \frac{1}{2}e^x (\sin x - \cos x).$$

Section 6.3 Exercises

In Exercises 1–4, evaluate the integral. Confirm your answer by differentiation.

1. $\int x \sin x \, dx$

2. $\int x^2 \cos x \, dx$

3. $\int y \ln y \, dy$

4. $\int \tan^{-1} y \, dy$

In Exercises 5–8, evaluate the integral. Support your answer by superimposing the graph of one of the antiderivatives on a slope field of the integrand.

5. $\int x \sec^2 x \, dx$

6. $\int \sin^{-1} \theta \, d\theta$

7. $\int t^2 \sin t \, dt$

8. $\int t \csc^2 t \, dt$

In Exercises 9–14, evaluate the integral.

9. $\int x^3 \ln x \, dx$

10. $\int x^4 e^{-x} \, dx$

11. $\int (x^2 - 5x)e^x \, dx$

12. $\int x^3 e^{-2x} \, dx$

13. $\int e^y \sin y \, dy$

14. $\int e^{-y} \cos y \, dy$

In Exercises 15–18, evaluate the integral analytically. Support your answer using NINT.

15. $\displaystyle\int_{0}^{\pi/2} x^2 \sin 2x \, dx$

16. $\displaystyle\int_{0}^{\pi/2} x^3 \cos 2x \, dx$

17. $\displaystyle\int_{-2}^{3} e^{2x} \cos 3x \, dx$

18. $\displaystyle\int_{-3}^{2} e^{-2x} \sin 2x \, dx$

In Exercises 19–22, solve the differential equation.

19. $\dfrac{dy}{dx} = x^2 \, e^{4x}$

20. $\dfrac{dy}{dx} = x^2 \ln x$

21. $\dfrac{dy}{d\theta} = \theta \sec^{-1} \theta, \quad \theta > 1$

22. $\dfrac{dy}{d\theta} = \theta \sec \theta \tan \theta$

23. *Finding Area* Find the area of the region enclosed by the x-axis and the curve $y = x \sin x$ for

  (a) $0 \le x \le \pi$,   **(b)** $\pi \le x \le 2\pi$,   **(c)** $0 \le x \le 2\pi$.

24. *Finding Area* Find the area of the region enclosed by the y-axis and the curves $y = x^2$ and $y = (x^2 + x + 1)e^{-x}$.

25. *Average Value* A retarding force, symbolized by the dashpot in the figure, slows the motion of the weighted spring so that the mass's position at time t is

$$y = 2e^{-t} \cos t, \quad t \ge 0.$$

Find the average value of y over the interval $0 \le t \le 2\pi$.

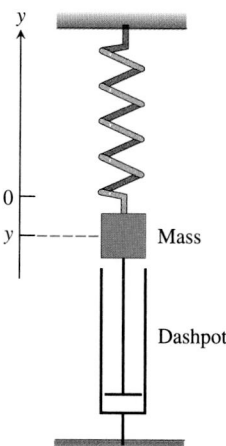

Exploration

26. Consider the integral $\int x^n e^x \, dx$. Use integration by parts to evaluate the integral if

  (a) $n = 1$.   **(b)** $n = 2$.   **(c)** $n = 3$.

  (d) Conjecture the value of the integral for any positive integer n.

  (e) Writing to Learn Give a convincing argument that your conjecture in (d) is true.

In Exercises 27–30, evaluate the integral by using a substitution prior to integration by parts.

27. $\displaystyle\int \sin \sqrt{x} \, dx$

28. $\displaystyle\int e^{\sqrt{3x + 9}} \, dx$

29. $\displaystyle\int x^7 e^{x^2} \, dx$

30. $\displaystyle\int \sin (\ln r) \, dr$

In Exercises 31–34, use integration by parts to establish the *reduction formula*.

31. $\displaystyle\int x^n \cos x \, dx = x^n \sin x - n \int x^{n-1} \sin x \, dx$

32. $\displaystyle\int x^n \sin x \, dx = -x^n \cos x + n \int x^{n-1} \cos x \, dx$

33. $\displaystyle\int x^n e^{ax} \, dx = \dfrac{x^n e^{ax}}{a} - \dfrac{n}{a} \int x^{n-1} e^{ax} \, dx, \quad a \neq 0$

34. $\displaystyle\int (\ln x)^n \, dx = x(\ln x)^n - n \int (\ln x)^{n-1} \, dx$

Extending the Ideas

35. *Integrating Inverse Functions* Assume that the function f has an inverse.

  (a) Show that $\int f^{-1}(x)dx = \int yf'(y) \, dy$. (*Hint:* Use the substitution $y = f^{-1}(x)$.)

  (b) Use integration by parts on the second integral in (a) to show that

$$\int f^{-1}(x) \, dx = \int yf'(y) \, dy = xf^{-1}(x) - \int f(y) \, dy.$$

36. *Integrating Inverse Functions* Assume that the function f has an inverse. Use integration by parts directly to show that

$$\int f^{-1}(x) \, dx = xf^{-1}(x) - \int x\left(\dfrac{d}{dx} f^{-1}(x)\right)dx.$$

In Exercises 37–40, evaluate the integral using

  (a) the technique of Exercise 35.

  (b) the technique of Exercise 36.

  (c) Show that the expressions (with $C = 0$) obtained in parts (a) and (b) are the same.

37. $\displaystyle\int \sin^{-1} x \, dx$

38. $\displaystyle\int \tan^{-1} x \, dx$

39. $\displaystyle\int \cos^{-1} x \, dx$

40. $\displaystyle\int \log_2 x \, dx$

6.4 Exponential Growth and Decay

Law of Exponential Change • Continuously Compounded Interest • Radioactivity • Newton's Law of Cooling • Resistance Proportional to Velocity

Law of Exponential Change

Suppose we are interested in a quantity y (population, radioactive element, money, whatever) that increases or decreases at a rate proportional to the amount present. If we also know the amount present at time $t = 0$, say y_0, we can find y as a function of t by solving the following initial value problem.

Differential equation: $\dfrac{dy}{dt} = ky$

Initial condition: $y = y_0$ when $t = 0$

If y is positive and increasing, then k is positive and the rate of growth is proportional to what has already been accumulated. If y is positive and decreasing, then k is negative and the rate of decay is proportional to the amount still left.

The constant function $y = 0$ is a solution of the differential equation but we usually aren't interested in that solution. To find nonzero solutions we separate the variables and integrate.

$$\frac{dy}{y} = k\,dt$$

$$\ln|y| = kt + C \qquad \text{Integrate.}$$

$$e^{\ln|y|} = e^{kt+C} \qquad \text{Exponentiate.}$$

$$|y| = e^C \bullet e^{kt} \qquad e^{\ln u} = u,\ e^{a+b} = e^a \bullet e^b$$

$$y = \pm e^C e^{kt} \qquad |y| = r \Rightarrow y = \pm r$$

$$y = Ae^{kt} \qquad \text{Let } A = \pm e^C.$$

By allowing A to take on the value 0 in addition to all the possible values of $\pm e^C$, we can include the solution $y = 0$. This solves the differential equation.

To solve the initial value problem we set $t = 0$ and $y = y_0$ and solve for A.

$$y_0 = Ae^{(k)(0)} = A$$

The solution of the initial value problem is $y = y_0 e^{kt}$.

Law of Exponential Change

If y changes at a rate proportional to the amount present ($dy/dt = ky$) and $y = y_0$ when $t = 0$, then

$$y = y_0 e^{kt},$$

where $k > 0$ represents growth and $k < 0$ represents decay.

The number k is the **rate constant** of the equation.

Continuously Compounded Interest

Suppose that A_0 dollars are invested at a fixed annual interest rate r (expressed as a decimal). If interest is added to the account k times a year, the amount of money present after t years is

$$A(t) = A_0 \left(1 + \frac{r}{k}\right)^{kt}.$$

The interest might be added ("compounded," bankers say) monthly ($k = 12$), weekly ($k = 52$), daily ($k = 365$), or even more frequently, by the hour or by the minute.

It can be shown that

$$\lim_{k \to \infty} A_0 \left(1 + \frac{r}{k}\right)^{kt} = A_0 e^{rt}.$$

We will see how this limit is evaluated in Section 8.1, Exercise 49.

If, instead of being added at discrete intervals, the interest is added continuously at a rate proportional to the amount in the account, we can model the growth of the account with the initial value problem.

$$\text{Differential equation:} \quad \frac{dA}{dt} = rA$$

$$\text{Initial condition:} \quad A(0) = A_0$$

The amount of money in the account after t years is then

$$A(t) = A_0 e^{rt}.$$

Interest paid according to this formula is said to be **compounded continuously.** The number r is the **continuous interest rate.**

Example 1 COMPOUNDING INTEREST CONTINUOUSLY

Suppose you deposit $800 in an account that pays 6.3% annual interest. How much will you have 8 years later if the interest is **(a)** compounded continuously? **(b)** compounded quarterly?

Solution Here $A_0 = 800$ and $r = 0.063$. The amount in the account to the nearest cent after 8 years is

(a) $A(8) = 800 e^{(0.063)(8)} = 1324.26.$

(b) $A(8) = 800 \left(1 + \dfrac{0.063}{4}\right)^{(4)(8)} = 1319.07.$

You might have expected to generate more than an additional $5.19 with interest compounded continuously.

Radioactivity

When an atom emits some of its mass as radiation, the remainder of the atom reforms to make an atom of some new element. This process of radiation and change is **radioactive decay,** and an element whose atoms go spontaneously through this process is **radioactive.** Radioactive carbon-14 decays into nitrogen. Radium, through a number of intervening radioactive steps, decays into lead.

For radium-226, which used to be painted on watch dials to make them glow at night (a dangerous practice for the painters, who licked their brush-tips), t is measured in years and $k = 4.3 \times 10^{-4}$. For radon-222 gas, t is measured in days and $k = 0.18$. The decay of radium in the earth's crust is the source of the radon we sometimes find in our basements.

Experiments have shown that at any given time the rate at which a radioactive element decays (as measured by the number of nuclei that change per unit of time) is approximately proportional to the number of radioactive nuclei present. Thus, the decay of a radioactive element is described by the equation $dy/dt = -ky$, $k > 0$. If y_0 is the number of radioactive nuclei present at time zero, the number still present at any later time t will be

$$y = y_0 e^{-kt}, \quad k > 0.$$

Convention

It is conventional to use $-k$ ($k > 0$) here instead of k ($k < 0$) to emphasize that y is decreasing.

The **half-life** of a radioactive element is the time required for half of the radioactive nuclei present in a sample to decay. Example 2 shows the surprising fact that the half-life is a constant that depends only on the radioactive substance and not on the number of radioactive nuclei present in the sample.

Example 2 FINDING HALF-LIFE

Find the half-life of a radioactive substance with decay equation $y = y_0 e^{-kt}$ and show that the half-life depends only on k.

Solution

Model

The half-life is the solution to the equation

$$y_0 e^{-kt} = \frac{1}{2} y_0.$$

Solve Algebraically

$$e^{-kt} = \frac{1}{2} \qquad \text{Divide by } y_0.$$

$$-kt = \ln \frac{1}{2} \qquad \text{Take ln of both sides.}$$

$$t = -\frac{1}{k} \ln \frac{1}{2} = \frac{\ln 2}{k} \qquad \ln \frac{1}{a} = -\ln a$$

Interpret

This value of t is the half-life of the element. It depends only on the value of k. Note that the number y_0 does not appear.

Carbon-14 Dating

The decay of radioactive elements can sometimes be used to date events from earth's past. The ages of rocks more than 2 billion years old have been measured by the extent of the radioactive decay of uranium (half-life 4.5 billion years!). In a living organism, the ratio of radioactive carbon, carbon-14, to ordinary carbon stays fairly constant during the lifetime of the organism, being approximately equal to the ratio in the organism's surroundings at the time. After the organism's death, however, no new carbon is ingested, and the proportion of carbon-14 decreases as the carbon-14 decays. It is possible to estimate the ages of fairly old organic remains by comparing the proportion of carbon-14 they contain with the proportion assumed to have been in the organism's environment at the time it lived. Archeologists have dated shells (which contain $CaCO_3$), seeds, and wooden artifacts this way. The estimate of 15,500 years for the age of the cave paintings at Lascaux, France, is based on carbon-14 dating. After generations of controversy, the Shroud of Turin, long believed by many to be the burial cloth of Christ, was shown by carbon-14 dating in 1988 to have been made after A.D. 1200.

Half-life

The **half-life** of a radioactive substance with rate constant k ($k > 0$) is

$$\text{half-life} = \frac{\ln 2}{k}.$$

Example 3 USING CARBON-14 DATING

Scientists who do carbon-14 dating use 5700 years for its half-life. Find the age of a sample in which 10% of the radioactive nuclei originally present have decayed.

Solution Because half-life $= 5700 = (\ln 2)/k$, we have $k = (\ln 2)/5700$.

Model

We need to find the value of t for which

$$y_0 e^{-kt} = 0.9 y_0 \qquad \text{or} \qquad e^{-kt} = 0.9,$$

where $k = (\ln 2)/5700$.

Solve Algebraically

$$e^{-kt} = 0.9$$

$$-kt = \ln 0.9 \quad \text{ln of both sides}$$

$$t = -\frac{1}{k}\ln 0.9 = -\frac{5700}{\ln 2}\ln 0.9 \approx 866$$

Interpret

The sample is about 866 years old.

Newton's Law of Cooling

Soup left in a cup cools to the temperature of the surrounding air. A hot silver ingot immersed in water cools to the temperature of the surrounding water. In situations like these, the rate at which an object's temperature is changing at any given time is roughly proportional to the difference between its temperature and the temperature of the surrounding medium. This observation is *Newton's Law of Cooling*, although it applies to warming as well, and there is an equation for it.

If T is the temperature of the object at time t, and T_s is the surrounding temperature, then

$$\frac{dT}{dt} = -k(T - T_s). \tag{1}$$

Since $dT = d(T - T_s)$, Equation 1 can be written as

$$\frac{d}{dt}(T - T_s) = -k(T - T_s).$$

Its solution, by the law of exponential change, is

$$T - T_s = (T_0 - T_s)e^{-kt},$$

where T_0 is the temperature at time $t = 0$. This equation also bears the name **Newton's Law of Cooling.**

Example 4   USING NEWTON'S LAW OF COOLING

A hard-boiled egg at 98°C is put in a pan under running 18°C water to cool. After 5 minutes, the egg's temperature is found to be 38°C. How much longer will it take the egg to reach 20°?

Solution

Model

Using Newton's Law of Cooling with $T_s = 18$ and $T_0 = 98$, we have

$$T - 18 = (98 - 18)e^{-kt} \quad \text{or} \quad T = 18 + 80e^{-kt}.$$

To find k we use the information that $T = 38$ when $t = 5$.

$$38 = 18 + 80e^{-5k}$$

$$e^{-5k} = \frac{1}{4}$$

$$-5k = \ln\frac{1}{4} = -\ln 4$$

$$k = \frac{1}{5}\ln 4$$

The egg's temperature at time t is $T = 18 + 80e^{-(0.2\ln 4)t}$.

J. Ernest Wilkins, Jr.

1923–

By the age of nineteen, J. Ernest Wilkins had earned a Ph.D. degree in Mathematics from the University of Chicago. He then taught, served on the Manhattan project (the goal of which was to build the first atomic bomb), and worked as a mathematician and physicist for several corporations. In 1970, Dr. Wilkins joined the faculty at Howard University and has served as head of the electrical engineering, physics, chemistry, and mathematics departments. He was one of those rare teachers who could make complicated topics easy to understand.

Intersection
X = 13.30482   Y = 20

[0, 20] by [10, 40]

Figure 6.8 The egg will reach 20°C about 13.3 min after being placed in the pan to cool. (Example 4)

Table 6.3 Experimental Data		
Time (sec)	T (°C)	$T - T_s$ (°C)
2	64.8	60.3
5	49.0	44.5
10	31.4	26.9
15	22.0	17.5
20	16.5	12.0
25	14.2	9.7
30	12.0	7.5

Solve Graphically

We can now use a grapher to find the time when the egg's temperature is 20°C. Figure 6.8 shows that the graphs of

$$y = 20 \quad \text{and} \quad y = T = 18 + 80e^{-(0.2 \ln 4)t}$$

intersect at about $t = 13.3$.

Interpret

The egg's temperature will reach 20°C in about 13.3 min after it is put in the pan under running water to cool. Because it took 5 min to reach 38°C, it will take slightly more than 8 additional minutes to reach 20°C.

The next example shows how to use exponential regression to fit a function to real data. A CBL™ temperature probe was used to collect the data.

Example 5 USING NEWTON'S LAW OF COOLING

A temperature probe (thermometer) is removed from a cup of coffee and placed in water that has a temperature of $T_s = 4.5$°C. Temperature readings T, as recorded in Table 6.3, are taken after 2 sec, 5 sec, and every 5 sec thereafter. Estimate

(a) the coffee's temperature at the time the temperature probe was removed.

(b) the time when the temperature probe reading will be 8°C.

Solution

Model

According to Newton's Law of Cooling, $T - T_s = (T_0 - T_s)e^{-kt}$, where $T_s = 4.5$ and T_0 is the temperature of the coffee (probe reading) at $t = 0$.

We use exponential regression to find that

$$T - 4.5 = 61.66(0.9277^t)$$

is a model for the $(t, T - T_s) = (t, T - 4.5)$ data.

Thus,

$$T = 4.5 + 61.66(0.9277^t)$$

is a model for the (t, T) data.

Figure 6.9a shows the graph of the model superimposed on a scatter plot of the (t, T) data.

(a) At time $t = 0$, when the probe was removed, the temperature was

$$T = 4.5 + 61.66(0.9277^0) \approx 66.16°C.$$

(b) Solve Graphically

Figure 6.9b shows that the graphs of

$$y = 8 \quad \text{and} \quad y = T = 4.5 + 61.66(0.9277^t)$$

intersect at about $t = 38$.

Interpret

The temperature of the coffee was about 66.2°C when the temperature probe was removed. The temperature probe will reach 8°C about 38 sec after it is removed from the coffee and placed in the water.

[0, 35] by [0, 70]

(a)

[0, 60] by [−20, 70]

(b)

Figure 6.9 (Example 5)

Resistance Proportional to Velocity

In some cases it is reasonable to assume that, other forces being absent, the resistance encountered by a moving object, such as a car coasting to a stop, is proportional to the object's velocity. The slower the object moves, the less its forward progress is resisted by the air through which it passes. To describe this in mathematical terms, we picture the object as a mass m moving along a coordinate line with position function s and velocity v at time t. The resisting force opposing the motion is

$$\text{Force} = \text{mass} \times \text{acceleration} = m\frac{dv}{dt}.$$

We can express the assumption that the resisting force is proportional to velocity by writing

$$m\frac{dv}{dt} = -kv \quad \text{or} \quad \frac{dv}{dt} = -\frac{k}{m}v \quad (k > 0).$$

This is a differential equation of exponential change. The solution to the differential equation with initial condition $v = v_0$ at $t = 0$ is

$$v = v_0 e^{-(k/m)t}. \tag{2}$$

In Exploration 1 we fix the values of v_0 and k and investigate what happens as the mass m increases.

Exploration 1 **Slowing Down More Slowly**

An object is traveling along a coordinate line with velocity $v = 100e^{-(0.5/m)t}$.

1. Graph v for $m = 1, 2, 4, 6$ in the window $[0, 20]$ by $[0, 120]$. Compare and describe the motion in each case.

2. Explain why $s(t) = \int_0^t v(u)\,du$ gives the distance traveled by the object at time t. What is $s(0)$?

3. Graph $s(t)$ for $m = 1, 2, 4, 6$ in the window $[0, 70]$ by $[0, 1500]$. Compare the total distances traveled.

Example 6 COASTING TO A STOP

For a 50-kg ice skater, the k in Equation 2 is about 2.5 kg/sec.

(a) How long will it take the skater to coast from 7 m/sec to 1 m/sec?

(b) How far will the skater coast before coming to a complete stop?

Solution

Model

If $t = 0$ represents the time when the skater's velocity is 7 m/sec, then $v_0 = 7$ and Equation 2 becomes

$$v = 7e^{-(2.5/50)t} = 7e^{-0.05t}.$$

(a) Solve Algebraically

We want the value of t when $v = 1$.

$$1 = 7e^{-0.05t}$$

$$e^{-0.05t} = \frac{1}{7}$$

$$-0.05t = \ln\frac{1}{7} = -\ln 7$$

$$t = \frac{\ln 7}{0.05} \approx 38.9 \text{ sec}$$

(b) Solve Analytically

The integral of velocity is distance, so

$$s = \int 7e^{-0.05t}\, dt = \frac{7e^{-0.05t}}{-0.05} + C = -140e^{-0.05t} + C.$$

We assume $s = 0$ when $t = 0$, so

$$0 = -140e^0 + C$$

$$C = 140.$$

Thus,

$$s = 140 - 140e^{-0.05t} = 140(1 - e^{-0.05t}).$$

Notice that the limit of $s(t)$ as $t \to \infty$ is 140.

Interpret

Mathematically, s never quite reaches 140 m. However, for all practical purposes, the skater will come to a complete stop after traveling 140 m.

In Exercise 31, you will show that the total distance coasted by a moving object encountering resistance proportional to its velocity is $v_0 m/k$. Further, you will show that if $s(0) = 0$, then

$$s(t) = \frac{v_0 m}{k}\left(1 - e^{-(k/m)t}\right). \tag{3}$$

Example 7   COASTING ON IN-LINE SKATES

The data in Table 6.4 were collected with a motion detector and a CBL™ by Valerie Sharritts, a mathematics teacher at St. Francis DeSales High School in Columbus, Ohio. The table shows the distance s (meters) coasted on in-line skates in t seconds by her daughter Ashley when she was 10 years old.

Table 6.4 Ashley Sharritts Skating Data

t (sec)	s (m)	t (sec)	s (m)	t (sec)	s (m)
0	0	2.24	3.05	4.48	4.77
0.16	0.31	2.40	3.22	4.64	4.82
0.32	0.57	2.56	3.38	4.80	4.84
0.48	0.80	2.72	3.52	4.96	4.86
0.64	1.05	2.88	3.67	5.12	4.88
0.80	1.28	3.04	3.82	5.28	4.89
0.96	1.50	3.20	3.96	5.44	4.90
1.12	1.72	3.36	4.08	5.60	4.90
1.28	1.93	3.52	4.18	5.76	4.91
1.44	2.09	3.68	4.31	5.92	4.90
1.60	2.30	3.84	4.41	6.08	4.91
1.76	2.53	4.00	4.52	6.24	4.90
1.92	2.73	4.16	4.63	6.40	4.91
2.08	2.89	4.32	4.69	6.56	4.91

Find a model for Ashley's position given by the data in Table 6.4 in the form of Equation 3. Her initial velocity was $v_0 = 2.75$ m/sec, her mass $m = 39.92$ kg (she weighed 88 lb), and her total coasting distance was 4.91 m.

Solution   We use the total coasting distance to determine k.

$$\frac{v_0 m}{k} = \text{coasting distance}$$

$$\frac{(2.75)(39.92)}{k} = 4.91$$

$$k \approx 22.36$$

Thus, the distance function is

$$s(t) = 4.91(1 - e^{-(22.36/39.92)t}).$$

Figure 6.10 shows the graph of $s(t)$ superimposed on a scatter plot of the position data in Table 6.4.

In Exercise 33 of this section and in Review Exercise 61, we will investigate the skating data of two friends of the Sharrittses: Kelly Schmitzer and Johnathon Krueger.

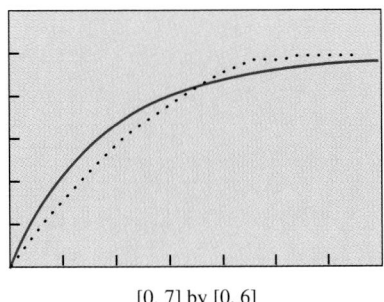

[0, 7] by [0, 6]

Figure 6.10 Notice how well the model obtained using mathematics fits the data obtained experimentally. (Example 7)

Quick Review 6.4

In Exercises 1 and 2, rewrite the equation in exponential form or logarithmic form.

1. $\ln a = b$

2. $e^c = d$

In Exercises 3–8, solve the equation.

3. $\ln (x + 3) = 2$

4. $100e^{2x} = 600$

5. $0.85^x = 2.5$

6. $2^{k+1} = 3^k$

7. $1.1^t = 10$

8. $e^{-2t} = \dfrac{1}{4}$

In Exercises 9 and 10, solve for y.

9. $\ln (y + 1) = 2x - 3$

10. $\ln |y + 2| = 3t - 1$

Section 6.4 Exercises

In Exercises 1–4, find the solution of the differential equation $dy/dt = ky$, k a constant, that satisfies the given conditions.

1. $k = 1.5$, $y(0) = 100$

2. $k = -0.5$, $y(0) = 200$

3. $y(0) = 50$, $y(5) = 100$

4. $y(0) = 60$, $y(10) = 30$

In Exercises 5–8, complete the table for an investment if interest is compounded continuously.

	Initial Deposit ($)	Annual Rate (%)	Doubling Time (yr)	Amount in 30 yr ($)
5.	1000	8.6		
6.	2000		15	
7.		5.25		2898.44
8.	1200			10,405.37

In Exercises 9 and 10, find the amount of time required for a $2000 investment to double if the annual interest rate r is compounded **(a)** annually, **(b)** monthly, **(c)** quarterly, and **(d)** continuously.

9. $r = 4.75\%$

10. $r = 8.25\%$

11. *Growth of Cholera Bacteria* Suppose that the cholera bacteria in a colony grows unchecked according to the Law of Exponential Change. The colony starts with 1 bacterium and doubles in number every half hour.

(a) How many bacteria will the colony contain at the end of 24 h?

(b) Writing to Learn Use (a) to explain why a person who feels well in the morning may be dangerously ill by evening even though, in an infected person, many bacteria are destroyed.

12. *Bacteria Growth* A colony of bacteria is grown under ideal conditions in a laboratory so that the population increases exponentially with time. At the end of 3 h there are 10,000 bacteria. At the end of 5 h there are 40,000 bacteria. How many bacteria were present initially?

13. *Radon-222* The decay equation for radon-222 gas is known to be $y = y_0 e^{-0.18t}$, with t in days. About how long will it take the amount of radon in a sealed sample of air to decay to 90% of its original value?

14. *Polonium-210* The number of radioactive atoms remaining after t days in a sample of polonium-210 that starts with y_0 radioactive atoms is $y = y_0 e^{-0.005t}$.

(a) Find the element's half-life.

(b) Your sample will not be useful to you after 95% of the radioactive nuclei present on the day the sample arrives have disintegrated. For about how many days after the sample arrives will you be able to use the polonium?

In Exercises 15 and 16, find the exponential function $y = y_0 e^{kt}$ whose graph passes through the two points.

15.

(2, 5)

16.

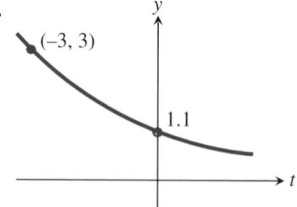

(-3, 3)

1.1

17. *Mean Life of Radioactive Nuclei* Physicists using the radioactive decay equation $y = y_0 e^{-kt}$ call the number $1/k$ the *mean life* of a radioactive nucleus. The mean life of a radon-222 nucleus is about $1/0.18 \approx 5.6$ days. The mean life of a carbon-14 nucleus is more than 8000 years. Show that 95% of the radioactive nuclei originally present in any sample will disintegrate within three mean lifetimes, that is, by time $t = 3/k$. Thus, the mean life of a nucleus gives a quick way to estimate how long the radioactivity of a sample will last.

18. *Finding the Original Temperature of a Beam* An aluminum beam was brought from the outside cold into a machine shop where the temperature was held at 65°F. After 10 min, the beam warmed to 35°F and after another 10 min its temperature was 50°F. Use Newton's Law of Cooling to estimate the beam's initial temperature.

19. *Cooling Soup* Suppose that a cup of soup cooled from 90°C to 60°C in 10 min in a room whose temperature was 20°C. Use Newton's Law of Cooling to answer the following questions.

(a) How much longer would it take the soup to cool to 35°C?

(b) Instead of being left to stand in the room, the cup of 90°C soup is put into a freezer whose temperature is −15°C. How long will it take the soup to cool from 90°C to 35°C?

20. *Cooling Silver* The temperature of an ingot of silver is 60°C above room temperature right now. Twenty minutes ago, it was 70°C above room temperature. How far above room temperature will the silver be

(a) 15 minutes from now?

(b) 2 hours from now?

(c) When will the silver be 10°C above room temperature?

21. *Dating Crater Lake* The charcoal from a tree killed in the volcanic eruption that formed Crater Lake in Oregon contained 44.5% of the carbon-14 found in living matter. About how old is Crater Lake?

22. *Carbon-14 Dating Measurement Sensitivity* To see the effect of a relatively small error in the estimate of the amount of carbon-14 in a sample being dated, answer the following questions about this hypothetical situation.

(a) A fossilized bone found in central Illinois in the year A.D. 2000 contains 17% of its original carbon-14 content. Estimate the year the animal died.

(b) Repeat part (a) assuming 18% instead of 17%.

(c) Repeat part (a) assuming 16% instead of 17%.

23. *Coasting Bicycle* A 66-kg cyclist on a 7-kg bicycle starts coasting on level ground at 9 m/sec. The k in Equation 2 is about 3.9 kg/sec.

(a) About how far will the cyclist coast before reaching a complete stop?

(b) How long will it take the cyclist's speed to drop to 1 m/sec?

24. *Coasting Battleship* Suppose an Iowa class battleship has mass around 51,000 metric tons (51,000,000 kg) and a k value in Equation 2 of about 59,000 kg/sec. Assume the ship loses power when it is moving at a speed of 9 m/sec.

(a) About how far will the ship coast before it is dead in the water?

(b) About how long will it take the ship's speed to drop to 1 m/sec?

25. *The Inversion of Sugar* The processing of raw sugar has an "inversion" step that changes the sugar's molecular structure. Once the process has begun, the rate of change of the amount of raw sugar is proportional to the amount of raw sugar remaining. If 1000 kg of raw sugar reduces to 800 kg of raw sugar during the first 10 h, how much raw sugar will remain after another 14 h?

26. *Oil Depletion* Suppose the amount of oil pumped from one of the canyon wells in Whittier, California, decreases at the continuous rate of 10% per year. When will the well's output fall to one-fifth of its present level?

27. *Atmospheric Pressure* Earth's atmospheric pressure p is often modeled by assuming that the rate dp/dh at which p changes with the altitude h above sea level is proportional to p. Suppose that the pressure at sea level is 1013 millibars (about 14.7 lb/in^2) and that the pressure at an altitude of 20 km is 90 millibars.

(a) Solve the initial value problem

Differential equation: $\dfrac{dp}{dh} = kp,$

Initial condition: $p = p_0$ when $h = 0$,

to express p in terms of h. Determine the values of p_0 and k from the given altitude-pressure data.

(b) What is the atmospheric pressure at $h = 50$ km?

(c) At what altitude does the pressure equal 900 millibars?

28. *First Order Chemical Reactions* In some chemical reactions the rate at which the amount of a substance changes with time is proportional to the amount present. For the change of δ-glucono lactone into gluconic acid, for example,

$$\frac{dy}{dt} = -0.6y$$

when y is measured in grams and t is measured in hours. If there are 100 grams of a δ-glucono lactone present when $t = 0$, how many grams will be left after the first hour?

29. *Discharging Capacitor Voltage* Suppose that electricity is draining from a capacitor at a rate proportional to the voltage V across its terminals and that, if t is measured in seconds,

$$\frac{dV}{dt} = -\frac{1}{40}V.$$

(a) Solve this differential equation for V, using V_0 to denote the value of V when $t = 0$.

(b) How long will it take the voltage to drop to 10% of its original value?

30. *John Napier's Answer* John Napier (1550–1617), the Scottish laird who invented logarithms, was the first person to answer the question, "What happens if you invest an amount of money at 100% yearly interest, compounded continuously?"

(a) **Writing to Learn** What does happen? Explain.

(b) How long does it take to triple your money?

(c) **Writing to Learn** How much can you earn in a year?

31. *Coasting Object* Assume that the resistance encountered by a moving object is proportional to the object's velocity so that its velocity is $v = v_0 e^{-(k/m)t}$, as given by Equation 2.

(a) Integrate the velocity function with respect to t to obtain the distance function s. Assume that $s(0) = 0$ and show that

$$s(t) = \frac{v_0 m}{k}\left(1 - e^{-(k/m)t}\right).$$

(b) Show that the total coasting distance traveled by the object as it coasts to a complete stop is $v_0 m/k$.

32. *Benjamin Franklin's Will* The Franklin Technical Institute of Boston owes its existence to a provision in a codicil to Benjamin Franklin's will. In part the codicil reads:

> I wish to be useful even after my Death, if possible, in forming and advancing other young men that may be serviceable to their Country in both Boston and Philadelphia. To this end I devote Two thousand Pounds Sterling, which I give, one thousand thereof to the Inhabitants of the Town of Boston in Massachusetts, and the other thousand to the inhabitants of the City of Philadelphia, in Trust and for the Uses, Interests and Purposes hereinafter mentioned and declared.

Franklin's plan was to lend money to young apprentices at 5% interest with the provision that each borrower should pay each year along

> ... with the yearly Interest, one tenth part of the Principal, which sums of Principal and Interest shall be again let to fresh Borrowers. ... If this plan is executed and succeeds as projected without interruption for one hundred Years, the Sum will then be one hundred and thirty-one thousand Pounds of which I would have the Managers of the Donation to the Inhabitants of the Town of Boston, then lay out at their discretion one hundred thousand Pounds in Public Works. ... The remaining thirty-one thousand Pounds, I would have continued to be let out on Interest in the manner above directed for another hundred Years. ... At the end of this second term if no unfortunate accident has prevented the operation the sum will be Four Millions and Sixty-one Thousand Pounds.

It was not always possible to find as many borrowers as Franklin had planned, but the managers of the trust did the best they could. At the end of 100 years from the receipt of the Franklin gift, in January 1894, the fund had grown from 1000 pounds to almost 90,000 pounds. In 100 years the original capital had multiplied about 90 times instead of the 131 times Franklin had imagined.

(a) What annual rate of interest, compounded continuously for 100 years, would have multiplied Benjamin Franklin's original capital by 90?

(b) In Benjamin Franklin's estimate that the original 1000 pounds would grow to 131,000 in 100 years, he was using an annual rate of 5% and compounding once each year. What rate of interest per year when compounded continuously for 100 years would multiply the original amount by 131?

33. *Coasting to a Stop* Table 6.5 shows the distance s (meters) coasted on in-line skates in terms of time t (seconds) by Kelly Schmitzer. Find a model for her position in the form of Equation 3 and superimpose its graph on a scatter plot of the data. Her initial velocity was $v_0 = 0.80$ m/sec, her mass $m = 49.90$ kg (110 lb), and her total coasting distance was 1.32 m.

Table 6.5 Kelly Schmitzer Skating Data

t (sec)	s (m)	t (sec)	s (m)	t (sec)	s (m)
0	0	1.5	0.89	3.1	1.30
0.1	0.07	1.7	0.97	3.3	1.31
0.3	0.22	1.9	1.05	3.5	1.32
0.5	0.36	2.1	1.11	3.7	1.32
0.7	0.49	2.3	1.17	3.9	1.32
0.9	0.60	2.5	1.22	4.1	1.32
1.1	0.71	2.7	1.25	4.3	1.32
1.3	0.81	2.9	1.28	4.5	1.32

34. *Temperature Experiment* A temperature probe is removed from a cup of coffee and placed in water whose temperature (T_s) is 10°C. The data in Table 6.6 were collected over the next 30 sec with a CBL™ temperature probe.

Table 6.6 Experimental Data

Time (sec)	T (°C)	$T - T_s$ (°C)
2	80.47	70.47
5	69.39	59.39
10	49.66	39.66
15	35.26	25.26
20	28.15	18.15
25	23.56	13.56
30	20.62	10.62

(a) Find an exponential regression equation for the $(t, T - T_s)$ data.

(b) Use the regression equation in (a) to find a model for the (t, T) data. Superimpose the graph of the model on a scatter plot of the (t, T) data.

(c) Estimate when the temperature probe will read 12°C.

(d) Estimate the coffee's temperature when the temperature probe was removed.

Explorations

35. *Newton's Law of Cooling* Let T be the temperature of an object at time t, T_s the surrounding temperature, and T_0 the value of T at time zero.

(a) Solve the equation

$$\frac{dT}{dt} = -k(T - T_s)$$

by separation of variables to derive Newton's Law of Cooling.

(b) Find $\lim_{t \to \infty} T$ and identify any horizontal asymptotes.

36. *Rules of 70 and 72* The rules state that it takes about $70/i$ or $72/i$ years for money to double at i percent, compounded continuously, using whichever of 70 or 72 is easier to divide by i.

(a) Show that it takes $t = (\ln 2)/r$ years for money to double if it is invested at annual interest rate r (in decimal form) compounded continuously.

(b) Graph the functions

$$y_1 = \frac{\ln 2}{r}, \quad y_2 = \frac{70}{100r}, \quad \text{and} \quad y_3 = \frac{72}{100r}$$

in the $[0, 0.1]$ by $[0, 100]$ viewing window.

(c) Writing to Learn Explain why these two rules of thumb for mental computation are reasonable.

(d) Use the rules to estimate how long it takes to double money at 5% compounded continuously.

(e) Invent a rule for estimating the number of years needed to triple your money. ■

Extending the Ideas

37. *Continuously Compounded Interest*

(a) Use tables to give a numerical argument that

$$\lim_{x \to \infty} \left(1 + \frac{1}{x}\right)^x = e.$$

Support your argument graphically.

(b) For several different values of r, give numerical and graphical evidence that

$$\lim_{x \to \infty} \left(1 + \frac{r}{x}\right)^x = e^r.$$

(c) Writing to Learn Explain why compounding interest over smaller and smaller periods of time leads to the concept of interest compounded continuously.

38. *Skydiving* If a body of mass m falling from rest under the action of gravity encounters an air resistance proportional to the square of the velocity, then the body's velocity $v(t)$ is modeled by the initial value problem

Differential equation: $\quad m\dfrac{dv}{dt} = mg - kv^2,$

Initial condition: $\quad v(0) = 0,$

where t represents time in seconds, g is the acceleration due to gravity, and k is a constant that depends on the body's aerodynamic properties and the density of the air. (We assume that the fall is short enough so that variation in the air's density will not affect the outcome.)

(a) Show that the function

$$v(t) = \sqrt{\frac{mg}{k}} \, \frac{e^{at} - e^{-at}}{e^{at} + e^{-at}},$$

where $a = \sqrt{gk/m}$, is a solution of the initial value problem.

(b) Find the body's limiting velocity, $\lim_{t \to \infty} v(t)$.

(c) For a 160-lb skydiver ($mg = 160$), and with time in seconds and distance in feet, a typical value for k is 0.005. What is the diver's limiting velocity in feet per second? in miles per hour?

Skydivers can vary their limiting velocities by changing the amount of body area opposing the fall. Their velocities can vary from 94 to 321 miles per hour.

6.5 Population Growth

Exponential Model • Logistic Growth Model • Logistic Regression

Exponential Model

Strictly speaking, the number of individuals in a population is a discontinuous function of time because it takes on only whole number values. However, one common way to model a population is with a differentiable function P growing at a rate proportional to the size of the population. Thus, for some constant k,

$$\frac{dP}{dt} = kP.$$

Notice that

$$\frac{dP/dt}{P} = k$$

is constant. This rate is called the **relative growth rate.** As we saw in Section 6.4, we can represent the population by the model $P = P_0 e^{kt}$, where P_0 is the size of the population at time $t = 0$.

In Table 1.7 of Section 1.3 we gave the world population for the years 1986 to 1991, which we repeat here as Table 6.7.

Table 6.7 World Population

Year	Population (millions)	Ratio
1986	4936	
1987	5023	$5023/4936 \approx 1.0176$
1988	5111	$5111/5023 \approx 1.0175$
1989	5201	$5201/5111 \approx 1.0176$
1990	5329	$5329/5201 \approx 1.0246$
1991	5422	$5422/5329 \approx 1.0175$

Source: Statistical Office of the United Nations, *Monthly Bulletin of Statistics,* 1991.

Example 1 PREDICTING WORLD POPULATION

Find an initial value problem model for world population and use it to predict the population in the year 2010. Graph the model and the data.

Solution We let $t = 0$ represent 1986, $t = 1$ represent 1987, and so forth. The year 2010 will be represented by $t = 24$.

If we approximate the ratios in Table 6.7 by 1.018, then the relative rate k of growth satisfies $e^k = 1.018$, so k is about 0.0178. We obtain the initial value problem

Differential equation: $\quad \dfrac{dP}{dt} = 0.0178P,$

Initial condition: $\quad P(0) = 4936.$

The solution to this initial value problem gives us the model $P = 4936e^{0.0178t}$.

[−1, 25] by [4500, 8000]

X = 24   Y = 7566.6867

Figure 6.11 Notice that the value of the model $P = 4936e^{0.0178t}$ is 7566.6867 when $t = 24$. (Example 1)

The year 2010 is represented by $t = 24$, so

$$P(24) \approx 7567.$$

Figure 6.11 shows the graph of the model superimposed on a scatter plot of the data.

Interpret

This model predicts the world population in the year 2010 to be about 7567 million, or 7.6 billion, slightly less than the prediction in Example 1 of Section 1.3.

Logistic Growth Model

The exponential model for population growth assumes unlimited growth. This is realistic only for a short period of time when the initial population is small. A more realistic assumption is that the relative growth rate is positive but decreases as the population increases due to environmental or economic factors. In other words there is a maximum population M, the **carrying capacity,** that the environment is capable of sustaining in the long run. If we assume the relative growth rate is proportional to $1 - (P/M)$ with positive proportionality constant k, then

$$\frac{dP/dt}{P} = k\left(1 - \frac{P}{M}\right) \quad \text{or} \quad \frac{dP}{dt} = \frac{k}{M}P(M - P). \tag{1}$$

The solution to this **logistic differential equation** is called the **logistic growth model.** Notice that the rate of growth is proportional to both P and $(M - P)$, where M is the assumed maximum population. If P were to exceed M, the growth rate would be negative (as $k > 0$, $M > 0$) and the population would be decreasing.

Example 2   MODELING BEAR POPULATION

A national park is known to be capable of supporting no more than 100 grizzly bears. Ten bears are in the park at present. We model the population with a logistic differential equation with $k = 0.1$.

(a) Draw and describe a slope field for the differential equation.

(b) Find a logistic growth model $P(t)$ for the population and draw its graph.

(c) When will the bear population reach 50?

Solution **(a)** The carrying capacity is 100, so $M = 100$. The model we seek is a solution to the following differential equation.

$$\frac{dP}{dt} = \frac{0.1}{100}P(100 - P) \quad \text{Eq. 1 with } k = 0.1, M = 100$$

$$= 0.001P(100 - P)$$

Figure 6.12 shows a slope field for this differential equation. There appears to be a horizontal asymptote at $P = 100$. The solution curves fall toward this level from above, and rise toward it from below.

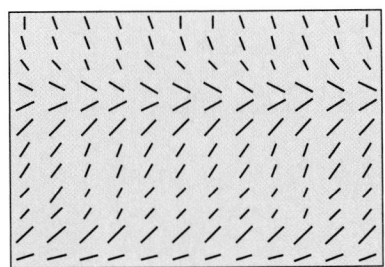

[0, 150] by [0, 150]

Figure 6.12 A slope field for the logistic differential equation

$$\frac{dP}{dt} = 0.001P(100 - P).$$

(Example 2)

Partial fractions are reviewed in more detail in Section 8.4.

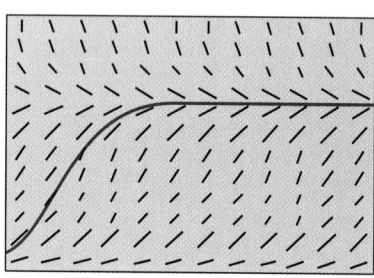

[0, 150] by [0, 150]

Figure 6.13 The graph of

$$P = \frac{100}{1 + 9e^{-0.1t}}$$

superimposed on a slope field for

$$\frac{dP}{dt} = 0.001P(100 - P).$$

(Example 2)

(b) We can assume that $t = 0$ when the bear population is 10, so $P(0) = 10$. The logistic growth model we seek is the solution to the following initial value problem.

Differential equation: $\quad \dfrac{dP}{dt} = 0.001P(100 - P)$

Initial condition: $\quad P(0) = 10$

To prepare for integration we rewrite the differential equation in the form

$$\frac{1}{P(100 - P)} \frac{dP}{dt} = 0.001.$$

Using partial fractions, we rewrite $1/(P(100 - P))$ as

$$\frac{1}{P(100 - P)} = \frac{1}{100}\left(\frac{1}{P} + \frac{1}{100 - P}\right).$$

Substituting this expression into the differential equation and multiplying both sides by 100 we obtain

$$\left(\frac{1}{P} + \frac{1}{100 - P}\right)\frac{dP}{dt} = 0.1$$

$$\ln|P| - \ln|100 - P| = 0.1t + C \qquad \text{Integrate with respect to } t.$$

$$\ln\left|\frac{P}{100 - P}\right| = 0.1t + C$$

$$\ln\left|\frac{100 - P}{P}\right| = -0.1t - C \qquad \ln\frac{a}{b} = -\ln\frac{b}{a}$$

$$\left|\frac{100 - P}{P}\right| = e^{-0.1t - C} \qquad \text{Exponentiate.}$$

$$\frac{100 - P}{P} = (\pm e^{-C})e^{-0.1t}$$

$$\frac{100}{P} - 1 = Ae^{-0.1t} \qquad \text{Let } A = \pm e^{-C}.$$

$$P = \frac{100}{1 + Ae^{-0.1t}}. \qquad \text{Solve for } P.$$

This is the general solution to the differential equation. When $t = 0$, $P = 10$, and we obtain

$$10 = \frac{100}{1 + Ae^0}$$

$$1 + A = 10$$

$$A = 9.$$

Thus, the logistic growth model is

$$P = \frac{100}{1 + 9e^{-0.1t}}.$$

Its graph (Figure 6.13) is superimposed on the slope field from Figure 6.12.

(c) For this model, the bear population will reach 50 when

$$50 = \frac{100}{1 + 9e^{-0.1t}}$$

$$1 + 9e^{-0.1t} = 2$$

$$e^{-0.1t} = \frac{1}{9}$$

$$e^{0.1t} = 9$$

$$t = \frac{\ln 9}{0.1} \approx 22 \text{ years.}$$

The solution of the general logistic differential equation

$$\frac{dP}{dt} = \frac{k}{M}P(M - P)$$

(Equation 1) can be obtained as in Example 2. In Exercise 25, we ask you to show that the solution is

$$P = \frac{M}{1 + Ae^{-kt}}.$$

The value of A is determined by an appropriate initial condition.

Logistic Regression

The logistic growth model is built in on many graphers as the logistic regression equation. It is not a good idea to use logistic regression with the data in Table 6.7, though, because the time interval is too short. In the next example we consider U.S. population data from 1880 to 1990.

Example 3 USING LOGISTIC REGRESSION

(a) Find the logistic regression equation for the U.S. population data in Table 6.8 and superimpose its graph on a scatter plot of the data.

(b) Use the model in (a) to find the carrying capacity predicted by the regression equation.

(c) Determine when the rate of growth predicted by the regression equation changes from increasing to decreasing. Estimate the population at this time.

Solution **(a)** We let $x = 0$ represent 1880, $x = 10$ represent 1890, and so forth. The logistic regression equation obtained from the grapher we use is approximately

$$y = \frac{540.7}{1 + 8.972e^{-0.01856x}}.$$

See Figure 6.14.

(b) The form of the logistic model allows us to read that the carrying capacity is 540.7 million. This number is also the limit of y as $x \to \infty$.

$$\lim_{x \to \infty} y = \lim_{x \to \infty} \frac{540.7}{1 + 8.972e^{-0.01856x}} = 540.7$$

Table 6.8 U.S. Population, 1880–1990

Year	Population (millions)
1880	50.2
1890	63.0
1900	76.0
1910	92.0
1920	105.7
1930	122.8
1940	131.7
1950	151.3
1960	179.3
1970	203.3
1980	226.5
1990	248.7

Source: The Statesman's Yearbook, 133rd ed. (London: The Macmillan Press, Ltd., 1996).

[0, 200] by [0, 600]

Figure 6.14 The graph of $y = 540.7/(1 + 8.972e^{-0.01856x})$ superimposed on a scatter plot of the data in Table 6.8. (Example 3)

[0, 200] by [–0.03, 0.03]

Figure 6.15 The graph of the second derivative of $y = 540.7/(1 + 8.972e^{-0.01856x})$ obtained using NDER twice. (Example 3)

(c) In the long run, the population will stabilize at 540.7 million. If the first derivative of a twice-differentiable function changes from increasing to decreasing, it must do so at a point of inflection. We look for a zero of y''. Figure 6.15 shows that $y'' = 0$ when $x \approx 118$, or in 1998. The population at this time, $y^{(118)}$, is predicted to be about 270 million.

Suppose we have reason to believe that the maximum population found in Example 3 is too high. Example 4 shows how we can use a projected value of the population to find a new regression model with smaller carrying capacity.

Example 4   USING LOGISTIC REGRESSION

Use the data in Table 6.8 and the estimate that the U.S. population will be 350 million in 2080 to find a new logistic growth model for the population. Graph the data and the model. What is the new carrying capacity?

Solution We add the year 2080 with population 350 million to the data in Table 6.8 and compute the new logistic regression model to be about

$$y = \frac{378}{1 + 6.48e^{-0.02232x}}.$$

The new carrying capacity is 378 million (Figure 6.16).

[0, 300] by [0, 400]

Figure 6.16 (Example 4)

Quick Review 6.5

In Exercises 1–8, let

$$f(x) = \frac{50}{1 + 5e^{-0.1x}}.$$

1. Find where f is continuous.

2. Find $\lim_{x \to \infty} f(x)$; $\lim_{x \to -\infty} f(x)$.

3. Find the horizontal asymptotes of the graph of f.

4. Find the domain of f' and the domain of f''.

5. Draw the graph of f and find its zeros.

6. Draw the graph of f' and find the intervals on which f is
(a) increasing, (b) decreasing.

7. Draw the graph of f'' and find where the graph of f is
(a) concave up, (b) concave down.

8. Find the points of inflection of the graph of f.

In Exercises 9 and 10, find the values of A and B to complete the partial fraction decomposition.

9. $\dfrac{x - 12}{x^2 - 4x} = \dfrac{A}{x} + \dfrac{B}{x - 4}$

10. $\dfrac{2x + 16}{x^2 + x - 6} = \dfrac{A}{x + 3} + \dfrac{B}{x - 2}$

Section 6.5 Exercises

In Exercises 1–4, a population is described.

(a) Write a differential equation for the population.

(b) Find a formula for the population P in terms of t.

(c) Superimpose the graph of the population function on a slope field for the differential equation.

1. The relative growth rate of Clairsville is 0.025 and its current population is $P(0) = 75{,}000$.

2. The relative growth rate of Blairsville is 1.9% and its current population is $P(0) = 110{,}000$.

3. logistic growth with $k = 0.05$, $M = 200$, and $P(0) = 10$

4. logistic growth with $k = 0.02$, $M = 150$, and $P(0) = 15$

In Exercises 5 and 6, find the relative growth rate of the population represented by the differential equation.

5. $\dfrac{dP}{dt} = -0.3P$ **6.** $\dfrac{dP}{dt} = 0.075P$

In Exercises 7 and 8, find k and the carrying capacity for the population represented by the logistic differential equation.

7. $\dfrac{dP}{dt} = 0.04P - 0.0004P^2$ **8.** $\dfrac{50}{P}\dfrac{dP}{dt} = 2 - \dfrac{P}{250}$

In Exercises 9–12, match the differential equation with its slope field.

9. $\dfrac{dy}{dx} = 0.065y$ **10.** $\dfrac{dy}{dx} = 0.06y\left(1 - \dfrac{y}{100}\right)$

11. $\dfrac{dy}{dx} = \dfrac{y}{x} - y$ **12.** $\dfrac{dy}{dx} = 0.06y\left(1 - \dfrac{y}{150}\right)$

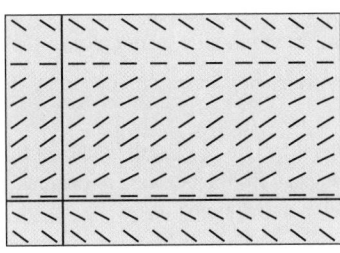

[–20, 100] by [–50, 200]

(a)

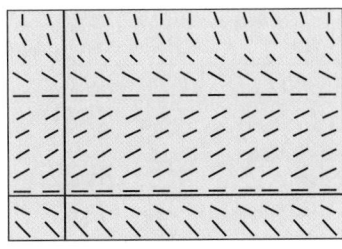

[–20, 100] by [–50, 200]

(b)

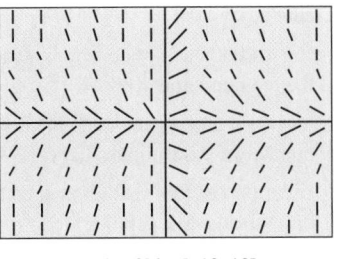

[–10, 10] by [–10, 10]

(c)

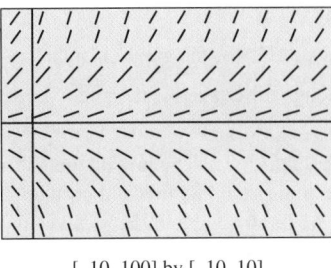

[–10, 100] by [–10, 10]

(d)

In Exercises 13 and 14, a population function is given.

(a) Show that the function is a solution of a logistic differential equation. Identify k and the carrying capacity.

(b) Writing to Learn Estimate $P(0)$. Explain its meaning in the context of the problem.

13. *Rabbit Population* A population of rabbits is given by the formula
$$P(t) = \frac{1000}{1 + e^{4.8 - 0.7t}},$$
where t is the number of months after a few rabbits are released.

14. *Spread of Measles* The number of students infected by measles in a certain school is given by the formula
$$P(t) = \frac{200}{1 + e^{5.3 - t}},$$
where t is the number of days after students are first exposed to an infected student.

15. *U.S. Population* The Museum of Science in Boston displays a running total of the U.S. population. On May 11, 1993, the total was increasing at the rate of 1 person every 14 sec. The displayed population figure at 3:45 P.M. that day was 257,313,431. Assume that the relative growth rate is constant.

(a) What is the relative growth rate per year (365 days)?

(b) At this rate, what will the U.S. population be at 3:45 P.M. Boston time on May 11, 2001?

16. *Spread of a Disease* Assume that the relative rate of spread of a certain disease is constant. Suppose also that in the course of any given year the number of cases of the disease is reduced by 20%.

(a) If there are 10,000 cases today, how many years will it take to reduce the number to 1000?

(b) How long will it take to eradicate the disease, that is, to reduce the number of cases to less than 1?

17. *Guppy Population* A 2000-gallon tank can support no more than 150 guppies. Six guppies are introduced into the tank. Assume that the rate of growth of the population is

$$\frac{dP}{dt} = 0.0015P(150 - P),$$

where time t is in weeks.

(a) Find a formula for the guppy population in terms of t.

(b) How long will it take for the guppy population to be 100? 125?

18. *Gorilla Population* A certain wild animal preserve can support no more than 250 lowland gorillas. Twenty-eight gorillas were known to be in the preserve in 1970. Assume that the rate of growth of the population is

$$\frac{dP}{dt} = 0.0004P(250 - P),$$

where time t is in years.

(a) Find a formula for the gorilla population in terms of t.

(b) How long will it take for the gorilla population to reach the carrying capacity of the preserve?

19. *Tooth Size* The analysis of tooth shrinkage by C. Loring Brace and colleagues at the University of Michigan's Museum of Anthropology indicates that human tooth size is continuing to decrease and that the evolutionary process did not come to a halt some 30,000 years ago as many scientists contend. In northern Europe, for example, tooth size reduction now has a constant relative rate of 1% per 1000 years.

(a) Find a model for northern European tooth size y in terms of time t in years.

(b) In about how many years will northern European tooth size be 90% of its present size?

(c) What will be northern European tooth size 20,000 years from now as a percentage of our present tooth size?
Source: LSA Magazine, Spring 1989, Vol. 12, p. 19, Ann Arbor, MI.

20. *Honeybee Population* A population of honeybees grows at an annual rate equal to 1/4 of the number present when there are no more than 10,000 bees. If there are more than 10,000 bees but fewer than 50,000 bees, the growth rate is equal to 1/12 of the number present. If there are 5000 bees now, when will there be 25,000 bees?

21. *Continuous Compounding* You have $1000 with which to open an account to which you plan also to add $1000 per year. All funds in the account will earn 10% annual interest compounded continuously. If the added deposits are also credited to your account continuously, the number of dollars x in your account at time t (years) will satisfy the initial value problem.

Differential equation: $\dfrac{dx}{dt} = 1000 + 0.10x$

Initial condition: $x(0) = 1000$

(a) Solve the initial value problem for x as a function of t.

(b) About how many years will it take for the amount in your account to reach $100,000?

22. *Continuous Price Discounting* To encourage buyers to place 100-unit orders, your firm's sales department applies a continuous discount that makes the unit price a function $p(x)$ of the number of units ordered. The discount decreases the price at the rate of $0.01 per unit ordered. The price per unit for a 100-unit order is $p(100) = \$20.09$.

(a) Find $p(x)$ by solving the following initial value problem.

Differential equation: $\dfrac{dp}{dx} = -\dfrac{1}{100}p$

Initial condition: $p(100) = 20.09$

(b) Find the unit price $p(10)$ for a 10-unit order and the unit price $p(90)$ for a 90-unit order.

(c) The sales department has asked you to find out whether it is discounting so much that the firm's revenue, $r(x) = x \cdot p(x)$, will actually be less for a 100-unit order than, say, for a 90-unit order. Reassure them by showing that r has its maximum value at $x = 100$.

In Exercises 23 and 24, use the census data provided.

(a) Find the logistic regression equation for the data and superimpose its graph on a scatter plot of the data.

(b) Find the carrying capacity predicted by the regression equation.

(c) Find when the rate of growth predicted by the regression equation changes from increasing to decreasing. Estimate the population at this time.

23. *State of New York* Let $x = 0$ represent 1910, $x = 20$ represent 1930, and so forth.

Table 6.9 New York Population

Year	Population (millions)
1910	9.1
1930	12.6
1960	16.8
1980	17.6
1990	18.0

Source: The Statesman's Yearbook, 133rd ed. (London: The Macmillan Press, Ltd., 1996).

24. *State of Florida* Let $x = 0$ represent 1950, $x = 10$ represent 1960, and so forth.

Table 6.10 Florida Population	
Year	Population (millions)
1950	2.8
1960	5.0
1970	6.8
1980	9.7
1990	12.9

Source: The Statesman's Yearbook, 133rd ed. (London: The Macmillan Press, Ltd., 1996).

25. *Logistic Differential Equation* Show that the solution of the differential equation (Equation 1)

$$\frac{dP}{dt} = \frac{k}{M}P(M - P) \quad \text{is} \quad P = \frac{M}{1 + Ae^{-kt}},$$

where A is an arbitrary constant.

26. *Tinkering with Carrying Capacity* Suppose you decide to estimate the population of Florida to be 16 million in the year 2020. Include this data point with the data in Table 6.10.

(a) Find a logistic regression equation for the new data.

(b) What is the new carrying capacity?

In Exercises 27–30, solve the initial value problem.

27. $\dfrac{dy}{dx} = (\cos x)\, e^{\sin x}, \quad y(0) = 0$

28. $\dfrac{dy}{dx} = -2(y - 3), \quad y(0) = 5$

29. $\dfrac{dy}{dx} = \dfrac{x}{y}, \quad y(0) = 2$ **30.** $\dfrac{dy}{dx} = y\sqrt{x}, \quad y(0) = 1$

Exploration

31. *Extinct Populations* One theory states that if the size of a population falls below a minimum m, the population will become extinct. This condition leads to the *extended* logistic differential equation

$$\frac{dP}{dt} = kP\left(1 - \frac{P}{M}\right)\left(1 - \frac{m}{P}\right)$$

$$= \frac{k}{M}(M - P)(P - m),$$

with $k > 0$ the proportionality constant and M the population maximum.

(a) Show that dP/dt is positive for $m < P < M$ and negative if $P < m$ or $P > M$.

(b) Let $m = 100$, $M = 1200$, and assume that $m < P < M$. Show that the differential equation can be rewritten in the form

$$\left[\frac{1}{1200 - P} + \frac{1}{P - 100}\right]\frac{dP}{dt} = \frac{11}{12}k.$$

Use a procedure similar to that used in Example 2 to solve this differential equation.

(c) Find the solution to (b) that satisfies $P(0) = 300$.

(d) Superimpose the graph of the solution in (c) with $k = 0.1$ on a slope field of the differential equation.

(e) Solve the general extended differential equation with the restriction $m < P < M$. ∎

Extending the Ideas

32. *Variable Inflation Rate* When the inflation rate k varies with time instead of being constant, the formula $p = p_0 e^{kt}$ no longer gives the solution of the differential equation $dp/dt = kp$.

(a) Show that the solution of the differential equation satisfying $p(0) = p_0$ is

$$p(t) = p_0 e^{\int_0^t k(u)\,du}.$$

(b) **Writing to Learn** Suppose that $p_0 = 100$ and $k(t) = 0.04/(1 + t)$, a rate of inflation that starts at 4% at time $t = 0$ but decreases steadily as the years pass. Find $p(9)$ and give a meaning to its value.

(c) Find the price of a \$100 item at $t = 9$ years if the annual rate of inflation is a constant 4%.

(d) Suppose that $p_0 = 100$ and $k(t) = 0.04 + 0.004t$, a rate of inflation that starts at 4% when $t = 0$ but increases steadily as the years pass. Find $p(9)$.

33. *Catastrophic Solution* Let k and P_0 be positive constants.

(a) Solve the initial value problem.

$$\frac{dP}{dt} = kP^2, \quad P(0) = P_0$$

(b) Show that the graph of the solution in (a) has a vertical asymptote at a positive value of t.

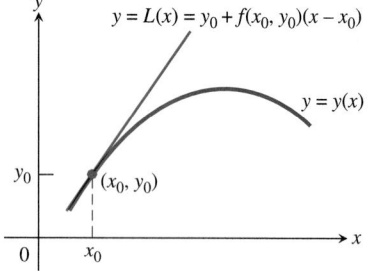

Figure 6.17 The linearization of $y = y(x)$ at $x = x_0$.

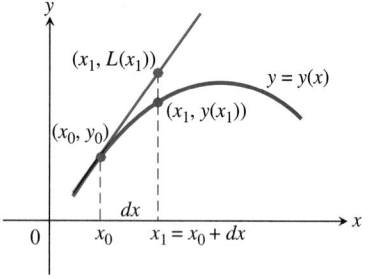

Figure 6.18 The first Euler step approximates $y(x_1)$ with $L(x_1)$.

6.6 Numerical Methods

Euler's Method • Numerical Solutions • Graphical Solutions • Improved Euler's Method

Euler's Method

If we are given a differential equation $dy/dx = f(x, y)$ and an initial condition $y(x_0) = y_0$, we can approximate the solution $y = y(x)$ by its linearization

$$L(x) = y(x_0) + y'(x_0)(x - x_0) \quad \text{or} \quad L(x) = y_0 + f(x_0, y_0)(x - x_0).$$

The function $L(x)$ will give a good approximation to the solution $y(x)$ in a short interval about x_0 (Figure 6.17). The basis of Euler's method is to patch together a string of linearizations to approximate the curve over a longer stretch. Here is how the method works.

We know the point (x_0, y_0) lies on the solution curve. Suppose we specify a new value for the independent variable to be $x_1 = x_0 + dx$. If the increment dx is small, then

$$y_1 = L(x_1) = y_0 + f(x_0, y_0) \, dx$$

is a good approximation to the exact solution value $y = y(x_1)$. So from the point (x_0, y_0), which lies *exactly* on the solution curve, we have obtained the point (x_1, y_1), which lies very close to the point $(x_1, y(x_1))$ on the solution curve (Figure 6.18).

Using the point (x_1, y_1) and the slope $f(x_1, y_1)$ of the solution curve through (x_1, y_1), we take a second step. Setting $x_2 = x_1 + dx$, we use the linearization of the solution curve through (x_1, y_1) to calculate

$$y_2 = y_1 + f(x_1, y_1) \, dx.$$

This gives the next approximation (x_2, y_2) to values along the solution curve $y = y(x)$ (Figure 6.19). Continuing in this fashion, we take a third step from the point (x_2, y_2) with slope $f(x_2, y_2)$ to obtain the third approximation

$$y_3 = y_2 + f(x_2, y_2) \, dx,$$

and so on. We are literally building an approximation to one of the solutions by following the direction of the slope field of the differential equation.

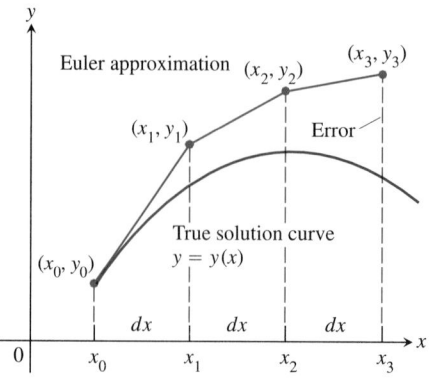

Figure 6.19 Three steps in the Euler approximation to the solution of the initial value problem $y' = f(x, y)$, $y(x_0) = y_0$. As we take more steps, the errors involved usually accumulate, but not in the exaggerated way shown here.

The steps in Figure 6.19 are drawn large to illustrate the construction process, so the approximation looks crude. In practice, dx would be small enough to make the red curve hug the blue one and give a good approximation throughout.

Example 1 USING EULER'S METHOD

Find the first three approximations y_1, y_2, y_3 using Euler's method for the initial value problem

$$y' = 1 + y, \quad y(0) = 1,$$

starting at $x_0 = 0$ with $dx = 0.1$.

Solution We have $x_0 = 0$, $y_0 = 1$, $x_1 = x_0 + dx = 0.1$, $x_2 = x_0 + 2dx = 0.2$, and $x_3 = x_0 + 3dx = 0.3$.

First:
$$\begin{aligned} y_1 &= y_0 + f(x_0, y_0)\, dx \\ &= y_0 + (1 + y_0)\, dx \\ &= 1 + (1 + 1)(0.1) = 1.2 \end{aligned}$$

Second:
$$\begin{aligned} y_2 &= y_1 + f(x_1, y_1)\, dx \\ &= y_1 + (1 + y_1)\, dx \\ &= 1.2 + (1 + 1.2)(0.1) = 1.42 \end{aligned}$$

Third:
$$\begin{aligned} y_3 &= y_2 + f(x_2, y_2)\, dx \\ &= y_2 + (1 + y_2)\, dx \\ &= 1.42 + (1 + 1.42)(0.1) = 1.662 \end{aligned}$$

Numerical Solutions

If we do not require or cannot immediately find an *exact* solution for an initial value problem $y' = f(x, y)$, $y(x_0) = y_0$, we can use a grapher to generate a table of approximate numerical values of y for values of x in an appropriate interval. Such a table is a **numerical solution** of the initial value problem and the method by which we generate the table is a **numerical method.**

The step-by-step process used in Example 1 can be continued easily. If we use equally spaced values for the independent variable in the table and generate n of them, we first set

$$x_1 = x_0 + dx,$$
$$x_2 = x_1 + dx,$$
$$\vdots$$
$$x_n = x_{n-1} + dx.$$

Then we calculate the approximations to the solution,

$$y_1 = y_0 + f(x_0, y_0)\, dx,$$
$$y_2 = y_1 + f(x_1, y_1)\, dx,$$
$$\vdots$$
$$y_n = y_{n-1} + f(x_{n-1}, y_{n-1})\, dx.$$

The number of steps n can be as large as we like, but errors can accumulate if n is too large.

This process can be easily programmed. We use the program EULERT (EULER Table) to generate numerical solutions to initial value problems. It allows us to choose n and the step size dx.

In Exercise 5, you will show that the exact solution to the initial value problem of Example 1 is $y = 2e^x - 1$. We will use this information in Example 2.

Example 2 INVESTIGATING ACCURACY OF EULER'S METHOD

Use Euler's Method to solve

$$y' = 1 + y, \quad y(0) = 1,$$

on the interval $0 \le x \le 1$, starting at $x_0 = 0$ and taking **(a)** $dx = 0.1$ and **(b)** $dx = 0.05$. Compare the approximations to the values of the exact solution $y = 2e^x - 1$.

Solution **(a)** We use EULERT to generate the approximate values in Table 6.11. The "error" column is obtained by subtracting the unrounded Euler values from the unrounded values found using the exact solution. All entries are then rounded to 4 decimal places.

Table 6.11 Euler Solution of $y' = 1 + y$, $y(0) = 1$, Step Size $dx = 0.1$

x	y (Euler)	y (exact)	Error
0	1	1	0
0.1	1.2	1.2103	0.0103
0.2	1.42	1.4428	0.0228
0.3	1.662	1.6997	0.0377
0.4	1.9282	1.9836	0.0554
0.5	2.2210	2.2974	0.0764
0.6	2.5431	2.6442	0.1011
0.7	2.8974	3.0275	0.1301
0.8	3.2872	3.4511	0.1639
0.9	3.7159	3.9192	0.2033
1.0	4.1875	4.4366	0.2491

By the time we reach $x = 1$ (after 10 steps), the error is about 5.6% of the exact solution.

(b) One way to try to reduce the error is to decrease the step size. Table 6.12 shows the results and their comparisons with the exact solutions when we decrease the step size to 0.05, doubling the number of steps to 20. As in Table 6.11, all computations are performed before rounding. This time when we reach $x = 1$ the relative error is only about 2.9%.

Table 6.12 Euler Solution of $y' = 1 + y$, $y(0) = 1$, Step Size $dx = 0.05$

x	y (Euler)	y (exact)	Error
0	1	1	0
0.05	1.1	1.1025	0.0025
0.10	1.205	1.2103	0.0053
0.15	1.3153	1.3237	0.0084
0.20	1.4310	1.4428	0.0118
0.25	1.5526	1.5681	0.0155
0.30	1.6802	1.6997	0.0195
0.35	1.8142	1.8381	0.0239
0.40	1.9549	1.9836	0.0287
0.45	2.1027	2.1366	0.0340
0.50	2.2578	2.2974	0.0397
0.55	2.4207	2.4665	0.0458
0.60	2.5917	2.6442	0.0525
0.65	2.7713	2.8311	0.0598
0.70	2.9599	3.0275	0.0676
0.75	3.1579	3.2340	0.0761
0.80	3.3657	3.4511	0.0853
0.85	3.5840	3.6793	0.0953
0.90	3.8132	3.9192	0.1060
0.95	4.0539	4.1714	0.1175
1.00	4.3066	4.4366	0.1300

It might be tempting to reduce the step size even further to obtain greater accuracy. However, each additional calculation not only requires additional grapher time but more importantly adds to the buildup of round-off errors due to the approximate representations of numbers inside the grapher.

The analysis of error and the investigation of methods to reduce it when making numerical calculations are important, but appropriate for a more advanced course. There are numerical methods more accurate than Euler's method, as you will see in your further study of differential equations.

Graphical Solutions

Investigating a graph is usually more instructive than analyzing a large data table. If we plot the data pairs in a numerical solution of an initial value problem we obtain a **graphical solution,** as shown in Example 3.

Example 3 VISUALIZING EULER'S APPROXIMATIONS

Figure 6.20 gives a visualization of the numerical solution shown as Table 6.11 by superimposing the graph of the exact solution on a scatter plot of the data points in the table.

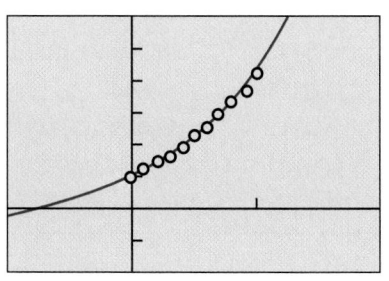

[–1, 2] by [–2, 6]

Figure 6.20 The graph of $y = 2e^x - 1$ superimposed on a scatter plot of the Euler approximations shown in Table 6.11. (Example 3)

Improved Euler's Method

We can improve on Euler's method by taking an average of two slopes. We first estimate y_n as in the original Euler method, but denote it by z_n. We then take the average of $f(x_{n-1}, y_{n-1})$ and $f(x_n, z_n)$ in place of $f(x_{n-1}, y_{n-1})$ in the next step. Thus,

$$z_n = y_{n-1} + f(x_{n-1}, y_{n-1})\, dx,$$

$$y_n = y_{n-1} + \left[\frac{f(x_{n-1}, y_{n-1}) + f(x_n, z_n)}{2} \right] dx.$$

We use the program IMPEULT (IMProved EULer Table) to perform the computations.

Example 4 INVESTIGATING THE ACCURACY OF IMPROVED EULER'S METHOD

Use the improved Euler's method to solve

$$y' = 1 + y, \quad y(0) = 1$$

on the interval $0 \le x \le 1$, starting at $x_0 = 0$ and taking $dx = 0.1$. Compare the approximations to the values of the exact solution $y = 2e^x - 1$.

Solution We use IMPEULT to generate the approximate values in Table 6.13. The "error" column is obtained by subtracting the unrounded improved Euler values from the unrounded values found using the exact solution. All entries are then rounded to 4 decimal places.

Table 6.13 Improved Euler Solution of $y' = 1 + y$, $y(0) = 1$, Step Size $dx = 0.1$

x	y (IMPEULT)	y (exact)	Error
0	1	1	0
0.1	1.21	1.2103	0.0003
0.2	1.4421	1.4428	0.0008
0.3	1.6985	1.6997	0.0013
0.4	1.9818	1.9836	0.0018
0.5	2.2949	2.2974	0.0025
0.6	2.6409	2.6442	0.0034
0.7	3.0231	3.0275	0.0044
0.8	3.4456	3.4511	0.0055
0.9	3.9124	3.9192	0.0068
1.0	4.4282	4.4366	0.0084

By the time we reach $x = 1$ (after 10 steps), the relative error is about 0.19%.

By comparing Tables 6.11 and 6.13, we see that the improved Euler's method is considerably more accurate than the regular Euler's method, at least for the initial value problem $y' = 1 + y$, $y(0) = 1$.

Instead of making a scatter plot of the Euler approximations, we can use the program EULERG (EULER Graph), a modification of EULERT, to plot the points from the Euler approximation directly to obtain a graphical solution.

Example 5 FINDING GRAPHICAL SOLUTIONS

Find a graphical solution for the initial value problem

$$y' = x - y, \quad y(0) = 1.$$

Compare with the graph of the exact solution, $y = x - 1 + 2e^{-x}$.

Solution Figure 6.21a shows the EULERG solution of the initial value problem starting with $x = 0$ with step size 0.1. Figure 6.21b shows the graph of the exact solution superimposed on the Euler graph.

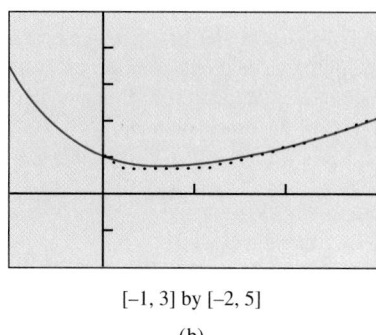

[−1, 3] by [−2, 5]　　　　　　　　　　　　[−1, 3] by [−2, 5]

(a)　　　　　　　　　　　　　　　　　　(b)

Figure 6.21 (b) The graph of the exact solution $y = x - 1 + 2e^{-x}$ of the initial value problem $y' = x - y, y(0) = 1$ agrees with (a) the graph of the Euler solution. (Example 5)

Quick Review 6.6

In Exercises 1 and 2, let $f(x) = x^3 - 3x$.

1. Find $f'(2)$.

2. Find the linearization $L(x)$ of f at $x = 2$.

In Exercises 3 and 4, let $f(x) = \tan x$.

3. Find $f'(\pi/4)$.

4. Find the linearization $L(x)$ of f at $x = \pi/4$.

In Exercises 5 and 6, let $f(x) = 0.1x^2 + (5/x)$.

5. Find $f'(4)$.

6. Find the linearization $L(x)$ of f at $x = 4$.

In Exercises 7–10, use the approximation $L(x) \approx f(x)$ found in Exercise 6. Find the (a) error and (b) percentage error in the approximation at $x = a$.

7. $a = 4.1$　　　　　　　　**8.** $a = 4.2$

9. $a = 4.5$　　　　　　　　**10.** $a = 3.5$

Section 6.6 Exercises

In Exercises 1–4, use differentiation and substitution to show that the function $y = f(x)$ is the exact solution of the initial value problem.

1. $y' = x - y, \quad y(0) = 1, \quad f(x) = x - 1 + 2e^{-x}$

2. $y' = x - y, \quad y(0) = -2, \quad f(x) = x - 1 - e^{-x}$

3. $y' = 2y + \sin x, \quad y(0) = 0, \quad f(x) = \dfrac{e^{2x} - 2 \sin x - \cos x}{5}$

4. $y' = y - e^{2x} + 1, \quad y(0) = -1, \quad f(x) = e^x - e^{2x} - 1$

In Exercises 5–8, use analytic methods to find the exact solution of the initial value problem.

5. $y' = 1 + y$, $y(0) = 1$ **6.** $y' = x(1 - y)$, $y(-2) = 0$

7. $y' = 2y(x + 1)$, $y(-2) = 2$

8. $y' = y^2(1 + 2x)$, $y(-1) = -1$

In Exercises 9 and 10, use Euler's method to solve the initial value problem on the interval $0 \le x \le 1$ starting at $x_0 = 0$ with $dx = 0.1$. Compare the approximations to the values of the exact solution.

9. $y' = 2y + \sin x$, $y(0) = 0$
(See Exercise 3 for the exact solution.)

10. $y' = x - y$, $y(0) = -2$
(See Exercise 2 for the exact solution.)

In Exercises 11 and 12, use the improved Euler's method to solve the initial value problem on the interval $-2 \le x \le 0$ starting at $x_0 = -2$ with $dx = 0.1$. Compare the approximations to the values of the exact solution.

11. $y' = 2y(x + 1)$, $y(-2) = 2$
(See Exercise 7 for the exact solution.)

12. $y' = x(1 - y)$, $y(-2) = 0$
(See Exercise 6 for the exact solution.)

In Exercises 13 and 14, *work in groups of two or three.* Use Euler's method and the improved Euler's method to solve the initial value problem on the interval $0 \le x \le 2$ starting at $x_0 = 0$ with $dx = 0.1$. Compare the accuracy of the two methods.

13. $y' = x - y$, $y(0) = 1$
(See Exercise 1 for the exact solution.)

14. $y' = y - e^{2x} + 1$, $y(0) = -1$
(See Exercise 4 for the exact solution.)

In Exercises 15 and 16, **(a)** find the exact solution of the initial value problem. Then compare the accuracy of the approximation to $y(x^*)$ using Euler's method starting at x_0 with step size **(b)** 0.2, **(c)** 0.1, and **(d)** 0.05.

15. $y' = 2y^2(x - 1)$, $y(2) = -1/2$, $x_0 = 2$, $x^* = 3$

16. $y' = y - 1$, $y(0) = 3$, $x_0 = 0$, $x^* = 1$

In Exercises 17 and 18, compare the accuracy of the approximation to $y(x^*)$ using the improved Euler's method starting at x_0 with step size **(a)** 0.2, **(b)** 0.1, and **(c)** 0.05.
(d) Writing to Learn Explain what happens to the error.

17. $y' = 2y^2(x - 1)$, $y(2) = -1/2$, $x_0 = 2$, $x^* = 3$
(See Exercise 15 for the exact solution.)

18. $y' = y - 1$, $y(0) = 3$, $x_0 = 0$, $x^* = 1$
(See Exercise 16 for the exact solution.)

In Exercises 19–22, superimpose the graph of the exact solution on a scatter plot of the Euler or improved Euler approximations from the given exercise.

19. Exercise 9   **20.** Exercise 10

21. Exercise 11   **22.** Exercise 12

In Exercises 23 and 24, use the indicated method to solve the initial value problem $y' = x + y$, $y(0) = 1$, on the interval $-1 \le x \le 0$ starting at $x_0 = 0$ with $dx = -0.1$. Compare the approximations to the values of the exact solution $y = 2e^x - x - 1$.

23. Euler's method   **24.** improved Euler's method

In Exercises 25–28, use the indicated method to solve the initial value problem graphically starting at $x_0 = 0$ with **(a)** $dx = 0.1$ and **(b)** $dx = 0.05$.

25. Euler's method, $y' = y + e^x - 2$, $y(0) = 2$

26. Euler's method, $y' = \cos(2x - y)$, $y(0) = 2$

27. improved Euler's method,

$$y' = y\left(\frac{1}{2} - \ln|y|\right), \quad y(0) = \frac{1}{3}$$

28. improved Euler's method, $y' = \sin(2x - y)$ $y(0) = 1$

29. Use Euler's method with $dx = 0.05$ to estimate $y(1)$ if $y' = y$ and $y(0) = 1$. What is the exact value of $y(1)$?

30. Use the improved Euler's method with $dx = 0.05$ to estimate $y(1)$ if $y' = 3y$ and $y(0) = 1$. What is the exact value of $y(1)$?

Extending the Ideas

The calculator programs RUNKUTG (G for "graph") and RUNKUTT (the second T for "table") solve initial value problems using a numerical method called the **Runge-Kutta method.** As implemented by the calculator, the method uses a weighted average of four intermediate calculations to go from each solution step to the next. With

$$k_1 = f(x_{n-1}, y_{n-1})\, dx,$$

$$k_2 = f\left(x_{n-1} + \frac{dx}{2}, y_{n-1} + \frac{k_1}{2}\right) dx,$$

$$k_3 = f\left(x_{n-1} + \frac{dx}{2}, y_{n-1} + \frac{k_2}{2}\right) dx,$$

$$k_4 = f(x_{n-1} + dx, y_{n-1} + k_3)\, dx,$$

the programs calculate y_n from y_{n-1} as

$$y_n = y_{n-1} + \frac{1}{6}(k_1 + 2k_2 + 2k_3 + k_4).$$

31. Use the Runge-Kutta program RUNKUTT to solve the initial value problem $y' = 1 + y$, $y(0) = 1$, on the interval $0 \le x \le 1$ starting at $x_0 = 0$ with $dx = 0.1$. Compare the approximations to the values of the exact solution $y = 2e^x - 1$.

32. Use the Runge-Kutta program RUNKUTG to solve the following initial value problems graphically starting at $x_0 = 0$ with $dx = 0.1$.

(a) $y' = x - y$, $y(0) = 1$ **(b)** $y' = x - y$, $y(0) = -2$

Calculus at Work

I have a Bachelor's and Master's degree in Aerospace Engineering from the University of California at Davis. I started my professional career as a Facility Engineer managing productivity and maintenance projects in the Unitary Project Wind Tunnel facility at NASA Ames Research Center. I used calculus and differential equations in fluid mechanic analyses of the tunnels. I then moved to the position of Test Manager, still using some fluid mechanics and other mechanical engineering analysis tools to solve problems. For example, the lift and drag forces acting on an airplane wing can be determined by integrating the known pressure distribution on the wing.

I am currently a NASA On-Site Systems Engineer for the Lunar Prospector spacecraft project, at Lockheed Martin Missiles and Space in Sunnyvale, California. Differential equations and integration are used to design some of the flight hardware for the spacecraft. I work on ensuring that the different systems of the spacecraft are adequately integrated together to meet the specified design requirements. This often means doing some analysis to determine if the systems will function properly and within the constraints of the space environment. Some of these analyses require use of differential equations and integration to determine the most exact results, within some margin of error.

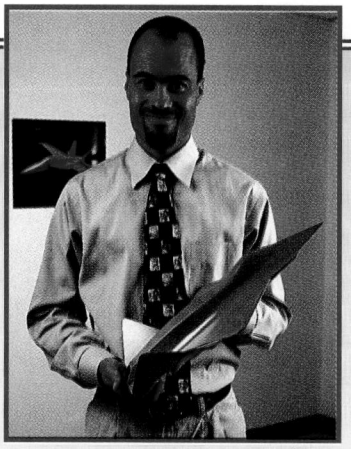

Ross Shaw
NASA Ames Research Center
Sunnyvale, CA

Chapter 6 Key Terms

Chapter 6 Review Exercises

In Exercises 1–8, evaluate the integral analytically. Then use NINT to support your result.

1. $\int_0^{\pi/3} \sec^2 \theta \, d\theta$

2. $\int_1^2 \left(x + \frac{1}{x^2} \right) dx$

3. $\int_0^1 \frac{36 \, dx}{(2x + 1)^3}$

4. $\int_{-1}^1 2x \sin (1 - x^2) \, dx$

5. $\int_0^{\pi/2} 5 \sin^{3/2} x \cos x \, dx$

6. $\int_{1/2}^4 \frac{x^2 + 3x}{x} \, dx$

7. $\int_0^{\pi/4} e^{\tan x} \sec^2 x \, dx$

8. $\int_1^e \frac{\sqrt{\ln r}}{r} \, dr$

In Exercises 9–20, evaluate the integral.

9. $\int \frac{\cos x}{2 - \sin x} \, dx$

10. $\int \frac{dx}{\sqrt[3]{3x + 4}}$

11. $\int \frac{t \, dt}{t^2 + 5}$

12. $\int \frac{1}{\theta^2} \sec \frac{1}{\theta} \tan \frac{1}{\theta} \, d\theta$

13. $\int \frac{\tan (\ln y)}{y} \, dy$

14. $\int e^x \sec (e^x) \, dx$

15. $\int \frac{dx}{x \ln x}$

16. $\int \frac{dt}{t\sqrt{t}}$

17. $\int x^3 \cos x \, dx$

18. $\int x^4 \ln x \, dx$

19. $\int e^{3x} \sin x \, dx$

20. $\int x^2 e^{-3x} \, dx$

In Exercises 21–28, solve the initial value problem analytically. Support your solution by overlaying its graph on a slope field of the differential equation.

21. $\frac{dy}{dx} = 1 + x + \frac{x^2}{2}, \quad y(0) = 1$

22. $\frac{dy}{dx} = \left(x + \frac{1}{x} \right)^2, \quad y(1) = 1$

23. $\frac{dy}{dt} = \frac{1}{t + 4}, \quad y(-3) = 2$

24. $\frac{dy}{d\theta} = \csc 2\theta \cot 2\theta, \quad y(\pi/4) = 1$

25. $\frac{d^2 y}{dx^2} = 2x - \frac{1}{x^2}, \quad x > 0, \quad y'(1) = 1, \quad y(1) = 0$

26. $\frac{d^3 r}{dt^3} = -\cos t, \quad r''(0) = r'(0) = r(0) = -1$

27. $\frac{dy}{dx} = y + 2, \quad y(0) = 2$

28. $\frac{dy}{dx} = (2x + 1)(y + 1), \quad y(-1) = 1$

In Exercises 29–32, assume that $1 - \sqrt{x}$ is an antiderivative of f and $x + 2$ is an antiderivative of g. Find the indefinite integral of the function.

29. $-f(x)$

30. $x + f(x)$

31. $2f(x) - g(x)$

32. $g(x) - 4$

In Exercises 33 and 34, match the indefinite integral with the graph of one of the antiderivatives of the integrand.

33. $\int \frac{\sin x}{x} \, dx$

34. $\int e^{-x^2} \, dx$

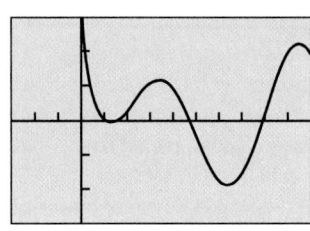

[−3, 10] by [−3, 3]

(a)

[−10, 10] by [−3, 3]

(b)

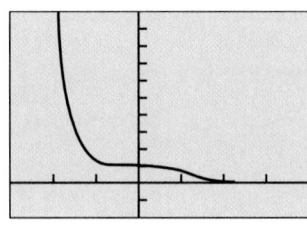

[−3, 4] by [−2, 10]

(c)

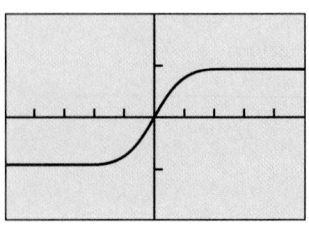

[−5, 5] by [−2, 2]

(d)

35. Writing to Learn The figure shows the graph of the function $y = f(x)$ that is the solution of one of the following initial value problems. Which one? How do you know?

i. $dy/dx = 2x, \quad y(1) = 0$

ii. $dy/dx = x^2, \quad y(1) = 1$

iii. $dy/dx = 2x + 2, \quad y(1) = 1$

iv. $dy/dx = 2x, \quad y(1) = 1$

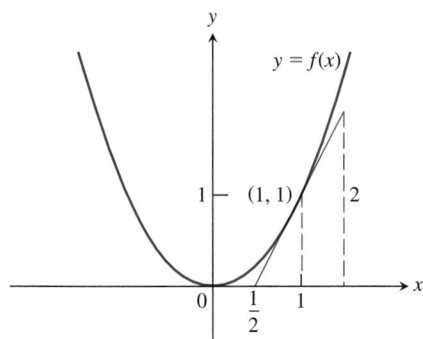

36. Writing to Learn Does the following initial value problem have a solution? Explain.

$$\frac{d^2y}{dx^2} = 0, \quad y'(0) = 1, \quad y(0) = 0$$

37. Moving Particle The acceleration of a particle moving along a coordinate line is

$$\frac{d^2s}{dt^2} = 2 + 6t \text{ m/sec}^2.$$

At $t = 0$ the velocity is 4 m/sec.

(a) Find the velocity as a function of time t.

(b) How far does the particle move during the first second of its trip, from $t = 0$ to $t = 1$?

38. Sketching Solutions Draw a possible graph for the function $y = f(x)$ with slope field given in the figure that satisfies the initial condition $y(0) = 0$.

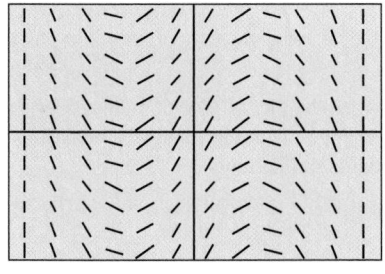

[–10, 10] by [–10, 10]

In Exercises 39 and 40, use the stated method to solve the initial value problem on the given interval starting at x_0 with $dx = 0.1$.

39. Euler; $y' = y + \cos x, \quad y(0) = 0; \quad 0 \le x \le 2; \quad x_0 = 0$

40. improved Euler; $y' = (2 - y)(2x + 3), \quad y(-3) = 1;$
$-3 \le x \le -1; \quad x_0 = -3$

In Exercises 41 and 42, use the stated method with $dx = 0.05$ to estimate $y(c)$ where y is the solution to the given initial value problem.

41. improved Euler; $c = 3; \quad \dfrac{dy}{dx} = \dfrac{x - 2y}{x + 1}, \quad y(0) = 1$

42. Euler; $c = 4; \quad \dfrac{dy}{dx} = \dfrac{x^2 - 2y + 1}{x}, \quad y(1) = 1$

In Exercises 43 and 44, use the stated method to solve the initial value problem graphically, starting at $x_0 = 0$ with **(a)** $dx = 0.1$ and **(b)** $dx = -0.1$.

43. Euler; $\dfrac{dy}{dx} = \dfrac{1}{e^{x+y+2}}, \quad y(0) = -2$

44. improved Euler; $\dfrac{dy}{dx} = -\dfrac{x^2 + y}{e^y + x}, \quad y(0) = 0$

45. *Californium-252* What costs $27 million per gram and can be used to treat brain cancer, analyze coal for its sulfur content, and detect explosives in luggage? The answer is californium-252, a radioactive isotope so rare that only about 8 g of it have been made in the western world since its discovery by Glenn Seaborg in 1950. The half-life of the isotope is 2.645 years—long enough for a useful service life and short enough to have a high radioactivity per unit mass. One microgram of the isotope releases 170 million neutrons per second.

(a) What is the value of k in the decay equation for this isotope?

(b) What is the isotope's mean life? (See Exercise 17, Section 6.4.)

46. *Cooling a Pie* A deep-dish apple pie, whose internal temperature was 220°F when removed from the oven, was set out on a 40°F breezy porch to cool. Fifteen minutes later, the pie's internal temperature was 180°F. How long did it take the pie to cool from there to 70°F?

47. *Finding Temperature* A pan of warm water (46°C) was put into a refrigerator. Ten minutes later, the water's temperature was 39°C; 10 minutes after that, it was 33°C. Use Newton's Law of Cooling to estimate how cold the refrigerator was.

48. *Art Forgery* A painting attributed to Vermeer (1632–1675), which should contain no more than 96.2% of its original carbon-14, contains 99.5% instead. About how old is the forgery?

49. *Carbon-14* What is the age of a sample of charcoal in which 90% of the carbon-14 that was originally present has decayed?

50. *Appreciation* A violin made in 1785 by John Betts, one of England's finest violin makers, cost $250 in 1924 and sold for $7500 in 1988. Assuming a constant relative rate of appreciation, what was that rate?

51. *Working Underwater* The intensity $L(x)$ of light x feet beneath the surface of the ocean satisfies the differential equation

$$\frac{dL}{dx} = -kL,$$

where k is a constant. As a diver you know from experience that diving to 18 ft in the Caribbean Sea cuts the intensity in half. You cannot work without artificial light when the intensity falls below a tenth of the surface value. About how deep can you expect to work without artificial light?

52. *Transport through a Cell Membrane* Under certain conditions, the result of the movement of a dissolved substance across a cell's membrane is described by the equation

$$\frac{dy}{dt} = k\frac{A}{V}(c - y).$$

In this equation, y is the concentration of the substance inside the cell, and dy/dt is the rate with which y changes over time. The letters k, A, V, and c stand for constants, k being the *permeability coefficient* (a property of the membrane), A the surface area of the membrane, V the cell's volume, and c the concentration of the substance outside the cell. The equation says that the rate at which the concentration changes within the cell is proportional to the difference between it and the outside concentration.

(a) Solve the equation for $y(t)$, using $y_0 = y(0)$.

(b) Find the steady-state concentration, $\lim_{t\to\infty} y(t)$.

53. *Logistic Equation* The spread of flu in a certain school is given by the formula

$$P(t) = \frac{150}{1 + e^{4.3-t}},$$

where t is the number of days after students are first exposed to infected students.

(a) Show that the function is a solution of a logistic differential equation. Identify k and the carrying capacity.

(b) **Writing to Learn** Estimate $P(0)$. Explain its meaning in the context of the problem.

(c) Estimate the number of days it will take for a total of 125 students to become infected.

54. *Confirming a Solution* Show that

$$y = \int_0^x \sin(t^2)\, dt + x^3 + x + 2$$

is the solution of the initial value problem.

Differential equation: $y'' = 2x\cos(x^2) + 6x$

Initial conditions: $y'(0) = 1, \quad y(0) = 2$

55. *Finding an Exact Solution* Use analytic methods to find the exact solution to

$$\frac{dP}{dt} = 0.002P\left(1 - \frac{P}{800}\right), \quad P(0) = 50.$$

56. *Supporting a Solution* Give two ways to provide graphical support for the integral formula

$$\int x^2 \ln x\, dx = \frac{x^3}{3}\ln x - \frac{x^3}{9} + C.$$

57. *Doubling Time* Find the amount of time required for $10,000 to double if the 6.3% annual interest is compounded **(a)** annually, **(b)** continuously.

58. *Constant of Integration* Let

$$f(x) = \int_0^x u(t)\, dt \quad \text{and} \quad g(x) = \int_3^x u(t)\, dt.$$

(a) Show that f and g are antiderivatives of $u(x)$.

(b) Find a constant C so that $f(x) = g(x) + C$.

59. *England and Wales Population* Table 6.14 gives the combined population of England and Wales for several years. Let $x = 0$ represent 1800, $x = 1$ represent 1801, and so forth.

Table 6.14 England and Wales Population

Year	Population (millions)	Year	Population (millions)
1801	8.9	1901	32.5
1811	10.2	1911	36.1
1821	12.0	1921	37.9
1831	13.9	1931	40.0
1841	15.9	1941	
1851	17.9	1951	43.8
1861	20.1	1961	46.1
1871	22.7	1971	48.7
1881	26.0	1981	49.0
1891	29.0	1991	49.2

Source: The Statesman's Yearbook, 133rd ed. (London: The Macmillan Press, Ltd., 1996).

(a) Find the logistic regression equation for the data and superimpose its graph on a scatter plot of the data.

(b) Find the carrying capacity predicted by the regression equation.

(c) Find when the rate of growth predicted by the regression equation changes from increasing to decreasing. Estimate the population at that time.

60. *Temperature Experiment* A temperature probe is removed from a cup of hot chocolate and placed in water whose temperature (T_s) is 0°C. The data in Table 6.15 were collected over the next 30 sec with a CBL™ temperature probe.

Table 6.15 Experimental Data

Time t (sec)	T (°C)
2	74.68
5	61.99
10	34.89
15	21.95
20	15.36
25	11.89
30	10.02

(a) Find an exponential regression equation for the (t, T) data. Superimpose its graph on a scatter plot of the data.

(b) Estimate when the temperature probe will read 40°C.

(c) Estimate the hot chocolate's temperature when the temperature probe was removed.

61. *Coasting to a Stop* Table 6.16 shows the distance s (meters) coasted on in-line skates in t seconds by Johnathon Krueger. Find a model for his position in the form of Equation 3 of Section 6.4 and superimpose its graph on a scatterplot of the data. His initial velocity was $v_0 = 0.86$ m/sec, his mass $m = 30.84$ kg (he weighed 68 lb), and his total coasting distance 0.97 m.

Table 6.16 Johnathon Krueger Skating Data

t (sec)	s (m)	t (sec)	s (m)	t (sec)	s (m)
0	0	0.93	0.61	1.86	0.93
0.13	0.08	1.06	0.68	2.00	0.94
0.27	0.19	1.20	0.74	2.13	0.95
0.40	0.28	1.33	0.79	2.26	0.96
0.53	0.36	1.46	0.83	2.39	0.96
0.67	0.45	1.60	0.87	2.53	0.97
0.80	0.53	1.73	0.90	2.66	0.97

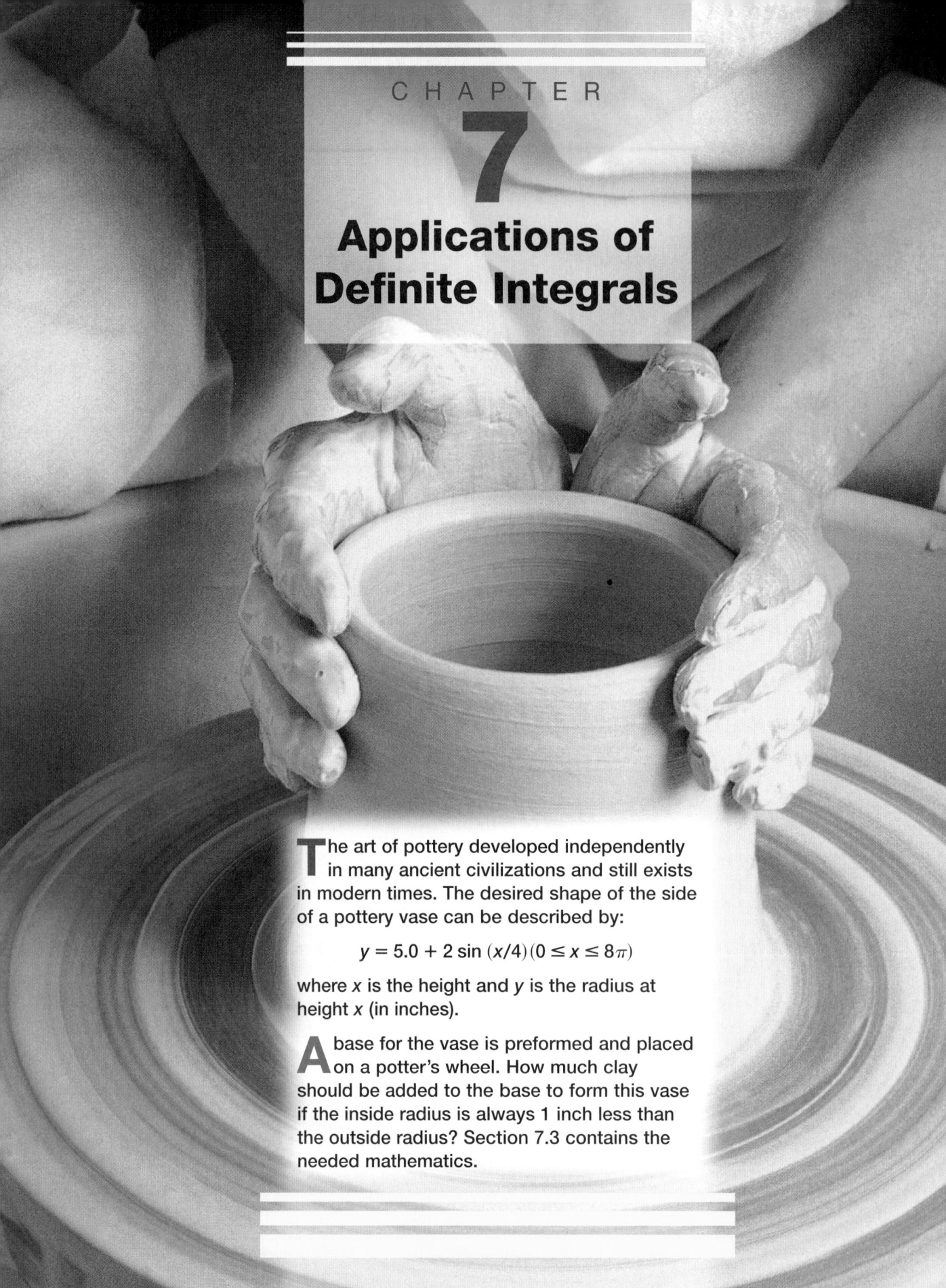

CHAPTER

7

Applications of Definite Integrals

The art of pottery developed independently in many ancient civilizations and still exists in modern times. The desired shape of the side of a pottery vase can be described by:

$$y = 5.0 + 2 \sin (x/4)\,(0 \le x \le 8\pi)$$

where x is the height and y is the radius at height x (in inches).

A base for the vase is preformed and placed on a potter's wheel. How much clay should be added to the base to form this vase if the inside radius is always 1 inch less than the outside radius? Section 7.3 contains the needed mathematics.

Chapter 7 Overview

By this point it should be apparent that finding the limits of Riemann sums is not just an intellectual exercise; it is a natural way to calculate mathematical or physical quantities that appear to be irregular when viewed as a whole, but which can be fragmented into regular pieces. We calculate values for the regular pieces using known formulas, then sum them to find a value for the irregular whole. This approach to problem solving was around for thousands of years before calculus came along, but it was tedious work and the more accurate you wanted to be the more tedious it became.

With calculus it became possible to get *exact* answers for these problems with almost no effort, because in the limit these sums became definite integrals and definite integrals could be evaluated with antiderivatives. With calculus, the challenge became one of fitting an integrable function to the situation at hand (the "modeling" step) and then finding an antiderivative for it.

Today we can finesse the antidifferentiation step (occasionally an insurmountable hurdle for our predecessors) with programs like NINT, but the modeling step is no less crucial. Ironically, it is the modeling step that is thousands of years old. Before either calculus or technology can be of assistance, we must still break down the irregular whole into regular parts and set up a function to be integrated. We have already seen how the process works with area, volume, and average value, for example. Now we will focus more closely on the underlying modeling step: how to set up the function to be integrated.

7.1 Integral as Net Change

Linear Motion Revisited • General Strategy • Consumption over Time • Net Change from Data • Work

Linear Motion Revisited

In many applications, the integral is viewed as net change over time. The classic example of this kind is distance traveled, a problem we discussed in Chapter 5.

Example 1   INTERPRETING A VELOCITY FUNCTION

Figure 7.1 shows the velocity

$$\frac{ds}{dt} = v(t) = t^2 - \frac{8}{(t+1)^2} \quad \frac{cm}{sec}$$

of a particle moving along a horizontal s-axis for $0 \le t \le 5$. Describe the motion.

Solution

Solve Graphically

The graph of v (Figure 7.1) starts with $v(0) = -8$, which we interpret as saying that the particle has an initial velocity of 8 cm/sec to the left. It slows to a halt at about $t = 1.25$ sec, after which it moves to the right ($v > 0$) with increasing speed, reaching a velocity of $v(5) \approx 24.8$ cm/sec at the end.

[0, 5] by [−10, 30]

Figure 7.1 The velocity function in Example 1.

Example 2 FINDING POSITION FROM DISPLACEMENT

Suppose the initial position of the particle in Example 1 is $s(0) = 9$. What is the particle's position at **(a)** $t = 1$ sec? **(b)** $t = 5$ sec?

Solution

Solve Analytically

(a) The position at $t = 1$ is the initial position $s(0)$ plus the displacement (the amount, Δs, that the position changed from $t = 0$ to $t = 1$). When velocity is constant during a motion, we can find the displacement (change in position) with the formula

$$\text{Displacement} = \text{rate of change} \times \text{time}.$$

But in our case the velocity varies, so we resort instead to partitioning the time interval $[0, 1]$ into subintervals of length Δt so short that the velocity is effectively constant on each subinterval. If t_k is any time in the kth sub-interval, the particle's velocity throughout that interval will be close to $v(t_k)$. The change in the particle's position during the brief time this constant velocity applies is

$$v(t_k)\,\Delta t. \quad \text{rate of change} \times \text{time}$$

If $v(t_k)$ is negative, the displacement is negative and the particle will move left. If $v(t_k)$ is positive, the particle will move right. The sum

$$\sum v(t_k)\,\Delta t$$

of all these small position changes approximates the displacement for the time interval $[0, 1]$.

The sum $\sum v(t_k)\,\Delta t$ is a Riemann sum for the continuous function $v(t)$ over $[0, 1]$. As the norms of the partitions go to zero, the approximations improve and the sums converge to the integral of v over $[0, 1]$, giving

$$\text{Displacement} = \int_0^1 v(t)\,dt$$

$$= \int_0^1 \left(t^2 - \frac{8}{(t+1)^2}\right) dt$$

$$= \left[\frac{t^3}{3} + \frac{8}{t+1}\right]_0^1$$

$$= \frac{1}{3} + \frac{8}{2} - 8 = -\frac{11}{3}.$$

During the first second of motion, the particle moves 11/3 cm to the left. It starts at $s(0) = 9$, so its position at $t = 1$ is

$$\text{New position} = \text{initial position} + \text{displacement} = 9 - \frac{11}{3} = \frac{16}{3}.$$

Reminder from Section 3.4

A change in position is a **displacement.** If $s(t)$ is a body's position at time t, the displacement over the time interval from t to $t + \Delta t$ is $s(t + \Delta t) - s(t)$. The displacement may be positive, negative, or zero, depending on the motion.

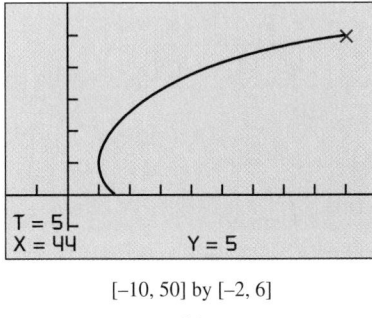

T = 1
X = 5.3333333   Y = 1

[-10, 50] by [-2, 6]

(a)

T = 5
X = 44   Y = 5

[-10, 50] by [-2, 6]

(b)

Figure 7.2 Using TRACE and the parametrization in Example 2 you can "see" the left and right motion of the particle.

(b) If we model the displacement from $t = 0$ to $t = 5$ in the same way, we arrive at

$$\text{Displacement} = \int_0^5 v(t)\, dt = \left[\frac{t^3}{3} + \frac{8}{t+1} \right]_0^5 = 35.$$

The motion has the net effect of displacing the particle 35 cm to the right of its starting point. The particle's final position is

$$\text{Final position} = \text{initial position} + \text{displacement}$$
$$= s(0) + 35 = 9 + 35 = 44.$$

Support Graphically

The position of the particle at any time t is given by

$$s(t) = \int_0^t \left[u^2 - \frac{8}{(u+1)^2} \right] du + 9,$$

because $s'(t) = v(t)$ and $s(0) = 9$. Figure 7.2 shows the graph of $s(t)$ given by the parametrization

$$x(t) = \text{NINT}\,(v(u), u, 0, t), \quad y(t) = t, \quad 0 \le t \le 5.$$

(a) Figure 7.2a supports that the position of the particle at $t = 1$ is 16/3.

(b) Figure 7.2b shows the position of the particle is 44 at $t = 5$. Therefore, the displacement is $44 - 9 = 35$.

The reason for our method in Example 2 was to illustrate the *modeling step* that will be used throughout this chapter. We can also solve Example 2 using the techniques of Chapter 6 as shown in Exploration 1.

Exploration 1   **Revisiting Example 2**

The velocity of a particle moving along a horizontal s-axis for $0 \le t \le 5$ is

$$\frac{ds}{dt} = t^2 - \frac{8}{(t+1)^2}.$$

1. Use the indefinite integral of ds/dt to find the solution of the initial value problem

$$\frac{ds}{dt} = t^2 - \frac{8}{(t+1)^2}, \quad s(0) = 9.$$

2. Determine the position of the particle at $t = 1$. Compare your answer with the answer to Example 2a.

3. Determine the position of the particle at $t = 5$. Compare your answer with the answer to Example 2b.

We know now that the particle in Example 1 was at $s(0) = 9$ at the beginning of the motion and at $s(5) = 44$ at the end. But it did not travel from 9 to 44 directly—it began its trip by moving to the left (Figure 7.2). How much distance did the particle actually travel? We find out in Example 3.

Example 3 CALCULATING TOTAL DISTANCE TRAVELED

Find the *total distance traveled* by the particle in Example 1.

Solution

Solve Analytically

We partition the time interval as in Example 2 but record every position shift as *positive* by taking absolute values. The Riemann sum approximating total distance traveled is

$$\sum |v(t_k)| \, \Delta t,$$

and we are led to the integral

$$\text{Total distance traveled} = \int_0^5 |v(t)| \, dt = \int_0^5 \left| t^2 - \frac{8}{(t+1)^2} \right| dt.$$

Evaluate Numerically

We have

$$\text{NINT}\left(\left| t^2 - \frac{8}{(t+1)^2} \right|, t, 0, 5 \right) \approx 42.59.$$

What we learn from Examples 2 and 3 is this: Integrating velocity gives displacement (net area between the velocity curve and the time axis). Integrating the *absolute value* of velocity gives total distance traveled (total area between the velocity curve and the time axis).

General Strategy

The idea of fragmenting net effects into finite sums of easily estimated small changes is not new. We used it in Section 5.1 to estimate cardiac output, volume, and air pollution. What *is* new is that we can now identify many of these sums as Riemann sums and express their limits as integrals. The advantages of doing so are twofold. First, we can evaluate one of these integrals to get an accurate result in less time than it takes to crank out even the crudest estimate from a finite sum. Second, the integral itself becomes a formula that enables us to solve similar problems without having to repeat the modeling step.

The strategy that we began in Section 5.1 and have continued here is the following:

> **Strategy for Modeling with Integrals**
>
> 1. *Approximate what you want to find as a Riemann sum* of values of a continuous function multiplied by interval lengths. If $f(x)$ is the function and $[a, b]$ the interval, and you partition the interval into subintervals of length Δx, the approximating sums will have the form $\sum f(c_k) \, \Delta x$ with c_k a point in the kth subinterval.
> 2. *Write a definite integral*, here $\int_a^b f(x) \, dx$, to express the limit of these sums as the norms of the partitions go to zero.
> 3. *Evaluate the integral* numerically or with an antiderivative.

Example 4 MODELING THE EFFECTS OF ACCELERATION

A car moving with initial velocity of 5 mph accelerates at the rate of

$$a(t) = 2.4t$$

mph per second for 8 seconds.

(a) How fast is the car going when the 8 seconds are up?

(b) How far did the car travel during those 8 seconds?

Solution

(a) We first model the effect of the acceleration on the car's velocity.

Step 1:

Approximate the net change in velocity as a Riemann sum. When acceleration is constant,

velocity change = acceleration × time applied.   rate of change × time

To apply this formula, we partition $[0, 8]$ into short subintervals of length Δt. On each subinterval the acceleration is nearly constant, so if t_k is any point in the kth subinterval, the change in velocity imparted by the acceleration in the subinterval is approximately

$$a(t_k) \, \Delta t \ \text{mph}. \qquad \frac{\text{mph}}{\text{sec}} \times \text{sec}$$

The net change in velocity for $0 \le t \le 8$ is approximately

$$\sum a(t_k) \, \Delta t \ \text{mph}.$$

Step 2:

Write a definite integral. The limit of these sums as the norms of the partitions go to zero is

$$\int_0^8 a(t) \, dt.$$

Step 3:

Evaluate the integral. Using an antiderivative, we have

$$\text{Net velocity change} = \int_0^8 2.4t \, dt = 1.2t^2 \Big]_0^8 = 76.8 \ \text{mph}.$$

So, how fast is the car going when the 8 seconds are up? Its initial velocity is 5 mph and the acceleration adds another 76.8 mph for a total of 81.8 mph.

(b) There is nothing special about the upper limit 8 in the preceding calculation. Applying the acceleration for any length of time t adds

$$\int_0^t 2.4u \, du \ \text{mph} \qquad u \text{ is just a dummy variable here.}$$

to the car's velocity, giving

$$v(t) = 5 + \int_0^t 2.4u \, du = 5 + 1.2t^2 \ \text{mph}.$$

The distance traveled from $t = 0$ to $t = 8$ sec is

$$\int_0^8 |v(t)|\ dt = \int_0^8 (5 + 1.2t^2)\ dt \qquad \text{Extension of Example 3}$$

$$= \left[5t + 0.4t^3\right]_0^8$$

$$= 244.8 \text{ mph} \times \text{seconds.}$$

Miles-per-hour second is not a distance unit that we normally work with! To convert to miles we multiply by hours/second $= 1/3600$, obtaining

$$244.8 \times \frac{1}{3600} = 0.068 \text{ mile.} \qquad \frac{\text{mi}}{\text{h}} \times \text{sec} \times \frac{\text{h}}{\text{sec}} = \text{mi}$$

The car traveled 0.068 mi during the 8 seconds of acceleration.

Consumption over Time

The integral is a natural tool to calculate net change and total accumulation of more quantities than just distance and velocity. Integrals can be used to calculate growth, decay, and, as in the next example, consumption. Whenever you want to find the cumulative effect of a varying rate of change, integrate it.

Example 5 POTATO CONSUMPTION

From 1970 to 1980, the rate of potato consumption in a particular country was $C(t) = 2.2 + 1.1^t$ millions of bushels per year, with t being years since the beginning of 1970. How many bushels were consumed from the beginning of 1972 to the end of 1973?

Solution We seek the cumulative effect of the consumption rate for $2 \le t \le 4$.

Step 1:

Riemann sum. We partition $[2, 4]$ into subintervals of length Δt and let t_k be a time in the kth subinterval. The amount consumed during this interval is approximately

$$C(t_k)\ \Delta t \text{ million bushels.}$$

The consumption for $2 \le t \le 4$ is approximately

$$\sum C(t_k)\ \Delta t \text{ million bushels.}$$

Step 2:

Definite integral. The amount consumed from $t = 2$ to $t = 4$ is the limit of these sums as the norms of the partitions go to zero.

$$\int_2^4 C(t)\ dt = \int_2^4 (2.2 + 1.1^t)\ dt \text{ million bushels}$$

Step 3:

Evaluate. Evaluating numerically, we obtain

$$\text{NINT } (2.2 + 1.1^t, t, 2, 4) \approx 7.066 \text{ million bushels.}$$

Net Change from Data

Many real applications begin with data, not a fully modeled function. In the next example, we are given data on the rate at which a pump operates in consecutive 5-minute intervals and asked to find the total amount pumped.

Example 6   FINDING GALLONS PUMPED FROM RATE DATA

A pump connected to a generator operates at a varying rate, depending on how much power is being drawn from the generator to operate other machinery. The rate (gallons per minute) at which the pump operates is recorded at 5-minute intervals for one hour as shown in Table 7.1. How many gallons were pumped during that hour?

Solution Let $R(t)$, $0 \le t \le 60$, be the pumping rate as a continuous function of time for the hour. We can partition the hour into short subintervals of length Δt on which the rate is nearly constant and form the sum $\sum R(t_k)\,\Delta t$ as an approximation to the amount pumped during the hour. This reveals the integral formula for the number of gallons pumped to be

$$\text{Gallons pumped} = \int_0^{60} R(t)\,dt.$$

We have no formula for R in this instance, but the 13 equally spaced values in Table 7.1 enable us to estimate the integral with the Trapezoidal Rule:

$$\int_0^{60} R(t)\,dt \approx \frac{60}{2\cdot 12}\left[58 + 2(60) + 2(65) + \cdots + 2(63) + 63\right]$$

$$= 3582.5.$$

The total amount pumped during the hour is about 3580 gal.

Table 7.1 Pumping Rates	
Time (min)	Rate (gal/min)
0	58
5	60
10	65
15	64
20	58
25	57
30	55
35	55
40	59
45	60
50	60
55	63
60	63

Work

In everyday life, *work* means an activity that requires muscular or mental effort. In science, the term refers specifically to a force acting on a body and the body's subsequent displacement. When a body moves a distance d along a straight line as a result of the action of a force of constant magnitude F in the direction of motion, the **work** done by the force is

$$W = Fd.$$

The equation $W = Fd$ is the **constant-force formula** for work.

The units of work are force × distance. In the metric system, the unit is the newton-meter, which, for historical reasons, is called a joule (see margin note). In the U.S. customary system, the most common unit of work is the **foot-pound.**

Hooke's Law for springs says that the force it takes to stretch or compress a spring x units from its natural (unstressed) length is a constant times x. In symbols,

$$F = kx,$$

where k, measured in force units per unit length, is a characteristic of the spring called the **force constant.**

Joules

The joule, abbreviated J and pronounced "jewel," is named after the English physicist James Prescott Joule (1818–1889). The defining equation is

1 joule = (1 newton)(1 meter).

In symbols, 1 J = 1 N · m.

It takes a force of about 1 N to lift an apple from a table. If you lift it 1 m you have done about 1 J of work on the apple. If you eat the apple, you will have consumed about 80 food calories, the heat equivalent of nearly 335,000 joules. If this energy were directly useful for mechanical work (it's not), it would enable you to lift 335,000 more apples up 1 m.

Example 7 A BIT OF WORK

It takes a force of 10 N to stretch a spring 2 m beyond its natural length. How much work is done in stretching the spring 4 m from its natural length?

Solution We let $F(x)$ represent the force in newtons required to stretch the spring x meters from its natural length. By Hooke's Law, $F(x) = kx$ for some constant k. We are told that

$$F(2) = 10 = k \cdot 2, \qquad \text{The force required to stretch the spring 2 m is 10 newtons.}$$

so $k = 5$ N/m and $F(x) = 5x$ for this particular spring.

We construct an integral for the work done in applying F over the interval from $x = 0$ to $x = 4$.

Step 1:

Riemann sum. We partition the interval into subintervals on each of which F is so nearly constant that we can apply the constant-force formula for work. If x_k is any point in the kth subinterval, the value of F throughout the interval is approximately $F(x_k) = 5x_k$. The work done by F across the interval is approximately $5x_k \, \Delta x$, where Δx is the length of the interval. The sum

$$\sum F(x_k) \, \Delta x = \sum 5x_k \, \Delta x$$

approximates the work done by F from $x = 0$ to $x = 4$.

Steps 2 and 3:

Integrate. The limit of these sums as the norms of the partitions go to zero is

Numerically, work is the area under the force graph.

$$\int_0^4 F(x) \, dx = \int_0^4 5x \, dx = 5\frac{x^2}{2} \Big]_0^4 = 40 \text{ N} \cdot \text{m}.$$

We will revisit work in Section 7.5.

Quick Review 7.1

In Exercises 1–10, find all values of x (if any) at which the function changes sign on the given interval. Sketch a number line graph of the interval, and indicate the sign of the function on each subinterval.

Example: $f(x) = x^2 - 1$ on $[-2, 3]$ changes sign at $x = \pm 1$.

$f(x)$

1. $\sin 2x$ on $[-3, 2]$

2. $x^2 - 3x + 2$ on $[-2, 4]$

3. $x^2 - 2x + 3$ on $[-4, 2]$

4. $2x^3 - 3x^2 + 1$ on $[-2, 2]$

5. $x \cos 2x$ on $[0, 4]$

6. xe^{-x} on $[0, \infty)$

7. $\dfrac{x}{x^2 + 1}$ on $[-5, 30]$

8. $\dfrac{x^2 - 2}{x^2 - 4}$ on $[-3, 3]$

9. $\sec(1 + \sqrt{1 - \sin^2 x})$ on $(-\infty, \infty)$

10. $\sin(1/x)$ on $[0.1, 0.2]$

Section 7.1 Exercises

In Exercises 1–8, the function $v(t)$ is the velocity in m/sec of a particle moving along the *x*-axis. Use analytic methods to do each of the following:

(a) Determine when the particle is moving to the right, to the left, and stopped.

(b) Find the particle's displacement for the given time interval.

(c) Find the total distance traveled by the particle.

1. $v(t) = 5 \cos t, \quad 0 \le t \le 2\pi$

2. $v(t) = 6 \sin 3t, \quad 0 \le t \le \pi/2$

3. $v(t) = 49 - 9.8t, \quad 0 \le t \le 10$

4. $v(t) = 6t^2 - 18t + 12, \quad 0 \le t \le 2$

5. $v(t) = 5 \sin^2 t \cos t, \quad 0 \le t \le 2\pi$

6. $v(t) = \sqrt{4 - t}, \quad 0 \le t \le 4$

7. $v(t) = e^{\sin t} \cos t, \quad 0 \le t \le 2\pi$

8. $v(t) = \dfrac{t}{1 + t^2}, \quad 0 \le t \le 3$

9. An automobile accelerates from rest at $1 + 3\sqrt{t}$ mph/sec for 9 seconds.

   (a) What is its velocity after 9 seconds?

   (b) How far does it travel in those 9 seconds?

10. A particle travels with velocity
$$v(t) = (t - 2) \sin t \text{ m/sec}$$
   for $0 \le t \le 4$ sec.

   (a) What is the particle's displacement?

   (b) What is the total distance traveled?

11. *Projectile* Recall that the acceleration due to Earth's gravity is 32 ft/sec^2. From ground level, a projectile is fired straight upward with velocity 90 feet per second.

   (a) What is its velocity after 3 seconds?

   (b) When does it hit the ground?

   (c) When it hits the ground, what is the net distance it has traveled?

   (d) When it hits the ground, what is the total distance it has traveled?

In Exercises 12–16, a particle moves along the *x*-axis (units in cm). Its initial position at $t = 0$ sec is $x(0) = 15$. The figure shows the graph of the particle's velocity $v(t)$. The numbers are the *areas* of the enclosed regions.

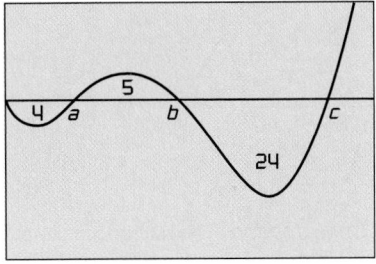

12. What is the particle's displacement between $t = 0$ and $t = c$?

13. What is the total distance traveled by the particle in the same time period?

14. Give the positions of the particle at times a, b, and c.

15. Approximately where does the particle achieve its greatest positive acceleration on the interval $[0, b]$?

16. Approximately where does the particle achieve its greatest positive acceleration on the interval $[0, c]$?

In Exercises 17–20, the graph of the velocity of a particle moving on the *x*-axis is given. The particle starts at $x = 2$ when $t = 0$.

(a) Find where the particle is at the end of the trip.

(b) Find the total distance traveled by the particle.

17.

18.

19.

20.

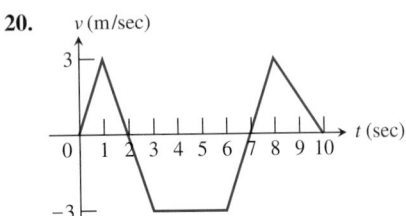

21. *U.S. Oil Consumption* The rate of consumption of oil in the United States during the 1980s (in billions of barrels per year) is modeled by the function $C = 27.08 \cdot e^{t/25}$, where t is the number of years after January 1, 1980. Find the total consumption of oil in the United States from January 1, 1980 to January 1, 1990.

22. *Home Electricity Use* The rate at which your home consumes electricity is measured in kilowatts. If your home consumes electricity at the rate of 1 kilowatt for 1 hour, you will be charged for 1 "kilowatt-hour" of electricity. Suppose that the average consumption rate for a certain home is modeled by the function $C(t) = 3.9 - 2.4 \sin (\pi t/12)$, where $C(t)$ is measured in kilowatts and t is the number of hours past midnight. Find the average daily consumption for this home, measured in kilowatt-hours.

23. *Population Density* Population density measures the number of people per square mile inhabiting a given living area. Washerton's population density, which decreases as you move away from the city center, can be approximated by the function $10,000(2 - r)$ at a distance r miles from the city center.

(a) If the population density approaches zero at the edge of the city, what is the city's radius?

(b) A thin ring around the center of the city has thickness Δr and radius r. If you straighten it out, it suggests a rectangular strip. Approximately what is its area?

(c) **Writing to Learn** Explain why the population of the ring in (b) is approximately

$$10,000(2 - r)(2\pi r) \, \Delta r.$$

(d) Estimate the total population of Washerton by setting up and evaluating a definite integral.

24. *Oil Flow* Oil flows through a cylindrical pipe of radius 3 inches, but friction from the pipe slows the flow toward the outer edge. The speed at which the oil flows at a distance r inches from the center is $8(10 - r^2)$ inches per second.

(a) In a plane cross section of the pipe, a thin ring with thickness Δr at a distance r inches from the center approximates a rectangular strip when you straighten it

out. What is the area of the strip (and hence the approximate area of the ring)?

(b) Explain why we know that oil passes through this ring at approximately $8(10 - r^2)(2\pi r) \, \Delta r$ cubic inches per second.

(c) Set up and evaluate a definite integral that will give the rate (in cubic inches per second) at which oil is flowing through the pipe.

25. Writing to Learn As a school project, Anna accompanies her mother on a trip to the grocery store and keeps a log of the car's speed at 10-second intervals. Explain how she can use the data to estimate the distance from her home to the store. What is the connection between this process and the definite integral?

In Exercises 26 and 27, *work in groups of two or three.*

26. *Bagel Sales* From 1985 to 1995, the Konigsberg Bakery noticed a consistent increase in annual sales of its bagels. The annual sales (in thousands of bagels) are shown below.

Year	Sales (thousands)
1985	5
1986	8.9
1987	16
1988	26.3
1989	39.8
1990	56.5
1991	76.4
1992	99.5
1993	125.8
1994	155.3
1995	188

(a) What was the total number of bagels sold over the 11-year period? (This is not a calculus question!)

(b) Use quadratic regression to model the annual bagel sales (in thousands) as a function $B(x)$, where x is the number of years after 1985.

(c) Integrate $B(x)$ over the interval $[0, 11]$ to find total bagel sales for the 11-year period.

(d) Explain graphically why the answer in (a) is smaller than the answer in (c).

27. *(Continuation of Exercise 26)*

(a) Integrate $B(x)$ over the interval $[-0.5, 10.5]$ to find total bagel sales for the 11-year period.

(b) Explain graphically why the answer in (a) is better than the answer in 26(c).

28. *Filling Milk Cartons* A machine fills milk cartons with milk at an approximately constant rate, but backups along the assembly line cause some variation. The rates (in cases per hour) are recorded at hourly intervals during a 10-hour period, from 8:00 A.M. to 6:00 P.M.

Time	Rate (cases/h)
8	120
9	110
10	115
11	115
12	119
1	120
2	120
3	115
4	112
5	110
6	121

Use the Trapezoidal Rule with $n = 10$ to determine approximately how many cases of milk were filled by the machine over the 10-hour period.

29. *Hooke's Law* A certain spring requires a force of 6 N to stretch it 3 cm beyond its natural length.

(a) What force would be required to stretch the string 9 cm beyond its natural length?

(b) What would be the work done in stretching the string 9 cm beyond its natural length?

30. *Hooke's Law* Hooke's Law also applies to *compressing* springs; that is, it requires a force of kx to compress a spring a distance x from its natural length. Suppose a 10,000-lb force compressed a spring from its natural length of 12 inches to a length of 11 inches. How much work was done in compressing the spring

(a) the first half-inch?   **(b)** the second half-inch?

Extending the Ideas

31. *Inflation* Although the economy is continuously changing, we analyze it with discrete measurements. The following table records the *annual* inflation rate as measured each month for 13 consecutive months. Use the Trapezoidal Rule with $n = 12$ to find the overall inflation rate for the year.

Month	Annual Rate
January	0.04
February	0.04
March	0.05
April	0.06
May	0.05
June	0.04
July	0.04
August	0.05
September	0.04
October	0.06
November	0.06
December	0.05
January	0.05

32. *Inflation Rate* The table below shows the *monthly* inflation rate (in *thousandths*) for energy prices for thirteen consecutive months. Use the Trapezoidal Rule with $n = 12$ to approximate the *annual* inflation rate for the 12-month period running from the middle of the first month to the middle of the last month.

Month	Monthly Rate (in thousandths)
January	3.6
February	4.0
March	3.1
April	2.8
May	2.8
June	3.2
July	3.3
August	3.1
September	3.2
October	3.4
November	3.4
December	3.9
January	4.0

33. *Center of Mass* Suppose we have a finite collection of masses in the coordinate plane, the mass m_k located at the point (x_k, y_k) as shown in the figure.

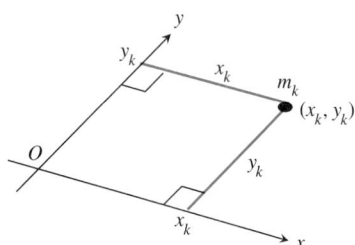

Each mass m_k has **moment $m_k y_k$ with respect to the x-axis** and **moment $m_k x_k$ about the y-axis.** The moments of the entire system with respect to the two axes are

$$\text{Moment about } x\text{-axis: } M_x = \sum m_k y_k,$$

$$\text{Moment about } y\text{-axis: } M_y = \sum m_k x_k.$$

The **center of mass** is $(\overline{x}, \overline{y})$ where

$$\overline{x} = \frac{M_y}{M} = \frac{\sum m_k x_k}{\sum m_k} \quad \text{and} \quad \overline{y} = \frac{M_x}{M} = \frac{\sum m_k y_k}{\sum m_k}.$$

Suppose we have a thin, flat plate occupying a region in the plane.

(a) Imagine the region cut into thin strips parallel to the y-axis. Show that

$$\overline{x} = \frac{\int x \, dm}{\int dm},$$

where $dm = \delta \, dA$, $\delta =$ density (mass per unit area), and $A =$ area of the region.

(b) Imagine the region cut into thin strips parallel to the x-axis. Show that

$$\overline{y} = \frac{\int y \, dm}{\int dm},$$

where $dm = \delta \, dA$, $\delta =$ density, and $A =$ area of the region.

In Exercises 34 and 35, use Exercise 33 to find the center of mass of the region with given density.

34. the region bounded by the parabola $y = x^2$ and the line $y = 4$ with constant density δ

35. the region bounded by the lines $y = x$, $y = -x$, $x = 2$ with constant density δ

7.2 Areas in the Plane

Area Between Curves • Area Enclosed by Intersecting Curves • Boundaries with Changing Functions • Integrating with Respect to y • Saving Time with Geometry Formulas

Area Between Curves

We know how to find the area of a region between a curve and the x-axis but many times we want to know the area of a region that is bounded above by one curve, $y = f(x)$, and below by another, $y = g(x)$ (Figure 7.3).

We find the area as an integral by applying the first two steps of the modeling strategy developed in Section 7.1.

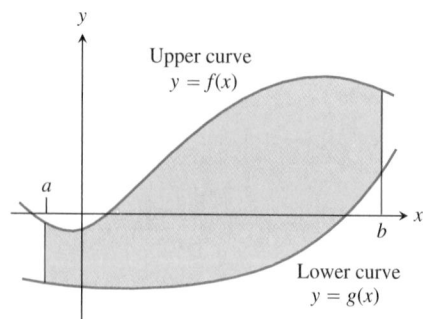

Figure 7.3 The region between $y = f(x)$ and $y = g(x)$ and the lines $x = a$ and $x = b$.

1. We partition the region into vertical strips of equal width Δx and approximate each strip with a rectangle with base parallel to $[a, b]$ (Figure 7.4). Each rectangle has area

$$[f(c_k) - g(c_k)] \, \Delta x$$

for some c_k in its respective subinterval (Figure 7.5). This expression will be nonnegative even if the region lies below the x-axis. We approximate the area of the region with the Riemann sum

$$\sum [f(c_k) - g(c_k)] \, \Delta x.$$

2. The limit of these sums as $\Delta x \to 0$ is

$$\int_a^b [f(x) - g(x)] \, dx.$$

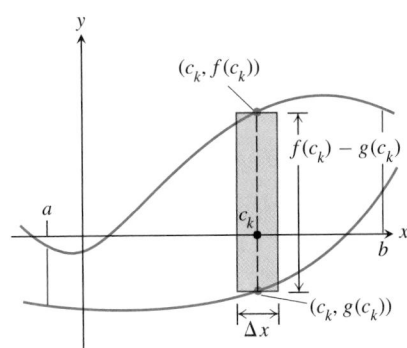

Figure 7.4 We approximate the region with rectangles perpendicular to the x-axis.

Figure 7.5 The area of a typical rectangle is $[f(c_k) - g(c_k)] \, \Delta x$.

This approach to finding area captures the properties of area, so it can serve as a definition.

Definition Area Between Curves

If f and g are continuous with $f(x) \geq g(x)$ throughout $[a, b]$, then the **area between the curves $y = f(x)$ and $y = g(x)$ from a to b** is the integral of $[f - g]$ from a to b,

$$A = \int_a^b [f(x) - g(x)] \, dx.$$

Example 1   APPLYING THE DEFINITION

Find the area of the region between $y = \sec^2 x$ and $y = \sin x$ from $x = 0$ to $x = \pi/4$.

Solution We graph the curves (Figure 7.6) to find their relative positions in the plane, and see that $y = \sec^2 x$ lies *above* $y = \sin x$ on $[0, \pi/4]$. The area is therefore

$$A = \int_0^{\pi/4} [\sec^2 x - \sin x] \, dx$$

$$= \Big[\tan x + \cos x \Big]_0^{\pi/4}$$

$$= \frac{\sqrt{2}}{2} \text{ units squared.}$$

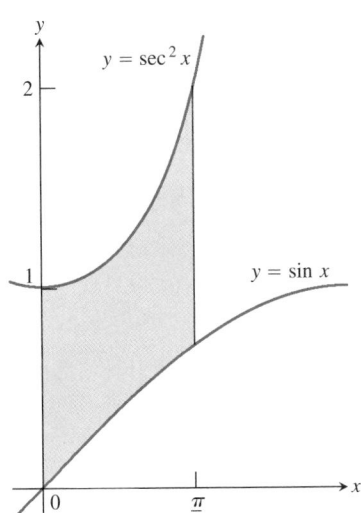

Figure 7.6 The region in Example 1.

$y_1 = 2k - k \sin kx$
$y_2 = k \sin kx$

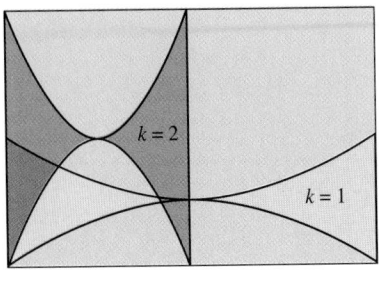

$[0, \pi]$ by $[0, 4]$

Figure 7.7 Two members of the family of butterfly-shaped regions described in Exploration 1.

Exploration 1 **A Family of Butterflies**

For each positive integer k, let A_k denote the area of the butterfly-shaped region enclosed between the graphs of $y = k \sin kx$ and $y = 2k - k \sin kx$ on the interval $[0, \pi/k]$. The regions for $k = 1$ and $k = 2$ are shown in Figure 7.7.

1. Find the areas of the two regions in Figure 7.7.
2. Make a conjecture about the areas A_k for $k \geq 3$.
3. Set up a definite integral that gives the area A_k. Can you make a simple u-substitution that will transform this integral into the definite integral that gives the area A_1?
4. What is $\lim_{k \to \infty} A_k$?
5. If P_k denotes the perimeter of the kth butterfly-shaped region, what is $\lim_{k \to \infty} P_k$? (You can answer this question without an explicit formula for P_k.)

Area Enclosed by Intersecting Curves

When a region is enclosed by intersecting curves, the intersection points give the limits of integration.

Example 2   AREA OF AN ENCLOSED REGION

Find the area of the region enclosed by the parabola $y = 2 - x^2$ and the line $y = -x$.

Solution We graph the curves to view the region (Figure 7.8).

The limits of integration are found by solving the equation

$$2 - x^2 = -x$$

either algebraically or by calculator. The solutions are $x = -1$ and $x = 2$.

Since the parabola lies above the line on $[-1, 2]$, the area integrand is $2 - x^2 - (-x)$.

$$A = \int_{-1}^{2} [2 - x^2 - (-x)] \, dx$$

$$= \left[2x - \frac{x^3}{3} + \frac{x^2}{2} \right]_{-1}^{2}$$

$$= \frac{9}{2} \text{ units squared}$$

$y_1 = 2 - x^2$
$y_2 = -x$

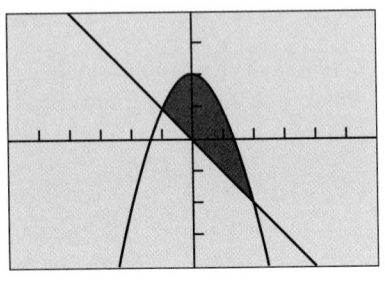

$[-6, 6]$ by $[-4, 4]$

Figure 7.8 The region in Example 2.

Example 3   USING A CALCULATOR

Find the area of the region enclosed by the graphs of $y = 2 \cos x$ and $y = x^2 - 1$.

Solution The region is shown in Figure 7.9.

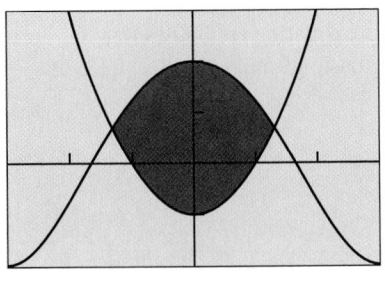

$y_1 = 2 \cos x$
$y_2 = x^2 - 1$

[–3, 3] by [–2, 3]

Figure 7.9 The region in Example 3.

Finding Intersections by Calculator

The coordinates of the points of intersection of two curves are sometimes needed for other calculations. To take advantage of the accuracy provided by calculators, use them to solve for the values and *store* the ones you want.

Using a calculator, we solve the equation

$$2 \cos x = x^2 - 1$$

to find the x-coordinates of the points where the curves intersect. These are the limits of integration. The solutions are $x = \pm 1.265423706$. We store the negative value as A and the positive value as B. The area is

$$\text{NINT}\,(2 \cos x - (x^2 - 1),\ x,\ A,\ B) \approx 4.994907788.$$

This is the final calculation, so we are now free to round. The area is about 4.99.

Boundaries with Changing Functions

If a boundary of a region is defined by more than one function, we can partition the region into subregions that correspond to the function changes and proceed as usual.

Example 4   FINDING AREA USING SUBREGIONS

Find the area of the region R in the first quadrant that is bounded above by $y = \sqrt{x}$ and below by the x-axis and the line $y = x - 2$.

Solution   The region is shown in Figure 7.10.

While it appears that no single integral can give the area of R (the bottom boundary is defined by two different curves), we can split the region at $x = 2$ into two regions A and B. The area of R can be found as the sum of the areas of A and B.

$$\text{Area of } R = \underbrace{\int_0^2 \sqrt{x}\, dx}_{\text{area of } A} + \underbrace{\int_2^4 [\sqrt{x} - (x - 2)]\, dx}_{\text{area of } B}$$

$$= \frac{2}{3} x^{3/2} \Big]_0^2 + \left[\frac{2}{3} x^{3/2} - \frac{x^2}{2} + 2x \right]_2^4$$

$$= \frac{10}{3} \text{ units squared}$$

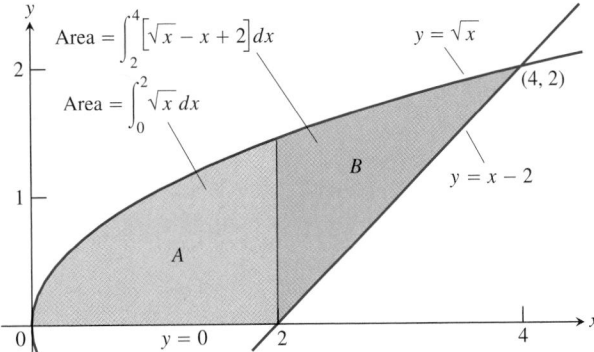

Figure 7.10 Region R split into subregions A and B. (Example 4)

Integrating with Respect to *y*

Sometimes the boundaries of a region are more easily described by functions of *y* than by functions of *x*. We can use approximating rectangles that are horizontal rather than vertical and the resulting basic formula has *y* in place of *x*.

For regions like these

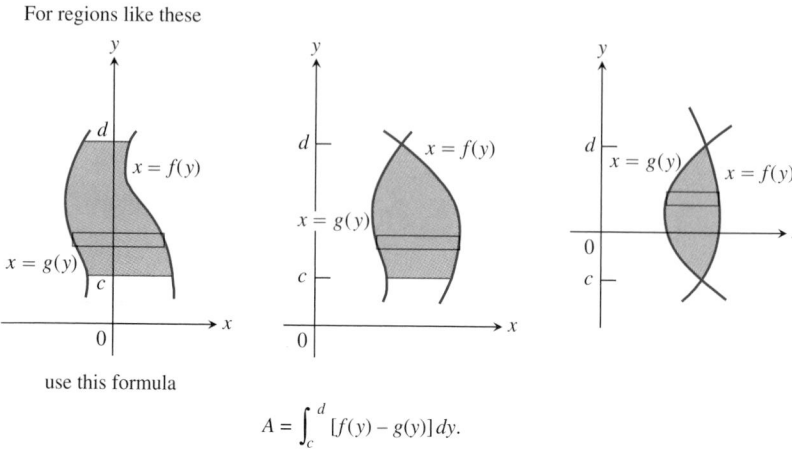

use this formula

$$A = \int_c^d [f(y) - g(y)]\, dy.$$

Example 5 INTEGRATING WITH RESPECT TO *y*

Find the area of the region in Example 4 by integrating with respect to *y*.

Solution We remarked in solving Example 4 that "it appears that no single integral can give the area of *R*," but notice how appearances change when we think of our rectangles being summed over *y*. The interval of integration is [0, 2], and the rectangles run between the same two curves on the entire interval. There is no need to split the region (Figure 7.11).

We need to solve for *x* in terms of *y* in both equations:

$$y = x - 2 \quad \text{becomes} \quad x = y + 2,$$
$$y = \sqrt{x} \quad \text{becomes} \quad x = y^2, \quad y \ge 0.$$

We must still be careful to subtract the lower number from the higher number when forming the integrand. In this case, the higher numbers are the higher *x*-values, which are on the line *x* = *y* + 2 because the line lies to the *right* of the parabola. So,

$$\text{Area of } R = \int_0^2 (y + 2 - y^2)\, dy$$

$$= \left[\frac{y^2}{2} + 2y - \frac{y^3}{3} \right]_0^2$$

$$= \frac{10}{3} \text{ units squared.}$$

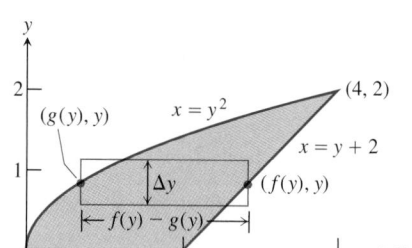

Figure 7.11 It takes two integrations to find the area of this region if we integrate with respect to *x*. It takes only one if we integrate with respect to *y*. (Example 5)

$y_1 = x^3, y_2 = \sqrt{x+2}, y_3 = -\sqrt{x+2}$

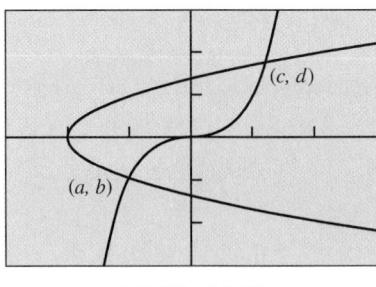

[–3, 3] by [–3, 3]

Figure 7.12 The region in Example 6.

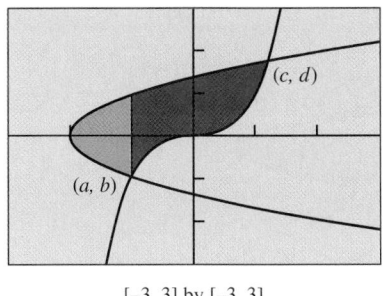

[–3, 3] by [–3, 3]

Figure 7.13 If we integrate with respect to x in Example 6, we must split the region at $x = a$.

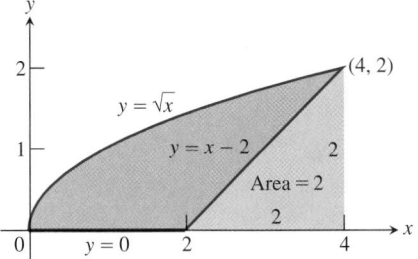

Figure 7.14 The area of the blue region is the area under the parabola $y = \sqrt{x}$ minus the area of the triangle. (Example 7)

Example 6   MAKING THE CHOICE

Find the area of the region enclosed by the graphs of $y = x^3$ and $x = y^2 - 2$.

Solution We can produce a graph of the region on a calculator by graphing the three curves $y = x^3$, $y = \sqrt{x+2}$, and $y = -\sqrt{x+2}$ (Figure 7.12).

This conveniently gives us all of our bounding curves as functions of x. If we integrate in terms of x, however, we need to split the region at $x = a$ (Figure 7.13).

On the other hand, we can integrate from $y = b$ to $y = d$ and handle the entire region at once. We solve the cubic for x in terms of y:

$$y = x^3 \quad \text{becomes} \quad x = y^{1/3}.$$

To find the limits of integration, we solve $y^{1/3} = y^2 - 2$. It is easy to see that the lower limit is $b = -1$, but a calculator is needed to find that the upper limit $d = 1.793003715$. We store this value as D.

The cubic lies to the right of the parabola, so

$$\text{Area} = \text{NINT}\,(y^{1/3} - (y^2 - 2),\, y,\, -1,\, \text{D}) = 4.214939673.$$

The area is about 4.21.

Saving Time with Geometry Formulas

Here is yet another way to handle Example 4.

Example 7   USING GEOMETRY

Find the area of the region in Example 4 by subtracting the area of the triangular region from the area under the square root curve.

Solution Figure 7.14 illustrates the strategy, which enables us to integrate with respect to x without splitting the region.

$$\text{Area} = \int_0^4 \sqrt{x}\, dx - \frac{1}{2}\,(2)(2)$$

$$= \frac{2}{3} x^{3/2} \Big]_0^4 - 2$$

$$= \frac{10}{3} \text{ units squared}$$

The moral behind Examples 4, 5, and 7 is that you often have options for finding the area of a region, some of which may be easier than others. You can integrate with respect to x or with respect to y, you can partition the region into subregions, and sometimes you can even use traditional geometry formulas. Sketch the region first and take a moment to determine the best way to proceed.

Quick Review 7.2

In Exercises 1–5, find the area between the *x*-axis and the graph of the given function over the given interval.

1. $y = \sin x$ over $[0, \pi]$

2. $y = e^{2x}$ over $[0, 1]$

3. $y = \sec^2 x$ over $[-\pi/4, \pi/4]$

4. $y = 4x - x^3$ over $[0, 2]$

5. $y = \sqrt{9 - x^2}$ over $[-3, 3]$

In Exercises 6–10, find the *x*- and *y*-coordinates of all points where the graphs of the given functions intersect. If the curves never intersect, write "NI."

6. $y = x^2 - 4x$ and $y = x + 6$

7. $y = e^x$ and $y = x + 1$

8. $y = x^2 - \pi x$ and $y = \sin x$

9. $y = \dfrac{2x}{x^2 + 1}$ and $y = x^3$

10. $y = \sin x$ and $y = x^3$

Section 7.2 Exercises

In Exercises 1–8, find the area of the shaded region analytically.

1.

2.

3.

4.

5.

NOT TO SCALE

6.

7.

8.

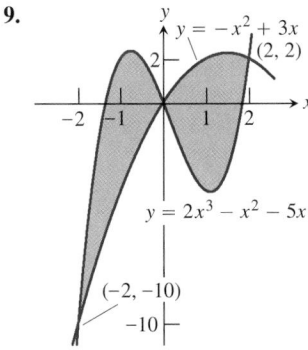

In Exercises 9 and 10, find the total shaded area.

9.

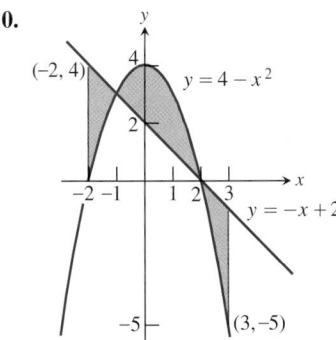

10.

In Exercises 11–30, find the area of the regions enclosed by the lines and curves.

11. $y = x^2 - 2$ and $y = 2$

12. $y = 2x - x^2$ and $y = -3$

13. $y = 7 - 2x^2$ and $y = x^2 + 4$

14. $y = x^4 - 4x^2 + 4$ and $y = x^2$

15. $y = x\sqrt{a^2 - x^2}$, $a > 0$,   and $y = 0$

16. $y = \sqrt{|x|}$ and $5y = x + 6$
   (How many intersection points are there?)

17. $y = |x^2 - 4|$ and $y = (x^2/2) + 4$

18. $x = y^2$ and $x = y + 2$

19. $y^2 - 4x = 4$ and $4x - y = 16$

20. $x - y^2 = 0$ and $x + 2y^2 = 3$

21. $x + y^2 = 0$ and $x + 3y^2 = 2$

22. $4x^2 + y = 4$ and $x^4 - y = 1$

23. $x + y^2 = 3$ and $4x + y^2 = 0$

24. $y = 2 \sin x$ and $y = \sin 2x$, $0 \le x \le \pi$

25. $y = 8 \cos x$ and $y = \sec^2 x$, $-\pi/3 \le x \le \pi/3$

26. $y = \cos(\pi x/2)$ and $y = 1 - x^2$

27. $y = \sin(\pi x/2)$ and $y = x$

28. $y = \sec^2 x$, $y = \tan^2 x$, $x = -\pi/4$, $x = \pi/4$

29. $x = \tan^2 y$ and $x = -\tan^2 y$, $-\pi/4 \le y \le \pi/4$

30. $x = 3 \sin y \sqrt{\cos y}$ and $x = 0$, $0 \le y \le \pi/2$

31. Find the area of the propeller-shaped region enclosed by the curve $x - y^3 = 0$ and the line $x - y = 0$.

32. Find the area of the region in the first quadrant bounded by the line $y = x$, the line $x = 2$, the curve $y = 1/x^2$, and the x-axis.

33. Find the area of the "triangular" region in the first quadrant bounded on the left by the y-axis and on the right by the curves $y = \sin x$ and $y = \cos x$.

34. Find the area of the region between the curve $y = 3 - x^2$ and the line $y = -1$ by integrating with respect to **(a)** x, **(b)** y.

35. The region bounded below by the parabola $y = x^2$ and above by the line $y = 4$ is to be partitioned into two subsections of equal area by cutting across it with the horizontal line $y = c$.

   (a) Sketch the region and draw a line $y = c$ across it that looks about right. In terms of c, what are the coordinates of the points where the line and parabola intersect? Add them to your figure.

   (b) Find c by integrating with respect to y. (This puts c in the limits of integration.)

   (c) Find c by integrating with respect to x. (This puts c into the integrand as well.)

36. Find the area of the region in the first quadrant bounded on the left by the y-axis, below by the line $y = x/4$, above left by the curve $y = 1 + \sqrt{x}$, and above right by the curve $y = 2/\sqrt{x}$.

In Exercises 37–40, *work in groups of two or three.*

37. The figure here shows triangle AOC inscribed in the region cut from the parabola $y = x^2$ by the line $y = a^2$. Find the limit of the ratio of the area of the triangle to the area of the parabolic region as a approaches zero.

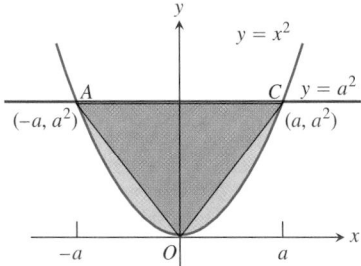

38. Suppose the area of the region between the graph of a positive continuous function f and the x-axis from $x = a$ to $x = b$ is 4 square units. Find the area between the curves $y = f(x)$ and $y = 2f(x)$ from $x = a$ to $x = b$.

39. Writing to Learn Which of the following integrals, if either, calculates the area of the shaded region shown here? Give reasons for your answer.

i. $\displaystyle\int_{-1}^{1} (x - (-x))\, dx = \int_{-1}^{1} 2x\, dx$

ii. $\displaystyle\int_{-1}^{1} (-x - (x))\, dx = \int_{-1}^{1} -2x\, dx$

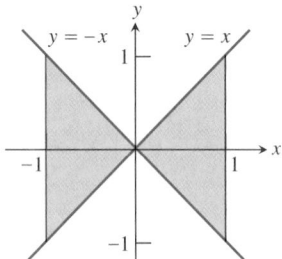

40. Writing to Learn Is the following statement true, sometimes true, or never true? The area of the region between the graphs of the continuous functions $y = f(x)$ and $y = g(x)$ and the vertical lines $x = a$ and $x = b$ $(a < b)$ is

$$\int_a^b [f(x) - g(x)]\, dx.$$

Give reasons for your answer.

41. Find the area of the propeller-shaped region enclosed between the graphs of

$$y = \frac{2x}{x^2 + 1} \quad \text{and} \quad y = x^3.$$

42. Find the area of the propeller-shaped region enclosed between the graphs of $y = \sin x$ and $y = x^3$.

43. Find the positive value of k such that the area of the region enclosed between the graph of $y = k \cos x$ and the graph of $y = kx^2$ is 2.

Exploration

44. *Area of Ellipse* *Work in groups of two or three.*

An ellipse with major axis of length $2a$ and minor axis of length $2b$ can be coordinatized with its center at the origin and its major axis horizontal, in which case it is defined by the equation

$$\frac{x^2}{a^2} + \frac{y^2}{b^2} = 1.$$

(a) Find the equations that define the upper and lower semiellipses as functions of x.

(b) Write an integral expression that gives the area of the ellipse.

(c) With your group, use NINT to find the areas of ellipses for various lengths of a and b.

(d) There is a simple formula for the area of an ellipse with major axis of length $2a$ and minor axis of length $2b$. Can you tell what it is from the areas you and your group have found?

(e) Work with your group to write a *proof* of this area formula by showing that it is the exact value of the integral expression in (b). ■

Extending the Ideas

45. *Cavalieri's Theorem* Bonaventura Cavalieri (1598–1647) discovered that if two plane regions can be arranged to lie over the same interval of the x-axis in such a way that they have identical vertical cross sections at every point (see figure), then the regions have the same area. Show that this theorem is true.

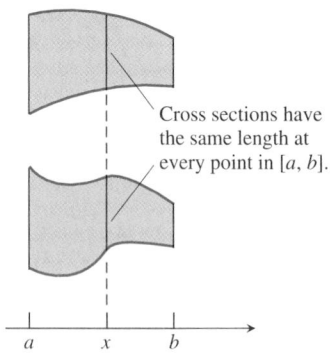

Cross sections have the same length at every point in $[a, b]$.

46. Find the area of the region enclosed by the curves

$$y = \frac{x}{x^2 + 1} \quad \text{and} \quad y = mx, \quad 0 < m < 1.$$

7.3 Volumes

Volume as an Integral • Square Cross Sections • Circular Cross Sections • Cylindrical Shells • Other Cross Sections

Volume as an Integral

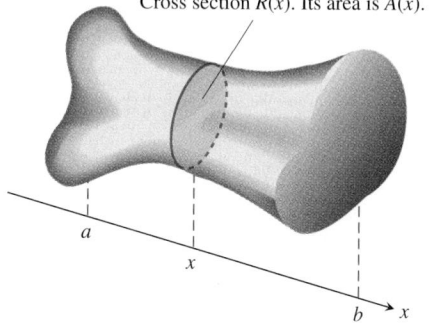

Cross section $R(x)$. Its area is $A(x)$.

Figure 7.15 The cross section of an arbitrary solid at point x.

In Section 5.1, Example 3, we estimated the volume of a sphere by partitioning it into thin slices that were nearly cylindrical and summing the cylinders' volumes using MRAM. MRAM sums are Riemann sums, and had we known how at the time, we could have continued on to express the volume of the sphere as a definite integral.

Starting the same way, we can now find the volumes of a great many solids by integration. Suppose we want to find the volume of a solid like the one in Figure 7.15. The cross section of the solid at each point x in the interval $[a, b]$ is a region $R(x)$ of area $A(x)$. If A is a continuous function of x, we can use it to define and calculate the volume of the solid as an integral in the following way.

We partition $[a, b]$ into subintervals of length Δx and slice the solid, as we would a loaf of bread, by planes perpendicular to the x-axis at the partition points. The kth slice, the one between the planes at x_{k-1} and x_k, has approximately the same volume as the cylinder between the two planes based on the region $R(x_k)$ (Figure 7.16).

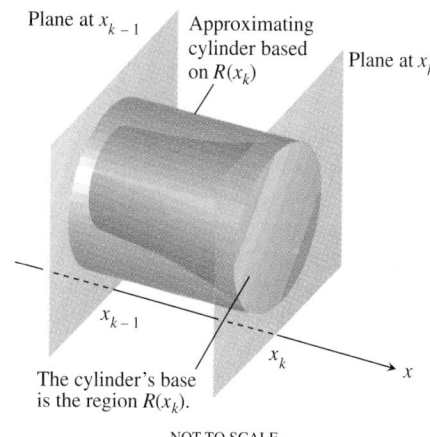

Plane at x_{k-1} Approximating cylinder based on $R(x_k)$ Plane at x_k

x_{k-1}

x_k

x

The cylinder's base is the region $R(x_k)$.

NOT TO SCALE

Figure 7.16 Enlarged view of the slice of the solid between the planes at x_{k-1} and x_k

The volume of the cylinder is

$$V_k = \text{base area} \times \text{height} = A(x_k) \times \Delta x.$$

The sum

$$\sum V_k = \sum A(x_k) \times \Delta x$$

approximates the volume of the solid.

This is a Riemann sum for $A(x)$ on $[a, b]$. We expect the approximations to improve as the norms of the partitions go to zero, so we define their limiting integral to be the *volume of the solid*.

> **Definition** Volume of a Solid
>
> The **volume** of a solid of known integrable cross section area $A(x)$ from $x = a$ to $x = b$ is the integral of A from a to b,
>
> $$V = \int_a^b A(x)\, dx.$$

To apply this formula, we proceed as follows.

> **How to Find Volume by the Method of Slicing**
>
> 1. Sketch the solid and a typical cross section.
> 2. Find a formula for $A(x)$.
> 3. Find the limits of integration.
> 4. Integrate $A(x)$ to find the volume.

Bonaventura Cavalieri (1598–1647)

Cavalieri, a student of Galileo, discovered that if two plane regions can be arranged to lie over the same interval of the x-axis in such a way that they have identical vertical cross sections at every point, then the regions have the same area. This theorem and a letter of recommendation from Galileo were enough to win Cavalieri a chair at the University of Bologna in 1629. The solid geometry version in Example 1, which Cavalieri never proved, was named after him by later geometers.

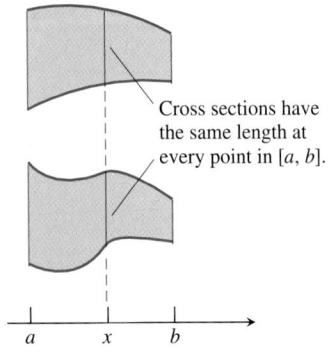

Cross sections have the same length at every point in $[a, b]$.

Example 1 CAVALIERI'S VOLUME THEOREM

Cavalieri's volume theorem says that solids with equal altitudes and identical cross section areas at each height have the same volume (Figure 7.17). This follows immediately from the definition of volume, because the cross section area function $A(x)$ and the interval $[a, b]$ are the same for both solids.

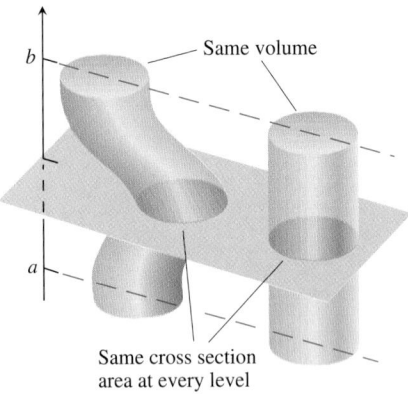

Same volume

Same cross section area at every level

Figure 7.17 *Cavalieri's volume theorem:* These solids have the same volume. You can illustrate this yourself with stacks of coins. (Example 1)

Square Cross Sections

Let us apply the volume formula to a solid with square cross sections.

Example 2   A SQUARE-BASED PYRAMID

A pyramid 3 m high has congruent triangular sides and a square base that is 3 m on each side. Each cross section of the pyramid parallel to the base is a square. Find the volume of the pyramid.

Solution We follow the steps for the method of slicing.

1. *Sketch.* We draw the pyramid with its vertex at the origin and its altitude along the interval $0 \le x \le 3$. We sketch a typical cross section at a point x between 0 and 3 (Figure 7.18).

2. *Find a formula for $A(x)$.* The cross section at x is a square x meters on a side, so

$$A(x) = x^2.$$

3. *Find the limits of integration.* The squares go from $x = 0$ to $x = 3$.

4. *Integrate to find the volume.*

$$V = \int_0^3 A(x)\, dx = \int_0^3 x^2 = \frac{x^3}{3}\Bigg]_0^3 = 9 \text{ m}^3$$

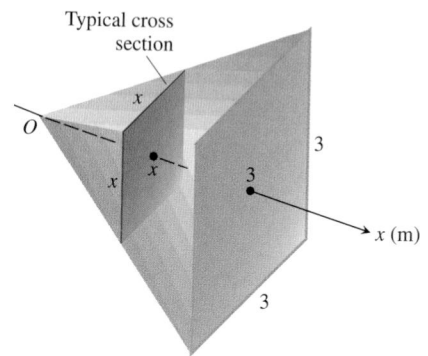

Figure 7.18 A cross section of the pyramid in Example 2.

Circular Cross Sections

The only thing that changes when the cross sections of a solid are circular is the formula for $A(x)$. Many such solids are **solids of revolution,** as in the next example.

Example 3   A SOLID OF REVOLUTION

The region between the graph of $f(x) = 2 + x \cos x$ and the x-axis over the interval $[-2, 2]$ is revolved about the x-axis to generate a solid. Find the volume of the solid.

Solution Revolving the region (Figure 7.19) about the x-axis generates the vase-shaped solid in Figure 7.20. The cross section at a typical point x is circular, with radius equal to $f(x)$. Its area is

$$A(x) = \pi (f(x))^2.$$

The volume of the solid is

$$V = \int_{-2}^2 A(x)\, dx$$

$$\approx \text{NINT}\left(\pi(2 + x \cos x)^2, x, -2, 2\right)$$

$$\approx 52.43 \text{ units cubed.}$$

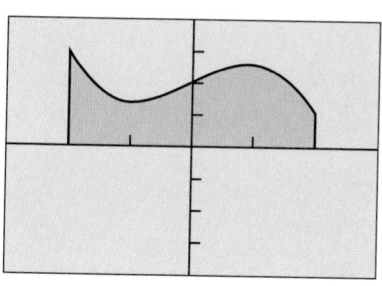

[−3, 3] by [−4, 4]

Figure 7.19 The region in Example 3.

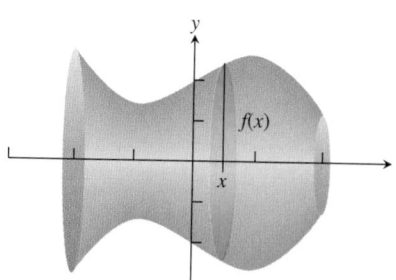

Figure 7.20 The region in Figure 7.19 is revolved about the x-axis to generate a solid. A typical cross section is circular, with radius $f(x) = 2 + x \cos x$. (Example 3)

Example 4 WASHER CROSS SECTIONS

The region in the first quadrant enclosed by the y-axis and the graphs of $y = \cos x$ and $y = \sin x$ is revolved about the x-axis to form a solid. Find its volume.

Solution The region is shown in Figure 7.21.

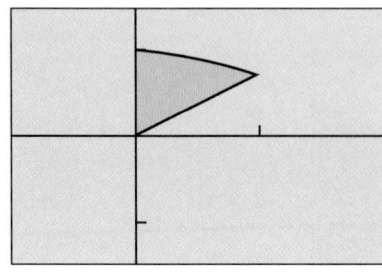

$[-\pi/4, \pi/2]$ by $[-1.5, 1.5]$

Figure 7.21 The region in Example 4.

We revolve it about the x-axis to generate a solid with a cone-shaped cavity in its center (Figure 7.22).

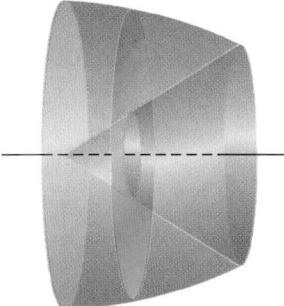

Figure 7.22 The solid generated by revolving the region in Figure 7.21 about the x-axis. A typical cross section is a washer: a circular region with a circular region cut out of its center. (Example 4)

This time each cross section perpendicular to the *axis of revolution* is a *washer,* a circular region with a circular region cut from its center. The area of a washer can be found by subtracting the inner area from the outer area (Figure 7.23).

CAUTION!

The area of a washer is $\pi R^2 - \pi r^2$, which you can simplify to $\pi(R^2 - r^2)$, but *not* to $\pi(R - r)^2$. No matter how tempting it is to make the latter simplification, it's wrong. Don't do it.

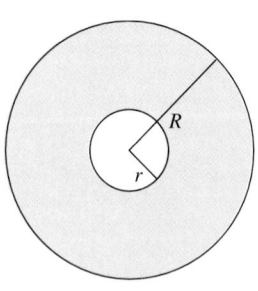

Figure 7.23 The area of a washer is $\pi R^2 - \pi r^2$. (Example 4)

In our region the cosine curve defines the outer radius, and the curves intersect at $x = \pi/4$. The volume is

$$V = \int_0^{\pi/4} \pi(\cos^2 x - \sin^2 x)\, dx$$

$$= \pi \int_0^{\pi/4} \cos 2x \, dx \qquad \text{identity: } \cos^2 x - \sin^2 x = \cos 2x$$

$$= \pi \left[\frac{\sin 2x}{2} \right]_0^{\pi/4} = \frac{\pi}{2} \text{ units cubed.}$$

We could have done the integration in Example 4 with NINT, but we wanted to demonstrate how a trigonometric identity can be useful under unexpected circumstances in calculus. The double-angle identity turned a difficult integrand into an easy one and enabled us to get an exact answer by antidifferentiation.

Cylindrical Shells

There is another way to find volumes of solids of rotation that can be useful when the axis of revolution is perpendicular to the axis containing the natural interval of integration. Instead of summing volumes of thin slices, we sum volumes of thin cylindrical shells that grow outward from the axis of revolution like tree rings.

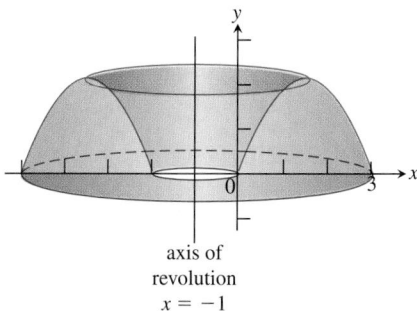

[–6, 4] by [–3, 3]

Figure 7.24 The graph of the region in Exploration 1, before revolution.

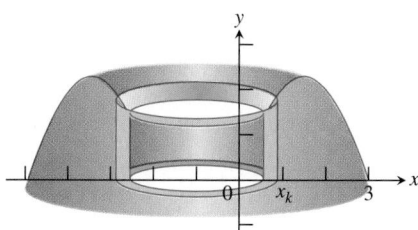

Figure 7.25 The region in Figure 7.24 is revolved about the line $x = -1$ to form a solid cake. The natural interval of integration is along the x-axis, perpendicular to the axis of revolution. (Exploration 1)

Figure 7.26 Cutting the cake into thin cylindrical slices, working from the inside out. Each slice occurs at some x_k between 0 and 3 and has thickness Δx. (Exploration 1)

Exploration 1   **Volume by Cylindrical Shells**

The region enclosed by the x-axis and the parabola $y = f(x) = 3x - x^2$ is revolved about the line $x = -1$ to generate the shape of a cake (Figures 7.24, 7.25). (Such a cake is often called a bundt cake.) What is the volume of the cake?

Integrating with respect to y would be awkward here, as it is not easy to get the original parabola in terms of y. (Try finding the volume by washers and you will soon see what we mean.) To integrate with respect to x, you can do the problem by *cylindrical shells,* which requires that you cut the cake in a rather unusual way.

1. Instead of cutting the usual wedge shape, cut a *cylindrical* slice by cutting straight down all the way around close to the inside hole. Then cut another cylindrical slice around the enlarged hole, then another, and so on. The radii of the cylinders gradually increase, and the heights of the cylinders follow the contour of the parabola: smaller to larger, then back to smaller (Figure 7.26). Each slice is sitting over a subinterval of the x-axis of length Δx. Its radius is approximately $(1 + x_k)$. What is its height?

2. If you unroll the cylinder at x_k and flatten it out, it becomes (essentially) a rectangular slab with thickness Δx. Show that the volume of the slab is approximately $2\pi(x_k + 1)(3x_k - x_k{}^2)\Delta x$.

3. $\sum 2\pi(x_k + 1)(3x_k - x_k{}^2)\Delta x$ is a Riemann sum. What is the limit of these Riemann sums as $\Delta x \to 0$?

4. Evaluate the integral you found in step 3 to find the volume of the cake!

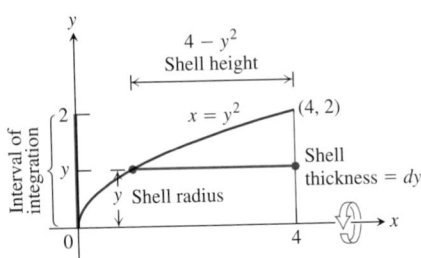

Figure 7.27 The region, shell dimensions, and interval of integration in Example 5.

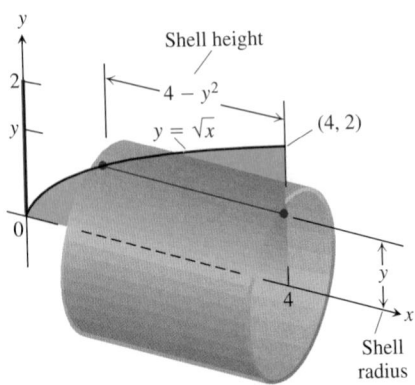

Figure 7.28 The shell swept out by the line segment in Figure 7.27.

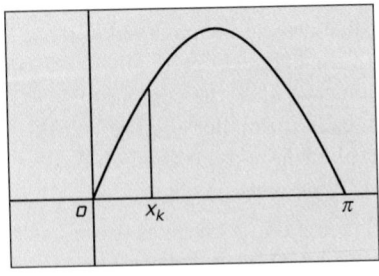

[−1, 3.5] by [−0.8, 2.2]

Figure 7.29 The base of the paperweight in Example 6. The segment perpendicular to the *x*-axis at x_k is the diameter of a semicircle that is perpendicular to the base.

Example 5   FINDING VOLUMES USING CYLINDRICAL SHELLS

The region bounded by the curve $y = \sqrt{x}$, the *x*-axis, and the line $x = 4$ is revolved about the *x*-axis to generate a solid. Find the volume of the solid.

Solution

1. Sketch the region and draw a line segment across it parallel to the axis of revolution (Figure 7.27). Label the segment's length (shell height) and distance from the axis of revolution (shell radius). The width of the segment is the shell thickness *dy*. (We drew the shell in Figure 7.28, but you need not do that.)

2. Identify the limits of integration: *y* runs from 0 to 2.

3. Integrate to find the volume.

$$V = \int_0^2 2\pi \left(\begin{array}{c} \text{shell} \\ \text{radius} \end{array} \right) \left(\begin{array}{c} \text{shell} \\ \text{height} \end{array} \right) dy$$

$$= \int_0^2 2\pi(y)(4 - y^2)\, dy = 8\pi$$

Other Cross Sections

The method of cross-section slicing can be used to find volumes of a wide variety of unusually shaped solids, so long as the cross sections have areas that we can describe with some formula. Admittedly, it does take a special artistic talent to *draw* some of these solids, but a crude picture is usually enough to suggest how to set up the integral.

Example 6   A MATHEMATICIAN'S PAPERWEIGHT

A mathematician has a paperweight made so that its base is the shape of the region between the *x*-axis and one arch of the curve $y = 2 \sin x$ (linear units in inches). Each cross section cut perpendicular to the *x*-axis (and hence to the *xy*-plane) is a semicircle whose diameter runs from the *x*-axis to the curve. (Think of the cross section as a semicircular fin sticking up out of the plane.) Find the volume of the paperweight.

Solution The paperweight is not easily drawn, but we know what it looks like. Its base is the region in Figure 7.29, and the cross sections perpendicular to the base are semicircular fins like those in Figure 7.30.

The semicircle at each point *x* has

$$\text{radius} = \frac{2 \sin x}{2} = \sin x \quad \text{and area} \quad A(x) = \frac{1}{2}\pi(\sin x)^2.$$

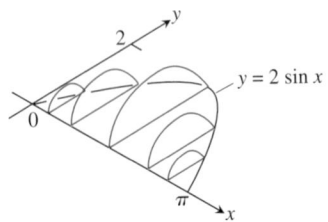

Figure 7.30 Cross sections perpendicular to the region in Figure 7.29 are semicircular. (Example 6)

The volume of the paperweight is

$$V = \int_0^{\pi} A(x)\, dx$$

$$= \frac{\pi}{2} \int_0^{\pi} (\sin x)^2\, dx$$

$$\approx \frac{\pi}{2} \, \text{NINT}\,((\sin x)^2, x, 0, \pi)$$

$$\approx \frac{\pi}{2}(1.570796327).$$

The number in parentheses looks like half of π, an observation that can be confirmed analytically, and which we support numerically by dividing by π to get 0.5. The volume of the paperweight is

$$\frac{\pi}{2} \cdot \frac{\pi}{2} = \frac{\pi^2}{4} \approx 2.47 \text{ in}^3.$$

Exploration 2 Surface Area

We know how to find the volume of a solid of revolution, but how would we find the *surface area*? As before, we partition the solid into thin slices, but now we wish to form a Riemann sum of approximations to *surface areas of slices* (rather than of volumes of slices).

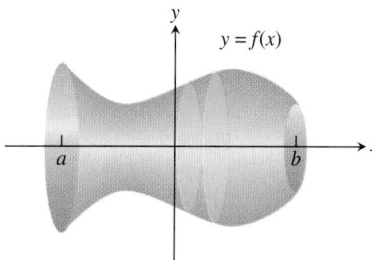

A typical slice has a surface area that can be approximated by $2\pi \cdot f(x) \cdot \Delta s$, where Δs is the tiny *slant height* of the slice. We will see in Section 7.4, when we study arc length, that $\Delta s = \sqrt{\Delta x^2 + \Delta y^2}$, and that this can be written as $\Delta s = \sqrt{1 + (f'(x_k))^2}\, \Delta x$.

Thus, the surface area is approximated by the Riemann sum

$$\sum_{k=1}^{n} 2\pi f(x_k) \sqrt{1 + (f'(x_k))^2}\, \Delta x.$$

1. Write the limit of the Riemann sums as a definite integral from a to b. When will the limit exist?

2. Use the formula from part 1 to find the surface area of the solid generated by revolving a single arch of the curve $y = \sin x$ about the x-axis.

3. The region enclosed by the graphs of $y^2 = x$ and $x = 4$ is revolved about the x-axis to form a solid. Find the surface area of the solid.

Quick Review 7.3

In Exercises 1–10, give a formula for the area of the plane region in terms of the single variable x.

1. a square with sides of length x

2. a square with diagonals of length x

3. a semicircle of radius x

4. a semicircle of diameter x

5. an equilateral triangle with sides of length x

6. an isosceles right triangle with legs of length x

7. an isosceles right triangle with hypotenuse x

8. an isosceles triangle with two sides of length $2x$ and one side of length x

9. a triangle with sides $3x$, $4x$, and $5x$

10. a regular hexagon with sides of length x

Section 7.3 Exercises

In Exercises 1 and 2, find a formula for the area $A(x)$ of the cross sections of the solid that are perpendicular to the x-axis.

1. The solid lies between planes perpendicular to the x-axis at $x = -1$ and $x = 1$. The cross sections perpendicular to the x-axis between these planes run from the semicircle $y = -\sqrt{1 - x^2}$ to the semicircle $y = \sqrt{1 - x^2}$.

(a) The cross sections are circular disks with diameters in the xy-plane.

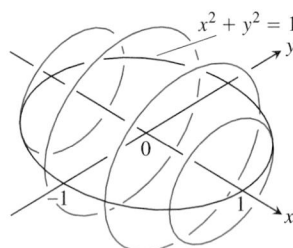

(b) The cross sections are squares with bases in the xy-plane.

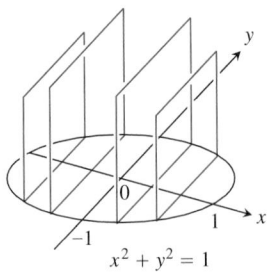

(c) The cross sections are squares with diagonals in the xy-plane. (The length of a square's diagonal is $\sqrt{2}$ times the length of its sides.)

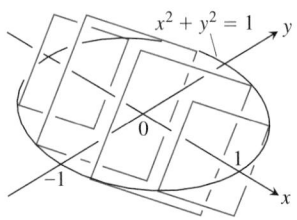

(d) The cross sections are equilateral triangles with bases in the xy-plane.

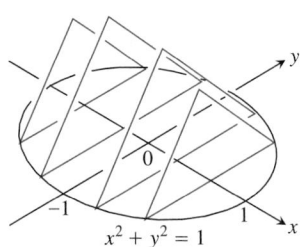

2. The solid lies between planes perpendicular to the x-axis at $x = 0$ and $x = 4$. The cross sections perpendicular to the x-axis between these planes run from $y = -\sqrt{x}$ to $y = \sqrt{x}$.

(a) The cross sections are circular disks with diameters in the xy-plane.

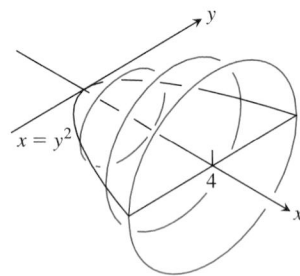

(b) The cross sections are squares with bases in the xy-plane.

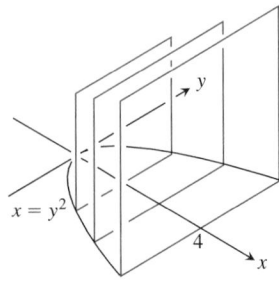

(c) The cross sections are squares with diagonals in the *xy*-plane.

(d) The cross sections are equilateral triangles with bases in the *xy*-plane.

In Exercises 3–10, find the volume of the solid analytically.

3. The solid lies between planes perpendicular to the *x*-axis at $x = 0$ and $x = 4$. The cross sections perpendicular to the axis on the interval $0 \le x \le 4$ are squares whose diagonals run from $y = -\sqrt{x}$ to $y = \sqrt{x}$.

4. The solid lies between planes perpendicular to the *x*-axis at $x = -1$ and $x = 1$. The cross sections perpendicular to the *x*-axis are circular disks whose diameters run from the parabola $y = x^2$ to the parabola $y = 2 - x^2$.

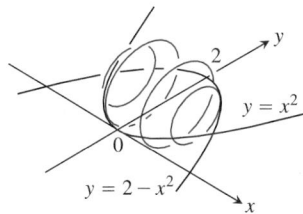

5. The solid lies between planes perpendicular to the *x*-axis at $x = -1$ and $x = 1$. The cross sections perpendicular to the *x*-axis between these planes are squares whose bases run from the semicircle $y = -\sqrt{1 - x^2}$ to the semicircle $y = \sqrt{1 - x^2}$.

6. The solid lies between planes perpendicular to the *x*-axis at $x = -1$ and $x = 1$. The cross sections perpendicular to the *x*-axis between these planes are squares whose diagonals run from the semicircle $y = -\sqrt{1 - x^2}$ to the semicircle $y = \sqrt{1 - x^2}$.

7. The base of a solid is the region between the curve $y = 2\sqrt{\sin x}$ and the interval $[0, \pi]$ on the *x*-axis. The cross sections perpendicular to the *x*-axis are

(a) equilateral triangles with bases running from the *x*-axis to the curve as shown in the figure.

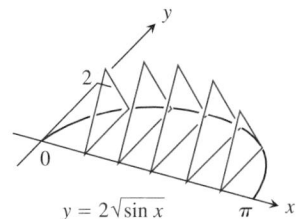

(b) squares with bases running from the *x*-axis to the curve.

8. The solid lies between planes perpendicular to the *x*-axis at $x = -\pi/3$ and $x = \pi/3$. The cross sections perpendicular to the *x*-axis are

(a) circular disks with diameters running from the curve $y = \tan x$ to the curve $y = \sec x$.

(b) squares whose bases run from the curve $y = \tan x$ to the curve $y = \sec x$.

9. The solid lies between planes perpendicular to the *y*-axis at $y = 0$ and $y = 2$. The cross sections perpendicular to the *y*-axis are circular disks with diameters running from the *y*-axis to the parabola $x = \sqrt{5}y^2$.

10. The base of the solid is the disk $x^2 + y^2 \le 1$. The cross sections by planes perpendicular to the *y*-axis between $y = -1$ and $y = 1$ are isosceles right triangles with one leg in the disk.

11. *A Twisted Solid* A square of side length *s* lies in a plane perpendicular to a line *L*. One vertex of the square lies on *L*. As this square moves a distance *h* along *L*, the square turns one revolution about *L* to generate a corkscrew-like column with square cross sections.

(a) Find the volume of the column.

(b) Writing to Learn What will the volume be if the square turns twice instead of once? Give reasons for your answer.

12. Writing to Learn A solid lies between planes perpendicular to the *x*-axis at $x = 0$ and $x = 12$. The cross sections by planes perpendicular to the *x*-axis are circular disks whose diameters run from the line $y = x/2$ to the line $y = x$ as shown in the figure. Explain why the solid has the same volume as a right circular cone with base radius 3 and height 12.

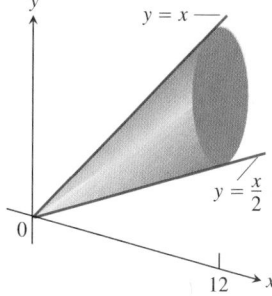

In Exercises 13–16, find the volume of the solid generated by revolving the shaded region about the given axis.

13. about the *x*-axis

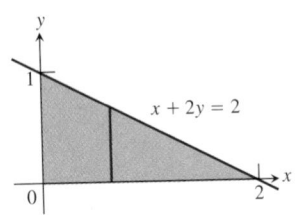
$x + 2y = 2$

14. about the *y*-axis

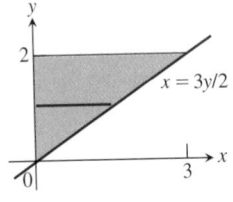
$x = 3y/2$

15. about the *y*-axis

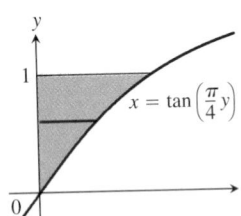
$x = \tan\left(\dfrac{\pi}{4}y\right)$

16. about the *x*-axis

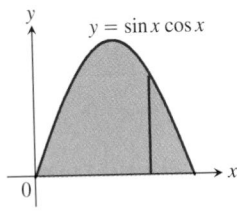
$y = \sin x \cos x$

In Exercises 17–26, find the volume of the solid generated by revolving the region bounded by the lines and curves about the *x*-axis.

17. $y = x^2$, $y = 0$, $x = 2$

18. $y = x^3$, $y = 0$, $x = 2$

19. $y = \sqrt{9 - x^2}$, $y = 0$

20. $y = x - x^2$, $y = 0$

21. $y = x$, $y = 1$, $x = 0$

22. $y = 2x$, $y = x$, $x = 1$

23. $y = x^2 + 1$, $y = x + 3$

24. $y = 4 - x^2$, $y = 2 - x$

25. $y = \sec x$, $y = \sqrt{2}$, $-\pi/4 \le x \le \pi/4$

26. $y = -\sqrt{x}$, $y = -2$, $x = 0$

In Exercises 27 and 28, find the volume of the solid generated by revolving the region about the given line.

27. the region in the first quadrant bounded above by the line $y = \sqrt{2}$, below by the curve $y = \sec x \tan x$, and on the left by the *y*-axis, about the line $y = \sqrt{2}$

28. the region in the first quadrant bounded above by the line $y = 2$, below by the curve $y = 2 \sin x$, $0 \le x \le \pi/2$, and on the left by the *y*-axis, about the line $y = 2$

In Exercises 29–34, find the volume of the solid generated by revolving the region about the *y*-axis.

29. the region enclosed by $x = \sqrt{5}y^2$, $x = 0$, $y = -1$, $y = 1$

30. the region enclosed by $x = y^{3/2}$, $x = 0$, $y = 2$

31. the region enclosed by the triangle with vertices $(1, 0)$, $(2, 1)$, and $(1, 1)$

32. the region enclosed by the triangle with vertices $(0, 1)$, $(1, 0)$, and $(1, 1)$

33. the region in the first quadrant bounded above by the parabola $y = x^2$, below by the *x*-axis, and on the right by the line $x = 2$

34. the region bounded above by the curve $y = \sqrt{x}$ and below by the line $y = x$

In Exercises 35–38, *work in groups of two or three.*

35. Find the volume of the solid generated by revolving the region bounded by $y = \sqrt{x}$ and the lines $y = 2$ and $x = 0$ about

(a) the *x*-axis.   (b) the *y*-axis.

(c) the line $y = 2$.   (d) the line $x = 4$.

36. Find the volume of the solid generated by revolving the triangular region bounded by the lines $y = 2x$, $y = 0$, and $x = 1$ about

(a) the line $x = 1$.   (b) the line $x = 2$.

37. Find the volume of the solid generated by revolving the region bounded by the parabola $y = x^2$ and the line $y = 1$ about

(a) the line $y = 1$.   (b) the line $y = 2$.

(c) the line $y = -1$.

38. By integration, find the volume of the solid generated by revolving the triangular region with vertices $(0, 0)$, $(b, 0)$, $(0, h)$ about

(a) the *x*-axis.   (b) the *y*-axis.

In Exercises 39–42, use the cylindrical shell method to find the volume of the solid generated by revolving the region bounded by the curves about the *y*-axis.

39. $y = x$, $y = -x/2$, $x = 2$

40. $y = x^2$, $y = 2 - x$, $x = 0$, for $x \ge 0$

41. $y = \sqrt{x}$, $y = 0$, $x = 4$

42. $y = 2x - 1$, $y = \sqrt{x}$, $x = 0$

In Exercises 43 and 44, use the cylindrical shell method to find the volume of the solid generated by revolving the shaded region about the indicated axis.

43. (a) the *x*-axis   (b) the line $y = 1$

(c) the line $y = 8/5$ (d) the line $y = -2/5$

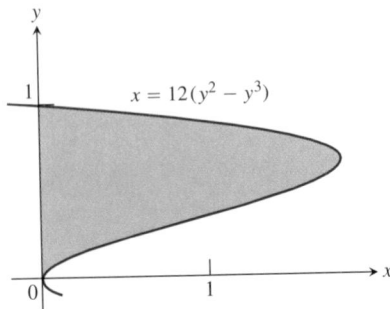
$x = 12(y^2 - y^3)$

44. (a) the *x*-axis   **(b)** the line $y = 2$

(c) the line $y = 5$ **(d)** the line $y = -5/8$

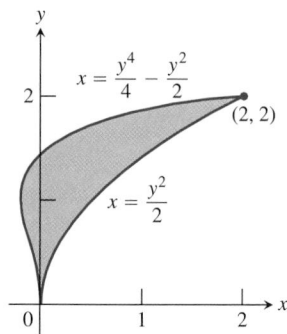

$$x = \frac{y^4}{4} - \frac{y^2}{2}$$

(2, 2)

$$x = \frac{y^2}{2}$$

45. Find the volume of the solid generated by revolving the region in the first quadrant bounded by $y = x^3$ and $y = 4x$ about **(a)** the *x*-axis, **(b)** the line $y = 8$.

46. Find the volume of the solid generated by revolving the region bounded by $y = 2x - x^2$ and $y = x$ about **(a)** the *y*-axis, **(b)** the line $x = 1$.

47. The region in the first quadrant that is bounded above by the curve $y = 1/\sqrt{x}$, on the left by the line $x = 1/4$, and below by the line $y = 1$ is revolved about the *y*-axis to generate a solid. Find the volume of the solid by **(a)** the washer method and **(b)** the cylindrical shell method.

48. Let $f(x) = \begin{cases} (\sin x)/x, & 0 < x \le \pi \\ 1, & x = 0. \end{cases}$

(a) Show that $x\,f(x) = \sin x, \quad 0 \le x \le \pi$.

(b) Find the volume of the solid generated by revolving the shaded region about the *y*-axis.

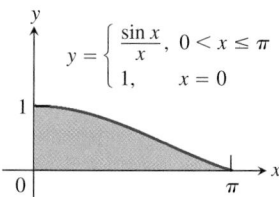

$$y = \begin{cases} \dfrac{\sin x}{x}, & 0 < x \le \pi \\ 1, & x = 0 \end{cases}$$

49. *Designing a Plumb Bob* Having been asked to design a brass plumb bob that will weigh in the neighborhood of 190 g, you decide to shape it like the solid of revolution shown here.

(a) Find the plumb bob's volume.

(b) If you specify a brass that weighs 8.5 g/cm^3, how much will the plumb bob weigh to the nearest gram?

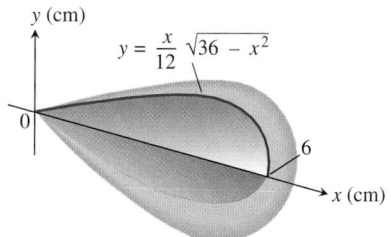

y (cm)

$$y = \frac{x}{12}\sqrt{36 - x^2}$$

6

x (cm)

Explorations

50. *Max-Min* The arch $y = \sin x, \ 0 \le x \le \pi,$ is revolved about the line $y = c, \ 0 \le c \le 1,$ to generate the solid in the figure.

(a) Find the value of *c* that minimizes the volume of the solid. What is the minimum volume?

(b) What value of *c* in $[0, 1]$ maximizes the volume of the solid?

(c) **Writing to Learn** Graph the solid's volume as a function of *c*, first for $0 \le c \le 1$ and then on a larger domain. What happens to the volume of the solid as *c* moves away from $[0, 1]$? Does this make sense physically? Give reasons for your answers.

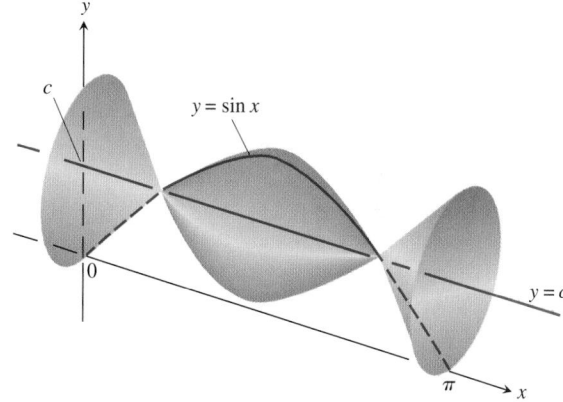

$y = \sin x$

$y = c$

51. *A Vase* We wish to estimate the volume of a flower vase using only a calculator, a string, and a ruler. We measure the height of the vase to be 6 inches. We then use the string and the ruler to find circumferences of the vase (in inches) at half-inch intervals. (We list them from the top down to correspond with the picture of the vase.)

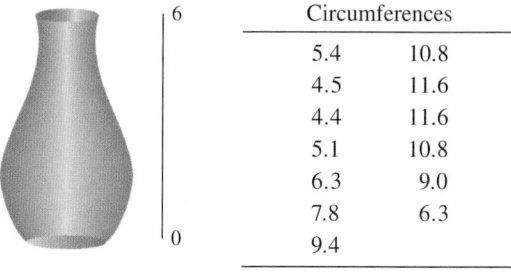

Circumferences	
5.4	10.8
4.5	11.6
4.4	11.6
5.1	10.8
6.3	9.0
7.8	6.3
9.4	

(a) Find the areas of the cross sections that correspond to the given circumferences.

(b) Express the volume of the vase as an integral with respect to *y* over the interval $[0, 6]$.

(c) Approximate the integral using the Trapezoidal Rule with $n = 12$. ∎

52. *Volume of a Bowl*   A bowl has a shape that can be generated by revolving the graph of $y = x^2/2$ between $y = 0$ and $y = 5$ about the y-axis.

(a) Find the volume of the bowl.

(b) If we fill the bowl with water at a constant rate of 3 cubic units per second, how fast will the water level in the bowl be rising when the water is 4 units deep?

53. *The Classical Bead Problem*   A round hole is drilled through the center of a spherical solid of radius r. The resulting cylindrical hole has height 4 cm.

(a) What is the volume of the solid that remains?

(b) What is unusual about the answer?

54. Writing to Learn   Explain how you could estimate the volume of a solid of revolution by measuring the shadow cast on a table parallel to its axis of revolution by a light shining directly above it.

55. *Same Volume about Each Axis*   The region in the first quadrant enclosed between the graph of $y = ax - x^2$ and the x-axis generates the same volume whether it is revolved about the x-axis or the y-axis. Find the value of a.

56. *(Continuation of Exploration 2)*   Let $x = g(y) > 0$ have a continuous first derivative on $[c, d]$. Show that the area of the surface generated by revolving the curve $x = g(y)$ about the y-axis is

$$S = \int_c^d 2\pi\, g(y)\, \sqrt{1 + (g'(y))^2}\, dy.$$

In Exercises 57–64, find the area of the surface generated by revolving the curve about the indicated axis.

57. $x = \sqrt{y}, \quad 0 \le y \le 2; \quad y$-axis

58. $x = y^3/3, \quad 0 \le y \le 1; \quad y$-axis

59. $x = y^{1/2} - (1/3)^{3/2}, \quad 1 \le y \le 3; \quad y$-axis

60. $x = \sqrt{2y - 1}, \quad (5/8) \le y \le 1; \quad y$-axis

61. $y = x^2, \quad 0 \le x \le 2; \quad x$-axis

62. $y = 3x - x^2, \quad 0 \le x \le 3; \quad x$-axis

63. $y = \sqrt{2x - x^2}, \quad 0.5 \le x \le 1.5; \quad x$-axis

64. $y = \sqrt{x + 1}, \quad 1 \le x \le 5; \quad x$-axis

Extending the Ideas

65. *Volume of a Hemisphere*   Derive the formula $V = (2/3)\,\pi R^3$ for the volume of a hemisphere of radius R by comparing its cross sections with the cross sections of a solid right circular cylinder of radius R and height R from which a solid right circular cone of base radius R and height R has been removed as suggested by the figure.

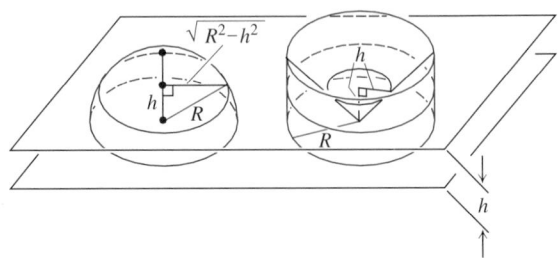

66. *Volume of a Torus*   The disk $x^2 + y^2 \le a^2$ is revolved about the line $x = b$ $(b > a)$ to generate a solid shaped like a doughnut, called a *torus*. Find its volume. (*Hint:* $\int_{-a}^{a} \sqrt{a^2 - y^2}\, dy = \pi a^2/2$, since it is the area of a semicircle of radius a.)

67. *Filling a Bowl*

(a) *Volume*   A hemispherical bowl of radius a contains water to a depth h. Find the volume of water in the bowl.

(b) *Related Rates*   Water runs into a sunken concrete hemispherical bowl of radius 5 m at a rate of 0.2 m³/sec. How fast is the water level in the bowl rising when the water is 4 m deep?

68. *Consistency of Volume Definitions*   The volume formulas in calculus are consistent with the standard formulas from geometry in the sense that they agree on objects to which both apply.

(a) As a case in point, show that if you revolve the region enclosed by the semicircle $y = \sqrt{a^2 - x^2}$ and the x-axis about the x-axis to generate a solid sphere, the calculus formula for volume at the beginning of the section will give $(4/3)\pi a^3$ for the volume just as it should.

(b) Use calculus to find the volume of a right circular cone of height h and base radius r.

7.4 Lengths of Curves

A Sine Wave • Length of a Smooth Curve • Vertical Tangents, Corners, and Cusps

Group Exploration

Later in this section we will use an integral to find the length of the sine wave with great precision. But there are ways to get good approximations without integrating. Take five minutes to come up with a written estimate of the curve's length. No fair looking ahead.

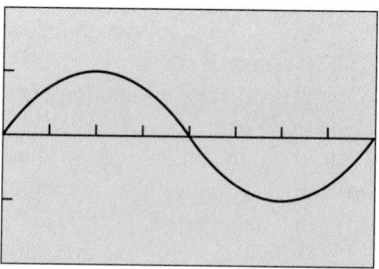

[0, 2π] by [−2, 2]

Figure 7.31 One wave of a sine curve has to be longer than 2π.

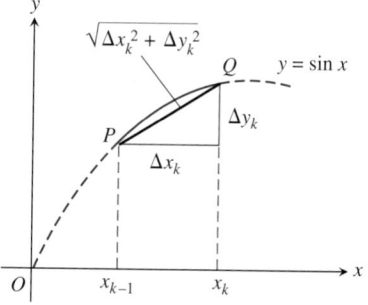

Figure 7.32 The line segment approximating the arc PQ of the sine curve above the subinterval $[x_{k-1}, x_k]$. (Example 1)

A Sine Wave

How long is a sine wave (Figure 7.31)?

The usual meaning of *wavelength* refers to the fundamental period, which for $y = \sin x$ is 2π. But how long is the curve itself? If you straightened it out like a piece of string along the positive x-axis with one end at 0, where would the other end be?

Example 1 THE LENGTH OF A SINE WAVE

What is the length of the curve $y = \sin x$ from $x = 0$ to $x = 2\pi$?

Solution We answer this question with integration, following our usual plan of breaking the whole into measurable parts. We partition $[0, 2\pi]$ into intervals so short that the pieces of curve (call them "arcs") lying directly above the intervals are nearly straight. That way, each arc is nearly the same as the line segment joining its two ends and we can take the length of the segment as an approximation to the length of the arc.

Figure 7.32 shows the segment approximating the arc above the subinterval $[x_{k-1}, x_k]$. The length of the segment is $\sqrt{\Delta x_k^2 + \Delta y_k^2}$. The sum

$$\sum \sqrt{\Delta x_k^2 + \Delta y_k^2}$$

over the entire partition approximates the length of the curve. All we need now is to find the limit of this sum as the norms of the partitions go to zero. That's the usual plan, but this time there is a problem. Do you see it?

The problem is that the sums as written are not Riemann sums. They do not have the form $\sum f(c_k) \Delta x$. We can rewrite them as Riemann sums if we multiply and divide each square root by Δx_k.

$$\sum \sqrt{\Delta x_k^2 + \Delta y_k^2} = \sum \frac{\sqrt{(\Delta x_k)^2 + (\Delta y_k)^2}}{\Delta x_k} \Delta x_k$$

$$= \sum \sqrt{1 + \left(\frac{\Delta y_k}{\Delta x_k}\right)^2} \Delta x_k$$

This is better, but we still need to write the last square root as a function evaluated at some c_k in the kth subinterval. For this, we call on the Mean Value Theorem for differentiable functions (Section 4.2), which says that since $\sin x$ is continuous on $[x_{k-1}, x_k]$ and is differentiable on (x_{k-1}, x_k) there is a point c_k in $[x_{k-1}, x_k]$ at which $\Delta y_k/\Delta x_k = \sin' c_k$ (Figure 7.33). That gives us

$$\sum \sqrt{1 + (\sin' c_k)^2} \, \Delta x_k,$$

which *is* a Riemann sum.

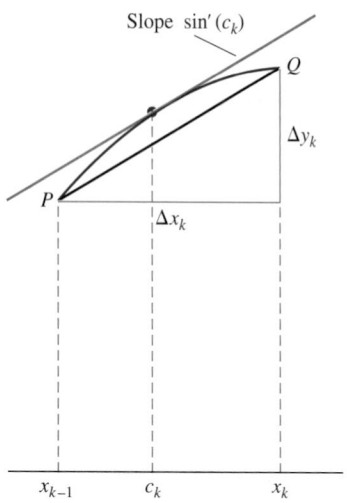

Figure 7.33 The portion of the sine curve above $[x_{k-1}, x_k]$. At some c_k in the interval, $\sin'(c_k) = \Delta y_k/\Delta x_k$, the slope of segment PQ. (Example 1)

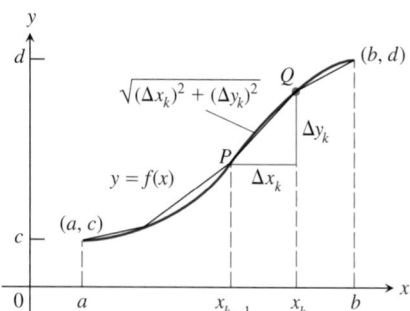

Figure 7.34 The graph of f, approximated by line segments.

Now we take the limit as the norms of the subdivisions go to zero and find that the length of one wave of the sine function is

$$\int_0^{2\pi} \sqrt{1 + (\sin' x)^2}\, dx = \int_0^{2\pi} \sqrt{1 + \cos^2 x}\, dx \approx 7.64. \quad \text{Using NINT}$$

How close was your estimate?

Length of a Smooth Curve

We are almost ready to define the length of a curve as a definite integral, using the procedure of Example 1. We first call attention to two properties of the sine function that came into play along the way.

We obviously used *differentiability* when we invoked the Mean Value Theorem to replace $\Delta y_k/\Delta x_k$ by $\sin'(c_k)$ for some c_k in the interval $[x_{k-1}, x_k]$. Less obviously, we used the continuity of the derivative of sine in passing from $\sum \sqrt{1 + (\sin'(c_k))^2}\, \Delta x_k$ to the Riemann integral. The requirement for finding the length of a curve by this method, then, is that the function have a continuous first derivative. We call this property **smoothness.** A function with a continuous first derivative is **smooth** and its graph is a **smooth curve.**

Let us review the process, this time with a general smooth function $f(x)$. Suppose the graph of f begins at the point (a, c) and ends at (b, d), as shown in Figure 7.34. We partition the interval $a \leq x \leq b$ into subintervals so short that the arcs of the curve above them are nearly straight. The length of the segment approximating the arc above the subinterval $[x_{k-1}, x_k]$ is $\sqrt{\Delta x_k^2 + \Delta y_k^2}$. The sum $\sum \sqrt{\Delta x_k^2 + \Delta y_k^2}$ approximates the length of the curve. We apply the Mean Value Theorem to f on each subinterval to rewrite the sum as a Riemann sum,

$$\sum \sqrt{\Delta x_k^2 + \Delta y_k^2} = \sum \sqrt{1 + \left(\frac{\Delta y_k}{\Delta x_k}\right)^2}\, \Delta x_k$$

$$= \sum \sqrt{1 + (f'(c_k))^2}\, \Delta x_k. \quad \begin{array}{l}\text{For some point}\\ c_k \text{ in } (x_{k-1}, x_k)\end{array}$$

Passing to the limit as the norms of the subdivisions go to zero gives the length of the curve as

$$L = \int_a^b \sqrt{1 + (f'(x))^2}\, dx = \int_a^b \sqrt{1 + \left(\frac{dy}{dx}\right)^2}\, dx.$$

We could as easily have transformed $\sum \sqrt{\Delta x_k^2 + \Delta y_k^2}$ into a Riemann sum by dividing and multiplying by Δy_k, giving a formula that involves x as a function of y (say, $x = g(y)$) on the interval $[c, d]$:

$$L \approx \sum \frac{\sqrt{(\Delta x_k)^2 + (\Delta y_k)^2}}{\Delta y_k}\, \Delta y_k = \sum \sqrt{1 + \left(\frac{\Delta x_k}{\Delta y_k}\right)^2}\, \Delta y_k$$

$$= \sum \sqrt{1 + (g'(c_k))^2}\, \Delta y_k. \quad \begin{array}{l}\text{For some } c_k\\ \text{in } (y_{k-1}, y_k)\end{array}$$

The limit of these sums, as the norms of the subdivisions go to zero, gives another reasonable way to calculate the curve's length,

$$L = \int_c^d \sqrt{1 + (g'(y))^2}\, dy = \int_c^d \sqrt{1 + \left(\frac{dx}{dy}\right)^2}\, dy.$$

Putting these two formulas together, we have the following definition for the length of a smooth curve.

Definition Arc Length: Length of a Smooth Curve

If a smooth curve begins at (a, c) and ends at (b, d), $a < b$, $c < d$, then the **length (arc length) of the curve** is

$$L = \int_a^b \sqrt{1 + \left(\frac{dy}{dx}\right)^2}\, dx \quad \text{if } y \text{ is a smooth function of } x \text{ on } [a, b];$$

$$L = \int_c^d \sqrt{1 + \left(\frac{dx}{dy}\right)^2}\, dy \quad \text{if } x \text{ is a smooth function of } y \text{ on } [c, d].$$

Example 2 APPLYING THE DEFINITION

Find the *exact* length of the curve

$$y = \frac{4\sqrt{2}}{3} x^{3/2} - 1 \quad \text{for} \quad 0 \le x \le 1.$$

Solution

$$\frac{dy}{dx} = \frac{4\sqrt{2}}{3} \cdot \frac{3}{2} x^{1/2} = 2\sqrt{2}\, x^{1/2},$$

which is continuous on $[0, 1]$. Therefore,

$$L = \int_0^1 \sqrt{1 + \left(\frac{dy}{dx}\right)^2}\, dx$$

$$= \int_0^1 \sqrt{1 + \left(2\sqrt{2}x^{1/2}\right)^2}\, dx$$

$$= \int_0^1 \sqrt{1 + 8x}\, dx$$

$$= \frac{2}{3} \cdot \frac{1}{8} (1 + 8x)^{3/2} \Big]_0^1$$

$$= \frac{13}{6}.$$

We asked for an exact length in Example 2 to take advantage of the rare opportunity it afforded of taking the antiderivative of an arc length integrand. When you add 1 to the square of the derivative of an arbitrary smooth function and then take the square root of that sum, the result is rarely antidifferentiable by reasonable methods. We know a few more functions that give "nice" integrands, but we are saving those for the exercises.

Vertical Tangents, Corners, and Cusps

Sometimes a curve has a vertical tangent, corner, or cusp where the derivative we need to work with is undefined. We can sometimes get around such a difficulty in ways illustrated by the following examples.

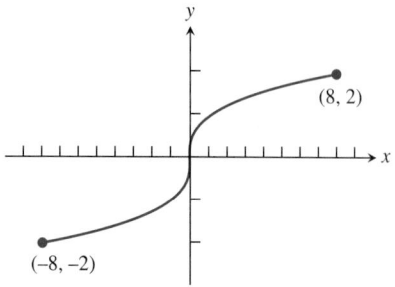

Figure 7.35 The graph of $y = x^{1/3}$ has a vertical tangent line at the origin where dy/dx does not exist. (Example 3)

Example 3 A VERTICAL TANGENT

Find the length of the curve $y = x^{1/3}$ between $(-8, -2)$ and $(8, 2)$.

Solution The derivative

$$\frac{dy}{dx} = \frac{1}{3}x^{-2/3} = \frac{1}{3x^{2/3}}$$

is not defined at $x = 0$. Graphically, there is a vertical tangent at $x = 0$ where the derivative becomes infinite (Figure 7.35). If we change to x as a function of y, the tangent at the origin will be horizontal (Figure 7.36) and the derivative will be zero instead of undefined. Solving $y = x^{1/3}$ for x gives $x = y^3$, and we have

$$L = \int_{-2}^{2} \sqrt{1 + \left(\frac{dx}{dy}\right)^2}\, dy = \int_{-2}^{2} \sqrt{1 + (3y^2)^2}\, dy \approx 17.26. \quad \text{Using NINT}$$

What happens if you fail to notice that dy/dx is undefined at $x = 0$ and ask your calculator to compute

$$\text{NINT}\left(\sqrt{1 + \left((1/3)\, x^{-2/3}\right)^2}, x, -8, 8\right)?$$

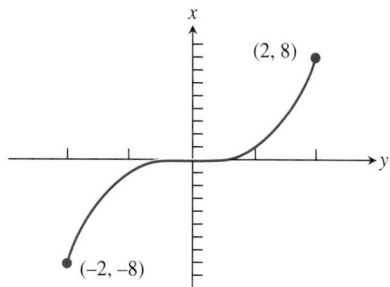

Figure 7.36 The curve in Figure 7.35 plotted with x as a function of y. The tangent at the origin is now horizontal. (Example 3)

This actually depends on your calculator. If, in the process of its calculations, it tries to evaluate the function at $x = 0$, then some sort of domain error will result. If it tries to find convergent Riemann sums near $x = 0$, it might get into a long, futile loop of computations that you will have to interrupt. Or it might actually produce an answer—in which case you hope it would be sufficiently bizarre for you to realize that it should not be trusted.

Example 4 GETTING AROUND A CORNER

Find the length of the curve $y = x^2 - 4|x| - x$ from $x = -4$ to $x = 4$.

Solution We should always be alert for abrupt slope changes when absolute value is involved. We graph the function to check (Figure 7.37).

There is clearly a corner at $x = 0$ where neither dy/dx nor dx/dy can exist. To find the length, we split the curve at $x = 0$ to write the function *without* absolute values:

$$x^2 - 4|x| - x = \begin{cases} x^2 + 3x & \text{if} \quad x < 0, \\ x^2 - 5x & \text{if} \quad x \geq 0. \end{cases}$$

Then,

$$L = \int_{-4}^{0} \sqrt{1 + (2x + 3)^2}\, dx + \int_{0}^{4} \sqrt{1 + (2x - 5)^2}\, dx$$

$$\approx 19.56. \quad \text{By NINT}$$

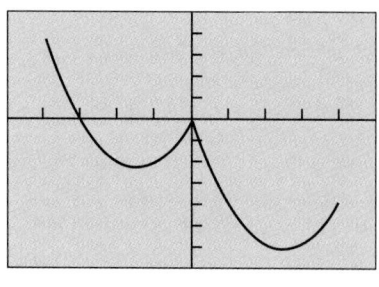

$[-5, 5]$ by $[-7, 5]$

Figure 7.37 The graph of
$$y = x^2 - 4|x| - x, \quad -4 \leq x \leq 4,$$
has a corner at x = 0 where neither dy/dx nor dx/dy exists. We find the lengths of the two smooth pieces and add them together. (Example 4)

Finally, cusps are handled the same way corners are: split the curve into smooth pieces and add the lengths of those pieces.

Quick Review 7.4

In Exercises 1–5, simplify the function.

1. $\sqrt{1 + 2x + x^2}$ on $[1, 5]$

2. $\sqrt{1 - x + \dfrac{x^2}{4}}$ on $[-3, -1]$

3. $\sqrt{1 + (\tan x)^2}$ on $[0, \pi/3]$

4. $\sqrt{1 + (x/4 - 1/x)^2}$ on $[4, 12]$

5. $\sqrt{1 + \cos 2x}$ on $[0, \pi/2]$

In Exercises 6–10, identify all values of x for which the function fails to be differentiable.

6. $f(x) = |x - 4|$

7. $f(x) = 5x^{2/3}$

8. $f(x) = \sqrt[5]{x + 3}$

9. $f(x) = \sqrt{x^2 - 4x + 4}$

10. $f(x) = 1 + \sqrt[3]{\sin x}$

Section 7.4 Exercises

In Exercises 1–10,

(a) set up an integral for the length of the curve;

(b) graph the curve to see what it looks like;

(c) use NINT to find the length of the curve.

1. $y = x^2$, $-1 \le x \le 2$

2. $y = \tan x$, $-\pi/3 \le x \le 0$

3. $x = \sin y$, $0 \le y \le \pi$

4. $x = \sqrt{1 - y^2}$, $-1/2 \le y \le 1/2$

5. $y^2 + 2y = 2x + 1$, from $(-1, -1)$ to $(7, 3)$

6. $y = \sin x - x \cos x$, $0 \le x \le \pi$

7. $y = \int_0^x \tan t \, dt$, $0 \le x \le \pi/6$

8. $x = \int_0^y \sqrt{\sec^2 t - 1} \, dt$, $-\pi/3 \le y \le \pi/4$

9. $y = \sec x$, $-\pi/3 \le x \le \pi/3$

10. $y = (e^x + e^{-x})/2$, $-3 \le x \le 3$

In Exercises 11–18, find the exact length of the curve analytically by antidifferentiation. You will need to simplify the integrand algebraically before finding an antiderivative.

11. $y = (1/3)(x^2 + 2)^{3/2}$ from $x = 0$ to $x = 3$

12. $y = x^{3/2}$ from $x = 0$ to $x = 4$

13. $x = (y^3/3) + 1/(4y)$ from $y = 1$ to $y = 3$
    (*Hint:* $1 + (dx/dy)^2$ is a perfect square.)

14. $x = (y^4/4) + 1/(8y^2)$ from $y = 1$ to $y = 2$
    (*Hint:* $1 + (dx/dy)^2$ is a perfect square.)

15. $x = (y^3/6) + 1/(2y)$ from $y = 1$ to $y = 2$
    (*Hint:* $1 + (dx/dy)^2$ is a perfect square.)

16. $y = (x^3/3) + x^2 + x + 1/(4x + 4)$, $0 \le x \le 2$

17. $x = \int_0^y \sqrt{\sec^4 t - 1} \, dt$, $-\pi/4 \le y \le \pi/4$

18. $y = \int_{-2}^x \sqrt{3t^4 - 1} \, dt$, $-2 \le x \le -1$

In Exercises 19 and 20, *work in groups of two or three.*

19. (a) Find a curve through the point $(1, 1)$ whose length integral is
$$L = \int_1^4 \sqrt{1 + \frac{1}{4x}} \, dx.$$

(b) **Writing to Learn** How many such curves are there? Give reasons for your answer.

20. (a) Find a curve through the point $(0, 1)$ whose length integral is
$$L = \int_1^2 \sqrt{1 + \frac{1}{y^4}} \, dy.$$

(b) **Writing to Learn** How many such curves are there? Give reasons for your answer.

21. Find the length of the curve
$$y = \int_0^x \sqrt{\cos 2t} \, dt$$
from $x = 0$ to $x = \pi/4$.

22. *The Length of an Astroid* The graph of the equation $x^{2/3} + y^{2/3} = 1$ is one of the family of curves called *astroids* (not "asteroids") because of their starlike appearance (see figure). Find the length of this particular astroid by finding the length of half the first quadrant portion, $y = (1 - x^{2/3})^{3/2}$, $\sqrt{2}/4 \le x \le 1$, and multiplying by 8.

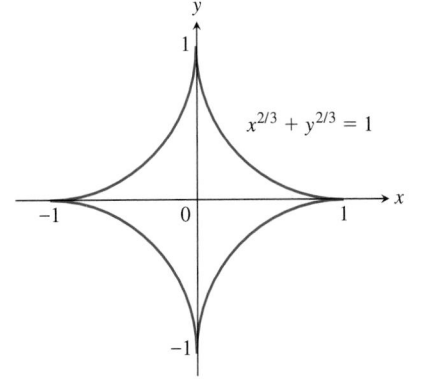

23. *Fabricating Metal Sheets* Your metal fabrication company is bidding for a contract to make sheets of corrugated steel roofing like the one shown here. The cross sections of the corrugated sheets are to conform to the curve

$$y = \sin\left(\frac{3\pi}{20}x\right), \quad 0 \le x \le 20 \text{ in.}$$

If the roofing is to be stamped from flat sheets by a process that does not stretch the material, how wide should the original material be? Give your answer to two decimal places.

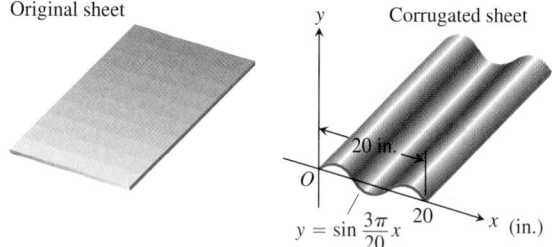

Original sheet   Corrugated sheet

$y = \sin\dfrac{3\pi}{20}x$

24. *Tunnel Construction* Your engineering firm is bidding for the contract to construct the tunnel shown here. The tunnel is 300 ft long and 50 ft wide at the base. The cross section is shaped like one arch of the curve $y = 25 \cos(\pi x/50)$. Upon completion, the tunnel's inside surface (excluding the roadway) will be treated with a waterproof sealer that costs \$1.75 per square foot to apply. How much will it cost to apply the sealer?

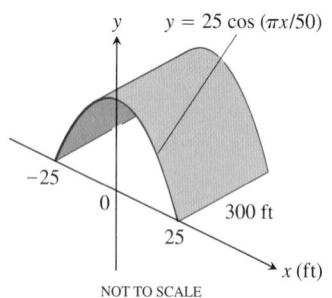

$y = 25 \cos(\pi x/50)$

300 ft

x (ft)

NOT TO SCALE

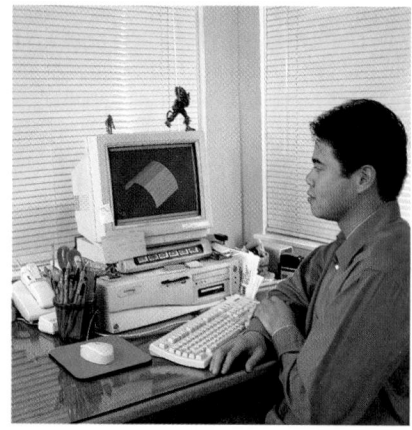

Exploration

25. *Modeling Running Tracks* Two lanes of a running track are modeled by the semiellipses as shown. The equation for lane 1 is $y = \sqrt{100 - 0.2x^2}$, and the equation for lane 2 is $y = \sqrt{150 - 0.2x^2}$. The starting point for lane 1 is at the negative x-intercept $(-\sqrt{500}, 0)$. The finish points for both lanes are the positive x-intercepts. Where should the starting point be placed on lane 2 so that the two lane lengths will be equal (running clockwise)?

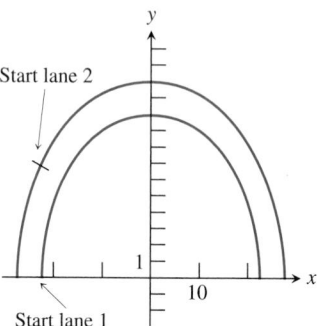

Start lane 2

10

Start lane 1

In Exercises 26 and 27, find the length of the curve.

26. $f(x) = x^{1/3} + x^{2/3}, \quad 0 \le x \le 2$

27. $f(x) = \dfrac{x-1}{4x^2+1}, \quad -\dfrac{1}{2} \le x \le 1$

In Exercises 28–30, find the length of the nonsmooth curve.

28. $y = x^3 + 5|x|$ from $x = -2$ to $x = 1$

29. $\sqrt{x} + \sqrt{y} = 1$

30. $y = \sqrt[4]{x}$ from $x = 0$ to $x = 16$

31. Writing to Learn Explain geometrically why it does not work to use short *horizontal* line segments to approximate the lengths of small arcs when we search for a Riemann sum that leads to the formula for arc length.

32. Writing to Learn A curve is totally contained inside the square with vertices $(0, 0)$, $(1, 0)$, $(1, 1)$, and $(0, 1)$. Is there any limit to the possible length of the curve? Explain.

Extending the Ideas

33. *Using Tangent Fins to Find Arc Length* Assume f is smooth on $[a, b]$ and partition the interval $[a, b]$ in the usual way. In each subinterval $[x_{k-1}, x_k]$ construct the *tangent fin* at the point $(x_{k-1}, f(x_{k-1}))$ as shown in the figure.

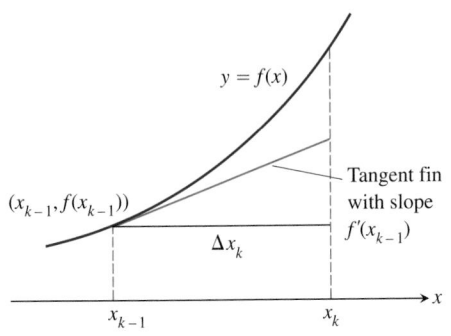

(a) Show that the length of the kth tangent fin over the interval $[x_{k-1}, x_k]$ equals

$$\sqrt{(\Delta x_k)^2 + (f'(x_{k-1})\Delta x_k)^2}.$$

(b) Show that

$$\lim_{n\to\infty} \sum_{k=1}^{n} (\text{length of } k\text{th tangent fin}) = \int_a^b \sqrt{1 + (f'(x))^2}\, dx,$$

which is the length L of the curve $y = f(x)$ from $x = a$ to $x = b$.

34. Is there a smooth curve $y = f(x)$ whose length over the interval $0 \le x \le a$ is always $a\sqrt{2}$? Give reasons for your answer.

7.5 Applications from Science and Statistics

Work Revisited • Fluid Force and Fluid Pressure • Normal Probabilities

Our goal in this section is to hint at the diversity of ways in which the definite integral can be used. The contexts may be new to you, but we will explain what you need to know as we go along.

Work Revisited

Recall from Section 7.1 that *work* is defined as force (in the direction of motion) times displacement. A familiar example is to move against the force of gravity to lift an object. The object has to move, incidentally, before "work" is done, no matter how tired you get *trying*.

4.4 newtons ≈ 1 lb

(1 newton)(1 meter) = 1 N • m = 1 Joule

Example 1 WORK DONE LIFTING

A leaky bucket weighs 22 newtons (N) empty. It is lifted from the ground at a constant rate to a point 20 m above the ground by a rope weighing 0.4 N/m. The bucket starts with 70 N (approximately 7.1 liters) of water, but it leaks at a constant rate and just finishes draining as the bucket reaches the top. Find the amount of work done

(a) lifting the bucket alone;

(b) lifting the water alone;

(c) lifting the rope alone;

(d) lifting the bucket, water, and rope together.

Figure 7.38 The work done by a constant 22-N force lifting a bucket 20 m is 440 N • m. (Example 1)

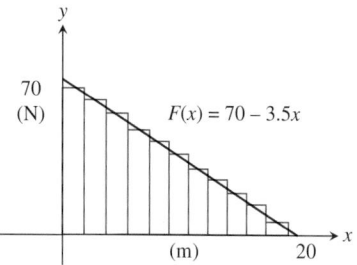

Figure 7.39 The force required to lift the water varies with distance but the work still corresponds to the area under the force graph. (Example 1)

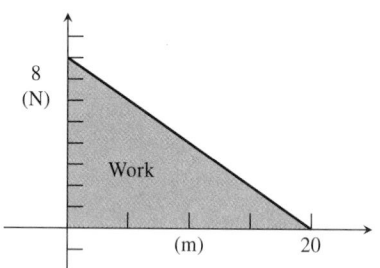

Figure 7.40 The work done lifting the rope to the top corresponds to the area of another triangle. (Example 1)

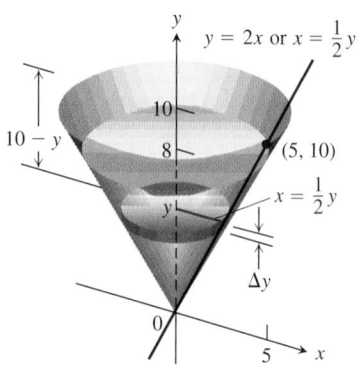

Figure 7.41 The conical tank in Example 2.

Solution

(a) *The bucket alone.* This is easy because the bucket's weight is constant. To lift it, you must exert a force of 22 N through the entire 20-meter interval.

$$\text{Work} = (22 \text{ N}) \times (20 \text{ m}) = 440 \text{ N} \cdot \text{m} = 440 \text{ J}$$

Figure 7.38 shows the graph of force vs. distance applied. The work corresponds to the area under the force graph.

(b) *The water alone.* The force needed to lift the water is equal to the water's weight, which decreases steadily from 70 N to 0 N over the 20-m lift. When the bucket is x m off the ground, the water weighs

$$F(x) = 70\left(\frac{20 - x}{20}\right) = 70\left(1 - \frac{x}{20}\right) = 70 - 3.5x \text{ N.}$$

original weight of water   proportion left at elevation x

The work done is (Figure 7.39)

$$W = \int_a^b F(x)\, dx$$

$$= \int_0^{20} (70 - 3.5x)\, dx = \left[70x - 1.75x^2\right]_0^{20} = 1400 - 700 = 700 \text{ J.}$$

(c) *The rope alone.* The force needed to lift the rope is also variable, starting at $(0.4)(20) = 8$ N when the bucket is on the ground and ending at 0 N when the bucket and rope are all at the top. As with the leaky bucket, the rate of decrease is constant. At elevation x meters, the $(20 - x)$ meters of rope still there to lift weigh $F(x) = (0.4)(20 - x)$ N. Figure 7.40 shows the graph of F. The work done lifting the rope is

$$\int_0^{20} F(x)\, dx = \int_0^{20} (0.4)(20 - x)\, dx$$

$$= \left[8x - 0.2x^2\right]_0^{20} = 160 - 80 = 80 \text{ N} \cdot \text{m} = 80 \text{ J.}$$

(d) *The bucket, water, and rope together.* The total work is

$$440 + 700 + 80 = 1220 \text{ J.}$$

Example 2   WORK DONE PUMPING

The conical tank in Figure 7.41 is filled to within 2 ft of the top with olive oil weighing 57 lb/ft³. How much work does it take to pump the oil to the rim of the tank?

Solution We imagine the oil partitioned into thin slabs by planes perpendicular to the y-axis at the points of a partition of the interval $[0, 8]$. (The 8 represents the top of the oil, not the top of the tank.)

The typical slab between the planes at y and $y + \Delta y$ has a volume of about

$$\Delta V = \pi(\text{radius})^2(\text{thickness}) = \pi\left(\frac{1}{2}y\right)^2 \Delta y = \frac{\pi}{4}y^2 \Delta y \text{ ft}^3.$$

The force $F(y)$ required to lift this slab is equal to its weight,

$$F(y) = 57 \, \Delta V = \frac{57\pi}{4} y^2 \, \Delta y \ \text{lb.} \quad \text{Weight} = \left(\begin{array}{c}\text{weight per} \\ \text{unit volume}\end{array}\right) \times \text{volume}$$

The distance through which $F(y)$ must act to lift this slab to the level of the rim of the cone is about $(10 - y)$ ft, so the work done lifting the slab is about

$$\Delta W = \frac{57\pi}{4}(10 - y)y^2 \, \Delta y \ \text{ft} \cdot \text{lb.}$$

The work done lifting all the slabs from $y = 0$ to $y = 8$ to the rim is approximately

$$W \approx \sum \frac{57\pi}{4}(10 - y) \, y^2 \, \Delta y \ \text{ft} \cdot \text{lb.}$$

This is a Riemann sum for the function $(57\pi/4)(10 - y)y^2$ on the interval from $y = 0$ to $y = 8$. The work of pumping the oil to the rim is the limit of these sums as the norms of the partitions go to zero.

$$W = \int_0^8 \frac{57\pi}{4}(10 - y)y^2 \, dy = \frac{57\pi}{4}\int_0^8 (10y^2 - y^3) \, dy$$

$$= \frac{57\pi}{4}\left[\frac{10y^3}{3} - \frac{y^4}{4}\right]_0^8 \approx 30{,}561 \ \text{ft} \cdot \text{lb}$$

Fluid Force and Fluid Pressure

We make dams thicker at the bottom than at the top (Figure 7.42) because the pressure against them increases with depth. It is a remarkable fact that the pressure at any point on a dam depends only on how far below the surface the point lies and not on how much water the dam is holding back. In any liquid, the **fluid pressure** p (force per unit area) at depth h is

$$p = wh, \quad \text{Dimensions check: } \frac{\text{lb}}{\text{ft}^2} = \frac{\text{lb}}{\text{ft}^3} \times \text{ft, for example}$$

where w is the *weight-density* (weight per unit volume) of the liquid.

Figure 7.42 To withstand the increasing pressure, dams are built thicker toward the bottom.

Typical Weight-densities (lb/ft³)

Gasoline	42
Mercury	849
Milk	64.5
Molasses	100
Seawater	64
Water	62.4

Example 3 THE GREAT MOLASSES FLOOD OF 1919

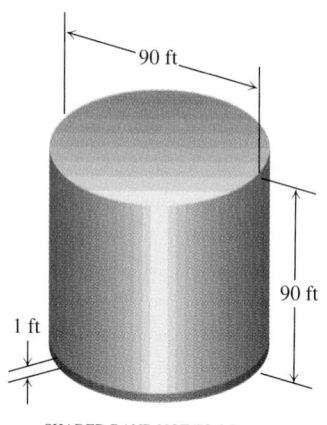

1 ft

SHADED BAND NOT TO SCALE

Figure 7.43 The molasses tank of Example 3.

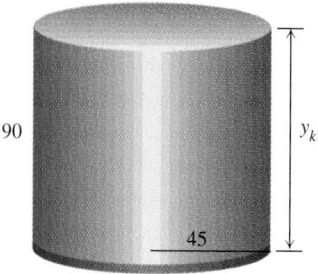

Figure 7.44 The 1-ft band at the bottom of the tank wall can be partitioned into thin strips on which the pressure is approximately constant. (Example 3)

At 1:00 P.M. on January 15, 1919 (an unseasonably warm day), a 90-ft-high, 90-foot-diameter cylindrical metal tank in which the Puritan Distilling Company stored molasses at the corner of Foster and Commercial streets in Boston's North End exploded. Molasses flooded the streets 30 feet deep, trapping pedestrians and horses, knocking down buildings, and oozing into homes. It was eventually tracked all over town and even made its way into the suburbs via trolley cars and people's shoes. It took weeks to clean up.

(a) Given that the tank was full of molasses weighing 100 lb/ft^3, what was the total force exerted by the molasses on the bottom of the tank at the time it ruptured?

(b) What was the total force against the bottom foot-wide band of the tank wall (Figure 7.43)?

Solution

(a) At the bottom of the tank, the molasses exerted a constant pressure of

$$p = wh = \left(100 \; \frac{\text{lb}}{\text{ft}^3}\right)\left(90 \; \text{ft}\right) = 9000 \; \frac{\text{lb}}{\text{ft}^2}.$$

Since the area of the base was $\pi(45)^2$, the total force on the base was

$$\left(9000 \; \frac{\text{lb}}{\text{ft}^2}\right)\left(2025 \; \pi \; \text{ft}^2\right) \approx 57{,}225{,}526 \; \text{lb}.$$

(b) We partition the band from depth 89 ft to depth 90 ft into narrower bands of width Δy and choose a depth y_k in each one. The pressure at this depth y_k is $p = wh = 100 \; y_k$ lb/ft^2 (Figure 7.44). The force against each narrow band is approximately

$$\text{pressure} \times \text{area} = (100 y_k)(90\pi \; \Delta y) = 9000\pi \; y_k \; \Delta y \; \text{lb}.$$

Adding the forces against all the bands in the partition and passing to the limit as the norms go to zero, we arrive at

$$F = \int_{89}^{90} 9000\pi y \, dy = 9000\pi \int_{89}^{90} y \, dy \approx 2{,}530{,}553 \; \text{lb}$$

for the force against the bottom foot of tank wall.

Normal Probabilities

Suppose you find an old clock in the attic. What is the probability that it has stopped somewhere between 2:00 and 5:00?

If you imagine time being measured continuously over a 12-hour interval, it is easy to conclude that the answer is 1/4 (since the interval from 2:00 to 5:00 contains one-fourth of the time), and that is correct. Mathematically, however, the situation is not quite that clear because both the 12-hour interval and the 3-hour interval contain an *infinite* number of times. In what sense does the ratio of one infinity to another infinity equal 1/4?

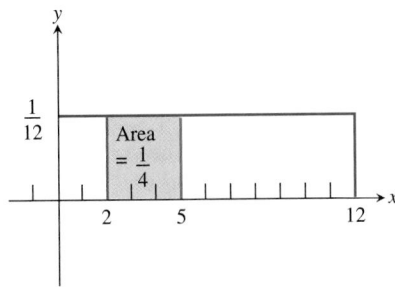

Figure 7.45 The probability that the clock has stopped between 2:00 and 5:00 can be represented as an area of 1/4. The rectangle over the entire interval has area 1.

Improper Integrals

More information about improper integrals like $\int_{-\infty}^{\infty} f(x)\,dx$ can be found in Section 8.3. (You will not need that information here.)

The easiest way to resolve that question is to look at area. We represent the total probability of the 12-hour interval as a rectangle of area 1 sitting above the interval (Figure 7.45).

Not only does it make perfect sense to say that the rectangle over the time interval [2, 5] has an area that is one-fourth the area of the total rectangle, the area actually *equals* 1/4, since the total rectangle has area 1. That is why mathematicians represent probabilities as areas, and that is where definite integrals enter the picture.

Definition   **Probability Density Function (pdf)**

A **probability density function** is a function $f(x)$ with domain all reals such that

$$f(x) \geq 0 \text{ for all } x \quad \text{and} \quad \int_{-\infty}^{\infty} f(x)\,dx = 1.$$

Then the probability associated with an interval $[a, b]$ is

$$\int_a^b f(x)\,dx.$$

Probabilities of events, such as the clock stopping between 2:00 and 5:00, are integrals of an appropriate pdf.

Example 4   PROBABILITY OF THE CLOCK STOPPING

Find the probability that the clock stopped between 2:00 and 5:00.

Solution   The pdf of the clock is

$$f(t) = \begin{cases} 1/12, & 0 \leq t \leq 12 \\ 0, & \text{otherwise.} \end{cases}$$

The probability that the clock stopped at some time t with $2 \leq t \leq 5$ is

$$\int_2^5 f(t)\,dt = \frac{1}{4}.$$

By far the most useful kind of pdf is the *normal* kind. ("Normal" here is a technical term, referring to a curve with the shape in Figure 7.46.) The **normal curve,** often called the "bell curve," is one of the most significant curves in applied mathematics because it enables us to describe entire populations based on the statistical measurements taken from a reasonably-sized sample. The measurements needed are the *mean* (μ) and the *standard deviation* (σ), which your calculators will approximate for you from the data. The symbols on the calculator will probably be $\bar{x}$ and s (see your *Owner's Manual*), but go ahead and use them as μ and σ, respectively. Once you have the numbers, you can find the curve by using the following remarkable formula discovered by Karl Friedrich Gauss.

$\int f(x)dx = .17287148$

Figure 7.46 A normal probability density function. The probability associated with the interval [a, b] is the area under the curve, as shown.

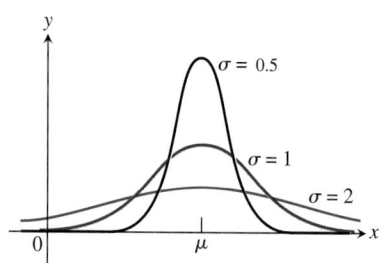

Figure 7.47 The 68-95-99.7 rule for normal distributions.

> **Definition   Normal Probability Density Function (pdf)**
>
> The **normal probability density function (Gaussian curve)** for a population with mean μ and standard deviation σ is
>
> $$f(x) = \frac{1}{\sigma\sqrt{2\pi}} e^{-(x-\mu)^2/(2\sigma^2)}.$$

The mean μ represents the average value of the variable x. The standard deviation σ measures the "scatter" around the mean. For a normal curve, the mean and standard deviation tell you where most of the probability lies. The rule of thumb, illustrated in Figure 7.47, is this:

> **The 68-95-99.7 Rule for Normal Distributions**
>
> Given a normal curve,
>
> • 68% of the area will lie within σ of the mean μ,
>
> • 95% of the area will lie within 2σ of the mean μ,
>
> • 99.7% of the area will lie within 3σ of the mean μ.

Even with the 68-95-99.7 rule, the area under the curve can spread quite a bit, depending on the size of σ. Figure 7.48 shows three normal pdfs with mean $\mu = 2$ and standard deviations equal to 0.5, 1, and 2.

Example 5   A TELEPHONE HELP LINE

Suppose a telephone help line takes a mean of 2 minutes to answer calls. If the standard deviation is $\sigma = 0.5$, then 68% of the calls are answered in the range of 1.5 to 2.5 minutes and 99.7% of the calls are answered in the range of 0.5 to 3.5 minutes. If $\sigma = 2$, then 68% of the calls are answered in the range of 0 (immediately) to 4 minutes. That is, you are just as likely to have a 4-minute wait as an immediate response.

Figure 7.48 Normal pdf curves with mean $\mu = 2$ and $\sigma = 0.5$, 1, and 2.

Example 6   WEIGHTS OF SPINACH BOXES

Suppose that frozen spinach boxes marked as "10 ounces" of spinach have a mean weight of 10.3 ounces and a standard deviation of 0.2 ounce.

(a) What percentage of *all* such spinach boxes can be expected to weigh between 10 and 11 ounces?

(b) What percentage would we expect to weigh less than 10 ounces?

(c) What is the probability that a box weighs *exactly* 10 ounces?

Solution Assuming that some person or machine is *trying* to pack 10 ounces of spinach into these boxes, we expect that most of the weights will be around 10, with probabilities tailing off for boxes being heavier or lighter. We expect, in other words, that a normal pdf will model these probabilities. First, we define $f(x)$ using the formula:

$$f(x) = \frac{1}{0.2\sqrt{2\pi}} e^{-(x-10.3)^2/(0.08)}.$$

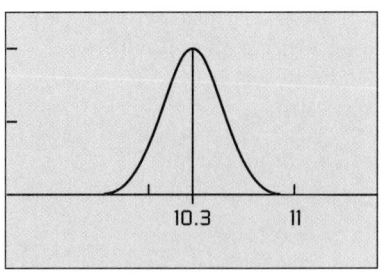

[9, 11.5] by [–1, 2.5]

Figure 7.49 The normal pdf for the spinach weights in Example 6. The mean is at the center.

The graph (Figure 7.49) has the look we are expecting.

(a) For an arbitrary box of this spinach, the probability that it weighs between 10 and 11 ounces is the area under the curve from 10 to 11, which is

$$\text{NINT}\,(f(x),\,x,\,10,\,11) \approx 0.933.$$

So without doing any more measuring, we can predict that about 93.3% of all such spinach boxes will weigh between 10 and 11 ounces.

(b) For the probability that a box weighs less than 10 ounces, we use the entire area under the curve to the left of $x = 10$. The curve actually approaches the x-axis as an asymptote, but you can see from the graph (Figure 7.49) that $f(x)$ approaches zero quite quickly. Indeed, $f(9)$ is only slightly larger than a billionth. So getting the area from 9 to 10 should do it:

$$\text{NINT}\,(f(x),\,x,\,9,\,10) \approx 0.067.$$

We would expect only about 6.7% of the boxes to weigh less than 10 ounces.

(c) This would be the integral from 10 to 10, which is zero. This zero probability might seem strange at first, but remember that we are assuming a continuous, unbroken interval of possible spinach weights, and 10 is but one of an infinite number of them.

Quick Review 7.5

In Exercises 1–5, find the definite integral by **(a)** antiderivatives and **(b)** using NINT.

1. $\displaystyle\int_0^1 e^{-x}\,dx$

2. $\displaystyle\int_0^1 e^x\,dx$

3. $\displaystyle\int_{\pi/4}^{\pi/2} \sin x\,dx$

4. $\displaystyle\int_0^3 (x^2 + 2)\,dx$

5. $\displaystyle\int_1^2 \frac{x^2}{x^3 + 1}\,dx$

In Exercises 6–10 find, but do not evaluate, the definite integral that is the limit as the norms of the partitions go to zero of the Riemann sums on the closed interval $[0, 7]$.

6. $\displaystyle\sum 2\pi(x_k + 2)(\sin x_k)\,\Delta x$

7. $\displaystyle\sum (1 - x_k^2)(2\pi x_k)\,\Delta x$

8. $\displaystyle\sum \pi(\cos x_k)^2\Delta x$

9. $\displaystyle\sum \pi\left(\frac{y_k}{2}\right)^2 (10 - y_k)\,\Delta y$

10. $\displaystyle\sum \frac{\sqrt{3}}{4}\,(\sin^2 x_k)\,\Delta x$

Section 7.5 Exercises

In Exercises 1–4, find the work done by the force of $F(x)$ newtons along the x-axis from $x = a$ meters to $x = b$ meters.

1. $F(x) = xe^{-x/3}, \quad a = 0, \quad b = 5$

2. $F(x) = x\sin(\pi x/4), \quad a = 0, \quad b = 3$

3. $F(x) = x\sqrt{9 - x^2}, \quad a = 0, \quad b = 3$

4. $F(x) = e^{\sin x}\cos x + 2, \quad a = 0, \quad b = 10$

5. *Leaky Bucket* The workers in Example 1 changed to a larger bucket that held 50 L (490 N) of water, but the new

bucket had an even larger leak so that it too was empty by the time it reached the top. Assuming the water leaked out at a steady rate, how much work was done lifting the water to a point 20 meters above the ground? (Do not include the rope and bucket.)

6. *Leaky Bucket* The bucket in Exercise 5 is hauled up more quickly so that there is still 10 L (98 N) of water left when the bucket reaches the top. How much work is done lifting the water this time? (Do not include the rope and bucket.)

7. *Leaky Sand Bag*   A bag of sand originally weighing 144 lb was lifted at a constant rate. As it rose, sand leaked out at a constant rate. The sand was half gone by the time the bag had been lifted 18 ft. How much work was done lifting the sand this far? (Neglect the weights of the bag and lifting equipment.)

8. *Stretching a Spring*   A spring has a natural length of 10 in. An 800-lb force stretches the spring to 14 in.

(a) Find the force constant.

(b) How much work is done in stretching the spring from 10 in. to 12 in.?

(c) How far beyond its natural length will a 1600-lb force stretch the spring?

9. *Subway Car Springs*   It takes a force of 21,714 lb to compress a coil spring assembly on a New York City Transit Authority subway car from its free height of 8 in. to its fully compressed height of 5 in.

(a) What is the assembly's force constant?

(b) How much work does it take to compress the assembly the first half inch? the second half inch? Answer to the nearest inch-pound.
(*Source:* Data courtesy of Bombardier, Inc., Mass Transit Division, for spring assemblies in subway cars delivered to the New York City Transit Authority from 1985 to 1987.)

10. *Bathroom Scale*   A bathroom scale is compressed 1/16 in. when a 150-lb person stands on it. Assuming the scale behaves like a spring that obeys Hooke's Law,

(a) how much does someone who compresses the scale 1/8 in. weigh?

(b) how much work is done in compressing the scale 1/8 in.?

11. *Hauling a Rope*   A mountain climber is about to haul up a 50-m length of hanging rope. How much work will it take if the rope weighs 0.624 N/m?

Exploration

12. *Compressing Gas*   Suppose that gas in a circular cylinder of cross section area A is being compressed by a piston (see figure).

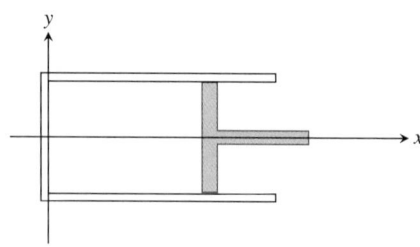

(a) If p is the pressure of the gas in pounds per square inch and V is the volume in cubic inches, show that the work done in compressing the gas from state (p_1, V_1) to state (p_2, V_2) is given by the equation

$$\text{Work} = \int_{(p_1,\, V_1)}^{(p_2,\, V_2)} p \, dV \quad \text{in.} \cdot \text{lb},$$

where the force against the piston is pA.

(b) Find the work done in compressing the gas from $V_1 = 243 \text{ in}^3$ to $V_2 = 32 \text{ in}^3$ if $p_1 = 50 \text{ lb/in}^3$ and p and V obey the gas law $pV^{1.4} = $ constant (for adiabatic processes). ∎

In Exercises 13–16, the vertical end of a tank containing water (blue shading) weighing 62.4 lb/ft³ has the given shape. *Work in groups of two or three.*

(a) **Writing to Learn**   Explain how to approximate the force against the end of the tank by a Riemann sum.

(b) Find the force as an integral and evaluate it.

13. semicircle

14. semiellipse

15. triangle

16. parabola

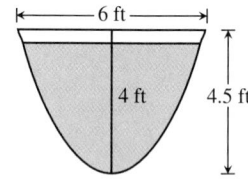

17. *Pumping Water* The rectangular tank shown here, with its top at ground level, is used to catch runoff water. Assume that the water weighs 62.4 lb/ft^3.

(a) How much work does it take to empty the tank by pumping the water back to ground level once the tank is full?

(b) If the water is pumped to ground level with a (5/11)-horsepower motor (work output 250 ft · lb/sec), how long will it take to empty the full tank (to the nearest minute)?

(c) Show that the pump in (b) will lower the water level 10 ft (halfway) during the first 25 min of pumping.

(d) *The Weight of Water* Because of differences in the strength of Earth's gravitational field, the weight of a cubic foot of water at sea level can vary from as little as 62.26 lb at the equator to as much as 62.59 lb near the poles, a variation of about 0.5%. A cubic foot of water that weighs 62.4 lb in Melbourne or New York City will weigh 62.5 lb in Juneau or Stockholm. What are the answers to (a) and (b) in a location where water weighs 62.26 lb/ft^3? 62.5 lb/ft^3?

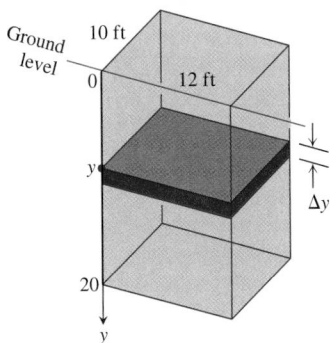

18. *Emptying a Tank* A vertical right cylindrical tank measures 30 ft high and 20 ft in diameter. It is full of kerosene weighing 51.2 lb/ft^3. How much work does it take to pump the kerosene to the level of the top of the tank?

19. *Writing to Learn* The cylindrical tank shown here is to be filled by pumping water from a lake 15 ft below the bottom of the tank. There are two ways to go about this. One is to pump the water through a hose attached to a valve in the bottom of the tank. The other is to attach the hose to the rim of the tank and let the water pour in. Which way will be faster? Give reasons for your answer.

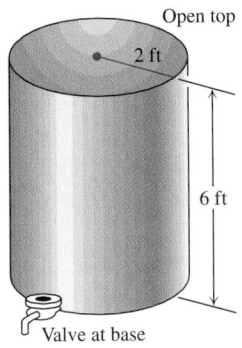

20. *Drinking a Milkshake* The truncated conical container shown here is full of strawberry milkshake that weighs (4/9) oz/in^3. As you can see, the container is 7 in. deep, 2.5 in. across at the base, and 3.5 in. across at the top (a standard size at Brigham's in Boston). The straw sticks up an inch above the top. About how much work does it take to drink the milkshake through the straw (neglecting friction)? Answer in inch-ounces.

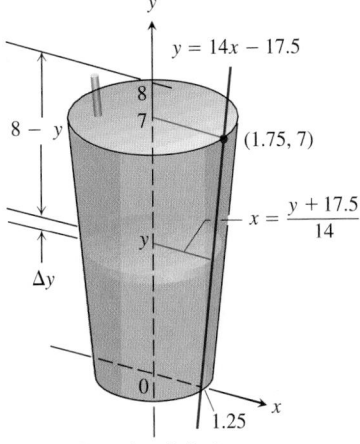

Dimensions in inches

21. *Revisiting Example 2* How much work will it take to pump the oil in Example 2 to a level 3 ft above the cone's rim?

22. *Pumping Milk* Suppose the conical tank in Example 2 contains milk weighing 64.5 lb/ft^3 instead of olive oil. How much work will it take to pump the contents to the rim?

23. **Writing to Learn** You are in charge of the evacuation and repair of the storage tank shown here. The tank is a hemisphere of radius 10 ft and is full of benzene weighing 56 lb/ft^3. A firm you contacted says it can empty the tank for 1/2 cent per foot-pound of work. Find the work required to empty the tank by pumping the benzene to an outlet 2 ft above the tank. If you have budgeted $5000 for the job, can you afford to hire the firm?

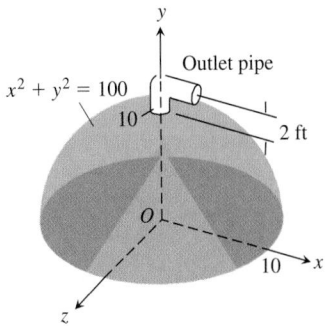

24. *Water Tower* Your town has decided to drill a well to increase its water supply. As the town engineer, you have determined that a water tower will be necessary to provide the pressure needed for distribution, and you have designed the system shown here. The water is to be pumped from a 300-ft well through a vertical 4-in. pipe into the base of a cylindrical tank 20 ft in diameter and 25 ft high. The base of the tank will be 60 ft above ground. The pump is a 3-hp pump, rated at 1650 ft · lb/sec. To the nearest hour, how long will it take to fill the tank the first time? (Include the time it takes to fill the pipe.) Assume water weighs 62.4 lb/ft^3.

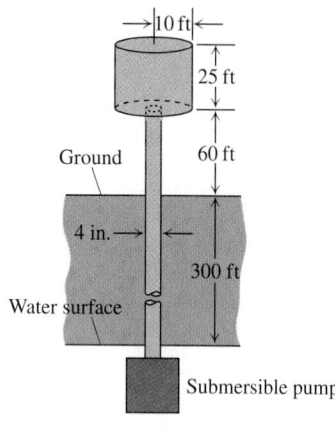

NOT TO SCALE

25. *Fish Tank* A rectangular freshwater fish tank with base 2 × 4 ft and height 2 ft (interior dimensions) is filled to within 2 in. of the top.

(a) Find the fluid force against each end of the tank.

(b) Suppose the tank is sealed and stood on end (without spilling) so that one of the square ends is the base. What does that do to the fluid forces on the rectangular sides?

26. *Milk Carton* A rectangular milk carton measures 3.75 in. by 3.75 in. at the base and is 7.75 in. tall. Find the force of the milk (weighing 64.5 lb/ft^3) on one side when the carton is full.

27. *Heights of Females* The mean height of an adult female in New York City is estimated to be 63.4 inches with a standard deviation of 3.2 inches. What proportion of the adult females in New York City are

(a) less than 63.4 inches tall?

(b) between 63 and 65 inches tall?

(c) taller than 6 feet?

(d) exactly 5 feet tall?

28. *Test Scores* The mean score on a national aptitude test is 498 with a standard deviation of 100 points.

(a) What percentage of the population has scores between 400 and 500?

(b) If we sample 300 test-takers at random, about how many should have scores above 700?

29. *Writing to Learn* Exercises 27 and 28 are subtly different, in that the heights in Exercise 27 are measured *continuously* and the scores in Exercise 28 are measured *discretely*. The discrete probabilities determine rectangles above the individual test scores, so that there actually is a nonzero probability

of scoring, say, 560. The rectangles would look like the figure below, and would have total area 1.

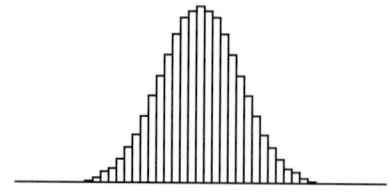

Explain why integration gives a good estimate for the probability, even in the discrete case.

30. **Writing to Learn** Suppose that $f(t)$ is the probability density function for the lifetime of a certain type of lightbulb where t is in hours. What is the meaning of the integral

$$\int_{100}^{800} f(t)\, dt?$$

Extending the Ideas

31. *Putting a Satellite into Orbit* The strength of Earth's gravitational field varies with the distance r from Earth's center, and the magnitude of the gravitational force experienced by a satellite of mass m during and after launch is

$$F(r) = \frac{mMG}{r^2}.$$

Here, $M = 5.975 \times 10^{24}$ kg is Earth's mass, $G = 6.6726 \times 10^{-11}$ N · m^{2}kg^{-2} is the *universal gravitational constant*, and r is measured in meters. The work it takes to lift a 1000-kg satellite from Earth's surface to a circular orbit 35,780 km above Earth's center is therefore given by the integral

$$\text{Work} = \int_{6,370,000}^{35,780,000} \frac{1000MG}{r^2}\, dr \text{ joules.}$$

The lower limit of integration is Earth's radius in meters at the launch site. Evaluate the integral. (This calculation does not take into account energy spent lifting the launch vehicle or energy spent bringing the satellite to orbit velocity.)

32. *Forcing Electrons Together* Two electrons r meters apart repel each other with a force of

$$F = \frac{23 \times 10^{-29}}{r^2} \text{ newton.}$$

(a) Suppose one electron is held fixed at the point $(1, 0)$ on the x-axis (units in meters). How much work does it take to move a second electron along the x-axis from the point $(-1, 0)$ to the origin?

(b) Suppose an electron is held fixed at each of the points $(-1, 0)$ and $(1, 0)$. How much work does it take to move a third electron along the x-axis from $(5, 0)$ to $(3, 0)$?

33. *Kinetic Energy* If a variable force of magnitude $F(x)$ moves a body of mass m along the x-axis from x_1 to x_2, the body's velocity v can be written as dx/dt (where t represents time). Use Newton's second law of motion, $F = m(dv/dt)$, and the Chain Rule

$$\frac{dv}{dt} = \frac{dv}{dx}\frac{dx}{dt} = v\frac{dv}{dx}$$

to show that the net work done by the force in moving the body from x_1 to x_2 is

$$W = \int_{x_1}^{x_2} F(x)\,dx = \frac{1}{2}mv_2{}^2 - \frac{1}{2}mv_1{}^2, \qquad (1)$$

where v_1 and v_2 are the body's velocities at x_1 and x_2. In physics the expression $(1/2)mv^2$ is the *kinetic energy* of the body moving with velocity v. Therefore, *the work done by the force equals the change in the body's kinetic energy*, and we can find the work by calculating this change.

Weight vs. Mass

Weight is the force that results from gravity pulling on a mass. The two are related by the equation in Newton's second law,

weight = mass × acceleration.

Thus,

newtons = kilograms × m/sec^2,

pounds = slugs × ft/sec^2.

To convert mass to weight, multiply by the acceleration of gravity. To convert weight to mass, divide by the acceleration of gravity.

In Exercises 34–40, use Equation 1 from Exercise 33.

34. *Tennis* A 2-oz tennis ball was served at 160 ft/sec (about 109 mph). How much work was done on the ball to make it go this fast?

35. *Baseball* How many foot-pounds of work does it take to throw a baseball 90 mph? A baseball weighs 5 oz = 0.3125 lb.

36. *Golf* A 1.6-oz golf ball is driven off the tee at a speed of 280 ft/sec (about 191 mph). How many foot-pounds of work are done getting the ball into the air?

37. *Tennis* During the match in which Pete Sampras won the 1990 U.S. Open men's tennis championship, Sampras hit a serve that was clocked at a phenomenal 124 mph. How much work did Sampras have to do on the 2-oz ball to get it to that speed?

38. *Football* A quarterback threw a 14.5-oz football 88 ft/sec (60 mph). How many foot-pounds of work were done on the ball to get it to that speed?

39. *Softball* How much work has to be performed on a 6.5-oz softball to pitch it at 132 ft/sec (90 mph)?

40. *A Ball Bearing* A 2-oz steel ball bearing is placed on a vertical spring whose force constant is $k = 18$ lb/ft. The spring is compressed 3 in. and released. About how high does the ball bearing go? (*Hint:* The kinetic (compression) energy, mgh, of a spring is $\frac{1}{2}ks^2$, where s is the distance the spring is compressed, m is the mass, g is the acceleration of gravity, and h is the height.)

Calculus at Work

I am working toward becoming an archeoastronomer and ethnoastronomer of Africa. I have a Bachelor's degree in Physics, a Master's degree in Astronomy, and a Ph.D. in Astronomy and Astrophysics. From 1988 to 1990 I was a member of the Peace Corps, and I taught mathematics to high school students in the Fiji Islands. Calculus is a required course in high schools there.

For my Ph.D. dissertation, I investigated the possibility of the birthrate of stars being related to the composition of star formation clouds. I collected data on the absorption of electromagnetic emissions emanating from these regions. The intensity of emissions graphed versus wavelength produces a flat curve with downward spikes at the characteristic wavelengths of the elements present. An estimate of the area between a spike and the flat curve results in a concentration in molecules/cm^2 of an element. This area is the difference in the integrals of the flat and spike curves. In particular, I was looking for a large concentration of water-ice, which increases the probability of planets forming in a region.

Currently, I am applying for two research grants. One will allow me to use the NASA infrared telescope on Mauna Kea to search for C_3S_2 in comets. The other will help me study the history of astronomy in Tunisia.

Jarita Holbrook
Los Angeles, CA

Chapter 7 Key Terms

arc length (p. 397)
area between curves (p. 375)
Cavalieri's theorems (p. 384)
center of mass (p. 374)
constant-force formula (p. 369)
cylindrical shells (p. 387)
displacement (p. 364)
fluid force (p. 403)
fluid pressure (p. 403)
foot-pound (p. 369)
force constant (p. 369)
Gaussian curve (p. 406)
Hooke's Law (p. 369)

inflation rate (p. 373)
joule (p. 369)
length of a curve (p. 397)
mean (p. 405)
moment (p. 374)
net change (p. 363)
newton (p. 369)
normal curve (p. 405)
normal pdf (p. 406)
probability density function (pdf)
   (p. 405)
68-95-99.7 rule (p. 406)
smooth curve (p. 396)

smooth function (p. 396)
solid of revolution (p. 385)
standard deviation (p. 405)
surface area (p. 389)
total distance traveled (p. 366)
universal gravitational constant (p. 410)
volume by cylindrical shells (p. 387)
volume by slicing (p. 384)
volume of a solid (p. 383)
weight-density (p. 403)
work (p. 369)

Chapter 7   Review Exercises

In Exercises 1–5, the application involves the accumulation of small changes over an interval to give the net change over that entire interval. Set up an integral to model the accumulation and evaluate it to answer the question.

1. A toy car slides down a ramp and coasts to a stop after 5 sec. Its velocity from $t = 0$ to $t = 5$ is modeled by $v(t) = t^2 - 0.2t^3$ ft/sec. How far does it travel?

2. The fuel consumption of a diesel motor between weekly maintenance periods is modeled by the function $c(t) = 4 + 0.001t^4$ gal/day, $0 \le t \le 7$. How many gallons does it consume in a week?

3. The number of billboards per mile along a 100-mile stretch of an interstate highway approaching a certain city is modeled by the function $B(x) = 21 - e^{0.03x}$, where x is the distance from the city in miles. About how many billboards are along that stretch of highway?

4. A 2-meter rod has a variable density modeled by the function $\rho(x) = 11 - 4x$ g/m, where x is the distance in meters from the base of the rod. What is the total mass of the rod?

5. The electrical power consumption (measured in kilowatts) at a factory t hours after midnight during a typical day is modeled by $E(t) = 300(2 - \cos(\pi t/12))$. How many kilowatt-hours of electrical energy does the company consume in a typical day?

In Exercises 6–19, find the area of the region enclosed by the lines and curves.

6. $y = x$, $y = 1/x^2$, $x = 2$

7. $y = x + 1$, $y = 3 - x^2$

8. $\sqrt{x} + \sqrt{y} = 1$, $x = 0$, $y = 0$

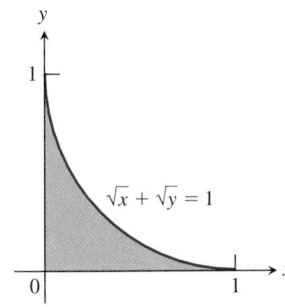

9. $x = 2y^2$, $x = 0$, $y = 3$

10. $4x = y^2 - 4$, $4x = y + 16$

11. $y = \sin x$, $y = x$, $x = \pi/4$

12. $y = 2 \sin x$, $y = \sin 2x$, $0 \le x \le \pi$

13. $y = \cos x$, $y = 4 - x^2$

14. $y = \sec^2 x$, $y = 3 - |x|$

15. *The Necklace* one of the smaller bead-shaped regions enclosed by the graphs of $y = 1 + \cos x$ and $y = 2 - \cos x$

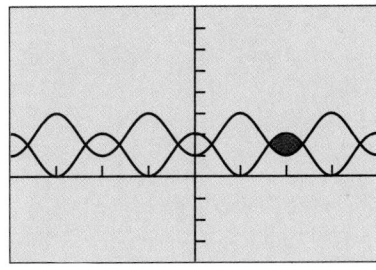

$[-4\pi, 4\pi]$ by $[-4, 8]$

16. one of the larger bead-shaped regions enclosed by the curves in Exercise 15

17. *The Bow Tie* the region enclosed by the graphs of

$$y = x^3 - x \quad \text{and} \quad y = \frac{x}{x^2 + 1}$$

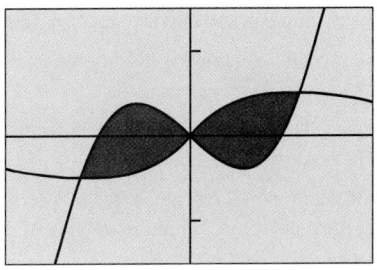

$[-2, 2]$ by $[-1.5, 1.5]$

18. *The Bell* the region enclosed by the graphs of

$$y = 3^{1-x^2} \quad \text{and} \quad y = \frac{x^2 - 3}{10}$$

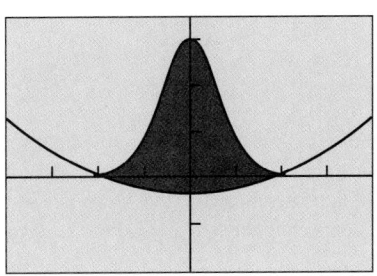

$[-4, 4]$ by $[-2, 3.5]$

19. *The Kissing Fish* the region enclosed between the graphs of $y = x \sin x$ and $y = -x \sin x$ over the interval $[-\pi, \pi]$

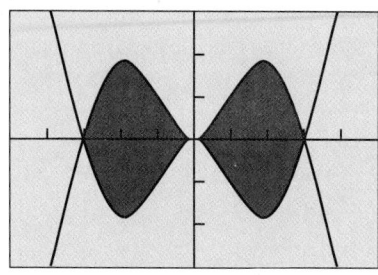

[–5, 5] by [–3, 3]

20. Find the volume of the solid generated by revolving the region bounded by the x-axis, the curve $y = 3x^4$, and the lines $x = -1$ and $x = 1$ about the x-axis.

21. Find the volume of the solid generated by revolving the region enclosed by the parabola $y^2 = 4x$ and the line $y = x$ about

   (a) the x-axis.　　**(b)** the y-axis.

   (c) the line $x = 4$.　　**(d)** the line $y = 4$.

22. The section of the parabola $y = x^2/2$ from $y = 0$ to $y = 2$ is revolved about the y-axis to form a bowl.

   (a) Find the volume of the bowl.

   (b) Find how much the bowl is holding when it is filled to a depth of k units $(0 < k < 2)$.

   (c) If the bowl is filled at a rate of 2 cubic units per second, how fast is the depth k increasing when $k = 1$?

23. The profile of a football resembles the ellipse shown here (all dimensions in inches). Find the volume of the football to the nearest cubic inch.

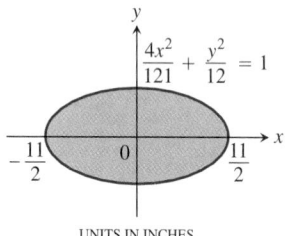

$$\frac{4x^2}{121} + \frac{y^2}{12} = 1$$

UNITS IN INCHES

24. The base of a solid is the region enclosed between the graphs of $y = \sin x$ and $y = -\sin x$ from $x = 0$ to $x = \pi$. Each cross section perpendicular to the x-axis is a semicircle with diameter connecting the two graphs. Find the volume of the solid.

25. The region enclosed by the graphs of $y = e^{x/2}$, $y = 1$, and $x = \ln 3$ is revolved about the x-axis. Find the volume of the solid generated.

26. A round hole of radius $\sqrt{3}$ feet is bored through the center of a sphere of radius 2 feet. Find the volume of the piece cut out.

27. Find the length of the arch of the parabola $y = 9 - x^2$ that lies above the x-axis.

28. Find the *perimeter* of the bow-tie-shaped region enclosed between the graphs of $y = x^3 - x$ and $y = x - x^3$.

29. A particle travels at 2 units per second along the curve $y = x^3 - 3x^2 + 2$. How long does it take to travel from the local maximum to the local minimum?

30. *Work in groups of two or three.* One of the following statements is true for all $k > 0$ and one is false. Which is which? Explain.

   (a) The graphs of $y = k \sin x$ and $y = \sin kx$ have the same length on the interval $[0, 2\pi]$.

   (b) The graph of $y = k \sin x$ is k times as long as the graph of $y = \sin x$ on the interval $[0, 2\pi]$.

31. Let $F(x) = \int_1^x \sqrt{t^4 - 1} \, dt$. Find the *exact* length of the graph of F from $x = 2$ to $x = 5$ without using a calculator.

32. *Rock Climbing* A rock climber is about to haul up 100 N (about 22.5 lb) of equipment that has been hanging beneath her on 40 m of rope weighing 0.8 N/m. How much work will it take to lift

   (a) the equipment?　　**(b)** the rope?

   (c) the rope and equipment together?

33. *Hauling Water* You drove an 800-gallon tank truck from the base of Mt. Washington to the summit and discovered on arrival that the tank was only half full. You had started out with a full tank of water, had climbed at a steady rate, and had taken 50 minutes to accomplish the 4750-ft elevation change. Assuming that the water leaked out at a steady rate, how much work was spent in carrying the water to the summit? Water weighs 8 lb/gal. (Do not count the work done getting you and the truck to the top.)

34. *Stretching a Spring* If a force of 80 N is required to hold a spring 0.3 m beyond its unstressed length, how much work does it take to stretch the spring this far? How much work does it take to stretch the spring an additional meter?

35. **Writing to Learn** It takes a lot more effort to roll a stone up a hill than to roll the stone down the hill, but the weight of the stone and the distance it covers are the same. Does this mean that the same amount of work is done? Explain.

36. *Emptying a Bowl* A hemispherical bowl with radius 8 inches is filled with punch (weighing 0.04 pound per cubic inch) to within 2 inches of the top. How much work is done emptying the bowl if the contents are pumped just high enough to get over the rim?

37. *Fluid Force* The vertical triangular plate shown below is the end plate of a feeding trough full of hog slop, weighing 80 pounds per cubic foot. What is the force against the plate?

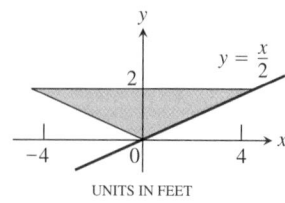

UNITS IN FEET

38. *Fluid Force* A standard olive oil can measures 5.75 in. by 3.5 in. by 10 in. Find the fluid force against the base and each side of the can when it is full. (Olive oil has a weight-density of 57 pounds per cubic foot.)

39. *Volume* A solid lies between planes perpendicular to the x-axis at $x = 0$ and at $x = 6$. The cross sections between the planes are squares whose bases run from the x-axis up to the curve $\sqrt{x} + \sqrt{y} = \sqrt{6}$. Find the volume of the solid.

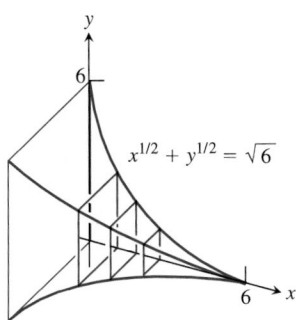

40. *Yellow Perch* A researcher measures the lengths of 3-year-old yellow perch in a fish hatchery and finds that they have a mean length of 17.2 cm with a standard deviation of 3.4 cm. What proportion of 3-year-old yellow perch raised under similar conditions can be expected to reach a length of 20 cm or more?

41. *Work in groups of two or three.* Using as large a sample of classmates as possible, measure the span of each person's fully stretched hand, from the tip of the pinky finger to the tip of the thumb. Based on the mean and standard deviation of your sample, what percentage of students your age would have a finger span of more than 10 inches?

42. *The 68-95-99.7 Rule* **(a)** Verify that for every normal pdf, the proportion of the population lying within one standard deviation of the mean is close to 68%. (*Hint:* Since it is the same for every pdf, you can simplify the function by assuming that $\mu = 0$ and $\sigma = 1$. Then integrate from -1 to 1.)

(b) Verify the two remaining parts of the rule.

43. Writing to Learn Explain why the area under the graph of a probability density function has to equal 1.

In Exercises 44–48, use the cylindrical shell method to find the volume of the solid generated by revolving the region bounded by the curves about the y-axis.

44. $y = 2x$, $\quad y = x/2$, $\quad x = 1$

45. $y = 1/x$, $\quad y = 0$, $\quad x = 1/2$, $\quad x = 2$

46. $y = \sin x$, $\quad y = 0$, $\quad 0 \le x \le \pi$

47. $y = x - 3$, $\quad y = x^2 - 3x$

48. the bell-shaped region in Exercise 18

49. *Bundt Cake* A bundt cake (see Exploration 1, Section 7.3) has a hole of radius 2 inches and an outer radius of 6 inches at the base. It is 5 inches high, and each cross-sectional slice is parabolic.

(a) Model a typical slice by finding the equation of the parabola with y-intercept 5 and x-intercepts ± 2.

(b) Revolve the parabolic region about an appropriate line to generate the bundt cake and find its volume.

50. *Finding a Function* Find a function f that has a continuous derivative on $(0, \infty)$ and that has both of the following properties.

i. The graph of f goes through the point $(1, 1)$.

ii. The length L of the curve from $(1, 1)$ to any point $(x, f(x))$ is given by the formula $L = \ln x + f(x) - 1$.

In Exercises 51 and 52, find the area of the surface generated by revolving the curve about the indicated axis.

51. $y = \tan x$, $\quad 0 \le x \le \pi/4$; $\quad x$-axis

52. $xy = 1$, $\quad 1 \le y \le 2$; $\quad y$-axis

CHAPTER

8

L'Hôpital's Rule, Improper Integrals, and Partial Fractions

NASA's Mars Pathfinder collected and transmitted scientific data and photographs back to Earth.

How much work must be done against gravity for the 2000-pound Pathfinder to escape Earth's gravity, that is, to be lifted an infinite distance above the surface of the Earth? Assume that the force due to gravity on an object of weight *w, r* miles from the center of the Earth is:

$$F = 16{,}000{,}000w/(r^2)\,(r \geq 4000) \text{ (in pounds)}.$$

The radius of the Earth is approximately 4000 miles. The concepts in Section 8.3 will help you solve this problem.

Chapter 8 Overview

In the late seventeenth century, John Bernoulli discovered a rule for calculating limits of fractions whose numerators and denominators both approach zero. The rule is known today as l'Hôpital's Rule, after Guillaume François Antoine de l'Hôpital (1661–1704), Marquis de St. Mesme, a French nobleman who wrote the first differential calculus text, where the rule first appeared in print. We will also use l'Hôpital's Rule to compare the rates at which functions of x grow as |x| becomes large. In the process, we introduce the little-oh and big-oh notation sometimes used to describe the results of these comparisons.

In Chapter 5 we saw how to evaluate definite integrals of continuous functions and bounded functions with a finite number of discontinuities on finite closed intervals. These ideas are extended to integrals where one or both limits of integration are infinite, and to integrals whose integrands become unbounded on the interval of integration. We also extend the use of partial fractions to integrals whose integrands are rational functions.

8.1 L'Hôpital's Rule

Indeterminate Form 0/0 • Indeterminate Forms ∞/∞, $\infty \cdot 0$, $\infty - \infty$ • Indeterminate Forms 1^∞, 0^0, ∞^0

Indeterminate Form 0/0

If functions $f(x)$ and $g(x)$ are both zero at $x = a$, then

$$\lim_{x \to a} \frac{f(x)}{g(x)}$$

cannot be found by substituting $x = a$. The substitution produces $0/0$, a meaningless expression known as an **indeterminate form.** Our experience so far has been that limits that lead to indeterminate forms may or may not be hard to find algebraically. It took a lot of analysis in Exercise 63 of Section 2.1 to find $\lim_{x \to 0}$ $(\sin x)/x$. But we have had remarkable success with the limit

$$f'(a) = \lim_{x \to a} \frac{f(x) - f(a)}{x - a},$$

from which we calculate derivatives and which always produces the equivalent of $0/0$. L'Hôpital's Rule enables us to draw on our success with derivatives to evaluate limits that otherwise lead to indeterminate forms.

Bernard A. Harris Jr.

1956–

Bernard Harris, M.D., became an astronaut in 1991. In 1995, as the Payload Commander on the STS-63 mission, he became the first African American to walk in space. During this mission, which included a rendezvous with the Russian Space Station, Mir, Harris traveled over 2.9 million miles. Dr. Harris left NASA in 1996 to become Vice President of Microgravity and Life Sciences for SPACEHAB Incorporated.

Theorem 1 L'Hôpital's Rule (First Form)

Suppose that $f(a) = g(a) = 0$, that $f'(a)$ and $g'(a)$ exist, and that $g'(a) \neq 0$. Then

$$\lim_{x \to a} \frac{f(x)}{g(x)} = \frac{f'(a)}{g'(a)}.$$

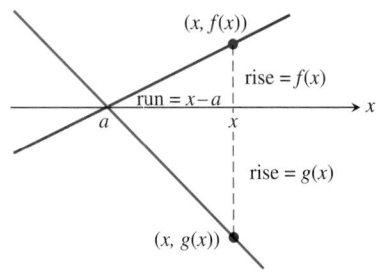

Figure 8.1 A zoom-in view of the graphs of the differentiable functions f and g at $x = a$. (Theorem 1)

Proof
Graphical Argument

If we zoom in on the graphs of f and g at $(a, f(a)) = (a, g(a)) = (a, 0)$, the graphs (Figure 8.1) appear to be straight lines because differentiable functions are locally linear. Let m_1 and m_2 be the slopes of the lines for f and g, respectively. Then for x near a,

$$\frac{f(x)}{g(x)} = \frac{\dfrac{f(x)}{x-a}}{\dfrac{g(x)}{x-a}} = \frac{m_1}{m_2}.$$

As $x \to a$, m_1 and m_2 approach $f'(a)$ and $g'(a)$, respectively. Therefore,

$$\lim_{x \to a} \frac{f(x)}{g(x)} = \lim_{x \to a} \frac{m_1}{m_2} = \frac{f'(a)}{g'(a)}.$$

Confirm Analytically

Working backwards from $f'(a)$ and $g'(a)$, which are themselves limits, we have

$$\frac{f'(a)}{g'(a)} = \frac{\displaystyle\lim_{x \to a} \frac{f(x) - f(a)}{x-a}}{\displaystyle\lim_{x \to a} \frac{g(x) - g(a)}{x-a}} = \lim_{x \to a} \frac{\dfrac{f(x) - f(a)}{x-a}}{\dfrac{g(x) - g(a)}{x-a}}$$

$$= \lim_{x \to a} \frac{f(x) - f(a)}{g(x) - g(a)} = \lim_{x \to a} \frac{f(x) - 0}{g(x) - 0} = \lim_{x \to a} \frac{f(x)}{g(x)}. \qquad \blacksquare$$

Example 1 USING L'HÔPITAL'S RULE

$$\lim_{x \to 0} \frac{\sqrt{1+x} - 1}{x} = \left.\frac{\dfrac{d}{dx}(\sqrt{1+x} - 1)}{\dfrac{d}{dx}(x)}\right|_{x=0} = \left.\frac{1/(2\sqrt{1+x})}{1}\right|_{x=0} = \frac{1}{2}$$

Example 2 WORKING WITH INDETERMINATE FORM 0/0

$$\lim_{x \to 0} \frac{1 - \cos x}{x + x^2} \qquad \frac{0}{0}$$

$$= \left.\frac{\sin x}{(1 + 2x)}\right|_{x=0} = \frac{0}{1} = 0$$

Sometimes after differentiation the new numerator and denominator both equal zero at $x = a$, as we will see in Example 3. In these cases we apply a stronger form of l'Hôpital's Rule.

Theorem 2 L'Hôpital's Rule (Stronger Form)

Suppose that $f(a) = g(a) = 0$, that f and g are differentiable on an open interval I containing a, and that $g'(x) \neq 0$ on I if $x \neq a$. Then

$$\lim_{x \to a} \frac{f(x)}{g(x)} = \lim_{x \to a} \frac{f'(x)}{g'(x)}.$$

Augustin-Louis Cauchy (1789–1857)

An engineer with a genius for mathematics and mathematical modeling, Cauchy created an early modeling of surface wave propagation that is now a classic in hydrodynamics. Cauchy (pronounced "CO-she") invented our notion of continuity and proved the Intermediate Value Theorem for continuous functions. He invented modern limit notation and was the first to prove the convergence of $(1 + 1/n)^n$. His mean value theorem, the subject of Exercise 55, is the key to proving the stronger form of l'Hôpital's Rule. His work advanced not only calculus and mathematical analysis, but also the fields of complex function theory, error theory, differential equations, and celestial mechanics.

When you apply l'Hôpital's Rule, look for a change from $0/0$ into something else. This is where the limit is revealed.

Example 3 APPLYING STRONGER FORM OF L'HÔPITAL'S RULE

$$\lim_{x \to 0} \frac{\sqrt{1+x} - 1 - x/2}{x^2} \qquad \frac{0}{0}$$

$$= \lim_{x \to 0} \frac{(1/2)(1+x)^{-1/2} - 1/2}{2x} \qquad \text{Still } \frac{0}{0}; \text{ differentiate again.}$$

$$= \lim_{x \to 0} \frac{-(1/4)(1+x)^{-3/2}}{2} = -\frac{1}{8} \qquad \text{Not } \frac{0}{0}; \text{ limit is found.}$$

L'Hôpital's Rule applies to one-sided limits as well.

Exploration 1 **Exploring L'Hôpital's Rule Graphically**

Consider the function $f(x) = \dfrac{\sin x}{x}$.

1. Use l'Hôpital's Rule to find $\lim_{x \to 0} f(x)$.
2. Let $y_1 = \sin x$, $y_2 = x$, $y_3 = y_1/y_2$, $y_4 = y_1'/y_2'$. Explain how graphing y_3 and y_4 in the same viewing window provides support for l'Hôpital's Rule in part 1.
3. Let $y_5 = y_3'$. Graph y_3, y_4, and y_5 in the same viewing window. Based on what you see in the viewing window, make a statement about what l'Hôpital's Rule does *not* say.

Example 4 USING L'HÔPITAL'S RULE WITH ONE-SIDED LIMITS

(a) $\displaystyle\lim_{x \to 0^+} \frac{\sin x}{x^2} \qquad \frac{0}{0}$

$\displaystyle = \lim_{x \to 0^+} \frac{\cos x}{2x} = \infty \qquad \frac{1}{0}$

(b) $\displaystyle\lim_{x \to 0^-} \frac{\sin x}{x^2} \qquad \frac{0}{0}$

$\displaystyle = \lim_{x \to 0^-} \frac{\cos x}{2x} = -\infty \qquad -\frac{1}{0}$

These results are supported by the graph in Figure 8.2.

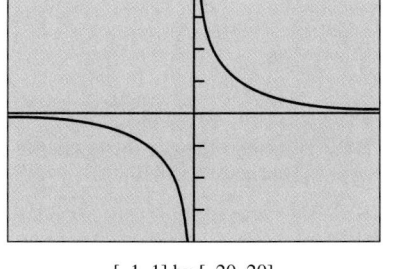

[−1, 1] by [−20, 20]

Figure 8.2 The graph of $f(x) = (\sin x)/x^2$. (Example 4)

When we reach a point where one of the derivatives approaches 0, as in Example 4, and the other does not, then the limit in question is 0 (if the numerator approaches 0) or infinity (if the denominator approaches 0).

Indeterminate Forms ∞/∞, $\infty \cdot 0$, $\infty - \infty$

A version of l'Hôpital's Rule also applies to quotients that lead to the indeterminate form ∞/∞. If $f(x)$ and $g(x)$ both approach infinity as $x \to a$, then

$$\lim_{x \to a} \frac{f(x)}{g(x)} = \lim_{x \to a} \frac{f'(x)}{g'(x)},$$

provided the latter limit exists. The a here (and in the indeterminate form $0/0$) may itself be finite or infinite, and may be an endpoint of the interval I of Theorem 2.

Example 5   WORKING WITH INDETERMINATE FORM ∞/∞

Find   **(a)** $\lim\limits_{x \to \pi/2} \dfrac{\sec x}{1 + \tan x}$ **(b)** $\lim\limits_{x \to \infty} \dfrac{\ln x}{2\sqrt{x}}$.

Solution

(a) Solve Analytically

The numerator and denominator are discontinuous at $x = \pi/2$, so we investigate the one-sided limits there. To apply l'Hôpital's Rule we can choose I to be any open interval with $x = \pi/2$ as an endpoint.

$$\lim_{x \to (\pi/2)^-} \frac{\sec x}{1 + \tan x} \qquad \frac{\infty}{\infty} \text{ from the left}$$

$$= \lim_{x \to (\pi/2)^-} \frac{\sec x \tan x}{\sec^2 x} = \lim_{x \to (\pi/2)^-} \sin x = 1$$

The right-hand limit is 1 also, with $(-\infty)/(-\infty)$ as the indeterminate form. Therefore, the two-sided limit is equal to 1.

Support Graphically

The graph of $(\sec x)/(1 + \tan x)$ in Figure 8.3 appears to pass right through the point $(\pi/2, 1)$.

(b) $\lim\limits_{x \to \infty} \dfrac{\ln x}{2\sqrt{x}} = \lim\limits_{x \to \infty} \dfrac{1/x}{1/\sqrt{x}} = \lim\limits_{x \to \infty} \dfrac{1}{\sqrt{x}} = 0$

We can sometimes handle the indeterminate forms $\infty \cdot 0$ and $\infty - \infty$ by using algebra to get $0/0$ or ∞/∞ instead. Here again we do not mean to suggest that there is a number $\infty \cdot 0$ or $\infty - \infty$ any more than we mean to suggest that there is a number $0/0$ or ∞/∞. These forms are not numbers but descriptions of function behavior.

Example 6   WORKING WITH INDETERMINATE FORM $\infty \cdot 0$

Find   **(a)** $\lim\limits_{x \to \infty} \left(x \sin \dfrac{1}{x} \right)$ **(b)** $\lim\limits_{x \to -\infty} \left(x \sin \dfrac{1}{x} \right)$.

Solution   Figure 8.4 suggests that the limits exist.

(a) $\lim\limits_{x \to \infty} \left(x \sin \dfrac{1}{x} \right)$ $\infty \cdot 0$

$$= \lim_{h \to 0^+} \left(\frac{1}{h} \sin h \right) \qquad \text{Let } h = 1/x.$$

$$= 1$$

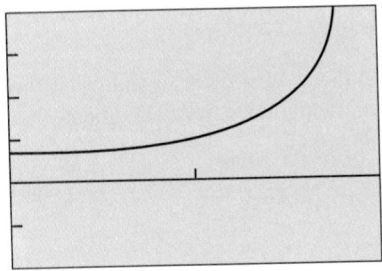

$[\pi/4, 3\pi/4]$ by $[-2, 4]$

Figure 8.3 The graph of $y = (\sec x)/(1 + \tan x)$. (Example 5)

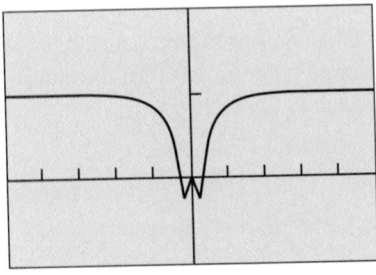

$[-5, 5]$ by $[-1, 2]$

Figure 8.4 The graph of $y = x \sin (1/x)$. (Example 6)

(b) Similarly,

$$\lim_{x \to -\infty} \left(x \sin \frac{1}{x} \right) = 1.$$

Example 7   WORKING WITH INDETERMINATE FORM ∞ − ∞

Find $\displaystyle \lim_{x \to 1} \left(\frac{1}{\ln x} - \frac{1}{x - 1} \right).$

Solution Combining the two fractions converts the indeterminate form ∞ − ∞ to 0/0, to which we can apply l'Hôpital's Rule.

$$\lim_{x \to 1} \left(\frac{1}{\ln x} - \frac{1}{x - 1} \right) \qquad \infty - \infty$$

$$= \lim_{x \to 1} \frac{x - 1 - \ln x}{(x - 1) \ln x} \qquad \text{Now } \frac{0}{0}$$

$$= \lim_{x \to 1} \frac{1 - 1/x}{\dfrac{x - 1}{x} + \ln x}$$

$$= \lim_{x \to 1} \frac{x - 1}{x \ln x + x - 1} \qquad \text{Still } \frac{0}{0}$$

$$= \lim_{x \to 1} \frac{1}{2 + \ln x}$$

$$= \frac{1}{2}$$

Interpret

The vertical distance between the graphs of $y = 1/(\ln x)$ and $y = 1/(x - 1)$ approaches half a unit as $x \to 1$ from either side (Figure 8.5).

[0, 2] by [−4, 5]

Figure 8.5 The graphs of $y = 1/(\ln x)$, $y = 1/(x - 1)$, and $y = 1/(\ln x) - 1/(x - 1)$. (Example 7)

Indeterminate Forms 1^{∞}, 0^0, ∞^0

Limits that lead to the indeterminate forms 1^{∞}, 0^0, and ∞^0 can sometimes be handled by taking logarithms first. We use l'Hôpital's Rule to find the limit of the logarithm and then exponentiate to reveal the original function's behavior.

Since $b = e^{\ln b}$ for every positive number b, we can write $f(x)$ as

$$f(x) = e^{\ln f(x)}$$

for any positive function $f(x)$.

$$\lim_{x \to a} \ln f(x) = L \quad \Rightarrow \quad \lim_{x \to a} f(x) = \lim_{x \to a} e^{\ln f(x)} = e^{L}$$

Here a can be finite or infinite.

In Section 1.3 we used graphs and tables to investigate the values of $f(x) = (1 + 1/x)^x$ as $x \to \infty$. Now we find this limit with l'Hôpital's Rule.

Example 8   WORKING WITH INDETERMINATE FORM 1^{∞}

Find $\displaystyle \lim_{x \to \infty} \left(1 + \frac{1}{x} \right)^x.$

Solution Let $f(x) = (1 + 1/x)^x$. Then taking logarithms of both sides converts the indeterminate form 1^∞ to $0/0$, to which we can apply l'Hôpital's Rule.

$$\ln f(x) = \ln \left(1 + \frac{1}{x}\right)^x = x \ln \left(1 + \frac{1}{x}\right) = \frac{\ln \left(1 + \frac{1}{x}\right)}{\frac{1}{x}}$$

We apply l'Hôpital's Rule to the last expression above.

$$\lim_{x \to \infty} \ln f(x) = \lim_{x \to \infty} \frac{\ln \left(1 + \frac{1}{x}\right)}{\frac{1}{x}} \qquad \frac{0}{0}$$

$$= \lim_{x \to \infty} \frac{\frac{1}{1 + \frac{1}{x}} \left(-\frac{1}{x^2}\right)}{-\frac{1}{x^2}} \qquad \text{Differentiate numerator and denominator.}$$

$$= \lim_{x \to \infty} \frac{1}{1 + \frac{1}{x}} = 1$$

Therefore,

$$\lim_{x \to \infty} \left(1 + \frac{1}{x}\right)^x = \lim_{x \to \infty} f(x) = \lim_{x \to \infty} e^{\ln f(x)} = e^1 = e.$$

Example 9 WORKING WITH INDETERMINATE FORM 0^0

Determine whether $\lim_{x \to 0^+} x^x$ exists and find its value if it does.

Solution

Investigate Graphically

Figure 8.6 suggests that the limit exists and has a value near 1.

Solve Analytically

The limit leads to the indeterminate form 0^0. To convert the problem to one involving $0/0$, we let $f(x) = x^x$ and take the logarithm of both sides.

$$\ln f(x) = x \ln x = \frac{\ln x}{1/x}$$

Applying l'Hôpital's Rule to $(\ln x)/(1/x)$ we obtain

$$\lim_{x \to 0^+} \ln f(x) = \lim_{x \to 0^+} \frac{\ln x}{1/x} \qquad \frac{-\infty}{\infty}$$

$$= \lim_{x \to 0^+} \frac{1/x}{-1/x^2} \qquad \text{Differentiate.}$$

$$= \lim_{x \to 0^+} (-x) = 0.$$

Therefore,

$$\lim_{x \to 0^+} x^x = \lim_{x \to 0^+} f(x) = \lim_{x \to 0^+} e^{\ln f(x)} = e^0 = 1.$$

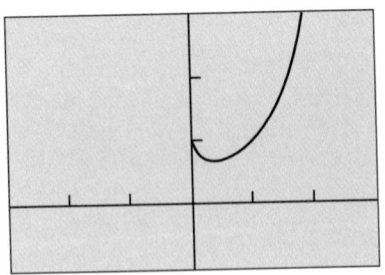

[−3, 3] by [−1, 3]

Figure 8.6 The graph of $y = x^x$. (Example 9)

Example 10 WORKING WITH INDETERMINATE FORM ∞^0

Find $\lim\limits_{x\to\infty} x^{1/x}$.

Solution Let $f(x) = x^{1/x}$. Then

$$\ln f(x) = \frac{\ln x}{x}.$$

Applying l'Hôpital's Rule to $\ln f(x)$ we obtain

$$\lim_{x\to\infty} \ln f(x) = \lim_{x\to\infty} \frac{\ln x}{x} \qquad \frac{\infty}{\infty}$$

$$= \lim_{x\to\infty} \frac{1/x}{1} \qquad \text{Differentiate.}$$

$$= \lim_{x\to\infty} \frac{1}{x} = 0.$$

Therefore,

$$\lim_{x\to\infty} x^{1/x} = \lim_{x\to\infty} f(x) = \lim_{x\to\infty} e^{\ln f(x)} = e^0 = 1.$$

Quick Review 8.1

In Exercises 1 and 2, use tables to estimate the value of the limit.

1. $\lim\limits_{x\to\infty} \left(1 + \dfrac{0.1}{x}\right)^x$

2. $\lim\limits_{x\to0^+} x^{1/(\ln x)}$

In Exercises 3–8, use graphs or tables to estimate the value of the limit.

3. $\lim\limits_{x\to0^-} \left(1 - \dfrac{1}{x}\right)^x$

4. $\lim\limits_{x\to-1^-} \left(1 + \dfrac{1}{x}\right)^x$

5. $\lim\limits_{t\to1} \dfrac{t-1}{\sqrt{t}-1}$

6. $\lim\limits_{x\to\infty} \dfrac{\sqrt{4x^2+1}}{x+1}$

7. $\lim\limits_{x\to0} \dfrac{\sin 3x}{x}$

8. $\lim\limits_{\theta\to\pi/2} \dfrac{\tan\theta}{2+\tan\theta}$

In Exercises 9 and 10, substitute $x = 1/h$ to express y as a function of h.

9. $y = x \sin \dfrac{1}{x}$

10. $y = \left(1 + \dfrac{1}{x}\right)^x$

Section 8.1 Exercises

In Exercises 1–4, estimate the limit graphically. Use l'Hôpital's Rule to confirm your estimate.

1. $\lim\limits_{x\to2} \dfrac{x-2}{x^2-4}$

2. $\lim\limits_{x\to0} \dfrac{\sin 5x}{x}$

3. $\lim\limits_{x\to0^+} \left(1 + \dfrac{1}{x}\right)^x$

4. $\lim\limits_{x\to\infty} \dfrac{5x^2-3x}{7x^2+1}$

In Exercises 5–8, use l'Hôpital's Rule to evaluate the limit. Support your answer graphically.

5. $\lim\limits_{x\to1} \dfrac{x^3-1}{4x^3-x-3}$

6. $\lim\limits_{x\to0} \dfrac{1-\cos x}{x^2}$

7. $\lim\limits_{x\to0^+} (e^x + x)^{1/x}$

8. $\lim\limits_{x\to\infty} \dfrac{2x^2+3x}{x^3+x+1}$

In Exercises 9 and 10, **(a)** complete the table and estimate the limit. **(b)** Use l'Hôpital's Rule to confirm your estimate.

9. $\lim\limits_{x\to\infty} f(x),\ f(x) = \dfrac{\ln x^5}{x}$

x	10	10^2	10^3	10^4	10^5
$f(x)$					

10. $\lim\limits_{x\to0^+} f(x),\ f(x) = \dfrac{x - \sin x}{x^3}$

x	10^0	10^{-1}	10^{-2}	10^{-3}	10^{-4}
$f(x)$					

In Exercises 11–14, use tables to estimate the limit. Confirm your estimate using l'Hôpital's Rule.

11. $\lim\limits_{\theta \to 0} \dfrac{\sin 3\theta}{\sin 4\theta}$

12. $\lim\limits_{t \to 0} \left(\dfrac{1}{\sin t} - \dfrac{1}{t} \right)$

13. $\lim\limits_{x \to \infty} (1 + x)^{1/x}$

14. $\lim\limits_{x \to \infty} \dfrac{x - 2x^2}{3x^2 + 5x}$

In Exercises 15–42, use l'Hôpital's Rule to evaluate the limit.

15. $\lim\limits_{\theta \to 0} \dfrac{\sin \theta^2}{\theta}$

16. $\lim\limits_{\theta \to \pi/2} \dfrac{1 - \sin \theta}{1 + \cos 2\theta}$

17. $\lim\limits_{t \to 0} \dfrac{\cos t - 1}{e^t - t - 1}$

18. $\lim\limits_{t \to 1} \dfrac{t - 1}{\ln t - \sin \pi t}$

19. $\lim\limits_{x \to \infty} \dfrac{\ln (x + 1)}{\log_2 x}$

20. $\lim\limits_{x \to \infty} \dfrac{\log_2 x}{\log_3 (x + 3)}$

21. $\lim\limits_{y \to 0^+} \dfrac{\ln (y^2 + 2y)}{\ln y}$

22. $\lim\limits_{y \to \pi/2} \left(\dfrac{\pi}{2} - y \right) \tan y$

23. $\lim\limits_{x \to 0^+} x \ln x$

24. $\lim\limits_{x \to \infty} x \tan \dfrac{1}{x}$

25. $\lim\limits_{x \to 0^+} (\csc x - \cot x + \cos x)$

26. $\lim\limits_{x \to \infty} (\ln 2x - \ln (x + 1))$

27. $\lim\limits_{x \to 0^+} (\ln x - \ln \sin x)$

28. $\lim\limits_{x \to 0^+} \left(\dfrac{1}{x} - \dfrac{1}{\sqrt{x}} \right)$

29. $\lim\limits_{x \to 0} (e^x + x)^{1/x}$

30. $\lim\limits_{x \to 0} \left(\dfrac{1}{x^2} \right)^x$

31. $\lim\limits_{x \to \pm\infty} \dfrac{3x - 5}{2x^2 - x + 2}$

32. $\lim\limits_{x \to 0} \dfrac{\sin 7x}{\tan 11x}$

33. $\lim\limits_{x \to \infty} (\ln x)^{1/x}$

34. $\lim\limits_{x \to \infty} (1 + 2x)^{1/(2 \ln x)}$

35. $\lim\limits_{x \to 1} (x^2 - 2x + 1)^{x-1}$

36. $\lim\limits_{x \to (\pi/2)^-} (\cos x)^{\cos x}$

37. $\lim\limits_{x \to 0^+} (1 + x)^{1/x}$

38. $\lim\limits_{x \to 1} x^{1/(x-1)}$

39. $\lim\limits_{x \to 0^+} (\sin x)^x$

40. $\lim\limits_{x \to 0^+} (\sin x)^{\tan x}$

41. $\lim\limits_{x \to 1^+} x^{1/(1-x)}$

42. $\lim\limits_{x \to \infty} \displaystyle\int_x^{2x} \dfrac{dt}{t}$

In Exercises 43 and 44, *work in groups of two or three.*

(a) Writing to Learn Explain why l'Hôpital's Rule does not help you to find the limit.

(b) Use a graph to estimate the limit.

(c) Evaluate the limit analytically using the techniques of Chapter 2.

43. $\lim\limits_{x \to \infty} \dfrac{\sqrt{9x + 1}}{\sqrt{x + 1}}$

44. $\lim\limits_{x \to \pi/2} \dfrac{\sec x}{\tan x}$

Explorations

45. Give an example of two differentiable functions f and g with $\lim\limits_{x \to 3} f(x) = \lim\limits_{x \to 3} g(x) = 0$ that satisfy the following.

(a) $\lim\limits_{x \to 3} \dfrac{f(x)}{g(x)} = 7$ **(b)** $\lim\limits_{x \to 3} \dfrac{f(x)}{g(x)} = 0$

(c) $\lim\limits_{x \to 3} \dfrac{f(x)}{g(x)} = \infty$

46. Give an example of two differentiable functions f and g with $\lim\limits_{x \to \infty} f(x) = \lim\limits_{x \to \infty} g(x) = \infty$ that satisfy the following.

(a) $\lim\limits_{x \to \infty} \dfrac{f(x)}{g(x)} = 3$ **(b)** $\lim\limits_{x \to \infty} \dfrac{f(x)}{g(x)} = 0$

(c) $\lim\limits_{x \to \infty} \dfrac{f(x)}{g(x)} = \infty$ ∎

47. *Continuous Extension* Find a value of c that makes the function

$$f(x) = \begin{cases} \dfrac{9x - 3 \sin 3x}{5x^3}, & x \neq 0 \\ c, & x = 0 \end{cases}$$

continuous at $x = 0$. Explain why your value of c works.

48. *Continuous Extension* Let $f(x) = |x|^x$, $x \neq 0$. Show that f has a removable discontinuity at $x = 0$ and extend the definition of f to $x = 0$ so that the extended function is continuous there.

49. *Interest Compounded Continuously*

(a) Show that $\lim\limits_{k \to \infty} A_0 \left(1 + \dfrac{r}{k} \right)^{kt} = A_0 e^{rt}$.

(b) Writing to Learn Explain how the limit in (a) connects interest compounded k times per year with interest compounded continuously.

50. *L'Hôpital's Rule* Let

$$f(x) = \begin{cases} x + 2, & x \neq 0 \\ 0, & x = 0 \end{cases} \quad \text{and} \quad g(x) = \begin{cases} x + 1, & x \neq 0 \\ 0, & x = 0. \end{cases}$$

(a) Show that

$$\lim\limits_{x \to 0} \dfrac{f'(x)}{g'(x)} = 1 \quad \text{but} \quad \lim\limits_{x \to 0} \dfrac{f(x)}{g(x)} = 2.$$

(b) Writing to Learn Explain why this does not contradict l'Hôpital's Rule.

51. *Solid of Revolution* Let $A(t)$ be the area of the region in the first quadrant enclosed by the coordinate axes, the curve $y = e^{-x}$, and the line $x = t > 0$ as shown in the figure. Let $V(t)$ be the volume of the solid generated by revolving the region about the x-axis. Find the following limits.

(a) $\lim\limits_{t \to \infty} A(t)$ **(b)** $\lim\limits_{t \to \infty} \dfrac{V(t)}{A(t)}$ **(c)** $\lim\limits_{t \to 0^+} \dfrac{V(t)}{A(t)}$

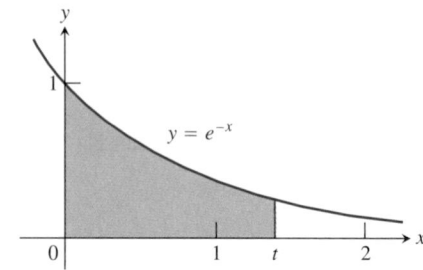

52. *L'Hôpital's Trap* Let $f(x) = \dfrac{1 - \cos x}{x + x^2}$.

(a) Use graphs or tables to estimate $\lim_{x \to 0} f(x)$.

(b) Find the error in the following incorrect application of L'Hôpital's Rule.

$$\lim_{x \to 0} \frac{1 - \cos x}{x + x^2} = \lim_{x \to 0} \frac{\sin x}{1 + 2x}$$
$$= \lim_{x \to 0} \frac{\cos x}{2}$$
$$= \frac{1}{2}$$

53. *Exponential Functions* **(a)** Use the equation

$$a^x = e^{x \ln a}$$

to find the domain of

$$f(x) = \left(1 + \frac{1}{x}\right)^x.$$

(b) Find $\lim_{x \to -1^-} f(x)$.

(c) Find $\lim_{x \to -\infty} f(x)$.

Extending the Ideas

54. *Grapher Precision* Let $f(x) = \dfrac{1 - \cos x^6}{x^{12}}$.

(a) Explain why some graphs of f may give false information about $\lim_{x \to 0} f(x)$. (*Hint:* Try the window $[-1, 1]$ by $[-0.5, 1]$.)

(b) Explain why tables may give false information about $\lim_{x \to 0} f(x)$. (*Hint:* Try tables with increments of 0.01.)

(c) Use l'Hôpital's Rule to find $\lim_{x \to 0} f(x)$.

(d) **Writing to Learn** This is an example of a function for which graphers do not have enough precision to give reliable information. Explain this statement in your own words.

55. *Cauchy's Mean Value Theorem* Suppose that functions f and g are continuous on $[a, b]$ and differentiable throughout (a, b) and suppose also that $g' \neq 0$ throughout (a, b). Then there exists a number c in (a, b) at which

$$\frac{f'(c)}{g'(c)} = \frac{f(b) - f(a)}{g(b) - g(a)}.$$

Find all values of c in (a, b) that satisfy this property for the following given functions and intervals.

(a) $f(x) = x^3 + 1$, $g(x) = x^2 - x$, $[a, b] = [-1, 1]$

(b) $f(x) = \cos x$, $g(x) = \sin x$, $[a, b] = [0, \pi/2]$

56. *Why 0^∞ and $0^{-\infty}$ Are Not Indeterminate Forms* Assume that $f(x)$ is nonnegative in an open interval containing c and $\lim_{x \to c} f(x) = 0$.

(a) If $\lim_{x \to c} g(x) = \infty$, show that $\lim_{x \to c} f(x)^{g(x)} = 0$.

(b) If $\lim_{x \to c} g(x) = -\infty$, show that $\lim_{x \to c} f(x)^{g(x)} = \infty$.

8.2 **Relative Rates of Growth**

Comparing Rates of Growth • Order and Oh-Notation • Sequential versus Binary Search

Comparing Rates of Growth

We restrict our attention to functions whose values eventually become and remain positive as $x \to \infty$.

The exponential function e^x grows so rapidly and the logarithm function $\ln x$ grows so slowly that they set standards by which we can judge the growth of other functions. The graphs (Figure 8.7) of e^x, $\ln x$, and x suggest how rapidly and slowly e^x and $\ln x$, respectively, grow in comparison to x.

In fact, all the functions a^x, $a > 1$, grow faster (eventually) than any power of x, even $x^{1,000,000}$ (Exercise 35), and hence faster (eventually) than any polynomial function.

To get a feeling for how rapidly the values of e^x grow with increasing x, think of graphing the function on a large blackboard, with the axes scaled in centimeters. At $x = 1$ cm, the graph is $e^1 \approx 3$ cm above the x-axis. At $x = 6$ cm, the graph is $e^6 \approx 403$ cm ≈ 4 m high. (It is about to go through the ceiling if it

$y = e^x$
$y = \ln x$
$y = x$

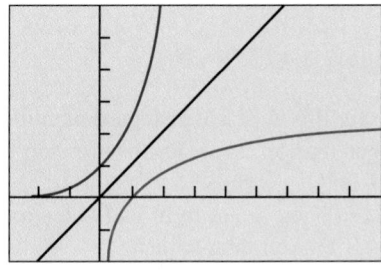

[–3, 9] by [–2, 6]

Figure 8.7 The graphs of $y = e^x$, $y = \ln x$, and $y = x$.

Grace Murray Hopper

1906–1992

Computer scientists use function comparisons like the ones in this section to measure the relative efficiencies of computer programs. The pioneering work of Rear Admiral Grace Murray Hopper in the field of computer technology led the navy, and the country, into the computer age. Hopper graduated from Yale in 1934 with a Ph.D. in Mathematics. During World War II she joined the navy and became director of a project that resulted in the development of COBOL, a computer language that enabled computers to "talk to one another." On September 6, 1997, the navy commissioned a multi-mission ship the "USS Hopper."

hasn't done so already.) At $x = 10$ cm, the graph is $e^{10} \approx 22{,}026$ cm ≈ 220 m high, higher than most buildings. At $x = 24$ cm, the graph is more than halfway to the moon, and at $x = 43$ cm from the origin, the graph is high enough to reach well past the sun's closest stellar neighbor, the red dwarf star Proxima Centauri:

$$e^{43} \approx 4.7 \times 10^{18} \text{ cm}$$

$$= 4.7 \times 10^{13} \text{ km}$$

$$\approx 1.57 \times 10^8 \text{ light-seconds} \qquad \text{Light travels about } 300{,}000 \text{ km/sec in a vacuum.}$$

$$\approx 5.0 \text{ light-years.}$$

The distance to Proxima Centauri is 4.2 light-years. Yet with $x = 43$ cm from the origin, the graph is still less than 2 feet to the right of the *y*-axis.

In contrast, the logarithm function $\ln x$ grows more slowly as $x \to \infty$ than any positive power of x, even $x^{1/1{,}000{,}000}$ (Exercise 37). Because $\ln x$ and e^x are inverse functions, the calculations above show that with axes scaled in centimeters, you have to go nearly 5 light-years out on the *x*-axis to find where the graph of $\ln x$ is even 43 cm high.

In fact, all the functions $\log_a x$, $a > 1$, grow slower (eventually) than any positive power of x.

These comparisons of exponential, polynomial, and logarithmic functions can be made precise by defining what it means for a function $f(x)$ to *grow faster* than another function $g(x)$ as $x \to \infty$.

Definitions Faster, Slower, Same-rate Growth as $x \to \infty$

Let $f(x)$ and $g(x)$ be positive for x sufficiently large.

1. f **grows faster** than g (and g **grows slower** than f) as $x \to \infty$ if

$$\lim_{x \to \infty} \frac{f(x)}{g(x)} = \infty, \quad \text{or, equivalently, if} \quad \lim_{x \to \infty} \frac{g(x)}{f(x)} = 0.$$

2. f and g **grow at the same rate** as $x \to \infty$ if

$$\lim_{x \to \infty} \frac{f(x)}{g(x)} = L \neq 0. \quad \textit{L finite and not zero}$$

According to these definitions, $y = 2x$ does not grow faster than $y = x$ as $x \to \infty$. The two functions grow at the same rate because

$$\lim_{x \to \infty} \frac{2x}{x} = \lim_{x \to \infty} 2 = 2,$$

which is a finite nonzero limit. The reason for this apparent disregard of common sense is that we want "f grows faster than g" to mean that for large *x*-values, g is negligible in comparison to f.

If $L = 1$ in part 2 of the definition, then f and g are right end behavior models for each other (Section 2.2). If f grows faster than g, then

$$\lim_{x \to \infty} \frac{f(x) + g(x)}{f(x)} = \lim_{x \to \infty} \left(1 + \frac{g(x)}{f(x)} \right) = 1 + 0 = 1,$$

so f is a right end behavior model for $f + g$. Thus, for large x-values, g can be ignored in the sum $f + g$. This explains why, for large x-values, we can ignore the terms

$$g(x) = a_{n-1}x^{n-1} + \cdots + a_0$$

in

$$f(x) = a_n x^n + a_{n-1}x^{n-1} + \cdots + a_0;$$

that is, why $a_n x^n$ is an end behavior model for

$$a_n x^n + a_{n-1}x^{n-1} + \cdots + a_0.$$

L'Hôpital's Rule can help us to compare rates of growth, as shown in Example 1.

Example 1   COMPARING e^x AND x^2 AS $x \to \infty$

The function e^x grows faster than x^2 as $x \to \infty$ because

$$\underbrace{\lim_{x \to \infty} \frac{e^x}{x^2}}_{\infty/\infty} = \underbrace{\lim_{x \to \infty} \frac{e^x}{2x}}_{\infty/\infty} = \lim_{x \to \infty} \frac{e^x}{2} = \infty. \qquad \text{Using l'Hôpital's Rule twice}$$

Exploration 1   **Comparing Rates of Growth as $x \to \infty$**

1. Show that $a^x, a > 1$, grows faster than x^2 as $x \to \infty$.
2. Show that 3^x grows faster than 2^x as $x \to \infty$.
3. If $a > b > 1$, show that a^x grows faster than b^x as $x \to \infty$.

Example 2   COMPARING ln x WITH x AND x^2 AS $x \to \infty$

Show that $\ln x$ grows slower than **(a)** x and **(b)** x^2 as $x \to \infty$.

Solution

(a) Solve Analytically

$$\lim_{x \to \infty} \frac{\ln x}{x} = \lim_{x \to \infty} \frac{1/x}{1} \qquad \text{l'Hôpital's Rule}$$

$$= \lim_{x \to \infty} \frac{1}{x} = 0$$

Support Graphically

Figure 8.8 suggests that the graph of the function $f(x) = (\ln x)/x$ drops dramatically toward the x-axis as x outstrips $\ln x$.

(b) $\lim\limits_{x \to \infty} \dfrac{\ln x}{x^2} = \lim\limits_{x \to \infty} \left(\dfrac{\ln x}{x} \cdot \dfrac{1}{x} \right) = 0 \cdot 0 = 0$

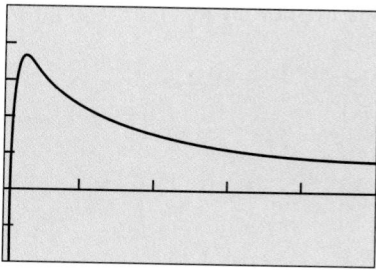

[0, 50] by [–0.2, 0.5]

Figure 8.8 The x-axis is a horizontal asymptote of the function $f(x) = (\ln x)/x$. (Example 2)

Example 3 COMPARING LOGARITHMIC FUNCTIONS AS $x \to \infty$

Let a and b be numbers greater than 1. Show that $\log_a x$ and $\log_b x$ grow at the same rate as $x \to \infty$.

Solution

$$\lim_{x \to \infty} \frac{\log_a x}{\log_b x} = \lim_{x \to \infty} \frac{\ln x / \ln a}{\ln x / \ln b} = \frac{\ln b}{\ln a}$$

The limiting value is finite and nonzero.

Growing at the same rate is a *transitive relation*.

Transitivity of Growing Rates

If f grows at the same rate as g as $x \to \infty$ and g grows at the same rate as h as $x \to \infty$, then f grows at the same rate as h as $x \to \infty$.

The reason is that

$$\lim_{x \to \infty} \frac{f}{g} = L \quad \text{and} \quad \lim_{x \to \infty} \frac{g}{h} = M$$

together imply that

$$\lim_{x \to \infty} \frac{f}{h} = \lim_{x \to \infty} \left(\frac{f}{g} \cdot \frac{g}{h} \right) = LM.$$

If L and M are finite and nonzero, then so is LM.

Example 4 GROWING AT THE SAME RATE AS $x \to \infty$

Show that $f(x) = \sqrt{x^2 + 5}$ and $g(x) = (2\sqrt{x} - 1)^2$ grow at the same rate as $x \to \infty$.

Solution

Solve Analytically

We show that f and g grow at the same rate by showing that they both grow at the same rate as $h(x) = x$.

$$\lim_{x \to \infty} \frac{\sqrt{x^2 + 5}}{x} = \lim_{x \to \infty} \sqrt{1 + \frac{5}{x^2}} = 1$$

$$\lim_{x \to \infty} \frac{(2\sqrt{x} - 1)^2}{x} = \lim_{x \to \infty} \left(\frac{2\sqrt{x} - 1}{\sqrt{x}} \right)^2 = \lim_{x \to \infty} \left(2 - \frac{1}{\sqrt{x}} \right)^2 = 4$$

Thus,

$$\lim_{x \to \infty} \frac{f}{g} = \lim_{x \to \infty} \left(\frac{f}{h} \cdot \frac{h}{g} \right) = 1 \cdot \frac{1}{4} = \frac{1}{4},$$

and f and g grow at the same rate as $x \to \infty$.

Support Graphically

The graph of $y = g/f$ in Figure 8.9 suggests that the quotient g/f is an increasing function with horizontal asymptote $y = 4$. This supports that f and g grow at the same rate.

[0, 100] by [−2, 5]

Figure 8.9 The graph of g/f appears to have the line $y = 4$ as a horizontal asymptote. (Example 4)

Order and Oh-Notation

Here we introduce the "little-oh" and "big-oh" notation invented by number theorists a hundred years ago and now commonplace in mathematical analysis and computer science.

Definition *f* of Smaller Order than *g*

Let *f* and *g* be positive for *x* sufficiently large. Then *f* is of **smaller order** than *g* as $x \to \infty$ if

$$\lim_{x \to \infty} \frac{f(x)}{g(x)} = 0.$$

We write $f = o(g)$ and say "*f* is little-oh of *g*."

Saying $f = o(g)$ as $x \to \infty$ is another way to say that *f* grows slower than *g* as $x \to \infty$.

Example 5   SHOWING LITTLE-OH ORDER

$\ln x = o(x)$ as $x \to \infty$ because $\displaystyle \lim_{x \to \infty} \frac{\ln x}{x} = 0.$

$x^2 = o(x^3 + 1)$ as $x \to \infty$ because $\displaystyle \lim_{x \to \infty} \frac{x^2}{x^3 + 1} = 0.$

Definition *f* of at Most the Order of *g*

Let $f(x)$ and $g(x)$ be positive for *x* sufficiently large. Then *f* is of **at most the order** of *g* as $x \to \infty$ if there is a positive integer *M* for which

$$\frac{f(x)}{g(x)} \le M$$

for *x* sufficiently large. We write $f = O(g)$ and say "*f* is big-oh of *g*."

Example 6   SHOWING BIG-OH ORDER

Show that $x + \sin x = O(x)$ as $x \to \infty$.

Solution

Solve Graphically

Figure 8.10 suggests that

$$\frac{x + \sin x}{x} \le 2 \quad \text{for} \quad x > 0.$$

Confirm Algebraically

For $x \ge 1$,

$$\frac{x + \sin x}{x} = 1 + \frac{\sin x}{x} \le 1 + 1 = 2, \quad \text{because} \quad \left| \frac{\sin x}{x} \right| \le \frac{1}{x} \le 1.$$

$[0, 4\pi]$ by $[-1, 3]$

Figure 8.10 $y = (x + \sin x)/x$ appears to have an absolute maximum of 2 in the interval $[0, \infty)$. (Example 6)

[0, 10] by [−1, 2]

Figure 8.11 The graph of $y = (e^x + x^2)/e^x$ suggests that this function has an absolute maximum less than 2 on the interval $[0, \infty)$. (Example 7)

Note

You would not use a sequential search method to find a word, but you might program a computer to search for a word using this technique.

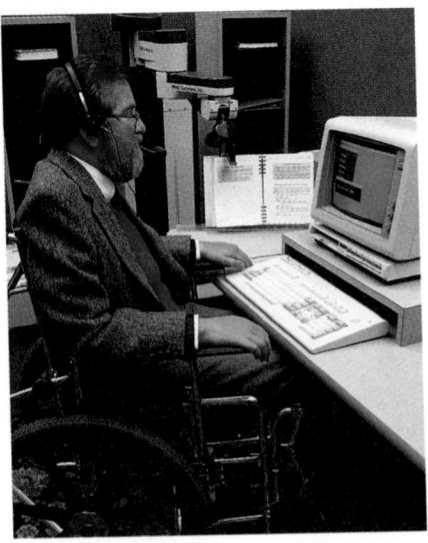

Figure 8.12 Computer scientists look for the most efficient algorithms when they program searches.

Example 7 SHOWING BIG-OH ORDER

Show that $e^x + x^2 = O(e^x)$ as $x \to \infty$.

Solution Because

$$\lim_{x \to \infty} \frac{e^x + x^2}{e^x} = 1,$$

we must have

$$\frac{e^x + x^2}{e^x} \le 2$$

for x sufficiently large. This is supported by the graph in Figure 8.11.

Sequential versus Binary Search

Computer scientists sometimes measure the efficiency of an algorithm by counting the number of steps a computer must take to make the algorithm do something (Figure 8.12). (Your graphing calculator works according to algorithms programmed into it.) There can be significant differences in how efficiently algorithms perform, even if they are designed to accomplish the same task. These differences are often described in big-oh notation. Here is an example.

Webster's *Third New International Dictionary* lists about 26,000 words that begin with the letter *a*. One way to look up a word, or to learn if it is not there, is to read through the list one word at a time until you either find the word or determine that it is not there. This **sequential search** method makes no particular use of the words' alphabetical arrangement. You are sure to get an answer, but it might take about 26,000 steps.

Another way to find the word or to learn that it is not there is to go straight to the middle of the list (give or take a few words). If you do not find the word, then go to the middle of the half that would contain it and forget about the half that would not. (You know which half would contain it because you know the list is ordered alphabetically.) This **binary search** method eliminates roughly 13,000 words in this first step. If you do not find the word on the second try, then jump to the middle of the half that would contain it. Continue this way until you have found the word or divided the list in half so many times that there are no words left. How many times do you have to divide the list to find the word or learn that it is not there? At most 15, because

$$\frac{26,000}{2^{15}} < 1.$$

This certainly beats a possible 26,000 steps.

For a list of length n, a sequential search algorithm takes on the order of n steps to find a word or determine that it is not in the list. A binary search takes on the order of $\log_2 n$ steps. The reason is that if $2^{m-1} < n \le 2^m$, then $m - 1 < \log_2 n \le m$, and the number of bisections required to narrow the list to one word will be at most m, the smallest integer greater than or equal to $\log_2 n$.

Big-oh notation provides a compact way to say all this.

To find an item in a list of length n,

    a sequential search takes $O(n)$ steps,

    a binary search takes $O(\log_2 n)$ steps.

In our example, there is a big difference between the two (26,000 versus 15), and the difference can only increase with n because n grows faster than $\log_2 n$ as $n \to \infty$.

Quick Review 8.2

In Exercises 1–4, evaluate the limit.

1. $\lim\limits_{x \to \infty} \dfrac{\ln x}{e^x}$

2. $\lim\limits_{x \to \infty} \dfrac{e^x}{x^3}$

3. $\lim\limits_{x \to -\infty} \dfrac{x^2}{e^{2x}}$

4. $\lim\limits_{x \to \infty} \dfrac{x^2}{e^{2x}}$

In Exercises 5 and 6, find an end behavior model (Section 2.2) for the function.

5. $f(x) = -3x^4 + 5x^3 - x + 1$

6. $f(x) = \dfrac{2x^3 - 3x + 1}{x + 2}$

In Exercises 7 and 8, show that g is a right end behavior model for f.

7. $g(x) = x, \quad f(x) = x + \ln x$

8. $g(x) = 2x, \quad f(x) = \sqrt{4x^2 + 5x}$

9. Let $f(x) = \dfrac{e^x + x^2}{e^x}$. Find the

    (a) local extreme values of f and where they occur.

    (b) intervals on which f is increasing.

    (c) intervals on which f is decreasing.

10. Let $f(x) = \dfrac{x + \sin x}{x}$.

    Find the absolute maximum value of f and where it occurs.

Section 8.2 Exercises

In Exercises 1–10, determine whether the function grows faster than e^x, at the same rate as e^x, or slower than e^x as $x \to \infty$.

1. $x^3 - 3x + 1$

2. $\sqrt{1 + x^4}$

3. 4^x

4. $(5/2)^x$

5. e^{x+1}

6. $x \ln x - x$

7. $e^{\cos x}$

8. xe^x

9. x^{1000}

10. $(e^x + e^{-x})/2$

In Exercises 11–16, determine whether the function grows faster than x^2, at the same rate as x^2, or slower than x^2 as $x \to \infty$.

11. $x^2 + 4x$

12. $x^3 + 3$

13. $15x + 3$

14. $\sqrt{x^4 + 5x}$

15. $\ln x$

16. 2^x

In Exercises 17–22, determine whether the function grows faster than $\ln x$, at the same rate as $\ln x$, or slower than $\ln x$ as $x \to \infty$.

17. $\log_2 x^2$

18. $\log \sqrt{x}$

19. $1/\sqrt{x}$

20. e^{-x}

21. $x - 2 \ln x$

22. $5 \ln x$

In Exercises 23 and 24, order the functions from slowest-growing to fastest-growing as $x \to \infty$.

23. $e^x, \quad x^x, \quad (\ln x)^x, \quad e^{x/2}$

24. $2^x, \quad x^2, \quad (\ln 2)^x, \quad e^x$

In Exercises 25–28, show that the three functions grow at the same rate as $x \to \infty$.

25. $f_1(x) = \sqrt{x}, \quad f_2(x) = \sqrt{10x + 1}, \quad f_3(x) = \sqrt{x + 1}$

26. $f_1(x) = x^2, \quad f_2(x) = \sqrt{x^4 + x}, \quad f_3(x) = \sqrt{x^4 - x^3}$

27. $f_1(x) = 3^x, \quad f_2(x) = \sqrt{9^x + 2^x}, \quad f_3(x) = \sqrt{9^x - 4^x}$

28. $f_1(x) = x^3, \quad f_2(x) = \dfrac{x^4 + 2x^2 - 1}{x + 1}, \quad f_3(x) = \dfrac{2x^5 - 1}{x^2 + 1}$

In Exercises 29 and 30, determine whether the statements are true or false as $x \to \infty$. Explain.

29. (a) $x = o(x)$ (b) $x = o(x + 5)$

   (c) $x = O(x + 5)$ (d) $x = O(2x)$

   (e) $e^x = o(e^{2x})$ (f) $x + \ln x = O(x)$

   (g) $\ln x = o(\ln 2x)$ (h) $\sqrt{x^2 + 5} = O(x)$

30. (a) $\dfrac{1}{x + 3} = O\left(\dfrac{1}{x}\right)$ (b) $\dfrac{1}{x} + \dfrac{1}{x^2} = O\left(\dfrac{1}{x}\right)$

   (c) $\dfrac{1}{x} - \dfrac{1}{x^2} = o\left(\dfrac{1}{x}\right)$ (d) $2 + \cos x = O(2)$

   (e) $e^x + x = O(e^x)$ (f) $x \ln x = o(x^2)$

   (g) $\ln(\ln x) = O(\ln x)$ (h) $\ln x = o(\ln(x^2 + 1))$

In Exercises 31–34, only one of the following is true.

   i. $f = o(g)$

   ii. $g = o(f)$

   iii. f and g grow at the same rate.

Use the given graph of f/g to determine which one is true.

31.
[0, 100] by [−1000, 10000]

32.
[0, 10] by [−0.5, 1]

33.
[0, 100] by [−1, 1.5]

34.
[0, 20] by [−1, 3]

In Exercises 35–37, *work in groups of two or three.*

35. *Comparing Exponential and Power Functions*

   (a) **Writing to Learn** Explain why e^x grows faster than x^n as $x \to \infty$ for any positive integer n, even $n = 1{,}000{,}000$. (*Hint:* What is the nth derivative of x^n?)

   (b) **Writing to Learn** Explain why $a^x, a > 1$, grows faster than x^n as $x \to \infty$ for any positive integer n.

36. *Comparing Exponential and Polynomial Functions*

   (a) **Writing to Learn** Show that e^x grows faster than any polynomial

$$a_n x^n + a_{n-1} x^{n-1} + \cdots + a_1 x + a_0, \ a_n > 0,$$

as $x \to \infty$. Explain.

(b) **Writing to Learn** Show that $a^x, a > 1$, grows faster than any polynomial

$$a_n x^n + a_{n-1} x^{n-1} + \cdots + a_1 x + a_0, \ a_n > 0,$$

as $x \to \infty$. Explain.

37. *Comparing Logarithm and Power Functions*

   (a) **Writing to Learn** Show that $\ln x$ grows slower than $x^{1/n}$ as $x \to \infty$ for any positive integer n, even $n = 1{,}000{,}000$. Explain.

   (b) **Writing to Learn** Show that for any number $a > 0$, $\ln x$ grows slower than x^a as $x \to \infty$. Explain.

38. *Comparing Logarithm and Polynomial Functions* Show that $\ln x$ grows slower than any nonconstant polynomial

$$a_n x^n + a_{n-1} x^{n-1} + \cdots + a_1 x + a_0, \ a_n > 0,$$

as $x \to \infty$.

39. *Search Algorithms* Suppose you have three different algorithms for solving the same problem and each algorithm provides for a number of steps that is of order of one of the functions listed here.

$$n \log_2 n, \quad n^{3/2}, \quad n(\log_2 n)^2$$

Which of the algorithms is likely the most efficient in the long run? Give reasons for your answer.

40. *Sequential and Binary Search* Suppose you are looking for an item in an ordered list one million items long. How many steps might it take to find the item with (a) a sequential search? (b) a binary search?

41. *Growing at the Same Rate* Show that if functions f and g grow at the same rate as $x \to \infty$, then $f = O(g)$ and $g = O(f)$.

42. *Growing at the Same Rate* Suppose that polynomials $p(x)$ and $q(x)$ grow at the same rate as $x \to \infty$. What can you conclude about

   (a) $\displaystyle \lim_{x \to \infty} \frac{p(x)}{q(x)}$?   (b) $\displaystyle \lim_{x \to -\infty} \frac{p(x)}{q(x)}$?

Explorations

43. Let

$$f(x) = a_n x^n + a_{n-1} x^{n-1} + \cdots + a_1 x + a_0$$

and

$$g(x) = b_m x^m + b_{m-1} x^{m-1} + \cdots + b_1 x + b_0$$

be any two polynomial functions with $a_n > 0$, $b_m > 0$.

   (a) Compare the rates of growth of x^5 and x^2 as $x \to \infty$.

   (b) Compare the rates of growth of $5x^3$ and $2x^3$ as $x \to \infty$.

   (c) If x^m grows faster than x^n as $x \to \infty$, what can you conclude about m and n?

   (d) If x^m grows at the same rate as x^n as $x \to \infty$, what can you conclude about m and n?

   (e) If $g(x)$ grows faster than $f(x)$ as $x \to \infty$, what can you conclude about their degrees?

   (f) If $g(x)$ grows at the same rate as $f(x)$ as $x \to \infty$, what can you conclude about their degrees?

44. *Simpson's Rule and Trapezoidal Rule (Section 5.5)*
The definitions in this section can be made more general by lifting the restriction that $x \to \infty$ and considering limits as $x \to a$ for any real number a.

(a) Write definitions for big-oh and little-oh as $x \to a$.

(b) Show that the error E_S in the Simpson's Rule approximation of a definite integral is $O(h^4)$ as $h \to 0$.

(c) Show that the error E_T in the Trapezoidal Rule approximation of a definite integral is $O(h^2)$ as $h \to 0$.

This gives another way to explain the relative accuracies of the two methods. ∎

Extending the Ideas

45. Suppose that the values of the functions $f(x)$ and $g(x)$ eventually become and remain negative as $x \to \infty$. We say that

i. *f* **decreases faster** than *g* as $x \to \infty$ if
$$\lim_{x \to \infty} \frac{f(x)}{g(x)} = \infty.$$

ii. *f* and *g* **decrease at the same rate** as $x \to \infty$ if
$$\lim_{x \to \infty} \frac{f(x)}{g(x)} = L \ne 0.$$

(a) Show that if *f* decreases faster than *g* as $x \to \infty$, then $|f|$ grows faster than $|g|$ as $x \to \infty$.

(b) Show that if *f* and *g* decrease at the same rate as $x \to \infty$, then $|f|$ and $|g|$ grow at the same rate as $x \to \infty$.

46. Suppose that the values of the functions $f(x)$ and $g(x)$ eventually become and remain positive as $x \to -\infty$. We say that

i. *f* **grows faster** than *g* as $x \to -\infty$ if
$$\lim_{x \to -\infty} \frac{f(x)}{g(x)} = \infty.$$

ii. *f* and *g* **grow at the same rate** as $x \to -\infty$ if
$$\lim_{x \to -\infty} \frac{f(x)}{g(x)} = L \ne 0.$$

(a) Show that if *f* grows faster than *g* as $x \to -\infty$, then $f(-x)$ grows faster than $g(-x)$ as $x \to \infty$.

(b) Show that if *f* and *g* grow at the same rate as $x \to -\infty$, then $f(-x)$ and $g(-x)$ grow at the same rate as $x \to \infty$.

8.3 Improper Integrals

Infinite Limits of Integration • The Integral $\int_1^\infty dx/x^p$ • Integrands with Infinite Discontinuities • Tests for Convergence and Divergence • Applications

Infinite Limits of Integration

Consider the infinite region that lies under the curve $y = e^{-x/2}$ in the first quadrant (Figure 8.13a). You might think this region has infinite area, but we will see that it is finite. Here is how we assign a value to the area. First we find the area $A(b)$ of the portion of the region that is bounded on the right by $x = b$ (Figure 8.13b).

$$A(b) = \int_0^b e^{-x/2}\, dx = -2e^{-x/2}\Big]_0^b = -2e^{-b/2} + 2$$

Then we find the limit of $A(b)$ as $b \to \infty$.

$$\lim_{b \to \infty} A(b) = \lim_{b \to \infty}\left(-2e^{-b/2} + 2\right) = 2$$

The area under the curve from 0 to ∞ is

$$\int_0^\infty e^{-x/2}\, dx = \lim_{b \to \infty} \int_0^b e^{-x/2}\, dx = 2.$$

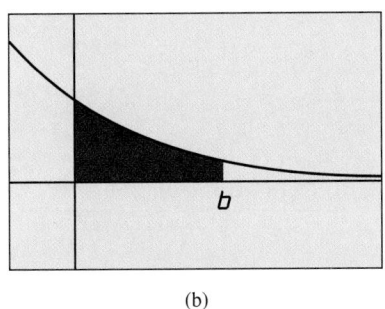

(a)

(b)

Figure 8.13 (a) The area in the first quadrant under the curve $y = e^{-x/2}$ is (b)
$$\lim_{b \to \infty} \int_0^b e^{-x/2}\, dx.$$

> **Definition** Improper Integrals with Infinite Integration Limits
>
> Integrals with infinite limits of integration are **improper integrals.**
>
> **1.** If $f(x)$ is continuous on $[a, \infty)$, then
>
> $$\int_a^\infty f(x)\, dx = \lim_{b \to \infty} \int_a^b f(x)\, dx.$$
>
> **2.** If $f(x)$ is continuous on $(-\infty, b]$, then
>
> $$\int_{-\infty}^b f(x)\, dx = \lim_{a \to -\infty} \int_a^b f(x)\, dx.$$
>
> **3.** If $f(x)$ is continuous on $(-\infty, \infty)$, then
>
> $$\int_{-\infty}^\infty f(x)\, dx = \int_{-\infty}^c f(x)\, dx + \int_c^\infty f(x)\, dx,$$
>
> where c is any real number.

In parts 1 and 2, if the limit is finite the improper integral **converges** and the limit is the **value** of the improper integral. If the limit fails to exist, the improper integral **diverges.** In part 3, the integral on the left-hand side of the equation **converges** if both improper integrals on the right-hand side converge, otherwise it **diverges** and has no value. It can be shown that the choice of c in part 3 is unimportant. We can evaluate or determine the convergence or divergence of $\int_{-\infty}^\infty f(x)\, dx$ with any convenient choice.

Example 1 EVALUATING AN IMPROPER INTEGRAL ON $[1, \infty)$

Does the improper integral $\displaystyle\int_1^\infty \frac{dx}{x}$ converge or diverge?

Solution

$$\int_1^\infty \frac{dx}{x} = \lim_{b \to \infty} \int_1^b \frac{dx}{x} \qquad \text{Definition}$$

$$= \lim_{b \to \infty} \ln x \Big]_1^b$$

$$= \lim_{b \to \infty} (\ln b - \ln 1) = \infty$$

Thus, the integral diverges.

Example 2 EVALUATING AN INTEGRAL ON $(-\infty, \infty)$

Evaluate $\displaystyle\int_{-\infty}^\infty \frac{dx}{1 + x^2}$.

Solution According to the definition (part 3) we can write

$$\int_{-\infty}^\infty \frac{dx}{1 + x^2} = \int_{-\infty}^0 \frac{dx}{1 + x^2} + \int_0^\infty \frac{dx}{1 + x^2}.$$

Next, we evaluate each improper integral on the right-hand side of the equation above.

$$\int_{-\infty}^{0} \frac{dx}{1 + x^2} = \lim_{a \to -\infty} \int_{a}^{0} \frac{dx}{1 + x^2}$$

$$= \lim_{a \to -\infty} \tan^{-1} x \Big]_{a}^{0}$$

$$= \lim_{a \to -\infty} (\tan^{-1} 0 - \tan^{-1} a) = 0 - \left(-\frac{\pi}{2}\right) = \frac{\pi}{2}$$

$$\int_{0}^{\infty} \frac{dx}{1 + x^2} = \lim_{b \to \infty} \int_{0}^{b} \frac{dx}{1 + x^2}$$

$$= \lim_{b \to \infty} \tan^{-1} x \Big]_{0}^{b}$$

$$= \lim_{b \to \infty} (\tan^{-1} b - \tan^{-1} 0) = \frac{\pi}{2} - 0 = \frac{\pi}{2}$$

Thus,

$$\int_{-\infty}^{\infty} \frac{dx}{1 + x^2} = \frac{\pi}{2} + \frac{\pi}{2} = \pi.$$

The Integral $\int_{1}^{\infty} \dfrac{dx}{x^p}$

The function $y = 1/x$ is the boundary between the convergent and divergent improper integrals with integrands of the form $y = 1/x^p$. Example 3 explains.

Example 3 DETERMINING CONVERGENCE

For what values of p does the integral $\int_{1}^{\infty} dx/x^p$ converge? When the integral does converge, what is its value?

Solution Example 1 shows that the integral diverges if $p = 1$. If $p \neq 1$,

$$\int_{1}^{b} \frac{dx}{x^p} = \frac{x^{-p+1}}{-p + 1} \bigg]_{1}^{b} = \frac{1}{1 - p}(b^{-p+1} - 1) = \frac{1}{1 - p}\left(\frac{1}{b^{p-1}} - 1\right).$$

Thus,

$$\int_{1}^{\infty} \frac{dx}{x^p} = \lim_{b \to \infty} \int_{1}^{b} \frac{dx}{x^p}$$

$$= \lim_{b \to \infty} \left[\frac{1}{1 - p}\left(\frac{1}{b^{p-1}} - 1\right)\right] = \begin{cases} \dfrac{1}{p - 1}, & p > 1 \\ \infty, & p < 1 \end{cases}$$

because

$$\lim_{b \to \infty} \frac{1}{b^{p-1}} = \begin{cases} 0, & p > 1 \\ \infty, & p < 1. \end{cases}$$

Therefore, the integral converges to the value $1/(p - 1)$ if $p > 1$. It diverges if $p \leq 1$ (see Example 1).

[0, 2] by [−1, 5]

(a)

[0, 2] by [−1, 5]

(b)

Figure 8.14 (a) The area under the curve $y = 1/\sqrt{x}$ from $x = 0$ to $x = 1$ is (b)

$$\lim_{a \to 0^+} \int_a^1 (1/\sqrt{x})\, dx.$$

Integrands with Infinite Discontinuities

Another type of improper integral arises when the integrand has a vertical asymptote — an infinite discontinuity — at a limit of integration or at some point between the limits of integration.

Consider the infinite region in the first quadrant that lies under the curve $y = 1/\sqrt{x}$ from $x = 0$ to $x = 1$ (Figure 8.14a). First we find the area of the portion from a to 1 (Figure 8.14b).

$$\int_a^1 \frac{dx}{\sqrt{x}} = 2\sqrt{x}\,\Big]_a^1 = 2 - 2\sqrt{a}$$

Then, we find the limit of this area as $a \to 0^+$.

$$\lim_{a \to 0^+} \int_a^1 \frac{dx}{\sqrt{x}} = \lim_{a \to 0^+} (2 - 2\sqrt{a}) = 2$$

The area under the curve from 0 to 1 is

$$\int_0^1 \frac{dx}{\sqrt{x}} = \lim_{a \to 0^+} \int_a^1 \frac{dx}{\sqrt{x}} = 2.$$

Definition Improper Integrals with Infinite Discontinuities

Integrals of functions that become infinite at a point within the interval of integration are **improper integrals.**

1. If $f(x)$ is continuous on $(a, b]$, then

$$\int_a^b f(x)\, dx = \lim_{c \to a^+} \int_c^b f(x)\, dx.$$

2. If $f(x)$ is continuous on $[a, b)$, then

$$\int_a^b f(x)\, dx = \lim_{c \to b^-} \int_a^c f(x)\, dx.$$

3. If $f(x)$ is continuous on $[a, c) \cup (c, b]$, then

$$\int_a^b f(x)\, dx = \int_a^c f(x)\, dx + \int_c^b f(x)\, dx.$$

In parts 1 and 2, if the limit is finite the improper integral **converges** and the limit is the **value** of the improper integral. If the limit fails to exist the improper integral **diverges**. In part 3, the integral on the left-hand side of the equation **converges** if both integrals on the right-hand side have values, otherwise it **diverges**.

Exploration 1 **Investigating** $\displaystyle \int_0^1 \frac{dx}{x^p}$

1. Explain why these integrals are improper if $p > 0$.

2. Show that the integral diverges if $p = 1$.

3. Show that the integral diverges if $p > 1$.

4. Show that the integral converges if $0 < p < 1$.

Example 4   INFINITE DISCONTINUITY AT AN INTERIOR POINT

Evaluate $\displaystyle\int_0^3 \frac{dx}{(x-1)^{2/3}}$.

Solution The integrand has a vertical asymptote at $x = 1$ and is continuous on $[0, 1)$ and $(1, 3]$. Thus, by part 3 of the definition above

$$\int_0^3 \frac{dx}{(x-1)^{2/3}} = \int_0^1 \frac{dx}{(x-1)^{2/3}} + \int_1^3 \frac{dx}{(x-1)^{2/3}}.$$

Next, we evaluate each improper integral on the right-hand side of this equation.

$$\int_0^1 \frac{dx}{(x-1)^{2/3}} = \lim_{c\to1^-} \int_0^c \frac{dx}{(x-1)^{2/3}}$$

$$= \lim_{c\to1^-} 3(x-1)^{1/3} \Big]_0^c$$

$$= \lim_{c\to1^-} [3(c-1)^{1/3} + 3] = 3$$

$$\int_1^3 \frac{dx}{(x-1)^{2/3}} = \lim_{c\to1^+} \int_c^3 \frac{dx}{(x-1)^{2/3}}$$

$$= \lim_{c\to1^+} 3(x-1)^{1/3} \Big]_c^3$$

$$= \lim_{c\to1^+} [3(3-1)^{1/3} - 3(c-1)^{1/3}] = 3\sqrt[3]{2}$$

We conclude that

$$\int_0^3 \frac{dx}{(x-1)^{2/3}} = 3 + 3\sqrt[3]{2}.$$

Example 5   INFINITE DISCONTINUITY AT AN INTERIOR POINT

Evaluate $\displaystyle\int_1^4 \frac{dx}{x-2}$.

Solution The integrand has an infinite discontinuity at $x = 2$ and is continuous on $[1, 2)$ and $(2, 4]$. Thus,

$$\int_1^4 \frac{dx}{x-2} = \int_1^2 \frac{dx}{x-2} + \int_2^4 \frac{dx}{x-2}.$$

Next, we evaluate each improper integral on the right-hand side.

$$\int_1^2 \frac{dx}{x-2} = \lim_{c\to2^-} \int_1^c \frac{dx}{x-2}$$

$$= \lim_{c\to2^-} \ln|x-2| \Big]_1^c$$

$$= \lim_{c\to2^-} (\ln|c-2| - \ln|-1|) = -\infty$$

There is no need to evaluate the second integral on the right-hand side as we can now conclude that the original integral diverges and has no value.

Tests for Convergence and Divergence

When we cannot evaluate an improper integral directly (often the case in practice) we first try to determine whether it converges or diverges. If the integral diverges, that's the end of the story. If it converges, we can then use numerical methods to approximate its value. The principal tests for convergence or divergence are the direct comparison test and the limit comparison test.

Example 6   INVESTIGATING CONVERGENCE

Does the integral $\int_1^\infty e^{-x^2}\,dx$ converge?

Solution

Solve Analytically

By definition,

$$\int_1^\infty e^{-x^2}\,dx = \lim_{b\to\infty}\int_1^b e^{-x^2}\,dx.$$

We cannot evaluate the latter integral directly because there is no simple formula for the antiderivative of e^{-x^2}. We must therefore determine its convergence or divergence some other way. Because $e^{-x^2} > 0$ for all x, $\int_1^b e^{-x^2}\,dx$ is an increasing function of b. Therefore, as $b\to\infty$, the integral either becomes infinite as $b\to\infty$ or it is bounded from above and is forced to converge (have a finite limit).

The two curves $y = e^{-x^2}$ and $y = e^{-x}$ intersect at $(1, e^{-1})$, and $0 < e^{-x^2} \le e^{-x}$ for $x \ge 1$ (Figure 8.15). Thus, for any $b > 1$,

$$0 < \int_1^b e^{-x^2}\,dx \le \int_1^b e^{-x}\,dx = -e^{-b} + e^{-1} < e^{-1} \approx 0.368. \qquad \text{Rounded up to be safe}$$

As an increasing function of b bounded above by 0.368, the integral $\int_1^\infty e^{-x^2}\,dx$ must converge. This does not tell us much about the value of the improper integral, however, except that it is positive and less than 0.368.

Support Graphically

The graph of NINT $(e^{-x^2}, x, 1, x)$ is shown in Figure 8.16. The value of the integral rises rapidly as x first moves away from 1 but changes little past $x = 3$. Values sampled along the curve suggest a limit of about 0.13940 as $x\to\infty$. (Exercise 51 shows how to confirm the accuracy of this estimate.)

The comparison of e^{-x^2} and e^{-x} in Example 6 is a special case of the following test.

Bounded Monotonic Functions

It can be shown that a monotonic function $f(x)$ that is bounded on an infinite interval (a, ∞) must have a finite limit as $x\to\infty$. In Example 6, this is applied to the function

$$f(b) = \int_1^b e^{-x^2}\,dx$$

as $b\to\infty$.

[0, 3] by [−0.5, 1.5]

Figure 8.15 The graph of $y = e^{-x^2}$ lies below the graph of $y = e^{-x}$ for $x > 1$. (Example 6)

$X = 19 \qquad Y = .13940279$

[0, 20] by [−0.1, 0.3]

Figure 8.16 The graph of NINT $(e^{-x^2}, x, 1, x)$. (Example 6)

Theorem 3   Direct Comparison Test

Let f and g be continuous on $[a, \infty)$ with $0 \le f(x) \le g(x)$ for all $x \ge a$. Then

1. $\displaystyle\int_a^\infty f(x)\,dx$ converges if $\displaystyle\int_a^\infty g(x)\,dx$ converges.

2. $\displaystyle\int_a^\infty g(x)\,dx$ diverges if $\displaystyle\int_a^\infty f(x)\,dx$ diverges.

Example 7 USING THE DIRECT COMPARISON TEST

(a) $\displaystyle\int_1^\infty \frac{\sin^2 x}{x^2}\,dx$ converges because

$$0 \le \frac{\sin^2 x}{x^2} \le \frac{1}{x^2} \quad \text{on} \quad [1, \infty) \quad \text{and} \quad \int_1^\infty \frac{1}{x^2}\,dx \text{ converges.} \qquad \text{Example 3}$$

(b) $\displaystyle\int_1^\infty \frac{1}{\sqrt{x^2 - 0.1}}\,dx$ diverges because

$$\frac{1}{\sqrt{x^2 - 0.1}} \ge \frac{1}{x} \quad \text{on} \quad [1, \infty) \quad \text{and} \quad \int_1^\infty \frac{1}{x}\,dx \text{ diverges.} \qquad \text{Example 1}$$

Theorem 4 Limit Comparison Test

If the positive functions f and g are continuous on $[a, \infty)$ and if

$$\lim_{x \to \infty} \frac{f(x)}{g(x)} = L, \quad 0 < L < \infty,$$

then

$$\int_a^\infty f(x)\,dx \quad \text{and} \quad \int_a^\infty g(x)\,dx \quad \text{both converge or both diverge.}$$

In the language of Section 8.2, Theorem 4 says that if two functions grow at the same rate as $x \to \infty$, then their integrals from a to ∞ behave alike; they both converge or both diverge. This does not mean that their integrals necessarily have the same value, however, as Example 8 shows.

Example 8 USING THE LIMIT COMPARISON TEST

Show that

$$\int_1^\infty \frac{dx}{1 + x^2}$$

converges by comparison with $\int_1^\infty (1/x^2)\,dx$. Find and compare the two integral values.

Solution The functions $f(x) = 1/x^2$ and $g(x) = 1/(1 + x^2)$ are positive and continuous on $[1, \infty)$. Also,

$$\lim_{x \to \infty} \frac{f(x)}{g(x)} = \lim_{x \to \infty} \frac{1/x^2}{1/(1 + x^2)} = \lim_{x \to \infty} \frac{1 + x^2}{x^2}$$

$$= \lim_{x \to \infty} \frac{2x}{2x} = 1. \quad \text{L'Hopital's Rule}$$

Thus, $\int_1^\infty (1/(1 + x^2))\,dx$ converges because $\int_1^\infty (1/x^2)\,dx$ converges (Example 3).

To evaluate the two integrals, we begin with the following from Example 3,

$$\int_1^\infty \frac{dx}{x^2} = \frac{1}{2 - 1} = 1.$$

By definition,

$$\int_1^\infty \frac{dx}{1+x^2} = \lim_{b\to\infty} \int_1^b \frac{dx}{1+x^2}$$

$$= \lim_{b\to\infty} \tan^{-1}x \Big]_1^b$$

$$= \lim_{b\to\infty} (\tan^{-1} b - \tan^{-1} 1) = \frac{\pi}{2} - \frac{\pi}{4} = \frac{\pi}{4}.$$

The integrals converge to the different values 1 and $\pi/4$.

Example 9   USING THE LIMIT COMPARISON TEST

Show that $\displaystyle\int_1^\infty \frac{1}{e^x - 5} \, dx$ converges.

Solution From Example 6 it is easy to see that $\int_1^\infty e^{-x} \, dx = \int_1^\infty (1/e^x) \, dx$ converges. Because

$$\lim_{x\to\infty} \frac{1/e^x}{1/(e^x - 5)} = \lim_{x\to\infty} \frac{e^x - 5}{e^x} = \lim_{x\to\infty} \left(1 - \frac{5}{e^x}\right) = 1,$$

the integral $\displaystyle\int_1^\infty \frac{3}{e^x - 5} \, dx$ *also* converges.

Applications

Example 10   FINDING THE VOLUME OF AN INFINITE SOLID

Find the volume of the solid obtained by revolving the curve $y = xe^{-x}$, $0 \le x < \infty$ about the *x*-axis.

Solution Figure 8.17 shows a portion of the region to be revolved about the *x*-axis. The area of a typical cross section of the solid is

$$\pi(\text{radius})^2 = \pi y^2 = \pi x^2 e^{-2x}.$$

The volume of the solid is

$$V = \pi \int_0^\infty x^2 e^{-2x} \, dx = \pi \lim_{b\to\infty} \int_0^b x^2 e^{-2x} \, dx.$$

Integrating by parts twice we obtain the following.

$$\int x^2 e^{-2x} \, dx = -\frac{x^2}{2} e^{-2x} + \int x e^{-2x} \, dx$$

$$\quad u = x^2, \, dv = e^{-2x} \, dx$$
$$\quad du = 2x \, dx, \, v = -\frac{1}{2} e^{-2x}$$

$$= -\frac{x^2}{2} e^{-2x} - \frac{x}{2} e^{-2x} + \frac{1}{2} \int e^{-2x} \, dx$$

$$\quad u = x, \, dv = e^{-2x} \, dx$$
$$\quad du = dx, \, v = -\frac{1}{2} e^{-2x}$$

$$= -\frac{x^2}{2} e^{-2x} - \frac{x}{2} e^{-2x} - \frac{1}{4} e^{-2x} + C$$

$$= -\frac{2x^2 + 2x + 1}{4e^{2x}} + C$$

[0, 5] by [–0.5, 1]

Figure 8.17 The graph of $y = xe^{-x}$. (Example 10)

Thus,

$$V = \pi \lim_{b \to \infty} \left[-\frac{2x^2 + 2x + 1}{4e^{2x}} \right]_0^b$$

$$= \pi \lim_{b \to \infty} \left[-\frac{2b^2 + 2b + 1}{4e^{2b}} + \frac{1}{4} \right] = \frac{\pi}{4},$$

and the volume of the solid is $\pi/4$.

Example 11 FINDING CIRCUMFERENCE

Use the arc length formula (Section 7.4) to show that the circumference of the circle $x^2 + y^2 = 4$ is 4π.

Solution One fourth of this circle is given by $y = \sqrt{4 - x^2}, 0 \le x \le 2$. Its arc length is

$$L = \int_0^2 \sqrt{1 + (y')^2} \, dx, \quad \text{where} \quad y' = -\frac{x}{\sqrt{4 - x^2}}.$$

The integral is improper because y' is not defined at $x = 2$. We evaluate it as a limit.

$$L = \int_0^2 \sqrt{1 + (y')^2} \, dx = \int_0^2 \sqrt{1 + \frac{x^2}{4 - x^2}} \, dx$$

$$= \int_0^2 \sqrt{\frac{4}{4 - x^2}} \, dx$$

$$= \lim_{b \to 2^-} \int_0^b \sqrt{\frac{4}{4 - x^2}} \, dx$$

$$= \lim_{b \to 2^-} \int_0^b \sqrt{\frac{1}{1 - (x/2)^2}} \, dx$$

$$= \lim_{b \to 2^-} 2 \sin^{-1} \frac{x}{2} \Big]_0^b$$

$$= \lim_{b \to 2^-} 2 \left[\sin^{-1} \frac{b}{2} - 0 \right] = \pi$$

The circumference of the quarter circle is π; the circumference of the circle is 4π.

Quick Review 8.3

In Exercises 1–4, evaluate the integral.

1. $\displaystyle\int_0^3 \frac{dx}{x+3}$

2. $\displaystyle\int_{-1}^1 \frac{x\,dx}{x^2+1}$

3. $\displaystyle\int \frac{dx}{x^2+4}$

4. $\displaystyle\int \frac{dx}{x^4}$

In Exercises 5 and 6, find the domain of the function.

5. $g(x) = \dfrac{1}{\sqrt{9-x^2}}$

6. $h(x) = \dfrac{1}{\sqrt{x-1}}$

In Exercises 7 and 8, confirm the inequality.

7. $\left|\dfrac{\cos x}{x^2}\right| \le \dfrac{1}{x^2}, \quad -\infty < x < \infty$

8. $\dfrac{1}{\sqrt{x^2-1}} \ge \dfrac{1}{x}, \quad x > 1$

In Exercises 9 and 10, show that the functions f and g grow at the same rate as $x \to \infty$.

9. $f(x) = 4e^x - 5, \quad g(x) = 3e^x + 7$

10. $f(x) = \sqrt{2x-1}, \quad g(x) = \sqrt{x+3}$

Section 8.3 Exercises

In Exercises 1–6, **(a)** state why the integral is improper or involves improper integrals. Then, **(b)** determine whether the integral converges or diverges, and **(c)** evaluate the integral if it converges.

1. $\displaystyle\int_0^\infty \frac{dx}{x^2+1}$

2. $\displaystyle\int_0^1 \frac{dx}{\sqrt{x}}$

3. $\displaystyle\int_{-8}^1 \frac{dx}{x^{1/3}}$

4. $\displaystyle\int_{-\infty}^\infty \frac{2x\,dx}{(x^2+1)^2}$

5. $\displaystyle\int_0^{\ln 2} x^{-2}e^{1/x}\,dx$

6. $\displaystyle\int_0^{\pi/2} \cot\theta\,d\theta$

In Exercises 7–26, evaluate the integral or state that it diverges.

7. $\displaystyle\int_1^\infty \frac{dx}{x^{1.001}}$

8. $\displaystyle\int_{-1}^1 \frac{dx}{x^{2/3}}$

9. $\displaystyle\int_0^4 \frac{dr}{\sqrt{4-r}}$

10. $\displaystyle\int_0^1 \frac{dr}{r^{0.999}}$

11. $\displaystyle\int_0^1 \frac{dx}{\sqrt{1-x^2}}$

12. $\displaystyle\int_{-\infty}^2 \frac{2\,dx}{x^2+4}$

13. $\displaystyle\int_{-\infty}^{-2} \frac{2\,dx}{x^2-1}$

14. $\displaystyle\int_2^\infty \frac{3\,dt}{t^2-t}$

15. $\displaystyle\int_0^1 \frac{\theta+1}{\sqrt{\theta^2+2\theta}}\,d\theta$

16. $\displaystyle\int_0^2 \frac{s+1}{\sqrt{4-s^2}}\,ds$

17. $\displaystyle\int_0^\infty \frac{dx}{(1+x)\sqrt{x}}$

18. $\displaystyle\int_1^\infty \frac{dx}{x\sqrt{x^2-1}}$

19. $\displaystyle\int_1^2 \frac{ds}{s\sqrt{s^2-1}}$

20. $\displaystyle\int_{-1}^\infty \frac{d\theta}{\theta^2+5\theta+6}$

21. $\displaystyle\int_0^\infty \frac{16\tan^{-1}x}{1+x^2}\,dx$

22. $\displaystyle\int_{-1}^4 \frac{dx}{\sqrt{|x|}}$

23. $\displaystyle\int_{-\infty}^0 \theta e^\theta\,d\theta$

24. $\displaystyle\int_0^\infty 2e^{-\theta}\sin\theta\,d\theta$

25. $\displaystyle\int_{-\infty}^\infty e^{-|x|}\,dx$

26. $\displaystyle\int_0^1 x\ln x\,dx$

In Exercises 27–46, use integration, the direct comparison test, or the limit comparison test to determine whether the integral converges or diverges.

27. $\displaystyle\int_0^{\pi/2} \tan\theta\,d\theta$

28. $\displaystyle\int_0^\pi \frac{\sin\theta\,d\theta}{\sqrt{\pi-\theta}}$

29. $\displaystyle\int_{-\infty}^\infty 2x\,e^{-x^2}\,dx$

30. $\displaystyle\int_0^4 \frac{e^{-\sqrt{x}}}{\sqrt{x}}\,dx$

31. $\displaystyle\int_0^\pi \frac{dt}{\sqrt{t}+\sin t}$

32. $\displaystyle\int_4^\infty \frac{dx}{\sqrt{x}-1}$

33. $\displaystyle\int_1^\infty \frac{dx}{x^3+1}$

34. $\displaystyle\int_0^2 \frac{dx}{1-x^2}$

35. $\displaystyle\int_0^2 \frac{dx}{1-x}$

36. $\displaystyle\int_{-1}^1 \ln|x|\,dx$

37. $\displaystyle\int_1^\infty \frac{d\theta}{1+e^\theta}$

38. $\displaystyle\int_2^\infty \frac{dx}{\sqrt{x^2-1}}$

39. $\displaystyle\int_1^\infty \frac{\sqrt{x+1}}{x^2}\,dx$

40. $\displaystyle\int_0^\infty \frac{dx}{\sqrt{x}}$

41. $\displaystyle\int_\pi^\infty \frac{2+\cos x}{x}\,dx$

42. $\displaystyle\int_\pi^\infty \frac{1+\sin x}{x^2}\,dx$

43. $\displaystyle\int_{-\infty}^\infty \frac{dx}{e^x+e^{-x}}$

44. $\displaystyle\int_{-\infty}^\infty \frac{dx}{\sqrt{x^4+1}}$

45. $\displaystyle\int_0^\infty \frac{dy}{(1+y^2)(1+\tan^{-1}y)}$

46. $\displaystyle\int_{-\infty}^\infty \frac{e^{-y}\,dy}{y^2+1}$

In Exercises 47 and 48, find the area of the region in the first quadrant that lies under the given curve.

47. $y = \dfrac{\ln x}{x^2}$ **48.** $y = \dfrac{\ln x}{x}$

Explorations

49. The Infinite Paint Can, or Gabriel's Horn Consider the region R in the first quadrant bounded above by $y = 1/x$ and on the left by $x = 1$. The region is revolved about the x-axis to form an infinite solid as shown in the figure.

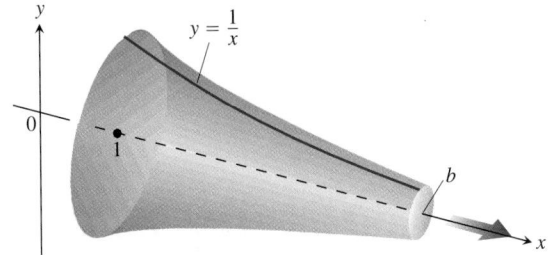

(a) Writing to Learn Explain how Example 1 shows that the region R has infinite area.

(b) Find the surface area of the solid.

(c) Find the volume of the solid.

(d) Why is Gabriel's horn sometimes described as a can that does not hold enough paint to cover its own outside surface?

50. Normal Probability Distribution Function In Section 7.5, we encountered the bell-shaped normal distribution curve that is the graph of

$$f(x) = \frac{1}{\sigma\sqrt{2\pi}}\, e^{-\frac{1}{2}\left(\frac{x-\mu}{\sigma}\right)^2},$$

the normal probability density function with mean μ and standard deviation σ. The number μ tells where the distribution is centered, and σ measures the "scatter" around the mean.

From the theory of probability, it is known that

$$\int_{-\infty}^{\infty} f(x)\, dx = 1.$$

In what follows, let $\mu = 0$ and $\sigma = 1$.

(a) Draw the graph of f. Find the intervals on which f is increasing, the intervals on which f is decreasing, and any local extreme values and where they occur.

(b) Evaluate $\displaystyle\int_{-n}^{n} f(x)\, dx$ for $n = 1, 2, 3$.

(c) Give a convincing argument that $\displaystyle\int_{-\infty}^{\infty} f(x)\, dx = 1.$

(*Hint:* Show that $0 < f(x) < e^{-x/2}$ for $x > 1$, and for $b > 1$,

$$\int_{b}^{\infty} e^{-x/2}\, dx \to 0 \quad \text{as} \quad b \to \infty.)$$

51. Approximating the Value of $\int_1^{\infty} e^{-x^2}\, dx$

(a) Show that $\displaystyle\int_{6}^{\infty} e^{-x^2}\, dx \le \int_{6}^{\infty} e^{-6x}\, dx < 4 \times 10^{-17}.$

(b) Writing to Learn Explain why

$$\int_{1}^{\infty} e^{-x^2}\, dx \approx \int_{1}^{6} e^{-x^2}\, dx$$

with error of at most 4×10^{-17}.

(c) Use the approximation in (b) to estimate the value of $\int_1^{\infty} e^{-x^2}\, dx$. Compare this estimate with the value displayed in Figure 8.16.

(d) Writing to Learn Explain why

$$\int_{0}^{\infty} e^{-x^2}\, dx \approx \int_{0}^{3} e^{-x^2}\, dx$$

with error of at most 0.000042.   ■

52. Work in groups of two or three.
(a) Show that if f is an even function and the necessary integrals exist, then

$$\int_{-\infty}^{\infty} f(x)\, dx = 2\int_{0}^{\infty} f(x)\, dx.$$

(b) Show that if f is odd and the necessary integrals exist, then

$$\int_{-\infty}^{\infty} f(x)\, dx = 0.$$

53. Writing to Learn

(a) Show that the integral $\displaystyle\int_{0}^{\infty} \frac{2x\, dx}{x^2 + 1}$ diverges.

(b) Explain why we can conclude from (a) that

$$\int_{-\infty}^{\infty} \frac{2x\, dx}{x^2 + 1} \quad \text{diverges.}$$

(c) Show that $\displaystyle\lim_{b \to \infty} \int_{-b}^{b} \frac{2x\, dx}{x^2 + 1} = 0.$

(d) Explain why the result in (c) does not contradict (b).

54. Finding Perimeter Find the perimeter of the 4-sided figure $x^{2/3} + y^{2/3} = 1$.

Extending the Ideas

55. Use properties of integrals to give a convincing argument that Theorem 3 is true.

56. Consider the integral

$$f(n + 1) = \int_0^\infty x^n e^{-x}\, dx$$

where $n \geq 0$.

(a) Show that $\int_0^\infty x^n e^{-x}\, dx$ converges for $n = 0, 1, 2$.

(b) Use integration by parts to show that $f(n + 1) = nf(n)$.

(c) Give a convincing argument that $\int_0^\infty x^n e^{-x}\, dx$ converges for all integers $n \geq 0$.

57. Let $f(x) = \int_0^x \dfrac{\sin t}{t}\, dt$.

(a) Use graphs and tables to investigate the values of $f(x)$ as $x \to \infty$.

(b) Does the integral $\int_0^\infty (\sin x)/x\, dx$ converge? Give a convincing argument.

58. (a) Show that we get the same value for the improper integral in Example 2 if we express

$$\int_{-\infty}^\infty \frac{dx}{1 + x^2} = \int_{-\infty}^1 \frac{dx}{1 + x^2} + \int_1^\infty \frac{dx}{1 + x^2},$$

and then evaluate these two integrals.

(b) Show that it doesn't matter what we choose for c in (Improper Integrals with Infinite Integration Limits, part 3)

$$\int_{-\infty}^\infty f(x)\, dx = \int_{-\infty}^c f(x)\, dx + \int_c^\infty f(x)\, dx.$$

8.4 Partial Fractions and Integral Tables

Partial Fractions • General Description of the Method • Integral Tables • Trigonometric Substitutions

Partial Fractions

In Example 2 of Section 6.5, we solved the logistic differential equation

$$\frac{dP}{dt} = 0.001P(100 - P)$$

by rewriting it as

$$\frac{100}{P(100 - P)}\, dP = 0.1\, dt, \quad \text{Variables separated}$$

expanding the fraction on the left into two basic fractions,

$$\frac{100}{P(100 - P)} = \frac{1}{P} + \frac{1}{100 - P},$$

and integrating both sides to find the solution

$$\ln |P| - \ln |100 - P| = 0.1t + C.$$

This expansion technique is the **method of partial fractions.** Any rational function can be written as a sum of basic fractions, called **partial fractions,** using the method of partial fractions. We can then integrate the rational function by integrating the sum of partial fractions instead.

Example 1 USING PARTIAL FRACTIONS

Use the method of partial fractions to evaluate $\displaystyle\int \frac{5x - 3}{x^2 - 2x - 3}\, dx$.

Solution First we factor the denominator: $x^2 - 2x - 3 = (x + 1)(x - 3)$. Then we determine the values of A and B so that

$$\frac{5x - 3}{x^2 - 2x - 3} = \frac{A}{x + 1} + \frac{B}{x - 3}.$$

Undetermined Coefficients

The A and B in the partial fraction decomposition are referred to as **undetermined coefficients.**

We proceed as follows:

$$5x - 3 = A(x - 3) + B(x + 1) \quad \text{Multiply by } (x + 1)(x - 3).$$

$$= (A + B)x - 3A + B. \quad \text{Combine terms.}$$

Equating coefficients, we obtain the following system of linear equations.

$$A + B = 5$$
$$-3A + B = -3$$

Solving these equations simultaneously yields $A = 2$ and $B = 3$. Therefore,

$$\int \frac{5x - 3}{x^2 - 2x - 3}\, dx = \int \frac{2}{x + 1}\, dx + \int \frac{3}{x - 3}\, dx$$

$$= 2 \ln |x + 1| + 3 \ln |x - 3| + C.$$

General Description of the Method

Success in writing a rational function $f(x)/g(x)$ as a sum of partial fractions depends on two things:

• *The degree of $f(x)$ must be less than the degree of $g(x)$.* That is, the fraction must be *proper*. If it isn't, divide $f(x)$ by $g(x)$ and work with the remainder term. See Example 4 of this section.

• *We must know the factors of $g(x)$.* In theory, any polynomial with real coefficients can be written as a product of real linear factors and real quadratic factors. In practice, the factors may be hard to find.

Here is how we find the partial fractions of a proper fraction $f(x)/g(x)$ when the factors of g are known.

Method of Partial Fractions ($f(x)/g(x)$ Proper)

1. Let $x - r$ be a linear factor of $g(x)$. Suppose $(x - r)^m$ is the highest power of $x - r$ that divides $g(x)$. Then, to this factor, assign the sum of the m partial fractions:

$$\frac{A_1}{x - r} + \frac{A_2}{(x - r)^2} + \cdots + \frac{A_m}{(x - r)^m}.$$

Do this for each distinct linear factor of $g(x)$.

2. Let $x^2 + px + q$ be a quadratic factor of $g(x)$. Suppose $(x^2 + px + q)^n$ is the highest power of this factor that divides $g(x)$. Then, to this factor, assign the sum of the n partial fractions:

$$\frac{B_1 x + C_1}{x^2 + px + q} + \frac{B_2 x + C_2}{(x^2 + px + q)^2} + \cdots + \frac{B_n x + C_n}{(x^2 + px + q)^n}.$$

Do this for each distinct quadratic factor of $g(x)$ that cannot be factored into linear factors with real coefficients.

3. Set the original fraction $f(x)/g(x)$ equal to the sum of all these partial fractions. Clear the resulting equation of fractions and arrange the terms in decreasing powers of x.

4. Equate the coefficients of corresponding powers of x and solve the resulting equations for the undetermined coefficients.

Example 2 USING A REPEATED LINEAR FACTOR

Express as a sum of partial fractions: $\dfrac{6x + 7}{(x + 2)^2}$.

Solution According to the description above, we must express the fraction as a sum of partial fractions with undetermined coefficients.

$$\frac{6x + 7}{(x + 2)^2} = \frac{A}{x + 2} + \frac{B}{(x + 2)^2}$$

$$6x + 7 = A(x + 2) + B \qquad \text{Multiply both sides by } (x + 2)^2.$$

$$= Ax + (2A + B) \qquad \text{Combine terms.}$$

Equating coefficients of like terms gives

$$A = 6 \quad \text{and} \quad 2A + B = 12 + B = 7, \quad \text{or}$$

$$A = 6 \quad \text{and} \quad B = -5.$$

Therefore,

$$\frac{6x + 7}{(x + 2)^2} = \frac{6}{x + 2} - \frac{5}{(x + 2)^2}.$$

Example 3 USING PARTIAL FRACTIONS

Evaluate $\displaystyle\int \frac{6x + 7}{(x + 2)^2}\,dx$.

Solution

$$\int \frac{6x + 7}{(x + 2)^2}\,dx = \int \left(\frac{6}{x + 2} - \frac{5}{(x + 2)^2} \right) dx \qquad \text{Example 2}$$

$$= 6 \int \frac{dx}{x + 2} - 5 \int (x + 2)^{-2}\,dx$$

$$= 6 \ln |x + 2| + 5(x + 2)^{-1} + C$$

Example 4 INTEGRATING AN IMPROPER FRACTION

Evaluate $\displaystyle\int \frac{2x^3 - 4x^2 - x - 3}{x^2 - 2x - 3}\,dx$.

Solution First we divide the denominator into the numerator to get a polynomial plus a proper fraction.

$$
\begin{array}{r}
2x \\
x^2 - 2x - 3 \overline{)\,2x^3 - 4x^2 -\ x - 3} \\
\underline{2x^3 - 4x^2 - 6x} \\
5x - 3
\end{array}
$$

Then we write the improper fraction as a polynomial plus a proper fraction.

$$\frac{2x^3 - 4x^2 - x - 3}{x^2 - 2x - 3} = 2x + \frac{5x - 3}{x^2 - 2x - 3}$$

Finally, using $\int 2x \, dx = x^2$ and Example 1 we obtain

$$\int \frac{2x^3 - 4x^2 - x - 3}{x^2 - 2x - 3} \, dx = \int 2x \, dx + \int \frac{5x - 3}{x^2 - 2x - 3} \, dx$$

$$= x^2 + 2 \ln |x + 1| + 3 \ln |x - 3| + C.$$

Example 5 SOLVING AN INITIAL VALUE PROBLEM

Find the solution to $dy/dx = 2xy(y^2 + 1)$ that satisfies $y(0) = 1$.

Solution Separating the variables, we rewrite the differential equation as

$$\frac{1}{y(y^2 + 1)} \, dy = 2x \, dx.$$

Integrating both sides gives

$$\int \frac{1}{y(y^2 + 1)} \, dy = \int 2x \, dx = x^2 + C_1.$$

We use partial fractions to rewrite the integrand on the left.

$$\frac{1}{y(y^2 + 1)} = \frac{A}{y} + \frac{By + C}{y^2 + 1}$$

$$1 = A(y^2 + 1) + (By + C)y \quad \text{Multiply by } y(y^2 + 1).$$

$$= (A + B)y^2 + Cy + A$$

Equating coefficients of like terms gives $A + B = 0$, $C = 0$, and $A = 1$. Solving these equations simultaneously, we find $A = 1$, $B = -1$, and $C = 0$.

Accordingly,

$$\int \frac{1}{y(y^2 + 1)} \, dy = \int \frac{1}{y} \, dy - \int \frac{y}{y^2 + 1} \, dy$$

$$= \ln |y| - \frac{1}{2} \ln (y^2 + 1) + C_2.$$

The solution to the differential equation is

$$\ln |y| - \frac{1}{2} \ln (y^2 + 1) = x^2 + C. \quad C = C_1 + C_2$$

Substituting $x = 0$, $y = 1$, we find

$$0 - \frac{1}{2} \ln 2 = C, \quad \text{or} \quad C = -\ln \sqrt{2}.$$

The solution to the initial value problem is

$$\ln |y| - \frac{1}{2} \ln (y^2 + 1) = x^2 - \ln \sqrt{2}.$$

Integral Tables

Every partial fraction that occurs in the decomposition of a rational function can be integrated using analytic methods. This means that we can find the indefinite integral of any rational function analytically. However, rarely do you need to use these techniques. You can usually use a CAS to evaluate the integral, or you can look up the integral in a table like the Brief Table of Integrals in Appendix A7. Example 6 uses this table to evaluate an integral.

Example 6   USING A TABLE OF INTEGRALS

Evaluate $\displaystyle \int \left(\frac{1}{x^2 + 1} + \frac{1}{x^2 - 2x + 5} \right) dx.$

Solution

$$\int \left(\frac{1}{x^2 + 1} + \frac{1}{x^2 - 2x + 5} \right) dx = \int \frac{1}{x^2 + 1}\, dx + \int \frac{1}{x^2 - 2x + 5}\, dx \quad (1)$$

Both integrals on the right-hand side of Equation 1 can be evaluated using Formula 16 from the Brief Table of Integrals.

$$\textbf{16.} \int \frac{dx}{a^2 + x^2} = \frac{1}{a} \tan^{-1} \frac{x}{a} + C$$

You probably remember

$$\int \frac{dx}{1 + x^2} = \tan^{-1} x + C,$$

but if you don't, you can obtain the result by setting $a = 1$ in Formula 16.

To find the value of the second integral, we need to complete the square:

$$x^2 - 2x + 5 = x^2 - 2x + 1 + 4 = (x - 1)^2 + 4.$$

Then,

$$\int \frac{dx}{x^2 - 2x + 5} = \int \frac{dx}{(x - 1)^2 + 4}$$

$$= \int \frac{du}{u^2 + 4} \qquad u = x - 1,\ du = dx$$

$$= \frac{1}{2} \tan^{-1} \frac{u}{2} + C \qquad \text{Formula 16 with } a = 2$$

$$= \frac{1}{2} \tan^{-1} \frac{x - 1}{2} + C.$$

Combining the two integrations gives

$$\int \left(\frac{1}{x^2 + 1} + \frac{1}{x^2 - 2x + 5} \right) dx = \tan^{-1} x + \frac{1}{2} \tan^{-1} \frac{x - 1}{2} + C.$$

The manipulation and substitution in Example 6 are typical of what is needed to use integral tables to evaluate integrals.

Trigonometric Substitutions

Trigonometric substitutions enable us to replace binomials of the form $a^2 + x^2$, $a^2 - x^2$, and $x^2 - a^2$ by single squared terms, and thereby transform a number of integrals into ones we can evaluate directly or find in a table of integrals.

Three of the most common substitutions are $x = a \tan \theta$, $x = a \sin \theta$, and $x = a \sec \theta$. They come from the reference right triangles in Figure 8.18.

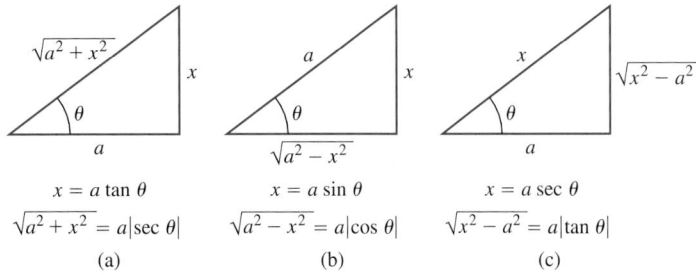

Figure 8.18 Reference triangles for trigonometric substitutions that change binomials into single squared terms.

With $x = a \tan \theta$,

$$a^2 + x^2 = a^2 + a^2 \tan^2 \theta = a^2(1 + \tan^2 \theta) = a^2 \sec^2 \theta.$$

With $x = a \sin \theta$,

$$a^2 - x^2 = a^2 - a^2 \sin^2 \theta = a^2(1 - \sin^2 \theta) = a^2 \cos^2 \theta.$$

With $x = a \sec \theta$,

$$x^2 - a^2 = a^2 \sec^2 \theta - a^2 = a^2(\sec^2 \theta - 1) = a^2 \tan^2 \theta.$$

Trigonometric Substitutions

1. $x = a \tan \theta$ replaces $a^2 + x^2$ with $a^2 \sec^2 \theta$

2. $x = a \sin \theta$ replaces $a^2 - x^2$ with $a^2 \cos^2 \theta$

3. $x = a \sec \theta$ replaces $x^2 - a^2$ with $a^2 \tan^2 \theta$

We want any substitution we use in an integration to be reversible so we can change back to the original variable afterward. For example, if $x = a \tan \theta$, we want to be able to set $\theta = \tan^{-1}(x/a)$ after the integration takes place. If $x = a \sin \theta$, we want to be able to set $\theta = \sin^{-1}(x/a)$ when we're done, and similarly for $x = a \sec \theta$.

For reversibility,

$$x = a \tan \theta \quad \text{requires} \quad \theta = \tan^{-1}\left(\frac{x}{a}\right) \quad \text{with} \quad -\frac{\pi}{2} < \theta < \frac{\pi}{2},$$

$$x = a \sin \theta \quad \text{requires} \quad \theta = \sin^{-1}\left(\frac{x}{a}\right) \quad \text{with} \quad -\frac{\pi}{2} \le \theta \le \frac{\pi}{2},$$

$$x = a \sec \theta \quad \text{requires} \quad \theta = \sec^{-1}\left(\frac{x}{a}\right) \quad \text{with} \quad \begin{cases} 0 \le \theta < \dfrac{\pi}{2}, & \dfrac{x}{a} \ge 1 \\[2mm] \dfrac{\pi}{2} < \theta \le \pi, & \dfrac{x}{a} \le -1. \end{cases}$$

Example 7 USING THE SUBSTITUTION $x = a \tan \theta$

Evaluate $\int \sqrt{4 + x^2}\, dx$.

Solution We set

$$x = 2 \tan \theta, \qquad dx = 2 \sec^2 \theta\, d\theta, \qquad -\frac{\pi}{2} < \theta < \frac{\pi}{2},$$

$$4 + x^2 = 4 + 4 \tan^2 \theta = 4(1 + \tan^2 \theta) = 4 \sec^2 \theta.$$

Then

$$\int \sqrt{4 + x^2}\, dx = \int \sqrt{4 \sec^2 \theta}\, (2 \sec^2 \theta)\, d\theta$$

$$= 4 \int |\sec \theta| \sec^2 \theta\, d\theta \qquad \sqrt{\sec^2 \theta} = |\sec \theta|$$

$$= 4 \int \sec^3 \theta\, d\theta. \qquad \sec \theta > 0 \text{ for } -\frac{\pi}{2} < \theta < \frac{\pi}{2}$$

Now we are in a position to use Formula 92 from the table of integrals, a *reduction formula*.

92. $\displaystyle\int \sec^n ax\, dx = \frac{\sec^{n-2} ax \tan ax}{a(n-1)} + \frac{n-2}{n-1} \int \sec^{n-2} ax\, dx, \quad n \neq 1$

Reduction Formulas

Formula 92 and the formulas in Section 6.3, Exercises 31–34, are called **reduction formulas** because they replace an integral containing some power of a function with an integral of the same form with the power reduced. By applying such a formula repeatedly, we can eventually express the original integral in terms of a power low enough to be evaluated directly.

Applying Formula 92 with $n = 3$, we obtain

$$\int \sec^3 \theta\, d\theta = \frac{\sec \theta \tan \theta}{2} + \frac{1}{2} \int \sec \theta\, d\theta \qquad n = 3, a = 1, x = \theta$$

$$= \frac{\sec \theta \tan \theta}{2} + \frac{1}{2} \ln |\sec \theta + \tan \theta| + C \qquad \text{Formula 88, } a = 1$$

$$= \frac{\sec \theta \tan \theta}{2} + \frac{1}{2} \ln (\sec \theta + \tan \theta) + C.$$

The last step is valid because

$$\sec \theta + \tan \theta = \frac{1 + \sin \theta}{\cos \theta} > 0 \qquad \text{for} \qquad -\frac{\pi}{2} < \theta < \frac{\pi}{2}.$$

Thus,

$$\int \sqrt{4 + x^2}\, dx = 4 \int \sec^3 \theta\, d\theta$$

$$= 2 \sec \theta \tan \theta + 2 \ln (\sec \theta + \tan \theta) + C$$

$$= 2 \cdot \frac{\sqrt{4 + x^2}}{2} \cdot \frac{x}{2} + 2 \ln \left(\frac{\sqrt{4 + x^2}}{2} + \frac{x}{2} \right) + C \qquad \text{Figure 8.18a, } a = 2$$

$$= \frac{x\sqrt{4 + x^2}}{2} + 2 \ln (\sqrt{4 + x^2} + x) + C.$$

Example 8   USING THE SUBSTITUTION $x = a \sin \theta$

Evaluate $\displaystyle\int \frac{x^3 \, dx}{\sqrt{9 - x^2}}$.

Solution We set

$$x = 3 \sin \theta, \quad dx = 3 \cos \theta \, d\theta, \quad -\frac{\pi}{2} \le \theta \le \frac{\pi}{2}$$

$$9 - x^2 = 9 - 9 \sin^2 \theta = 9(1 - \sin^2 \theta) = 9 \cos^2 \theta.$$

Then

$$\int \frac{x^3 \, dx}{\sqrt{9 - x^2}} = \int \frac{27 \sin^3 \theta \cdot 3 \cos \theta \, d\theta}{|3 \cos \theta|}$$

$$= 27 \int \sin^3 \theta \, d\theta \qquad \cos \theta > 0 \text{ for } -\frac{\pi}{2} < \theta < \frac{\pi}{2}$$

$$= 27 \int (1 - \cos^2 \theta) \sin \theta \, d\theta \qquad \sin^2 \theta = 1 - \cos^2 \theta$$

$$= -27 \cos \theta + 9 \cos^3 \theta + C$$

$$= -27 \cdot \frac{\sqrt{9 - x^2}}{3} + 9 \left(\frac{\sqrt{9 - x^2}}{3} \right)^3 + C \qquad \text{Figure 8.18b, } a = 3$$

$$= -9 \sqrt{9 - x^2} + \frac{(9 - x^2)^{3/2}}{3} + C.$$

Quick Review 8.4

In Exercises 1 and 2, find the simultaneous solution to the equations.

1. $\begin{aligned} 3A + B &= -5 \\ -2A + 3B &= 7 \end{aligned}$

2. $\begin{aligned} A + 2B - C &= 0 \\ 3A - B + 2C &= 1 \\ A + B + C &= 4 \end{aligned}$

In Exercises 3 and 4, use long division to express the improper fraction (degree of numerator $\ge$ degree of denominator) as a polynomial plus a proper fraction (degree of numerator $<$ degree of denominator).

3. $\dfrac{2x^3 - 5x^2 - 10x - 7}{x^2 - 3x - 4}$

4. $\dfrac{2x^2 + 11x + 6}{x^2 + 4x + 5}$

In Exercises 5 and 6, factor the polynomials as completely as possible.

5. $x^3 - 3x^2 + x - 3$

6. $y^4 - 5y^2 + 4$

In Exercises 7–10, use the least common denominator to combine the fractions.

7. $\dfrac{2}{x + 3} - \dfrac{3}{x - 2}$

8. $\dfrac{x - 1}{x^2 - 4x + 5} - \dfrac{2}{x + 5}$

9. $\dfrac{t - 1}{t^2 + 2} - \dfrac{3t + 4}{t^2 + 1}$

10. $\dfrac{2}{x - 1} - \dfrac{3}{(x - 1)^2} + \dfrac{1}{(x - 1)^3}$

Section 8.4 Exercises

In Exercises 1–6, express the rational function as a sum of partial fractions.

1. $\dfrac{5x - 7}{x^2 - 3x + 2}$

2. $\dfrac{2x + 2}{x^2 - 2x + 1}$

3. $\dfrac{t + 1}{t^2(t - 1)}$

4. $\dfrac{4}{s^3 - s^2 - 6s}$

5. $\dfrac{x^2 + 8}{x^2 - 5x + 6}$

6. $\dfrac{y^3 + 1}{y^2 + 4}$

In Exercises 7–26, use partial fractions to evaluate the integral.

7. $\displaystyle\int \dfrac{dx}{1 - x^2}$

8. $\displaystyle\int \dfrac{dx}{x^2 + 2x}$

9. $\displaystyle\int \dfrac{y \, dy}{y^2 - 2y - 3}$

10. $\displaystyle\int \dfrac{y + 4}{y^2 + y} \, dy$

11. $\displaystyle\int \dfrac{dt}{t^3 + t^2 - 2t}$

12. $\displaystyle\int \dfrac{t + 3}{2t^3 - 8t} \, dt$

13. $\displaystyle\int \dfrac{s^3 \, ds}{s^2 + 4}$

14. $\displaystyle\int \dfrac{s^4 + 2s}{s^2 + 1} \, ds$

15. $\displaystyle\int \dfrac{5x^2 \, dx}{x^2 + x + 1}$

16. $\displaystyle\int \dfrac{x^2 \, dx}{(x - 1)(x^2 + 2x + 1)}$

17. $\displaystyle\int \dfrac{dx}{(x^2 - 1)^2}$

18. $\displaystyle\int \dfrac{x + 4}{x^2 + 5x - 6} \, dx$

19. $\displaystyle\int \dfrac{2 \, dr}{r^2 - 2r + 2}$

20. $\displaystyle\int \dfrac{3 \, dr}{r^2 - 4r + 5}$

21. $\displaystyle\int \dfrac{x^2 - 2x - 2}{x^3 - 1} \, dx$

22. $\displaystyle\int \dfrac{x^2 - 4x + 4}{x^3 + 1} \, dx$

23. $\displaystyle\int \dfrac{3x^2 - 2x + 12}{(x^2 + 4)^2} \, dx$

24. $\displaystyle\int \dfrac{x^3 + 2x^2 + 2}{(x^2 + 1)^2} \, dx$

25. $\displaystyle\int_0^1 \dfrac{\theta}{\theta + 1} \, d\theta$

26. $\displaystyle\int_0^2 \dfrac{\theta^2}{\theta^2 + 1} \, d\theta$

In Exercises 27–30, solve the initial value problem.

27. $\dfrac{dy}{dx} = e^x (y^2 - y), \quad y(0) = 2$

28. $\dfrac{dy}{d\theta} = (y + 1)^2 \sin \theta, \quad y(\pi/2) = 0$

29. $\dfrac{dy}{dx} = \dfrac{1}{x^2 - 3x + 2}, \quad y(3) = 0$

30. $\dfrac{ds}{dt} = \dfrac{2s + 2}{t^2 + 2t}, \quad s(1) = 1$

In Exercises 31 and 32, *work in groups of two or three.*

31. (a) Writing to Learn Use Formula 18,
$$\int \frac{dx}{a^2 - x^2} = \frac{1}{2a} \ln \left| \frac{x + a}{x - a} \right| + C,$$
from the Brief Table of Integrals to evaluate
$$\int \frac{dx}{5 + 4x - x^2},$$
explaining each step.

(b) Use differentiation to confirm Formula 18.

32. (a) Writing to Learn Use Formula 17,
$$\int \frac{dx}{(a^2 + x^2)^2} = \frac{x}{2a^2(a^2 + x^2)} + \frac{1}{2a^3} \tan^{-1} \frac{x}{a} + C,$$
from the Brief Table of Integrals to evaluate
$$\int \frac{dx}{(x^2 - 2x + 2)^2},$$
explaining each step.

(b) Use differentiation to confirm Formula 17.

In Exercises 33 and 34, find the volume of the solid generated by revolving the shaded region about the indicated axis.

33. the *x*-axis

34. the *y*-axis

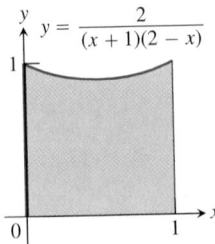

In Exercises 35–42, use trigonometric substitution and the table of integrals (as necessary) to evaluate the integral.

35. $\displaystyle\int \frac{dy}{\sqrt{9 + y^2}}$

36. $\displaystyle\int \sqrt{25 - t^2} \, dt$

37. $\displaystyle\int \frac{dx}{\sqrt{4x^2 - 49}}, \quad x > \frac{7}{2}$ **38.** $\displaystyle\int \frac{dx}{x^2\sqrt{x^2 + 1}}$

39. $\displaystyle\int \frac{x^3\,dx}{\sqrt{1 - x^2}}$ **40.** $\displaystyle\int \frac{2\,dx}{x^3\sqrt{x^2 - 1}}, \quad x > 1$

41. $\displaystyle\int \frac{\sqrt{16 - z^2}}{z}\,dz, \quad 0 < z < 4$ **42.** $\displaystyle\int \frac{8\,dw}{w^2\sqrt{4 - w^2}}$

In Exercises 43 and 44, solve the initial value problem.

43. $\sqrt{x^2 - 9}\,\dfrac{dy}{dx} = 1, \quad x > 3, \quad y(5) = \ln 3$

44. $(x^2 + 1)^2\,\dfrac{dy}{dx} = \sqrt{x^2 + 1}, \quad y(0) = 1$

45. Find the area of the region in the first quadrant that is enclosed by the coordinate axes and the curve $y = \sqrt{9 - x^2}/3$.

46. Find the volume of the solid generated by revolving about the x-axis the region in the first quadrant enclosed by the coordinate axes, the curve $y = 2/(1 + x^2)$, and the line $x = 1$.

Exploration

47. *Social Diffusion* Sociologists sometimes use the phrase "social diffusion" to describe the way information spreads through a population. The information might be a rumor, a cultural fad, or news about a technical innovation. In a sufficiently large population, the number of people x who have the information is treated as a differentiable function of time t, and the rate of diffusion, dx/dt, is assumed to be proportional to the number of people who have the information times the number of people who do not. This leads to the differential equation

$$\frac{dx}{dt} = kx(N - x),$$

where N is the size of the population.

Suppose t is time in days, $k = 1/250$, and two people start a rumor at time $t = 0$ in a population of $N = 1000$ people.

(a) Find x as a function of t.

(b) When will half the population have heard the rumor?

(c) Show that the rumor is spreading the fastest at the time found in (b).   ■

48. *Arc Length* Find the length of the curve $y = \ln (1 - x^2)$ from $x = 0$ to $x = 1/2$.

Extending the Ideas

49. The substitution

$$z = \tan \frac{x}{2}$$

(see figure) reduces the problem of integrating any rational function of $\sin x$ and $\cos x$ to a problem involving a rational function of z. This in turn can be solved by partial fractions.

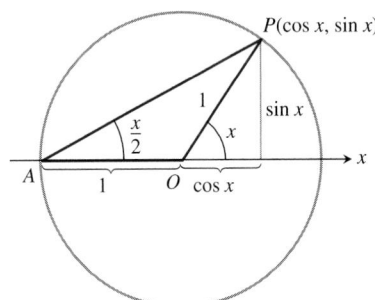

Show that each equation is true.

(a) $\tan \dfrac{x}{2} = \dfrac{\sin x}{1 + \cos x}$ **(b)** $\cos x = \dfrac{1 - z^2}{1 + z^2}$

(c) $\sin x = \dfrac{2z}{1 + z^2}$ **(d)** $dx = \dfrac{2\,dz}{1 + z^2}$

In Exercises 50–53, use the substitution $z = \tan (x/2)$ and the results in Exercise 49 to evaluate the integral.

50. $\displaystyle\int \frac{dx}{1 + \sin x}$ **51.** $\displaystyle\int \frac{dx}{1 - \cos x}$

52. $\displaystyle\int \frac{d\theta}{1 - \sin \theta}$ **53.** $\displaystyle\int \frac{dt}{1 + \sin t + \cos t}$

Chapter 8   Key Terms

at most the order (p. 429)
big-oh (p. 429)
binary search (p. 430)
convergence of improper integral (p. 434)
direct comparison test (p. 438)
divergence of improper integral (p. 434)
grows at the same rate (p. 426)
grows faster (p. 426)
grows slower (p. 426)

improper integral (pp. 434, 436)
indeterminate form (p. 417)
l'Hôpital's Rule, first form (p. 417)
l'Hôpital's Rule, stronger form (p. 418)
limit comparison test (p. 439)
little-oh (p. 429)
method of partial fractions (p. 444)
monotonic function (p. 438)
partial fractions (p. 444)

reduction formula (p. 450)
sequential search (p. 430)
smaller order (p. 429)
transitivity of growing rates (p. 428)
trigonometric substitution (p. 449)
undetermined coefficients (p. 444)
value of improper integral (p. 434)

Chapter 8   Review Exercises

In Exercises 1–14, find the limit.

1. $\lim\limits_{t \to 0} \dfrac{t - \ln(1 + 2t)}{t^2}$

2. $\lim\limits_{t \to 0} \dfrac{\tan 3t}{\tan 5t}$

3. $\lim\limits_{x \to 0} \dfrac{x \sin x}{1 - \cos x}$

4. $\lim\limits_{x \to 1} x^{1/(1-x)}$

5. $\lim\limits_{x \to \infty} x^{1/x}$

6. $\lim\limits_{x \to \infty} \left(1 + \dfrac{3}{x}\right)^x$

7. $\lim\limits_{r \to \infty} \dfrac{\cos r}{\ln r}$

8. $\lim\limits_{\theta \to \pi/2} \left(\theta - \dfrac{\pi}{2}\right)\sec \theta$

9. $\lim\limits_{x \to 1} \left(\dfrac{1}{x - 1} - \dfrac{1}{\ln x}\right)$

10. $\lim\limits_{x \to 0^+} \left(1 + \dfrac{1}{x}\right)^x$

11. $\lim\limits_{\theta \to 0^+} (\tan \theta)^\theta$

12. $\lim\limits_{\theta \to \infty} \theta^2 \sin\left(\dfrac{1}{\theta}\right)$

13. $\lim\limits_{x \to \infty} \dfrac{x^3 - 3x^2 + 1}{2x^2 + x - 3}$

14. $\lim\limits_{x \to \infty} \dfrac{3x^2 - x + 1}{x^4 - x^3 + 2}$

In Exercises 15–26, determine whether f grows faster than, slower than, or at the same rate as g as $x \to \infty$. Give reasons for your answer.

15. $f(x) = x, \quad g(x) = 5x$

16. $f(x) = \log_2 x, \quad g(x) = \log_3 x$

17. $f(x) = x, \quad g(x) = x + \dfrac{1}{x}$

18. $f(x) = \dfrac{x}{100}, \quad g(x) = xe^{-x}$

19. $f(x) = x, \quad g(x) = \tan^{-1} x$

20. $f(x) = \csc^{-1} x, \quad g(x) = \dfrac{1}{x}$

21. $f(x) = x^{\ln x}, \quad g(x) = x^{\log_2 x}$

22. $f(x) = 3^{-x}, \quad g(x) = 2^{-x}$

23. $f(x) = \ln 2x, \quad g(x) = \ln x^2$

24. $f(x) = 10x^3 + 2x^2, \quad g(x) = e^x$

25. $f(x) = \tan^{-1}\dfrac{1}{x}, \quad g(x) = \dfrac{1}{x}$

26. $f(x) = \sin^{-1}\dfrac{1}{x}, \quad g(x) = \dfrac{1}{x^2}$

In Exercises 27 and 28,

  (a) show that f has a removable discontinuity at $x = 0$.

  (b) define f at $x = 0$ so that it is continuous there.

27. $f(x) = \dfrac{2^{\sin x} - 1}{e^x - 1}$

28. $f(x) = x \ln x$

In Exercises 29–38, determine whether the statement is true or false. Give reasons for your answer.

29. $\dfrac{1}{x^2} + \dfrac{1}{x^4} = O\left(\dfrac{1}{x^2}\right)$

30. $\dfrac{1}{x^2} + \dfrac{1}{x^4} = O\left(\dfrac{1}{x^4}\right)$

31. $x = o(x + \ln x)$

32. $\ln(\ln x) = o(\ln x)$

33. $\tan^{-1} x = O(1)$

34. $\dfrac{1}{x^4} = O\left(\dfrac{1}{x^2} + \dfrac{1}{x^4}\right)$

35. $\dfrac{1}{x^4} = o\left(\dfrac{1}{x^2} + \dfrac{1}{x^4}\right)$

36. $\ln x = o(x + 1)$

37. $\ln 2x = O(\ln x)$

38. $\sec^{-1} x = O(1)$

In Exercises 39–48, evaluate the improper integral or state that it diverges.

39. $\displaystyle\int_0^3 \dfrac{dx}{\sqrt{9 - x^2}}$

40. $\displaystyle\int_0^1 \ln x \, dx$

41. $\displaystyle\int_{-1}^1 \dfrac{dy}{y^{2/3}}$

42. $\displaystyle\int_{-2}^0 \dfrac{d\theta}{(\theta + 1)^{3/5}}$

43. $\displaystyle\int_3^\infty \dfrac{2\,dx}{x^2 - 2x}$

44. $\displaystyle\int_1^\infty \dfrac{3x - 1}{4x^3 - x^2} \, dx$

45. $\displaystyle\int_0^\infty x^2 e^{-x} \, dx$

46. $\displaystyle\int_{-\infty}^0 x\, e^{3x} \, dx$

47. $\displaystyle\int_{-\infty}^\infty \dfrac{4t^3 + t - 1}{t^2(t - 1)(t^2 + 1)} \, dt$

48. $\displaystyle\int_{-\infty}^\infty \dfrac{4\,dx}{x^2 + 16}$

In Exercises 49–54, determine whether the improper integral converges or diverges. Give reasons for your answer.

49. $\displaystyle\int_0^\infty \dfrac{d\theta}{\sqrt{\theta^2 + 1}}$

50. $\displaystyle\int_0^\infty e^{-u} \cos u \, du$

51. $\displaystyle\int_1^\infty \dfrac{\ln z}{z} \, dz$

52. $\displaystyle\int_1^\infty \dfrac{e^{-t}}{\sqrt{t}} \, dt$

53. $\displaystyle\int_{-\infty}^\infty \dfrac{dx}{e^x + e^{-x}}$

54. $\displaystyle\int_{-\infty}^\infty \dfrac{dx}{x^2(1 + e^x)}$

In Exercises 55–60, evaluate the integral.

55. $\displaystyle\int \dfrac{2x + 1}{x^2 - 7x + 12} \, dx$

56. $\displaystyle\int \dfrac{8\,dx}{x^3(x + 2)}$

57. $\displaystyle\int \dfrac{3t^2 + 4t + 4}{t^3 + t} \, dt$

58. $\displaystyle\int \dfrac{dt}{t^4 + 4t^2 + 3}$

59. $\displaystyle\int \dfrac{x^3 + 1}{x^3 - x} \, dx$

60. $\displaystyle\int \dfrac{x^3 + 4x^2}{x^2 + 4x + 3} \, dx$

In Exercises 61 and 62, solve the initial value problem.

61. $\dfrac{dy}{dx} = 0.002y(500 - y), \quad y(0) = 20$

62. $\dfrac{dy}{dx} = \dfrac{y^2 + 1}{x + 1}, \quad y(0) = \pi/4$

In Exercises 63–66, use trigonometric substitution and the table of integrals (if necessary) to evaluate the integral.

63. $\displaystyle\int \dfrac{3\,dy}{\sqrt{1 + 9y^2}}$

64. $\displaystyle\int \sqrt{1 - 9t^2}\,dt$

65. $\displaystyle\int \dfrac{5\,dx}{\sqrt{25x^2 - 9}}, \quad x > \dfrac{3}{5}$

66. $\displaystyle\int \dfrac{4x^2\,dx}{(1 - x^2)^{3/2}}$

67. *Infinite Solid* The infinite region bounded by the coordinate axes and the curve $y = -\ln x$ in the first quadrant (see figure) is revolved about the x-axis to generate a solid. Find the volume of the solid.

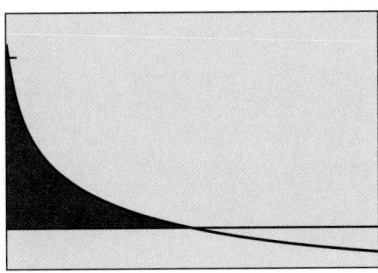

[0, 2] by [–1, 5]

68. *Infinite Region* Find the area of the region in the first quadrant under the curve $y = xe^{-x}$ (see figure).

[0, 5] by [–0.5, 1]

69. *Second Order Chemical Reactions* Many chemical reactions are the result of the interaction of two molecules that undergo a change to produce a new product. The rate of the reaction typically depends on the concentrations of the two kinds of molecules. If a is the amount of substance A and b is the amount of substance B at time $t = 0$, and if x is the amount of product at time t, then the rate of formation of the product may be given by the separable differential equation

$$\dfrac{dx}{dt} = k(a - x)(b - x),$$

where k is a constant for the reaction.

Solve this equation to obtain a relation between x and t **(a)** if $a = b$, and **(b)** if $a \neq b$. Assume in each case that $x = 0$ when $t = 0$.

CHAPTER

9
Infinite Series

One mathematical constant crucial to the analysis of the world is π. The p-series

$$\frac{\pi^2}{6} = 1 + \frac{1}{2^2} + \frac{1}{3^2} + \frac{1}{4^2} + \frac{1}{5^2} + \cdots$$

approximates the value of π. The error, or remainder, of such an approximation is the difference between the actual sum and the nth partial sum. For this p-series, the remainder is estimated by $R_n \leq 1/n$.

Shown here is a close-up of a high speed microprocessor chip. If a computer adds 1,000,000 terms of the p-series in one second, how many places of accuracy will it achieve in 24 hours? Section 9.5 provides a discussion of p-series.

Chapter 9 Overview

One consequence of the early and dramatic successes that scientists enjoyed when using calculus to explain natural phenomena was that there suddenly seemed to be no limits, so to speak, on how infinite processes might be exploited. There was still considerable mystery about "infinite sums" and "division by infinitely small quantities" in the years after Newton and Leibniz, but even mathematicians normally insistent on rigorous proof were inclined to throw caution to the wind while things were working. The result was a century of unprecedented progress in understanding the physical universe. (Moreover, we can note happily in retrospect, the proofs eventually followed.)

One infinite process that had puzzled mathematicians for centuries was the summing of infinite series. Sometimes an infinite series of terms added to a number, as in

$$\frac{1}{2} + \frac{1}{4} + \frac{1}{8} + \frac{1}{16} + \cdots = 1.$$

(You can see this by adding the areas in the "infinitely halved" unit square at the right.) But sometimes the infinite sum was infinite, as in

$$\frac{1}{1} + \frac{1}{2} + \frac{1}{3} + \frac{1}{4} + \frac{1}{5} + \cdots = \infty$$

(although this is far from obvious), and sometimes the infinite sum was impossible to pin down, as in

$$1 - 1 + 1 - 1 + 1 - 1 + \cdots$$

(Is it 0? Is it 1? Is it neither?).

Nonetheless, mathematicians like Gauss and Euler successfully used infinite series to derive previously inaccessible results. Laplace used infinite series to prove the stability of the solar system (although that does not stop some people from worrying about it today when they feel that "too many" planets have swung to the same side of the sun). It was years later that careful analysts like Cauchy developed the theoretical foundation for series computations, sending many mathematicians (including Laplace) back to their desks to verify their results.

Our approach in this chapter will be to discover the calculus of infinite series as the pioneers of calculus did: proceeding intuitively, accepting what works and rejecting what does not. Toward the end of the chapter we will return to the crucial question of convergence and take a careful look at it.

9.1 Power Series

Geometric Series • Representing Functions by Series • Differentiation and Integration • Identifying a Series

Geometric Series

The first thing to get straight about an infinite series is that it is not simply an example of addition. Addition of real numbers is a *binary* operation, meaning that we really add numbers two at a time. The only reason that $1 + 2 + 3$

makes sense as "addition" is that we can *group* the numbers and then add them two at a time. The associative property of addition guarantees that we get the same sum no matter how we group them:

$$1 + (2 + 3) = 1 + 5 = 6 \quad \text{and} \quad (1 + 2) + 3 = 3 + 3 = 6.$$

In short, a *finite sum* of real numbers always produces a real number (the result of a finite number of binary additions), but an *infinite sum* of real numbers is something else entirely. That is why we need the following definition.

Definition Infinite Series

An **infinite series** is an expression of the form

$$a_1 + a_2 + a_3 + \cdots + a_n + \cdots, \quad \text{or} \quad \sum_{k=1}^{\infty} a_k.$$

The numbers a_1, a_2, ... are the **terms** of the series; a_n is the **nth term.**

The **partial sums** of the series form a sequence

$$s_1 = a_1$$

$$s_2 = a_1 + a_2$$

$$s_3 = a_1 + a_2 + a_3$$

$$\vdots$$

$$s_n = \sum_{k=1}^{n} a_k$$

$$\vdots$$

of real numbers, each defined as a finite sum. If the sequence of partial sums has a limit S as $n \to \infty$, we say the series **converges** to the sum S, and we write

$$a_1 + a_2 + a_3 + \cdots + a_n + \cdots = \sum_{k=1}^{\infty} a_k = S.$$

Otherwise, we say the series **diverges.**

Example 1 IDENTIFYING A DIVERGENT SERIES

Does the series $1 - 1 + 1 - 1 + 1 - 1 + \cdots$ converge?

Solution You might be tempted to pair the terms as

$$(1 - 1) + (1 - 1) + (1 - 1) + \cdots.$$

That strategy, however, requires an *infinite* number of pairings, so it cannot be justified by the associative property of addition. This is an infinite series, not a finite sum, so if it has a sum it *has to be* the limit of its sequence of partial sums,

$$1, 0, 1, 0, 1, 0, 1, \ldots.$$

Since this sequence has no limit, the series has no sum. It diverges.

Example 2 IDENTIFYING A CONVERGENT SERIES

Does the series

$$\frac{3}{10} + \frac{3}{100} + \frac{3}{1000} + \cdots + \frac{3}{10^n} + \cdots$$

converge?

Solution Here is the sequence of partial sums, written in decimal form.

$$0.3, \; 0.33, \; 0.333, \; 0.3333, \; \ldots$$

This sequence has a limit $0.\overline{3}$, which we recognize as the fraction $1/3$. The series converges to the sum $1/3$.

The series in Example 2 is a **geometric series** because each term is obtained from its preceding term by multiplying by the same number r—in this case, $r = 1/10$. (The series of areas for the infinitely-halved square at the beginning of this chapter is also geometric.) The convergence of geometric series is one of the few infinite processes with which mathematicians were reasonably comfortable prior to calculus. You may have already seen the following result in a previous course.

The **geometric series**

$$a + ar + ar^2 + ar^3 + \cdots + ar^{n-1} + \cdots = \sum_{n=1}^{\infty} ar^{n-1}$$

converges to the sum $a/(1 - r)$ if $|r| < 1$, and diverges if $|r| \geq 1$.

This completely settles the issue for geometric series. We know which ones converge and which ones diverge, and for the convergent ones we know what the sums must be. The interval $-1 < r < 1$ is the **interval of convergence.**

Example 3 ANALYZING GEOMETRIC SERIES

Tell whether each series converges or diverges. If it converges, give its sum.

(a) $\displaystyle\sum_{n=1}^{\infty} 3\left(\frac{1}{2}\right)^{n-1}$

(b) $1 - \dfrac{1}{2} + \dfrac{1}{4} - \dfrac{1}{8} + \cdots + \left(-\dfrac{1}{2}\right)^{n-1} + \cdots$

(c) $\displaystyle\sum_{k=0}^{\infty} \left(\frac{3}{5}\right)^k$

(d) $\dfrac{\pi}{2} + \dfrac{\pi^2}{4} + \dfrac{\pi^3}{8} + \cdots$

Solution

(a) First term is $a = 3$ and $r = 1/2$. The series converges to

$$\frac{3}{1 - (1/2)} = 6.$$

(b) First term is $a = 1$ and $r = -1/2$. The series converges to

$$\frac{1}{1 - (-1/2)} = \frac{2}{3}.$$

(c) First term is $a = (3/5)^0 = 1$ and $r = 3/5$. The series converges to

$$\frac{1}{1 - (3/5)} = \frac{5}{2}.$$

(d) In this series, $r = \pi/2 > 1$. The series diverges.

We have hardly begun our study of infinite series, but knowing everything there is to know about the convergence and divergence of an *entire class* of series (geometric) is an impressive start. Like the Renaissance mathematicians, we are ready to explore where this might lead. We are ready to bring in x.

Representing Functions by Series

If $|x| < 1$, then the geometric series formula assures us that

$$1 + x + x^2 + x^3 + \cdots + x^n + \cdots = \frac{1}{1 - x}.$$

Consider this statement for a moment. The expression on the right defines a function whose domain is the set of all numbers $x \neq 1$. The expression on the left defines a function whose domain is the interval of convergence, $|x| < 1$. The equality is understood to hold only on this latter domain, where both sides of the equation are defined. On this domain, the series *represents* the function $1/(1 - x)$.

The partial sums of the infinite series on the left are all polynomials, so we can graph them (Figure 9.1). As expected, we see that the convergence is strong in the interval $(-1, 1)$ but breaks down when $|x| \geq 1$.

Partial Sums

$y = 1/(1-x)$

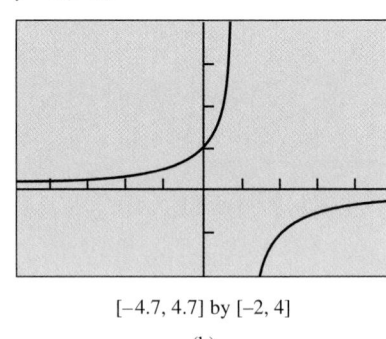

[−4.7, 4.7] by [−2, 4]　　　　[−4.7, 4.7] by [−2, 4]

(a)　　　　　　　　　(b)

Figure 9.1 (a) Partial sums converging to $1/(1 - x)$ on the interval $(0, 1)$. The partial sums graphed here are $1 + x + x^2$, $1 + x + x^2 + x^3$, and $1 + x + x^2 + x^3 + x^4$. Notice how they resemble the graph of (b) $1/(1 - x)$ on the interval $(-1, 1)$ but are not even close when $|x| \geq 1$.

The expression $\sum_{n=0}^{\infty} x^n$ is like a polynomial in that it is a sum of coefficients times powers of x, but polynomials have *finite* degrees and do not suffer from divergence for the wrong values of x. Just as an infinite series of numbers is not a mere sum, this series of powers of x is not a mere polynomial.

When we set $x = 0$ in the expression

$$\sum_{n=0}^{\infty} c_n x^n = c_0 + c_1 x + c_2 x^2 + \cdots + c_n x^n + \cdots,$$

we get c_0 on the right but $c_0 \cdot 0^0$ on the left. Since 0^0 is not a number, this is a slight flaw in the notation, which we agree to overlook. The same situation arises when we set

$$x = a \quad \text{in} \quad \sum_{n=0}^{\infty} c_n (x - a)^n.$$

In either case, we agree that the expression will equal c_0. (It really *should* equal c_0, so we are not compromising the mathematics; we are clarifying the notation we use to convey the mathematics.)

Definition Power Series

An expression of the form

$$\sum_{n=0}^{\infty} c_n x^n = c_0 + c_1 x + c_2 x^2 + \cdots + c_n x^n + \cdots$$

is a **power series centered at $x = 0$.** An expression of the form

$$\sum_{n=0}^{\infty} c_n (x - a)^n = c_0 + c_1 (x - a) + c_2 (x - a)^2 + \cdots + c_n (x - a)^n + \cdots$$

is a **power series centered at $x = a$.** The term $c_n (x - a)^n$ is the **nth term;** the number a is the **center.**

The geometric series

$$\sum_{n=0}^{\infty} x^n = 1 + x + x^2 + \cdots + x^n + \cdots$$

is a power series centered at $x = 0$. It converges on the interval $-1 < x < 1$, also centered at $x = 0$. This is typical behavior, as we will see in Section 9.4. A power series either converges for all x, converges on a finite interval with the same center as the series, or converges only at the center itself.

We have seen that the power series $\sum_{n=0}^{\infty} x^n$ represents the function $1/(1 - x)$ on the domain $(-1, 1)$. Can we find power series to represent other functions?

Exploration 1   Finding Power Series for Other Functions

Given that $1/(1 - x)$ is represented by the power series

$$1 + x + x^2 + \cdots + x^n + \cdots$$

on the interval $(-1, 1)$,

1. find a power series that represents $1/(1 + x)$ on $(-1, 1)$.

2. find a power series that represents $x/(1 + x)$ on $(-1, 1)$.

3. find a power series that represents $1/(1 - 2x)$ on $(-1/2, 1/2)$.

4. find a power series that represents

$$\frac{1}{x} = \frac{1}{1 + (x - 1)}$$

on $(0, 2)$.

Could you have found the intervals of convergence yourself?

5. Find a power series that represents

$$\frac{1}{3x} = \frac{1}{3} \cdot \left(\frac{1}{1 + (x - 1)} \right)$$

and give its interval of convergence.

Differentiation and Integration

So far we have only represented functions by power series that happen to be geometric. The partial sums that converge to those power series, however, are *polynomials,* and we can apply calculus to polynomials. It would seem logical that the calculus of polynomials (the first rules we encountered in Chapter 3) would also apply to power series.

Example 4 FINDING A POWER SERIES BY DIFFERENTIATION

Given that $1/(1-x)$ is represented by the power series

$$1 + x + x^2 + \cdots + x^n + \cdots$$

on the interval $(-1, 1)$, find a power series to represent $1/(1-x)^2$.

Solution Notice that $1/(1-x)^2$ is the derivative of $1/(1-x)$. To find the power series, we differentiate both sides of the equation

$$\frac{1}{1-x} = 1 + x + x^2 + x^3 + \cdots + x^n + \cdots.$$

$$\frac{d}{dx}\left(\frac{1}{1-x}\right) = \frac{d}{dx}(1 + x + x^2 + x^3 + \cdots + x^n + \cdots)$$

$$\frac{1}{(1-x)^2} = 1 + 2x + 3x^2 + 4x^3 + \cdots + nx^{n-1} + \cdots$$

What about the interval of convergence? Since the original series converges for $-1 < x < 1$, it would seem that the differentiated series ought to converge on the same open interval. Graphs (Figure 9.2) of the partial sums $1 + 2x + 3x^2$, $1 + 2x + 3x^2 + 4x^3$, and $1 + 2x + 3x^2 + 4x^3 + 5x^4$ suggest that this is the case (although such empirical evidence does not constitute a proof).

The basic theorem about differentiating power series is the following.

Partial Sums

[−4.7, 4.7] by [−2, 4]

(a)

$y = 1/(1-x)^2$

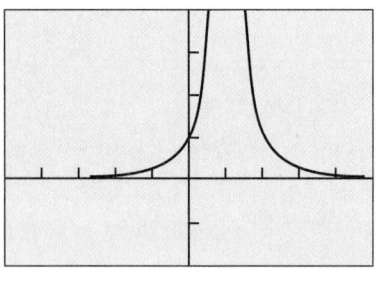

[−4.7, 4.7] by [−2, 4]

(b)

Figure 9.2 (a) The polynomial partial sums of the power series we derived for (b) $1/(1-x)^2$ seem to converge on the open interval $(-1, 1)$. (Example 4)

Theorem 1 Term-by-Term Differentiation

If $f(x) = \sum\limits_{n=0}^{\infty} c_n(x-a)^n = c_0 + c_1(x-a) + c_2(x-a)^2 + \cdots + c_n(x-a)^n + \cdots$

converges for $|x-a| < R$, then the series

$$\sum_{n=1}^{\infty} nc_n(x-a)^{n-1} = c_1 + 2c_2(x-a) + 3c_3(x-a)^2 + \cdots + nc_n(x-a)^{n-1} + \cdots,$$

obtained by differentiating the series for f term by term, converges for $|x-a| < R$ and represents $f'(x)$ on that interval. If the series for f converges for all x, then so does the series for f'.

Theorem 1 says that if a power series is differentiated term by term, the new series will converge on the same interval to the derivative of the function represented by the original series. This gives a way to generate new connections between functions and series.

Another way to reveal new connections between functions and series is by integration.

Example 5 FINDING A POWER SERIES BY INTEGRATION

Given that

$$\frac{1}{1+x} = 1 - x + x^2 - x^3 + \cdots + (-x)^n + \cdots, \qquad -1 < x < 1$$

(Exploration 1, part 1), find a power series to represent $\ln(1 + x)$.

Solution Recall that $1/(1 + x)$ is the derivative of $\ln(1 + x)$. We can therefore integrate the series for $1/(1 + x)$ to obtain a series for $\ln(1 + x)$ (no absolute value bars are necessary because $(1 + x)$ is positive for $-1 < x < 1$).

$$\frac{1}{1+x} = 1 - x + x^2 - x^3 + \cdots + (-x)^n + \cdots$$

$$= 1 - x + x^2 - x^3 + \cdots + (-1)^n x^n + \cdots$$

$$\int_0^x \frac{1}{1+t}\, dt = \int_0^x (1 - t + t^2 - t^3 + \cdots + (-1)^n t^n + \cdots)\, dt \qquad \text{\small\textit{t} is a dummy variable.}$$

$$\ln(1 + t)\Big]_0^x = \left[t - \frac{t^2}{2} + \frac{t^3}{3} - \frac{t^4}{4} + \cdots + (-1)^n \frac{t^{n+1}}{n+1} + \cdots \right]_0^x$$

$$\ln(1 + x) = x - \frac{x^2}{2} + \frac{x^3}{3} - \frac{x^4}{4} + \cdots + (-1)^n \frac{x^{n+1}}{n+1} + \cdots$$

It would seem logical for the new series to converge where the original series converges, on the open interval $(-1, 1)$. The graphs of the partial sums in Figure 9.3 support this idea.

Partial Sums

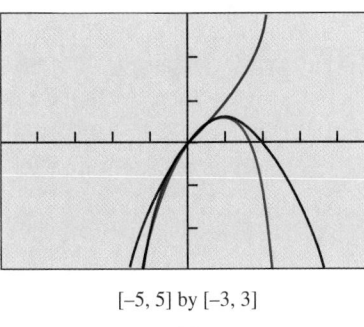

[−5, 5] by [−3, 3]

(a)

$y = \ln(1 + x)$

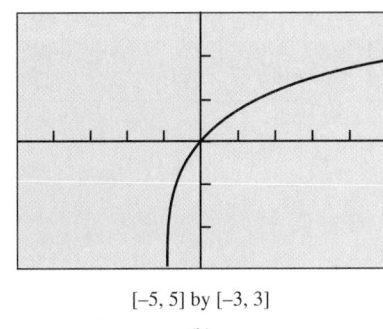

[−5, 5] by [−3, 3]

(b)

Figure 9.3 (a) The graphs of the partial sums

$$x - \frac{x^2}{2}, \qquad x - \frac{x^2}{2} + \frac{x^3}{3}, \qquad \text{and} \qquad x - \frac{x^2}{2} + \frac{x^3}{3} - \frac{x^4}{4}$$

closing in on (b) the graph of $\ln(1 + x)$ over the interval $(-1, 1)$. (Example 5)

The idea that the integrated series in Example 5 converges to $\ln(1+x)$ for all x between -1 and 1 is confirmed by the following theorem.

Some calculators have a *sequence mode* that enables you to generate a sequence of partial sums, but you can also do it with simple commands on the home screen. Try entering the two multiple-step commands shown on the first screen below.

```
0→N: 1→T
                        1
N+1→N: T+(-1)^N/(
N+1) →T
                      .5
■
```

```
              .6997694067
               .686611512
              .699598525
              .6867780122
               .69943624
               .68693624
               .699281919
```

If you are successful, then every time you hit ENTER, the calculator will display the next partial sum of the series

$$1 - \frac{1}{2} + \frac{1}{3} - \frac{1}{4} + \cdots + \frac{(-1)^n}{n+1} + \cdots.$$

The second screen shows the result of about 80 ENTERs. The sequence certainly seems to be converging to $\ln 2 = 0.6931471806\ldots$.

Theorem 2 Term-by-Term Integration

If $f(x) = \displaystyle\sum_{n=0}^{\infty} c_n(x-a)^n = c_0 + c_1(x-a) + c_2(x-a)^2 + \cdots + c_n(x-a)^n + \cdots$

converges for $|x-a| < R$, then the series

$$\sum_{n=0}^{\infty} c_n \frac{(x-a)^{n+1}}{n+1} = c_0(x-a) + c_1\frac{(x-a)^2}{2} + c_2\frac{(x-a)^3}{3} + \cdots + c_n\frac{(x-a)^{n+1}}{n+1} + \cdots,$$

obtained by integrating the series for f term by term, converges for $|x-a| < R$ and represents $\int_a^x f(t)\,dt$ on that interval. If the series for f converges for all x, then so does the series for the integral.

Theorem 2 says that if a power series is integrated term by term, the new series will converge on the same interval to the integral of the function represented by the original series.

There is still more to be learned from Example 5. The original equation

$$\frac{1}{1+x} = 1 - x + x^2 - x^3 + \cdots + (-x)^n + \cdots$$

clearly diverges at $x = 1$ (see Example 1). The behavior is not so apparent, however, for the new equation

$$\ln(1+x) = x - \frac{x^2}{2} + \frac{x^3}{3} - \frac{x^4}{4} + \cdots + (-1)^n\frac{x^{n+1}}{n+1} + \cdots.$$

If we let $x = 1$ on both sides of this equation, we get

$$\ln 2 = 1 - \frac{1}{2} + \frac{1}{3} - \frac{1}{4} + \cdots + \frac{(-1)^n}{n+1} + \cdots,$$

which looks like a reasonable statement. It looks even more reasonable if you look at the partial sums of the series and watch them converge toward $\ln 2$ (see margin note). It would appear that our new series converges at 1 despite the fact that we obtained it from a series that did not! This is all the more reason to take a careful look at convergence later. Meanwhile, we can enjoy the observation that we have created a series that apparently works better than we might have expected and better than Theorem 2 could guarantee.

Exploration 2 Finding a Power Series for $\tan^{-1} x$

1. Find a power series that represents $1/(1 + x^2)$ on $(-1, 1)$.
2. Use the technique of Example 5 to find a power series that represents $\tan^{-1} x$ on $(-1, 1)$.
3. Graph the first four partial sums. Do the graphs suggest convergence on the open interval $(-1, 1)$?
4. Do you think that the series for $\tan^{-1} x$ converges at $x = 1$? Can you support your answer with evidence?

Identifying a Series

So far we have been finding power series to represent functions. Let us now try to find the function that a given power series represents.

Exploration 3 A Series with a Curious Property

Define a function f by a power series as follows:

$$f(x) = 1 + x + \frac{x^2}{2!} + \frac{x^3}{3!} + \frac{x^4}{4!} + \cdots + \frac{x^n}{n!} + \cdots .$$

1. Find $f'(x)$.
2. Find $f(0)$.
3. What well-known function do you suppose f is?
4. Use your responses to parts 1 and 2 to set up an initial value problem that the function f must solve. You will need a differential equation and an initial condition.
5. Solve the initial value problem to prove your conjecture in part 3.
6. Graph the first three partial sums. What appears to be the interval of convergence?
7. Graph the next three partial sums. Did you underestimate the interval of convergence?

The correct answer to part 7 in Exploration 3 above is "yes," unless you had the keen insight (or reckless bravado) to answer "all real numbers" in part 6. We will prove the remarkable fact that this series converges *for all x* when we revisit the question of convergence of this series in Section 9.3, Example 4.

Quick Review 9.1

In Exercises 1 and 2, find the first four terms and the 30th term of the sequence

$$\{u_n\}_{n=1}^{\infty} = \{u_1, u_2, \dots, u_n, \dots\}.$$

1. $u_n = \dfrac{4}{n+2}$ **2.** $u_n = \dfrac{(-1)^n}{n}$

In Exercises 3 and 4, the sequences are geometric ($a_{n+1}/a_n = r$, a constant). Find

    (a) the common ratio r.   **(b)** the tenth term.

    (c) a rule for the nth term.

3. $\{2, 6, 18, 54, \dots\}$

4. $\{8, -4, 2, -1, \dots\}$

In Exercises 5–10,

    (a) graph the sequence $\{a_n\}$.

    (b) determine $\lim\limits_{n\to\infty} a_n$.

5. $a_n = \dfrac{1-n}{n^2}$ **6.** $a_n = \left(1 + \dfrac{1}{n}\right)^n$

7. $a_n = (-1)^n$ **8.** $a_n = \dfrac{1-2n}{1+2n}$

9. $a_n = 2 - \dfrac{1}{n}$ **10.** $a_n = \dfrac{\ln(n+1)}{n}$

Section 9.1 Exercises

1. Replace the $*$ with an expression that will generate the series

$$1 - \frac{1}{4} + \frac{1}{9} - \frac{1}{16} + \cdots.$$

    (a) $\displaystyle\sum_{n=1}^{\infty} (-1)^{n-1}\left(\dfrac{1}{*}\right)$ **(b)** $\displaystyle\sum_{n=0}^{\infty} (-1)^{n}\left(\dfrac{1}{*}\right)$

    (c) $\displaystyle\sum_{n=*}^{\infty} (-1)^{n}\left(\dfrac{-1}{(n-2)^2}\right)$

2. Write an expression for the nth term, a_n.

    (a) $\displaystyle\sum_{n=0}^{\infty} a_n = 1 + \frac{1}{3} + \frac{1}{9} + \frac{1}{27} + \frac{1}{81} + \cdots$

    (b) $\displaystyle\sum_{n=1}^{\infty} a_n = 1 - \frac{1}{2} + \frac{1}{3} - \frac{1}{4} + \frac{1}{5} - \cdots$

    (c) $\displaystyle\sum_{n=0}^{\infty} a_n = 5 + 0.5 + 0.05 + 0.005 + 0.0005 + \cdots$

In Exercises 3–6, tell whether the series is the same as

$$\sum_{n=1}^{\infty} \left(-\frac{1}{2}\right)^{n-1}.$$

3. $\displaystyle\sum_{n=1}^{\infty} -\left(\frac{1}{2}\right)^{n-1}$ **4.** $\displaystyle\sum_{n=0}^{\infty} \left(-\frac{1}{2}\right)^{n}$

5. $\displaystyle\sum_{n=0}^{\infty} (-1)^n\left(\frac{1}{2}\right)^{n}$ **6.** $\displaystyle\sum_{n=1}^{\infty} \frac{(-1)^n}{2^{n-1}}$

In Exercises 7–16, tell whether the series converges or diverges. If it converges, give its sum.

7. $1 + \dfrac{2}{3} + \left(\dfrac{2}{3}\right)^2 + \left(\dfrac{2}{3}\right)^3 + \cdots + \left(\dfrac{2}{3}\right)^n + \cdots$

8. $1 - 2 + 3 - 4 + 5 - \cdots + (-1)^n(n+1) + \cdots$

9. $\displaystyle\sum_{n=0}^{\infty} \left(\frac{5}{4}\right)\left(\frac{2}{3}\right)^{n}$ **10.** $\displaystyle\sum_{n=0}^{\infty} \left(\frac{2}{3}\right)\left(\frac{5}{4}\right)^{n}$

11. $\displaystyle\sum_{n=0}^{\infty} \cos(n\pi)$

12. $3 - 0.3 + 0.03 - 0.003 + 0.0003 - \cdots + 3(-0.1)^n + \cdots$

13. $\displaystyle\sum_{n=0}^{\infty} \sin^n\left(\frac{\pi}{4} + n\pi\right)$

14. $\dfrac{1}{2} + \dfrac{2}{3} + \dfrac{3}{4} + \dfrac{4}{5} + \cdots + \dfrac{n}{n+1} + \cdots$

15. $\displaystyle\sum_{n=1}^{\infty} \left(\frac{e}{\pi}\right)^{n}$ **16.** $\displaystyle\sum_{n=0}^{\infty} \frac{5^n}{6^{n+1}}$

In Exercises 17–20, find the interval of convergence and the function of x represented by the geometric series.

17. $\displaystyle\sum_{n=0}^{\infty} 2^n x^n$ **18.** $\displaystyle\sum_{n=0}^{\infty} (-1)^n (x+1)^n$

19. $\displaystyle\sum_{n=0}^{\infty} \left(-\frac{1}{2}\right)^n (x-3)^n$ **20.** $\displaystyle\sum_{n=0}^{\infty} 3\left(\frac{x-1}{2}\right)^n$

In Exercises 21 and 22, find the values of x for which the geometric series converges and find the function of x it represents.

21. $\displaystyle\sum_{n=0}^{\infty} \sin^n x$ **22.** $\displaystyle\sum_{n=0}^{\infty} \tan^n x$

23. Writing to Learn Each of the following series diverges in a slightly different way. Explain what is happening to the sequence of partial sums in each case.

(a) $\displaystyle\sum_{n=1}^{\infty} 2n$ **(b)** $\displaystyle\sum_{n=0}^{\infty} (-1)^n$ **(c)** $\displaystyle\sum_{n=1}^{\infty} (-1)^n (2n)$

24. Prove that $\displaystyle\sum_{n=0}^{\infty} \frac{e^{n\pi}}{\pi^{ne}}$ diverges.

25. Solve for x: $\displaystyle\sum_{n=0}^{\infty} x^n = 20.$

26. Writing to Learn Explain how it is possible, given any real number at all, to construct an infinite series of non-zero terms that converges to it.

27. Make up a geometric series $\sum ar^{n-1}$ that converges to the number 5 if

(a) $a = 2$ **(b)** $a = 13/2$

In Exercises 28 and 29, express the repeating decimal as a geometric series and find its sum.

28. $0.\overline{21}$ **29.** $0.\overline{234}$

In Exercises 30–35, express the number as the ratio of two integers.

30. $0.\overline{7} = 0.7777\ldots$

31. $0.\overline{d} = 0.dddd\ldots,$ where d is a digit

32. $0.0\overline{6} = 0.06666\ldots$

33. $1.\overline{414} = 1.414\ 414\ 414\ldots$

34. $1.24\overline{123} = 1.24\ 123\ 123\ 123\ldots$

35. $3.\overline{142857} = 3.142857\ 142857\ldots$

36. *Bouncing Ball* A ball is dropped from a height of 4 m. Each time it strikes the pavement after falling from a height of h m, it rebounds to a height of $0.6h$ m. Find the total distance the ball travels up and down.

37. *(Continuation of Exercise 36)* Find the total number of seconds that the ball in Exercise 36 travels. (*Hint:* A freely falling ball travels $4.9t^2$ meters in t seconds, so it will fall h meters in $\sqrt{h/4.9}$ seconds. Bouncing from ground to apex takes the same time as falling from apex to ground.)

38. *Summing Areas* The figure below shows the first five of an infinite sequence of squares. The outermost square has an area of 4 m². Each of the other squares is obtained by joining the midpoints of the sides of the preceding square. Find the sum of the areas of all the squares.

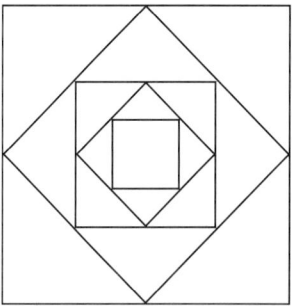

39. *Summing Areas* The accompanying figure shows the first three rows and part of the fourth row of a sequence of rows of semicircles. There are 2^n semicircles in the nth row, each of radius $1/(2^n)$. Find the sum of the areas of all the semicircles.

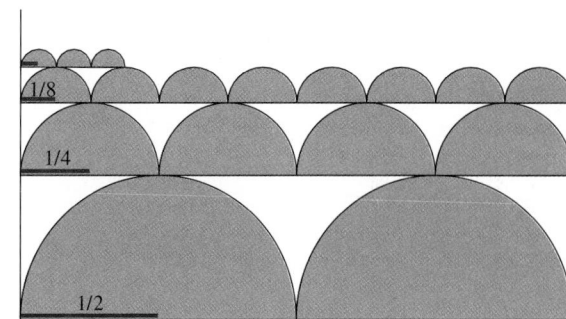

40. *Sum of a Finite Geometric Progression* Let a and r be real numbers with $r \neq 1$, and let

$$S = a + ar + ar^2 + ar^3 + \cdots + ar^{n-1}.$$

(a) Find $S - rS$.

(b) Use the result in (a) to show that $S = \dfrac{a - ar^n}{1 - r}.$

41. *Sum of a Convergent Geometric Series* Exercise 40 gives a formula for the nth partial sum of an infinite geometric series. Use this formula to show that $\sum_{n=1}^{\infty} ar^{n-1}$ diverges when $|r| \geq 1$ and converges to $a/(1 - r)$ when $|r| < 1$.

In Exercises 42–47, find a power series to represent the given function and identify its interval of convergence. When writing the power series, include a formula for the nth term.

42. $\dfrac{1}{1 + 3x}$

43. $\dfrac{x}{1 - 2x}$

44. $\dfrac{3}{1 - x^3}$

45. $\dfrac{1}{1 + (x - 4)}$

46. $\dfrac{1}{4x} = \dfrac{1}{4}\left(\dfrac{1}{1 + (x - 1)}\right)$

47. $\dfrac{1}{2 - x}$ (*Hint:* Rewrite $2 - x$.)

48. Find the value of b for which $1 + e^b + e^{2b} + e^{3b} + \cdots = 9$.

49. Let S be the series

$$\sum_{n=0}^{\infty}\left(\dfrac{t}{1 + t}\right)^n, \quad t \neq 0.$$

(a) Find the value to which S converges when $t = 1$.

(b) Determine all values of t for which S converges.

(c) Find all values of t that make the sum of the series S greater than 10.

Exploration

50. Let $f(t) = \dfrac{4}{1 + t^2}$ and $G(x) = \displaystyle\int_0^x f(t)\,dt$.

(a) Find the first four nonzero terms and the general term for a power series for $f(t)$ centered at $t = 0$.

(b) Find the first four nonzero terms and the general term for a power series for $G(x)$ centered at $x = 0$.

(c) Find the interval of convergence of the power series in (a).

(d) The interval of convergence of the power series in (b) is almost the same as the interval in (c), but includes two more numbers. What are the numbers? ∎

51. *A Series for ln x* Starting with the power series found for $1/x$ in Exploration 1, Part 4, find a power series for $\ln x$ centered at $x = 1$.

52. *Differentiation* Use differentiation to find a series for $f(x) = 2/(1 - x)^3$. What is the interval of convergence of your series?

53. *Intervals of Convergence* Work in groups of two or three. How much can the interval of convergence of a power series be changed by integration or differentiation? To be specific, suppose that the power series

$$f(x) = c_0 + c_1 x + c_2 x^2 + \cdots + c_n x^n + \cdots$$

converges for $-1 < x < 1$ and diverges for all other values of x.

(a) Writing to Learn Could the series obtained by integrating the series for f term by term possibly converge for $-2 < x < 2$? Explain. (*Hint:* Apply Theorem 1, not Theorem 2.)

(b) Writing to Learn Could the series obtained by differentiating the series for f term by term possibly converge for $-2 < x < 2$? Explain.

Extending the Ideas

The sequence $\{a_n\}$ **converges** to the number L if to every positive number ε there corresponds an integer N such that for all n,

$$n > N \quad \Rightarrow \quad |a_n - L| < \varepsilon.$$

L is the **limit** of the sequence and we write $\lim_{n\to\infty} a_n = L$. If no such number L exists, we say that $\{a_n\}$ **diverges.**

54. *Tail of a Sequence* Prove that if $\{a_n\}$ is a convergent sequence, then to every positive number ε there corresponds an integer N such that for all m and n,

$$m > N \quad \text{and} \quad n > N \quad \Rightarrow \quad |a_m - a_n| < \varepsilon.$$

(*Hint:* Let $\lim_{n\to\infty} a_n = L$. As the terms approach L, how far apart can they be?)

55. *Uniqueness of Limits* Prove that limits of sequences are unique. That is, show that if L_1 and L_2 are numbers such that $\lim_{n\to\infty} a_n = L_1$ and $\lim_{n\to\infty} a_n = L_2$, then $L_1 = L_2$.

56. *Limits and Subsequences* Prove that if two subsequences of a sequence $\{a_n\}$ have different limits $L_1 \neq L_2$, then $\{a_n\}$ diverges.

57. *Limits and Asymptotes*

(a) Show that the sequence with nth term $a_n = (3n + 1)/(n + 1)$ converges.

(b) If $\lim_{n\to\infty} a_n = L$, explain why $y = L$ is a horizontal asymptote of the graph of the function

$$f(x) = \dfrac{3x + 1}{x + 1}$$

obtained by replacing n by x in the nth term.

9.2 Taylor Series

Constructing a Series • Series for sin x and cos x • Beauty Bare • Maclaurin and Taylor Series • Combining Taylor Series • Table of Maclaurin Series

Constructing a Series

A comprehensive understanding of geometric series served us well in Section 9.1, enabling us to find power series to represent certain functions, and functions that are equivalent to certain power series (all of these equivalencies being subject to the condition of convergence). In this section we learn a more general technique for constructing power series, one that makes good use of the tools of calculus.

Let us start by constructing a polynomial.

Exploration 1 **Designing a Polynomial to Specifications**

Construct a polynomial $P(x) = a_0 + a_1x + a_2x^2 + a_3x^3 + a_4x^4$ with the following behavior at $x = 0$:

$$P(0) = 1,$$
$$P'(0) = 2,$$
$$P''(0) = 3,$$
$$P'''(0) = 4, \text{ and}$$
$$P^{(4)}(0) = 5.$$

This task might look difficult at first, but when you try it you will find that the predictability of differentiation when applied to polynomials makes it straightforward. (Be sure to check this out before you move on.)

There is nothing special about the number of derivatives in Exploration 1. We could have prescribed the value of the polynomial and its first n derivatives at $x = 0$ for any n and found a polynomial of degree at most n to match. Our plan now is to use the technique of Exploration 1 to construct polynomials that approximate functions by emulating their behavior at 0.

Example 1 APPROXIMATING ln $(1 + x)$ BY A POLYNOMIAL

Construct a polynomial $P(x) = a_0 + a_1x + a_2x^2 + a_3x^3 + a_4x^4$ that matches the behavior of $\ln(1 + x)$ at $x = 0$ through its first four derivatives. That is,

$$P(0) = \ln(1 + x) \quad \text{at } x = 0,$$
$$P'(0) = (\ln(1 + x))' \quad \text{at } x = 0,$$
$$P''(0) = (\ln(1 + x))'' \quad \text{at } x = 0,$$
$$P'''(0) = (\ln(1 + x))''' \quad \text{at } x = 0, \text{ and}$$
$$P^{(4)}(0) = (\ln(1 + x))^{(4)} \quad \text{at } x = 0.$$

Solution This is just like Exploration 1, except that first we need to find out what the numbers are.

$$P(0) = \ln (1 + x)\Big|_{x=0} = 0$$

$$P'(0) = \frac{1}{1 + x}\Big|_{x=0} = 1$$

$$P''(0) = -\frac{1}{(1 + x)^2}\Big|_{x=0} = -1$$

$$P'''(0) = \frac{2}{(1 + x)^3}\Big|_{x=0} = 2$$

$$P^{(4)}(0) = -\frac{6}{(1 + x)^4}\Big|_{x=0} = -6$$

In working through Exploration 1, you probably noticed that the coefficient of the term x^n in the polynomial we seek is $P^{(n)}(0)$ divided by $n!$. The polynomial is

$$P(x) = 0 + x - \frac{x^2}{2} + \frac{x^3}{3} - \frac{x^4}{4}.$$

We have just constructed the fourth order **Taylor polynomial** for the function $\ln (1 + x)$ at $x = 0$. You might recognize it as the beginning of the power series we discovered for $\ln (1 + x)$ in Example 5 of Section 9.1, when we came upon it by integrating a geometric series. If we keep going, of course, we will gradually reconstruct that entire series one term at a time, improving the approximation near $x = 0$ with every term we add. The series is called the **Taylor series** generated by the function $\ln (1 + x)$ at $x = 0$.

You might also recall Figure 9.3, which shows how the polynomial approximations converge nicely to $\ln (1 + x)$ near $x = 0$, but then gradually peel away from the curve as x gets farther away from 0 in either direction. Given that the coefficients are totally determined by specifying behavior at $x = 0$, that is exactly what we ought to expect.

Series for sin x and cos x

We can use the technique of Example 1 to construct Taylor series about $x = 0$ for any function, as long as we can keep taking derivatives there. Two functions that are particularly well-suited for this treatment are the sine and cosine.

Example 2   CONSTRUCTING A POWER SERIES FOR sin x

Construct the seventh order Taylor polynomial and the Taylor series for $\sin x$ at $x = 0$.

Solution We need to evaluate $\sin x$ and its first seven derivatives at $x = 0$. Fortunately, this is not hard to do.

$$\sin(0) = 0$$

$$\sin'(0) = \cos(0) = 1$$

$$\sin''(0) = -\sin(0) = 0$$

$$\sin'''(0) = -\cos(0) = -1$$

$$\sin^{(4)}(0) = \sin(0) = 0$$

$$\sin^{(5)}(0) = \cos(0) = 1$$

$$\vdots$$

The pattern $0, 1, 0, -1$ will keep repeating forever.

The unique seventh order Taylor polynomial that matches all these derivatives at $x = 0$ is

$$P_7(x) = 0 + 1x - 0x^2 - \frac{1}{3!}x^3 + 0x^4 + \frac{1}{5!}x^5 - 0x^6 - \frac{1}{7!}x^7$$

$$= x - \frac{x^3}{3!} + \frac{x^5}{5!} - \frac{x^7}{7!}.$$

P_7 is the seventh order Taylor polynomial for $\sin x$ at $x = 0$. (It also happens to be of seventh *degree,* but that does not always happen. For example, you can see that P_8 for $\sin x$ will be the same polynomial as P_7.)

To form the Taylor series, we just keep on going:

$$x - \frac{x^3}{3!} + \frac{x^5}{5!} - \frac{x^7}{7!} + \frac{x^9}{9!} - \cdots = \sum_{n=0}^{\infty} (-1)^n \frac{x^{2n+1}}{(2n+1)!}.$$

Exploration 2 **A Power Series for the Cosine**

Work in groups of two or three.

1. Construct the sixth order Taylor polynomial and the Taylor series at $x = 0$ for $\cos x$.

2. Compare your method for attacking part 1 with the methods of other groups. Did anyone find a shortcut?

Beauty Bare

Edna St. Vincent Millay, an early twentieth-century American poet, referring to the experience of simultaneously seeing and understanding the geometric underpinnings of nature, wrote "Euclid alone has looked on Beauty bare." In case you have never experienced that sort of reverie when gazing upon something geometric, we intend to give you that opportunity now.

In Example 2 we constructed a power series for $\sin x$ by matching the behavior of $\sin x$ at $x = 0$. Let us graph the first nine partial sums together with $y = \sin x$ to see how well we did (Figure 9.4).

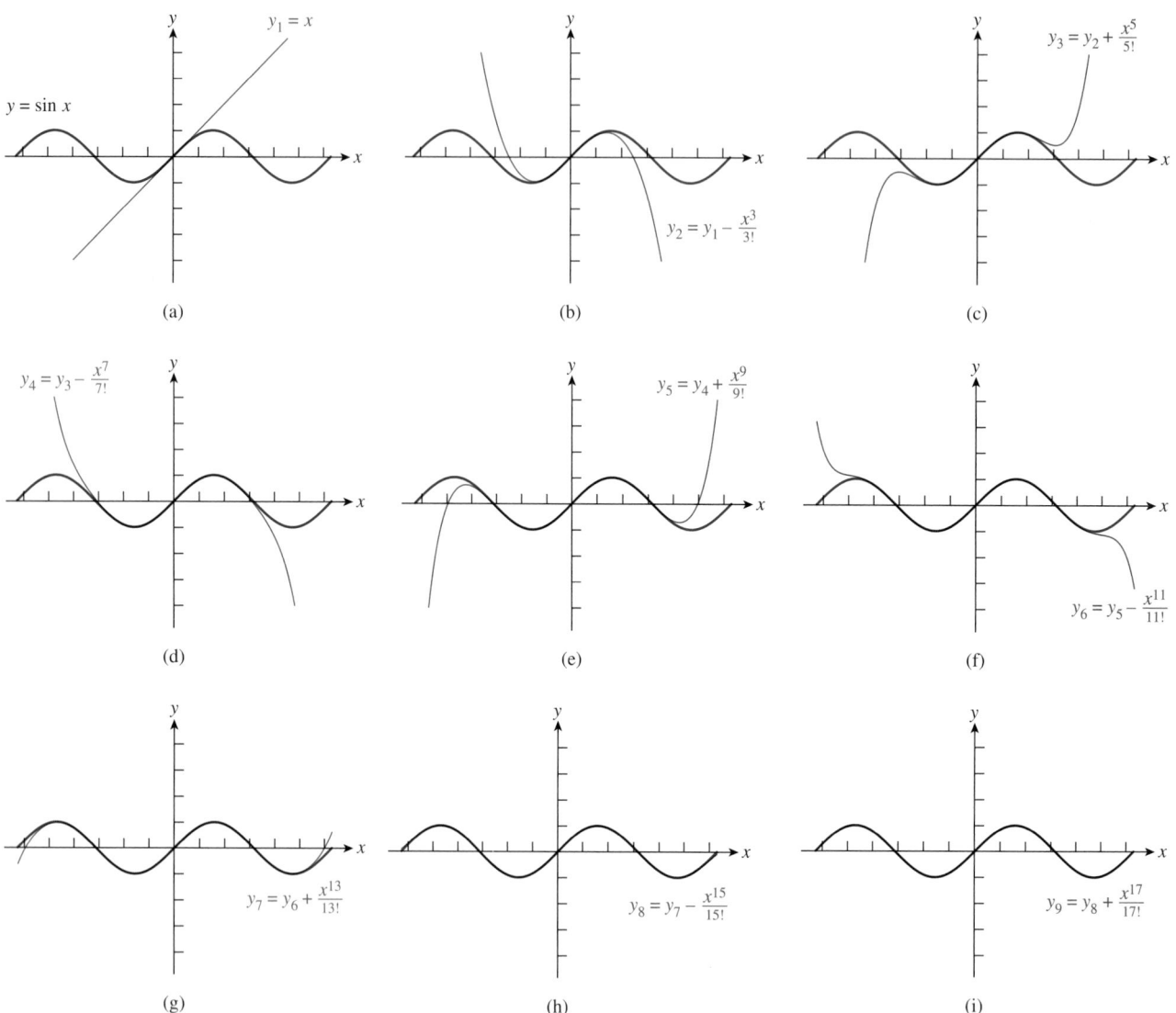

Figure 9.4 $y = \sin x$ and its nine Taylor polynomials $P_1, P_3, \ldots, P_{17}$ for $-2\pi \leq x \leq 2\pi$. Try graphing these functions in the window $[-2\pi, 2\pi]$ by $[-5, 5]$.

Behold what is occurring here! These polynomials were constructed to mimic the behavior of $\sin x$ near $x = 0$. The *only* information we used to construct the coefficients of these polynomials was information about the sine function and its derivatives at 0. Yet, somehow, the information at $x = 0$ is producing a series whose graph not only looks like sine near the origin, but appears to be a clone of the *entire* sine curve. This is no deception, either; we will show in Section 9.3, Example 3 that the Taylor series for $\sin x$ does, in fact, converge to $\sin x$ over the entire real line. We have managed to construct an entire function by knowing its behavior at a single point! (The same is true about the series for $\cos x$ found in Exploration 2.)

We still must remember that convergence is an infinite process. Even the one-billionth order Taylor polynomial begins to peel away from $\sin x$ as we move away from 0, although imperceptibly at first, and eventually becomes unbounded, as any polynomial must. Nonetheless, we can approximate the sine of *any* number to whatever accuracy we want if we just have the patience to work out enough terms of this series!

This kind of dramatic convergence does not occur for all Taylor series. The Taylor polynomials for $\ln (1 + x)$ do not converge outside the interval from -1 to 1, no matter how many terms we add.

Maclaurin and Taylor Series

If we generalize the steps we followed in constructing the coefficients of the power series in this section so far, we arrive at the following definition.

Who invented Taylor series?

Brook Taylor (1685–1731) did not invent Taylor series, and Maclaurin series were not developed by Colin Maclaurin (1698–1746). James Gregory was already working with Taylor series when Taylor was only a few years old, and he published the Maclaurin series for $\tan x$, $\sec x$, $\arctan x$, and $\text{arcsec } x$ ten years before Maclaurin was born. Nicolaus Mercator discovered the Maclaurin series for $\ln (1 + x)$ at about the same time.

Taylor was unaware of Gregory's work when he published his book *Methodus Incrementorum Directa et Inversa* in 1715, containing what we now call Taylor series. Maclaurin quoted Taylor's work in a calculus book he wrote in 1742. The book popularized series representations of functions and although Maclaurin never claimed to have discovered them, Taylor series centered at $x = 0$ became known as Maclaurin series. History evened things up in the end. Maclaurin, a brilliant mathematician, was the original discoverer of the rule for solving systems of equations that we call Cramer's rule.

> **Definition** Taylor Series Generated by *f* at *x* = 0 (Maclaurin Series)
>
> Let f be a function with derivatives of all orders throughout some open interval containing 0. Then the **Taylor series generated by** f at $x = 0$ is
>
> $$f(0) + f'(0)x + \frac{f''(0)}{2!}x^2 + \cdots + \frac{f^{(n)}(0)}{n!}x^n + \cdots = \sum_{k=0}^{\infty} \frac{f^{(k)}(0)}{k!}x^k.$$
>
> This series is also called the **Maclaurin series generated by** *f.* The partial sum
>
> $$P_n(x) = \sum_{k=0}^{n} \frac{f^{(k)}(0)}{k!}x^k$$
>
> is the **Taylor polynomial of order *n* for** *f* **at** *x* = 0.

We use $f^{(0)}$ to mean *f.* *Every* power series constructed in this way converges to the function f at $x = 0$, but we have seen that the convergence might well extend to an interval containing 0, or even to the entire real line. When this happens, the Taylor polynomials that form the partial sums of a Taylor series provide good approximations for f near 0.

Example 3 APPROXIMATING A FUNCTION NEAR 0

Find the fourth order Taylor polynomial that approximates $y = \cos 2x$ near $x = 0$.

Solution The polynomial we want is $P_4(x)$, the Taylor polynomial for $\cos 2x$ at $x = 0$. Before we go cranking out derivatives though, remember that we can use a known power series to generate another, as we did in Section 9.1. We know from Exploration 2 that

$$\cos x = 1 - \frac{x^2}{2!} + \frac{x^4}{4!} - \cdots + (-1)^n \frac{x^{2n}}{(2n)!} + \cdots.$$

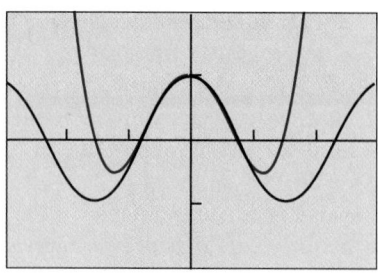

[–3, 3] by [–2, 2]

Figure 9.5 The graphs of $y = 1 - 2x^2 + (2/3)x^4$ and $y = \cos 2x$ near $x = 0$. (Example 3)

Therefore,

$$\cos 2x = 1 - \frac{(2x)^2}{2!} + \frac{(2x)^4}{4!} - \cdots + (-1)^n \frac{(2x)^{2n}}{(2n)!} + \cdots.$$

So,

$$P_4(x) = 1 - \frac{(2x)^2}{2!} + \frac{(2x)^4}{4!}$$
$$= 1 - 2x^2 + (2/3)x^4.$$

The graph in Figure 9.5 shows how well the polynomial approximates the cosine near $x = 0$.

These polynomial approximations can be useful in a variety of ways. For one thing, it is easy to do calculus with polynomials. For another thing, polynomials are built using only the two basic operations of addition and multiplication, so computers can handle them easily.

Exploration 3 | **Approximating sin 13**

How many terms of the series

$$\sum_{n=0}^{\infty} (-1)^n \frac{x^{2n+1}}{(2n+1)!}$$

are required to approximate sin 13 accurate to the third decimal place?

1. Find sin 13 on your calculator (radians, of course).

2. Enter these two multiple-step commands on your home screen. They will give you the first order and second order Taylor polynomial approximations for sin 13. Notice that the second order approximation, in particular, is not very good.

```
0 →N: 13→T
                          13
N+1 →N: T+(−1)^N*1
3^(2N+1)/(2N+1)!
→ T
                  −353.1666667
```

3. Continue to hit ENTER. Each time you will add one more term to the Taylor polynomial approximation. Be patient; things will get worse before they get better.

4. How many terms are required before the polynomial approximations stabilize in the thousandths place for $x = 13$?

Section 9.2 *Taylor Series* **475**

This strategy for approximation would be of limited practical value if we were restricted to power series at $x = 0$—but we are not. We can match a power series with f in the same way at *any* value $x = a$, provided we can take the derivatives. In fact, we can get a formula for doing that by simply "shifting horizontally" the formula we already have.

Definition Taylor Series Generated by *f* at *x* = *a*

Let f be a function with derivatives of all orders throughout some open interval containing a. Then the **Taylor series generated by f at $x = a$** is

$$f(a) + f'(a)(x - a) + \frac{f''(a)}{2!}(x - a)^2 + \cdots + \frac{f^{(n)}(a)}{n!}(x - a)^n + \cdots$$

$$= \sum_{k=0}^{\infty} \frac{f^{(k)}(a)}{k!}(x - a)^k.$$

The partial sum

$$P_n(x) = \sum_{k=0}^{n} \frac{f^{(k)}(a)}{k!}(x - a)^k$$

is the **Taylor polynomial of order n for f at $x = a$.**

[–1, 4] by [–10, 50]

Figure 9.6 The graphs of $y = e^x$ and $y = P_3(x)$ (the third order Taylor polynomial for e^x at $x = 2$). (Example 4)

Example 4 A TAYLOR SERIES AT $x = 2$

Find the Taylor series generated by $f(x) = e^x$ at $x = 2$.

Solution We first observe that $f(2) = f'(2) = f''(2) = \cdots = f^{(n)}(2) = e^2$. The series, therefore, is

$$e^x = e^2 + e^2(x - 2) + \frac{e^2}{2!}(x - 2)^2 + \cdots + \frac{e^2}{n!}(x - 2)^n + \cdots$$

$$= \sum_{k=0}^{\infty} \left(\frac{e^2}{k!}\right)(x - 2)^k.$$

We illustrate the convergence near $x = 2$ by sketching the graphs of $y = e^x$ and $y = P_3(x)$ in Figure 9.6.

Example 5 A TAYLOR POLYNOMIAL FOR A POLYNOMIAL

Find the third order Taylor polynomial for $f(x) = 2x^3 - 3x^2 + 4x - 5$

(a) at $x = 0$.　　**(b)** at $x = 1$.

Solution

(a) This is easy. This polynomial is already written in powers of x and is of degree three, so it is its own third order (and fourth order, etc.) Taylor polynomial at $x = 0$.

(b) This would also be easy if we could quickly rewrite the formula for f as a polynomial in powers of $x - 1$, but that would require some messy tinkering. Instead, we apply the Taylor series formula.

$$f(1) = 2x^3 - 3x^2 + 4x - 5\Big|_{x=1} = -2$$

$$f'(1) = 6x^2 - 6x + 4\Big|_{x=1} = 4$$

$$f''(1) = 12x - 6\Big|_{x=1} = 6$$

$$f'''(1) = 12$$

So,

$$P_3(x) = -2 + 4(x - 1) + \frac{6}{2!}(x - 1)^2 + \frac{12}{3!}(x - 1)^3$$
$$= 2(x - 1)^3 + 3(x - 1)^2 + 4(x - 1) - 2.$$

This polynomial function agrees with f at every value of x (as you can verify by multiplying it out) but it is written in powers of $(x - 1)$ instead of x.

Combining Taylor Series

On the intersection of their intervals of convergence, Taylor series can be added, subtracted, and multiplied by constants and powers of x, and the results are once again Taylor series. The Taylor series for $f(x) + g(x)$ is the sum of the Taylor series for $f(x)$ and the Taylor series for $g(x)$ because the nth derivative of $f + g$ is $f^{(n)} + g^{(n)}$, and so on. We can obtain the Maclaurin series for $(1 + \cos 2x)/2$ by substituting $2x$ in the Maclaurin series for $\cos x$, adding 1, and dividing the result by 2. The Maclaurin series for $\sin x + \cos x$ is the term-by-term sum of the series for $\sin x$ and $\cos x$. We obtain the Maclaurin series for $x \sin x$ by multiplying all the terms of the Maclaurin series for $\sin x$ by x.

Table of Maclaurin Series

We conclude the section by listing some of the most useful Maclaurin series, all of which have been derived in one way or another in the first two sections of this chapter. The exercises will ask you to use these series as basic building blocks for constructing other series (e.g., $\tan^{-1} x^2$ or $7xe^x$). We also list the intervals of convergence, although rigorous proofs of convergence are deferred until we develop convergence tests in Sections 9.4 and 9.5.

Maclaurin Series

$$\frac{1}{1-x} = 1 + x + x^2 + \cdots + x^n + \cdots = \sum_{n=0}^{\infty} x^n \quad (|x| < 1)$$

$$\frac{1}{1+x} = 1 - x + x^2 - \cdots + (-x)^n + \cdots = \sum_{n=0}^{\infty} (-1)^n x^n \quad (|x| < 1)$$

$$e^x = 1 + x + \frac{x^2}{2!} + \cdots + \frac{x^n}{n!} + \cdots = \sum_{n=0}^{\infty} \frac{x^n}{n!} \quad (\text{all real } x)$$

$$\sin x = x - \frac{x^3}{3!} + \frac{x^5}{5!} - \cdots + (-1)^n \frac{x^{2n+1}}{(2n+1)!} + \cdots$$

$$= \sum_{n=0}^{\infty} (-1)^n \frac{x^{2n+1}}{(2n+1)!} \quad (\text{all real } x)$$

$$\cos x = 1 - \frac{x^2}{2!} + \frac{x^4}{4!} - \cdots + (-1)^n \frac{x^{2n}}{(2n)!} + \cdots$$

$$= \sum_{n=0}^{\infty} (-1)^n \frac{x^{2n}}{(2n)!} \quad (\text{all real } x)$$

$$\ln(1+x) = x - \frac{x^2}{2} + \frac{x^3}{3} - \cdots + (-1)^{n-1} \frac{x^n}{n} + \cdots$$

$$= \sum_{n=1}^{\infty} (-1)^{n-1} \frac{x^n}{n} \quad (-1 < x \le 1)$$

$$\tan^{-1} x = x - \frac{x^3}{3} + \frac{x^5}{5} - \cdots + (-1)^n \frac{x^{2n+1}}{2n+1} + \cdots$$

$$= \sum_{n=0}^{\infty} (-1)^n \frac{x^{2n+1}}{2n+1} \quad (|x| \le 1)$$

Quick Review 9.2

In Exercises 1–5, find a formula for the nth derivative of the function.

1. e^{2x}

2. $\dfrac{1}{x-1}$

3. 3^x

4. $\ln x$

5. x^n

In Exercises 6–10, find dy/dx. (Assume that letters other than x represent constants.)

6. $y = \dfrac{x^n}{n!}$

7. $y = \dfrac{2^n(x-a)^n}{n!}$

8. $y = (-1)^n \dfrac{x^{2n+1}}{(2n+1)!}$

9. $y = \dfrac{(x+a)^{2n}}{(2n)!}$

10. $y = \dfrac{(1-x)^n}{n!}$

Section 9.2 Exercises

In Exercises 1–8, use the table of Maclaurin series on the preceding page. Construct the first three nonzero terms and the general term of the Maclaurin series generated by the function and give the interval of convergence.

1. $\sin 2x$

2. $\ln(1 - x)$

3. $\tan^{-1} x^2$

4. $7x e^x$

5. $\cos(x + 2)$
   (*Hint:* $\cos(x + 2) = (\cos 2)(\cos x) - (\sin 2)(\sin x)$)

6. $x^2 \cos x$

7. $\dfrac{x}{1 - x^3}$

8. e^{-2x}

In Exercises 9–12, find the Taylor polynomials of orders 0, 1, 2, and 3 generated by f at $x = a$.

9. $f(x) = \dfrac{1}{x}, \quad a = 2$

10. $f(x) = \sin x, \quad a = \pi/4$

11. $f(x) = \cos x, \quad a = \pi/4$

12. $f(x) = \sqrt{x}, \quad a = 4$

In Exercises 13–15, find the Taylor polynomial of order 3 generated by f

   (a) at $x = 0$;   **(b)** at $x = 1$.

13. $f(x) = x^3 - 2x + 4$

14. $f(x) = 2x^3 + x^2 + 3x - 8$ **15.** $f(x) = x^4$

16. Let f be a function that has derivatives of all orders for all real numbers. Assume $f(0) = 4$, $f'(0) = 5$, $f''(0) = -8$, and $f'''(0) = 6$.

   (a) Write the third order Taylor polynomial for f at $x = 0$ and use it to approximate $f(0.2)$.

   (b) Write the second order Taylor polynomial for f', the derivative of f, at $x = 0$ and use it to approximate $f'(0.2)$.

17. Let f be a function that has derivatives of all orders for all real numbers. Assume $f(1) = 4$, $f'(1) = -1$, $f''(1) = 3$, and $f'''(1) = 2$.

   (a) Write the third order Taylor polynomial for f at $x = 1$ and use it to approximate $f(1.2)$.

   (b) Write the second order Taylor polynomial for f', the derivative of f, at $x = 1$ and use it to approximate $f'(1.2)$.

18. The Maclaurin series for $f(x)$ is

$$f(x) = 1 + \frac{x}{2!} + \frac{x^2}{3!} + \frac{x^3}{4!} + \cdots + \frac{x^n}{(n + 1)!} + \cdots.$$

   (a) Find $f'(0)$ and $f^{(10)}(0)$.

   (b) Let $g(x) = xf(x)$. Write the Maclaurin series for $g(x)$, showing the first three nonzero terms and the general term.

   (c) Write $g(x)$ in terms of a familiar function without using series.

19. **(a)** Write the first three nonzero terms and the general term of the Taylor series generated by $e^{x/2}$ at $x = 0$.

   (b) Write the first three nonzero terms and the general term of the Taylor series at $x = 0$ for

$$g(x) = \frac{e^x - 1}{x}.$$

   (c) For the function g in part (b), find $g'(1)$ and use it to show that

$$\sum_{n=1}^{\infty} \frac{n}{(n + 1)!} = 1.$$

20. Let

$$f(t) = \frac{2}{1 - t^2} \quad \text{and} \quad G(x) = \int_0^x f(t)\, dt.$$

   (a) Find the first four terms and the general term for the Maclaurin series generated by f.

   (b) Find the first four nonzero terms and the Maclaurin series for G.

21. **(a)** Find the first four nonzero terms in the Taylor series generated by $f(x) = \sqrt{1 + x}$ at $x = 0$.

   (b) Use the results found in part (a) to find the first four nonzero terms in the Taylor series for $g(x) = \sqrt{1 + x^2}$ at $x = 0$.

   (c) Find the first four nonzero terms in the Taylor series at $x = 0$ for the function h such that $h'(x) = \sqrt{1 + x^2}$ and $h(0) = 5$.

22. Consider the power series

$$\sum_{n=0}^{\infty} a_n x^n, \quad \text{where } a_0 = 1 \text{ and } a_n = \left(\frac{3}{n}\right) a_{n-1} \text{ for } n \geq 1.$$

(This defines the coefficients *recursively*.)

   (a) Find the first four terms and the general term of the series.

   (b) What function f is represented by this power series?

   (c) Find the exact value of $f'(1)$.

23. Use the technique of Exploration 3 to determine the number of terms of the Maclaurin series for $\cos x$ that are needed to approximate the value of $\cos 18$ accurate to within 0.001 of the true value.

24. **Writing to Learn** Based on what you know about polynomial functions, explain why no Taylor polynomial of any order could actually equal $\sin x$.

25. Writing to Learn Your friend has memorized the Maclaurin series for both $\sin x$ and $\cos x$ but is having a hard time remembering which is which. Assuming that your friend knows the trigonometric functions well, what are some tips you could give that would help match $\sin x$ and $\cos x$ with their correct series?

26. What is the coefficient of x^5 in the Maclaurin series generated by $\sin 3x$?

27. What is the coefficient of $(x - 2)^3$ in the Taylor series generated by $\ln x$ at $x = 2$?

28. Writing to Learn Review the definition of the *linearization* of a differentiable function f at a in Chapter 4. What is the connection between the linearization of f and Taylor polynomials?

29. *Linearizations at Inflection Points*

(a) As the figure below suggests, linearizations fit particularly well at inflection points. As another example, graph *Newton's serpentine* $f(x) = 4x/(x^2 + 1)$ together with its linearizations at $x = 0$ and $x = \sqrt{3}$.

(b) Show that if the graph of a twice-differentiable function $f(x)$ has an inflection point at $x = a$, then the linearization of f at $x = a$ is also the second order Taylor polynomial of f at $x = a$. This explains why tangent lines fit so well at inflection points.

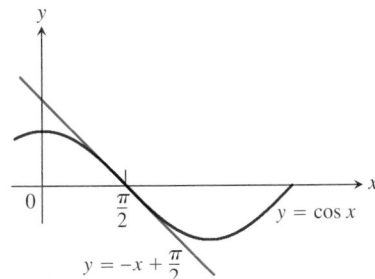

The graph of $f(x) = \cos x$ and its linearization at $\pi/2$. (Exercise 29)

30. According to the table of Maclaurin series, the power series

$$x - \frac{x^3}{3} + \frac{x^5}{5} - \cdots + (-1)^n \frac{x^{2n+1}}{2n + 1} + \cdots$$

converges at $x = \pm 1$. To what number does it converge when $x = 1$? To what number does it converge when $x = -1$?

Explorations

31. (a) Using the table of Maclaurin series, find a power series to represent $f(x) = (\sin x)/x$.

(b) The power series you found in part (a) is not quite a Maclaurin series for f, because f is technically not eligible to have a Maclaurin series. Why not?

(c) If we redefine f as follows, then the power series in (a) *will* be a Maclaurin series for f. What is the value of k?

$$f(x) = \begin{cases} \dfrac{\sin x}{x}, & x \neq 0, \\ k, & x = 0 \end{cases}$$

32. *Work in groups of two or three.* Find a function f whose Maclaurin series is

$$1x^1 + 2x^2 + 3x^3 + \cdots + nx^n + \cdots .$$ ■

Extending the Ideas

33. *The Binomial Series* Let $f(x) = (1 + x)^m$ for some nonzero constant m.

(a) Show that $f'''(x) = m(m - 1)(m - 2)(1 + x)^{m-3}$.

(b) Extend the result of part (a) to show that
$$f^{(k)}(0) = m(m - 1)(m - 2) \cdots (m - k + 1).$$

(c) Find the coefficient of x^k in the Maclaurin series generated by f.

(d) We define the symbol $\binom{m}{k}$ as follows:

$$\binom{m}{k} = \frac{m(m - 1)(m - 2) \cdots (m - k + 1)}{k!},$$

with the understanding that

$$\binom{m}{0} = 1 \quad \text{and} \quad \binom{m}{1} = m.$$

With this notation, show that the Maclaurin series generated by $f(x) = (1 + x)^m$ is

$$\sum_{k=0}^{\infty} \binom{m}{k} x^k.$$

This is called the **binomial series.**

34. *(Continuation of Exercise 33)* If m is a positive integer, explain why the Maclaurin series generated by f is a polynomial of degree m. (This means that

$$(1 + x)^m = \sum_{k=0}^{m} \binom{m}{k} x^k.$$

You may recognize this result as the **Binomial Theorem** from algebra.)

Taylor's Theorem

About Taylor Polynomials • The Remainder • Remainder
Estimation Theorem • Euler's Formula

About Taylor Polynomials

While there is a certain unspoiled beauty in the exactness of a convergent Taylor
series, it is the inexact Taylor polynomials that essentially do all the work. It is
satisfying to know, for example, that $\sin x$ can be found *exactly* by summing an
infinite Taylor series, but if we want to use that information to find $\sin 3$, we will
have to evaluate Taylor polynomials until we arrive at an *approximation* with
which we are satisfied. Even a computer must deal with finite sums.

Example 1 APPROXIMATING A FUNCTION TO SPECIFICATIONS

Find a Taylor polynomial that will serve as an adequate substitute for $\sin x$
on the interval $[-\pi, \pi]$.

Solution You do not have to be a professional mathematician to appreciate
the imprecision of this problem as written. We are simply unable to proceed
until someone decides what an "adequate" substitute is! We will revisit
this issue shortly, but for now let us accept the following clarification of
"adequate."

By "adequate," we mean that the polynomial should differ from $\sin x$ by
less than 0.0001 anywhere on the interval.

Now we have a clear mission: Choose $P_n(x)$ so that $|P_n(x) - \sin x| < 0.0001$
for every x in the interval $[-\pi, \pi]$. How do we do this?

Recall the nine graphs of the partial sums of the Maclaurin series for $\sin x$
in Section 9.2. They show that the approximations get worse as x moves
away from 0, suggesting that if we can make $|P_n(\pi) - \sin \pi| < 0.0001$,
then P_n will be adequate throughout the interval. Since $\sin \pi = 0$, this
means that we need to make $|P_n(\pi)| < 0.0001$.

We evaluate the partial sums at $x = \pi$, adding a term at a time, eventually
arriving at the following:

$$\pi - \pi^3/3! + \pi^{\wedge}5/5! - \pi$$
$$^{\wedge}7/7! + \pi^{\wedge}9/9! - \pi^{\wedge}1$$
$$1/11! + \pi^{\wedge}13/13!$$
$$2.114256749_{E}^{-}5$$

As graphical support that the polynomial $P_{13}(x)$ is adequate throughout
the interval, we graph the *absolute error* of the approximation, namely
$|P_{13}(x) - \sin x|$, in the window $[-\pi, \pi]$ by $[-0.00004, 0.00004]$
(Figure 9.7).

$[-\pi, \pi]$ by $[-0.00004, 0.00004]$

Figure 9.7 The graph shows that
$|P_{13}(x) - \sin x| < 0.00010$ throughout
the interval $[-\pi, \pi]$. (Example 1)

In practical terms, then, we would like to be able to use Taylor polynomials to approximate functions over the intervals of convergence of the Taylor series, and we would like to keep the error of the approximation within specified bounds. Since the error results from *truncating* the series down to a polynomial (that is, cutting it off after some number of terms), we call it the **truncation error.**

Example 2 TRUNCATION ERROR FOR A GEOMETRIC SERIES

Find a formula for the truncation error if we use $1 + x^2 + x^4 + x^6$ to approximate $1/(1 - x^2)$ over the interval $(-1, 1)$.

Solution We recognize this polynomial as the fourth partial sum of the geometric series for $1/(1 - x^2)$. Since this series converges to $1/(1 - x^2)$ on $(-1, 1)$, the truncation error is the absolute value of the part that we threw away, namely

$$\left| x^8 + x^{10} + \cdots + x^{2n} + \cdots \right|.$$

This is the absolute value of a geometric series with first term x^8 and $r = x^2$. Therefore,

$$\left| x^8 + x^{10} + \cdots + x^{2n} + \cdots \right| = \left| \frac{x^8}{1 - x^2} \right| = \frac{x^8}{1 - x^2}.$$

Figure 9.8 shows that the error is small near 0, but increases as x gets closer to 1 or -1.

$y = x^8/(1 - x^2)$

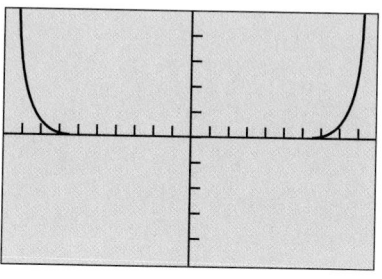

[–1, 1] by [–5, 5]

Figure 9.8 A graph of the truncation error on $(-1, 1)$ if $P_6(x)$ is used to approximate $1/(1 - x^2)$. (Example 2)

You can probably infer from our solution in Example 2 that the truncation error after 5 terms would be $x^{10}/(1 - x^2)$, and after n terms would be $x^{2n}/(1 - x^2)$. Figure 9.9 shows how these errors get closer to 0 on the interval $(-1, 1)$ as n gets larger, and that they still get worse as we approach -1 and 1.

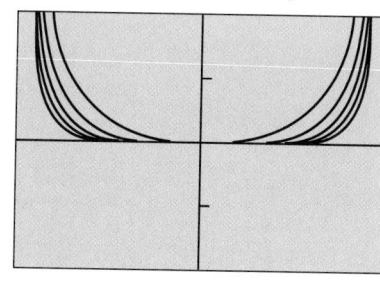

[–1, 1] by [–2, 2]

Figure 9.9 The truncation errors for $n = 2, 4, 6, 8, 10$, when we approximate $1/(1 - x^2)$ by its Taylor polynomials of higher and higher order. (The errors for the higher order polynomials are on the bottom.)

It was fortunate for our error analysis that this series was geometric, since the error was consequently a geometric series itself. This enabled us to write it as a (non-series) function and study it exactly. But how could we handle the error if we were to truncate a *nongeometric* series? That practical question sets the stage for Taylor's Theorem.

The Remainder

Every truncation splits a Taylor series into two equally significant pieces: the Taylor polynomial $P_n(x)$ that gives us the approximation, and the *remainder* $R_n(x)$ that tells us whether the approximation is any good. Taylor's Theorem is about both pieces.

Theorem 3 Taylor's Theorem with Remainder

If f has derivatives of all orders in an open interval I containing a, then for each positive integer n and for each x in I,

$$f(x) = f(a) + f'(a)(x - a) + \frac{f''(a)}{2!}(x - a)^2 + \cdots +$$

$$\frac{f^{(n)}(a)}{n!}(x - a)^n + R_n(x),$$

where

$$R_n(x) = \frac{f^{(n+1)}(c)}{(n + 1)!}(x - a)^{n+1}$$

for some c between a and x.

Pause for a moment to consider how remarkable this theorem is. If we wish to approximate f by a polynomial of degree n over an interval I, the theorem gives us both a formula for the *polynomial* and a formula for the *error* involved in using that approximation over the interval I.

The first equation in Taylor's Theorem is **Taylor's formula.** The function $R_n(x)$ is the **remainder of order n** or the **error term** for the approximation of f by $P_n(x)$ over I. It is also called the **Lagrange form** of the remainder, and bounds on $R_n(x)$ found using this form are **Lagrange error bounds.**

The introduction of $R_n(x)$ finally gives us a mathematically precise way to define what we mean when we say that a Taylor series converges to a function on an interval. If $R_n(x) \to 0$ as $n \to \infty$ for all x in I, we say that the Taylor series generated by f at $x = a$ **converges to f** on I, and we write

$$f(x) = \sum_{k=0}^{\infty} \frac{f^{(k)}(a)}{k!} (x - a)^k.$$

Example 3 PROVING CONVERGENCE OF A MACLAURIN SERIES

Prove that the series

$$\sum_{k=0}^{\infty} (-1)^k \frac{x^{2k+1}}{(2k + 1)!}$$

converges to $\sin x$ for all real x.

Solution We need to consider what happens to $R_n(x)$ as $n \to \infty$.

By Taylor's Theorem,

$$R_n(x) = \frac{f^{(n+1)}(c)}{(n + 1)!}(x - 0)^{n+1},$$

where $f^{(n+1)}(c)$ is the $(n + 1)$st derivative of $\sin x$ evaluated at some c between x and 0. This does not seem at first glance to give us much information, but *for this particular function* we can say something very significant about $f^{(n+1)}(c)$: it lies between -1 and 1 inclusive. Therefore, no matter what x is, we have

$$|R_n(x)| = \left| \frac{f^{(n+1)}(c)}{(n+1)!} (x-0)^{n+1} \right|$$

$$= \frac{|f^{(n+1)}(c)|}{(n+1)!} |x^{n+1}|$$

$$\leq \frac{1}{(n+1)!} |x^{n+1}| = \frac{|x|^{n+1}}{(n+1)!}.$$

What happens to $|x|^{n+1}/(n+1)!$ as $n \to \infty$? The numerator is a product of $n+1$ factors, all of them $|x|$. The denominator is a product of $n+1$ factors, the largest of which eventually exceed $|x|$ and keep on growing as $n \to \infty$. The factorial growth in the denominator, therefore, eventually outstrips the power growth in the numerator, and we have $|x|^{n+1}/(n+1)! \to 0$ for all x. This means that $R_n(x) \to 0$ for all x, which completes the proof.

Exploration 1 **Your Turn**

Modify the steps of the proof in Example 3 to prove that

$$\sum_{k=0}^{\infty} (-1)^k \frac{x^{2k}}{(2k)!}$$

converges to $\cos x$ for all real x.

Remainder Estimation Theorem

Notice that we were able to use the remainder formula in Taylor's Theorem to verify the convergence of two Taylor series to their generating functions ($\sin x$ and $\cos x$), and yet in neither case did we have to find an actual value for $f^{(n+1)}(c)$. Instead, we were able to put an *upper bound* on $|f^{(n+1)}(c)|$, which was enough to ensure that $R_n(x) \to 0$ for all x. This strategy is so convenient that we state it as a theorem for future reference.

Theorem 4 **Remainder Estimation Theorem**

If there are positive constants M and r such that $|f^{(n+1)}(t)| \leq Mr^{n+1}$ for all t between a and x, then the remainder $R_n(x)$ in Taylor's Theorem satisfies the inequality

$$|R_n(x)| \leq M \frac{r^{n+1}|x-a|^{n+1}}{(n+1)!}.$$

If these conditions hold for every n and all the other conditions of Taylor's Theorem are satisfied by f, then the series converges to $f(x)$.

It does not matter if M and r are huge; the important thing is that they do not get any *more huge* as $n \to \infty$. This allows the factorial growth to outstrip the power growth and thereby sweep $R_n(x)$ to zero.

Example 4 PROVING CONVERGENCE

Use the Remainder Estimation Theorem to prove that

$$e^x = \sum_{k=0}^{\infty} \frac{x^k}{k!}$$

for all real x.

Solution We have already seen that this is the Taylor series generated by e^x at $x = 0$, so all that remains is to verify that $R_n(x) \to 0$ for all x. By the Remainder Estimation Theorem, it suffices to find M and r such that $|f^{(n+1)}(t)| = e^t$ is bounded by Mr^{n+1} for t between 0 and an arbitrary x.

We know that e^t is an increasing function on any interval, so it reaches its maximum value at the right-hand endpoint. We can pick M to be that maximum value and simply let $r = 1$. If the interval is $[0, x]$, we let $M = e^x$; if the interval is $[x, 0]$, we let $M = e^0 = 1$. In either case, we have $e^t \leq M$ throughout the interval, and the Remainder Estimation Theorem guarantees convergence.

Example 5 ESTIMATING A REMAINDER

The approximation $\ln(1 + x) \approx x - (x^2/2)$ is used when x is small. Use the Remainder Estimation Theorem to get a bound for the maximum error when $|x| \leq 0.1$. Support the answer graphically.

Solution In the notation of the Remainder Estimation Theorem, $f(x) = \ln(1 + x)$, the polynomial is $P_2(x)$, and we need a bound for $|R_2(x)|$. On the interval $[-0.1, 0.1]$, the function $|f^{(3)}(t)| = 2/(1 + t)^3$ is strictly decreasing, achieving its maximum value at the left-hand endpoint, -0.1. We can therefore bound $|f^{(3)}(t)|$ by

$$M = \left| \frac{2}{(1 + (-0.1))^3} \right| = \frac{2000}{729}.$$

We can let $r = 1$.

By the Remainder Estimation Theorem, we may conclude that

$$|R_2(x)| \leq \frac{2000}{729} \cdot \frac{|x|^3}{3!} \leq \frac{2000}{729} \cdot \frac{|\pm 0.1|^3}{3!} < 4.6 \times 10^{-4}. \quad \text{Rounded up, to be safe}$$

Since $R_2(x) = \ln(1 + x) - (x - x^2/2)$, it is an easy matter to produce a graph to *observe* the behavior of the error on the interval $[-0.1, 0.1]$ (Figure 9.10).

The graph almost appears to have odd-function symmetry, but evaluation shows that $R_2(-0.1) \approx -3.605 \times 10^{-4}$ and $R_2(0.1) \approx 3.102 \times 10^{-4}$. The maximum *absolute* error on the interval is 3.605×10^{-4}, which is indeed less than the bound, 4.6×10^{-4}.

$y = \ln(1 + x) - (x - x^2/2)$

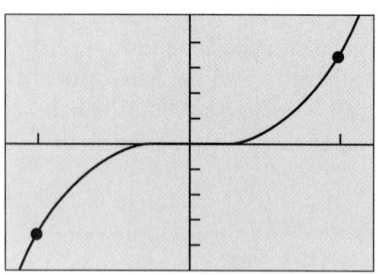

[−0.12, 0.12] by [−0.0005, 0.0005]

Figure 9.10 The graph of the error term $R_2(x)$ in Example 5. Maximum error for $|x| \leq 0.1$ occurs at the left-hand endpoint of the interval.

Euler's Formula

We have seen that $\sin x$, $\cos x$, and e^x equal their respective Maclaurin series for all real numbers x. It can also be shown that this is true for all *complex* numbers, although we would need to extend our concept of limit to know what

Srinivasa Ramanujan

1887–1920

Ramanujan, from southern India, wrote with amazing originality and depth on a wide range of topics in mathematics, including infinite series, prime and composite numbers, integers as the sum of squares, function theory, and combinatorics. His theorems have influenced medical research and statistical mechanics. One of his identities has been used by computer programmers to calculate the decimal expansion of pi to millions of digits. There are still areas of his work that have not been explored. Ramanujan was largely self taught and, although he worked with the British mathematician G. H. Hardy of Cambridge, he never graduated from college because he neglected his other studies for mathematics.

convergence would mean in that context. Accept for the moment that we can substitute complex numbers into these power series, and let us see where that might lead.

We mentioned at the beginning of the chapter that Leonhard Euler had derived some powerful results using infinite series. One of the most impressive was the surprisingly simple relationship he discovered that connects the exponential function e^x to the trigonometric functions $\sin x$ and $\cos x$. You do not need a deep understanding of complex numbers to understand what Euler did, but you do need to recall the powers of $i = \sqrt{-1}$.

$$i^1 = i$$
$$i^2 = -1$$
$$i^3 = -i$$
$$i^4 = 1$$
$$i^5 = i$$

etc.

Now try this exploration!

Exploration 2   **Euler's Formula**

Assume that e^x, $\cos x$, and $\sin x$ equal their Maclaurin series (as in the table in Section 9.2) for complex numbers as well as for real numbers.

1. Find the Maclaurin series for e^{ix}.

2. Use the result of part 1 and the Maclaurin series for $\cos x$ and $\sin x$ to prove that $e^{ix} = \cos x + i \sin x$. This equation is known as **Euler's formula.**

3. Use Euler's formula to prove that $e^{i\pi} + 1 = 0$. This beautiful equation, which brings together the five most celebrated numbers in mathematics in such a stunningly unexpected way, is also widely known as Euler's formula. (There are still others. The prolific Euler had more than his share.)

Quick Review 9.3

In Exercises 1–5, find the smallest number M that bounds $|f|$ from above on the interval I (that is, find the smallest M such that $|f(x)| \le M$ for all x in I).

1. $f(x) = 2 \cos (3x)$, $I = [-2\pi, 2\pi]$

2. $f(x) = x^2 + 3$, $I = [1, 2]$

3. $f(x) = 2^x$, $I = [-3, 0]$

4. $f(x) = \dfrac{x}{x^2 + 1}$, $I = [-2, 2]$

5. $f(x) = \begin{cases} 2 - x^2, & x \le 1, \\ 2x - 1, & x > 1, \end{cases}$ $I = [-3, 3]$

In Exercises 6–10, tell whether the function has derivatives of all orders at the given value of a.

6. $\dfrac{x}{x + 1}$, $a = 0$

7. $|x^2 - 4|$, $a = 2$

8. $\sin x + \cos x$, $a = \pi$

9. e^{-x}, $a = 0$

10. $x^{3/2}$, $a = 0$

Section 9.3 Exercises

In Exercises 1–5, find the Taylor polynomial of order four for the function at $x = 0$, and use it to approximate the value of the function at $x = 0.2$.

1. e^{-2x}

2. $\cos(\pi x/2)$

3. $5 \sin(-x)$

4. $\ln(1 + x^2)$

5. $(1 - x)^{-2}$

In Exercises 6–10, find the Maclaurin series for the function.

6. xe^x

7. $\sin x - x + \dfrac{x^3}{3!}$

8. $\cos^2 x \quad \left(= \dfrac{1 + \cos 2x}{2} \right)$

9. $\sin^2 x$

10. $\dfrac{x^2}{1 - 2x}$

11. For approximately what values of x can you replace $\sin x$ by $x - (x^3/6)$ with an error magnitude no greater than 5×10^{-4}? Give reasons for your answer.

12. If $\cos x$ is replaced by $1 - (x^2/2)$ and $|x| < 0.5$, what estimate can be made of the error? Does $1 - (x^2/2)$ tend to be too large or too small? Support your answer graphically.

13. How close is the approximation $\sin x \approx x$ when $|x| < 10^{-3}$? For which of these values of x is $x < \sin x$? Support your answer graphically.

14. The approximation $\sqrt{1 + x} \approx 1 + (x/2)$ is used when x is small. Estimate the maximum error when $|x| < 0.01$.

15. The approximation $e^x \approx 1 + x + (x^2/2)$ is used when x is small. Use the Remainder Estimation Theorem to estimate the error when $|x| < 0.1$.

16. *Hyperbolic sine and cosine* The hyperbolic sine and hyperbolic cosine functions, denoted sinh and cosh respectively, are defined as

$$\sinh x = \frac{e^x - e^{-x}}{2} \quad \text{and} \quad \cosh x = \frac{e^x + e^{-x}}{2}.$$

(Appendix A6 gives more information about hyperbolic functions.)

Find the Maclaurin series generated by $\sinh x$ and $\cosh x$.

17. *(Continuation of Exercise 16)* Use the Remainder Estimation Theorem to prove that $\cosh x$ equals its Maclaurin series for all real numbers x.

18. Writing to Learn Review the statement of the Mean Value Theorem (Section 4.2) and explain its relationship to Taylor's Theorem.

Quadratic Approximations Just as we call the Taylor polynomial of order 1 generated by f at $x = a$ the *linearization* of f at a, we call the Taylor polynomial of order 2 generated by f at $x = a$ the *quadratic approximation* of f at a.

In Exercises 19–23, find **(a)** the linearization and **(b)** the quadratic approximation of f at $x = 0$. Then **(c)** graph the function and its linear and quadratic approximations together around $x = 0$ and comment on how the graphs are related.

19. $f(x) = \ln(\cos x)$

20. $f(x) = e^{\sin x}$

21. $f(x) = 1/\sqrt{1 - x^2}$

22. $f(x) = \sec x$

23. $f(x) = \tan x$

24. Use the Taylor polynomial of order 2 to find the quadratic approximation of $f(x) = (1 + x)^k$ at $x = 0$ (k a constant). If $k = 3$, for approximately what values of x in the interval $[0, 1]$ will the magnitude of the error in the quadratic approximation be less than $1/100$?

25. *A Cubic Approximation of e^x* The approximation

$$e^x \approx 1 + x + \frac{x^2}{2} + \frac{x^3}{6}$$

is used on small intervals about the origin. Estimate the magnitude of the approximation error for $|x| \le 0.1$.

26. *A Cubic Approximation* Use the Taylor polynomial of order 3 to find the *cubic approximation* of $f(x) = 1/(1 - x)$ at $x = 0$. Give an upper bound for the magnitude of the approximation error for $|x| \le 0.1$.

27. Consider the initial value problem,

$$\frac{dy}{dx} = e^{-x^2} \quad \text{and} \quad y = 2 \quad \text{when} \quad x = 0.$$

(a) Can you find a formula for the function y that does not involve any integrals?

(b) Can you represent y by a power series?

(c) For what values of x does this power series actually equal the function y? Give a reason for your answer.

28. (a) Construct the Maclaurin series for $\ln(1 - x)$.

(b) Use this series and the series for $\ln(1 + x)$ to construct a Maclaurin series for

$$\ln \frac{1 + x}{1 - x}.$$

29. *Identifying Graphs* Which well-known functions are approximated on the interval $(-\pi/2, \pi/2)$ by the following Taylor polynomials?

(a) $x + \dfrac{x^3}{3} + \dfrac{2x^5}{15} + \dfrac{17x^7}{315} + \dfrac{62x^9}{2835}$

(b) $1 + \dfrac{x^2}{2} + \dfrac{5x^4}{24} + \dfrac{61x^6}{720} + \dfrac{277x^8}{8064}$

Explorations

30. *Work in groups of two or three.* Try to reinforce each other's ideas and verify your computations at each step.

(a) Use the identity

$$\sin^2 x = \frac{1}{2}(1 - \cos 2x)$$

to obtain the Maclaurin series for $\sin^2 x$.

(b) Differentiate this series to obtain the Maclaurin series for $2 \sin x \cos x$.

(c) Verify that this is the series for $\sin 2x$.

31. *Improving Approximations to* π

(a) Let P be an approximation of π accurate to n decimal places. Check with a calculator to see that $P + \sin P$ gives an approximation correct to $3n$ decimal places!

(b) Use the Remainder Estimation Theorem and the Maclaurin series for $\sin x$ to explain what is happening in part (a). (*Hint:* Let $P = \pi + x$, where x is the error of the estimate. Why should $(P + \sin P) - \pi$ be less than x^3?)

32. *Euler's Identities* Use Euler's formula to show that

(a) $\cos \theta = \dfrac{e^{i\theta} + e^{-i\theta}}{2}$, and

(b) $\sin \theta = \dfrac{e^{i\theta} - e^{-i\theta}}{2i}$.

Extending the Ideas

33. When a and b are real numbers, we define $e^{(a+ib)x}$ with the equation

$$e^{(a+ib)x} = e^{ax} \cdot e^{ibx} = e^{ax}(\cos bx + i \sin bx).$$

Differentiate the right-hand side of this equation to show that

$$\frac{d}{dx} e^{(a+ib)x} = (a + ib) e^{(a+ib)x}.$$

Thus, the familiar rule

$$\frac{d}{dx} e^{kx} = k e^{kx}$$

holds for complex values of k as well as for real values.

34. (*Continuation of Exercise 33*)

(a) Confirm the antiderivative formula

$$\int e^{(a+ib)x}\, dx = \frac{a - ib}{a^2 + b^2} e^{(a+ib)x} + C$$

by differentiating both sides. (In this case, $C = C_1 + iC_2$ is a complex constant of integration.)

(b) Two complex numbers $a + ib$ and $c + id$ are equal if and only if $a = c$ and $b = d$. Use this fact and the formula in (a) to evaluate $\int e^{ax} \cos bx\, dx$ and $\int e^{ax} \sin bx\, dx$.

| 9.4 | **Radius of Convergence** |

Convergence • *n*th-Term Test • Comparing Nonnegative Series • Ratio Test • Endpoint Convergence

Seki Kowa (1642–1708)

Child prodigy, brilliant mathematician, and inspirational teacher, Seki Kowa was born into a samurai warrior family in Fujioka, Kozuke, Japan, and adopted by the family of an accountant. Among his contributions were an improved method of solving higher-degree equations, the use of determinants in solving simultaneous equations, and a form of calculus known in Japan as *yenri*. It is difficult to know the full extent of his work because the samurai code demanded great modesty. Seki Kowa is credited with awakening in Japan a scientific spirit that continues to this day.

Convergence

Throughout our explorations of infinite series we stressed the importance of convergence. In terms of numbers, the difference between a convergent series and a divergent series could hardly be more stark: a convergent series is a number and may be treated as such; a divergent series is not a number and must not be treated as one.

Recall that the symbol "=" means many different things in mathematics.

1. $1 + 1 = 2$ signifies *equality of real numbers*. It is a true sentence.

2. $2(x - 3) = 2x - 6$ signifies *equivalent expressions*. It is a true sentence.

3. $x^2 + 3 = 7$ is an *equation*. It is an *open sentence,* because it can be true or false, depending on whether x is a solution to the equation.

4. $(x^2 - 1)/(x + 1) = x - 1$ is an *identity*. It is a true sentence (very much like (2)), but with the important qualification that x must be in the *domain of both expressions*. If either side of the equality is undefined, the sentence is meaningless. Substituting -1 into both sides of the equation in (3) gives a sentence that is mathematically false (i.e., $4 = 7$); substituting -1 into both sides of this identity gives a sentence that is meaningless.

$y = \dfrac{1}{1 + x^2}$

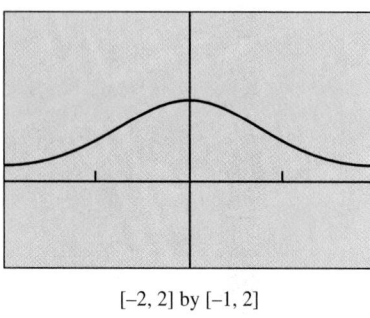

[−2, 2] by [−1, 2]

(a)

Partial Sums

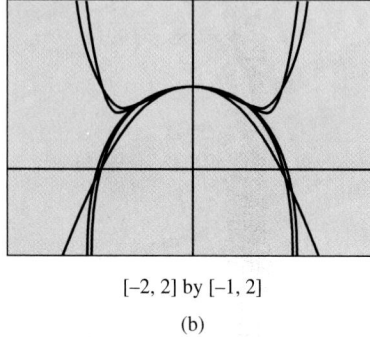

[−2, 2] by [−1, 2]

(b)

Figure 9.11 (a) The graph of $y = 1/(1 + x^2)$ and (b) the graphs of the Taylor polynomials $P_2(x)$, $P_4(x)$, $P_6(x)$, $P_8(x)$, and $P_{10}(x)$. The approximations become better and better, but only over the interval of convergence $(−1, 1)$. (Example 1)

Example 1 THE IMPORTANCE OF CONVERGENCE

Consider the sentence

$$\frac{1}{1 + x^2} = 1 - x^2 + x^4 - x^6 + \cdots + (-1)^n x^{2n} + \cdots .$$

For what values of x is this an identity?

Solution The function on the left has domain all real numbers. The function on the right can be viewed as a limit of Taylor polynomials. Each Taylor polynomial has domain all real numbers, but the polynomial values *converge* only when $|x| < 1$, so the *series* has domain $(−1, 1)$. If we graph the Taylor polynomials (Figure 9.11) we can see the dramatic convergence to $1/(1 + x^2)$ over the interval $(−1, 1)$. The divergence is just as dramatic for $|x| \geq 1$.

For values of x outside the interval, the statement in this example is meaningless. The Taylor series on the right diverges so it is not a number. The sentence is an identity for x in $(−1, 1)$.

As convincing as these graphs are, they do not *prove* convergence or divergence as $n \to \infty$. The series in Example 1 happens to be geometric, so we do have an analytic proof that it converges for $|x| < 1$ and diverges for $|x| \geq 1$, but for nongeometric series we do not have such undeniable assurance about convergence (yet).

In this section we develop a strategy for finding the interval of convergence of an arbitrary power series and backing it up with proof. We begin by noting that any power series of the form $\sum_{n=0}^{\infty} c_n(x - a)^n$ always converges at $x = a$, thus assuring us of at least one coordinate on the real number line where the series must converge. We have encountered power series that converge for all real numbers (the Maclaurin series for $\sin x$, $\cos x$, and e^x), and we have encountered power series like the series in Example 1 that converge only on a finite interval centered at a. A useful fact about power series is that those are the only possibilities, as the following theorem attests.

Theorem 5 The Convergence Theorem for Power Series

There are three possibilities for $\sum_{n=0}^{\infty} c_n(x - a)^n$ with respect to convergence:

1. There is a positive number R such that the series diverges for $|x - a| > R$ but converges for $|x - a| < R$. The series may or may not converge at either of the endpoints $x = a - R$ and $x = a + R$.

2. The series converges for every x $(R = \infty)$.

3. The series converges at $x = a$ and diverges elsewhere $(R = 0)$.

The number R is the **radius of convergence,** and the set of all values of x for which the series converges is the **interval of convergence.** The radius of convergence completely determines the interval of convergence if R is either zero or infinite. For $0 < R < \infty$, however, there remains the question of what happens at the endpoints of the interval. The table of Maclaurin series at the end of Section 9.2 includes intervals of convergence that are open, half-open, and closed.

We will learn how to find the radius of convergence first, and then we will settle the endpoint question in Section 9.5.

*n*th-Term Test

The most obvious requirement for convergence of a series is that the *n*th term must go to zero as $n \to \infty$. If the partial sums are approaching a limit *S*, then they also must be getting close to one another, so that for a convergent series $\sum a_n$,

$$\lim_{n \to \infty} a_n = \lim_{n \to \infty} (S_n - S_{n-1}) = S - S = 0.$$

This gives a handy test for divergence:

Theorem 6   The *n*th-Term Test for Divergence

$\sum_{n=1}^{\infty} a_n$ diverges if $\lim_{n \to \infty} a_n$ fails to exist or is different from zero.

Example 2   A SERIES WITH RADIUS OF CONVERGENCE 0

Find the radius of convergence of the series $\sum_{n=0}^{\infty} n! x^n$.

Solution If $|x| \geq 1$, the terms $n! x^n$ grow without bound as $n \to \infty$, $\lim_{n \to \infty} n! x^n$ fails to exist, and the series diverges by the *n*th-Term Test.

If $0 < |x| < 1$, then

$$\lim_{n \to \infty} n! x^n = \lim_{n \to \infty} \frac{n!}{(1/x)^n},$$

a quotient in which both numerator and denominator grow without bound. Recall, however, that factorial growth outstrips power growth, so again the limit fails to exist.

If $x = 0$, the series does converge. The radius of convergence is $R = 0$, and the interval of convergence is $\{0\}$.

Comparing Nonnegative Series

An effective way to show that a series $\sum a_n$ of nonnegative numbers *converges* is to compare it term by term with a known convergent series $\sum c_n$.

Theorem 7   The Direct Comparison Test

Let $\sum a_n$ be a series with no negative terms.

(a) $\sum a_n$ converges if there is a convergent series $\sum c_n$ with $a_n \leq c_n$ for all $n > N$, for some integer *N*.

(b) $\sum a_n$ diverges if there is a divergent series $\sum d_n$ of nonnegative terms with $a_n \geq d_n$ for all $n > N$, for some integer *N*.

If we can show that $\sum a_n$, $a_n \geq 0$ is eventually dominated by a convergent series, that will establish the convergence of $\sum a_n$. If we can show that $\sum a_n$ eventually dominates a divergent series of nonnegative terms, that will establish the divergence of $\sum a_n$.

We leave the proof to Exercises 42 and 43.

Example 3 PROVING CONVERGENCE BY COMPARISON

Prove that $\displaystyle\sum_{n=0}^{\infty} \frac{x^{2n}}{(n!)^2}$ converges for all real x.

Solution Let x be any real number. The series

$$\sum_{n=0}^{\infty} \frac{x^{2n}}{(n!)^2}$$

has no negative terms.

For any n, we have

$$\frac{x^{2n}}{(n!)^2} \le \frac{x^{2n}}{n!} = \frac{(x^2)^n}{n!}.$$

We recognize

$$\sum_{n=0}^{\infty} \frac{(x^2)^n}{n!}$$

as the Taylor series for e^{x^2}, which we know converges to e^{x^2} for all real numbers. Since the e^{x^2} series dominates

$$\sum_{n=0}^{\infty} \frac{x^{2n}}{(n!)^2}$$

term by term, the latter series must also converge for all real numbers by the Direct Comparison Test.

For the Direct Comparison Test to apply, the terms of the unknown series must be *nonnegative*. The fact that $\sum a_n$ is dominated by a convergent positive series means nothing if $\sum a_n$ diverges to $-\infty$. You might think that the requirement of nonnegativity would limit the usefulness of the Direct Comparison Test, but in practice this does not turn out to be the case. We can apply our test to $\sum |a_n|$ (which certainly has no negative terms); if $\sum |a_n|$ converges, then $\sum a_n$ converges.

Definition Absolute Convergence

If the series $\sum |a_n|$ of absolute values converges, then $\sum a_n$ **converges absolutely.**

Theorem 8 Absolute Convergence Implies Convergence

If $\sum |a_n|$ converges, then $\sum a_n$ converges.

Proof For each n,

$$-|a_n| \le a_n \le |a_n|, \quad \text{so} \quad 0 \le a_n + |a_n| \le 2\,|a_n|.$$

If $\sum |a_n|$ converges, then $\sum 2|a_n|$ converges, and by the Direct Comparison Test, the nonnegative series $\sum (a_n + |a_n|)$ converges. The equality $a_n = (a_n + |a_n|) - |a_n|$ now allows us to express $\sum a_n$ as the difference of two convergent series:

$$\sum a_n = \sum (a_n + |a_n| - |a_n|) = \sum (a_n + |a_n|) - \sum |a_n|.$$

Therefore, $\sum a_n$ converges. ∎

Example 4 USING ABSOLUTE CONVERGENCE

Show that

$$\sum_{n=0}^{\infty} \frac{(\sin x)^n}{n!}$$

converges for all x.

Solution Let x be any real number. The series

$$\sum_{n=0}^{\infty} \frac{|\sin x|^n}{n!}$$

has no negative terms, and it is term-by-term less than or equal to the series $\sum_{n=0}^{\infty} (1/n!)$, which we know converges to e. Therefore,

$$\sum_{n=0}^{\infty} \frac{|\sin x|^n}{n!}$$

converges by direct comparison. Since

$$\sum_{n=0}^{\infty} \frac{(\sin x)^n}{n!}$$

converges absolutely, it converges.

Ratio Test

Our strategy for finding the radius of convergence for an arbitrary power series will be to check for absolute convergence using a powerful test called the *Ratio Test*.

L'Hôpital's Rule is occasionally helpful in determining the limits that arise here.

Theorem 9 The Ratio Test

Let $\sum a_n$ be a series with positive terms, and with

$$\lim_{n \to \infty} \frac{a_{n+1}}{a_n} = L.$$

Then,

(a) the series *converges* if $L < 1$,

(b) the series *diverges* if $L > 1$,

(c) the test is *inconclusive* if $L = 1$.

Proof

(a) $L < 1$:

Choose some number r such that $L < r < 1$. Since

$$\lim_{n \to \infty} \frac{a_{n+1}}{a_n} = L,$$

we know that there is some N large enough so that a_{n+1}/a_n is arbitrarily close to L for all $n \geq N$. In particular, we can guarantee that for some N large enough, $(a_{n+1}/a_n) < r$ for all $n \geq N$. (See Figure 9.12.)

Figure 9.12 Since

$$\lim_{n \to \infty} \frac{a_{n+1}}{a_n} = L,$$

there is some N large enough so that a_{n+1}/a_n lies inside this open interval around L for all $n \geq N$. This guarantees that $a_{n+1}/a_n < r < 1$ for all $n \geq N$.

Thus,

$$\frac{a_{N+1}}{a_N} < r \quad \text{so} \quad a_{N+1} < ra_N$$

$$\frac{a_{N+2}}{a_{N+1}} < r \quad \text{so} \quad a_{N+2} < ra_{N+1} < r^2 a_N$$

$$\frac{a_{N+3}}{a_{N+2}} < r \quad \text{so} \quad a_{N+3} < ra_{N+2} < r^3 a_N$$

$$\vdots$$

This shows that for $n \geq N$ we can dominate $\Sigma\, a_n$ by $a_N (1 + r + r^2 + \cdots)$. Since $0 < r < 1$, this latter series is a convergent geometric series, and so $\Sigma\, a_n$ converges by the Direct Comparison Test.

(b) $L > 1$:

From some index M,

$$\frac{a_{n+1}}{a_n} > 1$$

for all $n \geq M$. In particular,

$$a_M < a_{M+1} < a_{M+2} < \cdots.$$

The terms of the series do not approach 0, so $\Sigma\, a_n$ diverges by the nth-Term Test.

(c) $L = 1$:

In Exploration 1 you will finish the proof by showing that the Ratio Test is inconclusive when $L = 1$. ∎

A Note on Absolute Convergence: The proof of the Ratio Test shows that the convergence of a power series inside its radius of convergence is *absolute* convergence, a stronger result than we first stated in Theorem 5. We will learn more about the distinction between convergence and absolute convergence in Section 9.5.

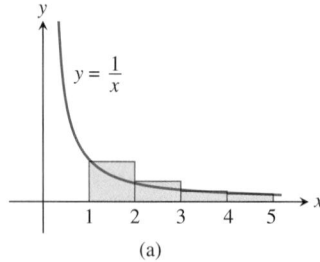

Figure 9.13 Find these areas. (Exploration 1)

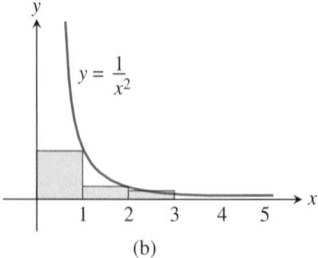

Figure 9.14 The areas of the rectangles form a series in each case. (Exploration 1)

Exploration 1 **Finishing the Proof of the Ratio Test**

Consider

$$\sum_{n=1}^{\infty} \frac{1}{n} \quad \text{and} \quad \sum_{n=1}^{\infty} \frac{1}{n^2}.$$

(We will refer to them hereafter in this exploration as $\Sigma\, 1/n$ and $\Sigma\, 1/n^2$.)

1. Show that the Ratio Test yields $L = 1$ for both series.
2. Use improper integrals to find the areas shaded in Figures 9.13a and 9.13b for $1 \leq x < \infty$.
3. Explain how Figure 9.14a shows that $\Sigma\, 1/n$ diverges, while Figure 9.14b shows that $\Sigma\, 1/n^2$ converges.
4. Explain how this proves the last part of the Ratio Test.

Example 5 FINDING THE RADIUS OF CONVERGENCE

Find the radius of convergence of

$$\sum_{n=0}^{\infty} \frac{nx^n}{10^n}.$$

Solution We check for absolute convergence using the Ratio Test.

$$\lim_{n \to \infty} \frac{|a_{n+1}|}{|a_n|} = \lim_{n \to \infty} \frac{(n+1)\,|x^{n+1}|}{10^{n+1}} \cdot \frac{10^n}{n|x^n|}$$

$$= \lim_{n \to \infty} \left(\frac{n+1}{n} \right) \frac{|x|}{10} = \frac{|x|}{10}$$

Setting $|x|/10 < 1$, we see that the series converges absolutely (and hence converges) for $-10 < x < 10$. The series diverges for $|x| > 10$, which means (by Theorem 5, the Convergence Theorem for Power Series) that it diverges for $x > 10$ and for $x < -10$. The radius of convergence is 10.

Endpoint Convergence

The Ratio Test, which is really a test for absolute convergence, establishes the radius of convergence for $\sum |c_n(x-a)^n|$. Theorem 5 guarantees that this is the same as the radius of convergence of $\sum c_n(x-a)^n$. Therefore, all that remains to be resolved about the convergence of an arbitrary power series is the question of convergence at the endpoints of the convergence interval when the radius of convergence is a finite, nonzero number.

Exploration 2 Revisiting a Maclaurin Series

For what values of x does the series

$$x - \frac{x^2}{2} + \frac{x^3}{3} - \cdots + (-1)^{n-1}\frac{x^n}{n} + \cdots$$

converge?

1. Apply the Ratio Test to determine the radius of convergence.
2. Substitute the left-hand endpoint of the interval into the power series. Use Figure 9.14a of Exploration 1 to help you decide whether the resulting series converges or diverges.
3. Substitute the right-hand endpoint of the interval into the power series. You should get

$$1 - \frac{1}{2} + \frac{1}{3} - \frac{1}{4} + \cdots + \frac{(-1)^{n-1}}{n} + \cdots.$$

   Chart the progress of the partial sums of this series geometrically on a number line as follows: Start at 0. Go forward 1. Go back 1/2. Go forward 1/3. Go back 1/4. Go forward 1/5, etc.

4. Does the series converge at the right-hand endpoint? Give a convincing argument based on your geometric journey in part 3.
5. Does the series converge *absolutely* at the right-hand endpoint?

The question of convergence of a power series at an endpoint is really a question about the convergence of a series of numbers. If the series is geometric with first term a and common ratio r, then the series converges to $a/(1 - r)$ if $|r| < 1$ and diverges if $|r| \geq 1$. Another type of series whose sums are easily found are **telescoping series,** as illustrated in Example 6.

Telescoping Series

We call the series in Example 6 a *telescoping series* because its partial sums collapse like an old handheld telescope.

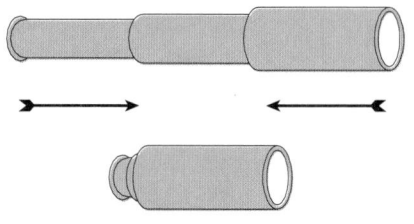

Example 6   SUMMING A TELESCOPING SERIES

Find the sum of $\displaystyle\sum_{n=1}^{\infty} \frac{1}{n(n + 1)}$.

Solution Use partial fractions to rewrite the nth term.

$$\frac{1}{n(n + 1)} = \frac{1}{n} - \frac{1}{n + 1}$$

We compute a few partial sums to find a general formula.

$$s_1 = 1 - \frac{1}{2}$$

$$s_2 = \left(1 - \frac{1}{2}\right) + \left(\frac{1}{2} - \frac{1}{3}\right) = 1 - \frac{1}{3}$$

$$s_3 = \left(1 - \frac{1}{2}\right) + \left(\frac{1}{2} - \frac{1}{3}\right) + \left(\frac{1}{3} - \frac{1}{4}\right) = 1 - \frac{1}{4}$$

We can see that, in general,

$$s_n = 1 - \frac{1}{n},$$

because all the terms between the first and last cancel when the parentheses are removed. Therefore, the sum of the series is

$$S = \lim_{n \to \infty} s_n = 1.$$

The final section of this chapter will formalize some of the strategies used in Exploration 2 and Example 6 and will develop additional tests that can be used to determine series behavior at endpoints.

Quick Review 9.4

In Exercises 1–5, find the limit of the expression as $n \to \infty$. Assume x remains fixed as n changes.

1. $\dfrac{n|x|}{n + 1}$

2. $\dfrac{n^2 |x - 3|}{n(n - 1)}$

3. $\dfrac{|x|^n}{n!}$

4. $\dfrac{(n + 1)^4 x^2}{(2n)^4}$

5. $\dfrac{|2x + 1|^{n+1} 2^n}{2^{n+1} |2x + 1|^n}$

In Exercises 6–10, let a_n be the nth term of the first and b_n the nth term of the second series. Find the smallest positive integer N for which $a_n > b_n$ for all $n \geq N$. Identify a_n and b_n.

6. $\sum 5n, \quad \sum n^2$

7. $\sum n^5, \quad \sum 5^n$

8. $\sum \ln n, \quad \sum \sqrt{n}$

9. $\sum \dfrac{1}{10^n}, \quad \sum \dfrac{1}{n!}$

10. $\sum \dfrac{1}{n^2}, \quad \sum n^{-3}$

Section 9.4 Exercises

In Exercises 1–16, determine the convergence or divergence of the series. Identify the test (or tests) you use. There may be more than one correct way to determine convergence or divergence of a given series.

1. $\sum_{n=1}^{\infty} \dfrac{n}{n+1}$

2. $\sum_{n=1}^{\infty} \dfrac{2^n}{n+1}$

3. $\sum_{n=1}^{\infty} \dfrac{n^2-1}{2^n}$

4. $\sum_{n=1}^{\infty} -\dfrac{1}{8^n}$

5. $\sum_{n=1}^{\infty} \dfrac{2^n}{3^n+1}$

6. $\sum_{n=1}^{\infty} n\sin\left(\dfrac{1}{n}\right)$

7. $\sum_{n=0}^{\infty} n^2 e^{-n}$

8. $\sum_{n=0}^{\infty} \dfrac{n^{10}}{10^n}$

9. $\sum_{n=1}^{\infty} \dfrac{(n+3)!}{3!\,n!\,3^n}$

10. $\sum_{n=1}^{\infty} \left(1+\dfrac{1}{n}\right)^n$

11. $\sum_{n=0}^{\infty} \dfrac{(-2)^n}{3^n}$

12. $\sum_{n=1}^{\infty} n!\,e^{-n}$

13. $\sum_{n=1}^{\infty} \dfrac{3^n}{n^3 2^n}$

14. $\sum_{n=1}^{\infty} \dfrac{n\ln n}{2^n}$

15. $\sum_{n=1}^{\infty} \dfrac{n!}{(2n+1)!}$

16. $\sum_{n=1}^{\infty} \dfrac{n!}{n^n}$ (*Hint:* If you do not recognize L, try recognizing the reciprocal of L.)

17. Give an example to show that the converse of the nth-Term Test is false. That is, $\sum a_n$ might diverge even though $\lim_{n\to\infty} a_n = 0$.

18. Find two convergent series $\sum a_n$ and $\sum b_n$ such that $\sum (a_n/b_n)$ diverges.

In Exercises 19–34, find the *radius* of convergence of the power series.

19. $\sum_{n=0}^{\infty} x^n$

20. $\sum_{n=0}^{\infty} (x+5)^n$

21. $\sum_{n=0}^{\infty} (-1)^n (4x+1)^n$

22. $\sum_{n=1}^{\infty} \dfrac{(3x-2)^n}{n}$

23. $\sum_{n=0}^{\infty} \dfrac{(x-2)^n}{10^n}$

24. $\sum_{n=0}^{\infty} \dfrac{nx^n}{n+2}$

25. $\sum_{n=1}^{\infty} \dfrac{x^n}{n\sqrt{n}\,3^n}$

26. $\sum_{n=0}^{\infty} \dfrac{x^{2n+1}}{n!}$

27. $\sum_{n=0}^{\infty} \dfrac{n(x+3)^n}{5^n}$

28. $\sum_{n=0}^{\infty} \dfrac{nx^n}{4^n(n^2+1)}$

29. $\sum_{n=0}^{\infty} \dfrac{\sqrt{n}\,x^n}{3^n}$

30. $\sum_{n=0}^{\infty} n!(x-4)^n$

31. $\sum_{n=0}^{\infty} (-2)^n(n+1)(x-1)^n$

32. $\sum_{n=1}^{\infty} \dfrac{(4x-5)^{2n+1}}{n^{3/2}}$

33. $\sum_{n=1}^{\infty} \dfrac{(x+\pi)^n}{\sqrt{n}}$

34. $\sum_{n=0}^{\infty} \dfrac{(x-\sqrt{2})^{2n+1}}{2^n}$

In Exercises 35–40, find the *interval* of convergence of the series and, within this interval, the sum of the series as a function of x.

35. $\sum_{n=0}^{\infty} \dfrac{(x-1)^{2n}}{4^n}$

36. $\sum_{n=0}^{\infty} \dfrac{(x+1)^{2n}}{9^n}$

37. $\sum_{n=0}^{\infty} \left(\dfrac{\sqrt{x}}{2}-1\right)^n$

38. $\sum_{n=0}^{\infty} (\ln x)^n$

39. $\sum_{n=0}^{\infty} \left(\dfrac{x^2-1}{3}\right)^n$

40. $\sum_{n=0}^{\infty} \left(\dfrac{\sin x}{2}\right)^n$

41. Writing to Learn We reviewed in Section 9.1 how to find the interval of convergence for the geometric series $\sum_{n=0}^{\infty} x^n$. Can we find the interval of convergence of a geometric series by using the Ratio Test? Explain.

Explorations

Nondecreasing Sequences As you already know, a nondecreasing (or increasing) function $f(x)$ that is bounded from above on an interval $[a, \infty)$ has a limit as $x\to\infty$ that is less than or equal to the bound. The same is true of sequences of numbers. If $s_1 \le s_2 \le s_3 \le \dots \le s_n \dots$ and there is a number M such that $|s_n| \le M$ for all n, then the sequence converges to a limit $S \le M$. You will need this fact as you work through Exercises **42** and **43**. *Work in groups of two or three.*

42. *Proof of the Direct Comparison Test, Part a* Let $\sum a_n$ be a series with no negative terms, and let $\sum c_n$ be a convergent series such that $a_n \le c_n$ for all $n \ge N$, for some integer N.

(a) Show that the partial sums of $\sum a_n$ are bounded above by

$$a_1 + \dots + a_N + \sum_{n=N+1}^{\infty} c_n.$$

(b) Explain why this shows that $\sum a_n$ must converge.

43. *Proof of the Direct Comparison Test, Part b* Let Σa_n be a series with no negative terms, and let Σd_n be a divergent series of nonnegative terms such that $a_n \geq d_n$ for all $n \geq N$, for some integer N.

(a) Show that the partial sums of Σd_n are bounded above by

$$d_1 + \cdots + d_N + \sum_{n=N+1}^{\infty} a_n.$$

(b) Explain why this leads to a contradiction if we assume that Σa_n converges.

44. *Work in groups of n students each.* Within your group, have each student make up a power series with radius of convergence equal to one of the numbers $1, 2, \ldots, n$. Then exchange series with another group and match the other group's series with the correct radii of convergence. ∎

In Exercises 45–51, find the sum of the telescoping series.

45. $\displaystyle\sum_{n=1}^{\infty} \frac{4}{(4n-3)(4n+1)}$

46. $\displaystyle\sum_{n=1}^{\infty} \frac{6}{(2n-1)(2n+1)}$

47. $\displaystyle\sum_{n=1}^{\infty} \frac{40n}{(2n-1)^2(2n+1)^2}$

48. $\displaystyle\sum_{n=1}^{\infty} \frac{2n+1}{n^2(n+1)^2}$

49. $\displaystyle\sum_{n=1}^{\infty} \left(\frac{1}{\sqrt{n}} - \frac{1}{\sqrt{n+1}} \right)$

50. $\displaystyle\sum_{n=1}^{\infty} \left(\frac{1}{\ln(n+2)} - \frac{1}{\ln(n+1)} \right)$

51. $\displaystyle\sum_{n=1}^{\infty} (\tan^{-1}(n) - \tan^{-1}(n+1))$

Extending the Ideas

52. We can show that the series

$$\sum_{n=0}^{\infty} \frac{n^2}{2^n}$$

converges by the Ratio Test, but what is its sum?

To find out, express $1/(1-x)$ as a geometric series. Differentiate both sides of the resulting equation with respect to x, multiply both sides of the result by x, differentiate again, multiply by x again, and set x equal to 1/2. What do you get? (*Source:* David E. Dobbs's letter to the editor, *Illinois Mathematics Teacher*, Vol. 33, Issue 4, 1982, p. 27.)

| **9.5** | **Testing Convergence at Endpoints** |

Integral Test • Harmonic Series and *p*-series • Comparison Tests • Alternating Series • Absolute and Conditional Convergence • Intervals of Convergence • A Word of Caution

Integral Test

In Exploration 1 of Section 9.4, you showed that $\Sigma 1/n$ *diverges* by modeling it as a sum of rectangle areas that contain the area under the curve $y = 1/x$ from 1 to ∞. You also showed that $\Sigma 1/n^2$ *converges* by modeling it as a sum of rectangle areas contained by the area under the curve $y = 1/x^2$ from 1 to ∞. This area-based convergence test in its general form is known as the *Integral Test*.

Theorem 10 The Integral Test

Let $\{a_n\}$ be a sequence of positive terms. Suppose that $a_n = f(n)$, where f is a continuous, positive, decreasing function of x for all $x \geq N$ (N a positive integer). Then the series $\sum_{n=N}^{\infty} a_n$ and the integral $\int_N^{\infty} f(x)\, dx$ either both converge or both diverge.

Proof We will illustrate the proof for $N = 1$ to keep the notation simple, but the illustration can be shifted horizontally to any value of N without affecting the logic of the proof.

The proof is entirely contained in these two pictures (Figure 9.15):

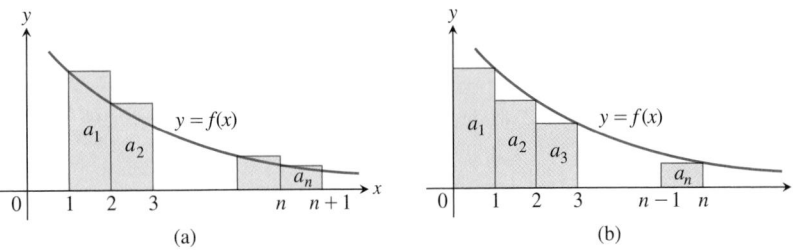

Figure 9.15 (a) The sum $a_1 + a_2 + \cdots + a_n$ provides an upper bound for $\int_1^{n+1} f(x)\, dx$. (b) The sum $a_2 + a_3 + \cdots + a_n$ provides a lower bound for $\int_1^n f(x)\, dx$. (Theorem 10)

We leave it to you (in Exercise 44) to supply the words. ∎

Example 1 APPLYING THE INTEGRAL TEST

Does $\displaystyle\sum_{n=1}^{\infty} \frac{1}{n\sqrt{n}}$ converge?

Solution The Integral Test applies because

$$f(x) = \frac{1}{x\sqrt{x}}$$

is a continuous, positive, decreasing function of x for $x > 1$.

We have

$$\int_1^{\infty} \frac{1}{x\sqrt{x}}\, dx = \lim_{k\to\infty} \int_1^k x^{-3/2}\, dx$$

$$= \lim_{k\to\infty} \left[-2x^{-1/2} \right]_1^k$$

$$= \lim_{k\to\infty} \left(-\frac{2}{\sqrt{k}} + 2 \right)$$

$$= 2.$$

Since the integral converges, so must the series.

Caution

The series and the integral in the Integral Test need not have the same value in the convergent case. Although the integral converges to 2 in Example 1, the series might have a quite different sum. If you use your calculator to compute or graph partial sums for the series, you can see that the 11th partial sum is already greater than 2. The *Technology Resource Manual* contains two programs, PARTSUMT, which displays partial sums in table form, and PARTSUMG, which displays partial sums graphically.

Harmonic Series and *p*-series

The Integral Test can be used to settle the question of convergence for any series of the form $\sum_{n=1}^{\infty} (1/n^p)$, p a real constant. (The series in Example 1 had this form, with $p = 3/2$.) Such a series is called a ***p*-series.**

Exploration 1 The *p*-Series Test

1. Use the Integral Test to prove that $\sum_{n=1}^{\infty} (1/n^p)$ converges if $p > 1$.

2. Use the Integral Test to prove that $\sum_{n=1}^{\infty} (1/n^p)$ diverges if $p < 1$.

3. Use the Integral Test to prove that $\sum_{n=1}^{\infty} (1/n^p)$ diverges if $p = 1$.

What is harmonic about the harmonic series?

The terms in the harmonic series correspond to the nodes on a vibrating string that produce multiples of the fundamental frequency. For example, 1/2 produces the harmonic that is twice the fundamental frequency, 1/3 produces a frequency that is three times the fundamental frequency, and so on. The fundamental frequency is the lowest note or pitch we hear when a string is plucked. (Figure 9.17)

(a)

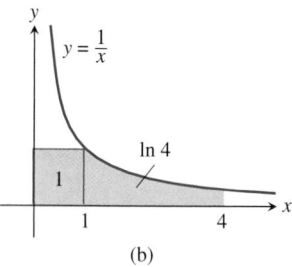

(b)

Figure 9.16 Finding an upper bound for one of the partial sums of the harmonic series. (Example 2)

The p-series with $p = 1$ is the **harmonic series,** and it is probably the most famous divergent series in mathematics. The p-Series Test shows that the harmonic series is just *barely* divergent; if we increase p to 1.000000001, for instance, the series converges!

The slowness with which the harmonic series approaches infinity is most impressive. Consider the following example.

Example 2 THE SLOW DIVERGENCE OF THE HARMONIC SERIES

Approximately how many terms of the harmonic series are required to form a partial sum larger than 20?

Solution Before you set your graphing calculator to the task of finding this number, you might want to estimate how long the calculation might take. The graphs tell the story (Figure 9.16).

Let H_n denote the nth partial sum of the harmonic series. Comparing the two graphs, we see that $H_4 < (1 + \ln 4)$ and (in general) that $H_n \le (1 + \ln n)$. If we wish H_n to be greater than 20, then

$$1 + \ln n > H_n > 20$$

$$1 + \ln n > 20$$

$$\ln n > 19$$

$$n > e^{19}.$$

The exact value of e^{19} rounds up to 178,482,301. It will take *at least* that many terms of the harmonic series to move the partial sums beyond 20. It would take your calculator several weeks to compute a partial sum of this many terms. Nonetheless, the harmonic series really does diverge!

Comparison Tests

The p-Series Test tells everything there is to know about the convergence or divergence of series of the form $\sum (1/n^p)$. This is admittedly a rather narrow class of series, but we can test many other kinds (including those in which the nth term is any rational function of n) by *comparing* them to p-series.

The Direct Comparison Test (Theorem 7, Section 9.4) is one method of comparison, but the *Limit Comparison Test* is another.

Theorem 11 The Limit Comparison Test (LCT)

Suppose that $a_n > 0$ and $b_n > 0$ for all $n \ge N$ (N a positive integer).

1. If $\lim\limits_{n \to \infty} \dfrac{a_n}{b_n} = c$, $0 < c < \infty$, then $\sum a_n$ and $\sum b_n$ both converge or both diverge.

2. If $\lim\limits_{n \to \infty} \dfrac{a_n}{b_n} = 0$ and $\sum b_n$ converges, then $\sum a_n$ converges.

3. If $\lim\limits_{n \to \infty} \dfrac{a_n}{b_n} = \infty$ and $\sum b_n$ diverges, then $\sum a_n$ diverges.

We omit the proof.

Example 3 USING THE LIMIT COMPARISON TEST

Determine whether the series converge or diverge.

(a) $\dfrac{3}{4} + \dfrac{5}{9} + \dfrac{7}{16} + \dfrac{9}{25} + \cdots = \displaystyle\sum_{n=1}^{\infty} \dfrac{2n+1}{(n+1)^2}$

(b) $\dfrac{1}{1} + \dfrac{1}{3} + \dfrac{1}{7} + \dfrac{1}{15} + \cdots = \displaystyle\sum_{n=1}^{\infty} \dfrac{1}{2^n - 1}$

(c) $\dfrac{8}{4} + \dfrac{11}{21} + \dfrac{14}{56} + \dfrac{17}{115} + \cdots = \displaystyle\sum_{n=2}^{\infty} \dfrac{3n+2}{n^3 - 2n}$

(d) $\sin 1 + \sin \dfrac{1}{2} + \sin \dfrac{1}{3} + \cdots = \displaystyle\sum_{n=1}^{\infty} \sin\left(\dfrac{1}{n}\right)$

Solution

(a) For n large,

$$\frac{2n+1}{(n+1)^2}$$

behaves like

$$\frac{2n}{n^2} = \frac{2}{n},$$

so we compare terms of the given series to terms of $\sum (1/n)$ and try the LCT.

$$\lim_{n\to\infty} \frac{a_n}{b_n} = \lim_{n\to\infty} \frac{(2n+1)/(n+1)^2}{1/n}$$

$$= \lim_{n\to\infty} \frac{2n+1}{(n+1)^2} \cdot \frac{n}{1}$$

$$= \lim_{n\to\infty} \frac{2n^2 + n}{n^2 + 2n + 1} = 2$$

Since the limit is positive and $\sum (1/n)$ diverges,

$$\sum_{n=1}^{\infty} \frac{2n+1}{(n+1)^2}$$

also diverges.

(b) For n large, $1/(2^n - 1)$ behaves like $1/2^n$, so we compare the given series to $\sum (1/2^n)$.

$$\lim_{n\to\infty} \frac{a_n}{b_n} = \lim_{n\to\infty} \frac{1}{2^n - 1} \cdot \frac{2^n}{1}$$

$$= \lim_{n\to\infty} \frac{2^n}{2^n - 1}$$

$$= \lim_{n\to\infty} \frac{1}{1 - (1/2^n)} = 1$$

Since $\sum (1/2^n)$ converges (geometric, $r = 1/2$), the LCT guarantees that

$$\sum_{n=1}^{\infty} \frac{1}{2^n - 1}$$

also converges.

Figure 9.17 On a guitar, the second harmonic note is produced when the finger is positioned halfway between the bridge and nut of the string while the string is plucked with the other hand.

(c) For n large,

$$\frac{3n + 2}{n^3 - 2n}$$

behaves like $3/n^2$, so we compare the given series to $\sum (1/n^2)$.

$$\lim_{n\to\infty} \frac{a_n}{b_n} = \lim_{n\to\infty} \frac{3n + 2}{n^3 - 2n} \cdot \frac{n^2}{1}$$

$$= \lim_{n\to\infty} \frac{3n^3 + 2n^2}{n^3 - 2n} = 3$$

Since $\sum (1/n^2)$ converges by the p-Series Test,

$$\sum_{n=2}^{\infty} \frac{3n + 2}{n^3 - 2n}$$

also converges (by the LCT).

(d) Recall that

$$\lim_{x\to 0} \frac{\sin x}{x} = 1,$$

so we try the LCT by comparing the given series to $\sum (1/n)$.

$$\lim_{n\to\infty} \frac{a_n}{b_n} = \lim_{n\to\infty} \frac{\sin (1/n)}{(1/n)} = 1$$

Since $\sum (1/n)$ diverges, $\sum_{n=1}^{\infty} \sin (1/n)$ also diverges.

As Example 3 suggests, applying the Limit Comparison Test has strong connections to analyzing end behavior in functions. In part (c) of Example 3, we could have reached the same conclusion if a_n had been *any* linear polynomial in n divided by *any* cubic polynomial in n, since any such rational function "in the end" will grow like $1/n^2$.

Alternating Series

A series in which the terms are alternately positive and negative is an **alternating series.**

Here are three examples.

$$1 - \frac{1}{2} + \frac{1}{3} - \frac{1}{4} + \frac{1}{5} - \cdots + \frac{(-1)^{n+1}}{n} + \cdots \tag{1}$$

$$-2 + 1 - \frac{1}{2} + \frac{1}{4} - \frac{1}{8} + \cdots + \frac{(-1)^n 4}{2^n} + \cdots \tag{2}$$

$$1 - 2 + 3 - 4 + 5 - 6 + \cdots + (-1)^{n+1} n + \cdots \tag{3}$$

Series 1, called the **alternating harmonic series,** converges, as we will see shortly. (You may have come to this conclusion already in Exploration 2 of Section 9.4.) Series 2, a geometric series with $a = -2$, $r = -1/2$, converges to $-2/[1 + (1/2)] = -4/3$. Series 3 diverges by the nth-Term Test.

We prove the convergence of the alternating harmonic series by applying the following test.

> **Theorem 12   The Alternating Series Test (Leibniz's Theorem)**
>
> The series
>
> $$\sum_{n=1}^{\infty} (-1)^{n+1}u_n = u_1 - u_2 + u_3 - u_4 + \cdots$$
>
> converges if all three of the following conditions are satisfied:
>
> **1.** each u_n is positive;
>
> **2.** $u_n \geq u_{n+1}$ for all $n \geq N$, for some integer N;
>
> **3.** $\lim_{n \to \infty} u_n \to 0$.

Figure 9.18 illustrates the convergence of the partial sums to their limit L.

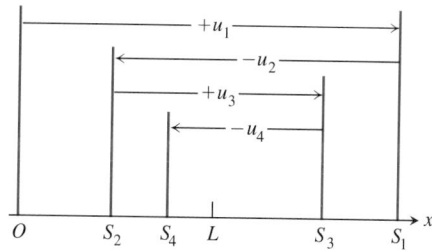

Figure 9.18 Closing in on the sum of a convergent alternating series. (Theorem 12)

The figure that proves the Alternating Series Test actually proves more than the *fact* of convergence; it also shows the *way* that an alternating series converges when it satisfies the conditions of the test. The partial sums keep "overshooting" the limit as they go back and forth on the number line, gradually closing in as the terms tend to zero. If we stop at the nth partial sum, we know that the next term $(\pm u_{n+1})$ will again cause us to overshoot the limit in the positive direction or negative direction, depending on the sign carried by u_{n+1}. This gives us a convenient bound for the truncation error, which we state as another theorem.

> **Theorem 13   The Alternating Series Estimation Theorem**
>
> If the alternating series $\sum_{n=1}^{\infty} (-1)^{n+1}u_n$ satisfies the conditions of Theorem 12, then the truncation error for the nth partial sum is less than u_{n+1} and has the same sign as the unused term.

Example 4   THE ALTERNATING HARMONIC SERIES

Prove that the alternating harmonic series is convergent, but not absolutely convergent. Find a bound for the truncation error after 99 terms.

Solution The terms are strictly alternating in sign and decrease in absolute value from the start:

$$1 > \frac{1}{2} > \frac{1}{3} > \cdots . \quad \text{Also,} \quad \frac{1}{n} \to 0.$$

A Note on the Error Bound

Theorem 13 does not give a *formula* for the truncation error, but a *bound* for the truncation error. The bound might be fairly conservative. For example, the first 99 terms of the alternating harmonic series add to about 0.6981721793, while the series itself has a sum of ln 2 ≈ 0.6931471806. That makes the actual truncation error very close to 0.005, about half the size of the bound of 0.01 given by Theorem 13.

By the Alternating Series Test,

$$\sum_{n=1}^{\infty} \frac{(-1)^{n+1}}{n}$$

converges.

On the other hand, the series $\sum_{n=1}^{\infty} (1/n)$ of absolute values is the harmonic series, which diverges, so the alternating harmonic series is not absolutely convergent.

The Alternating Series Estimation Theorem guarantees that the truncation error after 99 terms is less than $u_{99+1} = 1/(99 + 1) = 1/100$.

Absolute and Conditional Convergence

Because the alternating harmonic series is convergent but not absolutely convergent, we say it is **conditionally convergent** (or **converges conditionally**).

We take it for granted that we can rearrange the terms of a *finite* sum without affecting the sum. We can also rearrange a *finite number* of terms of an infinite series without affecting the sum. But if we rearrange an infinite number of terms of an infinite series, we can be sure of leaving the sum unaltered *only if it converges absolutely.*

Rearrangements of Absolutely Convergent Series

If $\sum a_n$ converges absolutely, and if $b_1, b_2, b_3, \ldots, b_n, \ldots$ is any rearrangement of the sequence $\{a_n\}$, then $\sum b_n$ converges absolutely and $\sum_{n=1}^{\infty} b_n = \sum_{n=1}^{\infty} a_n$.

On the other hand, consider this:

Rearrangements of Conditionally Convergent Series

If $\sum a_n$ converges conditionally, then the terms can be rearranged to form a divergent series. The terms can also be rearranged to form a series that converges to *any* preassigned sum.

This seems incredible, but it is a logical consequence of the definition of the sum as the *limit of the sequence of partial sums.* A conditionally convergent series consists of positive terms that sum to ∞ and negative terms that sum to −∞, so we can manipulate the partial sums to do virtually anything we wish. We illustrate the technique with the alternating harmonic series.

Example 5   REARRANGING THE ALTERNATING HARMONIC SERIES

Show how to rearrange the terms of

$$\sum_{n=1}^{\infty} \frac{(-1)^{n+1}}{n}$$

to form

(a) a divergent series;   **(b)** a series that converges to π.

Solution The series of positive terms,

$$1 + \frac{1}{3} + \frac{1}{5} + \cdots + \frac{1}{2n+1} + \cdots,$$

diverges to ∞, while the series of negative terms,

$$-\frac{1}{2} - \frac{1}{4} - \frac{1}{6} - \cdots - \frac{1}{2n} - \cdots,$$

diverges to $-\infty$. No matter what finite number of terms we use, the remaining positive terms or negative terms still diverge. So, we build our series as follows:

(a) Start by adding positive terms until the partial sum is greater than 1. Then add negative terms until the partial sum is less than -2. Then add positive terms until the sum is greater than 3. Then add negative terms until the sum is less than -4. Continue in this manner indefinitely, so that the sequence of partial sums swings arbitrarily far in both directions and hence diverges.

(b) Start by adding positive terms until the partial sum is greater than π. Then add negative terms until the partial sum is less than π. Then add positive terms until the sum is greater than π. Continue in this manner indefinitely, always closing in on π. Since the positive and negative terms of the original series both approach zero, the amount by which the partial sums exceed or fall short of π approaches zero.

Intervals of Convergence

Our purpose in this section has been to develop tests for convergence that can be used at the endpoints of the intervals of absolute convergence of power series. There are three possibilities at each endpoint: The series could diverge, it could converge absolutely, or it could converge conditionally.

How to Test a Power Series $\sum\limits_{n=0}^{\infty} c_n(x - a)^n$ for Convergence

1. Use the Ratio Test to find the values of x for which the series converges absolutely. Ordinarily, this is an open interval

$$a - R < x < a + R.$$

   In some instances, the series converges for all values of x. In rare cases, the series converges only at $x = a$.

2. If the interval of absolute convergence is finite, test for convergence or divergence at each endpoint. The Ratio Test fails at these points. Use a comparison test, the Integral Test, or the Alternating Series Test.

3. If the interval of absolute convergence is $a - R < x < a + R$, conclude that the series diverges (it does not even converge conditionally) for $|x - a| > R$, because for those values of x the nth term does not approach zero.

Example 6 FINDING INTERVALS OF CONVERGENCE

For what values of x do the following series converge?

(a) $\displaystyle\sum_{n=1}^{\infty} (-1)^{n+1}\frac{x^{2n}}{2n} = \frac{x^2}{2} - \frac{x^4}{4} + \frac{x^6}{6} - \cdots$

(b) $\displaystyle\sum_{n=0}^{\infty} \frac{(10x)^n}{n!} = 1 + 10x + \frac{100x^2}{2!} + \frac{1000x^3}{3!} + \cdots$

(c) $\displaystyle\sum_{n=0}^{\infty} n!(x + 1)^n = 1 + (x + 1) + 2!(x + 1)^2 + 3!(x + 1)^3 + \cdots$

(d) $\displaystyle\sum_{n=1}^{\infty} \frac{(x - 3)^n}{2n} = \frac{(x - 3)}{2} + \frac{(x - 3)^2}{4} + \frac{(x - 3)^3}{6} + \cdots$

Solution We apply the Ratio Test to find the interval of absolute convergence, then check the endpoints if they exist.

(a) $\displaystyle\lim_{n\to\infty} \left|\frac{u_{n+1}}{u_n}\right| = \lim_{n\to\infty} \frac{x^{2n+2}}{2n + 2} \cdot \frac{2n}{x^{2n}}$

$$= \lim_{n\to\infty} \left(\frac{2n}{2n + 2}\right)x^2 = x^2$$

The series converges absolutely for $x^2 < 1$, i.e., on the interval $(-1, 1)$. At $x = 1$, the series is

$$\sum \frac{(-1)^{n+1}}{2n},$$

which converges by the Alternating Series Test. (It is half the sum of the alternating harmonic series.) At $x = -1$, the series is the same as at $x = 1$, so it converges. The interval of convergence is $[-1, 1]$.

(b) $\displaystyle\lim_{n\to\infty} \left|\frac{u_{n+1}}{u_n}\right| = \lim_{n\to\infty} \frac{|10x|^{n+1}}{(n + 1)!} \cdot \frac{n!}{|10x|^n}$

$$= \lim_{n\to\infty} \frac{|10x|}{n + 1} = 0$$

The series converges absolutely for all x.

(c) $\displaystyle\lim_{n\to\infty} \left|\frac{u_{n+1}}{u_n}\right| = \lim_{n\to\infty} \frac{(n + 1)!|x + 1|^{n+1}}{n!|x + 1|^n}$

$$= \lim_{n\to\infty} (n + 1)|x + 1| = \begin{cases} \infty, & x \neq -1 \\ 0, & x = -1 \end{cases}$$

The series converges only at $x = -1$.

(d) $\displaystyle\lim_{n\to\infty} \left|\frac{u_{n+1}}{u_n}\right| = \lim_{n\to\infty} \frac{|x - 3|^{n+1}}{2n + 2} \cdot \frac{2n}{|x - 3|^n}$

$$= \lim_{n\to\infty} \left(\frac{2n}{2n + 2}\right)|x - 3| = |x - 3|$$

The series converges absolutely for $|x - 3| < 1$, i.e., on the interval $(2, 4)$. At $x = 2$, the series is $\sum (-1)^n/2n$, which converges by the Alternating Series Test. At $x = 4$, the series is $\sum 1/2n$, which diverges by limit comparison with the harmonic series. The interval of convergence is $[2, 4)$.

To facilitate testing convergence at endpoints we can use the following flowchart.

Procedure for Determining Convergence

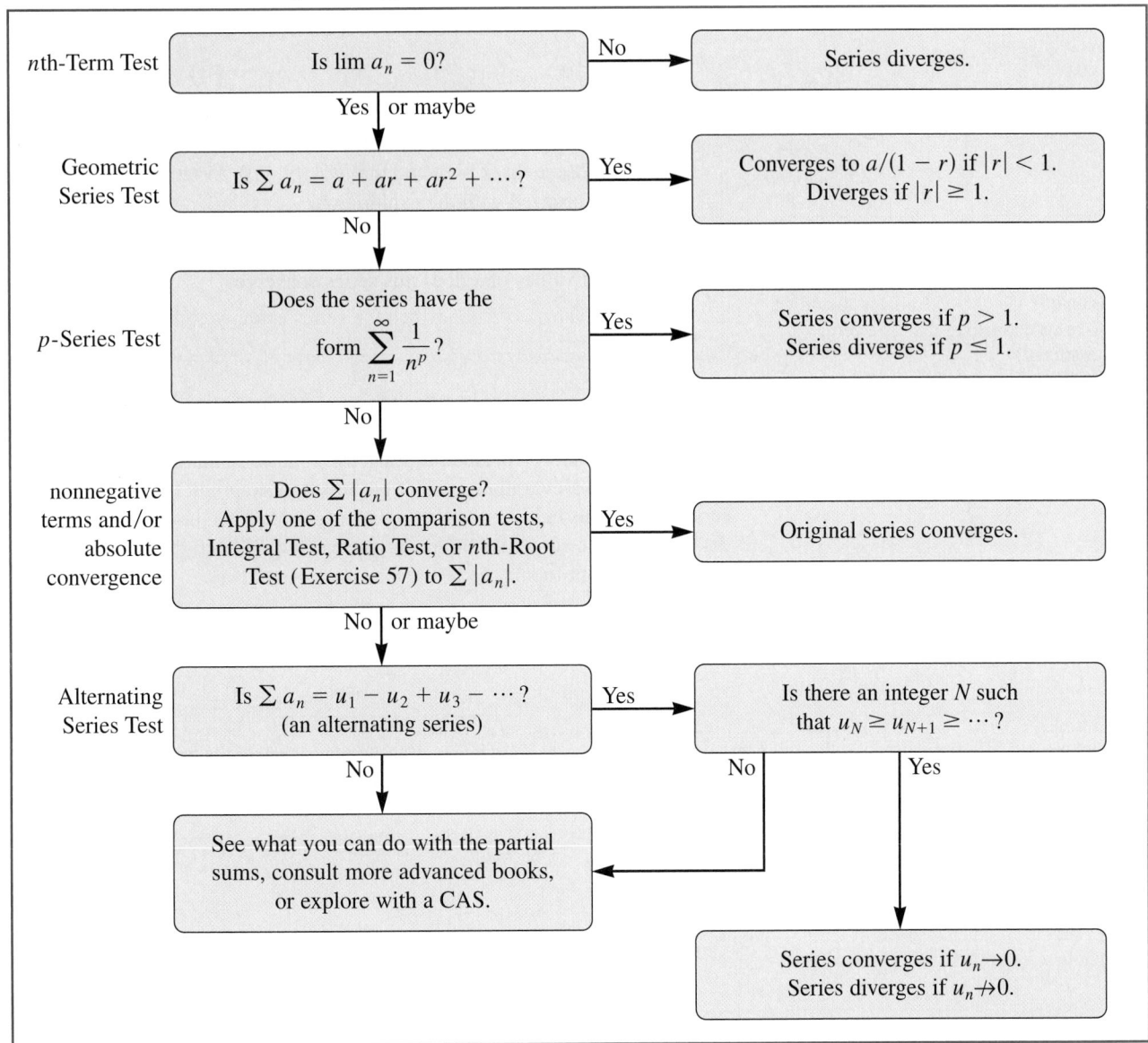

A Word of Caution

Although we can use the tests we have developed to find where a given power series converges, they do not tell us what function that power series is converging *to*. Even if the series is known to be a Maclaurin series generated by a function *f*, we cannot automatically conclude that the series converges *to the function f* on its interval of convergence. That is why it is so important to estimate the error.

For example, we can use the Ratio Test to show that the Maclaurin series for sin *x*, cos *x*, and e^x all converge absolutely for all real numbers. However, the reason we know that they converge to sin *x*, cos *x*, and e^x is that we used the Remainder Estimation Theorem to show that the respective truncation errors went to zero.

The following exploration shows what can happen with a strange function.

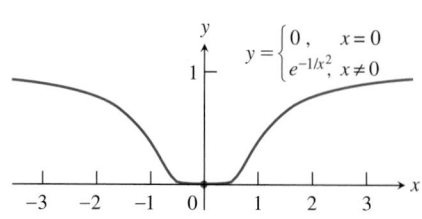

$$y = \begin{cases} 0, & x = 0 \\ e^{-1/x^2}, & x \neq 0 \end{cases}$$

Figure 9.19 The graph of the continuous extension of $y = e^{-1/x^2}$ is so flat at the origin that all of its derivatives there are zero. (Exploration 2)

Exploration 2 **The Maclaurin Series of a Strange Function**

Let $f(x) = \begin{cases} 0, & x = 0 \\ e^{-1/x^2}, & x \neq 0. \end{cases}$

It can be shown (although not easily) that f (Figure 9.19) has derivatives of all orders at $x = 0$ and that $f^{(n)}(0) = 0$ for all n. Use this fact as you proceed with the exploration.

1. Construct the Maclaurin series for f.
2. For what values of x does this series converge?
3. Find all values of x for which the series actually converges to $f(x)$.

If you are surprised by the behavior of the series in Exploration 2, remember that we identified it up front as a strange function. It was fortunate for the early history of calculus that the functions that modeled physical behavior in the Newtonian world were much more predictable, enabling the early theories to enjoy encouraging successes before they could be lost in detail. When the subtleties of convergence emerged later, the theory was prepared to confront them.

Quick Review 9.5

In Exercises 1–5, determine whether the improper integral converges or diverges. Give reasons for your answer. (You do not need to evaluate the integral.)

1. $\int_1^\infty \frac{1}{x^{4/3}} \, dx$

2. $\int_1^\infty \frac{x^2}{x^3 + 1} \, dx$

3. $\int_1^\infty \frac{\ln x}{x} \, dx$

4. $\int_1^\infty \frac{1 + \cos x}{x^2} \, dx$

5. $\int_1^\infty \frac{\sqrt{x}}{x + 1} \, dx$

In Exercises 6–10, determine whether the function is both positive and decreasing on some interval (N, ∞). (You do not need to identify N.)

6. $f(x) = \frac{3}{x}$

7. $f(x) = \frac{7x}{x^2 - 8}$

8. $f(x) = \frac{3 + x^2}{3 - x^2}$

9. $f(x) = \frac{\sin x}{x^5}$

10. $f(x) = \ln (1/x)$

Section 9.5 Exercises

In Exercises 1–16, determine whether the series converges or diverges. There may be more than one correct way to determine convergence or divergence of a given series.

1. $\sum_{n=1}^\infty \frac{5}{n + 1}$

2. $\sum_{n=1}^\infty \frac{3}{\sqrt{n}}$

3. $\sum_{n=2}^\infty \frac{\ln n}{n}$

4. $\sum_{n=1}^\infty \frac{1}{2n - 1}$

5. $\sum_{n=1}^\infty \frac{1}{(\ln 2)^n}$

6. $\sum_{n=1}^\infty \frac{1}{(\ln 3)^n}$

7. $\sum_{n=1}^\infty n \sin \left(\frac{1}{n} \right)$

8. $\sum_{n=0}^\infty \frac{e^n}{1 + e^{2n}}$

9. $\sum_{n=1}^\infty \frac{\sqrt{n}}{n^2 + 1}$

10. $\sum_{n=1}^\infty \frac{5n^3 - 3n}{n^2(n + 2)(n^2 + 5)}$

11. $\displaystyle\sum_{n=1}^{\infty} \frac{3^{n-1}+1}{3^n}$

12. $\displaystyle\sum_{n=2}^{\infty} (-1)^{n+1}\frac{1}{\ln n}$

13. $\displaystyle\sum_{n=1}^{\infty} (-1)^{n+1}\frac{10^n}{n^{10}}$

14. $\displaystyle\sum_{n=1}^{\infty} (-1)^{n+1}\frac{\sqrt{n}+1}{n+1}$

15. $\displaystyle\sum_{n=2}^{\infty} (-1)^{n+1}\frac{\ln n}{\ln n^2}$

16. $\displaystyle\sum_{n=1}^{\infty} \left(\frac{1}{n}-\frac{1}{n^2}\right)$

In Exercises 17–26, determine whether the series converges absolutely, converges conditionally, or diverges. Give reasons for your answer.

17. $\displaystyle\sum_{n=1}^{\infty} (-1)^{n+1}(0.1)^n$

18. $\displaystyle\sum_{n=1}^{\infty} (-1)^{n+1}\frac{1+n}{n^2}$

19. $\displaystyle\sum_{n=1}^{\infty} (-1)^n n^2\left(\frac{2}{3}\right)^n$

20. $\displaystyle\sum_{n=2}^{\infty} (-1)^{n+1}\frac{1}{n\ln n}$

21. $\displaystyle\sum_{n=1}^{\infty} (-1)^{n+1}\frac{n!}{2^n}$

22. $\displaystyle\sum_{n=1}^{\infty} (-1)^{n+1}\frac{\sin n}{n^2}$

23. $\displaystyle\sum_{n=1}^{\infty} \frac{(-1)^n}{1+\sqrt{n}}$

24. $\displaystyle\sum_{n=1}^{\infty} \frac{\cos n\pi}{n\sqrt{n}}$

25. $\displaystyle\sum_{n=1}^{\infty} \frac{\cos n\pi}{n}$

26. $\displaystyle\sum_{n=1}^{\infty} \frac{(-1)^n}{\sqrt{n}+\sqrt{n+1}}$

In Exercises 27–42, find **(a)** the *interval* of convergence of the series. For what values of x does the series converge **(b)** absolutely, **(c)** conditionally?

27. $\displaystyle\sum_{n=0}^{\infty} x^n$

28. $\displaystyle\sum_{n=0}^{\infty} (x+5)^n$

29. $\displaystyle\sum_{n=0}^{\infty} (-1)^n(4x+1)^n$

30. $\displaystyle\sum_{n=1}^{\infty} \frac{(3x-2)^n}{n}$

31. $\displaystyle\sum_{n=0}^{\infty} \frac{(x-2)^n}{10^n}$

32. $\displaystyle\sum_{n=0}^{\infty} \frac{nx^n}{n+2}$

33. $\displaystyle\sum_{n=1}^{\infty} \frac{x^n}{n\sqrt{n}\,3^n}$

34. $\displaystyle\sum_{n=0}^{\infty} \frac{x^{2n+1}}{n!}$

35. $\displaystyle\sum_{n=0}^{\infty} \frac{n(x+3)^n}{5^n}$

36. $\displaystyle\sum_{n=0}^{\infty} \frac{nx^n}{4^n(n^2+1)}$

37. $\displaystyle\sum_{n=0}^{\infty} \frac{\sqrt{n}\,x^n}{3^n}$

38. $\displaystyle\sum_{n=0}^{\infty} n!(x-4)^n$

39. $\displaystyle\sum_{n=0}^{\infty} (-2)^n(n+1)(x-1)^n$

40. $\displaystyle\sum_{n=1}^{\infty} \frac{(4x-5)^{2n+1}}{n^{3/2}}$

41. $\displaystyle\sum_{n=1}^{\infty} \frac{(x+\pi)^n}{\sqrt{n}}$

42. $\displaystyle\sum_{n=0}^{\infty} (\ln x)^n$

43. Not only do the figures in Example 2 show that the nth partial sum of the harmonic series is less than $1+\ln n$; they also show that it is *greater* than $\ln(n+1)$. Suppose you had started summing the harmonic series with $S_1 = 1$ at the time

the universe was formed, 13 billion years ago. If you had been able to add a term every *second* since then, about how large would your partial sum be today? (Assume a 365-day year.)

44. Writing to Learn Write out a proof of the Integral Test (Theorem 10) for $N = 1$, explaining what you see in Figure 9.15.

45. *(Continuation of Exercise 44)* Relabel the pictures for an arbitrary N and explain why the same conclusions about convergence can be drawn.

46. In each of the following cases, decide whether the infinite series converges. Justify your answer.

(a) $\displaystyle\sum_{k=1}^{\infty} \frac{1}{\sqrt{2k+7}}$

(b) $\displaystyle\sum_{k=1}^{\infty} \left(1+\frac{1}{k}\right)^k$

(c) $\displaystyle\sum_{k=1}^{\infty} \frac{\cos k}{k^2+\sqrt{k}}$

(d) $\displaystyle\sum_{k=3}^{\infty} \frac{18}{k(\ln k)}$

47. Construct a series that diverges more slowly than the harmonic series. Justify your answer.

48. Let $a_k = (-1)^{k+1}\int_0^{1/k} 6(kx)^2\,dx$.

(a) Evaluate a_k.

(b) Show that $\sum_{k=1}^{\infty} a_k$ converges.

(c) Show that

$$1 \le \sum_{k=1}^{\infty} a_k \le \frac{3}{2}.$$

49. (a) Determine whether the series

$$A = \sum_{n=1}^{\infty} \frac{n}{3n^2+1}$$

converges or diverges. Justify your answer.

(b) If S is the series formed by multiplying the nth term in A by the nth term in $\sum_{n=1}^{\infty}(3/n)$, write an expression using summation notation for S and determine whether S converges or diverges.

50. (a) Find the Taylor series generated by $f(x) = \ln(1+x)$ at $x = 0$. Include an expression for the general term.

(b) For what values of x does the series in part (a) converge?

(c) Use Theorem 13 to find a bound for the error in evaluating $\ln(3/2)$ by using only the first five nonzero terms of the series in part (a).

(d) Use the result found in part (a) to determine the logarithmic function whose Taylor series is

$$\sum_{n=1}^{\infty} \frac{(-1)^{n+1}x^{2n}}{2n}.$$

51. Determine all values of x for which the series

$$\sum_{k=0}^{\infty} \frac{2^k x^k}{\ln(k+2)}$$

converges. Justify your answer.

52. Consider the series $\displaystyle\sum_{n=2}^{\infty} \frac{1}{n^p \ln n}$, where $p \geq 0$.

(a) Show that the series converges for $p > 1$.

(b) **Writing to Learn** Determine whether the series converges or diverges for $p = 1$. Show your analysis.

(c) Show that the series diverges for $0 \leq p < 1$.

53. The Maclaurin series for $1/(1 + x)$ converges for $-1 < x < 1$, but when we integrate it term by term, the resulting series for $\ln |1 + x|$ converges for $-1 < x \leq 1$. Verify the convergence at $x = 1$.

54. The Maclaurin series for $1/(1 + x^2)$ converges for $-1 < x < 1$, but when we integrate it term by term, the resulting series for $\arctan x$ converges for $-1 \leq x \leq 1$. Verify the convergence at $x = 1$ and $x = -1$.

55. (a) The series

$$\frac{1}{3} - \frac{1}{2} + \frac{1}{9} - \frac{1}{4} + \frac{1}{27} - \frac{1}{8} + \cdots + \frac{1}{3^n} - \frac{1}{2^n} + \cdots$$

fails to satisfy one of the conditions of the Alternating Series Test. Which one?

(b) Find the sum of the series in (a).

Exploration

56. *Work in groups of n people.* Within your group, have each student construct a series that converges to one of the numbers $1, \ldots, n$. Then exchange your series with another group and try to figure out which number is matched with which series. ∎

Extending the Ideas

Here is a test called the *nth-Root Test*.

> **nth-Root Test** Let $\sum a_n$ be a series with $a_n \geq 0$ for $n \geq N$, and suppose that $\lim_{n \to \infty} \sqrt[n]{a_n} = L$. Then,
>
> (a) the series *converges* if $L < 1$,
>
> (b) the series *diverges* if $L > 1$ or L is infinite,
>
> (c) the test is *inconclusive* if $L = 1$.

57. Use the *nth*-Root Test and the fact that $\lim_{n \to \infty} \sqrt[n]{n} = 1$ to test the following series for convergence or divergence.

(a) $\displaystyle\sum_{n=1}^{\infty} \frac{n^2}{2^n}$ (b) $\displaystyle\sum_{n=1}^{\infty} \left(\frac{n}{2n - 1} \right)^n$

(c) $\displaystyle\sum_{n=1}^{\infty} a_n$, where $a_n = \begin{cases} n/2^n, & n \text{ is odd} \\ 1/2^n, & n \text{ is even} \end{cases}$

58. Use the *nth*-Root Test and whatever else you need to find the intervals of convergence of the following series.

(a) $\displaystyle\sum_{n=0}^{\infty} \frac{(x - 1)^n}{4^n}$ (b) $\displaystyle\sum_{n=1}^{\infty} \frac{(x - 2)^n}{n \cdot 3^n}$

(c) $\displaystyle\sum_{n=1}^{\infty} 2^n x^n$ (d) $\displaystyle\sum_{n=0}^{\infty} (\ln x)^n$

Chapter 9   Key Terms

In Exercises 1–16, find **(a)** the radius of convergence for the series and **(b)** its interval of convergence. Then identify the values of x for which the series converges **(c)** absolutely and **(d)** conditionally.

1. $\displaystyle\sum_{n=0}^{\infty} \frac{(-x)^n}{n!}$

2. $\displaystyle\sum_{n=1}^{\infty} \frac{(x+4)^n}{n3^n}$

3. $\displaystyle\sum_{n=0}^{\infty} \left(\frac{2}{3}\right)^n (x-1)^n$

4. $\displaystyle\sum_{n=1}^{\infty} \frac{(x-1)^{2n-2}}{(2n-1)!}$

5. $\displaystyle\sum_{n=1}^{\infty} \frac{(-1)^{n-1}(3x-1)^n}{n^2}$

6. $\displaystyle\sum_{n=0}^{\infty} (n+1)x^{3n}$

7. $\displaystyle\sum_{n=0}^{\infty} \frac{(n+1)(2x+1)^n}{(2n+1)2^n}$

8. $\displaystyle\sum_{n=1}^{\infty} \frac{x^n}{n^n}$

9. $\displaystyle\sum_{n=1}^{\infty} \frac{x^n}{\sqrt{n}}$

10. $\displaystyle\sum_{n=1}^{\infty} \frac{e^n}{n^e} x^n$

11. $\displaystyle\sum_{n=0}^{\infty} \frac{(n+1)x^{2n-1}}{3^n}$

12. $\displaystyle\sum_{n=0}^{\infty} \frac{(-1)^n(x-1)^{2n+1}}{2n+1}$

13. $\displaystyle\sum_{n=1}^{\infty} \frac{n!}{2^n} x^{2n}$

14. $\displaystyle\sum_{n=2}^{\infty} \frac{(10x)^n}{\ln n}$

15. $\displaystyle\sum_{n=0}^{\infty} (n+1)! x^n$

16. $\displaystyle\sum_{n=0}^{\infty} \left(\frac{x^2-1}{2}\right)^n$

In Exercises 17–22, the series is the value of the Maclaurin series of a function $f(x)$ at a particular point. What function and what point? What is the sum of the series?

17. $1 - \dfrac{1}{4} + \dfrac{1}{16} - \cdots + (-1)^n \dfrac{1}{4^n} + \cdots$

18. $\dfrac{2}{3} - \dfrac{4}{18} + \dfrac{8}{81} - \cdots + (-1)^{n-1} \dfrac{2^n}{n3^n} + \cdots$

19. $\pi - \dfrac{\pi^3}{3!} + \dfrac{\pi^5}{5!} - \cdots + (-1)^n \dfrac{\pi^{2n+1}}{(2n+1)!} + \cdots$

20. $1 - \dfrac{\pi^2}{9 \cdot 2!} + \dfrac{\pi^4}{81 \cdot 4!} - \cdots + (-1)^n \dfrac{\pi^{2n}}{3^{2n}(2n)!} + \cdots$

21. $1 + \ln 2 + \dfrac{(\ln 2)^2}{2!} + \cdots + \dfrac{(\ln 2)^n}{n!} + \cdots$

22. $\dfrac{1}{\sqrt{3}} - \dfrac{1}{9\sqrt{3}} + \dfrac{1}{45\sqrt{3}} - \cdots +$
$(-1)^{n-1} \dfrac{1}{(2n-1)(\sqrt{3})^{2n-1}} + \cdots$

In Exercises 23–36, find a Maclaurin series for the function.

23. $\dfrac{1}{1-6x}$

24. $\dfrac{1}{1+x^3}$

25. $x^9 - 2x^2 + 1$

26. $\dfrac{4x}{1-x}$

27. $\sin \pi x$

28. $-\sin \dfrac{2x}{3}$

29. $-x + \sin x$

30. $\dfrac{e^x + e^{-x}}{2}$

31. $\cos \sqrt{5x}$

32. $e^{(\pi x/2)}$

33. xe^{-x^2}

34. $\tan^{-1} 3x$

35. $\ln (1 - 2x)$

36. $x \ln (1 - x)$

In Exercises 37–40, find the first four nonzero terms and the general term of the Taylor series generated by f at $x = a$.

37. $f(x) = \dfrac{1}{3-x}$, $\quad a = 2$

38. $f(x) = x^3 - 2x^2 + 5$, $\quad a = -1$

39. $f(x) = \dfrac{1}{x}$, $\quad a = 3$
40. $f(x) = \sin x$, $\quad a = \pi$

In Exercises 41–52, determine if the series converges absolutely, converges conditionally, or diverges. Give reasons for your answer.

41. $\displaystyle\sum_{n=1}^{\infty} \frac{-5}{n}$

42. $\displaystyle\sum_{n=1}^{\infty} \frac{(-1)^n}{\sqrt{n}}$

43. $\displaystyle\sum_{n=1}^{\infty} \frac{\ln n}{n^3}$

44. $\displaystyle\sum_{n=1}^{\infty} \frac{n+1}{n!}$

45. $\displaystyle\sum_{n=1}^{\infty} \frac{(-1)^n}{\ln (n+1)}$

46. $\displaystyle\sum_{n=2}^{\infty} \frac{1}{n(\ln n)^2}$

47. $\displaystyle\sum_{n=1}^{\infty} \frac{(-3)^n}{n!}$

48. $\displaystyle\sum_{n=1}^{\infty} \frac{2^n 3^n}{n^n}$

49. $\displaystyle\sum_{n=1}^{\infty} \frac{(-1)^n(n^2+1)}{2n^2+n-1}$

50. $\displaystyle\sum_{n=1}^{\infty} \frac{1}{\sqrt{n(n+1)(n+2)}}$

51. $\displaystyle\sum_{n=2}^{\infty} \frac{1}{n\sqrt{n^2-1}}$

52. $\displaystyle\sum_{n=1}^{\infty} \left(\frac{n}{n+1}\right)^n$

In Exercises 53 and 54, find the sum of the series.

53. $\displaystyle\sum_{n=3}^{\infty} \frac{1}{(2n-3)(2n-1)}$

54. $\displaystyle\sum_{n=2}^{\infty} \frac{-2}{n(n+1)}$

55. Let f be a function that has derivatives of all orders for all real numbers. Assume that $f(3) = 1$, $f'(3) = 4$, $f''(3) = 6$, and $f'''(3) = 12$.

(a) Write the third order Taylor polynomial for f at $x = 3$ and use it to approximate $f(3.2)$.

(b) Write the second order Taylor polynomial for f' at $x = 3$ and use it to approximate $f'(2.7)$.

(c) Does the linearization of f underestimate or overestimate the values of $f(x)$ near $x = 3$? Justify your answer.

56. Let

$$P_4(x) = 7 - 3(x - 4) + 5(x - 4)^2 - 2(x - 4)^3 + 6(x - 4)^4$$

be the Taylor polynomial of order 4 for the function f at $x = 4$. Assume f has derivatives of all orders for all real numbers.

(a) Find $f(4)$ and $f'''(4)$.

(b) Write the second order Taylor polynomial for f' at $x = 4$ and use it to approximate $f'(4.3)$.

(c) Write the fourth order Taylor polynomial for $g(x) = \int_4^x f(t)\, dt$ at $x = 4$.

(d) Can the exact value of $f(3)$ be determined from the information given? Justify your answer.

57. (a) Write the first three nonzero terms and the general term of the Taylor series generated by $f(x) = 5 \sin(x/2)$ at $x = 0$.

(b) What is the interval of convergence for the series found in (a)? Show your method.

(c) Writing to Learn What is the minimum number of terms of the series in (a) needed to approximate $f(x)$ on the interval $(-2, 2)$ with an error not exceeding 0.1 in magnitude? Show your method.

58. Let $f(x) = 1/(1 - 2x)$.

(a) Write the first four terms and the general term of the Taylor series generated by $f(x)$ at $x = 0$.

(b) What is the interval of convergence for the series found in part (a)? Show your method.

(c) Find $f(-1/4)$. How many terms of the series are adequate for approximating $f(-1/4)$ with an error not exceeding one percent in magnitude? Justify your answer.

59. Let

$$f(x) = \sum_{n=1}^{\infty} \frac{x^n n^n}{n!}$$

for all x for which the series converges.

(a) Find the radius of convergence of this series.

(b) Use the first three terms of this series to approximate $f(-1/3)$.

(c) Estimate the error involved in the approximation in part (b). Justify your answer.

60. Let $f(x) = 1/(x - 2)$.

(a) Write the first four terms and the general term of the Taylor series generated by $f(x)$ at $x = 3$.

(b) Use the result from part (a) to find the first four terms and the general term of the series generated by $\ln|x - 2|$ at $x = 3$.

(c) Use the series in part (b) to compute a number that differs from $\ln(3/2)$ by less than 0.05. Justify your answer.

61. Let $f(x) = e^{-2x^2}$.

(a) Find the first four nonzero terms and the general term for the power series generated by $f(x)$ at $x = 0$.

(b) Find the interval of convergence of the series generated by $f(x)$ at $x = 0$. Show the analysis that leads to your conclusion.

(c) Writing to Learn Let g be the function defined by the sum of the first four nonzero terms of the series generated by $f(x)$. Show that $|f(x) - g(x)| < 0.02$ for $-0.6 \le x \le 0.6$.

62. (a) Find the Maclaurin series generated by $f(x) = x^2/(1 + x)$.

(b) Does the series converge at $x = 1$? Explain.

63. *Evaluating Nonelementary Integrals* Maclaurin series can be used to express nonelementary integrals in terms of series.

(a) Express $\int_0^x \sin t^2\, dt$ as a power series.

(b) According to the Alternating Series Estimation Theorem, how many terms of the series in (a) should you use to estimate $\int_0^1 \sin x^2\, dx$ with an error of less than 0.001?

(c) Use NINT to approximate $\int_0^1 \sin x^2\, dx$.

(d) How close to the answer in (c) do you get if you use four terms of the series in (a)?

64. *Estimating an Integral* Suppose you want a quick noncalculator estimate for the value of $\int_0^1 x^2 e^x\, dx$. There are several ways to get one.

(a) Use the Trapezoidal Rule with $n = 2$ to estimate $\int_0^1 x^2 e^x\, dx$.

(b) Write the first three nonzero terms of the Maclaurin series for $x^2 e^x$ to obtain the fourth order Maclaurin polynomial $P_4(x)$ for $x^2 e^x$. Use $\int_0^1 P_4(x)\, dx$ to obtain another estimate of $\int_0^1 x^2 e^x\, dx$.

(c) Writing to Learn The second derivative of $f(x) = x^2 e^x$ is positive for all $x > 0$. Explain why this enables you to conclude that the Trapezoidal Rule estimate obtained in (a) is too large.

(d) Writing to Learn All the derivatives of $f(x) = x^2 e^x$ are positive for $x > 0$. Explain why this enables you to conclude that all Maclaurin series approximations to $f(x)$ for x in $[0, 1]$ will be too small. (*Hint:* $f(x) = P_n(x) + R_n(x)$.)

(e) Use integration by parts to evaluate $\int_0^1 x^2 e^x\, dx$.

65. *Perpetuities* Suppose you want to give a favorite school or charity $1000 a year forever. This kind of gift is called a *perpetuity*. Assume you can earn 8% annually on your money, i.e., that a payment of a_n today will be worth $a_n (1.08)^n$ in n years.

(a) Show that the amount you must invest today to cover the nth $1000 payment in n years is $1000(1.08)^{-n}$.

(b) Construct an infinite series that gives the amount you must invest today to cover *all* the payments in the perpetuity.

(c) Show that the series in (b) converges and find its sum. This sum is called the *present value* of the perpetuity. What does it represent?

66. *(Continuation of Exercise 65)* Find the present value of a $1000-per-year perpetuity at 6% annual interest.

67. *Expected Payoff* How much would you expect to win playing the following game?

Toss a *fair* coin (heads and tails equally likely). Every time it comes up heads you win a dollar, but the game is over as soon as it comes up tails.

(a) The *expected payoff* of the game is computed by summing all possible payoffs times their respective probabilities. If the probability of tossing the first tail on the *n*th toss is $(1/2)^n$, express the expected payoff of this game as an infinite series.

(b) Differentiate both sides of

$$\frac{1}{1-x} = 1 + x + x^2 + \cdots + x^n + \cdots$$

to get a series for $1/(1-x)^2$.

(c) Use the series in (b) to get a series for $x^2/(1-x)^2$.

(d) Use the series in (c) to evaluate the expected payoff of the game.

68. *Punching out Triangles* This exercise refers to the "right side up" equilateral triangle with sides of length $2b$ in the accompanying figure. "Upside down" equilateral triangles are removed from the original triangle as the sequence of pictures suggests. The sum of the areas removed from the original triangle forms an infinite series.

(a) Find this infinite series.

(b) Find the sum of this infinite series and hence find the total area removed from the original triangle.

(c) Is every point on the original triangle removed? Explain why or why not.

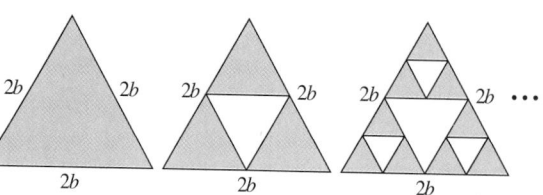

69. *Nicole Oresme's (pronounced "O-rem's") Theorem* Prove Nicole Oresme's theorem that

$$1 + \frac{1}{2}\cdot 2 + \frac{1}{4}\cdot 3 + \cdots + \frac{n}{2^{n-1}} + \cdots = 4.$$

(*Hint:* Differentiate both sides of the equation $1/(1-x) = 1 + \sum_{n=1}^{\infty} x^n$.)

70. (a) Show that

$$\sum_{n=1}^{\infty} \frac{n(n+1)}{x^n} = \frac{2x^2}{(x-1)^3}$$

for $|x| > 1$ by differentiating the identity

$$\sum_{n=1}^{\infty} x^{n+1} = \frac{x^2}{1-x}$$

twice, multiplying the result by x, and then replacing x by $1/x$.

(b) Use part (a) to find the real solution greater than 1 of the equation

$$x = \sum_{n=1}^{\infty} \frac{n(n+1)}{x^n}.$$

Calculus at Work

I attended the University of California at Los Angeles and received my B.S., M.S., and Ph.D. degrees in Geophysics and Space Physics. After graduating, I became a Research Physicist at SRI International in Menlo Park, California. There I studied the ionosphere, particularly from high latitudes, for the Sondre Stromfjord Incoherent Scatter Radar project.

I am currently an Associate Research Scientist at the University of Michigan, where I continue my upper atmospheric research. I study the lower thermosphere, which extends from 85 to 150 km above the earth. At this height, many physical processes interact in complex ways, making it difficult to understand the behavior of winds and temperatures and to observe atmospheric parameters. Atmospheric models are therefore vitally important in constructing a global perspective of lower thermospheric dynamics. The models use calculus in a number of ways. Polynomial expansions are used to express variables globally. Integrals are solved by summation over small steps in independent variables to determine the global wind and temperature fields. And complex processes, such as turbulence, are described by equations of motion and state. These equations are solved differentially, using series approximations where necessary.

Roberta M. Johnson
University of Michigan
Ann Arbor, MI

CHAPTER

10

Parametric, Vector, and Polar Functions

In 1935, air traffic control was conducted with a system of teletype machines, wall-sized blackboards, large table maps, and movable markers representing airplanes. Today's radar data processing includes an automatic display of aircraft identification, speed, altitude, and velocity vectors.

A DC-10 flying due west at 600 mph enters a region with a steady air current coming from the southwest at 100 mph. How should the pilot adjust the airplane's course and speed to maintain its original velocity vector? This type of problem is covered in Section 10.2.

Chapter 10 Overview

When a body travels in the xy-plane, the parametric equations $x = f(t)$ and $y = g(t)$ can be used to model the body's motion and path. In this chapter, we introduce the *vector* form of parametric equations, which allows us to track the positions of moving bodies with vectors, calculate the directions and magnitudes of their velocities and accelerations, and predict the effects of the forces we see working on them.

One of the principal applications of vector functions is the analysis of motion in space. Planetary motion is best described with polar coordinates (another of Newton's inventions, although James Bernoulli usually gets the credit because he published first), so we prepare for the future by investigating curves, derivatives, and integrals in this new coordinate system.

10.1 Parametric Functions

Derivatives • Parametric Formula for d^2y/dx^2 • Length of a Smooth Curve • Cycloids • Surface Area

Derivatives

If f and g are functions of t, then the curve given by the parametric equations $x = f(t)$, $y = g(t)$ can be treated as the graph of a function of the parameter t, because each value of t produces a unique point on the curve.

You may want to review the material on parametrization in Section 1.4.

> **Definition Derivative at a Point**
>
> A parametrized curve $x = f(t)$, $y = g(t)$, $a \le t \le b$, has a **derivative** at $t = t_0$ if f and g have derivatives at $t = t_0$.

The curve is **differentiable** if it is differentiable at every parameter value. The curve is **smooth** if f' and g' are continuous and not simultaneously zero.

Example 1 USING THE DEFINITION

The parametrization

$$x = \cos t, \quad y = \sin t, \quad 0 \le t \le 2\pi,$$

of the unit circle is differentiable at every t in the interval $[0, 2\pi]$ because $\cos t$ and $\sin t$ are differentiable in $[0, 2\pi]$. Thus, the unit circle is a differentiable curve. It is also smooth because $x' = -\sin t$ and $y' = \cos t$ are never zero at the same time.

Parametric Formula for d^2y/dx^2

A parametrized curve can yield one or more ways to define y as a function of x. In Section 3.6, we showed that at a point where both the curve and $y(x)$ are differentiable and $dx/dt \ne 0$, the derivatives dx/dt, dy/dt, and dy/dx are related by the formula

$$\frac{dy}{dx} = \frac{dy/dt}{dx/dt}. \tag{1}$$

This is the formula we use to express dy/dx as a function of t.

If the parametric equations define y as a twice-differentiable function of x, we can apply Equation 1 to the function $dy/dx = y'$ to calculate d^2y/dx^2 as a function of t:

$$\frac{d^2y}{dx^2} = \frac{d}{dx}(y') = \frac{dy'/dt}{dx/dt}. \qquad \text{Eq. 1 with } y' \text{ in place of } y$$

Parametric Formula for d^2y/dx^2

If the equations $x = f(t)$, $y = g(t)$ define y as a twice-differentiable function of x, then at any point where $dx/dt \neq 0$,

$$\frac{d^2y}{dx^2} = \frac{dy'/dt}{dx/dt}.$$

Finding d^2y/dx^2 in terms of t:

1. Express $y' = dy/dx$ in terms of t.
2. Find dy'/dt.
3. Divide dy'/dt by dx/dt.

Example 2 FINDING d^2y/dx^2 FOR A PARAMETRIZED CURVE

Find d^2y/dx^2 as a function of t if $x = t - t^2$, $y = t - t^3$.

Solution

Step 1: Express $y' = dy/dx$ in terms of t.

$$y' = \frac{dy}{dx} = \frac{dy/dt}{dx/dt} = \frac{1 - 3t^2}{1 - 2t}$$

Step 2: Differentiate y' with respect to t.

$$\frac{dy'}{dt} = \frac{d}{dt}\left(\frac{1 - 3t^2}{1 - 2t}\right) = \frac{2 - 6t + 6t^2}{(1 - 2t)^2} \qquad \text{Quotient Rule}$$

Step 3: Divide dy'/dt by dx/dt.

$$\frac{d^2y}{dx^2} = \frac{dy'/dt}{dx/dt} = \frac{(2 - 6t + 6t^2)/(1 - 2t)^2}{1 - 2t} = \frac{2 - 6t + 6t^2}{(1 - 2t)^3}$$

Length of a Smooth Curve

Line segment approximations similar to those in Section 7.4 lead to the following formula for the length of a smooth parametrized curve.

Length (Arc Length) of a Smooth Parametrized Curve

If a smooth curve $x = f(t)$, $y = g(t)$, $a \leq t \leq b$, is traversed exactly once as t increases from a to b, the curve's length is

$$L = \int_a^b \sqrt{\left(\frac{dx}{dt}\right)^2 + \left(\frac{dy}{dt}\right)^2}\, dt.$$

In Exercises 25 and 26, you will see how the length formulas in Section 7.4 are special cases of this parametric formula.

$x = \cos^3 t, \; y = \sin^3 t, \; 0 \le t \le 2\pi$

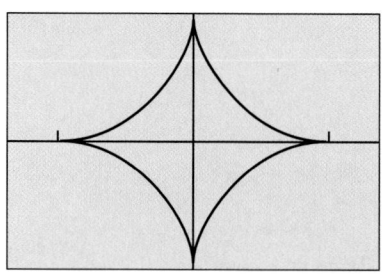

[−1.5, 1.5] by [−1, 1]

Figure 10.1 The astroid in Example 3.

Example 3 FINDING LENGTH

Find the length of the astroid (Figure 10.1)

$$x = \cos^3 t, \qquad y = \sin^3 t, \qquad 0 \le t \le 2\pi.$$

Solution

Solve Analytically

The curve is traced once as t goes from 0 to 2π. Because of the curve's symmetry with respect to the coordinate axes, its length is four times the length of the first quadrant portion. We have

$$\left(\frac{dx}{dt}\right)^2 = \left((3\cos^2 t)(-\sin t)\right)^2 = 9\cos^4 t \sin^2 t$$

$$\left(\frac{dy}{dt}\right)^2 = \left((3\sin^2 t)(\cos t)\right)^2 = 9\sin^4 t \cos^2 t$$

$$\sqrt{\left(\frac{dx}{dt}\right)^2 + \left(\frac{dy}{dt}\right)^2} = \sqrt{9\cos^2 t \sin^2 t \underbrace{(\cos^2 t + \sin^2 t)}_{1}}$$

$$= \sqrt{9\cos^2 t \sin^2 t}$$

$$= 3|\cos t \sin t|.$$

Thus, the length of the first quadrant portion of the curve is

$$\int_0^{\pi/2} 3|\cos t \sin t| \, dt = 3\int_0^{\pi/2} \cos t \sin t \, dt \quad \cos t \sin t \ge 0, \, 0 \le t \le \pi/2$$

$$= \frac{3}{2}\sin^2 t \Big]_0^{\pi/2} \qquad u = \sin t, \, du = \cos t \, dt$$

$$= \frac{3}{2}.$$

The length of the astroid is $4(3/2) = 6$.

Support Numerically

NINT $(3|\cos t \sin t|, t, 0, 2\pi) = 6$.

Cycloids

Suppose that a wheel of radius a rolls along a horizontal line without slipping (see Figure 10.2). The path traced by a point P on the wheel's edge is a **cycloid,** where P is originally at the origin.

Example 4 FINDING PARAMETRIC EQUATIONS FOR A CYCLOID

Find parametric equations for the path of the point P in Figure 10.2.

Solution We suppose that the wheel rolls to the right, P being at the origin when the turn angle t equals 0. Figure 10.2 shows the wheel after it has turned t radians. The base of the wheel is at distance at from the origin. The wheel's center is at (at, a), and the coordinates of P are

$$x = at + a\cos\theta, \qquad y = a + a\sin\theta.$$

Figure 10.2 The position of $P(x, y)$ on the edge of the wheel when the wheel has turned t radians. (Example 4)

Huygens's Clock

The problem with a pendulum clock whose bob swings in a circular arc is that the frequency of the swing depends on the amplitude of the swing. The wider the swing, the longer it takes the bob to return to center.

This does not happen if the bob can be made to swing in a cycloid. In 1673, Christiaan Huygens (1629–1695), the Dutch mathematician, physicist, and astronomer who discovered the rings of Saturn, designed a pendulum clock whose bob would swing in a cycloid. Driven by a need to make accurate determinations of longitude at sea, he hung the bob from a fine wire constrained by guards that caused it to draw up as it swung away from center. How were the guards shaped? They were cycloids, too.

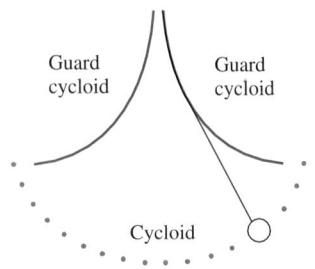

To express θ in terms of t, we observe that $t + \theta = 3\pi/2 + 2k\pi$ for some integer k, so

$$\theta = \frac{3\pi}{2} - t + 2k\pi.$$

Thus,

$$\cos \theta = \cos\left(\frac{3\pi}{2} - t + 2k\pi\right) = -\sin t,$$

$$\sin \theta = \sin\left(\frac{3\pi}{2} - t + 2k\pi\right) = -\cos t.$$

Therefore,

$$x = at - a \sin t = a(t - \sin t),$$

$$y = a - a \cos t = a(1 - \cos t).$$

Exploration 1 | Investigating Cycloids

Consider the cycloids with parametric equations

$$x = a(t - \sin t), \qquad y = a(1 - \cos t), \qquad a > 0.$$

1. Graph the equations for $a = 1, 2,$ and 3.
2. Find the x-intercepts.
3. Show that $y \geq 0$ for all t.
4. Explain why the arches of a cycloid are congruent.
5. What is the maximum value of y? Where is it attained?
6. Describe the graph of a cycloid.

Example 5 FINDING LENGTH

Find the length of one arch of the cycloid

$$x = a(t - \sin t), \qquad y = a(1 - \cos t), \qquad a > 0.$$

Solution Figure 10.3 shows the first arch of the cycloid and part of the next for $a = 1$. In Exploration 1 you found that the x-intercepts occur at t equal to multiples of 2π and that the arches are congruent.

The length of the first arch is

$$\int_0^{2\pi} \sqrt{\left(\frac{dx}{dt}\right)^2 + \left(\frac{dy}{dt}\right)^2}\, dt.$$

We have

$$\left(\frac{dx}{dt}\right)^2 = [a(1 - \cos t)]^2 = a^2(1 - 2\cos t + \cos^2 t)$$

$$\left(\frac{dy}{dt}\right)^2 = [a \sin t]^2 = a^2 \sin^2 t$$

$$\sqrt{\left(\frac{dx}{dt}\right)^2 + \left(\frac{dy}{dt}\right)^2} = a\sqrt{2 - 2\cos t}. \quad a > 0, \sin^2 t + \cos^2 t = 1$$

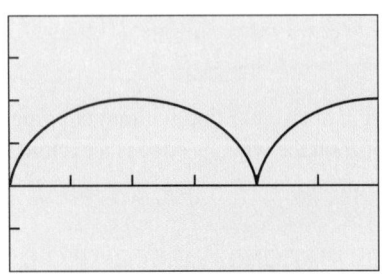

[0, 9π] by [−2, 4]

Figure 10.3 The graph of the cycloid $x = t - \sin t$, $y = 1 - \cos t$, $t \geq 0$. (Example 5)

Therefore,

$$\int_0^{2\pi} \sqrt{\left(\frac{dx}{dt}\right)^2 + \left(\frac{dy}{dt}\right)^2}\, dt = a \int_0^{2\pi} \sqrt{2 - 2\cos t}\, dt = 8a. \qquad \text{Using NINT}$$

The length of one arch of the cycloid is $8a$.

Surface Area

The following formulas give the area of the surface obtained by revolving the smooth parametrized curve $x = f(t)$, $y = g(t)$, $a \le t \le b$, about a coordinate axis.

Surface Area (from a Smooth Parametrized Curve)

If a smooth curve $x = f(t)$, $y = g(t)$, $a \le t \le b$, is traversed exactly once as t increases from a to b, then the areas of the surfaces generated by revolving the curve about the coordinate axes are as follows.

1. Revolution about the x-axis ($y \ge 0$):

$$S = \int_a^b 2\pi y \sqrt{\left(\frac{dx}{dt}\right)^2 + \left(\frac{dy}{dt}\right)^2}\, dt$$

2. Revolution about the y-axis ($x \ge 0$):

$$S = \int_a^b 2\pi x \sqrt{\left(\frac{dx}{dt}\right)^2 + \left(\frac{dy}{dt}\right)^2}\, dt$$

Example 6 FINDING SURFACE AREA

The standard parametrization of the circle of radius 1 centered at the point $(0, 1)$ in the xy-plane is

$$x = \cos t, \qquad y = 1 + \sin t, \qquad 0 \le t \le 2\pi.$$

Use this parametrization to find the area of the surface swept out by revolving the circle about the x-axis (Figure 10.4).

Solution We evaluate the formula

$$S = \int_a^b 2\pi y \sqrt{\left(\frac{dx}{dt}\right)^2 + \left(\frac{dy}{dt}\right)^2}\, dt.$$

$$S = \int_0^{2\pi} 2\pi(1 + \sin t)\sqrt{(-\sin t)^2 + (\cos t)^2}\, dt$$

$$= 2\pi \int_0^{2\pi} (1 + \sin t)\, dt \qquad \sin^2 t + \cos^2 t = 1$$

$$= 2\pi \left[t - \cos t\right]_0^{2\pi} = 4\pi^2$$

The surface area of the solid in Figure 10.4 is $4\pi^2$.

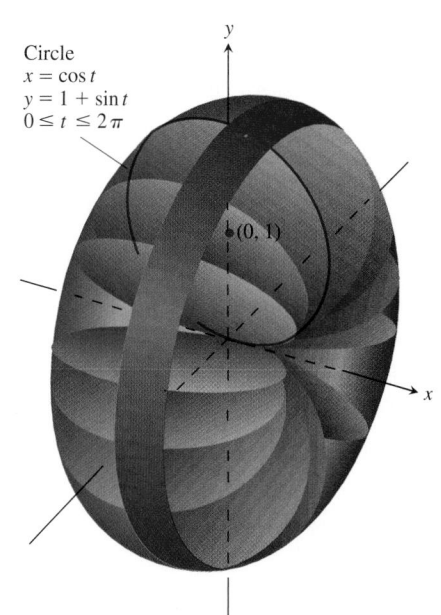

Circle
$x = \cos t$
$y = 1 + \sin t$
$0 \le t \le 2\pi$

$(0, 1)$

Figure 10.4 The surface in Example 6.

Quick Review 10.1

Exercises 1–4 refer to the parametrized curve $x = \cos t$, $y = \sin t$, $0 \le t \le 3\pi/2$.

1. Identify the initial point of the curve.

2. Identify the terminal point of the curve.

3. Find a Cartesian equation for a curve that contains the graph of the parametrized curve.

4. Which portion of the Cartesian graph in Exercise 3 is traced by the parametrized curve?

In Exercises 5 and 6, find a parametrization for the curve.

5. The portion of the curve $y = x^2 + 1$ between the points $(-1, 2)$ and $(3, 10)$.

6. $(x - 2)^2 + (y - 3)^2 = 4$

Exercises 7–9 refer to the parametrized curve $x = 2 \cos t$, $y = 3 \sin t$, $0 \le t \le 2\pi$.

7. Find dy/dx at the point where $t = 3\pi/4$.

8. Find an equation for the tangent at the point where $t = 3\pi/4$.

9. Find an equation for the normal at the point where $t = 3\pi/4$.

10. Find the length of the curve $y = x^{3/2}$ from $x = 0$ to $x = 3$.

Section 10.1 Exercises

In Exercises 1–6, find (a) dy/dx and (b) d^2y/dx^2 in terms of t.

1. $x = 4 \sin t$, $y = 2 \cos t$
2. $x = \cos t$, $y = \sqrt{3} \cos t$
3. $x = -\sqrt{t + 1}$, $y = \sqrt{3t}$
4. $x = 1/t$, $y = -2 + \ln t$
5. $x = t^2 - 3t$, $y = t^3$
6. $x = t^2 + t$, $y = t^2 - t$

In Exercises 7–10, find the points at which the tangent to the curve is (a) horizontal, (b) vertical.

7. $x = 2 + \cos t$, $y = -1 + \sin t$
8. $x = \sec t$, $y = \tan t$
9. $x = 2 - t$, $y = t^3 - 4t$
10. $x = -2 + 3 \cos t$, $y = 1 + 3 \sin t$

In Exercises 11–16, find the length of the curve.

11. $x = \cos t$, $y = t + \sin t$, $0 \le t \le \pi$
12. $x = \dfrac{(2t + 3)^{3/2}}{3}$, $y = t + \dfrac{t^2}{2}$, $0 \le t \le 3$
13. $x = \dfrac{1}{3}t^3$, $y = \dfrac{1}{2}t^2$, $0 \le t \le 1$
14. $x = 8 \cos t + 8t \sin t$, $y = 8 \sin t - 8t \cos t$, $0 \le t \le \pi/2$
15. $x = \ln(\sec t + \tan t) - \sin t$, $y = \cos t$, $0 \le t \le \pi/3$
16. $x = e^t - t^2$, $y = t + e^{-t}$, $-1 \le t \le 2$

In Exercises 17–20, find the area of the surface generated by revolving the curve about the indicated axis.

17. $x = \cos t$, $y = 2 + \sin t$, $0 \le t \le 2\pi$; x-axis
18. $x = (2/3)t^{3/2}$, $y = 2\sqrt{t}$, $0 \le t \le 2$; y-axis
19. $x = t + 1$, $y = t^2 + 2$, $0 \le t \le 3$; y-axis
20. $x = \ln(\sec t + \tan t) - \sin t$, $y = \cos t$, $0 \le t \le \pi/3$; x-axis

Explorations

21. *A Frustum of a Cone* The line segment joining the points $(0, 1)$ and $(2, 2)$ is revolved about the x-axis to generate a frustum of a cone.

   (a) Find a parametrization for the line segment joining the two points.

   (b) Find the surface area of the frustum using the parametric equations in (a).

   (c) Find the (lateral) surface area using the geometry formula

   $$\text{Area} = \pi(r_1 + r_2)(\text{slant height}).$$

22. *A Cone* The line segment joining the origin to the point (h, r) is revolved about the x-axis to generate a cone of height h and base radius r.

   (a) **Writing to Learn** Explain why

   $$x = ht, \quad y = rt, \quad 0 \le t \le 1$$

   gives a parametrization of the line segment joining the two points.

   (b) Find the cone's (lateral) surface area using the parametrization in (a).

   (c) Use the geometry formula

   $$\text{Area} = \pi(r)(\text{slant height})$$

   to compute the surface area of the cone. ∎

23. *Length is Independent of Parametrization* To illustrate the fact that the numbers we get for length do not usually depend on the way we parametrize our curves, calculate the length of the semicircle $y = \sqrt{1 - x^2}$ with these two different parametrizations.

(a) $x = \cos 2t, \quad y = \sin 2t, \quad 0 \le t \le \pi/2$

(b) $x = \sin \pi t, \quad y = \cos \pi t, \quad -1/2 \le t \le 1/2$

24. *Perimeter of an Ellipse* Find the length of the ellipse

$$x = 3 \cos t, \quad y = 4 \sin t, \quad 0 \le t \le 2\pi.$$

25. *Cartesian Length Formula* The graph of a function $y = f(x)$ over an interval $[a, b]$ automatically has the parametrization

$$x = x, \quad y = f(x), \quad a \le x \le b.$$

The parameter in this case is x itself. Show that for this parametrization, the length formula

$$L = \int_a^b \sqrt{\left(\frac{dx}{dt}\right)^2 + \left(\frac{dy}{dt}\right)^2} \, dt$$

reduces to the Cartesian formula

$$L = \int_a^b \sqrt{1 + \left(\frac{dy}{dx}\right)^2} \, dx$$

derived in Section 7.4.

26. *(Continuation of Exercise 25)* Show that the Cartesian formula

$$L = \int_c^d \sqrt{1 + \left(\frac{dx}{dy}\right)^2} \, dy$$

for the length of the curve $x = g(y), \ c \le y \le d,$ from Section 7.4 is a special case of the parametric length formula

$$L = \int_a^b \sqrt{\left(\frac{dx}{dt}\right)^2 + \left(\frac{dy}{dt}\right)^2} \, dt.$$

27. *Finding a Midpoint* A particle travels from $t = 0$ to $t = 4$ along the curve

$$x = \frac{t^2}{2}, \quad y = \frac{1}{3}(2t + 1)^{3/2}, \quad 0 \le t \le 4.$$

At what point (x, y) has the particle covered half of the length of the curve?

28. *Finding Surface Area* Find the area of the surface generated by revolving the curve

$$x = 3 \sin t, \quad y = 5 + 3 \sin 2t, \quad 0 \le t \le 2\pi$$

about the y-axis.

29. *Finding Surface Area* Find the area of the surface generated by revolving the curve

$$x = 5 + 3 \sin t, \quad y = 3 \sin 2t, \quad 0 \le t \le 2\pi$$

about the x-axis.

30. *Cycloid* Find the area of the surface generated by revolving one arch of the cycloid

$$x = a(t - \sin t), \quad y = a(1 - \cos t)$$

about the x-axis.

Exercises 31 and 32 refer to the region bounded by the x-axis and one arch of the cycloid

$$x = a(t - \sin t), \quad y = a(1 - \cos t)$$

that is shaded in the figure.

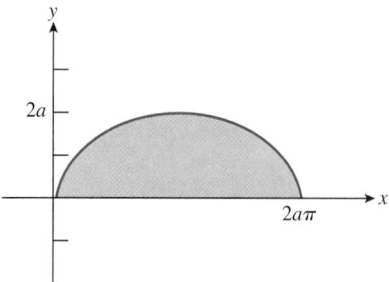

31. Find the area of the shaded region. (*Hint:* $dx = (dx/dt) \, dt$)

32. Find the volume swept out by revolving the region about the x-axis. (*Hint:* $dV = \pi y^2 \, dx = \pi y^2 (dx/dt) \, dt$)

33. *Involute of a Circle* Work in groups of two or three. If a string wound around a fixed circle is unwound while being held taut in the plane of the circle, its end P traces an *involute* of the circle as suggested by the diagram below. In the diagram, the circle is the unit circle in the xy-plane, and the initial position of the tracing point is the point $(1, 0)$ on the x-axis. The unwound portion of the string is tangent to the circle at Q, and t is the radian measure of the angle from the positive x-axis to the segment OQ.

(a) Derive parametric equations for the involute by expressing the coordinates x and y of P in terms of t for $t \ge 0$.

(b) Find the length of the involute for $0 \le t \le 2\pi$.

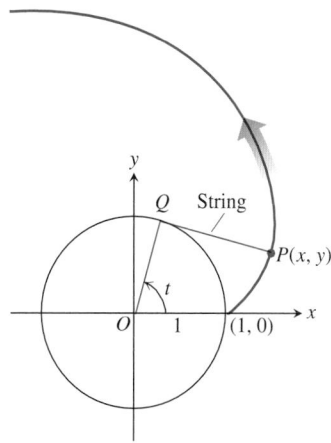

34. *(Continuation of Exercise 33)* Repeat Exercise 33 using the circle of radius a centered at the origin, $x^2 + y^2 = a^2$.

In Exercises 35–38, a projectile is launched over horizontal ground at an angle θ with the horizontal and with initial velocity v_0 ft/sec. Its path is given by the parametric equations

$$x = (v_0 \cos \theta)t, \quad y = (v_0 \sin \theta)t - 16t^2.$$

(a) Find the length of the path traveled by the projectile.

(b) Estimate the maximum height of the projectile.

35. $\theta = 20°, \quad v_0 = 150$ **36.** $\theta = 30°, \quad v_0 = 150$

37. $\theta = 60°, \quad v_0 = 150$ **38.** $\theta = 90°, \quad v_0 = 150$

Extending the Ideas

39. The function $y = f(x)$ is nonnegative on the interval $[a, b]$. Use the parametrization

$$x = x, \quad y = f(x), \quad a \le x \le b,$$

to show that the area of the surface generated by revolving the curve $y = f(x)$ from $x = a$ to $x = b$ about the x-axis is

$$S = \int_a^b 2\pi f(x) \sqrt{1 + \left(\frac{dy}{dx}\right)^2} \, dx.$$

In Exercises 40–42, use the formula of Exercise 39 to find the area of the surface generated by revolving the curve $y = f(x)$ from $x = a$ to $x = b$ about the x-axis.

40. $f(x) = e^x, \quad a = 0, \quad b = 3$

41. $f(x) = 1/x, \quad a = 1, \quad b = 4$

42. $f(x) = 2^x + 2^{-x}, \quad a = -2, \quad b = 2$

10.2 Vectors in the Plane

Component Form • Zero Vector • Vector Operations • Angle Between Vectors • Applications

Component Form

Some things we measure are determined by their *magnitudes*. To record mass, length, or time, for example, we need only write down a real number and name an appropriate unit of measure. These are *scalar quantities,* and the associated real numbers are **scalars.**

A quantity such as force, displacement, or velocity has direction as well as magnitude and is represented by a **directed line segment** (Figure 10.5). The arrow points in the direction of the action and its length gives the magnitude of the action in terms of a suitably chosen unit. The directed line segment $\overrightarrow{AB}$ has **initial point** A and **terminal point** B; its **length** is denoted by $|\overrightarrow{AB}|$. Directed line segments that have the same length and direction are **equivalent.**

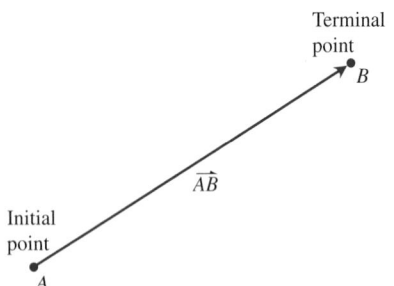

Figure 10.5 The directed line segment $\overrightarrow{AB}$.

Definitions Vector, Equal Vectors

A **vector** in the plane is represented by a directed line segment. Two vectors are **equal** (or **the same**) if they have the same length and direction.

Thus, the arrows we use when we draw vectors are understood to represent the same vector if they have the same length, are parallel, and point in the same direction (Figure 10.6).

In textbooks, vectors are usually written in lowercase, boldface letters, for example **u**, **v**, and **w**. In handwritten form, it is customary to draw small arrows above the letters, for example $\vec{u}$, $\vec{v}$, and $\vec{w}$.

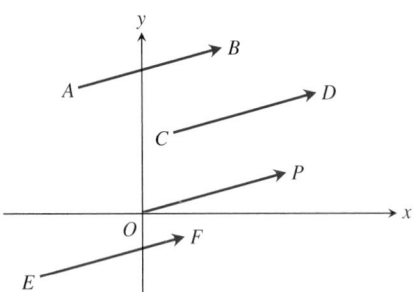

Figure 10.6 The four arrows (directed line segments) shown here have the same length and direction. They therefore represent the same vector, and we write $\overrightarrow{AB} = \overrightarrow{CD} = \overrightarrow{OP} = \overrightarrow{EF}$.

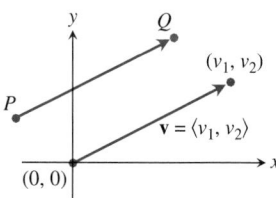

Figure 10.7 Two equal vectors. (Example 1)

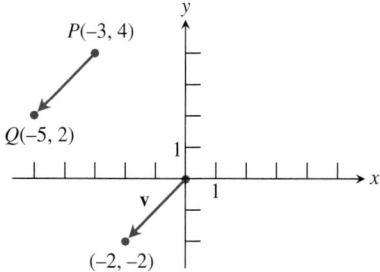

Figure 10.8 The standard position of a vector.

Example 1   SHOWING VECTORS ARE EQUAL

Let $A = (0, 0)$, $B = (3, 4)$, $C = (-4, 2)$, and $D = (-1, 6)$. Show that the vectors $\mathbf{u} = \overrightarrow{AB}$ and $\mathbf{v} = \overrightarrow{CD}$ are equal.

Solution We need to show that $\mathbf{u}$ and $\mathbf{v}$ have the same length and direction (Figure 10.7). We use the distance formula to find their lengths.

$$|\mathbf{u}| = |\overrightarrow{AB}| = \sqrt{(3 - 0)^2 + (4 - 0)^2} = 5$$

$$|\mathbf{v}| = |\overrightarrow{CD}| = \sqrt{(-1 - (-4))^2 + (6 - 2)^2} = 5$$

Next we calculate the slopes of the two line segments.

$$\text{slope of } \overrightarrow{AB} = \frac{4 - 0}{3 - 0} = \frac{4}{3}, \quad \text{slope of } \overrightarrow{CD} = \frac{6 - 2}{-1 - (-4)} = \frac{4}{3}$$

The line segments have the same direction because they are parallel and directed toward the upper right. Therefore, $\mathbf{u} = \mathbf{v}$ because they have the same length and direction.

Let $\mathbf{v} = \overrightarrow{PQ}$. There is one directed line segment equivalent to $\overrightarrow{PQ}$ whose initial point is the origin (Figure 10.8). It is the representative of $\mathbf{v}$ in **standard position** and is the vector we normally use to represent $\mathbf{v}$.

Definition   Component Form of a Vector

If $\mathbf{v}$ is a vector in the plane equal to the vector with initial point $(0, 0)$ and terminal point (v_1, v_2), then the **component form** of $\mathbf{v}$ is

$$\mathbf{v} = \langle v_1, v_2 \rangle.$$

The coordinates v_1 and v_2 are the **components** of $\mathbf{v}$. The **magnitude (length)** of $\mathbf{v} = \langle v_1, v_2 \rangle$ is

$$|\mathbf{v}| = \sqrt{v_1{}^2 + v_2{}^2}.$$

Two vectors $\langle a, b \rangle$ and $\langle c, d \rangle$ are equal if and only if $a = c$ and $b = d$. The vector $\langle a, b \rangle$ is called the **position vector** of the point (a, b).

Example 2   FINDING COMPONENT FORM

Find the **(a)** component form and **(b)** length of the vector with initial point $P = (-3, 4)$ and terminal point $Q = (-5, 2)$.

Solution

(a) We can move the initial point $(-3, 4)$ of the vector $\overrightarrow{PQ}$ to the origin by translating the vector 3 units right and 4 units down (Figure 10.9). Thus, Q translates to $(-2, -2)$ and the component form of $\overrightarrow{PQ}$ is

$$\mathbf{v} = \langle -2, -2 \rangle.$$

(b) The length of $\mathbf{v} = \overrightarrow{PQ}$ is

$$|\mathbf{v}| = \sqrt{(-2)^2 + (-2)^2} = 2\sqrt{2}.$$

Figure 10.9 The vectors $\overrightarrow{PQ}$ and $\mathbf{v}$ are equal. (Example 2)

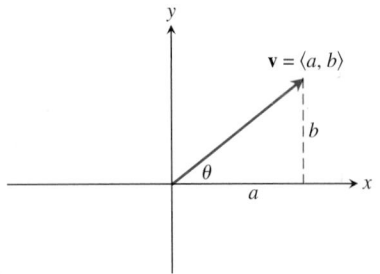

Figure 10.10 The vector $\mathbf{v} = \langle a, b \rangle$ has slope b/a.

Example 2 can be generalized in the following way. The component form of the vector with initial point $P = (p_1, p_2)$ and terminal point $Q = (q_1, q_2)$ is

$$\mathbf{v} = \langle v_1, v_2 \rangle = \langle q_1 - p_1, q_2 - p_2 \rangle,$$

and its length is

$$|\mathbf{v}| = \sqrt{v_1^2 + v_2^2} = \sqrt{(q_1 - p_1)^2 + (q_2 - p_2)^2}.$$

If $|\mathbf{v}| = 1$, then $\mathbf{v}$ is a **unit vector.**

The **slope** of a nonvertical vector is the slope shared by the lines parallel to the vector. Thus, if $a \neq 0$, the vector $\mathbf{v} = \langle a, b \rangle$ has slope b/a (Figure 10.10), and $a = |\mathbf{v}| \cos \theta$, $b = |\mathbf{v}| \sin \theta$.

Example 3 FINDING COMPONENT FORM

Find the component form of the vector $\mathbf{v}$ of length 3 that makes an angle of $40°$ with the positive x-axis.

Solution From Figure 10.10 and the discussion preceding this example, we have

$$\mathbf{v} = \langle 3 \cos 40°, 3 \sin 40° \rangle \approx \langle 2.298, 1.928 \rangle.$$

Zero Vector

It is convenient to define $\langle 0, 0 \rangle$ as the **zero vector,**

$$\mathbf{0} = \langle 0, 0 \rangle.$$

The zero vector has length $|\mathbf{0}| = \sqrt{0^2 + 0^2} = 0$ and is the only vector with no direction.

Vector Operations

Two principal operations involving vectors are *vector addition* and *scalar multiplication.*

Definitions Vector Operations

Let $\mathbf{u} = \langle u_1, u_2 \rangle$, $\mathbf{v} = \langle v_1, v_2 \rangle$ be vectors with k a scalar (real number).

Addition: $\quad \mathbf{u} + \mathbf{v} = \langle u_1, u_2 \rangle + \langle v_1, v_2 \rangle = \langle u_1 + v_1, u_2 + v_2 \rangle$

Subtraction: $\quad \mathbf{u} - \mathbf{v} = \langle u_1, u_2 \rangle - \langle v_1, v_2 \rangle = \langle u_1 - v_1, u_2 - v_2 \rangle$

Scalar multiplication: $k\mathbf{u} = \langle ku_1, ku_2 \rangle$

Negative (opposite): $\quad -\mathbf{u} = (-1)\mathbf{u} = \langle -u_1, -u_2 \rangle$

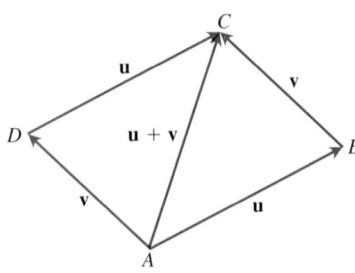

Figure 10.11 The parallelogram law of addition.

Two vectors $\mathbf{u}$ and $\mathbf{v}$ may be added geometrically by placing the initial point of one at the terminal point of the other as shown in Figure 10.11. The sum, called the **resultant vector,** is the diagonal of the resulting parallelogram. This geometric description of addition is sometimes called the **parallelogram law** of addition. In Exercise 33, you will show that the parallelogram law agrees with the component form of addition.

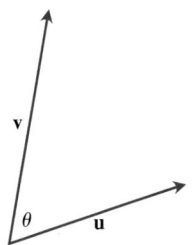

Figure 10.12 Scalar multiples of **u**.

The product $k\mathbf{u}$ of the scalar k and the vector $\mathbf{u}$ can be obtained geometrically by a stretch or shrink of $\mathbf{u}$ by the factor k when $k > 0$ (Figure 10.12). If $k < 0$, then there is a stretch or shrink by the factor $|k|$, and the direction of $k\mathbf{u}$ is opposite that of $\mathbf{u}$.

Example 4 PERFORMING OPERATIONS ON VECTORS

Let $\mathbf{u} = \langle -1, 3 \rangle$ and $\mathbf{v} = \langle 4, 7 \rangle$. Find

(a) $2\mathbf{u} + 3\mathbf{v}$,   (b) $\mathbf{u} - \mathbf{v}$,   (c) $\left| \dfrac{1}{2}\mathbf{u} \right|$.

Solution

(a) $2\mathbf{u} + 3\mathbf{v} = 2\langle -1, 3 \rangle + 3\langle 4, 7 \rangle$

$\qquad = \langle 2(-1) + 3(4), 2(3) + 3(7) \rangle = \langle 10, 27 \rangle$

(b) $\mathbf{u} - \mathbf{v} = \langle -1, 3 \rangle - \langle 4, 7 \rangle$

$\qquad = \langle -1 - 4, 3 - 7 \rangle = \langle -5, -4 \rangle$

(c) $\left| \dfrac{1}{2}\mathbf{u} \right| = \left| \left\langle -\dfrac{1}{2}, \dfrac{3}{2} \right\rangle \right| = \sqrt{\left(-\dfrac{1}{2} \right)^2 + \left(\dfrac{3}{2} \right)^2} = \dfrac{1}{2}\sqrt{10}$

Vector operations have many of the properties of ordinary arithmetic.

Properties of Vector Operations

Let $\mathbf{u}$, $\mathbf{v}$, $\mathbf{w}$ be vectors and a, b be scalars.

1. $\mathbf{u} + \mathbf{v} = \mathbf{v} + \mathbf{u}$

2. $(\mathbf{u} + \mathbf{v}) + \mathbf{w} = \mathbf{u} + (\mathbf{v} + \mathbf{w})$

3. $\mathbf{u} + \mathbf{0} = \mathbf{u}$

4. $\mathbf{u} + (-\mathbf{u}) = \mathbf{0}$

5. $0\mathbf{u} = \mathbf{0}$

6. $1\mathbf{u} = \mathbf{u}$

7. $a(b\mathbf{u}) = (ab)\mathbf{u}$

8. $a(\mathbf{u} + \mathbf{v}) = a\mathbf{u} + a\mathbf{v}$

9. $(a + b)\mathbf{u} = a\mathbf{u} + b\mathbf{u}$

Angle Between Vectors

When two nonzero vectors $\mathbf{u}$ and $\mathbf{v}$ are placed so their initial points coincide, they form an angle θ of measure $0 \le \theta \le \pi$ (Figure 10.13). This angle is the **angle between** $\mathbf{u}$ and $\mathbf{v}$.

Theorem 1 gives a formula we can use to determine the angle between two vectors.

Figure 10.13 The angle between **u** and **v**.

Theorem 1 Angle Between Two Vectors

The angle θ between two nonzero vectors $\mathbf{u} = \langle u_1, u_2 \rangle$ and $\mathbf{v} = \langle v_1, v_2 \rangle$ is given by

$$\theta = \cos^{-1} \frac{u_1 v_1 + u_2 v_2}{|\mathbf{u}||\mathbf{v}|}.$$

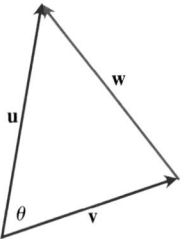

Figure 10.14 The parallelogram law of addition of vectors gives $\mathbf{w} = \mathbf{u} - \mathbf{v}$.

Proof   Applying the law of cosines to the triangle in Figure 10.14 we find

$$|\mathbf{w}|^2 = |\mathbf{u}|^2 + |\mathbf{v}|^2 - 2|\mathbf{u}||\mathbf{v}| \cos \theta$$

$$2|\mathbf{u}||\mathbf{v}| \cos \theta = |\mathbf{u}|^2 + |\mathbf{v}|^2 - |\mathbf{w}|^2.$$

Because $\mathbf{w} = \mathbf{u} - \mathbf{v}$, the component form of $\mathbf{w}$ is $\langle u_1 - v_1, u_2 - v_2 \rangle$. Thus,

$$|\mathbf{u}|^2 = \left(\sqrt{u_1^2 + u_2^2} \right)^2 = u_1^2 + u_2^2,$$

$$|\mathbf{v}|^2 = \left(\sqrt{v_1^2 + v_2^2} \right)^2 = v_1^2 + v_2^2,$$

$$|\mathbf{w}|^2 = \left(\sqrt{(u_1 - v_1)^2 + (u_2 - v_2)^2} \right)^2 = (u_1 - v_1)^2 + (u_2 - v_2)^2$$

$$= (u_1^2 - 2u_1 v_1 + v_1^2) + (u_2^2 - 2u_2 v_2 + v_2^2),$$

and

$$|\mathbf{u}|^2 + |\mathbf{v}|^2 - |\mathbf{w}|^2 = 2(u_1 v_1 + u_2 v_2).$$

Therefore,

$$2|\mathbf{u}||\mathbf{v}| \cos \theta = |\mathbf{u}|^2 + |\mathbf{v}|^2 - |\mathbf{w}|^2 = 2(u_1 v_1 + u_2 v_2)$$

$$|\mathbf{u}||\mathbf{v}| \cos \theta = u_1 v_1 + u_2 v_2$$

$$\cos \theta = \frac{u_1 v_1 + u_2 v_2}{|\mathbf{u}||\mathbf{v}|}.$$

So

$$\theta = \cos^{-1} \left(\frac{u_1 v_1 + u_2 v_2}{|\mathbf{u}||\mathbf{v}|} \right). \qquad \blacksquare$$

Definition   Dot Product (Inner Product)

The **dot product** (or **inner product**) $\mathbf{u} \cdot \mathbf{v}$ ("**u** dot **v**") of vectors $\mathbf{u} = \langle u_1, u_2 \rangle$ and $\mathbf{v} = \langle v_1, v_2 \rangle$ is the number

$$\mathbf{u} \cdot \mathbf{v} = |\mathbf{u}||\mathbf{v}| \cos \theta = u_1 v_1 + u_2 v_2.$$

We can use the dot product to rewrite the formula in Theorem 1 for finding the angle between two vectors.

Corollary   Angle Between Two Vectors

The angle between nonzero vectors $\mathbf{u}$ and $\mathbf{v}$ is

$$\theta = \cos^{-1} \left(\frac{\mathbf{u} \cdot \mathbf{v}}{|\mathbf{u}||\mathbf{v}|} \right). \qquad (1)$$

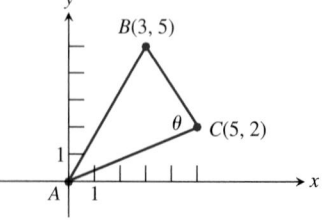

Figure 10.15 The triangle in Example 5.

Example 5   FINDING AN ANGLE OF A TRIANGLE

Find the angle θ in the triangle ABC determined by the vertices $A = (0, 0)$, $B = (3, 5)$, and $C = (5, 2)$ (Figure 10.15).

Solution   The angle θ is the angle between the vectors $\overrightarrow{CA}$ and $\overrightarrow{CB}$. The component forms of these two vectors are

$$\overrightarrow{CA} = \langle -5, -2 \rangle \quad \text{and} \quad \overrightarrow{CB} = \langle -2, 3 \rangle.$$

We use Equation 1.

$$\vec{CA} \cdot \vec{CB} = (-5)(-2) + (-2)(3) = 4$$

$$|\vec{CA}| = \sqrt{(-5)^2 + (-2)^2} = \sqrt{29}$$

$$|\vec{CB}| = \sqrt{(-2)^2 + (3)^2} = \sqrt{13}$$

Thus,

$$\theta = \cos^{-1}\left(\frac{\vec{CA} \cdot \vec{CB}}{|\vec{CA}||\vec{CB}|}\right)$$

$$= \cos^{-1}\left(\frac{4}{(\sqrt{29})(\sqrt{13})}\right)$$

$$\approx 78.1° \text{ or } 1.36 \text{ radians.}$$

Applications

Suppose the motion of a particle in the plane is represented by parametric equations. The tangent line, suitably directed, models the *direction* of the motion at the point of tangency (Figure 10.16).

A vector is **tangent** or **normal** to a curve at a point P if it is parallel or normal, respectively, to the line that is tangent to the curve at P. Example 6 shows how to find such vectors.

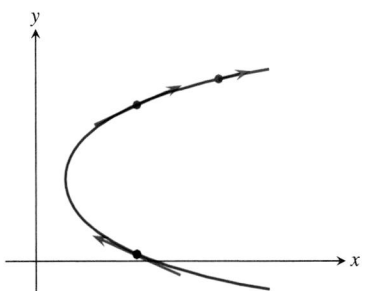

Figure 10.16 The tangent line models the direction of a moving particle at a given point.

Example 6 FINDING VECTORS TANGENT AND NORMAL TO A CURVE

Find unit vectors tangent and normal to the parametrized curve

$$x = \frac{t}{2} + 1, \quad y = \sqrt{t} + 1, \quad t \geq 0,$$

at the point where $t = 4$.

Solution Since $x = y = 3$ when $t = 4$, we seek the unit vectors that are parallel and normal to the curve's tangent line at $(3, 3)$ (Figure 10.17). The slope of the tangent is

$$\frac{dy}{dx} = \frac{dy/dt}{dx/dt} = \frac{(1/2)t^{-1/2}}{1/2}\bigg|_{t=4} = \frac{1}{2}.$$

The vector $\mathbf{v} = \langle 2, 1\rangle$ has slope 1/2, as does every nonzero multiple of $\mathbf{v}$. To find a multiple of $\mathbf{v}$ that is a unit vector, we divide $\mathbf{v}$ by its length

$$|\mathbf{v}| = \sqrt{2^2 + 1^2} = \sqrt{5},$$

obtaining

$$\mathbf{u} = \frac{\mathbf{v}}{|\mathbf{v}|} = \left\langle \frac{2}{\sqrt{5}}, \frac{1}{\sqrt{5}}\right\rangle.$$

The vector $\mathbf{u}$ has length 1 and is tangent to the curve at $(3, 3)$ because it has the same direction as $\mathbf{v}$. Of course,

$$-\mathbf{u} = \left\langle -\frac{2}{\sqrt{5}}, -\frac{1}{\sqrt{5}}\right\rangle,$$

which points in the opposite direction, is also tangent to the curve at $(3, 3)$. Unless there are additional requirements, either $\mathbf{u}$ or $-\mathbf{u}$ is acceptable.

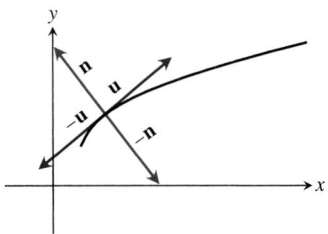

Figure 10.17 Unit tangent and normal vectors at the point $(3, 3)$ on the curve $x = (1/2)t + 1$, $y = \sqrt{t} + 1$, $t \geq 0$. (Example 6)

To find unit vectors normal to the curve at (3, 3), we look for unit vectors whose slopes are the negative reciprocal of the slope of **u**. This is quickly done by interchanging the components of **u** and changing the sign of one of them. We obtain

$$\mathbf{n} = \left\langle -\frac{1}{\sqrt{5}}, \frac{2}{\sqrt{5}} \right\rangle \quad \text{and} \quad -\mathbf{n} = \left\langle \frac{1}{\sqrt{5}}, -\frac{2}{\sqrt{5}} \right\rangle.$$

Again, either one will do. The vectors have opposite directions but both are normal to the curve at (3, 3) and have length 1.

An important application of vectors occurs in navigation.

Example 7 FINDING GROUND SPEED AND DIRECTION

A Boeing® 727® airplane, flying due east at 500 mph in still air, encounters a 70-mph tail wind acting in the direction 60° north of east. The airplane holds its compass heading due east but, because of the wind, acquires a new ground speed and direction. What are they?

Solution If **u** = the velocity of the airplane alone and **v** = the velocity of the tail wind, then $|\mathbf{u}| = 500$ and $|\mathbf{v}| = 70$ (Figure 10.18).

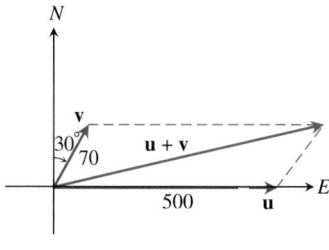

NOT TO SCALE

Figure 10.18 Vectors representing the velocities of the airplane and tail wind in Example 7.

We need to find the magnitude and direction of the *resultant vector* **u** + **v**. If we let the positive *x*-axis represent east and the positive *y*-axis represent north, then the component forms of **u** and **v** are

$$\mathbf{u} = \langle 500, 0 \rangle \quad \text{and} \quad \mathbf{v} = \langle 70 \cos 60°, 70 \sin 60° \rangle = \langle 35, 35\sqrt{3} \rangle.$$

Therefore,

$$\mathbf{u} + \mathbf{v} = \langle 535, 35\sqrt{3} \rangle,$$

$$|\mathbf{u} + \mathbf{v}| = \sqrt{535^2 + (35\sqrt{3})^2} \approx 538.4,$$

and

$$\theta = \tan^{-1} \frac{35\sqrt{3}}{535} \approx 6.5°.$$

Interpret

The new ground speed of the airplane is about 538.4 mph, and its new direction is about 6.5° north of east.

Daniel James, Jr.

1920-1978

Daniel "Chappie" James, one of the first African American fighter pilots in the U.S. armed forces, was also the first African American to receive the rank of four-star general. From 1975 to his retirement in 1978, James served as commander of the North American Air Defense Command.

Quick Review 10.2

In Exercises 1–3, let $P = (1, 2)$ and $Q = (5, 3)$.

1. Find the distance between the points P and Q.

2. Find the slope of line segment PQ.

3. If $R = (3, b)$, determine b so that segments PQ and RQ are perpendicular.

In Exercises 4 and 5, determine the missing coordinate so that the four points form a parallelogram *ABCD*.

4. $A = (0, 0)$, $B = (1, 3)$, $C = (5, 3)$, $D = (a, 0)$

5. $A = (1, 1)$, $B = (3, 5)$, $C = (8, b)$, $D = (6, 2)$

In Exercises 6–8, find the measure of the angle θ in **(a)** degrees and **(b)** radians.

6. $\theta = \cos^{-1}(-1/2)$

7. $\theta = \sin^{-1}(-1/2)$

8. $\theta = \tan^{-1}(-1)$

9. Use the law of cosines to find the value of c in the triangle.

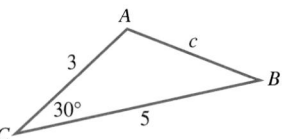

10. Use the law of cosines to find the value of θ in the triangle.

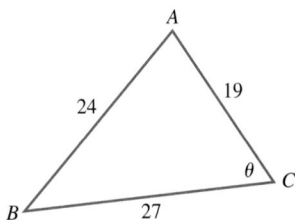

Section 10.2 Exercises

In Exercises 1–8, let $\mathbf{u} = \langle 3, -2 \rangle$ and $\mathbf{v} = \langle -2, 5 \rangle$. Find the **(a)** component form and **(b)** magnitude of the vector.

1. $3\mathbf{u}$

2. $-2\mathbf{v}$

3. $\mathbf{u} + \mathbf{v}$

4. $\mathbf{u} - \mathbf{v}$

5. $2\mathbf{u} - 3\mathbf{v}$

6. $-2\mathbf{u} + 5\mathbf{v}$

7. $\dfrac{3}{5}\mathbf{u} + \dfrac{4}{5}\mathbf{v}$

8. $-\dfrac{5}{13}\mathbf{u} + \dfrac{12}{13}\mathbf{v}$

In Exercises 9–16, find the component form of the vector.

9. the vector $\overrightarrow{PQ}$, where $P = (1, 3)$ and $Q = (2, -1)$

10. the vector $\overrightarrow{OP}$ where O is the origin and P is the midpoint of segment RS, where $R = (2, -1)$ and $S = (-4, 3)$

11. the vector from the point $A = (2, 3)$ to the origin

12. the sum of $\overrightarrow{AB}$ and $\overrightarrow{CD}$, where $A = (1, -1)$, $B = (2, 0)$, $C = (-1, 3)$, and $D = (-2, 2)$

13. the unit vector that makes an angle $\theta = 2\pi/3$ with the positive x-axis

14. the unit vector that makes an angle $\theta = -3\pi/4$ with the positive x-axis

15. the unit vector obtained by rotating the vector $\langle 0, 1 \rangle$ 120° counterclockwise about the origin

16. the unit vector obtained by rotating the vector $\langle 1, 0 \rangle$ 135° counterclockwise about the origin

In Exercises 17 and 18, copy vectors **u**, **v**, and **w** head to tail as needed to sketch the indicated vector.

17.

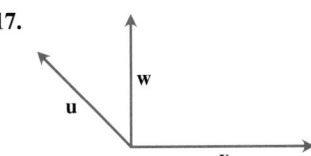

(a) u + v   **(b) u + v + w**

(c) u − v   **(d) u − w**

18.

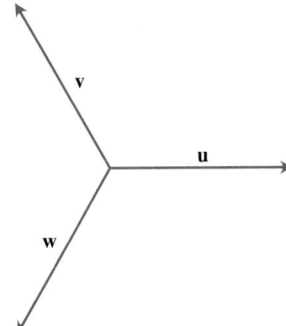

(a) u − v   **(b) u − v + w**

(c) 2u − v   **(d) u + v + w**

In Exercises 19–22, find a unit vector in the direction of the given vector.

19. $\langle 3, 4 \rangle$

20. $\langle 4, -3 \rangle$

21. $\langle -15, 8 \rangle$

22. $\langle -5, -2 \rangle$

In Exercises 23–26, find the unit vectors (four vectors in all) that are tangent and normal to the curve at the given point.

23. $x = \sqrt{t} + 1, \quad y = t + 1 + 2\sqrt{t}, \quad t = 1$

24. $x = \ln(t - 1), \quad y = t - 1, \quad t = 3$

25. $x = 4 \cos t, \quad y = 5 \sin t, \quad t = \pi/3$

26. $x = 3 \cos t, \quad y = 3 \sin t, \quad t = -\pi/4$

27. *Triangle* Find the measures of the angles of the triangle whose vertices are $A = (-1, 0)$, $B = (2, 1)$, and $C = (1, -2)$.

28. *Rectangle* Find the measures of the angles between the diagonals of the rectangle whose vertices are $A = (1, 0)$, $B = (0, 3)$, $C = (3, 4)$, and $D = (4, 1)$.

In Exercises 29–34, *work in groups of two or three.*

29. *Distributive Property* Use components to show that for any vectors **u**, **v**, and **w**,

(a) $\mathbf{u} \cdot (\mathbf{v} + \mathbf{w}) = \mathbf{u} \cdot \mathbf{v} + \mathbf{u} \cdot \mathbf{w}$

(b) $(\mathbf{u} + \mathbf{v}) \cdot \mathbf{w} = \mathbf{u} \cdot \mathbf{w} + \mathbf{v} \cdot \mathbf{w}$

30. *Inner Product and Length* Show that $\mathbf{u} \cdot \mathbf{u} = |\mathbf{u}|^2$ for any vector **u**.

31. *Inner Product and Length* Show that $(\mathbf{u} + \mathbf{v}) \cdot (\mathbf{u} - \mathbf{v}) = |\mathbf{u}|^2 - |\mathbf{v}|^2$ for any two vectors **u** and **v**.

32. *Test for Orthogonality* If **u** and **v** are nonzero vectors, show that $\mathbf{u} \cdot \mathbf{v} = 0$ if and only if **u** and **v** are **orthogonal** (perpendicular).

33. *Addition* Show that geometric addition of vectors agrees with component-form addition. (*Hint:* Place Figure 10.11 in a coordinate plane and write component forms for the vectors.)

34. *Subtraction*

(a) Give a geometric version of subtraction of vectors.

(b) Show that your geometric version of subtraction agrees with component-form subtraction. (*Hint:* See Exercise 33.)

Explorations

35. *Location on a Line Segment* Let M be a point on the line segment PQ.

(a) If M is the midpoint of the segment PQ, show that

$$\overrightarrow{OM} = \frac{1}{2}\overrightarrow{OP} + \frac{1}{2}\overrightarrow{OQ}.$$

(b) If M is one third of the way from P to Q, find a formula for the position vector $\overrightarrow{OM}$ in terms of the position vectors $\overrightarrow{OP}$ and $\overrightarrow{OQ}$.

(c) If M is two thirds of the way from P to Q, find a formula for the position vector $\overrightarrow{OM}$ in terms of the position vectors $\overrightarrow{OP}$ and $\overrightarrow{OQ}$.

(d) **Writing to Learn** If M is between P and Q on the segment PQ, state a formula for the position vector $\overrightarrow{OM}$ in terms of the position vectors $\overrightarrow{OP}$ and $\overrightarrow{OQ}$. Then prove your statement.

36. *Circle* Suppose that AB is the diameter of a circle with center O and that C is a point on one of the two arcs joining A and B, as shown in the figure. Show that $\overrightarrow{CA}$ and $\overrightarrow{CB}$ are orthogonal (perpendicular).

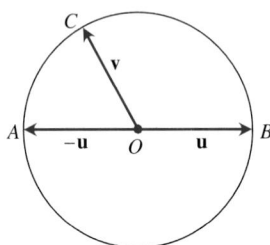

37. *Rhombus* Show that the diagonals of a rhombus (parallelogram with all sides of equal length) are perpendicular.

38. *Square* Show that squares are the only rectangles with perpendicular diagonals.

39. *Parallelogram* Prove the fact, often exploited by carpenters, that a parallelogram is a rectangle if and only if the diagonals are equal in length. ∎

40. *Parallelogram* Show that the indicated diagonal of the parallelogram determined by vectors **u** and **v** bisects the angle between **u** and **v** if $|\mathbf{u}| = |\mathbf{v}|$.

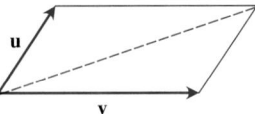

41. **Writing to Learn** What can you say about the slopes of the (nonzero) vectors **v** and $-\mathbf{v}$?

42. **Writing to Learn** If $|\mathbf{v}| = 0$, what can you say about the vector **v**?

43. *Velocity* An airplane is flying in the direction 25° west of north at 800 km/h. Find the component form of the velocity of the airplane, assuming that the positive x-axis represents due east and the positive y-axis represents due north.

44. *Velocity* An airplane is flying in the direction 10° east of south at 600 km/h. Find the component form of the velocity of the airplane, assuming that the positive x-axis represents due east and the positive y-axis represents due north.

45. *Navigation* An airplane, flying in the direction 20° east of north at 325 mph in still air, encounters a 40-mph tail wind acting in the direction 40° west of north. The airplane maintains its compass heading but, because of the wind, acquires a new ground speed and direction. What are they?

46. *Inclined Plane* Suppose a box is being towed up an inclined plane as shown in the figure. Find the force **w** needed to make the component of the force parallel to the inclined plane equal to 2.5 lb.

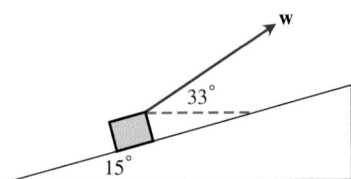

47. *Combining Forces* Juana and Diego Gonzales, ages six and four respectively, own a strong and stubborn puppy named Corporal. It is so hard to take Corporal for a walk that they devise a scheme to use two leashes. If Juana and Diego pull with forces of 23 lb and 18 lb at the angles shown in the figure, how hard is Corporal pulling if the puppy holds the children at a standstill?

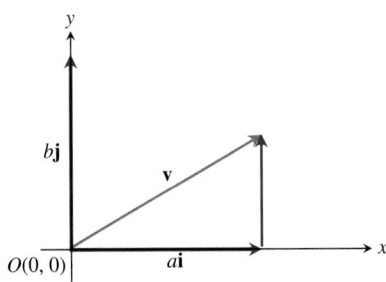

48. *Locating Points* A bird flies from its nest 7 km in the direction northeast, where it stops to rest on a tree. It then flies 8 km in the direction 30° south of west to land atop a telephone pole. Place an xy-coordinate system so that the origin is the bird's nest, the x-axis points east, and the y-axis points north.

 (a) At what point is the tree located?

 (b) At what point is the telephone pole located?

In Exercises 49 and 50, show that the vectors $\overrightarrow{AB}$ and $\overrightarrow{CD}$ are equal.

49. $A = (0, 0)$, $B = (-3, 4)$, $C = (4, 1)$, $D = (1, 5)$

50. $A = (-4, 3)$, $B = (-2, -2)$, $C = (1, 1)$, $D = (3, -4)$

Extending the Ideas

51. *Vector Operation Properties* Use the component form for vectors to prove the nine properties of vector operations listed in the text.

52. *Linear Combinations of Vectors* Write the vector $\langle 3, 4 \rangle$ as the sum of a vector parallel to $\mathbf{u} = \langle 1, 1 \rangle$ and a vector perpendicular to $\mathbf{u}$.

53. *Equations for Lines* Write an equation for the line in the plane that passes through the point $(-2, 1)$ and is

 (a) parallel to the vector $\langle 1, -1 \rangle$.

 (b) perpendicular to the vector $\langle 1, -1 \rangle$.

54. *Angle Between Lines* Find the acute angle between the lines $3x - 4y = 3$, $x - y = 7$.

10.3 # Vector-valued Functions

Standard Unit Vectors • Planar Curves • Limits and Continuity • Derivatives and Motion • Differentiation Rules • Integrals

Standard Unit Vectors

Any vector $\mathbf{v} = \langle a, b \rangle$ in the plane can be written as a *linear combination* of the two **standard unit vectors**

$$\mathbf{i} = \langle 1, 0 \rangle \quad \text{and} \quad \mathbf{j} = \langle 0, 1 \rangle$$

as follows:

$$\mathbf{v} = \langle a, b \rangle = \langle a, 0 \rangle + \langle 0, b \rangle = a\langle 1, 0 \rangle + b\langle 0, 1 \rangle$$
$$= a\mathbf{i} + b\mathbf{j}.$$

The vector $\mathbf{v}$ is a **linear combination** of the vectors $\mathbf{i}$ and $\mathbf{j}$; the scalar a is the **horizontal component** of $\mathbf{v}$ and the scalar b is the **vertical component** of $\mathbf{v}$ (Figure 10.19).

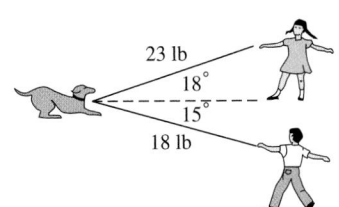

Figure 10.19 $\mathbf{v}$ is a linear combination of $\mathbf{i}$ and $\mathbf{j}$.

Example 1   EXPRESSING VECTORS AS LINEAR COMBINATIONS OF i AND j

Let $P = (-1, 5)$ and $Q = (3, 2)$. Write the vector $\mathbf{v} = \overrightarrow{PQ}$ as a linear combination of $\mathbf{i}$ and $\mathbf{j}$.

Solution The component form of $\mathbf{v}$ is $\langle 3 - (-1), 2 - 5 \rangle = \langle 4, -3 \rangle$. Thus,

$$\mathbf{v} = \langle 4, -3 \rangle = 4\mathbf{i} + (-3)\mathbf{j} = 4\mathbf{i} - 3\mathbf{j}.$$

Planar Curves

When a particle moves through the plane during a time interval I, we think of the particle's coordinates as functions defined on I:

$$x = f(t), \quad y = g(t), \quad t \in I. \tag{1}$$

The points $(x, y) = (f(t), g(t))$, $t \in I$, make up the curve in the plane that is the particle's **path.** The equations and interval in Equation 1 parametrize the curve. The vector

$$\mathbf{r}(t) = \overrightarrow{OP} = \langle f(t), g(t) \rangle = f(t)\mathbf{i} + g(t)\mathbf{j} \tag{2}$$

from the origin to the particle's **position** $P(f(t), g(t))$ at time t is the particle's **position vector.** The functions f and g are the **component functions (components)** of the position vector. We think of the particle's path as the **curve traced by r** during the time interval I.

Equation 2 defines $\mathbf{r}$ as a *vector function* of the real variable t on the interval I. More generally, a **vector function** or **vector-valued function** on a domain D is a rule that assigns a vector in the plane to each element in D. The curve traced by a vector function is its **graph.**

We refer to real-valued functions as **scalar functions** to distinguish them from vector functions. The components of $\mathbf{r}$ are scalar functions of t. When we define a vector-valued function by giving its component functions, we assume the vector function's domain to be the common domain of the components.

Example 2   GRAPHING AN ARCHIMEDES SPIRAL

Graph the vector function

$$\mathbf{r}(t) = (t \cos t)\mathbf{i} + (t \sin t)\mathbf{j}, \quad t \geq 0.$$

Solution We can graph the vector function parametrically using

$$x = t \cos t, \quad y = t \sin t, \quad t \geq 0.$$

The curve (Figure 10.20) is a spiral that winds around the origin, getting farther away from the origin as t increases. (Enter the equations into your calculator with $t\text{Max} = 50$ and watch the curve develop.)

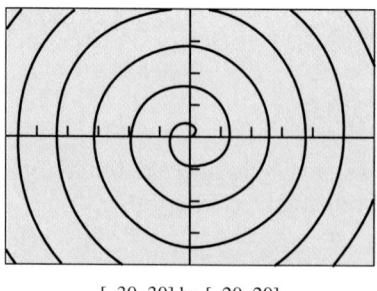

[–30, 30] by [–20, 20]

Figure 10.20 The graph of $\mathbf{r}(t) = (t \cos t)\mathbf{i} + (t \sin t)\mathbf{j}$, $t \geq 0$, is the curve $x = t \cos t$, $y = t \sin t$, $t \geq 0$. (Example 2)

Limits and Continuity

We define limits of vector functions in terms of their scalar components.

Definition Limit

Let $\mathbf{r}(t) = f(t)\mathbf{i} + g(t)\mathbf{j}$. If

$$\lim_{t \to c} f(t) = L_1 \quad \text{and} \quad \lim_{t \to c} g(t) = L_2,$$

then the **limit** of $\mathbf{r}(t)$ as t approaches c is

$$\lim_{t \to c} \mathbf{r}(t) = \mathbf{L} = L_1\mathbf{i} + L_2\mathbf{j}.$$

Example 3   FINDING A LIMIT OF A VECTOR FUNCTION

If $\mathbf{r}(t) = (\cos t)\mathbf{i} + (\sin t)\mathbf{j}$, then

$$\lim_{t \to \pi/4} \mathbf{r}(t) = \left(\lim_{t \to \pi/4} \cos t \right)\mathbf{i} + \left(\lim_{t \to \pi/4} \sin t \right)\mathbf{j} = \frac{\sqrt{2}}{2}\mathbf{i} + \frac{\sqrt{2}}{2}\mathbf{j}.$$

We define continuity for vector functions in the same way we define continuity for scalar functions.

Definition Continuity at a Point

A vector function $\mathbf{r}(t)$ is **continuous at a point** $t = c$ in its domain if

$$\lim_{t \to c} \mathbf{r}(t) = \mathbf{r}(c).$$

A vector function $\mathbf{r}(t)$ is **continuous** if it is continuous at every point in its domain. Since limits of vector functions are defined in terms of components we have the following test for continuity.

Component Test for Continuity at a Point

The vector function $\mathbf{r}(t) = f(t)\mathbf{i} + g(t)\mathbf{j}$ is continuous at $t = c$ if and only if f and g are continuous at $t = c$.

Example 4   FINDING POINTS OF CONTINUITY & DISCONTINUITY

(a) The function
$$\mathbf{r}(t) = (t \cos t)\mathbf{i} + (t \sin t)\mathbf{j}$$
is continuous everywhere because the component functions $t \cos t$ and $t \sin t$ are continuous everywhere.

(b) The function
$$\mathbf{r}(t) = (1/t)\mathbf{i} + (\sin t)\mathbf{j}$$
is not continuous at $t = 0$ because the first component is not continuous at $t = 0$. It is a continuous function, however, because it is continuous on its domain, the set of all nonzero real numbers.

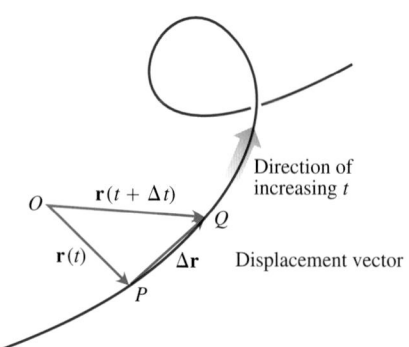

Figure 10.21 Between time t and time $t + \Delta t$, the particle moving along the path shown here undergoes the displacement $\overrightarrow{PQ} = \Delta\mathbf{r}$. The vector sum $\mathbf{r}(t) + \Delta\mathbf{r}$ gives the new position, $\mathbf{r}(t + \Delta t)$.

Derivatives and Motion

Suppose that $\mathbf{r}(t) = f(t)\mathbf{i} + g(t)\mathbf{j}$ is the position vector of a particle moving along a curve in the plane and that f and g are differentiable functions of t. Then (see Figure 10.21) the difference between the particle's positions at time $t + \Delta t$ and time t is $\Delta\mathbf{r} = \mathbf{r}(t + \Delta t) - \mathbf{r}(t)$. In terms of components,

$$
\begin{aligned}
\Delta\mathbf{r} &= \mathbf{r}(t + \Delta t) - \mathbf{r}(t) \\
&= [f(t + \Delta t)\mathbf{i} + g(t + \Delta t)\mathbf{j}] - [f(t)\mathbf{i} + g(t)\mathbf{j}] \\
&= [f(t + \Delta t) - f(t)]\mathbf{i} + [g(t + \Delta t) - g(t)]\mathbf{j}.
\end{aligned}
$$

As Δt approaches zero, three things seem to happen simultaneously. First, Q approaches P along the curve. Second, the secant line PQ seems to approach a limiting position tangent to the curve at P. Third, the quotient $\Delta\mathbf{r}/\Delta t$ approaches the limit

$$
\begin{aligned}
\lim_{\Delta t \to 0} \frac{\Delta\mathbf{r}}{\Delta t} &= \left[\lim_{\Delta t \to 0} \frac{f(t + \Delta t) - f(t)}{\Delta t}\right]\mathbf{i} + \left[\lim_{\Delta t \to 0} \frac{g(t + \Delta t) - g(t)}{\Delta t}\right]\mathbf{j} \\
&= \left[\frac{df}{dt}\right]\mathbf{i} + \left[\frac{dg}{dt}\right]\mathbf{j}.
\end{aligned}
$$

We are therefore led by past experience to the following definition.

Definition Derivative at a Point

The vector function $\mathbf{r}(t) = f(t)\mathbf{i} + g(t)\mathbf{j}$ has a **derivative (is differentiable) at t** if f and g have derivatives at t. The derivative is the vector

$$
\frac{d\mathbf{r}}{dt} = \lim_{\Delta t \to 0} \frac{\mathbf{r}(t + \Delta t) - \mathbf{r}(t)}{\Delta t} = \frac{df}{dt}\mathbf{i} + \frac{dg}{dt}\mathbf{j}.
$$

A vector function $\mathbf{r}$ is **differentiable** if it is differentiable at every point of its domain. The curve traced by $\mathbf{r}$ is **smooth** if $d\mathbf{r}/dt$ is continuous and never $\mathbf{0}$, that is, if f and g have continuous first derivatives that are not simultaneously 0. On a smooth curve there are no sharp corners or cusps.

The vector $d\mathbf{r}/dt$, when different from $\mathbf{0}$, is also a vector *tangent* to the curve. The **tangent line** to the curve at a point $P = (f(a), g(a))$ is defined to be the line through P parallel to $d\mathbf{r}/dt$ at $t = a$.

A curve that is made up of a finite number of smooth curves pieced together in a continuous fashion is **piecewise smooth** (Figure 10.22).

Look once again at Figure 10.21. We drew the figure for Δt positive, so $\Delta\mathbf{r}$ points forward, in the direction of the motion. The vector $\Delta\mathbf{r}/\Delta t$ (not shown), having the same direction as $\Delta\mathbf{r}$, points forward also. Had Δt been negative, $\Delta\mathbf{r}$ would have pointed backwards, against the direction of motion. The quotient $\Delta\mathbf{r}/\Delta t$, however, being a negative scalar multiple of $\Delta\mathbf{r}$, would have once again pointed forward. No matter how $\Delta\mathbf{r}$ points, $\Delta\mathbf{r}/\Delta t$ points forward and we expect the vector $d\mathbf{r}/dt = \lim_{\Delta t \to 0} \Delta\mathbf{r}/\Delta t$, when different from $\mathbf{0}$, to do the same. This means that the derivative $d\mathbf{r}/dt$ is just what we want for modeling a particle's velocity. It points in the direction of motion and gives the rate of change of position with respect to time. For a smooth curve, the velocity is never zero; the particle does not stop or reverse direction.

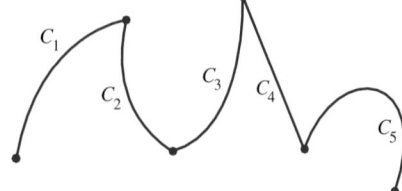

Figure 10.22 A piecewise smooth curve made by connecting five smooth curves end to end in continuous fashion.

Definitions Velocity, Speed, Acceleration, Direction of Motion

If **r** is the position vector of a particle moving along a smooth curve in the plane, then at any time t,

1. $\mathbf{v}(t) = \dfrac{d\mathbf{r}}{dt}$ is the particle's **velocity vector** and is tangent to the curve.

2. $|\mathbf{v}(t)|$, the magnitude of **v**, is the particle's **speed.**

3. $\mathbf{a}(t) = \dfrac{d\mathbf{v}}{dt} = \dfrac{d^2\mathbf{r}}{dt^2}$, the derivative of velocity and the second derivative of position, is the particle's **acceleration vector.**

4. $\dfrac{\mathbf{v}}{|\mathbf{v}|}$, a unit vector, is the **direction of motion.**

We can express the velocity of a moving particle as the product of its speed and direction.

$$\text{velocity} = |\mathbf{v}|\left(\frac{\mathbf{v}}{|\mathbf{v}|}\right) = (\text{speed})(\text{direction})$$

Example 5 STUDYING MOTION

The vector $\mathbf{r}(t) = (3\cos t)\mathbf{i} + (3\sin t)\mathbf{j}$ gives the position of a moving particle at time t. Find

(a) the velocity and acceleration vectors.

(b) the velocity, acceleration, speed, and direction of motion at $t = \pi/4$.

(c) $\mathbf{v} \cdot \mathbf{a}$. Interpret this result graphically.

Solution

(a) $\mathbf{v} = \dfrac{d\mathbf{r}}{dt} = (-3\sin t)\mathbf{i} + (3\cos t)\mathbf{j}$

$\mathbf{a} = \dfrac{d\mathbf{v}}{dt} = (-3\cos t)\mathbf{i} - (3\sin t)\mathbf{j}$

(b) At $t = \pi/4$, the particle's velocity and acceleration are

velocity: $\quad \mathbf{v}\left(\dfrac{\pi}{4}\right) = \left(-3\sin\dfrac{\pi}{4}\right)\mathbf{i} + \left(3\cos\dfrac{\pi}{4}\right)\mathbf{j} = -\dfrac{3}{\sqrt{2}}\mathbf{i} + \dfrac{3}{\sqrt{2}}\mathbf{j};$

acceleration: $\quad \mathbf{a}\left(\dfrac{\pi}{4}\right) = \left(-3\cos\dfrac{\pi}{4}\right)\mathbf{i} - \left(3\sin\dfrac{\pi}{4}\right)\mathbf{j} = -\dfrac{3}{\sqrt{2}}\mathbf{i} - \dfrac{3}{\sqrt{2}}\mathbf{j}.$

Its speed and direction are

speed: $\quad \left|\mathbf{v}\left(\dfrac{\pi}{4}\right)\right| = \sqrt{\left(\dfrac{-3}{\sqrt{2}}\right)^2 + \left(\dfrac{3}{\sqrt{2}}\right)^2} = 3;$

direction: $\quad \dfrac{\mathbf{v}(\pi/4)}{|\mathbf{v}(\pi/4)|} = \dfrac{-3/\sqrt{2}}{3}\mathbf{i} + \dfrac{3/\sqrt{2}}{3}\mathbf{j} = -\dfrac{1}{\sqrt{2}}\mathbf{i} + \dfrac{1}{\sqrt{2}}\mathbf{j}.$

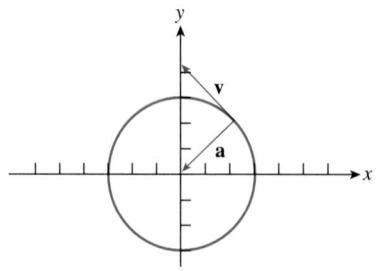

Figure 10.23 At $t = \pi/4$, the velocity vector $-(3/\sqrt{2})\mathbf{i} + (3/\sqrt{2})\mathbf{j}$ is tangent to the circle, and the acceleration vector $-(3/\sqrt{2})\mathbf{i} - (3/\sqrt{2})\mathbf{j}$ is perpendicular to the tangent, pointing toward the center of the circle. (Example 5)

(c) $\mathbf{v} \cdot \mathbf{a} = 9 \sin t \cos t - 9 \sin t \cos t = 0$

Thus, $\mathbf{v}$ and $\mathbf{a}$ are perpendicular for all values of t (Exercise 32, Section 10.2). Figure 10.23 shows the path, a circle with radius 3 centered at the origin, and the velocity and acceleration vectors at $t = \pi/4$.

Example 6 STUDYING MOTION

The vector $\mathbf{r}(t) = (2t^3 - 3t^2)\mathbf{i} + (t^3 - 12t)\mathbf{j}$ gives the position of a moving particle at time t.

(a) Write an equation for the line tangent to the path of the particle at the point where $t = -1$.

(b) Find the coordinates of each point on the path where the horizontal component of the velocity is 0.

Solution

(a) $\mathbf{v}(t) = \dfrac{d\mathbf{r}}{dt} = (6t^2 - 6t)\mathbf{i} + (3t^2 - 12)\mathbf{j}$

At $t = -1$, $\mathbf{r}(-1) = -5\mathbf{i} + 11\mathbf{j}$ and $\mathbf{v}(-1) = 12\mathbf{i} - 9\mathbf{j}$. Thus, we want the equation of the line through $(-5, 11)$ with slope $-9/12 = -3/4$.

$$y - 11 = -\frac{3}{4}(x + 5) \quad \text{or} \quad y = -\frac{3}{4}x + \frac{29}{4}$$

(b) The horizontal component of the velocity is $6t^2 - 6t$. It equals 0 when $t = 0$ and $t = 1$. The point corresponding to $t = 0$ is the origin $(0, 0)$; the point corresponding to $t = 1$ is $(-1, -11)$.

Differentiation Rules

Because the derivatives of vector functions are computed component by component, the rules for differentiating vector functions have the same form as the rules for differentiating scalar functions.

Differentiation Rules for Vector Functions

Let $\mathbf{u}$ and $\mathbf{v}$ be differentiable vector functions of t, and $\mathbf{C}$ a constant vector.

1. *Constant Function Rule:* $\dfrac{d}{dt}\mathbf{C} = \mathbf{0}$

2. *Scalar Multiple Rules:* $\dfrac{d}{dt}(c\mathbf{u}) = c\dfrac{d\mathbf{u}}{dt}$ c any scalar

$\dfrac{d}{dt}(f\mathbf{u}) = \dfrac{df}{dt}\mathbf{u} + f\dfrac{d\mathbf{u}}{dt}$ f any differentiable scalar function

3. *Sum Rule:* $\dfrac{d}{dt}(\mathbf{u} + \mathbf{v}) = \dfrac{d\mathbf{u}}{dt} + \dfrac{d\mathbf{v}}{dt}$

4. *Difference Rule:* $\dfrac{d}{dt}(\mathbf{u} - \mathbf{v}) = \dfrac{d\mathbf{u}}{dt} - \dfrac{d\mathbf{v}}{dt}$

5. *Dot Product Rule:* $\dfrac{d}{dt}(\mathbf{u} \cdot \mathbf{v}) = \dfrac{d\mathbf{u}}{dt} \cdot \mathbf{v} + \mathbf{u} \cdot \dfrac{d\mathbf{v}}{dt}$

6. *Chain Rule:* $\dfrac{d\mathbf{r}}{ds} = \dfrac{d\mathbf{r}}{dt}\dfrac{dt}{ds}$ $\mathbf{r}$ a differentiable function of t, t a differentiable function of s

We will prove the Dot Product Rule but leave the rest for exercises.

Proof of Rule 5 Suppose

$$\mathbf{u} = u_1(t)\mathbf{i} + u_2(t)\mathbf{j} \quad \text{and} \quad \mathbf{v} = v_1(t)\mathbf{i} + v_2(t)\mathbf{j}.$$

Then

$$\frac{d}{dt}(\mathbf{u} \cdot \mathbf{v}) = \frac{d}{dt}(u_1 v_1 + u_2 v_2)$$

$$= u_1' v_1 + u_2' v_2 + u_1 v_1' + u_2 v_2'$$

$$= \quad \mathbf{u}' \cdot \mathbf{v} \quad + \quad \mathbf{u} \cdot \mathbf{v}'. \qquad \blacksquare$$

Integrals

A differentiable vector function $\mathbf{R}(t)$ is an **antiderivative** of a vector function $\mathbf{r}(t)$ on an interval I if $d\mathbf{R}/dt = \mathbf{r}$ at each point t of I. If $\mathbf{R}$ is an antiderivative of $\mathbf{r}$ on I, it can be shown, working one component at a time, that every antiderivative of $\mathbf{r}$ on I has the form $\mathbf{R} + \mathbf{C}$ for some constant vector $\mathbf{C}$ (Exercise 37).

Definition Indefinite Integral

The **indefinite integral** of $\mathbf{r}$ with respect to t is the set of all antiderivatives of $\mathbf{r}$, denoted by $\int \mathbf{r}(t)\, dt$. If $\mathbf{R}$ is any antiderivative of $\mathbf{r}$, then

$$\int \mathbf{r}(t)dt = \mathbf{R}(t) + \mathbf{C}.$$

The usual arithmetic rules for indefinite integrals apply.

Example 7 FINDING ANTIDERIVATIVES

$$\int ((\cos t)\mathbf{i} - 2t\mathbf{j})\, dt = \left(\int \cos t\, dt\right)\mathbf{i} - \left(\int 2t\, dt\right)\mathbf{j} \qquad (3)$$

$$= (\sin t + C_1)\mathbf{i} - (t^2 + C_2)\mathbf{j} \qquad (4)$$

$$= (\sin t)\mathbf{i} - t^2\mathbf{j} + \mathbf{C} \qquad \mathbf{C} = C_1\mathbf{i} + C_2\mathbf{j}$$

As with integration of scalar functions, we recommend that you skip the steps in Equations 3 and 4 and go directly to the final form. Find an antiderivative for each component and add a constant vector at the end.

As with derivatives and indefinite integrals, definite integrals of vector functions are calculated component by component.

Definition Definite Integral

If the components of $\mathbf{r}(t) = f(t)\mathbf{i} + g(t)\mathbf{j}$ are integrable on $[a, b]$, then so is $\mathbf{r}$, and the **definite integral** of $\mathbf{r}$ from a to b is

$$\int_a^b \mathbf{r}(t)\, dt = \left(\int_a^b f(t)\, dt\right)\mathbf{i} + \left(\int_a^b g(t)\, dt\right)\mathbf{j}.$$

Example 8 EVALUATING DEFINITE INTEGRALS

$$\int_0^\pi ((\cos t)\mathbf{i} - 2t\mathbf{j})\, dt = \left(\int_0^\pi \cos t\, dt\right)\mathbf{i} - \left(\int_0^\pi 2t\, dt\right)\mathbf{j}$$

$$= \left(\sin t\Big]_0^\pi\right)\mathbf{i} - \left(t^2\Big]_0^\pi\right)\mathbf{j} = 0\mathbf{i} - \pi^2\mathbf{j} = -\pi^2\mathbf{j}$$

Example 9 FINDING A PATH

The velocity vector of a particle moving in the plane (scaled in meters) is

$$\frac{d\mathbf{r}}{dt} = \frac{1}{t+1}\mathbf{i} + 2t\mathbf{j}, \quad t \ge 0.$$

(a) Find the particle's position as a vector function of t if $\mathbf{r} = (\ln 2)\mathbf{i}$ when $t = 1$.

(b) Find the distance the particle travels from $t = 0$ to $t = 2$.

Solution

(a) $\mathbf{r} = \left(\int \dfrac{dt}{t+1}\right)\mathbf{i} + \left(\int 2t\, dt\right)\mathbf{j} = (\ln (t+1))\mathbf{i} + t^2\mathbf{j} + \mathbf{C}$

$$\mathbf{r}(1) = (\ln 2)\mathbf{i} + \mathbf{j} + \mathbf{C} = (\ln 2)\mathbf{i}$$

Thus, $\mathbf{C} = -\mathbf{j}$ and

$$\mathbf{r} = (\ln (t+1))\mathbf{i} + (t^2 - 1)\mathbf{j}.$$

(b) The parametrization

$$x = \ln (t+1), \quad y = t^2 - 1, \quad 0 \le t \le 2$$

is smooth, and because x and y are increasing functions of t the path is traversed exactly once as t increases from 0 to 2 (Figure 10.24). The length is

$$L = \int_0^2 \sqrt{\left(\frac{dx}{dt}\right)^2 + \left(\frac{dy}{dt}\right)^2}\, dt = \int_0^2 \sqrt{\left(\frac{1}{t+1}\right)^2 + (2t)^2}\, dt \approx 4.34 \text{ m}.$$

$x = \ln (t + 1), y = t^2 - 1$

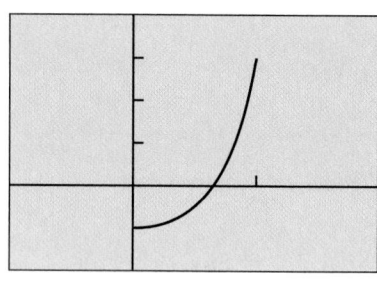

[−1, 2] by [−2, 4]

Figure 10.24. The path of the particle in Example 9 for $0 \le t \le 2$.

Quick Review 10.3

In Exercises 1 and 2, let $f(x) = \sqrt{4 - x^2}$.

1. Write an equation for the line tangent to f at $x = 1$.

2. Write an equation for the line normal to f at $x = 1$.

Exercises 3–6 refer to the curve defined parametrically by $x = 4 \cos t$, $y = 5 \sin t$.

3. Find the slope of the tangent to the curve at $t = \pi/2$.

4. Find the slope of the tangent to the curve at $t = \pi$.

5. Write an equation for the tangent to the curve at $t = \pi/6$.

6. Write an equation for the line normal to the curve at $t = \pi/6$.

7. Evaluate $\displaystyle\lim_{x \to 2} \frac{x - 2}{x^2 - 4}$.

8. Find the length of the curve $y = 3x - x^2$ from $x = 0$ to $x = 2$.

9. Find the length of the curve

$$x = t \sin t, \quad y = t \cos t, \quad 0 \le t \le 2.$$

10. Find y if $dy/dx = xe^x$ and $y(0) = 2$.

Section 10.3 Exercises

In Exercises 1–4, express each vector as a linear combination of **i** and **j**.

1. the vector $\overrightarrow{PQ}$, where $P = (-1, 4)$ and $Q = (5, 1)$

2. the vector from the point $P = (3, -4)$ to the origin

3. the vectors **(a)** $\overrightarrow{AB} + \overrightarrow{CD}$ and **(b)** $\overrightarrow{AB} - \overrightarrow{CD}$, where $A = (-3, 0)$, $B = (0, 2)$, $C = (4, 0)$, and $D = (0, -3)$

4. the vectors **(a)** $\mathbf{u} + \mathbf{v}$, **(b)** $\mathbf{u} - \mathbf{v}$, **(c)** $3\mathbf{u}$, **(d)** $2\mathbf{u} - 3\mathbf{v}$, where $\mathbf{u} = 5\mathbf{i} - 2\mathbf{j}$ and $\mathbf{v} = 3\mathbf{i} + 4\mathbf{j}$

In Exercises 5–8, $\mathbf{r}(t)$ is the position vector of a particle in the plane at time t.

(a) Draw the graph of the path of the particle.

(b) Find the velocity and acceleration vectors.

(c) Find the particle's speed and direction of motion at the given value of t.

(d) Write the particle's velocity at that time as the product of its speed and direction.

5. $\mathbf{r}(t) = (2 \cos t)\mathbf{i} + (3 \sin t)\mathbf{j}, \quad t = \pi/2$

6. $\mathbf{r}(t) = (\cos 2t)\mathbf{i} + (2 \sin t)\mathbf{j}, \quad t = 0$

7. $\mathbf{r}(t) = (\sec t)\mathbf{i} + (\tan t)\mathbf{j}, \quad t = \pi/6$

8. $\mathbf{r}(t) = (2 \ln (t + 1))\mathbf{i} + (t^2)\mathbf{j}, \quad t = 1$

In Exercises 9 and 10, find an equation for the line that is **(a)** tangent and **(b)** normal to the curve $\mathbf{r}(t)$ at the point determined by the given value of t.

9. $\mathbf{r}(t) = (\sin t)\mathbf{i} + (t^2 - \cos t)\mathbf{j}, \quad t = 0$

10. $\mathbf{r}(t) = (2 \cos t - 3)\mathbf{i} + (3 \sin t + 1)\mathbf{j}, \quad t = \pi/4$

In Exercises 11–14, evaluate the integral.

11. $\displaystyle\int_1^2 [(6 - 6t)\mathbf{i} + 3\sqrt{t}\,\mathbf{j}]\, dt$

12. $\displaystyle\int_{-\pi/4}^{\pi/4} [(\sin t)\mathbf{i} + (1 + \cos t)\mathbf{j}]\, dt$

13. $\displaystyle\int [(\sec t \tan t)\mathbf{i} + (\tan t)\mathbf{j}]\, dt$

14. $\displaystyle\int \left[\frac{1}{t}\mathbf{i} + \frac{1}{5 - t}\mathbf{j}\right] dt$

In Exercises 15–18, solve the initial value problem for $\mathbf{r}$ as a vector function of t.

15. $\dfrac{d\mathbf{r}}{dt} = \dfrac{3}{2}(t + 1)^{1/2}\mathbf{i} + e^{-t}\mathbf{j}, \quad \mathbf{r}(0) = \mathbf{0}$

16. $\dfrac{d\mathbf{r}}{dt} = (t^3 + 4t)\mathbf{i} + t\mathbf{j}, \quad \mathbf{r}(0) = \mathbf{i} + \mathbf{j}$

17. $\dfrac{d^2\mathbf{r}}{dt^2} = -32\mathbf{j}, \quad \mathbf{r}(0) = 100\mathbf{i}, \quad \left.\dfrac{d\mathbf{r}}{dt}\right|_{t=0} = 8\mathbf{i} + 8\mathbf{j}$

18. $\dfrac{d^2\mathbf{r}}{dt^2} = -\mathbf{i} - \mathbf{j}, \quad \mathbf{r}(0) = 10\mathbf{i} + 10\mathbf{j}, \quad \left.\dfrac{d\mathbf{r}}{dt}\right|_{t=0} = \mathbf{0}$

In Exercises 19–22, $\mathbf{r}(t)$ is the position vector of a particle in the plane at time t. Find the time, or times, in the given time interval when the velocity and acceleration vectors are perpendicular.

19. $\mathbf{r}(t) = (t - \sin t)\mathbf{i} + (1 - \cos t)\mathbf{j}, \quad 0 \le t \le 2\pi$

20. $\mathbf{r}(t) = (\sin t)\mathbf{i} + t\mathbf{j}, \quad t \ge 0$

21. $\mathbf{r}(t) = (3 \cos t)\mathbf{i} + (4 \sin t)\mathbf{j}, \quad t \ge 0$

22. $\mathbf{r}(t) = (5 \cos t)\mathbf{i} + (5 \sin t)\mathbf{j}, \quad t \ge 0$

In Exercises 23 and 24, $\mathbf{r}(t)$ is the position vector of a particle in the plane at time t. Find the angle between the velocity and acceleration vectors at the given value of t.

23. $\mathbf{r}(t) = (2 \cos t)\mathbf{i} + (\sin t)\mathbf{j}, \quad t = \pi/4$

24. $\mathbf{r}(t) = (3t + 1)\mathbf{i} + (t^2)\mathbf{j}, \quad t = 0$

In Exercises 25 and 26, **(a)** evaluate the limit, and **(b)** find the values of t for which the vector function is continuous and **(c)** discontinuous.

25. $\displaystyle\lim_{t \to 3} \left[t\mathbf{i} + \frac{t^2 - 9}{t^2 + 3t}\mathbf{j}\right]$

26. $\displaystyle\lim_{t \to 0} \left[\frac{\sin 2t}{t}\mathbf{i} + (\ln (t + 1))\mathbf{j}\right]$

27. *Finding Distance Traveled* The position of a particle in the plane at time t is given by

$$\mathbf{r}(t) = (1 - \cos t)\mathbf{i} + (t - \sin t)\mathbf{j}.$$

Find the distance the particle travels along the path from $t = 0$ to $t = 2\pi/3$.

28. Let C be the path traced by

$$\mathbf{r}(t) = \left(\frac{1}{4}e^{4t} - t\right)\mathbf{i} + (e^{2t})\mathbf{j}, \quad 0 \le t \le 2.$$

(a) Find the initial and terminal point of C.

(b) Find the length of C.

(c) Find the area of the surface generated by revolving C about the y-axis.

In Exercises 29–32, *work in groups of two or three.*

29. The position of a particle is given by

$$\mathbf{r}(t) = (\sin t)\mathbf{i} + (\cos 2t)\mathbf{j}.$$

(a) Find the velocity vector for the particle.

(b) For what values of t in the interval $0 \le t \le 2\pi$ is $d\mathbf{r}/dt$ equal to $\mathbf{0}$?

(c) Find a Cartesian equation for a curve that contains the particle's path. What portion of the graph of the Cartesian equation is traced by the particle? Describe the motion as t increases from 0 to 2π.

30. *Revisiting Example 6* The position of a particle is given by $\mathbf{r}(t) = (2t^3 - 3t^2)\mathbf{i} + (t^3 - 12t)\mathbf{j}$.

(a) Find dy/dx in terms of t.

(b) Writing to Learn Find the x- and y-coordinates for each critical point of the path (point where dy/dx is zero or does not exist). Does the path have a vertical or horizontal tangent at the critical point? Explain.

31. *Finding a Position Vector* At time $t = 0$, a particle is located at the point $(1, 2)$. It travels in a straight line to the point $(4, 1)$, has speed 2 at $(1, 2)$, and constant acceleration $3\mathbf{i} - \mathbf{j}$. Find an equation for the position vector $\mathbf{r}(t)$ of the particle at time t.

32. The path of a particle for $t > 0$ is given by

$$\mathbf{r}(t) = \left(t + \frac{2}{t}\right)\mathbf{i} + (3t^2)\mathbf{j}.$$

(a) Find the coordinates of each point on the path where the horizontal component of the velocity of the particle is zero.

(b) Find dy/dx when $t = 1$.

(c) Find d^2y/dx^2 when $y = 12$.

Explorations

33. *Flying a Kite* The position of a kite is given by

$$\mathbf{r}(t) = \frac{t}{8}\mathbf{i} - \frac{3}{64}t(t - 160)\mathbf{j},$$

where $t \geq 0$ is measured in seconds and distance is measured in meters.

(a) How long is the kite above ground?

(b) How high is the kite at $t = 40$ sec?

(c) At what rate is the kite's altitude increasing at $t = 40$ sec?

(d) At what time does the kite start to lose altitude?

34. *Colliding Particles* The paths of two particles for $t \geq 0$ are given by

$$\mathbf{r}_1(t) = (t - 3)\mathbf{i} + (t - 3)^2\mathbf{j},$$

$$\mathbf{r}_2(t) = \left(\frac{3t}{2} - 4\right)\mathbf{i} + \left(\frac{3t}{2} - 2\right)\mathbf{j}.$$

(a) Determine the exact time(s) at which the particles collide.

(b) Find the direction of motion of each particle at the time(s) of collision.

35. *A Satellite in Circular Orbit* A satellite of mass m is moving at a constant speed v around a planet of mass M in a circular orbit of radius r_0, as measured from the planet's center of mass. Determine the satellite's orbital period T (the time to complete one full orbit), as follows:

(a) Coordinatize the orbital plane by placing the origin at the planet's center of mass, with the satellite on the x-axis at $t = 0$ and moving counterclockwise, as in the accompanying figure.

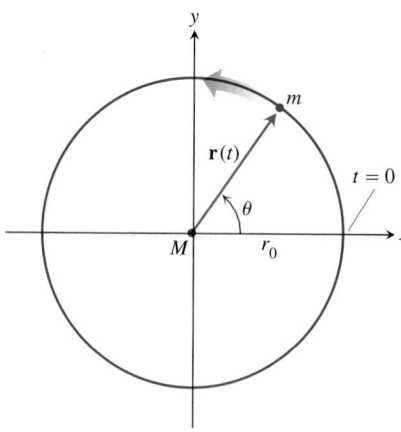

Let $\mathbf{r}(t)$ be the satellite's position vector at time t. Show that $\theta = vt/r_0$ and hence that

$$\mathbf{r}(t) = \left(r_0 \cos \frac{vt}{r_0}\right)\mathbf{i} + \left(r_0 \sin \frac{vt}{r_0}\right)\mathbf{j}.$$

(b) Find the acceleration of the satellite.

(c) According to Newton's law of gravitation, the gravitational force exerted on the satellite by the planet is directed toward the origin and is given by

$$\mathbf{F} = \left(-\frac{GmM}{r_0^2}\right)\frac{\mathbf{r}}{r_0},$$

where G is the universal constant of gravitation. Using Newton's second law, $\mathbf{F} = m\mathbf{a}$, show that $v^2 = GM/r_0$.

(d) Show that the orbital period T satisfies $vT = 2\pi r_0$.

(e) From parts (c) and (d), deduce that

$$T^2 = \frac{4\pi^2}{GM}r_0^3;$$

that is, the square of the period of a satellite in circular orbit is proportional to the cube of the radius from the orbital center. ∎

36. *Revisiting Example 9* Find a Cartesian equation for a curve that contains the parametrized curve

$$x = \ln(t + 1), \quad y = t^2 - 1, \quad 0 \leq t < \infty.$$

What portion of the graph of the Cartesian equation is traced by the parametrized curve?

37. *Antiderivatives of Vector Functions*

(a) Use Corollary 3 in Section 4.2 (a consequence of the Mean Value Theorem for scalar functions) to show that two vector functions $\mathbf{R}_1(t)$ and $\mathbf{R}_2(t)$ that have identical derivatives on an interval I differ by a constant vector value throughout I.

(b) Use the result in (a) to show that if $\mathbf{R}(t)$ is any antiderivative of $\mathbf{r}(t)$ on I, then every other antiderivative of $\mathbf{r}(t)$ on I equals $\mathbf{R}(t) + \mathbf{C}$ for some constant vector $\mathbf{C}$.

38. *Constant Length* Let **v** be a differentiable vector function of t. Show that if $\mathbf{v} \cdot (d\mathbf{v}/dt) = 0$ for all t, then $|\mathbf{v}|$ is constant.

Extending the Ideas

39. *Constant Function Rule* Prove that if **u** is the vector function with the constant value **C**, then $d\mathbf{u}/dt = \mathbf{0}$.

40. *Scalar Multiple Rules*

   (a) Prove that if **u** is a differentiable function of t and c is any real number, then

$$\frac{d(c\mathbf{u})}{dt} = c\frac{d\mathbf{u}}{dt}.$$

   (b) Prove that if **u** is a differentiable function of t and f is a differentiable scalar function of t, then

$$\frac{d(f\mathbf{u})}{dt} = \frac{df}{dt}\mathbf{u} + f\frac{d\mathbf{u}}{dt}.$$

41. *Sum and Difference Rules* Prove that if **u** and **v** are differentiable functions of t, then

   (a) $\dfrac{d}{dt}(\mathbf{u} + \mathbf{v}) = \dfrac{d\mathbf{u}}{dt} + \dfrac{d\mathbf{v}}{dt}.$

   (b) $\dfrac{d}{dt}(\mathbf{u} - \mathbf{v}) = \dfrac{d\mathbf{u}}{dt} - \dfrac{d\mathbf{v}}{dt}.$

42. *Chain Rule* Prove that if $\mathbf{r}(t) = f(t)\mathbf{i} + g(t)\mathbf{j}$ is a differentiable vector function of t and t is a differentiable scalar function of s, then

$$\frac{d\mathbf{r}}{ds} = \frac{d\mathbf{r}}{dt} \cdot \frac{dt}{ds}.$$

43. *Differentiable Vector Functions Are Continuous* Show that if $\mathbf{r}(t) = f(t)\mathbf{i} + g(t)\mathbf{j}$ is differentiable at $t = c$, then **r** is continuous at c as well.

44. *Integration Properties* Establish the following properties of integrable vector functions.

   (a) *Constant Scalar Multiple Rule:*

$$\int_a^b k\mathbf{r}(t)\,dt = k\int_a^b \mathbf{r}(t)\,dt$$

   for any scalar constant k.

   (b) *Sum and Difference Rules:*

$$\int_a^b (\mathbf{r}_1(t) \pm \mathbf{r}_2(t))\,dt = \int_a^b \mathbf{r}_1(t)\,dt \pm \int_a^b \mathbf{r}_2(t)\,dt$$

   (c) *Constant Vector Multiple Rule:*

$$\int_a^b \mathbf{C} \cdot \mathbf{r}(t)\,dt = \mathbf{C} \cdot \int_a^b \mathbf{r}(t)\,dt$$

   for any constant vector **C**.

45. *Fundamental Theorem of Calculus* The Fundamental Theorem of Calculus for scalar functions of a real variable holds for vector functions of a real variable as well.

   (a) Prove this by using the theorem for scalar functions to show that if a vector function $\mathbf{r}(t)$ is continuous for $a \leq t \leq b$, then

$$\frac{d}{dt}\int_a^t \mathbf{r}(q)\,dq = \mathbf{r}(t)$$

   at every point t of $[a, b]$.

   (b) Use the conclusion in part (b) of Exercise 37 to show that if **R** is any antiderivative of **r** on $[a, b]$ then

$$\int_a^b \mathbf{r}(t)\,dt = \mathbf{R}(b) - \mathbf{R}(a).$$

10.4 **Modeling Projectile Motion**

Ideal Projectile Motion • Height, Flight Time, and Range • Ideal Trajectories Are Parabolic • Firing from (x_0, y_0) • Projectile Motion with Wind Gusts • Projectile Motion with Air Resistance

Ideal Projectile Motion

To derive equations for *ideal* projectile motion, we assume that the projectile behaves like a particle moving in a vertical coordinate plane and that the only force acting on the projectile during its flight is the constant force of gravity, which always points straight down. In practice, none of these assumptions really holds. The ground moves beneath the projectile as the earth turns, the air creates a frictional force that varies with the projectile's speed and altitude, and the force (vector) of gravity changes magnitude and direction as the projectile moves along. All this must be taken into account by applying corrections to the predictions of the ideal equations we are about to derive.

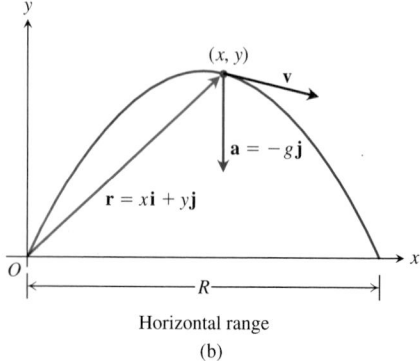

Figure 10.25 (a) Position, velocity, acceleration due to gravity, and launch angle at $t = 0$. (b) Position, velocity, and acceleration at a later time t.

We assume the projectile is launched from the origin at time $t = 0$ into the first quadrant with an initial velocity $\mathbf{v}_0$ (Figure 10.25). If $\mathbf{v}_0$ makes an angle α with the horizontal, then

$$\mathbf{v}_0 = (|\mathbf{v}_0| \cos \alpha)\mathbf{i} + (|\mathbf{v}_0| \sin \alpha)\mathbf{j}.$$

If we use the simpler notation v_0 for the initial speed $|\mathbf{v}_0|$, then

$$\mathbf{v}_0 = (v_0 \cos \alpha)\mathbf{i} + (v_0 \sin \alpha)\mathbf{j}. \tag{1}$$

The projectile's initial position is

$$\mathbf{r}_0 = 0\mathbf{i} + 0\mathbf{j} = \mathbf{0}. \tag{2}$$

Newton's second law of motion says that the force acting on the projectile is equal to the projectile's mass m times its acceleration, or $m(d^2\mathbf{r}/dt^2)$ if $\mathbf{r}$ is the projectile's position vector and t is time. If the force is solely the gravitational force $-mg\mathbf{j}$, then

$$m\frac{d^2\mathbf{r}}{dt^2} = -mg\mathbf{j} \quad \text{and} \quad \frac{d^2\mathbf{r}}{dt^2} = -g\mathbf{j}.$$

We find $\mathbf{r}$ as a function of t by solving the following initial value problem.

Differential equation: $\quad \dfrac{d^2\mathbf{r}}{dt^2} = -g\mathbf{j}$

Initial conditions: $\quad \mathbf{r} = \mathbf{r}_0 \quad \text{and} \quad \dfrac{d\mathbf{r}}{dt} = \mathbf{v}_0 \quad \text{when} \quad t = 0$

The first integration gives

$$\frac{d\mathbf{r}}{dt} = -(gt)\mathbf{j} + \mathbf{v}_0.$$

A second integration gives

$$\mathbf{r} = -\frac{1}{2}gt^2\mathbf{j} + \mathbf{v}_0 t + \mathbf{r}_0. \tag{3}$$

Substituting the values of $\mathbf{v}_0$ and $\mathbf{r}_0$ from Equations 1 and 2 gives

$$\mathbf{r} = -\frac{1}{2}gt^2\mathbf{j} + \underbrace{(v_0 \cos \alpha)t\mathbf{i} + (v_0 \sin \alpha)t\mathbf{j}}_{\mathbf{v}_0 t} + \mathbf{0}$$

or

$$\mathbf{r} = (v_0 \cos \alpha)t\mathbf{i} + \left((v_0 \sin \alpha)t - \frac{1}{2}gt^2\right)\mathbf{j}. \tag{4}$$

Equation 4 is the **vector equation** for ideal projectile motion. The angle α is the projectile's **launch angle (firing angle, angle of elevation)**, and v_0, as we said before, is the projectile's **initial speed.**

Equation 4 gives us a pair of scalar equations,

$$x = (v_0 \cos \alpha)t, \quad y = (v_0 \sin \alpha)t - \frac{1}{2}gt^2, \tag{5}$$

the **parametric equations** for ideal projectile motion. If time is measured in seconds and distance in meters, g is 9.8 m/sec^2 and Equations 5 give x and y in meters. With feet in place of meters, g is 32 ft/sec^2 and Equations 5 give x and y in feet.

Example 1 FIRING AN IDEAL PROJECTILE

A projectile is fired from the origin over horizontal ground at an initial speed of 500 m/sec and a launch angle of 60°. Where will the projectile be 10 sec later?

Solution We use Equations 5 with $v_0 = 500$, $\alpha = 60°$, $g = 9.8$, and $t = 10$ to find the projectile's coordinates 10 sec after firing.

$$x = (v_0 \cos \alpha)t = (500)\left(\frac{1}{2}\right)10 = 2500$$

$$y = (v_0 \sin \alpha)t - \frac{1}{2}gt^2$$

$$= (500)\left(\frac{\sqrt{3}}{2}\right)10 - \left(\frac{1}{2}\right)(9.8)10^2 \approx 3840.13$$

Interpret

Ten seconds after firing, the projectile is about 3840 m in the air and 2500 m downrange.

Height, Flight Time, and Range

Equations 5 enable us to answer most questions about the ideal motion for a projectile fired from the origin.

The projectile reaches its highest point when its vertical velocity component is zero, that is, when

$$\frac{dy}{dt} = v_0 \sin \alpha - gt = 0, \quad \text{or} \quad t = \frac{v_0 \sin \alpha}{g}.$$

For this value of t, the value of y is

$$y_{\max} = (v_0 \sin \alpha)\left(\frac{v_0 \sin \alpha}{g}\right) - \frac{1}{2}g\left(\frac{v_0 \sin \alpha}{g}\right)^2 = \frac{(v_0 \sin \alpha)^2}{2g}.$$

To find when the projectile lands when fired over horizontal ground, we set y equal to zero in Equations 5 and solve for t.

$$(v_0 \sin \alpha)t - \frac{1}{2}gt^2 = 0$$

$$t\left(v_0 \sin \alpha - \frac{1}{2}gt\right) = 0$$

$$t = 0, \quad t = \frac{2v_0 \sin \alpha}{g}$$

Since 0 is the time the projectile is fired, $(2v_0 \sin \alpha)/g$ must be the time when the projectile strikes the ground.

To find the projectile's **range** R, the distance from the origin to the point of impact on horizontal ground, we find the value of x when $t = (2v_0 \sin \alpha)/g$.

$$x = (v_0 \cos \alpha)t$$

$$R = (v_0 \cos \alpha)\left(\frac{2v_0 \sin \alpha}{g}\right) = \frac{v_0^2}{g}(2 \sin \alpha \cos \alpha) = \frac{v_0^2}{g}\sin 2\alpha$$

The range is largest when $\sin 2\alpha = 1$ or $\alpha = 45°$.

Height, Flight Time, and Range for Ideal Projectile Motion

For ideal projectile motion when an object is launched from the origin over a horizontal surface with initial speed v_0 and launch angle α:

$$\text{Maximum height:} \qquad y_{\max} = \frac{(v_0 \sin \alpha)^2}{2g} \qquad\qquad (6)$$

$$\text{Flight time:} \qquad t = \frac{2v_0 \sin \alpha}{g} \qquad\qquad (7)$$

$$\text{Range:} \qquad R = \frac{v_0^2}{g} \sin 2\alpha \qquad\qquad (8)$$

Example 2 INVESTIGATING IDEAL PROJECTILE MOTION

Find the maximum height, flight time, and range of a projectile fired from the origin over horizontal ground at an initial speed of 500 m/sec and a launch angle of 60° (same projectile as Example 1). Then graph the path of the projectile.

Solution

Maximum height (Equation 6): $y_{\max} = \dfrac{(v_0 \sin \alpha)^2}{2g}$

$$= \frac{(500 \sin 60°)^2}{2(9.8)} \approx 9566.33 \text{ m}$$

Flight time (Equation 7): $t = \dfrac{2v_0 \sin \alpha}{g}$

$$= \frac{2(500) \sin 60°}{9.8} \approx 88.37 \text{ sec}$$

Range (Equation 8): $R = \dfrac{v_0^2}{g} \sin 2\alpha$

$$= \frac{(500)^2 \sin 120°}{9.8} \approx 22,092.48 \text{ m}$$

Equations 5 give parametric equations for the path of the projectile:

$$x = (v_0 \cos \alpha)t = (500 \cos 60°)t = 250t,$$

$$y = (v_0 \sin \alpha)t - \frac{1}{2}gt^2$$

$$= (500 \sin 60°)t - \frac{1}{2}(9.8)t^2 = (250\sqrt{3})t - 4.9t^2.$$

A graph of these equations (the path of the projectile) is shown in Figure 10.26.

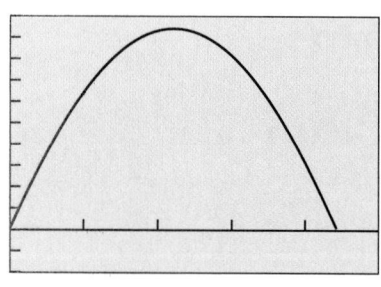

[0, 25000] by [−2000, 10000]

Figure 10.26 The graph of the parametric equations $x = 250t$, $y = (250\sqrt{3})t - 4.9t^2$ for $0 \le t \le 88.4$. (Example 2)

Ideal Trajectories Are Parabolic

It is often claimed that water from a hose traces a parabola in the air, but anyone who looks closely enough will see this is not so. The air slows the water down, and its forward progress is too slow at the end to keep pace with the rate at which it falls.

What is really being claimed is that ideal projectiles move along parabolas, and this we can see from Equations 5. If we substitute $t = x/(v_0 \cos \alpha)$ from the first equation into the second, we obtain the Cartesian-coordinate equation

$$y = -\left(\frac{g}{2v_0{}^2 \cos^2 \alpha}\right)x^2 + (\tan \alpha)x.$$

This equation has the form $y = ax^2 + bx$, so its graph is a parabola.

Firing from (x_0, y_0)

If we fire our ideal projectile from the point (x_0, y_0) instead of the origin (Figure 10.27), the equations that replace Equations 5 are

$$x = x_0 + (v_0 \cos \alpha)t, \qquad y = y_0 + (v_0 \sin \alpha)t - \frac{1}{2}gt^2, \qquad (9)$$

as you will be asked to show in Exercise 19.

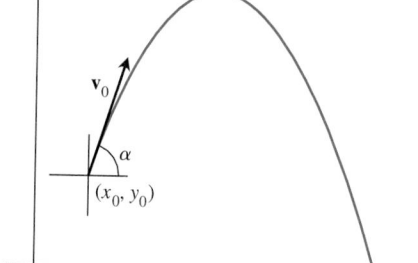

Figure 10.27 The path of a projectile fired from (x_0, y_0) with an initial velocity $\mathbf{v}_0$ at an angle of α degrees with the horizontal.

Example 3 FIRING A FLAMING ARROW

To open the 1992 Summer Olympics in Barcelona, bronze medalist archer Antonio Rebollo lit the Olympic torch with a flaming arrow (Figure 10.28). Suppose that Rebollo shot the arrow at a height of 6 ft above ground level 30 yd from the 70-ft-high cauldron, and he wanted the arrow to reach maximum height exactly 4 ft above the center of the cauldron (Figure 10.29).

Figure 10.28 Spanish archer Antonio Rebollo lights the Olympic torch in Barcelona with a flaming arrow.

(a) Express $y_{\max}$ in terms of the initial speed v_0 and firing angle α.

(b) Use $y_{\max} = 74$ ft (Figure 10.29 on the next page) and the result from part (a) to find the value of $v_0 \sin \alpha$.

(c) Find the value of $v_0 \cos \alpha$.

(d) Find the initial firing angle of the arrow.

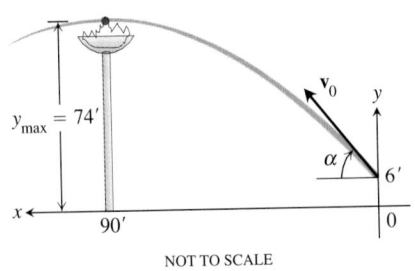

$y_{max} = 74'$

$90'$

$6'$

NOT TO SCALE

Figure 10.29 Ideal path of the arrow that lit the Olympic torch. (Example 3)

Solution

(a) We use a coordinate system in which the positive x-axis lies along the ground toward the left (to match the second photograph in Figure 10.28) and the coordinates of the flaming arrow at $t = 0$ are $x_0 = 0$ and $y_0 = 6$ (Figure 10.29). We have

$$y = y_0 + (v_0 \sin \alpha)t - \frac{1}{2}gt^2 \quad \text{Eqs. 9}$$

$$= 6 + (v_0 \sin \alpha)t - \frac{1}{2}gt^2. \quad y_0 = 6$$

We find the time when the arrow reaches its highest point by setting $dy/dt = 0$ and solving for t, obtaining

$$t = \frac{v_0 \sin \alpha}{g}.$$

For this value of t, the value of y is

$$y_{max} = 6 + (v_0 \sin \alpha)\left(\frac{v_0 \sin \alpha}{g}\right) - \frac{1}{2}g\left(\frac{v_0 \sin \alpha}{g}\right)^2$$

$$= 6 + \frac{(v_0 \sin \alpha)^2}{2g}.$$

(b) Using $y_{max} = 74$ and $g = 32$, we see from the preceding equation in part (a) that

$$74 = 6 + \frac{(v_0 \sin \alpha)^2}{2(32)}$$

or

$$v_0 \sin \alpha = \sqrt{(68)(64)}.$$

(c) When the arrow reaches y_{max}, the horizontal distance traveled to the center of the cauldron is $x = 90$ ft. We substitute the time to reach y_{max} from part (a) and the horizontal distance $x = 90$ ft into Equations 9 to obtain

$$x = x_0 + (v_0 \cos \alpha)t \qquad \text{Eqs. 9}$$

$$90 = 0 + (v_0 \cos \alpha)t \qquad x = 90, \ x_0 = 0$$

$$= (v_0 \cos \alpha)\left(\frac{v_0 \sin \alpha}{g}\right). \qquad t = (v_0 \sin \alpha)/g$$

Solving this equation for $v_0 \cos \alpha$ and using $g = 32$ and the result from part (b), we have

$$v_0 \cos \alpha = \frac{90g}{v_0 \sin \alpha} = \frac{(90)(32)}{\sqrt{(68)(64)}}.$$

(d) Parts (b) and (c) together tell us that

$$\tan \alpha = \frac{v_0 \sin \alpha}{v_0 \cos \alpha} = \frac{(\sqrt{(68)(64)})^2}{(90)(32)} = \frac{68}{45}$$

or

$$\alpha = \tan^{-1}\left(\frac{68}{45}\right) \approx 56.5°.$$

This is Rebollo's firing angle.

Projectile Motion with Wind Gusts

We begin with an example that shows how to account for a second force acting on a projectile. We also assume that the path of the baseball in Example 4 lies in a vertical plane.

Example 4 HITTING A BASEBALL

A baseball is hit when it is 3 ft above the ground. It leaves the bat with initial speed of 152 ft/sec, making an angle of 20° with the horizontal. At the instant the ball is hit, an instantaneous gust of wind blows in the horizontal direction directly opposite the direction the ball is taking toward the outfield, adding a component of $-8.8\mathbf{i}$ (ft/sec) to the ball's initial velocity (8.8 ft/sec = 6 mph).

(a) Find the vector equation (position vector) and the parametric equations for the path of the baseball.

(b) How high does the baseball go, and when does it reach maximum height?

(c) Assuming the ball is not caught, find its range and flight time.

Solution

Solve Analytically

(a) Using Equation 1 and accounting for the gust of wind, the initial velocity of the baseball is

$$\mathbf{v}_0 = (v_0 \cos \alpha)\mathbf{i} + (v_0 \sin \alpha)\mathbf{j} - 8.8\mathbf{i}$$
$$= (152 \cos 20°)\mathbf{i} + (152 \sin 20°)\mathbf{j} - (8.8)\mathbf{i}$$
$$= (152 \cos 20° - 8.8)\mathbf{i} + (152 \sin 20°)\mathbf{j}.$$

The initial position is $\mathbf{r}_0 = 0\mathbf{i} + 3\mathbf{j}$.

Substituting the values of $\mathbf{v}_0$ and $\mathbf{r}_0$ into Equation 3 gives the position vector of the baseball.

$$\mathbf{r} = -\frac{1}{2}gt^2\mathbf{j} + \mathbf{v}_0 t + \mathbf{r}_0$$
$$= -16t^2\mathbf{j} + (152 \cos 20° - 8.8)t\mathbf{i} + (152 \sin 20°)t\mathbf{j} + 3\mathbf{j}$$
$$= (152 \cos 20° - 8.8)t\mathbf{i} + (3 + (152 \sin 20°)t - 16t^2)\mathbf{j}$$

The corresponding parametric equations are

$$x = (152 \cos 20° - 8.8)t, \quad y = 3 + (152 \sin 20°)t - 16t^2.$$

Solve Graphically

(b) Figure 10.30a shows the path of the baseball as the graph of the parametric equations in part (a). Tracing along the path, we find that the maximum height is about 45.2 ft, and it is reached in about 1.6 sec.

(c) The x-axis represents the ground, so the baseball hits the ground at the x-intercept (Figure 10.30b). Continuing to trace, we see that the horizontal range is about 442.3 ft, and the flight time about 3.3 sec.

$x = (152 \cos 20 - 8.8)t$
$y = 3 + (152 \sin 20)t - 16t^2$

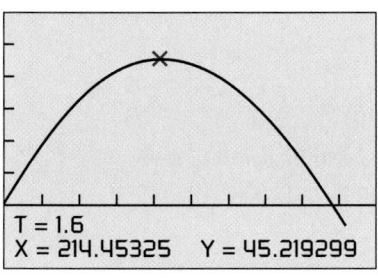

T = 1.6
X = 214.45325 Y = 45.219299

[0, 500] by [−20, 60]

(a)

T = 3.3
X = 442.30982 Y = .31730389

[0, 500] by [−20, 60]

(b)

Figure 10.30 The (a) maximum height and (b) horizontal range of the baseball of Example 4.

Exploration 1 **Hitting a Home Run**

Suppose the baseball in Example 4 is hit toward the right-centerfield fence, 400 ft from the batter.

1. If the fence is 15 ft high, is the hit a home run (does it clear the fence in flight)? Simulate the fence and the flight of the baseball.

2. Find the range and the flight time if the angle at which the ball comes off the bat is **(a)** 25°, **(b)** 30°, **(c)** 45°.

3. Suppose there is no gust of wind. Simulate this flight and the fence. Is this hit a home run?

4. Repeat part 2 for the flight in part 3.

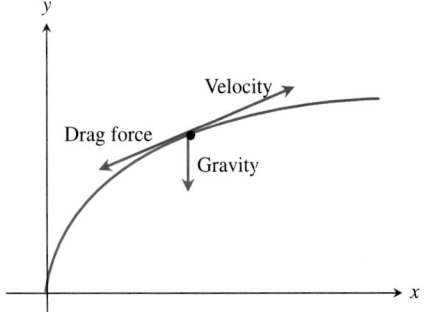

Figure 10.31 Projectile motion with air resistance.

Projectile Motion with Air Resistance

The main force affecting the motion of a projectile, other than gravity, is air resistance. This slowing down force is **drag force** and it acts in a direction *opposite* to the velocity of the projectile (Figure 10.31).

For some high speed projectiles, the drag force is proportional to different powers of the speed over different velocity ranges. For projectiles moving through the air at relatively low speeds, however, the drag force is (very nearly) proportional to the speed (to the first power) and so is called **linear.** Here is what happens for linear drag force.

Projectile Motion with Linear Drag

Equations for the motion of a projectile with linear drag force launched from the origin over a horizontal surface at $t = 0$:

Vector form:

$$\mathbf{r} = \frac{v_0}{k}(1 - e^{-kt})(\cos \alpha)\mathbf{i} +$$
$$\left[\frac{v_0}{k}(1 - e^{-kt})(\sin \alpha) + \frac{g}{k^2}(1 - kt - e^{-kt}) \right]\mathbf{j} \qquad (10)$$

Parametric form:

$$x = \frac{v_0}{k}(1 - e^{-kt}) \cos \alpha,$$

$$y = \frac{v_0}{k}(1 - e^{-kt})(\sin \alpha) + \frac{g}{k^2}(1 - kt - e^{-kt}) \qquad (11)$$

where the **drag coefficient** k is a positive constant representing resistance due to air density, v_0 and α are the projectile's initial speed and launch angle, and g is the acceleration of gravity.

Proof Because drag is linear, directly proportional to the particle's speed, and acts in the direction opposite to the velocity vector, we have the following initial value problem to solve for $\mathbf{r} = x\mathbf{i} + y\mathbf{j}$.

Differential equation: $\dfrac{d^2\mathbf{r}}{dt^2} = -g\mathbf{j} - k\mathbf{v} = -g\mathbf{j} - k\dfrac{d\mathbf{r}}{dt}$ (12)

Initial conditions: $\mathbf{r}(0) = \mathbf{0}$

$$\dfrac{d\mathbf{r}}{dt}\bigg|_{t=0} = \mathbf{v}_0 = (v_0 \cos \alpha)\mathbf{i} + (v_0 \sin \alpha)\mathbf{j}$$

Substituting $\mathbf{r} = x\mathbf{i} + y\mathbf{j}$ into Equation 12 gives

$$\dfrac{d^2\mathbf{r}}{dt^2} = -g\mathbf{j} - k\left(\dfrac{dx}{dt}\mathbf{i} + \dfrac{dy}{dt}\mathbf{j}\right)$$

$$\dfrac{d^2x}{dt^2}\mathbf{i} + \dfrac{d^2y}{dt^2}\mathbf{j} = -k\dfrac{dx}{dt}\mathbf{i} - \left(g + k\dfrac{dy}{dt}\right)\mathbf{j}. \qquad \dfrac{d^2\mathbf{r}}{dt^2} = \dfrac{d^2x}{dt^2}\mathbf{i} + \dfrac{d^2y}{dt^2}\mathbf{j}$$

Comparing components of the two sides, we have

$$\dfrac{d^2x}{dt^2} = -k\dfrac{dx}{dt} \tag{13}$$

and

$$\dfrac{d^2y}{dt^2} = -g - k\dfrac{dy}{dt}. \tag{14}$$

We integrate both sides of Equation 13 with respect to t and use $x = 0$ and $dx/dt = v_0 \cos \alpha$ when $t = 0$ to obtain

$$\dfrac{dx}{dt} = -kx + v_0 \cos \alpha. \tag{15}$$

To solve Equation 15 for x, we first separate the variables (Section 6.2).

$$\dfrac{dx}{v_0 \cos \alpha - kx} = dt$$

Integrating both sides, we obtain

$$-\dfrac{1}{k} \ln |v_0 \cos \alpha - kx| = t + C$$

$$\ln |v_0 \cos \alpha - kx| = -kt + C$$

$$|v_0 \cos \alpha - kx| = Ce^{-kt}$$

$$v_0 \cos \alpha - kx = Ce^{-kt}$$

$$v_0 \cos \alpha - kx = (v_0 \cos \alpha)e^{-kt} \qquad x(0) = 0 \Rightarrow C = v_0 \cos \alpha$$

$$x = \dfrac{v_0}{k}(1 - e^{-kt})(\cos \alpha),$$

the x-component of the parametric form.

To solve Equation 14, we set $w = dy/dt$ and separate the variables.

$$\dfrac{dw}{dt} = -g - kw$$

$$\dfrac{dw}{-g - kw} = dt$$

$$-\dfrac{1}{k} \ln |-g - kw| = t + C$$

$$\ln |-g - kw| = -kt + C$$

$$-g - kw = Ce^{-kt} \tag{16}$$

Now when $t = 0$,

$$w = \frac{dy}{dt} = v_0 \sin \alpha,$$

so

$$C = -g - kv_0 \sin \alpha.$$

Equation 16 becomes

$$-g - kw = (-g - kv_0 \sin \alpha)e^{-kt}$$

$$-g - k\frac{dy}{dt} = (-g - kv_0 \sin \alpha)e^{-kt}. \quad w = dy/dt \quad (17)$$

Integrating both sides of Equation 17 with respect to t, solving for the constant of integration, and then solving for y, we obtain

$$y = \frac{v_0}{k}(1 - e^{-kt})(\sin \alpha) + \frac{g}{k^2}(1 - kt - e^{-kt}),$$

the y-component of the parametric form. ∎

Exploration 2 **Hitting a Baseball—Linear Drag**

A baseball is hit as described in Example 4, but this time with linear drag to slow it down instead of wind (a very calm day). For the temperature and humidity on game day, it has been established that the drag coefficient k is 0.05.

1. Find and graph the parametric equations for the path of the baseball.

2. How high does the baseball go, and when does it reach maximum height?

3. Assuming that the ball is not caught, find the range and flight time of the baseball.

Quick Review 10.4

In Exercises 1 and 2, find the component form of the vector **v**.

1.

2.

In Exercises 3–6, let $f(x) = 2x^2 + 11x - 40$ and $g(x) = 20x - x^2$.

3. Find the x- and y-intercepts of the graph of f.

4. Find the coordinates of the vertex of the graph of f.

5. Find the x- and y-intercepts of the graph of g.

6. Find the coordinates of the vertex of the graph of g.

In Exercises 7 and 8, solve the initial value problem.

7. $\frac{dy}{dx} = \sin x$, $y(\pi/2) = 2$

8. $\frac{d^2y}{dt^2} = 2t$, $y(-1) = 5$, $y'(-1) = 4$

In Exercises 9 and 10, use separation of variables to solve the initial value problem.

9. $\frac{dy}{dt} = 16 - y$, $y(0) = 20$

10. $\frac{dy}{dx} = (4 - 2y)x$, $y(0) = 1$

Section 10.4 Exercises

Projectile flights in the following exercises are to be treated as ideal unless stated otherwise. All launch angles are assumed to be measured from the horizontal. All projectiles are assumed to be launched from the origin over a horizontal surface unless stated otherwise.

1. *Travel Time* A projectile is fired at a speed of 840 m/sec at an angle of 60°. How long will it take to get 21 km downrange?

2. *Finding Muzzle Speed* Find the muzzle speed of a gun whose maximum range is 24.5 km if the angle of elevation is 45°.

3. *Flight Time and Height* A projectile is fired with an initial speed of 500 m/sec at an angle of elevation of 45°.

   (a) When and how far away will the projectile strike?

   (b) How high overhead will the projectile be when it is 5 km downrange?

   (c) What is the greatest height reached by the projectile?

4. *Throwing a Baseball* A baseball is thrown from the stands 32 ft above the field at an angle of 30° up from the horizontal. When and how far away will the ball strike the ground if its initial speed is 32 ft/sec?

5. *Shot Put* An athlete puts a 16-lb shot at an angle of 45° to the horizontal from 6.5 ft above the ground at an initial speed of 44 ft/sec as suggested in the figure. How long after launch and how far from the inner edge of the stopboard does the shot land?

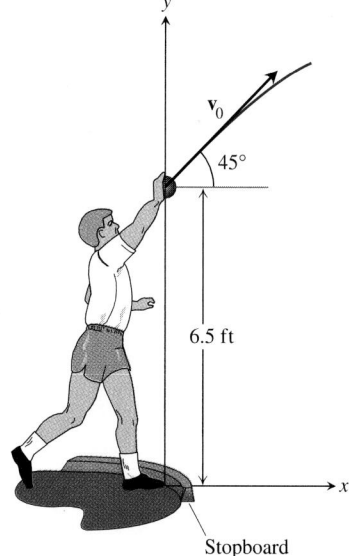

6. *(Continuation of Exercise 5)* Because of its initial elevation, the shot in Exercise 5 would have gone slightly farther if it had been launched at a 40° angle. How much farther? Answer in inches.

7. *Firing Golf Balls* A spring gun at ground level fires a golf ball at an angle of 45°. The ball lands 10 m away.

   (a) What was the ball's initial speed?

   (b) For the same initial speed, find the two firing angles that make the range 6 m.

8. *Beaming Electrons* An electron in a TV tube is beamed horizontally at a speed of 5×10^6 m/sec toward the face of the tube 40 cm away. About how far will the electron drop before it hits?

9. *Finding Golf Ball Speed* Laboratory tests designed to find how far golf balls of different hardness go when hit with a driver showed that a 100-compression ball hit with a club-head speed of 100 mph at a launch angle of 9° carried 248.8 yd. What was the launch speed of the ball? (It was more than 100 mph. At the same time the club head was moving forward, the compressed ball was kicking away from the club face, adding to the ball's forward speed.)

10. **Writing to Learn** A *human cannonball* is to be fired with an initial speed of $v_0 = 80\sqrt{10}/3$ ft/sec. The circus performer (of the right caliber, naturally) hopes to land on a special cushion located 200 ft downrange at the same height as the muzzle of the cannon. The circus is being held in a large room with a flat ceiling 75 ft higher than the muzzle. Can the performer be fired to the cushion without striking the ceiling? If so, what should the cannon's angle of elevation be?

11. **Writing to Learn** A golf ball leaves the ground at a 30° angle at a speed of 90 ft/sec. Will it clear the top of a 30-ft tree that is in the way, 135 ft down the fairway? Explain.

12. *Elevated Green* A golf ball is hit with an initial speed of 116 ft/sec at an angle of elevation of 45° from the tee to a green that is elevated 45 ft above the tee as shown in the diagram. Assuming that the pin, 369 ft downrange, does not get in the way, where will the ball land in relation to the pin?

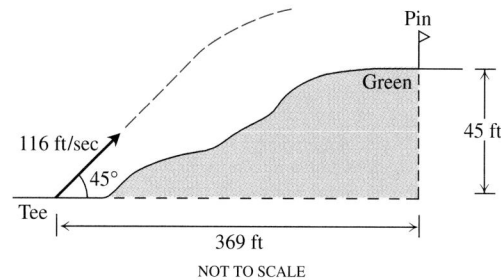

13. *The Green Monster* A baseball hit by a Boston Red Sox player at a 20° angle from 3 ft above the ground just cleared the left end of the "Green Monster," the left-field wall in Fenway Park (see figure). This wall is 37 ft high and 315 ft from home plate.

    (a) What was the initial speed of the ball?

    (b) How long did it take the ball to reach the wall?

14. *Equal Range Firing Angles* Show that a projectile fired at an angle of α degrees, $0 < \alpha < 90$, has the same range as a projectile fired at the same speed at an angle of $(90 - \alpha)$ degrees. (In models that take air resistance into account, this symmetry is lost.)

15. *Equal Range Firing Angles* What two angles of elevation will enable a projectile to reach a target 16 km downrange on the same level as the gun if the projectile's initial speed is 400 m/sec?

16. *Range and Height vs Speed*

    (a) Show that doubling a projectile's initial speed at a given launch angle multiplies its range by 4.

    (b) By about what percentage should you increase the initial speed to double the height and range?

17. *Shot Put* In Moscow in 1987, Natalya Lisouskaya set a women's world record by putting an 8-lb 13-oz shot 73 ft 10 in. Assuming that she launched the shot at a 40° angle to the horizontal from 6.5 ft above the ground, what was the shot's initial speed?

18. *Height vs Time* Show that a projectile attains three-quarters of its maximum height in half the time it takes to reach the maximum height.

19. *Firing from* (x_0, y_0) Derive the equations
$$x = x_0 + (v_0\cos\alpha)t,$$
$$y = y_0 + (v_0\sin\alpha)t - \frac{1}{2}gt^2,$$
(Equations 9 in the text) by solving the following initial value problem for a vector **r** in the plane.

Differential equation: $\dfrac{d^2\mathbf{r}}{dt^2} = -g\mathbf{j}$

Initial conditions: $\mathbf{r}(0) = x_0\mathbf{i} + y_0\mathbf{j}$
$$\frac{d\mathbf{r}}{dt}(0) = (v_0\cos\alpha)\mathbf{i} + (v_0\sin\alpha)\mathbf{j}$$

20. *Flaming Arrow* Using the firing angle found in Example 3, find the speed at which the flaming arrow left Rebollo's bow. See Figure 10.29.

21. *Flaming Arrow* The cauldron in Example 3 is 12 ft in diameter. Using Equations 9 and Example 3c, find how long it takes the flaming arrow to cover the horizontal distance to the rim. How high is the arrow at this time?

22. **Writing to Learn** Describe the path of a projectile given by Equations 5 when $\alpha = 90°$.

In Exercises 23–25, *work in groups of two or three.*

23. *Model Train* The multiflash photograph below shows a model train engine moving at a constant speed on a straight horizontal track. As the engine moved along, a marble was fired into the air by a spring in the engine's smokestack. The marble, which continued to move with the same forward speed as the engine, rejoined the engine 1 sec after it was fired. Measure the angle the marble's path made with the horizontal and use the information to find how high the marble went and how fast the engine was moving.

24. **Writing to Learn** The figure shows an experiment with two marbles. Marble A was launched toward marble B with launch angle α and initial speed v_0. At the same instant, marble B was released to fall from rest at $R\tan\alpha$ units directly above a spot R units downrange from A. The marbles were found to collide regardless of the value of v_0. Was this mere coincidence, or must this happen? Give reasons for your answer.

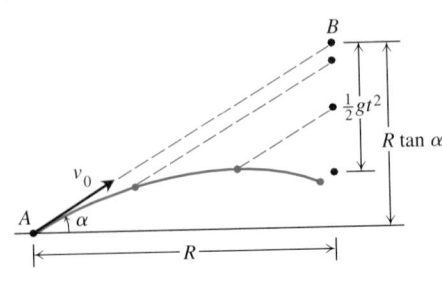

25. *Launching Downhill* An ideal projectile is launched straight down an inclined plane as shown in the figure.

(a) Show that the greatest downhill range is achieved when the initial velocity vector bisects angle *AOR*.

(b) If the projectile were fired uphill instead of down, what launch angle would maximize its range? Give reasons for your answer.

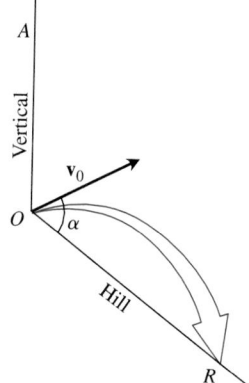

26. *Hitting a Baseball* A baseball is hit when it is 2.5 ft above the ground. It leaves the bat with an initial velocity of 145 ft/sec at a launch angle of 23°. At the instant the ball is hit, an instantaneous gust of wind blows against the ball, adding a component of $-14\mathbf{i}$ (ft/sec) to the ball's initial velocity. A 15-ft-high fence is 300 ft from home plate in the direction of the flight.

(a) Find vector and parametric forms for the path of the baseball.

(b) How high does the baseball go, and when does it reach maximum height?

(c) Find the range and flight time of the baseball, assuming the ball is not caught.

(d) When is the baseball 20 ft high? How far (ground distance) is the baseball from home plate at that height?

(e) **Writing to Learn** Has the batter hit a home run? Explain.

Explorations

27. *Volleyball* A volleyball is hit when it is 4 ft above the ground and 12 ft from a 6-ft-high net. It leaves the point of impact with an initial velocity of 35 ft/sec at an angle of 27° and slips by the opposing team untouched.

(a) Find vector and parametric forms for the path of the volleyball.

(b) How high does the volleyball go, and when does it reach maximum height?

(c) Find its range and flight time.

(d) When is the volleyball 7 ft above the ground? How far (ground distance) is the volleyball from where it will land?

(e) **Writing to Learn** Suppose the net is raised to 8 ft. Does this change things? Explain.

28. *Hitting a Baseball* Consider the baseball problem of Exploration 2 again. Assume a drag coefficient of 0.12.

(a) Find vector and parametric forms for the path of the baseball.

(b) How high does the baseball go, and when does it reach maximum height?

(c) Find the range and flight time of the baseball.

(d) When is the baseball 30 ft high? How far (ground distance) is the baseball from home plate at that height?

(e) **Writing to Learn** A 10-ft-high outfield fence is 340 ft from home plate in the direction of the flight of the baseball. The outfielder can jump and catch any ball up to 11 ft off the ground to stop it from going over the fence. Has the batter hit a home run?

29. *Hitting a Baseball* Consider the baseball problem of Exploration 2 again. This time assume a drag coefficient of 0.08 *and* an instantaneous gust of wind that adds a component of $-17.6\mathbf{i}$ (ft/sec) to the initial velocity at the instant the baseball is hit.

(a) Find vector and parametric forms for the path of the baseball.

(b) How high does the baseball go, and when does it reach maximum height?

(c) Find the range and flight time of the baseball.

(d) When is the baseball 35 ft high? How far (ground distance) is the baseball from home plate at that height?

(e) **Writing to Learn** A 20-ft-high outfield fence is 380 ft from home plate in the direction of the flight of the baseball. Has the batter hit a home run? If "yes," what change in the horizontal component of the ball's initial velocity would have kept the ball in the park? If "no," what change would have allowed it to be a home run?

30. *Hitting a Baseball* Consider the baseball problem of Exploration 2 again. Assume drag coefficients of 0.01, 0.02, 0.10, 0.15, 0.20, and 0.25.

(a) Graph the paths of the baseballs in the same viewing window.

(b) In each case, how high does the baseball go, and when does it reach maximum height?

(c) In each case, find the range and flight time of the baseball.

(d) **Writing to Learn** As $k \to 0$, show that the right-hand members of Equations 11 approach the right-hand members of Equations 5. Explain why this makes sense.

Extending the Ideas

31. *Where Trajectories Crest* For a projectile fired from the ground at launch angle α with initial speed v_0, consider α as a variable and v_0 as a fixed constant. For each α, $0 < \alpha < \pi/2$, we obtain a parabolic trajectory as shown in the figure. Show that the points in the plane that give the maximum heights of these parabolic trajectories all lie on the ellipse

$$x^2 + 4\left(y - \frac{v_0^2}{4g}\right)^2 = \frac{v_0^4}{4g^2},$$

where $x \geq 0$.

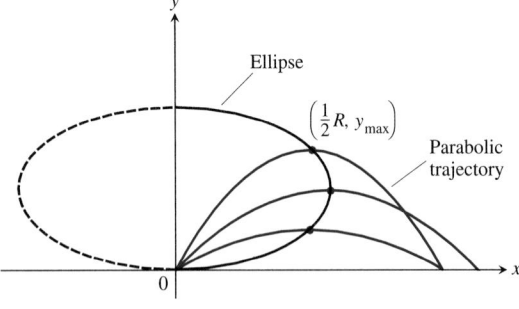

32. *The Linear Drag Equation* Show directly that Equation 10 is the solution to this initial value problem.

Differential equation: $\dfrac{d^2\mathbf{r}}{dt^2} = -g\mathbf{j} - k\mathbf{v} = -g\mathbf{j} - k\dfrac{d\mathbf{r}}{dt}$

Initial conditions: $\quad \mathbf{r}(0) = \mathbf{0}$

$$\left.\frac{d\mathbf{r}}{dt}\right|_{t=0} = \mathbf{v}_0 = (v_0\cos\alpha)\mathbf{i} + (v_0\sin\alpha)\mathbf{j}$$

10.5 Polar Coordinates and Polar Graphs

Polar Coordinates • Polar Graphing • Relating Polar and Cartesian Coordinates

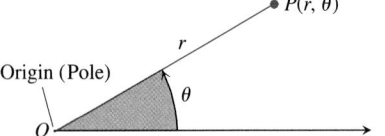

Figure 10.32 To define polar coordinates for the plane, we start with an origin, called the pole, and an initial ray.

Polar Coordinates

To define polar coordinates, we first fix an **origin** O (called the **pole**) and an **initial ray** from O (Figure 10.32). Then each point P can be located by assigning to it a **polar coordinate pair** (r, θ) in which r gives the directed distance from O to P and θ gives the directed angle from the initial ray to ray OP.

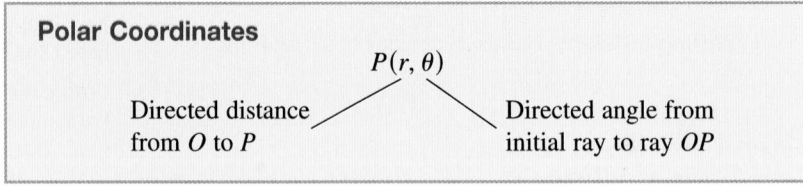

As in trigonometry, θ is positive when measured counterclockwise and negative when measured clockwise. The angle associated with a given point is not unique. For instance, the point 2 units from the origin along the ray $\theta = \pi/6$ has polar coordinates $r = 2$, $\theta = \pi/6$. It also has coordinates $r = 2$, $\theta = -11\pi/6$ (Figure 10.33).

There are also occasions when we wish to allow r to be negative. This is why we use *directed* distance in defining $P(r, \theta)$. The point $P(2, 7\pi/6)$ can be reached by turning $7\pi/6$ radians counterclockwise from the initial ray and going forward 2 units (Figure 10.35 on the next page). It can also be reached by turning $\pi/6$ radians counterclockwise from the initial ray and going *backwards* 2 units. So the point also has polar coordinates $r = -2$, $\theta = \pi/6$.

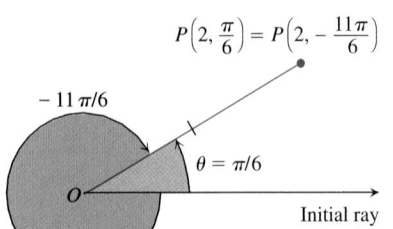

Figure 10.33 Polar coordinates are not unique.

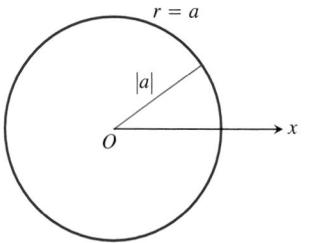

Figure 10.34 The polar equation for this circle is $r = a$.

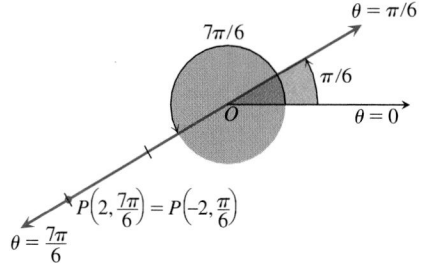

Figure 10.35 Polar coordinates can have negative r values.

Polar Graphing

If we hold r fixed at a constant value $a \neq 0$, the point $P(r, \theta)$ will lie $|a|$ units from the origin O. As θ varies over any interval of length 2π, P traces a circle of radius $|a|$ centered at O (Figure 10.34).

If we hold θ fixed at a constant value $\theta = \alpha$ and let r vary between $-\infty$ and ∞, the point $P(r, \theta)$ traces the line through O that makes an angle of measure α with the initial ray.

Equation	Polar Graph
$r = a$	Circle of radius $|a|$ centered at O
$\theta = \alpha$	Line through O making an angle α with the initial ray

Example 1 FINDING POLAR EQUATIONS FOR GRAPHS

(a) $r = 1$ and $r = -1$ are equations for the circle of radius 1 centered at O.

(b) $\theta = \pi/6$, $\theta = 7\pi/6$, and $\theta = -5\pi/6$ are equations for the line in Figure 10.35.

Equations of the form $r = a$ and $\theta = \alpha$ can be combined to define regions, segments, and rays.

Example 2 IDENTIFYING GRAPHS

Graph the set of points whose polar coordinates satisfy the given conditions.

(a) $1 \leq r \leq 2$ and $0 \leq \theta \leq \dfrac{\pi}{2}$

(b) $-3 \leq r \leq 2$ and $\theta = \dfrac{\pi}{4}$

(c) $r \leq 0$ and $\theta = \dfrac{\pi}{4}$

(d) $\dfrac{2\pi}{3} \leq \theta \leq \dfrac{5\pi}{6}$ (no restriction on r)

Solution The graphs are shown in Figure 10.36.

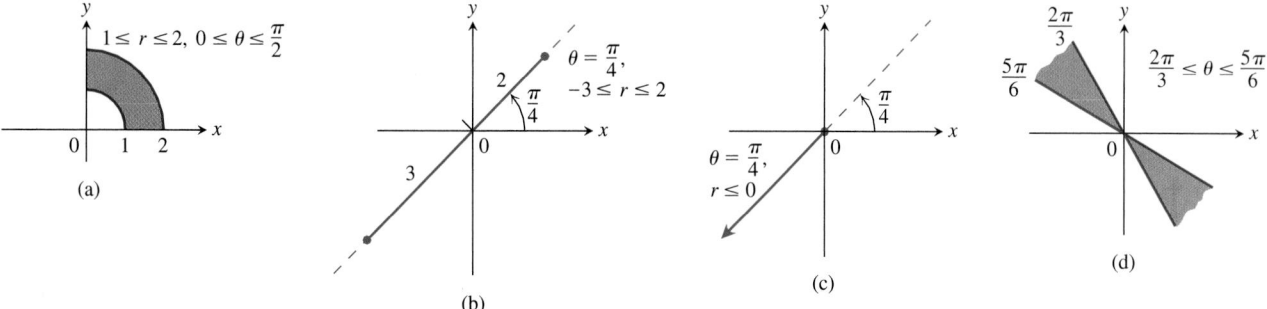

Figure 10.36 The graphs of typical inequalities in r and θ. (Example 2)

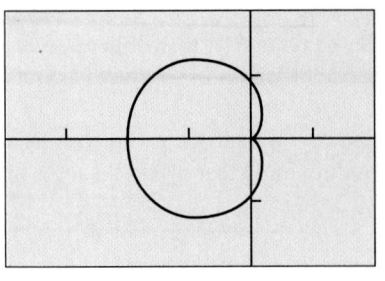

[−4, 2] by [−2, 2]

Figure 10.37 The graph of the cardioid $r = 1 - \cos \theta$ (calculator in polar mode). (Example 3)

The curve in Figure 10.37 is called a **cardioid** because of its heart-like shape.

Example 3 USING A POLAR GRAPHER

Draw the graph of $r = 1 - \cos \theta$. What is the shortest length a θ-interval can have and still produce the graph?

Solution Figure 10.37 shows the graph of

$$r = 1 - \cos \theta, \quad 0 \le \theta \le 2\pi.$$

Any interval of length 2π will produce the graph. Any interval shorter than 2π will leave part of the cardioid untraced.

The graph in Figure 10.37 appears to be symmetric about the x-axis. Figure 10.38 illustrates the standard polar coordinate tests for symmetry.

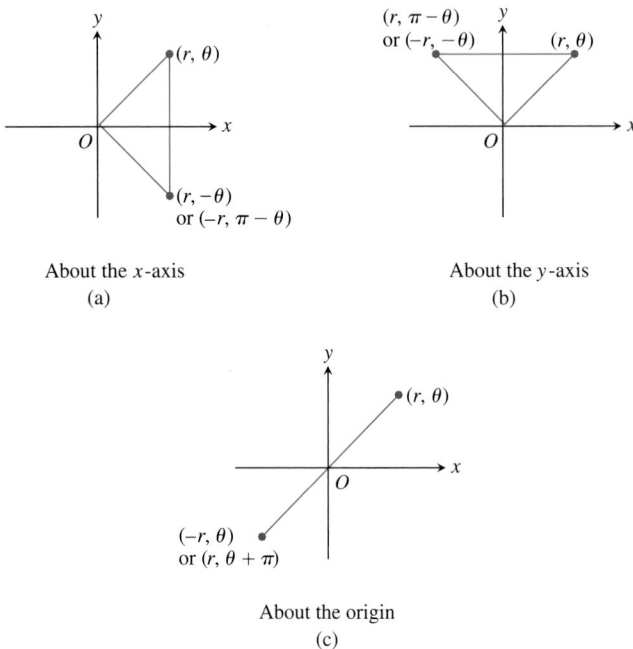

Figure 10.38 Three tests for symmetry.

Symmetry Tests for Polar Graphs

1. **Symmetry about the x-axis:** If the point (r, θ) lies on the graph, the point $(r, -\theta)$ or $(-r, \pi - \theta)$ lies on the graph (Figure 10.38a).

2. **Symmetry about the y-axis:** If the point (r, θ) lies on the graph, the point $(r, \pi - \theta)$ or $(-r, -\theta)$ lies on the graph (Figure 10.38b).

3. **Symmetry about the origin:** If the point (r, θ) lies on the graph, the point $(-r, \theta)$ or $(r, \theta + \pi)$ lies on the graph (Figure 10.38c).

If we replace θ by $-\theta$ in the equation $r = 1 - \cos \theta$ of the cardioid of Figure 10.37 we obtain the same equation

$$r = 1 - \cos(-\theta) = 1 - \cos \theta.$$

Thus, if point (r, θ) lies on the graph of $r = 1 - \cos \theta$, so does the point $(r, -\theta)$. This confirms our conjecture that the graph in Figure 10.37 is symmetric about the x-axis.

Exploration 1 **Investigating Polar Graphs**

Let $r^2 = 4 \cos \theta$.

1. Graph $r^2 = 4 \cos \theta$ in a square viewing window using $r_1 = 2\sqrt{\cos \theta}$ and $r_2 = -2\sqrt{\cos \theta}$.
2. Explain why the origin is part of the graph.
3. What is the shortest length a θ-interval can have and still produce the graph?
4. Show algebraically that the graph is symmetric about the x-axis, the y-axis, and the origin.

Relating Polar and Cartesian Coordinates

When we use both polar and Cartesian coordinates in a plane, we place the two origins together and take the polar initial ray as the positive x-axis. The ray $\theta = \pi/2$, $r > 0$, becomes the positive y-axis (Figure 10.39). The two coordinate systems are then related by the following equations.

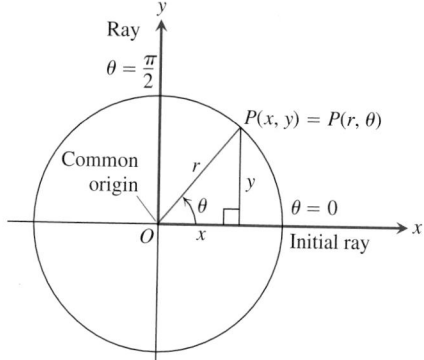

Figure 10.39 The usual way to relate polar and Cartesian coordinates.

Equations Relating Polar and Cartesian Coordinates

$$x = r \cos \theta, \quad y = r \sin \theta, \quad x^2 + y^2 = r^2, \quad \frac{y}{x} = \tan \theta$$

We use these equations and algebra (sometimes a lot of it!) to rewrite polar equations in Cartesian form and vice versa.

Example 4 EQUIVALENT EQUATIONS

Polar equation	Cartesian equivalent
$r \cos \theta = 2$	$x = 2$
$r^2 \cos \theta \sin \theta = 4$	$xy = 4$
$r^2 \cos^2 \theta - r^2 \sin^2 \theta = 1$	$x^2 - y^2 = 1$
$r = 1 + 2r \cos \theta$	$y^2 - 3x^2 - 4x - 1 = 0$
$r = 1 - \cos \theta$	$x^4 + y^4 + 2x^2y^2 + 2x^3 + 2xy^2 - y^2 = 0$

With some curves, we are better off with polar coordinates; with others, we aren't.

Example 5 CONVERTING CARTESIAN TO POLAR

Find a polar equation for the circle $x^2 + (y - 3)^2 = 9$ (Figure 10.40a). Support graphically.

Solution

$$x^2 + y^2 - 6y + 9 = 9 \quad \text{Expand } (y - 3)^2.$$

$$x^2 + y^2 - 6y = 0$$

$$r^2 - 6r \sin \theta = 0 \quad x^2 + y^2 = r^2, \quad y = r \sin \theta$$

$$r = 0 \quad \text{or} \quad r - 6 \sin \theta = 0$$

The equation $r = 6 \sin \theta$ includes the possibility that $r = 0$, as supported by its graph in Figure 10.40b.

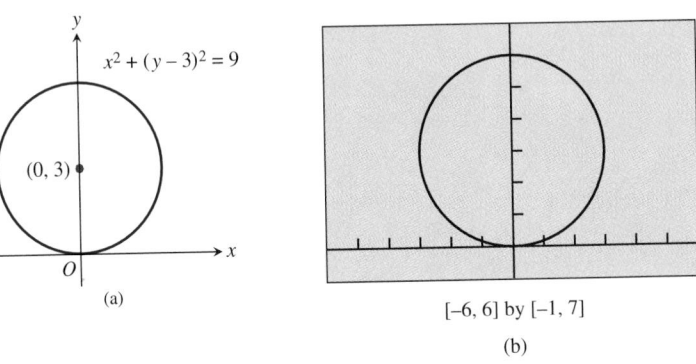

(a)

[–6, 6] by [–1, 7]

(b)

Figure 10.40 The graphs of (a) the circle $x^2 + (y - 3)^2 = 9$, and (b) the polar equation $r = 6 \sin \theta$. (Example 5)

Example 6 CONVERTING POLAR TO CARTESIAN

Find a Cartesian equivalent for the polar equation. Identify the graph.

(a) $r^2 = 4r \cos \theta$ **(b)** $r = \dfrac{4}{2 \cos \theta - \sin \theta}$

Solution

(a)

$$r^2 = 4r \cos \theta$$

$$x^2 + y^2 = 4x \qquad r^2 = x^2 + y^2, \quad r \cos \theta = x$$

$$x^2 - 4x + y^2 = 0$$

$$x^2 - 4x + 4 + y^2 = 4 \qquad \text{Completing the square}$$

$$(x - 2)^2 + y^2 = 4$$

The graph of the equivalent Cartesian equation $(x - 2)^2 + y^2 = 4$ is a circle with radius 2 and center $(2, 0)$.

(b)

$$r = \frac{4}{2\cos\theta - \sin\theta}$$

$$r(2\cos\theta - \sin\theta) = 4$$

$$2r\cos\theta - r\sin\theta = 4$$

$$2x - y = 4 \qquad r\cos\theta = x, \quad r\sin\theta = y$$

$$y = 2x - 4$$

The graph of the equivalent Cartesian equation $y = 2x - 4$ is a line with slope 2 and y-intercept -4.

Exploration 2 **Graphing Rose Curves**

The polar curves $r = a\cos n\theta$ and $r = a\sin n\theta$, where n is an integer and $|n| > 1$, are *rose curves*.

1. Graph $r = 2\cos n\theta$ for $n = \pm 2, \pm 4, \pm 6$. Describe the curves.

2. What is the shortest length a θ-interval can have and still produce the graphs in (1)?

3. Based on your observations in (1), describe the graph of $r = 2\cos n\theta$ when n is a nonzero even integer.

4. Graph $r = 2\cos n\theta$ for $n = \pm 3, \pm 5, \pm 7$. Describe the curves.

5. What is the shortest length a θ-interval can have and still produce the graphs in (4)?

6. Based on your observations in (4), describe the graph of $r = 2\cos n\theta$ when n is an odd integer different from ± 1.

Quick Review 10.5

In Exercises 1–3, write a Cartesian equation for the curve.

1. the line through the points $(-2, 4)$ and $(3, -1)$

2. the circle with center $(0, 0)$ and radius 3

3. the circle with center $(-2, 4)$ and radius 2

In Exercises 4–7, determine whether the curve is symmetric about **(a)** the x-axis, **(b)** the y-axis, or **(c)** the origin.

4. $y = x^3 - x$

5. $y = x^2 - x$

6. $y = \cos x$

7. $4x^2 + 9y^2 = 36$

In Exercises 8 and 9, explain how to use a function grapher to graph the curve.

8. $y^2 = x - 2$

9. $x^2 + 3y^2 = 4$

10. Use completing the square to find the center and radius of the circle $x^2 + y^2 - 4x + 6y + 9 = 0$.

Section 10.5 Exercises

In Exercises 1 and 2, determine which polar coordinate pairs name the same point.

1. (a) $(3, 0)$ (b) $(-3, 0)$ (c) $(2, 2\pi/3)$

    (d) $(2, 7\pi/3)$ (e) $(-3, \pi)$ (f) $(2, \pi/3)$

    (g) $(-3, 2\pi)$ (h) $(-2, -\pi/3)$

2. (a) $(-2, \pi/3)$ (b) $(2, -\pi/3)$ (c) (r, θ)

    (d) $(r, \theta + \pi)$ (e) $(-r, \theta)$ (f) $(2, -2\pi/3)$

    (g) $(-r, \theta + \pi)$ (h) $(-2, 2\pi/3)$

In Exercises 3 and 4, plot the points with the given polar coordinates and find their Cartesian coordinates.

3. (a) $(\sqrt{2}, \pi/4)$ (b) $(1, 0)$

    (c) $(0, \pi/2)$ (d) $(-\sqrt{2}, \pi/4)$

4. (a) $(-3, 5\pi/6)$ (b) $(5, \tan^{-1}(4/3))$

    (c) $(-1, 7\pi)$ (d) $(2\sqrt{3}, 2\pi/3)$

In Exercises 5 and 6, plot the points with the given Cartesian coordinates and find two sets of polar coordinates for each.

5. (a) $(-1, 1)$ (b) $(1, -\sqrt{3})$

    (c) $(0, 3)$ (d) $(-1, 0)$

6. (a) $(-\sqrt{3}, -1)$ (b) $(3, 4)$

    (c) $(0, -2)$ (d) $(2, 0)$

In Exercises 7–18, graph the set of points whose polar coordinates satisfy the given equations and inequalities.

7. $r = 2$ **8.** $0 \le r \le 2$

9. $r \ge 1$ **10.** $0 \le \theta \le \pi/6, \quad r \ge 0$

11. $\theta = 2\pi/3, \quad r \le -2$ **12.** $\theta = \pi/3, \quad -1 \le r \le 3$

13. $0 \le \theta \le \pi, \quad r = 1$ **14.** $0 \le \theta \le \pi, \quad r = -1$

15. $\theta = \pi/2, \quad r \le 0$

16. $\pi/4 \le \theta \le 3\pi/4, \quad 0 \le r \le 1$

17. $-\pi/4 \le \theta \le \pi/4, \quad -1 \le r \le 1$

18. $0 \le \theta \le \pi/2, \quad 1 \le |r| \le 2$

In Exercises 19–36, replace the polar equation by an equivalent Cartesian equation. Then identify or describe the graph.

19. $r \sin \theta = 0$ **20.** $r \cos \theta = 0$

21. $r = 4 \csc \theta$ **22.** $r = -3 \sec \theta$

23. $r \cos \theta + r \sin \theta = 1$ **24.** $r^2 = 1$

25. $r^2 = 4r \sin \theta$ **26.** $r = \dfrac{5}{\sin \theta - 2 \cos \theta}$

27. $r^2 \sin 2\theta = 2$ **28.** $r = \cot \theta \csc \theta$

29. $r = \csc \theta \; e^{r \cos \theta}$ **30.** $\cos^2 \theta = \sin^2 \theta$

31. $r \sin \theta = \ln r + \ln \cos \theta$ **32.** $r^2 + 2r^2 \cos \theta \sin \theta = 1$

33. $r^2 = -4r \cos \theta$ **34.** $r = 8 \sin \theta$

35. $r = 2 \cos \theta + 2 \sin \theta$ **36.** $r \sin\left(\theta + \dfrac{\pi}{6}\right) = 2$

In Exercises 37–48, replace the Cartesian equation by an equivalent polar equation. Support graphically.

37. $x = 7$ **38.** $y = 1$

39. $x = y$ **40.** $x - y = 3$

41. $x^2 + y^2 = 4$ **42.** $x^2 - y^2 = 1$

43. $\dfrac{x^2}{9} + \dfrac{y^2}{4} = 1$ **44.** $xy = 2$

45. $y^2 = 4x$ **46.** $x^2 + xy + y^2 = 1$

47. $x^2 + (y - 2)^2 = 4$

48. $(x - 3)^2 + (y + 1)^2 = 4$

In Exercises 49–58, (a) graph the polar curve. (b) What is the shortest length a θ-interval can have and still produce the graph?

49. $r = 1 + \cos \theta$ **50.** $r = 2 - 2 \cos \theta$

51. $r^2 = -\sin 2\theta$ **52.** $r = 1 - \sin \theta$

53. $r = 1 - 2 \sin 3\theta$ **54.** $r = \sin (\theta/2)$

55. $r = \theta$ **56.** $r = 1 + \sin \theta$

57. $r = 2 \cos 3\theta$ **58.** $r = 1 + 2 \sin \theta$

In Exercises 59–62, *work in groups of two or three.* Determine the symmetries of the curve.

59. $r^2 = 4 \cos 2\theta$ **60.** $r^2 = 4 \sin 2\theta$

61. $r = 2 + \sin \theta$ **62.** $r^2 = -\cos 2\theta$

63. Writing to Learn

(a) Explain why every vertical line in the plane has a polar equation of the form $r = a \sec \theta$.

(b) Find an analogous polar equation for horizontal lines. Give reasons for your answer.

Explorations

64. *Rose Curves* Let $r = 2 \sin n\theta$.

(a) Graph $r = 2 \sin n\theta$ for $n = \pm 2, \pm 4, \pm 6$. Describe the curves.

(b) What is the smallest length a θ-interval can have and still produce the graphs in (a)?

(c) Based on your observations in (a), describe the graph of $r = 2 \sin n\theta$ when n is a nonzero even integer.

(d) Graph $r = 2 \sin n\theta$ for $n = \pm 3, \pm 5, \pm 7$. Describe the curves.

(e) What is the smallest length a θ-interval can have and still produce the graphs in (d)?

(f) Based on your observations in (d), describe the graph of $r = 2 \sin n\theta$ when n is an odd integer different from ± 1.

65. *Relating Polar Equations to Parametric Equations* Let $r = f(\theta)$ be a polar curve.

(a) Writing to Learn Explain why

$$x = f(t) \cos t, \quad y = f(t) \sin t$$

are parametric equations for the curve.

(b) Use (a) to write parametric equations for the circle $r = 3$. Support your answer by graphing the parametric equations.

(c) Repeat (b) with $r = 1 - \cos \theta$.

(d) Repeat (b) with $r = 3 \sin 2\theta$.

66. *Rotating Curves* Let $r_1(\theta) = 3(1 - \cos \theta)$ and $r_2(\theta) = r_1(\theta - \alpha)$.

(a) Graph r_2 for $\alpha = \pi/6, \pi/4, \pi/3,$ and $\pi/2$ and compare with the graph of r_1.

(b) Graph r_2 for $\alpha = -\pi/6, -\pi/4, -\pi/3,$ and $-\pi/2$ and compare with the graph of r_1.

(c) Based on your observations in (a) and (b), describe the relationship between the graphs of $r_1 = f(\theta)$ and $r_2 = f(\theta - \alpha)$.

Extending the Ideas

67. *Distance Formula* Show that the distance between two points (r_1, θ_1) and (r_2, θ_2) in polar coordinates is

$$d = \sqrt{r_1{}^2 + r_2{}^2 - 2r_1 r_2 \cos (\theta_1 - \theta_2)}.$$

68. Let $r = \dfrac{2}{1 + k \cos \theta}$.

(a) Graph r in a square viewing window for $k = 0.1, 0.3, 0.5, 0.7,$ and 0.9. Describe the graphs.

(b) Based on your observations in (a), conjecture what happens to the graphs for $0 < k < 1$ and $k \to 0^+$.

69. Let $r = \dfrac{2}{1 + k \cos \theta}$.

(a) Graph r in a square viewing window for $k = 1.1, 1.3, 1.5, 1.7,$ and 1.9. Describe the graphs.

(b) Based on your observations in (a), conjecture what happens to the graphs for $k > 1$ and $k \to 1^+$.

70. Let $r = \dfrac{k}{1 + \cos \theta}$.

(a) Graph r in a square viewing window for $k = 1, 3, 5, 7,$ and 9. Describe the graphs.

(b) Based on your observations in (a), conjecture what happens to the graphs for $k > 0$ and $k \to 0^+$.

10.6 Calculus of Polar Curves

Slope • Area in the Plane • Length of a Curve • Area of a Surface of Revolution

Slope

The slope of the tangent line to a polar curve $r = f(\theta)$ is given by dy/dx, not by $r' = df/d\theta$. To see why, think of the graph of f as the graph of the parametric equations (see Section 10.5, Exercise 65)

$$x = r \cos \theta = f(\theta) \cos \theta, \quad y = r \sin \theta = f(\theta) \sin \theta.$$

If f is a differentiable function of θ, then so are x and y and, when $dx/d\theta \neq 0$, we can calculate dy/dx from the parametric formula

$$
\begin{aligned}
\frac{dy}{dx} &= \frac{dy/d\theta}{dx/d\theta} \qquad \text{\small Section 10.1, Eq. 1 with } t = \theta \\[2mm]
&= \frac{\dfrac{d}{d\theta}(f(\theta) \sin \theta)}{\dfrac{d}{d\theta}(f(\theta) \cos \theta)} \\[2mm]
&= \frac{\dfrac{df}{d\theta} \sin \theta + f(\theta) \cos \theta}{\dfrac{df}{d\theta} \cos \theta - f(\theta) \sin \theta} . \qquad \text{\small Product Rule for derivatives}
\end{aligned}
$$

Slope of the Curve $r = f(\theta)$

$$\frac{dy}{dx}\bigg|_{(r,\,\theta)} = \frac{dy/d\theta}{dx/d\theta} = \frac{f'(\theta)\sin\theta + f(\theta)\cos\theta}{f'(\theta)\cos\theta - f(\theta)\sin\theta}, \tag{1}$$

provided $dx/d\theta \neq 0$ at (r, θ).

We can see from Equation 1 and its derivation that the curve $r = f(\theta)$ has a

1. horizontal tangent at a point where $dy/d\theta = 0$ and $dx/d\theta \neq 0$,

2. vertical tangent at a point where $dx/d\theta = 0$ and $dy/d\theta \neq 0$.

If both derivatives are zero, no conclusion can be drawn without further investigation, as illustrated in Example 1.

Example 1   FINDING HORIZONTAL AND VERTICAL TANGENTS

Find the horizontal and vertical tangents to the cardioid $r = 1 - \cos\theta$, $0 \le \theta \le 2\pi$.

Solution The graph in Figure 10.41 suggests that there are at least two horizontal and three vertical tangents.

The parametric form of the equation is

$$x = r\cos\theta = (1 - \cos\theta)\cos\theta = \cos\theta - \cos^2\theta,$$

$$y = r\sin\theta = (1 - \cos\theta)\sin\theta = \sin\theta - \cos\theta\sin\theta.$$

We need to find the zeros of $dy/d\theta$ and $dx/d\theta$.

(a) Zeros of $dy/d\theta$ in $[0, 2\pi]$:

$$\frac{dy}{d\theta} = \cos\theta + \sin^2\theta - \cos^2\theta = \cos\theta + (1 - \cos^2\theta) - \cos^2\theta$$

$$= 1 + \cos\theta - 2\cos^2\theta = (1 + 2\cos\theta)(1 - \cos\theta)$$

Now,

$$1 - \cos\theta = 0 \;\Rightarrow\; \theta = 0, 2\pi,$$

$$1 + 2\cos\theta = 0 \;\Rightarrow\; \theta = 2\pi/3, 4\pi/3.$$

Thus, $dy/d\theta = 0$ in $0 \le \theta \le 2\pi$ if $\theta = 0, 2\pi/3, 4\pi/3$, or 2π.

(b) Zeros of $dx/d\theta$ in $[0, 2\pi]$:

$$\frac{dx}{d\theta} = -\sin\theta + 2\cos\theta\sin\theta = (2\cos\theta - 1)\sin\theta$$

Now,

$$2\cos\theta - 1 = 0 \;\Rightarrow\; \theta = \pi/3, 5\pi/3,$$

$$\sin\theta = 0 \;\Rightarrow\; \theta = 0, \pi, 2\pi.$$

Thus, $dx/d\theta = 0$ in $0 \le \theta \le 2\pi$ if $\theta = 0, \pi/3, \pi, 5\pi/3$, or 2π.

We can now see that there are horizontal tangents ($dy/d\theta = 0$, $dx/d\theta \neq 0$) at the points where $\theta = 2\pi/3$ and $4\pi/3$, and vertical tangents ($dx/d\theta = 0$, $dy/d\theta \neq 0$) at the points where $\theta = \pi/3$, π, and $5\pi/3$.

$r = 1 - \cos\theta, \; 0 \le \theta \le 2\pi$

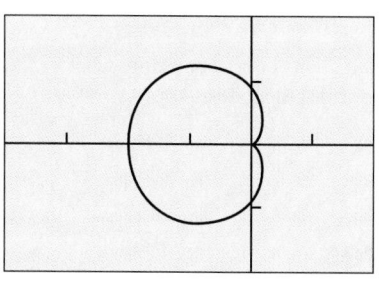

[–4, 2] by [–2, 2]

Figure 10.41 Where are the horizontal and vertical tangents to this cardioid? (Example 1)

At the points where $\theta = 0$ or 2π, the right side of Equation 1 takes the form 0/0. We can use l'Hôpital's rule (Section 8.1) to see that

$$\lim_{\theta \to 0, 2\pi} \frac{dy/d\theta}{dx/d\theta} = \lim_{\theta \to 0, 2\pi} \frac{1 + \cos\theta - 2\cos^2\theta}{2\cos\theta\sin\theta - \sin\theta}$$

$$= \lim_{\theta \to 0, 2\pi} \frac{-\sin\theta + 4\cos\theta\sin\theta}{2\cos^2\theta - 2\sin^2\theta - \cos\theta} = \frac{0}{1} = 0.$$

The curve has a horizontal tangent at the point where $\theta = 0$ or 2π. Summarizing, we have

horizontal tangents at $(0, 0) = (0, 2\pi)$, $(1.5, 2\pi/3)$, $(1.5, 4\pi/3)$;

vertical tangents at $(0.5, \pi/3)$, $(2, \pi)$, $(0.5, 5\pi/3)$.

If the curve $r = f(\theta)$ passes through the origin at $\theta = \theta_0$, then $f(\theta_0) = 0$, and Equation 1 gives

$$\left.\frac{dy}{dx}\right|_{(0, \theta_0)} = \frac{f'(\theta_0)\sin\theta_0}{f'(\theta_0)\cos\theta_0} = \tan\theta_0,$$

provided $f'(\theta_0) \neq 0$ (not the case in Example 1). The reason we say "slope at $(0, \theta_0)$" and not just "slope at the origin" is that a polar curve may pass through the origin more than once, with different slopes at different θ-values.

Example 2 FINDING TANGENT LINES AT THE POLE (ORIGIN)

Find the lines tangent to the rose curve

$$r = f(\theta) = 2\sin 3\theta, \qquad 0 \le \theta \le \pi,$$

at the pole.

Solution $f(\theta)$ is zero when $\theta = 0$, $\pi/3$, $2\pi/3$, and π. The derivative $f'(\theta) = 6\cos 3\theta$ is not zero at these four values of θ. Thus, this curve has tangent lines at the pole (Figure 10.42) with slopes $\tan 0 = \tan\pi = 0$, $\tan(\pi/3) = \sqrt{3}$, and $\tan(2\pi/3) = -\sqrt{3}$. The three corresponding tangent lines are $y = 0$, $y = \sqrt{3}x$, and $y = -\sqrt{3}x$.

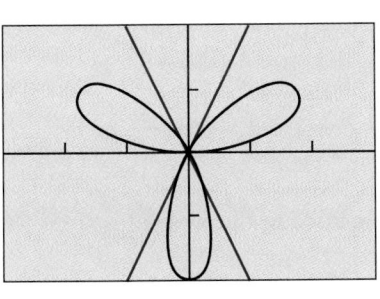

[–3, 3] by [–2, 2]

Figure 10.42 The three tangent lines to $r = f(\theta) = 2\sin 3\theta$, $0 \le \theta \le \pi$, are $y = 0$, $y = \sqrt{3}x$, and $y = -\sqrt{3}x$. (Example 2)

Area in the Plane

The region OTS in Figure 10.43 is bounded by the rays $\theta = \alpha$ and $\theta = \beta$ and the curve $r = f(\theta)$. We approximate the region with n nonoverlapping circular sectors based on a partition P of angle TOS. The typical sector has radius $r_k = f(\theta_k)$ and central angle of radian measure $\Delta\theta_k$. Its area is

$$A_k = \frac{1}{2}r_k^2 \Delta\theta_k = \frac{1}{2}(f(\theta_k))^2 \Delta\theta_k.$$

The area of the region OTS is approximately

$$\sum_{k=1}^{n} A_k = \sum_{k=1}^{n} \frac{1}{2}(f(\theta_k))^2 \Delta\theta.$$

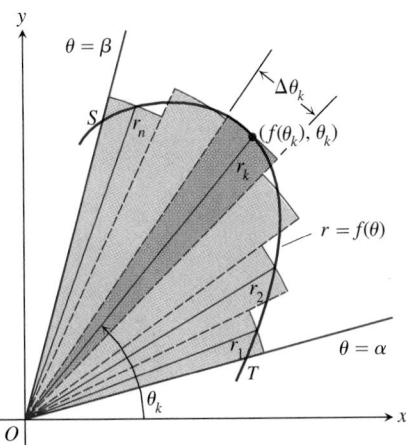

Figure 10.43 To derive a formula for the area of the region OTS, we approximate the region with thin circular sectors.

If f is continuous, we expect the approximations to improve as $\|P\| \to 0$, and we are led to the following formula for the region's area:

$$A = \lim_{\|P\| \to 0} \sum_{k=1}^{n} \frac{1}{2}(f(\theta_k))^2 \, \Delta\theta$$

$$= \int_{\alpha}^{\beta} \frac{1}{2}(f(\theta))^2 \, d\theta.$$

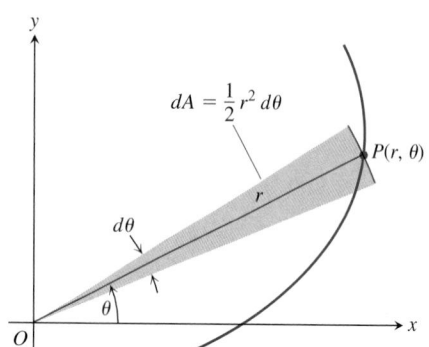

Figure 10.44 The area differential dA.

Area in Polar Coordinates

The **area** of the region **between the origin and the curve** $r = f(\theta)$, $\alpha \le \theta \le \beta$, is

$$A = \int_{\alpha}^{\beta} \frac{1}{2} r^2 \, d\theta.$$

This is the integral of the **area differential** (Figure 10.44),

$$dA = \frac{1}{2} r^2 \, d\theta.$$

Example 3 FINDING AREA

Find the area of the region in the plane enclosed by the cardioid $r = 2(1 + \cos\theta)$.

Solution We graph the cardioid (Figure 10.45) and determine that the *radius OP* sweeps out the region exactly once as θ runs from 0 to 2π.

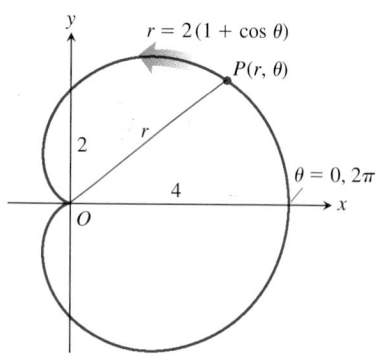

Figure 10.45 The cardioid in Example 3.

Solve Analytically

The area is therefore

$$\int_{\theta=0}^{\theta=2\pi} \frac{1}{2} r^2 \, d\theta = \int_{0}^{2\pi} \frac{1}{2} \cdot 4(1 + \cos\theta)^2 \, d\theta$$

$$= \int_{0}^{2\pi} 2(1 + 2\cos\theta + \cos^2\theta) \, d\theta$$

$$= \int_{0}^{2\pi} \left(2 + 4\cos\theta + 2\frac{1 + \cos 2\theta}{2}\right) d\theta$$

$$= \int_{0}^{2\pi} (3 + 4\cos\theta + \cos 2\theta) \, d\theta$$

$$= \left[3\theta + 4\sin\theta + \frac{\sin 2\theta}{2}\right]_{0}^{2\pi} = 6\pi - 0 = 6\pi.$$

Support Numerically

NINT $(2(1 + \cos\theta)^2, \theta, 0, 2\pi) = 18.84955592$, which agrees with 6π to eight decimal places.

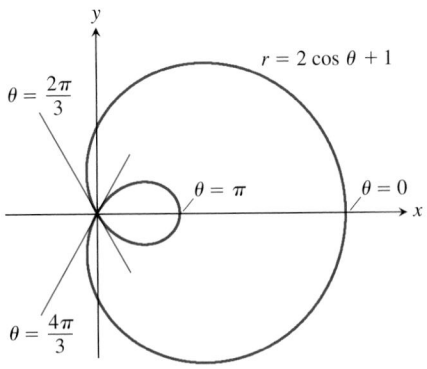

Figure 10.46 The limaçon in Example 4. ("Limaçon (pronounced LEE-ma-sahn) is an old French word for *snail*.)

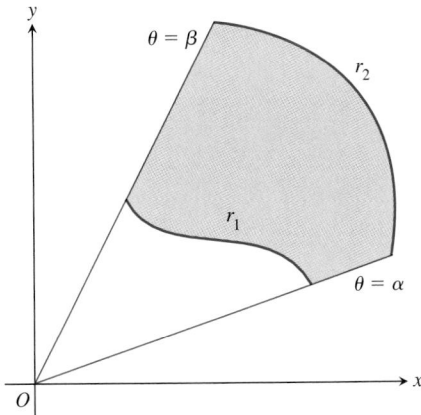

Figure 10.47 The area of the shaded region is calculated by subtracting the area of the region between r_1 and the origin from the area of the region between r_2 and the origin.

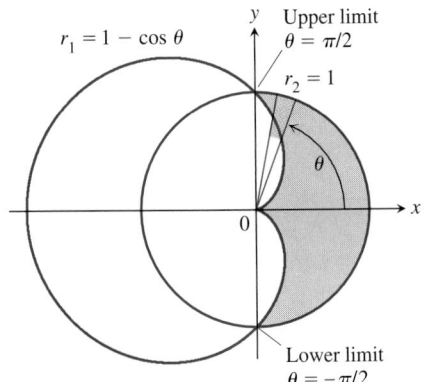

Figure 10.48 The region and limits of integration in Example 5.

Example 4 FINDING AREA

Find the area inside the smaller loop of the limaçon $r = 2 \cos \theta + 1$.

Solution After watching the grapher generate the curve (Figure 10.46), we see that the smaller loop is traced by the point (r, θ) as θ increases from $\theta = 2\pi/3$ to $\theta = 4\pi/3$. The area we seek is

$$A = \int_{2\pi/3}^{4\pi/3} \frac{1}{2} r^2 \, d\theta = \frac{1}{2} \int_{2\pi/3}^{4\pi/3} (2 \cos \theta + 1)^2 \, d\theta.$$

Solve Numerically

$$\frac{1}{2} \, \text{NINT} \, ((2 \cos \theta + 1)^2, \, \theta, \, 2\pi/3, \, 4\pi/3) \approx 0.544.$$

To find the area of a region like the one in Figure 10.47, which lies between two polar curves $r_1 = r_1(\theta)$ and $r_2 = r_2(\theta)$ from $\theta = \alpha$ to $\theta = \beta$, we subtract the integral of $(1/2)r_1^2$ from the integral of $(1/2)r_2^2$. This leads to the following formula.

> **Area Between Polar Curves**
>
> The area of the region $0 \le r_1(\theta) \le r_2(\theta)$, $\alpha \le \theta \le \beta$, is
>
> $$A = \int_\alpha^\beta \frac{1}{2} r_2^2 \, d\theta - \int_\alpha^\beta \frac{1}{2} r_1^2 \, d\theta = \int_\alpha^\beta \frac{1}{2} (r_2^2 - r_1^2) \, d\theta. \qquad (2)$$

Example 5 FINDING AREA BETWEEN CURVES

Find the area of the region that lies inside the circle $r = 1$ and outside the cardioid $r = 1 - \cos \theta$.

Solution The region is shown in Figure 10.48. The outer curve is $r_2 = 1$, the inner curve is $r_1 = 1 - \cos \theta$, and θ runs from $-\pi/2$ to $\pi/2$. The area, from Equation 2, is

$$A = \int_{-\pi/2}^{\pi/2} \frac{1}{2} (r_2^2 - r_1^2) \, d\theta$$

$$= 2 \int_0^{\pi/2} \frac{1}{2} (r_2^2 - r_1^2) \, d\theta \qquad \text{Symmetry}$$

$$= \int_0^{\pi/2} (1 - (1 - 2 \cos \theta + \cos^2 \theta)) \, d\theta$$

$$= \int_0^{\pi/2} (2 \cos \theta - \cos^2 \theta) \, d\theta \approx 1.215. \qquad \text{Using NINT}$$

In case you are interested, the exact value is $2 - \pi/4$.

Length of a Curve

We can obtain a polar coordinate formula for the length of a curve $r = f(\theta)$, $\alpha \le \theta \le \beta$, by parametrizing the curve as

$$x = r\cos\theta = f(\theta)\cos\theta, \quad y = r\sin\theta = f(\theta)\sin\theta, \quad \alpha \le \theta \le \beta. \quad (3)$$

The parametric length formula from Section 10.1 then gives the length as

$$L = \int_\alpha^\beta \sqrt{\left(\frac{dx}{d\theta}\right)^2 + \left(\frac{dy}{d\theta}\right)^2}\, d\theta.$$

This equation becomes

$$L = \int_\alpha^\beta \sqrt{r^2 + \left(\frac{dr}{d\theta}\right)^2}\, d\theta$$

when Equations 3 are substituted for x and y (Exercise 43).

> ### Length of a Polar Curve
>
> If $r = f(\theta)$ has a continuous first derivative for $\alpha \le \theta \le \beta$ and if the point $P(r, \theta)$ traces the curve $r = f(\theta)$ exactly once as θ runs from α to β, then the length of the curve is
>
> $$L = \int_\alpha^\beta \sqrt{r^2 + \left(\frac{dr}{d\theta}\right)^2}\, d\theta. \quad (4)$$

Example 6   FINDING LENGTH OF A CARDIOID

Find the length of the cardioid $r = 1 - \cos\theta$.

Solution The graph is shown in Figure 10.49. The point $P(r, \theta)$ traces the curve once counterclockwise as θ runs from 0 to 2π, so these are the values we take for α and β.

Since $r = 1 - \cos\theta$, $dr/d\theta = \sin\theta$, and we have

$$r^2 + \left(\frac{dr}{d\theta}\right)^2 = (1 - \cos\theta)^2 + (\sin\theta)^2$$

$$= 1 - 2\cos\theta + \underbrace{\cos^2\theta + \sin^2\theta}_{1} = 2 - 2\cos\theta.$$

Therefore,

$$L = \int_\alpha^\beta \sqrt{r^2 + \left(\frac{dr}{d\theta}\right)^2}\, d\theta$$

$$= \int_0^{2\pi} \sqrt{2 - 2\cos\theta}\, d\theta \approx 8. \quad \text{Using NINT}$$

This is, in fact, the exact value of the integral.

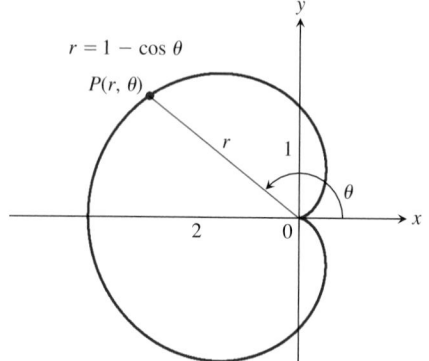

Figure 10.49 Example 6 calculates the length of this cardioid.

Area of a Surface of Revolution

To derive polar coordinate formulas for the area of a surface of revolution, we parametrize the curve $r = f(\theta)$, $\alpha \leq \theta \leq \beta$, with Equations 3 and apply the parametric surface area equations in Section 10.1.

Area of a Surface of Revolution

If $r = f(\theta)$ has a continuous first derivative for $\alpha \leq \theta \leq \beta$ and if the point $P(r, \theta)$ traces the curve $r = f(\theta)$ exactly once as θ runs from α to β, then the areas of the surfaces generated by revolving the curve about the x- and y-axes are given by the following formulas.

Revolution about the x-axis ($y \geq 0$):

$$S = \int_\alpha^\beta 2\pi r \sin\theta \sqrt{r^2 + \left(\frac{dr}{d\theta}\right)^2}\, d\theta \qquad (5)$$

Revolution about the y-axis ($x \geq 0$):

$$S = \int_\alpha^\beta 2\pi r \cos\theta \sqrt{r^2 + \left(\frac{dr}{d\theta}\right)^2}\, d\theta \qquad (6)$$

Example 7 FINDING SURFACE AREA

Find the area of the surface generated by revolving the right-hand loop of the lemniscate $r^2 = \cos 2\theta$ about the y-axis.

Solution The lemniscate is shown in Figure 10.50a. You can use a grapher to see that the equation $r_1 = \sqrt{\cos 2\theta}$ produces the entire graph. Using $r_1 = \sqrt{\cos 2\theta}$, the point $P(r, \theta)$ traces the right-hand loop of the curve once counterclockwise as θ runs from $-\pi/4$ to $\pi/4$, so these are the values we take for α and β.

We evaluate the area integrand in Equation 6 in stages. Since $r_1 = \sqrt{\cos 2\theta}$,

$$\frac{dr_1}{d\theta} = \frac{1}{2}(\cos 2\theta)^{-1/2}(-2 \sin 2\theta) = -\frac{\sin 2\theta}{\sqrt{\cos 2\theta}},$$

and

$$r_1^2 + \left(\frac{dr_1}{d\theta}\right)^2 = \cos 2\theta + \frac{\sin^2 2\theta}{\cos 2\theta}$$

$$= \frac{\cos^2 2\theta + \sin^2 2\theta}{\cos 2\theta} = \frac{1}{\cos 2\theta}.$$

Equation 6 becomes

$$S = \int_\alpha^\beta 2\pi r_1 \cos\theta \sqrt{r_1^2 + \left(\frac{dr_1}{d\theta}\right)^2}\, d\theta$$

$$= \int_{-\pi/4}^{\pi/4} 2\pi \sqrt{\cos 2\theta}\, \cos\theta \sqrt{\frac{1}{\cos 2\theta}}\, d\theta$$

$$= 2\pi \int_{-\pi/4}^{\pi/4} \cos\theta\, d\theta = 2\pi \Big[\sin\theta\Big]_{-\pi/4}^{\pi/4}$$

$$= 2\pi\sqrt{2} \approx 8.886.$$

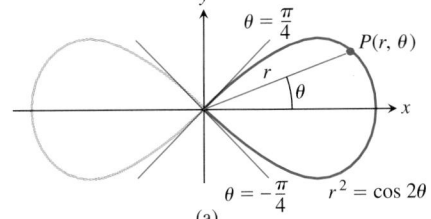

$\theta = \frac{\pi}{4}$ $P(r, \theta)$ r θ

$\theta = -\frac{\pi}{4}$ $r^2 = \cos 2\theta$

(a)

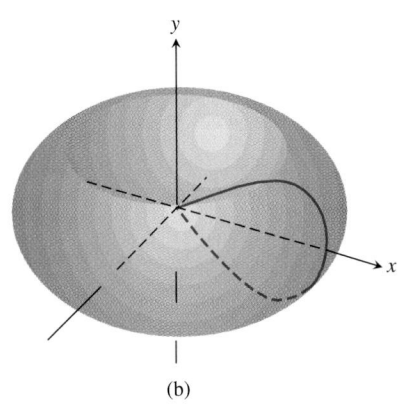

(b)

Figure 10.50 (a) The right-hand half of a lemniscate is (b) revolved about the y-axis to generate a surface whose area is calculated in Example 7.

Quick Review 10.6

Exercises 1–5 refer to the parametrized curve

$$x = 3 \cos t, \quad y = 5 \sin t, \quad 0 \le t \le 2\pi.$$

1. Find dy/dx.

2. Find the slope of the curve at $t = 2$.

3. Find the points on the curve where the slope is zero.

4. Find the points on the curve where the slope is not defined.

5. Find the length of the curve from $t = 0$ to $t = \pi$.

In Exercises 6–8, describe the portion of the graph of the polar curve $r = 1 + 2 \cos \theta$ traced in the given θ-interval.

6. $0 \le \theta \le \dfrac{2\pi}{3}$

7. $\dfrac{2\pi}{3} \le \theta \le \dfrac{4\pi}{3}$

8. $\dfrac{4\pi}{3} \le \theta \le 2\pi$

9. Find the area in the first quadrant under the curve $y = 6x - x^2$.

10. Find the area of the region enclosed by the curves $y = 2 \sin x$ and $y = x^2 - 2x + 1$.

Section 10.6 Exercises

In Exercises 1–4, find the slope of the curve at each indicated point.

1. $r = -1 + \sin \theta, \ \theta = 0, \pi$

2. $r = \cos 2\theta, \ \theta = 0, \pm\pi/2, \pi$

3. $r = 2 - 3 \sin \theta$

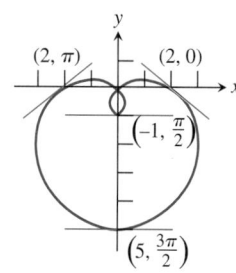

4. $r = 3(1 - \cos \theta)$

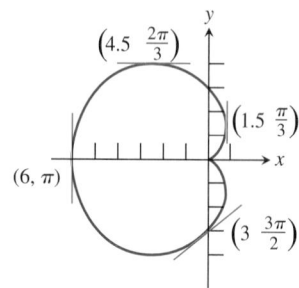

In Exercises 5–8, find the tangent lines at the pole.

5. $r = 3 \cos \theta, \quad 0 \le \theta \le 2\pi$

6. $r = 2 \cos 3\theta, \quad 0 \le \theta \le \pi$

7. $r = \sin 5\theta, \quad 0 \le \theta \le \pi$

8. $r = 2 \sin 2\theta, \quad 0 \le \theta \le 2\pi$

In Exercises 9–12, *work in groups of two or three.* Find equations for the horizontal and vertical tangent lines to the curve.

9. $r = -1 + \sin \theta, \quad 0 \le \theta \le 2\pi$

10. $r = 1 + \cos \theta, \quad 0 \le \theta \le 2\pi$

11. $r = 2 \sin \theta, \quad 0 \le \theta \le \pi$

12. $r = 3 - 4 \cos \theta, \quad 0 \le \theta \le 2\pi$

In Exercises 13–28, find the area of the region.

13. inside the oval limaçon $r = 4 + 2 \cos \theta$

14. inside the cardioid $r = a(1 + \cos \theta), \ a > 0$

15. inside the lemniscate $r^2 = 2a^2 \cos 2\theta, \ a > 0$

16. inside one leaf of the four-leaved rose $r = \cos 2\theta$

17. inside one loop of the lemniscate $r^2 = 4 \sin 2\theta$

18. inside the six-leaved rose $r^2 = 2 \sin 3\theta$

19. shared by the circles $r = 2 \cos \theta$ and $r = 2 \sin \theta$

20. shared by the circles $r = 1$ and $r = 2 \sin \theta$

21. shared by the circle $r = 2$ and the cardioid $r = 2(1 - \cos \theta)$

22. shared by the cardioids $r = 2(1 + \cos \theta)$ and $r = 2(1 - \cos \theta)$

23. inside the circle $r = 3a \cos \theta$ and outside the cardioid $r = a(1 + \cos \theta), \ a > 0$

24. inside the lemniscate $r^2 = 6 \cos 2\theta$ and outside the circle $r = \sqrt{3}$

25. inside the circle $r = 2$ and outside the cardioid $r = 2(1 - \sin \theta)$

26. (a) inside the outer loop of the limaçon $r = 2 \cos \theta + 1$ (see Figure 10.46)

(b) inside the outer loop and outside the inner loop of the limaçon $r = 2 \cos \theta + 1$

27. inside the circle $r = 6$ above the line $r = 3 \csc \theta$

28. inside the lemniscate $r^2 = 6 \cos 2\theta$ to the right of the line $r = (3/2) \sec \theta$

29. (a) Find the area of the shaded region.

(b) Writing to Learn It looks as if the graph of $r = \tan \theta$, $-\pi/2 < \theta < \pi/2$, could be asymptotic to the lines $x = 1$ and $x = -1$. Is it? Give reasons for your answer.

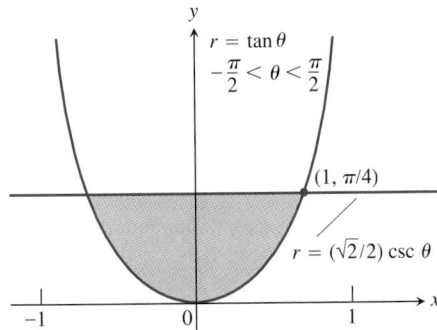

30. Writing to Learn The area of the region that lies inside the cardioid $r = \cos \theta + 1$ and outside the circle $r = \cos \theta$ is not

$$\frac{1}{2} \int_0^{2\pi} [(\cos \theta + 1)^2 - \cos^2 \theta] \, d\theta = \pi.$$

Why not? What is the area? Explain.

In Exercises 31–38, find the length of the curve.

31. the spiral $r = \theta^2$, $0 \le \theta \le \sqrt{5}$

32. the spiral $r = e^\theta/\sqrt{2}$, $0 \le \theta \le \pi$

33. the cardioid $r = 1 + \cos \theta$

34. the curve $r = a \sin^2 (\theta/2)$, $0 \le \theta \le \pi$, $a > 0$

35. the parabolic segment $r = 6/(1 + \cos \theta)$, $0 \le \theta \le \pi/2$

36. the parabolic segment $r = 2/(1 - \cos \theta)$, $\pi/2 \le \theta \le \pi$

37. the curve $r = \cos^3 (\theta/3)$, $0 \le \theta \le \pi/4$

38. the curve $r = \sqrt{1 + \sin 2\theta}$, $0 \le \theta \le \pi\sqrt{2}$

In Exercises 39–42, find the area of the surface generated by revolving the curve about the indicated axis.

39. $r = \sqrt{\cos 2\theta}$, $0 \le \theta \le \pi/4$, *y*-axis

40. $r = \sqrt{2} e^{\theta/2}$, $0 \le \theta \le \pi/2$, *x*-axis

41. $r^2 = \cos 2\theta$, *x*-axis

42. $r = 2a \cos \theta$, $a > 0$, *y*-axis

43. *Length of a Polar Curve* Assuming that the necessary derivatives are continuous, show how the substitutions

$$x = f(\theta) \cos \theta, \quad y = f(\theta) \sin \theta$$

(Equations 3 in the text) transform

$$L = \int_\alpha^\beta \sqrt{\left(\frac{dx}{d\theta}\right)^2 + \left(\frac{dy}{d\theta}\right)^2} \, d\theta$$

into

$$L = \int_\alpha^\beta \sqrt{r^2 + \left(\frac{dr}{d\theta}\right)^2} \, d\theta.$$

44. *Average Value* If f is continuous, the average value of the polar coordinate r over the curve $r = f(\theta)$, $\alpha \le \theta \le \beta$, with respect to θ is

$$r_{av} = \frac{1}{\beta - \alpha} \int_\alpha^\beta f(\theta) \, d\theta.$$

Use this formula to find the average value of r with respect to θ over the following curves ($a > 0$).

(a) the cardioid $r = a(1 - \cos \theta)$

(b) the circle $r = a$

(c) the circle $r = a \cos \theta$, $-\pi/2 \le \theta \le \pi/2$

45. Writing to Learn Can anything be said about the relative lengths of the curves

$$r = f(\theta), \quad \alpha \le \theta \le \beta,$$

and

$$r = 2f(\theta), \quad \alpha \le \theta \le \beta?$$

Give reasons for your answer.

46. Writing to Learn The curves

$$r = f(\theta), \quad \alpha \le \theta \le \beta,$$

and

$$r = 2f(\theta), \quad \alpha \le \theta \le \beta,$$

are revolved about the *x*-axis to generate surfaces. Can anything be said about the relative areas of these surfaces? Give reasons for your answer.

Explorations

47. *Videocassette Tape Length* The length of a tape wound onto a take-up reel as shown in the figure is

$$L = \int_0^\alpha \sqrt{r^2 + \left(\frac{b}{2\pi}\right)^2} \, d\theta,$$

where b is the tape thickness and

$$r = r_0 + \left(\frac{\alpha}{2\pi}\right) b$$

is the radius of the tape on the take-up reel. The initial radius of the tape on the take-up reel is r_0, and α is the angle in radians through which the wheel has turned.

(a) Simulate the tape accumulating on the take-up reel using polar graphing with $r_0 = 1.75$ cm and $b = 0.06$ cm.

(b) Confirm the formula for L analytically.

(c) Determine the length of tape on the take-up reel if the reel has turned through an angle of 80π with $r_0 = 1.75$ cm and $b = 0.06$ cm.

(d) Assume that b is very small in comparison to r at any time. Show analytically that

$$L_a = \int_0^\alpha r\,d\theta$$

is an excellent approximation to the exact value of L.

(e) For the values given in part (c), compare L_a with L.

48. *(Continuation of Exercise 47)* Let n be the number of complete turns the take-up reel has made.

(a) Find a formula for n in terms of L, the tape length.

(b) When a VCR operates, the tape moves past the heads at a constant speed. Describe the speed of the take-up reel as time progresses.

(c) Suppose that the VCR tape counter is the number n of complete turns of the take-up reel. Describe the counter values as a function of time t. ■

Extending the Ideas

Centroids of Polar Regions (You might find it helpful to review Exercises 33–35 in Section 7.1 before you attempt Exercises 49 and 50.)

When thin flat plates are manufactured from material of constant density, the location of their centers of mass depends only on how the plates are shaped. Congruent plates of aluminum and steel, for instance, have their centers of mass in the same location. The location is thus a feature of the geometry of the plate and not of the material of which the plate is made. In such cases, engineers call the center of mass the **centroid** of the shape, as in "the centroid of a parabolic plate, triangle, or fan-shaped region."

Since the centroid of a triangle is located on each median, two-thirds of the way from the vertex to the opposite base, the lever arm for the moment about the x-axis of the thin triangular region in Figure 10.51 is about $(2/3)r \sin \theta$. Similarly, the lever arm for the moment of the triangular region about the y-axis is about $(2/3)r \cos \theta$.

These approximations improve as $\Delta\theta \to 0$ and lead to the following formulas for the coordinates $(\bar{x}, \bar{y})$ of the centroid of region *AOB:*

$$\bar{x} = \frac{\displaystyle\int_\alpha^\beta \frac{2}{3}r \cos \theta \cdot \frac{1}{2}r^2\,d\theta}{\displaystyle\int_\alpha^\beta \frac{1}{2}r^2\,d\theta} = \frac{\displaystyle\frac{2}{3}\int_\alpha^\beta r^3 \cos \theta\,d\theta}{\displaystyle\int_\alpha^\beta r^2\,d\theta},$$

$$\bar{y} = \frac{\displaystyle\int_\alpha^\beta \frac{2}{3}r \sin \theta \cdot \frac{1}{2}r^2\,d\theta}{\displaystyle\int_\alpha^\beta \frac{1}{2}r^2\,d\theta} = \frac{\displaystyle\frac{2}{3}\int_\alpha^\beta r^3 \sin \theta\,d\theta}{\displaystyle\int_\alpha^\beta r^2\,d\theta},$$

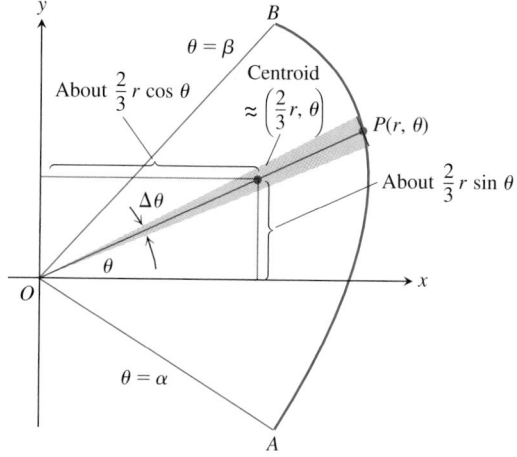

Figure 10.51 The moment of the thin triangular sector about the x-axis is approximately

$$\frac{2}{3}r \sin \theta\,dA = \frac{2}{3}r \sin \theta \cdot \frac{1}{2}r^2\,d\theta = \frac{1}{3}r^3 \sin \theta\,d\theta.$$

49. Find the centroid of the region enclosed by the cardioid $r = a(1 + \cos \theta)$.

50. Find the centroid of the semicircular region $0 \le r \le a$, $0 \le \theta \le \pi$.

Chapter 10 Key Terms

acceleration vector (p. 533)
angle between vectors (p. 523)
antiderivative of a vector function (p. 535)
Archimedes spiral (p. 530)
area between polar curves (p. 563)
area differential (p. 562)
area in polar coordinates (p. 562)
cardioid (p. 554)
centroid (p. 568)
component form of a vector (p. 521)
component functions (p. 530)
component test for continuity (p. 531)
components of a position vector (p. 530)
components of a vector (p. 521)
continuity of vector function at a point
   (p. 531)
continuous vector function (p. 531)
curve traced by vector function (p. 530)
cycloid (p. 515)
definite integral of vector function (p. 535)
derivative of parametric function at a point
   (p. 513)
derivative of vector function at a point
   (p. 532)
difference of vectors (p. 522)
differentiable parametrized curve (p. 513)
differentiable vector function (p. 532)
differentiation rules (p. 534)
directed line segment (p. 520)
direction of motion (p. 533)
distributive property of vectors (p. 528)
dot product of vectors (p. 524)
drag coefficient (p. 546)
drag force (p. 546)
equal vectors (p. 520)
equivalent directed line segments (p. 520)

graph of a vector function (p. 530)
horizontal component of a vector (p. 529)
Huygens's clock (p. 516)
ideal projectile motion (p. 539)
indefinite integral of vector function
   (p. 535)
initial point of directed line segment
   (p. 520)
initial speed of a projectile (p. 540)
inner product of vectors (p. 524)
integration properties (p. 539)
launch angle (p. 540)
length of directed line segment (p. 520)
length of polar curve (p. 564)
length of smooth parametrized curve
   (p. 514)
length of vector (p. 520)
limit of vector function (p. 531)
linear combination of vectors (p. 529)
linear drag (p. 546)
magnitude of a vector (p. 521)
maximum height of a projectile (p. 542)
negative of a vector (p. 522)
normal vector (p. 525)
opposite of a vector (p. 522)
orthogonal vectors (p. 528)
parallelogram law (p. 522)
parametric equations for ideal projectile
   motion (p. 540)
parametric formula for d^2y/dx^2 (p. 514)
particle's path (p. 530)
particle's position (p. 530)
particle's position vector (p. 530)
piecewise smooth curve (p. 532)
polar coordinates (p. 552)
polar graphing (p. 553)

pole (p. 552)
position vector (p. 521)
properties of vectors (p. 523)
projectile's range (p. 541)
projectile's flight path (p. 542)
resultant vector (p. 522)
scalar (p. 520)
scalar function (p. 530)
scalar multiple of a vector (p. 522)
slope of a polar curve (p. 560)
slope of a vector (p. 522)
smooth parametrized curve (p. 514)
smooth vector curve (p. 532)
speed of a particle (p. 533)
standard position of a vector (p. 521)
standard unit vectors (p. 529)
sum of vectors (p. 522)
surface area of revolution for a
   parametrized curve (p. 517)
surface area of revolution for a
   polar curve (p. 565)
symmetry tests for polar graphs (p. 554)
tangent line to a vector curve (p. 532)
tangent vector (p. 525)
terminal point of directed line segment
   (p. 520)
unit vector (p. 522)
vector (p. 520)
vector equation for ideal projectile motion
   (p. 540)
vector (vector-valued) function (p. 530)
velocity vector (p. 533)
vertical component of a vector (p. 529)
zero vector (p. 522)

Chapter 10 Review Exercises

In Exercises 1–4, let $\mathbf{u} = \langle -3, 4 \rangle$ and $\mathbf{v} = \langle 2, -5 \rangle$. Find
(a) the component form of the vector and **(b)** its magnitude.

1. $3\mathbf{u} - 4\mathbf{v}$

2. $\mathbf{u} + \mathbf{v}$

3. $-2\mathbf{u}$

4. $5\mathbf{v}$

In Exercises 5–8, find the component form of the vector.

5. the vector obtained by rotating $\langle 0, 1 \rangle$ through an angle
   of $2\pi/3$ radians

6. the unit vector that makes an angle of $\pi/6$ radian with the
   positive x-axis

7. the vector 2 units long in the direction $4\mathbf{i} - \mathbf{j}$

8. the vector 5 units long in the direction opposite to the
   direction of $(3/5)\mathbf{i} + (4/5)\mathbf{j}$

In Exercises 9 and 10, **(a)** find an equation for the tangent to
the curve at the point corresponding to the given value of t,
and **(b)** find the value of d^2y/dx^2 at this point.

9. $x = (1/2)\tan t, \quad y = (1/2)\sec t; \quad t = \pi/3$

10. $x = 1 + 1/t^2, \quad y = 1 - 3/t; \quad t = 2$

In Exercises 11–14, find the points at which the tangent to the curve is **(a)** horizontal; **(b)** vertical.

11. $x = (1/2)\tan t, \quad y = (1/2)\sec t$

12. $x = -2\cos t, \quad y = 2\sin t$

13. $x = -\cos t, \quad y = \cos^2 t$

14. $x = 4\cos t, \quad y = 9\sin t$

In Exercises 15 and 16, graph the set of points whose polar coordinates satisfy the inequality.

15. $0 \le r \le 6\cos\theta$

16. $-4\sin\theta \le r \le 0$

In Exercises 17–20, **(a)** graph the polar curve. **(b)** What is the smallest length θ-interval that will produce the graph?

17. $r = \cos 2\theta$

18. $r\cos\theta = 1$

19. $r^2 = \sin 2\theta$

20. $r = -\sin\theta$

In Exercises 21 and 22, *work in groups of two or three.* Find the tangent lines at the pole.

21. $r = \cos 2\theta, \quad 0 \le \theta \le 2\pi$

22. $r = 1 + \cos 2\theta, \quad 0 \le \theta \le 2\pi$

In Exercises 23 and 24, *work in groups of two or three.* Find equations for the horizontal and vertical tangent lines to the curve.

23. $r = 1 - \cos(\theta/2), \quad 0 \le \theta \le 4\pi$

24. $r = 2(1 - \sin\theta), \quad 0 \le \theta \le 2\pi$

25. Find equations for the lines that are tangent to the tips of the petals of the four-leaved rose $r = \sin 2\theta$.

26. Find equations for the lines that are tangent to the cardioid $r = 1 + \sin\theta$ at the points where it crosses the x-axis.

In Exercises 27–30, replace the polar equation by an equivalent Cartesian equation. Then identify or describe the graph.

27. $r\cos\theta = r\sin\theta$

28. $r = 3\cos\theta$

29. $r = 4\tan\theta\sec\theta$

30. $r\cos(\theta + \pi/3) = 2\sqrt{3}$

In Exercises 31–34, replace the Cartesian equation by an equivalent polar equation.

31. $x^2 + y^2 + 5y = 0$

32. $x^2 + y^2 - 2y = 0$

33. $x^2 + 4y^2 = 16$

34. $(x + 2)^2 + (y - 5)^2 = 16$

In Exercises 35–42, find the length of the curve.

35. $x = e^{2t} - t/8, \quad y = e^t, \quad 0 \le t \le \ln 2$

36. the loop of the curve $x = t^2, \; y = (t^3/3) - t$, that starts at $t = -\sqrt{3}$ and ends at $t = \sqrt{3}$ as shown in the figure

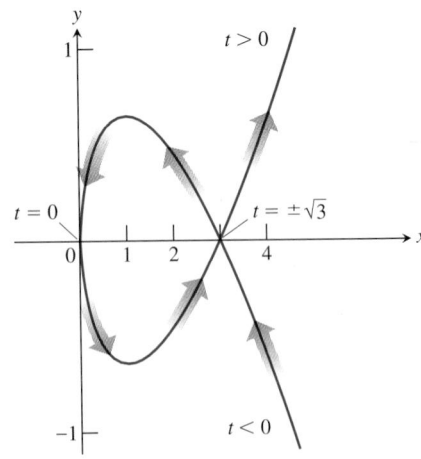

37. $r = -1 + \cos\theta, \quad 0 \le \theta \le 2\pi$

38. $r = 2\sin\theta + 2\cos\theta, \quad 0 \le \theta \le \pi/2$

39. $r = 8\sin^3(\theta/3), \quad 0 \le \theta \le \pi/4$

40. $r = \sqrt{1 + \cos 2\theta}, \quad -\pi/2 \le \theta \le \pi/2$

41. $\mathbf{r} = (2\cos t)\mathbf{i} + (t^2)\mathbf{j}, \quad 0 \le t \le \pi/2$

42. $\mathbf{r} = (3\sin t)\mathbf{i} + (2t^{3/2})\mathbf{j}, \quad 0 \le t \le 3$

In Exercises 43–46, find the area of the region.

43. enclosed by the limaçon $r = 2 - \cos\theta$

44. enclosed by one leaf of the three-leaved rose $r = \sin 3\theta$

45. inside the "figure eight" $r = 1 + \cos 2\theta$ and outside the circle $r = 1$

46. inside the cardioid $r = 2(1 + \sin\theta)$ and outside the circle $r = 2\sin\theta$

In Exercises 47–50, find the area of the surface generated by revolving the curve about the indicated axis.

47. $x = t^2/2, \quad y = 2t, \quad 0 \le t \le \sqrt{5}; \quad x$-axis

48. $x = t^2 + 1/(2t), \quad y = 4t, \quad 1/\sqrt{2} \le t \le 1; \quad y$-axis

49. $r = \sqrt{\cos 2\theta}, \quad 0 \le \theta \le \pi/4; \quad x$-axis

50. $r^2 = \sin 2\theta; \quad y$-axis

In Exercises 51 and 52, $\mathbf{r}(t)$ is the position vector of a particle in the plane at time t.

(a) Find the velocity and acceleration vectors.

(b) Find the speed at the given value of t.

(c) Find the angle between the velocity and acceleration vectors at the given value of t.

51. $\mathbf{r}(t) = (4\cos t)\mathbf{i} + (\sqrt{2}\sin t)\mathbf{j}, \quad t = \pi/4$

52. $\mathbf{r}(t) = (\sqrt{3}\sec t)\mathbf{i} + (\sqrt{3}\tan t)\mathbf{j}, \quad t = 0$

53. The position of a particle in the plane at time t is

$$\mathbf{r} = \frac{1}{\sqrt{1+t^2}}\mathbf{i} + \frac{t}{\sqrt{1+t^2}}\mathbf{j}.$$

Find the particle's greatest speed.

54. Writing to Learn Suppose that

$$\mathbf{r}(t) = (e^t\cos t)\mathbf{i} + (e^t\sin t)\mathbf{j}.$$

Show that the angle between $\mathbf{r}$ and the acceleration vector $\mathbf{a}$ never changes. What is the angle?

In Exercises 55 and 56, evaluate the integral.

55. $\displaystyle\int_0^1 [(3 + 6t)\mathbf{i} + (6\pi\cos\pi t)\mathbf{j}]\,dt$

56. $\displaystyle\int_e^{e^2} \left[\left(\frac{2\ln t}{t}\right)\mathbf{i} + \left(\frac{1}{t\ln t}\right)\mathbf{j}\right]dt$

In Exercises 57–60, solve the initial value problem.

57. $\dfrac{d\mathbf{r}}{dt} = -(\sin t)\mathbf{i} + (\cos t)\mathbf{j}, \quad \mathbf{r}(0) = \mathbf{j}$

58. $\dfrac{d\mathbf{r}}{dt} = \dfrac{1}{t^2+1}\mathbf{i} + \dfrac{t}{\sqrt{t^2+1}}\mathbf{j}, \quad \mathbf{r}(0) = \mathbf{i} + \mathbf{j}$

59. $\dfrac{d^2\mathbf{r}}{dt^2} = 2\mathbf{j}, \quad \dfrac{d\mathbf{r}}{dt}\bigg|_{t=0} = \mathbf{0}, \quad \mathbf{r}(0) = \mathbf{i}$

60. $\dfrac{d^2\mathbf{r}}{dt^2} = -2\mathbf{i} - 2\mathbf{j}, \quad \dfrac{d\mathbf{r}}{dt}\bigg|_{t=1} = 4\mathbf{i}, \quad \mathbf{r}(1) = 3\mathbf{i} + 3\mathbf{j}$

61. Particle Motion A particle moves in the plane in such a manner that its coordinates at time t are

$$x = 3\cos\frac{\pi}{4}t, \quad y = 5\sin\frac{\pi}{4}t.$$

(a) Find the length of the velocity vector at $t = 3$.

(b) Find the x- and y-components of the acceleration of the particle at $t = 3$.

(c) Find a single equation in x and y for the path of the particle.

62. Solid of Revolution Let C be the curve

$$x = \frac{t-2}{2}, \quad y = t\left(\frac{10-t}{2}\right), \quad 0 \le t \le 10.$$

Let R be the region bounded by C and the x-axis.

(a) Find the length of C.

(b) Find the volume of the solid generated by revolving R about the x-axis.

(c) Find the surface area of the solid generated by revolving R about the x-axis.

63. Particle Motion At time t, $0 \le t \le 4$, the position of a particle moving along a path in the plane is given by the parametric equations

$$x = e^t\cos t, \quad y = e^t\sin t.$$

(a) Find the slope of the path of the particle at time $t = \pi$.

(b) Find the speed of the particle when $t = 3$.

(c) Find the distance traveled by the particle along the path from $t = 0$ to $t = 3$.

64. Particle Motion The position of a particle at any time $t \ge 0$ is given by

$$x(t) = t^2 - 2, \quad y(t) = \frac{2}{5}t^3.$$

(a) Find the magnitude of the velocity vector at $t = 4$.

(b) Find the total distance traveled by the particle from $t = 0$ to $t = 4$.

(c) Find dy/dx as a function of x.

65. Navigation An airplane, flying in the direction $80°$ east of north at 540 mph in still air, encounters a 55-mph tail wind acting in the direction $100°$ east of north. The airplane holds its compass heading but, because of the wind, acquires a different ground speed and direction. What are they?

66. Combining Forces A force of 120 lb pulls up on an object at an angle of $20°$ with the horizontal. A second force of 300 lb pulls down on the object at an angle of $-5°$. Find the direction and length of the resultant force vector.

67. Shot Put A shot leaves the thrower's hand 6.5 ft above the ground at a $45°$ angle at 44 ft/sec. Where is it 3 sec later?

68. Javelin A javelin leaves the thrower's hand 7 ft above the ground at a $45°$ angle at 80 ft/sec. How high does it go?

69. Rolling Wheel *Work in groups of two or three.* A circular wheel with radius 1 ft and center C rolls to the right along the x-axis at a half-turn per second (see figure). At time t seconds, the position vector of the point P on the wheel's circumference is

$$\mathbf{r}(t) = (\pi t - \sin\pi t)\mathbf{i} + (1 - \cos\pi t)\mathbf{j}.$$

(a) Graph the curve traced by P during the interval $0 \le t \le 3$.

(b) Find velocity and acceleration vectors $\mathbf{v}$ and $\mathbf{a}$ at $t = 0$, 1, 2, and 3.

(c) Writing to Learn At any given time, what is the forward speed of the topmost point of the wheel? of C? Give reasons for your answers.

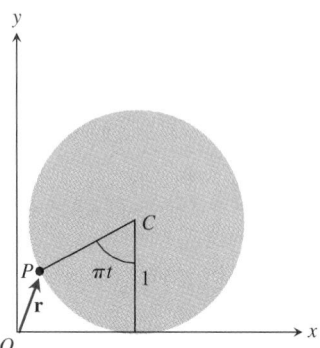

70. *The Dictator* The Civil War mortar Dictator weighed so much (17,120 lb) that it had to be mounted on a railroad car. It had a 13-in. bore and used a 20-lb powder charge to fire a 200-lb shell. The mortar was made by Mr. Charles Knapp in his ironworks in Pittsburgh, Pennsylvania, and was used by the Union army in 1864 in the siege of Petersburg, Virginia. How far did it shoot? Here we have a difference of opinion. The ordnance manual claimed 4325 yd, while field officers claimed 4752 yd. Assuming a 45° firing angle, what muzzle speeds are involved here?

71. *World's Record for Popping a Champagne Cork*

(a) Until 1988, the world's record for popping a champagne cork, 109 ft 6 in., was held by Captain Michael Hill of the British Royal Artillery (of course). Assuming Captain Hill held the bottle at ground level at a 45° angle and the cork behaved like an ideal projectile, how fast was the cork going as it left the bottle?

(b) A new world record, 177 ft 9 in., was set on June 5, 1988, by Prof. Emeritus Heinrich of Rensselaer Polytechnic Institute, firing from 4 ft above ground level at a 45° angle

at the Woodbury Vineyards Winery, New York. Assuming an ideal trajectory, what was the cork's initial speed?

72. *Javelin* In Potsdam in 1988, Petra Felke of (then) East Germany set a women's world record by throwing a javelin 262 ft 5 in.

(a) Assuming that Felke launched the javelin at a 40° angle to the horizontal from 6.5 ft above the ground, what was the javelin's initial speed?

(b) How high did the javelin go?

73. *Synchronous Curves* By eliminating α from the ideal projectile equations

$$x = (v_0 \cos \alpha)t, \quad y = (v_0 \sin \alpha)t - \frac{1}{2}gt^2,$$

show that $x^2 + (y + gt^2/2)^2 = v_0^2 t^2$. This shows that projectiles launched simultaneously from the origin at the same initial speed will, at any given instant, all lie on the circle of radius $v_0 t$ centered at $(0, -gt^2/2)$, regardless of their launch angle. These circles are the *synchronous curves* of the launching.

74. *Hitting a Baseball* A baseball is hit when it is 4 ft above the ground. It leaves the bat with an initial velocity of 155 ft/sec, making an angle of 18° with the horizontal. At the instant the ball is hit, an instantaneous 11.7 ft/sec gust of wind blows in the horizontal direction against the ball, adding a component of $-11.7\mathbf{i}$ to the ball's initial velocity. A 10-foot-high fence is 380 ft from home plate in the direction of the flight.

(a) Find vector and parametric forms for the path of the baseball.

(b) How high does the baseball go, and when does it reach maximum height?

(c) Find the range and flight time of the baseball.

(d) When is the baseball 25 ft high? How far (ground distance) is the baseball from home plate at that height?

(e) Writing to Learn Has the batter hit a home run? Explain.

75. *(Continuation of Exercise 74)* Consider the baseball problem of Exercise 74 again. This time, assume a linear drag model with a drag coefficient of 0.09.

(a) Find vector and parametric forms for the path of the baseball.

(b) How high does the baseball go, and when does it reach maximum height?

(c) Find the range and flight time of the baseball.

(d) When is the baseball 30 ft high? How far (ground distance) is the baseball from home plate at that height?

(e) Has the batter hit a home run? If "yes," find a drag coefficient that would have prevented a home run. If "no," find a drag coefficient that would have allowed the hit to be a home run.

76. *Parallelogram* The accompanying figure shows parallelogram *ABCD* and the midpoint *P* of diagonal *BD*.

    (a) Express $\overrightarrow{BD}$ in terms of $\overrightarrow{AB}$ and $\overrightarrow{AD}$.

    (b) Express $\overrightarrow{AP}$ in terms of $\overrightarrow{AB}$ and $\overrightarrow{AD}$.

(c) Prove that *P* is also the midpoint of diagonal *AC*.

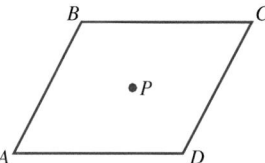

77. *Archimedes Spirals* The graph of an equation of the form $r = a\theta$, where *a* is a nonzero constant, is called an Archimedes spiral. Is there anything special about the widths between the successive turns of such a spiral?

CHAPTER

11

Vectors and Analytic Geometry in Space

Long before sled dog racing became a formal sport, Alaskan natives trained sled dogs to pull loads and help with chores. During the 1925 "Great Race of Mercy to Nome," sled dogs carried diphtheria antitoxin serum 674 miles in less than six days, saving many lives.

A child on a sled is pulled from point $(0, 0, 0)$ to point $(12, 16, 0)$ by two dogs. The forces exerted by the dogs are given by $7\mathbf{i} + 8\mathbf{j} + \mathbf{k}$ and $6\mathbf{i} + 9\mathbf{j} + 2\mathbf{k}$. If distances are in feet and forces are in pounds, find the amount of work done by the dogs. Section 11.2 covers the mathematics you need to answer this question.

Chapter 11 Overview

This chapter introduces three-dimensional vector operations and coordinate systems. We define distance, practice with the arithmetic of vectors in space, and make connections between sets of points and equations and inequalities. Just as the coordinate plane is the natural place to study functions of a single variable, coordinate space is the place to study functions of two variables (or more). We establish coordinates in space by adding a third axis that measures distance above and below the *xy*-plane.

11.1 Cartesian (Rectangular) Coordinates and Vectors in Space

Cartesian Coordinates • Vectors in Space • Magnitude • Zero and Unit Vectors • Magnitude and Direction • Distance and Spheres in Space • Midpoints of Line Segments

Cartesian Coordinates

To locate points in space, we use three mutually perpendicular coordinate axes, arranged as in Figure 11.1. The *x*-, *y*-, and *z*-axes shown there make a **right-handed coordinate frame:** When you hold your right hand so that the fingers curl from the positive *x*-axis toward the positive *y*-axis, your thumb points along the positive *z*-axis.

The **Cartesian coordinates** (x, y, z) **of a point** P in space are the numbers at which the planes through P perpendicular to the axes cut the axes.

Points on the *x*-axis have *y*- and *z*-coordinates equal to zero. That is, they have coordinates of the form $(x, 0, 0)$. Similarly, points on the *y*-axis have coordinates of the form $(0, y, 0)$. Points on the *z*-axis have coordinates of the form $(0, 0, z)$.

The points in a plane perpendicular to the *x*-axis all have the same *x*-coordinate, this being the number at which that plane cuts the *x*-axis. The *y*- and *z*-coordinates can be any numbers. Similarly, the points in a plane perpendicular to the *y*-axis have a common *y*-coordinate and the points in a plane perpendicular to the *z*-axis have a common *z*-coordinate. To write equations for these planes, we name the common coordinate's value. The plane $x = 2$ is the plane perpendicular to the *x*-axis at $x = 2$. The plane $y = 3$ is the plane perpendicular to the *y*-axis at $y = 3$. The plane $z = 5$ is the plane perpendicular to the *z*-axis at $z = 5$. Figure 11.2 shows the planes $x = 2$, $y = 3$, and $z = 5$, together with their intersection point $(2, 3, 5)$.

The planes $x = 2$ and $y = 3$ in Figure 11.2 intersect in a line parallel to the *z*-axis. This line is described by the *pair* of equations $x = 2$, $y = 3$. A point (x, y, z) lies on the line if and only if $x = 2$ and $y = 3$. Similarly, the line of intersection of the planes $y = 3$ and $z = 5$ is described by the equation pair $y = 3$, $z = 5$. This line runs parallel to the *x*-axis. The line of intersection of the planes $x = 2$ and $z = 5$, parallel to the *y*-axis, is described by the equation pair $x = 2$, $z = 5$.

The three planes determined by the coordinate axes are the ***xy*-plane,** whose standard equation is $z = 0$; the ***yz*-plane,** whose standard equation is

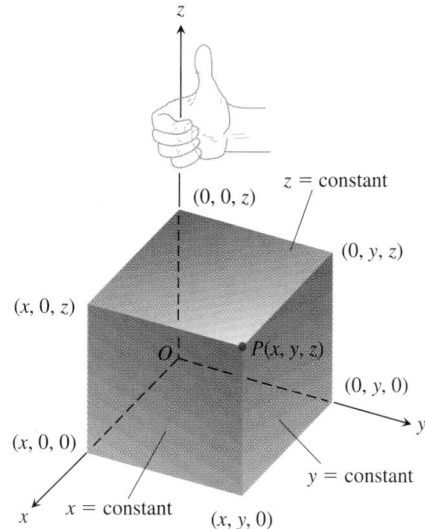

Figure 11.1 The Cartesian coordinate system is right-handed.

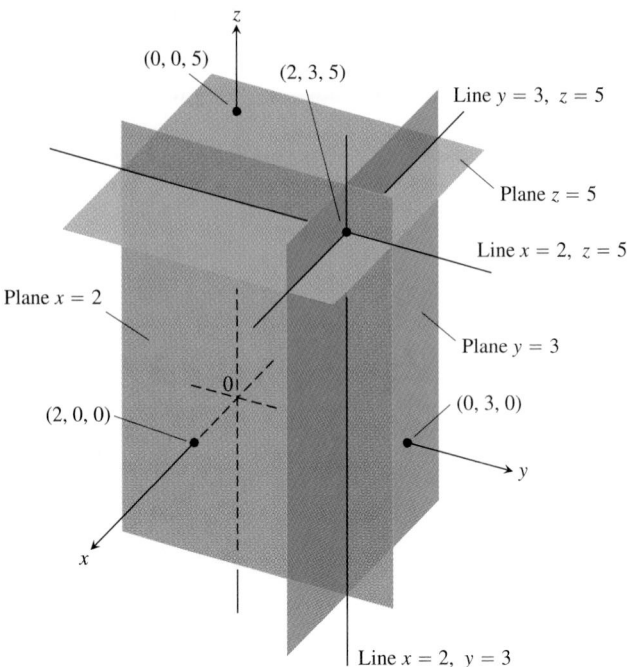

Figure 11.2 The planes $x = 2$, $y = 3$, and $z = 5$ determine three lines through the point $(2, 3, 5)$.

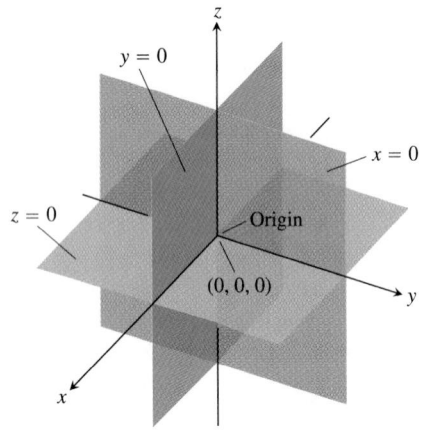

Figure 11.3 The planes $x = 0$, $y = 0$, and $z = 0$ divide space into eight octants.

$x = 0$; and the **xz-plane,** whose standard equation is $y = 0$. They meet at the **origin** $(0, 0, 0)$ (Figure 11.3).

The three **coordinate planes** $x = 0$, $y = 0$, and $z = 0$ divide space into eight cells called **octants.** The octant in which the point coordinates are all nonnegative is the **first octant;** there is no conventional numbering for the other seven octants.

Cartesian coordinates for space are also called **rectangular coordinates** because the axes that define them meet at right angles.

In the following examples, we match coordinate equations and inequalities with the sets of points they define in space.

Example 1 INTERPRETING EQUATIONS AND INEQUALITIES

(a) $z \geq 0$ — The half-space consisting of the points on and above the xy-plane.

(b) $x = -3$ — The plane perpendicular to the x-axis at $x = -3$. This plane parallels the yz-plane and lies 3 units behind it.

(c) $z = 0, x \leq 0, y \geq 0$ — The second quadrant of the xy-plane.

(d) $x \geq 0, y \geq 0, z \geq 0$ — The first octant.

(e) $-1 \leq y \leq 1$ — The slab between the planes $y = -1$ and $y = 1$, including the planes.

(f) $y = -2, z = 2$ — The line in which the planes $y = -2$ and $z = 2$ intersect. Alternatively, the line through the point $(0, -2, 2)$ parallel to the x-axis.

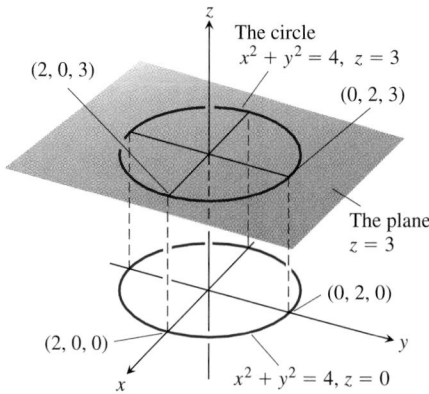

Figure 11.4 The circle $x^2 + y^2 = 4$, $z = 3$. (Example 2)

Recall that two vectors are *equal* (or the *same*) if they have the same length and direction.

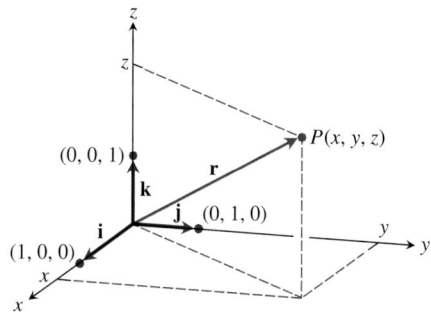

Figure 11.5 The position vector of a point in space.

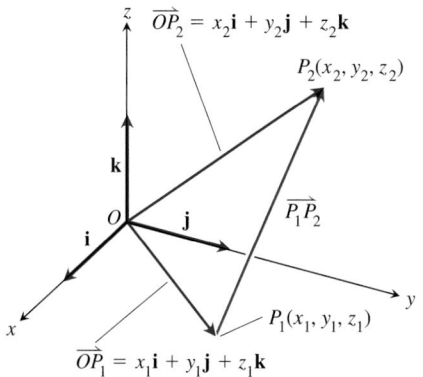

Figure 11.6 The vector from P_1 to P_2 is $\overrightarrow{P_1P_2} = (x_2 - x_1)\mathbf{i} + (y_2 - y_1)\mathbf{j} + (z_2 - z_1)\mathbf{k}$. (Example 3)

Example 2 GRAPHING EQUATIONS

What points $P(x, y, z)$ satisfy the equations

$$x^2 + y^2 = 4 \qquad \text{and} \qquad z = 3?$$

Solution The points lie in the horizontal plane $z = 3$ and, in this plane, make up the circle $x^2 + y^2 = 4$. We call this set of points "the circle $x^2 + y^2 = 4$ in the plane $z = 3$" or, more simply, "the circle $x^2 + y^2 = 4$, $z = 3$" (Figure 11.4).

Vectors in Space

The sets of equivalent directed line segments that we use to represent forces, displacements, and velocities are called vectors, just as in the plane (see Section 10.2).

The vectors represented by the directed line segments from the origin to the points $(1, 0, 0)$, $(0, 1, 0)$, and $(0, 0, 1)$ are the **standard unit vectors i, j,** and **k** (Figure 11.5). The **position vector r** from the origin O to the typical point $P(x, y, z)$ is

$$\mathbf{r} = \overrightarrow{OP} = x\mathbf{i} + y\mathbf{j} + z\mathbf{k}.$$

The definitions of addition, subtraction, and scalar multiplication are the same as in the plane.

Definitions Vector Operations

Let $\mathbf{A} = a_1\mathbf{i} + a_2\mathbf{j} + a_3\mathbf{k}$ and $\mathbf{B} = b_1\mathbf{i} + b_2\mathbf{j} + b_3\mathbf{k}$ be vectors with k a scalar (real number).

Addition: $\qquad\qquad \mathbf{A} + \mathbf{B} = (a_1 + b_1)\mathbf{i} + (a_2 + b_2)\mathbf{j} + (a_3 + b_3)\mathbf{k}$

Subtraction: $\qquad\quad \mathbf{A} - \mathbf{B} = (a_1 - b_1)\mathbf{i} + (a_2 - b_2)\mathbf{j} + (a_3 - b_3)\mathbf{k}$

Scalar multiplication: $k\mathbf{A} = (ka_1)\mathbf{i} + (ka_2)\mathbf{j} + (ka_3)\mathbf{k}$

Example 3 FINDING THE VECTOR FROM ONE POINT TO ANOTHER

Show that the vector from $P_1(x_1, y_1, z_1)$ to $P_2(x_2, y_2, z_2)$ is

$$\overrightarrow{P_1P_2} = (x_2 - x_1)\mathbf{i} + (y_2 - y_1)\mathbf{j} + (z_2 - z_1)\mathbf{k}.$$

Solution Since $\overrightarrow{OP_1} + \overrightarrow{P_1P_2} = \overrightarrow{OP_2}$ (Figure 11.6), we have

$$\begin{aligned}
\overrightarrow{P_1P_2} &= \overrightarrow{OP_2} - \overrightarrow{OP_1} \\
&= (x_2\mathbf{i} + y_2\mathbf{j} + z_2\mathbf{k}) - (x_1\mathbf{i} + y_1\mathbf{j} + z_1\mathbf{k}) \\
&= (x_2 - x_1)\mathbf{i} + (y_2 - y_1)\mathbf{j} + (z_2 - z_1)\mathbf{k}.
\end{aligned}$$

Magnitude

As always, the important features of a vector are its magnitude and direction. We find a formula for the magnitude (length) of $a_1\mathbf{i} + a_2\mathbf{j} + a_3\mathbf{k}$ by applying

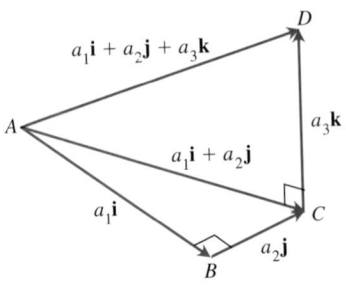

Figure 11.7 The length of $\overrightarrow{AC}$ is $\sqrt{a_1^2 + a_2^2}$ and the length of $\overrightarrow{AD}$ is $\sqrt{a_1^2 + a_2^2 + a_3^2}$.

the Pythagorean Theorem to the right triangles in Figure 11.7. From triangle *ABC*,

$$\left|\overrightarrow{AC}\right| = \sqrt{a_1^2 + a_2^2},$$

and from triangle *ACD*,

$$|a_1\mathbf{i} + a_2\mathbf{j} + a_3\mathbf{k}| = \left|\overrightarrow{AD}\right| = \sqrt{\left|\overrightarrow{AC}\right|^2 + \left|\overrightarrow{CD}\right|^2} = \sqrt{a_1^2 + a_2^2 + a_3^2}.$$

Definition Magnitude (Length) of a Vector

The **magnitude (length)** of $\mathbf{A} = a_1\mathbf{i} + a_2\mathbf{j} + a_3\mathbf{k}$ is

$$|\mathbf{A}| = |a_1\mathbf{i} + a_2\mathbf{j} + a_3\mathbf{k}| = \sqrt{a_1^2 + a_2^2 + a_3^2}.$$

Zero and Unit Vectors

The **zero vector** in space is the vector $\mathbf{0} = 0\mathbf{i} + 0\mathbf{j} + 0\mathbf{k}$. As in the plane, $\mathbf{0}$ has length zero and no direction.

A **unit vector** in space is a vector of length 1. The equations

$$|\mathbf{i}| = |1\mathbf{i} + 0\mathbf{j} + 0\mathbf{k}| = \sqrt{1^2 + 0^2 + 0^2} = 1,$$

$$|\mathbf{j}| = |0\mathbf{i} + 1\mathbf{j} + 0\mathbf{k}| = \sqrt{0^2 + 1^2 + 0^2} = 1,$$

$$|\mathbf{k}| = |0\mathbf{i} + 0\mathbf{j} + 1\mathbf{k}| = \sqrt{0^2 + 0^2 + 1^2} = 1,$$

show that the standard unit vectors are indeed unit vectors.

Magnitude and Direction

If $\mathbf{A} \neq \mathbf{0}$, then $\mathbf{A}/|\mathbf{A}|$ is a unit vector in the direction of $\mathbf{A}$. We can use the equation

$$\mathbf{A} = |\mathbf{A}|\frac{\mathbf{A}}{|\mathbf{A}|}$$

to express $\mathbf{A}$ as a product of its magnitude and direction.

Example 4 WRITING A VECTOR AS LENGTH TIMES DIRECTION

Express $\mathbf{A} = \mathbf{i} - 2\mathbf{j} + 3\mathbf{k}$ as a product of its magnitude and direction.

Solution $\mathbf{A}/|\mathbf{A}|$ is a unit vector in the direction of $\mathbf{A}$. The magnitude of $\mathbf{A}$ is

$$|\mathbf{A}| = \sqrt{(1)^2 + (-2)^2 + (3)^2} = \sqrt{1 + 4 + 9} = \sqrt{14}.$$

Thus,

$$\mathbf{A} = |\mathbf{A}| \cdot \frac{\mathbf{A}}{|\mathbf{A}|}$$

$$= \sqrt{14} \cdot \frac{\mathbf{i} - 2\mathbf{j} + 3\mathbf{k}}{\sqrt{14}}$$

$$= \sqrt{14} \cdot \left(\frac{1}{\sqrt{14}}\mathbf{i} - \frac{2}{\sqrt{14}}\mathbf{j} + \frac{3}{\sqrt{14}}\mathbf{k}\right)$$

$$= (\text{length of } \mathbf{A}) \cdot (\text{direction of } \mathbf{A}).$$

Example 5 FINDING A UNIT VECTOR

Find a unit vector $\mathbf{u}$ in the direction of the vector from $P_1(1, 0, 1)$ to $P_2(3, 2, 0)$.

Solution We divide $\overrightarrow{P_1P_2}$ by its length:

$$\overrightarrow{P_1P_2} = (3 - 1)\mathbf{i} + (2 - 0)\mathbf{j} + (0 - 1)\mathbf{k} = 2\mathbf{i} + 2\mathbf{j} - \mathbf{k}$$

$$\left|\overrightarrow{P_1P_2}\right| = \sqrt{(2)^2 + (2)^2 + (-1)^2} = \sqrt{4 + 4 + 1} = \sqrt{9} = 3$$

$$\mathbf{u} = \frac{\overrightarrow{P_1P_2}}{\left|\overrightarrow{P_1P_2}\right|} := \frac{2\mathbf{i} + 2\mathbf{j} - \mathbf{k}}{3} = \frac{2}{3}\mathbf{i} + \frac{2}{3}\mathbf{j} - \frac{1}{3}\mathbf{k}.$$

Example 6 FINDING A VECTOR WITH GIVEN MAGNITUDE AND DIRECTION

Find a vector 6 units long in the direction of $\mathbf{A} = 2\mathbf{i} + 2\mathbf{j} - \mathbf{k}$.

Solution The vector we want is

$$6\frac{\mathbf{A}}{|\mathbf{A}|} = 6\frac{2\mathbf{i} + 2\mathbf{j} - \mathbf{k}}{\sqrt{2^2 + 2^2 + (-1)^2}} = 6\frac{2\mathbf{i} + 2\mathbf{j} - \mathbf{k}}{3} = 4\mathbf{i} + 4\mathbf{j} - 2\mathbf{k}.$$

Distance and Spheres in Space

The distance between two points P_1 and P_2 in space is the length of $\overrightarrow{P_1P_2}$.

Distance Between Points

The distance between $P_1(x_1, y_1, z_1)$ and $P_2(x_2, y_2, z_2)$ is

$$\left|\overrightarrow{P_1P_2}\right| = \sqrt{(x_2 - x_1)^2 + (y_2 - y_1)^2 + (z_2 - z_1)^2}.$$

Figure 11.8 The sphere of radius a centered at (x_0, y_0, z_0).

Example 7 FINDING DISTANCE BETWEEN POINTS

The distance between $P_1(2, 1, 5)$ and $P_2(-2, 3, 0)$ is

$$\left|\overrightarrow{P_1P_2}\right| = \sqrt{(-2 - 2)^2 + (3 - 1)^2 + (0 - 5)^2}$$
$$= \sqrt{16 + 4 + 25} = \sqrt{45} \approx 6.708.$$

We use the distance formula to write equations for spheres in space (Figure 11.8). A *sphere* is the set of all points $P(x, y, z)$ in space a fixed distance from a fixed point $P_0(x_0, y_0, z_0)$. A point $P(x, y, z)$ lies on the sphere of radius a centered at $P_0(x_0, y_0, z_0)$ precisely when $\left|\overrightarrow{P_0P}\right| = a$ or

$$(x - x_0)^2 + (y - y_0)^2 + (z - z_0)^2 = a^2.$$

Standard Equation for the Sphere of Radius a and Center (x_0, y_0, z_0)

$$(x - x_0)^2 + (y - y_0)^2 + (z - z_0)^2 = a^2$$

Drawing Lesson

How to Draw Three-Dimensional Objects to Look Three-Dimensional

1. *Break lines.* When one line passes behind another, break it to show that it doesn't touch and that part of it is hidden.

Intersecting

CD behind *AB*

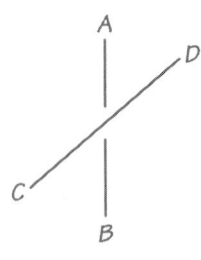

AB behind *CD*

2. Make the angle between the positive *x*-axis and the positive *y*-axis large enough.

This

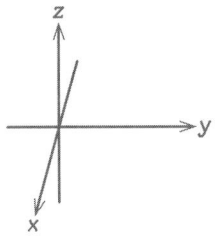

Not this

3. Draw planes parallel to the coordinate planes as if they were rectangles with sides parallel to the coordinate axes.

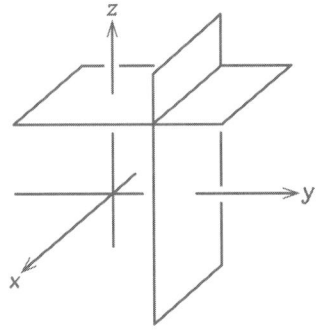

4. Dash or omit hidden portions of lines. Don't let the line touch the boundary of the parallelogram that represents the plane, unless the line lies in the plane.

Line below plane

Line above plane

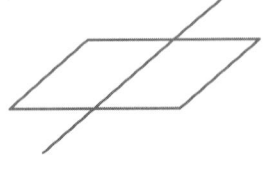

Line *in* plane

5. *Spheres.* Draw the sphere first (outline and equator); draw axes, if any, later. Use line breaks and dashed lines.

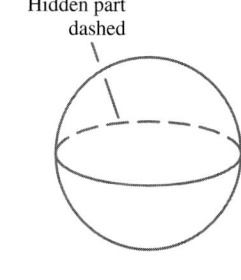

Hidden part dashed

Sphere first

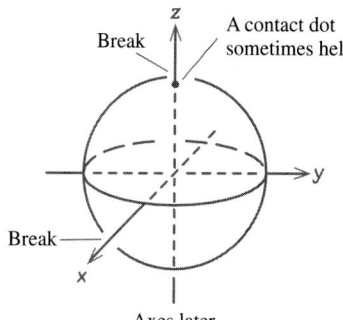

Break

A contact dot sometimes helps

Break

Axes later

6. A general rule for perspective: Draw the object as if it lies some distance away, below, and to the left.

Advice ignored

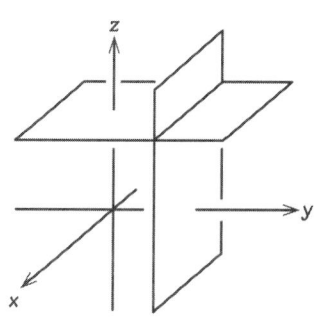

Advice followed

7. To draw a plane that crosses all three coordinate axes, follow the steps shown here. (a) Sketch the axes and mark the intercepts. (b) Connect the intercepts to form two sides of a parallelogram. (c) Complete the parallelogram and enlarge it by drawing lines parallel to its sides. (d) Darken the exposed parts, break hidden lines, and, if desired, dash hidden portions of the axes. You may wish to erase the smaller parallelogram at this point.

(a)

(b)

(c)

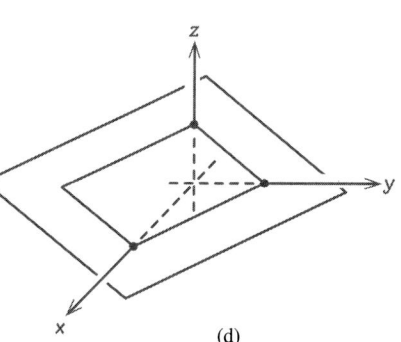

(d)

Example 8 FINDING THE CENTER AND RADIUS OF A SPHERE

Find the center and radius of the sphere

$$x^2 + y^2 + z^2 + 3x - 4z + 1 = 0.$$

Solution We find the center and radius of a sphere the way we find the center and radius of a circle: Complete the squares on the x-, y-, and z-terms as necessary and write each quadratic as a squared liner expression. Then, from the equation in standard form, read off the center and radius. For the sphere here we have

$$x^2 + y^2 + z^2 + 3x - 4z + 1 = 0$$

$$(x^2 + 3x) + y^2 + (z^2 - 4z) = -1$$

$$\left(x^2 + 3x + \left(\frac{3}{2}\right)^2\right) + y^2 + \left(z^2 - 4z + \left(\frac{-4}{2}\right)^2\right) = -1 + \left(\frac{3}{2}\right)^2 + \left(\frac{-4}{2}\right)^2$$

$$\left(x + \frac{3}{2}\right)^2 + y^2 + (z - 2)^2 = -1 + \frac{9}{4} + 4 = \frac{21}{4}.$$

From this standard form we read that $x_0 = -3/2$, $y_0 = 0$, $z_0 = 2$, and $a = \sqrt{21}/2$. The center is $(-3/2, 0, 2)$. The radius is $\sqrt{21}/2$.

In Example 9 we describe sets that are bounded by spheres or portions of spheres.

Example 9 INTERPRETING EQUATIONS AND INEQUALITIES

(a) $x^2 + y^2 + z^2 < 4$ The interior of the sphere $x^2 + y^2 + z^2 = 4$.

(b) $x^2 + y^2 + z^2 \leq 4$ The solid ball bounded by the sphere $x^2 + y^2 + z^2 = 4$. Alternatively, the sphere $x^2 + y^2 + z^2 = 4$ together with its interior.

(c) $x^2 + y^2 + z^2 > 4$ The exterior of the sphere $x^2 + y^2 + z^2 = 4$.

(d) $x^2 + y^2 + z^2 = 4, z \leq 0$ The lower hemisphere cut from the sphere $x^2 + y^2 + z^2 = 4$ by the xy-plane (the plane $z = 0$).

Midpoints of Line Segments

Just as in the plane, the coordinates of the midpoints of a line segment are found by averaging the coordinates of the endpoints of the segment.

Midpoint

The midpoint M of the line segment joining points $P_1(x_1, y_1, z_1)$ and $P_2(x_2, y_2, z_2)$ is the point

$$\left(\frac{x_1 + x_2}{2}, \frac{y_1 + y_2}{2}, \frac{z_1 + z_2}{2}\right).$$

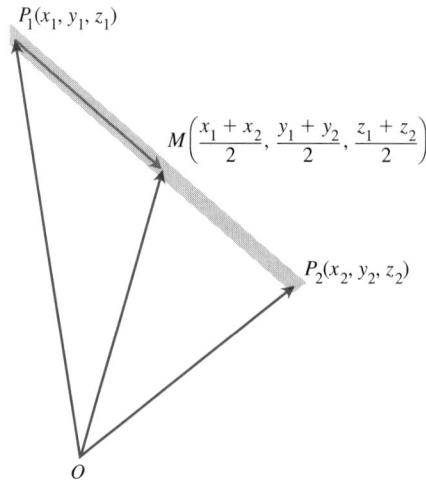

Figure 11.9 The coordinates of the midpoint are the averages of the coordinates of P_1 and P_2.

To see why, observe (Figure 11.9) that

$$\overrightarrow{OM} = \overrightarrow{OP_1} + \frac{1}{2}\left(\overrightarrow{P_1P_2}\right) = \overrightarrow{OP_1} + \frac{1}{2}\left(\overrightarrow{OP_2} - \overrightarrow{OP_1}\right)$$

$$= \frac{1}{2}\left(\overrightarrow{OP_1} + \overrightarrow{OP_2}\right)$$

$$= \frac{x_1 + x_2}{2}\mathbf{i} + \frac{y_1 + y_2}{2}\mathbf{j} + \frac{z_1 + z_2}{2}\mathbf{k}.$$

Example 10 FINDING A MIDPOINT

The midpoint of the line segment joining $P_1(3, -2, 0)$ and $P_2(7, 4, 4)$ is

$$\left(\frac{3 + 7}{2}, \frac{-2 + 4}{2}, \frac{0 + 4}{2}\right) = (5, 1, 2).$$

Quick Review 11.1

In Exercises 1 and 2, let $P_1(-2, 3)$ and $P_2(5, 2)$ be two points in the xy-plane.

1. Find the distance between the points P_1 and P_2.

2. Find the midpoint of the line segment P_1P_2.

In Exercises 3–5, let $\mathbf{u} = \langle 2, 3 \rangle = 2\mathbf{i} + 3\mathbf{j}$ be a vector in the xy-plane.

3. Find the magnitude of $\mathbf{u}$.

4. Find a unit vector in the direction of $\mathbf{u}$.

5. Find a vector 3 units long in the direction of $-\mathbf{u}$.

In Exercises 6 and 7, give a geometric description of the set of points in the xy-plane whose coordinates satisfy the equation or inequality.

6. $\dfrac{x^2}{4} + \dfrac{y^2}{9} = 1$ **7.** $\dfrac{x^2}{4} + \dfrac{y^2}{9} \leq 1$

8. Write an equation for the circle in the xy-plane with center $(2, -3)$ and radius 3 units.

9. Find the center and radius of the circle in the xy-plane: $x^2 + y^2 + 6x - 2y = -6$.

10. Find the vector in the xy-plane from the point $P_1(-2, 3)$ to the point $P_2(3, -1)$.

Section 11.1 Exercises

In Exercises 1–10, give a geometric description of the set of points in space whose coordinates satisfy the pair of equations.

1. $x = 2, y = 3$ **2.** $x = -1, z = 0$

3. $y = 0, z = 0$ **4.** $x = 1, y = 0$

5. $x^2 + y^2 = 4, z = -2$ **6.** $x^2 + z^2 = 4, y = 0$

7. $y^2 + z^2 = 1, x = 0$

8. $x^2 + y^2 + z^2 = 25, y = -4$.

9. $x^2 + y^2 + (z + 3)^2 = 25, z = 0$

10. $x^2 + (y - 1)^2 + z^2 = 4, y = 0$

In Exercises 11–16, describe the set of points in space whose coordinates satisfy the given inequalities or combination of equations and inequalities.

11. (a) $x \geq 0, y \geq 0, z = 0$

  (b) $x \geq 0, y \leq 0, z = 0$

12. (a) $0 \leq x \leq 1$

  (b) $0 \leq x \leq 1, 0 \leq y \leq 1$

  (c) $0 \leq x \leq 1, 0 \leq y \leq 1, 0 \leq z \leq 1$

13. (a) $x^2 + y^2 + z^2 \leq 1$

  (b) $x^2 + y^2 + z^2 > 1$

14. (a) $x^2 + y^2 \leq 1, z = 0$

  (b) $x^2 + y^2 \leq 1, z = 3$

  (c) $x^2 + y^2 \leq 1$, no restriction on z

15. (a) $x^2 + y^2 + z^2 = 1$, $z \geq 0$

   (b) $x^2 + y^2 + z^2 \leq 1$, $z \geq 0$

16. (a) $x = y$, $z = 0$

   (b) $x = y$, no restriction on z

In Exercises 17–26, describe the given set with a single equation or a pair of equations.

17. The plane perpendicular to the

   (a) x-axis at $(3, 0, 0)$ **(b)** y-axis at $(0, -1, 0)$

   (c) z-axis at $(0, 0, -2)$

18. The plane through the point $(3, -1, 2)$ perpendicular to the

   (a) x-axis   **(b)** y-axis   **(c)** z-axis

19. The plane through the point $(3, -1, 1)$ parallel to the

   (a) xy-plane   **(b)** yz-plane   **(c)** xz-plane

20. The circle of radius 2 centered at $(0, 0, 0)$ and lying in the

   (a) xy-plane   **(b)** yz-plane   **(c)** xz-plane

21. The circle of radius 2 centered at $(0, 2, 0)$ and lying in the

   (a) xy-plane   **(b)** yz-plane   **(c)** plane $y = 2$

22. The circle of radius 1 centered at $(-3, 4, 1)$ and lying in a plane parallel to the

   (a) xy-plane   **(b)** yz-plane   **(c)** xz-plane

23. The line through the point $(1, 3, -1)$ parallel to the

   (a) x-axis   **(b)** y-axis   **(c)** z-axis

24. The set of points in space equidistant from the origin and the point $(0, 2, 0)$

25. The circle in which the plane through the point $(1, 1, 3)$ perpendicular to the z-axis meets the sphere of radius 5 centered at the origin

26. The set of points in space that lie 2 units from the point $(0, 0, 1)$ and, at the same time, 2 units from the point $(0, 0, -1)$

In Exercises 27–32, write inequalities to describe the set.

27. The slab bounded by the planes $z = 0$ and $z = 1$ (planes included)

28. The solid cube in the first octant bounded by the coordinate planes and the planes $x = 2$, $y = 2$, and $z = 2$

29. The half-space consisting of the points on and below the xy-plane

30. The upper hemisphere of the sphere of radius 1 centered at the origin

31. The **(a)** interior and **(b)** exterior of the sphere of radius 1 centered at the point $(1, 1, 1)$

32. The closed region bounded by the spheres of radius 1 and radius 2 centered at the origin. (*Closed* means the spheres are to be included. Had we wanted the spheres left out, we would have asked for the *open* region bounded by the

spheres. This is analogous to the way we use *closed* and *open* to described intervals: *Closed* means endpoints included, *open* means endpoints left out. Closed sets include boundaries; open sets leave them out.)

In Exercises 33–38, express the vector as a product of its length and direction.

33. $2\mathbf{i} + \mathbf{j} - 2\mathbf{k}$

34. $9\mathbf{i} - 2\mathbf{j} + 6\mathbf{k}$

35. $5\mathbf{k}$

36. $\dfrac{3}{5}\mathbf{i} + \dfrac{4}{5}\mathbf{k}$

37. $\dfrac{1}{\sqrt{6}}\mathbf{i} - \dfrac{1}{\sqrt{6}}\mathbf{j} - \dfrac{1}{\sqrt{6}}\mathbf{k}$

38. $\dfrac{\mathbf{i}}{\sqrt{3}} + \dfrac{\mathbf{j}}{\sqrt{3}} + \dfrac{\mathbf{k}}{\sqrt{3}}$

In Exercises 39 and 40, find the vector whose length and direction are given. Try to do the calculation mentally.

39. Length / **Direction**

   (a) 2 / $\mathbf{i}$

   (b) $\sqrt{3}$ / $-\mathbf{k}$

   (c) $\dfrac{1}{2}$ / $\dfrac{3}{5}\mathbf{j} + \dfrac{4}{5}\mathbf{k}$

   (d) 7 / $\dfrac{6}{7}\mathbf{i} - \dfrac{2}{7}\mathbf{j} + \dfrac{3}{7}\mathbf{k}$

40. Length / **Direction**

   (a) 7 / $-\mathbf{j}$

   (b) $\sqrt{2}$ / $-\dfrac{3}{5}\mathbf{i} - \dfrac{4}{5}\mathbf{k}$

   (c) $\dfrac{13}{12}$ / $\dfrac{3}{13}\mathbf{i} - \dfrac{4}{13}\mathbf{j} - \dfrac{12}{13}\mathbf{k}$

   (d) $a > 0$ / $\dfrac{1}{\sqrt{2}}\mathbf{i} + \dfrac{1}{\sqrt{3}}\mathbf{j} - \dfrac{1}{\sqrt{6}}\mathbf{k}$

41. Find a vector of magnitude 7 in the direction of $\mathbf{A} = 12\mathbf{i} - 5\mathbf{k}$.

42. Find a vector of magnitude 3 in the direction opposite to the direction of $\mathbf{A} = (1/2)\mathbf{i} - (1/2)\mathbf{j} - (1/2)\mathbf{k}$.

In Exercises 43–46, find

   (a) the distance between the points P_1 and P_2,

   (b) the direction of $\overrightarrow{P_1 P_2}$,

   (c) the midpoint of line segment $P_1 P_2$.

43. $P_1(-1, 1, 5)$, $P_2(2, 5, 0)$

44. $P_1(1, 4, 5)$, $P_2(4, -2, 7)$

45. $P_1(3, 4, 5)$, $P_2(2, 3, 4)$

46. $P_1(0, 0, 0)$, $P_2(2, -2, -2)$

In Exercises 47 and 48, find the center and radius of the sphere.

47. $(x + 2)^2 + y^2 + (z - 2)^2 = 8$

48. $x^2 + \left(y + \dfrac{1}{3}\right)^2 + \left(z - \dfrac{1}{3}\right)^2 = \dfrac{29}{9}$

In Exercises 49 and 50, find an equation for the sphere with given center and radius.

49. Center: (1, 2, 3), radius: $\sqrt{14}$

50. Center: (0, −1, 5), radius: 2

In Exercises 51–54, find the center and radius of the sphere.

51. $x^2 + y^2 + z^2 + 4x - 4z = 0$

52. $x^2 + y^2 + z^2 - 6y + 8z = 0$

53. $2x^2 + 2y^2 + 2z^2 + x + y + z = 9$

54. $3x^2 + 3y^2 + 3z^2 + 2y - 2z = 9$

In Exercises 55–58, *work in groups of two or three.*

55. If $\overrightarrow{AB} = \mathbf{i} + 4\mathbf{j} - 2\mathbf{k}$ and B is the point (5, 1, 3), find A.

56. If $\overrightarrow{AB} = -7\mathbf{i} + 3\mathbf{j} + 8\mathbf{k}$ and A is the point $(-2, -3, 6)$, find B.

57. Find a formula for the distance from the point $P(x, y, z)$ to the

    (a) *x*-axis   **(b)** *y*-axis   **(c)** *z*-axis

58. Find a formula for the distance from the point $P(x, y, z)$ to the

    (a) *xy*-plane   **(b)** *yz*-plane   **(c)** *xz*-plane

59. Writing to Learn Let $ABCD$ be a general, not necessarily planar, quadrilateral in space. Show that the two segments joining the midpoints of opposite sides of $ABCD$ bisect each other.

60. Writing to Learn Suppose A, B, and C are vertices of a triangle and a, b, and c are, respectively, the midpoints of the opposite sides.

Show that $\overrightarrow{Aa} + \overrightarrow{Bb} + \overrightarrow{Cc} = 0$.

61. Using only vectors, show that $P_1 P_2 P_3 P_4$ form a parallelogram where $P_1 = (0, 0, 0)$, $P_2 = (1, 2, 3)$, $P_3 = (3, 5, 7)$, and $P_4 = (2, 3, 4)$.

Explorations

62. *Cross Sections of Sphere*

    (a) Find the cross sections of the sphere

$$x^2 + y^2 + z^2 = 16$$

formed by its intersection with the planes $z = 0$, $z = -2$, and $z = 3$.

    (b) Graph the three curves in part (a) in a two-dimensional coordinate system.

    (c) Show how to set up your grapher to graph the cross section of the sphere for any value of z.

    (d) For each value of z, what is the area A_z of the cross section?

    (e) How is $\int_{-4}^{4} A_z \, dz$ related to the sphere?

63. *Cross Sections of Sphere*

    (a) Find the cross sections of the sphere

$$x^2 + y^2 + z^2 = 9$$

formed by its intersection with the planes $x = -1$, $x = 0$, and $x = 2$.

    (b) Graph the three curves in part (a) in a two-dimensional coordinate system.

    (c) Show how to set up your grapher to graph the cross section of the sphere for any value of x.

    (d) For each value of x, what is the area A_x of the cross section?

    (e) How is $\int_{-3}^{3} A_x \, dx$ related to the sphere? ■

Extending The Ideas

64. *Medians of a Triangle* Suppose A, B, and C are the corners of the thin triangular plate of constant density shown here.

    (a) Find the vector from C to the midpoint M of side AB.

    (b) Find the vector from C to the point that lies two-thirds of the way from C to M on the median CM.

    (c) Find the coordinates of the point in which the medians of $\triangle ABC$ intersect. (This point is the plate's center of mass (c.m.).)

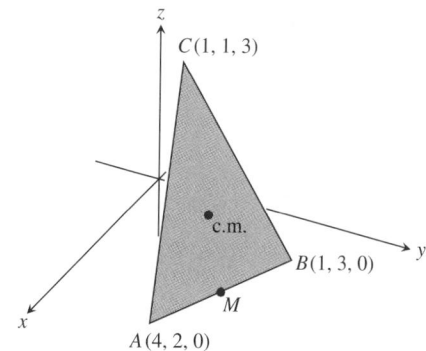

65. *Medians of a Triangle* Find the vector from the origin to the point of intersection of the medians of the triangle whose vertices are $A(1, -1, 2)$, $B(2, 1, 3)$, and $C(-1, 2, -1)$.

66. *Regular Polygons* Vectors are drawn from the center of a regular *n*-sided polygon in the plane to the vertices of the polygon. Show that the sum of the vectors is zero. (*Hint:* What happens to the sum if you rotate the polygon about its center?)

11.2   Dot Products

Component Form • Properties of the Dot Product • Perpendicular (Orthogonal) Vectors and Projections • Writing a Vector as a Sum of Orthogonal Vectors • Work

Component Form

The dot product (or inner product) of two vectors in space is defined in the same way as for vectors in the plane (see Section 10.2). When two nonzero vectors $\mathbf{A}$ and $\mathbf{B}$ are placed so their initial points coincide, they form an angle θ of measure $0 \leq \theta \leq \pi$.

Definition   Dot Product (Inner Product)

The **dot product** (or **inner product**) $\mathbf{A} \cdot \mathbf{B}$ ("$\mathbf{A}$ dot $\mathbf{B}$") of vectors $\mathbf{A}$ and $\mathbf{B}$ is the number

$$\mathbf{A} \cdot \mathbf{B} = |\mathbf{A}||\mathbf{B}| \cos \theta,$$

where θ is the angle between $\mathbf{A}$ and $\mathbf{B}$.

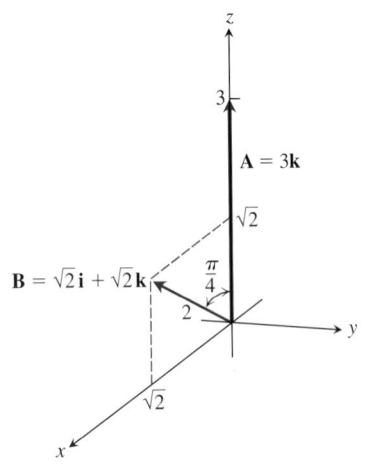

Figure 11.10 The vectors in Example 1.

Example 1   COMPUTING A DOT PRODUCT

If $\mathbf{A} = 3\mathbf{k}$ and $\mathbf{B} = \sqrt{2}\mathbf{i} + \sqrt{2}\mathbf{k}$ (Figure 11.10), then

$$\mathbf{A} \cdot \mathbf{B} = |\mathbf{A}||\mathbf{B}| \cos \theta = (3)(2) \cos \frac{\pi}{4} = 6 \cdot \frac{\sqrt{2}}{2} = 3\sqrt{2}.$$

In Section 10.2, we also proved (see Theorem 1) that the dot product can be expressed using the components of the vectors. The same proof yields the following formula.

If $\mathbf{A} = a_1\mathbf{i} + a_2\mathbf{j} + a_3\mathbf{k}$ and $\mathbf{B} = b_1\mathbf{i} + b_2\mathbf{j} + b_3\mathbf{k}$, then

$$\mathbf{A} \cdot \mathbf{B} = a_1b_1 + a_2b_2 + a_3b_3.$$

Solving for θ in the definition of dot product gives a formula for finding angles between vectors.

The angle between two nonzero vectors $\mathbf{A}$ and $\mathbf{B}$ is

$$\theta = \cos^{-1}\left(\frac{\mathbf{A} \cdot \mathbf{B}}{|\mathbf{A}||\mathbf{B}|}\right).$$

Example 2 FINDING THE ANGLE BETWEEN VECTORS

Find the angle between $\mathbf{A} = \mathbf{i} - 2\mathbf{j} - 2\mathbf{k}$ and $\mathbf{B} = 6\mathbf{i} + 3\mathbf{j} + 2\mathbf{k}$.

Solution Using the formula above, we have

$$\mathbf{A} \cdot \mathbf{B} = (1)(6) + (-2)(3) + (-2)(2) = -4$$

$$|\mathbf{A}| = \sqrt{(1)^2 + (-2)^2 + (-2)^2} = \sqrt{9} = 3$$

$$|\mathbf{B}| = \sqrt{(6)^2 + (3)^2 + (2)^2} = \sqrt{49} = 7$$

$$\theta = \cos^{-1}\left(\frac{\mathbf{A} \cdot \mathbf{B}}{|\mathbf{A}||\mathbf{B}|}\right)$$

$$= \cos^{-1}\left(\frac{-4}{(3)(7)}\right) \approx 1.762 \text{ radians}$$

Properties of the Dot Product

We can use the component form of the dot product to establish the following properties (see Exercise 45).

Properties of the Dot Product

Let $\mathbf{A}$, $\mathbf{B}$, $\mathbf{C}$, and $\mathbf{D}$ be vectors and c any number (scalar).

1. $\mathbf{A} \cdot \mathbf{B} = \mathbf{B} \cdot \mathbf{A}$
2. $(c\mathbf{A}) \cdot \mathbf{B} = \mathbf{A} \cdot (c\mathbf{B}) = c(\mathbf{A} \cdot \mathbf{B})$
3. $\mathbf{A} \cdot (\mathbf{B} + \mathbf{C}) = \mathbf{A} \cdot \mathbf{B} + \mathbf{A} \cdot \mathbf{C}$
4. $(\mathbf{A} + \mathbf{B}) \cdot \mathbf{C} = \mathbf{A} \cdot \mathbf{C} + \mathbf{B} \cdot \mathbf{C}$
5. $(\mathbf{A} + \mathbf{B}) \cdot (\mathbf{C} + \mathbf{D}) = \mathbf{A} \cdot \mathbf{C} + \mathbf{A} \cdot \mathbf{D} + \mathbf{B} \cdot \mathbf{C} + \mathbf{B} \cdot \mathbf{D}$
6. $\mathbf{A} \cdot \mathbf{A} = |\mathbf{A}|^2$

Perpendicular (Orthogonal) Vectors and Projections

Two nonzero vectors $\mathbf{A}$ and $\mathbf{B}$ are **perpendicular** or **orthogonal** if the angle between them is $\pi/2$. For such vectors, we automatically have $\mathbf{A} \cdot \mathbf{B} = 0$ because $\cos(\pi/2) = 0$. The converse is also true. If $\mathbf{A}$ and $\mathbf{B}$ are nonzero vectors with $\mathbf{A} \cdot \mathbf{B} = |\mathbf{A}||\mathbf{B}| \cos \theta = 0$, then $\cos \theta = 0$ and $\theta = \cos^{-1} 0 = \pi/2$.

Theorem 1 Perpendicular Vectors

Nonzero vectors $\mathbf{A}$ and $\mathbf{B}$ are perpendicular (orthogonal) if and only if $\mathbf{A} \cdot \mathbf{B} = 0$.

Example 3 SHOWING VECTORS PERPENDICULAR

The nonzero vectors $\mathbf{A} = 3\mathbf{i} - 2\mathbf{j} + \mathbf{k}$ and $\mathbf{B} = 2\mathbf{j} + 4\mathbf{k}$ are orthogonal because

$$\mathbf{A} \cdot \mathbf{B} = (3)(0) + (-2)(2) + (1)(4) = 0.$$

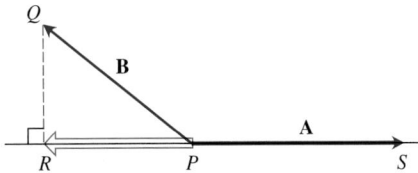

Figure 11.11 The vector projection of **B** onto **A**.

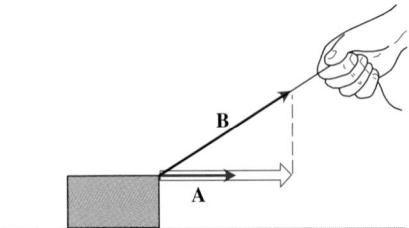

Figure 11.12 If we pull on a box with force **B**, the effective force in the direction of **A** is the vector projection of **B** onto **A**.

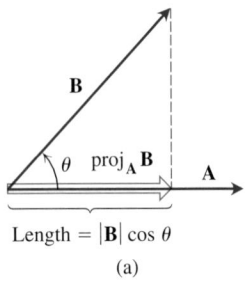

Length = $|\mathbf{B}| \cos \theta$

(a)

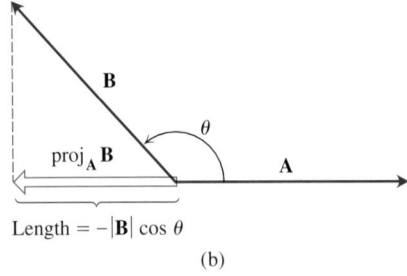

Length = $-|\mathbf{B}| \cos \theta$

(b)

Figure 11.13 The length of $\mathrm{proj}_\mathbf{A}\mathbf{B}$ is (a) $|\mathbf{B}| \cos \theta$ if $\cos \theta \geq 0$ and (b) $-|\mathbf{B}| \cos \theta$ if $\cos \theta < 0$.

The **vector projection** of $\mathbf{B} = \overrightarrow{PQ}$ onto a nonzero vector $\mathbf{A} = \overrightarrow{PS}$ (Figure 11.11) is the vector $\overrightarrow{PR}$ determined by dropping a perpendicular from Q to the line PS. The notation for this vector is

$$\mathrm{proj}_\mathbf{A}\mathbf{B} \quad \text{(``the vector projection of } \mathbf{B} \text{ onto } \mathbf{A}\text{'')}.$$

If **B** represents a force, then $\mathrm{proj}_\mathbf{A}\mathbf{B}$ represents the effective force in the direction of **A** (Figure 11.12).

If the angle θ between **A** and **B** is acute, $\mathrm{proj}_\mathbf{A}\mathbf{B}$ has length $|\mathbf{B}| \cos \theta$ and direction $\mathbf{A}/|\mathbf{A}|$ (Figure 11.13). If θ is obtuse, $\cos \theta < 0$ and $\mathrm{proj}_\mathbf{A}\mathbf{B}$ has length $-|\mathbf{B}| \cos \theta$ and direction $-\mathbf{A}/|\mathbf{A}|$. In either case,

$$\mathrm{proj}_\mathbf{A}\mathbf{B} = \left(|\mathbf{B}| \cos \theta \right) \frac{\mathbf{A}}{|\mathbf{A}|}$$

$$= \left(\frac{\mathbf{A} \cdot \mathbf{B}}{|\mathbf{A}|} \right) \frac{\mathbf{A}}{|\mathbf{A}|} \qquad |\mathbf{B}| \cos \theta = \frac{|\mathbf{A}||\mathbf{B}| \cos \theta}{|\mathbf{A}|} = \frac{\mathbf{A} \cdot \mathbf{B}}{|\mathbf{A}|}$$

$$= \left(\mathbf{B} \cdot \frac{\mathbf{A}}{|\mathbf{A}|} \right) \frac{\mathbf{A}}{|\mathbf{A}|} \qquad \begin{array}{l} \text{This is } \mathbf{B} \text{ ``dotted'' with the direction} \\ \text{of } \mathbf{A}, \text{ times the direction of } \mathbf{A}. \end{array}$$

> **Projection of B onto A**
>
> $$\mathrm{proj}_\mathbf{A}\mathbf{B} = \left(\mathbf{B} \cdot \frac{\mathbf{A}}{|\mathbf{A}|} \right) \frac{\mathbf{A}}{|\mathbf{A}|} = \left(\frac{\mathbf{B} \cdot \mathbf{A}}{\mathbf{A} \cdot \mathbf{A}} \right) \mathbf{A} \qquad (1)$$

The number $|\mathbf{B}| \cos \theta$ is the **scalar component of B in the direction of A.** Since

$$|\mathbf{B}| \cos \theta = \mathbf{B} \cdot \frac{\mathbf{A}}{|\mathbf{A}|}, \qquad (2)$$

we can find the scalar component by "dotting" **B** with the direction of **A**. Equation (1) says that the vector projection of **B** onto **A** is the scalar component of **B** in the direction of **A** times the direction of **A**.

Although the first part of Equation (1) describes the effect of **B** in the direction of **A,** the second part is better for calculations because it avoids square roots.

Example 4 USING THE PROJECTION FORMULA (EQ. (1))

Find the vector projection of $\mathbf{B} = 6\mathbf{i} + 3\mathbf{j} + 2\mathbf{k}$ onto $\mathbf{A} = \mathbf{i} - 2\mathbf{j} - 2\mathbf{k}$ and the scalar component of **B** in the direction of **A**.

Solution   We find $\mathrm{proj}_\mathbf{A}\mathbf{B}$ from Equation (1):

$$\mathrm{proj}_\mathbf{A}\mathbf{B} = \frac{\mathbf{B} \cdot \mathbf{A}}{\mathbf{A} \cdot \mathbf{A}} \mathbf{A} = \frac{6 - 6 - 4}{1 + 4 + 4} (\mathbf{i} - 2\mathbf{j} - 2\mathbf{k}) = -\frac{4}{9}\mathbf{i} + \frac{8}{9}\mathbf{j} + \frac{8}{9}\mathbf{k}.$$

We find the scalar component of **B** in the direction of **A** from Equation (2):

$$|\mathbf{B}|\cos\theta = \mathbf{B} \cdot \frac{\mathbf{A}}{|\mathbf{A}|} = (6\mathbf{i} + 3\mathbf{j} + 2\mathbf{k}) \cdot \left(\frac{1}{3}\mathbf{i} - \frac{2}{3}\mathbf{j} - \frac{2}{3}\mathbf{k}\right)$$

$$= 2 - 2 - \frac{4}{3} = -\frac{4}{3}.$$

Exploration 1 **Proving a Theorem in Solid Geometry**

1. Let P and Q be two endpoints of a diameter of a sphere centered at the origin. If P has coordinates (a, b, c), find the coordinates of Q.
2. Let $X = (x, y, z)$ be any other point on the sphere. Use the equation of a sphere to show that $x^2 + y^2 + z^2 = a^2 + b^2 + c^2$.
3. Find the components of vectors $\overrightarrow{PX}$ and $\overrightarrow{QX}$.
4. Compute the dot product of $\overrightarrow{PX}$ and $\overrightarrow{QX}$ and use the equation in step 2 to show that it is zero.
5. What can you conclude about angle PXQ? What geometric theorem have you proved?

Writing a Vector as a Sum of Orthogonal Vectors

In mechanics, we often need to express a vector **B** as a sum of a vector parallel to a vector **A** and a vector orthogonal to **A**. We can accomplish this with the equation

$$\mathbf{B} = \text{proj}_{\mathbf{A}}\mathbf{B} + (\mathbf{B} - \text{proj}_{\mathbf{A}}\mathbf{B}),$$

shown in Figure 11.14.

Figure 11.14 Writing **B** as the sum of vectors parallel and orthogonal to **A.**

Writing B as a Vector Parallel to A Plus a Vector Orthogonal to A

$$\mathbf{B} = \text{proj}_{\mathbf{A}}\mathbf{B} + (\mathbf{B} - \text{proj}_{\mathbf{A}}\mathbf{B})$$

$$= \underbrace{\left(\frac{\mathbf{B} \cdot \mathbf{A}}{\mathbf{A} \cdot \mathbf{A}}\right)\mathbf{A}}_{\text{parallel to } \mathbf{A}} + \underbrace{\left(\mathbf{B} - \left(\frac{\mathbf{B} \cdot \mathbf{A}}{\mathbf{A} \cdot \mathbf{A}}\right)\mathbf{A}\right)}_{\text{orthogonal to } \mathbf{A}} \qquad (3)$$

Example 5   DECOMPOSING A VECTOR

Express $\mathbf{B} = 2\mathbf{i} + \mathbf{j} - 3\mathbf{k}$ as the sum of a vector parallel to $\mathbf{A} = 3\mathbf{i} - \mathbf{j}$ and a vector orthogonal to **A.**

Solution   We use Equation (3). With

$$\mathbf{A} \cdot \mathbf{B} = 6 - 1 = 5 \quad \text{and} \quad \mathbf{A} \cdot \mathbf{A} = 9 + 1 = 10,$$

Equation (3) gives

$$\mathbf{B} = \frac{\mathbf{B} \cdot \mathbf{A}}{\mathbf{A} \cdot \mathbf{A}}\mathbf{A} + \left(\mathbf{B} - \frac{\mathbf{B} \cdot \mathbf{A}}{\mathbf{A} \cdot \mathbf{A}}\mathbf{A}\right)$$

$$= \frac{5}{10}(3\mathbf{i} - \mathbf{j}) + \left(2\mathbf{i} + \mathbf{j} - 3\mathbf{k} - \frac{5}{10}(3\mathbf{i} - \mathbf{j})\right)$$

$$= \left(\frac{3}{2}\mathbf{i} - \frac{1}{2}\mathbf{j}\right) + \left(\frac{1}{2}\mathbf{i} + \frac{3}{2}\mathbf{j} - 3\mathbf{k}\right).$$

To check this result, notice that the first vector is parallel to **A** because it is (1/2)**A**. The second vector in the sum is orthogonal to **A** because

$$\left(\frac{1}{2}\mathbf{i} + \frac{3}{2}\mathbf{j} - 3\mathbf{k}\right) \cdot (3\mathbf{i} - \mathbf{j}) = \frac{3}{2} - \frac{3}{2} = 0.$$

Work

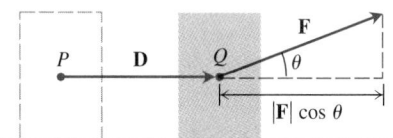

Figure 11.15 The work done by a constant force **F** during a displacement **D** is $(|\mathbf{F}| \cos \theta)|\mathbf{D}|$.

In Section 7.1, we calculated the work done by a constant force of magnitude F in moving an object through a distance d as $W = Fd$. That formula holds only if the force is directed along the line of motion. If a force **F** moving an object through a displacement $\mathbf{D} = \overrightarrow{PQ}$ has some other direction, the work is performed by the component of **F** in the direction of **D**. If θ is the angle between **F** and **D** (Figure 11.15), then

$$\text{Work} = \left(\begin{array}{c}\text{scalar component of } \mathbf{F}\\ \text{in the direction of } \mathbf{D}\end{array}\right)(\text{length of } \mathbf{D})$$

$$= (|\mathbf{F}| \cos \theta)|\mathbf{D}|$$

$$= \mathbf{F} \cdot \mathbf{D}$$

Definition Work

The **work** done by a constant force **F** acting through a displacement $\mathbf{D} = \overrightarrow{PQ}$ is

$$W = \mathbf{F} \cdot \mathbf{D} = |\mathbf{F}||\mathbf{D}| \cos \theta,$$

where θ is the angle between **F** and **D**.

Example 6   FINDING WORK DONE BY A FORCE

The standard units of work are the foot-pound and the newton-meter, both force-distance units. The newton-meter is usually called a *joule*. For more about the notion of work and the units involved, read pages 369 and 370.

If $|\mathbf{F}| = 40$ N (newtons), $|\mathbf{D}| = 3$ m, and $\theta = 60°$, the work done by **F** in acting from P to Q is

$$\text{Work} = |\mathbf{F}||\mathbf{D}| \cos \theta = (40)(3) \cos 60°$$

$$= (120)(1/2) = 60 \text{ J (joules)}.$$

We will encounter more interesting work problems in Chapter 15 when we can find the work done by a variable force along a path in space.

Quick Review 11.2

In Exercises 1–3, let $A = i + j$ and $B = i - j$.

1. Find $A \cdot B$.

2. Find the angle between A and B.

3. Find $A \cdot i$.

In Exercises 4 and 5, write an equation for the line in the *xy*-plane that passes through the point $(3, 2)$ and satisfies the given condition.

4. Parallel to $i - j$.

5. Perpendicular to $i - j$.

In Exercises 6 and 7, let $y = \sin x$.

6. Find the two unit vectors that are tangent to the curve at $x = \pi/6$.

7. Find the two unit vectors that are normal to the curve at $x = \pi/6$.

In Exercises 8 and 9, let $x = 5 \cos t$ and $y = 3 \sin t$.

8. Find the two unit vectors that are tangent to the curve at $t = 3\pi/4$.

9. Find the two unit vectors that are normal to the curve at $t = 3\pi/4$.

10. Find the work done by a force of 50 N moving along the *x*-axis from $x = -2$ to $x = 3$.

Section 11.2 Exercises

In Exercises 1–6, find

    (a) $A \cdot B$, $|A|$, $|B|$;

    (b) the angle between A and B;

    (c) the scalar component of B in the direction of A;

    (d) the vector $\text{proj}_A B$.

1. $A = 2i - 4j + \sqrt{5}k$, $B = -2i + 4j - \sqrt{5}k$

2. $A = (3/5)i + (4/5)k$, $B = 5i + 12j$

3. $A = 10i + 11j - 2k$, $B = 3j + 4k$

4. $A = 2i + 10j - 11k$, $B = 2i + 2j + k$

5. $A = 5j - 3k$, $B = i + j + k$

6. $A = -i + j$, $B = \sqrt{2}i + \sqrt{3}j + 2k$

In Exercises 7–9, write B as the sum of a vector parallel to A and a vector orthogonal to A.

7. $B = 3j + 4k$, $A = i + j$

8. $B = j + k$, $A = i + j$

9. $B = 8i + 4j - 12k$, $A = i + 2j - k$

10. *Sum of Vectors* $B = i + (j + k)$ is already the sum of a vector parallel to i and a vector orthogonal to i. If you use Equation (3) with $A = i$, do you get $B_1 = i$ and $B_2 = j + k$? Try it and find out.

11. Writing to Learn *Orthogonal Sums and Differences* In the accompanying figure, it looks as if $v_1 + v_2$ and $v_1 - v_2$ are orthogonal. Is this mere coincidence, or are there circumstances under which we may expect the sum of two vectors to be orthogonal to their differences? Give reasons for your answer.

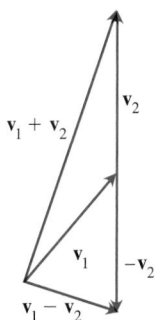

12. *Work in Groups of Two or Three* The accompanying figure shows a pyramid *OABCD* with a square base whose sides are 1 unit long. The pyramid's height is also 1 unit, and the point *D* stands directly above the midpoint of the diagonal *OB*. Find the angle between $\overrightarrow{OB}$ and $\overrightarrow{OD}$.

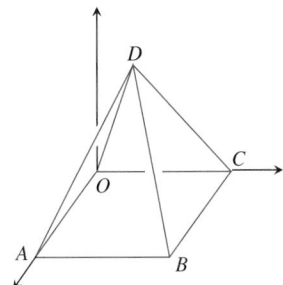

Exploration

13. *Direction Angles and Direction Cosines*   The **direction angles** α, β, and γ of a vector $\mathbf{v} = a\mathbf{i} + b\mathbf{j} + c\mathbf{k}$ are defined as follows:

α is the angle between $\mathbf{v}$ and the positive x-axis $(0 \le \alpha \le \pi)$,

β is the angle between $\mathbf{v}$ and the positive y-axis $(0 \le \beta \le \pi)$,

γ is the angle between $\mathbf{v}$ and the positive z-axis $(0 \le \gamma \le \pi)$.

(a) Show that

$$\cos \alpha = \frac{a}{|\mathbf{v}|}, \qquad \cos \beta = \frac{b}{|\mathbf{v}|}, \qquad \cos \gamma = \frac{c}{|\mathbf{v}|},$$

and $\cos^2 \alpha + \cos^2 \beta + \cos^2 \gamma = 1$. These cosines are the direction cosines of $\mathbf{v}$.

(b) Show that if $\mathbf{v} = a\mathbf{i} + b\mathbf{j} + c\mathbf{k}$ is a unit vector, then a, b, and c are the direction cosines of $\mathbf{v}$.

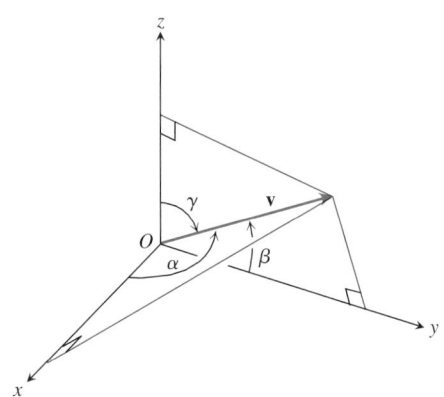

14. *Interior Angles of a Triangle*   Find the interior angles of the triangle ABC whose vertices are $A(-1, 0, 2)$, $B(2, 1, -1)$, and $C(1, -2, 2)$.

15. *Diagonal of a Cube*   Find the angle between the diagonal of a cube and the diagonal of one of its faces. (*Hint:* Use a cube whose edges represent $\mathbf{i}$, $\mathbf{j}$, and $\mathbf{k}$.)

16. *Grade*   A water main is to be constructed with a 20% grade in the north direction and a 10% grade in the east direction as shown in the figure. Determine the angle θ required in the water main for the turn from north to east.

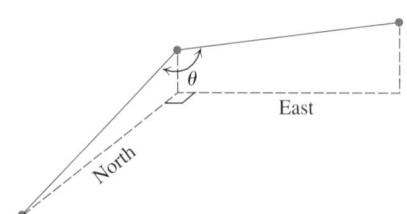

17. *Orthogonal Unit Vectors*   If $\mathbf{u}_1$ and $\mathbf{u}_2$ are orthogonal unit vectors and $\mathbf{v} = a\mathbf{u}_1 + b\mathbf{u}_2$, find $\mathbf{v} \cdot \mathbf{u}_1$.

18. **Writing to Learn** *Cancellation in Dot Product*   In real-number multiplication, if $ab_1 = ab_2$ and a is not zero, we

can cancel the a and conclude that $b_1 = b_2$. Does the same rule hold for vector multiplication: If $\mathbf{A} \cdot \mathbf{B}_1 = \mathbf{A} \cdot \mathbf{B}_2$ and $\mathbf{A} \ne \mathbf{0}$, can you conclude that $\mathbf{B}_1 = \mathbf{B}_2$? Give reasons for your answer.

19. *Linear Combinations of Orthogonal Vectors*   Suppose $\mathbf{A}$, $\mathbf{B}$, and $\mathbf{C}$ are mutually orthogonal vectors. Let $\mathbf{D} = 5\mathbf{A} - 6\mathbf{B} + 3\mathbf{C}$.

(a) If $\mathbf{A}$, $\mathbf{B}$, and $\mathbf{C}$ are unit vectors, find $|\mathbf{D}|$, the magnitude of $\mathbf{D}$.

(b) If $|\mathbf{A}| = 2$, $|\mathbf{B}| = 3$, and $|\mathbf{C}| = 4$, find $|\mathbf{D}|$.

20. *Linear Combinations of Orthogonal Vectors*   Suppose $\mathbf{A}$, $\mathbf{B}$, and $\mathbf{C}$ are mutually orthogonal vectors. If $\mathbf{D}$ is a vector such that $\mathbf{D} = \alpha\mathbf{A} + \beta\mathbf{B} + \gamma\mathbf{C}$ where α, β, and γ are scalars, prove that $\alpha = \mathbf{D} \cdot \mathbf{A}$, $\beta = \mathbf{D} \cdot \mathbf{B}$, and $\gamma = \mathbf{D} \cdot \mathbf{C}$.

21. *Work Along a Line in Space*   Find the work done by a force $\mathbf{F} = 5\mathbf{k}$ (magnitude 5 N) in moving an object along the line from the origin to the point $(1, 1, 1)$ (distance in meters).

22. *The Locomotive Big Boy*   The Union Pacific's *Big Boy* locomotive could pull 6000-ton trains with a tractive effort (pull) of 602,148 N (135,368 lb). At this level of effort, about how much work did *Big Boy* do on the (approximately straight) 605-km journey from San Francisco to Los Angeles?

23. *Work Sliding a Crate*   How much work does it take to slide a crate 20 m along a loading dock by pulling on it with a 200-N force at an angle of 30° from the horizontal?

24. *Wind at Work*   The wind passing over a boat's sail exerted a 1000-lb magnitude force $\mathbf{F}$ as shown here. How much work did the wind perform in moving the boat forward 1 mi? Answer in foot-pounds.

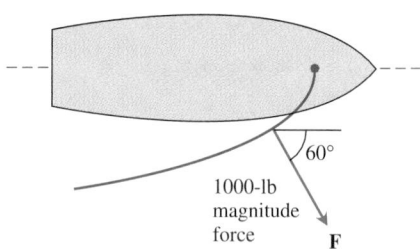

25. *Vector Perpendicular to a Line*   Show that the vector $\mathbf{v} = a\mathbf{i} + b\mathbf{j}$ is perpendicular to the line $ax + by = c$ by establishing that the slope of any line segment representing $\mathbf{v}$ is the negative reciprocal of the slope of the line.

26. *Vector Parallel to a Line*   Show that the vector $\mathbf{v} = a\mathbf{i} + b\mathbf{j}$ is parallel to the line $bx - ay = c$ by establishing that the slope of any line segment representing $\mathbf{v}$ is the same as the slope of the line.

In Exercises 27–30, use the result of Exercise 25 to find an equation for the line through P perpendicular to $\mathbf{v}$. Then sketch the line. Include $\mathbf{v}$ in your sketch *as a vector starting at the origin*.

27. $P(2, 1)$, $\mathbf{v} = \mathbf{i} + 2\mathbf{j}$

28. $P(-1, 2)$, $\mathbf{v} = -2\mathbf{i} - \mathbf{j}$

29. $P(-2, -7)$, $\mathbf{v} = -2\mathbf{i} + \mathbf{j}$

30. $P(11, 10)$, $\mathbf{v} = 2\mathbf{i} - 3\mathbf{j}$

In Exercises 31–34, use the result of Exercise 26 to find an equation for the line through P parallel to $\mathbf{v}$. Then sketch the line. Include $\mathbf{v}$ in your sketch *as a vector starting at the origin.*

31. $P(-2, 1)$, $\mathbf{v} = \mathbf{i} - \mathbf{j}$

32. $P(0, -2)$, $\mathbf{v} = 2\mathbf{i} + 3\mathbf{j}$

33. $P(1, 2)$, $\mathbf{v} = -\mathbf{i} - 2\mathbf{j}$

34. $P(1, 3)$, $\mathbf{v} = 3\mathbf{i} - 2\mathbf{j}$

In Exercises 35–40, *work in groups of two or three*. The **acute angle between intersecting lines in the plane** that do not cross at right angles is the same as the angle determined by vectors normal to the lines or by vectors parallel to the lines as shown in the figure. Use this fact and the results of Exercises 25 and 26 to find the acute angle between the given lines.

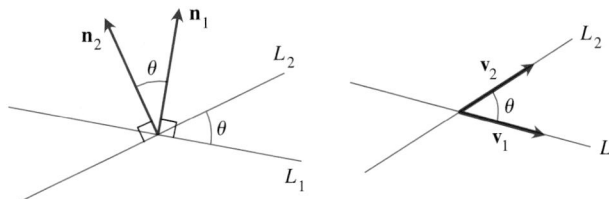

35. $3x + y = 5$, $2x - y = 4$

36. $y = \sqrt{3}x - 1$, $y = -\sqrt{3}x + 2$

37. $\sqrt{3}x - y = -2$, $x - \sqrt{3}y = 1$

38. $x + \sqrt{3}y = 1$, $(1 - \sqrt{3})x + (1 + \sqrt{3})y = 8$

39. $3x - 4y = 3$, $x - y = 7$

40. $12x + 5y = 1$, $2x - 2y = 3$

In Exercises 41–44, find the angle between the two differentiable curves at each point of intersection. The **angle between**

differentiable curves at a point of intersection is the (acute) angle between the curves' tangent lines at this point.

41. $y = \dfrac{3}{2} - x^2$, $y = x^2$

42. $x = \dfrac{3}{4} - y^2$, $x = y^2 - \dfrac{3}{4}$

43. $y = x^3$, $x = y^2$

44. $y = -x^2$, $y = \sqrt[3]{x}$

45. *Properties of the Dot Product* Let $\mathbf{A}$, $\mathbf{B}$, $\mathbf{C}$, and $\mathbf{D}$ be vectors and c any scalar. Use the component form of the dot product to establish the following properties.

   (a) $\mathbf{A} \cdot \mathbf{B} = \mathbf{B} \cdot \mathbf{A}$

   (b) $(c\mathbf{A}) \cdot \mathbf{B} = \mathbf{A} \cdot (c\mathbf{B}) = c(\mathbf{A} \cdot \mathbf{B})$

   (c) $\mathbf{A} \cdot (\mathbf{B} + \mathbf{C}) = \mathbf{A} \cdot \mathbf{B} + \mathbf{A} \cdot \mathbf{C}$

   (d) $(\mathbf{A} + \mathbf{B}) \cdot \mathbf{C} = \mathbf{A} \cdot \mathbf{C} + \mathbf{B} \cdot \mathbf{C}$

   (e) $(\mathbf{A} + \mathbf{B}) \cdot (\mathbf{C} + \mathbf{D}) = \mathbf{A} \cdot \mathbf{C} + \mathbf{A} \cdot \mathbf{D} + \mathbf{B} \cdot \mathbf{C} + \mathbf{B} \cdot \mathbf{D}$

   (f) $\mathbf{A} \cdot \mathbf{A} = |\mathbf{A}|^2$

Extending the Ideas

46. (a) Use the equation $\mathbf{u} \cdot \mathbf{v} = |\mathbf{u}||\mathbf{v}| \cos \theta$ to show that the inequality $|\mathbf{u} \cdot \mathbf{v}| \le |\mathbf{u}||\mathbf{v}|$ holds for any vectors $\mathbf{u}$ and $\mathbf{v}$.

   (b) Writing to Learn Under what circumstances, if any, does $|\mathbf{u} \cdot \mathbf{v}|$ equal $|\mathbf{u}||\mathbf{v}|$? Give reasons for your answer.

47. Copy the axes and vector shown here. Then shade in the points (x, y) for which $(x\mathbf{i} + y\mathbf{j}) \cdot \mathbf{v} \le 0$. Justify your answer.

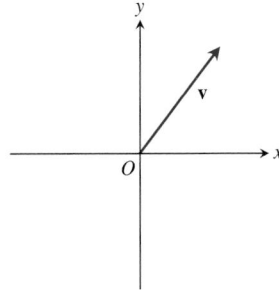

11.3 # Cross Products

Definition of Cross Product • Are Cross Products Commutative? • $|\mathbf{A} \times \mathbf{B}|$ Is the Area of a Parallelogram • Torque • Associative and Distributive Laws • Determinant Formula for $\mathbf{A} \times \mathbf{B}$ • Triple Scalar or Box Product

Definition of Cross Product

In studying lines in the plane, we needed to describe how a line was tilting. We did so with the notions of slope and angle of inclination. In space, we need to be able to describe how a plane is tilting. We accomplish this by multiplying

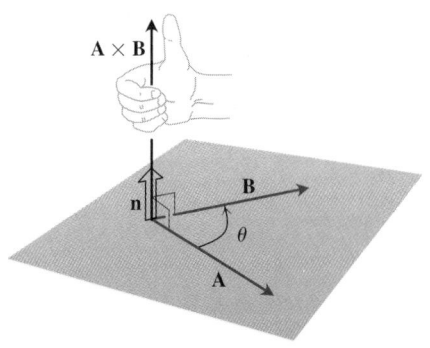

Figure 11.16 The construction of $\mathbf{A} \times \mathbf{B}$.

two vectors in the plane together to get a third vector perpendicular to the plane. The direction of this third vector tells us the "inclination" of the plane. The product we use to multiply the vectors together is the *vector* or *cross product*.

We start with two nonzero vectors $\mathbf{A}$ and $\mathbf{B}$ in space. If $\mathbf{A}$ and $\mathbf{B}$ are not parallel, they determine a plane. We select a unit vector $\mathbf{n}$ perpendicular to the plane by the **right-hand rule.** This means we choose $\mathbf{n}$ to be the unit (normal) vector that points the way your right thumb points when your fingers curl through the angle θ from $\mathbf{A}$ to $\mathbf{B}$ (Figure 11.16).

Definition Vector Product $\mathbf{A} \times \mathbf{B}$

The **vector product $\mathbf{A} \times \mathbf{B}$** ("A cross B") **of A and B** is the vector

$$\mathbf{A} \times \mathbf{B} = (|\mathbf{A}||\mathbf{B}| \sin \theta)\mathbf{n}.$$

We sometimes call the dot product a **scalar product** to distinguish it from the vector product. The vector $\mathbf{A} \times \mathbf{B}$ is orthogonal to both $\mathbf{A}$ and $\mathbf{B}$ because it is a scalar multiple of $\mathbf{n}$. The vector product of $\mathbf{A}$ and $\mathbf{B}$ is often called the **cross product** of $\mathbf{A}$ and $\mathbf{B}$ because of the cross in the notation $\mathbf{A} \times \mathbf{B}$.

Since the sines of 0 and π are both zero, it makes sense to define the cross product of two parallel nonzero vectors to be $\mathbf{0}$.

If one or both of $\mathbf{A}$ and $\mathbf{B}$ are zero, we also define $\mathbf{A} \times \mathbf{B}$ to be zero. This way, the cross product of two vectors $\mathbf{A}$ and $\mathbf{B}$ is zero if and only if $\mathbf{A}$ and $\mathbf{B}$ are parallel or one or both of them are zero.

Nonzero vectors $\mathbf{A}$ and $\mathbf{B}$ are parallel if and only if $\mathbf{A} \times \mathbf{B} = \mathbf{0}$.

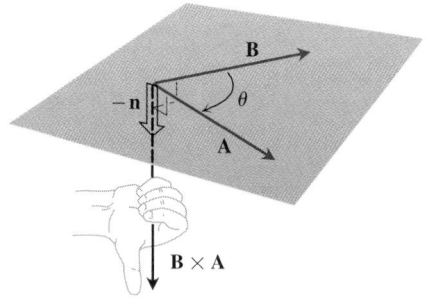

Figure 11.17 The construction of $\mathbf{B} \times \mathbf{A}$.

Are Cross Products Commutative?

Reversing the order of the factors in a nonzero cross product reverses the direction of the product. When the fingers of our right hand curl through the angle θ from $\mathbf{B}$ to $\mathbf{A}$, our thumb points the opposite way and the unit vector we choose in forming $\mathbf{B} \times \mathbf{A}$ is the negative of the one we choose in forming $\mathbf{A} \times \mathbf{B}$ (Figure 11.17). Thus, for all vectors $\mathbf{A}$ and $\mathbf{B}$,

---

$$\mathbf{B} \times \mathbf{A} = -(\mathbf{A} \times \mathbf{B}).$$

Unlike the dot product, the cross product is not commutative.

When we apply the definition to calculate the pairwise cross products of $\mathbf{i}, \mathbf{j}$, and $\mathbf{k}$, we find (Figure 11.18)

$$\mathbf{i} \times \mathbf{j} = -(\mathbf{j} \times \mathbf{i}) = \mathbf{k}$$
$$\mathbf{j} \times \mathbf{k} = -(\mathbf{k} \times \mathbf{j}) = \mathbf{i}$$
$$\mathbf{k} \times \mathbf{i} = -(\mathbf{i} \times \mathbf{k}) = \mathbf{j}$$

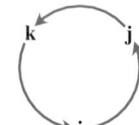

Diagram for recalling these products.

Figure 11.18 The pairwise cross products of $\mathbf{i}, \mathbf{j}$, and $\mathbf{k}$.

and

$$\mathbf{i} \times \mathbf{i} = \mathbf{j} \times \mathbf{j} = \mathbf{k} \times \mathbf{k} = \mathbf{0}.$$

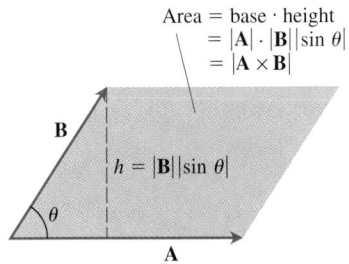

Area = base · height
= $|\mathbf{A}| \cdot |\mathbf{B}||\sin\theta|$
= $|\mathbf{A} \times \mathbf{B}|$

$h = |\mathbf{B}||\sin\theta|$

Figure 11.19 The parallelogram determined by **A** and **B**.

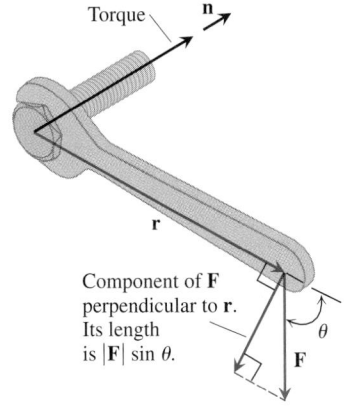

Figure 11.20 The torque vector describes the tendency of the force **F** to drive the bolt forward.

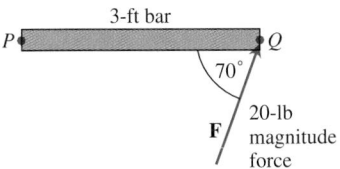

Figure 11.21 The figure for Example 1.

$|\mathbf{A} \times \mathbf{B}|$ Is the Area of a Parallelogram

Because **n** is a unit vector, the magnitude of $\mathbf{A} \times \mathbf{B}$ is

$$|\mathbf{A} \times \mathbf{B}| = |\mathbf{A}||\mathbf{B}||\sin\theta||\mathbf{n}| = |\mathbf{A}||\mathbf{B}||\sin\theta|.$$

This is the area of the parallelogram determined by **A** and **B** (Figure 11.19), $|\mathbf{A}|$ being the base of the parallelogram and $|\mathbf{B}||\sin\theta|$ the height.

Torque

When we tighten a bolt by applying a force **F** to a wrench (Figure 11.20), the torque we produce acts along the axis of the bolt to drive the bolt forward. The magnitude of the torque depends on how far out on the wrench the force is applied and on how much of the force is perpendicular to the wrench at the point of application. The number we use to measure the torque's magnitude is the product of the length of the lever arm **r** and the scalar component of **F** perpendicular to **r**. The usual units are foot-pounds and Newton-meters, the same units as work.

In the notation of Figure 11.20,

$$\text{Magnitude of torque vector} = |\mathbf{r}||\mathbf{F}|\sin\theta,$$

or $|\mathbf{r} \times \mathbf{F}|$. If we let **n** be a unit vector along the axis of the bolt in the direction of the torque, then a complete description of the torque vector is $\mathbf{r} \times \mathbf{F}$, or

$$\text{Torque vector} = (|\mathbf{r}||\mathbf{F}|\sin\theta)\mathbf{n}.$$

Recall that we defined $\mathbf{A} \times \mathbf{B}$ to be **0** when **A** and **B** are parallel. This is consistent with the torque interpretation as well. If the force **F** in Figure 11.20 is parallel to the wrench, meaning that we are trying to turn the bolt by pushing or pulling along the line of the wrench's handle, the torque produced is zero.

Example 1 FINDING THE MAGNITUDE OF A TORQUE

The magnitude of the torque exerted by the force **F** about the pivot point P in Figure 11.21 is

$$\left|\overrightarrow{PQ} \times \mathbf{F}\right| = \left|\overrightarrow{PQ}\right||\mathbf{F}|\sin 70°$$

$$\approx (3)(20)(0.94) = 56.4 \text{ ft-lb}$$

Associative and Distributive Laws

As a rule, cross-product multiplication is *not associative* because $(\mathbf{A} \times \mathbf{B}) \times \mathbf{C}$ lies in the plane of **A** and **B** whereas $\mathbf{A} \times (\mathbf{B} \times \mathbf{C})$ lies in the plane of **B** and **C**. However, the following properties do hold.

Properties of the Cross Product
Scalar Distributive Laws

$$(r\mathbf{A}) \times (s\mathbf{B}) = (rs)(\mathbf{A} \times \mathbf{B}) \tag{1}$$

$$(-\mathbf{A}) \times \mathbf{B} = \mathbf{A} \times (-\mathbf{B}) = -(\mathbf{A} \times \mathbf{B}) \tag{2}$$

Vector Distributive Laws

$$\mathbf{A} \times (\mathbf{B} + \mathbf{C}) = \mathbf{A} \times \mathbf{B} + \mathbf{A} \times \mathbf{C} \tag{3}$$

$$(\mathbf{B} + \mathbf{C}) \times \mathbf{A} = \mathbf{B} \times \mathbf{A} + \mathbf{C} \times \mathbf{A} \tag{4}$$

Equation (1) can be verified by applying the definition of cross product to both sides of the equation and comparing the results. Equation (2) is a special case of Equation (1). We will skip the proof of Equation (3) because of the complexity of the geometry involved. Equation (4) follows from Equation (3) by multiplying both sides of Equation (3) by -1 and reversing the orders of the products.

Determinant Formula for A $\times$ B

Our next objective is to calculate $\mathbf{A} \times \mathbf{B}$ from the components of $\mathbf{A}$ and $\mathbf{B}$ relative to a Cartesian coordinate system.

Suppose

$$\mathbf{A} = a_1\mathbf{i} + a_2\mathbf{j} + a_3\mathbf{k}, \quad \mathbf{B} = b_1\mathbf{i} + b_2\mathbf{j} + b_3\mathbf{k}.$$

Then the distributive laws and the rules for multiplying $\mathbf{i}$, $\mathbf{j}$, and $\mathbf{k}$ tell us that

$$\begin{aligned}
\mathbf{A} \times \mathbf{B} &= (a_1\mathbf{i} + a_2\mathbf{j} + a_3\mathbf{k}) \times (b_1\mathbf{i} + b_2\mathbf{j} + b_3\mathbf{k}) \\
&= a_1b_1\mathbf{i} \times \mathbf{i} + a_1b_2\mathbf{i} \times \mathbf{j} + a_1b_3\mathbf{i} \times \mathbf{k} \\
&\quad + a_2b_1\mathbf{j} \times \mathbf{i} + a_2b_2\mathbf{j} \times \mathbf{j} + a_2b_3\mathbf{j} \times \mathbf{k} \\
&\quad + a_3b_1\mathbf{k} \times \mathbf{i} + a_3b_2\mathbf{k} \times \mathbf{j} + a_3b_3\mathbf{k} \times \mathbf{k} \\
&= (a_2b_3 - a_3b_2)\mathbf{i} - (a_1b_3 - a_3b_1)\mathbf{j} + (a_1b_2 - a_2b_1)\mathbf{k}.
\end{aligned}$$

The terms in the last line are the same as the terms in the expansion of the symbolic determinant

$$\begin{vmatrix} \mathbf{i} & \mathbf{j} & \mathbf{k} \\ a_1 & a_2 & a_3 \\ b_1 & b_2 & b_3 \end{vmatrix}.$$

We therefore have the following formula.

Determinants

LS(For more information, see Appendix 8.)

$$\begin{vmatrix} a & b \\ c & d \end{vmatrix} = ad - bc$$

Example

$$\begin{aligned}
\begin{vmatrix} 2 & 1 \\ -4 & 3 \end{vmatrix} &= (2)(3) - (1)(-4) \\
&= 6 + 4 = 10 \quad \blacksquare
\end{aligned}$$

$$\begin{vmatrix} a_1 & a_2 & a_3 \\ b_1 & b_2 & b_3 \\ c_1 & c_2 & c_3 \end{vmatrix}$$

$$= a_1 \begin{vmatrix} b_2 & b_3 \\ c_2 & c_3 \end{vmatrix} - a_2 \begin{vmatrix} b_1 & b_3 \\ c_1 & c_3 \end{vmatrix} + a_3 \begin{vmatrix} b_1 & b_2 \\ c_1 & c_2 \end{vmatrix}$$

Example

$$\begin{vmatrix} -5 & 3 & 1 \\ 2 & 1 & 1 \\ -4 & 3 & 1 \end{vmatrix}$$

$$= (-5) \begin{vmatrix} 1 & 1 \\ 3 & 1 \end{vmatrix} - (3) \begin{vmatrix} 2 & 1 \\ -4 & 1 \end{vmatrix}$$

$$+ (1) \begin{vmatrix} 2 & 1 \\ -4 & 3 \end{vmatrix}$$

$$= -5(1 - 3) - 3(2 + 4) + 1(6 + 4)$$

$$= 10 - 18 + 10 = 2 \quad \blacksquare$$

Evaluating Cross Products Using Determinants

If $\mathbf{A} = a_1\mathbf{i} + a_2\mathbf{j} + a_3\mathbf{k}$ and $\mathbf{B} = b_1\mathbf{i} + b_2\mathbf{j} + b_3\mathbf{k}$, then

$$\mathbf{A} \times \mathbf{B} = \begin{vmatrix} \mathbf{i} & \mathbf{j} & \mathbf{k} \\ a_1 & a_2 & a_3 \\ b_1 & b_2 & b_3 \end{vmatrix}.$$

Example 2   COMPUTING CROSS PRODUCTS WITH DETERMINANTS

Find $\mathbf{A} \times \mathbf{B}$ and $\mathbf{B} \times \mathbf{A}$ if

$$\mathbf{A} = 2\mathbf{i} + \mathbf{j} + \mathbf{k}, \qquad \mathbf{B} = -4\mathbf{i} + 3\mathbf{j} + \mathbf{k}.$$

Solution

$$\mathbf{A} \times \mathbf{B} = \begin{vmatrix} \mathbf{i} & \mathbf{j} & \mathbf{k} \\ 2 & 1 & 1 \\ -4 & 3 & 1 \end{vmatrix} = \begin{vmatrix} 1 & 1 \\ 3 & 1 \end{vmatrix} \mathbf{i} - \begin{vmatrix} 2 & 1 \\ -4 & 1 \end{vmatrix} \mathbf{j} + \begin{vmatrix} 2 & 1 \\ -4 & 3 \end{vmatrix} \mathbf{k}$$

$$= -2\mathbf{i} - 6\mathbf{j} + 10\mathbf{k}$$

$$\mathbf{B} \times \mathbf{A} = -(\mathbf{A} \times \mathbf{B}) = 2\mathbf{i} + 6\mathbf{j} - 10\mathbf{k}$$

Example 3 FINDING VECTORS PERPENDICULAR TO A PLANE

Find a vector perpendicular to the plane of $P(1, -1, 0)$, $Q(2, 1, -1)$, and $R(-1, 1, 2)$.

Solution   The vector $\overrightarrow{PQ} \times \overrightarrow{PR}$ is perpendicular to the plane because it is perpendicular to both vectors. In terms of components,

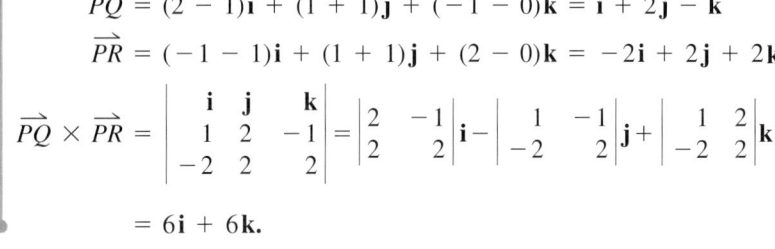

$$\overrightarrow{PQ} = (2 - 1)\mathbf{i} + (1 + 1)\mathbf{j} + (-1 - 0)\mathbf{k} = \mathbf{i} + 2\mathbf{j} - \mathbf{k}$$
$$\overrightarrow{PR} = (-1 - 1)\mathbf{i} + (1 + 1)\mathbf{j} + (2 - 0)\mathbf{k} = -2\mathbf{i} + 2\mathbf{j} + 2\mathbf{k}$$

$$\overrightarrow{PQ} \times \overrightarrow{PR} = \begin{vmatrix} \mathbf{i} & \mathbf{j} & \mathbf{k} \\ 1 & 2 & -1 \\ -2 & 2 & 2 \end{vmatrix} = \begin{vmatrix} 2 & -1 \\ 2 & 2 \end{vmatrix} \mathbf{i} - \begin{vmatrix} 1 & -1 \\ -2 & 2 \end{vmatrix} \mathbf{j} + \begin{vmatrix} 1 & 2 \\ -2 & 2 \end{vmatrix} \mathbf{k}$$

$$= 6\mathbf{i} + 6\mathbf{k}.$$

Example 4 FINDING THE AREA OF A TRIANGLE

Find the area of the triangle with vertices $P(1, -1, 0)$, $Q(2, 1, -1)$, and $R(-1, 1, 2)$ (Figure 11.22).

Solution   The area of the parallelogram determined by P, Q, and R is

$$\left| \overrightarrow{PQ} \times \overrightarrow{PR} \right| = |6\mathbf{i} + 6\mathbf{k}|$$
$$= \sqrt{6^2 + 6^2} = 6\sqrt{2}.$$

The triangle's area is half of this, or $3\sqrt{2}$ units squared.

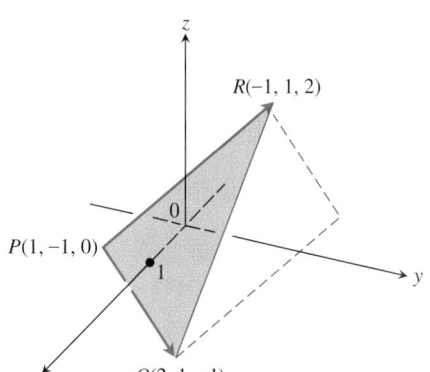

Figure 11.22 The area of the triangle PQR is half of $\left| \overrightarrow{PQ} \times \overrightarrow{PR} \right|$. (Example 4)

Example 5 FINDING A UNIT NORMAL TO A PLANE

Find a unit vector perpendicular to the plane containing $P(1, -1, 0)$, $Q(2, 1, -1)$, and $R(-1, 1, 2)$.

Solution   Since $\overrightarrow{PQ} \times \overrightarrow{PR}$ is perpendicular to the plane, its direction $\mathbf{n}$ is a unit vector perpendicular to the plane. Taking values from Examples 3 and 4, we have

$$\mathbf{n} = \frac{\overrightarrow{PQ} \times \overrightarrow{PR}}{\left| \overrightarrow{PQ} \times \overrightarrow{PR} \right|} = \frac{6\mathbf{i} + 6\mathbf{k}}{6\sqrt{2}} = \frac{1}{\sqrt{2}} \mathbf{i} + \frac{1}{\sqrt{2}} \mathbf{k}.$$

Triple Scalar or Box Product

The product $(\mathbf{A} \times \mathbf{B}) \cdot \mathbf{C}$ is the **triple scalar product** of $\mathbf{A}$, $\mathbf{B}$, and $\mathbf{C}$ (in that order). As you can see from the formula

$$|(\mathbf{A} \times \mathbf{B}) \cdot \mathbf{C}| = |\mathbf{A} \times \mathbf{B}||\mathbf{C}||\cos \theta|,$$

the absolute value of the product is the volume of the parallelepiped (parallelogram-sided box) determined by $\mathbf{A}$, $\mathbf{B}$, and $\mathbf{C}$ (Figure 11.23). The number $|\mathbf{A} \times \mathbf{B}|$ is the area of the base parallelogram. The number $|\mathbf{C}||\cos \theta|$ is the parallelepiped's height. Because of this geometry, $(\mathbf{A} \times \mathbf{B}) \cdot \mathbf{C}$ is also called the **box product** of $\mathbf{A}$, $\mathbf{B}$, and $\mathbf{C}$.

By treating the planes of $\mathbf{B}$ and $\mathbf{C}$ and of $\mathbf{C}$ and $\mathbf{A}$ as the base planes of the parallelepiped determined by $\mathbf{A}$, $\mathbf{B}$, and $\mathbf{C}$, we see that

$$(\mathbf{A} \times \mathbf{B}) \cdot \mathbf{C} = (\mathbf{B} \times \mathbf{C}) \cdot \mathbf{A} = (\mathbf{C} \times \mathbf{A}) \cdot \mathbf{B}.$$

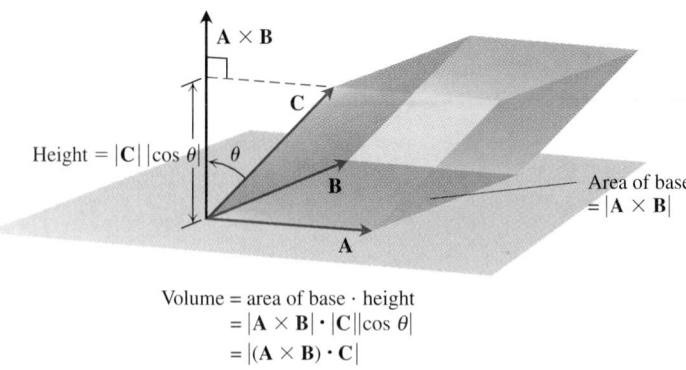

$$\text{Volume} = \text{area of base} \cdot \text{height}$$
$$= |\mathbf{A} \times \mathbf{B}| \cdot |\mathbf{C}||\cos \theta|$$
$$= |(\mathbf{A} \times \mathbf{B}) \cdot \mathbf{C}|$$

Figure 11.23 The number $|(\mathbf{A} \times \mathbf{B}) \cdot \mathbf{C}|$ is the volume of a parallelepiped.

Triple Scalar Product

The dot and cross may be interchanged in a triple scalar product without altering its value.

Since the dot product is commutative, the equation above also gives

$$(\mathbf{A} \times \mathbf{B}) \cdot \mathbf{C} = \mathbf{A} \cdot (\mathbf{B} \times \mathbf{C}).$$

The triple scalar product can be evaluated as a determinant:

$$\mathbf{A} \cdot (\mathbf{B} \times \mathbf{C}) = \mathbf{A} \cdot \left[\begin{vmatrix} b_2 & b_3 \\ c_2 & c_3 \end{vmatrix} \mathbf{i} - \begin{vmatrix} b_1 & b_3 \\ c_1 & c_3 \end{vmatrix} \mathbf{j} + \begin{vmatrix} b_1 & b_2 \\ c_1 & c_2 \end{vmatrix} \mathbf{k} \right]$$

$$= a_1 \begin{vmatrix} b_2 & b_3 \\ c_2 & c_3 \end{vmatrix} - a_2 \begin{vmatrix} b_1 & b_3 \\ c_1 & c_3 \end{vmatrix} + a_3 \begin{vmatrix} b_1 & b_2 \\ c_1 & c_2 \end{vmatrix}$$

$$= \begin{vmatrix} a_1 & a_2 & a_3 \\ b_1 & b_2 & b_3 \\ c_1 & c_2 & c_3 \end{vmatrix}$$

Triple Scalar Product

$$\mathbf{A} \cdot (\mathbf{B} \times \mathbf{C}) = (\mathbf{A} \times \mathbf{B}) \cdot \mathbf{C} = \begin{vmatrix} a_1 & a_2 & a_3 \\ b_1 & b_2 & b_3 \\ c_1 & c_2 & c_3 \end{vmatrix} \tag{5}$$

Example 6 FINDING THE VOLUME OF A PARALLELEPIPED

Find the volume of the box (parallelepiped) determined by $\mathbf{A} = \mathbf{i} + 2\mathbf{j} - \mathbf{k}$, $\mathbf{B} = -2\mathbf{i} + 3\mathbf{k}$, and $\mathbf{C} = 7\mathbf{j} - 4\mathbf{k}$.

Solution   Using a calculator, we find

$$\mathbf{A} \cdot (\mathbf{B} \times \mathbf{C}) = \begin{vmatrix} 1 & 2 & -1 \\ -2 & 0 & 3 \\ 0 & 7 & -4 \end{vmatrix} = -23.$$

The volume is $|\mathbf{A} \cdot (\mathbf{B} \times \mathbf{C})| = 23$ units cubed.

Quick Review 11.3

In Exercises 1–3, consider the points $A(1, 1)$, $B(3, 4)$, $C(7, 6)$, and $D(5, 3)$.

1. Show that the figure $ABCD$ is a parallelogram.

2. Find the area of $ABCD$.

3. Find the measure of the interior angles of $ABCD$.

In Exercises 4 and 5, solve the equation.

4. $\sin 2x = 0$, $0 \le x \le 2\pi$

5. $\cos 2x = 0$, $0 \le x \le 2\pi$

In Exercises 6–10, evaluate the determinants.

6. $\begin{vmatrix} 2 & 3 \\ -1 & 1 \end{vmatrix}$

7. $\begin{vmatrix} -3 & 4 \\ 0 & 1 \end{vmatrix}$

8. $\begin{vmatrix} 2 & 3 & -1 \\ -1 & 2 & 4 \\ 1 & 0 & 3 \end{vmatrix}$

9. $\begin{vmatrix} -1 & 0 & 1 \\ 1 & 1 & 1 \\ 0 & 1 & -1 \end{vmatrix}$

10. $\begin{vmatrix} 2 & -1 & 3 \\ 4 & 0 & -2 \\ -6 & 1 & -1 \end{vmatrix}$

Section 11.3 Exercises

In Exercises 1–8, find the length and direction (when defined) of $\mathbf{A} \times \mathbf{B}$ and $\mathbf{B} \times \mathbf{A}$.

1. $\mathbf{A} = 2\mathbf{i} - 2\mathbf{j} - \mathbf{k}$, $\mathbf{B} = \mathbf{i} - \mathbf{k}$

2. $\mathbf{A} = 2\mathbf{i} - 3\mathbf{j}$, $\mathbf{B} = -\mathbf{i} + \mathbf{j}$

3. $\mathbf{A} = 2\mathbf{i} - 2\mathbf{j} + 4\mathbf{k}$, $\mathbf{B} = -\mathbf{i} + \mathbf{j} - 2\mathbf{k}$

4. $\mathbf{A} = \mathbf{i} + \mathbf{j} - \mathbf{k}$, $\mathbf{B} = 0$

5. $\mathbf{A} = 2\mathbf{i}$, $\mathbf{B} = -3\mathbf{j}$

6. $\mathbf{A} = \mathbf{i} \times \mathbf{j}$, $\mathbf{B} = \mathbf{j} \times \mathbf{k}$

7. $\mathbf{A} = -8\mathbf{i} - 2\mathbf{j} - 4\mathbf{k}$, $\mathbf{B} = 2\mathbf{i} + 2\mathbf{j} + \mathbf{k}$

8. $\mathbf{A} = \dfrac{3}{2}\mathbf{i} - \dfrac{1}{2}\mathbf{j} + \mathbf{k}$, $\mathbf{B} = \mathbf{i} + \mathbf{j} + \mathbf{k}$

In Exercises 9–14, sketch the coordinate axes and then include the vectors $\mathbf{A}$, $\mathbf{B}$, and $\mathbf{A} \times \mathbf{B}$ as vectors starting at the origin.

9. $\mathbf{A} = \mathbf{i}$, $\mathbf{B} = \mathbf{j}$

10. $\mathbf{A} = \mathbf{i} - \mathbf{k}$, $\mathbf{B} = \mathbf{j}$

11. $\mathbf{A} = \mathbf{i} - \mathbf{k}$, $\mathbf{B} = \mathbf{j} + \mathbf{k}$

12. $\mathbf{A} = 2\mathbf{i} - \mathbf{j}$, $\mathbf{B} = \mathbf{i} + 2\mathbf{j}$

13. $\mathbf{A} = \mathbf{i} + \mathbf{j}$, $\mathbf{B} = \mathbf{i} - \mathbf{j}$

14. $\mathbf{A} = \mathbf{j} + 2\mathbf{k}$, $\mathbf{B} = \mathbf{i}$

In Exercises 15–18, find

   (a) the area of the triangle determined by the points P, Q, and R.

   (b) a unit vector perpendicular to plane PQR.

15. $P(1, -1, 2)$, $Q(2, 0, -1)$, $R(0, 2, 1)$

16. $P(1, 1, 1)$, $Q(2, 1, 3)$, $R(3, -1, 1)$

17. $P(2, -2, 1)$, $Q(3, -1, 2)$, $R(3, -1, 1)$

18. $P(-2, 2, 0)$, $Q(0, 1, -1)$, $R(-1, 2, -2)$

19. Writing to Learn Let $\mathbf{A} = 5\mathbf{i} - \mathbf{j} + \mathbf{k}$, $\mathbf{B} = \mathbf{j} - 5\mathbf{k}$, and $\mathbf{C} = -15\mathbf{i} + 3\mathbf{j} - 3\mathbf{k}$. Which vectors, if any, are **(a)** perpendicular, **(b)** parallel? Give reasons for your answers.

20. Writing to Learn Let $\mathbf{A} = \mathbf{i} + 2\mathbf{j} - \mathbf{k}$, $\mathbf{B} = -\mathbf{i} + \mathbf{j} + \mathbf{k}$, $\mathbf{C} = \mathbf{i} + \mathbf{k}$, and $\mathbf{D} = -(\pi/2)\mathbf{i} - \pi\mathbf{j} + (\pi/2)\mathbf{k}$. Which vectors, if any, are **(a)** perpendicular, **(b)** parallel? Give reasons for your answers.

In Exercises 21 and 22, *work in groups of two or three.* Find the magnitude of the torque exerted by $\mathbf{F}$ on the bolt at P if $\left|\overrightarrow{PQ}\right| = 8$ in. and $|\mathbf{F}| = 30$ lb. Answer in foot-pounds.

21.   **22.**

In Exercises 23–26, verify that $(\mathbf{A} \times \mathbf{B}) \cdot \mathbf{C} = (\mathbf{B} \times \mathbf{C}) \cdot \mathbf{A} = (\mathbf{C} \times \mathbf{A}) \cdot \mathbf{B}$ and find the volume of the parallelepiped (box) determined by $\mathbf{A}$, $\mathbf{B}$, and $\mathbf{C}$.

	A	**B**	**C**
23.	$2\mathbf{i}$	$2\mathbf{j}$	$2\mathbf{k}$
24.	$\mathbf{i} - \mathbf{j} + \mathbf{k}$	$2\mathbf{i} + \mathbf{j} - 2\mathbf{k}$	$-\mathbf{i} + 2\mathbf{j} - \mathbf{k}$
25.	$2\mathbf{i} + \mathbf{j}$	$2\mathbf{i} - \mathbf{j} + \mathbf{k}$	$\mathbf{i} + 2\mathbf{k}$
26.	$\mathbf{i} + \mathbf{j} - 2\mathbf{k}$	$-\mathbf{i} - \mathbf{k}$	$2\mathbf{i} + 4\mathbf{j} - 2\mathbf{k}$

Explorations

27. Which of the following are *always true* and which are *not always true?* Give reasons for your answers.

   (a) $|\mathbf{A}| = \sqrt{\mathbf{A} \cdot \mathbf{A}}$

   (b) $\mathbf{A} \cdot \mathbf{A} = |\mathbf{A}|$

   (c) $\mathbf{A} \times \mathbf{0} = \mathbf{0} \times \mathbf{A} = \mathbf{0}$

   (d) $\mathbf{A} \times (-\mathbf{A}) = \mathbf{0}$

   (e) $\mathbf{A} \times \mathbf{B} = \mathbf{B} \times \mathbf{A}$

   (f) $\mathbf{A} \times (\mathbf{B} + \mathbf{C}) = \mathbf{A} \times \mathbf{B} + \mathbf{A} \times \mathbf{C}$

   (g) $(\mathbf{A} \times \mathbf{B}) \cdot \mathbf{B} = 0$

   (h) $(\mathbf{A} \times \mathbf{B}) \cdot \mathbf{C} = \mathbf{A} \cdot (\mathbf{B} \times \mathbf{C})$

28. Which of the following are *always true* and which are *not always true?* Give reasons for your answers.

(a) $\mathbf{A} \cdot \mathbf{B} = \mathbf{B} \cdot \mathbf{A}$ (b) $\mathbf{A} \times \mathbf{B} = -(\mathbf{B} \times \mathbf{A})$

(c) $(-\mathbf{A}) \times \mathbf{B} = -(\mathbf{A} \times \mathbf{B})$

(d) $(c\mathbf{A}) \cdot \mathbf{B} = \mathbf{A} \cdot (c\mathbf{B}) = c(\mathbf{A} \cdot \mathbf{B})$, c any number

(e) $c(\mathbf{A} \times \mathbf{B}) = (c\mathbf{A}) \times \mathbf{B} = \mathbf{A} \times (c\mathbf{B})$, c any number

(f) $\mathbf{A} \cdot \mathbf{A} = |\mathbf{A}|^2$ (g) $(\mathbf{A} \times \mathbf{A}) \cdot \mathbf{A} = 0$

(h) $(\mathbf{A} \times \mathbf{B}) \cdot \mathbf{A} = \mathbf{B} \cdot (\mathbf{A} \times \mathbf{B})$

29. Given nonzero vectors **A**, **B**, and **C**, use dot-product and cross-product notation, as appropriate, to describe the following.

(a) The vector projection of **A** onto **B**

(b) A vector orthogonal to **A** and **B**

(c) A vector orthogonal to $\mathbf{A} \times \mathbf{B}$ and **C**

(d) The volume of the parallelepiped determined by **A**, **B**, and **C**

30. Given nonzero vectors **A**, **B**, and **C**, use dot-product and cross-product notation to describe the following.

(a) A vector orthogonal to $\mathbf{A} \times \mathbf{B}$ and $\mathbf{A} \times \mathbf{C}$

(b) A vector orthogonal to $\mathbf{A} + \mathbf{B}$ and $\mathbf{A} - \mathbf{B}$

(c) A vector of length $|\mathbf{A}|$ in the direction of **B**

(d) The area of the parallelogram determined by **A** and **C** ∎

31. Writing to Learn Let **A**, **B**, and **C** be vectors. Which of the following make sense, and which do not? Give reasons for your answers.

(a) $(\mathbf{A} \times \mathbf{B}) \cdot \mathbf{C}$ (b) $\mathbf{A} \times (\mathbf{B} \cdot \mathbf{C})$

(c) $\mathbf{A} \times (\mathbf{B} \times \mathbf{C})$ (d) $\mathbf{A} \cdot (\mathbf{B} \cdot \mathbf{C})$

32. Writing to Learn Show that except in degenerate cases, $(\mathbf{A} \times \mathbf{B}) \times \mathbf{C}$ lies in the plane of **A** and **B** whereas $\mathbf{A} \times (\mathbf{B} \times \mathbf{C})$ lies in the plane of **B** and **C**. What are the degenerate cases?

In Exercises 33–36, find the area of the parallelogram whose vertices are given.

33. $A(1, 0)$, $B(0, 1)$, $C(-1, 0)$, $D(0, -1)$

34. $A(0, 0)$, $B(7, 3)$, $C(9, 8)$, $D(2, 5)$

35. $A(-1, 2)$, $B(2, 0)$, $C(7, 1)$, $D(4, 3)$

36. $A(-6, 0)$, $B(1, -4)$, $C(3, 1)$, $D(-4, 5)$

In Exercises 37–40, find the area of the triangle whose vertices are given.

37. $A(0, 0)$, $B(-2, 3)$, $C(3, 1)$

38. $A(-1, -1)$, $B(3, 3)$, $C(2, 1)$

39. $A(-5, 3)$, $B(1, -2)$, $C(6, -2)$

40. $A(-6, 0)$, $B(10, -5)$, $C(-2, 4)$

Extending the Ideas

41. Writing to Learn Find a formula for the area of the triangle in the *xy*-plane with vertices at $(0, 0)$, (a_1, a_2), and (b_1, b_2). Explain your work.

42. Find a concise formula for the area of a triangle in the *xy*-plane with vertices (a_1, a_2), (b_1, b_2), and (c_1, c_2).

43. *Cancellation in Cross Products* If $\mathbf{A} \times \mathbf{B} = \mathbf{A} \times \mathbf{C}$ and $\mathbf{A} \neq \mathbf{0}$, then does $\mathbf{B} = \mathbf{C}$? Give reasons for your answer.

44. *Double Cancellation* If $\mathbf{A} \neq \mathbf{0}$ and if $\mathbf{A} \times \mathbf{B} = \mathbf{A} \times \mathbf{C}$ and $\mathbf{A} \cdot \mathbf{B} = \mathbf{A} \cdot \mathbf{C}$, then does $\mathbf{B} = \mathbf{C}$? Give reasons for your answer.

11.4   Lines and Planes in Space

Lines and Line Segments in Space   •   Equations for Planes in Space
•   Lines of Intersection

Lines and Line Segments in Space

Suppose L is a line in space passing through a point $P_0(x_0, y_0, z_0)$ parallel to a vector $\mathbf{v} = A\mathbf{i} + B\mathbf{j} + C\mathbf{k}$. Then L is the set of all points $P(x, y, z)$ for which $\overrightarrow{P_0P}$ is parallel to **v** (Figure 11.24). That is, P lies on L if and only if $\overrightarrow{P_0P}$ is a scalar multiple of **v**.

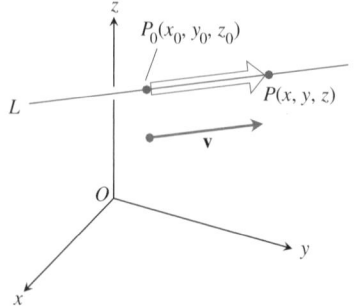

Figure 11.24 A point P lies on the line through P_0 parallel to **v** if and only if $\overrightarrow{P_0P}$ is a scalar multiple of **v**.

> **Vector Equation for a Line**
>
> A **vector equation for the line through $P_0(x_0, y_0, z_0)$ parallel to v** is
> $$\overrightarrow{P_0P} = t\mathbf{v}, \qquad -\infty < t < \infty. \qquad (1)$$

Equating the corresponding components of the two sides of Equation (1) gives three scalar equations involving the parameter t:

$$(x - x_0)\mathbf{i} + (y - y_0)\mathbf{j} + (z - z_0)\mathbf{k} = t(A\mathbf{i} + B\mathbf{j} + C\mathbf{k}) \quad \text{Eq. (1) expanded}$$

$$x - x_0 = tA, \qquad y - y_0 = tB, \qquad z - z_0 = tC. \quad \text{Components equated}$$

When rearranged, these equations give the standard parametrization of the line for the parameter interval $-\infty < t < \infty$.

Parametric Equations for a Line

The standard parametrization of the line through $P_0(x_0, y_0, z_0)$ parallel to $\mathbf{v} = A\mathbf{i} + B\mathbf{j} + C\mathbf{k}$ is

$$x = x_0 + tA, \qquad y = y_0 + tB, \qquad z = z_0 + tC, \qquad -\infty < t < \infty. \qquad (2)$$

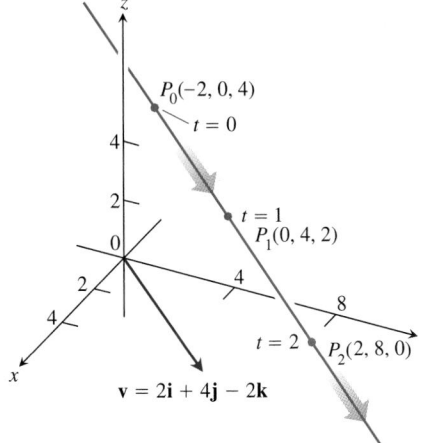

Figure 11.25 Selected points and parameter values on the line of Example 1. The arrows show the direction of increasing t.

Just as in the plane, parametrizations are not unique.

Example 1   PARAMETRIZING A LINE PASSING THROUGH A POINT PARALLEL TO A VECTOR

Find parametric equations for the line through $(-2, 0, 4)$ parallel to $\mathbf{v} = 2\mathbf{i} + 4\mathbf{j} - 2\mathbf{k}$ (Figure 11.25).

Solution   With $P_0(x_0, y_0, z_0)$ equal to $(-2, 0, 4)$ and $A\mathbf{i} + B\mathbf{j} + C\mathbf{k}$ equal to $2\mathbf{i} + 4\mathbf{j} - 2\mathbf{k}$, Equations (2) become

$$x = -2 + 2t, \qquad y = 4t, \qquad z = 4 - 2t.$$

Example 2   PARAMETRIZING A LINE THROUGH TWO POINTS

Find parametric equations for the line through $P(-3, 2, -3)$ and $Q(1, -1, 4)$.

Solution   The vector

$$\overrightarrow{PQ} = \big(1 - (-3)\big)\mathbf{i} + \big(-1 - 2\big)\mathbf{j} + \big(4 - (-3)\big)\mathbf{k}$$
$$= 4\mathbf{i} - 3\mathbf{j} + 7\mathbf{k}$$

is parallel to the line, and Equations (2) with $(x_0, y_0, z_0) = (-3, 2, -3)$ give

$$x = -3 + 4t, \qquad y = 2 - 3t, \qquad z = -3 + 7t.$$

We could have chosen $Q(1, -1, 4)$ as the "base point" and written

$$x = 1 + 4t, \qquad y = -1 - 3t, \qquad z = 4 + 7t.$$

These equations serve as well as the first; they simply place you at a different point for a given value of t.

To parametrize a line segment joining two points, we first parametrize the line through the points just as we did in the plane. We then find the t-values for the endpoints and restrict t to lie in the closed interval bounded by these values. The line equations together with this added restriction parametrize the segment, as illustrated in Example 3.

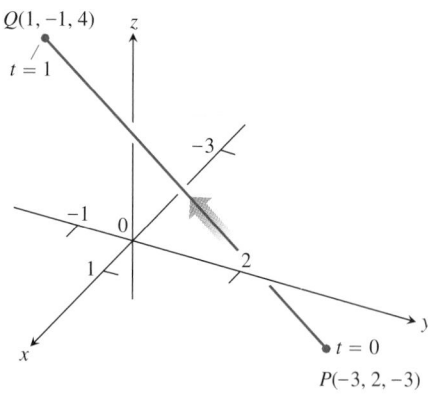

Figure 11.26 Example 3 derives a parametrization of line segment *PQ*. The arrow shows the direction of increasing *t*.

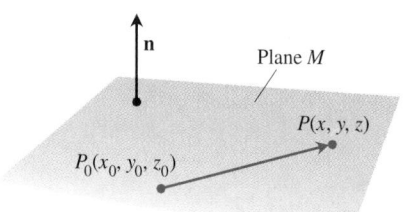

Figure 11.27 The standard equation for a plane in space is defined in terms of a vector normal to the plane: A point *P* lies in the plane through P_0 normal to **n** if and only if $\mathbf{n} \cdot \overrightarrow{P_0P} = 0$.

Example 3 PARAMETRIZING A LINE SEGMENT

Parametrize the line segment joining the points $P(-3, 2, -3)$ and $Q(1, -1, 4)$ (Figure 11.26).

Solution We begin with equations for the line through *P* and *Q*, taking them, in this case, from Example 2:

$$x = -3 + 4t, \qquad y = 2 - 3t, \qquad z = -3 + 7t.$$

We observe that the point

$$(x, y, z) = (-3 + 4t, 2 - 3t, -3 + 7t)$$

passes through $P(-3, 2, -3)$ at $t = 0$ and $Q(1, -1, 4)$ at $t = 1$. We add the restriction $0 \leq t \leq 1$ to parametrize the segment:

$$x = -3 + 4t, \qquad y = 2 - 3t, \qquad z = -3 + 7t, \qquad 0 \leq t \leq 1.$$

Equations for Planes in Space

Suppose plane *M* passes through a point $P_0(x_0, y_0, z_0)$ and is normal (perpendicular) to the nonzero vector $\mathbf{n} = A\mathbf{i} + B\mathbf{j} + C\mathbf{k}$. Then *M* is the set of all points $P(x, y, z)$ for which $\overrightarrow{P_0P}$ is orthogonal to **n** (Figure 11.27). That is, *P* lies on *M* if and only if $\mathbf{n} \cdot \overrightarrow{P_0P} = 0$. This equation is equivalent to

$$(A\mathbf{i} + B\mathbf{j} + C\mathbf{k}) \cdot [(x - x_0)\mathbf{i} + (y - y_0)\mathbf{j} + (z - z_0)\mathbf{k}] = 0$$

or

$$A(x - x_0) + B(y - y_0) + C(z - z_0) = 0.$$

Equations for a Plane

The **plane through $P_0(x_0, y_0, z_0)$ normal to $\mathbf{n} = A\mathbf{i} + B\mathbf{j} + C\mathbf{k}$** has

Vector equation:	$\mathbf{n} \cdot \overrightarrow{P_0P} = 0$
Component equation:	$A(x - x_0) + B(y - y_0) + C(z - z_0) = 0$
Component equation simplified:	$Ax + By + Cz = D$, where $D = Ax_0 + By_0 + Cz_0$

Example 4 FINDING AN EQUATION FOR A PLANE

Find an equation for the plane through $P_0(-3, 0, 7)$ perpendicular to $\mathbf{n} = 5\mathbf{i} + 2\mathbf{j} - \mathbf{k}$.

Solution The component equation is

$$5(x - (-3)) + 2(y - 0) + (-1)(z - 7) = 0.$$

Simplifying, we obtain

$$5x + 15 + 2y - z + 7 = 0$$
$$5x + 2y - z = -22.$$

$A\mathbf{i} + B\mathbf{j} + C\mathbf{k}$ is normal to the plane $Ax + By + Cz = D$.

Notice in Example 4 how the components of $\mathbf{n} = 5\mathbf{i} + 2\mathbf{j} - \mathbf{k}$ became the coefficients of x, y, and z in the equation $5x + 2y - z = -22$, and $D = Ax_0 + By_0 + Cz_0 = -22$.

Exploration 1 **Finding an Equation for the Plane Through Three Points**

Two points in the xy-plane determine a line. Given the coordinates of the two points, we can easily determine the corresponding linear equation $ax + by = c$. Similarly, three points in xyz-space determine a plane with a linear equation $ax + by + cz = d$. Given the coordinates of three points, we can use vectors to find the equation, as in this exploration.

1. Given the three points $A(0, 0, 1)$, $B(2, 0, 0)$, and $C(0, 3, 0)$, find the vectors $\overrightarrow{AB}$ and $\overrightarrow{AC}$ in component form.

2. Use the cross product to find a vector perpendicular to both $\overrightarrow{AB}$ and $\overrightarrow{AC}$. Call it **v.**

3. Find an equation of the plane through A perpendicular to **v.**

Example 5 FINDING THE INTERSECTION OF A LINE AND A PLANE

Find the point where the line

$$x = \frac{8}{3} + 2t, \qquad y = -2t, \qquad z = 1 + t$$

intersects the plane $3x + 2y + 6z = 6$.

Solution The point

$$\left(\frac{8}{3} + 2t, -2t, 1 + t \right)$$

lies in the plane if its coordinates satisfy the equation of the plane; that is, if

$$3 \left(\frac{8}{3} + 2t \right) + 2(-2t) + 6(1 + t) = 6,$$

or

$$8 + 6t - 4t + 6 + 6t = 6$$
$$8t = -8$$
$$t = -1.$$

The point of intersection is

$$(x, y, z)\Big|_{t=-1} = \left(\frac{8}{3} - 2, 2, 1 - 1 \right) = \left(\frac{2}{3}, 2, 0 \right)$$

Lines of Intersection

Two planes that are not parallel intersect in a line.

Example 6 FINDING A VECTOR PARALLEL TO THE LINE OF INTERSECTION OF TWO PLANES

Find a vector parallel to the line of intersection of the planes $3x - 6y - 2z = 15$ and $2x + y - 2z = 5$.

Solution The line of intersection of two planes is perpendicular to the planes' normal vectors $\mathbf{n}_1$ and $\mathbf{n}_2$ (Figure 11.28) and therefore parallel to $\mathbf{n}_1 \times \mathbf{n}_2$. Turning this around, $\mathbf{n}_1 \times \mathbf{n}_2$ is a vector parallel to the planes' line of intersection. In our case,

$$\mathbf{n}_1 \times \mathbf{n}_2 = \begin{vmatrix} \mathbf{i} & \mathbf{j} & \mathbf{k} \\ 3 & -6 & -2 \\ 2 & 1 & -2 \end{vmatrix} = 14\mathbf{i} + 2\mathbf{j} + 15\mathbf{k}.$$

Any nonzero scalar multiple of $\mathbf{n}_1 \times \mathbf{n}_2$ will do as well.

Figure 11.28 How the line of intersection of two planes is related to the planes' normal vectors. (Example 6)

Example 7 PARAMETRIZING THE LINE OF INTERSECTION OF TWO PLANES

Find parametric equations for the line in which the planes $3x - 6y - 2z = 15$ and $2x + y - 2z = 5$ intersect.

Solution We find a vector parallel to the line and a point on the line and use Equations (2).

Example 6 identifies $\mathbf{v} = 14\mathbf{i} + 2\mathbf{j} + 15\mathbf{k}$ as a vector parallel to the line. To find a point on the line, we can take any point common to the two planes. Substituting $z = 0$ in the planes' equations and solving for x and y simultaneously identifies one of the points as $(3, -1, 0)$. The line is

$$x = 3 + 14t, \qquad y = -1 + 2t, \qquad z = 15t.$$

Quick Review 11.4

1. Which of the following points are solutions of the equation $2x - y + 3z = -3$?

(a) $(1, 2, -1)$ (b) $(3, 1, 2)$

(c) $(-1, 3, -4)$ (d) $(5, 0, 4)$

In Exercises 2–4, find a parametrization for the curve.

2. The line through $(-2, -3)$ and $(3, 1)$

3. The line segment with endpoints $(-2, -3)$ and $(3, 1)$

4. The line through $(1, 2)$ and parallel to the line $2x + 3y = 1$

In Exercises 5 and 6, solve the system of equations.

5. $x - 2y = -1$ 6. $3x + 2y = 1$
    $2x + \ y = 2$ $4x - 3y = -1$

7. Find a unit vector in the direction of $\mathbf{v} = -2\mathbf{i} + \mathbf{j} + 2\mathbf{k}$.

In Exercises 8–10, let $\mathbf{v}_1 = 3\mathbf{i} - 2\mathbf{j} + \mathbf{k}$ and $\mathbf{v}_2 = -\mathbf{i} + \mathbf{j} - 2\mathbf{k}$.

8. Find $\mathbf{u} = \mathbf{v}_1 \times \mathbf{v}_2$ and $\mathbf{w} = \mathbf{v}_2 \times \mathbf{v}_1$.

9. Show that $\mathbf{u} \cdot \mathbf{v}_1 = \mathbf{w} \cdot \mathbf{v}_1 = 0$.

10. Show that $\mathbf{u} \cdot \mathbf{v}_2 = \mathbf{w} \cdot \mathbf{v}_2 = 0$.

Section 11.4 Exercises

In Exercises 1–10, find parametric equations for the line.

1. The line through the point $P(3, -4, -1)$ parallel to the vector $\mathbf{i} + \mathbf{j} + \mathbf{k}$

2. The line through $P(1, 2, -1)$ and $Q(-1, 0, 1)$

3. The line through $P(-2, 0, 3)$ and $Q(3, 5, -2)$

4. The line through the origin parallel to the vector $2\mathbf{j} + \mathbf{k}$

5. The line through the point $(3, -2, 1)$ parallel to the line $x = 1 + 2t,\ y = 2 - t,\ z = 3t$

6. The line through $(1, 1, 1)$ parallel to the z-axis

7. The line through $(2, 4, 5)$ perpendicular to the plane $3x + 7y - 5z = 21$

8. The line through $(0, -7, 0)$ perpendicular to the plane $x + 2y + 2z = 13$

9. The line through $(2, 3, 0)$ perpendicular to the vectors $\mathbf{A} = \mathbf{i} + 2\mathbf{j} + 3\mathbf{k}$ and $\mathbf{B} = 3\mathbf{i} + 4\mathbf{j} + 5\mathbf{k}$

10. The x-axis

In Exercises 11–14, find parametrizations for the line segment joining the points. Draw coordinate axes and sketch each segment, indicating the direction of increasing t for your parametrization.

11. $(0, 0, 0)$, $(1, 1, 3/2)$

12. $(1, 0, 0)$, $(1, 1, 0)$

13. $(0, 1, 1)$, $(0, -1, 1)$

14. $(1, 0, -1)$, $(0, 3, 0)$

In Exercises 15–20, find an equation for the plane.

15. The plane through $P_0(0, 2, -1)$ normal to $\mathbf{n} = 3\mathbf{i} - 2\mathbf{j} - \mathbf{k}$

16. The plane through $(1, -1, 3)$ parallel to the plane $3x + y + z = 7$

17. The plane through $(1, 1, -1)$, $(2, 0, 2)$, and $(0, -2, 1)$

18. The plane through $(2, 4, 5)$, $(1, 5, 7)$, and $(-1, 6, 8)$

19. The plane through $P_0(2, 4, 5)$ perpendicular to the line $x = 5 + t,\ y = 1 + 3t,\ z = 4t$

20. The plane through $A(1, -2, 1)$ perpendicular to the vector from the origin to A

21. Find the point of intersection of the lines $x = 2t + 1,\ y = 3t + 2,\ z = 4t + 3$, and $x = s + 2,\ y = 2s + 4,\ z = -4s - 1$, and then find the plane determined by these lines.

22. Find the point of intersection of the lines $x = t,\ y = -t + 2,\ z = t + 1$, and $x = 2s + 2,\ y = s + 3,\ z = 5s + 6$, and then find the plane determined by these lines.

In Exercises 23 and 24, find the plane determined by the intersecting lines.

23. $L_1: x = -1 + t,\ y = 2 + t,\ z = 1 - t,\ -\infty < t < \infty$

$L_2: x = 1 - 4s,\ y = 1 + 2s,\ z = 2 - 2s,\ -\infty < s < \infty$

24. $L_1: x = t,\ y = 3 - 3t,\ z = -2 - t,\ -\infty < t < \infty$

$L_2: x = 1 + s,\ y = 4 + s,\ z = -1 + s,\ -\infty < s < \infty$

25. Find a plane through $P_0(2, 1, -1)$ and perpendicular to the line of intersection of the planes $2x + y - z = 3$, $x + 2y + z = 2$.

26. Find a plane through the points $P_1(1, 2, 3)$, $P_2(3, 2, 1)$ and perpendicular to the plane $4x - y + 2z = 7$.

Exploration

27. *Distance from a Point to a Line* Follow these steps to find the distance from a point S to a line that passes through a point P parallel to a vector $\mathbf{v}$ as shown in the figure.

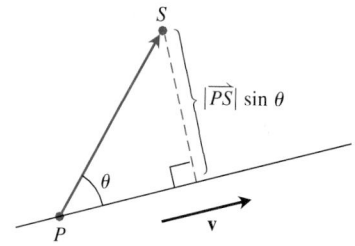

(a) Show that the length of the component of $\overrightarrow{PS}$ normal to the line is $\left|\overrightarrow{PS}\right| \sin \theta$.

(b) Show that the distance d from S to the line through P parallel to $\mathbf{v}$ is

$$d = \frac{\left|\overrightarrow{PS} \times \mathbf{v}\right|}{|\mathbf{v}|}.$$

In Exercises 28–30, use the result of Exercise 27 to find the distance from the point to the line.

28. $(0, 0, 0)$; $x = 5 + 3t,\ y = 5 + 4t,\ z = -3 - 5t$

29. $(2, 1, 3)$; $x = 2 + 2t,\ y = 1 + 6t,\ z = 3$

30. $(3, -1, 4)$; $x = 4 - t,\ y = 3 + 2t,\ z = -5 + 3t$

Explorations

31. *Distance from a Point to a Plane* Follow these steps to find the distance d from a point S to a plane $Ax + By + Cz = D$.

(a) Find a point P on the plane.

(b) Find $\overrightarrow{PS}$.

(c) Show that the distance is

$$d = \left|\overrightarrow{PS} \cdot \frac{\mathbf{n}}{|\mathbf{n}|}\right|,$$

where $\mathbf{n} = A\mathbf{i} + B\mathbf{j} + C\mathbf{k}$.

In Exercises 32–34, use the result of Exercise 31 to find the distance from the point to the plane.

32. $(2, -3, 4)$, $x + 2y + 2z = 13$

33. $(0, 1, 1)$, $4y + 3z = -12$

34. $(0, -1, 0)$, $2x + y + 2z = 4$

35. Find the distance from the plane $x + 2y + 6z = 1$ to the plane $x + 2y + 6z = 10$.

36. Find the distance from the line $x = 2 + t$, $y = 1 + t$, $z = -(1/2) - (1/2)t$ to the plane $x + 2y + 6z = 10$.

Exploration

37. *Angle Between Planes* The **angle between two intersecting planes** is defined to be the (acute) angle determined by the normal vectors as shown in the figure.

(a) If $\mathbf{n}_1$ and $\mathbf{n}_2$ are the normals to two planes, show that the angle between the planes is

$$\theta = \cos^{-1}\left(\frac{|\mathbf{n}_1 \cdot \mathbf{n}_2|}{|\mathbf{n}_1||\mathbf{n}_2|}\right).$$

(b) Show that the angle between the planes $3x - 6y - 2z = 15$ and $2x + y - 2z = 5$ is about 1.38 radians.

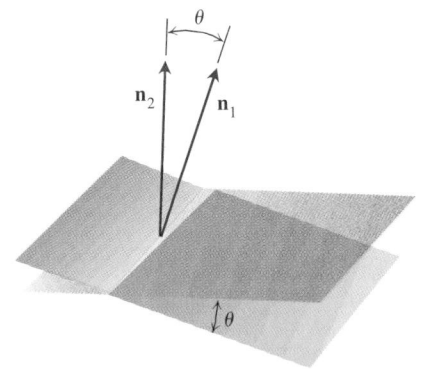

In Exercises 38–40, use the result of Exercise 37 to find the acute angle between the planes.

38. $x + y + z = 1$, $z = 0$

39. $2x + 2y - z = 3$, $x + 2y + z = 2$

40. $4y + 3z = -12$, $3x + 2y + 6z = 6$

In Exercises 41–44, find the point in which the line meets the plane.

41. $x = 1 - t$, $y = 3t$, $z = 1 + t$; $2x - y + 3z = 6$

42. $x = 2$, $y = 3 + 2t$, $z = -2 - 2t$; $6x + 3y - 4z = -12$

43. $x = 1 + 2t$, $y = 1 + 5t$, $z = 3t$; $x + y + z = 2$

44. $x = -1 + 3t$, $y = -2$, $z = 5t$; $2x - 3z = 7$

In Exercises 45–48, find a parametrization for the line in which the planes intersect.

45. $x + y + z = 1$, $x + y = 2$

46. $3x - 6y - 2z = 3$, $2x + y - 2z = 2$

47. $x - 2y + 4z = 2$, $x + y - 2z = 5$

48. $5x - 2y = 11$, $4y - 5z = -17$

In Exercises 49 and 50, *work in groups of two or three*. Given two lines in space, either they are parallel, or they intersect, or they are skew (imagine, for example, the flight paths of two airplanes in the sky). Determine whether the lines, taken two at a time, are parallel, intersect, or are skew. If they intersect, find the point of intersection.

49. L_1: $x = 3 + 2t$, $y = -1 + 4t$, $z = 2 - t$, $-\infty < t < \infty$

L_2: $x = 1 + 4s$, $y = 1 + 2s$, $z = -3 + 4s$, $-\infty < s < \infty$

L_3: $x = 3 + 2r$, $y = 2 + r$, $z = -2 + 2r$, $-\infty < r < \infty$

50. L_1: $x = 1 + 2t$, $y = -1 - t$, $z = 3t$, $-\infty < t < \infty$

L_2: $x = 2 - s$, $y = 3s$, $z = 1 + s$, $-\infty < s < \infty$

L_3: $x = 5 + 2r$, $y = 1 - r$, $z = 8 + 3r$, $-\infty < r < \infty$

51. Use Equations (2) to generate a parametrization of the line through $P_1(2, -4, 7)$ parallel to $\mathbf{v}_1 = 2\mathbf{i} - \mathbf{j} + 3\mathbf{k}$. Then generate another parametrization of the line using the point $P_2(-2, -2, 1)$ and the vector $\mathbf{v}_2 = -\mathbf{i} + (1/2)\mathbf{j} - (3/2)\mathbf{k}$.

52. Use the component form to generate an equation for the plane through $P_1(4, 1, 5)$ normal to $\mathbf{n}_1 = \mathbf{i} - 2\mathbf{j} + \mathbf{k}$. Then generate another equation for the same plane using the point $P_2(3, -2, 0)$ and the normal vector $\mathbf{n}_2 = -\sqrt{2}\mathbf{i} + 2\sqrt{2}\mathbf{j} - \sqrt{2}\mathbf{k}$.

53. Writing to Learn Find the points in which the line $x = 1 + 2t$, $y = -1 - t$, $z = 3t$ meets the coordinate planes. Describe the reasoning behind your answer.

54. Writing to Learn Find equations for the line in the plane $z = 3$ that makes an angle of $\pi/6$ rad with $\mathbf{i}$ and an angle of $\pi/3$ rad with $\mathbf{j}$. Describe the reasoning behind your answer.

55. Writing to Learn Is the line $x = 1 - 2t$, $y = 2 + 5t$, $z = -3t$ parallel to the plane $2x + y - z = 8$? Give reasons for your answer.

56. Writing to Learn How can you tell when two planes $A_1x + B_1y + C_1z = D_1$ and $A_2x + B_2y + C_2z = D_2$ are parallel? Perpendicular? Give reasons for your answer.

57. Find two different planes whose intersection is the line $x = 1 + t$, $y = 2 - t$, $z = 3 + 2t$. Write equations for each plane in the form $Ax + By + Cz = D$.

58. Writing to Learn Find a plane through the origin that meets the plane M: $2x + 3y + z = 12$ in a right angle. How do you know that your plane is perpendicular to M?

59. Writing to Learn For any nonzero numbers a, b, and c, the graph of $(x/a) + (y/b) + (z/c) = 1$ is a plane. Which planes have an equation of this form?

60. Writing to Learn Suppose L_1 and L_2 are disjoint (nonintersecting) nonparallel lines. Is it possible for a nonzero vector to be perpendicular to both L_1 and L_2? Give reasons for your answer.

Extending the Ideas

61. *Perspective in Computer Graphics* In computer graphics and perspective drawing, we need to represent objects seen by the eye in space as images on a two-dimensional plane. Suppose the eye is at $E(x_0, 0, 0)$ as shown here and we want to represent a point $P_1(x_1, y_1, z_1)$ as a point on the yz-plane. We do this by projecting P_1 onto the plane with a ray from E. The point P_1 will be portrayed as the point $P(0, y, z)$. The problem for us as graphics designers is to find y and z given E and P_1.

(a) Write a vector equation that holds between $\overrightarrow{EP}$ and $\overrightarrow{EP_1}$. Use the equation to express y and z in terms of x_0, x_1, y_1, and z_1.

(b) Test the formula obtained for y and z in part (a) by investigating their behavior at $x_1 = 0$ and $x_1 = x_0$ and by seeing what happens as $x_0 \to \infty$.

62. *Hidden Lines* Here is another typical problem in computer graphics. Your eye is at $(4, 0, 0)$. You are looking at a

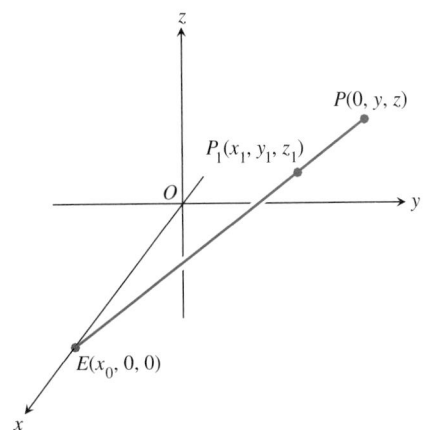

triangular plate whose vertices are $(1, 0, 1)$, $(1, 1, 0)$, and $(-2, 2, 2)$. The line segment from $(1, 0, 0)$ to $(0, 2, 2)$ passes through the plate. What portion of the line segment is hidden from your view by the plate?

11.5 Cylinders and Cylindrical Coordinates

Cylinders • Cylindrical Coordinates

Cylinders

We normally think of a circular cylinder as something like a tin can. When we do geometry with coordinates in space, however, we take a more general view. Cylinders no longer have tops or bottoms, and their sides extend infinitely in both directions. They are surfaces made of parallel lines passing through a circle. Once we take this point of view, it is easy to generalize still further. Why does the curve through which the parallel lines pass have to be a circle? Why can't it be a curve of any kind? Indeed, it can. There can be elliptical cylinders (generating curve an ellipse), parabolic cylinders, hyperbolic cylinders, and so on.

> **Definitions** Cylinder and Generating Curve
>
> A **cylinder** is a surface composed of the lines that pass through a given plane curve parallel to a given line in space. The curve is a **generating curve** for the cylinder. See Figure 11.29.

Figure 11.29 A cylinder and generating curve.

Generating curve (in the yz-plane)

Lines through generating curve parallel to x-axis

Example 1 FINDING AN EQUATION FOR A PARABOLIC CYLINDER

Find an equation for the cylinder composed of the lines that pass through the curve $y = x^2$, $z = 0$ parallel to the z-axis (Figure 11.30).

Solution If the point $P_0(x_0, x_0{}^2, 0)$ lies on the parabola in the xy-plane, then the point $Q(x_0, x_0{}^2, z)$ will lie on the cylinder because it lies on the line $x = x_0$, $y = x_0{}^2$ through P_0 parallel to the z-axis. Conversely, any point $Q(x_0, x_0{}^2, z)$ whose y-coordinate is the square of its x-coordinate will lie on the cylinder because it lies on the line $x = x_0$, $y = x_0{}^2$ through P_0 parallel

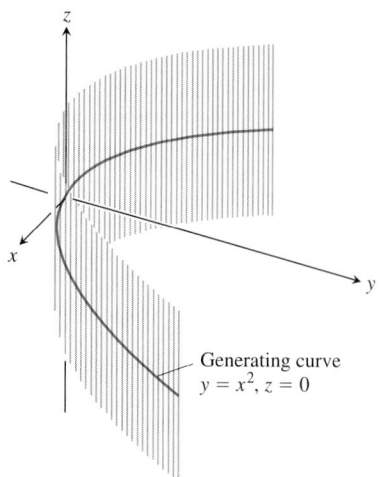

Figure 11.30 The cylinder of lines through the parabola $y = x^2$ in the xy-plane parallel to the z-axis. (Example 1)

to the z-axis. Regardless of the value of their z-coordinate, the points on the surface are the points whose x- and y-coordinates satisfy the equation $y = x^2$ (Figure 11.31). This is the equation we seek.

Example 2   SKETCHING CYLINDERS

Identify and sketch the cylinders (a) $x^2 + y^2 = 1$ and (b) $z = y^2$.

Solution

(a) The cylinder $x^2 + y^2 = 1$: Since the equation places no restriction on z, the cylinder consists of all the points in space lying directly above or below points of the unit circle in the xy-plane. The cylinder is the circular cylinder of radius 1 with axis the z-axis. See the drawing lesson on page 609 for how to sketch it.

(b) The cylinder $z = y^2$: This time there is no restriction on the coordinate x. The cylinder consists of all the points lying on lines through the parabola $z = y^2$ in the yz-plane parallel to the x-axis, the points whose y- and z-coordinates satisfy the equation $z = y^2$ regardless of the value of x. (See the drawing lesson on page 609.) The graph in Figure 11.32 supports this conclusion.

As the examples suggest, any curve $f(x, y) = c$ in the xy-plane defines a cylinder parallel to the z-axis whose equation is also $f(x, y) = c$. Similarly, an equation in x and z defines a cylinder parallel to the y-axis, and an equation in y and z defines a cylinder parallel to the x-axis. See Figures 11.33 and 11.34.

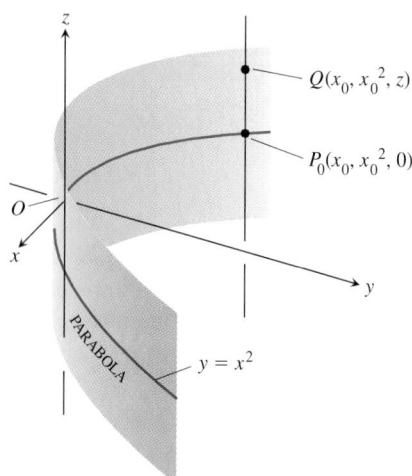

Figure 11.31 The cylinder $y = x^2$.

Figure 11.32 $z = y^2$. (Example 2)

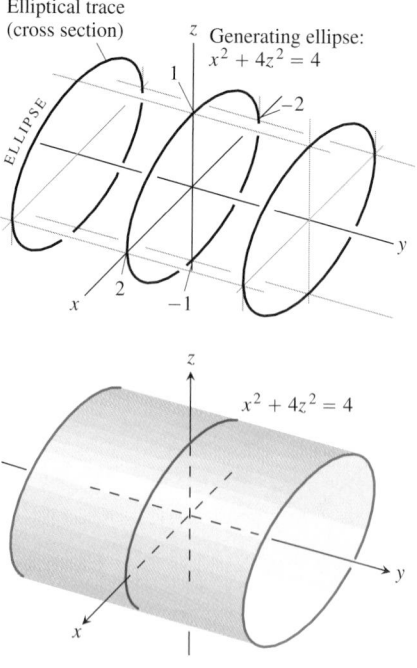

Figure 11.33 The elliptical cylinder $x^2 + 4z^2 = 4$ is made of lines parallel to the y-axis through the ellipse $x^2 + 4z^2 = 4$ in the xz-plane. The cross sections or "traces" of the cylinder in planes perpendicular to the y-axis are ellipses congruent to the generating ellipse. The cylinder extends along the entire y-axis.

Drawing Lesson

How to Draw Cylinders Parallel to the Coordinate Axes

$x^2 + y^2 = 1$ $z = y^2$

1. Sketch all three coordinate axes *very lightly*.

    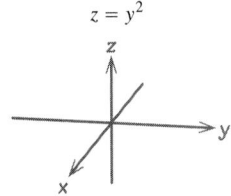

2. Sketch the trace of the cylinder in the coordinate plane of the two variables that appear in the cylinder's equation. Sketch *very lightly*.

    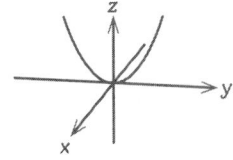

3. Sketch traces in parallel planes on either side (again, *lightly*).

    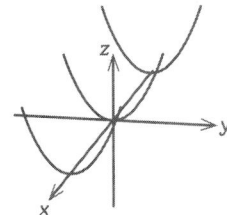

4. Add parallel outer edges to give the shape definition.

    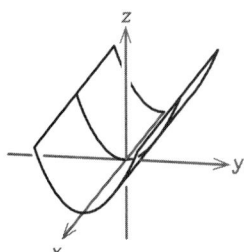

5. If more definition is required, darken the parts of the lines that are exposed to view. Leave the hidden parts light. Use line breaks when you can.

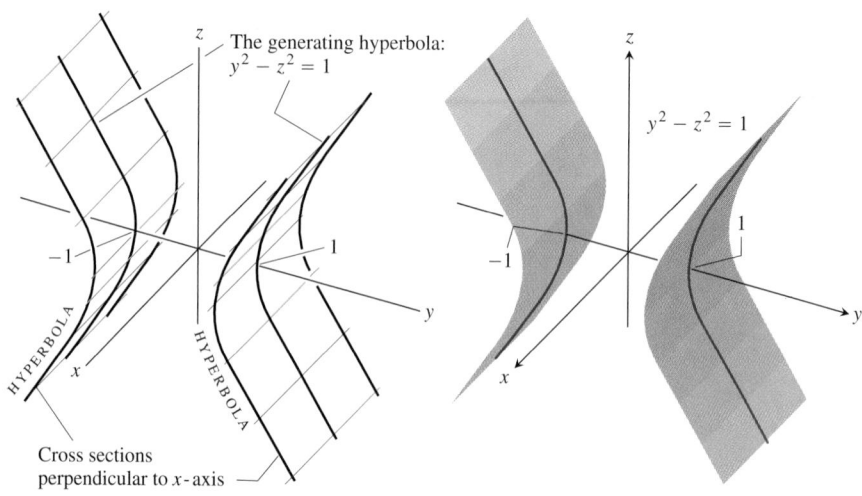

Figure 11.34 The hyperbolic cylinder $y^2 - z^2 = 1$ is made of lines parallel to the x-axis through the hyperbola $y^2 - z^2 = 1$ in the yz-plane. The cross sections perpendicular to the x-axis are hyperbolas congruent to the generating hyperbola.

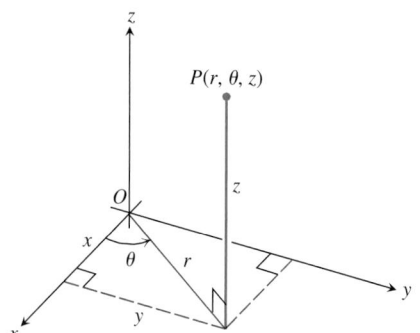

Figure 11.35 The cylindrical coordinates of the point P are r, θ, and z.

Cylindrical Coordinates

We obtain cylindrical coordinates for space by replacing the Cartesian coordinates in the xy-plane by polar coordinates. This assigns to every point in space one or more coordinate triples of the form (r, θ, z), as shown in Figure 11.35.

Definition Cylindrical Coordinates

Cylindrical coordinates represent a point P in space by ordered triples (r, θ, z) in which

1. r and θ are polar coordinates for the vertical projection of P on the xy-plane, and

2. z is the rectangular vertical coordinate.

The values of x, y, r, and θ are related by the usual equations.

Equations Relating Rectangular and Cylindrical Coordinates

$$x = r \cos \theta, \qquad y = r \sin \theta, \qquad z = z$$
$$r^2 = x^2 + y^2 \qquad \tan \theta = y/x \qquad\qquad (1)$$

In cylindrical coordinates, the equation $r = a$, $a > 0$ describes not just a circle of radius a in the plane but an entire cylinder about the z-axis (see Figure

11.36). The simplicity of this equation is one reason for the popularity of cylindrical coordinates. Another is the efficiency cylindrical coordinates bring to descriptions of planetary motion (Section 12.4).

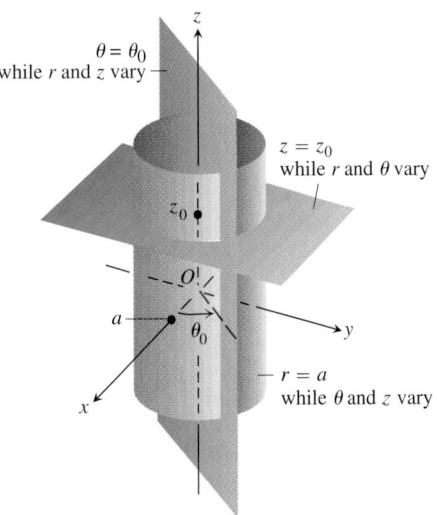

Figure 11.36 Constant-coordinate equations in cylindrical coordinates define cylinders and planes.

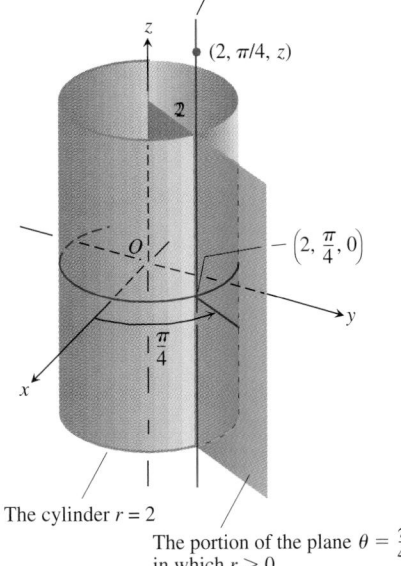

Along this line, z varies while r and θ have the constant values $r = 2$ and $\theta = \pi/4$.

The cylinder $r = 2$

The portion of the plane $\theta = \dfrac{\pi}{4}$ in which $r > 0$

Figure 11.37 The points whose first two cylindrical coordinates are $r = 2$ and $\theta = \pi/4$ form a line parallel to the z-axis. (Example 3)

Example 3 IDENTIFYING A VERTICAL LINE

What points satisfy the equations

$$r = 2, \qquad \theta = \frac{\pi}{4}?$$

Solution These points make up the line in which the cylinder $r = 2$ cuts the portion of the plane $\theta = \pi/4$ where r is positive (Figure 11.37). This is the line through the point $(2, \pi/4, 0)$ parallel to the z-axis.

Exploration 1 **Easy as Pi**

1. A point has cylindrical coordinates (π, π, π). What are its rectangular coordinates?

2. A point has rectangular coordinates (π, π, π). What are its cylindrical coordinates?

3. Describe the graph of the equation $r = \pi$.

4. Describe the graph of the equation $\theta = \pi$.

5. Describe the graph formed by the intersection of the graphs in parts 3 and 4.

6. Describe the graph of all points satisfying the simultaneous equations $r = \pi$ and $\theta = \pi$.

7. Why are the answers to parts 5 and 6 different?

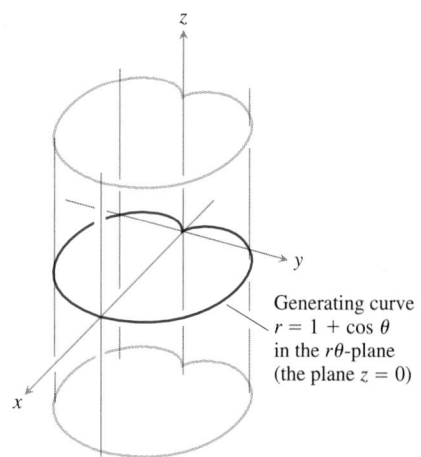

Generating curve
$r = 1 + \cos \theta$
in the $r\theta$-plane
(the plane $z = 0$)

Figure 11.38 The equation $r = 1 + \cos \theta$ defines a cylinder in space whose cross sections perpendicular to the z-axis are cardioids. (Example 4)

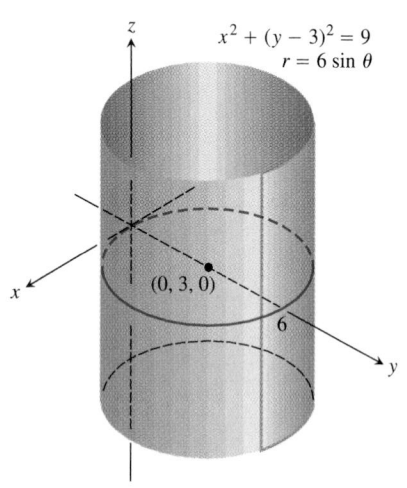

$x^2 + (y - 3)^2 = 9$
$r = 6 \sin \theta$

$(0, 3, 0)$

6

Figure 11.39 The cylinder in Example 6.

Example 4   SKETCHING A SURFACE IN CYLINDRICAL COORDINATES

Sketch the surface $r = 1 + \cos \theta$.

Solution   The equation involves only r and θ; the variable z can take on any value. Therefore, the surface is a cylinder of lines that pass through the cardioid $r = 1 + \cos \theta$ in the $r\theta$-plane parallel to the z-axis. The rules for sketching the cylinder are the same as before: Sketch the x-, y-, and z-axes, draw a few cross sections, connect the cross sections with parallel lines, and darken the exposed parts (Figure 11.38).

Example 5   CHANGING FROM CYLINDRICAL TO CARTESIAN COORDINATES

Find a Cartesian equation for the cylinder $r = 3/2$.

Solution   Using Equations (1), we have

$$r = \frac{3}{2}$$

$$\sqrt{x^2 + y^2} = \frac{3}{2} \qquad \text{Substitute } \sqrt{x^2 + y^2} \text{ for } r.$$

$$4x^2 + 4y^2 = 9 \qquad \text{Square and mulitiply both sides by 4.}$$

Example 6   CHANGING FROM CARTESIAN TO CYLINDRICAL COORDINATES

Find a cylindrical-coordinate equation for the cylinder $x^2 + (y - 3)^2 = 9$ (Figure 11.39).

Solution   The equation for the cylinder in cylindrical coordinates is the same as the polar equation for the cylinder's trace (cross section) in the xy-plane, namely $r = 6 \sin \theta$ (see Section 10.5, Example 5).

Quick Review 11.5

In Exercises 1–4, graph the equation in the xy-plane and identify the graph.

1. (a) $x^2 + 2x + y^2 = 0$ (b) $x^2 + y^2 - 4y = 0$

2. (a) $x^2 + 2y^2 = 8$ (b) $4x^2 + y^2 = 36$

3. (a) $xy = 4$ (b) $x^2 - 4y^2 = 36$

4. (a) $y = x^2 - 1$ (b) $x = 4 - y^2$

In Exercises 5–8, find a polar equation for the circle. Then graph the circle in the polar-coordinate plane.

5. $x^2 + y^2 = 9$

6. $(x - 1)^2 + y^2 = 1$

7. $x^2 + (y + 2)^2 = 4$

8. $(x + 3)^2 + y^2 = 9$

In Exercises 9 and 10, graph the equation in the polar-coordinate plane and identify the graph.

9. (a) $r = 1 + \cos \theta$ **(b)** $r = \sin \theta - 1$

10. (a) $r = 2 \csc \theta$ **(b)** $r = -\sec \theta$

Section 11.5 Exercises

In Exercises 1 and 2, match the equation with the cylinder it defines. What kinds of cylinders are these?

1. $9y^2 + z^2 = 16$ **2.** $x^2 + 2z^2 = 8$

(a)

(b)
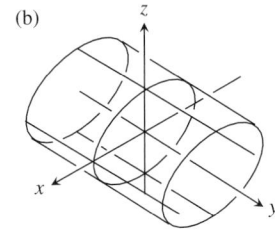

In Exercises 3–12, *work in groups of two or three* to sketch the surface.

3. $x^2 + y^2 = 4$ **4.** $x^2 + z^2 = 4$

5. $z^2 - y^2 = 1$ **6.** $x = y^2$

7. $x^2 + 4z^2 = 16$ **8.** $4x^2 + y^2 = 36$

9. $16x^2 + 4y^2 = 1$ **10.** $x = 4 - y^2$

11. $z^2 + 4y^2 = 9$ **12.** $x^2 - 4y^2 = 1$

In Exercises 13–16, match the equation with its graph.

13. $z = y^2 - 1$ **14.** $yz = 1$

15. $x^2 z = 1$ **16.** $z = 1 - x^2$

(a)

(b)

(c)

(d)

In Exercises 17–24, the coordinates of a point in space are given in one of two coordinate systems. Find the coordinates of the point in the other system. There may be more than one answer because points in cylindrical coordinates can have more than one coordinate triple.

	Rectangular (x, y, z)	Cylindrical (r, θ, z)
17.	$(0, 0, 0)$	
18.	$(1, 0, 0)$	
19.	$(0, 1, 0)$	
20.	$(0, 0, 1)$	
21.		$(1, 0, 0)$
22.		$(\sqrt{2}, 0, 1)$
23.		$(1, \pi/2, 1)$
24.		$(-1, \pi/2, -1)$

In Exercises 25–34, translate the equation from the given coordinate system (rectangular, cylindrical) into an equation in the other system. Then identify the figure in space defined by the equation or by the equation and inequality.

25. $r = 0$ **26.** $x^2 + y^2 = 5$

27. $z = 0$ **28.** $z = -2$

29. $x^2 + y^2 + z^2 = 4$ **30.** $x^2 + y^2 + \left(z - \dfrac{1}{2}\right)^2 = \dfrac{1}{4}$

31. $r = \csc \theta$ **32.** $r = -3 \sec \theta$

33. $x^2 + y^2 + (z-1)^2 = 1, z \le 1$

34. $r^2 + z^2 = 4, z \le -\sqrt{2}$

In Exercises 35–38, describe the set of points whose cylindrical coordinates satisfy the given equation. Sketch each set. *Work in groups of two or three.*

35. $r = -2 \sin \theta$ **36.** $r = 2 \cos \theta$

37. $r = 1 - \cos \theta$ **38.** $r = 1 + \sin \theta$

Extending the Ideas

39. Find the Cartesian coordinates of the center of the sphere

$$r^2 + z^2 = 4r \cos \theta + 6r \sin \theta + 2z.$$

40. Find an equation in the form $r = f(\theta)$ in cylindrical coordinates for the plane $ax + by = c, c \ne 0$.

41. Writing to Learn *Vertical Planes* Show that planes perpendicular to the x-axis have equations of the form $r = a \sec \theta$ in cylindrical coordinates.

42. Writing to Learn *Vertical Planes* Show that planes perpendicular to the y-axis have equations of the form $r = b \csc \theta$ in cylindrical coordinates.

43. Writing to Learn *Symmetry* What symmetry will you find in a surface that has an equation of the form $r = f(z)$ in cylindrical coordinates?

44. Describe the surface $y - \sqrt{4 - z^2} = 0$.

CAS Explorations

In Exercises 45–48, plot the surface over the indicated domain. If you can, rotate the surface into different viewing positions. Then describe the surface in words.

45. $z = y^2, -2 \le x \le 2, -0.5 \le y \le 2$

46. $z = 1 - y^2, -2 \le x \le 2, -2 \le y \le 2$

47. $z = \sin x, -2\pi \le x \le 2\pi, -2\pi \le y \le 2\pi$

48. $z = x^2 + 2y^2$ over

(a) $-3 \le x \le 3, -3 \le y \le 3$

(b) $-1 \le x \le 1, -2 \le y \le 3$

(c) $-2 \le x \le 2, -2 \le y \le 2$

(d) $-2 \le x \le 2, -1 \le y \le 1$

In Exercise 49, perform the following steps.

(a) Solve the equation for r. Choose the expression for the positive square root and simplify it.

(b) Plot r as a function of θ and z. Use the specified ranges of the variables for your plot.

(c) From your plot in (b), estimate the center and radius of the sphere to the nearest integer value.

(d) Convert your equation to Cartesian coordinates. Simplify the result. You may need to complete the final simplification by hand to obtain an equation for the sphere in the form

$$(x - x_0)^2 + (y - y_0)^2 + (z - z_0)^2 = a^2.$$

Compare the coordinates of the center given by this form with your estimates in (c).

(e) Plot the implicit equation obtained in (d). How does it compare with the plot produced in (b)? Can you explain any discrepancies from the way in which a computer grapher plots surfaces?

49. $r^2 + z^2 = 2r(\cos \theta + \sin \theta) + 2, \pi/4 \le \theta \le 9\pi/4, -2 \le z \le 2$

11.6 Quadric Surfaces

Definition of Quadric Surface • Ellipsoids • Paraboloids • Cones • Hyperboloids • Liquid Mirror Telescopes

Definition of Quadric Surface

A **quadric surface** is the graph in space of a second-degree equation in x, y, and z. The most general form is

$$Ax^2 + By^2 + Cz^2 + Dxy + Eyz + Fxz + Gx + Hy + Jz + K = 0,$$

where A, B, C, and so on are constants, but the equation can be simplified by translation and rotation, as in the two-dimensional case in Appendix A5.3. We will study only the simpler equations. Although the definition of cylinder did not require it, the cylinders in Figures 11.31–11.34 were also quadric surfaces.

We now examine ellipsoids (these include spheres), paraboloids, cones, and hyperboloids.

Ellipsoids

Example 1   GRAPHING ELLIPSOIDS

The **ellipsoid**

$$\frac{x^2}{a^2} + \frac{y^2}{b^2} + \frac{z^2}{c^2} = 1$$

(Figure 11.40) cuts the coordinate axes at $(\pm a, 0, 0)$, $(0, \pm b, 0)$, $(0, 0, \pm c)$. It lies within the rectangular box defined by the inequalities $|x| \le a$, $|y| \le b$, and $|z| \le c$. The surface is symmetric with respect to each of the coordinate planes because the variables in the defining equation are squared.

The curves in which the three coordinate planes cut the surface are ellipses. For example,

$$\frac{x^2}{a^2} + \frac{y^2}{b^2} = 1 \qquad \text{when } z = 0.$$

The section cut from the surface by the plane $z = z_0$, $|z_0| < c$, is the ellipse

$$\frac{x^2}{a^2\left(1 - (z_0/c)^2\right)} + \frac{y^2}{b^2\left(1 - (z_0/c)^2\right)} = 1.$$

If any two of the semiaxes a, b, and c are equal, the surface is an **ellipsoid of revolution.** If all three are equal, the surface is a sphere.

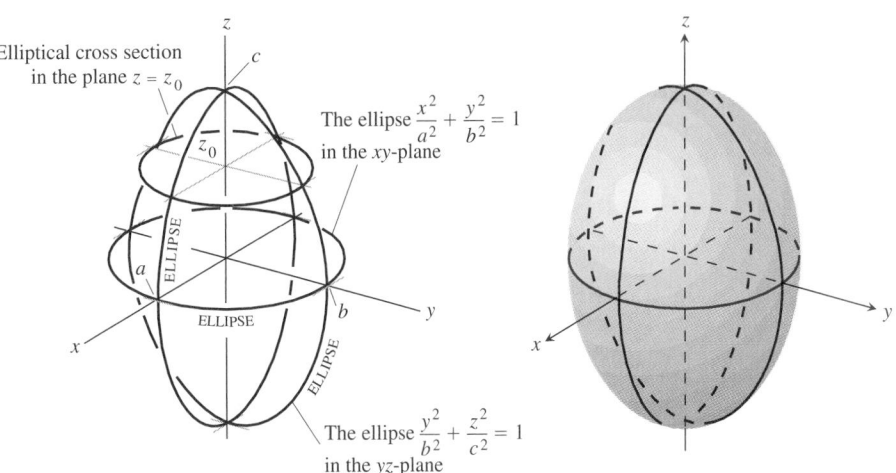

Figure 11.40 The ellipsoid $x^2/a^2 + y^2/b^2 + z^2/c^2 = 1$ in Example 1.

A CAS three-dimensional graphing program can help in visualizing surfaces in space. Many systems can rotate the figure so you can see it as if it were a physical model you could turn in your hand. Hidden-line algorithms (see Exercise 62, Section 11.4) are used to block out portions of the surface that you would not see from your current viewing angle. Most three-dimensional computer graphing programs allow the graph to be displayed in a box as illustrated with the graph of the ellipsoid in Figure 11.41.

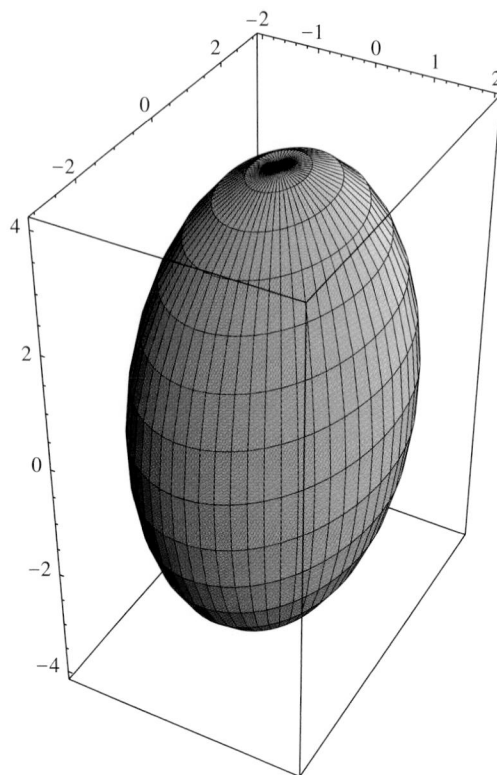

Figure 11.41 The graph of the ellipsoid $x^2/4 + y^2/9 + z^2/16 = 1$ drawn in a box using Mathematica.

Paraboloids

Example 2 GRAPHING AN ELLIPTIC PARABOLOID

The **elliptic paraboloid**

$$\frac{x^2}{a^2} + \frac{y^2}{b^2} = \frac{z}{c}$$

is symmetric with respect to the planes $x = 0$ and $y = 0$ (Figure 11.42). The only intercept on the axes is the origin. Except for this point, the surface lies above or entirely below the xy-plane, depending on the sign of c. The sections cut by the coordinate planes are

$$x = 0: \qquad \text{the parabola } z = \frac{c}{b^2}y^2$$

$$y = 0: \qquad \text{the parabola } z = \frac{c}{a^2}x^2$$

$$z = 0: \qquad \text{the point } (0, 0, 0) \qquad .$$

Each plane $z = z_0$ above the xy-plane cuts the surface in the ellipse

$$\frac{x^2}{a^2} + \frac{y^2}{b^2} = \frac{z_0}{c}.$$

Figure 11.42 The elliptic paraboloid $x^2/a^2 + y^2/b^2 = z/c$ in Example 2, shown for $c > 0$. The cross sections perpendicular to the z-axis above the xy-plane are ellipses. The cross sections in the planes that contain the z-axis are parabolas.

The parabola $z = \dfrac{c}{a^2}x^2$ in the xz-plane

The ellipse $\dfrac{x^2}{a^2} + \dfrac{y^2}{b^2} = 1$ in the plane $z = c$

$z = c$

PARABOLA

ELLIPSE

The parabola $z = \dfrac{c}{b^2}y^2$ in the yz-plane

Example 3   GRAPHING A CIRCULAR PARABOLOID

The **circular paraboloid** or **paraboloid of revolution**

$$\frac{x^2}{a^2} + \frac{y^2}{a^2} = \frac{z}{c}$$

is obtained by taking $b = a$ in the equation for the elliptic paraboloid of Example 2. The cross sections of the surface by planes perpendicular to the z-axis are circles centered on the z-axis. The cross sections by planes containing the z-axis are congruent parabolas with a common focus at the point $(0, 0, a^2/(4c))$.

Shapes cut from circular paraboloids are used for antennas in radio telescopes, satellite trackers, and microwave radio links (Figure 11.43).

Cones

Example 4   GRAPHING AN ELLIPTIC CONE

The **elliptic cone**

$$\frac{x^2}{a^2} + \frac{y^2}{b^2} = \frac{z^2}{c^2}$$

is symmetric with respect to the three coordinate planes (Figure 11.44). The sections cut by the coordinate planes are

$$x = 0: \qquad \text{the lines } z = \pm\frac{c}{b}y$$

$$y = 0: \qquad \text{the lines } z = \pm\frac{c}{a}x$$

$$z = 0 \qquad \text{the point } (0, 0, 0).$$

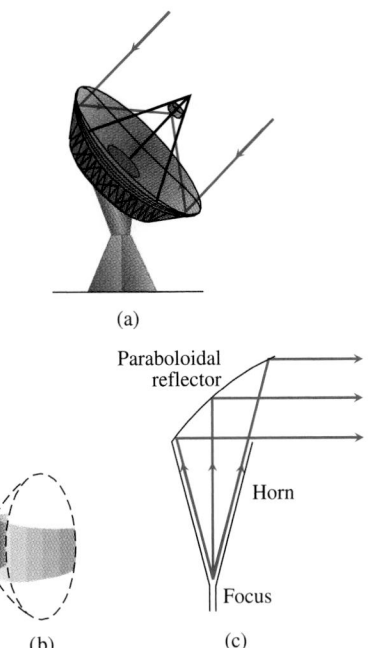

(a)

Paraboloidal reflector

Horn

Focus

(b)   (c)

Figure 11.43 Many antennas are shaped like pieces of paraboloids of revolution. (a) Radio telescopes use the same principles as optical telescopes. (b) A "rectangular-cut" radar reflector. (c) The profile of a horn antenna in a microwave radio link.

The sections cut by the planes $z = z_0$ above and below the xy-plane are ellipses whose centers lie on the z-axis and whose vertices lie on the above lines.

If $a = b$, the cone is a **right circular cone.**

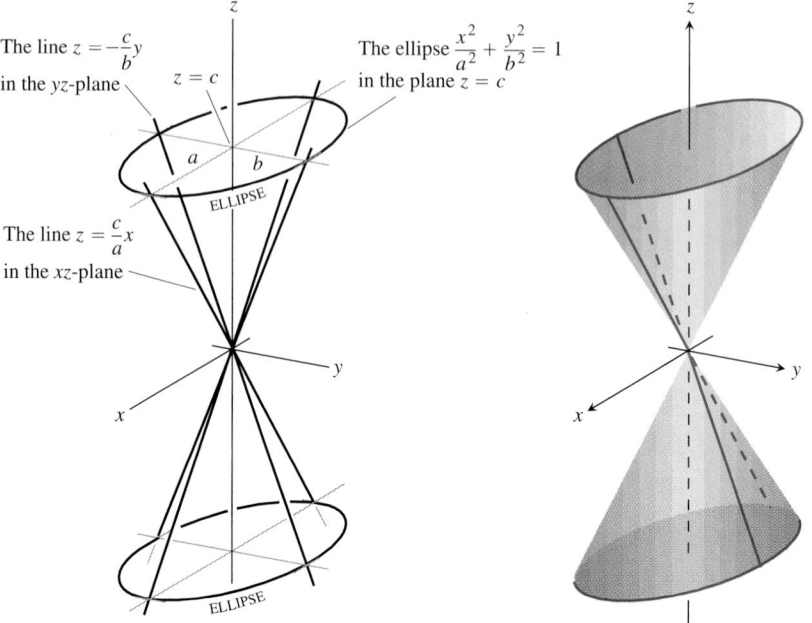

The line $z = -\dfrac{c}{b}y$ in the yz-plane

$z = c$

The ellipse $\dfrac{x^2}{a^2} + \dfrac{y^2}{b^2} = 1$ in the plane $z = c$

ELLIPSE

The line $z = \dfrac{c}{a}x$ in the xz-plane

ELLIPSE

Figure 11.44 The elliptic cone $x^2/a^2 + y^2/b^2 = z^2/c^2$ in Example 4. Planes perpendicular to the z-axis cut the cone in ellipses above and below the xy-plane. Vertical planes that contain the z-axis cut it in a pair of intersecting lines.

Hyperboloids

Example 5 GRAPHING A HYPERBOLOID OF ONE SHEET

The **hyperboloid of one sheet**

$$\frac{x^2}{a^2} + \frac{y^2}{b^2} - \frac{z^2}{c^2} = 1$$

is symmetric with respect to each of the three coordinate planes (Figure 11.45). The sections cut by the coordinate planes are

$$x = 0: \qquad \text{the hyperbola} \quad \frac{y^2}{b^2} - \frac{z^2}{c^2} = 1$$

$$y = 0: \qquad \text{the hyperbola} \quad \frac{x^2}{a^2} - \frac{z^2}{c^2} = 1$$

$$z = 0: \qquad \text{the ellipse} \quad \frac{x^2}{a^2} - \frac{y^2}{b^2} = 1$$

The plane $z = z_0$ cuts the surface in an ellipse with center on the z-axis and vertices on one of the above hyperbolas.

The surface is connected, meaning that it is possible to travel from one point on it to any other without leaving the surface. For this reason, it is said to have *one sheet,* in contrast to the hyperboloid in the next example, which has two sheets.

If $a = b$, the hyperboloid is a surface of revolution.

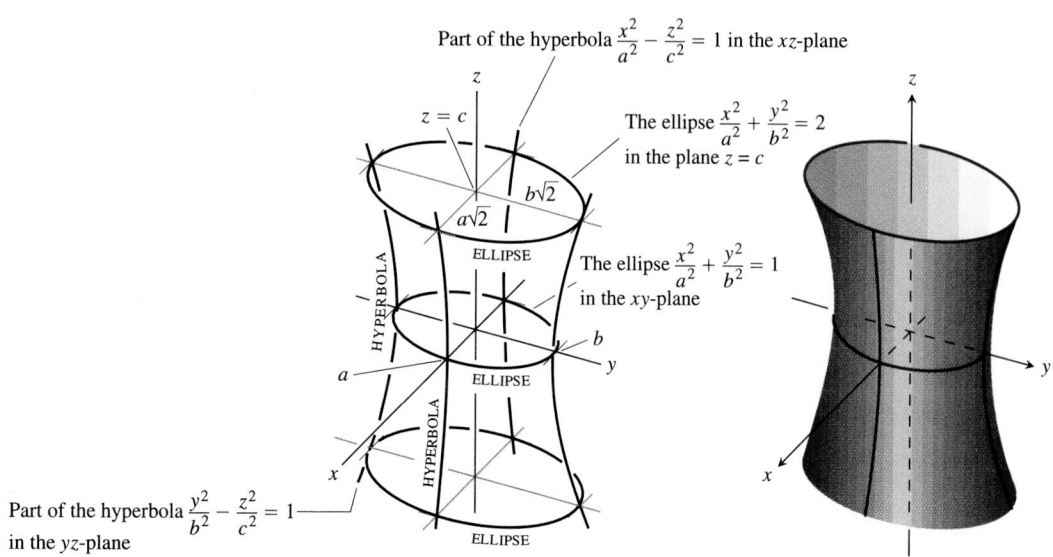

Part of the hyperbola $\dfrac{x^2}{a^2} - \dfrac{z^2}{c^2} = 1$ in the xz-plane

$z = c$

The ellipse $\dfrac{x^2}{a^2} + \dfrac{y^2}{b^2} = 2$ in the plane $z = c$

$b\sqrt{2}$

$a\sqrt{2}$

ELLIPSE

The ellipse $\dfrac{x^2}{a^2} + \dfrac{y^2}{b^2} = 1$ in the xy-plane

HYPERBOLA

b

a

ELLIPSE

y

HYPERBOLA

Part of the hyperbola $\dfrac{y^2}{b^2} - \dfrac{z^2}{c^2} = 1$ in the yz-plane

ELLIPSE

Figure 11.45 The hyperboloid $x^2/a^2 + y^2/b^2 - z^2/c^2 = 1$ in Example 5. Planes perpendicular to the z-axis cut it in ellipses. Vertical planes containing the z-axis cut it in hyperbolas.

Example 6   GRAPHING A HYPERBOLOID OF TWO SHEETS

The **hyperboloid of two sheets**

$$\frac{z^2}{c^2} - \frac{x^2}{a^2} - \frac{y^2}{b^2} = 1$$

is symmetric with respect to the three coordinate planes (Figure 11.46). The plane $z = 0$ does not intersect the surface; in fact, for a horizontal plane to intersect the surface, we must have $|z| \ge c$. The hyperbolic sections

$$x = 0: \qquad \frac{z^2}{c^2} - \frac{y^2}{b^2} = 1$$

$$y = 0: \qquad \frac{z^2}{c^2} - \frac{x^2}{a^2} = 1$$

have their vertices and foci on the z-axis. The surface is separated into two portions, one above the plane $z = c$ and the other below the plane $z = -c$. This accounts for its name.

The equation for the hyperboloid of one sheet and the equation for the hyperboloid of two sheets have different numbers of negative terms. The number in each case is the same as the number of sheets of the hyperboloid. If we replace the 1 on the right side of either equation by 0, we obtain the equation

$$\frac{x^2}{a^2} + \frac{y^2}{b^2} = \frac{z^2}{c^2}$$

for an elliptic cone. The hyperboloids are asymptotic to this cone (Figure 11.47) in the same way that the hyperbolas

$$\frac{x^2}{a^2} - \frac{y^2}{b^2} = \pm 1$$

are asymptotic to the lines

$$\frac{x^2}{a^2} - \frac{y^2}{b^2} = 0$$

in the xy-plane.

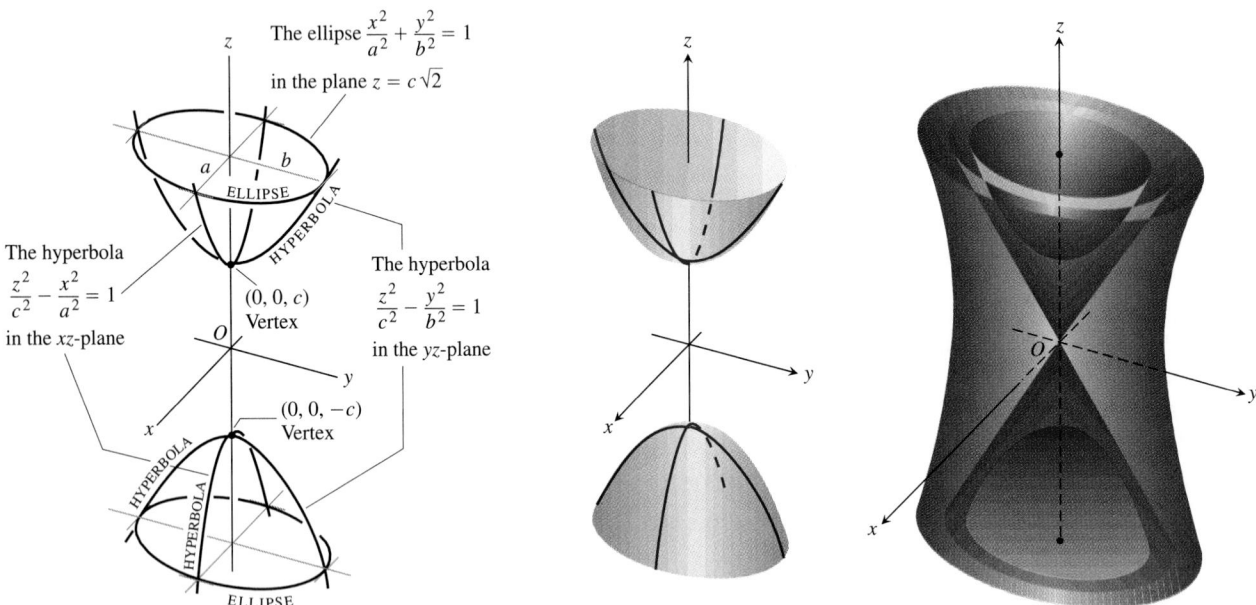

Figure 11.46 The hyperboloid $z^2/c^2 - x^2/a^2 - y^2/b^2 = 1$ in Example 6. Planes perpendicular to the z-axis above and below the vertices cut it in ellipses. Vertical planes containing the z-axis cut it into hyperbolas.

Figure 11.47 Both hyperboloids are asymptotic to the cone. (Example 6)

Example 7 GRAPHING A HYPERBOLIC PARABOLOID

The **hyperbolic paraboloid**

$$\frac{y^2}{b^2} - \frac{x^2}{a^2} = \frac{z}{c}, \qquad c > 0$$

has symmetry with respect to the planes $x = 0$ and $y = 0$ (Figure 11.48). The sections in these planes are

$$x = 0: \qquad \text{the parabola} \quad z = \frac{c}{b^2}y^2$$

$$y = 0: \qquad \text{the parabola} \quad z = -\frac{c}{a^2}x^2.$$

In the plane $x = 0$, the parabola opens upward from the origin. The parabola in the plane $y = 0$ opens downward.

If we cut the surface by a plane $z = z_0 > 0$, the section is a hyperbola

$$\frac{y^2}{b^2} - \frac{x^2}{a^2} = \frac{z_0}{c},$$

with its focal axis parallel to the y-axis and its vertices on the parabola

$$z = \frac{c}{b^2} y^2.$$

If z_0 is negative, the focal axis is parallel to the x-axis and the vertices lie on the hyperbola

$$z = -\frac{c}{a^2} x^2.$$

Near the origin, the surface is shaped like a saddle. To a person traveling along the surface in the yz-plane, the origin looks like a minimum. To a person traveling in the xz-plane, the origin looks like a maximum. Such a point is a **minimax** or **saddle point** of a surface.

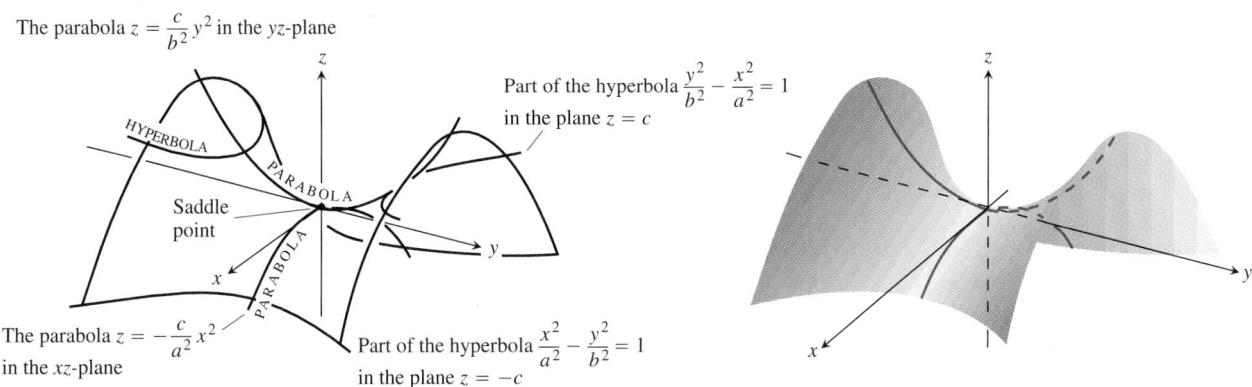

Figure 11.48 The hyperbolic paraboloid $y^2/b^2 - x^2/a^2 = z/c$, $c > 0$ in Example 7. The cross sections in planes perpendicular to the z-axis above and below the xy-plane are hyperbolas. The cross sections in planes perpendicular to the other axes are parabolas.

Exploration 1 **Identifying Shapes of Familiar Objects**

Match each of the familiar objects on the left with its identifying shape of the right. In each case, try to come up with a few other examples of familiar objects having the same identifying shape. (Some shapes are more commonly found than others.)

1. Jelly bean

2. Reflector in a flashlight

3. Nuclear plant cooling tower

4. Saddle

A. Circular paraboloid

B. Ellipsoid of revolution

C. Hyperbolic paraboloid

D. Hyperboloid of revolution

Drawing Lesson

How to Draw Quadric Surfaces

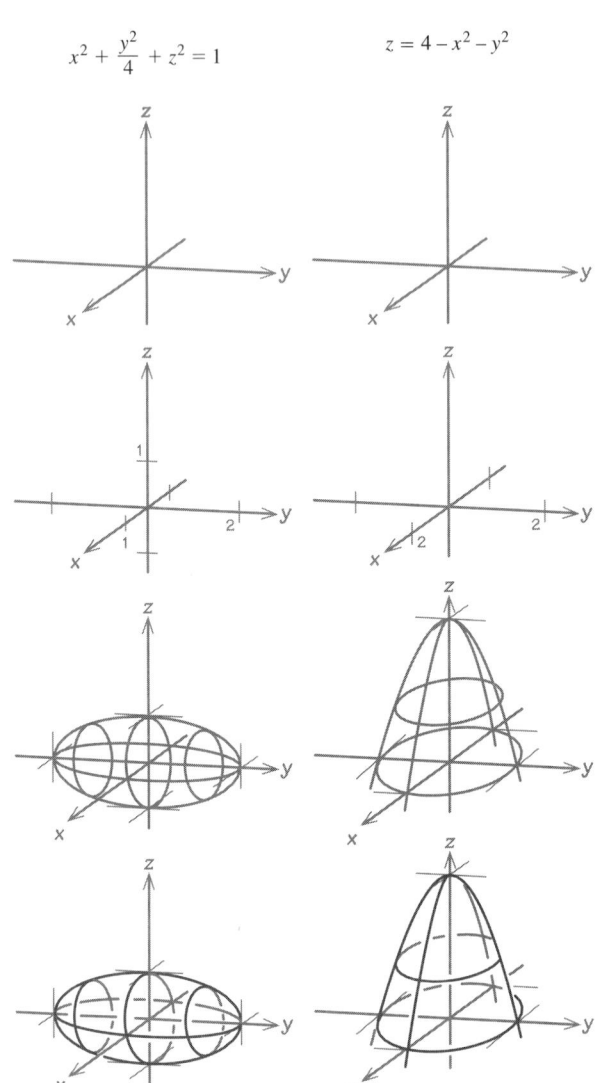

$$x^2 + \frac{y^2}{4} + z^2 = 1$$

$$z = 4 - x^2 - y^2$$

1. Lightly sketch the three coordinate axes.

2. Decide on a scale and mark the intercepts on the axes.

3. Sketch cross sections in the coordinate planes and in a few parallel planes, but don't clutter the picture. Use tangent lines as guides.

4. If more is required, darken the parts exposed to view. Leave the rest light. Use line breaks when you can.

Liquid Mirror Telescopes

When a circular pan of liquid is rotated about its vertical axis, the surface of the liquid does not stay flat. Instead, it assumes the shape of a paraboloid of revolution, exactly what is needed for the primary mirror of a reflecting telescope. At the beginning of the twentieth century, attempts to make reliable mirrors with revolving mercury failed because of surface ripples and focus losses caused by variations in the speed of rotation. Today these difficulties can be overcome with synchronous motors driven by oscillator-stabilized power supplies and checked against constant clocks.

Using the same idea, astronomers at the Steward Observatory's Mirror Laboratory in Tucson, Arizona, have used a large heated spinning turntable to cast borosilicate glass blanks for lightweight mirrors. Spincast mirrors cost less than traditionally cast mirrors and can be made larger. Their shorter focal lengths also allow them to be installed in compact telescope frames that are less expensive to house and less likely to flex in strong winds.

Quick Review 11.6

In Exercises 1 and 2, consider the ellipse

$$\frac{x^2}{4} + \frac{y^2}{9} = 1.$$

1. Solve for y and give the two functions that can be entered into a function grapher to produce the graph of the ellipse.

2. Give parametric equations that will produce the graph of the ellipse.

In Exercises 3 and 4, consider the hyperbola

$$\frac{x^2}{4} - \frac{y^2}{9} = 1$$

3. Solve for y and give the two functions that can be entered into a function grapher to produce the graph of the hyperbola.

4. Give parametric equations that will produce the graph of the hyperbola.

In Exercises 5 and 6, find the x- and y-intercepts of the graph of the equation.

5. $\dfrac{x^2}{9} + \dfrac{y^2}{4} = 1$ **6.** $\dfrac{x^2}{9} - \dfrac{y^2}{4} = 1$

In Exercises 7–10, determine whether the graph of the equation is symmetric with respect to the **(a)** x-axis, **(b)** y-axis, **(c)** origin.

7. $\dfrac{x^2}{25} + \dfrac{y^2}{16} = 1$ **8.** $\dfrac{y^2}{25} - \dfrac{x^2}{4} = 1$

9. $y = \dfrac{3}{2}x^2$ **10.** $x = \dfrac{4}{3}y^2$

Section 11.6 Exercises

In Exercises 1–10, match the equation with the surface it defines. Also, identify each surface by type (paraboloid, ellipsoid, etc.). The surfaces are labeled (a)–(j).

1. $x^2 + y^2 + 4z^2 = 10$ **2.** $z^2 + 4y^2 - 4x^2 = 4$

3. $y^2 + z^2 = x^2$ **4.** $x = y^2 - z^2$

5. $x = -y^2 - z^2$ **6.** $z^2 + x^2 - y^2 = 1$

7. $x = z^2 - y^2$ **8.** $z = -4x^2 - y^2$

9. $x^2 + 4z^2 = y^2$ **10.** $9x^2 + 4y^2 + 2z^2 = 36$

(a)

(b)

(c)

(d)

(e)

(f)

(g)

(h)

(i)

(j)
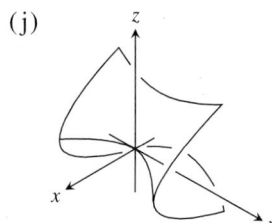

In Exercises 11–14, sketch the ellipsoid.

11. $9x^2 + y^2 + z^2 = 9$ **12.** $4x^2 + 4y^2 + z^2 = 16$

13. $4x^2 + 9y^2 + 4z^2 = 36$

14. $9x^2 + 4y^2 + 36z^2 = 36$

In Exercises 15–20, sketch the paraboloid.

15. $z = x^2 + 4y^2$ **16.** $z = x^2 + 9y^2$

17. $z = 8 - x^2 - y^2$ **18.** $z = 18 - x^2 - 9y^2$

19. $x = 4 - 4y^2 - z^2$ **20.** $y = 1 - x^2 - z^2$

In Exercises 21–24, sketch the cone.

21. $x^2 + y^2 = z^2$ **22.** $y^2 + z^2 = x^2$

23. $4x^2 + 9z^2 = 9y^2$ **24.** $9x^2 + 4y^2 = 36z^2$

In Exercises 25–32, sketch the hyperboloid.

25. $x^2 + y^2 - z^2 = 1$ **26.** $y^2 + z^2 - x^2 = 1$

27. $\dfrac{y^2}{4} + \dfrac{z^2}{9} - \dfrac{x^2}{4} = 1$ **28.** $\dfrac{x^2}{4} + \dfrac{y^2}{4} - \dfrac{z^2}{9} = 1$

29. $z^2 - x^2 - y^2 = 1$ **30.** $\dfrac{y^2}{4} - \dfrac{x^2}{4} - z^2 = 1$

31. $x^2 - y^2 - \dfrac{z^2}{4} = 1$ **32.** $\dfrac{x^2}{4} - y^2 - \dfrac{z^2}{4} = 1$

In Exercises 33 and 34, sketch the hyperbolic paraboloid.

33. $y^2 - x^2 = z$ **34.** $x^2 - y^2 = z$

In Exercises 35–50, sketch the surface.

35. $x^2 + y^2 + z^2 = 4$ **36.** $4x^2 + 4y^2 = z^2$

37. $z = 1 + y^2 - x^2$ **38.** $y^2 - z^2 = 6$

39. $y = -(x^2 + z^2)$ **40.** $z^2 - 4x^2 - 4y^2 = 4$

41. $x^2 + y^2 - z^2 = 4$ **42.** $x = 6 - y^2$

43. $x^2 + z^2 = 2$ **44.** $4x^2 + 4y^2 + z^2 = 4$

45. $16y^2 + 9z^2 = 4x^2$ **46.** $z = x^2 - y^2 - 1$

47. $z = -(x^2 + y^2)$ **48.** $y^2 - x^2 - z^2 = 1$

49. $36x^2 + 9y^2 + 4z^2 = 36$

50. $9x^2 + 16y^2 = 4z^2$

Explorations

51. *Volumes of Ellipsoids* **(a)** Express the area of the cross section cut from the ellipsoid

$$x^2 + \frac{y^2}{4} + \frac{z^2}{9} = 1$$

by the plane $z = c$ as a function of c. (The area of an ellipse with semiaxes a and b is (πab).)

(b) Use slices perpendicular to the z-axis to find the volume of the ellipsoid in (a).

(c) Now find the volume of the ellipsoid

$$\frac{x^2}{a^2} + \frac{y^2}{b^2} + \frac{z^2}{c^2} = 1.$$

Does your formula give the volume of a sphere of radius a if $a = b = c$?

52. *Volume of a Barrel* The barrel shown here is shaped like an ellipsoid with equal pieces cut from the ends by planes perpendicular to the z-axis. The cross sections perpendicular to the z-axis are circular. The barrel is $2h$ units high, its midsection radius is R, and its end radii are both r. Find a formula for the barrel's volume. Then check two things. First, suppose the sides of the barrel are straightened to turn the barrel into a cylinder of radius R and height $2h$. Does your formula give the cylinder's volume? Second, suppose $r = 0$ and $h = R$ so the barrel is a sphere. Does your formula give the sphere's volume?

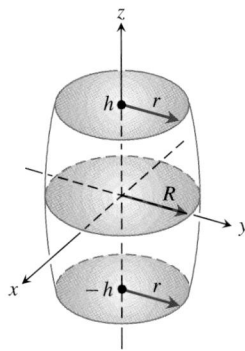

53. Show that the volume of the segment cut from the paraboloid

$$\frac{x^2}{a^2} + \frac{y^2}{b^2} = \frac{z}{c}$$

by the plane $z = h$ equals half the segment's base times its altitude. (Figure 11.42 shows the segment for the special case $h = c$.)

54. (a) Find the volume of the solid bounded by the hyperboloid

$$\frac{x^2}{a^2} + \frac{y^2}{b^2} - \frac{z^2}{c^2} = 1$$

and the planes $z = 0$ and $z = h$, $h > 0$.

(b) Express your answer in (a) in terms of h and the areas A_0 and A_h of the regions cut by the hyperboloid from the planes $z = 0$ and $z = h$.

(c) Show that the volume in (a) is also given by the formula

$$V = \frac{h}{6}(A_0 + 4A_m + A_h),$$

where A_m is the area of the region cut by the hyperboloid from the plane $z = h/2$.

55. Writing to Learn Every time we found the trace of a quadric surface in a plane parallel to one of the coordinate planes, it turned out to be a conic section. Was this mere coincidence? Did it have to happen? Give reasons for your answer.

56. Writing to Learn Suppose you intersect a quadric surface with a plane that is *not* parallel to one of the coordinate planes. What will the trace in the plane be like? Give reasons for your answer.

Extending the Ideas

57. If the hyperbolic paraboloid

$$\frac{y^2}{b^2} - \frac{x^2}{a^2} = \frac{z}{c}$$

is cut by the plane $y = y_1$, the resulting curve is a parabola. Find its vertex and focus.

58. Writing to Learn Suppose you set $z = 0$ in the equation

$$Ax^2 + By^2 + Cz^2 + Dxy + Eyz$$
$$+ \ Fxz + Gx + Hy + Jz + K = 0$$

to obtain a curve in the xy-plane. What will the curve be like? Give reasons for your answer.

CAS Explorations

In Exercises 59–63, use a CAS to plot the surface. Identify the type of quadric surface from your graph.

59. $\dfrac{x^2}{9} + \dfrac{y^2}{36} = 1 - \dfrac{z^2}{25}$

60. $\dfrac{x^2}{9} - \dfrac{z^2}{9} = 1 - \dfrac{y^2}{16}$

61. $5x^2 = z^2 - 3y^2$

62. $\dfrac{y^2}{16} = 1 - \dfrac{x^2}{9} + z$

63. $\dfrac{x^2}{9} - 1 = \dfrac{y^2}{16} + \dfrac{z^2}{2}$

Chapter 11 Key Terms

Chapter 11 Review Exercises

In Exercises 1 and 2, express the vector as a product of its length and direction.

1. $2\mathbf{i} - 3\mathbf{j} + 6\mathbf{k}$

2. $\mathbf{i} + 2\mathbf{j} - \mathbf{k}$

3. Find a vector 2 units long in the direction of $\mathbf{A} = 4\mathbf{i} - \mathbf{j} + 4\mathbf{k}$.

4. The points O, A, B, C, D, and E are vertices of the rectangular box shown here. Express $\overrightarrow{OD}$ and $\overrightarrow{OE}$ in terms of $\overrightarrow{OA}$, $\overrightarrow{OB}$, and $\overrightarrow{OC}$.

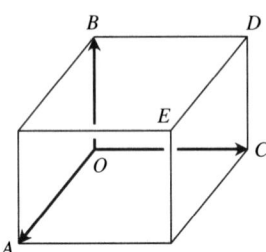

5. Copy the vectors $\mathbf{u}$ and $\mathbf{v}$ and sketch the vector projection of $\mathbf{v}$ onto $\mathbf{u}$.

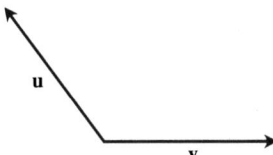

6. Express vectors $\mathbf{a}$, $\mathbf{b}$, and $\mathbf{c}$ in terms of $\mathbf{u}$ and $\mathbf{v}$.

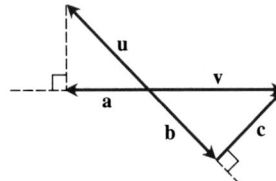

In Exercises 7 and 8, find $|\mathbf{A}|$, $|\mathbf{B}|$, $\mathbf{A} \cdot \mathbf{B}$, $\mathbf{B} \cdot \mathbf{A}$, $\mathbf{A} \times \mathbf{B}$, $\mathbf{B} \times \mathbf{A}$, $|\mathbf{A} \times \mathbf{B}|$, the angle θ between $\mathbf{A}$ and $\mathbf{B}$, the scalar component of $\mathbf{B}$ in the direction of $\mathbf{A}$, and the vector projection of $\mathbf{B}$ onto $\mathbf{A}$.

7. $\mathbf{A} = \mathbf{i} + \mathbf{j}$
$\mathbf{B} = 2\mathbf{i} + \mathbf{j} - 2\mathbf{k}$

8. $\mathbf{A} = \mathbf{i} - \mathbf{j} + 2\mathbf{k}$
$\mathbf{B} = -\mathbf{i} - \mathbf{k}$

In Exercises 9 and 10, write $\mathbf{B}$ as a sum of a vector parallel to $\mathbf{A}$ and a vector orthogonal to $\mathbf{A}$

9. $\mathbf{A} = 2\mathbf{i} + \mathbf{j} - \mathbf{k}$
$\mathbf{B} = \mathbf{i} + \mathbf{j} - 5\mathbf{k}$

10. $\mathbf{A} = \mathbf{i} - 2\mathbf{j}$
$\mathbf{B} = \mathbf{i} + \mathbf{j} + \mathbf{k}$

In Exercises 11 and 12, draw coordinate axes and then sketch $\mathbf{A}$, $\mathbf{B}$, and $\mathbf{A} \times \mathbf{B}$ as vectors at the origin.

11. $\mathbf{A} = \mathbf{i}, \mathbf{B} = \mathbf{i} + \mathbf{j}$

12. $\mathbf{A} = \mathbf{i} - \mathbf{j}, \mathbf{B} = \mathbf{i} + \mathbf{j}$

In Exercises 13 and 14, find the unit vectors that are tangent and normal to the curve at point P.

13. $y = \tan x, \ P(\pi/4, 1)$

14. $x^2 + y^2 = 25, \ P(3, 4)$

15. For any vectors $\mathbf{A}$ and $\mathbf{B}$, show that $|\mathbf{A} + \mathbf{B}|^2 + |\mathbf{A} - \mathbf{B}|^2 = 2|\mathbf{A}|^2 + 2|\mathbf{B}|^2$.

Exploration

16. Let ABC be the triangle determined by vectors $\mathbf{u}$ and $\mathbf{v}$.

(a) Express the area of $\triangle ABC$ in terms of $\mathbf{u}$ and $\mathbf{v}$.

(b) Express the triangle's altitude h in terms of $\mathbf{u}$ and $\mathbf{v}$.

(c) Find the area and altitude if $\mathbf{u} = \mathbf{i} - \mathbf{j} + \mathbf{k}$ and $\mathbf{v} = 2\mathbf{i} + \mathbf{k}$.

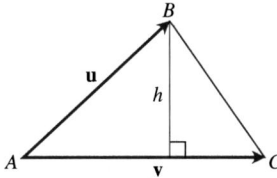

17. If $|\mathbf{v}| = 2$, $|\mathbf{w}| = 3$, and the angle between $\mathbf{v}$ and $\mathbf{w}$ is $\pi/3$, find $|\mathbf{v} - 2\mathbf{w}|$.

18. For what value or values of a will the vectors $\mathbf{u} = 2\mathbf{i} + 4\mathbf{j} - 5\mathbf{k}$ and $\mathbf{v} = -4\mathbf{i} - 8\mathbf{j} + a\mathbf{k}$ be parallel?

In Exercises 19 and 20, find **(a)** the area of the parallelogram determined by vectors $\mathbf{A}$ and $\mathbf{B}$, **(b)** the volume of the parallelepiped determined by the vectors $\mathbf{A}$, $\mathbf{B}$, and $\mathbf{C}$.

19. $\mathbf{A} = \mathbf{i} + \mathbf{j} - \mathbf{k}, \mathbf{B} = 2\mathbf{i} + \mathbf{j} + \mathbf{k}, \mathbf{C} = -\mathbf{i} - 2\mathbf{j} + 3\mathbf{k}$

20. $\mathbf{A} = \mathbf{i} + \mathbf{j}, \mathbf{B} = \mathbf{j}, \mathbf{C} = \mathbf{i} + \mathbf{j} + \mathbf{k}$

21. Writing to Learn Suppose $\mathbf{n}$ is normal to a plane and $\mathbf{v}$ is parallel to the plane. Describe how you would find a vector $\mathbf{u}$ that is both perpendicular to $\mathbf{v}$ and parallel to the plane.

22. Find a vector in the plane parallel to the line $ax + by = c$.

In Exercises 23 and 24, find the distance from the point to the line.

21. $(2, 2, 0)$; $x = -t, y = t, z = -1 + t$

24. $(0, 4, 1)$; $x = 2 + t, y = 2 + t, z = t$

25. Parametrize the line that passes through the point $(1, 2, 3)$ parallel to the vector $\mathbf{v} = -3\mathbf{i} + 7\mathbf{k}$.

26. Parametrize the line segment joining the points $P(1, 2, 0)$ and $Q(1, 3, -1)$.

In Exercises 27 and 28, find the distance from the point to the plane.

27. $(6, 0, -6)$, $\quad x - y = 4$

28. $(3, 0, 10)$, $\quad 2x + 3y + z = 2$

29. Find an equation for the plane that passes through the point $(3, -2, 1)$ normal to the vector $\mathbf{n} = 2\mathbf{i} + \mathbf{j} + \mathbf{k}$.

30. Find an equation for the plane that passes through the point $(-1, 6, 0)$ perpendicular to the line $x = -1 + t$, $y = 6 - 2t, z = 3t$.

In Exercises 31 and 32, find an equation for the plane through points P, Q, and R.

31. $P(1, -1, 2), Q(2, 1, 3), R(-1, 2, -1)$

32. $P(1, 0, 0), Q(0, 1, 1), R(0, 0, 1)$

33. Find the points in which the line $x = 1 + 2t, y = -1 - t$, $z = 3t$ meets the three coordinate planes.

34. Find the point in which the line through the origin perpendicular to the plane $2x - y - z = 4$ meets the plane $3x - 5y + 2z = 6$.

In Exercises 35 and 36, find the acute angle between the planes.

35. $x = 7, x + y \sqrt{2} z = -3$

36. $x + y = 1, y + z = 1$

37. Find parametric equations for the line in which the planes $x + 2y + z = 1$ and $x - y + 2z = -8$ intersect.

38. Show that the line in which the planes $x + 2y - 2z = 5$ and $5x - 2y - z = 0$ intersect is parallel to the line $x = -3 + 2t, y = 3t, z = 1 + 4t$.

39. The planes $3x + 6z = 1$ and $2x + 2y - z = 3$ intersect in a line.

(a) Show that the planes are orthogonal.

(b) Find equations for the line of intersection.

40. Find an equation for the plane that passes through the point $(1, 2, 3)$ parallel to $\mathbf{u} = 2\mathbf{i} + 3\mathbf{j} + \mathbf{k}$ and $\mathbf{v} = \mathbf{i} - \mathbf{j} + 2\mathbf{k}$.

41. Writing to Learn Is $\mathbf{v} = 2\mathbf{i} - 4\mathbf{j} + \mathbf{k}$ related in any special way to the plane $2x + y = 5$? Give reasons for your answer.

42. The equation $\mathbf{n} \cdot \overrightarrow{P_0 P} = 0$ represents the plane through P_0 normal to $\mathbf{n}$. What set does the inequality $\mathbf{n} \cdot \overrightarrow{P_0 P} > 0$ represent?

43. Find the distance from the point $P(1, 4, 0)$ to the plane through $A(0, 0, 0)$, $B(2, 0, -1)$ and $C(2, -1, 0)$.

44. Find the distance from the point $(2, 2, 3)$ to the plane $2x + 3y + 5z = 0$.

45. Find a vector parallel to the plane $2x - y - z = 4$ and orthogonal to $\mathbf{i} + \mathbf{j} + \mathbf{k}$.

46. Find a unit vector orthogonal to $\mathbf{A}$ in the plane of $\mathbf{B}$ and $\mathbf{C}$ if $\mathbf{A} = 2\mathbf{i} - \mathbf{j} + \mathbf{k}, \mathbf{B} = \mathbf{i} + 2\mathbf{j} + \mathbf{k}$, and $\mathbf{C} = \mathbf{i} + \mathbf{j} - 2\mathbf{k}$.

47. Find a vector of magnitude 2 parallel to the line of intersection of the planes $x + 2y + z - 1 = 0$ and $x - y + 2z + 7 = 0$.

48. Find the point in which the line through the origin perpendicular to the plane $2x - y - z = 4$ meets the plane $3x - 5y + 2z = 6$.

49. Find the point in which the line through $P(3, 2, 1)$ normal to the plane $2x - y + 2z = -2$ meets the plane.

50. What angle does the line of intersection of the planes $2x + y - z = 0$ and $x + y + 2z = 0$ make with the positive x-axis?

51. The line
$$L: x = 3 + 2t, \qquad y = 2t, \qquad z = t$$
intersects the plane $x + 3y - z = -4$ in a point P. Find the coordinates of P and find equations for the line through P perpendicular to L.

52. Show that for every real number k the plane $x - 2y + z + 3 + k(2x - y - z + 1) = 0$ contains the line of intersection of the planes $x - 2y + z + 3 = 0$ and $2x - y - z + 1 = 0$.

53. Find an equation for the plane through $A(-2, 0, -3)$ and $B(1, -2, 1)$ that lies parallel to the line through $C(-2, -13/5, 26/5)$ and $D(16/5, -13/5, 0)$.

54. Writing to Learn Is the line $x = 1 + 2t, y = -2 + 3t$, $z = -5t$ related in any way to the plane $-4x - 6y + 10z = 9$? Give reasons for your answer.

In Exercises 55 and 56, *work in groups of two or three.*

55. Which of the following are equations for the plane through the points $P(1, 1, -1)$, $Q(3, 0, 2)$, and $R(-2, 1, 0)$?

(a) $(2\mathbf{i} - 3\mathbf{j} + 3\mathbf{k}) \cdot ((x + 2)\mathbf{i} + (y - 1)\mathbf{j} + z\mathbf{k}) = 0$

(b) $x = 3 - t, y = -11t, z = 2 - 3t$

(c) $(x + 2) + 11(y - 1) = 3z$

(d) $(2\mathbf{i} - 3\mathbf{j} + 3\mathbf{k}) \times ((x + 2)\mathbf{i} + (y - 1)\mathbf{j} + z\mathbf{k}) = \mathbf{0}$

(e) $((2\mathbf{i} - \mathbf{j} + 3\mathbf{k}) \times (-3\mathbf{i} + \mathbf{k})) \cdot ((x + 2)\mathbf{i}(y - 1)\mathbf{j} + z\mathbf{k}) = 0$

56. The parallelogram shown here has vertices at $A(2, -1, 4)$, $B(1, 0, -1)$, $C(1, 2, 3)$, and D. Find

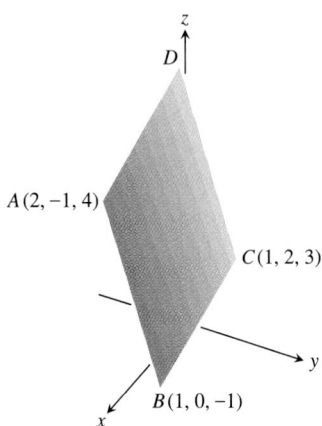

(a) the coordinates of D.

(b) the cosine of the interior angle at B.

(c) the vector projection of $\vec{BA}$ onto $\vec{BC}$.

(d) the area of the parallelogram.

(e) an equation for the plane of the parallelogram.

(f) the areas of the orthogonal projections of the parallelogram on the three coordinate planes.

57. *Distance Between Lines* Find the distance between the line L_1 through the points $A(1, 0, -1)$ and $B(-1, 1, 0)$ and the line L_2 through the points $C(3, 1, -1)$ and $D(4, 5, -2)$. The distance is to be measured along the line perpendicular to the two lines. First find a vector $\mathbf{n}$ perpendicular to both lines. Then project $\vec{AC}$ onto $\mathbf{n}$.

58. *(Continuation of Exercise 57)* Find the distance between the line through $A(4, 0, 2)$ and $B(2, 4, 1)$ and the line through $C(1, 3, 2)$ and $D(2, 2, 4)$.

In Exercises 59 and 60, the equation defines sets both in the plane and in three-dimensional space. Identify both sets.

59. $x = 0$ **60.** $x + y = 1$

61. Find the center and radius of the sphere
$$2x^2 + 2y^2 + 2z^2 - 12x + 4y - 8z + 10 = 0.$$

62. *Work Pushing a Car* Find the work done in pushing a car 250 m with a force of magnitude 160 N directed at an angle

of $\pi/6$ rad downward from the horizontal against the back of the car as shown in the figure.

63. *Torque on a Spark Plug* The operator's manual for the Toro® 21-in. lawnmower says "tighten the spark plug to 15 ft-lb (20.4 N · m)." If you are installing the plug with a 10.5-in. socket wrench that places the center of your hand 9 in. from the axis of the spark plug as shown in the figure, about how hard should you pull? Answer in pounds.

64. In the figure here, D is the midpoint of side AB of triangle ABC and E is one-third of the way between C and B. Use vectors to prove that F is the midpoint of line segment CD.

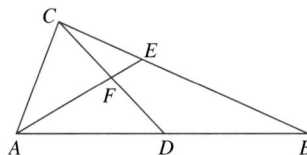

65. *Determinant Equations for Planes* **(a)** Show that
$$\begin{vmatrix} x_1 - x & y_1 - y & z_1 - z \\ x_2 - x & y_2 - y & z_2 - z \\ x_3 - x & y_3 - y & z_3 - z \end{vmatrix} = 0$$

is an equation for the plane through the three noncollinear points $P_1(x_1, y_1, z_1)$, $P_2(x_2, y_2, z_2)$, and $P_3(x_3, y_3, z_3)$.

(b) What set of points in space is described by the equation
$$\begin{vmatrix} x & y & z & 1 \\ x_1 & y_1 & z_1 & 1 \\ x_2 & y_2 & z_2 & 1 \\ x_3 & y_3 & z_3 & 1 \end{vmatrix} = 0?$$

66. *Distance from a Point to a Line* Use vectors to prove that the distance from $P_1(x_1, y_1)$ to the line $ax + by = c$ is
$$d = \frac{|ax_1 + by_1 - c|}{\sqrt{a^2 + b^2}}.$$

In Exercises 67–70, the coordinates of a point in space are given in one of two coordinate systems. Find the coordinates of the point in the other system. There may be more than one answer because points in cylindrical coordinates can have more than one coordinate triple.

Rectangular (x, y, z)	**Cylindrical** (r, θ, z)
67.	$(1, 0, 0)$
68.	$(1, \pi/2, 0)$
69. $(-1, 0, -1)$	
70. $(0, -1, 1)$	

In Exercises 71–74, translate the equation from the given coordinate system (rectangular, cylindrical) into the other system. Identify the set of points in space defined by the equation.

71. $z = 2$

72. $x^2 + y^2 + (z - 3)^2 = 9$

73. $r = 7 \sin \theta$

74. $r = 4 \cos \theta$

In Exercises 75–84, sketch the surface.

75. $x^2 + y^2 + z^2 = 4$

76. $x^2 + (y - 1)^2 + z^2 = 1$

77. $x^2 + z^2 = y^2$

78. $4y^2 + z^2 - 4x^2 = 4$

79. $y^2 - x^2 - z^2 = 1$

80. $z = x^2 + y^2 + 1$

81. $z^2 - \dfrac{x^2}{4} - y^2 = 1$

82. $9x^2 + 4y^2 + z^2 = 36$

83. $4x^2 + 9z^2 = y^2$

84. $z = 4x^2 + y^2 - 4$

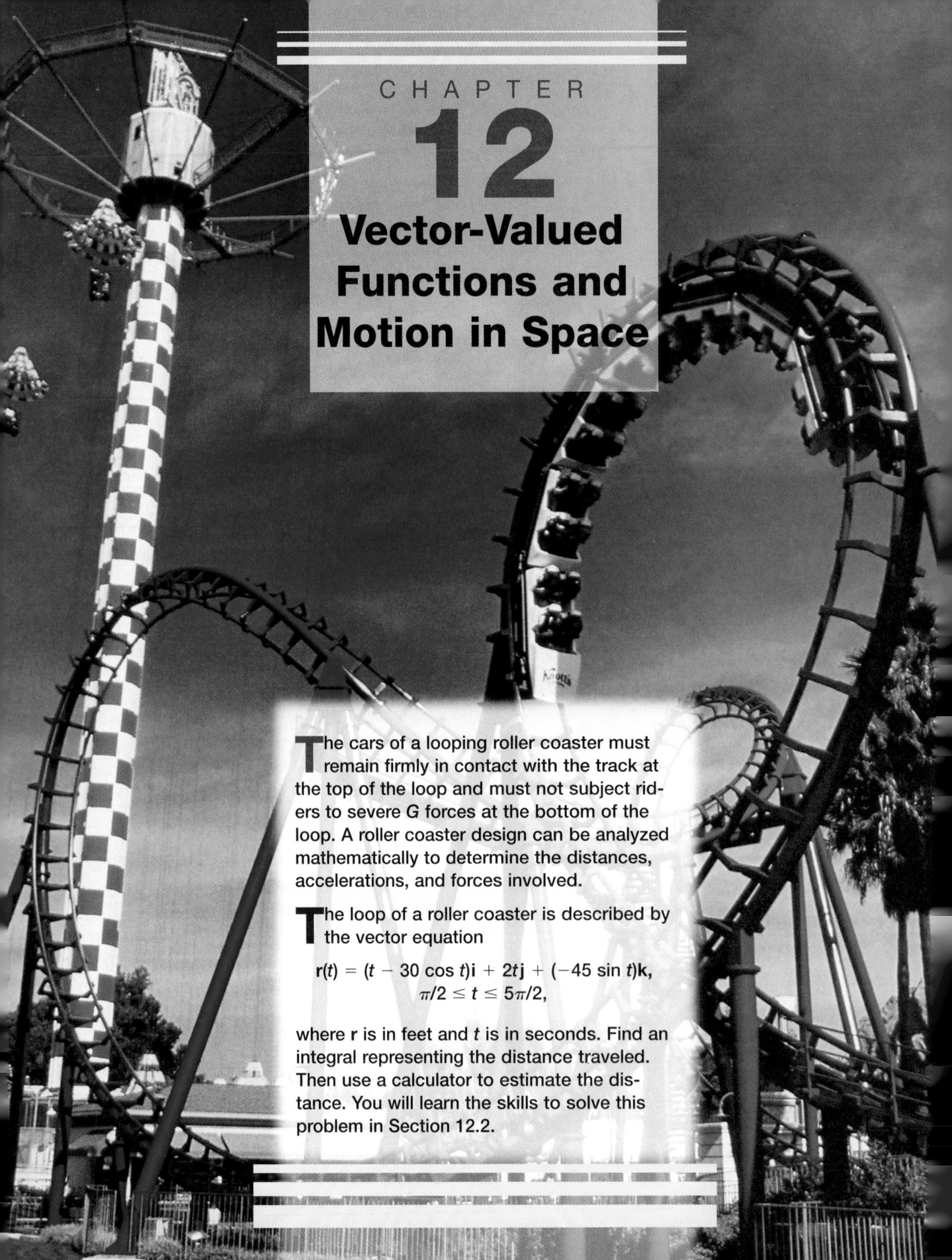

CHAPTER

12
Vector-Valued Functions and Motion in Space

The cars of a looping roller coaster must remain firmly in contact with the track at the top of the loop and must not subject riders to severe *G* forces at the bottom of the loop. A roller coaster design can be analyzed mathematically to determine the distances, accelerations, and forces involved.

The loop of a roller coaster is described by the vector equation

$$\mathbf{r}(t) = (t - 30 \cos t)\mathbf{i} + 2t\mathbf{j} + (-45 \sin t)\mathbf{k},$$
$$\pi/2 \le t \le 5\pi/2,$$

where **r** is in feet and *t* is in seconds. Find an integral representing the distance traveled. Then use a calculator to estimate the distance. You will learn the skills to solve this problem in Section 12.2.

Chapter 12 Overview

When a body travels through space, the equations $x = f(t)$, $y = g(t)$, and $z = h(t)$ that give the body's coordinates as functions of time serve as parametric equations for the body's motion and path. With vector notation, we can condense these into a single equation $\mathbf{r}(t) = f(t)\mathbf{i} + g(t)\mathbf{j} + h(t)\mathbf{k}$ that gives the body's position as a vector function of time.

In this chapter, we show how to use calculus to study the paths, velocities, and accelerations of moving bodies. As we go along, we see how our work answers the standard questions about the paths and motions of planets and satellites. In the final section, we use our new vector calculus to derive Kepler's laws of planetary motion from Newton's laws of motion and gravitation.

12.1 Vector-Valued Functions and Space Curves

Space Curves • Limits and Continuity • Derivatives and Motion on Smooth Curves • Differentiation Rules • Vector Functions of Constant Length • Integrals of Vector Functions

Space Curves

To track a particle moving in space, we run a vector $\mathbf{r}$ from the origin to the particle (Figure 12.1) and study the changes in $\mathbf{r}$. When a particle moves through space during a time interval I, we think of the particle's coordinates as functions defined on I:

$$x = f(t), \qquad y = g(t), \qquad z = h(t), \qquad t \in I. \tag{1}$$

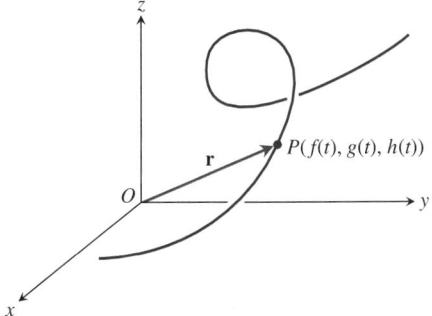

Figure 12.1 The position vector $\mathbf{r}(t) = \overrightarrow{OP}$ of a particle moving through space is a function of time.

The points $(x, y, z) = (f(t), g(t), h(t))$, $t \in I$, make up the **curve** in space that is the particle's **path.** The equations and interval in Equation (1) **parametrize** the curve. The vector

$$\mathbf{r}(t) = \overrightarrow{OP} = f(t)\mathbf{i} + g(t)\mathbf{j} + h(t)\mathbf{k} \tag{2}$$

from the origin to the particle's **position** $P(f(t), g(t), h(t))$ at time t is the particle's **position vector.** The functions f, g, and h are the **component functions (components)** of the position vector. We think of the particle's path as the **curve traced by $\mathbf{r}$** (the terminal point of $\mathbf{r}$) during the time interval I.

Equations (1) and (2) define $\mathbf{r}$ as a vector function of the real variable t on the interval I. More generally, a **vector function** or **vector-valued function** on a domain set D is a rule that assigns a vector in space to each element in D. For now, the domains will be intervals of real numbers. Later, in Chapter 15, the domains will be regions in the plane or space. Vector functions will then be called "vector fields."

We refer to real-valued functions as **scalar functions** to distinguish them from vector functions. The components of $\mathbf{r}$ are scalar functions of t. When we define a vector-valued function by giving its component functions, we assume the vector function's domain to be the common domain of the components. You will find this material to be very similar to the material on planar curves in Section 10.3.

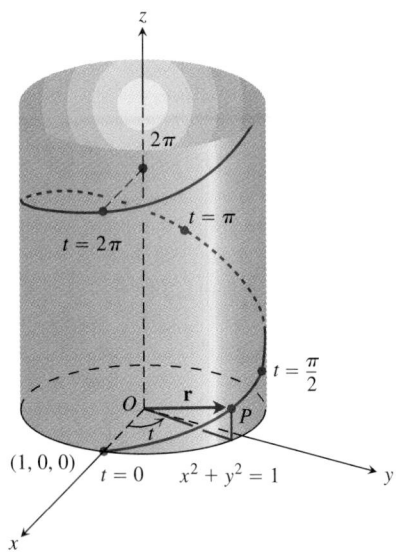

Figure 12.2 The upper half of the helix $\mathbf{r}(t) = (\cos t)\mathbf{i} + (\sin t)\mathbf{j} + t\mathbf{k}$. (Example 1)

Example 1 GRAPHING A HELIX

Graph the vector function

$$\mathbf{r}(t) = (\cos t)\mathbf{i} + (\sin t)\mathbf{j} + t\mathbf{k}.$$

Solution The vector function $\mathbf{r}$ is defined for all real values of t. The curve traced by $\mathbf{r}$ is a **helix** (from an old Greek word for "spiral") that winds around the circular cylinder $x^2 + y^2 = 1$ (Figure 12.2). The curve lies on the cylinder because the $\mathbf{i}$- and $\mathbf{j}$-components of $\mathbf{r}$, being the x- and y-coordinates of the tip of $\mathbf{r}$, satisfy the cylinder's equation:

$$x^2 + y^2 = (\cos t)^2 + (\sin t)^2 = 1.$$

The curve rises as the $\mathbf{k}$-component $z = t$ increases. Each time t increases by 2π, the curve completes one turn around the cylinder. The equations

$$x = \cos t, \qquad y = \sin t, \qquad z = t$$

parametrize the helix, the interval $-\infty < t < \infty$ being understood. You will find more helices in Figure 12.3.

Figure 12.4 shows the graph of the vector function of Exercise 12, $\mathbf{r}(t) = (2 \cos 3t)\mathbf{i} + (e^{-t})\mathbf{j} + (2 \sin 3t)\mathbf{k}$.

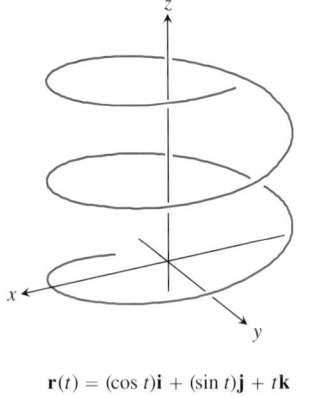

$$\mathbf{r}(t) = (\cos t)\mathbf{i} + (\sin t)\mathbf{j} + t\mathbf{k}$$

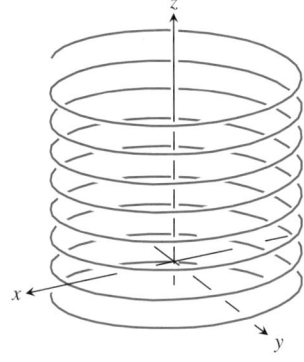

$$\mathbf{r} = (\cos t)\mathbf{i} + (\sin t)\mathbf{j} + 0.3t\mathbf{k}$$

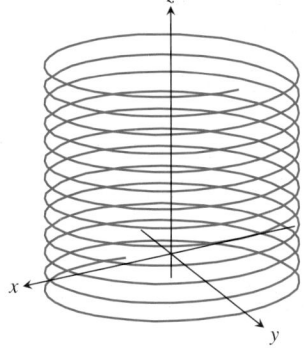

$$\mathbf{r}(t) = (\cos 5t)\mathbf{i} + (\sin 5t)\mathbf{j} + t\mathbf{k}$$

(Generated by Mathematica)

Figure 12.3 Helices drawn by computer.

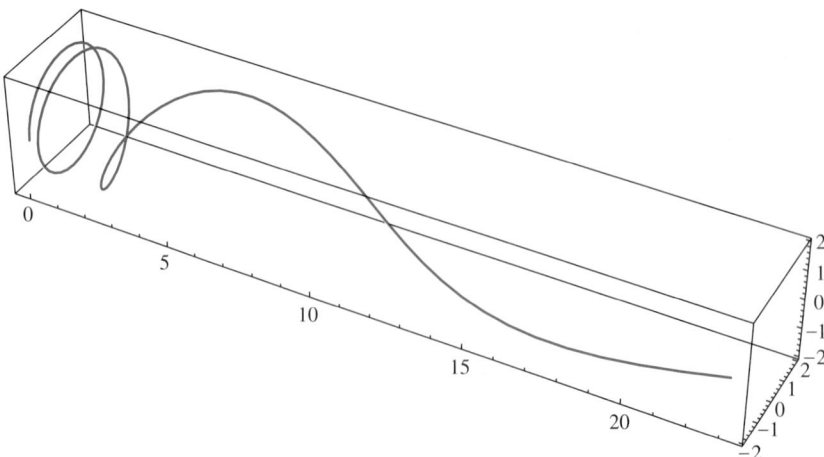

Figure 12.4 The graph of $\mathbf{r}(t) = (2 \cos 3t)\mathbf{i} + (e^{-t})\mathbf{j} + (2 \sin 3t)\mathbf{k}$ obtained parametrically for $-\pi \le t \le \pi$ using Mathematica.

> ### Exploration 1   How Vector-Valued Functions in Space Ought to Work
>
> The rules for vector-valued functions in space and vector-valued functions in the plane (Section 10.3) are so similar you can probably figure them out for yourself. How many of these can you answer before we give you the definitions?
>
> 1. If $\mathbf{r}(t) = (t^2 + 3)\mathbf{i} + (\cos t)\mathbf{j} + (\sin t/t)\mathbf{k}$, what is $\lim_{t \to 0} \mathbf{r}(t)$?
> 2. Is $\mathbf{g}(t) = (e^{3t+1})\mathbf{i} + (|t|)\mathbf{j} + (\tan^2 t)\mathbf{k}$ continuous at $t = 0$? Is it differentiable there?
> 3. Find $d\mathbf{r}/dt$ if $\mathbf{r}(t) = (t^3)\mathbf{i} + (\ln(t + 3))\mathbf{j} + (\tan t)\mathbf{k}$.
> 4. Find $\displaystyle\int_0^4 \mathbf{r}(t)\, dt$ if $\mathbf{r}(t) = (t^2)\mathbf{i} + (\sin \pi t)\mathbf{j} + (\sqrt{t})\mathbf{k}$.
> 5. A particle moves through space with position vector at time t given by $\mathbf{r}(t) = (2t^3)\mathbf{i} + (\sin t)\mathbf{j} + (e^{2t})\mathbf{k}$. Find the velocity vector.
> 6. Find the acceleration vector for the particle in part (5).
> 7. Find the speed and direction of motion of the particle in (5) at time $t = 0$.
> 8. If $\mathbf{u}$ and $\mathbf{v}$ are vectors in three-dimensional space, what is the rule for finding $(d/dt)\,(\mathbf{u} \cdot \mathbf{v})$?
> 9. If $\mathbf{r}$ is a differentiable function of t and t is a differentiable function of s, what would the Chain Rule say about $d\mathbf{r}/ds$?
> 10. We did not consider the cross product of two-dimensional vectors, so we have no two-dimensional formula for $(d/dt)\,(\mathbf{u} \times \mathbf{v})$. Nonetheless, what do you suppose the three-dimensional formula for the derivative of a cross product would be?

Limits and Continuity

The way we define limits of vector-valued functions is similar to the way we define limits of real-valued functions.

> ### Definition   Limit
>
> Let $\mathbf{r}(t) = f(t)\mathbf{i} + g(t)\mathbf{j} + h(t)\mathbf{k}$ be a vector function and $\mathbf{L}$ a vector. We say that $\mathbf{r}$ has **limit** $\mathbf{L}$ as t approaches t_0 and write
>
> $$\lim_{t \to t_0} \mathbf{r}(t) = \mathbf{L}$$
>
> if, for every number $\varepsilon > 0$, there exists a corresponding $\delta > 0$ such that for all t
>
> $$0 < |t - t_0| < \delta \implies |\mathbf{r}(t) - \mathbf{L}| < \varepsilon.$$

If $\mathbf{L} = L_1\mathbf{i} + L_2\mathbf{j} + L_3\mathbf{k}$, then, just as in Section 10.3, $\lim_{t \to t_0} \mathbf{r}(t) = \mathbf{L}$ precisely when

$$\lim_{t \to t_0} f(t) = L_1, \qquad \lim_{t \to t_0} g(t) = L_2, \qquad \lim_{t \to t_0} h(t) = L_3.$$

The equation

$$\lim_{t \to t_0} \mathbf{r}(t) = \left(\lim_{t \to t_0} f(t) \right) \mathbf{i} + \left(\lim_{t \to t_0} g(t) \right) \mathbf{j} + \left(\lim_{t \to t_0} h(t) \right) \mathbf{k}$$

provides a practical way to calculate limits of vector functions.

Example 2 FINDING LIMITS OF VECTOR FUNCTIONS

If $\mathbf{r}(t) = (\cos t)\mathbf{i} + (\sin t)\mathbf{j} + t\mathbf{k}$, then

$$\lim_{t \to \pi/4} \mathbf{r}(t) = \left(\lim_{t \to \pi/4} \cos t \right) \mathbf{i} + \left(\lim_{t \to \pi/4} \sin t \right) \mathbf{j} + \left(\lim_{t \to \pi/4} t \right) \mathbf{k}$$

$$= \frac{\sqrt{2}}{2} \mathbf{i} + \frac{\sqrt{2}}{2} \mathbf{j} + \frac{\pi}{4} \mathbf{k}.$$

Figure 12.5 Checking the result of Example 2 on a graphing calculator. Notice that the displayed vectors omit the commas.

Limits can be checked with a CAS. For example, the vector $\mathbf{r}(t)$ of Example 2 is entered as [cos (t), sin (t), t] on the advanced graphing calculator we use. Figure 12.5 shows what Example 2 looks like on this calculator.

We define continuity for vector functions the same way we define continuity for scalar functions.

Definition Continuity at a Point

A vector function $\mathbf{r}(t)$ is **continuous at a point $t = t_0$** in its domain if

$$\lim_{t \to t_0} \mathbf{r}(t) = \mathbf{r}(t_0).$$

A vector function $\mathbf{r}(t)$ is **continuous** if it is continuous at every point in its domain. Since limits can be expressed in terms of components, we can test vector functions for continuity by examining their components (Exercise 42).

Component Test for Continuity at a Point

The vector function $\mathbf{r}(t) = f(t)\mathbf{i} + g(t)\mathbf{j} + h(t)\mathbf{k}$ is continuous at $t = t_0$ if and only if f, g, and h are continuous at t_0.

Example 3 FINDING POINTS OF CONTINUITY AND DISCONTINUITY

(a) The function

$$\mathbf{r}(t) = (\cos t)\mathbf{i} + (\sin t)\mathbf{j} + t\mathbf{k}$$

is continuous because $\cos t$, $\sin t$, and t are continuous.

(b) The function

$$\mathbf{g}(t) = (\cos t)\mathbf{i} + (\sin t)\mathbf{j} + (\text{int } t)\mathbf{k}$$

is discontinuous (not continuous) at every integer.

Derivatives and Motion on Smooth Curves

Suppose $\mathbf{r}(t) = f(t)\mathbf{i} + g(t)\mathbf{j} + h(t)\mathbf{k}$ is the position vector of a particle moving along a curve in space and f, g, and h are differentiable functions of t. Then (see Figure 12.6) the difference between the particle's positions at time $t + \Delta t$ and time t is $\Delta \mathbf{r} = \mathbf{r}(t + \Delta t) - \mathbf{r}(t)$. Using the same derivation as on page 532 with one more component here, we are led to the following definition.

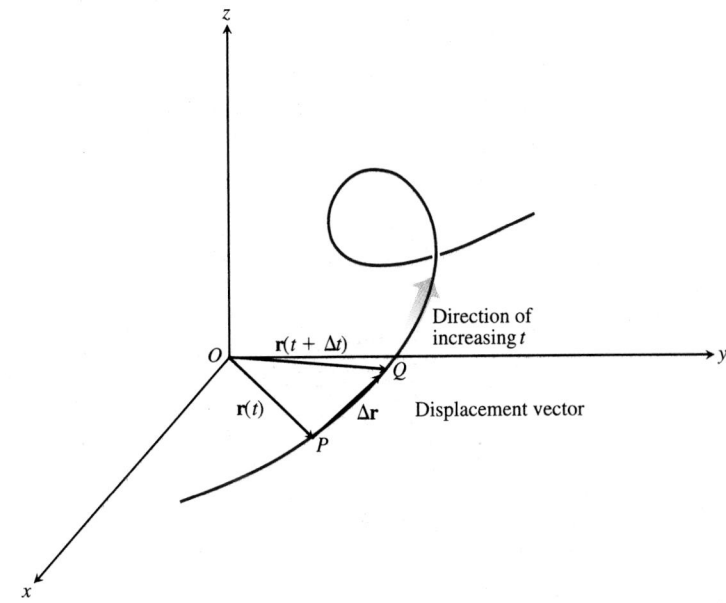

Figure 12.6 Between time t and time $t + \Delta t$, the particle moving along the path shown here undergoes the displacement $\overrightarrow{PQ} = \Delta \mathbf{r}.$ The vector sum $\mathbf{r}(t) + \Delta \mathbf{r}$ gives the new position, $\mathbf{r}(t + \Delta t)$.

Definition Derivative at a Point

The vector function $\mathbf{r}(t) = f(t)\mathbf{i} + g(t)\mathbf{j} + h(t)\mathbf{k}$ is **differentiable at** $t = t_0$ if f, g, and h are differentiable at t_0. The **derivative** is the vector

$$\frac{d\mathbf{r}}{dt} = \lim_{\Delta t \to 0} \frac{\mathbf{r}(t + \Delta t) - \mathbf{r}(t)}{\Delta t} = \frac{df}{dt}\mathbf{i} + \frac{dg}{dt}\mathbf{j} + \frac{dh}{dt}\mathbf{k}.$$

A vector function $\mathbf{r}$ is **differentiable** if it is differentiable at every point of its domain. The curve traced by $\mathbf{r}$ is **smooth** if $d\mathbf{r}/dt$ is continuous and never **0**, that is, if f, g, and h have continuous first derivatives that are not simultaneously 0.

The vector $d\mathbf{r}/dt$, when different from **0**, is also a vector tangent to the curve. The **tangent line** to the curve at a point $(f(t_0), g(t_0), h(t_0))$ is defined to be the line through the point parallel to $d\mathbf{r}/dt$ at $t = t_0$. We require $d\mathbf{r}/dt \neq \mathbf{0}$ for a smooth curve to make sure the curve has a continuously turning tangent at each point. On a smooth curve there are no sharp corners or cusps.

A curve that is made up of a finite number of smooth curves pieced together in a continuous fashion is **piecewise smooth** (Figure 12.7).

Just as we did in Section 10.3, we may conclude that the derivative $d\mathbf{r}/dt$ is just what we want for modeling a particle's velocity. It points in the direction of motion and gives the rate of change of position with respect to time. For a smooth curve, the velocity is never zero; the particle does not stop or reverse direction.

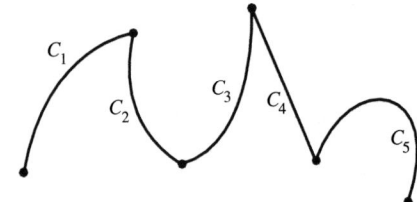

Figure 12.7 A piecewise smooth curve made up of five smooth curves connected end to end in continuous fashion.

Definitions Velocity, Speed, Acceleration, Direction of Motion

If **r** is the position vector of a particle moving along a smooth curve in space, then at any time t,

1. $\mathbf{v}(t) = \dfrac{d\mathbf{r}}{dt}$, the derivative of position, is the particle's **velocity vector** and is tangent to the curve.

2. $|\mathbf{v}(t)|$, the magnitude of **v,** is the particle's **speed.**

3. $\mathbf{a}(t) = \dfrac{d\mathbf{v}}{dt} = \dfrac{d^2\mathbf{r}}{dt^2}$, the derivative of velocity and the second derivative of position, is the particle's **acceleration vector.**

4. $\dfrac{\mathbf{v}}{|\mathbf{v}|}$, a unit vector, is the **direction of motion.**

We can express the velocity of a smoothly moving particle as the product of its speed and direction.

$$\text{Velocity} = |\mathbf{v}|\left(\frac{\mathbf{v}}{|\mathbf{v}|}\right) = (\text{speed})(\text{direction})$$

Example 4   STUDYING MOTION

The vector $\mathbf{r}(t) = (3\cos t)\mathbf{i} + (3\sin t)\mathbf{j} + t^2\mathbf{k}$ gives the position of a moving body at time t. Find

(a) the velocity and acceleration vectors.

(b) the body's speed and direction at $t = 2$.

(c) at what times the body's velocity and acceleration are orthogonal.

Solution

(a)
$$\mathbf{r} = (3\cos t)\mathbf{i} + (3\sin t)\mathbf{j} + t^2\mathbf{k}$$

$$\mathbf{v} = \frac{d\mathbf{r}}{dt} = -(3\sin t)\mathbf{i} + (3\cos t)\mathbf{j} + 2t\mathbf{k}$$

$$\mathbf{a} = \frac{d^2\mathbf{r}}{dt^2} = -(3\cos t)\mathbf{i} - (3\sin t)\mathbf{j} + 2\mathbf{k}$$

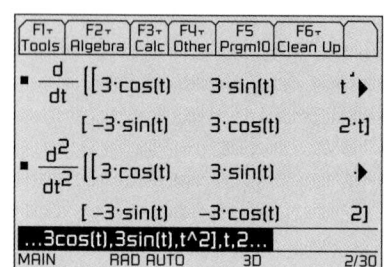

Figure 12.8 The velocity and acceleration of Example 4.

We can also check derivative computations with our advanced graphing calculator. Figure 12.8 shows **v** and **a** computed on our graphing calculator.

(b) At $t = 2$, the body's speed and direction are

$$\text{Speed: } |\mathbf{v}(2)| = \sqrt{(-3\sin 2)^2 + (3\cos 2)^2 + (4)^2} = 5,$$

$$\text{Direction: } \frac{\mathbf{v}(2)}{|\mathbf{v}(2)|} = -\left(\frac{3}{5}\sin 2\right)\mathbf{i} + \left(\frac{3}{5}\cos 2\right)\mathbf{j} + \frac{4}{5}\mathbf{k}.$$

(c) To find the times when $\mathbf{v}$ and $\mathbf{a}$ are orthogonal, we look for values of t for which

$$\mathbf{v} \cdot \mathbf{a} = 9 \sin t \cos t - 9 \cos t \sin t + 4t = 0.$$

The only value is $t = 0$.

Differentiation Rules

Because the derivatives of vector functions may be computed component by component, the rules for differentiating vector functions have the same form as the rules for differentiating scalar functions.

Differentiation Rules for Vector Functions

Let $\mathbf{u}$ and $\mathbf{v}$ be differentiable vector functions of t, $\mathbf{C}$ any constant vector, c any scalar, and f any differentiable scalar function of t.

1. Constant Function Rule: $\dfrac{d}{dt} \mathbf{C} = \mathbf{0}$

2. Scalar Multiple Rules: $\dfrac{d}{dt}(c\mathbf{u}) = c\dfrac{d\mathbf{u}}{dt}$

$\dfrac{d}{dt}(f\mathbf{u}) = \dfrac{df}{dt}\mathbf{u} + f\dfrac{d\mathbf{u}}{dt}$

3. Sum Rule: $\dfrac{d}{dt}(\mathbf{u} + \mathbf{v}) = \dfrac{d\mathbf{u}}{dt} + \dfrac{d\mathbf{v}}{dt}$

4. Difference Rule: $\dfrac{d}{dt}(\mathbf{u} - \mathbf{v}) = \dfrac{d\mathbf{u}}{dt} - \dfrac{d\mathbf{v}}{dt}$

5. Dot-Product Rule: $\dfrac{d}{dt}(\mathbf{u} \cdot \mathbf{v}) = \dfrac{d\mathbf{u}}{dt} \cdot \mathbf{v} + \mathbf{u} \cdot \dfrac{d\mathbf{v}}{dt}$

6. Cross-Product Rule: $\dfrac{d}{dt}(\mathbf{u} \times \mathbf{v}) = \dfrac{d\mathbf{u}}{dt} \times \mathbf{v} + \mathbf{u} \times \dfrac{d\mathbf{v}}{dt}$

7. Chain Rule (Short Form):   If $\mathbf{r}$ is a differentiable function of t and t is a differentiable function of s, then

$$\frac{d\mathbf{r}}{ds} = \frac{d\mathbf{r}}{dt}\frac{dt}{ds}.$$

As an algebraic convenience, we sometimes write the product of a scalar c and a vector $\mathbf{v}$ as $\mathbf{v}c$ instead of $c\mathbf{v}$. This permits us, for instance, to write the Chain Rule in a familiar form:

$$\frac{d\mathbf{r}}{ds} = \frac{d\mathbf{r}}{dt}\frac{dt}{ds}.$$

The proof of the Dot-Product Rule is very similar (one more component) to the proof on page 535. We will prove the Cross-Product Rule and the Chain Rule but leave the rest of the rules as exercises.

Proof of the Cross-Product Rule   We model the proof after the proof of the Product Rule for scalar functions (see page 116). According to the definition of derivative,

$$\frac{d}{dt}(\mathbf{u} \times \mathbf{v}) = \lim_{h \to 0} \frac{\mathbf{u}(t + h) \times \mathbf{v}(t + h) - \mathbf{u}(t) \times \mathbf{v}(t)}{h}.$$

To change this fraction into an equivalent one that contains the difference quotients for the derivatives of **u** and **v**, we subtract and add $\mathbf{u}(t) \times \mathbf{v}(t + h)$ in the numerator. Then

$$\frac{d}{dt}(\mathbf{u} \times \mathbf{v})$$

$$= \lim_{h \to 0} \frac{\mathbf{u}(t + h) \times \mathbf{v}(t + h) - \mathbf{u}(t) \times \mathbf{v}(t + h) + \mathbf{u}(t) \times \mathbf{v}(t + h) - \mathbf{u}(t) \times \mathbf{v}(t)}{h}$$

$$= \lim_{h \to 0} \left[\frac{\mathbf{u}(t + h) - \mathbf{u}(t)}{h} \times \mathbf{v}(t + h) + \mathbf{u}(t) \times \frac{\mathbf{v}(t + h) - \mathbf{v}(t)}{h} \right]$$

$$= \lim_{h \to 0} \frac{\mathbf{u}(t + h) - \mathbf{u}(t)}{h} \times \lim_{h \to 0} \mathbf{v}(t + h) + \lim_{h \to 0} \mathbf{u}(t) \times \lim_{h \to 0} \frac{\mathbf{v}(t + h) - \mathbf{v}(t)}{h}.$$

The last equality holds because the limit of the cross product of two vector functions is the cross product of their limits if the latter exist (Exercise 43). As h approaches zero, $\mathbf{v}(t + h)$ approaches $\mathbf{v}(t)$ because **v**, being differentiable at t, is continuous at t (Exercise 44). The two fractions approach the values of $d\mathbf{u}/dt$ and $d\mathbf{v}/dt$ at t. In short,

$$\frac{d}{dt}(\mathbf{u} \times \mathbf{v}) = \frac{d\mathbf{u}}{dt} \times \mathbf{v} + \mathbf{u} \times \frac{d\mathbf{v}}{dt}.$$

Proof of the Chain Rule Suppose $\mathbf{r}(t) = f(t)\mathbf{i} + g(t)\mathbf{j} + h(t)\mathbf{k}$ is a differentiable vector function of t and t is a differentiable scalar function of some other variable s. Then f, g, and h are differentiable functions of s, and the Chain Rule for differentiable real-valued functions gives

$$\frac{d\mathbf{r}}{ds} = \frac{df}{ds}\mathbf{i} + \frac{dg}{ds}\mathbf{j} + \frac{dh}{ds}\mathbf{k}$$

$$= \frac{df}{dt}\frac{dt}{ds}\mathbf{i} + \frac{dg}{dt}\frac{dt}{ds}\mathbf{j} + \frac{dh}{dt}\frac{dt}{ds}\mathbf{k}$$

$$= \left(\frac{df}{dt}\mathbf{i} + \frac{dg}{dt}\mathbf{j} + \frac{dh}{dt}\mathbf{k} \right) \frac{dt}{ds}$$

$$= \frac{d\mathbf{r}}{dt}\frac{dt}{ds}. \qquad \blacksquare$$

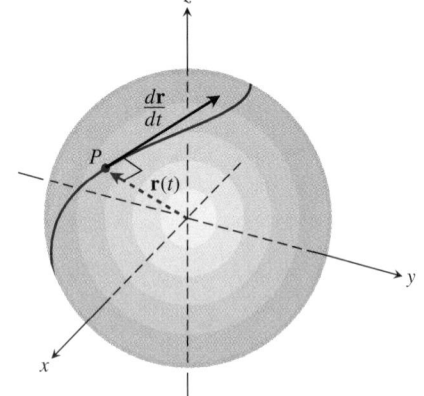

Figure 12.9 If a particle moves on a sphere in such a way that its position **r** is a differentiable function of time, then $\mathbf{r} \cdot (d\mathbf{r}/dt) = 0$.

Vector Functions of Constant Length

When we track a particle moving on a sphere centered at the origin (Figure 12.9), the position vector has a constant length equal to the radius of the sphere. The velocity vector $d\mathbf{r}/dt$, tangent to the path of motion, is tangent to the sphere and hence perpendicular to **r**. This is always the case for a differentiable vector function of constant length: The vector and its first derivative are orthogonal. With the length constant, the change in the function is a change in direction only, and direction changes take place at right angles.

We will use the observation in Equation (3) repeatedly in Section 12.3.

If **u** is a differentiable vector function of t of constant length, then

$$\mathbf{u} \cdot \frac{d\mathbf{u}}{dt} = 0. \qquad (3)$$

To see why Equation (3) holds, suppose $\mathbf{u}$ is a differentiable function of t and $|\mathbf{u}|$ is constant. Then $\mathbf{u} \cdot \mathbf{u} = |\mathbf{u}^2|$ is constant and we may differentiate both sides of this equation to get

$$\frac{d}{dt}(\mathbf{u} \cdot \mathbf{u}) = \frac{d}{dt}(\text{constant}) = 0$$

$$\frac{d\mathbf{u}}{dt} \cdot \mathbf{u} + \mathbf{u} \cdot \frac{d\mathbf{u}}{dt} = 0 \qquad \text{Dot-Product Rule with } \mathbf{v} = \mathbf{u}$$

$$2\mathbf{u} \cdot \frac{d\mathbf{u}}{dt} = 0 \qquad \text{Dot multiplication is commutative}$$

$$\mathbf{u} \cdot \frac{d\mathbf{u}}{dt} = 0.$$

Example 5 SUPPORTING EQUATION (3)

Show that $\mathbf{u}(t) = (\sin t)\mathbf{i} + (\cos t)\mathbf{j} + \sqrt{3}\,\mathbf{k}$ has constant length and is orthogonal to its derivative.

Solution

$$\mathbf{u}(t) = (\sin t)\mathbf{i} + (\cos t)\mathbf{j} + \sqrt{3}\,\mathbf{k}$$

$$|\mathbf{u}(t)| = \sqrt{(\sin t)^2 + (\cos t)^2 + (\sqrt{3})^2} = \sqrt{1 + 3} = 2$$

$$\frac{d\mathbf{u}}{dt} = (\cos t)\mathbf{i} - (\sin t)\mathbf{j}$$

$$\mathbf{u} \cdot \frac{d\mathbf{u}}{dt} = \sin t \cos t - \sin t \cos t = 0$$

Integrals of Vector Functions

A differentiable vector function $\mathbf{R}(t)$ is an **antiderivative** of a vector function $\mathbf{r}(t)$ on an interval I if $d\mathbf{R}/dt = \mathbf{r}$ at each point of I. If $\mathbf{R}$ is an antiderivative of $\mathbf{r}$ on I, it can be shown, working one component at a time, that every antiderivative of $\mathbf{r}$ on I has the form $\mathbf{R} + \mathbf{C}$ for some constant vector $\mathbf{C}$ (Exercise 47).

> **Definition Indefinite Integral**
>
> The **indefinite integral** of $\mathbf{r}$ with respect to t is the set of all antiderivatives of $\mathbf{r}$, denoted by $\int \mathbf{r}(t)\, dt$. If $\mathbf{R}$ is any antiderivative of $\mathbf{r}$, then
>
> $$\int \mathbf{r}(t)\, dt = \mathbf{R}(t) + \mathbf{C}.$$

The usual arithmetic rules for indefinite integrals apply.

Example 6 FINDING ANTIDERIVATIVES

We can check evaluation of integrals (both definite and indefinite) with our advanced graphing calculator.

$$\int ((\cos t)\mathbf{i} + \mathbf{j} - 2t\mathbf{k})\, dt = \left(\int \cos t\, dt\right)\mathbf{i} + \left(\int dt\right)\mathbf{j} - \left(\int 2t\, dt\right)\mathbf{k} \quad (4)$$

$$= (\sin t + C_1)\mathbf{i} + (t + C_2)\mathbf{j} - (t^2 + C_3)\mathbf{k} \quad (5)$$

$$= (\sin t)\mathbf{i} + t\mathbf{j} - t^2\mathbf{k} + \mathbf{C} \quad \mathbf{C} = C_1\mathbf{i} + C_2\mathbf{j} - C_3\mathbf{k}$$

As in the integration of scalar functions, we recommend that you skip the steps in Equations (4) and (5) and go directly to the final form. Find an antiderivative for each component and add a constant vector at the end.

Definite integrals of vector functions are also defined in terms of components.

Definition Definite Integrals

If the components of $\mathbf{r}(t) = f(t)\mathbf{i} + g(t)\mathbf{j} + h(t)\mathbf{k}$ are integrable over $[a, b]$, then so is $\mathbf{r}$, and the **definite integral** of $\mathbf{r}$ from a to b is

$$\int_a^b \mathbf{r}(t)\, dt = \left(\int_a^b f(t)\, dt \right) \mathbf{i} + \left(\int_a^b g(t)\, dt \right) \mathbf{j} + \left(\int_a^b h(t)\, dt \right) \mathbf{k}.$$

Example 7   EVALUATING DEFINITE INTEGRALS

$$\int_0^\pi \left((\cos t)\mathbf{i} + \mathbf{j} - 2t\mathbf{k} \right) dt = \left(\int_0^\pi \cos t\, dt \right) \mathbf{i} + \left(\int_0^\pi dt \right) \mathbf{j} - \left(\int_0^\pi 2t\, dt \right) \mathbf{k}$$

$$= \left(\sin t \,\big]_0^\pi \right)\mathbf{i} + \left(t \,\big]_0^\pi \right)\mathbf{j} - \left(t^2 \,\big]_0^\pi \right)\mathbf{k}$$

$$= (0 - 0)\mathbf{i} + (\pi - 0)\mathbf{j} - (\pi^2 - 0^2)\mathbf{k}$$

$$= \pi\mathbf{j} - \pi^2\mathbf{k}$$

Example 8   FINDING A PARTICLE'S POSITION FUNCTION

The velocity of a particle moving in space is

$$\frac{d\mathbf{r}}{dt} = (\cos t)\mathbf{i} - (\sin t)\mathbf{j} + \mathbf{k}.$$

Find the particle's position function as a function of t if $\mathbf{r} = 2\mathbf{i} + \mathbf{k}$ when $t = 0$.

Solution   Our goal is to solve the initial value problem that consists of

The differential equation: $\dfrac{d\mathbf{r}}{dt} = (\cos t)\mathbf{i} - (\sin t)\mathbf{j} + \mathbf{k}$

The initial condition: $\mathbf{r}(0) = 2\mathbf{i} + \mathbf{k}.$

Integrating both sides of the differential equation with respect to t gives

$$\mathbf{r}(t) = (\sin t)\mathbf{i} + (\cos t)\mathbf{j} + t\mathbf{k} + \mathbf{C}.$$

The initial condition determines the value of $\mathbf{C}$:

$$(\sin 0)\mathbf{i} + (\cos 0)\mathbf{j} + (0)\mathbf{k} + \mathbf{C} = 2\mathbf{i} + \mathbf{k} \qquad \mathbf{r}(0) = 2\mathbf{i} + \mathbf{k}$$

$$\mathbf{j} + \mathbf{C} = 2\mathbf{i} + \mathbf{k}$$

$$\mathbf{C} = 2\mathbf{i} - \mathbf{j} + \mathbf{k}.$$

The particle's position as a function of t is

$$\mathbf{r}(t) = (\sin t + 2)\mathbf{i} + (\cos t - 1)\mathbf{j} + (t + 1)\mathbf{k}.$$

To check (always a good idea), we can see from this formula that

$$\frac{d\mathbf{r}}{dt} = (\cos t + 0)\mathbf{i} + (-\sin t - 0)\mathbf{j} + (1 + 0)\mathbf{k}$$

$$= (\cos t)\mathbf{i} - (\sin t)\mathbf{j} + \mathbf{k}$$

and

$$\mathbf{r}(0) = (\sin 0 + 2)\mathbf{i} + (\cos 0 - 1)\mathbf{j} + (0 + 1)\mathbf{k}$$

$$= 2\mathbf{i} + \mathbf{k}.$$

Quick Review 12.1

In Exercises 1–4, let $f(x) = \text{int } x$, the greatest integer function.

1. Find $\lim\limits_{x \to -2^-} \text{int } x$.

2. Find $\lim\limits_{x \to -2^+} \text{int } x$.

3. At what values of x is f continuous?

4. Find the points of discontinuity of f.

In Exercises 5 and 6, let $f(x) = (\cot x)/x$.

5. Find an equation of the line tangent to the graph of f at $x = \pi/4$.

6. Find an equation of the line normal to the graph of f at $x = \pi/4$.

7. Find $\dfrac{d^2y}{dx^2}$ for $y = \dfrac{2x + 1}{x + 3}$.

8. Evaluate $\displaystyle\int \frac{dx}{1 + x}$.

9. Evaluate $\displaystyle\int_0^2 \frac{dx}{1 + x}$.

10. Solve the initial value problem

$$\frac{dy}{dx} = \cos x + \sin x, \quad y\left(\frac{\pi}{2}\right) = 1.$$

Section 12.1 Exercises

In Exercises 1–4, $\mathbf{r}(t)$ is the position of a particle in the xy-plane at time t. Find an equation in x and y whose graph is the path of the particle. Then find the particle's velocity and acceleration vectors at the given value of t.

1. $\mathbf{r}(t) = (t + 1)\mathbf{i} + (t^2 - 1)\mathbf{j}, \qquad t = 1$

2. $\mathbf{r}(t) = (t^2 + 1)\mathbf{i} + (2t - 1)\mathbf{j}, \qquad t = 1/2$

3. $\mathbf{r}(t) = e^t\mathbf{i} + \dfrac{2}{9}e^{2t}\mathbf{j}, \qquad t = \ln 3$

4. $\mathbf{r}(t) = (\cos 2t)\mathbf{i} + (3 \sin 2t)\mathbf{j}, \qquad t = 0$

In Exercises 5–8, $\mathbf{r}(t)$ is the position vector of a particle moving along a curve in the xy-plane. Find the particle's velocity and acceleration vectors at the stated times and sketch them as vectors on the curve.

5. *Motion on the Circle* $x^2 + y^2 = 1$

$\mathbf{r}(t) = (\sin t)\mathbf{i} + (\cos t)\mathbf{j}; \qquad t = \pi/4$ and $\pi/2$

6. *Motion on the Circle* $x^2 + y^2 = 16$

$\mathbf{r}(t) = \left(4\cos\dfrac{t}{2}\right)\mathbf{i} + \left(4\sin\dfrac{t}{2}\right)\mathbf{j}; \qquad t = \pi$ and $3\pi/2$

7. *Motion on the Cycloid* $x = t - \sin t, y = 1 - \cos t$

$\mathbf{r}(t) = (t - \sin t)\mathbf{i} + (1 - \cos t)\mathbf{j}: \qquad t = \pi$ and $3\pi/2$

8. *Motion on the Parabola* $y = x^2 + 1$

$\mathbf{r}(t) = t\mathbf{i} + (t^2 + 1)\mathbf{j}; \qquad t = -1, 0$, and 1

In Exercises 9–12, $\mathbf{r}(t)$ is the position of a particle in space at time t. **(a)** Find the particle's velocity and acceleration vectors.

Then **(b)** find the particle's speed and direction of motion at the given value of *t*. **(c)** Write the particle's velocity at that time as the product of its speed and direction.

9. $\mathbf{r}(t) = (t + 1)\mathbf{i} + (t^2 - 1)\mathbf{j} + 2t\mathbf{k}, \quad t = 1$

10. $\mathbf{r}(t) = (2 \cos t)\mathbf{i} + (3 \sin t)\mathbf{j} + 4t\mathbf{k}, \quad t = \pi/2$

11. $\mathbf{r}(t) = (\sec t)\mathbf{i} + (\tan t)\mathbf{j} + \dfrac{4}{3}t\mathbf{k}, \quad t = \pi/6$

12. $\mathbf{r}(t) = (2 \cos 3t)\mathbf{i} + (e^{-t})\mathbf{j} + (2 \sin 3t)\mathbf{k}, \quad t = 0$

In Exercises 13–16, $\mathbf{r}(t)$ is the position of a particle in space at time *t*. Find the angle between the velocity and acceleration vectors at time $t = 0$.

13. $\mathbf{r}(t) = (3t + 1)\mathbf{i} + \sqrt{3}t\mathbf{j} + t^2\mathbf{k}$

14. $\mathbf{r}(t) = \left(\dfrac{\sqrt{2}}{2}t\right)\mathbf{i} + \left(\dfrac{\sqrt{2}}{2}t - 16t^2\right)\mathbf{j}$

15. $\mathbf{r}(t) = (\ln(t^2 + 1))\mathbf{i} + (\tan^{-1} t)\mathbf{j} + \sqrt{t^2 + 1}\mathbf{k}$

16. $\mathbf{r}(t) = \dfrac{4}{9}(1 + t)^{3/2}\mathbf{i} + \dfrac{4}{9}(1 - t)^{3/2}\mathbf{j} + \dfrac{1}{3}t\mathbf{k}$

In Exercises 17 and 18, $\mathbf{r}(t)$ is the position vector of a particle in space at time *t*. Find the time or times in the given time interval when the velocity and acceleration vectors are orthogonal.

17. $\mathbf{r}(t) = (t - \sin t)\mathbf{i} + (1 - \cos t)\mathbf{j} + t\mathbf{k}; \quad 0 \le t \le 2\pi$

18. $\mathbf{r}(t) = (\sin t)\mathbf{i} + t\mathbf{j} + (\cos t)\mathbf{k}, \quad t \ge 0$

In Exercises 19–22, evaluate the integral.

19. $\displaystyle\int_0^1 [t^3\mathbf{i} + 7\mathbf{j} + (t + 1)\mathbf{k}]\, dt$

20. $\displaystyle\int_0^{\pi/3} [(\sec t \tan t)\mathbf{i} + (\tan t)\mathbf{j} + (2 \sin t \cos t)\mathbf{k}]\, dt$

21. $\displaystyle\int_1^4 \left[\dfrac{1}{t}\mathbf{i} + \dfrac{1}{5 - t}\mathbf{j} + \dfrac{1}{2t}\mathbf{k}\right] dt$

22. $\displaystyle\int_0^1 \left[\dfrac{2}{\sqrt{1 - t^2}}\mathbf{i} + \dfrac{\sqrt{3}}{1 + t^2}\mathbf{k}\right] dt$

In Exercises 23–26, solve the initial value problem for **r** as a vector function of *t*.

23. Differential equation: $\dfrac{d\mathbf{r}}{dt} = \dfrac{3}{2}(t + 1)^{1/2}\mathbf{i} + e^{-t}\mathbf{j} + \dfrac{1}{t + 1}\mathbf{k}$

Initial condition: $\mathbf{r}(0) = \mathbf{k}$

24. Differential equation: $\dfrac{d\mathbf{r}}{dt} = (t^3 + 4t)\mathbf{i} + t\mathbf{j} + 2t^2\mathbf{k}$

Initial condition: $\mathbf{r}(0) = \mathbf{i} + \mathbf{j}$

25. Differential equation: $\dfrac{d^2\mathbf{r}}{dt^2} = -32\mathbf{k}$

Initial conditions: $\mathbf{r}(0) = 100\mathbf{k}$ and $\left.\dfrac{d\mathbf{r}}{dt}\right|_{t=0} = 8\mathbf{i} + 8\mathbf{j}$

26. Differential equation: $\dfrac{d^2\mathbf{r}}{dt^2} = -(\mathbf{i} + \mathbf{j} + \mathbf{k})$

Initial conditions: $\mathbf{r}(0) = 10\mathbf{i} + 10\mathbf{j} + 10\mathbf{k}$ and $\left.\dfrac{d\mathbf{r}}{dt}\right|_{t=0} = \mathbf{0}$

In Exercises 27–30, find parametric equations for the line that is tangent to the given curve at the given parameter value $t = t_0$.

27. $\mathbf{r}(t) = (\sin t)\mathbf{i} + (t^2 - \cos t)\mathbf{j} + e^t\mathbf{k}, \quad t_0 = 0$

28. $\mathbf{r}(t) = (2 \sin t)\mathbf{i} + (2 \cos t)\mathbf{j} + 5t\mathbf{k}, \quad t_0 = 4\pi$

29. $\mathbf{r}(t) = (a \sin t)\mathbf{i} + (a \cos t)\mathbf{j} + bt\mathbf{k}, \quad t_0 = 2\pi$

30. $\mathbf{r}(t) = (\cos t)\mathbf{i} + (\sin t)\mathbf{j} + (\sin 2t)\mathbf{k}, \quad t_0 = \dfrac{\pi}{2}$

Exploration

31. *Motion on Circular Paths* Each of the following equations (a)–(e) describes the motion of a particle having the same path, namely the unit circle $x^2 + y^2 = 1$. Although the path of each particle is the same, the behavior, or "dynamics," of each particle is different. For each particle, answer the following questions.

  (i) Does the particle have constant speed? If so, what is its constant speed?

  (ii) Is the particle's acceleration vector always orthogonal to its velocity vector?

  (iii) Does the particle move clockwise or counterclockwise around the circle?

  (iv) Does the particle begin at the point (1, 0)?

  (a) $\mathbf{r}(t) = (\cos t)\mathbf{i} + (\sin t)\mathbf{j}, \quad t \ge 0$

  (b) $\mathbf{r}(t) = \cos(2t)\mathbf{i} + \sin(2t)\mathbf{j}, \quad t \ge 0$

  (c) $\mathbf{r}(t) = \cos\left(t - \dfrac{\pi}{2}\right)\mathbf{i} + \sin\left(t - \dfrac{\pi}{2}\right)\mathbf{j}, \quad t \ge 0$

  (d) $\mathbf{r}(t) = (\cos t)\mathbf{i} - (\sin t)\mathbf{j}, \quad t \ge 0$

  (e) $\mathbf{r}(t) = \cos(t^2)\mathbf{i} + \sin(t^2)\mathbf{j}, \quad t \ge 0$

32. *Motion Along a Circular Path* Show that the vector-valued function

$$\mathbf{r}(t) = (2\mathbf{i} + 2\mathbf{j} + \mathbf{k}) + \cos t\left(\dfrac{1}{\sqrt{2}}\mathbf{i} - \dfrac{1}{\sqrt{2}}\mathbf{j}\right)$$
$$+ \sin t\left(\dfrac{1}{\sqrt{3}}\mathbf{i} + \dfrac{1}{\sqrt{3}}\mathbf{j} + \dfrac{1}{\sqrt{3}}\mathbf{k}\right)$$

describes the motion of a particle moving on the circle of radius 1 centered at the point (2, 2, 1) and lying in the plane $x + y - 2z = 2$.

33. *Motion Along a Line* At time $t = 0$, a particle is located at the point (1, 2, 3). It travels in a straight line toward the point (4, 1, 4) and has speed 2 at (1, 2, 3) and constant acceleration $3\mathbf{i} - \mathbf{j} + \mathbf{k}$. Find an equation for the position vector $\mathbf{r}(t)$ of the particle at time t.

34. *Motion Along a Line* A particle traveling in a straight line is located at the point $(1, -1, 2)$ and has speed 2 at time $t = 0$. The particle moves toward the point $(3, 0, 3)$ with constant acceleration $2\mathbf{i} + \mathbf{j} + \mathbf{k}$. Find its position vector $\mathbf{r}(t)$ at time t.

35. *Motion Along a Parabola* A particle moves along the top of the parabola $y^2 = 2x$ in the xy-plane from left to right at a constant speed of 5 units per second. Find the velocity of the particle as it moves through the point $(2, 2)$.

36. *Motion Along a Cycloid* A particle moves on a cycloid in the xy-plane in such a way that its position at time t is

$$\mathbf{r}(t) = (t - \sin t)\mathbf{i} + (1 - \cos t)\mathbf{j}.$$

Find the maximum and minimum values of $|\mathbf{v}|$ and $|\mathbf{a}|$. (*Hint:* Find the extreme values of $|\mathbf{v}|^2$ and $|\mathbf{a}|^2$ first and take square roots later.)

37. *Motion Along an Ellipse* A particle moves around the ellipse $(y/3)^3 + (z/2)^2 = 1$ in the yz-plane in such a way that its position at time t is

$$\mathbf{r}(t) = (3 \cos t)\mathbf{j} + (2 \sin t)\mathbf{k}.$$

Find the maximum and minimum values of $|\mathbf{v}|$ and $|\mathbf{a}|$. (See the hint in Exercise 36.)

38. *Orthogonality Implies Constant Length* Let $\mathbf{v}$ be a differentiable vector function of t. Show that if $\mathbf{v} \cdot (d\mathbf{v}/dt) = 0$ for all t, then $|\mathbf{v}|$ is constant.

39. *Constant Function Rule* Prove that if $\mathbf{u}$ is the vector function with the constant value $\mathbf{C}$, then $d\mathbf{u}/dt = \mathbf{0}$.

40. *Scalar Multiple Rules*

(a) Prove that if $\mathbf{u}$ is a differentiable function of t and c is any real number, then

$$\frac{d(c\mathbf{u})}{dt} = c\frac{d\mathbf{u}}{dt}.$$

(b) Prove that if $\mathbf{u}$ is a differentiable function of t and f is a differentiable scalar function of t, then

$$\frac{d}{dt}(f\mathbf{u}) = \frac{df}{dt}\mathbf{u} + f\frac{d\mathbf{u}}{dt}.$$

41. *Sum and Difference Rules* Prove that if $\mathbf{u}$ and $\mathbf{v}$ are differentiable functions of t, then

$$\frac{d}{dt}(\mathbf{u} + \mathbf{v}) = \frac{d\mathbf{u}}{dt} + \frac{d\mathbf{v}}{dt}$$

and

$$\frac{d}{dt}(\mathbf{u} - \mathbf{v}) = \frac{d\mathbf{u}}{dt} - \frac{d\mathbf{v}}{dt}.$$

42. *Component Test for Continuity* Show that the vector function $\mathbf{r}$ defined by the rule

$$\mathbf{r}(t) = f(t)\mathbf{i} + g(t)\mathbf{j} + h(t)\mathbf{k}$$

is continuous at $t = t_0$ if and only if f, g, and h are continuous at t_0.

43. *Limits of Cross Products* Suppose that

$$\mathbf{r}_1(t) = f_1(t)\mathbf{i} + g_1(t)\mathbf{j} + h_1(t)\mathbf{k}, \mathbf{r}_2(t) = f_2(t)\mathbf{i} + g_2(t)\mathbf{j} + h_2(t)\mathbf{k},$$

$\lim_{t \to t_0} \mathbf{r}_1(t) = \mathbf{A}$, and $\lim_{t \to t_0} \mathbf{r}_2(t) = \mathbf{B}$. Use the determinant formula for cross products and the Limit Product Rule for scalar functions to show that

$$\lim_{t \to t_0}(\mathbf{r}_1(t) \times \mathbf{r}_2(t)) = \mathbf{A} \times \mathbf{B}.$$

44. *Differentiability Implies Continuity* Show that if $\mathbf{r}(t) = f(t)\mathbf{i} + g(t)\mathbf{j} + h(t)\mathbf{k}$ is differentiable at $t = t_0$, then it is continuous at t_0 as well.

45. *Properties of Integrable Vector Functions* Establish the following properties of integrable vector functions.

(a) The *Constant Scalar Multiple Rule:*

$$\int_a^b k\mathbf{r}(t)\, dt = k\int_a^b \mathbf{r}(t)\, dt \quad \text{(any scalar } k\text{)}$$

The Rule for Negatives,

$$\int_a^b \left(-\mathbf{r}(t)\right) dt = -\int_a^b \mathbf{r}(t)\, dt$$

is obtained by taking $k = -1$.

(b) The *Sum and Difference Rules:*

$$\int_a^b \left(\mathbf{r}_1(t) \pm \mathbf{r}_2(t)\right) dt = \int_a^b \mathbf{r}_1(t)\, dt \pm \int_a^b \mathbf{r}_2(t)\, dt$$

(c) The *Constant Vector Multiple Rules:*

$$\int_a^b \mathbf{C} \cdot \mathbf{r}(t)\, dt = \mathbf{C} \cdot \int_a^b \mathbf{r}(t)\, dt \quad \text{(any constant vector } \mathbf{C}\text{)}$$

and

$$\int_a^b \mathbf{C} \times \mathbf{r}(t)\, dt = \mathbf{C} \times \int_a^b \mathbf{r}(t)\, dt \quad \text{(any constant vector } \mathbf{C}\text{)}$$

46. *Products of Scalar and Vector Functions* Suppose the scalar function $u(t)$ and the vector function $\mathbf{r}(t)$ are both defined for $a \leq t \leq b$.

(a) Show that $u\mathbf{r}$ is continuous on $[a, b]$ if u and $\mathbf{r}$ are continuous on $[a, b]$.

(b) If u and $\mathbf{r}$ are both differentiable on $[a, b]$, show that $u\mathbf{r}$ is differentiable on $[a, b]$ and that

$$\frac{d}{dt}(u\mathbf{r}) = u\frac{d\mathbf{r}}{dt} + \mathbf{r}\frac{du}{dt}.$$

47. *Antiderivatives of Vector Functions*

(a) Use Corollary 2 of the Mean Value Theorem for derivatives of scalar functions (p. 189) to show that if two vector functions $\mathbf{R}_1(t)$ and $\mathbf{R}_2(t)$ have identical derivatives on an interval I, then the functions differ by a constant vector value throughout I.

(b) Use the result in (a) to show that if $\mathbf{R}(t)$ is any antiderivative of $\mathbf{r}(t)$ on I, then every antiderivative of $\mathbf{r}$ on I equals $\mathbf{R}(t) + \mathbf{C}$ for some constant vector $\mathbf{C}$.

48. *Fundamental Theorem of Calculus*

The Fundamental Theorem of Calculus for scalar functions of a real variable holds for vector functions of a real variable as well. Prove this by using the theorem for scalar functions to show first that if a vector function $\mathbf{r}(t)$ is continuous for $a \le t \le b$, then

$$\frac{d}{dt} \int_a^b \mathbf{r}(\tau)\, d\tau = \mathbf{r}(t)$$

at every point τ of $[a, b]$. Then use the conclusion in part (b) of Exercise 47 to show that if $\mathbf{R}$ is any antiderivative of $\mathbf{r}$ on $[a, b]$, then

$$\int_a^b \mathbf{r}(t)\, dt = \mathbf{R}(b) - \mathbf{R}(a).$$

49. **Writing to Learn**

Explain why the motion in Exercises 5–8 takes place on the curves indicated in each exercise.

Extending the Ideas

50. *Derivatives of Triple Scalar Products*

(a) Show that if $\mathbf{u}$, $\mathbf{v}$, and $\mathbf{w}$ are differentiable vector functions of t, then

$$\frac{d}{dt}(\mathbf{u} \cdot \mathbf{v} \times \mathbf{w}) = \frac{d\mathbf{u}}{dt} \cdot \mathbf{v} \times \mathbf{w} + \mathbf{u} \cdot \frac{d\mathbf{v}}{dt} \times \mathbf{w} + \mathbf{u} \cdot \mathbf{v} \times \frac{d\mathbf{w}}{dt}.$$

(b) Show that the equation in part (a) is equivalent to

$$\frac{d}{dt} \begin{vmatrix} u_1 & u_2 & u_3 \\ v_1 & v_2 & v_3 \\ w_1 & w_2 & w_3 \end{vmatrix} = \begin{vmatrix} \frac{du_1}{dt} & \frac{du_2}{dt} & \frac{du_3}{dt} \\ v_1 & v_2 & v_3 \\ w_1 & w_2 & w_3 \end{vmatrix}$$

$$+ \begin{vmatrix} u_1 & u_2 & u_3 \\ \frac{dv_1}{dt} & \frac{dv_2}{dt} & \frac{dv_3}{dt} \\ w_1 & w_2 & w_3 \end{vmatrix} + \begin{vmatrix} u_1 & u_2 & u_3 \\ v_1 & v_2 & v_3 \\ \frac{dw_1}{dt} & \frac{dw_2}{dt} & \frac{dw_3}{dt} \end{vmatrix}$$

This equation says that the derivative of a 3 by 3 determinant of differentiable functions is the sum of the three determinants obtained from the original by differentiating one row at a time. The result extends to determinants of any order.

51. *(Continuation of Exercise 50)*

Suppose $\mathbf{r}(t) = f(t)\mathbf{i} + g(t)\mathbf{j} + h(t)\mathbf{k}$ and f, g, and h have derivatives through order three. Use the results of Exercise 50 to show that

$$\frac{d}{dt}\left(\mathbf{r} \cdot \frac{d\mathbf{r}}{dt} \times \frac{d^2\mathbf{r}}{dt^2} \right) = \mathbf{r} \cdot \left(\frac{d\mathbf{r}}{dt} \times \frac{d^3\mathbf{r}}{dt^3} \right).$$

(Hint: Differentiate on the left and look for vectors whose products are zero.)

CAS Explorations

Tangent Lines

In Exercises 52–55, perform the following steps.

(a) Use a CAS to plot the space curve traced by the position vector $\mathbf{r}$.

(b) Find the components of the velocity vector $d\mathbf{r}/dt$.

(c) Evaluate $d\mathbf{r}/dt$ at the given point t_0 and find parametric equations for the line tangent to the curve at $\mathbf{r}(t_0)$.

(d) Use a CAS to plot the tangent line together with the curve over the given interval.

52. $\mathbf{r}(t) = (\sin t - t \cos t)\mathbf{i} + (\cos t + t \sin t)\mathbf{j} + t^2\, \mathbf{k}$, $0 \le t \le 6\pi$, $t_0 = 3\pi/2$

53. $\mathbf{r}(t) = \sqrt{2}\, t\mathbf{i} + e^t\mathbf{j} + e^{-t}\mathbf{k}$, $\quad -2 \le t \le 3$, $\quad t_0 = 1$

54. $\mathbf{r}(t) = (\sin 2t)\mathbf{i} + (\ln(1 + t))\mathbf{j} + t\mathbf{k}$, $\quad 0 \le t \le 4\pi$, $t_0 = \pi/4$

55. $\mathbf{r}(t) = (\ln(t^2 + 2))\mathbf{i} + (\tan^{-1} 3t)\mathbf{j} + \sqrt{t^2 + 1}\, \mathbf{k}$, $-3 \le t \le 5$, $\quad t_0 = 3$

Helices

In Exercises 56 and 57, you will use a CAS to explore graphically the behavior of the helix

$$\mathbf{r}(t) = (\cos at)\mathbf{i} + (\sin at)\mathbf{j} + bt\mathbf{k}$$

as you change the values of the constants a and b.

56. Writing to Learn Set $b = 1$. Plot the helix $\mathbf{r}(t)$ together with the tangent line to the curve at $t = 3\pi/2$ for $a = 1, 2, 4$, and 6 over the interval $0 \le t \le 4\pi$. Describe in your own words what happens to the graph of the helix and the position of the tangent line as a increases through these positive values.

57. Writing to Learn Set $a = 1$. Plot the helix $\mathbf{r}(t)$ together with the tangent line to the curve at $t = 3/2$ for $b = 1/4, 1/2$, 2, and 4 over the interval $0 \le t \le 4\pi$. Describe in your own words what happens to the graph of the helix and the position of the tangent line as b increases through these positive values.

Arc Length and the Unit Tangent Vector T

Arc Length • Speed on a Smooth Curve • The Unit Tangent Vector **T**

Arc Length

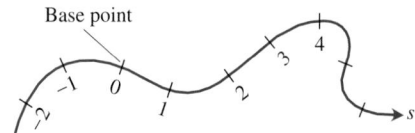

Figure 12.10 Smooth curves can be scaled like number lines, the coordinate of each point being its directed distance from a preselected base point.

One special feature of smooth curves is that they have a measurable length. This enables us to locate points on these curves by giving their directed distances along the curve from some **base point,** the way we locate points on coordinate axes by giving their directed distances from the origin (Figure 12.10). Time is the natural parameter for describing a moving body's velocity and acceleration, but *s*, the arc length, is the natural parameter for studying a curve's shape. Both parameters appear in analyses of space flight.

To measure distance along a smooth curve in space, we add a *z*-term to the formula we use for curves in the plane.

> **Definition** Arc Length: Length of a Smooth Curve
> The **length** of a smooth curve $\mathbf{r}(t) = f(t)\mathbf{i} + g(t)\mathbf{j} + h(t)\mathbf{k}, a \leq t \leq b,$ that is traced exactly once as *t* increases from $t = a$ to $t = b$ is
>
> $$L = \int_a^b \sqrt{\left(\frac{df}{dt}\right)^2 + \left(\frac{dg}{dt}\right)^2 + \left(\frac{dh}{dt}\right)^2}\, dt$$
>
> $$= \int_a^b \sqrt{\left(\frac{dx}{dt}\right)^2 + \left(\frac{dy}{dt}\right)^2 + \left(\frac{dz}{dt}\right)^2}\, dt \qquad (1)$$
>
> $$= \int_a^b |\mathbf{v}|\, dt$$

Just as for plane curves, we can calculate the length of a curve in space from any convenient parametrization that meets the stated conditions. We omit the proof.

Example 1 APPLYING THE DEFINITION OF LENGTH

Find the length of one turn of the helix $\mathbf{r}(t) = (\cos t)\mathbf{i} + (\sin t)\mathbf{j} + t\mathbf{k}$.

Solution The helix makes one full turn as *t* runs from 0 to 2π (Figure 12.11). The length of this portion of the curve is

$$L = \int_a^b |\mathbf{v}|\, dt = \int_0^{2\pi} \sqrt{(-\sin t)^2 + (\cos t)^2 + (1)^2}\, dt$$

$$= \int_0^{2\pi} \sqrt{2}\, dt = 2\pi\sqrt{2} \text{ units.}$$

This is $\sqrt{2}$ times the length of the circle in the *xy*-plane over which the helix stands.

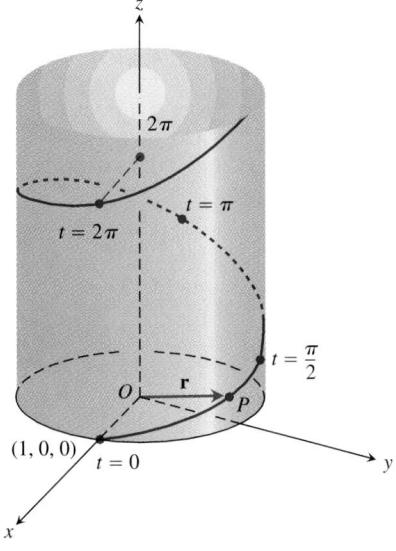

Figure 12.11 The helix $\mathbf{r}(t) = (\cos t)\mathbf{i} + (\sin t)\mathbf{j} + t\mathbf{k}$. (Example 1)

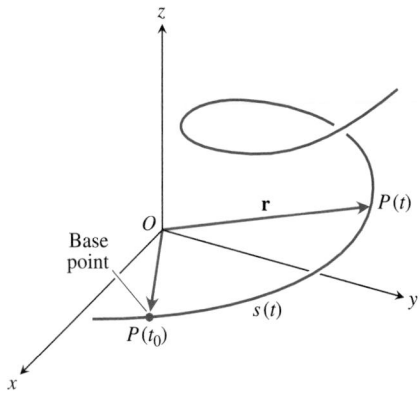

Figure 12.12 The directed distance along the curve from $P(t_0)$ to any point is $P(t)$ is
$$s(t) = \int_{t_0}^{t} |\mathbf{v}(\tau)| \, d\tau.$$

We use the Greek letter τ ("tau") as the variable of integration in the integral for s because the letter t is already in use as the upper limit.

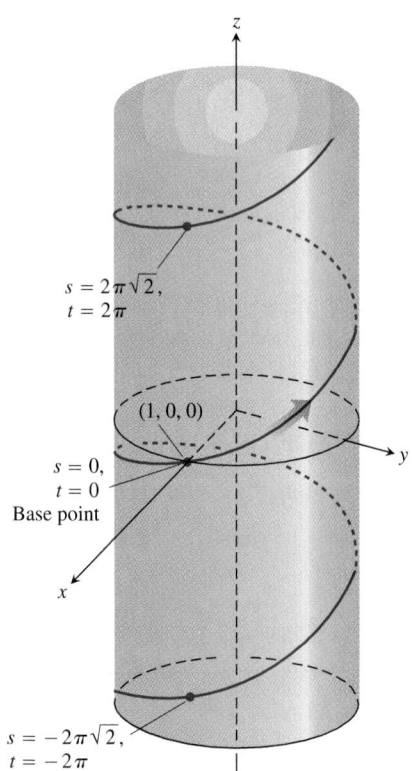

Figure 12.13 Arc length parameter values on the helix $\mathbf{r}(t) = (\cos t)\mathbf{i} + (\sin t)\mathbf{j} + t\mathbf{k}$. (Example 2)

If we choose a base point $P(t_0)$ on a smooth curve C parametrized by t, each value of t determines a point $P(t) = (x(t), y(t), z(t))$ on C and a "directed distance"

$$s(t) = \int_{t_0}^{t} |\mathbf{v}(\tau)| \, d\tau,$$

measured along C from the base point (Figure 12.12). If $t > t_0$, $s(t)$ is the distance from $P(t_0)$ to $P(t)$. If $t < t_0$, $s(t)$ is the negative of the distance. Each value of s determines a point on C and this parametrizes C with respect to s. We call s an **arc length parameter** for the curve. The parameter's value increases in the direction of increasing t.

> **Arc Length Parameter with Base Point $P(t_0)$ on a Smooth Curve**
>
> $$s(t) = \int_{t_0}^{t} \sqrt{[x'(\tau)]^2 + [y'(\tau)]^2 + [z'(\tau)]^2} \, d\tau = \int_{t_0}^{t} |\mathbf{v}(\tau)| \, d\tau \quad (2)$$

Example 2 FINDING AN ARC LENGTH PARAMETER

If $t_0 = 0$, the arc length parameter along the helix

$$\mathbf{r}(t) = (\cos t)\mathbf{i} + (\sin t)\mathbf{j} + t\mathbf{k}$$

from t_0 to t is

$$s(t) = \int_{t_0}^{t} |\mathbf{v}(\tau)| \, d\tau \qquad \text{Eq. (2)}$$

$$= \int_{0}^{t} \sqrt{2} \, d\tau \qquad \text{Value from Example 1}$$

$$= \sqrt{2}\,t.$$

Thus, $s(2\pi) = 2\pi\sqrt{2}$, $s(-2\pi) = -2\pi\sqrt{2}$, and so on (Figure 12.13).

Example 3 FINDING DIRECTED DISTANCE ALONG A LINE

Show that if $\mathbf{u} = u_1\mathbf{i} + u_2\mathbf{j} + u_3\mathbf{k}$ is a unit vector, then the directed distance along the line L given by

$$\mathbf{r}(t) = (x_0 + tu_1)\mathbf{i} + (y_0 + tu_2)\mathbf{j} + (z_0 + tu_3)\mathbf{k}$$

from the point $P_0(x_0, y_0, z_0)$ where $t = 0$ is t itself (Figure 12.14).

Solution

$$\mathbf{v} = \frac{d}{dt}(x_0 + tu_1)\mathbf{i} + \frac{d}{dt}(y_0 + tu_2)\mathbf{j} + \frac{d}{dt}(z_0 + tu_3)\mathbf{k}$$

$$= u_1\mathbf{i} + u_2\mathbf{j} + u_3\mathbf{k} = \mathbf{u},$$

so

$$s(t) = \int_{0}^{t} |\mathbf{v}| \, d\tau = \int_{0}^{t} |\mathbf{u}| \, d\tau = \int_{0}^{t} 1 \, d\tau = t.$$

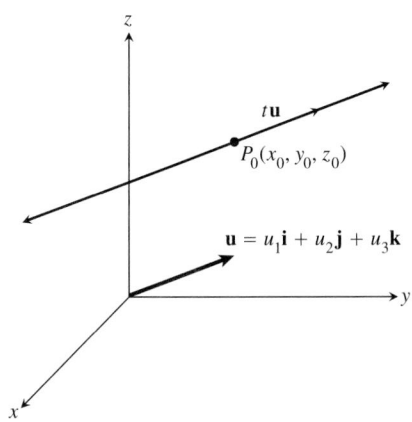

Figure 12.14 The figure for Example 3.

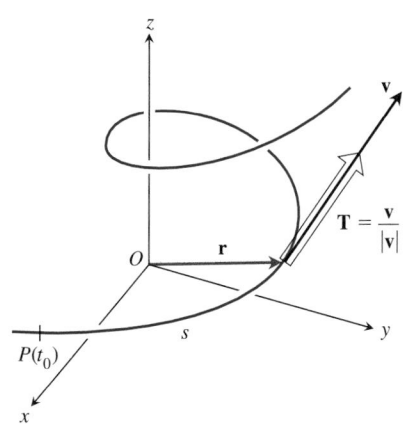

Figure 12.15 We find the unit tangent vector **T** by dividing **v** by $|\mathbf{v}|$.

Speed on a Smooth Curve

Since the derivatives beneath the radical in Equation (2) are continuous (the curve is smooth), the Fundamental Theorem of Calculus tells us that s is a differentiable function of t with derivative

$$\frac{ds}{dt} = |\mathbf{v}(t)|. \tag{3}$$

As we expect, the speed with which the particle moves along its path is the magnitude of **v**.

Although the base point $P(t_0)$ plays a role in defining s in Equation (2), it plays no role in Equation (3). The rate at which a moving particle covers distance along its path has nothing to do with how far away the base point is.

Notice also that $ds/dt > 0$ since $|\mathbf{v}(t)|$ is never zero for a smooth curve. We see once again that s is an increasing function of t.

The Unit Tangent Vector T

Since $ds/dt > 0$ for the curves we are considering, s is one-to-one and has an inverse that gives t as a differentiable function of s (Section 3.8). The derivative of the inverse is

$$\frac{dt}{ds} = \frac{1}{ds/dt} = \frac{1}{|\mathbf{v}|}.$$

This makes **r** a differentiable function of s whose derivative can be calculated with the Chain Rule to be

$$\frac{d\mathbf{r}}{ds} = \frac{d\mathbf{r}}{dt}\frac{dt}{ds} = \mathbf{v}\frac{1}{|\mathbf{v}|} = \frac{\mathbf{v}}{|\mathbf{v}|}.$$

This equation says that $d\mathbf{r}/ds$ is a unit vector in the direction of **v**. The vector $d\mathbf{r}/ds$ is the unit tangent vector of the curve traced by **r** and denoted by **T** (Figure 12.15).

Definition Unit Tangent Vector

The **unit tangent vector** of a smooth curve $\mathbf{r}(t)$ is

$$\mathbf{T} = \frac{d\mathbf{r}}{ds} = \frac{d\mathbf{r}/dt}{ds/dt} = \frac{\mathbf{v}}{|\mathbf{v}|}. \tag{4}$$

The unit tangent vector **T** is a differentiable function of t whenever **v** is a differentiable function of t. As we see in Section 12.3, **T** is one of three unit vectors in a traveling reference frame that is used to describe the motion of space vehicles and other bodies moving in three dimensions.

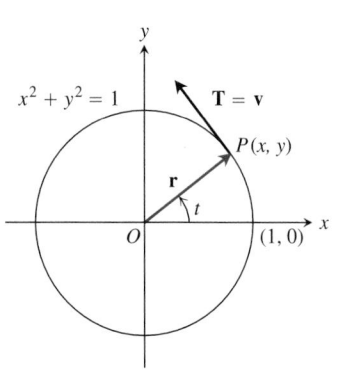

Figure 12.16 The motion $\mathbf{r}(t) = (\cos t)\mathbf{i} + (\sin t)\mathbf{j}$. (Example 5)

Example 4   FINDING THE UNIT TANGENT VECTOR T

Find the unit tangent vector of the helix $\mathbf{r}(t) = (\cos t)\mathbf{i} + (\sin t)\mathbf{j} + t\mathbf{k}.$

Solution

$$\mathbf{v} = (-\sin t)\mathbf{i} + (\cos t)\mathbf{j} + \mathbf{k}$$

$$|\mathbf{v}| = \sqrt{(-\sin t)^2 + (\cos t)^2 + (1)^2} = \sqrt{2}$$

$$\mathbf{T} = \frac{\mathbf{v}}{|\mathbf{v}|} = -\frac{\sin t}{\sqrt{2}}\mathbf{i} + \frac{\cos t}{\sqrt{2}}\mathbf{j} + \frac{1}{\sqrt{2}}\mathbf{k}$$

Example 5   FINDING THE UNIT TANGENT VECTOR T

For the counterclockwise motion $\mathbf{r}(t) = (\cos t)\mathbf{i} + (\sin t)\mathbf{j}$ around the unit circle,

$$\mathbf{v} = (-\sin t)\mathbf{i} + (\cos t)\mathbf{j}$$

is already a unit vector, so $\mathbf{T} = \mathbf{v}$ (Figure 12.16).

Quick Review 12.2

In Exercises 1–4, find an equation for the tangent line to the curve $x^2 + y^2 = 9$ at the indicated point.

1. $(3, 0)$ **2.** $(0, 3)$

3. $\left(-\dfrac{3}{\sqrt{2}}, \dfrac{3}{\sqrt{2}}\right)$ **4.** $\left(\dfrac{3}{\sqrt{2}}, \dfrac{3}{\sqrt{2}}\right)$

5. Let $x = 2\cos t$ and $y = 3\sin t$. Find two unit vectors

   (a) tangent to the curve at $t = \pi/3$.

   (b) normal to the curve at $t = \pi/3$.

6. Find the length of the curve $y = \sin x$ from $x = 0$ to $x = \pi/2$.

7. Find the length of the curve $x = \cos^3 t$, $y = \sin^3 t$, $0 \le t \le \pi$.

In Exercises 8–10, find dy/dx.

8. $y = \displaystyle\int_2^x t \sin t \, dt$

9. $y = \displaystyle\int_0^{x^2} \cos(t^2) \, dt$

10. $y = \displaystyle\int_{\cos x}^{\sin x} \ln(1 + t^2) \, dt$

Section 12.2 Exercises

In Exercises 1–8, find **(a)** the curve's unit tangent vector. Also, find **(b)** the length of the indicated portion of the curve.

1. $\mathbf{r}(t) = (2\cos t)\mathbf{i} + (2\sin t)\mathbf{j} + \sqrt{5}t\mathbf{k}$, $0 \le t \le \pi$

2. $\mathbf{r}(t) = (6\sin 2t)\mathbf{i} + (6\cos 2t)\mathbf{j} + 5t\mathbf{k}$, $0 \le t \le \pi$

3. $\mathbf{r}(t) = t\mathbf{i} + (2/3)t^{3/2}\mathbf{k}$, $0 \le t \le 8$

4. $\mathbf{r}(t) = (2 + t)\mathbf{i} - (t + 1)\mathbf{j} + t\mathbf{k}$, $0 \le t \le 3$

5. $\mathbf{r}(t) = (\cos^3 t)\mathbf{j} + (\sin^3 t)\mathbf{k}$, $0 \le t \le \pi/2$

6. $\mathbf{r}(t) = 6t^3\mathbf{i} - 2t^3\mathbf{j} - 3t^3\mathbf{k}$, $1 \le t \le 2$

7. $\mathbf{r}(t) = (t\cos t)\mathbf{i} + (t\sin t)\mathbf{j} + (2\sqrt{2}/3)t^{3/2}\mathbf{k}$, $0 \le t \le \pi$

8. $\mathbf{r}(t) = (t\sin t + \cos t)\mathbf{i} + (t\cos t - \sin t)\mathbf{j}$, $\sqrt{2} \le t \le 2$

9. Find the point on the curve

$$\mathbf{r}(t) = (5\sin t)\mathbf{i} + (5\cos t)\mathbf{j} + 12t\mathbf{k}$$

at a distance 26π units along the curve from the origin in the direction of increasing arc length.

10. Find the point on the curve

$$\mathbf{r}(t) = (12\sin t)\mathbf{i} - (12\cos t)\mathbf{j} + 5t\mathbf{k}$$

at a distance 13π units along the curve from the origin in the direction opposite to the direction of increasing arc length.

In Exercises 11–14, find **(a)** the arc length parameter along the curve from the point where $t = 0$ by evaluating the integral

$$s = \int_0^t |\mathbf{v}(\tau)| \, d\tau.$$

Then find **(b)** the length of the indicated portion of the curve.

11. $\mathbf{r}(t) = (4 \cos t)\mathbf{i} + (4 \sin t)\mathbf{j} + 3t\mathbf{k}, \qquad 0 \le t \le \pi/2$

12. $\mathbf{r}(t) = (\cos t + t \sin t)\mathbf{i} + (\sin t - t \cos t)\mathbf{j}, \qquad \pi/2 \le t \le \pi$

13. $\mathbf{r}(t) = (e^t \cos t)\mathbf{i} + (e^t \sin t)\mathbf{j} + e^t\mathbf{k}, \qquad -\ln 4 \le t \le 0$

14. $\mathbf{r}(t) = (1 + 2t)\mathbf{i} + (1 + 3t)\mathbf{j} + (6 - 6t)\mathbf{k}, \qquad -1 \le t \le 0$

15. Find the length of the curve
$$\mathbf{r}(t) = (\sqrt{2}\,t)\mathbf{i} + (\sqrt{2}\,t)\mathbf{j} + (1 - t^2)\mathbf{k}$$
from $(0, 0, 1)$ to $(\sqrt{2}, \sqrt{2}, 0)$.

16. Writing to Learn The length $2\pi\sqrt{2}$ of the turn of the helix of Example 1 is also the length of the diagonal of a square 2π units on a side. Show how to obtain this square by cutting away and flattening a portion of the cylinder around which the helix winds.

Exercises 17–20 refer to the curve $\mathbf{r}(t) = (\cos t)\mathbf{i} + (\sin t)\mathbf{j} + (1 - \cos t)\mathbf{k}, \quad 0 \le t \le 2\pi$.

17. Show that the curve is an ellipse by showing that it is the intersection of a right circular cylinder and a plane. Find equations for the cylinder and plane.

18. *(Continuation of Exercise 17)* Sketch the ellipse on the cylinder. Add to your sketch the unit tangent vectors at $t = 0, \pi/2, \pi,$ and $3\pi/2$.

19. *(Continuation of Exercises 17 and 18)* Show that the acceleration vector always lies parallel to the plane of the

ellipse (orthogonal to a vector normal to the plane). Thus, if you draw the acceleration as a vector attached to the ellipse, it will lie in the plane of the ellipse. Add the acceleration vectors for $t = 0, \pi/2, \pi,$ and $3\pi/2$ to the sketch you made in Exercise 18.

20. Write and evaluate an integral for the length of the ellipse.

21. Writing to Learn Explain how to use a two-dimensional grapher to graph the curves in Exercises 3, 5, and 12.

Exploration

22. *Length Is Independent of Parametrization* To illustrate that the length of a smooth space curve does not depend on the parametrization you use to compute it, calculate the length of one turn of the helix in Example 1 with each of the following parametrizations.

(a) $\mathbf{r}(t) = (\cos 4t)\mathbf{i} + (\sin 4t)\mathbf{j} + 4t\mathbf{k}, \quad 0 \le t \le \pi/2$

(b) $\mathbf{r}(t) = [\cos (t/2)]\mathbf{i} + [\sin (t/2)]\mathbf{j} + (t/2)\mathbf{k}, \quad 0 \le t \le 4\pi$

(c) $\mathbf{r}(t) = (\cos t)\mathbf{i} - (\sin t)\mathbf{j} - t\mathbf{k}, \quad -2\pi \le t \le 0$ ∎

Extending the Ideas

23. *Arc Length in Cylindrical Coordinates*

(a) Show that when you express $ds^2 = dx^2 + dy^2 + dz^2$ in terms of cylindrical coordinates, you get
$ds^2 = dr^2 + r^2 \, d\theta^2 + dz^2$.

(b) Interpret this result geometrically in terms of the edges and diagonal of a box. Sketch the box.

(c) Use the result in (a) to find the length of the curve
$r = e^\theta, z = e^\theta, 0 \le \theta \le \ln 8$.

12.3 Curvature, Torsion, and the TNB Frame

Curvature of a Plane Curve • The Principal Unit Normal Vector for Plane Curves • Circle of Curvature and Radius of Curvature • Curvature and Normal Vectors for Space Curves • Torsion and the Binormal Vector • Tangential and Normal Components of Acceleration • Formulas for Computing Curvature and Torsion

Figure 12.17 As P moves along the curve in the direction of increasing arc length, the unit tangent vector turns. The value of $|d\mathbf{T}/ds|$ is the *curvature* of the curve at P.

Curvature of a Plane Curve

As a particle moves along a smooth curve in the plane, $\mathbf{T} = d\mathbf{r}/ds$ turns as the curve bends. Since $\mathbf{T}$ is a unit vector, its length remains constant and only its direction changes as the particle moves along the curve. The rate at which the direction changes per unit of length along the curve is the *curvature* (Figure 12.17). The traditional symbol for the curvature is the Greek letter κ ("kappa").

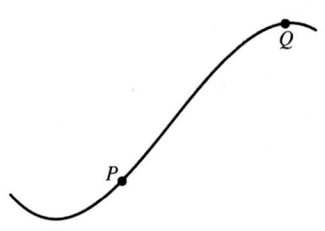

Figure 12.18 The curvature at Q is greater than the curvature at P.

> **Definition** Curvature
>
> If **T** is the unit tangent vector of a smooth curve, the **curvature** function of the curve is
>
> $$\kappa = \left| \frac{d\mathbf{T}}{ds} \right|.$$

The curvature measures how sharply a curve bends at a point. If $|d\mathbf{T}/ds|$ is large, **T** turns sharply as the particle passes through P and the curvature at P is large. If $|d\mathbf{T}/ds|$ is close to zero, **T** turns more slowly and the curvature at P is smaller. In Figure 12.18, the curvature at Q is greater than the curvature at P.

Testing the definition, we see in Examples 1 and 2 that the curvature is constant for straight lines and circles.

Example 1   FINDING THE CURVATURE OF A STRAIGHT LINE

Show that the curvature of a straight line is zero.

Solution   On a straight line, the unit tangent vector **T** always points in the same direction, so its components are constants. Therefore, $|d\mathbf{T}/ds| = |\mathbf{0}| = 0$ (Figure 12.19).

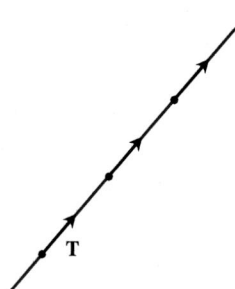

Figure 12.19 Along a straight line, **T** always points in the same direction. The curvature, $|d\mathbf{T}/ds|$, is zero. (Example 1)

Example 2   FINDING THE CURVATURE OF A CIRCLE

Show that the curvature of a circle of radius a is $1/a$.

Solution   To see why (Figure 12.20), start with the parametrization

$$\mathbf{r}(\theta) = (a \cos \theta)\mathbf{i} + (a \sin \theta)\mathbf{j}$$

and substitute $\theta = s/a$ to parametrize in terms of arc length s:

$$\mathbf{r} = \left(a \cos \frac{s}{a} \right) \mathbf{i} + \left(a \sin \frac{s}{a} \right) \mathbf{j}.$$

Then

$$\mathbf{T} = \frac{d\mathbf{r}}{ds} = \left(-\sin \frac{s}{a} \right) \mathbf{i} + \left(\cos \frac{s}{a} \right) \mathbf{j}$$

and

$$\frac{d\mathbf{T}}{ds} = \left(-\frac{1}{a} \cos \frac{s}{a} \right) \mathbf{i} - \left(\frac{1}{a} \sin \frac{s}{a} \right) \mathbf{j}.$$

Hence, for any value of s,

$$\kappa = \left| \frac{d\mathbf{T}}{ds} \right|$$

$$= \sqrt{ \frac{1}{a^2} \cos^2 \left(\frac{s}{a} \right) + \frac{1}{a^2} \sin^2 \left(\frac{s}{a} \right) }$$

$$= \frac{1}{\sqrt{a^2}} = \frac{1}{|a|} = \frac{1}{a}. \qquad \text{Since } a > 0, |a| = a.$$

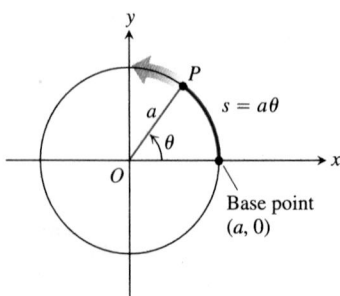

Figure 12.20 The point P has coordinates $(a \cos \theta, a \sin \theta) = (a \cos(s/a), a \sin (s/a))$. (Example 2)

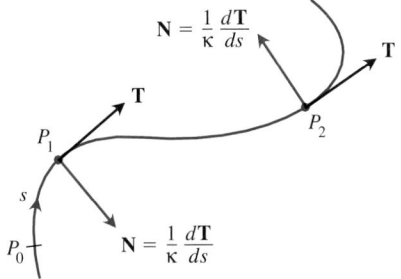

Figure 12.21 The vector $d\mathbf{T}/ds$, normal to the curve, always points in the direction in which $\mathbf{T}$ is turning. The vector $\mathbf{N}$ is the direction of $d\mathbf{T}/ds$.

The Principal Unit Normal Vector for Plane Curves

Since $\mathbf{T}$ has constant length, the vector $d\mathbf{T}/ds$ is orthogonal to $\mathbf{T}$ (Section 12.1). Therefore, if at a point where $d\mathbf{T}/ds \neq \mathbf{0}$, we divide $d\mathbf{T}/ds$ by its length κ, we obtain a *unit* vector orthogonal to $\mathbf{T}$ (Figure 12.21).

> **Definition Principal Unit Normal Vector**
>
> At a point where $\kappa \neq 0$, the **principal unit normal vector** for a curve in the plane is
>
> $$\mathbf{N} = \frac{1}{\kappa}\frac{d\mathbf{T}}{ds}.$$

The vector $d\mathbf{T}/ds$ points in the direction in which $\mathbf{T}$ turns as the curve bends. Therefore, if we face in the direction of increasing arc length, the vector $d\mathbf{T}/ds$ points to the right if $\mathbf{T}$ turns clockwise and toward the left if $\mathbf{T}$ turns counterclockwise. In other words, the principal normal vector $\mathbf{N}$ will point toward the concave side of the curve (Figure 12.21). Exercise 10 illustrates what happens at a point where the concavity changes and $\mathbf{T}$ momentarily stops turning.

Because the arc length parameter for a smooth curve $\mathbf{r}(t) = f(t)\mathbf{i} + g(t)\mathbf{j}$ is defined with ds/dt positive, $ds/dt = |ds/dt|$, and the Chain Rule gives

$$\mathbf{N} = \frac{d\mathbf{T}/ds}{|d\mathbf{T}/ds|}$$

$$= \frac{(d\mathbf{T}/dt)(dt/ds)}{|d\mathbf{T}/dt||dt/ds|}$$

$$= \frac{d\mathbf{T}/dt}{|d\mathbf{T}/dt|}. \tag{1}$$

This formula enables us to find $\mathbf{N}$ without having to find κ and s first.

Example 3 FINDING $\mathbf{T}$ AND $\mathbf{N}$

Find $\mathbf{T}$ and $\mathbf{N}$ for the circular motion $\mathbf{r}(t) = (\cos 2t)\mathbf{i} + (\sin 2t)\mathbf{j}$.

Solution We first find $\mathbf{T}$:

$$\mathbf{v} = -(2\sin 2t)\mathbf{i} + (2\cos 2t)\mathbf{j},$$

$$|\mathbf{v}| = \sqrt{4\sin^2 2t + 4\cos^2 2t} = 2,$$

$$\mathbf{T} = \frac{\mathbf{v}}{|\mathbf{v}|} = -(\sin 2t)\mathbf{i} + (\cos 2t)\mathbf{j}.$$

From this we find

$$\frac{d\mathbf{T}}{dt} = -(2\cos 2t)\mathbf{i} - (2\sin 2t)\mathbf{j},$$

$$\left|\frac{d\mathbf{T}}{dt}\right| = \sqrt{4\cos^2 2t + 4\sin^2 2t} = 2,$$

and

$$\mathbf{N} = \frac{d\mathbf{T}/dt}{|d\mathbf{T}/dt|} \qquad \text{Eq. (1)}$$

$$= -(\cos 2t)\mathbf{i} - (\sin 2t)\mathbf{j}.$$

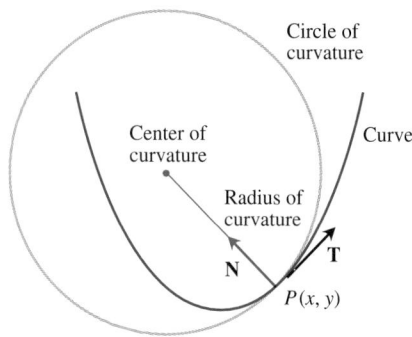

Figure 12.22 The osculating circle at $P(x, y)$ lies toward the inner side of the curve.

Circle of Curvature and Radius of Curvature

The **circle of curvature** or **osculating circle** at a point P on a plane curve where $\kappa \neq 0$ is the circle in the plane of the curve that

1. is tangent to the curve at P (has the same tangent line the curve has);

2. has the same curvature the curve has at P; and

3. lies toward the concave or inner side of the curve (as in Figure 12.22).

The **radius of curvature** of the curve at P is the radius of the circle of curvature, which, according to Example 2, is

$$\text{Radius of curvature} = \rho = \frac{1}{\kappa}. \qquad (2)$$

To find ρ, we find κ and take the reciprocal. The **center of curvature** of the curve at P is the center of the circle of curvature.

Curvature and Normal Vectors for Space Curves

Just as it does for a curve in the plane, the arc length parameter s gives the unit tangent vector $\mathbf{T} = d\mathbf{r}/ds$ for a smooth curve in space. We again define the curvature to be

$$\kappa = \left| \frac{d\mathbf{T}}{ds} \right|. \qquad (3)$$

The vector $d\mathbf{T}/ds$ is orthogonal to $\mathbf{T}$, and we define the principal unit normal to be

$$\mathbf{N} = \frac{1}{\kappa} \frac{d\mathbf{T}}{ds} = \frac{d\mathbf{T}/dt}{|d\mathbf{T}/dt|}. \qquad (4)$$

Example 4 FINDING CURVATURE ON A HELIX

Find the curvature for the helix (Figure 12.23)
$$\mathbf{r}(t) = (a \cos t)\mathbf{i} + (a \sin t)\mathbf{j} + bt\mathbf{k}, \qquad a, b \geq 0, \qquad a^2 + b^2 \neq 0.$$

Solution We calculate $\mathbf{T}$ from the velocity vector $\mathbf{v}$:

$$\mathbf{v} = -(a \sin t)\mathbf{i} + (a \cos t)\mathbf{j} + b\mathbf{k},$$

$$|\mathbf{v}| = \sqrt{a^2 \sin^2 t + a^2 \cos^2 t + b^2} = \sqrt{a^2 + b^2},$$

$$\mathbf{T} = \frac{\mathbf{v}}{|\mathbf{v}|} = \frac{1}{\sqrt{a^2 + b^2}} [-(a \sin t)\mathbf{i} + (a \cos t)\mathbf{j} + b\mathbf{k}].$$

Then, we find $d\mathbf{T}/ds$ as

$$\frac{d\mathbf{T}}{ds} = \frac{d\mathbf{T}}{dt} \frac{dt}{ds} \qquad \text{Chain Rule}$$

$$= \frac{d\mathbf{T}}{dt} \cdot \frac{1}{|\mathbf{v}|} \qquad \frac{ds}{dt} = |\mathbf{v}|, \text{ so } \frac{dt}{ds} = \frac{1}{|\mathbf{v}|}$$

$$= \frac{1}{\sqrt{a^2 + b^2}} [-(a \cos t)\mathbf{i} - (a \sin t)\mathbf{j}] \cdot \left(\frac{1}{\sqrt{a^2 + b^2}} \right)$$

$$= \frac{a}{a^2 + b^2} [-(\cos t)\mathbf{i} - (\sin t)\mathbf{j}].$$

Figure 12.23 The helix $\mathbf{r}(t) = (a \cos t)\mathbf{i} + (a \sin t)\mathbf{j} + bt\mathbf{k}$ drawn with a and b positive and $t \geq 0$. (Example 4)

Therefore,

$$\kappa = \left| \frac{d\mathbf{T}}{ds} \right| \qquad\qquad \text{Eq. (3)}$$

$$= \frac{a}{a^2 + b^2} \left| -(\cos t)\mathbf{i} - (\sin t)\mathbf{j} \right|$$

$$= \frac{a}{a^2 + b^2} \sqrt{(\cos t)^2 + (\sin t)^2} = \frac{a}{a^2 + b^2}.$$

From this equation we see that increasing b for fixed a decreases the curvature. Decreasing a for fixed b eventually decreases the curvature as well. Stretching a spring tends to straighten it.

If $b = 0$, the helix reduces to a circle of radius a and its curvature reduces to $1/a$, as it should. If $a = 0$, the helix becomes the z-axis, and its curvature reduces to 0, again as it should.

Example 5   FINDING THE PRINCIPAL UNIT NORMAL VECTOR **N**

Find **N** for the helix in Example 4.

Solution   We have

$$\frac{d\mathbf{T}}{dt} = -\frac{1}{\sqrt{a^2 + b^2}} \left[(a \cos t)\mathbf{i} + (a \sin t)\mathbf{j} \right], \qquad \text{Example 4}$$

$$\left| \frac{d\mathbf{T}}{dt} \right| = \frac{1}{\sqrt{a^2 + b^2}} \sqrt{a^2 \cos^2 t + a^2 \sin^2 t} = \frac{a}{\sqrt{a^2 + b^2}},$$

$$\mathbf{N} = \frac{d\mathbf{T}/dt}{|d\mathbf{T}/dt|} \qquad\qquad \text{Eq. (4)}$$

$$= -\frac{\sqrt{a^2 + b^2}}{a} \cdot \frac{1}{\sqrt{a^2 + b^2}} \left[(a \cos t)\mathbf{i} + (a \sin t)\mathbf{j} \right]$$

$$= -(\cos t)\mathbf{i} - (\sin t)\mathbf{j}.$$

Torsion and the Binormal Vector

The **binormal vector** of a curve in space is $\mathbf{B} = \mathbf{T} \times \mathbf{N}$, a unit vector orthogonal to both **T** and **N** (Figure 12.24). Together **T**, **N**, and **B** define a moving right-handed vector frame that always travels with a body moving along a curve in space. It is the **Frenet** ("fre-nay") **frame** (after Jean-Frédéric Frenet, 1816–1900), or the **TNB frame.** This vector frame plays a significant role in calculating the flight paths of space vehicles.

How does $d\mathbf{B}/ds$ behave in relation to **T**, **N**, and **B**? From the rule for differentiating a cross product, we have

$$\frac{d\mathbf{B}}{ds} = \frac{d\mathbf{T}}{ds} \times \mathbf{N} + \mathbf{T} \times \frac{d\mathbf{N}}{ds}.$$

Since **N** is the direction of $d\mathbf{T}/ds$, $(d\mathbf{T}/ds) \times \mathbf{N} = \mathbf{0}$ and

$$\frac{d\mathbf{B}}{ds} = \mathbf{0} + \mathbf{T} \times \frac{d\mathbf{N}}{ds} = \mathbf{T} \times \frac{d\mathbf{N}}{ds}.$$

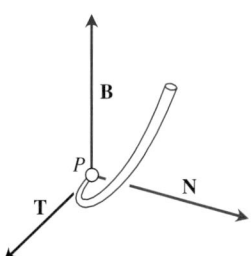

Figure 12.24 The vectors **T**, **N**, and **B** (in that order) make a right-handed frame of mutually orthogonal unit vectors in space.

From this we see that $d\mathbf{B}/ds$ is orthogonal to $\mathbf{T}$ since a cross product is orthogonal to its factors.

Since $d\mathbf{B}/ds$ is also orthogonal to $\mathbf{B}$ (the latter has constant length), it follows that $d\mathbf{B}/ds$ is orthogonal to the plane of $\mathbf{B}$ and $\mathbf{T}$. In other words, $d\mathbf{B}/ds$ is parallel to $\mathbf{N}$, so $d\mathbf{B}/ds$ is a scalar multiple of $\mathbf{N}$. In symbols,

$$\frac{d\mathbf{B}}{ds} = -\tau\mathbf{N}.$$

The minus sign in this equation is traditional. The scalar τ is the *torsion* along the curve. Notice that

$$\frac{d\mathbf{B}}{ds}\cdot\mathbf{N} = -\tau\mathbf{N}\cdot\mathbf{N} = -\tau(1) = -\tau,$$

so that

$$\tau = -\frac{d\mathbf{B}}{ds}\cdot\mathbf{N}.$$

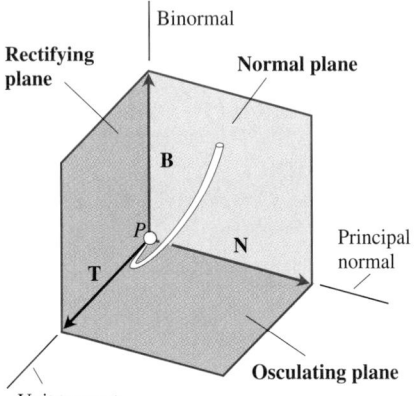

Figure 12.25 The names of the three planes determined by $\mathbf{T}$, $\mathbf{N}$, and $\mathbf{B}$.

> **Definition Torsion**
>
> Let $\mathbf{B} = \mathbf{T}\times\mathbf{N}$. The **torsion** function of a smooth curve is
>
> $$\tau = -\frac{d\mathbf{B}}{ds}\cdot\mathbf{N}.$$

Unlike the curvature κ, which is never negative, the torsion τ may be positive, negative, or zero.

The three planes determined by $\mathbf{T}$, $\mathbf{N}$, and $\mathbf{B}$ are shown in Figure 12.25. The curvature $\kappa = |d\mathbf{T}/ds|$ can be thought of as the rate at which the normal plane turns as the point P moves along the curve. Similarly, the torsion $\tau = -(d\mathbf{B}/ds)\cdot\mathbf{N}$ is the rate at which the osculating plane turns about $\mathbf{T}$ as P moves along the curve. Torsion measures how the curve twists.

Tangential and Normal Components of Acceleration

When a body is accelerated by gravity, brakes, a combination of rocket motors, or whatever, we usually want to know how much of the acceleration acts to move the body straight ahead in the direction of motion, in the tangential direction $\mathbf{T}$. We can find out if we use the Chain Rule to rewrite $\mathbf{v}$ as

$$\mathbf{v} = \frac{d\mathbf{r}}{ds} = \frac{d\mathbf{r}}{ds}\frac{ds}{dt} = \mathbf{T}\frac{ds}{dt}$$

and differentiate both ends of this string of equations to get

$$\mathbf{a} = \frac{d\mathbf{v}}{dt} = \frac{d}{dt}\left(\mathbf{T}\frac{ds}{dt}\right) = \frac{d^2s}{dt^2}\mathbf{T} + \frac{ds}{dt}\frac{d\mathbf{T}}{dt}$$

$$= \frac{d^2s}{dt^2}\mathbf{T} + \frac{ds}{dt}\left(\frac{d\mathbf{T}}{ds}\frac{ds}{dt}\right) = \frac{d^2s}{dt^2}\mathbf{T} + \frac{ds}{dt}\left(\kappa\mathbf{N}\frac{ds}{dt}\right)$$

$$= \frac{d^2s}{dt^2}\mathbf{T} + \kappa\left(\frac{ds}{dt}\right)^2\mathbf{N}.$$

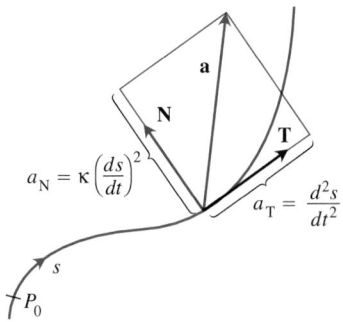

Figure 12.26 The tangential and normal components of acceleration. The acceleration **a** always lies in the plane of **T** and **N**, orthogonal to **B**.

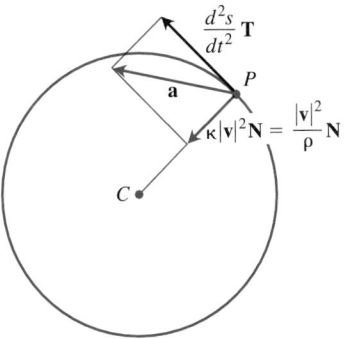

Figure 12.27 The tangential and normal components of the acceleration of a body that is speeding up as it moves counterclockwise around a circle of radius ρ.

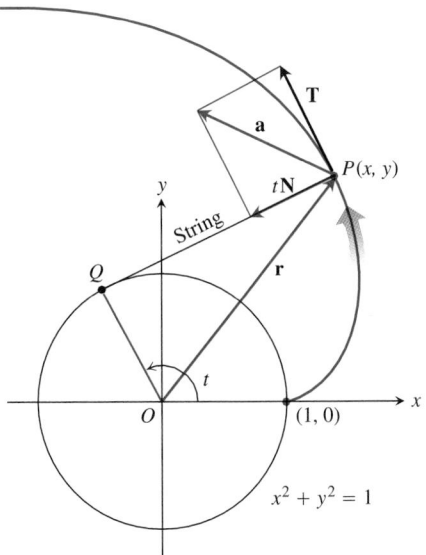

Figure 12.28 The tangential and normal components of the acceleration of the motion $\mathbf{r}(t) = (\cos t + t \sin t)\mathbf{i} +$ $(\sin t - t \cos t)\mathbf{j}$, for $t > 0$. (Example 6)

Definition Tangential and Normal Components of Acceleration

$$\mathbf{a} = a_T\mathbf{T} + a_N\mathbf{N}, \tag{5}$$

where

$$a_T = \frac{d^2s}{dt^2} = \frac{d}{dt}|\mathbf{v}| \quad \text{and} \quad a_N = \kappa\left(\frac{ds}{dt}\right)^2 = \kappa|\mathbf{v}^2| \tag{6}$$

are the **tangential** and **normal** scalar components of acceleration.

Equation (5) is remarkable in that **B** does not appear. No matter how the path of the moving body we are watching may appear to twist and turn in space, the acceleration **a** *always lies in the plane of* **T** and **N** orthogonal to **B**. The equation also tells exactly how much of the acceleration takes place tangent to the motion (d^2s/dt^2) and how much takes place normal to the motion $[\kappa(ds/dt)^2]$ (Figure 12.26).

What other information can we read from Equation (5)? By definition, the acceleration **a** is the rate of change of velocity **v**, and in general, both length and direction of **v** changes as a body moves along its path. The tangential component of acceleration a_T measures the rate of change of the *length* of **v** (that is, the change in speed). The normal component of acceleration a_N measures the rate of change of the *direction* of **v**.

Notice that the normal scalar component of the acceleration is the curvature times the *square* of the speed. This explains why you have to hold on when your car makes a sharp (large κ), high-speed (large $|\mathbf{v}|$) turn. If you double the speed of your car, you will experience four times the normal component of acceleration for the same curvature.

If a body moves in a circle at a constant speed, d^2s/dt^2 is zero and all the acceleration points along **N** toward the circle's center. If the body is speeding up or slowing down, **a** has a nonzero tangential component (Figure 12.27).

To calculate a_N, we usually use the formula $a_N = \sqrt{|\mathbf{a}|^2 - a_T^2}$, which comes from solving the equation $|\mathbf{a}|^2 = \mathbf{a} \cdot \mathbf{a} = a_T^2 + a_N^2$ for a_N. With this formula we find a_N without having to calculate κ first.

$$a_N = \sqrt{|\mathbf{a}|^2 - a_T^2} \tag{7}$$

Example 6 FINDING AN ACCELERATION VECTOR

Without finding **T** and **N**, write the acceleration of the motion

$$\mathbf{r}(t) = (\cos t + t \sin t)\mathbf{i} + (\sin t - t \cos t)\mathbf{j}, \quad t > 0$$

in the form $\mathbf{a} = a_T\mathbf{T} + a_N\mathbf{N}$. (The path of motion is the involute of the circle in Figure 12.28.)

Solution We use the first of Equations (6) to find a_T:

$$\mathbf{v} = \frac{d\mathbf{r}}{dt} = (t \cos t)\mathbf{i} + (t \sin t)\mathbf{j},$$

$$|\mathbf{v}| = \sqrt{t^2 \cos^2 t + t^2 \sin^2 t} = \sqrt{t^2} = |t| = t, \quad t > 0$$

$$a_T = \frac{d}{dt}|\mathbf{v}| = 1. \qquad \text{Eq. (6)}$$

Knowing a_T, we use Equation (7) to find a_N:

$$\mathbf{a} = (\cos t - t \sin t)\mathbf{i} + (\sin t + t \cos t)\mathbf{j},$$

$$|\mathbf{a}|^2 = t^2 + 1, \qquad \text{After some algebra}$$

$$a_N = \sqrt{|\mathbf{a}|^2 - a_T{}^2} \qquad \text{Eq. (7)}$$

$$= \sqrt{(t^2 + 1) - (1)} = \sqrt{t^2} = t.$$

We then use Equation (5) to find **a**:

$$\mathbf{a} = a_T\mathbf{T} + a_N\mathbf{N} = (1)\mathbf{T} + t\mathbf{N} \quad \mathbf{T} + t\mathbf{N}.$$

See Figure 12.28.

Formulas for Computing Curvature and Torsion

We now present some easy-to-use formulas for computing the curvature and torsion of a smooth curve. From Equation (5), we have

$$\mathbf{v} \times \mathbf{a} = \left(\frac{ds}{dt}\mathbf{T}\right) \times \left[\frac{d^2s}{dt^2}\mathbf{T} + \kappa\left(\frac{ds}{dt}\right)^2\mathbf{N}\right] \qquad \begin{array}{l}\text{From Section 12.2,}\\ \text{Eqs. (3) and (4),}\\ |\mathbf{v}|\,\mathbf{T} = (ds/dt)\mathbf{T}.\end{array}$$

$$= \left(\frac{ds}{dt}\frac{d^2s}{dt^2}\right)(\mathbf{T} \times \mathbf{T}) + \kappa\left(\frac{ds}{dt}\right)^3(\mathbf{T} \times \mathbf{N})$$

$$= \kappa\left(\frac{ds}{dt}\right)^3\mathbf{B} \qquad \mathbf{T} \times \mathbf{T} = \mathbf{0} \text{ and } \mathbf{T} \times \mathbf{N} = \mathbf{B}$$

It follows that

$$|\mathbf{v} \times \mathbf{a}| = \kappa\left|\frac{ds}{dt}\right|^3|\mathbf{B}| = \kappa|\mathbf{v}|^3. \qquad \frac{ds}{dt} = |\mathbf{v}| \text{ and } |\mathbf{B}| = 1$$

Solving for κ gives the following formula.

A Vector Formula for Curvature

$$\kappa = \frac{|\mathbf{v} \times \mathbf{a}|}{|\mathbf{v}|^3} \tag{8}$$

Equation (8) calculates the curvature, a geometric property of the curve, from the velocity and acceleration of any representation of the curve in which $|\mathbf{v}|$ is different from zero. Take a moment to think about how remarkable this really is: From any formula for motion along a curve, no matter how variable the motion may be (as long as **v** is never zero), we can calculate a physical property of the curve that seems to have nothing to do with the way the curve is traversed.

The most widely used formula for torsion, derived in more advanced texts, is

Newton's Dot Notation for Derivatives

The dots in Equation (9) denote differentiation with respect to t, one derivative for each dot. Thus, $\dot{x}$ ("x dot") means dx/dt, $\ddot{x}$ ("x double dot") means d^2x/dt^2, and $\dddot{x}$ ("x triple dot") means d^3x/dt^3. Similarly, $\dot{y} = dy/dt$, and so on.

$$\tau = \frac{\begin{vmatrix} \dot{x} & \dot{y} & \dot{z} \\ \ddot{x} & \ddot{y} & \ddot{z} \\ \dddot{x} & \dddot{y} & \dddot{z} \end{vmatrix}}{|\mathbf{v} \times \mathbf{a}|^2} \qquad (\text{if } \mathbf{v} \times \mathbf{a} \neq \mathbf{0}). \tag{9}$$

This formula calculates the torsion directly from the derivatives of the component functions $x = f(t)$, $y = g(t)$, $z = h(t)$ that make up **r**. The determinant's first row comes from **v**, the second row from **a**, and the third from $\dot{\mathbf{a}} = d\mathbf{a}/dt$.

Example 7 FINDING CURVATURE AND TORSION

Use Equations (8) and (9) to find κ and τ for the helix

$$\mathbf{r}(t) = (a \cos t)\mathbf{i} + (a \sin t)\mathbf{j} + bt\mathbf{k}, \qquad a, b \geq 0, \qquad a^2 + b^2 \neq 0.$$

Solution We calculate the curvature with Equation (8):

$$\mathbf{v} = -(a \sin t)\mathbf{i} + (a \cos t)\mathbf{j} + b\mathbf{k},$$

$$\mathbf{a} = -(a \cos t)\mathbf{i} - (a \sin t)\mathbf{j},$$

$$\mathbf{v} \times \mathbf{a} = \begin{vmatrix} \mathbf{i} & \mathbf{j} & \mathbf{k} \\ -a \sin t & a \cos t & b \\ -a \cos t & -a \sin t & 0 \end{vmatrix}$$

$$= (ab \sin t)\mathbf{i} - (ab \cos t)\mathbf{j} + a^2\mathbf{k},$$

$$\kappa = \frac{|\mathbf{v} \times \mathbf{a}|}{|\mathbf{v}|^3} = \frac{\sqrt{a^2 b^2 + a^4}}{(a^2 + b^2)^{3/2}} = \frac{a\sqrt{a^2 + b^2}}{(a^2 + b^2)^{3/2}} = \frac{a}{a^2 + b^2}. \qquad (10)$$

Notice that Equation (10) agrees with the result in Example 4, where we calculated the curvature directly from its definition.

To evaluate Equation (9) for the torsion, we find the entries in the determinant by differentiating **r** with respect to t. We already have **v** and **a**, and

$$\dot{\mathbf{a}} = \frac{d\mathbf{a}}{dt} = (a \sin t)\mathbf{i} - (a \cos t)\mathbf{j}.$$

Hence,

$$\tau = \frac{\begin{vmatrix} \dot{x} & \dot{y} & \dot{z} \\ \ddot{x} & \ddot{y} & \ddot{z} \\ \dddot{x} & \dddot{y} & \dddot{z} \end{vmatrix}}{|\mathbf{v} \times \mathbf{a}|^2} = \frac{\begin{vmatrix} -a \sin t & a \cos t & b \\ -a \cos t & -a \sin t & 0 \\ a \sin t & -a \cos t & 0 \end{vmatrix}}{(a\sqrt{a^2 + b^2})^2} \quad \text{Value of } |\mathbf{v} \times \mathbf{a}| \text{ from Eq. (10)}$$

$$= \frac{b(a^2 \cos^2 t + a^2 \sin^2 t)}{a^2(a^2 + b^2)}$$

$$= \frac{b}{a^2 + b^2}. \qquad (11)$$

From Equation (11), we see that the torsion of a helix about a circular cylinder is constant. In fact, constant curvature and constant torsion characterize the helix among all curves in space.

The DNA molecule, the basic building block of life forms, is designed in the form of two helices winding around each other, a little like the rungs and sides of a twisted rope ladder (Figure 12.29). Not only is the space occupied by the DNA molecule very much smaller than it would be if it were unraveled, but when the molecule is damaged the imperfect piece can be snipped out by a kind of molecular scissors (because the curvature and torsion functions are constant) and the DNA made right again.

Figure 12.29 The helical shape of a DNA molecule is characterized by its constant curvature and torsion.

Formulas for Curves in Space

Unit tangent vector:
$$\mathbf{T} = \frac{\mathbf{v}}{|\mathbf{v}|}$$

Principal unit normal vector:
$$\mathbf{N} = \frac{d\mathbf{T}/dt}{|d\mathbf{T}/dt|}$$

Binormal vector:
$$\mathbf{B} = \mathbf{T} \times \mathbf{N}$$

Curvature:
$$\kappa = \left|\frac{d\mathbf{T}}{ds}\right| = \frac{|\mathbf{v} \times \mathbf{a}|}{|\mathbf{v}|^3}$$

Torsion:
$$\tau = -\frac{d\mathbf{B}}{ds} \cdot \mathbf{N} = \frac{\begin{vmatrix} \dot{x} & \dot{y} & \dot{z} \\ \ddot{x} & \ddot{y} & \ddot{z} \\ \dddot{x} & \dddot{y} & \dddot{z} \end{vmatrix}}{|\mathbf{v} \times \mathbf{a}|^2}$$

Tangential and normal scalar components of acceleration:
$$\mathbf{a} = a_T\mathbf{T} + a_N\mathbf{N}$$
$$a_T = \frac{d}{dt}|\mathbf{v}|$$
$$a_N = \kappa|\mathbf{v}|^2 = \sqrt{|\mathbf{a}|^2 - a_T^2}$$

Quick Review 12.3

In Exercises 1–5, let $\mathbf{u} = \dfrac{\mathbf{i} + \mathbf{j}}{\sqrt{2}}$, $\mathbf{v} = \dfrac{\mathbf{i} - \mathbf{j}}{\sqrt{2}}$, and $\mathbf{w} = 5\mathbf{i} - \mathbf{j}$.

1. Show that $\mathbf{u} \cdot \mathbf{v} = 0$.

2. Show that $\mathbf{w} = 2\sqrt{2}\mathbf{u} + 3\sqrt{2}\mathbf{v}$.

3. Show that $\mathbf{w} \cdot \mathbf{u} = 2\sqrt{2}$.

4. Show that $\mathbf{w} \cdot \mathbf{v} = 3\sqrt{2}$.

5. If $\mathbf{w} = a\mathbf{u} + b\mathbf{v}$, show that $a = \mathbf{w} \cdot \mathbf{u}$ and $b = \mathbf{w} \cdot \mathbf{v}$.

In Exercises 6–8, $\mathbf{r}(t) = (\sin t)\mathbf{i} + t\mathbf{j}$ is the position vector of a particle in the plane at time t.

6. Find $\mathbf{v}$. **7.** Find $\mathbf{a}$.

8. Find the times t when $\mathbf{v} \cdot \mathbf{a} = 0$ and neither $\mathbf{v}$ nor $\mathbf{a}$ are zero. What are $\mathbf{v}$ and $\mathbf{a}$ and for these values of t?

In Exercises 9 and 10, use determinants to find $\mathbf{u} \times \mathbf{v}$.

9. $\mathbf{u} = 2\mathbf{i} - 3\mathbf{j} - \mathbf{k}$, $\mathbf{v} = -2\mathbf{i} + 7\mathbf{j} - 3\mathbf{k}$

10. $\mathbf{u} = -\mathbf{i} + \mathbf{k}$, $\mathbf{v} = 3\mathbf{j} - 4\mathbf{k}$

Section 12.3 Exercises

In Exercises 1–4, find $\mathbf{T}$, $\mathbf{N}$, and κ for the plane curve.

1. $\mathbf{r}(t) = t\mathbf{i} + (\ln \cos t)\mathbf{j}$, $\quad -\pi/2 < t < \pi/2$

2. $\mathbf{r}(t) = (\ln \sec t)\mathbf{i} + t\mathbf{j}$, $\quad -\pi/2 < t < \pi/2$

3. $\mathbf{r}(t) = (2t + 3)\mathbf{i} + (5 - t^2)\mathbf{j}$

4. $\mathbf{r}(t) = (\cos t + t \sin t)\mathbf{i} + (\sin t - t \cos t)$, $\quad t > 0$

In Exercises 5 and 6, write $\mathbf{a}$ in the form $\mathbf{a} = a_T\mathbf{T} + a_N\mathbf{N}$ without finding $\mathbf{T}$ and $\mathbf{N}$.

5. $\mathbf{r}(t) = (2t + 3)\mathbf{i} + (t^2 - 1)\mathbf{j}$

6. $\mathbf{r}(t) = \ln(t^2 + 1)\mathbf{i} + (t - 2 \tan^{-1} t)\mathbf{j}$

Explorations

7. *A Formula for the Curvature of the Graph of a Function in the xy-Plane* The graph $y = f(x)$ in the xy-plane automatically has the parametrization $x = x$, $y = f(x)$, and the vector formula $\mathbf{r}(x) = x\mathbf{i} + f(x)\mathbf{j}$.

(a) Use this formula to show that if f is a twice differentiable function of x, then

$$\kappa(x) = \frac{|f''(x)|}{[1 + (f'(x))^2]^{3/2}}.$$

(b) Use the formula for κ in (a) to find the curvature of $y = \ln(\cos x)$, $-\pi/2 < x < \pi/2$. Compare your answer with the answer in Exercise 1.

(c) Show that the curvature is zero at a point of inflection.

8. *A Formula for the Curvature of a Parametrized Plane Curve* Let $\mathbf{r}(t) = f(t)\mathbf{i} + g(t)\mathbf{j}$ be a smooth curve defined by the twice-differentiable functions $x = f(t)$ and $y = g(t)$.

(a) Show that the curvature of $\mathbf{r}$ is given by the formula

$$\kappa = \frac{|\dot{x}\ddot{y} - \dot{y}\ddot{x}|}{(\dot{x}^2 + \dot{y}^2)^{3/2}}.$$

(b) Apply the formula in (a) to find the curvature of $\mathbf{r}(t) = t\mathbf{i} + (\ln \sin t)\mathbf{j}$, $0 < t < \pi$.

9. *Normals to Plane Curves*

(a) Show that $\mathbf{n}(t) = -g'(t)\mathbf{i} + f'(t)\mathbf{j}$ and $-\mathbf{n}(t) = g'(t)\mathbf{i} - f'(t)\mathbf{j}$ are both normal to the curve $\mathbf{r}(t) = f(t)\mathbf{i} + g(t)\mathbf{j}$ at the point $(f(t), g(t))$.

To obtain $\mathbf{N}$ for a particular plane curve, we can choose the one of $\mathbf{n}$ or $-\mathbf{n}$ from part (a) that points toward the concave side of the curve, and make it into a unit vector. (See Figure 12.21.) Apply this method to find $\mathbf{N}$ for the following curves.

(b) $\mathbf{r}(t) = t\mathbf{i} + e^{2t}\mathbf{j}$

(c) $\mathbf{r}(t) = \sqrt{4 - t^2}\,\mathbf{i} + t\mathbf{j}$, $-2 \le t \le 2$

10. *(Continuation of Exercise 9)*

(a) Use the method of Exercise 9 to find $\mathbf{N}$ for the curve $\mathbf{r}(t) = t\mathbf{i} + (1/3)t^3\mathbf{j}$ when $t < 0$; when $t > 0$.

(b) Writing to Learn Calculate

$$\mathbf{N} = \frac{d\mathbf{T}/dt}{|d\mathbf{T}/dt|}, \qquad t \ne 0,$$

for the curve in (a). Does $\mathbf{N}$ exist at $t = 0$? Graph the curve and explain what happens to $\mathbf{N}$ as t passes from negative to positive values. ■

In Exercises 11–16, find $\mathbf{T}$, $\mathbf{N}$, $\mathbf{B}$, κ, and τ for the space curve.

11. $\mathbf{r}(t) = (3 \sin t)\mathbf{i} + (3 \cos t)\mathbf{j} + 4t\mathbf{k}$

12. $\mathbf{r}(t) = (\cos t + t \sin t)\mathbf{i} + (\sin t - t \cos t)\mathbf{j} + 3\mathbf{k}$, $t > 0$

13. $\mathbf{r}(t) = (e^t \cos t)\mathbf{i} + (e^t \sin t)\mathbf{j} + 2\mathbf{k}$

14. $\mathbf{r}(t) = (6 \sin 2t)\mathbf{i} + (6 \cos 2t)\mathbf{j} + 5t\mathbf{k}$

15. $\mathbf{r}(t) = (t^3/3)\mathbf{i} + (t^2/2)\mathbf{j}$, $t > 0$

16. $\mathbf{r}(t) = (\cos^3 t)\mathbf{i} + (\sin^3 t)\mathbf{j}$, $0 < t < \pi/2$

In Exercises 17 and 18, write $\mathbf{a}$ in the form $\mathbf{a} = a_T\mathbf{T} + a_N\mathbf{N}$ without finding $\mathbf{T}$ and $\mathbf{N}$.

17. $\mathbf{r}(t) = (a \cos t)\mathbf{i} + (a \sin t)\mathbf{j} + bt\mathbf{k}$

18. $\mathbf{r}(t) = (1 + 3t)\mathbf{i} + (t - 2)\mathbf{j} - 3t\mathbf{k}$

In Exercises 19–22, write $\mathbf{a}$ in the form $\mathbf{a} = a_T\mathbf{T} + a_N\mathbf{N}$ at the given value of t without finding $\mathbf{T}$ and $\mathbf{N}$.

19. $\mathbf{r}(t) = (t + 1)\mathbf{i} + 2t\mathbf{j} + t^2\mathbf{k}$, $t = 1$

20. $\mathbf{r}(t) = (t \cos t)\mathbf{i} + (t \sin t)\mathbf{j} + t^2\mathbf{k}$, $t = 0$

21. $\mathbf{r}(t) = t^2\mathbf{i} + \left(t + \frac{1}{3}t^3\right)\mathbf{j} + \left(t - \frac{1}{3}t^3\right)\mathbf{k}$, $t = 0$

22. $\mathbf{r}(t) = (e^t \cos t)\mathbf{i} + (e^t \sin t)\mathbf{j} + \sqrt{2}e^t\mathbf{k}$, $t = 0$

In Exercises 23 and 24, find $\mathbf{r}$, $\mathbf{T}$, $\mathbf{N}$, and $\mathbf{B}$ at the given value of t. Then find equations for the osculating, normal, and rectifying planes at that value of t (see Figure 12.25 for definitions).

23. $\mathbf{r}(t) = (\cos t)\mathbf{i} + (\sin t)\mathbf{j} - \mathbf{k}$, $t = \pi/4$

24. $\mathbf{r}(t) = (\cos t)\mathbf{i} + (\sin t)\mathbf{j} + t\mathbf{k}$, $t = 0$

25. Writing to Learn The speedometer on your car reads a steady 35 mph. Could you be accelerating? Explain.

26. Writing to Learn Can anything be said about the acceleration of a particle that is moving at a constant speed? Give reasons for your answer.

27. Writing to Learn Can anything be said about the speed of a particle whose acceleration is always orthogonal to its velocity? Give reasons for your answer.

28. *Motion Along a Parabola* An object of mass m travels along the parabola $y = x^2$ with a constant speed of 10 units/sec. What is the force on the object due to its acceleration at $(0, 0)$? At $(2^{1/2}, 2)$? Write your answers in terms of $\mathbf{i}$ and $\mathbf{j}$. (Remember Newton's law, $\mathbf{F} = m\mathbf{a}$.)

29. *Curvature in Physics* **Writing to Learn** The following is a quotation from an article in *The American Mathematical Monthly*, titled "Curvature in the Eighties" by Robert Osserman (October 1990, page 731):

Curvature also plays a key role in physics. The magnitude of a force required to move an object at constant speed along a curved path is, according to Newton's laws, a constant multiple of the curvature of the trajectories.

Explain mathematically why the second sentence of the quotation is true.

30. *What happens if $a_N = 0$?* Show that a moving particle will move in a straight line if the normal component of its acceleration is zero.

31. *Curvature Extremes* Show that the parabola $y = ax^2$, $a \ne 0$, has its largest curvature at its vertex and has no minimum curvature. (*Note:* Since the curvature of a curve remains the same if the curve is translated or rotated, this result is true for any parabola.)

32. *Curvature Extremes* Show that the ellipse $x = a \cos t$, $y = b \sin t$, $a > b > 0$, has its largest curvature on its major axis and its smallest curvature on its minor axis. (As in Exercise 31, the same is true for any ellipse.)

33. *Maximizing the Curvature of a Helix* In Example 4, we found the curvature of the helix $\mathbf{r}(t) = (a \cos t)\mathbf{i} + (a \sin t)\mathbf{j} + bt\mathbf{k}$ $(a, b \geq 0)$ to be $\kappa = a/(a^2 + b^2)$. What is the largest value κ can have for a given value of b? Give reasons for your answer.

34. *A Useful Shortcut to Curvature* If you already know $|a_N|$ and $|\mathbf{v}|$, then the formula $a_N = \kappa|\mathbf{v}|^2$ gives a convenient way to find the curvature. Use it to find the curvature and radius of curvature of the curve $\mathbf{r}(t) = (\cos t + t \sin t)\mathbf{i} + (\sin t - t \cos t)\mathbf{j}$, $t > 0$. (Take a_N and $|\mathbf{v}|$ from Example 6.)

35. *Curvature and Torsion for a Line* Show that κ and τ are both zero for the line $\mathbf{r}(t) = (x_0 + At)\mathbf{i} + (y_0 + Bt)\mathbf{j} + (z_0 + Ct)\mathbf{k}$.

36. *Total Curvature* We find the **total curvature** of the portion of a smooth curve that runs from $s = s_0$ to $s = s_1 > s_0$ by integrating κ from s_0 to s_1. If the curve has some other parameter, say t, then the total curvature is

$$K = \int_{s_0}^{s_1} \kappa \, ds = \int_{t_0}^{t_1} \kappa \frac{ds}{dt} \, dt = \int_{t_0}^{t_1} \kappa|\mathbf{v}| \, dt,$$

where t_0 and t_1 correspond to s_0 and s_1. Find the total curvature of the portion on the helix $\mathbf{r}(t) = (3 \cos t)\mathbf{i} + (3 \sin t)\mathbf{j} + t\mathbf{k}$, $0 \leq t \leq 4\pi$.

37. *(Continuation of Exercise 36)* Find the total curvatures of the following curves.

(a) The involute of the unit circle: $\mathbf{r}(t) = (\cos t + t \sin t)\mathbf{i} + (\sin t - t \cos t)\mathbf{j}$, $a \leq t \leq b$ $(a > 0)$. (Exercise 34 gives a convenient way to find κ. Use values from Example 6.)

(b) The parabola $y = x^2$, $-\infty < x < \infty$

38. *Circle of Curvature*

(a) Find an equation for the circle of curvature of the curve $\mathbf{r}(t) = t\mathbf{i} + (\sin t)\mathbf{j}$ at the point $(\pi/2, 1)$. (The curve parametrizes the graph of $y = \sin x$ in the xy-plane.)

(b) Find an equation for the circle of curvature of the curve $\mathbf{r}(t) = (2 \ln t)\mathbf{i} - [t + (1/t)]\mathbf{j}$, $e^{-2} \leq t \leq e^2$, at the point $(0, -2)$, where $t = 1$.

39. **Writing to Learn** What can be said about the torsion of a (sufficiently differentiable) plane curve $\mathbf{r}(t) = f(t)\mathbf{i} + g(t)\mathbf{j}$? Give reasons for your answer.

40. *The Torsion of a Helix* In Example 7, we found the torsion of the helix $\mathbf{r}(t) = (a \cos t)\mathbf{i} + (a \sin t)\mathbf{j} + bt\mathbf{k}$, $a, b \geq 0$ to be $\tau = b/(a^2 + b^2)$. What is the largest value τ can have for a given value of a? Give reasons for your answer.

Explorations

Plotting Curvature Functions

In Exercises 41–44, find the curvature of the plane curve $y = f(x)$ using the formula

$$\kappa(x) = \frac{|f''(x)|}{[1 + (f'(x))^2]^{3/2}}$$

derived in Exercise 7. Then graph $f(x)$ together with $\kappa(x)$ over the given interval. You will find some surprises.

41. $y = x^2$, $\quad -2 \leq x \leq 2$

42. $y = x^4/4$, $\quad -2 \leq x \leq 2$

43. $y = \sin x$, $\quad 0 \leq x \leq 2\pi$

44. $y = e^x$, $\quad -1 \leq x \leq 2$ ∎

Extending the Ideas

45. *Differentiable Curves with Zero Torsion Lie in Planes* That a sufficiently differentiable curve with zero torsion lies in a plane is a special case of the fact that a particle whose velocity remains perpendicular to a fixed vector $\mathbf{C}$ moves in a plane perpendicular to $\mathbf{C}$. This, in turn, can be viewed as the solution of the following problem in calculus.

Suppose $\mathbf{r}(t) = f(t)\mathbf{i} + g(t)\mathbf{j} + h(t)\mathbf{k}$ is twice differentiable for all t in an interval $[a, b]$, $\mathbf{r} = \mathbf{0}$ when $t = a$, and $\mathbf{v} \cdot \mathbf{k} = 0$ for all t in $[a, b]$. Then $h(t) = 0$ for all t in $[a, b]$.

Solve this problem. (*Hint:* Start with $\mathbf{a} = d^2\mathbf{r}/dt^2$ and apply the initial conditions in reverse order.)

46. *A Formula That Calculates τ from $\mathbf{B}$ and $\mathbf{v}$* If we start with the definition $\tau = -(d\mathbf{B}/ds) \cdot \mathbf{N}$ and apply the Chain Rule to rewrite $d\mathbf{B}/ds$ as

$$\frac{d\mathbf{B}}{ds} = \frac{d\mathbf{B}}{dt}\frac{dt}{ds} = \frac{d\mathbf{B}}{dt}\frac{1}{|\mathbf{v}|},$$

we arrive at the formula

$$\tau = -\frac{1}{|\mathbf{v}|}\left(\frac{d\mathbf{B}}{dt} \cdot \mathbf{N}\right).$$

The advantage of this formula over Equation (9) is that it is easier to derive and state. The disadvantage is that it can take a lot of work to evaluate with pencil and paper. Use the new formula to find the torsion of the helix in Example 7.

CAS Explorations

Osculating Circles In Exercises 47–50, you will explore the osculating circle at a point P on a plane curve where $\kappa \neq 0$. Perform the following steps.

(a) Plot the plane curve over the specified interval to see what it looks like.

(b) Calculate the curvature κ of the curve at the given value t_0 using the appropriate formula from Exercise 8.

(c) Find the unit normal vector $\mathbf{N}$ at t_0. Notice that the signs of the components of $\mathbf{N}$ depend on whether the unit tangent vector $\mathbf{T}$ is turning clockwise or counterclockwise at $t = t_0$. (See Exercise 9.)

(d) If $\mathbf{C} = a\mathbf{i} + b\mathbf{j}$ is the vector from the origin to the center (a, b) of the osculating circle, find the center $\mathbf{C}$ from the vector equation

$$\mathbf{C} = \mathbf{r}(t_0) + \frac{1}{\kappa(t_0)}\mathbf{N}(t_0).$$

The point $P(x_0, y_0)$ on the curve is given by the position vector $\mathbf{r}(t_0)$.

(e) Graph the equation $(x - a)^2 + (y - b)^2 = 1/\kappa^2$ of the osculating circle. Then graph the curve and osculating circle together. You may need to experiment with the size of the viewing window, but be sure it is square.

47. $\mathbf{r}(t) = (3 \cos t)\mathbf{i} + (5 \sin t)\mathbf{j}, \qquad 0 \le t \le 2\pi, \qquad t_0 = \pi/4$

48. $\mathbf{r}(t) = (\cos^3 t)\mathbf{i} + (\sin^3 t)\mathbf{j}, \qquad 0 \le t \le 2\pi, \qquad t_0 = \pi/4$

49. $\mathbf{r}(t) = t^2\mathbf{i} + (t^3 - 3t)\mathbf{j}, \qquad -4 \le t \le 4, \qquad t_0 = 3/5$

50. $\mathbf{r}(t) = (t^3 - 2t^2 - t)\mathbf{i} + \dfrac{3t}{\sqrt{1+t^2}}\mathbf{j}, \qquad -2 \le t \le 5,$
$t_0 = 1$

12.4 Planetary Motion and Satellites

Motion in Polar and Cylindrical Coordinates • Planets Move in Planes • Coordinates and Initial Conditions • Kepler's First Law (The Conic Section Law) • Kepler's Second Law (The Equal Area Law) • Proof of Kepler's First Law • Kepler's Third Law (The Time–Distance Law) • Orbit Data

Motion in Polar and Cylindrical Coordinates

When a particle moves along a curve in the polar coordinate plane, we express its position, velocity, and acceleration in terms of the moving unit vectors

$$\mathbf{u}_r = (\cos \theta)\mathbf{i} + (\sin \theta)\mathbf{j}, \qquad \mathbf{u}_\theta = -(\sin \theta)\mathbf{i} + (\cos \theta)\mathbf{j}, \qquad (1)$$

shown in Figure 12.30. The vector $\mathbf{u}_r$ points along the position vector $\overrightarrow{OP}$, so $\mathbf{r} = r\mathbf{u}_r$. The vector $\mathbf{u}_\theta$ orthogonal to $\mathbf{u}_r$ points in the direction of increasing θ.

We find from Equations (1) that

$$\frac{d\mathbf{u}_r}{d\theta} = -(\sin \theta)\mathbf{i} + (\cos \theta)\mathbf{j} = \mathbf{u}_\theta,$$

$$\frac{d\mathbf{u}_r}{d\theta} = -(\cos \theta)\mathbf{i} - (\sin \theta)\mathbf{j} = -\mathbf{u}_r. \qquad (2)$$

When we differentiate $\mathbf{u}_r$ and $\mathbf{u}_\theta$ with respect to t to find how they change with time, the Chain Rule gives

$$\dot{\mathbf{u}}_r = \frac{d\mathbf{u}_r}{d\theta}\dot{\theta} = \dot{\theta}\mathbf{u}_\theta, \qquad \dot{\mathbf{u}}_\theta = \frac{d\mathbf{u}_\theta}{d\theta}\dot{\theta} = -\dot{\theta}\mathbf{u}_r. \qquad (3)$$

Hence,

$$\mathbf{v} = \dot{\mathbf{r}} = \frac{d}{dt}(r\mathbf{u}_r) = \dot{r}\mathbf{u}_r + r\dot{\mathbf{u}}_r = \dot{r}\mathbf{u}_r + r\dot{\theta}\mathbf{u}_\theta. \qquad (4)$$

See Figure 12.31.

The acceleration is

$$\mathbf{a} = \dot{\mathbf{v}} = (\ddot{r}\mathbf{u}_r + \dot{r}\dot{\mathbf{u}}_r) + (\dot{r}\dot{\theta}\mathbf{u}_\theta + r\ddot{\theta}\mathbf{u}_\theta + r\dot{\theta}\dot{\mathbf{u}}_\theta). \qquad (5)$$

When Equations (3) are used to evaluate $\dot{\mathbf{u}}_r$ and $\dot{\mathbf{u}}_\theta$ and the components are separated, the equation for acceleration becomes

$$\mathbf{a} = (\ddot{r} - r\dot{\theta}^2)\mathbf{u}_r + (r\ddot{\theta} + 2\dot{r}\dot{\theta})\mathbf{u}_\theta. \qquad (6)$$

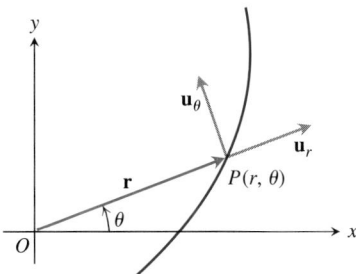

Figure 12.30 The length of $\mathbf{r}$ is the positive polar coordinate r of the point P. Thus, $\mathbf{u}_r$, which is $\mathbf{r}/|\mathbf{r}|$, is also $\mathbf{r}/r$.

As in the previous section, we use Newton's dot notation for time derivatives to keep the formulas as simple as we can: $\dot{\mathbf{u}}_r$ means $d\mathbf{u}_r/dt$, $\dot{\theta}$ means $d\theta/dt$, and so on.

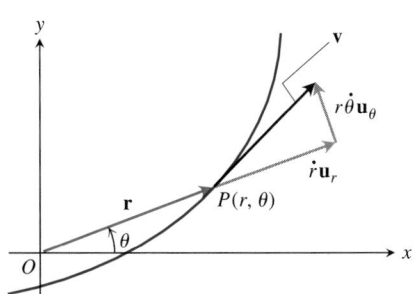

Figure 12.31 In polar coordinates, the velocity vector is $\mathbf{v} = \dot{r}\mathbf{u}_r + r\dot{\theta}\mathbf{u}_\theta$.

Notice that $|\mathbf{r}| \neq r$ if $z \neq 0$.

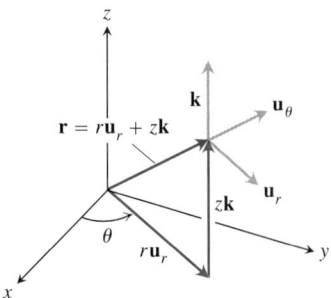

Figure 12.32 Position vector and basic unit vectors in cylindrical coordinates.

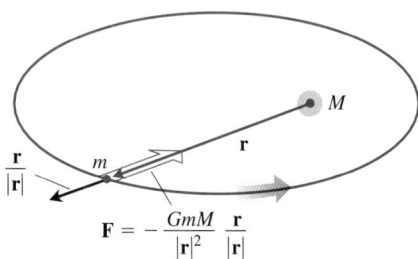

Figure 12.33 The force of gravity is directed along the line joining the centers of mass.

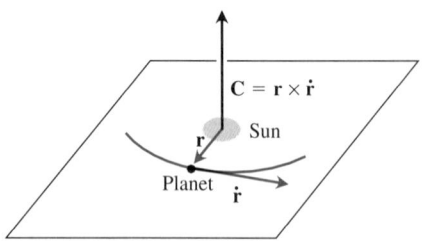

Figure 12.34 A planet that obeys Newton's laws of gravitation and motion travels in the plane through its sun's center of mass perpendicular to $\mathbf{C} = \mathbf{r} \times \dot{\mathbf{r}}$.

To extend these equations of motion to space, we add $z\mathbf{k}$ to the right-hand side of the equation $\mathbf{r} = r\mathbf{u}_r$. Then, in cylindrical coordinates,

$$\mathbf{r} = r\mathbf{u}_r + z\mathbf{k}$$
$$\mathbf{v} = \dot{r}\mathbf{u}_r + r\dot{\theta}\mathbf{u}_\theta + \dot{z}\mathbf{k} \tag{7}$$
$$\mathbf{a} = (\ddot{r} - r\dot{\theta}^2)\mathbf{u}_r + (r\ddot{\theta} + 2\dot{r}\dot{\theta})\mathbf{u}_\theta + \ddot{z}\mathbf{k}.$$

The vectors $\mathbf{u}_r$, $\mathbf{u}_\theta$, and $\mathbf{k}$ make a right-handed frame (Figure 12.32) in which

$$\mathbf{u}_r \times \mathbf{u}_\theta = \mathbf{k}, \qquad \mathbf{u}_\theta \times \mathbf{k} = \mathbf{u}_r, \qquad \mathbf{k} \times \mathbf{u}_r = \mathbf{u}_\theta. \tag{8}$$

Planets Move in Planes

Newton's Law of Gravitation says that if $\mathbf{r}$ is the radius vector from the center of a sun of mass M to the center of a planet of mass m, then the force $\mathbf{F}$ of the gravitational attraction between the planet and sun is

$$\mathbf{F} = -\frac{GmM}{|\mathbf{r}|^2} \frac{\mathbf{r}}{|\mathbf{r}|} \tag{9}$$

(Figure 12.33). The number G is the **universal gravitational constant.** If we measure mass in kilograms, force in newtons, and distance in meters, G is about 6.6720×10^{-11} Nm²kg⁻².

Combining Equation (9) with Newton's second law, $\mathbf{F} = m\ddot{\mathbf{r}}$, for the force acting on the planet gives

$$m\ddot{\mathbf{r}} = -\frac{GmM}{|\mathbf{r}|^2} \frac{\mathbf{r}}{|\mathbf{r}|},$$

$$\ddot{\mathbf{r}} = -\frac{GM}{|\mathbf{r}|^2} \frac{\mathbf{r}}{|\mathbf{r}|}. \tag{10}$$

The planet is accelerated toward its sun's center at all times.

Equation (10) says that $\ddot{\mathbf{r}}$ is a scalar multiple of $\mathbf{r}$ at every value of t, so that

$$\mathbf{r} \times \ddot{\mathbf{r}} = \mathbf{0}. \tag{11}$$

A routine calculation shows $\mathbf{r} \times \ddot{\mathbf{r}}$ to be the derivative of $\mathbf{r} \times \dot{\mathbf{r}}$:

$$\frac{d}{dt}(\mathbf{r} \times \dot{\mathbf{r}}) = \underbrace{\dot{\mathbf{r}} \times \dot{\mathbf{r}}}_{\mathbf{0}} + \mathbf{r} \times \ddot{\mathbf{r}} = \mathbf{r} \times \ddot{\mathbf{r}}. \tag{12}$$

Hence Equation (11) is equivalent to

$$\frac{d}{dt}(\mathbf{r} \times \dot{\mathbf{r}}) = \mathbf{0}, \tag{13}$$

which integrates to

$$\mathbf{r} \times \dot{\mathbf{r}} = \mathbf{C} \tag{14}$$

for some constant $\mathbf{C}$.

Equation (14) tells us that $\mathbf{r}$ and $\dot{\mathbf{r}}$ always lie in a plane perpendicular to $\mathbf{C}$. Hence, the planet moves in a fixed plane through the center of its sun (Figure 12.34).

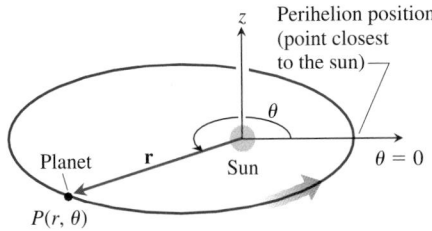

Figure 12.35 The coordinate system for planetary motion. The motion is counterclockwise when viewed from above, as it is here, and $\dot\theta > 0$.

Coordinates and Initial Conditions

We now introduce cylindrical coordinates in a way that places the origin at the sun's center of mass and makes the plane of the planet's motion the polar coordinate plane. This makes **r** the planet's polar coordinate position vector and makes $|\mathbf{r}|$ equal to r and $\mathbf{r}/|\mathbf{r}|$ equal to $\mathbf{u}_r$. We also position the z-axis in a way that makes **k** the direction of **C**. Thus, **k** has the same right-hand relation to $\mathbf{r} \times \dot{\mathbf{r}}$ that **C** does, and the planet's motion is counterclockwise when viewed from the positive z-axis. This makes θ increase with t, so that $\dot\theta > 0$ for all t. Finally, we rotate the polar coordinate plane about the z-axis, if necessary, to make the initial ray coincide with the direction **r** has when the planet is closest to the sun. This runs the initial ray through the planet's **perihelion position** (Figure 12.35).

If we measure time so that $t = 0$ at perihelion, we have the following initial conditions for the planet's motion.

1. $r = r_0$, the minimum radius, when $t = 0$.

2. $\dot{r} = 0$ when $t = 0$ (because r has a minimum value then).

3. $\theta = 0$ when $t = 0$.

4. $|\mathbf{v}| = v_0$ when $t = 0$.

Since

$$
\begin{aligned}
v_0 &= |\mathbf{v}|_{t=0} \\
&= |\dot{r}\mathbf{u}_r + r\dot\theta\mathbf{u}_\theta|_{t=0} \quad &\text{Eq. (4)} \\
&= |r\dot\theta\mathbf{u}_\theta|_{t=0} \quad &\dot{r} = 0 \text{ when } t = 0 \\
&= (|r\dot\theta||\mathbf{u}_\theta|)_{t=0} \\
&= |r\dot\theta|_{t=0} \quad &|\mathbf{u}_\theta| = 1 \\
&= (r\dot\theta)_{t=0} \quad &r \text{ and } \dot\theta \text{ both positive}
\end{aligned}
$$

we also know that

5. $r\dot\theta = v_0$ when $t = 0$.

Kepler's First Law (The Conic Section Law)

Kepler's first law says that a planet's path is a conic section with the sun at one focus. The eccentricity of the conic is

$$ e = \frac{r_0 v_0^2}{GM} - 1 \tag{15} $$

and the polar equation is

$$ r = \frac{(1 + e)r_0}{1 + e \cos\theta} \tag{16} $$

(Appendix A5, Section 4).

The derivation uses Kepler's second law, so we will state and prove the second law before proving the first law.

Kepler's Second Law (The Equal Area Law)

Kepler's second law says that the radius vector from the sun to a planet (the vector **r** in our model) sweeps out equal areas in equal times (Figure

Multiplying through by 2 and integrating with respect to r gives

$$p^2 = (\dot{r})^2 = -\frac{r_0^2 v_0^2}{r^2} + \frac{2GM}{r} + C_1. \tag{24}$$

The initial conditions that $r = r_0$ and $\dot{r} = 0$ when $t = 0$ determine the value of C_1 to be

$$C_1 = v_0^2 - \frac{2GM}{r_0}.$$

Accordingly, Equation (24), after a suitable rearrangement, becomes

$$\dot{r}^2 = v_0^2 \left(1 - \frac{r_0^2}{r^2}\right) + 2GM \left(\frac{1}{r} - \frac{1}{r_0}\right). \tag{25}$$

The effect of going from Equation (21) to Equation (25) has been to replace a second-order differential equation in r by a first-order differential equation in r. Our goal is still to express r in terms of θ, so we now bring θ back into the picture. To accomplish this, we divide both sides of Equation (25) by the squares of the corresponding sides of Equation (19), $r^2 \dot{\theta} = r_0 v_0$, and use the fact that $\dot{r}/\dot{\theta} = (dr/dt)/(d\theta/dt) = dr/d\theta$ to get

$$\frac{1}{r^4}\left(\frac{dr}{d\theta}\right)^2 = \frac{1}{r_0^2} - \frac{1}{r^2} + \frac{2GM}{r_0^2 v_0^2}\left(\frac{1}{r} - \frac{1}{r_0}\right) \tag{26}$$

$$= \frac{1}{r_0^2} - \frac{1}{r^2} + 2h\left(\frac{1}{r} - \frac{1}{r_0}\right). \qquad h = \frac{GM}{r_0^2 v_0^2}$$

To simplify further, we substitute

$$u = \frac{1}{r}, \; u_0 = \frac{1}{r_0}, \; \frac{du}{d\theta} = -\frac{1}{r^2}\frac{dr}{d\theta}, \; \left(\frac{du}{d\theta}\right)^2 = \frac{1}{r^4}\left(\frac{dr}{d\theta}\right)^2,$$

obtaining

$$\left(\frac{du}{d\theta}\right)^2 = u_0^2 - u^2 + 2hu - 2hu_0 = (u_0 - h)^2 - (u - h)^2 \tag{27}$$

$$\frac{du}{d\theta} = \pm\sqrt{(u_0 - h)^2 + (u - h)^2} \tag{28}$$

Which sign do we take? We know that $\dot{\theta} = r_0 v_0/r^2$ is positive. Also, r starts from a minimum value at $t = 0$, so it cannot immediately decrease, and $\dot{r} \geq 0$, at least for early positive values of t. Therefore,

$$\frac{dr}{d\theta} = \frac{\dot{r}}{\dot{\theta}} \geq 0 \quad \text{and} \quad \frac{du}{d\theta} = -\frac{1}{r^2}\frac{dr}{d\theta} \leq 0.$$

The correct sign for Equation (28) is the negative sign. With this determined, we rearrange Equation (28) and integrate both sides with respect to θ:

$$\frac{-1}{\sqrt{(u_0 - h)^2 - (u - h)^2}}\frac{du}{d\theta} = 1 \tag{29}$$

$$\cos^{-1}\left(\frac{u - h}{u_0 - h}\right) = \theta + C_2.$$

The constant C_2 is zero because $u = u_0$ when $\theta = 0$ and $\cos^{-1}(1) = 0$. Therefore,

$$\frac{u - h}{u_0 - h} = \cos\theta$$

and

$$\frac{1}{r} = u = h + (u_0 - h)\cos\theta. \tag{30}$$

A few more algebraic maneuvers produce the final equation

$$r = \frac{(1 + e)r_0}{1 + e\cos\theta}, \tag{31}$$

where

$$e = \frac{1}{r_0 h} - 1 = \frac{r_0 v_0{}^2}{GM} - 1. \tag{32}$$

Together, Equations (31) and (32) say that the path of the planet is a conic section with one focus at the sun and with eccentricity $(r_0 v_0{}^2)/(GM) - 1$. This is the modern formulation of Kepler's first law.

Kepler's Third Law (The Time–Distance Law)

The time T it takes a planet to go around its sun once is the planet's **orbital period.** *Kepler's third law* says that T and the orbit's semimajor axis a are related by the equation

$$\frac{T^2}{a^3} = \frac{4\pi^2}{GM}. \tag{33}$$

Since the right-hand side of this equation is constant within a given solar system, the ratio of T^2 to a^3 *is the same for every planet in the system.*

Kepler's third law is the starting point for working out the size of our solar system. It allows the semimajor axis of each planetary orbit to be expressed in astronomical units, Earth's semimajor axis being one unit. The distance between any two planets at any time can then be predicted in astronomical units. All that remains is to find one of these distances in kilometers, which can be done by bouncing radar waves off Venus, for example. The astronomical unit is now known, after a series of such measurements, to be 149,597,870 km.

We derive Kepler's third law by combining two formulas for the area enclosed by the planet's elliptical orbit:

Formula 1:   Area $= \pi a b$ The geometry formula in which *a* is the semimajor axis and *b* is the semiminor axis.

Formula 2:   Area $= \displaystyle\int_0^T dA$

$$= \int_0^T \frac{1}{2} r_0 v_0 \, dt \qquad \text{Eq. (20)}$$

$$= \frac{1}{2} T r_0 v_0.$$

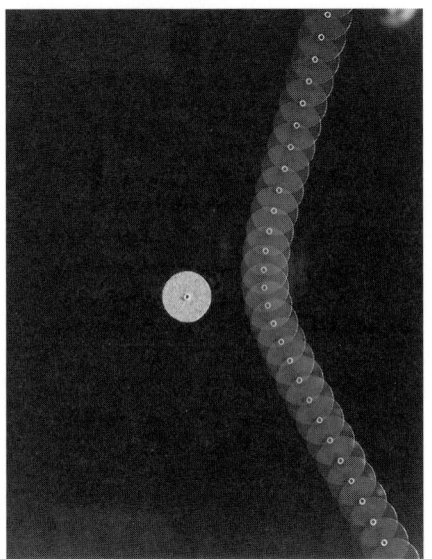

Figure 12.37 This multiflash photograph shows a body being deflected by an inverse square law force. It moves along a hyperbola.

Equating these gives

$$T = \frac{2\pi ab}{r_0 v_0} = \frac{2\pi a^2}{r_0 v_0}\sqrt{1 - e^2}. \qquad \text{For any ellipse,} \quad b = a\sqrt{1 - e^2}. \quad (34)$$

It remains only to express a and e in terms of r_0, v_0, G, and M. Equation (32) does this for e. For a, we observe that setting θ equal to π in Equation (31) gives

$$r_{\max} = r_0 \frac{1 + e}{1 - e}.$$

Hence,

$$2a = r_0 + r_{\max} = \frac{2r_0}{1 - e} = \frac{2r_0 GM}{2GM - r_0 v_0^2}. \qquad (35)$$

Squaring both sides of Equation (34) and substituting the results of Equations (32) and (35) now produces Kepler's third law (Exercise 15).

Orbit Data

Although Kepler discovered his laws empirically and stated them only for the six planets known at the time, the modern derivations of Kepler's laws show that they apply to any body driven by a force that obeys an inverse square law. They apply to Halley's comet and the asteroid Icarus. They apply to the moon's orbit about Earth, and they applied to the orbit of the spacecraft *Apollo 8* about the moon. They also applied to the air puck shown in Figure 12.37 being deflected by an inverse square law force; its path is a hyperbola. Charged particles fired at the nuclei of atoms scatter along hyperbolic paths.

Tables 12.1 to 12.3 give additional data for planetary orbits and for orbits of seven of Earth's artificial satellites (Figure 12.38). *Vanguard I* sent back data that revealed differences between the levels of Earth's oceans and provided the first determination of the precise locations of some of the more isolated Pacific islands. The data also verified that the gravitation of the sun and moon would affect the orbits of Earth's satellites and that solar radiation could exert enough pressure to deform an orbit.

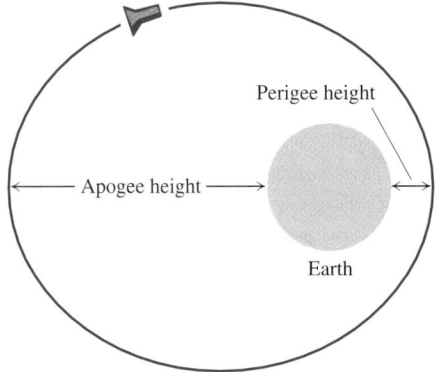

Figure 12.38 The orbit of an Earth satellite: $2a$ = diameter of Earth + perigee height + apogee height.

Table 12.1 Values of a, e, and T for the Major Planets

Planet	Semimajor axis $a^\dagger$	Eccentricity e	Period T
Mercury	57.95	0.2056	87.967 days
Venus	108.11	0.0068	224.701 days
Earth	149.57	0.0167	365.256 days
Mars	227.84	0.0934	1.8808 years
Jupiter	778.14	0.0484	11.8613 years
Saturn	1427.0	0.0543	29.4568 years
Uranus	2870.3	0.0460	84.0081 years
Neptune	4499.9	0.0082	164.784 years
Pluto	5909	0.2481	248.35 years

†Millions of kilometers

Table 12.2 Data on Earth's Satellites

Name	Launch date	Time or expected time aloft	Mass at launch (kg)	Period (min)	Perigee height (km)	Apogee height (km)	Semimajor axis a (km)	Eccentricity
Sputnik 1	Oct. 1957	57.6 days	83.6	96.2	215	939	6,955	0.052
Vanguard 1	March 1958	300 years	1.47	138.5	649	4,340	8,872	0.208
Syncom 3	Aug. 1964	$>10^6$ years	39	1436.2	35,718	35,903	42,189	0.002
Skylab 4	Nov. 1973	84.06 days	13,980	93.11	422	437	6,808	0.001
Tiros 11	Oct. 1978	500 years	734	102.12	850	866	7,236	0.001
GOES 4	Sept. 1980	$>10^6$ years	627	1436.2	35,776	35,800	42,166	0.0003
Intelsat 5	Dec 1980	$>10^6$ years	1,928	1417.67	35,143	35,707	41,803	0.007

Syncom 3 is one of a series of U.S. Department of Defense telecommunications satellites. *Tiros II* (for "television infrared observation satellite") is one of a series of weather satellites. *GOES 4* (for "geostationary operational environmental satellite") is one of a series of satellites designed to gather information about Earth's atmosphere. Its orbital period, 1436.2 minutes, is nearly the same as Earth's rotational period of 1436.1 minutes, and its orbit is nearly circular ($e = 0.0003$). *Intelsat 5* is a heavy-capacity commercial telecommunications satellite.

Table 12.3 Numerical Data

Universal gravitational constant: $G = 6.6720 \times 10^{-11} \text{ Nm}^2\text{kg}^{-2}$
(When you use this value of G in a calculation, remember to express force in newtons, distance in meters, mass in kilograms, and time in seconds.)

Sun's mass: 1.99×10^{30} kg

Earth's mass: 5.975×10^{24} kg

Equatorial radius of Earth: 6378.533 km

Polar radius of Earth: 6356.912 km

Earth's rotational period: 1436.1 min

Earth's orbital period: 1 year = 365.256 days

Quick Review 12.4

In Exercises 1–3, find the eccentricity of the conic section.

1. $4x^2 + 9y^2 = 36$ **2.** $12x^2 - 3y^2 = 24$

3. $y^2 + 4y - x + 6 = 0$

In Exercises 4–6, find a polar equation for the conic with focus at (0, 0) and given eccentricity and directrix.

4. $e = 1$, $x = -2$ **5.** $e = 5/4$, $x = 4$

6. $e = 2/3$, $x = -5$

In Exercises 7 and 8, find the endpoints of the major and minor axes of the ellipse.

7. $\dfrac{x^2}{36} + \dfrac{y^2}{64} = 1$ **8.** $\dfrac{x^2}{49} + \dfrac{y^2}{16} = 1$

In Exercises 9 and 10, solve the separable differential equation.

9. $\dfrac{dy}{dx} = 2(x + y^2 x)$ **10.** $(y + 1)\dfrac{dy}{dx} = y(x - 1)$

Section 12.4 Exercises

Reminder: When a calculation involves the universal gravitational constant G, express force in newtons, distance in meters, mass in kilograms, and time in seconds.

1. *Period of Skylab 4* Because the orbit of *Skylab 4* had a semimajor axis of $a = 6808$ km, Kepler's third law with M equal to Earth's mass should give the period. Calculate it. Compare your result with the value in Table 12.2.

2. *Earth's Velocity at Perihelion* Earth's distance from the sun at perihelion is approximately 149,577,000 km, and the eccentricity of Earth's orbit about the sun is 0.0167. Find the velocity v_0 of Earth in its orbit at perihelion. (Use Equation (15).)

3. *Semimajor Axis of Proton I* In July 1965, the USSR launched *Proton I*, weighing 12,200 kg (at launch), with a perigee height of 183 km, an apogee height of 589 km, and a period of 92.25 min. Using the relevant data for the mass of Earth and the gravitational constant G, find the semimajor axis a of the orbit from Equation (33). Compare your answer with the number you get by adding the perigee and apogee heights to the diameter of Earth.

4. *Semimajor Axis of Viking I* The *Viking I* orbiter, which surveyed Mars from August 1975 to June 1976, had a period of 1639 min. Use this and the fact that the mass of Mars is 6.418×10^{23} kg to find the semimajor axis of the *Viking I* orbit.

5. *Average Diameter of Mars* The *Viking I* orbiter was 1499 km from the surface of Mars at its closest point and 35,800 km from the surface at its farthest point. Use this information together with the value you obtained in Exercise 4 to estimate the average diameter of Mars.

6. *Period of Viking 2* The *Viking 2* orbiter, which surveyed Mars from September 1975 to August 1976, moved in an ellipse whose semimajor axis was 22,030 km. What was the orbital period? (Express your answer in minutes.)

Exploration

7. *Geosynchronous Orbits* Several satellites in Earth's equatorial plane have nearly circular orbits whose periods are the same as Earth's rotational period. Such orbits are **geosynchronous** or **geostationary** because they hold the satellite over the same spot on Earth's surface.

(a) Writing to Learn Approximately what is the semimajor axis of a geosynchronous orbit? Give reasons for your answer.

(b) About how high is a geosynchronous orbit above Earth's surface?

(c) Which of the satellites in Table 12.2 have (nearly) geosynchronous orbits?   ■

8. Writing to Learn *Semimajor Axis of Mars's Orbit* The mass of Mars is 6.418×10^{23} kg. If a satellite of Mars is to hold a stationary orbit (have the same period as the period of Mars's rotation, which is 1477.4 min), what must the semimajor axis of its orbit be? Give reasons for your answer.

9. *Distance from Earth to the Moon* The period of the moon's revolution about Earth is 2.36055×10^6 sec. About how far away is the moon?

10. *Finding Satellite Speed* A satellite moves around Earth in a circular orbit. Express the satellite's speed as a function of the orbit's radius.

11. *Finding T^2/a^3* If T is measured in seconds and a in meters, what is the value of T^2/a^3 for planets in our solar system? For satellites orbiting Earth? For satellites orbiting the moon? (The moon's mass is 7.354×10^{22} kg.)

12. *Type of Orbit* For what values of v_0 in Equation (15) is the orbit in Equation (16) a circle? An ellipse? A parabola? A hyperbola?

13. *Circular Orbits* Show that a planet in a circular orbit moves with a constant speed. (*Hint:* This is a consequence of one of Kepler's laws.)

14. Writing to Learn *The Area Derivative in Kepler's Second Law* Suppose **r** is the position vector of a particle moving along a plane curve and *dA/dt* is the rate at which the vector sweeps out area. Without introducing coordinates, and assuming the necessary derivatives exist, give a geometric argument based on increments and limits for the validity of the equation

$$\frac{dA}{dt} = \frac{1}{2}|\mathbf{r} \times \dot{\mathbf{r}}|.$$

15. Writing to Learn *Kepler's Third Law* Complete the derivation of Kepler's third law (the part following Equation (34)).

16. *Trouble in River City—The Problem with the Geocentric View* Two planets *A*, and *B* are orbiting their sun in circular orbits, with *A* being the inner planet and *B* being farther away from the sun. Suppose the positions of *A* and *B* at time *t* are

$$\mathbf{r}_A(t) = 2\cos(2\pi t)\mathbf{i} + 2\sin(2\pi t)\mathbf{j}$$

and

$$\mathbf{r}_B(t) = 3\cos(\pi t)\mathbf{i} + 3\sin(\pi t)\mathbf{j},$$

respectively, where the sun is assumed to be located at the origin and distance is measured in astronomical units. (Notice that planet *A* moves faster than planet *B*.)

The people on planet *A* regard their planet, not the sun, as the center of their planetary system (their solar system).

(a) Using planet *A* as the origin of a new coordinate system, give parametric equations for the location of planet *B* at time *t*. Write your answer in terms of cos(π*t*) and sin(π*t*).

(b) Using planet *A* as the origin, graph the path of planet *B*.

This exercise illustrates the difficulty that people before Kepler's time, with an Earth-centered or geocentric (planet *A*) view of our solar system, had in understanding the motions of the planets (planet *B* = Mars). See D. G. Saari's article in *The American Mathematical Monthly*, Vol. 97, Feb. 1990, pp. 105–119.

Extending the Ideas

17. Writing to Learn *Earth Orbit* Kepler discovered that the path of Earth around the sun is an ellipse with the sun at one of the foci. Let **r**(*t*) be the position vector from the center of the sun to the center of Earth at time *t*. Let **w** be the vector from Earth's South Pole to North Pole. It is known that **w** is constant and not orthogonal to the plane of the ellipse (Earth's axis is tilted). In terms of **r**(*t*) and **w,** give the mathematical meaning of (a) perihelion, (b) **aphelion** (position farthest from the sun), (c) equinox, (d) summer solstice, (e) winter solstice.

Chapter 12 Review Exercises

In Exercises 1 and 2, graph the curve and sketch its velocity and acceleration vectors at the given value(s) of t. Then write **a** in the form $\mathbf{a} = a_T\mathbf{T} + a_N\mathbf{N}$ without finding **T** and **N**, and find the value of κ at the given value(s) of t.

1. $\mathbf{r}(t) = (4 \cos t)\mathbf{i} + (\sqrt{2} \sin t)\mathbf{j}$, $\quad t = 0$, $\quad t = \pi/4$

2. $\mathbf{r}(t) = (\sqrt{3} \sec t)\mathbf{i} + (\sqrt{3} \tan t)\mathbf{j}$, $\quad t = 0$

3. *Highest Speed* The position of a particle in the plane at time t is

$$\mathbf{r} = \frac{1}{\sqrt{1 + t^2}}\mathbf{i} + \frac{t}{\sqrt{1 + t^2}}\mathbf{j}.$$

Find the particle's highest speed.

4. *Constant Angle Between Position and Acceleration* Suppose $\mathbf{r}(t) = (e^t \cos t)\mathbf{i} + (e^t \sin t)\mathbf{j}$. Show that the angle between **r** and **a** never changes. What is the angle?

5. *Finding Curvature* At point P, the velocity and acceleration of a particle moving in the plane are $\mathbf{v} = 3\mathbf{i} + 4\mathbf{j}$ and $\mathbf{a} = 5\mathbf{i} + 15\mathbf{j}$. Find the curvature of the particle's path at P.

6. *Finding Where the Curvature is Greatest* Find the point on the curve $y = e^x$ where the curvature is greatest.

7. *Which Direction Does It Go?* A particle moves around the unit circle in the xy-plane. Its position at time t is $\mathbf{r} = x\mathbf{i} + y\mathbf{j}$, where x and y are differentiable functions of t. Find dy/dt if $\mathbf{v} \cdot \mathbf{i} = y$. Is the motion clockwise, or counterclockwise?

8. *Sending a Message through a Tube* You send a message through a pneumatic tube that follows the curve $9y = x^3$ (distance in meters). At the point $(3, 3)$, $\mathbf{v} \cdot \mathbf{i} = 4$ and $\mathbf{a} \cdot \mathbf{i} = -2$. Find the values of $\mathbf{v} \cdot \mathbf{j}$ and $\mathbf{a} \cdot \mathbf{j}$ at $(3, 3)$.

In Exercises 9 and 10, solve the initial value problem for **r** as a vector function of t.

9. Differential equation: $\quad \dfrac{d\mathbf{r}}{dt} = -t\mathbf{i} - t\mathbf{j} - t\mathbf{k}$

Initial condition: $\quad \mathbf{r}(0) = \mathbf{i} + 2\mathbf{j} + 3\mathbf{k}$

10. Differential equation: $\quad \dfrac{d\mathbf{r}}{dt} = (180t)\mathbf{i} + (180t - 16t^2)\mathbf{j}$

Initial condition: $\quad \mathbf{r}(0) = 100\mathbf{j}$

11. *Velocity Orthogonal to Position* A particle moves in the plane so that its velocity and position vectors are always orthogonal. Show that the particle moves in a circle centered at the origin.

12. *Radius of Curvature* Show that the radius of curvature of a twice-differentiable plane curve $\mathbf{r}(t) = f(t)\mathbf{i} + g(t)\mathbf{j}$ is given by the formula

$$\rho = \frac{\dot{x}^2 + \dot{y}^2}{\sqrt{\ddot{x}^2 + \ddot{y}^2 - \dot{s}^2}}, \qquad \text{where } \dot{s} = \frac{d}{dt}\sqrt{\dot{x}^2 + \dot{y}^2}.$$

In Exercises 13 and 14, find the length of the curve.

13. $\mathbf{r}(t) = (2 \cos t)\mathbf{i} + (2 \sin t)\mathbf{j} + t^2\mathbf{k}$, $\quad 0 \le t \le \pi/4$

14. $\mathbf{r}(t) = (3 \cos t)\mathbf{i} + (3 \sin t)\mathbf{j} + 2t^{3/2}\mathbf{k}$, $\quad 0 \le t \le 3$

In Exercises 15–18, find **T**, **N**, **B**, κ, and τ at the given value of t.

15. $\mathbf{r}(t) = \dfrac{4}{9}(1 + t)^{3/2}\,\mathbf{i} + \dfrac{4}{9}(1 - t)^{3/2}\,\mathbf{j} + \dfrac{1}{3}t\mathbf{k}$, $\quad t = 0$

16. $\mathbf{r}(t) = (e^t \sin 2t)\mathbf{i} + (e^t \cos 2t)\mathbf{j} + 2e^t\mathbf{k}$, $\quad t = 0$

17. $\mathbf{r}(t) = t\mathbf{i} + \dfrac{1}{2}e^{2t}\,\mathbf{j}$, $\quad t = \ln 2$

18. $\mathbf{r}(t) = \dfrac{3}{2}(e^{2t} + e^{-2t})\mathbf{i} + \dfrac{3}{2}(e^{2t} - e^{-2t})\mathbf{j} + 6t\mathbf{k}$, $\quad t = \ln 2$

In Exercises 19 and 20, write **a** in the form $\mathbf{a} = a_T\mathbf{T} + a_N\mathbf{N}$ at $t = 0$ without finding **T** and **N**.

19. $\mathbf{r}(t) = (2 + 3t + 3t^2)\mathbf{i} + (4t + 4t^2)\mathbf{j} - (6 \cos t)\mathbf{k}$

20. $\mathbf{r}(t) = (2 + t)\mathbf{i} + (t + 2t^2)\mathbf{j} + (1 + t^2)\mathbf{k}$

21. *Finding Curvature* Express the curvature of the curve

$$\mathbf{r}(t) = \left(\int_0^t \cos\left(\frac{1}{2}\pi\theta^2\right) d\theta\right)\mathbf{i} + \left(\int_0^t \sin\left(\frac{1}{2}\pi\theta^2\right) d\theta\right)\mathbf{j}$$

as a function of the directed distance s measured along the curve from the origin (see figure).

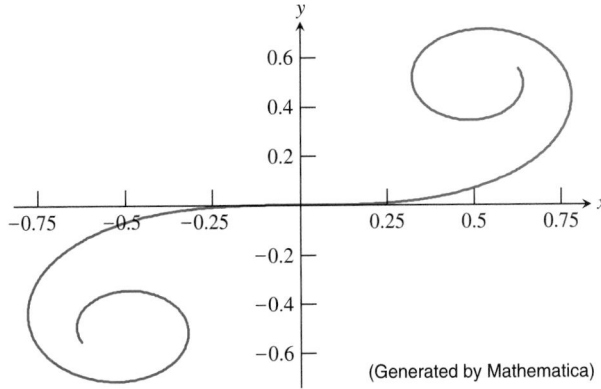

(Generated by Mathematica)

22. *An Alternative Definition of Curvature in the Plane* An alternative definition gives the curvature of a sufficiently differentiable plane curve to be $|d\phi/ds|$, where ϕ is the angle between $\mathbf{T}$ and $\mathbf{i}$ (see figure, part (a)). Part (b) of the figure shows the distance s measured counterclockwise around the circle $x^2 + y^2 = a^2$ from the point $(a, 0)$ to a point P, along with the angle ϕ at P. Calculate the circle's curvature using the alternative definition.

(a)

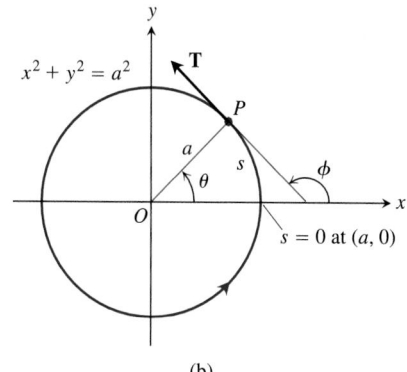

(b)

23. Find $\mathbf{T}$, $\mathbf{N}$, $\mathbf{B}$, κ, and τ as functions of t if
$\mathbf{r}(t) = (\sin t)\mathbf{i} + (\sqrt{2} \cos t)\mathbf{j} + (\sin t)\mathbf{k}$.

24. *Velocity and Acceleration Perpendicular* At what times in the interval $0 \leq t \leq \pi$ are the velocity and acceleration vectors of the motion $\mathbf{r}(t) = \mathbf{i} + (5 \cos t)\mathbf{j} + (3 \sin t)\mathbf{k}$ orthogonal?

25. The position of a particle moving in space at time $t \geq 0$ is

$$\mathbf{r}(t) = 2\mathbf{i} + \left(4 \sin \frac{t}{2}\right)\mathbf{j} + \left(3 - \frac{t}{\pi}\right)\mathbf{k}.$$

Find the first time $\mathbf{r}$ is orthogonal to the vector $\mathbf{i} - \mathbf{j}$.

26. *Finding Frenet-Frame Planes* Find equations for the osculating, normal, and rectifying planes of the curve $\mathbf{r}(t) = t\mathbf{i} + t^2\mathbf{j} + t^3\mathbf{k}$ at the point $(1, 1, 1)$.

27. *Finding Parametric Equations for a Line* Find parametric equations for the line that is tangent to the curve
$\mathbf{r}(t) = e^t\mathbf{i} + (\sin t)\mathbf{j} + \ln(1 - t)\mathbf{k}$ at $t = 0$.

28. *Finding Parametric Equations for a Line* Find parametric equations for the line tangent to the helix $\mathbf{r}(t) = (\sqrt{2} \cos t)\mathbf{i} + (\sqrt{2} \sin t)\mathbf{j} + t\mathbf{k}$ at the point where $t = \pi/4$.

29. *The View from Skylab 4* What percentage of Earth's surface area could astronauts see when *Skylab 4* was at its apogee height, 437 km above the surface? To find out, model the visible surface as the surface generated by revolving the circular arc *GT*, shown here, about the y-axis. Then carry out the following steps.

(a) Use similar triangles in the figure to show that $y_0/6380 = 6380/(6380 + 437)$. Solve for y_0.

(b) To four significant digits, calculate the visible area as

$$VA = \int_{y_0}^{6380} 2\pi x \sqrt{1 + \left(\frac{dx}{dy}\right)^2}\, dy.$$

(c) Express the result in (b) as a percentage of Earth's surface area.

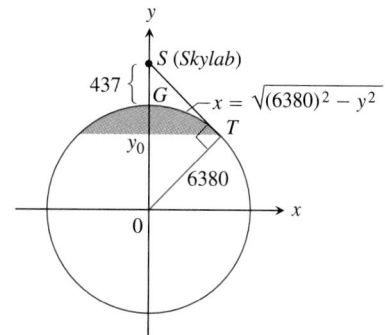

30. *Orbit Equation* Deduce from the orbit equation

$$r = \frac{(1 + e)r_0}{1 + e \cos \theta}$$

that a planet is closest to its sun when $\theta = 0$ and show that $r = r_0$ at that time.

31. *A Kepler Equation* The problem of locating a planet in its orbit at a given time and date eventually leads to solving "Kepler" equations of the form

$$f(x) = x - 1 - \frac{1}{2} \sin x = 0.$$

(a) Show that this particular equation has a solution between $x = 0$ and $x = 2$.

(b) Find the solution to as many decimal places as you can.

32. In Section 12.4, we found the velocity of a particle moving in the plane to be

$$\mathbf{v} = \dot{x}\mathbf{i} + \dot{y}\mathbf{j} = \dot{r}\mathbf{u}_r + r\dot{\theta}\mathbf{u}_\theta.$$

(a) Express $\dot{x}$ and $\dot{y}$ in terms of $\dot{r}$ and $r\dot{\theta}$ by evaluating the dot products $\mathbf{v} \cdot \mathbf{i}$ and $\mathbf{v} \cdot \mathbf{j}.$

(b) Express $\dot{r}$ and $r\dot{\theta}$ in terms of $\dot{x}$ and $\dot{y}$ by evaluating the dot products $\mathbf{v} \cdot \mathbf{u}_r$ and $\mathbf{v} \cdot \mathbf{u}_\theta.$

33. *The Curvature of Polar Curves* Express the curvature of a twice-differentiable curve $r = f(\theta)$ in the polar coordinate plane in terms of r and its derivatives.

34. *The Voyage of the Beetle* A slender rod through the origin of the polar coordinate plane rotates (in the plane) about the origin at the rate of 3 rad/min. A beetle starting from the point $(2, 0)$ crawls along the rod toward the origin at the rate of 1 in./min.

(a) Find the beetle's acceleration and velocity in polar form when it is halfway to (1 in. from) the origin.

(b) To the nearest tenth of an inch, what will be the length of the path the beetle has traveled by the time it reaches the origin?

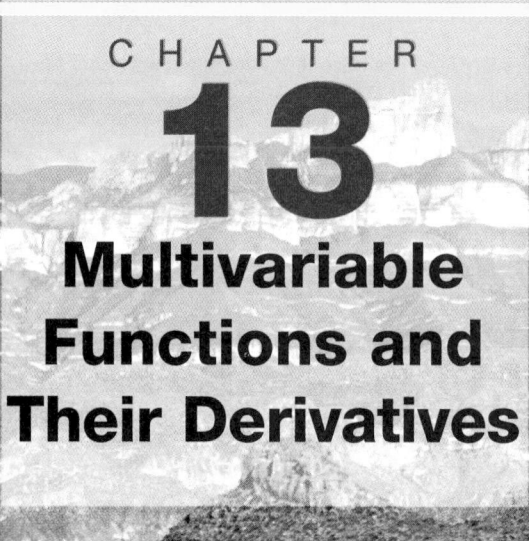

CHAPTER

13

Multivariable Functions and Their Derivatives

The appearance of a region's landscape is important to its human inhabitants. It is part of their cultural identity and affects the health of their environment and economy. Natural resource planners use 3-D modeling to produce realistic models of current land-scapes and to simulate the effects of changes on them.

Using 3-D modeling software, a natural resource planner creates a mountain range with the function $z = f(x, y) = -x^4 - y^4 - 14x^2y^2 + 32x^2 + 128y^2$. Find the high-est elevation in the mountain range. You will explore problems of this type in Section 13.7.

Chapter 13 Overview

Functions with two or more independent variables appear in science more often than functions of a single variable, and their calculus is even richer. Their derivatives are more varied and more interesting because of the different ways in which the variables can interact, and their integrals lead to a greater variety of applications. The studies of probability, statistics, fluid dynamics, and electricity, to mention only a few, all lead in natural ways to functions of more than one variable. The mathematics of these functions is one of the finest achievements in science.

As we see in this chapter, the rules of calculus remain essentially the same as we move into higher dimensions. We do not have to reinvent the theory. In higher dimensions the points representing independent variables can move around in a variety of directions instead of just along a line, but this is easily handled with the vector notation of the previous chapters.

13.1 Functions of Several Variables

Functions and Variables • Domains and Regions • Graphs and Level Curves of Functions of Two Variables • Contour Lines • Level Surfaces of Functions of Three Variables • Computer Graphing • An interesting Application (Optional)

Functions and Variables

Many functions depend on more than one independent variable. The function $V = \pi r^2 h$ calculates the volume of a right circular cylinder from its radius and height. The function $f(x, y) = x^2 + y^2$ calculates the height of the paraboloid $z = x^2 + y^2$ above the point $P(x, y)$ from the two coordinates of P.

The domains of real-valued functions of several independent real variables are sets of ordered pairs (triples, quadruples, whatever) of real numbers. The ranges are sets of real numbers of the kind we have worked with all along.

Definitions Multivariable Functions

Suppose D is a set of n-tuples of real numbers $(x_1, x_2, \ldots, x_n)$. A **real-valued function f on D** is a rule that assigns a real number

$$w = f(x_1, x_2, \ldots, x_n)$$

to each element in D. The set D is the function's **domain.** The set of w-values taken on by f is the function's **range.** The symbol w is the **dependent variable** of f, and f is a function of the n **independent variables** x_1 to x_n. The x's are the function's **input variables;** w is the function's **output variable.**

If f is a function of two independent variables, we usually call the independent variables x and y and picture the domain of f as a region in the xy-plane. If f is a function of three independent variables, we call the variables $x, y,$ and z and picture the domain as a region in space.

In applications, we tend to use letters that remind us of what the variables stand for. To say that the volume of a right circular cylinder is a function of its radius and height, we might write $V = f(r, h)$. To be more specific, we might replace the notation $f(r, h)$ by the formula that calculates the value of V from the values of r and h and write $V = \pi r^2 h$. In either case, r and h would be the independent variables and V the dependent variable of the function.

As usual, we evaluate functions defined by formulas by substituting the values of the independent variables in the formula and calculating the corresponding value of the dependent variable.

Example 1   FINDING DISTANCE FROM THE ORIGIN IN SPACE

When we use rectangular coordinates in three-dimensional space, the distance of a point (x, y, z) from the origin is given by the function $D(x, y, z) = \sqrt{x^2 + y^2 + z^2}$. The value of D at the point $(3, 0, 4)$ is $D(3, 0, 4) = \sqrt{3^2 + 0^2 + 4^2} = \sqrt{25} = 5$.

Domains and Regions

In defining functions of more than one variable, we follow the usual practice of excluding inputs that lead to complex numbers or division by zero. If $f(x, y) = \sqrt{y - x^2}$, we do not allow y to be less than x^2. If $f(x, y) = 1/(xy)$, we do not allow x or y to be zero. The domains of functions are otherwise assumed to be the largest sets for which the defining rules generate real numbers.

Exploration 1 **Finding Domains and Ranges**

For each of the following functions of *two* variables, the *domain* is given. Fill in the *range:*

Function	Domain	Range
$w = \sqrt{y - x^2}$	$y \geq x^2$	
$w = \dfrac{1}{xy}$	$xy \neq 0$	
$w = x \ln y$	Half-plane $y > 0$	
$w = \sqrt{1 - (x^2 + y^2)}$	Disk $x^2 + y^2 \leq 1$	

For each of the following functions of *three* variables, the *range* is given. Fill in the *domain:*

Function	Domain	Range
$w = \sqrt{x^2 + y^2 + z^2}$		$(0, \infty)$
$w = \dfrac{1}{x^2 + y^2 + z^2}$		$(0, \infty)$
$w = \sin xyz$		$[-1, 1]$
$w = \sqrt{4 - x^2 - y^2 - z^2}$		$[0, 2]$

(a) Interior point

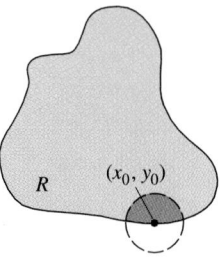

(b) Boundary point

Figure 13.1 Interior points and boundary points of a plane region R. An interior point is necessarily a point of R. A boundary point of R need not belong to R.

The domains of functions defined on portions of the plane can have interior points and boundary points just the way the domains of functions defined on intervals of the real line can.

Definitions Interior, Boundary, Open and Closed in the Plane

A point (x_0, y_0) in a region (set) R in the xy-plane is an **interior point** of R if it is the center of a disk that lies entirely in R (Figure 13.1a). A point (x_0, y_0) is a **boundary point** of R if every disk centered at (x_0, y_0) contains points that lie outside of R as well as points that lie inside R (Figure 13.1b). The boundary point itself need not belong to R.

The interior points of a region, as a set, make up the **interior** of the region. The region's boundary points make up its **boundary.** A region is **open** if it consists entirely of interior points. A region is **closed** if it contains all its boundary points (Figure 13.2).

As with intervals of real numbers, some regions in the plane are neither open nor closed. If you start with the open disk in Figure 13.2 and add to it some but not all of its boundary points, the resulting set is neither open nor closed. The boundary points that *are* there keep the set from being open. The absence of the remaining boundary points keeps the set from being closed.

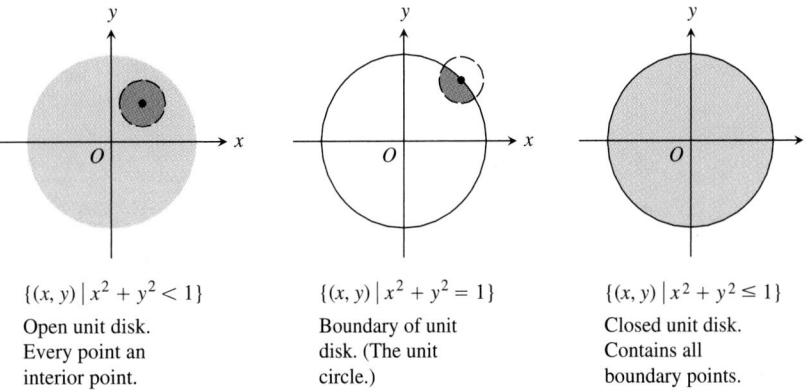

$\{(x, y) \mid x^2 + y^2 < 1\}$
Open unit disk.
Every point an
interior point.

$\{(x, y) \mid x^2 + y^2 = 1\}$
Boundary of unit
disk. (The unit
circle.)

$\{(x, y) \mid x^2 + y^2 \le 1\}$
Closed unit disk.
Contains all
boundary points.

Figure 13.2 Interior points and boundary points of the unit disk in the plane.

Definitions Bounded and Unbounded Regions in the Plane

A region in the plane is **bounded** if it lies inside a disk of finite radius. A region is **unbounded** if it is not bounded.

Examples of *bounded* sets in the plane include line segments, triangles, interiors of triangles, rectangles, circles, and disks. Examples of *unbounded* sets in the plane include lines, coordinate axes, the graphs of functions defined on infinite intervals, quadrants, half-planes, and the plane itself.

Example 2 DESCRIBING THE DOMAIN OF A FUNCTION OF TWO VARIABLES

Describe the domain of the function $f(x, y) = \sqrt{y - x^2}$.

Solution

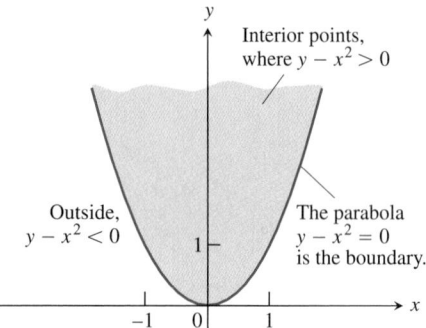

Figure 13.3 The domain of $f(x, y) = \sqrt{y - x^2}$ consists of the shaded region and its bounding parabola $y = x^2$.

Since f is defined only where $y - x^2 \geq 0$, the domain is the closed, unbounded region shown in Figure 13.3. The parabola $y = x^2$ is the boundary of the domain. The points above the parabola make up the domain's interior.

The definitions of interior, boundary, open, closed, bounded, and unbounded for regions in space are similar to those for regions in the plane. To accommodate the extra dimension, we use solid spheres instead of disks.

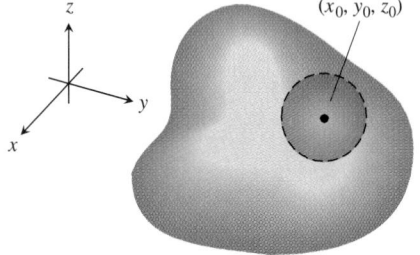

(a) Interior point

(b) Boundary point

Figure 13.4 Interior points and boundary points of a region in space.

Definitions Interior, Boundary, Open and Closed in Space

A point (x_0, y_0, z_0) in a region R in space is an **interior point** of R if it is the center of a solid sphere that lies entirely in R (Figure 13.4a). A point (x_0, y_0, z_0) is a **boundary point** of R if every sphere centered at (x_0, y_0, z_0) encloses points that lie outside of R as well as points that lie inside R (Figure 13.4b). The **interior** of R is the set of interior points of R. The **boundary** of R is the set of boundary points of R.

A region is **open** if it consists entirely of interior points. A region is **closed** if it contains its entire boundary.

A region in space is **bounded** if it lies inside a sphere of finite radius. A region in space is **unbounded** if it is not bounded.

Examples of *open* sets in space include the interior of a sphere, the open half-space $z > 0$, the first octant (where x, y, and z are all positive), and space itself.

Examples of *closed* sets in space include lines, planes, the closed half-space $z \geq 0$, the first octant together with its bounding planes, and space itself (since it has no boundary points).

A solid sphere with part of its boundary removed or a solid cube with one missing face, edge, or corner point would be *neither open nor closed*.

Graphs and Level Curves of Functions of Two Variables

There are two standard ways to picture the values of a function $f(x, y)$. One is to draw and label curves in the domain on which f has a constant value (each of which is a curve in the xy-plane). The other is to sketch the surface $z = f(x, y)$ in space.

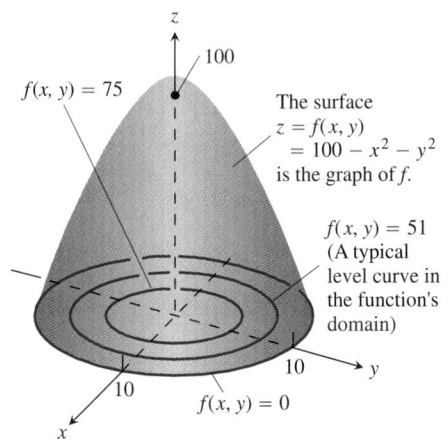

Figure 13.5 The graph and selected level curves of the function $f(x, y) = 100 - x^2 - y^2$.

> ### Definitions Level Curve, Graph, Surface (Functions of Two Variables)
>
> The set of points in the plane where a function $f(x, y)$ has a constant value $f(x, y) = c$ is a **level curve** of f. The set of points $(x, y, f(x, y))$ in space for (x, y) in the domain of f is called the **graph** of f. The graph of f is also called the **surface** $z = f(x, y)$.

Example 3 GRAPHING A FUNCTION OF TWO VARIABLES

Graph the function $f(x, y) = 100 - x^2 - y^2$ in space. Also plot the level curves $f(x, y) = 0, f(x, y) = 51$, and $f(x, y) = 75$ in the domain of f in the plane.

Solution   The domain of f is the entire xy-plane. The range of f is the set of real numbers less than or equal to 100. The graph is the paraboloid $z = 100 - x^2 - y^2$, a portion of which is shown in Figure 13.5.

The level curve $f(x, y) = 0$ is the set of points in the xy-plane at which

$$f(x, y) = 100 - x^2 - y^2 = 0, \quad \text{or} \quad x^2 + y^2 = 100,$$

which is the circle of radius 10 centered at the origin (Figure 13.5). Similarly, the level curves $f(x, y) = 51$ and $f(x, y) = 75$ are the circles

$$f(x, y) = 100 - x^2 - y^2 = 51, \quad \text{or} \quad x^2 + y^2 = 49,$$

$$f(x, y) = 100 - x^2 - y^2 = 75, \quad \text{or} \quad x^2 + y^2 = 25.$$

The level curve $f(x, y) = 100$ consists of the origin alone. (It is still called a level curve.)

Contour Lines

The curve in space in which the plane $z = c$ cuts a surface $z = f(x, y)$ is made up of the points that represent the function value $f(x, y) = c$. It is called the **contour line** $f(x, y) = c$ to distinguish it from the level curve $f(x, y) = c$ in the domain of f. Figure 13.6 shows the contour line $f(x, y) = 75$ on the surface $z = 100 - x^2 - y^2$ defined by the function $f(x, y) = 100 - x^2 - y^2$. The contour line lies 75 units directly above the circle $x^2 + y^2 = 25$, which is the level curve $f(x, y) = 75$ in the function's domain.

Not everyone makes this distinction between level curves and contour lines, but the distinction can be discerned from context. For example, on most maps the curves that represent constant elevation (height above sea level) are called contours, not level curves, even though they are all drawn in the same plane (Figure 13.7).

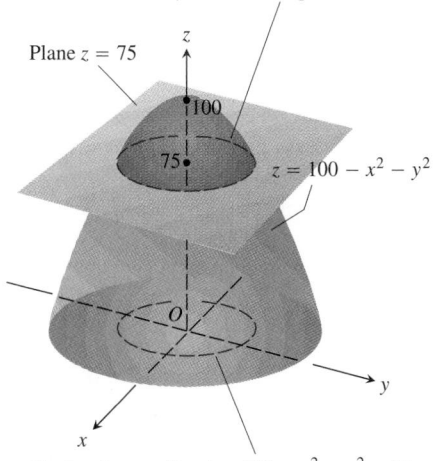

The contour line $f(x, y) = 100 - x^2 - y^2 = 75$ is the circle $x^2 + y^2 = 25$ in the plane $z = 75$.

The level curve $f(x, y) = 100 - x^2 - y^2 = 75$ is the circle $x^2 + y^2 = 25$ in the xy-plane.

Figure 13.6 The graph of $f(x, y) = 100 - x^2 - y^2$ and its intersection with the plane $z = 75$.

Figure 13.7 Contours on Mt. Greylock in western Massachusetts.

Level Surfaces of Functions of Three Variables

In the plane, the points where a function of two independent variables has a constant value $f(x, y) = c$ make a curve in the function's domain. In space, the points where a function of three independent variables has a constant value $f(x, y, z) = c$ make a surface in the function's domain.

> **Definition Level Surface**
>
> The set of points (x, y, z) in space where a function of three independent variables has a constant value $f(x, y, z) = c$ is a **level surface** of f.

Since the graphs of functions of three variables consist of points $(x, y, z, f(x, y, z))$ lying in a four-dimensional space, we cannot sketch them effectively in our three-dimensional frame of reference. We can see how the function behaves, however, by looking at its three-dimensional level surfaces.

Example 4 DESCRIBING LEVEL SURFACES OF A FUNCTION OF THREE VARIABLES

Describe the level surfaces of the function $f(x, y, z) = \sqrt{x^2 + y^2 + z^2}$.

Solution The value of f is the distance from the origin to the point (x, y, z). Each level surface $\sqrt{x^2 + y^2 + z^2} = c$, $c > 0$, is a sphere of radius c centered at the origin. Figure 13.8 shows a cutaway view of three of these spheres. The level surface $\sqrt{x^2 + y^2 + z^2} = 0$ consists of the origin alone.

We are not graphing the function here; we are looking at level surfaces in the function's three-dimensional domain. The level surfaces show how the function's values change as we move through its domain. If we remain on a sphere of radius c centered at the origin, the function maintains a constant value, namely c. If we move from one sphere to another, the function's value changes. As we move away from the origin, it increases; as we

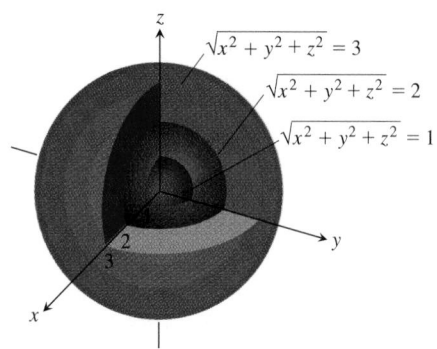

Figure 13.8 The level surfaces of $f(x, y, z) = \sqrt{x^2 + y^2 + z^2}$ are concentric spheres.

move toward the origin, it decreases. The way the values change depends on the direction we take. The relationship of change in value to direction is important and will be revisited in Section 13.7.

Computer Graphing

Three-dimensional graphing programs for computers make it possible to graph functions of two variables with only a few keystrokes. We can often get information more quickly from a graph than from a formula.

Some graphing calculators have three-dimensional graphing features. At this time, they are not as robust as computer graphing programs. They are slower, are not able to graph parametrically, and do not do a good job with contour plots. For these reasons, and except for an occasional example, we will stick with computer graphing programs for three-dimensional graphing.

Just as in two dimensions, graphing in three dimensions requires *viewing* skills.

Example 5 MODELING TEMPERATURE BENEATH THE EARTH'S SURFACE

The temperature beneath the earth's surface is a function of x, the depth beneath the surface, and t, the time of the year. If we measure x in feet and t in days elapsed from the average date of the yearly surface high, we can model the variation in temperature with the function

$$w = \cos(1.7 \times 10^{-2}t - 0.2x)e^{-0.2x}.$$

(The temperature at 0 ft is scaled to vary from $+1$ to -1, so that the variation at x feet can be interpreted as a fraction of the variation at the surface.)

(a) Sketch the two-dimensional graphs of one year's temperature variation at depths of 0, 3, and 9 ft.

(b) Interpret these graphs to say what happens to the variation of temperature as depth increases.

(c) Figure 13.9 shows a computer-generated graph of w as a function of x and t. Interpret this graph to say what happens to the variation of temperature as depth increases.

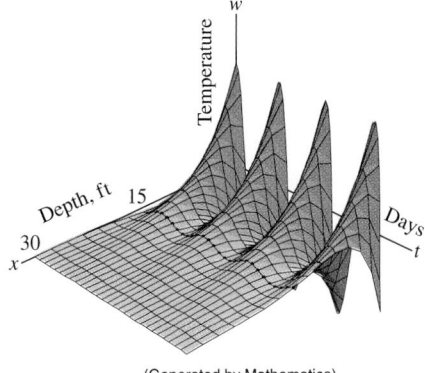

(Generated by Mathematica)

Figure 13.9 This computer-generated graph of $w = \cos(1.7 \times 10^{-2}t - 0.2x)e^{-0.2x}$ shows the seasonal variation of the temperature below ground as a fraction of surface temperature. At $x = 15$ ft, the variation is only 5% of the variation at the surface. At $x = 30$ ft, the variation is less than 0.25% of the surface variation. (Adapted from art provided by Norton Starr for G. C. Berresford, "Differential Equations and Root Cellars," *The UMAP Journal*, Vol. 2, No. 3 [1981], pp. 53–75.)

(a)

(b)

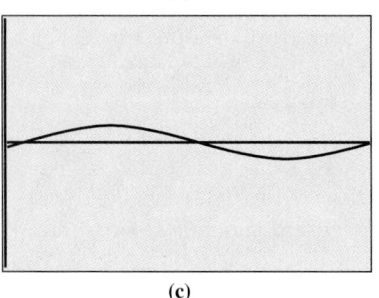

(c)

Figure 13.10 These graphs show the variation in ground temperature as a function of time, at depths of (a) 0 ft, (b) 3 ft, and (c) 9 ft. The graphs are scaled to show variation from the mean as a percentage of the variation ± 1 at the surface (graph (a)). There is less temperature variation at greater depths. Also, the temperature cycle undergoes a phase shift to the right as depth increases, indicating that the temperature cycle below the surface lags behind the cycle at ground level. (Example 5)

Solution

(a) Figure 13.10 shows the graphs of $w = \cos(1.7 \times 10^{-2}t - 0.2x)e^{-0.2x}$ for $x = 0$, $x = 3$, and $x = 9$. Each curve shows w as a function of t in the window [0, 365] by [−1.2, 1.2].

(b) From these graphs, it would appear that the variation is greatest at the surface, decreasing in amplitude as the depth increases. It also appears that the wave is undergoing a gradual phase shift to the right as depth increases, indicating that variation lags increasingly behind surface variation as the depth increases.

(c) The computer-generated graph (Figure 13.9) shows how w varies with time and depth simultaneously. The variation in temperature (vertical amplitude in the figure) decreases dramatically as depth increases, decreasing to about 5% of surface variation at 15 ft and to negligible variation at 30 ft. The graph also shows that the temperature 15 ft below the surface is about a half a year out of phase with the surface temperature. (Follow the lines perpendicular to the 15-ft line and see how they go from peaks at 15 ft to valleys at 0 ft.) When the temperature is lowest at the surface (late January, say), it is at its highest 15 ft below. Fifteen feet below the ground, the seasons are reversed!

An Interesting Application (Optional)

Exploration 2 illustrates one way to visualize the complex zeros of any single valued function $f(x)$. You will need a CAS for this exploration.

Exploration 2 **Visualizing the Complex Zeros of a Function**

Let $f(x)$ be a function of the single variable x. For the complex number $x + yi$, define $g(x, y) = |f(x + yi)|$. The graph of g is sometimes called the *complex modulus surface of f*.

1. Explain why the zeros of f occur where the surface g touches or crosses the xy-plane.

2. Let $f(x) = x^3 + 1$. Use a CAS to show that

$$g(x, y) = \sqrt{x^6 + 3x^4y^2 + 2x^3 + 3x^2y^4 - 6xy^2 + y^6 + 1}.$$

3. Find the zeros of f (real and nonreal) algebraically. (*Hint:* Factor f.)

4. Use the graph of g to estimate the zeros of f graphically. Compare your graph with the graph in Figure 13.11a. We used the viewing window $-1.5 \le x \le 1.5$, $-1.5 \le y \le 1.5$, $-1 \le z \le 3$. The graph suggests that the surface looks like it has three "teeth" that touch the xy-plane. Of course, we cannot "see" exactly where they touch.

5. Improve your estimate in (4) using contour plots of g. We used Figure 13.11b to "see" the real zero of -1 along the x-axis and estimate that the other two complex zeros are near $0.5 \pm 0.75i$. Figure 13.11c shows a zoom-in view in the first quadrant that with trace allows us to estimate this zero as $0.5 + 0.857i$. Compare with your answer in (3).

(a)

(b)

(c)

Figure 13.11 The graph of (a) $g(x, y) = |f(x + yi)|$ in $-1.5 \le x \le 1.5$, $-1.5 \le y \le 1.5$, $-1 \le z \le 3$, together with its (b) contour plot in $-1.5 \le x \le 1.5$, $-1.5 \le y \le 1.5$, and the (c) zoom-in contour plot in $0 \le x \le 1.5$, $0 \le y \le 1.5$.

Quick Review 13.1

Find a formula for each of the following functions of more than one independent variable:

1. The volume V of a right circular cone with radius r and height h

2. The total surface area S of a closed cylindrical container with radius r and height h

3. The time t required for a car traveling x miles per hour to travel y feet

4. The altitude h to the base b of an isosceles triangle with legs of length a

5. The area A of a triangle in which sides of length a and b meet at a 20° angle

6. The volume of the solid generated by revolving the region enclosed by the graphs of $y^2 = ax$ and the line $x = b$ (where a and b are positive) about the x-axis

7. The volume V enclosed when an x-inch-by-y-inch rectangle has a pair of opposite sides taped together to form a cylinder (two possible formulas)

8. The volume V of a cylinder of height h inscribed inside a sphere of radius r

9. The area A of an isosceles triangle with legs of length a and vertex angle θ

10. The linear velocity v of a carousel rider on a horse k feet from the axis of rotation, if the carousel makes m revolutions per minute

Section 13.1 Exercises

In Exercises 1–8, find the function's domain and range and determine whether the domain is bounded or unbounded.

1. $f(x, y) = y - x$
2. $f(x, y) = \sqrt{y - x}$
3. $f(x, y) = \ln(x^2 + y^2)$
4. $f(x, y) = \cos(x^2 + y^2)$
5. $f(x, y) = xy$
6. $f(x, y) = y/x^2$
7. $f(x, y) = 1/\sqrt{16 - x^2 - y^2}$
8. $f(x, y) = \sqrt{9 - x^2 - y^2}$

In Exercises 9–16, display the values of the functions in two ways: (**a**) by sketching the surface $z = f(x, y)$ and (**b**) by drawing an assortment of level curves in the function's domain. Label each level curve you draw with its function value.

9. $f(x, y) = y^2$
10. $f(x, y) = 4 - y^2$
11. $f(x, y) = x^2 + y^2$
12. $f(x, y) = \sqrt{x^2 + y^2}$
13. $f(x, y) = -(x^2 + y^2)$
14. $f(x, y) = 4 - x^2 - y^2$

15. $f(x, y) = 4x^2 + y^2$
16. $f(x, y) = 4x^2 + y^2 + 1$

Exercises 17–22 show level curves for the functions graphed in (a)–(f). Match each set of level curves with the appropriate function.

17.

18.

19.

20.

21.

22.

(a)

$z = (\cos x)(\cos y) \, e^{-\sqrt{x^2 + y^2}/4}$

(b)

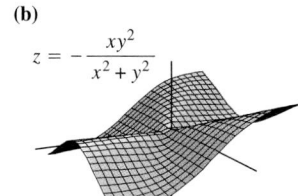

$z = -\dfrac{xy^2}{x^2 + y^2}$

(c)

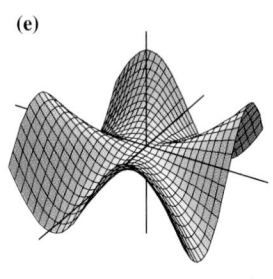

$z = \dfrac{1}{(4x^2 + y^2)}$

(d)

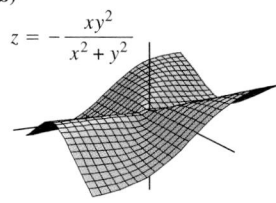

$z = e^{-y} \cos x$

(e)

$z = \dfrac{xy(x^2 - y^2)}{(x^2 + y^2)}$

(f)

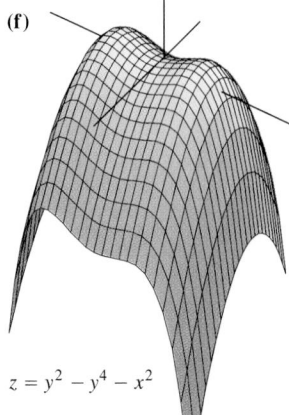

$z = y^2 - y^4 - x^2$

In Exercises 23–30, sketch a typical level surface for the function.

23. $f(x, y, z) = x^2 + y^2 + z^2$

24. $f(x, y, z) = \ln(x^2 + y^2 + z^2)$

25. $f(x, y, z) = x + z$ **26.** $f(x, y, z) = z$

27. $f(x, y, z) = x^2 + y^2$ **28.** $f(x, y, z) = y^2 + z^2$

29. $f(x, y, z) = z - x^2 - y^2$

30. $f(x, y, z) = \dfrac{x^2}{25} + \dfrac{y^2}{16} + \dfrac{z^2}{9}$

In Exercises 31–34, find an equation for the level curve of the function $f(x, y)$ that passes through the given point.

31. $f(x, y) = 16 - x^2 - y^2,$ $(2\sqrt{2}, \sqrt{2})$

32. $f(x, y) = \sqrt{x^2 - 1},$ $(1, 0)$

33. $f(x, y) = \displaystyle\int_x^y \dfrac{dt}{1 + t^2},$ $(-\sqrt{2}, \sqrt{2})$

34. $f(x, y) = \displaystyle\sum_{n=0}^{\infty} \left(\dfrac{x}{y}\right)^n,$ $(1, 2)$

In Exercises 35–38, find an equation for the level surface of the function through the given point.

35. $f(x, y, z) = \sqrt{x} - y - \ln z,$ $(3, -1, 1)$

36. $f(x, y, z) = \ln(x^2 + y + z^2),$ $(-1, 2, 1)$

37. $g(x, y, z) = \displaystyle\sum_{n=0}^{\infty} \dfrac{(x + y)^n}{n! z^n},$ $(\ln 2, \ln 4, 3)$

38. $g(x, y, z) = \displaystyle\int_x^y \dfrac{d\theta}{\sqrt{1 - \theta^2}} + \int_{\sqrt{2}}^z \dfrac{dt}{t\sqrt{t^2 - 1}},$ $(0, 1/2, 2)$

39. Writing to Learn *Defining the Graphs of Functions of n Independent Variables* As you know, the graph of a real-valued function of a single variable is a set in two-coordinate space. The graph of a real-valued function of two independent variables is a set in three-coordinate space. The graph of a real-valued function of three independent variables is a set in four-coordinate space. How would you define the graph of a real-valued function $f(x_1, x_2, x_3, x_4)$ of four independent real variables? How would you define the graph of a real-valued function $f(x_1, x_2, x_3, \ldots, x_n)$ of n independent variables?

40. Writing to Learn *Reading a Topographic Map* The "contour lines" on a topographic map (as in Figure 13.7) are really level curves of the function $h(x, y)$ that gives the height above sea level of a point with coordinates (x, y) on the map. The level curves are drawn for h values at regular intervals (for example, 20 ft).

(a) How can you spot a mountain on a topographic map? Explain.

(b) How can you spot a level area on a topographic map? Explain.

(c) Why do rivers and streams usually run perpendicular to contour lines?

(d) What geological formation would be indicated by a series of straight contour lines very close together? Explain.

Explorations

41. *Maximum Value of a Function on a Line in Space* Does the function $f(x, y, z) = xyz$ have a maximum value on the line $x = 20 - t$, $y = t$, $z = 20$? If so, what is it? Give reasons for your answer. (*Hint:* Along the line, $w = f(x, y, z)$ is a differentiable function of the single variable t.)

42. *Minimum Value of a Function on a Line in Space* Does the function $f(x, y, z) = xy - z$ have a minimum value on the line $x = t - 1$, $y = t - 2$, $z = t + 7$? If so, what is it? Give reasons for your answer. (*Hint:* Along the line, $w = f(x, y, z)$ is a differentiable function of the single variable t.) ∎

Extending the Ideas

43. *The Concorde's Sonic Booms* The width w of the region in which people on the ground hear the Concorde's sonic boom directly, not reflected from a layer of the atmosphere, is a function of

T = air temperature at ground level (in degrees Kelvin),

h = the Concorde's altitude (in kilometers), and

d = the vertical temperature gradient (temperature drop in degrees Kelvin per km).

The formula for w is

$$w = 4(Th/d)^{\frac{1}{2}}.$$

See Figure 13.12.

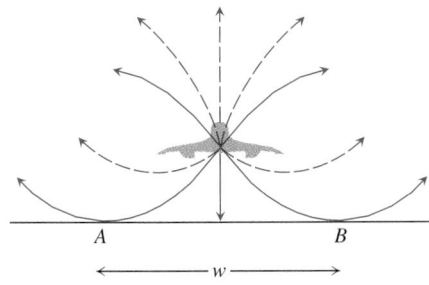

Sonic boom carpet

Figure 13.12 Sound waves from the Concorde bend as the temperature changes above and below the altitude at which the plane flies. The sonic boom carpet is the region on the ground that receives shock waves directly from the plane, not reflected from the atmosphere or diffracted along the ground. The carpet is determined by the grazing rays striking the ground from the point directly under the plane. (Exercise 43)

The Washington, D.C.–bound Concorde approaches the United States from Europe on a course that takes it south of Nantucket Island at an altitude of 16.8 km. If the surface temperature is 290 K and the vertical temperature gradient is 5 K/km, how many kilometers south of Nantucket must the plane be flown to keep its sonic boom carpet away from the island? (From N. K. Balachandra, W. L. Donn, and D. H. Rind, "Concorde Sonic Booms as an Atmospheric Probe," *Science,* July 1, 1977, Vol. 197, pp. 47–49.)

CAS Explorations

In Exercises 44–47, perform the following steps for the given function.

(a) Plot the surface over the given rectangle.

(b) Plot several level curves in the rectangle.

(c) Plot the level curve of f through the given point.

44. $f(x, y) = x \sin \dfrac{y}{2} + y \sin 2x$, $\quad 0 \le x \le 5\pi$, $0 \le y \le 5\pi$, $\quad P(3\pi, 3\pi)$

45. $f(x, y) = (\sin x)(\cos y)e^{\sqrt{x^2 + y^2}/8}$, $\quad 0 \le x \le 5\pi$, $0 \le y \le 5\pi$, $\quad P(4\pi, 4\pi)$

46. $f(x, y) = \sin(x + 2 \cos y)$, $\quad -2\pi \le x \le 2\pi$, $-2\pi \le y \le 2\pi$, $\quad P(\pi, \pi)$

47. $f(x, y) = e^{(x^{0.1} - y)} \sin(x^2 + y^2)$, $\quad 0 \le x \le 2\pi$, $-2\pi \le y \le \pi$, $\quad P(\pi, -\pi)$

In Exercises 48–51, plot the level surfaces.

48. $4 \ln (x^2 + y^2 + z^2) = 1$

49. $x^2 + z^2 = 1$

50. $x + y^2 - 3z^2 = 1$

51. $\sin \left(\dfrac{x}{2}\right) - (\cos y)\sqrt{x^2 + z^2} = 2$

Just as you describe curves in the plane parametrically with a pair of equations $x = f(t)$, $y = g(t)$ defined on some parameter interval I, you can sometimes describe surfaces in space with a triple of equations $x = f(u, v)$, $y = g(u, v)$, $z = h(u, v)$ defined on some parameter rectangle $a \le u \le b$, $c \le v \le d$. Many computer algebra systems permit you to plot such surfaces in *parametric mode.* (Parametrized surfaces are discussed in detail in Section 15.6.)

In Exercises 52–55, plot the surfaces. Also plot several level curves in the xy-plane.

52. $x = u \cos v$, $\quad y = u \sin v$, $\quad z = u$, $\quad 0 \le u \le 2$, $0 \le v \le 2\pi$

53. $x = u \cos v$, $\quad y = u \sin v$, $\quad z = v$, $\quad 0 \le u \le 2$, $0 \le v \le 2\pi$

54. $x = (2 + \cos u) \cos v$, $\quad y = (2 + \cos u) \sin v$, $z = \sin u$, $\quad 0 \le u \le 2\pi$, $\quad 0 \le v \le 2\pi$

55. $x = 2 \cos u \cos v$, $\quad y = 2 \cos u \sin v$, $\quad z = 2 \sin u$, $0 \le u \le 2\pi$, $\quad 0 \le v \le \pi$

In Exercises 56 and 57, find all the zeros of the function.

56. $f(x) = x^5 - 1$ $\qquad\qquad$ **57.** $f(x) = x^4 + 1$

Limits and Continuity in Higher Dimensions

Limits • Continuity • The Max-Min Property for Continuous Functions

Limits

If the values of a real-valued function $f(x, y)$ lie close to a fixed real number L for all points (x, y) sufficiently close to the point (x_0, y_0) but not equal to (x_0, y_0), we say that L is the limit of f as (x, y) approaches (x_0, y_0). In symbols, we write

$$\lim_{(x,y)\to(x_0,y_0)} f(x, y) = L,$$

and we say, "the limit of f as (x, y) approaches (x_0, y_0) equals L." This is like the limit of a function of one variable, except that two independent variables are involved instead of one, complicating the issue of "closeness." If (x_0, y_0) is an interior point of f's domain, (x, y) can approach (x_0, y_0) from any direction. To take the added dimension into account, we adjust the definition from Section 2.1 to measure distances in the domain with the two-dimensional distance formula rather than simple absolute value.

Definition Limit of a Function of Two Independent Variables

The function f **has limit** L as (x, y) approaches (x_0, y_0) if, given any positive number ε, there is a positive number δ such that for all (x, y) in the domain of f,

$$0 < \sqrt{(x - x_0)^2 + (y - y_0)^2} < \delta \;\Rightarrow\; |f(x, y) - L| < \varepsilon.$$

We write $\lim_{(x,y)\to(x_0,y_0)} f(x, y) = L.$

Closeness in *both* dimensions in the *xy*-plane implies closeness in *each* dimension (and vice versa), so we can also use the following, equivalent, definition of limit.

Definition Limit of a Function of Two Independent Variables

The function f **has limit** L as (x, y) approaches (x_0, y_0) if, given any positive number ε, there is a positive number δ such that for all (x, y) in the domain of f,

$$0 < |x - x_0| < \delta \quad \text{and} \quad 0 < |y - y_0| < \delta \;\Rightarrow\; |f(x, y) - L| < \varepsilon.$$

We write $\lim_{(x,y)\to(x_0,y_0)} f(x, y) = L.$

These definitions apply to boundary points as well as interior points of the domain of f. The only requirement is that the point (x, y) remain in the domain at all times.

It can be shown, as for functions of a single variable, that

$$\lim_{(x,y) \to (x_0, y_0)} x = x_0$$

$$\lim_{(x,y) \to (x_0, y_0)} y = y_0$$

$$\lim_{(x,y) \to (x_0, y_0)} k = k \qquad \text{(any number } k\text{).}$$

It can also be shown that the limit of the sum of two functions is the sum of their limits (when they both exist), with similar results for the limits of their differences, products, constant multiples, quotients, and powers.

Theorem 1 Properties of Limits of Functions of Two Variables

If L, M, and k are real numbers and

$$\lim_{(x,y) \to (x_0, y_0)} f(x, y) = L \qquad \text{and} \qquad \lim_{(x,y) \to (x_0, y_0)} g(x, y) = M,$$

then

1. *Sum Rule:* $\qquad\qquad\qquad \lim\limits_{(x,y) \to (x_0, y_0)} [f(x, y) + g(x, y)] = L + M$

2. *Difference Rule:* $\qquad\quad \lim\limits_{(x,y) \to (x_0, y_0)} [f(x, y) - g(x, y)] = L - M$

3. *Product Rule:* $\qquad\qquad \lim\limits_{(x,y) \to (x_0, y_0)} [f(x, y) \cdot g(x, y)] = L \cdot M$

4. *Constant-Multiple-Rule:* $\quad \lim\limits_{(x,y) \to (x_0, y_0)} k \cdot f(x, y) = k \cdot L$

$$\text{(any number } k\text{)}$$

5. *Quotient Rule:* $\qquad\qquad \lim\limits_{(x,y) \to (x_0, y_0)} \dfrac{f(x, y)}{g(x, y)} = \dfrac{L}{M} \qquad \text{if } M \neq 0$

6. *Power Rule:* $\qquad\qquad$ Let $a > 0$. Then

$$\lim_{(x,y) \to (x_0, y_0)} [f(x, y)]^a = L^a,$$

provided L^a is a real number.

When we apply Theorem 1 to polynomials and rational functions, we obtain the useful result that the limits of these functions as $(x, y) \to (x_0, y_0)$ can be calculated by evaluating the functions at (x_0, y_0). The only requirement is that the rational functions be defined at (x_0, y_0).

Example 1 CALCULATING LIMITS OF RATIONAL FUNCTIONS

Find **(a)** $\lim\limits_{(x,y) \to (3,-4)} (x^2 + y^2)$ and **(b)** $\lim\limits_{(x,y) \to (0,1)} \dfrac{x - xy + 3}{x^2 y + 5xy - y^3}$.

Solution

(a) $\lim\limits_{(x,y)\to(3,-4)} (x^2 + y^2) = (3)^2 + (-4)^2 = 25$

(b) $\lim\limits_{(x,y)\to(0,1)} \dfrac{x - xy + 3}{x^2y + 5xy - y^3} = \dfrac{0 - 0\cdot1 + 3}{0^2\cdot1 + 5\cdot0\cdot1 - 1^3} = -3$

As in the single-variable case, it is the *continuity* of polynomial and rational functions over their domains that enables us to compute these limits so easily.

Example 2   CALCULATING LIMITS

Find

$$\lim_{(x,y)\to(0,0)} \frac{x^2 - xy}{\sqrt{x} - \sqrt{y}}.$$

Solution   Since the denominator $\sqrt{x} - \sqrt{y}$ approaches 0 as $(x, y) \to (0, 0)$, we cannot use the Quotient Rule from Theorem 1. However, if we multiply numerator and denominator by $\sqrt{x} + \sqrt{y}$, we produce an equivalent fraction (equivalent for $(x, y) \neq (0, 0)$) whose limit we *can* find:

$$\lim_{(x,y)\to(0,0)} \frac{x^2 - xy}{\sqrt{x} - \sqrt{y}} = \lim_{(x,y)\to(0,0)} \frac{(x^2 - xy)(\sqrt{x} + \sqrt{y})}{(\sqrt{x} - \sqrt{y})(\sqrt{x} + \sqrt{y})}$$

$$= \lim_{(x,y)\to(0,0)} \frac{x(x - y)(\sqrt{x} + \sqrt{y})}{x - y} \qquad \text{Algebra}$$

$$= \lim_{(x,y)\to(0,0)} x(\sqrt{x} + \sqrt{y}) = 0 \qquad \begin{array}{l}\text{Cancel the}\\ \text{factor } (x - y).\end{array}$$

We can cancel the factor $(x - y)$ in Example 2 because the path $y = x$ (along which $x - y = 0$) is *not* in the domain of the function

$$\frac{x^2 - xy}{\sqrt{x} - \sqrt{y}}.$$

Continuity

The definition of continuity for functions of two variables is essentially the same as for functions of a single variable.

Definitions Continuity at a Point, Continuity

A function $f(x, y)$ is **continuous at the point (x_0, y_0)** if

1. f is defined at (x_0, y_0).
2. $\lim\limits_{(x,y)\to(x_0,y_0)} f(x, y)$ exists.
3. $\lim\limits_{(x,y)\to(x_0,y_0)} f(x, y) = f(x_0, y_0)$.

A function is **continuous** if it is continuous at every point of its domain.

As with the definition of limit, the definition of continuity applies at boundary points as well as interior points of the domain of f. The only requirement is that the point (x, y) remain in the domain at all times.

As you may have guessed, one consequence of Theorem 1 is that algebraic combinations of continuous functions are continuous at every point at

(Generated by Mathematica)

(a)

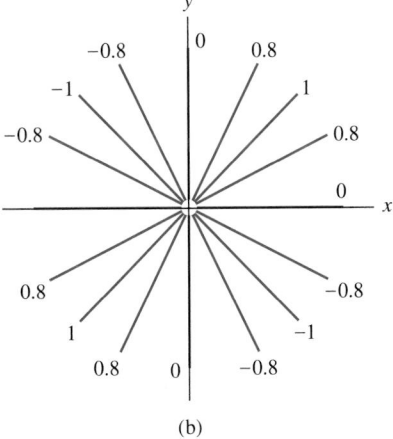

(b)

Figure 13.13 (a) The graph of

$$f(x, y) = \begin{cases} \dfrac{2xy}{x^2 + y^2}, & (x, y) \neq (0, 0) \\ 0, & (x, y) = (0, 0) \end{cases}$$

is continuous at every point except the origin. (b) The level curve of *f*.

Figure 13.14 Read the expression in the figure as the limit of $f(x, y)$ as $x \to 0$ with the restriction that $y = mx$. The vertical bar translates to "with the restriction that."

which all the functions involved are defined. Hence, sums, differences, products, constant multiples, quotients, and powers of continuous functions are continuous where defined. In particular, polynomials and rational functions of two variables are continuous at every point at which they are defined.

If $z = f(x, y)$ is a continuous function of x and y and if $w = g(z)$ is a continuous function of z, then the composite $w = g(f(x, y))$ is continuous. Thus,

$$e^{x-y}, \qquad \cos\left(\frac{xy}{x^2 + 1}\right), \qquad \ln(1 + x^2 y^2)$$

are all continuous at every point (x, y).

As with functions of a single variable, the general rule is that composites of continuous functions are continuous. The only requirement is that each function be continuous where it is applied.

Example 3 SHOWING THAT A FUNCTION HAS A SINGLE DISCONTINUITY

Show that

$$f(x, y) = \begin{cases} \dfrac{2xy}{x^2 + y^2}, & (x, y) \neq (0, 0) \\ 0, & (x, y) = (0, 0) \end{cases}$$

is continuous at every point except the origin (Figure 13.13).

Solution The function f is continuous at any point $(x, y) \neq (0, 0)$ because its values are then given by a rational function of x and y.

At $(0, 0)$, the value of f is defined, but f, we claim, has no limit as $(x, y) \to (0, 0)$. The reason is that different paths of approach to the origin can lead to different limits, as we will now see.

For every value of m, the function f has a constant value on the "punctured" line $y = mx$, $x \neq 0$, because

$$f(x, y)\Big|_{y=mx} = \frac{2xy}{x^2 + y^2}\bigg|_{y=mx} = \frac{2x(mx)}{x^2 + (mx)^2} = \frac{2m}{1 + m^2}.$$

Therefore, f has this number as its limit as (x, y) approaches $(0, 0)$ along the line:

$$\lim_{\substack{\text{along } y=mx \\ (x,y)\to(0,0)}} f(x, y) = \lim_{(x,y)\to(0,0)} \left[f(x, y)\big|_{y=mx} \right] = \frac{2m}{1 + m^2}.$$

The limit changes with m, however. There is therefore no single number we may call the limit of f as (x, y) approaches the origin. The limit fails to exist, and the function is not continuous.

We can support the result of Example 3 with CAS as suggested by Figure 13.14.

Example 3 illustrates an important point about the limits of functions of two variables (or even more variables, for that matter). For a limit to exist at a point, the limit must be the same along every approach path. If we ever find paths with different limits, we know that the function has no limit at the point they approach.

(Generated by Mathematica)

(a)

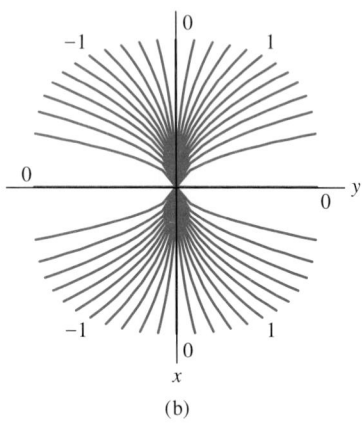

(b)

Figure 13.15 (a) The graph of $f(x, y) = 2x^2y/(x^4 + y^2)$. As the graph suggests and the level-curve values in (b) confirm,
$\lim\limits_{(x,y)\to(0,0)} f(x, y)$ does not exist.

> ### The Two-Path Test for Discontinuity
>
> If a function $f(x, y)$ has different limits along two different paths as (x, y) approaches (x_0, y_0), then $\lim\limits_{(x,y)\to(x_0,y_0)} f(x, y)$ does not exist.

Exploration 1 **Exploring Pathways to the Origin**

Does the function

$$f(x, y) = \frac{2x^2y}{x^4 + y^2}$$

have a limit as (x, y) approaches $(0, 0)$?

(a) Find $f(x, y)\big|_{y=x^2}$ and compute $\lim\limits_{\substack{(x,y)\to(0,0)\\ \text{along } y=x^2}} f(x, y)$.

(b) Find $f(x, y)\big|_{y=-x^2}$ and compute $\lim\limits_{\substack{(x,y)\to(0,0)\\ \text{along } y=-x^2}} f(x, y)$.

(c) Explain why we can conclude that $\lim\limits_{(x,y)\to(0,0)} f(x, y)$ does not exist.

(d) Show that the values of

$$f(x, y) = \frac{2x^2y}{x^4 + y^2}$$

are *constant* along any parabola of the form $y = kx^2$. Explain why such parabolas are level curves of f.

(e) Several level curves of f are graphed in Figure 13.15. How could we conclude from this graph that $\lim\limits_{(x,y)\to(0,0)} f(x, y)$ does not exist?

The language we use in Exploration 1 showing that f has no limit at the origin might seem contradictory; one could easily argue that f has, in fact, many limits there. But that is just the point. There is no path-independent limit; therefore, by our definition, $\lim\limits_{(x,y)\to(0,0)} f(x, y)$ does not exist. It is our translating this formal statement into the more colloquial "has no limit" that creates the apparent contradiction. The mathematics is fine; the problem arises in how we tend to talk about it. In the crunch, we need the formality to keep things straight.

The Max-Min Property for Continuous Functions

We have seen that a function of a single variable that is continuous throughout a closed, bounded interval takes on an absolute maximum value and an absolute minimum value at least once in that interval. The same is true of a function $z = f(x, y)$ that is continuous on a closed, bounded set R in the plane (like a line segment, a disk, or a filled-in triangle). The function takes on an absolute maximum value at some point in R and an absolute minimum value at some point in R.

The max-min property and other results of this section hold for functions of three or more variables. A continuous function $w = f(x, y, z)$, for example, must take on absolute maximum and minimum values on any closed, bounded set (solid ball or cube, spherical shell, rectangular solid) on which it is defined.

We will learn how to find these extreme values in Section 13.7, but first we need to know about derivatives in higher dimensions, which is the topic of the next section.

Quick Review 13.2

In Exercises 1–5, find the limit.

1. $\lim\limits_{x \to 2} \dfrac{3x^2 - 7x + 5}{x^3 + x^2 - 9}$

2. $\lim\limits_{x \to 0} \sqrt{3 + \sec x}$

3. $\lim\limits_{x \to -4} \dfrac{x + 5}{\sqrt{x^2 + 9}}$

4. $\lim\limits_{x \to 3} \dfrac{2x^2 - 7x + 3}{x^2 - 9}$

5. $\lim\limits_{x \to \pi} \dfrac{\sin x}{x - \pi}$

In Exercises 6–10, find the points at which f is continuous.

6. $f(x) = \dfrac{\sin 4x}{x^2 - 4}$

7. $f(x) = (x^2 + 1) \ln x$

8. $f(x) = \dfrac{|2x + 3|}{2x + 3}$

9. $f(x) = \begin{cases} \dfrac{\sin 2x}{\sin x}, & x \neq 0 \\ 0, & x = 0 \end{cases}$

10. $f(x) = \dfrac{x^2 - 3x - 10}{2 + \sin x}$

Section 13.2 Exercises

In Exercises 1–12, find the limit.

1. $\lim\limits_{(x,y) \to (0,0)} \dfrac{3x^2 - y^2 + 5}{x^2 + y^2 + 2}$

2. $\lim\limits_{(x,y) \to (1,1)} \ln|1 + x^2 y^2|$

3. $\lim\limits_{(x,y) \to (0,\,\ln 2)} e^{x-y}$

4. $\lim\limits_{(x,y) \to (0,4)} \dfrac{x}{\sqrt{y}}$

5. $\lim\limits_{(x,y,z) \to (1,3,4)} \sqrt{x^2 + y^2 + z^2 - 1}$

6. $\lim\limits_{(x,y,z) \to (1,2,6)} \left(\dfrac{1}{x} + \dfrac{1}{y} + \dfrac{1}{z}\right)$

7. $\lim\limits_{(x,y) \to (0,\pi/4)} \sec x \tan y$

8. $\lim\limits_{(x,y) \to (0,0)} \cos \dfrac{x^2 + y^2}{x + y + 1}$

9. $\lim\limits_{(x,y) \to (1,1)} \cos \sqrt[3]{|xy| - 1}$

10. $\lim\limits_{(x,y) \to (1,0)} \dfrac{x \sin y}{x^2 + 1}$

11. $\lim\limits_{(x,y) \to (0,0)} \dfrac{e^y \sin x}{x}$

12. $\lim\limits_{(x,y) \to (0,0)} \tan^{-1}\left(1/\sqrt{x^2 + y^2}\right)$

In Exercises 13–20, find the limit by rewriting the fraction first.

13. $\lim\limits_{\substack{(x,y) \to (1,1) \\ x \neq y}} \dfrac{x^2 - 2xy + y^2}{x - y}$

14. $\lim\limits_{\substack{(x,y) \to (1,1) \\ x \neq y}} \dfrac{x^2 - y^2}{x - y}$

15. $\lim\limits_{\substack{(x,y) \to (1,1) \\ x \neq 1}} \dfrac{xy - y - 2x + 2}{x - 1}$

16. $\lim\limits_{\substack{(x,y) \to (2,-4) \\ y \neq -4,\, x \neq x^2}} \dfrac{y + 4}{x^2 y - xy + 4x^2 - 4x}$

17. $\lim\limits_{\substack{(x,y) \to (0,0) \\ x \neq y}} \dfrac{x - y + 2\sqrt{x} - 2\sqrt{y}}{\sqrt{x} - \sqrt{y}}$

18. $\lim\limits_{\substack{(x,y) \to (2,2) \\ x+y \neq 4}} \dfrac{x + y - 4}{\sqrt{x + y} - 2}$

19. $\lim\limits_{\substack{(x,y) \to (2,0) \\ 2x-y \neq 4}} \dfrac{\sqrt{2x - y} - 2}{2x - y - 4}$

20. $\lim\limits_{\substack{(x,y) \to (4,3) \\ x \neq y+1}} \dfrac{\sqrt{x} - \sqrt{y+1}}{x - y - 1}$

In Exercises 21–26, find the limit. In each exercise, P stands for the point (x, y, z).

21. $\lim\limits_{P \to (2,3,-6)} \sqrt{x^2 + y^2 + z^2}$

22. $\lim\limits_{P \to (0,-2,0)} \ln \sqrt{x^2 + y^2 + z^2}$

23. $\lim\limits_{P \to (3,3,0)} (\sin^2 x + \cos^2 y + \sec^2 z)$

24. $\lim\limits_{P\to(\pi,0,3)} ze^{-2y}\cos 2x$

25. $\lim\limits_{P\to(-1/4,\pi/2,2)} \tan^{-1} xyz$

26. $\lim\limits_{P\to(1,-1,-1)} \dfrac{2xy + yz}{x^2 + z^2}$

In Exercises 27–30, at what points (x, y) in the plane is the function continuous?

27. (a) $f(x, y) = \sin(x + y)$

    (b) $f(x, y) = \ln(x^2 + y^2)$

28. (a) $f(x, y) = \dfrac{x + y}{x - y}$

    (b) $f(x, y) = \dfrac{y}{x^2 + 1}$

29. (a) $g(x, y) = \sin\dfrac{1}{xy}$

    (b) $g(x, y) = \dfrac{x + y}{2 + \cos x}$

30. (a) $g(x, y) = \dfrac{x^2 + y^2}{x^2 - 3x + 2}$

    (b) $g(x, y) = \dfrac{1}{x^2 - y}$

In Exercises 31–34, at what points (x, y, z) in space is the function continuous?

31. (a) $f(x, y, z) = x^2 + y^2 - 2z^2$

    (b) $f(x, y, z) = \sqrt{x^2 + y^2 - 1}$

32. (a) $f(x, y, z) = \ln xyz$

    (b) $f(x, y, z) = e^{x+y}\cos z$

33. (a) $f(x, y, z) = xy \sin(1/z)$

    (b) $f(x, y, z) = \dfrac{1}{x^2 + y^2 + z^2 - 1}$

34. (a) $f(x, y, z) = \dfrac{1}{|x| + |y| + |z|}$

    (b) $f(x, y, z) = \dfrac{1}{|xy| + |z|}$

In Exercises 35–42, show that the function has no limit as (x, y) approaches $(0, 0)$ by considering different paths of approach.

35. $f(x, y) = -\dfrac{x}{\sqrt{x^2 + y^2}}$

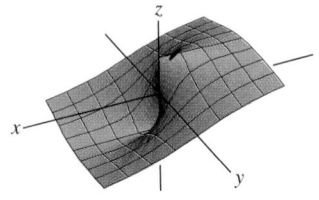

(Generated by Mathematica)

36. $f(x, y) = \dfrac{x^4}{x^4 + y^2}$

(Generated by Mathematica)

37. $f(x, y) = \dfrac{x^4 - y^2}{x^4 + y^2}$

38. $f(x, y) = \dfrac{xy}{|xy|}$

39. $g(x, y) = \dfrac{x - y}{x + y}$

40. $g(x, y) = \dfrac{x + y}{x - y}$

41. $h(x, y) = \dfrac{x^2 + y}{y}$

42. $h(x, y) = \dfrac{x^2}{x^2 - y}$

43. Writing to Learn If $\lim\limits_{(x,y)\to(x_0,y_0)} f(x, y) = L$, must f be defined at (x_0, y_0)? Give reasons for your answer.

44. Writing to Learn If $f(x_0, y_0) = 3$, what can you say about $\lim\limits_{(x,y)\to(x_0,y_0)} f(x, y)$ if f is continuous at (x_0, y_0)? If f is not continuous at (x_0, y_0)? Give reasons for your answers.

Explorations

45. *(Continuation of Example 3)*

(a) Reread Example 3. Then substitute $m = \tan\theta$ into the formula

$$f(x, y)\big|_{y=mx} = \dfrac{2m}{1 + m^2}$$

and simplify the result to show how the value of f varies with the line's angle of inclination.

(b) Use the formula you obtained in (a) to show that the limit of f as $(x, y) \to (0, 0)$ along the line $y = mx$ varies from -1 to 1 depending on the angle of approach.

46. *Removing a Discontinuity in the Plane* Define $f(0, 0)$ in a way that extends

$$f(x, y) = xy\,\dfrac{x^2 - y^2}{x^2 + y^2}$$

to be continuous at the origin. ■

Extending the Ideas

The *Sandwich Theorem* for functions of two variables states that if $g(x, y) \le f(x, y) \le h(x, y)$ for all $(x, y) \ne (x_0, y_0)$ in a disk centered at (x_0, y_0) and if g and h have the same finite limit L as $(x, y) \to (x_0, y_0)$, then $\lim\limits_{(x,y)\to(x_0,y_0)} f(x, y) = L$. In Exercises 47–50, use this result to support your answers to the question.

47. Does knowing that

$$1 - \dfrac{x^2 y^2}{3} < \dfrac{\tan^{-1} xy}{xy} < 1$$

tell you anything about

$$\lim_{(x,y)\to(0,0)} \frac{\tan^{-1} xy}{xy}?$$

Give reasons for your answer.

48. Does knowing that

$$2|xy| - \frac{x^2 y^2}{6} < 4 - 4\cos\sqrt{|xy|} < 2|xy|$$

tell you anything about

$$\lim_{(x,y)\to(0,0)} \frac{4 - 4\cos\sqrt{|xy|}}{|xy|}?$$

Give reasons for your answer.

49. Does knowing that

$$\left| \sin\frac{1}{x} \right| \le 1$$

tell you anything about

$$\lim_{(x,y)\to(0,0)} y \sin\frac{1}{x}?$$

Give reasons for your answer.

50. Does knowing that

$$\left| \cos\frac{1}{y} \right| \le 1$$

tell you anything about

$$\lim_{(x,y)\to(0,0)} x \cos\frac{1}{y}?$$

Give reasons for your answer.

CAS Explorations

51. Explore the graphs of the four functions whose limits you considered in Exercises 47–50. Try to find a view that supports your results in the previous exercises.

13.3 Partial Derivatives

Definitions and Notation • Calculations • Functions of More Than Two Variables • Partial Derivatives and Continuity • Second-Order Partial Derivatives • The Mixed Derivative Theorem • Partial Derivatives of Still Higher Order

Definitions and Notation

When we hold all but one of the independent variables of a function constant and differentiate with respect to that one variable, we get a "partial" derivative. For example, let (x_0, y_0) be some point in the domain of the function $f(x, y)$, and suppose we hold the value of y fixed at y_0. The vertical plane $y = y_0$ will cut the surface $z = f(x, y)$ in the curve $z = f(x, y_0)$ (Figure 13.16). This curve

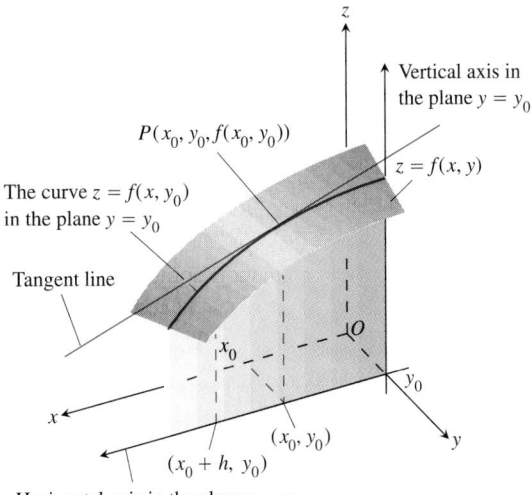

Figure 13.16 The intersection of the plane $y = y_0$ with the surface $z = f(x, y)$, viewed from a point above the first quadrant of the xy-plane.

is the graph of the function $z = f(x, y_0)$ in the plane $y = y_0$. The horizontal coordinate in this plane is x; the vertical coordinate is z.

We define the partial derivative of f with respect to x at the point (x_0, y_0) as the ordinary derivative of $f(x, y_0)$ with respect to x at the point $x = x_0$.

Definition Partial Derivative with Respect to x

The **partial derivative of $f(x, y)$ with respect to x** at the point (x_0, y_0) is

$$\frac{\partial f}{\partial x}\bigg|_{(x_0,y_0)} = \frac{d}{dx} f(x, y_0)\bigg|_{(x_0,y_0)} = \lim_{h \to 0} \frac{f(x_0 + h, y_0) - f(x_0, y_0)}{h},$$

provided the limit exists.

The stylized "∂" (similar to the lowercase Greek letter "δ" used in the limit definition in Section 2.1) is just another kind of "d." It is convenient to have this distinguishable way of extending the Leibniz differential notation into a multivariable context.

The slope of the curve $z = f(x, y_0)$ at the point $P(x_0, y_0, f(x_0, y_0))$ in the plane $y = y_0$ is the value of the partial derivative of f with respect to x at (x_0, y_0). The tangent line to the curve at P is the line in the plane $y = y_0$ that passes through P with this slope. The partial derivative $\partial f/\partial x$ at (x_0, y_0) gives the rate of change of f with respect to x when y is held fixed at the value y_0. Geometrically, this is the rate of change of f in the direction of $\mathbf{i}$ at (x_0, y_0).

The notation that we use for a partial derivative depends on what aspect of the derivative we want to emphasize:

$\frac{\partial f}{\partial x}(x_0, y_0)$ or $f_x(x_0, y_0)$ — Partial derivative of f with respect to "x at (x_0, y_0)" or "f sub x at (x_0, y_0)." Convenient for stressing the point (x_0, y_0).

$\frac{\partial z}{\partial x}\big|_{(x_0,y_0)}$ — "Partial derivative of z with respect to x at (x_0, y_0)." Common in science and engineering when you are dealing with variables and do not mention the function explicitly.

f_x, $\frac{\partial f}{\partial x}$, z_x, or $\frac{\partial z}{\partial x}$ — "Partial derivative of f (or z) with respect to x." Convenient when you are treating the partial derivative as a function in its own right.

The definition of the partial derivative of $f(x, y)$ with respect to y at (x_0, y_0) is similar to the definition of the partial derivative of f with respect to x. We hold x fixed at the value x_0 and take the ordinary derivative of $f(x_0, y)$ with respect to y at y_0.

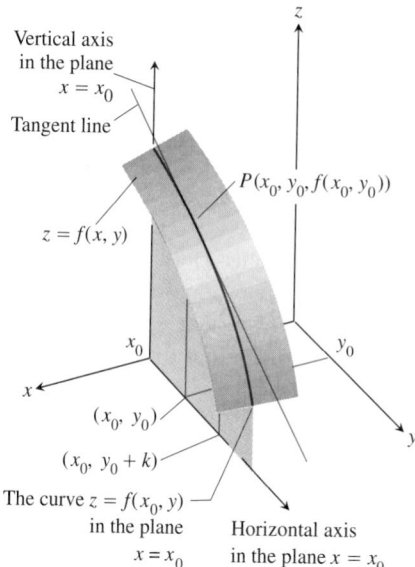

Figure 13.17 The intersection of the plane $x = x_0$ with the surface $z = f(x, y)$, viewed from above the first quadrant of the xy-plane.

Definition Partial Derivative with Respect to y

The **partial derivative of $f(x, y)$ with respect to y** at the point (x_0, y_0) is

$$\frac{\partial f}{\partial y}\bigg|_{(x_0,y_0)} = \frac{d}{dy}f(x_0, y)\bigg|_{(x_0,y_0)} = \lim_{h\to 0}\frac{f(x_0, y_0 + h) - f(x_0, y_0)}{h},$$

provided the limit exists.

The slope of the curve $z = f(x_0, y)$ at the point $P(x_0, y_0, f(x_0, y_0))$ in the plane $x = x_0$ (Figure 13.17) is the value of the partial derivative of f with respect to y at (x_0, y_0). The tangent line to the curve at P is the line in the plane $x = x_0$ that passes through P with this slope. The partial derivative $\partial f/\partial y$ at (x_0, y_0) gives the rate of change of f with respect to y when x is held fixed at the value x_0. Geometrically, this is the rate of change of f in the direction of $\mathbf{j}$ at (x_0, y_0).

The partial derivative with respect to y is denoted the same way as the partial derivative with respect to x:

$$\frac{\partial f}{dy}(x_0, y_0), \qquad f_y(x_0, y_0), \qquad \frac{\partial f}{\partial y}, \qquad f_y.$$

Notice that we now have two tangent lines associated with the surface $z = f(x, y)$ at the point $P(x_0, y_0, f(x_0, y_0))$. (See Figure 13.18.) Is the plane they determine tangent to the surface at P? It would be nice if it were, but we have to learn more about partial derivatives before we can find out.

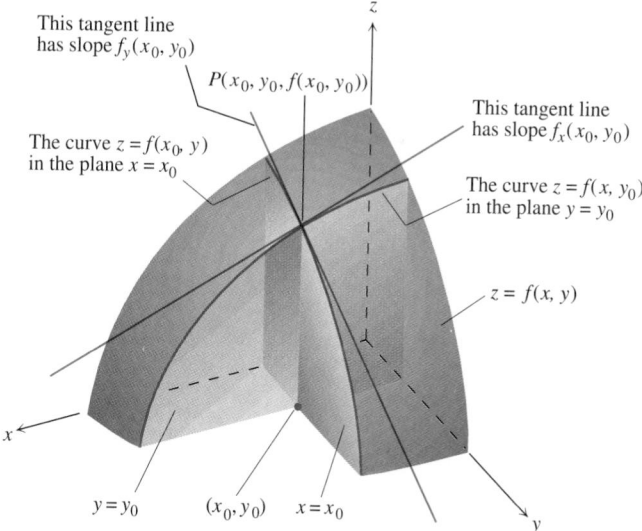

Figure 13.18 Figures 13.14 and 13.15 combined. The tangent lines at the point $P(x_0, y_0, f(x_0, y_0))$ determine a plane that appears (at least in this picture) to be tangent to the surface.

Calculations

The definitions of $\partial f/\partial x$ and $\partial f/\partial y$ give two different ways to differentiate f at a point: with respect to x in the usual way while treating y as constant and with respect to y in the usual way while treating x as constant. Can one partial derivative be predicted from the other? Before we calculate some partial derivatives, Exploration 1 anticipates the answer to that question.

Exploration 1 **An Intersection of Two Hiking Paths**

On a piece of cardboard or stiff paper, draw two straight lines to represent hiking paths that cross in the woods. Hikers *A* and *B* are headed toward the intersection along different paths. Draw and label a point on each path to represent *A* and *B*, and indicate each hiker's direction of travel with an arrow. Notice that you can change the slope of the "terrain" by tilting the cardboard in various directions.

1. Tilt the cardboard so that both hikers will be hiking uphill when they meet.
2. Tilt the cardboard so that both hikers will be hiking downhill when they meet.
3. Tilt the cardboard so that *A* will be hiking uphill and *B* downhill when they meet.
4. Tilt the cardboard so that *A* will be hiking uphill and *B* will be on a level path when they meet.
5. What do you conclude about the likelihood of being able to predict one partial derivative from knowing the other?

Example 1 FINDING PARTIAL DERIVATIVES AT A POINT

If $f(x, y) = x^2 + 3xy + y - 1$, find the values of $\partial f/\partial x$ and $\partial f/\partial y$ at the point $(4, -5)$.

Solution To find $\partial f/\partial x$, we regard y as fixed and differentiate with respect to x:

$$\frac{\partial f}{\partial x} = \frac{\partial}{\partial x}(x^2 + 3xy + y - 1) = 2x + 3y + 0 - 0 = 2x + 3y.$$

The value of $\partial f/\partial x$ at $(4, -5)$ is $2(4) + 3(-5) = -7$.

To find $\partial f/\partial y$, we regard x as fixed and differentiate with respect to y:

$$\frac{\partial f}{\partial y} = \frac{\partial}{\partial y}(x^2 + 3xy + y - 1) = 0 + 3x + 1 - 0 = 3x + 1.$$

The value of $\partial f/\partial y$ at $(4, -5)$ is $3(4) + 1 = 13$.

Partial Derivatives with NDER

Most calculators that find numerical derivatives (see Section 3.2) require you to specify the variable of differentiation. Any other variables will be given their previously stored values. This means that NDER will actually compute partial derivatives. Just store the values of x_0 and y_0 into the respective registers for x and y, and then differentiate with respect to x at x_0 or with respect to y at y_0. Computers and calculators with CAS (computer algebra systems) can compute partial derivatives numerically or symbolically, using almost the same commands as in the single-variable case.

Example 2 FINDING A PARTIAL DERIVATIVE AS A FUNCTION

Find $\partial f/\partial y$ if $f(x, y) = y \sin xy$.

Solution We treat x as a constant and f as a product of y and $\sin xy$:

$$\frac{\partial f}{\partial y} = \frac{\partial}{\partial y}(y \sin xy) = y \cdot \frac{\partial}{\partial y}(\sin xy) + \sin xy \cdot \frac{\partial}{\partial y}(y)$$

$$= y \cos xy \cdot \frac{\partial}{\partial y}(xy) + \sin xy = xy \cos xy + \sin xy.$$

Example 3   PARTIAL DERIVATIVES MAY BE DIFFERENT FUNCTIONS

Find $\partial f/\partial x$ and $\partial f/\partial y$ if

$$f(x, y) = \frac{2y}{y + \cos x}.$$

Solution   We apply the Quotient Rule, first with y held constant:

$$f_x = \frac{\partial}{\partial x}\left(\frac{2y}{y + \cos x}\right) = \frac{(y + \cos x)\left(\dfrac{\partial}{\partial x}\right)(2y) - 2y\left(\dfrac{\partial}{\partial x}\right)(y + \cos x)}{(y + \cos x)^2}$$

$$= \frac{(y + \cos x)(0) - 2y(-\sin x)}{(y + \cos x)^2} = \frac{2y \sin x}{(y + \cos x)^2}.$$

With x held constant, we get

$$f_y = \frac{\partial}{\partial y}\left(\frac{2y}{y + \cos x}\right) = \frac{(y + \cos x)\left(\dfrac{\partial}{\partial y}\right)(2y) - 2y\left(\dfrac{\partial}{\partial y}\right)(y + \cos x)}{(y + \cos x)^2}$$

$$= \frac{(y + \cos x)(2) - 2y(1)}{(y + \cos x)^2} = \frac{2 \cos x}{(y + \cos x)^2}.$$

Implicit differentiation works for partial derivatives the way it works for ordinary derivatives, as the next example illustrates.

Example 4   IMPLICIT PARTIAL DIFFERENTIATION

Find $\partial z/\partial x$ if the equation

$$yz - \ln z = x + y$$

defines z as a function of two independent variables x and y and the partial derivative exists.

Solution   We differentiate both sides of the equation with respect to x, holding y constant and treating z as a differentiable function of x:

$$\frac{\partial}{\partial x}(yz) - \frac{\partial}{\partial x}(\ln z) = \frac{\partial x}{\partial x} + \frac{\partial y}{\partial x}$$

$$y\frac{\partial z}{\partial x} - \frac{1}{z}\frac{\partial z}{\partial x} = 1 + 0 \qquad \text{With } y \text{ constant,}$$
$$\qquad\qquad\qquad\qquad\qquad\qquad \frac{\partial}{\partial x}(yz) = y\frac{\partial z}{\partial x}.$$

$$\left(y - \frac{1}{z}\right)\frac{\partial z}{\partial x} = 1$$

$$\frac{\partial z}{\partial x} = \frac{z}{yz - 1}.$$

Functions of More Than Two Variables

The definitions of the partial derivatives of functions of three or more variables are like the definitions for the functions of two variables. They are ordinary derivatives with respect to one variable, taken while the other independent variables are held constant.

Example 5   A FUNCTION OF THREE VARIABLES

If x, y, and z are independent variables and $f(x, y, z) = x \sin(y + 3z)$, find $\partial f / \partial z$.

Solution

$$\frac{\partial f}{\partial z} = \frac{\partial}{\partial z}\left[x \sin(y + 3z)\right] = x\frac{\partial}{\partial z}\sin(y + 3z)$$

$$= x \cos(y + 3z)\frac{\partial}{\partial z}(y + 3z) = 3x \cos(y + 3z)$$

Example 6   ELECTRICAL RESISTORS IN PARALLEL

If resistors of R_1, R_2, and R_3 ohms are connected in parallel to make an R-ohm resistor, the value of R can be found from the equation

$$\frac{1}{R} = \frac{1}{R_1} + \frac{1}{R_2} + \frac{1}{R_3} \qquad \text{(Figure 13.19)}.$$

Find the value of $\partial R / \partial R_2$ when $R_1 = 30$, $R_2 = 45$, and $R_3 = 90$ ohms.

Solution   To find $\partial R / \partial R_2$, we treat R_1 and R_3 as constants and differentiate with respect to R_2:

$$\frac{\partial}{\partial R_2}\left(\frac{1}{R}\right) = \frac{\partial}{\partial R_2}\left(\frac{1}{R_1} + \frac{1}{R_2} + \frac{1}{R_3}\right)$$

$$-\frac{1}{R^2}\frac{\partial R}{\partial R_2} = 0 - \frac{1}{R_2{}^2} + 0$$

$$\frac{\partial R}{\partial R_2} = \frac{R^2}{R_2{}^2} = \left(\frac{R}{R_2}\right)^2.$$

When $R_1 = 30$, $R_2 = 45$, and $R_3 = 90$,

$$\frac{1}{R} = \frac{1}{30} + \frac{1}{45} + \frac{1}{90} = \frac{1}{15},$$

so $R = 15$ and

$$\frac{\partial R}{\partial R_2} = \left(\frac{15}{45}\right)^2 = \frac{1}{9}.$$

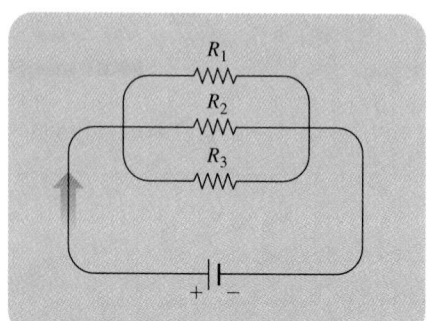

Figure 13.19 Resistors arranged this way are said to be connected in parallel (Example 6). Each resistor lets a portion of the current through. Their combined resistance R is calculated with the formula $1/R = 1/R_1 + 1/R_2 + 1/R_3$.

Partial Derivatives and Continuity

Since the existence of the derivative at a point always implies continuity in the single-variable case, you might be surprised to learn that a function $f(x, y)$ can have partial derivatives with respect to both x and y at a point and yet *fail* to be continuous there. The following example illustrates what can happen.

Example 7 $\partial f / \partial x$ AND $\partial f / \partial y$ EXIST BUT f IS DISCONTINUOUS

Let

$$f(x, y) = \begin{cases} 0 & \text{if } xy \neq 0, \\ 1 & \text{if } xy = 0 \end{cases} \qquad \text{(Figure 13.20)}.$$

(a) Find the limit of f as (x, y) approaches $(0, 0)$ along the line $y = x$.

(b) Prove that f is not continuous at the origin.

(c) Show that both partial derivatives $\partial f / \partial x$ and $\partial f / \partial y$ exist at the origin.

$$z = \begin{cases} 0, & xy \neq 0 \\ 1, & xy = 0 \end{cases}$$

Figure 13.20 The graph of

$$f(x, y) = \begin{cases} 0 & \text{if } xy \neq 0, \\ 1 & \text{if } xy = 0 \end{cases}$$

consists of the lines L_1 and L_2 in the plane $z = 1$ and the four open quadrants of the xy-plane one unit below. The function has partial derivatives at the origin but is not continuous there.

Solution

(a) Since $f(x, y)$ is constantly zero along the line $y = x$ except at the origin, we have

$$\lim_{(x,y) \to (0,0)} f(x, y) \bigg|_{y=x} = \lim_{(x,y) \to (0,0)} \bigg|_{y=x} 0 = 0.$$

(b) Since $f(0, 0) = 1$, the limit in (a) proves f cannot be continuous at $(0, 0)$.

(c) To find $\partial f/\partial x$ at $(0, 0)$, we hold y fixed at $y = 0$. Then $f(x, y) = 1$ for all x, and the graph of f is the line L_1 in Figure 13.20. The slope of this line at any x is $\partial f/\partial x = 0$. In particular, $\partial f/\partial x = 0$ at $(0, 0)$. Similarly, $\partial f/\partial y$ is the slope of line L_2 at any y, so $\partial f/\partial y = 0$ at $(0, 0)$.

Example 7 notwithstanding, it is still true in higher dimensions that *differentiability* at a point implies continuity. What Example 7 suggests is that we need a stronger requirement for differentiability in higher dimensions than the mere existence of the partial derivatives. We define differentiability for functions of two variables in the next section, at which time we revisit the connection to continuity.

Second-order Partial Derivatives

When we differentiate a function $f(x, y)$ twice, we produce its second-order derivatives. These derivatives are usually denoted by

$\dfrac{\partial^2 f}{\partial x^2}$	"d squared f d x squared"	or	f_{xx}	"f sub x x"
$\dfrac{\partial^2 f}{\partial y^2}$	"d squared f d y squared"		f_{yy}	"f sub y y"
$\dfrac{\partial^2 f}{\partial x \partial y}$	"d squared f d x d y"		f_{yx}	"f sub y x"
$\dfrac{\partial^2 f}{\partial y \partial x}$	"d squared f d y d x"		f_{xy}	"f sub x y"

The defining equations are

$$\frac{\partial^2 f}{\partial x^2} = \frac{\partial}{\partial x}\left(\frac{\partial f}{\partial x}\right) \quad \text{or} \quad f_{xx} = (f_x)_x,$$

$$\frac{\partial^2 f}{\partial x \partial y} = \frac{\partial}{\partial x}\left(\frac{\partial f}{\partial y}\right) \quad \text{or} \quad f_{yx} = (f_y)_x,$$

and so on. Notice the order in which the derivatives are taken: both $\partial^2 f/\partial x \partial y$ and f_{yx} mean to differentiate first with respect to y, then with respect to x.

Example 8 FINDING SECOND-ORDER PARTIAL DERIVATIVES

If $f(x, y) = x \cos y + ye^x$, find $\partial^2 f/\partial x^2$, $\partial^2 f/\partial y \partial x$, $\partial^2 f/\partial y^2$, and $\partial^2 f/\partial x \partial y$.

Solution

$$\frac{\partial f}{\partial x} = \frac{\partial}{\partial x}(x \cos y + ye^x)$$
$$= \cos y + ye^x.$$

So

$$\frac{\partial^2 f}{\partial x^2} = \frac{\partial}{\partial x}\left(\frac{\partial f}{\partial x}\right) = ye^x$$

$$\frac{\partial^2 f}{\partial y \partial x} = \frac{\partial}{\partial y}\left(\frac{\partial f}{\partial x}\right) = -\sin y + e^x.$$

$$\frac{\partial f}{\partial y} = \frac{\partial}{\partial y}(x \cos y + ye^x)$$
$$= -x \sin y + e^x.$$

So

$$\frac{\partial^2 f}{\partial y^2} = \frac{\partial}{\partial y}\left(\frac{\partial f}{\partial y}\right) = -x \cos y$$

$$\frac{\partial^2 f}{\partial x \partial y} = \frac{\partial}{\partial x}\left(\frac{\partial f}{\partial y}\right) = -\sin y + e^x$$

The Mixed Derivative Theorem

You probably noticed in Example 8 that the "mixed" second-order partial derivatives

$$\frac{\partial^2 f}{\partial y \partial x} \quad \text{and} \quad \frac{\partial^2 f}{\partial x \partial y}$$

were equal. This was no mere coincidence; they have to be equal whenever all five of the functions $f, f_x, f_y, f_{xy},$ and f_{yx} are continuous. That is the assertion of the following theorem, also known as *Euler's Theorem.*

Theorem 2 **The Mixed Derivative Theorem**

If $f(x, y)$ and its partial derivatives $f_x, f_y, f_{xy},$ and f_{yx} are defined throughout an open region containing a point (a, b) and are all continuous at (a, b), then

$$f_{xy}(a, b) = f_{yx}(a, b).$$

Theorem 2 says that to calculate a mixed second-order partial derivative we may normally differentiate in either order. This can work to our advantage, as shown in the following example.

Example 9 CHOOSING THE ORDER OF DIFFERENTIATION

Find $\partial^2 w / \partial x \partial y$ if $w = xy + e^y/(y^2 + 1)$.

Solution The symbol $\partial^2 w / \partial x \partial y$ tells us to differentiate first with respect to y and then respect to x. If we postpone the differentiation with respect to y and differentiate first with respect to x, however, we get the answer more quickly. In two steps,

$$\frac{\partial w}{\partial x} = y \quad \text{and} \quad \frac{\partial^2 w}{\partial y \partial x} = 1.$$

We are in for significantly more work if we differentiate first with respect to y. (Just try it.)

Partial Derivatives of Still Higher Order

Although we deal mostly with first- and second-order partial derivatives, because they appear the most frequently in applications, there is no theoretical limit to how many times we can differentiate a function as long as the derivatives involved exist. Thus, we get third- and fourth-order derivatives denoted by symbols like

$$\frac{\partial^3 f}{\partial x \partial y^2} = f_{yyx},$$

$$\frac{\partial^4 f}{\partial x^2 \partial y^2} = f_{yyxx},$$

and so on. As with second-order derivatives, the order of differentiation is immaterial as long as the function and its derivatives through the order in question are all continuous.

Quick Review 13.3

In Exercises 1–10, y is given as a function of a *single* variable x. The number k is a constant. Find dy/dx and d^2y/dx^2.

1. $y = kx^2 + 5x$

2. $y = \sin kx$

3. $y = 20e^{kx}$

4. $y = (x + k)e^x$

5. $y = \tan^{-1}(x\sqrt{k})$

6. $y = \ln|x - k| + 3x$

7. $y = k^x$

8. $y = e^k \cos(kx)$

9. $y = \dfrac{x}{x - k}$

10. $y = \dfrac{k}{k - x}$

Section 13.3 Exercises

In Exercises 1–18, find $\partial f/\partial x$ and $\partial f/\partial y$.

1. $f(x, y) = 2x$

2. $f(x, y) = 2x - 3y - 4$

3. $f(x, y) = x(y - 1)$

4. $f(x, y) = x^2 + y^2$

5. $f(x, y) = (x^2 - 1)(y + 2)$

6. $f(x, y) = 5xy - 7x^2 - y^2 + 3x - 6y + 2$

7. $f(x, y) = \dfrac{1}{x + y}$

8. $f(x, y) = (2x - 3y)^3$

9. $f(x, y) = \sqrt{x^2 + y^2}$

10. $f(x, y) = \dfrac{x}{x^2 + y^2}$

11. $f(x, y) = \dfrac{x + y}{xy - 1}$

12. $f(x, y) = \sin(x + y)$

13. $f(x, y) = e^x \sin(y + 1)$

14. $f(x, y) = \tan^{-1}\left(\dfrac{y}{x}\right)$

15. $f(x, y) = e^{(x+y+1)}$

16. $f(x, y) = \ln(x + y)$

17. $f(x, y) = \log_y x$

18. $f(x, y) = \displaystyle\int_x^y g(t)\, dt$ (g continuous for all t)

In Exercises 19–24, find f_x, f_y, and f_z.

19. $f(x, y, z) = xy + yz + xz$

20. $f(x, y, z) = (x^2 + y^2 + z^2)^{-1/2}$

21. $f(x, y, z) = 1 + y^2 + 2z^2$

22. $f(x, y, z) = \tan(x + y + z)$

23. $f(x, y, z) = \ln(x + 2y + 3z)$

24. $f(x, y, z) = \sin(x + yz)$

In Exercises 25–30, find the partial derivative of the function with respect to each variable.

25. $f(t, \alpha) = \cos(2\pi t - \alpha)$

26. $g(u, v) = v^2 e^{(2u/v)}$

27. $h(\rho, \phi, \theta) = \rho \sin \phi \cos \theta$

28. $g(r, \theta, z) = r(1 - \cos \theta) - z$

29. *Work Done by Heart* (Section 4.5, Exercise 47)

$$W(P, V, \delta, v, g) = PV + \dfrac{V\delta v^2}{2g}$$

30. *Wilson Lot Size Formula* (Section 4.4, Exercise 50)

$$A(c, h, k, m, q) = \dfrac{km}{q} + cm + \dfrac{hq}{2}$$

In Exercises 31–36, find the second-order partial derivatives f_{xx}, f_{yy}, f_{xy}, and f_{yx} for the function.

31. $f(x, y) = x + y + xy$

32. $f(x, y) = \sin xy$

33. $f(x, y) = x^2 y + \cos y + y \sin x$

34. $f(x, y) = xe^y + y + 1$

35. $f(x, y) = \ln(x + y)$

36. $f(x, y) = \tan^{-1}(y/x)$

In Exercises 37–40, verify that $w_{xy} = w_{yx}$.

37. $w = \ln(2x + 3y)$

38. $w = e^x + x \ln y + y \ln x$

39. $w = xy^2 + x^2 y^3 + x^3 y^4$

40. $w = x \sin y + y \sin x + xy$

Explorations (Order of Differentiation)

Exercises 41 and 42 should be done without writing anything down.

41. Which order of differentiation will calculate f_{xy} faster: x first or y first?

(a) $f(x, y) = x \sin y + e^y$

(b) $f(x, y) = 1/x$

(c) $f(x, y) = y + (x/y)$

(d) $f(x, y) = y + x^2 y + 4y^3 - \ln(y^2 + 1)$

(e) $f(x, y) = x^2 + 5xy + \sin x + 7e^x$

(f) $f(x, y) = x \ln xy$

42. The fifth-order partial derivative $\partial^5 f/\partial x^2 \partial y^3$ is zero for each of the following functions. To show this as quickly as possible, which variable would you differentiate with respect to first: x or y?

(a) $f(x, y) = y^2 x^4 e^x + 2$

(b) $f(x, y) = y^2 + y(\sin x - x^4)$

(c) $f(x, y) = x^2 + 5xy + \sin x + 7e^x$

(d) $f(x, y) = xe^{y^2/2}$

43. Assume that the equation

$$x^3z + z^3x - 2yz = 0$$

defines z as a differentiable function of the two independent variables x and y. Find the value of $\partial z/\partial x$ at the point $(x, y, z) = (1, 1, 1)$.

44. Assume that the equation

$$xz + y \ln x - 9/z = 0$$

defines x as a differentiable function of the two independent variables y and z. Find the value of $\partial x/\partial z$ at the point $(x, y, z) = (1, -1, 3)$.

45. Writing to Learn Let $w = f(x, y, z)$ be a function of three independent variables, and write the formal definition of the partial derivative $\partial f/\partial z$ at (x_0, y_0, z_0). Use this definition to find $\partial f/\partial z$ at $(1, 2, 3)$ for $f(x, y, z) = x^2yz^2$.

46. Writing to Learn Let $w = f(x, y, z)$ be a function of three independent variables, and write the formal definition of the partial derivative $\partial f/\partial y$ at (x_0, y_0, z_0). Use this definition to find $\partial f/\partial y$ at $(-1, 0, 3)$ for $f(x, y, z) = -2xy^2 + yz^2$.

Extending the Ideas

47. *Laplace Equations* The **three-dimensional Laplace equation**

$$\frac{\partial^2 f}{\partial x^2} + \frac{\partial^2 f}{\partial y^2} + \frac{\partial^2 f}{\partial z^2} = 0$$

is satisfied by steady-state temperature distributions $T = f(x, y, z)$ in space, by gravitational potentials, and by electrostatic potentials. The **two-dimensional Laplace equation**

$$\frac{\partial^2 f}{\partial x^2} + \frac{\partial^2 f}{\partial y^2} = 0,$$

obtained by dropping the $\partial^2 f/\partial z^2$ term, describes potentials and steady-state temperature distributions in a plane (Figure 13.21). Show that each of the following functions satisfies a Laplace equation.

(a) $f(x, y, z) = x^2 + y^2 - 2z^2$

(b) $f(x, y, z) = 2z^3 - 3(x^2 + y^2)z$

(c) $f(x, y) = e^{-2y} \cos 2x$

(d) $f(x, y) = \ln \sqrt{x^2 + y^2}$

(e) $f(x, y, z) = (x^2 + y^2 + z^2)^{-1/2}$

(f) $f(x, y, z) = e^{3x+4y} \cos 5z$

48. *The One-Dimensional Wave Equation* If we stand on an ocean shore and take a snapshot of the waves, the picture shows a regular pattern of peaks and valleys in an instant of time (Figure 13.22). Show that the following functions are all solutions of the wave equation.

(a) $w = \sin(x + ct)$ (b) $w = \cos(2x + 2ct)$

(c) $w = \sin(x + cy) + \cos(2x + 2ct)$

(d) $w = \ln(2x + 2ct)$ (e) $w = \tan(2x - 2ct)$

(f) $w = 5 \cos(3x + 3ct) + e^{x+ct}$

(a)

$$\frac{\partial^2 f}{\partial x^2} + \frac{\partial^2 f}{\partial y^2} = 0$$

(b)

$$\frac{\partial^2 f}{\partial x^2} + \frac{\partial^2 f}{\partial y^2} + \frac{\partial^2 f}{\partial z^2} = 0$$

Boundary temperatures controlled

Figure 13.21 Steady-state temperature distributions in planes and solids satisfy Laplace equations. The plane (a) may be treated as a thin slice of the solid (b) perpendicular to the z-axis.

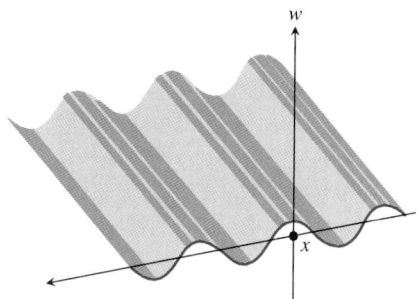

Figure 13.22 Waves in water at an instant in time. As time passes,

$$\frac{\partial^2 w}{\partial t^2} = c^2 \frac{\partial^2 w}{\partial x^2}.$$

We see periodic vertical motion in space with respect to distance. If we stand in the water, we can feel the rise and fall of the water as the waves go by. We see periodic vertical motion in time. In physics, this beautiful symmetry is expressed by the **one-dimensional wave equation**

$$\frac{\partial^2 w}{\partial t^2} = c^2 \frac{\partial^2 w}{\partial x^2},$$

where w is the wave height, x is the distance variable, and c is the velocity with which the waves are propagated.

In our example, x is the distance across the ocean's surface, but in other applications, x might be the distance along a vibrating string, distance through the air (sound waves), or distance through space (light waves). The number c varies with the medium and type of wave.

13.4 Differentiability, Linearization, and Differentials

Differentiability • Linearization of a Function of Two Variables •
Accuracy of the Standard Linear Approximation • Predicting
Change with Differentials • Absolute, Relative, and Percentage
Change • Functions of More Than Two Variables

Differentiability

Surprising as it may seem, the starting point for differentiability is not
Fermat's difference quotient, but rather the idea of increment. As Example 7 in
Section 13.3 demonstrates, differentiability in higher dimensions is complicated because a function $z = f(x, y)$ can actually be *discontinuous* at a point
where its first partial derivatives exist. This paradox arises because limits of
functions of more than one independent variable cannot be considered one
variable at a time as "partial limits." What we need, informally, is a criterion
that will guarantee infintesimal smallness in all the variables at once. We find
it in the Increment Theorem from advanced calculus.

**Theorem 3 The Increment Theorem for Functions
of Two Variables**

Suppose the first partial derivatives of $f(x, y)$ are defined throughout
an open region R containing the point (x_0, y_0) and f_x and f_y are continuous at (x_0, y_0). Then the change

$$\Delta z = f(x_0 + \Delta x, y_0 + \Delta y) - f(x_0, y_0)$$

in the value of f that results from moving from (x_0, y_0) to another point
$(x_0 + \Delta x, y_0 + \Delta y)$ in R satisfies an equation of the form

$$\Delta z = f_x(x_0, y_0)\, \Delta x + f_y(x_0, y_0)\, \Delta y + \varepsilon_1\, \Delta x + \varepsilon_2\, \Delta y,$$

in which $\varepsilon_1, \varepsilon_2 \to 0$ as $\Delta x, \Delta y \to 0$.

The condition on Δz at the end of the Increment Theorem (which generalizes to n dimensions with n epsilons) is exactly the condition that we will
require for differentiability. Even without seeing the proof of the Increment
Theorem, you can probably appreciate how the combined effect of all those
factors approaching zero will guarantee impressive smallness in all the independent variables at once.

Definition Differentiability of a Function of Two Variables

A function $z = f(x, y)$ is **differentiable at** (x_0, y_0) if $f_x(x_0, y_0)$ and
$f_y(x_0, y_0)$ exist and Δz satisfies an equation of the form

$$\Delta z = f_x(x_0, y_0)\, \Delta x + f_y(x_0, y_0)\, \Delta y + \varepsilon_1\, \Delta x + \varepsilon_2\, \Delta y,$$

in which $\varepsilon_1, \varepsilon_2 \to 0$ as $\Delta x, \Delta y \to 0$. We call f **differentiable** if it is
differentiable at every point in its domain.

In light of this definition, we have the immediate corollary of Theorem 3 that a function is differentiable if its first partial derivatives are *continuous*.

> **Corollary of Theorem 3** Continuity of Partial Derivatives Implies Differentiability
>
> If the partial derivatives f_x and f_y of a function $f(x, y)$ are continuous throughout an open region R, then f is differentiable at every point of R.

As we can see from Theorems 3 and 4, a function $f(x, y)$ must be continuous at a point (x_0, y_0) if its partial derivatives f_x and f_y are continuous throughout an open region containing (x_0, y_0). But remember that it is still possible for a function of two variables to be discontinuous at a point where its first partial derivatives exist. We saw this in Section 13.3, Example 7. Existence alone is not enough.

If $z = f(x, y)$ is differentiable, then the definition of differentiability assures that $\Delta z = f(x_0 + \Delta x, y_0 + \Delta y) - f(x_0, y_0)$ approaches 0 as Δx and Δy approach 0. This tells us that a function of two variables is continuous at every point where it is differentiable.

> **Theorem 4** Differentiability Implies Continuity
>
> If a function $f(x, y)$ is differentiable at (x_0, y_0), then f is continuous at (x_0, y_0).

Linearization of a Function of Two Variables

Functions of two variables can be complicated, but sometimes we can replace them with simpler ones that give the accuracy required for specific applications without being so hard to work with. We do this in a way that is similar to the way we find linear replacements for functions of a single variable (Section 4.5).

Suppose the function we wish to replace is $z = f(x, y)$ and that we want the replacement to be effective near a point (x_0, y_0) at which we know the values of f, f_x, and f_y and at which f is differentiable (Figure 13.23).

Since f is differentiable,

$$\Delta z = f_x(x_0, y_0)\,\Delta x + f_y(x_0, y_0)\,\Delta y + \varepsilon_1\,\Delta x + \varepsilon_2\,\Delta y,$$

where $\varepsilon_1, \varepsilon_2 \to 0$ as $\Delta x, \Delta y \to 0$.

Rewriting Δz, Δx, and Δy, we have

$$f(x, y) - f(x_0, y_0) = f_x(x_0, y_0)(x - x_0) + f_y(x_0, y_0)(y - y_0) + \varepsilon_1\,\Delta x + \varepsilon_2\,\Delta y.$$

If the increments Δx and Δy are small, the products $\varepsilon_1\,\Delta x$ and $\varepsilon_2\,\Delta y$ will eventually be smaller still and we will have

$$f(x, y) \approx \underbrace{f(x_0, y_0) + f_x(x_0, y_0)(x - x_0) + f_y(x_0, y_0)(y - y_0)}_{L\,(x,y)}.$$

In other words, as long as Δx and Δy are small, f will have approximately the same value as the linear function L. If f is hard to use and if our work can tolerate the error involved, we can safely replace f by L.

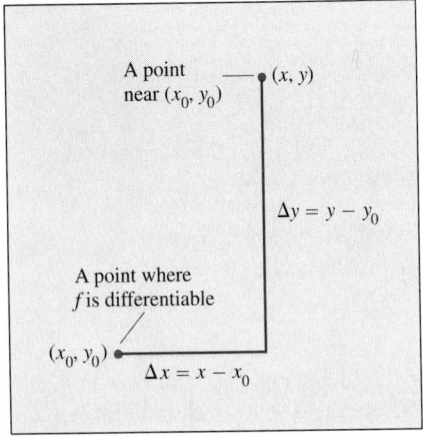

Figure 13.23 If f is differentiable at (x_0, y_0), then the value of f at any point (x, y) nearby is approximately
$f(x_0, y_0) + f_x(x_0, y_0)\,\Delta x + f_y(x_0, y_0)\,\Delta y.$

> **Definitions** Linearization, Standard Linear Approximation
>
> The **linearization** of a function $f(x, y)$ at a point (x_0, y_0) where f is differentiable is the function
>
> $$L(x, y) = f(x_0, y_0) + f_x(x_0, y_0)(x - x_0) + f_y(x_0, y_0)(y - y_0).$$
>
> The approximation
>
> $$f(x, y) \approx L(x, y)$$
>
> is the **standard linear approximation** of f at (x_0, y_0).

In Section 13.6, we see that the plane $z = L(x, y)$ is tangent to the surface $z = f(x, y)$ at the point (x_0, y_0). Thus, the linearization of a function of two variables is a tangent-*plane* approximation in the same way that the linearization of a function of a single variable is a tangent-*line* approximation.

Example 1 FINDING A LINEARIZATION

Find the linearization of $f(x, y) = x^2 - xy + \dfrac{1}{2}y^2 + 3$ at the point $(3, 2)$.

Solution   We first evaluate f, f_x, and f_y at the point $(3, 2)$.

$$f(3, 2) = \left(x^2 - xy + \frac{1}{2}y^2 + 3 \right)\bigg|_{(3,2)} = 8,$$

$$f_x(3, 2) = \frac{\partial}{\partial x}\left(x^2 - xy + \frac{1}{2}y^2 + 3 \right)\bigg|_{(3,2)} = (2x - y)|_{(3,2)} = 4,$$

$$f_y(3, 2) = \frac{\partial}{\partial y}\left(x^2 - xy + \frac{1}{2}y^2 + 3 \right)\bigg|_{(3,2)} = (-x + y)|_{(3,2)} = -1,$$

getting

$$L(x, y) = f(x_0, y_0) + f_x(x_0, y_0)(x - x_0) + f_y(x_0, y_0)(y - y_0)$$
$$= 8 + (4)(x - 3) + (-1)(y - 2) = 4x - y - 2.$$

Accuracy of the Standard Linear Approximation

Suppose $L(x, y)$ is the linearization of a differentiable function $f(x, y)$ at (x_0, y_0) and we use L to approximate f at points (x, y) close to (x_0, y_0). How accurate can we expect the approximation to be? As you might expect, the closeness of the approximation depends on three things:

1. the closeness of x to x_0,

2. the closeness of y to y_0, and

3. the "curviness" of f near (x_0, y_0), as measured by the magnitude of the second partial derivatives.

In fact, if we can find a common upper bound M for $|f_{xx}|$, $|f_{yy}|$, and $|f_{xy}|$ on a rectangle R containing (x_0, y_0) (Figure 13.24), then we can bound the error throughout R by using a simple formula. The formula is derived from Taylor's Theorem (Section 9.3), but we will omit the derivation here.

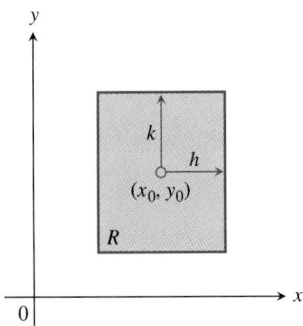

Figure 13.24 The rectangular region R: $|x - x_0| \le h$, $|y - y_0| \le k$ in the xy-plane. On this kind of region, we can find a useful error bound for the standard linear approximation.

The Error in the Standard Linear Approximation

If f has continuous first and second partial derivatives throughout an open set containing a rectangle R centered at (x_0, y_0) and if M is an upper bound for the values of $|f_{xx}|$, $|f_{yy}|$, and $|f_{xy}|$ on R, then the error $E(x, y)$ incurred in replacing $f(x, y)$ on R by its linearization

$$L(x, y) = f(x_0, y_0) + f_x(x_0, y_0)(x - x_0) + f_y(x_0, y_0)(y - y_0)$$

satisfies the inequality

$$|E(x, y)| \le \frac{1}{2} M(|x - x_0| + |y - y_0|)^2.$$

When we need to make $|E(x, y)|$ small for a given M, we just make $|x - x_0|$ and $|y - y_0|$ small.

Example 2 BOUNDING THE ERROR IN EXAMPLE 1

In Example 1, we found the linearization of $f(x, y) = x^2 - xy + \frac{1}{2}y^2 + 3$ at the point $(3, 2)$ to be $L(x, y) = 4x - y - 2$. Find an upper bound for the error in the approximation $f(x, y) \approx L(x, y)$ over the rectangle

$$R: |x - 3| \le 0.1, \qquad |y - 2| \le 0.1.$$

Express the upper bound as a percentage of $f(3, 2)$, the value of f at the center of the rectangle.

Solution We use the inequality $|E(x, y)| \le \frac{1}{2} M(|x - x_0| + |y - y_0|)^2$.

To find a suitable value for M, we calculate f_{xx}, f_{yy}, and f_{xy}, finding, after some routine differentiation, that all three derivatives are constant. In terms of magnitude,

$$|f_{xx}| = |2| = 2, \qquad |f_{xy}| = |-1| = 1, \qquad |f_{yy}| = |1| = 1.$$

The largest of these is 2, so we may safely take M to be 2. With $(x_0, y_0) = (3, 2)$, we know that, throughout R,

$$|E(x, y)| \le \frac{1}{2}(2)(|x - 3| + |y - 2|)^2 = (|x - 3| + |y - 2|)^2.$$

Finally, since $|x - 3| \le 0.1$ and $|y - 2| \le 0.1$ on R, we have

$$|E(x, y)| \le (0.1 + 0.1)^2 = 0.04.$$

As a percentage of $f(3, 2) = 8$, the error is no greater than $0.04/8 = 0.5\%$. As long as (x, y) stays in R, the approximation $f(x, y) \approx L(x, y)$ will be in error by no more than 0.04, which is one-half of one percent of the value of f at the center of R.

Predicting Change with Differentials

Suppose we know the value of a differentiable function $f(x, y)$ and its first partial derivatives at a point (x_0, y_0) and we want to predict how much the value of

f will change if we move to a point $(x_0 + \Delta x, y_0 + \Delta y)$ nearby. If Δx and Δy are small, f and its linearization at (x_0, y_0) will change by nearly the same amount, so the change in L will give a practical estimate of the change in f. (We used the differential in the same way in Section 4.5.)

The change in f is

$$\Delta f = f(x_0 + \Delta x, y_0 + \Delta y) - f(x_0, y_0).$$

This formula is usually as hard to work with as the formula for f. On the other hand, the change in L is

$$\Delta L = L(x_0 + \Delta x, y_0 + \Delta y) - L(x_0, y_0)$$

$$= f(x_0, y_0) + f_x(x_0, y_0)\,\Delta x + f_y(x_0, y_0)\,\Delta y - L(x_0, y_0) \quad \text{Definition of linearization}$$

$$= f_x(x_0, y_0)\,\Delta x + f_y(x_0, y_0)\,\Delta y. \quad \begin{array}{l}\text{Since}\\ L(x_0, y_0) = f(x_0, y_0)\end{array}$$

This is just a known constant times Δx plus a known constant times Δy. Moving to differential notation, we use the more suggestive notation df for ΔL and write

$$df = f_x(x_0, y_0)\,dx + f_y(x_0, y_0)\,dy,$$

in which df denotes the change in the linearization that results from the changes dx and dy in x and y. As usual, we call dx and dy differentials of x and y. We call df the corresponding *total* differential of f.

Definition **Total Differential**

If we move from (x_0, y_0) to a point $(x_0 + dx, y_0 + dy)$ nearby, the resulting change in the linearization of f is

$$df = f_x(x_0, y_0)\,dx + f_y(x_0, y_0)\,dy.$$

This change in the linearization of f is called the **total differential of f.**

Exploration 1 **Sensitivity to Change**

Your company manufactures right circular cylindrical molasses storage tanks that are 25 ft high with a radius of 5 ft. How sensitive is the volume of such a tank to small variations in height and radius (Figure 13.25)?

1. Express the volume $V(r, h)$ of the tank as a function of radius r and height h.

2. Find dV at $(5, 25)$ in terms of dr and dh.

3. As a quality control engineer concerned with being sure the tanks have the correct volume, would you be more concerned with variations in their radii or variations in their heights? Explain.

4. Suppose the values of r and h were reversed and your company were producing shorter, wider cylinders of radius 25 and height 5. Would this affect your answer to part 3? Why?

5. Is there a ratio of height to radius that would make the volume of the tank equally sensitive to changes in either variable? If so, what is it?

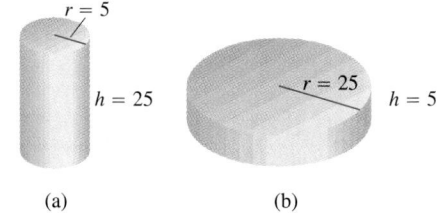

(a)   (b)

Figure 13.25 The volume of cylinder (a) is more sensitive to a small change in r than it is to an equally small change in h. The volume of cylinder (b) is more sensitive to small changes in h than it is to small changes in r. (Exploration 1)

Solution   We write the given conditions in terms of relative errors for the sake of computation:

$$\left|\frac{dr}{r}\right| \le 0.02 \qquad \text{and} \qquad \left|\frac{dh}{h}\right| \le 0.005.$$

Since

$$\frac{dV}{V} = \frac{2\pi rh\, dr + \pi r^2\, dh}{\pi r^2 h} = \frac{2dr}{r} + \frac{dh}{h},$$

we have

$$\left|\frac{dV}{V}\right| = \left|2\frac{dr}{r} + \frac{dh}{h}\right|$$

$$\le \left|2\frac{dr}{r}\right| + \left|\frac{dh}{h}\right| \qquad \text{Property of asbolute value}$$

$$\le 2(0.02) + 0.005 = 0.045.$$

We estimate the error in the volume calculation to be at most 4.5%.

How accurately do we have to measure r and h to have a reasonable chance of calculating $V = \pi r^2 h$ with an error of, say, less than 2%? Questions like this can be difficult because there is usually no single right answer. Since

$$\frac{dV}{V} = 2\frac{dr}{r} + \frac{dh}{h},$$

we see that dV/V is controlled by a combination of dr/r and dh/h. If we can measure h with great accuracy, we might come out all right even if we are sloppy about measuring r. On the other hand, our measurement of h might have so large a dh that the resulting dV/V would be too crude an estimate of $\Delta V/V$ to be useful even if dr were zero.

What we do in such cases is look for a reasonable square about the measured values (r_0, h_0) in which V will not vary by more than the allowed amount from $V_0 = \pi r_0^2 h_0$.

Example 5   CONTROLLING THE ERROR

Find a reasonable square about the point $(r_0, h_0) = (5, 12)$ in which the value of $V = \pi r^2 h$ will not vary from its value at $(5, 12)$ by more than ± 0.1.

Solution   We approximate the variation ΔV by the differential

$$dV = 2\pi r_0 h_0\, dr + \pi r_0^2\, dh = 2\pi(5)(12)\, dr + \pi(5)^2\, dh$$
$$= 120\pi\, dr + 25\pi\, dh.$$

Since the region to which we are restricting our attention is a square (Figure 13.26), we may set $dh = dr$ to get

$$dV = 120\pi\, dr + 25\pi\, dr = 145\pi\, dr.$$

We wish to choose dr small enough to ensure that $|dV| \le 0.1$. Writing dV in terms of dr, we have

$$|dV| = |145\pi\, dr| \le 0.1.$$

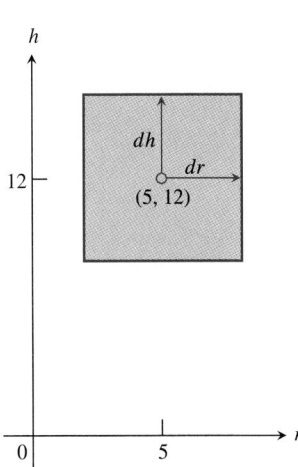

Figure 13.26 A small square about the point (5, 12) in the *rh*-plane. (Example 5)

Solving, we find a corresponding upper bound for *dr*:

$$|dr| \leq \frac{0.1}{145\pi} \approx 2.1 \times 10^{-4}$$
Rounding down to make sure that *dr* will not accidentally be too big

With $dh = dr$, then, the square we want is described by the inequalities

$$|r - 5| \leq 2.1 \times 10^{-4}, \qquad |h - 12| \leq 2.1 \times 10^{-4}.$$

As long as (r, h) stays in this square, we may expect $|dV|$ to be less than or equal to 0.1, and we may expect $|\Delta V|$ to be approximately the same size.

Functions of More Than Two Variables

Analogous results hold for differentiable functions of more than two variables; we summarize a few of them here.

1. The **linearization** of $f(x, y, z)$ at a point $P_0(x_0, y_0, z_0)$ is
$$L(x, y, z) = f(P_0) + f_x(P_0)(x - x_0) + f_y(P_0)(y - y_0) + f_z(P_0)(z - z_0).$$

2. Suppose R is a closed rectangular solid centered at P_0 and lying in an open region on which the second partial derivatives of f are continuous. Also suppose $|f_{xx}|, |f_{yy}|, |f_{zz}|, |f_{xy}|, |f_{xz}|$, and $|f_{yz}|$ are all less than or equal to M throughout R. Then the **error** $E(x, y, z) = f(x, y, z) - L(x, y, z)$ in the approximation of f by L is bounded throughout R by the inequality

$$|E| \leq \frac{1}{2} M(|x - x_0| + |y - y_0| + |z - z_0|)^2.$$

3. If the second partial derivatives of f are continuous and if x, y, and z change from x_0, y_0, and z_0 by small amounts dx, dy, and dz, the **total differential**

$$df = f_x(P_0)\, dx + f_y(P_0)\, dy + f_z(P_0)\, dz$$

gives a good approximation of the resulting change in f.

Example 6   LINEARIZING A FUNCTION OF THREE VARIABLES

Find the linearization $L(x, y, z)$ of the function $f(x, y, z) = x^2 - xy + 3 \sin z$ at the point $(2, 1, 0)$, and find an upper bound for the error incurred when L is used to approximate f on the rectangle R:

$$|x - 2| \leq 0.01, \qquad |y - 1| \leq 0.02, \qquad |z| \leq 0.01.$$

Solution   Differentiation and evaluation gives

$$f(2, 1, 0) = 2, \qquad f_x(2, 1, 0) = 3, \qquad f_y(2, 1, 0) = -2, \qquad f_z(2, 1, 0) = 3.$$

Thus,

$$L(x, y, z) = 2 + 3(x - 2) + (-2)(y - 1) + 3(z - 0) = 3x - 2y + 3z - 2.$$

Since

$$f_{xx} = 2, \qquad f_{yy} = 0, \qquad f_{zz} = -3 \sin z,$$
$$f_{xy} = -1, \qquad f_{xz} = 0, \qquad f_{yz} = 0,$$

we may safely take M in the error bound formula to be 3, the maximum possible value of $|-3 \sin z|$. Hence

$$|E| \le \frac{1}{2} \, 3(0.01 + 0.02 + 0.01)^2 = 0.0024.$$

The error will be no greater than 0.0024.

Example 7 FINDING THE SAG IN UNIFORMLY LOADED BEAMS

A horizontal rectangular beam, supported at both ends, will sag when subjected to a uniform load (constant weight per meter of beam length). The amount S of sag (Figure 13.27) is calculated with the formula

$$S = C \frac{px^4}{wh^3}.$$

In this equation,

p = the load (newtons per meter of beam length),

x = the length between supports (meters),

w = the width of the beam (meters),

h = the height of the beam (meters),

C = a constant that depends on the units of measurement and the material from which the beam is made.

Find dS for a beam 4 m long, 10 cm wide, and 20 cm high that is subjected to a load of 100 N/m (Figure 13.28). What conclusions can be drawn about the beam from the expression for dS?

Solution   Since S is a function of the four independent variables p, x, w, and h, its total differential is

$$dS = S_p \, dp + S_x \, dx + S_w \, dw + S_h \, dh.$$

When we write this out for a particular set of values p_0, x_0, w_0, and h_0 and simplify the result, we find that

$$dS = S_0 \left(\frac{dp}{p_0} + \frac{4 \, dx}{x_0} - \frac{dw}{w_0} - \frac{3 \, dh}{h_0} \right),$$

where

$$S_0 = S(p_0, x_0, w_0, h_0) = C \frac{p_0 x_0^4}{w_0 h_0^3}.$$

If $p_0 = 100$ N/m, $x_0 = 4$ m, $w_0 = 0.1$ m, and $h_0 = 0.2$ m, then

$$dS = S_0 \left(\frac{dp}{100} + dx - 10 \, dw - 15 \, dh \right).$$

Here is what we can learn from this equation for dS:

Since dp and dx appear with positive coefficients, increases in p (load) and x (length) will increase the sag. But dw and dh appear with negative coefficients, so increases in w (width) and h (height) will decrease the sag (make the beam stiffer). The sag is not very sensitive to changes in load because the coefficient of dp is 1/100. The magnitude of the coefficient of dh is greater than the magnitude of the coefficient of dw. Making the beam 1 cm higher will therefore decrease the sag more than making the beam 1 cm wider.

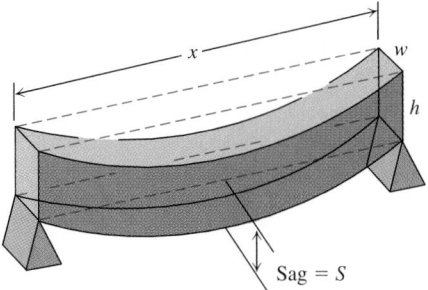

Figure 13.27 A beam supported at its two ends before and after loading. Example 7 shows how the sag S is related to the weight of the load and the dimensions of the beam.

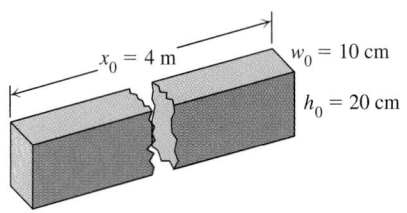

Figure 13.28 The dimensions of the beam in Example 7.

Quick Review 13.4

Let $f(x) = x - \sin 2x + 1$. (Notice that f is a function of a single independent variable.)

1. Find the linearization of f at $(0, 1)$.

2. Use the linearization to estimate $f(0.2)$.

3. With a calculator, find the actual error in the estimate in Exercise 2.

4. Find the relative error in Exercise 2.

5. Find the percentage error in Exercise 2.

6. Using a graphing calculator, find the largest open interval $(-a, a)$ on which the linearization approximates f with an error less than 0.02.

7. Write an equation relating the differentials df and dx.

8. Use the differential formula in Exercise 7 to estimate the change in f when x changes from π to $\pi + 1/2$.

9. Looking at the graph of f near $x = \pi$, why would you expect the estimate in Exercise 8 to be fairly accurate?

10. Looking at the graph of f near $x = 1/2$, why would you expect the differential formula in Exercise 7 to be less accurate in estimating the change in f when x changes from 1/2 to 1?

Section 13.4 Exercises

In Exercises 1–6, find the linearization $L(x, y)$ of the function f at (a) the origin and (b) the given point.

1. $f(x, y) = x^2 + y^2 + 1$, $(1, 1)$

2. $f(x, y) = (x + y + 2)^2$, $(1, 2)$

3. $f(x, y) = 3x - 4y + 5$, $(1, 1)$

4. $f(x, y) = x^3 y^4$, $(1, 1)$

5. $f(x, y) = e^x \cos y$, $(0, \pi/2)$

6. $f(x, y) = e^{2y-x}$, $(1, 2)$

In Exercises 7–10, find the linearization $L(x, y)$ of the function $f(x, y)$ at the given point. Then use the technique of Example 2 to find an upper bound for the magnitude $|E|$ of the error in the approximation $f(x, y) \approx L(x, y)$ over the rectangle R. (Use $|\cos y| \le 1$ and $|\sin y| \le 1$ when estimating E in Exercises 8 and 9.)

7. $f(x, y) = x^2 - 3xy + 5$ $(2, 1)$
   $R: |x - 2| \le 0.1$, $|y - 1| \le 0.1$

8. $f(x, y) = 1 + y + x \cos y$ $(0, 0)$
   $R: |x| \le 0.2$, $|y| \le 0.2$

9. $f(x, y) = e^x \cos y$ $(0, 0)$
   $R: |x| \le 0.1$, $|y| \le 0.1$
   (Use $e^x \le 1.11$ in estimating E.)

10. $f(x, y) = \ln x + \ln y$ $(1, 1)$
    $R: |x - 1| \le 0.2$, $|y - 1| \le 0.2$

11. **Writing to Learn** You plan to calculate the area of a long, thin rectangle from measurements of its length and width. Which dimension should you measure more carefully? Give reasons for your answer.

12. **Writing to Learn**
    (a) Around the point $(1, 0)$, is $f(x, y) = x^2(y + 1)$ more sensitive to changes in x or to changes in y? Give reasons for your answer.
    (b) What ratio of dx to dy will make df equal zero at $(1, 0)$?

13. Suppose T is to be found from the formula $T = x(e^y + e^{-y})$, where x and y are found to be 2 and $\ln 2$, respectively, with maximum possible errors of magnitudes $|dx| = 0.1$ and $|dy| = 0.02$. Estimate the maximum possible error in the computed value of T.

14. About how accurately may $V = \pi r^2 h$ be calculated from measurements of r and h that are in error by 1%?

15. If $r = 5.0$ cm and $h = 12.0$ cm to the nearest millimeter, what should we expect to be the maximum percentage error in calculating $V = \pi r^2 h$?

16. To estimate the volume of a cylinder of radius about 2 m and height about 3 m, about how accurately should the radius and height be measured so that the error in the volume estimate will not exceed 0.1 m³? Assume that the possible error dr in measuring r is equal to the possible error dh in measuring h.

17. Give a reasonable square centered at $(1, 1)$ over which the value of $f(x, y) = x^3 y^4$ will not vary by more than ± 0.1.

18. *Variation in Electrical Resistance* The resistance R produced by wiring resistors of R_1 and R_2 ohms in parallel (Figure 13.29) can be calculated from the formula

$$\frac{1}{R} = \frac{1}{R_1} + \frac{1}{R_2}.$$

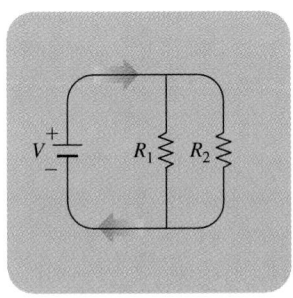

Figure 13.29 The circuit in Exercises 18 and 19.

(a) Show that

$$dR = \left(\frac{R}{R_1}\right)^2 dR_1 + \left(\frac{R}{R_2}\right)^2 dR_2$$

(b) You have designed a two-resistor circuit like the one in Figure 13.29 to have resistances of $R_1 = 100$ ohms and $R_2 = 400$ ohms, but there is always some variation in manufacturing, so the resistors received by your firm will probably not have these exact values. Will the values of R be more sensitive to variations in R_1 or to variations in R_2? Give reasons for your answer.

19. *(Continuation of Exercise 18)* In another circuit like the one in Figure 13.29, you plan to change R_1 from 20 to 20.1 ohms and R_2 from 25 to 24.9 ohms. By what percentage will this change R?

20. *Error Carry-over in Coordinate Changes*

(a) If $x = 3 \pm 0.01$ and $y = 4 \pm 0.01$, as shown here, with approximately what accuracy can you calculate the polar coordinates r and θ of the point $P(x, y)$ from the formulas $r^2 = x^2 + y^2$ and $\theta = \tan^{-1}(y/x)$? Express your estimates as percentage changes of the values that r and θ have at the point $(3, 4)$.

(b) At the point $(3, 4)$, are the values of r and θ more sensitive to changes in x or to changes in y? Give reasons for your answer.

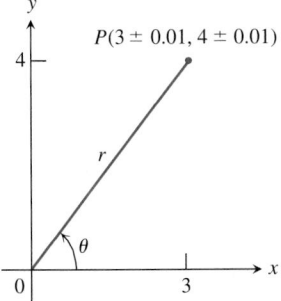

In Exercises 21–26, find the linearization $L(x, y, z)$ of the given function at each of the given points.

21. $f(x, y, z) = xy + yz + xz$ at
   (a) $(1, 1, 1)$ **(b)** $(1, 0, 0)$ **(c)** $(0, 0, 0)$

22. $f(x, y, z) = x^2 + y^2 + z^2$ at
   (a) $(1, 1, 1)$ **(b)** $(0, 1, 0)$ **(c)** $(1, 0, 0)$

23. $f(x, y, z) = \sqrt{x^2 + y^2 + z^2}$ at
   (a) $(1, 0, 0)$ **(b)** $(1, 1, 0)$ **(c)** $(1, 2, 2)$

24. $f(x, y, z) = (\sin xy)/z$ at
   (a) $(\pi/2, 1, 1)$ **(b)** $(2, 0, 1)$

25. $f(x, y, z) = e^x + \cos(y + z)$ at
   (a) $(0, 0, 0)$ **(b)** $\left(0, \frac{\pi}{2}, 0\right)$ **(c)** $\left(0, \frac{\pi}{4}, \frac{\pi}{4}\right)$

26. $f(x, y, z) = \tan^{-1}(xyz)$ at
   (a) $(1, 0, 0)$ **(b)** $(1, 1, 0)$ **(c)** $(1, 1, 1)$

In Exercises 27–30, find the linearization $L(x, y, z)$ of the function $f(x, y, z)$ at the given point. Then use the technique of Example 6 to find an upper bound for the magnitude $|E|$ of the error in the approximation $f(x, y, z) \approx L(x, y, z)$ over the rectangle R.

27. $f(x, y, z) = xz - 3yz + 2$ $(1, 1, 2)$
   $R: |x - 1| \leq 0.01,$ $|y - 1| \leq 0.01,$ $|z - 2| \leq 0.02$

28. $f(x, y, z) = x^2 + xy + yz + \dfrac{z^2}{4}$ $(1, 1, 2)$
   $R: |x - 1| \leq 0.01,$ $|y - 1| \leq 0.01,$ $|z - 2| \leq 0.08$

29. $f(x, y, z) = xy + 2yz - 3xz$ $(1, 1, 0)$
   $R: |x - 1| \leq 0.01,$ $|y - 1| \leq 0.01,$ $|z| \leq 0.01$

30. $f(x, y, z) = \sqrt{2} \cos x \sin(y + z)$ $(0, 0, \pi/4)$
   $R: |x| \leq 0.01,$ $|y| \leq 0.01,$ $|z - \pi/4| \leq 0.01$

31. *The Sagging Beam Revisited* The beam of Example 7 is tipped on its side so that $h = 0.1$ m and $w = 0.2$ m.

   (a) What is the value of dS now?

   (b) Compare the sensitivity of the newly positioned beam to a small change in height with its sensitivity to an equally small change in width.

32. Estimate how strongly simultaneous errors of 2% in a, b, and c might affect the calculation of the product $p(a, b, c) = abc$.

33. Suppose $u = xe^y + y \sin z$ and x, y, and z can be measured with maximum possible errors of ± 0.2, ± 0.6, and $\pm \pi/180$, respectively. Estimate the maximum possible error in calculating u from the measured values $x = 2$, $y = \ln 3$, and $z = \pi/2$.

34. **Writing to Learn** Does a function $f(x, y)$ with continuous first partial derivatives throughout an open region R have to be continuous on R? Give reasons for your answer.

35. **Writing to Learn** If a function $f(x, y)$ has continuous second partial derivatives throughout an open region R, must the first-order partial derivatives of *f* be continuous on *R?* Give reasons for your answer.

Explorations

36. *Designing a Box* Estimate how much wood it takes to make a hollow rectangular box whose inside measurements are 5 ft long by 3 ft wide by 2 ft deep if the box is to be made of 1/2-in.-thick lumber and is to have no top.

37. *Designing a Soda Can* A standard 12-ounce soda can is essentially a cylinder of radius $r = 1$ in. and height $h = 5$ in.

(a) At these dimensions, how sensitive is the can's volume to a small change in the radius versus a small change in the height?

(b) Could you design a soda can that *appears* to hold more soda but in fact holds the same 12 ounces? What might its dimensions be? (There is more than one correct answer.)

38. *Surveying a Triangular Field* The area of a triangle is $(1/2)\,ab \sin C$, where *a* and *b* are the lengths of two sides of the triangle and *C* is the measure of the included angle. In surveying a triangular plot, you have measured *a*, *b*, and *C* to be 150 ft, 200 ft, and 60° respectively. By about how much could your area calculation be in error if your values of *a* and *b* are off by half a foot each and your measurement of *C* is off by 2°? (See the figure.) (*Note:* Be sure to use radians.)

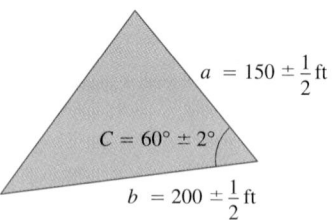

$a = 150 \pm \frac{1}{2}$ ft

$C = 60° \pm 2°$

$b = 200 \pm \frac{1}{2}$ ft

Extending the Ideas

39. *Determinant of a Matrix* If $|a|$ is much greater than $|b|$, $|c|$, and $|d|$, is the value of the determinant

$$f(a, b, c, d) = \begin{vmatrix} a & b \\ c & d \end{vmatrix}$$

more sensitive to small changes in *a*, *b*, *c*, or *d?* Give reasons for your answer.

40. *The Wilson Lot Size Formula* The Wilson Lot Size Formula in economics says that the most economical quantity *Q* of goods (radios, shoes, brooms, whatever) for a store to order is given by the formula $Q = \sqrt{2KM/h}$, where *K* is the cost of placing the order, *M* is the number of items sold per week, and *h* is the weekly holding cost for each item (cost of space, utilities, security, and so on). To which of the variables *K*, *M*, and *h* is *Q* most sensitive near the point $(K_0, M_0, h_0) = (2, 20, 0.05)$? Give reasons for your answers.

13.5 The Chain Rule

Composite Functions in Higher Dimensions • Functions of Two Variables • Functions of Three Variables • Functions Defined on Surfaces • Implicit Differentiation Improved • Functions of Many Variables

Composite Functions in Higher Dimensions

When we are interested in the temperature $w = f(x, y, z)$ at points along a curve $x = g(t)$, $y = h(t)$, $z = k(t)$ in space or in the pressure or density along a path through a gas or fluid, we may think of *f* as a function of the single variable *t*. For each value of *t*, the temperature at the point $(g(t), h(t), k(t))$ is the value of the composite function $f(g(t), h(t), k(t))$. If we then wish to know the rate at which *f* changes with respect to *t* along the path, we only have to differentiate this composite function with respect to *t*, provided, of course, the derivative exists.

Sometimes we can find the derivative by substituting the formulas for *g*, *h*, and *k* into the formula for *f* and differentiating directly with respect to *t*. Often, however, we have to work with functions whose formulas are too complicated for convenient substitution or for which formulas are not readily available. To find a function's derivatives under circumstances like these (and to get further insight into how calculus works in higher dimensions), we use a multivariable form of the Chain Rule. The form that the rule takes depends on how many variables are involved, but, except for the presence of additional variables, it works just like the Chain Rule in Section 3.6.

Functions of Two Variables

In Section 3.6, we studied what happened when $w = f(x)$ was a differentiable function of x and $x = g(t)$ was a differentiable function of t. This made w a differentiable function of t, and the Chain Rule said that dw/dt could be calculated with the formula

$$\frac{dw}{dt} = \frac{dw}{dx}\frac{dx}{dt}.$$

The analogous formula for a function $w = f(x, y)$ is given in Theorem 5.

Theorem 5 Chain Rule for Functions of Two Variables

If $w = f(x, y)$ is differentiable and x and y are differentiable functions of t, then w is a differentiable function of t and

$$\frac{dw}{dt} = \frac{\partial f}{\partial x}\frac{dx}{dt} + \frac{\partial f}{\partial y}\frac{dy}{dt}.$$

Proof of Theorem 5 The proof consists of showing that if x and y are differentiable at $t = t_0$, then w is differentiable at t_0 and

$$\left(\frac{dw}{dt}\right)_{t_0} = \left(\frac{\partial w}{\partial x}\right)_{P_0}\left(\frac{dx}{dt}\right)_{t_0} + \left(\frac{\partial w}{\partial y}\right)_{P_0}\left(\frac{dy}{dt}\right)_{t_0},$$

where $P_0 = (x(t_0), y(t_0))$.

Let Δx, Δy, and Δw be the increments that result from changing t from t_0 to $t_0 + \Delta t$. Since f is differentiable (remember the definition in Section 13.4),

$$\Delta w = \left(\frac{\partial w}{\partial x}\right)_{P_0}\Delta x + \left(\frac{\partial w}{\partial y}\right)_{P_0}\Delta y + \varepsilon_1\,\Delta x + \varepsilon_2\,\Delta y,$$

where $\varepsilon_1, \varepsilon_2 \to 0$ as $\Delta x, \Delta y \to 0$. To find dw/dt, we divide this equation through by Δt and let Δt approach zero. The division gives

$$\frac{\Delta w}{\Delta t} = \left(\frac{\partial w}{\partial x}\right)_{P_0}\frac{\Delta x}{\Delta t} + \left(\frac{\partial w}{\partial y}\right)_{P_0}\frac{\Delta y}{\Delta t} + \varepsilon_1\frac{\Delta x}{\Delta t} + \varepsilon_2\frac{\Delta y}{\Delta t}.$$

Letting Δt approach zero gives

$$\left(\frac{dw}{dt}\right)_{t_0} = \lim_{\Delta t \to 0}\frac{\Delta w}{\Delta t}$$

$$= \left(\frac{\partial w}{\partial x}\right)_{P_0}\left(\frac{dx}{dt}\right)_{t_0} + \left(\frac{\partial w}{\partial y}\right)_{P_0}\left(\frac{dy}{dt}\right)_{t_0} + 0 \cdot \left(\frac{dx}{dt}\right)_{t_0} + 0 \cdot \left(\frac{dy}{dt}\right)_{t_0}$$

$$= \left(\frac{\partial w}{\partial x}\right)_{P_0}\left(\frac{dx}{dt}\right)_{t_0} + \left(\frac{\partial w}{\partial y}\right)_{P_0}\left(\frac{dy}{dt}\right)_{t_0}. \qquad \blacksquare$$

The **tree diagram** in the margin provides a convenient way to remember the Chain Rule for functions of two variables. Notice that both partial derivatives and single-variable derivatives are involved. From the diagram you see that when $t = t_0$, the single-variable derivatives dx/dt and dy/dt are evaluated at t_0, whereas the partial derivatives $\partial w/\partial x$ and $\partial w/\partial y$ are evaluated at the point (x_0, y_0) determined by evaluating the differentiable functions x and y at t_0. The "true" independent variable is t, whereas x and y are **intermediate variables** controlled by t. The "true" dependent variable is w.

One way to remember the Chain Rule is to picture the diagram below. To find dw/dt, start at w and read down each route to t, multiplying derivatives along the way. Then add the products.

Chain Rule

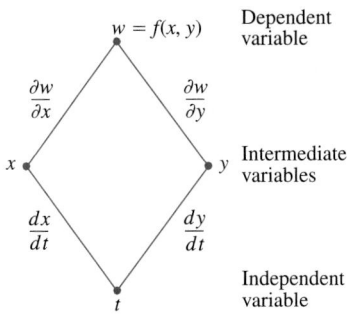

$$\frac{dw}{dt} = \frac{\partial w}{\partial x}\frac{dx}{dt} + \frac{\partial w}{\partial y}\frac{dy}{dt}$$

A more precise notation for the Chain Rule shows how the partial and single-variable derivatives involved are evaluated:

$$\frac{dw}{dt}(t_0) = \frac{\partial f}{\partial x}(x_0, y_0) \cdot \frac{dx}{dt}(t_0) + \frac{\partial f}{\partial y}(x_0, y_0) \cdot \frac{dy}{dt}(t_0).$$

Example 1   APPLYING THE CHAIN RULE

Use the Chain Rule to find the derivative of $w = xy$ with respect to t along the path $x = \cos t$, $y = \sin t$. What is the derivative's value at $t = \pi/2$?

Solution   We apply the Chain Rule to find dw/dt as follows:

$$\frac{dw}{dt} = \frac{\partial w}{\partial x}\frac{dx}{dt} + \frac{\partial w}{\partial y}\frac{dy}{dt}$$

$$= \frac{\partial(xy)}{\partial x} \cdot \frac{d}{dt}(\cos t) + \frac{\partial(xy)}{\partial y} \cdot \frac{d}{dt}(\sin t)$$

$$= (y)(-\sin t) + (x)(\cos t)$$

$$= (\sin t)(-\sin t) + (\cos t)(\cos t)$$

$$= -\sin^2 t + \cos^2 t$$

$$= \cos(2t)$$

At the given value of t,

$$\left(\frac{dw}{dt}\right)_{t=\frac{\pi}{2}} = \cos\left(2 \cdot \frac{\pi}{2}\right) = -1.$$

Figure 13.30 shows CAS support for the computation of dw/dt in Example 1 because $2\cos^2 t - 1 = \cos(2t)$.

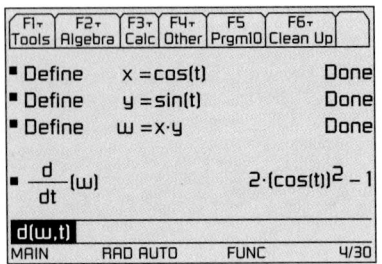

Figure 13.30 We have used the "Define" command to express x, y, and therefore w as functions of t. (Example 1)

Exploration 1   **Bypassing the Chain Rule**

Example 1 was intended to illustrate how the Chain Rule works for a function of two variables, but notice that the output of the process was the derivative of w with respect to the single variable t. Could we have done this as a single-variable problem and avoided confronting the multivariable nature of w altogether?

1. Write the function w of Example 1 explicitly as a function of t.
2. Find dw/dt using the techniques of Chapter 3. Do you get the same function as in the solution to Example 1?
3. Find dw/dt along the same (x, y) path if $w = x^2 e^y$. Again, find it in two ways: by using the Chain Rule for functions of two variables and by writing w explicitly as a function of t.
4. Do you think that the two methods will always lead to the same solution?

Exploration 1 suggests that the "new" Chain Rule leads to the same dw/dt that we could have found using the rules of Chapter 3. That is an important property of the new rule: It works! Remember that when we differentiated an

Chain Rule

Here we have three routes from w to t instead of two, but finding dw/dt is still the same. Read down each route, multiplying derivatives along the way; then add.

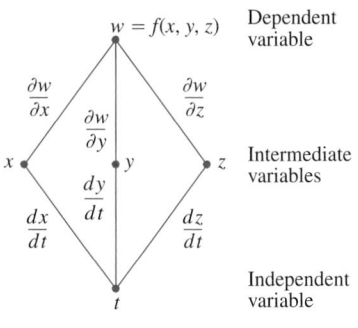

Chain Rule

$$\frac{dw}{dt} = \frac{\partial w}{\partial x}\frac{dx}{dt} + \frac{\partial w}{\partial y}\frac{dy}{dt} + \frac{\partial w}{\partial z}\frac{dz}{dt}$$

expression like $x^2 e^y$ with respect to t in Chapter 3, we were not thinking of it as a function of two independent variables controlled by t (even though it was). As we develop the rules for multivariable derivatives, we must expect them to be consistent with our previous results.

Functions of Three Variables

You can probably predict the Chain Rule for functions of three variables, as it only involves adding the expected third term to the two-variable formula.

> **Theorem 6 Chain Rule for Functions of Three Variables**
>
> If $w = f(x, y, z)$ is differentiable and x, y, and z are differentiable functions of t, then w is a differentiable function of t and
>
> $$\frac{dw}{dt} = \frac{\partial f}{\partial x}\frac{dx}{dt} + \frac{\partial f}{\partial y}\frac{dy}{dt} + \frac{\partial f}{\partial z}\frac{dz}{dt}.$$

The proof is identical to the proof of Theorem 5, except that there is a third intermediate variable to be dealt with every step of the way. There is also a similar diagram (in the margin) for remembering the new equation, this one with three routes from w to t.

Example 2 CHANGES IN A FUNCTION'S VALUES ALONG A HELIX

Find dw/dt if $w = xy + z$, $x = \cos t$, $y = \sin t$, and $z = t$ (Figure 13.31). What is the derivative's value at $t = 0$?

Solution

$$\frac{dw}{dt} = \frac{\partial w}{\partial x}\frac{dx}{dt} + \frac{\partial w}{\partial y}\frac{dy}{dt} + \frac{\partial w}{\partial z}\frac{dz}{dt}$$

$$= (y)(-\sin t) + (x)(\cos t) + (1)(1)$$

$$= (\sin t)(-\sin t) + (\cos t)(\cos t) + 1$$

$$= -\sin^2 t + \cos^2 t + 1$$

$$= 1 + \cos 2t$$

$$\left(\frac{dw}{dt}\right)_{t=0} = 1 + \cos(0) = 2$$

Functions Defined on Surfaces

If we are interested in the temperature $w = f(x, y, z)$ at points (x, y, z) on a globe in space, we might prefer to think of x, y, and z as functions of the variables r and s that give the points' longitudes and latitudes. If $x = g(r, s)$, $y = h(r, s)$, and $z = k(r, s)$, we could then express the temperature as a function of r and s with the composite function

$$w = f(g(r, s), h(r, s), k(r, s)).$$

Under the right conditions, w would have partial derivatives with respect to both r and s that could be calculated in the following way.

The helix
$$\mathbf{r} = (\cos t)\mathbf{i} + (\sin t)\mathbf{j} + t\mathbf{k}$$

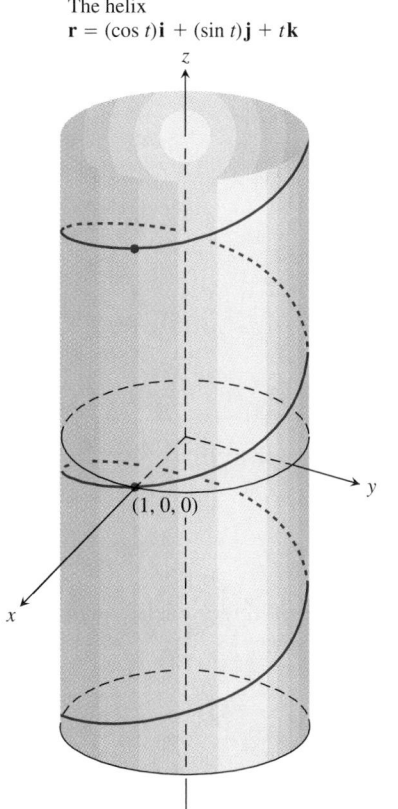

Figure 13.31 Example 2 shows how the values of $w = xy + z$ vary with t along this helix.

Dependent variable

Intermediate variables

Independent variables

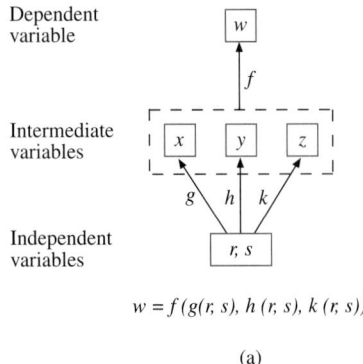

$w = f(g(r, s), h(r, s), k(r, s))$

(a)

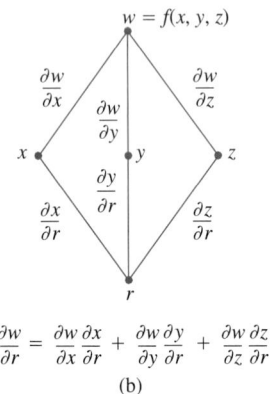

$$\frac{\partial w}{\partial r} = \frac{\partial w}{\partial x}\frac{\partial x}{\partial r} + \frac{\partial w}{\partial y}\frac{\partial y}{\partial r} + \frac{\partial w}{\partial z}\frac{\partial z}{\partial r}$$

(b)

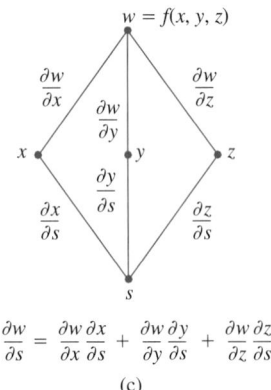

$$\frac{\partial w}{\partial s} = \frac{\partial w}{\partial x}\frac{\partial x}{\partial s} + \frac{\partial w}{\partial y}\frac{\partial y}{\partial s} + \frac{\partial w}{\partial z}\frac{\partial z}{\partial s}$$

(c)

Figure 13.32 Composite function and tree diagrams to accompany Theorem 7.

> **Theorem 7 Chain Rule for Two Independent Variables and Three Intermediate Variables**
>
> Suppose $w = f(x, y, z)$, $x = g(r, s)$, $y = h(r, s)$, and $z = k(r, s)$. If all four functions are differentiable, then w has partial derivatives with respect to r and s, given by the formulas
>
> $$\frac{\partial w}{\partial r} = \frac{\partial w}{\partial x}\frac{\partial x}{\partial r} + \frac{\partial w}{\partial y}\frac{\partial y}{\partial r} + \frac{\partial w}{\partial z}\frac{\partial z}{\partial r},$$
>
> $$\frac{\partial w}{\partial s} = \frac{\partial w}{\partial x}\frac{\partial x}{\partial s} + \frac{\partial w}{\partial y}\frac{\partial y}{\partial s} + \frac{\partial w}{\partial z}\frac{\partial z}{\partial s}.$$

The first of these equations can be derived from the Chain Rule in Theorem 6 by holding s fixed and treating r as t. The second can be derived in the same way, holding r fixed and treating s as t. The tree diagrams for both equations are shown in Figure 13.32.

Example 3 PARTIAL DERIVATIVES USING THEOREM 7

Express $\partial w/\partial r$ and $\partial w/\partial s$ in terms of r and s if

$$w = x + 2y + z^2, \qquad x = \frac{r}{s}, \qquad y = r^2 + \ln s, \qquad z = 2r.$$

Solution

$$\frac{\partial w}{\partial r} = \frac{\partial w}{\partial x}\frac{\partial x}{\partial r} + \frac{\partial w}{\partial y}\frac{\partial y}{\partial r} + \frac{\partial w}{\partial z}\frac{\partial z}{\partial r}$$

$$= (1)\left(\frac{1}{s}\right) + (2)(2r) + (2z)(2)$$

$$= \frac{1}{s} + 4r + (4r)(2)$$

$$= \frac{1}{s} + 12r$$

$$\frac{\partial w}{\partial s} = \frac{\partial w}{\partial x}\frac{\partial x}{\partial s} + \frac{\partial w}{\partial y}\frac{\partial y}{\partial s} + \frac{\partial w}{\partial z}\frac{\partial z}{\partial s}$$

$$= (1)\left(-\frac{r}{s^2}\right) + (2)\left(\frac{1}{s}\right) + (2z)(0)$$

$$= \frac{2}{s} - \frac{r}{s^2}$$

If f is a function of two variables instead of three, each equation in Theorem 7 becomes correspondingly one term shorter.

> If $w = f(x, y)$, $x = g(r, s)$, and $y = h(r, s)$, then
>
> $$\frac{\partial w}{\partial r} = \frac{\partial w}{\partial x}\frac{\partial x}{\partial r} + \frac{\partial w}{\partial y}\frac{\partial y}{\partial r} \qquad \text{and} \qquad \frac{\partial w}{\partial s} = \frac{\partial w}{\partial x}\frac{\partial x}{\partial s} + \frac{\partial w}{\partial y}\frac{\partial y}{\partial s}.$$

Figure 13.33 shows the tree diagram for the first of these equations. The diagram for the second equation is similar: Just replace r with s.

Chain Rule

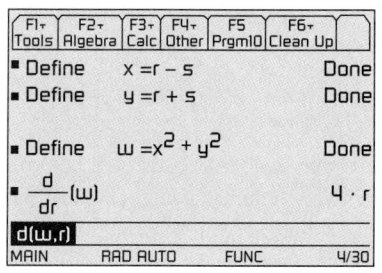

Figure 13.33 Tree diagram for the equation

$$\frac{\partial w}{\partial r} = \frac{\partial w}{\partial x}\frac{\partial x}{\partial r} + \frac{\partial w}{\partial y}\frac{\partial y}{\partial r}.$$

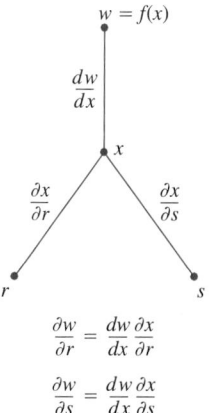

Figure 13.34 This figure illustrates how to use a CAS to write *x*, *y*, and therefore *w* as function of *r* and *s*. (Example 4)

Chain Rule

$$w = f(x)$$

$$\frac{\partial w}{\partial r} = \frac{dw}{dx}\frac{\partial x}{\partial r}$$

$$\frac{\partial w}{\partial s} = \frac{dw}{dx}\frac{\partial x}{\partial s}$$

Figure 13.35 Tree diagram for differentiating *f* as a composite function of *r* and *s* with one intermediate variable.

Example 4 MORE PARTIAL DERIVATIVES

Express $\partial w/\partial r$ and $\partial w/\partial s$ in terms of *r* and *s* if

$$w = x^2 + y^2, \qquad x = r - s, \qquad y = r + s.$$

Solution

$$\frac{\partial w}{\partial r} = \frac{\partial w}{\partial x}\frac{\partial x}{\partial r} + \frac{\partial w}{\partial y}\frac{\partial y}{\partial r}$$

$$= (2x)(1) + (2y)(1)$$

$$= 2(r - s) + 2(r + s)$$

$$= 4r$$

$$\frac{\partial w}{\partial s} = \frac{\partial w}{\partial x}\frac{\partial x}{\partial s} + \frac{\partial w}{\partial y}\frac{\partial y}{\partial s}$$

$$= (2x)(-1) + (2y)(1)$$

$$= -2(r - s) + 2(r + s)$$

$$= 4s$$

Figure 13.34 shows CAS support for the computation of $\partial w/\partial r$ in Example 4. If you have a CAS available, try providing support for $\partial w/\partial s$.

If *f* is a function of *x* alone, our equations become even simpler.

If $w = f(x)$ and $x = g(r, s)$, then

$$\frac{\partial w}{\partial r} = \frac{dw}{dx}\frac{\partial x}{\partial r} \qquad \text{and} \qquad \frac{\partial w}{\partial s} = \frac{dw}{dx}\frac{\partial x}{\partial s}.$$

Notice that, in this case, we can use the ordinary (single-variable) derivative, dw/dx. The tree diagram is shown in Figure 13.35.

Implicit Differentiation Improved

Believe it or not, the two-variable Chain Rule in Theorem 5 leads to a formula that takes most of the work out of implicit differentiation (Section 3.7). Suppose:

1. the function $F(x, y)$ is differentiable, and

2. the equation $F(x, y) = 0$ defines *y* implicitly as a differentiable function of *x*, say $y = h(x)$.

Since $w = F(x, y) = 0$, the derivative dw/dx must be zero. Computing the derivative from the Chain Rule (tree diagram in Figure 13.36), we find

$$0 = \frac{dw}{dx} = F_x\frac{dx}{dx} + F_y\frac{dy}{dx}$$

$$= F_x \cdot 1 + F_y \cdot \frac{dy}{dx}.$$

If $F_y = \partial w/\partial y \neq 0$, we can solve this equation for dy/dx to get

$$\frac{dy}{dx} = -\frac{F_x}{F_y}.$$

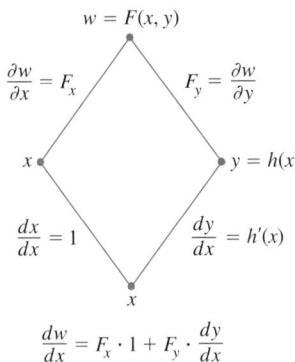

$w = F(x, y)$

$\dfrac{\partial w}{\partial x} = F_x$ $F_y = \dfrac{\partial w}{\partial y}$

x $y = h(x)$

$\dfrac{dx}{dx} = 1$ $\dfrac{dy}{dx} = h'(x)$

x

$\dfrac{dw}{dx} = F_x \cdot 1 + F_y \cdot \dfrac{dy}{dx}$

Figure 13.36 Tree diagram for differentiating $w = F(x, y)$ with respect to x. Setting $dw/dx = 0$ leads to a simple computational formula for implicit differentiation. (Theorem 8)

This relationship gives a surprisingly simple shortcut to finding derivatives of implicitly defined functions, which we state here as a theorem.

Theorem 8 A Formula for Implicit Differentiation

Suppose $F(x, y)$ is differentiable and the equation $F(x, y) = 0$ defines y implicitly as a differentiable function of x. Then, at any point where $F_y \neq 0$,

$$\frac{dy}{dx} = -\frac{F_x}{F_y}.$$

Example 5 SPEEDY IMPLICIT DIFFERENTIATION

Use Theorem 8 to find dy/dx if $x^2 - xy + y^2 = 7$.

Solution   Take $F(x, y) = x^2 - xy + y^2 - 7$. Then

$$\frac{dy}{dx} = -\frac{F_x}{F_y} = -\frac{2x - y}{-x + 2y} = \frac{y - 2x}{2y - x}.$$

This calculation is significantly shorter than the single-variable calculation with which we found dy/dx in Section 3.7, Example 4.

Functions of Many Variables

We have seen several different forms of the Chain Rule in this section, but you do not have to memorize them all if you can see them as special cases of the same general formula. When solving particular problems, it may help to draw the appropriate tree diagram, placing the dependent variable at the top, the intermediate variables in the middle, and the selected independent variable at the bottom. To find the derivative of the dependent variable with respect to the selected independent variable, start at the dependent variable and read down all the different routes to the independent variable, calculating and multiplying the derivatives along each route. Then add the products you found along the different routes.

In general, suppose $w = f(x, y, \ldots, v)$ is a differentiable function of the variables $x, y, \ldots, v$ (a finite set) and the $x, y, \ldots, v$ are differentiable functions of the variables $p, q, \ldots, t$ (another finite set). Then w is a differentiable function of the variables p through t, and the partial derivatives of w with respect to these variables are given by equations of the form

$$\frac{\partial w}{\partial p} = \frac{\partial w}{\partial x}\frac{\partial x}{\partial p} + \frac{\partial w}{\partial y}\frac{\partial y}{\partial p} + \cdots + \frac{\partial w}{\partial v}\frac{\partial v}{\partial p}.$$

The other equations are obtained by replacing p by $q, \ldots, t$, one at a time.

Another way to remember this equation is to think of the right-hand side as the dot product of two vectors with components

$$\left(\frac{\partial w}{\partial x}, \frac{\partial w}{\partial y}, \ldots, \frac{\partial w}{\partial v} \right) \quad \text{and} \quad \left(\frac{\partial x}{\partial p}, \frac{\partial y}{\partial p}, \ldots, \frac{\partial v}{\partial p} \right).$$

Derivatives of w with respect to the intermediate variables

Derivatives of the intermediate variables with respect to the selected independent variable

Quick Review 13.5

In Exercises 1–5, find $\partial f/\partial x$ and $\partial f/\partial y$.

1. $f(x, y) = xy + \sin y$

2. $f(x, y) = x^3 - y^2 + 5$

3. $f(x, y) = \tan(x + y)$

4. $f(x, y) = e^{xy} + 10$

5. $f(x, y) = x \sin y - y \cos x$

In Exercises 6–10, find dy/dt at the given value of t.

6. $y = \cos x,$ $\quad x = t^2,$ $\quad t = \sqrt{\pi}$

7. $y = \tan x,$ $\quad x = e^{t/2},$ $\quad t = 2 \ln \pi$

8. $y = \ln x,$ $\quad x = \cos^2 t,$ $\quad t = \pi/4$

9. $ty - 3t^2 = 5,$ $\quad t = 1$

10. $\tan y = x,$ $\quad t = \sqrt[3]{x},$ $\quad t = \sqrt{2}$

Section 13.5 Exercises

In Exercises 1–6, **(a)** express dw/dt as a function of t, both by using the Chain Rule and by expressing w in terms of t and differentiating directly with respect to t. Then **(b)** evaluate dw/dt at the given value of t.

1. $w = x^2 + y^2,$ $\quad x = \cos t,$ $\quad y = \sin t;$ $\quad t = \pi$

2. $w = x^2 + y^2,$ $\quad x = \cos t + \sin t,$ $\quad y = \cos t - \sin t;$ $\quad t = 0$

3. $w = \dfrac{x}{z} + \dfrac{y}{z},$ $\quad x = \cos^2 t,$ $\quad y = \sin^2 t,$ $\quad z = 1/t;$ $\quad t = 3$

4. $w = \ln(x^2 + y^2 + z^2),$ $\quad x = \cos t,$ $\quad y = \sin t,$ $z = 4\sqrt{t};$ $\quad t = 3$

5. $w = 2ye^x - \ln z,$ $\quad x = \ln(t^2 + 1),$ $\quad y = \tan^{-1} t,$ $z = e^t;$ $\quad t = 1$

6. $w = z - \sin xy,$ $\quad x = t,$ $\quad y = \ln t,$ $\quad z = e^{t-1};$ $\quad t = 1$

In Exercises 7 and 8, **(a)** express $\partial z/\partial r$ and $\partial z/\partial \theta$ as functions of r and θ both by using the Chain Rule and by expressing w directly in terms of r and θ before differentiating. Then **(b)** evaluate $\partial z/\partial r$ and $\partial z/\partial \theta$ at the given point (r, θ).

7. $z = 4e^x \ln y,$ $\quad x = \ln(r \cos \theta),$ $\quad y = r \sin \theta;$ $(r, \theta) = (2, \pi/4)$

8. $z = \tan^{-1}(x/y),$ $\quad x = r \cos \theta,$ $\quad y = r \sin \theta;$ $(r, \theta) = (1.3, \pi/6)$

In Exercises 9 and 10, **(a)** express $\partial w/\partial u$ and $\partial w/\partial v$ as functions of u and v both by using the Chain Rule and by expressing w in terms of u and v before differentiating. Then **(b)** evaluate $\partial w/\partial u$ and $\partial w/\partial v$ at the given point (u, v).

9. $w = xy + yz + xz,$ $\quad x = u + v,$ $\quad y = u - v,$ $z = uv;$ $\quad (u, v) = (1/2, 1)$

10. $w = \ln(x^2 + y^2 + z^2),$ $\quad x = ue^v \sin u,$ $\quad y = ue^v \cos u,$ $z = ue^v;$ $\quad (u, v) = (-2, 0)$

In Exercises 11 and 12, **(a)** express $\partial u/\partial x$, $\partial u/\partial y$, and $\partial u/\partial z$ as functions of x, y, and z both by using the Chain Rule and by expressing u directly in terms of x, y, and z before

differentiating. Then **(b)** evaluate $\partial u/\partial x$, $\partial u/\partial y$, and $\partial u/\partial z$ at the given point (x, y, z).

11. $u = \dfrac{p - q}{q - r},$ $\quad p = x + y + z,$ $\quad q = x - y + z,$ $r = x + y - z;$ $\quad (x, y, z) = (\sqrt{3}, 2, 1)$

12. $u = e^{qr} \sin^{-1} p,$ $\quad p = \sin x,$ $\quad q = z^2 \ln y,$ $\quad r = 1/z;$ $(x, y, z) = (\pi/4, 1/2, -1/2)$

In Exercises 13–24, draw a tree diagram and write a Chain Rule formula for each derivative.

13. $\dfrac{dz}{dt}$ for $z = f(x, y),$ $\quad x = g(t),$ $\quad y = h(t)$

14. $\dfrac{dz}{dt}$ for $z = f(u, v, w),$ $\quad u = g(t),$ $\quad v = h(t),$ $w = k(t)$

15. $\dfrac{\partial w}{\partial u}$ and $\dfrac{\partial w}{\partial v}$ for $w = h(x, y, z),$ $\quad x = f(u, v),$ $y = g(u, v),$ $\quad z = k(u, v)$

16. $\dfrac{\partial w}{\partial x}$ and $\dfrac{\partial w}{\partial y}$ for $w = f(r, s, t),$ $\quad r = g(x, y),$ $s = h(x, y),$ $\quad t = k(x, y)$

17. $\dfrac{\partial w}{\partial u}$ and $\dfrac{\partial w}{\partial v}$ for $w = g(x, y),$ $\quad x = h(u, v),$ $\quad y = k(u, v)$

18. $\dfrac{\partial w}{\partial x}$ and $\dfrac{\partial w}{\partial y}$ for $w = g(u, v),$ $\quad u = h(x, y),$ $\quad v = k(x, y)$

19. $\dfrac{\partial z}{\partial t}$ and $\dfrac{\partial z}{\partial s}$ for $z = f(x, y),$ $\quad x = g(t, s),$ $\quad y = h(t, s)$

20. $\dfrac{\partial y}{\partial r}$ for $y = f(u),$ $\quad u = g(r, s)$

21. $\dfrac{\partial w}{\partial s}$ and $\dfrac{\partial w}{\partial t}$ for $w = g(u),$ $\quad u = h(s, t)$

22. $\dfrac{\partial w}{\partial p}$ for $w = f(x, y, z, v),$ $\quad x = g(p, q),$ $\quad y = h(p, q),$ $z = j(p, q),$ $\quad v = k(p, q)$

23. $\dfrac{\partial w}{\partial r}$ and $\dfrac{\partial w}{\partial s}$ for $w = f(x, y),$ $\quad x = g(r),$ $\quad y = h(s)$

24. $\dfrac{\partial w}{\partial s}$ for $w = g(x, y),$ $\quad x = h(r, s, t),$ $\quad y = k(r, s, t)$

In Exercises 25–28, assume that the equation defines y as a differentiable function of x and use Theorem 8 to find the value of dy/dx at the given point.

25. $x^3 - 2y^2 + xy = 0,$ $\quad$ (1, 1)

26. $xy + y^2 - 3x - 3 = 0,$ $\quad$ (−1, 1)

27. $x^2 + xy + y^2 - 7 = 0,$ $\quad$ (1, 2)

28. $xe^y + \sin xy + y - \ln 2 = 0,$ $\quad$ (0, ln 2)

In Exercises 29–34, find the specified partial derivatives.

29. $\partial w/\partial r$ when $r = 1, s = -1,$ if $w = (x + y + z)^2,$
$x = r - s, y = \cos(r + s), z = \sin(r + s)$

30. $\partial w/\partial v$ when $u = -1, v = 2,$ if $w = xy + \ln z,$
$x = v^2/u, y = u + v, z = \cos u$

31. $\partial w/\partial v$ when $u = 0, v = 0,$ if $w = x^2 + (y/x),$
$x = u - 2v + 1, y = 2u + v - 2$

32. $\partial z/\partial u$ when $u = 0, v = 1,$ if $z = \sin xy + x \sin y,$
$x = u^2 + v^2, y = uv$

33. $\partial z/\partial u$ and $\partial z/\partial v$ when $u = \ln 2, v = 1,$
if $z = 5 \tan^{-1} x, x = e^u + \ln v$

34. $\partial z/\partial u$ and $\partial z/\partial v$ when $u = 1, v = -2,$
if $z = \ln q, q = \sqrt{v} + 3 \tan^{-1} u$

35. *A Function Defined on a Curve in Space* Let $w = x^2 e^{2y}$ $\cos 3z$. Find the value of dw/dt at the point $(1, \ln 2, 0)$ on the curve $x = \cos t, y = \ln(t + 2), z = t$.

36. *Temperature on a Circle* Let $T = f(x, y)$ be the temperature at the point (x, y) on the circle defined parametrically by

$$x = \cos t, \qquad y = \sin t, \qquad 0 \le t \le 2\pi,$$

and suppose

$$\frac{\partial T}{\partial x} = 8x - 4y, \qquad \frac{\partial T}{\partial y} = 8y - 4x.$$

(a) Find where the maximum and minimum temperatures on the circle occur by examining the derivatives dT/dt and d^2T/dt^2.

(b) Suppose $T = 4x^2 - 4xy + 4y^2$. Find the maximum and minimum values of T on the circle.

37. *Temperature on an Ellipse* Let $T = g(x, y)$ be the temperature at the point (x, y) on the ellipse defined parametrically by

$$x = 2\sqrt{2} \cos t, \qquad y = \sqrt{2} \sin t, \qquad 0 \le t \le 2\pi,$$

and suppose

$$\frac{\partial T}{\partial x} = y, \qquad \frac{\partial T}{\partial y} = x.$$

(a) Locate the maximum and minimum temperatures on the ellipse by examining dT/dt and d^2T/dt^2.

(b) Suppose $T = xy - 2$. Find the maximum and minimum values of T on the ellipse.

38. Writing to Learn *Narrowing the Search* Suppose the partial derivatives of a function $f(x, y, z)$ at points on the helix $x = \cos t, y = \sin t, z = t$ are

$$f_x = \cos t, \qquad f_y = \sin t, \qquad f_z = t^2 + t - 2.$$

Explain how we can conclude that the points on the curve where f can take on extreme values are
$(\cos(-2), \sin(-2), -2)$ and $(\cos 1, \sin 1, 1)$.

Explorations

39. If $f(u, v, w)$ is differentiable and $u = x - y, v = y - z,$ and $w = z - x,$ show that

$$\frac{\partial f}{\partial x} + \frac{\partial f}{\partial y} + \frac{\partial f}{\partial z} = 0.$$

40. *Changes in an Electric Circuit* The voltage V in a circuit that satisfies the law $V = IR$ is slowly dropping as the battery wears out. At the same time, the resistance R is increasing as the resistor heats up. Use the equation

$$\frac{dV}{dt} = \frac{\partial V}{\partial I}\frac{dI}{dt} + \frac{\partial V}{\partial R}\frac{dR}{dt}$$

to find how the current is changing at the instant when $R = 600$ ohms, $I = 0.04$ amp, $dR/dt = 0.5$ ohms/sec, and $dV/dt = -0.01$ volt/sec.

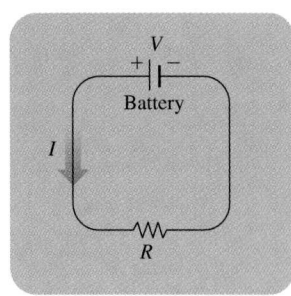

41. *The Effects of Changing the Dimensions of a Box* The lengths a, b, and c of the edges of a rectangular box are changing with time. At the instant in question, $a = 1$ m, $b = 2$ m, $c = 3$ m, $da/dt = db/dt = 1$ m/sec, and $dc/dt = -3$ m/sec. At what rates are the box's volume V and surface area S changing at that instant? Are the box's interior diagonals increasing in length, or decreasing? How can you tell?

42. *Laplace Equations* Show that if $w = f(u, v)$ satisfies the Laplace equation $f_{uu} + f_{vv} = 0$ and if $u = (x^2 - y^2)/2$ and $v = xy$, then w satisfies the Laplace equation $w_{xx} + w_{yy} = 0$.

43. *Laplace Equations* Let $w = f(u) + g(v)$, where $u = x + iy$ and $v = x - iy$ and $i = \sqrt{-1}$. Show that w satisfies the Laplace equation $w_{xx} + w_{yy} = 0$ if all the necessary functions are differentiable. ■

Extending the Ideas

44. *Extending Implicit Differentiation to Three Variables* Theorem 8 can be generalized to functions of three variables and even more. The three-variable version goes like this: If the equation $F(x, y, z) = 0$ determines x as a differentiable function of x and y, then, at points where $F_z \neq 0$,

$$\frac{\partial z}{\partial x} = -\frac{F_x}{F_z} \quad \text{and} \quad \frac{\partial z}{\partial y} = -\frac{F_y}{F_z}.$$

Use these equations to find the values of $\partial z / \partial x$ and $\partial z / \partial y$ at the points given:

(a) $z^3 - xy + yz + y^3 - 2 = 0,$ $(1, 1, 1)$

(b) $\dfrac{1}{x} + \dfrac{1}{y} + \dfrac{1}{z} = 1,$ $(2, 3, 6)$

(c) $\sin(x + y) + \sin(y + z) + \sin(x + z) = 0,$ (π, π, π)

(d) $xe^y + ye^z + 2 \ln x - 2 - 3 \ln 2 = 0,$ $(1, \ln 2, \ln 3)$

45. *Polar Coordinates* Suppose we substitute polar coordinates $x = r \cos \theta$ and $y = r \sin \theta$ into a differentiable function $w = f(x, y)$.

(a) Show that

$$\frac{\partial w}{\partial r} = f_x \cos \theta + f_y \sin \theta$$

and

$$\frac{1}{r} \frac{\partial w}{\partial \theta} = -f_x \sin \theta + f_y \cos \theta.$$

(b) Solve the equations in (a) simultaneously to express f_x and f_y in terms of $\partial w / \partial r$ and $\partial w / \partial \theta$.

(c) Show that

$$(f_x)^2 + (f_y)^2 = \left(\frac{\partial w}{\partial r}\right)^2 + \frac{1}{r^2}\left(\frac{\partial w}{\partial \theta}\right)^2.$$

46. *Differentiating Integrals* Under mild continuity restrictions, it is true that if

$$F(x) = \int_a^b g(t, x)\, dt,$$

then

$$F'(x) = \int_a^b g_x(t, x)\, dt.$$

Using this fact and the Chain Rule, we can find the derivative of

$$F(x) = \int_a^{f(x)} g(t, x)\, dt$$

by letting

$$G(u, x) = \int_a^b g(t, x)\, dt,$$

where $u = f(x)$. Find the derivatives of the following functions defined by integrals.

(a) $F(x) = \displaystyle\int_0^{x^2} \sqrt{t^4 + x^3}\, dt$

(b) $F(x) = \displaystyle\int_{x^2}^1 \sqrt{t^3 + x^2}\, dt$

13.6 Directional Derivatives, Gradient Vectors, and Tangent Planes

Directional Derivatives in the Plane • Geometric Interpretation of the Directional Derivative • Calculation • Properties of Directional Derivatives • Gradients and Tangents to Level Curves • Functions of Three Variables • Tangent Planes and Normal Lines • Planes Tangent to a Surface $z = f(x, y)$ • Increments and Distance • Algebra Rules for Gradients

Directional Derivatives in the Plane

Suppose the function $f(x, y)$ is defined throughout a region R in the xy-plane, $P_0(x_0, y_0)$ is a point in R, and $\mathbf{u} = u_1\mathbf{i} + u_2\mathbf{j}$ is a unit vector. Then the equations

$$x = x_0 + su_1, \qquad y = y_0 + su_2$$

parametrize the line through P_0 parallel to $\mathbf{u}$. The parameter s measures arc length from P_0 in the direction of $\mathbf{u}$. We find the rate of change of f at P_0 in the direction of $\mathbf{u}$ by calculating df/ds at P_0 (Figure 13.37).

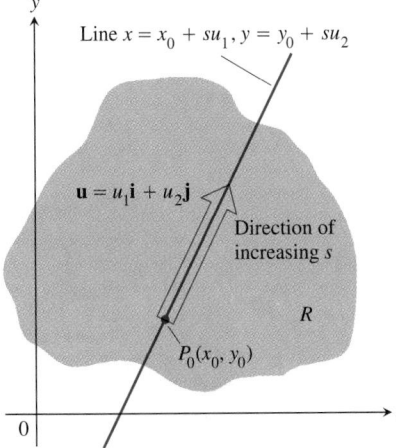

Figure 13.37 The rate of change of f in the direction of $\mathbf{u}$ at a point P_0 is the rate at which f changes along this line at P_0.

> **Definition** Directional Derivative
>
> The **derivative of f at $P_0(x_0, y_0)$ in the direction of the unit vector $\mathbf{u} = u_1\mathbf{i} + u_2\mathbf{j}$** is the number
>
> $$\left(\frac{df}{ds}\right)_{\mathbf{u}, P_0} = \lim_{s \to 0} \frac{f(x_0 + su_1, y_0 + su_2) - f(x_0, y_0)}{s}, \qquad (1)$$
>
> provided the limit exists.

The directional derivative is also denoted by

$$(D_{\mathbf{u}}f)_{P_0}. \qquad \text{"The derivative of f at P_0 in the direction of $\mathbf{u}$"}$$

Example 1 FINDING A DIRECTIONAL DERIVATIVE

Find the derivative of $f(x, y) = x^2 + xy$ at $P_0(1, 2)$ in the direction of the unit vector $\mathbf{u} = (1/\sqrt{2})\mathbf{i} + (1/\sqrt{2})\mathbf{j}$.

Solution

$$\left(\frac{df}{ds}\right)_{\mathbf{u}, P_0} = \lim_{s \to 0} \frac{f(x_0 + su_1, y_0 + su_2) - f(x_0, y_0)}{s} \qquad \text{Eq. (1)}$$

$$= \lim_{s \to 0} \frac{f\left(1 + s \cdot \dfrac{1}{\sqrt{2}}, \ 2 + s \cdot \dfrac{1}{\sqrt{2}}\right) - f(1, 2)}{s}$$

$$= \lim_{s \to 0} \frac{\left(1 + \dfrac{s}{\sqrt{2}}\right)^2 + \left(1 + \dfrac{s}{\sqrt{2}}\right)\left(2 + \dfrac{s}{\sqrt{2}}\right) - (1^2 + 1 \cdot 2)}{s}$$

$$= \lim_{s \to 0} \frac{\left(1 + \dfrac{2s}{\sqrt{2}} + \dfrac{s^2}{2}\right) + \left(2 + \dfrac{3s}{\sqrt{2}} + \dfrac{s^2}{2}\right) - 3}{s}$$

$$= \lim_{s \to 0} \frac{\dfrac{5s}{\sqrt{2}} + s^2}{s} = \lim_{s \to 0} \left(\frac{5}{\sqrt{2}} + s\right) = \frac{5}{\sqrt{2}}.$$

The rate of change of $f(x, y) = x^2 + xy$ at $P_0(1, 2)$ in the direction of $\mathbf{u} = (1/\sqrt{2})\mathbf{i} + (1/\sqrt{2})\mathbf{j}$ is $5/\sqrt{2}$.

Geometric Interpretation of the Directional Derivative

The equation $z = f(x, y)$ represents a surface S in space. If $z_0 = f(x_0, y_0)$, then the point $P(x_0, y_0, z_0)$ lies on S. The vertical plane that passes through P and $P_0(x_0, y_0)$ parallel to $\mathbf{u}$ intersects S in a curve C (Figure 13.38). The rate of change of f in the direction of $\mathbf{u}$ is the slope of the tangent to C at P.

Notice that when $\mathbf{u} = \mathbf{i}$, the directional derivative at P_0 is $\partial f/\partial x$ evaluated at (x_0, y_0). When $\mathbf{u} = \mathbf{j}$, the directional derivative at P_0 is $\partial f/\partial y$ evaluated at (x_0, y_0). The directional derivative generalizes the two partial derivatives. We can now ask for the rate of change of f in any direction $\mathbf{u}$, not just the directions $\mathbf{i}$ and $\mathbf{j}$.

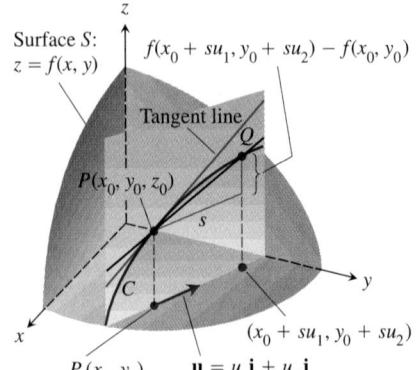

Figure 13.38 The slope of the curve C at P_0 is

$$\lim_{Q \to P} \text{slope}\,(PQ)$$

$$= \lim_{s \to 0} \frac{f(x_0 + su_1, y_0 + su_2) - f(x_0, y_0)}{s}$$

$$= \left(\frac{df}{ds}\right)_{\mathbf{u}, P_0}.$$

Calculation

As you know, it is rarely convenient to calculate a derivative directly from its definition as a limit, and the directional derivative is no exception. We can develop a more efficient formula in the following way. We begin with the line

$$x = x_0 + su_1, \qquad y = y_0 + su_2, \tag{2}$$

through $P_0(x_0, y_0)$, parametrized with the arc length parameter s increasing in the direction of the unit vector $\mathbf{u} = u_1\mathbf{i} + u_2\mathbf{j}$. Then

$$\left(\frac{df}{ds}\right)_{\mathbf{u},P_0} = \left(\frac{\partial f}{\partial x}\right)_{P_0}\frac{dx}{ds} + \left(\frac{\partial f}{\partial y}\right)_{P_0}\frac{dy}{ds} \qquad \text{Chain Rule}$$

$$= \left(\frac{\partial f}{\partial x}\right)_{P_0} \cdot u_1 + \left(\frac{\partial f}{\partial y}\right)_{P_0} \cdot u_2 \qquad \begin{array}{l}\text{From Eqs. (2), } dx/ds = \\ u_1 \text{ and } dy/ds = u_2.\end{array}$$

$$= \underbrace{\left[\left(\frac{\partial f}{\partial x}\right)_{P_0}\mathbf{i} + \left(\frac{\partial f}{\partial y}\right)_{P_0}\mathbf{j}\right]}_{\text{gradient of } f \text{ at } P_0} \bullet \underbrace{[u_1\mathbf{i} + u_2\mathbf{j}]}_{\text{direction } \mathbf{u}} \tag{3}$$

> **Definition** Gradient Vector or Gradient
>
> The **gradient vector (gradient)** of $f(x, y)$ at a point $P_0(x_0, y_0)$ is the vector
>
> $$\nabla f = \frac{\partial f}{\partial x}\mathbf{i} + \frac{\partial f}{\partial y}\mathbf{j}$$
>
> obtained by evaluating the partial derivatives of f at P_0.

The notation ∇f is read "grad f" as well as "gradient of f" and "del f." The symbol ∇ by itself is read "del." Another notation for the gradient is grad f, read the way it is written.

Equation (3) says that the derivative of f in the direction of $\mathbf{u}$ at P_0 is the dot product of $\mathbf{u}$ with the gradient of f at P_0.

> **Theorem 9** The Directional Derivative is a Scalar Product
>
> If the partial derivatives of $f(x, y)$ are defined at $P_0(x_0, y_0)$, then
>
> $$\left(\frac{df}{ds}\right)_{\mathbf{u},P_0} = (\nabla f)_{P_0} \bullet \mathbf{u}, \tag{4}$$
>
> the scalar product of the gradient f at P_0 and $\mathbf{u}$.

Example 2 FINDING THE DIRECTIONAL DERIVATIVE

Find the derivative of $f(x, y) = xe^y + \cos(xy)$ at the point $(2, 0)$ in the direction of $\mathbf{A} = 3\mathbf{i} - 4\mathbf{j}$.

Solution The direction of $\mathbf{A}$ is obtained by dividing $\mathbf{A}$ by its length

$$\mathbf{u} = \frac{\mathbf{A}}{|\mathbf{A}|} = \frac{\mathbf{A}}{5} = \frac{3}{5}\mathbf{i} - \frac{4}{5}\mathbf{j}.$$

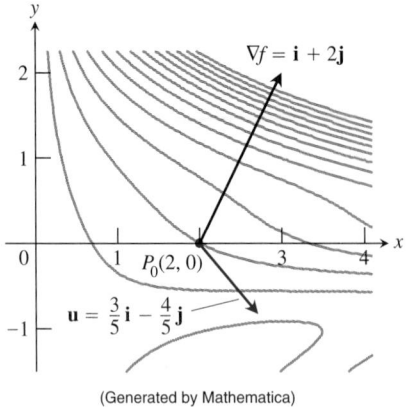

(Generated by Mathematica)

Figure 13.39 It is customary to picture ∇f as a vector in the domain of f. In the case of $f(x, y) = xe^y + \cos(xy)$, the domain is the entire plane. The rate at which f changes in the direction $\mathbf{u} = (3/5)\mathbf{i} - (4/5)\mathbf{j}$ at P_0 is $\nabla f \cdot \mathbf{u} = -1$. (Example 2)

The partial derivatives of f at $(2, 0)$ are

$$f_x(2, 0) = \left(e^y - y \sin(xy)\right)_{(2,0)} = e^0 - 0 = 1,$$
$$f_y(2, 0) = \left(xe^y - x \sin(xy)\right)_{(2,0)} = 2e^0 - 2 \cdot 0 = 2.$$

The gradient of f at $(2, 0)$ is

$$\nabla f|_{(2,0)} = f_x(2, 0)\ \mathbf{i} + f_y(2, 0)\ \mathbf{j} = \mathbf{i} + 2\mathbf{j}$$

(Figure 13.39). The derivative of f at $(2, 0)$ in the direction of $\mathbf{A}$ is therefore

$$(D_\mathbf{u}f)|_{(2,0)} = \nabla f|_{(2,0)} \cdot \mathbf{u} \qquad\qquad \text{Eq. (4)}$$

$$= (\mathbf{i} + 2\mathbf{j}) \cdot \left(\frac{3}{5}\mathbf{i} - \frac{4}{5}\mathbf{j}\right) = \frac{3}{5} - \frac{8}{5} = -1.$$

Properties of Directional Derivatives

Evaluating the dot product in the formula

$$D_\mathbf{u}f = \nabla f \cdot \mathbf{u} = |\nabla f||\mathbf{u}|\cos\theta = |\nabla f|\cos\theta,$$

where θ is the angle between the vectors $\mathbf{u}$ and ∇f, reveals the following properties.

Properties of the Directional Derivative
$$D_\mathbf{u}f = \nabla f \cdot \mathbf{u} = |\nabla f|\cos\theta$$

1. The function f increases most rapidly when $\cos\theta = 1$ or when $\mathbf{u}$ is the direction of ∇f. That is, at each point P in its domain, f increases most rapidly in the direction of the gradient vector ∇f at P. The derivative in this direction is
$$D_\mathbf{u}f = |\nabla f| \cos(0) = |\nabla f|.$$

2. Similarly, f decreases most rapidly in the direction of $-\nabla f$. The derivative in this direction is $D_\mathbf{u}f = |\nabla f| \cos(\pi) = -|\nabla f|$.

3. Any direction $\mathbf{u}$ orthogonal to the gradient is a direction of zero change in f because θ then equals $\pi/2$ and
$$D_\mathbf{u}f = |\nabla f| \cos(\pi/2) = |\nabla f| \cdot 0 = 0.$$

As we will discuss later, these properties hold in three dimensions as well as two.

Example 3 FINDING DIRECTIONS OF MAXIMAL, MINIMAL, AND ZERO CHANGE

Find the directions in which $f(x, y) = (x^2/2) + (y^2/2)$ **(a)** increases most rapidly and **(b)** decreases most rapidly at the point $(1, 1)$. **(c)** What are the directions of zero change in f at $(1, 1)$?

Solution

(a) The function increases most rapidly in the direction of ∇f at $(1, 1)$. The gradient is

$$(\nabla f)_{(1,1)} = (x\mathbf{i} + y\mathbf{j})_{(1,1)} = \mathbf{i} + \mathbf{j}.$$

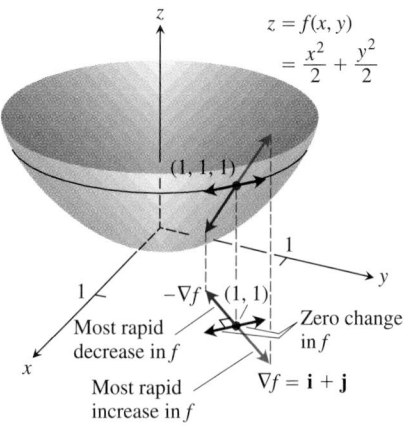

Figure 13.40 The direction in which $f(x, y) = (x^2/2) + (y^2/2)$ increases most rapidly at $(1, 1)$ is the direction of $\nabla f_{(1,1)} = \mathbf{i} + \mathbf{j}$. It corresponds to the direction of steepest ascent on the surface at $(1, 1, 1)$.

Its direction is

$$\mathbf{u} = \frac{\mathbf{i} + \mathbf{j}}{|\mathbf{i} + \mathbf{j}|} = \frac{\mathbf{i} + \mathbf{j}}{\sqrt{(1)^2 + (1)^2}} = \frac{1}{\sqrt{2}}\mathbf{i} + \frac{1}{\sqrt{2}}\mathbf{j}.$$

(b) The function decreases most rapidly in the direction of $-\nabla f$ at $(1, 1)$, which is

$$-\mathbf{u} = -\frac{1}{\sqrt{2}}\mathbf{i} - \frac{1}{\sqrt{2}}\mathbf{j}.$$

(c) The directions of zero change at $(1, 1)$ are the directions orthogonal to ∇f:

$$\mathbf{n} = -\frac{1}{\sqrt{2}}\mathbf{i} + \frac{1}{\sqrt{2}}\mathbf{j} \quad \text{and} \quad -\mathbf{n} = \frac{1}{\sqrt{2}}\mathbf{i} - \frac{1}{\sqrt{2}}\mathbf{j}.$$

See Figure 13.40.

Gradients and Tangents to Level Curves

If a differentiable function $f(x, y)$ has a constant value c along a smooth curve $\mathbf{r} = g(t)\mathbf{i} + h(t)\mathbf{j}$ (making the curve a level curve of f), then $f(g(t), h(t)) = c$. Differentiating both sides of this equation with respect to t leads to the equations

$$\frac{d}{dt}f\big(g(t), h(t)\big) = \frac{d}{dt}(c),$$

$$\frac{\partial f}{\partial x}\frac{dg}{dt} + \frac{\partial f}{\partial y}\frac{dh}{dt} = 0, \qquad \text{Chain Rule}$$

$$\underbrace{\left(\frac{\partial f}{\partial x}\mathbf{i} + \frac{\partial f}{\partial y}\mathbf{j}\right)}_{\nabla f} \cdot \underbrace{\left(\frac{dg}{dt}\mathbf{i} + \frac{dh}{dt}\mathbf{j}\right)}_{\frac{d\mathbf{r}}{dt}} = 0. \tag{5}$$

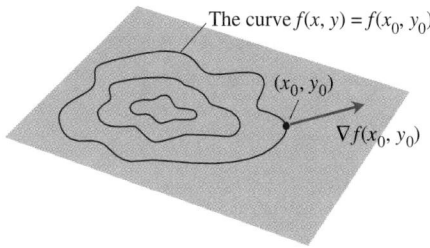

Figure 13.41 The gradient of a differentiable function of two variables at a point is always normal to the function's level curve through the point.

Equation (5) says that ∇f is normal to the tangent vector $d\mathbf{r}/dt$, so it is normal to the curve.

At every point (x_0, y_0) in the domain of $f(x, y)$, the gradient of f is normal to the level curve through (x_0, y_0) (Figure 13.41). This observation enables us to find equations for tangent lines to level curves. They are the lines normal to the gradients. The line through a point $P_0(x_0, y_0)$ normal to a vector $\mathbf{N} = A\mathbf{i} + B\mathbf{j}$ has the equation

$$A(x - x_0) + B(y - y_0) = 0$$

(Exercise 59). If $\mathbf{N}$ is the gradient $(\nabla f)_{(x_0, y_0)} = f_x(x_0, y_0)\mathbf{i} + f_y(x_0, y_0)\mathbf{j}$, the equation becomes

$$f_x(x_0, y_0)(x - x_0) + f_y(x_0, y_0)(y - y_0) = 0. \tag{6}$$

Example 4 FINDING AN EQUATION FOR A TANGENT LINE

Find an equation for the line tangent to the ellipse

$$\frac{x^2}{4} + y^2 = 2$$

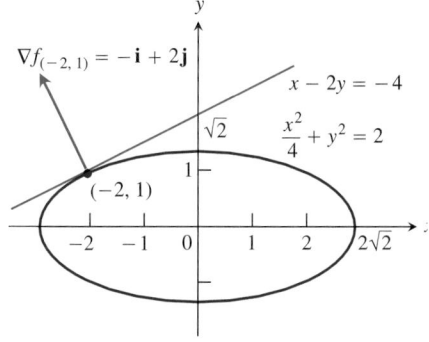

Figure 13.42 We can find the tangent to the ellipse $(x^2/4) + y^2 = 2$ by treating the ellipse as a level curve of the function $f(x, y) = (x^2/4) + y^2$. (Example 4)

(Figure 13.42) at the point $(-2, 1)$.

Solution   The ellipse is a level curve of the function

$$f(x, y) = \frac{x^2}{4} + y^2.$$

The gradient of f at $(-2, 1)$ is

$$\nabla f|_{(-2,1)} = \left(\frac{x}{2}\mathbf{i} + 2y\mathbf{j}\right)_{(-2,1)} = -\mathbf{i} + 2\mathbf{j}.$$

The tangent (from Equation 6) is the line

$$(-1)(x + 2) + (2)(y - 1) = 0$$
$$x - 2y = -4.$$

Functions of Three Variables

We obtain three-variable formulas by adding the z-terms to the two-variable formulas. For a differentiable function $f(x, y, z)$ and a unit vector $\mathbf{u} = u_1\mathbf{i} + u_2\mathbf{j} + u_3\mathbf{k}$ in space, we have

$$\nabla f = \frac{\partial f}{\partial x}\mathbf{i} + \frac{\partial f}{\partial y}\mathbf{j} + \frac{\partial f}{\partial z}\mathbf{k}$$

and

$$D_{\mathbf{u}}f = \nabla f \cdot \mathbf{u} = \frac{\partial f}{\partial x}u_1 + \frac{\partial f}{\partial y}u_2 + \frac{\partial f}{\partial z}u_3.$$

The directional derivative can once again be written in the form

$$D_{\mathbf{u}}f = \nabla f \cdot \mathbf{u} = |\nabla f||\mathbf{u}|\cos \theta = |\nabla f|\cos \theta,$$

so the properties listed earlier for functions of two variables continue to hold. At any given point, f increases most rapidly in the direction of ∇f and decreases most rapidly in the direction of $-\nabla f$. In any direction orthogonal to ∇f, the derivative is zero.

Example 5   FINDING DIRECTIONS OF MAXIMAL, MINIMAL, AND ZERO CHANGE

(a) Find the derivative of $f(x, y, z) = x^3 - xy^2 - z$ at $P_0(1, 1, 0)$ in the direction of $\mathbf{A} = 2\mathbf{i} - 3\mathbf{j} + 6\mathbf{k}$.

(b) In what directions does f change most rapidly at P_0, and what are the rates of change in these directions?

Solution

(a) The direction of $\mathbf{A}$ is obtained by dividing $\mathbf{A}$ by its length:

$$|\mathbf{A}| = \sqrt{(2)^2 + (-3)^2 + (6)^2} = \sqrt{49} = 7$$

$$\mathbf{u} = \frac{\mathbf{A}}{|\mathbf{A}|} = \frac{2}{7}\mathbf{i} - \frac{3}{7}\mathbf{j} + \frac{6}{7}\mathbf{k}.$$

The partial derivatives of f at P_0 are

$$f_x = 3x^2 - y^2|_{(1,1,0)} = 2, \qquad f_y = -2xy|_{(1,1,0)} = -2, \qquad f_z = -1|_{(1,1,0)} = -1.$$

The gradient of f at P_0 is

$$\nabla f|_{(1,1,0)} = 2\mathbf{i} - 2\mathbf{j} - \mathbf{k}.$$

The derivative of f at P_0 in the direction of $\mathbf{A}$ is therefore

$$(D_{\mathbf{u}}f)\big|_{(1,1,0)} = \nabla f\big|_{(1,1,0)} \cdot \mathbf{u} = (2\mathbf{i} - 2\mathbf{j} - \mathbf{k}) \cdot \left(\frac{2}{7}\mathbf{i} - \frac{3}{7}\mathbf{j} + \frac{6}{7}\mathbf{k}\right)$$

$$= \frac{4}{7} + \frac{6}{7} - \frac{6}{7} = \frac{4}{7}.$$

(b) The function increases most rapidly in the direction of $\nabla f = 2\mathbf{i} - 2\mathbf{j} - \mathbf{k}$ and decreases most rapidly in the direction of $-\nabla f$. The rates of change in the directions are, respectively,

$$|\nabla f| = \sqrt{(2)^2 + (-2)^2 + (-1)^2} = \sqrt{9} = 3 \qquad \text{and} \qquad -|\nabla f| = -3.$$

Tangent Planes and Normal Lines

If $\mathbf{r} = g(t)\mathbf{i} + h(t)\mathbf{j} + k(t)\mathbf{k}$ is a smooth curve on the level surface $f(x, y, z) = c$ of a differentiable function f, then $f(g(t), h(t), k(t)) = c$. Differentiating both sides of this equation with respect to t leads to

$$\frac{d}{dt}f\big(g(t), h(t), k(t)\big) = \frac{d}{dt}(c),$$

$$\frac{\partial f}{\partial x}\frac{dg}{dt} + \frac{\partial f}{\partial y}\frac{dh}{dt} + \frac{\partial f}{\partial z}\frac{dk}{dt} = 0, \qquad \text{Chain Rule}$$

$$\underbrace{\left(\frac{\partial f}{\partial x}\mathbf{i} + \frac{\partial f}{\partial y}\mathbf{j} + \frac{\partial f}{\partial z}\mathbf{k}\right)}_{\nabla f} \cdot \underbrace{\left(\frac{dg}{dt}\mathbf{i} + \frac{dh}{dt}\mathbf{j} + \frac{dk}{dt}\mathbf{k}\right)}_{d\mathbf{r}/dt} = 0. \qquad (7)$$

At every point on the curve, ∇f is orthogonal to the curve's velocity vector.

Now let us restrict our attention to the curves that pass through a fixed point P_0 on the surface (Figure 13.43). All the velocity vectors at P_0 are orthogonal to ∇f at P_0, so the curve's tangent lines all lie in the plane through P_0 normal to ∇f. This is the *tangent plane* of the surface at P_0. The line through P_0 perpendicular to the plane is the surface's *normal line* at P_0.

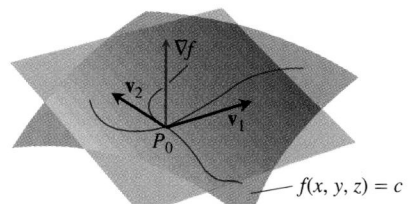

Figure 13.43 ∇f is orthogonal to the velocity vector of every smooth curve in the surface through P_0. The velocity vectors at P_0 therefore lie in a common plane, the tangent plane at P_0.

Definitions Tangent Plane and Normal Line

The **tangent plane** at the point $P_0(x_0, y_0, z_0)$ on the level surface $f(x, y, z) = c$ is the plane through P_0 normal to $\nabla f\big|_{P_0}$. The **normal line** of the surface at P_0 is the line through P_0 parallel to $\nabla f\big|_{P_0}$.

Thus, from Section 11.4, the tangent plane and normal line, respectively, have the following equations:

$$f_x(P_0)(x - x_0) + f_y(P_0)(y - y_0) + f_z(P_0)(z - z_0) = 0, \qquad (8)$$

$$x = x_0 + f_x(P_0)t, \qquad y = y_0 + f_y(P_0)t, \qquad z = z_0 + f_z(P_0)t. \qquad (9)$$

Example 6 FINDING TANGENT PLANE AND NORMAL LINE

Find the tangent plane and normal line of the surface

$$f(x, y, z) = x^2 + y^2 + z - 9 = 0 \qquad \text{A circular paraboloid}$$

at the point $P_0(1, 2, 4)$.

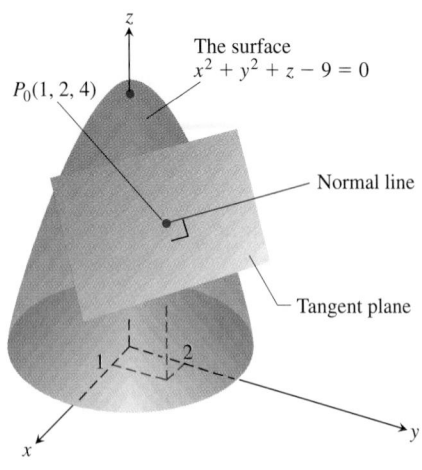

Figure 13.44 The tangent plane and normal line to the surface $x^2 + y^2 + z - 9 = 0$ at $P_0(1, 2, 4)$. (Example 6)

Solution The surface is shown in Figure 13.44.

The tangent plane is the plane through P_0 perpendicular to the gradient of f at P_0. The gradient is

$$\nabla f\big|_{P_0} = (2x\mathbf{i} + 2y\mathbf{j} + \mathbf{k})_{(1,2,4)} = 2\mathbf{i} + 4\mathbf{j} + \mathbf{k}.$$

The plane is therefore the plane

$$2(x - 1) + 4(y - 2) + (z - 4) = 0, \qquad \text{or} \qquad 2x + 4y + z = 14.$$

The line normal to the surface at P_0 is

$$x = 1 + 2t, \qquad y = 2 + 4t, \qquad z = 4 + t.$$

Example 7   FINDING PARAMETRIC EQUATIONS FOR A LINE TANGENT TO A CURVE

The surfaces

$$f(x, y, z) = x^2 + y^2 - 2 = 0 \qquad \text{A cylinder}$$

and

$$g(x, y, z) = x + z - 4 = 0 \qquad \text{A plane}$$

meet in an ellipse (Figure 13.45). Find parametric equations for the line tangent to the ellipse at the point $P_0(1, 1, 3)$.

Solution The tangent line is orthogonal to both ∇f and ∇g at P_0 and is therefore parallel to $\mathbf{v} = \nabla f \times \nabla g$. The components of $\mathbf{v}$ and the coordinates of P_0 give us equations for the line. We have

$$\nabla f_{(1,1,3)} = (2x\mathbf{i} + 2y\mathbf{j})_{(1,1,3)} = 2\mathbf{i} + 2\mathbf{j},$$

$$\nabla g_{(1,1,3)} = (\mathbf{i} + \mathbf{k})_{(1,1,3)} = \mathbf{i} + \mathbf{k},$$

$$\mathbf{v} = (2\mathbf{i} + 2\mathbf{j}) \times (\mathbf{i} + \mathbf{k}) = \begin{vmatrix} \mathbf{i} & \mathbf{j} & \mathbf{k} \\ 2 & 2 & 0 \\ 1 & 0 & 1 \end{vmatrix} = 2\mathbf{i} - 2\mathbf{j} - 2\mathbf{k}.$$

The line is

$$x = 1 + 2t, \qquad y = 1 - 2t, \qquad z = 3 - 2t.$$

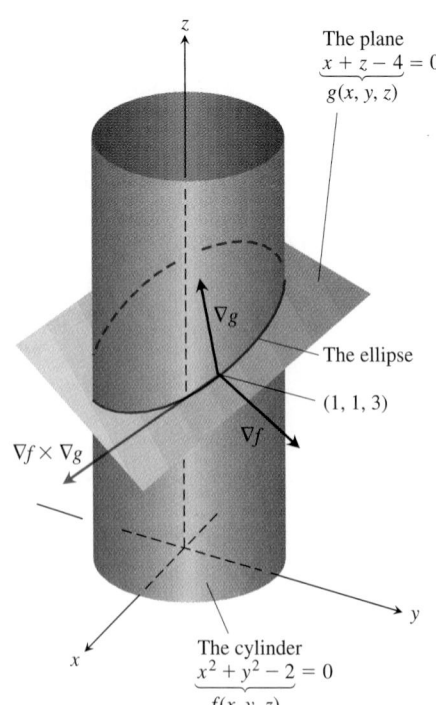

Figure 13.45 The cylinder $f(x, y, z) = x^2 + y^2 - 2 = 0$ and the plane $g(x, y, z) = x + z - 4 = 0$ intersect in an ellipse. (Example 7)

Planes Tangent to a Surface *z* = *f*(*x, y*)

To find an equation for the plane tangent to a surface $z = f(x, y)$ at a point $P_0(x_0, y_0, z_0)$, where $z_0 = f(x_0, y_0)$, we first observe that the equation $z = f(x, y)$ is equivalent to $f(x, y) - z = 0$. The surface $z = f(x, y)$ is therefore the zero level surface of the function $F(x, y, z) = f(x, y) - z$. The partial derivatives of F are

$$F_x = \frac{\partial}{\partial x}\big(f(x, y) - z\big) = f_x - 0 = f_x,$$

$$F_y = \frac{\partial}{\partial y}\big(f(x, y) - z\big) = f_y - 0 = f_y,$$

$$F_z = \frac{\partial}{\partial z}\big(f(x, y) - z\big) = 0 - 1 = -1.$$

The formula

$$F_x(P_0)(x - x_0) + F_y(P_0)(y - y_0) + F_z(P_0)(z - z_0) = 0 \qquad \text{Eq. (8) restated for } F(x, y, z)$$

for the plane tangent to the level surface at P_0 therefore reduces to

$$f_x(x_0, y_0)(x - x_0) + f_y(x_0, y_0)(y - y_0) - (z - z_0) = 0.$$

Plane Tangent to a Surface $z = f(x, y)$ at $(x_0, y_0, f(x_0, y_0))$

The plane tangent to the surface $z = f(x, y)$ at the point
$P(x_0, y_0, z_0) = (x_0, y_0, f(x_0, y_0))$ is

$$f_x(x_0, y_0)(x - x_0) + f_y(x_0, y_0)(y - y_0) - (z - z_0) = 0. \qquad (10)$$

Example 8 FINDING A PLANE TANGENT TO A SURFACE $z = f(x, y)$

Find the plane tangent to the surface $z = x \cos y - ye^x$ at $(0, 0, 0)$.

Solution We calculate the partial derivatives of $f(x, y) = x \cos y - ye^x$ and use Equation (10):

$$f_x(0, 0) = \left(\cos y - ye^x\right)_{(0, 0)} = 1 - 0 \cdot 1 = 1,$$

$$f_y(0, 0) = (-x \sin y - e^x)_{(0, 0)} = 0 - 1 = -1.$$

The tangent plane is therefore

$$1 \cdot (x - 0) - 1 \cdot (y - 0) - (z - 0) = 0, \qquad \text{Eq. (10)}$$

or

$$x - y - z = 0.$$

Increments and Distance

The directional derivative plays the role of an ordinary derivative when we want to estimate how much the value of a function f will change if we move a small distance ds from a point P_0 to another point nearby. If f were a function of a single variable, we would have

$$df = f'(P_0) \, ds. \qquad \text{Ordinary derivative} \times \text{increment}$$

For a function of two or more variables, we use the formula

$$df = \left(\nabla f|_{P_0} \cdot \mathbf{u}\right) ds, \qquad \text{Directional derivative} \times \text{increment}$$

where $\mathbf{u}$ is the direction of the motion away from P_0.

Estimating the Change in f in a Direction u

To estimate the change in the value of a function f when we move a small distance ds from a point P_0 in a particular direction $\mathbf{u}$, use the formula

$$df = \underbrace{\left(\nabla f|_{P_0} \cdot \mathbf{u}\right)}_{\substack{\text{directional} \\ \text{derivative}}} \cdot \underbrace{ds.}_{\substack{\text{distance} \\ \text{increment}}}$$

Example 9   ESTIMATING CHANGE

Estimate how much the value of $f(x, y, z) = xe^y + yz$ will change if the point $P(x, y, z)$ moves 0.1 units from $P_0(2, 0, 0)$ straight toward $P_1(4, 1, -2)$.

Solution   We first find the derivative of f in the direction of

$$\overrightarrow{P_0P_1} = 2\mathbf{i} + \mathbf{j} - 2\mathbf{k}.$$

The direction of this vector is

$$\mathbf{u} = \frac{\overrightarrow{P_0P_1}}{\left|\overrightarrow{P_0P}\right|} = \frac{\overrightarrow{P_0P}}{3} = \frac{2}{3}\mathbf{i} + \frac{1}{3}\mathbf{j} - \frac{2}{3}\mathbf{k}.$$

The gradient of f at P_0 is

$$\nabla f\big|_{(2,0,0)} = (e^y\mathbf{i} + (xe^y + z)\mathbf{j} + y\mathbf{k})\big|_{(2,0,0)} = \mathbf{i} + 2\mathbf{j}.$$

Therefore,

$$\nabla f\big|_{P_0} \cdot \mathbf{u} = (\mathbf{i} + 2\mathbf{j}) \cdot \left(\frac{2}{3}\mathbf{i} + \frac{1}{3}\mathbf{j} - \frac{2}{3}\mathbf{k}\right) = \frac{2}{3} + \frac{2}{3} = \frac{4}{3}.$$

The change in f that results from moving $ds = 0.1$ unit away from P_0 in the direction of $\mathbf{u}$ is approximately

$$df = \left(\nabla f\big|_{P_0} \cdot \mathbf{u}\right)(ds) = \left(\frac{4}{3}\right)(0.1) \approx 0.13.$$

Algebra Rules for Gradients

If we know the gradients of two functions f and g, we automatically know the gradients of their constant multiples, sum, difference, product, and quotient.

These rules have the same form as the corresponding rules for derivatives, as they should (Exercise 65).

Algebra Rules for Gradients

1. *Constant Multiple Rule:* $\nabla(kf) = k\nabla f$ (any number k)

2. *Sum Rule:* $\nabla(f + g) = \nabla f + \nabla g$

3. *Difference Rule:* $\nabla(f - g) = \nabla f - \nabla g$

4. *Product Rule:* $\nabla(fg) = f\nabla g + g\nabla f$

5. *Quotient Rule:* $\nabla\left(\dfrac{f}{g}\right) = \dfrac{g\nabla f - f\nabla g}{g^2}$

Example 10   ILLUSTRATING THE GRADIENT RULES

We illustrate the rules with $f(x, y, z) = x - y$, $g(x, y, z) = z$.

Solution   With $\nabla f = \mathbf{i} - \mathbf{j}$ and $\nabla g = \mathbf{k}$, we have

1. $\nabla(2f) = \nabla(2x - 2y) = 2\mathbf{i} - 2\mathbf{j} = 2\nabla f.$

2. $\nabla(f + g) = \nabla(x - y + z) = \mathbf{i} - \mathbf{j} + \mathbf{k} = \nabla f + \nabla g.$

3. $\nabla(f - g) = \nabla(x - y - z) = \mathbf{i} - \mathbf{j} - \mathbf{k} = \nabla f - \nabla g.$

4. $\nabla(fg) = \nabla(xz - yz) = z\mathbf{i} - z\mathbf{j} + (x - y)\mathbf{k} = g\nabla f + f\nabla g.$

$$5. \ \nabla\left(\frac{f}{g}\right) = \nabla\left(\frac{x-y}{z}\right)$$

$$= \frac{\partial}{\partial x}\left(\frac{x-y}{z}\right)\mathbf{i} + \frac{\partial}{\partial y}\left(\frac{x-y}{z}\right)\mathbf{j} + \frac{\partial}{\partial z}\left(\frac{x-y}{z}\right)\mathbf{k}$$

$$= \frac{1}{z}\mathbf{i} - \frac{1}{z}\mathbf{j} - \frac{x-y}{z^2}\mathbf{k}$$

$$= \frac{z\mathbf{i} - z\mathbf{j} - (x-y)\mathbf{k}}{z^2} = \frac{g\nabla f - f\nabla g}{g^2}.$$

Quick Review 13.6

In Exercises 1 and 2, let $\mathbf{u} = 2\mathbf{i} - \mathbf{j} + \mathbf{k}$.

1. Find the length of $\mathbf{u}$.

2. Find the direction of $\mathbf{u}$.

In Exercises 3–6, let $\mathbf{u} = 3\mathbf{i} - \mathbf{j}$ and $\mathbf{v} = -\mathbf{i} + 2\mathbf{j} - \mathbf{k}$.

3. Find $\mathbf{u} \cdot \mathbf{v}$.

4. Find $\mathbf{u} \times \mathbf{v}$.

5. Find a vector perpendicular to the plane determined by $\mathbf{u}$ and $\mathbf{v}$.

6. Find the direction of $\mathbf{u} \times \mathbf{v}$.

7. Find parametric equations for the line through the points $(-1, 2, -1)$ and $(3, 2, 0)$.

8. Find parametric equations for the line through the point $(2, -1, 1)$ and parallel to the vector $3\mathbf{i} - 2\mathbf{j} + \mathbf{k}$.

9. Find an equation for the line through the point $(2, 0, -2)$ perpendicular to the plane $x - 2y + 3z = 1$.

10. Find an equation for the plane through the point $(1, 0, -2)$ normal to the vector $\mathbf{n} = -\mathbf{i} + 3\mathbf{j} - 2\mathbf{k}$.

Section 13.6 Exercises

In Exercises 1–4, find the gradient of the function at the given point. Then sketch the gradient together with the level curve that passes through the point.

1. $f(x, y) = y - x$, $(2, 1)$

2. $f(x, y) = \ln(x^2 + y^2)$, $(1, 1)$

3. $g(x, y) = y - x^2$, $(-1, 0)$

4. $g(x, y) = \frac{x^2}{2} - \frac{y^2}{2}$, $(\sqrt{2}, 1)$

In Exercises 5–8, find ∇f at the given point.

5. $f(x, y, z) = x^2 + y^2 - 2z^2 + z \ln x$, $(1, 1, 1)$

6. $f(x, y, z) = 2z^3 - 3(x^2 + y^2)z + \tan^{-1} xz$, $(1, 1, 1)$

7. $f(x, y, z) = (x^2 + y^2 + z^2)^{-1/2} + \ln(xyz)$, $(-1, 2, -2)$

8. $f(x, y, z) = e^{x+y} \cos z + (y + 1) \sin^{-1} x$, $(0, 0, \pi/6)$

In Exercises 9–16, find the derivative of the function at P_0 in the direction of $\mathbf{A}$.

9. $f(x, y) = 2xy - 3y^2$, $P_0(5, 5)$, $\mathbf{A} = 4\mathbf{i} + 3\mathbf{j}$

10. $f(x, y) = 2x^2 + y^2$, $P_0(-1, 1)$, $\mathbf{A} = 3\mathbf{i} - 4\mathbf{j}$

11. $g(x, y) = x - (y^2/x) + \sqrt{3} \sec^{-1}(2xy)$, $P_0(1, 1)$, $\mathbf{A} = 12\mathbf{i} + 5\mathbf{j}$

12. $h(x, y) = \tan^{-1}(y/x) + \sqrt{3} \sin^{-1}(xy/2)$, $P_0(1, 1)$, $\mathbf{A} = 3\mathbf{i} - 2\mathbf{j}$

13. $f(x, y, z) = xy + yz + zx$, $P_0(1, -1, 2)$, $\mathbf{A} = 3\mathbf{i} + 6\mathbf{j} - 2\mathbf{k}$

14. $f(x, y, z) = x^2 + 2y^2 - 3z^2$, $P_0(1, 1, 1)$, $\mathbf{A} = \mathbf{i} + \mathbf{j} + \mathbf{k}$

15. $g(x, y, z) = 3e^x \cos yz$, $P_0(0, 0, 0)$, $\mathbf{A} = 2\mathbf{i} + \mathbf{j} - 2\mathbf{k}$

16. $h(x, y, z) = \cos xy + e^{yz} + \ln zx$, $P_0(1, 0, 1/2)$, $\mathbf{A} = \mathbf{i} + 2\mathbf{j} + 2\mathbf{k}$

In Exercises 17–22, find the directions in which the function increases and decreases most rapidly at P_0. Then find the derivative of the function in these directions.

17. $f(x, y) = x^2 + xy + y^2$, $P_0(-1, 1)$

18. $f(x, y) = x^2 y + e^{xy} \sin y$, $P_0(1, 0)$

19. $f(x, y, z) = (x/y) - yz$, $P_0(4, 1, 1)$

20. $g(x, y, z) = xe^y + z^2$, $P_0(1, \ln 2, 1/2)$

21. $f(x, y, z) = \ln xy + \ln yz + \ln xz$, $P_0(1, 1, 1)$

22. $h(x, y, z) = \ln(x^2 + y^2 - 1) + y + 6z$, $P_0(1, 1, 0)$

23. *Estimating Change* By about how much will

$$f(x, y, z) = \ln \sqrt{x^2 + y^2 + z^2}$$

change if the point $P(x, y, z)$ moves from $P_0(3, 4, 12)$ a distance of $ds = 0.1$ units in the direction of $3\mathbf{i} + 6\mathbf{j} - 2\mathbf{k}$?

24. *Estimating Change* By about how much will

$$f(x, y, z) = e^x \cos yz$$

change as the point $P(x, y, z)$ moves from the origin a distance of $ds = 0.1$ units in the direction of $2\mathbf{i} + 2\mathbf{j} - 2\mathbf{k}$?

25. *Estimating Change* By about how much will

$$g(x, y, z) = x + x \cos z - y \sin z + y$$

change if the point $P(x, y, z)$ moves from $P_0(2, -1, 0)$ a distance of $ds = 0.2$ units toward the point $P_1(0, 1, 2)$?

26. *Estimating Change* By about how much will

$$h(x, y, z) = \cos(\pi xy) + xz^2$$

change if the point $P(x, y, z)$ moves from $P_0(-1, -1, -1)$ a distance of $ds = 0.1$ units toward the origin?

In Exercises 27–34, find an equation for the **(a)** tangent plane and **(b)** normal line at the point P_0 on the given surface.

27. $x^2 + y^2 + z^2 = 3$, $\quad P_0(1, 1, 1)$

28. $x^2 + y^2 - z^2 = 18$, $\quad P_0(3, 5, -4)$

29. $2z - x^2 = 0$, $\quad P_0(2, 0, 2)$

30. $x^2 + 2xy - y^2 + z^2 = 7$, $\quad P_0(1, -1, 3)$

31. $\cos \pi x - x^2 y + e^{xz} + yz = 4$, $\quad P_0(0, 1, 2)$

32. $x^2 - xy - y^2 - z = 0$, $\quad P_0(1, 1, -1)$

33. $x + y + z = 1$, $\quad P_0(0, 1, 0)$

34. $x^2 + y^2 - 2xy - x + 3y - z = -4$, $\quad P_0(2, -3, 18)$

In Exercises 35–38, find an equation for the plane that is tangent to the given surface at the given point.

35. $z = \ln(x^2 + y^2)$, $\quad (1, 0, 0)$

36. $z = e^{-(x^2+y^2)}$, $\quad (0, 0, 1)$

37. $z = \sqrt{y - x}$, $\quad (1, 2, 1)$

38. $z = 4x^2 + y^2$, $\quad (1, 1, 5)$

In Exercises 39–42, sketch the curve $f(x, y) = c$ together with ∇f and the tangent line at the given point. Then write an equation for the tangent line.

39. $x^2 + y^2 = 4$, $\quad (\sqrt{2}, \sqrt{2})$

40. $x^2 - y = 1$, $\quad (\sqrt{2}, 1)$

41. $xy = -4$, $\quad (2, -2)$

42. $x^2 - xy + y^2 = 7$, $\quad (-1, 2)$ $\quad$ (These are the curve and point in Section 3.7, Example 4.)

In Exercises 43–48, *work in groups of two or three* to find parametric equations for the line tangent to the curve of intersection of the surfaces at the given point.

43. Surfaces: $x + y^2 + 2z = 4$, $\quad x = 1$
Point: $(1, 1, 1)$

44. Surfaces: $xyz = 1$, $\quad x^2 + 2y^2 + 3z^2 = 6$
Point: $(1, 1, 1)$

45. Surfaces: $x^2 + 2y + 2z = 4$, $\quad y = 1$
Point: $(1, 1, 1/2)$

46. Surfaces: $x + y^2 + z = 2$, $\quad y = 1$
Point: $(1/2, 1, 1/2)$

47. Surfaces: $x^3 + 3x^2 y^2 + y^3 + 4xy - z^2 = 0$,
$x^2 + y^2 + z^2 = 11$
Point: $(1, 1, 3)$

48. Surfaces: $x^2 + y^2 = 4$, $\quad x^2 + y^2 - z = 0$
Point: $(\sqrt{2}, \sqrt{2}, 4)$

49. In what directions is the derivative of $f(x, y) = xy + y^2$ at $P(3, 2)$ equal to zero?

50. In what two directions is the derivative of $f(x, y) = (x^2 - y^2)/(x^2 + y^2)$ at $P(1, 1)$ equal to zero?

51. Writing to Learn Is there a direction $\mathbf{A}$ in which the rate of change of $f(x, y) = x^2 - 3xy + 4y^2$ at $P(1, 2)$ equals 14? Give reasons for your answer.

52. Writing to Learn Is there a direction $\mathbf{A}$ in which the rate of change of the temperature function $T(x, y, z) = 2xy - yz$ (temperature in degrees Celsius, distance in meters) at $P(1, -1, 1)$ is $-3°$C/m? Give reasons for your answer.

53. Writing to Learn The derivative of $f(x, y)$ at $P_0(1, 2)$ in the direction of $\mathbf{i} + \mathbf{j}$ is $2\sqrt{2}$ and in the direction of $-2\mathbf{j}$ is -3. What is the derivative of f in the direction of $-\mathbf{i} - 2\mathbf{j}$? Give reasons for your answer.

54. The derivative of $f(x, y, z)$ at a point P is greatest in the direction of $\mathbf{A} = \mathbf{i} + \mathbf{j} - \mathbf{k}$. In this direction the value of the derivative is $2\sqrt{3}$.

(a) Writing to Learn What is ∇f at P? Give reasons for your answer.

(b) What is the derivative of f at P in the direction of $\mathbf{i} + \mathbf{j}$?

Explorations

55. *Temperature Change Along a Circle* Suppose the Celsius temperature at the point (x, y) in the xy-plane is $T(x, y) = x \sin 2y$ and distance in the xy-plane is measured in meters. A particle is moving *clockwise* around the circle of radius 1 m centered at the origin at the constant rate of 2 m/sec.

(a) How fast is the temperature experienced by the particle changing in °C/m at the point $P(1/2, \sqrt{3}/2)$?

(b) How fast is the temperature experienced by the particle changing in degrees Celsius per second at P?

56. *Change Along the Involute of a Circle* Find the derivative of $f(x, y) = x^2 + y^2$ in the direction of the unit tangent vector of the curve $\mathbf{r}(t) = (\cos t + t \sin t)\mathbf{i} + (\sin t - t \cos t)\mathbf{j}$, $t > 0$ (Figure 13.46).

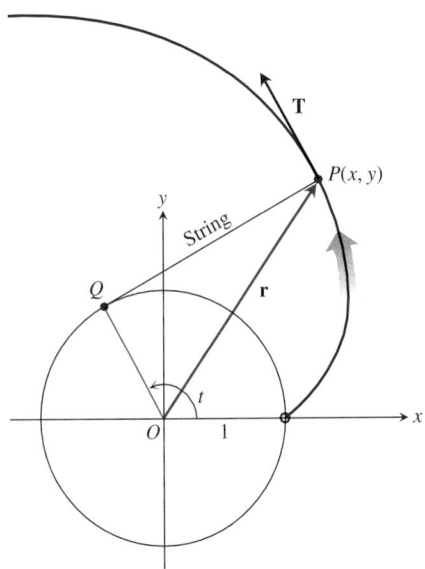

Figure 13.46 If you move along the involute of the unit circle, covering distance along the curve at a constant rate, your distance from the origin will increase at a constant rate as well. (This is how to interpret the result of your calculation in Exercise 56.)

57. *Change Along a Helix* Find the derivative of $f(x, y, z) = x^2 + y^2 + z^2$ in the direction of the unit tangent vector of the helix $\mathbf{r}(t) = (\cos t)\mathbf{i} + (\sin t)\mathbf{j} + t\mathbf{k}$ at the points where $t = -\pi/4$, 0, and $\pi/4$. The function f gives the square of the distance from a point $P(x, y, z)$ on the helix to the origin. The derivatives calculated here give the rates at which the square of the distance is changing with respect to t as P moves through the points where $t = -\pi/4$, 0, and $\pi/4$.

58. *Changing Temperature* The Celsius temperature in a region in space is given by $T(x, y, z) = 2x^2 - xyz$. A particle is moving in this region and its position at time t is given by $x = 2t^2$, $y = 3t$, $z = -t^2$, where time is measured in seconds and distance in meters.

(a) How fast is the temperature experienced by the particle changing in degrees Celsius per meter when the particle is at the point $P(8, 6, -4)$?

(b) How fast is the temperature experienced by the particle changing in degrees Celsius per second at P? ∎

59. *Lines in the xy-Plane* Show that $A(x - x_0) + B(y - y_0) = 0$ is an equation for the line in the xy-plane through the point (x_0, y_0) normal to the vector $\mathbf{N} = A\mathbf{i} + B\mathbf{j}$.

60. *Normal Curves and Tangent Curves* A curve is **normal** to a surface $f(x, y, z) = c$ at a point of intersection if the curve's velocity vector is a scalar multiple of ∇f at the point. The curve is **tangent** to the surface at a point of intersection if its velocity vector is orthogonal to ∇f there.

(a) Show that the curve
$$\mathbf{r}(t) = \sqrt{t}\,\mathbf{i} + \sqrt{t}\,\mathbf{j} - \frac{1}{4}(t + 3)\mathbf{k}$$
is normal to the surface $x^2 + y^2 - z = 3$ when $t = 1$.

(b) Show that the curve
$$\mathbf{r}(t) = \sqrt{t}\,\mathbf{i} + \sqrt{t}\,\mathbf{j} + (2t - 1)\mathbf{k}$$
is tangent to the surface $x^2 + y^2 - z = 1$ when $t = 1$.

Extending the Ideas

61. *Another Way to See Why Gradients Are Normal to Level Curves* Suppose a differentiable function $f(x, y)$ has a constant value c along the differentiable curve $x = g(t)$, $y = h(t)$ for all values of t. Differentiate both sides of the equation $f(g(t), h(t)) = c$ with respect to t to show that ∇f is normal to the curve's tangent vector at every point.

62. *The Linearization of f(x,y) Is a Tangent-Plane Approximation* Show that the tangent plane at the point $P_0(x_0, y_0, f(x_0, y_0))$ on the surface $z = f(x, y)$ defined by a differentiable function f is the plane
$$f_x(x_0, y_0)(x - x_0) + f_y(x_0, y_0)(y - y_0) \\ - (z - f(x_0, y_0)) = 0$$
or
$$z = f(x_0, y_0) + f_x(x_0, y_0)(x - x_0) + f_y(x_0, y_0)(y - y_0).$$
Thus, the tangent plane at P_0 is the graph of the linearization of f at P_0 (Figure 13.47).

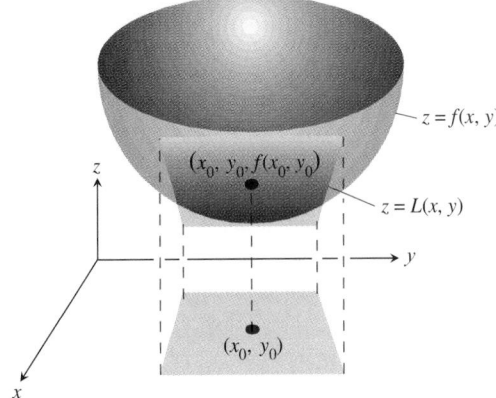

Figure 13.47 The graph of a function $z = f(x, y)$ and its linearization at a point (x_0, y_0). The plane defined by L is tangent to the surface at the point above the point (x_0, y_0). This furnishes a geometric explanation of why the values of L lie close to those of f in the immediate neighborhood of (x_0, y_0). (Exercise 62)

63. *Directional Derivatives and Scalar Components* How is the derivative of a differentiable function $f(x, y, z)$ at a point P_0 in the direction of a unit vector $\mathbf{u}$ related to the scalar component of $(\nabla f)_{P_0}$ in the direction of $\mathbf{u}$? Give reasons for your answer.

64. *Directional Derivatives and Partial Derivatives*   Assuming that the necessary derivatives of $f(x, y, z)$ are defined, how are $D_\mathbf{i} f$, $D_\mathbf{j} f$, and $D_\mathbf{k} f$ related to f_x, f_y, and f_z? Give reasons for your answer.

65. *The Algebra Rules for Gradients*   Given a constant k and the gradients

$$\nabla f = \frac{\partial f}{\partial x}\mathbf{i} + \frac{\partial f}{\partial y}\mathbf{j} + \frac{\partial f}{\partial z}\mathbf{k}$$

and

$$\nabla g = \frac{\partial g}{\partial x}\mathbf{i} + \frac{\partial g}{\partial y}\mathbf{j} + \frac{\partial g}{\partial z}\mathbf{k},$$

use the scalar equations

$$\frac{\partial}{\partial x}(kf) = k\frac{\partial f}{\partial x},$$

$$\frac{\partial}{\partial x}(f \pm g) = \frac{\partial f}{\partial x} \pm \frac{\partial g}{\partial x},$$

$$\frac{\partial}{\partial x}(fg) = f\frac{\partial g}{\partial x} + g\frac{\partial f}{\partial x},$$

$$\frac{\partial}{\partial x}\left(\frac{f}{g}\right) = \frac{g\frac{\partial f}{\partial x} - f\frac{\partial g}{\partial x}}{g^2},$$

and so on, to establish the following rules:

(a) $\nabla(kf) = k\,\nabla f$

(b) $\nabla(f + g) = \nabla f + \nabla g$

(c) $\nabla(f - g) = \nabla f - \nabla g$

(d) $\nabla(fg) = f\,\nabla g + g\,\nabla f$

(e) $\nabla\left(\dfrac{f}{g}\right) = \dfrac{g\nabla f - f\nabla g}{g^2}$

CAS Explorations

66. Draw a graph of the surface and its tangent plane determined in

(a) Exercise 29.

(b) Exercise 32.

(c) Exercise 34.

13.7 Extreme Values and Saddle Points

Behavior on Closed Bounded Regions • Derivative Tests for Local Extreme Values • Absolute Extrema on Closed Bounded Regions • Limitations of Theorem 10 • Summary

Behavior on Closed Bounded Regions

As we saw when we were working with functions of a single variable, differentiable functions were just what we wanted for modeling optimization problems. Because these functions were continuous, we knew that on finite closed intervals they did assume both maximum and minimum values. Because they were differentiable, we knew that they would assume these values only at domain endpoints or at interior domain points where the first derivative vanished. Occasionally, we encountered functions that failed to be differentiable at one or more interior domain points and we had to add these points to the list to be investigated as well.

We also saw that the condition that $f'(c) = 0$ did not always signal the presence of an extreme value. At such a point c, the graph might have an inflection point instead of a local maximum or minimum. The graph might rise as it approached c from the left, level off at c, then rise again as it left c. Or, it might fall toward c, level off at c, and then resume falling. That is, the graph might cross its tangent line at $x = c$.

Functions of two variables exhibit similar behavior. As we mentioned in Section 13.2, continuous functions of two variables assume extreme values on closed, bounded domains (see Figures 13.48 and 13.49). As we see in this section, we can narrow the search for these extreme values by examining the functions' first partial derivatives. A function of two variables can assume extreme values only at domain boundary points or at interior domain points where both first partial derivatives are zero or where one or both of the first partial derivatives fails to exist.

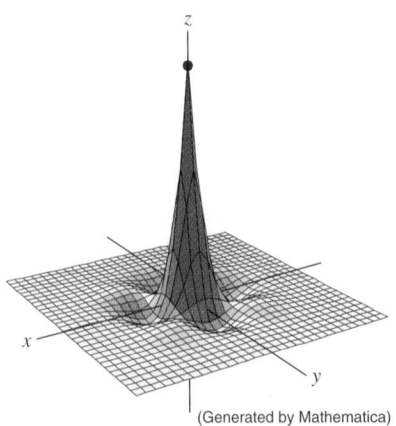

(Generated by Mathematica)

Figure 13.48   The function $z = (\cos x) \cdot (\cos y)e^{-\sqrt{x^2+y^2}}$ has a maximum value of 1 and a minimum value of about -0.067 on the square region $|x| \leq 3\pi/2$, $|y| \leq 3\pi/2$.

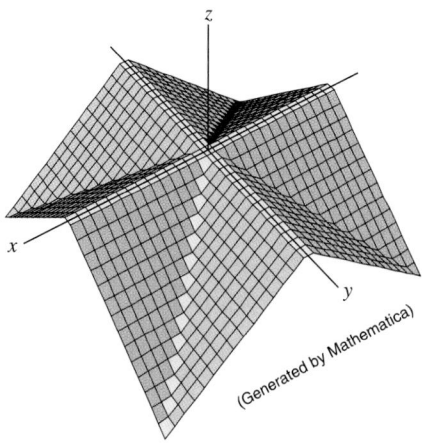

Figure 13.49 The "roof surface" $z = (1/2)(\||x| - |y|\| - |x| - |y|)$, $|x| \le a$, $|y| \le a$, viewed from the point $(10, 15, 20)$. The defining function has a maximum value of 0 and a minimum value of $-a$.

Once again, the vanishing of derivatives at an interior point (a, b) does not always signal the presence of an extreme value. Right above (a, b) the surface that is the graph of the function might be shaped like a saddle and cross its tangent plane there. (We will say more about saddle points when the time comes.)

Derivative Tests for Local Extreme Values

For functions of two independent variables, we define local maximum and minimum values in the following way.

Definitions Local Maximum and Local Minimum

Let $f(x, y)$ be defined on a region R containing the point (a, b). Then

1. $f(a, b)$ is a **local maximum** value of f if $f(a, b) \ge f(x, y)$ for all domain points (x, y) in an open disk centered at (a, b).
2. $f(a, b)$ is a **local minimum** value of f if $f(a, b) \le f(x, y)$ for all domain points (x, y) in an open disk centered at (a, b).

Local maxima correspond to mountain peaks on the surface $z = f(x, y)$ and local minima correspond to valley bottoms (Figure 13.50). At such points the tangent planes, when they exist, are horizontal. Local extrema are also called **relative extrema**.

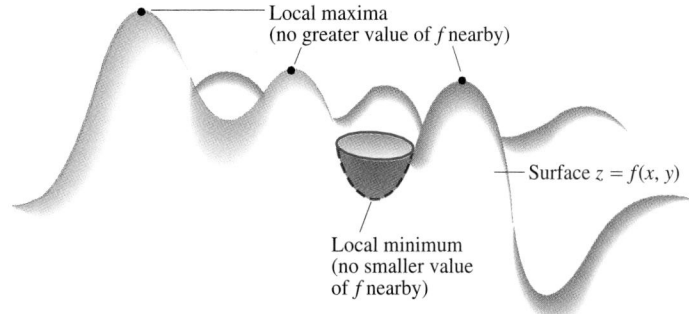

Local maxima
(no greater value of f nearby)

Surface $z = f(x, y)$

Local minimum
(no smaller value
of f nearby)

Figure 13.50 A local maximum is a mountain peak, and a local minimum is a valley low.

As with functions of a single variable, the key to identifying the local extrema is a first derivative test.

Theorem 10 First Derivative Test for Local Extreme Values

If $f(x, y)$ has a local maximum or minimum value at an interior point (a, b) of its domain and if the first partial derivatives exist there, then $f_x(a, b) = 0$ and $f_y(a, b) = 0$.

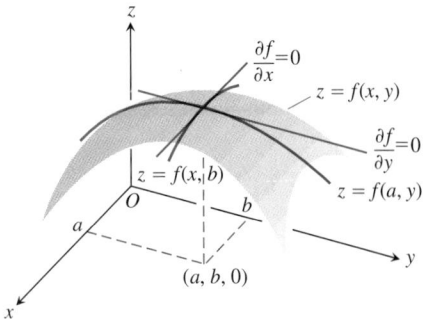

Figure 13.51 The maximum of f occurs at $x = a$, $y = b$.

Proof of Theorem 10 Suppose f has a local maximum at an interior point (a, b) of its domain. Then

1. $x = a$ is an interior point of the domain of the curve $z = f(x, b)$ in which the plane $y = b$ cuts the surface $z = f(x, y)$ (Figure 13.51).

2. The function $z = f(x, b)$ is a differentiable function of x at $x = a$ (the derivative is $f_x(a, b)$).

3. The function $z = f(x, b)$ has a local maximum value at $x = a$.

4. The value of the derivative of $z = f(x, b)$ at $x = a$ is therefore zero (Theorem 2, Section 4.1). Since this derivative is $f_x(a, b)$, we conclude that $f_x(a, b) = 0$.

A similar argument with the function $z = f(a, y)$ shows that $f_y(a, b) = 0$.

This proves the theorem for local maximum values. The proof for local minimum values is left as Exercise 36. ∎

If we substitute the values $f_x(a, b) = 0$ and $f_y(a, b) = 0$ into the equation

$$f_x(a, b)(x - a) + f_y(a, b)(y - b) - (z - f(a, b)) = 0$$

for the tangent plane to the surface $z = f(x, y)$ at (a, b), the equation reduces to

$$0 \cdot (x - a) + 0 \cdot (y - b) - z + f(a, b) = 0$$

or

$$z = f(a, b).$$

Thus, Theorem 10 says that the surface does indeed have a horizontal tangent plane at a local extremum, provided there is a tangent plane there.

As in the single-variable case, Theorem 10 says that the only places a function $f(x, y)$ can ever have an extreme value are

1. interior points where $f_x = f_y = 0$,

2. interior points where one or both of f_x and f_y do not exist, and

3. boundary points of the function's domain.

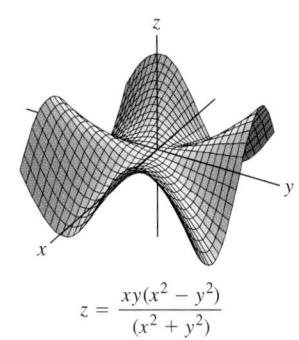

$$z = \frac{xy(x^2 - y^2)}{(x^2 + y^2)}$$

Definition Critical Point

An interior point of the domain of a function $f(x, y)$ where both f_x and f_y are zero or where one or both of f_x and f_y do not exist is a **critical point** of f.

Thus, the only points where a function $f(x, y)$ can assume extreme values are critical points and boundary points. As with differentiable functions of a single variable, not every critical point gives rise to a local extremum. A differentiable function of two variables might have a *saddle point* instead.

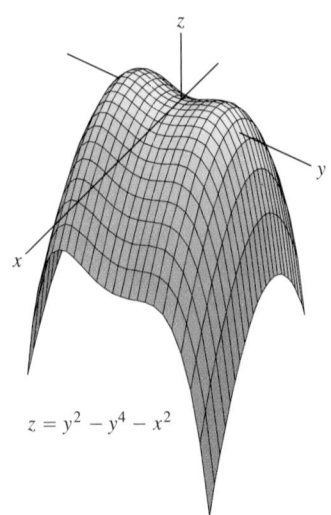

$$z = y^2 - y^4 - x^2$$

(Generated by Mathematica)

Figure 13.52 Saddle points at the origin.

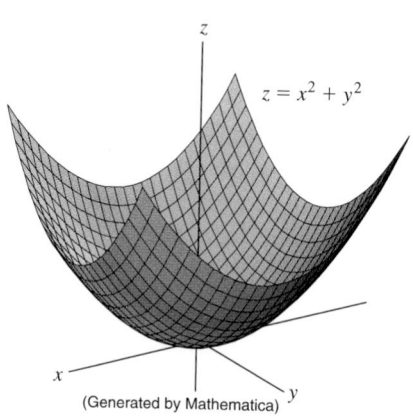

Figure 13.53 The graph of the function $f(x, y) = x^2 + y^2$ is the paraboloid $z = x^2 + y^2$. The function has only one critical point, the origin, which gives rise to a local minimum value of 0. (Example 1)

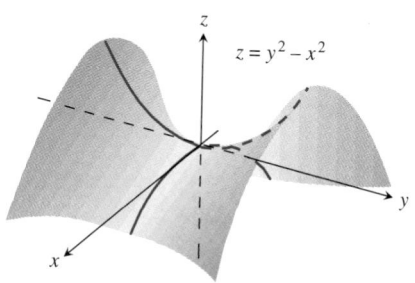

Figure 13.54 The origin is a saddle point of the function $f(x, y) = y^2 - x^2$. There are no local extreme values. (Example 2)

Definition Saddle Point

A differentiable function $f(x, y)$ has a **saddle point** at a critical point (a, b) if in every open disk centered at (a, b) there are domain points (x, y), where $f(x, y) > f(a, b)$, and domain points (x, y), where $f(x, y) < f(a, b)$. The corresponding point $(a, b, f(a, b))$ on the surface $z = f(x, y)$ is a saddle point of the surface (Figure 13.52).

Example 1 FINDING LOCAL EXTREME VALUES

Find the local extreme values of $f(x, y) = x^2 + y^2$.

Solution   The domain of f is the entire plane (so there are no boundary points), and the partial derivatives $f_x = 2x$ and $f_y = 2y$ exist everywhere. Therefore, local extreme can occur only where

$$f_x = 2x = 0 \quad \text{and} \quad f_y = 2y = 0.$$

The only possibility is the origin, where the value of f is zero. Since f is never negative, we see that the origin gives a local minimum (Figure 13.53).

Example 2 IDENTIFYING A SADDLE POINT

Find the local extreme values (if any) of $f(x, y) = y^2 - x^2$.

Solution   The domain of f is the entire plane (so there are no boundary points), and the partial derivatives $f_x = -2x$ and $f_y = 2y$ exist everywhere. Therefore, local extrema can occur only at the origin $(0, 0)$. Along the positive x-axis, however, f has the value $f(x, 0) = -x^2 < 0$; along the positive y-axis, f has the value $f(0, y) = y^2 > 0$. Therefore, every open disk in the xy-plane centered at $(0, 0)$ contains points where the function is positive and points where it is negative. The function has a saddle point at the origin (Figure 13.54) instead of a local extreme value. We conclude that the function has no local extreme values.

That $f_x = f_y = 0$ at an interior point (a, b) of R does not guarantee a local extreme value at (a, b). If f and its first and second partial derivatives are continuous on R, however, we may be able to learn more from the following theorem, whose proof will be omitted.

Theorem 11 Second Derivative Test for Local Extreme Values

Suppose $f(x, y)$ and its first and second partial derivatives are continuous throughout a disk centered at (a, b) and $f_x(a, b) = f_y(a, b) = 0$. Then

(i) f has a **local maximum** at (a, b) if $f_{xx} < 0$ and $f_{xx}f_{yy} - f_{xy}^2 > 0$ at (a, b);

(ii) f has a **local minimum** at (a, b) if $f_{xx} > 0$ and $f_{xx}f_{yy} - f_{xy}^2 > 0$ at (a, b);

(iii) f has a **saddle point** at (a, b) if $f_{xx}f_{yy} - f_{xy}^2 < 0$ at (a, b).

(iv) **The test is inconclusive** at (a, b) if $f_{xx}f_{yy} - f_{xy}^2 = 0$ at (a, b). In this case, we must find some other way to determine the behavior of f at (a, b).

The expression $f_{xx}f_{yy} - f_{xy}{}^2$ is the **discriminant** of f. It is sometimes easier to remember the determinant form,

$$f_{xx}f_{yy} - f_{xy}{}^2 = \begin{vmatrix} f_{xx} & f_{xy} \\ f_{xy} & f_{yy} \end{vmatrix}.$$

Theorem 11 says that if the discriminant is positive at the point (a, b), then the surface curves the same way in all directions: downward if $f_{xx} < 0$, giving rise to a local maximum, and upward if $f_{xx} > 0$, giving a local minimum. On the other hand, if the discriminant is negative at (a, b), then the surface curves up in some directions and down in others, so we have a saddle point.

Example 3 IDENTIFYING LOCAL EXTREME VALUES

Find the local extreme values of $f(x, y) = xy - x^2 - y^2 - 2x - 2y + 4$.

Solution The function is defined and differentiable for all x and y and its domain has no boundary points. The function has extreme values only at the points where f_x and f_y are simultaneously zero. This leads to

$$f_x = y - 2x - 2 = 0, \qquad f_y = x - 2y - 2 = 0,$$

or

$$x = y = -2.$$

Therefore, the point $(-2, -2)$ is the only point where f may take on an extreme value. To see if it does so, we calculate

$$f_{xx} = -2, \qquad f_{yy} = -2, \qquad f_{xy} = 1.$$

The discriminant of f at $(a, b) = (-2, -2)$ is

$$f_{xx}f_{yy} - f_{xy}{}^2 = (-2)(-2) - (1)^2 = 3.$$

The combination

$$f_{xx} < 0 \qquad \text{and} \qquad f_{xx}f_{yy} - f_{xy}{}^2 > 0$$

tells us that f has a local maximum at $(-2, -2)$. The value of f at this point is $f(-2, -2) = 8$.

Example 4 SEARCHING FOR LOCAL EXTREME VALUES

Find the local extreme values of $f(x, y) = xy$.

Solution Since f is differentiable everywhere (Figure 13.55), it can assume extreme values only where

$$f_x = y = 0 \qquad \text{and} \qquad f_y = x = 0.$$

Thus, the origin is the only point where f might have an extreme value. To see what happens there, we calculate

$$f_{xx} = 0, \qquad f_{yy} = 0, \qquad f_{xy} = 1.$$

The discriminant,

$$f_{xx}f_{yy} - f_{xy}{}^2 = -1,$$

is negative. Therefore, the function has a saddle point at $(0, 0)$. We conclude that $f(x, y) = xy$ has no local extreme values.

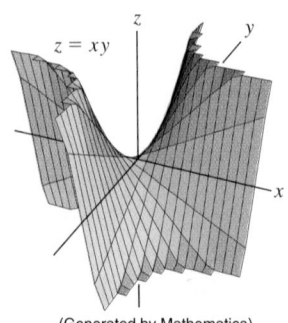

$z = xy$

(Generated by Mathematica)

Figure 13.55 The surface $z = xy$ has a saddle point at the origin. (Example 4)

Absolute Extrema on Closed Bounded Regions

We organize the search for the absolute extrema of a continuous function $f(x, y)$ on a closed and bounded region R into three steps.

Step 1: *List the interior points* of R where f may have local maxima and minima and evaluate f at these points. These are the points where $f_x = f_y = 0$ or where one or both of f_x and f_y fail to exist (the critical points of f).

Step 2: *List the boundary points* of R where f has local maxima and minima and evaluate f at these points. We show how to do this shortly.

Step 3: *Look through the two lists* for the maximum and minimum values of f. These will be the absolute maximum and minimum values of f on R. Since absolute maxima and minima are also local maxima and minima, the absolute maximum and minimum values of f already appear somewhere in the lists made in steps 1 and 2. We have only to glance at the lists to see what they are.

Figure 13.56 This triangular plate is the domain of the function in Example 5.

Example 5   FINDING ABSOLUTE EXTREMA

Find the absolute maximum and minimum values of

$$f(x, y) = 2 + 2x + 2y - x^2 - y^2$$

on the triangular plate in the first quadrant bounded by the lines $x = 0$, $y = 0$, $y = 9 - x$.

Solution   Since f is differentiable, the only places where f can assume these values are points inside the triangle (Figure 13.56), where $f_x = f_y = 0$ and points on the boundary.

Interior points. For these, we have

$$f_x = 2 - 2x = 0, \qquad f_y = 2 - 2y = 0,$$

yielding the single point $(x, y) = (1, 1)$. The value of f there is

$$f(1, 1) = 4.$$

Boundary points. We take the triangle one side at a time:

1. On the segment OA, $y = 0$. The function

$$f(x, y) = f(x, 0) = 2 + 2x - x^2$$

may now be regarded as a function of x defined on the closed interval $0 \le x \le 9$. Its extreme values (we know from Chapter 4) may occur at the endpoints

$$x = 0, \qquad \text{where} \qquad f(0, 0) = 2,$$
$$x = 9, \qquad \text{where} \qquad f(9, 0) = 2 + 18 - 81 = -61,$$

and at the interior points where $f'(x, 0) = 2 - 2x = 0$. The only interior point where $f'(x, 0) = 0$ is $x = 1$, where

$$f(x, 0) = f(1, 0) = 3.$$

2. On the segment OB, $x = 0$ and

$$f(x, y) = f(0, y) = 2 + 2y - y^2.$$

We know from the symmetry of f in x and y and from the analysis we just carried out that the candidates on this segment are

$$f(0, 0) = 2, \qquad f(0, 9) = -61, \qquad f(0, 1) = 3.$$

3. We have already accounted for the values of f at the endpoints of AB, so we need only look at the interior points of AB. With $y = 9 - x$, we have

$$f(x, y) = 2 + 2x + 2(9 - x) - x^2 - (9 - x)^2 = -61 + 18x - 2x^2.$$

Setting $f'(x, 9 - x) = 18 - 4x = 0$ gives

$$x = \frac{18}{4} = \frac{9}{2}.$$

At this value of x,

$$y = 9 - \frac{9}{2} = \frac{9}{2} \qquad \text{and} \qquad f(x, y) = f\left(\frac{9}{2}, \frac{9}{2}\right) = -\frac{41}{2}.$$

Looking through the two lists. The candidates are: $4, 2, -61, 3, -(41/2)$. The maximum is 4, which f assumes at $(1, 1)$. The minimum is -61, which f assumes at $(0, 9)$ and $(9, 0)$.

Limitations of Theorem 10

Despite the power of Theorem 10, we urge you to remember its limitations. It does not apply to boundary points of a function's domain, where it is possible for a function to have extreme values along with nonzero derivatives, and it does not apply to points where either f_x or f_y fails to exist.

Summary

Summary of Max-Min Tests

The extreme values of $f(x, y)$ can occur only at

  (i) boundary points of the domain of f and

  (ii) critical points (interior points where $f_x = f_y = 0$ or points where f_x or f_y fails to exist).

If the first- and second-order partial derivatives of f are continuous throughout a disk centered at a point (a, b) and if $f_x(a, b) = f_y(a, b) = 0$, you may be able to classify $f(a, b)$ with the *second derivative test:*

  (i) $f_{xx} < 0$ and $f_{xx}f_{yy} - f_{xy}^2 > 0$ at $(a, b) \Rightarrow$ **local maximum,**

  (ii) $f_{xx} > 0$ and $f_{xx}f_{yy} - f_{xy}^2 > 0$ at $(a, b) \Rightarrow$ **local minimum,**

  (iii) $f_{xx}f_{yy} - f_{xy}^2 < 0$ at $(a, b) \Rightarrow$ **saddle point,**

  (iv) $f_{xx}f_{yy} - f_{xy}^2 = 0$ at $(a, b) \Rightarrow$ **test is inconclusive.**

Quick Review 13.7

In Exercises 1–4, let $f(x) = x/(x^2 + 1)$

1. Find the critical points of f.

2. Find the local minimum values of f and where they occur.

3. Find the local maximum values of f and where they occur.

4. Find the absolute extrema of f and where they occur.

In Exercises 5–8, let $f(x) = 3 + 2x - x^2$, $-2 \le x \le 5$.

5. Find the critical points of f.

6. Find the local minimum values of f and where they occur.

7. Find the local maximum values of f and where they occur.

8. Find the absolute extrema of f and where they occur.

In Exercises 9 and 10, solve the system of equations.

9. $-2x + y = 2$
   $x - 2y = 2$

10. $3x - 4y = 5$
    $2x + y = 1$

Section 13.7 Exercises

In Exercises 1–18, find all the local maxima, local minima, and saddle points of the function.

1. $f(x, y) = x^2 + xy + y^2 + 3x - 3y + 4$

2. $f(x, y) = 2xy - 5x^2 - 2y^2 + 4x + 4y - 4$

3. $f(x, y) = x^2 + xy + 3x + 2y + 5$

4. $f(x, y) = 2xy - x^2 - 2y^2 + 3x + 4$

5. $f(x, y) = 5xy - 7x^2 + 3x - 6y + 2$

6. $f(x, y) = 3x^2 + 6xy + 7y^2 - 2x + 4y$

7. $f(x, y) = 2x^2 + 3xy + 4y^2 - 5x + 2y$

8. $f(x, y) = x^2 - 2xy + 2y^2 - 2x + 2y + 1$

9. $f(x, y) = x^2 - y^2 - 2x + 4y + 6$

10. $f(x, y) = x^3 + 3xy + y^3$

11. $f(x, y) = 6x^2 - 2x^3 + 3y^2 + 6xy$

12. $f(x, y) = x^3 + y^3 + 3x^2 - 3y^2 - 8$

13. $f(x, y) = 4xy - x^4 - y^4$

14. $f(x, y) = x^4 + y^4 + 4xy$

15. $f(x, y) = \dfrac{1}{x^2 + y^2 - 1}$

16. $f(x, y) = \dfrac{1}{x} + xy + \dfrac{1}{y}$

17. $f(x, y) = y \sin x$ 18. $f(x, y) = e^{2x} \cos y$

In Exercises 19–26, find the absolute maxima and minima of the function on the given domain.

19. $f(x, y) = 2x^2 - 4x + y^2 - 4y + 1$ on the closed triangular plate bounded by the lines $x = 0$, $y = 2$, $y = 2x$ in the first quadrant

20. $f(x, y) = x^2 + y^2$ on the closed triangular plate bounded by the lines $x = 0$, $y = 0$, $y + 2x = 2$ in the first quadrant

21. $T(x, y) = x^2 + xy + y^2 - 6x + 2$ on the rectangular plate $0 \le x \le 5$, $-3 \le y \le 0$

22. $f(x, y) = 48xy - 32x^3 - 24y^2$ on the rectangular plate $0 \le x \le 1, 0 \le y \le 1$

23. $f(x, y) = (4x - x^2) \cos y$ on the rectangular plate $1 \le x \le 3$, $-\pi/4 \le y \le \pi/4$ (Figure 13.57)

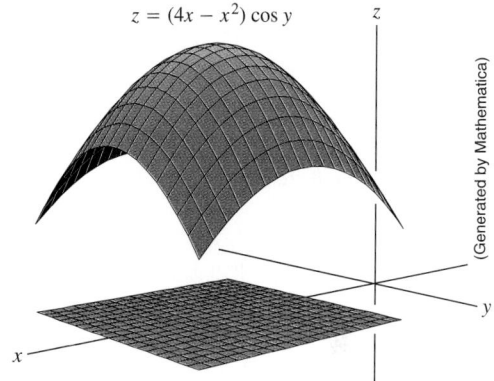

$z = (4x - x^2) \cos y$

(Generated by Mathematica)

Figure 13.57 The function and domain in Exercise 23.

24. $f(x, y) = 4x - 8xy + 2y + 1$ on the triangular plate bounded by the lines $x = 0$, $y = 0$, $x + y = 1$ in the first quadrant

25. $f(x, y) = x^3 + y^2$ on the closed disk $x^2 + y^2 \le 1$

26. $f(x, y) = 3x^2 + x + y^2 - 1$ on the closed disk $x^2 + y^2 \le 4$

27. *Maximizing an Integral* Find two numbers a and b with $a \le b$ such that

$$\int_a^b (6 - x - x^2) \, dx$$

has its largest value.

28. *Maximizing an Integral* Find two numbers a and b with $a \le b$ such that

$$\int_a^b (24 - 2x - x^2)^{1/3} \, dx$$

has its largest value.

29. *Temperature Extremes* The flat circular plate in Figure 13.58 has the shape of the region $x^2 + y^2 \leq 1$. The plate, including the boundary where $x^2 + y^2 = 1$, is heated so that the temperature at the point (x, y) is

$$T(x, y) = x^2 + 2y^2 - x.$$

Find the temperatures at the hottest and coldest points on the plate.

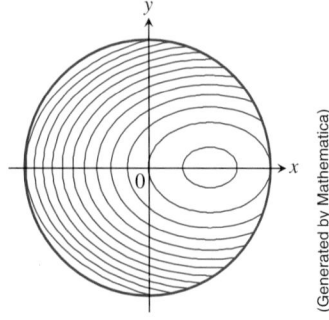

(Generated by Mathematica)

Figure 13.58 Curves of constant temperature are called **isotherms**. The figure shows isotherms of the temperature function $T(x, y) = x^2 + 2y^2 - x$ on the disk $x^2 + y^2 \leq 1$ in the xy-plane. Exercise 29 asks you to locate the extreme temperatures.

30. Find the critical points of

$$f(x, y) = xy + 2x - \ln x^2 y$$

in the open first quadrant ($x > 0$, $y > 0$) and show that f takes on a minimum there (Figure 13.59).

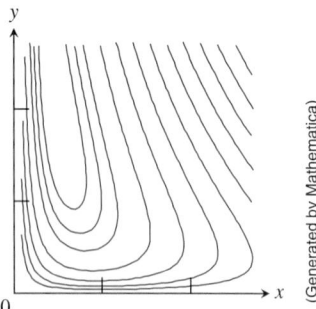

(Generated by Mathematica)

Figure 13.59 The function $f(x, y) = xy + 2x - \ln x^2 y$ (selected level curves shown here) takes on a minimum value somewhere in the open first quadrant $x > 0$, $y > 0$. (Exercise 30)

31. Writing to Learn Find the maxima, minima, and saddle points of $f(x, y)$, if any, given that

(a) $f_x = 2x - 4y$ and $f_y = 2y - 4x$.

(b) $f_x = 2x - 2$ and $f_y = 2y - 4$.

(c) $f_x = 9x^2 - 9$ and $f_y = 2y + 4$.

Describe your reasoning in each case.

32. Writing to Learn *When the Second Derivative Test is Inconclusive* The discriminant $f_{xx}f_{yy} - f_{xy}^2$ is zero at the origin for each of the following functions, so the second derivative test fails there. Determine whether the function has a local maximum, a local minimum, or neither at the origin by imagining what the surface $z = f(x, y)$ looks like. Describe your reasoning in each case.

(a) $f(x, y) = x^2y^2$ **(b)** $f(x, y) = 1 - x^2y^2$

(c) $f(x, y) = xy^2$ **(d)** $f(x, y) = x^3y^2$

(e) $f(x, y) = x^3y^3$ **(f)** $f(x, y) = x^4y^4$

33. Show that $(0, 0)$ is a critical point of $f(x, y) = x^2 + kxy + y^2$ no matter what value the constant k has. (*Hint:* Consider two cases: $k = 0$ and $k \neq 0$.)

34. Writing to Learn For what values of the constant k does the second derivative test guarantee that $f(x, y) = x^2 + kxy + y^2$ will have a saddle point at $(0, 0)$? A local minimum at $(0, 0)$? For what values of k is the second derivative test inconclusive? Give reasons for your answers.

In Exercises 35–40, *work in groups of two or three* to solve the problem.

35. (a) Writing to Learn If $f_x(a, b) = f_y(a, b) = 0$, must f have a local maximum or minimum value at (a, b)? Give reasons for your answer.

(b) Writing to Learn Can you conclude anything about $f(a, b)$ if f and its first and second partial derivatives are continuous throughout a disk centered at (a, b) and also $f_{xx}(a, b)$ and $f_{yy}(a, b)$ differ in sign? Give reasons for your answer.

36. Using the proof of Theorem 10 given in the text for the case in which f has a local maximum at (a, b), prove the theorem for the case in which f has a local minimum at (a, b).

37. Among all the points on the graph of $z = 10 - x^2 - y^2$ that lie above the plane $x + 2y + 3z = 0$, find the point farthest from the plane.

38. Find the point on the graph of $z = x^2 + y^2 + 10$ nearest the plane $x + 2y - z = 0$.

39. Writing to Learn The function $f(x, y) = x + y$ fails to have an absolute maximum value in the closed first quadrant $x \geq 0$ and $y \geq 0$. Does this contradict the discussion on finding absolute extrema given in the text? Give reasons for your answer.

40. Consider the function

$$f(x, y) = x^2 + y^2 + 2xy - x - y + 1$$

over the square $0 \leq x \leq 1$ and $0 \leq y \leq 1$.

(a) Show that f has an absolute minimum along the line segment $2x + 2y = 1$ in this square. What is the absolute minimum value?

(b) Find the absolute maximum value of f over the square.

Explorations

Extreme Values on Parametrized Curves To find the extreme values of a function $f(x, y)$ on a curve $x = x(t)$, $y = y(t)$, we treat f as a function of the single variable t and use the Chain Rule to find where df/dt is zero. As in any other single-variable case, the extreme values of f are then found among the values at the

(i) critical points (points where df/dt is zero or fails to exist), and

(ii) endpoints of the parameter domain.

In Exercises 41–44, find the absolute maximum and minimum values of the functions on the given curves.

41. Functions:

   (a) $f(x, y) = x + y$ (b) $g(x, y) = xy$

   (c) $h(x, y) = 2x^2 + y^2$

   Curves:

   (i)   The semicircle $x^2 + y^2 = 4$, $y \geq 0$

   (ii) The quarter circle $x^2 + y^2 = 4$, $x \geq 0$, $y \geq 0$

   Use the parametric equations $x = 2 \cos t$, $y = 2 \sin t$.

42. Functions:

   (a) $f(x, y) = 2x + 3y$ (b) $g(x, y) = xy$

   (c) $h(x, y) = x^2 + 3y^2$

   Curves:

   (i) The semi-ellipse $(x^2/9) + (y^2/4) = 1$, $y \geq 0$

   (ii) The quarter ellipse $(x^2/9) + (y^2/4) = 1$, $x \geq 0$, $y \geq 0$

   Use the parametric equations $x = 3 \cos t$, $y = 2 \sin t$.

43. Function: $f(x, y) = xy$

   Curves:

   (i) The line $x = 2t$, $y = t + 1$

   (ii) The line segment $x = 2t$, $y = t + 1$, $-1 \leq t \leq 0$

   (iii) The line segment $x = 2t$, $y = t + 1$, $0 \leq t \leq 1$

44. Functions:

   (a) $f(x, y) = x^2 + y^2$ (b) $g(x, y) = \dfrac{1}{x^2 + y^2}$

   Curves:

   (i) The line $x = t$, $y = 2 - 2t$

   (ii) The line segment $x = t$, $y = 2 - 2t$, $0 \leq t \leq 1$ ∎

Extending the Ideas

CAS Explorations

In Exercises 45–50, you will use a CAS to identify the local extrema of the function. Perform the following steps:

(a) Plot the function over the given rectangle.

(b) Plot some level curves in the rectangle.

(c) Calculate the function's first partial derivatives and find the critical points. How do the critical points relate to the level curves plotted in (b)? Which critical points, if any, appear to give a saddle point? Give reasons for your answer.

45. $f(x, y) = x^2 + y^3 - 3xy$, $-5 \leq x \leq 5$, $-5 \leq y \leq 5$

46. $f(x, y) = x^3 - 3xy^2 + y^2$, $-2 \leq x \leq 2$, $-2 \leq y \leq 2$

47. $f(x, y) = x^4 + y^2 - 8x^2 - 6y + 16$, $-3 \leq x \leq 3$, $-6 \leq y \leq 6$

48. $f(x, y) = 2x^4 + y^4 - 2x^2 - 2y^2 + 3$, $-3/2 \leq x \leq 3/2$, $-3/2 \leq y \leq 3/2$

49. $f(x, y) = 5x^6 + 18x^5 - 30x^4 + 30xy^2 - 120x^3$, $-4 \leq x \leq 3$, $-2 \leq y \leq 2$

50. $f(x, y) = \begin{cases} x^5 \ln(x^2 + y^2), & (x, y) \neq (0, 0) \\ 0, & (x, y) = (0, 0) \end{cases}$,

   $-2 \leq x \leq 2$, $-2 \leq y \leq 2$

51. *Least Squares and Regression Lines* When we try to fit a line $y = mx + b$ to a set of numerical data points (x_1, y_1), $(x_2, y_2), \ldots, (x_n, y_n)$ (Figure 13.60) we usually choose the line that minimizes the sum of the squares of the vertical distances from the points to the line. In theory, this means finding the values of m and b that minimize the value of the function

$$w = (mx_1 + b - y_1)^2 + \cdots + (mx_n + b - y_n)^2.$$

Use the first and second derivative tests to show that these values are

$$m = \frac{\left(\sum x_x\right)\left(\sum y_k\right) - n \sum x_k y_k}{\left(\sum x_k\right)^2 - n \sum x_k^2},$$

$$b = \frac{1}{n}\left(\sum y_k - m \sum x_k\right).$$

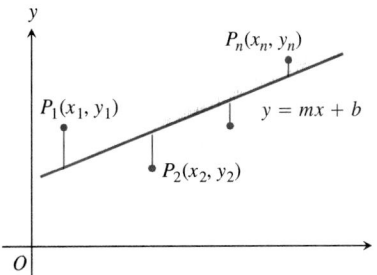

Figure 13.60 To fit a line to noncollinear points, we choose the line that minimizes the sum of the squares of the deviations.

13.8 Lagrange Multipliers

Constrained Maxima and Minima • The Method of Lagrange
Multipliers • Lagrange Multipliers with Two Constraints

Constrained Maxima and Minima

$z = 49 - x^2 - y^2$

Free maximum

Constrained
maximum

$x + 3y - 10 = 0$
constraint on
x and y

Figure 13.61 The function
$f(x, y) = 49 - x^2 - y^2$, subject to the constraint $g(x, y) = x + 3y - 10 = 0$.

As we saw in Section 13.7, we sometimes need to find the extreme values of
a function whose domain is constrained to lie within some particular subset of
the plane, such as a disk or a closed triangular region. But, as Figure 13.61
suggests, a function may be subject to other kinds of constraints as well.

Example 1   FINDING A MINIMUM WITH CONSTRAINT

Find the point $P(x, y, z)$ closest to the origin on the plane
$2x + y - z - 5 = 0$.

Solution   The problem asks us to find the minimum value of the function

$$\left| \overrightarrow{OP} \right| = \sqrt{(x - 0)^2 + (y - 0)^2 + (z - 0)^2}$$
$$= \sqrt{x^2 + y^2 + z^2}$$

subject to the constraint that

$$2x + y - z - 5 = 0.$$

Since $\left| \overrightarrow{OP} \right|$ has a minimum value wherever the function

$$f(x, y, z) = x^2 + y^2 + z^2$$

has a minimum value, we may solve the problem by finding the minimum
value of $f(x, y, z)$ subject to the constraint $2x + y - z - 5 = 0$ (thus avoid-
ing square roots). If we regard x and y as the independent variables in this
equation and write z as

$$z = 2x + y - 5,$$

the problem reduces to one of finding the points (x, y) at which the function

$$h(x, y) = f(x, y, 2x + y - 5) = x^2 + y^2 + (2x + y - 5)^2$$

has its minimum value or values. Since the domain of h is the entire
xy-plane, the first derivative test of Section 13.7 tells us that any minima
that h might have must occur at points where

$$h_x = 2x + 2(2x + y - 5)(2) = 0, \qquad h_y = 2y + 2(2x + y - 5) = 0.$$

This leads to

$$10x + 4y = 20, \qquad 4x + 4y = 10,$$

and the solution

$$x = \frac{5}{3}, \qquad y = \frac{5}{6}.$$

We may apply a geometric argument together with the second derivative test to show that these values minimize h. The z-coordinate of the corresponding point on the plane $z = 2x + y - 5$ is

$$z = 2\left(\frac{5}{3}\right) + \frac{5}{6} - 5 = -\frac{5}{6}.$$

Therefore, the point we seek is

Closest point: $\quad P\left(\dfrac{5}{3}, \dfrac{5}{6}, -\dfrac{5}{6}\right).$

The distance from P to the origin is $5/\sqrt{6} \approx 2.04$.

Attempts to solve a constrained maximum or minimum problem by substitution, as we might call the method of Example 1, do not always go smoothly. This is one of the reasons for learning the new method of this section.

Example 2 FINDING A MINIMUM WITH CONSTRAINT

Find the points closest to the origin on the hyperbolic cylinder

$$x^2 - z^2 - 1 = 0.$$

Solution 1 The cylinder is shown in Figure 13.62. We seek the points on the cylinder closest to the origin. These are the points whose coordinates minimize the value of the function

$$f(x, y, z) = x^2 + y^2 + z^2 \qquad \text{Square of the distance}$$

subject to the constraint that $x^2 - z^2 - 1 = 0$. If we regard x and y as independent variables, the constraint equation is

$$z^2 = x^2 - 1,$$

and the values of $f(x, y, z) = x^2 + y^2 + z^2$ on the cylinder are given by the function

$$h(x, y) = x^2 + y^2 + (x^2 - 1) = 2x^2 + y^2 - 1.$$

To find the points on the cylinder whose coordinates minimize f, we look for the points in the xy-plane whose coordinates minimize h. The only extreme value of h occurs where

$$h_x = 4x = 0 \qquad \text{and} \qquad h_y = 2y = 0,$$

that is, at the point $(0, 0)$. But now we're in trouble—there are no points on the cylinder where both x and y are zero. What went wrong?

What happened was that the first derivative test found (as it should have) the point *in the domain of h* where h has a minimum value. We, on the other hand, want the points *on the cylinder* where h has a minimum value. Although the domain of h is the entire xy-plane, the domain from which we can select the first two coordinates of the points (x, y, z) on the cylinder is restricted to the "shadow" of the cylinder on the xy-plane; it does not include the band between the lines $x = -1$ and $x = 1$ (Figure 13.63).

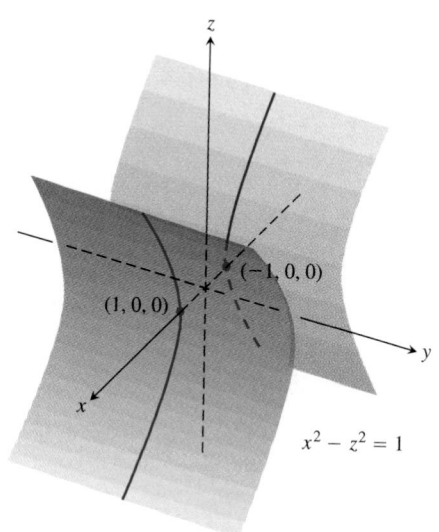

Figure 13.62 The hyperbolic cylinder $x^2 - z^2 - 1 = 0$ in Example 2.

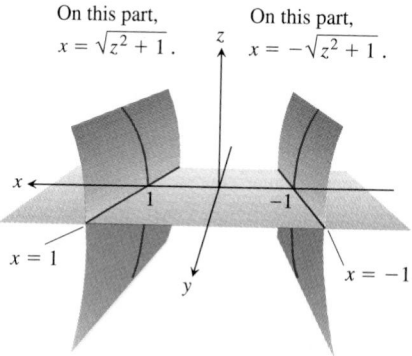

Figure 13.63 The region in the xy-plane from which the first two coordinates of the points (x, y, z) on the hyperbolic cylinder $x^2 - z^2 = 1$ are selected excludes the band $-1 < x < 1$ in the xy-plane.

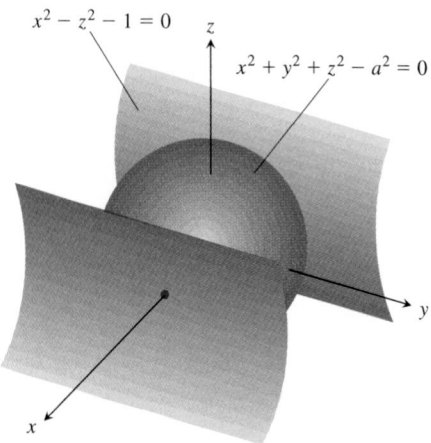

$x^2 - z^2 - 1 = 0$
$x^2 + y^2 + z^2 - a^2 = 0$

Figure 13.64 A sphere expanding like a soap bubble centered at the origin until it just touches the hyperbolic cylinder $x^2 - z^2 - 1 = 0$. See Solution 2 of Example 2.

We can avoid this problem if we treat y and z as independent variables (instead of x and y) and express x in terms of y and z as

$$x^2 = z^2 + 1.$$

With this substitution, $f(x, y, z) = x^2 + y^2 + z^2$ becomes

$$k(y, z) = (z^2 + 1) + y^2 + z^2 = 1 + y^2 + 2z^2$$

and we look for the points where k takes on its smallest value. The domain of k in the yz-plane now matches the domain from which we select the y- and z-coordinates of the points (x, y, z) on the cylinder. Hence, the points that minimize k in the plane will have corresponding points on the cylinder. The smallest values of k occur where

$$k_y = 2y = 0 \quad \text{and} \quad k_z = 4z = 0,$$

or where $y = z = 0$. This leads to

$$x^2 = z^2 + 1 = 1, \qquad x = \pm 1.$$

The corresponding points on the cylinder are $(\pm 1, 0, 0)$. We can see from the inequality

$$k(y, z) = 1 + y^2 + 2z^2 \geq 1$$

that the points $(\pm 1, 0, 0)$ give a minimum value for k. We can also see that the minimum distance from the origin to a point on the cylinder is 1 unit.

Solution 2   Another way to find the points on the cylinder closest to the origin is to imagine a small sphere centered at the origin expanding like a soap bubble until it just touches the cylinder (Figure 13.64). At each point of contact, the cylinder and sphere have the same tangent plane and normal line. Therefore, if the sphere and cylinder are represented as the level surfaces obtained by setting

$$f(x, y, z) = x^2 + y^2 + z^2 - a^2 \quad \text{and} \quad g(x, y, z) = x^2 - z^2 - 1$$

equal to 0, then the gradients ∇f and ∇g will be parallel where the surfaces touch. At any point of contact we should therefore be able to find a scalar λ ("lambda") such that

$$\nabla f = \lambda \nabla g,$$

or

$$2x\mathbf{i} + 2y\mathbf{j} + 2z\mathbf{k} = \lambda(2x\mathbf{i} - 2z\mathbf{k}).$$

Thus, the coordinates x, y, and z of any point of tangency will have to satisfy the three scalar equations

$$2x = 2\lambda x, \qquad 2y = 0, \qquad 2z = -2\lambda z. \tag{1}$$

For what values of λ will a point (x, y, z) whose coordinates satisfy the equations in (1) also lie on the surface $x^2 - z^2 - 1 = 0$? To answer this question, we remember that no point on the surface has a zero x-coordinate and conclude that $x \neq 0$ in the first equation in (1). Thus, $2x = 2\lambda x$ only if

$$2 = 2\lambda, \qquad \text{or} \qquad \lambda = 1.$$

For $\lambda = 1$, the equation $2z = -2\lambda z$ becomes $2z = -2z$. If this equation is to be satisfied as well, z must be zero. Since $y = 0$ also (from the equation

$2y = 0$), we conclude that the points we seek all have coordinates of the form

$$(x, 0, 0).$$

What points on the surface $x^2 - z^2 = 1$ have coordinates of this form? The answer is the points $(x, 0, 0)$ for which

$$x^2 - (0)^2 = 1, \qquad x^2 = 1, \qquad \text{or} \qquad x = \pm 1.$$

The points on the cylinder closest to the origin are the points $(\pm 1, 0, 0)$.

The Method of Lagrange Multipliers

In Solution 2 of Example 2, we solved the problem by the **method of Lagrange multipliers.** In general terms, the method says that the extreme values of a function $f(x, y, z)$ whose variables are subject to a constraint $g(x, y, z) = 0$ are to be found on the surface $g = 0$ at the points where

$$\nabla f = \lambda \nabla g$$

for some scalar λ (called a **Lagrange multiplier**).

To explore the method further and see why it works, we first make the following observation, which we state as a theorem.

Theorem 12 The Orthogonal Gradient Theorem

Suppose $f(x, y, z)$ is differentiable in a region whose interior contains a smooth curve

$$C: \mathbf{r} = g(t)\mathbf{i} + h(t)\mathbf{j} + k(t)\mathbf{k}.$$

If P_0 is a point on C where f has a local maximum or minimum relative to its values on C, then ∇f is orthogonal to C at P_0. At every point on the curve, $\nabla f \cdot \mathbf{v} = 0$.

Proof of Theorem 12 We show that ∇f is orthogonal to the curve's velocity vector at P_0. The values of f on C are given by the composite $f(g(t), h(t), k(t))$, whose derivative with respect to t is

$$\frac{df}{dt} = \frac{\partial f}{\partial x}\frac{dg}{dt} + \frac{\partial f}{\partial y}\frac{dh}{dt} + \frac{\partial f}{\partial z}\frac{dk}{dt} = \nabla f \cdot \mathbf{v}.$$

At any point P_0 where f has a local maximum or minimum relative to its values on the curve, $df/dt = 0$, so

$$\nabla f \cdot \mathbf{v} = 0. \qquad \blacksquare$$

By dropping the z-terms in Theorem 12, we obtain a similar result for functions of two variables.

Corollary of Theorem 12

At the points on a smooth curve $\mathbf{r} = g(t)\mathbf{i} + h(t)\mathbf{j}$ where a differentiable function $f(x, y)$ takes on its local maxima and minimum relative to its values on the curve, $\nabla f \cdot \mathbf{v} = 0$.

Theorem 12 is the key to the method of Lagrange multipliers. Suppose $f(x, y, z)$ and $g(x, y, z)$ are differentiable and P_0 is a point on the surface $g(x, y, z) = 0$ where f has a local maximum or minimum value relative to its other values on the surface. Then f takes on a local maximum or minimum at P_0 relative to its values on every differentiable curve through P_0 on the surface $g(x, y, z) = 0$. Therefore, ∇f is orthogonal to the velocity vector of every such differentiable curve through P_0. But so is ∇g (because ∇g is orthogonal to the level surface $g = 0$, as we saw in Section 13.6). Therefore, at P_0, ∇f is some scalar multiple λ of ∇g.

The Method of Lagrange Multipliers

Suppose $f(x, y, z)$ and $g(x, y, z)$ are differentiable. To find the local maximum and minimum values of f subject to the constraint $g(x, y, z) = 0$, find the values of x, y, z, and λ that simultaneously satisfy the equations

$$\nabla f = \lambda \nabla g \qquad \text{and} \qquad g(x, y, z) = 0.$$

For functions of two independent variables, the appropriate equations are

$$\nabla f = \lambda \nabla g \qquad \text{and} \qquad g(x, y) = 0.$$

Example 3 USING THE METHOD OF LAGRANGE MULTIPLIERS

Find the greatest and smallest values that the function

$$f(x, y) = xy$$

takes on the ellipse (Figure 13.65)

$$\frac{x^2}{8} + \frac{y^2}{2} = 1.$$

Solution We want to find the extreme values of $f(x, y) = xy$ subject to the constraint

$$g(x, y) = \frac{x^2}{8} + \frac{y^2}{2} - 1 = 0.$$

To do so, we first find the values of x, y, and λ for which

$$\nabla f = \lambda \nabla g \qquad \text{and} \qquad g(x, y) = 0.$$

The gradient equation gives

$$y\mathbf{i} + x\mathbf{j} = \frac{\lambda}{4} x\mathbf{i} + \lambda y\mathbf{j},$$

from which we find

$$y = \frac{\lambda}{4} x, \qquad x = \lambda y, \qquad \text{and} \qquad y = \frac{\lambda}{4}(\lambda y) = \frac{\lambda^2}{4} y,$$

so that $y = 0$ or $\lambda = \pm 2$. We now consider these two cases.

Case 1: If $y = 0$, then $x = y = 0$. But $(0, 0)$ is not on the ellipse. Hence, $y \neq 0$.

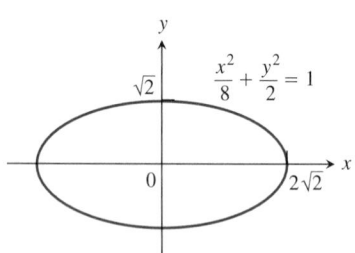

Figure 13.65 Example 3 shows how to find the largest and smallest values of the product xy on this ellipse.

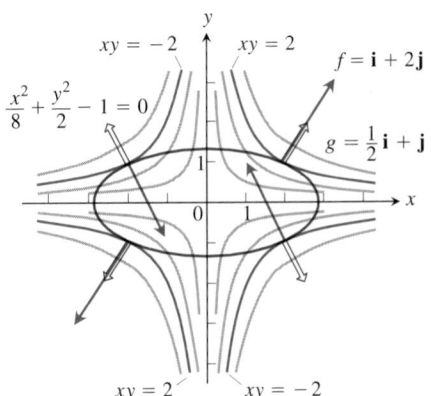

$xy = -2$ $xy = 2$

$f = \mathbf{i} + 2\mathbf{j}$

$\dfrac{x^2}{8} + \dfrac{y^2}{2} - 1 = 0$

$g = \dfrac{1}{2}\mathbf{i} + \mathbf{j}$

$xy = 2$ $xy = -2$

Figure 13.66 When subjected to the constraint $g(x, y) = x^2/8 + y^2/2 - 1 = 0$, the function $f(x, y) = xy$ takes on extreme values at the four points $(\pm 2, \pm 1)$. These are the points on the ellipse where ∇f (red) is a scalar multiple of ∇g (blue). (Example 3)

Case 2: If $y \neq 0$, then $\lambda = \pm 2$ and $x = \pm 2y$. Substituting this in the equation $g(x, y) = 0$ gives

$$\frac{(\pm 2y)^2}{8} + \frac{y^2}{2} = 1, \qquad y^2 + y^2 = 2, \qquad \text{and} \qquad y = \pm 1.$$

The function $f(x, y) = xy$ therefore takes on its extreme values on the ellipse at the four points $(\pm 2, 1)$, $(\pm 2, -1)$. The extreme values are $xy = 2$ and $xy = -2$.

The Geometry of the Solution The level curves of the function $f(x, y) = xy$ are the hyperbolas $xy = c$ (Figure 13.66). The further the hyperbolas lie from the origin, the larger the absolute value of f. We want to find the extreme values of $f(x, y)$, given that the point (x, y) also lies on the ellipse $x^2 + 4y^2 = 8$. Which hyperbolas intersecting the ellipse lie farthest from the origin? The hyperbolas that just graze the ellipse, the ones that are tangent to it. At these points, any vector normal to the hyperbola is normal to the ellipse, so $\nabla f = y\mathbf{i} + x\mathbf{j}$ is a multiple ($\lambda = \pm 2$) of $\nabla g = (x/4)\mathbf{i} + y\mathbf{j}$. At the point $(2, 1)$, for example,

$$\nabla f = \mathbf{i} + 2\mathbf{j}, \qquad \nabla g = \frac{1}{2}\mathbf{i} + \mathbf{j}, \qquad \text{and} \qquad \nabla f = 2\nabla g.$$

At the point $(-2, 1)$,

$$\nabla f = \mathbf{i} - 2\mathbf{j}, \qquad \nabla g = -\frac{1}{2}\mathbf{i} + \mathbf{j}, \qquad \text{and} \qquad \nabla f = -2\nabla g.$$

Example 4 FINDING EXTREME FUNCTION VALUES
ON A CIRCLE

Find the maximum and minimum values of the function $f(x, y) = 3x + 4y$ on the circle $x^2 + y^2 = 1$.

Solution We model this as a Lagrange multiplier problem with

$$f(x, y) = 3x + 4y, \qquad g(x, y) = x^2 + y^2 - 1$$

and look for the values of x, y, and λ that satisfy the equations

$$\nabla f = \lambda \nabla g: \quad 3\mathbf{i} + 4\mathbf{j} = 2x\lambda\mathbf{i} + 2y\lambda\mathbf{j},$$

$$g(x, y) = 0: \quad x^2 + y^2 - 1 = 0.$$

The gradient equation implies that $\lambda \neq 0$ and gives

$$x = \frac{3}{2\lambda}, \qquad y = \frac{2}{\lambda}.$$

These equations tell us, among other things, that x and y have the same sign. With these values for x and y, the equation $g(x, y) = 0$ gives

$$\left(\frac{3}{2\lambda}\right)^2 + \left(\frac{2}{\lambda}\right)^2 - 1 = 0,$$

so

$$\frac{9}{4\lambda^2} + \frac{4}{\lambda^2} = 1, \qquad 9 + 16 = 4\lambda^2, \qquad 4\lambda^2 = 25, \qquad \text{and} \qquad \lambda = \pm\frac{5}{2}.$$

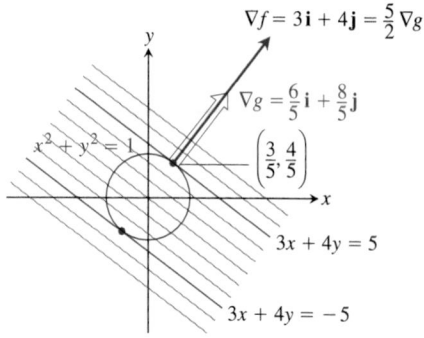

Figure 13.67 The function $f(x, y) = 3x + 4y$ takes on its largest value on the unit circle $g(x, y) = x^2 + y^2 - 1 = 0$ at the point $(3/5, 4/5)$ and its smallest value at the point $(-3/5, -4/5)$ (Example 4). At each point, ∇f is a scalar multiple of ∇g. The figure shows the gradients at the first point but not the second.

Thus,

$$x = \frac{3}{2\lambda} = \pm\frac{3}{5}, \quad y = \frac{2}{\lambda} = \pm\frac{4}{5},$$

and $f(x, y) = 3x + 4y$ has extreme values at $(x, y) = \pm(3/5, 4/5)$.

By calculating the value of $3x + 4y$ at the points $\pm(3/5, 4/5)$, we see that its maximum and minimum values on the circle $x^2 + y^2 = 1$ are

$$3\left(\frac{3}{5}\right) + 4\left(\frac{4}{5}\right) = \frac{25}{5} = 5 \quad \text{and} \quad 3\left(-\frac{3}{5}\right) + 4\left(-\frac{4}{5}\right) = -\frac{25}{5} = -5.$$

The Geometry of the Solution (Figure 13.67)   The level curves of $f(x, y) = 3x + 4y$ are the lines $3x + 4y = c$. The further the lines lie from the origin, the larger the absolute value of f. We want to find the extreme values of $f(x, y)$ given that the point (x, y) also lies on the circle $x^2 + y^2 = 1$. Which lines intersecting the circle lie farthest from the origin? The lines tangent to the circle. At the points of tangency, any vector normal to the line is normal to the circle, so the gradient $\nabla f = 3\mathbf{i} + 4\mathbf{j}$ is a multiple ($\lambda = \pm 5/2$) of the gradient $\nabla g = 2x\mathbf{i} + 2y\mathbf{j}$. At the point $(3/5, 4/5)$, for example,

$$\nabla f = 3\mathbf{i} + 4\mathbf{j}, \qquad \nabla g = \frac{6}{5}\mathbf{i} + \frac{8}{5}\mathbf{j}, \qquad \text{and} \qquad \nabla f = \frac{5}{2}\nabla g.$$

Lagrange Multipliers with Two Constraints

Many problems require us to find the extreme values of a differentiable function $f(x, y, z)$ whose variables are subject to two constraints. If the constraints are

$$g_1(x, y, z) = 0 \qquad \text{and} \qquad g_2(x, y, z) = 0$$

and g_1 and g_2 are differentiable, with ∇g_1 not parallel to ∇g_2, we find the constrained local maxima and minima of f by introducing two Lagrange multipliers λ and μ (mu, pronounced "mew"). That is, we locate points $P(x, y, z)$ where f takes on its constrained extreme values by finding the values of x, y, z, λ, and μ that simultaneously satisfy the equations

$$\nabla f = \lambda \nabla g_1 + \mu \nabla g_2, \qquad g_1(x, y, z) = 0, \qquad g_2(x, y, z) = 0. \qquad (2)$$

The equations in (2) have a nice geometric interpretation. The surfaces $g_1 = 0$ and $g_2 = 0$ (usually) intersect in a smooth curve, say C (Figure 13.68), and along this curve we seek the points where f has local maximum and minimum values relative to its other values on the curve. These are the points where ∇f is normal to C, as we saw in Theorem 12. But ∇g_1 and ∇g_2 are also normal to C at these points because C lies in the surfaces $g_1 = 0$ and $g_2 = 0$. Therefore, ∇f lies in the plane determined by ∇g_1 and ∇g_2, which means that $\nabla f = \lambda \nabla g_1 + \mu \nabla g_2$ for some λ and μ. Since the points we seek also lie in both surfaces, their coordinates must satisfy the equations $g_1(x, y, z) = 0$ and $g_2(x, y, z) = 0$, which are the remaining requirements in Equation (2).

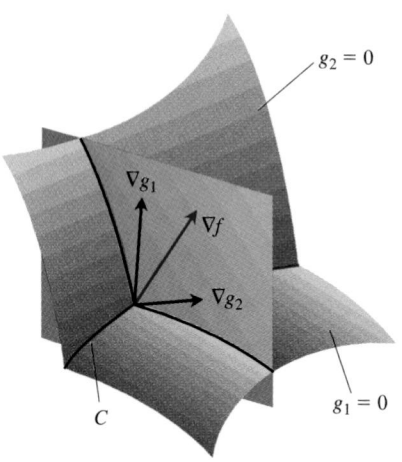

Figure 13.68 The vectors ∇g_1 and ∇g_2 lie in a plane perpendicular to the curve C because ∇g_1 is normal to the surface $g_1 = 0$ and ∇g_2 is normal to the surface $g_2 = 0$.

Example 5   FINDING EXTREMES OF DISTANCE ON AN ELLIPSE

The plane $x + y + z = 1$ cuts the cylinder $x^2 + y^2 = 1$ in an ellipse (Figure 13.69). Find the points on the ellipse that lie closest to and farthest from the origin.

Solution   We find the extreme values of

$$f(x, y, z) = x^2 + y^2 + z^2$$

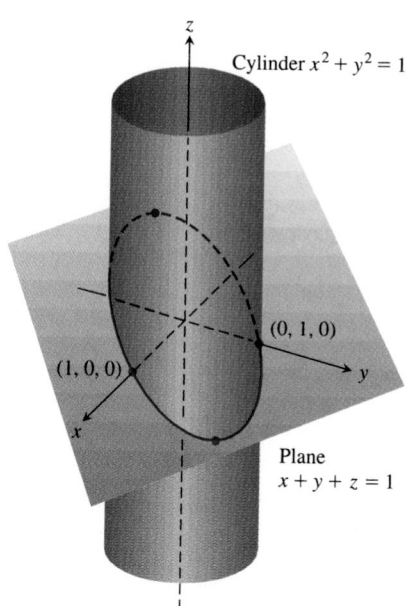

Figure 13.69 On the ellipse where the plane and cylinder meet, what are the points closest to and farthest from the origin? (Example 5)

(the square of the distance from (x, y, z) to the origin) subject to the constraints

$$g_1(x, y, z) = x^2 + y^2 - 1 = 0 \qquad (3)$$

$$g_2(x, y, z) = x + y + z - 1 = 0. \qquad (4)$$

The gradient equation in (2) gives

$$\nabla f = \lambda \nabla g_1 + \mu \nabla g_2 \qquad \text{Eq. (2)}$$

$$2x\mathbf{i} + 2y\mathbf{j} + 2z\mathbf{k} = \lambda(2x\mathbf{i} + 2y\mathbf{j}) + \mu(\mathbf{i} + \mathbf{j} + \mathbf{k})$$

$$2x\mathbf{i} + 2y\mathbf{j} + 2z\mathbf{k} = (2\lambda x + \mu)\mathbf{i} + (2\lambda y + \mu)\mathbf{j} + \mu\mathbf{k}$$

or

$$2x = 2\lambda x + \mu, \qquad 2y = 2\lambda y + \mu, \qquad 2z = \mu. \qquad (5)$$

The scalar equations in (5) yield

$$2x = 2\lambda x + 2z \quad \Rightarrow \quad (1 - \lambda)x = z,$$
$$2y = 2\lambda y + 2z \quad \Rightarrow \quad (1 - \lambda)y = z. \qquad (6)$$

Equations (6) are satisfied simultaneously if either $\lambda = 1$ and $z = 0$ or $\lambda \neq 1$ and $x = y = z/(1 - \lambda)$.

If $z = 0$, then solving Equations (3) and (4) simultaneously to find the corresponding points on the ellipse gives two points: $(1, 0, 0)$ and $(0, 1, 0)$. This makes sense when you look at Figure 13.69.

If $x = y$, then Equations (3) and (4) give

$$x^2 + x^2 - 1 = 0 \qquad\qquad x + x + z - 1 = 0$$

$$2x^2 = 1 \qquad\qquad\qquad z = 1 - 2x$$

$$x = \pm \frac{\sqrt{2}}{2} \qquad\qquad\qquad z = 1 \mp \sqrt{2}.$$

The corresponding points on the ellipse are

$$P_1 = \left(\frac{\sqrt{2}}{2}, \frac{\sqrt{2}}{2}, 1 - \sqrt{2} \right) \quad \text{and} \quad P_2 = \left(-\frac{\sqrt{2}}{2}, -\frac{\sqrt{2}}{2}, 1 + \sqrt{2} \right).$$

We need to be careful here, however. Although P_1 and P_2 both give local maxima of f on the ellipse, P_2 is farther from the origin than P_1.

The points on the ellipse closest to the origin are $(1, 0, 0)$ and $(0, 1, 0)$. The point on the ellipse farthest from the origin is P_2.

Quick Review 13.8

In Exercises 1–4, find all the local maxima and local minima of the function.

1. $f(x, y) = 2xy - 5x^2 - 2y^2 + 4x - 4$

2. $f(x, y) = 3 + 2x + 2y - 2x^2 - 2xy - y^2$

3. $f(x, y) = x^2 + 3xy + 3y^2 - 6x + 3y - 6$

4. $f(x, y) = 8x^3 + y^3 + 6xy$

In Exercises 5–7, find del (the gradient) of the function.

5. $g(y, z) = 2y^2 - 3z^2$

6. $k(x, y) = 3x^3y - 2xy^3 + \sin xy$

7. $h(x, y, z) = 3x - 2y + 2z^2$

In Exercises 8–10, let $\mathbf{u} = 2\mathbf{i} - 3\mathbf{j} + 4\mathbf{k}$, $\mathbf{v} = \mathbf{i} - 2\mathbf{j} - 2\mathbf{k}$, and $\mathbf{w} = -7\mathbf{i} + 14\mathbf{j} + 14\mathbf{k}$. Decide whether the two vectors listed are parallel, perpendicular, or neither.

8. $\mathbf{u}$ and $\mathbf{v}$

9. $\mathbf{u}$ and $\mathbf{w}$

10. $\mathbf{v}$ and $\mathbf{w}$

Section 13.8 Exercises

In Exercises 1–16, the problem involves a function of two variables with one constraint.

1. Find the points on the ellipse $x^2 + 2y^2 = 1$ where $f(x, y) = xy$ has its extreme values.

2. Find the extreme values of $f(x, y) = xy$ subject to the constraint $g(x, y) = x^2 + y^2 - 10 = 0$.

3. Find the maximum value of $f(x, y) = 49 - x^2 - y^2$ on the line $x + 3y = 10$ (Figure 13.61).

4. Find the local extreme values of $f(x, y) = x^2y$ on the line $x + y = 3$.

5. Find the points on the curve $xy^2 = 54$ nearest the origin.

6. Find the points on the curve $x^2y = 2$ nearest the origin.

7. **Writing to Learn** Use the method of Lagrange multipliers to find

   (a) the minimum value of $x + y$, subject to the constraints $xy = 16, x > 0, y > 0$;

   (b) the maximum value of xy, subject to the constraint $x + y = 16$.

   Comment on the geometry of each solution.

8. Find the points on the curve $x^2 + xy + y^2 = 1$ in the xy-plane that lie nearest to and farthest from the origin.

9. *Best Cylindrical Can* Find the dimensions of the closed right circular cylindrical can of smallest surface area whose volume is 16π cm^3.

10. *Cylinder in a Sphere* Find the radius and height of the open right circular cylinder of largest surface area that can be inscribed in a sphere of radius a. What is the largest surface area?

11. *Rectangle of Greatest Area in an Ellipse* Use the method of Lagrange multipliers to find the dimensions of the rectangle of greatest area that can be inscribed in the ellipse $x^2/16 + y^2/9 = 1$ with sides parallel to the coordinate axes.

12. *Rectangle of Greatest Perimeter in an Ellipse* Find the dimensions of the rectangle of largest perimeter that can be inscribed in the ellipse $x^2/a^2 + y^2/b^2 = 1$ with sides parallel to the coordinate axes. What is the largest perimeter?

13. Find the maximum and minimum values of $x^2 + y^2$ subject to the constraint $x^2 - 2x + y^2 - 4y = 0$.

14. Find the maximum and minimum values of $3x - y + 6$ subject to the constraint $x^2 + y^2 = 4$.

15. *Ant on a Metal Plate* The temperature at a point (x, y) on a metal plate is $T(x, y) = 4x^2 - 4xy + y^2$. An ant on the plate walks around the circle of radius 5 centered at the origin. What are the highest and lowest temperatures encountered by the ant?

16. *Storing LPG* Your firm has been asked to design a storage tank for liquid petroleum gas. The customer's specifications call for a cylindrical tank with hemispherical ends, and the tank is to hold 8000 m^3 of gas. The customer also wants to use the smallest amount of material possible in building the tank. What radius and height do you recommend for the cylindrical portion of the tank?

In Exercises 17–30, the problem involves a function of three variables with one constraint.

17. *Closest Point* Find the point on the plane $x + 2y + 3z = 13$ closest to the point $(1, 1, 1)$.

18. *Farthest Point* Find the point on the sphere $x^2 + y^2 + z^2 = 4$ which is farthest from the point $(1, -1, 1)$.

19. *Minimum Distance* Find the minimum distance from the surface $x^2 - y^2 - z^2 = 1$ to the origin.

20. *Nearest Point* Find the point on the surface $z = xy + 1$ nearest the origin.

21. *Closest Point* Find the points on the surface $z^2 = xy + 4$ closest to the origin.

22. *Closest Point* Find the point(s) on the surface $xyz = 1$ closest to the origin.

23. *Extreme Values on a Sphere* Find the maximum and minimum values of

$$f(x, y, z) = x - 2y + 5z$$

on the sphere $x^2 + y^2 + z^2 = 30$.

24. *Extreme Values on a Sphere* Find the points on the sphere $x^2 + y^2 + z^2 = 25$ where $f(x, y, z) = x + 2y + 3z$ has its maximum and minimum values.

25. *Smallest Sum* Find three real numbers whose sum is 9 and the sum of whose squares is as small as possible.

26. *Largest Product* Find the largest product the positive numbers x, y, and z can have if $x + y + z^2 = 16$.

27. *Largest Box in a Sphere* Find the dimensions of the closed rectangular box with maximum volume that can be inscribed in the unit sphere.

28. *Largest Volume* Find the largest volume a closed rectangular box can have if it lies in the first octant with three faces in the coordinate planes and a vertex on the plane $x/a + y/b + z/c = 1$, where $a > 0$, $b > 0$, and $c > 0$.

29. *Hottest Point on a Space Probe* A space probe in the shape of the ellipsoid

$$4x^2 + y^2 + 4z^2 = 16$$

enters the earth's atmosphere and its surface begins to heat. After 1 hour, the temperature at the point (x, y, z) on the probe's surface is

$$T(x, y, z) = 8x^2 + 4yz - 16z + 600.$$

Find the hottest point on the probe's surface.

30. *Temperature Extremes on a Sphere* Suppose the Celsius temperature at the point (x, y, z) on the sphere $x^2 + y^2 + z^2 = 1$ is $T = 400xyz^2$. Locate the highest and lowest temperatures on the sphere.

Explorations

31. Maximizing a Utility Function. An Example from Economics In economics, the usefulness or utility of amounts x and y of two capital goods G_1 and G_2 is sometimes measured by a function $U(x, y)$. For example, G_1 and G_2 might be two chemicals a pharmaceutical company needs to have on hand and $U(x, y)$ the gain from manufacturing a product whose synthesis requires different amounts of the chemicals depending on the process used. If G_1 costs a dollars per kilogram, G_2 costs b dollars per kilogram, and the total amount allocated for the purchase of G_1 and G_2 together is c dollars, then the company's managers want to maximize $U(x, y)$ given that $ax + by = c$. Thus, they need to solve a typical Lagrange multiplier problem.

Suppose

$$U(x, y) = xy + 2x$$

and the equation $ax + by = c$ simplifies to

$$2x + y = 30.$$

Find the maximum value of U and the corresponding values of x and y subject to this latter constraint.

32. Placing a Radio Telescope on a Newly Discovered Planet You are in charge of erecting a radio telescope on a newly discovered planet. To minimize interference, you want to place it where the magnetic field of the planet is weakest. The planet is spherical, with a radius of 6 units. Based on a coordinate system whose origin is at the center of the planet, the strength of the magnetic field is given by $M(x, y, z) = 6x - y^2 + xz + 60$. Where should you locate the radio telescope? ∎

In Exercises 33–40, the problem involves two constraints.

33. Maximize the function $f(x, y, z) = x^2 + 2y - z^2$ subject to the constraints $2x - y = 0$ and $y + z = 0$.

34. Maximize the function $f(x, y, z) = x^2 + y^2 + z^2$ subject to the constraints $x + 2y + 3z = 6$ and $x + 3y + 9z = 9$.

35. Find the point closest to the origin on the line of intersection of the planes $y + 2z = 12$ and $x + y = 6$.

36. Find the maximum value that $f(x, y, z) = x^2 + 2y - z^2$ can have on the line of intersection of the planes $2x - y = 0$ and $y + z = 0$.

37. Find the extreme values of $f(x, y, z) = x^2yz + 1$ on the intersection of the plane $z = 1$ with the sphere $x^2 + y^2 + z^2 = 10$.

38. (a) Find the maximum value of $w = xyz$ on the line of intersection of the two planes $x + y + z = 40$ and $x + y - z = 0$.

(b) Writing to Learn Give a geometric argument to support your claim that you have found a maximum, and not a minimum, value of w.

39. Find the extreme values of the function $f(x, y, z) = xy + z^2$ on the circle in which the plane $y - x = 0$ intersects the sphere $x^2 + y^2 + z^2 = 4$.

40. Find the point closest to the origin on the curve of intersection of the plane $2y + 4z = 5$ and the cone $z^2 = 4x^2 + 4y^2$.

Extending the Ideas

41. The Condition $\nabla f = \lambda \nabla g$ Is Not Sufficient Although $\nabla f = \lambda \nabla g$ is a necessary condition for the occurrence of an extreme value of $f(x, y)$ subject to the condition $g(x, y) = 0$, it does not in itself guarantee that an extreme value exists. As a case in point, try using the method of Lagrange multipliers to find a maximum value of $f(x, y) = x + y$ subject to the constraint that $xy = 16$. The method will identify the two points $(4, 4)$ and $(-4, -4)$ as candidates for the location of extreme values. Yet the sum $(x + y)$ has no maximum value on the hyperbola $xy = 16$. The farther you go from the origin on this hyperbola in the first quadrant, the larger the sum $f(x, y) = x + y$ becomes.

42. A Least Squares Plane The plane $z = Ax + By + C$ is to be "fitted" to the following points (x_k, y_k, z_k):

$$(0, 0, 0), \quad (0, 1, 1), \quad (1, 1, 1), \quad (1, 0, -1).$$

Find the values of A, B, and C that minimize

$$\sum_{k=1}^{4} (Ax_k + By_k + C - z_k)^2,$$

the sum of the squares of the deviations.

43. Geometric vs. Arithmetic Means

(a) Show that the maximum value of $a^2b^2c^2$ on a sphere of radius r centered at the origin of a Cartesian abc-coordinate system is $(r^2/3)^3$.

(b) Using part (a), show that for nonnegative numbers a, b, and c,

$$(abc)^{1/3} \leq \frac{a + b + c}{3}.$$

That is, the *geometric mean* of three numbers is less than or equal to their *arithmetic mean*.

44. Let $a_1, a_2, \ldots, a_n$ be n positive numbers. Find the maximum of $\sum_{i=1}^{n} a_i x_i$ subject to the constraint $\sum_{i=1}^{n} x_i^2 = 1$.

CAS Explorations

In Exercises 45–50, perform the following steps implementing the method of Lagrange multipliers for finding constrained extrema:

(a) Form the function $h = f - \lambda_1 g_1 - \lambda_2 g_2$, where f is the function to optimize subject to the constraints $g_1 = 0$ and $g_2 = 0$.

(b) Determine all the first partial derivatives of h, including the partials with respect to λ_1 and λ_2, and set them equal to 0.

(c) Solve the system of equations found in (b) for all the unknowns, including λ_1 and λ_2.

(d) Evaluate f at each of the solution points found in (c) and select the extreme value subject to the constraints asked for in the exercise.

45. Minimize $f(x, y, z) = xy + yz$ subject to the constraints $x^2 + y^2 - 2 = 0$ and $x^2 + z^2 - 2 = 0$.

46. Minimize $f(x, y, z) = xyz$ subject to the constraints $x^2 + y^2 - 1 = 0$ and $x - z = 0$.

47. Maximize $f(x, y, z) = x^2 + y^2 + z^2$ subject to the constraints $2y + 4z - 5 = 0$ and $4x^2 + 4y^2 - z^2 = 0$.

48. Minimize $f(x, y, z) = x^2 + y^2 + z^2$ subject to the constraints $x^2 - xy + y^2 - z^2 - 1 = 0$ and $x^2 + y^2 - 1 = 0$.

49. Minimize $f(x, y, z, w) = x^2 + y^2 + z^2 + w^2$ subject to the constraints $2x - y + z - w - 1 = 0$ and $x + y - z + w - 1 = 0$.

50. Determine the distance from the line $y = x + 1$ to the parabola $y^2 = x$. (*Hint*: Let (x, y) be a point on the line and (w, z) a point on the parabola. You want to minimize $(x - w)^2 + (y - z)^2$.)

Chapter 13 Key Terms

absolute change (p. 708)
algebra rules for gradients (p. 732)
boundary of a region in space (p. 678)
boundary of a region in the plane (p. 677)
boundary point of a region in space (p. 678)
boundary point of a region in the plane (p. 677)
bounded region in the plane (p. 677)
Chain Rule for functions of three variables (p. 717)
Chain Rule for functions of two variables (p. 715)
closed region in space (p. 678)
closed region in the plane (p. 677)
continuity at a point (p. 688)
continuous function (p. 688)
contour line (p. 679)
critical point (p. 738)
del (p. 736)
dependent variable (p. 675)
differentiable at a point (p. 703)
differentiable function (p. 703)
directional derivative (p. 723)
discriminant of $f(x, y)$ (p. 740)
domain of a multivariable function (p. 675)

error in an approximation (p. 706)
first derivative test for local extrema (p. 737)
formula for implicit differentiation (p. 719)
gradient (p. 725)
gradient vector (p. 725)
graph of $f(x, y)$ (p. 679)
Increment Theorem (p. 703)
independent variables (p. 675)
input variables (p. 675)
interior of a region in space (p. 678)
interior point of a region in the plane (p. 677)
interior of a region in the plane (p. 677)
interior point of a region in space (p. 678)
Lagrange multiplier (p. 749)
Lagrange multiplier method (p. 749)
Laplace equation (p. 702)
level curve (p. 679)
level surface (p. 680)
limit of multivariable function (p. 686)
linearization of a function (p. 705)
local maximum of a multivariable function (p. 737)

local minimum of a multivariable function (p. 737)
Mixed Derivative Theorem (p. 700)
normal line at a point on a surface (p. 729)
open region in space (p. 678)
open region in the plane (p. 677)
Orthogonal Gradient Theorem (p. 749)
output variable (p. 675)
partial derivative of a function (p. 694)
percentage change (p. 708)
range of a multivariable function (p. 675)
real-valued multivariable function (p. 675)
relative change (p. 737)
saddle point (p. 739)
second derivative test for a local extrema (p. 739)
second-order partial derivatives (p. 699)
standard linear approximation (p. 705)
surface $z = f(x, y)$ (p. 679)
tangent plane at a point on a surface (p. 729)
total differential (p. 710)
two-path test for discontinuity (p. 690)
unbounded region in the plane (p. 677)
wave equation (p. 702)

Chapter 13 Review Exercises

In Exercises 1–4, find the domain and range of the function and identify its level curves. Sketch a typical level curve.

1. $f(x, y) = 9x^2 + y^2$

2. $f(x, y) = e^{x+y}$

3. $g(x, y) = \dfrac{1}{xy}$

4. $g(x, y) = \sqrt{x^2 - y}$

In Exercise 5–8, find the domain and range of the function and identify its level surfaces. Sketch a typical level surface.

5. $f(x, y, z) = x^2 + y^2 - z$

6. $g(x, y, z) = x^2 + 4y^2 + 9z^2$

7. $h(x, y, z) = \dfrac{1}{x^2 + y^2 + z^2}$

8. $k(x, y, z) = \dfrac{1}{x^2 + y^2 + z^2 + 1}$

In Exercises 9–14, find the limit.

9. $\displaystyle\lim_{(x, y) \to (\pi, \ln 2)} e^y \cos x$

10. $\displaystyle\lim_{(x, y) \to (0, 0)} \dfrac{2 + y}{x + \cos y}$

11. $\displaystyle\lim_{\substack{(x, y) \to (1, 1) \\ x \neq y}} \dfrac{x - y}{x^2 - y^2}$

12. $\displaystyle\lim_{(x - y) \to (1, 1)} \dfrac{x^3 y^3 - 1}{xy - 1}$

13. $\displaystyle\lim_{P \to (1, -1, e)} \ln |x + y + z|$

14. $\displaystyle\lim_{P \to (1, -1, -1)} \tan^{-1}(x + y + z)$

In Exercises 15 and 16, show that the limit does not exist by considering different paths of approach.

15. $\lim\limits_{\substack{(x, y)\to(0,0)\\ y\neq x^2}} \dfrac{y}{x^2 - y}$

16. $\lim\limits_{\substack{(x, y)\to(0,0)\\ xy\neq 0}} \dfrac{x^2 + y^2}{xy}$

17. (a) Writing to Learn Let $f(x, y) = (x^2 - y^2)/(x^2 + y^2)$ for $(x, y) \neq (0, 0)$. Is it possible to define $f(0, 0)$ in a way that makes f continuous at the origin? Why?

(b) Writing to Learn Let

$$f(x, y) = \begin{cases} \dfrac{\sin (x - y)}{|x| + |y|}, & |x| + |y| \neq 0 \\ 0, & (x, y) = (0, 0). \end{cases}$$

Is f continuous at the origin? Why?

18. Let

$$f(r, \theta) = \begin{cases} \dfrac{\sin 6r}{6r}, & r \neq 0 \\ 1, & r = 0, \end{cases}$$

where r and θ are polar coordinates. Find

(a) $\lim\limits_{r\to0} f(r, \theta)$ **(b)** $f_r(0, 0)$

(c) $f_\theta(r, \theta)$, $r \neq 0$

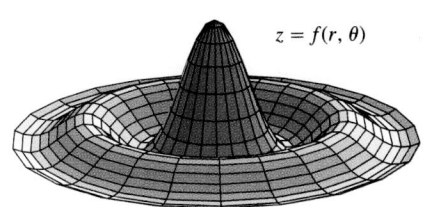

$z = f(r, \theta)$

(Generated by Mathematica)

In Exercises 19–24, find the partial derivatives of the function with respect to each variable.

19. $g(r, \theta) = r \cos \theta + r \sin \theta$

20. $f(x, y) = \dfrac{1}{2} \ln (x^2 + y^2) + \tan^{-1} \dfrac{y}{x}$

21. $f(R_1, R_2, R_3) = \dfrac{1}{R_1} + \dfrac{1}{R_2} + \dfrac{1}{R_3}$

22. $h(x, y, z) = \sin (2\pi x + y - 3z)$

23. $P(n, R, T, V) = \dfrac{nRT}{V}$ (the Ideal Gas Law)

24. $f(r, l, T, w) = \dfrac{1}{2rl} \sqrt{\dfrac{T}{\pi w}}$

In Exercises 25–28, find the second-order partial derivatives of the function.

25. $g(x, y) = y + \dfrac{x}{y}$

26. $g(x, y) = e^x + y \sin x$

27. $f(x, y) = x + xy - 5x^3 + \ln(x^2 + 1)$

28. $f(x, y) = y^2 - 3xy + \cos y + 7e^y$

In Exercises 29 and 30, find the linearization $L(x, y)$ of the function $f(x, y)$ at the point P_0. Then find an upper bound for the magnitude of the error E in the approximation $f(x, y) \approx L(x, y)$ over the rectangle R.

29. $f(x, y) = \sin x \cos y$, $P_0(\pi/4, \pi/4)$

$R: \left| x - \dfrac{\pi}{4} \right| \leq 0.1, \quad \left| y - \dfrac{\pi}{4} \right| \leq 0.1$

30. $f(x, y) = xy - 3y^2 + 2$, $P_0(1, 1)$

$R: |x - 1| \leq 0.1, \quad |y - 1| \leq 0.2$

In Exercises 31 and 32, find the linearization of the function at the given points.

31. $f(x, y, z) = xy + 2yz - 3xz$ at $(1, 0, 0)$ and $(1, 1, 0)$

32. $f(x, y, z) = \sqrt{2} \cos x \sin(y + z)$ at $(0, 0, \pi/4)$ and $(\pi/4, \pi/4, 0)$

33. Writing to Learn You plan to calculate the volume inside a stretch of pipeline that is about 36 in. in diameter and 1 mi long. With which measurement should you be more careful, the length or the diameter? Why?

34. Writing to Learn Near the point $(1, 2)$, is $f(x, y) = x^2 - xy + y^2 - 3$ more sensitive to changes in x, or to changes in y? How do you know?

35. *Current, Voltage, and Resistance* Suppose the current I (amperes) in an electrical circuit is related to the voltage V (volts) and the resistance R (ohms) by the equation $I = V/R$. If the voltage drops from 24 to 23 volts and the resistance drops from 100 to 80 ohms, will I increase or decrease? By about how much? Express the changes in V and R and the estimated change in I as percentages of their original values.

36. If $a = 10$ cm and $b = 16$ cm to the nearest millimeter, what should you expect the maximum percentage error to be in the calculated area πab of the ellipse $x^2/a^2 + y^2/b^2 = 1$?

37. Let $y = uv$ and $z = u + v$, where u and v are positive independent variables.

(a) If u is measured with an error of 2% and v with an error of 3%, about what is the percentage error in the calculated value of y?

(b) Show that the percentage error in the calculated value of z is less than the percentage error in the value of y.

38. *Cardiac Index* To make different people comparable in studies of cardiac output (Section 4.6, Exercise 25), researchers divide the measured cardiac output by the body surface area to find the *cardiac index* C:

$$C = \dfrac{\text{cardiac output}}{\text{body surface area}}.$$

The body surface area B is calculated with the formula

$$B = 71.84 w^{0.425} h^{0.725},$$

which gives B in square centimeters when the weight w is measured in kilograms and the height h in centimeters. You are about to calculate the cardiac index of a person with the following measurements:

Cardiac output:	7 L/min
Weight:	70 kg
Height:	180 cm

Which will have a greater effect on the calculation, a 1-kg error in measuring the weight or a 1-cm error in measuring the height?

39. Find dw/dt at $t = 0$ if $w = \sin(xy + \pi)$, $x = e^t$, and $y = \ln(t + 1)$.

40. Find $\partial w/\partial u$ and $\partial w/\partial v$ when $u = v = 0$ if $w = \ln \sqrt{1 + x^2} - \tan^{-1} x$ and $x = 2e^u \cos v$.

41. Find the value of the derivative of $f(x, y, z) = xy + yz + xz$ with respect to t on the curve $x = \cos t$, $y = \sin t$, $z = \cos 2t$ at $t = 1$.

42. Show that if $w = f(s)$ is any differentiable function of s and if $s = y + 5x$, then

$$\frac{\partial w}{\partial x} - 5\frac{\partial w}{\partial y} = 0.$$

In Exercises 43 and 44, assume that the equation defines y as a differentiable function of x and find the value of dy/dx at point P.

43. $1 - x - y^2 - \sin xy = 0$, $\quad P(0, 1)$

44. $2xy + e^{x+y} - 2 = 0$, $\quad P(0, \ln 2)$

In Exercises 45–48, find the directions in which f increases and decreases most rapidly at P_0 and find the derivative of f in each direction. Also, find the derivative of f at P_0 in the direction of the vector **A**.

45. $f(x, y) = \cos x \cos y$, $\quad P_0(\pi/4, \pi/4)$, $\quad \mathbf{A} = 3\mathbf{i} + 4\mathbf{j}$

46. $f(x, y) = x^2 e^{-2y}$, $\quad P_0(1, 0)$, $\quad \mathbf{A} = \mathbf{i} + \mathbf{j}$

47. $f(x, y, z) = \ln(2x + 3y + 6z)$, $\quad P_0(-1, -1, 1)$, $\mathbf{A} = 2\mathbf{i} + 3\mathbf{j} + 6\mathbf{k}$

48. $f(x, y, z) = x^2 + 3xy - z^2 + 2y + z + 4$, $\quad P_0(0, 0, 0)$, $\mathbf{A} = \mathbf{i} + \mathbf{j} + \mathbf{k}$

49. Find the derivative of $f(x, y, z) = xyz$ in the direction of the velocity vector of the helix

$$\mathbf{r}(t) = (\cos 3t)\mathbf{i} + (\sin 3t)\mathbf{j} + 3t\mathbf{k}$$

at $t = \pi/3$.

50. What is the largest value that the directional derivative of $f(x, y, z) = xyz$ can have at the point $(1, 1, 1)$?

51. At the point $(1, 2)$ the function $f(x, y)$ has a derivative of 2 in the direction toward $(2, 2)$ and derivative of -2 in the direction toward $(1, 1)$.

(a) Find $f_x(1, 2)$ and $f_y(1, 2)$.

(b) Find the derivative of f at $(1, 2)$ in the direction toward the point $(4, 6)$.

52. Which of the following statements are true if $f(x, y)$ is differentiable at (x_0, y_0)?

(a) If **u** is a unit vector, the derivative of f at (x_0, y_0) in the direction of **u** is $(f_x(x_0, y_0)\mathbf{i} + f_y(x_0, y_0)\mathbf{j}) \cdot \mathbf{u}.$

(b) The derivative of f at (x_0, y_0) in the direction of **u** is a vector.

(c) The directional derivative of f at (x_0, y_0) has its greatest value in the direction of ∇f.

(d) At (x_0, y_0), vector ∇f is normal to the curve $f(x, y) = f(x_0, y_0)$.

In Exercises 53 and 54, sketch the surface $f(x, y, z) = c$ together with ∇f at the given points.

53. $x^2 + y + z^2 = 0$; $\quad (0, -1, \pm1)$, $\quad (0, 0, 0)$

54. $y^2 + z^2 = 4$; $\quad (2, \pm2, 0)$, $\quad (2, 0, \pm2)$

In Exercises 55 and 56, find an equation for the plane tangent to the level surface $f(x, y, z) = c$ at the point P_0. Also, find parametric equations for the line that is normal to the surface at P_0.

55. $x^2 - y - 5z = 0$, $\quad P_0(2, -1, 1)$

56. $x^2 + y^2 + z = 4$, $\quad P_0(1, 1, 2)$

In Exercises 57 and 58, find an equation for the plane tangent to the surface $z = f(x, y)$ at the given point.

57. $z = \ln(x^2 + y^2)$, $\quad (0, 1, 0)$

58. $z = 1/(x^2 + y^2)$, $\quad (1, 1, 1/2)$

In Exercises 59 and 60, find equations for the lines that are tangent and normal to the level curve $f(x, y) = c$ at the point P_0. Then sketch the lines and level curve together with ∇f at P_0.

59. $y - \sin x = 1$, $\quad P_0(\pi, 1)$

60. $\dfrac{y^2}{2} - \dfrac{x^2}{2} = \dfrac{3}{2}$, $\quad P_0(1, 2)$

In Exercises 61 and 62, find parametric equations for the line that is tangent to the curve of intersection of the surfaces at the given point.

61. Surfaces: $\quad x^2 + 2y + 2z = 4$, $\quad y = 1$

Point: $\quad (1, 1, 1/2)$

62. Surfaces: $\quad x + y^2 + z = 2$, $\quad y = 1$

Point: $\quad (1/2, 1, 1/2)$

In Exercises 63–68, find the local maxima, local minima, and saddle points of the function. Find the function's values at these points.

63. $f(x, y) = x^2 - xy + y^2 + 2x + 2y - 4$

64. $f(x, y) = 5x^2 + 4xy - 2y^2 + 4x - 4y$

65. $f(x, y) = 2x^3 + 3xy + 2y^3$

66. $f(x, y) = x^3 + y^3 - 3xy + 15$

67. $f(x, y) = x^3 + y^3 + 3x^2 - 3y^2$

68. $f(x, y) = x^4 - 8x^2 + 3y^2 - 6y$

In Exercises 69–76, find the absolute maximum and minimum values of f on the region R.

69. $f(x, y) = x^2 + xy + y^2 - 3x + 3y$

R: The triangular region cut from the first quadrant by the line $x + y = 4$

70. $f(x, y) = x^2 - y^2 - 2x + 4y + 1$

R: The rectangular region in the first quadrant bounded by the coordinates axes and the lines $x = 4$ and $y = 2$

71. $f(x, y) = y^2 - xy - 3y + 2x$

R: The square region enclosed by the lines $x = \pm 2$ and $y = \pm 2$

72. $f(x, y) = 2x + 2y - x^2 - y^2$

R: The square bounded by the coordinate axes and the lines $x = 2$, $y = 2$ in the first quadrant

73. $f(x, y) = x^2 - y^2 - 2x + 4y$

R: The triangular region bounded below by the x-axis, above by the line $y = x + 2$, and on the right by the line $x = 2$

74. $f(x, y) = 4xy - x^4 - y^4 + 16$

R: The triangular region bounded below by the line $y = -2$, above by the line $y = x$, and on the right by the line $x = 2$

75. $f(x, y) = x^3 + y^3 + 3x^2 - 3y^2$

R: The square region enclosed by the lines $x = \pm 1$ and $y = \pm 1$

76. $f(x, y) = x^3 + 3xy + y^3 + 1$

R: The square region enclosed by the lines $x = \pm 1$ and $y = \pm 1$

77. Find the extreme values of $f(x, y) = x^3 + y^2$ on the circle $x^2 + y^2 = 1$.

78. Find the extreme values of $f(x, y) = xy$ on the circle $x^2 + y^2 = 1$.

79. Find the extreme values of $f(x, y) = x^2 + 3y^2 + 2y$ on the unit disk $x^2 + y^2 \le 1$.

80. Find the extreme values of $f(x, y) = x^2 + y^2 - 3x - xy$ on the disk $x^2 + y^2 \le 9$.

81. Find the extreme values of $f(x, y, z) = x - y + z$ on the unit sphere $x^2 + y^2 + z^2 = 1$.

82. Find the points on the surface $z^2 - xy = 4$ closest to the origin.

83. A closed rectangular box is to have volume V cm^3. The cost of the material used in the box is a cents/cm^2 for top and bottom, b cents/cm^2 for front and back, and c cents/cm^2 for the remaining sides. What dimensions minimize the total cost of materials?

84. Find the plane $x/a + y/b + z/c = 1$ that passes through the point $(2, 1, 2)$ and cuts off the least volume from the first octant.

85. Find the extreme values of $f(x, y, z) = x(y + z)$ on the curve of intersection of the right circular cylinder $x^2 + y^2 = 1$ and the hyperbolic cylinder $xz = 1$.

86. Find the point closest to the origin on the curve of intersection of the plane $x + y + z = 1$ and the cone $z^2 = 2x^2 + 2y^2$.

87. Let $w = f(r, \theta)$, $r = \sqrt{x^2 + y^2}$, and $\theta = \tan^{-1}(y/x)$. Find $\partial w/\partial x$ and $\partial w/\partial y$ and express your answers in terms of r and θ.

88. Let $z = f(u,v)$, $u = ax + by$, and $v = ax - by$. Express z_x and z_y in terms of f_u, f_v, and the constants a and b.

89. If $w = \ln(x^2 + y^2 + 2z)$, $x = r + s$, $y = r - s$, and $z = 2rs$, find w_r and w_s by the Chain Rule. Then check your answer another way.

90. The equations $e^u \cos v - x = 0$ and $e^u \sin v - y = 0$ define u and v as differentiable functions of x and y. Show that the angle between the vectors

$$\frac{\partial u}{\partial x}\mathbf{i} + \frac{\partial u}{\partial y}\mathbf{j} \quad \text{and} \quad \frac{\partial v}{\partial x}\mathbf{i} + \frac{\partial v}{\partial y}\mathbf{j}$$

is constant.

91. Find the points on the surface

$$(y + z)^2 + (z - x)^2 = 16$$

where the normal line is parallel to the yz-plane.

92. Find the points on the surface

$$xy + yz + zx - x - z^2 = 0$$

where the tangent plane is parallel to the xy-plane.

93. Suppose $\nabla f(x, y, z)$ is always parallel to the position vector $x\mathbf{i} + y\mathbf{j} + z\mathbf{k}$. Show that $f(0, 0, a) = f(0, 0, -a)$ for any a.

94. Show that the directional derivative of

$$f(x, y, z) = \sqrt{x^2 + y^2 + z^2}$$

at the origin equals 1 in any direction but that f has no gradient vector at the origin.

95. Show that the line normal to the surface $xy + z = 2$ at the point $(1, 1, 1)$ passes through the origin.

96. (a) Sketch the surface $x^2 - y^2 + z^2 = 4$.

(b) Find a vector normal to the surface at $(2, -3, 3)$. Add the vector to your sketch.

(c) Find the equations for the tangent plane and normal line at $(2, -3, 3)$.

Magnetic compasses are required equipment in most aircraft, yet they are adversely influenced by vibration, by other metal and electrical components, and during turns and acceleration. Most aircraft are also equipped with a directional gyroscope, which resists changes in direction while the aircraft rotates around it.

One directional gyroscope has a disk with a radius of 6 cm, thickness of 0.5 cm, and a density of 8.9 gm/cm^3, suspended on a weightless axle. If the gyroscope spins at an angular velocity of 1800 radians per second, what is its kinetic energy? Section 14.2 covers the necessary mathematics to solve this problem.

Chapter 14 Overview

The problems we can solve by integrating functions of two and three variables are similar to the problems solved by single-variable integration, but more general. As in the previous chapter, we can perform the necessary calculations by drawing on our experience with functions of a single variable.

14.1 Double Integrals

Double Integrals over Rectangles • Properties of Double Integrals • Double Integrals as Volumes • Fubini's Theorem for Calculating Double Integrals • Double Integrals over Bounded Nonrectangular Regions • Finding Limits of Integration

Double Integrals over Rectangles

Suppose $f(x, y)$ is defined on a rectangular region R given by

$$R: \quad a \leq x \leq b, \quad c \leq y \leq d.$$

We imagine R to be covered by a network of lines parallel to the x- and y-axes (Figure 14.1). These lines divide R into small pieces of area $\Delta A = \Delta x \, \Delta y$. We number these in some order $\Delta A_1, \Delta A_2, \ldots, \Delta A_n$, choose a point (x_k, y_k) in each piece ΔA_k, and form the sum

$$S_n = \sum_{k=1}^{n} f(x_k, y_k) \, \Delta A_k. \tag{1}$$

If f is continuous throughout R, then, as we refine the mesh width to make both Δx and Δy go to zero, the sums in Equation (1) approach a limit, the **double integral** of f over R. The notation for it is

$$\iint\limits_R f(x, y) \, dA \qquad \text{or} \qquad \iint\limits_R f(x, y) \, dx \, dy.$$

Thus,

$$\iint\limits_R f(x, y) \, dA = \lim_{\Delta A \to 0} \sum_{k=1}^{n} f(x_k, y_k) \, \Delta A_k. \tag{2}$$

As with functions of a single variable, the sums approach this limit no matter how the intervals $[a, b]$ and $[c, d]$ that determine R are partitioned, as long as the norms of the partitions both go to zero. The limit in Equation (2) is also independent of the order in which the areas ΔA_k are numbered and independent of the choice of the point (x_k, y_k) within each ΔA_k. The values of the individual approximating sums S_n depend on these choices, but the sums approach the same limit in the end. The proof of the existence and uniqueness of this limit for a continuous function f is given in more advanced texts. The continuity of f is a sufficient condition for the existence of the double integral, but not a necessary one. The limit in question exists for many discontinuous functions as well.

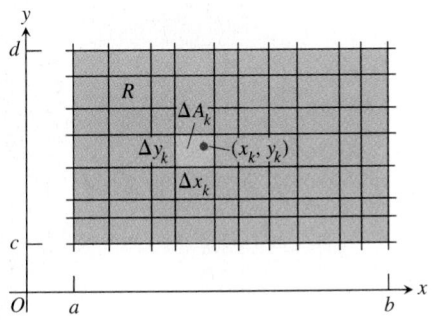

Figure 14.1 Rectangular grid partitioning the region R into small rectangles of area $\Delta A_k = \Delta x_k \, \Delta y_k$.

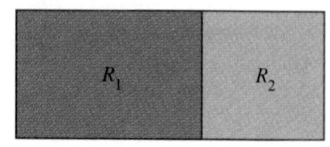

$$\iint\limits_{R_1 \cup R_2} f(x, y)\, dA = \iint\limits_{R_1} f(x, y)\, dA + \iint\limits_{R_2} f(x, y)\, dA$$

Figure 14.2 Double integrals have the same kind of domain additivity property that single integrals have.

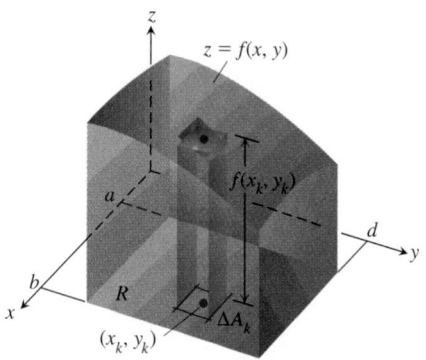

Figure 14.3 Approximating solids with rectangular prisms leads us to define the volumes of more general prisms as double integrals. The volume of the prism shown here is the double integral of $f(x, y)$ over the base region R.

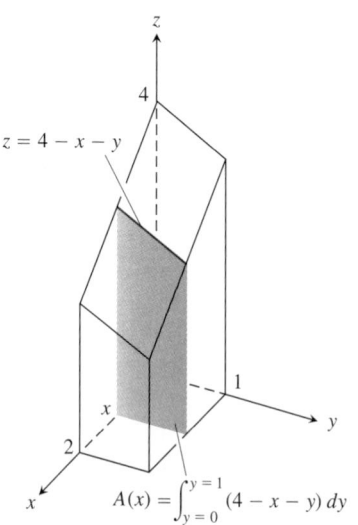

Figure 14.4 To obtain the cross-section area $A(x)$, we hold x fixed and integrate with respect to y.

Properties of Double Integrals

Like single integrals, double integrals of continuous functions have algebraic properties that are useful in computations and applications.

Properties of Double Integrals

1. *Constant Multiple:* $\displaystyle \iint\limits_{R} kf(x, y)\, dA = k \iint\limits_{R} f(x, y)\, dA$

   (any number k)

2. *Sum and Difference:*

$$\iint\limits_{R} (f(x, y) \pm g(x, y))\, dA = \iint\limits_{R} f(x, y)\, dA \pm \iint\limits_{R} g(x, y)\, dA$$

3. *Domination:*

   (a) $\displaystyle \iint\limits_{R} f(x, y)\, dA \geq 0 \qquad$ if $\qquad f(x, y) \geq 0$ on R

   (b) $\displaystyle \iint\limits_{R} f(x, y)\, dA \geq \iint\limits_{R} g(x, y)\, dA \qquad$ if $\qquad f(x, y) \geq g(x, y)$ on R

4. *Additivity:* $\displaystyle \iint\limits_{R} f(x, y)\, dA = \iint\limits_{R_1} f(x, y)\, dA + \iint\limits_{R_2} f(x, y)\, dA$

   if R is the union of two nonoverlapping rectangles R_1 and R_2 (Figure 14.2).

Double Integrals as Volumes

When $f(x, y)$ is positive, we may interpret the double integral of f over a rectangular region R as the volume of the solid prism bounded below by R and above by the surface $z = f(x, y)$ (Figure 14.3). Each term $f(x_k, y_k)\, \Delta A_k$ in the sum $S_n = \sum f(x_k, y_k)\, \Delta A_k$ is the volume of a vertical rectangular prism that approximates the volume of the portion of the solid that stands directly above the base ΔA_k. The sum S_n thus approximates what we want to call the *total volume of the solid*. We define this volume to be

$$\text{Volume} = \lim S_n = \iint\limits_{R} f(x, y)\, dA. \tag{3}$$

As you might expect, this more general method of calculating volume agrees with the methods in Chapter 7, but we will not prove this here.

Fubini's Theorem for Calculating Double Integrals

Suppose we wish to calculate the volume under the plane $z = 4 - x - y$ over the rectangular region R: $0 \leq x \leq 2$, $0 \leq y \leq 1$ in the xy-plane. If we apply the method of slicing from Section 7.3, with slices perpendicular to the x-axis (Figure 14.4), then the volume is

$$\int_{x=0}^{x=2} A(x)\, dx, \tag{4}$$

where $A(x)$ is the cross-section area at x. For each value of x we may calculate $A(x)$ as the integral

$$A(x) = \int_{y=0}^{y=1} (4 - x - y) \, dy, \tag{5}$$

which is the area under the curve $z = 4 - x - y$ in the plane of the cross section at x. In calculating $A(x)$, x is held fixed and the integration takes place with respect to y. Combining Equations (4) and (5), we see that the volume of the entire solid is

$$\text{Volume} = \int_{x=0}^{x=2} A(x) \, dx = \int_{x=0}^{x=2} \left(\int_{y=0}^{y=1} (4 - x - y) \, dy \right) dx$$

$$= \int_{x=0}^{x=2} \left[4y - xy - \frac{y^2}{2} \right]_{y=0}^{y=1} dx = \int_{x=0}^{x=2} \left(\frac{7}{2} - x \right) dx \tag{6}$$

$$= \left[\frac{7}{2} x - \frac{x^2}{2} \right]_0^2 = 5 \text{ units cubed.}$$

If we had just wanted to write instructions for calculating the volume, without carrying out any of the integrations, we could write

$$\text{Volume} = \int_0^2 \int_0^1 (4 - x - y) \, dy \, dx.$$

The expression on the right, an **iterated** or **repeated integral,** says that the volume is obtained by integrating $4 - x - y$ with respect to y from $y = 0$ to $y = 1$, holding x fixed, and then integrating the resulting expression in x with respect to x from $x = 0$ to $x = 2$.

What would have happened if we had calculated the volume by slicing with planes perpendicular to the y-axis (Figure 14.5)? As a function of y, the typical cross-section area is

$$A(y) = \int_{x=0}^{x=2} (4 - x - y) \, dx = \left[4x - \frac{x^2}{2} - xy \right]_{x=0}^{x=2} = 6 - 2y. \tag{7}$$

The volume of the entire solid is therefore

$$\text{Volume} = \int_{y=0}^{y=1} A(y) \, dy = \int_{y=0}^{y=1} (6 - 2y) \, dy = \left[6y - y^2 \right]_0^1 = 5 \text{ cubic units,}$$

in agreement with our earlier calculation.

Again, we may give instructions for calculating the volume as an iterated integral by writing

$$\text{Volume} = \int_0^1 \int_0^2 (4 - x - y) \, dx \, dy.$$

The expression on the right says we can find the volume by integrating $4 - x - y$ with respect to x from $x = 0$ to $x = 2$ (as in Equation (7)) and integrating the result with respect to y from $y = 0$ to $y = 1$. In this iterated integral, the order of integration is first x and then y, the reverse of the order in Equation (6).

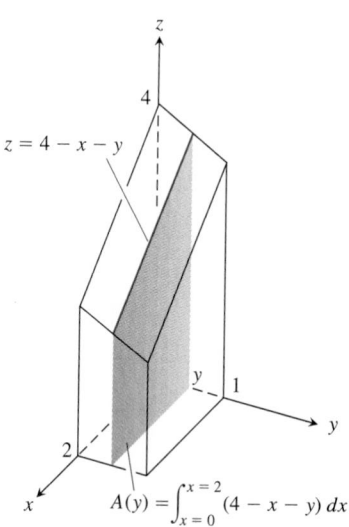

$$A(y) = \int_{x=0}^{x=2} (4 - x - y) \, dx$$

Figure 14.5 To obtain the cross-section area $A(y)$, we hold y fixed and integrate with respect to x.

What do these two volume calculations with iterated integrals have to do with the double integral

$$\iint_R (4 - x - y)\, dA$$

over the rectangle $R: 0 \leq x \leq 2, 0 \leq y \leq 1$? The answer is they both give the value of the double integral. A theorem published in 1907 by Guido Fubini (1879–1943) says that the double integral of any continuous function over a rectangle can be calculated as an iterated integral in either order of integration. (Fubini proved his theorem in greater generality, but this is how it translates into what we are doing now.)

Theorem 1 Fubini's Theorem (First Form)

If $f(x, y)$ is continuous on the rectangular region $R: a \leq x \leq b,$ $c \leq y \leq d$, then

$$\iint_R f(x, y)\, dA = \int_c^d \int_a^b f(x, y)\, dx\, dy = \int_a^b \int_c^d f(x, y)\, dy\, dx.$$

Fubini's theorem says that double integrals over rectangles can be calculated as iterated integrals. Thus, we can evaluate a double integral by integrating with respect to one variable at a time.

Fubini's theorem also says that we may calculate the double integral by integrating in *either* order, a genuine convenience, as we will see in Example 3. In particular, when we calculate a volume by slicing, we may use planes either perpendicular to the x-axis or to the y-axis.

Example 1 EVALUATING A DOUBLE INTEGRAL

Calculate $\iint_R f(x, y)\, dA$ for $f(x, y) = 1 - 6x^2y$ and $R: 0 \leq x \leq 2,$ $-1 \leq y \leq 1.$

Solution By Fubini's theorem,

$$\iint_R f(x, y)\, dA = \int_{-1}^1 \int_0^2 (1 - 6x^2y)\, dx\, dy = \int_{-1}^1 [x - 2x^3y]_{x=0}^{x=2}\, dy$$

$$= \int_{-1}^1 (2 - 16y)\, dy = [2y - 8y^2]_{-1}^1 = 4.$$

Reversing the order of integration gives the same answer:

$$\int_0^2 \int_{-1}^1 (1 - 6x^2y)\, dy\, dx = \int_0^2 [y - 3x^2y^2]_{y=-1}^{y=1}\, dx$$

$$= \int_0^2 [(1 - 3x^2) - (-1 - 3x^2)]\, dx$$

$$= \int_0^2 2\, dx = 4 \text{ cubic units.}$$

(a)

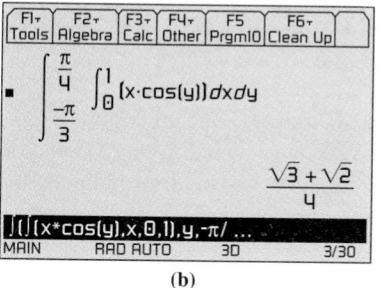

(b)

Figure 14.6 The value of $\iint x^2 y \, dx \, dy$

(a) and the value of $\int_{-\pi/3}^{\pi/4} \int_0^1 x \cos y \, dx \, dy$ (b).

Technology *Multiple Integration* Most computer algebra systems (CAS) can calculate both multiple and iterated integrals. The typical procedure is to apply the CAS integrate command in nested iterations according to the order of integration you specify. Figure 14.6 shows how one new advanced graphing calculator with CAS features calculates

$$\iint x^2 y \, dx \, dy \qquad \text{and} \qquad \int_{-\pi/3}^{\pi/4} \int_0^1 x \cos y \, dx \, dy.$$

If a CAS cannot produce an exact value for a definite integral, it can usually find an approximate value numerically.

Double Integrals over Bounded Nonrectangular Regions

To define the double integral of a function $f(x, y)$ over a bounded nonrectangular region, like the one shown in Figure 14.7, we again imagine R to be covered by a rectangular grid, but we include in the partial sum only the small pieces of area $\Delta A = \Delta x \, \Delta y$ that lie entirely within the region (shaded in the figure). We number the pieces in some order, choose an arbitrary point (x_k, y_k) in each ΔA_k, and form the sum

$$S_n = \sum_{k=1}^n f(x_k, y_k) \, \Delta A_k.$$

The only difference between this sum and the one in Equation (1) for rectangular regions is that now the areas ΔA_k may not cover all R. As the mesh becomes increasingly fine and the number of terms in S_n increases, however, more and more of R is included. If f is continuous and the boundary of R is made from the graphs of a finite number of continuous functions of x and/or continuous functions of y joined end to end, then the sums S_n will have a limit as the norms of the partitions that define the rectangular grid independently approach zero. The limit is the **double integral** of f over R:

$$\iint_R f(x, y) \, dA = \lim_{\Delta A \to 0} \sum f(x_k, y_k) \, \Delta A_k.$$

This limit may also exist under less restrictive circumstances.

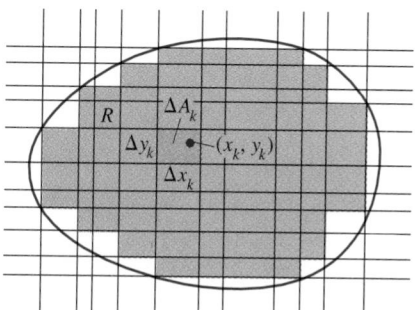

Figure 14.7 A rectangular grid partitioning a bounded nonrectangular region into cells.

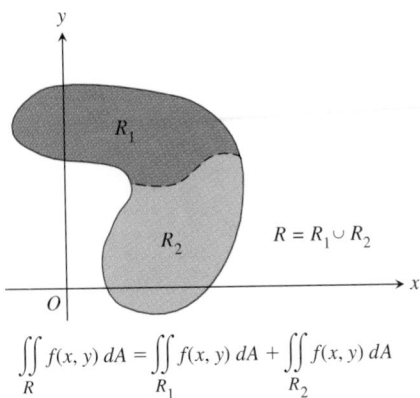

$$\iint_R f(x, y)\, dA = \iint_{R_1} f(x, y)\, dA + \iint_{R_2} f(x, y)\, dA$$

Figure 14.8 The additivity property for rectangular regions holds for regions bounded by continuous curves.

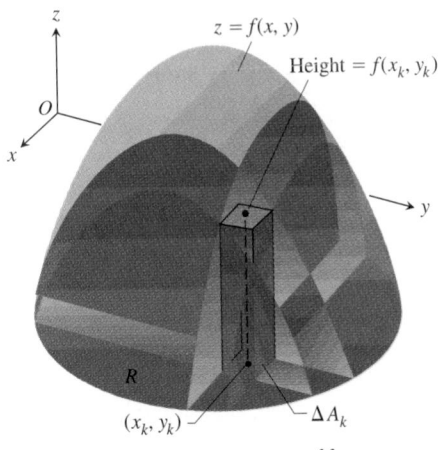

$$\text{Volume} = \lim \Sigma f(x_k, y_k)\, \Delta A_k = \iint_R f(x, y)\, dA$$

Figure 14.9 We define the volume of solids with curved bases the same way we define the volumes of solids with rectangular bases.

Double integrals of continuous functions over nonrectangular regions have the same algebraic properties as integrals over rectangular regions. The domain additivity property corresponding to property 4 says that if R is decomposed into nonoverlapping regions R_1 and R_2 with boundaries that are again made of a finite number of line segments or smooth curves (see Figure 14.8 for an example), then

$$\iint_R f(x, y)\, dA = \iint_{R_1} f(x, y)\, dA + \iint_{R_2} f(x, y)\, dA.$$

If $f(x, y)$ is positive and continuous over R (Figure 14.9), we define the volume of the solid region between R and the surface $z = f(x, y)$ to be $\iint_R f(x, y)\, dA$, as before.

If R is a region like the one shown in the xy-plane in Figure 14.10, bounded "above" and "below" by the curves $y = g_2(x)$ and $y = g_1(x)$ and on the sides by the lines $x = a$, $x = b$, we may again calculate the volume by the method of slicing. We first calculate the cross-section area

$$A(x) = \int_{y=g_1(x)}^{y=g_2(x)} f(x, y)\, dy$$

and then integrate $A(x)$ from $x = a$ to $x = b$ to get the volume as an iterated integral:

$$V = \int_a^b A(x)\, dx = \int_a^b \int_{g_1(x)}^{g_2(x)} f(x, y)\, dy\, dx. \tag{8}$$

Similarly, if R is a region like the one shown in Figure 14.11, bounded by the curves $x = h_2(y)$ and $x = h_1(y)$ and the lines $y = c$ and $y = d$, then the volume calculated by slicing is given by the iterated integral

$$\text{Volume} = \int_c^d \int_{h_1(y)}^{h_2(y)} f(x, y)\, dx\, dy. \tag{9}$$

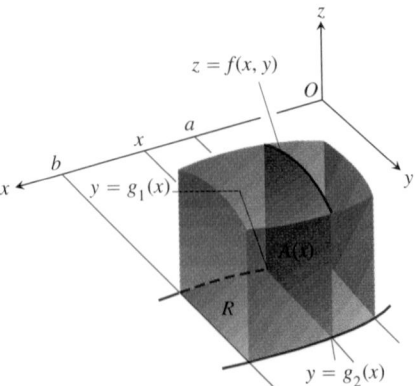

Figure 14.10 The area of the vertical slice shown here is $A(x) = \int_{g_1(x)}^{g_2(x)} f(x, y)\, dy$. To calculate the volume of the solid, we integrate this area from $x = a$ to $x = b$.

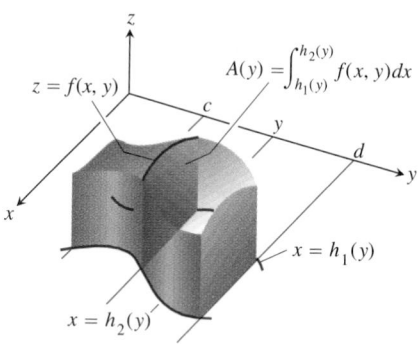

Figure 14.11 The volume of the solid shown here is

$$\int_c^d A(y)\, dy = \int_c^d \int_{h_1(y)}^{h_2(y)} f(x, y)\, dx\, dy.$$

That the iterated integrals in Equations (8) and (9) both give the volume that we define to be the double integral of f over R is a consequence of the following stronger form of Fubini's theorem.

Theorem 2 Fubini's Theorem (Stronger Form)

Let $f(x, y)$ be continuous on a region R.

1. If R is defined by $a \le x \le b$, $g_1(x) \le y \le g_2(x)$, with g_1 and g_2 continuous on $[a, b]$, then

$$\iint_R f(x, y)\, dA = \int_a^b \int_{g_1(x)}^{g_2(x)} f(x, y)\, dy\, dx.$$

2. If R is defined by $c \le y \le d$, $h_1(y) \le x \le h_2(y)$, with h_1 and h_2 continuous on $[c, d]$, then

$$\iint_R f(x, y)\, dA = \int_c^d \int_{h_1(y)}^{h_2(y)} f(x, y)\, dx\, dy.$$

Example 2 FINDING VOLUME

Find the volume of the prism whose base is the triangle in the xy-plane bounded by the x-axis and the lines $y = x$ and $x = 1$ and whose top lies in the plane $z = f(x, y) = 3 - x - y$.

Solution   See Figure 14.12. For any x between 0 and 1, y may vary from $y = 0$ to $y = x$ (Figure 14.12b). Hence,

$$V = \int_0^1 \int_0^x (3 - x - y)\, dy\, dx = \int_0^1 \left[3y - xy - \frac{y^2}{2} \right]_{y=0}^{y=x} dx$$

$$= \int_0^1 \left(3x - \frac{3x^2}{2} \right) dx = \left[\frac{3x^2}{2} - \frac{x^3}{2} \right]_{x=0}^{x=1} = 1 \text{ cubic unit.}$$

When the order of integration is reversed (Figure 14.12c), the integral for the volume is

$$V = \int_0^1 \int_y^1 (3 - x - y)\, dx\, dy = \int_0^1 \left[3x - \frac{x^2}{2} - xy \right]_{x=y}^{x=1} dy$$

$$= \int_0^1 \left(3 - \frac{1}{2} - y - 3y + \frac{y^2}{2} + y^2 \right) dy$$

$$= \int_0^1 \left[\frac{5}{2} - 4y + \frac{3}{2} y^2 \right] dy = \left[\frac{5}{2} y - 2y^2 + \frac{y^3}{2} \right]_{y=0}^{y=1} = 1 \text{ cubic unit.}$$

The two integrals are equal, as they should be.

(a)

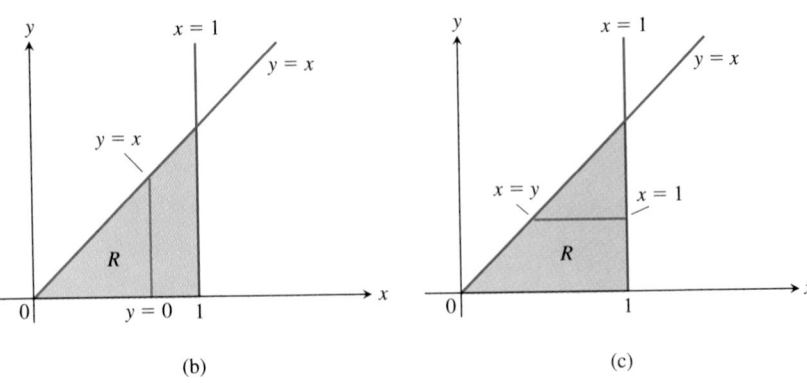

(b)

(c)

Figure 14.12 (a) Prism with a triangular base in the xy-plane. The volume of this prism is defined as a double integral over R. To evaluate it as an iterated integral, we may integrate first with respect to y and then with respect to x, or the other way around (Example 2). (b) Integration limits of $\int_{x=0}^{x=1} \int_{y=0}^{y=x} f(x, y)\, dy\, dx.$ If we integrate first with respect to y, we integrate along a vertical line through R and then integrate from left to right to include all the vertical lines in R.

(c) Integration limits of $\int_{y=0}^{y=1} \int_{x=y}^{x=1} f(x, y)\, dx\, dy.$ If we integrate first with respect to x, we integrate along a horizontal line through R and then integrate from bottom to top to include all the horizontal lines in R.

Although Fubini's theorem assures us that a double integral may be calculated as an iterated integral in either order of integration, the value of one integral may be easier to find than the value of the other. The next example shows how this can happen.

Example 3   EVALUATING A DOUBLE INTEGRAL

Calculate

$$\iint_R \frac{\sin x}{x}\, dA,$$

where R is the triangle in the xy-plane bounded by the x-axis, the line $y = x$, and the line $x = 1$.

Solution   The region of integration is shown in Figure 14.13. If we integrate first with respect to y and then with respect to x, we find

$$\int_0^1 \left(\int_0^x \frac{\sin x}{x}\, dy \right) dx = \int_0^1 \left(y\, \frac{\sin x}{x} \right]_{y=0}^{y=x} \right) dx = \int_0^1 \sin x\, dx$$

$$= -\cos(1) + 1 \approx 0.46 \text{ units cubed.}$$

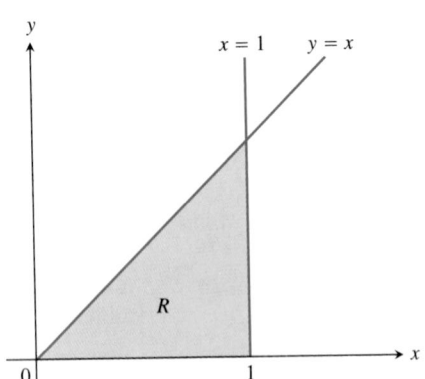

Figure 14.13 The region of integration in Example 3.

If we reverse the order of integration and attempt to calculate (by hand)

$$\int_0^1 \int_y^1 \frac{\sin x}{x} \, dx \, dy,$$

we are stopped because $\int ((\sin x)/x) \, dx$ cannot be expressed in terms of elementary functions. (A CAS will, however, produce an approximation to the answer.)

There is no general rule for predicting which order of integration will be the good one for computation without technology in circumstances like these, so don't worry about how to start. Just forge ahead, and if the order you first choose doesn't work, try the other.

Finding Limits of Integration

The most difficult part of evaluating a double integral can be finding the limits of integration. Fortunately, there is a good procedure to follow.

Procedure for Finding Limits of Integration

A. To evaluate $\iint_R dA$ over a region R, integrating first with respect to y and then with respect to x, take the following steps:

            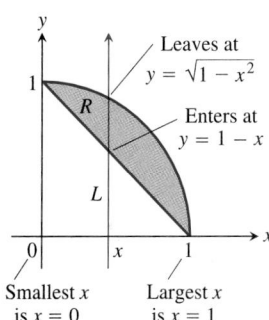

1. *A sketch.* Sketch the region of integration and label the bounding curves.

2. *The y-limits of integration.* Imagine a vertical line L cutting through R in the direction of increasing y. Mark the y-values where L enters and leaves. These are the y-limits of integration.

3. *The x-limits of integration.* Choose x-limits that include all the vertical lines through R. The integral is

$$\iint_R f(x, y) \, dA =$$

$$\int_{x=0}^{x=1} \int_{y=1-x}^{y=\sqrt{1-x^2}} f(x, y) \, dy \, dx.$$

B. To evaluate the same double integral as an iterated integral with the order of integration reversed, use horizontal lines instead of vertical lines. The integral is

$$\iint_R f(x, y) \, dA = \int_0^1 \int_{1-y}^{\sqrt{1-y^2}} f(x, y) \, dx \, dy.$$

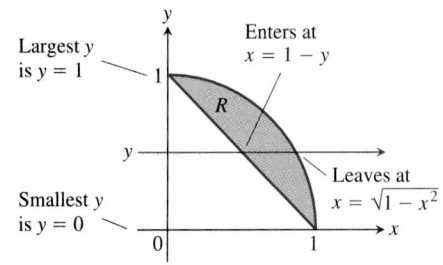

Example 4 REVERSING ORDER OF INTEGRATION

Sketch the region of integration for the integral

$$\int_0^2 \int_{x^2}^{2x} (4x + 2)\, dy\, dx$$

and write an equivalent integral with the order of integration reversed.

Solution The region of integration is given by the inequalities $x^2 \le y \le 2x$ and $0 \le x \le 2$. It is therefore the region bounded by the curves $y = x^2$ and $y = 2x$ between $x = 0$ and $x = 2$ (Figure 14.14a).

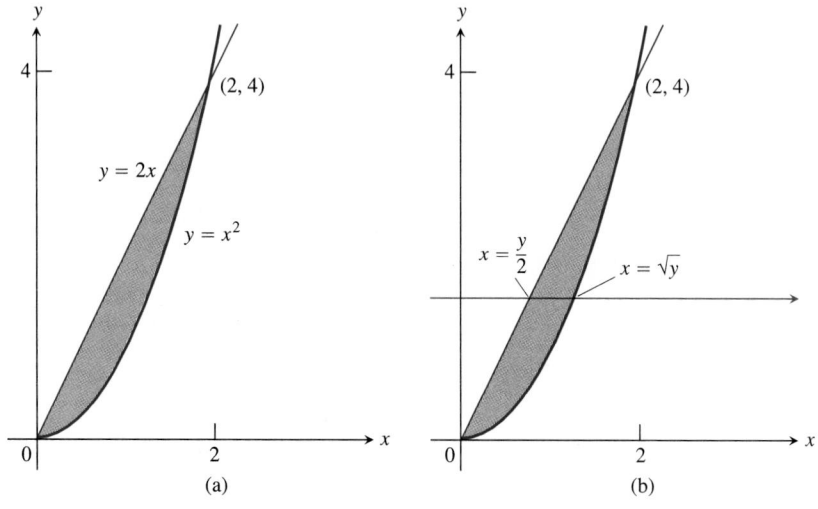

Figure 14.14 Figure for Example 4.

To find limits for integrating in the reverse order, we imagine a horizontal line passing from left to right through the region. It enters at $x = y/2$ and leaves at $x = \sqrt{y}$. To include all such lines, we let y run from $y = 0$ to $y = 4$ (Figure 14.14b). The integral is

$$\int_0^4 \int_{y/2}^{\sqrt{y}} (4x + 2)\, dx\, dy.$$

• The common value of these integrals is 8.

Quick Review 14.1

In Exercises 1 and 2, express the limit as a definite integral.

1. $\displaystyle\lim_{\|P\| \to 0} \sum_{k=1}^{n} c_k{}^3 \Delta x_k$, where P is any partition of $[0, 3]$

2. $\displaystyle\lim_{\|P\| \to 0} \sum_{k=1}^{n} \sqrt[3]{2 - c_k{}^2}\, \Delta x_k$, where P is any partition of $[-1, 1]$

In Exercises 3–8, evaluate the integral.

3. $\displaystyle\int e^{\sin x} \cos x\, dx$

4. $\displaystyle\int \frac{2}{1 - x}\, dx$

5. $\displaystyle\int \left(e^x - \frac{5}{x^2} \right) dx$

6. $\displaystyle\int_0^{\pi/4} \sec x \tan x\, dx$

7. $\displaystyle\int_{-2}^{1} (1 + |x|)\, dx$

8. $\displaystyle\int_{-1}^{1} \frac{\sin x}{x}\, dx$

In Exercises 9 and 10, find the volume of the solid.

9. The solid lies between the planes perpendicular to the x-axis at $x = 0$ and $x = 2$. The cross sections perpendicular to the x-axis on the interval $0 \le x \le 2$ are squares whose diagonals run from $y = -\sqrt{x}$ to $y = \sqrt{x}$.

10. The solid lies between planes perpendicular to the x-axis at $x = -2$ and $x = 2$. The cross sections perpendicular to the x-axis are circular disks whose diameters run from the parabola $y = x^2$ to the parabola $y = 8 - x^2$.

Section 14.1 Exercises

In Exercises 1–10, sketch the region of integration and evaluate the integral.

1. $\displaystyle\int_0^3 \int_0^2 (4 - y^2) \, dy \, dx$

2. $\displaystyle\int_0^3 \int_{-2}^0 (x^2 y - 2xy) \, dy \, dx$

3. $\displaystyle\int_{-1}^0 \int_{-1}^1 (x + y + 1) \, dx \, dy$

4. $\displaystyle\int_\pi^{2\pi} \int_0^\pi (\sin x + \cos y) \, dx \, dy$

5. $\displaystyle\int_0^\pi \int_0^x x \sin y \, dy \, dx$

6. $\displaystyle\int_0^\pi \int_0^{\sin x} y \, dy \, dx$

7. $\displaystyle\int_1^{\ln 8} \int_0^{\ln y} e^{x+y} \, dx \, dy$

8. $\displaystyle\int_1^2 \int_y^{y^2} dx \, dy$

9. $\displaystyle\int_0^1 \int_0^{y^2} 3y^3 e^{xy} \, dx \, dy$

10. $\displaystyle\int_1^4 \int_0^{\sqrt{x}} \frac{3}{2} e^{y/\sqrt{x}} \, dy \, dx$

In Exercises 11–16, integrate f over the given region.

11. $f(x, y) = x/y$ over the region in the first quadrant bounded by the lines $y = x$, $y = 2x$, $x = 1$, $x = 2$

12. $f(x, y) = 1/(xy)$ over the square $1 \le x \le 2$, $1 \le y \le 2$

13. $f(x, y) = x^2 + y^2$ over the triangular region with vertices $(0, 0)$, $(1, 0)$, and $(0, 1)$

14. $f(x, y) = y \cos xy$ over the rectangle $0 \le x \le \pi$, $0 \le y \le 1$

15. $f(u, v) = v - \sqrt{u}$ over the triangular region cut from the first quadrant of the uv-plane by the line $u + v = 1$

16. $f(s, t) = e^s \ln t$ over the region in the first quadrant of the st-plane that lies above the curve $s = \ln t$ from $t = 1$ to $t = 2$

In Exercises 17–20, an integral over a region in a Cartesian coordinate plane is given. Sketch the region and evaluate the integral.

17. $\displaystyle\int_{-2}^0 \int_v^{-v} 2 \, dp \, dv$ (the pv-plane)

18. $\displaystyle\int_0^1 \int_0^{\sqrt{1-s^2}} 8t \, dt \, ds$ (the st-plane)

19. $\displaystyle\int_{-\pi/3}^{\pi/3} \int_0^{\sec t} 3 \cos t \, du \, dt$ (the tu-plane)

20. $\displaystyle\int_0^3 \int_{-2}^{4-2u} \frac{4 - 2u}{v^2} \, dv \, du$ (the uv-plane)

In Exercises 21–30, sketch the region of integration and write an equivalent double integral with the order of integration reversed.

21. $\displaystyle\int_0^1 \int_2^{4-2x} dy \, dx$

22. $\displaystyle\int_0^2 \int_{y-2}^0 dx \, dy$

23. $\displaystyle\int_0^1 \int_y^{\sqrt{y}} dx \, dy$

24. $\displaystyle\int_0^1 \int_{1-x}^{1-x^2} dy \, dx$

25. $\displaystyle\int_0^1 \int_1^{e^x} dy \, dx$

26. $\displaystyle\int_0^{\ln 2} \int_{e^y}^2 dx \, dy$

27. $\displaystyle\int_0^{3/2} \int_0^{9-4x^2} 16x \, dy \, dx$

28. $\displaystyle\int_0^2 \int_0^{4-y^2} y \, dx \, dy$

29. $\displaystyle\int_0^1 \int_{-\sqrt{1-y^2}}^{\sqrt{1-y^2}} 3y \, dx \, dy$

30. $\displaystyle\int_0^2 \int_{-\sqrt{4-x^2}}^{\sqrt{4-x^2}} 6x \, dy \, dx$

In Exercises 31–40, sketch the region of integration, determine an order of integration that will enable you to evaluate the integral analytically, and evaluate the integral.

31. $\displaystyle\int_0^\pi \int_x^\pi \frac{\sin y}{y} \, dy \, dx$

32. $\displaystyle\int_0^2 \int_x^2 2y^2 \sin xy \, dy \, dx$

33. $\displaystyle\int_0^1 \int_y^1 x^2 e^{xy} \, dx \, dy$

34. $\displaystyle\int_0^2 \int_0^{4-x^2} \frac{xe^{2y}}{4 - y} \, dy \, dx$

35. $\displaystyle\int_0^{2\sqrt{\ln 3}} \int_{y/2}^{\sqrt{\ln 3}} e^{x^2} \, dx \, dy$

36. $\displaystyle\int_0^3 \int_{\sqrt{x/3}}^1 e^{y^3} \, dy \, dx$

37. $\displaystyle\int_0^{1/16} \int_{y^{1/4}}^{1/2} \cos(16\pi x^5) \, dx \, dy$

38. $\displaystyle\int_0^8 \int_{\sqrt[3]{x}}^2 \frac{dy \, dx}{y^4 + 1}$

39. $\iint\limits_{R} (y - 2x^2) \, dA$, where R is the region inside the square

$|x| + |y| = 1$

40. $\iint\limits_{R} xy \, dA$, where R is the region bounded by the lines

$y = x$, $y = 2x$, and $x + y = 2$

In Exercises 41–50, find the volume of the solid.

41. The region that lies under the paraboloid $z = x^2 + y^2$ and above the triangle enclosed by the lines $y = x$, $x = 0$, and $x + y = 2$ in the xy-plane

42. The solid that is bounded above by the cylinder $z = x^2$ and below by the region enclosed by the parabola $y = 2 - x^2$ and the line $y = x$ in the xy-plane

43. The solid whose base is the region in the xy-plane that is bounded by the parabola $y = 4 - x^2$ and the line $y = 3x$, while the top of the solid is bounded by the plane $z = x + 4$

44. The solid in the first octant bounded by the coordinate planes, the cylinder $x^2 + y^2 = 4$, and the plane $z + y = 3$

45. The solid in the first octant bounded by the coordinate planes, the plane $x = 3$, and the parabolic cylinder $z = 4 - y^2$

46. The solid cut from the first octant by the surface $z = 4 - x^2 - y$

47. The wedge cut from the first octant by the cylinder $z = 12 - 3y^2$ and the plane $x + y = 2$

48. The solid cut from the square column $|x| + |y| \leq 1$ by the planes $z = 0$ and $3x + z = 3$

49. The solid that is bounded on the front and back by the planes $x = 2$ and $x = 1$, on the sides by the cylinders $y = \pm 1/x$, and above by the planes $z = x + 1$ and $z = 0$

50. The solid that is bounded on the front and back by the planes $x = \pm \pi/3$, on the sides by the cylinders $y = \pm \sec x$, above by the cylinder $z = 1 + y^2$, and below by the xy-plane

In Exercises 51–54, *work in groups of two or three* to evaluate the improper integrals as iterated integrals.

51. $\displaystyle\int_{1}^{\infty} \int_{e^{-x}}^{1} \frac{1}{x^3 y} \, dy \, dx$

52. $\displaystyle\int_{-1}^{1} \int_{-1/\sqrt{1-x^2}}^{1/\sqrt{1-x^2}} (2y + 1) \, dy \, dx$

53. $\displaystyle\int_{-\infty}^{\infty} \int_{-\infty}^{\infty} \frac{1}{(x^2 + 1)(y^2 + 1)} \, dx \, dy$

54. $\displaystyle\int_{0}^{\infty} \int_{0}^{\infty} xe^{-(x+2y)} \, dx \, dy$

Explorations

In Exercises 55 and 56, approximate the double integral of $f(x, y)$ over the region R partitioned by the given vertical lines $x = a$ and horizontal lines $y = c$. In each subrectangle, use (x_k, y_k) as indicated for your approximation.

$$\iint\limits_{R} f(x, y) \, dA \approx \sum_{k=1}^{n} f(x_k, y_k) \, \Delta A_k$$

55. $f(x, y) = x + y$ over the region R bounded above by the semicircle $y = \sqrt{1 - x^2}$ and below by the x-axis, using the partition $x = -1, -1/2, 0, 1/4, 1/2, 1$ and $y = 0, 1/2, 1$ with (x_k, y_k) the lower left corner in the kth subrectangle

56. $f(x, y) = x + 2y$ over the region R inside the circle $(x - 2)^2 + (y - 3)^2 = 1$ using the partition $x = 1, 3/2, 2, 5/2, 3$ and $y = 2, 5/2, 3, 7/2, 4$ with (x_k, y_k) the center (centroid) in the kth subrectangle ∎

57. Integrate $f(x, y) = \sqrt{4 - x^2}$ over the smaller sector cut from the disk $x^2 + y^2 \leq 4$ by the rays $\theta = \pi/6$ and $\theta = \pi/2$.

58. Integrate $f(x, y) = 1/[(x^2 - x)(y - 1)^{2/3}]$ over the infinite rectangle $2 \leq x < \infty$, $0 \leq y \leq 2$.

59. A solid right (noncircular) cylinder has its base R in the xy-plane and is bounded above by the paraboloid $z = x^2 + y^2$. The cylinder's volume is

$$V = \int_{0}^{1} \int_{0}^{y} (x^2 + y^2) \, dx \, dy + \int_{1}^{2} \int_{0}^{2-y} (x^2 + y^2) \, dx \, dy.$$

Sketch the base region R and express the cylinder's volume as a single iterated integral with the order of integration reversed. Then evaluate the integral to find the volume.

60. Evaluate the integral

$$\int_{0}^{2} (\tan^{-1} \pi x - \tan^{-1} x) \, dx.$$

(*Hint:* Write the integrand as an integral.)

61. Writing to Learn What region R in the xy-plane maximizes the value of

$$\iint\limits_{R} (4 - x^2 - 2y^2) \, dA?$$

Give reasons for your answer.

62. Writing to Learn What region R in the xy-plane minimizes the value of

$$\iint\limits_{R} (x^2 + y^2 - 9) \, dA?$$

Give reasons for your answer.

63. Writing to Learn Is it all right to evaluate the integral of a continuous function $f(x, y)$ over a rectangular region in the xy-plane and get different answers depending on the order of integration? Give reasons for your answer.

64. Writing to Learn How would you evaluate the double integral of a continuous function $f(x, y)$ over the region R in the xy-plane enclosed by the triangle with vertices $(0, 1)$, $(2, 0)$, and $(1, 2)$? Give reasons for your answer.

Extending the Ideas

65. Prove that

$$\int_{-\infty}^{\infty} \int_{-\infty}^{\infty} e^{-x^2 - y^2} \, dx \, dy = \lim_{b \to \infty} \int_{-b}^{b} \int_{-b}^{b} e^{-x^2 - y^2} \, dx \, dy$$

$$= 4 \left(\int_{0}^{\infty} e^{-x^2} \, dx \right)^2.$$

66. Evaluate the improper integral

$$\int_{0}^{1} \int_{0}^{3} \frac{x^2}{(y-1)^{2/3}} \, dy \, dx.$$

CAS Explorations

In Exercises 67–70, evaluate the integrals.

67. $\displaystyle\int_{1}^{3} \int_{1}^{x} \frac{1}{xy} \, dy \, dx$

68. $\displaystyle\int_{0}^{1} \int_{0}^{1} e^{-(x^2 + y^2)} \, dy \, dx$

69. $\displaystyle\int_{0}^{1} \int_{0}^{1} \tan^{-1} xy \, dy \, dx$

70. $\displaystyle\int_{-1}^{1} \int_{0}^{\sqrt{1-x^2}} 3\sqrt{1 - x^2 - y^2} \, dy \, dx$

14.2 Areas, Moments, and Centers of Mass

Areas of Bounded Regions in the Plane • Average Value • First and Second Moments and Centers of Mass • Centroids of Geometric Figures

Areas of Bounded Regions in the Plane

If we take $f(x, y) = 1$ in the definition of the double integral over a region R in the preceding section, the partial sums reduce to

$$S_n = \sum_{k=1}^{n} f(x_k, y_k) \, \Delta A_k = \sum_{k=1}^{n} \Delta A_k. \tag{1}$$

This approximates what we would like to call the area of R. As Δx and Δy approach zero, the coverage of R by the ΔA_k's (Figure 14.15) becomes increasingly complete, and we define the *area of R* to be the limit

$$\text{Area} = \lim_{n \to \infty} \sum_{k=1}^{n} \Delta A_k = \iint_R dA. \tag{2}$$

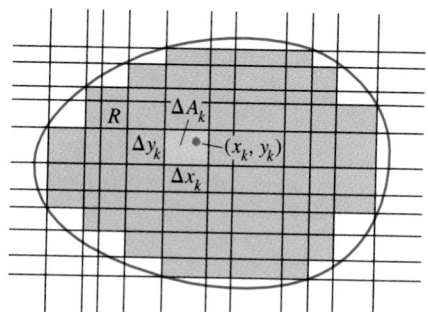

Figure 14.15 The first step in defining the area of a region is to partition the interior of the region into cells.

Definition Area

The **area** of a closed, bounded plane region R is

$$A = \iint_R dA. \tag{3}$$

As with the other definitions in this chapter, the definition here applies to a greater variety of regions than does the earlier single-variable definition of area, but it agrees with the earlier definition on regions to which they both apply.

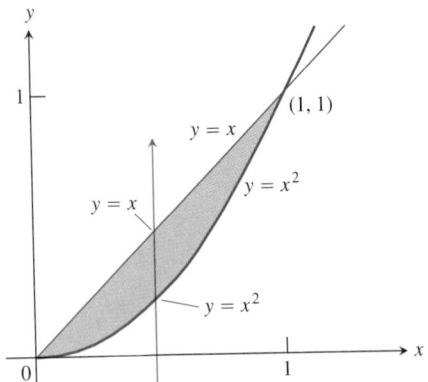

Figure 14.16 The area of the region between the parabola and the line in Example 1 is $\int_0^1 \int_{x^2}^x dy\, dx$.

To evaluate the integral in (3), we integrate the constant function $f(x, y) = 1$ over R.

Example 1   FINDING AREA

Find the area of the region R bounded by $y = x$ and $y = x^2$ in the first quadrant.

Solution   We sketch the region (Figure 14.16) and calculate the area as

$$A = \int_0^1 \int_{x^2}^x dy\, dx = \int_0^1 [y]_{x^2}^x\, dx = \int_0^1 (x - x^2)\, dx$$

$$= \left[\frac{x^2}{2} - \frac{x^3}{3}\right]_0^1 = \frac{1}{6} \text{ square units}$$

Example 2   FINDING AREA

Find the area of the region R enclosed by the parabola $y = x^2$ and the line $y = x + 2$.

Solution   If we divide R into the regions R_1 and R_2 shown in Figure 14.17a, we may calculate the area as

$$A = \iint_{R_1} dA + \iint_{R_2} dA = \int_0^1 \int_{-\sqrt{y}}^{\sqrt{y}} dx\, dy + \int_1^4 \int_{y-2}^{\sqrt{y}} dx\, dy.$$

On the other hand, reversing the order of integration (Figure 14.17b) gives

$$A = \int_{-1}^2 \int_{x^2}^{x+2} dy\, dx.$$

This result is simpler and is the only one we would bother to write down in practice. The area is

$$A = \int_{-1}^2 [y]_{x^2}^{x+2}\, dx = \int_{-1}^2 (x + 2 - x^2)\, dx$$

$$= \left[\frac{x^2}{2} + 2x - \frac{x^3}{3}\right]_{-1}^2 = \frac{9}{2} \text{ square units}.$$

(a)

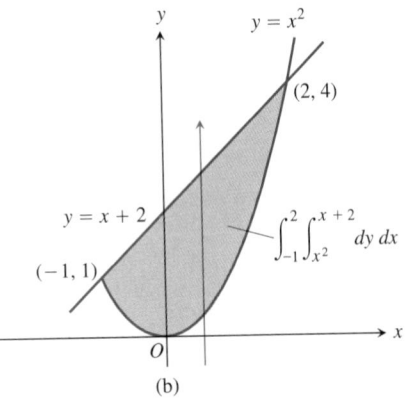

(b)

Figure 14.17 Calculating the area takes (a) two double integrals if the first integration is with respect to x, but (b) only one if the first integration is with respect to y. (Example 2)

Average Value

The average value of an integrable function of a single variable on a closed interval is the integral of the function over the interval divided by the length of the interval. For an integrable function of two variables defined on a closed and bounded region that has measurable area, the *average value* is the integral over the region divided by the area of the region. If f is the function and R the region, then

$$\textbf{Average value} \text{ of } f \text{ over } R = \frac{1}{\text{area of } R} \iint_R f\, dA. \qquad (4)$$

If f is the area density of a thin plate covering R, then the double integral of f over R divided by the area of R is the plate's **average density** in units of mass per unit area. If $f(x, y)$ is the distance from the point (x, y) to a fixed point P, then the average value of f over R is the **average distance** of points in R from P.

Example 3 FINDING AVERAGE VALUE

Find the average value of $f(x, y) = x \cos xy$ over the rectangle $R: 0 \le x \le \pi,\ 0 \le y \le 1$.

Solution The value of the integral of f over R is

$$\int_0^\pi \int_0^1 x \cos xy\, dy\, dx = \int_0^\pi \left[\sin xy\right]_{y=0}^{y=1} dx$$

$$= \int_0^\pi (\sin x - 0)\, dx = -\cos x\big]_0^\pi = 1 + 1 = 2.$$

The area of R is π. The average value of f over R is $2/\pi$.

First and Second Moments and Centers of Mass

To find the moments and centers of mass of thin sheets and plates, we use formulas similar to those in Exercises 7.1. The main difference is that now, with double integrals, we can accommodate a greater variety of shapes and density functions. The formulas are given in Table 14.1. The examples that follow show how the formulas are used.

The mathematical difference between the **first moments** M_x and M_y and the **moments of inertia,** or **second moments,** I_x and I_y is that the second moments use the *squares* of the "lever-arm" distances x and y.

The moment I_0 is also called the **polar moment** of inertia about the origin. It is calculated by integrating the density $\delta(x, y)$ (mass per unit area) times $r^2 = x^2 + y^2$, the square of the distance from a representative point (x, y) to the origin. Notice that $I_0 = I_x + I_y$; once we find any two of these, we get the third automatically. The moment I_0 is sometimes called I_z, for moment of inertia about the z-axis. The identity $I_z = I_x + I_y$ is then called the **Perpendicular Axis Theorem**.

The **radius of gyration** R_x is defined by the equation

$$I_x = MR_x^2.$$

It tells how far from the x-axis the entire mass of the plate might be concentrated to give the same I_x. The radius of gyration gives a convenient way to express the moment of inertia in terms of a mass and a length. The radii R_y and R_0 are defined in a similar way, with

$$I_y = MR_y^2 \qquad \text{and} \qquad I_0 = MR_0^2.$$

We take square roots to get the formulas in Table 14.1.

Why the interest in moments of inertia? A body's first moments tell us about balance and about the torque the body exerts about different axes in a gravitational field, but if the body is a rotating shaft, we are more likely to be interested in how much energy is stored in the shaft or about how much energy

Table 14.1 Mass and Moment Formulas for Thin Plates Covering Regions in the *xy*-Plane

Density: $\delta(x, y)$

Mass: $M = \iint \delta(x, y)\, dA$

First moments: $M_x = \iint y\delta(x, y)\, dA, \qquad M_y = \iint x\delta(x, y)\, dA$

Center of mass: $\bar{x} = \dfrac{M_y}{M}, \qquad \bar{y} = \dfrac{M_x}{M}$

Moments of inertia (second moments):

About the *x*-axis: $I_x = \iint y^2\delta(x, y)\, dA$

About the *y*-axis: $I_y = \iint x^2\delta(x, y)\, dA$

About the origin (**polar moment**):

$$I_0 = \iint (x^2 + y^2)\delta(x, y)\, dA = I_x + I_y$$

About a line *L*: $\quad I_L = \iint r^2(x, y)\,\delta(x, y)\, dA, \qquad$ where $r(x, y) = $ distance from (x, y) to L

Radii of gyration:

About the *x*-axis: $R_x = \sqrt{I_x/M}$

About the *y*-axis: $R_y = \sqrt{I_y/M}$

About the origin: $R_0 = \sqrt{I_0/M}$

it will take to accelerate the shaft to a particular angular velocity. This is where the second moment or moment of inertia comes in.

Think of partitioning the shaft into small blocks of mass Δm_k and let r_k denote the distance from the *k*th block's center of mass to the axis of rotation (Figure 14.18). If the shaft rotates at an angular velocity of $\omega = d\theta/dt$

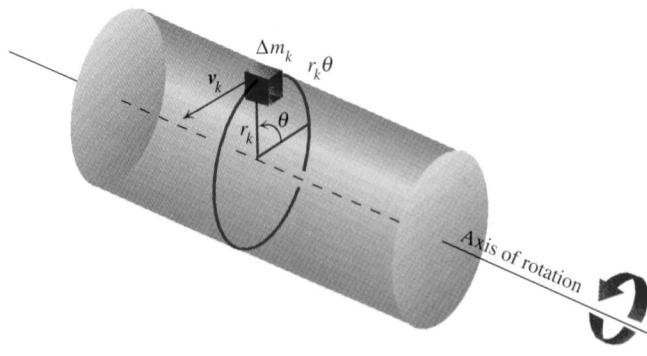

Figure 14.18 To find an integral for the amount of energy stored in a rotating shaft, we first imagine the shaft to be partitioned into small blocks. Each block has its own kinetic energy. We add the contributions of the individual blocks to find the kinetic energy of the shaft.

radians per second, the block's center of mass will trace its orbit at a linear speed of

$$v_k = \frac{d}{dt}(r_k\theta) = r_k\frac{d\theta}{dt} = r_k\omega. \tag{5}$$

The block's kinetic energy will be approximately

$$\frac{1}{2}\Delta m_k v_k^2 = \frac{1}{2}\Delta m_k(r_k\omega)^2 = \frac{1}{2}\omega^2 r_k^2 \Delta m_k. \tag{6}$$

The kinetic energy of the shaft will be approximately

$$\sum \frac{1}{2}\omega^2 r_k^2 \Delta m_k. \tag{7}$$

The integral approached by these sums as the shaft is partitioned into smaller and smaller blocks gives the shaft's kinetic energy:

$$KE_{shaft} = \int \frac{1}{2}\omega^2 r^2\, dm = \frac{1}{2}\omega^2 \int r^2\, dm. \tag{8}$$

The factor

$$I = \int r^2\, dm \tag{9}$$

is the moment of inertia of the shaft about its axis of rotation, and we see from Equation (8) that the shaft's kinetic energy is

$$KE_{shaft} = \frac{1}{2}I\omega^2. \tag{10}$$

To start a shaft of inertial moment I rotating at an angular velocity ω, we need to provide a kinetic energy of $KE = (1/2)I\omega^2$. To stop the shaft, we have to take this amount of energy back out. To start a locomotive with mass m and accelerate it to a linear velocity v, we need to provide a kinetic energy of $KE = (1/2)mv^2$. To stop the locomotive, we have to remove this amount of energy. The shaft's moment of inertia is analogous to the locomotive's mass. What makes the locomotive hard to start or stop is its mass. What makes the shaft hard to start or stop is its moment of inertia. The moment of inertia takes into account not only the mass but also its distribution.

The moment of inertia also plays a role in determining how much a horizontal metal beam will bend under a load. The stiffness of the beam is a constant times I, the polar moment of inertia of a typical cross section of the beam perpendicular to the beam's longitudinal axis. The greater the value of I, the stiffer the beam and the less it will bend under a given load. This is why we use I beams instead of beams whose cross sections are square. The flanges at the top and bottom of the beam hold most of the beam's mass away from the longitudinal axis to maximize the value of I (Figure 14.19).

Beam A

Axis

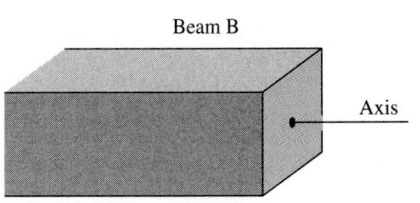

Beam B

Axis

Figure 14.19 The greater the polar moment of inertia of the cross section of a beam about the beam's longitudinal axis, the stiffer the beam. Beams A and B have the same cross-section area, but A is stiffer.

If you want to see the moment of inertia at work, try the following experiment.

1. Tape two coins to the ends of a pencil and twiddle the pencil about the center of mass. The moment of inertia accounts for the resistance you feel each time you change the direction of motion.

2. Now move the coins an equal distance toward the center of mass and twiddle the pencil again. The system has the same mass and the same center of mass but now offers less resistance to the changes in motion. The moment of inertia has been reduced.

The moment of inertia is what gives a baseball bat, golf club, or tennis racket its "feel." Tennis rackets that weigh the same, look the same, and have identical centers of mass will feel and behave differently if their masses are not distributed the same way.

First moments are "balancing" moments.
Second moments are "turning" moments.

Example 4 APPLYING FORMULAS FROM TABLE 14.1

A thin plate covers the triangular region bounded by the x-axis and the lines $x = 1$ and $y = 2x$ in the first quadrant. The plate's density at the point (x, y) is $\delta(x, y) = 6x + 6y + 6$. Find the plate's mass, first moments, center of mass, moments of inertia, and radii of gyration about the coordinate axes.

Solution   We sketch the plate and put in enough detail to determine the limits of integration for the integrals we have to evaluate (Figure 14.20).

The plate's mass is

$$M = \int_0^1 \int_0^{2x} \delta(x, y) \, dy \, dx = \int_0^1 \int_0^{2x} (6x + 6y + 6) \, dy \, dx$$

$$= \int_0^1 \left[6xy + 3y^2 + 6y \right]_{y=0}^{y=2x} dx$$

$$= \int_0^1 (24x^2 + 12x) \, dx = \left[8x^3 + 6x^2 \right]_0^1 = 14.$$

The first moment about the x-axis is

$$M_x = \int_0^1 \int_0^{2x} y\delta(x, y) \, dy \, dx = \int_0^1 \int_0^{2x} (6xy + 6y^2 + 6y) \, dy \, dx$$

$$= \int_0^1 \left[3xy^2 + 2y^3 + 3y^2 \right]_{y=0}^{y=2x} dx = \int_0^1 (28x^3 + 12x^2) \, dx$$

$$= \left[7x^4 + 4x^3 \right]_0^1 = 11.$$

A similar calculation gives

$$M_y = \int_0^1 \int_0^{2x} x\delta(x, y) \, dy \, dx = 10.$$

Figure 14.20 The triangular region covered by the plate in Example 4.

The coordinates of the center of mass are therefore

$$\bar{x} = \frac{M_y}{M} = \frac{10}{14} = \frac{5}{7}, \qquad \bar{y} = \frac{M_x}{M} = \frac{11}{14}.$$

The moment of inertia about the x-axis is

$$I_x = \int_0^1 \int_0^{2x} y^2 \delta(x, y)\, dy\, dx = \int_0^1 \int_0^{2x} (6xy^2 + 6y^3 + 6y^2)\, dy\, dx$$

$$= \int_0^1 \left[2xy^3 + \frac{3}{2} y^4 + 2y^3 \right]_{y=0}^{y=2x} dx = \int_0^1 (40x^4 + 16x^3)\, dx$$

$$= [8x^5 + 4x^4]_0^1 = 12.$$

Similarly, the moment of inertia about the y-axis is

$$I_y = \int_0^1 \int_0^{2x} x^2 \delta(x, y)\, dy\, dx = \frac{39}{5}.$$

Since we know I_x and I_y, we do not need to evaluate an integral to find I_0; we can use the equation $I_0 = I_x + I_y$ instead:

$$I_0 = 12 + \frac{39}{5} = \frac{99}{5}.$$

The three radii of gyration are

$$R_x = \sqrt{I_x/M} = \sqrt{12/14} = \sqrt{6/7} \approx 0.93,$$

$$R_y = \sqrt{I_y/M} = \sqrt{\left(\frac{39}{5}\right)/14} = \sqrt{39/70} \approx 0.75,$$

$$R_0 = \sqrt{I_0/M} = \sqrt{\left(\frac{99}{5}\right)/14} = \sqrt{99/70} \approx 1.19.$$

Centroids of Geometric Figures

When the density of an object is constant, it cancels out of the numerator and denominator of the formulas for $\bar{x}$ and $\bar{y}$. As far as $\bar{x}$ and $\bar{y}$ are concerned, δ might as well be 1. Thus, when δ is constant, the location of the center of mass becomes a feature of the object's shape and not of the material of which it is made. In such cases, engineers may call the center of mass the **centroid** of the shape. To find a centroid, we set δ equal to 1 and proceed to find $\bar{x}$ and $\bar{y}$ as before, by dividing first moments by masses.

Example 5   FINDING THE CENTROID OF A REGION

Find the centroid of the region in the first quadrant that is bounded above by the line $y = x$ and below by the parabola $y = x^2$.

Solution   We sketch the region and include enough detail to determine the limits of integration (Figure 14.21.). We then set δ equal to 1 and evaluate the appropriate formulas from Table 14.1:

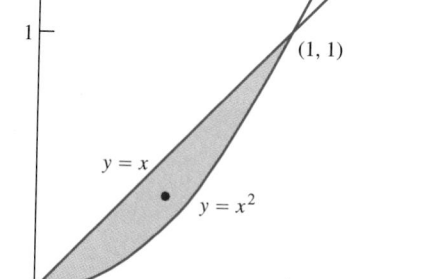

Figure 14.21 Example 5 finds the centroid of the region shown here.

$$M = \int_0^1 \int_{x^2}^x dy\, dx = \int_0^1 [y]_{y=x^2}^{y=x}\, dx = \int_0^1 (x - x^2)\, dx$$

$$= \left[\frac{x^2}{2} - \frac{x^3}{3}\right]_0^1 = \frac{1}{6},$$

$$M_x = \int_0^1 \int_{x^2}^x y\, dy\, dx = \int_0^1 \left[\frac{y^2}{2}\right]_{y=x^2}^{y=x}\, dx$$

$$= \int_0^1 \left(\frac{x^2}{2} - \frac{x^4}{2}\right) dx = \left[\frac{x^3}{6} - \frac{x^5}{10}\right]_0^1 = \frac{1}{15},$$

$$M_y = \int_0^1 \int_{x^2}^x x\, dy\, dx = \int_0^1 [xy]_{y=x^2}^{y=x}\, dx$$

$$= \int_0^1 (x^2 - x^3)\, dx = \left[\frac{x^3}{3} - \frac{x^4}{4}\right]_0^1 = \frac{1}{12}.$$

From these values of M, M_x, and M_y, we find

$$\bar{x} = \frac{M_y}{M} = \frac{1/12}{1/6} = \frac{1}{2} \quad \text{and} \quad \bar{y} = \frac{M_x}{M} = \frac{1/15}{1/6} = \frac{2}{5}.$$

The centroid is the point $\left(\frac{1}{2}, \frac{2}{5}\right)$.

Quick Review 14.2

In Exercises 1 and 2, find the area of the region between the *x*-axis and the graph of the given function over the given interval.

1. $y = e^x$ over $[0, 1]$

2. $y = \cos x$ over $[0, \pi/2]$

In Exercises 3 and 4, consider the region between the curve $y = 2 - x^2$ and the line $y = -2$.

3. Find the area of the region by integrating with respect to *x*.

4. Find the area of the region by integrating with respect to *y*.

5. Find the average value of $f(x) = \sin x$ on $[0, \pi]$.

6. Find the area of the region enclosed by $y = x^2 - 3$ and $y = 1$.

Suppose we have a finite collection of masses in the coordinate plane, the mass m_k located at the point (x_k, y_k) as shown in the figure.

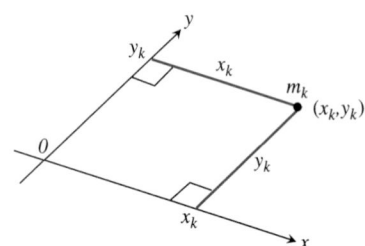

Each mass m_k has **moment** $m_k y_k$ **about the *x*-axis** and **moment** $m_k x_k$ **about the *y*-axis**. The moments of the entire system about the coordinate axes are

Moment about *x*-axis: $M_x = \sum m_k y_k$,

Moment about *y*-axis: $M_y = \sum m_k x_k$.

The **center of mass** is $(\bar{x}, \bar{y})$, where

$$\bar{x} = \frac{M_y}{M} = \frac{\sum m_k x_k}{\sum m_k} \quad \text{and} \quad \bar{y} = \frac{M_x}{M} = \frac{\sum m_k y_k}{\sum m_k}.$$

Suppose we have a thin, flat plate occupying a region in the plane.

7. Imagine the region cut into thin strips parallel to the *y*-axis. Show that

$$\bar{x} = \frac{\displaystyle\int x\, dm}{\displaystyle\int dm},$$

where $dm = \delta\, dA$, δ = density (mass per unit area), and A = area of the region.

8. Imagine the region cut into thin strips parallel to the *x*-axis. Show that

$$\bar{y} = \frac{\int y \, dm}{\int dm},$$

where $dm = \delta \, dA$, δ = density, and A = area of the region.

In Exercises 9 and 10, use the results of Exercises 7 and 8 to find the center of mass of the region with given density.

9. The region bounded by the parabola $y = x^2$ and the line $y = 4$ with constant density δ

10. The region bounded by the lines $y = x$, $y = -x$, $x = 2$ with constant density δ

Section 14.2 Exercises

In Exercises 1–8, sketch the region bounded by the given lines and curves. Then express the region's area as an iterated double integral and evaluate the integral.

1. The coordinate axes and the line $x + y = 2$

2. The lines $x = 0$, $y = 2x$, and $y = 4$

3. The parabola $x = -y^2$ and the line $y = x + 2$

4. The parabola $x = y - y^2$ and the line $y = -x$

5. The curve $y = e^x$ and the lines $y = 0$, $x = 0$, and $x = \ln 2$

6. The curves $y = \ln x$ and $y = 2 \ln x$ and the line $x = e$, in the first quadrant

7. The parabolas $x = y^2$ and $x = 2y - y^2$

8. The parabolas $x = y^2 - 1$ and $x = 2y^2 - 2$

In Exercises 9–14, the integral or sum of integrals gives the area of a region in the *xy*-plane. Sketch each region, label each bounding curve with its equation, and give the coordinates of the points where the curves intersect. Then find the area of the region.

9. $\int_0^6 \int_{y^2/3}^{2y} dx \, dy$

10. $\int_0^3 \int_{-x}^{x(2-x)} dy \, dx$

11. $\int_0^{\pi/4} \int_{\sin x}^{\cos x} dy \, dx$

12. $\int_{-1}^2 \int_{y^2}^{y+2} dx \, dy$

13. $\int_{-1}^0 \int_{-2x}^{1-x} dy \, dx + \int_0^2 \int_{-x/2}^{1-x} dy \, dx$

14. $\int_0^2 \int_{x^2-4}^0 dy \, dx + \int_0^4 \int_0^{\sqrt{x}} dy \, dx$

15. Find the average value of $f(x, y) = \sin(x + y)$ over

(a) the rectangle $0 \le x \le \pi$, $0 \le y \le \pi$.

(b) the rectangle $0 \le x \le \pi$, $0 \le y \le \pi/2$.

16. Which do you think will be larger, the average value of $f(x, y) = xy$ over the square $0 \le x \le 1$, $0 \le y \le 1$, or the average value of f over the quarter circle $x^2 + y^2 \le 1$ in the first quadrant? Calculate them to find out.

17. Find the average height of the paraboloid $z = x^2 + y^2$ over the square $0 \le x \le 2$, $0 \le y \le 2$.

18. Find the average value of $f(x, y) = 1/(xy)$ over the square $\ln 2 \le x \le 2 \ln 2$, $\ln 2 \le y \le 2 \ln 2$.

In Exercises 19–30, the density of the region is constant.

19. Find the center of mass of a thin plate of density $\delta = 3$ bounded by the lines $x = 0$, $y = x$, and the parabola $y = 2 - x^2$ in the first quadrant.

20. Find the moments of inertia and radii of gyration about the coordinate axes of a thin rectangular plate of constant density δ bounded by the lines $x = 3$ and $y = 3$ in the first quadrant.

21. Find the centroid of the region in the first quadrant bounded by the *x*-axis, the parabola $y^2 = 2x$, and the line $x + y = 4$.

22. Find the centroid of the triangular region cut from the first quadrant by the line $x + y = 3$.

23. Find the centroid of the semicircular region bounded by the *x*-axis and the curve $y = \sqrt{1 - x^2}$.

24. The area of the region in the first quadrant bounded by the parabola $y = 6x - x^2$ and the line $y = x$ is 125/6 square units. Find the centroid.

25. Find the centroid of the region cut from the first quadrant by the circle $x^2 + y^2 = a^2$.

26. Find the moment of inertia about the *x*-axis of a thin plate of density $\delta = 1$ bounded by the circle $x^2 + y^2 = 4$. Then use your result to find I_y and I_0 for the plate.

27. Find the centroid of the region between the *x*-axis and the arch $y = \sin x$, $0 \le x \le \pi$.

28. Find the moment of inertia with respect to the *y*-axis of a thin sheet of constant density $\delta = 1$ bounded by the curve $y = (\sin^2 x)/x^2$ and the interval $\pi \le x \le 2\pi$ of the *x*-axis.

29. *Centroid of an Infinite Region* Find the centroid of the infinite region in the second quadrant enclosed by the coordinate axes and the curve $y = e^x$. (Use improper integrals in the mass-moment formulas.)

30. *First Moment of an Infinite Plate* Find the first moment about the *y*-axis of a thin plate of density $\delta(x, y) = 1$ covering the infinite region under the curve $y = e^{-x^2/2}$ in the first quadrant.

In Exercises 31–40, the density of the region varies throughout the region.

31. Find the moment of inertia and radius of gyration about the *x*-axis of a thin plate bounded by the parabola $x = y - y^2$ and the line $x + y = 0$ if $\delta(x, y) = x + y$.

32. Find the mass of a thin plate occupying the smaller region cut from the ellipse $x^2 + 4y^2 = 12$ by the parabola $x = 4y^2$ if $\delta(x, y) = 5x$.

33. Find the center of mass of a thin triangular plate bounded by the y-axis and the lines $y = x$ and $y = 2 - x$ if $\delta(x, y) = 6x + 3y + 3$.

34. Find the center of mass and moment of inertia about the x-axis of a thin plate bounded by the curves $x = y^2$ and $x = 2y - y^2$ if the density at the point (x, y) is $\delta(x, y) = y + 1$.

35. Find the center of mass and moment of inertia and radius of gyration about the y-axis of a thin rectangular plate cut from the first quadrant by the lines $x = 6$ and $y = 1$ if $\delta(x, y) = x + y + 1$.

36. Find the center of mass and the moment of inertia and radius of gyration about the y-axis of a thin plate bounded by the line $y = 1$ and the parabola $y = x^2$ if the density is $\delta(x, y) = y + 1$.

37. Find the center of mass and the moment of inertia and radius of gyration about the y-axis of a thin plate bounded by the x-axis, the lines $x = \pm 1$, and the parabola $y = x^2$ if $\delta(x, y) = 7y + 1$.

38. Find the center of mass and the moment of inertia and radius of gyration about the x-axis of a thin rectangular plate bounded by the lines $x = 0$, $x = 20$, $y = -1$, and $y = 1$ if $\delta(x, y) = 1 + (x/20)$.

39. Find the center of mass, the moments of inertia and radii of gyration about the coordinate axes, and the polar moment of inertia and radius of gyration of a thin triangular plate bounded by the lines $y = x$, $y = -x$, and $y = 1$ if $\delta(x, y) = y + 1$.

40. Repeat Exercise 39 for $\delta(x, y) = 3x^2 + 1$.

In Exercises 41 and 42, *work in groups of two or three* to solve the problem.

41. *Number of Bacteria in a Rectangle* If $f(x, y) = (10,000e^y)/(1 + |x|/2)$ represents the "population density" of a certain bacteria on the xy-plane, where x and y are measured in centimeters, find the total population of bacteria within the rectangle $-5 \leq x \leq 5$ and $-2 \leq y \leq 0$.

42. *Number of People in a Region* If $f(x, y) = 100(y + 1)$ represents the population density of a planar region on Earth, where x and y are measured in miles, find the number of people in the region bounded by the curves $x = y^2$ and $x = 2y - y^2$.

Explorations

43. *Appliance Design* When we design an appliance, one of the concerns is how difficult the appliance will be to tip over. When tipped, it will right itself as long as its center of mass lies on the correct side of the *fulcrum*, the point on which the appliance is riding as it tips. Suppose the profile of an appliance of approximate constant density is parabolic, like

an old-fashioned radio. It fills the region $0 \leq y \leq a(1 - x^2)$, $-1 \leq x \leq 1$, in the xy-plane (Figure 14.22). What values of a will guarantee that the appliance will have to be tipped more than 45° to fall over?

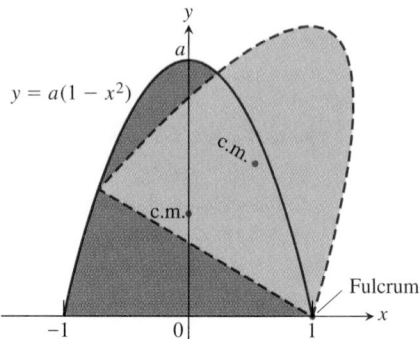

Figure 14.22 The profile of the appliance in Exercise 43.

44. *Minimizing a Moment of Inertia* A rectangular plate of constant density $\delta(x, y) = 1$ occupies the region bounded by the lines $x = 4$ and $y = 2$ in the first quadrant. The moment of inertia I_a of the rectangle about the line $y = a$ is given by the integral

$$I_a = \int_0^4 \int_0^2 (y - a)^2 \, dy \, dx.$$

Find the value of a that minimizes I_a. ∎

45. *Centroid of an Infinite Region* Find the centroid of the infinite region in the xy-plane bounded by the curves $y = 1/\sqrt{1 - x^2}$, $y = -1/\sqrt{1 - x^2}$, and the lines $x = 0$, $x = 1$.

46. *Radii of Gyration* Find the radius of gyration of a slender rod of constant linear density δ gm/cm and length L cm with respect to an axis

(a) through the rod's center of mass perpendicular to the rod's axis.

(b) perpendicular to the rod's axis at one end of the rod.

47. A thin plate of constant density δ occupies the region R in the xy-plane bounded by the curves $x = y^2$ and $x = 2y - y^2$ (see Exercise 34).

(a) Find δ such that the plate has the same mass as the plate in Exercise 34.

(b) *Average Density* Compare the value of δ found in part (a) with the average value of $\delta(x, y) = y + 1$ over R.

48. **Writing to Learn** *Average Temperature in Texas* According to the *Texas Almanac*, Texas has 254 counties and a National Weather Service station in each county. Assume that at time t_0 each of the 254 weather stations recorded the local temperature. Find a formula that would give a reasonable approximation to the average temperature in Texas at time t_0. Your answer should involve information that is readily available in the *Texas Almanac*.

Extending the Ideas

The Parallel Axis Theorem

Let $L_{c.m.}$ be a line in the xy-plane that runs through the center of mass of a thin plate of mass m covering a region in the plane. Let L be a line in the plane parallel to and h units away from $L_{c.m.}$. The **Parallel Axis Theorem** says that under these conditions, the moments of inertia I_L and $L_{c.m.}$ of the plate about L and $L_{c.m.}$ satisfy the equation

$$I_L = I_{c.m.} + mh^2. \qquad (1)$$

This equation gives a quick way to calculate one moment when the other moment and the mass are known.

49. *Proof of the Parallel Axis Theorem*

(a) Show that the first moment of a thin flat plate about any line in the plane of the plate through the plate's center of mass is zero. (*Hint:* Place the center of mass at the origin with the line along the y-axis. What does the formula $\bar{x} = M_y/M$ then tell you?)

(b) Use the result in (a) to derive the Parallel Axis Theorem. Assume that the plate is coordinatized in a way that makes $L_{c.m.}$ the y-axis and L the line $x = h$. Then expand the integrand of the integral for I_L to rewrite the integral as the sum of integrals whose values you recognize.

50. *Applying the Parallel Axis Theorem*

(a) Use the Parallel Axis Theorem and the results of Example 4 to find the moments of inertia of the plate in Example 4 about the vertical and horizontal lines through the plate's center of mass.

(b) Use the results in (a) to find the plate's moments of inertia about the lines $x = 1$ and $y = 2$.

Pappus's Formula

Suppose m_1 and m_2 are the masses of thin plates P_1 and P_2 that cover nonoverlapping regions in the xy-plane. Let $\mathbf{c}_1$ and $\mathbf{c}_2$ be the vectors from the origin to the respective centers of mass of P_1 and P_2. Then the center of mass of the union $P_1 \cup P_2$ of the two plates is determined by the vector

$$\mathbf{c} = \frac{m_1\mathbf{c}_1 + m_2\mathbf{c}_2}{m_1 + m_2}. \qquad (2)$$

Equation (2) is known as **Pappus's formula**. For more than two nonoverlapping plates, as long as their number is finite, the formula generalizes to

$$\mathbf{c} = \frac{m_1\mathbf{c}_1 + m_2\mathbf{c}_2 + \cdots + m_n\mathbf{c}_n}{m_1 + m_2 + \cdots + m_n}. \qquad (3)$$

This formula is especially useful for finding the centroid of a plate of irregular shape that is made up of pieces of constant density whose centroids we know from geometry. We find the centroid of each piece and apply Equation (3) to find the centroid of the plate.

51. Derive Pappus's formula (Equation (2)). (*Hint:* Sketch the plates as regions in the first quadrant and label their centers of mass as $(\bar{x}_1, \bar{y}_1)$ and $(\bar{x}_2, \bar{y}_2)$. What are the moments of $P_1 \cup P_2$ about the coordinate axes?)

52. Use Equation (2) and mathematical induction to show that Equation (3) holds for any positive integer $n > 2$.

53. Let A, B, and C be the shapes indicated in Figure 14.23a. Use Pappus's formula to find the centroid of

(a) $A \cup B$ (b) $A \cup C$

(c) $B \cup C$ (d) $A \cup B \cup C$

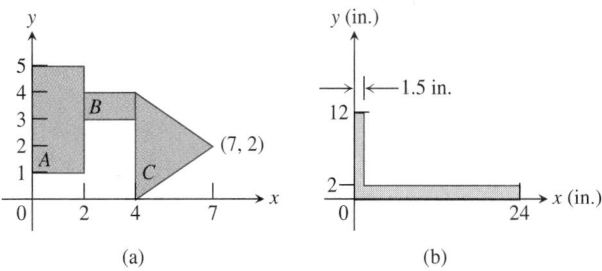

(a)   (b)

Figure 14.23 The figures for Exercises 53 and 54.

54. *Center of Mass of a Carpenter's Square* Locate the center of mass of the carpenter's square in Figure 14.23b.

55. **Writing to Learn** *Designing a Two-Dimensional Ice Cream Cone* An isosceles triangle T has base $2a$ and altitude h. The base lies along the diameter of a semicircular disk D of radius a so that the two together make a shape resembling an ice cream cone. What relation must hold between a and h to place the centroid of $T \cup D$ on the common boundary of T and D? inside T?

56. **Writing to Learn** *Shape Design* An isosceles triangle T of altitude h has as its base one side of a square Q whose edges have length s. (The square and triangle do not overlap.) What relation must hold between h and s to place the centroid of $T \cup Q$ on the base of the triangle? Compare your answer with the answer to Exercise 55.

14.3   Double Integrals in Polar Form

Integrals in Polar Coordinates • Finding Limits of Integration •
Changing Cartesian Integrals into Polar Integrals

Integrals in Polar Coordinates

When we defined the double integral of a function over a region R in the xy-plane, we began by cutting R into rectangles whose sides were parallel to the coordinate axes. These were the natural shapes to use because their sides have either constant x-values or constant y-values. In polar coordinates, the natural shapes are "polar rectangles" whose sides have constant r- and θ-values.

Suppose a function $f(r, \theta)$ is defined over a region R that is bounded by the rays $\theta = \alpha$ and $\theta = \beta$ and by the continuous curves $r = g_1(\theta)$ and $r = g_2(\theta)$. Also suppose $0 \le g_1(\theta) \le g_2(\theta) \le a$ for every value of θ between α and β. Then R lies in the fan-shaped region Q defined by the inequalities $0 \le r \le a$ and $\alpha \le \theta \le \beta$. See Figure 14.24.

We cover Q by a grid of circular arcs and rays. The arcs are cut from circles centered at the origin, with radii $\Delta r, 2\,\Delta r, \ldots, m\,\Delta r$, where $\Delta r = a/m$. The rays are given by

$$\theta = \alpha, \theta = \alpha + \Delta\theta, \theta = \alpha + 2\Delta\theta, \ldots, \theta = \alpha + m'\,\Delta\theta = \beta,$$

where $\Delta\theta = (\beta - \alpha)/m'$. The arcs and rays partition Q into small patches called "polar rectangles."

We number the polar rectangles that lie inside R (the order does not matter), calling their areas $\Delta A_1, \Delta A_2, \ldots, \Delta A_n$.

We let (r_k, θ_k) be the center of the polar rectangle whose area is ΔA_k. By "center" we mean the point that lies halfway between the circular arcs on the ray that bisects the arcs. We then form the sum

$$S_n = \sum_{k=1}^{n} f(r_k, \theta_k)\,\Delta A_k. \tag{1}$$

If f is continuous throughout R, this sum will approach a limit as we refine the grid to make Δr and $\Delta\theta$ go to zero. This limit is the **double integral of f over R.** In symbols,

$$\lim_{n \to \infty} S_n = \iint\limits_{R} f(r, \theta)\, dA.$$

Figure 14.24 The region R: $g_1(\theta) \le r \le g_2(\theta)$, $\alpha \le \theta \le \beta$ is contained in the fan-shaped region Q: $0 \le r \le a$, $\alpha \le \theta \le \beta$. The partition of Q by circular arcs and rays induces a partition of R.

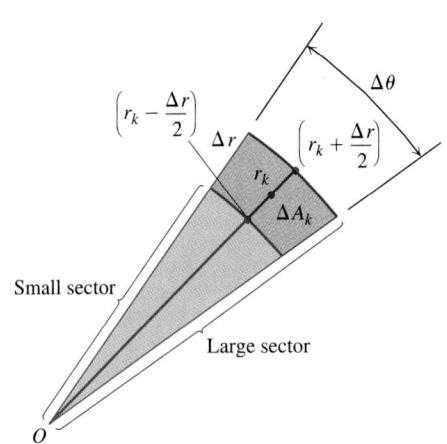

Figure 14.25 The observation that

$$\Delta A_k = \begin{pmatrix} \text{area of} \\ \text{large sector} \end{pmatrix} - \begin{pmatrix} \text{area of} \\ \text{small sector} \end{pmatrix}$$

leads to the formula $\Delta A_k = r_k \, \Delta r \, \Delta \theta$. The text explains why.

To see how to evaluate this limit, we first write the sum S_n in a way that expresses ΔA_k in terms of Δr and $\Delta \theta$. The radius of the inner arc bounding ΔA_k is $r_k - (\Delta r/2)$ (Figure 14.25). The radius of the outer arc is $r_k + (\Delta r/2)$. The areas of the circular sectors subtended by these arcs at the origin are

$$\text{Inner radius: } \frac{1}{2}\left(r_k - \frac{\Delta r}{2}\right)^2 \Delta \theta \qquad \text{Outer radius: } \frac{1}{2}\left(r_k + \frac{\Delta r}{2}\right)^2 \Delta \theta. \qquad (2)$$

Therefore,

$$\Delta A_k = \text{area of large sector} - \text{area of small sector}$$

$$= \frac{\Delta \theta}{2}\left[\left(r_k + \frac{\Delta r}{2}\right)^2 - \left(r_k - \frac{\Delta r}{2}\right)^2\right] = \frac{\Delta \theta}{2}(2r_k \, \Delta r) = r_k \, \Delta r \, \Delta \theta.$$

Combining this result with Equation (1) gives

$$S_n = \sum_{k=1}^{n} f(r_k, \theta_k) r_k \, \Delta r \, \Delta \theta. \qquad (3)$$

A version of Fubini's theorem now says that the limit approached by these sums can be evaluated by repeated single integration with respect to r and θ as

$$\iint_R f(r, \theta) \, dA = \int_{\theta=\alpha}^{\theta=\beta} \int_{r=g_1(\theta)}^{r=g_2(\theta)} f(r, \theta) r \, dr \, d\theta. \qquad (4)$$

Finding Limits of Integration

The procedure for finding limits of integration in rectangular coordinates also works for polar coordinates.

How to Integrate in Polar Coordinates

To evaluate $\iint_R f(r, \theta) \, dA$ over a region R in polar coordinates, integrating first with respect to r and then with respect to θ, take the following steps.

1. *A sketch.* Sketch the region and label the bounding curves.

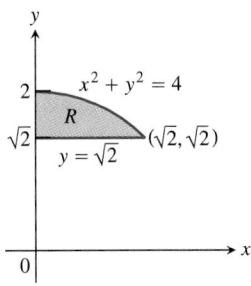

2. *The r-limits of integration.* Imagine a ray L from the origin cutting through R in the direction of increasing r. Mark the r-values where L enters and leaves R. These are the r-limits of integration. They usually depend on the angle θ that L makes with the positive x-axis.

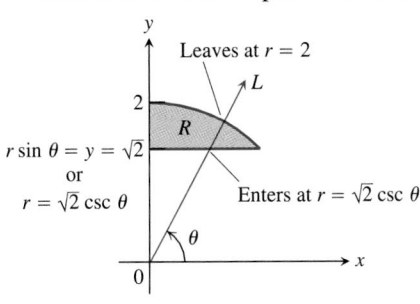

3. *The θ-limits of integration.* Find the smallest and largest θ-values that bound R. These are the θ-limits of integration.

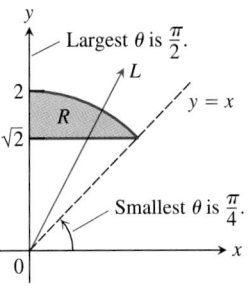

The integral is

$$\iint_R f(r, \theta) \, dA = \int_{\theta=\pi/4}^{\theta=\pi/2} \int_{r=\sqrt{2}\csc\theta}^{r=2} f(r, \theta) \, r \, dr \, d\theta.$$

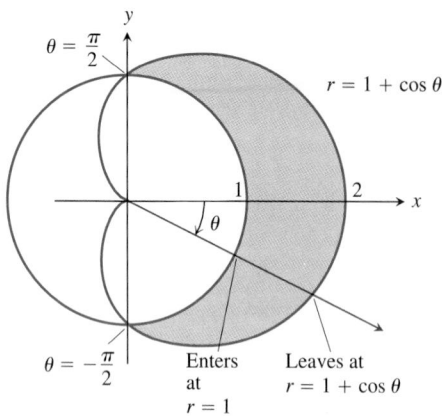

Figure 14.26 The sketch for Example 1.

Example 1 FINDING LIMITS OF INTEGRATION

Find the limits of integration for integrating $f(r, \theta)$ over the region R that lies inside the cardioid $r = 1 + \cos \theta$ and outside the circle $r = 1$.

Solution

Step 1: *A sketch.* We sketch the region and label the bounding curves (Figure 14.26)

Step 2: *The r-limits of integration.* A typical ray from the origin enters R where $r = 1$ and leaves where $r = 1 + \cos \theta$.

Step 3: *The θ-limits of integration.* The rays from the origin that intersect R run from $\theta = -\pi/2$ to $\theta = \pi/2$. The integral is

$$\int_{-\pi/2}^{\pi/2} \int_{1}^{1+\cos \theta} f(r, \theta) r \, dr \, d\theta.$$

If $f(r, \theta)$ is the constant function whose value is 1, then the integral of f over R is the area of R.

Area in Polar Coordinates

The area of a closed and bounded region R in the polar coordinate plane is

$$A = \iint_R r \, dr \, d\theta. \qquad (5)$$

As you might expect, this formula for area is consistent with all earlier formulas, although we will not prove this fact.

Example 2 FINDING AREA IN POLAR COORDINATES

Find the area enclosed by the lemniscate $r^2 = 4 \cos 2\theta$.

Solution We graph the lemniscate to determine the limits of integration (Figure 14.27) and see that the total area is four times the first-quadrant portion.

$$A = 4 \int_0^{\pi/4} \int_0^{\sqrt{4 \cos 2\theta}} r \, dr \, d\theta = 4 \int_0^{\pi/4} \left[\frac{r^2}{2} \right]_{r=0}^{r=\sqrt{4 \cos 2\theta}} d\theta$$

$$= 4 \int_0^{\pi/4} 2 \cos 2\theta \, d\theta = 4 \sin 2\theta\big]_0^{\pi/4} = 4.$$

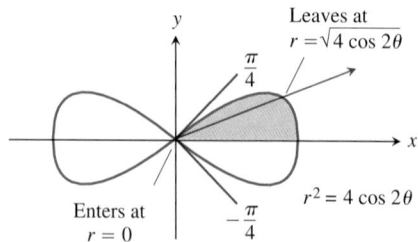

Figure 14.27 To integrate over the shaded region, we run r from 0 to $\sqrt{4 \cos 2\theta}$ and θ from 0 to $\pi/4$. (Example 2)

Changing Cartesian Integrals into Polar Integrals

The procedure for changing a Cartesian integral $\int_R f(x, y) \, dx \, dy$ into a polar integral has two steps.

Step 1: Substitute $x = r \cos \theta$ and $y = r \sin \theta$, and replace $dx \, dy$ in the Cartesian integral by $r \, dr \, d\theta$.

Step 2: Supply polar limits of integration for the boundary of R.

The Cartesian integral then becomes

$$\iint\limits_{R} f(x, y)\, dx\, dy = \iint\limits_{G} f(r \cos \theta, r \sin \theta) r\, dr\, d\theta, \tag{6}$$

where G denotes the region of integration in polar coordinates. This is like the substitution method in Chapter 6 except that there are now two variables to substitute for instead of one. Notice that $dx\, dy$ is not replaced by $dr\, d\theta$ but by $r\, dr\, d\theta$. We will see why in Section 14.7.

Example 3 CHANGING CARTESIAN INTEGRALS TO POLAR

Find the polar moment of inertia about the origin of a thin plate of density $\delta(x, y) = 1$ bounded by the quarter circle $x^2 + y^2 = 1$ in the first quadrant.

Solution We sketch the plate to determine the limits of integration (Figure 14.28).

In Cartesian coordinates, the polar moment is the value of the integral

$$\int_0^1 \int_0^{\sqrt{1-x^2}} (x^2 + y^2)\, dy\, dx.$$

Integration with respect to y gives

$$\int_0^1 \left(x^2 \sqrt{1 - x^2} + \frac{(1 - x^2)^{3/2}}{3} \right) dx,$$

an integral difficult to evaluate without tables or technology.

Things go better if we change the original integral to polar coordinates. Substituting $x = r \cos \theta$, $y = r \sin \theta$ and replacing $dx\, dy$ by $r\, dr\, d\theta$, we get

$$\int_0^1 \int_0^{\sqrt{1-x^2}} (x^2 + y^2)\, dy\, dx = \int_0^{\pi/2} \int_0^1 (r^2) r\, dr\, d\theta$$

$$= \int_0^{\pi/2} \left[\frac{r^4}{4} \right]_{r=0}^{r=1} d\theta = \int_0^{\pi/2} \frac{1}{4}\, d\theta = \frac{\pi}{8}.$$

Why is the polar coordinate transformation so effective here? One reason is that $x^2 + y^2$ simplifies to r^2. Another is that the limits of integration become constants.

Example 4 EVALUATING INTEGRALS USING POLAR COORDINATES

Evaluate

$$\iint\limits_{R} e^{x^2 + y^2}\, dy\, dx,$$

where R is the semicircular region bounded by the x-axis and the curve $y = \sqrt{1 - x^2}$ (Figure 14.29).

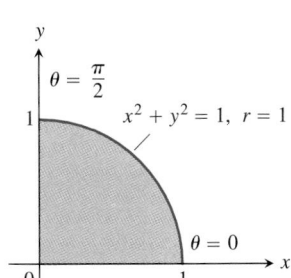

Figure 14.28 In polar coordinates, this region is described by simple inequalities: $0 \le r \le 1$ and $0 \le \theta \le \pi/2$. (Example 3)

Figure 14.29 The semicircular region in Example 4 is the region $0 \le r \le 1$, $0 \le \theta \le \pi$.

Solution In Cartesian coordinates, the integral in question is a nonelementary integral and there is no direct way to integrate $e^{x^2+y^2}$ analytically with respect to either x or y. We could use a CAS to get an approximation (2.699) to its value. Let's try polar coordinates. Substituting $x = r\cos\theta$, $y = r\sin\theta$ and replacing $dy\,dx$ by $r\,dr\,d\theta$ enables us to evaluate the integral as

$$\iint\limits_{R} e^{x^2+y^2}\,dy\,dx = \int_0^{\pi}\int_0^1 e^{r^2}r\,dr\,d\theta = \int_0^{\pi}\left[\frac{1}{2}e^{r^2}\right]_0^1 d\theta$$

$$= \int_0^{\pi}\frac{1}{2}(e-1)\,d\theta = \frac{\pi}{2}(e-1).$$

The r in the $r\,dr\,d\theta$ was just what we needed to integrate e^{r^2} analytically. Without it, we would have been stuck, as we were at the beginning.

Quick Review 14.3

In Exercises 1–3, replace the Cartesian equation by an equivalent polar equation.

1. $x^2 + y^2 = 9$ **2.** $x = 3$

3. $x^2 + (y + 2)^2 = 4$

In Exercises 4 and 5, replace the polar equation by an equivalent Cartesian equation.

4. $r^2 = 4r\cos\theta$ **5.** $r = \sec\theta$

In Exercises 6 and 7, use polar coordinate inequalities to describe the region of the plane.

6. The portion of $x^2 + y^2 \le 4$ in the second quadrant and not on the coordinate axes.

7. The portion of $x^2 + y^2 \le 4$ in the third and fourth quadrants and not on the coordinate axes.

In Exercises 8–10, find the area of the region.

8. Inside the cardioid $r = 2(1 + \cos\theta)$

9. Inside the lemniscate $r^2 = 2\cos 2\theta$

10. Shared by the circle $r = 2$ and the cardioid $r = 2(1 + \cos\theta)$

Section 14.3 Exercises

In Exercises 1–16, change the Cartesian integral into an equivalent polar integral. Then evaluate the polar integral.

1. $\displaystyle\int_{-1}^{1}\int_0^{\sqrt{1-x^2}} dy\,dx$

2. $\displaystyle\int_{-1}^{1}\int_{-\sqrt{1-x^2}}^{\sqrt{1-x^2}} dy\,dx$

3. $\displaystyle\int_0^1\int_0^{\sqrt{1-y^2}} (x^2 + y^2)\,dx\,dy$

4. $\displaystyle\int_{-1}^{1}\int_{-\sqrt{1-y^2}}^{\sqrt{1-y^2}} (x^2 + y^2)\,dx\,dy$

5. $\displaystyle\int_{-a}^{a}\int_{-\sqrt{a^2-x^2}}^{\sqrt{a^2-x^2}} dy\,dx$

6. $\displaystyle\int_0^2\int_0^{\sqrt{4-y^2}} (x^2 + y^2)\,dx\,dy$

7. $\displaystyle\int_0^6\int_0^y x\,dx\,dy$

8. $\displaystyle\int_0^2\int_0^x y\,dy\,dx$

9. $\displaystyle\int_{-1}^{0}\int_{-\sqrt{1-x^2}}^{0} \frac{2}{1 + \sqrt{x^2 + y^2}}\,dy\,dx$

10. $\displaystyle\int_{-1}^{1}\int_{-\sqrt{1-y^2}}^{0} \frac{4\sqrt{x^2 + y^2}}{1 + x^2 + y^2}\,dx\,dy$

11. $\displaystyle\int_0^{\ln 2}\int_0^{\sqrt{(\ln 2)^2-y^2}} e^{\sqrt{x^2+y^2}}\,dx\,dy$

12. $\displaystyle\int_0^1\int_0^{\sqrt{1-x^2}} e^{-(x^2+y^2)}\,dy\,dx$

13. $\displaystyle\int_0^2\int_0^{\sqrt{1-(x-1)^2}} \frac{x + y}{x^2 + y^2}\,dy\,dx$

14. $\displaystyle\int_0^2\int_{-\sqrt{1-(y-1)^2}}^{0} xy^2\,dx\,dy$

15. $\displaystyle\int_{-1}^{1}\int_{-\sqrt{1-y^2}}^{\sqrt{1-y^2}} \ln(x^2 + y^2 + 1)\,dx\,dy$

16. $\displaystyle\int_{-1}^{1}\int_{-\sqrt{1-x^2}}^{\sqrt{1-x^2}} \frac{2}{(1 + x^2 + y^2)^2}\,dy\,dx$

In Exercises 17–22, find the area of the region.

17. The region cut from the first quadrant by the curve $r = 2(2 - \sin 2\theta)^{1/2}$

18. The region that lies inside the cardioid $r = 1 + \cos\theta$ and outside the circle $r = 1$

19. The region enclosed by one leaf of the rose $r = 12\cos 3\theta$

20. The region enclosed by the positive x-axis and spiral $r = 4\theta/3$, $0 \le \theta \le 2\pi$ (the region looks like a snail shell)

21. The region cut from the first quadrant by the cardioid $r = 1 + \sin\theta$

22. The region common to the interiors of the cardioids $r = 1 + \cos\theta$ and $r = 1 - \cos\theta$

23. *First Moment of a Plate* Find the first moment about the x-axis of a thin plate of constant density $\delta(x, y) = 3$, bounded below by the x-axis and above by the cardioid $r = 1 - \cos\theta$.

24. *Inertial and Polar Moments of a Disk* Find the moment of inertia about the x-axis and the polar moment of inertia about the origin of a thin disk bounded by the circle $x^2 + y^2 = a^2$ if the disk's density at the point (x, y) is $\delta(x, y) = k(x^2 + y^2)$, k a constant.

25. *Mass* Find the mass of a thin plate covering the region outside the circle $r = 3$ and inside the circle $r = 6\sin\theta$ if the plate's density function is $\delta(x, y) = 1/r$.

26. *Polar Moment of a Cardioid Region Outside a Circle* Find the polar moment of inertia about the origin of a thin plate covering the region that lies inside the cardioid $r = 1 - \cos\theta$ and outside the circle $r = 1$ if the plate's density function is $\delta(x, y) = 1/r^2$.

27. *Centroid of a Cardioid Region* Find the centroid of the region enclosed by the cardioid $r = 1 + \cos\theta$.

28. *Polar Moment of a Cardioid Region* Find the polar moment of inertia about the origin of a thin plate enclosed by the cardioid $r = 1 + \cos\theta$ if the plate's density function is $\delta(x, y) = 1$.

29. *Average Height of a Surface* Find the average height of the hemisphere $z = \sqrt{a^2 - x^2 - y^2}$ above the disk $x^2 + y^2 \le a^2$ in the xy-plane.

30. *Average Height of a Surface* Find the average height of the (single) cone $z = \sqrt{x^2 + y^2}$ above the disk $x^2 + y^2 \le a^2$ in the xy-plane.

31. *Average Distance from the Origin* Find the average distance from a point $P(x, y)$ in the disk $x^2 + y^2 \le a^2$ to the origin.

32. *Average Squared Distance* Find the average value of the *square* of the distance from the point $P(x, y)$ in the disk $x^2 + y^2 \le 1$ to the boundary point $A(1, 0)$.

33. Integrate $f(x, y) = [\ln(x^2 + y^2)]/\sqrt{x^2 + y^2}$ over the region $1 \le x^2 + y^2 \le e$.

34. Integrate $f(x, y) = [\ln(x^2 + y^2)]/(x^2 + y^2)$ over the region $1 \le x^2 + y^2 \le e$.

35. *Volume* The region that lies inside the cardioid $r = 1 + \cos\theta$ and outside the circle $r = 1$ is the base of a solid right cylinder. The top of the cylinder lies in the plane $z = x$. Find the cylinder's volume.

36. *Volume* The region enclosed by the lemniscate $r^2 = 2\cos 2\theta$ is the base of a solid right cylinder whose top is bounded by the sphere $z = \sqrt{2 - r^2}$. Find the cylinder's volume.

Explorations

37. *The Error Function erf(x)*

(a) The usual way to evaluate the improper integral

$$I = \int_0^\infty e^{-x^2}\, dx \text{ is first to calculate its square:}$$

$$I^2 = \left(\int_0^\infty e^{-x^2}\, dx\right)\left(\int_0^\infty e^{-y^2}\, dy\right) = \int_0^\infty \int_0^\infty e^{-(x^2 + y^2)}\, dx\, dy.$$

Evaluate the last integral using polar coordinates and solve the resulting equation for I.

(b) Evaluate

$$\lim_{x \to \infty} \operatorname{erf}(x) = \lim_{x \to \infty} \int_0^x \frac{2e^{-t^2}}{\sqrt{\pi}}\, dt.$$

38. Evaluate the integral

$$\int_0^\infty \int_0^\infty \frac{1}{(1 + x^2 + y^2)^2}\, dx\, dy. \qquad \blacksquare$$

In Exercises 39–41, *work in groups of two or three* to solve the problem.

39. **Writing to Learn** Integrate the function $f(x, y) = 1/(1 - x^2 - y^2)$ over the disk $x^2 + y^2 \le 3/4$. Does the integral of $f(x, y)$ over the disk $x^2 + y^2 \le 1$ exist? Give reasons for your answer.

40. Use the double integral in polar coordinates to derive the formula

$$A = \int_\alpha^\beta \frac{1}{2} r^2\, d\theta$$

for the area of the fan-shaped region between the origin and polar curve $r = f(\theta)$, $\alpha \le \theta \le \beta$.

41. *Average Squared Distance* Let P_0 be a point inside a circle of radius a and let h denote the distance from P_0 to the center of the circle. Let d denote the distance from an arbitrary point P to P_0. Find the average value of d^2 over the region enclosed by the circle. (*Hint:* Simplify your work by placing the center of the circle at the origin and P_0 on the x-axis.)

Extending the Ideas

CAS Explorations

In Exercises 42–45, perform the following steps.

   (a) Plot the Cartesian region of integration in the *xy*-plane.

   (b) Change each boundary curve of the Cartesian region in (a) to its polar representation by solving its Cartesian equation for r and θ.

   (c) Plot the polar region of integration in the $r\theta$-plane.

(d) Change the integrand from Cartesian to polar coordinates. Determine the limits of integration from your plot in (c) and evaluate the polar integral.

42. $\displaystyle\int_0^1 \int_x^1 \frac{y}{x^2 + y^2}\, dy\, dx$ **43.** $\displaystyle\int_0^1 \int_0^{x/2} \frac{x}{x^2 + y^2}\, dy\, dx$

44. $\displaystyle\int_0^1 \int_{-y/3}^{y/3} \frac{y}{\sqrt{x^2 + y^2}}\, dx\, dy$

45. $\displaystyle\int_0^1 \int_y^{2-y} \sqrt{x + y}\, dx\, dy$

14.4 Triple Integrals in Rectangular Coordinates

Triple Integrals • Properties of Triple Integrals • Volume of a Region in Space • Finding Limits of Integration • Average Value of a Function in Space

Triple Integrals

If $F(x, y, z)$ is a function defined on a closed bounded region D in space—the region occupied by a solid ball, for example, or a lump of clay—then the integral of F over D may be defined in the following way. We partition a rectangular region containing D into rectangular cells by planes parallel to the coordinate planes (Figure 14.30). We number the cells that lie inside D from 1 to n in some order, a typical cell having dimensions Δx_k by Δy_k by Δz_k and volume ΔV_k. We choose a point (x_k, y_k, z_k) in each cell and form the sum

$$S_n = \sum_{k=1}^{n} F(x_k, y_k, z_k)\, \Delta V_k. \tag{1}$$

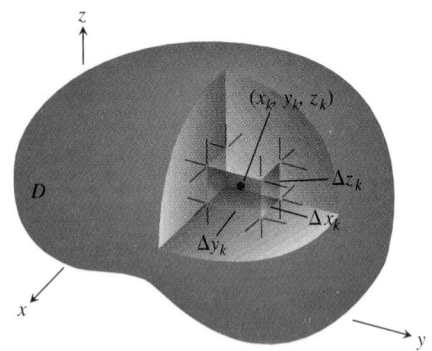

Figure 14.30 Partitioning a solid with rectangular cells of volume ΔV_k.

If F is continuous and the bounding surface of D is made up of smooth surfaces joined along continuous curves, then as Δx_k, Δy_k, and Δz_k approach zero independently, the sums S_n approach a limit

$$\lim_{n \to \infty} S_n = \iiint_D F(x, y, z)\, dV. \tag{2}$$

This limit is the **triple integral of F over D.** The limit also exists for some discontinuous functions.

Properties of Triple Integrals

Triple integrals of continuous functions have the same algebraic properties as double and single integrals.

Properties of Triple Integrals

1. *Constant Multiple:* $\iiint_D kF\,dV = k\iiint_D F\,dV$ (any number k)

2. *Sum and Difference:*
$$\iiint_D (F \pm G)\,dV = \iiint_D F\,dV \pm \iiint_D G\,dV$$

3. *Domination:*

(a) $\iiint_D F\,dV \geq 0$ if $F \geq 0$ on D

(b) $\iiint_D F\,dV \geq \iiint_D G\,dV$ if $F \geq G$ on D

4. *Additivity:*
$$\iiint_D F\,dV = \iiint_{D_1} F\,dV + \iiint_{D_2} F\,dV + \cdots + \iiint_{D_n} F\,dV$$

if D is the union of a finite number of nonoverlapping cells D_1, $D_2, \ldots, D_n$.

Volume of a Region in Space

If F is the constant function whose value is 1, then the sums in Equation (1) reduce to

$$S_n = \sum_{k=1}^{n} F(x_k, y_k, z_k)\,\Delta V_k = \sum_{k=1}^{n} \Delta V_k. \tag{3}$$

As Δx, Δy, and Δz approach zero, the cells ΔV_k become smaller and more numerous and fill up more and more of D. We therefore define the volume of D to be the triple integral

$$\lim_{n\to\infty} \sum_{k=1}^{n} \Delta V_k = \iiint_D dV.$$

Definition Volume

The **volume** of a closed, bounded region in space is

$$V = \iiint_D dV. \tag{4}$$

As we will see in a moment, this integral enables us to calculate the volumes of solids enclosed by curved surfaces.

Finding Limits of Integration

We evaluate triple integrals by applying a three-dimensional version of Fubini's theorem to evaluate them by repeated single integrations. As with double integrals, there is a geometric procedure for finding the limits of integration.

How to Find Limits of Integration in Triple Integrals

To evaluate

$$\iiint_D F(x, y, z)\, dV$$

over a region D, integrating first with respect to z, then with respect to y, finally with x, take the following steps.

1. *A sketch.* Sketch the region D along with its "shadow" R (vertical projection) in the xy-plane. Label the upper and lower bounding surfaces of D and the upper and lower bounding curves of R.

2. *The z-limits of integration.* Draw a line M passing through a typical point (x, y) in R parallel to the z-axis. As z increases, M enters D at $z = f_1(x, y)$ and leaves at $z = f_2(x, y)$. These are the z-limits of integration.

3. *The y-limits of integration.* Draw a line L through (x, y) parallel to the y-axis. As y increases, L enters R at $y = g_1(x)$ and leaves at $y = g_2(x)$. These are the y-limits of integration.

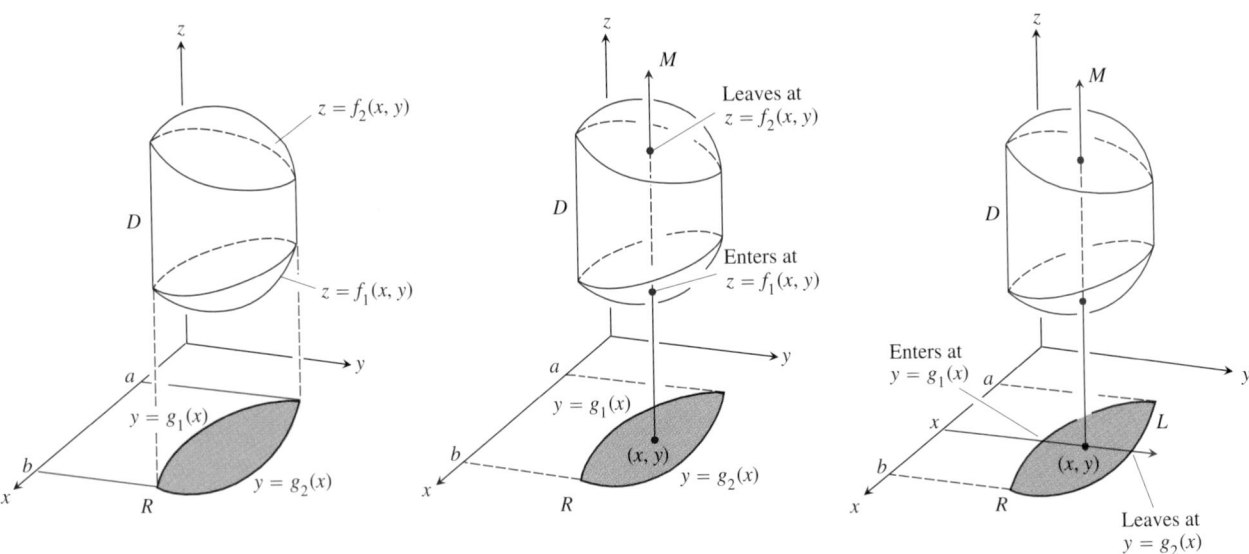

4. *The x-limits of integration.* Choose x-limits that include all lines through R parallel to the y-axis ($x = a$ and $x = b$ in the preceding figure). These are the x-limits of integration. The integral is

$$\int_{x=a}^{x=b} \int_{y=g_1(x)}^{y=g_2(x)} \int_{z=f_1(x,y)}^{z=f_2(x,y)} F(x, y, z)\, dz\, dy\, dx.$$

Follow similar procedures if you change the order of integration. The "shadow" of region D lies in the plane of the last two variables with respect to which the iterated integration takes place.

Example 1   FINDING VOLUME

Find the volume of the region D enclosed by the surfaces $z = x^2 + 3y^2$ and $z = 8 - x^2 - y^2$.

Solution   The volume is

$$V = \iiint_D dz\, dy\, dx,$$

the integral of $F(x, y, z) = 1$ over D. To find the limits of integration for evaluating the integral, we take these steps.

Step 1: *A sketch.* The surfaces (Figure 14.31) intersect on the elliptical cylinder $x^2 + 3y^2 = 8 - x^2 - y^2$ or $x^2 + 2y^2 = 4$. The boundary of the region R, the projection of D onto the xy-plane, is an ellipse with the same equation: $x^2 + 2y^2 = 4$. The "upper" boundary of R is the curve $y = \sqrt{(4 - x^2)/2}$. The lower boundary is the curve $y = -\sqrt{(4 - x^2)/2}$.

Step 2: *The z-limits of integration.* The line M passing through a typical point (x, y) in R parallel to the z-axis enters D at $z = x^2 + 3y^2$ and leaves at $z = 8 - x^2 - y^2$.

Step 3: *The y-limits of integration.* The line L through (x, y) parallel to the y-axis enters R at $y = -\sqrt{(4 - x^2)/2}$ and leaves at $y = \sqrt{(4 - x^2)/2}$.

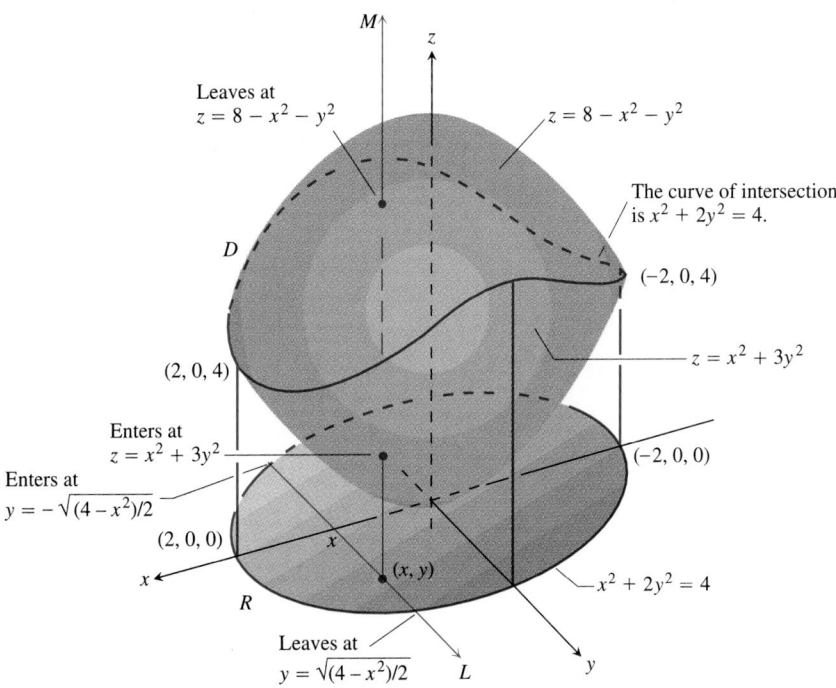

Figure 14.31 The volume of the region enclosed by these two paraboloids is calculated in Example 1.

Step 4: *The x-limits of integration.* As L sweeps across R, the values of x vary from $x = -2$ at $(-2, 0, 0)$ to $x = 2$ at $(2, 0, 0)$. The volume of D is

$$V = \iiint_D dz \, dy \, dx$$

$$= \int_{-2}^{2} \int_{-\sqrt{(4-x^2)/2}}^{\sqrt{(4-x^2)/2}} \int_{x^2+3y^2}^{8-x^2-y^2} dz \, dy \, dx$$

$$= \int_{-2}^{2} \int_{-\sqrt{(4-x^2)/2}}^{\sqrt{(4-x^2)/2}} (8 - 2x^2 - 4y^2) \, dy \, dx$$

$$= \int_{-2}^{2} \left[(8 - 2x^2)y - \frac{4}{3} y^3 \right]_{y=-\sqrt{(4-x^2)/2}}^{y=\sqrt{(4-x^2)/2}} dx$$

$$= \int_{-2}^{2} \left(2(8 - 2x^2)\sqrt{\frac{4 - x^2}{2}} - \frac{8}{3} \left(\frac{4 - x^2}{2} \right)^{3/2} \right) dx$$

$$= \int_{-2}^{2} \left[8 \left(\frac{4 - x^2}{2} \right)^{3/2} - \frac{8}{3} \left(\frac{4 - x^2}{2} \right)^{3/2} \right] = \frac{4\sqrt{2}}{3} \int_{-2}^{2} (4 - x^2)^{3/2} \, dx$$

$$= 8\pi\sqrt{2} \text{ units cubed} \qquad \text{After integration with the substitution } x = 2 \sin u$$

In the next example, we project D onto the xz-plane instead of the xy-plane.

Example 2 FINDING LIMITS OF INTEGRATION

Set up the limits of integration for evaluating the triple integral of a function $F(x, y, z)$ over the tetrahedron D with vertices $(0, 0, 0)$, $(1, 1, 0)$, $(0, 1, 0)$, and $(0, 1, 1)$.

Solution

Step 1: *A sketch.* We sketch D along with its "shadow" R in the xz-plane (Figure 14.32). The upper (right-hand) bounding surface of D lies in the plane $y = 1$. The lower (left-hand) bounding surface lies in the plane $y = x + z$. The upper boundary of R is the line $z = 1 - x$. The lower boundary is the line $z = 0$.

Step 2: *The y-limits of integration.* The line through a typical point (x, z) in R parallel to the y-axis enters D at $y = x + z$ and leaves at $y = 1$.

Step 3: *The z-limits of integration.* The line L through (x, z) parallel to the z-axis enters R at $z = 0$ and leaves at $z = 1 - x$.

Step 4: *The x-limits of integration.* As L sweeps across R, the value of x varies from $x = 0$ to $x = 1$. The integral is

$$\int_0^1 \int_0^{1-x} \int_{x+z}^1 F(x, y, z) \, dy \, dz \, dx.$$

As we know, there are sometimes (but not always) two different orders in which the single integrations for evaluating a double integral may be worked. For triple integrals, there could be as many as *six*.

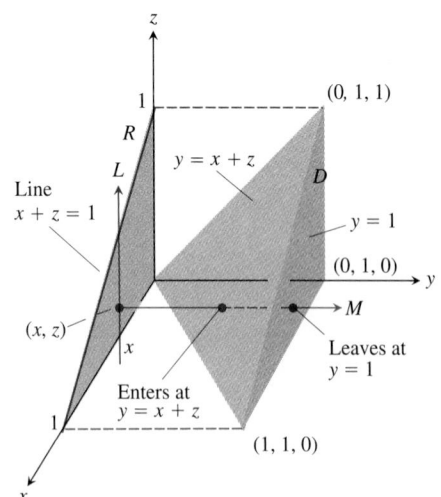

Figure 14.32 The tetrahedron in Example 2.

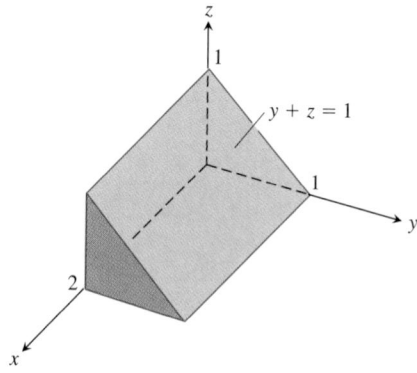

Figure 14.33 Example 3 gives six different iterated triple integrals for the volume of this prism.

Example 3 FINDING VOLUME

Each of the following integrals gives the volume of the solid shown in Figure 14.33.

(a) $\displaystyle\int_0^1 \int_0^{1-z} \int_0^2 dx \, dy \, dz$ **(b)** $\displaystyle\int_0^1 \int_0^{1-y} \int_0^2 dx \, dz \, dy$

(c) $\displaystyle\int_0^1 \int_0^2 \int_0^{1-z} dy \, dx \, dz$ **(d)** $\displaystyle\int_0^2 \int_0^1 \int_0^{1-z} dy \, dz \, dx$

(e) $\displaystyle\int_0^1 \int_0^2 \int_0^{1-y} dz \, dx \, dy$ **(f)** $\displaystyle\int_0^2 \int_0^1 \int_0^{1-y} dz \, dy \, dx$

Average Value of a Function in Space

The average value of a function F over a region D in space is defined by the formula

$$\textbf{Average value of } F \text{ over } D = \frac{1}{\text{volume of } D} \iiint_D F \, dV. \qquad (5)$$

For example, if $F(x, y, z) = \sqrt{x^2 + y^2 + z^2}$, then the average value of F over D is the average distance of points in D form the origin. If $F(x, y, z)$ is the density of a solid that occupies a region D in space, then the average value of F over D is the average density of the solid in units of mass per unit volume.

Example 4 FINDING AVERAGE VALUE

Find the average value of $F(x, y, z) = xyz$ over the cube bounded by the coordinate planes and the planes $x = 2$, $y = 2$, and $z = 2$ in the first octant.

Solution We sketch the cube with enough detail to show the limits of integration (Figure 14.34). We then use Equation (5) to calculate the average value of F over the cube.

The volume of the cube is $(2)(2)(2) = 8$. The value of the integral of F over the cube is

$$\int_0^2 \int_0^2 \int_0^2 xyz \, dx \, dy \, dz = \int_0^2 \int_0^2 \left[\frac{x^2}{2} yz \right]_{x=0}^{x=2} dy \, dz = \int_0^2 \int_0^2 2yz \, dy \, dz$$

$$= \int_0^2 [y^2 z]_{y=0}^{y=2} dz = \int_0^2 4z \, dz = [2z^2]_0^2 = 8.$$

With these values, Equation (5) gives

$$\text{Average value of } xyz \text{ over the cube} = \frac{1}{\text{volume}} \iiint_{\text{cube}} xyz \, dV$$

$$= \left(\frac{1}{8}\right)(8) = 1.$$

In evaluating the integral, we choose the order dx, dy, dz, but any of the other five possible orders would have done as well.

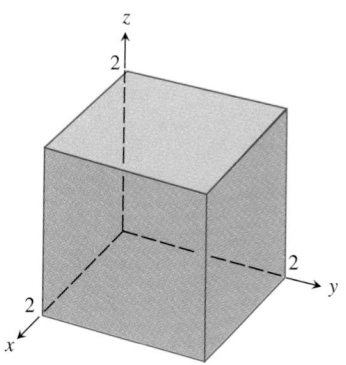

Figure 14.34 The region of integration in Example 4.

Quick Review 14.4

In Exercises 1 and 2, assume $\int_0^3 f(x)\, dx = 2$. Evaluate the integral.

1. $\int_0^3 2f(x)\, dx$

2. $\int_0^3 (x + f(x))\, dx$

In Exercises 3 and 4, explain, without evaluating the integral, why the inequality is a true statement.

3. $\int_0^3 (x^2 + 4)\, dx \geq 0$

4. $\int_1^4 x(4 - x)\, dx \geq \int_1^4 xe^{-x}\, dx$

In Exercises 5–8, let $\int_a^b f(x)\, dx = -2$, $\int_a^b g(x)\, dx = 3$,

$\int_b^c f(x)\, dx = 4$, and $\int_b^c g(x)\, dx = -1$. Evaluate the integral.

5. $\int_a^c (f(x) - g(x))\, dx$

6. $\int_a^c (f(x) + g(x))\, dx$

7. $\int_a^c (2f(x) + 3g(x))\, dx$

8. $\int_a^c (rf(x) + sg(x))\, dx$ (r, s constants)

9. Find the average value of $f(x) = \sin x$ over the interval $[0, \pi]$.

10. Find the average value of $f(x, y) = \cos(x + y)$ over the rectangle $0 \leq x \leq \pi$, $0 \leq y \leq \pi$.

Section 14.4 Exercises

1. Find the common value of the integrals in Example 3.

2. Write the six different iterated triple integrals for the volume of the rectangular solid in the first octant bounded by the coordinate planes and the planes $x = 1$, $y = 2$, and $z = 3$. Evaluate one of the integrals.

3. Write six different iterated triple integrals for the volume of the tetrahedron cut from the first octant by the plane $6x + 3y + 2z = 6$. Evaluate one of the integrals.

4. Write six different iterated triple integrals for the volume of the region in the first octant enclosed by the cylinder $x^2 + z^2 = 4$ and the plane $y = 3$. Evaluate one of the integrals.

5. Let D be the region bounded by the paraboloids $z = 8 - x^2 - y^2$ and $z = x^2 + y^2$. Write six different triple iterated integrals for the volume of D. Evaluate one of the integrals.

6. Let D be the region bounded by the paraboloid $z = x^2 + y^2$ and the plane $z = 2y$. Write triple iterated integrals in the order $dz\, dx\, dy$ and $dz\, dy\, dx$ that give the volume of D. Do not evaluate either integral.

In Exercises 7–20, evaluate the integral.

7. $\int_0^1 \int_0^1 \int_0^1 (x^2 + y^2 + z^2)\, dz\, dy\, dx$

8. $\int_0^{\sqrt{2}} \int_0^{3y} \int_{x^2+3y^2}^{8-x^2-y^2} dz\, dx\, dy$

9. $\int_1^e \int_1^e \int_1^e \frac{1}{xyz}\, dx\, dy\, dz$

10. $\int_0^1 \int_0^{3-3x} \int_0^{3-3x-y} dz\, dy\, dx$

11. $\int_0^1 \int_0^{\pi} \int_0^{\pi} y \sin z\, dx\, dy\, dz$

12. $\int_{-1}^1 \int_{-1}^1 \int_{-1}^1 (x + y + z)\, dy\, dx\, dz$

13. $\int_0^3 \int_0^{\sqrt{9-x^2}} \int_0^{\sqrt{9-x^2}} dz\, dy\, dx$

14. $\int_0^2 \int_{-\sqrt{4-y^2}}^{\sqrt{4-y^2}} \int_0^{2x+y} dz\, dx\, dy$

15. $\int_0^1 \int_0^{2-x} \int_0^{2-x-y} dz\, dy\, dx$

16. $\int_0^1 \int_0^{1-x^2} \int_3^{4-x^2-y} x\, dz\, dy\, dx$

17. $\int_0^{\pi} \int_0^{\pi} \int_0^{\pi} \cos(u + v + w)\, du\, dv\, dw$ (uvw-space)

18. $\int_1^e \int_1^e \int_1^e \ln r \ln s \ln t\, dt\, dr\, ds$ (rst-space)

19. $\int_0^{\pi/4} \int_0^{\ln \sec v} \int_{-\infty}^{2t} e^x\, dx\, dt\, dv$ (tvx-space)

20. $\int_0^7 \int_0^2 \int_0^{\sqrt{4-q^2}} \frac{q}{r+1}\, dp\, dq\, dr$ (pqr-space)

21. Here is the region of integration of the integral

$$\int_{-1}^{1} \int_{x^2}^{1} \int_{0}^{1-y} dz\, dy\, dx.$$

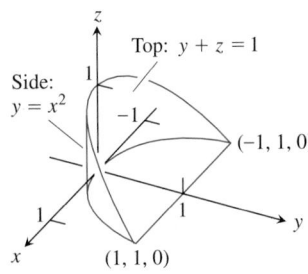

Top: $y + z = 1$
Side: $y = x^2$
-1
$(-1, 1, 0)$
$(1, 1, 0)$

Rewrite the integral as an equivalent iterated integral in the order

(a) $dy\, dz\, dx$ (b) $dy\, dx\, dz$

(c) $dx\, dy\, dz$ (d) $dx\, dz\, dy$

(e) $dz\, dx\, dy$

22. Here is the region of integration of the integral

$$\int_{0}^{1} \int_{-1}^{0} \int_{0}^{y^2} dz\, dy\, dx.$$

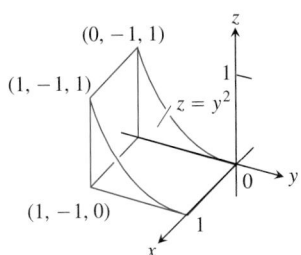

$(0, -1, 1)$
$(1, -1, 1)$
$z = y^2$
$(1, -1, 0)$

Rewrite the integral as an equivalent iterated integral in the order

(a) $dy\, dz\, dx$ (b) $dy\, dx\, dz$

(c) $dx\, dy\, dz$ (d) $dx\, dz\, dy$

(e) $dz\, dx\, dy$

In Exercises 23–36, find the volume of the region.

23. The region between the cylinder $z = y^2$ and the xy-plane that is bounded by the planes $x = 0$, $x = 1$, $y = -1$, $y = 1$

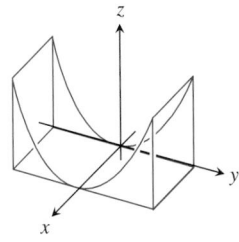

24. The region in the first octant bounded by the coordinate planes and the planes $x + z = 1$, $y + 2z = 2$

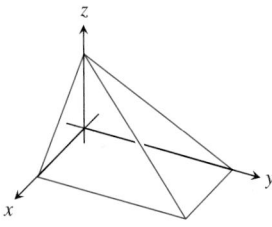

25. The region in the first octant bounded by the coordinate planes, the plane $y + z = 2$, and the cylinder $x = 4 - y^2$

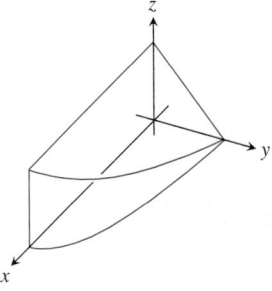

26. The wedge cut from the cylinder $x^2 + y^2 = 1$ by the planes $z = -y$ and $z = 0$

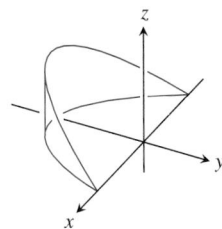

27. The tetrahedron in the first octant bounded by the coordinate planes and the plane $x + y/2 + z/3 = 1$

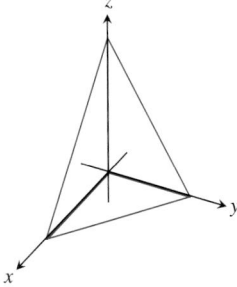

28. The region in the first octant bounded by the coordinate planes, the plane $y = 1 - x$, and the surface $z = \cos(\pi x/2)$, $0 \le x \le 1$

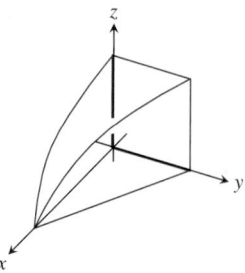

29. The region common to the interiors of the cylinders $x^2 + y^2 = 1$ and $x^2 + z^2 = 1$

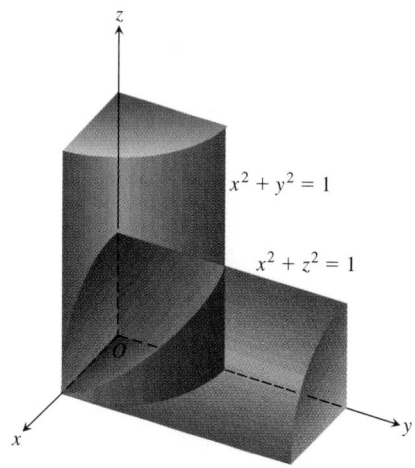

$x^2 + y^2 = 1$

$x^2 + z^2 = 1$

Figure 14.35 One-eighth of the region common to the cylinders $x^2 + y^2 = 1$ and $x^2 + z^2 = 1$ in Exercise 29.

30. The region in the first octant bounded by the coordinate planes and the surface $z = 4 - x^2 - y^2$

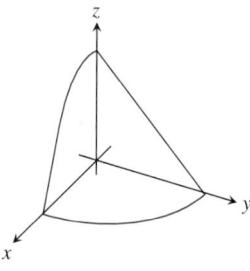

31. The region in the first octant bounded by the coordinate planes, the plane $x + y = 4$, and the cylinder $y^2 + 4z^2 = 16$

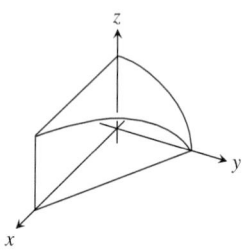

32. The region cut from the cylinder $x^2 + y^2 = 4$ by the plane $z = 0$ and the plane $x + z = 3$

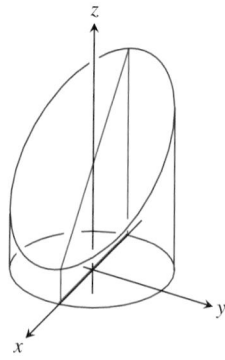

33. The region between the planes $x + y + 2z = 2$ and $2x + 2y + z = 4$ in the first octant

34. The finite region bounded by the planes $z = x$, $x + z = 8$, $z = y$, $y = 8$, and $z = 0$

35. The region cut from the solid elliptical cylinder $x^2 + 4y^2 \le 4$ by the xy-plane and the plane $z = x + 2$

36. The region bounded in back by the plane $x = 0$, on the front and sides by the parabolic cylinder $x = 1 - y^2$, on the top by the paraboloid $z = x^2 + y^2$, and on the bottom by the xy-plane

In Exercises 37–40, find the average value of $F(x, y, z)$ over the given region.

37. $F(x, y, z) = x^2 + 9$ over the cube in the first octant bounded by the coordinate planes and the planes $x = 2$, $y = 2$, and $z = 2$

38. $F(x, y, z) = x + y - z$ over the rectangular solid in the first octant bounded by the coordinate planes and the planes $x = 1$, $y = 1$, and $z = 2$

39. $F(x, y, z) = x^2 + y^2 + z^2$ over the cube in the first octant bounded by the coordinate planes and the planes $x = 1$, $y = 1$, and $z = 1$

40. $F(x, y, z) = xyz$ over the cube in the first octant bounded by the coordinate planes and the planes $x = 2$, $y = 2$, and $z = 2$

In Exercises 41–44, *work in groups of two or three* to evaluate the integrals by changing the order of integration in an appropriate way.

41. $\displaystyle\int_0^4 \int_0^1 \int_{2y}^2 \frac{4\cos(x^2)}{2\sqrt{z}}\, dx\, dy\, dz$

42. $\displaystyle\int_0^1 \int_0^1 \int_{x^2}^1 12xze^{zy^2}\, dy\, dx\, dz$

43. $\displaystyle\int_0^1 \int_{\sqrt[3]{z}}^1 \int_0^{\ln 3} \frac{\pi e^{2x}\sin \pi y^2}{y^2}\, dx\, dy\, dz$

44. $\displaystyle\int_0^2 \int_0^{4-x^2} \int_0^x \frac{\sin 2z}{4 - z}\, dy\, dz\, dx$

Explorations

45. Solve for a:

$$\int_0^1 \int_0^{4-a-x^2} \int_a^{4-x^2-y} dz\, dy\, dx = \frac{4}{15}.$$

46. For what value of c is the volume of the ellipsoid $x^2 + (y/2)^2 + (z/c)^2 = 1$ equal to 8π?

47. Writing to Learn *Minimizing the Value of an Integral* What domain D in space minimizes the value of the integral

$$\iiint_D (4x^2 + 4y^2 + z^2 - 4)\, dV?$$

Give reasons for your answer.

48. Writing to Learn *Maximizing the Value of an Integral* What domain D in space maximizes the value of the integral

$$\iiint_D (1 - x^2 - y^2 - z^2)\, dV?$$

Give reasons for your answer.

Extending the Ideas

CAS Explorations

In Exercises 49–52, evaluate the triple integral of $F(x, y, z)$ over the specified solid region.

49. $F(x, y, z) = x^2 y^2 z$ over the solid cylinder bounded by $x^2 + y^2 = 1$ and the planes $z = 0$ and $z = 1$

50. $F(x, y, z) = |xyz|$ over the solid bounded below by the paraboloid $z = x^2 + y^2$ and above by the plane $z = 1$

51. $F(x, y, z) = z/(x^2 + y^2 + z^2)^{3/2}$ over the solid bounded below by the cone $z = \sqrt{x^2 + y^2}$ and above by the plane $z = 1$

52. $F(x, y, z) = x^4 + y^2 + z^2$ over the solid sphere $x^2 + y^2 + z^2 \le 1$

| 14.5 | **Masses and Moments in Three Dimensions** |

Masses and Moments

Masses and Moments

If $\delta(x, y)$ is the density of an object occupying a region D in space (mass per unit volume), the integral of δ over D gives the mass of the object. To see why, imagine partitioning the object into n mass elements like the one in Figure 14.36. The object's mass is the limit

$$M = \lim_{n \to \infty} \sum_{k=1}^n \Delta m_k = \lim_{n \to \infty} \sum_{k=1}^n \delta(x_k, y_{yk}, z_k)\, \Delta V_k = \iiint_D \delta(x, y, z)\, dV. \quad (1)$$

If $r(x, y, z)$ is the distance from the point (x, y, z) in D to a line L, then the moment of inertia of the mass $\Delta m_k = \delta(x_k, y_k, z_k)\, \Delta V_k$ about the line L (shown in Figure 14.36) is approximately $\Delta I_k = r^2(x_k, y_k, z_k)\, \Delta m_k$. The moment of inertia of the entire object about L is

$$I_L = \lim_{n \to \infty} \sum_{k=1}^n \Delta I_k = \lim_{n \to \infty} \sum r^2(x_k, y_k, z_k)\, \delta(x_k, y_k, z_k)\, \Delta V_k = \iiint_D r^2 \delta\, dV.$$

If L is the x-axis, then $r^2 = y^2 + z^2$ (Figure 14.37) and

$$I_x = \iiint_D (y^2 + z^2)\, \delta\, dV.$$

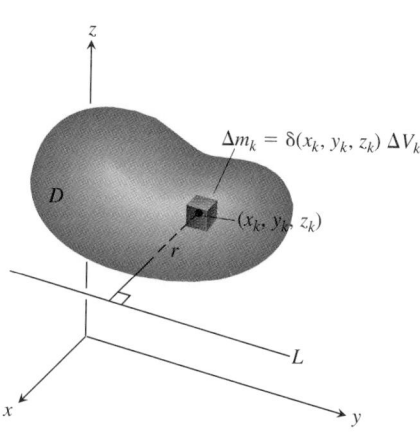

Figure 14.36 To define an object's mass and moment of inertia about a line, we first imagine it to be partitioned into a finite number of mass elements Δm_k.

$\Delta m_k = \delta(x_k, y_k, z_k)\, \Delta V_k$

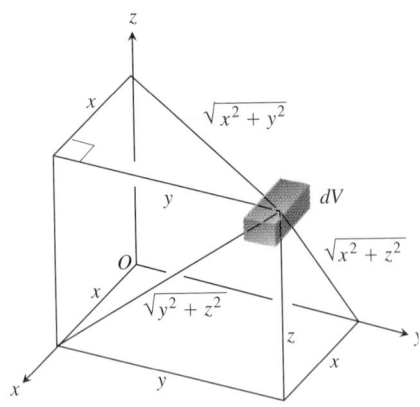

Figure 14.37 Distances from dV to the coordinate planes and axes.

Similarly,

$$I_y = \iiint_D (x^2 + z^2)\, \delta\, dV \qquad \text{and} \qquad I_z = \iiint_D (x^2 + y^2)\delta\, dV.$$

These and other useful formulas are summarized in Table 14.2.

Table 14.2 Mass and Moment Formulas for Objects in Space

Mass: $M = \iiint_D \delta\, dV \qquad (\delta = \text{density})$

First moments about the coordinate planes:

$$M_{yz} = \iiint_D x\delta\, dV, \qquad M_{xz} = \iiint_D y\delta\, dV, \qquad M_{xy} = \iiint_D z\delta\, dV$$

Center of mass:

$$\bar{x} = \frac{M_{yz}}{M}, \qquad \bar{y} = \frac{M_{xz}}{M}, \qquad \bar{z} = \frac{M_{xy}}{M}$$

Moments of inertia (second moments):

$$I_x = \iiint_D (y^2 + z^2)\delta\, dV$$

$$I_y = \iiint_D (x^2 + z^2)\delta\, dV$$

$$I_z = \iiint_D (x^2 + y^2)\delta\, dV$$

Moment of inertia about a line L:

$$I_L = \iiint_D r^2\delta\, dV \qquad \text{(where } r(x, y, z) = \text{distance from points}$$
$$(x, y, z) \text{ to line } L)$$

Radius of gyration about a line L:

$$R_L = \sqrt{I_L/M}$$

Example 1 FINDING MOMENTS OF INERTIA

Find I_x, I_y, and I_z for the rectangular solid of constant density δ shown in Figure 14.38.

Solution The preceding formula for I_x gives

$$I_x = \int_{-c/2}^{c/2} \int_{-b/2}^{b/2} \int_{-a/2}^{a/2} (y^2 + z^2)\delta\, dx\, dy\, dz. \tag{2}$$

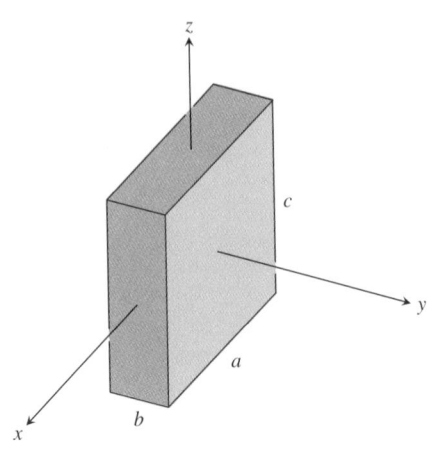

Figure 14.38 Example 1 calculates I_x, I_y, and I_z for the block shown here. The origin lies at the center of the block.

We can avoid some of the work of integration by observing that $(y^2 + z^2)\delta$ is an even function of x, y, *and* z and therefore

$$I_x = 8 \int_0^{c/2} \int_0^{b/2} \int_0^{a/2} (y^2 + z^2)\delta \, dx \, dy \, dz = 4a\delta \int_0^{c/2} \int_0^{b/2} (y^2 + z^2) \, dy \, dz$$

$$= 4a\delta \int_0^{c/2} \left[\frac{y^3}{3} + z^2 y \right]_{y=0}^{y=b/2} dz$$

$$= 4a\delta \int_0^{c/2} \left(\frac{b^3}{24} + \frac{z^2 b}{2} \right) dz$$

$$= 4a\delta \left(\frac{b^3 c}{48} + \frac{c^3 b}{48} \right) = \frac{abc\delta}{12} (b^2 + c^2)$$

$$= \frac{M}{12} (b^2 + c^2). \qquad\qquad M = abc\,\delta$$

Similarly,

$$I_y = \frac{M}{12} (a^2 + c^2) \qquad \text{and} \qquad I_z = \frac{M}{12} (a^2 + b^2).$$

Example 2 FINDING CENTER OF MASS

Find the center of mass of a solid of constant density δ bounded below by the disk $R\colon x^2 + y^2 \le 4$ in the plane $z = 0$ and above by the paraboloid $z = 4 - x^2 - y^2$ (Figure 14.39).

Solution By symmetry, $\bar{x} = \bar{y} = 0$. To find $\bar{z}$, we first calculate

$$M_{xy} = \iint_R \int_{z=0}^{z=4-x^2-y^2} z\delta \, dz \, dy \, dx = \iint_R \left[\frac{z^2}{2} \right]_{z=0}^{z=4-x^2-y^2} \delta \, dy \, dx$$

$$= \frac{\delta}{2} \iint_R (4 - x^2 - y^2)^2 \, dy \, dx$$

$$= \frac{\delta}{2} \int_0^{2\pi} \int_0^2 (4 - r^2)^2 r \, dr \, d\theta \qquad \text{Polar coordinates}$$

$$= \frac{\delta}{2} \int_0^{2\pi} \left[-\frac{1}{6}(4 - r^2)^3 \right]_{r=0}^{r=2} d\theta = \frac{16\delta}{3} \int_0^{2\pi} d\theta = \frac{32\pi\delta}{3}.$$

A similar calculation gives

$$M = \iint_R \int_0^{4-x^2-y^2} \delta \, dz \, dy \, dx = 8\pi\delta.$$

Therefore $\bar{z} = (M_{xy}/M) = 4/3$, and the center of mass is $(\bar{x}, \bar{y}, \bar{z}) = (0, 0, 4/3)$.

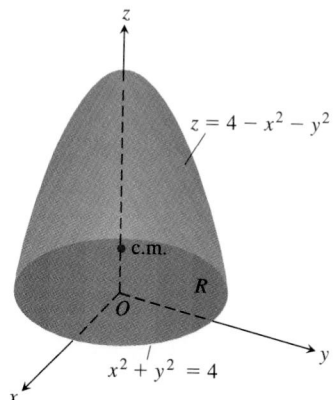

Figure 14.39 Example 2 finds the center of mass of this solid.

When the density of a solid object is constant (as in Examples 1 and 2), the center of mass is the **centroid** of the object (as was the case for two-dimensional shapes in Section 14.2).

Quick Review 14.5

In Exercises 1–10, a thin plate of density $\delta = 2$ occupies the region in the xy-plane bounded by the lines $x = 0$, $y = 3x$, and the parabola $y = 4 - x^2$ in the first quadrant. Find the plate's

1. mass.

2. moment about the x-axis.

3. moment about the y-axis.

4. center of mass.

5. moment of inertia about the x-axis.

6. moment of inertia about the y-axis.

7. moment of inertia about the origin.

8. radius of gyration about the x-axis.

9. radius of gyration about the y-axis.

10. radius of gyration about the origin.

Section 14.5 Exercises

In Exercises 1–12, the solid has constant density $\delta = 1$.

1. Evaluate the integral for I_x in Equation (2) in the text directly to show that the shortcut in Example 1 gives the same answer. Use the results in Example 1 to find the radius of gyration of the rectangular solid about each coordinate axis.

2. The coordinate axes in the figure shown here run through the centroid of a solid wedge parallel to the labeled edges. Find I_x, I_y, and I_z, if $a = b = 6$ and $c = 4$.

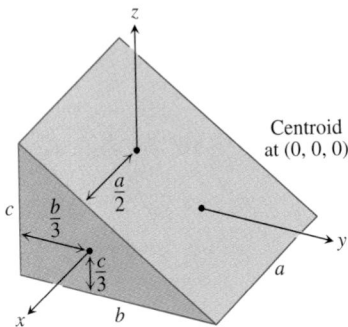

3. Find the moments of inertia of the rectangular solid shown here with respect to its edges by calculating I_x, I_y, and I_z.

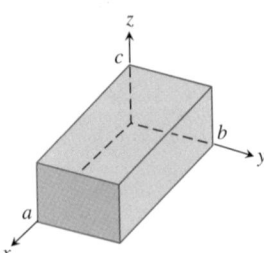

4. (a) Find the centroid and the moments of inertia I_x, I_y, and I_z of the tetrahedron whose vertices are the points $(0, 0, 0)$, $(1, 0, 0)$, $(0, 1, 0)$, and $(0, 0, 1)$.

(b) Find the radius of gyration of the tetrahedron about the x-axis. Compare it with the distance from the centroid to the x-axis.

5. A solid "trough" is bounded below by the surface $z = 4y^2$, above by the plane $z = 4$, and on the ends by the planes $x = 1$ and $x = -1$. Find the center of mass and the moments of inertia with respect to the three axes.

6. A solid is bounded below by the plane $z = 0$, on the sides by the elliptical cylinder $x^2 + 4y^2 = 4$, and above by the plane $z = 2 - x$ (see the figure).

(a) Find $\bar{x}$ and $\bar{y}$.

(b) Evaluate the integral

$$M_{xy} = \int_{-2}^{2} \int_{-(1/2)\sqrt{4-x^2}}^{(1/2)\sqrt{4-x^2}} \int_{0}^{2-x} z \, dz \, dy \, dx,$$

using integral tables or technology to carry out the final integration with respect to x. Then divide M_{xy} by M to verify that $\bar{z} = 5/4$.

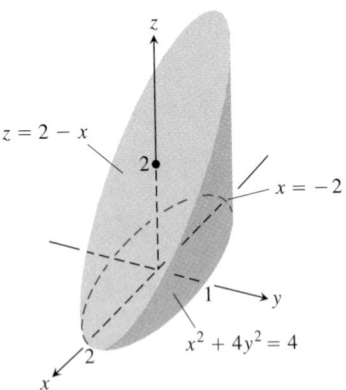

7. (a) Find the center of mass of a solid bounded below by the paraboloid $z = x^2 + y^2$ and above by the plane $z = 4$.

(b) Find the plane $z = c$ that divides the solid into two parts of equal volume. This plane does not pass through the center of mass.

8. A solid cube, 2 units on a side, is bounded by the planes $x = \pm 1$, $z = \pm 1$, $y = 3$, and $y = 5$. Find the center of mass and the moments of inertia and radii of gyration about the coordinate axes.

9. A wedge like the one in Exercise 2 has $a = 4$, $b = 6$, and $c = 3$. Make a quick sketch to check for yourself that the square of the distance from a typical point (x, y, z) of the wedge to the line L: $z = 0$, $y = 6$ is $r^2 = (y - 6)^2 + z^2$. Then calculate the moment of inertia and radius of gyration of the wedge about L.

10. A wedge like the one in Exercise 2 has $a = 4$, $b = 6$, and $c = 3$. Make a quick sketch to check for yourself that the square of the distance from a typical point (x, y, z) of the wedge to the line L: $x = 4$, $y = 0$ is $r^2 = (x - 4)^2 + y^2$. Then calculate the moment of inertia and radius of gyration of the wedge about L.

11. A solid like the one in Exercise 3 has $a = 4$, $b = 2$, and $c = 1$. Make a quick sketch to check for yourself that the square of the distance from a typical point (x, y, z) of the wedge to the line L: $y = 2$, $z = 0$ is $r^2 = (y - 2)^2 + z^2$. Then find the moment of inertia and radius of gyration of the solid about L.

12. A solid like the one in Exercise 3 has $a = 4$, $b = 2$, and $c = 1$. Make a quick sketch to check for yourself that the square of the distance from a typical point (x, y, z) of the wedge to the line L: $x = 4$, $y = 0$ is $r^2 = (x - 4)^2 + y^2$. Then find the moment of inertia and radius of gyration of the solid about L.

In Exercises 13–18, the solid has variable density. In Exercises 13 and 14, find **(a)** the mass of the solid and **(b)** the center of mass.

13. A solid region in the first octant is bounded by the coordinate planes and the plane $x + y + z = 2$. The density of the solid is $\delta(x, y, z) = 2x$.

14. A solid in the first octant is bounded by the planes $y = 0$ and $z = 0$ and by the surfaces $z = 4 - x^2$ and $x = y^2$ (see the figure). Its density function is $\delta(x, y, z) = kxy$.

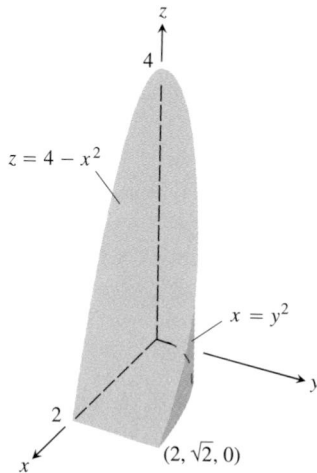

In Exercises 15 and 16, find

(a) the mass of the solid,

(b) the center of mass,

(c) the moments of inertia about the coordinate axes,

(d) the radii of gyration about the coordinate axes.

15. A solid cube in the first octant is bounded by the coordinate planes and by the planes $x = 1$, $y = 1$, and $z = 1$. The density of the cube is $\delta(x, y, z) = x + y + z + 1$.

16. A wedge like the one in Exercise 2 has dimensions $a = 2$, $b = 6$, and $c = 3$. The density is $\delta(x, y, z) = x + 1$. Notice that if the density is constant, the center of mass will be $(0, 0, 0)$.

17. Find the mass of the solid bounded by the planes $x + z = 1$, $x - z = -1$, $y = 0$ and the surface $y = \sqrt{z}$. The density of the solid is $\delta(x, y, z) = 2y + 5$.

18. Find the mass of the solid region bounded by the parabolic surfaces $z = 16 - 2x^2 - 2y^2$ and $z = 2x^2 + 2y^2$ if the density of the solid is $\delta(x, y, z) = \sqrt{x^2 + y^2}$.

Explorations

In Exercises 19 and 20, calculate the following.

(a) Writing to Learn The amount of work done by (constant) gravity g in moving the liquid filling the container to the xy-plane. (*Hint:* Partition the liquid into small volume elements ΔV_i and find the work done (approximately) by gravity on each element. Summation and passage to the limit gives a triple integral to evaluate.)

(b) The work done by gravity in moving the center of mass down to the xy-plane.

19. The container is a cubical box in the first octant bounded by the coordinate planes and the planes $x = 1$, $y = 1$, and $z = 1$. The density of the liquid filling the box is $\delta(x, y, z) = x + y + z + 1$ (refer to Exercise 15).

20. The container is in the shape of the region bounded by $y = 0$, $z = 0$, $z = 4 - x^2$, and $x = y^2$. The density of the liquid filling the region is $\delta(x, y, z) = kxy$ (see Exercise 14). ∎

Extending the Ideas

The Parallel Axis Theorem
The Parallel Axis Theorem (Exercises 14.2) holds in three dimensions as well as in two. Let $L_{c.m.}$ be a line through the center of mass of a body of mass m and let L be a parallel line h units away from $L_{c.m.}$. The **Parallel Axis Theorem** says that the moments of inertia $I_{c.m.}$ and I_L of the body about $L_{c.m.}$ and L satisfy the equation

$$I_L = I_{c.m.} + mh^2. \qquad (1)$$

As in the two-dimensional case, the theorem gives a quick way to calculate one moment when the other moment and the mass are known.

21. *Proof of the Parallel Axis Theorem*

(a) Show that the first moment of a body in space about any plane through the body's center of mass is zero. (*Hint:* Place the body's center of mass at the origin and let the plane be the yz-plane. What does the formula $\bar{x} = M_{yz}/M$ then tell you?)

(b) To prove the Parallel Axis Theorem, place the body with its center of mass at the origin, with the line $L_{c.m.}$ along the z-axis and the line L perpendicular to the xy-plane at the point $(h, 0, 0)$. Let D be the region of space occupied by the body. Then, in the notation of the figure,

$$I_L = \int \int \int |\mathbf{v} - h\mathbf{i}|^2 \, dm. \tag{2}$$

Expand the integrand in this integral and complete the proof.

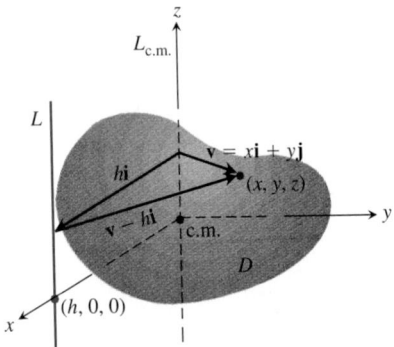

22. *Using the Parallel Axis Theorem* The moment of inertia about a diameter of a solid sphere of constant density and radius a is $(2/5)ma^2$, where m is the mass of the sphere. Find the moment of inertia about a line tangent to the sphere.

23. *Using the Parallel Axis Theorem* The moment of inertia of the solid in Exercise 3 about the z-axis is $I_z = abc(a^2 + b^2)/3$.

(a) Use Equation (1) to find the moment of inertia and radius of gyration of the solid about the line parallel to the z-axis through the solid's center of mass.

(b) Use Equation (1) and the result in (a) to find the moment of inertia and radius of gyration of the solid about the line $x = 0$, $y = 2b$.

24. *Using the Parallel Axis Theorem* If $a = b = 6$ and $c = 4$, the moment of inertia of the solid wedge in Exercise 2 about the x-axis is $I_x = 208$. Find the moment of inertia of the wedge about the line $y = 4$, $z = -4/3$ (the edge of the wedge's narrow end).

Pappus's Formula

Pappus's formula (Exercises 14.2) holds in three dimensions as well as in two. Suppose bodies B_1 and B_2 of mass m_1 and m_2, respectively, occupy nonoverlapping regions in space and $\mathbf{c}_1$ and $\mathbf{c}_2$ are the vectors from the origin to the bodies' respective centers of mass. Then the center of mass of the union $B_1 \cup B_2$ of the two bodies is determined by the vector

$$\mathbf{c} = \frac{m_1\mathbf{c}_1 + m_2\mathbf{c}_2}{m_1 + m_2}. \tag{3}$$

As before, this formula is **Pappus's formula**. As in the two-dimensional case, the formula generalizes to

$$\mathbf{c} = \frac{m_1\mathbf{c}_1 + m_2\mathbf{c}_2 + \cdots + m_n\mathbf{c}_n}{m_1 + m_2 + \cdots + m_n} \tag{4}$$

for n bodies.

25. *Deriving Pappus's Formula* Derive Pappus's formula (Equation (3)). (*Hint:* Sketch B_1 and B_2 as nonoverlapping regions in the first octant and label their centers of mass $(\bar{x}_1, \bar{y}_1, \bar{z}_1)$ and $(\bar{x}_2, \bar{y}_2, \bar{z}_2)$. Express the moments of $B_1 \cup B_2$ about the coordinate planes in terms of the masses m_1 and m_2 and the coordinates of these centers.)

26. *Applying Pappus's Formula* The figure shows a solid made from three rectangular solids of constant density $\delta = 1$. Use Pappus's formula to find the center of mass of

(a) $A \cup B$. (b) $A \cup C$.

(c) $B \cup C$. (d) $A \cup B \cup C$.

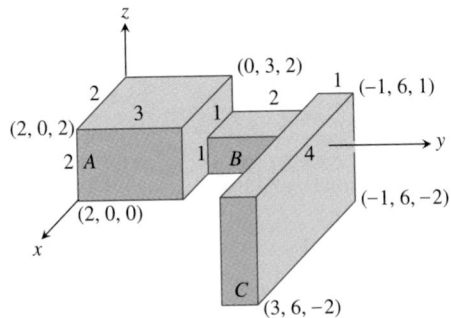

27. (a) **Writing to Learn** *Designing an Ice Cream Cone* Suppose a solid right circular cone C of base radius a and altitude h is constructed on the circular base of a solid hemisphere S of radius a so that the union of the two solids resembles an ice cream cone. The centroid of a solid cone lies one-fourth of the way from the base toward the vertex. The centroid of a solid hemisphere lies three-eighths of the way from the base to the top. What relation must hold between h and a to place the centroid of $C \cup S$ in the common base of the two solids?

(b) If you have not already done so, answer the analogous question about a triangle and a semicircle (Section 14.2, Exercise 55). The answers are not the same.

28. **Writing to Learn** *Shape Design* A solid pyramid P with height h and four congruent sides is built with its base as one face of a solid cube C whose edges have length s. The centroid of a solid pyramid lies one-fourth of the way from the base toward the vertex. What relation must hold between h and s to place the centroid of $P \cup C$ in the base of the pyramid? Compare your answer with the answer to Exercise 27. Also compare it to the answer to Exercise 56 in Section 14.2.

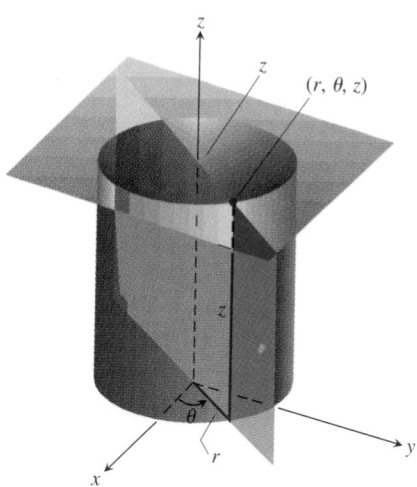

Figure 14.40 Cylindrical coordinates and typical surfaces of constant coordinate value.

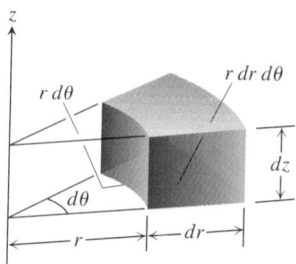

Figure 14.41 The volume element in cylindrical coordinates is $dV = dz\, r\, dr\, d\theta$.

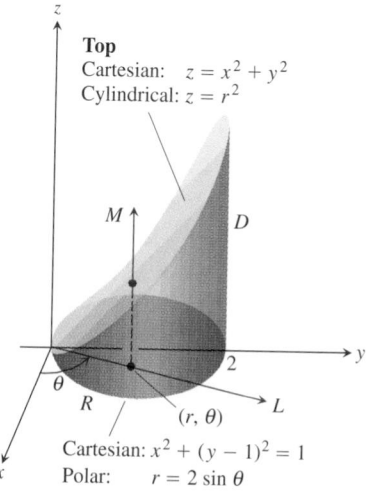

Cartesian: $x^2 + (y-1)^2 = 1$
Polar: $r = 2\sin\theta$

Figure 14.42 The figure for Example 1.

Triple Integrals in Cylindrical and Spherical Coordinates

14.6

Integration in Cylindrical Coordinates • Spherical Coordinates •
Integration in Spherical Coordinates

Integration in Cylindrical Coordinates

Cylindrical coordinates (Figure 14.40) are good for describing cylinders whose axes run along the z-axis and planes that either contain the z-axis or lie perpendicular to the z-axis. As we saw in Section 11.5, surfaces like these have equations of constant coordinate value:

$$r = 4, \qquad \text{Cylinder, radius 4, axis the } z\text{-axis}$$

$$\theta = \frac{\pi}{3}, \qquad \text{Plane containing the } z\text{-axis}$$

$$z = 2. \qquad \text{Plane perpendicular to the } z\text{-axis}$$

The volume element for subdividing a region in space with cylindrical coordinates is

$$dV = dz\, r\, dr\, d\theta \tag{1}$$

(Figure 14.41). Triple integrals in cylindrical coordinates are then evaluated as iterated integrals, as in the following example.

EXAMPLE 1 FINDING LIMITS OF INTEGRATION IN CYLINDRICAL COORDINATES

Find the limits of integration in cylindrical coordinates for integrating a function $f(r, \theta, z)$ over the region D bounded below by the plane $z = 0$, laterally by the circular cylinder $x^2 + (y-1)^2 = 1$, and above by the paraboloid $z = x^2 + y^2$.

Solution

Step 1: *A sketch* (Figure 14.42). The base of D is also the region's projection R on the xy-plane. The boundary of R is the circle $x^2 + (y-1)^2 = 1$. Its polar coordinate equation is

$$x^2 + (y-1)^2 = 1$$
$$x^2 + y^2 - 2y + 1 = 1$$
$$r^2 - 2r\sin\theta = 0$$
$$r = 2\sin\theta.$$

Step 2: *The z-limits of integration.* A line M through a typical point (r, θ) in R parallel to the z-axis enters D at $z = 0$ and leaves at $z = x^2 + y^2 = r^2$.

Step 3: *The r-limits of integration.* A ray L through (r, θ) from the origin enters R at $r = 0$ and leaves at $r = 2\sin\theta$.

Step 4: *The θ-limits of integration.* As L sweeps across R, the angle θ it makes with the positive x-axis runs from $\theta = 0$ to $\theta = \pi$. The integral is

$$\iiint_D f(r, \theta, z)\, dV = \int_0^\pi \int_0^{2\sin\theta} \int_0^{r^2} f(r, \theta, z)\, dz\, r\, dr\, d\theta.$$

Example 1 illustrates a good procedure for finding limits of integration in cylindrical coordinates. The procedure is summarized below.

How to Integrate in Cylindrical Coordinates

To evaluate

$$\iiint_D f(r, \theta, z)\, dV$$

over a region D in space in cylindrical coordinates, integrating first with respect to z, then with respect to r, and finally with respect to θ, take the following steps.

1. *A sketch.* Sketch the region D along with its projection R on the xy-plane. Label the surfaces and curves that bound D and R.

2. *The z-limits of integration.* Draw a line M through a typical point (r, θ) of R parallel to the z-axis. As z increases, M enters D at $z = g_1(r, \theta)$ and leaves at $z = g_2(r, \theta)$. These are the z-limits of integration.

3. *The r-limits of integration.* Draw a ray L through (r, θ) from the origin. The ray enters R at $r = h_1(\theta)$ and leaves at $r = h_2(\theta)$. These are the r-limits of integration.

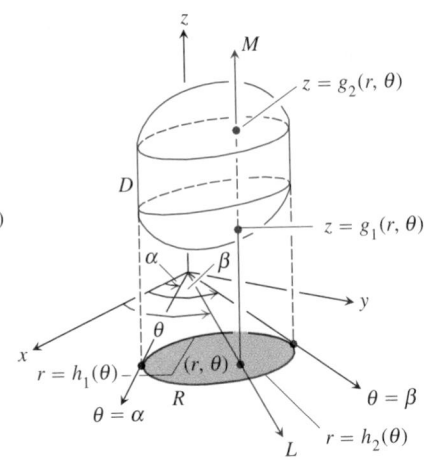

4. *The θ-limits of integration.* As L sweeps across R, the angle θ it makes with the positive x-axis runs from $\theta = \alpha$ to $\theta = \beta$. These are the θ-limits of integration. The integral is

$$\iiint_D f(r, \theta, z)\, dV = \int_{\theta=\alpha}^{\theta=\beta} \int_{r=h_1(\theta)}^{r=h_2(\theta)} \int_{z=g_1(r,\theta)}^{z=g_2(r,\theta)} f(r, \theta, z)\, dz\, r\, dr\, d\theta.$$

Example 2 FINDING A CENTROID

Find the centroid ($\delta = 1$) of the solid enclosed by the cylinder $x^2 + y^2 = 4$, bounded above by the paraboloid $z = x^2 + y^2$ and below by the xy-plane.

Solution

Step 1: *A sketch.* We sketch the solid, bounded above by the paraboloid $z = r^2$ and below by the plane $z = 0$ (Figure 14.43). Its base R is the disk $|r| \le 2$ in the xy-plane.

The solid's centroid $(\bar{x}, \bar{y}, \bar{z})$ lies on its axis of symmetry, here the z-axis. This makes $\bar{x} = \bar{y} = 0$. To find $\bar{z}$, we divide the first moment M_{xy} by the mass M.

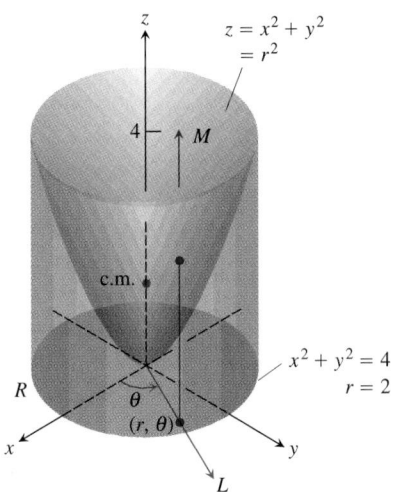

Figure 14.43 Example 2 shows how to find the centroid of this solid.

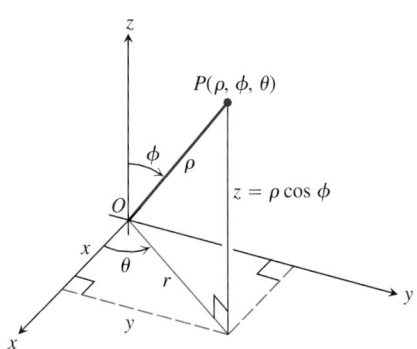

Figure 14.44 The spherical coordinates ρ, ϕ, and θ and their relation to x, y, z, and r. From the figure, we read the relations $r = \rho \sin \phi$, $z = \rho \cos \phi$, $x = r \cos \theta$, $y = r \sin \theta$.

A few books give spherical coordinates in the order (ρ, θ, ϕ), with θ and ϕ reversed. In some cases, you may also find r being used for ρ. Watch out for this when you read elsewhere.

To find the limits of integration for the mass and moment integrals, we continue with the four basic steps. We completed step 1 with our initial sketch. The remaining steps give the limits of integration.

Step 2: *The z-limits.* A line M through a typical point (r, θ) in the base parallel to the z-axis enters the solid at $z = 0$ and leaves at $z = r^2$.

Step 3: *The r-limits.* A ray L through (r, θ) from the origin enters R at $r = 0$ and leaves at $r = 2$.

Step 4: *The θ-limits.* As L sweeps over the base like a clock hand moving backward, the angle θ it makes with the positive x-axis runs from $\theta = 0$ to $\theta = 2\pi$. The value of M_{xy} is

$$M_{xy} = \int_0^{2\pi} \int_0^2 \int_0^{r^2} z\, dz\, r\, dr\, d\theta = \int_0^{2\pi} \int_0^2 \left[\frac{z^2}{2} \right]_0^{r^2} r\, dr\, d\theta$$

$$= \int_0^{2\pi} \int_0^2 \frac{r^5}{2}\, dr\, d\theta = \int_0^{2\pi} \left[\frac{r^6}{12} \right]_0^2 d\theta = \int_0^{2\pi} \frac{16}{3}\, d\theta = \frac{32\pi}{3}.$$

The value of M is

$$M = \int_0^{2\pi} \int_0^2 \int_0^{r^2} dz\, r\, dr\, d\theta = \int_0^{2\pi} \int_0^2 [z]_0^{r^2}\, r\, dr\, d\theta$$

$$= \int_0^{2\pi} \int_0^2 r^3\, dr\, d\theta = \int_0^{2\pi} \left[\frac{r^4}{4} \right]_0^2 d\theta = \int_0^{2\pi} 4\, d\theta = 8\pi.$$

Therefore,

$$\bar{z} = \frac{M_{xy}}{M} = \frac{32\pi}{3} \frac{1}{8\pi} = \frac{4}{3},$$

and the centroid is (0, 0, 4/3). Notice that the centroid lies outside of the solid.

Spherical Coordinates

Spherical coordinates locate points in space with two angles and a distance, as shown in Figure 14.44.

The first coordinate, $\rho = \left| \overrightarrow{OP} \right|$, is the point's distance from the origin. Unlike r, *the variable ρ is never negative.* The second coordinate, ϕ, is the angle $\overrightarrow{OP}$ makes with the positive z-axis. It is required to lie in the interval $[0, \pi]$. The third coordinate is the angle θ as measured in cylindrical coordinates.

Definition Spherical Coordinates

Spherical coordinates represent a point P in space by ordered triples (ρ, ϕ, θ) in which

1. ρ is the distance from P to the origin,
2. ϕ is the angle $\overrightarrow{OP}$ makes with the positive z-axis ($0 \le \phi \le \pi$), and
3. θ is the angle from cylindrical coordinates.

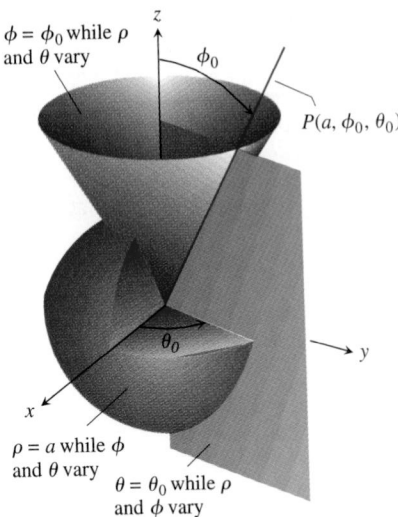

$\phi = \phi_0$ while ρ and θ vary

ϕ_0

$P(a, \phi_0, \theta_0)$

θ_0

$\rho = a$ while ϕ and θ vary

$\theta = \theta_0$ while ρ and ϕ vary

Figure 14.45 Constant-coordinate equations in spherical coordinates yield spheres, single cones, and half-planes.

The equation $\rho = a$ describes the sphere of radius a centered at the origin (Figure 14.45). The equation $\phi = \phi_0$ describes a single cone whose vertex lies at the origin and whose axis lies along the z-axis. (We broaden our interpretation to include the xy-plane as the cone $\phi = \pi/2$.) If ϕ_0 is greater than $\pi/2$, the cone $\phi = \phi_0$ opens downward.

Equations Relating Spherical Coordinates to Cartesian and Cylindrical Coordinates

$$r = \rho \sin \phi, \qquad x = r \cos \theta = \rho \sin \phi \cos \theta,$$

$$z = \rho \cos \phi, \qquad y = r \sin \theta = \rho \sin \phi \sin \theta, \qquad (3)$$

$$\rho = \sqrt{x^2 + y^2 + z^2} = \sqrt{r^2 + z^2}$$

Example 3   CONVERTING CARTESIAN TO SPHERICAL

Find a spherical coordinate equation for the sphere $x^2 + y^2 + (z - 1)^2 = 1$.

Solution   We use Equations (3) to substitute for x, y, and z:

$$x^2 + y^2 + (z - 1)^2 = 1$$

$$\rho^2 \sin^2 \phi \cos^2 \theta + \rho^2 \sin^2 \phi \sin^2 \theta + (\rho \cos \phi - 1)^2 = 1 \qquad \text{Eqs. (3)}$$

$$\rho^2 \sin^2 \phi \underbrace{(\cos^2 \theta + \sin^2 \theta)}_{1} + \rho^2 \cos^2 \phi - 2\rho \cos \phi + 1 = 1$$

$$\rho^2 \underbrace{(\sin^2 \phi + \cos^2 \phi)}_{1} = 2\rho \cos \phi$$

$$\rho^2 = 2\rho \cos \phi$$

$$\rho = 2 \cos \phi.$$

See Figure 14.46.

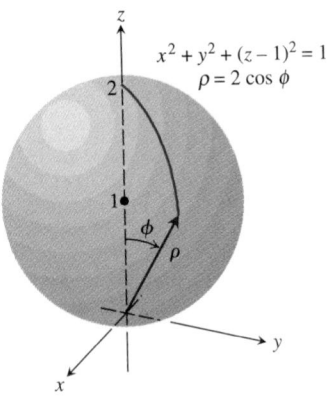

$x^2 + y^2 + (z - 1)^2 = 1$
$\rho = 2 \cos \phi$

ϕ

ρ

Figure 14.46 The sphere in Example 3.

Example 4   CONVERTING CARTESIAN TO SPHERICAL

Find a spherical coordinate equation for the cone $z = \sqrt{x^2 + y^2}$ (Figure 14.47).

Solution 1   Use geometry. The cone is symmetric with respect to the z-axis and cuts the first quadrant of the yz-plane along the line $z = y$. The angle between the cone and the positive z-axis is therefore $\pi/4$ radians. The cone consists of the points whose spherical coordinates have ϕ equal to $\pi/4$, so its equation is $\phi = \pi/4$.

Solution 2   Use algebra. If we use Equations (3) to substitute for x, y, and z, we obtain the same result:

$$z = \sqrt{x^2 + y^2}$$

$$\rho \cos \phi = \sqrt{\rho^2 \sin^2 \phi} \qquad \text{Example 3 details}$$

$$\rho \cos \phi = \rho \sin \phi \qquad \rho \geq 0, \ \sin \phi \geq 0$$

$$\cos \phi = \sin \phi$$

$$\phi = \frac{\pi}{4}. \qquad 0 \leq \phi \leq \pi$$

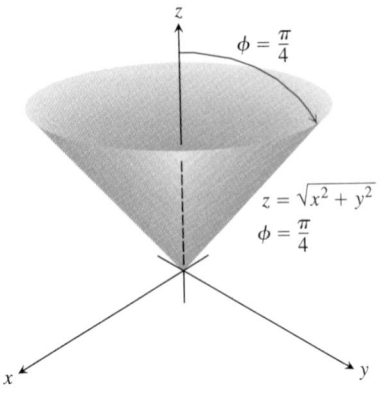

$\phi = \frac{\pi}{4}$

$z = \sqrt{x^2 + y^2}$
$\phi = \frac{\pi}{4}$

Figure 14.47 The cone in Example 4.

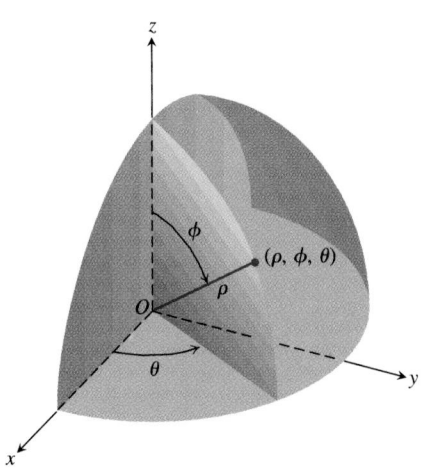

Figure 14.48 Spherical coordinates are measured with a distance and two angles.

Integration in Spherical Coordinates

Spherical coordinates (Figure 14.48) are good for describing spheres centered at the origin, half-planes hinged along the z-axis, and single-napped cones whose vertices lie at the origin and whose axes lie along the z-axis (see Example 4). Surfaces like these have equations of constant coordinate value:

$$\rho = 4, \qquad \text{Sphere, radius 4, center at the origin}$$

$$\phi = \frac{\pi}{3}, \qquad \text{Cone opening up from the origin, making an} \\ \text{angle of } \pi/3 \text{ radians with the positive } z\text{-axis}$$

$$\theta = \frac{\pi}{3}. \qquad \text{Half-plane, hinged along the } z\text{-axis, making an} \\ \text{angle of } \pi/3 \text{ radians with the positive } x\text{-axis}$$

The volume element in spherical coordinates is the volume of a **spherical wedge** defined by the differentials $d\rho$, $d\phi$, and $d\theta$ (Figure 14.49). The wedge is approximately a rectangular box with one side a circular arc of length $\rho\,d\phi$, another side a circular arc of length $\rho \sin\phi\, d\theta$, and thickness $d\rho$. Therefore, the volume element in spherical coordinates is

$$dV = \rho^2 \sin\phi\, d\rho\, d\phi\, d\theta, \tag{4}$$

and triple integrals take the form

$$\iiint F(\rho, \phi, \theta)\, dV = \iiint F(\rho, \phi, \theta)\rho^2 \sin\phi\, d\rho\, d\phi\, d\theta. \tag{5}$$

To evaluate these integrals, we usually integrate first with respect to ρ. The procedure for finding limits of integration follows. We restrict our attention to integrating over domains that are solids of revolution about the z-axis (or portions thereof) and for which the limits for θ and ϕ are constant.

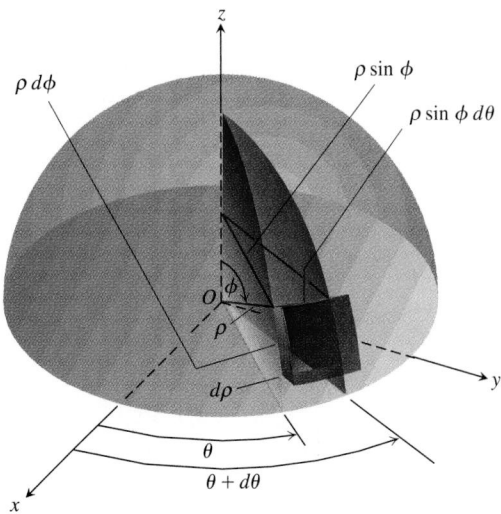

Figure 14.49 The volume element in spherical coordinates is

$$dV = d\rho \cdot \rho\, d\phi \cdot \rho \sin\phi\, d\theta$$
$$= \rho^2 \sin\phi\, d\rho\, d\phi\, d\theta.$$

How to Integrate in Spherical Coordinates

To evaluate

$$\iiint_D f(\rho, \phi, \theta)\, dV$$

over a region D in space in spherical coordinates, integrating first with respect to ρ, then with respect to ϕ, and finally with respect to θ, take the following steps.

1. *A sketch.* Sketch the region D along with its projection R on the xy-plane. Label the surfaces that bound D.

2. *The ρ-limits of integration.* Draw a ray M from the origin making an angle ϕ with the positive z-axis. Also draw the projection of M on the xy-plane (call the projection L). The ray L makes an angle θ with the positive x-axis. As ρ increases, M enters D at $\rho = g_1(\phi, \theta)$ and leaves at $\rho = g_2(\phi, \theta)$. These are the ρ-limits of integration.

3. *The ϕ-limits of integration.* For any given θ, the angle ϕ that M makes with the z-axis runs from $\phi = \phi_{min}$ to $\phi = \phi_{max}$. These are the ϕ-limits of integration.

4. *The θ-limits of integration.* The ray L sweeps over R as θ runs from α to β. These are the θ-limits of integration. The integral is

$$\iiint_D f(\rho, \phi, \theta)\, dV = \int_{\theta=\alpha}^{\theta=\beta} \int_{\phi=\phi_{min}}^{\phi=\phi_{max}} \int_{\rho=g_1(\phi,\theta)}^{\rho=g_2(\phi,\theta)} f(\rho, \phi, \theta)\, \rho^2 \sin\phi \, d\rho\, d\phi\, d\theta. \qquad (6)$$

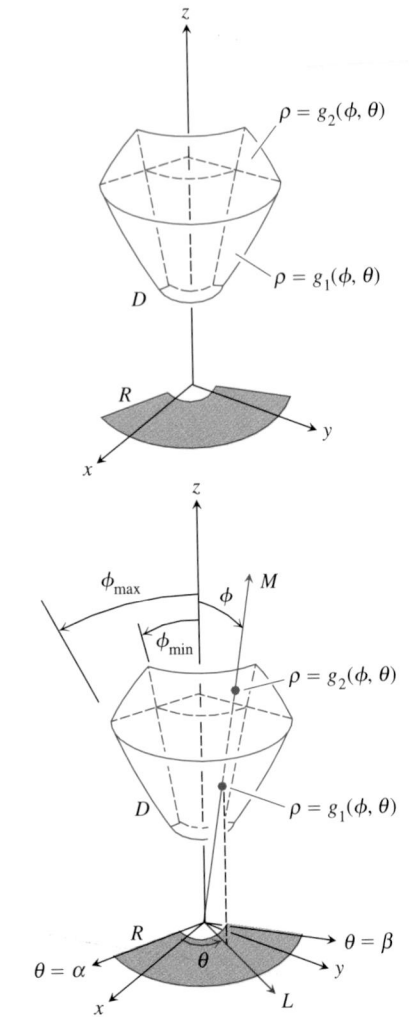

Example 5   FINDING VOLUME IN SPHERICAL COORDINATES

Find the volume of the upper region D cut from the solid sphere $\rho \le 1$ by the cone $\phi = \pi/3$.

Solution   The volume is $\iiint_D \rho^2 \sin\phi \, d\rho\, d\phi\, d\theta$, the integral of $f(\rho, \phi, \theta) = 1$ over D.

To find the limits of integration for evaluating the integral, we take the following steps.

Step 1: *A sketch.* We sketch D and its projection R on the xy-plane (Figure 14.50).

Step 2: *The ρ-limits of integration.* We draw a ray M from the origin making an angle ϕ with the positive z-axis. We also draw L, the projection of M on the xy-plane, along with the angle θ that L makes with the positive x-axis. Ray M enters D at $\rho = 0$ and leaves at $\rho = 1$.

Step 3: *The ϕ-limits of integration.* The cone $\phi = \pi/3$ makes an angle of $\pi/3$ with the positive z-axis. For any given θ, the angle ϕ can run from $\phi = 0$ to $\phi = \pi/3$.

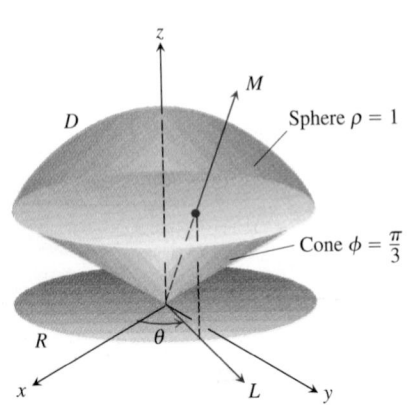

Figure 14.50 The solid in Example 5.

Step 4: *The θ-limits of integration.* The ray L sweeps over R as θ runs from 0 to 2π. The volume is

$$V = \iiint_D \rho^2 \sin\phi \, d\rho \, d\phi \, d\theta = \int_0^{2\pi}\int_0^{\pi/3}\int_0^1 \rho^2 \sin\phi \, d\rho \, d\phi \, d\theta$$

$$= \int_0^{2\pi}\int_0^{\pi/3}\left[\frac{\rho^3}{3}\right]_0^1 \sin\phi \, d\phi \, d\theta = \int_0^{2\pi}\int_0^{\pi/3}\frac{1}{3}\sin\phi \, d\phi \, d\theta$$

$$= \int_0^{2\pi}\left[-\frac{1}{3}\cos\phi\right]_0^{\pi/3} d\theta = \int_0^{2\pi}\left(-\frac{1}{6}+\frac{1}{3}\right) d\theta = \frac{\pi}{3} \text{ units cubed.}$$

Example 6 FINDING A MOMENT OF INERTIA

A solid of constant density $\delta = 1$ occupies the region D in Example 5. Find the solid's moment of inertia about the z-axis.

Solution In rectangular coordinates, the moment is

$$I_z = \iiint (x^2 + y^2) \, dV.$$

In spherical coordinates,

$$x^2 + y^2 = (\rho \sin\phi \cos\theta)^2 + (\rho \sin\phi \sin\theta)^2 = \rho^2 \sin^2\phi.$$

Hence,

$$I_z = \iiint (\rho^2 \sin^2\phi)\rho^2 \sin\phi \, d\rho \, d\phi \, d\theta = \iiint \rho^4 \sin^3\phi \, d\rho \, d\phi \, d\theta.$$

For the region in Example 5, this becomes

$$I_z = \int_0^{2\pi}\int_0^{\pi/3}\int_0^1 \rho^4\sin^3\phi \, d\rho \, d\phi \, d\theta = \int_0^{2\pi}\int_0^{\pi/3}\left[\frac{\rho^5}{5}\right]_0^1 \sin^3\phi \, d\phi \, d\theta$$

$$= \frac{1}{5}\int_0^{2\pi}\int_0^{\pi/3}(1-\cos^2\phi)\sin\phi \, d\phi \, d\theta = \frac{1}{5}\int_0^{2\pi}\left[-\cos\phi + \frac{\cos^3\phi}{3}\right]_0^{\pi/3} d\theta$$

$$= \frac{1}{5}\int_0^{2\pi}\left(-\frac{1}{2}+1+\frac{1}{24}-\frac{1}{3}\right) d\theta = \frac{1}{5}\int_0^{2\pi}\frac{5}{24}d\theta = \frac{\pi}{12}.$$

Coordinate Conversion Formulas

Cylindrical to Rectangular	Spherical to Rectangular	Spherical to Cylindrical
$x = r\cos\theta$	$x = \rho\sin\phi\cos\theta$	$r = \rho\sin\phi$
$y = r\sin\theta$	$y = \rho\sin\phi\sin\theta$	$z = \rho\cos\phi$
$z = z$	$z = \rho\cos\phi$	$\theta = \theta$

Corresponding volume elements

$$dV = dx \, dy \, dz$$
$$= dz \, r \, dr \, d\theta$$
$$= \rho^2 \sin\phi \, d\rho \, d\phi \, d\theta$$

Quick Review 14.6

1. Translate $z = \sqrt{x^2 + y^2}$, $1 \le z \le 2$, into cylindrical coordinates.

2. Translate $z^2 - r^2 = 1$ into rectangular coordinates.

In Exercises 3 and 4, describe the set of points whose cylindrical coordinates satisfy the given equation.

3. $r = 2 \sin \theta$ **4.** $r = 1 + \cos \theta$

In Exercises 5–7, find the trace of the surface $z = x^2 + y^2$ in the given plane.

5. xz-plane   **6.** yz-plane

7. xy-plane

In Exercises 8–10, find the trace of the surface $z = \sqrt{x^2 + y^2}$ in the given plane.

8. xz-plane   **9.** yz-plane

10. plane $z = 2$

Section 14.6 Exercises

In Exercises 1–6, the coordinates of a point in space are given in one of two coordinate systems. Find the coordinates of the point in the other system. There may be more than one answer because points in spherical coordinates can have more that one coordinate triple.

Rectangular (x, y, z)	Spherical (ρ, ϕ, θ)
1. $(0, 0, 0)$	
2. $(1, 0, 0)$	
3. $(0, 0, 1)$	
4.	$(\sqrt{3}, \pi/3, -\pi/2)$
5.	$(2\sqrt{2}, \pi/2, 3\pi/2)$
6.	$(\sqrt{2}, \pi, \pi)$

In Exercises 7–20, translate the equation or the equation and inequality from the given coordinate system (rectangular, cylindrical, spherical) into equations and inequalities in the other two systems. Then identify the figure in space defined by the equation or by the equation and inequality.

7. $r = 0$ **8.** $x^2 + y^2 = 5$

9. $z = 0$ **10.** $z = -2$

11. $z = \sqrt{x^2 + y^2}$, $z \le 1$ **12.** $\tan^2 \phi = 1$

13. $\rho \sin \phi \cos \theta = 0$ **14.** $x^2 + y^2 + z^2 = 4$

15. $\rho = 5 \cos \phi$ **16.** $\rho = 9 \csc \phi$

17. $x^2 + y^2 + (z - 1)^2 = 1$, $z \le 1$

18. $\rho = 3$, $\pi/3 \le \phi \le 2\pi/3$

19. $\phi = 3\pi/4$, $0 \le \rho \le \sqrt{2}$

20. $\phi = \pi/2$, $0 \le \rho \le \sqrt{7}$

21. Describe the set of points in space whose spherical coordinates satisfy the equation $\rho = 1 - \cos \phi$. Sketch the graph of the equation.

22. *Horizontal Planes in Cylindrical and Spherical Coordinates*

(a) Show that the plane whose equation is $z = c$ ($c \ne 0$) in rectangular and cylindrical coordinates has the equation $\rho = c \sec \phi$ in spherical coordinates.

(b) Find an equation for the xy-plane in spherical coordinates.

23. *Vertical Circular Cylinders in Spherical Coordinates* Find an equation of the form $\rho = f(\phi)$ for the cylinder $x^2 + y^2 = a^2$.

24. **Writing to Learn** *Symmetry* What symmetry will you find in a surface that has an equation of the form $\rho = f(\phi)$ in spherical coordinates? Give reasons for your answer.

In Exercises 25–28, evaluate the cylindrical coordinate integral.

25. $\displaystyle\int_0^{2\pi} \int_0^1 \int_r^{\sqrt{2-r^2}} dz\, r\, dr\, d\theta$

26. $\displaystyle\int_0^{\pi} \int_0^{\theta/\pi} \int_{-\sqrt{4-r^2}}^{3\sqrt{4-r^2}} z\, dz\, r\, dr\, d\theta$

27. $\displaystyle\int_0^{2\pi} \int_0^1 \int_r^{1/\sqrt{2-r^2}} 3\, dz\, r\, dr\, d\theta$

28. $\displaystyle\int_0^{2\pi} \int_0^1 \int_{-1/2}^{1/2} (r^2 \sin^2 \theta + z^2)\, dz\, r\, dr\, d\theta$

The integrals we have seen so far suggest that there are preferred orders of integration for cylindrical coordinates, but other orders usually work well and are occasionally easier to evaluate. In Exercises 29–32, evaluate the integral.

29. $\displaystyle\int_0^{2\pi}\int_0^3\int_0^{z/3} r^3\, dr\, dz\, d\theta$

30. $\displaystyle\int_{-1}^1\int_0^{2\pi}\int_0^{1+\cos\theta} 4r\, dr\, d\theta\, dz$

31. $\displaystyle\int_0^1\int_0^{\sqrt{z}}\int_0^{2\pi} (r^2\cos^2\theta + z^2)r\, d\theta\, dr\, dz$

32. $\displaystyle\int_0^2\int_{r-2}^{\sqrt{4-r^2}}\int_0^{2\pi} (r\sin\theta + 1)r\, d\theta\, dz\, dr$

In Exercises 33–36, *work in groups of two or three* to solve the problem.

33. Let D be the region bounded below by the plane $z = 0$, above by the sphere $x^2 + y^2 + z^2 = 4$, and on the sides by the cylinder $x^2 + y^2 = 1$. Set up the triple integrals in cylindrical coordinates that give the volume of D using the following orders of integration.

(a) $dz\, dr\, d\theta$ (b) $dr\, dz\, d\theta$

(c) $d\theta\, dz\, dr$

34. Let D be the region bounded below by the cone $z = \sqrt{x^2 + y^2}$ and above by the paraboloid $z = 2 - x^2 - y^2$. Set up the triple integrals in cylindrical coordinates that give the volume of D using the following orders of integration.

(a) $dz\, dr\, d\theta$ (b) $dr\, dz\, d\theta$

(c) $d\theta\, dz\, dr$

35. Give the limits of integration for evaluating the integral

$$\int\int\int f(r, \theta, z)\, dz\, r\, dr\, d\theta$$

as an iterated integral over the region that is bounded below by the plane $z = 0$, on the side by the cylinder $r = \cos\theta$, and on top by the paraboloid $z = 3r^2$.

36. Convert the integral

$$\int_{-1}^1\int_0^{\sqrt{1-y^2}}\int_0^x (x^2 + y^2)\, dz\, dx\, dy$$

to an equivalent integral in cylindrical coordinates and evaluate the result.

In Exercises 37–42, set up the iterated integral for evaluating $\displaystyle\int\int\int_D f(r, \theta, z)\, dz\, r\, dr\, d\theta$ over the given region D.

37. D is the right circular cylinder whose base is the circle $r = 2\sin\theta$ in the xy-plane and whose top lies in the plane $z = 4 - y$ as illustrated in the figure.

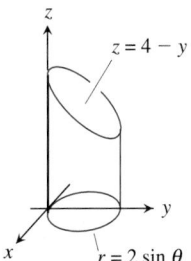

38. D is the right circular cylinder whose base is the circle $r = 3\cos\theta$ and whose top lies in the plane $z = 5 - x$ as illustrated in the figure.

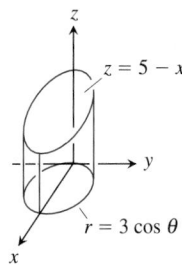

39. D is the solid right cylinder whose base is the region in the xy-plane that lies inside the cardioid $r = 1 + \cos\theta$ and outside the circle $r = 1$ and whose top lies in the plane $z = 4$ as illustrated in the figure.

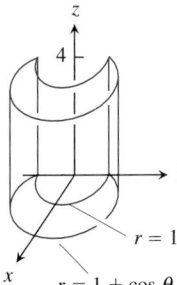

40. *D* is the solid right cylinder whose base is the region between the circles $r = \cos\theta$ and $r = 2\cos\theta$ and whose top lies in the plane $z = 3 - y$ as illustrated in the figure.

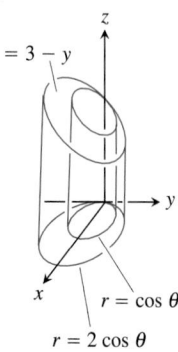

$z = 3 - y$

$r = \cos\theta$

$r = 2\cos\theta$

41. *D* is the prism whose base is the triangle in the *xy*-plane bounded by the *x*-axis and the lines $y = x$ and $x = 1$ and whose top lies in the plane $z = 2 - y$ as illustrated in the figure.

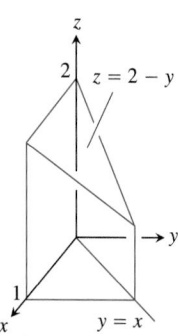

$z = 2 - y$

$y = x$

42. *D* is the prism whose base is the triangle in the *xy*-plane bounded by the *y*-axis and the lines $y = x$ and $y = 1$ and whose top lies in the plane $z = 2 - x$ as illustrated in the figure.

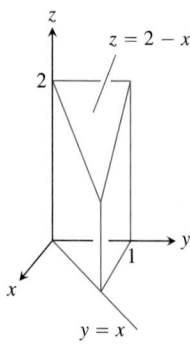

$z = 2 - x$

$y = x$

In Exercises 43–46, evaluate the spherical coordinate integral.

43. $\displaystyle\int_0^\pi \int_0^\pi \int_0^{2\sin\phi} \rho^2 \sin\phi \, d\rho \, d\phi \, d\theta$

44. $\displaystyle\int_0^{2\pi} \int_0^{\pi/4} \int_0^2 (\rho\cos\phi)\rho^2 \sin\phi \, d\rho \, d\phi \, d\theta$

45. $\displaystyle\int_0^{2\pi} \int_0^{\pi/3} \int_{\sec\phi}^2 3\rho^2 \sin\phi \, d\rho \, d\phi \, d\theta$

46. $\displaystyle\int_0^{2\pi} \int_0^{\pi/4} \int_0^{\sec\phi} (\rho\cos\phi)\rho^2 \sin\phi \, d\rho \, d\phi \, d\theta$

While $d\rho \, d\phi \, d\theta$ is the usual order of integration, other orders are possible and occasionally easier to evaluate. In Exercises 47–50, evaluate the integral.

47. $\displaystyle\int_0^2 \int_{-\pi}^0 \int_{\pi/4}^{\pi/2} \rho^3 \sin 2\phi \, d\phi \, d\theta \, d\rho$

48. $\displaystyle\int_{\pi/6}^{\pi/3} \int_{\csc\phi}^{2\csc\phi} \int_0^{2\pi} \rho^2 \sin\phi \, d\theta \, d\rho \, d\phi$

49. $\displaystyle\int_0^1 \int_0^\pi \int_0^{\pi/4} 12\rho \sin^3\phi \, d\phi \, d\theta \, d\rho$

50. $\displaystyle\int_{\pi/6}^{\pi/2} \int_{-\pi/2}^{\pi/2} \int_{\csc\phi}^2 5\rho^4 \sin^3\phi \, d\rho \, d\theta \, d\phi$

51. Let *D* be the region in Exercise 33. Set up the triple integrals in spherical coordinates that give the volume of *D* using the following orders of integration.

(a) $d\rho \, d\phi \, d\theta$ (b) $d\phi \, d\rho \, d\theta$

52. Let *D* be the region bounded below by the cone $z = \sqrt{x^2 + y^2}$ and above by the plane $z = 1$. Set up the triple integrals in spherical coordinates that give the volume of *D* using the following orders of integration.

(a) $d\rho \, d\phi \, d\theta$ (b) $d\phi \, d\rho \, d\theta$

In Exercises 53–58, (a) find the spherical coordinate limits for the integral that calculates the volume of the given solid and (b) then evaluate the integral.

53. The solid between the sphere $\rho = \cos\phi$ and the hemisphere $\rho = 2$, $z \geq 0$, as illustrated in the figure.

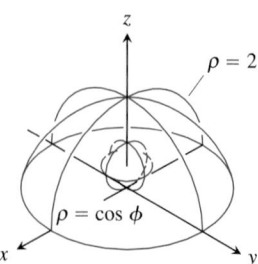

$\rho = 2$

$\rho = \cos\phi$

54. The solid bounded below by the hemisphere $\rho = 1$, $z \geq 0$, and above by the cardioid of revolution $\rho = 1 + \cos \phi$ as illustrated in the figure.

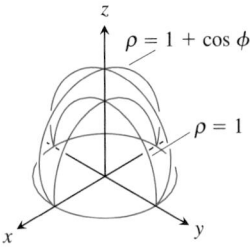

55. The solid enclosed by the cardioid of revolution $\rho = 1 - \cos \phi$.

56. The upper portion cut from the solid in Exercise 55 by the *xy*-plane.

57. The solid bounded below by the sphere $\rho = 2 \cos \phi$ and above by the cone $z = \sqrt{x^2 + y^2}$ as illustrated in the figure.

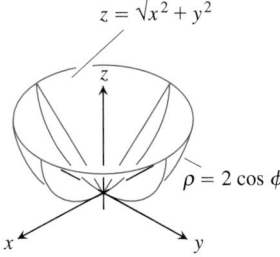

58. The solid bounded below by the *xy*-plane, on the sides by the sphere $\rho = 2$, and above by the cone $\phi = \pi/3$ as illustrated in the figure.

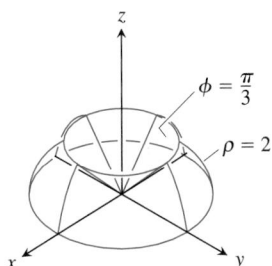

Explorations

59. Set up triple integrals for the volume of the sphere $\rho = 2$ in **(a)** spherical, **(b)** cylindrical, and **(c)** rectangular coordinates.

60. Let D be the region in the first octant that is bounded below by the cone $\phi = \pi/4$ and above by the sphere $\rho = 3$. Express the volume of D as an iterated triple integral in **(a)** cylindrical and **(b)** spherical coordinates. Then **(c)** find V.

61. Let D be the smaller cap cut from a solid sphere of radius 2 units by a plane 1 unit from the center of the sphere. Express the volume of D as an iterated triple integral in **(a)** spherical, **(b)** cylindrical, and **(c)** rectangular coordinates. Then **(d)** find the volume by evaluating one of the three triple integrals.

62. Express the moment of inertia I_z of the solid hemisphere $x^2 + y^2 + z^2 \leq 1$, $z \geq 0$, as an iterated integral in **(a)** cylindrical and **(b)** spherical coordinates. Then **(c)** find I_z. ∎

In Exercises 63–68, find the volume of the solid shown in the figure.

63.

64.

65.

66.

67.

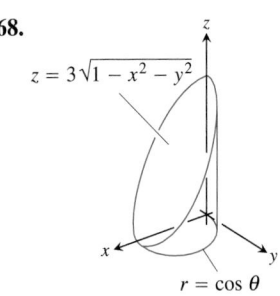

68.

In Exercises 69–76, find the volume of the region.

69. The portion of the solid sphere $\rho \leq a$ that lies between the cones $\phi = \pi/3$ and $\phi = 2\pi/3$

70. The smaller region cut from the solid sphere $\rho \leq 2$ by the plane $z = 1$

71. The solid enclosed by the cone $z = \sqrt{x^2 + y^2}$ between the planes $z = 1$ and $z = 2$

72. The region bounded below by the plane $z = 0$, laterally by the cylinder $x^2 + y^2 = 1$, and above by the paraboloid $z = x^2 + y^2$

73. The solid cut from the thick-walled cylinder $1 \leq x^2 + y^2 \leq 2$ by the cones $z = \pm\sqrt{x^2 + y^2}$

74. The region that lies inside the sphere $x^2 + y^2 + z^2 = 2$ and outside the cylinder $x^2 + y^2 = 1$

75. The region enclosed by the cylinder $x^2 + y^2 = 4$ and the planes $z = 0$ and $y + z = 4$

76. The region cut from the solid cylinder $x^2 + y^2 \leq 1$ by the sphere $x^2 + y^2 + z^2 = 4$

77. *Average Value* Find the average value of the function $f(r, \theta, z) = r$ over the region bounded by the cylinder $r = 1$ between the planes $z = -1$ and $z = 1$.

78. *Average Value* Find the average value of the function $f(\rho, \phi, \theta) = \rho$ over the solid ball $\rho \leq 1$.

79. *Center of Mass* A solid of constant density is bounded below by the plane $z = 0$, above by the cone $z = r, r \geq 0$, and on the sides by the cylinder $r = 1$. Find the center of mass.

80. *Centroid* Find the centroid of the region in the first octant that is bounded above by the cone $z = \sqrt{x^2 + y^2}$, below by the plane $z = 0$, and on the sides by the cylinder $x^2 + y^2 = 4$ and the planes $x = 0$ and $y = 0$.

81. *Centroid* Find the centroid of the solid in Exercise 58.

82. *Centroid* Find the centroid of the solid bounded above by the sphere $\rho = a$ and below by the cone $\phi = \pi/4$.

83. *Centroid* Find the centroid of the region that is bounded above by the surface $z = \sqrt{r}$, on the sides by the cylinder $r = 4$, and below by the xy-plane.

84. *Thick-Walled Cylinder* Find the moment of inertia and radius of gyration about the z-axis of a thick-walled right circular cylinder bounded on the inside by the cylinder $r = 1$, on the outside by the cylinder $r = 2$, and on the top and bottom by the planes $z = 4$ and $z = 0$. (Take $\delta = 1$.)

85. *Solid Sphere, Constant Density* Find the moment of inertia of a solid sphere of radius a about a diameter. (Take $\delta = 1$.)

86. *Solid Sphere, Variable Densities* A solid sphere is bounded by the sphere $\rho = a$. Find the moment of inertia and radius of gyration about the z-axis if the density is

(a) $\delta(\rho, \phi, \theta) = \rho^2$

(b) $\delta(\rho, \phi, \theta) = r = \rho \sin \phi$.

Extending the Ideas

87. *Atmospheric Mass* A spherical planet of radius R has an atmosphere whose density is $\mu = \mu_0 e^{-ch}$, where h is the altitude above the surface of the planet, μ_0 is the density at sea level, and c is a positive constant. Find the mass of the planet's atmosphere.

88. *Density at the Center of a Planet* A planet is in the shape of a sphere of radius R and total mass M with spherically symmetric density distribution that increases linearly as one approaches its center. What is the density at the center of this planet if the density at its edge (surface) is taken to be zero?

14.7 # Substitutions in Multiple Integrals

Substitutions in Double Integrals • Substitutions in Triple Integrals

Substitutions in Double Integrals

The polar coordinate substitution of Section 14.3 is a special case of a more general substitution method for double integrals, a method that pictures changes in variables as transformations of regions. As always, the goal of substitution is the transformation of difficult or unfamiliar integrals into integrals we can recognize or handle more easily.

Suppose a region G in the uv-plane is transformed one-to-one into the region R in the xy-plane by equations of the form

$$x = g(u, v), \qquad y = h(u, v),$$

as suggested in Figure 14.51. R is the **image** of G under the transformation, and G is the **preimage** of R. Any function $f(x, y)$ defined on R can be thought of as a function $f(g(u, v), h(u, v))$ defined on G as well. How is the integral of $f(x, y)$ over R related to the integral of $f(g(u, v), h(u, v))$ over G?

The answer is: If g, h, and f have continuous partial derivatives and $J(u, v)$ (to be discussed in a moment) is zero only at isolated points, if at all, then

Notice the "Reversed" Order

The transforming equations $x = g(u, v)$ and $y = h(u, v)$ go from G to R, but we use them to change an integral over R into an integral over G.

$$\iint_R f(x, y) \, dx \, dy = \iint_G f(g(u, v), h(u, v))|J(u, v)| \, du \, dv. \qquad (1)$$

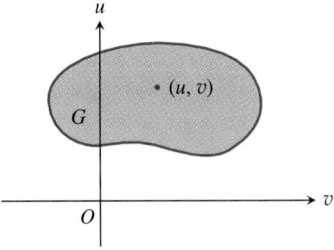

Cartesian-plane

$$x = g(u, v)$$
$$y = h(u, v)$$

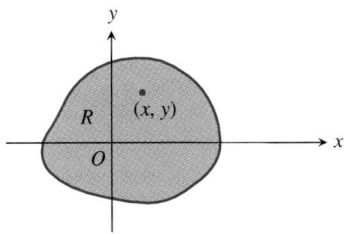

Cartesian xy-plane

Figure 14.51 The equations $x = g(u, v)$ and $y = h(u, v)$ allow us to change an integral over a region R in the xy-plane into an integral over a region G in the uv-plane.

The factor $J(u, v)$, whose absolute value appears in Equation (1), is the *Jacobian* of the coordinate transformation, named after the mathematician Carl Jacobi.

Definition Jacobian Determinant or Jacobian

The **Jacobian determinant** or **Jacobian** of the coordinate transformation $x = g(u, v), y = h(u, v)$ is

$$J(u, v) = \begin{vmatrix} \dfrac{\partial x}{\partial u} & \dfrac{\partial x}{\partial v} \\ \dfrac{\partial y}{\partial u} & \dfrac{\partial y}{\partial v} \end{vmatrix} = \frac{\partial x}{\partial u}\frac{\partial y}{\partial v} - \frac{\partial y}{\partial u}\frac{\partial x}{\partial v}. \qquad (2)$$

The Jacobian is also denoted by

$$J(u, v) = \frac{\partial(x, y)}{\partial(u, v)}$$

to help remember how the determinant in Equation (2) is constructed from the partial derivatives of x and y. The derivation of Equation (1) is intricate and properly belongs to a course in advanced calculus. We will not give the derivation here.

For polar coordinates, we have r and θ in place of u and v. With $x = r \cos \theta$ and $y = r \sin \theta$, the Jacobian is

$$J(r, \theta) = \begin{vmatrix} \dfrac{\partial x}{\partial r} & \dfrac{\partial x}{\partial \theta} \\ \dfrac{\partial y}{\partial r} & \dfrac{\partial y}{\partial \theta} \end{vmatrix} = \begin{vmatrix} \cos \theta & -r \sin \theta \\ \sin \theta & r \cos \theta \end{vmatrix} = r(\cos^2 \theta + \sin^2 \theta) = r.$$

Hence, Equation (1) becomes

$$\iint\limits_{R} f(x, y)\, dx\, dy = \iint\limits_{G} f(r \cos \theta, r \sin \theta)|r|\, dr\, d\theta$$

$$= \iint\limits_{G} f(r \cos \theta, r \sin \theta) r\, dr\, d\theta, \qquad \text{If } r \geq 0 \ (3)$$

which is Equation (6) in Section 14.3.

Figure 14.52 shows how the equations $x = r \cos \theta, y = r \sin \theta$ transform the rectangle $G: 0 \leq r \leq 1, 0 \leq \theta \leq \pi/2$ into the quarter circle R bounded by $x^2 + y^2 = 1$ in the first quadrant of the xy-plane.

Notice that the integral on the right-hand side of Equation (3) is not the integral of $f(r \cos \theta, r \sin \theta)$ over a region in the polar coordinate plane. It is the integral of the product of $f(r \cos \theta, r \sin \theta)$ and r over a region G in the *Cartesian $r\theta$-plane*.

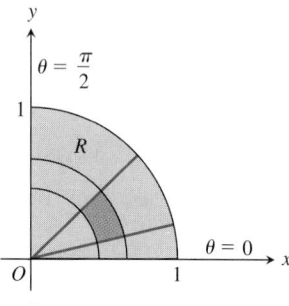

Figure 14.52 The equations $x = r \cos \theta$, $y = r \sin \theta$ transform G into R.

Here is an example of another substitution.

Example 1 APPLYING TRANSFORMATIONS TO INTEGRATE

Evaluate

$$\int_0^4 \int_{x=y/2}^{x=(y/2)+1} \frac{2x - y}{2} \, dx \, dy$$

by applying the transformation

$$u = \frac{2x - y}{2}, \qquad v = \frac{y}{2} \qquad (4)$$

and integrating over an appropriate region in the uv-plane.

Solution   We sketch the region R of integration in the xy-plane and identify its boundaries (Figure 14.53).

To apply Equation (1), we need to find the corresponding uv-region G and the Jacobian of the transformation. To find them, we first solve Equations (4) for x and y in terms of u and v. Routine algebra gives

$$x = u + v, \qquad y = 2v. \qquad (5)$$

We then find the boundaries of G by substituting these expressions into the equations for the boundaries of R (Figure 14.53).

xy-equations for the boundary of R	Corresponding uv-equations for the boundary of G	Simplified uv-equations
$x = y/2$	$u + v = 2v/2 = v$	$u = 0$
$x = (y/2) + 1$	$u + v = (2v/2) + 1 = v + 1$	$u = 1$
$y = 0$	$2v = 0$	$v = 0$
$y = 4$	$2v = 4$	$v = 2$

The Jacobian of the transformation (again from Equations (5)) is

$$J(u, v) = \begin{vmatrix} \dfrac{\partial x}{\partial u} & \dfrac{\partial x}{\partial v} \\ \dfrac{\partial y}{\partial u} & \dfrac{\partial y}{\partial v} \end{vmatrix} = \begin{vmatrix} \dfrac{\partial}{\partial u}(u + v) & \dfrac{\partial}{\partial v}(u + v) \\ \dfrac{\partial}{\partial u}(2v) & \dfrac{\partial}{\partial v}(2v) \end{vmatrix} = \begin{vmatrix} 1 & 1 \\ 0 & 2 \end{vmatrix} = 2.$$

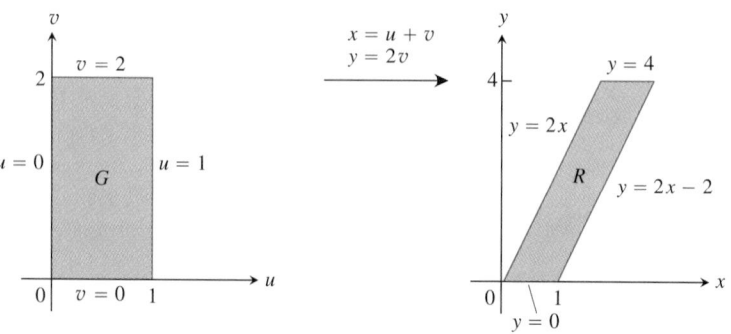

Figure 14.53 The equations $x = u + v$ and $y = 2v$ transform G into R. Reversing the transformation by the equations $u = (2x - y)/2$ and $v = y/2$ transforms R into G. (Example 1)

We now have everything we need to apply Equation (1):

$$\int_0^4 \int_{x=y/2}^{x=(y/2)+1} \frac{2x - y}{2} \, dx \, dy$$

$$= \int_{v=0}^{v=2} \int_{u=0}^{u=1} u \left| J(u, v) \right| \, du \, dv$$

$$= \int_0^2 \int_0^1 (u)(2) \, du \, dv = \int_0^2 \left[u^2 \right]_0^1 \, dv = \int_0^2 \, dv = 2.$$

Example 2 APPLYING TRANSFORMATIONS TO INTEGRATE

Evaluate $\displaystyle\int_0^1 \int_0^{1-x} \sqrt{x + y}\,(y - 2x)^2 \, dy \, dx.$

Solution We sketch the region R of integration in the xy-plane and identify its boundaries (Figure 14.54). The integrand suggests the transformation $u = x + y$ and $v = y - 2x$. Routine algebra produces x and y as functions of u and v:

$$x = \frac{u}{3} - \frac{v}{3}, \qquad y = \frac{2u}{3} + \frac{v}{3}. \tag{6}$$

From Equations (6) we can find the boundaries of the uv-region G (Figure 14.54).

xy-equations for the boundary of R	Corresponding uv-equations for the boundary of G	Simplified uv-equations
$x + y = 1$	$\left(\dfrac{u}{3} - \dfrac{v}{3}\right) + \left(\dfrac{2u}{3} + \dfrac{v}{3}\right) = 1$	$u = 1$
$x = 0$	$\dfrac{u}{3} - \dfrac{v}{3} = 0$	$v = u$
$y = 0$	$\dfrac{2u}{3} + \dfrac{v}{3} = 0$	$v = -2u$

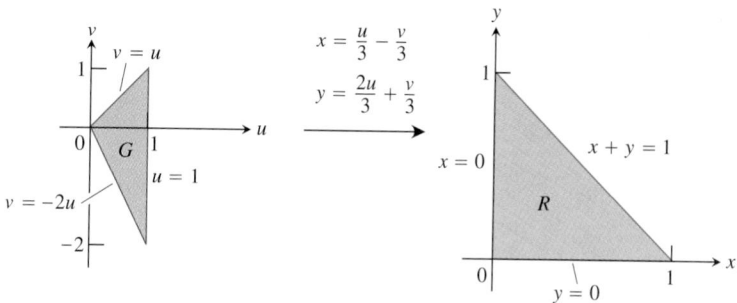

Figure 14.54 The equations $x = (u/3) - (v/3)$ and $y = (2u/3) + (v/3)$ transform G into R. Reversing the transformations by the equations $u = x + y$ and $v = y - 2x$ transforms R into G. (Example 2)

The Jacobian of the transformation in Equation (6) is

$$J(u, v) = \begin{vmatrix} \dfrac{\partial x}{\partial u} & \dfrac{\partial x}{\partial v} \\[2mm] \dfrac{\partial y}{\partial u} & \dfrac{\partial y}{\partial v} \end{vmatrix} = \begin{vmatrix} \dfrac{1}{3} & -\dfrac{1}{3} \\[2mm] \dfrac{2}{3} & \dfrac{1}{3} \end{vmatrix} = \dfrac{1}{3}.$$

Applying Equation (1), we evaluate the integral:

$$\int_0^1 \int_0^{1-x} \sqrt{x + y}\,(y - 2x)^2 \, dy \, dx = \int_{u=0}^{u=1} \int_{v=-2u}^{v=u} u^{1/2} v^2 |J(u, v)| \, dv \, du$$

$$= \int_0^1 \int_{-2u}^u u^{1/2} v^2 \left(\frac{1}{3}\right) dv \, du$$

$$= \frac{1}{3} \int_0^1 u^{1/2} \left[\frac{v^3}{3}\right]_{v=-2u}^{v=u} du$$

$$= \frac{1}{9} \int_0^1 u^{1/2} (u^3 + 8u^3) \, du$$

$$= \int_0^1 u^{7/2} \, du = \frac{2}{9} u^{9/2} \Big|_0^1 = \frac{2}{9}.$$

Substitutions in Triple Integrals

The cylindrical and spherical coordinate substitutions in Section 14.6 are special cases of a substitution method that pictures changes of variables in triple integrals as transformations of three-dimensional regions. The method is like the method for double integrals except that now we work in three dimensions instead of two.

Suppose a region G in uvw-space is transformed one-to-one into the region D in xyz-space by differentiable equations of the form

$$x = g(u, v, w), \qquad y = h(u, v, w), \qquad z = k(u, v, w),$$

as suggested in Figure 14.55. Then any function $F(x, y, z)$ defined on D can be thought of as a function

$$F(g(u, v, w), h(u, v, w), k(u, v\ w)) = H(u, v, w)$$

defined on G. If g, h, and k have continuous first partial derivatives, then the integral of $F(x, y, z)$ over D is related to the integral of $H(u, v\ w)$ over G by the equation

$$\iiint\limits_D F(x, y, z) \, dx \, dy \, dz = \iiint\limits_G H(u, v, w)|J(u, v, w)| \, du \, dv \, dw. \qquad (7)$$

The factor $J(u, v\ w)$, whose absolute value appears in this equation, is the **Jacobian determinant**

$$J(u, v, w) = \begin{vmatrix} \dfrac{\partial x}{\partial u} & \dfrac{\partial x}{\partial v} & \dfrac{\partial x}{\partial w} \\[2mm] \dfrac{\partial y}{\partial u} & \dfrac{\partial y}{\partial v} & \dfrac{\partial y}{\partial w} \\[2mm] \dfrac{\partial z}{\partial u} & \dfrac{\partial z}{\partial v} & \dfrac{\partial z}{\partial w} \end{vmatrix} = \dfrac{\partial(x, y, z)}{\partial(u, v, w)}. \qquad (8)$$

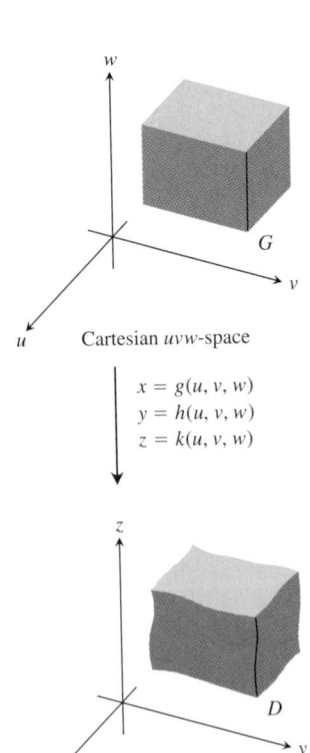

Cartesian uvw-space

$$x = g(u, v, w)$$
$$y = h(u, v, w)$$
$$z = k(u, v, w)$$

Cartesian xyz-space

Figure 14.55 The equations $x = g(u, v, w)$, $y = h(u, v, w)$, $z = k(u, v, w)$ allow us to change an integral over a region D in Cartesian xyz-space into an integral over a region G in Cartesian uvw-space.

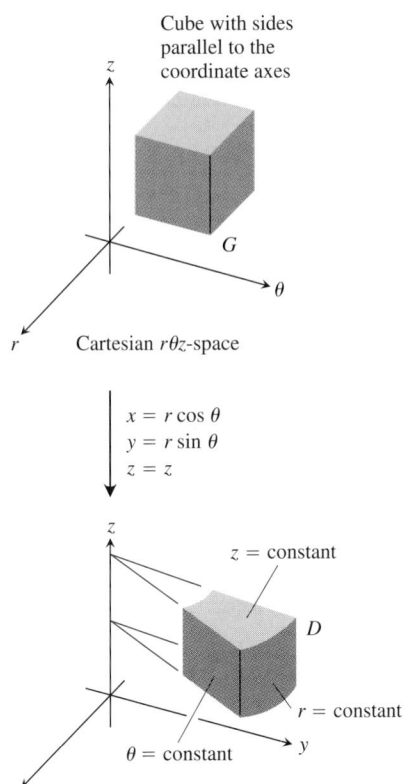

Cube with sides parallel to the coordinate axes

G

Cartesian $r\theta z$-space

$x = r \cos \theta$
$y = r \sin \theta$
$z = z$

$z = $ constant

D

$r = $ constant

$\theta = $ constant

Cartesian xyz-space

Figure 14.56 The equations $x = r \cos \theta$, $y = r \sin \theta$, and $z = z$ transform G into D.

As in the two-dimensional case, the derivation of the change-of-variable formula in Equation (7) is complicated, and we will not go into it here.

For cylindrical coordinates, we have r, θ, and z in place of u, v, and w. The transformation from Cartesian $r\theta z$-space to Cartesian xyz-space is given by the equations

$$x = r \cos \theta, \qquad y = r \sin \theta, \qquad z = z$$

(Figure 14.56). The Jacobian of the transformation is

$$J(r, \theta, z) = \begin{vmatrix} \dfrac{\partial x}{\partial r} & \dfrac{\partial x}{\partial \theta} & \dfrac{\partial x}{\partial z} \\[2mm] \dfrac{\partial y}{\partial r} & \dfrac{\partial y}{\partial \theta} & \dfrac{\partial y}{\partial z} \\[2mm] \dfrac{\partial z}{\partial r} & \dfrac{\partial z}{\partial \theta} & \dfrac{\partial z}{\partial z} \end{vmatrix} = \begin{vmatrix} \cos \theta & -r \sin \theta & 0 \\ \sin \theta & r \cos \theta & 0 \\ 0 & 0 & 1 \end{vmatrix}$$

$$= r \cos^2 \theta + r \sin^2 \theta = r.$$

The corresponding version of Equation (7) is

$$\iiint_D F(x, y, z) \, dx \, dy \, dz = \iiint_G H(r, \theta, z)|r| \, dr \, d\theta \, dz. \tag{9}$$

We can drop the absolute value signs whenever $r \geq 0$.

For spherical coordinates, ρ, ϕ, and θ take the place of u, v, and w. The transformation from Cartesian $\rho\phi\theta$-space to Cartesian xyz-space is given by

$$x = \rho \sin \phi \cos \theta, \qquad y = \rho \sin \phi \sin \theta, \qquad z = \rho \cos \phi$$

(Figure 14.57). The Jacobian of the transformation is

$$J(\rho, \phi, \theta) = \begin{vmatrix} \dfrac{\partial x}{\partial \rho} & \dfrac{\partial x}{\partial \phi} & \dfrac{\partial x}{\partial \theta} \\[2mm] \dfrac{\partial y}{\partial \rho} & \dfrac{\partial y}{\partial \phi} & \dfrac{\partial y}{\partial \theta} \\[2mm] \dfrac{\partial z}{\partial \rho} & \dfrac{\partial z}{\partial \phi} & \dfrac{\partial z}{\partial \theta} \end{vmatrix} = \rho^2 \sin \phi \tag{10}$$

(Exercise 17). The corresponding version of Equation (7) is

$$\iiint_D F(x, y, z) \, dx \, dy \, dz = \iiint_G H(\rho, \phi, \theta)\left|\rho^2 \sin \phi\right| \, d\rho \, d\phi \, d\theta. \tag{11}$$

We can drop the absolute value signs because $\sin \phi$ is never negative.

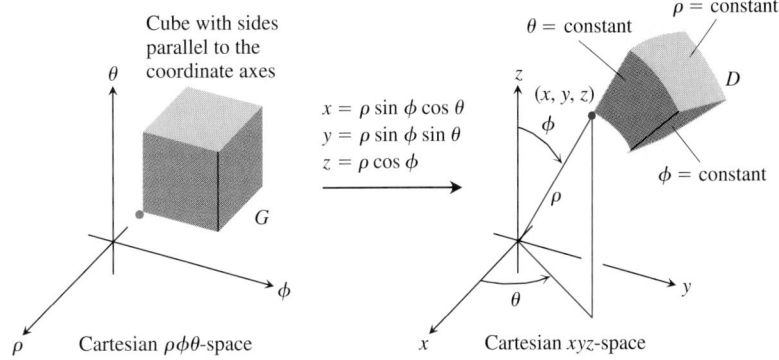

Cube with sides parallel to the coordinate axes

θ

$x = \rho \sin \phi \cos \theta$
$y = \rho \sin \phi \sin \theta$
$z = \rho \cos \phi$

G

ϕ

ρ Cartesian $\rho\phi\theta$-space

$\rho = $ constant

$\theta = $ constant

D

(x, y, z)

ϕ

$\phi = $ constant

ρ

θ

y

x Cartesian xyz-space

Figure 14.57 The equations $x = \rho \sin \phi \cos \theta$, $y = \rho \sin \phi \sin \theta$, and $z = \rho \cos \phi$ transform G into D.

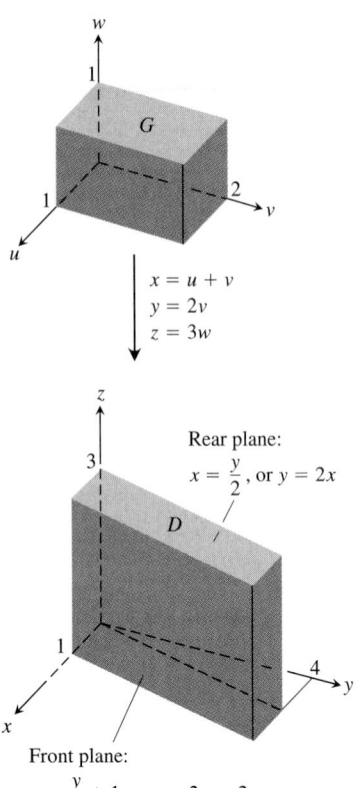

Figure 14.58 The equations $x = u + v$, $y = 2v$, and $z = 3w$ transform G into D. Reversing the transformation by the equations $u = (2x - y)/2$, $v = y/2$, and $w = z/3$ transforms D into G. (Example 3)

Carl Gustav Jacob Jacobi

Jacobi (1804–1851), one of nineteenth-century Germany's most accomplished scientists, developed the theory of determinants and transformations into a powerful tool for evaluating multiple integrals and solving differential equations. He also applied transformation methods to study nonelementary integrals like the ones that arise in the calculation of arc length. Like Euler, Jacobi was a prolific writer and an even more prolific calculator and worked in a variety of mathematical and applied fields.

Here is an example of another substitution.

Example 3   APPLYING TRANSFORMATIONS TO INTEGRATE

Evaluate

$$\int_0^3 \int_0^4 \int_{x=y/2}^{x=(y/2)+1} \left(\frac{2x - y}{2} + \frac{z}{3} \right) dx\, dy\, dz$$

by applying the transformation

$$u = (2x - y)/2, \qquad v = y/2, \qquad w = z/3 \tag{12}$$

and integrating over an appropriate region in uvw-space.

Solution   We sketch the region D of integration in xyz-space and identify its boundaries (Figure 14.58). In this case, the bounding surfaces are planes.

   To apply Equation (7), we need to find the corresponding uvw-region G and the Jacobian of the transformation. To find them, we first solve Equations (12) for x, y, and z in terms of u, v, and w. Routine algebra gives

$$x = u + v, \qquad y = 2v, \qquad z = 3w. \tag{13}$$

We then find the boundaries of G by substituting these expressions into the equations for the boundaries of D:

xyz-equations for the boundary of D	Corresponding uvw-equations for the boundary of G	Simplified uvw-equations
$x = y/2$	$u + v = 2v/2 = v$	$u = 0$
$x = (y/2) + 1$	$u + v = (2v/2) + 1 = v + 1$	$u = 1$
$y = 0$	$2v = 0$	$v = 0$
$y = 4$	$2v = 4$	$v = 2$
$z = 0$	$3w = 0$	$w = 0$
$z = 3$	$3w = 3$	$w = 1$

The Jacobian of the transformation, again from Equations (13), is

$$J(u, v, w) = \begin{vmatrix} \dfrac{\partial x}{\partial u} & \dfrac{\partial x}{\partial v} & \dfrac{\partial x}{\partial w} \\[2mm] \dfrac{\partial y}{\partial u} & \dfrac{\partial y}{\partial v} & \dfrac{\partial y}{\partial w} \\[2mm] \dfrac{\partial z}{\partial u} & \dfrac{\partial z}{\partial v} & \dfrac{\partial z}{\partial w} \end{vmatrix} = \begin{vmatrix} 1 & 1 & 0 \\ 0 & 2 & 0 \\ 0 & 0 & 3 \end{vmatrix} = 6.$$

We now have everything we need to apply Equation (7):

$$\int_0^3 \int_0^4 \int_{x=y/2}^{x=(y/2)+1} \left(\frac{2x - y}{2} + \frac{z}{3} \right) dx\, dy\, dz$$

$$= \int_0^1 \int_0^2 \int_0^1 (u + w)|J(u, v, w)|\, du\, dv\, dw$$

$$= \int_0^1 \int_0^2 \int_0^1 (u + w)(6)\, du\, dv\, dw = 6 \int_0^1 \int_0^2 \left[\frac{u^2}{2} + uw \right]_0^1 dv\, dw$$

$$= 6 \int_0^1 \int_0^2 \left(\frac{1}{2} + w\right) dv \, dw = 6 \int_0^1 \left[\frac{v}{2} + vw\right]_0^2 dw = 6 \int_0^1 (1 + 2w) \, dw$$

$$= 6[w + w^2]_0^1 = 12.$$

Quick Review 14.7

In Exercises 1 and 2, find the value of the determinant.

1. $\begin{vmatrix} 2 & 5 \\ -1 & 6 \end{vmatrix}$

2. $\begin{vmatrix} -3 & 4 & 7 \\ 2 & 0 & 3 \\ 4 & -1 & 2 \end{vmatrix}$

3. Solve for x and y:

$$2x - 3y = 7$$
$$x + 2y = 0$$

4. Solve for x, y, and z:

$$3x - y + 2z = -1$$
$$x + 2y - 3z = -9$$
$$2x + 3y - 4z = -13$$

In Exercises 5–7, solve for x and y in terms of u and v.

5. $3x + 2y = u$

$$x - y = v$$

6. $2x - y = 2u$

$$y = 2v$$

7. $x + y = u$

$$y - 2x = v$$

In Exercises 8 and 9, solve for x, y, and z in terms of u, v, and w.

8. $3x - 2y + z = u + 1$

$$2x + y - z = v - 1$$
$$x + 2y - 3z = w$$

9. $2x - y = 2u$

$$y = 2v$$
$$z = 3w$$

10. Let $x = u \cos(vw)$ and find

$$\frac{\partial x}{\partial u}, \frac{\partial x}{\partial v}, \frac{\partial x}{\partial w}.$$

Section 14.7 Exercises

In Exercises 1–4, *work in groups of two or three* to solve the problem.

1. **(a)** Solve the system

$$u = x - y, \quad v = 2x + y$$

for x and y in terms of u and v. Then find the value of the Jacobian $\partial(x, y)/\partial(u, v)$.

(b) Find the image under the transformation $u = x - y$, $v = 2x + y$ of the triangular region with vertices $(0, 0)$, $(1, 1)$, and $(1, -2)$ in the xy-plane. Sketch the transformed region in the uv-plane.

2. **(a)** Solve the system

$$u = x + 2y, \quad v = x - y$$

for x and y in terms of u and v. Then find the value of the Jacobian $\partial(x, y)/\partial(u, v)$.

(b) Find the image under the transformation $u = x + 2y$, $v = x - y$ of the triangular region in the xy-plane bounded by the lines $y = 0$, $y = x$, and $x + 2y = 2$. Sketch the transformed region in the uv-plane.

3. **(a)** Solve the system

$$u = 3x + 2y, \quad v = x + 4y$$

for x and y in terms of u and v. Then find the value of the Jacobian $\partial(x, y)/\partial(u, v)$.

(b) Find the image under the transformation $u = 3x + 2y$, $v = x + 4y$ of the triangular region in the xy-plane bounded by the x-axis, the y-axis, and the line $x + y = 1$. Sketch the transformed region in the uv-plane.

4. **(a)** Solve the system

$$u = 2x - 3y, \quad v = -x + y$$

for x and y in terms of u and v. Then find the value of the Jacobian $\partial(x, y)/\partial(u, v)$.

(b) Find the image under the transformation $u = 2x - 3y$, $v = -x + y$ of the parallelogram R in the xy-plane with boundaries $x = -3$, $x = 0$, $y = x$, and $y = x + 1$. Sketch the transformed region in the uv-plane.

5. Find the Jacobian $\partial(x, y)/\partial(u, v)$ for the transformation

(a) $x = u \cos v$, $\quad y = u \sin v$.

(b) $x = u \sin v$, $\quad y = u \cos v$.

6. Find the Jacobian $\partial(x, y, z)/\partial(u, v, w)$ for the transformation

(a) $x = u \cos v$, $\quad y = u \sin v$, $\quad z = w$.

(b) $x = 2u - 1$, $\quad y = 3v - 4$, $\quad z = \frac{1}{2}(w - 4)$.

7. Evaluate the integral

$$\int_0^4 \int_{x=y/2}^{x=(y/2)+1} \frac{2x - y}{2} \, dx \, dy$$

from Example 1 directly by integration with respect to x and y to confirm that its value is 2.

8. Use the transformation in Exercise 1 to evaluate the integral

$$\iint_R (2x^2 - xy - y^2) \, dx \, dy$$

for the region R in the first quadrant bounded by the lines $y = -2x + 4$, $y = -2x + 7$, $y = x - 2$, and $y = x + 1$.

9. Use the transformation in Exercise 3 to evaluate the integral

$$\iint_R (3x^2 + 14xy + 8y^2) \, dx \, dy$$

for the region R in the first quadrant bounded by the lines

$$y = -\frac{3}{2}x + 1, \qquad y = -\frac{3}{2}x + 3,$$

$$y = -\frac{1}{4}x, \qquad \text{and} \quad y = -\frac{1}{4}x + 1.$$

10. Use the transformation and parallelogram R in Exercise 4 to evaluate the integral

$$\iint_R 2(x - y) \, dx \, dy.$$

11. Let R be the region in the first quadrant of the xy-plane bounded by the hyperbolas $xy = 1$, $xy = 9$ and the lines $y = x$, $y = 4x$. Use the transformation $x = u/v$, $y = uv$ with $u > 0$ and $v > 0$ to rewrite

$$\iint_R \left(\sqrt{\frac{y}{x}} + \sqrt{xy} \right) dx \, dy$$

as an integral over an appropriate region G in the uv-plane. Then evaluate the uv-integral over G.

12. **(a)** Find the Jacobian of the transformation $x = u$, $y = uv$ and sketch the region G: $1 \leq u \leq 2$, $1 \leq uv \leq 2$ in the uv-plane.

(b) Then use Equation (1) to transform the integral

$$\int_1^2 \int_1^2 \frac{y}{x} \, dy \, dx$$

into an integral over G and evaluate both integrals.

Explorations

13. *Polar Moment of an Elliptical Plate* A thin plate of constant density covers the region bounded by the ellipse $x^2/a^2 + y^2/b^2 = 1$, $a > 0$, $b > 0$, in the xy-plane. Find the polar moment of the plate about the origin. (*Hint:* Use the transformation $x = ar \cos \theta$, $y = br \sin \theta$.)

14. *Finding the Area of an Ellipse* The area πab of the ellipse $x^2/a^2 + y^2/b^2 = 1$ can be found by integrating the function $f(x, y) = 1$ over the region bounded by the ellipse in the xy-plane. Evaluating the integral directly requires a trigonometric substitution. An easier way to evaluate the integral is to use the transformation $x = au$, $y = bv$ and evaluate the transformed integral over the disk G: $u^2 + v^2 \leq 1$ in the uv-plane. Find the area this way. ∎

15. Use the transformation in Exercise 2 to evaluate the integral

$$\int_0^{2/3} \int_y^{2-2y} (x + 2y)e^{(y-x)} \, dx \, dy$$

by first writing it as an integral over a region G in the uv-plane.

16. Use the transformation $x = u + (1/2)v$, $y = v$ to evaluate the integral

$$\int_0^2 \int_{y/2}^{(y+4)/2} y^3(2x - y)e^{(2x-y)^2} \, dx \, dy$$

by first writing it as an integral over a region in the uv-plane.

17. Evaluate the determinant in Equation (10) to show that the Jacobian of the transformation from Cartesian $\rho\phi\theta$-space to Cartesian xyz-space is $\rho^2 \sin \phi$.

18. Evaluate the integral in Example 3 by integrating with respect to x, y, and z.

19. *Volume of an Ellipsoid* Find the volume of the ellipsoid

$$\frac{x^2}{a^2} + \frac{y^2}{b^2} + \frac{z^2}{c^2} = 1.$$

(*Hint:* Let $x = au$, $y = bv$, and $z = cw$. Then find the volume of an appropriate region in uvw-space.)

20. Evaluate

$$\iiint |xyz| \, dx \, dy \, dz$$

over the solid ellipsoid

$$\frac{x^2}{a^2} + \frac{y^2}{b^2} + \frac{z^2}{c^2} \leq 1.$$

(*Hint:* Let $x = au$, $y = bv$, and $z = cw$. Then find the volume of an appropriate region in uvw-space.)

21. Let D be the region in xyz-space defined by the inequalities

$$1 \leq x \leq 2, \qquad 0 \leq xy \leq 2, \qquad 0 \leq z \leq 1.$$

Evaluate

$$\iiint_D (x^2y + 3xyz) \, dx \, dy \, dz$$

by applying the transformation

$$u = x, \qquad v = xy, \qquad w = 3z$$

and integrating over an appropriate region G in uvw-space.

22. *Centroid of a Solid Semi-ellipsoid* Assuming the result that the centroid of a solid hemisphere lies on the axis of symmetry three-eighths of the way from the base toward the top, show, by transforming the appropriate integrals, that the centroid of a solid semi-ellipsoid $(x^2/a^2) + (y^2/b^2) + (z^2/c^2) \le 1$, $z \ge 0$, lies on the z-axis three-eighths of the way from the base toward the top. (You can do this without evaluating any of the integrals.)

Extending the Ideas

23. *Substitutions in Single Integrals* How can substitutions in single definite integrals be viewed as transformations of regions? What is the Jacobian in such a case? Illustrate with an example.

24. *Cylindrical Shells* In Section 7.3, we learned how to find the volume of a solid of revolution using the shell method, namely if the region between the curve $y = f(x)$ and the x-axis from a to b $(0 < a < b)$ is revolved about the y-axis the volume of the resulting solid is $\int_a^b 2\pi x f(x)\, dx$. Prove that finding volumes by using triple integrals gives the same result. (*Hint:* Use cylindrical coordinates with the roles of y and z changed.)

Chapter 14 Key Terms

area of a region (p. 773)
average density (p. 775)
average distance (p. 775)
average value of a function in space (p. 795)
average value of a function in the plane (p. 774)
center of mass (p. 775)
centroid (p. 779)
double integral of f over R (p. 765)
first moment (p. 775)

Fubini's theorems (p. 762)
image (p. 816)
iterated integral (p. 763)
Jacobian (p. 817)
Jacobian determinant (p. 817)
mass (p. 800)
moment of inertia (p. 775)
Pappus's formula (p. 783)
Parallel Axis Theorem (p. 783)
Perpendicular Axis Theorem (p. 775)

polar moment (p. 775)
preimage (p. 816)
radius of gyration (p. 775)
repeated integral (p. 763)
second moment (p. 775)
spherical coordinates (p. 807)
spherical wedge (p. 809)
triple integral of F over D (p. 790)
volume element (p. 810)
volume of a region (p. 791)

Chapter 14 Review Exercises

In Exercises 1–4, sketch the region of integration and evaluate the double integral analytically.

1. $\displaystyle \int_1^{10} \int_0^{1/y} y e^{xy} \, dx \, dy$

2. $\displaystyle \int_0^1 \int_0^{x^3} e^{y/x} \, dy \, dx$

3. $\displaystyle \int_0^{3/2} \int_{-\sqrt{9-4t^2}}^{\sqrt{9-4t^2}} t \, ds \, dt$

4. $\displaystyle \int_0^1 \int_{\sqrt{y}}^{2-\sqrt{y}} xy \, dx \, dy$

In Exercises 5–8, sketch the region of integration and write an equivalent integral with the order of integration reversed. Then evaluate both integrals analytically.

5. $\displaystyle \int_0^4 \int_{-\sqrt{4-y}}^{(y-4)/2} dx \, dy$

6. $\displaystyle \int_0^1 \int_{x^2}^{x} \sqrt{x} \, dy \, dx$

7. $\displaystyle \int_0^{3/2} \int_{-\sqrt{9-4y^2}}^{\sqrt{9-4y^2}} y \, dx \, dy$

8. $\displaystyle \int_0^2 \int_0^{4-x^2} 2x \, dy \, dx$

In Exercises 9–12, evaluate the integral analytically.

9. $\displaystyle \int_0^1 \int_{2y}^{2} 4\cos(x^2) \, dx \, dy$

10. $\displaystyle \int_0^2 \int_{y/2}^{1} e^{x^2} \, dx \, dy$

11. $\displaystyle \int_0^8 \int_{\sqrt[3]{x}}^{2} \frac{dy \, dx}{y^4 + 1}$

12. $\displaystyle \int_0^1 \int_{\sqrt[3]{y}}^{1} \frac{2\pi \sin \pi x^2}{x^2} \, dx \, dy$

13. *Area* Find the area of the region enclosed by the line $y = 2x + 4$ and the parabola $y = 4 - x^2$ in the xy-plane.

14. *Area* Find the area of the "triangular" region in the xy-plane that is bounded on the right by the parabola $y = x^2$, on the left by the line $x + y = 2$, and above by the line $y = 4$.

15. *Volume* Find the volume under the paraboloid $z = x^2 + y^2$ above the triangle enclosed by the lines $y = x$, $x = 0$, and $x + y = 2$ in the xy-plane.

16. *Volume* Find the volume under the parabolic cylinder $z = x^2$ above the region enclosed by the parabola $y = 6 - x^2$ and the line $y = x$ in the xy-plane.

Average Value In Exercises 17 and 18, find the average value of $f(x, y) = xy$ over the given region.

17. The square bounded by the lines $x = 1$, $y = 1$ in the first quadrant

18. The quarter circle $x^2 + y^2 \le 1$ in the first quadrant

19. *Centroid* Find the centroid of the "triangular" region bounded by the lines $x = 2$, $y = 2$ and the hyperbola $xy = 2$ in the xy-plane.

20. *Centroid* Find the centroid of the region between the parabola $x + y^2 - 2y = 0$ and the line $x + 2y = 0$ in the xy-plane.

21. *Polar Moment* Find the polar moment of inertia about the origin of a thin triangular plate of constant density $\delta = 3$, bounded by the y-axis and the lines $y = 2x$ and $y = 4$ in the xy-plane.

22. *Polar Moment* Find the polar moment of inertia about the center of a thin rectangular sheet of constant density $\delta = 1$ bounded by the lines

(a) $x = \pm 2$, $y = \pm 1$ in the xy-plane.

(b) $x = \pm a$, $y = \pm b$ in the xy-plane.

(*Hint:* Find I_x. Then use the formula for I_x to find I_y and add the two to find I_0.)

23. *Inertial Moment and Radius of Gyration* Find the moment of inertia and radius of gyration about the x-axis of a thin plate of constant density δ covering the triangle with vertices $(0, 0)$, $(3, 0)$, and $(3, 2)$ in the xy-plane.

24. *Plate with Variable Density* Find the center of mass and the moments of inertia and radii of gyration about the coordinate axes of a thin plate bounded by the line $y = x$ and the parabola $y = x^2$ in the xy-plane if the density is $\delta(x, y) = x + 1$.

25. *Plate with Variable Density* Find the mass and first moments about the coordinate axes of a thin square plate bounded by the lines $x = \pm 1$, $y = \pm 1$ in the xy-plane if the density is $\delta(x, y) = x^2 + y^2 + 1/3$.

26. *Triangles with the Same M_x and R_x* Find the moment of inertia and radius of gyration about the x-axis of a thin triangular plate of constant density δ whose base lies along the interval $[0, b]$ on the x-axis and whose vertex lies on the line $y = h$ above the x-axis. As you will see, it does not matter where on the line this vertex lies. All such triangles have the same moment of inertia and radius of gyration.

In Exercises 27 and 28, evaluate the integral by changing to polar coordinates.

27. $\displaystyle \int_{-1}^{1} \int_{-\sqrt{1-x^2}}^{\sqrt{1-x^2}} \frac{2 \, dy \, dx}{(1 + x^2 + y^2)^2}$

28. $\displaystyle \int_{-1}^{1} \int_{-\sqrt{1-y^2}}^{\sqrt{1-y^2}} \ln(x^2 + y^2 + 1) \, dx \, dy$

29. *Centroid in Polar Coordinates* Find the centroid of the region in the polar coordinate plane defined by the inequalities $0 \le r \le 3$ and $-\pi/3 \le \theta \le \pi/3$.

30. *Centroid in Polar Coordinates* Find the centroid of the region in the first quadrant bounded by the rays $\theta = 0$ and $\theta = \pi/2$ and the circles $r = 1$ and $r = 3$.

31. *Centroid in Polar Coordinates*

(a) Find the centroid of the region in the polar coordinate plane that lies inside the cardioid $r = 1 + \cos \theta$ and outside the circle $r = 1$.

(b) Sketch the region and show the centroid in your sketch.

32. **(a)** **Writing to Learn** Find the centroid of the plane region defined by the polar coordinate inequalities $0 \le r \le a$, $-\alpha \le \theta \le \alpha$ ($0 < \alpha \le \pi$). How does the centroid move as $\alpha \to \pi^-$?

(b) Sketch the region for $\alpha = 5\pi/6$ and show the centroid in your sketch.

33. Integrate the function $f(x, y) = 1/(1 + x^2 + y^2)^2$ over the region enclosed by one loop of the lemniscate $(x^2 + y^2)^2 - (x^2 - y^2) = 0$.

34. Integrate $f(x, y) = 1/(1 + x^2 + y^2)^2$ over

(a) the triangle with vertices $(0, 0)$, $(1, 0)$, $(1, \sqrt{3})$.

(b) the first quadrant of the xy-plane.

In Exercises 35–38, evaluate the integral analytically.

35. $\displaystyle \int_0^\pi \int_0^\pi \int_0^\pi \cos(x + y + z) \, dx \, dy \, dz$

36. $\displaystyle \int_{\ln 6}^{\ln 7} \int_0^{\ln 2} \int_{\ln 4}^{\ln 5} e^{(x+y+z)} \, dz \, dy \, dx$

37. $\displaystyle \int_0^1 \int_0^{x^2} \int_0^{x+y} (2x - y - z) \, dz \, dy \, dx$

38. $\displaystyle \int_1^e \int_1^x \int_0^z \frac{2y}{z^3} \, dy \, dz \, dx$

39. *Volume* Find the volume of the wedge-shaped region enclosed on the side by the cylinder $x = -\cos y$, $-\pi/2 \le y \le \pi/2$, on the top by the plane $z = -2x$, and below by the xy-plane.

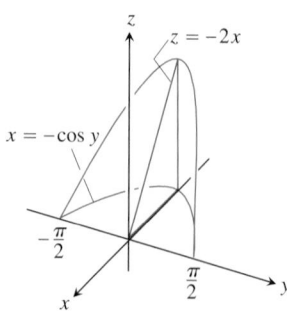

40. *Volume* Find the volume of the solid that is bounded above by the cylinder $z = 4 - x^2$, on the sides by the cylinder $x^2 + y^2 = 4$, and below by the xy-plane.

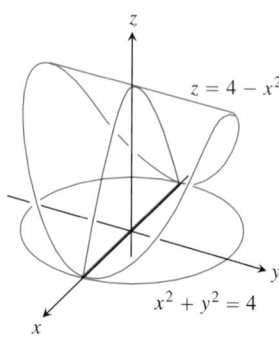

41. *Average Value (Rectangular Coordinates)* Find the average value of $f(x, y, z) = 30xz\sqrt{x^2 + y}$ over the rectangular solid in the first octant bounded by the coordinate planes and the planes $x = 1$, $y = 3$, $z = 1$.

42. *Average Value (Cylindrical Coordinates)* Find the average value of ρ over the solid sphere $\rho \leq a$ (spherical coordinates).

Coordinate Conversions In Exercises 43–46, translate the equation from the given coordinate system (rectangular, cylindrical, spherical) into the other two systems. Identify the set of points defined by the equation.

43. $x^2 + y^2 + (z + 1)^2 = 1$

44. $\rho \cos \phi + \rho^2 \sin^2 \phi = 1$

45. $r = 4 \cos \theta$ **46.** $\phi = 3\pi/4$

47. Convert

$$\int_0^{2\pi} \int_0^{\sqrt{2}} \int_r^{\sqrt{4-r^2}} 3 \, dz \, r \, dr \, d\theta, \qquad r \geq 0$$

to **(a)** rectangular coordinates with the order of integration $dz \, dx \, dy$ and **(b)** spherical coordinates. Then **(c)** evaluate one of the integrals.

48. (a) Convert to cylindrical coordinates. Then **(b)** evaluate the new integral.

$$\int_0^1 \int_{-\sqrt{1-x^2}}^{\sqrt{1-x^2}} \int_{-(x^2+y^2)}^{(x^2+y^2)} 21xy^2 \, dz \, dy \, dx$$

49. (a) Convert to spherical coordinates. Then **(b)** evaluate the new integral.

$$\int_{-1}^1 \int_{-\sqrt{1-x^2}}^{\sqrt{1-x^2}} \int_{\sqrt{x^2+y^2}}^1 dz \, dy \, dx$$

50. Write an iterated triple integral for the integral of $f(x, y, z) = 6 + 4y$ over the region in the first octant bounded by the cone $z = \sqrt{x^2 + y^2}$, the cylinder $x^2 + y^2 = 1$, and the coordinate planes in **(a)** rectangular coordinates, **(b)** cylindrical coordinates, and **(c)** spherical coordinates. Then **(d)** find the integral of f by evaluating one of the triple integrals.

51. Set up an integral in rectangular coordinates equivalent to the integral

$$\int_0^{\pi/2} \int_1^{\sqrt{3}} \int_1^{\sqrt{4-r^2}} r^3 \sin \theta \cos \theta z^2 \, dz \, dr \, d\theta.$$

Arrange the order of integration to be z first, then y, then x.

52. The volume of a solid is

$$\int_0^2 \int_0^{\sqrt{2x-x^2}} \int_{-\sqrt{4-x^2-y^2}}^{\sqrt{4-x^2-y^2}} dz \, dy \, dx.$$

(a) Describe the solid by giving equations for the surfaces that form its boundary.

(b) Convert the integral to cylindrical coordinates but do not evaluate the integral.

53. *Equivalent Triple Integrals for Volume* Let D be the smaller spherical cap cut from a solid ball of radius 2 by a plane 1 unit from the center of the sphere. Express the volume of D as an iterated triple integral in **(a)** rectangular, **(b)** cylindrical, and **(c)** spherical coordinates. *Do not evaluate the integrals.*

54. *Equivalent Triple Integrals for I_z* Express the moment of inertia I_z of the solid hemisphere bounded below by the plane $z = 0$ and above by the sphere $x^2 + y^2 + z^2 = 1$ as an iterated integral in **(a)** rectangular, **(b)** cylindrical, and **(c)** spherical coordinates. *Do not evaluate the integrals.*

55. *Spherical vs. Cylindrical Coordinates* Triple integrals involving spherical shapes do not always require spherical coordinates for convenient evaluation. Some calculations may be accomplished more easily with cylindrical coordinates. As a case in point, find the volume of the region bounded above by the sphere $x^2 + y^2 + z^2 = 8$ and below by the plane $z = 2$ by using **(a)** cylindrical coordinates and **(b)** spherical coordinates.

56. *Calculating I_z in Spherical Coordinates* Find the moment of inertia about the z-axis of a solid of constant density $\delta = 1$ that is bounded above by the sphere $\rho = 2$ and below by the cone $\phi = \pi/3$ (spherical coordinates).

57. *Moment of Inertia of a Thick Sphere* Find the moment of inertia of a solid of constant density δ bounded by two concentric spheres of radii a and b ($a < b$) about a diameter.

58. *Moment of Inertia of an Apple* Find the moment of inertia about the z-axis of a solid of density $\delta = 1$ enclosed by the spherical coordinate surface $\rho = 1 - \cos \phi$ as illustrated in the figure.

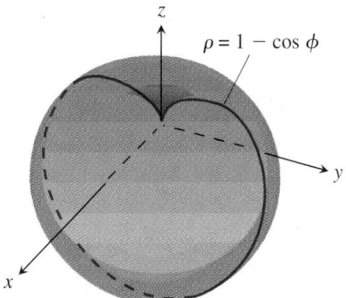

In Exercises 59 and 60, *work in groups of two or three* to solve the problem.

59. Show that if $u = x - y$ and $v = y$, then

$$\int_0^{\infty} \int_0^x e^{-sx} f(x - y, y) \, dy \, dx = \int_0^{\infty} \int_0^{\infty} e^{-s(u+v)} f(u, v) \, du \, dv.$$

60. Writing to Learn What relationship must hold between the constants a, b, and c to make

$$\int_{-\infty}^{\infty} \int_{-\infty}^{\infty} e^{-(ax^2+2bxy+cy^2)} \, dx \, dy = 1?$$

(*Hint:* Let $s = \alpha x + \beta y$ and $t = \gamma x + \delta y$, where $(\alpha\delta - \beta\gamma)^2 = ac - b^2$. Then $ax^2 + 2bxy + cy^2 = s^2 + t^2$.)

In addition to its use as a sweetener, honey has been used as a medicine, an aphrodisiac, and a wound dressing. Honey reacts with the water molecules in wounds, causing death by dehydration for infection-causing bacteria. It also produces hydrogen peroxide through enzymatic reactions, enhancing its antibacterial properties.

In a bottling facility, honey is forced through a long cylindrical tube of radius 0.5 cm. Its velocity function is given by $\mathbf{v} = 5(0.25 - r^2)\mathbf{k}$ cm/sec, where $r = \sqrt{x^2 + y^2}$ is the distance (in centimeters) from the center of the tube. Find the output of the tube in cubic centimeters per second. This problem can be solved using the ideas in Section 15.5.

Chapter 15 Overview

In this chapter we encounter Green's Theorem, perhaps the most powerful theorem of calculus. It is deep and surprising and has far-reaching consequences. In pure mathematics, it ranks in importance with the Fundamental Theorem of Calculus. In applied mathematics, the generalizations of Green's Theorem provide the foundation for the great vector integral theorems of electricity and magnetism.

The mathematics of this chapter is the mathematics engineers and physicists use to model fluid flow, design underwater transmission cables, explain the distribution of heat in stars, and calculate the work it takes to put satellites into orbit.

15.1 Line Integrals

Definitions and Notation • Evaluation for Smooth Curves •
Additivity • Mass and Moment Calculations

Definitions and Notation

Suppose $f(x, y, z)$ is a function whose domain contains the curve $\mathbf{r}(t) = g(t)\mathbf{i} + h(t)\mathbf{j} + k(t)\mathbf{k}, a \leq t \leq b$. The values of f along the curve are given by the composite function $f(g(t), h(t), k(t))$. If we integrate this composite with respect to arc length from $t = a$ to $t = b$, we calculate the **line integral** of f over the curve.

To define this integral, we partition the curve into a finite number of subarcs (Figure 15.1). The typical subarc has length Δs_k. In each subarc, we choose a point (x_k, y_k, z_k) and form the sum

$$S_n = \sum_{k=1}^{n} f(x_k, y_k, z_k)\, \Delta s_k. \tag{1}$$

If f is continuous and the functions g, h, and k have continuous first derivatives, then the sums in Equation (1) approach a limit as n increases and the lengths Δs_k approach zero. This limit is the **(line) integral of f over the curve from a to b.** If the curve is denoted by a single letter, C for example, the notation for the integral is

$$\int_C f(x, y, z)\, ds. \qquad \text{“The integral of } f \text{ over } C\text{”} \tag{2}$$

Figure 15.1 The curve $\mathbf{r}(t) = g(t)\mathbf{i} + h(t)\mathbf{j} + k(t)\mathbf{k}$, partitioned into small arcs from $t = a$ to $t = b$. The length of a typical subarc is Δs_k.

Evaluation for Smooth Curves

If $\mathbf{r}(t)$ is smooth for $a \leq t \leq b$ ($\mathbf{v} = d\mathbf{r}/dt$ is continuous and never $\mathbf{0}$), we can use the equation

$$s(t) = \int_a^t |\mathbf{v}(\tau)|\, d\tau \qquad \text{Eq. (2) of Section 12.2 with } t_0 = a$$

to express ds in Equation (2) as $ds = |\mathbf{v}(t)|\, dt$. A theorem from advanced calculus says we can then evaluate the integral of f over C as

$$\int_C f(x, y, z)\, ds = \int_a^b f\big(g(t), h(t), k(t)\big) |\mathbf{v}(t)|\, dt.$$

This formula evaluates the integral correctly no matter what parametrization we use, as long as the parametrization is smooth.

> **How to Evaluate a Line Integral**
>
> To integrate a continuous function $f(x, y, z)$ over a curve C:
>
> **1.** Find a smooth parametrization of C,
> $$\mathbf{r}(t) = g(t)\mathbf{i} + h(t)\mathbf{j} + k(t)\mathbf{k}, \qquad a \le t \le b.$$
>
> **2.** Evaluate the integral as
> $$\int_C f(x, y, z)\, ds = \int_a^b f\big(g(t), h(t), k(t)\big)|\mathbf{v}(t)|\, dt. \qquad (3)$$

If f has the constant value 1, then the integral of f over C gives the length of C.

Example 1 EVALUATING A LINE INTEGRAL

Integrate $f(x, y, z) = x - 3y^2 + z$ over the line segment C joining the origin and the point $(1, 1, 1)$ (Figure 15.2).

Solution   We choose the simplest parametrization we can think of:
$$\mathbf{r}(t) = t\mathbf{i} + t\mathbf{j} + t\mathbf{k}, \qquad 0 \le t \le 1.$$

The components have continuous first derivatives and $|\mathbf{v}(t)| = \sqrt{1^2 + 1^2 + 1^2} = \sqrt{3}$ is never 0, so the parametrization is smooth. The integral of f over C is

$$\int_C f(x, y, z)\, ds = \int_0^1 f(t, t, t) \cdot (\sqrt{3})\, dt \qquad \text{Eq. (3)}$$

$$= \int_0^1 (t - 3t^2 + t)\sqrt{3}\, dt$$

$$= \sqrt{3} \int_0^1 (2t - 3t^2)\, dt = \sqrt{3}\,[t^2 - t^3]_0^1 = 0.$$

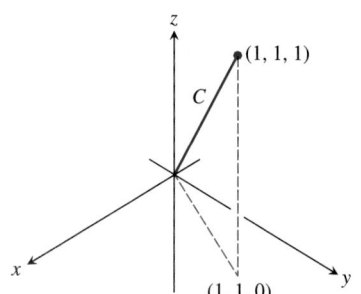

Figure 15.2 The integration path in Example 1.

Additivity

If a curve C is made by joining a finite number of curves $C_1, C_2, \ldots, C_n$ end to end, then the integral of a function over C is the sum of the integrals over the curves that make it up:

$$\int_C f\, ds = \int_{C_1} f\, ds + \int_{C_2} f\, ds + \cdots + \int_{C_n} f\, ds. \qquad (4)$$

Example 2 ILLUSTRATING EQUATION (4)

Figure 15.3 shows another path from the origin to $(1, 1, 1)$, the union of line segments C_1 and C_2. Integrate $f(x, y, z) = x - 3y^2 + z$ over $C_1 \cup C_2$.

Solution   We choose the simplest parametrizations for C_1 and C_2 we can think of, checking the lengths of the velocity vectors as we go along:

$$C_1\!: \mathbf{r}(t) = t\mathbf{i} + t\mathbf{j}, \qquad 0 \le t \le 1; \qquad |\mathbf{v}| = \sqrt{1^2 + 1^2} = \sqrt{2},$$

$$C_2\!: \mathbf{r}(t) = \mathbf{i} + \mathbf{j} + t\mathbf{k}, \qquad 0 \le t \le 1; \qquad |\mathbf{v}| = \sqrt{0^2 + 0^2 + 1^2} = 1.$$

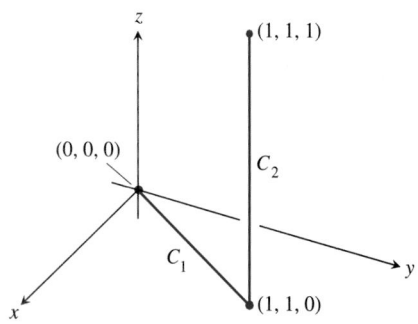

Figure 15.3 The path of integration in Example 2.

With these parametrizations we find

$$\int_{C_1 \cup C_2} f(x, y, z) \, ds = \int_{C_1} f(x, y, z) \, ds + \int_{C_2} f(x, y, z) \, ds \qquad \text{Eq. (4)}$$

$$= \int_0^1 f(t, t, 0)\sqrt{2} \, dt + \int_0^1 f(1, 1, t)(1) \, dt \qquad \text{Eq. (3)}$$

$$= \int_0^1 (t - 3t^2 + 0)\sqrt{2} \, dt + \int_0^1 (1 - 3 + t)(1) \, dt$$

$$= \sqrt{2}\left[\frac{t^2}{2} - t^3\right]_0^1 + \left[\frac{t^2}{2} - 2t\right]_0^1 = -\frac{\sqrt{2}}{2} - \frac{3}{2}.$$

Integral Evaluation

The integrals with respect to *t* in Examples 1 and 2 were simple enough to do by hand. More complicated integrals might require an integral evaluator like the one in your graphing calculator.

Notice three things about the integrations in Examples 1 and 2. First, as soon as the components of the appropriate curve were substituted into the formula for *f*, the integration became a standard integration with respect to *t*. Second, the integral of *f* over $C_1 \cup C_2$ was obtained by integrating *f* over each section of the path and adding the results. Third, the integrals of *f* over *C* and $C_1 \cup C_2$ had different values. For most functions, the value of the integral along a path joining two points changes if you change the path between them. For a special class of functions, however, the value remains the same, as we see in Section 15.3.

Mass and Moment Calculations

We treat coil springs and wires like masses distributed along smooth curves in space. The distribution is described by a continuous density function $\delta(x, y, z)$ (mass per unit length). The spring's or wire's mass, center of mass, and moments are then calculated with the formulas in Table 15.1. The formulas also apply to thin rods.

Table 15.1 Mass and Moment Formulas for Coil Springs, Thin Rods, and Wires Lying Along a Smooth Curve C in Space

Mass: $M = \displaystyle\int_C \delta(x, y, z) \, ds$

First moments about the coordinate planes:

$$M_{yz} = \int_C x\delta \, ds, \qquad M_{xz} = \int_C y\delta \, ds, \qquad M_{xy} = \int_C z\delta \, ds$$

Coordinates of the center of mass:

$$\bar{x} = M_{yz}/M, \qquad \bar{y} = M_{xz}/M, \qquad \bar{z} = M_{xy}/M$$

Moments of inertia (second moments):

$$I_x = \int_C (y^2 + z^2)\delta \, ds, \qquad I_y = \int_C (x^2 + z^2)\delta \, ds$$

$$I_z = \int_C (x^2 + y^2)\delta \, ds, \qquad I_L = \int_C r^2\delta \, ds$$

$r(x, y, z) = $ distance from point (x, y, z) to line *L*

Radius of gyration about a line *L*: $R_L = \sqrt{I_L/M}$

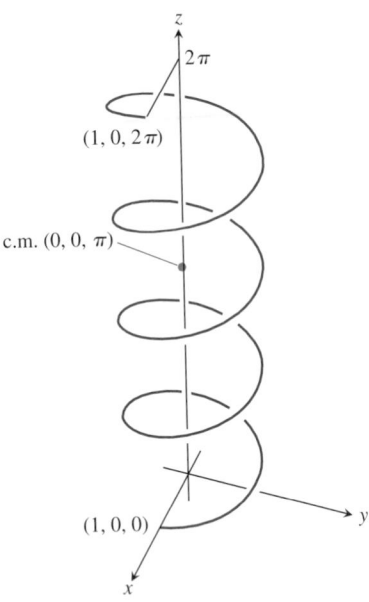

Figure 15.4 The helical spring in Example 3.

Example 3 FINDING MASS, CENTER OF MASS, MOMENT OF INERTIA, AND RADIUS OF GYRATION

A coil spring lies along the helix

$$\mathbf{r}(t) = (\cos 4t)\mathbf{i} + (\sin 4t)\mathbf{j} + t\mathbf{k}, \qquad 0 \le t \le 2\pi.$$

The spring's density is a constant, $\delta = 1$. Find the spring's mass and center of mass, and its moment of inertia and radius of gyration about the z-axis.

Solution We sketch the spring (Figure 15.4). The symmetry and constant density put the center of mass at the point $(0, 0, \pi)$ on the z-axis.

For the remaining calculations, we first find $|\mathbf{v}(t)|$:

$$|\mathbf{v}(t)| = \sqrt{\left(\frac{dx}{dt}\right)^2 + \left(\frac{dy}{dt}\right)^2 + \left(\frac{dz}{dt}\right)^2}$$

$$= \sqrt{(-4\sin 4t)^2 + (4\cos 4t)^2 + (1)^2} = \sqrt{17}.$$

We then evaluate the formulas from Table 15.1 using Equation (3):

$$M = \int_{\text{Helix}} \delta \, ds = \int_0^{2\pi} (1)\sqrt{17} \, dt = 2\pi\sqrt{17}$$

$$I_z = \int_{\text{Helix}} (x^2 + y^2)\delta \, ds = \int_0^{2\pi} (\cos^2 4t + \sin^2 4t)(1)\sqrt{17} \, dt$$

$$= \int_0^{2\pi} \sqrt{17} \, dt = 2\pi\sqrt{17}$$

$$R_z = \sqrt{I_z/M} = \sqrt{2\pi\sqrt{17}/(2\pi\sqrt{17})} = 1.$$

Notice that the radius of gyration about the z-axis is the radius of the cylinder around which the helix winds.

A Balancing Trick

The center of mass of the arch in Example 4 lies outside the arch. This is the principle behind a neat dinner-table trick. If you overlap the prongs of two forks and insert a half-dollar part way to hold the forks together, you will be able to hold the forks out in space by balancing the coin on the rim of a glass.

Example 4 FINDING A CENTER OF MASS

A slender metal arch, denser at the bottom than top, lies along the semicircle $y^2 + z^2 = 1$, $z \ge 0$, in the yz-plane (Figure 15.5). Find the center of the arch's mass if the density at the point (x, y, z) on the arch is $\delta(x, y, z) = 2 - z$.

Solution We know $\bar{x} = 0$ and $\bar{y} = 0$ because the arch lies in the yz-plane with its mass distributed symmetrically about the z-axis. To find $\bar{z}$, we parametrize the semicircle as

$$\mathbf{r}(t) = (\cos t)\mathbf{j} + (\sin t)\mathbf{k}, \qquad 0 \le t \le \pi.$$

For this parametrization,

$$|\mathbf{v}(t)| = \sqrt{\left(\frac{dx}{dt}\right)^2 + \left(\frac{dy}{dt}\right)^2 + \left(\frac{dz}{dt}\right)^2} = \sqrt{(0)^2 + (-\sin t)^2 + (\cos t)^2} = 1,$$

so $ds = vt$. The formulas in Table 15.1 then give

$$M = \int_C \delta \, ds = \int_C (2 - z) \, ds = \int_0^\pi (2 - \sin t) \, dt = 2\pi - 2$$

$$M_{xy} = \int_C z\delta \, ds = \int_C z(2 - z) \, ds = \int_0^\pi (\sin t)(2 - \sin t) \, dt$$

$y^2 + z^2 = 1, z \geq 0$

$$= \int_0^{\pi} (2 \sin t - \sin^2 t) \, dt = \frac{8 - \pi}{2}$$

$$\bar{z} = \frac{M_{xy}}{M} = \frac{8 - \pi}{2} \cdot \frac{1}{2\pi - 2} = \frac{8 - \pi}{4\pi - 4} \approx 0.57.$$

● With $\bar{z}$ to the nearest hundredth, the center of mass is (0, 0, 0.57).

Figure 15.5 Example 4 shows how to find the center of mass of a circular arch of variable density.

Quick Review 15.1

In Exercises 1–5, evaluate the integrals analytically.

1. $\int \cos 3x \, dx$

2. $\int \sin 2x \, dx$

3. $\int (x^2 - x^3 + 2) \, dx$

4. $\int_{-1}^{2} (x^2 - x^3 + 2) \, dx$

5. $\int_0^{\pi} \cos 3x \, dx$

In Exercises 6–10, parametrize the line segment with the given endpoints.

6. (0, 0), (2, 3)

7. (0, 0), (a, b)

8. (−2, 1), (3, −2)

9. (0, 0, 0), (2, 3, 4)

10. (−1, 1, 2), (3, 3, −2)

Section 15.1 Exercises

In Exercises 1–8, match the vector equation with the graphs in Figure 15.6.

(a)

(b)

(c)

(d)

(e)

(f)

(g)

(h)
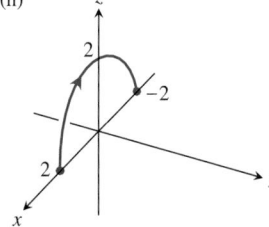

Figure 15.6 The graphs for Exercises 1–8.

1. $\mathbf{r}(t) = t\mathbf{i} + (1 - t)\mathbf{j}, \qquad 0 \leq t \leq 1$

2. $\mathbf{r}(t) = \mathbf{i} + \mathbf{j} + t\mathbf{k}, \qquad -1 \leq t \leq 1$

3. $\mathbf{r}(t) = (2 \cos t)\mathbf{i} + (2 \sin t)\mathbf{j}, \qquad 0 \leq t \leq 2\pi$

4. $\mathbf{r}(t) = t\mathbf{i}, \qquad -1 \leq t \leq 1$

5. $\mathbf{r}(t) = t\mathbf{i} + t\mathbf{j} + t\mathbf{k}, \qquad 0 \leq t \leq 2$

6. $\mathbf{r}(t) = t\mathbf{j} + (2 - 2t)\mathbf{k}, \qquad 0 \leq t \leq 1$

7. $\mathbf{r}(t) = (t^2 - 1)\mathbf{j} + 2t\mathbf{k}, \qquad -1 \leq t \leq 1$

8. $\mathbf{r}(t) = (2 \cos t)\mathbf{i} + (2 \sin t)\mathbf{k}, \qquad 0 \leq t \leq \pi$

9. Evaluate $\int_C (x + y)\, ds$, where C is the straight-line segment $x = t$, $y = (1 - t)$, $z = 0$, from $(0, 1, 0)$ to $(1, 0, 0)$.

10. Evaluate $\int_C (x - y + z - 2)\, ds$, where C is the straight-line segment $x = t$, $y = (1 - t)$, $z = 1$, from $(0, 1, 1)$ to $(1, 0, 1)$.

11. Evaluate $\int_C (xy + y + z)\, ds$ along the curve $\mathbf{r}(t) = 2t\mathbf{i} + t\mathbf{j} + (2 - 2t)\mathbf{k}, 0 \le t \le 1$.

12. Evaluate $\int_C \sqrt{x^2 + y^2}\, ds$ along the curve $\mathbf{r}(t) = (4 \cos t)\mathbf{i} + (4 \sin t)\mathbf{j} + 3t\mathbf{k}, -2\pi \le t \le 2\pi$.

13. Find the line integral of $f(x, y, z) = x + y + z$ over the straight-line segment from $(1, 2, 3)$ to $(0, -1, 1)$.

14. Find the line integral of $f(x, y, z) = \sqrt{3}/(x^2 + y^2 + z^2)$ over the curve $\mathbf{r}(t) = t\mathbf{i} + t\mathbf{j} + t\mathbf{k}, 1 \le t \le \infty$.

Explorations

15. Integrate $f(x, y, z) = x + \sqrt{y} - z^2$ over the path from $(0, 0, 0)$ to $(1, 1, 1)$ (Figure 15.7a) given by

$C_1\colon \mathbf{r}(t) = t\mathbf{i} + t^2\mathbf{j}, \quad 0 \le t \le 1,$
$C_2\colon \mathbf{r}(t) = \mathbf{i} + \mathbf{j} + t\mathbf{k}, \quad 0 \le t \le 1.$

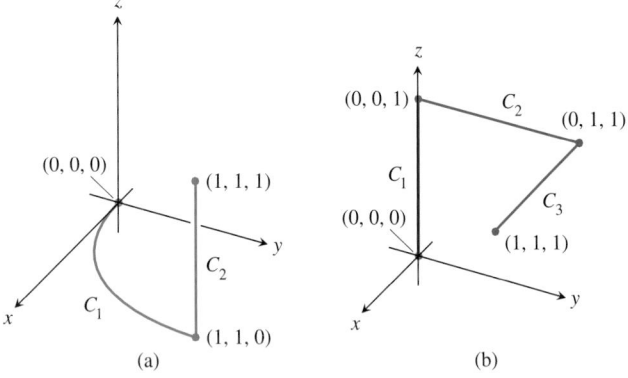

Figure 15.7 The paths of integration for Exercises 15 and 16.

16. Integrate $f(x, y, z) = x + \sqrt{y} - z^2$ over the path from $(0, 0, 0)$ to $(1, 1, 1)$ (Figure 15.7b) given by

$C_1\colon \mathbf{r}(t) = t\mathbf{k}, \quad 0 \le t \le 1,$
$C_2\colon \mathbf{r}(t) = t\mathbf{j} + \mathbf{k}, \quad 0 \le t \le 1,$
$C_3\colon \mathbf{r}(t) = t\mathbf{i} + \mathbf{j} + \mathbf{k}, \quad 0 \le t \le 1.$ ∎

17. Integrate

$$f(x, y, z) = \frac{x + y + z}{x^2 + y^2 + z^2}$$

over the path $\mathbf{r}(t) = t\mathbf{i} + t\mathbf{j} + t\mathbf{k}, \quad 0 < a \le t \le b.$

18. Integrate

$$f(x, y, z) = -\sqrt{x^2 + z^2}$$

over the circle $\mathbf{r}(t) = (a \cos t)\mathbf{j} + (a \sin t)\mathbf{k}, \quad 0 \le t \le 2\pi.$

In Exercises 19–22, integrate f over the given curve in the xy-plane.

19. $f(x, y) = x^3/y$, $\quad C\colon y = x^2/2$, $\quad 0 \le x \le 2$

20. $f(x, y) = \dfrac{x + y^2}{\sqrt{1 + x^2}}$, $\quad C\colon y = \dfrac{x^2}{2}$, from $(1, 1/2)$ to $(0, 0)$

21. $f(x, y) = x + y$, $\quad C\colon x^2 + y^2 = 4$ in the first quadrant from $(2, 0)$ to $(0, 2)$

22. $f(x, y) = x^2 - y$, $\quad C\colon x^2 + y^2 = 4$ in the first quadrant from $(0, 2)$ to $(\sqrt{2}, \sqrt{2})$

23. *Mass of a Wire* Find the mass of a wire that lies along the curve $\mathbf{r}(t) = (t^2 - 1)\mathbf{j} + 2t\mathbf{k}, 0 \le t \le 1$, if the density is $\delta = (3/2)t$.

24. *Center of Mass of a Curved Wire* A wire of density $\delta(x, y, z) = 15\sqrt{y + 2}$ lies along the curve $\mathbf{r}(t) = (t^2 - 1)\mathbf{j} + 2t\mathbf{k}, -1 \le t \le 1$. Find its center of mass. Then sketch the curve and center of mass together.

25. *Wire with Different Density Functions* Find the mass of a thin wire lying along the curve $\mathbf{r}(t) = \sqrt{2}t\mathbf{i} + \sqrt{2}t\mathbf{j} + (4 - t^2)\mathbf{k}, 0 \le t \le 1$, if the density is **(a)** $\delta = 3t$, **(b)** $\delta = 1$.

26. *Wire with Variable Density* Find the center of mass of a thin wire lying along the curve $\mathbf{r}(t) = t\mathbf{i} + 2t\mathbf{j} + (2/3)t^{3/2}\mathbf{k}$, $0 \le t \le 2$, if the density is $\delta = 3\sqrt{5 + t}$.

27. *A Wire Hoop* A circular wire hoop of constant density δ lies along the circle $x^2 + y^2 = a^2$ in the xy-plane. Find the hoop's moment of inertia and radius of gyration about the z-axis.

28. *A Slender Rod* A slender rod of constant density lies along the line segment $\mathbf{r}(t) = t\mathbf{j} + (2 - 2t)\mathbf{k}, 0 \le t \le 1$, in the yz-plane. Find the moments of inertia and radii of gyration of the rod about the three coordinate axes.

In Exercises 29 and 30, *work in groups of two or three* to solve the problem.

29. Writing to Learn *Two Springs of Constant Density* A spring of constant density δ lies along the helix $\mathbf{r}(t) = (\cos t)\mathbf{i} + (\sin t)\mathbf{j} + t\mathbf{k}, 0 \le t \le 2\pi$.

(a) Find I_z and R_z.

(b) Suppose you have another spring of constant density δ that is twice as long as the spring in (a) and lies along the helix for $0 \le t \le 4\pi$. Do you expect I_z and R_z for the longer spring to be the same as those for the shorter one, or should they be different? Check your prediction by calculating I_z and R_z for the longer spring.

30. *Wire of Constant Density* A wire of constant density $\delta = 1$ lies along the curve $\mathbf{r}(t) = (t \cos t)\mathbf{i} + (t \sin t)\mathbf{j} + (2\sqrt{2}/3)t^{3/2}\mathbf{k}, 0 \le t \le 1$. Find $\bar{z}$, I_z, and R_z.

31. *The Arch in Example 4* Find I_x and R_x for the arch in Example 4.

32. *Wire with Variable Density* Find the center of mass, and the moments of inertia and radii of gyration about the coordinate axes of a thin wire lying along the curve

$$\mathbf{r}(t) = t\mathbf{i} + \frac{2\sqrt{2}}{3}t^{3/2}\mathbf{j} + \frac{t^2}{2}\mathbf{k}, \quad 0 \le t \le 2,$$

if the density is $\delta = 1/(t + 1)$.

Extending the Ideas

CAS Explorations

In Exercises 33–36, perform the following steps to evaluate the line integral:

(a) Find $ds = |\mathbf{v}(t)| \, dt$ for the path $\mathbf{r}(t) = g(t)\mathbf{i} + h(t)\mathbf{j} + k(t)\mathbf{k}.$

(b) Express the integrand $f(g(t), h(t), k(t))|\mathbf{v}(t)|$ as a function of the parameter t.

(c) Evaluate $\int_C f \, ds$ using Equation (3) in the text.

33. $f(x, y, z) = \sqrt{1 + 30x^2 + 10y}$; $\mathbf{r}(t) = t\mathbf{i} + t^2\mathbf{j} + 3t^2\mathbf{k},$
   $0 \le t \le 2$

34. $f(x, y, z) = \sqrt{1 + x^3 + 5y^3}$; $\mathbf{r}(t) = t\mathbf{i} + \dfrac{1}{3}t^2\mathbf{j} + \sqrt{t}\,\mathbf{k},$
   $0 \le t \le 2$

35. $f(x, y, z) = x\sqrt{y} - 3z^2$; $\mathbf{r}(t) = \cos 2t\,\mathbf{i} + \sin 2t\,\mathbf{j} + 5t\mathbf{k},$
   $0 \le t \le \pi/2$

36. $f(x, y, z) = \left(1 + \dfrac{9}{4}z^{1/3}\right)^{1/4}$;
   $\mathbf{r}(t) = \cos 2t\,\mathbf{i} + \sin 2t\,\mathbf{j} + t^{5/2}\mathbf{k},$ $0 \le t \le 2\pi$

Figure 15.8 The velocity vectors $\mathbf{v}(t)$ of a projectile's motion make a vector field along the trajectory.

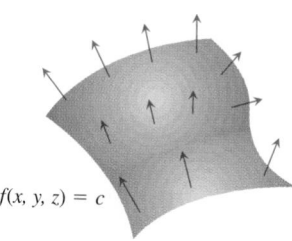

$f(x, y, z) = c$

Figure 15.9 The field of gradient vectors ∇f on a surface $f(x, y, z) = c$.

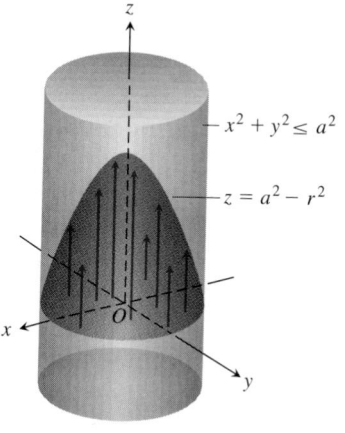

$x^2 + y^2 \le a^2$

$z = a^2 - r^2$

15.2 Vector Fields, Work, Circulation, and Flux

Vector Fields • Gradient Fields • Work Done by a Force over a Curve in Space • Notation and Evaluation • Flow Integrals and Circulation • Flux Across a Plane Curve

Vector Fields

A **vector field** on a domain in the plane or in space is a function that assigns a vector to each point in the domain. In a typical field of three-dimensional vectors has a formula like

$$\mathbf{F}(x, y, z) = M(x, y, z)\mathbf{i} + N(x, y, z)\mathbf{j} + P(x, y, z)\mathbf{k}.$$

The field is **continuous** if the **component functions** M, N, and P are continuous, **differentiable** if M, N, and P are differentiable, and so on. A typical field of two-dimensional vectors has a formula like

$$\mathbf{F}(x, y) = M(x, y)\mathbf{i} + N(x, y)\mathbf{j}.$$

If we attach a projectile's velocity vector to each point of the projectile's trajectory in the plane of motion, we have a two-dimensional field defined along the trajectory. If we attach the gradient vector of a scalar function to each point of a level surface of the function, we have a three-dimensional field on the surface. If we attach the velocity vector to each point of a flowing fluid, we have a three-dimensional field defined on a region in space. These and other fields are illustrated in Figures 15.8 through 15.16. Some of the illustrations give formulas for the fields as well.

Figure 15.10 The flow of fluid in a long cylindrical pipe. The vectors $\mathbf{v} = (a^2 - r^2)\mathbf{k}$ inside the cylinder that have their bases in the xy-plane have their tips on the paraboloid $z = a^2 - r^2$.

Figure 15.11 Velocity vectors of a flow around an airfoil in a wind tunnel. The streamlines were made visible by kerosene smoke. (Adapted from *NCFMF Book of Film Notes,* 1974, MIT Press with Education Development Center, Inc., Newton, Massachusetts.)

Figure 15.12 Streamlines in a contracting channel. The water speeds up as the channel narrows and the velocity vectors increase in length. (Adapted from *NCFMF Book of Film Notes,* 1974, MIT Press with Education Development Center, Inc., Newton, Massachusetts.)

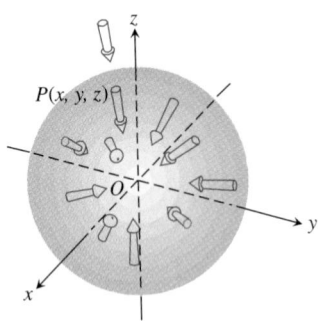

Figure 15.13 Vectors in the gravitational field

$$\mathbf{F} = -\frac{GM(x\mathbf{i} + y\mathbf{j} + z\mathbf{k})}{(x^2 + y^2 + z^2)^{3/2}}.$$

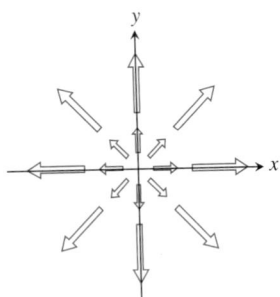

Figure 15.14 The radial field $\mathbf{F} = x\mathbf{i} + y\mathbf{j}$ of position vectors of points in the plane. Notice the convention that an arrow is drawn with its tail, not its head, at the point where $\mathbf{F}$ is evaluated.

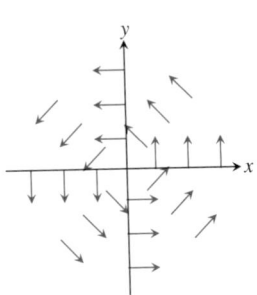

Figure 15.15 The circumferential or "spin" field of unit vectors

$$\mathbf{F} = (-y\mathbf{i} + x\mathbf{j})/(x^2 + y^2)^{1/2}$$

in the plane. The field is not defined at the origin.

Figure 15.16 NASA's *Seasat* used radar during a three-day period in September 1978 to take 350,000 wind measurements over the world's oceans. The arrows show wind direction; their length and the color contouring indicate speed. Notice the heavy storm south of Greenland.

To sketch the fields that had formulas, we picked a representative selection of domain points and sketched the vectors attached to them. Notice the convention that the arrows representing the vectors are drawn with their tails, not their heads, at the points where the vector functions are evaluated. This is different from the way we drew the position vectors of the planets and projectiles in Chapters 10 and 11, with their tails at the origin and their heads at the planet's and projectile's location.

Gradient Fields

> **Definition** Gradient Field
>
> The **gradient field** of a differentiable function $f(x, y, z)$ is the field of gradient vectors
>
> $$\nabla f = \frac{\partial f}{\partial x}\mathbf{i} + \frac{\partial f}{\partial y}\mathbf{j} + \frac{\partial f}{\partial z}\mathbf{k}.$$

Example 1 FINDING GRADIENT FIELDS

Find the gradient field of $f(x, y, z) = xyz$.

Solution   The gradient field of f is the field $\mathbf{F} = \nabla f = yz\mathbf{i} + xz\mathbf{j} + xy\mathbf{k}.$

As we see in Section 15.3, gradient fields play a special role in engineering, mathematics, and physics.

Work Done by a Force over a Curve in Space

Suppose the vector field

$$\mathbf{F} = M(x, y, z)\mathbf{i} + N(x, y, z)\mathbf{j} + P(x, y, z)\mathbf{k}$$

represents a force throughout a region in space (it might be the force of gravity or an electromagnetic force of some kind) and

$$\mathbf{r}(t) = g(t)\mathbf{i} + h(t)\mathbf{j} + k(t)\mathbf{k}, \qquad a \le t \le b,$$

is a smooth curve in the region. Then the integral of $\mathbf{F} \cdot \mathbf{T}$, the scalar component of $\mathbf{F}$ in the direction of the curve's unit tangent vector, over the curve is the *work done by* $\mathbf{F}$ *over the curve* from a to b (Figure 15.17).

> **Definition** Work
>
> The **work** done by a force $\mathbf{F} = M(x, y, z)\mathbf{i} + N(x, y, z)\mathbf{j} + P(x, y, z)\mathbf{k}$ over a smooth curve $\mathbf{r}(t) = g(t)\mathbf{i} + h(t)\mathbf{j} + k(t)\mathbf{k}$ from $t = a$ to $t = b$ is
>
> $$W = \int_{t=a}^{t=b} \mathbf{F} \cdot \mathbf{T} \, ds. \qquad (1)$$

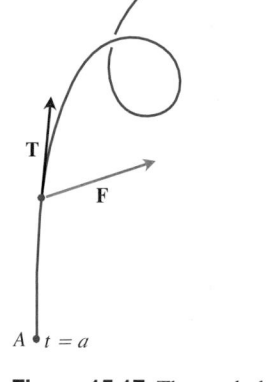

Figure 15.17 The work done by a continuous field $\mathbf{F}$ over a smooth path $\mathbf{r} = g(t)\mathbf{i} + h(t)\mathbf{j} + k(t)\mathbf{k}$ from A to B is the integral of $\mathbf{F} \cdot \mathbf{T}$ over the path from $t = a$ to $t = b$.

We motivate Equation (1) with the same kind of reasoning we used in Section 7.5 to derive the formula $W = \int_{a}^{b} F(x) \, dx$ for the work done by a continuous force of magnitude $F(x)$ directed along an interval of the x-axis. We divide the curve into short segments, apply the constant-force-times-distance formula for work to approximate the work over each curved segment, add the results to approximate the work over the entire curve, and calculate the work as the limit of the approximating sums as the segments become shorter and more numerous. To find exactly what the limiting integral should be, we partition the parameter interval $I = [a, b]$ in the usual way and choose a point c_k in each subinterval $[t_k, t_{k+1}]$. The partition of I determines ("induces," we say) a partition of the curve, with the point P_k being the tip of the position vector $\mathbf{r}$ at $t = t_k$ and Δs_k being the length of the curve segment $P_k P_{k+1}$ (Figure 15.18).

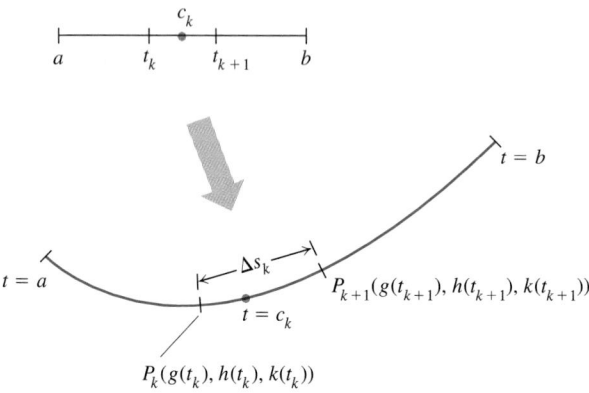

Figure 15.18 Each partition interval $a \le t \le b$ induces a partition of the curve $\mathbf{r} = g(t)\mathbf{i} + h(t)\mathbf{j} + k(t)\mathbf{k}$.

If $\mathbf{F}_k$ denotes the value of $\mathbf{F}$ at the point on the curve corresponding to $t = c_k$ and $\mathbf{T}_k$ denotes the curve's tangent vector at this point, then $\mathbf{F}_k \cdot \mathbf{T}_k$ is the scalar component of $\mathbf{F}$ in the direction of $\mathbf{T}$ at $t = c_k$ (Figure 15.19). The work done by $\mathbf{F}$ along the curve segment $P_k P_{k+1}$ will be approximately

$$\begin{pmatrix} \text{Force component in} \\ \text{direction of motion} \end{pmatrix} \times \begin{pmatrix} \text{distance} \\ \text{applied} \end{pmatrix} = \mathbf{F}_k \cdot \mathbf{T}_k \, \Delta s_k.$$

The work done by $\mathbf{F}$ along the curve from $t = a$ to $t = b$ will be approximately

$$\sum_{k=1}^{n} \mathbf{F}_k \cdot \mathbf{T}_k \, \Delta s_k.$$

As the norm of the partition of $[a, b]$ approaches zero, the norm of the induced partition of the curve approaches zero and these sums approach the line integral

$$\int_{t=a}^{t=b} \mathbf{F} \cdot \mathbf{T} \, ds.$$

The sign of the number we calculate with this integral depends on the direction in which the curve is traversed as t increases. If we reverse the direction of motion, then we reverse the direction of $\mathbf{T}$ and change the sign of $\mathbf{F} \cdot \mathbf{T}$ and its integral.

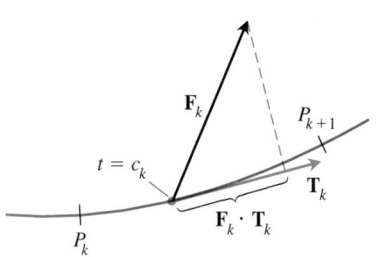

Figure 15.19 An enlarged view of the curve segment $P_k P_{k+1}$ in Figure 15.18, showing the force vector and unit tangent vector at the point on the curve where $t = c_k$.

Notation and Evaluation

Table 15.2 shows six ways to write the work integral in Equation (1). Despite their variety, the formulas in Table 15.2 are all evaluated the same way.

Table 15.2 Different Ways to Write the Work Integral

$$W = \int_{t=a}^{t=b} \mathbf{F} \cdot \mathbf{T}\, ds \qquad \text{The definition}$$

$$= \int_{t=a}^{t=b} \mathbf{F} \cdot d\mathbf{r} \qquad \text{Compact differential form}$$

$$= \int_{a}^{b} \mathbf{F} \cdot \frac{d\mathbf{r}}{dt}\, dt \qquad \begin{array}{l}\text{Expanded to include } dt; \\ \text{emphasizes the parameter } t \text{ and} \\ \text{velocity vector } d\mathbf{r}/dt\end{array}$$

$$= \int_{a}^{b} \left(M \frac{dg}{dt} + N \frac{dh}{dt} + P \frac{dk}{dt}\right) dt \qquad \begin{array}{l}\text{Emphasizes the component} \\ \text{functions}\end{array}$$

$$= \int_{a}^{b} \left(M \frac{dx}{dt} + N \frac{dy}{dt} + P \frac{dz}{dt}\right) dt \qquad \text{Abbreviates the components of } \mathbf{r}$$

$$= \int_{a}^{b} M\, dx + N\, dy + P\, dz \qquad \begin{array}{l}dt\text{'s canceled; the most common} \\ \text{form}\end{array}$$

How to Evaluate a Work Integral

To evaluate the work integral, take these steps:

1. Evaluate $\mathbf{F}$ on the curve as a function of the parameter t.

2. Find $d\mathbf{r}/dt$.

3. Dot $\mathbf{F}$ with $d\mathbf{r}/dt$.

4. Integrate the resulting scalar function from $t = a$ to $t = b$.

Example 2   FINDING WORK DONE BY A FORCE

Find the work done by $\mathbf{F} = (y - x^2)\mathbf{i} + (z - y^2)\mathbf{j} + (x - z^2)\mathbf{k}$ over the curve $\mathbf{r}(t) = t\mathbf{i} + t^2\mathbf{j} + t^3\mathbf{k},\ 0 \le t \le 1$, from $(0, 0, 0)$ to $(1, 1, 1)$ (Figure 15.20).

Solution

Step 1: *Evaluate* $\mathbf{F}$ *on the curve.*

$$\mathbf{F} = (y - x^2)\mathbf{i} + (z - y^2)\mathbf{j} + (x - z^2)\mathbf{k}$$
$$= \underbrace{(t^2 - t^2)}_{0}\mathbf{i} + (t^3 - t^4)\mathbf{j} + (t - t^6)\mathbf{k}$$

Step 2: *Find* $d\mathbf{r}/dt$.

$$\frac{d\mathbf{r}}{dt} = \frac{d}{dt}(t\mathbf{i} + t^2\mathbf{j} + t^3\mathbf{k}) = \mathbf{i} + 2t\mathbf{j} + 3t^2\mathbf{k}$$

Step 3: *Dot* $\mathbf{F}$ *with* $d\mathbf{r}/dt$.

$$\mathbf{F} \cdot \frac{d\mathbf{r}}{dt} = \left[(t^3 - t^4)\mathbf{j} + (t - t^6)\mathbf{k}\right] \cdot (\mathbf{i} + 2t\mathbf{j} + 3t^2\mathbf{k})$$

$$= (t^3 - t^4)(2t) + (t - t^6)(3t^2) = 2t^4 - 2t^5 + 3t^3 - 3t^8$$

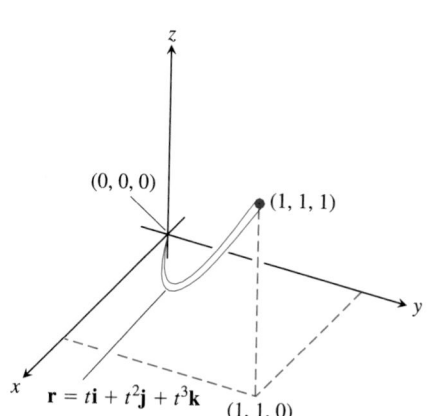

z

$(0, 0, 0)$ $(1, 1, 1)$

y

x $\mathbf{r} = t\mathbf{i} + t^2\mathbf{j} + t^3\mathbf{k}$ $(1, 1, 0)$

Figure 15.20 The curve in Example 2.

In Step 4 of Example 2, you can use your grapher to approximate the value of the integral or a CAS to find the exact value shown.

Step 4: *Integrate from $t = 0$ to $t = 1$.*

$$\text{Work} = \int_0^1 (2t^4 - 2t^5 + 3t^3 - 3t^8)\, dt = \frac{29}{60}$$

Flow Integrals and Circulation

Instead of being a force field, suppose $\mathbf{F} = M\mathbf{i} + N\mathbf{j} + P\mathbf{k}$ represents the velocity field of a fluid flowing through a region in space (a tidal basin or the turbine chamber of a hydroelectric generator, for example). Under these circumstances, the integral of $\mathbf{F} \cdot \mathbf{T}$ along a curve in the region gives the *fluid's flow along the curve.*

> **Definitions** Flow, Flow Integral, and Circulation
>
> If $\mathbf{r}(t) = g(t)\mathbf{i} + h(t)\mathbf{j} + k(t)\mathbf{k}$, $a \le t \le b$, is a smooth curve in the domain of a continuous velocity field $\mathbf{F} = M(x, y, z)\mathbf{i} + N(x, y, z)\mathbf{j} + P(x, y, z)\mathbf{k}$, the **flow** along the curve from $t = a$ to $t = b$ is the integral of $\mathbf{F} \cdot \mathbf{T}$ over the curve from a to b:
>
> $$\text{Flow} = \int_a^b \mathbf{F} \cdot \mathbf{T}\, ds. \tag{2}$$
>
> The integral in this case is a **flow integral.** If the curve is a closed loop, then the flow is the **circulation** around the curve.

We evaluate flow integrals the same way we evaluate work integrals.

Example 3 FINDING THE FLOW

A fluid's velocity field is $\mathbf{F} = x\mathbf{i} + z\mathbf{j} + y\mathbf{k}$. Find the flow along the helix $\mathbf{r}(t) = (\cos t)\mathbf{i} + (\sin t)\mathbf{j} + t\mathbf{k}$, $0 \le t \le \pi/2$.

Solution

Step 1: *Evaluate $\mathbf{F}$ on the curve.*

$$\mathbf{F} = x\mathbf{i} + z\mathbf{j} + y\mathbf{k} = (\cos t)\mathbf{i} + t\mathbf{j} + (\sin t)\mathbf{k}$$

Step 2: *Find $d\mathbf{r}/dt$.*

$$\frac{d\mathbf{r}}{dt} = (-\sin t)\mathbf{i} + (\cos t)\mathbf{j} + \mathbf{k}$$

Step 3: *Find $\mathbf{F} \cdot (d\mathbf{r}/dt)$.*

$$\mathbf{F} \cdot \frac{d\mathbf{r}}{dt} = (\cos t)(-\sin t) + (t)(\cos t) + (\sin t)(1)$$

$$= -\sin t \cos t + t \cos t + \sin t$$

Step 4: *Integrate from $t = a$ to $t = b$.*

$$\text{Flow} = \int_{t=a}^{t=b} \mathbf{F} \cdot \frac{d\mathbf{r}}{dt}\, dt = \int_0^{\pi/2} (-\sin t \cos t + t \cos t + \sin t)\, dt$$

$$= \frac{\pi}{2} - \frac{1}{2}$$

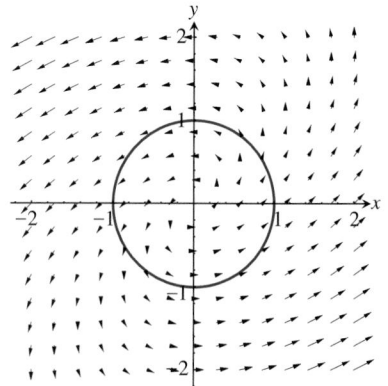

Figure 15.21 This figure suggests that the projections of the vectors of **F** that start on the circle onto the tangent vectors to the circle point in the direction of increasing t. This supports the positive value for circulation found in Example 4.

Example 4   FINDING CIRCULATION

Find the circulation of the field $\mathbf{F} = (x - y)\mathbf{i} + x\mathbf{j}$ around the circle $\mathbf{r}(t) = (\cos t)\mathbf{i} + (\sin t)\mathbf{j}, 0 \le t \le 2\pi$.

Solution

1. On the circle, $\mathbf{F} = (x - y)\mathbf{i} + x\mathbf{j} = (\cos t - \sin t)\mathbf{i} + (\cos t)\mathbf{j}.$

2. $\dfrac{d\mathbf{r}}{dt} = (-\sin t)\mathbf{i} + (\cos t)\mathbf{j}$

3. $\mathbf{F} \cdot \dfrac{d\mathbf{r}}{dt} = -\sin t \cos t + \underbrace{\sin^2 t + \cos^2 t}_{1}$

4. Circulation $= \displaystyle\int_0^{2\pi} \mathbf{F} \cdot \frac{d\mathbf{r}}{dt}\, dt = \int_0^{2\pi} (1 - \sin t \cos t)\, dt$

$$= \left[t - \frac{\sin^2 t}{2} \right]_0^{2\pi} = 2\pi$$

Figure 15.21 shows the circulation vector field **F** together with the unit circle $\mathbf{r}(t)$.

Flux Across a Plane Curve

To find the rate at which a fluid is entering or leaving a region enclosed by a smooth curve C in the xy-plane, we calculate the line integral over C of $\mathbf{F} \cdot \mathbf{n}$, the scalar component of the fluid's velocity field in the direction of the curve's outward-pointing normal vector. The value of this integral is the *flux of* $\mathbf{F}$ *across* C. *Flux* is Latin for *flow,* but many flux calculations involve no motion at all. If **F** were an electric field or a magnetic field, for instance, the integral of $\mathbf{F} \cdot \mathbf{n}$ would still be called the flux of the field across C.

Definition   Flux

If C is a smooth closed curve in the domain of a continuous vector field $\mathbf{F} = M(x, y)\mathbf{i} + N(x, y)\mathbf{j}$ in the plane and if **n** is the outward-pointing unit normal vector on C, then the **flux** of **F** across C is given by the following line integral:

$$\text{Flux of } \mathbf{F} \text{ across } C = \int_C \mathbf{F} \cdot \mathbf{n}\, ds. \tag{3}$$

Notice the difference between flux and circulation. The flux of **F** across C is the line integral with respect to arc length of $\mathbf{F} \cdot \mathbf{n}$, the scalar component of **F** in the direction of the outward normal. The circulation of **F** around C is the line integral with respect to arc length of $\mathbf{F} \cdot \mathbf{T}$, the scalar component of **F** in the direction of the unit tangent vector. Flux is the integral of the normal component of **F**; circulation is the integral of the tangential component of **F**.

To evaluate the integral in Equation (3), we begin with a parametrization

$$x = g(t), \qquad y = h(t), \qquad a \le t \le b,$$

that traces the curve C exactly once as t increases from a to b. We can find the outward unit normal vector **n** by crossing the curve's unit tangent vector **T**

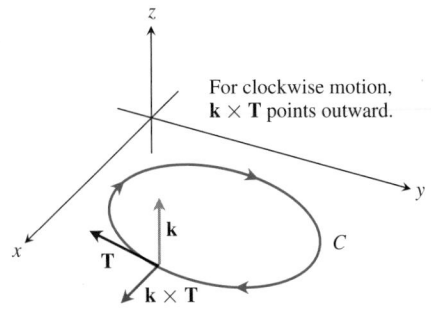

For clockwise motion, $\mathbf{k} \times \mathbf{T}$ points outward.

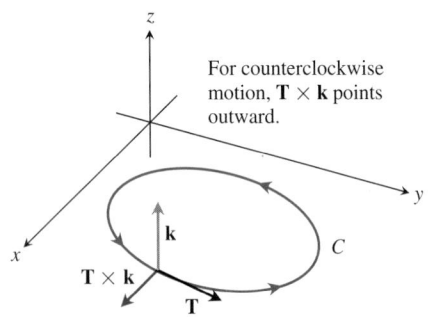

For counterclockwise motion, $\mathbf{T} \times \mathbf{k}$ points outward.

Figure 15.22 To find an outward unit normal vector for a smooth curve C in the xy-plane that is traversed counterclockwise as t increases, we take $\mathbf{n} = \mathbf{T} \times \mathbf{k}$.

with the vector $\mathbf{k}$, but which order do we choose, $\mathbf{T} \times \mathbf{k}$ or $\mathbf{k} \times \mathbf{T}$? Which one points outward? It depends on which way C is traversed as the parameter t increases. If the motion is clockwise, then $\mathbf{k} \times \mathbf{T}$ points outward; if the motion is counterclockwise, then $\mathbf{T} \times \mathbf{k}$ points outward (Figure 15.22). The usual choice is $\mathbf{n} = \mathbf{T} \times \mathbf{k}$, the choice that assumes counterclockwise motion. Thus, although the value of the arc length integral in the definition of flux in Equation (3) does not depend on which way C is traversed, the formulas we are about to derive for *evaluating* the integral in Equation (3) will assume counterclockwise motion.

In terms of components,

$$\mathbf{n} = \mathbf{T} \times \mathbf{k} = \left(\frac{dx}{ds}\mathbf{i} + \frac{dy}{ds}\mathbf{j}\right) \times \mathbf{k} = \frac{dy}{ds}\mathbf{i} - \frac{dx}{ds}\mathbf{j}.$$

If $\mathbf{F} = M(x, y)\mathbf{i} + N(x, y)\mathbf{j}$, then

$$\mathbf{F} \cdot \mathbf{n} = M(x, y)\frac{dy}{ds} - N(x, y)\frac{dx}{ds}.$$

Hence,

$$\int_C \mathbf{F} \cdot \mathbf{n}\, ds = \int_C \left(M\frac{dy}{ds} - N\frac{dx}{ds}\right) ds = \oint_C M\, dy - N\, dx.$$

We put a directed circle $\circlearrowleft$ on the last integral as a reminder that the integration around the closed curve C is to be in the counterclockwise direction. To evaluate this integral, we express M, dy, N, and dx in terms of t and integrate from $t = a$ to $t = b$. We do not need to know either $\mathbf{n}$ or ds to find the flux.

> **Formula for Calculating Flux Across a Smooth Closed Plane Curve**
>
> $$\text{(Flux of } \mathbf{F} = M\mathbf{i} + N\mathbf{j} \text{ across } C) = \oint_C M\, dy - N\, dx \qquad (4)$$
>
> The integral can be evaluated from any smooth parametrization $x = g(t)$, $y = h(t)$, $a \le t \le b$, that traces C counterclockwise exactly once.

Example 5   FINDING FLUX ACROSS A CIRCLE

Find the flux of $\mathbf{F} = (x - y)\mathbf{i} + x\mathbf{j}$ across the circle $x^2 + y^2 = 1$ in the xy-plane.

Solution   The parametrization $\mathbf{r}(t) = (\cos t)\mathbf{i} + (\sin t)\mathbf{j}, 0 \le t \le 2\pi$, traces the circle counterclockwise exactly once. We can therefore use this parametrization in Equation (4). With

$$M = x - y = \cos t - \sin t, \qquad dy = d(\sin t) = \cos t\, dt$$
$$N = x = \cos t, \qquad dx = d(\cos t) = -\sin t\, dt,$$

we find

$$\text{Flux} = \oint_C M\, dy - N\, dx = \int_0^{2\pi} (\cos^2 t - \sin t \cos t + \cos t \sin t)\, dt \quad \text{Eq. (4)}$$
$$= \int_0^{2\pi} \cos^2 t\, dt = \int_0^{2\pi} \frac{1 + \cos 2t}{2} = \pi.$$

The flux of **F** across the circle is π. Since the answer is positive, the net flow across the curve is outward. A net inward flow would have given a negative flux.

Quick Review 15.2

1. Let $f(x, y) = x \sin y + y \cos x$. Find $\partial f/\partial x$ and $\partial f/\partial y$.

2. Let $f(x, y, z) = x \sin (yz) + xyz$. Find $\partial f/\partial x$, $\partial f/\partial y$, and $\partial f/\partial z$.

In Exercises 3 and 4, find dy.

3. $y = x^3 - 2 \cos x$ **4.** $y = x \sin x$

5. Let $x = 2 \cos t$ and $y = 3 \sin t$. Find dy and dx.

6. Let $\mathbf{u} = t\mathbf{i} + t^2\mathbf{j} + (\cos t)\mathbf{k}$ and $\mathbf{v} = (\cos t)\mathbf{i} + (\sin t)\mathbf{j} - t\mathbf{k}$. Compute $\mathbf{u} \cdot \mathbf{v}$.

In Exercises 7–10, identify whether the circle is being traced clockwise or counterclockwise.

7. $x = 2 \cos t$, $y = 2 \sin t$, $0 \le t \le 2\pi$

8. $x = -2 \cos t$, $y = 2 \sin t$, $0 \le t \le 2\pi$

9. $x = -2 \cos t$, $y = -2 \sin t$, $0 \le t \le 2\pi$

10. $x = 2 \sin t$, $y = 2 \cos t$, $0 \le t \le 2\pi$

Section 15.2 Exercises

In Exercises 1–4, find the gradient field of the function.

1. $f(x, y, z) = (x^2 + y^2 + z^2)^{-1/2}$

2. $f(x, y, z) = \ln\sqrt{x^2 + y^2 + z^2}$

3. $g(x, y, z) = e^z - \ln(x^2 + y^2)$

4. $g(x, y, z) = xy + yz + xz$

5. Give a formula $\mathbf{F} = M(x, y)\mathbf{i} + N(x, y)\mathbf{j}$ for the vector field in the plane that has the property that **F** points toward the origin with magnitude inversely proportional to the square of the distance from to the origin. (The field is not defined at (0, 0).)

6. Give a formula $\mathbf{F} = M(x, y)\mathbf{i} + N(x, y)\mathbf{j}$ for the vector field in the plane that has the property that $\mathbf{F} = \mathbf{0}$ at (0, 0) and that, at any other point (a, b), **F** is tangent to the circle $x^2 + y^2 = a^2 + b^2$ and points in the clockwise direction with magnitude $|\mathbf{F}| = \sqrt{a^2 + b^2}$.

In Exercises 7–12, *work in groups of two or three* to find the work done by force **F** from (0, 0, 0) to (1, 1, 1) over each of the following paths (Figure 15.23).

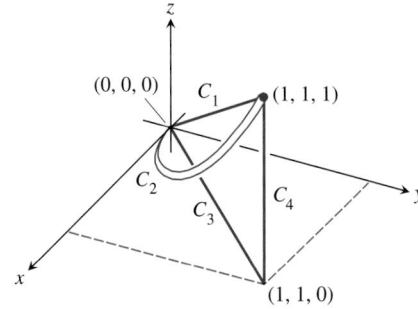

Figure 15.23 The paths from (0, 0, 0) to (1, 1, 1).

(a) The straight-line path C_1: $\mathbf{r}(t) = t\mathbf{i} + t\mathbf{j} + t\mathbf{k}, 0 \le t \le 1$

(b) The curved path C_2: $\mathbf{r}(t) = t\mathbf{i} + t^2\mathbf{j} + t^4\mathbf{k}, 0 \le t \le 1$

(c) The path $C_3 \cup C_4$ consisting of the line segment from (0, 0, 0) to (1, 1, 0) followed by the segment from (1, 1, 0) to (1, 1, 1)

7. $\mathbf{F} = 3y\mathbf{i} + 2x\mathbf{j} + 4z\mathbf{k}$ **8.** $\mathbf{F} = [1/(x^2 + 1)]\mathbf{j}$

9. $\mathbf{F} = \sqrt{z}\mathbf{i} - 2x\mathbf{j} + \sqrt{y}\mathbf{k}$ **10.** $\mathbf{F} = xy\mathbf{i} + yz\mathbf{j} + xz\mathbf{k}$

11. $\mathbf{F} = (3x^2 - 3x)\mathbf{i} + 3z\mathbf{j} + \mathbf{k}$

12. $\mathbf{F} = (y + z)\mathbf{i} + (z + x)\mathbf{j} + (x + y)\mathbf{k}$

In Exercises 13–16, find the work done by **F** over the curve in the direction of increasing t.

13. $\mathbf{F} = xy\mathbf{i} + y\mathbf{j} - yz\mathbf{k}$
$\mathbf{r}(t) = t\mathbf{i} + t^2\mathbf{j} + t\mathbf{k}$, $0 \le t \le 1$

14. $\mathbf{F} = 2y\mathbf{i} + 3x\mathbf{j} + (x + y)\mathbf{k}$
$\mathbf{r}(t) = (\cos t)\mathbf{i} + (\sin t)\mathbf{j} + (t/6)\mathbf{k}$, $0 \le t \le 2\pi$

15. $\mathbf{F} = z\mathbf{i} + x\mathbf{j} + y\mathbf{k}$
$\mathbf{r}(t) = (\sin t)\mathbf{i} + (\cos t)\mathbf{j} + t\mathbf{k}$, $0 \le t \le 2\pi$

16. $\mathbf{F} = 6z\mathbf{i} + y^2\mathbf{j} + 12x\mathbf{k}$
$\mathbf{r}(t) = (\sin t)\mathbf{i} + (\cos t)\mathbf{j} + (t/6)\mathbf{k}$, $0 \le t \le 2\pi$

17. Evaluate $\displaystyle\int_C xy\, dx + (x + y)\, dy$ along the curve $y = x^2$ in the xy-plane from $(-1, 1)$ to $(2, 4)$.

18. Evaluate $\displaystyle\int_C (x - y)\, dx + (x + y)\, dy$ counterclockwise around the triangle in the xy-plane with vertices (0, 0), (1, 0), and (0, 1).

19. Evaluate $\int_C \mathbf{F} \cdot \mathbf{T} \, ds$ for the vector field $\mathbf{F} = x^2\mathbf{i} - y\mathbf{j}$ along the curve $x = y^2$ from (4, 2) to (1, −1) in the *xy*-plane.

20. Evaluate $\int_C \mathbf{F} \cdot d\mathbf{r}$ for the vector field $\mathbf{F} = y\mathbf{i} - x\mathbf{j}$ counterclockwise along the unit circle $x^2 + y^2 = 1$ from (1, 0) to (0, 1) in the *xy*-plane.

21. *Work* Find the work done by the force $\mathbf{F} = xy\mathbf{i} + (y - x)\mathbf{j}$ over the straight line from (1, 1) to (2, 3) in the *xy*-plane.

22. *Work* Find the work done by the gradient of $f(x, y) = (x + y)^2$ counterclockwise around the circle $x^2 + y^2 = 4$ from (2, 0) to itself in the *xy*-plane.

23. *Circulation and Flux* Find the circulation and flux of the fields

$$\mathbf{F}_1 = x\mathbf{i} + y\mathbf{j} \quad \text{and} \quad \mathbf{F}_2 = -y\mathbf{i} + x\mathbf{j}$$

around and across each of the following curves in the *xy*-plane.

(a) The circle $\mathbf{r}(t) = (\cos t)\mathbf{i} + (\sin t)\mathbf{j}, 0 \le t \le 2\pi$

(b) The ellipse $\mathbf{r}(t) = (\cos t)\mathbf{i} + (4 \sin t)\mathbf{j}, 0 \le t \le 2\pi$

24. *Flux Across a Circle* Find the flux of the fields
$\mathbf{F}_1 = 2x\mathbf{i} - 3y\mathbf{j} \quad \text{and} \quad \mathbf{F}_2 = -2x\mathbf{i} + (x - y)\mathbf{j}$
across the circle $\mathbf{r}(t) = (a \cos t)\mathbf{i} + (a \sin t)\mathbf{j}$,
$0 \le t \le 2\pi$ in the *xy*-plane.

In Exercises 25–28, find the circulation and flux of the field $\mathbf{F}$ around and across the closed semicircular path in the *xy*-plane that consists of the semicircular arch $\mathbf{r}_1(t) = (a \cos t)\mathbf{i} + (a \sin t)\mathbf{j}$, $0 \le t \le \pi$, followed by the line segment $\mathbf{r}_2(t) = t\mathbf{i}, -a \le t \le a$.

25. $\mathbf{F} = x\mathbf{i} + y\mathbf{j}$ **26.** $\mathbf{F} = x^2\mathbf{i} + y^2\mathbf{j}$

27. $\mathbf{F} = -y\mathbf{i} + x\mathbf{j}$ **28.** $\mathbf{F} = -y^2\mathbf{i} + x^2\mathbf{j}$

29. *Flow* Evaluate the flow integral of the velocity field $\mathbf{F} = (x + y)\mathbf{i} - (x^2 + y^2)\mathbf{j}$ along each of the following paths from (1, 0) to (−1, 0) in the *xy*-plane.

(a) The upper half of the circle $x^2 + y^2 = 1$

(b) The line segment from (1, 0) to (−1, 0)

(c) The line segment from (1, 0) to (0, −1) followed by the line segment from (0, −1) to (−1, 0)

30. *Flux* Find the flux of the field in Exercise 29 outward across the triangle with vertices (1, 0), (0, 1), (−1, 0).

31. *Spin Field* Draw the spin field

$$\mathbf{F} = -\frac{y}{\sqrt{x^2 + y^2}}\mathbf{i} + \frac{x}{\sqrt{x^2 + y^2}}\mathbf{j}$$

(see Figure 15.15) along with its horizontal and vertical components at a representative assortment of points on the circle $x^2 + y^2 = 4$ in the *xy*-plane.

32. *Radial Field* Draw the radial field

$$\mathbf{F} = x\mathbf{i} + y\mathbf{j}$$

(see Figure 15.14) along with its horizontal and vertical components at a representative assortment of points on the circle $x^2 + y^2 = 1$ in the *xy*-plane.

33. *A Field of Tangent Vectors*

(a) Find a field $\mathbf{G} = P(x, y)\mathbf{i} + Q(x, y)\mathbf{j}$ in the *xy*-plane with the property that at any point $(a, b) \ne (0, 0)$, $\mathbf{G}$ is a vector of magnitude $\sqrt{a^2 + b^2}$ tangent to the circle $x^2 + y^2 = a^2 + b^2$ and pointing in the counterclockwise direction. (The field is undefined at (0, 0).)

(b) Writing to Learn How is $\mathbf{G}$ related to the spin field $\mathbf{F}$ in Figure 15.15?

34. *A Field of Tangent Vectors*

(a) Find a field $\mathbf{G} = P(x, y)\mathbf{i} + Q(x, y)\mathbf{j}$ in the *xy*-plane with the property that at any point $(a, b) \ne (0, 0)$, $\mathbf{G}$ is a unit vector tangent to the circle $x^2 + y^2 = a^2 + b^2$ and pointing in the clockwise direction.

(b) Writing to Learn How is $\mathbf{G}$ related to the spin field $\mathbf{F}$ in Figure 15.15?

35. *Unit Vectors Pointing Toward the Origin* Find a field $\mathbf{F} = M(x, y)\mathbf{i} + N(x, y)\mathbf{j}$ in the *xy*-plane with the property that at each point $(x, y) \ne (0, 0)$, $\mathbf{F}$ is a unit vector pointing toward the origin. (The field is undefined at (0, 0).)

36. *Two "Central" Fields* Find a field $\mathbf{F} = M(x, y)\mathbf{i} + N(x, y)\mathbf{j}$ in the *xy*-plane with the property that at each point $(x, y) \ne (0, 0)$, $\mathbf{F}$ points toward the origin and $|\mathbf{F}|$ is **(a)** the distance from (x, y) to the origin and **(b)** inversely proportional to the distance from (x, y) to the origin. (The field is undefined at (0, 0).)

Flow In Exercises 37–40, $\mathbf{F}$ is the velocity field of a fluid flowing through a region in space. Find the flow along the given curve in the direction of increasing *t*.

37. $\mathbf{F} = -4xy\mathbf{i} + 8y\mathbf{j} + 2\mathbf{k}$

$\mathbf{r}(t) = t\mathbf{i} + t^2\mathbf{j} + \mathbf{k}, \quad 0 \le t \le 2$

38. $\mathbf{F} = x^2\mathbf{i} + yz\mathbf{j} + y^2\mathbf{k}$

$\mathbf{r}(t) = 3t\mathbf{j} + 4t\mathbf{k}, \quad 0 \le t \le 1$

39. $\mathbf{F} = (x - z)\mathbf{i} + x\mathbf{k}$

$\mathbf{r}(t) = (\cos t)\mathbf{i} + (\sin t)\mathbf{k}, \quad 0 \le t \le \pi$

40. $\mathbf{F} = -y\mathbf{i} + x\mathbf{j} + 2\mathbf{k}$

$\mathbf{r}(t) = (-2 \cos t)\mathbf{i} + (2 \sin t)\mathbf{j} + 2t\mathbf{k}, \quad 0 \le t \le 2\pi$

41. *Circulation* Find the circulation of $\mathbf{F} = 2x\mathbf{i} + 2z\mathbf{j} + 2y\mathbf{k}$ around the closed path consisting of the following three curves traversed in the direction of increasing *t*:

$C_1: \mathbf{r}(t) = (\cos t)\mathbf{i} + (\sin t)\mathbf{j} + t\mathbf{k}, \quad 0 \le t \le \pi/2$

$C_2: \mathbf{r}(t) = \mathbf{j} + (\pi/2)(1 - t)\mathbf{k}, \quad 0 \le t \le 1$

$C_3: \mathbf{r}(t) = t\mathbf{i} + (1 - t)\mathbf{j}, \quad 0 \le t \le 1$

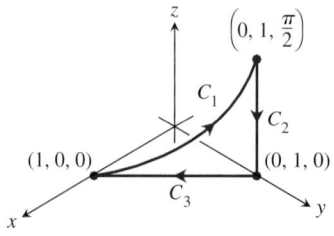

In Exercises 42–44, *work in groups of two or three* to solve the problem.

42. Writing to Learn *Zero Circulation* Let C be the ellipse in which the plane $2x + 3y - z = 0$ meets the cylinder $x^2 + y^2 = 12$. Show, without evaluating either line integral directly, that the circulation of the field $\mathbf{F} = x\mathbf{i} + y\mathbf{j} + z\mathbf{k}$ around C in either direction is zero.

43. *Flow Along a Curve* The field $\mathbf{F} = xy\mathbf{i} + y\mathbf{j} - yz\mathbf{k}$ is the velocity field of a flow in space. Find the flow from $(0, 0, 0)$ to $(1, 1, 1)$ along the curve of intersection of the cylinder $y = x^2$ and the plane $z = x$.

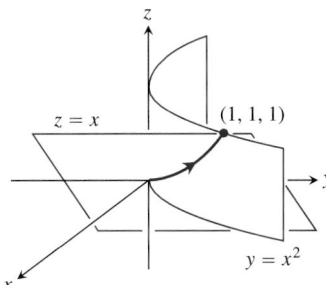

44. *Flow of a Gradient Field* Find the flow of the field $\mathbf{F} = \nabla(xy^2z^3)$

(a) once around the curve in Exercise 42, clockwise as viewed from above.

(b) along the line segment from $(1, 1, 1)$ to $(2, 1, -1)$.

Extending the Ideas

45. Writing to Learn *Work and Area* Suppose $f(t)$ is differentiable and positive for $a \le t \le b$. Let C be the path $\mathbf{r}(t) = t\mathbf{i} + f(t)\mathbf{j}$, and $\mathbf{F} = y\mathbf{i}$. Is there any relation between the value of the work integral

$$\int_C \mathbf{F} \cdot d\mathbf{r}$$

and the area of the region bounded by the t-axis, the graph of f, and the lines $t = a$ and $t = b$? Give reasons for your answer.

46. *Work Done by a Force Moving a Particle* A particle in the xy-plane moves along the smooth curve $y = f(x)$ from $(a, f(a))$ to $(b, f(b))$. The force moving the particle has constant magnitude k and always points away from the origin. Show how the work done by the force is

$$\int \mathbf{F} \cdot \mathbf{T} \, ds = k[(b^2 + (f(b))^2)^{1/2} - (a^2 + (f(a))^2)^{1/2}] \quad \blacksquare$$

CAS Explorations

In Exercises 47–52, perform the following steps for finding the work done by force $\mathbf{F}$ over the given path:

(a) Find $d\mathbf{r}$ for the path $\mathbf{r}(t) = g(t)\mathbf{i} + h(t)\mathbf{j} + k(t)\mathbf{k}$.

(b) Evaluate the force $\mathbf{F}$ along the path.

(c) Evaluate $\displaystyle\int_C \mathbf{F} \cdot d\mathbf{r}$.

47. $\mathbf{F} = xy^6\mathbf{i} + 3x(xy^5 + 2)\mathbf{j}$
$\mathbf{r}(t) = (2 \cos t)\mathbf{i} + (\sin t)\mathbf{j}, \qquad 0 \le t \le 2\pi$

48. $\mathbf{F} = \dfrac{3}{1 + x^2}\mathbf{i} + \dfrac{2}{1 + y^2}\mathbf{j}$
$\mathbf{r}(t) = (\cos t)\mathbf{i} + (\sin t)\mathbf{j}, \qquad 0 \le t \le \pi$

49. $\mathbf{F} = (y + yz \cos xyz)\mathbf{i} + (x^2 + xz \cos xyz)\mathbf{j} + (z + xy \cos xyz)\mathbf{k}$
$\mathbf{r}(t) = (2 \cos t)\mathbf{i} + (3 \sin t)\mathbf{j} + \mathbf{k}, \qquad 0 \le t \le 2\pi$

50. $\mathbf{F} = 2xy\mathbf{i} - y^2\mathbf{j} + ze^x\mathbf{k}$
$\mathbf{r}(t) = -t\mathbf{i} + \sqrt{t}\mathbf{j} + 3t\mathbf{k}, \qquad 1 \le t \le 4$

51. $\mathbf{F} = (2y + \sin x)\mathbf{i} + (z^2 + (1/3) \cos y)\mathbf{j} + x^4\mathbf{k}$
$\mathbf{r}(t) = (\sin t)\mathbf{i} + (\cos t)\mathbf{j} + (\sin 2t)\mathbf{k}, \qquad -\pi/2 \le t \le \pi/2$

52. $\mathbf{F} = (x^2y)\mathbf{i} + \dfrac{1}{3}x^3\mathbf{j} + xy\mathbf{k}$
$\mathbf{r}(t) = (\cos t)\mathbf{i} + (\sin t)\mathbf{j} + (2 \sin^2 t - 1)\mathbf{k}, \qquad 0 \le t \le 2\pi$

53. Graph the vector field and curve given in

(a) Exercise 37.

(b) Exercise 39.

54. Graph the circulation field $\mathbf{F}$ and curve C_1 given in Exercise 41.

15.3 Path Independence, Potential Functions, and Conservative Fields

Path Independence • Assumptions Now in Effect • Line Integrals in Conservative Fields • Finding Potentials for Conservative Fields • Exact Differential Forms

Path Independence

If A and B are two points in an open region D in space, the work $\displaystyle\int \mathbf{F} \cdot d\mathbf{r}$ done in moving a particle from A and B by a field defined on D usually depends on the path taken. For some special fields, however, the integral's value is the same for all paths from A to B. If this is true for all points A and B in D, we say that the *integral* $\int \mathbf{F} \cdot d\mathbf{r}$ *is path independent in D* and that $\mathbf{F}$ *is conservative on D.*

Definitions Path Independence and Conservative Field

Let $\mathbf{F}$ be a field defined on an open region D in space, and suppose for any two points A and B in D the work $\int_A^B \mathbf{F} \cdot d\mathbf{r}$ done in moving from A to B is the same for all paths from A to B. Then the integral $\int \mathbf{F} \cdot d\mathbf{r}$ is **path independent in D** and the field $\mathbf{F}$ is **conservative on D.**

Conservative

The word **conservative** comes from physics, where it refers to fields in which the principle of conservation of energy holds (it does, in conservative fields).

Under conditions normally met in practice, a field $\mathbf{F}$ is conservative if and only if it is the gradient field of a scalar function f — that is, if and only if $\mathbf{F} = \nabla f$ for some f. Such a function f is a *potential function* for $\mathbf{F}$.

Definition Potential Function

If $\mathbf{F}$ is a field defined on D and $\mathbf{F} = \nabla f$ for some scalar function f on D, then f is a **potential function** for $\mathbf{F}$ on D.

An **electric potential** is a scalar function whose gradient field is an electric field, a **gravitational potential** is a scalar function whose gradient field is a gravitational field, and so on. As we will see in a moment, once we have found a potential function f for a field $\mathbf{F}$, we can evaluate all the work integrals in the domain of $\mathbf{F}$ by calculating the potential difference between two end points of the path along which the work is done

$$\int_A^B \mathbf{F} \cdot d\mathbf{r} = \int_A^B \nabla f \cdot d\mathbf{r} = f(B) - f(A). \tag{1}$$

If you think of ∇f for functions of several variables as being something like the derivative f' for functions of a single variable, then you see that Equation (1) is the vector-calculus analogue of the Fundamental Theorem of Calculus formula

$$\int_a^b f'(x)\, dx = f(b) - f(a).$$

Naturally, we need to impose conditions on the curves, fields, and domains to make Equation (1) and its implications hold.

Assumptions Now in Effect

We assume that all curves are *piecewise smooth*—that is, made up of finitely many smooth pieces connected end to end—as discussed in Section 10.3. We also assume that the components of **F** have continuous first partial derivatives. When $\mathbf{F} = \nabla f$, this continuity requirement guarantees that the mixed second derivatives of the potential function f are equal, a result we will find revealing in studying conservative fields **F**.

We assume D to be a *connected open* region in space. Open means that every point in D is the center of a ball that lies entirely in D, and **connected,** in an open region, means that every point can be connected to every other point by a smooth curve lying inside the region.

Line Integrals in Conservative Fields

Here is the theorum that says the value of an integral along a path in a conservative field depends only on the location of the endpoints of the path and not on the specific path taken.

Theorem 1 The Fundamental Theorem of Line Integrals

1. Let $\mathbf{F} = M\mathbf{i} + N\mathbf{j} + P\mathbf{k}$ be a vector field whose components are continuous throughout an open connected region D in space. Then there exists a differentiable function f such that

$$\mathbf{F} = \nabla f = \frac{\partial f}{\partial x}\mathbf{i} + \frac{\partial f}{\partial y}\mathbf{j} + \frac{\partial f}{\partial z}\mathbf{k}$$

if and only if for all points A and B in D the value of $\displaystyle\int_A^B \mathbf{F}\cdot d\mathbf{r}$ is independent of the path joining A to B in D.

2. If the integral is independent of the path from A to B, its value is

$$\int_A^B \mathbf{F}\cdot d\mathbf{r} = f(B) - f(A).$$

Proof That $\mathbf{F} = \nabla f$ Implies Path Independence of the Integral Suppose A and B are two points in D and that $C: \mathbf{r}(t) = g(t)\mathbf{i} + h(t)\mathbf{j} + k(t)\mathbf{k}$, $a \le t \le b$, is a smooth curve in D joining A and B. Along the curve, f is a differentiable function of t and

$$\frac{df}{dt} = \frac{\partial f}{\partial x}\frac{dx}{dt} + \frac{\partial f}{\partial y}\frac{dy}{dt} + \frac{\partial f}{\partial z}\frac{dz}{dt} \qquad \text{Chain Rule}$$

$$= \nabla f \cdot \left(\frac{dx}{dt}\mathbf{i} + \frac{dy}{dt}\mathbf{j} + \frac{dz}{dt}\mathbf{k}\right) = \nabla f \cdot \frac{d\mathbf{r}}{dt} = \mathbf{F}\cdot\frac{d\mathbf{r}}{dt} \qquad \begin{array}{l}\text{Because} \\ \mathbf{F} = \nabla f\end{array} \quad (2)$$

Therefore,

$$\int_C \mathbf{F}\cdot d\mathbf{r} = \int_{t=a}^{t=b} \mathbf{F}\cdot\frac{d\mathbf{r}}{dt}\,dt = \int_a^b \frac{df}{dt}\,dt \qquad \text{Eq. (2)}$$

$$= f\big(g(t), h(t), k(t)\big)\big]_a^b = f(B) - f(A).$$

Thus, the value of the work integral depends only on the values of f at A and B and not on the path in between. This proves Part 2 as well as the forward implication in Part 1. We omit the more technical proof of the reverse implication. ∎

Example 1 FINDING WORK DONE BY A CONSERVATIVE FIELD

Find the work done by the conservative field

$$\mathbf{F} = yz\mathbf{i} + xz\mathbf{j} + xy\mathbf{k} = \nabla(xyz)$$

along any smooth curve C joining the point $(-1, 3, 9)$ to $(1, 6, -4)$.

Solution   With $f(x, y, z) = xyz$, we have

$$\int_A^B \mathbf{F} \cdot d\mathbf{r} = \int_A^B \nabla f \cdot d\mathbf{r} \qquad \text{F} = \nabla f$$

$$= f(B) - f(A) \qquad \text{Fundamental Theorem, Part 2}$$

$$= xyz\Big|_{(1,6,-4)} - xyz\Big|_{(-1,3,9)}$$

$$= (1)(6)(-4) - (-1)(3)(9)$$

$$= -24 + 27 = 3.$$

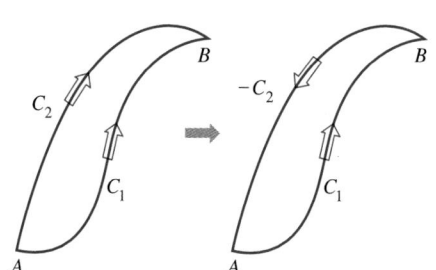

Figure 15.24 If we have two paths from A to B, one of them can be reversed to make a loop.

Theorem 2 Closed-Loop Property of Conservative Fields

Under the assumptions now in effect, the following statements are equivalent:

1. $\mathbf{F} \cdot d\mathbf{r} = 0$ around every closed loop in D.
2. The field $\mathbf{F}$ is conservative on D.

Proof That (1) ⇒ (2)   We want to show that for any two points A and B in D the integral of $\mathbf{F} \cdot d\mathbf{r}$ has the same value over any two paths C_1 and C_2 from A to B. We reverse the direction on C_2 to make a path $-C_2$ from B to A (Figure 15.24). Together C_1 and $-C_2$ make a closed loop C, and

$$\int_{C_1} \mathbf{F} \cdot d\mathbf{r} - \int_{C_2} \mathbf{F} \cdot d\mathbf{r} = \int_{C_1} \mathbf{F} \cdot d\mathbf{r} + \int_{-C_2} \mathbf{F} \cdot d\mathbf{r} = \int_C \mathbf{F} \cdot d\mathbf{r} = 0.$$

Thus, the integrals over C_1 and C_2 give the same value.

Proof That (2) ⇒ (1)   We want to show that the integral of $\mathbf{F} \cdot d\mathbf{r}$ is zero over any closed loop C. We pick two points A and B on C and use them to break C into two pieces: C_1 from A to B followed by C_2 from B back to A (Figure 15.25). Then

$$\oint_C \mathbf{F} \cdot d\mathbf{r} = \int_{C_1} \mathbf{F} \cdot d\mathbf{r} + \int_{C_2} \mathbf{F} \cdot d\mathbf{r} = \int_A^B \mathbf{F} \cdot d\mathbf{r} - \int_A^B \mathbf{F} \cdot d\mathbf{r} = 0. \quad ∎$$

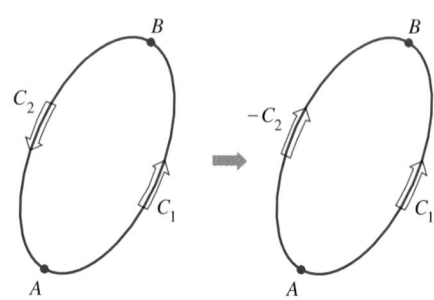

Figure 15.25 If A and B lie on a loop, we can reverse part of the loop to make two paths from A to B.

The following diagram summarizes the results of Theorems 1 and 2.

$$\mathbf{F} = \nabla f \text{ on } D \quad \overset{\text{Theorem 1}}{\Longleftrightarrow} \quad \begin{matrix} \text{F conservative} \\ \text{on } D \end{matrix} \quad \overset{\text{Theorem 2}}{\Longleftrightarrow} \quad \begin{matrix} \oint_C \mathbf{F} \cdot d\mathbf{r} = 0 \\ \text{over any closed} \\ \text{path in } D \end{matrix}$$

Now that we see how convenient it is to evaluate the line integrals in conservative fields, two questions remain:

1. How do we know when a given field **F** is conservative?

2. If **F** is in fact conservative, how do we find a potential function for **F**?

Finding Potential Functions for Conservative Fields

The test for a field's being conservative is the following.

Component Test for Conservative Fields

Let $\mathbf{F} = M(x, y, z)\mathbf{i} + N(x, y, z)\mathbf{j} + P(x, y, z)\mathbf{k}$ be a field whose component functions have continuous first partial derivatives. Then, **F** is conservative if and only if

$$\frac{\partial P}{\partial y} = \frac{\partial N}{\partial z}, \qquad \frac{\partial M}{\partial z} = \frac{\partial P}{\partial x}, \qquad \text{and} \qquad \frac{\partial N}{\partial x} = \frac{\partial M}{\partial y}. \tag{3}$$

Proof   We show that Equations (3) must hold if **F** is conservative. There is a potential function f such that

$$\mathbf{F} = M\mathbf{i} + N\mathbf{j} + P\mathbf{k} = \frac{\partial f}{\partial x}\mathbf{i} + \frac{\partial f}{\partial y}\mathbf{j} + \frac{\partial f}{\partial z}\mathbf{k}.$$

Hence,

$$\frac{\partial P}{\partial y} = \frac{\partial}{\partial y}\left(\frac{\partial f}{\partial z}\right) = \frac{\partial^2 f}{\partial y\,\partial z}$$

$$= \frac{\partial^2 f}{\partial z\,\partial y} \qquad \begin{array}{l}\text{Continuity implies that}\\ \text{the mixed partial}\\ \text{derivatives are equal.}\end{array}$$

$$= \frac{\partial}{\partial z}\left(\frac{\partial f}{\partial y}\right) = \frac{\partial N}{\partial z}.$$

The other two equations in (3) are proved similarly.

The second half of the proof, that Equations (3) imply that **F** is conservative, is a consequence of Stokes's Theorem taken up in Section 15.7.   ∎

Once we know **F** is conservative, we usually want to find a potential function for **F**. This requires solving the equation $\nabla f = \mathbf{F}$ or

$$\frac{\partial f}{\partial x}\mathbf{i} + \frac{\partial f}{\partial y}\mathbf{j} + \frac{\partial f}{\partial z}\mathbf{k} = M\mathbf{i} + N\mathbf{j} + P\mathbf{k}$$

for f. We accomplish this by integrating the three equations

$$\frac{\partial f}{\partial x} = M, \qquad \frac{\partial f}{\partial y} = N, \qquad \frac{\partial f}{\partial z} = P.$$

Example 2   FINDING A POTENTIAL FUNCTION

Show that $\mathbf{F} = (e^x \cos y + yz)\mathbf{i} + (xz - e^x \sin y)\mathbf{j} + (xy + z)\mathbf{k}$ is conservative and find a potential function for it.

Solution   We apply the test in Equations (3) to

$$M = e^x \cos y + yz, \qquad N = xz - e^x \sin y, \qquad P = xy + z$$

and calculate

$$\frac{\partial P}{\partial y} = x = \frac{\partial N}{\partial z}, \qquad \frac{\partial M}{\partial z} = y = \frac{\partial P}{\partial x}, \qquad \frac{\partial N}{\partial x} = z - e^x \sin y = \frac{\partial M}{\partial y}.$$

Together, these equations tell us that there is a function f with $\nabla f = \mathbf{F}$.

We find f by integrating the equations

$$\frac{\partial f}{\partial x} = e^x \cos y + yz, \qquad \frac{\partial f}{\partial y} = xz - e^x \sin y, \qquad \frac{\partial f}{\partial z} = xy + z. \quad (4)$$

We integrate the first equation with respect to x, holding y and z fixed, to get

$$f(x, y, z) = e^x \cos y + xyz + g(y, z).$$

We write the constant of integration as a function of y and z because its value may change if y and z change. We then calculate $\partial f/\partial y$ from this equation and match it with the expression for $\partial f/\partial y$ in Equations (4). This gives

$$-e^x \sin y + xz + \frac{\partial g}{\partial y} = xz - e^x \sin y,$$

so $\partial g/\partial y = 0$. Therefore, g is a function of z alone, and

$$f(x, y, z) = e^x \cos y + xyz + h(z).$$

We now calculate $\partial f/\partial z$ from this equation and match it to the formula for $\partial f/\partial z$ in Equations (4). This gives

$$xy + \frac{dh}{dz} = xy + z, \qquad \text{or} \qquad \frac{dh}{dz} = z,$$

so

$$h(z) = \frac{z^2}{2} + C.$$

Hence,

$$f(x, y, z) = e^x \cos y + xyz + \frac{z^2}{2} + C.$$

We have found infinitely many potential functions for $\mathbf{F}$, one for each value of C.

Example 3   SHOWING A FIELD IS NOT CONSERVATIVE

Show that $\mathbf{F} = (2x - 3)\mathbf{i} - z\mathbf{j} + (\cos z)\mathbf{k}$ is not conservative.

Solution   We apply the component test in Equations (3) and find right away that

$$\frac{\partial P}{\partial y} = \frac{\partial}{\partial y}(\cos z) = 0, \qquad \frac{\partial N}{\partial z} = \frac{\partial}{\partial z}(-z) = -1.$$

The two are unequal, so $\mathbf{F}$ is not conservative. No further testing is required.

Exact Differential Forms

As we see in the next section and again later on, it is often convenient to express work and circulation integrals in the "differential" form

$$\int_A^B M\,dx + N\,dy + P\,dz$$

mentioned in Section 15.2. Such integrals are relatively easy to evaluate if $M\,dx + N\,dy + P\,dz$ is the differential of a function f. For then

$$\int_A^B M\,dx + N\,dy + P\,dz = \int_A^B \frac{\partial f}{\partial x}\,dx + \frac{\partial f}{\partial y}\,dy + \frac{\partial f}{\partial z}\,dz$$

$$= \int_A^B \nabla f \cdot d\mathbf{r}$$

$$= f(B) - f(A). \qquad \text{Theorem 1}$$

Thus,

$$\int_A^B df = f(B) - f(A),$$

just as with differentiable functions of a single variable.

Definitions **Differential Form and Exact Differential Form**

The form $M(x, y, z)\,dx + N(x, y, z)\,dy + P(x, y, z)\,dz$ is a **differential form**. A differential form is **exact** on a domain D in space if

$$M\,dx + N\,dy + P\,dz = \frac{\partial f}{\partial x}\,dx + \frac{\partial f}{\partial y}\,dy + \frac{\partial f}{\partial z}\,dz = df$$

for some (scalar) function f throughout D.

Notice that if $M\,dx + N\,dy + P\,dz = df$ on D, then $\mathbf{F} = M\mathbf{i} + N\mathbf{j} + P\mathbf{k}$ is the gradient field of f on D. Conversely, if $\mathbf{F} = \nabla f$, then the form $M\,dx + N\,dy + P\,dz$ is exact. The test for the form's being exact is therefore the same as the test for $\mathbf{F}$'s being conservative.

Component Test for Exactness of $M\,dx + N\,dy + P\,dz$

The differential form $M\,dx + N\,dy + P\,dz$ is exact if and only if

$$\frac{\partial P}{\partial y} = \frac{\partial N}{\partial z}, \qquad \frac{\partial M}{\partial z} = \frac{\partial P}{\partial x}, \qquad \text{and} \qquad \frac{\partial N}{\partial x} = \frac{\partial M}{\partial y}. \qquad (5)$$

This is equivalent to saying that the field $\mathbf{F} = M\mathbf{i} + N\mathbf{j} + P\mathbf{k}$ is conservative.

Example 4 SHOWING A DIFFERENTIAL FORM IS EXACT

Show that $y\,dx + x\,dy + 4\,dz$ is exact and evaluate the integral

$$\int_{(1,1,1)}^{(2,3,-1)} y\,dx + x\,dy + 4\,dz$$

over the line segment from $(1, 1, 1)$ to $(2, 3, -1)$.

Solution We let $M = y$, $N = x$, $P = 4$ and apply the test of Equations (5):

$$\frac{\partial P}{\partial y} = 0 = \frac{\partial N}{\partial z}, \qquad \frac{\partial M}{\partial z} = 0 = \frac{\partial P}{\partial x}, \qquad \frac{\partial N}{\partial x} = 1 = \frac{\partial M}{\partial y}.$$

These equalities tell us that $y\,dx + x\,dy + 4\,dz$ is exact, so

$$y\,dx + x\,dy + 4\,dz = df$$

for some function f, and the integral's value is $f(2, 3, -1) - f(1, 1, 1)$.
We find f up to a constant by integrating the equations

$$\frac{\partial f}{\partial x} = y, \qquad \frac{\partial f}{\partial y} = x, \qquad \frac{\partial f}{\partial z} = 4. \tag{6}$$

From the first equation we get

$$f(x, y, z) = xy + g(y, z).$$

The second equation tells us that

$$\frac{\partial f}{\partial y} = x + \frac{\partial g}{\partial y} = x, \qquad \text{or} \qquad \frac{\partial g}{\partial y} = 0.$$

Hence, g is a function of z alone, and

$$f(x, y, z) = xy + h(z).$$

The third of Equations (6) tells us that

$$\frac{\partial f}{\partial z} = 0 + \frac{dh}{dz} = 4, \qquad \text{or} \qquad h(z) = 4z + C.$$

Therefore,

$$f(x, y, z) = xy + 4z + C.$$

The value of the integral is

$$f(2, 3, -1) - f(1, 1, 1) = 2 + C - (5 + C) = -3.$$

Quick Review 15.3

In Exercises 1 and 2, compute the indicated value of the $f(x, y, z) = e^{xyz} + \sin(xy + xz)$.

1. $f(0, 1, 2)$

2. $f(1, \pi/2, 0)$

In Exercises 3–7, find the indicated partial derivative of $f(x, y, z) = \cos^{-1}(xyz)$.

3. $\dfrac{\partial f}{\partial x}$

4. $\dfrac{\partial f}{\partial y}$

5. $\dfrac{\partial f}{\partial z}$

6. $\dfrac{\partial^2 f}{\partial y\, \partial x}$

9. $\dfrac{\partial f}{\partial y} = 2ye^{x^2+y^2+z^2} + xz + 2y$

7. $\dfrac{\partial^2 f}{\partial x\, \partial z}$

10. $\dfrac{\partial f}{\partial z} = 2ze^{x^2+y^2+z^2} + xy + z$

In Exercises 8–10, let $f(x, y, z) = e^{x^2+y^2+z^2} + xyz + g(x, y, z)$.
Determine the most general function g that satisfies the given
partial differential equation.

8. $\dfrac{\partial f}{\partial x} = 2xe^{x^2+y^2+z^2} + yz$

Section 15.3 Exercises

In Exercises 1–6, determine whether the field is conservative or
not conservative.

1. $\mathbf{F} = yz\mathbf{i} + xz\mathbf{j} + xy\mathbf{k}$

2. $\mathbf{F} = (y \sin z)\mathbf{i} + (x \sin z)\mathbf{j} + (xy \cos z)\mathbf{k}$

3. $\mathbf{F} = y\mathbf{i} + (x + z)\mathbf{j} - y\mathbf{k}$

4. $\mathbf{F} = -y\mathbf{i} + x\mathbf{j}$

5. $\mathbf{F} = (z + y)\mathbf{i} + z\mathbf{j} + (y + x)\mathbf{k}$

6. $\mathbf{F} = (e^x \cos y)\mathbf{i} - (e^x \sin y)\mathbf{j} + z\mathbf{k}$

In Exercises 7–12, find a potential function f for the field $\mathbf{F}$.

7. $\mathbf{F} = 2x\mathbf{i} + 3y\mathbf{j} + 4z\mathbf{k}$

8. $\mathbf{F} = (y + z)\mathbf{i} + (x + z)\mathbf{j} + (x + y)\mathbf{k}$

9. $\mathbf{F} = e^{y+2z}(\mathbf{i} + x\mathbf{j} + 2x\mathbf{k})$

10. $\mathbf{F} = (y \sin z)\mathbf{i} + (x \sin z)\mathbf{j} + (xy \cos z)\mathbf{k}$

11. $\mathbf{F} = (\ln x + \sec^2(x + y))\mathbf{i} +$
$\left(\sec^2 (x + y) + \dfrac{y}{y^2 + z^2}\right)\mathbf{j} + \dfrac{z}{y^2 + z^2}\mathbf{k}$

12. $\mathbf{F} = \dfrac{y}{1 + x^2y^2}\mathbf{i} + \left(\dfrac{x}{1 + x^2y^2} + \dfrac{z}{\sqrt{1 - y^2z^2}}\right)\mathbf{j}$
$+ \left(\dfrac{y}{\sqrt{1 - y^2z^2}} + \dfrac{1}{z}\right)\mathbf{k}$

In Exercises 13–22, show that the differential form in the
integral is exact. Then evaluate the integral.

13. $\displaystyle\int_{(0,0,0)}^{(2,3,-6)} 2x\, dx + 2y\, dy + 2z\, dz$

14. $\displaystyle\int_{(1,1,2)}^{(3,5,0)} yz\, dx + xz\, dy + xy\, dz$

15. $\displaystyle\int_{(0,0,0)}^{(1,2,3)} 2xy\, dx + (x^2 - z^2)\, dy - 2yz\, dz$

16. $\displaystyle\int_{(0,0,0)}^{(3,3,1)} 2x\, dx - y^2\, dy - \dfrac{4}{1 + z^2}\, dz$

17. $\displaystyle\int_{(1,0,0)}^{(0,1,1)} \sin y \cos x\, dx + \cos y \sin x\, dy + dz$

18. $\displaystyle\int_{(0,2,1)}^{(1,\pi/2,2)} 2 \cos y\, dx + \left(\dfrac{1}{y} - 2x \sin y\right) dy + \dfrac{1}{z}\, dz$

19. $\displaystyle\int_{(1,1,1)}^{(1,2,3)} 3x^2\, dx + \dfrac{z^2}{y}\, dy + 2z \ln y\, dz$

20. $\displaystyle\int_{(1,2,1)}^{(2,1,1)} (2x \ln y - yz)\, dx + \left(\dfrac{x^2}{y} - xz\right) dy - xy\, dz$

21. $\displaystyle\int_{(1,1,1)}^{(2,2,2)} \dfrac{1}{y}\, dx + \left(\dfrac{1}{z} - \dfrac{x}{y^2}\right) dy - \dfrac{y}{z^2}\, dz$

22. $\displaystyle\int_{(-1,-1,-1)}^{(2,2,2)} \dfrac{2x\, dx + 2y\, dy + 2z\, dz}{x^2 + y^2 + z^2}$

In Exercises 23 and 24, *work in groups of two or three* to solve
the problem.

23. Evaluate the integral
$$\int_{(1,1,1)}^{(2,3,-1)} y\, dx + x\, dy + 4\, dz$$
from Example 4 by finding parametric equations for the line
segment from $(1, 1, 1)$ to $(2, 3, -1)$ and evaluating the line
integral of $\mathbf{F} = y\mathbf{i} + x\mathbf{j} + 4\mathbf{k}$ along the segment. Since $\mathbf{F}$ is
conservative, the integral is independent of the path.

24. Evaluate $\displaystyle\int_C x^2\, dx + yz\, dy + (y^2/2)\, dz$ along the line
segment C joining $(0, 0, 0)$ to $(0, 3, 4)$.

Writing to Learn In Exercises 25 and 26, show that the value
of the integral does not depend on the path taken from A to B.

25. $\displaystyle\int_A^B z^2\, dx + 2y\, dy + 2xz\, dz$

26. $\int_A^B \dfrac{x\,dx + y\,dy + z\,dz}{\sqrt{x^2 + y^2 + z^2}}$

In Exercises 27 and 28, express **F** in the form ∇f.

27. $\mathbf{F} = \dfrac{2x}{y}\mathbf{i} + \left(\dfrac{1 - x^2}{y^2}\right)\mathbf{j}$

28. $\mathbf{F} = (e^x \ln y)\mathbf{i} + \left(\dfrac{e^x}{y} + \sin z\right)\mathbf{j} + (y \cos z)\mathbf{k}$

29. *Work Along Different Paths* Find the work done by

$$\mathbf{F} = (x^2 + y)\mathbf{i} + (y^2 + x)\mathbf{j} + ze^z\mathbf{k}$$

over the following paths from $(1, 0, 0)$ to $(1, 0, 1)$.

(a) The line segment $x = 1$, $y = 0$, $0 \le z \le 1$

(b) The helix

$$\mathbf{r}(t) = (\cos t)\mathbf{i} + (\sin t)\mathbf{j} + \left(\dfrac{t}{2\pi}\right)\mathbf{k}, \qquad 0 \le t \le 2\pi$$

(c) The x-axis from $(1, 0, 0)$ to $(0, 0, 0)$ followed by the parabola $z = x^2$, $y = 0$ from $(0, 0, 0)$ to $(1, 0, 1)$

30. *Work Along Different Paths* Find the work done by

$$\mathbf{F} = e^{yz}\mathbf{i} + (xze^{yz} + z \cos y)\mathbf{j} + (xye^{yz} + \sin y)\mathbf{k}$$

over the following paths from $(1, 0, 1)$ to $(1, \pi/2, 0)$.

(a) The line segment $x = 1$, $y = \pi t/2$, $z = 1 - t$, $0 \le t \le 1$

(b) The line segment from $(1, 0, 1)$ to the origin followed by the line segment from the origin to $(1, \pi/2, 0)$

(c) The line segment from $(1, 0, 1)$ to $(1, 0, 0)$, followed by the x-axis from $(1, 0, 0)$ to the origin, followed by the parabola $y = \pi x^2/2$, $z = 0$

Explorations

31. *Evaluating a Work Integral Two Ways* Let $\mathbf{F} = \nabla(x^3 y^2)$ and let C be the path in the xy-plane from $(-1, 1)$ to $(1, 1)$ that consists of the line segment from $(-1, 1)$ to $(0, 0)$ followed by the line segment from $(0, 0)$ to $(1, 1)$. Evaluate $\int_C \mathbf{F} \cdot d\mathbf{r}$ in the following two ways.

(a) Find parametrizations for the segments that make up C and evaluate the integral directly.

(b) Use the fact that $f(x, y) = x^3 y^2$ is a potential function for **F**.

32. *Same Integrand, Different Paths* Evaluate

$$\int_C 2x \cos y\,dx - x^2 \sin y\,dy$$

along the following paths C in the xy-plane.

(a) The parabola $y = (x - 1)^2$ from $(1, 0)$ to $(0, 1)$

(b) The line segment from $(-1, \pi)$ to $(1, 0)$

(c) The x-axis from $(-1, 0)$ to $(1, 0)$

(d) The astroid $\mathbf{r}(t) = (\cos^3 t)\mathbf{i} + (\sin^3 t)\mathbf{j}$, $0 \le t \le 2\pi$, counterclockwise from $(1, 0)$ back to $(1, 0)$ ∎

33. *A Gravitational Field* Find a potential function for the gravitational field

$$\mathbf{F} = -GmM \dfrac{x\mathbf{i} + y\mathbf{j} + z\mathbf{k}}{(x^2 + y^2 + z^2)^{3/2}}$$

$(G, m, \text{ and } M \text{ are constants})$.

34. (*Continuation of Exercise 33*) Let P_1 and P_2 be points in space at distances s_1 and s_2 from the origin. Show that the work done by the gravitational field in Exercise 33 in moving a particle from P_1 to P_2 is the quantity

$$GmM\left(\dfrac{1}{s_2} - \dfrac{1}{s_1}\right).$$

35. **(a)** How are the constants a, b, and c related if the following differential form is exact?

$$(ay^2 + 2czx)\,dx + y(bx + cz)\,dy + (ay^2 + cx^2)\,dz$$

(b) For what values of b and c will

$$\mathbf{F} = (y^2 + 2czx)\mathbf{i} + y(bx + cy)\mathbf{j} + (y^2 + cx^2)\mathbf{k}$$

be a gradient field?

36. Suppose $\mathbf{F} = \nabla f$ is a conservative vector field and

$$g(x, y, z) = \int_{(0, 0, 0)}^{(x, y, z)} \mathbf{F} \cdot d\mathbf{r}.$$

Show that $\nabla g = \mathbf{F}$.

37. *Writing to Learn Path of Least Work* You have been asked to find the path along which a force field **F** will perform the least work in moving a particle between two locations. A quick calculation on your part shows **F** to be conservative. How should you respond? Give reasons for your answer.

38. *Writing to Learn A Revealing Experiment* By experiment, you find that a force field **F** performs only half as much work in moving an object along path C_1 from A to B as it does in moving the object along path C_2 from A to B. What can you conclude about **F**? Give reasons for your answer.

Extending the Ideas

39. *Work Done by Gravity on a String* A string lies along the circle $x^2 + y^2 = 4$ from $(2, 0)$ to $(0, 2)$ in the first quadrant. The density of the string is $\rho(x, y) = xy$.

(a) Partition the string into a finite number of subarcs to show that the work done by gravity to move the string straight down to the x-axis is given by

$$\text{Work} = \lim_{n \to \infty} \sum_{k=1}^{n} gx_k y_k^2 \,\Delta s_k = \int_C gxy^2 \,ds,$$

where g is the universal gravitational constant.

(b) Find the total work done by evaluating the line integral in part (a).

(c) Show that the total work done equals the work required to move the string's center of mass $(\bar{x}, \bar{y})$ straight down to the x-axis.

CAS Exploration

40. Draw the force field **F** and curve $\mathbf{r}(t)$ given in Exercise 29 (b).

15.4 Green's Theorem in the Plane

Flux Density at a Point: Divergence • Circulation Density at a Point: The **k**-Component of Curl • Two Forms for Green's Theorem • Mathematical Assumptions • Using Green's Theorem to Evaluate Line Integrals • Proof of Green's Theorem for Special Regions • Extending the Proof to Other Regions

Flux Density at a Point: Divergence

Figure 15.26 The rectangle for defining the flux density (divergence) of a vector field at a point (x, y).

We need two new ideas for Green's Theorem. The first is the idea of the *flux density* of a vector field at a point, which in mathematics is the *divergence* of the vector field. We obtain it in the following way.

Suppose $\mathbf{F}(x, y) = M(x, y)\mathbf{i} + N(x, y)\mathbf{j}$ is the vector velocity of a fluid flow in the plane and the first partial derivatives of M and N are continuous at each point of a region R. Let (x, y) be a point in R, and let A be a small rectangle with one corner at (x, y) that, along with its interior, lies entirely in R (Figure 15.26). The sides of the rectangle, parallel to the coordinate axes, have lengths of Δx and Δy. The rate at which fluid leaves the rectangle across the bottom edge is approximately

$$\mathbf{F}(x, y) \cdot (-\mathbf{j}) \, \Delta x = -N(x, y) \, \Delta x.$$

This is the scalar component of the velocity at (x, y) in the direction of the outward normal times the length of the segment. If the velocity is in meters per second, for example, the exit rate will be in meters per second times meters or square meters per second. The rates at which the fluid crosses the other three sides in the directions of their outward normals can be estimated in a similar way. All told, we have

Top: $\mathbf{F}(x, y + \Delta y) \cdot \mathbf{j} \, \Delta x = N(x, y + \Delta y) \, \Delta x$

Bottom: $\mathbf{F}(x, y) \cdot (-\mathbf{j}) \, \Delta x = -N(x, y) \, \Delta x$

Right: $\mathbf{F}(x + \Delta x, y) \cdot \mathbf{i} \, \Delta y = M(x + \Delta x, y) \, \Delta y$

Left: $\mathbf{F}(x, y) \cdot (-\mathbf{i}) \, \Delta y = -M(x, y) \, \Delta y.$

Combining opposite pairs gives

Top and bottom: $(N(x, y + \Delta y) - N(x, y)) \, \Delta x \approx \left(\dfrac{\partial N}{\partial y} \Delta y \right) \Delta x$

Right and left: $(M(x + \Delta x, y) - M(x, y)) \, \Delta y \approx \left(\dfrac{\partial M}{\partial x} \Delta x \right) \Delta y.$

Adding these last two approximations gives

$$\text{Flux across rectangle boundary} \approx \left(\frac{\partial M}{\partial x} + \frac{\partial N}{\partial y} \right) \Delta x \, \Delta y.$$

We now divide by $\Delta x \, \Delta y$ to estimate the total flux per unit area or *flux density* for the rectangle:

$$\frac{\text{Flux across rectangular boundary}}{\text{Rectangle area}} \approx \left(\frac{\partial M}{\partial x} + \frac{\partial N}{\partial y} \right).$$

Finally, we let Δx and Δy approach zero to define the *flux density* of $\mathbf{F}$ at the point (x, y).

In mathematics, the *flux density* is the *divergence* of $\mathbf{F}$. The symbol for it is div $\mathbf{F}$, pronounced "divergence of $\mathbf{F}$" or "div $\mathbf{F}$."

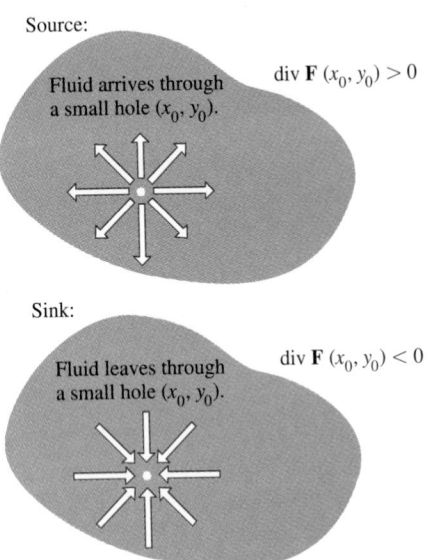

Source:

Fluid arrives through a small hole (x_0, y_0).

div $\mathbf{F}\,(x_0, y_0) > 0$

Sink:

Fluid leaves through a small hole (x_0, y_0).

div $\mathbf{F}\,(x_0, y_0) < 0$

Definition Flux Density or Divergence

The **flux density** or **divergence** of a vector field $\mathbf{F} = M\mathbf{i} + N\mathbf{j}$ at the point (x, y) is

$$\text{div } \mathbf{F} = \frac{\partial M}{\partial x} + \frac{\partial N}{\partial y}. \tag{1}$$

Intuitively, if water were flowing into a region through a small hole at the point (x_0, y_0), the lines of flow would diverge there (hence the name) and, since water would be flowing out of a small rectangle about (x_0, y_0), the divergence of $\mathbf{F}$ at (x_0, y_0) would be positive. If the water were draining out instead of flowing in, the divergence would be negative. See Figure 15.27.

Figure 15.27 In the flow of an incompressible fluid across a plane region, the divergence is positive at a "source," a point where fluid enters the system, and negative at a "sink," a point where the fluid leaves the system.

Example 1   FINDING DIVERGENCE

Find the divergence of $\mathbf{F}(x, y) = (x^2 - y)\mathbf{i} + (xy - y^2)\mathbf{j}$.

Solution   We use the formula in Equation (1):

$$\text{div } \mathbf{F} = \frac{\partial M}{\partial x} + \frac{\partial N}{\partial y} = \frac{\partial}{\partial x}(x^2 - y) + \frac{\partial}{\partial y}(xy - y^2)$$

$$= 2x + x - 2y = 3x - 2y.$$

Circulation Density at a Point: The k-Component of Curl

The second of the two new ideas we need for Green's Theorem is the idea of *circulation density* of a vector field $\mathbf{F}$ at a point. To obtain it, we return to the velocity field

$$\mathbf{F}(x, y) = M(x, y)\mathbf{i} + N(x, y)\mathbf{j}$$

and the rectangle A. The rectangle is redrawn here as Figure 15.28.

The counterclockwise circulation of $\mathbf{F}$ around the boundary of A is the sum of flow rates along the sides. For the bottom edge, the flow rate is approximately

$$\mathbf{F}(x, y) \cdot \mathbf{i} \, \Delta x = M(x, y) \, \Delta x.$$

This is the scalar component of the velocity $\mathbf{F}(x, y)$ in the direction of the tangent vector $\mathbf{i}$ times the length of the segment. The rates of flow along the other sides in the counterclockwise direction are expressed in a similar way. In all, we have

Top: $\qquad \mathbf{F}(x, y + \Delta y) \cdot (-\mathbf{i}) \, \Delta x = -M(x, y + \Delta y) \, \Delta x$

Bottom: $\qquad\qquad \mathbf{F}(x, y) \cdot \mathbf{i} \, \Delta x = M(x, y) \, \Delta x$

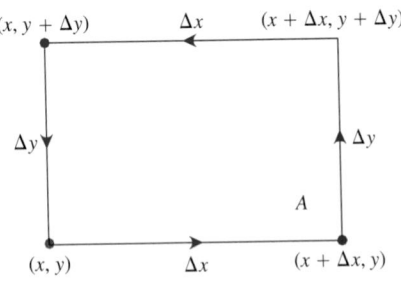

$(x, y + \Delta y)$ $\qquad \Delta x \qquad$ $(x + \Delta x, y + \Delta y)$

Δy $\qquad\qquad\qquad\qquad\qquad$ Δy

A

(x, y) $\qquad \Delta x \qquad$ $(x + \Delta x, y)$

Figure 15.28 The rectangle for defining the circulation density (curl) of a vector field at a point (x, y).

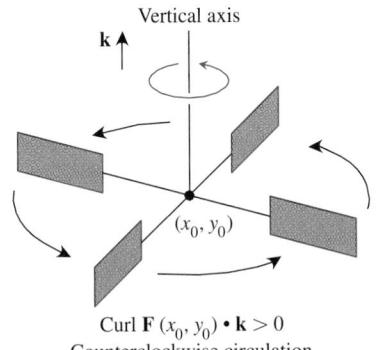

Curl $\mathbf{F}\,(x_0, y_0) \cdot \mathbf{k} > 0$
Counterclockwise circulation

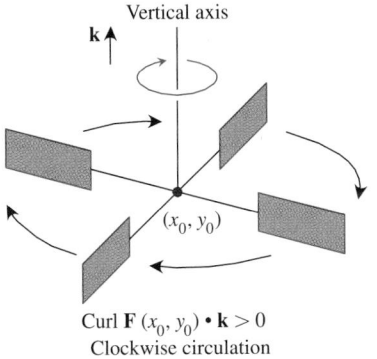

Curl $\mathbf{F}\,(x_0, y_0) \cdot \mathbf{k} > 0$
Clockwise circulation

Figure 15.29 In the flow of an incompressible fluid over a plane region, the **k**-component of the curl measures the rate of the fluid's rotation at a point. The **k**-component is positive at points where the rotation is counterclockwise and negative where the rotation is clockwise.

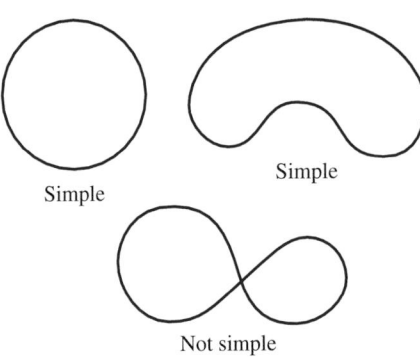

Figure 15.30 In Proving Green's Theorem, we distinguish between two kinds of closed curves, simple and not simple. Simple curves do not cross themselves. A circle is simple; a figure eight is not.

Right: $\mathbf{F}(x + \Delta x, y) \cdot \mathbf{j}\,\Delta y = N(x + \Delta x, y)\,\Delta y$

Left: $\mathbf{F}(x, y) \cdot (-\mathbf{j})\,\Delta y = -N(x, y)\,\Delta y.$

We add opposite pairs to get

Top and bottom: $-(M(x, y + \Delta y) - M(x, y))\,\Delta x \approx -\left(\dfrac{\partial M}{\partial y}\,\Delta y\right)\Delta x$

Right and left: $(N(x + \Delta x, y) - N(x, y))\,\Delta y \approx \left(\dfrac{\partial N}{\partial x}\,\Delta x\right)\Delta y.$

Adding these last two approximations and dividing by $\Delta x\,\Delta y$ gives an estimate of the *circulation density* for the rectangle:

$$\frac{\text{Circulation around rectangle}}{\text{Rectangle area}} \approx \frac{\partial N}{\partial x} - \frac{\partial M}{\partial y}.$$

Finally, we let Δx and Δy approach zero to define the circulation density of $\mathbf{F}$ at the point (x, y).

The positive orientation for circulation in the xy-plane is counterclockwise around a vertical axis looking downward on the plane from the tip of the (vertical) unit vector $\mathbf{k}$ (Figure 15.29). The general density just defined is actually the $\mathbf{k}$-component of a more general circulation-density vector we investigate in Section 15.7, called *curl* $\mathbf{F}$, the *curl* of the vector field $\mathbf{F}$. For Green's Theorem, we need only the $\mathbf{k}$-component of this vector.

Definition k-Component of Circulation Density or Curl

The **k**-component of the circulation density or curl of a vector field $\mathbf{F} = M\mathbf{i} + N\mathbf{j}$ at the point (x, y) is the scalar

$$(\text{curl } \mathbf{F}) \cdot \mathbf{k} = \frac{\partial N}{\partial x} - \frac{\partial M}{\partial y} \tag{2}$$

If water is moving about a region in the xy-plane in a thin layer, then the **k**-component of the circulation, or curl, at a point (x_0, y_0) gives a way to measure how fast and in what direction a small paddle wheel will spin if it is put into the water at (x_0, y_0) with its axis perpendicular to the plane, parallel to the unit vector $\mathbf{k}$ (Figure 15.29).

Example 2 FINDING THE k-COMPONENT OF CURL

Find the **k**-component of the curl for $\mathbf{F}(x, y) = (x^2 - y)\mathbf{i} + (xy - y^2)\mathbf{j}.$

Solution   We use the formula in Equation (2):

$$(\text{curl } \mathbf{F}) \cdot \mathbf{k} = \frac{\partial N}{\partial x} - \frac{\partial M}{\partial y} = \frac{\partial}{\partial x}(xy - y^2) - \frac{\partial}{\partial y}(x^2 - y) = y + 1.$$

Two Forms for Green's Theorem

In one form, Green's Theorem says that under suitable conditions the outward flux of a vector field across a simple closed curve in the plane (Figure 15.30) equals the double integral of the divergence of the field over the region enclosed by the curve. Recall the formulas for flux in Equations (3) and (4) in Section 15.2.

> ### Theorem 3 Green's Theorem (Flux-Divergence or Normal Form)
>
> The outward flux of a field $\mathbf{F} = M\mathbf{i} + N\mathbf{j}$ across a simple closed curve C equals the double integral of div $\mathbf{F}$ over the region R enclosed by C.
>
> $$\underset{\text{outward flux}}{\underbrace{\oint_C \mathbf{F} \cdot \mathbf{n}\, ds}} = \oint_C M\, dy - N\, dx = \underset{\text{divergence integral}}{\underbrace{\iint_R \left(\frac{\partial M}{\partial x} + \frac{\partial N}{\partial y} \right) dx\, dy}} \qquad (3)$$

In another form, Green's Theorem says that the counterclockwise circulation of a vector field around a simple closed curve is the double integral of the $\mathbf{k}$-component of the curl of the field over the region enclosed by the curve.

For a two-dimensional field $\mathbf{F} = M\mathbf{i} + N\mathbf{j}$, the integral in Equation (2), Section 15.2, for circulation takes the equivalent form

$$\oint_C \mathbf{F} \cdot \mathbf{T}\, ds = \oint_C M\, dx + N\, dy.$$

> ### Theorem 4 Green's Theorem (Circulation-Curl or Tangential Form)
>
> The counterclockwise circulation of a field $\mathbf{F} = M\mathbf{i} + N\mathbf{j}$ around a simple closed curve C in the plane equals the double integral of (curl $\mathbf{F}$) $\cdot$ $\mathbf{k}$ over the region R enclosed by C.
>
> $$\underset{\substack{\text{counterclockwise}\\\text{circulation}}}{\underbrace{\oint_C \mathbf{F} \cdot \mathbf{T}\, ds}} = \oint_C M\, dx + N\, dy = \underset{\text{curl integral}}{\underbrace{\iint_R \left(\frac{\partial N}{\partial x} - \frac{\partial M}{\partial y} \right) dx\, dy}} \qquad (4)$$

The two forms of Green's Theorem are equivalent. We do not need to prove them both. Applying Equation (3) to the field $\mathbf{G}_1 = N\mathbf{i} - M\mathbf{j}$ gives Equation (4), and applying Equation (4) to $\mathbf{G}_2 = -N\mathbf{i} + M\mathbf{j}$ gives Equation (3).

Mathematical Assumptions

We need two kinds of assumptions for Green's Theorem to hold. First, we need conditions on M and N to ensure the existence of the integrals. The usual assumptions are that M, N, and their first partial derivatives are continuous at every point of some open region containing C and R. Second, we need geometric conditions on the curve C. It must be simple, closed, and made up of pieces along which we can integrate M and N. The usual assumptions are that C is piecewise smooth. The proof we give for Green's Theorem, however, assumes things about the shape of R as well. You can find proofs that are less restrictive in more advanced texts. First, let's look at some examples.

Example 3   VERIFYING GREEN'S THEOREM

Verify both forms of Green's Theorem for the field $\mathbf{F}(x, y) = (x - y)\mathbf{i} + x\mathbf{j}$ and the region R bounded by the unit circle

$$C: \quad \mathbf{r}(t) = (\cos t)\mathbf{i} + (\sin t)\mathbf{j}, \qquad 0 \le t \le 2\pi.$$

Solution We first express all functions, derivatives, and differentials in terms of t:

$$M = \cos t - \sin t, \qquad dx = d(\cos t) = -\sin t \, dt,$$

$$N = \cos t, \qquad dy = d(\sin t) = \cos t \, dt,$$

$$\frac{\partial M}{\partial x} = 1, \qquad \frac{\partial M}{\partial y} = -1, \qquad \frac{\partial N}{\partial x} = 1, \qquad \frac{\partial N}{\partial y} = 0.$$

The two sides of Equation (3) are

$$\oint_C M \, dy - N \, dx = \int_{t=0}^{t=2\pi} (\cos t - \sin t)(\cos t \, dt) - (\cos t)(-\sin t \, dt)$$

$$= \int_0^{2\pi} \cos^2 t \, dt = \pi,$$

$$\iint_R \left(\frac{\partial M}{\partial x} + \frac{\partial N}{\partial y}\right) dx \, dy = \iint_R (1 + 0) \, dx \, dy$$

$$= \iint_R dx \, dy = \text{area of unit circle} = \pi.$$

The two sides of Equation (4) are

$$\oint_C M \, dx + N \, dy = \int_{t=0}^{t=2\pi} (\cos t - \sin t)(-\sin t \, dt) + (\cos t)(\cos t \, dt)$$

$$= \int_0^{2\pi} (-\sin t \cos t + 1) \, dt = 2\pi,$$

$$\iint_R \left(\frac{\partial N}{\partial x} - \frac{\partial M}{\partial y}\right) dx \, dy = \iint_R \left(1 - (-1)\right) dx \, dy = 2\iint_R dx \, dy = 2\pi.$$

Using Green's Theorem to Evaluate Line Integrals

If we construct a closed curve C by piecing a number of different curves end to end, the process of evaluating a line integral over C can be lengthy because there are so many different integrals to evaluate. However, if C bounds a region R to which Green's Theorem applies, we can use Green's Theorem to change the line integral around C into one double integral over R.

Example 4 EVALUATING A LINE INTEGRAL USING GREEN'S THEOREM

Evaluate the integral

$$\oint_C xy \, dy - y^2 \, dx,$$

where C is the square cut from the first quadrant by the lines $x = 1$ and $y = 1$.

The Green of Green's Theorem

The Green of Green's Theorem was George Green (1793–1841), a self-taught scientist in Nottingham, England. Green's work on the mathematical foundations of gravitation, electricity, and magnetism was published privately in 1828 in a short book entitled *An Essay on the Application of Mathematical Analysis to Electricity and Magnetism.* The book sold all of 52 copies (fewer than 100 were printed), the copies going mostly to Green's patrons and personal friends. A few weeks before Green's death in 1841, Sir William Thomson noticed a reference to Green's book and in 1845 was finally able to locate a copy. Excited by what he read, Thomson shared Green's ideas with other scientists and had the book republished in a series of journal articles. Green's mathematics provided the foundation on which Thomson, Stokes, Rayleigh, and Maxwell built the present-day theory of electromagnetism.

Solution We can use either form of Green's Theorem to change the line integral into a double integral over the square.

1. *With Equation (3):* Taking $M = xy$, $N = y^2$, and C and R as the square's boundary and interior gives

$$\oint_C xy\,dy - y^2\,dx = \iint_R (y + 2y)\,dx\,dy = \int_0^1 \int_0^1 3y\,dx\,dy$$

$$= \int_0^1 [3xy]_{x=0}^{x=1}\,dy = \int_0^1 3y\,dy = \frac{3}{2}y^2\Big]_0^1 = \frac{3}{2}.$$

2. *With Equation (4):* Taking $M = -y^2$ and $N = xy$ gives the same result:

$$\oint_C -y^2\,dx + xy\,dy = \iint_R (y - (-2y))\,dx\,dy = \frac{3}{2}.$$

Example 5 FINDING OUTWARD FLUX

Calculate the outward flux of the field $\mathbf{F}(x, y) = x\mathbf{i} + y^2\mathbf{j}$ across the square bounded by the lines $x = \pm 1$ and $y = \pm 1$.

Solution Calculating the flux with a line integral would take four integrations, one for each side of the square. With Green's theorem, we can change the line integral to one double integral. With $M = x$, $N = y^2$, C the square, and R the square's interior, we have

$$\text{Flux} = \oint_C \mathbf{F} \cdot \mathbf{n}\,ds = \oint_C M\,dy - N\,dx$$

$$= \iint_R \left(\frac{\partial M}{\partial x} + \frac{\partial N}{\partial y}\right) dx\,dy$$

$$= \int_{-1}^1 \int_{-1}^1 (1 + 2y)\,dx\,dy = \int_{-1}^1 [x + 2xy]_{x=-1}^{x=1}\,dy$$

$$= \int_{-1}^1 (2 + 4y)\,dy = [2y + 2y^2]_{-1}^1 = 4.$$

Proof of Green's Theorem for Special Regions

Let C be a smooth simple closed curve in the xy-plane with the property that lines parallel to the axes cut it in no more than two points. Let R be the region enclosed by C, and suppose M, N, and their first partial derivatives are continuous at every point of some open region containing C and R. We want to prove the circulation-curl form of Green's Theorem,

$$\oint_C M\,dx + N\,dy = \iint_R \left(\frac{\partial N}{\partial x} - \frac{\partial M}{\partial y}\right) dx\,dy. \tag{5}$$

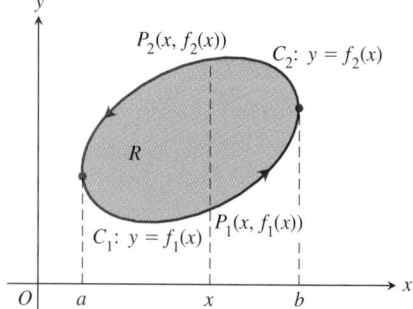

Figure 15.31 The boundary curve C is made up of C_1, the graph of $y = f_1(x)$, and C_2, the graph of $y = f_2(x)$.

Figure 15.31 shows C made up of two directed parts:

$$C_1: \quad y = f_1(x), \qquad a \le x \le b, \qquad C_2: \quad y = f_2(x), \qquad b \ge x \ge a.$$

For any x between a and b, we can integrate $\partial M/\partial y$ with respect to y from $y = f_1(x)$ to $y = f_2(x)$ and obtain

$$\int_{f_1(x)}^{f_2(x)} \frac{\partial M}{\partial y} \, dy = M(x, y) \Big]_{y=f_1(x)}^{y=f_2(x)} = M(x, f_2(x)) - M(x, f_1(x)).$$

We can then integrate this with respect to x from a to b:

$$\int_a^b \int_{f_1(x)}^{f_2(x)} \frac{\partial M}{\partial y} \, dy \, dx = \int_a^b \left[M(x, f_2(x)) - M(x, f_1(x)) \right] dx$$

$$= -\int_b^a M(x, f_2(x)) \, dx - \int_a^b M(x, f_1(x)) \, dx$$

$$= -\int_{C_2} M \, dx - \int_{C_1} M \, dx$$

$$= -\oint_C M \, dx.$$

Therefore,

$$\oint_C M \, dx = \iint_R \left(-\frac{\partial M}{\partial y} \right) dx \, dy. \qquad (6)$$

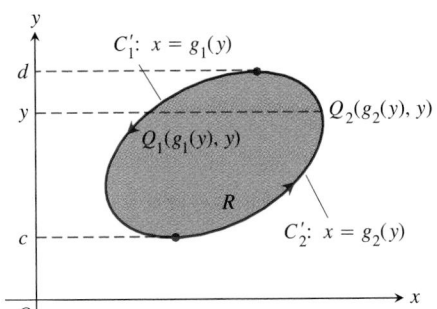

Figure 15.32 The boundary curve C is made up of C_1', the graph of $x = g_1(y)$, and C_2', the graph of $x = g_2(y)$.

Equation (6) is half the result we need for Equation (5). We derive the other half by integrating $\partial N/\partial x$ first with respect to x and then with respect to y, as suggested by Figure 15.32. This shows the curve C of Figure 15.31 decomposed into the two directed parts

$$C_1': \quad x = g_1(y), \quad d \geq y \geq c, \quad \text{and} \quad C_2': \quad x = g_2(y), \quad c \leq y \leq d.$$

The result of this double integration is

$$\oint_C N \, dy = \iint_R \frac{\partial N}{\partial x} \, dx \, dy. \qquad (7)$$

Combining Equations (6) and (7) gives Equation (5). This concludes the proof.

Extending the Proof to Other Regions

The argument we just gave does not apply directly to the rectangular region in Figure 15.33 because the lines $x = a$, $x = b$, $y = c$, and $y = d$ meet the region's boundary in more than two points. If we divide the boundary C into four directed line segments, however,

$$C_1: \quad y = c, \quad a \leq x \leq b, \qquad C_2: \quad x = b, \quad c \leq y \leq d,$$

$$C_3: \quad y = d, \quad b \geq x \geq a, \qquad C_4: \quad x = a, \quad d \geq y \geq c,$$

we can modify the argument in the following way.

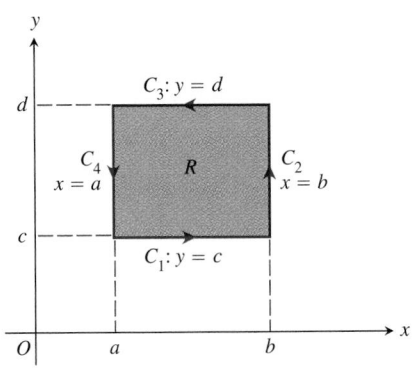

Figure 15.33 To prove Green's Theorem for a rectangle, we divide the boundary into four directed line segments.

(a)

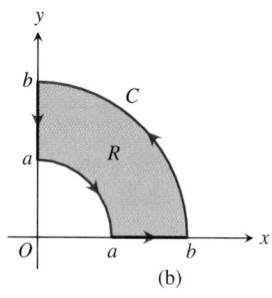

(b)

Figure 15.34 Other regions to which Green's Theorem applies.

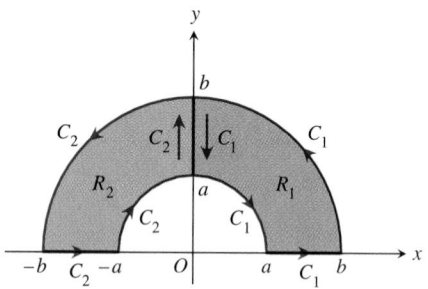

Figure 15.35 A region R that combines regions R_1 and R_2.

Proceeding as in the proof of Equation (7), we have

$$\int_c^d \int_a^b \frac{\partial N}{\partial x}\, dx\, dy = \int_c^d (N(b, y) - N(a, y))\, dy$$

$$= \int_c^d N(b, y)\, dy + \int_d^c N(a, y)\, dy \qquad (8)$$

$$= \int_{C_2} N\, dy + \int_{C_4} N\, dy.$$

Because y is constant along C_1 and C_3, $\int_{C_1} N\, dy = \int_{C_3} N\, dy = 0$, so we can add $\int_{C_1} N\, dy + \int_{C_3} N\, dy$ to the right-hand side of Equation (8) without changing the equality. Doing so, we have

$$\int_c^d \int_a^b \frac{\partial N}{\partial x}\, dx\, dy = \oint_C N\, dy. \qquad (9)$$

Similarly, we can show that

$$\int_a^b \int_c^d \frac{\partial M}{\partial y}\, dy\, dx = -\oint_C M\, dx. \qquad (10)$$

Subtracting Equation (10) from Equation (9), we again arrive at

$$\oint_C M\, dx + N\, dy = \iint_R \left(\frac{\partial N}{\partial x} - \frac{\partial M}{\partial y} \right) dx\, dy.$$

Regions like those in Figure 15.34 can be handled with no greater difficulty. Equation (5) still applies. It also applies to the horseshoe-shaped region R shown in Figure 15.35, as we see by putting together the regions R_1 and R_2 and their boundaries. Green's Theorem applies to C_1, R_1, and to C_2, R_2, yielding

$$\int_{C_1} M\, dx + N\, dy = \iint_{R_1} \left(\frac{\partial N}{\partial x} - \frac{\partial M}{\partial y} \right) dx\, dy,$$

$$\int_{C_2} M\, dx + N\, dy = \iint_{R_2} \left(\frac{\partial N}{\partial x} - \frac{\partial M}{\partial y} \right) dx\, dy.$$

When we add these two equations, the line integral along the y-axis from b to a for C_1 cancels the integral over the same segment but in the opposite direction for C_2. Hence,

$$\oint_C M\, dx + N\, dy = \iint_R \left(\frac{\partial N}{\partial x} - \frac{\partial M}{\partial y} \right) dx\, dy,$$

where C consists of the two segments of the x-axis from $-b$ to $-a$ and from a to b and of the two semicircles and where R is the region inside C.

The device of adding line integrals over separate boundaries to build up an integral over a single boundary can be extended to any finite number of sub-

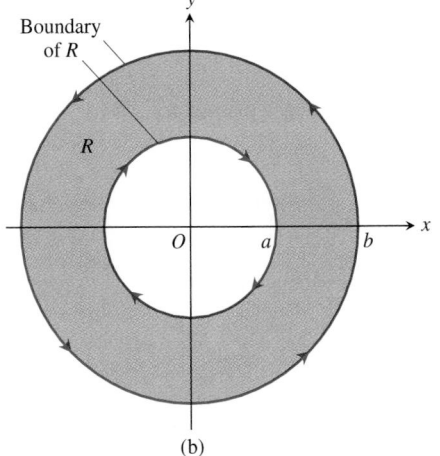

Figure 15.36 The annular region R combines four smaller regions. In polar coordinates, $r = a$ for the inner circle, $r = b$ for the outer circle, and $a \le r \le b$ for the region itself.

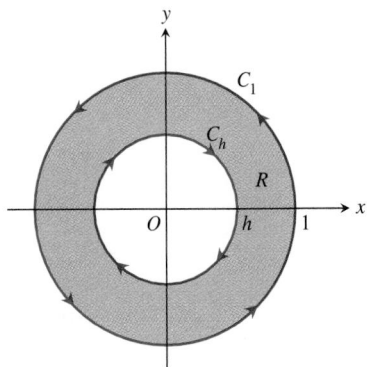

Figure 15.37 Green's Theorem may be applied to the annular region R by integrating along the boundaries as shown. (Example 6)

regions. In Figure 15.36a, let C_1 be the boundary, oriented counterclockwise, of the region R_1 in the first quadrant. Similarly for the other three quadrants: C_i is the boundary of the region R_i, $i = 1, 2, 3, 4$. By Green's Theorem,

$$\oint_{C_i} M\,dx + N\,dy = \iint_{R_i} \left(\frac{\partial N}{\partial x} - \frac{\partial M}{\partial y} \right) dx\,dy. \tag{11}$$

We add Equations (11) for $i = 1, 2, 3, 4$ and get (Figure 15.36b):

$$\oint_{r=b} (M\,dx + N\,dy) + \oint_{r=a} (M\,dx + N\,dy) = \iint_{a \le r \le b} \left(\frac{\partial N}{\partial x} - \frac{\partial M}{\partial y} \right) dx\,dy. \tag{12}$$

Equation (12) says that the double integral of $(\partial N/\partial x) - (\partial M/\partial y)$ over the annular ring R equals the line integral of $M\,dx + N\,dy$ over the complete boundary of R in the direction that keeps R on our left as we progress (Figure 15.36b).

Example 6 VERIFYING GREEN'S THEOREM

Verify the circulation form of Green's Theorem (Equation (4)) on the annular ring R: $h^2 \le x^2 + y^2 \le 1$, $0 < h < 1$ (Figure 15.37), if

$$M = \frac{-y}{x^2 + y^2}, \qquad N = \frac{x}{x^2 + y^2}.$$

Solution The boundary of R consists of the circle

$$C_1: \quad x = \cos t, \qquad y = \sin t, \qquad 0 \le t \le 2\pi,$$

traversed counterclockwise as t increases and the circle

$$C_h: \quad x = h\cos\theta, \qquad y = -h\sin\theta, \qquad 0 \le \theta \le 2\pi,$$

traversed clockwise as θ increases. The functions M and N and their partial derivatives are continuous throughout R. Moreover,

$$\frac{\partial M}{\partial y} = \frac{(x^2 + y^2)(-1) + y(2y)}{(x^2 + y^2)^2}$$

$$= \frac{y^2 - x^2}{(x^2 + y^2)^2} = \frac{\partial N}{\partial x},$$

so

$$\iint_R \left(\frac{\partial N}{\partial x} - \frac{\partial M}{\partial y} \right) dx\,dy = \iint_R 0\,dx\,dy = 0.$$

The integral of $M\,dx + N\,dy$ over the boundary of R is

$$\int_C M\,dx + N\,dy = \oint_{C_1} \frac{x\,dy - y\,dx}{x^2 + y^2} + \oint_{C_h} \frac{x\,dy - y\,dx}{x^2 + y^2}$$

$$= \int_0^{2\pi} (\cos^2 t + \sin^2 t)\,dt - \int_0^{2\pi} \frac{h^2(\cos^2\theta + \sin^2\theta)}{h^2}\,d\theta$$

$$= 2\pi - 2\pi = 0.$$

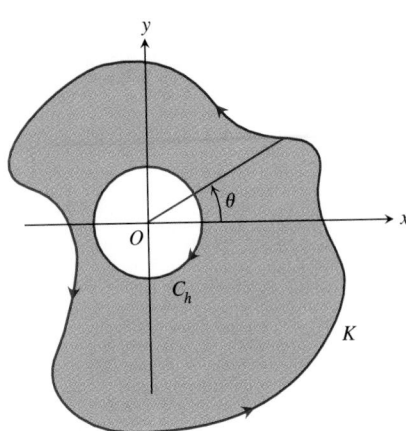

Figure 15.38 The region bounded by the circle C_h and the curve K.

The functions M and N in Example 6 are discontinuous at $(0, 0)$, so we cannot apply Green's Theorem to the circle C_1 and the region inside it. We must exclude the origin. We do so by excluding the points inside C_h.

We could replace the circle C_1 in Example 6 by an ellipse or any other simple closed curve K surrounding C_h (Figure 15.38). The result would still be

$$\oint_K (M\,dx + N\,dy) + \oint_{C_h} (M\,dx + N\,dy) = \iint_R \left(\frac{\partial N}{\partial x} - \frac{\partial M}{\partial y}\right) dy\,dx = 0,$$

which leads to the surprising conclusion that

$$\oint_K (M\,dx + N\,dy) = 2\pi$$

for any such curve K. We can explain this result by changing to polar coordinates. With

$$x = r\cos\theta, \qquad\qquad y = r\sin\theta,$$
$$dx = -r\sin\theta\,d\theta + \cos\theta\,dr, \qquad dy = r\cos\theta\,d\theta + \sin\theta\,dr,$$

we have

$$\frac{x\,dy - y\,dx}{x^2 + y^2} = \frac{r^2(\cos^2\theta + \sin^2\theta)\,d\theta}{r^2} = d\theta,$$

and θ increases by 2π as we traverse K once counterclockwise.

Quick Review 15.4

In Exercises 1–4, parametrize the given straight line segment.

1. C_1 from $(0, 0)$ to $(1, 0)$

2. C_2 from $(1, 0)$ to $(1, 1)$

3. C_3 from $(1, 1)$ to $(0, 1)$

4. C_4 from $(0, 1)$ to $(0, 0)$

In Exercises 5–8, evaluate the line integral $\int_C xy\,dy - y^2\,dx$ for the given curve.

5. $C = C_1$ of Exercise 1

6. $C = C_2$ of Exercise 2

7. $C = C_3$ of Exercise 3

8. $C = C_4$ of Exercise 4

In Exercises 9 and 10, let $M(x, y) = -y/(x^2 + y^2)$.

9. Find $\partial M/\partial x$.   **10.** Find $\partial M/\partial y$.

Section 15.4 Exercises

In Exercises 1–4, verify Green's Theorem by evaluating both sides of Equations (3) and (4) for the field $\mathbf{F} = M\mathbf{i} + N\mathbf{j}$. Take the domains of integration in each case to be the disk R: $x^2 + y^2 \leq a^2$ and its bounding circle C: $\mathbf{r} = (a\cos t)\mathbf{i} + (a\sin t)\mathbf{j}, 0 \leq t \leq 2\pi$.

1. $\mathbf{F} = -y\mathbf{i} + x\mathbf{j}$ **2.** $\mathbf{F} = y\mathbf{i}$

3. $\mathbf{F} = 2x\mathbf{i} - 3y\mathbf{j}$ **4.** $\mathbf{F} = -x^2 y\mathbf{i} + xy^2\mathbf{j}$

In Exercises 5–10, use Green's theorem to find the counterclockwise circulation and outward flux for the field $\mathbf{F}$ and curve C.

5. $\mathbf{F} = (x - y)\mathbf{i} + (y - x)\mathbf{j}$
     C: The square bounded by $x = 0, x = 1, y = 0, y = 1$

6. $\mathbf{F} = (x^2 + 4y)\mathbf{i} + (x + y^2)\mathbf{j}$
  C: The square bounded by $x = 0$, $x = 1$, $y = 0$, $y = 1$

7. $\mathbf{F} = (y^2 - x^2)\mathbf{i} + (x^2 + y^2)\mathbf{j}$
  C: The triangle bounded by $y = 0$, $x = 3$, and $y = x$

8. $\mathbf{F} = (x + y)\mathbf{i} + (x^2 + y^2)\mathbf{j}$
  C: The triangle bounded by $y = 0$, $x = 1$, and $y = x$

9. $\mathbf{F} = (x + e^x \sin y)\mathbf{i} + (x + e^x \cos y)\mathbf{j}$
  C: The right-hand loop of the lemniscate $r^2 = \cos 2\theta$

10. $\mathbf{F} = \left(\tan^{-1}\dfrac{y}{x}\right)\mathbf{i} + \ln(x^2 + y^2)\mathbf{j}$

  C: The boundary of the region defined by the polar
    coordinate inequalities $1 \le r \le 2$, $0 \le \theta \le \pi$

11. Find the counterclockwise circulation and outward flux of
  the field $\mathbf{F} = xy\mathbf{i} + y^2\mathbf{j}$ around and over the boundary of the
  region enclosed by the curves $y = x^2$ and $y = x$ in the first
  quadrant.

12. Find the counterclockwise circulation and the outward flux
  of the field $\mathbf{F} = (-\sin y)\mathbf{i} + (x \cos y)\mathbf{j}$ around and over the
  square cut from the first quadrant by the lines $x = \pi/2$ and
  $y = \pi/2$.

13. Find the outward flux of the field

$$\mathbf{F} = \left(3xy - \dfrac{x}{1 + y^2}\right)\mathbf{i} + (e^x + \tan^{-1} y)\mathbf{j}$$

  across the cardioid $r = a(1 + \cos \theta)$, $a > 0$.

14. Find the counterclockwise circulation of

$$\mathbf{F} = (y + e^x \ln y)\mathbf{i} + (e^x/y)\mathbf{j}$$

  around the boundary of the region that is bounded above by
  the curve $y = 3 - x^2$ and below by the curve $y = x^4 + 1$.

In Exercises 15 and 16, find the work done by $\mathbf{F}$ in moving a
particle once counterclockwise around the given curve.

15. $\mathbf{F} = 2xy^3\mathbf{i} + 4x^2y^2\mathbf{j}$
  C: The boundary of the "triangular" region in the first
  quadrant enclosed by the *x*-axis, the line $x = 1$, and the
  curve $y = x^3$

16. $\mathbf{F} = (4x - 2y)\mathbf{i} + (2x - 4y)\mathbf{j}$
  C: The circle $(x - 2)^2 + (y - 2)^2 = 4$

In Exercises 17–20, use Green's Theorem to evaluate the
integral.

17. $\oint_C (y^2\, dx + x^2\, dy)$
  C: The triangle bounded by $x = 0$, $x + y = 1$, $y = 0$

18. $\oint_C (3y\, dx + 2x\, dy)$
  C: The boundary of $0 \le x \le \pi$, $0 \le y \le \sin x$

19. $\oint_C (6y + x)\, dx + (y + 2x)\, dy$
  C: The circle $(x - 2)^2 + (y - 3)^2 = 4$

20. $\oint_C (2x + y^2)\, dx + (2xy + 3y)\, dy$
  C: Any simple closed curve in the plane for which Green's
  Theorem holds

Explorations

If a simple closed curve *C* in the plane and the region *R* it
encloses satisfy the hypotheses of Green's Theorem, the area of
R is given by

Green's Theorem Area Formula

$$\text{Area of } R = \dfrac{1}{2}\oint_C x\, dy - y\, dx \qquad (13)$$

---

21. (a) Prove Equation (13). (*Hint:* Use Equation (11).)

  (b) Use Equation (13) to find the area of the region enclosed
  by the circle $\mathbf{r}(t) = (a \cos t)\mathbf{i} + (a \sin t)\mathbf{j}$, $0 \le t \le 2\pi$.

In Exercises 22–24, use Equation (13) to find the area of the
region enclosed by the curve.

22. The ellipse $\mathbf{r}(t) = (a \cos t)\mathbf{i} + (b \sin t)\mathbf{j}$, $\quad 0 \le t \le 2\pi$

23. The astroid $\mathbf{r}(t) = (\cos^3 t)\mathbf{i} + (\sin^3 t)\mathbf{j}$, $\quad 0 \le t \le 2\pi$

24. The curve $\mathbf{r}(t) = t^2\mathbf{i} + \left(\dfrac{t^3}{3} - t\right)\mathbf{j}$, $\quad -\sqrt{3} \le t \le \sqrt{3}$ ∎

25. Let *C* be the boundary of a region on which Green's
  Theorem holds. Use Green's Theorem to calculate

  (a) $\oint_C f(x)\, dx + g(y)\, dy$.

  (b) $\oint_C ky\, dx + hx\, dy$ $\quad$ (*k* and *h* constants.)

26. *Integral Dependent Only On Area* Show that the value of

$$\oint_C xy^2\, dx + (x^2y + 2x)\, dy$$

  around any square depends only on the area of the square
  and not on its location in the plane.

27. Writing to Learn What is special about the integral

$$\oint_C 4x^3y\, dx + x^4\, dy?$$

  Give reasons for your answer.

28. Writing to Learn What is special about the integral

$$\oint_C -y^3\, dx + x^3\, dy?$$

  Give reasons for your answer.

29. *Area as a Line Integral* Show that if R is a region in the plane bounded by a piecewise smooth simple closed curve C, then

$$\text{Area of } R = \oint_C x \, dy = -\oint_C y \, dx.$$

30. *Definite Integral as Line Integral* Suppose a nonnegative function $y = f(x)$ has a continuous first derivative on $[a, b]$. Let C be the boundary of the region in the xy-plane that is bounded below by the x-axis, above by the graph of f, and on the sides by the lines $x = a$ and $x = b$. Show that

$$\int_a^b f(x) \, dx = -\oint_C y \, dx.$$

31. *Area and the Centroid*

(a) Let A be the area and $\bar{x}$ the x-coordinate of the centroid of a region R that is bounded by a piecewise smooth simple closed curve C in the xy-plane. Show that

$$\frac{1}{2}\oint_C x^2 \, dy = -\oint_C xy \, dy = \frac{1}{3}\oint_C x^2 \, dy - xy \, dx = A\bar{x}.$$

(b) What is the analogous statement for A and $\bar{y}$?

32. *Moment of Inertia* Let I_y be the moment of inertia about the y-axis of the region in Exercise 31. Show that

$$\frac{1}{3}\oint_C x^3 \, dy = -\oint_C x^2 y \, dx = \frac{1}{4}\oint_C x^3 \, dy - x^2 y \, dx = I_y.$$

In Exercises 33–36, *work in groups of two or three to solve the problem.*

33. *Green's Theorem and Laplace's Equation* Assuming that all the necessary derivatives exist and are continuous, show that if $f(x, y)$ satisfies the **Laplace equation**

$$\frac{\partial^2 f}{\partial x^2} + \frac{\partial^2 f}{\partial y^2} = 0,$$

then

$$\oint_C \frac{\partial f}{\partial y} \, dx - \frac{\partial f}{\partial x} \, dy = 0$$

for all closed curves to which Green's Theorem applies. (The converse is also true: If the line integral is always zero, then f satisfies the Laplace equation.)

34. *Maximizing Work* Among all smooth simple closed curves in the plane, oriented counterclockwise, find the one along which the work done by

$$\mathbf{F} = \left(\frac{1}{4}x^2 y + \frac{1}{3}y^3\right)\mathbf{i} + x\mathbf{j}$$

is greatest. (*Hint:* Where is (curl $\mathbf{F}$) • $\mathbf{k}$ positive?)

35. *Regions with Many Holes* Green's Theorem holds for a region R with any finite number of holes as long as the

bounding curves are smooth, simple, and closed and we integrate over each component of the boundary in the direction that keeps R on our immediate left as we go along (Figure 15.39).

(a) Let $f(x, y) = \ln(x^2 + y^2)$ and let C be the circle $x^2 + y^2 = a^2$. Evaluate the flux integral

$$\oint_C \nabla f \cdot \mathbf{n} \, ds.$$

(b) Let K be an arbitrary smooth simple closed curve in the plane that does not pass through $(0, 0)$. Use Green's Theorem to show that

$$\oint_K \nabla f \cdot \mathbf{n} \, ds$$

has two possible values, depending on whether $(0, 0)$ lies inside K or outside K.

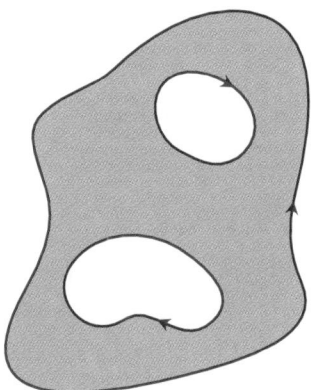

Figure 15.39 Green's Theorem holds for regions with more than one hole. (Exercise 35)

36. *Bendixson's Criterion* The **streamlines** of a planar fluid flow are the smooth curves traced by the fluid's individual particles. The vectors $\mathbf{F} = M(x, y)\mathbf{i} + N(x, y)\mathbf{j}$ of the flow's velocity field are the tangent vectors of the streamlines. Show that if the flow takes place over a *simply connected* region R (no holes or missing points) and that if $M_x + N_y \neq 0$ throughout R, then none of the streamlines in R is closed. In other words, no particle of fluid ever has a closed trajectory in R. The criterion is $M_x + N_y \neq 0$ **Bendixson's criterion** for the nonexistence of closed trajectories.

37. Establish Equation (7) to finish the proof of the special case of Green's Theorem.

38. Establish Equation (10) to complete the argument for the extension of Green's Theorem.

39. Writing to Learn *Curl Components of Conservative Fields* Can anything be said about the curl component of a conservative two-dimensional vector field? Give reasons for your answer.

40. Writing to Learn *The Circulation of a Conservative Field* Does Green's Theorem give any information about the circulation of a conservative field? Does this agree with anything else you know? Give reasons for your answer.

Extending the Ideas

CAS Explorations

In Exercises 41–44, perform the following steps:

(a) Plot C in the xy-plane.

(b) Determine the integrand $\partial N/\partial x - \partial M/\partial y$ for the curl form of Green's Theorem.

(c) Determine the (double integral) limits of integration from your plot in (a) and evaluate the curl integral for the counterclockwise circulation of the field $\mathbf{F}$ around the curve C.

41. $\mathbf{F} = (2x - y)\mathbf{i} + (x + 3y)\mathbf{j}$
C: The ellipse $x^2 + 4y^2 = 4$

42. $\mathbf{F} = (2x^3 - y^3)\mathbf{i} + (x^3 + y^3)\mathbf{j}$
C: The ellipse $\dfrac{x^2}{4} + \dfrac{y^2}{9} = 1$

43. $\mathbf{F} = x^{-1}e^y\mathbf{i} + (e^y \ln x + 2x)\mathbf{j}$
C: The boundary of the region defined by $y = 1 + x^4$ (below) and $y = 2$ (above)

44. $\mathbf{F} = xe^y\mathbf{i} + 4x^2 \ln y\mathbf{j}$
C: The triangle with vertices $(0, 0)$, $(2, 0)$, and $(0, 4)$

15.5 Surface Area and Surface Integrals

Surface Area • A Practical Formula • Surface Integrals • Algebraic Properties: The Surface Area Differential • Orientation • Surface Integral for Flux • Moments and Masses of Thin Shells

Surface Area

Figure 15.40 shows a surface S lying above its "shadow" region R in a plane beneath it. The surface is defined by the equation $f(x, y, z) = c$. If the surface is **smooth** (∇f is continuous and never vanishes on S), we can define and calculate its area as a double integral over R.

The first step in defining the area of S is to partition the region R into small rectangles ΔA_k of the kind we would use if we were defining an integral over R. Directly above each ΔA_k lies a patch of surface $\Delta \sigma_k$ that we may approximate with a portion ΔP_k of the tangent plane. To be specific, we suppose ΔP_k is a portion of the plane that is tangent to the surface at the point $T_k(x_k, y_k, z_k)$ directly above the back corner C_k of ΔA_k. If the tangent plane is parallel to R, then ΔP_k will be congruent to ΔA_k. Otherwise, it will be a parallelogram whose area is somewhat larger than the area of ΔA_k.

Figure 15.41 gives a magnified view of $\Delta \sigma_k$ and ΔP_k, showing the gradient vector $\nabla f(x_k, y_k, z_k)$ at T_k and a unit vector $\mathbf{p}$ that is normal to R. The figure also shows the angle γ_k between ∇f and $\mathbf{p}$. The other vectors in the picture, $\mathbf{u}_k$ and $\mathbf{v}_k$, lie along the edges of the patch ΔP_k in the tangent plane. Thus, both $\mathbf{u}_k \times \mathbf{v}_k$ and ∇f are normal to the tangent plane.

We now need from advanced vector geometry that $|(\mathbf{u}_k \times \mathbf{v}_k) \cdot \mathbf{p}|$ is the area of the projection of the parallelogram determined by $\mathbf{u}_k$ and $\mathbf{v}_k$ onto any plane normal to $\mathbf{p}$. In our case, this translates into the statement

$$|(\mathbf{u}_k \times \mathbf{v}_k) \cdot \mathbf{p}| = \Delta A_k. \tag{1}$$

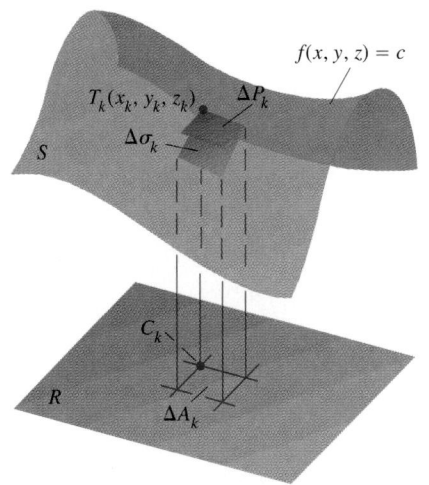

Figure 15.40 A surface S and its vertical projection onto a plane beneath it. You can think of R as the shadow of S on the plane. The tangent plate ΔP_k approximates the surface patch $\Delta \sigma_k$ above ΔA_k.

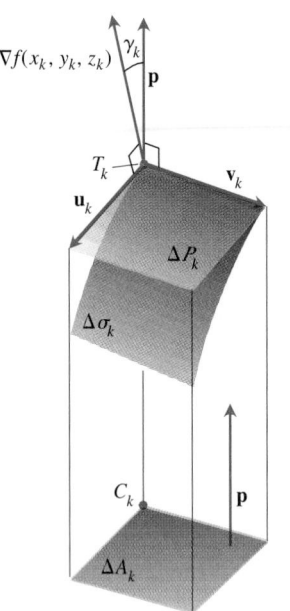

$\nabla f(x_k, y_k, z_k)$

Figure 15.41 Magnified view from Figure 15.40. The vector $\mathbf{u}_k \times \mathbf{v}_k$ (not shown) is parallel to the vector ∇f because both vectors are normal to the plane of ΔP_k.

Now, $|\mathbf{u}_k \times \mathbf{v}_k|$ itself is the area ΔP_k (standard fact about cross products), so Equation (1) becomes

$$\underbrace{|\mathbf{u}_k \times \mathbf{v}_k|}_{\Delta P_k} \underbrace{|\mathbf{p}|}_{1} \underbrace{|\cos(\text{angle between } \mathbf{u}_k \times \mathbf{v}_k \text{ and } \mathbf{p})|}_{\substack{\text{same as } |\cos \gamma_k| \text{ because} \\ \nabla f \text{ and } \mathbf{u}_k \times \mathbf{v}_k \text{ are both} \\ \text{normal to the tangent plane}}} = \Delta A_k \tag{2}$$

or

$$\Delta P_k |\cos \gamma_k| = \Delta A_k$$

or

$$\Delta P_k = \frac{\Delta A_k}{|\cos \gamma_k|},$$

provided $\cos \gamma_k \neq 0$. We will have $\cos \gamma_k \neq 0$ as long as ∇f is not parallel to the ground plane and $\nabla f \cdot \mathbf{p} \neq 0$.

Since the patches ΔP_k approximate the surface patches $\Delta \sigma_k$ that fit together to make S, the sum

$$\sum \Delta P_k = \sum \frac{\Delta A_k}{|\cos \gamma_k|} \tag{3}$$

looks like an approximation of what we might like to call the surface area of S. It also looks as if the approximation would improve if we refined the partition of R. In fact, the sums on the right-hand side of Equation (3) are approximating sums for the double integral

$$\iint_R \frac{1}{|\cos \gamma_k|} \, dA. \tag{4}$$

We therefore define the **area** of S to be the value of this integral whenever it exists.

A Practical Formula

For any surface $f(x, y, z) = c$, we have $|\nabla f \cdot \mathbf{p}| = |\nabla f||\mathbf{p}||\cos \gamma|$, so

$$\frac{1}{|\cos \gamma|} = \frac{|\nabla f|}{|\nabla f \cdot \mathbf{p}|}.$$

This combines with Equation (4) to give a practical formula for area.

Formula for Surface Area

The area of the surface $f(x, y, z) = c$ over a closed and bounded plane region R is

$$\text{Surface area} = \iint_R \frac{|\nabla f|}{|\nabla f \cdot \mathbf{p}|} \, dA, \tag{5}$$

where $\mathbf{p}$ is a unit vector normal to R and $\nabla f \cdot \mathbf{p} \neq 0$.

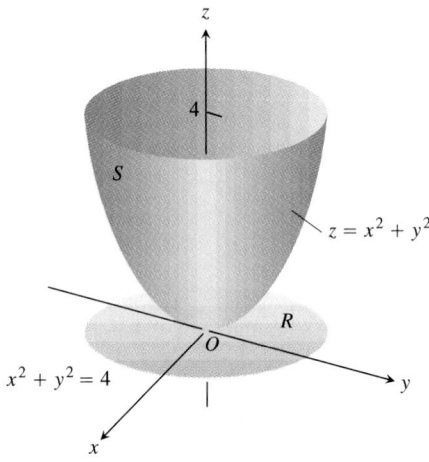

Figure 15.42 The area of this parabolic surface is calculated in Example 1.

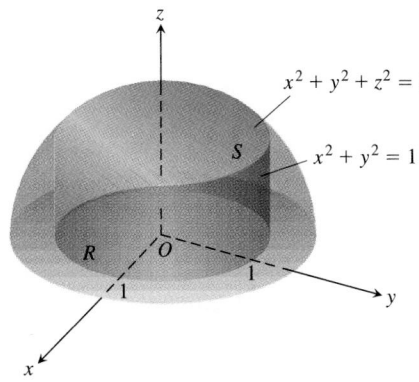

Figure 15.43 The cap cut from the hemisphere by the cylinder projects vertically onto the disk $R: x^2 + y^2 \le 1$. (Example 2)

Thus, the area is the double integral over R of the magnitude of ∇f divided by the magnitude of the scalar component of ∇f normal to R.

We reached Equation (5) under the assumptions that $\nabla f \cdot \mathbf{p} \ne 0$ throughout R and that ∇f is continuous. Whenever the integral exists, however, we define its value to be the area of the portion of the surface $f(x, y, z) = c$ that lies over R.

Example 1 FINDING SURFACE AREA

Find the area of the surface cut from the bottom of the paraboloid $x^2 + y^2 - z = 0$ by the plane $z = 4$.

Solution We sketch the surface S and the region R below it in the xy-plane (Figure 15.42). The surface S is part of the level surface $f(x, y, z) = x^2 + y^2 - z = 0$, and R is the disk $x^2 + y^2 \le 4$ in the xy-plane. To get a unit normal to the plane of R, we can take $\mathbf{p} = \mathbf{k}.$

At any point (x, y, z) on the surface, we have

$$f(x, y, z) = x^2 + y^2 - z$$

$$\nabla f = 2x\mathbf{i} + 2y\mathbf{j} - \mathbf{k}$$

$$|\nabla f| = \sqrt{(2x)^2 + (2y)^2 + (-1)^2}$$

$$= \sqrt{4x^2 + 4y^2 + 1}$$

$$|\nabla f \cdot \mathbf{p}| = |\nabla f \cdot \mathbf{k}| = |-1| = 1.$$

In the region R, $dA = dx\, dy$. Therefore,

$$\text{Surface area} = \iint_R \frac{|\nabla f|}{|\nabla f \cdot \mathbf{p}|}\, dA \qquad \text{Eq. (5)}$$

$$= \iint_{x^2+y^2\le 4} \sqrt{4x^2 + 4y^2 + 1}\, dx\, dy$$

$$= \int_0^{2\pi} \int_0^2 \sqrt{4r^2 + 1}\, r\, dr\, d\theta \qquad \begin{array}{l}\text{Polar}\\ \text{coordinates}\end{array}$$

$$= \int_0^{2\pi} \left[\frac{1}{12}(4r^2 + 1)^{3/2}\right]_0^2 d\theta$$

$$= \frac{\pi}{6}(17^{3/2} - 1) \approx 36.177 \text{ units squared.}$$

Example 2 FINDING SURFACE AREA

Find the area of the cap cut from the hemisphere $x^2 + y^2 + z^2 = 2$, $z \ge 0$, by the cylinder $x^2 + y^2 = 1$ (Figure 15.43).

Solution The cap S is part of the level surface $f(x, y, z) = x^2 + y^2 + z^2 = 2$. It projects one-to-one onto the disk $R: x^2 + y^2 \le 1$ in the xy-plane. The vector $\mathbf{p} = \mathbf{k}$ is normal to the plane of R.

At any point on the surface,

$$f(x, y, z) = x^2 + y^2 + z^2$$

$$\nabla f = 2x\mathbf{i} + 2y\mathbf{j} + 2z\mathbf{k}$$

$$|\nabla f| = 2\sqrt{x^2 + y^2 + z^2} = 2\sqrt{2} \qquad \text{Because } x^2 + y^2 + z^2 = 2$$
$$\text{at points of } S$$

$$|\nabla f \cdot \mathbf{p}| = |2z| = 2z.$$

Therefore,

$$\text{Surface area} = \iint_R \frac{|\nabla f|}{|\nabla f \cdot \mathbf{p}|}\, dA = \iint_R \frac{2\sqrt{2}}{2z}\, dA = \sqrt{2} \iint_R \frac{dA}{z}. \qquad (6)$$

What do we do about the z?

Since z is the z-coordinate of a point on the sphere, we can express it in terms of x and y as

$$z = \sqrt{2 - x^2 - y^2}.$$

With this substitution, we continue the work of Equation (6):

$$\text{Surface area} = \sqrt{2} \iint_R \frac{dA}{z} = \sqrt{2} \iint_R \frac{dA}{\sqrt{2 - x^2 - y^2}}$$

$$= \sqrt{2} \int_0^{2\pi} \int_0^1 \frac{r\, dr\, d\theta}{\sqrt{2 - r^2}} \qquad \text{Polar coordinates}$$

$$= \sqrt{2} \int_0^{2\pi} \left[-(2 - r^2)^{1/2} \right]_{r=0}^{r=1} d\theta$$

$$= 2\pi(2 - \sqrt{2}).$$

Surface Integrals

We now show how to integrate a function over a surface, using the ideas just developed for calculating surface area.

Suppose, for example, we have an electrical charge distributed over a surface $f(x, y, z) = c$ like the one in Figure 15.44 and we know the charge density $g(x, y, z)$ (charge per unit area) at each point on S. Then we may calculate the total charge on S in the following way.

We partition the shadow region R on the ground plane beneath S into small rectangles of the kind we would use if we were defining the area of S. Then directly above each ΔA_k lies a patch of surface $\Delta \sigma_k$ that we approximate with a parallelogram-shaped portion of tangent plane, ΔP_k.

Up to this point the construction proceeds as in the definition of surface area, but now we take one additional step: We evaluate g at (x_k, y_k, z_k) and then approximate the total charge on the surface patch $\Delta \sigma_k$ by the product $g(x_k, y_k, z_k)\, \Delta P_k$. The rationale is that when the partition of R is sufficiently fine, the value of g throughout $\Delta \sigma_k$ is nearly constant and ΔP_k is nearly the same as $\Delta \sigma_k$. The total charge over S is then approximated by the sum

$$\text{Total charge} \approx \sum g(x_k, y_k, z_k)\, \Delta P_k = \sum g(x_k, y_k, z_k) \frac{\Delta A_k}{|\cos \gamma_k|}. \qquad (7)$$

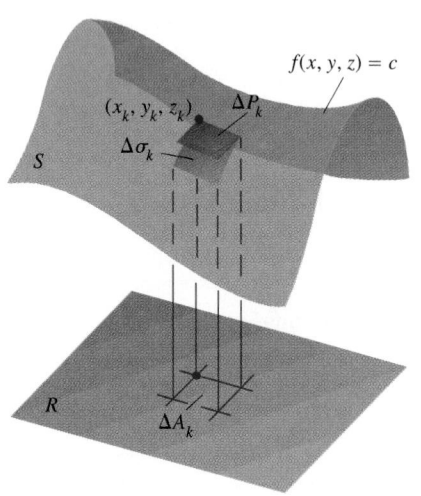

Figure 15.44 If we know how an electrical charge is distributed over a surface, we can find the total charge with a surface integral.

If f, the function defining the surface S, and its first partial derivatives are continuous and if g is continuous over S, then the sums on the right-hand side of Equation (7) approach the limit

$$\iint_R g(x, y, z) \frac{dA}{|\cos \gamma_k|} = \iint_R g(x, y, z) \frac{|\nabla f|}{|\nabla f \cdot \mathbf{p}|} dA \qquad (8)$$

as the partition of R is refined in the usual way. This limit is the *integral of g over the surface S* and is calculated as a double integral over R. The value of the integral is the total charge on the surface S.

As you might expect, the formula in Equation (8) defines the integral of *any* function g over the surface S as long as the integral exists.

Definition Integral of g over S and Surface Integral

If R is the shadow region of a surface S defined by the equation $f(x, y, z) = c$ and g is a continuous function defined at the points of S, then the **integral of g over S** is the integral

$$\iint_R g(x, y, z) \frac{|\nabla f|}{|\nabla f \cdot \mathbf{p}|} dA, \qquad (9)$$

where $\mathbf{p}$ is a unit vector normal to R and $\nabla f \cdot \mathbf{p} \neq 0$. The integral itself is a **surface integral.**

The integral in Equation (9) takes on different meanings in different applications. If g has the constant value 1, the integral gives the area of S. If g gives the mass density of a thin shell of material modeled by S, the integral gives the mass of the shell.

Algebraic Properties; The Surface Area Differential

By writing $d\sigma$ for $(|\nabla f|/|\nabla f \cdot \mathbf{p}|) \, dA$, we can abbreviate the integral in Equation (9).

Surface Area Differential and Differential Form for Surface Integrals

$$d\sigma = \frac{|\nabla f|}{|\nabla f \cdot \mathbf{p}|} dA \qquad\qquad \iint_S g \, d\sigma \qquad (10)$$

surface area differential   differential formula for surface integrals

---

Surface integrals behave like other double integrals, the integral of the sum of two functions being the sum of their integrals and so on. The domain additivity property takes the form

$$\iint_S g \, d\sigma = \iint_{S_1} g \, d\sigma + \iint_{S_2} g \, d\sigma + \cdots + \iint_{S_n} g \, d\sigma.$$

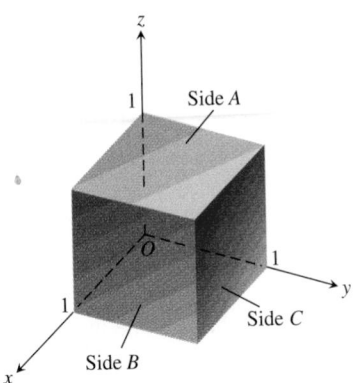

Figure 15.45 To integrate a function over the surface of a cube, we integrate over each face and add the results. (Example 3)

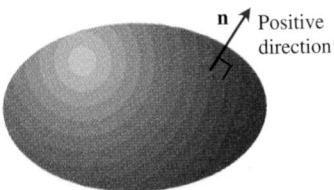

Figure 15.46 Smooth closed surfaces in space are orientable. The outward unit normal vector defines the positive direction at each point.

Figure 15.47 To make a Möbius band, take a rectangular strip of paper *abcd*, give the end *bc* a single twist, and paste the ends of the strip together to match *a* with *c* and *b* with *d*. The Möbius band is a nonorientable or one-sided surface.

The idea is that if S is partitioned by smooth curves into a finite number of nonoverlapping smooth patches (i.e., if S is **piecewise smooth**), then the integral over S is the sum of the integrals over the patches. Thus, the integral of a function over the surface of a cube is the sum of the integrals over the faces of the cube. We integrate over a turtle shell of welded plates by integrating one plate at a time and adding the results.

Example 3 INTEGRATING OVER A SURFACE

Integrate $g(x, y, z) = xyz$ over the surface of the cube cut from the first octant by the planes $x = 1$, $y = 1$, and $z = 1$ (Figure 15.45).

Solution We integrate xyz over each of the six sides and add the results. Since $xyz = 0$ on the sides that lie in the coordinate planes, the integral over the surface of the cube reduces to

$$\iint_{\substack{\text{cube}\\ \text{surface}}} xyz\, d\sigma = \iint_{\text{side } A} xyz\, d\sigma + \iint_{\text{side } B} xyz\, d\sigma + \iint_{\text{side } C} xyz\, d\sigma.$$

Side A is the surface $f(x, y, z) = z = 1$ over the square region R_{xy}: $0 \le x \le 1$, $0 \le y \le 1$, in the xy-plane. For this surface and region,

$$\mathbf{p} = \mathbf{k}, \qquad \nabla f = \mathbf{k}, \qquad |\nabla f| = 1, \qquad |\nabla f \cdot \mathbf{p}| = |\mathbf{k} \cdot \mathbf{k}| = 1,$$

$$d\sigma = \frac{|\nabla f|}{|\nabla f \cdot \mathbf{p}|}\, dA = \frac{1}{1}\, dx\, dy = dx\, dy,$$

$$xyz = xy(1) = xy,$$

and

$$\iint_{\text{side } A} xyz\, d\sigma = \iint_{R_{xy}} xy\, dx\, dy = \int_0^1 \int_0^1 xy\, dx\, dy = \int_0^1 \frac{y}{2}\, dy = \frac{1}{4}.$$

Symmetry tells us that the integrals of xyz over sides B and C are also 1/4. Hence,

$$\iint_{\substack{\text{cube}\\ \text{surface}}} xyz\, d\sigma = \frac{1}{4} + \frac{1}{4} + \frac{1}{4} = \frac{3}{4}.$$

Orientation

A smooth surface S is **orientable** or **two-sided** if it is possible to define a field $\mathbf{n}$ of unit normal vectors on S that varies continuously with position. Any patch or subportion of an orientable surface is orientable. Spheres and other smooth closed surfaces in space (smooth surfaces that enclose solids) are orientable. By convention, we choose $\mathbf{n}$ on a closed surface to point outward.

Once $\mathbf{n}$ has been chosen, we have **oriented** the surface, and the surface together with its normal field is an **oriented surface**. The vector $\mathbf{n}$ at any point is the **positive direction** at that point (Figure 15.46).

The Möbius band in Figure 15.47 is not orientable. No matter where you start to construct a continuous unit normal field (shown as the shaft of a thumbtack in the figure), moving the vector continuously around the surface in the manner shown will return it to the starting point with a direction opposite to the one it had when it started out. The vector at that point cannot point both ways and yet it must if the field is to be continuous. We conclude that no such field exists.

Surface Integral for Flux

Suppose **F** is a continuous vector field defined over an oriented surface S and **n** is the chosen unit normal field on the surface. The integral of **F** • **n** over S is the *flux across* S in the positive direction. Thus, the flux is the integral over S of the scalar component of **F** in the direction of **n.**

Definition Flux

The **flux** of a three-dimensional vector field **F** across an oriented surface S in the direction of **n** is given by the formula

$$\text{Flux} = \iint_S \mathbf{F} \cdot \mathbf{n} \, d\sigma. \tag{11}$$

The definition is analogous to the flux of a two-dimensional field **F** across a plane curve C. In the plane (Section 15.2), the flux is

$$\int_C \mathbf{F} \cdot \mathbf{n} \, ds,$$

the integral of the scalar component of **F** normal to the curve.

If **F** is the velocity field of a three-dimensional fluid flow, then the flux of **F** across S is the net rate at which fluid is crossing S in the chosen positive direction. We discuss such flows in more detail in Section 15.7.

If S is part of a level surface $g(x, y, z) = c$, then **n** may be taken to be one of the two fields

$$\mathbf{n} = \pm \frac{\nabla g}{|\nabla g|}, \tag{12}$$

depending on which one gives the preferred direction. The corresponding flux is

$$\begin{aligned}
\text{Flux} &= \iint_S \mathbf{F} \cdot \mathbf{n} \, d\sigma && \text{Eq. (11)} \\[2mm]
&= \iint_R \left(\mathbf{F} \cdot \frac{\pm \nabla g}{|\nabla g|} \right) \frac{|\nabla g|}{|\nabla g \cdot \mathbf{p}|} \, dA && \text{Eqs. (12) and (10)} \\[2mm]
&= \iint_R \mathbf{F} \cdot \frac{\pm \nabla g}{|\nabla g \cdot \mathbf{p}|} \, dA && \text{(13)}
\end{aligned}$$

Example 4   FINDING FLUX

Find the flux of $\mathbf{F} = yz\mathbf{j} + z^2\mathbf{k}$ outward through the surface S cut from the cylinder $y^2 + z^2 = 1$, $z \geq 0$, by the planes $x = 0$ and $x = 1$.

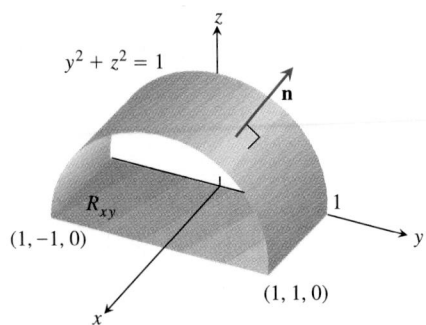

Figure 15.48 Example 4 calculates the flux of a vector field outward through this surface. The area of the shadow region R_{xy} is 2.

Solution The outward normal field on S (Figure 15.48) may be calculated from the gradient of $g(x, y, z) = y^2 + z^2$ to be

$$\mathbf{n} = +\frac{\nabla g}{|\nabla g|} = \frac{2y\mathbf{j} + 2z\mathbf{k}}{\sqrt{4y^2 + 4z^2}} = \frac{2y\mathbf{j} + 2z\mathbf{k}}{2} = y\mathbf{j} + z\mathbf{k}.$$

With $\mathbf{p} = \mathbf{k}$, we also have

$$d\sigma = \frac{|\nabla g|}{|\nabla g \cdot \mathbf{k}|} \, dA = \frac{2}{|2z|} \, dA = \frac{1}{z} \, dA. \qquad \begin{matrix} |z| = z \text{ because} \\ z \geq 0 \text{ on } S \end{matrix}$$

The value of $\mathbf{F} \cdot \mathbf{n}$ on the surface is given by the formula

$$
\begin{aligned}
\mathbf{F} \cdot \mathbf{n} &= (yz\mathbf{j} + z^2\mathbf{k}) \cdot (y\mathbf{j} + z\mathbf{k}) \\
&= y^2z + z^3 = z(y^2 + z^2) \\
&= z \qquad y^2 + z^2 = 1 \text{ on } S
\end{aligned}
$$

Therefore, the flux of $\mathbf{F}$ outward through S is

$$\iint_S \mathbf{F} \cdot \mathbf{n} \, d\sigma = \iint_S (z)\left(\frac{1}{z} \, dA\right) = \iint_{R_{xy}} dA = \text{area}(R_{xy}) = 2.$$

Moments and Masses of Thin Shells

Thin shells of material like bowls, metal drums, and domes are modeled with surfaces. Their moments and masses are calculated with the formulas in Table 15.3.

Table 15.3 Mass and Moment Formulas for Very Thin Shells

Mass: $M = \iint_S \delta(x, y, z) \, d\sigma$ $\qquad$ ($\delta(x, y, z) = $ density at (x, y, z), mass per unit area)

First moments about the coordinate planes:

$$M_{yz} = \iint_S x\delta \, d\sigma, \qquad M_{xz} = \iint_S y\delta \, d\sigma, \qquad M_{xy} = \iint_S z\delta \, d\sigma$$

Coordinates of center of mass:

$$\bar{x} = M_{yz}/M, \qquad \bar{y} = M_{xz}/M, \qquad \bar{z} = M_{xy}/M$$

Moments of inertia:

$$I_x = \iint_S (y^2 + z^2)\delta \, d\sigma, \qquad I_y = \iint_S (x^2 + z^2)\delta \, d\sigma,$$

$$I_z = \iint_S (x^2 + y^2)\delta \, d\sigma, \qquad I_L = \iint_S r^2 \delta \, d\sigma,$$

$r(x, y, z) = $ distance from point (x, y, z) to line L

Radius of gyration about a line L: $R_L = \sqrt{I_L/M}$

Example 5 FINDING CENTER OF MASS

Find the center of mass of a thin hemispherical shell of radius a and constant density δ.

Solution We model the shell with the hemisphere

$$f(x, y, z) = x^2 + y^2 + z^2 = a^2, \qquad z \geq 0$$

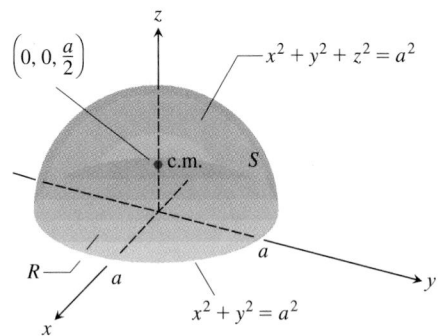

Figure 15.49 The center of mass of a thin hemispherical shell of constant density lies on the axis of symmetry halfway from the base to the top. (Example 5)

(Figure 15.49). The symmetry of the surface about the z-axis tells us that $\bar{x} = \bar{y} = 0$. It remains only to find $\bar{z}$ from the formula $\bar{z} = M_{xy}/M$.

The mass of the shell is

$$M = \iint_S \delta \, d\sigma = \delta \iint_S d\sigma = (\delta)(\text{area of } S) = 2\pi a^2 \delta.$$

To evaluate the integral for M_{xy}, we take $\mathbf{p} = \mathbf{k}$ and calculate

$$|\nabla f| = |2x\mathbf{i} + 2y\mathbf{j} + 2z\mathbf{k}| = 2\sqrt{x^2 + y^2 + z^2} = 2a,$$

$$|\nabla f \cdot \mathbf{p}| = |\nabla f \cdot \mathbf{k}| = |2z| = 2z,$$

$$d\sigma = \frac{|\nabla f|}{|\nabla f \cdot \mathbf{p}|} \, dA = \frac{a}{z} \, dA.$$

Then

$$M_{xy} = \iint_S z\delta \, d\sigma = \delta \iint_R z\frac{a}{z} \, dA = \delta a \iint_R dA = \delta a(\pi a^2) = \delta \pi a^3,$$

$$\bar{z} = \frac{M_{xy}}{M} = \frac{\pi a^3 \delta}{2\pi a^2 \delta} = \frac{a}{2}.$$

The shell's center of mass is the point $(0, 0, a/2)$.

Quick Review 15.5

In Exercises 1–5, let $z = f(x, y) = \sqrt{4 - x^2 - y^2}$ and P_0 be the point $(1, 1, \sqrt{2})$.

1. Evaluate f_x at P_0.   **2.** Evaluate f_y at P_0.

3. Find a vector perpendicular to the surface $z = f(x, y)$ at P_0.

4. Find an equation for the plane tangent to the surface $z = f(x, y)$ at P_0.

5. Find an equation for the normal line to the surface $z = f(x, y)$ at P_0.

In Exercises 6–10, evaluate the integral.

6. $\int_0^{2\pi} \int_0^2 \sqrt{4r^2 + 1} \, r \, dr \, d\theta$

7. $\int_0^{2\pi} \int_0^1 \frac{r \, dr \, d\theta}{\sqrt{2 - r^2}}$ **8.** $\int_0^2 \int_0^{\sqrt{4-x^2}} dy \, dx$

9. $\int_0^2 \int_0^2 e^{x+y} \, dy \, dx$ **10.** $\int_0^2 \int_0^{y/2} y \sin(xy) \, dx \, dy$

Section 15.5 Exercises

In Exercises 1–12, find the area of the surface.

1. The surface cut from the paraboloid $x^2 + y^2 - z = 0$ by the plane $z = 2$

2. The band cut from the paraboloid $x^2 + y^2 - z = 0$ by the planes $z = 2$ and $z = 6$

3. The region cut from the plane $x + 2y + 2z = 5$ by the cylinder whose walls are $x = y^2$ and $x = 2 - y^2$

4. The portion of the surface $x^2 - 2z = 0$ that lies above the triangle bounded by the lines $x = \sqrt{3}$, $y = 0$, and $y = x$ in the xy-plane

5. The portion of the surface $x^2 - 2y - 2z = 0$ that lies above the triangle bounded by the lines $x = 2$, $y = 0$, and $y = 3x$ in the xy-plane

6. The cap cut from the sphere $x^2 + y^2 + z^2 = 2$ by the cone $z = \sqrt{x^2 + y^2}$

7. The ellipse cut from the plane $z = cx$ by the cylinder $x^2 + y^2 = 1$

8. The upper portion of the cylinder $x^2 + z^2 = 1$ that lies between the planes $x = \pm 1/2$ and $y = \pm 1/2$

9. The portion of the paraboloid $x = 4 - y^2 - z^2$ that lies above the ring $1 \le y^2 + z^2 \le 4$ in the yz-plane

10. The surface cut from the paraboloid $x^2 + y + z^2 = 2$ by the plane $y = 0$

11. The portion of the surface $x^2 - 2 \ln x + \sqrt{15} y - z = 0$ above the square $R: 1 \le x \le 2, 0 \le y \le 1$, in the xy-plane

12. The portion of the surface $2x^{3/2} + 2y^{3/2} - 3z = 0$ above the square $R: 0 \le x \le 1, 0 \le y \le 1$, in the xy-plane

In Exercises 13–18, integrate g over the surface.

13. $g(x, y, z) = x + y + z$ over the surface of the cube cut from the first octant by the planes $x = a, y = a, z = a$

14. $g(x, y, z) = y + z$ over the surface of the wedge in the first octant bounded by the coordinate planes and the planes $x = 2$ and $y + z = 1$

15. $g(x, y, z) = xyz$ over the surface of the rectangular solid cut from the first octant by the planes $x = a, y = b$, and $z = c$

16. $g(x, y, z) = xyz$ over the surface of the rectangular solid bounded by the planes $x = \pm a, y = \pm b$, and $z = \pm c$

17. $g(x, y, z) = x + y + z$ over the portion of the plane $2x + 2y + z = 2$ that lies in the first octant.

18. $g(x, y, z) = x\sqrt{y^2 + 4}$ over the surface cut from the parabolic cylinder $y^2 + 4z = 16$ by the planes $x = 0, x = 1$, and $z = 0$

In Exercises 19 and 20, find the flux of the field $\mathbf{F}$ across the portion of the given surface in the specified direction.

19. $\mathbf{F}(x, y, z) = -\mathbf{i} + 2\mathbf{j} + 3\mathbf{k}$

    S: rectangular surface $z = 0$, $\quad 0 \le x \le 2$, $\quad 0 \le y \le 3$, direction $\mathbf{k}$

20. $\mathbf{F}(x, y, z) = yx^2\mathbf{i} - 2\mathbf{j} + xz\mathbf{k}$

    S: rectangular surface $y = 0$, $\quad -1 \le x \le 2$, $\quad 2 \le z \le 7$, direction $-\mathbf{j}$

In Exercises 21–26, find the flux of the field $\mathbf{F}$ across the portion of the sphere $x^2 + y^2 + z^2 = a^2$ in the first octant in the direction away from the origin.

21. $\mathbf{F}(x, y, z) = z\mathbf{k}$

22. $\mathbf{F}(x, y, z) = -y\mathbf{i} + x\mathbf{j}$

23. $\mathbf{F}(x, y, z) = y\mathbf{i} - x\mathbf{j} + \mathbf{k}$

24. $\mathbf{F}(x, y, z) = zx\mathbf{i} + zy\mathbf{j} + z^2\mathbf{k}$

25. $\mathbf{F}(x, y, z) = x\mathbf{i} + y\mathbf{j} + z\mathbf{k}$

26. $\mathbf{F}(x, y, z) = \dfrac{x\mathbf{i} + y\mathbf{j} + z\mathbf{k}}{\sqrt{x^2 + y^2 + z^2}}$

27. *Flux* Find the flux of the field $\mathbf{F}(x, y, z) = z^2\mathbf{i} + x\mathbf{j} - 3z\mathbf{k}$ upward through the surface cut from the parabolic cylinder $z = 4 - y^2$ by the planes $x = 0, x = 1$, and $z = 0$.

28. *Flux* Find the flux of the field $\mathbf{F}(x, y, z) = 4x\mathbf{i} + 4y\mathbf{j} + 2\mathbf{k}$ outward (away from the z-axis) through the surface cut from the bottom of the paraboloid $z = x^2 + y^2$ by the plane $z = 1$.

29. *Flux* Let S be the portion of the cylinder $y = e^x$ in the first octant that projects parallel to the x-axis onto the rectangle $R_{yz}: 1 \le y \le 2, 0 \le z \le 1$ in the yz-plane (Figure 15.50). Let $\mathbf{n}$ be the unit vector normal to S that points away from the yz-plane. Find the flux of the field $\mathbf{F}(x, y, z) = -2\mathbf{i} + 2y\mathbf{j} + z\mathbf{k}$ across S in the direction of $\mathbf{n}$.

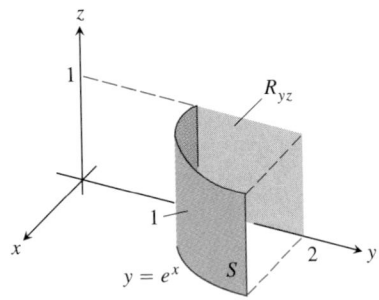

Figure 15.50 The surface and region in Exercise 29.

30. *Flux* Let S be the portion of the cylinder $y = \ln x$ in the first octant whose projection parallel to the y-axis onto the xz-plane is the rectangle $R_{xz}: 1 \le x \le e, 0 \le z \le 1$. Let $\mathbf{n}$ be the unit vector normal to S that points away from the xz-plane. Find the flux of $\mathbf{F} = 2y\mathbf{j} + z\mathbf{k}$ through S in the direction of $\mathbf{n}$.

31. *Flux* Find the outward flux of the field $\mathbf{F} = 2xy\mathbf{i} + 2yz\mathbf{j} + 2xz\mathbf{k}$ across the surface of the cube cut from the first octant by the planes $x = a, y = a, z = a$.

32. *Flux* Find the outward flux of the field $\mathbf{F} = xz\mathbf{i} + yz\mathbf{j} + \mathbf{k}$ across the surface of the upper cap cut from the solid sphere $x^2 + y^2 + z^2 \le 25$ by the plane $z = 3$.

In Exercises 33–36, *work in groups of two or three* to solve the problem.

33. *Centroid* Find the centroid of the portion of the sphere $x^2 + y^2 + z^2 = a^2$ that lies in the first octant.

34. *Centroid* Find the centroid of the surface cut from the cylinder $y^2 + z^2 = 9, z \ge 0$, by the planes $x = 0$ and $x = 3$ (resembles the surface in Example 4).

35. *Thin Shell of Constant Density* Find the center of mass and the moment of inertia and radius of gyration about the z-axis of a thin shell of constant density δ cut from the cone $x^2 + y^2 - z^2 = 0$ by the planes $z = 1$ and $z = 2$.

36. *Conical Surface of Constant Density* Find the moment of inertia about the z-axis of a thin shell of constant density δ cut from the cone $4x^2 + 4y^2 - z^2 = 0$, $z \geq 0$, by the circular cylinder $x^2 + y^2 = 2x$ (Figure 15.51).

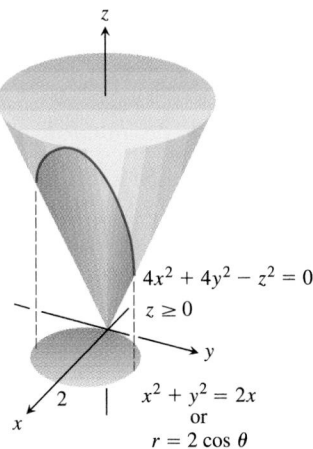

$$4x^2 + 4y^2 - z^2 = 0$$
$$z \geq 0$$
$$x^2 + y^2 = 2x$$
$$\text{or}$$
$$r = 2 \cos \theta$$

Figure 15.51 The surface in Exercises 36.

Explorations

37. *Spherical Shells*

(a) Find the moment of inertia about a diameter of a thin spherical shell of radius a and constant density δ. (Work with a hemispherical shell and double the result.)

(b) Use the Parallel Axis Theorem (Exercises 14.5) and the result in (a) to find the moment of inertia about a line tangent to the shell.

38. *Cones with and without Ice Cream*

(a) Find the centroid of the lateral surface of a solid cone of base radius a and height h (cone surface minus the base).

(b) Use Pappus's formula (Exercises 14.5) and the result in (a) to find the centroid of the complete surface of a solid cone (side plus base).

(c) **Writing to Learn** A cone of radius a and height h is joined to a hemisphere of radius a to make a surface S that resembles an ice cream cone. Use Pappus's formula and the results in (a) and Example 5 to find the centroid of S. How high does the cone have to be to place the centroid in the plane shared by the bases of the hemisphere and cone? ∎

Extending the Ideas

Specialized Formulas for Surface Area If S is the surface defined by a function $z = f(x, y)$ that has continuous first partial derivatives throughout a region R_{xy} in the xy-plane (Figure 15.52), then S is also the level surface $F(x, y, z) = 0$ of the function $F(x, y, z) = f(x, y) - z$. Taking the unit normal to R_{xy} to be $\mathbf{p} = \mathbf{k}$ then gives

$$|\nabla f| = |f_x \mathbf{i} + f_y \mathbf{j} - \mathbf{k}| = \sqrt{f_x^2 + f_y^2 + 1},$$
$$|\nabla f \cdot \mathbf{p}| = |(f_x \mathbf{i} + f_y \mathbf{j} - \mathbf{k}) \cdot \mathbf{k}| = |-1| = 1,$$

and

$$A = \iint\limits_{R_{xy}} \frac{|\nabla f|}{|\nabla f \cdot \mathbf{p}|} \, dA = \iint\limits_{R_{xy}} \sqrt{f_x^2 + f_y^2 + 1} \, dx \, dy. \quad (14)$$

Similarly, the area of a smooth surface $x = f(y, z)$ over a region R_{yz} in the yz-plane is

$$A = \iint\limits_{R_{yz}} \sqrt{f_y^2 + f_z^2 + 1} \, dy \, dz, \quad (15)$$

and the area of a smooth surface $y = f(x, z)$ over a region R_{xz} in the xz-plane is

$$A = \iint\limits_{R_{xz}} \sqrt{f_x^2 + f_z^2 + 1} \, dx \, dz. \quad (16)$$

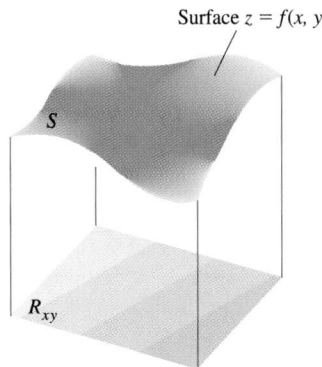

Surface $z = f(x, y)$

Figure 15.52 For a surface $z = f(x, y)$, the surface area formula in Equation (5) takes the form

$$A = \iint\limits_{R_{xy}} \sqrt{f_x^2 + f_y^2 + 1} \, dx \, dy.$$

In Exercises 39–44, use Equations (14)–(16) to find the areas of the surfaces.

39. The surface cut from the bottom of the paraboloid $z = x^2 + y^2$ by the plane $z = 3$

40. The surface cut from the "nose" of the paraboloid $x = 1 - y^2 - z^2$ by the yz-plane

41. The portion of the cone $z = \sqrt{x^2 + y^2}$ that lies over the region between the circle $x^2 + y^2 = 1$ and the ellipse $9x^2 + 4y^2 = 36$ in the xy-plane. (*Hint:* Use formulas from geometry to find the area of the region.)

42. The triangle cut from the plane $2x + 6y + 3z = 6$ by the bounding planes of the first octant. Calculate the area three ways, once with each formula.

43. The surface in the first octant cut from the cylinder $y = (2/3)z^{3/2}$ by the planes $x = 1$ and $y = 16/3$

44. The portion of the plane $y + z = 4$ that lies above the region cut from the first quadrant of the xz-plane by the parabola $x = 4 - z^2$

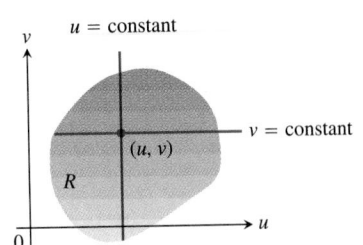

Figure 15.53 A parametrized surface.

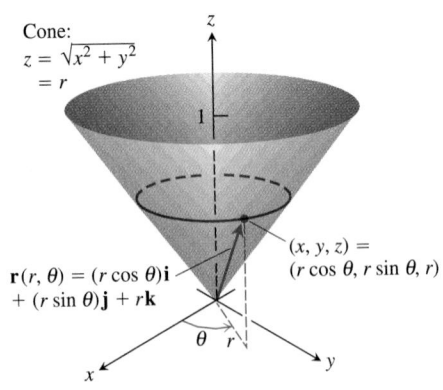

Figure 15.54 The cone in Example 1.

Parametrized Surfaces

Parametrizations of Surfaces • Surface Area • Surface Integrals

Parametrizations of Surfaces

Let

$$\mathbf{r}(u, v) = f(u, v)\mathbf{i} + g(u, v)\mathbf{j} + h(u, v)\mathbf{k} \tag{1}$$

be a continuous vector function that is defined on a region R in the uv-plane and one-to-one on the interior of R (Figure 15.53). The range of $\mathbf{r}$ is the **surface S** defined or traced by $\mathbf{r}$. Equation (1) and the domain R together constitute a **parametrization** of the surface. The variables u and v are the **parameters,** and R is the **parameter domain.** To simplify our discussion, we will take R to be a rectangle defined by inequalities of the form $a \le u \le b$, $c \le v \le d$. The requirement that $\mathbf{r}$ be one-to-one on the interior of R ensures that S does not cross itself. Notice that Equation (1) is the vector equivalent of *three* parametric equations:

$$x = f(u, v), \qquad y = g(u, v), \qquad z = h(u, v).$$

Example 1 PARAMETRIZING A CONE

Find a parametrization of the cone $z = \sqrt{x^2 + y^2}$, $0 \le z \le 1$.

Solution Here, cylindrical coordinates provide everything we need. A typical point (x, y, z) on the cone (Figure 15.54) has $x = r \cos \theta$, $y = r \sin \theta$, and $z = \sqrt{x^2 + y^2} = r$, with $0 \le r \le 1$ and $0 \le \theta \le 2\pi$. Taking $u = r$ and $v = \theta$ in Equation (1) gives the parametrization

$$\mathbf{r}(r, \theta) = (r \cos \theta)\mathbf{i} + (r \sin \theta)\mathbf{j} + r\mathbf{k}, \qquad 0 \le r \le 1, \qquad 0 \le \theta \le 2\pi.$$

Example 2 PARAMETRIZING A SPHERE

Find a parametrization of the sphere $x^2 + y^2 + z^2 = a^2$.

Solution Spherical coordinates provide what we need. A typical point (x, y, z) on the sphere (Figure 15.55) has $x = a \sin \phi \cos \theta$, $y = a \sin \phi \sin \theta$, and $z = a \cos \phi$, $0 \le \phi \le \pi$, $0 \le \theta \le 2\pi$. Taking $u = \phi$ and $v = \theta$ in Equation (1) gives the parametrization

$$\mathbf{r}(\phi, \theta) = (a \sin \phi \cos \theta)\mathbf{i} + (a \sin \phi \sin \theta)\mathbf{j} + (a \cos \phi)\mathbf{k},$$
$$0 \le \phi \le \pi, \qquad 0 \le \theta \le 2\pi.$$

Example 3 PARAMETRIZING A CYLINDER

Find a parametrization of the cylinder $x^2 + (y - 3)^2 = 9$, $0 \le z \le 5$.

Solution In cylindrical coordinates, a point (x, y, z) has $x = r \cos \theta$, $y = r \sin \theta$, and $z = z$. For points on the cylinder $x^2 + (y - 3)^2 = 9$ (Figure 15.56), $r = 6 \sin \theta$, $0 \le \theta \le \pi$ (Section 11.5, Example 6). A typical point on the cylinder therefore has

$$x = r \cos \theta = 6 \sin \theta \cos \theta = 3 \sin 2\theta,$$
$$y = r \sin \theta = 6 \sin^2 \theta,$$
$$z = z.$$

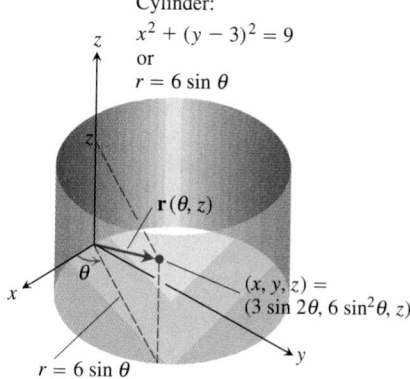

$(x, y, z) = (a \sin \phi \cos \theta, a \sin \phi \sin \theta, a \cos \phi)$

Figure 15.55 The sphere in Example 2.

Cylinder:
$x^2 + (y - 3)^2 = 9$
or
$r = 6 \sin \theta$

$\mathbf{r}(\theta, z)$

$(x, y, z) = (3 \sin 2\theta, 6 \sin^2\theta, z)$

$r = 6 \sin \theta$

Figure 15.56 The cylinder in Example 3.

Taking $u = \theta$ and $v = z$ in Equation (1) gives the parametrization

$\mathbf{r}(\theta, z) = (3 \sin 2\theta)\mathbf{i} + (6 \sin^2 \theta)\mathbf{j} + z\mathbf{k}, \qquad 0 \le \theta \le \pi, \qquad 0 \le z \le 5.$

Surface Area

Our goal is to find a double integral for calculating the area of a curved surface S based on the parametrization

$\mathbf{r}(u, v) = f(u, v)\mathbf{i} + g(u, v)\mathbf{j} + h(u, v)\mathbf{k}, \qquad a \le u \le b, \qquad c \le v \le d.$

We need to assume that S is smooth enough for the construction we are about to carry out. The definition of smoothness involves the partial derivatives of $\mathbf{r}$ with respect to u and v:

$$\mathbf{r}_u = \frac{\partial \mathbf{r}}{\partial u} = \frac{\partial f}{\partial u}\mathbf{i} + \frac{\partial g}{\partial u}\mathbf{j} + \frac{\partial h}{\partial u}\mathbf{k},$$

$$\mathbf{r}_v = \frac{\partial \mathbf{r}}{\partial v} = \frac{\partial f}{\partial v}\mathbf{i} + \frac{\partial g}{\partial v}\mathbf{j} + \frac{\partial h}{\partial v}\mathbf{k}.$$

Definition Smooth Parametrized Surface

A parametrized surface $\mathbf{r}(u, v) = f(u, v)\mathbf{i} + g(u, v)\mathbf{j} + h(u, v)\mathbf{k}$ is **smooth** if $\mathbf{r}_u$ and $\mathbf{r}_v$ are continuous and $\mathbf{r}_u \times \mathbf{r}_v$ is never zero on the parameter domain.

Now consider a small rectangle in ΔA_{uv} in R with sides on the lines $u = u_0$, $u = u_0 + \Delta u$, $v = v_0$, and $v = v_0 + \Delta v$ (Figure 15.57). Each side of ΔA_{uv} maps to a curve on the surface S, and together these four curves bound a "curved area element" $\Delta \sigma_{uv}$. In the notation of the figure, the side $v = v_0$ maps to curve C_1, the side $u = u_0$ maps to C_2, and their common vertex (u_0, v_0) maps to P_0. Figure 15.58 shows an enlarged view of $\Delta \sigma_{uv}$. The vector $\mathbf{r}_u(u_0, v_0)$ is tangent to C_1 at P_0. Likewise, $\mathbf{r}_v(u_0, v_0)$ is tangent to C_2 at P_0. The cross product $\mathbf{r}_u \times \mathbf{r}_v$ is normal to the surface at P_0. (Here is where we begin to use the assumption that S is smooth. We want to be sure that $\mathbf{r}_u \times \mathbf{r}_v \ne \mathbf{0}$.)

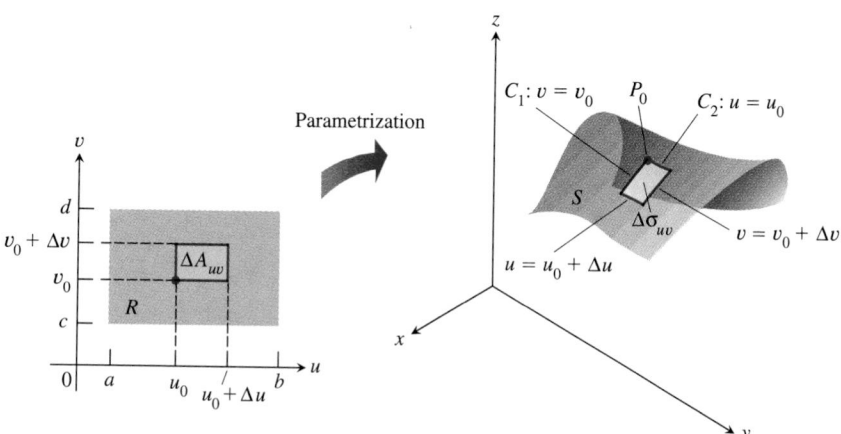

Figure 15.57 A rectangular area element ΔA_{uv} in the uv-plane maps onto a curved area element $\Delta \sigma_{uv}$ on S.

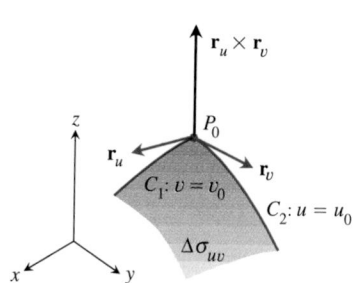

Figure 15.58 A magnified view of a surface area element $\Delta \sigma_{uv}$.

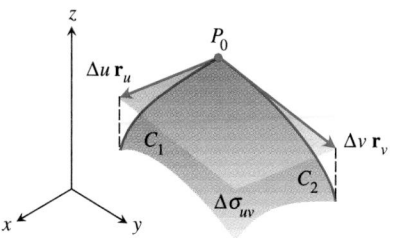

Figure 15.59 The parallelogram determined by the vectors $\Delta u \mathbf{r}_u$ and $\Delta v \mathbf{r}_v$ approximates the surface area element $\Delta \sigma_{uv}$.

We next approximate the surface element $\Delta \sigma_{uv}$ by the parallelogram on the tangent plane whose sides are determined by the vectors $\Delta u \mathbf{r}_u$ and $\Delta v \mathbf{r}_v$ (Figure 15.59). The area of the parallelogram is

$$\left| \Delta u \mathbf{r}_u \times \Delta v \mathbf{r}_v \right| = \left| \mathbf{r}_u \times \mathbf{r}_v \right| \Delta u \, \Delta v. \tag{2}$$

A partition of the region R in the uv-plane by rectangular regions ΔA_{uv} generates a partition of the surface S into surface area elements $\Delta \sigma_{uv}$. We approximate the area of each surface element $\Delta \sigma_{uv}$ by the parallelogram area in Equation (2) and sum these areas together to obtain an approximation of the area of S:

$$\sum_u \sum_v \left| \mathbf{r}_u \times \mathbf{r}_v \right| \Delta u \, \Delta v. \tag{3}$$

As Δu and Δv approach zero independently, the continuity of $\mathbf{r}_u$ and $\mathbf{r}_v$ guarantees that the sum in Equation (3) approaches the double integral $\int_c^d \int_a^b \left| \mathbf{r}_u \times \mathbf{r}_v \right| du \, dv$. This double integral gives the area of the surface S.

Parametric Formula for Surface Area

The area of the smooth surface

$$\mathbf{r}(u, v) = f(u, v)\mathbf{i} + g(u, v)\mathbf{j} + h(u, v)\mathbf{k}, \quad a \leq u \leq b, \quad c \leq v \leq d$$

is

$$A = \int_c^d \int_a^b \left| \mathbf{r}_u \times \mathbf{r}_v \right| du \, dv. \tag{4}$$

As in Section 15.5, we can abbreviate the integral in Equation (4) by writing $d\sigma$ for $\left| \mathbf{r}_u \times \mathbf{r}_v \right| du \, dv$.

Surface Area Differential and Differential Formula for Surface Area

$$d\sigma = \left| \mathbf{r}_u \times \mathbf{r}_v \right| du \, dv \qquad \iint_S d\sigma \tag{5}$$

surface area differential

differential formula for surface area

Example 4   FINDING SURFACE AREA (CONE)

Find the surface area of the cone in Example 1 (Figure 15.54).

Solution   In Example 1, we found the parametrization

$$\mathbf{r}(r, \theta) = (r \cos \theta)\mathbf{i} + (r \sin \theta)\mathbf{j} + r\mathbf{k}, \quad 0 \leq r \leq 1, \quad 0 \leq \theta \leq 2\pi.$$

To apply Equation (4) we first find $\mathbf{r}_r \times \mathbf{r}_\theta$:

$$\mathbf{r}_r \times \mathbf{r}_\theta = \begin{vmatrix} \mathbf{i} & \mathbf{j} & \mathbf{k} \\ \cos \theta & \sin \theta & 1 \\ -r \sin \theta & r \cos \theta & 0 \end{vmatrix}$$

$$= -(r \cos \theta)\mathbf{i} - (r \sin \theta)\mathbf{j} + \underbrace{(r \cos^2 \theta + r \sin^2 \theta)}_{r}\mathbf{k}.$$

Thus, $|\mathbf{r}_r \times \mathbf{r}_\theta| = \sqrt{r^2 \cos^2 \theta + r^2 \sin^2 \theta + r^2} = \sqrt{2r^2} = \sqrt{2}\, r$. The area of the cone is

$$A = \int_0^{2\pi} \int_0^1 |\mathbf{r}_r \times \mathbf{r}_\theta|\, dr\, d\theta \qquad\qquad \text{Eq. (4) with } u = r,\, v = \theta$$

$$= \int_0^{2\pi} \int_0^1 \sqrt{2}\, r\, dr\, d\theta = \int_0^{2\pi} \frac{\sqrt{2}}{2}\, d\theta = \pi\sqrt{2} \text{ units squared.}$$

Example 5 FINDING SURFACE AREA (SPHERE)

Find the surface area of a sphere of radius a.

Solution   We use the parametrization from Example 2:

$$\mathbf{r}(\phi, \theta) = (a \sin \phi \cos \theta)\mathbf{i} + (a \sin \phi \sin \theta)\mathbf{j} + (a \cos \phi)\mathbf{k},$$
$$0 \le \phi \le \pi, \qquad 0 \le \theta \le 2\pi.$$

For $\mathbf{r}_\phi \times \mathbf{r}_\theta$ we get

$$\mathbf{r}_\phi \times \mathbf{r}_\theta = \begin{vmatrix} \mathbf{i} & \mathbf{j} & \mathbf{k} \\ a \cos \phi \cos \theta & a \cos \phi \sin \theta & -a \sin \phi \\ -a \sin \phi \sin \theta & a \sin \phi \cos \theta & 0 \end{vmatrix}$$

$$= (a^2 \sin^2 \phi \cos \theta)\mathbf{i} + (a^2 \sin^2 \phi \sin \theta)\mathbf{j} + (a^2 \sin \phi \cos \phi)\mathbf{k}.$$

Thus,

$$|\mathbf{r}_\phi \times \mathbf{r}_\theta| = \sqrt{a^4 \sin^4 \phi \cos^2 \theta + a^4 \sin^4 \phi \sin^2 \theta + a^4 \sin^2 \phi \cos^2 \phi}$$

$$= \sqrt{a^4 \sin^4 \phi + a^4 \sin^2 \phi \cos^2 \phi}$$

$$= \sqrt{a^4 \sin^2 \phi(\sin^2 \phi + \cos^2 \phi)}$$

$$= a^2 \sqrt{\sin^2 \phi} = a^2 \sin \phi$$

since $\sin \phi \ge 0$ for $0 \le \phi \le \pi$. Therefore, the area of the sphere is

$$A = \int_0^{2\pi} \int_0^\pi a^2 \sin \phi\, d\phi\, d\theta$$

$$= \int_0^{2\pi} [-a^2 \cos \phi]_0^\pi\, d\theta = \int_0^{2\pi} 2a^2\, d\theta = 4\pi a^2 \text{ units squared.}$$

Surface Integrals

Having found the formula for calculating the area of a parametrized surface, we can now integrate a function over the surface using the parametrized form.

Definition Integral of G(x, y, z) over S

If S is a smooth surface defined parametrically as

$$\mathbf{r}(u, v) = f(u, v)\mathbf{i} + g(u, v)\mathbf{j} + h(u, v)\mathbf{k}, \qquad a \le u \le b, \qquad c \le v \le d,$$

and $G(x, y, z)$ is a continuous function defined on S, then the **integral of G over S** is

$$\iint\limits_S G(x, y, z)\, d\sigma = \int_c^d \int_a^b G(f(u, v), g(u, v), h(u, v)) |\mathbf{r}_u \times \mathbf{r}_v|\, du\, dv.$$

Example 6 INTEGRATING OVER A SURFACE PARAMETRICALLY

Integrate $G(x, y, z) = x^2$ over the cone $z = \sqrt{x^2 + y^2}, 0 \le z \le 1$.

Solution Continuing the work in Examples 1 and 4, we have $|\mathbf{r}_r \times \mathbf{r}_\theta| = \sqrt{2}\, r$ and

$$\iint_S x^2\, d\sigma = \int_0^{2\pi}\int_0^1 (r^2\, \cos^2\theta)(\sqrt{2}\, r)\, dr\, d\theta$$

$$= \sqrt{2}\int_0^{2\pi}\int_0^1 r^3\, \cos^2\theta\, dr\, d\theta$$

$$= \frac{\sqrt{2}}{4}\int_0^{2\pi} \cos^2\theta\, d\theta = \frac{\sqrt{2}}{4}\left[\frac{\theta}{2} + \frac{1}{4}\sin 2\theta\right]_0^{2\pi} = \frac{\pi\sqrt{2}}{4}.$$

Example 7 FINDING FLUX

Find the flux of $\mathbf{F} = yz\mathbf{i} + x\mathbf{j} - z^2\mathbf{k}$ outward through the parabolic cylinder $y = x^2, 0 \le x \le 1, 0 \le z \le 4$ (Figure 15.60).

Solution On the surface we have $x = x$, $y = x^2$, and $z = z$, so we automatically have the parametrization $\mathbf{r}(x, z) = x\mathbf{i} + x^2\mathbf{j} + z\mathbf{k}, 0 \le x \le 1$, $0 \le z \le 4$. The cross product of tangent vectors is

$$\mathbf{r}_x \times \mathbf{r}_z = \begin{vmatrix} \mathbf{i} & \mathbf{j} & \mathbf{k} \\ 1 & 2x & 0 \\ 0 & 0 & 1 \end{vmatrix} = 2x\mathbf{i} - \mathbf{j}.$$

The unit normal pointing outward from the surface is

$$\mathbf{n} = \frac{\mathbf{r}_x \times \mathbf{r}_z}{|\mathbf{r}_x \times \mathbf{r}_z|} = \frac{2x\mathbf{i} - \mathbf{j}}{\sqrt{4x^2 + 1}}.$$

On the surface, $y = x^2$, so the vector field is

$$\mathbf{F} = yz\mathbf{i} + x\mathbf{j} - z^2\mathbf{k} = x^2 z\mathbf{i} + x\mathbf{j} - z^2\mathbf{k}.$$

Thus,

$$\mathbf{F} \cdot \mathbf{n} = \frac{1}{\sqrt{4x^2 + 1}}\left((x^2 z)(2x) + (x)(-1) + (-z^2)(0)\right)$$

$$= \frac{2x^3 z - x}{\sqrt{4x^2 + 1}}.$$

The flux of $\mathbf{F}$ outward through the surface is

$$\iint_S \mathbf{F} \cdot \mathbf{n}\, d\sigma = \int_0^4\int_0^1 \frac{2x^3 z - x}{\sqrt{4x^2 + 1}}\, |\mathbf{r}_x \times \mathbf{r}_z|\, dx\, dz$$

$$= \int_0^4\int_0^1 \frac{2x^3 z - x}{\sqrt{4x^2 + 1}}\sqrt{4x^2 + 1}\, dx\, dz$$

$$= \int_0^4\int_0^1 (2x^3 z - x)\, dx\, dz = \int_0^4 \left[\frac{1}{2}x^4 z - \frac{1}{2}x^2\right]_{x=0}^{x=1} dz$$

$$= \int_0^4 \frac{1}{2}(z - 1)\, dz = \frac{1}{4}(z - 1)^2\Big]_0^4$$

$$= 2.$$

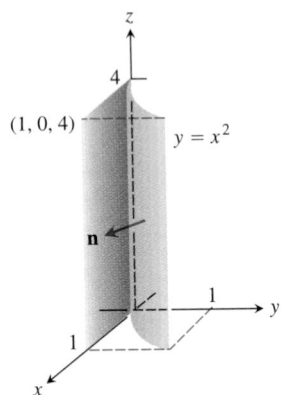

Figure 15.60 The parabolic surface in Example 7.

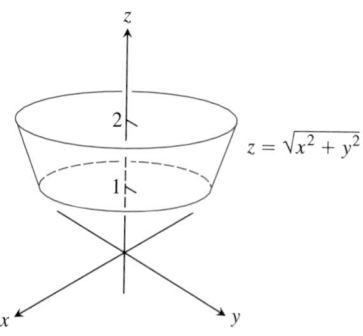

Figure 15.61 The cone frustum in Example 8.

Example 8 FINDING CENTER OF MASS

Find the center of mass of a thin shell of constant density δ cut from the cone $z = \sqrt{x^2 + y^2}$ by the planes $z = 1$ and $z = 2$ (Figure 15.61).

Solution The symmetry of the surface about the z-axis tells us that $\bar{x} = \bar{y} = 0$. We find $\bar{z} = M_{xy}/M$. Working as in Examples 1 and 4 we have

$$\mathbf{r}(r, \theta) = (r \cos \theta)\mathbf{i} + (r \sin \theta)\mathbf{j} + r\mathbf{k}, \qquad 1 \le r \le 2, \qquad 0 \le \theta \le 2\pi,$$

and

$$|\mathbf{r}_r \times \mathbf{r}_\theta| = \sqrt{2}\,r.$$

Therefore,

$$M = \iint\limits_S \delta\, d\sigma = \int_0^{2\pi}\int_1^2 \delta\sqrt{2}\,r\,dr\,d\theta$$

$$= \delta\sqrt{2}\int_0^{2\pi}\left[\frac{r^2}{2}\right]_1^2 d\theta = \frac{3}{2}\delta\sqrt{2}\int_0^{2\pi} d\theta$$

$$= 3\pi\delta\sqrt{2},$$

$$M_{xy} = \iint\limits_S \delta z\, d\sigma = \int_0^{2\pi}\int_1^2 \delta r\sqrt{2}\,r\,dr\,d\theta$$

$$= \delta\sqrt{2}\int_0^{2\pi}\int_1^2 r^2\,dr\,d\theta = \delta\sqrt{2}\int_0^{2\pi}\left[\frac{r^3}{3}\right]_1^2 d\theta$$

$$= \frac{7}{3}\delta\sqrt{2}\int_0^{2\pi} d\theta = \frac{14}{3}\pi\delta\sqrt{2}$$

$$\bar{z} = \frac{M_{xy}}{M} = \frac{14\pi\delta\sqrt{2}}{3(3\pi\delta\sqrt{2})} = \frac{14}{9}.$$

The shell's center of mass is the point $(0, 0, 14/9)$.

Quick Review 15.6

In Exercises 1 and 2, the rectangular coordinates of P are $(1, 1, 1)$.

1. Find cylindrical coordinates for P.

2. Find spherical coordinates for P.

In Exercises 3 and 4, consider the rectangular equation $x^2 + (y + 2)^2 = 4$.

3. Describe the surface defined by the equation.

4. Find equivalent equations for the surface in (**a**) cylindrical and (**b**) spherical coordinates.

In Exercises 5 and 6, consider the rectangular equation $x^2 + y^2 + z^2 = 9$.

5. Describe the surface defined by the equation.

6. Find equivalent equations for the surface in (**a**) cylindrical and (**b**) spherical coordinates.

In Exercises 7 and 8, consider the cylindrical equation $r = 2 \csc \theta$.

7. Describe the surface defined by the equation.

8. Find equivalent equations for the surface in (**a**) rectangular and (**b**) spherical coordinates.

In Exercises 9 and 10, let $\mathbf{u} = 2\mathbf{i} - 3\mathbf{j} + \mathbf{k}$ and $\mathbf{v} = -2\mathbf{i} + 4\mathbf{j} - 2\mathbf{k}$. Use determinants to find the vector.

9. $\mathbf{u} \times \mathbf{v}$

10. $(\mathbf{u} + \mathbf{v}) \times (\mathbf{u} - \mathbf{v})$

Section 15.6 Exercises

In Exercises 1–16, find a parametrization of the surface. (There are many correct ways to do these, so your answers may not be the same as those in the back of the book.)

1. *Parabolic Bowl* The paraboloid $z = x^2 + y^2, z \leq 4$

2. *Inverted Parabolic Bowl* The paraboloid $z = 9 - x^2 - y^2$, $z \geq 0$

3. *Conical Triangle* The first-octant portion of the cone $z = \sqrt{x^2 + y^2}/2$ between the planes $z = 0$ and $z = 3$

4. *Cone Frustum* The portion of the cone $z = 2\sqrt{x^2 + y^2}$ between the planes $z = 2$ and $z = 4$

5. *Spherical Cap* The cap cut from the sphere $x^2 + y^2 + z^2 = 9$ by the cone $z = \sqrt{x^2 + y^2}$

6. *Spherical Triangle* The portion of the sphere $x^2 + y^2 + z^2 = 4$ in the first octant between the xy-plane and the cone $z = \sqrt{x^2 + y^2}$

7. *Spherical Band* The portion of the sphere $x^2 + y^2 + z^2 = 3$ between the planes and $z = \sqrt{3}/2$ and $z = -\sqrt{3}/2$

8. *Spherical Mushroom Cap* The upper portion cut from the sphere $x^2 + y^2 + z^2 = 8$ by the plane $z = -2$

9. *Parabolic Reflector* The surface cut from the parabolic cylinder $z = 4 - y^2$ by the planes $x = 0$, $x = 2$, and $z = 0$

10. *Parabolic Reflector* The surface cut from the parabolic cylinder $y = x^2$ by the planes $z = 0$, $z = 3$, and $y = 2$

11. *Circular Cylindrical Band* The portion of the cylinder $y^2 + z^2 = 9$ between the planes $x = 0$ and $x = 3$

12. *Half of a Circular Cylindrical Band* The portion of the cylinder $x^2 + z^2 = 4$ above the xy-plane between the planes $y = -2$ and $y = 2$

13. *Tilted Plane Inside a Cylinder* The portion of the plane $x + y + z = 1$
    (a) inside the cylinder $x^2 + y^2 = 9$
    (b) inside the cylinder $y^2 + z^2 = 9$

14. *Tilted Plane Inside a Cylinder* The portion of the plane $x - y + 2z = 2$
    (a) inside the cylinder $x^2 + z^2 = 3$
    (b) inside the cylinder $y^2 + z^2 = 2$

15. *Circular Cylindrical Band* The portion of the cylinder $(x - 2)^2 + z^2 = 4$ between the planes $y = 0$ and $y = 3$

16. *Circular Cylindrical Band* The portion of the cylinder $y^2 + (z - 5)^2 = 25$ between the planes $x = 0$ and $x = 10$

In Exercises 17–26, use a parametrization to express the area of the surface as a double integral. Then evaluate the integral. (There are many correct ways to set up the integrals, so your integrals may not be the same as those in the back of the book. They should have the same values, however.)

17. *Tilted Plane Inside a Cylinder* The portion of the plane $y + 2z = 2$ inside the cylinder $x^2 + y^2 = 1$

18. *Tilted Plane Inside a Cylinder* The portion of the plane $z = -x$ inside the cylinder $x^2 + y^2 = 4$

19. *Cone Frustum* The portion of the cone $z = 2\sqrt{x^2 + y^2}$ between the planes $z = 2$ and $z = 6$

20. *Cone Frustum* The portion of the cone $z = \sqrt{x^2 + y^2}/3$ between the planes $z = 1$ and $z = 4/3$

21. *Circular Cylindrical Band* The portion of the cylinder $x^2 + y^2 = 1$ between the planes $z = 1$ and $z = 4$

22. *Circular Cylindrical Band* The portion of the cylinder $x^2 + z^2 = 10$ between the planes $y = -1$ and $y = 1$

23. *Parabolic Cap* The cap cut from the paraboloid $z = 2 - x^2 - y^2$ by the cone $z = \sqrt{x^2 + y^2}$

24. *Parabolic Band* The portion of the paraboloid $z = x^2 + y^2$ between the planes $z = 1$ and $z = 4$

25. *Inverted Spherical Mushroom Cap* The lower portion cut from the sphere $x^2 + y^2 + z^2 = 2$ by the cone $z = \sqrt{x^2 + y^2}$

26. *Spherical Band* The portion of the sphere $x^2 + y^2 + z^2 = 4$ between the planes $z = -1$ and $z = \sqrt{3}$

In Exercises 27–34, integrate the given function over the given surface.

27. *Parabolic Cylinder* $G(x, y, z) = x$, over the parabolic cylinder $y = x^2, 0 \leq x \leq 2, 0 \leq z \leq 3$

28. *Circular Cylinder* $G(x, y, z) = z$, over the cylindrical surface $y^2 + z^2 = 4, z \geq 0, 1 \leq x \leq 4$

29. *Unit Sphere* $G(x, y, z) = x^2$, over the unit sphere $x^2 + y^2 + z^2 = 1$

30. *Hemisphere* $G(x, y, z) = z^2$, over the hemisphere $x^2 + y^2 + z^2 = a^2, z \geq 0$

31. *Rhombus* $F(x, y, z) = z$, over the portion of the plane $x + y + z = 4$ that lies above the square $0 \leq x \leq 1$, $0 \leq y \leq 1$, in the xy-plane

32. *Cone* $F(x, y, z) = z - x$, over the cone $z = \sqrt{x^2 + y^2}$, $0 \leq z \leq 1$

33. *Parabolic Dome* $H(x, y, z) = x^2\sqrt{5 - 4z}$, over the parabolic dome $z = 1 - x^2 - y^2, z \geq 0$

34. *Spherical Cap* $H(x, y, z) = yz$, over the part of the sphere $x^2 + y^2 + z^2 = 4$ that lies above the cone $z = \sqrt{x^2 + y^2}$

In Exercises 35–44, use a parametrization to find the flux $\iint_S \mathbf{F} \cdot \mathbf{n} \, d\sigma$ across the surface in the given direction.

35. $\mathbf{F} = z^2\mathbf{i} + x\mathbf{j} - 3z\mathbf{k}$ outward (normal away from the x-axis) through the surface cut from the parabolic cylinder $z = 4 - y^2$ by the planes $x = 0$, $x = 1$, and $z = 0$

36. $\mathbf{F} = x^2\mathbf{j} - xz\mathbf{k}$ outward (normal away from the yz-plane) through the surface cut from the parabolic cylinder $y = x^2$, $-1 \leq x \leq 1$, by the planes $z = 0$ and $z = 2$

37. $\mathbf{F} = z\mathbf{k}$ across the portion of the sphere $x^2 + y^2 + z^2 = a^2$ in the first octant in the direction away from the origin

38. $\mathbf{F} = x\mathbf{i} + y\mathbf{j} + z\mathbf{k}$ across the sphere $x^2 + y^2 + z^2 = a^2$ in the direction away from the origin

39. $\mathbf{F} = 2xy\mathbf{i} + 2yz\mathbf{j} + 2xz\mathbf{k}$ upward across the portion of the plane $x + y + z = 2a$ that lies above the square $0 \leq x \leq a$, $0 \leq y \leq a$, in the xy-plane

40. $\mathbf{F} = x\mathbf{i} + y\mathbf{j} + z\mathbf{k}$ outward through the portion of the cylinder $x^2 + y^2 = 1$ cut by the planes $z = 0$ and $z = a$

41. $\mathbf{F} = xy\mathbf{i} - z\mathbf{k}$ outward (normal away from the z-axis) through the cone $z = \sqrt{x^2 + y^2}$, $0 \leq z \leq 1$

42. $\mathbf{F} = y^2\mathbf{i} + xz\mathbf{j} - \mathbf{k}$ outward (normal away from the z-axis) through the cone $z = 2\sqrt{x^2 + y^2}$, $0 \leq z \leq 2$

43. $\mathbf{F} = -x\mathbf{i} - y\mathbf{j} + z^2\mathbf{k}$ outward (normal away from the z-axis) through the portion of the cone $z = \sqrt{x^2 + y^2}$ between the planes $z = 1$ and $z = 2$

44. $\mathbf{F} = 4x\mathbf{i} + 4y\mathbf{j} + 2\mathbf{k}$ outward (normal away from the z-axis) through the surface cut from the bottom of the paraboloid $z = x^2 + y^2$ by the plane $z = 1$

In Exercises 45–48, *work in groups of two or three* to find the appropriate center, moment, or radius of gyration.

45. *A Quarter Dome* Find the centroid of the portion of the sphere $x^2 + y^2 + z^2 = a^2$ that lies in the first octant.

46. *A Thin Shell* Find the center of mass and the moment of inertia and radius of gyration about the z-axis of a thin shell of constant density δ cut from the cone $x^2 + y^2 - z^2 = 0$ by the planes $z = 1$ and $z = 2$.

47. *A Spherical Shell* Find the moment of inertia about the z-axis of a thin spherical shell $x^2 + y^2 + z^2 = a^2$ of constant density δ.

48. *A Conical Shell* Find the moment of inertia about the z-axis of a thin conical shell $z = \sqrt{x^2 + y^2}$, $0 \leq z \leq 1$, of constant density δ.

Explorations

Tangent Planes The **tangent plane** at a point

$$P_0(f(u_0, v_0), g(u_0, v_0), h(u_0, v_0))$$

on a parametrized surface

$$\mathbf{r}(u, v) = f(u, v)\mathbf{i} + g(u, v)\mathbf{j} + h(u, v)\mathbf{k}$$

is the plane through P_0 normal to the vector $\mathbf{r}_u(u_0, v_0) \times \mathbf{r}_v(u_0, v_0)$, which is the cross product of the tangent vectors $\mathbf{r}_u(u_0, v_0)$ and $\mathbf{r}_v(u_0, v_0)$ at P_0.

In Exercises 49–52, find an equation for the plane that is tangent to the surface at the given point P_0. Then find a Cartesian equation for the surface and sketch the surface and tangent plane together.

49. *Cone* The cone $\mathbf{r}(r, \theta) = (r \cos \theta)\mathbf{i} + (r \sin \theta)\mathbf{j} + r\mathbf{k}$, $r \geq 0$, $0 \leq \theta \leq 2\pi$, at the point $P_0(\sqrt{2}, \sqrt{2}, 2)$ corresponding to $(r, \theta) = (2, \pi/4)$

50. *Hemisphere* The hemisphere surface
$$\mathbf{r}(\phi, \theta) = (4 \sin \phi \cos \theta)\mathbf{i} + (4 \sin \phi \sin \theta)\mathbf{j} + (4 \cos \phi)\mathbf{k},$$
$0 \leq \phi \leq \pi/2$, $0 \leq \theta \leq 2\pi$, at the point $P_0(\sqrt{2}, \sqrt{2}, 2\sqrt{3})$ corresponding to $(\phi, \theta) = (\pi/6, \pi/4)$

51. *Circular Cylinder* The circular cylinder
$$\mathbf{r}(\theta, z) = (3 \sin 2\theta)\mathbf{i} + (6 \sin^2 \theta)\mathbf{j} + z\mathbf{k},$$
$0 \leq \theta \leq \pi$, at the point $P_0(3\sqrt{3}/2, 9/2, 0)$ corresponding to $(\theta, z) = (\pi/3, 0)$. (See Example 3.)

52. *Parabolic Cylinder* The parabolic cylinder surface
$$\mathbf{r}(x, y) = x\mathbf{i} + y\mathbf{j} - x^2\mathbf{k}, \quad -\infty < x < \infty, \ -\infty < y < \infty,$$
at the point $P_0(1, 2, -1)$ corresponding to $(x, y) = (1, 2)$ ∎

Extending the Ideas

53. *Torus of Revolution*

(a) A **torus of revolution** (doughnut) is the surface obtained by rotating a circle C in the xz-plane about the z-axis in space. If the radius of C is $r > 0$ and the center is $(R, 0, 0)$, $R > r$, show that a parametrization of the torus is

$$\mathbf{r}(u, v) = ((R + r \cos u) \cos v)\mathbf{i} + ((R + r \cos u) \sin v)\mathbf{j} + (r \sin u)\mathbf{k},$$

where $0 \leq u \leq 2\pi$ and $0 \leq v \leq 2\pi$ are the angles in Figure 15.62.

(b) Show that the surface area of the torus is
$$A = 4\pi^2 Rr.$$

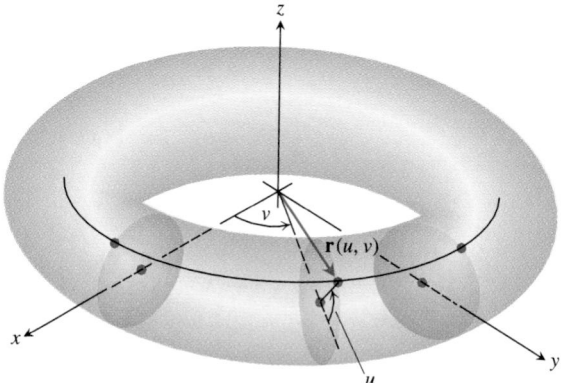

Figure 15.62 The torus surface in Exercise 53.

54. *Parametrizing a Surface of Revolution* Suppose the parametrized curve C: $(f(u), g(u))$, $a \leq u \leq b$, $g(u) \geq 0$, is revolved about the x-axis.

(a) Show that

$$\mathbf{r}(u, v) = f(u)\mathbf{i} + (g(u) \cos v)\mathbf{j} + (g(u) \sin v)\mathbf{k}$$

is a parametrization of the resulting surface of revolution, where $0 \leq v \leq 2\pi$ is the angle from the xy-plane to the point $\mathbf{r}(u, v)$ on the surface. (See the accompanying figure.) Notice that $f(u)$ measures distance *along* the axis of revolution and $g(u)$ measures distance *from* the axis of revolution

(b) Find a parametrization for the surface obtained by revolving the curve $x = y^2$, $y \geq 0$, about the x-axis.

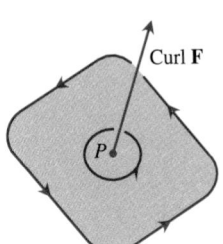

Figure 15.63 The circulation vector at a point P in a plane in a three dimensional fluid flow. Notice its right-hand relation to the circulation line.

15.7 Stokes's Theorem

Circulation Density: Curl • Stokes's Theorem • Getting the Most from a Paddle Wheel • Proof of Stokes's Theorem for Polyhedral Surfaces • Stokes's Theorem for Surfaces with Holes • An Important Identity • Conservative Fields and Stokes's Theorem

Circulation Density: Curl

As we saw in Section 15.4, the $\mathbf{k}$-component of the circulation density or curl of a two-dimensional field $\mathbf{F} = M\mathbf{i} + N\mathbf{j}$ at a point (x, y) is the scalar quantity $(\partial N/\partial x - \partial M/\partial y)$. In three dimensions, the circulation around a point P in a plane is described by a vector normal to the plane of the circulation (Figure 15.63), pointing in the direction that gives it a right-hand relation to the circulation line. The length of the vector gives the rate of the fluid's rotation, which usually varies as the circulation plane is tilted about P. It turns out (Stokes's Theorem) that the vector of greatest circulation in a flow with velocity field $\mathbf{F} = M\mathbf{i} + N\mathbf{j} + P\mathbf{k}$ is

$$\text{curl } \mathbf{F} = \left(\frac{\partial P}{\partial y} - \frac{\partial N}{\partial z}\right)\mathbf{i} + \left(\frac{\partial M}{\partial z} - \frac{\partial P}{\partial x}\right)\mathbf{j} + \left(\frac{\partial N}{\partial x} - \frac{\partial M}{\partial y}\right)\mathbf{k}. \quad (1)$$

Notice that (curl $\mathbf{F}$) $\cdot$ $\mathbf{k}$ equals $\partial N/\partial x - \partial M/\partial y$, consistent with the definition in Section 15.4 for $\mathbf{F} = M\mathbf{i} + N\mathbf{j}$.

The formula for curl $\mathbf{F}$ in Equation (1) is usually written using the symbolic operator

$$\nabla = \mathbf{i}\frac{\partial}{\partial x} + \mathbf{j}\frac{\partial}{\partial y} + \mathbf{k}\frac{\partial}{\partial z}. \quad (2)$$

The symbol ∇ is pronounced "del." The curl of $\mathbf{F}$ can be found by calculating $\nabla \times \mathbf{F}$:

$$\nabla \times \mathbf{F} = \begin{vmatrix} \mathbf{i} & \mathbf{j} & \mathbf{k} \\ \dfrac{\partial}{\partial x} & \dfrac{\partial}{\partial y} & \dfrac{\partial}{\partial z} \\ M & N & P \end{vmatrix}$$

$$= \left(\frac{\partial P}{\partial y} - \frac{\partial N}{\partial z}\right)\mathbf{i} + \left(\frac{\partial M}{\partial z} - \frac{\partial P}{\partial x}\right)\mathbf{j} + \left(\frac{\partial N}{\partial x} - \frac{\partial M}{\partial y}\right)\mathbf{k} \quad (3)$$

$$= \text{curl } \mathbf{F}.$$

George Gabriel Stokes

Sir George Gabriel Stokes (1819–1903), one of the most influential scientific figures of his century, was Lucasian Professor of Mathematics at Cambridge University from 1849 until his death in 1903. His theoretical and experimental investigations covered hydrodynamics, elasticity, light, gravity, sound, heat, meteorology, and solar physics. He left electricity and magnetism to his friend William Thomson, Baron Kelvin of Largs. It is another one of those delightful quirks of history that the theorem we call Stokes's Theorem isn't his theorem at all. He learned of it from Thomson in 1850 and a few years later included it among the questions on an examination he wrote for the Smith Prize. It has been known as Stokes's Theorem ever since. As usual, things have balanced out. Stokes was the original discover of the principles of spectrum analysis that we now credit to Bunsen and Kirchhoff.

$$\text{curl } \mathbf{F} = \nabla \times \mathbf{F} \qquad (4)$$

Example 1 FINDING CURL

Find the curl of $\mathbf{F} = (x^2 - y)\mathbf{i} + 4z\mathbf{j} + x^2\mathbf{k}.$

Solution

$$\text{curl } \mathbf{F} = \nabla \times \mathbf{F} \qquad \text{Eq. (4)}$$

$$= \begin{vmatrix} \mathbf{i} & \mathbf{j} & \mathbf{k} \\ \dfrac{\partial}{\partial x} & \dfrac{\partial}{\partial y} & \dfrac{\partial}{\partial z} \\ x^2 - y & 4z & x^2 \end{vmatrix}$$

$$= \left(\frac{\partial}{\partial y}(x^2) - \frac{\partial}{\partial z}(4z) \right)\mathbf{i} - \left(\frac{\partial}{\partial x}(x^2) - \frac{\partial}{\partial z}(x^2 - y) \right)\mathbf{j}$$

$$+ \left(\frac{\partial}{\partial x}(4z) - \frac{\partial}{\partial y}(x^2 - y) \right)\mathbf{k}$$

$$= (0 - 4)\mathbf{i} - (2x - 0)\mathbf{j} + (0 + 1)\mathbf{k}$$

$$= -4\mathbf{i} - 2x\mathbf{j} + \mathbf{k}$$

As we will see, the operator ∇ has a number of other applications. For instance, when applied to a scalar function $f(x, y, z)$, it gives the gradient of f:

$$\nabla f = \frac{\partial f}{\partial x}\mathbf{i} + \frac{\partial f}{\partial y}\mathbf{j} + \frac{\partial f}{\partial z}\mathbf{k}.$$

This may now be read as "del f" as well as "grad f."

Stokes's Theorem

Stokes's Theorem says that, under conditions normally met in practice, the circulation of a vector field around the boundary of an oriented surface in space in the direction counterclockwise with respect to the surface's unit normal vector field $\mathbf{n}$ (Figure 15.64) equals the integral of the normal component of the curl of the field over the surface.

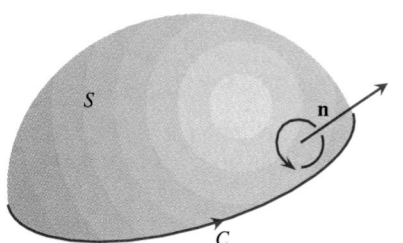

Figure 15.64 The orientation of the bounding curve C gives it a right-handed relation to the normal field $\mathbf{n}$.

Theorem 5 Stokes's Theorem

The circulation of $\mathbf{F} = M\mathbf{i} + N\mathbf{j} + P\mathbf{k}$ around the boundary C of an oriented surface S in the direction counterclockwise with respect to the surface's unit normal vector $\mathbf{n}$ equals the integral of $\nabla \times \mathbf{F} \cdot \mathbf{n}$ over S.

$$\underbrace{\oint_C \mathbf{F} \cdot d\mathbf{r}}_{\substack{\text{counterclockwise} \\ \text{circulation}}} = \underbrace{\iint_S \nabla \times \mathbf{F} \cdot \mathbf{n}\, d\sigma}_{\text{curl integral}} \qquad (5)$$

Notice from Equation (5) that if two different oriented surfaces S_1 and S_2 have the same boundary C, then their curl integrals are equal:

$$\iint_{S_1} \nabla \times \mathbf{F} \cdot \mathbf{n}_1 \, d\sigma = \iint_{S_2} \nabla \times \mathbf{F} \cdot \mathbf{n}_2 \, d\sigma.$$

Both curl integrals equal the counterclockwise circulation integral on the left side of Equation (5) as long as the unit normal vectors $\mathbf{n}_1$ and $\mathbf{n}_2$ correctly orient the surfaces.

Naturally, we need some mathematical restrictions on $\mathbf{F}$, C, and S to ensure the existence of the integrals in Stokes's equation. The usual restrictions are that all the functions and derivatives be continuous.

If C is a curve in the xy-plane, oriented counterclockwise, and R is the region in the xy-plane bounded by C, then $d\sigma = dx \, dy$ and

$$(\nabla \times \mathbf{F}) \cdot \mathbf{n} = (\nabla \times \mathbf{F}) \cdot \mathbf{k} = \left(\frac{\partial N}{\partial x} - \frac{\partial M}{\partial y} \right). \tag{6}$$

Under these conditions, Stokes's equation becomes

$$\oint_C \mathbf{F} \cdot d\mathbf{r} = \iint_R \left(\frac{\partial N}{\partial x} - \frac{\partial M}{\partial y} \right) dx \, dy,$$

which is the circulation-curl form of the equation in Green's theorem. Conversely, by reversing these steps we can rewrite the circulation-curl form of Green's Theorem for two-dimensional fields in del notation as

$$\oint_C \mathbf{F} \cdot d\mathbf{r} = \iint_R \nabla \times \mathbf{F} \cdot \mathbf{k} \, dA. \tag{7}$$

See Figure 15.65.

Green:

Stokes:

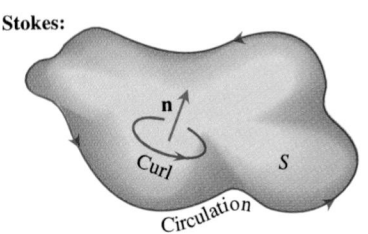

Figure 15.65 Green's Theorem vs. Stokes's Theorem.

Example 2 VERIFYING STOKES'S EQUATION

Evaluate Equation (5) for the hemisphere S: $x^2 + y^2 + z^2 = 9$, $z \geq 0$, its bounding circle C: $x^2 + y^2 = 9$, $z = 0$, and the field $\mathbf{F} = y\mathbf{i} - x\mathbf{j}$.

Solution   We calculate the counterclockwise circulation around C (as viewed from above) using the parametrization $\mathbf{r}(\theta) = (3 \cos \theta)\mathbf{i} + (3 \sin \theta)\mathbf{j}$, $0 \leq \theta \leq 2\pi$:

$$d\mathbf{r} = (-3 \sin \theta \, d\theta)\mathbf{i} + (3 \cos \theta \, d\theta)\mathbf{j}$$

$$\mathbf{F} = y\mathbf{i} - x\mathbf{j} = (3 \sin \theta)\mathbf{i} - (3 \cos \theta)\mathbf{j}$$

$$\mathbf{F} \cdot d\mathbf{r} = -9 \sin^2 \theta \, d\theta - 9 \cos^2 \theta \, d\theta = -9 \, d\theta$$

$$\oint_C \mathbf{F} \cdot d\mathbf{r} = \int_0^{2\pi} -9 \, d\theta = -18\pi.$$

For the curl integral of **F,** we have

$$\nabla \times \mathbf{F} = \left(\frac{\partial P}{\partial y} - \frac{\partial N}{\partial z}\right)\mathbf{i} + \left(\frac{\partial M}{\partial z} - \frac{\partial P}{\partial x}\right)\mathbf{j} + \left(\frac{\partial N}{\partial x} - \frac{\partial M}{\partial y}\right)\mathbf{k}$$

$$= (0 - 0)\mathbf{i} + (0 - 0)\mathbf{j} + (-1 - 1)\mathbf{k} = -2\mathbf{k},$$

$$\mathbf{n} = \frac{x\mathbf{i} + y\mathbf{j} + z\mathbf{k}}{\sqrt{x^2 + y^2 + z^2}} = \frac{x\mathbf{i} + y\mathbf{j} + z\mathbf{k}}{3}, \qquad \text{Outer unit normal}$$

$$d\sigma = \frac{3}{z}\, dA, \qquad\qquad \text{Section 15.5,}$$
Section 15.5, Example 5, with $a = 3$

$$\nabla \times \mathbf{F} \cdot \mathbf{n} = -\frac{2z}{3}\frac{3}{3}\, dA = -dA,$$

and

$$\iint_S \nabla \times \mathbf{F} \cdot \mathbf{n}\, d\sigma = \iint_{x^2 + y^2 \le 9} -2\, dA = -18\pi.$$

The circulation around the circle equals the integral of the curl over the hemisphere, as it should.

Example 3   FINDING CIRCULATION

Find the circulation of the field $\mathbf{F} = (x^2 - y)\mathbf{i} + 4z\mathbf{j} + x^2\mathbf{k}$ around the curve C in which the plane $z = 2$ meets the cone $z = \sqrt{x^2 + y^2}$, counterclockwise as viewed from above (Figure 15.66).

Solution   Stokes's Theorem enables us to find the circulation by integrating over the surface of the cone. Traversing C in the counterclockwise direction viewed from above corresponds to taking the *inner* normal **n** to the cone (which has a positive z-component).

We parametrize the cone as

$$\mathbf{r}(r, \theta) = (r \cos \theta)\mathbf{i} + (r \sin \theta)\mathbf{j} + r\mathbf{k}, \qquad 0 \le r \le 2, \qquad 0 \le \theta \le 2\pi.$$

We then have

$$\mathbf{n} = \frac{\mathbf{r}_r \times \mathbf{r}_\theta}{|\mathbf{r}_r \times \mathbf{r}_\theta|} = \frac{-(r \cos \theta)\mathbf{i} - (r \sin \theta)\mathbf{j} + r\mathbf{k}}{r\sqrt{2}} \qquad \text{Section 15.6, Example 4}$$

$$= \frac{1}{\sqrt{2}}(-(\cos \theta)\mathbf{i} - (\sin \theta)\mathbf{j} + \mathbf{k}),$$

$$d\sigma = r\sqrt{2}\, dr\, d\theta, \qquad \text{Section 15.6, Example 4}$$

$$\nabla \times \mathbf{F} = -4\mathbf{i} - 2x\mathbf{j} + \mathbf{k} \qquad \text{Example 1}$$

$$= -4\mathbf{i} - 2(r \cos \theta)\mathbf{j} + \mathbf{k}. \qquad x = r \cos \theta$$

Accordingly,

$$\nabla \times \mathbf{F} \cdot \mathbf{n} = \frac{1}{\sqrt{2}}(4 \cos \theta + 2r \cos \theta \sin \theta + 1)$$

$$= \frac{1}{\sqrt{2}}(4 \cos \theta + r \sin 2\theta + 1),$$

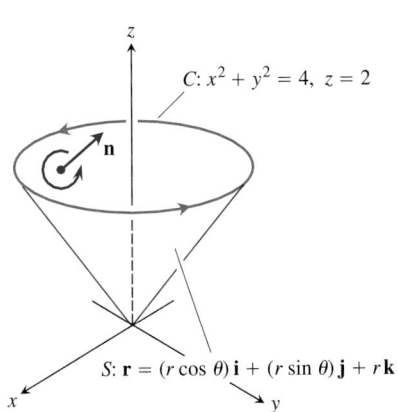

C: $x^2 + y^2 = 4$, $z = 2$

n

S: $\mathbf{r} = (r \cos \theta)\mathbf{i} + (r \sin \theta)\mathbf{j} + r\mathbf{k}$

Figure 15.66 The curve C and cone S in Example 3.

and the circulation is

$$\oint_C \mathbf{F} \cdot d\mathbf{r} = \iint_S \nabla \times \mathbf{F} \cdot \mathbf{n} \, d\sigma \qquad \text{Stokes's Theorem}$$

$$= \int_0^{2\pi} \int_0^2 \frac{1}{\sqrt{2}} (4 \cos \theta + r \sin 2\theta + 1)(r\sqrt{2} \, dr \, d\theta) = 4\pi.$$

Getting the Most from a Paddle Wheel

Suppose $\mathbf{v}(x, y, z)$ is the velocity of a moving fluid whose density at (x, y, z) is $\delta(x, y, z)$, and let $\mathbf{F} = \delta \mathbf{v}$. Then

$$\oint_C \mathbf{F} \cdot d\mathbf{r}$$

is the circulation of the fluid around the closed curve C. By Stokes's Theorem, the circulation is equal to the flux of $\nabla \times \mathbf{F}$ through a surface spanning C:

$$\oint_C \mathbf{F} \cdot d\mathbf{r} = \iint_S \nabla \times \mathbf{F} \cdot \mathbf{n} \, d\sigma.$$

Suppose we fix a point Q in the domain of $\mathbf{F}$ and a direction $\mathbf{u}$ at Q. Let C be a circle of radius ρ, with center at Q, whose plane is normal to $\mathbf{u}$. If $\nabla \times \mathbf{F}$ is continuous at Q, then the average value of the $\mathbf{u}$-component of $\nabla \times \mathbf{F}$ over the circular disk S bounded by C approaches the $\mathbf{u}$-component of $\nabla \times \mathbf{F}$ at Q as $\rho \to 0$:

$$(\nabla \times \mathbf{F} \cdot \mathbf{u})_Q = \lim_{\rho \to 0} \frac{1}{\pi\rho^2} \iint_S \nabla \times \mathbf{F} \cdot \mathbf{u} \, d\sigma. \tag{8}$$

If we replace the double integral in Equation (8) by the circulation, we get

$$(\nabla \times \mathbf{F} \cdot \mathbf{u})_Q = \lim_{\rho \to 0} \frac{1}{\pi\rho^2} \oint_C \mathbf{F} \cdot d\mathbf{r}. \tag{9}$$

The left-hand side of Equation (9) has its maximum value when $\mathbf{u}$ is the direction of $\nabla \times \mathbf{F}$. When ρ is small, the limit on the right-hand side of Equation (9) is approximately

$$\frac{1}{\pi\rho^2} \oint_C \mathbf{F} \cdot d\mathbf{r},$$

which is the circulation around C divided by the area of the disk (the circulation density). Suppose a small paddle wheel of radius ρ is introduced into the fluid at Q, with its axle directed along $\mathbf{u}$. The circulation of the fluid around C will affect the rate of spin of the paddle wheel. The wheel will spin fastest when the circulation integral is maximized; therefore, it will spin fastest when the axle of the paddle wheel points in the direction of $\nabla \times \mathbf{F}$ (Figure 15.67).

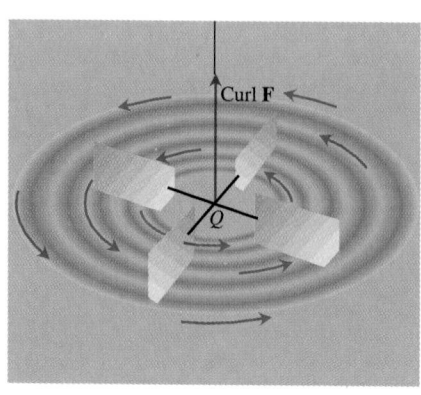

Figure 15.67 The paddle interpretation of curl **F**.

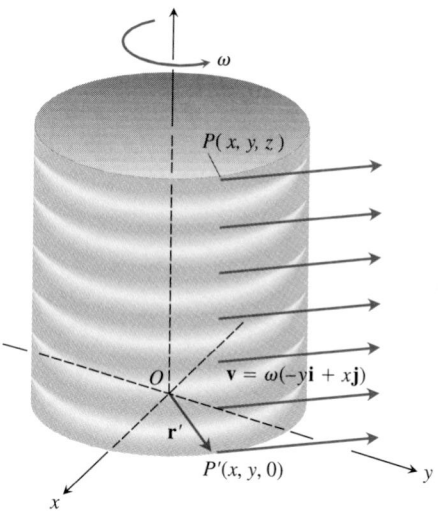

Figure 15.68 A steady rotational flow parallel to the xy-plane, with constant angular velocity ω in the positive (counterclockwise) direction.

Example 4 RELATING $\nabla \times \mathbf{F}$ TO CIRCULATION DENSITY

A fluid of constant density rotates around the z-axis with velocity $\mathbf{v} = \omega(-y\mathbf{i} + x\mathbf{j})$, where ω is a positive constant called the *angular velocity* of the rotation (Figure 15.68). If $\mathbf{F} = \mathbf{v}$, find $\nabla \times \mathbf{F}$ and relate it to the circulation density.

Solution With $\mathbf{F} = \mathbf{v} = -\omega y\mathbf{i} + \omega x\mathbf{j}$,

$$\nabla \times \mathbf{F} = \left(\frac{\partial P}{\partial y} - \frac{\partial N}{\partial z}\right)\mathbf{i} + \left(\frac{\partial M}{\partial z} - \frac{\partial P}{\partial x}\right)\mathbf{j} + \left(\frac{\partial N}{\partial x} - \frac{\partial M}{\partial y}\right)\mathbf{k}$$

$$= (0 - 0)\mathbf{i} + (0 - 0)\mathbf{j} + (\omega - (-\omega))\mathbf{k} = 2\omega\mathbf{k}.$$

By Stokes's Theorem, the circulation of $\mathbf{F}$ around a circle C of radius ρ bounding a disk S in a plane normal to $\nabla \times \mathbf{F}$, say the xy-plane, is

$$\oint_C \mathbf{F} \cdot d\mathbf{r} = \iint_S \nabla \times \mathbf{F} \cdot \mathbf{n} \, d\sigma = \iint_S 2\omega\mathbf{k} \cdot \mathbf{k} \, dx \, dy = (2\omega)(\pi\rho^2).$$

Thus,

$$(\nabla \times \mathbf{F}) \cdot \mathbf{k} = 2\omega = \frac{1}{\pi\rho^2} \oint_C \mathbf{F} \cdot d\mathbf{r},$$

in agreement with Equation (9) with $\mathbf{u} = \mathbf{k}$.

Example 5 APPLYING STOKES'S THEOREM

Use Stokes's Theorem to evaluate $\int_C \mathbf{F} \cdot d\mathbf{r}$, if $\mathbf{F} = xz\mathbf{i} + xy\mathbf{j} + 3xz\mathbf{k}$ and C is the boundary of the portion of the plane $2x + y + z = 2$ in the first octant, traversed counterclockwise as viewed from above (Figure 15.69).

Solution The plane is the level surface $f(x, y, z) = 2$ of the function $f(x, y, z) = 2x + y + z$. The unit normal vector

$$\mathbf{n} = \frac{\nabla f}{|\nabla f|} = \frac{2\mathbf{i} + \mathbf{j} + \mathbf{k}}{|2\mathbf{i} + \mathbf{j} + \mathbf{k}|} = \frac{1}{\sqrt{6}}(2\mathbf{i} + \mathbf{j} + \mathbf{k})$$

is consistent with the counterclockwise motion around C. To apply Stokes's Theorem, we find

$$\text{curl } \mathbf{F} = \nabla \times \mathbf{F} = \begin{vmatrix} \mathbf{i} & \mathbf{j} & \mathbf{k} \\ \dfrac{\partial}{\partial x} & \dfrac{\partial}{\partial y} & \dfrac{\partial}{\partial z} \\ xz & xy & 3xz \end{vmatrix} = (x - 3z)\mathbf{j} + y\mathbf{k}.$$

On the plane, z equals $2 - 2x - y$, so

$$\nabla \times \mathbf{F} = (x - 3(2 - 2x - y))\mathbf{j} + y\mathbf{k} = (7x + 3y - 6)\mathbf{j} + y\mathbf{k}$$

and

$$\nabla \times \mathbf{F} \cdot \mathbf{n} = \frac{1}{\sqrt{6}}(7x + 3y - 6 + y) = \frac{1}{\sqrt{6}}(7x + 4y - 6).$$

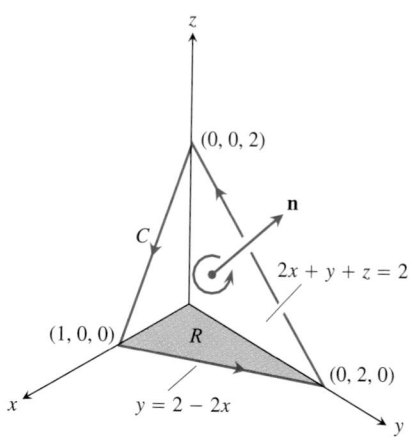

Figure 15.69 The planar surface in Example 5.

The surface area element is

$$d\sigma = \frac{|\nabla f|}{|\nabla f \cdot \mathbf{k}|} \, dA = \frac{\sqrt{6}}{1} \, dx \, dy.$$

The circulation is $\oint_C \mathbf{F} \cdot d\mathbf{r} = \iint_S \nabla \times \mathbf{F} \cdot \mathbf{n} \, d\sigma$ Stokes's Theorem

$$= \int_0^1 \int_0^{2-2x} \frac{1}{\sqrt{6}} (7x + 4y - 6)\sqrt{6} \, dy \, dx$$

$$= \int_0^1 \int_0^{2-2x} (7x + 4y - 6) \, dy \, dx = -1.$$

Proof of Stokes's Theorem for Polyhedral Surfaces

Let S be a polyhedral surface consisting of a finite number of plane regions. (Think of one of Buckminster Fuller's geodesic domes.) We apply Green's theorem to each separate panel of S. There are two types of panels:

1. those that are surrounded on all sides by other panels and

2. those that have one or more edges that are not adjacent to other panels.

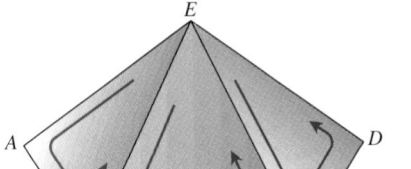

Figure 15.70 Part of a polyhedral surface.

The boundary Δ of S consists of edges of panels of the second type that are not adjacent to other panels. In Figure 15.70, the triangles EAB, BCE, and CDE represent a part of S, with $ABCD$ part of the boundary Δ. Applying Green's Theorem to the three triangles in turn and adding the results, we get

$$\left(\oint_{EAB} + \oint_{BCE} + \oint_{CDE} \right) \mathbf{F} \cdot d\mathbf{r} = \left(\iint_{EAB} + \iint_{BCE} + \iint_{CDE} \right) \nabla \times \mathbf{F} \cdot \mathbf{n} \, d\sigma. \quad (10)$$

The three line integrals on the left-hand side of Equation (10) combine into a single line integral taken around the periphery $ABCDE$ because the integrals along interior segments cancel in pairs. For example, the integral along segment BE in triangle ABE is opposite in sign to the integral along the same segment in triangle EBC. The same holds for segment CE. Hence, Equation (10) reduces to

$$\oint_{ABCDE} \mathbf{F} \cdot d\mathbf{r} = \iint_{ABCDE} \nabla \times \mathbf{F} \cdot \mathbf{n} \, d\sigma.$$

When we apply Green's Theorem to all the panels and add the results, we get

$$\oint_\Delta \mathbf{F} \cdot d\mathbf{r} = \iint_S \nabla \times \mathbf{F} \cdot \mathbf{n} \, d\sigma. \quad (11)$$

This is Stokes's Theorem for a polyhedral surface S. You can find proofs for more general surfaces in advanced calculus texts.

Stokes's Theorem for Surfaces with Holes

Stokes's Theorem can be extended to an oriented surface S that has one or more holes (Figure 15.71), in a way analogous to the extension of Green's Theorem: The surface integral over S of the normal component of $\nabla \times \mathbf{F}$

Figure 15.71 Stokes's Theorem also holds for oriented surfaces with holes.

Connected and simply connected.

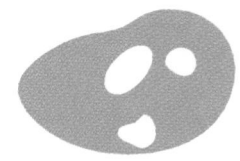

Connected but not simply connected.

Connected and simply connected.

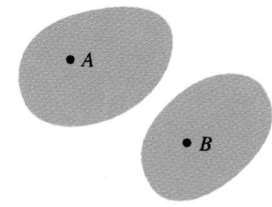

Simply connected but not connected.
No path from A to B lies entirely in the region.

Figure 15.72 Connectivity and simple connectivity are not the same. Neither implies the other, as these pictures of plane regions illustrate. To make three dimensional regions with these properties, thicken the plane regions into cylinders.

equals the sum of the line integrals around all the boundary curves of the tangential component of **F,** where the curves are to be traced in the direction induced by the orientation of S.

An Important Identity

The following identity arises frequently in mathematics and the physical sciences.

$$\text{curl grad } f = \mathbf{0} \qquad \text{or} \qquad \nabla \times \nabla f = \mathbf{0} \qquad (12)$$

This identity holds for any function $f(x, y, z)$ whose second partial derivatives are continuous. The proof goes like this:

$$\nabla \times \nabla f = \begin{vmatrix} \mathbf{i} & \mathbf{j} & \mathbf{k} \\ \dfrac{\partial}{\partial x} & \dfrac{\partial}{\partial y} & \dfrac{\partial}{\partial z} \\ \dfrac{\partial f}{\partial x} & \dfrac{\partial f}{\partial y} & \dfrac{\partial f}{\partial z} \end{vmatrix} = (f_{zy} - f_{yz})\mathbf{i} - (f_{zx} - f_{xz})\mathbf{j} + (f_{yx} - f_{xy})\mathbf{k}.$$

If the second partial derivatives are continuous, the mixed second derivatives in parentheses are equal (The Mixed Derivative Theorem, Section 13.3) and the vector is zero.

Conservative Fields and Stokes's Theorem

In Section 15.3, we found that saying that a field **F** is conservative in an open region D in space is equivalent to saying that the integral of **F** around every closed loop in D is zero. This, in turn, is equivalent in *simply connected* open regions to saying that $\nabla \times \mathbf{F} = \mathbf{0}$. A region D is **simply connected** if every closed path in D can be contracted to a point in D without ever leaving D. If D consisted of space with a line removed, for example, D would not be simply connected. There would be no way to contract a loop around the line to a point without leaving D. On the other hand, space itself is simply connected (Figure 15.72).

Theorem 6 Zero Curl Implies Path Independence

If $\nabla \times \mathbf{F} = \mathbf{0}$ at every point of a simply connected open region D in space, then on any piecewise smooth closed path C in D,

$$\oint_C \mathbf{F} \cdot d\mathbf{r} = 0.$$

Sketch of a Proof Theorem 6 is usually proved in two steps. The first step is for simple closed curves. A theorem from topology, a branch of advanced mathematics, states that every differentiable simple closed curve C in a simply connected open region D is the boundary of a smooth two-sided surface S that also lies in D. Hence, by Stokes's Theorem,

$$\oint_C \mathbf{F} \cdot d\mathbf{r} = \iint_S \nabla \times \mathbf{F} \cdot \mathbf{n} \, d\sigma = 0.$$

The second step is for curves that cross themselves, like the one in Figure 15.73. The idea is to break these into simple loops spanned by orientable surfaces, apply Stokes's Theorem one loop at a time, and add the results. ∎

The following diagram summarizes the results for conservative fields defined on connected, simply connected open regions. In such regions, under the conditions imposed on **F** and *C* in Section 15.3, the properties in the diagram's four corners are equivalent.

Figure 15.73 In a simply connected open region in space, differentiable curves that cross themselves can be divided into loops to which Stokes's Theorem applies.

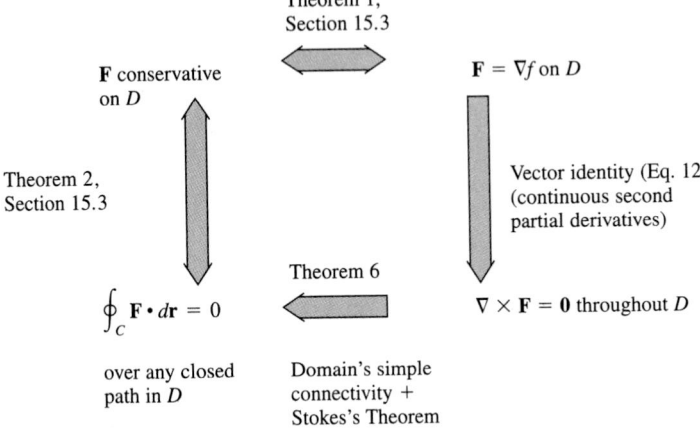

Quick Review 15.7

In Exercises 1–6, let $f(x, y, z) = e^{-xyz} \ln(x + y + z)$. Compute the indicated partial derivative.

1. f_x

2. f_z

3. f_{xy}

4. f_{yx}

5. f_{yz}

6. f_{zy}

In Exercises 7 and 8, let $\mathbf{u} = x\mathbf{i} + y\mathbf{j} - z\mathbf{k}$ and $\mathbf{v} = a\mathbf{i} + b\mathbf{j} + c\mathbf{k}$. Use determinants to compute the vector.

7. $\mathbf{u} \times \mathbf{v}$

8. $\mathbf{v} \times \mathbf{u}$

9. Find ∇f if $f(x, y, z) = x \sin (y + z)$.

10. Solve for *x*:

$$\begin{vmatrix} 1 & x & x^2 \\ 2 & 3 & -1 \\ 1 & 0 & 1 \end{vmatrix} = 0.$$

Section 15.7 Exercises

In Exercises 1–6, use the surface integral in Stokes's Theorem to calculate the circulation of the field **F** around the curve *C* in the indicated direction.

1. $\mathbf{F} = x^2\mathbf{i} + 2x\mathbf{j} + z^2\mathbf{k}$

C: The ellipse $4x^2 + y^2 = 4$ in the *xy*-plane, counterclockwise when viewed from above

2. $\mathbf{F} = 2y\mathbf{i} + 3x\mathbf{j} - z^2\mathbf{k}$

C: The circle $x^2 + y^2 = 9$ in the *xy*-plane, counterclockwise when viewed from above

3. $\mathbf{F} = y\mathbf{i} + xz\mathbf{j} + x^2\mathbf{k}$

C: The boundary of the triangle cut from the plane $x + y + z = 1$ by the first octant, counterclockwise when viewed from above

4. $\mathbf{F} = (y^2 + z^2)\mathbf{i} + (x^2 + z^2)\mathbf{j} + (x^2 + y^2)\mathbf{k}$

C: The boundary of the triangle cut from the plane $x + y + z = 1$ by the first octant, counterclockwise when viewed from above

5. $\mathbf{F} = (y^2 + z^2)\mathbf{i} + (x^2 + y^2)\mathbf{j} + (x^2 + y^2)\mathbf{k}$

   C: The square bounded by the lines $x = \pm 1$ and $y = \pm 1$ in the *xy*-plane, counterclockwise when viewed from above

6. $\mathbf{F} = x^2 y^3 \mathbf{i} + \mathbf{j} + z\mathbf{k}$

   C: The intersection of the cylinder $x^2 + y^2 = 4$ and the hemisphere $x^2 + y^2 + z^2 = 16, z \geq 0$

7. Let $\mathbf{n}$ be the outer unit normal of the elliptical shell

   $$S: 4x^2 + 9y^2 + 36z^2 = 36, \qquad z \geq 0,$$

   and let

   $$\mathbf{F} = y\mathbf{i} + x^2\mathbf{j} + (x^2 + y^4)^{3/2} \sin e^{\sqrt{xyz}} \mathbf{k}.$$

   Find the value of

   $$\iint_S \nabla \times \mathbf{F} \cdot \mathbf{n} \, d\sigma.$$

   (*Hint:* One parametrization of the ellipse at the base of the shell is $x = 3 \cos t, y = 2 \sin t, 0 \leq t \leq 2\pi$.)

8. Let $\mathbf{n}$ be the outer normal (normal away from the origin) of the parabolic shell

   $$S: 4x^2 + y + z^2 = 4, \qquad y \geq 0,$$

   and let

   $$\mathbf{F} = \left(-z + \frac{1}{2 + x}\right)\mathbf{i} + (\tan^{-1} y)\mathbf{j} + \left(x + \frac{1}{4 + z}\right)\mathbf{k}.$$

   Find the value of

   $$\iint_S \nabla \times \mathbf{F} \cdot \mathbf{n} \, d\sigma.$$

9. Let S be the cylinder $x^2 + y^2 = a^2, 0 \leq z \leq h$, together with its top, $x^2 + y^2 \leq a^2, z = h$. Let $\mathbf{F} = -y\mathbf{i} + x\mathbf{j} + x^2\mathbf{k}$. Use Stokes's Theorem to calculate the flux of $\nabla \times \mathbf{F}$ outward through S.

10. Evaluate

   $$\iint_S \nabla \times (y\mathbf{i}) \cdot \mathbf{n} \, d\sigma,$$

   where S is the hemisphere $x^2 + y^2 + z^2 = 1, z \geq 0$.

11. Show that

   $$\iint_S \nabla \times \mathbf{F} \cdot \mathbf{n} \, d\sigma$$

   has the same value for all oriented surfaces that span C and that induce the same positive direction on C.

12. Writing to Learn Let $\mathbf{F}$ be a differentiable vector field defined on a region containing a smooth closed oriented surface S and its interior. Let $\mathbf{n}$ be the unit normal vector field on S. Suppose S is the union of two surfaces S_1 and S_2 joined along a smooth simple closed curve C. Can anything be said about

   $$\iint_S \nabla \times \mathbf{F} \cdot \mathbf{n} \, d\sigma?$$

   Give reasons for your answer.

In Exercises 13–18, use the surface integral in Stokes's Theorem to calculate the flux of the curl of the field $\mathbf{F}$ across the surface S in the direction of the outward unit normal $\mathbf{n}$.

13. $\mathbf{F} = 2z\mathbf{i} + 3x\mathbf{j} + 5y\mathbf{k}$

   S: $\mathbf{r}(r, \theta) = (r \cos \theta)\mathbf{i} + (r \sin \theta)\mathbf{j} + (4 - r^2)\mathbf{k},$
   $\qquad 0 \leq r \leq 2, \qquad 0 \leq \theta \leq 2\pi$

14. $\mathbf{F} = (y - z)\mathbf{i} + (z - x)\mathbf{j} + (x + z)\mathbf{k}$

   S: $\mathbf{r}(r, \theta) = (r \cos \theta)\mathbf{i} + (r \sin \theta)\mathbf{j} + (9 - r^2)\mathbf{k},$
   $\qquad 0 \leq r \leq 3, \qquad 0 \leq \theta \leq 2\pi$

15. $\mathbf{F} = x^2 y\mathbf{i} + 2y^3 z\mathbf{j} + 3z\mathbf{k}$

   S: $\mathbf{r}(r, \theta) = (r \cos \theta)\mathbf{i} + (r \sin \theta)\mathbf{j} + r\mathbf{k}, \qquad 0 \leq r \leq 1,$
   $\qquad 0 \leq \theta \leq 2\pi$

16. $\mathbf{F} = (x - y)\mathbf{i} + (y - z)\mathbf{j} + (z - x)\mathbf{k}$

   S: $\mathbf{r}(r, \theta) = (r \cos \theta)\mathbf{i} + (r \sin \theta)\mathbf{j} + (5 - r)\mathbf{k},$
   $\qquad 0 \leq r \leq 5, \qquad 0 \leq \theta \leq 2\pi$

17. $\mathbf{F} = 3y\mathbf{i} + (5 - 2x)\mathbf{j} + (z^2 - 2)\mathbf{k}$

   S: $\mathbf{r}(r, \theta) = (\sqrt{3} \sin \phi \cos \theta)\mathbf{i} + (\sqrt{3} \sin \phi \sin \theta)\mathbf{j} +$
   $\qquad (\sqrt{3} \cos \phi)\mathbf{k}, \qquad 0 \leq \phi \leq \pi/2, \qquad 0 \leq \theta \leq 2\pi$

18. $\mathbf{F} = y^2\mathbf{i} + z^2\mathbf{j} + x\mathbf{k}$

   S: $\mathbf{r}(\phi, \theta) = (2 \sin \phi \cos \theta)\mathbf{i} + (2 \sin \phi \sin \theta)\mathbf{j} + (2 \cos \phi)\mathbf{k},$
   $\qquad 0 \leq \phi \leq \pi/2, \qquad 0 \leq \theta \leq 2\pi$

Explorations

19. Use the identity $\nabla \times \nabla f = \mathbf{0}$ (Equation (12) in the text) and Stokes's Theorem to show that the circulation of the following fields around the boundary of any smooth orientable surface in space are zero.

   (a) $\mathbf{F} = 2x\mathbf{i} + 2y\mathbf{j} + 2z\mathbf{k}$

   (b) $\mathbf{F} = \nabla(xy^2 z^3)$

   (c) $\mathbf{F} = \nabla \times (x\mathbf{i} + y\mathbf{j} + z\mathbf{k})$

   (d) $\mathbf{F} = \nabla f$

20. Let $f(x, y, z) = (x^2 + y^2 + z^2)^{-1/2}$. Show that the clockwise circulation of the field $\mathbf{F} = \nabla f$ around the circle $x^2 + y^2 = a^2$ in the *xy*-plane is zero

   (a) by taking $\mathbf{r} = (a \cos t)\mathbf{i} + (a \sin t)\mathbf{j}, 0 \leq t \leq 2\pi$, and integrating $\mathbf{F} \cdot d\mathbf{r}$ over the circle.

   (b) by applying Stokes's Theorem. $\blacksquare$

In Exercises 21–24, *work in groups of two or three* to solve the problem.

21. Let C be a simple closed smooth curve in the plane $2x + 2y + z = 2$, oriented as shown here. Show that

   $$\oint_C 2y \, dx + 3z \, dy - x \, dz$$

depends only on the area of the region enclosed by C and not on the position or shape of C.

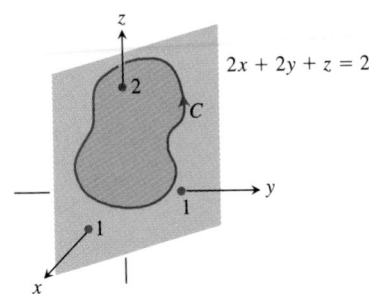

22. Show that if $\mathbf{F} = x\mathbf{i} + y\mathbf{j} + z\mathbf{k}$, then $\nabla \times \mathbf{F} = \mathbf{0}$.

23. Find a vector field with twice-differentiable components whose curl is $x\mathbf{i} + y\mathbf{j} + z\mathbf{k}$ or prove that no such field exists.

24. Writing to Learn Does Stokes's Theorem say anything about circulation in a field whose curl is zero? Give reasons for your answer.

Extending the Ideas

25. Let R be a region in the xy-plane that is bounded by a piecewise smooth simple closed curve C, and suppose the moments of inertia of R about the x- and y-axes are known to be I_x and I_y. Evaluate the integral

$$\oint_C \nabla (r^4) \cdot \mathbf{n}\, ds,$$

where $r = \sqrt{x^2 + y^2}$, in terms of I_x and I_y.

26. *Zero Curl, Yet Not Conservative* Show that the curl of

$$\mathbf{F} = \frac{-y}{x^2 + y^2}\mathbf{i} + \frac{x}{x^2 + y^2}\mathbf{j} + z\mathbf{k}$$

is zero but that

$$\oint_C \mathbf{F} \cdot d\mathbf{r}$$

is not zero if C is the circle $x^2 + y^2 = 1$ in the xy-plane. (Theorem 6 does not apply here because the domain of $\mathbf{F}$ is not simply connected. The field $\mathbf{F}$ is not defined along the z-axis so there is no way to contract C to a point without leaving the domain of $\mathbf{F}$.)

15.8 The Divergence Theorem and a Unified Theory

Divergence in Three Dimensions • The Divergence Theorem • Proof of the Divergence Theorem for Special Regions • The Divergence Theorem for Other Regions • Gauss's Law: One of the Four Great Laws of Electromagnetic Theory • Continuity Equation of Hydrodynamics • Unifying the Integral Theorems

Divergence in Three Dimensions

The **divergence** of a vector field

$$\mathbf{F} = M(x, y, z)\mathbf{i} + N(x, y, z)\mathbf{j} + P(x, y, z)\mathbf{k}$$

is the scalar function

$$\text{div } \mathbf{F} = \nabla \cdot \mathbf{F} = \frac{\partial M}{\partial x} + \frac{\partial N}{\partial y} + \frac{\partial P}{\partial z}. \tag{1}$$

The symbol "div $\mathbf{F}$" is read as "divergence of $\mathbf{F}$" or "div $\mathbf{F}$." The notation $\nabla \cdot \mathbf{F}$ is read "del dot $\mathbf{F}$."

Div $\mathbf{F}$ has the same physical interpretation in three dimensions that it does in two. If $\mathbf{F}$ is the velocity field of a fluid flow, the value of div $\mathbf{F}$ at a point (x, y, z) is the rate at which fluid is being piped in or drained away at (x, y, z). The divergence is the flux per unit volume or flux density at the point.

Divergence Theorem

Mikhail Vassilievich Ostrogradsky (1801–1862) was the first mathematician to publish a proof of the Divergence Theorem. Upon being denied his degree at Kharkhov University by the minister for religious affairs and national education (for atheism), Ostrogradsky left Russia for Paris in 1822, attracted by the presence of Laplace, Legendre, Fourier, Poisson, and Cauchy. While working on the theory of heat in the mid-1820s, he formulated the Divergence Theorem as a tool for converting volume integrals to surface integrals.

Carl Friedrich Gauss (1777–1855) had already proved the theorem while working on the theory of gravitation, but his notebooks were not to be published until many years later. (The theorem is sometimes called Gauss's Theorem.) The list of Gauss's accomplishments in science and mathematics is truly astonishing, ranging from the co-invention of the electric telegraph with Wilhelm Weber in 1833 to the development of a wonderfully accurate theory of planetary orbits and to work in non-Euclidean geometry that later became fundamental to Einstein's general theory of relativity.

Example 1 FINDING DIVERGENCE

Find the divergence of $\mathbf{F} = 2xz\mathbf{i} - xy\mathbf{j} - z\mathbf{k}$.

Solution   The divergence of $\mathbf{F}$ is

$$\nabla \cdot \mathbf{F} = \frac{\partial}{\partial x}(2xz) + \frac{\partial}{\partial y}(-xy) + \frac{\partial}{\partial z}(-z) = 2z - x - 1.$$

The Divergence Theorem

The Divergence Theorem says that under suitable conditions the outward flux of a vector field across a closed surface (oriented outward) equals the triple integral of the divergence of the field over the region enclosed by the surface.

Theorem 7 Divergence Theorem

The flux of a vector field $\mathbf{F} = M\mathbf{i} + N\mathbf{j} + P\mathbf{k}$ across a closed oriented surface S in the direction of the surface's outward unit normal field $\mathbf{n}$ equals the integral of $\nabla \cdot \mathbf{F}$ over the region D enclosed by the surface:

$$\underset{\substack{S \\ \text{outward} \\ \text{flux}}}{\iint} \mathbf{F} \cdot \mathbf{n}\, d\sigma = \underset{\substack{D \\ \text{divergence} \\ \text{integral}}}{\iiint} \nabla \cdot \mathbf{F}\, dV. \qquad (2)$$

Example 2 VERIFYING THE DIVERGENCE THEOREM

Evaluate both sides of Equation (2) for the field $\mathbf{F} = x\mathbf{i} + y\mathbf{j} + z\mathbf{k}$ over the sphere $x^2 + y^2 + z^2 = a^2$.

Solution   The outer normal to S, calculated from the gradient of $f(x, y, z) = x^2 + y^2 + z^2 - a^2$, is

$$\mathbf{n} = \frac{2(x\mathbf{i} + y\mathbf{j} + z\mathbf{k})}{\sqrt{4(x^2 + y^2 + z^2)}} = \frac{x\mathbf{i} + y\mathbf{j} + z\mathbf{k}}{a}$$

Hence,

$$\mathbf{F} \cdot \mathbf{n}\, d\sigma = \frac{x^2 + y^2 + z^2}{a}\, d\sigma = \frac{a^2}{a}\, d\sigma = a\, d\sigma$$

because $x^2 + y^2 + z^2 = a^2$ on the surface. Therefore,

$$\iint_S \mathbf{F} \cdot \mathbf{n}\, d\sigma = \iint_S a\, d\sigma = a(4\pi a^2) 4\pi a^3.$$

The divergence of $\mathbf{F}$ is

$$\nabla \cdot \mathbf{F} = \frac{\partial}{\partial x}(x) + \frac{\partial}{\partial y}(y) + \frac{\partial}{\partial z}(z) = 3,$$

so

$$\iiint_D \nabla \cdot \mathbf{F}\, dV = \iiint_D 3\, dV = 3\left(\frac{4}{3}\pi a^3\right) = 4\pi a^3.$$

Example 3 FINDING FLUX

Find the flux of $\mathbf{F} = xy\mathbf{i} + yz\mathbf{j} + xz\mathbf{k}$ outward through the surface of the cube cut from the first octant by the planes $x = 1$, $y = 1$, and $z = 1$.

Solution Instead of calculating the flux as a sum of six separate integrals, one for each face of the cube, we can calculate the flux by integrating the divergence

$$\nabla \cdot \mathbf{F} = \frac{\partial}{\partial x}(xy) + \frac{\partial}{\partial y}(yz) + \frac{\partial}{\partial z}(xz) = y + z + x$$

over the cube's interior:

$$\text{Flux} = \iint_{\substack{S \\ \text{cube} \\ \text{surface}}} \mathbf{F} \cdot \mathbf{n} \, d\sigma = \iiint_{\substack{D \\ \text{cube} \\ \text{interior}}} \nabla \cdot \mathbf{F} \, dV \qquad \text{Divergence Theorem}$$

$$= \int_0^1 \int_0^1 \int_0^1 (x + y + z) \, dx \, dy \, dz = \frac{3}{2} \qquad \text{Routine integration}$$

Proof of the Divergence Theorem for Special Regions

To prove the Divergence Theorem, we assume that the components of $\mathbf{F}$ have continuous first partial derivatives. We also assume that D is a convex region with no holes or bubbles, such as a solid sphere, cube, or ellipsoid, and that S is a piecewise smooth surface. In addition, we assume that any line perpendicular to the xy-plane at an interior point of the region R_{xy} that is the projection of D on the xy-plane intersects the surface S in exactly two points, producing surfaces

$$S_1: \quad z = f_1(x, y), \qquad (x, y) \text{ in } R_{xy}$$
$$S_2: \quad z = f_2(x, y), \qquad (x, y) \text{ in } R_{xy}$$

with $f_1 \leq f_2$. We make similar assumptions about the projection of D onto the other two coordinate planes. See Figure 15.74.

The components of the unit normal vector $\mathbf{n} = n_1\mathbf{i} + n_2\mathbf{j} + n_3\mathbf{k}$ are the cosines of the angles α, β, and γ that $\mathbf{n}$ makes with $\mathbf{i}$, $\mathbf{j}$, and $\mathbf{k}$ (Figure 15.75). This is true because all the vectors involved are unit vectors. We have

$$n_1 = \mathbf{n} \cdot \mathbf{i} = |\mathbf{n}||\mathbf{i}| \cos \alpha = \cos \alpha,$$
$$n_2 = \mathbf{n} \cdot \mathbf{j} = |\mathbf{n}||\mathbf{j}| \cos \beta = \cos \beta,$$
$$n_3 = \mathbf{n} \cdot \mathbf{k} = |\mathbf{n}||\mathbf{k}| \cos \gamma = \cos \gamma.$$

Thus,

$$\mathbf{n} = (\cos \alpha)\mathbf{i} + (\cos \beta)\mathbf{j} + (\cos \gamma)\mathbf{k}$$

and

$$\mathbf{F} \cdot \mathbf{n} = M \cos \alpha + N \cos \beta + P \cos \gamma.$$

In component form, the Divergence Theorem states that

$$\iint_S (M \cos \alpha + N \cos \beta + P \cos \gamma) \, d\sigma = \iiint_D \left(\frac{\partial M}{\partial x} + \frac{\partial N}{\partial y} + \frac{\partial P}{\partial z} \right) dx \, dy \, dz.$$

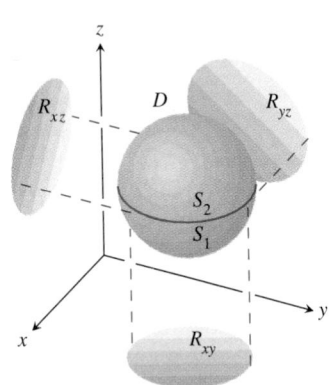

Figure 15.74 We first prove the Divergence Theorem for the kind of three-dimensional region shown here. We then extend the theorem to other regions.

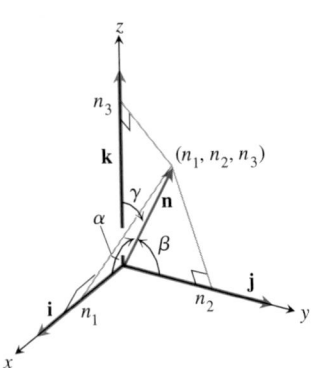

Figure 15.75 The scalar components of a unit normal vector $\mathbf{n}$ are the cosines of the angles α, β, and γ that it makes with $\mathbf{i}$, $\mathbf{j}$, and $\mathbf{k}$.

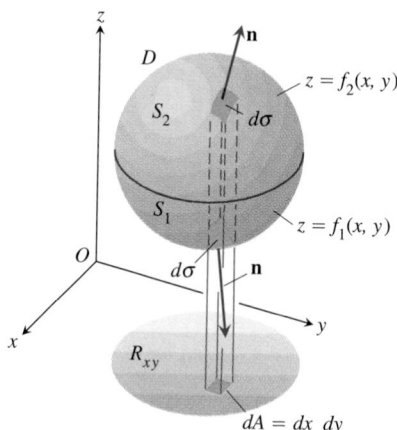

Figure 15.76 The three-dimensional region D enclosed by the surfaces S_1 and S_2 shown here projects vertically onto a two-dimensional region R_{xy} in the xy-plane.

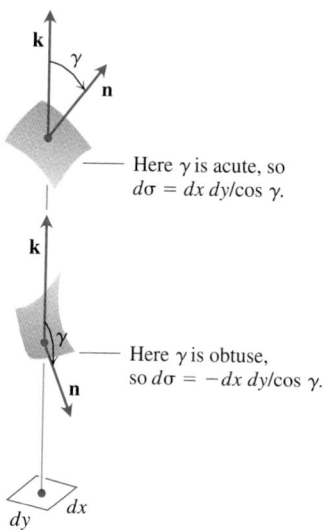

Figure 15.77 An enlarged view of the area patches in Figure 15.75. The relations $d\sigma = \pm dx\, dy/\cos \gamma$ are derived in Section 15.5.

We prove the theorem by proving the three equalities:

$$\iint\limits_{S} M \cos \alpha \, d\sigma = \iiint\limits_{D} \frac{\partial M}{\partial x} dx\, dy\, dz, \tag{3}$$

$$\iint\limits_{S} N \cos \beta \, d\sigma = \iiint\limits_{D} \frac{\partial N}{\partial y} dx\, dy\, dz, \tag{4}$$

$$\iint\limits_{S} P \cos \gamma \, d\sigma = \iiint\limits_{D} \frac{\partial P}{\partial z} dx\, dy\, dz. \tag{5}$$

We prove Equation (5) by converting the surface integral on the left to a double integral over the projection R_{xy} of D on the xy-plane (Figure 15.76). The surface S consists of an upper part S_2 whose equation is $z = f_2(x, y)$ and a lower part S_1 whose equation is $z = f_1(x, y)$. On S_2, the outer normal $\mathbf{n}$ has a positive $\mathbf{k}$-component and

$$\cos \gamma \, d\sigma = dx\, dy \qquad \text{because} \qquad d\sigma = \frac{dA}{|\cos \gamma|} = \frac{dx\, dy}{\cos \gamma}.$$

See Figure 15.77. On S_1, the outer normal $\mathbf{n}$ has a negative $\mathbf{k}$-component and

$$\cos \gamma \, d\sigma = -dx\, dy.$$

Therefore,

$$\iint\limits_{S} P \cos \gamma \, d\sigma = \iint\limits_{S_2} P \cos \gamma \, d\sigma + \iint\limits_{S_1} P \cos \gamma \, d\sigma$$

$$= \iint\limits_{R_{xy}} P(x, y, f_2(x, y)) \, dx\, dy - \iint\limits_{R_{xy}} P(x, y, f_1(x, y)) \, dx\, dy$$

$$= \iint\limits_{R_{xy}} (P(x, y, f_2(x, y)) - P(x, y, f_1(x, y))) \, dx\, dy$$

$$= \iint\limits_{R_{xy}} \left[\int_{f_1(x,y)}^{f_2(x,y)} \frac{\partial P}{\partial z} dz \right] dx\, dy = \iiint\limits_{D} \frac{\partial P}{\partial z} dz\, dx\, dy.$$

This proves Equation (5).

The proofs for Equations (3) and (4) follow the same pattern, or just permute x, y, z; M, N, P; α, β, γ, in order, and get those results from Equation (5).

The Divergence Theorem for Other Regions

The Divergence Theorem can be extended to regions that can be partitioned into a finite number of simple regions of the type just discussed and to regions that can be defined as limits of simpler regions in certain ways. For example, suppose D is the region between two concentric spheres and $\mathbf{F}$ has continuously differentiable components throughout D and on the bounding surfaces. Split D by an equatorial plane and apply the Divergence Theorem to each half

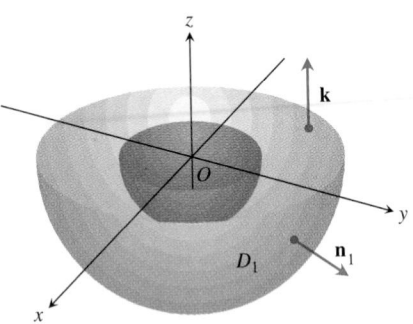

Figure 15.78 The lower half of the solid region between two concentric spheres.

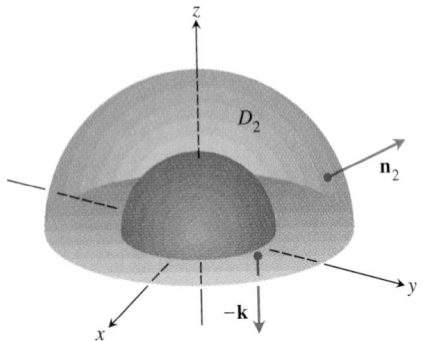

Figure 15.79 The upper half of the solid region between two concentric spheres.

separately. The bottom half, D_1, is shown in Figure 15.78. The surface S_1 that bounds D_1 consists of an outer hemisphere, a plane washer-shaped base, and an inner hemisphere. The Divergence Theorem says that

$$\iint_{S_1} \mathbf{F} \cdot \mathbf{n}_1 \, d\sigma_1 = \iiint_{D_1} \nabla \cdot \mathbf{F} \, dV_1. \tag{6}$$

The unit normal $\mathbf{n}_1$ that points outward from D_1 points away from the origin along the outer surface, equals $\mathbf{k}$ along the flat base, and points toward the origin along the inner surface. Next apply the Divergence Theorem to D_2, as shown in Figure 15.79:

$$\iint_{S_2} \mathbf{F} \cdot \mathbf{n}_2 \, d\sigma_2 = \iiint_{D_2} \nabla \cdot \mathbf{F} \, dV_2. \tag{7}$$

As we follow $\mathbf{n}_2$ over S_2, pointing outward from D_2, we see that $\mathbf{n}_2$ equals $-\mathbf{k}$ along the washer-shaped base in the xy-plane, points away from the origin on the outer sphere, and points toward the origin on the inner sphere. When we add Equations (6) and (7), the integrals over the flat base cancel because of the opposite signs of $\mathbf{n}_1$ and $\mathbf{n}_2$. We thus arrive at the result

$$\iint_{S} \mathbf{F} \cdot \mathbf{n} \, d\sigma_2 = \iiint_{D} \nabla \cdot \mathbf{F} \, dV,$$

with D the region between the spheres, S the boundary of D consisting of two spheres, and $\mathbf{n}$ the unit normal to S directed outward from D.

Example 4   FINDING OUTWARD FLUX

Find the net outward flux of the field

$$\mathbf{F} = \frac{x\mathbf{i} + y\mathbf{j} + z\mathbf{k}}{\rho^3}, \qquad \rho = \sqrt{x^2 + y^2 + z^2}$$

across the boundary of the region D: $0 < a^2 \le x^2 + y^2 + z^2 \le b^2$.

Solution   The flux can be calculated by integrating $\nabla \cdot \mathbf{F}$ over D. We have

$$\frac{\partial \rho}{\partial x} = \frac{1}{2}(x^2 + y^2 + z^2)^{-1/2}(2x) = \frac{x}{\rho}$$

and

$$\frac{\partial M}{\partial x} = \frac{\partial}{\partial x}(x\rho^{-3}) = \rho^{-3} - 3x\rho^{-4}\frac{\partial \rho}{\partial x} = \frac{1}{\rho^3} - \frac{3x^2}{\rho^5}.$$

Similarly,

$$\frac{\partial N}{\partial y} = \frac{1}{\rho^3} - \frac{3y^2}{\rho^5} \qquad \text{and} \qquad \frac{\partial P}{\partial z} = \frac{1}{\rho^3} - \frac{3z^2}{\rho^5}.$$

Hence,

$$\text{div } \mathbf{F} = \frac{3}{\rho^3} - \frac{3}{\rho^5}(x^2 + y^2 + z^2) = \frac{3}{\rho^3} - \frac{3\rho^2}{\rho^5} = 0$$

and

$$\iiint_D \nabla \cdot \mathbf{F}\, dV = 0.$$

So the integral of $\nabla \cdot \mathbf{F}$ over D is zero and the net outward flux across the boundary of D is zero, but there is more to learn from this example. The flux leaving D across the inner sphere S_a is the negative of the flux leaving D across the outer sphere S_b (because the sum of these fluxes is zero). Hence, the flux of $\mathbf{F}$ across S_a in the direction away from the origin equals the flux of $\mathbf{F}$ across S_b in the direction away from the origin. Thus, the flux of $\mathbf{F}$ across a sphere centered at the origin is independent of the radius of the sphere. What is this flux?

To find it, we evaluate the flux integral directly. The outward unit normal on the sphere of radius a is

$$\mathbf{n} = \frac{x\mathbf{i} + y\mathbf{j} + z\mathbf{k}}{\sqrt{x^2 + y^2 + z^2}} = \frac{x\mathbf{i} + y\mathbf{j} + z\mathbf{k}}{a}.$$

Hence, on the sphere,

$$\mathbf{F} \cdot \mathbf{n} = \frac{x\mathbf{i} + y\mathbf{j} + z\mathbf{k}}{a^3} \cdot \frac{x\mathbf{i} + y\mathbf{j} + z\mathbf{k}}{a} = \frac{x^2 + y^2 + z^2}{a^4} = \frac{a^2}{a^4} = \frac{1}{a^2}$$

and

$$\iint_{S_a} \mathbf{F} \cdot \mathbf{n}\, d\sigma = \frac{1}{a^2} \iint_{S_a} d\sigma = \frac{1}{a^2}(4\pi a^2) = 4\pi.$$

The outward flux of $\mathbf{F}$ across any sphere centered at the origin is 4π.

Gauss's Law: One of the Four Great Laws of Electromagnetic Theory

There is more to be learned from Example 4. In electromagnetic theory, the electric field created by a point charge q located at the origin is the inverse square field

$$\mathbf{E}(x, y, z) = \frac{1}{4\pi\varepsilon_0} \frac{q}{|\mathbf{r}|^2}\left(\frac{\mathbf{r}}{|\mathbf{r}|}\right) = \frac{q}{4\pi\varepsilon_0} \frac{\mathbf{r}}{|\mathbf{r}|^3} = \frac{q}{4\pi\varepsilon_0} \frac{x\mathbf{i} + y\mathbf{j} + z\mathbf{k}}{\rho^3},$$

where ε_0 is a physical constant, $\mathbf{r}$ is the position vector of the point (x, y, z), and $\rho = |\mathbf{r}| = \sqrt{x^2 + y^2 + z^2}$. In the notation of Example 4,

$$\mathbf{E} = \frac{q}{4\pi\varepsilon_0} \mathbf{F}.$$

The calculations in Example 4 show that the outward flux of $\mathbf{E}$ across any sphere centered at the origin is q/ε_0, but this result is not confined to spheres. The outward flux of $\mathbf{E}$ across any closed surface S that encloses the origin (and to which the Divergence Theorem applies) is also q/ε_0. To see why, we have only to imagine a large sphere S_a centered at the origin and enclosing the surface S. Since

$$\nabla \cdot \mathbf{E} = \nabla \cdot \frac{q}{4\pi\varepsilon_0} \mathbf{F} = \frac{q}{4\pi\varepsilon_0} \nabla \cdot \mathbf{F} = 0$$

when $\rho > 0$, the integral of $\nabla \cdot \mathbf{E}$ over the region D between S and S_a is zero. Hence, by the Divergence Theorem,

$$\iint_{\substack{\text{boundary} \\ \text{of } D}} \mathbf{E} \cdot \mathbf{n} \, d\sigma = 0,$$

and the flux of $\mathbf{E}$ across S in the direction away from the origin must be the same as the flux of $\mathbf{E}$ across S_a in the direction away from the origin, which is $4\pi q$. This statement, called *Gauss's law*, also applies to charge distributions that are more general than the one assumed here, as you will see in nearly any physics text.

$$\text{Guass's Law: } \iint_S \mathbf{E} \cdot \mathbf{n} \, d\sigma = \frac{q}{\varepsilon_0}$$

Continuity Equation of Hydrodynamics

Let D be a region in space bounded by a closed oriented surface S. If $\mathbf{v}(x, y, z)$ is the velocity field of a fluid flowing smoothly through D, $\delta = \delta(t, x, y, z)$ is the fluid's density at (x, y, z) at time t, and $\mathbf{F} = \delta \mathbf{v}$, then the **continuity equation** of hydrodynamics states that

$$\nabla \cdot \mathbf{F} + \frac{\partial \delta}{\partial t} = 0.$$

If the functions involved here have continuous first partial derivatives, the equation evolves naturally from the Divergence Theorem, as we will now see. First, the integral

$$\iint_S \mathbf{F} \cdot \mathbf{n} \, d\sigma$$

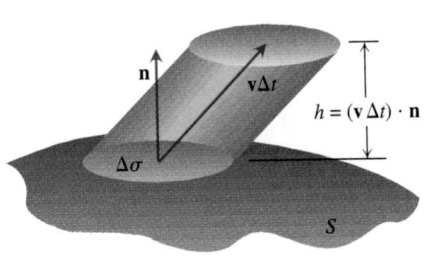

Figure 15.80 The fluid that flows upward through the patch $\Delta\sigma$ in a short time Δt fills a "cylinder" whose volume is approximately base $\times$ height $\mathbf{v} \cdot \mathbf{n} \, \Delta\sigma \, \Delta t$.

is the rate at which mass leaves D across S (leaves because $\mathbf{n}$ is the outer normal). To see why, consider a patch of area $\Delta\sigma$ on the surface (Figure 15.80). In a short time interval Δt, the volume ΔV of fluid that flows across the patch is approximately equal to the volume of a cylinder with base area $\Delta\sigma$ and height $(\mathbf{v} \, \Delta t) \cdot \mathbf{n}$, where $\mathbf{v}$ is a velocity vector rooted at a point of the patch:

$$\Delta V \approx \mathbf{v} \cdot \mathbf{n} \, \Delta\sigma \, \Delta t.$$

The mass of this volume of fluid is about

$$\Delta m \approx \delta \, \mathbf{v} \cdot \mathbf{n} \, \Delta\sigma \, \Delta t,$$

so the rate at which mass is flowing out of D across the patch is about

$$\frac{\Delta m}{\Delta t} \approx \delta \, \mathbf{v} \cdot \mathbf{n} \, \Delta\sigma.$$

This leads to the approximation

$$\sum \frac{\Delta m}{\Delta t} \approx \sum \delta \, \mathbf{v} \cdot \mathbf{n} \, \Delta\sigma$$

as an estimate of the average rate at which mass flows across S. Finally, letting $\Delta\sigma \to 0$ and $\Delta t \to 0$ gives the instantaneous rate at which mass leaves D across S as

$$\frac{dm}{dt} = \iint_S \delta\, \mathbf{v} \cdot \mathbf{n}\, d\sigma,$$

which for our particular flow is

$$\frac{dm}{dt} = \iint_S \mathbf{F} \cdot \mathbf{n}\, d\sigma.$$

Now let B be a solid sphere centered at a point Q in the flow. The average value of $\nabla \cdot \mathbf{F}$ over B is

$$\frac{1}{\text{volume of } B} \iiint_B \nabla \cdot \mathbf{F}\, dV.$$

It is a consequence of the continuity of the divergence that $\nabla \cdot \mathbf{F}$ actually takes on this value at some point P in B. Thus,

$$(\nabla \cdot \mathbf{F})_P = \frac{1}{\text{volume of } B} \iiint_B \nabla \cdot \mathbf{F}\, dV = \frac{\displaystyle\iint_S \mathbf{F} \cdot \mathbf{n}\, d\sigma}{\text{volume of } B}$$

$$= \frac{\text{rate at which mass leaves } B \text{ across its surfaces}}{\text{volume of } B}. \qquad (8)$$

The fraction on the right describes the decrease in mass per unit volume.

Now let the radius of B approach zero while the center Q stays fixed. The left-hand side of Equation (8) converges to $(\nabla \cdot \mathbf{F})_Q$, the right side to $(-\partial\delta/\partial t)_Q$. The equality of these two limits is the continuity equation

$$\nabla \cdot \mathbf{F} = -\frac{\partial\delta}{\partial t}.$$

The continuity equation "explains" $\nabla \cdot \mathbf{F}$: The divergence of $\mathbf{F}$ at a point is the rate at which the density of the fluid is decreasing there.

The Divergence Theorem

$$\iint_S \mathbf{F} \cdot \mathbf{n}\, d\sigma = \iiint_D \nabla \cdot \mathbf{F}\, dV$$

now says that the net decrease in density of the fluid in region D is accounted for by the mass transported across the surface S. In this context, the theorem becomes a statement about conservation of mass.

Unifying the Integral Theorems

If we think of a two-dimensional field $\mathbf{F} = M(x, y)\mathbf{i} + N(x, y)\mathbf{j}$ as a three-dimensional field whose $\mathbf{k}$-component is zero, then $\nabla \cdot \mathbf{F} = (\partial M/\partial x) + (\partial N/\partial y)$ and the normal form of Green's Theorem can be written as

$$\oint_C \mathbf{F} \cdot \mathbf{n}\, ds = \iint_R \left(\frac{\partial M}{\partial x} + \frac{\partial N}{\partial y}\right) dx\, dy = \iint_R \nabla \cdot \mathbf{F}\, dA.$$

Similarly, $\nabla \times \mathbf{F} \cdot \mathbf{k} = (\partial N/\partial x) - (\partial M/\partial y)$, so the tangential form of Green's Theorem can be written as

$$\oint_C \mathbf{F} \cdot d\mathbf{r} = \iint_R \left(\frac{\partial N}{\partial x} - \frac{\partial M}{\partial y} \right) dx\, dy = \iint_R \nabla \times \mathbf{F} \cdot \mathbf{k}\, dA.$$

With the equations of Green's Theorem now in del notation, we can see their relationship to the equations in Stokes's Theorem and the Divergence Theorem.

Green's Theorem and its Generalization to Three Dimensions

Normal form of Green's Theorem:
$$\oint_C \mathbf{F} \cdot \mathbf{n}\, ds = \iint_R \nabla \cdot \mathbf{F}\, dA$$

Divergence Theorem:
$$\iint_S \mathbf{F} \cdot \mathbf{n}\, d\sigma = \iiint_D \nabla \cdot \mathbf{F}\, dV$$

Tangential form of Green's Theorem:
$$\oint_C \mathbf{F} \cdot d\mathbf{r} = \iint_R \nabla \cdot \mathbf{F} \cdot \mathbf{k}\, dA$$

Stokes's Theorem:
$$\oint_C \mathbf{F} \cdot d\mathbf{r} = \iint_S \nabla \cdot \mathbf{F} \cdot \mathbf{n}\, d\sigma$$

Stokes's Theorem generalizes the tangential (curl) form of Green's Theorem from a flat surface in the plane to a surface in three-dimensional space. In each theorem, the integral of the normal component of curl **F** over the interior of the surface equals the circulation of **F** around the boundary.

Likewise, the Divergence Theorem generalizes the normal (flux) form of Green's Theorem from a two-dimensional region in the plane to a three-dimensional region in space. In each case, the integral of $\nabla \cdot \mathbf{F}$ over the interior of the region equals the total flux of the field across the boundary.

There is still more to be learned here. All these results can be viewed as forms of a *single fundamental theorem*. Think back to the Fundamental Theorem of Calculus in Section 5.4. It says that if $f(x)$ is differentiable on $[a, b]$, then

$$\int_a^b \frac{df}{dx}\, dx = f(b) - f(a).$$

Figure 15.81 The outward unit normals at the boundary of $[a, b]$ in one dimensional space.

If we let $\mathbf{F} = f(x)\mathbf{i}$ throughout $[a, b]$, then $(df/dx) = \nabla \cdot \mathbf{F}$. If we define the unit vector **n** normal to the boundary of $[a, b]$ to be **i** at b and $-\mathbf{i}$ at a (Figure 15.81), then

$$f(b) - f(a) = f(b)\mathbf{i} \cdot (\mathbf{i}) + f(a)\mathbf{i} \cdot (-\mathbf{i})$$

$$= \mathbf{F}(b) \cdot \mathbf{n} + \mathbf{F}(a) \cdot \mathbf{n}$$

$$= \text{total outward flux of } \mathbf{F} \text{ across the boundary of } [a, b].$$

The Fundamental Theorem now says that

$$\int_{[a,b]} \nabla \cdot \mathbf{F} \, dx = \text{total outward flux of } \mathbf{F} \text{ across the boundary.}$$

The Fundamental Theorem of Calculus, the flux form of Green's Theorem, and the Divergence Theorem all say that the integral of the differential operator $\nabla \cdot$ operating on a field $\mathbf{F}$ over a region equals the sum of the normal field components over the boundary of the region.

Stokes's Theorem and the circulation form of Green's Theorem say that, when things are properly oriented, the integral of the normal component of the curl operating on a field over a region equals the sum of the tangential field components on the boundary of the region.

The beauty of these interpretations is the observance of a marvelous underlying principle, which we might state as follows.

> The integral of a differential operator acting on a field over a region equals the sum of the field components appropriate to the operator over the boundary of the region.

Quick Review 15.8

In Exercises 1–4, evaluate the integral.

1. $\int_0^2 \int_0^{2-x} \int_0^{2-x-y} dz \, dy \, dx$

2. $\int_0^2 \int_0^{2-x} \int_0^{2-x-y} x \, dz \, dy \, dx$

3. $\int_0^2 \int_0^{2-x} \int_0^{2-x-y} y \, dz \, dy \, dx$

4. $\int_0^2 \int_0^{2-x} \int_0^{2-x-y} z \, dz \, dy \, dx$

In Exercises 5–8, rewrite the integral $\int_0^1 \int_0^1 \int_0^{y^2} dz \, dy \, dx$ as an iterated integral in the given order.

5. $dy \, dz \, dx$ **6.** $dy \, dx \, dz$

7. $dx \, dy \, dz$ **8.** $dx \, dz \, dy$

In Exercises 9 and 10, find the angle that the vector $\mathbf{i} + \mathbf{j} + \mathbf{k}$ makes with the indicated axis.

9. x-axis   **10.** y-axis

Section 15.8 Exercises

In Exercises 1–4, find the divergence of the field.

1. The spin field in Figure 15.15

2. The radial field in Figure 15.14

3. The gravitational field in Figure 15.13

4. The velocity field in Figure 15.10

In Exercises 5–16, use the Divergence Theorem to find the outward flux of $\mathbf{F}$ across the boundary of the region D.

5. $\mathbf{F} = (y - x)\mathbf{i} + (z - y)\mathbf{j} + (y - x)\mathbf{k}$

    D: The cube bounded by the planes $x = \pm 1$, $y = \pm 1$, and $z = \pm 1$

6. $\mathbf{F} = x^2\mathbf{i} + y^2\mathbf{j} + z^2\mathbf{k}$

    (a) D: The cube cut from the first octant by the planes $x = 1$, $y = 1$, and $z = 1$

    (b) D: The cube bounded by the planes $x = \pm 1$, $y = \pm 1$, and $z = \pm 1$

    (c) D: The region cut from the solid cylinder $x^2 + y^2 \leq 4$ by the planes $z = 0$ and $z = 1$

7. $\mathbf{F} = y\mathbf{i} + xy\mathbf{j} - z\mathbf{k}$

    D: The region inside the solid cylinder $x^2 + y^2 \leq 4$ between the planes $z = 0$ and the paraboloid $z = x^2 + y^2$

8. $\mathbf{F} = x^2\mathbf{i} + xz\mathbf{j} + 3z\mathbf{k}$

   D: the solid sphere $x^2 + y^2 + z^2 \leq 4$

9. $\mathbf{F} = x^2\mathbf{i} - 2xy\mathbf{j} + 3xz\mathbf{k}$

   D: The region cut from the first octant by the sphere $x^2 + y^2 + z^2 = 4$

10. $\mathbf{F} = (6x^2 + 2xy)\mathbf{i} + (2y + x^2z)\mathbf{j} + 4x^2y^3\mathbf{k}$

    D: The region cut from the first octant by the cylinder $x^2 + y^2 = 4$ and the plane $z = 3$

11. $\mathbf{F} = 2xz\mathbf{i} - xy\mathbf{j} - z^2\mathbf{k}$

    D: The wedge cut from the first octant by the plane $y + z = 4$ and the elliptical cylinder $4x^2 + y^2 = 16$

12. $\mathbf{F} = x^3\mathbf{i} + y^3\mathbf{j} + z^3\mathbf{k}$

    D: The solid sphere $x^2 + y^2 + z^2 \leq a^2$

13. $\mathbf{F} = \sqrt{x^2 + y^2 + z^2}\,(x\mathbf{i} + y\mathbf{j} + z\mathbf{k})$

    D: The region $1 \leq x^2 + y^2 + z^2 \leq 2$

14. $\mathbf{F} = (x\mathbf{i} + y\mathbf{j} + z\mathbf{k})/\sqrt{x^2 + y^2 + z^2}$

    D: The region $1 \leq x^2 + y^2 + z^2 \leq 4$

15. $\mathbf{F} = (5x^3 + 12xy^2)\mathbf{i} + (y^3 + e^y \sin z)\mathbf{j} + (5z^3 + e^y \cos z)\mathbf{k}$

    D: The solid region between the spheres $x^2 + y^2 + z^2 = 1$ and $x^2 + y^2 + z^2 = 2$

16. $\mathbf{F} = \ln(x^2 + y^2)\mathbf{i} - \left(\dfrac{2z}{x} \tan^{-1}\dfrac{y}{x}\right)\mathbf{j} + z\sqrt{x^2 + y^2}\,\mathbf{k}$

    D: The thick-walled cylinder $1 \leq x^2 + y^2 \leq 2,\ -1 \leq z \leq 2$

Explorations

17. *div (curl G) = 0*

    (a) Show that if the necessary partial derivatives of the components of the field $\mathbf{G} = M\mathbf{i} + N\mathbf{j} + P\mathbf{k}$ are continuous, then $\nabla \cdot \nabla \times \mathbf{G} = 0$.

    (b) Writing to Learn What, if anything, can you conclude about the flux of the field $\nabla \times \mathbf{G}$ across a closed surface? Give reasons for your answer.

18. *Identities* Let $\mathbf{F}_1$ and $\mathbf{F}_2$ be differentiable vector fields, and let a and b be arbitrary real constants. Verify the following identities.

    (a) $\nabla \cdot (a\mathbf{F}_1 + b\mathbf{F}_2) = a\nabla \cdot \mathbf{F}_1 + b\nabla \cdot \mathbf{F}_2$

    (b) $\nabla \times (a\mathbf{F}_1 + b\mathbf{F}_2) = a\nabla \times \mathbf{F}_1 + b\nabla \times \mathbf{F}_2$

    (c) $\nabla \cdot (\mathbf{F}_1 \times \mathbf{F}_2) = \mathbf{F}_2 \cdot \nabla \times \mathbf{F}_1 - \mathbf{F}_1 \cdot \nabla \times \mathbf{F}_2$

19. *Identities* Let $\mathbf{F}$ be a differentiable vector field and let $g(x, y, z)$ be a differentiable scalar function. Verify the following identities.

    (a) $\nabla \cdot (g\mathbf{F}) = g\nabla \cdot \mathbf{F} + \nabla g \cdot \mathbf{F}$

(b) $\nabla \times (g\mathbf{F}) = g\nabla \times \mathbf{F} + \nabla g \times \mathbf{F}$

20. *Identities* If $\mathbf{F} = M\mathbf{i} + N\mathbf{j} + P\mathbf{k}$ is a differentiable vector field, we define the notation $\mathbf{F} \cdot \nabla$ to mean

$$M\frac{\partial}{\partial x} + N\frac{\partial}{\partial y} + P\frac{\partial}{\partial z}.$$

For differentiable vector fields $\mathbf{F}_1$ and $\mathbf{F}_2$ verify the following identities.

   (a) $\nabla \times (\mathbf{F}_1 \times \mathbf{F}_2) = (\mathbf{F}_2 \cdot \nabla)\mathbf{F}_1 - (\mathbf{F}_1 \cdot \nabla)\mathbf{F}_2 + (\nabla \cdot \mathbf{F}_2)\mathbf{F}_1 - (\nabla \cdot \mathbf{F}_1)\mathbf{F}_2$

   (b) $\nabla(\mathbf{F}_1 \cdot \mathbf{F}_2) = (\mathbf{F}_1 \cdot \nabla)\mathbf{F}_2 + (\mathbf{F}_2 \cdot \nabla)\mathbf{F}_1 + \mathbf{F}_1 \times (\nabla \times \mathbf{F}_2) + \mathbf{F}_2 \times (\nabla \times \mathbf{F}_1)$ ∎

In Exercises 21–24, *work in groups of two or three* to solve the problem.

21. **Writing to Learn** Let $\mathbf{F}$ be a field whose components have continuous first partial derivatives throughout a portion of space containing a region D bounded by a smooth closed surface S. If $|\mathbf{F}| \leq 1$, can any bound be placed on the size of

$$\iiint_D \nabla \cdot \mathbf{F}\ dV?$$

Give reasons for your answer.

22. **Writing to Learn** The base of the closed cubelike surface shown here is the unit square in the xy-plane. The four sides lie in the planes $x = 0$, $x = 1$, $y = 0$, and $y = 1$. The top is an arbitrary smooth surface whose identity is unknown. Let $\mathbf{F} = x\mathbf{i} - 2y\mathbf{j} + (z + 3)\mathbf{k}$, and suppose the outward flux of $\mathbf{F}$ through side A is 1 and through side B is -3. Can you

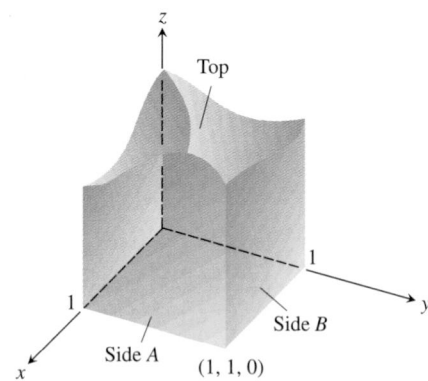

conclude anything about the outward flux through the top? Give reasons for your answer.

23. **(a)** Show that the flux of the position vector field $\mathbf{F} = x\mathbf{i} + y\mathbf{j} + z\mathbf{k}$ outward through a smooth closed surface S is three times the volume of the region enclosed by the surface.

(b) Let **n** be the outward unit normal vector field on S. Show that it is not possible for **F** to be orthogonal to **n** at every point of S.

24. Among all rectangular solids defined by the inequalities $0 \le x \le a, 0 \le y \le b, 0 \le z \le 1$, find the one for which the total flux of $\mathbf{F} = (-x^2 - 4xy)\mathbf{i} - 6yz\mathbf{j} + 12z\mathbf{k}$ outward through the six sides is greatest. What is the greatest flux?

25. Let $\mathbf{F} = x\mathbf{i} + y\mathbf{j} + z\mathbf{k}$, and suppose the surface S and region D satisfy the hypotheses of the Divergence Theorem. Show that the volume of D is given by the formula

$$\text{Volume of } D = \frac{1}{3} \iint_S \mathbf{F} \cdot \mathbf{n} \, d\sigma.$$

26. Show that the outward flux of a constant vector field $\mathbf{F} = \mathbf{C}$ across any closed surface to which the Divergence Theorem applies is zero.

27. *Harmonic Functions* A function $f(x, y, z)$ is **harmonic** in a region D in space if it satisfies the Laplace equation

$$\nabla^2 f = \nabla \cdot \nabla f = \frac{\partial^2 f}{\partial x^2} + \frac{\partial^2 f}{\partial y^2} + \frac{\partial^2 f}{\partial z^2} = 0$$

throughout D.

(a) Suppose f is harmonic throughout a bounded region D enclosed by a smooth surface S and **n** is the chosen unit normal vector on S. Show that the integral over S of $\nabla f \cdot \mathbf{n}$, the derivative of f in the direction of **n**, is zero.

(b) Show that if f is harmonic on D, then

$$\iint_S f \nabla f \cdot \mathbf{n} \, d\sigma = \iiint_D |\nabla f|^2 \, dV.$$

28. Let S be the surface of the portion of the solid sphere $x^2 + y^2 + z^2 \le a^2$ that lies in the first octant and let $f(x, y, z) = \ln\sqrt{x^2 + y^2 + z^2}$. Calculate

$$\iint_S \nabla f \cdot \mathbf{n} \, d\sigma.$$

($\nabla f \cdot \mathbf{n}$ is the derivative of f in the direction of **n**.)

29. *Green's First Formula* Suppose f and g are scalar functions with continuous first- and second-order partial derivatives throughout a closed region D that is bounded by a piecewise smooth surface S. Show that

$$\iint_S f \nabla g \cdot \mathbf{n} \, d\sigma = \iiint_D (f\nabla^2 g + \nabla f \cdot \nabla g) \, dV. \qquad (9)$$

Equation (9) is **Green's first formula**. (*Hint:* Apply the Divergence Theorem to the field $\mathbf{F} = f \nabla g$.)

30. *Green's Second Formula* (*Continuation of Exercise 29*) Interchange f and g in Equation (9) to obtain a similar formula. Then subtract this formula from Equation (9) to show that

$$\iint_S (f\nabla g - g\nabla f) \cdot \mathbf{n} \, d\sigma = \iiint_D (f\nabla^2 g - g\nabla^2 f) \, dV. \quad (10)$$

This equation is **Green's second formula.**

Extending the Ideas

31. *Conservation of Mass* Let $\mathbf{v}(t, x, y, z)$ be a continuously differentiable vector field over the region D in space and let $p(t, x, y, z)$ be a continuously differentiable scalar function. The variable t represents the time domain. The Law of Conservation of Mass asserts that

$$\frac{d}{dt} \iiint_D p(t, x, y, z) \, dV = -\iint_S p\mathbf{v} \cdot \mathbf{n} \, d\sigma,$$

where S is the surface enclosing D.

(a) Give a physical interpretation of the conservation of mass law if **v** is a velocity flow field and p represents the density of the fluid at point (x, y, z) at time t.

(b) Use the Divergence Theorem and Leibniz's rule,

$$\frac{d}{dt} \iiint_D p(t, x, y, z) \, dV = \iiint_D \frac{\partial p}{\partial t} \, dV,$$

to show that the Law of Conservation of Mass is equivalent to the continuity equation,

$$\nabla \cdot p\mathbf{v} + \frac{\partial p}{\partial t} = 0.$$

(In the first term, $\nabla \cdot p\mathbf{v}$, the variable t is held fixed, and in the second term, $\partial p/\partial t$, it is assumed that the point (x, y, z) in D is held fixed.)

32. *General Diffusion Equation* Let $T(t, x, y, z)$ be a function with continuous second derivatives giving the temperature at time t at the point (x, y, z) of a solid occupying a region D in space. If the solid's specific heat and mass density are denoted by the constants c and ρ respectively, the quantity $c\rho T$ is the solid's **heat energy per unit volume.**

(a) Explain why $-\nabla T$ points in the direction of heat flow.

(b) Let $-k \nabla T$ denote the **energy flux vector.** (Here the constant k is the **conductivity.**) Assuming the Law of Conservation of Mass with $-k \nabla T = \mathbf{v}$ and $c\rho T = p$ in Exercise 31, derive the **diffusion** (heat) **equation**

Chapter 15 Key Terms

Chapter 15 Review Exercises

In Exercises 1–4, evaluate the line integral over the indicated paths.

1. Figure 15.82 shows two polygonal paths in space joining the origin to the point $(1, 1, 1)$. Integrate
$f(x, y, z) = 2x - 3y^2 - 2z + 3$ over each path.

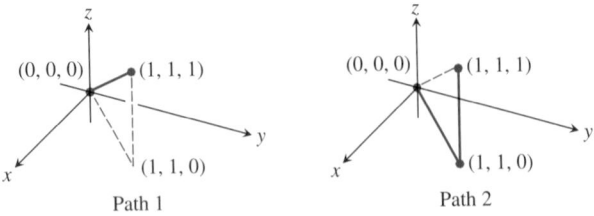

Figure 15.82 The paths in Exercise 1.

2. Figure 15.83 shows three polygonal paths joining the origin to the point $(1, 1, 1)$. Integrate $f(x, y, z) = x^2 + y - z$ over each path.

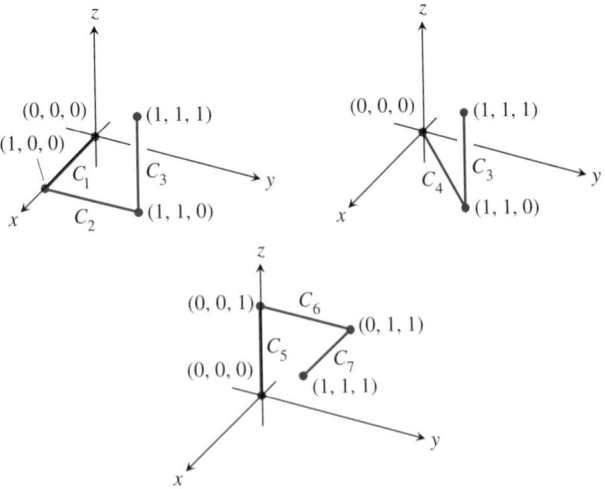

Figure 15.83 The paths in Exercise 2.

3. Integrate $f(x, y, z) = \sqrt{x^2 + z^2}$ over the circle
$\mathbf{r}(t) = (a \cos t)\mathbf{j} + (a \sin t)\mathbf{k},\ 0 \le t \le 2\pi.$

4. Integrate $f(x, y, z) = \sqrt{x^2 + z^2}$ over the curve $\mathbf{r}(t) =$
$(\cos t + t \sin t)\mathbf{i} + (\sin t - t \cos t)\mathbf{j},\ 0 \le t \le \sqrt{3}.$

In Exercises 5 and 6, evaluate the integral.

5. $\displaystyle\int_{(-1,1,1)}^{(4,-3,0)} \frac{dx + dy + dz}{\sqrt{x + y + z}}$

6. $\displaystyle\int_{(1,1,1)}^{(10,3,3)} dx - \sqrt{\frac{z}{y}}\, dy - \sqrt{\frac{y}{z}}\, dz$

7. Integrate $\mathbf{F} = (y \sin z)\mathbf{i} + (x \sin z)\mathbf{j} + (xy \cos z)\mathbf{k}$ around
the circle cut from the sphere $x^2 + y^2 + z^2 = 5$ by the
plane $z = -1$, clockwise as viewed from above.

8. Integrate $\mathbf{F} = 3x^2 y\mathbf{i} + (x^3 + 1)\mathbf{j} + 9z^2\mathbf{k}$ around the circle
cut from the sphere $x^2 + y^2 + z^2 = 9$ by the plane $x = 2$.

In Exercises 9 and 10, evaluate the line integral.

9. $\displaystyle\int_C 8x \sin y\, dx - 8y \cos x\, dy$

C is the square cut from the first quadrant by the lines
$x = \pi/2$ and $y = \pi/2$.

10. $\displaystyle\int_C y^2\, dx + x^2\, dy$

C is the circle $x^2 + y^2 = 4$.

11. Find the area of the elliptical region cut from the plane
$x + y + z = 1$ by the cylinder $x^2 + y^2 = 1$.

12. Find the area of the cap cut from the paraboloid
$y^2 + z^2 = 3x$ by the plane $x = 1$.

13. Find the area of the cap cut from the top of the sphere
$x^2 + y^2 + z^2 = 1$ by the plane $z = \sqrt{2}/2$.

14. (a) Find the area of the surface cut from the hemisphere
$x^2 + y^2 + z^2 = 4$, $z \ge 0$, by the cylinder $x^2 + y^2 = 2x$.

(b) Find the area of the portion of the cylinder that lies
inside the hemisphere. (*Hint:* Project onto the xz-plane. Or,
evaluate the integral $\displaystyle\int h\, ds$, where h is the altitude of the

cylinder and ds is the element of arc length on the circle
$x^2 + y^2 = 2x$ in the xy-plane.)

15. Find the area of the triangle in which the plane $(x/a) +$
$(y/b) + (z/c) = 1$ $(a, b, c > 0)$ intersects the first octant.
Check your answer with an appropriate vector calculation.

16. Integrate

(a) $g(x, y, z) = \dfrac{yz}{\sqrt{4y^2 + 1}}$

(b) $g(x, y, z) = \dfrac{z}{\sqrt{4y^2 + 1}}$

over the surface cut from the parabolic cylinder $y^2 - z = 1$
by the planes $x = 0$, $x = 3$, and $z = 0$.

17. Integrate $g(x, y, z) = x^4 y(y^2 + z^2)$ over the portion of the
cylinder $y^2 + z^2 = 25$ that lies in the first octant between
the planes $x = 0$ and $x = 1$ and above the plane $z = 3$.

18. *Area of Wyoming* The state of Wyoming is bounded by the
meridians $111°3'$ and $104°3'$ west longitude and by the
circles $41°$ and $45°$ north latitude. Assuming that Earth is a
sphere of radius $R = 3959$ mi, find the area of Wyoming.

In Exercises 19–24, find a parametrization for the surface.
(There are many ways to do these, so your answers may not be
the same as those in the back of the book.)

19. The portion of the sphere $x^2 + y^2 + z^2 = 36$ between the
planes $z = -3$ and $z = 3\sqrt{3}$

20. The portion of the paraboloid $z = -(x^2 + y^2)/2$ above the
plane $z = -2$

21. The cone $z = 1 + \sqrt{x^2 + y^2}$, $z \le 3$

22. The portion of the plane $4x + 2y + 4z = 12$ that lies above
the square $0 \le x \le 2$, $0 \le y \le 2$ in the first quadrant

23. The portion of the paraboloid $y = 2(x^2 + z^2)$, $y \le 2$, that
lies above the xy-plane

24. The portion of the hemisphere $x^2 + y^2 + z^2 = 10$, $y \ge 0$, in
the first octant

25. Find the area of the surface $\mathbf{r}(u, v) = (u + v)\mathbf{i} + (u - v)\mathbf{j} + v\mathbf{k}$, $0 \le u \le 1$, $0 \le v \le 1$.

26. Integrate $f(x, y, z) = xy - z^2$ over the surface in Exercise 25.

27. *Area of a Helicoid* Find the surface area of the helicoid
$\mathbf{r}(r, \theta) = r \cos \theta\mathbf{i} + r \sin \theta\mathbf{j} + \theta\mathbf{k}$, $0 \le \theta \le 2\pi$ and
$0 \le r \le 1$, in the accompanying figure.

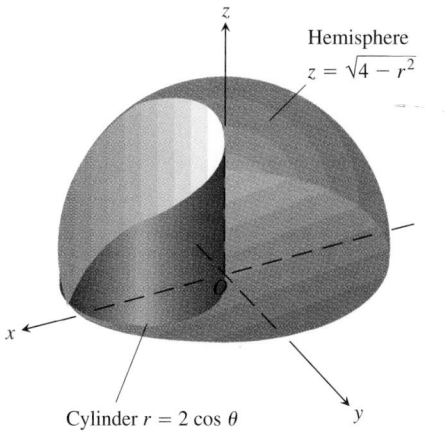

Hemisphere
$z = \sqrt{4 - r^2}$

Cylinder $r = 2 \cos \theta$

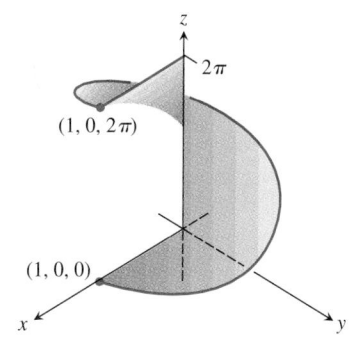

2π

$(1, 0, 2\pi)$

$(1, 0, 0)$

28. *Integrating over a Helicoid* Evaluate the integral

$$\iint_S \sqrt{x^2 + y^2 + 1}\, d\sigma,$$ where S is the helicoid in Exercise 27.

In Exercises 29–32, identify whether the field is conservative or not conservative.

29. $\mathbf{F} = x\mathbf{i} + y\mathbf{j} + z\mathbf{k}$

30. $\mathbf{F} = (x\mathbf{i} + y\mathbf{j} + z\mathbf{k})/(x^2 + y^2 + z^2)^{3/2}$

31. $\mathbf{F} = xe^y\mathbf{i} + ye^x\mathbf{j} + ze^x\mathbf{k}$

32. $\mathbf{F} = (\mathbf{i} + z\mathbf{j} + y\mathbf{k})/(x + yz)$

In Exercises 33 and 34, find a potential function for the field.

33. $\mathbf{F} = 2\mathbf{i} + (2y + z)\mathbf{j} + (y + 1)\mathbf{k}$

34. $\mathbf{F} = (z \cos xz)\mathbf{i} + e^y\mathbf{j} + (x \cos xz)\mathbf{k}$

In Exercises 35 and 36, find the work done by the field along the paths from (0, 0, 0) to (1, 1, 1) in Figure 15.82.

35. $\mathbf{F} = 2xy\mathbf{i} + \mathbf{j} + x^2\mathbf{k}$ **36.** $\mathbf{F} = 2xy\mathbf{i} + x^2\mathbf{j} + \mathbf{k}$

In Exercises 37–40, *work in groups of two or three* to solve the problem.

37. *Finding Work Two Ways* Find the work done by

$$\mathbf{F} = \frac{x\mathbf{i} + y\mathbf{j}}{(x^2 + y^2)^{3/2}}$$

over the plane curve $\mathbf{r}(t) = (e^t \cos t)\mathbf{i} + (e^t \sin t)\mathbf{j}$ from the point (1, 0) to the point $(e^{2\pi}, 0)$ in two ways:

(a) by using the parametrization of the curve to evaluate the work integral.

(b) by evaluating a potential function for $\mathbf{F}$.

38. *Flows* Find the flow of the field $\mathbf{F} = \nabla(x^2ze^y)$

(a) once around the ellipse C in which the plane $x + y + z = 1$ intersects the cylinder $x^2 + z^2 = 25$, clockwise as viewed from the positive y-axis.

(b) along the curved boundary of the helicoid in Exercise 27 from (1, 0, 0) to (1, 0, 2π).

39. *Flows* Suppose $\mathbf{F}(x, y) = (x + y)\mathbf{i} - (x^2 + y^2)\mathbf{j}$ is the velocity field of a fluid flowing across the xy-plane. Find the flow along each of the following paths from (1, 0) to (−1, 0).

(a) The upper half of the circle $x^2 + y^2 = 1$

(b) The line segment from (1, 0) to (−1, 0)

(c) The line segment from (1, 0) to (−1, 0) followed by the line segment from (0, −1) to (−1, 0)

40. *Circulation* Find the circulation of $\mathbf{F} = 2x\mathbf{i} + 2z\mathbf{j} + 2y\mathbf{k}$ along the closed path consisting of the helix $\mathbf{r}_1(t) = (\cos t)\mathbf{i} + (\sin t)\mathbf{j} + t\mathbf{k}, 0 \le t \le \pi/2$, followed by the line segments $\mathbf{r}_2(t) = \mathbf{j} + (\pi/2)(1 - t)\mathbf{k}, 0 \le t \le 1$, and $\mathbf{r}_3(t) = t\mathbf{i} + (1 - t)\mathbf{j}, 0 \le t \le 1$.

In Exercises 41 and 42, use the surface integral in Stokes's Theorem to find the circulation of the field $\mathbf{F}$ around the curve C in the indicated direction.

41. $\mathbf{F} = y^2\mathbf{i} - y\mathbf{j} + 3z^2\mathbf{k}$

C: The ellipse in which the plane $2x + 6y - 3z = 6$ meets the cylinder $x^2 + y^2 = 1$, counterclockwise as viewed from above

42. $\mathbf{F} = (x^2 + y)\mathbf{i} + (x + y)\mathbf{j} + (4y^2 - z)\mathbf{k}$

C: The circle in which the plane $z = -y$ meets the sphere $x^2 + y^2 + z^2 = 4$, counterclockwise as viewed from above

43. *Mass of a Wire* Find the mass of a thin wire lying along the curve $\mathbf{r}(t) = \sqrt{2}\, t\mathbf{i} + \sqrt{2}\, t\mathbf{j} + (4 - t^2)\mathbf{k}, 0 \le t \le 1$, if the density at t is **(a)** $\delta = 3t$ and **(b)** $\delta = 1$.

44. *Center of Mass of a Curved Wire* Find the center of mass of a thin wire lying along the curve $\mathbf{r}(t) = t\mathbf{i} + 2t\mathbf{j} + (2/3)t^{2/3}\mathbf{k}, 0 \le t \le 2$, if the density at t is $\delta = 3\sqrt{5 + t}$.

45. *Center of Mass, Moments, and Radii of Gyration* Find the center of mass and the moments of inertia and radii of gyration about the coordinate axes of a thin wire lying along the curve

$$\mathbf{r}(t) = t\mathbf{i} + \frac{2\sqrt{2}}{3} t^{3/2}\mathbf{j} + \frac{t^2}{2}\mathbf{k}, 0 \le t \le 2,$$

if the density is $\delta = 1/(t + 1)$.

46. *A Metal Arch* A slender metal arch lies along the semicircle $y = \sqrt{a^2 - x^2}$ in the xy-plane. The density at the point (x, y) on the arch is $\delta(x, y) = 2a - y$. Find the center of mass.

47. *Wire of Constant Density* A wire of constant density $\delta = 1$ lies along the curve $\mathbf{r}(t) = (e^t \cos t)\mathbf{i} + (e^t \sin t)\mathbf{j} + e^t\mathbf{k}, 0 \le t \le \ln 2$. Find $\bar{z}, I_z$, and R_z.

48. *Wire of Constant Density* Find the mass and center of mass of a wire of constant density δ that lies along the helix $\mathbf{r}(t) = (2 \sin t)\mathbf{i} + (2 \cos t)\mathbf{j} + 3t\mathbf{k}, 0 \le t \le 2\pi$.

49. *Thin Shell of Variable Density* Find I_z, R_z, and the center of mass of a thin shell of density $\delta(x, y, z) = z$ cut from the upper portion of the sphere $x^2 + y^2 + z^2 = 25$ by the plane $z = 3$.

50. *Moment of Inertia of a Cube about One Edge* Find the moment of inertia about the z-axis of the surface of the cube cut from the first octant by the planes $x = 1, y = 1$, and $z = 1$ if the density is $\delta = 1$.

In Exercises 51 and 52, use Green's Theorem to find the counterclockwise circulation and outward flux for the given field and curve.

51. $\mathbf{F} = (2xy + x)\mathbf{i} + (xy - y)\mathbf{j}$

C: The square bounded by $x = 0, x = 1, y = 0, y = 1$

52. $\mathbf{F} = (y - 6x^2)\mathbf{i} + (x + y^2)\mathbf{j}$

C: The triangle made by the lines $y = 0, y = x$, and $x = 1$.

53. Show that

$$\oint_C \ln x \sin y \, dy - \frac{\cos y}{x} \, dx = 0$$

for any closed curve C to which Green's Theorem applies.

54. (a) *Outward Flux and Area* Show that the outward flux of the position vector field $\mathbf{F} = x\mathbf{i} + y\mathbf{j}$ across any closed curve to which Green's Theorem applies is twice the area of the region enclosed by the curve.

(b) Writing to Learn Let $\mathbf{n}$ be the outward unit normal vector to a closed curve to which Green's Theorem applies. Show that it is not possible for $\mathbf{F} = x\mathbf{i} + y\mathbf{j}$ to be orthogonal to $\mathbf{n}$ at every point of C.

In Exercises 55–58, find the outward flux of $\mathbf{F}$ across the boundary of D.

55. $\mathbf{F} = 2xy\mathbf{i} + 2yz\mathbf{j} + 2xz\mathbf{k}$

D: The cube cut from the first octant by the planes $x = 1$, $y = 1$, $z = 1$

56. $\mathbf{F} = xz\mathbf{i} + yz\mathbf{j} + \mathbf{k}$

D: The entire surface of the upper cap cut from the solid sphere $x^2 + y^2 + z^2 \le 25$ by the plane $z = 3$

57. $\mathbf{F} = -2x\mathbf{i} - 3y\mathbf{j} + z\mathbf{k}$

D: The upper region cut from the solid sphere $x^2 + y^2 + z^2 \le 2$ by the paraboloid $z = x^2 + y^2$

58. $\mathbf{F} = (6x + y)\mathbf{i} - (x + z)\mathbf{j} + 4yz\mathbf{k}$

D: The region in the first octant bounded by the cone $z = \sqrt{x^2 + y^2}$, the cylinder $x^2 + y^2 = 1$, and the coordinate planes

59. *Flux* Let S be the surface that is bounded on the left by the hemisphere $x^2 + y^2 + z^2 = a^2$, $y \le 0$, in the middle by the cylinder $x^2 + z^2 = a^2$, $0 \le y \le a$, and on the right by the plane $y = a$. Find the flux of the field $\mathbf{F} = y\mathbf{i} + z\mathbf{j} + x\mathbf{k}$ outward across S.

60. *Flux* Find the outward flux of the field $\mathbf{F} = 3xz^2\mathbf{i} + y\mathbf{j} - z^3\mathbf{k}$ across the surface of the solid in the first octant that is bounded by the cylinder $x^2 + 4y^2 = 16$ and the planes $y = 2x$, $x = 0$, and $z = 0$.

61. *Flux* Use the Divergence Theorem to find the flux of $\mathbf{F} = xy^2\mathbf{i} + x^2y\mathbf{j} + y\mathbf{k}$ outward through the surface of the region enclosed by the cylinder $x^2 + y^2 = 1$ and the planes $z = 1$ and $z = -1$.

62. *Finding Flux Two Ways* Find the flux of $\mathbf{F} = (3z + 1)\mathbf{k}$ upward across the hemisphere $x^2 + y^2 + z^2 = a^2$, $z \ge 0$ **(a)** with the Divergence Theorem and **(b)** by evaluating the flux integral directly.

Cumulative Review Exercises: Chapters 1–10

In Exercises 1–8, determine the limit.

1. $\lim\limits_{x \to 1} \dfrac{2x^2 - x - 1}{x^2 + x - 12}$

2. $\lim\limits_{x \to 0} \dfrac{\sin 3x}{4x}$

3. $\lim\limits_{x \to 0} \dfrac{\dfrac{1}{x+1} - 1}{x}$

4. $\lim\limits_{x \to \infty} \dfrac{x + e^x}{x - e^x}$

5. $\lim\limits_{t \to 0} \dfrac{t(1 - \cos t)}{t - \sin t}$

6. $\lim\limits_{x \to 0^+} \dfrac{\ln(e^x - 1)}{\ln x}$

7. $\lim\limits_{x \to 0} (e^x + x)^{1/x}$

8. $\lim\limits_{x \to 0} \left(\dfrac{3x + 1}{x} - \dfrac{1}{\sin x} \right)$

9. Let $f(x) = \begin{cases} 2x - x^2, & x \le 1 \\ 2 - x, & x > 1. \end{cases}$

   (a) Find $\lim_{x \to 1^-} f(x)$.

   (b) Find $\lim_{x \to 1^+} f(x)$.

   (c) Find $\lim_{x \to 1} f(x)$.

   (d) Is f continuous at $x = 1$?

   (e) Is f differentiable at $x = 1$?

10. Find all the points of discontinuity of
$$f(x) = \sqrt{\dfrac{1}{4 - x^2}}.$$

11. Identify all horizontal and vertical asymptotes of
$$y = \dfrac{\cos x}{2x^2 - x}.$$

12. Sketch a possible graph for a function $y = f(x)$ that satisfies:
$$\lim\limits_{x \to 2^-} f(x) = \infty, \quad \lim\limits_{x \to 2^+} f(x) = -1,$$
$$\lim\limits_{x \to -\infty} f(x) = -3, \quad \lim\limits_{x \to \infty} f(x) = 3.$$

13. Find the average rate of change of the function $f(x) = \sqrt{x + 4}$ over the interval $[0, 5]$.

In Exercises 14–28, find dy/dx.

14. $y = \dfrac{x + 1}{x - 2}$

15. $y = \cos(\sqrt{1 - 3x})$

16. $y = \sin x \tan x$

17. $y = \ln(x^2 + 1)$

18. $y = e^{x^2 - x}$

19. $y = x^2 \tan^{-1} x$

20. $y = x^{-3} e^x$

21. $y = \left(\dfrac{\csc x}{1 + \cos x} \right)^3$

22. $y = \cos^{-1} x - \cot^{-1} x$

23. $\cos(xy) + y^2 - \ln x = 1$

24. $y = \sqrt{|x|}$

25. $x = 1 + \cos t, \quad y = 1 - \sin t$

26. $y = (\cos x)^x, \quad -\dfrac{\pi}{2} < x < \dfrac{\pi}{2}$

27. $y = \displaystyle\int_0^x \sqrt{1 + t^3} \, dt$

28. $y = \displaystyle\int_{2x}^{x^2} \sin t \, dt$

29. Find d^2y/dx^2 if $y^2 + 2y = \sec x$.

30. Suppose u and v are differentiable functions of x and that $u(0) = 2$, $u'(0) = -1$, $v(0) = -3$, and $v'(0) = 3$. Find
$$\dfrac{d}{dx} \left(\dfrac{u}{1 + v} \right) \bigg|_{x=0}.$$

31. A particle moves along the x-axis with its position at time t in seconds given by $x = t^3 - 6t^2 + 9t$, $0 \le t \le 5$, in meters.

   (a) Determine the velocity and acceleration of the particle at time t.

   (b) When is the particle at rest?

   (c) When is the particle moving to the right? left?

   (d) What is the velocity when the acceleration is zero?

In Exercises 32–36, find an equation for **(a)** the tangent line and **(b)** the normal line to the curve at the indicated point.

32. $y = 2x^3 - 6x^2 + 4x - 1$ at $x = 1$

33. $y = x \cos x$ at $x = \pi/3$

34. $\dfrac{x^2}{4} + \dfrac{y^2}{9} = 1$ at $\left(1, \dfrac{3\sqrt{3}}{2} \right)$

35. $x = 2 \cos t$, $y = 3 \sin t$, at $t = \pi/3$

36. $\mathbf{r}(t) = (\sec t)\mathbf{i} + (\tan t)\mathbf{j}$, at $t = \pi/4$

37. Sketch the graph of a continuous function f with
$$f(3) = 1 \quad \text{and} \quad f'(x) = \begin{cases} -1, & x < 3 \\ 2, & x > 3. \end{cases}$$

38. The graph of the function f over the interval $[-2, 3]$ is given. At what domain points does f appear to be

   (a) differentiable?

   (b) continuous but not differentiable?

   (c) neither continuous nor differentiable?

   (d) Identify any extreme values and where they occur.

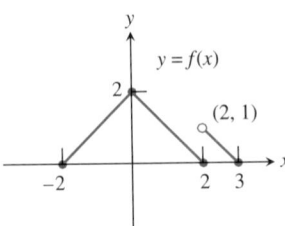

39. A driver handed in a ticket at a toll booth showing that in 1.5 h he had covered 111 mi on a toll road with speed limit of 65 mph. The driver was cited for speeding. Why?

40. Assume that f is continuous and differentiable on the interval $[-2, 2]$. The table gives some values of f'.

x	$f'(x)$	x	$f'(x)$
-2	-11	0.25	3.63
-1.75	-8.38	0.5	4
-1.5	-6	0.75	4.13
-1.25	-3.88	1	4
-1	-2	1.25	3.63
-0.75	-0.38	1.5	3
-0.5	1	1.75	2.13
-0.25	2.13	2	1
0	3		

(a) Estimate where f is increasing, decreasing, and has local extrema.

(b) Find a quadratic regression equation for the data in the table and superimpose its graph on a scatter plot of the data.

(c) Use the equation in (b) to find a formula for f that satisfies $f(0) = 1$.

41. Find the function f with $f'(x) = 2x - 3 + \sin x$ whose graph passes through the point $P(0, -2)$.

42. Suppose $f(x) = x^2\sqrt{4 - x^2}$. Find the intervals on which the graph of f is **(a)** increasing, **(b)** decreasing, **(c)** concave up, **(d)** concave down. Then find any **(e)** local extreme values and where they occur, and **(f)** any inflection points.

43. A function f is continuous on its domain $[-1, 3]$, $f(-1) = 1$, $f(3) = -2$, and f' and f'' have the following properties.

x	$-1 < x < 1$	$x = 1$	$1 < x < 2$	$x = 2$	$2 < x < 3$
f'	$+$	0	$-$	does not exist	$-$
f''	$-$	-1	$-$	does not exist	$+$

(a) Find where all absolute extrema of f occur.

(b) Find where the points of inflection of f occur.

(c) Sketch a possible graph of f.

44. A rectangle with base on the x-axis is to be inscribed under the upper half of the ellipse
$$\frac{x^2}{16} + \frac{y^2}{4} = 1.$$
What are the dimensions of the rectangle with largest area, and what is the largest area?

45. Find the linearization of $f(x) = \sec x$ at $x = \pi/4$.

46. The edge of a cube is measured as 8 cm with an error of 1%. The cube's volume is to be calculated from this measurement. Estimate the percentage error in the volume calculation.

47. A dinghy is pulled toward a dock by a rope from the bow through a ring on the dock 5 ft above the bow as shown in the figure. The rope is hauled in at the rate of 1.5 ft/sec.

(a) How fast is the boat approaching the dock when 8 ft of rope are out?

(b) At what rate is the angle θ changing at that moment?

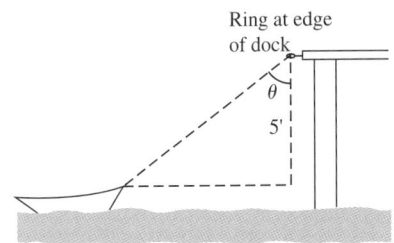

Ring at edge of dock

θ

$5'$

48. Coffee is draining from a conical filter into a cylindrical coffeepot at the rate of 9 in.^{3}/min as suggested in the figure.

(a) How fast is the level in the pot rising when the coffee in the cone is 5 in. deep?

(b) How fast is the level in the cone falling at that moment?

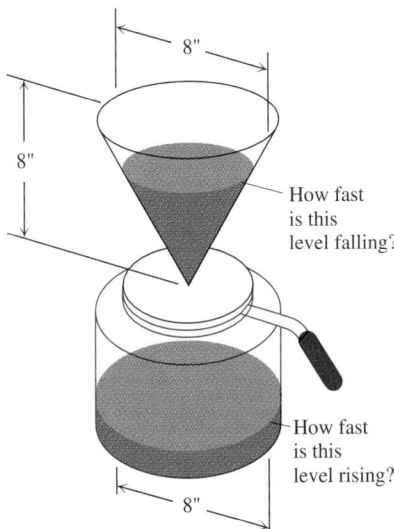

$8''$

$8''$

How fast is this level falling?

How fast is this level rising?

$8''$

49. The table below shows the velocity of a model train engine moving along a track for 10 sec. Estimate the distance traveled by the engine, using 10 subintervals of length 1 with **(a)** LRAM and **(b)** RRAM.

Time (sec)	Velocity (in./sec)	Time (sec)	Velocity (in./sec)
0	0	6	28.8
1	1.8	7	29.4
2	6.4	8	25.6
3	12.6	9	16.2
4	19.2	10	0
5	25.0		

In Exercises 50–61, evaluate the integral analytically.

50. $\displaystyle\int_{-2}^{1} |x|\, dx$

51. $\displaystyle\int_{-2}^{2} \sqrt{4 - x^2}\, dx$

52. $\displaystyle\int_{1}^{3} \left(x^2 + \frac{1}{x}\right) dx$

53. $\displaystyle\int_{0}^{\pi/4} \sec^2 x\, dx$

54. $\displaystyle\int_{1}^{4} \frac{2 + \sqrt{x}}{\sqrt{x}}\, dx$

55. $\displaystyle\int_{e}^{2e} \frac{dx}{x(\ln x)^2}$

56. $\displaystyle\int_{1}^{3} [(3 - 2t)\mathbf{i} + (1/t)\mathbf{j}]\, dt$

57. $\displaystyle\int e^x \cot^2 (e^x + 1)\, dx$

58. $\displaystyle\int \frac{ds}{s^2 + 4}$

59. $\displaystyle\int \frac{\sin (x - 3)}{\cos^3 (x - 3)}\, dx$

60. $\displaystyle\int e^{-x} \cos 2x\, dx$

61. $\displaystyle\int \frac{x + 2}{x^2 - 5x - 6}\, dx$

62. A rectangular swimming pool is 25 ft wide and 40 ft long. The table below shows the depth $h(x)$ of the water at 5-ft intervals from one end of the pool to the other. Estimate the volume of the water in the pool using the Trapezoidal Rule with $n = 8$.

Position (ft) x	Depth (ft) $h(x)$	Position (ft) x	Depth (ft) $h(x)$
0	3	25	10.7
5	8.3	30	9.9
10	9.9	35	8.3
15	10.7	40	3
20	11		

In Exercises 63 and 64, solve the initial value problem.

63. $\dfrac{dy}{dt} = (t + 1)^{-2} + e^{-2t}, \quad y(0) = 2$

64. $\dfrac{d^2y}{d\theta^2} = \sin 2\theta - \cos \theta, \quad y(\pi/2) = y'(\pi/2) = 0$

65. Evaluate $\int x^2 \sin x\, dx$. Support your answer by superimposing the graph of one of the antiderivatives on a slope field of the integrand.

66. Evaluate $\int xe^x\, dx$. Confirm your answer by differentiation.

67. A colony of bacteria is grown under ideal conditions in a laboratory so that the population increases exponentially with time. At the end of 2 h there are 6,000 bacteria. At the end of 5 h there are 10,000 bacteria.

(a) Find a formula for the number of bacteria present at any time t.

(b) How many bacteria were present initially?

68. The temperature of an ingot of silver is 50°C above room temperature right now. Fifteen minutes ago, it was 65°C above room temperature.

(a) How far above room temperature will the silver be 2 hours from now?

(b) When will the silver be 5°C above room temperature?

In Exercises 69 and 70, solve the differential equation.

69. $\dfrac{dy}{dx} = 0.08y \left(1 - \dfrac{y}{500}\right)$

70. $\dfrac{dy}{dx} = (y - 4)(x + 3)$

71. Use Euler's method to solve the initial value problem
$$y' = y + \cos x, \quad y(0) = 0,$$
on the interval $0 \le x \le 1$ with $dx = 0.1$.

In Exercises 72–75, find the area of the region enclosed by the curves.

72. $y = \sin 2x, \quad y = 0, \quad x = -\pi, \quad x = \pi$

73. $y = 5 - x^2, \quad y = x^2 - 3$

74. $x = y^2 - 3, \quad y = x - 2$

75. $r = 3(1 + \cos \theta)$

In Exercises 76 and 77, find the volume of the solid generated by revolving the region bounded by the curves about the indicated axis.

76. $y = x^3/2, \quad y = 0, \quad x = -1, \quad x = 1; \quad x\text{-axis}$

77. $y = 4x - x^2, \quad y = 0; \quad y\text{-axis}$

78. Find the average value of $\sqrt{\sin x}$ on the interval $[0, \pi]$.

In Exercises 79–81, find the length of the curve.

79. $y = \tan x, \quad -\pi/4 \le x \le \pi/4$

80. $x = \sin t, \quad y = t + \cos t, \quad -\pi/2 \le t \le \pi/2$

81. $r = \theta, \quad 0 \le \theta \le \pi$

In Exercises 82–84, find the area of the surface generated by revolving the curve about the indicated axis.

82. $y = e^{-x/2}, \quad 0 \le x \le 2; \quad x\text{-axis}$

83. $x = \sin t, \quad y = t + \cos t, \quad 0 \le t \le \pi/2; \quad y\text{-axis}$

84. $r = \theta, \quad \pi/2 \le \theta \le \pi; \quad x\text{-axis}$

85. A solid lies between planes perpendicular to the x-axis at $x = 0$ and $x = 1$. The cross sections of the solid perpendicular to the x-axis between these planes are circular disks with diameters running from the parabola $y = x^2$ to the parabola $y = \sqrt{x}$. Find the volume of the solid.

86. Find the volume of the solid generated by revolving about the x-axis the region bounded by $y = 2 \tan x, \quad y = 0,$ $x = -\pi/4,$ and $x = \pi/4$. (The region lies in the first and third quadrants and resembles a bow tie.)

87. A force of 200 N will stretch a garage door spring 0.8 m beyond its unstressed length.

(a) How far will a 300-N force stretch the spring from its unstressed length?

(b) In (a), how much work was done in stretching the spring that far?

88. A right circular conical tank, point down, with top radius 5 ft and height 10 ft is filled with a liquid whose weight-density is 60 lb/ft^3.

(a) To the nearest foot-pound, how much work will it take to pump the liquid to a point 2 ft above the tank?

(b) If the pump is driven by a motor rated at 275 ft · lb/sec (1/2-hp), about how long will it take it to empty the tank?

89. You plan to store mercury (weight density 849 lb/ft^3) in a vertical right circular cylindrical tank of inside radius 1 ft whose interior side wall can withstand a total fluid force of 40,000 lb. About how many cubic feet of mercury can you store at any one time?

90. Does $f(x) = \ln x$ grow faster than, at the same rate as, or slower than $g(x) = \sqrt{x}$ as $x \to \infty$?

In Exercises 91–96, determine whether the integral converges or diverges.

91. $\int_3^\infty \dfrac{dt}{t^2 - 4}$

92. $\int_2^\infty \dfrac{dx}{\ln x}$

93. $\int_{-\infty}^\infty e^{-|x|}\, dx$

94. $\int_0^1 \dfrac{4r\, dr}{\sqrt{1 - r^2}}$

95. $\int_0^{10} \dfrac{dx}{1 - x}$

96. $\int_0^2 \dfrac{dx}{\sqrt[3]{x - 1}}$

97. Find a power series to represent

$$\frac{1}{1 + 2x}$$

and identify its interval of convergence.

98. (a) Find a power series for

$$F(x) = \int_0^x \cos(t^2)\, dt.$$

(b) What is the interval of convergence of the series? Explain.

99. Find the Maclaurin series generated by $\ln(2 + 2x)$. What is its interval of convergence?

100. Find the Taylor series generated by $\sin x$ at $x = 2\pi$.

101. Find a polynomial that you know will approximate e^{-x} throughout the interval $[0, 1]$ with an error of magnitude less than 10^{-3}. Explain.

102. Find the Taylor series generated by $f(x) = \sqrt[3]{1 + x}$ at $x = 0$ and identify its radius of convergence.

In Exercises 103–106, determine whether the series converges or diverges.

103. $\sum_{n=0}^\infty \dfrac{2}{3^n}$

104. $\sum_{n=1}^\infty \dfrac{2}{\sqrt{n}}$

105. $\sum_{n=0}^\infty \dfrac{(-1)^n}{n + 1}$

106. $\sum_{n=0}^\infty \dfrac{3^n}{n!}$

In Exercises 107 and 108, **(a)** find the radius and interval of convergence. For what values of x is the convergence **(b)** absolute? **(c)** conditonal?

107. $\sum_{n=1}^\infty \dfrac{(-1)^n (x + 2)^n}{n}$

108. $\sum_{n=2}^\infty \dfrac{x^n}{n(\ln n)^2}$

109. Find the unit vector in the direction of $\langle 2, -3 \rangle$.

110. Find the component form of the unit vector that makes an angle of $\pi/3$ with the positive x-axis.

111. Find the unit vectors (four vectors in all) that are tangent and normal to the curve $x = 4 \sin t$, $y = 3 \cos t$, at $t = 3\pi/4$.

112. The position of a particle in the plane is given by

$$\mathbf{r}(t) = (1 - \sin t)\mathbf{i} + (t - \cos t)\mathbf{j}.$$

(a) Find the velocity and acceleration of the particle.

(b) Find the distance the particle travels along the path from $t = \pi/2$ to $t = 3\pi/2$.

113. A golf ball leaves the ground at a 45° angle at a speed of 100 ft/sec. Will it clear the top of a 35-ft tree 130 ft away? Explain.

114. Replace the polar equation $r \cos \theta - r \sin \theta = 2$ by an equivalent Cartesian equation. Then identify the graph.

115. Graph the polar curve $r = 1 + 2 \sin \theta$. What is the shortest length a θ-interval can have and still produce the graph?

116. Find equations for the horizontal and vertical tangents to the curve $r = 1 - \cos \theta$, $0 \le \theta \le 2\pi$.

Cumulative Review Exercises: Chapters 11–15

1. Let $\mathbf{A} = 2\mathbf{i} - 2\mathbf{j} + \mathbf{k}$ and $\mathbf{B} = \mathbf{i} + \mathbf{j} - \mathbf{k}$. Find

   (a) $|\mathbf{A}|$ (b) $|\mathbf{B}|$ (c) $\mathbf{A} \cdot \mathbf{B}$

   (d) $\mathbf{B} \cdot \mathbf{A}$ (e) $\mathbf{A} \times \mathbf{B}$ (f) $\mathbf{B} \times \mathbf{A}$

   (g) the angle between $\mathbf{A}$ and $\mathbf{B}$

   (h) the vector projection of $\mathbf{B}$ onto $\mathbf{A}$

2. For what values of a will the vectors $\mathbf{u} = a\mathbf{i} + \mathbf{j} + 3\mathbf{k}$ and $\mathbf{v} = a\mathbf{i} - \mathbf{j} - \mathbf{k}$ be orthogonal?

3. Let $\mathbf{A} = 2\mathbf{i} + \mathbf{j}$, $\mathbf{B} = \mathbf{i} - 2\mathbf{j}$, and $\mathbf{C} = \mathbf{i} + \mathbf{j} + \mathbf{k}$.

   Find

   (a) the area of the parallelogram determined by $\mathbf{A}$ and $\mathbf{B}$.

   (b) the volume of the parallelepiped determined by $\mathbf{A}$, $\mathbf{B}$, and $\mathbf{C}$.

4. Write the vector $\mathbf{B} = 2\mathbf{i} + 3\mathbf{j} - \mathbf{k}$ as a sum of a vector parallel to $\mathbf{A} = \mathbf{i} + \mathbf{j} + \mathbf{k}$ and a vector orthogonal to $\mathbf{A}$.

5. Parametrize the line that passes through the point $(-2, 1, -1)$ parallel to the vector $\mathbf{v} = \mathbf{i} - 2\mathbf{j} + 3\mathbf{k}$.

6. Parametrize the line segment determined by the points $P(2, 1, -1)$ and $Q(1, 2, 4)$.

7. Find the distance from the point $(2, 0, 1)$ to the line $x = 1 + t, y = 2t, z = -2 + t$

8. Find the distance from the point $(-1, 2, 3)$ to the plane $2x + 3y + z = 2$.

9. Find an equation for the plane that passes through the point $(-1, -2, 3)$ normal to the vector $\mathbf{n} = 3\mathbf{i} - \mathbf{j} + 2\mathbf{k}$.

10. Find an equation for the plane that passes through the point $(-2, 3, 4)$ perpendicular to the line $x = -2 + 3t$, $y = 3 - 2t, z = 4 + 2t$.

11. Find an equation for the plane through the points $P(1, 1, 0)$, $Q(0, 1, 1)$, and $R(1, 2, 1)$.

12. Find parametric equations for the line in which the planes $2x - y + z = 1$ and $x - 2y + z = 2$ intersect.

13. Find the acute angle between the planes $x + y - z = 1$ and $2x - y + 3z = 2$.

14. The equation $x - 2y = 1$ defines sets both in the plane and in three-dimensional space. Identify both sets.

15. Translate the cylindrical coordinate equation $r = 4 \sin \theta$ into rectangular coordinates. Identify the set of points defined by the equation.

In Exercises 16–18, sketch the surface.

16. $z = y^2 - 1$ 17. $x^2 + y^2 - 16z^2 = 16$

18. $4z^2 - x^2 - y^2 = 4$

19. Solve the following initial value problem for $\mathbf{r}$ as a vector function of t.

    Differential equation: $\dfrac{d\mathbf{r}}{dt} = e^t\mathbf{i} + (\cos t)\mathbf{j} - 3t^2\mathbf{k}$

    Initial condition: $\mathbf{r}(0) = 2\mathbf{i} + 3\mathbf{j}$

20. Find the length of the curve $\mathbf{r}(t) = (\cos t)\mathbf{i} + (\sin t)\mathbf{j} + t^2\mathbf{k}$, $0 \le t \le \pi$.

21. Let $\mathbf{r}(t) = (e^t \sin t)\mathbf{i} + (e^t \cos t)\mathbf{j} + e^t\mathbf{k}$. Find $\mathbf{T}$, $\mathbf{N}$, $\mathbf{B}$, κ, and τ at $t = 0$.

22. Let $\mathbf{r}(t) = (1 + 2t^3)\mathbf{i} + (t + t^2)\mathbf{j} - (2 \sin t)\mathbf{k}$. Write $\mathbf{a}$ in the form $a_\mathbf{T}\mathbf{T} + a_\mathbf{N}\mathbf{N}$ at $t = 0$.

23. At what times in the interval $0 \le t \le 2\pi$ are the velocity and acceleration vectors of the motion $\mathbf{r}(t) = t\mathbf{i} + (2 \sin t)\mathbf{j} + (3 \cos t)\mathbf{k}$ orthogonal?

24. Find equations for the normal, rectifying, and osculating planes of the curve $\mathbf{r}(t) = t\mathbf{i} - t^2\mathbf{j} + 2t^3\mathbf{k}$ at the point $(1, -1, 2)$.

25. *Rowing Across a River* A straight river is 100 m wide (see figure). A rowboat leaves the far shore at time $t = 0$. The person in the boat rows at a rate of 20 m/min, always toward the near shore. The velocity of the river at (x, y) is

    $$\mathbf{v} = \left(-\frac{1}{250}(y - 50)^2 + 10\right)\mathbf{i} \text{ m/min}, \qquad 0 < y < 100.$$

    (a) Given that $\mathbf{r}(0) = 0\mathbf{i} + 100\mathbf{j}$, what is the position of the boat at time t?

    (b) How far downstream will the boat land on the near shore?

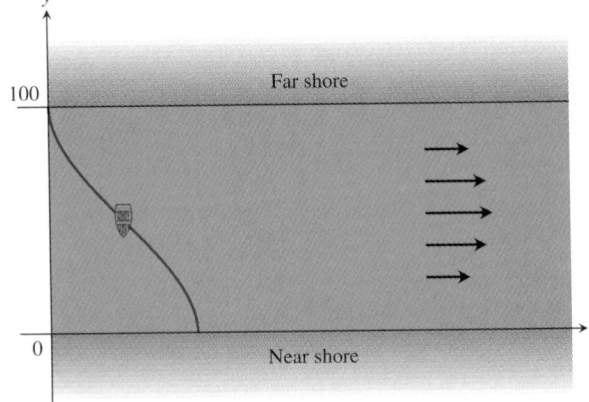

26. *A Radio Controlled Toy Boat* A straight river is 20 m wide (see figure). The velocity of the river at (x, y) is

$$\mathbf{v} = -\frac{3x(20 - x)}{100}\mathbf{j} \text{ m/min}, \qquad 0 \le x \le 20.$$

A radio controlled toy boat leaves the shore at $(0, 0)$ and travels through the water with a constant velocity. It arrives at the opposite shore at $(20, 0)$. The speed of the boat is always $\sqrt{20}$ m/min.

(a) Find the velocity of the boat.

(b) Find the location of the boat at time t.

(c) Sketch the path of the boat.

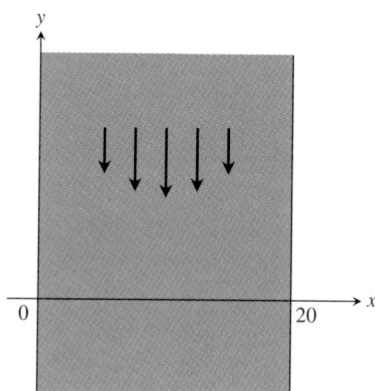

27. The line through the origin and the point $A(1, 1, 1)$ is the axis of a rigid body rotating with a constant angular speed of 32 rad/sec (see figure). The rotation appears to be clockwise when we look toward the origin from A. Find the velocity $\mathbf{v}$ of the point of the body that is at the position $B(1, 3, 2)$.

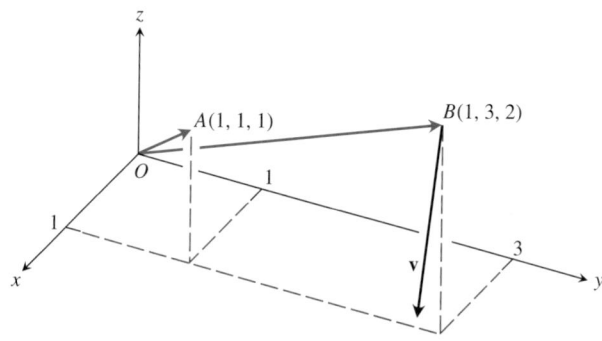

28. *Reconstructing a Quadric Surface* The curve

$$\mathbf{r}(t) = (2\sqrt{t} \cos t)\mathbf{i} + (3\sqrt{t} \sin t)\mathbf{j} + \sqrt{1 - t}\,\mathbf{k},$$
$0 \le t \le 1$, lies on a quadric surface. Describe the surface and find an equation for it.

29. *Sliding Down a Helix* A frictionless particle P, starting from rest at time $t = 0$ at the point $(a, 0, 0)$, slides down the helix

$$\mathbf{r}(\theta) = (a \cos \theta)\mathbf{i} + (a \sin \theta)\mathbf{j} + b\theta\mathbf{k}, \qquad a, b > 0$$

under the influence of gravity, as in the figure. The θ in this equation is the cylindrical coordinate θ and the helix is the curve $r = a$, $z = b\theta$, $\theta \ge 0$, in cylindrical coordinates. We assume θ to be a differentiable function of t for the motion. The law of conservation of energy tells us that the particle's speed after it has fallen straight down a distance z is $\sqrt{2gz}$, where g is the constant acceleration of gravity.

(a) Find the angular velocity $d\theta/dt$ when $\theta = 2\pi$.

(b) Express the particle's θ- and z-coordinates as function of t.

(c) Express the tangential and normal components of the velocity $d\mathbf{r}/dt$ and acceleration $d^2\mathbf{r}/dt^2$ as functions of t. Does the acceleration have any nonzero component in the direction of the binormal vector $\mathbf{B}$?

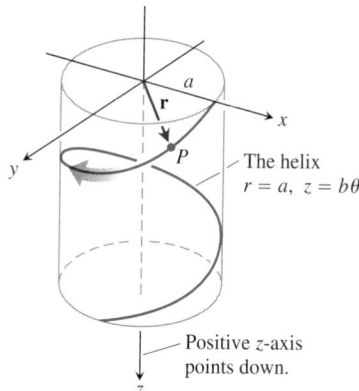

30. *Conserving Angular Momentum* Let $\mathbf{r}(t)$ denote the position in space of a moving object at time t. Suppose the force acting on the object at time t is

$$\mathbf{F}(t) = -\frac{c}{|\mathbf{r}(t)|^3}\mathbf{r}(t),$$

where c is a constant. In physics, the **angular momentum** of an object at time t is defined to be $\mathbf{L}(t) = \mathbf{r}(t) \times m\mathbf{v}(t)$, where m is the mass of the object and $\mathbf{v}(t)$ is the velocity. Prove that angular momentum is a conserved quantity; that is, prove that $\mathbf{L}(t)$ is a constant vector, independent of time. Remember Newton's law, $\mathbf{F} = m\mathbf{a}$.

In Exercises 31 and 32, find the limit, if it exists, or show that it does not exist.

31. $\displaystyle\lim_{\substack{(x,y)\to(1,2) \\ y\ne 2x}} \frac{2x - y}{4x^2 - y^2}$

32. $\displaystyle\lim_{(x,y)\to(0,0)} \frac{y^2}{x^2 + y^2}$

33. Writing to Learn Let

$$f(x, y) = \frac{2 \sin (x^2 + y^2)}{x^2 + y^2}$$

for $(x, y) \ne (0, 0)$. Is it possible to define $f(0, 0)$ in a way that makes f continuous at the origin? Explain.

34. Let $f(x, y, z) = xe^{-(x^2+y^2+z^2)}$. Find f_x, f_y, f_z, and f_{xy}.

In Exercises 35 and 36, find the linearization of the function at the given point.

35. $f(x, y) = e^x \sin y$ at $(0, \pi/6)$

36. $f(x, y, z) = \sin^{-1}(xyz)$ at $(1, \sqrt{2}, 1/2)$

37. Find the linearization $L(x, y)$ of the function
$f(x, y) = xy^2 + y \cos(x - 1)$ at $P(1, 2)$. Then find an upper bound for the magnitude $|E|$ of the error in the approximation $f(x, y) \approx L(x, y)$ over the rectangle $|x - 1| \leq 0.1, |y - 2| \leq 0.1$.

38. Find dw/dt at $t = 1$ if $w = xe^y + y \sin z - \cos z$, $x = 2\sqrt{t}, y = t - 1 + \ln t, z = \pi t$.

39. Find $\partial w/\partial r$ and $\partial w/\partial s$ when $r = \pi$ and $s = 0$ if $w = \sin(2x - y), x = r + \sin s, y = rs$.

40. Find a function $w = f(x, y)$ whose first partial derivatives are $\partial w/\partial x = 1 + e^x \cos y$ and $\partial w/\partial y = 2y - e^x \sin y$ and whose value at the point $(\ln 2, 0)$ is $\ln 2$.

41. Suppose f and g are functions of x and y such that

$$\frac{\partial f}{\partial y} = \frac{\partial g}{\partial x} \quad \text{and} \quad \frac{\partial f}{\partial x} = \frac{\partial g}{\partial y},$$

and suppose

$$\frac{\partial f}{\partial x} = 0,$$

$f(1, 2) = g(1, 2) = 5$, and $f(0, 0) = 4$. Find $f(x, y)$ and $g(x, y)$.

42. Find the derivative of $f(x, y, z) = xyz$ in the direction of the velocity vector of the helix
$\mathbf{r}(t) = (\cos 2t)\mathbf{i} + (\sin 2t)\mathbf{j} + 2t\mathbf{k}$ at $t = \pi/2$.

43. *Maximizing Temperature Change* Let S be the surface that is the graph of $f(x, y) = 10 - x^2 - y^2$. Suppose the temperature in space at each point (x, y, z) is $T(x, y, z) = x^2y + y^2z + 4x + 14y + z$.

(a) Among all the possible directions tangential to the surface S at the point $(0, 0, 10)$, which direction will make the rate of change of temperature at $(0, 0, 10)$ a maximum?

(b) Which direction tangential to S at the point $(1, 1, 8)$ will make the rate of change of temperature there a maximum?

44. Find an equation for the plane tangent to the surface $z = \sin(x + y)$ at the point $(1, -1, 0)$.

45. Find the local maxima, local minima, and saddle points of $f(x, y) = 2x^3 + 2y^3 - 9x^2 + 3y^2 - 12y$. Find the function's values at these points.

46. Find the absolute maximum and minimum values of $f(x, y) = x^2 - xy + y^2 + 1$ on the closed triangular plate in the first quadrant bounded by the lines $x = 0, y = 4$, $y = x$.

47. Find the extreme values of $f(x, y) = x^2 + y^3$ on the circle $x^2 + y^2 = 1$.

48. Find the extreme values of $f(x, y, z) = y(x + z)$ on the curve of intersection of the right circular cylinder $x^2 + y^2 = 1$ and the hyperbolic cylinder $yz = 1$.

49. Find the maximum value of $f(x, y) = 6xye^{-(2x+3y)}$ in the closed first quadrant (includes the nonnegative axes).

50. *Smallest Tetrahedron* Find the minimum volume for a tetrahedron bounded by the planes $x = 0, y = 0, z = 0$ and a plane tangent to the ellipsoid

$$\frac{x^2}{a^2} + \frac{y^2}{b^2} + \frac{z^2}{c^2} = 1$$

at a point in the first octant.

51. *Reassessing the Order of Integration* Sketch the region of integration of the integral

$$\int_0^1 \int_{x^2}^x f(x, y) \, dy \, dx$$

and write an equivalent iterated integral with the order of integration reversed.

In Exercises 52–55, evaluate the integral.

52. $\int_2^5 \int_0^{(3/2)\sqrt{x^2-4}} y \, dy \, dx$

53. $\int_0^2 \int_0^{x^2} \int_0^{x+y} (x + 2y - z) \, dz \, dy \, dx$

54. $\int_0^{2\pi} \int_0^3 \int_{r^2/3}^{\sqrt{18-r^2}} dz \, r \, dr \, d\theta$

55. $\int_0^{2\pi} \int_0^{\pi} \int_0^{(1-\cos\phi)/2} \rho^2 \sin\phi \, d\rho \, d\phi \, d\theta$

56. Find the area of the region bounded by the curves $y^2 = x$ and $x + y = 2$.

57. *Volume of a Sand Pile* The base of a sand pile covers the region in the xy-plane that is bounded by the parabola $x^2 + y = 6$ and the line $y = x$. The height of the sand above the point (x, y) is x^2. Express the volume of the sand as (a) a double integral and (b) a triple integral. Then (c) find the volume.

58. Find the volume of the region bounded above by the paraboloid $z = 3 - x^2 - y^2$ and below by the paraboloid $z = 2x^2 + 2y^2$.

59. Find the volume of the region bounded by the spherical coordinate surface $\rho = 2 \sin \phi$ (see figure).

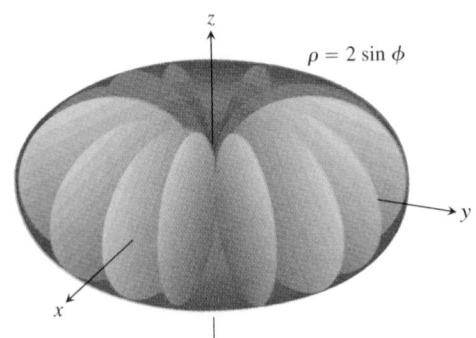

$\rho = 2 \sin \phi$

60. *Coring a Sphere* A circular cylindrical hole is bored through a solid sphere, the axis of the hole being a diameter of the sphere. The volume of the remaining solid is

$$V = 2 \int_0^{2\pi} \int_0^{\sqrt{3}} \int_1^{\sqrt{4-z^2}} r \, dr \, dz \, d\theta.$$

(a) Find the radius of the hole and the radius of the sphere.

(b) Evaluate the integral.

61. *Minimizing a Polar Moment* A thin plate of constant density is to occupy the triangular region in the first quadrant of the xy-plane having vertices $(0, 0)$, $(a, 0)$, and $(a, 1/a)$. What value of a will minimize the plate's polar moment of inertia about the origin?

62. *Polar Moment of a Triangular Plate* Find the polar moment of inertia about the origin of a thin triangular plate of constant density $\delta = 3$ bounded by the y-axis and the lines $y = 2x$ and $y = 4$ in the xy-plane.

63. *Centroid of a Polar Region* Find the centroid of the region in the polar coordinate plane that lies inside the cardioid $r = 1 + \cos\theta$ and outside the circle $r = 1$.

64. *Radii of Gyration of a Polar Region* Find the radii of gyration about the x- and y-axes of a thin plate of density $\delta = 1$ enclosed by one loop of the lemniscate $r^2 = 2a^2 \cos 2\theta$.

65. *Physical Properties of a Solid Cone* A solid is bounded below by the cone $z = \sqrt{x^2 + y^2}$ and above by the plane $z = 1$. Find the center of mass and the moment of inertia and radius of gyration about the z-axis if the density is **(a)** $\delta(r, \theta, z) = z$ and **(b)** $\delta(r, \theta, z) = z^2$.

66. *Averaging a Directional Derivative* Let $D_\mathbf{u}f$ denote the derivative of $f(x, y) = (x^2 + y^2)/2$ in the direction of the unit vector $\mathbf{u} = u_1\mathbf{i} + u_2\mathbf{j}$.

(a) Find the average value of $D_\mathbf{u}f$ over the triangular region cut from the first quadrant by the line $x + y = 1$.

(b) Show in general that the average value of $D_\mathbf{u}f$ over a region in the xy-plane is the value of $D_\mathbf{u}f$ at the centroid of the region.

67. *Finding Electrical Charge* The electrical charge distribution on a circular plate of radius R meters is $\sigma(r, \theta) = kr(1 - \sin\theta)$ coulomb/m^2 (k a constant). Integrate σ over the plate to find the total charge Q.

68. *A Parabolic Rain Gauge* A bowl is in the shape of the graph of $z = x^2 + y^2$ from $z = 0$ to $z = 10$ in. You plan to calibrate the bowl to make it into a rain gauge. What height in the bowl would correspond to 1 in. of rain? 3 in. of rain?

69. *An Infinite Half-cylinder* Let D be the interior of the infinite right circular half-cylinder of radius 1 with its single-end face suspended 1 unit above the origin and its axis the ray from $(0, 0, 1)$ to ∞ on the z-axis. Use cylindrical coordinates to evaluate $\iiint_D z(r^2 + z^2)^{-5/2} \, dV.$

70. Integrate $f(x, y, z) = x^2 - 3y + z^2$ over the line segment joining the origin and the point $(1, 1, 2)$.

71. Integrate $f(x, y, z) = x + y + z$ over the curve $\mathbf{r}(t) = (2 \cos t)\mathbf{i} + (2 \sin t)\mathbf{j} + 3t\mathbf{k}, 0 \le t \le 2\pi.$

72. Evaluate $\int_C y \, dy - (y - x) \, dx$, where C is the circle $x^2 + y^2 = 1.$

73. *Circulation* Find the circulation and flux of $\mathbf{F} = x^2\mathbf{i} + y^2\mathbf{j}$ around and across the curve $\mathbf{r}(t) = (3 \cos t)\mathbf{i} + (4 \sin t)\mathbf{j}, 0 \le t \le \pi.$

74. *Work* Find the work done by the force $\mathbf{F} = xy\mathbf{i} + (y - x)\mathbf{j}$ counterclockwise around the circle $x^2 + y^2 = 4$ from $(2, 0)$ to itself.

75. *Finding a Potential Function* Show that the following field is conservative and find a potential function for it.

$$\mathbf{F} = (\cos(x + z) + 2xyz)\mathbf{i} + \left(\frac{2y}{y^2 + z^2} + x^2z\right)\mathbf{j}$$

$$+ \left(\frac{2z}{y^2 + z^2} + x^2y\right)\mathbf{k}$$

76. *Integrating an Exact Differential Form* Show that the differential form in the integral is exact. Then evaluate the integral.

$$\int_{(1,1,1)}^{(2,3,4)} (yz + 2x) \, dx + (xz + 2y) \, dy + (xy + 2z) \, dz.$$

77. *Circulation and Flux from Green's Theorem* Use Green's Theorem to find the counterclockwise circulation and outward flux for the field $\mathbf{F} = (y^2 + 2x)\mathbf{i} + (x^2 + 2y)\mathbf{j}$ around and across the triangle determined by $y = 0, x = 2, y = 2x.$

78. *Evaluating a Line Integral with Green's Theorem* Use Green's Theorem to evaluate the integral

$$\oint_C (yx^2 \, dx + xy^2 \, dy), \text{ where } C \text{ is the boundary of}$$

$-\pi/2 \le x \le \pi/2, 0 \le y \le \cos x.$

79. *Area of a Spherical Cap* Find the area of the cap cut from the sphere $x^2 + y^2 + z^2 = 4$ by the cone $z = \sqrt{x^2 + y^2}.$

80. *Integrating over a Surface* Integrate $f(x, y, z) = xyz$ over the surface of the rectangular solid cut from the first octant by the planes $x = 2, y = 3$, and $z = 5.$

81. **(a)** *Area of a Parametrized Surface* Find the area of the surface $\mathbf{r}(\phi, \theta) = (3 \sin\phi \cos\theta)\mathbf{i} + (3 \sin\phi \sin\theta)\mathbf{j} + (3 \cos\phi)\mathbf{k}, \pi/3 \le \phi \le 2\pi/3, 0 \le \theta \, 2\pi.$

(b) *Integrating over the Surface* Integrate $f(x, y, z) = x^2 + y^2 - z^2$ over the surface in part (a).

82. (a) *Verifying a Parametrization* Using the angles θ and ϕ as defined in spherical coordinates, show that

$$\mathbf{r}(\theta, \phi) = (a \cos \theta \cos \phi)\mathbf{i} + (b \sin \theta \cos \phi)\mathbf{j} + (c \sin \phi)\mathbf{k}$$

is a parametrization of the ellipsoid

$$\frac{x^2}{z^2} + \frac{y^2}{b^2} + \frac{z^2}{c^2} = 1.$$

(b) *Setting up a Surface Area Integral* Write an integral for the surface area of the ellipsoid, but do not evaluate the integral.

83. Find all points (a, b, c) on the sphere $x^2 + y^2 + z^2 = R^2$, where the vector field $\mathbf{F} = yz^2\mathbf{i} + xz^2\mathbf{j} + 2xyz\mathbf{k}$ is normal to the surface and $\mathbf{F}(a, b, c) \neq \mathbf{0}$.

84. *Mass of a Shell with Variable Density* Find the mass of a spherical shell of radius R such that at each point (x, y, z) on the surface the mass density $\delta(x, y, z)$ is its distance to some fixed point (a, b, c) on the surface.

85. *Mass of a Helicoid with Variable Density* Find the mass of a helicoid $\mathbf{r}(r, \theta) = (r \cos \theta)\mathbf{i} + (r \sin \theta)\mathbf{j} + \theta\mathbf{k}$, $0 \leq r \leq 1, 0 \leq \theta \leq 2\pi$, if the density function is

$$\delta(x, y, z) = 2\sqrt{x^2 + y^2}.$$

86. *Rectangle of Least Outward Flux* Among all rectangular regions $0 \leq x \leq a, 0 \leq y \leq b$, find the one for which the total outward flux of $\mathbf{F} = (x^2 + 4xy)\mathbf{i} - 6y\mathbf{j}$ across the four sides is least. What is the least flux?

87. *Circle of Greatest Circulation* Find an equation for the plane through the origin such that the circulation of the flow field $\mathbf{F} = z\mathbf{i} + x\mathbf{j} + y\mathbf{k}$ around the circle of intersection of the plane with the sphere $x^2 + y^2 + z^2 = 4$ is a maximum.

A P P E N D I C E S

Formulas from Precalculus Mathematics

Algebra • Geometry • Trigonometry

Algebra

1. Laws of Exponents

$$a^m a^n = a^{m+n}, \quad (ab)^m = a^m b^m, \quad (a^m)^n = a^{mn}, \quad a^{m/n} = \sqrt[n]{a^m}$$

If $a \neq 0$, $\qquad \dfrac{a^m}{a^n} = a^{m-n}, \quad a^0 = 1, \quad a^{-m} = \dfrac{1}{a^m}$

2. Zero Division by zero is not defined.

If $a \neq 0$: $\qquad \dfrac{0}{a} = 0, \quad a^0 = 1, \quad 0^a = 0$

For any number a: $\qquad a \cdot 0 = 0 \cdot a = 0$

3. Fractions

$$\frac{a}{b} + \frac{c}{d} = \frac{ad + bc}{bd}, \quad \frac{a}{b} \cdot \frac{c}{d} = \frac{ac}{bd}, \quad \frac{a/b}{c/d} = \frac{a}{b} \cdot \frac{d}{c}, \quad \frac{-a}{b} = -\frac{a}{b} = \frac{a}{-b},$$

$$\frac{(a/b) + (c/d)}{(e/f) + (g/h)} = \frac{(a/b) + (c/d)}{(e/f) + (g/h)} \cdot \frac{bdfh}{bdfh} = \frac{(ad + bc)fh}{(eh + fg)bd}$$

4. The Binomial Theorem
For any positive integer n,

$$(a + b)^n = a^n + na^{n-1}b + \frac{n(n-1)}{1 \cdot 2} a^{n-2} b^2$$

$$+ \frac{n(n-1)(n-2)}{1 \cdot 2 \cdot 3} a^{n-3} b^3 + \cdots + nab^{n-1} + b^n.$$

For instance, $\qquad (a + b)^1 = a + b,$

$$(a + b)^2 = a^2 + 2ab + b^2,$$

$$(a + b)^3 = a^3 + 3a^2 b + 3ab^2 + b^3,$$

$$(a + b)^4 = a^4 + 4a^3 b + 6a^2 b^2 + 4ab^3 + b^4.$$

5. Differences of Like Integer Powers, $n > 1$

$$a^n - b^n = (a - b)(a^{n-1} + a^{n-2}b + a^{n-3}b^2 + \cdots + ab^{n-2} + b^{n-1})$$

For instance, $\qquad a^2 - b^2 = (a - b)(a + b),$

$$a^3 - b^3 = (a - b)(a^2 + ab + b^2),$$

$$a^4 - b^4 = (a - b)(a^3 + a^2 b + ab^2 + b^3).$$

6. Completing the Square

If $a \neq 0$, we can rewrite the quadratic $ax^2 + bx + c$ in the form $au^2 + C$ by a process called completing the square:

$$ax^2 + bx + c = a\left(x^2 + \frac{b}{a}x\right) + c \qquad \text{Factor } a \text{ from the first two terms.}$$

$$= a\left(x^2 + \frac{b}{a}x + \frac{b^2}{4a^2} - \frac{b^2}{4a^2}\right) + c \qquad \text{Add and subtract the square of half the coefficient of } x.$$

$$= a\left(x^2 + \frac{b}{a}x + \frac{b^2}{4a^2}\right) + a\left(-\frac{b^2}{4a^2}\right) + c \qquad \text{Bring out the } -b^2/(4a^2).$$

$$= a\underbrace{\left(x^2 + \frac{b}{a}x + \frac{b^2}{4a^2}\right)}_{\text{This is } \left(x + \frac{b}{2a}\right)^2.} + \underbrace{c - \frac{b^2}{4a}}_{\substack{\text{Call this} \\ \text{part } C.}}$$

$$= au^2 + C \qquad\qquad u = x + \frac{b}{2a}$$

7. The Quadratic Formula

By completing the square on the first two terms of the equation

$$ax^2 + bx + c = 0$$

and solving the resulting equation for x (details omitted), we obtain

$$x = \frac{-b \pm \sqrt{b^2 - 4ac}}{2a}.$$

This equation is the **quadratic formula.**

The solutions of the equation $2x^2 + 3x - 1 = 0$ are

$$x = \frac{-3 \pm \sqrt{(3)^2 - 4(2)(-1)}}{2(2)} = \frac{-3 \pm \sqrt{9 + 8}}{4}$$

or

$$x = \frac{-3 + \sqrt{17}}{4} \quad \text{and} \quad x = \frac{-3 - \sqrt{17}}{4}.$$

Geometry

(A = area, B = area of base, C = circumference, h = height, S = lateral area or surface area, V = volume)

1. Triangle

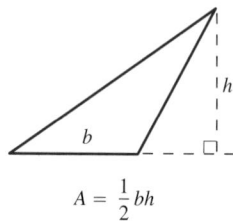

$$A = \frac{1}{2}bh$$

2. Similar Triangles

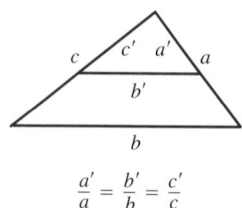

$$\frac{a'}{a} = \frac{b'}{b} = \frac{c'}{c}$$

3. Pythagorean Theorem

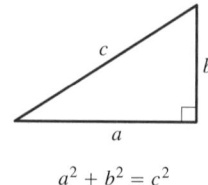

$$a^2 + b^2 = c^2$$

4. Parallelogram

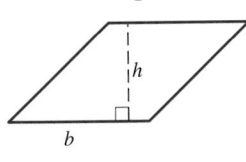

$$A = bh$$

5. Trapezoid

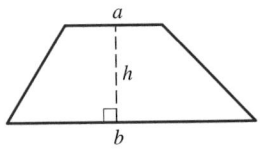

$$A = \frac{1}{2}(a + b)h$$

6. Circle

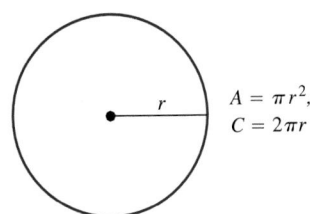

$$A = \pi r^2,$$
$$C = 2\pi r$$

7. Any Cylinder or Prism with Parallel Bases

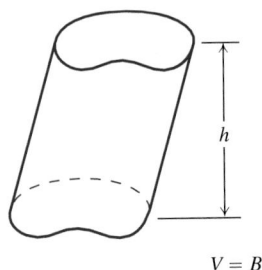

$$V = Bh$$

8. Right Circular Cylinder

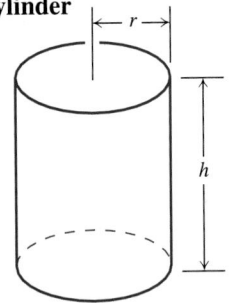

$$V = \pi r^2 h, \quad S = 2\pi rh$$

9. Any Cone or Pyramid

$$V = \frac{1}{3}Bh$$

10. Right Circular Cone

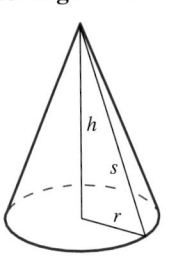

$$V = \frac{1}{3}\pi r^2 h, \quad S = \pi rs$$

11. Sphere

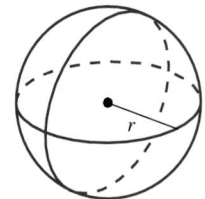

$$V = \frac{4}{3}\pi r^3, \quad S = 4\pi r^2$$

Trigonometry

1. Definitions of Fundamental Identities

Sine: $\quad \sin \theta = \dfrac{y}{r} = \dfrac{1}{\csc \theta}$

Cosine: $\quad \cos \theta = \dfrac{x}{r} = \dfrac{1}{\sec \theta}$

Tangent: $\quad \tan \theta = \dfrac{y}{x} = \dfrac{1}{\cot \theta}$

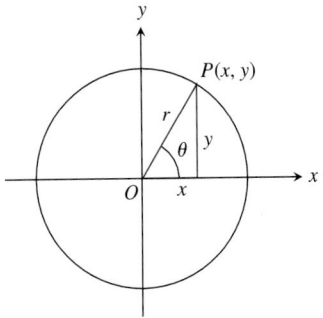

2. Identities

$$\sin(-\theta) = -\sin\theta, \quad \cos(-\theta) = \cos\theta$$

$$\sin^2\theta + \cos^2\theta = 1, \quad \sec^2\theta = 1 + \tan^2\theta, \quad \csc^2\theta = 1 + \cot^2\theta$$

$$\sin 2\theta = 2\sin\theta\cos\theta, \quad \cos 2\theta = \cos^2\theta - \sin^2\theta$$

$$\cos^2\theta = \frac{1 + \cos 2\theta}{2}, \quad \sin^2\theta = \frac{1 - \cos 2\theta}{2}$$

$$\sin(A + B) = \sin A \cos B + \cos A \sin B$$

$$\sin(A - B) = \sin A \cos B - \cos A \sin B$$

$$\cos(A + B) = \cos A \cos B - \sin A \sin B$$

$$\cos(A - B) = \cos A \cos B + \sin A \sin B$$

$$\tan(A + B) = \frac{\tan A + \tan B}{1 - \tan A \tan B}$$

$$\tan(A - B) = \frac{\tan A - \tan B}{1 + \tan A \tan B}$$

$$\sin\left(A - \frac{\pi}{2}\right) = -\cos A, \quad \cos\left(A - \frac{\pi}{2}\right) = \sin A$$

$$\sin\left(A + \frac{\pi}{2}\right) = \cos A, \quad \cos\left(A + \frac{\pi}{2}\right) = -\sin A$$

$$\sin A \sin B = \frac{1}{2}\cos(A - B) - \frac{1}{2}\cos(A + B)$$

$$\cos A \cos B = \frac{1}{2}\cos(A - B) + \frac{1}{2}\cos(A + B)$$

$$\sin A \cos B = \frac{1}{2}\sin(A - B) + \frac{1}{2}\sin(A + B)$$

$$\sin A + \sin B = 2\sin\frac{1}{2}(A + B)\cos\frac{1}{2}(A - B)$$

$$\sin A - \sin B = 2\cos\frac{1}{2}(A + B)\sin\frac{1}{2}(A - B)$$

$$\cos A + \cos B = 2\cos\frac{1}{2}(A + B)\cos\frac{1}{2}(A - B)$$

$$\cos A - \cos B = -2\sin\frac{1}{2}(A + B)\sin\frac{1}{2}(A - B)$$

3. Common Reference Triangles

4. Angles and Sides of a Triangle

Law of cosines: $c^2 = a^2 + b^2 - 2ab \cos C$

Law of sines: $\dfrac{\sin A}{a} = \dfrac{\sin B}{b} = \dfrac{\sin C}{c}$

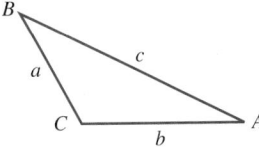

$$\text{Area} = \frac{1}{2}bc \sin A = \frac{1}{2}ac \sin B = \frac{1}{2}ab \sin C$$

A2   Mathematical Induction

Mathematical Induction Principle • Other Starting Integers

Mathematical Induction Principle

Many formulas, like

$$1 + 2 + \cdots + n = \frac{n(n + 1)}{2},$$

can be shown to hold for every positive integer n by applying an axiom called the *mathematical induction principle*. A proof that uses this axiom is a *proof by mathematical induction* or a *proof by induction*.

The steps in proving a formula by induction are the following.

Step 1: Check that the formula holds for $n = 1$.

Step 2: Prove that if the formula holds for any positive integer $n = k$, then it also holds for the next integer, $n = (k + 1)$.

Once these steps are completed (the axiom says), we know that the formula holds for all positive integers n. By step 1 it holds for $n = 1$. By step 2 it holds for $n = 2$, and therefore by step 2 also for $n = 3$, and by step 2 again for $n = 4$, and so on. If the first domino falls, and the kth domino always knocks over the $(k + 1)$st when it falls, all the dominoes fall.

From another point of view, suppose we have a sequence of statements S_1, $S_2, \ldots, S_n, \ldots$, one for each positive integer. Suppose we can show that assuming any one of the statements to be true implies that the next statement in line is true. Suppose that we can also show that S_1 is true. Then we may conclude that the statements are true from S_1 on.

Example 1 SUM OF THE FIRST n POSITIVE INTEGERS

Show that for every positive integer n,

$$1 + 2 + \cdots + n = \frac{n(n+1)}{2}.$$

Solution We accomplish the proof by carrying out the two steps.

Step 1: The formula holds for $n = 1$ because

$$1 = \frac{1(1+1)}{2}.$$

Step 2: If the formula holds for $n = k$, does it hold for $n = (k+1)$? The answer is yes, and here's why: If

$$1 + 2 + \cdots + k = \frac{k(k+1)}{2},$$

then

$$1 + 2 + \cdots + k + (k+1) = \frac{k(k+1)}{2} + (k+1)$$

$$= \frac{k^2 + k + 2k + 2}{2}$$

$$= \frac{(k+1)(k+2)}{2} = \frac{(k+1)((k+1)+1)}{2}.$$

The last expression in this string of equalities is the expression $n(n+1)/2$ for $n = (k+1)$.

The mathematical induction priniciple now guarantees the original formula for all positive integers n. All *we* have to do is carry out steps 1 and 2. The mathematical induction principle does the rest.

Example 2 SUMS OF POWERS OF 1/2

Show that for all positive integers n,

$$\frac{1}{2^1} + \frac{1}{2^2} + \cdots + \frac{1}{2^n} = 1 - \frac{1}{2^n}.$$

Solution We accomplish the proof by carrying out the two steps of mathematical induction.

Step 1: The formula holds for $n = 1$ because

$$\frac{1}{2^1} = 1 - \frac{1}{2^1}.$$

Step 2: If

$$\frac{1}{2^1} + \frac{1}{2^2} + \cdots + \frac{1}{2^k} = 1 - \frac{1}{2^k},$$

then

$$\frac{1}{2^1} + \frac{1}{2^2} + \cdots + \frac{1}{2^k} + \frac{1}{2^{k+1}} = 1 - \frac{1}{2^k} + \frac{1}{2^{k+1}}$$

$$= 1 - \frac{1 \cdot 2}{2^k \cdot 2} + \frac{1}{2^{k+1}}$$

$$= 1 - \frac{2}{2^{k+1}} + \frac{1}{2^{k+1}}$$

$$= 1 - \frac{1}{2^{k+1}}.$$

Thus, the original formula holds for $n = (k + 1)$ whenever it holds for $n = k$.

With these steps verified, the mathematical induction principle now guarantees the formula for every positive integer n.

Other Starting Integers

Instead of starting at $n = 1$, some induction arguments start at another integer. The steps for such an argument are as follows.

Step 1: Check that the formula holds for $n = n_1$ (the first appropriate integer).

Step 2: Prove that if the formula holds for any integer $n = k \geq n_1$, then it also holds for $n = (k + 1)$.

Once these steps are completed, the mathematical induction principle guarantees the formula for all $n \geq n_1$.

Example 3 FACTORIAL EXCEEDING EXPONENTIAL

Show that $n! > 3^n$ if n is large enough.

Solution How large is large enough? We experiment:

n	1	2	3	4	5	6	7
$n!$	1	2	6	24	120	720	5040
3^n	3	9	27	81	243	729	2187

It looks as if $n! > 3^n$ for $n \geq 7$. To be sure, we apply mathematical induction. We take $n_1 = 7$ in step 1 and try for step 2.

Suppose $k! > 3^k$ for some $k \geq 7$. Then

$$(k + 1)! = (k + 1)(k!) > (k + 1)3^k > 7 \cdot 3^k > 3^{k+1}.$$

Thus, for $k \geq 7$,

$$k! > 3^k \quad \Rightarrow \quad (k + 1)! > 3^{k+1}.$$

The mathematical induction principle now guarantees $n! > 3^n$ for all $n \geq 7$.

Section A2 Exercises

1. *General Triangle Inequality* Assuming that the triangle inequality $|a + b| \le |a| + |b|$ holds for any two numbers a and b, show that

$$|x_1 + x_2 + \cdots + x_n| \le |x_1| + |x_2| + \cdots + |x_n|$$

for any n numbers.

2. *Partial Sums of Geometric Series* Show that if $r \ne 1$, then

$$1 + r + r^2 + \cdots + r^n = \frac{1 - r^{n+1}}{1 - r}$$

for every positive integer n.

3. *Positive Integer Power Rule* Use the Product Rule,

$$\frac{d}{dx}(uv) = u\frac{dv}{dx} + v\frac{du}{dx},$$

and the fact that

$$\frac{d}{dx}(x) = 1$$

to show that

$$\frac{d}{dx}(x^n) = nx^{n-1}$$

for every positive integer n.

4. *Products into Sums* Suppose that a function $f(x)$ has the property that $f(x_1 x_2) = f(x_1) + f(x_2)$ for any two positive numbers x_1 and x_2. Show that

$$f(x_1 x_2 \ldots x_n) = f(x_1) + f(x_2) + \cdots + f(x_n)$$

for the product of any n positive numbers $x_1, x_2, \ldots, x_n$.

5. Show that

$$\frac{2}{3^1} + \frac{2}{3^2} + \cdots + \frac{2}{3^n} = 1 - \frac{1}{3^n}$$

for all positive integers n.

6. Show that $n! > n^3$ if n is large enough.

7. Show that $2^n > n^2$ if n is large enough.

8. Show that $2^n \ge 1/8$ for $n \ge -3$.

9. *Sums of Squares* Show that the sum of the squares of the first n positive integers is

$$\frac{n\left(n + \frac{1}{2}\right)(n + 1)}{3}.$$

10. *Sums of Cubes* Show that the sum of the cubes of the first n positive integers is $(n(n + 1)/2)^2$.

11. *Rules for Finite Sums* Show that the following finite sum rules hold for every positive integer n.

(a) $\displaystyle\sum_{k=1}^{n} (a_k + b_k) = \sum_{k=1}^{n} a_k + \sum_{k=1}^{n} b_k$

(b) $\displaystyle\sum_{k=1}^{n} (a_k - b_k) = \sum_{k=1}^{n} a_k - \sum_{k=1}^{n} b_k$

(c) $\displaystyle\sum_{k=1}^{n} ca_k = c \cdot \sum_{k=1}^{n} a_k$ (Any number c)

(d) $\displaystyle\sum_{k=1}^{n} a_k = n \cdot c$, if a_k has the constant value c.

12. *Absolute Values* Show that $|x^n| = |x|^n$ for every positive integer n and every real number x.

A3 Using the Limit Definition

Limit Definition • Finding Deltas for Given Epsilons • Proving Limit Theorems

Limit Definition

We begin by setting the stage for the definition of limit. Recall that the limit of f of x as x approaches c equals L ($\lim_{x \to c} f(x) = L$) means that the values $f(x)$ of the function f approach or equal L as the values of x approach (but do not equal) c. Suppose we are watching the values of a function $f(x)$ as x approaches c (without taking on the value of c itself). Certainly we want to be able to say that $f(x)$ stays within one-tenth of a unit of L as soon as x stays within some distance δ of c (Figure A3.1). But that in itself is not enough, because as x continues on its course toward c, what is to prevent $f(x)$ from jittering about within the interval from $L - 1/10$ to $L + 1/10$ without tending toward L?

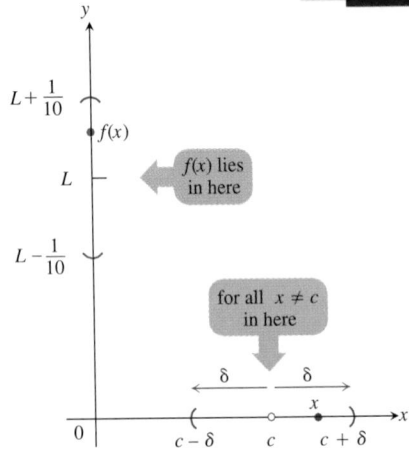

Figure A3.1 A preliminary stage in the development of the definition of limit.

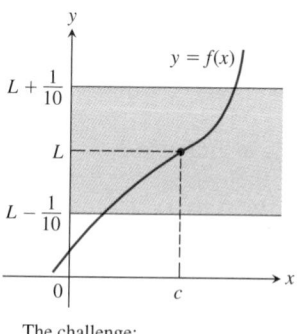

The challenge:
Make $|f(x) - L| < \varepsilon = \frac{1}{10}$

Figure A3.2 The first of a possibly endless sequence of challenges.

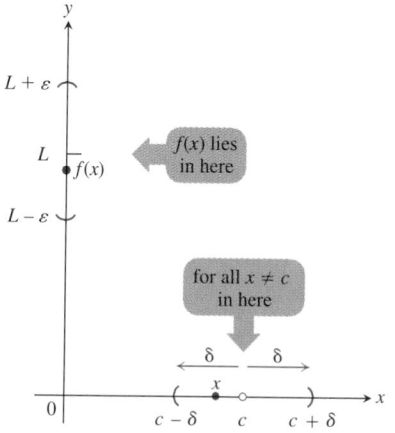

Figure A3.3 The relation of the δ and ε in the definition of limit.

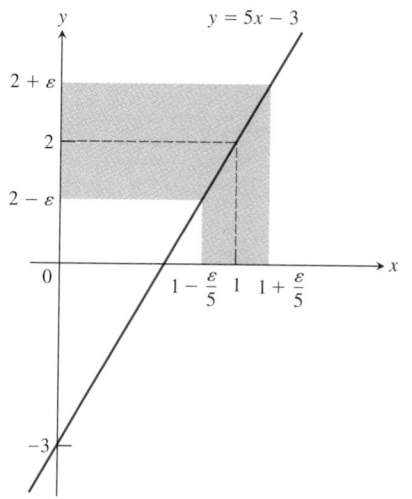

Figure A3.4 If $f(x) = 5x - 3$, then $0 < |x - 1| < \varepsilon/5$ guarantees that $|f(x) - 2| < \varepsilon$. (Example 1)

We can insist that $f(x)$ stay within 1/100 or 1/1000 or 1/100,000 of L. Each time, we find a new δ-interval about c so that keeping x within that interval keeps $f(x)$ within $\varepsilon = 1/100$ or 1/1000 or 1/100,000 of L. And each time the possibility exists that c jitters away from L at the last minute.

Figure A3.2 illustrates the problem. You can think of this as a quarrel between a skeptic and a scholar. The skeptic presents ε-challenges to prove that the limit does not exist or, more precisely, that there is room for doubt, and the scholar answers every challenge with a δ-interval around c.

How do we stop this seemingly endless sequence of challenges and responses? By proving that for every ε-distance that the challenger can produce, we can find, calculate, or conjure a matching δ-distance that keeps x "close enough" to c to keep $f(x)$ within that distance of L (Figure A3.3).

The following definition that we made in Section 2.1 provides a mathematical way to say that the closer x gets to c, the closer $f(x)$ must get to L.

Definition Limit

Let c and L be real numbers. The function f **has limit L as x approaches c** if, given any positive number ε, there is a positive number δ such that for all x

$$0 < |x - c| < \delta \quad \Rightarrow \quad |f(x) - L| < \varepsilon.$$

We write

$$\lim_{x \to c} f(x) = L.$$

Finding Deltas for Given Epsilons

From our work in Chapter 2 we know that $\lim_{x \to 1}(5x - 3) = 2$. In Example 1, we confirm this result using the definition of limit.

Example 1 USING THE DEFINITION OF LIMIT

Show that $\lim_{x \to 1}(5x - 3) = 2$.

Solution Set $c = 1$, $f(x) = 5x - 3$, and $L = 2$ in the definition of limit. For any given $\varepsilon > 0$ we have to find a suitable $\delta > 0$ so that if $x \neq 1$ and x is within distance δ of $c = 1$, that is, if

$$0 < |x - 1| < \delta,$$

then $f(x)$ is within distance ε of $L = 2$, that is,

$$|f(x) - 2| < \varepsilon.$$

We find δ by working backwards from the ε-inequality:

$$|(5x - 3) - 2| = |5x - 5| < \varepsilon$$

$$5|x - 1| < \varepsilon$$

$$|x - 1| < \varepsilon/5$$

Thus we can take $\delta = \varepsilon/5$ (Figure A3.4). If $0 < |x - 1| < \delta = \varepsilon/5$, then

$$|(5x - 3) - 2| = |5x - 5|$$

$$= 5|x - 1| < 5(\varepsilon/5) = \varepsilon.$$

This proves that $\lim_{x \to 1}(5x - 3) = 2$.

Intersection
X = .49751244   Y = 2.01

[0.48, 0.52] by [1.98, 2.02]
(a)

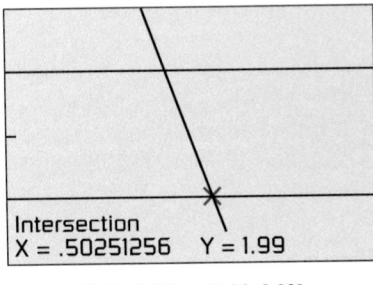

Intersection
X = .50251256   Y = 1.99

[0.48, 0.52] by [1.98, 2.02]
(b)

Figure A3.5 We can see from the two graphs that if $0.498 < x < 0.502$, then $1.99 < f(x) < 2.01$. (Example 2)

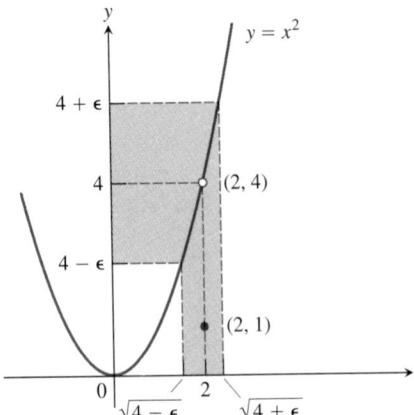

Figure A3.6 The function in Example 3.

The value of $\delta = \varepsilon/5$ is not the only value that will make $0 < |x - 1| < \delta$ imply $|f(x) - 2| = |5x - 5| < \varepsilon$ in Example 1. Any smaller positive δ will do as well. The definition does not ask for a "best" positive δ, just one that will work.

We can use graphs to find a δ for a specific ε as in Example 2.

Example 2   FINDING A δ GRAPHICALLY

For the limit $\lim_{x \to 0.5}(1/x) = 2$, find a δ that works for $\varepsilon = 0.01$. That is, find a $\delta > 0$ such that for all x

$$0 < |x - 0.5| < \delta \quad \Rightarrow \quad |f(x) - 2| < 0.01.$$

Solution Here $f(x) = 1/x$, $c = 0.5$, and $L = 2$. Figure A3.5 shows the graphs of f and the two horizontal lines

$$y = L - \varepsilon = 2 - 0.01 = 1.99 \quad \text{and} \quad y = L + \varepsilon = 2 + 0.01 = 2.01.$$

Figure A3.5a shows that the graph of f intersects the horizontal line $y = 2.01$ at about $(0.49751244, 2.01)$, and Figure A3.5b shows that the graph of f intersects the horizontal line $y = 1.99$ at about $(0.50251256, 1.99)$. It follows that

$$0 < |x - 0.5| < 0.002 \quad \Rightarrow \quad |f(x) - 2| < 0.01.$$

Thus, $\delta = 0.002$ works.

Example 3   FINDING δ ALGEBRAICALLY

Prove that $\lim_{x \to 2} f(x) = 4$ if

$$f(x) = \begin{cases} x^2, & x \neq 2 \\ 1, & x = 2. \end{cases}$$

Solution Our task is to show that given $\varepsilon > 0$ there exists a $\delta > 0$ such that for all x

$$0 < |x - 2| < \delta \quad \Rightarrow \quad |f(x) - 4| < \varepsilon.$$

Step 1: Solve the inequality $|f(x) - 4| < \varepsilon$ to find an open interval about $c = 2$ on which the inequality holds for all $x \neq c$.

For $x \neq c = 2$, we have $f(x) = x^2$, and the inequality to solve is $|x^2 - 4| < \varepsilon$:

$$|x^2 - 4| < \varepsilon$$

$$-\varepsilon < x^2 - 4 < \varepsilon$$

$$4 - \varepsilon < x^2 < 4 + \varepsilon$$

$$\sqrt{4 - \varepsilon} < |x| < \sqrt{4 + \varepsilon} \qquad \text{Assume } \varepsilon < 4.$$

$$\sqrt{4 - \varepsilon} < x < \sqrt{4 + \varepsilon} \qquad \begin{array}{l}\text{An open interval about 2}\\\text{that solves the inequality}\end{array}$$

The inequality $|f(x) - 4| < \varepsilon$ holds for all $x \neq 2$ in the open interval $(\sqrt{4 - \varepsilon}, \sqrt{4 + \varepsilon})$ (Figure A3.6).

Step 2: Find a value of $\delta > 0$ that places the *centered* interval $(2 - \delta, 2 + \delta)$ inside the open interval $(\sqrt{4 - \varepsilon}, \sqrt{4 + \varepsilon})$.

Take δ to be the distance from $c = 2$ to the nearer endpoint of $(\sqrt{4 - \varepsilon}, \sqrt{4 + \varepsilon})$. In other words, take

$$\delta = \min \{2 - \sqrt{4 - \varepsilon}, \sqrt{4 + \varepsilon} - 2\},$$

the *minimum* (the smaller) of the two numbers $2 - \sqrt{4 - \varepsilon}$ and $\sqrt{4 + \varepsilon} - 2$. If δ has this or any smaller positive value, the inequality

$$0 < |x - 2| < \delta$$

will automatically place x between $\sqrt{4 - \varepsilon}$ and $\sqrt{4 + \varepsilon}$ to make

$$|f(x) - 4| < \varepsilon.$$

For all x,

$$0 < |x - 2| < \delta \quad \Rightarrow \quad |f(x) - 4| < \varepsilon.$$

This completes the proof.

Why was it all right to assume $\varepsilon < 4$ in Example 3? Because, in finding a δ such that, for all x, $0 < |x - 2| < \delta$ implied $|f(x) - 4| < \varepsilon < 4$, we found a δ that would work for any larger ε as well.

Finally, notice the freedom we gained in letting

$$\delta = \min \{2 - \sqrt{4 - \varepsilon}, \sqrt{4 + \varepsilon} - 2\}.$$

We did not have to spend time deciding which, if either, number was the smaller of the two. We just let δ represent the smaller and went on to finish the argument.

Proving Limit Theorems

We use the limit definition to prove parts 1, 3, and 5 of Theorem 1 (Properties of Limits) from Section 2.1.

Theorem 1 Properties of Limits

If L, M, c, and k are real numbers and

$$\lim_{x \to c} f(x) = L \quad \text{and} \quad \lim_{x \to c} g(x) = M, \quad \text{then}$$

1. *Sum Rule:* $\qquad\qquad\qquad \lim_{x \to c} (f(x) + g(x)) = L + M$

2. *Difference Rule:* $\qquad\quad\; \lim_{x \to c} (f(x) - g(x)) = L - M$

3. *Product Rule:* $\qquad\qquad\;\; \lim_{x \to c} (f(x) \cdot g(x)) = L \cdot M$

4. *Constant Multiple Rule:* $\quad \lim_{x \to c} k \cdot f(x) = k \cdot L$

5. *Quotient Rule:* $\qquad\qquad\; \lim_{x \to c} \dfrac{f(x)}{g(x)} = \dfrac{L}{M}, \quad M \neq 0$

6. *Power Rule:* $\qquad\qquad\quad$ If r and s are integers, $s \neq 0$, then

$$\lim_{x \to c} (f(x))^{r/s} = L^{r/s}$$

provided $L^{r/s}$ is a real number.

Proof of the Limit Sum Rule We need to show that for any $\varepsilon > 0$, there is a $\delta > 0$ such that for all x in the common domain D of f and g,

$$0 < |x - c| < \delta \quad \Rightarrow \quad |f(x) + g(x) - (L + M)| < \varepsilon.$$

Regrouping terms, we get

$$|f(x) + g(x) - (L + M)| = |(f(x) - L) + (g(x) - M)|$$
$$\leq |f(x) - L| + |g(x) - M|. \quad |a + b| \leq |a| + |b|$$

Here we have applied the triangle inequality, which states that for all real numbers a and b, $|a + b| \leq |a| + |b|$. Since $\lim_{x \to c} f(x) = L$, there exists a number $\delta_1 > 0$ such that for all x in D

$$0 < |x - c| < \delta_1 \quad \Rightarrow \quad |f(x) - L| < \varepsilon/2.$$

Similarly, since $\lim_{x \to c} g(x) = M$, there exists a number $\delta_2 > 0$ such that for all x in D

$$0 < |x - c| < \delta_2 \quad \Rightarrow \quad |g(x) - M| < \varepsilon/2.$$

Let $\delta = \min \{\delta_1, \delta_2\}$, the smaller of δ_1 and δ_2. If $0 < |x - c| < \delta$ then

$$0 < |x - c| < \delta_1, \quad \text{so} \quad |f(x) - L| < \varepsilon/2,$$

and

$$0 < |x - c| < \delta_2, \quad \text{so} \quad |g(x) - M| < \varepsilon/2.$$

Therefore, $|f(x) + g(x) - (L + M)| < \dfrac{\varepsilon}{2} + \dfrac{\varepsilon}{2} = \varepsilon.$

This shows that $\lim_{x \to c} (f(x) + g(x)) = L + M$. ∎

Proof of the Limit Product Rule We show that for any $\varepsilon > 0$, there is a $\delta > 0$ such that for all x in the common domain D of f and g,

$$0 < |x - c| < \delta \quad \Rightarrow \quad |f(x)g(x) - LM| < \varepsilon.$$

Write $f(x)$ and $g(x)$ as $f(x) = L + (f(x) - L)$, $g(x) = M + (g(x) - M)$.

Multiply these expressions together and subtract LM:

$$f(x) \cdot g(x) - LM = (L + (f(x) - L))(M + (g(x) - M)) - LM$$
$$= LM + L(g(x) - M) + M(f(x) - L) + (f(x) - L)(g(x) - M) - LM \quad (1)$$
$$= L(g(x) - M) + M(f(x) - L) + (f(x) - L)(g(x) - M)$$

Since f and g have limits L and M as $x \to c$, there exist positive numbers $\delta_1, \delta_2, \delta_3$, and δ_4 such that for all x in D

$$0 < |x - c| < \delta_1 \quad \Rightarrow \quad |f(x) - L| < \sqrt{\varepsilon/3}$$
$$0 < |x - c| < \delta_2 \quad \Rightarrow \quad |g(x) - M| < \sqrt{\varepsilon/3}$$
$$0 < |x - c| < \delta_3 \quad \Rightarrow \quad |f(x) - L| < \frac{\varepsilon}{3(1 + |M|)} \quad (2)$$
$$0 < |x - c| < \delta_4 \quad \Rightarrow \quad |g(x) - M| < \frac{\varepsilon}{3(1 + |L|)}.$$

If we take δ to be the smallest of the numbers δ_1 through δ_4, the inequalities on the right-hand side of (2) will hold simultaneously for $0 < |x - c| < \delta$. Then, applying the triangle inequality to Equation 1, we have for all x in D, $0 < |x - c| < \delta$ implies

$$|f(x) \cdot g(x) - LM|$$

$$\leq |L||g(x) - M| + |M||f(x) - L| + |f(x) - L||g(x) - M|$$

$$\leq (1 + |L|)|g(x) - M| + (1 + |M|)|f(x) - L| + |f(x) - L||g(x) - M|$$

$$\leq \frac{\varepsilon}{3} + \frac{\varepsilon}{3} + \sqrt{\frac{\varepsilon}{3}}\sqrt{\frac{\varepsilon}{3}} = \varepsilon. \quad \text{Values from (2)}$$

This completes the proof of the Limit Product Rule. ∎

Proof of the Limit Quotient Rule We show that $\lim_{x \to c} (1/g(x)) = 1/M$. We can then conclude that

$$\lim_{x \to c} \frac{f(x)}{g(x)} = \lim_{x \to c} \left(f(x) \cdot \frac{1}{g(x)} \right) = \lim_{x \to c} f(x) \cdot \lim_{x \to c} \frac{1}{g(x)} = L \cdot \frac{1}{M} = \frac{L}{M}$$

by the Limit Product Rule.

Let $\varepsilon > 0$ be given. To show that $\lim_{x \to c} (1/g(x)) = 1/M$, we need to show that there exists a $\delta > 0$ such that for all x

$$0 < |x - c| < \delta \quad \Rightarrow \quad \left| \frac{1}{g(x)} - \frac{1}{M} \right| < \varepsilon.$$

Since $|M| > 0$, there exists a positive number δ_1 such that for all x

$$0 < |x - c| < \delta_1 \quad \Rightarrow \quad |g(x) - M| < \frac{|M|}{2}. \tag{3}$$

For any numbers A and B it can be shown that

$$|A| - |B| \leq |A - B| \quad \text{and} \quad |B| - |A| \leq |A - B|,$$

from which it follows that $||A| - |B|| \leq |A - B|$. With $A = g(x)$ and $B = M$, this becomes

$$||g(x)| - |M|| \leq |g(x) - M|,$$

which can be combined with the inequality on the right in (3) to get, in turn,

$$||g(x)| - |M|| < \frac{|M|}{2}$$

$$-\frac{|M|}{2} < |g(x)| - |M| < \frac{|M|}{2}$$

$$\frac{|M|}{2} < |g(x)| < \frac{3|M|}{2}$$

$$\frac{1}{|g(x)|} < \frac{2}{|M|} < \frac{3}{|g(x)|}. \quad \text{Multiply by } 2/(|M||g(x)|). \tag{4}$$

Therefore, $0 < |x - c| < \delta$ implies that

$$\left| \frac{1}{g(x)} - \frac{1}{M} \right| = \left| \frac{M - g(x)}{Mg(x)} \right| \le \frac{1}{|M|} \cdot \frac{1}{|g(x)|} \cdot |M - g(x)|$$

$$< \frac{1}{|M|} \cdot \frac{2}{|M|} \cdot |M - g(x)|. \quad \text{Inequality (4)} \quad (5)$$

Since $(1/2)|M|^2 \varepsilon > 0$, there exists a number $\delta_2 > 0$ such that for all x in D

$$0 < |x - c| < \delta_2 \quad \Rightarrow \quad |M - g(x)| < \frac{\varepsilon}{2}|M|^2. \qquad (6)$$

If we take δ to be the smaller of δ_1 and δ_2, the conclusions in (5) and (6) both hold for all x such that $0 < |x - c| < \delta$. Combining these conclusions gives

$$0 < |x - c| < \delta_2 \quad \Rightarrow \quad \left| \frac{1}{g(x)} - \frac{1}{M} \right| < \varepsilon.$$

This completes the proof of the Limit Quotient Rule. ∎

The last proof we give is of the Sandwich Theorem (Theorem 4) of Section 2.1.

Theorem 4 The Sandwich Theorem

If $g(x) \le f(x) \le h(x)$ for all $x \ne c$ in some interval about c, and

$$\lim_{x \to c} g(x) = \lim_{x \to c} h(x) = L,$$

then

$$\lim_{x \to c} f(x) = L.$$

Proof for Right-hand Limits Suppose that $\lim_{x \to c^+} g(x) = \lim_{x \to c^+} h(x) = L$. Then for any $\varepsilon > 0$ there exists a $\delta > 0$ such that for all x the inequality $c < x < c + \delta$ implies

$$L - \varepsilon < g(x) < L + \varepsilon \quad \text{and} \quad L - \varepsilon < h(x) < L + \varepsilon.$$

These inequalities combine with the inequality $g(x) \le f(x) \le h(x)$ to give

$$L - \varepsilon < g(x) \le f(x) \le h(x) < L + \varepsilon,$$

$$L - \varepsilon < f(x) < L + \varepsilon,$$

$$-\varepsilon < f(x) - L < \varepsilon.$$

Thus, for all x, the inequality $c < x < c + \delta$ implies $|f(x) - L| < \varepsilon$. Therefore, $\lim_{x \to c^+} f(x) = L$.

Proof for Left-hand Limits Suppose that $\lim_{x \to c^-} g(x) = \lim_{x \to c^-} h(x) = L$. Then for any $\varepsilon > 0$ there exists a $\delta > 0$ such that for all x the inequality $c - \delta < x < c$ implies

$$L - \varepsilon < g(x) < L + \varepsilon \quad \text{and} \quad L - \varepsilon < h(x) < L + \varepsilon.$$

We conclude as before that for all x, $c - \delta < x < c$ implies $|f(x) - L| < \varepsilon$. Therefore, $\lim_{x \to c^-} f(x) = L$.

Proof for Two-sided Limits If $\lim_{x \to c} g(x) = \lim_{x \to c} h(x) = L$, then $g(x)$ and $h(x)$ both approach L as $x \to c^+$ and $x \to c^-$; so $\lim_{x \to c^+} f(x) = L$ and $\lim_{x \to c^-} f(x) = L$. Hence $\lim_{x \to c} f(x)$ exists and equals L. ∎

Section A3 Exercises

In Exercises 1 and 2, sketch the interval (a, b) on the x-axis with the point c inside. Then find a value of $\delta > 0$ such that for all x,

$$0 < |x - c| < \delta \;\Rightarrow\; a < x < b.$$

1. $a = 4/9, \quad b = 4/7, \quad c = 1/2$

2. $a = 2.7591, \quad b = 3.2391, \quad c = 3$

In Exercises 3 and 4, use the graph to find a $\delta > 0$ such that for all x $0 < |x - c| < \delta \;\Rightarrow\; |f(x) - L| < \varepsilon$.

3.

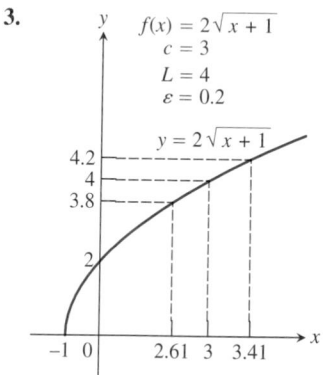

$f(x) = 2\sqrt{x + 1}$
$c = 3$
$L = 4$
$\varepsilon = 0.2$

$y = 2\sqrt{x + 1}$

NOT TO SCALE

4.

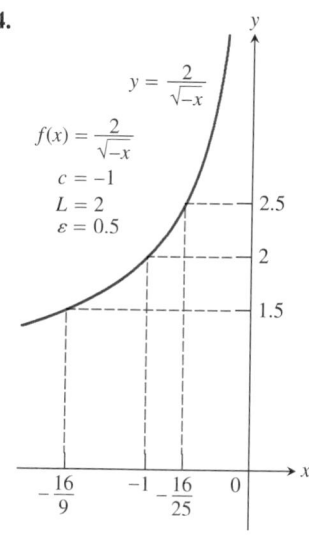

$y = \dfrac{2}{\sqrt{-x}}$

$f(x) = \dfrac{2}{\sqrt{-x}}$
$c = -1$
$L = 2$
$\varepsilon = 0.5$

Exercises 5–8 give a function $f(x)$ and numbers L, c, and ε. Find an open interval about c on which the inequality $|f(x) - L| < \varepsilon$ holds. Then give a value for $\delta > 0$ such that for all x satisfying $0 < |x - c| < \delta$ the inequality $|f(x) - L| < \varepsilon$ holds. Use algebra to find your answers.

5. $f(x) = 2x - 2, \quad L = -6, \quad c = -2, \quad \varepsilon = 0.02$

6. $f(x) = \sqrt{x + 1}, \quad L = 1, \quad c = 0, \quad \varepsilon = 0.1$

7. $f(x) = \sqrt{19 - x}, \quad L = 3, \quad c = 10, \quad \varepsilon = 1$

8. $f(x) = x^2, \quad L = 4, \quad c = -2, \quad \varepsilon = 0.5$

Exercises 9–12 give a function $f(x)$, a point c, and a positive number ε. (a) Find $L = \lim_{x \to c} f(x)$. Then (b) find a number $\delta > 0$ such that for all x

$$0 < |x - c| < \delta \;\Rightarrow\; |f(x) - L| < \varepsilon.$$

9. $f(x) = \dfrac{x^2 + 6x + 5}{x + 5}, \quad c = -5, \quad \varepsilon = 0.05$

10. $f(x) = \begin{cases} 4 - 2x, & x < 1, \\ 6x - 4, & x \geq 1, \end{cases} \quad c = 1, \quad \varepsilon = 0.5$

11. $f(x) = \sin x, \quad c = 1, \quad \varepsilon = 0.01$

12. $f(x) = \dfrac{x}{x^2 - 4}, \quad c = -1, \quad \varepsilon = 0.1$

In Exercises 13 and 14, use the definition of limit to prove the limit statement.

13. $\lim\limits_{x \to 1} f(x) = 1 \quad$ if $\quad f(x) = \begin{cases} x^2, & x \neq 1 \\ 2, & x = 1 \end{cases}$

14. $\lim\limits_{x \to \sqrt{3}} \dfrac{1}{x^2} = \dfrac{1}{3}$

15. *Relating to Limits* Given $\varepsilon > 0$, (a) find an interval $I = (5, 5 + \delta)$, $\delta > 0$, such that if x lies in I, then $\sqrt{x - 5} < \varepsilon$. (b) What limit is being verified?

16. *Relating to Limits* Given $\varepsilon > 0$, (a) find an interval $I = (4 - \delta, 4)$, $\delta > 0$, such that if x lies in I, then $\sqrt{4 - x} < \varepsilon$. (b) What limit is being verified?

17. Prove the Constant Multiple Rule for limits.

18. Prove the Difference Rule for limits.

19. *Generalized Limit Sum Rule* Suppose that functions $f_1(x)$, $f_2(x)$, and $f_3(x)$ have limits L_1, L_2, and L_3, respectively, as $x \to c$. Show that their sum has limit $L_1 + L_2 + L_3$. Use mathematical induction (Appendix A2) to generalize this result to the sum of any finite number of functions.

20. *Generalized Limit Product Rule* Use mathematical induction and the Limit Product Rule in Theorem 1 to show that if functions $f_1(x)$, $f_2(x)$, ..., $f_n(x)$ have limits L_1, L_2, ..., L_n, respectively, as $x \to c$, then

$$\lim\limits_{x \to c} (f_1(x) \cdot f_2(x) \cdot \cdots \cdot f_n(x)) = L_1 \cdot L_2 \cdot \cdots \cdot L_n.$$

21. *Positive Integer Power Rule* Use the fact that $\lim_{x \to c} x = c$ and the result of Exercise 20 to show that $\lim_{x \to c} x^n = c^n$ for any integer $n > 1$.

22. *Limits of Polynomials* Use the fact that $\lim_{x \to c} k = k$ for any number k together with the results of Exercises 19 and 21 to show that $\lim_{x \to c} f(x) = f(c)$ for any polynomial function

$$f(x) = a_n x^n + a_{n-1} x^{n-1} + \cdots + a_1 x + a_0.$$

23. *Limits of Rational Functions* Use Theorem 1 and the result of Exercise 22 to show that if $f(x)$ and $g(x)$ are polynomial functions and $g(c) \neq 0$, then

$$\lim_{x \to c} \frac{f(x)}{g(x)} = \frac{f(c)}{g(c)}.$$

24. *Composites of Continuous Functions* Figure A3.7 gives the diagram for a proof that the composite of two continuous functions is continuous. Reconstruct the proof from the diagram. The statement to be proved is this: If f is continuous at $x = c$ and g is continuous at $f(c)$, then $g \circ f$ is continuous at c.

Assume that c is an interior point of the domain of f and that $f(c)$ is an interior point of the domain of g. This will make the limits involved two-sided. (The argument for the cases that involve one-sided limits are similar.)

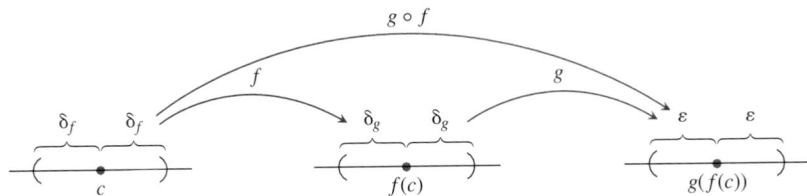

Figure A3.7 The continuity of composites holds for any finite number of functions. The only requirement is that each function be continuous where it is applied. Here, f is to be continuous at c and g at $f(c)$.

A4   Proof of the Chain Rule

Error in the Approximation $\Delta f \approx df$ •   The Proof

Error in the Approximation $\Delta f \approx df$

Let $f(x)$ be differentiable at $x = a$ and suppose that Δx is an increment of x. We know that the differential $df = f'(a)\Delta x$ is an approximation for the change $\Delta f = (f(a + \Delta x) - f(a))$ in f as x changes from a to $(a + \Delta x)$. How well does df approximate Δf?

We measure the approximation error by subtracting df from Δf:

$$\text{Approximation error} = \Delta f - df = \Delta f - f'(a)\Delta x$$

$$= \underbrace{f(a + \Delta x) - f(a)}_{\Delta f} - f'(a)\Delta x$$

$$= \underbrace{\left(\frac{f(a + \Delta x) - f(a)}{\Delta x} - f'(a) \right)}_{\text{Call this part } \varepsilon.} \cdot \Delta x$$

$$= \varepsilon \cdot \Delta x$$

As $\Delta x \to 0$, the difference quotient $(f(a + \Delta x) - f(a))/\Delta x$ approaches $f'(a)$ (remember the definition of $f'(a)$), so the quantity in parentheses becomes a very small number (which is why we called it ε). In fact, $\varepsilon \to 0$ as $\Delta x \to 0$. When Δx is small, the **approximation error** $\varepsilon\Delta x$ is smaller still.

$$\underbrace{\Delta f}_{\substack{\text{true} \\ \text{change}}} = \underbrace{f'(a)\,\Delta x}_{\substack{\text{estimated} \\ \text{change}}} + \underbrace{\varepsilon\,\Delta x}_{\text{error}} \tag{1}$$

The Proof

Our goal is to show that if $f(u)$ is a differentiable function of u and $u = g(x)$ is a differentiable function of x, then the composite $y = f(g(x))$ is a differentiable function of x. More precisely, if g is differentiable at a and f is differentiable at $g(a)$, then the composite is differentiable at a and

$$\frac{dy}{dx}\bigg|_{x=a} = f'(g(a)) \cdot g'(a).$$

Let Δx be an increment in x and let Δu and Δy be the corresponding increments in u and y. As you can see in Figure A4.1,

$$\frac{dy}{dx}\bigg|_{x=a} = \lim_{\Delta x \to 0} \frac{\Delta y}{\Delta x},$$

so our goal is to show that the limit is $f'(g(a)) \cdot g'(a)$.

By Equation 1,

$$\Delta u = g'(a)\Delta x + \varepsilon_1 \Delta x = (g'(a) + \varepsilon_1)\Delta x,$$

where $\varepsilon_1 \to 0$ as $\Delta x \to 0$. Similarly, since f is differentiable at $g(a)$,

$$\Delta y = f'(g(a))\Delta u + \varepsilon_2 \Delta u = (f'(g(a)) + \varepsilon_2)\Delta u,$$

where $\varepsilon_2 \to 0$ as $\Delta u \to 0$. Notice also that $\Delta u \to 0$ as $\Delta x \to 0$. Combining the equations for Δu and Δy gives $\Delta y = (f'(g(a)) + \varepsilon_2)(g'(a) + \varepsilon_1)\Delta x$, so

$$\frac{\Delta y}{\Delta x} = f'(g(a))g'(a) + \varepsilon_2 g'(a) + f'(g(a))\varepsilon_1 + \varepsilon_2 \varepsilon_1.$$

Since ε_1 and ε_2 go to zero as Δx goes to zero, three of the four terms on the right vanish in the limit, leaving

$$\lim_{\Delta x \to 0} \frac{\Delta y}{\Delta x} = f'(g(a))g'(a).$$

This concludes the proof. ∎

Figure A4.1 The graph of y as a function of x. The derivative of y with respect to x at $x = a$ is $\lim_{\Delta x \to 0} (\Delta y / \Delta x)$.

A5 Conic Sections

Overview

Conic sections are the paths traveled by planets, satellites, and other bodies (even electrons) whose motions are driven by inverse-square forces. Once we know that the path of a moving body is a conic section, we immediately have information about the body's velocity and the force that drives it. In this appendix, we study the connections between conic sections and quadratic equations and classify conic sections by eccentricity (Pluto's orbit is highly eccentric while Earth's is nearly circular).

A5.1 Conic Sections and Quadratic Equations

Circles • Interiors and Exteriors of Circles • Parabolas • Ellipses • Axes of an Ellipse • Hyperbolas • Asymptotes and Drawing • Reflective Properties • Other Applications

Circles

The Greeks of Plato's time defined conic sections as the curves formed by cutting through a double cone with a plane (Figure A5.1). Today, we define conic sections with the distance function in the coordinate plane.

Definition Circle

A **circle** is the set of points in a plane whose distance from a given fixed point in the plane is constant. The fixed point is the **center** of the circle; the constant distance is the **radius.**

If $a > 0$, the equation $x^2 + y^2 = a^2$ represents all the points (x, y) in the plane whose distance from the origin is

$$\sqrt{(x - 0)^2 + (y - 0)^2} = \sqrt{x^2 + y^2} = \sqrt{a^2} = a.$$

These are the points of the circle of radius a centered at the origin. If we shift the circle to place its center at the point (h, k), its equation becomes $(x - h)^2 + (y - k)^2 = a^2$.

Circle of Radius a Centered at (h, k)

$$(x - h)^2 + (y - k)^2 = a^2$$

Example 1   FINDING CENTER AND RADIUS

Find the center and radius of the circle

$$x^2 + y^2 + 4x - 6y - 3 = 0.$$

Solution We convert the equation to standard form by completing the squares in x and y:

$x^2 + y^2 + 4x - 6y - 3 = 0$ — Start with the given equation.

$(x^2 + 4x) + (y^2 - 6y) = 3$ — Gather terms. Move the constant to the right-hand side.

$\left(x^2 + 4x + \left(\frac{4}{2}\right)^2\right) + \left(y^2 - 6y + \left(\frac{-6}{2}\right)^2\right) = 3 + \left(\frac{4}{2}\right)^2 + \left(\frac{-6}{2}\right)^2$ — Add the square of half the coefficient of x to each side of the equation. Do the same for y. The parenthetical expressions on the left-hand side are now perfect squares.

$(x^2 + 4x + 4) + (y^2 - 6y + 9) = 3 + 4 + 9$

$(x + 2)^2 + (y - 3)^2 = 16$ — Write each quadratic as a squared linear expression.

With the equation now in standard form, we read off the center's coordinates and the radius: $(h, k) = (-2, 3)$ and $a = 4$.

Circle: plane perpendicular
to cone axis

Ellipse

Parabola: plane parallel
to side of cone

(a)

Hyperbola: plane
parallel to cone axis

Point: plane through
cone vertex only

Single line: plane
tangent to cone

Pair of intersecting lines

(b)

Figure A5.1 The standard conic sections (a) are the curves in which a plane cuts a double cone. Hyperbolas come in two parts, called *branches*. The point and lines obtained by passing the plane through the cone's vertex (b) are *degenerate* conic sections.

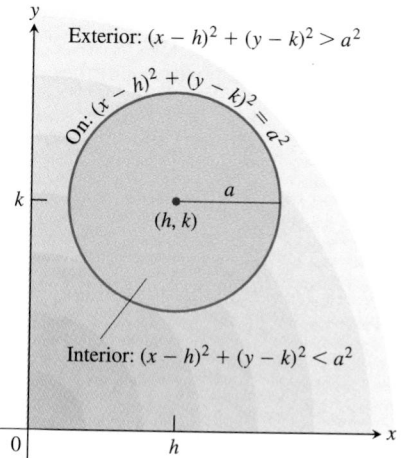

Figure A5.2 The interior and exterior of the circle $(x - h)^2 + (y - k)^2 = a^2$.

Interiors and Exteriors of Circles

The points that lie inside the circle $(x - h)^2 + (y - k)^2 = a^2$ are the points less than a units from (h, k). They satisfy the inequality

$$(x - h)^2 + (y - k)^2 < a^2.$$

They make up the region we call the **interior** of the circle (Figure A5.2).

The circle's **exterior** consists of the points that lie more than a units from (h, k). These points satisfy the inequality

$$(x - h)^2 + (y - k)^2 > a^2.$$

Example 2 INTERPRETING INEQUALITIES

Inequality	Region
$x^2 + y^2 < 1$	Interior of the unit circle
$x^2 + y^2 \leq 1$	Unit circle plus its interior
$x^2 + y^2 > 1$	Exterior of the unit circle
$x^2 + y^2 \geq 1$	Unit circle plus its exterior

Parabolas

> **Definition Parabola**
>
> A set that consists of all the points in a plane equidistant from a given fixed point and a given fixed line in the plane is a **parabola.** The fixed point is the **focus** of the parabola. The fixed line is the **directrix.**

If the focus F lies on the directrix L, the parabola is the line through F perpendicular to L. We consider this to be a degenerate case and assume henceforth that F does not lie on L.

A parabola has its simplest equation when its focus and directrix straddle one of the coordinate axes. For example, suppose that the focus lies at the point $F(0, p)$ on the positive y-axis and that the directrix is the line $y = -p$ (Figure A5.3). In the notation of the figure, a point $P(x, y)$ lies on the parabola if and only if $PF = PQ$. From the distance formula,

$$PF = \sqrt{(x - 0)^2 + (y - p)^2} = \sqrt{x^2 + (y - p)^2}$$
$$PQ = \sqrt{(x - x)^2 + (y - (-p))^2} = \sqrt{(y + p)^2}.$$

When we equate these expressions, square, and simplify, we get

$$y = \frac{x^2}{4p} \quad \text{or} \quad x^2 = 4py. \quad \text{Standard form} \tag{1}$$

These equations reveal the parabola's symmetry about the y-axis. We call the y-axis the **axis** of the parabola (short for "axis of symmetry").

The point where a parabola crosses its axis is the **vertex.** The vertex of the parabola $x^2 = 4py$ lies at the origin (Figure A5.3). The positive number p is the parabola's **focal length.**

If the parabola opens downward, with its focus at $(0, -p)$ and its directrix the line $y = p$, Equations 1 become

$$y = -\frac{x^2}{4p} \quad \text{or} \quad x^2 = -4py$$

(Figure A5.4). We obtain similar equations for parabolas opening to the right or to the left (Figure A5.5 and Table A5.1).

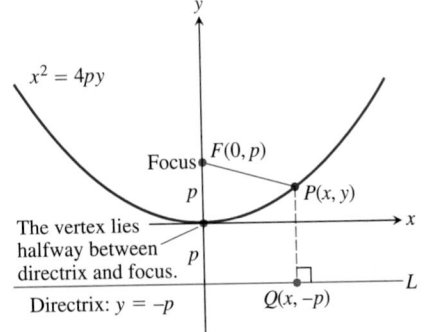

Figure A5.3 The parabola $x^2 = 4py$.

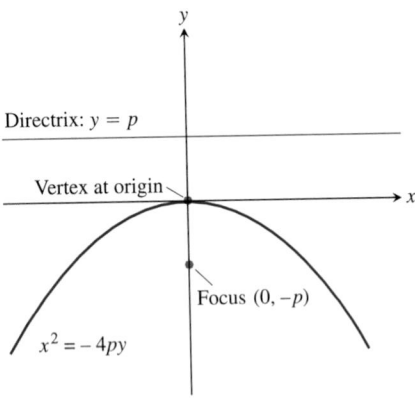

Figure A5.4 The parabola $x^2 = -4py$.

Figure A5.5 (a) The parabola $y^2 = 4px$. (b) The parabola $y^2 = -4px$.

Table A5.1 Standard-form equations for parabolas with vertices at the origin ($p > 0$)

Equation	Focus	Directrix	Axis	Opens
$x^2 = 4py$	$(0, p)$	$y = -p$	y-axis	Up
$x^2 = -4py$	$(0, -p)$	$y = p$	y-axis	Down
$y^2 = 4px$	$(p, 0)$	$x = -p$	x-axis	To the right
$y^2 = -4px$	$(-p, 0)$	$x = p$	x-axis	To the left

Example 3   FINDING FOCUS AND DIRECTRIX

Find the focus and directrix of the parabola $y^2 = 10x$.

Solution We find the value of p in the standard equation $y^2 = 4px$:

$$4p = 10, \quad \text{so} \quad p = \frac{10}{4} = \frac{5}{2}.$$

Then we find the focus and directrix for this value of p:

$$\text{Focus:} \qquad (p, 0) = \left(\frac{5}{2}, 0\right)$$

$$\text{Directrix:} \quad x = -p \quad \text{or} \quad x = -\frac{5}{2}.$$

Ellipses

Definition Ellipse

An **ellipse** is the set of points in a plane whose distances from two fixed points in the plane have a constant sum. The fixed points are the **foci** of the ellipse. The line through the foci is the **focal axis.** The point on the axis halfway between the foci is the **center.** The points where the focal axis and ellipse cross are the **vertices** (Figure A5.6).

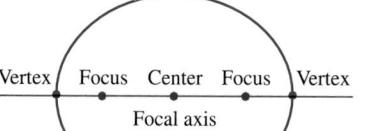

Figure A5.6 Points on the focal axis of an ellipse.

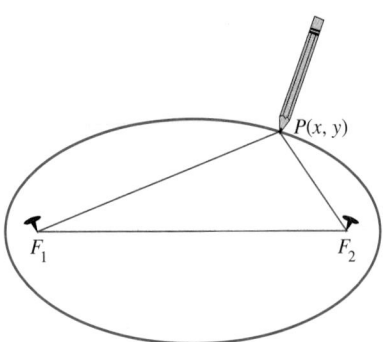

Figure A5.7 How to draw an ellipse.

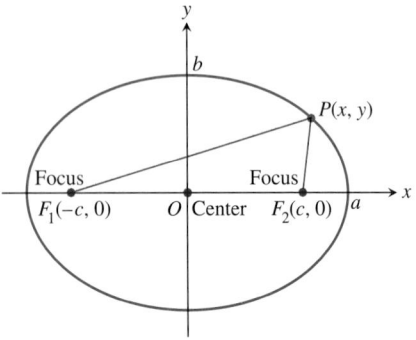

Figure A5.8 The ellipse defined by the equation $PF_1 + PF_2 = 2a$ is the graph of the equation $(x^2/a^2) + (y^2/b^2) = 1$.

The quickest way to construct an ellipse uses the definition. Put a loop of string around two tacks F_1 and F_2, pull the string taut with a pencil point P, and move the pencil around to trace a closed curve (Figure A5.7). The curve is an ellipse because the sum $PF_1 + PF_2$, being the length of the loop minus the distance between the tacks, remains constant. The ellipse's foci lie at F_1 and F_2.

If the foci are $F_1(-c, 0)$ and $F_2(c, 0)$ (Figure A5.8) and $PF_1 + PF_2$ is denoted by $2a$, then the coordinates of a point P on the ellipse satisfy the equation

$$\sqrt{(x + c)^2 + y^2} + \sqrt{(x - c)^2 + y^2} = 2a.$$

To simplify this equation, we move the second radical to the right-hand side, square, isolate the remaining radical, and square again, obtaining

$$\frac{x^2}{a^2} + \frac{y^2}{a^2 - c^2} = 1. \tag{2}$$

Since $PF_1 + PF_2$ is greater than the length F_1F_2 (triangle inequality for triangle PF_1F_2), the number $2a$ is greater than $2c$. Accordingly, $a > c$ and the number $a^2 - c^2$ in Equation 2 is positive.

The algebraic steps leading to Equation 2 can be reversed to show that every point P whose coordinates satisfy an equation of this form with $0 < c < a$ also satisfies the equation $PF_1 + PF_2 = 2a$. A point therefore lies on the ellipse if and only if its coordinates satisfy Equation 2.

If

$$b = \sqrt{a^2 - c^2}, \tag{3}$$

then $a^2 - c^2 = b^2$ and Equation 2 takes the form

$$\frac{x^2}{a^2} + \frac{y^2}{b^2} = 1. \tag{4}$$

Equation 4 reveals that this ellipse is symmetric with respect to the origin and both coordinate axes. It lies inside the rectangle bounded by the lines $x = \pm a$ and $y = \pm b$. It crosses the axes at the points $(\pm a, 0)$ and $(0, \pm b)$. The tangents at these points are perpendicular to the axes because

$$\frac{dy}{dx} = -\frac{b^2 x}{a^2 y} \qquad \text{Obtained from Eq. 4 by implicit differentiation}$$

is zero if $x = 0$ and infinite if $y = 0$.

Axes of an Ellipse

The **major axis** of the ellipse in Equation 4 is the line segment of length $2a$ joining the points $(\pm a, 0)$. The **minor axis** is the line segment of length $2b$ joining the points $(0, \pm b)$. The number a itself is the **semimajor axis,** the number b the **semiminor axis.** The number c found from Equation 3 as

$$c = \sqrt{a^2 - b^2},$$

is the **center-to-focus** distance of the ellipse.

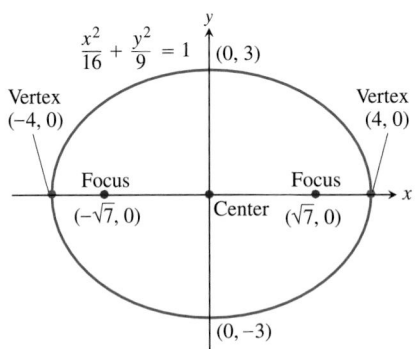

$$\frac{x^2}{16} + \frac{y^2}{9} = 1 \quad (0, 3)$$

Vertex (−4, 0)   Vertex (4, 0)

Focus (−√7, 0)   Center   Focus (√7, 0)

(0, −3)

Figure A5.9 Major axis horizontal. (Example 4)

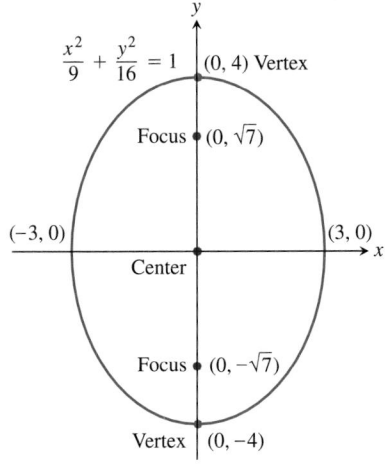

$$\frac{x^2}{9} + \frac{y^2}{16} = 1 \quad (0, 4) \text{ Vertex}$$

Focus (0, √7)

(−3, 0)   (3, 0)

Center

Focus (0, −√7)

Vertex (0, −4)

Figure A5.10 Major axis vertical. (Example 5)

Example 4 MAJOR AXIS HORIZONTAL

The ellipse

$$\frac{x^2}{16} + \frac{y^2}{9} = 1 \tag{5}$$

(Figure A5.9) has

Semimajor axis: $a = \sqrt{16} = 4$

Semiminor axis: $b = \sqrt{9} = 3$

Center-to-focus distance: $c = \sqrt{16 - 9} = \sqrt{7}$

Foci: $(\pm c, 0) = (\pm\sqrt{7}, 0)$

Vertices: $(\pm a, 0) = (\pm 4, 0)$

Center: $(0, 0).$

Example 5 MAJOR AXIS VERTICAL

The ellipse

$$\frac{x^2}{9} + \frac{y^2}{16} = 1, \tag{6}$$

obtained by interchanging x and y in Equation 5, has its major axis vertical instead of horizontal (Figure A5.10). With a^2 still equal to 16 and b^2 equal to 9, we have

Semimajor axis: $a = \sqrt{16} = 4$

Semiminor axis: $b = \sqrt{9} = 3$

Center-to-focus distance: $c = \sqrt{16 - 9} = \sqrt{7}$

Foci: $(0, \pm c) = (0, \pm\sqrt{7})$

Vertices: $(0, \pm a) = (0, \pm 4)$

Center: $(0, 0).$

There is never any cause for confusion in analyzing equations like (5) and (6). We simply find the intercepts on the coordinate axes; then we know which way the major axis runs because it is the longer of the two axes. The center always lies at the origin and the foci lie on the major axis.

> ### Standard-Form Equations for Ellipses Centered at the Origin
>
> *Foci on the x-axis:* $\dfrac{x^2}{a^2} + \dfrac{y^2}{b^2} = 1 \quad (a > b)$
>
>   Center-to-focus distance: $c = \sqrt{a^2 - b^2}$
>   Foci: $(\pm c, 0)$
>   Vertices: $(\pm a, 0)$
>
> *Foci on the y-axis:* $\dfrac{x^2}{b^2} + \dfrac{y^2}{a^2} = 1 \quad (a > b)$
>
>   Center-to-focus distance: $c = \sqrt{a^2 - b^2}$
>   Foci: $(0, \pm c)$
>   Vertices: $(0, \pm a)$
>
> In each case, a is the semimajor axis and b is the semiminor axis.

Hyperbolas

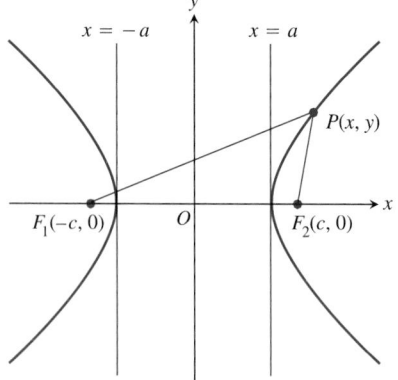

$x = -a$ $x = a$

$P(x, y)$

$F_1(-c, 0)$ O $F_2(c, 0)$

Figure A5.11 Hyperbolas have two branches. For points on the right-hand branch of the hyperbola shown here, $PF_1 - PF_2 = 2a$. For points on the left-hand branch, $PF_2 - PF_1 = 2a$.

> **Definition Hyperbola**
>
> A **hyperbola** is the set of points in a plane whose distances from two fixed points in the plane have a constant difference. The two fixed points are the **foci** of the hyperbola.

If the foci are $F_1(-c, 0)$ and $F_2(c, 0)$ (Figure A5.11) and the constant difference is $2a$, then a point (x, y) lies on the hyperbola if and only if

$$\sqrt{(x + c)^2 + y^2} - \sqrt{(x - c)^2 + y^2} = \pm 2a. \qquad (7)$$

To simplify this equation, we move the second radical to the right-hand side, square, isolate the remaining radical, and square again, obtaining

$$\frac{x^2}{a^2} + \frac{y^2}{a^2 - c^2} = 1. \qquad (8)$$

So far, this looks just like the equation for an ellipse. But now $a^2 - c^2$ is negative because $2a$, being the difference of two sides of triangle PF_1F_2 is less than $2c$, the third side.

The algebraic steps leading to Equation 8 can be reversed to show that every point P whose coordinates satisfy an equation of this form with $0 < a < c$ also satisfies Equation 7. A point therefore lies on the hyperbola if and only if its coordinates satisfy Equation 8.

If we let b denote the positive square root of $c^2 - a^2$,

$$b = \sqrt{c^2 - a^2}, \qquad (9)$$

then $a^2 - c^2 = -b^2$ and Equation 8 takes the more compact form

$$\frac{x^2}{a^2} - \frac{y^2}{b^2} = 1. \qquad (10)$$

The differences between Equation 10 and the equation for an ellipse (Equation 4) are the minus sign and the new relation

$$c^2 = a^2 + b^2. \quad \text{From Eq. 9}$$

Like the ellipse, the hyperbola is symmetric with respect to the origin and coordinate axes. It crosses the x-axis at the points $(\pm a, 0)$. The tangents at these points are vertical because

$$\frac{dy}{dx} = \frac{b^2 x}{a^2 y} \quad \begin{array}{l}\text{Obtained from Eq. 10 by} \\ \text{implicit differentiation}\end{array}$$

is infinite when $y = 0$. The hyperbola has no y-intercepts; in fact, no part of the curve lies between the lines $x = -a$ and $x = a$.

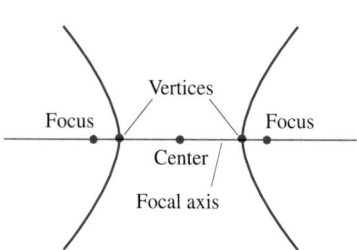

Focus   Vertices   Focus

Center

Focal axis

Figure A5.12 Points on the focal axis of a hyperbola.

> **Definition Parts of a Hyperbola**
>
> The line through the foci of a hyperbola is the **focal axis**. The point on the axis halfway between the foci is the hyperbola's **center.** The points where the focal axis and hyperbola cross are the **vertices** (Figure A5.12).

Asymptotes and Drawing

The hyperbola

$$\frac{x^2}{a^2} - \frac{y^2}{b^2} = 1 \tag{11}$$

has two asymptotes, the lines

$$y = \pm\frac{b}{a}x.$$

The asymptotes give the guidance we need to draw hyperbolas quickly. (See the drawing lesson.) The fastest way to find the equations of the asymptotes is to replace the 1 in Equation 11 by 0 and solve the new equation for *y:*

$$\underbrace{\frac{x^2}{a^2} - \frac{y^2}{b^2} = 1}_{\text{hyperbola}} \quad \Rightarrow \quad \underbrace{\frac{x^2}{a^2} - \frac{y^2}{b^2} = 0}_{\text{0 for 1}} \quad \Rightarrow \quad \underbrace{y = \pm\frac{b}{a}x.}_{\text{asymptotes}}$$

Standard-Form Equations for Hyperbolas Centered at the Origin

Foci on the x-axis: $\dfrac{x^2}{a^2} - \dfrac{y^2}{b^2} = 1$

Center-to-focus distance: $c = \sqrt{a^2 + b^2}$

Foci: $(\pm c, 0)$

Vertices: $(\pm a, 0)$

Asymptotes: $\dfrac{x^2}{a^2} - \dfrac{y^2}{b^2} = 0$ or $y = \pm\dfrac{b}{a}x$

Foci on the y-axis: $\dfrac{y^2}{a^2} - \dfrac{x^2}{b^2} = 1$

Center-to-focus distance: $c = \sqrt{a^2 + b^2}$

Foci: $(0, \pm c)$

Vertices: $(0, \pm a)$

Asymptotes: $\dfrac{y^2}{a^2} - \dfrac{x^2}{b^2} = 0$ or $y = \pm\dfrac{a}{b}x$

Notice the difference in the asymptote equations (*b/a* in the first, *a/b* in the second).

Drawing Lesson

How to Graph the Hyperbola $\dfrac{x^2}{a^2} - \dfrac{y^2}{b^2} = 1$

1. Mark the points $(\pm a, 0)$ and $(0, \pm b)$ with line segments and complete the rectangle they determine.

2. Sketch the asymptotes by extending the rectangle's diagonals.

3. Use the rectangle and asymptotes to guide your drawing.

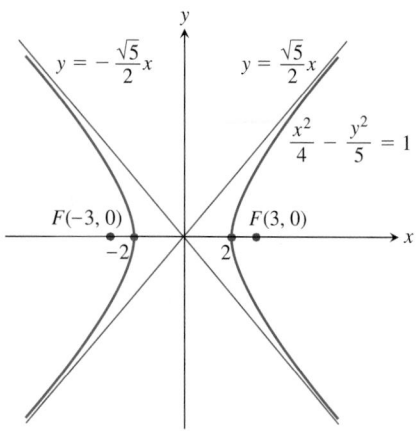

Figure A5.13 The hyperbola in Example 6.

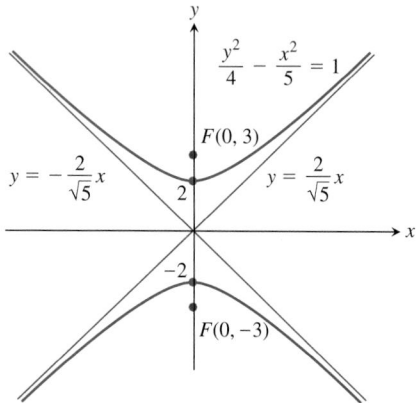

Figure A5.14 The hyperbola in Example 7.

Example 6 FOCI ON THE x-AXIS

The equation

$$\frac{x^2}{4} - \frac{y^2}{5} = 1 \tag{12}$$

is Equation 10 with $a^2 = 4$ and $b^2 = 5$ (Figure A5.13). We have

Center-to-focus distance: $\quad c = \sqrt{a^2 + b^2} = \sqrt{4 + 5} = 3$

Foci: $\quad (\pm c, 0) = (\pm 3, 0)$

Vertices: $\quad (\pm a, 0) = (\pm 2, 0)$

Center: $\quad (0, 0)$

Asymptotes: $\quad \dfrac{x^2}{4} - \dfrac{y^2}{5} = 0 \quad$ or $\quad y = \pm \dfrac{\sqrt{5}}{2}x.$

Example 7 FOCI ON THE y-AXIS

The hyperbola

$$\frac{y^2}{4} - \frac{x^2}{5} = 1,$$

obtained by interchanging x and y in Equation 12, has its vertices on the y-axis instead of the x-axis (Figure A5.14). With a^2 still equal to 4 and b^2 equal to 5, we have

Center-to-focus distance: $\quad c = \sqrt{a^2 + b^2} = \sqrt{4 + 5} = 3$

Foci: $\quad (0, \pm c) = (0, \pm 3)$

Vertices: $\quad (0, \pm a) = (0, \pm 2)$

Center: $\quad (0, 0)$

Asymptotes: $\quad \dfrac{y^2}{4} - \dfrac{x^2}{5} = 0 \quad$ or $\quad y = \pm \dfrac{2}{\sqrt{5}}x.$

Reflective Properties

The chief applications of parabolas involve their use as reflectors of light and radio waves. Rays originating at a parabola's focus are reflected out of the parabola parallel to the parabola's axis (Figure A5.15). This property is used by flashlight, headlight, and spotlight reflectors and by microwave broadcast

Figure A5.15 Two of the many uses of parabolic reflectors.

antennas to direct radiation from point sources into narrow beams. Conversely, electromagnetic waves arriving parallel to a parabolic reflector's axis are directed toward the reflector's focus. This property is used to intensify signals picked up by radio telescopes and television satellite dishes, to focus arriving light in telescopes, and to concentrate sunlight in solar heaters.

If an ellipse is revolved about its major axis to generate a surface (a surface called an *ellipsoid*) and the interior is silvered to produce a mirror, light from one focus will be reflected to the other focus (Figure A5.16). Ellipsoids reflect sound the same way, and this property is used to construct *whispering galleries,* rooms in which a person standing at one focus can hear a whisper from the other focus. Statuary Hall in the U.S. Capitol building is a whispering gallery. Ellipsoids also appear in instruments used to study aircraft noise in wind tunnels (sound at one focus can be received at the other focus with relatively little interference from other sources).

Light directed toward one focus of a hyperbolic mirror is reflected toward the other focus. This property of hyperbolas is combined with the reflective properties of parabolas and ellipses in designing modern telescopes. In Figure A5.17 starlight reflects off a primary parabolic mirror toward the mirror's focus F_P. It is then reflected by a small hyperbolic mirror, whose focus is $F_H = F_P$, toward the second focus of the hyperbola, $F_E = F_H$. Since this focus is shared by an ellipse, the light is reflected by the elliptical mirror to the ellipse's second focus to be seen by an observer.

As past experience with NASA's Hubble space telescope shows, the mirrors have to be nearly perfect to focus properly. The aberration that caused the malfunction in Hubble's primary mirror (now corrected with additional mirrors) amounted to about half a wavelength of visible light, no more than 1/50 the width of a human hair.

Other Applications

Water pipes are sometimes designed with elliptical cross sections to allow for expansion when the water freezes. The triggering mechanisms in some lasers are elliptical, and stones on a beach become more and more elliptical as they are ground down by waves. There are also applications of ellipses to fossil formation. The ellipsolith, once thought to be a separate species, is now known to be an elliptically deformed nautilus.

Hyperbolic paths arise in Einstein's theory of relativity and form the basis for the (unrelated) LORAN radio navigation system. (LORAN is short for "long range navigation.") Hyperbolas also form the basis for a new system the Burlington Northern Railroad developed for using synchronized electronic signals from satellites to track freight trains. Computers aboard Burlington Northern locomotives in Minnesota can track trains to within one mile per hour of their speed and to within feet of their actual location.

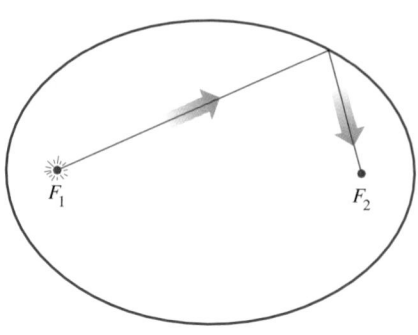

Figure A5.16 An elliptical mirror (shown here in profile) reflects light from one focus to the other.

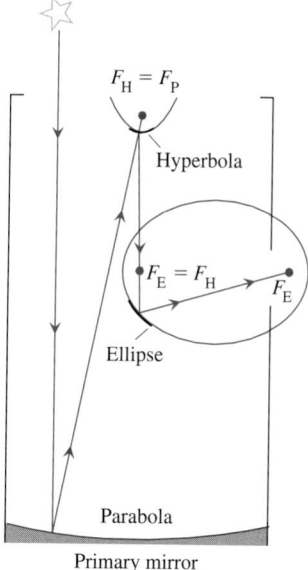

Figure A5.17 Schematic drawing of a reflecting telescope.

Section A5.1 Exercises

In Exercises 1 and 2, find an equation for the circle with center $C(h, k)$ and radius a. Sketch the circle in the xy-plane. Label the circle's center and x- and y-intercepts (if any) with their coordinate pairs.

1. $C(0, 2), \quad a = 2$

2. $C(-1, 5), \quad a = \sqrt{10}$

In Exercises 3 and 4, find the center and radius of the circle. Then sketch the circle.

3. $x^2 + y^2 + 4x - 4y + 4 = 0$ **4.** $x^2 + y^2 - 4x + 4y = 0$

In Exercises 5 and 6, describe the regions defined by the inequalities and pairs of inequalities.

5. $(x - 1)^2 + y^2 \le 4$

6. $x^2 + y^2 > 1, \quad x^2 + y^2 < 4$

Match the parabolas in Exercises 7–10 with the following equations:

$$x^2 = 2y, \quad x^2 = -6y, \quad y^2 = 8x, \quad y^2 = -4x.$$

Then find the parabola's focus and directrix.

7.

8.

9.

10.
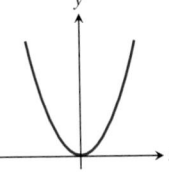

Match each conic section in Exercises 11–14 with one of these equations:

$$\frac{x^2}{4} + \frac{y^2}{9} = 1, \quad \frac{x^2}{2} + y^2 = 1,$$

$$\frac{y^2}{4} - x^2 = 1, \quad \frac{x^2}{4} - \frac{y^2}{9} = 1.$$

Then find the conic section's foci and vertices. If the conic section is a hyperbola, find its asymptotes as well.

11.

12.

13.

14.
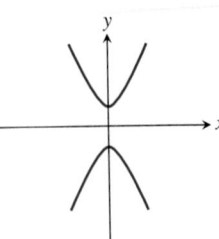

Exercises 15 and 16 give equations of parabolas. Find each parabola's focus and directrix. Then sketch the parabola. Include the focus and directrix in your sketch.

15. $y^2 = 12x$

16. $y = 4x^2$

Exercises 17 and 18 give equations for ellipses. Put each equation in standard form. Then sketch the ellipse. Include the foci in your sketch.

17. $16x^2 + 25y^2 = 400$

18. $3x^2 + 2y^2 = 6$

Exercises 19 and 20 give information about the foci and vertices of ellipses centered at the origin of the xy-plane. In each case, find the ellipse's standard-form equation from the given information.

19. Foci: $(\pm\sqrt{2}, 0)$
   Vertices: $(\pm 2, 0)$

20. Foci: $(0, \pm 4)$
   Vertices: $(0, \pm 5)$

Exercises 21 and 22 give equations for hyperbolas. Put each equation in standard form and find the hyperbola's asymptotes. Then sketch the hyperbola. Include the asymptotes and foci in your sketch.

21. $x^2 - y^2 = 1$

22. $8y^2 - 2x^2 = 16$

Exercises 23 and 24 give information about the foci, vertices, and asymptotes of hyperbolas centered at the origin of the xy-plane. In each case, find the hyperbola's standard-form equation from the information given.

23. Foci: $(0, \pm\sqrt{2})$
   Asymptotes: $y = \pm x$

24. Vertices: $(\pm 3, 0)$
   Asymptotes: $y = \pm\frac{4}{3}x$

25. The parabola $y^2 = 8x$ is shifted down 2 units and right 1 unit to generate the parabola $(y + 2)^2 = 8(x - 1)$.
   (a) Find the new parabola's vertex, focus, and directrix.
   (b) Plot the new vertex, focus, and directrix, and sketch in the parabola.

26. The ellipse $x^2/16 + y^2/9 = 1$ is shifted 4 units to the right and 3 units up to generate the ellipse

$$\frac{(x - 4)^2}{16} + \frac{(y - 3)^2}{9} = 1.$$

   (a) Find the foci, vertices, and center of the new ellipse.
   (b) Plot the new foci, vertices, and center, and sketch in the new ellipse.

27. The hyperbola $x^2/16 - y^2/9 = 1$ is shifted 2 units to the right to generate the hyperbola

$$\frac{(x-2)^2}{16} - \frac{y^2}{9} = 1.$$

(a) Find the center, foci, vertices, and asymptotes of the new hyperbola. **(b)** Plot the new center, foci, vertices, and asymptotes, and sketch in the hyperbola.

Exercises 28–31 give equations for conic sections and tell how many units up or down and to the right or left each is to be shifted. Find an equation for the new conic section and find the new vertices, foci, directrices, center, and asymptotes, as appropriate.

28. $y^2 = 4x,$ left 2, down 3

29. $\dfrac{x^2}{6} + \dfrac{y^2}{9} = 1,$ left 2, down 1

30. $\dfrac{x^2}{4} - \dfrac{y^2}{5} = 1,$ right 2, up 2

31. $y^2 - x^2 = 1,$ left 1, up 1

Find the center, foci, vertices, asymptotes, and radius, as appropriate, of each conic section in Exercises 32–36.

32. $x^2 + 4x + y^2 = 12$

33. $2x^2 + 2y^2 - 28x + 12y + 114 = 0$

34. $x^2 + 2x + 4y - 3 = 0$

35. $x^2 + 5y^2 + 4x = 1$

36. $x^2 - y^2 - 2x + 4y = 4$

37. *Archimedes' Formula for the Volume of a Parabolic Solid* The region enclosed by the parabola $y = (4h/b^2)x^2$ and the line $y = h$ is revolved about the y-axis to generate the solid shown here. Show that the volume of the solid is $3/2$ the volume of the corresponding cone.

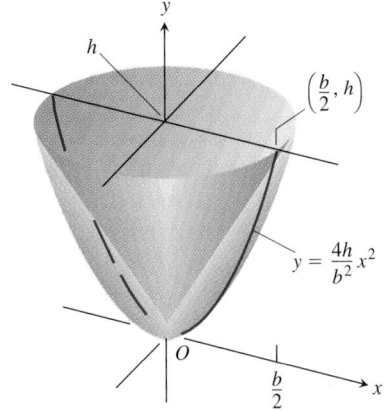

38. *Comparing Volumes* If lines are drawn parallel to the coordinate axes through a point P on the parabola $y^2 = kx$, $k > 0$, the parabola partitions the rectangular region bounded by these lines and the coordinate axes into two smaller regions, A and B.

(a) If the two smaller regions are revolved about the y-axis, show that they generate solids whose volumes have the ratio 4:1.

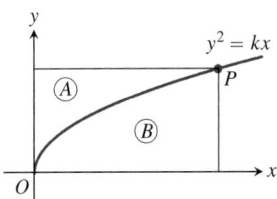

(b) What is the ratio of the volumes of the solids generated by revolving the regions about the x-axis?

39. *Perpendicular Tangents* Show that the tangents to the curve $y^2 = 4px$ from any point on the line $x = -p$ are perpendicular.

40. *Maximizing Area* Find the dimensions of the rectangle of largest area that can be inscribed in the ellipse $x^2 + 4y^2 = 4$ with its sides parallel to the coordinate axes. What is the area of the rectangle?

41. *Volume* Find the volume of the solid generated by revolving the region enclosed by the ellipse $9x^2 + 4y^2 = 36$ about the **(a)** x-axis, **(b)** y-axis.

42. *Volume* The "triangular" region in the first quadrant bounded by the x-axis, the line $x = 4$, and the hyperbola $9x^2 - 4y^2 = 36$ is revolved about the x-axis to generate a solid. Find the volume of the solid.

43. *Volume* The region bounded on the left by the y-axis, on the right by the hyperbola $x^2 - y^2 = 1$, and above and below by the lines $y = \pm 3$ is revolved about the y-axis to generate a solid. Find the volume of the solid.

Extending the Ideas

44. *Suspension Bridge Cables* The suspension bridge cable shown here supports a uniform load of w pounds per horizontal foot. It can be shown that if H is the horizontal tension of the cable at the origin, then the curve of the cable satisfies the equation

$$\frac{dy}{dx} = \frac{w}{H}x.$$

Show that the cable hangs in a parabola by solving this differential equation subject to the initial condition that $y = 0$ when $x = 0$.

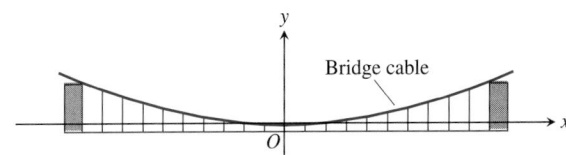

45. *Ripple Tank* Circular waves were made by touching the surface of a ripple tank, first at a point A and shortly thereafter at a nearby point B. As the waves expanded, their points of intersection appeared to trace a hyperbola. Did they really do that? To find out, we can model the waves with circles in the plane centered at nearby points labeled A and B as in the accompanying figure.

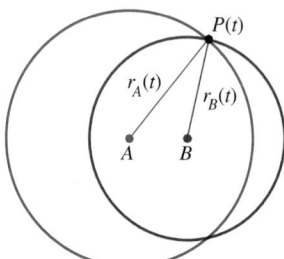

At time t, the point P is $r_A(t)$ units from A and $r_B(t)$ units from B. Since the radii of the circles increase at a constant rate, the rate at which the waves are traveling is

$$\frac{dr_A}{dt} = \frac{dr_B}{dt}.$$

Conclude from this equation that $r_A - r_B$ has a constant value, so that P must lie on a hyperbola with foci at A and B.

46. *How the Astronomer Kepler Used String to Draw Parabolas* Kepler's method for drawing a parabola (with more modern tools) requires a string the length of a T square and a table whose edge can serve as the parabola's directrix. Pin one end of the string to the point where you want the focus to be and the other end to the upper end of the T square. Then, holding the string taut against the T square with a pencil, slide the T square along the table's edge. As the T square moves, the pencil will trace a parabola. Why?

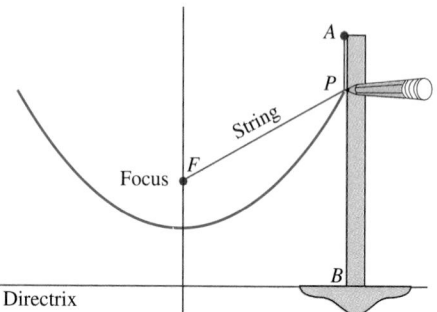

A5.2 Classifying Conic Sections by Eccentricity

Ellipses and Orbits • Hyperbolas • Focus-Directrix Equation

Ellipses and Orbits

We now associate with each conic section a number called the eccentricity. The eccentricity tells whether the conic is a circle, ellipse, parabola, or hyperbola, and, in the case of ellipses and hyperbolas, describes the conic's proportions. We begin with the ellipse.

Although the center-to-focus distance c does not appear in the equation

$$\frac{x^2}{a^2} + \frac{y^2}{b^2} = 1, \quad (a > b)$$

for an ellipse, we can still determine c from the equation $c = \sqrt{a^2 - b^2}$. If we fix a and vary c over the interval $0 \le c \le a$, the resulting ellipses will vary in shape (Figure A5.18). They are circles if $c = 0$ (so that $a = b$) and flatten as c increases. If $c = a$, the foci and vertices overlap and the ellipse degenerates into a line segment.

We use the ratio of c to a to describe the various shapes the ellipse can take. We call this ratio the ellipse's eccentricity.

$c = 0$ $e = 0$

$F_1 = F_2$

$c = \dfrac{4a}{5}$ $e = \dfrac{4}{5}$

F_1 F_2

F_1 $c = a$ $e = 1$ F_2

Figure A5.18 The ellipse changes from a circle to a line segment as c increases from 0 to a.

> **Definition** Eccentricity of Ellipse
>
> The **eccentricity** of the ellipse $x^2/a^2 + y^2/b^2 = 1$ $(a > b)$ is
>
> $$e = \frac{c}{a} = \frac{\sqrt{a^2 - b^2}}{a}.$$

The planets in the solar system revolve around the sun in elliptical orbits with the sun at one focus. Most of the orbits are nearly circular, as can be seen from the eccentricities in Table A5.2. Pluto has a fairly eccentric orbit, with $e = 0.25$, as does Mercury, with $e = 0.21$. Other members of the solar system have orbits that are even more eccentric. Icarus, an asteroid about 1 mile wide that revolves around the sun every 409 Earth days, has an orbital eccentricity of 0.83 (Figure A5.19).

Table A5.2 Eccentricities of planetary orbits	
Mercury	0.21
Venus	0.01
Earth	0.02
Mars	0.09
Jupiter	0.05
Saturn	0.06
Uranus	0.05
Neptune	0.01
Pluto	0.25

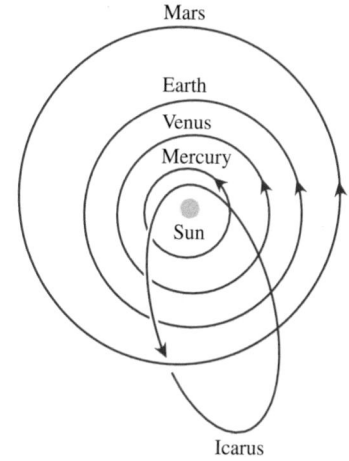

Figure A5.19 The orbit of the asteroid Icarus is highly eccentric. Earth's orbit is so nearly circular that its foci lie inside the sun.

Halley's comet

Edmund Halley (1656–1742; pronounced "*haw*-ley"), British biologist, geologist, sea captain, pirate, spy, Antarctic voyager, astronomer, adviser on fortifications, company founder and director, and the author of the first actuarial mortality tables, was also the mathematician who pushed and harried Newton into writing his *Principia*. Despite these accomplishments, Halley is known today chiefly as the man who calculated the orbit of the great comet of 1682: "wherefore if according to what we have already said [the comet] should return again about the year 1758, candid posterity will not refuse to acknowledge that this was first discovered by an Englishman." Indeed, candid posterity did not refuse — ever since the comet's return in 1758, it has been known as Halley's comet.

Last seen rounding the sun during the winter and spring of 1985–1986, the comet is due to return in the year 2062. The comet has made about 2000 cycles so far with about the same number to go before the sun erodes it away.

Example 1 FINDING THE ECCENTRICITY OF AN ORBIT

The orbit of Halley's comet is an ellipse 36.18 astronomical units long by 9.12 astronomical units wide. (One *astronomical unit* [AU] is 149,597,870 km, the semimajor axis of Earth's orbit.) Its eccentricity is

$$e = \frac{\sqrt{a^2 - b^2}}{a} = \frac{\sqrt{(36.18/2)^2 - (9.12/2)^2}}{(1/2)(36.18)}$$

$$= \frac{\sqrt{(18.09)^2 - (4.56)^2}}{18.09} \approx 0.97.$$

Example 2 LOCATING VERTICES

Locate the vertices of an ellipse of eccentricity 0.8 whose foci lie at the points $(0, \pm 7)$.

Solution Since $e = c/a$, the vertices are the points $(0, \pm a)$ where

$$a = \frac{c}{e} = \frac{7}{0.8} = 8.75,$$

or $(0, \pm 8.75)$.

Whereas a parabola has one focus and one directrix, each ellipse has two foci and two **directrices.** These are the lines perpendicular to the major axis at distances $\pm a/e$ from the center. The parabola has the property that

$$PF = 1 \cdot PD \tag{1}$$

for any point P on it, where F is the focus and D is the point nearest P on the directrix. For an ellipse, it can be shown that the equations that replace (1) are

$$PF_1 = e \cdot PD_1, \qquad PF_2 = e \cdot PD_2. \tag{2}$$

Here, e is the eccentricity, P is any point on the ellipse, F_1 and F_2 are the foci, and D_1 and D_2 are the points on the directrices nearest P (Figure A5.20).

In each equation in (2) the directrix and focus must correspond; that is, if we use the distance from P to F_1, we must also use the distance from P to the directrix at the same end of the ellipse. The directrix $x = -a/e$ corresponds to $F_1(-c, 0)$, and the directrix $x = a/e$ corresponds to $F_2(c, 0)$.

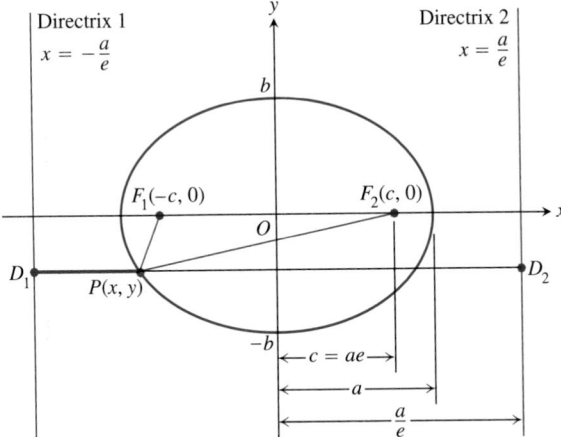

Figure A5.20 The foci and directrices of the ellipse $x^2/a^2 + y^2/b^2 = 1$. Directrix 1 corresponds to focus F_1, and directrix 2 to focus F_2.

Hyperbolas

The eccentricity of a hyperbola is also $e = c/a$, only in this case c equals $\sqrt{a^2 + b^2}$ instead of $\sqrt{a^2 - b^2}$. In contrast to the eccentricity of an ellipse, the eccentricity of a hyperbola is always greater than 1.

Definition Eccentricity of Hyperbola

The **eccentricity** of the hyperbola $x^2/a^2 - y^2/b^2 = 1$ is

$$e = \frac{c}{a} = \frac{\sqrt{a^2 + b^2}}{a}.$$

In both ellipse and hyperbola, the eccentricity is the ratio of the distance between the foci to the distance between the vertices (because $c/a = 2c/2a$).

$$\text{Eccentricity} = \frac{\text{distance between foci}}{\text{distance between vertices}}$$

In an ellipse, the foci are closer together than the vertices and the ratio is less than 1. In a hyperbola, the foci are farther apart than the vertices and the ratio is greater than 1.

Example 3 FINDING ECCENTRICITY

Find the eccentricity of the hyperbola $9x^2 - 16y^2 = 144$.

Solution We divide both sides of the hyperbola's equation by 144 to put it in standard form, obtaining

$$\frac{9x^2}{144} - \frac{16y^2}{144} = 1 \quad \text{or} \quad \frac{x^2}{16} - \frac{y^2}{9} = 1.$$

With $a^2 = 16$ and $b^2 = 9$, we find that $c = \sqrt{a^2 + b^2} = \sqrt{16 + 9} = 5$, so

$$e = \frac{c}{a} = \frac{5}{4}.$$

As with the ellipse, it can be shown that the lines $x = \pm a/e$ act as **directrices** for the hyperbola and that

$$PF_1 = e \cdot PD_1 \quad \text{and} \quad PF_2 = e \cdot PD_2. \tag{3}$$

Here P is any point on the hyperbola, F_1 and F_2 are the foci, and D_1 and D_2 are the points nearest P on the directrices (Figure A5.21).

Focus-Directrix Equation

To complete the picture, we define the eccentricity of a parabola to be $e = 1$. Equations 1–3 then have the common form $PF = e \cdot PD$.

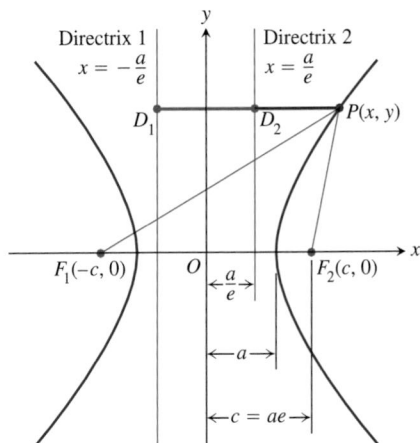

Figure A5.21 The foci and directrices of the hyperbola $x^2/a^2 - y^2/b^2 = 1$. No matter where P lies on the hyperbola, $PF_1 = e \cdot PD_1$, and $PF_2 = e \cdot PD_2$.

> **Definition** Eccentricity of Parabola
> The **eccentricity** of a parabola is $e = 1$.

The **focus-directrix equation** $PF = e \cdot PD$ unites the parabola, ellipse, and hyperbola in the following way. Suppose that the distance PF of a point P from a fixed point F (the focus) is a constant multiple of its distance from a fixed line (the directrix). That is, suppose

$$PF = e \cdot PD, \tag{4}$$

where e is the constant of proportionality. Then the path traced by P is

(a) a *parabola* if $e = 1$,

(b) an *ellipse* of eccentricity e if $e < 1$, and

(c) a *hyperbola* of eccentricity e if $e > 1$.

Equation 4 may not look like much to get excited about. There are no coordinates in it and when we try to translate it into coordinate form it translates in different ways, depending on the size of e. At least, that is what happens in Cartesian coordinates. However, in polar coordinates, the equation

$PF = e \cdot PD$ translates into a single equation regardless of the value of e, an equation so simple that it has been the equation of choice of astronomers and space scientists for nearly 300 years.

Given the focus and corresponding directrix of a hyperbola centered at the origin and with foci on the x-axis, we can use the dimensions shown in Figure A5.21 to find e. Knowing e, we can derive a Cartesian equation for the hyperbola from the equation $PF = e \cdot PD$, as in the next example. We can find equations for ellipses centered at the origin and with foci on the x-axis in a similar way, using the dimensions shown in Figure A5.20.

Example 4 USING FOCUS AND DIRECTRIX

Find a Cartesian equation for the hyperbola centered at the origin that has a focus at $(3, 0)$ and the line $x = 1$ as the corresponding directrix.

Solution We first use the dimensions shown in Figure A5.21 to find the hyperbola's eccentricity. The focus is

$$(c, 0) = (3, 0), \quad \text{so} \quad c = 3.$$

The directrix is the line

$$x = \frac{a}{e} = 1, \quad \text{so} \quad a = e.$$

When combined with the equation $e = c/a$ that defines eccentricity, these results give

$$e = \frac{c}{a} = \frac{3}{e}, \quad \text{so} \quad e^2 = 3 \quad \text{and} \quad e = \sqrt{3}.$$

Knowing e, we can now derive the equation we want from the equation $PF = e \cdot PD$. In the notation of Figure A5.22, we have

$$PF = e \cdot PD \qquad \text{Eq. 4}$$
$$\sqrt{(x - 3)^2 + (y - 0)^2} = \sqrt{3}\,|x - 1| \qquad e = \sqrt{3}$$
$$x^2 - 6x + 9 + y^2 = 3(x^2 - 2x + 1)$$
$$2x^2 - y^2 = 6$$
$$\frac{x^2}{3} - \frac{y^2}{6} = 1.$$

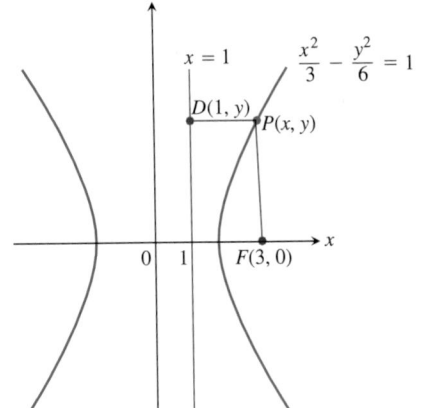

Figure A5.22 The hyperbola in Example 4.

Section A5.2 Exercises

In Exercises 1–4, find the eccentricity, foci, and directrices of the ellipse.

1. $16x^2 + 25y^2 = 400$

2. $2x^2 + y^2 = 2$

3. $3x^2 + 2y^2 = 6$

4. $6x^2 + 9y^2 = 54$

Exercises 5–8 give the foci or vertices and the eccentricities of ellipses centered at the origin of the xy-plane. In each case, find the ellipse's standard-form equation.

5. Foci: $(0, \pm 3)$
Eccentricity: 0.5

6. Foci: $(\pm 8, 0)$
Eccentricity: 0.2

7. Vertices: $(\pm 10, 0)$
Eccentricity: 0.24

8. Vertices: $(0, \pm 70)$
Eccentricity: 0.1

Exercises 9 and 10 give foci and corresponding directrices of ellipses centered at the origin of the xy-plane. In each case, use the dimensions in Figure A5.20 to find the eccentricity of the ellipse. Then find the ellipse's standard-form equation.

9. Focus: $(\sqrt{5}, 0)$
Directrix: $x = \dfrac{9}{\sqrt{5}}$

10. Focus: $(-4, 0)$
Directrix: $x = -16$

11. Draw an ellipse of eccentricity 4/5. Explain your procedure.

12. Draw the orbit of Pluto (eccentricity 0.25) to scale. Explain your procedure.

13. The endpoints of the major and minor axes of an ellipse are $(1, 1)$, $(3, 4)$, $(1, 7)$, and $(-1, 4)$. Sketch the ellipse, give its equation in standard form, and find its foci, eccentricity, and directrices.

14. Find an equation for the ellipse of eccentricity 2/3 that has the line $x = 9$ as a directrix and the point $(4, 0)$ as the corresponding focus.

In Exercises 15–18, find the eccentricity, foci, and directrices of the hyperbola.

15. $9x^2 - 16y^2 = 144$ **16.** $y^2 - x^2 = 8$

17. $8x^2 - 2y^2 = 16$ **18.** $8y^2 - 2x^2 = 16$

Exercises 19 and 20 give the eccentricities and the vertices or foci of hyperbolas centered at the origin of the xy-plane. In each case, find the hyperbola's standard-form equation.

19. Eccentricity: 3   **20.** Eccentricity: 3
     Vertices: $(0, \pm 1)$ Foci: $(\pm 3, 0)$

Exercises 21 and 22 give foci and corresponding directrices of hyperbolas centered at the origin of the xy-plane. In each case, find the hyperbola's eccentricity. Then find the hyperbola's standard-form equation.

21. Focus: $(4, 0)$ **22.** Focus: $(-2, 0)$
     Directrix: $x = 2$ Directrix: $x = -\dfrac{1}{2}$

23. A hyperbola of eccentricity 3/2 has one focus at $(1, -3)$. The corresponding directrix is the line $y = 2$. Find an equation for the hyperbola.

Explorations

24. *The Effect of Eccentricity on a Hyperbola's Shape* What happens to the graph of a hyperbola as its eccentricity increases? To find out, rewrite the equation $x^2/a^2 - y^2/b^2 = 1$ in terms of a and e instead of a and b. Graph the hyperbola for various values of e and describe what you find.

25. *Determining Constants* What values of the constants a, b, and c make the ellipse

$$4x^2 + y^2 + ax + by + c = 0$$

lie tangent to the x-axis at the origin and pass through the point $(-1, 2)$? What is the eccentricity of the ellipse? ■

Extending the Ideas

26. *The Reflective Property of Ellipses* An ellipse is revolved about its major axis to generate an ellipsoid. The inner surface of the ellipsoid is silvered to make a mirror. Show that a ray of light emanating from one focus will be reflected to the other focus. Sound waves also follow such paths, and this property is used in constructing "whispering galleries." (*Hint:* Place the ellipse in standard position in the xy-plane and show that the lines from a point P on the ellipse to the two foci make congruent angles with the tangent to the ellipse at P.)

27. *The Reflective Property of Hyperbolas* Show that a ray of light directed toward one focus of a hyperbolic mirror, as in the accompanying figure, is reflected toward the other focus. (*Hint:* Show that the tangent to the hyperbola at P bisects the angle made by segments PF_1 and PF_2.)

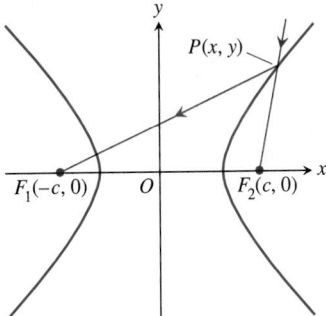

28. *A Confocal Ellipse and Hyperbola* Show that an ellipse and a hyperbola that have the same foci A and B, as in the accompanying figure, cross at right angles at their points of intersection. [*Hint:* A ray of light from focus A that met the hyperbola at P would be reflected from the hyperbola as if it came directly from B (Exercise 27). The same ray would be reflected off the ellipse to pass through B (Exercise 26).]

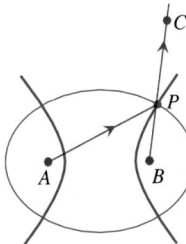

A5.3 Quadratic Equations and Rotations

Quadratic Curves • Cross Product Term • Rotating Axes to
Eliminate *Bxy* • Possible Graphs of Quadratic Equations •
Discriminant Test • Technology Application

Quadratic Curves

In this section, we examine one of the most amazing results in analytic geometry, which is that the Cartesian graph of any equation

$$Ax^2 + Bxy + Cy^2 + Dx + Ey + F = 0, \qquad (1)$$

in which A, B, and C are not all zero, is nearly always a conic section. The exceptions are the cases in which there is no graph at all or the graph consists of two parallel lines. It is conventional to call all graphs of Equation 1, curved or not, **quadratic curves.**

Cross Product Term

You may have noticed that the term Bxy did not appear in the equations for the conic sections in Section A5.1. This happened because the axes of the conic sections ran parallel to (in fact, coincided with) the coordinate axes.

To see what happens when the parallelism is absent, let us write an equation for a hyperbola with $a = 3$ and foci at $F_1(-3, -3)$ and $F_2(3, 3)$ (Figure A5.23). The equation $|PF_1 - PF_2| = 2a$ becomes $|PF_1 - PF_2| = 2(3) = 6$ and

$$\sqrt{(x + 3)^2 + (y + 3)^2} - \sqrt{(x - 3)^2 + (y - 3)^2} = \pm 6.$$

When we transpose one radical, square, solve for the remaining radical and square again, the equation reduces to

$$2xy = 9, \qquad (2)$$

a case of Equation 1 in which the cross product term is present. The asymptotes of the hyperbola in Equation 2 are the x- and y-axes, and the focal axis makes an angle of $\pi/4$ radians with the positive x-axis. As in this example, the cross product term is present in Equation 1 only when the axes of the conic are tilted.

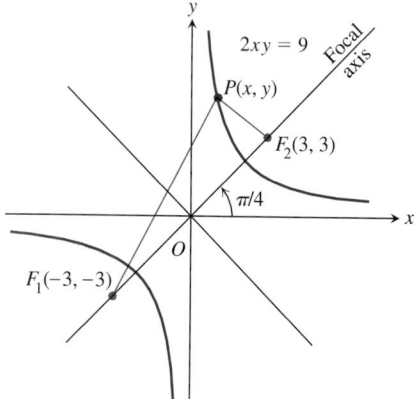

Figure A5.23 The focal axis of the hyperbola $2xy = 9$ makes an angle of $\pi/4$ radians with the positive x-axis.

Rotating Axes to Eliminate *Bxy*

To eliminate the xy-term from the equation of a conic, we rotate the coordinate axes to eliminate the "tilt" in the axes of the conic. The equations for the rotations we use are derived in the following way. In the notation of Figure A5.24, which shows a counterclockwise rotation about the origin through an angle α,

$$x = OM = OP \cos(\theta + \alpha) = OP \cos\theta \cos\alpha - OP \sin\theta \sin\alpha$$

$$y = MP = OP \sin(\theta + \alpha) = OP \cos\theta \sin\alpha + OP \sin\theta \cos\alpha. \qquad (3)$$

Since

$$OP \cos\theta = OM' = x'$$

and

$$OP \sin\theta = M'P = y',$$

the equations in (3) reduce to the following.

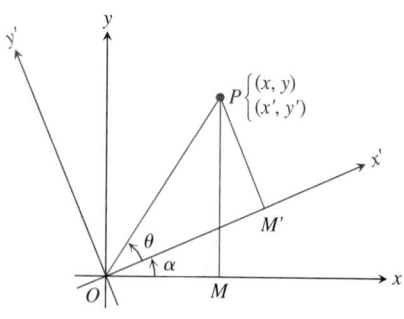

Figure A5.24 A counterclockwise rotation through angle α about the origin.

Equations for Rotating Coordinate Axes

$$x = x' \cos \alpha - y' \sin \alpha$$

$$y = x' \sin \alpha + y' \cos \alpha$$

(4)

Example 1 CHANGING AN EQUATION

The x- and y-axes are rotated through an angle of $\pi/4$ radians about the origin. Find an equation for the hyperbola $2xy = 9$ in the new coordinates.

Solution Since $\cos \pi/4 = \sin \pi/4 = 1/\sqrt{2}$, we substitute

$$x = \frac{x' - y'}{\sqrt{2}}, \qquad y = \frac{x' + y'}{\sqrt{2}}$$

from Equations 4 into the equation $2xy = 9$, obtaining

$$2\left(\frac{x' - y'}{\sqrt{2}}\right)\left(\frac{x' + y'}{\sqrt{2}}\right) = 9$$

$$x'^2 - y'^2 = 9$$

$$\frac{x'^2}{9} - \frac{y'^2}{9} = 1.$$

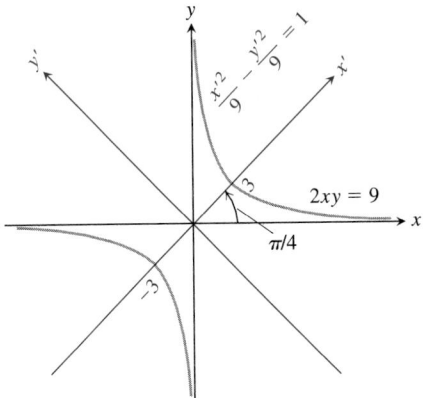

Figure A5.25 The hyperbola in Example 1 (x' and y' are the new coordinates).

See Figure A5.25.

If we apply Equations 4 to the quadratic Equation 1, we obtain a new quadratic equation

$$A'x'^2 + B'x'y' + C'y'^2 + D'x' + E'y' + F' = 0. \tag{5}$$

The new and old coefficients are related by the equations

$$A' = A \cos^2 \alpha + B \cos \alpha \sin \alpha + C \sin^2 \alpha$$

$$B' = B \cos 2\alpha + (C - A) \sin 2\alpha$$

$$C' = A \sin^2 \alpha - B \sin \alpha \cos \alpha + C \cos^2 \alpha \tag{6}$$

$$D' = D \cos \alpha + E \sin \alpha$$

$$E' = -D \sin \alpha + E \cos \alpha$$

$$F' = F.$$

Equations 6 show, among other things, that if we start with an equation for a curve in which the cross product term is present ($B \neq 0$), we can find a rotation angle α that produces an equation in which no cross product term appears ($B' = 0$). To find α, we set $B' = 0$ in the second equation in (6) and solve the resulting equation,

$$B \cos 2\alpha + (C - A) \sin 2\alpha = 0,$$

for α. In practice, this means determining α from one of the two equations.

$$\cot 2\alpha = \frac{A - C}{B} \qquad \text{or} \qquad \tan 2\alpha = \frac{B}{A - C}. \tag{7}$$

Example 2 ELIMINATING A CROSS PRODUCT TERM

The coordinate axes are to be rotated through an angle α to produce an equation for the curve

$$2x^2 + \sqrt{3}xy + y^2 - 10 = 0$$

that has no cross product term. Find a suitable α and the corresponding new equation. Identify the curve.

Solution The equation $2x^2 + \sqrt{3}xy + y^2 - 10 = 0$ has $A = 2$, $B = \sqrt{3}$, and $C = 1$. We substitute these values into Equation 7 to find α:

$$\cot 2\alpha = \frac{A - C}{B} = \frac{2 - 1}{\sqrt{3}} = \frac{1}{\sqrt{3}}.$$

From the right triangle in Figure A5.26, we see that one appropriate choice of angle is $2\alpha = \pi/3$, so we take $\alpha = \pi/6$. Substituting $\alpha = \pi/6$, $A = 2$, $B = \sqrt{3}$, $C = 1$, $D = E = 0$, and $F = -10$ into Equations 6 gives

$$A' = \frac{5}{2}, \quad B' = 0, \quad C' = \frac{1}{2}, \quad D' = E' = 0, \quad F' = -10.$$

Equation 5 then gives

$$\frac{5}{2}x'^2 + \frac{1}{2}y'^2 - 10 = 0, \quad \text{or} \quad \frac{x'^2}{4} + \frac{y'^2}{20} = 1.$$

The curve is an ellipse with foci on the new y'-axis (Figure A5.27).

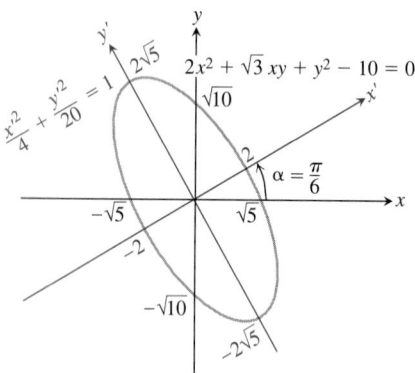

Figure A5.26 This triangle identifies $2\alpha = \cot^{-1}(1/\sqrt{3})$ as $\pi/3$. (Example 2)

Figure A5.27 The conic section in Example 2.

Possible Graphs of Quadratic Equations

We now return to the graph of the general quadratic equation.

Since axes can always be rotated to eliminate the cross product term, there is no loss of generality in assuming that this has been done and that the equation has the form

$$Ax^2 + Cy^2 + Dx + Ey + F = 0. \tag{8}$$

Equation 8 represents

(a) a *circle* if $A = C \neq 0$ (special cases: the graph is a point or there is no graph at all);

(b) a *parabola* if Equation 8 is quadratic in one variable and linear in the other;

(c) an *ellipse* if A and C are both positive or both negative (special cases: circles, a single point, or no graph at all);

(d) a *hyperbola* if A and C have opposite signs (special case: a pair of intersecting lines);

(e) a *straight line* if A and C are zero and at least one of D and E is different from zero;

(f) *one or two straight lines* if the left-hand side of Equation 8 can be factored into the product of two linear factors.

See Table A5.3 (on page 960) for examples.

Discriminant Test

We do not need to eliminate the xy-term from the equation

$$Ax^2 + Bxy + Cy^2 + Dx + Ey + F = 0 \qquad (9)$$

to tell what kind of conic section the equation represents. If this is the only information we want, we can apply the following test instead.

As we have seen, if $B \neq 0$, then rotating the coordinate axes through an angle α that satisfies the equation

$$\cot 2\alpha = \frac{A - C}{B} \qquad (10)$$

will change Equation 9 into an equivalent form

$$A'x'^2 + C'y'^2 + D'x' + E'y' + F' = 0 \qquad (11)$$

without a cross product term.

Now, the graph of Equation 11 is a (real or degenerate)

(a) *parabola* if A' or $C' = 0$; that is, if $A'C' = 0$;

(b) *ellipse* if A' and C' have the same sign; that is, if $A'C' > 0$;

(c) *hyperbola* if A' and C' have opposite signs; that is, if $A'C' < 0$.

It can also be verified from Equations 6 that for any rotation of axes,

$$B^2 - 4AC = B'^2 - 4A'C'. \qquad (12)$$

This means that the quantity $B^2 - 4AC$ is not changed by a rotation. But when we rotate through the angle α given by Equation 10, B' becomes zero, so

$$B^2 - 4AC = -4A'C'.$$

Since the curve is a parabola if $A'C' = 0$, an ellipse if $A'C' > 0$, and a hyperbola if $A'C' < 0$, the curve must be a parabola if $B^2 - 4AC = 0$, an ellipse if $B^2 - 4AC < 0$, and a hyperbola if $B^2 - 4AC > 0$. The number $B^2 - 4AC$ is called the **discriminant** of Equation 9.

Discriminant Test

With the understanding that occasional degenerate cases may arise, the quadratic curve $Ax^2 + Bxy + Cy^2 + Dx + Ey + F = 0$ is

(a) a **parabola** if $B^2 - 4AC = 0$,

(b) an **ellipse** if $B^2 - 4AC < 0$,

(c) a **hyperbola** if $B^2 - 4AC > 0$.

Example 3   APPLYING THE DISCRIMINANT TEST

(a) $3x^2 - 6xy + 3y^2 + 2x - 7 = 0$ represents a parabola because

$$B^2 - 4AC = (-6)^2 - 4 \cdot 3 \cdot 3 = 36 - 36 = 0.$$

(b) $x^2 + xy + y^2 - 1 = 0$ represents an ellipse because

$$B^2 - 4AC = (1)^2 - 4 \cdot 1 \cdot 1 = -3 < 0.$$

(c) $xy - y^2 - 5y + 1 = 0$ represents a hyperbola because

$$B^2 - 4AC = (1)^2 - 4(0)(-1) = 1 > 0.$$

Technology Application

How Some Calculators Use Rotations to Evaluate Sines and Cosines
Some calculators use rotations to calculate sines and cosines of arbitrary angles. The procedure goes something like this: The calculator has, stored,

1. ten angles or so, say

$$\alpha_1 = \sin^{-1}(10^{-1}), \quad \alpha_2 = \sin^{-1}(10^{-2}), \quad \ldots, \quad \alpha_{10} = \sin^{-1}(10^{-10}),$$

and

2. twenty numbers, the sines and cosines of the angles $\alpha_1, \alpha_2, \ldots, \alpha_{10}$.

To calculate the sine and cosine of an arbitrary angle θ, we enter θ (in radians) into the calculator. The calculator substracts or adds multiples of 2π to θ to replace θ by the angle between 0 and 2π that has the same sine and cosine as θ (we continue to call the angle θ). The calculator then "writes" θ as a sum of multiples of α_1 (as many as possible without overshooting) plus multiples of α_2 (again, as many as possible), and so on, working its way to α_{10}. This gives

$$\theta \approx m_1\alpha_1 + m_2\alpha_2 + \cdots + m_{10}\alpha_{10}.$$

The calculator then rotates the point $(1, 0)$ through m_1 copies of α_1 (through α_1, m_1 times in succession), plus m_2 copies of α_2, and so on, finishing off with m_{10} copies of α_{10} (Figure A5.28). The coordinates of the final position of $(1, 0)$ on the unit circle are the values the calculator gives for $(\cos\theta, \sin\theta)$.

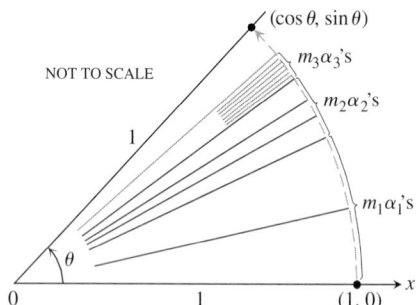

Figure A5.28 To calculate the sine and cosine of an angle θ between 0 and 2π, the calculator rotates the point $(1, 0)$ to an appropriate location on the unit circle and displays the resulting coordinates.

Table A5.3 Examples of quadratic curves

$$Ax^2 + Bxy + Cy^2 + Dx + Ey + F = 0$$

	A	B	C	D	E	F	Equation	Remarks
Circle	1		1			−4	$x^2 + y^2 = 4$	$A = C; F < 0$
Parabola			1	−9			$y^2 = 9x$	Quadratic in y, linear in x
Ellipse	4		9			−36	$4x^2 + 9y^2 = 36$	A, C have same sign, $A \neq C; F < 0$
Hyperbola	1		−1			−1	$x^2 - y^2 = 1$	A, C have opposite signs
One line (still a conic section)	1						$x^2 = 0$	y-axis
Intersecting lines (still a conic section)		1		1	−1	−1	$xy + x - y - 1 = 0$	Factors to $(x - 1)(y + 1) = 0$, so $x = 1, y = -1$
Parallel lines (not a conic section)	1			−3		2	$x^2 - 3x + 2 = 0$	Factors to $(x - 1)(x - 2) = 0$, so $x = 1, x = 2$
Point	1		1				$x^2 + y^2 = 0$	The origin
No graph	1					1	$x^2 = -1$	No graph

Section A5.3 Exercises

Use the discriminant $B^2 - 4AC$ to decide whether the equations in Exercises 1–16 represent parabolas, ellipses, or hyperbolas.

1. $x^2 - 3xy + y^2 - x = 0$

2. $3x^2 - 18xy + 27y^2 - 5x + 7y = -4$

3. $3x^2 - 7xy + \sqrt{17}y^2 = 1$

4. $2x^2 - \sqrt{15}xy + 2y^2 + x + y = 0$

5. $x^2 + 2xy + y^2 + 2x - y + 2 = 0$

6. $2x^2 - y^2 + 4xy - 2x + 3y = 6$

7. $x^2 + 4xy + 4y^2 - 3x = 6$

8. $x^2 + y^2 + 3x - 2y = 10$

9. $xy + y^2 - 3x = 5$

10. $3x^2 + 6xy + 3y^2 - 4x + 5y = 12$

11. $3x^2 - 5xy + 2y^2 - 7x - 14y = -1$

12. $2x^2 - 4.9xy + 3y^2 - 4x = 7$

13. $x^2 - 3xy + 3y^2 + 6y = 7$

14. $25x^2 + 21xy + 4y^2 - 350x = 0$

15. $6x^2 + 3xy + 2y^2 + 17y + 2 = 0$

16. $3x^2 + 12xy + 12y^2 + 435x - 9y + 72 = 0$

In Exercises 17–26, rotate the coordinate axes to change the given equation into an equation that has no cross product (xy) term. Then identify the graph of the equation. (The new equations will vary with the size and direction of the rotation you use.)

17. $xy = 2$ **18.** $x^2 + xy + y^2 = 1$

19. $3x^2 + 2\sqrt{3}xy + y^2 - 8x + 8\sqrt{3}y = 0$

20. $x^2 - \sqrt{3}xy + 2y^2 = 1$

21. $x^2 - 2xy + y^2 = 2$

22. $3x^2 - 2\sqrt{3}xy + y^2 = 1$

23. $\sqrt{2}x^2 + 2\sqrt{2}xy + \sqrt{2}y^2 - 8x + 8y = 0$

24. $xy - y - x + 1 = 0$

25. $3x^2 + 2xy + 3y^2 = 19$

26. $3x^2 + 4\sqrt{3}xy - y^2 = 7$

27. Find the sine and cosine of an angle through which the coordinate axes can be rotated to eliminate the cross product term from the equation

$$14x^2 + 16xy + 2y^2 - 10x + 26{,}370y - 17 = 0.$$

Do not carry out the rotation.

28. Find the sine and cosine of an angle through which the coordinate axes can be rotated to eliminate the cross product term from the equation

$$4x^2 - 4xy + y^2 - 8\sqrt{5}x - 16\sqrt{5}y = 0.$$

Do not carry out the rotation.

The conic sections in Exercises 17–26 were chosen to have rotation angles that were "nice" in the sense that once we knew $\cot 2\alpha$ or $\tan 2\alpha$ we could identify 2α and find $\sin \alpha$ and $\cos \alpha$ from familiar triangles. The conic sections encountered in practice may not have such nice rotation angles, and we may have to use a calculator to determine α from the value of $\cot 2\alpha$ or $\tan 2\alpha$.

In Exercises 29–34, use a calculator to find an angle α through which the coordinate axes can be rotated to change the given equation into a quadratic equation that has no cross product term. Then find $\sin \alpha$ and $\cos \alpha$ to two decimal places and use Equations 6 to find the coefficients of the new equation to the nearest decimal place. In each case, say whether the conic section is an ellipse, hyperbola, or parabola.

29. $x^2 - xy + 3y^2 + x - y - 3 = 0$

30. $2x^2 + xy - 3y^2 + 3x - 7 = 0$

31. $x^2 - 4xy + 4y^2 - 5 = 0$

32. $2x^2 - 12xy + 18y^2 - 49 = 0$

33. $3x^2 + 5xy + 2y^2 - 8y - 1 = 0$

34. $2x^2 + 7xy + 9y^2 + 20x - 86 = 0$

35. *The Hyperbola* $xy = a$ The hyperbola $xy = 1$ is one of many hyperbolas of the form $xy = a$ that appear in science and mathematics.

   (a) Rotate the coordinate axes through an angle of $45°$ to change the equation $xy = 1$ into an equation with no xy-term. What is the new equation?

   (b) Do the same for the equation $xy = a$.

36. **Writing to Learn** Can anything be said about the graph of the equation $Ax^2 + Bxy + Cy^2 + Dx + Ey + F = 0$ if $AC < 0$? Give reasons for your answer.

37. **Writing to Learn** Does any nondegenerate conic section $Ax^2 + Bxy + Cy^2 + Dx + Ey + F = 0$ have all of the following properties?

   (a) It is symmetric with respect to the origin.

   (b) It passes through the point $(1, 0)$.

   (c) It is tangent to the line $y = 1$ at the point $(-2, 1)$.

   Give reasons for your answer.

38. *When A = C* Show that rotating the axes through an angle of $\pi/4$ radians will eliminate the xy-term from Equation 1 whenever $A = C$.

Explorations

39. *90° Rotations* What effect does a 90° rotation about the origin have on the equations of the following conic sections? Give the new equation in each case.

(a) The ellipse $x^2/a^2 + y^2/b^2 = 1$ $(a > b)$

(b) The hyperbola $x^2/a^2 - y^2/b^2 = 1$

(c) The circle $x^2 + y^2 = a^2$

(d) The line $y = mx$

(e) The line $y = mx + b$

40. *180° Rotations* What effect does a 180° rotation about the origin have on the equations of the following conic sections? Give the new equation in each case.

(a) The ellipse $x^2/a^2 + y^2/b^2 = 1$ $(a > b)$

(b) The hyperbola $x^2/a^2 - y^2/b^2 = 1$

(c) The circle $x^2 + y^2 = a^2$

(d) The line $y = mx$

(e) The line $y = mx + b$ ∎

41. *Identifying a Conic Section*

(a) What kind of conic section is the curve $xy + 2x - y = 0$?

(b) Solve the equation $xy + 2x - y = 0$ for y and sketch the curve as the graph of a rational function of x.

(c) Find equations for the lines parallel to the line $y = -2x$ that are normal to the curve. Add the lines to your sketch.

42. *Sign of AC* Prove or find counterexamples to the following statements about the graph of $Ax^2 + Bxy + Cy^2 + Dx + Ey + F = 0$.

(a) If $AC > 0$, the graph is an ellipse.

(b) If $AC > 0$, the graph is a hyperbola.

(c) If $AC < 0$, the graph is a hyperbola.

Extending the Ideas

43. *Degenerate Conic Section*

(a) Decide whether the equation

$$x^2 + 4xy + 4y^2 + 6x + 12y + 9 = 0$$

represents an ellipse, a parabola, or a hyperbola.

(b) Show that the graph of the equation in (a) is the line $2y = -x - 3$.

44. *Degenerate Conic Section*

(a) Decide whether the conic section with equation

$$9x^2 + 6xy + y^2 - 12x - 4y + 4 = 0$$

represents a parabola, an ellipse, or a hyperbola.

(b) Show that the graph of the equation in (a) is the line $y = -3x + 2$.

45. *A Nice Area Formula for Ellipses* When $B^2 - 4AC$ is negative, the equation

$$Ax^2 + Bxy + Cy^2 = 1$$

represents an ellipse. If the ellipse's semi-axes are a and b, its area is πab (a standard formula). Show that the area is also $2\pi/\sqrt{4AC - B^2}$. (*Hint:* Rotate the coordinate axes to eliminate the xy-term and apply Equation 12 to the new equation.)

46. *Other Rotation Invariants* We describe the fact that $B'^2 - 4A'C'$ equals $B^2 - 4AC$ after a rotation about the origin by saying that the discriminant of a quadratic equation is an **invariant** of the equation. Use Equations 6 to show that the numbers **(a)** $A + C$ and **(b)** $D^2 + E^2$ are also invariants, in the sense that

$$A' + C' = A + C \quad \text{and} \quad D'^2 + E'^2 = D^2 + E^2.$$

We can use these equalities to check against numerical errors when we rotate axes. They can also be helpful in shortening the work required to find values for the new coefficients.

A5.4 Polar Equations for Conic Sections

The Problem with Rectangular Coordinates • Lines • Circles • Ellipses, Parabolas, and Hyperbolas from One Simple Equation

The Problem with Rectangular Coordinates

One problem with using rectangular coordinates in astronomy and astronautical engineering is that in rectangular coordinates we need three different kinds of equations for modeling the paths along which celestial objects move when subject to the inverse-square force of gravitation. We need one kind of equation for objects whose paths are ellipses (planetary orbits), another kind for objects that trace hyperbolas (certain comets and other one-time visitors from outside the solar system), and still another for parabolas (not seen in practice, perhaps, but useful to the theory). With a shift to polar coordinates, however, we find that all these different orbits can be described with one simple equa-

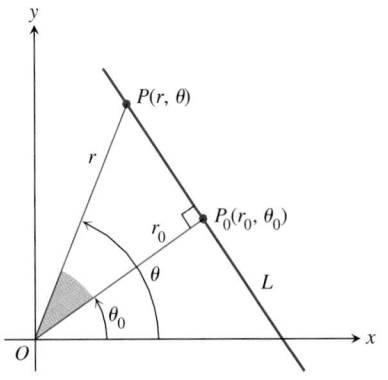

A5.29 We obtain a polar equation for line L by reading the relation $r_0/r = \cos(\theta - \theta_0)$ from triangle OP_0P.

tion (which is why polar and cylindrical coordinates are used so extensively today in the study of motion in space). We develop that equation here.

Since it is also useful from time to time to be able to produce polar equations for lines, we begin our treatment with them.

Lines

Suppose the perpendicular from the origin to line L meets L at the point $P_0(r_0, \theta_0)$, with $r_0 \geq 0$ (Figure A5.29). Then, if $P(r, \theta)$ is any other point on L, the points P, P_0, and O are the vertices of a right triangle, from which we can read the relation

$$\frac{r_0}{r} = \cos(\theta - \theta_0)$$

or

$$r \cos(\theta - \theta_0) = r_0.$$

The Standard Polar Equation for Lines

If the point $P_0(r_0, \theta_0)$ is the foot of the perpendicular from the origin to the line L and if $r_0 \geq 0$, then an equation for L is

$$r \cos(\theta - \theta_0) = r_0. \tag{1}$$

Example 1   APPLYING EQUATION (1)

Find an equation for the line in Figure A5.30.

Solution   From the figure, we see that the foot of the perpendicular from the origin to the line is the point $(2, \pi/3)$. With $\theta = \pi/3$ and $r_0 = 2$, Equation (1) becomes

$$r \cos\left(\theta - \frac{\pi}{3}\right) = 2.$$

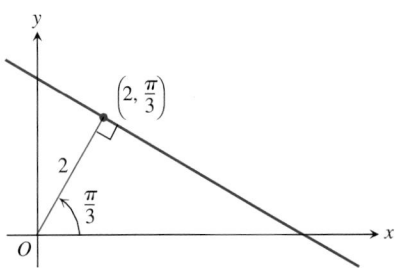

A5.30 The standard polar equation of this line is

$$r \cos\left(\theta - \frac{\pi}{3}\right) = 2.$$

(Example 1)

Example 2   CONVERTING FROM POLAR TO CARTESIAN

Use the identity $\cos(A - B) = \cos A \cos B + \sin A \sin B$ to find a Cartesian equation for the line in Example 1.

Solution
$$r \cos\left(\theta - \frac{\pi}{3}\right) = 2$$

$$r\left(\cos\theta \cos\frac{\pi}{3} + \sin\theta \sin\frac{\pi}{3}\right) = 2$$

$$\frac{1}{2} r \cos\theta + \frac{\sqrt{3}}{2} r \sin\theta = 2$$

$$\frac{1}{2} x + \frac{\sqrt{3}}{2} y = 2$$

$$x + \sqrt{3}y = 4$$

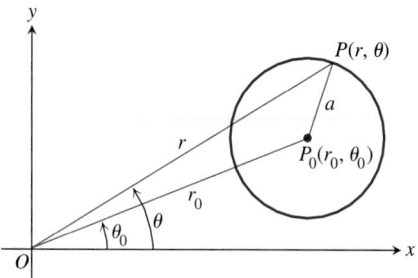

A5.31 We can get a polar equation for this circle by applying the Law of Cosines to triangle OP_0P.

Circles

To find a polar equation for the circle of radius a centered at $P_0(r_0, \theta_0)$, we let $P(r, \theta)$ be a point on the circle and apply the Law of Cosines to triangle OP_0P (Figure A5.31). This gives

$$a^2 = r_0^2 + r^2 - 2r_0 r \cos (\theta - \theta_0). \tag{2}$$

If the circle passes through the origin, then $r_0 = a$ and Equation (2) simplifies to

$$a^2 = a^2 + r^2 - 2ar \cos (\theta - \theta_0) \qquad \text{Eq. (2) with } r_0 = a$$

$$r^2 = 2ar \cos (\theta - \theta_0)$$

$$r = 2a \cos (\theta - \theta_0). \tag{3}$$

If the circle's center lies on the positive x-axis, $\theta_0 = 0$ and Equation (3) becomes

$$r = 2a \cos \theta. \tag{4}$$

If the center lies on the positive y-axis, $\theta = \pi/2$, $\cos (\theta - \pi/2) = \sin \theta$, and Equation (3) becomes

$$r = 2a \sin \theta. \tag{5}$$

Equations for circles through the origin centered on the negative x- and y-axes can be obtained from Equations (4) and (5) by replacing r with $-r$.

Polar Equations for Circles Through the Origin Centered On the x- and y-Axes, Radius a

Example 3   CIRCLES THROUGH THE ORIGIN

Radius	Center (polar coordinates)	Equation
3	$(3, 0)$	$r = 6 \cos \theta$
2	$(2, \pi/2)$	$r = 4 \sin \theta$
1/2	$(-1/2, 0)$	$r = -\cos \theta$
1	$(-1, \pi/2)$	$r = -2 \sin \theta$

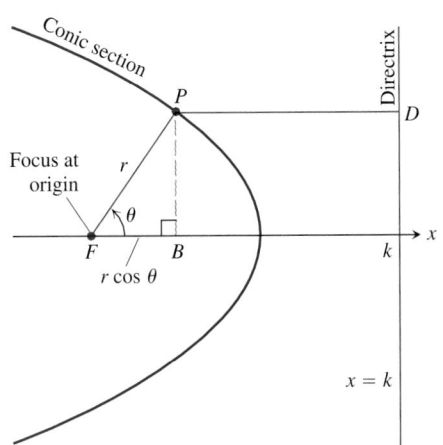

A5.32 If a conic section is put in this position, then $PF = r$ and $PD = k - r \cos \theta$.

Ellipses, Parabolas, and Hyperbolas from One Simple Equation

To find polar equation for ellipses, parabolas, and hyperbolas, we place one focus at the origin and the corresponding directrix to the right of the origin along the vertical line $x = k$ (Figure A5.32). This makes

$$PF = r$$

and

$$PD = k - FB = k - r \cos \theta.$$

The conic's focus–directrix equation $PF = e \cdot PD$ then becomes

$$r = e(k - r \cos \theta),$$

which can be solved for r to obtain

$$r = \frac{ke}{1 + e \cos \theta}. \tag{6}$$

Equation (6) represents an ellipse if $0 < e < 1$, a parabola if $e = 1$, and a hyperbola if $e > 1$. And there we have it: ellipses, parabolas, and hyperbolas from one simple equation.

Example 4 CONICS FROM EQUATION (6)

$$e = \frac{1}{2}: \qquad \text{ellipse} \qquad r = \frac{k}{2 + \cos \theta}$$

$$e = 1: \qquad \text{parabola} \qquad r = \frac{k}{1 + \cos \theta}$$

$$e = 2: \qquad \text{hyperbola} \qquad r = \frac{2k}{1 + 2 \cos \theta}$$

You may see variations of Equation (6) from time to time, depending on the location of the directrix. If the directrix is the line $x = -k$ to the left of the origin (the origin is still a focus), we replace Equation (6) by

$$r = \frac{ke}{1 - e \cos \theta}.$$

The denominator now has a $(-)$ instead of a $(+)$. If the directrix is either of the lines $y = k$ or $y = -k$, the equations we get have sines in them instead of cosines, as shown in Table A5.4.

Table A5.4 Equations for conic sections (e > 0)

A.

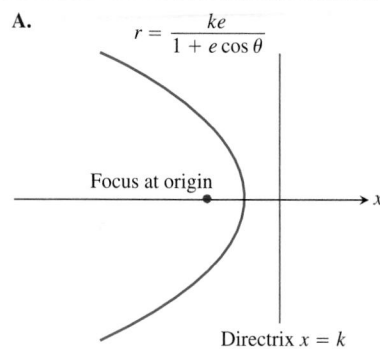

$$r = \frac{ke}{1 + e \cos \theta}$$

Focus at origin

Directrix $x = k$

B.

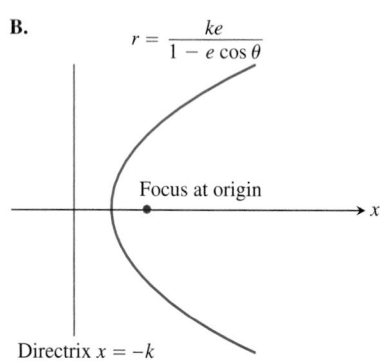

$$r = \frac{ke}{1 - e \cos \theta}$$

Focus at origin

Directrix $x = -k$

C.

$$r = \frac{ke}{1 + e \sin \theta}$$

Directrix $y = k$

Focus at origin

D.

$$r = \frac{ke}{1 - e \sin \theta}$$

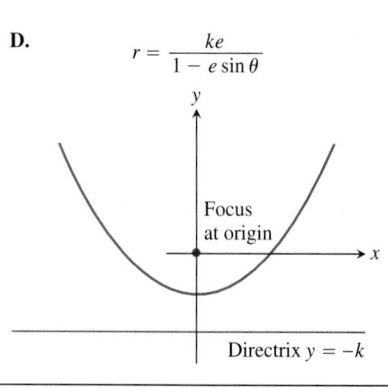

Focus at origin

Directrix $y = -k$

Example 5   FINDING AN EQUATION FOR A HYPERBOLA

Find an equation for the hyperbola with eccentricity 3/2 and directrix $x = 2$.

Solution   We use Equation (A) in Table A5.4 with $k = 2$ and $e = 3/2$ to get

$$r = \frac{2(3/2)}{1 + (3/2) \cos \theta} \quad \text{or} \quad r = \frac{6}{2 + 3 \cos \theta}.$$

Example 6   FINDING THE DIRECTRIX OF A PARABOLA

Find the directrix of the parabola

$$r = \frac{25}{10 + 10 \cos \theta}.$$

Solution   We divide the numerator and denominator by 10 to put the equation in standard form:

$$r = \frac{5/2}{1 + \cos \theta}.$$

This is the equation

$$r = \frac{ke}{1 + e \cos \theta}.$$

with $k = 5/2$ and $e = 1$. The equation of the directrix is $x = 5/2$.

From the ellipse diagram in Figure A5.33, we see that k is related to the eccentricity e and the semimajor axis a by the equation

$$k = \frac{a}{e} - ea. \tag{7}$$

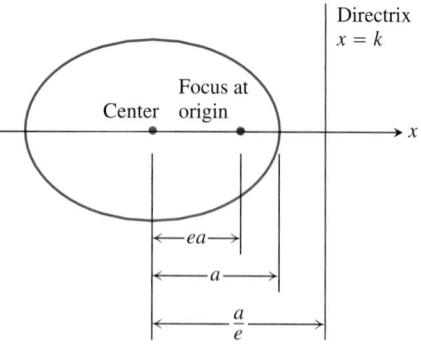

Directrix $x = k$

Focus at origin

Center

ea

a

$\frac{a}{e}$

A5.33 In an ellipse with semimajor axis a, the focus–directrix distance is $k = (a/e) - ea$, so $ke = a(1 - e^2)$.

From this, we find that $ke = a(1 - e^2)$. Replacing ke in Equation (6) by $a(1 - e^2)$ gives the standard polar equation for an ellipse.

Ellipse with Eccentricity *e* and Semimajor Axis *a*

$$r = \frac{a(1 - e^2)}{1 + e \cos \theta} \tag{8}$$

Notice that when $e = 0$, Equation (8) becomes $r = a$, which represents a circle. Equation (8) is the starting point for calculating planetary orbits.

Example 7   FINDING A PLANETARY ORBIT

Find a polar equation for an ellipse with semimajor axis 39.44 AU (astronomical units) and eccentricity 0.25. This is the approximate size of Pluto's orbit around the sun.

Solution   We use Equation (8) with $a = 39.44$ and $e = 0.25$ to find

$$r = \frac{39.44(1 - (0.25)^2)}{1 + 0.25 \cos \theta} = \frac{147.9}{4 + \cos \theta}.$$

At its point of closest approach (perihelion), Pluto is

$$r = \frac{147.9}{4 + 1} = 29.58 \text{ AU}$$

from the sun. At its most distant point (aphelion), Pluto is

$$r = \frac{147.9}{4 - 1} = 49.3 \text{ AU}$$

from the sun (Figure A5.34).

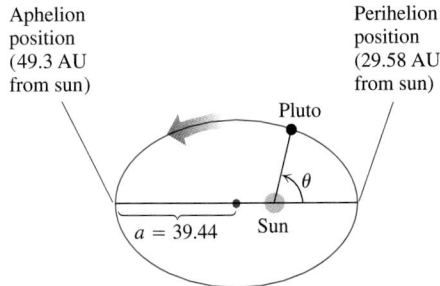

Aphelion position (49.3 AU from sun)

Perihelion position (29.58 AU from sun)

Pluto

θ

$a = 39.44$ Sun

A5.34 The orbit of Pluto. (Example 7)

Example 8   FINDING A FOCUS-TO-DIRECTRIX DISTANCE

Find the distance from one focus of the ellipse in Example 7 to the associated directrix.

Solution   We use Equation (7) with $a = 39.44$ and $e = 0.25$ to find

$$k = 39.44 \left(\frac{1}{0.25} - 0.25 \right) = 147.9 \text{ AU}.$$

Section A5.4 Exercises

In Exercises 1–4, find polar and Cartesian equations for the line.

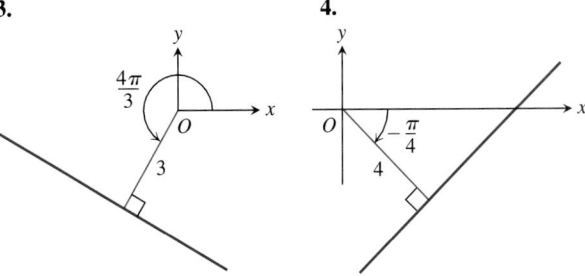

1.

y

5

$\pi/6$

O x

2.

y

2

$\frac{3\pi}{4}$

O x

3.

y

$\frac{4\pi}{3}$

O x

3

4.

y

O $-\frac{\pi}{4}$ x

4

In Exercises 5–8, sketch the line and find a Cartesian equation for it.

5. $r \cos\left(\theta - \dfrac{\pi}{4}\right) = \sqrt{2}$ **6.** $r \cos\left(\theta + \dfrac{3\pi}{4}\right) = 1$

7. $r \cos\left(\theta - \dfrac{2\pi}{3}\right) = 3$ **8.** $r \cos\left(\theta + \dfrac{\pi}{3}\right) = 2$

In Exercises 9–12, find a polar equation in the form $r \cos(\theta - \theta_0) = r_0$ for the line.

9. $\sqrt{2}x + \sqrt{2}y = 6$ **10.** $\sqrt{3}x - y = 1$

11. $y = -5$ **12.** $x = -4$

In Exercises 13–16, find a polar equation for the circle.

13.

Radius = 4

14.

Radius = 1

15.

Radius = $\sqrt{2}$

16.

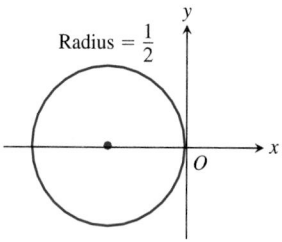
Radius = $\dfrac{1}{2}$

In Exercises 17–20, sketch the circle. Give polar coordinates for the center and identify the radius.

17. $r = 4 \cos\theta$ **18.** $r = 6 \sin\theta$

19. $r = -2 \cos\theta$ **20.** $r = -8 \sin\theta$

In Exercises 21–24, find a polar equation for the circle. Sketch the circle in the coordinate plane and label it with both its Cartesian and polar equations.

21. $(x - 6)^2 + y^2 = 36$

22. $x^2 + (y - 5)^2 = 25$

23. $x^2 + 2x + y^2 = 0$

24. $x^2 + y^2 + y = 0$

Exercises 25–32 give the eccentricities of conic sections with one focus at the origin, along with the directrix corresponding to that focus. Find a polar equation for each conic section.

25. $e = 1$, $x = 2$ **26.** $e = 1$, $y = 2$

27. $e = 5$, $y = -6$ **28.** $e = 2$, $x = 4$

29. $e = 1/2$, $x = 1$ **30.** $e = 1/4$, $x = -2$

31. $e = 1/5$, $y = -10$ **32.** $e = 1/3$, $y = 6$.

In Exercises 33–40, sketch the parabola or ellipse. Include the directrix that corresponds to the focus at the origin. Label the vertices with appropriate polar coordinates. Label the center of the ellipse as well. Support with a grapher.

33. $r = \dfrac{1}{1 + \cos\theta}$ **34.** $r = \dfrac{6}{2 + \cos\theta}$

35. $r = \dfrac{25}{10 - 5\cos\theta}$ **36.** $r = \dfrac{4}{2 - 2\cos\theta}$

37. $r = \dfrac{400}{16 + 8\sin\theta}$ **38.** $r = \dfrac{12}{3 + 3\sin\theta}$

39. $r = \dfrac{8}{2 - 2\sin\theta}$ **40.** $r = \dfrac{4}{2 - \sin\theta}$

41. (a) Find Cartesian equations for the curves $r = 4 \sin\theta$ and $r = \sqrt{3}\sec\theta$.

(b) Sketch the curves together and label their points of intersection in both Cartesian and polar coordinates.

42. Repeat Exercise 41 for $r = 8 \cos\theta$ and $r = 2 \sec\theta$.

43. Find a polar equation for the parabola with focus (0, 0) and directrix $r \cos\theta = 4$.

44. Find a polar equation for the parabola with focus (0, 0) and directrix $r \cos(\theta - \pi/2) = 2$.

Explorations

45. Writing to Learn In the same viewing window, graph the conics

$$r = \dfrac{e}{1 + e\cos\theta}$$

for $e = 0.1, 0.2, 0.3, 0.4, 0.5, 0.6, 0.7, 0.8,$ and 0.9. Explain the effects of the parameter e.

46. Writing to Learn In the same viewing window, graph the conics

$$r = \dfrac{e}{1 + e\cos\theta}$$

for $e = 1.1, 1.2, 1.3, 1.4, 1.5, 2, 2.5, 3,$ and 10. Explain the effect of the parameter e.

47. Writing to Learn In the same viewing window, graph the conics

$$r = \dfrac{k}{1 + \cos\theta}$$

for $k = -8, -4, -3, -2, -1, 0, 1, 2, 3, 4,$ and 8. Explain the effect of the parameter k.

48. Writing to Learn In the same viewing window, graph the conics

$$r = \frac{k}{1 - \sin \theta}$$

for $k = -8, -4, -3, -2, -1, 0, 1, 2, 3, 4,$ and 8. Explain the effect of the parameter k. ∎

Extending the Ideas

Graphing Inequalities In Exercises 49 and 50, sketch the region defined by the inequality.

49. $0 \le r \le 2 \cos \theta$ **50.** $-3 \cos \theta \le r \le 0$

51. *Perihelion and Aphelion.* A planet travels about its sun in an ellipse whose semimajor axis has length a.

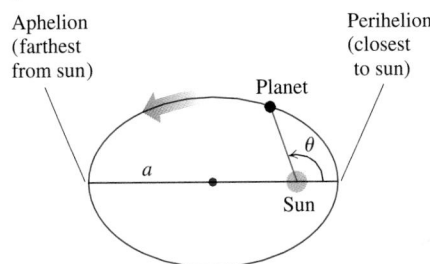

(a) Writing to Learn Show that $r = a(1 - e)$ when the planet is closest to the sun and that $r = a(1 + e)$ when the planet is farthest from the sun.

(b) Use the data in Table 5.5 to find how close each planet in our solar system comes to the sun and how far away each planet gets from the sun.

52. *Planetary Orbits.* In Example 7, we drew the orbit of Pluto.

(a) Use the data in Table A5.5 to find polar equations for the orbits of the other planets. Draw the orbits in the same viewing window for Mercury through Mars. Then draw the orbits in the same viewing window for Jupiter through Pluto. Explain why it was necessary to use two viewing windows.

(b) Study the orbits of Neptune and Pluto together. The right side of the viewing window suggests an astronomical event that occurred on January 21, 1979 and March 14, 1999, and will not happen again until September 2226. The behavior that better identifies the event is hidden. Make a conjecture about what the event is. Zoom in to help you decide that the event is indeed possible.

Table A5.5 Semimajor axes and eccentricities of the planets in our solar system

Planet	Semimajor axis (astronomical units)	Eccentricity
Mercury	0.3871	0.2056
Venus	0.7233	0.0068
Earth	1.000	0.0167
Mars	1.524	0.0934
Jupiter	5.203	0.0484
Saturn	9.539	0.0543
Uranus	19.18	0.0460
Neptune	30.06	0.0082
Pluto	39.44	0.2481

53. Writing to Learn *The Space Engineer's Formula for Eccentricity* The space engineer's formula for the eccentricity of an elliptical orbit is

$$e = \frac{r_{max} - r_{min}}{r_{max} + r_{min}},$$

where r is the distance from the space vehicle to the attracting focus of the ellipse along which it travels. Why does the formula work?

(b) Writing to Learn *Drawing the Orbit of Mercury* You have a string with a knot in each end that can be pinned to a drawing board. The string is 10 in. long from the center of one knot to the center of the other. How far apart should the pins be to use the method illustrated in Figure A5.7 (Section A5.1) to draw an ellipse of eccentricity 0.2? The resulting ellipse would resemble the orbit of Mercury.

54. *Halley's Comet* (See Section A5.2, Example 1.)

(a) *Orbit Equation* Write an equation for the orbit of Halley's comet in a coordinate system in which the sun lies at the origin and the other focus lies on the negative x-axis, scaled in astronomical units.

(b) *Perihelion* How close does the comet come to the sun in astronomical units? In kilometers?

(c) *Aphelion* What is the farthest the comet gets from the sun in astronomical units? In kilometers?

A6   # Hyperbolic Functions

Background • Definitions • Identities • Derivatives and Integrals • Inverse Hyperbolic Functions • Identities for $\text{sech}^{-1} x$, $\text{csch}^{-1} x$, $\coth^{-1} x$ • Derivatives of Inverse Hyperbolic Functions; Associated Integrals

Background

Suspension cables like those of the Golden Gate Bridge, which support a constant load per horizontal foot, hang in parabolas (Section A5.1, Exercise 44). Cables like power line cables, which hang freely, hang in curves called hyperbolic cosine curves.

Besides describing the shapes of hanging cables, hyperbolic functions describe the motions of waves in elastic solids, the temperature distributions in metal cooling fins, and the motions of falling bodies that encounter air resistance proportional to the square of the velocity. If a hanging cable were turned upside down (without changing shape) to form an arch, the internal forces, then reversed, would once again be in equilibrium, making the inverted hyperbolic cosine curve the ideal shape for a self-standing arch. The center line of the Gateway Arch to the West in St. Louis follows a hyperbolic cosine curve.

Definitions

The hyperbolic cosine and sine functions are defined by the first two equations in Table A6.1. The table also defines the hyperbolic tangent, cotangent, secant, and cosecant. As we will see, the hyperbolic functions bear a number of similarities to trigonometric functions after which they are named.

Pronouncing "cosh" and "sinh"

"Cosh" is often pronounced "kosh," rhyming with "gosh" or "gauche." "Sinh" is pronounced as if spelled "cinch" or "shine."

Table A6.1 The six basic hyperbolic functions

Hyperbolic cosine of x:	$\cosh x = \dfrac{e^x + e^{-x}}{2}$
Hyperbolic sine of x:	$\sinh x = \dfrac{e^x - e^{-x}}{2}$
Hyperbolic tangent:	$\tanh x = \dfrac{\sinh x}{\cosh x} = \dfrac{e^x - e^{-x}}{e^x + e^{-x}}$
Hyperbolic cotangent:	$\coth x = \dfrac{\cosh x}{\sinh x} = \dfrac{e^x + e^{-x}}{e^x - e^{-x}}$
Hyperbolic secant:	$\operatorname{sech} x = \dfrac{1}{\cosh x} = \dfrac{2}{e^x + e^{-x}}$
Hyperbolic cosecant:	$\operatorname{csch} x = \dfrac{1}{\sinh x} = \dfrac{2}{e^x - e^{-x}}$

See Figure A6.1 for graphs.

Table A6.2 Identities for hyperbolic functions

$$\sinh 2x = 2 \sinh x \cosh x$$

$$\cosh 2x = \cosh^2 x + \sinh^2 x$$

$$\cosh^2 x = \frac{\cosh 2x + 1}{2}$$

$$\sinh^2 x = \frac{\cosh 2x - 1}{2}$$

$$\cosh^2 x - \sinh^2 x = 1$$

$$\tanh^2 x = 1 - \operatorname{sech}^2 x$$

$$\coth^2 x = 1 + \operatorname{csch}^2 x$$

Identities

Hyperbolic functions satisfy the identities in Table A6.2. Except for differences in sign, these are identities we already know for trigonometric functions.

Derivatives and Integrals

The six hyperbolic functions, being rational combinations of the differentiable functions e^x and e^{-x}, have derivatives at every point at which they are defined (Table A6.3 on the following page). Again, there are similarities with trigonometric functions. The derivative formulas in Table A6.3 lead to the integral formulas seen there.

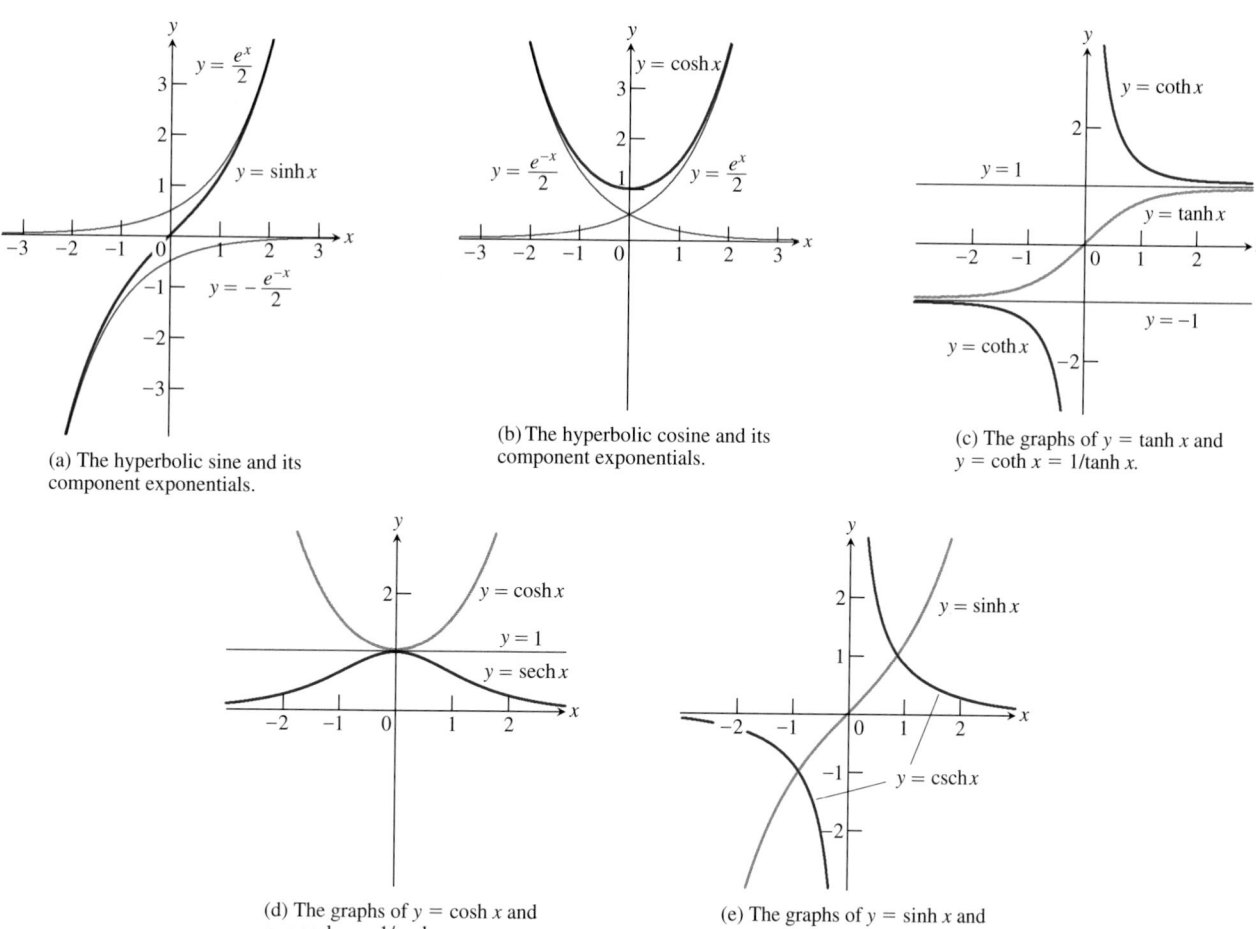

(a) The hyperbolic sine and its component exponentials.

(b) The hyperbolic cosine and its component exponentials.

(c) The graphs of $y = \tanh x$ and $y = \coth x = 1/\tanh x$.

(d) The graphs of $y = \cosh x$ and $y = \operatorname{sech} x = 1/\cosh x$.

(e) The graphs of $y = \sinh x$ and $y = \operatorname{csch} x = 1/\sinh x$.

Figure A6.1 The graphs of the six hyperbolic functions.

Table A6.3 Derivatives and companion integrals

$\dfrac{d}{dx}(\sinh u) = \cosh u\,\dfrac{du}{dx}$	$\displaystyle\int \cosh u\,du = \sinh u + C$
$\dfrac{d}{dx}(\cosh u) = \sinh u\,\dfrac{du}{dx}$	$\displaystyle\int \sinh u\,du = \cosh u + C$
$\dfrac{d}{dx}(\tanh u) = \operatorname{sech}^2 u\,\dfrac{du}{dx}$	$\displaystyle\int \operatorname{sech}^2 u\,du = \tanh u + C$
$\dfrac{d}{dx}(\coth u) = -\operatorname{csch}^2 u\,\dfrac{du}{dx}$	$\displaystyle\int \operatorname{csch}^2 u\,du = -\coth u + C$
$\dfrac{d}{dx}(\operatorname{sech} u) = -\operatorname{sech} u \tanh u\,\dfrac{du}{dx}$	$\displaystyle\int \operatorname{sech} u \tanh u\,du = -\operatorname{sech} u + C$
$\dfrac{d}{dx}(\operatorname{csch} u) = -\operatorname{csch} u \coth u\,\dfrac{du}{dx}$	$\displaystyle\int \operatorname{csch} u \coth u\,du = -\operatorname{csch} u + C$

Example 1   FINDING A DERIVATIVE

$$\frac{d}{dt}(\tanh \sqrt{1+t^2}) = \text{sech}^2 \sqrt{1+t^2} \cdot \frac{d}{dt}(\sqrt{1+t^2})$$

$$= \frac{t}{\sqrt{1+t^2}} \text{sech}^2 \sqrt{1+t^2}$$

Example 2   INTEGRATING A HYPERBOLIC COTANGENT

$$\int \coth 5x \, dx = \int \frac{\cosh 5x}{\sinh 5x} \, dx = \frac{1}{5} \int \frac{du}{u} \qquad \begin{array}{l} u = \sinh 5x, \\ du = 5 \cosh 5x \, dx \end{array}$$

$$= \frac{1}{5} \ln |u| + C = \frac{1}{5} \ln |\sinh 5x| + C$$

Example 3   USING AN IDENTITY TO INTEGRATE

Evaluate $\displaystyle\int_0^1 \sinh^2 x \, dx$.

Solution

Solve Numerically

To five decimal places,

$$\text{NINT}((\sinh x)^2, x, 0, 1) = 0.40672.$$

Confirm Analytically

$$\int_0^1 \sinh^2 x \, dx = \int_0^1 \frac{\cosh 2x - 1}{2} \, dx \qquad \text{Table A6.2}$$

$$= \frac{1}{2} \int_0^1 (\cosh 2x - 1) \, dx = \frac{1}{2} \left[\frac{\sinh 2x}{2} - x \right]_0^1$$

$$= \frac{\sinh 2}{4} - \frac{1}{2} \approx 0.40672$$

Inverse Hyperbolic Functions

We use the inverses of the six basic hyperbolic functions in integration. Since $d(\sinh x)/dx = \cosh x > 0$, the hyperbolic sine is an increasing function of x. We denote its inverse by

$$y = \sinh^{-1} x.$$

For every value of x in the interval $-\infty < x < \infty$, the value of $y = \sinh^{-1} x$ is the number whose hyperbolic sine is x (Figure A6.2a).

The function $y = \cosh x$ is not one-to-one, as we can see from the graph in Figure A6.1. But the restricted function $y = \cosh x$, $x \geq 0$, is one-to-one and therefore has an inverse, denoted by

$$y = \cosh^{-1} x.$$

For every value of $x \geq 1$, $y = \cosh^{-1} x$ is the number in the interval $0 \leq y < \infty$ whose hyperbolic cosine is x (Figure A6.2b).

Like $y = \cosh x$, the function $y = \text{sech } x = 1/\cosh x$ fails to be one-to-one, but its restriction to nonnegative values of x does have an inverse, denoted by

$$y = \text{sech}^{-1} x.$$

(a)

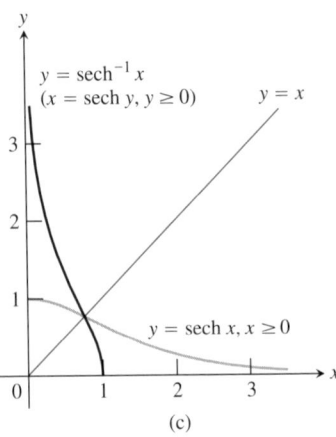

(b)

(c)

Figure A6.2 The graphs of the inverse hyperbolic sine, cosine, and secant of x. Notice the symmetries about the line $y = x$.

For every value of x in the interval $(0, 1]$, $y = \text{sech}^{-1} x$ is the nonnegative number whose hyperbolic secant is x (Figure A6.2c).

The hyperbolic tangent, cotangent, and cosecant are one-to-one on their domains and therefore have inverses, denoted by

$$y = \tanh^{-1} x, \quad y = \coth^{-1} x, \quad y = \text{csch}^{-1} x$$

(Figure A6.3).

Exploration 1 **Viewing Inverses**

Let $x_1(t) = t,$ $\quad y_1(t) = t,$

$\qquad x_2(t) = t,$ $\quad y_2(t) = 1/\cosh t,$

$\qquad x_3(t) = y_2(t), \quad y_3(t) = x_2(t).$

1. Graph the parametric equations simultaneously in a square viewing window that contains $0 \le x \le 6$, $0 \le y \le 4$. Set $t\text{Min} = 0$, $t\text{Max} = 6$, and t-step $= 0.05$. Explain what you see. Explain the domain of each function.

2. Let $x_4(t) = t$, $y_4(t) = \cosh^{-1}(1/t)$. Graph and compare (x_3, y_3) and (y_4, x_4). Predict what you should see, and explain what you do see.

Table A6.4 Identities for inverse hyperbolic functions

$$\text{sech}^{-1} x = \cosh^{-1} \frac{1}{x}$$

$$\text{csch}^{-1} x = \sinh^{-1} \frac{1}{x}$$

$$\coth^{-1} x = \tanh^{-1} \frac{1}{x}$$

Identities for $\text{sech}^{-1} x$, $\text{csch}^{-1} x$, $\coth^{-1} x$

We use the identities in Table A6.4 to calculate the values of $\text{sech}^{-1} x$, $\text{csch}^{-1} x$, and $\coth^{-1} x$ on calculators that give only $\cosh^{-1} x$, $\sinh^{-1} x$, and $\tanh^{-1} x$.

Derivatives of Inverse Hyperbolic Functions; Associated Integrals

The chief use of inverse hyperbolic functions lies in integrations that reverse the derivative formulas in Table A6.5.

(a)

(b)

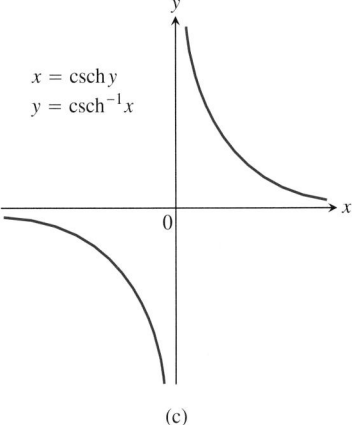

(c)

Figure A6.3 The graphs of the inverse hyperbolic tangent, cotangent, and cosecant of x.

Table A6.5 Derivatives of inverse hyperbolic functions

$$\frac{d(\sinh^{-1} u)}{dx} = \frac{1}{\sqrt{1 + u^2}} \frac{du}{dx}$$

$$\frac{d(\cosh^{-1} u)}{dx} = \frac{1}{\sqrt{u^2 - 1}} \frac{du}{dx}, \quad u > 1$$

$$\frac{d(\tanh^{-1} u)}{dx} = \frac{1}{1 - u^2} \frac{du}{dx}, \quad |u| < 1$$

$$\frac{d(\coth^{-1} u)}{dx} = \frac{1}{1 - u^2} \frac{du}{dx}, \quad |u| > 1$$

$$\frac{d(\operatorname{sech}^{-1} u)}{dx} = \frac{-du/dx}{u\sqrt{1 - u^2}}, \quad 0 < u < 1$$

$$\frac{d(\operatorname{csch}^{-1} u)}{dx} = \frac{-du/dx}{|u|\sqrt{1 + u^2}}, \quad u \neq 0$$

The restrictions $|u| < 1$ and $|u| > 1$ on the derivative formulas for $\tanh^{-1} u$ and $\coth^{-1} u$ come from the natural restrictions on the values of these functions. (See Figures A6.3a and b.) The distinction between $|u| < 1$ and $|u| > 1$ becomes important when we convert the derivative formulas into integral formulas. If $|u| < 1$, the integral of $1/(1 - u^2)$ is $\tanh^{-1} u + C$. If $|u| > 1$, the integral is $\coth^{-1} u + C$.

With appropriate substitutions, the derivative formulas in Table A6.5 lead to the integral formulas in Table A6.6.

Table A6.6 Integrals leading to inverse hyperbolic functions

$$1. \int \frac{du}{\sqrt{a^2 + u^2}} = \sinh^{-1}\left(\frac{u}{a}\right) + C, \quad a > 0$$

$$2. \int \frac{du}{\sqrt{u^2 - a^2}} = \cosh^{-1}\left(\frac{u}{a}\right) + C, \quad u > a > 0$$

$$3. \int \frac{du}{a^2 - u^2} = \begin{cases} \frac{1}{a} \tanh^{-1}\left(\frac{u}{a}\right) + C & \text{if } u^2 < a^2 \\ \frac{1}{a} \coth^{-1}\left(\frac{u}{a}\right) + C & \text{if } u^2 > a^2 \end{cases}$$

$$4. \int \frac{du}{u\sqrt{a^2 - u^2}} = -\frac{1}{a} \operatorname{sech}^{-1}\left(\frac{u}{a}\right) + C, \quad 0 < u < a$$

$$5. \int \frac{du}{u\sqrt{a^2 + u^2}} = -\frac{1}{a} \operatorname{csch}^{-1}\left|\frac{u}{a}\right| + C, \quad u \neq 0$$

Example 4   USING TABLE A6.6

Evaluate $\displaystyle\int_0^1 \frac{2\,dx}{\sqrt{3+4x^2}}$.

Solution

Solve Analytically

The indefinite integral is

$$\int \frac{2\,dx}{\sqrt{3+4x^2}} = \int \frac{du}{\sqrt{a^2+u^2}} \qquad u=2x,\ du=2\,dx,\ a=\sqrt{3}$$

$$= \sinh^{-1}\left(\frac{u}{a}\right) + C \qquad \text{Formula from Table A6.6}$$

$$= \sinh^{-1}\left(\frac{2x}{\sqrt{3}}\right) + C.$$

Therefore,

$$\int_0^1 \frac{2\,dx}{\sqrt{3+4x^2}} = \sinh^{-1}\left(\frac{2x}{\sqrt{3}}\right)\Bigg]_0^1 = \sinh^{-1}\left(\frac{2}{\sqrt{3}}\right) - \sinh^{-1}(0)$$

$$= \sinh^{-1}\left(\frac{2}{\sqrt{3}}\right) - 0 \approx 0.98665.$$

Support Numerically

To five decimal places,

$$\text{NINT}\,(2/\sqrt{3+4x^2}, x, 0, 1) = 0.98665.$$

Section A6 Exercises

In Exercises 1–4, find the values of the remaining five hyperbolic functions.

1. $\sinh x = -\dfrac{3}{4}$

2. $\sinh x = \dfrac{4}{3}$

3. $\cosh x = \dfrac{17}{15}, \quad x > 0$

4. $\cosh x = \dfrac{13}{5}, \quad x > 0$

In Exercises 5–10, rewrite the expression in terms of exponentials and simplify the results as much as you can. Support your answers graphically.

5. $2\cosh(\ln x)$

6. $\sinh(2\ln x)$

7. $\cosh 5x + \sinh 5x$

8. $\cosh 3x - \sinh 3x$

9. $(\sinh x + \cosh x)^4$

10. $\ln(\cosh x + \sinh x) + \ln(\cosh x - \sinh x)$

11. Use the identities

$$\sinh(x+y) = \sinh x \cosh y + \cosh x \sinh y$$
$$\cosh(x+y) = \cosh x \cosh y + \sinh x \sinh y$$

to show that

(a) $\sinh 2x = 2\sinh x \cosh x$;

(b) $\cosh 2x = \cosh^2 x + \sinh^2 x$.

12. Use the definitions of $\cosh x$ and $\sinh x$ to show that

$$\cosh^2 x - \sinh^2 x = 1.$$

In Exercises 13–24, find the derivative of y with respect to the appropriate variable.

13. $y = 6\sinh\dfrac{x}{3}$

14. $y = \dfrac{1}{2}\sinh(2x+1)$

15. $y = 2\sqrt{t}\tanh\sqrt{t}$

16. $y = t^2\tanh\dfrac{1}{t}$

17. $y = \ln (\sinh z)$　　　　**18.** $y = \ln (\cosh z)$

19. $y = \operatorname{sech} \theta (1 - \ln \operatorname{sech} \theta)$　　**20.** $y = \operatorname{csch} \theta (1 - \ln \operatorname{csch} \theta)$

21. $y = \ln \cosh x - \dfrac{1}{2} \tanh^2 x$　　**22.** $y = \ln \sinh x - \dfrac{1}{2} \coth^2 x$

23. $y = (x^2 + 1) \operatorname{sech} (\ln x)$　(*Hint:* Before differentiating, express in terms of exponentials and simplify.)

24. $y = (4x^2 - 1) \operatorname{csch} (\ln 2x)$

In Exercises 25–36, find the derivative of y with respect to the appropriate variable.

25. $y = \sinh^{-1} \sqrt{x}$　　　　**26.** $y = \cosh^{-1} (2\sqrt{x + 1})$

27. $y = (1 - \theta) \tanh^{-1} \theta$

28. $y = (\theta^2 + 2\theta) \tanh^{-1} (\theta + 1)$

29. $y = (1 - t) \coth^{-1} \sqrt{t}$

30. $y = (1 - t^2) \coth^{-1} t$

31. $y = \cos^{-1} x - x \operatorname{sech}^{-1} x$

32. $y = \ln x + \sqrt{1 - x^2} \operatorname{sech}^{-1} x$

33. $y = \operatorname{csch}^{-1} \left(\dfrac{1}{2}\right)^{\theta}$　　　**34.** $y = \operatorname{csch}^{-1} 2^{\theta}$

35. $y = \sinh^{-1} (\tan x)$

36. $y = \cosh^{-1} (\sec x), \quad 0 < x < \pi/2$

Verify the integration formulas in Exercises 37–40.

37. (a) $\displaystyle\int \operatorname{sech} x \, dx = \tan^{-1} (\sinh x) + C$

(b) $\displaystyle\int \operatorname{sech} x \, dx = \sin^{-1} (\tanh x) + C$

38. $\displaystyle\int x \operatorname{sech}^{-1} x \, dx = \dfrac{x^2}{2} \operatorname{sech}^{-1} x - \dfrac{1}{2}\sqrt{1 - x^2} + C$

39. $\displaystyle\int x \coth^{-1} x \, dx = \dfrac{x^2 - 1}{2} \coth^{-1} x + \dfrac{x}{2} + C$

40. $\displaystyle\int \tanh^{-1} x \, dx = x \tanh^{-1} x + \dfrac{1}{2} \ln (1 - x^2) + C$

Evaluate the integrals in Exercises 41–50.

41. $\displaystyle\int \sinh 2x \, dx$　　　　**42.** $\displaystyle\int \sinh \dfrac{x}{5} \, dx$

43. $\displaystyle\int 6 \cosh \left(\dfrac{x}{2} - \ln 3\right) dx$　　**44.** $\displaystyle\int 4 \cosh (3x - \ln 2) \, dx$

45. $\displaystyle\int \tanh \dfrac{x}{7} \, dx$　　　**46.** $\displaystyle\int \coth \dfrac{\theta}{\sqrt{3}} \, d\theta$

47. $\displaystyle\int \operatorname{sech}^2 \left(x - \dfrac{1}{2}\right) dx$　　**48.** $\displaystyle\int \operatorname{csch}^2 (5 - x) \, dx$

49. $\displaystyle\int \dfrac{\operatorname{sech} \sqrt{t} \tanh \sqrt{t} \, dt}{\sqrt{t}}$

50. $\displaystyle\int \dfrac{\operatorname{csch} (\ln t) \coth (\ln t) \, dt}{t}$

Evaluate the integrals in Exercises 51–60 analytically and support with NINT.

51. $\displaystyle\int_{\ln 2}^{\ln 4} \coth x \, dx$　　　**52.** $\displaystyle\int_{0}^{\ln 2} \tanh 2x \, dx$

53. $\displaystyle\int_{-\ln 4}^{-\ln 2} 2e^{\theta} \cosh \theta \, d\theta$　　**54.** $\displaystyle\int_{0}^{\ln 2} 4e^{-\theta} \sinh \theta \, d\theta$

55. $\displaystyle\int_{-\pi/4}^{\pi/4} \cosh (\tan \theta) \sec^2 \theta \, d\theta$

56. $\displaystyle\int_{0}^{\pi/2} 2 \sinh (\sin \theta) \cos \theta \, d\theta$

57. $\displaystyle\int_{1}^{2} \dfrac{\cosh (\ln t)}{t} \, dt$　　**58.** $\displaystyle\int_{1}^{4} \dfrac{8 \cosh \sqrt{x}}{\sqrt{x}} \, dx$

59. $\displaystyle\int_{-\ln 2}^{0} \cosh^2 \left(\dfrac{x}{2}\right) dx$　　**60.** $\displaystyle\int_{0}^{\ln 10} 4 \sinh^2 \left(\dfrac{x}{2}\right) dx$

In Exercises 61 and 62, find the volume of the solid generated by revolving the shaded region about the x-axis.

61.

62.

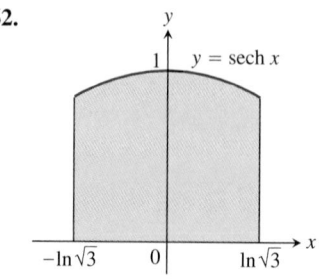

63. Find the volume of the solid generated by revolving the shaded region about the line $y = 1$.

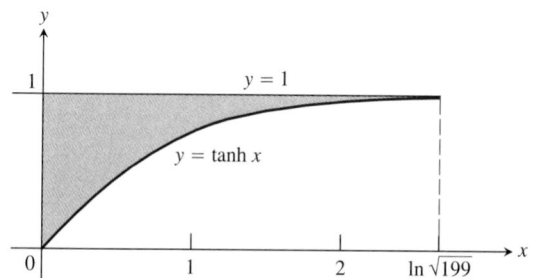

64. (a) Find the length of the curve $y = (1/2) \cosh 2x$, $0 \le x \le \ln \sqrt{5}$.

(b) Find the length of the curve $y = (1/a) \cosh ax$, $0 \le x \le b$.

Extending the Ideas

65. *Even-Odd Decompositions*

(a) Show that if a function f is defined on an interval symmetric about the origin (so that f is defined at $-x$ whenever it is defined at x), then

$$f(x) = \frac{f(x) + f(-x)}{2} + \frac{f(x) - f(-x)}{2}. \qquad (1)$$

Then show that

$$\frac{f(x) + f(-x)}{2} \quad \text{is even}$$

and

$$\frac{f(x) - f(-x)}{2} \quad \text{is odd}.$$

(b) In Equation 1, set $f(x) = e^x$. Identify the even and odd parts of f.

66. Writing to Learn *(Continuation of Exercise 65)* Equation 1 in Exercise 65 simplifies considerably if f itself is **(a)** even or **(b)** odd. What are the new equations? Explain.

67. *Skydiving* If a body of mass m falling from rest under the action of gravity encounters an air resistance proportional to the square of the velocity, then the body's velocity t seconds into the fall satisfies the differential equation

$$m\frac{dv}{dt} = mg - kv^2,$$

where k is a constant that depends on the body's aerodynamic properties and the density of the air. (We assume that the fall is short enough so that variation in the air's density will not affect the outcome.)

Show that

$$v = \sqrt{\frac{mg}{k}} \tanh\left(\sqrt{\frac{gk}{m}}\, t\right)$$

satisfies the differential equation and the initial condition that $v = 0$ when $t = 0$.

68. *Accelerations Whose Magnitudes Are Proportional to Displacement* Suppose that the position of a body moving along a coordinate line at time t is

(a) $s = a \cos kt + b \sin kt$,

(b) $s = a \cosh kt + b \sinh kt$.

Show in both cases that the acceleration d^2s/dt^2 is proportional to s but that in the first case it is directed toward the origin while in the second case it is directed away from the origin.

69. *Tractor Trailers and the Tractrix* When a tractor trailer turns into a cross street or driveway, its rear wheels follow a curve like the one shown here. (This is why the rear wheels sometimes ride up over the curb.)

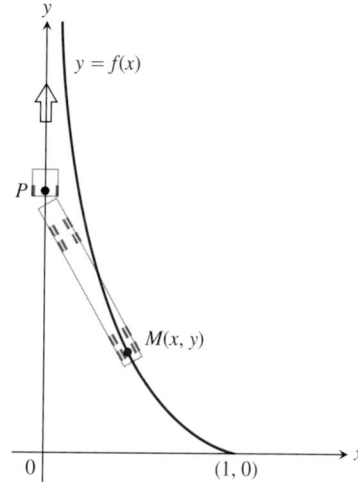

We can find an equation for the curve if we picture the rear wheels as a mass M at the point $(1, 0)$ on the x-axis attached by a rod of unit length to a point P representing the cab at the origin. As P moves up the y-axis, it drags M along behind it. The curve traced by M, called a *tractrix* from the Latin word *tractum* for "drag," can be shown to be the graph of the function $y = f(x)$ that solves the initial value problem

Differential equation: $\dfrac{dy}{dx} = -\dfrac{1}{x\sqrt{1 - x^2}} + \dfrac{x}{\sqrt{1 - x^2}}$,

Initial condition: $y = 0$ when $x = 1$.

Solve the initial value problem to find an equation for the curve. (You need an inverse hyperbolic function.)

70. *A Minimal Surface* Find the area of the surface swept out by revolving the curve $y = 4 \cosh(x/4)$, $-\ln 16 \le x \le \ln 81$, about the x-axis.

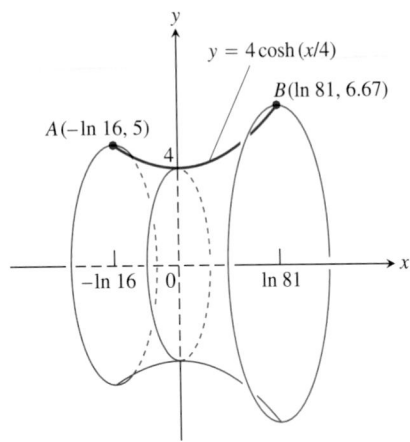

It can be shown that, of all continuously differentiable curves joining points A and B in the figure, the curve $y = 4 \cosh(x/4)$ generates the surface of least area. If you made a rigid wire frame of the end-circles through A and B and dipped them in a soap-film solution, the surface spanning the circles would be the one generated by the curve.

71. *Hanging Cables* Show that the function $y = a \cosh(x/a)$ solves the initial value problem

$$y'' = (1/a)\sqrt{1 + (y')^2}, \quad y'(0) = 0, \quad y(0) = a.$$

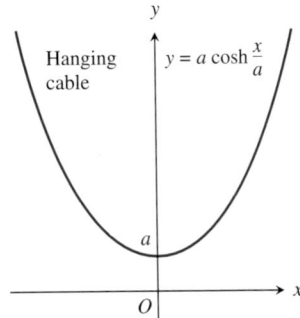

By analyzing the forces on hanging cables, we can show that the curves they hang in always satisfy the differential equation and initial conditions given here. That is how we know that hanging cables hang in hyperbolic cosines.

72. *The Hyperbolic in Hyperbolic Functions* In case you are wondering where the name *hyperbolic* comes from, here is the answer: Just as $x = \cos u$ and $y = \sin u$ are identified with points (x, y) on the unit circle, the functions $x = \cosh u$ and $y = \sinh u$ are identified with points (x, y) on the right-hand branch of the unit hyperbola $x^2 - y^2 = 1$ (Figure A6.4).

Another analogy between hyperbolic and circular functions is that the variable u in the coordinates $(\cosh u, \sinh u)$ for the points of the right-hand branch of the hyperbola $x^2 - y^2 = 1$ is twice the area of the sector AOP pictured in Figure A6.5. To see why, carry out the following steps.

(a) Let $A(u)$ be the area of sector AOP. Show that

$$A(u) = \frac{1}{2}\cosh u \sinh u - \int_1^{\cosh u} \sqrt{x^2 - 1}\, dx.$$

(b) Differentiate both sides of the equation in (a) with respect to u to show that

$$A'(u) = \frac{1}{2}.$$

(c) Solve the equation in (b) for $A(u)$. What is the value of $A(0)$? What is the value of the constant of integration C in your solution? With C determined, what does your solution say about the relationship of u to $A(u)$?

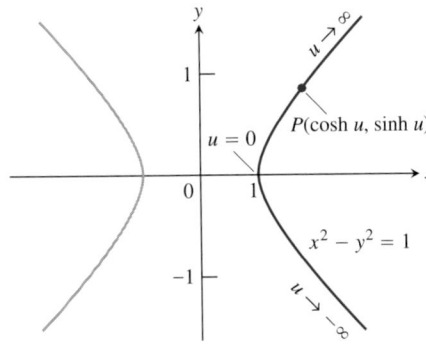

Figure A6.4 Since $\cosh^2 u - \sinh^2 u = 1$, the point $(\cosh u, \sinh u)$ lies on the right-hand branch of the hyperbola $x^2 - y^2 = 1$ for every value of u (Exercise 72).

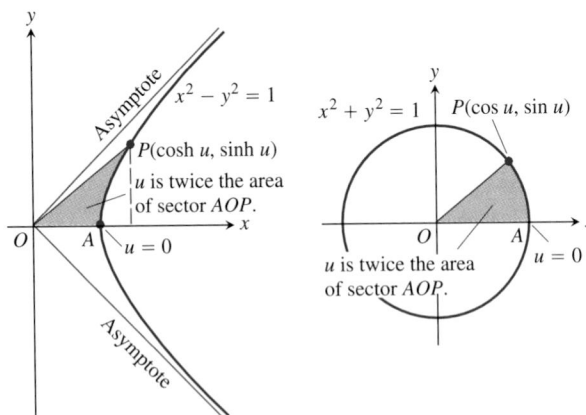

Figure A6.5 One of the analogies between hyperbolic and circular functions is revealed by these two diagrams (Exercise 72).

A7   A Brief Table of Integrals

The formulas below are stated in terms of constants a, b, c, m, n, and so on. These constants can usually assume any real value and need not be integers. Occasional limitations on their values are stated with the formulas. Formula 5 requires $n \neq -1$, for example, and Formula 11 requires $n \neq -2$. The formulas also assume that the constants do not take on values that require dividing by zero or taking even roots of negative numbers. For example, Formula 8 assumes $a \neq 0$, and Formula 13(a) cannot be used unless b is negative.

1. $\displaystyle \int u \, dv = uv - \int v \, du$

2. $\displaystyle \int a^u \, du = \frac{a^u}{\ln a} + C, \quad a \neq 1, \quad a > 0$

3. $\displaystyle \int \cos u \, du = \sin u + C$

4. $\displaystyle \int \sin u \, du = -\cos u + C$

5. $\displaystyle \int (ax + b)^n \, dx = \frac{(ax + b)^{n+1}}{a(n + 1)} + C, \quad n \neq -1$

6. $\displaystyle \int (ax + b)^{-1} \, dx = \frac{1}{a} \ln |ax + b| + C$

7. $\displaystyle \int x(ax + b)^n \, dx = \frac{(ax + b)^{n+1}}{a^2} \left[\frac{ax + b}{n + 2} - \frac{b}{n + 1} \right] + C, \quad n \neq -1, -2$

8. $\displaystyle \int x(ax + b)^{-1} \, dx = \frac{x}{a} - \frac{b}{a^2} \ln |ax + b| + C$

9. $\displaystyle \int x(ax + b)^{-2} \, dx = \frac{1}{a^2} \left[\ln |ax + b| + \frac{b}{ax + b} \right] + C$

10. $\displaystyle \int \frac{dx}{x(ax + b)} = \frac{1}{b} \ln \left| \frac{x}{ax + b} \right| + C$

11. $\displaystyle \int (\sqrt{ax + b})^n \, dx = \frac{2}{a} \frac{(\sqrt{ax + b})^{n+2}}{n + 2} + C, \quad n \neq -2$

12. $\displaystyle \int \frac{\sqrt{ax + b}}{x} \, dx = 2\sqrt{ax + b} + b \int \frac{dx}{x\sqrt{ax + b}}$

13. (a) $\displaystyle \int \frac{dx}{x\sqrt{ax + b}} = \frac{2}{\sqrt{-b}} \tan^{-1} \sqrt{\frac{ax + b}{-b}} + C, \quad \text{if } b < 0$

(b) $\displaystyle \int \frac{dx}{x\sqrt{ax + b}} = \frac{1}{\sqrt{b}} \ln \left| \frac{\sqrt{ax + b} - \sqrt{b}}{\sqrt{ax + b} + \sqrt{b}} \right| + C, \quad \text{if } b > 0$

14. $\displaystyle \int \frac{\sqrt{ax + b}}{x^2} \, dx = -\frac{\sqrt{ax + b}}{x} + \frac{a}{2} \int \frac{dx}{x\sqrt{ax + b}} + C$

15. $\displaystyle \int \frac{dx}{x^2\sqrt{ax + b}} = -\frac{\sqrt{ax + b}}{bx} - \frac{a}{2b} \int \frac{dx}{x\sqrt{ax + b}} + C$

16. $\displaystyle \int \frac{dx}{a^2 + x^2} = \frac{1}{a} \tan^{-1} \frac{x}{a} + C$

17. $\displaystyle \int \frac{dx}{(a^2 + x^2)^2} = \frac{x}{2a^2(a^2 + x^2)} + \frac{1}{2a^3} \tan^{-1} \frac{x}{a} + C$

18. $\displaystyle \int \frac{dx}{a^2 - x^2} = \frac{1}{2a} \ln \left| \frac{x + a}{x - a} \right| + C$

19. $\displaystyle \int \frac{dx}{(a^2 - x^2)^2} = \frac{x}{2a^2(a^2 - x^2)} + \frac{1}{2a^2} \int \frac{dx}{a^2 - x^2}$

20. $\displaystyle\int \frac{dx}{\sqrt{a^2 + x^2}} = \sinh^{-1}\frac{x}{a} + C = \ln\left(x + \sqrt{a^2 + x^2}\right) + C$

21. $\displaystyle\int \sqrt{a^2 + x^2}\, dx = \frac{x}{2}\sqrt{a^2 + x^2} + \frac{a^2}{2}\ln\left(x + \sqrt{a^2 + x^2}\right) + C$

22. $\displaystyle\int x^2\sqrt{a^2 + x^2}\, dx = \frac{x}{8}(a^2 + 2x^2)\sqrt{a^2 + x^2} - \frac{a^4}{8}\ln\left(x + \sqrt{a^2 + x^2}\right) + C$

23. $\displaystyle\int \frac{\sqrt{a^2 + x^2}}{x}\, dx = \sqrt{a^2 + x^2} - a\ln\left|\frac{a + \sqrt{a^2 + x^2}}{x}\right| + C$

24. $\displaystyle\int \frac{\sqrt{a^2 + x^2}}{x^2}\, dx = \ln\left(x + \sqrt{a^2 + x^2}\right) - \frac{\sqrt{a^2 + x^2}}{x} + C$

25. $\displaystyle\int \frac{x^2}{\sqrt{a^2 + x^2}}\, dx = -\frac{a^2}{2}\ln\left(x + \sqrt{a^2 + x^2}\right) + \frac{x\sqrt{a^2 + x^2}}{2} + C$

26. $\displaystyle\int \frac{dx}{x\sqrt{a^2 + x^2}} = -\frac{1}{a}\ln\left|\frac{a + \sqrt{a^2 + x^2}}{x}\right| + C$
   27. $\displaystyle\int \frac{dx}{x^2\sqrt{a^2 + x^2}} = -\frac{\sqrt{a^2 + x^2}}{a^2 x} + C$

28. $\displaystyle\int \frac{dx}{\sqrt{a^2 - x^2}} = \sin^{-1}\frac{x}{a} + C$
   29. $\displaystyle\int \sqrt{a^2 - x^2}\, dx = \frac{x}{2}\sqrt{a^2 - x^2} + \frac{a^2}{2}\sin^{-1}\frac{x}{a} + C$

30. $\displaystyle\int x^2\sqrt{a^2 - x^2}\, dx = \frac{a^4}{8}\sin^{-1}\frac{x}{a} - \frac{1}{8}x\sqrt{a^2 - x^2}(a^2 - 2x^2) + C$

31. $\displaystyle\int \frac{\sqrt{a^2 - x^2}}{x}\, dx = \sqrt{a^2 - x^2} - a\ln\left|\frac{a + \sqrt{a^2 - x^2}}{x}\right| + C$

32. $\displaystyle\int \frac{\sqrt{a^2 - x^2}}{x^2}\, dx = -\sin^{-1}\frac{x}{a} - \frac{\sqrt{a^2 - x^2}}{x} + C$
   33. $\displaystyle\int \frac{x^2}{\sqrt{a^2 - x^2}}\, dx = \frac{a^2}{2}\sin^{-1}\frac{x}{a} - \frac{1}{2}x\sqrt{a^2 - x^2} + C$

34. $\displaystyle\int \frac{dx}{x\sqrt{a^2 - x^2}} = -\frac{1}{a}\ln\left|\frac{a + \sqrt{a^2 - x^2}}{x}\right| + C$
   35. $\displaystyle\int \frac{dx}{x^2\sqrt{a^2 - x^2}} = -\frac{\sqrt{a^2 - x^2}}{a^2 x} + C$

36. $\displaystyle\int \frac{dx}{\sqrt{x^2 - a^2}} = \cosh^{-1}\frac{x}{a} + C = \ln\left|x + \sqrt{x^2 - a^2}\right| + C$

37. $\displaystyle\int \sqrt{x^2 - a^2}\, dx = \frac{x}{2}\sqrt{x^2 - a^2} - \frac{a^2}{2}\ln\left|x + \sqrt{x^2 - a^2}\right| + C$

38. $\displaystyle\int \left(\sqrt{x^2 - a^2}\right)^n dx = \frac{x\left(\sqrt{x^2 - a^2}\right)^n}{n + 1} - \frac{na^2}{n + 1}\int \left(\sqrt{x^2 - a^2}\right)^{n-2} dx, \quad n \neq -1$

39. $\displaystyle\int \frac{dx}{\left(\sqrt{x^2 - a^2}\right)^n} = \frac{x\left(\sqrt{x^2 - a^2}\right)^{2-n}}{(2 - n)a^2} - \frac{n - 3}{(n - 2)a^2}\int \frac{dx}{\left(\sqrt{x^2 - a^2}\right)^{n-2}}, \quad n \neq 2$

40. $\displaystyle\int x\left(\sqrt{x^2-a^2}\right)^n dx = \frac{\left(\sqrt{x^2-a^2}\right)^{n+2}}{n+2} + C, \quad n \neq -2$

41. $\displaystyle\int x^2\sqrt{x^2-a^2}\,dx = \frac{x}{8}(2x^2-a^2)\sqrt{x^2-a^2} - \frac{a^4}{8}\ln\left|x+\sqrt{x^2-a^2}\right| + C$

42. $\displaystyle\int \frac{\sqrt{x^2-a^2}}{x}\,dx = \sqrt{x^2-a^2} - a\sec^{-1}\left|\frac{x}{a}\right| + C$

43. $\displaystyle\int \frac{\sqrt{x^2-a^2}}{x^2}\,dx = \ln\left|x+\sqrt{x^2-a^2}\right| - \frac{\sqrt{x^2-a^2}}{x} + C$

44. $\displaystyle\int \frac{x^2}{\sqrt{x^2-a^2}}\,dx = \frac{a^2}{2}\ln\left|x+\sqrt{x^2-a^2}\right| + \frac{x}{2}\sqrt{x^2-a^2} + C$

45. $\displaystyle\int \frac{dx}{x\sqrt{x^2-a^2}} = \frac{1}{a}\sec^{-1}\left|\frac{x}{a}\right| + C = \frac{1}{a}\cos^{-1}\left|\frac{a}{x}\right| + C$

46. $\displaystyle\int \frac{dx}{x^2\sqrt{x^2-a^2}} = \frac{\sqrt{x^2-a^2}}{a^2x} + C$
    47. $\displaystyle\int \frac{dx}{\sqrt{2ax-x^2}} = \sin^{-1}\left(\frac{x-a}{a}\right) + C$

48. $\displaystyle\int \sqrt{2ax-x^2}\,dx = \frac{x-a}{2}\sqrt{2ax-x^2} + \frac{a^2}{2}\sin^{-1}\left(\frac{x-a}{a}\right) + C$

49. $\displaystyle\int \left(\sqrt{2ax-x^2}\right)^n dx = \frac{(x-a)\left(\sqrt{2ax-x^2}\right)^n}{n+1} + \frac{na^2}{n+1}\int\left(\sqrt{2ax-x^2}\right)^{n-2}dx$

50. $\displaystyle\int \frac{dx}{\left(\sqrt{2ax-x^2}\right)^n} = \frac{(x-a)\left(\sqrt{2ax-x^2}\right)^{2-n}}{(n-2)a^2} + \frac{(n-3)}{(n-2)a^2}\int\frac{dx}{\left(\sqrt{2ax-x^2}\right)^{n-2}}$

51. $\displaystyle\int x\sqrt{2ax-x^2}\,dx = \frac{(x+a)(2x-3a)\sqrt{2ax-x^2}}{6} + \frac{a^3}{2}\sin^{-1}\left(\frac{x-a}{a}\right) + C$

52. $\displaystyle\int \frac{\sqrt{2ax-x^2}}{x}\,dx = \sqrt{2ax-x^2} + a\sin^{-1}\left(\frac{x-a}{a}\right) + C$

53. $\displaystyle\int \frac{\sqrt{2ax-x^2}}{x^2}\,dx = -2\sqrt{\frac{2a-x}{x}} - \sin^{-1}\left(\frac{x-a}{a}\right) + C$

54. $\displaystyle\int \frac{x\,dx}{\sqrt{2ax-x^2}} = a\sin^{-1}\left(\frac{x-a}{a}\right) - \sqrt{2ax-x^2} + C$
    55. $\displaystyle\int \frac{dx}{x\sqrt{2ax-x^2}} = -\frac{1}{a}\sqrt{\frac{2a-x}{x}} + C$

56. $\displaystyle\int \sin ax\,dx = -\frac{1}{a}\cos ax + C$
    57. $\displaystyle\int \cos ax\,dx = \frac{1}{a}\sin ax + C$

58. $\displaystyle\int \sin^2 ax\,dx = \frac{x}{2} - \frac{\sin 2ax}{4a} + C$
    59. $\displaystyle\int \cos^2 ax\,dx = \frac{x}{2} + \frac{\sin 2ax}{4a} + C$

60. $\displaystyle\int \sin^n ax\, dx = -\frac{\sin^{n-1} ax \cos ax}{na} + \frac{n-1}{n}\int \sin^{n-2} ax\, dx$

61. $\displaystyle\int \cos^n ax\, dx = \frac{\cos^{n-1} ax \sin ax}{na} + \frac{n-1}{n}\int \cos^{n-2} ax\, dx$

62. (a) $\displaystyle\int \sin ax \cos bx\, dx = -\frac{\cos (a+b)x}{2(a+b)} - \frac{\cos (a-b)x}{2(a-b)} + C, \quad a^2 \neq b^2$

(b) $\displaystyle\int \sin ax \sin bx\, dx = \frac{\sin (a-b)x}{2(a-b)} - \frac{\sin (a+b)x}{2(a+b)} + C, \quad a^2 \neq b^2$

(c) $\displaystyle\int \cos ax \cos bx\, dx = \frac{\sin (a-b)x}{2(a-b)} + \frac{\sin (a+b)x}{2(a+b)} + C, \quad a^2 \neq b^2$

63. $\displaystyle\int \sin ax \cos ax\, dx = -\frac{\cos 2ax}{4a} + C$
  64. $\displaystyle\int \sin^n ax \cos ax\, dx = \frac{\sin^{n+1} ax}{(n+1)a} + C, \quad n \neq -1$

65. $\displaystyle\int \frac{\cos ax}{\sin ax}\, dx = \frac{1}{a} \ln |\sin ax| + C$

66. $\displaystyle\int \cos^n ax \sin ax\, dx = -\frac{\cos^{n+1} ax}{(n+1)a} + C, \quad n \neq -1$
  67. $\displaystyle\int \frac{\sin ax}{\cos ax}\, dx = -\frac{1}{a} \ln |\cos ax| + C$

68. $\displaystyle\int \sin^n ax \cos^m ax\, dx = -\frac{\sin^{n-1} ax \cos^{m+1} ax}{a(m+n)} + \frac{n-1}{m+n}\int \sin^{n-2} ax \cos^m ax\, dx, \quad n \neq -m \quad (\text{If } n = -m, \text{ use No. 86.})$

69. $\displaystyle\int \sin^n ax \cos^m ax\, dx = \frac{\sin^{n+1} ax \cos^{m-1} ax}{a(m+n)} + \frac{m-1}{m+n}\int \sin^n ax \cos^{m-2} ax\, dx, \quad m \neq -n \quad (\text{If } m = -n, \text{ use No. 87.})$

70. $\displaystyle\int \frac{dx}{b + c \sin ax} = \frac{-2}{a\sqrt{b^2 - c^2}} \tan^{-1}\left[\sqrt{\frac{b-c}{b+c}} \tan\left(\frac{\pi}{4} - \frac{ax}{2}\right)\right] + C, \quad b^2 > c^2$

71. $\displaystyle\int \frac{dx}{b + c \sin ax} = \frac{-1}{a\sqrt{c^2 - b^2}} \ln\left|\frac{c + b \sin ax + \sqrt{c^2 - b^2}\cos ax}{b + c \sin ax}\right| + C, \quad b^2 < c^2$

72. $\displaystyle\int \frac{dx}{1 + \sin ax} = -\frac{1}{a} \tan\left(\frac{\pi}{4} - \frac{ax}{2}\right) + C$
  73. $\displaystyle\int \frac{dx}{1 - \sin ax} = \frac{1}{a} \tan\left(\frac{\pi}{4} + \frac{ax}{2}\right) + C$

74. $\displaystyle\int \frac{dx}{b + c \cos ax} = \frac{2}{a\sqrt{b^2 - c^2}} \tan^{-1}\left[\sqrt{\frac{b-c}{b+c}} \tan\frac{ax}{2}\right] + C, \quad b^2 > c^2$

75. $\displaystyle\int \frac{dx}{b + c \cos ax} = \frac{1}{a\sqrt{c^2 - b^2}} \ln\left|\frac{c + b \cos ax + \sqrt{c^2 - b^2}\sin ax}{b + c \cos ax}\right| + C, \quad b^2 < c^2$

76. $\displaystyle\int \frac{dx}{1 + \cos ax} = \frac{1}{a} \tan\frac{ax}{2} + C$
  77. $\displaystyle\int \frac{dx}{1 - \cos ax} = -\frac{1}{a} \cot\frac{ax}{2} + C$

78. $\displaystyle\int x \sin ax \, dx = \frac{1}{a^2} \sin ax - \frac{x}{a} \cos ax + C$

79. $\displaystyle\int x \cos ax \, dx = \frac{1}{a^2} \cos ax + \frac{x}{a} \sin ax + C$

80. $\displaystyle\int x^n \sin ax \, dx = -\frac{x^n}{a} \cos ax + \frac{n}{a}\int x^{n-1} \cos ax \, dx$

81. $\displaystyle\int x^n \cos ax \, dx = \frac{x^n}{a} \sin ax - \frac{n}{a}\int x^{n-1} \sin ax \, dx$

82. $\displaystyle\int \tan ax \, dx = \frac{1}{a} \ln |\sec ax| + C$

83. $\displaystyle\int \cot ax \, dx = \frac{1}{a} \ln |\sin ax| + C$

84. $\displaystyle\int \tan^2 ax \, dx = \frac{1}{a} \tan ax - x + C$

85. $\displaystyle\int \cot^2 ax \, dx = -\frac{1}{a} \cot ax - x + C$

86. $\displaystyle\int \tan^n ax \, dx = \frac{\tan^{n-1} ax}{a(n-1)} - \int \tan^{n-2} ax \, dx, \quad n \neq 1$

87. $\displaystyle\int \cot^n ax \, dx = -\frac{\cot^{n-1} ax}{a(n-1)} - \int \cot^{n-2} ax \, dx, \quad n \neq 1$

88. $\displaystyle\int \sec ax \, dx = \frac{1}{a} \ln |\sec ax + \tan ax| + C$

89. $\displaystyle\int \csc ax \, dx = -\frac{1}{a} \ln |\csc ax + \cot ax| + C$

90. $\displaystyle\int \sec^2 ax \, dx = \frac{1}{a} \tan ax + C$

91. $\displaystyle\int \csc^2 ax \, dx = -\frac{1}{a} \cot ax + C$

92. $\displaystyle\int \sec^n ax \, dx = \frac{\sec^{n-2} ax \tan ax}{a(n-1)} + \frac{n-2}{n-1}\int \sec^{n-2} ax \, dx, \quad n \neq 1$

93. $\displaystyle\int \csc^n ax \, dx = -\frac{\csc^{n-2} ax \cot ax}{a(n-1)} + \frac{n-2}{n-1}\int \csc^{n-2} ax \, dx, \quad n \neq 1$

94. $\displaystyle\int \sec^n ax \tan ax \, dx = \frac{\sec^n ax}{na} + C, \quad n \neq 0$

95. $\displaystyle\int \csc^n ax \cot ax \, dx = -\frac{\csc^n ax}{na} + C, \quad n \neq 0$

96. $\displaystyle\int \sin^{-1} ax \, dx = x \sin^{-1} ax + \frac{1}{a}\sqrt{1 - a^2 x^2} + C$

97. $\displaystyle\int \cos^{-1} ax \, dx = x \cos^{-1} ax - \frac{1}{a}\sqrt{1 - a^2 x^2} + C$

98. $\displaystyle\int \tan^{-1} ax \, dx = x \tan^{-1} ax - \frac{1}{2a} \ln (1 + a^2 x^2) + C$

99. $\displaystyle\int x^n \sin^{-1} ax \, dx = \frac{x^{n+1}}{n+1} \sin^{-1} ax - \frac{a}{n+1}\int \frac{x^{n+1} \, dx}{\sqrt{1 - a^2 x^2}}, \quad n \neq -1$

100. $\displaystyle\int x^n \cos^{-1} ax \, dx = \frac{x^{n+1}}{n+1} \cos^{-1} ax + \frac{a}{n+1}\int \frac{x^{n+1} \, dx}{\sqrt{1 - a^2 x^2}}, \quad n \neq -1$

101. $\displaystyle\int x^n \tan^{-1} ax \, dx = \frac{x^{n+1}}{n+1} \tan^{-1} ax - \frac{a}{n+1}\int \frac{x^{n+1} \, dx}{1 + a^2 x^2}, \quad n \neq -1$

102. $\displaystyle\int e^{ax} \, dx = \frac{1}{a} e^{ax} + C$

103. $\displaystyle\int b^{ax} \, dx = \frac{1}{a} \frac{b^{ax}}{\ln b} + C, \quad b > 0, \quad b \neq 1$

104. $\int xe^{ax}\,dx = \dfrac{e^{ax}}{a^2}(ax-1) + C$

105. $\int x^n e^{ax}\,dx = \dfrac{1}{a}x^n e^{ax} - \dfrac{n}{a}\int x^{n-1}e^{ax}\,dx$

106. $\int x^n b^{ax}\,dx = \dfrac{x^n b^{ax}}{a\ln b} - \dfrac{n}{a\ln b}\int x^{n-1}b^{ax}\,dx, \quad b > 0, \quad b \ne 1$

107. $\int e^{ax}\sin bx\,dx = \dfrac{e^{ax}}{a^2+b^2}(a\sin bx - b\cos bx) + C$

108. $\int e^{ax}\cos bx\,dx = \dfrac{e^{ax}}{a^2+b^2}(a\cos bx + b\sin bx) + C$

109. $\int \ln ax\,dx = x\ln ax - x + C$

110. $\int x^n(\ln ax)^m\,dx = \dfrac{x^{n+1}(\ln ax)^m}{n+1} - \dfrac{m}{n+1}\int x^n(\ln ax)^{m-1}\,dx, \quad n \ne -1$

111. $\int x^{-1}(\ln ax)^m\,dx = \dfrac{(\ln ax)^{m+1}}{m+1} + C, \quad m \ne -1$

112. $\int \dfrac{dx}{x\ln ax} = \ln|\ln ax| + C$

113. $\int \sinh ax\,dx = \dfrac{1}{a}\cosh ax + C$

114. $\int \cosh ax\,dx = \dfrac{1}{a}\sinh ax + C$

115. $\int \sinh^2 ax\,dx = \dfrac{\sinh 2ax}{4a} - \dfrac{x}{2} + C$

116. $\int \cosh^2 ax\,dx = \dfrac{\sinh 2ax}{4a} + \dfrac{x}{2} + C$

117. $\int \sinh^n ax\,dx = \dfrac{\sinh^{n-1} ax\cosh ax}{na} - \dfrac{n-1}{n}\int \sinh^{n-2} ax\,dx, \quad n \ne 0$

118. $\int \cosh^n ax\,dx = \dfrac{\cosh^{n-1} ax\sinh ax}{na} + \dfrac{n-1}{n}\int \cosh^{n-2} ax\,dx, \quad n \ne 0$

119. $\int x\sinh ax\,dx = \dfrac{x}{a}\cosh ax - \dfrac{1}{a^2}\sinh ax + C$

120. $\int x\cosh ax\,dx = \dfrac{x}{a}\sinh ax - \dfrac{1}{a^2}\cosh ax + C$

121. $\int x^n\sinh ax\,dx = \dfrac{x^n}{a}\cosh ax - \dfrac{n}{a}\int x^{n-1}\cosh ax\,dx$

122. $\int x^n\cosh ax\,dx = \dfrac{x^n}{a}\sinh ax - \dfrac{n}{a}\int x^{n-1}\sinh ax\,dx$

123. $\int \tanh ax\,dx = \dfrac{1}{a}\ln(\cosh ax) + C$

124. $\int \coth ax\,dx = \dfrac{1}{a}\ln|\sinh ax| + C$

125. $\int \tanh^2 ax\,dx = x - \dfrac{1}{a}\tanh ax + C$

126. $\int \coth^2 ax\,dx = x - \dfrac{1}{a}\coth ax + C$

127. $\int \tanh^n ax\,dx = -\dfrac{\tanh^{n-1} ax}{(n-1)a} + \int \tanh^{n-2} ax\,dx, \quad n \ne 1$

128. $\int \coth^n ax\,dx = -\dfrac{\coth^{n-1} ax}{(n-1)a} + \int \coth^{n-2} ax\,dx, \quad n \ne 1$

129. $\int \operatorname{sech} ax\,dx = \dfrac{1}{a}\sin^{-1}(\tanh ax) + C$

130. $\int \operatorname{csch} ax\,dx = \dfrac{1}{a}\ln\left|\tanh\dfrac{ax}{2}\right| + C$

131. $\int \text{sech}^2 ax \, dx = \frac{1}{a} \tanh ax + C$

132. $\int \text{csch}^2 ax \, dx = -\frac{1}{a} \coth ax + C$

133. $\int \text{sech}^n ax \, dx = \frac{\text{sech}^{n-2} ax \tanh ax}{(n-1)a} + \frac{n-2}{n-1} \int \text{sech}^{n-2} ax \, dx, \quad n \neq 1$

134. $\int \text{csch}^n ax \, dx = -\frac{\text{csch}^{n-2} ax \coth ax}{(n-1)a} - \frac{n-2}{n-1} \int \text{csch}^{n-2} ax \, dx, \quad n \neq 1$

135. $\int \text{sech}^n ax \tanh ax \, dx = -\frac{\text{sech}^n ax}{na} + C, \quad n \neq 0$

136. $\int \text{csch}^n ax \coth ax \, dx = -\frac{\text{csch}^n ax}{na} + C, \quad n \neq 0$

137. $\int e^{ax} \sinh bx \, dx = \frac{e^{ax}}{2} \left[\frac{e^{bx}}{a+b} - \frac{e^{-bx}}{a-b} \right] + C, \quad a^2 \neq b^2$

138. $\int e^{ax} \cosh bx \, dx = \frac{e^{ax}}{2} \left[\frac{e^{bx}}{a+b} + \frac{e^{-bx}}{a-b} \right] + C, \quad a^2 \neq b^2$

139. $\int_0^\infty x^{n-1} e^{-x} \, dx = (n-1)!, \quad n > 0$

140. $\int_0^\infty e^{-ax^2} \, dx = \frac{1}{2} \sqrt{\frac{\pi}{a}}, \quad a > 0$

141. $\int_0^{\pi/2} \sin^n x \, dx = \int_0^{\pi/2} \cos^n x \, dx = \begin{cases} \dfrac{1 \cdot 3 \cdot 5 \cdots (n-1)}{2 \cdot 4 \cdot 6 \cdots n} \cdot \dfrac{\pi}{2}, & \text{if } n \text{ is an even integer} \geq 2 \\ \dfrac{2 \cdot 4 \cdot 6 \cdots (n-1)}{3 \cdot 5 \cdot 7 \cdots n}, & \text{if } n \text{ is an odd integer} \geq 3 \end{cases}$

A8 Determinants and Cramer's Rule

Matrices • Determinants • Minors and Cofactors • Expanding by Columns or by Other Rows • Useful Facts about Determinants • Cramer's Rule

Matrices

A rectangular array of numbers like

$$A = \begin{bmatrix} 2 & 1 & 3 \\ 1 & 0 & -2 \end{bmatrix}$$

is called a **matrix.** We call A a 2 by 3 matrix because it has two rows and three columns. An m by n matrix has m rows and n columns, and the **entry** or **element** (number) in the ith row and jth column is denoted by a_{ij} (Figure A8.1). The matrix

$$A = \begin{bmatrix} 2 & 1 & 3 \\ 1 & 0 & -2 \end{bmatrix}$$

has

$$a_{11} = 2, \quad a_{12} = 1, \quad a_{13} = 3,$$
$$a_{21} = 1, \quad a_{22} = 0, \quad a_{23} = -2.$$

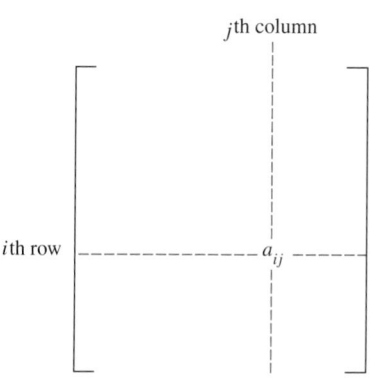

Figure A8.1 The location of entry a_{ij}.

A matrix with the same number of rows as columns is a **square matrix.** It is a **matrix of order n** if the number of rows and columns is n.

Determinants

The vertical bar in the notation $|a_{ij}|$ does not mean absolute value.

With each square matrix A we associate a number det A or $|a_{ij}|$, called the **determinant** of A, calculated from the entries of A in the following way. For $n = 1$ and $n = 2$, we define

$$\det[a] = a, \tag{1}$$

$$\det \begin{bmatrix} a_{11} & a_{12} \\ a_{21} & a_{22} \end{bmatrix} = a_{11}a_{22} - a_{21}a_{12}. \tag{2}$$

For a matrix of order 3, we define

$$\det A = \det \begin{bmatrix} a_{11} & a_{12} & a_{13} \\ a_{21} & a_{22} & a_{23} \\ a_{31} & a_{32} & a_{33} \end{bmatrix} = \begin{matrix} \text{sum of all signed products} \\ \text{of the form } \pm a_{1i}\,a_{2j}\,a_{3k}, \end{matrix} \tag{3}$$

where i, j, k is a permutation of 1, 2, 3 in some order. There are $3! = 6$ such permutations, so there are six terms in the sum. The sign is positive when the index of the permutation is even and negative when the index is odd.

Definition Index of a Permutation

Given any permutation of the numbers $1, 2, 3, \ldots, n$, denote the permutation by $i_1, i_2, i_3, \ldots, i_n$. In this arrangement, some of the numbers following i_1 may be less than i_1, and the number of these is called the **number of inversions** in the arrangement pertaining to i_1. Likewise, there are a number of inversions pertaining to each of the other i's; it is the number of indices that come after that particular i in the arrangement and are less than it. The **index** of the permutation is the sum of all the numbers of inversions pertaining to the separate indices.

Example 1 FINDING THE INDEX OF A PERMUTATION

Find the index of the permutation

$$5 \quad 3 \quad 1 \quad 2 \quad 4$$

of the numbers 1, 2, 3, 4, and 5.

Solution   The permutation

$$5 \quad 3 \quad 1 \quad 2 \quad 4$$

has four inversions pertaining to the first element, 5, two inversions pertaining to the second element, 3, and no further inversions, so the index is $4 + 2 = 6$.

The following table shows the permutations of 1, 2, 3, the index of each permutation, and the signed product in the determinant of Equation (3).

Permutation	Index	Signed product
1 2 3	0	$+a_{11}a_{22}a_{33}$
1 3 2	1	$-a_{11}a_{23}a_{32}$
2 1 3	1	$-a_{12}a_{21}a_{33}$
2 3 1	2	$+a_{12}a_{23}a_{31}$
3 1 2	2	$+a_{13}a_{21}a_{32}$
3 2 1	3	$-a_{13}a_{22}a_{31}$

The sum of the six signed products is

$$a_{11}(a_{22}a_{33} - a_{23}a_{32}) - a_{12}(a_{21}a_{33} - a_{23}a_{31}) + a_{13}(a_{21}a_{32} - a_{22}a_{31})$$

$$= a_{11}\begin{vmatrix} a_{22} & a_{23} \\ a_{32} & a_{33} \end{vmatrix} - a_{12}\begin{vmatrix} a_{21} & a_{23} \\ a_{31} & a_{33} \end{vmatrix} + a_{13}\begin{vmatrix} a_{21} & a_{22} \\ a_{31} & a_{32} \end{vmatrix} = \begin{vmatrix} a_{11} & a_{12} & a_{13} \\ a_{21} & a_{22} & a_{23} \\ a_{31} & a_{32} & a_{33} \end{vmatrix}.$$

The formula

$$\begin{vmatrix} a_{11} & a_{12} & a_{13} \\ a_{21} & a_{22} & a_{23} \\ a_{31} & a_{32} & a_{33} \end{vmatrix} = a_{11}\begin{vmatrix} a_{22} & a_{23} \\ a_{32} & a_{33} \end{vmatrix} - a_{12}\begin{vmatrix} a_{21} & a_{23} \\ a_{31} & a_{33} \end{vmatrix} + a_{13}\begin{vmatrix} a_{21} & a_{22} \\ a_{31} & a_{32} \end{vmatrix} \quad (4)$$

reduces the calculation of a 3 by 3 determinant to the calculation of three 2 by 2 determinants.

Many people prefer to remember the following scheme for calculating the six signed products in the determinant of a 3 by 3 matrix:

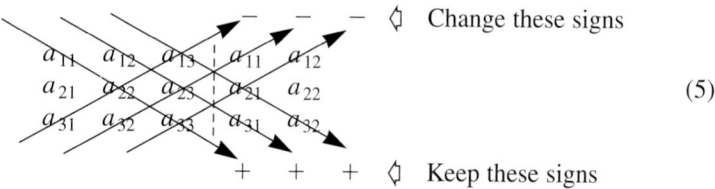

$$(5)$$

Minors and Cofactors

The second-order determinants on the right-hand side of Equation (4) are the **minors** (short for "minor determinants") of the entries they multiply. Thus,

$$\begin{vmatrix} a_{22} & a_{23} \\ a_{32} & a_{33} \end{vmatrix} \text{ is the minor of } a_{11}, \qquad \begin{vmatrix} a_{21} & a_{23} \\ a_{31} & a_{33} \end{vmatrix} \text{ is the minor of } a_{12},$$

and so on. The minor of the element a_{ij} in a matrix A is the determinant of the matrix that remains after we delete the row and column containing a_{ij}:

$$\begin{vmatrix} a_{11} & a_{12} & a_{13} \\ a_{21} & a_{22} & a_{23} \\ a_{31} & a_{32} & a_{33} \end{vmatrix}. \qquad \text{The minor of } a_{22} \text{ is } \begin{vmatrix} a_{11} & a_{13} \\ a_{31} & a_{33} \end{vmatrix}.$$

$$\begin{vmatrix} a_{11} & a_{12} & a_{13} \\ a_{21} & a_{22} & a_{23} \\ a_{31} & a_{32} & a_{33} \end{vmatrix}. \qquad \text{The minor of } a_{23} \text{ is } \begin{vmatrix} a_{11} & a_{12} \\ a_{31} & a_{32} \end{vmatrix}.$$

The **cofactor** A_{ij} of a_{ij} is $(-1)^{i+j}$ times the minor of a_{ij}. Thus,

$$A_{22} = (-1)^{2+2} \begin{vmatrix} a_{11} & a_{13} \\ a_{31} & a_{33} \end{vmatrix} = \begin{vmatrix} a_{11} & a_{13} \\ a_{31} & a_{33} \end{vmatrix},$$

$$A_{23} = (-1)^{2+3} \begin{vmatrix} a_{11} & a_{12} \\ a_{31} & a_{32} \end{vmatrix} = - \begin{vmatrix} a_{11} & a_{12} \\ a_{31} & a_{32} \end{vmatrix}.$$

The factor $(-1)^{i+j}$ changes the sign of the minor when $i + j$ is odd. There is a checkerboard pattern for remembering these changes:

$$\begin{matrix} + & - & + \\ - & + & - \\ + & - & + \end{matrix}$$

In the upper left corner, $i = 1$, $j = 1$, and $(-1)^{1+1} = +1$. In going from any cell to an adjacent cell in the same row or column, we change i by 1 or j by 1, but not both, so we change the exponent from even to odd or from odd to even, which changes the sign from + to − or from − to +.

When we rewrite Equation (4) in terms of cofactors we get

$$\det A = a_{11}A_{11} + a_{12}A_{12} + a_{13}A_{13}. \tag{6}$$

Example 2   FINDING A DETERMINANT TWO WAYS

Find the determinant of

$$A = \begin{vmatrix} 2 & 1 & 3 \\ 3 & -1 & -2 \\ 2 & 3 & 1 \end{vmatrix}.$$

Solution   Using Equation (6): The cofactors are

$$A_{11} = (-1)^{1+1} \begin{vmatrix} -1 & -2 \\ 3 & 1 \end{vmatrix}, \qquad A_{12} = (-1)^{1+2} \begin{vmatrix} 3 & -2 \\ 2 & 1 \end{vmatrix},$$

$$A_{13} = (-1)^{1+3} \begin{vmatrix} 3 & -1 \\ 2 & 3 \end{vmatrix}.$$

To find $\det A$, we multiply each element of the first row of A by its cofactor and add:

$$\det A = 2 \begin{vmatrix} -1 & -2 \\ 3 & 1 \end{vmatrix} + (-1) \begin{vmatrix} 3 & -2 \\ 2 & 1 \end{vmatrix} + 3 \begin{vmatrix} 3 & -1 \\ 2 & 3 \end{vmatrix}$$

$$= 2(-1 + 6) - 1(3 + 4) + 3(9 + 2) = 10 - 7 + 33 = 36.$$

Solution   Using the scheme in Equation (5): From Equation (5) we find

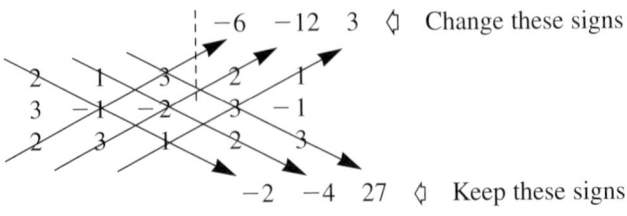

$$\det A = -(-6) - (-12) - 3 + (-2) + (-4) + 27 = 36.$$

Expanding by Columns or by Other Rows

The determinant of a square matrix can be calculated from the cofactors of any row or any column.

If we were to expand the determinant in Example 2 by cofactors according to elements of its third column, say, we would still get 36:

$$+3\begin{vmatrix} 3 & -1 \\ 2 & 3 \end{vmatrix} - (-2)\begin{vmatrix} 2 & 1 \\ 2 & 3 \end{vmatrix} + 1\begin{vmatrix} 2 & 1 \\ 3 & -1 \end{vmatrix}$$

$$= 3(9+2) + 2(6-2) + 1(-2-3) = 33 + 8 - 5 = 36.$$

Useful Facts about Determinants

Fact 1: If two rows (or columns) are identical, then the determinant is zero.

Fact 2: Interchanging two rows (or columns) changes the sign of the determinant.

Fact 3: The determinant is the sum of the products of the elements of the ith row (or column) by their cofactors, for any i.

Fact 4: The determinant of the transpose of a matrix is the same as the determinant of the original matrix. (The **transpose** of a matrix is obtained by writing the rows as columns. The first row becomes the first column, the second row the second column, and so on.)

Fact 5: Multiplying each element of some row (or column) by a constant c multiplies the determinant by c.

Fact 6: If all elements above the main diagonal (or all below it) are zero, then the determinant is the product of the elements on the main diagonal. (The **main diagonal** is the diagonal from upper left to lower right.)

Example 3   ILLUSTRATING FACT 6

$$\begin{vmatrix} 3 & 4 & 7 \\ 0 & -2 & 5 \\ 0 & 0 & 5 \end{vmatrix} = (3)(-2)(5) = -30$$

Fact 7: If the elements of any row are multiplied by the cofactors of the corresponding elements of a different row and these products are summed, then the sum is zero.

Example 4   ILLUSTRATING FACT 7

If A_{11}, A_{12}, A_{13} are the cofactors of the elements of the first row of $A = (a_{ij})$, then the sums

$$a_{21}A_{11} + a_{22}A_{12} + a_{23}A_{13}$$

(elements of second row times cofactors of elements of first row) and

$$a_{31}A_{11} + a_{32}A_{12} + a_{33}A_{13}$$

are both zero.

Fact 8: If the elements of any column are multiplied by the cofactors of the corresponding elements of a different column and these products are summed, then the sum is zero.

Fact 9: If each element of a row is multiplied by a constant c and the results added to a different row, then the determinant is not changed. A similar result holds for columns.

Example 5   ADDING A MULTIPLE OF ONE ROW TO ANOTHER ROW

If we start with

$$A = \begin{bmatrix} 2 & 1 & 3 \\ 3 & -1 & -2 \\ 2 & 3 & 1 \end{bmatrix}$$

and add -2 times row 1 to row 2 (subtract 2 times row 1 from row 2), we get

$$B = \begin{bmatrix} 2 & 1 & 3 \\ -1 & -3 & -8 \\ 2 & 3 & 1 \end{bmatrix}.$$

Since det $A = 36$ (Example 2), we should find that det $B = 36$ as well. Indeed we do, as the following calculation shows:

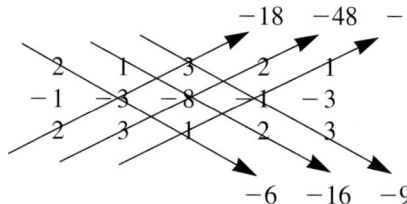

$$\det B = -(-18) - (-48) - (-1) + (-6) + (-16) + (-9)$$

$$= 18 + 48 + 1 - 6 - 16 - 9 = 67 - 31 = 36.$$

Example 6   TRANSFORMING A FOURTH-ORDER DETERMINANT TO MAKE IT EASIER TO EVALUATE

Evaluate the fourth-order determinant

$$D = \begin{vmatrix} 1 & -2 & 3 & 1 \\ 2 & 1 & 0 & 2 \\ -1 & 2 & 1 & -2 \\ 0 & 1 & 2 & 1 \end{vmatrix}.$$

Solution   We subtract 2 times row 1 from row 2 and add row 1 to row 3 to get

$$D = \begin{vmatrix} 1 & -2 & 3 & 1 \\ 0 & 5 & -6 & 0 \\ 0 & 0 & 4 & -1 \\ 0 & 1 & 2 & 1 \end{vmatrix}.$$

We then multiply the elements of the first column by their cofactors to get

$$D = (1) \begin{vmatrix} 5 & -6 & 0 \\ 0 & 4 & -1 \\ 1 & 2 & 1 \end{vmatrix} = 5(4 + 2) - (-6)(0 + 1) + 0 = 36.$$

Cramer's Rule

If the determinant $D = \det A = \begin{vmatrix} a_{11} & a_{12} \\ a_{21} & a_{22} \end{vmatrix} = 0$, then the system

$$a_{11}x + a_{12}y = b_1$$
$$a_{21}x + a_{22}y = b_2 \qquad (7)$$

has either infinitely many solutions or no solution at all. The system

$$x + y = 0,$$
$$2x + 2y = 0$$

whose determinant is

$$D = \begin{vmatrix} 1 & 1 \\ 2 & 2 \end{vmatrix} = 2 - 2 = 0$$

has infinitely many solutions. We can find an x to match any given y. The system

$$x + y = 0,$$
$$2x + 2y = 2$$

has no solution. If $x + y = 0$, then $2x + 2y = 2(x + y)$ cannot be 2.

If $D \neq 0$, then system (7) has a unique solution, and **Cramer's Rule** states that it may be found from the formulas

$$x = \frac{\begin{vmatrix} b_1 & a_{12} \\ b_2 & a_{22} \end{vmatrix}}{D}, \qquad y = \frac{\begin{vmatrix} a_{11} & b_1 \\ a_{21} & b_2 \end{vmatrix}}{D}. \qquad (8)$$

The numerator in the formula for x comes from replacing the first column in A (the x-column) by the column of constants b_1 and b_2 (the b-column). Replacing the y-column by the b-column gives the numerator of the y-solution.

Example 7 APPLYING CRAMER'S RULE

Solve the system

$$3x - y = 9$$
$$x + 2y = -4.$$

Solution We use Equations (8). The determinant of the coefficient matrix is

$$D = \begin{vmatrix} 3 & -1 \\ 1 & 2 \end{vmatrix} = 6 + 1 = 7.$$

Hence,

$$x = \frac{\begin{vmatrix} 9 & -1 \\ -4 & 2 \end{vmatrix}}{D} = \frac{18 - 4}{7} = \frac{14}{7} = 2,$$

$$y = \frac{\begin{vmatrix} 3 & 9 \\ 1 & -4 \end{vmatrix}}{D} = \frac{-12 - 9}{7} = \frac{-21}{7} = -3.$$

Systems of three equations in three unknowns work the same way. If

$$D = \det A = \begin{vmatrix} a_{11} & a_{12} & a_{13} \\ a_{21} & a_{22} & a_{23} \\ a_{31} & a_{32} & a_{33} \end{vmatrix} = 0,$$

the system

$$a_{11}x + a_{12}y + a_{13}z = b_1,$$
$$a_{21}x + a_{22}y + a_{23}z = b_2, \qquad (9)$$
$$a_{31}x + a_{32}y + a_{33}z = b_3,$$

has either infinitely many solutions or no solution at all. If $D \neq 0$, then the system has a unique solution, given by Cramer's Rule:

$$x = \frac{1}{D} \begin{vmatrix} b_1 & a_{12} & a_{13} \\ b_2 & a_{22} & a_{23} \\ b_3 & a_{32} & a_{33} \end{vmatrix}, \qquad y = \frac{1}{D} \begin{vmatrix} a_{11} & b_1 & a_{13} \\ a_{21} & b_2 & a_{23} \\ a_{31} & b_3 & a_{33} \end{vmatrix}.$$

$$z = \frac{1}{D} \begin{vmatrix} a_{11} & a_{12} & b_1 \\ a_{21} & a_{22} & b_2 \\ a_{31} & a_{32} & b_3 \end{vmatrix}.$$

The pattern continues in higher dimensions.

Section A8 Exercises

In Exercises 1–4, evaluate the determinant.

1. $\begin{vmatrix} 2 & 3 & 1 \\ 4 & 5 & 2 \\ 1 & 2 & 3 \end{vmatrix}$

2. $\begin{vmatrix} 2 & -1 & -2 \\ -1 & 2 & 1 \\ 3 & 0 & -3 \end{vmatrix}$

3. $\begin{vmatrix} 1 & 2 & 3 & 4 \\ 0 & 1 & 2 & 3 \\ 0 & 0 & 2 & 1 \\ 0 & 0 & 3 & 2 \end{vmatrix}$

4. $\begin{vmatrix} 1 & -1 & 2 & 3 \\ 2 & 1 & 2 & 6 \\ 1 & 0 & 2 & 3 \\ -2 & 2 & 0 & -5 \end{vmatrix}$

In Exercises 5–8, evaluate the determinant by expanding according to the cofactors of (a) the third row and (b) the second column.

5. $\begin{vmatrix} 2 & -1 & 2 \\ 1 & 0 & 3 \\ 0 & 2 & 1 \end{vmatrix}$

6. $\begin{vmatrix} 1 & 0 & -1 \\ 0 & 2 & -2 \\ 2 & 0 & 1 \end{vmatrix}$

7. $\begin{vmatrix} 1 & 1 & 0 & 0 \\ 0 & 0 & -2 & 1 \\ 0 & -1 & 0 & 7 \\ 3 & 0 & 2 & 1 \end{vmatrix}$

8. $\begin{vmatrix} 0 & 1 & 0 & 0 \\ 0 & 1 & 1 & 0 \\ 1 & 1 & 1 & 1 \\ 1 & 1 & 0 & 0 \end{vmatrix}$

In Exercises 9–16, solve the system of equations by Cramer's Rule.

9. $x + 8y = 4$
$3x - y = -13$

10. $2x + 3y = 5$
$3x - y = 2$

11. $4x - 3y = 6$
$3x - 2y = 5$

12. $x + y + z = 2$
$2x - y + z = 0$
$x + 2y - z = 4$

13. $2x + y - z = 2$
$x - y + z = 7$
$2x + 2y + z = 4$

14. $2x - 4y = 6$
$x + y + z = 1$
$5y + 7z = 10$

15. $x - z = 3$
$2y - 2z = 2$
$2x + z = 3$

16. $x_1 + x_2 - x_3 + x_4 = 2$
$x_1 - x_2 + x_3 + x_4 = -1$
$x_1 + x_2 + x_3 - x_4 = 2$
$x_1 + x_3 + x_4 = -1$

Explorations

17. Find values of h and k for which the system

$$2x + hy = 8,$$
$$x + 3y = k$$

has **(a)** infinitely many solutions and **(b)** no solution at all.

18. For what value of x will

$$\begin{vmatrix} x & x & 1 \\ 2 & 0 & 5 \\ 6 & 7 & 1 \end{vmatrix} = 0?$$

■

Extending the Ideas

19. **Writing to Learn** Suppose u, v, and w are twice-differentiable functions of x that satisfy the relation $au + bv + cw = 0$, where a, b, and c are constants, not all zero. Show that

$$\begin{vmatrix} u & v & w \\ u' & v' & w' \\ u'' & v'' & w'' \end{vmatrix} = 0.$$

20. *Partial Fractions.* Expanding the quotient

$$\frac{ax + b}{(x - r_1)(x - r_2)}$$

by partial fractions calls for finding the values of C and D that make the equation

$$\frac{ax + b}{(x - r_1)(x - r_2)} = \frac{C}{x - r_1} + \frac{D}{x - r_2}$$

hold for all x.

(a) Find a system of linear equations that determines C and D.

(b) Writing to Learn Under what circumstances does the system of equations in part (a) have a unique solution? That is, when is the determinant of the coefficient matrix of the system different from zero?

G L O S S A R Y

Absolute convergence: If the series $\Sigma|a_n|$ of absolute values converges, then Σa_n is said to converge absolutely.   p. 502

Absolute error: $|$true value $-$ approximate value$|$ p. 221

Absolute maximum: The function f has an absolute maximum value $f(c)$ at a point c in its domain D if and only if $f(x) \leq f(c)$ for all x in D.   pp. 177, 741.

Absolute minimum: The function f has an absolute minimum value $f(c)$ at a point c in its domain D if and only if $f(x) \geq f(c)$ for all x in D.   pp. 177, 741

Absolute, relative, and percentage change: As we move from $x = a$ to a nearby point $a + dx$, we can describe the corresponding change in the value of a function $f(x)$ three ways:

Absolute change:	$\Delta f = f(a + dx) - f(a)$
Relative change:	$\Delta f/f(a)$
Percentage change:	$(\Delta f/f(a)) \times 100$.

Similar definitions hold for functions of more than one variable pp. 226, 708

Absolute value function: The function $f(x) = |x|$. See also *Absolute value of a number.*   p. 15

Absolute value of a number: The absolute value of x is

$$|x| = \begin{cases} -x, & x < 0 \\ x, & x \geq 0. \end{cases}$$

Acceleration: The derivative of a velocity function with respect to time.   p. 125

Acceleration vector: If $\mathbf{r}$ is the position vector of a particle moving along a smooth curve in the plane, then at any time t,

$$\mathbf{a} = \frac{d^2\mathbf{r}}{dt^2}$$

is the particle's acceleration vector.   pp. 533, 636

Algebraic function: A function $y = f(x)$ that satisfies an equation of the form $P_n y^n + \cdots + P_1 y + P_0 = 0$ in which the P's are polynomials in x with rational coefficients. The function $y = 1/\sqrt{x + 1}$ is algebraic, for example, because it satisfies the equation $(x + 1)y^2 - 1 = 0$. Here, $P_2 = x + 1$, $P_1 = 0$, and $P_0 = -1$. All polynomials and rational functions are algebraic.

Alternating harmonic series: The series

$$\sum_{n=1}^{\infty} \frac{(-1)^{n-1}}{n}.$$

p. 500

Alternating series: A series in which the terms are alternately positive and negative.   p. 500

Amplitude: Of a periodic function $f(x)$ continuous for all real x, the number

$$\frac{(\text{absolute max of } f) - (\text{absolute min of } f)}{2};$$

of the sine function

$$f(x) = A \sin\left[\frac{2\pi}{B}(x - C)\right] + D,$$

is the number $|A|$.

Angle between curves: At a point of intersection of two differentiable curves, the angle between their tangent lines at the point of intersection.

Angle between vectors u and v: The angle

$$\theta = \cos^{-1}\left(\frac{\mathbf{u} \cdot \mathbf{v}}{|\mathbf{u}||\mathbf{v}|}\right).$$

p. 523

Antiderivative: A function $F(x)$ is an antiderivative of a function $f(x)$ if $F'(x) = f(x)$ for all x in the domain of f. A vector function $\mathbf{R}(t)$ is an antiderivative of a vector function $\mathbf{r}(t)$ on an internal I if $d\mathbf{R}/dt = \mathbf{r}(t)$ at each point of I.   pp. 190, 535, 639

Antidifferentiation: The process of finding an antiderivative. p. 190

Arbitrary constant: See *Constant of integration.*

Arc length: For a smooth plane curve $y = f(x)$, $a \leq x \leq b$, the number

$$L = \int_a^b \sqrt{1 + \left(\frac{dy}{dx}\right)^2}\, dx.$$

For a smooth space curve $\mathbf{r}(t) = f(t) = \mathbf{i} + g(t)\mathbf{j} + h(t)\mathbf{k}$, $a \leq t \leq b$, the number

$$L = \int_a^b \sqrt{\left(\frac{df}{dt}\right)^2 + \left(\frac{dg}{dt}\right)^2 + \left(\frac{dh}{dt}\right)^2}\, dt.$$

pp. 397, 645

Arc length parameter: The signed distance s measured from a base point on a smooth curve C that determines a point on the curve.   p. 646

Arccosine or inverse cosine function: The inverse of the cosine function with restricted domain $[0, \pi]$.   p. 46

Arcsine or inverse sine function: The inverse of the sine function with restricted domain $[-\pi/2, \pi/2]$.   p. 46

Arctangent or inverse tangent function: The inverse of the tangent function with restricted domain $(-\pi/2, \pi/2)$.   p. 46

Area of a closed bounded plane region *R*:

$$A = \iint\limits_R dA,$$

or, in polar coordinates,

$$A = \iint\limits_R r\, dr\, d\theta.$$

pp. 773, 786

Asymptote: The line $y = b$ is a horizontal asymptote of the graph of a function $y = f(x)$ if either

$$\lim_{x \to \infty} f(x) = b \qquad \text{or} \qquad \lim_{x \to -\infty} f(x) = b.$$

The line $x = a$ is a vertical asymptote of the graph of a function $y = f(x)$ if either

$$\lim_{x \to a^+} f(x) = \pm\infty \qquad \text{or} \qquad \lim_{x \to a^-} f(x) = \pm\infty.$$

p. 65

Average rate of change of a quantity over a period of time: The amount of change divided by the time it takes. p. 86

Average value
of a continuous function *f* on [*a*, *b*]:

$$\frac{1}{b-a} \int_a^b f(x)\, dx.$$

p. 271

of a function on a region D in space:

$$\frac{1}{\text{volume of } D} \iiint\limits_D F(\mathbf{x}, \mathbf{y}, \mathbf{z})\, dV$$

p. 795

of a function on a region R in the plane:

$$\frac{1}{\text{area of } R} \iint\limits_R f(\mathbf{x}, \mathbf{y})\, dA.$$

pp. 774, 795

Average velocity: Displacement (change in position) divided by time traveled. p. 123

Axis of revolution: The line about which a solid of revolution is generated. See *Solid of revolution.*

Base (exponential and logarithm): See *Exponential function with base a; Logarithm function with base a.*

Basic trigonometric functions: When an angle of measure θ is placed in standard position in the coordinate plane and (x, y) is the point in which its terminal ray intersects a circle with center $(0, 0)$ and radius r, the values of the six basic trigonometric functions of θ are

Sine: $\sin \theta = \dfrac{y}{r}$ Cosecant: $\csc \theta = \dfrac{r}{y}$

Cosine: $\cos \theta = \dfrac{x}{r}$ Secant: $\sec \theta = \dfrac{r}{x}$

Tangent: $\tan \theta = \dfrac{y}{x}$ Cotangent: $\cot \theta = \dfrac{x}{y}$.

p. 41

Big-oh notation: If functions f and g are positive for x sufficiently large, then f is big-oh of g (written $f = O(g)$) if there exists a positive integer M such that

$$\frac{f(x)}{g(x)} \le M$$

for x sufficiently large. p. 429

Binomial series: The Maclaurin series for $f(x) = (1 + x)^m$. p. 479

Binormal vector of a curve in space: The vector $\mathbf{B} = \mathbf{T} \times \mathbf{N}$, where $\mathbf{T}$ and $\mathbf{N}$ are the unit tangent and normal vectors. p. 653

Boundary point of a region R in the plane: A point (x_0, y_0) such that every disk centered at (x_0, y_0) contains points that lie outside of R as well as points that lie in R. The region's boundary points make up its *boundary.* p. 677

Boundary point of a region D in space: A point (x_0, y_0, z_0) such that every sphere centered at (x_0, y_0, z_0) contains points that lie outside of D as well as points that lie in D. The region's boundary points make up its *boundary.* p. 677

Bounded: A function f is bounded on a given domain if there are numbers m and M such that $m \le f(x) \le M$ for any x in the domain of f. The number m is a lower bound and the number M is an upper bound for f. p. 266

Bounded (from) above: A function f is bounded (from) above on a given domain if there is a number M such that $f(x) \le M$ for all x in the domain. p. 270

Bounded (from) below: A function f is bounded (from) below on a given domain if there is a number m such that $m \le f(x)$ for all x in the domain. p. 270

Bounded region: In the plane, a region that lies inside a disk of finite radius. In space, a region that lies inside a sphere of finite radius. p. 677

Cardioid: Any of the heart-shaped polar curves $r = a(1 - \cos \theta)$ or $r = a(1 - \sin \theta)$, $a \ne 0$ a constant. p. 554

Cartesian coordinates of a point in space: The ordered triple (x, y, z). p. 575

Center of curvature: See *Circle of curvature.*

Center of a power series: See *Power series.*

Centroid: The center of mass of a shape with constant density. p. 779

Chain Rule: If $y = f(u)$ is differentiable at the point $u = g(x)$ and g is differentiable at x, then the composite function $(f \circ g) = f(g(x))$ is differentiable at x, and

$$(f \circ g)'(x) = f'(g(x)) \cdot g'(x), \qquad \text{or} \qquad \frac{dy}{dx} = \frac{dy}{du} \cdot \frac{du}{dx},$$

where dy/du is evaluated at $u = g(x)$. p. 142

Chain Rule for differentiating vector functions (short form): If $\mathbf{r}$ is a differentiable function of t and t is a differentiable function of s, then

$$\frac{d\mathbf{r}}{ds} = \frac{d\mathbf{r}}{dt}\frac{dt}{ds}.$$

p. 637

Chain Rule for functions of three variables: If $w = f(x, y, z)$ is differentiable and x, y, and z are differentiable functions of t, then w is a differentiable function of t and

$$\frac{dw}{dt} = \frac{\partial f}{\partial x}\frac{dx}{dt} + \frac{\partial f}{\partial y}\frac{dy}{dt} + \frac{\partial f}{\partial z}\frac{dz}{dt}.$$

There are similar rules for functions of other numbers of variables. pp. 715, 717

Circular functions: The functions $\cos x$ and $\sin x$ with reference to their function values corresponding to points $(\cos x, \sin x)$, x in radians, on the unit circle. More generally, the six basic trigonometric functions. p. 41

Circle of curvature: At a point P on a plane curve where the curvature κ is not zero, the circle that

1. is tangent to the curve at P,

2. has the same curvature the curve has at P, and

3. lies toward the concave or inner side of the curve.

The *radius of curvature* of the curve at P is $\rho = 1/\kappa$, the reciprocal of the curvature at P. The *center of curvature* is the center of the circle of curvature. p. 652

Circular paraboloid: A surface generated by revolving a paraboloid in space about its axis. A typical equation (axis the z-axis) is

$$\frac{x^2}{a^2} + \frac{y^2}{a^2} = \frac{z}{c}.$$

p. 617

Circulation around a curve: The flow along a closed loop. p. 840

Circulation density: See *Curl*.

Closed interval [a, b]: The set of all real numbers x with $a \leq x \leq b$. p. 11

Closed region in the plane or space: A region that contains all its boundary points. pp. 677, 678

Comparison Test (Direct, for improper integrals): For f and g continuous on $[a, \infty)$ with $0 \leq f(x) \leq g(x)$ for all $x \geq a$,

(1) $\int_a^\infty f(x)\,dx$ converges if $\int_a^\infty g(x)\,dx$ converges;

(2) $\int_a^\infty g(x)\,dx$ diverges if $\int_a^\infty f(x)\,dx$ diverges. p. 438

Comparison Test (Direct, for infinite series): Let $\sum a_n$ be a series with no negative terms. Then

(1) $\sum a_n$ converges if there is a convergent series $\sum c_n$ with $a_n \leq c_n$ for all $n > N$ for some integer N;

(2) $\sum a_n$ diverges if there is a divergent series $\sum d_n$ of nonnegative terms with $a_n \geq d_n$ for all $n > N$ for some integer N. p. 489

Comparison Test (Limit, for improper integrals): If the positive functions f and g are continuous on $[a, \infty)$ and if

$$\lim_{x \to \infty} \frac{f(x)}{g(x)} = L, \qquad 0 < L < \infty,$$

then $\int_a^\infty f(x)\,dx$ and $\int_a^\infty g(x)\,dx$ both converge or both diverge. p. 439

Comparison Test (Limit, for infinite series): Suppose that $a_n > 0$ and $b_n > 0$ for all $n \geq N$, N an integer.

(1) if

$$\lim_{n \to \infty} \frac{a_n}{b_n} = c, \qquad 0 < c < \infty,$$

then $\sum a_n$ and $\sum b_n$ both converge or both diverge;

(2) if

$$\lim_{n \to \infty} \frac{a_n}{b_n} = 0$$

and $\sum b_n$ converges, then $\sum a_n$ converges;

(3) if

$$\lim_{n \to \infty} \frac{a_n}{b_n} = \infty$$

and $\sum b_n$ diverges, then $\sum a_n$ diverges. p. 498

Complex number: An expression of the form $a + bi$ where a and b are real numbers and i is a symbol for $\sqrt{-1}$.

Component equation for a plane: The component equation for the plane through $P_0(x_0, y_0, z_0)$ normal to $\mathbf{n} = A\mathbf{i} + B\mathbf{j} + C\mathbf{k}$ is $A(x - x_0) + B(y - y_0) + C(z - z_0) = 0$ or $Ax + By + Cz = D$, where $D = Ax_0 + By_0 + Cz_0$. p. 602

Component functions: Of a parametrized curve $x = f(t)$, $y = g(t)$, $t \in I$, the functions f and g. Of a vector function $\mathbf{r}(t) = f(t)\mathbf{i} + g(t)\mathbf{j}$, the functions f and g. Similarly in three dimensions.

Components of a vector: See *Component form of a vector*.

Component test for conservative fields: If $\mathbf{F} = M\mathbf{i} + N\mathbf{j} + P\mathbf{k}$ is a field whose component functions have continuous first partial derivatives, then $\mathbf{F}$ is conservative if and only if

$$\frac{\partial P}{\partial y} = \frac{\partial N}{\partial z}, \qquad \frac{\partial M}{\partial z} = \frac{\partial P}{\partial x}, \qquad \text{and} \qquad \frac{\partial N}{\partial x} = \frac{\partial M}{\partial y}.$$

p. 849

Composite function $f \circ g$: The function $(f \circ g)(x) = f(g(x))$. p. 16

Concave down: The graph of a differentiable function $y = f(x)$ is concave down on an open interval I if y' is decreasing on I. p. 197

Concave up: The graph of a differentiable function $y = f(x)$ is concave up on an open interval I if y' is increasing on I. p. 197

Conditional convergence: An infinite series is conditionally convergent if it is convergent but not absolutely convergent. p. 502

Conic section law: See *Kepler's first law.*

Connected open region: An open region in which each point can be connected to any other point by a smooth curve lying in the region. p. 847

Conservative field: A vector field $\mathbf{F}$ in an open region D in which the work done in moving from any one point to another is the same for all paths between those two points. The integral $\int \mathbf{F} \cdot d\mathbf{r}$ of such a field is said to be *path independent* in D. p. 846

Constant function: A function that assigns the same value to every element in its domain.

Constant of integration: The arbitrary constant C in $\int f(x)\, dx = F(x) + C$, where F is any antiderivative of f. p. 306

Continuity at an endpoint: A function $f(x)$ is continuous at a left endpoint a of its domain if $\lim_{x \to a^+} f(x) = f(a)$. The function is continuous at a right endpoint b of its domain if $\lim_{x \to b^-} f(x) = f(b)$. p. 74

Continuity at an interior point: A function $f(x)$ is continuous at an interior point c of its domain if $\lim_{x \to c} f(x) = f(c)$. p. 73

Continuity equation of hydrodynamics: If $\mathbf{v}$ is velocity field of a fluid flowing smoothly through a region D in space and $\delta(t, x, y, z)$ is the fluid's density at the point $P(x, y, z)$ at time t, then at each point of D,

$$\nabla \cdot (\delta \mathbf{v}) = -\frac{\partial \delta}{\partial t}.$$

p. 902

Continuity of vector functions: A vector function $\mathbf{r}(t)$ is continuous at $t = c$ if $\lim_{t \to c} \mathbf{r}(t) = \mathbf{r}(c)$ (use appropriate one-sided limits at endpoints). In terms of components, $\mathbf{r}(t) = f(t)\mathbf{i} + g(t)\mathbf{j} + h(t)\mathbf{k}$ is continuous at $t = c$ if and only if f, g and h are continuous at $t = c$. pp. 531, 634

Continuity on an interval: A function is continuous on an interval if and only if it is continuous at each point of the interval. p. 81

Continuous extension of a function f: A function identical to f except that it may be continuous at one or more points where f is not. p. 77

Continuous function: A function that is continuous at each point of its domain. pp. 77, 634, 688, 835

Contour line: The curve in space in which the plane $z = c$ cuts a surface $z = f(x, y)$. p. 679

Convergent improper integral: An improper integral whose related limit(s) is (are) finite. p. 434

Convergent sequence: A sequence that has a limit. See *Limit at infinity.*

Convergent series: An infinite series whose sequence of partial sums has a finite limit. p. 458

Coordinate planes: The planes $x = 0$, $y = 0$, and $z = 0$ that divide space into eight **octants.** The octant in which all coordinates are nonnegative is called the *first octant.* p. 576

Cosecant function: See *Basic trigonometric functions.*

Cosine function: See *Basic trigonometric functions.*

Cotangent function: See *Basic trigonometric functions.*

Critical point: A point (value) in the interior of the domain of a function $f(x)$ at which $f' = 0$ or f' does not exist. An interior point of the domain of a function $f(x, y)$ where both first partial derivatives are zero or where one or both of the first partial derivatives do not exist. pp. 180, 738

Cross-product rule for differentiating vector functions:

$$\frac{d}{dt}(\mathbf{u} \times v) = \mathbf{u} \times \frac{d\mathbf{v}}{dt} + \frac{d\mathbf{u}}{dt} \times \mathbf{v}$$

p. 637

Cross-section area: The area of a cross section of a solid. p. 383

Cross (vector) product: The cross (vector) product $\mathbf{A} \times \mathbf{B}$ of two nonparallel vectors $\mathbf{A}$ and $\mathbf{B}$ is the vector $\mathbf{A} \times \mathbf{B} = (|\mathbf{A}||\mathbf{B}| \sin \theta)\mathbf{n}$, where $\mathbf{n}$ is a unit vector normal to the plane determined by $\mathbf{A}$ and $\mathbf{B}$. p. 594

Curl: In a three-dimensional vector field $\mathbf{F}$, curl $\mathbf{F} = \nabla \times \mathbf{F}$. pp. 856, 886

Curvature: The curvature function of a smooth curve with unit tangent vector $\mathbf{T}$ is $\kappa = \left| \dfrac{d\mathbf{T}}{ds} \right|$. p. 650

Curve traced by a vector function: The curve traced by $\mathbf{r}(t) = f(t)\mathbf{i} + g(t)\mathbf{j} + h(t)\mathbf{k}$, $t \in I$, is the parametrized curve $x = f(t)$, $y = g(t)$, $z = h(t)$, $t \in I$. pp. 530, 631

Cylinder: A surface composed of the lines that pass through a given plane curve parallel to a given line in space. The curve is a *generating curve* for the cylinder. p. 607

Cylindrical coordinates of a point: Ordered triples (r, θ, z) in which r and θ are polar coordinates for the vertical projection of P on the xy-plane and z is the standard rectangular vertical coordinate. p. 610

Decay models: See *Exponential growth and decay.*

Decreasing function: Let f be a function defined on an interval I. Then f decreases on I if, for any two points x_1 and x_2 in I,

$$x_1 < x_2 \implies f(x_1) > f(x_2).$$

p. 188

Decreasing on an interval: See *Decreasing function.*

Definite integral of a vector function $\mathbf{r}(t) = f(t)\mathbf{i} + g(t)\mathbf{j} + h(t)\mathbf{k}$ from a to b: The vector

$$\int_a^b \mathbf{r}(t)\, dt = \left(\int_a^b f(t)\, dt \right)\mathbf{i} + \left(\int_a^b g(t)\, dt \right)\mathbf{j} + \left(\int_a^b h(t)\, dt \right)\mathbf{k}$$

pp. 535, 640

Definite integral of *f* over an interval [*a*, *b*]: For any partition P of [*a*, *b*], let a number c_k be chosen arbitrarily in each subinterval $[x_{k-1}, x_k]$ and let $\Delta x_k = x_k - x_{k-1}$. If there exists a number I such that

$$\lim_{\|P\| \to 0} \sum_{k=1}^{n} f(c_k)\, \Delta x_k = I$$

no matter how P and the c_k's are chosen, then I is the definite integral of *f* over [*a*, *b*].   p. 260

Delta notation (Δ): See *Increment*.

Density: See *Mass*.

Density function: See *Mass*.

Derivative of a function *f* at a point *a*:

$$\lim_{h \to 0} \frac{f(a + h) - f(a)}{h},$$

provided the limit exists.   p. 95.

Derivative of a function *f* with respect to *x*: The function f' whose value at x is

$$\lim_{h \to 0} \frac{f(x + h) - f(x)}{h},$$

provided the limit exists. Alternatively,

$$\lim_{x \to a} \frac{f(x) - f(a)}{x - a},$$

provided the limit exists.   p. 96

Derivative of a vector function at a point: The vector function $\mathbf{r}(t) = f(t)\mathbf{i} + g(t)\mathbf{j} + h(t)\mathbf{k}$ is differentiable at $t = t_0$ if f, g, and h are differentiable at t_0. The derivative is the vector

$$\frac{df}{dt}\mathbf{i} + \frac{dg}{dt}\mathbf{j} + \frac{dh}{dt}\mathbf{k}.$$

p. 635

Determinant formula for A × B: If $\mathbf{A} = a_1\mathbf{i} + a_2\mathbf{j} + a_3\mathbf{k}$ and $\mathbf{B} = b_1\mathbf{i} + b_2\mathbf{j} + b_3\mathbf{k}$, then

$$\mathbf{A} \times \mathbf{B} = \begin{vmatrix} \mathbf{i} & \mathbf{j} & \mathbf{k} \\ a_1 & a_2 & a_3 \\ b_1 & b_2 & b_3 \end{vmatrix}.$$

p. 595

Difference rule for differentiating vector functions:

$$\frac{d}{dt}(\mathbf{u} - \mathbf{v}) = \frac{d\mathbf{u}}{dt} - \frac{d\mathbf{v}}{dt}.$$

p. 637

Difference of vectors:

$$\mathbf{v} - \mathbf{u} = \langle v_1, v_2 \rangle - \langle u_1, u_2 \rangle = \langle v_1 - u_1, v_2 - u_2 \rangle$$

Similarly in three dimensions

pp. 522, 577

Difference quotient of the function *f* at *a*:

$$\frac{f(a + h) - f(a)}{h}.$$

Alternatively,

$$\frac{f(x) - f(a)}{x - a}.$$

p. 86

Differentiability: If $f'(x)$ exists, the function *f* is differentiable at x. A function that is differentiable at every point of its domain is a differentiable function.   p. 95

Differentiable curve: The graph of a differentiable function; also, a parametrized curve whose component functions are differentiable at every parameter value.   pp. 95, 513, 703

Differentiable vector function: $\mathbf{r}(t) = f(t)\mathbf{i} + g(t)\mathbf{j} + h(t)\mathbf{k}$ is differentiable at t if f, g and h are differentiable at t, and is differentiable if f, g and h are differentiable. Its derivative is the vector

$$\frac{d\mathbf{r}}{dt} = \frac{df}{dt}\mathbf{i} + \frac{dg}{dt}\mathbf{j} + \frac{dh}{dt}\mathbf{k}$$

p. 635

Differential: If $y = f(x)$ is a differentiable function, then the differential dx is an independent variable and the differential dy is $dy = f'(x)\, dx$.   p. 224

Differential calculus: The branch of mathematics that deals with derivatives.   p. 95

Differential equation: An equation containing a derivative, p. 303

Differential form: $M(x, y, z)\, dx + N(x, y, z)\, dy + P(x, y, z)\, dz$ is a differential form. A differential form is *exact* on a domain D in space if

$$M dx + N dy + P dz = \frac{\partial f}{\partial x}\, dx + \frac{\partial f}{\partial y}\, dy + \frac{\partial f}{\partial z}\, dz = df$$

for some (scalar) function *f* throughout D.   p. 851

Differentiation: The process of taking a derivative.   p. 95

Directional derivative: The rate of change $D_{\mathbf{u}}f = \nabla f \cdot \mathbf{u}$ of a function $f(x, y, z)$ at a point P_0 in the direction of a unit vector $\mathbf{u} = u_1\mathbf{i} + u_2\mathbf{j} + u_3\mathbf{j}$.   p. 726

Direction field: See *Slope field*.

Direction of a nonzero vector A: The vector $\mathbf{A}/|\mathbf{A}|$.

Direction of motion: $\mathbf{v}/|\mathbf{v}|$, where $\mathbf{v}$ is the (nonzero) velocity vector of the motion.   p. 533

Discontinuity: If a function *f* is not continuous at a point c, then c is a point of discontinuity of *f*.   p. 75

Disk method: A method for finding the volume of a solid of revolution by evaluating $\int_a^b A(x)\, dx$, where $A(x)$ is the area of the disk cut by a cross section of the solid perpendicular to the axis of revolution at x.   p. 384

Distance between points in space: The distance between $P_1(x_1, y_1, z_1)$ and $P_2(x_2, y_2, z_2)$ is

$$|\overrightarrow{P_1P_2}| = \sqrt{(x_2 - x_1)^2 + (y_2 - y_1)^2 + (z_2 - z_1)^2}.$$

p. 579

Distance traveled (from velocity): The integral of the absolute value of a velocity function with respect to time.   p. 298

Divergence of a three-dimensional vector field: div $\mathbf{F} = \nabla \cdot \mathbf{F}$.   p. 896

Divergence of a two-dimensional vector field: Given a vector field $\mathbf{F} = M\mathbf{i} + N\mathbf{j}$,

$$\text{div } \mathbf{F} = \frac{\partial M}{\partial x} + \frac{\partial N}{\partial y}.$$

Also called the *flux density of* $\mathbf{F}$.   p. 856

Divergence Theorem: Under suitable conditions, the outward flux of a vector field across a closed surface (oriented outward) equals the triple integral of the divergence of the field over the region enclosed by the surface.   p. 897

Divergent improper integral: An improper integral for which at least one of the defining limits does not exist.   p. 434

Divergent sequence: An infinite sequence that has no limit as $n \to \infty$.   p. 468

Divergent series: An infinite series whose sequence of partial sums diverges.   p. 458

Domain of a function: See *Function*.

Domination: A function of f dominates a function g on a domain D if $f(x) \geq g(x)$ for all x in D. A sequence $\{a_n\}$ dominates a sequence $\{b_n\}$ if $a_n \geq b_n$ for all n. An infinite series $\sum a_n$ dominates an infinite series $\sum b_n$ if $a_n \geq b_n$ for all n.

Dot (inner) product: Of vectors $\mathbf{u} = \langle u_1, u_2 \rangle$ and $\mathbf{v} = \langle v_1, v_2 \rangle$, the number $\mathbf{u} \cdot \mathbf{v} = u_1v_1 + u_2v_2$. Of vectors $\mathbf{A} = a_1\mathbf{i} + a_2\mathbf{j} + a_3\mathbf{k}$ and $\mathbf{B} = b_1\mathbf{i} + b_2\mathbf{j} + b_3\mathbf{k}$, the number $\mathbf{A} \cdot \mathbf{B} = a_1b_1 + a_2b_2 + a_3b_3$. It can also be written as $|\mathbf{A}||\mathbf{B}| \cos \theta$, where θ is the angle between $\mathbf{A}$ and $\mathbf{B}$.   pp. 524, 586

Dot product rule for differentiating vector functions:

$$\frac{d}{dt}(\mathbf{u} \cdot \mathbf{v}) = \mathbf{u} \cdot \frac{d\mathbf{v}}{dt} + \frac{d\mathbf{u}}{dt} \cdot \mathbf{v}$$

p. 637

Double integral of f over a region R in the plane:

$$\iint_R f(x, y) \, dA = \lim_{\Delta A \to 0} \sum_{k=1}^{n} f(x_k, y_k) \, \Delta A_k,$$

or, in polar form,

$$\iint_R f(r, \theta) \, dA = \lim_{n \to \infty} \sum_{k=1}^{n} f(r_k, \theta_k) \, \Delta A_k.$$

p. 761, 784

Dummy variable of integration: In $\int_a^b f(x) \, dx$, the variable x. It could be any other letter without changing the value of the integral.   p. 262

$\dfrac{dy}{dx}$**:** The derivative of y with respect to x.   p. 97

e (the number): To nine decimal places, $e = 2.718281828$. More formally,

$$e = \lim_{x \to \infty} \left(1 + \frac{1}{x}\right)^x.$$

p. 23

Ellipsoid: A surface defined by an equation of the form

$$\frac{x^2}{a^2} + \frac{y^2}{b^2} + \frac{z^2}{c^2} = 1.$$

p. 615

Ellipsoid of revolution: An ellipsoid in which any two of the semiaxes a, b, and c are equal.   p. 615

Elliptic cone: A surface defined by an equation of the form

$$\frac{x^2}{a^2} + \frac{y^2}{b^2} = \frac{z^2}{c^2}.$$

p. 617

Elliptic paraboloid: A surface defined by an equation of the form

$$\frac{x^2}{a^2} + \frac{y^2}{b^2} = \frac{z}{c}.$$

p. 616

End behavior model: A function g is

1. a right end behavior model for f if and only if

$$\lim_{x \to \infty} \frac{f(x)}{g(x)} = 1,$$

2. a left end behavior model for f if and only if

$$\lim_{x \to -\infty} \frac{f(x)}{g(x)} = 1,$$

3. an end behavior model for f if it is both a left and a right end behavior model for f.   p. 70

Equal area law: See *Kepler's second law*

Euler's method: A method using linearizations to approximate the solution of an initial value problem.   p. 350

Exact differential form: See *Differential form*.

Even function: A function f for which $f(-x) = f(x)$ for every x in the domain of f.   p. 13

Exponential change: See *Law of exponential change*.

Exponential function with base a: The function $f(x) = a^x$, $a > 0$ and $a \neq 1$.   p. 20

Exponential growth and decay: Growth and decay modeled by the functions $y = k \cdot a^x$, $k > 0$, with $a > 1$ for growth and $0 < a < 1$ for decay.   p. 20

Extreme value: See *Extremum*.

Extremum: A maximum or minimum value (extreme value) of a function on a set. See also *Absolute maximum; Absolute minimum; Local maximum; Local minimum.*

First derivative test for extreme values, multivariable: If $f(x, y)$ has a local maximum or minimum value at an interior point (a, b) of its domain and if the first partial derivatives exist there, then $f_x(a, b) = 0$ and $f_y(a, b) = 0$.   p. 737

First Derivative Test (for local extrema): For a continuous function f.

1. if f' changes sign from positive to negative at a critical point c, then f has a local maximum value at c;

2. if f' changes sign from negative to positive at a critical point c, then f has a local minimum value at c;

3. if f' does not change sign at a critical point c, then f has no local extreme value at c;

4. if $f' < 0$ ($f' > 0$) for $x > a$ where a is a left endpoint in the domain of f, then f has a local maximum (minimum) value at a;

5. if $f' < 0$ ($f' > 0$) for $x < b$ where b is a right endpoint in the domain of f, then f has a local minimum (maximum) value at b. p. 195

First moments:

For a thin plate covering a region R in the xy-plane, about the x- and y- axis:

$$M_x = \iint_R y\delta(x, y)\, dA, \qquad M_y = \iint_R x\delta(x, y)\, dA.$$

For a solid object occupying a region D in space, about the yz-, xz-, xy-planes,

$$M_{yz} = \iiint_D x\delta\, dV, \quad M_{xz} = \iiint_D y\delta\, dV, \quad M_{xy} = \iiint_D z\delta\, dV.$$

For coil springs, thin rods, and wires lying along a smooth curve C in space,

$$M_{yz} = \int_C x\delta\, ds, \qquad M_{xz} = \int_C y\delta\, ds, \qquad M_{xy} = \int_C z\delta\, ds.$$

For a thin shell modeled by a surface S:

$$M_{yz} = \iint_S x\delta\, d\sigma, \quad M_{xz} = \iint_S y\delta\, d\sigma, \quad M_{xy} = \iint_S z\delta\, d\sigma.$$

pp. 776, 800, 831, 874

First octant: See *Coordinate planes.*

Flow along a curve: The integral with respect to arc length over the curve of the dot product of a fluid's velocity field and the curve's unit tangent vector.   p. 840

Flow integral: Flow $= \int_a^b \mathbf{F} \cdot \mathbf{T}\, ds$.   p. 840

Flux across a smooth closed curve C: $\int_C \mathbf{F} \cdot \mathbf{n}\, ds$, where $\mathbf{n}$ is the outward-pointing unit normal vector on C.   p. 842

Flux across a surface S in the direction of the chosen unit normal field n: $\iint_S \mathbf{F} \cdot \mathbf{n}\, d\sigma$.   p. 841

Flux density: See *Divergence.*

Free fall equation: When air resistance is absent or insignificant and the only force acting on a falling body is the force of gravity, we call the way the body falls *free fall*. In a free fall short enough for the acceleration of gravity to be assumed constant, call it g, the position of a body released to fall from position s_0 at time $t = 0$ with velocity v_0 is modeled by the equation $s(t) = (1/2)gt^2 + v_0 t + s_0$.   p. 125

Frenet (TNB) frame: See TNB *frame.*

Frequency of a periodic function: The reciprocal of the period of the function, or the number of cycles or periods per unit time. The function $\sin x$, with x in seconds, has period 2π seconds and completes $1/(2\pi)$ cycles per second (has frequency $1/(2\pi)$). p. 43

Fubini's theorem: Under favorable conditions (normally met in practice), the double integral of a continuous function over a region in the plane can be evaluated as an iterated integral. pp. 762, 767

Function: A rule that assigns a unique element in a set R to each element in a set D. The set D is the *domain* of the function. The set of elements assigned from R is the *range* of the function. pp. 10, 631, 675, 835

Fundamental Theorem of Calculus, Part 1: If f is continuous on $[a, b]$, then the function $F(x) = \int_a^x f(t)\, dt$ has a derivative with respect to x at every point in $[a, b]$ and

$$\frac{dF}{dx} = \frac{d}{dx} \int_a^x f(t)\, dt = f(x).$$

p. 277

Fundamental Theorem of Calculus, Part 2: If f is continuous on $[a, b]$, and F is any antiderivative of f on $[a, b]$, then

$$\int_a^b f(x)\, dx = F(b) - F(a).$$

p. 282

Fundamental Theorem of Line Integrals: If $\mathbf{F}$ is a continuous vector field throughout an open connected region D in space, then there exists a differentiable function f such that

$$\mathbf{F} = \nabla f = \frac{\partial f}{\partial x}\mathbf{i} + \frac{\partial f}{\partial y}\mathbf{j} + \frac{\partial f}{\partial z}\mathbf{k}$$

if and only if for all points A and B in D the value of $\int_A^B \mathbf{F} \cdot d\mathbf{r}$ is independent of the path joining A to B in D. If the integral is independent of the path from A to B, its value is

$$\int_A^B \mathbf{F} \cdot d\mathbf{r} = f(B) - f(A). \quad \text{p. 847}$$

Gaussian curve: See *Normal curve.*

Gauss's law: The total electric flux out of any closed surface is proportional to the total electric charge inside the surface.

General linear equation: $Ax + By = C$ (A and B not both 0).

Generating curve for a cylinder: See *Cylinder.*

Geometric sequence: A sequence of the form $a, ar, ar^2, \ldots,$ $ar^n, \ldots,$ in which each term after the first is obtained from its preceding term by multiplying by the same number r. The number r is the *common ratio* of the sequence.

Geometric series: A series of the form

$$a + ar + ar^2 + \cdots + ar^n + \cdots, \quad \text{or} \quad \sum_{n=1}^{\infty} ar^{n-1},$$

in which each term after the first is obtained from its preceding term by multiplying by the same number r. The number r is the *common ratio* of the series. p. 459

Global maximum: See *Absolute maximum.*

Global minimum: See *Absolute minimum.*

Gradient field: The field of gradient vectors

$$\nabla f = \frac{\partial f}{\partial x}\mathbf{i} + \frac{\partial f}{\partial y}\mathbf{j} + \frac{\partial f}{\partial z}\mathbf{k}$$

of a differentiable function $f(x, y, z)$. p. 837

Gradient vector (gradient) of $f(x, y, z)$ at P_0: The vector

$$\nabla f = \frac{\partial f}{\partial x}\mathbf{i} + \frac{\partial f}{\partial y}\mathbf{j} + \frac{\partial f}{\partial z}\mathbf{k}$$

obtained by evaluating the partial derivatives of f at a point P_0. pp. 725, 727

Graph of a function $f(x)$: The set of points (x, y) in the coordinate plane whose coordinates are the input–output pairs of the function. p. 11

Graph of a function $f(x, y)$: The set of all points $(x, y, f(x, y))$ in space for (x, y) in the domain of f. The graph is also called the *surface* $z = f(x, y)$. p. 13

Gravitational constant: See *Universal gravitation constant.*

Green's Theorem (normal form): Under suitable conditions, the outward flux of a vector field across a simple closed curve in the plane equals the double integral of the divergence of the field over the region enclosed by the curve. p. 858

Green's Theorem (tangential form): Under suitable conditions, the counterclockwise circulation of a vector field around a simple closed curve is the double integral of the $\mathbf{k}$-component of the curl of the field over the region enclosed by the curve. p. 858

Growth models: See *Exponential growth and decay; Logistic growth.*

Growth rate: See *Relative growth rate.*

Half-life of a radioactive element: The time required for half of the radioactive nuclei present in a sample to decay. p. 21

Harmonic series: The series

$$1 + \frac{1}{2} + \frac{1}{3} + \frac{1}{4} + \cdots + \frac{1}{n} + \cdots = \sum_{n=1}^{\infty} \frac{1}{n}.$$

p. 498

Helix: A spiral with equally spaced coils that winds around a circular cylinder. p. 632

Hooke's Law: When a force is applied to stretch or compress a spring, the magnitude F of the force in the direction of motion is proportional to the distance x that the spring is stretched or compressed. In symbols, $F \sim x$ or $F = kx$, where k is the constant of proportionality. This relationship is *Hooke's Law.* If an elastic material is stretched too far, it becomes distorted and will not return to its original state. The distance beyond which distortion occurs is the material's *elastic limit.* Hooke's Law holds only as long as the material is not stretched past its elastic limit. p. 369

Horizontal line: In the Cartesian coordinate plane, a line parallel to the x-axis.

Hyperbolic paraboloid: A surface defined by an equation of the form

$$\frac{y^2}{b^2} - \frac{x^2}{a^2} = \frac{z}{c}, c > 0.$$

p. 620

Hyperboloid of one sheet: A surface defined by an equation of the form

$$\frac{x^2}{a^2} + \frac{y^2}{b^2} - \frac{z^2}{c^2} = 1.$$

p. 618

Hyperboloid of two sheets: A surface defined by an equation of the form

$$\frac{z^2}{c^2} - \frac{x^2}{a^2} - \frac{y^2}{b^2} = 1.$$

p. 619

Imaginary number: A complex number of the form $0 + bi$. See *Complex number.*

Implicit differentiation: A process for finding dy/dx when y is implicitly defined as a differentiable function of x by an equation of the form $f(x, y) = 0$. p. 149

Improper integral: An integral on an infinite interval or on a finite interval containing one or more points of infinite discontinuity of the integrand. Its value is found as a limit or sum of limits. p. 433

Increasing function: Let f be a function defined on an interval I. Then f increases on I if, for any two points x_1 and x_2 in I,

$$x_1 < x_2 \Rightarrow f(x_1) < f(x_2).$$

p. 188

Increasing on an interval: See *Increasing function.*

Increment: If coordinates change from (x_1, y_1) to (x_2, y_2), the increments in the coordinates are $\Delta x = x_2 - x_1$ and $\Delta y = y_2 - y_1$. The symbols Δx and Δy are read "delta x" and "delta y." p. 1

Indefinite integral of a function f: The set of all antiderivatives of f, denoted by $\int f(x)\,dx.$ p. 306

Indefinite integral of a vector function $\mathbf{r}(t)$: The set of all antiderivatives of $\mathbf{r}$, denoted by $\int \mathbf{r}(t)\,dt.$ pp. 535, 639

Indeterminate form: A nonnumeric expression of the form $0/0$, ∞/∞, $0 \cdot \infty$, $\infty - \infty$, 1^∞, ∞^0 or 0^0 obtained when trying substitution to evaluate a limit. The expression reveals nothing about the limit, but does suggest that l'Hôpital's Rule may be applied to help find the limit. p. 417

Infinite discontinuity: A point of discontinuity where one or both of the one-sided limits are infinite. p. 76

Infinite limit: If the values of a function $f(x)$ outgrow all positive bounds as x approaches a finite number a, we say

$$\lim_{x \to a} f(x) = \infty,$$

If the values of f become large and negative, exceeding all negative bounds as $x \to a$, we say

$$\lim_{x \to a} f(x) = -\infty,$$

p. 68

Infinite series: An expression of the form

$$a_1 + a_2 + a_3 + \cdots + a_n + \cdots, \quad \text{or} \quad \sum_{k=1}^{\infty} a_k.$$

The numbers $a_1, a_2, \ldots$ are the *terms* or the series; a_n is the nth term. The *partial sums* of the series form a sequence

$$s_1 = a_1$$
$$s_2 = a_1 + a_2$$
$$s_3 = a_1 + a_2 + a_3$$
$$\vdots$$
$$s_n = a_1 + a_2 + a_3 + \cdots + a_n$$
$$\vdots$$

of numbers, each defined as a finite sum. If the sequence of partial sums has a limit S as $n \to \infty$, the series *converges* to the *sum S*. A series that fails to converge *diverges.* p. 458

Inflection point: A point where the graph of a function has a tangent line and the concavity changes. p. 198

Initial condition: See *Initial value problem.*

Initial value problem: For a first-order differential equation, the problem of finding the solution that has a particular value at a

given point. The condition that the solution have this value at the point is the *initial condition* of the problem. For a second-order differential equation, the problem of finding the solution given its value and the value of its first derivative at a point (*two* initial conditions). p. 303

Inner product: See *Dot product.*

Instantaneous rate of change of f with respect to x at a: The derivative

$$f'(a) = \lim_{h \to 0} \frac{f(a + h) - f(a)}{h},$$

provided the limit exists. p. 87

Instantaneous velocity: The derivative of a position function with respect to time. p. 124

Integrable function on $[a, b]$: A function for which the definite integral over $[a, b]$ exists. p. 260

Integral calculus: The branch of calculus that deals with integrals. p. 247

Integral over a curve: See *Line integral.*

Integral over a surface: See *Surface integral.*

Integrand: $f(x)$ in $\int f(x)\,dx$ or in $\int_a^b f(x)\,dx.$ p. 261

Integration: The evaluation of a definite integral, an indefinite integral, or an improper integral. p. 261

Integration by partial fractions: A method for integrating a rational function by writing it as a sum of proper fractions, called *partial fractions*, whose denominators are positive integer powers of linear or quadratic polynominals. p. 444

Integration by parts: A method of integration in which $\int u\,dv$ is rewritten as $uv - \int v\,du.$ p. 323

Integration by substitution: A method of integration in which $\int f(g(x)) \cdot g'(x)\,dx$ is rewritten as $\int f(u)\,du$ by substituting $u = g(x)$ and $du = g'(x)\,dx.$ p. 318

Interior point of a region in space: A point (x_0, y_0, z_0) in a region R that lies at the center of a solid sphere that lies entirely in R. The interior points of the region, as a set, make up the *interior of the region.* p. 677

Interior point of a region in the plane: A point (x_0, y_0) in a region R that lies at the center of a disk that lies entirely in R. The interior points of the region, as a set, make up the *interior of the region.* p. 677

Intermediate Value Theorem for Continuous Functions: A function $y = f(x)$ that is continuous on a closed interval $[a, b]$ takes on every y-value between $f(a)$ and $f(b)$. p. 79

Intermediate Value Theorem for Derivatives: If a and b are any two points in an interval on which f is differentiable, then f' takes on every value between $f'(a)$ and $f'(b)$. p. 110

Interval: A subset of the number line formed by any of the following: (1) two points and the points in between, (2) only the points in between two points, (3) the points in between two points and one of the two points, (4) one point and the points to one side of it, (5) only the points to one side of a given point. The real line is also considered to be an interval.   p. 11

Interval of convergence: The interval of x-values for which a power series converges. See *Power series.*

Inverse function f^{-1}: The function obtained by reversing the ordered pairs of a one-to-one function f.   p. 33

Iterated integral: A multiple integral that is evaluated by repeated single-variable integrations.   pp. 763, 786, 752

Jacobian determinant (Jacobian): The Jacobian determinant of the coordinate transformation $x = g(u, v)$, $y = h(u, v)$ is

$$J(u, v) = \begin{vmatrix} \dfrac{\partial x}{\partial u} & \dfrac{\partial x}{\partial v} \\ \dfrac{\partial y}{\partial u} & \dfrac{\partial y}{\partial v} \end{vmatrix} = \frac{\partial x}{\partial u}\frac{\partial y}{\partial v} - \frac{\partial y}{\partial u}\frac{\partial x}{\partial v}.$$

p. 817

Jerk: The derivative of an acceleration function with respect to time.   p. 137

Jump discontinuity: A point of discontinuity where the one-sided limits exist but have different values. At such a point, the function jumps from one value to another.   p. 76

Kepler's first law (the Conic Section Law): A planet's path about its sun is a conic section with the sun at one focus. p. 663

Kepler's second law (the Equal Area Law): The radius vector from a sun to one of its planets sweeps out equal areas in equal times.   p. 663

Kepler's third law (the Time Distance Law): A planet's orbital period T about a sun of mass M is related to the orbit's semimajor axis a by the equation $T^2/a^3 = 4\pi^2/GM$ where G is the universal gravitational constant.   p. 666

Kilogram: See *Mass.*

Lagrange error bound: A bound for truncation error obtained from the Lagrange form of the remainder for Taylor series. p. 482

Lagrange form of the remainder: The formula

$$\frac{f^{n+1}(c)}{(n+1)!}(x-a)^{n+1}$$

for the remainder in Taylor's Theorem.   p. 482

Lagrange multiplier method: A method for finding extreme values of a function $f(x, y, z)$ whose variables are subject to a constraint $g(x, y, z) = 0$. The extreme values of f are found at points on the surface $g(x, y, z) = 0$ where ∇f is a scalar multiple λ of ∇g. The multiplier λ in the equation $\nabla f = \lambda \nabla g$ is called a **Lagrange multiplier,** after Joseph Louis Lagrange (1736–1813).

Law of exponential change: If a quantity y changes at a rate proportional to the amount present ($dy/dt = ky$) and $y = y_0$ when $t = 0$, then

$$y = y_0 e^{kt},$$

where $k > 0$ represents growth and $k < 0$ represents decay. The number k is the *rate constant.*   p. 330

Left end behavior model: See *End behavior model.*

Left-hand derivative: The derivative defined by a left-hand limit.   p. 100

Left-hand limit: The limit of f as x approaches c from the left, or $\lim_{x \to c^-} f(x)$.   p. 60

Level curve: A set of points in the plane where a function $f(x, y)$ has a constant value $f(x, y) = c$.   p. 680

L'Hôpital's Rule: Suppose that $f(a) = g(a) = 0$, that f and g are differentiable on an open interval I containing a, and that $g'(x) \neq 0$ on I if $x \neq a$. Then

$$\lim_{x \to a} \frac{f(x)}{g(x)} = \lim_{x \to a} \frac{f'(x)}{g'(x)}$$

provided the limit exists (or is $\pm \infty$). L'Hôpital's Rule also applies to quotients that lead to ∞/∞. If $f(x)$ and $g(x)$ both approach ∞ as $x \to a$, then

$$\lim_{x \to a} \frac{f(x)}{g(x)} = \lim_{x \to a} \frac{f'(x)}{g'(x)},$$

provided the latter limit exists (or is $\pm \infty$). In this case, a may itself be either finite or infinite.   p. 417

Limit: The function f has limit L as x approaches c if, given any positive number ε, there exists a positive number δ such that for all x,

$$0 < |x - c| < \delta \implies |f(x) - L| < \varepsilon.$$

This is represented as $\lim_{x \to c} f(x) = L$.   p. 57

Limit at infinity: The function f has limit L as x appoaches ∞ if, given any positive number ε, there exists a positive number N such that for all $x > N$, $|f(x) - L| < \varepsilon$. This is represented as $\lim_{x \to \infty} f(x) = L$.

The function f has limit L as x approaches $-\infty$ if, given any positive number ε, there is a negative number N such that for all x with $x < N$, $|f(x) - L| < \varepsilon$. This is represented as $\lim_{x \to -\infty} f(x) = L$.

The sequence $f(n) = x_n$ has limit L, if $\lim_{n \to \infty} f(n) = L$. p. 468

Limit of a sequence: See *Limit at infinity.*

Limit of a vector function: If $\lim_{t \to c} f(t) = L_1$, $\lim_{t \to c} g(t) = L_2$, $\lim_{t \to c} h(t) = L_3$, and $\mathbf{r}(t) = f(t)\mathbf{i} + g(t)\mathbf{j} + h(t)\mathbf{k}$, then $\lim_{t \to c} \mathbf{r}(t) = L_1\mathbf{i} + L_2\mathbf{j} + L_2\mathbf{k}$.

Limit of a function of two independent variables: The function $f(x, y)$ has limit L as (x, y) approaches (x_0, y_0) if, given any positive number ε, there exist a positive number δ such that for all (x, y) in the domain of f, $0 < \sqrt{(x - x_0)^2 + (y - y_0)^2} < \delta \implies |f(x, y) - L| < \varepsilon$. We write $\lim_{(x, y) \to (x_0, y_0)} f(x, y) = L$. Similar definitions hold for functions if three and more independent variables.   p. 686

Limits of integration: a and b in $\displaystyle\int_a^b f(x)\,dx$.   p. 261

Linear approximation (standard) of f at a: The approximation $f(x) \approx L(x)$, where $L(x)$ is the linearization of f at a.   pp. 220, 705, 710

Linear combination of vectors u and v: Any vector $a\mathbf{u} + b\mathbf{v}$, where a and b are numbers, is a linear combination of $\mathbf{u}$ and $\mathbf{v}$. p. 529

Linear equation: See *General linear equation*.

Linear function: A function that can be expressed in the form $f(x) = mx + b$.

Linearization of f at a: The approximating function $L(x) = f(a) + f'(a)(x - a)$ when f is differentiable at $x = a$. p. 221

Little-oh notation: If functions f and g are positive for x sufficiently large, then f is little-oh of g (written $f = o(g)$) if $\lim_{x \to \infty}(f(x)/g(x)) = 0$. p. 429

Local extrema: See *Local maximum; Local minimum*.

Local linearity: If a function $f(x)$ is differentiable at $x = a$, then, close to a, its graph resembles the tangent line at a. p. 107

Local maximum: The function f has a local maximum value $f(c)$ at a point c in the interior of its domain if and only if $f(x) \leq f(c)$ for all x in some open interval containing c. The function has a local maximum value at an endpoint c if the inequality holds for all x in some half-open domain interval containing c. For a function $f(x, y)$ defined on a region R containing the point (a, b), $f(a, b)$ is a local maximum if $f(a, b) \geq f(x, y)$ for all domain points (x, y) is an open disk centered at (a, b). pp. 179, 737

Local minimum: The function f has a local minimum value $f(c)$ at a point c in the interior of its domain if and only if $f(x) \geq f(c)$ for all x in some open interval containing c. The function has a local minimum value at an endpoint c if the inequality holds for all x in some half-open domain interval containing c. For a function $f(x, y)$ defined on a region R containing the point (a, b), $f(a, b)$ is a local minimum if $f(a, b) \leq f(x, y)$ for all domain points (x, y) is an open disk centered at (a, b). pp. 179, 737

Logarithm function with base a: The function $y = \log_a x$ that is the inverse of the exponential function $y = a^x$. $a > 0$, $a \neq 1$. p. 36

Logarithmic differentiation: The process of taking the natural logarithm of both sides of an equation, differentiating, and then solving for the desired derivative. p. 169

Logistic curve: A solution curve of the logistic differential equation. It describes population growth that begins slowly when the population is small, speeds up as the number of reproducing individuals increases and nutrients are still plentiful, and slows down again as the population reaches the carrying capacity of its environment. p. 199

Logistic differential equation:

$$\frac{dP}{dt} = \frac{k}{M} P(M - P)$$

where P is current population, t is time, M is the carrying capacity of the environment, and k is a positive proportionality constant. p. 343

Logistic growth model: The solution

$$P = \frac{M}{1 + Ae^{-kt}},$$

to the logistic differential equation, A an arbitrary constant. This model assumes that the relative growth rate of a population is positive, but decreases as the population increases due to environmental and economic factors. See *Logistic differential equation*. p. 343

Lower bound: See *Bounded*.

LRAM: Left-hand endpoint rectangular approximation method. The method of approximating a definite integral over an interval using the function values at the left-hand endpoints of the subintervals determined by a partition. p. 249

Maclaurin series: See *Taylor series*.

Magnitude: Of a number, its absolute value; of a vector, its length.

Mass: When a body is acted upon by no force or by a zero resultant force (the applied forces cancel), it moves with a constant velocity and zero acceleration. But what happens when the resultant force is not zero? If the body is initially at rest, it starts to move. If it is initially moving, the force may speed it up, slow it down, or change direction of its velocity. In each case, the body undergoes an acceleration (velocity change). We want to know the relation of the acceleration to the force, and that is what Newton's second law of motion, $F = ma$, is about.

Experiments with pucks on air-hockey tables show that if we apply a horizontal force at constant magnitude, the body's velocity changes at a constant rate during the time the force is applied. If we double the force, the velocity changes at twice the rate. If we halved the force, the velocity changes at half the rate. For a given body, the ratio F/a of the force to the acceleration it produces is constant, regardless of the magnitude of the force. We call this ratio the **mass** of the body. If we denote the mass by m, then $m = F/a$, or $F = ma$. If a large force is needed to give a body a certain acceleration (think of pushing a car), the mass is large. When only a small force is needed to give a body the same acceleration (think of pushing a bicycle), the mass is small. The mass is a quantitative measure of the body's inertia, its tendency to remain at rest or to keep moving once it is set in motion.

The SI unit of mass is the **kilogram,** officially defined to be the mass of a block of platinum-iridium alloy kept in a vault near Paris, France. This standard kilogram and the equation $F = ma$ are used to define the **newton,** one newton being the amount of force that gives an acceleration of one meter per second squared to a body with a mass of one kilogram. It takes a force of about one newton to lift an apple from a table.

In many bodies, the mass is not evenly distributed but varies from place to place. Snow in the mountains packs down during the winter to become heavier at the bottom than at the surface. We describe this situation by saying that the snow becomes denser as you go down, that its **density** (mass per unit volume) is greater at the bottom than at the top. (The densities of metal plates are measured in units of mass per unit area. The densities of wires, springs, and slender rods are measured in units of mass per unit length.)

The function that gives a body's density at each point is called its **density function**. Once we know this function, we can calculate the body's mass by integrating the function over the region occupied by the body. You will find formulas for doing so on pages 776, 800, 831, and 874.

Maximum: See *Absolute maximum; Local maximum.*

Mean value: See *Average value of a continuous function on* [*a, b*].

Mean Value Theorem for Definite Integrals: If *f* is continuous on [*a, b*], then at some point *c* in [*a, b*],

$$f(c) = \frac{1}{b - a} \int_a^b f(x)\, dx.$$

p. 272

Mean Value Theorem for Derivatives: If $y = f(x)$ is continuous at every point of the closed interval [*a, b*] and differentiable at every point of its interior (*a, b*), then there is at least one point *c* in (*a, b*) at which

$$f'(c) = \frac{f(b) - f(a)}{b - a}.$$

p. 186

Midpoint of a line segment: In space, the midpoint of the line segment joining points $P_1(x_1, y_1, z_1)$ and $P_2(x_2, y_2, z_2)$ is the point

$$\left(\frac{x_1 + x_2}{2}, \frac{y_1 + y_2}{2}, \frac{z_1 + z_2}{2} \right).$$

p. 582

Minimum: See *Absolute minimum; Local minimum.*

Mixed derivative theorem (Euler's Theorem): If f_x, f_y, f_{xy}, and f_{yx} are defined throughout an open region containing a point (*a, b*) and are all continuous at (*a, b*), then $f_{xy}(a, b) = f_{yx}(a, b)$. p. 700

Moment of inertia (second moment): For thin plates covering a region *R* in the *xy*-plane:

About the *x*-axis:

$$I_x = \iint_R y^2 \delta(x, y)\, dA.$$

About the *y*-axis:

$$I_y = \iint_R x^2(x, y)\, dA.$$

About the origin (polar moment):

$$I_0 = \iint_R (x^2 + y^2) \delta(x, y)\, dA = I_x + I_y.$$

About a line *L*:

$$I_L = \iint_R r^2(x, y) \delta(x, y)\, dA,$$

where $r(x, y)$ is the distance from (*x, y*) to *L*.
There are similar formulas for moments of inertia of solids, curves, and thin shells in space. pp. 776, 800, 831, 874

Monotonic (monotone) function: A function *f* is *monotonic increasing* on a domain *D* if

$$x_1 < x_2 \Rightarrow f(x_1) < f(x_2)$$

for all x_1 and x_2 in *D*. The function is *monotonic nondecreasing* on *D* if

$$x_1 < x_2 \Rightarrow f(x_1) \le f(x_2)$$

for all x_1 and x_2 in *D*. Similarly, *f* is *monotonic decreasing* on a domain *D* if

$$x_1 < x_2 \Rightarrow f(x_1) > f(x_2)$$

for all x_1 and x_2 in *D* and *monotonic nonincreasing* on *D* if

$$x_1 < x_2 \Rightarrow f(x_1) \ge f(x_2)$$

for all x_1 and x_2 in *D*. p. 188

Monotonic sequence: See *Monotonic function.*

MRAM: Midpoint rectangular approximation method. The method of approximating a definite integral over an interval using the function values at the midpoints of the subintervals determined by a partition. p. 249

Multivariable function: If *D* is a set of *n*-tuples $(x_1, x_2, \ldots, x_n)$, a *real-valued function* $w = f(x_1, x_2, \ldots, x_n)$ that assigns a real number to each element in *D* is called a multivariable function with domain *D*. The range is the set of *w*-values taken on by *f*. The dependent or output variable is *w*, and x_1 to x_n are the independent or input variables. p. 675

Natural logarithm: *a* is the natural logarithm of *b* if and only if $b = e^a$. The natural logarithm function $y = \ln x$ is the inverse of the exponential function $y = e^x$. p. 36

NDER(*f(x)*, *a*): The numerical derivative of *f* at $x = a$. p. 108

Newton (unit of force): pp. 369, 411. Also, see *Mass.*

Newton's dot notation: $\dot{x}$ means dx/dt, $\ddot{x}$ means d^2x/dt^2, $\dddot{x}$ means d^3x/dt^3, and so on. p. 656

Newton's second law (of motion): Most of the motion we observe undergoes changes that are the result of one or more applied forces. The overall net force, whether it be from a single force or a combination of forces, produces acceleration. The relationship of acceleration to force and mass is described by Newton's second law, which says that the acceleration of an object is directly proportional to the net force and inversely proportional to the object's mass:

$$\text{Acceleration} \sim \frac{\text{net force}}{\text{mass}} \qquad \text{or} \qquad a \sim \frac{F}{m}.$$

With appropriate units, the proportionality may be expressed as an exact equation: $F = ma$. To take direction into account as well as magnitude, we use the vector notation $\mathbf{F} = m\mathbf{a}$.   p. 540

NINT($f(x)$, x, a, b): The numerical integral of f with respect to x, from $x = a$ to $x = b$.   p. 281

Nonremovable discontinuity: A discontinuity that is not removable. See *Removable discontinuity*.

Normal component of acceleration:

$$a_N = \kappa \left(\frac{ds}{dt}\right)^2 = \kappa |\mathbf{v}|^2.$$

p. 655

Normal curve: The graph of a normal probability density function.   p. 405

Normal line at point on surface: The line through P_0 parallel to $\nabla f|_{P_0}$.   p. 729

Normal line to a curve at a point of the curve: The line perpendicular to the tangent at that point.   p. 86

Normal plane: The plane determined by $\mathbf{N}$ and $\mathbf{B}$.   p. 654

Normal probability density function: The normal probability density function for a population with *mean* μ and *standard deviation* σ is

$$f(x) = \frac{1}{\sigma\sqrt{2\pi}} e^{-(x-\mu)^2/(2\sigma^2)}.$$

The mean μ represents the average value of the variable x. The standard deviation σ measures the "scatter" around the mean. p. 405

Normal vector: To a differentiable curve at a point P, a vector perpendicular to the tangent to the curve at P. To a curve traced by a vector function at a point P where the velocity is not zero, a vector perpendicular to the velocity vector at P.   p. 525

Norm of a partition: The longest subinterval length, denoted $\|P\|$, for a partition P.   p. 260

Numerical derivative: An approximation of the derivative of a function using a numerical algorithm.   p. 108

Numerical integration: Approximating the integral of a function using a numerical algorithm.   p. 280

Numerical method: A method for generating a numerical solution of a problem. For example, a method for estimating the value of a definite integral, for estimating the zeros of a function or solutions of an equation, or for estimating values of the function that solves an initial value problem.   p. 351

Numerical solution: Of an equation $f(x) = 0$, an estimate of one or more of its roots; of an initial value problem, a table of estimated values of the solution function.   p. 351

Octants: See *Coordinate planes*.

Odd function: A function f for which $f(-x) = -f(x)$ for every x in the domain of f.   p. 13

One-sided limit: See *Left-hand limit; Right-hand limit*.

One-to-one function: A function f for which $f(a) \neq f(b)$ whenever $a \neq b$.   p. 32

Open interval (a, b): All numbers x with $a < x < b$.   p. 11

Open region in plane or space: A region that consists entirely of interior points.   p. 677

Optimization: In an application, maximizing or minimizing some aspect of the system being modeled.   p. 206

Orbital period: The time it takes a planet to go around its sun once.   p. 666

Order of a derivative: If y is a function of x, $y' = dy/dx$ is the first-order, or first, derivative of y with respect to x; $y'' = d^2y/dx^2$ is the second-order, or second, derivative of y; $y^{(n)} = dy^{(n-1)}/dx$ is the nth-order, or nth, derivative of y.   p. 119

Order of a differential equation: The order of the highest-order derivative in the equation.   p. 303

Orientable surface: A smooth surface S is orientable or two-sided if it is possible to define a field $\mathbf{n}$ of unit normal vectors on S that varies continuously with position. Once $\mathbf{n}$ is chosen, the surface is *oriented* and the surface together with its normal field is an *oriented surface*. The vector $\mathbf{n}$ at any point is the *positive direction* at that point.   p. 872

Origin: The point $(0, 0)$ in the Cartesian coordinate plane; the point $(0, 0)$ in the polar coordinate plane. The point $(0, 0, 0)$ in three-dimensional space.   pp. 552, 576

Orthogonal curves: See *Perpendicular curves*.

Orthogonal Gradient Theorem: If $f(x, y, z)$ is differentiable in a region whose interior contains a smooth curve C and P_0 is a point of C where f has a local maximum or minimum relative to its values on C, then ∇f is orthogonal (normal) to C at P_0. p. 749

Orthogonal vectors: Vectors making a $90°$ angle.   pp. 528, 587

Oscillating discontinuity: A point near which the function values oscillate too much for the function to have a limit.   p. 76

Osculating circle: See *Circle of curvature*.

Osculating plane: The plane determined by $\mathbf{N}$ and $\mathbf{T}$.   p. 654

Paraboloid of revolution: See *Circular paraboloid*.

Parallel curves: In the plane, curves that differ from one another by a vertical translation (shift).   p. 103

Parameter: See *Parametric equations*.

Parameter domain: See *Parametrization of a surface*.

Parameter interval: See *Parametric equations*.

Parametric equations: If x and y are given as functions $x = f(t)$, $y = g(t)$, over an interval of t-values, then the set of points $(x, y) = (f(t), g(t))$ defined by these equations is a *parametric curve*, the equations are *parametric equations* for the curve, the variable t is the *parameter* for the curve, and the interval of allowable t-values is the *parameter interval*.   p. 26

Parametric equations for a line: The standard parametrization of the line through $P_0(x_0, y_0, z_0)$ parallel to the vector $\mathbf{v} = A\mathbf{i} + B\mathbf{j} + C\mathbf{k}$ is $x = x_0 + tA$, $y = y_0 + tB$, $z = z_0 + tC$, $-\infty < t < \infty$. p. 601

Parametric formula for surface area: See *Surface area.*

Parametrization of a curve: The parametric equations and parameter interval describing a curve. pp. 27, 631

Parametrization of a surface: If $\mathbf{r}(u, v) = f(u, v)\mathbf{i} + g(u, v)\mathbf{j} + h(u, v)\mathbf{k}$ is a continuous vector function that is defined on a region R in the uv-plane and one-to-one on the interior of R, then the range of $\mathbf{r}$ is the *surface S* defined or traced by $\mathbf{r}$. The equation and the domain R together constitute a *parametrization* of the surface. The variables u and v are the *parameters*, and R is the *parameter domain.* p. 878

Partial derivative of a function: The partial derivative of $f(x, y)$ with respect to x at the point (x_0, y_0) is

$$\left.\frac{\partial f}{\partial x}\right|_{(x_0, y_0)} = \left.\frac{d}{dx} f(x, y_0)\right|_{(x_0, y_0)} = \lim_{h \to 0} \frac{f(x_0 + h, y_0) - f(x_0, y_0)}{h},$$

provided the limit exists. The partial derivative with respect to y has a similar definition. p. 694

Partial fractions: See *Integration by partial fractions.*

Partial sum: See *Infinite series.*

Particular solution: The unique solution of a differential equation satisfying given initial condition or conditions. p. 303

Partition of an interval [a, b]: A set

$$\{x_0 = a, x_1, x_2, \ldots, x_n = b\}$$

of points in [a, b] in order from left to right. p. 259

Path independent integral: See *Conservative field.*

Percentage change: See *Absolute, relative, and percentage change.*

Percentage error:

$$\frac{|\text{approximate value} - \text{exact value}|}{|\text{exact value}|} \times 100.$$

Perihelion position: The position of a planet in which it is closest to its sun. p. 663

Periodic function: A function f for which there is a positive number p such that $f(x + p) = f(x)$ for every value of x. p. 43

Period of a periodic function f: The smallest positive number p for which $f(x + p) = f(x)$ for every value of x. See *Periodic function.*

Perpendicular Axis Theorem: The moment of inertia about the z-axis can be calculated as I_z, $I_z = I_x + I_y$. p. 775

Perpendicular (orthogonal) curves: Two curves are said to be perpendicular (orthogonal) at a point of intersection if their tangents at that point are perpendicular. p. 148

Perpendicular vectors: See *Orthogonal vectors.*

Piecewise-defined function: A function that is defined by applying different formulas to different parts of its domain. p. 15

Piecewise smooth curve: A curve made from a finite number of smooth curves pieced together in a continuous fashion. p. 635

Polar coordinates: Each point P in the polar coordinate plane has polar coordinates (r, θ) where r gives the directed distance from the origin O to P and θ gives a directed angle from the initial ray to ray OP. p. 552

Pole: The polar coordinate origin. p. 552

Polynomial: An expression of the form

$$a_n x^n + a_{n-1} x^{n-1} + \cdots + a_1 x + a_0.$$

Position function: A function s that gives the position $s(t)$ of a body on a coordinate axis at time t. p. 86

Position vector: Of a point (a, b) in the plane, the vector $\langle a, b \rangle$. Of a point $P(x, y, z)$ in space, the vector $\mathbf{r} = \overrightarrow{OP} = x\mathbf{i} + y\mathbf{j} + z\mathbf{k}$. Of a particle $P(f(t), g(t), h(t))$ moving along a parametrized curve $x = f(t)$, $y = g(t)$, $z = h(t)$ in space, the vector $\mathbf{r} = \overrightarrow{OP} = f(t)\mathbf{i} + g(t)\mathbf{k} + h(t)\mathbf{k}$. pp. 540, 577, 631

Potential function: If $\mathbf{F}$ is a vector field defined on a region D, and if $\mathbf{F} = \nabla f$ for some scalar function f on D, then f is a potential function for $\mathbf{F}$. p. 846

Power series: An expression of the form

$$c_0 + c_1 x + c_2 x^2 + \cdots + c_n x^n + \cdots = \sum_{n=0}^{\infty} c_n x^n$$

is a power series centered at $x = 0$. An expression of the form $c_0 + c_1(x - a) + c_2(x - a)^2 + \cdots$

$$+ c_n(x - a)^n + \cdots = \sum_{n=0}^{\infty} c_n(x - a)^n$$

is a power series centered at $x = a$. The number a is the *center* of the series. p. 461

Prime notation ($f'(x)$): If $y = f(x)$, then both y' and $f'(x)$ denote the derivative of the function with respect to x. p. 97

Principal unit normal vector N: At a point on a smooth curve where $\kappa \neq 0$,

$$\mathbf{N} = \frac{1}{\kappa} \frac{d\mathbf{T}}{ds}.$$

p. 651

Probability: See *Probability density function.*

Probability density function (pdf): A function $f(x)$ such that $f(x) \geq 0$ for all x and $\int_{-\infty}^{\infty} f(x)\, dx = 1$. The *probability* associated with the interval [a, b] is $\int_a^b f(x)\, dx$. p. 405

Product Rule: The product of two differentiable functions u and v is differentiable, and

$$\frac{d}{dx}(uv) = u\frac{dv}{dx} + v\frac{du}{dx},$$

p. 115

p-series: A series of the form

$$\sum_{n=1}^{\infty} \frac{1}{n^p}, \qquad p \text{ a nonzero constant.}$$

p. 497

Quadric surface: The graph in space of a second-degree equation in x, y, z. The most general form is

$$Ax^2 + By^2 + Cz^2 + Dxy + Eyz + Fxz + Gx + Hy + Jz + K = 0.$$

Quotient Rule: At a point where $v \neq 0$, the quotient $y = u/v$ of two differentiable functions is differentiable, and

$$\frac{d}{dx}\left(\frac{u}{v}\right) = \frac{v\frac{du}{dx} - u\frac{dv}{dx}}{v^2}.$$

p. 117

Radian measure: If a central angle of a circle of radius r intercepts an arc of length s on the circle, then the radian measure of the angle is s/r.   p. 41

Radius of convergence: In general, the positive number R for which the power series

$$\sum_{n=0}^{\infty} c_n(x - a)^n$$

converges when $|x - a| < R$ and diverges when $|x - a| > R$. If the series converges only for $x = a$, then $R = 0$. If the series converges for every x, then $R = \infty$.   p. 488

Radius of curvature: See *Circle of curvature.*

Radius of gyration: The radius of gyration R_L of an object of mass M about a line L is defined by the equation $I_L = MR_L^2$. It tells how far from L the entire mass might be concentrated to give the same moment of inertia I_L. The radius of gyration gives a convenient way to express the moment of inertia in terms of a mass and a length. See pages 776, 800, 831, and 874 for formulas.

Range of a function: See *Function.*

Rate constant: See *Law of exponential change.*

Rate of change: See *Average rate of change of a quantity over a period of time; Instantaneous rate of change of f with respect to x at a.*

Rational function: A function that can be expressed as the quotient of two polynomial functions.

Ratio Test: Let $\sum a_n$ be a series with positive terms and with

$$\lim_{n \to \infty} \frac{a_{n-1}}{a_n} = L.$$

Then (1) the series converges if $L < 1$, (2) the series diverges of $L > 1$, (3) the test is inconclusive if $L = 1$.   p. 491

Rectifying plane: The plane determined by $\mathbf{B}$ and $\mathbf{T}$.   p. 654

Regular partition: A partition in which consecutive points are equally spaced.   p. 260

Related rates: An equation involving two or more variables that are differentiable functions of time can be used to find an equation that relates the rates at which the variables change.   p. 232

Relative change: See *Absolute, relative, and percentage change.*

Relative extrema: Same as *Local extrema.*

Relative growth rate: For a population modeled by a differentiable function $P(t)$ giving the number of individuals at time t, the quotient $(dP/dt)/P$.   p. 342

Relative maximum: Same as *Local maximum.*

Relative minimum: Same as *Local minimum.*

Remainder of order n: The remainder

$$R_n(x) = \frac{f^{(n+1)}(c)}{(n + 1)!}(x - a)^{n+1}$$

in Taylor's Theorem.   p. 482

Removable discontinuity: A discontinuity c of the function f for which $f(c)$ can be (re)defined so that $\lim_{x \to c} f(x) = f(c)$.   p. 76

Repeated integral: See *Iterated integral.*

Resultant vector: The vector that results from adding or subtracting two vectors.   p. 522

Riemann sum: A sum of the form

$$\sum_{k=1}^{n} f(c_k) \cdot \Delta x_k$$

where f is a continuous function on a closed interval $[a, b]$; and c_k is some point in, and Δx_k the length of the kth subinterval in some partition of $[a, b]$.   p. 258

Right end behavior model: See *End behavior model.*

Right-hand derivative: The derivative defined by a right-hand limit.   p. 100

Right-handed coordinate frame: A three-dimensional coordinate system with x-, y-, and z-axes such that when the right hand is held so that the fingers curl from the positive x-axis toward the positive y-axis, the thumb points along the positive z-axis.   p. 575

Right-hand limit: The limit of f as x approaches c from the right, or $\lim_{x \to c^+} f(x)$.   p. 60

Right-hand rule: The selection of a unit normal vector $\mathbf{n}$ to a plane determined by two vectors $\mathbf{A}$ and $\mathbf{B}$ in space such that $\mathbf{n}$ points the way a right thumb points when the fingers are curled through the angle from $\mathbf{A}$ to $\mathbf{B}$.   p. 594

Root of an equation: See *Zero of a function.*

Roundoff error: Error due to rounding.   p. 115

RRAM: Right-hand endpoint rectangular approximation method. The method of approximating a definite integral over an interval using the function values at the right-hand endpoints of the subintervals determined by a partition.   p. 249

Saddle point: A point on a surface that can look like a local minimum from one perspective and a local maximum from another perspective. It is sometimes called a *minimax* of the surface. For a surface $z = f(x, y)$, there is a saddle point at a critical point (a, b) if in every open disk centered at (a, b) there are domain points at which $f(x, y) > f(a, b)$ and domain points at which $f(x, y) < f(a, b)$.   pp. 621, 739

Scalar: In the context of vectors, scalars are real numbers that behave like scaling factors.   p. 520

Scalar component of B in the direction of A: The dot product of **B** with the direction of **A**.   p. 588

Scalar function: Name used for real-valued functions to distinguish them from vector-valued functions.   p. 530

Secant function: See *Basic trigonometric functions.*

Secant line to a curve: A line through two points on the curve. p. 82

Second derivative: If y is a function of x and $y' = dy/dx$ is the first derivative of y with respect to x, then $dy'/dx = y'' = d^2y/dx^2$ is the second derivative of y with respect to x. p. 119

Second derivative test for extreme values for functions of two variables: If $f(x, y)$ and its first and second partial derivatives are continuous throughout a disk centered at (a, b) and the first partial derivatives of f at (a, b) are 0, then
(1) f has a local maximum at (a, b) if $f_{xx} < 0$ and $f_{xx} f_{yy} - f_{xy}^2 > 0$ at (a, b);
(2) f has a local minimum at (a, b) if $f_{xx} > 0$ and $f_{xx} f_{yy} - f_{xy}^2 > 0$ at (a, b);
(3) f has a saddle point at (a, b) if $f_{xx} f_{yy} - f_{xy}^2 < 0$ at (a, b); and
(4) the test is inconclusive at (a, b) if $f_{xx} f_{yy} - f_{xy}^2 = 0$ at (a, b).   p. 739

Second derivative test (for local extrema): If $f'(c) = 0$ and $f''(c) < 0$, then f has a local maximum at $x = c$. If $f'(c) = 0$ and $f''(c) > 0$, then f has a local minimum at $x = c$.   p. 200

Second moment: See *Moment of inertia.*

Second-order partial derivatives: Derivatives obtained from a function by two successive partial differentiations.

Separable differential equation: A differential equation $y' = f(x, y)$ in which f can be expressed as a product of a function of x and a function of y.   p. 320

Separating the variables: For a separable differential equation $y' = f(x, y)$, the process of combining all the y-terms with y' on one side of the equation and putting all the x-terms on the other side.   p. 320

Sequence: A function whose domain is the set of positive integers.

Sigma notation: Notation using the Greek letter capital sigma, $\sum$, for writing lengthy sums in compact form.   p. 258

Simple harmonic motion: Periodic motion, like the vertical motion of a weight bobbing up and down at the end of a spring, that can be modeled with a sinusoidal position function.   p. 136

Simply connected region: A region D in which every closed path in D can be contracted to a point in D without ever leaving the region.   p. 893

Sine function: See *Basic trigonometric functions.*

Slope field: A slope field for the first order differential equation $dy/dx = f(x, y)$ is a plot of short line segments with slopes $f(x, y)$ at a lattice of points (x, y) in the domain of f.   p. 305

Slope of a curve: The slope of a curve $y = f(x)$ at the point $(a, f(a))$ is $f'(a)$, provided f is differentiable at a.   p. 85

Slope of a nonvertical vector: The common slope of the lines parallel to the vector.   p. 522

Smooth curve: The graph of a smooth function.   p. 396

Smooth function: A real-valued function $y = f(x)$ with a continuous first derivative. A vector function $\mathbf{r}(t)$ whose first derivative is continuous and never zero.   p. 396

Smooth parametrized curve: A plane curve $x = f(t)$, $y = g(t)$ for which the derivatives f' and g' are continuous and not simultaneously zero. A space curve $x = f(t)$, $y = g(t)$, $z = h(t)$ for which f', g' and h' are continuous and not simultaneously zero.   pp. 513, 635

Smooth parametrized surface: A parametrized surface $\mathbf{r}(u, v) = f(u, v)\mathbf{i} + g(u, v)\mathbf{j} + h(u, v)\mathbf{k}$ whose partial derivatives $\mathbf{r}_u$ and $\mathbf{r}_v$ are continuous and whose cross-product $\mathbf{r}_u \times \mathbf{r}_v$ is never zero on the parameter domain.   p. 879

Smooth surface: A surface $f(x, y, z) = c$ on which ∇f is continuous and never vanishes.   p. 867

Smooth vector curve: Curve traced by a smooth vector function. See *Smooth function.*

Solid of revolution: A solid generated by revolving a plane region about a line in the plane.   p. 385

Solution of a differential equation: Any function that together with its derivatives satisfies the equation. When we find all such functions, we have *solved* the differential equation.   p. 303

Space curve: When a particle $P(x, y, z)$ moves through space, we think of the particle's coordinates as functions $x = f(t)$, $y = g(t)$, $z = h(t)$ defined on some interval I. The curve traced by P as the parameter t runs through I is called a space curve.

Speed: The absolute value or magnitude of velocity.   p. 124

Spherical coordinates: Coordinates of the point P in the form (ρ, ϕ, θ), where ρ is the distance from the origin to the point, ϕ is the angle $\overline{OP}$ makes with the positive z-axis ($0 \le \phi \le \pi$), and θ is the angle from cylindrical coordinates.   p. 807

Spherical wedge: The three-dimensional figure, shaped like a slice of pie, defined by spherical coordinate differentials $d\rho$, $d\phi$, and $d\theta$.   p. 809

Standard linear approximation: The standard linear approximation of f at (x_0, y_0) is the function $L(x, y) = f(x_0, y_0) + f_x(x_0, y_0)(x - x_0) + f_y(x_0, y_0)(y - y_0)$.

Standard equation for a sphere: The standard equation for a sphere of radius a and center (x_0, y_0, z_0) is $(x - x_0)^2 + (y - y_0)^2 + (z - z_0)^2 = a^2$.

Standard position of a vector: The representative vector with initial point at the origin.   p. 521

Standard unit vectors: In the plane, the vectors $\mathbf{i} = \langle 1, 0 \rangle$, $\mathbf{j} = \langle 0, 1 \rangle$. In space, the vectors $\mathbf{i}$, $\mathbf{j}$, and $\mathbf{k}$ defined by the directed line segments from the origin to the points $(1, 0, 0)$, $(0, 1, 0)$, and $(0, 0, 1)$, respectively.   pp. 529, 577

Stokes's Theorem: The circulation of a vector field around the boundary of an oriented surface in space in the direction counterclockwise with respect to the surface's unit normal vector field equals the integral of the normal component of the curl of the field over the surface.   p. 886

Sum of a series: See *Infinite series.*

Surface: See *Parametrization of a surface* and *Graph of a function of two variables.*

Surface area

Gradient formula: The area of the surface $f(x, y, z) = c$ defined over a closed and bounded plane region R is

$$\iint\limits_{R} \frac{|\nabla f|}{|\nabla f \cdot \mathbf{p}|} \, dA,$$

where $\mathbf{p}$ is a unit vector normal to R and $\nabla f \cdot \mathbf{p} \neq 0$.

Parametric formula: The area of the smooth surface $\mathbf{r}(u, v) = f(u, v)\mathbf{i} + g(u, v)\mathbf{j} + h(u, v)\mathbf{k}$, $a \leq u \leq b$, $c \leq v \leq d$, is

$$A = \int_{c}^{d} \int_{a}^{b} |\mathbf{r}_u \times \mathbf{r}_v| \, du \, dv.$$

pp. 868, 880

Surface area differential:

$$d\sigma = \frac{|\nabla f|}{|\nabla f \cdot \mathbf{p}|} \, dA,$$

or, for a parametrized surface,

$$d\sigma = |\mathbf{r}_u \times \mathbf{r}_v| \, du \, dv.$$

pp. 871, 880

Surface integral: The integral of a function g over a surface S. pp. 871, 881

Symmetric difference quotient: The quotient

$$\frac{f(a + h) - f(a - h)}{2h}$$

that a graphing calculator uses to calculate NDER($f(x)$, a), the numerical derivative of f at $x = a$.   p. 108

Symmetry: For a plane curve to have

1. symmetry about the x-axis, the point (x, y) must lie on the curve if and only if $(x, -y)$ lies on the curve;

2. symmetry about the y-axis, (x, y) must lie on the curve if and only if $(-x, y)$ lies on the curve;

3. symmetry about the origin, (x, y) must lie on the curve if and only if $(-x, -y)$ lies on the curve.

p. 14

Tabular integration: A time-saving way to organize the work of repeated integrations by parts.   p. 327

Tangent function: See *Basic trigonometric functions.*

Tangent line: To the graph of a function $y = f(x)$ at a point $x = a$ where f' exists, the line through $(a, f(a))$ with slope $f'(a)$. To a parametrized curve at a point P where $dx/dt \neq 0$, the line through P with slope equal to the value of $(dy/dt)/(dx/dt)$ at P. To a curve traced by a vector function $\mathbf{r}(t)$ at a point P where $\mathbf{v} \neq \mathbf{0}$, the line through P parallel to $\mathbf{v}$.   p. 83

Tangent plane at a point on a surface: The plane tangent to a surface $f(x, y, z) = c$ at point P_0 is the plane through P_0 normal to $\nabla f|_{P_0}$.   p. 729

Tangential component of acceleration:

$$a_{\mathbf{T}} = \frac{d^2 s}{dt^2} = \frac{d}{dt} |\mathbf{v}|.$$

p. 655

Tangent vector to a differentiable curve at a point P: A vector that is parallel to the line tangent to the curve at P.   p. 532

Taylor polynomial: Let f be a function with derivatives through order n throughout some open interval containing 0. Then

$$P_n(x) = \sum_{k=0}^{n} \frac{f^{(k)}(0)}{k!} x^k$$

is the Taylor polynomial of order n for f at $x = 0$. If f is a function with derivatives through order n throughout some open interval containing the point $x = a$, then

$$P_n(x) = \sum_{k=0}^{n} \frac{f^{(k)}(a)}{k!} (x - a)^k$$

is the Taylor polynomial of order n for f at $x = a$.   p. 470

Taylor series: Let f be a function with derivatives of all orders throughout some open interval containing 0. Then

$$f(0) + f'(0)x + \frac{f''(0)}{2!} x^2 + \cdots + \frac{f^{(n)}(0)}{n!} x^n + \cdots,$$

$$\text{or} \quad \sum_{k=0}^{\infty} \frac{f^{(k)}(0)}{k!} x^k,$$

is the Taylor series generated by f at $x = 0$. This series is also called the *Maclaurin series* generated by f.

   If f is a function with derivatives of all orders throughout some open interval containing the point $x = a$, then

$$f(a) + f'(a)(x - a) + \frac{f''(a)}{2!} (x - a)^2 + \cdots$$

$$+ \frac{f^{(n)}(a)}{n!} (x - a)^n + \cdots,$$

$$\text{or} \quad \sum_{k=0}^{\infty} \frac{f^{(k)}(a)}{k!} (x - a)^k,$$

is the Taylor series generated by f at $x = a$.   p. 470

Term of a sequence or series: For the sequence

$$a_1, a_2, a_3, \ldots, a_n, \ldots,$$

or for the series

$$a_1 + a_2 + a_3 + \cdots + a_n + \cdots,$$

a_n is the nth term.   p. 458

Time Distance Law: See *Kepler's third law.*

TNB (Frenet) frame: The right-handed frame made by the mutually orthogonal unit vectors **T**, **N**, and **B** (in that order).

Torque: When we turn a threaded bolt with a wrench, the resulting force *along the axis of that bolt* is called a torque.

Torsion: The torsion function of a smooth curve is

$$\tau = -\frac{d\mathbf{B}}{ds} \cdot \mathbf{N}.$$

p. 654

Total differential: The change $df = f_x(x_0, y_0)\, dx + f_y(x_0, y_0)\, dy$ in the value of the linearization of $f(x, y)$ when the point of evaluation moves from (x_0, y_0) to $(x_0 + dx, y_0 + dy)$. Analogous definitions hold for functions of more than two independent variables. p. 707

Transcendental function: A function that is not algebraic. (See *Algebraic function.*) The six basic trigonometric functions are transcendental as are the inverse trigonometric functions and the exponential and logarithmic functions studied in this book.

Trapezoidal Rule: To approximate $\int_a^b f(x)\, dx$, use

$$T = \frac{h}{2}(y_0 + 2y_1 + 2y_2 + \cdots + 2y_{n-1} + y_n),$$

where $[a, b]$ is partitioned into n subintervals of equal length $h = (b - a)/n$ and y_i is the value of f at each partition point x_i. p. 290

Trigonometric function: See *Basic trigonometric functions.*

Triple integral of $F(x, y, z)$ over D:

$$\lim_{n \to \infty} \sum_{k=1}^{n} F(x_k, y_k, z_k)\, \Delta V_k = \iiint_D F(x, y, z)\, dV.$$

p. 790

Triple scalar (box) product: A product of the form $(\mathbf{A} \times \mathbf{B}) \cdot \mathbf{C}$, where $\mathbf{A}$, $\mathbf{B}$, $\mathbf{C}$ are vectors in space. p. 597

Truncation error: The error incurred in using a finite partial sum to estimate the sum of an infinite series. p. 481

Two-sided limit: Limit at an interior point of a function's domain. See *Limit.*

Two sided surface: See *Orientable surface.*

Unbounded region: A region that is not bounded.

Unit circle: The circle of radius 1 centered at the origin.

Unit tangent vector T: Every smooth curve $\mathbf{r} = f(t)\mathbf{i} + g(t)\mathbf{j} + h(t)\mathbf{k}$ has a unit tangent vector **T** defined by the equation

$$\mathbf{T} = \frac{d\mathbf{r}}{ds} = \frac{d\mathbf{r}/dt}{ds/dt} = \frac{\mathbf{v}}{|\mathbf{v}|}.$$

p. 647

Unit vector: A vector with magnitude 1. Also see *standard unit vectors.*

Universal gravitational constant: The constant $G = 6.6720 \times 10^{-11}\ \text{Nm}^2\ \text{kg}^{-2}$ in Newton's Law of Gravitation. p. 662

Upper bound: See *Bounded.*

u-substitution: See *Integration by substitution.*

Value of an improper integral: See *Improper integral.*

Variable of integration: In $\int f(x)\, dx$ or $\int_a^b f(x)\, dx$, the variable x. p. 261

Vector (cross) product: See *Cross (vector) product.*

Vector equation for a line: The vector equation for the line through $P_0(x_0, y_0, z_0)$ parallel to **v** is $\overrightarrow{PP_0} = t\mathbf{v}$, $-\infty < t < \infty$. p. 600

Vector equation for a plane: A point P lies in the plane through P_0 normal to **n** if and only if $\mathbf{n} \cdot \overrightarrow{PP_0} = 0$. p. 602

Vector field: A function on a domain in the plane or in space that assigns a vector to each point in the domain. The field is continuous if its component functions are continuous, differentiable if its component functions are differentiable, and so on. p. 835

Vector in the plane: A directed line segment in the plane with the understanding that two such vectors are equal if they have the same length and direction. p. 520

Vector projection: A vector projection of a vector $\mathbf{B} = \overrightarrow{PQ}$ onto a nonzero vector $\mathbf{A} = \overrightarrow{PS}$ is the vector $\overrightarrow{PR}$ determined by dropping a perpendicular from Q to the point R on the line PS. If **B** represents a force, then $\overrightarrow{PR}$ represents the effective force in the direction of A. p. 587

Vector-valued function on a domain D: A rule that assigns a vector in space to each element of D. p. 631

Vector (vector-valued) function: The equation $\mathbf{r}(t) = f(t)\mathbf{i} + g(t)\mathbf{j} + h(t)\mathbf{k}$, $t \in I$, defines **r** as a vector function of t on the interval I. See also *Vector field.* pp. 530, 631, 835

Velocity: The rate of change of position with respect to time. See also *Average velocity; Instantaneous velocity; Velocity vector.* p. 124

Velocity vector: If **r** is the position vector of a particle moving along a smooth curve in the plane or in space, then at any time t, $\mathbf{v}(t) = d\mathbf{r}/dt$ is the particle's velocity vector. pp. 533, 636

Vertical line: In the Cartesian coordinate plane, a line parallel to the y-axis.

Viewing window: On a graphing calculator, the portion of the coordinate plane displayed on the screen.

Volume by slicing: A method for finding the volume of a solid by evaluating $\int_a^b A(x)\, dx$, where $A(x)$ (assumed integrable) is the solid cross-section area at x. p. 384

Work: In everyday life, *work* means an activity that requires muscular or mental effort. In science, the term refers specifically to a force **F** acting on a body and the body's subsequent displacement.

The work done by a force **F** along a smooth path C in space is the integral of $\mathbf{F} \cdot \mathbf{T}$, the effective force in the direction of motion at each point, over the length of the path: $W = \int_C \mathbf{F} \cdot \mathbf{T} \, ds$
This formula simplifies to $W = \mathbf{F} \cdot \mathbf{D}$ for a constant force **F** acting through a displacement $\mathbf{D} = \overrightarrow{PQ}$. It simplifies still further to $W = Fd$ for a force of constant magnitude F acting in the direction of the displacement over a distance d. pp. 369, 582, 835

x-intercept: The x-coordinate of the point where a plane curve intersects the x-axis.   p. 8

xy-plane: The plane in space whose standard equation is $z = 0$. p. 575

xz-plane: The plane in space whose standard equation is $y = 0$. p. 576

y-intercept: The y-coordinate of the point where a plane curve intersects the y-axis.   p. 4

yz-plane: The plane in space whose standard equation is $x = 0$. p. 575

Zero of a function: A solution of the equation $f(x) = 0$ is a zero of the function f or a *root* of the equation.

Zero vector: In the plane, the vector $\mathbf{0} = 0\mathbf{i} + 0\mathbf{j}$. In space, the vector $\mathbf{0} = 0\mathbf{i} + 0\mathbf{j} + 0\mathbf{k}$.   pp. 522, 580

Chapter 1

Section 1.1 (pp. 1–9)

Quick Review

1. -2

2. -1

3. -1

4. $\dfrac{5}{4}$

5. (a) Yes **(b)** No

6. (a) Yes **(b)** No

7. $\sqrt{2}$

8. $\dfrac{5}{3}$

9. $y = \dfrac{4}{3}x - \dfrac{7}{3}$

10. $y = \dfrac{2}{5}x - \dfrac{3}{5}$

Exercises

1. $\Delta x = -2, \Delta y = -3$ **3.** $\Delta x = -5, \Delta y = 0$

5. (a) and **(c)** **(b)** 3

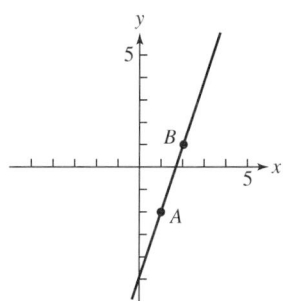

7. (a) and **(c)** **(b)** 0

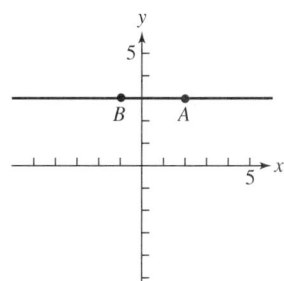

9. (a) $x = 2$ **(b)** $y = 3$

11. (a) $x = 0$ **(b)** $y = -\sqrt{2}$

13. $y = 1(x - 1) + 1$ **15.** $y = 2(x - 0) + 3$

17. $3x - 2y = 0$ **19.** $x = -2$

21. $y = 3x - 2$ **23.** $y = -\dfrac{1}{2}x - 3$

25. $y = \dfrac{5}{2}x$

27. (a) $-\dfrac{3}{4}$ **(b)** 3

(c)

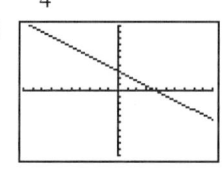

$[-10, 10]$ by $[-10, 10]$

29. (a) $-\dfrac{4}{3}$ **(b)** 4

(c)

$[-10, 10]$ by $[-10, 10]$

31. (a) $y = -x$ **(b)** $y = x$

33. (a) $x = -2$ **(b)** $y = 4$

35. $m = \dfrac{7}{2}, b = -\dfrac{3}{2}$ **37.** $y = -1$

39. (a) $y = 0.680x + 9.013$

(b) The slope is 0.68. It represents the approximate average weight gain in pounds per month.

(c)

$[15, 45]$ by $[15, 45]$

(d) 29 pounds

41. $y = 1(x - 3) + 4$
$y = x - 3 + 4$
$y = x + 1$, which is the same equation.

43. (a) $k = 2$ **(b)** $k = -2$

45. 5.97 atmospheres ($k = 0.0994$)

47. (a) $y = 5632x - 11,080,280$

(b) The rate at which the median price is increasing in dollars per year

(c) $y = 2732x - 5,362,360$

(d) In the Northeast

49. The coordinates of the three missing vertices are (5, 2), (−1, 4) and (−1, −2).

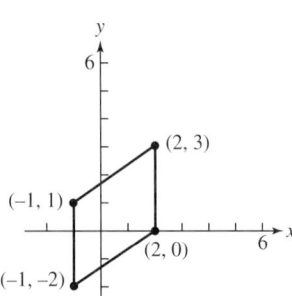

51. $y = -\dfrac{3}{4}(x - 3) + 4$ or $y = -\dfrac{3}{4}x + \dfrac{25}{4}$

Section 1.2 (pp. 9–19)

Quick Review

1. $[-2, \infty)$ **2.** $(-\infty, 0) \cup (2, \infty)$
3. $[-1, 7]$ **4.** $(-\infty, -3] \cup [7, \infty)$
5. $(-4, 4)$ **6.** $[-3, 3]$
7. Translate the graph of f 2 units left and 3 units downward.
8. Translate the graph of f 5 units right and 2 units upward.
9. (a) $x = -3, 3$ **(b)** No real solution
10. (a) $x = -\dfrac{1}{5}$ **(b)** No solution
11. (a) $x = 9$ **(b)** $x = -6$
12. (a) $x = -7$ **(b)** $x = 28$

Exercises

1. $A = \dfrac{\pi d^2}{4}$ **3.** $S = 6e^2$

5. (a) All reals   **(b)** $(-\infty, 4]$
(c)

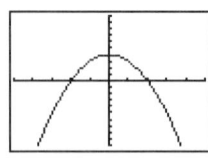

$[-5, 5]$ by $[-10, 10]$

(d) Symmetric about y-axis

7. (a) $[1, \infty)$ **(b)** $[2, \infty)$
(c)

$[-3, 10]$ by $[-3, 10]$

(d) None

9. (a) $(-\infty, 3]$ **(b)** $[0, \infty)$
(c)

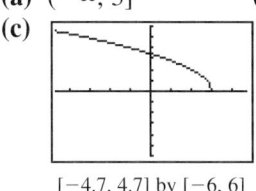

$[-4.7, 4.7]$ by $[-6, 6]$

(d) None

11. (a) All reals   **(b)** All reals
(c)

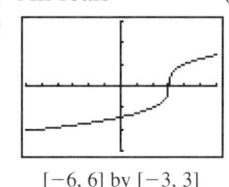

$[-6, 6]$ by $[-3, 3]$

(d) None

13. (a) $(-\infty, 0]$ **(b)** $[0, \infty)$
(c)

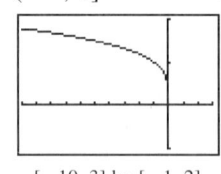

$[-10, 3]$ by $[-1, 2]$

(d) None

15. (a) $[-2, 2]$ **(b)** $[0, 2]$
(c)

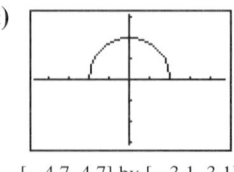

$[-4.7, 4.7]$ by $[-3.1, 3.1]$

(d) Symmetric about y-axis

17. (a) $(-\infty, 0) \cup (0, \infty)$ **(b)** $(1, \infty)$
(c)

[−4, 4] by [−1, 5]

(d) Symmetric about y-axis
19. Even   **21.** Neither
23. Even   **25.** Odd
27. Neither
29. (a)

[−9.4, 9.4] by [−6.2, 6.2]

(b) All reals   **(c)** $(-\infty, 2]$
31. (a)

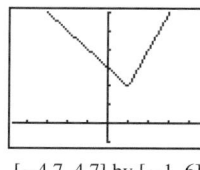

[−4.7, 4.7] by [−1, 6]

(b) All reals   **(c)** $[2, \infty)$
33. (a)

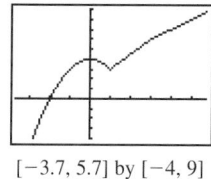

[−3.7, 5.7] by [−4, 9]

(b) All reals   **(c)** All reals
35. Because if the vertical line test holds, then for each x-coordinate, there is at most one y-coordinate giving a point on the curve. This y-coordinate would correspond to the value assigned to the x-coordinate. Since there's only one y-coordinate, the assignment would be unique.
37. No   **39.** Yes
41. $f(x) = \begin{cases} x, & 0 \le x \le 1 \\ 2 - x, & 1 < x \le 2 \end{cases}$

43. $f(x) = \begin{cases} 2 - x, & 0 < x \le 2 \\ \dfrac{5}{3} - \dfrac{x}{3}, & 2 < x \le 5 \end{cases}$

45. $f(x) = \begin{cases} -x, & -1 \le x < 0 \\ 1, & 0 < x \le 1 \\ \dfrac{3}{2} - \dfrac{x}{2}, & 1 < x < 3 \end{cases}$

47. $f(x) = \begin{cases} 0, & 0 \le x \le \dfrac{T}{2} \\ \dfrac{2}{T}x - 1, & \dfrac{T}{2} < x \le T \end{cases}$

49. (a) $x^2 + 2$ **(b)** $x^2 + 10x + 22$
(c) 2   **(d)** 22
(e) -2 **(f)** $x + 10$
51. (a) For $f \circ g$:

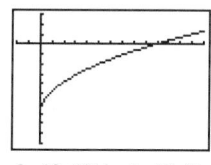

[−10, 70] by [−10, 3]

Domain: $[0, \infty)$; Range: $[-7, \infty)$
For $g \circ f$:

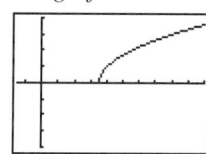

[−3, 20] by [−4, 4]

Domain: $[7, \infty)$; Range: $[0, \infty)$
(b) $(f \circ g)(x) = \sqrt{x} - 7$; $(g \circ f)(x) = \sqrt{x - 7}$
53. (a) For $f \circ g$:

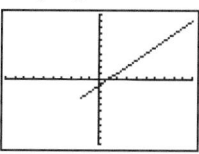

[−10, 10] by [−10, 10]

Domain: $[-2, \infty)$; Range: $[-3, \infty)$
For $g \circ f$:

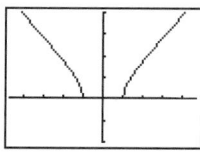

[−4.7, 4.7] by [−2, 4]

Domain: $(-\infty, -1] \cup [1, \infty)$; Range: $[0, \infty)$

53. continued

(b) $(f \circ g)(x) = (\sqrt{x+2})^2 - 3$

$= x - 1, x \geq -2$

$(g \circ f)(x) = \sqrt{x^2 - 1}$

55. Domain: $(-\infty, -2) \cup (2, \infty)$; Range: $(0, \infty)$

57. Domain: $(-\infty, -3) \cup (-3, 3) \cup (3, \infty)$

Range: $(-\infty, 0) \cup \left[\frac{2}{\sqrt[3]{9}}, \infty\right)$

59. (a)

(b)

61. (a)

(b)

63. (a) $g(x) = x^2$ **(b)** $g(x) = \frac{1}{x-1}$

(c) $f(x) = \frac{1}{x}$

(d) $f(x) = x^2$

(Note that the domain of the composite is $[0, \infty)$).

65. (a) Because the circumference of the original circle was 8π and a piece of length x was removed.

(b) $r = \dfrac{8\pi - x}{2\pi} = 4 - \dfrac{x}{2\pi}$

(c) $h = \sqrt{16 - r^2} = \dfrac{\sqrt{16\pi x - x^2}}{2\pi}$

(d) $V = \dfrac{1}{3}\pi r^2 h = \dfrac{(8\pi - x)^2 \sqrt{16\pi x - x^2}}{24\pi^2}$

67. (a)

[−3, 3] by [−1, 3]

(b) Domain of y_1: $[0, \infty)$
Domain of y_2: $(-\infty, 1]$
Domain of y_3: $[0, 1]$

(c) The results for $y_1 - y_2$, $y_2 - y_1$, and $y_1 \cdot y_2$ are the same as for $y_1 + y_2$ above.

Domain of $\dfrac{y_1}{y_2}$: $[0, 1)$

Domain of $\dfrac{y_2}{y_1}$: $(0, 1]$

(d) The domain of a sum, difference, or product of two functions is the intersection of their domains.
The domain of a quotient of two functions is the intersection of their domains with any zeros of the denominator removed.

Section 1.3 (pp. 20–26)

Quick Review

1. 2.924

2. 4.729

3. 0.192

4. 2.5713

5. 1.8882

6. ±1.0383

7. $630.58

8. $1201.16

9. $x^{-18}y^{-5} = \dfrac{1}{x^{18}y^5}$

10. $a^2b^{-1}c^{-6} = \dfrac{a^2}{bc^6}$

Exercises

1. (a)

3. (e)

5. (b)

7.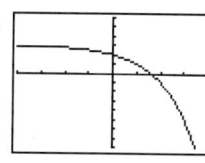

[−4, 4] by [−8, 6]

Domain: All reals
Range: $(−\infty, 3)$
x-intercept: ≈ 1.585
y-intercept: 2

9.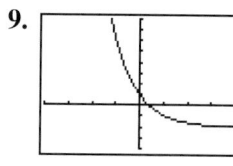

[−4, 4] by [−4, 8]

Domain: All reals
Range: $(−2, \infty)$
x-intercept: ≈ 0.405
y-intercept: 1

11. 3^{4x}

13. 2^{-6x}

15. $x \approx 2.3219$

17. $x \approx −0.6309$

19.

x	y	Δy
1	−1	
2	1	2
3	3	2
4	5	2

21.

x	y	Δy
1	1	
2	4	3
3	9	5
4	16	7

23. After 19 years

25. (a) $A(t) = 6.6\left(\dfrac{1}{2}\right)^{t/14}$

(b) About 38.1145 days later

27. ≈ 11.433 years

29. ≈ 11.090 years

31. ≈ 19.108 years

33. $2^{48} \approx 2.815 \times 10^{14}$

35. Since $\Delta x = 1$, the corresponding value of Δy is equal to the slope of the line. If the changes in x are constant for a linear function, then the corresponding changes in y are constant as well.

37. (a) Regression equation:
$P(x) = 6.033(1.030)^x$, where $x = 0$ represents 1900

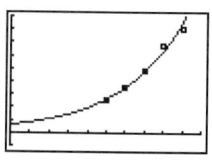

[0, 100] by [−10, 90]

(b) Approximately 6.03 million, which is not very close to the actual population

(c) The annual rate of growth is approximately 3%.

39. 7609.7 million

41. $a = 3$, $k = 1.5$

Section 1.4 (pp. 26–31)

Quick Review

1. $y = -\dfrac{5}{3}x + \dfrac{29}{3}$

2. $y = -4$

3. $x = 2$

4. x-intercepts: $x = -3$ and $x = 3$
y-intercepts: $y = -4$ and $y = 4$

5. x-intercepts: $x = -4$ and $x = 4$
y-intercepts: None

6. x-intercept: $x = -1$
y-intercepts: $y = -\dfrac{1}{\sqrt{2}}$ and $y = \dfrac{1}{\sqrt{2}}$

7. (a) Yes **(b)** No **(c)** Yes

8. (a) Yes **(b)** Yes **(c)** No

9. (a) $t = \dfrac{-2x - 5}{3}$ **(b)** $t = \dfrac{3y + 1}{2}$

10. (a) $a \geq 0$ **(b)** All reals **(c)** All reals

Exercises

1. Graph (c).
Window: $[-4, 4]$ by $[-3, 3]$, $0 \leq t \leq 2\pi$

3. Graph (d).
Window: $[-10, 10]$ by $[-10, 10]$, $0 \leq t \leq 2\pi$

5. (a) The resulting graph appears to be the right half of a hyperbola in the first and fourth quadrants. The parameter a determines the x-intercept. The parameter b determines the shape of the hyperbola. If b is smaller, the graph has less steep slopes and appears "sharper." If b is larger, the slopes are steeper and the graph appears more "blunt."

(b) This appears to be the left half of the same hyperbola.

(c) Because both $\sec t$ and $\tan t$ are discontinuous at these points. This might cause the grapher to include extraneous lines (the asymptotes to the hyperbola) in its graph.

(d) $\left(\dfrac{x}{a}\right)^2 - \left(\dfrac{y}{b}\right)^2 = (\sec t)^2 - (\tan t)^2 = 1$
by a standard trigonometric identity.

(e) This changes the orientation of the hyperbola. In this case, b determines the y-intercept of the hyperbola, and a determines the shape. The parameter interval $\left(-\dfrac{\pi}{2}, \dfrac{\pi}{2}\right)$ gives the upper half of the hyperbola. The parameter interval $\left(\dfrac{\pi}{2}, \dfrac{3\pi}{2}\right)$ gives the lower half. The same values of t cause discontinuities and may add extraneous lines to the graph.

7. (a)

[−3, 3] by [−2, 2]

Initial point: $(1, 0)$
Terminal point: $(−1, 0)$
(b) $x^2 + y^2 = 1$; upper half (or $y = \sqrt{1 − x^2}$; all)

9. (a)

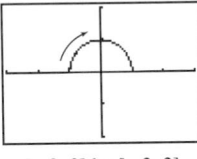

[−3, 3] by [−2, 2]

Initial point: $(1, 0)$
Terminal point: $(0, 1)$
(b) $x^2 + y^2 = 1$; upper half (or $y = \sqrt{1 − x^2}$; all)

11. (a)

[−4.7, 4.7] by [−3.1, 3.1]

Initial point: $(0, 2)$
Terminal point: $(0, −2)$
(b) $\left(\dfrac{x}{4}\right)^2 + \left(\dfrac{y}{2}\right)^2 = 1$;
right half (or $x = 2\sqrt{4 − y^2}$; all)

13. (a)

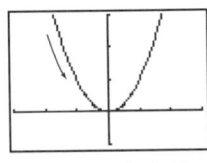

[−3, 3] by [−1, 3]

No initial or terminal point
(b) $y = x^2$; all

15. (a)

[−1, 5] by [−1, 3]

Initial point: $(0, 0)$
Terminal point: None
(b) $y = \sqrt{x}$; all (or $x = y^2$; upper half)

17. (a)

[−3, 3] by [−2, 2]

No initial or terminal point
(b) $x^2 − y^2 = 1$; left branch (or $x = −\sqrt{y^2 + 1}$;
all)

19. (a)

[−9, 9] by [−6, 6]

No initial or terminal point
(b) $y = 2x + 3$; all

21. (a)

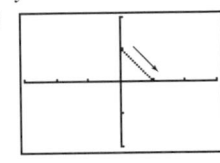

[−3, 3] by [−2, 2]

Initial point: $(0, 1)$
Terminal point: $(1, 0)$
(b) $y = −x + 1$; $(0, 1)$ to $(1, 0)$

23. (a)

[−6, 6] by [−2, 6]

Initial point: $(4, 0)$
Terminal point: None
(b) $y = −x + 4$; $x \le 4$

25. (a)

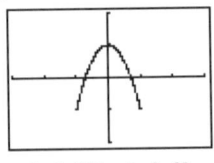

[−3, 3] by [−2, 2]

The curve is traced and retraced in both
directions, and there is no initial or terminal
point.
(b) $y = −2x^2 + 1$; $−1 \le x \le 1$

27. Possible answer:
$x = −1 + 5t, y = −3 + 4t, 0 \le t \le 1$

29. Possible answer:
$x = t^2 + 1, y = t, t \le 0$

31. Possible answer:
$x = 2 − 3t, y = 3 − 4t, t \ge 0$

33. $1 < t < 3$
35. $-5 \le t < -3$
37. Possible answer: $x = t, y = t^2 + 2t + 2, t > 0$
39. Possible answers:
   (a) $x = a \cos t, y = -a \sin t, 0 \le t \le 2\pi$
   (b) $x = a \cos t, y = a \sin t, 0 \le t \le 2\pi$
   (c) $x = a \cos t, y = -a \sin t, 0 \le t \le 4\pi$
   (d) $x = a \cos t, y = a \sin t, 0 \le t \le 4\pi$
41. $x = 2 \cot t, y = 2 \sin^2 t, 0 < t < \pi$

Section 1.5 (pp. 32–40)

Quick Review

1. 1   **2.** 5
3. $x^{2/3}$ **4.** $(x - 1)^{2/3} + 1$
5. Possible answer: $x = t, y = \dfrac{1}{t - 1}, t \ge 2$
6. Possible answer: $x = t, y = t, t < -3$
7. $(4, 5)$ **8.** $\left(\dfrac{8}{3}, -3\right) \approx (2.67, -3)$
9. (a) $(1.58, 3)$ **(b)** No intersection
10. (a) $(-1.39, 4)$ **(b)** No intersection

Exercises

1. No   **3.** Yes   **5.** Yes
7. Yes   **9.** No   **11.** No
13. $f^{-1}(x) = \dfrac{x - 3}{2}$
15. $f^{-1}(x) = (x + 1)^{1/3}$ or $\sqrt[3]{x + 1}$
17. $f^{-1}(x) = -x^{1/2}$ or $-\sqrt{x}$
19. $f^{-1}(x) = 2 - (-x)^{1/2}$ or $2 - \sqrt{-x}$
21. $f^{-1}(x) = \dfrac{1}{x^{1/2}}$ or $\dfrac{1}{\sqrt{x}}$
23. $f^{-1}(x) = \dfrac{1 - 3x}{x - 2}$
25.   **27.**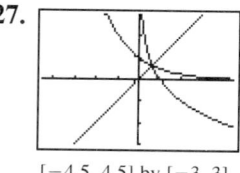
    $[-6, 6]$ by $[-4, 4]$ $[-4.5, 4.5]$ by $[-3, 3]$
29.   **31.**
    $[-4.5, 4.5]$ by $[-3, 3]$ $[-3, 3]$ by $[-2, 2]$

33. 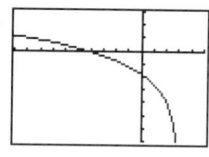   **35.**
    $[-10, 5]$ by $[-7, 3]$ $[-3, 6]$ by $[-2, 4]$
    Domain: $(-\infty, 3)$;   Domain: $(-1, \infty)$;
    Range: all reals   Range: all reals
37. $t = \dfrac{\ln 2}{\ln 1.045} \approx 15.75$
39. $x = \ln\left(\dfrac{3 \pm \sqrt{5}}{2}\right) \approx -0.96$ or 0.96
41. $y = e^{2t+4}$
43. $f^{-1}(x) = \log_2\left(\dfrac{x}{100 - x}\right)$
45. (a) $f(f(x)) = \sqrt{1 - (f(x))^2}$
           $= \sqrt{1 - (1 - x^2)}$
           $= \sqrt{x^2}$
           $= x$, since $x \ge 0$
   (b) $f(f(x)) = f\left(\dfrac{1}{x}\right) = \dfrac{1}{1/x} = x$ for all $x \ne 0$
47. About 14.936 years. (If the interest is only paid annually, it will take 15 years.)
49. (a) $y = -2539.852 + 636.896 \ln x$
   (b) 209.94 million metric tons
   (c) When $x \approx 101.08$ or 2001
51. (a) Suppose that $f(x_1) = f(x_2)$. Then
          $mx_1 + b = mx_2 + b$, which gives $x_1 = x_2$
          since $m \ne 0$.
   (b) $f^{-1}(x) = \dfrac{x - b}{m}$; the slopes are reciprocals.
   (c) They are also parallel lines with nonzero slope.
   (d) They are also perpendicular lines with nonzero slopes.
53. If the graph of $f(x)$ passes the horizontal line test, so will the graph of $g(x) = -f(x)$ since it's the same graph reflected about the x-axis.
55. (a) Domain: All reals
      Range: If $a > 0$, then (d, ∞)
               If $a < 0$, then $(-\infty, d)$
   (b) Domain: (c, ∞); Range: All reals

Section 1.6 (pp. 41–51)

Quick Review

1. $60°$ **2.** $-\left(\dfrac{450}{\pi}\right)° \approx -143.24°$
3. $-\dfrac{2\pi}{9}$ **4.** $\dfrac{\pi}{4}$

5. $x \approx 0.6435$, $x \approx 2.4981$

6. $x \approx 1.9823$, $x \approx 4.3009$

7. $x \approx 0.7854 \left(\text{or } \dfrac{\pi}{4}\right)$, $x \approx 3.9270 \left(\text{or } \dfrac{5\pi}{4}\right)$

8. $f(-x) = 2(-x)^2 - 3 = 2x^2 - 3 = f(x)$

The graph is symmetric about the y-axis because if a point (a, b) is on the graph, then so is the point $(-a, b)$.

9. $f(-x) = (-x)^3 - 3(-x) = -x^3 + 3x$
$= -(x^3 - 3x) = -f(x)$

The graph is symmetric about the origin because if a point (a, b) is on the graph, then so is the point $(-a, -b)$.

10. $x \geq 0$

Exercises

1. $\dfrac{5\pi}{4}$

3. $\dfrac{1}{2}$ radian or $\approx 28.65°$

5. Possible answers are:
  (a) $[0, 4\pi]$ by $[-3, 3]$
  (b) $[0, 4\pi]$ by $[-3, 3]$
  (c) $[0, 2\pi]$ by $[-3, 3]$

7. $\dfrac{\pi}{6}$ radian or $30°$

9. ≈ -1.3734 radians or $-78.6901°$

11. (a) π (b) 1.5
  (c) $[-2\pi, 2\pi]$ by $[-2, 2]$

13. (a) π (b) 3
  (c) $[-2\pi, 2\pi]$ by $[-4, 4]$

15. (a) 6 (b) 4
  (c) $[-3, 3]$ by $[-5, 5]$

17. (a) $\dfrac{2\pi}{3}$ (b) $x \neq \dfrac{k\pi}{3}$, for integers k
  (c) $(-\infty, -5] \cup [1, \infty)$
  (d)

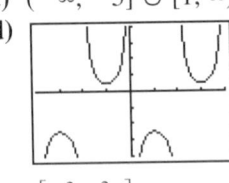

$\left[-\dfrac{2\pi}{3}, \dfrac{2\pi}{3}\right]$ by $[-8, 8]$

19. (a) $\dfrac{\pi}{3}$
  (b) $x \neq \dfrac{k\pi}{6}$, for odd integers k
  (c) All reals

(d)

$\left[-\dfrac{\pi}{2}, \dfrac{\pi}{2}\right]$ by $[-8, 8]$

21. $\cos \theta = \dfrac{15}{17}$ $\sin \theta = \dfrac{8}{17}$ $\tan \theta = \dfrac{8}{15}$

$\sec \theta = \dfrac{17}{15}$ $\csc \theta = \dfrac{17}{8}$ $\cot \theta = \dfrac{15}{8}$

23. $\cos \theta = -\dfrac{3}{5}$ $\sin \theta = \dfrac{4}{5}$ $\tan \theta = -\dfrac{4}{3}$

$\sec \theta = -\dfrac{5}{3}$ $\csc \theta = \dfrac{5}{4}$ $\cot \theta = -\dfrac{3}{4}$

25. $x \approx 1.190$ and $x \approx 4.332$

27. $x = \dfrac{\pi}{6}$ and $x = \dfrac{5\pi}{6}$

29. $x = \dfrac{7\pi}{6} + 2k\pi$ and $x = \dfrac{11\pi}{6} + 2k\pi$, k any integer

31. $\dfrac{\sqrt{72}}{11} \approx 0.771$

33. (a) $y = 1.543 \sin (2468.635x - 0.494) + 0.438$

$[0, 0.01]$ by $[-2.5, 2.5]$

  (b) Frequency $= 392.9$, so it must be a "G."

35. (a) 37 (b) 365
  (c) 101 (d) 25

37. (a) $\cot (-x) = \dfrac{\cos (-x)}{\sin (-x)} = \dfrac{\cos (x)}{-\sin (x)} = -\cot (x)$

  (b) Assume that f is even and g is odd.
    Then $\dfrac{f(-x)}{g(-x)} = \dfrac{f(x)}{-g(x)} = -\dfrac{f(x)}{g(x)}$ so $\dfrac{f}{g}$ is odd.
    The situation is similar for $\dfrac{g}{f}$.

39. Assume that f is even and g is odd.
  Then $f(-x)\, g(-x) = f(x)[-g(x)] = -f(x)g(x)$
  so fg is odd.

41. (a) $y = 3.0014 \sin (0.9996x + 2.0012) + 2.9999$
  (b) $y = 3 \sin (x + 2) + 3$

43. (a) $\sqrt{2} \sin \left(ax + \dfrac{\pi}{4}\right)$
  (b) See part (a).
  (c) It works.

(d) $\sin\left(ax + \dfrac{\pi}{4}\right)$

$= \sin(ax) \cdot \dfrac{1}{\sqrt{2}} + \cos(ax) \cdot \dfrac{1}{\sqrt{2}}$

$= \dfrac{1}{\sqrt{2}}(\sin ax + \cos ax)$

So, $\sin(ax) + \cos(ax) = \sqrt{2}\sin\left(ax + \dfrac{\pi}{4}\right)$.

45. Since $\sin(x)$ has period 2π,

$(\sin(x + 2\pi))^3 = (\sin(x))^3$. This function has period 2π. A graph shows that no smaller number works for the period.

47. One possible graph:

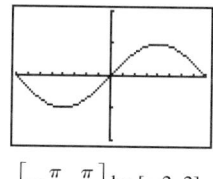

$\left[-\dfrac{\pi}{60}, \dfrac{\pi}{60}\right]$ by $[-2, 2]$

Chapter Review (pp. 52–53)

1. $y = 3x - 9$

2. $y = -\dfrac{1}{2}x + \dfrac{3}{2}$

3. $x = 0$

4. $y = -2x$

5. $y = 2$

6. $y = -\dfrac{2}{5}x + \dfrac{21}{5}$

7. $y = -3x + 3$

8. $y = 2x - 5$

9. $y = -\dfrac{4}{3}x - \dfrac{20}{3}$

10. $y = -\dfrac{5}{3}x - \dfrac{19}{3}$

11. $y = \dfrac{2}{3}x + \dfrac{8}{3}$

12. $y = \dfrac{5}{3}x - 5$

13. $y = -\dfrac{1}{2}x + 3$

14. $y = -\dfrac{2}{7}x - \dfrac{6}{7}$

15. Origin

16. *y*-axis

17. Neither

18. *y*-axis

19. Even

20. Odd

21. Even

22. Odd

23. Odd

24. Neither

25. Neither

26. Even

27. (a) Domain: all reals **(b)** Range: $[-2, \infty)$
(c)

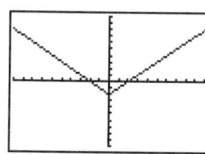

$[-10, 10]$ by $[-10, 10]$

28. (a) Domain: $(-\infty, 1]$ **(b)** Range: $[-2, \infty)$

(c)

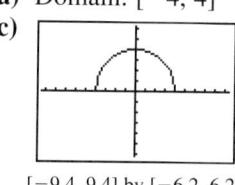

$[-9.4, 9.4]$ by $[-3, 3]$

29. (a) Domain: $[-4, 4]$ **(b)** Range: $[0, 4]$
(c)

$[-9.4, 9.4]$ by $[-6.2, 6.2]$

30. (a) Domain: all reals **(b)** Range: $(1, \infty)$
(c)

$[-6, 6]$ by $[-4, 20]$

31. (a) Domain: all reals **(b)** Range: $(-3, \infty)$
(c)

$[-4, 4]$ by $[-5, 15]$

32. (a) Domain: $x \neq \dfrac{k\pi}{4}$, for odd integers k

(b) Range: all reals
(c)

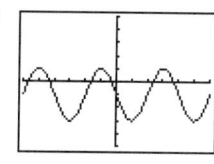

$\left[-\dfrac{\pi}{2}, \dfrac{\pi}{2}\right]$ by $[-8, 8]$

33. (a) Domain: all reals **(b)** Range: $[-3, 1]$
(c)

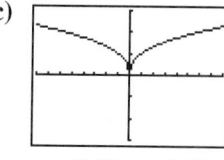

$[-\pi, \pi]$ by $[-5, 5]$

34. (a) Domain: all reals **(b)** Range: $[0, \infty)$
(c)

$[-8, 8]$ by $[-3, 3]$

35. (a) Domain: $(3, \infty)$ **(b)** Range: all reals
(c)
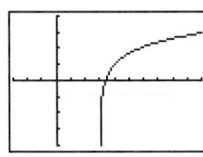
$[-3, 10]$ by $[-4, 4]$

36. (a) Domain: all reals **(b)** Range: all reals
(c)
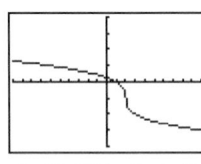
$[-10, 10]$ by $[-4, 4]$

37. (a) Domain: $[-4, 4]$ **(b)** Range: $[0, 2]$
(c)
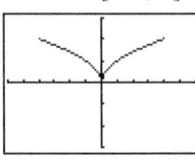
$[-6, 6]$ by $[-3, 3]$

38. (a) Domain: $[-2, 2]$ **(b)** Range: $[-1, 1]$
(c)
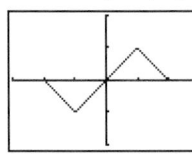
$[-3, 3]$ by $[-2, 2]$

39. $f(x) = \begin{cases} 1 - x, & 0 \le x < 1 \\ 2 - x, & 1 \le x \le 2 \end{cases}$

40. $f(x) = \begin{cases} \dfrac{5x}{2}, & 0 \le x < 2 \\ -\dfrac{5}{2}x + 10, & 2 \le x \le 4 \end{cases}$

41. (a) 1
(b) $\dfrac{1}{\sqrt{2.5}} \left(= \sqrt{\dfrac{2}{5}} \right)$
(c) $x, x \ne 0$
(d) $\dfrac{1}{\sqrt{1/\sqrt{x+2} + 2}}$

42. (a) 2 **(b)** 1 **(c)** x
(d) $\sqrt[3]{\sqrt[3]{x + 1} + 1}$

43. (a) $(f \circ g)(x) = -x, x \ge -2$
$(g \circ f)(x) = \sqrt{4 - x^2}$
(b) Domain $(f \circ g)$: $[-2, \infty)$
Domain $(g \circ f)$: $[-2, 2]$
(c) Range $(f \circ g)$: $(-\infty, 2]$
Range $(g \circ f)$: $[0, 2]$

44. (a) $(f \circ g)(x) = \sqrt[4]{1 - x}$
$(g \circ f)(x) = \sqrt{1 - \sqrt{x}}$
(b) Domain $(f \circ g)$: $(-\infty, 1]$
Domain $(g \circ f)$: $[0, 1]$
(c) Range $(f \circ g)$: $[0, \infty)$
Range $(g \circ f)$: $[0, 1]$

45. (a)
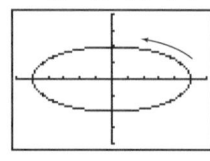
$[-6, 6]$ by $[-4, 4]$
Initial point: $(5, 0)$
Terminal point: $(5, 0)$
(b) $\left(\dfrac{x}{5}\right)^2 + \left(\dfrac{y}{2}\right)^2 = 1$; all

46. (a)
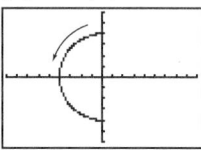
$[-9, 9]$ by $[-6, 6]$
Initial point: $(0, 4)$
Terminal point: $(0, -4)$
(b) $x^2 + y^2 = 16$; left half

47. (a)
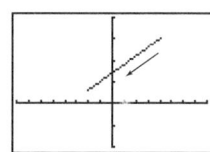
$[-8, 8]$ by $[-10, 20]$
Initial point: $(4, 15)$
Terminal point: $(-2, 3)$
(b) $y = 2x + 7$; from $(4, 15)$ to $(-2, 3)$

48. (a)
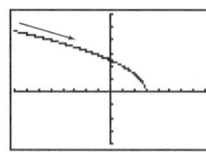
$[-8, 8]$ by $[-4, 6]$
Initial point: None
Terminal point: $(3, 0)$
(b) $y = \sqrt{6 - 2x}$; all

49. Possible answer:
$x = -2 + 6t, y = 5 - 2t, 0 \le t \le 1$
50. Possible answer:
$x = -3 + 7t, y = -2 + t, -\infty < t < \infty$
51. Possible answer:
$x = 2 - 3t, y = 5 - 5t, 0 \le t$

52. Possible answer:
$x = t, y = t(t - 4), t \leq 2$

53. (a) $f^{-1}(x) = \dfrac{2 - x}{3}$

(b)
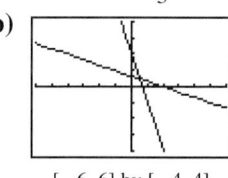
$[-6, 6]$ by $[-4, 4]$

54. (a) $f^{-1}(x) = \sqrt{x} - 2$

(b)

$[-6, 12]$ by $[-4, 8]$

55. ≈ 0.6435 radians or $36.8699°$

56. ≈ -1.1607 radians or $-66.5014°$

57. $\cos \theta = \dfrac{3}{7}$ $\quad \sin \theta = \dfrac{\sqrt{40}}{7}$ $\quad \tan \theta = \dfrac{\sqrt{40}}{3}$

$\sec \theta = \dfrac{7}{3}$ $\quad \csc \theta = \dfrac{7}{\sqrt{40}}$ $\quad \cot \theta = \dfrac{3}{\sqrt{40}}$

58. (a) $x \approx 3.3430$ and $x \approx 6.0818$

(b) $x \approx 3.3430 + 2k\pi$ and $x \approx 6.0818 + 2k\pi$, k any integer

59. $x = -5 \ln 4$

60. (a)

(b)

61. (a)

(b)
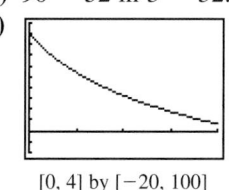

62. (a) $V = 100,000 - 10,000x, 0 \leq x \leq 10$

(b) After 4.5 years

63. (a) 90 units

(b) $90 - 52 \ln 3 \approx 32.8722$ units

(c)

$[0, 4]$ by $[-20, 100]$

64. After $\dfrac{\ln (10/3)}{\ln 1.08} \approx 15.6439$ years

(If the bank only pays interest at the end of the year, it will take 16 years.)

65. (a) $N = 4 \cdot 2^t$

    (b) 4 days: 64; one week: 512

    (c) After $\dfrac{\ln 500}{\ln 2} \approx 8.9658$ days, or after nearly 9 days.

    (d) Because it suggests the number of guppies will continue to double indefinitely and become arbitrarily large, which is impossible due to the finite size of the tank and the oxygen supply in the water.

66. (a) $y = 20.627x + 338.622$

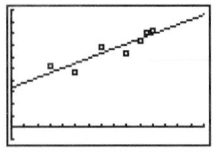

$[0, 30]$ by $[-100, 1000]$

    (b) Approximately 957

    (c) Slope is 20.627. It represents the approximate arrival increase in number of doctorates earned by Hispanic Americans per year.

67. (a) $y = 14.60175 \cdot 1.00232^x$

    (b) Sometime during the year 2132 (when $x \approx 232$)

    (c) 0.232%

Chapter 2

Section 2.1 (pp. 55–65)

Quick Review

1. 0

2. $\dfrac{11}{12}$

3. 0

4. $\dfrac{1}{3}$

5. $-4 < x < 4$

6. $-c^2 < x < c^2$

7. $-1 < x < 5$

8. $c - d^2 < x < c + d^2$

9. $x - 6$

10. $\dfrac{x}{x + 1}$

Exercises

1. (a) 3   **(b)** -2
    (c) No limit   **(d)** 1

3. (a) -4 **(b)** -4
    (c) -4 **(d)** -4

5. (a) 4   **(b)** -3
    (c) No limit   **(d)** 4

7. $-\dfrac{3}{2}$ **9.** -15 **11.** 0

13. 4   **15.** 1

17. Expression not defined at $x = -2$. There is no limit.

19. Expression not defined at $x = 0$. There is no limit.

21. $\dfrac{1}{2}$ **23.** $-\dfrac{1}{2}$

25. 12   **27.** -1 **29.** 0

31. (a) True   **(b)** True   **(c)** False
    (d) True   **(e)** True   **(f)** True
    (g) False   **(h)** False   **(i)** False
    (j) False

33. (c)   **35.** (d)   **37.** 0

39. 0   **41.** 1

43. (a) 6   **(b)** 0
    (c) 9   **(d)** -3

45. (a)

$[-3, 6]$ by $[-1, 5]$

    (b) Right-hand: 2
    Left-hand: 1

    (c) No, because the two one-sided limits are different.

47. (a)

$[-5, 5]$ by $[-4, 8]$

    (b) Right-hand: 4
    Left-hand: no limit

    (c) No, because the left-hand limit doesn't exist.

49. (a)

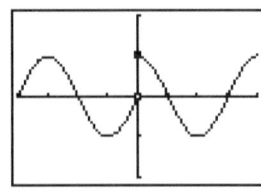

$[-2\pi, 2\pi]$ by $[-2, 2]$

    (b) $(-2\pi, 0) \cup (0, 2\pi)$

    (c) $c = 2\pi$ **(d)** $c = -2\pi$

51. (a)

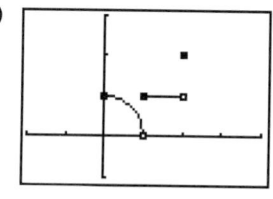

[−2, 4] by [−1, 3]

(b) $(0, 1) \cup (1, 2)$ (c) $c = 2$
(d) $c = 0$

53. 0   **55.** 0

57. (a) 14.7 m/sec   **(b)** 29.4 m/sec

59. (a)

x	−0.1	−0.01	−0.001	−0.0001
$f(x)$	−0.054402	−0.005064	−0.000827	−0.000031

(b)

x	0.1	0.01	0.001	0.0001
$f(x)$	−0.054402	−0.005064	−0.000827	−0.000031

The limit appears to be 0.

61. (a)

x	−0.1	−0.01	−0.001	−0.0001
$f(x)$	2.0567	2.2763	2.2999	2.3023

(b)

x	0.1	0.01	0.001	0.0001
$f(x)$	2.5893	2.3293	2.3052	2.3029

The limit appears to be approximately 2.3.

63. (a) Because the right-hand limit at zero depends only on the values of the function for positive x-values near zero.

(b) Use: area of triangle $= \left(\dfrac{1}{2}\right)$(base)(height)

area of circular sector $= \dfrac{\text{(angle)(radius)}^2}{2}$

(c) This is how the areas of the three regions compare.

(d) Multiply by 2 and divide by $\sin \theta$.

(e) Take reciprocals, remembering that all of the values involved are positive.

(f) The limits for $\cos \theta$ and 1 are both equal to 1. Since $\dfrac{\sin \theta}{\theta}$ is between them, it must also have a limit of 1.

(g) $\dfrac{\sin (-\theta)}{-\theta} = \dfrac{-\sin (\theta)}{-\theta} = \dfrac{\sin (\theta)}{\theta}$

(h) If the function is symmetric about the y-axis, and the right-hand limit at zero is 1, then the left-hand limit at zero must also be 1.

(i) The two one-sided limits both exist and are equal to 1.

65. (a) $f\left(\dfrac{\pi}{6}\right) = \dfrac{1}{2}$

(b) One possible answer: $a = 0.305, b = 0.775$
(c) One possible answer: $a = 0.513, b = 0.535$

Section 2.2 (pp. 65–73)

Quick Review

1. $j^{-1}(x) = \dfrac{x + 3}{2}$

[−12, 12] by [−8, 8]

2. $f^{-1}(x) = \ln (x)$

[−6, 6] by [−4, 4]

3. $f^{-1}(x) = \tan (x), -\dfrac{\pi}{2} < x < \dfrac{\pi}{2}$

[−6, 6] by [−4, 4]

4. $f^{-1}(x) = \cot (x), 0 < x < \pi$

[−6, 6] by [−4, 4]

5. $q(x) = \dfrac{2}{3}$

$r(x) = -3x^2 - \left(\dfrac{5}{3}\right)x + \dfrac{7}{3}$

6. $q(x) = 2x^2 + 2x + 1$
$r(x) = -x^2 - x - 2$

7. (a) $f(-x) = \cos x$ **(b)** $f\left(\dfrac{1}{x}\right) = \cos\left(\dfrac{1}{x}\right)$

8. (a) $f(-x) = e^x$ **(b)** $f\left(\dfrac{1}{x}\right) = e^{-1/x}$

9. (a) $f(-x) = -\dfrac{\ln (-x)}{x}$

(b) $f\left(\dfrac{1}{x}\right) = -x \ln x$

10. (a) $f(-x) = \left(x + \dfrac{1}{x}\right) \sin x$

(b) $f\left(\dfrac{1}{x}\right) = \left(\dfrac{1}{x} + x\right) \sin\left(\dfrac{1}{x}\right)$

Exercises

1. (a) 1   **(b)** 1   **(c)** $y = 1$

3. (a) 0   **(b)** $-\infty$ **(c)** $y = 0$

5. (a) 3   **(b)** -3

   (c) $y = 3, y = -3$

7. (a) 1   **(b)** -1

   (c) $y = 1, y = -1$

9. ∞ **11.** $-\infty$

13. 0   **15.** ∞

17. (a) $x = -2, x = 2$

   (b) Left-hand limit at -2 is ∞.

      Right-hand limit at -2 is $-\infty$.

      Left-hand limit at 2 is $-\infty$.

      Right-hand limit at 2 is ∞.

19. (a) $x = -1$

   (b) Left-hand limit at -1 is $-\infty$.

      Right-hand limit at -1 is ∞.

21. (a) $x = k\pi$, k any integer

   (b) At each vertical asymptote:

      Left-hand limit is $-\infty$.

      Right-hand limit is ∞.

23. Both are 1   **25.** Both are 1

27. Both are 0   **29. (a)**

31. (d)

33. (a) $3x^2$ **(b)** None

35. (a) $\dfrac{1}{2x}$ **(b)** $y = 0$

37. (a) $4x^2$ **(b)** None

39. (a) e^x **(b)** $-2x$

41. (a) x **(b)** x

43. At ∞: ∞ **45.** At ∞: 0

    At $-\infty$: 0   At $-\infty$: 0

47. (a) 0   **(b)** -1

   (c) $-\infty$ **(d)** -1

49. One possible answer:

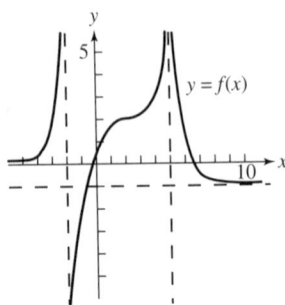

51. $\dfrac{f_1(x)/f_2(x)}{g_1(x)/g_2(x)} = \dfrac{f_1(x)/g_1(x)}{f_2(x)/g_2(x)}$

As x goes to infinity, $\dfrac{f_1}{g_1}$ and $\dfrac{f_2}{g_2}$ both approach 1.

Therefore, using the above equation, $\dfrac{f_1/f_2}{g_1/g_2}$ must

also approach 1.

53. (a) Using 1980 as $x = 0$:

    $y = -2.2316x^3 + 54.7134x^2 - 351.0933x$

        $+ 733.2224$

[0, 20] by [0, 800]

   (b) Again using 1980 as $x = 0$:

    $y = 1.458561x^4 - 60.5740x^3 + 905.8877x^2 -$

        $5706.0943x + 12967.6288$

[0, 20] by [0, 800]

   (c) Cubic: approximately -2256 dollars

      Quartic: approximately 9979 dollars

   (d) Cubic: End behavior model is $-2.2316x^3$.

              This model predicts that the grants

              will become negative by 1996.

      Quartic: End behavior model is $1.458561x^4$.

              This model predicts that the size of

              the grants will grow very rapidly

              after 1995.

    Neither of these seem reasonable. There is no

    reason to expect the grants to disappear

    (become negative) based on the data.

    Similarly, the data give no indication that a

    period of rapid growth is about to occur.

55. (a) This follows from $x - 1 < \text{int } x \le x$ which is

    true for all x. Dividing by x gives the result.

   (b) 1   **(c)** 1

57. This is because as x approaches infinity, $\sin x$

continues to oscillate between 1 and -1 and

doesn't approach any given real number.

59. Limit $= \ln(10)$,

    since $\dfrac{\ln x}{\log x} = \dfrac{\ln x}{\ln x/\ln 10} = \ln 10$.

Section 2.3 (pp. 73–81)

Quick Review

1. 2

2. (a) -2 **(b)** -1
   (c) No limit   **(d)** -1

3. (a) 1   **(b)** 2
   (c) No limit   **(d)** 2

4. $(f \circ g)(x) = \dfrac{x + 2}{6x + 1}, x \neq 0$

$(g \circ f)(x) = \dfrac{3x + 4}{2x - 1}, x \neq -5$

5. $g(x) = \sin x, x \geq 0$
$(f \circ g)(x) = \sin^2 x, x \geq 0$

6. $f(x) = \dfrac{1}{x^2} + 1, x > 0$

$(f \circ g)(x) = \dfrac{x}{x - 1}, x > 1$

7. $x = \dfrac{1}{2}, -5$ **8.** $x \approx 0.453$

9. $x = 1$ **10.** Any c in $[1, 2)$

Exercises

1. $x = -2$, infinite discontinuity

3. None

5. All points not in the domain, i.e., all $x < -\dfrac{3}{2}$

7. $x = 0$, jump discontinuity

9. $x = 0$, infinite discontinuity

11. (a) Yes   **(b)** Yes
   (c) Yes   **(d)** Yes

13. (a) No   **(b)** No   **15.** 0

17. No, because the right-hand and left-hand limits are not the same at zero.

19. (a) $x = 2$
   (b) Not removable, the one-sided limits are different.

21. (a) $x = 1$
   (b) Not removable, it's an infinite discontinuity.

23. (a) All points not in the domain along with $x = 0, 1$
   (b) $x = 0$ is a removable discontinuity, assign $f(0) = 0$.
     $x = 1$ is not removable, the two-sided limits are different.

25. $y = x - 3$

27. $y = \begin{cases} \dfrac{\sin x}{x}, & x \neq 0 \\ 1, & x = 0 \end{cases}$ **29.** $y = \sqrt{x} + 2$

31. Assume $y = x$, constant functions, and the square root function are continuous.
Use the sum, composite, and quotient theorems.
Domain: $(-2, \infty)$

33. Assume $y = x$ and the absolute value function are continuous.
Use the product, constant multiple, difference, and composite theorems.
Domain: $(-\infty, \infty)$

35. Possible answer

37. Possible answer

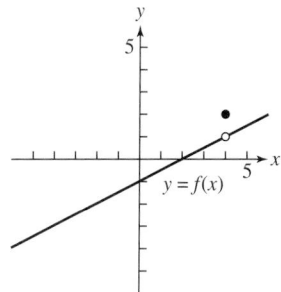

39. $x \approx -0.724$ and $x \approx 1.221$

41. $a = \dfrac{4}{3}$

43. (b)

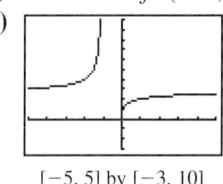

[0, 4.8] by [35000, 45000]

Continuous at all points in the domain $[0, 5)$ except at $t = 1, 2, 3, 4$.

45. (a) Domain of f: $(-\infty, -1) \cup (0, \infty)$
   (b)

[−5, 5] by [−3, 10]

   (c) Because f is undefined there due to division by 0.

45. continued
  (d) $x = 0$: removable, right-hand limit is 0
    $x = -1$: not removable, infinite discontinuity
  (e) 2.718 or e
47. Suppose not. Then f would be negative somewhere in the interval and positive somewhere else in the interval. So, by the Intermediate Value Theorem, it would have to be zero somewhere in the interval, which contradicts the hypothesis.
49. For any real number a, the limit of this function as x approaches a cannot exist. This is because as x approaches a, the values of the function will continually oscillate between 0 and 1.

Section 2.4 (pp. 82–90)

Quick Review

1. $\Delta x = 8, \Delta y = 3$
2. $\Delta x = a - 1, \Delta y = b - 3$
3. Slope $= -\dfrac{4}{7}$ **4.** Slope $= \dfrac{2}{3}$
5. $y = \dfrac{3}{2}x + 6$ **6.** $y = -\dfrac{7}{3}x + \dfrac{25}{3}$
7. $y = -\dfrac{3}{4}x + \dfrac{19}{4}$ **8.** $y = \dfrac{4}{3}x + \dfrac{8}{3}$
9. $y = -\dfrac{2}{3}x + \dfrac{7}{3}$ **10.** $b = \dfrac{19}{3}$

Exercises

1. (a) 19   **(b)** 1
3. (a) $\dfrac{1 - e^{-2}}{2} \approx 0.432$ **(b)** $\dfrac{e^3 - e}{2} \approx 8.684$
5. (a) $-\dfrac{4}{\pi} \approx -1.273$ **(b)** $-\dfrac{3\sqrt{3}}{\pi} \approx -1.654$
7. Using $Q_1 = (10, 225)$, $Q_2 = (14, 375)$, $Q_3 = (16.5, 475)$, $Q_4 = (18, 550)$, and $P = (20, 650)$

  (a)

Secant	Slope
PQ_1	43
PQ_2	46
PQ_3	50
PQ_4	50

  Units are meters/second
  (b) Approximately 50 m/sec
9. (a) -4 **(b)** $y = -4x - 4$
  (c) $y = \dfrac{1}{4}x + \dfrac{9}{2}$

(d)

$[-8, 7]$ by $[-1, 9]$

11. (a) -1 **(b)** $y = -x + 3$
  (c) $y = x - 1$
  (d)

$[-4.7, 4.7]$ by $[-3.1, 3.1]$

13. (a) 1   **(b)** -1
15. No. Slope from the left is -2; slope from the right is 2. The two-sided limit of the difference quotient doesn't exist.
17. Yes. The slope is $-\dfrac{1}{4}$.
19. (a) $2a$
  (b) The slope of the tangent steadily increases as a increases.
21. (a) $-\dfrac{1}{(a - 1)^2}$
  (b) The slope of the tangent is always negative. The tangents are very steep near $x = 1$ and nearly horizontal as a moves away from the origin.
23. 19.6 m/sec   **25.** 6π in²/in.
27. 3.72 m/sec   **29.** $(-2, -5)$
31. (a) At $x = 0$: $y = -x - 1$
    At $x = 2$: $y = -x + 3$
  (b) At $x = 0$: $y = x - 1$
    At $x = 2$: $y = x - 1$
33. (a) 0.3 billion dollars per year
  (b) 0.5 billion dollars per year
  (c) $y = 0.0571x^2 - 0.1514x + 1.3943$

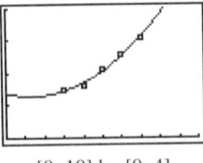
$[0, 10]$ by $[0, 4]$

  (d) 1993 to 1995: 0.31 billion dollars per year
    1995 to 1997: 0.53 billion dollars per year
  (e) 0.65 billion dollars per year
35. (a) $\dfrac{e^{1+h} - e}{h}$
  (b) Limit ≈ 2.718
  (c) They're about the same.
  (d) Yes, it has a tangent whose slope is about e.

37. No **39.** Yes

41. This function has a tangent with slope zero at the origin. It is sandwiched between two functions, $y = x^2$ and $y = -x^2$, both of which have slope zero at the origin.

Looking at the difference quotient,
$$-h \le \frac{f(0 + h) - f(0)}{h} \le h, \text{ so}$$
the Sandwich Theorem tells us that the limit is 0.

43. Slope ≈ 0.540

Chapter Review (pp. 91–93)

1. -15

2. $\dfrac{5}{21}$

3. No limit

4. No limit

5. $-\dfrac{1}{4}$

6. $\dfrac{2}{5}$

7. $+\infty, -\infty$

8. $\dfrac{1}{2}$

9. 2

10. 0

11. 6

12. 5

13. 0

14. 1

15. Limit exists

16. Limit exists

17. Limit exists

18. Doesn't exist

19. Limit exists

20. Limit exists

21. Yes

22. No

23. No

24. Yes

25. (a) 1 **(b)** 1.5 **(c)** No
   (d) g is discontinuous at $x = 3$ (and points not in domain).
   (e) Yes, can remove discontinuity at $x = 3$ by assigning the value 1 to $g(3)$.

26. (a) 1.5 **(b)** 0
   (c) 0 **(d)** No
   (e) k is discontinuous at $x = 1$ (and points not in domain).
   (f) Discontinuity at $x = 1$ is not removable because the two one-sided limits are different.

27. (a) Vertical Asymp.: $x = -2$
   (b) Left-hand limit $= -\infty$
       Right-hand limit $= \infty$

28. (a) Vertical Asymp.: $x = 0$ and $x = -2$
   (b) At $x = 0$:
       Left-hand limit $= -\infty$
       Right-hand limit $= -\infty$
       At $x = -2$:
       Left-hand limit $= \infty$
       Right-hand limit $= -\infty$

29. (a) At $x = -1$:
       Left-hand limit $= 1$
       Right-hand limit $= 1$
       At $x = 0$:
       Left-hand limit $= 0$
       Right-hand limit $= 0$
       At $x = 1$:
       Left-hand limit $= -1$
       Right-hand limit $= 1$
   (b) At $x = -1$:
       Yes, the limit is 1.
       At $x = 0$:
       Yes, the limit is 0.
       At $x = 1$:
       No, the limit doesn't exist because the two one-sided limits are different.
   (c) At $x = -1$:
       Continuous because $f(-1) = $ the limit.
       At $x = 0$:
       Discontinuous because $f(0) \ne$ the limit.
       At $x = 1$:
       Discontinuous because limit doesn't exist.

30. (a) Left-hand limit $= 3$
       Right-hand limit $= -3$
   (b) No, because the two one-sided limits are different.
   (c) Every place except for $x = 1$
   (d) At $x = 1$

31. $x = -2$ and $x = 2$

32. There are no points of discontinuity.

33. (a) $\dfrac{2}{x}$ **(b)** $y = 0$ (x-axis)

34. (a) 2 **(b)** $y = 2$

35. (a) x^2 **(b)** None

36. (a) x **(b)** None

37. (a) e^x **(b)** x

38. (a) $\ln|x|$ **(b)** $\ln|x|$

39. $k = 8$ **40.** $k = \dfrac{1}{2}$

41. One possible answer:

42. One possible answer:

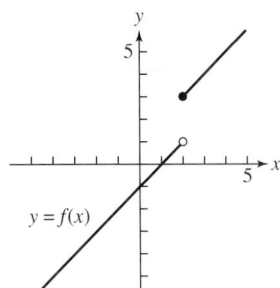

43. $\dfrac{2}{\pi}$

44. $\dfrac{2}{3}\pi aH$

45. $12a$

46. $2a - 1$

47. (a) -1

   (b) $y = -x - 1$

   (c) $y = x - 3$

48. $\left(\dfrac{3}{2}, -\dfrac{9}{4}\right)$

49. (a) 25. Perhaps this is the number of bears placed in the reserve when it was established.

   (b) 200

   (c) Perhaps this is the maximum number of bears which the reserve can support due to limitations of food, space, or other resources. Or, perhaps the number is capped at 200 and excess bears are moved to other locations.

50. (a)

$$f(x) = \begin{cases} 3.20 - 1.35 \cdot \text{int}\,(-x + 1), & 0 < x \le 20 \\ 0, & x = 0 \end{cases}$$

   (b)

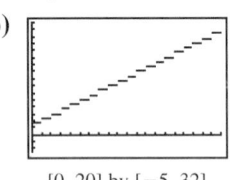

   [0, 20] by [−5, 32]

   f is discontinuous at integer values of x: 0, 1, 2, ..., 19

51. (a) Cubic:
$$y = -1.644x^3 + 42.981x^2 - 254.369x + 300.232$$
   Quartic:
$$y = 2.009x^4 - 102.081x^3 + 1884.997x^2 - 14918.180x + 43004.464$$

   (b) Cubic: $-1.644x^3$, predicts spending will go to 0.
   Quartic: $2.009x^4$, predicts spending will go to ∞.

52. $\displaystyle\lim_{x \to c} f(x) = \dfrac{3}{2}, \lim_{x \to c} g(x) = \dfrac{1}{2}$

53. (a)

   [3, 12] by [−2, 24]

   (b)

Year of Q	Slope of PQ
1995	3.48
1996	3.825
1997	4.1
1998	4.45
1999	4.9

   (c) Approximately 5 billion dollars per year

   (d) $y = 0.3214x^2 - 1.3471x + 1.3857$
   Predicted rate of change in 2000 is 5.081 billion dollars per year.

Chapter 3

Section 3.1 (pp. 95–104)

Quick Review

1. 4

2. $\dfrac{5}{2}$

3. -1

4. 8

5. 0

6. $(-\infty, 0]$ and $[2, \infty)$

7. $\displaystyle\lim_{x \to 1^+} f(x) = 0; \lim_{x \to 1^-} f(x) = 3$

8. 0

9. No, the two one-sided limits are different.

10. No. f is discontinuous at $x = 1$ because the limit doesn't exist there.

Exercises

1. (a) $y = 5x - 7$

   (b) $y = -\dfrac{1}{5}x + \dfrac{17}{5}$

3. $-\dfrac{1}{9}$

5. $\dfrac{dy}{dx} = 7$

7. (b)

9. (d)

11. $\frac{dy}{dx} = 4x - 13$, tangent line is $y = -x - 13$

13. (ii)

15. (a) Sometime around April 1. The rate then is approximately $\frac{1}{6}$ hour per day.

(b) Yes. Jan. 1 and July 1

(c) Positive: Jan. 1. through July 1
Negative: July 1 through Dec. 31

17. (a) 0 and 0

(b) 1700 and 1300

19. Graph of derivative:

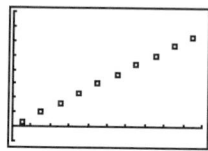

[0, 10] by [−10, 80]

(a) The speed of the skier

(b) Feet per second

(c) Approximately $D = 6.65t$

21.

[−π, π] by [−1.5, 1.5]

Cosine could be the derivative of sine. The values of cosine are positive where sine is increasing, zero where sine has horizontal tangents, and negative where sine is decreasing.

23. $\lim\limits_{h \to 0^+} \frac{f(0 + h) - f(0)}{h} = \lim\limits_{h \to 0^+} \frac{\sqrt{h}}{h} = \lim\limits_{h \to 0^+} \frac{1}{\sqrt{h}} = \infty$

Thus, the right-hand derivative at 0 does not exist.

25.

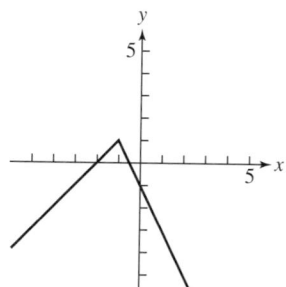

27. The y-intercept is $b - a$.

29. (a) 0.992   **(b)** 0.008

(c) If P is the answer to (b), then the probability of a shared birthday when there are four people is $1 - (1 - P)\frac{362}{365} \approx 0.016$.

(d) No. Clearly, February 29th is a much less likely birth date. Furthermore, census data do not support the assumption that the other 365 birth dates are equally likely. However, this simplifying assumption may still give us some insight into this problem even if the calculated probabilities aren't completely accurate.

Section 3.2 (pp. 105–112)

Quick Review

1. Yes   **2.** No   **3.** Yes

4. Yes   **5.** No   **6.** All reals

7. $[0, \infty)$ **8.** $[3, \infty)$ **9.** 3.2

10. 5

Exercises

1. Left-hand derivative = 0
Right-hand derivative = 1

3. Left-hand derivative = $\frac{1}{2}$
Right-hand derivative = 2

5. (a) All points in $[-3, 2]$
(b) None   **(c)** None

7. (a) All points in $[-3, 3]$ except $x = 0$
(b) None   **(c)** $x = 0$

9. (a) All points in $[-1, 2]$ except $x = 0$
(b) $x = 0$ **(c)** None

11. Discontinuity   **13.** Corner

15. Corner

17. All reals except $x = -1, 5$

19. All reals except $x = 0$

21. All reals except $x = 3$

23. (a) $x = 0$ is not in their domains, or, they are both discontinuous at $x = 0$.

(b) For $\frac{1}{x}$: 1,000,000

For $\frac{1}{x^2}$: 0

(c) It returns an incorrect response because even though these functions are not defined at $x = 0$, they are defined at $x = \pm 0.001$.

25.

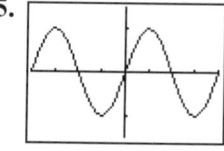

[−2π, 2π] by [−1.5, 1.5]

$\frac{dy}{dx} = \sin x$

27.

$[-2\pi, 2\pi]$ by $[-4, 4]$

$\dfrac{dy}{dx} = \tan x$

29. (a) $a + b = 2$

(b) $a = -3$ and $b = 5$

31. (a) Note that $-x \le x \sin \dfrac{1}{x} \le x$, for all x,

so $\lim\limits_{x \to 0} \left(x \sin \dfrac{1}{x} \right) = 0$ by the Sandwich

Theorem. Therefore, f is continuous at $x = 0$.

(b) $\dfrac{f(0 + h) - f(0)}{h} = \dfrac{h \sin \dfrac{1}{h} - 0}{h} = \sin \dfrac{1}{h}$

(c) The limit does not exist because $\sin \dfrac{1}{h}$ oscillates between -1 and 1 an infinite number of times arbitrarily close to $h = 0$.

(d) No

(e) $\dfrac{g(0 + h) - g(0)}{h} = \dfrac{h^2 \sin \dfrac{1}{h} - 0}{h}$

$= h \sin \dfrac{1}{h}$

As noted in part (a), the limit of this as x approaches zero is 0, so $g'(0) = 0$.

Section 3.3 (pp. 112–121)

Quick Review

1. $x + x^2 - 2x^{-1} - 2$ **2.** $x + x^{-1}$

3. $3x^2 - 2x^{-1} + 5x^{-2}$ **4.** $\dfrac{3}{2}x^2 - x + 2x^{-2}$

5. $x^{-3} + x^{-1} + 2x^{-2} + 2$ **6.** $x^2 + x$

7. Root: $x \approx 1.173$, $500x^6 \approx 1305$

Root: $x \approx 2.394$, $500x^6 \approx 94{,}212$

8. (a) 7   **(b)** 7

    (c) 7   **(d)** 0

9. (a) 0   **(b)** 0

    (c) 0

10. (a) $f'(x) = \dfrac{1}{\pi}$ **(b)** $f'(x) = -\pi x^{-2}$

Exercises

1. $\dfrac{dy}{dx} = -2x$, $\dfrac{d^2y}{dx^2} = -2$

3. $\dfrac{dy}{dx} = 2$, $\dfrac{d^2y}{dx^2} = 0$

5. $\dfrac{dy}{dx} = x^2 + x + 1$, $\dfrac{d^2y}{dx^2} = 2x + 1$

7. $\dfrac{dy}{dx} = 4x^3 - 21x^2 + 4x$, $\dfrac{d^2y}{dx^2} = 12x^2 - 42x + 4$

9. $\dfrac{dy}{dx} = -8x^{-3} - 8$, $\dfrac{d^2y}{dx^2} = 24x^{-4}$

11. (a) $3x^2 + 2x + 1$

    (b) $3x^2 + 2x + 1$

13. $-\dfrac{19}{(3x - 2)^2}$ **15.** $\dfrac{3}{x^4}$

17. $\dfrac{x^4 + 2x}{(1 - x^3)^2}$ **19.** $\dfrac{12 - 6x^2}{(x^2 - 3x + 2)^2}$

21. $\dfrac{d}{dx}(c \cdot f(x)) = c \cdot \dfrac{d}{dx} f(x) + f(x) \cdot \dfrac{d}{dx} c$

$= c \cdot \dfrac{d}{dx} f(x) + 0 = c \cdot \dfrac{d}{dx} f(x)$

23. (a) 13   **(b)** -7

    (c) $\dfrac{7}{25}$ **(d)** 20

25. (iii)   **27.** $y = -\dfrac{1}{9}x + \dfrac{29}{9}$

29. $(-1, 27)$ and $(2, 0)$

31. At $(0, 0)$: $y = 4x$

At $(1, 2)$: $y = 2$

33. $-\dfrac{nRT}{(V - nb)^2} + \dfrac{2an^2}{V^3}$ **35.** $\dfrac{dR}{dM} = CM - M^2$

37. If the radius of a sphere is changed by a very small amount Δr, the change in the volume can be thought of as $(4\pi r^2)(\Delta r)$, which means that the change in the volume divided by the change in the radius is just $4\pi r^2$.

39. It is going down approximately 20 cents per year. (rate ≈ -0.201 dollars/year)

Section 3.4 (pp. 122–133)

Quick Review

1. Downward   **2.** y-intercept $= -256$

3. x-intercepts $= 2, 8$ **4.** $(-\infty, 144]$

5. $(5, 144)$ **6.** $x = 3, 7$

7. $x = \dfrac{15}{8}$ **8.** $(-\infty, 5)$

9. 64   **10.** -32

Exercises

1. $3s^2$

3. (a) $\text{vel}(t) = 24 - 1.6t$, $\text{accel}(t) = -1.6$

    (b) 15 seconds   **(c)** 180 meters

    (d) About 4.393 seconds

    (e) 30 seconds

5. About 29.388 meters

7. For the moon:

$x_1(t) = 3(t < 160) + 3.1(t \geq 160)$

$y_1(t) = 832t - 2.6t^2$

t-values: 0 to 320

window: [0, 6] by [−10,000, 70,000]

For the earth:

$x_1(t) = 3(t < 26) + 3.1(t \geq 26)$

$y_1(t) = 832t - 16t^2$

t-values: 0 to 52

window: [0, 6] by [−1000, 11,000]

9. At the end of 10 minutes: 8000 gallons/minute
Average over first 10 minutes:
10,000 gallons/minute

11. (a)

[0, 50] by [−500, 2200]

The values of *x* which make sense are the whole numbers, $x \geq 0$.

(b) $\dfrac{2000}{(x + 1)^2}$

(c) Approximately $55.56

(d) The limit is 0. This means that as *x* gets large, one reaches a point where very little extra revenue can be expected from selling more desks.

13. At $t = 1$: 0 m/sec; At $t = 2$: 1 m/sec

15. (a)

[0, 200] by [−2, 12]

(b) $x \geq 0$ (whole numbers)

(c)

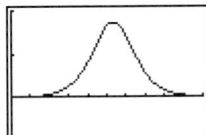

[0, 200] by [−0.1, 0.2]

P seems to be relatively sensitive to changes in *x* between approximately $x = 60$ and $x = 160$.

(d) The maximum occurs when $x \approx 106.44$. Since *x* must be an integer, $P(106) \approx 4.924$ thousand dollars or $4924.

(e) $13 per package sold, $165 per package sold, $118 per package sold, $31 per package sold, $6 per package sold, $P'(300) \approx 0$ (on the order of 10^{-6}, or $0.001 per package sold)

(f) The limit is 10. Maximum possible profit is $10,000 monthly.

(g) Yes. In order to sell more and more packages, the company might need to lower the price to a point where they won't make any additional profit.

17. Possible answer

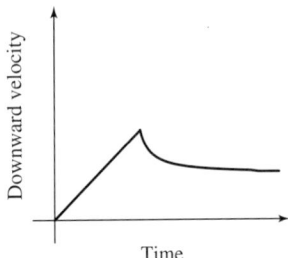

19. At $t \approx 2.83$

21. (a)

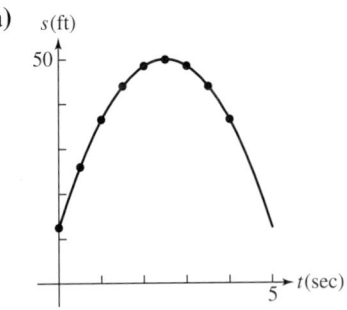

(b) $s'(1) = 18$, $s'(2.5) = 0$, $s'(3.5) = -12$

23. (a) At $t = 2$ and $t = 7$

(b) Between $t = 3$ and $t = 6$

(c)

(d)

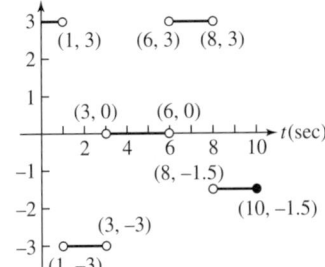

25. (a) Move forward: $0 \le t < 1$ and $5 < t < 7$
move backward: $1 < t < 5$
speed up: $1 < t < 2$ and $5 < t < 6$
slow down: $0 \le t < 1$, $3 < t < 5$,
and $6 < t < 7$

(b) Positive: $3 < t < 6$
negative: $0 \le t < 2$ and $6 < t < 7$
zero: $2 < t < 3$ and $7 < t \le 9$

(c) At $t = 0$ and $2 < t < 3$

(d) $7 < t \le 9$

27. (a) $\dfrac{4}{7}$ of a second.
Average velocity = 280 cm/sec

(b) Velocity = 560 cm/sec;
acceleration = 980 cm/sec^2

(c) About 28 flashes per second

29. Graph C is position, graph B is velocity, and
graph A is acceleration.
B is the derivative of C because it is negative and
zero where C is decreasing and has horizontal
tangents, respectively.
A is the derivative of B because it is positive,
negative, and zero where B is increasing,
decreasing, and has horizontal tangents,
respectively.

31. (a) 16π cubic feet of volume per foot of radius
(b) By about 11.092 cubic feet

33. Exit velocity ≈ 348.712 ft/sec ≈ 237.758 mi/h

35. Since profit = revenue − cost, using Rule 4 (the
"difference rule"), and taking derivatives, we see
that marginal profit
= marginal revenue − marginal cost.

37. (a) Assume that f is even. Then,

$$f'(-x) = \lim_{h \to 0} \frac{f(-x + h) - f(-x)}{h}$$

$$= \lim_{h \to 0} \frac{f(x - h) - f(x)}{h},$$

and substituting $k = -h$,

$$= \lim_{k \to 0} \frac{f(x + k) - f(x)}{-k}$$

$$= -\lim_{k \to 0} \frac{f(x + k) - f(x)}{k} = -f'(x)$$

So, f' is an odd function.

(b) Assume that f is odd. Then,

$$f'(-x) = \lim_{h \to 0} \frac{f(-x + h) - f(-x)}{h}$$

$$= \lim_{h \to 0} \frac{-f(x - h) + f(x)}{h}, \text{ and}$$

substituting $k = -h$,

$$= \lim_{k \to 0} \frac{-f(x + k) + f(x)}{-k}$$

$$= \lim_{k \to 0} \frac{f(x + k) - f(x)}{k} = f'(x)$$

So, f' is an even function.

Section 3.5 (pp. 134–141)

Quick Review

1. $\dfrac{3\pi}{4} \approx 2.356$ **2.** $\left(\dfrac{306}{\pi}\right)^{\circ} \approx 97.403^{\circ}$

3. $\dfrac{\sqrt{3}}{2}$

4. Domain: all reals; range: $[-1, 1]$

5. Domain: $x \ne k\dfrac{\pi}{2}$, where k is an odd integer;
range: all reals

6. 0 **7.** $\pm\dfrac{1}{\sqrt{2}}$

8. Multiply by $\dfrac{1 + \cos h}{1 + \cos h}$ and use
$1 - \cos^2 h = \sin^2 h$.

9. $y = 12x - 35$ **10.** 12

Exercises

1. $1 + \sin x$ **3.** $-\dfrac{1}{x^2} + 5 \cos x$

5. $-x^2 \cos x - 2x \sin x$ **7.** $4 \sec x \tan x$

9. $-\dfrac{\csc^2 x}{(1 + \cot x)^2} = -\dfrac{1}{(\sin x + \cos x)^2}$

11. $y = -x + \pi + 3$

13. Approximately $y = -8.063x + 25.460$

15. (a) $\dfrac{d}{dx} \tan x = \dfrac{d}{dx} \dfrac{\sin x}{\cos x}$

$$= \dfrac{(\cos x)(\cos x) - (\sin x)(-\sin x)}{}$$

$$= \dfrac{\cos^2 x + \sin^2 x}{\cos^2 x}$$

$$= \dfrac{1}{\cos^2 x} = \sec^2 x$$

(b) $\dfrac{d}{dx}\sec x = \dfrac{d}{dx}\dfrac{1}{\cos x}$

$\quad = \dfrac{(\cos x)(0) - (1)(-\sin x)}{(\cos x)^2}$

$\quad = \dfrac{\sin x}{(\cos x)^2} = \sec x \tan x$

17. $\dfrac{d}{dx}\sec x = \sec x \tan x$ which is 0 at $x = 0$, so the slope of the tangent line is 0.

$\dfrac{d}{dx}\cos x = -\sin x$ which is 0 at $x = 0$, so the slope of the tangent line is 0.

19. Tangent: $y = -x + \dfrac{\pi}{4} + 1$

Normal: $y = x + 1 - \dfrac{\pi}{4}$

21. (a) $y = -x + \dfrac{\pi}{2} + 2$

(b) $y = 4 - \sqrt{3}$

23. (a) Velocity: $-2\cos t$ m/sec
Speed: $|2\cos t|$ m/sec
Accel.: $2\sin t$ m/sec^2
Jerk: $2\cos t$ m/sec^3

(b) Velocity: $-\sqrt{2}$ m/sec
Speed: $\sqrt{2}$ m/sec
Accel.: $\sqrt{2}$ m/sec^2
Jerk: $\sqrt{2}$ m/sec^3

(c) The body starts at 2, goes to 0 and then oscillates between 0 and 4.
Speed: *Greatest* when $\cos t = \pm 1$ (or $t = k\pi$), at the center of the interval of motion.
Zero when $\cos t = 0$ $\left(\text{or } t = \dfrac{k\pi}{2}, k \text{ odd}\right)$, at the endpoints of the interval of motion.

Acceleration: *Greatest* (in magnitude) when $\sin t = \pm 1$ $\left(\text{or } t = \dfrac{k\pi}{2}, k \text{ odd}\right)$
Zero when $\sin t = 0$ (or $t = k\pi$)

Jerk: *Greatest* (in magnitude) when $\cos t = \pm 1$ (or $t = k\pi$)
Zero when $\cos t = 0$ $\left(\text{or } t = \dfrac{k\pi}{2}, k \text{ odd}\right)$

25. (a) The limit is $\dfrac{\pi}{180}$ because this is the conversion factor for changing from degrees to radians.

(b) This limit is still 0.

(c) $\dfrac{d}{dx}\sin x = \dfrac{\pi}{180}\cos x$

(d) $\dfrac{d}{dx}\cos x = -\dfrac{\pi}{180}\sin x$

(e) $\dfrac{d^2}{dx^2}\sin x = -\dfrac{\pi^2}{180^2}\sin x$

$\dfrac{d^3}{dx^3}\sin x = -\dfrac{\pi^3}{180^3}\cos x$

$\dfrac{d^2}{dx^2}\cos x = -\dfrac{\pi^2}{180^2}\cos x$

$\dfrac{d^3}{dx^3}\cos x = \dfrac{\pi^3}{180^3}\sin x$

27. $y'' = \dfrac{2 + 2\theta \tan \theta}{\cos^2 \theta} = \dfrac{2\cos \theta + 2\theta \sin \theta}{\cos^3 \theta}$

29. $\sin x$ **31.** $y = x$

33. $\dfrac{d}{dx}\sin 2x = \dfrac{d}{dx}2\sin x \cos x$

$\quad = 2[(\sin x)(-\sin x) + (\cos x)(\cos x)]$

$\quad = 2[\cos^2 x - \sin^2 x] = 2\cos 2x$

35. $\lim\limits_{h\to 0}\dfrac{\cos h - 1}{h} = \lim\limits_{h\to 0}\dfrac{(\cos h - 1)(\cos h + 1)}{h}$

$\quad = \lim\limits_{h\to 0}\dfrac{\cos^2 h - 1}{h(\cos h + 1)}$

$\quad = \lim\limits_{h\to 0}\dfrac{-\sin^2 h}{h(\cos h + 1)}$

$\quad = -\left(\lim\limits_{h\to 0}\dfrac{\sin h}{h}\right)\left(\lim\limits_{h\to 0}\dfrac{\sin h}{\cos h + 1}\right)$

$\quad = -(1)\left(\dfrac{0}{2}\right) = 0$

Section 3.6 (pp. 141–149)

Quick Review

1. $\sin(x^2 + 1)$ **2.** $\sin(49x^2 + 1)$
3. $49x^2 + 1$ **4.** $7x^2 + 7$
5. $\sin\dfrac{x^2 + 1}{7x}$ **6.** $g(f(x))$
7. $g(h(f(x)))$ **8.** $h(g(f(x)))$
9. $f(h(h(x)))$ **10.** $f(g(h(x)))$

Exercises

1. $3\cos(3x + 1)$ **3.** $-\sqrt{3}\sin(\sqrt{3}x)$
5. $\dfrac{10}{x^2}\csc^2\left(\dfrac{2}{x}\right)$ **7.** $-\sin(\sin x)\cos x$
9. $-2(x + \sqrt{x})^{-3}\left(1 + \dfrac{1}{2\sqrt{x}}\right)$
11. $-5\sin^{-6}x\cos x + 3\cos^2 x \sin x$
13. $4\sin^3 x\sec^2 4x + 3\sin^2 x\cos x\tan 4x$
15. $-3(2x + 1)^{-3/2}$
17. $6\sin(3x - 2)\cos(3x - 2) = 3\sin(6x - 4)$

19. $-42(1 + \cos^2 7x)^2 \cos 7x \sin 7x$

21. $3 \sin\left(\dfrac{\pi}{2} - 3t\right)$

23. $\dfrac{4}{\pi} \cos 3t - \dfrac{4}{\pi} \sin 5t$

25. $-\sec^2 (2 - \theta)$

27. $\dfrac{\theta \cos \theta + \sin \theta}{2\sqrt{\theta} \sin \theta}$

29. $2 \sec^2 x \tan x$

31. $18 \csc^2 (3x - 1) \cot (3x - 1)$

33. $\dfrac{5}{2}$ **35.** $-\dfrac{\pi}{4}$ **37.** 0

39. (a) $-6 \sin (6x + 2)$ **(b)** $-6 \sin (6x + 2)$

41. $y = -x + 2\sqrt{2}$ **43.** $y = -\dfrac{1}{2}x - \dfrac{1}{2}$

45. $y = x + \dfrac{1}{4}$

47. $y = \sqrt{3}x + 2 - \dfrac{\pi}{\sqrt{3}}$

49. (a) $\dfrac{\cos t}{2t + 1}$

(b) $\dfrac{d}{dt}\left(\dfrac{dy}{dx}\right) = -\dfrac{(2t + 1)(\sin t) + 2 \cos t}{}$

(c) $\dfrac{d}{dx}\left(\dfrac{dy}{dx}\right) = -\dfrac{(2t + 1)(\sin t) + 2 \cos t}{(2t + 1)^3}$

(d) part (c)

51. 5   **53.** $\dfrac{1}{2}$

55. Tangent: $y = \pi x - \pi + 2$;

Normal: $y = -\dfrac{1}{\pi}x + \dfrac{1}{\pi} + 2$

57. (a) 1   **(b)** 6   **(c)** 1

(d) $-\dfrac{1}{9}$ **(e)** $-\dfrac{40}{3}$ **(f)** -6

(g) $-\dfrac{4}{9}$

59. Because the symbols $\dfrac{dy}{dx}$, $\dfrac{dy}{du}$, and $\dfrac{du}{dx}$ are not fractions. The individual symbols dy, du, and dx do not have numerical values.

61. (a) On the 101st day (April 11th)
(b) About 0.637 degrees per day

63. Acceleration $= \dfrac{dv}{dt} = \dfrac{dv}{ds}\dfrac{ds}{dt} = \dfrac{dv}{ds}v$

$= \dfrac{k}{2\sqrt{s}} (k\sqrt{s}) = \dfrac{k^2}{2}$

65. Acceleration $= \dfrac{dv}{dt} = \dfrac{d f(x)}{dt}$

$= \left[\dfrac{d f(x)}{dx}\right]\left[\dfrac{dx}{dt}\right]$

$= f'(x)f(x)$

67. No, this does not contradict the Chain Rule. The Chain Rule states that if two functions are differentiable at the appropriate points, then their composite must also be differentiable.

69. As $h \to 0$, the second curve (the difference quotient) approaches the first $y = 2 \cos 2x$. This is because $2 \cos 2x$ is the derivative of $\sin 2x$, and the second curve is the difference quotient used to define the derivative of $\sin 2x$. As $h \to 0$, the difference quotient expression should be approaching the derivative.

71. (a) Let $f(x) = |x|$.

Then $\dfrac{d}{dx}|u| = \dfrac{d}{dx} f(u) = f'(u)\dfrac{du}{dx}$

$= f'(u)u' = \dfrac{u}{|u|} u'.$

The derivative of the absolute value function is $+1$ for positive values, -1 for negative values, and undefined at 0. So $f'(u)$ should be $+1$ when $u > 0$ and -1 when $u < 0$. But this is exactly how the expression $\dfrac{u}{|u|}$ evaluates.

(b) $f'(x) = \dfrac{(2x)(x^2 - 9)}{|x^2 - 9|}$

$g'(x) = |x| \cos x + \dfrac{x \sin x}{|x|}$

Section 3.7 (pp. 149–157)

Quick Review

1. $y_1 = \sqrt{x}$, $y_2 = -\sqrt{x}$

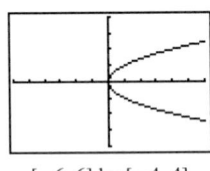

$[-6, 6]$ by $[-4, 4]$

2. $y_1 = \dfrac{2}{3}\sqrt{9 - x^2}$, $y_2 = -\dfrac{2}{3}\sqrt{9 - x^2}$

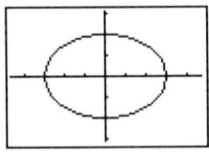

$[-4.7, 4.7]$ by $[-3.1, 3.1]$

3. $y_1 = \dfrac{x}{2}$, $y_2 = -\dfrac{x}{2}$

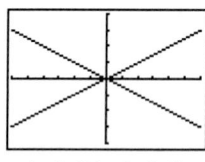

$[-6, 6]$ by $[-4, 4]$

4. $y_1 = \sqrt{9 - x^2}$, $y_2 = -\sqrt{9 - x^2}$

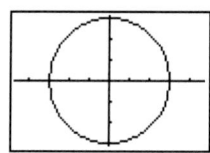

$[-4.7, 4.7]$ by $[-3.1, 3.1]$

5. $y_1 = \sqrt{2x + 3 - x^2}$, $y_2 = -\sqrt{2x + 3 - x^2}$

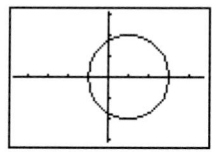

$[-4.7, 4.7]$ by $[-3.1, 3.1]$

6. $y' = \dfrac{4x - y + 2xy}{x^2}$

7. $y' = \dfrac{y + x \cos x}{\sin x - x}$

8. $y' = \dfrac{xy^2}{x^2 - y + x}$

9. $x^{3/2} - x^{5/6}$

10. $x^{-1/2} + x^{-5/6}$

Exercises

1. $\dfrac{9}{4}x^{5/4}$

3. $\dfrac{1}{3}x^{-2/3}$

5. $-(2x + 5)^{-3/2}$

7. $x^2(x^2 + 1)^{-1/2} + (x^2 + 1)^{1/2}$

9. $-\dfrac{2xy + y^2}{2xy + x^2}$

11. $\dfrac{1}{y(x + 1)^2}$

13. $-\dfrac{1}{4}(1 - x^{1/2})^{-1/2}x^{-1/2}$ **15.** $-\dfrac{9}{2}(\csc x)^{3/2} \cot x$

17. $\cos^2 y$

19. $-\dfrac{1}{x}\cos^2(xy) - \dfrac{y}{x}$

21. (b), (c), and (d)

23. $\dfrac{dy}{dx} = -\dfrac{x}{y}$

$\dfrac{d^2y}{dx^2} = -\dfrac{(x^2 + y^2)}{y^3} = -\dfrac{1}{y^3}$

25. $\dfrac{dy}{dx} = \dfrac{x + 1}{y}$

$\dfrac{d^2y}{dx^2} = \dfrac{y^2 - (x + 1)^2}{y^3} = -\dfrac{1}{y^3}$

27. (a) $y = \dfrac{7}{4}x - \dfrac{1}{2}$ **(b)** $y = -\dfrac{4}{7}x + \dfrac{29}{7}$

29. (a) $y = 3x + 6$ **(b)** $y = -\dfrac{1}{3}x + \dfrac{8}{3}$

31. (a) $y = \dfrac{6}{7}x + \dfrac{6}{7}$ **(b)** $y = -\dfrac{7}{6}x - \dfrac{7}{6}$

33. (a) $y = -\dfrac{\pi}{2}x + \pi$ **(b)** $y = \dfrac{2}{\pi}x - \dfrac{2}{\pi} + \dfrac{\pi}{2}$

35. (a) $y = 2\pi x - 2\pi$ **(b)** $y = -\dfrac{x}{2\pi} + \dfrac{1}{2\pi}$

37. (a) At $\left(\dfrac{\sqrt{3}}{4}, \dfrac{\sqrt{3}}{2}\right)$: Slope $= -1$;

at $\left(\dfrac{\sqrt{3}}{4}, \dfrac{1}{2}\right)$: Slope $= \sqrt{3}$

(b)

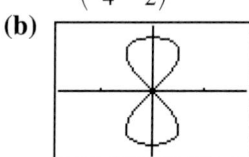

$[-1.8, 1.8]$ by $[-1.2, 1.2]$

Parameter interval:
$-1 \le t \le 1$

39. (a) $(-1)^3(1)^2 = \cos(\pi)$ is true since both sides equal: -1.

(b) The slope is $\dfrac{3}{2}$.

41. The points are $(\pm\sqrt{7}, 0)$.

$\dfrac{dy}{dx} = -\dfrac{2x + y}{2y + x}$

At both points, $\dfrac{dy}{dx} = -2$

43. First curve: $\dfrac{dy}{dx} = -\dfrac{2x}{3y}$

second curve: $\dfrac{dy}{dx} = \dfrac{3x^2}{2y}$

At $(1, 1)$, the slopes are $-\dfrac{2}{3}$ and $\dfrac{3}{2}$ respectively.

At $(1, -1)$, the slopes are $\dfrac{2}{3}$ and $-\dfrac{3}{2}$ respectively.

In both cases, the tangents are perpendicular.

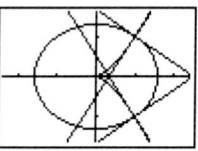

$[-2.4, 2.4]$ by $[-1.6, 1.6]$

45. Acceleration $= \dfrac{dv}{dt} = 4(s - t)^{-1/2}(v - 1)$
$= 32$ ft/sec^2

47. (a) At $(4, 2)$: $\dfrac{5}{4}$;

at $(2, 4)$: $\dfrac{4}{5}$

(b) At $(3\sqrt[3]{2}, 3\sqrt[3]{4}) \approx (3.780, 4.762)$

(c) At $(3\sqrt[3]{4}, 3\sqrt[3]{2}) \approx (4.762, 3.780)$

49. At $(-1, -1)$: $y = -2x - 3$;

at $(3, -3)$: $y = -2x + 3$

51. (a) $\dfrac{dy}{dx} = -\dfrac{b^2 x}{a^2 y}$

The tangent line is $y - y_1 = -\dfrac{b^2 x_1}{a^2 y_1}(x - x_1)$.

This gives: $a^2 y_1 y - a^2 y_1{}^2 = -b^2 x_1 x + b^2 x_1{}^2$,

$a^2 y_1 y + b^2 x_1 x = a^2 y_1{}^2 + b^2 x_1{}^2$.

But $a^2 y_1{}^2 + b^2 x_1{}^2 = a^2 b^2$ since (x_1, y_1) is on the ellipse.

Therefore, $a^2 y_1 y + b^2 x_1 x = a^2 b^2$, and

dividing by $a^2 b^2$ gives $\dfrac{x_1 x}{a^2} + \dfrac{y_1 y}{b^2} = 1$.

(b) $\dfrac{x_1 x}{a^2} - \dfrac{y_1 y}{b^2} = 1$.

Section 3.8 (pp. 157–163)

Quick Review

1. Domain: $[-1, 1]$

Range: $\left[-\dfrac{\pi}{2}, \dfrac{\pi}{2}\right]$

At 1: $\dfrac{\pi}{2}$

2. Domain: $[-1, 1]$

Range: $[0, \pi]$

At 1: 0

3. Domain: all reals

Range: $\left(-\dfrac{\pi}{2}, \dfrac{\pi}{2}\right)$

At 1: $\dfrac{\pi}{4}$

4. Domain: $(-\infty, -1] \cup [1, \infty)$

Range: $\left[0, \dfrac{\pi}{2}\right) \cup \left(\dfrac{\pi}{2}, \pi\right]$

At 1: 0

5. Domain: all reals

Range: all reals

At 1: 1

6. $f^{-1}(x) = \dfrac{x + 8}{3}$

7. $f^{-1}(x) = x^3 - 5$

8. $f^{-1}(x) = \dfrac{8}{x}$

9. $f^{-1}(x) = \dfrac{2}{3 - x}$

10. $f^{-1}(x) = 3 \tan x$, $-\dfrac{\pi}{2} < x < \dfrac{\pi}{2}$

Exercises

1. $-\dfrac{2x}{\sqrt{1 - x^4}}$

3. $\dfrac{\sqrt{2}}{\sqrt{1 - 2t^2}}$

5. $\dfrac{1}{|2s + 1|\sqrt{s^2 + s}}$

7. $-\dfrac{2}{(x^2 + 1)\sqrt{x^2 + 2}}$

9. $-\dfrac{1}{\sqrt{1 - t^2}}$

11. $-\dfrac{1}{2\sqrt{t}(t + 1)}$

13. $-\dfrac{2s^2}{\sqrt{1 - s^2}}$

15. $0, x > 1$

17. $\sin^{-1} x$

19. (a) $y = 2x - \dfrac{\pi}{2} + 1$ **(b)** $y = \dfrac{1}{2}x - \dfrac{1}{2} + \dfrac{\pi}{4}$

21. (a) $f'(x) = 3 - \sin x$ and $f'(x) \neq 0$. So f has a differentiable inverse by Theorem 3.

(b) $f(0) = 1, f'(0) = 3$

(c) $f^{-1}(1) = 0, (f^{-1})'(1) = \dfrac{1}{3}$

23. (a) $v(t) = \dfrac{dx}{dt} = \dfrac{1}{1 + t^2}$ which is always positive.

(b) $a(t) = \dfrac{dv}{dt} = -\dfrac{2t}{(1 + t^2)^2}$ which is always negative.

(c) $\dfrac{\pi}{2}$

25. $\dfrac{d}{dx} \cot^{-1} x = \dfrac{d}{dx}\left(\dfrac{\pi}{2} - \tan^{-1} x\right)$

$= 0 - \dfrac{d}{dx} \tan^{-1} x$

$= -\dfrac{1}{1 + x^2}$

27. (a) $y = \dfrac{\pi}{2}$ **(b)** $y = -\dfrac{\pi}{2}$

(c) None

29. (a) $y = \dfrac{\pi}{2}$ **(b)** $y = \dfrac{\pi}{2}$

(c) None

31. (a) None   **(b)** None

(c) None

33. (a)

$\alpha = \cos^{-1} x, \beta = \sin^{-1} x$

So $\dfrac{\pi}{2} = \alpha + \beta = \cos^{-1} x + \sin^{-1} x$.

(b)

$$\alpha = \tan^{-1} x, \beta = \cot^{-1} x$$

So $\dfrac{\pi}{2} = \alpha + \beta = \tan^{-1} x + \cot^{-1} x.$

(c)

$$\alpha = \sec^{-1} x, \beta = \csc^{-1} x$$

So $\dfrac{\pi}{2} = \alpha + \beta = \sec^{-1} x + \csc^{-1} x.$

35. If s is the length of a side of the square, and let α,

β, γ denote the angles labeled $\tan^{-1} 1$, $\tan^{-1} 2$,

and $\tan^{-1} 3$, respectively.

$\tan \alpha = \dfrac{s}{s} = 1$, so $\alpha = \tan^{-1} 1$ and

$\tan \beta = \dfrac{s}{\frac{s}{2}} = 2$, so $\beta = \tan^{-1} 2.$

$$\gamma = \pi - \alpha - \beta = \pi - \tan^{-1} 1 - \tan^{-1} 2$$
$$= \tan^{-1} 3.$$

Section 3.9 (pp. 163–171)

Quick Review

1. $\dfrac{\ln 8}{\ln 5}$ **2.** $e^{x \ln 7}$

3. $\tan x$ **4.** $\ln (x - 2)$

5. $3x - 15$ **6.** $\dfrac{5}{4}$

7. $\ln (4x^4)$ **8.** $x = \dfrac{\ln 19}{\ln 3} \approx 2.68$

9. $x = \dfrac{\ln 18 - \ln (\ln 5)}{\ln 5} \approx 1.50$

10. $x = \dfrac{\ln 3}{\ln 2 - \ln 3} \approx -2.71$

Exercises

1. $2e^x$ **3.** $-e^{-x}$

5. $\dfrac{2}{3} e^{2x/3}$ **7.** $e^2 - e^x$

9. $\dfrac{e^{\sqrt{x}}}{2\sqrt{x}}$ **11.** $\pi x^{\pi - 1}$

13. $-\sqrt{2} x^{-\sqrt{2} - 1}$ **15.** $8^x \ln 8$

17. $-3^{\csc x} (\ln 3)(\csc x \cot x)$

19. $\dfrac{2 x^{\ln x} \ln x}{x}$ **21.** $\dfrac{2}{x}$

23. $-\dfrac{1}{x}, x > 0$ **25.** $\dfrac{1}{x + 2}, x > -2$

27. $\dfrac{\sin x}{2 - \cos x}$ **29.** $\dfrac{1}{x \ln x}$

31. $\dfrac{2}{x \ln 4} = \dfrac{1}{x \ln 2}$ **33.** $\dfrac{3}{(3x + 1) \ln 2}, x > -\dfrac{1}{3}$

35. $-\dfrac{1}{x \ln 2}, x > 0$ **37.** $\dfrac{1}{x}, x > 0$

39. $\dfrac{1}{\ln 10}$ **41.** $y = ex$

43. $(\sin x)^x [x \cot x + \ln (\sin x)]$

45.
$$\left(\dfrac{(x - 3)^4 (x^2 + 1)}{(2x + 5)^3} \right)^{1/5} \left(\dfrac{4}{5(x - 3)} + \dfrac{2x}{5(x^2 + 1)} - \dfrac{6}{5(2x + 5)} \right)$$

47. rate ≈ 0.098 grams/day

49. (a) $\ln 2$ **(b)** $f'(0) = \lim\limits_{h \to 0} \dfrac{2^h - 1}{h}$

       (c) $\ln 2$ **(d)** $\ln 7$

51. (a) The graph of y_4 is a horizontal line at $y = a$.

       (b) The graph of y_3 is a horizontal line at $y = \ln a$.

       (c) $\dfrac{d}{dx} a^x = a^x$ if and only if $y_3 = \dfrac{y_2}{y_1} = 1$.

          So if $y_3 = \ln a$, then $\dfrac{d}{dx} a^x$ will equal a^x if and only if $\ln a = 1$, or $a = e$.

       (d) $y_2 = \dfrac{d}{dx} a^x = a^x \ln a$. This will equal $y_1 = a^x$ if and only if $\ln a = 1$, or $a = e$.

53. (a) $y = \dfrac{1}{e} x$

       (b) Because the graph of $\ln x$ lies below the graph of the line for all positive $x \neq e$.

       (c) Multiplying by e, $e(\ln x) < x$, or $\ln x^e < x$.

       (d) Exponentiate both sides of the inequality in (c).

       (e) Let $x = \pi$ to see that $\pi^e < e^\pi$.

Chapter Review (pp. 172–175)

1. $5x^4 - \dfrac{x}{4} + \dfrac{1}{4}$ **2.** $-21x^2 + 21x^6$

3. $-2 \cos^2 x + 2 \sin^2 x = 2 \cos 2x$

4. $-\dfrac{4}{(2x - 1)^2}$ **5.** $2 \sin (1 - 2t)$

6. $\dfrac{2}{t^2} \csc^2 \dfrac{2}{t}$ **7.** $\dfrac{1}{2\sqrt{x}} - \dfrac{1}{2x^{3/2}}$

8. $\dfrac{3x + 1}{\sqrt{2x + 1}}$

9. $3 \sec (1 + 3\theta) \tan (1 + 3\theta)$

10. $-4\theta \tan (3 - \theta^2) \sec^2 (3 - \theta^2)$

11. $-5x^2 \csc 5x \cot 5x + 2x \csc 5x$

12. $\dfrac{1}{2x}$, $x > 0$

13. $\dfrac{e^x}{1 + e^x}$

14. $-xe^{-x} + e^{-x}$

15. e

16. $\cot x$, where x is in an interval of the form $(k\pi, (k+1)\pi)$, k even

17. $-\dfrac{1}{\cos^{-1} x \sqrt{1 - x^2}}$

18. $\dfrac{2}{\theta \ln 2}$

19. $\dfrac{1}{(t - 7) \ln 5}$, $t > 7$

20. $-8^{-t} \ln 8$

21. $\dfrac{2 (\ln x) x^{\ln x}}{x}$

22. $\dfrac{(2 \cdot 2^x)[x^3 \ln 2 + x \ln 2 + 1]}{(x^2 + 1)^{3/2}}$ or

$\dfrac{(2x)2^x}{\sqrt{x^2 + 1}}\left(\dfrac{1}{x} + \ln 2 + \dfrac{x}{x^2 + 1}\right)$

23. $\dfrac{e^{\tan^{-1}}}{1 + x^2}$

24. $-\dfrac{u}{\sqrt{u^2 - u^4}} = -\dfrac{u}{|u|\sqrt{1 - u^2}}$

25. $\dfrac{t}{|t|\sqrt{t^2 - 1}} + \sec^{-1} t - \dfrac{1}{2t}$

26. $-\dfrac{2 + 2t^2}{1 + 4t^2} + 2t \cot^{-1} 2t$

27. $\cos^{-1} z$

28. $-\dfrac{1}{x} + \dfrac{\csc^{-1}\sqrt{x}}{\sqrt{x - 1}}$

29. $-\dfrac{\sin x}{|\sin x|} = -\text{sign} (\sin x)$, $x \neq \dfrac{\pi}{2}, \pi, \dfrac{3\pi}{2}$;

or $\begin{cases} -1, & 0 \leq x < \pi, & x \neq \dfrac{\pi}{2} \\ 1, & \pi < x \leq 2\pi & x \neq \dfrac{3\pi}{2} \end{cases}$

30. $2\left(\dfrac{1 + \sin \theta}{1 - \cos \theta}\right)\left(\dfrac{\cos \theta - \sin \theta - 1}{(1 - \cos \theta)^2}\right)$

31. For all $x \neq 0$

32. For all real x

33. For all $x < 1$

34. For all $x \neq \dfrac{7}{2}$

35. $-\dfrac{y + 2}{x + 3}$

36. $-\dfrac{1}{3}(xy)^{-1/5}$

37. $-\dfrac{y}{x}$ or $-\dfrac{1}{x^2}$

38. $\dfrac{1}{2y(x + 1)^2}$

39. $-\dfrac{2x}{y^5}$

40. $-\dfrac{1 + 2xy^2}{x^4 y^3}$

41. $-2\dfrac{(3y^2 + 1)^2 \cos x + 12y \sin^2 x}{(3y^2 + 1)^3}$

42. $\dfrac{2}{3}x^{-4/3}y^{1/3} + \dfrac{2}{3}x^{-5/3}y^{2/3}$

43. $y' = 2x^3 - 3x - 1$,
$y'' = 6x^2 - 3$,
$y''' = 12x$,
$y^{(4)} = 12$, and the rest are all zero.

44. $y' = \dfrac{x^4}{24}$,

$y'' = \dfrac{x^3}{6}$,

$y''' = \dfrac{x^2}{2}$,

$y^{(4)} = x$,

$y^{(5)} = 1$, and the rest are all zero.

45. (a) $y = \dfrac{2}{\sqrt{3}}x - \sqrt{3}$

(b) $y = -\dfrac{\sqrt{3}}{2}x + \dfrac{5\sqrt{3}}{2}$

46. (a) $y = -x + \dfrac{\pi}{2} + 2$

(b) $y = x - \dfrac{\pi}{2} + 2$

47. (a) $y = -\dfrac{1}{4}x + \dfrac{9}{4}$ **(b)** $y = 4x - 2$

48. (a) $y = -\dfrac{5}{4}x + 6$ **(b)** $y = \dfrac{4}{5}x - \dfrac{11}{5}$

49. $y = x - 2\sqrt{2}$ **50.** $y = \dfrac{4}{3}x + 4\sqrt{2}$

51. $y = \dfrac{10}{3}x - 5\sqrt{3}$

52. $y = (1 + \sqrt{2})x - \sqrt{2} - 1 - \dfrac{\pi}{4}$

or $y \approx 2.414x - 3.200$

53. (a)

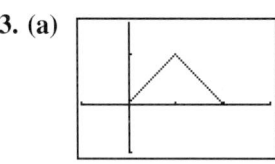

[−1, 3] by [−1, 5/3]

(b) Yes, because both of the one-sided limits as $x \to 1$ are equal to $f(1) = 1$.

(c) No, because the left-hand derivative at $x = 1$ is $+1$ and the right-hand derivative at $x = 1$ is -1.

54. (a) The function is continuous for all values of m, because the right-hand limit as $x \to 0$ is equal to $f(0) = 0$ for any value of m.

(b) The left-hand derivative at $x = 0$ is 2, and the right-hand derivative at $x = 0$ is m, so in order for the function to be differentiable at $x = 0$, m must be 2.

55. (a) For all $x \neq 0$ **(b)** At $x = 0$
(c) Nowhere

56. (a) For all x **(b)** Nowhere
(c) Nowhere

57. (a) $[-1, 0) \cup (0, 4]$ **(b)** At $x = 0$
(c) Nowhere in its domain

58. (a) $[-2, 0) \cup (0, 2]$ **(b)** Nowhere
   (c) Nowhere in its domain

59.

60.

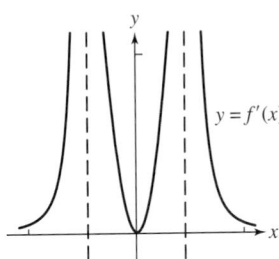

61. (a) iii   **(b)** i
   (c) ii

62.

63.

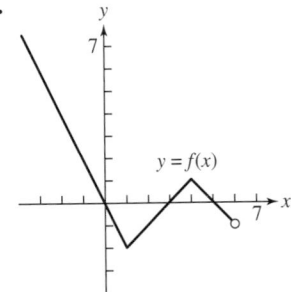

64. Answer is **D**: **i** and **iii** only could be true

65. (a)

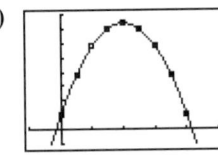

$[-1, 5]$ by $[-10, 80]$

(b)

t interval	avg. vel.
[0, 0.5]	56
[0.5, 1]	40
[1, 1.5]	24
[1.5, 2]	8
[2, 2.5]	−8
[2.5, 3]	−24
[3, 3.5]	−40
[3.5, 4]	−56

(c)

$[-1, 5]$ by $[-80, 80]$

(d) Average velocity is a good approximation to velocity.

66. (a) $-\dfrac{13}{10}$ **(b)** $-\dfrac{1}{3}$ **(c)** $\dfrac{1}{10}$

   (d) -1 **(e)** $-\dfrac{2}{3}$ **(f)** -12

67. (a) 5   **(b)** 0   **(c)** 8
   (d) 2   **(e)** 6   **(f)** -1

68. $\sqrt{3}$ **69.** $-\dfrac{1}{6}$

70. (a) One possible answer:
   $$x(t) = 10 \cos\left(t + \frac{\pi}{4}\right)$$
   $$y(t) = 1$$
   (b) $5\sqrt{2}$ **(c)** $s = -10$ and $s = 10$
   (d) At $t = \dfrac{\pi}{4}$:
   velocity $= -10$
   speed $= 10$
   acceleration $= 0$

71. (a) $\dfrac{ds}{dt} = 64 - 32t$
   $$\dfrac{d^2s}{dt^2} = -32$$
   (b) 2 sec   **(c)** 64 ft/sec
   (d) $\dfrac{64}{5.2} \approx 12.3$ sec;
   $s\left(\dfrac{64}{5.2}\right) \approx 393.8$ ft

72. (a) $\dfrac{4}{7}$ sec; 280 cm/sec
   (b) 560 cm/sec; 980 cm/sec^2

73. $\pi(20x - x^2)$

74. (a) $r(x) = \left(3 - \dfrac{x}{40}\right)^2 x = 9x - \dfrac{3}{20}x^2 + \dfrac{1}{1600}x^3$

(b) 40 people; $4.00

(c) One possible answer:
Probably not, since the company charges less overall for 60 passengers than it does for 40 passengers.

75. (a) -0.6 km/sec

(b) $\dfrac{18}{\pi} \approx 5.73$ revolutions/min

76. Yes

77. $y'(r) = -\dfrac{1}{2r^2 l}\sqrt{\dfrac{T}{\pi d}}$, so increasing r decreases the frequency.

$y'(l) = -\dfrac{1}{2rl^2}\sqrt{\dfrac{T}{\pi d}}$, so increasing l decreases the frequency.

$y'(d) = -\dfrac{1}{4rl}\sqrt{\dfrac{T}{\pi d^3}}$, so increasing d decreases the frequency.

$y'(T) = \dfrac{1}{4rl\sqrt{\pi Td}}$, so increasing T increases the frequency.

78. (a) $P(0) \approx 1.339$, so initially, one student was infected

(b) 200

(c) After 5 days, when the rate is 50 students/day

79. (a) $x \neq k\dfrac{\pi}{4}$, where k is an odd integer

(b) $\left(-\dfrac{\pi}{2}, \dfrac{\pi}{2}\right)$

(c) Where it's not defined, at $x = k\dfrac{\pi}{4}$, k an odd integer

(d) It has period $\dfrac{\pi}{2}$ and continues to repeat the pattern seen in this window.

80. $-\dfrac{1}{3\sqrt{3}}$

Chapter 4

Section 4.1 (pp. 177–185)

Quick Review

1. $\dfrac{-1}{2\sqrt{4 - x}}$

2. $\dfrac{3}{4}x^{-1/4}$

3. $\dfrac{2x}{(9 - x^2)^{3/2}}$

4. $\dfrac{-2x}{3(x^2 - 1)^{4/3}}$

5. $\dfrac{2x}{x^2 + 1}$

6. $-\dfrac{\sin(\ln x)}{x}$

7. $2e^{2x}$

8. 1

9. ∞

10. ∞

11. (a) 1   **(b)** 1   **(c)** Undefined

12. (a) $x \neq 2$

(b) $f'(x) = \begin{cases} 3x^2 - 2, x < 2 \\ 1, x > 2 \end{cases}$

Exercises

1. Maximum at $x = b$, minimum at $x = c_2$;
Extreme Value Theorem applies, and both the max and min exist.

3. Maximum at $x = c$, no minimum;
Extreme Value Theorem doesn't apply, since the function isn't defined on a closed interval.

5. Maximum at $x = c$, minimum at $x = a$;
Extreme Value Theorem doesn't apply, since the function isn't continuous.

7. Local minimum at $(-1, 0)$, local maximum at $(1, 0)$

9. Maximum at $(0, 5)$

11. Maximum value is $\dfrac{1}{4} + \ln 4$ at $x = 4$;
minimum value is 1 at $x = 1$;
local maximum at $\left(\dfrac{1}{2}, 2 - \ln 2\right)$

13. Maximum value is $\ln 4$ at $x = 3$;
minimum value is 0 at $x = 0$.

15. Maximum value is 1 at $x = \dfrac{\pi}{4}$;
minimum value is -1 at $x = \dfrac{5\pi}{4}$;
local minimum at $\left(0, \dfrac{1}{\sqrt{2}}\right)$;
local maximum at $\left(\dfrac{7\pi}{4}, 0\right)$

17. Maximum value is $3^{2/5}$ at $x = -3$;
minimum value is 0 at $x = 0$.

19. Minimum value is 1 at $x = 2$.

21. Local maximum at $(-2, 17)$;
local minimum at $\left(\dfrac{4}{3}, -\dfrac{41}{27}\right)$

23. Minimum value is 0 at $x = -1$ and $x = 1$.

25. Minimum value is 1 at $x = 0$.

27. Maximum value is 2 at $x = 1$;
minimum value is 0 at $x = -1$ and $x = 3$.

29. Maximum value is $\dfrac{1}{2}$ at $x = 1$;
minimum value is $-\dfrac{1}{2}$ at $x = -1$.

31. Maximum value is 11 at $x = 5$;
minimum value is 5 on the interval $[-3, 2]$;
local maximum at $(-5, 9)$

33. Maximum value is 5 on the interval $[3, \infty)$;
minimum value is -5 on the interval $(-\infty, -2]$.

35. (a) No
   (b) The derivative is defined and nonzero for
   $x \neq 2$. Also, $f(2) = 0$, and $f(x) > 0$ for all
   $x \neq 2$.
   (c) No, because $(-\infty, \infty)$ is not a closed interval.
   (d) The answers are the same as (a) and (b) with
   2 replaced by a.

37.

crit. pt.	derivative	extremum	value
$x = -\dfrac{4}{5}$	0	local max	$\dfrac{12}{25}10^{1/3} \approx 1.034$
$x = 0$	undefined	local min	0

39.

crit. pt.	derivative	extremum	value
$x = -2$	undefined	local max	0
$x = -\sqrt{2}$	0	minimum	-2
$x = \sqrt{2}$	0	maximum	2
$x = 2$	undefined	local min	0

41.

crit. pt.	derivative	extremum	value
$x = 1$	undefined	minimum	2

43.

crit. pt.	derivative	extremum	value
$x = -1$	0	maximum	5
$x = 1$	undefined	local min	1
$x = 3$	0	maximum	5

45. (c)   **47.** (d)

49. (a) Maximum value is 144 at $x = 2$.
   (b) The largest volume of the box is 144 cubic
   units and it occurs when $x = 2$.

51. (a) $f'(x) = 3ax^2 + 2bx + c$ is a quadratic, so it
can have 0, 1, or 2 zeros, which would be the
critical points of f. Examples:

[−3, 3] by [−5, 5]
The function $f(x) = x^3 - 3x$ has two critical
points at $x = -1$ and $x = 1$.

[−3, 3] by [−5, 5]
The function $f(x) = x^3 - 1$ has one critical
point at $x = 0$.

[−3, 3] by [−5, 5]
The function $f(x) = x^3 + x$ has no critical
points.

   (b) Two or none

53. (a)

[−0.1, 0.6] by [−1.5, 1.5]
$f(0) = 0$ is not a local extreme value because
in any open interval containing $x = 0$, there
are infinitely many points where $f(x) = 1$ and
where $f(x) = -1$.

   (b) One possible answer, on the interval $[0, 1]$:

$$f(x) = \begin{cases} (1 - x) \cos \dfrac{1}{1 - x}, & 0 \leq x < 1 \\ 0, & x = 1 \end{cases}$$

This function has no local extreme value at
$x = 1$. Note that it is continuous on $[0, 1]$.

Section 4.2 (pp. 186–194)

Quick Review

1. $(-\sqrt{3}, \sqrt{3})$
2. $(-\infty, -\sqrt{2}) \cup (\sqrt{2}, \infty)$
3. $[-2, 2]$

4. For all x in its domain, or, $[-2, 2]$

5. On $(-2, 2)$ **6.** $x \neq \pm 1$

7. For all x in its domain, or, for all $x \neq \pm 1$

8. For all x in its domain, or, for all $x \neq \pm 1$

9. $C = 3$ **10.** $C = -4$

Exercises

1. (a) Local maximum at $\left(\dfrac{5}{2}, \dfrac{25}{4}\right)$

(b) On $\left(-\infty, \dfrac{5}{2}\right]$ **(c)** On $\left[\dfrac{5}{2}, \infty\right)$

3. (a) None   **(b)** None

(c) On $(-\infty, 0)$ and $(0, \infty)$

5. (a) None   **(b)** On $(-\infty, \infty)$

(c) None

7. (a) Local maximum at $(-2, 4)$

(b) None   **(c)** On $[-2, \infty)$

9. (a) Local maximum at
 $\approx (2.67, 3.08)$;
 local minimum at $(4, 0)$

(b) On $\left(-\infty, \dfrac{8}{3}\right]$ **(c)** On $\left[\dfrac{8}{3}, 4\right]$

11. (a) Local maximum at $\left(-2, \dfrac{1}{4}\right)$;

 local minimum at $\left(2, -\dfrac{1}{4}\right)$

(b) On $(-\infty, -2]$ and $[2, \infty)$

(c) On $[-2, 2]$

13. (a) Local maximum at $\approx (-1.126, -0.036)$;
 local minimum at $\approx (0.559, -2.639)$

(b) On $(-\infty, -1.126]$ and $[0.559, \infty)$

(c) On $[-1.126, 0.559]$

15. (a) f is continuous on $[0, 1]$ and differentiable on
 $(0, 1)$.

(b) $c = \dfrac{1}{2}$

17. (a) f is continuous on $[-1, 1]$ and differentiable
 on $[-1, 1]$.

(b) $c \approx \pm 0.771$

19. (a) $y = \dfrac{5}{2}$ **(b)** $y = 2$

21. (a) Not differentiable at $x = 0$

(b)

$[-1, 1]$ by $[-1, 1]$

(c) $c = \pm 3^{-3/2} \approx \pm 0.192$

23. (a) Not differentiable at $x = 0$

(b)

$[-1, 1]$ by $[-1, 2]$

(c) There are none

25. $\dfrac{x^2}{2} + C$ **27.** $x^3 - x^2 + x + C$

29. $e^x + C$ **31.** $\dfrac{1}{x} + \dfrac{1}{2}, x > 0$

33. $\ln (x + 2) + 3$

35. Possible answers:

(a)

$[-2, 4]$ by $[-2, 4]$

(b)

$[-1, 4]$ by $[0, 3.5]$

(c)

$[-1, 4]$ by $[0, 3.5]$

37. One possible answer:

$[-3, 3]$ by $[-15, 15]$

39. Because the trucker's average speed was 79.5
 mph, and by the Mean Value Theorem, the
 trucker must have been going that speed at least
 once during the trip.

41. Because its average speed was approximately
 7.667 knots, and by the Mean Value Theorem, it
 must have been going that speed at least once
 during the trip.

43. (a) 48 m/sec   **(b)** 720 meters

(c) After about 27.604 seconds, and it will be
 going about 48.166 m/sec

45. Because the function is not continuous on $[0, 1]$.

47. $f(x)$ must be zero at least once between a and b by the Intermediate Value Theorem.
Now suppose that $f(x)$ is zero twice between a and b. Then by the Mean Value Theorem, $f'(x)$ would have to be zero at least once between the two zeros of $f(x)$, but this can't be true since we are given that $f'(x) \neq 0$ on this interval. Therefore, $f(x)$ is zero once and only once between a and b.

49. Let $f(x) = x + \ln(x + 1)$. Then $f(x)$ is continuous and differentiable everywhere on $[0, 3]$.
$$f'(x) = 1 + \frac{1}{x + 1}, \text{ which is never zero on } [0, 3].$$
Now $f(0) = 0$, so $x = 0$ is one solution of the equation. If there were a second solution, $f(x)$ would be zero twice in $[0, 3]$, and by the Mean Value Theorem, $f'(x)$ would have to be zero somewhere between the two zeros of $f(x)$. But this can't happen, since $f'(x)$ is never zero on $[0, 3]$. Therefore, $f(x) = 0$ has exactly one solution in the interval $[0, 3]$.

51. **(a)** Increasing: $[-2, -1.3]$ and $[1.3, 2]$;
decreasing: $[-1.3, 1.3]$;
local max: $x \approx -1.3$
local min: $x \approx 1.3$
(b) Regression equation: $y = 3x^2 - 5$

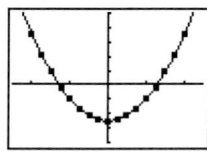

$[-2.5, 2.5]$ by $[-8, 10]$
(c) $f(x) = x^3 - 5x$

53. $\dfrac{f(b) - f(a)}{b - a} = \dfrac{\frac{1}{b} - \frac{1}{a}}{b - a} = -\dfrac{1}{ab}$

$f'(c) = -\dfrac{1}{c^2}$, so $-\dfrac{1}{c^2} = -\dfrac{1}{ab}$ and $c^2 = ab$.

Thus, $c = \sqrt{ab}$.

55. By the Mean Value Theorem,
$\sin b - \sin a = (\cos c)(b - a)$ for some c between a and b. Taking the absolute value of both sides and using $|\cos c| \leq 1$ gives the result.

57. Let $f(x)$ be a monotonic function defined on an interval D. For any two values in D, we may let x_1 be the smaller value and let x_2 be the larger value, so $x_1 < x_2$. Then either $f(x_1) < f(x_2)$ (if f is increasing), or $f(x_1) > f(x_2)$ (if f is decreasing), which means $f(x_1) \neq f(x_2)$. Therefore, f is one-to-one.

Section 4.3 (pp. 194–206)

Quick Review

1. $(-3, 3)$ **2.** $(-2, 0) \cup (2, \infty)$
3. f: all reals   **4.** f: all reals
   f': all reals f': $x \neq 0$
5. f: $x \neq 2$ **6.** f: all reals
   f': $x \neq 2$ f': $x \neq 0$
7. $y = 0$ **8.** $y = 0$
9. $y = 0$ and $y = 200$ **10.** $y = 0$ and $y = 375$

Exercises

1. **(a)** Zero: $x = \pm 1$;
positive: $(-\infty, -1)$ and $(1, \infty)$;
negative: $(-1, 1)$
(b) Zero: $x = 0$;
positive: $(0, \infty)$;
negative: $(-\infty, 0)$
3. **(a)** $(-\infty, -2]$ and $[0, 2]$
(b) $[-2, 0]$ and $[2, \infty)$
(c) Local maxima: $x = -2$ and $x = 2$;
local minimum: $x = 0$
5. **(a)** $[0, 1]$, $[3, 4]$, and $[5.5, 6]$
(b) $[1, 3]$ and $[4, 5.5]$
(c) Local maxima: $x = 1$, $x = 4$
(if f is continuous at $x = 4$), and $x = 6$;
local minima: $x = 0$, $x = 3$, and $x = 5.5$
7. **(a)** $\left[\dfrac{1}{2}, \infty\right)$ **(b)** $\left(-\infty, \dfrac{1}{2}\right]$
(c) $(-\infty, \infty)$ **(d)** Nowhere
(e) Local minimum at $\left(\dfrac{1}{2}, -\dfrac{5}{4}\right)$
(f) None
9. **(a)** $[-1, 0]$ and $[1, \infty)$
(b) $(-\infty, -1]$ and $[0, 1]$
(c) $\left(-\infty, -\dfrac{1}{\sqrt{3}}\right)$ and $\left(\dfrac{1}{\sqrt{3}}, \infty\right)$
(d) $\left(-\dfrac{1}{\sqrt{3}}, \dfrac{1}{\sqrt{3}}\right)$
(e) Local maximum: $(0, 1)$;
local minima: $(-1, -1)$ and $(1, -1)$
(f) $\left(\pm\dfrac{1}{\sqrt{3}}, -\dfrac{1}{9}\right)$
11. **(a)** $[-2, 2]$
(b) $[-\sqrt{8}, -2]$ and $[2, \sqrt{8}]$
(c) $(-\sqrt{8}, 0)$ **(d)** $(0, \sqrt{8})$
(e) Local maxima: $(-\sqrt{8}, 0)$ and $(2, 4)$;
local minima: $(-2, -4)$ and $(\sqrt{8}, 0)$
(f) $(0, 0)$

13. (a) $(-\infty, -2]$ and $\left[-\dfrac{3}{2}, \infty\right)$

(b) $\left[-2, -\dfrac{3}{2}\right]$ **(c)** $\left(-\dfrac{7}{4}, \infty\right)$

(d) $\left(-\infty, -\dfrac{7}{4}\right)$

(e) Local maximum: $(-2, -40)$;

     local minimum: $\left(-\dfrac{3}{2}, -\dfrac{161}{4}\right)$

(f) $\left(-\dfrac{7}{4}, -\dfrac{321}{8}\right)$

15. (a) $(-\infty, \infty)$ **(b)** None

   (c) $(-\infty, 0)$ **(d)** $(0, \infty)$

   (e) None   **(f)** $(0, 3)$

17. (a) $(-\infty, \infty)$ **(b)** None

   (c) $(-\infty, 5 \ln 3) \approx (-\infty, 5.49)$

   (d) $(5 \ln 3, \infty) \approx (5.49, \infty)$

   (e) None

   (f) $\left(5 \ln 3, \dfrac{5}{2}\right) \approx (5.49, 2.50)$

19. (a) $(-\infty, 1)$ **(b)** $[1, \infty)$

   (c) None   **(d)** $(1, \infty)$

   (e) None   **(f)** None

21. (a) $(-\infty, -\sqrt{2}\,]$ and $[\sqrt{2}, \infty)$

   (b) $[-\sqrt{2}, 0)$ and $(0, \sqrt{2}\,]$

   (c) $(0, \infty)$ **(d)** $(-\infty, 0)$

   (e) Local maximum:

     $(-\sqrt{2}, -\sqrt{2e}) \approx (-1.41, -2.33)$;

     local minimum: $(\sqrt{2}, \sqrt{2e}) \approx (1.41, 2.33)$

   (f) None

23. (a) $(-\infty, \infty)$ **(b)** None

   (c) $(-\infty, 0)$ **(d)** $(0, \infty)$

   (e) None   **(f)** $(0, 0)$

25. (a) $[1, \infty)$ **(b)** $(-\infty, 1]$

   (c) $(-\infty, -2)$ and $(0, \infty)$

   (d) $(-2, 0)$

   (e) Local minimum: $(1, -3)$

   (f) $\approx (-2, 7.56)$ and $(0, 0)$

27. (a) Approximately $[0.15, 1.40]$ and $[2.45, \infty)$

   (b) Approximately $(-\infty, 0.15]$, $[1.40, 2)$, and $(2, 2.45]$

   (c) $(-\infty, 1)$ and $(2, \infty)$

   (d) $(1, 2)$

   (e) Local maximum: $\approx (1.40, 1.29)$; local minima: $\approx (0.15, 0.48)$ and $(2.45, 9.22)$

   (f) $(1, 1)$

29. (a) None   **(b)** At $x = 2$

   (c) At $x = 1$ and $x = \dfrac{5}{3}$

31.

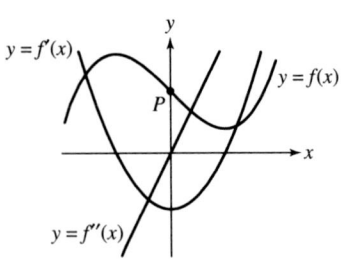

33. (a) Absolute maximum at $(1, 2)$; absolute minimum at $(3, -2)$

   (b) None

   (c) One possible answer:

35.

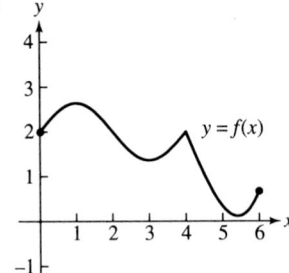

37. (a) $v(t) = 2t - 4$ **(b)** $a(t) = 2$

   (c) It begins at position 3 moving in a negative direction. It moves to position -1 when $t = 2$, and then changes direction, moving in a positive direction thereafter.

39. (a) $v(t) = 3t^2 - 3$

   (b) $a(t) = 6t$

   (c) It begins at position 3 moving in a negative direction. It moves to position 1 when $t = 1$, and then changes direction, moving in a positive direction thereafter.

41. (a) $t = 2.2, 6, 9.8$ **(b)** $t = 4, 8, 11$

43. No. f must have a horizontal tangent line at that point, but it could be increasing (or decreasing) on both sides of the point, and there would be no local extremum.

45. One possible answer:

47. One possible answer:

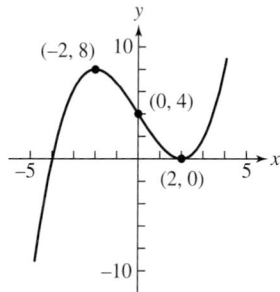

49. (a) Regression equation:
$$y = \frac{2161.4541}{1 + 28.1336e^{-0.8627x}}$$

[0, 8] by [−400, 2300]

(b) At approximately $x = 3.868$ (late in 1996), when the sales are about 1081 million dollars/year

(c) 2161.45 million dollars/year

51. (a) $f'(x) = \dfrac{abce^{bx}}{(e^{bx} + a)^2}$, so the sign of $f'(x)$ is the same as the sign of the product abc.

(b) $f''(x) = -\dfrac{ab^2ce^{bx}(e^{bx} - a)}{(e^{bx} + a)^3}$. Since $a > 0$, this changes sign when $x = \dfrac{\ln a}{b}$ due to the $e^{bx} - a$ factor in the numerator, and there is a point of inflection at that location.

Section 4.4 (pp. 206–220)

Quick Review

1. None
2. Local maximum: $(-2, 17)$;
local minimum: $(1, -10)$

3. $\dfrac{200\pi}{3}$ cm^3

4. $r \approx 4.01$ cm and $h \approx 19.82$ cm, or,
$r \approx 7.13$ cm and $h \approx 6.26$ cm

5. $-\sin \alpha$ **6.** $\cos \alpha$
7. $\sin \alpha$ **8.** $-\cos \alpha$
9. $x = 1$ and $y = \sqrt{3}$, or, $x = -1$ and $y = -\sqrt{3}$
10. $x = 0$ and $y = 3$, or, $x = -\dfrac{24}{13}$ and $y = \dfrac{15}{13}$

Exercises

1. (a) As large as possible: 0 and 20;
as small as possible: 10 and 10

(b) As large as possible: $\dfrac{79}{4}$ and $\dfrac{1}{4}$;
as small as possible: 0 and 20

3. Smallest perimeter = 16 in., dimensions are 4 in. by 4 in.

5. (a) $y = 1 - x$ **(b)** $A(x) = 2x(1 - x)$
(c) Largest area $= \dfrac{1}{2}$, dimensions are 1 by $\dfrac{1}{2}$

7. Largest volume is $\dfrac{2450}{27} \approx 90.74$ in^3;
dimensions: $\dfrac{5}{3}$ in. by $\dfrac{14}{3}$ in. by $\dfrac{35}{3}$ in.

9. Largest area $= 80{,}000$ m^2;
dimensions: 200 m (perpendicular to river) by 400 m (parallel to river)

11. (a) 10 ft by 10 ft by 5 ft
(b) Assume that the weight is minimized when the total area of the bottom and the 4 sides is minimized.

13. 18 in. high by 9 in. wide

15. $\theta = \dfrac{\pi}{2}$ **17.** $\dfrac{8}{\pi}$ to 1

19. (a) $V(x) = 2x(24 - 2x)(18 - 2x)$
(b) Domain: $(0, 9)$

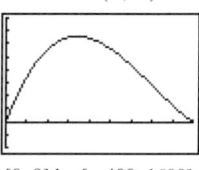

[0, 9] by [−400, 1600]

(c) Maximum volume ≈ 1309.95 in^3 when $x \approx 3.39$ in.

(d) $V'(x) = 24x^2 - 336x + 864$, so the critical point is at $x = 7 - \sqrt{13}$, which confirms the result in (c).

(e) $x = 2$ in. or $x = 5$ in.

19. continued

(f) The dimensions of the resulting box are $2x$ in., $(24 - 2x)$ in., and $(18 - 2x)$ in. Each of these measurements must be positive, so that gives the domain of $(0, 9)$.

21. Dimensions: width ≈ 3.44, height ≈ 2.61; maximum area ≈ 8.98

23. (a) At $x = 1$

(b)

a	b	A
0.1	3.71	0.33
0.2	2.86	0.44
0.3	2.36	0.46
0.4	2.02	0.43
0.5	1.76	0.38
0.6	1.55	0.31
0.7	1.38	0.23
0.8	1.23	0.15
0.9	1.11	0.08
1.0	1.00	0.00

(c)

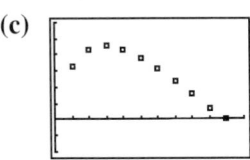

[0, 1.1] by [−0.2, 0.6]

(d) Quadratic:
$$A \approx -0.91a^2 + 0.54a + 0.34$$

[−0.5, 1.5] by [−0.2, 0.6]

Cubic:
$$A \approx 1.74a^3 - 3.78a^2 + 1.86a + 0.19$$

[−0.5, 1.5] by [−0.2, 0.6]

Quartic:
$$A \approx -1.92a^4 + 5.96a^3 - 6.87a^2 + 2.71a + 0.12$$

[−0.5, 1.5] by [−0.2, 0.6]

(e) Quadratic: $A \approx 0.42$; cubic: $A \approx 0.45$; quartic: $A \approx 0.46$

25. 18 in. by 18 in. by 36 in.

27. Radius $= \sqrt{2}$ m, height $= 1$ m, volume $\dfrac{2\pi}{3}$ m^3

29. $f'(x) = \dfrac{2x^3 - a}{x^2}$, so the only sign change in $f'(x)$ occurs at $x = \left(\dfrac{a}{2}\right)^{1/3}$, where the sign changes from negative to positive. This means there is a local minimum at that point, and there are no local maxima.

31. $\dfrac{32\pi}{3}$ cubic units

33. (a) 6 in. wide by $6\sqrt{3}$ in. deep

(b)

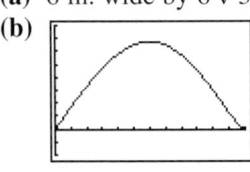

[0, 12] by [−2000, 8000]
$$y = x(144 - x^2)^{3/2}$$

(c)

[0, 12] by [−2000, 8000]
$$y = x^3(144 - x^2)^{1/2}$$

Changing the value of k changes the maximum stiffness, but not the dimensions of the stiffest beam. The graphs for different values of k look the same except that the vertical scale is different.

35. $2\sqrt{2}$ amps

37. The minimum distance is 2.

39. (a) Because $f(x)$ is periodic with period 2π.

(b) No. It has an absolute minimum at the point $(\pi, 0)$.

41. (a) At $t = \dfrac{\pi}{3}$ sec and at $t = \dfrac{4\pi}{3}$ sec

(b) The maximum distance between particles is 1 m.

(c) Near $t = \dfrac{\pi}{3}$ sec and near $t = \dfrac{4\pi}{3}$ sec

43. (a) Answers will vary.

(b) $x = \dfrac{51}{8} = 6.375$ in.

(c) Minimum length ≈ 11.04 in.

45. $x = \dfrac{c + 100}{2} = 50 + \dfrac{c}{2}$

47. The rate v is maximum when $x = \dfrac{a}{2}$.

The rate then is $\dfrac{ka^2}{4}$.

49. 67 people

51. $p(x) = 6x - (x^3 - 6x^2 + 15x),\ x \geq 0$. This function has its maximum value at the points $(0, 0)$ and $(3, 0)$.

53. **(a)** $y'(0) = 0$ **(b)** $y'(-L) = 0$

(c) $y(0) = 0$, so $d = 0$. $y'(0) = 0$, so $c = 0$.

Then $y(-L) = -aL^3 + bL^2 = H$ and

$y'(-L) = 3aL^2 - 2bL = 0$.

Solving, $a = 2\dfrac{H}{L^3}$ and $b = 3\dfrac{H}{L^2}$, which gives the equation shown.

55. **(a)** The x- and y-intercepts of the line through R and T are $x - \dfrac{a}{f'(x)}$ and $a - xf'(x)$ respectively.

The area of the triangle is the product of these two values.

(b) Domain: $(0, 10)$

[0, 10] by [−100, 1000]

The vertical asymptotes at $x = 0$ and $x = 10$ correspond to horizontal or vertical tangent lines, which do not form triangles.

(c) Height = 15, which is 3 times the y-coordinate of the center of the ellipse.

(d) Part (a) remains unchanged.

The domain is $(0, C)$ and the graph is similar.

The minimum area occurs when $x^2 = \dfrac{3C^2}{4}$.

From this, it follows that the triangle has minimum area when its height is $3B$.

Section 4.5 (pp. 220–232)

Quick Review

1. $2x \cos(x^2 + 1)$

2. $\dfrac{1 - \cos x - (x + 1)\sin x}{(x + 1)^2}$

3. $x \approx -0.567$ **4.** $x \approx -0.322$

5. $y = x + 1$ **6.** $y = 2ex + e + 1$

7. **(a)** $x = -1$

(b) $x = -\dfrac{e + 1}{2e} \approx -0.684$

8.

x	$f(x)$	$g(x)$
0.7	−1.457	−1.7
0.8	−1.688	−1.8
0.9	−1.871	−1.9
1.0	−2	−2
1.1	−2.069	−2.1
1.2	−2.072	−2.2
1.3	−2.003	−2.3

9.

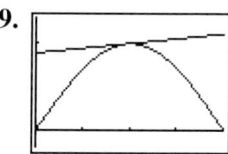

[0, π] by [−0.2, 1.3]

10.

[−1, 7] by [−2, 2]

Exercises

1. **(a)** $L(x) = 10x - 13$

(b) Differs from the true value in absolute value by less than 10^{-1}

3. **(a)** $L(x) = 2$

(b) Differs from the true value in absolute value by less than 10^{-2}

5. **(a)** $L(x) = x - \pi$

(b) Differs from the true value in absolute value by less than 10^{-3}

7. $f(0) = 1$. Also, $f'(x) = k(1 + x)^{k-1}$, so $f'(0) = k$. This means the linearization at $x = 0$ is $L(x) = 1 + kx$.

9. The linearization is $1 + \dfrac{3x}{2}$. It is the sum of the two individual linearizations.

11. Center $= -1,\ L(x) = -5$

13. Center $= 1,\ L(x) = \dfrac{x}{4} + \dfrac{1}{4}$, or

Center $= 1.5,\ L(x) = \dfrac{4x}{25} + \dfrac{9}{25}$

15. $x \approx 0.682328$

17. $x \approx 0.386237,\ 1.961569$

19. (a) $dy = (3x^2 - 3)\, dx$
  (b) $dy = 0.45$ at the given values

21. (a) $dy = (2x \ln x + x)\, dx$
  (b) $dy = 0.01$ at the
  given values

23. (a) $dy = (\cos x)\, e^{\sin x}\, dx$
  (b) $dy = 0.1$ at the given values

25. (a) $dy = \dfrac{dx}{(x+1)^2}$
  (b) $dy = 0.01$ at the given values

27. (a) 0.21 **(b)** 0.2 **(c)** 0.01

29. (a) $-\dfrac{2}{11}$ **(b)** $-\dfrac{1}{5}$ **(c)** $\dfrac{1}{55}$

31. $4\pi a^2\, dr$ **33.** $3a^2\, dx$ **35.** $2\pi ah\, dr$

37. (a) $x+1$ **(b)** $f(0.1) \approx 1.1$
  (c) The actual value is less than 1.1, since the derivative is decreasing over the interval $[0, 0.1]$.

39. The diameter grew $\dfrac{2}{\pi} \approx 0.6366$ in.
  The cross section area grew about 10 in².

41. The side should be measured to within 1%.
43. The angle should be measured to within 0.76%.
  This is also ± 0.01 radian or $\approx \pm 0.57$ degree.
45. (a) Within 0.5%   **(b)** Within 5%
47. About 37.87 to 1
49. If $f'(x_1) \ne 0$, then x_2 and all later approximations are equal to x_1.

51. $x_2 = -2,\ x_3 = 4,\ x_4 = -8,$ and $x_5 = 16;$
  $|x_n| = 2^{n-1}.$

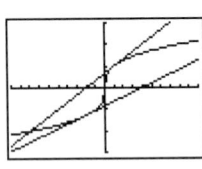

[-10, 10] by [-3, 3]

53. Just multiply the corresponding derivative formulas by dx.
55. $g(a) = c$, so if $E(a) = 0$, then $g(a) = f(a)$ and $c = f(a)$. Then
  $E(x) = f(x) - g(x) = f(x) - f(a) - m(x-a)$.
  Thus, $\dfrac{E(x)}{x-a} = \dfrac{f(x) - f(a)}{x-a} - m.$
  $\displaystyle\lim_{x \to a} \dfrac{f(x) - f(a)}{x-a} = f'(a)$, so if the limit of $\dfrac{E(x)}{x-a}$ is zero, then $m = f'(a)$ and $g(x) = L(x)$.

Section 4.6 (pp. 232–241)

Quick Review

1. $\sqrt{74}$ **2.** $\sqrt{a^2 + b^2}$

3. $\dfrac{1 - 2y}{2x + 2y - 1}$ **4.** $-\dfrac{y + \sin y}{x + x \cos y}$

5. $2x \cos^2 y$ **6.** $2x + 2y - 1$

7. One possible answer:
  $x = -2 + 6t,\ y = 1 - 4t,\ 0 \le t \le 1.$

8. One possible answer:
  $x = 5t,\ y = -4 + 4t,\ 0 \le t \le 1.$

9. One possible answer:
  $\dfrac{\pi}{2} \le t \le \dfrac{3\pi}{2}$

10. One possible answer:
  $\dfrac{3\pi}{2} \le t \le 2\pi$

Exercises

1. $\dfrac{dA}{dt} = 2\pi r \dfrac{dr}{dt}$

3. (a) $\dfrac{dV}{dt} = \pi r^2 \dfrac{dh}{dt}$

  (b) $\dfrac{dV}{dt} = 2\pi r h \dfrac{dr}{dt}$

  (c) $\dfrac{dV}{dt} = \pi r^2 \dfrac{dh}{dt} + 2\pi r h \dfrac{dr}{dt}$

5. $\dfrac{ds}{dt} = \dfrac{x \dfrac{dx}{dt} + y \dfrac{dy}{dt} + z \dfrac{dz}{dt}}{\sqrt{x^2 + y^2 + z^2}}$

7. (a) 1 volt/sec
  (b) $-\dfrac{1}{3}$ amp/sec
  (c) $\dfrac{dV}{dt} = I \dfrac{dR}{dt} + R \dfrac{dI}{dt}$
  (d) $\dfrac{dR}{dt} = \dfrac{3}{2}$ ohms/sec. R is increasing since $\dfrac{dR}{dt}$ is positive.

9. (a) $\dfrac{dA}{dt} = 14$ cm²/sec   **(b)** $\dfrac{dP}{dt} = 0$ cm/sec
  (c) $\dfrac{dD}{dt} = -\dfrac{14}{13}$ cm/sec
  (d) The area is increasing, because its derivative is positive.
  The perimeter is not changing, because its derivative is zero.
  The diagonal length is decreasing, because its derivative is negative.

11. $\dfrac{dx}{dt} = \dfrac{3000}{\sqrt{51}}$ mph ≈ 420.08 mph

13. (a) 12 ft/sec   **(b)** $-\dfrac{119}{2}$ ft²/sec

(c) -1 radian/sec

15. $\dfrac{19\pi}{2500} \approx 0.0239$ in³/min

17. (a) $\dfrac{32}{9\pi} \approx 1.13$ cm/min

(b) $-\dfrac{80}{3\pi} \approx -8.49$ cm/min

19. $V = \dfrac{4}{3}\pi r^3$, so $\dfrac{dV}{dt} = 4\pi r^2 \dfrac{dr}{dt}$. But $S = 4\pi r^2$, so we

are given that $\dfrac{dV}{dt} = kS = 4k\pi r^2$. Substituting,

$4k\pi r^2 = 4\pi r^2 \dfrac{dr}{dt}$ which gives $\dfrac{dr}{dt} = k$.

21. (a) $\dfrac{5}{2}$ ft/sec   **(b)** $-\dfrac{3}{20}$ radian/sec

23. (a) $\dfrac{dc}{dt} = 0.3, \dfrac{dr}{dt} = 0.9, \dfrac{dp}{dt} = 0.6$

(b) $\dfrac{dc}{dt} = -1.5625, \dfrac{dr}{dt} = 3.5, \dfrac{dp}{dt} = 5.0625$

25. $\dfrac{dy}{dt} = \dfrac{466}{1681} \approx 0.277$ L/min²

27. $\dfrac{2}{5}$ radian/sec   **29.** -3 ft/sec

31. In front: 2 radians/sec;

Half second later: 1 radian/sec

33. 80 mph   **35.** -6 deg/sec

37. (a) $\dfrac{24}{5}$ cm/sec   **(b)** 0 cm/sec

(c) $-\dfrac{1200}{160,801} \approx -0.00746$ cm/sec

39. (a) The point being plotted would correspond to a point on the edge of the wheel as the wheel turns.

(b) One possible answer:
$\theta = 16\pi t$, where t is in seconds.

(c) Assuming counterclockwise motion, the rates are as follows.

$\theta = \dfrac{\pi}{4}: \dfrac{dx}{dt} \approx -71.086$ ft/sec

$\dfrac{dy}{dt} \approx 71.086$ ft/sec

$\theta = \dfrac{\pi}{2}: \dfrac{dx}{dt} \approx -100.531$ ft/sec

$\dfrac{dy}{dt} = 0$ ft/sec

$\theta = \pi: \dfrac{dx}{dt} = 0$ ft/sec

$\dfrac{dy}{dt} \approx -100.531$ ft/sec

41. (a) 9% per year
(b) Increasing at 1% per year

Chapter Review (pp. 242–245)

1. Maximum: $\dfrac{4\sqrt{6}}{9}$ at $x = \dfrac{4}{3}$;
minimum: -4 at $x = -2$

2. No global extrema
3. (a) $[-1, 0)$ and $[1, \infty)$
(b) $(-\infty, -1]$ and $(0, 1]$
(c) $(-\infty, 0)$ and $(0, \infty)$
(d) None
(e) Local minima at $(1, e)$ and $(-1, e)$
(f) None
4. (a) $[-\sqrt{2}, \sqrt{2}]$
(b) $[-2, -\sqrt{2}]$ and $[\sqrt{2}, 2]$
(c) $(-2, 0)$ **(d)** $(0, 2)$
(e) Local max: $(-2, 0)$ and $(\sqrt{2}, 2)$;
local min: $(2, 0)$ and $(-\sqrt{2}, -2)$
(f) $(0, 0)$
5. (a) Approximately $(-\infty, 0.385]$
(b) Approximately $[0.385, \infty)$
(c) None   **(d)** $(-\infty, \infty)$
(e) Local maximum at $\approx (0.385, 1.215)$
(f) None
6. (a) $[1, \infty)$ **(b)** $(-\infty, 1]$
(c) $(-\infty, \infty)$ **(d)** None
(e) Local minimum at $(1, 0)$
(f) None
7. (a) $[0, 1)$ **(b)** $(-1, 0]$
(c) $(-1, 1)$ **(d)** None
(e) Local minimum at $(0, 1)$
(f) None
8. (a) $(-\infty, -2^{-1/3}] \approx (-\infty, -0.794]$
(b) $[-2^{-1/3}, 1) \approx [-0.794, 1)$ and $(1, \infty)$
(c) $(-\infty, -2^{1/3}) \approx (-\infty, -1.260)$ and $(1, \infty)$
(d) $(-1.260, 1)$
(e) Local maximum at
$\left(-2^{-1/3}, \dfrac{2}{3} \cdot 2^{-1/3}\right) \approx (-0.794, 0.529)$
(f) $\left(-2^{1/3}, \dfrac{1}{3} \cdot 2^{1/3}\right) \approx (-1.260, 0.420)$
9. (a) None   **(b)** $[-1, 1]$
(c) $(-1, 0)$ **(d)** $(0, 1)$
(e) Local maximum at $(-1, \pi)$;
local minimum at $(1, 0)$
(f) $\left(0, \dfrac{\pi}{2}\right)$

10. (a) $[-\sqrt{3}, \sqrt{3}]$

   (b) $(-\infty, -\sqrt{3}]$ and $[\sqrt{3}, \infty)$

   (c) Approximately $(-2.584, -0.706)$
   and $(3.290, \infty)$

   (d) Approximately $(-\infty, -2.584)$
   and $(-0.706, 3.290)$

   (e) Local maximum at
   $$\left(\sqrt{3}, \frac{\sqrt{3}-1}{4}\right) \approx (1.732, 0.183);$$
   local minimum at
   $$\left(-\sqrt{3}, \frac{-\sqrt{3}-1}{4}\right) \approx (-1.732, -0.683)$$

   (f) $\approx (-2.584, -0.573), (-0.706, -0.338),$ and
   $(3.290, 0.161)$

11. (a) $(0, 2]$ **(b)** $[-2, 0)$

   (c) None

   (d) $(-2, 0)$ and $(0, 2)$

   (e) Local maxima at $(-2, \ln 2)$ and $(2, \ln 2)$

   (f) None

12. (a) Approximately $[0, 0.176], \left[0.994, \frac{\pi}{2}\right],$
   $[2.148, 2.965], \left[3.834, \frac{3\pi}{2}\right],$ and $\left[5.591, 2\pi\right]$

   (b) Approximately $[0.176, 0.994], \left[\frac{\pi}{2}, 2.148\right],$
   $[2.965, 3.834],$ and $\left[\frac{3\pi}{2}, 5.591\right]$

   (c) Approximately $(0.542, 1.266), (1.876, 2.600),$
   $(3.425, 4.281),$ and $(5.144, 6.000)$

   (d) Approximately $(0, 0.542), (1.266, 1.876),$
   $(2.600, 3.425), (4.281, 5.144),$ and
   $(6.000, 2\pi)$

   (e) Local maxima at $\approx (0.176, 1.266), \left(\frac{\pi}{2}, 0\right)$
   and $(2.965, 1.266), \left(\frac{3\pi}{2}, 2\right),$ and $(2\pi, 1);$
   local minima at $\approx (0, 1), (0.994, -0.513),$
   $(2.148, -0.513), (3.834, -1.806),$ and
   $(5.591, -1.806)$

   Note that the local extrema at $x \approx 3.834,$
   $x = \frac{3\pi}{2},$ and $x \approx 5.591$ are also absolute
   extrema.

   (f) $\approx (0.542, 0.437), (1.266, -0.267),$
   $(1.876, -0.267), (2.600, 0.437),$
   $(3.425, -0.329), (4.281, 0.120),$
   $(5.144, 0.120),$ and $(6.000, -0.329)$

13. (a) $\left(0, \dfrac{2}{\sqrt{3}}\right]$

   (b) $(-\infty, 0]$ and $\left[\dfrac{2}{\sqrt{3}}, \infty\right)$

   (c) $(-\infty, 0)$ **(d)** $(0, \infty)$

   (e) Local maximum at
   $$\left(\frac{2}{\sqrt{3}}, \frac{16}{3\sqrt{3}}\right) \approx (1.155, 3.079)$$

   (f) None

14. (a) Approximately $[-0.578, 1.692]$

   (b) Approximately $(-\infty, -0.578]$ and $[1.692, \infty)$

   (c) Approximately $(-\infty, 1.079)$

   (d) Approximately $(1.079, \infty)$

   (e) Local maximum at $\approx (1.692, 20.517);$
   local minimum at $\approx (-0.578, 0.972)$

   (f) $\approx (1.079, 13.601)$

15. (a) $\left[0, \dfrac{8}{9}\right]$

   (b) $(-\infty, 0]$ and $\left[\dfrac{8}{9}, \infty\right)$

   (c) $\left(-\infty, -\dfrac{2}{9}\right)$

   (d) $\left(-\dfrac{2}{9}, 0\right)$ and $(0, \infty)$

   (e) Local maximum at $\approx (0.889, 1.011);$
   local minimum at $(0, 0)$

   (f) $\approx \left(\dfrac{2}{9}, 0.667\right)$

16. (a) Approximately $(-\infty, 0.215]$

   (b) Approximately $[0.215, 2)$ and $(2, \infty)$

   (c) Approximately $(2, 3.710)$

   (d) $(-\infty, 2)$ and approximately $(3.710, \infty)$

   (e) Local maximum at $\approx (0.215, -2.417)$

   (f) $\approx (3.710, -3.420)$

17. (a) None   **(b)** At $x = -1$

   (c) At $x = 0$ and $x = 2$

18. (a) At $x = -1$ **(b)** At $x = 2$

   (c) At $x = \dfrac{1}{2}$

19. $f(x) = -\dfrac{1}{4}x^{-4} - e^{-x} + C$

20. $f(x) = \sec x + C$

21. $f(x) = 2 \ln x + \dfrac{1}{3}x^3 + x + C$

22. $f(x) = \dfrac{2}{3}x^{3/2} + 2x^{1/2} + C$

23. $f(x) = -\cos x + \sin x + 2$

24. $f(x) = \dfrac{3}{4}x^{4/3} + \dfrac{x^3}{3} + \dfrac{x^2}{2} + x - \dfrac{31}{12}$

25. $s(t) = 4.9t^2 + 5t + 10$

26. $s(t) = 16t^2 + 20t + 5$

27. $L(x) = 2x + \dfrac{\pi}{2} - 1$

28. $L(x) = \sqrt{2}x - \dfrac{\pi\sqrt{2}}{4} + \sqrt{2}$

29. $L(x) = -x + 1$ **30.** $L(x) = 2x + 1$

31. Global minimum value of $\dfrac{1}{2}$ at $x = 2$

32. (a) T **(b)** P

33. (a) $(0, 2]$ **(b)** $[-3, 0)$
     (c) Local maxima at $(-3, 1)$ and $(2, 3)$

34. The 24th day

35.

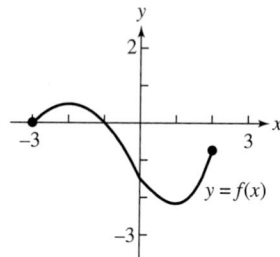

36. (a) Absolute minimum is -2 at $x = 1$;
     absolute maximum is 3 at $x = 3$
     (b) None
     (c)

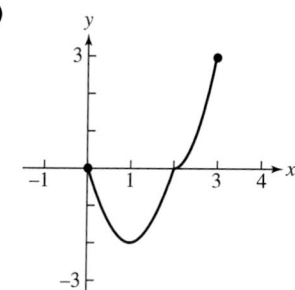

37. (a) $f(x)$ is continuous on $[0.5, 3]$ and differentiable on $(0.5, 3)$.
     (b) $c \approx 1.579$ **(c)** $y \approx 1.457x - 1.075$
     (d) $y \approx 1.457x - 1.579$

38. (a) $v(t) = -3t^2 - 6t + 4$
     (b) $a(t) = -6t - 6$
     (c) The particle starts at position 3 moving in the positive direction, but decelerating. At approximately $t = 0.528$, it reaches position 4.128 and changes direction, beginning to move in the negative direction. After that, it continues to accelerate while moving in the negative direction.

39. (a) $L(x) = -1$
     (b) Using the linearization, $f(0.1) \approx -1$
     (c) Greater than the approximation in (b), since $f'(x)$ is actually positive over the interval $(0, 0.1)$ and the estimate is based on the derivative being 0.

40. (a) $dy = (2x - x^2)e^{-x}\, dx$
     (b) $dy \approx 0.00368$

41. (a) $y = \dfrac{2701.73}{1 + 17.28e^{-0.36x}}$

[0, 20] by [−300, 2800]

     (b) In 1998. There are approximately 1351 million transactions in that year.
     (c) Approximately 2702 million transactions per year

42. $x \approx 0.828361$ **43.** 1200 m/sec

44. 1162.5 m   **45.** $r = 25$ ft and $s = 50$ ft

46. 54 square units

47. Base is 6 ft by 6 ft, height $= 3$ ft

48. Base is 4 ft by 4 ft, height $= 2$ ft

49. Height $= 2$, radius $= \sqrt{2}$

50. $r = h = 4$ ft

51. (a) $V(x) = x(15 - 2x)(5 - x)$
     (b) $0 < x < 5$

[0, 5] by [−10, 70]

     (c) Maximum volume ≈ 66.019 in^3 when $x \approx 1.962$ in.
     (d) $V'(x) = 6x^2 - 50x + 75$ which is zero at $x = \dfrac{25 - 5\sqrt{7}}{6} \approx 1.962$.

52. 29.925 square units

53. $x = \dfrac{48}{\sqrt{7}} \approx 18.142$ mi and $y = \dfrac{36}{\sqrt{7}} \approx 13.607$ mi

54. $x = 100$ m and $r = \dfrac{100}{\pi}$ m

55. 276 grade A and 553 grade B tires

56. (a) 0.765 units
     (b) When $t = \dfrac{7\pi}{8} \approx 2.749$
     (plus multiples of π if they keep going)

57. Dimensions: base is 6 in. by 12 in., height $= 2$ in.; maximum volume $= 144$ in^3

58. -40 m^2/sec   **59.** 5 m/sec

60. Increasing 1 cm/min   **61.** $\dfrac{dx}{dt} = 4$ units/second

62. (a) $h = \dfrac{5r}{2}$ (b) $\dfrac{125}{144\pi} \approx 0.276$ ft/min

63. 5 radians/sec

64. Not enough speed. Duck!

65. $dV \approx \dfrac{2\pi ah}{3}\, dr$

66. (a) Within 1%   (b) Within 3%

67. (a) Within 4%   (b) Within 8%
     (c) Within 12%

68. Height $= 14$ feet, estimated error $= \pm\dfrac{2}{45}$ feet

69. $\dfrac{dy}{dx} = 2 \sin x \cos x - 3.$
Since $\sin x$ and $\cos x$ are both between 1 and -1,
$2 \sin x \cos x$ is never greater than 2, and
therefore $\dfrac{dy}{dx} \le 2 - 3 = -1$ for all values of x.

Chapter 5

Section 5.1 (pp. 247–257)

Quick Review

1. 400 miles   **2.** 144 miles
3. 100 ft/sec $\approx$ 68.18 mph   **4.** 9.46×10^{12} km
5. 28 miles   **6.** 1200 gallons
7. $-3°$ **8.** 25,920,000 ft^3
9. 17,500   **10.** 176,400 times

Exercises

1. (a)

   (b)

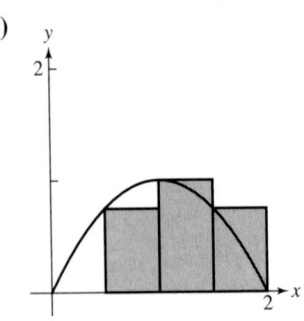

     LRAM $= 1.25$

3.

n	LRAM$_n$	MRAM$_n$	RRAM$_n$
10	1.32	1.34	1.32
50	1.3328	1.3336	1.3328
100	1.3332	1.3334	1.3332
500	1.333328	1.333336	1.333328

5. 13.5   **7.** 0.8821

9. $\approx$ 44.8, 6.7 L/min

11. (a) 5220 m   (b) 4920 m

13. (a) 0.969 mi
    (b) 0.006 h = 21.6 sec
       116 mph

15.

n	error	% error
10	2.61799	0.5
20	0.65450	0.125
40	0.16362	3.12×10^{-2}
80	0.04091	7.8×10^{-3}
160	0.01023	2×10^{-3}

17. (a) $S_8 \approx 120.95132$ (b) 10%
    Underestimate

19. (a) 15,465 ft^3 (b) 16,515 ft^3

21. 39.26991

23. (a) 240 ft/sec
    (b) 1520 ft with RRAM and $n = 5$

25. (a) Upper: 60.9 tons;
       lower: 46.8 tons
    (b) By the end of October

27. (a) 2   (b) $2\sqrt{2} \approx 2.828$
    (c) $8 \sin\left(\dfrac{\pi}{8}\right) \approx 3.061$
    (d) Each area is less than the area of the circle,
       π. As n increases, the polygon area
       approaches π.

29. RRAM$_n f =$ LRAM$_n f + f(x_n)\Delta x - f(x_0)\Delta x$
Since $f(a) = f(b)$, or $f(x_0) = f(x_n)$, we have
RRAM$_n f =$ LRAM$_n f$.

Section 5.2 (pp. 258–268)

Quick Review

1. 55 **2.** 20 **3.** 5500

4. $\displaystyle\sum_{k=1}^{99} k$ **5.** $\displaystyle\sum_{k=0}^{25} 2k$ **6.** $\displaystyle\sum_{k=1}^{500} 3k^2$

7. $\displaystyle\sum_{x=1}^{50} (2x^2 + 3x)$ **8.** $\displaystyle\sum_{k=0}^{20} x^k$

9. $\displaystyle\sum_{k=0}^{n} (-1)^k = 0$ if n is odd.

10. $\displaystyle\sum_{k=0}^{n} (-1)^k = 1$ if n is even.

Exercises

1. $\displaystyle\int_0^2 x^2\,dx$ **3.** $\displaystyle\int_1^4 \frac{1}{x}\,dx$

5. $\displaystyle\int_0^1 \sqrt{4 - x^2}\,dx$ **7.** 15

9. -480 **11.** 2.75 **13.** 21

15. $\dfrac{9\pi}{2}$ **17.** $\dfrac{5}{2}$ **19.** 3

21. $\dfrac{3\pi^2}{2}$ **23.** $\dfrac{1}{2}b^2$ **25.** $b^2 - a^2$

27. $\dfrac{3}{2}a^2$ **29.** 0 **31.** $\dfrac{1}{4}$

33. $\dfrac{3}{4}$ **35.** $\dfrac{1}{2}$ **37.** $-\dfrac{3}{4}$

39. 0.990501 **41.** $\dfrac{32}{3}$

43. (a) 0 **(b)** 1

45. (a) -1 **(b)** $-\dfrac{7}{2}$

47. (a) $f \to +\infty$

(b) Using right endpoints we have
$$\int_0^1 \frac{1}{x^2}\,dx = \lim_{n\to\infty} \sum_{k=1}^{n} \frac{1}{n}\frac{n^2}{k^2}$$
$$= \lim_{n\to\infty} \sum_{k=1}^{n} \frac{n}{k^2} = \lim_{n\to\infty} n\left[1 + \frac{1}{2^2} + \cdots + \frac{1}{n^2}\right].$$
$$n\left(1 + \frac{1}{2^2} + \cdots + \frac{1}{n^2}\right) > n \text{ and } n \to \infty$$
so $n\left(1 + \dfrac{1}{2^2} + \cdots + \dfrac{1}{n^2}\right) \to \infty$.

Section 5.3 (pp. 268–276)

Quick Review

1. $\sin x$ **2.** $\cos x$ **3.** $\tan x$

4. $\cot x$ **5.** $\sec x$ **6.** $\ln (x)$

7. x^n **8.** $-\dfrac{2^x \ln 2}{(2^x + 1)^2}$

9. $xe^x + e^x$ **10.** $\dfrac{1}{x^2 + 1}$

Exercises

1. (a) 0 **(b)** -8 **(c)** -12
  (d) 10 **(e)** -2 **(f)** 16

3. (a) 5 **(b)** $5\sqrt{3}$
  (c) -5 **(d)** -5

5. (a) 4 **(b)** -4

7. -14 **9.** 1

11. -1 **13.** $\dfrac{\pi}{2}$

15. $e^2 - 1 \approx 6.389$ **17.** $\dfrac{16}{3}$

19. $\dfrac{19}{3}$

21. 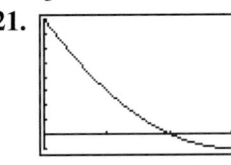 **(a)** 6 **(b)** $\dfrac{22}{3}$

$[0, 3]$ by $[-1, 8]$

23. 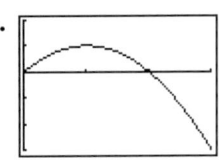 **(a)** 0 **(b)** $\dfrac{8}{3}$

$[0, 3]$ by $[-3, 2]$

25. 0, at $x = 1$ **27.** -2, at $x = \dfrac{1}{\sqrt{3}}$

29. $\dfrac{3}{2}$ **31.** 0

33. $\dfrac{1}{2} \le \displaystyle\int_0^1 \frac{1}{1 + x^4}\,dx \le 1$

35. $0 \le \displaystyle\int_0^1 \sin (x^2)\,dx \le \sin (1) < 1$

37. $0 \le \min f(b - a) \le \displaystyle\int_a^b f(x)\,dx$

39. $av(f) = \dfrac{1}{b - a}\displaystyle\int_a^b f(x)\,dx$ implies $\displaystyle\int_a^b f(x)\,dx$
  $= av(f)(b - a) = \displaystyle\int_a^b av(f)\,dx$

41. Avg rate $= \dfrac{\text{total amount released}}{\text{total time}}$
  $= \dfrac{2000 \text{ m}^3}{100 \text{ min} + 50 \text{ min}} = 13\dfrac{1}{3} \text{ m}^3/\text{min}$

43. $\dfrac{7}{6}$ **45.** $k \approx 2.39838$

Section 5.4 (pp. 277–288)

Quick Review

1. $2x \cos x^2$ **2.** $2 \sin x \cos x$

3. 0 **4.** 0

5. $2^x \ln 2$ **6.** $\dfrac{1}{2\sqrt{x}}$

7. $\dfrac{-x \sin x - \cos x}{x^2}$ **8.** $-\cot(t)$

9. $\dfrac{y+1}{2y-x}$ **10.** $\dfrac{1}{3x}$

Exercises

1. $5 - \ln 6 \approx 3.208$ **3.** 1

5. $\dfrac{5}{2}$ **7.** 2

9. $2\sqrt{3}$ **11.** 0

13. $\dfrac{8}{3}$ **15.** $\dfrac{5}{2}$

17. $\dfrac{1}{2}$

19. (a) No, $f(x) = \dfrac{x^2 - 1}{x + 1}$ is discontinuous at $x = -1$.

    (b) $-\dfrac{5}{2} \cdot f(x)$ is bounded with only one discontinuity. Split it up at $x = -1$, or use area.

21. (a) No, $f(x) = \tan x$ is discontinuous at $x = \dfrac{\pi}{2}$ and $x = \dfrac{3\pi}{2}$.

    (b) No, $\displaystyle\int_0^b \tan x \, dx \to \infty$ as $b \to \dfrac{\pi^-}{2}$.

23. (a) No, $f(x) = \dfrac{\sin x}{x}$ is discontinuous at $x = 0$.

    (b) ≈ 2.55. Area is finite. $\dfrac{\sin x}{x}$ is bounded with only one discontinuity.

25. $\dfrac{5}{6}$ **27.** π

29. ≈ 3.802 **31.** ≈ 0.914

33. $x \approx 0.699$ **35.** $-\dfrac{3}{2}$

37. $\sqrt{1 + x^2}$ **39.** $\dfrac{\sin x}{2\sqrt{x}}$

41. $3x^2 \cos(2x^3) - 2x \cos(2x^2)$

43. (d)   **45.** (b)

47. $x = a$ **49.** $L(x) = 2 + 10x$

51. $\displaystyle\int_0^{\pi/k} \sin kx \, dx = \dfrac{2}{k}$

53. (a) 0

    (b) H is increasing on $[0,6]$ where $H'(x) = f(x) > 0$.

    (c) H is concave up on $(9,12)$ where $H''(x) = f'(x) > 0$.

    (d) $H(12) = \displaystyle\int_0^{12} f(t) \, dt > 0$ because there is more area above the x-axis than below for $y = f(x)$.

    (e) $x = 6$ since $H'(6) = f(6) = 0$ and $H''(6) = f'(6) < 0$.

    (f) $x = 0$ since $H(x) > 0$ on $(0, 12]$.

55. (a) $s'(3) = f(3) = 0$

    (b) $s''(3) = f'(3) > 0$

    (c) $s(3) = \displaystyle\int_0^3 f(x) \, dx = -\dfrac{1}{2}(3)(6) = -9$ units

    (d) $s(t) = 0$ at $t = 6$ sec because $\displaystyle\int_0^6 f(x) \, dx = 0$

    (e) $s''(t) = f'(t) = 0$ at $t = 7$ sec

    (f) $0 < t < 3$: $s < 0$, $s' < 0 \Rightarrow$ away
      $3 < t < 6$: $s < 0$, $s' > 0 \Rightarrow$ toward
      $t > 6$: $s > 0$, $s' > 0 \Rightarrow$ away

    (g) Positive side

57. (a) $\$9$ **(b)** $\$10$

59. (a) 300 drums   **(b)** $\$6.00$

61. Using area, $\displaystyle\int_0^x f(t) \, dt = -\int_{-x}^0 f(t) \, dt = \int_0^{-x} f(t) \, dt$

63. f odd $\to \displaystyle\int_0^x f(t) \, dt$ is even, but $\dfrac{d}{dx} \displaystyle\int_0^x f(t) \, dt = f(x)$ so f is the derivative of an even function.

Similarly for f even.

Section 5.5 (pp. 289–297)

Quick Review

1. Concave down   **2.** Concave up
3. Concave down   **4.** Concave down
5. Concave up   **6.** Concave down
7. Concave up   **8.** Concave up
9. Concave down   **10.** Concave down

Exercises

1. (a) 2   **(b)** Exact   **(c)** 2
3. (a) 4.25   **(b)** Over   **(c)** 4
5. (a) 5.146   **(b)** Under   **(c)** $\dfrac{16}{3}$
7. 15,990 ft^3 **9.** 0.9785 mi

11. The average of the 13 discrete temperatures gives equal weight to the low values at the end.

13. $S_{50} = 3.1379$, $S_{100} = 3.14029$

15. $S_{50} = 1.37066$, $S_{100} = 1.37066$ using $a = 0.0001$ as lower limit

$S_{50} = 1.37076$, $S_{100} = 1.37076$ using $a = 0.000000001$ as lower limit

17. (a) $T_{10} = 1.983523538$, $T_{100} = 1.999835504$
$T_{1000} = 1.999998355$

(b)

n	$|E_T|$
10	$0.016476462 = 1.6476462 \times 10^{-2}$
100	1.64496×10^{-4}
1000	1.645×10^{-6}

(c) $\left|E_{T_{10n}}\right| \approx 10^{-2} \times \left|E_{T_n}\right|$

(d) $\left|E_{T_n}\right| \leq \dfrac{\pi^3 M}{12n^2}$. $\left|E_{T_{10n}}\right| \leq \dfrac{\pi^3 M}{12(10n)^2}$
$= \dfrac{\pi^3 M}{12n^2} \times 10^{-2}$

19. (a) $f''(x) = 2\cos(x^2) - 4x^2 \sin(x^2)$

(b)

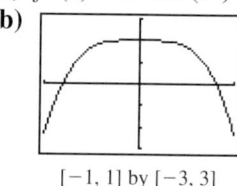

[−1, 1] by [−3, 3]

(c) The graph shows that $-3 \leq f''(x) \leq 2$ for $-1 \leq x \leq 1$.

(d) $|E_T| \leq \dfrac{1 - (-1)}{12}(h^2)(3) = \dfrac{h^2}{2}$

(e) $|E_T| \leq \dfrac{h^2}{2} \leq \dfrac{0.1^2}{2} < 0.1$

(f) $n \geq 20$

21. $466.67\ \text{in}^2$

23. Each quantity is equal to
$\dfrac{h}{2}(y_0 + 2y_1 + 2y_2 + \cdots + 2y_{n-1} + y_n)$.

Chapter Review (pp. 298–301)

1.

2. 3.75

3. 4.125

4. 3.75

5. 3.75

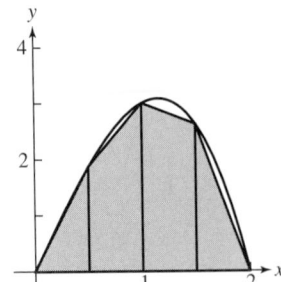

6. 4

7.

n	$LRAM_n$	$MRAM_n$	$RRAM_n$
10	1.78204	1.60321	1.46204
20	1.69262	1.60785	1.53262
30	1.66419	1.60873	1.55752
50	1.64195	1.60918	1.57795
100	1.62557	1.60937	1.59357
1000	1.61104	1.60944	1.60784

8. $\ln 5$

9. (a) True **(b)** True **(c)** False

10. (a) $V = \lim\limits_{n \to \infty} \sum\limits_{i=1}^{n} \pi \sin^2 (m_i) \, \Delta x$

(b) 4.9348

11. (a) 26.5 m

(b)

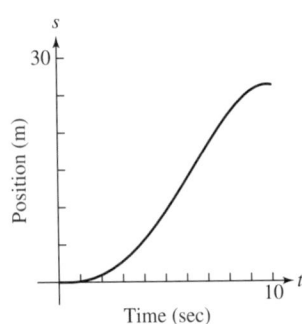

12. (a) $\int_0^{10} x^3 \, dx$ **(b)** $\int_0^{10} x \sin x \, dx$

(c) $\int_0^{10} x(3x - 2)^2 \, dx$ **(d)** $\int_0^{10} (1 + x^2)^{-1} \, dx$

(e) $\int_0^{10} \pi\left(9 - \sin^2 \dfrac{\pi x}{10}\right) dx$

13. 10 **14.** 2 **15.** 20

16. 42 **17.** $\dfrac{\sqrt{2}}{2}$ **18.** 16

19. 3 **20.** 2 **21.** 2

22. 1 **23.** $\sqrt{3}$ **24.** 1

25. 8 **26.** 2 **27.** -1

28. 0 **29.** $2 \ln 3$ **30.** π

31. 40 **32.** 64π

33. (a) Upper $= 4.392$ L; **(b)** 4.2 L
lower $= 4.008$ L

34. (a) Lower $= 87.15$ ft; **(b)** 95.1 ft
upper $= 103.05$ ft

35. One possible answer:
The dx is important because it corresponds to actual physical quantity Δx in a Riemann sum. Without the Δx, our integral approximations would be way off.

36. $\dfrac{16}{3}$

37. $0 \le \sqrt{1 + \sin^2 x} \le \sqrt{2}$

38. (a) $\dfrac{4}{3}$ **(b)** $\dfrac{2}{3}a^{3/2}$

39. $\sqrt{2 + \cos^3 x}$ **40.** $14x\sqrt{2 + \cos^3 (7x^2)}$

41. $\dfrac{-6}{3 + x^4}$ **42.** $\dfrac{2}{4x^2 + 1} - \dfrac{1}{x^2 + 1}$

43. \$230

44. $av(I) = 4800$ cases;
average holding cost $=$ \$192.00 per day

45. $x = 1.63052$ or $x = -3.09131$

46. (a) True **(b)** True **(c)** True
(d) False **(e)** True **(f)** False
(g) True

47. $F(1) - F(0)$ **48.** $y = \int_5^x \dfrac{\sin t}{t} \, dt + 3$

49. Use the fact that $y' = 2x + \dfrac{1}{x}$.

50. (b); $\dfrac{dy}{dx} = 2x \to y = x^2 + c.\ y(1) = 4 \to c = 3$

51. (a) ≈ 2.42 gal **(b)** ≈ 24.83 mpg

52. (a) 6,144 ft **(b)** 4,296 ft **(c)** B

53. (a) $h(y_1 + y_3) + 2(2hy_2) = h(y_1 + 4y_2 + y_3)$

(b) Each expression is equal to
$$\dfrac{1}{3}[h(y_0 + 4y_1 + y_2) + h(y_2 + 4y_3 + y_4)$$
$$+ \cdots + h(y_{2n-2} + 4y_{2n-1} + y_{2n}).]$$

54. (a) 0 **(b)** -1
(c) $-\pi$ **(d)** $x = 1$
(e) $y = 2x + 2 - \pi$ **(f)** $x = -1, x = 2$
(g) $[-2\pi, 0]$

55. (a) $NINT(e^{-x^2/2}, x, -10, 10) \approx 2.506628275$
$NINT(e^{-x^2/2}, x, -20, 20) \approx 2.506628275$
(b) The area is $\sqrt{2\pi}$.

56. ≈ 1500 yd^3

57. (a) $(V^2)_{av} = \dfrac{(V_{max})^2}{2}$ **(b)** 339 volts

$$V_{rms} = \dfrac{V_{max}}{\sqrt{2}}$$

Chapter 6

Section 6.1 (pp. 303–315)

Quick Review

1. $106.00   **2.** $106.14
3. $106.17   **4.** $106.18

5. $3 \cos 3x$ **6.** $\dfrac{5}{2} \sec^2 \dfrac{5}{2}x$

7. $2Ce^{2x}$ **8.** $\dfrac{1}{x+2}$

9. **10.**

[0.01, 5] by [−3, 3]   [−5, −0.01] by [−3, 3]
The graphs appear   The graphs appear
to be the same.   to be the same.

Exercises

1. $\dfrac{x^3}{3} - x^2 + x + C$ **3.** $\dfrac{x^3}{3} - \dfrac{8}{3}x^{3/2} + C$

5. $\dfrac{e^{4x}}{4} + C$ **7.** $\dfrac{x^6}{6} - 3x^2 + 3x + C$

9. $2e^{t/2} + \dfrac{5}{t} + C$ **11.** $\dfrac{x^4}{4} + \dfrac{1}{2x^2} + C$

13. $x^{1/3} + C$ **15.** $\sin\left(\dfrac{\pi}{2}x\right) + C$

17. $2 \ln |x - 1| + \ln |x| + C$

19. $\tan 5r + C$ **21.** $\dfrac{x}{2} + \dfrac{\sin 2x}{4} + C$

23. $\tan \theta - \theta + C$
25. **(a)** Graph (b)
     (b) The slope is always positive, so (a) and (c)
     can be ruled out.
27. $y = x^2 - x - 2$ **29.** $y = \tan x - 2$
31. $y = 3x^3 - 2x^2 + 5x + 10$
33. $y = -2e^{-t} + 1$ **35.** $y = -\sin \theta + \theta - 3$
37. $y = \dfrac{1}{2} \ln |t| + \dfrac{5}{4}t^2 - \dfrac{1}{4}$ **39.** $s = 4.9t^2 + 5t + 10$
41. $s = 16t^2 + 20t$
43.

[−2, 2] by [−3, 3]
45–47. The derivative of the right hand side of the
       equation is equal to the integrand on the left
       hand side.

49. **(a)** $y = \dfrac{x^2}{2} + \dfrac{1}{x} + \dfrac{1}{2}, x > 0$

     (b) $y = \dfrac{x^2}{2} + \dfrac{1}{x} + \dfrac{3}{2}, x < 0$

     (c) $y' = \begin{cases} x - \dfrac{1}{x^2}, & x < 0 \\ x - \dfrac{1}{x^2}, & x > 0 \end{cases}$

     (d) $C_1 = \dfrac{3}{2}, C_2 = \dfrac{1}{2}$

     (e) $C_1 = \dfrac{1}{2}, C_2 = -\dfrac{7}{2}$

51. $c(x) = x^3 - 6x^2 + 15x + 400$

53. **(a)** $s = \dfrac{-kt^2}{2} + 88t$ **(b)** $t = \dfrac{88}{k}$

     (c) $k = 16$ ft/sec^2

55. ≈ 1.240 sec

57. $h = \left[-\dfrac{125}{48\pi}t + 10^{5/2}\right]^{2/5}$

     $V = \dfrac{4\pi}{75}\left[-\dfrac{125}{48\pi}t + 10^{5/2}\right]^{6/5}$

59. **(a)** $y = 1200e^{0.0625t}$ **(b)** $t = \dfrac{\ln 3}{0.0625} \approx 17.6$ yr

61. **(a)** $\displaystyle\int_0^x te^t\, dt + C$ **(b)** $\displaystyle\int_0^x te^t\, dt + 1$

63.

[−6, 6] by [−4, 4]
The concavity of each solution curve indicates the
sign of y''.

65.

[−3, 3] by [−4, 10]
The concavity of each solution curve indicates the
sign of y''.

67. **(a)** $\dfrac{d}{dx}(\ln x + C) = \dfrac{1}{x}$ for $x > 0$

     (b) $\dfrac{d}{dx}(\ln (-x) + C) = \dfrac{1}{x}$ for $x < 0$

     (c) $\dfrac{d}{dx} \ln |x| = \dfrac{1}{x}$ for all x except 0.

     (d) $\dfrac{dy}{dx} = \dfrac{1}{x}$ for all x except 0.

Section 6.2 (pp. 315–323)

Quick Review

1. $\dfrac{32}{5}$ **2.** $\dfrac{16}{3}$

3. 3^x **4.** 3^x

5. $4(x^3 - 2x^2 + 3)^3(3x^2 - 4x)$

6. $8 \sin (4x - 5) \cos (4x - 5)$

7. $-\tan x$ **8.** $\cot x$

9. $\sec x$ **10.** $-\csc x$

Exercises

1. $-\dfrac{1}{3} \cos 3x + C$ **3.** $\dfrac{1}{2} \sec 2x + C$

5. $\dfrac{1}{3} \tan^{-1}\left(\dfrac{x}{3}\right) + C$ **7.** $\dfrac{2}{3}\left(1 - \cos \dfrac{t}{2}\right)^3 + C$

9. $\dfrac{1}{1 - x} + C$ **11.** $\dfrac{2}{3}(\tan x)^{3/2} + C$

13. $\ln (\ln 6)$ **15.** $\dfrac{1}{3} \sin (3z + 4) + C$

17. $\dfrac{1}{7}(\ln x)^7 + C$ **19.** $\dfrac{3}{4} \sin (s^{4/3} - 8) + C$

21. $\dfrac{1}{2} \sec (2t + 1) + C$ **23.** 0

25. $\dfrac{1}{2} \ln 5 \approx 0.805$

27. $\dfrac{1}{3} \ln \left|\sec (3x)\right| + C = -\dfrac{1}{3} \ln \left|\cos (3x)\right| + C$

29. $\ln \left|\sec x + \tan x\right| + C$ **31.** $\dfrac{14}{3}$

33. $-\dfrac{1}{2}$ **35.** $\dfrac{10}{3}$

37. $2\sqrt{3}$ **39.** $y = Ce^{(1/2)x^2 + 2x} - 5$

41. $y = -\ln (C - e^{\sin x})$ **43.** $y = \dfrac{1}{x^2 + 3}$

45. (a) $\dfrac{d}{dx}\left(\dfrac{2}{3}(x + 1)^{3/2} + C\right) = \sqrt{x + 1}$

(b) Because $\dfrac{dy_1}{dx} = \sqrt{x + 1}$ and $\dfrac{dy_2}{dx} = \sqrt{x + 1}$

(c) $4\dfrac{2}{3}$

(d) $C = y_1 - y_2$

$$= \int_0^x \sqrt{x + 1}\, dx - \int_3^x \sqrt{x + 1}\, dx$$

$$= \int_0^x \sqrt{x + 1}\, dx + \int_x^3 \sqrt{x + 1}\, dx$$

$$= \int_0^3 \sqrt{x + 1}\, dx$$

47. (a) $\dfrac{1}{2}\sqrt{10} - \dfrac{3}{2} \approx 0.081$

(b) $\dfrac{1}{2}\sqrt{10} - \dfrac{3}{2} \approx 0.081$

49. Show $\dfrac{dy}{dx} = \tan x$ and $y(3) = 5$.

51. (a) $\sin^2 x + C$ **(b)** $-\cos^2 x + C$

(c) $-\dfrac{1}{2} \cos 2x + C$

(d) The derivative of each expression is $2 \sin x \cos x$.

Section 6.3 (pp. 323–329)

Quick Review

1. $2x^3 \cos 2x + 3x^2 \sin 2x$

2. $\dfrac{3e^{2x}}{3x + 1} + 2e^{2x} \ln (3x+1)$

3. $\dfrac{2}{1 + 4x^2}$ **4.** $\dfrac{1}{\sqrt{1 - (x + 3)^2}}$

5. $x = \dfrac{1}{3}\tan y$ **6.** $x = \cos y - 1$

7. $\dfrac{2}{\pi}$ **8.** $y = \dfrac{1}{2}e^{2x} + C$

9. $y = \dfrac{1}{2}x^2 - \cos x + 3$

10. $\dfrac{d}{dx}\left[\dfrac{1}{2}e^x(\sin x - \cos x)\right] = e^x \sin x$

Exercises

1. $-x \cos x + \sin x + C$ **3.** $\dfrac{1}{2}y^2 \ln y - \dfrac{1}{4}y^2 + C$

5. $x \tan x + \ln \left|\cos x\right| + C$

7. $(2 - t^2) \cos t + 2t \sin t + C$

9. $\dfrac{1}{4}x^4 \ln x - \dfrac{1}{16}x^4 + C$ **11.** $(x^2 - 7x + 7)e^x + C$

13. $\dfrac{1}{2}e^y(\sin y - \cos y) + C$

15. $\dfrac{\pi^2}{8} - \dfrac{1}{2} \approx 0.734$

17. $\dfrac{1}{13}[e^6(2 \cos 9 + 3 \sin 9)$
$- e^{-4}(2 \cos 6 - 3 \sin 6)] \approx -18.186$

19. $y = \left(\dfrac{x^2}{4} - \dfrac{x}{8} + \dfrac{1}{32}\right)e^{4x} + C$

21. $y = \dfrac{\theta^2}{2}\sec^{-1} \theta - \dfrac{1}{2}\sqrt{\theta^2 - 1} + C$

23. (a) π **(b)** 3π **(c)** 4π

25. $\dfrac{1 - e^{-2\pi}}{2\pi} \approx 0.159$

27. $-2(\sqrt{x} \cos \sqrt{x} - \sin \sqrt{x}) + C$

29. $\dfrac{(x^6 - 3x^4 + 6x^2 - 6)e^{x^2}}{2} + C$

31. $u = x^n$, $dv = \cos x\, dx$ **33.** $u = x^n$, $dv = e^{ax}\, dx$

35. (a) Let $y = f^{-1}(x)$. Then $x = f(y)$,
so $dx = f'(y)\, dy$. Substitute directly.
(b) $u = y$, $dv = f'(y)\, dy$

37. (a) $\displaystyle\int \sin^{-1} x\, dx = x \sin^{-1} x + \cos(\sin^{-1} x) + C$

(b) $\displaystyle\int \sin^{-1} x\, dx = x \sin^{-1} x + \sqrt{1 - x^2} + C$

(c) $\cos(\sin^{-1} x) = \sqrt{1 - x^2}$

39. (a) $\displaystyle\int \cos^{-1} x\, dx = x \cos^{-1} x - \sin(\cos^{-1} x) + C$

(b) $\displaystyle\int \cos^{-1} x\, dx = x \cos^{-1} x - \sqrt{1 - x^2} + C$

(c) $\sin(\cos^{-1} x) = \sqrt{1 - x^2}$

Section 6.4 (pp. 330–341)

Quick Review

1. $a = e^b$ **2.** $c = \ln(d)$

3. $x = e^2 - 3$ **4.** $x = \dfrac{1}{2}\ln 6$

5. $x = \dfrac{\ln 2.5}{\ln 0.85} \approx -5.638$ **6.** $k = \dfrac{\ln 2}{\ln 3 - \ln 2} \approx 1.710$

7. $t = \dfrac{\ln 10}{\ln 1.1} \approx 24.159$ **8.** $t = \dfrac{1}{2}\ln 4 = \ln 2$

9. $y = -1 + e^{2x+3}$ **10.** $y = -2 \pm e^{3t-1}$

Exercises

Most of the numerical answers in this section are approximations.

1. $y(t) = 100e^{1.5t}$ **3.** $y(t) = 50e^{(0.2 \ln 2)t}$
5. 8.06 yr doubling time; \$13,197.10 in 30 yr
7. \$600 initially; 13.2 yr doubling time
9. (a) 14.94 yr **(b)** 14.62 yr
(c) 14.68 yr **(d)** 14.59 yr
11. (a) 2.8×10^{14} bacteria
(b) The bacteria reproduce fast enough that even if many are destroyed there are enough left to make the person sick.
13. 0.585 days **15.** $y \approx 2e^{0.4581t}$
17. $y = y_0 e^{-kt} = y_0 e^{-k(3/k)} = y_0 e^{-3} < 0.05 y_0$
19. (a) 17.53 minutes longer
(b) 13.26 minutes
21. 6658 years
23. (a) 168.5 meters **(b)** 41.13 seconds
25. 585.4 kg
27. (a) $p = 1013e^{-0.121h}$ **(b)** 2.383 millibars
(c) 0.977 km
29. (a) $V = V_0 e^{-t/40}$ **(b)** 92.1 seconds
31. (b) $\displaystyle\lim_{t\to\infty} s(t) = \dfrac{v_0 m}{k}$

33. $s(t) = 1.32(1 - e^{-0.606t})$
35. (b) $T = T_s$ is a horizontal asymptote.

37. (a)

x	$\left(1 + \dfrac{1}{x}\right)^x$
10	2.5937
100	2.7048
1000	2.7169
10,000	2.7181
100,000	2.7183

$e \approx 2.7183$

(c) As we compound more times the increment of time between compounding approaches 0. Continuous compounding is based on an instantaneous rate of change which is a limit of average rates as the increment in time approaches 0.

Section 6.5 (pp. 342–349)

Quick Review

1. All real numbers

2. $\displaystyle\lim_{x\to+\infty} f(x) = 50$; $\displaystyle\lim_{x\to-\infty} f(x) = 0$

3. $y = 0$ and $y = 50$ **4.** All real numbers

5.

[−30, 70] by [−10, 60]

no zeros

6.

[−30, 70] by [−0.5, 2]

(a) $(-\infty, \infty)$

(b) None

7.

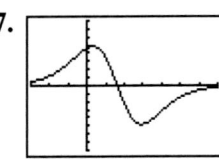

[−30, 70] by [−0.08, 0.08]

(a) $(-\infty, 10 \ln 5) \approx (-\infty, 16.094)$

(b) $(10 \ln 5, \infty) \approx (16.094, \infty)$

8. $(10 \ln 5, 25) \approx (16.094, 25)$

9. $A = 3$, $B = -2$ **10.** $A = -2$, $B = 4$

Exercises

1. (a) $\dfrac{dP}{dt} = 0.025P$ **(b)** $P(t) = 75{,}000e^{0.025t}$

1. continued

(c)

[0, 100] by [0, 1,000,000]

3. (a) $\dfrac{dP}{dt} = 0.00025P(200 - P)$

(b) $P(t) = \dfrac{200}{1 + 19e^{-0.05t}}$

(c)

[0, 100] by [0, 250]

5. -30%

7. $k = 0.04; M = 100$

9. (d)

11. (c)

13. (a) $k = 0.7; M = 1000$

(b) $P(0) \approx 8$; Initially there are 8 rabbits.

15. (a) 0.875% **(b)** $275{,}980{,}017$

17. (a) $P(t) = \dfrac{150}{1 + 24e^{-0.225t}}$

(b) About 17.21 weeks; 21.28 weeks

19. (a) $y = y_0 e^{kt}$ where $k = \ln 0.99^{(1/1000)} \approx -0.00001$

(b) $\approx 10{,}536$ yrs   **(c)** $\approx 81.9\%$

21. (a) $x = 11{,}000e^{0.1t} - 10{,}000$

(b) ≈ 23 yrs

23. (a) $y = \dfrac{18.70}{1 + 1.075e^{-0.0422x}}$

[0, 100] by [0, 20]

(b) 18.7 million

(c) $x \approx 1.7$ (year 1912);
population ≈ 9.35 million

25. Separate variables and rewrite $\dfrac{M}{P(M - P)}$ as
$\dfrac{1}{P} + \dfrac{1}{M - P}$ in order to integrate.

27. $y = e^{\sin x} - 1$ **29.** $y = \sqrt{x^2 + 4}$

31. (a) $\dfrac{dP}{dt}$ has the same sign as $(M - P)(P - m)$.

(b) $P(t) = \dfrac{1200Ae^{11kt/12} + 100}{1 + Ae^{11kt/12}}$

(c) $P(t) = \dfrac{300(8e^{11kt/12} + 3)}{9 + 2e^{11kt/12}}$

(d)

[0, 75] by [0, 1500]

(e) $P(t) = \dfrac{AMe^{(M-m)kt/M} + m}{}$ where

$A = \dfrac{P(0) - m}{M - P(0)}$

33. (a) $P(t) = \dfrac{P_0}{1 - kP_0 t}$

(b) Vertical asymptote at $t = \dfrac{1}{kP_0}$

Section 6.6 (pp. 350–356)

Quick Review

1. 9

2. $L(x) = 9x - 16$

3. 2

4. $L(x) = 2x - \dfrac{\pi}{2} + 1$

5. 0.4875

6. $y = 0.4875x + 0.9$

7. (a) 0.001762   **(b)** 0.061%

8. (a) 0.006976   **(b)** 0.236%

9. (a) 0.042361   **(b)** 1.351%

10. (a) 0.047321   **(b)** 1.783%

Exercises

1. $f'(x) = 1 - 2e^{-x} = x - f(x)$ and $f(0) = 1$.

3. $f'(x) = \dfrac{1}{5}(2e^{2x} - 2\cos x + \sin x) = 2f(x) + \sin x$
and $f(0) = 0$.

5. $y = 2e^x - 1$ **7.** $y = 2e^{x^2 + 2x}$

15. (a) $y = \dfrac{-1}{x^2 - 2x + 2}$, $y(3) = -0.2$

(b) -0.1851, error ≈ 0.0149

(c) -0.1929, error ≈ 0.0071

(d) -0.1965, error ≈ 0.0035

17. (a) -0.2024, error ≈ 0.0024

(b) -0.2005, error ≈ 0.0005

(c) -0.2001, error ≈ 0.0001

(d) As the step size decreases, the accuracy of
the method increases and the error decreases.

19.

[−0.1, 1.1] by [−0.13, 0.88]

21.

[−2.2, 0.2] by [−0.2, 2.2]

23.

x	y (Euler)	y (exact)	Error
0	1	1.0	0
-0.1	0.9000	0.9097	0.0097
-0.2	0.8200	0.8375	0.0175
-0.3	0.7580	0.7816	0.0236
-0.4	0.7122	0.7406	0.0284
-0.5	0.6810	0.7131	0.0321
-0.6	0.6629	0.6976	0.0347
-0.7	0.6566	0.6932	0.0366
-0.8	0.6609	0.6987	0.0377
-0.9	0.6748	0.7131	0.0383
-1.0	0.6974	0.7358	0.0384

25. (a)   (b)

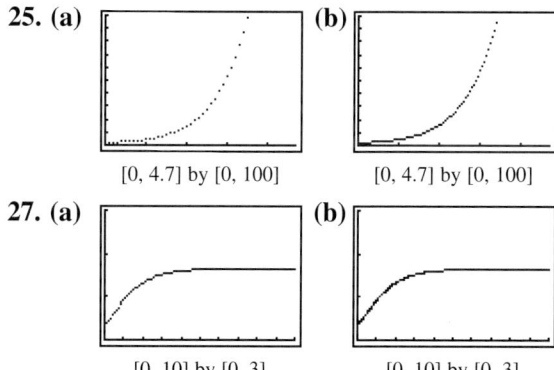

[0, 4.7] by [0, 100]   [0, 4.7] by [0, 100]

27. (a)   (b)

[0, 10] by [0, 3]   [0, 10] by [0, 3]

29. 2.6533, e

Chapter Review (pp. 358–361)

1. $\sqrt{3}$

2. 2

3. 8

4. 0

5. 2

6. $\dfrac{147}{8}$

7. $e - 1$

8. $\dfrac{2}{3}$

9. $-\ln|2 - \sin x| + C$

10. $\dfrac{1}{2}(3x + 4)^{2/3} + C$

11. $\dfrac{1}{2}\ln(t^2 + 5) + C$

12. $-\sec\dfrac{1}{\theta} + C$

13. $-\ln|\cos(\ln y)| + C$

14. $\ln|\sec(e^x) + \tan(e^x)| + C$

15. $\ln|\ln x| + C$

16. $\dfrac{-2}{\sqrt{t}} + C$

17. $x^3 \sin x + 3x^2 \cos x - 6x \sin x - 6 \cos x + C$

18. $\dfrac{x^5 \ln x}{5} - \dfrac{x^5}{25} + C$

19. $\left(\dfrac{3 \sin x}{10} - \dfrac{\cos x}{10}\right)e^{3x} + C$

20. $\left(-\dfrac{x^2}{3} - \dfrac{2x}{9} - \dfrac{2}{27}\right)e^{-3x} + C$

21. $y = \dfrac{x^3}{6} + \dfrac{x^2}{2} + x + 1$ **22.** $y = \dfrac{x^3}{3} + 2x - \dfrac{1}{x} - \dfrac{1}{3}$

23. $y = \ln(t + 4) + 2$ **24.** $y = -\dfrac{1}{2}\csc 2\theta + \dfrac{3}{2}$

25. $y = \dfrac{x^3}{3} + \ln x - x + \dfrac{2}{3}$

26. $r = \sin t - \dfrac{t^2}{2} - 2t - 1$

27. $y = 4e^x - 2$ **28.** $y = 2e^{x^2 + x} - 1$

29. $-1 + \sqrt{x} + C$ or $\sqrt{x} + C$

30. $\dfrac{x^2}{2} + 1 - \sqrt{x} + C$ or $\dfrac{x^2}{2} - \sqrt{x} + C$

31. $-2\sqrt{x} - x + C$

32. $2 - 3x + C$ or $-3x + C$

33. (b)   **34.** (d)

35. **iv**, since the given graph looks like $y = x^2$, which satisfies $\dfrac{dy}{dx} = 2x$ and $y(1) = 1$.

36. Yes, $y = x$ is a solution.

37. (a) $v = 2t + 3t^2 + 4$ (b) 6 m

38.

[−10, 10] by [−10, 10]

39.

x	y
0	0
0.1	0.1000
0.2	0.2095
0.3	0.3285
0.4	0.4568
0.5	0.5946
0.6	0.7418
0.7	0.8986
0.8	1.0649
0.9	1.2411
1.0	1.4273
1.1	1.6241
1.2	1.8319
1.3	2.0513
1.4	2.2832
1.5	2.5285
1.6	2.7884
1.7	3.0643
1.8	3.3579
1.9	3.6709
2.0	4.0057

40.

x	y
-3	1
-2.9	0.6680
-2.8	0.2599
-2.7	-0.2294
-2.6	-0.8011
-2.5	-1.4509
-2.4	-2.1687
-2.3	-2.9374
-2.2	-3.7333
-2.1	-4.5268
-2.0	-5.2840
-1.9	-5.9686
-1.8	-6.5456
-1.7	-6.9831
-1.6	-7.2562
-1.5	-7.3488
-1.4	-7.2553
-1.3	-6.9813
-1.2	-6.5430
-1.1	-5.9655
-1.0	-5.2805

41. 0.9063   **42.** 4.4974

43. (a) **(b)**

[−0.2, 4.5] by [−2.5, 0.5]   [−1, 0.2] by [−10, 2]

44. (a) 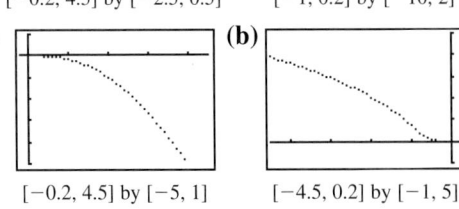 **(b)**

[−0.2, 4.5] by [−5, 1]   [−4.5, 0.2] by [−1, 5]

45. (a) $k \approx 0.262059$ **(b)** About 3.81593 years
46. About 92 minutes   **47.** $-3°C$
48. About 41.2 years
49. About 18,935 years old
50. About 5.3%   **51.** About 59.8 ft
52. (a) $y = c + (y_0 - c)e^{-(kA/V)t}$
   (b) c
53. (a) $k = 1$; carrying capacity $= 150$
   (b) ≈ 2; Initially there were 2 infected students.
   (c) About 6 days
54. Use the Fundamental Theorem of Calculus to obtain $y' = \sin(x^2) + 3x^2 + 1$. Then differentiate again and also verify the initial conditions.
55. $P = \dfrac{800}{1 + 15e^{-0.002t}}$

56. Method 1—Compare graph of $y_1 = x^2 \ln x$ with
$y_2 = \text{NDER}\left(\dfrac{x^3 \ln x}{3} - \dfrac{x^3}{9}\right)$.

Method 2—Compare graph of $y_1 = \text{NINT}(x^2 \ln x)$
with $y_2 = \dfrac{x^3 \ln x}{3} - \dfrac{x^3}{9}$.

57. (a) About 11.3 years **(b)** About 11 years

58. (a) $\dfrac{d}{dx}\displaystyle\int_0^x u(t)\,dt = u(x)$

   $\dfrac{d}{dx}\displaystyle\int_3^x u(t)\,dt = u(x)$

   (b) $C = \displaystyle\int_0^3 u(t)\,dt$

59. (a) $y = \dfrac{56.0716}{1 + 5.894e^{-0.0205x}}$

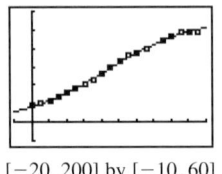

[−20, 200] by [−10, 60]

   (b) ≈ 56.0716 million **(c)** ≈ 1887, ≈ 28.0 million

60. (a) $T = 79.961(0.9273)^t$

[−1, 33] by [−5, 90]

   (b) About 9.2 sec   **(c)** About 79.96°C
61. $s = 0.97(1 - e^{-0.8866t})$
   A graph of the model is shown superimposed on a graph of the data.

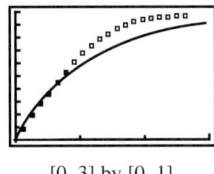

[0, 3] by [0, 1]

Chapter 7

Section 7.1 (pp. 363–374)

Quick Review

1. Changes sign at $-\dfrac{\pi}{2}, 0, \dfrac{\pi}{2}$

2. Changes sign at 1, 2

3. Always positive

4. Changes sign at $-\dfrac{1}{2}$

5. Changes sign at $\dfrac{\pi}{4}, \dfrac{3\pi}{4}, \dfrac{5\pi}{4}$

6. Always positive

7. Changes sign at 0

8. Changes sign at $-2, -\sqrt{2}, \sqrt{2}, 2$

9. Changes sign at $0.9633 + k\pi$
$$2.1783 + k\pi$$
where k is an integer

10. Changes sign at $\dfrac{1}{3\pi}, \dfrac{1}{2\pi}$

Exercises

1. (a) Right: $0 \le t < \dfrac{\pi}{2} \cup \dfrac{3\pi}{2} < t \le 2\pi$

Left: $\dfrac{\pi}{2} < t < \dfrac{3\pi}{2}$

Stopped: $t = \dfrac{\pi}{2}, \dfrac{3\pi}{2}$

(b) 0 **(c)** 20

3. (a) Right: $0 \le t < 5$

Left: $5 < t \le 10$

Stopped: $t = 5$

(b) 0 **(c)** 245

5. (a) Right: $0 < t < \dfrac{\pi}{2} \cup \dfrac{3\pi}{2} < t < 2\pi$

Left: $\dfrac{\pi}{2} < t < \pi, \pi < t < \dfrac{3\pi}{2}$

Stopped: $t = 0, \dfrac{\pi}{2}, \pi, \dfrac{3\pi}{2}, 2\pi$

(b) 0 **(c)** $\dfrac{20}{3}$

7. (a) Right: $0 \le t < \dfrac{\pi}{2} \cup \dfrac{3\pi}{2} < t \le 2\pi$

Left: $\dfrac{\pi}{2} < t < \dfrac{3\pi}{2}$

Stopped: $t = \dfrac{\pi}{2}, \dfrac{3\pi}{2}$

(b) 0 **(c)** $2e - \dfrac{2}{e} \approx 4.7$

9. (a) 63 mph **(b)** 344.52 feet

11. (a) -6 ft/sec **(b)** 5.625 sec

(c) 0 **(d)** 253.125 feet

13. 33 **15.** $t = a$

17. (a) 6 **(b)** 4 meters

19. (a) 5 **(b)** 7 meters

21. ≈ 332.965 billion barrels

23. (a) 2 miles **(b)** $2\pi r \Delta r$

(c) Population = Population density × Area

(d) Approximately 83,776

25. One possible answer:

Plot the speeds vs. time. Connect the points and find the area under the line graph. The definite integral also gives the area under the curve.

27. (a) 798.97 thousand

(b) The answer in (a) corresponds to the area of rectangles whose heights are the actual sales ("midpoint rectangles"). Part of each rectangle is above the curve and part is below.

29. (a) 18 N **(b)** 81 N · cm

31. 0.04875

33. (a, b) Take $dm = \delta\, dA$ as m_k and letting $dA \to 0$, $k \to \infty$ in the center of mass equations.

35. $\bar{x} = \dfrac{4}{3}, \bar{y} = 0$

Section 7.2 (pp. 374–382)

Quick Review

1. 2 **2.** $\dfrac{1}{2}(e^2 - 1) \approx 3.195$

3. 2 **4.** 4

5. $\dfrac{9\pi}{2}$ **6.** $(6, 12); (-1, 5)$

7. $(0, 1)$ **8.** $(0, 0); (\pi, 0)$

9. $(-1, -1); (0, 0); (1, 1)$

10. $(-0.9286, -0.8008); (0, 0); (0.9286, 0.8008)$

Exercises

1. $\dfrac{\pi}{2}$ **3.** $\dfrac{1}{12}$ **5.** $\dfrac{128}{15}$

7. $\dfrac{5}{6}$ **9.** 16 **11.** $10\dfrac{2}{3}$

13. 4 **15.** $\dfrac{2}{3}a^3$ **17.** $21\dfrac{1}{3}$

19. $30\dfrac{3}{8}$ **21.** $\dfrac{8}{3}$ **23.** 8

25. $6\sqrt{3}$ **27.** $\dfrac{4 - \pi}{\pi} \approx 0.273$

29. $4 - \pi \approx 0.858$ **31.** $\dfrac{1}{2}$

33. $\sqrt{2} - 1 \approx 0.414$

35. (a) $(-\sqrt{c}, c); (\sqrt{c}, c)$

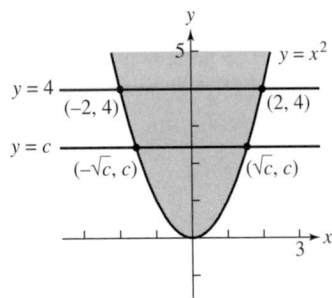

(b) $\int_0^c \sqrt{y}\, dy = \int_c^4 \sqrt{y}\, dy \Rightarrow c = 2^{4/3}$

(c) $\int_0^{\sqrt{c}} (c - x^2)\, dx$

$= (4 - c)\sqrt{c} + \int_{\sqrt{c}}^2 (4 - x^2)\, dx \Rightarrow c = 2^{4/3}$

37. $\dfrac{3}{4}$ **39.** Neither; both are zero

41. $\ln 4 - \dfrac{1}{2} \approx 0.886$ **43.** $k = 1.8269$

45. Since $f(x) - g(x)$ is the same for each region where $f(x)$ and $g(x)$ represent the upper and lower edges, area $= \int_a^b [f(x) - g(x)]\, dx$ will be the same for each.

Section 7.3 (pp. 383–394)

Quick Review

1. x^2 **2.** $\dfrac{x^2}{2}$ **3.** $\dfrac{\pi x^2}{2}$

4. $\dfrac{\pi x^2}{8}$ **5.** $\dfrac{\sqrt{3}}{4}x^2$ **6.** $\dfrac{x^2}{2}$

7. $\dfrac{x^2}{4}$ **8.** $\dfrac{\sqrt{15}}{4}x^2$ **9.** $6x^2$

10. $\dfrac{3\sqrt{3}}{2}x^2$

Exercises

1. (a) $\pi(1 - x^2)$ **(b)** $4(1 - x^2)$

(c) $2(1 - x^2)$ **(d)** $\sqrt{3}(1 - x^2)$

3. 16 **5.** $\dfrac{16}{3}$

7. (a) $2\sqrt{3}$ **(b)** 8

9. 8π

11. (a) s^2h **(b)** s^2h

13. $\dfrac{2}{3}\pi$ **15.** $4 - \pi$

17. $\dfrac{32\pi}{5}$ **19.** 36π

21. $\dfrac{2}{3}\pi$ **23.** $\dfrac{117\pi}{5}$

25. $\pi^2 - 2\pi$ **27.** 2.301

29. 2π **31.** $\dfrac{4}{3}\pi$

33. 8π

35. (a) 8π **(b)** $\dfrac{32\pi}{5}$

(c) $\dfrac{8\pi}{3}$ **(d)** $\dfrac{224\pi}{15}$

37. (a) $\dfrac{16\pi}{15}$ **(b)** $\dfrac{56\pi}{15}$

(c) $\dfrac{64\pi}{15}$

39. 8π **41.** $\dfrac{128\pi}{5}$

43. (a) $\dfrac{6\pi}{5}$ **(b)** $\dfrac{4\pi}{5}$

(c) 2π **(d)** 2π

45. (a) $\dfrac{512\pi}{21}$ **(b)** $\dfrac{832\pi}{21}$

47. (a) $\dfrac{11\pi}{48}$ **(b)** $\dfrac{11\pi}{48}$

49. (a) $\dfrac{36\pi}{5}$ cm^3 **(b)** 192.3 g

51. (a) 2.3, 1.6, 1.5, 2.1, 3.2, 4.8, 7.0, 9.3, 10.7, 10.7, 9.3, 6.4, 3.2

(b) $\dfrac{1}{4\pi} \int_0^6 (C(y))^2\, dy$ **(c)** ≈ 34.7 in^3

53. (a) $\dfrac{32\pi}{3}$

(b) The answer is independent of r.

55. 5 **57.** ≈ 13.614 **59.** ≈ 16.110

61. ≈ 53.226 **63.** ≈ 6.283

65. Hemisphere cross sectional area:

$\pi(\sqrt{R^2 - h^2})^2 = A_1$

Right circular cylinder with cone removed cross sectional area: $\pi R^2 - \pi h^2 = A_2$

Since $A_1 = A_2$, the two volumes are equal by Cavalieri's theorem. Thus,

volume of hemisphere

$=$ volume of cylinder $-$ volume of cone

$= \pi R^3 - \dfrac{1}{3}\pi R^3 = \dfrac{2}{3}\pi R^3.$

67. (a) $\dfrac{\pi h^2(3a - h)}{3}$ **(b)** $\dfrac{1}{120\pi}$ m/sec

Section 7.4 (pp. 395–401)

Quick Review

1. $x + 1$

2. $\dfrac{2 - x}{2}$

3. $\sec x$

4. $\dfrac{x^2 + 4}{4x}$

5. $\sqrt{2} \cos x$

6. 4

7. 0

8. -3

9. 2

10. $k\pi$, k any integer

Exercises

1. (a) $\displaystyle\int_{-1}^{2} \sqrt{1 + 4x^2}\, dx$

(b)

$[-1, 2]$ by $[-1, 5]$

(c) ≈ 6.126

3. (a) $\displaystyle\int_{0}^{\pi} \sqrt{1 + \cos^2 y}\, dy$

(b)

$[-1, 2]$ by $[-1, 4]$

(c) ≈ 3.820

5. (a) $\displaystyle\int_{-1}^{7} \sqrt{1 + \dfrac{1}{2x + 2}}\, dx$

(b)

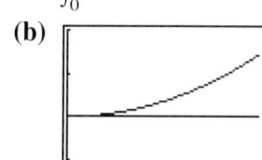

$[-1, 7]$ by $[-2, 4]$

(c) ≈ 9.294

7. (a) $\displaystyle\int_{0}^{\pi/6} \sqrt{1 + \tan^2 x}\, dx$

(b)

$\left[0, \dfrac{\pi}{6}\right]$ by $[-0.1, 0.2]$

(c) ≈ 0.549

9. (a) $\displaystyle\int_{-\pi/3}^{\pi/3} \sqrt{1 + \sec^2 x \tan^2 x}\, dx$

(b)

$\left[-\dfrac{\pi}{3}, \dfrac{\pi}{3}\right]$ by $[-1, 3]$

(c) ≈ 3.139

11. 12

13. $\dfrac{53}{6}$

15. $\dfrac{17}{12}$

17. 2

19. (a) $y = \sqrt{x}$

(b) Only one. We know the derivative of the function and the value of the function at one value of x.

21. 1

23. ≈ 21.07 inches

25. $(-19.909, 8.41)$

27. 2.1089

29. ≈ 1.623

31. Because the limit of the sum $\sum \Delta x_k$ as the norm of the partition goes to zero will always be the length $(b - a)$ of the interval (a, b).

33. (a) The fin is the hypotenuse of a right triangle with leg lengths Δx_k and
$$\dfrac{df}{dx}\bigg|_{x=x_{k-1}} \Delta x_k = f'(x_{k-1})\, \Delta x_k.$$

(b) $\displaystyle\lim_{n\to\infty} \sum_{k=1}^{n} \sqrt{(\Delta x_k)^2 + (f'(x_{k-1})\, \Delta x_k)^2}$

$$= \lim_{n\to\infty} \sum_{k=1}^{n} \Delta x_k \sqrt{1 + (f'(x_{k-1}))^2}$$

$$\int_{a}^{b} \sqrt{1 + (f'(x))^2}\, dx$$

Section 7.5 (pp. 401–411)

Quick Review

1. $1 - \dfrac{1}{e}$

2. $e - 1$

3. $\dfrac{\sqrt{2}}{2}$

4. 15

5. $\dfrac{1}{3} \ln\left(\dfrac{9}{2}\right)$

6. $\displaystyle\int_{0}^{7} 2\pi(x + 2) \sin x\, dx$

7. $\displaystyle\int_{0}^{7} (1 - x^2)(2\pi x)\, dx$

8. $\displaystyle\int_{0}^{7} \pi \cos^2 x\, dx$

9. $\displaystyle\int_{0}^{7} \pi\left(\dfrac{y}{2}\right)^2 (10 - y)\, dy$

10. $\displaystyle\int_{0}^{7} \dfrac{\sqrt{3}}{4} \sin^2 x\, dx$

Exercises

1. $\approx$4.4670 J   **3.** 9 J

5. 4900 J   **7.** 1944 ft-lb

9. (a) 7238 lb/in.
   (b) $\approx$905 in.-lb, $\approx$2714 in.-lb

11. 780 J   **13.** 1123.2 lb

15. 3705 lb

17. (a) 1,497,600 ft-lb   **(b)** $\approx$100 min
   (d) 1,494,240 ft-lb, $\approx$100 min
      1,500,000 ft-lb, 100 min

19. Through valve: 84,687.3 ft-lb
   Over the rim: 98,801.8 ft-lb
   Through a hose attached to a valve in the bottom
   is faster, because it takes more time to do more
   work.

21. 53,482.5 ft-lb   **23.** $\approx$967,611 ft-lb, yes

25. (a) $\approx$209.73 lb
   (b) $\approx$838.93 lb; the fluid force doubles

27. (a) 0.5 (50%)   **(b)** $\approx$0.24 (24%)
   (c) $\approx$0.0036 (0.36%)
   (d) 0 if we assume a continuous distribution;
      $\approx$0.071 between 59.5 in. and 60.5 in.

29. Integration is a good approximation to the area.

31. 5.1446×10^{10} J

33. $F = m\dfrac{dv}{dt} = mv\dfrac{dv}{dx}$, so $w = \displaystyle\int_{x_1}^{x_2} F(x)\, dx$

$$= \int_{x_1}^{x_2} mv\frac{dv}{dx}\, dx = \int_{v_1}^{v_2} mv\, dv = \frac{1}{2}mv_2^2 - \frac{1}{2}mv_1^2$$

35. $\approx$85.1 ft-lb   **37.** 64.6 ft-lb

39. 110.6 ft-lb

Chapter Review (pp. 413–415)

1. 10.417 ft   **2.** 31.361 gal   **3.** 1464

4. 14 g   **5.** 14,400   **6.** 1

7. $\dfrac{9}{2}$ **8.** $\dfrac{1}{6}$ **9.** 18

10. 30.375   **11.** 0.0155   **12.** 4

13. 8.9023   **14.** 2.1043

15. $2\sqrt{3} - \dfrac{2}{3}\pi \approx 1.370$ **16.** $2\sqrt{3} + \dfrac{4}{3}\pi \approx 7.653$

17. 1.2956   **18.** 5.7312

19. 4π **20.** 2π

21. (a) $\dfrac{32\pi}{3}$ **(b)** $\dfrac{128\pi}{15}$

   (c) $\dfrac{64\pi}{5}$ **(d)** $\dfrac{32\pi}{3}$

22. (a) 4π **(b)** πk^2 **(c)** $\dfrac{1}{\pi}$

23. 276 in^3 **24.** $\dfrac{\pi^2}{4}$

25. $\pi(2 - \ln 3)$ **26.** $\dfrac{28\pi}{3}$ ft^3

27. 19.4942   **28.** 5.2454

29. 2.296 sec   **30. (a)** is true

31. 39

32. (a) 4000 J   **(b)** 640 J
   (c) 4640 J

33. 22,800,000 ft-lb   **34.** 12 J, 213.3 J

35. No, the work going uphill is positive, but the
   work going downhill is negative.

36. 113.097 in.-lb   **37.** 426.67 lbs

38. 6.6385 lb,   5.7726 lb,   9.4835 lb
      base   front   sides
                  back

39. 14.4   **40.** 0.2051 (20.5%)

41. Answer will vary.

42. (a) 0.6827 (68.27%)   **(b)** 0.9545 (95.45%)
   (c) 0.9973 (99.73%)

43. The probability that the variable has some value
   in the range of all possible values is 1.

44. π **45.** 3π

46. $2\pi^2$ **47.** $\dfrac{16\pi}{3}$

48. 9.7717

49. (a) $y = 5 - \dfrac{5}{4}x^2$ **(b)** 335.1032 in^3

50. $f(x) = \dfrac{x^2 - 2\ln x + 3}{4}$ **51.** $\approx$3.84

52. $\approx$5.02

Chapter 8

Section 8.1 (pp. 417–425)

Quick Review

1. 1.1052   **2.** 2.7183   **3.** 1

4. ∞ **5.** 2   **6.** 2

7. 3   **8.** 1

9. $y = \dfrac{\sin h}{h}$ **10.** $y = (1 + h)^{1/h}$

Exercises

1. $\dfrac{1}{4}$ **3.** 1

5. $\dfrac{3}{11}$ **7.** e^2

9. (a)

x	10	10^2	10^3	10^4	10^5
$f(x)$	1.1513	0.2303	0.0354	0.0046	0.00058

Estimated limit = 0

(b) Note $\ln x^5 = 5 \ln x$.

$$\lim_{x \to \infty} \frac{5 \ln x}{x} = \lim_{x \to \infty} \frac{\frac{5}{x}}{1} = \frac{0}{1} = 0$$

11. $\dfrac{3}{4}$ **13.** 1 **15.** 0

17. -1 **19.** $\ln 2$ **21.** 1

23. 0 **25.** 1 **27.** 0

29. e^2 **31.** 0 **33.** 1

35. 1 **37.** e **39.** 1

41. e^{-1}

43. (a) L'Hôpital's rule does not help because applying l'Hôpital's rule to this quotient essentially "inverts" the problem by interchanging the numerator and denominator. It is still essentially the same problem and one is no closer to a solution. Applying l'Hôpital's rule a second time returns to the original problem.

(b, c) 3

45. Possible answers:

(a) $f(x) = 7(x - 3)$, $g(x) = x - 3$

(b) $f(x) = (x - 3)^2$, $g(x) = x - 3$

(c) $f(x) = x - 3$, $g(x) = (x - 3)^3$

47. $c = \dfrac{27}{10}$, because this is the limit of $f(x)$ as x approaches 0.

49. (a) $\ln \left(1 + \dfrac{r}{k}\right)^k = k \ln \left(1 + \dfrac{r}{k}\right)$. And, as $k \to \infty$,

$$\lim_{k \to \infty} k \ln \left(1 + \frac{r}{k}\right) = \lim_{k \to \infty} \frac{\ln \left(1 + \frac{r}{k}\right)}{\frac{1}{k}}$$

$$= \lim_{k \to \infty} \frac{\frac{-r}{k^2}\left(1 + \frac{r}{k}\right)}{\frac{-1}{k^2}}.$$

$$= \lim_{k \to \infty} \frac{r}{1 + \frac{r}{k}} = r.$$

Therefore, $\lim_{k \to \infty} \left(1 + \dfrac{r}{k}\right)^k = e^r$.

Hence, $\lim_{k \to \infty} A_0 \left(1 + \dfrac{r}{k}\right)^{kt} = A_0 e^{rt}$.

(b) Part (a) shows that as the number of compoundings per year increases toward infinity, the limit of interest compounded k times per year is interest compounded continuously.

51. (a) 1 **(b)** $\dfrac{\pi}{2}$ **(c)** π

53. (a) $(-\infty, -1) \cup (0, \infty)$

(b) ∞ **(c)** e

55. (a) $c = \dfrac{1}{3}$ **(b)** $c = \dfrac{\pi}{4}$

Section 8.2 (pp. 425–433)

Quick Review

1. 0 **2.** ∞ **3.** ∞

4. 0 **5.** $-3x^4$ **6.** $2x^2$

7. $\lim_{x \to \infty} \dfrac{f(x)}{g(x)} = \lim_{x \to \infty} \left(1 + \dfrac{\ln x}{x}\right) = 1 + 0 = 1$

8. $\lim_{x \to \infty} \dfrac{f(x)}{g(x)} = \lim_{x \to \infty} \sqrt{1 + \dfrac{5}{4x}} = 1$

9. (a) Local minimum at $(0, 1)$
Local maximum at $\approx (2, 1.541)$

(b) $[0, 2]$ **(c)** $(-\infty, 0]$ and $[2, \infty)$

10. f doesn't have an absolute maximum value. The values are always less than 2 and the values get arbitrarily close to 2 near $x = 0$, but the function is undefined at $x = 0$.

Exercises

1. Slower **3.** Faster **5.** Same rate

7. Slower **9.** Slower **11.** Same rate

13. Slower **15.** Slower **17.** Same rate

19. Slower **21.** Faster

23. $e^{x/2}$, e^x, $(\ln x)^x$, x^x

25. $\lim_{x \to \infty} \dfrac{f_2(x)}{f_1(x)} = \sqrt{10}$ and $\lim_{x \to \infty} \dfrac{f_3(x)}{f_1(x)} = 1$, so f_2 and f_3 also grow at the same rate.

27. $\lim_{x \to \infty} \dfrac{f_2(x)}{f_1(x)} = 1$ and $\lim_{x \to \infty} \dfrac{f_3(x)}{f_1(x)} = 1$, so f_2 and f_3 also grow at the same rate.

29. (a) False **(b)** False **(c)** True

(d) True **(e)** True **(f)** True

(g) False **(h)** True

31. $g = o(f)$

33. f and g grow at the same rate.

35. (a) The nth derivative of x^n is $n!$, which is a constant. Therefore n applications of l'Hôpital's rule give

$$\lim_{x \to \infty} \frac{e^x}{x^n} = \cdots = \lim_{x \to \infty} \frac{e^x}{n!} = \infty.$$

(b) In this case, n applications of l'Hôpital's rule give $\lim_{x \to \infty} \dfrac{a^x}{x^n} = \cdots = \lim_{x \to \infty} \dfrac{a^x (\ln a)^n}{n!} = \infty$

37. (a) $\displaystyle\lim_{x \to \infty} \frac{\ln x}{x^{1/n}} = \lim_{x \to \infty} \frac{\frac{1}{x}}{x^{(1/n)-1}\frac{1}{n}} = \lim_{x \to \infty} \frac{n}{x^{1/n}} = 0$

(b) $\displaystyle\lim_{x \to \infty} \frac{\ln x}{x^a} = \lim_{x \to \infty} \frac{\frac{1}{x}}{ax^{a-1}} = \lim_{x \to \infty} \frac{1}{ax^a} = 0$

39. The one which is $O(n \log_2 n)$ is likely the most efficient, because of the three given functions, it grows the most slowly as $n \to \infty$.

41. This is the case because if $\displaystyle\lim_{x \to \infty} \frac{f(x)}{g(x)} = L$ where L is a nonzero finite real number, then for sufficiently large x, it must be the case that $\dfrac{f(x)}{g(x)} < L + 1 \le M$, for some integer M. Similarly for $g = O(f)$

43. (a) x^5 grows faster than x^2.
(b) They grow at the same rate.
(c) $m > n$ **(d)** $m = n$
(e) $m > n$ (or, degree of $g >$ degree of f)
(f) $m = n$ (or, degree of $g =$ degree of f)

45. (a) and (b) both follow from the fact that if f and g are negative, then

$$\lim_{x \to \infty} \frac{|f(x)|}{|g(x)|} = \lim_{x \to \infty} \frac{-f(x)}{-g(x)} = \lim_{x \to \infty} \frac{f(x)}{g(x)}$$

Section 8.3 (pp. 433–444)

Quick Review

1. $\ln 2$ **2.** 0
3. $\dfrac{1}{2} \tan^{-1} \dfrac{x}{2} + C$ **4.** $-\dfrac{1}{3} x^{-3} + C$
5. $(-3, 3)$ **6.** $(1, \infty)$
7. Because $-1 \le \cos x \le 1$ for all x
8. Because $\sqrt{x^2 - 1} < \sqrt{x^2} = x$ for $x > 1$
9. $\displaystyle\lim_{x \to \infty} \frac{4e^x - 5}{3e^x + 7} = \frac{4}{3}$ **10.** $\displaystyle\lim_{x \to \infty} \frac{\sqrt{2x - 1}}{\sqrt{x + 3}} = \sqrt{2}$

Exercises

1. (a) Because of an infinite limit of integration
(b) Converges   **(c)** $\dfrac{\pi}{2}$
3. (a) Because the integrand has an infinite discontinuity at $x = 0$
(b) Converges   **(c)** $-\dfrac{9}{2}$
5. (a) Because the integrand has an infinite discontinuity at $x = 0$
(b) Diverges   **(c)** No value
7. 1000   **9.** 4   **11.** $\dfrac{\pi}{2}$
13. $\ln 3$ **15.** $\sqrt{3}$ **17.** π
19. $\dfrac{\pi}{3}$ **21.** $2\pi^2$ **23.** -1
25. 2   **27.** Diverges   **29.** Converges
31. Converges   **33.** Converges   **35.** Diverges
37. Converges   **39.** Converges   **41.** Diverges
43. Converges   **45.** Converges   **47.** 1
49. (a) The integral in Example 1 gives the area of the region.
(b) ∞ **(c)** π
(d) Gabriel's horn has finite volume so it could only hold a finite amount of paint, but it has infinite surface area so it would presumably require an infinite amount of paint to cover itself.

51.(a) For $x \ge 6$, $x^2 \ge 6x$, and therefore, $e^{-x^2} \le e^{-6x}$. The inequality for the integrals follows. The value of the second integral is $\dfrac{e^{-36}}{6}$, which is less than 4×10^{-17}.
(b) The error in the estimate is the integral over the interval $[6, \infty)$, and we have shown that it is bounded by 4×10^{-17} in part (a).
(c) 0.13940279264 (This agrees with Figure 8.16.)
(d) $\displaystyle\int_0^\infty e^{-x^2}\, dx = \int_0^3 e^{-x^2}\, dx + \int_3^\infty e^{-x^2}\, dx$. The error in the approximation is $\displaystyle\int_3^\infty e^{-x^2}\, dx \le \int_3^\infty e^{-3x}\, dx \le 0.000042$.

53. (a) It is divergent because as $x \to \infty$, $\displaystyle\lim_{x \to \infty} \ln (x^2 + 1) = \infty$.
(b) *Both* the integral over $[0, \infty)$ and the integral over $(-\infty, 0]$ must converge in order for the integral over $(-\infty, \infty)$ to converge.

(c) Since this is an odd function, the integral over any interval of the form $[-b, b]$ equals 0. Therefore the limit as $b \to \infty$ is 0.

(d) Because the determination of convergence is not made using the method in part (c). In order for the integral to converge, there must be finite areas in both directions (toward ∞ and toward $-\infty$). In this case, there are infinite areas in both directions, but when one computes the integral over an interval $[-b, b]$, there is cancellation which gives 0 as the result.

55. From the properties of integrals, for any $b > a$,

$$\int_a^b f(x) \le \int_a^b g(x)\, dx.$$

If the infinite integral of g converges, then taking the limit in the above inequality as $b \to \infty$ shows that the infinite integral of f is bounded above by the infinite integral of g. Therefore, the infinite integral of f must be finite and it converges. If the infinite integral of f diverges, it must grow to infinity. So taking the limit in the above inequality as $b \to \infty$ shows that the infinite integral of g must also diverge to infinity.

57. (a) Although the values oscillate a bit, they appear to be approaching a limit of approximately 1.57.

(b) Yes, it converges.

Section 8.4 (pp. 444–453)

Quick Review

1. $A = -2$, $B = 1$

2. $A = -1$, $B = 2$, $C = 3$

3. $2x + 1 + \dfrac{x - 3}{x^2 - 3x - 4}$

4. $2 + \dfrac{3x - 4}{x^2 + 4x + 5}$ **5.** $(x - 3)(x^2 + 1)$

6. $(y - 2)(y + 2)(y - 1)(y + 1)$

7. $-\dfrac{x + 13}{x^2 + x - 6}$ **8.** $\dfrac{-x^2 + 12x - 15}{(x + 5)(x^2 - 4x + 5)}$

9. $-\dfrac{2t^3 + 5t^2 + 5t + 9}{(t^2 + 2)(t^2 + 1)}$ **10.** $\dfrac{2x^2 - 7x + 6}{(x - 1)^3}$

Exercises

1. $\dfrac{2}{x - 1} + \dfrac{3}{x - 2}$ **3.** $\dfrac{2}{t - 1} - \dfrac{2}{t} - \dfrac{1}{t^2}$

5. $1 - \dfrac{12}{x - 2} + \dfrac{17}{x - 3}$

7. $\dfrac{1}{2} \ln |x + 1| - \dfrac{1}{2} \ln |x - 1| + C$

9. $\dfrac{1}{4} \ln |y + 1| + \dfrac{3}{4} \ln |y - 3| + C$

11. $\dfrac{1}{6} \ln |t + 2| + \dfrac{1}{3} \ln |t - 1| - \dfrac{1}{2} \ln |t| + C$

13. $\dfrac{s^2}{2} - 2 \ln (s^2 + 4) + C$

15. $5x - \dfrac{5}{2} \ln (x^2 + x + 1) - \dfrac{5}{\sqrt{3}} \tan^{-1} \left(\dfrac{2x + 1}{\sqrt{3}} \right) + C$

17. $\dfrac{1}{4} \ln |x + 1| - \dfrac{\frac{1}{4}}{x + 1} - \dfrac{1}{4} \ln |x - 1| - \dfrac{\frac{1}{4}}{x - 1} + C$

19. $2 \tan^{-1} (r - 1) + C$

21. $\ln |x^2 + x + 1| - \ln |x - 1| + C$

23. $\dfrac{1}{x^2 + 4} + \dfrac{3}{2} \tan^{-1} \dfrac{x}{2} + C$

25. $1 - \ln 2$

27. $\ln |y - 1| - \ln |y| = e^x - 1 - \ln 2$

29. $y = \ln |x - 2| - \ln |x - 1| + \ln 2$

31. (a) Use $5 + 4x - x^2 = 9 - (x - 2)^2$, substitute $u = x - 2$, then use Formula 18 with $x = u$ and $a = 3$.

The integral is $\dfrac{1}{6} \ln \left| \dfrac{x + 1}{x - 5} \right| + C$.

(b) Rewrite as $\dfrac{1}{2a} \ln |x + a| - \ln |x - a|$. Differentiating gives $\dfrac{1}{2a} \left(\dfrac{1}{x + a} - \dfrac{1}{x - a} \right)$ which equals $\dfrac{1}{a^2 - x^2}$.

33. $6\pi \ln 5$

35. $\ln \left| \sqrt{9 + y^2} + y \right| + C$

37. $\dfrac{1}{2} \ln \left| 2x + \sqrt{4x^2 - 49} \right| + C$

39. $-\dfrac{x^2 \sqrt{1 - x^2}}{3} - \dfrac{2}{3} \sqrt{1 - x^2} + C$

41. $\sqrt{16 - z^2} - 4 \ln \left| \dfrac{4}{z} + \dfrac{\sqrt{16 - z^2}}{z} \right| + C$

43. $y = \ln \left| \dfrac{x}{3} + \dfrac{\sqrt{x^2 - 9}}{3} \right|$ **45.** ≈ 2.356

47. (a) $x(t) = \dfrac{1000 e^{4t}}{e^{4t} + 499}$ or

$x(t) = \dfrac{1000}{1 + 499 e^{-4t}}$

(b) After $t = \dfrac{\ln 499}{4} \approx 1.553$ days

(c) Since $\dfrac{dx}{dt} = kx(1000 - x)$, $\dfrac{dx}{dt}$ will have its maximum value where $x(1000 - x)$ is greatest, which is at $x = 500$.

49. (a) This can be seen geometrically in the figure.

(b) Using part (a), substitute $z = \dfrac{\sin x}{1 + \cos x}$ and then obtain a trigonometric identity.

Or, use the trigonometric identity
$$\dfrac{1 - \tan^2 \theta}{1 + \tan^2 \theta} = \cos 2\theta \text{ with } \theta = \dfrac{x}{2}.$$

(c) Using part (a), substitute $z = \dfrac{\sin x}{1 + \cos x}$ and then obtain a trigonometric identity.

Or, use the trigonometric identity
$$\dfrac{2 \tan \theta}{1 + \tan^2 \theta} = \sin 2\theta \text{ with } \theta = \dfrac{x}{2}.$$

(d)
$$dz = \left(\sec^2 \dfrac{x}{2} \right) \dfrac{1}{2} \, dx$$
$$= \left(1 + \tan^2 \dfrac{x}{2} \right) \dfrac{1}{2} \, dx$$
$$= \left(1 + z^2 \right) \dfrac{1}{2} \, dx, \text{ then solve for } dx.$$

51. $-\dfrac{1}{\tan \frac{x}{2}} + C$ **53.** $\ln \left| 1 + \tan \dfrac{t}{2} \right| + C$

Chapter Review (pp. 454–455)

Exercises

1. The limit doesn't exist. **2.** $\dfrac{3}{5}$

3. 2 **4.** $\dfrac{1}{e}$ **5.** 1

6. e^3 **7.** 0 **8.** -1

9. $-\dfrac{1}{2}$ **10.** 1 **11.** 1

12. ∞ **13.** ∞ **14.** 0

15. Same rate, because $\lim\limits_{x \to \infty} \dfrac{f(x)}{g(x)} = \dfrac{1}{5}$

16. Same rate, because $\lim\limits_{x \to \infty} \dfrac{f(x)}{g(x)} = \dfrac{\ln 3}{\ln 2}$

17. Same rate, because $\lim\limits_{x \to \infty} \dfrac{f(x)}{g(x)} = 1$

18. Faster, because $\lim\limits_{x \to \infty} \dfrac{f(x)}{g(x)} = \infty$

19. Faster, because $\lim\limits_{x \to \infty} \dfrac{f(x)}{g(x)} = \infty$

20. Same rate, because $\lim\limits_{x \to \infty} \dfrac{f(x)}{g(x)} = 1$

21. Slower, because $\lim\limits_{x \to \infty} \dfrac{f(x)}{g(x)} = 0$

22. Slower, because $\lim\limits_{x \to \infty} \dfrac{f(x)}{g(x)} = 0$

23. Same rate, because $\lim\limits_{x \to \infty} \dfrac{f(x)}{g(x)} = \dfrac{1}{2}$

24. Slower, because $\lim\limits_{x \to \infty} \dfrac{f(x)}{g(x)} = 0$

25. Same rate, because $\lim\limits_{x \to \infty} \dfrac{f(x)}{g(x)} = 1$

26. Faster, because $\lim\limits_{x \to \infty} \dfrac{f(x)}{g(x)} = \infty$

27. (a) $\lim\limits_{x \to 0} f(x) = \ln 2$ **(b)** Define $f(0) = \ln 2$

28. (a) $\lim\limits_{x \to 0^+} f(x) = 0$ **(b)** Define $f(0) = 0$

29. True, $\lim\limits_{x \to \infty} \dfrac{\frac{1}{x^2} + \frac{1}{x^4}}{\frac{1}{x^2}} = 1$

30. False, $\lim\limits_{x \to \infty} \dfrac{\frac{1}{x^2} + \frac{1}{x^4}}{\frac{1}{x^4}} = \infty$

31. False, $\lim\limits_{x \to \infty} \dfrac{x}{x + \ln x} = 1$

32. True, $\lim\limits_{x \to \infty} \dfrac{\ln (\ln x)}{\ln x} = 0$

33. True, $\lim\limits_{x \to \infty} \dfrac{\tan^{-1} x}{1} = \dfrac{\pi}{2}$

34. True, $\lim\limits_{x \to \infty} \dfrac{\frac{1}{x^4}}{\frac{1}{x^2} + \frac{1}{x^4}} = 0$

35. True, $\lim\limits_{x \to \infty} \dfrac{\frac{1}{x^4}}{\frac{1}{x^2} + \frac{1}{x^4}} = 0$

36. True, $\lim\limits_{x \to \infty} \dfrac{\ln x}{x + 1} = 0$ **37.** True, $\lim\limits_{x \to \infty} \dfrac{\ln 2x}{\ln x} = 1$

38. True, $\lim\limits_{x \to \infty} \dfrac{\sec^{-1} x}{1} = \dfrac{\pi}{2}$ **39.** $\dfrac{\pi}{2}$

40. -1 **41.** 6 **42.** 0

43. $\ln 3$ **44.** $\ln \dfrac{3}{4} + 1$ **45.** 2

46. $-\dfrac{1}{9}$ **47.** Diverges **48.** π

49. Diverges, by the limit comparison test, comparing with $\dfrac{1}{\theta}$ or directly from the antiderivative

50. Converges; directly from the antiderivative, value $\dfrac{1}{2}$

51. Diverges; by the direct comparison test with $\dfrac{1}{z}$ or directly from the antiderivative.

52. Converges; by the direct comparison test with e^{-t}

53. Converges; by the direct comparison test with e^{-x} on $[0, \infty)$ and with e^{x} on $(-\infty, 0]$ or directly from the antiderivative, value $\dfrac{\pi}{2}$.

54. Diverges; the problem is near $x = 0$. Compare

with $\dfrac{1}{4x^2}$ there.

For $0 < x \le 1$, $1 + e^x \le 1 + e < 4$ and

$\dfrac{1}{x^2(1 + e^x)} \ge \dfrac{1}{4x^2}$.

55. $9 \ln |x - 4| - 7 \ln |x - 3| + C$

56. $\ln |x| - \ln |x + 2| + \dfrac{2}{x} - \dfrac{2}{x^2} + C$

57. $4 \ln |t| - \dfrac{1}{2} \ln (t^2 + 1) + 4 \tan^{-1} t + C$

58. $\dfrac{1}{2} \tan^{-1} t - \dfrac{1}{2\sqrt{3}} \tan^{-1} \dfrac{t}{\sqrt{3}} + C$

59. $x + \ln |x - 1| - \ln |x| + C$

60. $\dfrac{x^2}{2} + \dfrac{3}{2} \ln |x + 1| - \dfrac{9}{2} \ln |x + 3| + C$

61. $y = \dfrac{500}{1 + 24e^{-x}}$

62. $y = \tan \left(\ln |x + 1| + \tan^{-1} \dfrac{\pi}{4} \right)$

63. $\ln \left| \sqrt{1 + 9y^2} + 3y \right| + C$

64. $\dfrac{1}{6} \sin^{-1} 3t + \dfrac{1}{2} t \sqrt{1 - 9t^2} + C$

65. $\ln \left| 5x + \sqrt{25x^2 - 9} \right| + C$

66. $\dfrac{4x}{\sqrt{1 - x^2}} - 4 \sin^{-1} x + C$

67. 2π **68.** 1

69. (a) $x = a - \dfrac{1}{kt + \dfrac{1}{a}}$ **(b)** $x = \dfrac{ab(e^{akt} - e^{bkt})}{ae^{akt} - be^{bkt}}$

Chapter 9

Section 9.1 (pp. 457–468)

Quick Review

1. $\dfrac{4}{3}$, 1, $\dfrac{4}{5}$, $\dfrac{2}{3}$, $\dfrac{1}{8}$

2. -1, $\dfrac{1}{2}$, $-\dfrac{1}{3}$, $\dfrac{1}{4}$, $\dfrac{1}{30}$

3. (a) 3 **(b)** $39{,}366$
   (c) $a_n = 2(3^{n-1})$

4. (a) $-\dfrac{1}{2}$ **(b)** $-\dfrac{1}{64}$

   (c) $a_n = 8 \left(-\dfrac{1}{2} \right)^{n-1} = 8(-0.5)^{n-1}$

5. (a)

[0, 25] by [−0.5, 0.5]

   (b) 0

6. (a)

[0, 23.5] by [−1, 4]

   (b) e

7. (a)

[0, 23.5] by [−2, 2]

   (b) The limit does not exist.

8. (a)

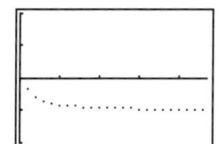

[0, 23.5] by [−2, 2]

   (b) -1

9. (a)

[0, 23,5] by [−1, 3]

   (b) 2

10. (a)

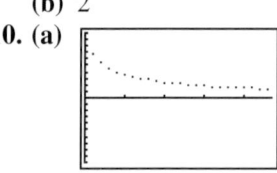

[0, 23.5] by [−1, 1]

   (b) 0

Exercises

1. (a) $* = n^2$ **(b)** $* = (n + 1)^2$
(c) $* = 3$

3. Different   **5.** Same

7. Converges; sum $= 3$ **9.** Converges; sum $= \dfrac{15}{4}$

11. Diverges
13. Converges; sum $= 2 - \sqrt{2}$

15. Converges; sum $= \dfrac{e}{\pi - e}$

17. Interval: $-\dfrac{1}{2} < x < \dfrac{1}{2}$; function: $\dfrac{1}{1 - 2x}$

19. Interval: $1 < x < 5$; function: $\dfrac{2}{x - 1}$

21. Converges for all values of x except odd integer multiples of $\dfrac{\pi}{2}$; function: $\dfrac{1}{1 - \sin x}$

23. (a) The partial sums tend toward infinity.
(b) The partial sums are alternately 1 and 0.
(c) The partial sums alternate between positive and negative while their magnitude increases toward infinity.

25. $x = \dfrac{19}{20}$

27. Assuming the series begins at $n = 1$:
(a) $\displaystyle\sum_{n=1}^{\infty} 2\left(\dfrac{3}{5}\right)^{n-1}$
(b) $\displaystyle\sum_{n=1}^{\infty} \dfrac{13}{2}\left(-\dfrac{3}{10}\right)^{n-1}$

29. Let $a = \dfrac{234}{1000}$ and $r = \dfrac{1}{1000}$, giving
$(0.234) + (0.234)(0.001) + (0.234)(0.001)^2 + (0.234)(0.001)^3 + \cdots$
So the sum is $\dfrac{26}{111}$.

31. $\dfrac{d}{9}$ **33.** $\dfrac{157}{111}$

35. $\dfrac{22}{7}$ **37.** ≈ 7.113 seconds

39. $\dfrac{\pi}{2}$

41. For $r \neq 1$, the result follows from:
If $|r| < 1$, $r^n \to 0$ as $n \to \infty$, and
if $|r| > 1$ or $r = -1$, r^n has no finite limit as $n \to \infty$.
When $r = 1$, the nth partial sum is na, which goes to $\pm\infty$.

43. Series: $x + 2x^2 + 4x^3 + \cdots + 2^{n-1}x^n + \cdots$
Interval: $-\dfrac{1}{2} < x < \dfrac{1}{2}$

45. Series: $1 - (x - 4) + (x - 4)^2 - (x - 4)^3 + \cdots + (-1)^n(x - 4)^n + \cdots$
Interval: $3 < x < 5$

47. One possible series:
$1 + (x - 1) + (x - 1)^2 + \cdots + (x - 1)^n + \cdots$
Interval: $0 < x < 2$

49. (a) 2   **(b)** $t > -\dfrac{1}{2}$
(c) $t > 9$

51. $(x - 1) - \dfrac{(x - 1)^2}{2} + \dfrac{(x - 1)^3}{3} - \cdots + \dfrac{(-1)^{n-1}(x - 1)^n}{n} + \cdots$

53. (a) No, because if you differentiate it again, you would have the original series for f, but by Theorem 1, that would have to converge for $-2 < x < 2$, which contradicts the assumption that the original series converges only for $-1 < x < 1$.
(b) No, because if you integrate it again, you would have the original series for f, but by Theorem 2, that would have to converge for $-2 < x < 2$, which contradicts the assumption that the original series converges only for $-1 < x < 1$.

55. Given an $\epsilon > 0$, by definition of convergence there corresponds an N such that for all $n < N$, $|L_1 - a_n| < \epsilon$ and $|L_2 - a_n| < \epsilon$. Now
$|L_2 - L_1| = |L_2 - a_n + a_n - L_1|$
$\leq |L_2 - a_n| + |a_n - L_1| < \epsilon + \epsilon = 2\epsilon.$
$|L_2 - L_1| < 2\epsilon$ says that the difference between two fixed values is smaller than any positive number 2ϵ. The only nonnegative number smaller than every positive number is 0, so $|L_2 - L_1| = 0$ or $L_1 = L_2$.

57. (a) $\displaystyle\lim_{n\to\infty} \dfrac{3n + 1}{n + 1} = 3$
(b) The line $y = 3$ is a horizontal asymptote of the graph of the function $f(x) = \dfrac{3x + 1}{x + 1}$, which means $\lim_{x\to\infty} f(x) = 3$. Because $f(n) = a_n$ for all positive integers n, it follows that $\lim_{n\to\infty} a_n$ must also be 3.

Section 9.2 (pp. 469–479)

Quick Review

1. $2^n e^{2x}$

2. $(-1)^n n!(x-1)^{-(n+1)}$

3. $3^x (\ln 3)^n$

4. $(-1)^{n-1}(n-1)! x^{-n}$

5. $n!$

6. $\dfrac{x^{n-1}}{(n-1)!}$

7. $\dfrac{2^n (x-a)^{n-1}}{(n-1)!}$

8. $\dfrac{(-1)^n x^{2n}}{(2n)!}$

9. $\dfrac{(x+a)^{2n-1}}{(2n-1)!}$

10. $-\dfrac{(1-x)^{n-1}}{(n-1)!}$

Exercises

1. $2x - \dfrac{4x^3}{3} + \dfrac{4x^5}{15} - \cdots + (-1)^n \dfrac{(2x)^{2n+1}}{(2n+1)!} + \cdots$
convergent for all real x

3. $x^2 - \dfrac{x^6}{3} + \dfrac{x^{10}}{5} - \cdots + (-1)^n \dfrac{x^{4n+2}}{(2n+1)} + \cdots$
convergent for $-1 \le x \le 1$

5. $(\cos 2) - (\sin 2)x - \dfrac{(\cos 2) x^2}{2} + \cdots$
$+ \dfrac{(-1)^A B x^n}{n!} + \cdots$, where $A = \text{int}\left(\dfrac{n+1}{2}\right)$, and
B is cos 2 if n is even and sin 2 if n is odd.

Alternately, the general term may be written as
$\left[\dfrac{1}{n!} \cos\left(2 + \dfrac{n\pi}{2}\right)\right] x^n$. The series converges for all
real x.

7. $x + x^4 + x^7 + \cdots + x^{3n+1} + \cdots$
convergent for $-1 < x < 1$

9. $P_0(x) = \dfrac{1}{2}$

$P_1(x) = \dfrac{1}{2} - \dfrac{x-2}{4}$

$P_2(x) = \dfrac{1}{2} - \dfrac{x-2}{4} + \dfrac{(x-2)^2}{8}$

$P_3(x) = \dfrac{1}{2} - \dfrac{x-2}{4} + \dfrac{(x-2)^2}{8} - \dfrac{(x-2)^3}{16}$

11. $P_0(x) = \dfrac{\sqrt{2}}{2}$

$P_1(x) = \dfrac{\sqrt{2}}{2} - \left(\dfrac{\sqrt{2}}{2}\right)\left(x - \dfrac{\pi}{4}\right)$

$P_2(x) = \dfrac{\sqrt{2}}{2} - \left(\dfrac{\sqrt{2}}{2}\right)\left(x - \dfrac{\pi}{4}\right) - \left(\dfrac{\sqrt{2}}{4}\right)\left(x - \dfrac{\pi}{4}\right)^2$

$P_3(x) = \dfrac{\sqrt{2}}{2} - \left(\dfrac{\sqrt{2}}{2}\right)\left(x - \dfrac{\pi}{4}\right) - \left(\dfrac{\sqrt{2}}{4}\right)\left(x - \dfrac{\pi}{4}\right)^2$
$+ \left(\dfrac{\sqrt{2}}{12}\right)\left(x - \dfrac{\pi}{4}\right)^3$

13. (a) $4 - 2x + x^3$
(b) $3 + (x-1) + 3(x-1)^2 + (x-1)^3$

15. (a) 0
(b) $1 + 4(x-1) + 6(x-1)^2 + 4(x-1)^3$

17. (a) $P_3(x) = 4 - (x-1) + \dfrac{3}{2}(x-1)^2 + \dfrac{1}{3}(x-1)^3$
$f(1.2) \approx P_3(1.2) \approx 3.863$
(b) For f', $P_2(x) = -1 + 3(x-1) + (x-1)^2$
$f'(1.2) \approx P_2(1.2) = -0.36$

19. (a) $1 + \dfrac{x}{2} + \dfrac{x^2}{8} + \cdots + \dfrac{x^n}{2^n \cdot n!} + \cdots$
(b) $1 + \dfrac{x}{2!} + \dfrac{x^2}{3!} + \cdots + \dfrac{x^n}{(n+1)!} + \cdots$
(c) $g'(1) = 1$ and from the series,
$g'(1) = \dfrac{1}{2!} + \dfrac{2}{3!} + \dfrac{3}{4!} + \cdots + \dfrac{n}{(n+1)!} + \cdots$

21. (a) $1 + \dfrac{x}{2} - \dfrac{x^2}{8} + \dfrac{x^3}{16}$
(b) $1 + \dfrac{x^2}{2} - \dfrac{x^4}{8} + \dfrac{x^6}{16}$
(c) $5 + x + \dfrac{x^3}{6} - \dfrac{x^5}{40}$

23. 27 terms (or, up to and including the 52nd degree term)

25. (1) sin x is odd and cos x is even
(2) sin $0 = 0$ and cos $0 = 1$

27. $\dfrac{1}{24}$

29. (a)

$[-2, 4]$ by $[-3, 3]$

(b) $f''(a)$ must be 0 because of the inflection point, so the second degree term in the Taylor series of f at $x = a$ is zero.

31. (a) $1 - \dfrac{x^2}{3!} + \dfrac{x^4}{5!} - \cdots + \dfrac{(-1)^n x^{2n}}{(2n+1)!} + \cdots$
(b) Because f is undefined at $x = 0$.
(c) $k = 1$

33. (a) Just differentiate 3 times.
(b) Differentiate k times and plug in $x = 0$.
(c) $\dfrac{m(m-1)(m-2) \cdots (m-k+1)}{k!}$
(d) $f(0) = 1, f'(0) = m$, and we're done by part (c).

Section 9.3 (pp. 480–487)

Quick Review

1. 2   **2.** 7

3. 1   **4.** $\dfrac{1}{2}$

5. 7   **6.** Yes
7. No   **8.** Yes
9. Yes   **10.** No

Exercises

1. $1 - 2x + 2x^2 - \dfrac{4}{3}x^3 + \dfrac{2}{3}x^4;\ f(0.2) \approx 0.6704$

3. $-5x + \dfrac{5}{6}x^3;\ f(0.2) \approx -0.9933$

5. $1 + 2x + 3x^2 + 4x^3 + 5x^4;\ f(0.2) \approx 1.56$

7. $\dfrac{x^5}{5!} - \dfrac{x^7}{7!} + \dfrac{x^9}{9!} - \cdots + (-1)^n\dfrac{x^{2n+5}}{(2n+5)!} + \cdots$

9. $x^2 - \dfrac{x^4}{3} + \dfrac{2x^6}{45} - \cdots + (-1)^n\dfrac{2^{2n+1}x^{2n+2}}{(2n+2)!} + \cdots$

11. Using the theorem, $-0.56 < x < 0.56$
Graphically, $-0.57 < x < 0.57$

13. $|\text{Error}| < 1.67 \times 10^{-10}$

$x < \sin x$ for negative values of x.

15. $|\text{Error}| < 1.842 \times 10^{-4}$

17. All of the derivatives of $\cosh x$ are either $\cosh x$ or $\sinh x$. For any real x, $\cosh x$ and $\sinh x$ are both bounded by $e^{|x|}$. So for any real x, let $M = e^{|x|}$ and $r = 1$ in the Remainder Estimation Theorem. It follows that the series converges to $\cosh x$ for all real values of x.

19. (a) 0   **(b)** $-\dfrac{x^2}{2}$

(c) The graphs of the linear and quadratic approximations fit the graph of the function near $x = 0$.

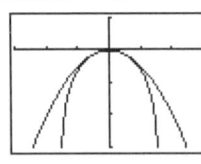

[-3, 3] by [-3, 1]

21. (a) 1   **(b)** $1 + \dfrac{x^2}{2}$

(c) The graphs of the linear and quadratic approximations fit the graph of the function near $x = 0$.

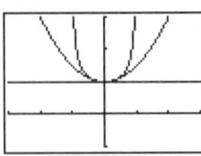

[-3, 3] by [-1, 3]

23. (a) x **(b)** x

(c) The graphs of the linear and quadratic approximations fit the graph of the function near $x = 0$.

[-3, 3] by [-2, 2]

25. $|\text{Error}| < 4.61 \times 10^{-6}$, by Remainder Estimation Theorem (actual maximum error is $\approx 4.251 \times 10^{-6}$)

27. (a) No

(b) Yes. $2 + x - \dfrac{x^3}{3} + \dfrac{x^5}{10} - \cdots$

$+ \dfrac{(-1)^{n+1}x^{2n-1}}{[(2n-1)(n-1)!]} + \cdots$

(c) For all real values of x. This is assured by Theorem 2 of Section 9.1, because the series for e^{-x^2} converges for all real values of x.

29. (a) $\tan x$ **(b)** $\sec x$

31. (a) It works.

(b) Let $P = \pi + x$ where x is the error in the original estimate. Then

$$P + \sin P = (\pi + x) + \sin(\pi + x)$$
$$= \pi + x - \sin x$$

But by the Remainder Theorem,
$|x - \sin x| < \dfrac{|x|^3}{6}$. Therefore, the difference between the new estimate $P + \sin P$ and π is less than $\dfrac{|x|^3}{6}$.

33. The derivative is
$(ae^{ax})(\cos bx + i \sin bx)$
$\quad + (e^{ax})(-b \sin bx + ib \cos bx)$
$= a[e^{ax}(\cos bx + i \sin bx)]$
$\quad + ib[e^{ax}(\cos bx + i \sin bx)]$
$= (a + ib)e^{(a+ib)x}$.

Section 9.4 (pp. 487–496)

Quick Review

1. $|x|$

2. $|x - 3|$

3. 0

4. $\dfrac{x^2}{16}$

5. $\dfrac{|2x + 1|}{2}$

6. $a_n = n^2, b_n = 5n, N = 6$

7. $a_n = 5^n, b_n = n^5, N = 6$

8. $a_n = \sqrt{n}, b_n = \ln n, N = 1$

9. $a_n = \dfrac{1}{10^n}, b^n = \dfrac{1}{n!}, N = 25$

10. $a_n = \dfrac{1}{n^2}, b_n = n^{-3}, N = 2$

Exercises

1. Diverges (*n*th-Term Test)

3. Converges (Ratio Test)

5. Converges (Ratio Test, Direct Comparison Test)

7. Converges (Ratio Test)

9. Converges (Ratio Test)

11. Converges (geometric series)

13. Diverges (*n*th-Term Test, Ratio Test)

15. Converges (Ratio Test)

17. One possible answer:

$\sum \dfrac{1}{n}$ diverges (see Exploration 1 in this section) even though $\lim\limits_{n\to\infty} \dfrac{1}{n} = 0$.

19. 1

21. $\dfrac{1}{4}$

23. 10

25. 3

27. 5

29. 3

31. $\dfrac{1}{2}$

33. 1

35. Interval: $-1 < x < 3$

Function: $-\dfrac{4}{x^2 - 2x - 3}$

37. Interval: $0 < x < 16$

Function: $\dfrac{2}{4 - \sqrt{x}}$

39. Interval: $-2 < x < 2$

Function: $\dfrac{3}{4 - x^2}$

41. Almost, but the ratio test won't determine whether there is convergence or divergence at the endpoints of the interval.

43. (a) For $k \le N$, it's obvious that

$$d_1 + \cdots + d_k$$
$$\le d_1 + \cdots + d_N + \sum_{n=N+1}^{\infty} a_n$$

For all $k > N$,

$$d_1 + \cdots + d_k$$
$$= d_1 + \cdots + d_N + d_{N+1} \quad + \cdots + d_k$$
$$\le d_1 + \cdots + d_N + a_{N+1} + \cdots + a_k$$
$$\le d_1 + \cdots + d_N + \sum_{n=N+1}^{\infty} a_n.$$

(b) If $\sum a_n$ converged, that would imply that $\sum d_n$ was also convergent.

45. 1

47. 5

49. 1

51. $-\dfrac{\pi}{4}$

Section 9.5 (pp. 496–508)

Quick Review

1. Converges, $p > 1$

2. Diverges, limit comparison test with integral of $\dfrac{1}{x}$

3. Diverges, comparison test with integral of $\dfrac{1}{x}$

4. Converges, comparison test with integral of $\dfrac{2}{x^2}$

5. Diverges, limit comparison test with integral of $\dfrac{1}{\sqrt{x}}$

6. Yes

7. Yes

8. No

9. No

10. No

Exercises

1. Diverges

3. Diverges

5. Diverges

7. Diverges

9. Converges

11. Diverges

13. Diverges

15. Diverges

17. Converges absolutely

19. Converges absolutely

21. Diverges

23. Converges conditionally

25. Converges conditionally

27. (a) $(-1, 1)$ **(b)** $(-1, 1)$ **(c)** None

29. (a) $\left(-\dfrac{1}{2}, 0\right)$ **(b)** $\left(-\dfrac{1}{2}, 0\right)$ **(c)** None

31. (a) $(-8, 12)$ **(b)** $(-8, 12)$ **(c)** None

33. (a) $[-3, 3]$ **(b)** $[-3, 3]$ **(c)** None

35. (a) $(-8, 2)$ **(b)** $(-8, 2)$ **(c)** None

37. (a) $(-3, 3)$ **(b)** $(-3, 3)$ **(c)** None

39. (a) $\left(\frac{1}{2}, \frac{3}{2}\right)$ **(b)** $\left(\frac{1}{2}, \frac{3}{2}\right)$ **(c)** None

41. (a) $[-\pi - 1, -\pi + 1)$

(b) $(-\pi - 1, -\pi + 1)$
(c) At $x = -\pi - 1$

43. $40.554 < \text{sum} < 41.555$

45.

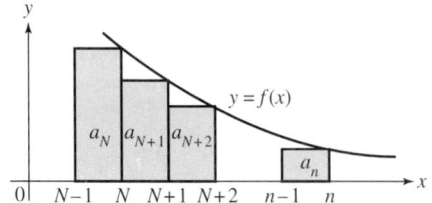

Comparing areas in the figures, we have for all

$$n \geq N, \int_{N}^{n+1} f(x)\, dx < a_N + \cdots + a_n < a_1$$
$$+ \int_{N}^{n} f(x)\, dx.$$

If the integral diverges, it must go to infinity, and the first inequality forces the partial sums of the series to go to infinity as well, so the series is divergent.
If the integral converges, then the second inequality puts an upper bound on the partial sums of the series, and since they are a nondecreasing sequence, they must converge to a finite sum for the series.

47. One possible answer: $\sum \dfrac{1}{n \ln n}$

This series diverges by the Integral Test, but its

partial sums are roughly $\ln (\ln n)$, so they are

much smaller than the partial sums for the

harmonic series, which are about $\ln n$.

49. (a) Diverges

(b) $S = \displaystyle\sum_{n=1}^{\infty} \frac{3n}{3n^3 + n} = \sum_{n=1}^{\infty} \frac{3}{3n^2 + 1}$ which

converges.

51. Convergent for $-\dfrac{1}{2} \leq x < \dfrac{1}{2}$.

Use the Ratio Test, Direct Comparison Test, and

Alternating Series Test.

53. Use the Alternating Series Test.
55. (a) It fails to satisfy $u_n \geq u_{n+1}$ for all $n \geq N$.

(b) The sum is $-\dfrac{1}{2}$.

57. (a) Converges   **(b)** Converges
(c) Converges

Chapter Review (pp. 509–511)

1. (a) ∞ **(b)** All real numbers
(c) All real numbers   **(d)** None
2. (a) 3   **(b)** $[-7, -1)$
(c) $(-7, -1)$ **(d)** At $x = -7$
3. (a) $\dfrac{3}{2}$ **(b)** $\left(-\dfrac{1}{2}, \dfrac{5}{2}\right)$

(c) $\left(-\dfrac{1}{2}, \dfrac{5}{2}\right)$ **(d)** None
4. (a) ∞ **(b)** All real numbers
(c) All real numbers   **(d)** None
5. (a) $\dfrac{1}{3}$ **(b)** $\left[0, \dfrac{2}{3}\right]$

(c) $\left[0, \dfrac{2}{3}\right]$ **(d)** None
6. (a) 1   **(b)** $(-1, 1)$
(c) $(-1, 1)$ **(d)** None
7. (a) 1   **(b)** $\left(-\dfrac{3}{2}, \dfrac{1}{2}\right)$

(c) $\left(-\dfrac{3}{2}, \dfrac{1}{2}\right)$ **(d)** None
8. (a) ∞ **(b)** All real numbers
(c) All real numbers   **(d)** None
9. (a) 1   **(b)** $[-1, 1)$
(c) $(-1, 1)$ **(d)** At $x = -1$
10. (a) $\dfrac{1}{e}$ **(b)** $\left[-\dfrac{1}{e}, \dfrac{1}{e}\right]$

(c) $\left[-\dfrac{1}{e}, \dfrac{1}{e}\right]$ **(d)** None
11. (a) $\sqrt{3}$ **(b)** $(-\sqrt{3}, \sqrt{3})$
(c) $(-\sqrt{3}, \sqrt{3})$ **(d)** None
12. (a) 1   **(b)** $[0, 2]$
(c) $(0, 2)$ **(d)** At $x = 0$ and $x = 2$
13. (a) 0   **(b)** $x = 0$ only
(c) $x = 0$ **(d)** None
14. (a) $\dfrac{1}{10}$ **(b)** $\left[-\dfrac{1}{10}, \dfrac{1}{10}\right)$

(c) $\left(-\dfrac{1}{10}, \dfrac{1}{10}\right)$ **(d)** At $x = -\dfrac{1}{10}$
15. (a) 0   **(b)** $x = 0$ only
(c) $x = 0$ **(d)** None
16. (a) $\sqrt{3}$ **(b)** $(-\sqrt{3}, \sqrt{3})$
(c) $(-\sqrt{3}, \sqrt{3})$ **(d)** None
17. $f(x) = \dfrac{1}{1 + x}$ evaluated at $x = \dfrac{1}{4}$. Sum $= \dfrac{4}{5}$.

18. $f(x) = \ln (1 + x)$ evaluated at $x = \dfrac{2}{3}$.
Sum $= \ln \left(\dfrac{5}{3}\right)$.

19. $f(x) = \sin x$ evaluated at $x = \pi$. Sum $= 0$.

20. $f(x) = \cos x$ evaluated at $x = \dfrac{\pi}{3}$. Sum $= \dfrac{1}{2}$.

21. $f(x) = e^x$ evaluated at $x = \ln 2$. Sum $= 2$.

22. $f(x) = \tan^{-1} x$ evaluated at $x = \dfrac{1}{\sqrt{3}}$. Sum $= \dfrac{\pi}{6}$.

23. $1 + 6x + 36x^2 + \cdots + (6x)^n + \cdots$

24. $1 - x^3 + x^6 - \cdots + (-1)^n x^{3n} + \cdots$

25. $1 - 2x^2 + x^9$

26. $4x + 4x^2 + 4x^3 + \cdots + 4x^{n+1} + \cdots$

27. $\pi x - \dfrac{(\pi x)^3}{3!} + \dfrac{(\pi x)^5}{5!} - \cdots + (-1)^n \dfrac{(\pi x)^{2n+1}}{(2n+1)!} + \cdots$

28. $-\dfrac{2x}{3} + \dfrac{4x^3}{81} - \dfrac{4x^5}{3645} + \cdots$
$+ \dfrac{(-1)^{n+1}}{(2n+1)!}\left(\dfrac{2x}{3}\right)^{2n+1} + \cdots$

29. $-\dfrac{x^3}{3!} + \dfrac{x^5}{5!} - \dfrac{x^7}{7!} + \cdots + (-1)^n \dfrac{x^{2n+1}}{(2n+1)!} + \cdots$

30. $1 + \dfrac{x^2}{2!} + \dfrac{x^4}{4!} + \cdots + \dfrac{x^{2n}}{(2n)!} + \cdots$

31. $1 - \dfrac{5x}{2!} + \dfrac{(5x)^2}{4!} - \cdots + (-1)^n \dfrac{(5x)^n}{(2n)!} + \cdots$

32. $1 + \dfrac{\pi x}{2} + \dfrac{\pi^2 x^2}{8} + \cdots + \dfrac{1}{n!}\left(\dfrac{\pi x}{2}\right)^n + \cdots$

33. $x - x^3 + \dfrac{x^5}{2!} - \dfrac{x^7}{3!} + \cdots + (-1)^n \dfrac{x^{2n+1}}{n!} + \cdots$

34. $3x - \dfrac{(3x)^3}{3} + \dfrac{(3x)^5}{5} - \cdots + (-1)^n \dfrac{(3x)^{2n+1}}{2n+1} + \cdots$

35. $-2x - 2x^2 - \dfrac{8x^3}{3} - \cdots - \dfrac{(2x)^n}{n} - \cdots$

36. $-x^2 - \dfrac{x^3}{2} - \dfrac{x^4}{3} - \cdots - \dfrac{x^{n+1}}{n} - \cdots$

37. $1 + (x-2) + (x-2)^2 + (x-2)^3$
$+ \cdots (x-2)^n + \cdots$

38. $2 + 7(x+1) - 5(x+1)^2 + (x+1)^3$

(Finite. general term $= 0$)

39. $\dfrac{1}{3} - \dfrac{x-3}{9} + \dfrac{(x-3)^2}{27} - \dfrac{(x-3)^3}{81} + \cdots$
$+ (-1)^n \dfrac{(x-3)^n}{3^{n+1}} + \cdots$

40. $-(x-\pi) + \dfrac{(x-\pi)^3}{3!} - \dfrac{(x-\pi)^5}{5!} + \dfrac{(x-\pi)^7}{7!} - \cdots$
$+ (-1)^{n+1} \dfrac{(x-\pi)^{2n+1}}{(2n+1)!} + \cdots$

41. Diverges. It's just -5 times the harmonic series.

42. Converges conditionally. Alternating Series Test and $p = \dfrac{1}{2}$.

43. Converges absolutely. Compare to $\dfrac{1}{n^2}$.

44. Converges absolutely. Ratio Test

45. Converges conditionally. Alternating Series Test and compare to $\dfrac{1}{n}$.

46. Converges absolutely. Integral Test

47. Converges absolutely. Ratio Test

48. Converges absolutely. nth-Root Test or Ratio Test

49. Diverges. nth-Term Test for Divergence

50. Converges absolutely. Compare to $\dfrac{1}{n^{3/2}}$.

51. Converges absolutely. Limit Comparison Test with $\dfrac{1}{n^2}$.

52. Diverges. nth-Term Test for Divergence

53. $\dfrac{1}{6}$ 54. -1

55. (a) $P_3(x) = 1 + 4(x-3) + 3(x-3)^2 + 2(x-3)^3$
$f(3.2) \approx P_3(3.2) = 1.936$
(b) For f: $P_2(x) = 4 + 6(x-3) + 6(x-3)^2$
$f'(2.7) \approx P_2(2.7) = 2.74$
(c) It underestimates the values, since the graph of f is concave up near $x = 3$.

56. (a) $f(4) = 7$ and $f'''(4) = -12$
(b) For f: $P_2(x) = -3 + 10(x-4) - 6(x-4)^2$
$f'(4.3) \approx P_2(4.3) = -0.54$
(c) $7(x-4) - \dfrac{3}{2}(x-4)^2 + \dfrac{5}{3}(x-4)^3 - \dfrac{1}{2}(x-4)^4$
(d) No. One would need the entire Taylor series for $f(x)$, and it would have to converge to $f(x)$ at $x = 3$.

57. (a) $\dfrac{5x}{2} - \dfrac{5x^3}{48} + \dfrac{x^5}{768} - \cdots$
$+ (-1)^n \dfrac{5}{(2n+1)!}\left(\dfrac{x}{2}\right)^n + \cdots$

(b) All real numbers. Use the Ratio Test.

(c) Note that the absolute value of $f^{(n)}(x)$ is bounded by $\dfrac{5}{2^n}$ for all x and all $n = 1, 2, 3, \cdots$
So if $-2 < x < 2$, the truncation error using P_n is bounded by $\dfrac{5}{2^{n+1}} \cdot \dfrac{2^{n+1}}{(n+1)!} = \dfrac{5}{(n+1)!}$.
To make this less than 0.1 requires $n \geq 4$.

So, two nonzero terms (up through degree 4) are needed.

58. (a) $1 + 2x + 4x^2 + 8x^3 + \cdots + (2x)^n + \cdots$

(b) $\left(-\dfrac{1}{2}, \dfrac{1}{2}\right)$. The series for $\dfrac{1}{1-t}$ is known to converge for $-1 < t < 1$, so by substituting $t = 2x$, we find the resulting series converges for $-1 < 2x < 1$.

(c) Possible answer: $f\left(-\dfrac{1}{4}\right) = \dfrac{2}{3}$, so one percent is approximately 0.0067. It takes 7 terms (up through degree 6). This can be found by trial and error. Also, the series is an alternating series for $x = -\dfrac{1}{4}$. If you use the Alternating Series Estimation Theorem, it shows that 8 terms (up through degree 7) are sufficient. It is also a geometric series, and one could use the remainder formula for a geometric series to estimate the number of terms needed.

59. (a) $\dfrac{1}{e}$ **(b)** $-\dfrac{5}{18} \approx -0.278$

(c) By the Alternating Series Estimation Theorem, the error is bounded by the size of the next term, which is $\dfrac{32}{243}$, or about 0.132.

60. (a) $1 - (x-3) + (x-3)^2 - (x-3)^3 + \cdots + (-1)^n(x-3)^n + \cdots$

(b) $(x-3) - \dfrac{(x-3)^2}{2} + \dfrac{(x-3)^3}{3} - \dfrac{(x-3)^4}{4} + \cdots + (-1)^n\dfrac{(x-3)^{n+1}}{n+1} + \cdots$

(c) Evaluate at $x = 3.5$. This is an alternating series. By the Alternating Series Estimation Theorem, since the size of the third term is $\dfrac{1}{24} < 0.05$, the first two terms will suffice. The estimate for $\ln\left(\dfrac{3}{2}\right)$ is 0.375.

61. (a) $1 - 2x^2 + 2x^4 - \dfrac{4x^6}{3} + \cdots + (-1)^n\dfrac{2^n x^{2n}}{n!} + \cdots$

(b) All real numbers. Use the Ratio Test.

(c) This is an alternating series. The difference will be bounded by the magnitude of the fifth term, which is $\dfrac{(2x^2)^4}{4!} = \dfrac{2x^8}{3}$. Since $-0.6 \le x \le 0.6$, this term is less than $\dfrac{2(0.6)^8}{3}$ which is less than 0.02.

62. (a) $x^2 - x^3 + x^4 - x^5 + \cdots + (-1)^n x^n + \cdots$

(b) No. The partial sums form the sequence 1, 0, 1, 0, 1, 0, ... which has no limit.

63. (a) $\dfrac{x^3}{3} - \dfrac{x^7}{7(3!)} + \dfrac{x^{11}}{11(5!)} + \cdots + \dfrac{(-1)^n x^{4n+3}}{(4n+3)(2n+1)!} + \cdots$

(b) The first two nonzero terms suffice (through degree 7).

(c) 0.31026830 **(d)** Within 1.5×10^{-7}

64. (a) 0.88566 **(b)** $\dfrac{41}{60} \approx 0.68333$

(c) Since f is concave up, the trapezoids used to estimate the area lie above the curve, and the estimate is too large.

(d) Since all the derivatives are positive (and $x > 0$), the remainder, $R_n(x)$, must be positive. This means that $P_n(x)$ is smaller than $f(x)$.

(e) $e - 2 \approx 0.71828$

65. (a) Because $[\$1000(1.08)^{-n}](1.08)^n = \$1000$ will be available after n years.

(b) Assume that the first payment goes to the charity at the end of the first year.
$1000(1.08)^{-1} + 1000(1.08)^{-2} + 1000(1.08)^{-3} + \cdots$

(c) This is a geometric series with sum equal to $\$12,500$. This represents the amount which must be invested today in order to completely fund the perpetuity forever.

66. $\$16,666.67$ [Again, assuming first payment at end of year.]

67. (a) $0\left(\dfrac{1}{2}\right) + 1\left(\dfrac{1}{2}\right)^2 + 2\left(\dfrac{1}{2}\right)^3 + 3\left(\dfrac{1}{2}\right)^4 + \cdots$

(b) $1 + 2x + 3x^2 + 4x^3 + \cdots$

(c) $\dfrac{x^2}{(1-x)^2} = x^2 + 2x^3 + 3x^4 + 4x^5 + \cdots$

(d) The expected payoff of the game is $\$1$.

68. (a) $\dfrac{b^2\sqrt{3}}{4} + \dfrac{3b^2\sqrt{3}}{4^2} + \dfrac{3^2 b^2\sqrt{3}}{4^3} + \cdots$

(b) $b^2\sqrt{3}$

(c) No, not every point is removed. But the remaining points are "isolated" enough that there are no regions and hence no area remaining.

69. $\dfrac{1}{(1-x)^2} = 1 + 2x + 3x^2 + 4x^3 + 5x^4 + \cdots$
Substitute $x = \dfrac{1}{2}$ to get the desired result.

70. (b) Solve $x = \dfrac{2x^2}{(x-1)^3}$. $x \approx 2.769$.

Chapter 10

Section 10.1 (pp. 513–520)

Quick Review

1. $(1, 0)$
2. $(0, -1)$
3. $x^2 + y^2 = 1$
4. The portion in the first three quadrants
5. $x = t$, $y = t^2 + 1$, $-1 \le t \le 3$
6. $x = 2\cos t + 2$, $y = 2\sin t + 3$, $0 \le t \le 2\pi$
7. $\dfrac{3}{2}$
8. $y = \dfrac{3}{2}x + 3\sqrt{2}$
9. $y = -\dfrac{2}{3}x + \dfrac{5\sqrt{2}}{6}$
10. $\dfrac{31^{3/2} - 8}{27}$

Exercises

1. (a) $-\dfrac{1}{2}\tan t$
   (b) $-\dfrac{1}{8}\sec^3 t$

3. (a) $-\sqrt{3 + \dfrac{3}{t}}$
   (b) $-\dfrac{\sqrt{3}}{t^{3/2}}$

5. (a) $\dfrac{3t^2}{2t - 3}$
   (b) $\dfrac{6t^2 - 18t}{(2t - 3)^3}$

7. (a) $(2, 0)$ and $(2, -2)$
   (b) $(1, -1)$ and $(3, -1)$

9. (a) At $t = \pm\dfrac{2}{\sqrt{3}}$, or $\approx (0.845, -3.079)$ and
   $(3.155, 3.079)$

   (b) Nowhere

11. 4
13. $\dfrac{2\sqrt{2} - 1}{3} \approx 0.609$

15. $\ln 2$
17. $8\pi^2$

19. ≈ 178.561

21. (a) $x(t) = 2t$, $y(t) = t + 1$, $0 \le t \le 1$
    (b) $3\pi\sqrt{5}$
    (c) $3\pi\sqrt{5}$

23. (a) π
    (b) π

25. Just substitute x for t and note that $\dfrac{dx}{dx} = 1$.

27. At $t = \sqrt{13} - 1$, or $\approx (3.394, 5.160)$
29. ≈ 144.513
31. $3\pi a^2$

33. (a) $x = \cos t + t\sin t$, $y = \sin t - t\cos t$
    (b) $2\pi^2$

35. (a) ≈ 461.749 ft
    (b) ≈ 41.125 ft

37. (a) ≈ 840.421 ft
    (b) $\dfrac{16{,}875}{64} \approx 263.672$ ft

39. Just substitute x for t and note that $\dfrac{dx}{dx} = 1$.

41. ≈ 9.417

Section 10.2 (pp. 520–529)

Quick Review

1. $\sqrt{17}$
2. $\dfrac{1}{4}$
3. $b = 11$
4. $a = 4$
5. $b = 6$
6. (a) $120°$
   (b) $\dfrac{2\pi}{3}$
7. (a) $-30°$
   (b) $-\dfrac{\pi}{6}$
8. (a) $-45°$
   (b) $-\dfrac{\pi}{4}$
9. $c \approx 2.832$
10. $\theta \approx 1.046$ radians or 59.935 degrees

Exercises

1. (a) $\langle 9, -6 \rangle$
   (b) $3\sqrt{13}$
3. (a) $\langle 1, 3 \rangle$
   (b) $\sqrt{10}$
5. (a) $\langle 12, -19 \rangle$
   (b) $\sqrt{505}$
7. (a) $\left\langle \dfrac{1}{5}, \dfrac{14}{5} \right\rangle$
   (b) $\dfrac{\sqrt{197}}{5}$
9. $\langle 1, -4 \rangle$
11. $\langle -2, -3 \rangle$
13. $\left\langle -\dfrac{1}{2}, \dfrac{\sqrt{3}}{2} \right\rangle$
15. $\left\langle -\dfrac{\sqrt{3}}{2}, -\dfrac{1}{2} \right\rangle$

17. The vector **v** is horizontal and 1 in. long. The vectors **u** and **w** are $\dfrac{11}{16}$ in. long. **w** is vertical and **u** makes a $45°$ angle with the horizontal. All vectors must be drawn to scale.

    (a)

    (b)

17. continued

(c)

(d)

19. $\left(\dfrac{3}{5}, \dfrac{4}{5}\right)$ **21.** $\left(-\dfrac{15}{17}, \dfrac{8}{17}\right)$

23. Tangent: $\pm\left(\dfrac{1}{\sqrt{17}}, \dfrac{4}{\sqrt{17}}\right)$

Normal: $\pm\left(\dfrac{4}{\sqrt{17}}, -\dfrac{1}{\sqrt{17}}\right)$

25. Tangent: $\pm\left(-\dfrac{12}{\sqrt{219}}, \dfrac{5}{\sqrt{73}}\right) \approx \pm\langle -0.811, 0.585\rangle$

Normal: $\pm\left(\dfrac{5}{\sqrt{73}}, \dfrac{12}{\sqrt{219}}\right) \approx \langle 0.585, 0.811\rangle$

27. Angle at $A = \cos^{-1}\left(\dfrac{1}{\sqrt{5}}\right) \approx 63.435$ degrees

Angle at $B = \cos^{-1}\left(\dfrac{3}{5}\right) \approx 53.130$ degrees

Angle at $C = \cos^{-1}\left(\dfrac{1}{\sqrt{5}}\right) \approx 63.435$ degrees

29. (a) Both equal $u_1(v_1 + w_1) + u_2(v_2 + w_2)$.
   (b) Both equal $(u_1 + v_1)w_1 + (u_2 + v_2)w_2$.

31. $(\mathbf{u} + \mathbf{v}) \cdot (\mathbf{u} - \mathbf{v})$
$= (u_1 + v_1)(u_1 - v_1) + (u_2 + v_2)(u_2 - v_2)$
$= u_1{}^2 - v_1{}^2 + u_2{}^2 - v_2{}^2$
$= (u_1{}^2 + u_2{}^2) - (v_1{}^2 + v_2{}^2)$
$= |\mathbf{u}|^2 - |\mathbf{v}|^2$

33.

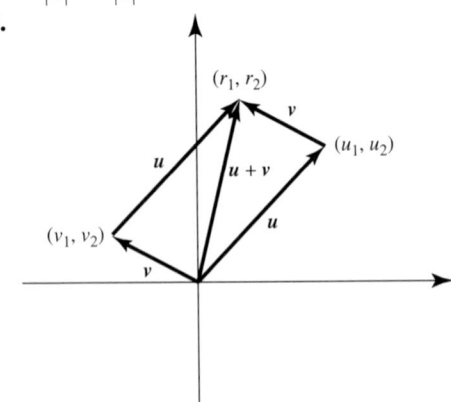

$r_1 - v_1 = u_1 \rightarrow r_1 = u_1 + v_1$
$r_2 - v_2 = u_2 \rightarrow r_2 = u_2 + v_2$

35. (a) Let $P = (a, b)$ and $Q = (c, d)$. Then

$$\left(\dfrac{1}{2}\right)\overrightarrow{OP} + \left(\dfrac{1}{2}\right)\overrightarrow{OQ} = \left(\dfrac{1}{2}\right)\langle a, b\rangle + \left(\dfrac{1}{2}\right)\langle c, d\rangle$$
$$= \left(\dfrac{(a + c)}{2}, \dfrac{(b + d)}{2}\right) = \overrightarrow{OM}$$

(b) $\overrightarrow{OM} = \left(\dfrac{2}{3}\right)\overrightarrow{OP} + \left(\dfrac{1}{3}\right)\overrightarrow{OQ}$

(c) $\overrightarrow{OM} = \left(\dfrac{1}{3}\right)\overrightarrow{OP} + \left(\dfrac{2}{3}\right)\overrightarrow{OQ}$

(d) Possible answer:

M is a fraction of the way from P to Q. Let d be this fraction. Then

$$\overrightarrow{OM} = d\overrightarrow{OQ} + (1 - d)\overrightarrow{OP}.$$

Proof: $\overrightarrow{PM} = d\overrightarrow{PQ}$ and $\overrightarrow{MQ} = (1 - d)\overrightarrow{PQ}$, so

$\overrightarrow{PQ} = \dfrac{1}{d}\overrightarrow{PM}$ and $\overrightarrow{PQ} = \dfrac{1}{1 - d}\overrightarrow{MQ}$.

Therefore, $\dfrac{1}{d}\overrightarrow{PM} = \dfrac{1}{1 - d}\overrightarrow{MQ}$.

But $\overrightarrow{PM} = \overrightarrow{OM} - \overrightarrow{OP}$ and $\overrightarrow{MQ} = \overrightarrow{OQ} - \overrightarrow{OM}$,

so $\dfrac{1}{d}\overrightarrow{OM} - \dfrac{1}{d}\overrightarrow{OP} = \dfrac{1}{1 - d}\overrightarrow{OQ} - \dfrac{1}{1 - d}\overrightarrow{OM}$.

Therefore,

$$\dfrac{1}{d}\overrightarrow{OM} + \dfrac{1}{1 - d}\overrightarrow{OM} = \dfrac{1}{d}\overrightarrow{OP} + \dfrac{1}{1 - d}\overrightarrow{OQ}.$$
$$\Rightarrow \overrightarrow{OM}\left(\dfrac{1}{d(1 - d)}\right) = \dfrac{1}{d}\overrightarrow{OP} + \dfrac{1}{1 - d}\overrightarrow{OQ}$$
$$\Rightarrow \overrightarrow{OM} = (1 - d)\overrightarrow{OP} + d\overrightarrow{OQ}.$$

37. Two adjacent sides of the rhombus can be given by two vectors of the same length, $\mathbf{u}$ and $\mathbf{v}$. Then the diagonals of the rhombus are $(\mathbf{u} + \mathbf{v})$ and $(\mathbf{u} - \mathbf{v})$. These two vectors are orthogonal since $(\mathbf{u} + \mathbf{v}) \cdot (\mathbf{u} - \mathbf{v}) = |\mathbf{u}|^2 - |\mathbf{v}|^2 = 0$.

39. Let two adjacent sides of the parallelogram be given by two vectors $\mathbf{u}$ and $\mathbf{v}$. The diagonals are then $(\mathbf{u} + \mathbf{v})$ and $(\mathbf{u} - \mathbf{v})$. So the lengths of the diagonals satisfy
$|\mathbf{u} + \mathbf{v}|^2 = (\mathbf{u} + \mathbf{v}) \cdot (\mathbf{u} + \mathbf{v})$
$= |\mathbf{u}|^2 + 2\mathbf{u} \cdot \mathbf{v} + |\mathbf{v}|^2$
and $|\mathbf{u} - \mathbf{v}|^2 = (\mathbf{u} - \mathbf{v}) \cdot (\mathbf{u} - \mathbf{v})$
$= |\mathbf{u}|^2 - 2\mathbf{u} \cdot \mathbf{v} + |\mathbf{v}|^2$.
The two lengths will be the same if and only if $\mathbf{u} \cdot \mathbf{v} = 0$, which means that $\mathbf{u}$ and $\mathbf{v}$ are perpendicular and the parallelogram is a rectangle.

41. The slopes are the same.

43. $\approx \langle -338.095, 725.046 \rangle$

45. Speed ≈ 346.735 mph
direction ≈ 14.266 degrees east of north

47. ≈ 39.337 lb

49. $\overrightarrow{AB} = \langle -3, 4 \rangle = \overrightarrow{CD}$

51. $\mathbf{u} = \langle u_1, u_2 \rangle$, $\mathbf{v} = \langle v_1, v_2 \rangle$, $\mathbf{w} = \langle w_1, w_2 \rangle$

(i) $\mathbf{u} + \mathbf{v} = \langle u_1 + v_1, u_2 + v_2 \rangle$
$= \langle v_1 + u_1, v_2 + u_2 \rangle = \mathbf{v} + \mathbf{u}$

(ii) $(\mathbf{u} + \mathbf{v}) + \mathbf{w}$
$= \langle u_1 + v_1, u_2 + v_2 \rangle + \langle w_1, w_2 \rangle$
$= \langle (u_1 + v_1) + w_1, (u_2 + v_2) + w_2 \rangle$
$= \langle u_1 + (v_1 + w_1), u_2 + (v_2 + w_2) \rangle$
$= \mathbf{u} + (\mathbf{v} + \mathbf{w})$

(iii) $\mathbf{u} + \mathbf{0} = \langle u_1, u_2 \rangle + \langle 0, 0 \rangle = \langle u_1 + 0, u_2 + 0 \rangle$
$= \langle u_1, u_2 \rangle = \mathbf{u}$

(iv) $\mathbf{u} + (-\mathbf{u}) = \langle u_1, u_2 \rangle + \langle -u_1, -u_2 \rangle$
$= \langle u_1 - u_1, u_2 - u_2 \rangle = \langle 0, 0 \rangle = \mathbf{0}$

(v) $0\mathbf{u} = 0\langle u_1, u_2 \rangle = \langle 0u_1, 0u_2 \rangle = \langle 0, 0 \rangle = \mathbf{0}$

(vi) $1\mathbf{u} = 1\langle u_1, u_2 \rangle = \langle 1u_1, 1u_2 \rangle = \langle u_1, u_2 \rangle = \mathbf{u}$

(vii) $a(b\mathbf{u}) = a(b \langle u_1, u_2 \rangle) = a\langle bu_1, bu_2 \rangle$
$= \langle abu_1, abu_2 \rangle = ab\langle u_1, u_2 \rangle = (ab)\mathbf{u}$

(viii) $a(\mathbf{u} + \mathbf{v}) = a\langle u_1 + v_1, u_2 + v_2 \rangle$
$= \langle au_1 + av_1, au_2 + av_2 \rangle$
$= \langle au_1, au_2 \rangle + \langle av_1, av_2 \rangle$
$= a\langle u_1, u_2 \rangle + a\langle v_1, v_2 \rangle$
$= a\mathbf{u} + a\mathbf{v}$

(ix) $(a + b)\mathbf{u} = (a + b)\langle u_1, u_2 \rangle$
$= \langle (a + b)u_1, (a + b)u_2 \rangle$
$= \langle au_1 + bu_1, au_2 + bu_2 \rangle$
$= \langle au_1, au_2 \rangle + \langle bu_1, bu_2 \rangle = a\mathbf{u} + b\mathbf{u}$

53. (a) $y = -x - 1$ **(b)** $y = x + 3$

Section 10.3 (pp. 529–539)

Quick Review

1. $y = \left(-\dfrac{1}{\sqrt{3}} \right)x + \dfrac{4}{\sqrt{3}}$, or approximately
$y = -0.577x + 2.309$

2. $y = \sqrt{3}x$ **3.** 0

4. Undefined; vertical tangent

5. $y = \left(-\dfrac{5\sqrt{3}}{4} \right)x + 10$ **6.** $y = \left(\dfrac{4\sqrt{3}}{15} \right)x + \dfrac{9}{10}$

7. $\dfrac{1}{4}$ **8.** ≈ 3.400

9. ≈ 2.958 **10.** $y = xe^x - e^x + 3$

Exercises

1. $6\mathbf{i} - 3\mathbf{j}$

3. (a) $-\mathbf{i} - \mathbf{j}$ **(b)** $7\mathbf{i} + 5\mathbf{j}$

5. (a)

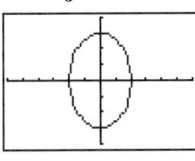

[−6, 6] by [−4, 4]

(b) $\mathbf{v}(t) = (-2 \sin t)\mathbf{i} + (3 \cos t)\mathbf{j}$
$\mathbf{a}(t) = (-2 \cos t)\mathbf{i} + (-3 \sin t)\mathbf{j}$

(c) Speed $= 2$
direction $= \langle -1, 0 \rangle$

(d) Velocity $= 2\langle -1, 0 \rangle$

7. (a)

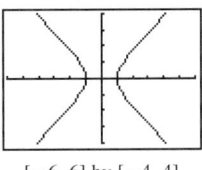

[−6, 6] by [−4, 4]

(b) $\mathbf{v}(t) = (\sec t \tan t)\mathbf{i} + (\sec^2 t)\mathbf{j}$
$\mathbf{a}(t) = (\sec t \tan^2 t + \sec^3 t)\mathbf{i} + (2 \sec^2 t \tan t)\mathbf{j}$

(c) Speed $= \dfrac{2\sqrt{5}}{3}$
direction $= \left\langle \dfrac{1}{\sqrt{5}}, \dfrac{2}{\sqrt{5}} \right\rangle$

(d) Velocity $= \left(\dfrac{2\sqrt{5}}{3} \right) \left\langle \dfrac{1}{\sqrt{5}}, \dfrac{2}{\sqrt{5}} \right\rangle$

9. (a) $y = -1$ **(b)** $x = 0$

11. $-3\mathbf{i} + (4\sqrt{2} - 2)\mathbf{j}$

13. $(\sec t)\mathbf{i} + (\ln |\sec t|)\mathbf{j} + \mathbf{C}$

15. $\mathbf{r}(t) = ((t + 1)^{3/2} - 1)\mathbf{i} - (e^{-t} - 1)\mathbf{j}$

17. $\mathbf{r}(t) = (8t + 100)\mathbf{i} + (-16t^2 + 8t)\mathbf{j}$

19. $t = 0, \pi, 2\pi$ **21.** $t = $ all multiples of $\dfrac{\pi}{2}$

23. $\cos^{-1}\left(\dfrac{3}{5} \right) \approx 53.130$ degrees

25. (a) $3\mathbf{i}$ **(b)** $t \neq 0, -3$ **(c)** $t = 0, -3$

27. 2

29. (a) $\mathbf{v}(t) = (\cos t)\mathbf{i} - (2 \sin 2t)\mathbf{j}$

(b) $t = \dfrac{\pi}{2}, \dfrac{3\pi}{2}$

(c) $y = 1 - 2x^2, -1 \leq x \leq 1$. The particle starts at $(0, 1)$, goes to $(1, -1)$, then goes to $(-1, -1)$, and then goes to $(0, 1)$, tracing the curve twice.

31. $\mathbf{r}(t) = \left(\dfrac{3}{2}t^2 + \dfrac{3\sqrt{10}}{5}t + 1 \right)\mathbf{i} + \left(-\dfrac{1}{2}t^2 - \dfrac{\sqrt{10}}{5}t + 2 \right)\mathbf{j}$

33. (a) 160 seconds **(b)** 225 m

(c) $\dfrac{15}{4}$ meters per second

(d) At $t = 80$ seconds

35. (a) Referring to the figure, look at the circular arc from the point where $t = 0$ to the point "m".
On one hand, this arc has length given by $(r_0\theta)$, but it also has length given by (vt). Setting those two quantities equal gives the result.

(b) $\mathbf{a}(t) = -\dfrac{v^2}{r_0}\left[\left(\cos\dfrac{vt}{r_0}\right)\mathbf{i} + \left(\sin\dfrac{vt}{r_0}\right)\mathbf{j}\right]$

(c) From part (b) above, $\mathbf{a}(t) = -\left(\dfrac{v}{r_0}\right)^2 \mathbf{r}(t)$.
So, by Newton's second law, $\mathbf{F} = -m\left(\dfrac{v}{r_0}\right)^2 \mathbf{r}$.
Substituting for $\mathbf{F}$ in the law of gravitation gives the result.

(d) Set $\dfrac{vT}{r_0} = 2\pi$ and solve for vT.

(e) Substitute $\dfrac{2\pi r_0}{T}$ for v in $v^2 = \dfrac{GM}{r_0}$ and solve for T^2.

37. (a) Apply the Corollary to each component separately.

(b) Follows immediately from (a) since any two anti-derivatives of $\mathbf{r}(t)$ must have identical derivatives, namely $\mathbf{r}(t)$.

39. Let $\mathbf{C} = \langle C_1, C_2 \rangle$. $\dfrac{d\mathbf{C}}{dt} = \left\langle \dfrac{dC_1}{dt}, \dfrac{dC_2}{dt} \right\rangle = \langle 0, 0 \rangle$.

41. $\mathbf{u} = \langle u_1, u_2 \rangle$, $\mathbf{v} = \langle v_1, v_2 \rangle$

(a) $\dfrac{d}{dt}(\mathbf{u} + \mathbf{v}) = \dfrac{d}{dt}(\langle u_1 + v_1, u_2 + v_2 \rangle)$

$= \left\langle \dfrac{d}{dt}(u_1 + v_1), \dfrac{d}{dt}(u_2 + v_2) \right\rangle$

$= \langle u_1' + v_1', u_2' + v_2' \rangle$

$= \langle u_1', u_2' \rangle + \langle v_1', v_2' \rangle = \dfrac{d\mathbf{u}}{dt} + \dfrac{d\mathbf{v}}{dt}$

(b) $\dfrac{d}{dt}(\mathbf{u} - \mathbf{v}) = \dfrac{d}{dt}(\langle u_1 - v_1, u_2 - v_2 \rangle)$

$= \left\langle \dfrac{d}{dt}(u_1 - v_1), \dfrac{d}{dt}(u_2 - v_2) \right\rangle$

$= \langle u_1' - v_1', u_2' - v_2' \rangle$

$= \langle u_1', u_2' \rangle - \langle v_1', v_2' \rangle$

$= \dfrac{d\mathbf{u}}{dt} - \dfrac{d\mathbf{v}}{dt}$

43. $f(t)$ and $g(t)$ differentiable at $c \Rightarrow f(t)$ and $g(t)$ continuous at $c \Rightarrow \mathbf{r}(t) = f(t)\mathbf{i} + g(t)\mathbf{j}$ is continuous at c.

45. (a) Let $\mathbf{r}(t) = f(t)\mathbf{i} + g(t)\mathbf{j}$. Then

$\dfrac{d}{dt}\displaystyle\int_a^t \mathbf{r}(q)\,dq = \dfrac{d}{dt}\int_a^t [f(q)\mathbf{i} + g(q)\mathbf{j}]\,dq$

$= \dfrac{d}{dt}\left[\left(\int_a^t f(q)\,dq\right)\mathbf{i} + \left(\int_a^t g(q)\,dq\right)\mathbf{j}\right]$

$= \left(\dfrac{d}{dt}\int_a^t f(q)\,dq\right)\mathbf{i} + \left(\dfrac{d}{dt}\int_a^t g(q)\,dq\right)\mathbf{j}$

$= f(t)\mathbf{i} + g(t)\mathbf{j} = \mathbf{r}(t)$.

(b) Let $\mathbf{S}(t) = \displaystyle\int_a^t \mathbf{r}(q)\,dq$. Then part (a) shows that $\mathbf{S}(t)$ is an antiderivative of $\mathbf{r}(t)$. Let $\mathbf{R}(t)$ be any antiderivative of $\mathbf{r}(t)$. Then according to 37(b), $\mathbf{S}(t) = \mathbf{R}(t) + \mathbf{C}$.

Letting $t = a$, we have

$\mathbf{0} = \mathbf{S}(a) = \mathbf{R}(a) + \mathbf{C}$. Therefore,

$\mathbf{C} = -\mathbf{R}(a)$ and $\mathbf{S}(t) = \mathbf{R}(t) - \mathbf{R}(a)$.

The result follows by letting $t = b$.

Section 10.4 (pp. 539–552)

Quick Review

1. x-component $= 50\cos 25° \approx 45.315$
y-component $= 50\sin 25° \approx 21.131$

2. x-component $= 80\cos 120° = -40$
y-component $= 80\sin 120° = 40\sqrt{3} \approx 69.282$

3. x-intercepts: $\left(\dfrac{5}{2}, 0\right)$ and $(-8, 0)$
y-intercept: $(0, -40)$

4. $\left(-\dfrac{11}{4}, -\dfrac{441}{8}\right)$

5. x-intercepts: $(0, 0)$ and $(20, 0)$
y-intercept: $(0, 0)$

6. $(10, 100)$

7. $y = -\cos x + 2$

8. $y = \dfrac{t^3}{3} + 3t + \dfrac{25}{3}$

9. $y = 16 + 4e^{-t}$

10. $y = 2 - e^{-x^2}$

Exercises

1. 50 seconds

3. (a) After ≈ 72.154 seconds, ≈ 25.510 km downrange

(b) 4020 m **(c)** ≈ 6377.55 m

5. After ≈ 2.135 seconds, ≈ 66.421 feet from the stopboard

7. (a) $7\sqrt{2} \approx 9.899$ m/sec
  (b) ≈ 18.435 or 71.565 degrees

9. ≈ 278.016 ft/sec, or ≈ 189.556 mph

11. No. When it has travelled 135 ft in the horizontal direction, it is only about 29.942 feet above the ground.

13. (a) ≈ 149.307 ft/sec **(b)** ≈ 2.245 seconds

15. ≈ 39.261 and 50.739 degrees

17. ≈ 46.597 ft/sec

19. Integrating, $\dfrac{d}{dt}\mathbf{r}(t) = c_1\mathbf{i} + (-gt + c_2)\mathbf{j}$.
The initial condition on the velocity gives

$c_1 = v_0 \cos \alpha$ and $c_2 = v_0 \sin \alpha$. Integrating again,

$$\mathbf{r}(t) = ((v_0 \cos \alpha)t + c_3)\mathbf{i}$$
$$+ \left(-\frac{gt^2}{2} + (v_0 \sin \alpha)t + c_4\right)\mathbf{j}.$$

The initial condition on the position gives

$c_3 = x_0$ and $c_4 = y_0$.

21. It takes about 1.924 seconds. The arrow passes about 3.698 feet above each rim. (It's 73.698 feet above the ground.)

23. Angle $\approx 62°$
  Maximum height $= 4$ feet (independent of the measured angle)
  Speed of engine ≈ 8.507 ft/sec (changes with the angle)

25. (a)

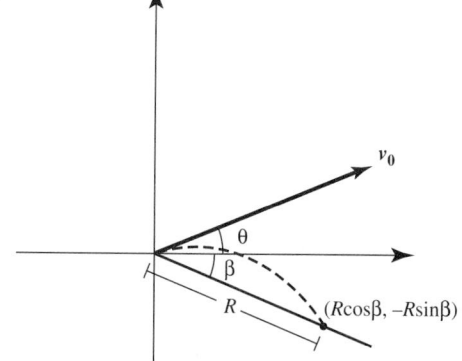

$$x = (v_0 \cos \theta)t$$
$$y = (v_0 \sin \theta)t = \frac{1}{2}gt^2$$

$$x = R \cos \beta \Rightarrow R \cos \beta = (v_0 \cos \theta)t$$

$$\Rightarrow t = \frac{R \cos \beta}{v_0 \cos \theta}. \text{ Then } y = -R \sin \beta$$

$$\Rightarrow -R \sin \beta = \frac{(v_0 \sin \theta) R \cos \beta}{v_0 \cos \theta} - \frac{g}{2}\frac{R^2 \cos^2 \beta}{v_0^2 \cos^2 \theta}$$

$$\Rightarrow R = \frac{2v_0^2}{g \cos^2 \beta} \cos \theta \sin (\theta + \beta).$$

Let $f(\theta) = \cos \theta \sin (\theta + \beta)$.
$f'(\theta) = \cos \theta \cos (\theta + \beta) - \sin \theta \sin (\theta + \beta)$
$f'(\theta) = 0 \Rightarrow \tan \theta \tan (\theta + \beta) = 1$
$\Rightarrow \tan \theta = \cot (\theta + \beta)$
$\Rightarrow \theta + \beta = 90° - \theta$. Note that $f''(\theta) < 0$, so
R is maximum when $\theta + \beta = 90° - \theta$.

(b)

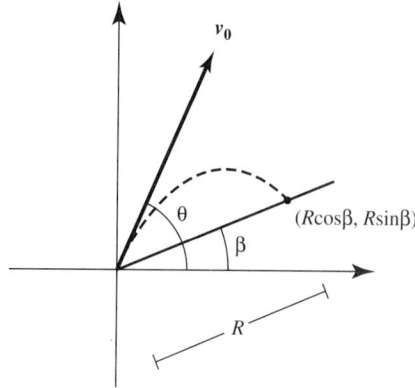

$R = \dfrac{2v_0^2}{g \cos^2 \beta} \cos \theta \sin (\theta - \beta)$ is maximum
when $\tan \theta = \cot (\theta - \beta)$,

so $\theta - \beta = 90° - \theta$.

The initial velocity vector bisects the angle between the hill and the vertical for max range.

27. (a) (Assuming that "x" is zero at the point of impact.)
  $\mathbf{r}(t) = (x(t))\mathbf{i} + (y(t))\mathbf{j}$, where
  $x(t) = (35 \cos 27°)t$ and
  $y(t) = 4 + (35 \sin 27°)t - 16t^2$.

  (b) At $t \approx 0.497$ seconds, it reaches its maximum height of about 7.945 feet.

  (c) Range ≈ 37.460 feet
  flight time ≈ 1.201 seconds

  (d) At $t \approx 0.254$ and $t \approx 0.740$ seconds, when it is ≈ 29.554 and ≈ 14.396 feet from where it will land.

  (e) Yes. It changes things because the ball won't clear the net.

29. (a) $\mathbf{r}(t) = (x(t))\mathbf{i} + (y(t))\mathbf{j}$, where

$$x(t) = \left(\frac{1}{0.08}\right)(1 - e^{-0.08t})(152 \cos 20° - 17.6) \text{ and}$$

$$y(t) = 3 + \left(\frac{152}{0.08}\right)(1 - e^{-0.08t})(\sin 20°)$$
$$+ \left(\frac{32}{0.08^2}\right)(1 - 0.08t - e^{-0.08t})$$

29. continued

   (b) At $t \approx 1.527$ seconds it reaches its maximum height of about 41.893 feet.

   (c) Range ≈ 351.734 feet
   Flight time ≈ 3.181 seconds

   (d) At $t \approx 0.877$ and $t \approx 2.190$ seconds, when it is about 106.028 and 251.530 feet from home plate.

   (e) No. The wind gust would need to be greater than 12.846 ft/sec in the direction of the hit in order for the ball to clear the fence for a home run.

31. The points in question are $(x, y) = \left(\dfrac{R}{2}, y_{max}\right)$. So,

$x = \dfrac{v_0^2 \sin \alpha \cos \alpha}{g}$, and $y = \dfrac{(v_0 \sin \alpha)^2}{2g}$.

Substituting these into the given equation for the ellipse yields an identity.

Section 10.5 (pp. 552–559)

Quick Review

1. $y = -x + 2$ **2.** $x^2 + y^2 = 9$

3. $(x + 2)^2 + (y - 4)^2 = 4$

4. (a) No   **(b)** No   **(c)** Yes

5. (a) No   **(b)** No   **(c)** No

6. (a) No   **(b)** Yes   **(c)** No

7. (a) Yes   **(b)** Yes   **(c)** Yes

8. Graph $y = (x - 2)^{1/2}$ and $y = -(x - 2)^{1/2}$

9. Graph $y = \left(\dfrac{4 - x^2}{3}\right)^{1/2}$ and $y = -\left(\dfrac{4 - x^2}{3}\right)^{1/2}$

10. $(x - 2)^2 + (y + 3)^2 = 4$, center $= (2, -3)$, radius $= 2$

Exercises

1. (a) and **(e)** are the same.

   (b) and **(g)** are the same.

   (c) and **(h)** are the same.

   (d) and **(f)** are the same.

3.

[−3, 3] by [−2, 2]

(a) $(1, 1)$ **(b)** $(1, 0)$

(c) $(0, 0)$ **(d)** $(-1, -1)$

5.

[−6, 6] by [−4, 4]

(a) $\left(\sqrt{2}, \dfrac{3\pi}{4}\right)$ or $\left(\sqrt{2}, -\dfrac{5\pi}{4}\right)$

(b) $\left(2, -\dfrac{\pi}{3}\right)$ or $\left(-2, \dfrac{2\pi}{3}\right)$

(c) $\left(3, \dfrac{\pi}{2}\right)$ or $\left(3, \dfrac{5\pi}{2}\right)$

(d) $(1, \pi)$ or $(-1, 0)$

7.   **9.**

          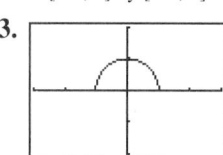

[−6, 6] by [−4, 4]   [−3, 3] by [−2, 2]

11.   **13.**

[−9, 9] by [−6, 6]   [−3, 3] by [−2, 2]

15.   **17.**

[−1.8, 1.8] by [−1.2, 1.2]

19. $y = 0$, the x-axis

21. $y = 4$, a horizontal line

23. $x + y = 1$, a line (slope $= -1$, y-int. $= 1$)

25. $x^2 + y^2 = 4y$, a circle (center $= (0, 2)$, radius $= 2$)

27. $xy = 1 \left(\text{or, } y = \dfrac{1}{x}\right)$, a hyperbola

29. $y = e^x$, the exponential curve

31. $y = \ln x$, the logarithmic curve

33. $x^2 + y^2 = -4x$, a circle (center $= (-2, 0)$, radius $= 2$)

35. $x^2 + y^2 = 2x + 2y$, a circle (center $= (1, 1)$, radius $= \sqrt{2}$)

37. $r \cos \theta = 7$ **39.** $\theta = \dfrac{\pi}{4}$

41. $r^2 = 4$ or $r = 2$

43. $r^2(4 \cos^2 \theta + 9 \sin^2 \theta) = 36$

45. $r \sin^2 \theta = 4 \cos \theta$ **47.** $r = 4 \sin \theta$

49. (a)

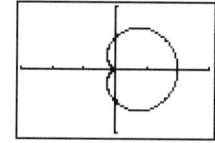

$[-3, 3]$ by $[-2, 2]$

(b) Length of interval $= 2\pi$

51. (a)

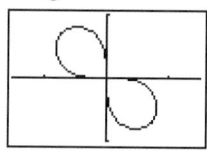

$[-1.5, 1.5]$ by $[-1, 1]$

(b) Length of interval $= \dfrac{\pi}{2}$

53. (a)

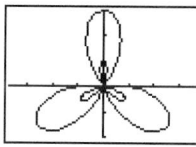

$[-3.75, 3.75]$ by $[-2, 3]$

(b) Length of interval $= 2\pi$

55. (a)

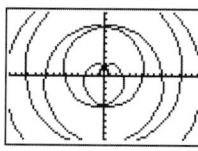

$[-15, 15]$ by $[-10, 10]$

(b) Required interval $= (-\infty, \infty)$

57. (a)

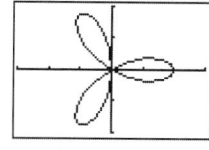

$[-3, 3]$ by $[-2, 2]$

(b) Length of interval $= \pi$

59. x-axis, y-axis, origin   **61.** y-axis

63. (a) Because $r = a \sec \theta$ is equivalent to
$r \cos \theta = a$ which is equivalent to the
cartesian equation $x = a$.

(b) $r = a \csc \theta$ is equivalent to $y = a$.

65. (a) We have $x = r \cos \theta$ and $y = r \sin \theta$. By
taking $t = \theta$, we have $r = f(t)$,
so $x = f(t) \cos t$ and $y = f(t) \sin t$.

(b) $x = 3 \cos t, y = 3 \sin t$

(c) $x = (1 - \cos t) \cos t, y = (1 - \cos t) \sin t$

(d) $x = (3 \sin 2t) \cos t, y = (3 \sin 2t) \sin t$

67. $d = [(x_2 - x_1)^2 + (y_2 - y_1)^2]^{1/2}$

$= [(r_2 \cos \theta_2 - r_1 \cos \theta_1)^2$

$\quad + (r_2 \sin \theta_2 - r_1 \sin \theta_1)^2]^{1/2}$

and then simplify using trigonometric identities.

69. (a)

$[-5, 25]$ by $[-10, 10]$

The graphs are hyperbolas.

(b) As $k \to 1^+$, the right branch of the hyperbola
goes to infinity and "disappears". The left
branch approaches the parabola $y^2 = 4 - 4x$.

Section 10.6 (pp. 559–568)

Quick Review

1. $-\dfrac{5}{3} \cot t$ **2.** $-\dfrac{5}{3} \cot 2 \approx 0.763$

3. $(0, 5)$ and $(0, -5)$ **4.** $(3, 0)$ and $(-3, 0)$

5. ≈ 12.763

6. The upper half of the outer loop

7. The inner loop

8. The lower half of the outer loop

9. 36   **10.** ≈ 2.403

Exercises

1. At $\theta = 0$: -1
At $\theta = \pi$: 1

3. At $(2, 0)$: $-\dfrac{2}{3}$

At $\left(-1, \dfrac{\pi}{2}\right)$: 0

At $(2, \pi)$: $\dfrac{2}{3}$

At $\left(5, \dfrac{3\pi}{2}\right)$: 0

5. $\theta = \dfrac{\pi}{2}$ $[x = 0]$

7. $\theta = 0$ $[y = 0]$

$\theta = \dfrac{\pi}{5}$ $\left[y = \left(\tan \dfrac{\pi}{5}\right)x\right]$

$\theta = \dfrac{2\pi}{5}$ $\left[y = \left(\tan \dfrac{2\pi}{5}\right)x\right]$

$\theta = \dfrac{3\pi}{5}$ $\left[y = \left(\tan \dfrac{3\pi}{5}\right)x\right]$

$\theta = \dfrac{4\pi}{5}$ $\left[y = \left(\tan \dfrac{4\pi}{5}\right)x\right]$

9. Horizontal at: $\left(-\dfrac{1}{2}, \dfrac{\pi}{6}\right)$ $\left[y = -\dfrac{1}{4}\right]$,

$\left(-\dfrac{1}{2}, \dfrac{5\pi}{6}\right)$ $\left[y = -\dfrac{1}{4}\right]$,

$\left(-2, \dfrac{3\pi}{2}\right)$ $[y = 2]$

Vertical at: $\left(0, \dfrac{\pi}{2}\right)$ $[x = 0]$,

$\left(-\dfrac{3}{2}, \dfrac{7\pi}{6}\right)$ $\left[x = \dfrac{3\sqrt{3}}{4}\right]$,

$\left(-1.5, \dfrac{11\pi}{6}\right)$ $\left[x = -\dfrac{3\sqrt{3}}{4}\right]$

11. Horizontal at: $(0, 0)$ $[y = 0]$,

$\left(2, \dfrac{\pi}{2}\right)$ $[y = 2]$,

$(0, \pi)$ $[y = 0]$

Vertical at: $\left(\sqrt{2}, \dfrac{\pi}{4}\right)$ $[x = 1]$,

$\left(\sqrt{2}, \dfrac{3\pi}{4}\right)$ $[x = -1]$

13. 18π **15.** $2a^2$ **17.** 2

19. $\dfrac{\pi}{2} - 1$ **21.** $5\pi - 8$ **23.** $a^2\pi$

25. $8 - \pi$ **27.** $12\pi - 9\sqrt{3}$

29. (a) $\dfrac{3}{2} - \dfrac{\pi}{4}$

(b) Yes. $x = \tan\theta\cos\theta \Rightarrow x = \sin\theta$

$y = \tan\theta\sin\theta \Rightarrow y = \dfrac{\sin^2\theta}{\cos\theta}$

$\displaystyle\lim_{\theta\to-\pi/2^+} x = -1, \quad \lim_{\theta\to-\pi/2^+} = +\infty$

$\displaystyle\lim_{\theta\to\pi/2^-} x = 1, \quad \lim_{\theta\to\pi/2^-} y = +\infty$

31. $\dfrac{19}{3}$ **33.** 8

35. ≈ 6.887 **37.** $\dfrac{\pi + 3}{8}$

39. $\pi\sqrt{2} \approx 4.443$ **41.** $(4 - 2\sqrt{2})\pi \approx 3.681$

43. $\left(\dfrac{dx}{d\theta}\right)^2 + \left(\dfrac{dy}{d\theta}\right)^2$

$= (f'(\theta)\cos\theta - f(\theta)\sin\theta)^2$
$\quad + (f'(\theta)\sin\theta + f(\theta)\cos\theta)^2$

$= (f'(\theta)\cos\theta)^2 + (f(\theta)\sin\theta)^2 + (f'(\theta)\sin\theta)^2$
$\quad + (f(\theta)\cos\theta)^2$

$= (f(\theta))^2(\cos^2\theta + \sin^2\theta)$
$\quad + (f'(\theta))^2(\cos^2\theta + \sin^2\theta)$

$= (f(\theta))^2 + (f'(\theta))^2 = r^2 + \left(\dfrac{dr}{d\theta}\right)^2$

45. If $g(\theta) = 2f(\theta)$, then
$(g(\theta)^2 + g'(\theta)^2)^{1/2} = 2(f(\theta)^2 + f'(\theta)^2)^{1/2}$ so the
length of g is 2 times the length of f.

47. (a) Let $r = 1.75 + \dfrac{0.06\theta}{2\pi}, \; 0 \le \theta \le 12\pi$.

(b) Since $\dfrac{dr}{d\theta} = \dfrac{b}{2\pi}$, this is just equation (4) for
the length of the curve.

(c) ≈ 741.420 cm, or, ≈ 7.414 m

(d) $\left(\left(r^2 + \dfrac{b}{2\pi}\right)^2\right)^{1/2} = r\left(\left(1 + \dfrac{b}{2\pi r}\right)^2\right)^{1/2} \approx r$ since

$\left(\dfrac{b}{2\pi r}\right)^2$ is a very small quantity squared.

(e) $L \approx 741.420$ cm (from part (c)),
$L_a \approx 741.416$ cm

49. $\left(\dfrac{5a}{6}, 0\right)$

Chapter Review (pp. 569–572)

1. (a) $\langle -17, 32 \rangle$ **(b)** $\sqrt{1313}$
2. (a) $\langle -1, -1 \rangle$ **(b)** $\sqrt{2}$
3. (a) $\langle 6, -8 \rangle$ **(b)** 10
4. (a) $\langle 10, -25 \rangle$ **(b)** $\sqrt{725} = 5\sqrt{29}$
5. $\left(-\dfrac{\sqrt{3}}{2}, -\dfrac{1}{2}\right)$ [assuming counterclockwise]
6. $\left(\dfrac{\sqrt{3}}{2}, \dfrac{1}{2}\right)$ **7.** $\left(\dfrac{8}{\sqrt{17}}, -\dfrac{2}{\sqrt{17}}\right)$
8. $\langle -3, -4 \rangle$
9. (a) $y = \dfrac{\sqrt{3}}{2}x + \dfrac{1}{4}$ **(b)** $\dfrac{1}{4}$
10. (a) $y = -3x + \dfrac{13}{4}$ **(b)** 6
11. (a) $\left(0, \dfrac{1}{2}\right)$ and $\left(0, -\dfrac{1}{2}\right)$
(b) Nowhere
12. (a) $(0, 2)$ and $(0, -2)$ **(b)** $(-2, 0)$ and $(2, 0)$
13. (a) $(0, 0)$ **(b)** Nowhere
14. (a) $(0, 9)$ and $(0, -9)$ **(b)** $(-4, 0)$ and $(4, 0)$
15.
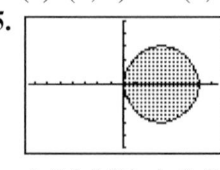
$[-7.5, 7.5]$ by $[-5, 5]$
16.
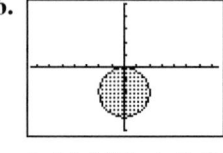
$[-7.5, 7.5]$ by $[-5, 5]$

17. (a)

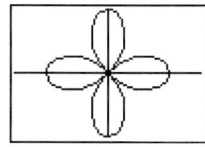

$[-1.5, 1.5]$ by $[-1, 1]$

(b) 2π

18. (a)

$[-3, 3]$ by $[-2, 2]$

(b) π

19. (a)

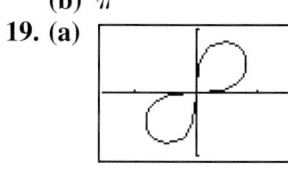

$[-1.5, 1.5]$ by $[-1, 1]$

(b) $\dfrac{\pi}{2}$

20. (a)

$[-1.5, 1.5]$ by $[-1, 1]$

(b) π

21. Tangent lines at $\theta = \dfrac{\pi}{4}, \dfrac{3\pi}{4}, \dfrac{5\pi}{4}$, and $\dfrac{7\pi}{4}$
Cartesian equations are $y = \pm x$.

22. Tangent lines at $\theta = \dfrac{\pi}{2}$ and $\dfrac{3\pi}{2}$
Cartesian equation is $x = 0$.

23. Horizontal: $y = 0$, $y \approx \pm 0.443$, $y \approx \pm 1.739$
Vertical: $x = 2$, $x \approx 0.067$, $x \approx -1.104$

24. Horizontal: $y = \dfrac{1}{2}$, $y = -4$
Vertical: $x = 0$, $x \approx \pm 2.598$

25. $y = \pm x + \sqrt{2}$ and $y = \pm x - \sqrt{2}$
26. $y = x - 1$ and $y = -x - 1$
27. $x = y$, a line
28. $x^2 + y^2 = 3x$,
a circle $\left(\text{center} = \left(\dfrac{3}{2}, 0\right), \text{radius} = \dfrac{3}{2}\right)$

29. $x^2 = 4y$, a parabola
30. $x - \sqrt{3}y = 4\sqrt{3}$ or $y = \dfrac{x}{\sqrt{3}} - 4$, a line
31. $r = -5 \sin \theta$ **32.** $r = 2 \sin \theta$

33. $r^2 \cos^2 \theta + 4r^2 \sin^2 \theta = 16$, or,
$r^2 = \dfrac{16}{\cos^2 \theta + 4 \sin^2 \theta}$

34. $(r \cos \theta + 2)^2 + (r \sin \theta - 5)^2 = 16$
35. $\dfrac{\ln 2 + 24}{8} \approx 3.087$ **36.** $4\sqrt{3}$
37. 8   **38.** $\pi\sqrt{2}$
39. $\pi - 3$ **40.** $\pi\sqrt{2}$
41. ≈ 3.183 **42.** ≈ 12.363
43. $\dfrac{9\pi}{2}$ **44.** $\dfrac{\pi}{12}$
45. $\dfrac{\pi}{4} + 2$ **46.** 5π
47. $\dfrac{76\pi}{3}$ **48.** ≈ 10.110
49. $\pi(2 - \sqrt{2}) \approx 1.840$ **50.** $4\pi \approx 12.566$
51. (a) $\mathbf{v}(t) = (-4 \sin t)\mathbf{i} + (\sqrt{2} \cos t)\mathbf{j}$
$\mathbf{a}(t) = (-4 \cos t)\mathbf{i} + (-\sqrt{2} \sin t)\mathbf{j}$
(b) 3
(c) $\cos^{-1} \dfrac{7}{9} \approx 38.942$ degrees
52. (a) $\mathbf{v}(t) = (\sqrt{3} \sec t \tan t)\mathbf{i} + (\sqrt{3} \sec^2 t)\mathbf{j}$
$\mathbf{a}(t) = \sqrt{3}(\sec t \tan^2 t + \sec^3 t)\mathbf{i}$
$+ (2\sqrt{3} \sec^2 t \tan t)\mathbf{j}$
(b) $\sqrt{3}$ **(c)** 90 degrees
53. 1
54. $\mathbf{a}(t) = (-2 e^t \sin t)\mathbf{i} + (2 e^t \cos t)\mathbf{j}$, and
$\mathbf{r}(t) \cdot \mathbf{a}(t) = 0$ for all t. The angle between $\mathbf{r}$ and
$\mathbf{a}$ is always 90 degrees.
55. $6\mathbf{i}$ **56.** $3\mathbf{i} + (\ln 2)\mathbf{j}$
57. $\mathbf{r}(t) = (\cos t - 1)\mathbf{i} + (\sin t + 1)\mathbf{j}$
58. $\mathbf{r}(t) = (\tan^{-1} t + 1)\mathbf{i} + ((t^2 + 1)^{1/2})\mathbf{j}$
59. $\mathbf{r}(t) = \mathbf{i} + t^2\mathbf{j}$
60. $\mathbf{r}(t) = (-t^2 + 6t - 2)\mathbf{i} + (-t^2 + 2t + 2)\mathbf{j}$
61. (a) $\dfrac{\pi\sqrt{34}}{4\sqrt{2}} = \dfrac{\pi\sqrt{17}}{4} \approx 3.238$
(b) x-component: $\dfrac{3\pi^2}{16\sqrt{2}}$
y-component: $-\dfrac{5\pi^2}{16\sqrt{2}}$
(c) $\dfrac{x^2}{9} + \dfrac{y^2}{25} = 1$
62. (a) ≈ 25.874
(b) Volume $= \dfrac{1250\pi}{3}$
(c) Area ≈ 1040.728
63. (a) 1   **(b)** $e^3\sqrt{2}$
(c) $(e^3 - 1)\sqrt{2}$

64. (a) $\dfrac{104}{5}$ **(b)** $\dfrac{4144}{135}$

(c) $\dfrac{dy}{dx} = \dfrac{3}{5}\sqrt{x+2}$

65. Speed ≈ 591.982 mph

Direction ≈ 8.179 degrees north of east

66. Direction ≈ 2.073 degrees

Magnitude ≈ 411.891 lbs

67. It hits the ground ≈ 2.135 seconds later, approximately 66.421 feet from where it left the thrower's hand. Assuming it doesn't bounce or roll, it will still be there 3 seconds after it was thrown.

68. 57 feet

69. (a)

[−2, 12] by [−2, 4]

(b) $\mathbf{v}(0) = \langle 0, 0 \rangle$ $\mathbf{v}(1) = \langle 2\pi, 0 \rangle$
$\mathbf{a}(0) = \langle 0, \pi^2 \rangle$ $\mathbf{a}(1) = \langle 0, -\pi^2 \rangle$

$\mathbf{v}(2) = \langle 0, 0 \rangle$ $\mathbf{v}(3) = \langle 2\pi, 0 \rangle$
$\mathbf{a}(2) = \langle 0, \pi^2 \rangle$ $\mathbf{a}(3) = \langle 0, -\pi^2 \rangle$

(c) Topmost point: 2π ft/sec

center of wheel: π ft/sec

Reasons: Since the wheel rolls half a circumference, or π feet every second, the center of the wheel will move π feet every second. Since the rim of the wheel is turning at a rate of π ft/sec about the center, the velocity of the topmost point relative to the center is π ft/sec, giving it a total velocity of 2π ft/sec.

70. For 4325 yds: $v \approx 644.360$ ft/sec

For 4752 yds: $v \approx 675.420$ ft/sec

71. (a) ≈ 59.195 ft/sec   **(b)** ≈ 74.584 ft/sec

72. (a) ≈ 91.01 ft/sec   **(b)** 59.97 ft

73. We have $x = (v_0 t) \cos \alpha$ and

$y + \dfrac{gt^2}{2} = (v_0 t) \sin \alpha$. Squaring and adding gives

$x^2 + \left(y + \dfrac{gt^2}{2} \right)^2 = (v_0 t)^2 (\cos^2 \alpha + \sin^2 \alpha) = v_0^2 t^2$.

74. (a) $\mathbf{r}(t) = (155 \cos 18° - 11.7)t\mathbf{i}$
$+ (4 + 155 \sin 18°t - 16t^2)\mathbf{j}$

$x(t) = (155 \cos 18° - 11.7)t$
$y(t) = 4 + 155 \sin 18°t - 16t^2$

(b) At ≈ 1.497 seconds, it reaches a maximum height of ≈ 39.847 ft.

(c) Range ≈ 417.307 ft

Flight time ≈ 3.075 secs

(d) At times $t \approx 0.534$ and $t \approx 2.460$ seconds, when it is ≈ 72.406 and ≈ 333.867 feet from home plate.

(e) Yes, the batter has hit a home run. When the ball is 380 feet from home plate (at $t \approx 2.800$ seconds), it is approximately 12.673 feet off the ground and therefore clears the fence by at least two feet.

75. (a)
$$\mathbf{r}(t) = \left[(155 \cos 18° - 11.7)\left(\frac{1}{0.09} \right)(1 - e^{-0.09t}) \right] \mathbf{i}$$
$$+ \left[4 + \left(\frac{155 \sin 18°}{0.09} \right)(1 - e^{-0.09t}) \right.$$
$$\left. + \frac{32}{0.09^2}(1 - 0.09t - e^{-0.09t}) \right] \mathbf{j}$$

$$x(t) = (155 \cos 18° - 11.7)\left(\frac{1}{0.09} \right)(1 - e^{-0.09t})$$

$$y(t) = 4 + \left(\frac{155 \sin 18°}{0.09} \right)(1 - e^{-0.09t})$$
$$+ \frac{32}{0.09^2}(1 - 0.09t - e^{-0.09t})$$

(b) At ≈ 1.404 seconds, it reaches a maximum height of ≈ 36.921 feet.

(c) Range ≈ 352.520 ft

Flight time ≈ 2.959 secs

(d) At times $t \approx 0.753$ and $t \approx 2.068$ seconds, when it is ≈ 98.799 and ≈ 256.138 feet from home plate

(e) No, the batter has not hit a home run. If the drag coefficient k is less than ≈ 0.011, the hit will be a home run.

76. (a) $\overrightarrow{BD} = \overrightarrow{AD} - \overrightarrow{AB}$

(b) $\overrightarrow{AP} = \overrightarrow{AB} + \dfrac{1}{2}\overrightarrow{BD} = \dfrac{1}{2}\overrightarrow{AB} + \dfrac{1}{2}\overrightarrow{AD}$

(c) $\overrightarrow{AC} = \overrightarrow{AB} + \overrightarrow{AD}$, so by part (b), $\overrightarrow{AP} = \dfrac{1}{2}\overrightarrow{AC}$.

77. The widths between the successive turns are constant and are given by $2\pi a$.

Chapter 11

Section 11.1 (pp. 575–585)

Quick Review

1. $5\sqrt{2}$

2. $(3/2, 5/2)$

3. $\sqrt{13}$

4. $\dfrac{2}{\sqrt{13}}\mathbf{i} + \dfrac{3}{\sqrt{13}}\mathbf{j}$

5. $-\dfrac{6}{\sqrt{13}}\mathbf{i} - \dfrac{9}{\sqrt{13}}\mathbf{j}$

6. Ellipse, center $(0, 0)$, x-intercepts: $(\pm 2, 0)$, vertices: $(0, \pm 3)$

7. Ellipse plus interior, center $(0, 0)$, x-intercepts: $(\pm 2, 0)$, vertices: $(0, \pm 3)$

8. $(x - 2)^2 + (y + 3)^2 = 9$

9. Center: $(-3, 1)$, raidus $= 2$

10. $5\mathbf{i} - 4\mathbf{j}$

Exercises

1. The line through $(2, 3, 0)$ parallel to the z-axis

3. The x-axis

5. The circle $x^2 + y^2 = 4$ in the plane $z = -2$

7. The circle $y^2 + z^2 = 1$ in the yz-plane

9. The circle $x^2 + y^2 = 16$ in the xy-plane

11. (a) The first quadrant of the xy-plane
    (b) The fourth quadrant of the xy-plane

13. (a) The ball of radius 1 centered at the origin
    (b) All points at distance greater than 1 unit from the origin

15. (a) The upper hemisphere of radius 1 centered at the origin
    (b) The solid upper hemisphere of radius 1 centered at the origin

17. (a) $x = 3$ (b) $y = -1$ (c) $z = -2$

19. (a) $z = 1$ (b) $x = 3$ (c) $y = -1$

21. (a) $x^2 + (y - 2)^2 = 4, z = 0$
    (b) $(y - 2)^2 + z^2 = 4, x = 0$
    (c) $x^2 + z^2 = 4, y = 2$

23. (a) $y = 3, z = -1$ (b) $x = 1, z = -1$
    (c) $x = 1, y = 3$

25. $x^2 + y^2 + z^2 = 25, z = 3$

27. $0 \le z \le 1$

29. $z \le 0$

31. (a) $(x - 1)^2 + (y - 1)^2 + (z - 1)^2 < 1$
    (b) $(x - 1)^2 + (y - 1)^2 + (z - 1)^2 > 1$

33. $2\mathbf{i} + \mathbf{j} - 2\mathbf{k} = 3\left(\dfrac{2}{3}\mathbf{i} + \dfrac{1}{3}\mathbf{j} - \dfrac{2}{3}\mathbf{k}\right)$

35. $5\mathbf{k} = 5(\mathbf{k})$

37. $\dfrac{1}{\sqrt{6}}\mathbf{i} - \dfrac{1}{\sqrt{6}}\mathbf{j} - \dfrac{1}{\sqrt{6}}\mathbf{k} = \sqrt{\dfrac{1}{2}}\left(\dfrac{1}{\sqrt{3}}\mathbf{i} - \dfrac{1}{\sqrt{3}}\mathbf{j} - \dfrac{1}{\sqrt{3}}\mathbf{k}\right)$

39. (a) $2\mathbf{i}$ (b) $-\sqrt{3}\mathbf{k}$
    (c) $\dfrac{3}{10}\mathbf{j} + \dfrac{2}{5}\mathbf{k}$ (d) $6\mathbf{i} - 2\mathbf{j} + 3\mathbf{k}$

41. $\dfrac{7}{13}(12\mathbf{i} - 5\mathbf{k})$

43. (a) $5\sqrt{2}$
    (b) $\dfrac{3}{5\sqrt{2}}\mathbf{i} + \dfrac{4}{5\sqrt{2}}\mathbf{j} - \dfrac{1}{\sqrt{2}}\mathbf{k}$
    (c) $\left(\dfrac{1}{2}, 3, \dfrac{5}{2}\right)$

45. (a) $\sqrt{3}$
    (b) $-\dfrac{1}{\sqrt{3}}\mathbf{i} - \dfrac{1}{\sqrt{3}}\mathbf{j} - \dfrac{1}{\sqrt{3}}\mathbf{k}$
    (c) $\left(\dfrac{5}{2}, \dfrac{7}{2}, \dfrac{9}{2}\right)$

47. Center $(-2, 0, 2)$, radius $2\sqrt{2}$

49. $(x - 1)^2 + (y - 2)^2 + (z - 3)^2 = 14$

51. Center $(-2, 0, 2)$, radius $\sqrt{8}$

53. Center $\left(-\dfrac{1}{4}, -\dfrac{1}{4}, -\dfrac{1}{4}\right)$, radius $\dfrac{5\sqrt{3}}{4}$

55. $A = (4, -3, 5)$

57. (a) $\sqrt{y^2 + z^2}$ (b) $\sqrt{x^2 + z^2}$
    (c) $\sqrt{x^2 + y^2}$

59. Let $A = (a_1, a_2, a_3)$, $B = (b_1, b_2, b_3)$, $C = (c_1, c_2, c_3)$, and $D = (d_1, d_2, d_3)$. The midpoints of the sides are

$$M_{AB} = \left(\dfrac{a_1 + b_1}{2}, \dfrac{a_2 + b_2}{2}, \dfrac{a_3 + b_3}{2}\right),$$

$$M_{BC} = \left(\dfrac{b_1 + c_1}{2}, \dfrac{b_2 + c_2}{2}, \dfrac{b_3 + c_3}{2}\right),$$

$$M_{CD} = \left(\dfrac{c_1 + d_1}{2}, \dfrac{c_2 + d_2}{2}, \dfrac{c_3 + d_3}{2}\right),$$

$$M_{AD} = \left(\dfrac{a_1 + d_1}{2}, \dfrac{a_2 + d_2}{2}, \dfrac{a_3 + d_3}{2}\right),$$

The midpoint of the segment joining M_{AB} and M_{CD} and the midpoint of the segment joining M_{BC} and M_{AD} have coordinates $\dfrac{a_i + b_i + c_i + d_i}{4}$ for $i = 1, 2, 3$.

61. $\overrightarrow{P_1P_2} = \mathbf{i} + 2\mathbf{j} + 3\mathbf{k} = \overrightarrow{P_4P_3}$ and
$\overrightarrow{P_2P_3} = 2\mathbf{i} + 3\mathbf{j} + 4\mathbf{k} = \overrightarrow{P_1P_4}$. As vectors the opposite sides are represented by equal vectors. Therefore, they are parallel and have the same length, and the figure is a parallelogram.

63. (a) $x = -1$: $y^2 + z^2 = 8$, $x = 0$: $y^2 + z^2 = 9$, $x = 2$: $y^2 + z^2 = 5$

(b)

[−4.7, 4.7] by [−3.1, 3.1]

(c) If $0 \le |c| \le 3$, then for $x = c$ graph $r = \sqrt{9 - c^2}$ in polar coordinate mode. If $|c| > 3$, then there is no graph.
(d) If $|x| \le 3$, then $A_x = \pi(9 - x^2)$.
(e) It is the volume of the sphere.

65. $\dfrac{2}{3}\mathbf{i} + \dfrac{2}{3}\mathbf{j} + \dfrac{4}{3}\mathbf{k}$

Section 11.2 (pp. 586–593)

Quick Review

1. 0

2. $\pi/2$

3. 1

4. $x + y = 5$

5. $x - y = 1$

6. $\pm\left(\dfrac{2}{\sqrt{7}}\mathbf{i} + \sqrt{\dfrac{3}{7}}\mathbf{j}\right)$

7. $\pm\left(\sqrt{\dfrac{3}{7}}\mathbf{i} - \dfrac{2}{\sqrt{7}}\mathbf{j}\right)$

8. $\pm\left(\dfrac{5}{\sqrt{34}}\mathbf{i} + \dfrac{3}{\sqrt{34}}\mathbf{j}\right)$

9. $\pm\left(\dfrac{3}{\sqrt{34}}\mathbf{i} - \dfrac{5}{\sqrt{34}}\mathbf{j}\right)$

10. $250\ \text{N} \cdot \text{m}$, assuming distance in meters (m)

Exercises

1. (a) $\mathbf{A} \cdot \mathbf{B} = -25$, $|\mathbf{A}| = 5$, $|\mathbf{B}| = 5$
(b) π **(c)** -5
(d) $-2\mathbf{i} + 4\mathbf{j} - \sqrt{5}\mathbf{k}$

3. (a) $\mathbf{A} \cdot \mathbf{B} = 25$, $|\mathbf{A}| = 15$, $|\mathbf{B}| = 5$
(b) $\cos^{-1}(1/3) \approx 1.23$ rad
(c) $5/3$ **(d)** $\dfrac{1}{9}(10\mathbf{i} + 11\mathbf{j} - 2\mathbf{k})$

5. (a) $\mathbf{A} \cdot \mathbf{B} = 2$, $|\mathbf{A}| = \sqrt{34}$, $|\mathbf{B}| = \sqrt{3}$
(b) $\cos^{-1}(2/\sqrt{102}) \approx 1.37$ rad
(c) $2/\sqrt{34}$ **(d)** $\dfrac{1}{17}(5\mathbf{j} - 3\mathbf{k})$

7. $\mathbf{B} = \left(\dfrac{3}{2}\mathbf{i} + \dfrac{3}{2}\mathbf{j}\right) + \left(-\dfrac{3}{2}\mathbf{i} + \dfrac{3}{2}\mathbf{j} + 4\mathbf{k}\right)$

9. $\mathbf{B} = \left(\dfrac{14}{3}\mathbf{i} + \dfrac{28}{3}\mathbf{j} - \dfrac{14}{3}\mathbf{k}\right) + \left(\dfrac{10}{3}\mathbf{i} - \dfrac{16}{3}\mathbf{j} - \dfrac{22}{3}\mathbf{k}\right)$

11. The sum of two vectors is *always* orthogonal to their difference: $(\mathbf{v}_1 + \mathbf{v}_2) \cdot (\mathbf{v}_1 - \mathbf{v}_2) = 0$.

15. $\theta = \cos^{-1}(2/\sqrt{6}) \approx 0.62$ rad $\approx 35.26°$

17. a

19. (a) $|\mathbf{D}| = \sqrt{70}$ **(b)** $|\mathbf{D}| = \sqrt{568}$

21. $5\ \text{N} \cdot \text{m}$ or 5 J

23. $3464.10\ \text{N} \cdot \text{m} = 3464.10$ J

25. Slope of $\mathbf{v} = a\mathbf{i} + b\mathbf{j}$ is b/a, slope of $ax + by = c$ is $-a/b$

27. $x + 2y = 4$

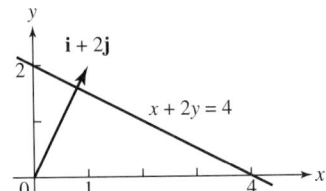

29. $-2x + y = -3$

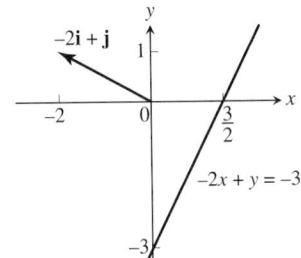

31. $x + y = -1$

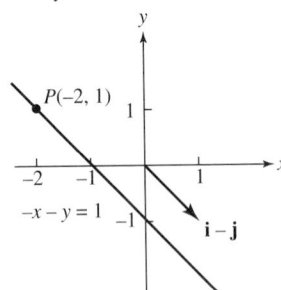

33. $2x - y = 0$

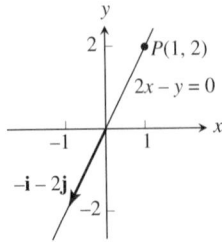

35. $\pi/4$ **37.** $\pi/6$

39. $\cos^{-1}\left(\dfrac{7}{5\sqrt{2}}\right) \approx 0.14$ rad

41. $\pi/3$ and $2\pi/3$ at each point

43. The angle at $(0, 0)$ is $\pi/2$; the angles at $(1, 1)$ are $\pi/4$ and $3\pi/4$.

47. $(x\mathbf{i} + y\mathbf{j}) \cdot \mathbf{v} = |x\mathbf{i} + y\mathbf{j}||\mathbf{v}|\cos\theta \le 0$ when $\pi/2 \le \theta \le \pi$. This means (x, y) has to be a point whose position vector makes an angle with $\mathbf{v}$ that is a right angle or bigger.

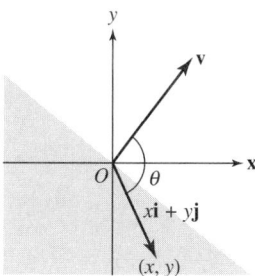

Section 11.3 (pp. 593–600)

Quick Review

1. Slope of AB = slope of $DC = 3/2$, slope of AD = slope of $BC = 1/2$, so opposite sides are parallel

2. 8

3. $\angle A$ and $\angle C$: $\cos^{-1}\left(\dfrac{14}{\sqrt{20}\sqrt{13}}\right) \approx 29.74°$

$\angle B$ and $\angle D$: $\cos^{-1}\left(\dfrac{14}{\sqrt{20}\sqrt{13}}\right) \approx 150.26°$

4. $x = 0, \pi/2, \pi, 3\pi/2, 2\pi$

5. $x = \pi/4, 3\pi/4, 5\pi/4, 7\pi/4$

6. 5 **7.** -3

8. 35 **9.** 3

10. 0

Exercises

1. $\mathbf{A} \times \mathbf{B}$: length = 3, direction = $\dfrac{2}{3}\mathbf{i} + \dfrac{1}{3}\mathbf{j} + \dfrac{2}{3}\mathbf{k}$

$\mathbf{B} \times \mathbf{A}$: length = 3, direction = $-\dfrac{2}{3}\mathbf{i} - \dfrac{1}{3}\mathbf{j} - \dfrac{2}{3}\mathbf{k}$

3. $\mathbf{A} \times \mathbf{B}$: length = 0, has no direction
$\mathbf{B} \times \mathbf{A}$: length = 0, has no direction

5. $\mathbf{A} \times \mathbf{B}$: length = 6, direction = $-\mathbf{k}$
$\mathbf{B} \times \mathbf{A}$: length = 6, direction = $\mathbf{k}$

7. $\mathbf{A} \times \mathbf{B}$: length = $6\sqrt{5}$, direction = $\dfrac{1}{\sqrt{5}}\mathbf{i} - \dfrac{2}{\sqrt{5}}\mathbf{k}$

$\mathbf{B} \times \mathbf{A}$: length = $6\sqrt{5}$,

direction = $-\dfrac{1}{\sqrt{5}}\mathbf{i} + \dfrac{2}{\sqrt{5}}\mathbf{k}$

9. $\mathbf{A} \times \mathbf{B} = \mathbf{k}$

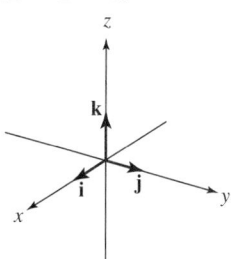

11. $\mathbf{A} \times \mathbf{B} = \mathbf{i} - \mathbf{j} + \mathbf{k}$

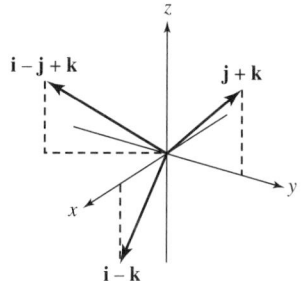

13. $\mathbf{A} \times \mathbf{B} = -2\mathbf{k}$

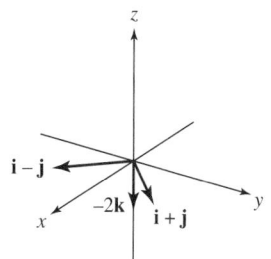

15. (a) $2\sqrt{6}$ **(b)** $\pm\dfrac{1}{\sqrt{6}}(2\mathbf{i} + \mathbf{j} + \mathbf{k})$

17. (a) $\sqrt{2}/2$ **(b)** $\pm\dfrac{1}{\sqrt{2}}(\mathbf{i} - \mathbf{j})$

19. (a) None
(b) $\mathbf{A}$ and $\mathbf{C}$ are parallel because $\mathbf{A} \times \mathbf{C} = \mathbf{0}$.

21. $10\sqrt{3}$ ft $\cdot$ lb

23. 8 **25.** 7

27. (a) True, $|\mathbf{A}| = \sqrt{a_1{}^2 + a_2{}^2 + a_3{}^2} = \sqrt{\mathbf{A} \cdot \mathbf{A}}$

(b) Not always true, $\mathbf{A} \cdot \mathbf{A} = |\mathbf{A}|^2$

(c) True, $\mathbf{A} \times \mathbf{0} = \begin{vmatrix} \mathbf{i} & \mathbf{j} & \mathbf{k} \\ a_1 & a_2 & a_3 \\ 0 & 0 & 0 \end{vmatrix} = \mathbf{0}$

(d) True, $\mathbf{A} \times (-\mathbf{A}) = \begin{vmatrix} \mathbf{i} & \mathbf{j} & \mathbf{k} \\ a_1 & a_2 & a_3 \\ -a_1 & -a_2 & -a_3 \end{vmatrix} = \mathbf{0}$

(e) Not always true, $\mathbf{i} \times \mathbf{j} = \mathbf{k} \ne -\mathbf{k} = \mathbf{j} \times \mathbf{i}$ for example

(f) True, Equation (3)

(g) True, $(\mathbf{A} \times \mathbf{B}) \cdot \mathbf{B} = \mathbf{A} \cdot (\mathbf{B} \times \mathbf{B})$
$$= \mathbf{A} \cdot \mathbf{0} = 0$$

(h) True, Equation (5)

29. (a) $\text{proj}_\mathbf{B}\mathbf{A} = \left(\dfrac{\mathbf{A} \cdot \mathbf{B}}{\mathbf{B} \cdot \mathbf{B}}\right)\mathbf{B}$

(b) $\pm(\mathbf{A} \times \mathbf{B})$ **(c)** $\pm(\mathbf{A} \times \mathbf{B}) \times \mathbf{C}$
(d) $|(\mathbf{A} \times \mathbf{B}) \cdot \mathbf{C}|$

31. (a) Yes, $\mathbf{A} \times \mathbf{B}$ and $\mathbf{C}$ are both vectors
(b) No, $\mathbf{A}$ is a vector but $\mathbf{B} \cdot \mathbf{C}$ is a scalar
(c) Yes, $\mathbf{A}$ and $\mathbf{B} \times \mathbf{C}$ are both vectors
(d) No, $\mathbf{A}$ is a vector but $\mathbf{B} \cdot \mathbf{C}$ is a scalar

33. 2 **35.** 13
37. 11/2 **39.** 25/2

41. $\dfrac{1}{2}|\mathbf{A} \times \mathbf{B}| = \pm\dfrac{1}{2}\begin{vmatrix} a_1 & a_2 \\ b_1 & b_2 \end{vmatrix}$. The applicable sign
$(+)$ is the acute angle from $\mathbf{A}$ to $\mathbf{B}$ runs counterclockwise in the xy-plane and $(-)$ if it runs clockwise, because the area must be a nonnegative number.

43. No. For example, $\mathbf{i} + \mathbf{j} \neq -\mathbf{i} + \mathbf{j}$, but
$\mathbf{i} \times (\mathbf{i} + \mathbf{j}) = \mathbf{i} \times (-\mathbf{i} + \mathbf{j}) = \mathbf{k}$.

Section 11.4 (pp. 600–607)

Quick Review

1. (a) Yes **(b)** No
(c) No **(d)** No
2. $x = -2 + 5t,\ y = -3 + 4t,\ -\infty < t < \infty$
3. $x = -2 + 5t,\ y = -3 + 4t,\ 0 \leq t \leq 1$
4. $x = 1 + 3t,\ y = 2 - 2t,\ -\infty < t < \infty$
5. $x = 3/5,\ y = 4/5$ **6.** $x = 1/17,\ y = 7/17$

7. $-\dfrac{2}{3}\mathbf{i} + \dfrac{1}{3}\mathbf{j} + \dfrac{2}{3}\mathbf{k}$

8. $\mathbf{u} = 3\mathbf{i} + 5\mathbf{j} + \mathbf{k},\ \mathbf{w} = -3\mathbf{i} - 5\mathbf{j} - \mathbf{k}$
9. $\mathbf{u} \cdot \mathbf{v}_1 = 9 - 10 + 1 = 0,$
$\mathbf{w} \cdot \mathbf{v}_1 = -9 + 10 - 1 = 0$
10. $\mathbf{u} \cdot \mathbf{v}_2 = -3 + 5 - 2 = 0,$
$\mathbf{w} \cdot \mathbf{v}_2 = 3 - 5 + 2 = 0$

Exercises

1. $x = 3 + t,\ y = -4 + t,\ z = -1 + t$
3. $x = -2 + 5t,\ y = 5t,\ z = 3 - 5t$
5. $x = 3 + 2t,\ y = -2 - t,\ z = 1 + 3t$
7. $x = 2 + 3t,\ y = 4 + 7t,\ z = 5 - 5t$
9. $x = 2 - 2t,\ y = 3 + 4y,\ z = -2t$

11. $x = t,\ y = t,\ z = \dfrac{3}{2}t,\ 0 \leq t \leq 1$

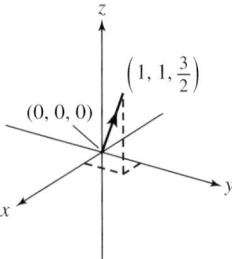

13. $x = 0,\ y = 1 - 2t,\ z = 1,\ 0 \leq t \leq 1$

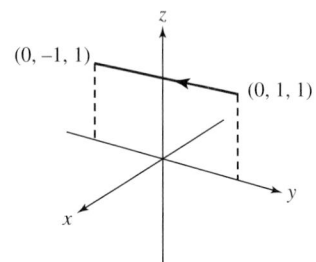

15. $3x - 2y - z = -3$ **17.** $7x - 5y - 4z = 6$
19. $x + 3y + 4z = 34$
21. $(1, 2, 3);\ -20x + 12y + z = 7$
23. $y + z = 3$
25. $x - y + z = 0$
29. 0
33. 19/5 **35.** $9/\sqrt{41}$

39. $\cos^{-1}\left(\dfrac{5}{3\sqrt{6}}\right) \approx 0.82$ rad

41. $\left(\dfrac{3}{2},\ -\dfrac{3}{2},\ \dfrac{1}{2}\right)$

43. $(1, 1, 0)$
45. $x = 1 - t,\ y = 1 + t,\ z = -1$
47. $x = 4,\ y = 3 + 6t,\ z = 1 + 3t$
49. L_1 and L_2 intersect at $(5, 3, 1)$, L_2 and L_3 are parallel, L_1 and L_3 are skew
51. $x = 2 + 2t,\ y = -4 - t,\ z = 7 + 3t;$
$x = -2 - t,\ y = -2 + \dfrac{1}{2}t,\ z = 1 - \dfrac{3}{2}t$
53. yz-plane: $(0, -1/2, -3/2)$ (set $x = 0$); xz-plane: $(-1, 0, -3)$ (set $y = 0$); xy-plane: $(1, -1, 0)$ (set $z = 0$)
55. No, because the line and plane intersect in the point $(-1, 7, -3)$
57. There are many possible answers. $x + y = 3$ and $2y + z = 7$ are two such planes.
59. All planes except those through the origin or parallel to a coordinate axis

61. (a) $\overrightarrow{EP} = c\overrightarrow{EP_1}$; $y = cy_1$ and $z = cz_1$
 (b) At $x_1 = 0$: $c = 1$, $y = y_1$, $z = z_1$;
 at $x_1 = x_0$: $x_0 = 0$, $y = 0$, $z = 0$;

$$\lim_{x_0 \to \infty} c = \lim_{x_0 \to \infty} \frac{-x_0}{x_1 - x_0} = \lim_{x_0 \to \infty} \frac{-1}{-1} = 1,$$

so $c \to 1$ and $y \to y_1$ and $z \to z_1$

Section 11.5 (pp. 607–614)

Quick Review

1. (a) Circle, center $= (-1, 0)$, radius $= 1$

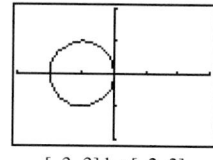

[−3, 3] by [−2, 2]

 (b) Circle, center $= (0, 2)$, radius $= 2$

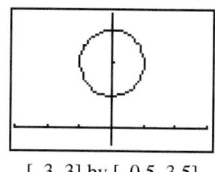

[−3, 3] by [−0.5, 3.5]

2. (a) Ellipse, center $= (0, 0)$, vertices: $\left(\pm\sqrt{2}, 0\right)$, y-intercepts: $(0, \pm 2)$

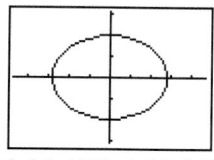

[−4.7, 4.7] by [−3.1, 3.1]

 (b) Ellipse, center $= (0, 0)$, x-intercepts: $(\pm 3, 0)$, vertices: $(0, \pm 6)$

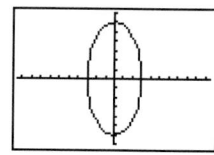

[−10.5, 10.5] by [−7, 7]

3. (a) Hyperbola, center $= (0, 0)$

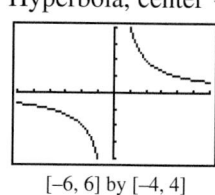

[−6, 6] by [−4, 4]

 (b) Hyperbola, center $= (0, 0)$, vertices $(\pm 6, 0)$

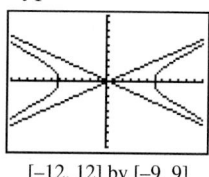

[−12, 12] by [−9, 9]

4. (a) Parabola, x-intercepts: $(\pm 1, 0)$, vertex: $(0, -1)$

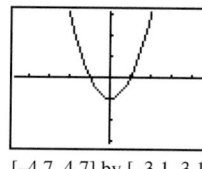

[−4.7, 4.7] by [−3.1, 3.1]

 (b) Parabola, vertex: $(4, 0)$, y-intercepts: $(0, \pm 2)$

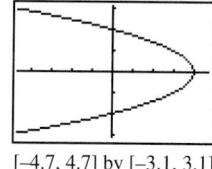

[−4.7, 4.7] by [−3.1, 3.1]

5. $r = 3$

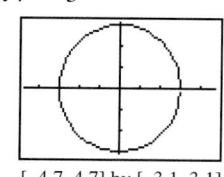

[−4.7, 4.7] by [−3.1, 3.1]

6. $r = 2 \cos \theta$

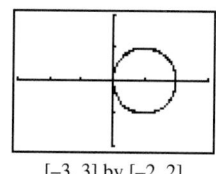

[−3, 3] by [−2, 2]

7. $r = -4 \sin \theta$

[−4.7, 4.7] by [−5, 1.2]

8. $r = -6 \cos \theta$

[−6, 6] by [−4, 4]

9. (a) Cardioid

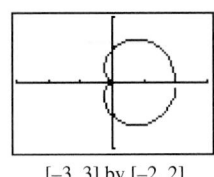

[–3, 3] by [–2, 2]

(b) Cardioid

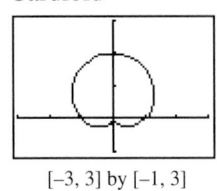

[–3, 3] by [–1, 3]

10. (a) Horizontal line $y = 2$

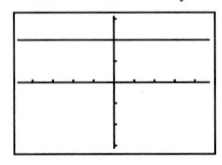

[–4.7, 4.7] by [–3.1, 3.1]

(b) Vertical line $x = -1$

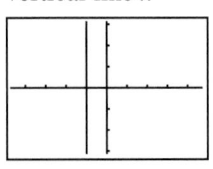

[–4.7, 4.7] by [–3.1, 3.1]

Exercises

1. a, elliptical cylinder

3.

5.

7.

9.

11.

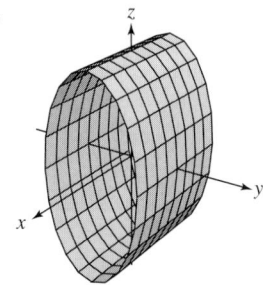

13. d   **15.** a
17. (0, 0, 0)   **19.** (1, π/2, 0)
21. (1, 0, 0)   **23.** (0, 1, 1)
25. $x^2 + y^2 = 0$, the z-axis
27. $z = 0$, the xy-plane
29. $r^2 + z^2 = 4$, a sphere of radius 2 centered at the origin
31. $y = 1$, the plane $y = 1$

33. $r^2 + z^2 = 2z$, $z \leq 1$, the lower half (hemisphere) of the sphere of radius 1 centered at $(0, 0, 1)$ (rectangular)

35. Right circular cylinder parallel to the z-axis

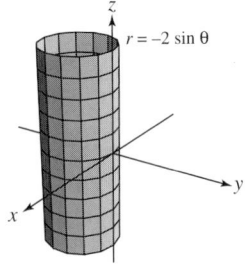

$r = -2 \sin \theta$

37. Cylinder parallel to the z-axis

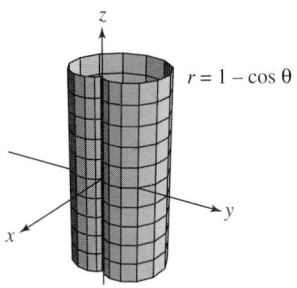

$r = 1 - \cos \theta$

39. $(2, 3, 1)$

43. The surface is symmetric with respect to the z-axis.

45. Parabolic cylinder

47. Sinusoidal cylinder

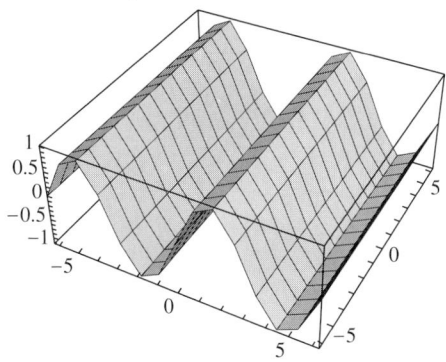

Section 11.6 (pp. 614–623)

Quick Review

1. $y = \pm 3 \sqrt{1 - \dfrac{x^2}{4}}$; $y_1 = 3 \sqrt{1 - \dfrac{x^2}{4}}$;

$y_2 = -3 \sqrt{1 - \dfrac{x^2}{4}}$

2. $x = 2 \cos t$, $y = 3 \sin t$, $0 \leq t \leq 2\pi$

3. $y = \pm 3 \sqrt{\dfrac{x^2}{4} - 1}$

4. $x = 2 \sec t$, $y = 3 \tan t$, $0 \leq t \leq 2\pi$

5. $(\pm 3, 0)$, $(0, \pm 2)$ **6.** $(\pm 3, 0)$

7. (a) Yes   **(b)** Yes   **(c)** Yes

8. (a) Yes   **(b)** Yes   **(c)** Yes

9. (a) No   **(b)** Yes   **(c)** No

10. (a) Yes   **(b)** No   **(c)** No

Exercises

1. b, ellipsoid   **3.** e, cone

5. c, paraboloid   **7.** i, hyperbolic paraboloid

9. f, cone

11.

13.

15.

17.

19.

21.

23.

25.

27.

29.

31.

33.

35.

37.

39.

41.

43.

45.

47.

49.
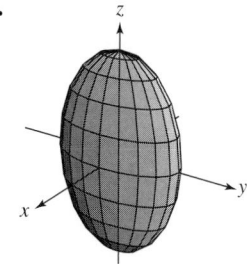

51. (a) $\dfrac{2\pi(9 - c^2)}{9}$ **(b)** 8π

(c) $\dfrac{4\pi abc}{3}$, yes $V = \dfrac{4\pi r^3}{3}$

53. Volume of segment: $V = \dfrac{\pi abh^2}{c}$; area of elliptic

base when $z = h$: $A = \dfrac{\pi abh}{c}$; $V = \dfrac{1}{2} Ah$

57. Vertex: $\left(0, y_1, \dfrac{cy_1{}^2}{b^2}\right)$, focus: $\left(0, y_1, \dfrac{cy_1{}^2}{b^2} - \dfrac{a^2}{4c}\right)$

Chapter 11 Review Exercises (pp. 626–629)

1. $2\mathbf{i} - 3\mathbf{j} + 6\mathbf{k} = 7\left(\dfrac{2}{7}\mathbf{i} - \dfrac{3}{7}\mathbf{j} + \dfrac{6}{7}\mathbf{k}\right)$

2. $\mathbf{i} + 2\mathbf{j} - \mathbf{k} = \sqrt{6}\left(\dfrac{1}{\sqrt{6}}\mathbf{i} + \dfrac{2}{\sqrt{6}}\mathbf{j} - \dfrac{1}{\sqrt{6}}\mathbf{k}\right)$

3. $\dfrac{8}{\sqrt{33}}\mathbf{i} - \dfrac{2}{\sqrt{33}}\mathbf{j} + \dfrac{8}{\sqrt{33}}\mathbf{k}$

4. $\overrightarrow{OD} = \overrightarrow{OC} + \overrightarrow{OB}$, $\overrightarrow{OE} = \overrightarrow{OC} + \overrightarrow{OA} + \overrightarrow{OB}$

5.

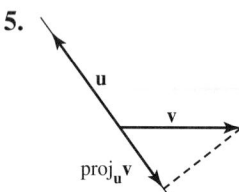

6. $a = proj_v u$, $b = proj_u v$, $c = v - proj_u v$

7. $|A| = \sqrt{2}$, $|B| = 3$, $A \cdot B = 3$, $B \cdot A = 3$,

$A \times B = -2i + 2j - k$, $B \times A = 2i - 2j + k$,

$|A \times B| = 3$, $\theta = \pi/4$, $|B| \cos \theta = 3/\sqrt{2}$,

$proj_A B = \dfrac{3}{2}(i + j)$

8. $|A| = \sqrt{6}$, $|B| = \sqrt{2}$, $A \cdot B = -3$, $B \cdot A = -3$,

$A \times B = -i - j + k$, $B \times A = i + j - k$,

$|A \times B| = \sqrt{3}$, $\theta = 5\pi/6$, $|B| \cos \theta = -\sqrt{3/2}$,

$proj_A B = -\dfrac{1}{2}(i + j + k)$

9. $B = \dfrac{4}{3}(2i + j - k) - \dfrac{1}{3}(5i + j + 11k)$

10. $B = -\dfrac{1}{5}(i - 2j) + \left(\dfrac{6}{5}i + \dfrac{3}{5}j + k\right)$

11. $A \times B = k$

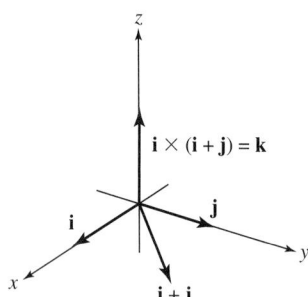

12. $A \times B = 2k$

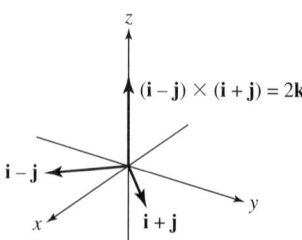

13. Tangent: $\pm\left(\dfrac{1}{\sqrt{5}}i + \dfrac{2}{\sqrt{5}}j\right)$,

normal: $\pm\left(-\dfrac{2}{\sqrt{5}}i + \dfrac{1}{\sqrt{5}}j\right)$

14. Tangent: $\pm\dfrac{1}{5}(4i - 3j)$,

normal: $\pm\dfrac{1}{5}(3i + 4j)$

16. (a) $\dfrac{1}{2}|u \times v|$ (b) $h = \dfrac{|u \times v|}{|v|}$

(c) Area: $\sqrt{6}/2$, h: $\sqrt{6/5}$

17. $2\sqrt{7}$ **18.** $a = 10$

19. (a) $\sqrt{14}$ (b) 1

20. (a) 1   (b) 1

21. $n \times v$ or $v \times n$ **22.** $bi - aj$

23. $\sqrt{78}/3$ **24.** $\sqrt{78}/3$

25. $x = 1 - 3t$, $y = 2$, $z = 3 + 7t$

26. $x = 1$, $y = 2 + t$, $z = -t$; $0 \le t \le 1$

27. $\sqrt{2}$ **28.** $\sqrt{14}$

29. $2x + y - z = 3$ **30.** $x - 2y + 3z = -13$

31. $-9x + y + 7z = 4$ **32.** $x + y + z = 1$

33. $x = 0$: $(0, -1/2, -3/2)$, $y = 0$: $(-1, 0, -3)$, $z = 0$: $(1, -1, 0)$

34. $(4/3, -2/3, -2/3)$ **35.** $\pi/3$

36. $\pi/3$

37. $x = -5 + 5t$, $y = 3 - t$, $z = -3t$

39. (a) Normals: $n_1 = 3i + 6k$, $n_2 = 2i + 2j - k$, and $n_1 \cdot n_2 = 0$

(b) $x = -12t$, $y = \dfrac{19}{12} + 15t$, and $z = \dfrac{1}{6} + 6t$

40. $7x - 3y - 5z = -14$

42. The half-space of points lying on one side of the plane in the direction of n

43. 3   **44.** $25/\sqrt{38}$

45. $-3j + 3k$ **46.** $\dfrac{1}{\sqrt{14}}(-2i - 3j + k)$

47. $\dfrac{2}{\sqrt{35}}(5i - j - 3k)$ **48.** $(4/3, -2/3, -2/3)$

49. $(11/9, 26/9, -7/9)$ **50.** $\cos^{-1}\left(\dfrac{3}{\sqrt{35}}\right) \approx 59.5°$

51. $(1, -2, 1)$; $x = 1 - 5t$, $y = -2 + 3t$, and $z = -1 + 4t$

53. $2x + 7y + 2z + 10 = 0$

55. (a) No   (b) No   (c) No
(d) No   (e) Yes

56. (a) $(2, 1, 8)$ (b) $\cos \theta = 3/\sqrt{15}$

(c) $\dfrac{9}{5}(j + 2k)$ (d) $6\sqrt{6}$

(e) $7x + 2y - z = 8$
(f) yz-plane: 14, xz-plane: 4, xy-plane: 2

57. $11/\sqrt{107}$ **58.** $12/\sqrt{62}$

59. The y-axis in the xy-plane; the yz-plane in three-dimensional space

60. The line $x + y = 1$ in the xy-plane; the plane $x + y = 1$ in three-dimensional space

61. Center: $(3, -1, 2)$, radius: $3\sqrt{2}$

62. $20,000\sqrt{3} \approx 34,641$ J   **63.** 20 lb

65. (b) All points in the plane determined by
$P_1(x_1, y_1, z_1)$, $P_2(x_2, y_2, z_2)$, and $P_3(x_3, y_3, z_3)$

67. $(1, 0, 0)$ **68.** $(0, 1, 0)$

69. $(1, \pi, -1)$ **70.** $(1, 3\pi/2, 1)$

71. $z = 2$, a plane parallel to the xy-plane

72. $r^2 + z^2 = 6z$, a sphere of radius 3 centered at $(0, 0, 3)$ (rectangular)

73. $x^2 + \left(y - \dfrac{7}{2}\right)^2 = \dfrac{49}{4}$, a circular cylinder parallel to the z-axis

74. $(x - 2)^2 + y^2 = 4$, a circular cylinder parallel to the z-axis

75.

76.

77.

78.

79.

80.   **81.**

82.

83.

84.

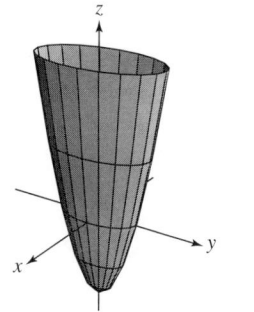

Chapter 12

Section 12.1 (pp. 631–644)

Quick Review

1. -3 **2.** -2

3. At all noninteger values of x

4. At every integer

5. $y - \dfrac{4}{\pi} = -\dfrac{8(\pi + 2)}{\pi^2}\left(x - \dfrac{\pi}{4}\right)$

6. $y - \dfrac{4}{\pi} = \dfrac{\pi^2}{8(\pi + 2)}\left(x - \dfrac{\pi}{4}\right)$

7. $-\dfrac{10}{(x + 3)^3}$ **8.** $\ln(|x + 1|) + C$

9. $\ln 3$ **10.** $y = \sin x - \cos x$

Exercises

1. $y = x^2 - 2x$, $\mathbf{v} = \mathbf{i} + 2\mathbf{j}$, $\mathbf{a} = 2\mathbf{j}$

3. $y = \dfrac{2}{9}x^2$, $\mathbf{v} = 3\mathbf{i} + 4\mathbf{j}$, $\mathbf{a} = 3\mathbf{i} + 8\mathbf{j}$

5. $\mathbf{v}\left(\dfrac{\pi}{4}\right) = \dfrac{\sqrt{2}}{2}\mathbf{i} - \dfrac{\sqrt{2}}{2}\mathbf{j}$, $\mathbf{a}\left(\dfrac{\pi}{4}\right) = -\dfrac{\sqrt{2}}{2}\mathbf{i} - \dfrac{\sqrt{2}}{2}\mathbf{j}$,

$\mathbf{v}\left(\dfrac{\pi}{2}\right) = -\mathbf{j}$, $\mathbf{a}\left(\dfrac{\pi}{2}\right) = -\mathbf{i}$

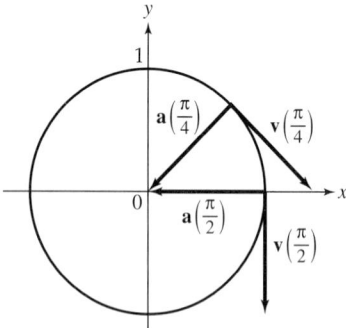

7. $\mathbf{v}(\pi) = 2\mathbf{i}$, $\mathbf{a}(\pi) = -\mathbf{j}$, $\mathbf{v}\left(\dfrac{3\pi}{2}\right) = \mathbf{i} - \mathbf{j}$,

$\mathbf{a}\left(\dfrac{3\pi}{2}\right) = -\mathbf{i}$

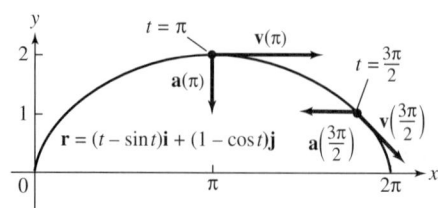

9. (a) $\mathbf{v} = \mathbf{i} + 2t\mathbf{j} + 2\mathbf{k}$, $\mathbf{a} = 2\mathbf{j}$

(b) Speed: 3; direction: $\dfrac{1}{3}\mathbf{i} + \dfrac{2}{3}\mathbf{j} + \dfrac{2}{3}\mathbf{k}$

(c) $\mathbf{v}(1) = 3\left(\dfrac{1}{3}\mathbf{i} + \dfrac{2}{3}\mathbf{j} + \dfrac{2}{3}\mathbf{k}\right)$

11. (a) $\mathbf{v} = (\sec t \tan t)\mathbf{i} + (\sec^2 t)\mathbf{j} + \dfrac{4}{3}\mathbf{k}$,

$\mathbf{a} = (\sec t \tan^2 t + \sec^3 t)\mathbf{i} + (2\sec^2 t \tan t)\mathbf{j}$

(b) Speed: 2; direction: $\dfrac{1}{3}\mathbf{i} + \dfrac{2}{3}\mathbf{j} + \dfrac{2}{3}\mathbf{k}$

(c) $\mathbf{v}\left(\dfrac{\pi}{6}\right) = 2\left(\dfrac{1}{3}\mathbf{i} + \dfrac{2}{3}\mathbf{j} + \dfrac{2}{3}\mathbf{k}\right)$

13. $\pi/2$

15. $\pi/2$

17. $t = 0$, π, or 2π

19. $\dfrac{1}{4}\mathbf{i} + 7\mathbf{j} + \dfrac{3}{2}\mathbf{k}$

21. $(\ln 4)\mathbf{i} + (\ln 4)\mathbf{j} + (\ln 2)\mathbf{k}$

23. $\mathbf{r} = [(t + 1)^{3/2} - 1]\mathbf{i} + (1 - e^{-t})\mathbf{j} + [1 + \ln(t + 1)]\mathbf{k}$

25. $\mathbf{r} = 8t\mathbf{i} + 8t\mathbf{j} + (100 - 16t^2)\mathbf{k}$

27. $x = t$, $y = -1$, $z = 1 + t$

29. $x = at$, $y = a$, $z = 2\pi b + bt$

31. (a) (i) Constant speed 1 **(ii)** Yes
    (iii) Counterclockwise **(iv)** Yes

(b) (i) Constant speed 2 **(ii)** Yes
    (iii) Counterclockwise **(iv)** Yes

(c) (i) Constant speed 1 **(ii)** Yes
    (iii) Counterclockwise **(iv)** No, starts at
    $(0, -1)$

(d) (i) Constant speed 1 **(ii)** Yes
    (iii) Clockwise **(iv)** Yes

(e) (i) Variable speed **(ii)** No
    (iii) Counterclockwise **(iv)** Yes

33. $\mathbf{r}(t) = \left(\dfrac{1}{2}t^2 + \dfrac{2}{\sqrt{11}}t\right)(3\mathbf{i} - \mathbf{j} + \mathbf{k}) + (\mathbf{i} + 2\mathbf{j} + 3\mathbf{k})$

35. $\mathbf{v} = 2\sqrt{5}\mathbf{i} + \sqrt{5}\mathbf{j}$

37. max $|\mathbf{v}| = 3$, min $|\mathbf{v}| = 2$, max $|\mathbf{a}| = 3$, min $|\mathbf{a}| = 2$

49. Show that the components of $\mathbf{r}(t)$ satisfy the corresponding equation.

Section 12.2 (pp. 645–649)

Quick Review

1. $x = 3$ **2.** $y = 3$
3. $y = x + 3\sqrt{2}$ **4.** $y = -x + 3\sqrt{2}$

5. (a) $\pm\left(\dfrac{2\sqrt{7}}{7}\mathbf{i} - \dfrac{\sqrt{21}}{7}\mathbf{j}\right)$

(b) $\pm\left(\dfrac{\sqrt{21}}{7}\mathbf{i} + \dfrac{2\sqrt{7}}{7}\mathbf{j}\right)$

6. ≈ 1.91 **7.** 3
8. $x \sin x$ **9.** $2x \cos(x^4)$
10. $(\cos x) \ln(1 + \sin^2 x) + (\sin x) \ln(1 + \cos^2 x)$

Exercises

1. (a) $\left(-\dfrac{2}{3}\sin t\right)\mathbf{i} + \left(\dfrac{2}{3}\cos t\right)\mathbf{j} + \dfrac{\sqrt{5}}{3}\mathbf{k}$

(b) 3π

3. (a) $\dfrac{1}{\sqrt{1+t}}\mathbf{i} + \dfrac{\sqrt{t}}{\sqrt{1+t}}\mathbf{k}$ **(b)** 52/3

5. (a) $(-\cos t)\mathbf{j} + (\sin t)\mathbf{k}$ **(b)** 3/2

7. (a) $\left(\dfrac{\cos t - \sin t}{t+1}\right)\mathbf{i} + \left(\dfrac{\sin t + t \cos t}{t+1}\right)\mathbf{j}$

$+ \left(\dfrac{\sqrt{2t}}{t+1}\right)\mathbf{k}$

(b) $\dfrac{\pi^2}{2} + \pi$

9. $(5, 0, 24\pi)$
11. (a) $s = 5t$ **(b)** $5\pi/2$
13. (a) $s = \sqrt{3}e^t - \sqrt{3}$ **(b)** $3\sqrt{3}/4$
15. $\sqrt{2} + \ln(1 + \sqrt{2}) \approx 2.30$
17. Cylinder: $x^2 + y^2 = 1$, plane: $x + z = 1$
19.

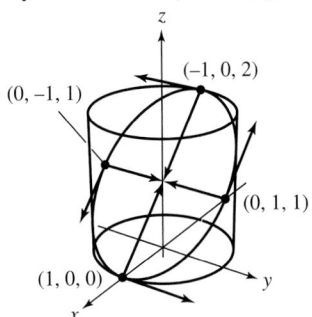

21. Graph parametrically. For example, use $x = t$, $y = (2/3)t^{3/2}$, $0 \le t \le 8$ for Exercise 3.

23. (b)

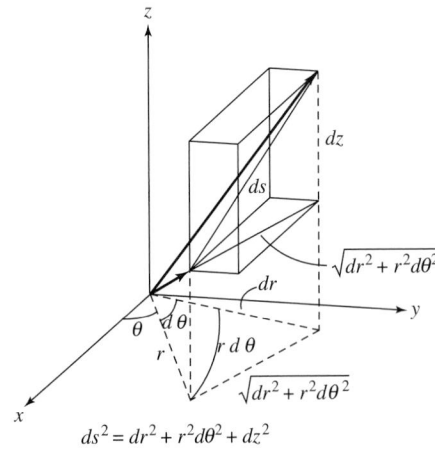

$$ds^2 = dr^2 + r^2 d\theta^2 + dz^2$$

(c) $7\sqrt{3}$

Section 12.3 (pp. 649–661)

Quick Review

6. $(\cos t)\mathbf{i} + \mathbf{j}$ **7.** $-(\sin t)\mathbf{i}$

8. $t = (2k + 1)\dfrac{\pi}{2}$: $\mathbf{v} = \mathbf{j}$, $\mathbf{a} = \begin{cases} -\mathbf{i}, & \text{if } k \text{ is even,} \\ \mathbf{i}, & \text{if } k \text{ is odd,} \end{cases}$

9. $16\mathbf{i} + 8\mathbf{j} + 8\mathbf{k}$ **10.** $-3\mathbf{i} - 4\mathbf{j} - 3\mathbf{k}$

Exercises

1. $\mathbf{T} = (\cos t)\mathbf{i} - (\sin t)\mathbf{j}$, $\mathbf{N} = -(\sin t)\mathbf{i} - (\cos t)\mathbf{j}$, $\kappa = \cos t$

3. $\mathbf{T} = \dfrac{1}{\sqrt{1+t^2}}\mathbf{i} - \dfrac{t}{\sqrt{1+t^2}}\mathbf{j}$,

$\mathbf{N} = \dfrac{-t}{\sqrt{1+t^2}}\mathbf{i} - \dfrac{1}{\sqrt{1+t^2}}\mathbf{j}$, $\kappa = \dfrac{1}{2}(1 + t^2)^{-3/2}$

5. $\mathbf{a} = \dfrac{2t}{\sqrt{1+t^2}}\mathbf{T} + \dfrac{2}{\sqrt{1+t^2}}\mathbf{N}$

7. (b) $\kappa = \cos t$
9. (a) Form the dot product with the tangent vector $\mathbf{v}$.

(b) $\mathbf{N} = \dfrac{-2e^{2t}}{\sqrt{1+4e^{4t}}}\mathbf{i} + \dfrac{1}{\sqrt{1+4e^{4t}}}\mathbf{j}$

(c) $\mathbf{N} = -\dfrac{1}{2}\left(\sqrt{4-t^2}\,\mathbf{i} + t\mathbf{j}\right)$

11. $\mathbf{T} = \left(\dfrac{3}{5}\cos t\right)\mathbf{i} - \left(\dfrac{3}{5}\sin t\right)\mathbf{j} + \dfrac{4}{5}\mathbf{k}$,

$\mathbf{N} = -(\sin t)\mathbf{i} - (\cos t)\mathbf{j}$,

$\mathbf{B} = \left(\dfrac{4}{5}\cos t\right)\mathbf{i} - \left(\dfrac{4}{5}\sin t\right)\mathbf{j} - \dfrac{3}{5}\mathbf{k}$,

$\kappa = 3/25$, $\tau = -4/25$

13. $T = \left(\dfrac{\cos t - \sin t}{\sqrt{2}}\right)\mathbf{i} + \left(\dfrac{\sin t + \cos t}{\sqrt{2}}\right)\mathbf{j}$,

$N = -\left(\dfrac{\cos t + \sin t}{\sqrt{2}}\right)\mathbf{i} + \left(\dfrac{-\sin t + \cos t}{\sqrt{2}}\right)\mathbf{j}$,

$B = \mathbf{k}, \kappa = \dfrac{1}{e^t\sqrt{2}}, \tau = 0$

15. $T = \dfrac{t}{\sqrt{t^2 + t}}\mathbf{i} + \dfrac{1}{\sqrt{t^2 + t}}\mathbf{j}$,

$N = \dfrac{1}{\sqrt{t^2 + 1}}\mathbf{i} - \dfrac{t}{\sqrt{t^2 + 1}}\mathbf{j}$,

$B = -\mathbf{k}, \kappa = \dfrac{1}{t(t^2 + 1)^{3/2}}, \tau = 0$

17. $\mathbf{a} = (0)\mathbf{T} + |a|\mathbf{N} = |a|\mathbf{N}$

19. $\mathbf{a}(1) = \dfrac{4}{3}\mathbf{T} + \dfrac{2\sqrt{5}}{3}\mathbf{N}$

21. $\mathbf{a}(0) = (0)\mathbf{T} + 2\mathbf{N} = 2\mathbf{N}$

23. $\mathbf{r}\left(\dfrac{\pi}{4}\right) = \dfrac{1}{\sqrt{2}}\mathbf{i} + \dfrac{1}{\sqrt{2}}\mathbf{j} - \mathbf{k}$,

$T\left(\dfrac{\pi}{4}\right) = -\dfrac{\sqrt{2}}{2}\mathbf{i} + \dfrac{\sqrt{2}}{2}\mathbf{j}$,

$N\left(\dfrac{\pi}{4}\right) = -\dfrac{\sqrt{2}}{2}\mathbf{i} - \dfrac{\sqrt{2}}{2}\mathbf{j}, B\left(\dfrac{\pi}{4}\right) = \mathbf{k}$, osculating

plane: $z = -1$, normal plane: $-x + y = 0$, rectifying plane: $x + y = \sqrt{2}$

25. Yes

27. Speed is constant.

33. $\kappa(b) = \dfrac{1}{2b}$

37. (a) $b - a$ **(b)** π

39. Torsion is zero.

41. $\kappa = \dfrac{2}{(1 + 4x^2)^{3/2}}$

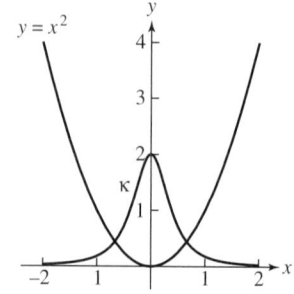

43. $\kappa = \dfrac{|\sin x|}{(1 + \cos^2 x)^{3/2}}$

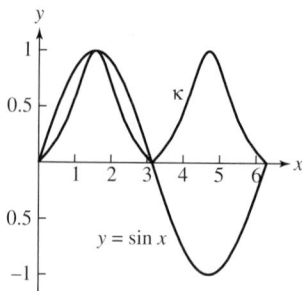

Section 12.4 (pp. 661–670)

Quick Review

1. $\sqrt{5}/3$ **2.** $\sqrt{5}$

3. 1   **4.** $r = \dfrac{2}{1 - \cos \theta}$

5. $r = \dfrac{20}{4 + 5 \cos \theta}$ **6.** $r = \dfrac{10}{3 - 2 \cos \theta}$

7. Major: $(0, \pm 8)$, minor: $(\pm 6, 0)$

8. Major: $(\pm 7, 0)$, minor: $(0, \pm 4)$

9. $\tan^{-1} y = x^2 + C$

10. $y + \ln (|y|) = \dfrac{x^2}{2} - x + C$

Exercises

1. $T \approx 93.2$ min

3. $a \approx 6763$ km; 6765 km

5. 6501 km

7. (a) 42,167 km   **(b)** 35,788 km
   (c) *Syncom 3, GOES 4,* and *Intelsat 5*

9. 376,821 km

11. Solar system: $\approx 2.97 \times 10^{-19}$ sec²/m³;
Earth: 0.903×10^{-14} sec²/m³;
Moon: 8.046×10^{-12} sec²/m³

17. (i) When $|\mathbf{r}(t)|$ is a minimum
   (ii) When $|\mathbf{r}(t)|$ is a maximum
   (iii) When $\mathbf{r}(t) \cdot \mathbf{w} = 0$
   (iv) When angles between $\mathbf{r}(t)$ and $\mathbf{w}$ is a maximum
   (v) When angles between $\mathbf{r}(t)$ and $\mathbf{w}$ is a minimum

Chapter Review (pp. 671–673)

1. $\mathbf{v}(0) = \sqrt{2}\,\mathbf{j}$, $\mathbf{a}(0) = -4\mathbf{i}$, $\mathbf{v}\left(\dfrac{\pi}{4}\right) = -2\sqrt{2}\mathbf{i} + \mathbf{j}$,

$\mathbf{a}\left(\dfrac{\pi}{4}\right) = -2\sqrt{2}\mathbf{i} - \mathbf{j}$; at $t = 0$: $\mathbf{a} = 4\mathbf{N}$,

$\kappa = 2$; at $t = \dfrac{\pi}{4}$: $\mathbf{a} = \dfrac{7}{3}\mathbf{T} + \dfrac{4\sqrt{2}}{3}\mathbf{N}$, $\kappa = \dfrac{4\sqrt{2}}{27}$

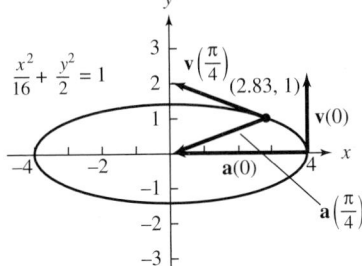

2. $\mathbf{v}(0) = \sqrt{3}\mathbf{j}$, $\mathbf{a}(0) = \sqrt{3}\mathbf{i}$; at $t = 0$: $\mathbf{a} = \sqrt{3}\mathbf{N}$,
$\kappa = 1/\sqrt{3}$

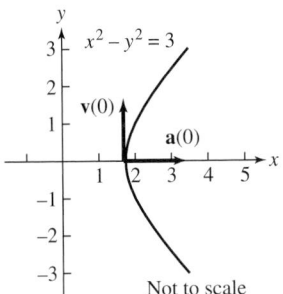

3. $|\mathbf{v}| = 1$ at $t = 0$ **4.** $\pi/2$

5. $\kappa = 1/5$ **6.** $\left(-\ln\sqrt{2}, \dfrac{1}{\sqrt{2}}\right)$

7. $\dfrac{dy}{dt} = -x$, clockwise

8. $\mathbf{v}\cdot\mathbf{j} = 12$, $\mathbf{a}\cdot\mathbf{j} = 26$

9. $\mathbf{r} = \left(-\dfrac{t^2}{2} + 1\right)\mathbf{i} + \left(-\dfrac{t^2}{2} + 2\right)\mathbf{j} + \left(-\dfrac{t^2}{2} + 3\right)\mathbf{k}$

10. $\mathbf{r} = 90t^2\mathbf{i} + \left(90t^2 - \dfrac{16}{3}t^3 + 100\right)\mathbf{j}$

13. $\dfrac{\pi}{4}\sqrt{1 + \dfrac{\pi^2}{16}} + \ln\left(\dfrac{\pi}{4} + \sqrt{1 + \dfrac{\pi^2}{16}}\right) \approx 1.72$

14. 14

15. $\mathbf{T}(0) = \dfrac{2}{3}\mathbf{i} - \dfrac{2}{3}\mathbf{j} + \dfrac{1}{3}\mathbf{k}$, $\mathbf{N}(0) = \dfrac{1}{\sqrt{2}}\mathbf{i} + \dfrac{1}{\sqrt{2}}\mathbf{j}$,

$\mathbf{B}(0) = -\dfrac{1}{3\sqrt{2}}\mathbf{i} + \dfrac{1}{3\sqrt{2}}\mathbf{j} + \dfrac{4}{3\sqrt{2}}\mathbf{k}$,

$\kappa(0) = \dfrac{\sqrt{2}}{3}$, $\tau(0) = 1/6$

16. $\mathbf{T}(0) = \dfrac{2}{3}\mathbf{i} + \dfrac{1}{3}\mathbf{j} + \dfrac{2}{3}\mathbf{k}$, $\mathbf{N}(0) = \dfrac{1}{\sqrt{5}}\mathbf{i} - \dfrac{2}{\sqrt{5}}\mathbf{j}$,

$\mathbf{B}(0) = \dfrac{4}{3\sqrt{5}}\mathbf{i} + \dfrac{2}{3\sqrt{5}}\mathbf{j} - \dfrac{5}{3\sqrt{5}}\mathbf{k}$,

$\kappa(0) = \dfrac{2\sqrt{5}}{9}$, $\tau(0) = -4/9$

17. $\mathbf{T}(\ln 2) = \dfrac{1}{\sqrt{17}}\mathbf{i} + \dfrac{4}{\sqrt{17}}\mathbf{j}$,

$\mathbf{N}(\ln 2) = -\dfrac{4}{\sqrt{17}}\mathbf{i} + \dfrac{1}{\sqrt{17}}\mathbf{j}$, $\mathbf{B}(\ln 2) = \mathbf{k}$,

$\kappa(\ln 2) = \dfrac{8}{17\sqrt{17}}$, $\tau(\ln 2) = 0$

18. $\mathbf{T}(\ln 2) = \dfrac{15}{17\sqrt{2}}\mathbf{i} + \dfrac{1}{\sqrt{2}}\mathbf{j} + \dfrac{8}{17\sqrt{2}}\mathbf{k}$,

$\mathbf{N}(\ln 2) = \dfrac{8}{17}\mathbf{i} - \dfrac{15}{17}\mathbf{k}$,

$\mathbf{B}(\ln 2) - \dfrac{15}{17\sqrt{2}}\mathbf{i} + \dfrac{1}{\sqrt{2}}\mathbf{j} - \dfrac{8}{17\sqrt{2}}\mathbf{k}$,

$\kappa(\ln 2) = 32/867$, $\tau(\ln 2) = 32/867$

19. $\mathbf{a}(0) = 10\mathbf{T} + 6\mathbf{N}$
20. $\mathbf{a}(0) = 2\sqrt{2}\mathbf{T} + 2\sqrt{3}\mathbf{N}$
21. $\kappa = \pi s$ **22.** $\kappa = 1/a$

23. $\mathbf{T} = \left(\dfrac{1}{\sqrt{2}}\cos t\right)\mathbf{i} - (\sin t)\mathbf{j} + \left(\dfrac{1}{\sqrt{2}}\cos t\right)\mathbf{k}$,

$\mathbf{N} = \left(-\dfrac{1}{\sqrt{2}}\sin t\right)\mathbf{i} - (\cos t)\mathbf{j} - \left(\dfrac{1}{\sqrt{2}}\sin t\right)\mathbf{k}$,

$\mathbf{B} = \dfrac{1}{\sqrt{2}}\mathbf{i} - \dfrac{1}{\sqrt{2}}\mathbf{k}$, $\kappa = 1/\sqrt{2}$, $\tau = 0$

24. $t = 0, \pi/2$, or π **25.** $t = \pi/3$
26. Osculating plane: $3x - 3y + z = 1$,
normal plane: $x + 2y + 3z = 6$,
rectifying plane: $11x + 8y - 9z = 10$
27. $x = 1 + t$, $y = t$, $z = -t$
28. $x = 1 - t$, $y = 1 + t$, $z = \dfrac{\pi}{4} + t$
29. (a) $y_0 \approx 5971$ km (b) 1.639×10^7 km^2
(c) 3.21%
31. (b) ≈ 1.498701134
32. (a) $\dot{x} = \dot{r}\cos\theta - r\dot{\theta}\sin\theta$, $\dot{y} = \dot{r}\sin\theta + r\dot{\theta}\cos\theta$
(b) $\dot{r} = \dot{x}\cos\theta + \dot{y}\sin\theta$, $r\dot{\theta} = -\dot{x}\sin\theta + \dot{y}\cos\theta$
33. $\kappa = \dfrac{f^2 - ff'' + 2(f')^2}{[(f')^2 + f^2]^{3/2}}$

34. (a) $\mathbf{v}(1) = -\mathbf{u}_r + 3\mathbf{u}_\theta$, $\mathbf{a}(1) = -9\mathbf{u}_r - 6\mathbf{u}_\theta$
(b) 6.5 in.

Chapter 13

Section 13.1 (pp. 675–685)

Quick Review

1. $V = \dfrac{1}{3}\pi r^2 h$

2. $S = 2\pi rh + 2\pi r^2$

3. $t = \dfrac{y}{5280x}$

4. $h = \sqrt{a^2 - \dfrac{b^2}{4}}$

5. $A = \dfrac{1}{2}ab\sin 20°$

6. $V = \dfrac{1}{2}\pi ab^2$

7. $V = \dfrac{x^2 y}{4\pi}$ or $V = \dfrac{xy^2}{4\pi}$

8. $V = \dfrac{4\pi r^2 h - \pi h^3}{4} = \dfrac{\pi h}{4}(4r^2 - h^2)$

9. $A = \dfrac{1}{2}a^2 \sin \theta$

10. $v = 2\pi mk$ ft/min

Exercises

1. Domain: all points in the xy-plane,
range: all real numbers,
domain unbounded

3. Domain: all $(x, y) \neq (0, 0)$,
range: all real numbers,
domain unbounded

5. Domain: all points in the xy-plane,
range: all real numbers,
domain unbounded

7. Domain: all (x, y) satisfying $x^2 + y^2 < 16$,
range: $z \geq 1/4$,
domain bounded

9. (a)   **(b)**

11. (a)   **(b)**

13. (a)   **(b)**

15. (a)   **(b)**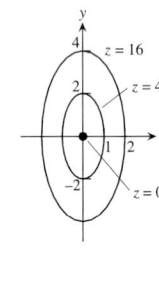

17. f

19. a

21. d

23.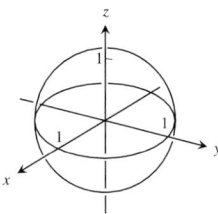

$f(x, y, z) = x^2 + y^2 + z^2 = 1$

25.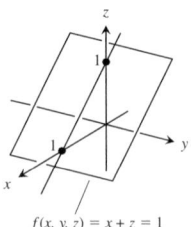

$f(x, y, z) = x + z = 1$

27.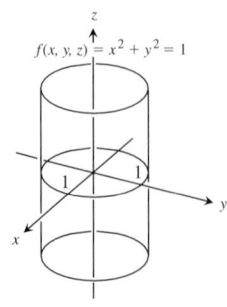

$f(x, y, z) = x^2 + y^2 = 1$

29.

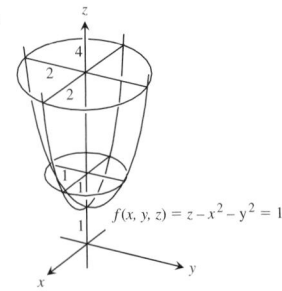

$f(x, y, z) = z - x^2 - y^2 = 1$

31. $x^2 + y^2 = 10$

33. $\tan^{-1} y - \tan^{-1} x = 2 \tan^{-1} \sqrt{2}$

35. $\sqrt{x - y} - \ln z = 2$

37. $\dfrac{x + y}{z} = \ln 2$

41. 2000

43. ≈ 63 km

57. $\dfrac{\sqrt{2}}{2} + \dfrac{\sqrt{2}}{2} i \approx 0.71 + 0.71i,$

$\dfrac{\sqrt{2}}{2} - \dfrac{\sqrt{2}}{2} i \approx 0.71 - 0.71i,$

$-\dfrac{\sqrt{2}}{2} + \dfrac{\sqrt{2}}{2} i \approx -0.71 + 0.71i,$

$-\dfrac{\sqrt{2}}{2} - \dfrac{\sqrt{2}}{2} i \approx -0.71 - 0.71i$

Section 13.2 (pp. 685–693)

Quick Review

1. 1 **2.** 2

3. 1/5 **4.** 5/6

5. −1 **6.** All $x \neq \pm 2$

7. All $x > 0$ **8.** All $x \neq -3/2$

9. All $x \neq 0$ **10.** All real numbers

Exercises

1. 5/2

3. 1/2

5. 5

7. 1

9. 1

11. 1

13. 0

15. −1

17. 2

19. 1/4

21. 7

23. 2

25. $\tan^{-1}\left(-\dfrac{\pi}{4}\right) \approx -0.67$

27. **(a)** All (x, y) **(b)** All $(x, y) \neq (0, 0)$

29. **(a)** All (x, y) except where $x = 0$ or $y = 0$
  (b) All (x, y)

31. **(a)** All (x, y, z)
  (b) All (x, y, z) except the interior of the cylinder
  $x^2 + y^2 = 1$

33. **(a)** All (x, y, z) where $z \neq 0$
  (b) All (x, y, z) where $x^2 + y^2 + z^2 \neq 1$

35. Consider paths along $y = x, x > 0$, and along
  $y = x, x < 0$

37. Consider the path $y = kx^2$, k a constant

39. Consider the paths $y = kx, k \neq -1$

41. Consider the path $y = kx^2, k \neq 0$

43. No

45. **(a)** $f(x, y)\big|_{y=mx} = \sin 2\theta$ where $\tan \theta = m$

47. The limit is 1.

49. The limit is 0.

Section 13.3 (pp. 693–702)

Quick Review

1. $\dfrac{dy}{dx} = 2kx + 5, \dfrac{d^2y}{dx^2} = 2k$

2. $\dfrac{dy}{dx} = k \cos kx, \dfrac{d^2y}{dx^2} = -k^2 \sin kx$

3. $\dfrac{dy}{dx} = 20ke^{kx}, \dfrac{d^2y}{dx^2} = 20k^2e^{kx}$

4. $\dfrac{dy}{dx} = (x + k + 1)e^x, \dfrac{d^2y}{dx^2} = (x + k + 2)e^x$

5. $\dfrac{dy}{dx} = \dfrac{\sqrt{k}}{1 + kx^2}, \dfrac{d^2y}{dx^2} = -\dfrac{2k^{3/2}x}{\left(1 + kx^2\right)^2}$

6. $\dfrac{dy}{dx} = \dfrac{1}{x - k} + 3, \dfrac{d^2y}{dx^2} = -\dfrac{1}{(x - k)^2}$

7. $\dfrac{dy}{dx} = k^x \ln k, \dfrac{d^2y}{dx^2} = k^x (\ln k)^2$

8. $\dfrac{dy}{dx} = -ke^k \sin (kx), \dfrac{d^2y}{dx^2} = -k^2e^k \cos (kx)$

9. $\dfrac{dy}{dx} = -\dfrac{k}{(x - k)^2}, \dfrac{d^2y}{dx^2} = \dfrac{2k}{(x - k)^3}$

10. $\dfrac{dy}{dx} = \dfrac{k}{(k - x)^2}, \dfrac{d^2y}{dx^2} = \dfrac{2k}{(k - x)^3}$

Exercises

1. $\dfrac{\partial f}{\partial x} = 2, \dfrac{\partial f}{\partial y} = 0$

3. $\dfrac{\partial f}{\partial x} = y - 1, \dfrac{\partial f}{\partial y} = x$

5. $\dfrac{\partial f}{\partial x} = 2x(y + 2), \dfrac{\partial f}{\partial y} = x^2 - 1$

7. $\dfrac{\partial f}{\partial x} = -\dfrac{1}{(x + y)^2}, \dfrac{\partial f}{\partial y} = -\dfrac{1}{(x + y)^2}$

9. $\dfrac{\partial f}{\partial x} = \dfrac{x}{\sqrt{x^2 + y^2}}, \dfrac{\partial f}{\partial y} = \dfrac{x}{\sqrt{y^2 + y^2}}$

11. $\dfrac{\partial f}{\partial x} = \dfrac{-y^2 - 1}{(xy - 1)^2}, \dfrac{\partial f}{\partial y} = \dfrac{-x^2 - 1}{(xy - 1)^2}$

13. $\dfrac{\partial f}{\partial x} = e^x \sin (y + 1), \dfrac{\partial f}{\partial y} = e^x \cos (y + 1)$

15. $\dfrac{\partial f}{\partial x} = e^{(x+y+1)}, \dfrac{\partial f}{\partial y} = e^{(x+y+1)}$

17. $\dfrac{\partial f}{\partial x} = \dfrac{1}{x \ln y}, \dfrac{\partial f}{\partial y} = \dfrac{-\ln x}{y(\ln y)^2}$

19. $f_x = y + z, f_y = x + z, f_z = y + x$

21. $f_x = 0, f_y = 2y, f_z = 4z$

23. $f_x = \dfrac{1}{x + 2y + 3z}, f_y = \dfrac{2}{x + 2y + 3z},$

$f_z = \dfrac{3}{x + 2y + 3z}$

25. $\dfrac{\partial f}{\partial t} = -2\pi \sin (2\pi t - \alpha), \dfrac{\partial f}{\partial \alpha} = \sin (2\pi t - \alpha)$

27. $\dfrac{\partial h}{\partial \rho} = \sin \phi \cos \theta, \dfrac{\partial h}{\partial \phi} = \rho \cos \phi \cos \theta,$

$\dfrac{\partial h}{\partial \theta} = -\rho \sin \phi \sin \theta$

29. $W_P = V, W_V = P + \dfrac{\delta v^2}{2g}, W_\delta = \dfrac{Vv^2}{2g},$

$W_v = \dfrac{V\delta v}{g}, W_g = -\dfrac{V\delta v^2}{2g^2}$

31. $f_{xx} = 0, f_{yy} = 0, f_{xy} = f_{yx} = 1$

33. $f_{xx} = 2y - y \sin x, f_{yy} = -\cos y,$
$f_{xy} = f_{yx} = 2x + \cos x$

35. $f_{xx} = -\dfrac{1}{(x + y)^2}, f_{yy} = -\dfrac{1}{(x + y)^2},$

$f_{xy} = f_{yx} = -\dfrac{1}{(x + y)^2}$

37. $w_{xy} = w_{yx} = -\dfrac{6}{(2x + 3y)^2}$

39. $w_{xy} = w_{yx} = 2y + 6xy^2 + 12x^2y^3$

41. (a) x first   **(b)** y first
(c) x first   **(d)** x first
(e) y first   **(f)** y first

43. -2

45. $\dfrac{\partial f}{\partial z}(x_0, y_0, z_0)$

$= \lim\limits_{h \to 0} \dfrac{f(x_0, y_0, z_0 + h) - f(x_0, y_0, z_0)}{h}; 12$

Section 13.4 (pp. 703–714)

Quick Review

1. $y = 1 - x$ **2.** 0.8
3. 0.01058   **4.** 0.013
5. 1.3%
6. Graph the error $y = |x - \sin (2x) + 1 - (1 - x)|$, along with the line $y = 0.02$ in $[-0.3, 0.3]$ by $[-0.01, 0.03]$.

Interval: $\approx (-0.2476, 0.2476)$
7. $df = (1 - 2 \cos (2x)) dx$
8. $-1/2$
9. The graph is quite straight near $x = \pi$, so the linearization at $x = \pi$ should be quite close to the graph of f.
10. The graph is turning around near $x = 1/2$, so a straight-line approximation will be less accurate.

Exercises

1. (a) $L(x, y) = 1$
(b) $L(x, y) = 2x + 2y - 1$
3. (a) $L(x, y) = 3x - 4y + 5$
(b) $L(x, y) = 3x - 4y + 5$
5. (a) $L(x, y) = x + 1$
(b) $L(x, y) = -y + \dfrac{\pi}{2}$

7. $L(x, y) = x - 6y + 7; |E(x, y)| \le 0.06$
9. $L(x, y) = x + 1; |E(x, y)| \le 0.0222$
11. width (smaller dimension)   **13.** 0.31
15. $\pm 2.415\%$
17. $|x - 1| \le 0.014$ and $|y - 1| \le 0.014$
19. $\approx 0.1\%$
21. (a) $L(x, y, z) = 2x + 2y + 2z - 3$
(b) $L(x, y, z) = y + z$
(c) $L(x, y, z) = 0$
23. (a) $L(x, y, z) = x$
(b) $L(x, y, z) = \dfrac{1}{\sqrt{2}}x + \dfrac{1}{\sqrt{2}}y$
(c) $L(x, y, z) = \dfrac{1}{3}x + \dfrac{2}{3}y + \dfrac{2}{3}z$
25. (a) $L(x, y, z) = 2 + x$
(b) $L(x, y, z) = x - y - z + \dfrac{\pi}{2} + 1$
(c) $L(x, y, z) = x - y - z + \dfrac{\pi}{2} + 1$

27. $L(x, y, z) = 2x - 6y - 2z + 6$;
$|E(x, y, z)| \leq 0.0024$

29. $L(x, y, z) = x + y - z - 1$; $|E(x, y, z)| \leq 0.00135$

31. (a) $dS = S_0 \left(\dfrac{1}{100} \, dp + dx - 5dw - 30dh \right)$

  (b) More sensitive to a change in height

33. 4.8

35. Yes

37. (a) The volume is about 10 times more sensitive to a change in r.

  (b) One solution is $dh = 1.5$, $dr = -0.15$, $h = 0.85$ in., $r = 0.85$ in.

39. d

Section 13.5 (pp. 714–723)

Quick Review

1. $\dfrac{\partial f}{\partial x} = y;\ \dfrac{\partial f}{\partial y} = x + \cos y$

2. $\dfrac{\partial f}{\partial x} = 3x^2;\ \dfrac{\partial f}{\partial y} = -2y$

3. $\dfrac{\partial f}{\partial x} = \sec^2 (x + y);\ \dfrac{\partial f}{\partial y} = \sec^2 (x + y)$

4. $\dfrac{\partial f}{\partial x} = ye^{xy};\ \dfrac{\partial f}{\partial y} = xe^{xy}$

5. $\dfrac{\partial f}{\partial x} = \sin y + y \sin x;\ \dfrac{\partial f}{\partial y} = x \cos y - \cos x$

6. 0   **7.** $\pi/2$

8. -2 **9.** -2

10. 2/3

Exercises

1. (a) $\dfrac{dw}{dt} = 0$ **(b)** $\dfrac{dw}{dt}(\pi) = 0$

3. (a) $\dfrac{dw}{dt} = 1$ **(b)** $\dfrac{dw}{dt}(3) = 1$

5. (a) $\dfrac{dw}{dt} = 4t \tan^{-1} t + 1$

  (b) $\dfrac{dw}{dt}(1) = \pi + 1$

7. (a) $\dfrac{\partial z}{\partial r} = (4 \cos \theta) \ln (r \sin \theta) + 4 \cos \theta$,

  $\dfrac{\partial z}{\partial \theta} = (-4r \sin \theta) \ln (r \sin \theta) + \dfrac{4r \cos^2 \theta}{\sin \theta}$

  (b) At $(2, \pi/4)$: $\dfrac{\partial z}{\partial r} = \sqrt{2} (\ln 2 + 2)$,

  $\dfrac{\partial z}{\partial \theta} = -2\sqrt{2} \ln 2 + 4\sqrt{2}$

9. (a) $\dfrac{\partial w}{\partial u} = 2u + 4uv,\ \dfrac{\partial w}{\partial v} = -2v + 2u^2$

  (b) At $(1/2, 1)$: $\dfrac{\partial w}{\partial u} = 3,\ \dfrac{\partial w}{\partial v} = -3/2$

11. (a) $\dfrac{\partial u}{\partial x} = 0,\ \dfrac{\partial u}{\partial y} = \dfrac{z}{(z - y)^2},\ \dfrac{\partial u}{\partial z} = -\dfrac{y}{(z - y)^2}$

  (b) At $(\sqrt{3}, 2, 1)$: $\dfrac{\partial u}{\partial x} = 0,\ \dfrac{\partial u}{\partial y} = 1,\ \dfrac{\partial u}{\partial z} = -2$

13. $\dfrac{dz}{dt} = \dfrac{\partial z}{\partial x} \dfrac{dx}{dt} + \dfrac{\partial z}{\partial y} \dfrac{dy}{dt}$

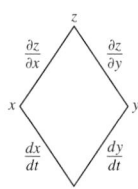

15. $\dfrac{\partial w}{\partial u} = \dfrac{\partial w}{\partial x} \dfrac{\partial x}{\partial u} + \dfrac{\partial w}{\partial y} \dfrac{\partial y}{\partial u} + \dfrac{\partial w}{\partial z} \dfrac{\partial z}{\partial u}$

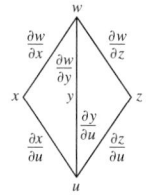

$\dfrac{\partial w}{\partial v} = \dfrac{\partial w}{\partial x} \dfrac{\partial x}{\partial v} + \dfrac{\partial w}{\partial y} \dfrac{\partial y}{\partial v} + \dfrac{\partial w}{\partial z} \dfrac{\partial z}{\partial v}$

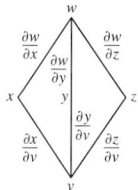

17. $\dfrac{\partial w}{\partial u} = \dfrac{\partial w}{\partial x} \dfrac{\partial x}{\partial u} + \dfrac{\partial w}{\partial y} \dfrac{\partial y}{\partial u}$

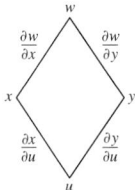

$\dfrac{\partial w}{\partial v} = \dfrac{\partial w}{\partial x} \dfrac{\partial x}{\partial v} + \dfrac{\partial w}{\partial y} \dfrac{\partial y}{\partial v}$

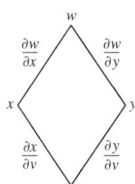

19. $\dfrac{\partial z}{\partial t} = \dfrac{\partial z}{\partial x}\dfrac{\partial x}{\partial t} + \dfrac{\partial z}{\partial y}\dfrac{\partial y}{\partial t}$

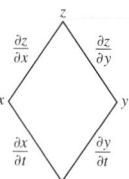

$\dfrac{\partial z}{\partial s} = \dfrac{\partial z}{\partial x}\dfrac{\partial x}{\partial s} + \dfrac{\partial z}{\partial y}\dfrac{\partial y}{\partial s}$

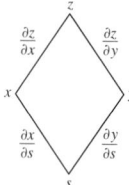

21. $\dfrac{\partial w}{\partial s} = \dfrac{dw}{du}\dfrac{\partial u}{\partial s}$

$\begin{array}{c} w \\ \Big| \frac{dw}{du} \\ u \\ \Big| \frac{\partial u}{\partial s} \\ s \end{array}$

$\dfrac{\partial w}{\partial t} = \dfrac{dw}{du}\dfrac{\partial u}{\partial t}$

$\begin{array}{c} w \\ \Big| \frac{dw}{du} \\ u \\ \Big| \frac{\partial u}{\partial t} \\ t \end{array}$

23. $\dfrac{\partial w}{\partial r} = \dfrac{\partial w}{\partial x}\dfrac{dx}{dr} + \dfrac{\partial w}{\partial y}\dfrac{dy}{dr} = \dfrac{\partial w}{\partial x}\dfrac{dx}{dr}$ since $\dfrac{dy}{dr} = 0$

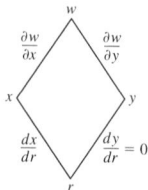

$\dfrac{\partial w}{\partial s} = \dfrac{\partial w}{\partial y}\dfrac{dy}{ds}$

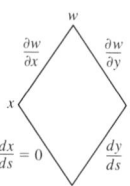

25. 4/3

27. −4/5

29. 12

31. −7

33. $\partial z/\partial u = 2$, $\partial z/v = 1$

35. 4

37. (a) Minimum: $(-2, 1)$, $(2, -1)$
maximum: $(2, 1)$, $(-2, -1)$
(b) Maximum 0, minimum −4

41. Volume is increasing at 3 m³/sec, surface area is not changing, diagonals are decreasing in length.

45. (b) $f_x = (\cos\,\theta)\dfrac{\partial w}{\partial r} - \left(\dfrac{\sin\theta}{r}\right)\dfrac{\partial w}{\partial\theta}$,

$f_y = (\sin\theta)\dfrac{\partial w}{\partial r} + \left(\dfrac{\cos\,\theta}{r}\right)\dfrac{\partial w}{\partial\theta}$

Section 13.6 (pp. 723–736)

Quick Review

1. $\sqrt{6}$

2. $\dfrac{2}{\sqrt{6}}\mathbf{i} - \dfrac{1}{\sqrt{6}}\mathbf{j} + \dfrac{1}{\sqrt{6}}\mathbf{k}$

3. −5

4. $\mathbf{i} + 3\mathbf{j} + 5\mathbf{k}$

5. $\mathbf{u} \times \mathbf{v} = \mathbf{i} + 3\mathbf{j} + 5\mathbf{k}$

6. $\dfrac{1}{\sqrt{35}}\mathbf{i} + \dfrac{3}{\sqrt{35}}\mathbf{j} + \dfrac{5}{\sqrt{35}}\mathbf{k}$

7. $x = -1 + 4t$, $y = 2$, $z = -1 + t$

8. $x = 2 + 3t$, $y = -1 - 2t$, $z = 1 + t$

9. $x = 2 + t$, $y = -2t$, $z = -2 + 3t$

10. $x - 3y + 2z = -3$

Exercises

1. $\nabla f = -\mathbf{i} + \mathbf{j}$, level curve: $y - x = -1$

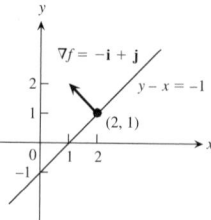

3. $\nabla g = 2\mathbf{i} + \mathbf{j}$, level curve: $y - x^2 = -1$

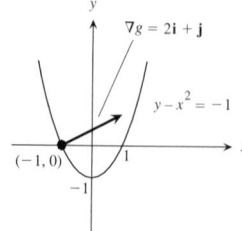

5. $3\mathbf{i} + 2\mathbf{j} - 4\mathbf{k}$

7. $-\dfrac{26}{27}\mathbf{i} + \dfrac{23}{54}\mathbf{j} - \dfrac{23}{54}\mathbf{k}$

9. -4

11. $31/13$

13. 3

15. 2

17. Increases: $\mathbf{u} = -\dfrac{1}{\sqrt{2}}\mathbf{i} + \dfrac{1}{\sqrt{2}}\mathbf{j},\ (D_{\mathbf{u}}f)_{P_0} = \sqrt{2}$;

decreases: $-\mathbf{u} = \dfrac{1}{\sqrt{2}}\mathbf{i} - \dfrac{1}{\sqrt{2}}\mathbf{j},(D_{-\mathbf{u}}f)_{P_0} = -\sqrt{2}$

19. Increases: $\mathbf{u} = \dfrac{1}{3\sqrt{3}}\mathbf{i} - \dfrac{5}{3\sqrt{3}}\mathbf{j} - \dfrac{1}{3\sqrt{3}}\mathbf{k}$,

$(D_{\mathbf{u}}f)_{P_0} = 3\sqrt{3}$;

decreases: $-\mathbf{u} = -\dfrac{1}{3\sqrt{3}}\mathbf{i} + \dfrac{5}{3\sqrt{3}}\mathbf{j} + \dfrac{1}{3\sqrt{3}}\mathbf{k}$,

$(D_{-\mathbf{u}}f)_{P_0} = -\sqrt{2}$

21. Increases: $\mathbf{u} = \dfrac{1}{\sqrt{3}}\mathbf{i} + \dfrac{1}{\sqrt{3}}\mathbf{j} + \dfrac{1}{\sqrt{3}}\mathbf{k}$,

$(D_{\mathbf{u}}f)_{P_0} = 2\sqrt{3}$;

decreases: $-\mathbf{u} = -\dfrac{1}{\sqrt{3}}\mathbf{i} - \dfrac{1}{\sqrt{3}}\mathbf{j} - \dfrac{1}{\sqrt{3}}\mathbf{k}$,

$(D_{-\mathbf{u}}f)_{P_0} = -2\sqrt{3}$

23. 0.000760

25. 0

27. (a) $x + y + z = 3$
  (b) $x = 1 + 2t,\ y = 1 + 2t,\ z = 1 + 2t$

29. (a) $2x - z = 2$
  (b) $x = 2 - 4t,\ y = 0,\ z = 2 + 2t$

31. (a) $2x + 2y + z = 4$
  (b) $x = 2t,\ y = 1 + 2t,\ z = 2 + t$

33. (a) $x + y + z = 1$
  (b) $x = t,\ y = 1 + t,\ z = t$

35. $2x - z = 2$

37. $x - y + 2z = 1$

39.

41.

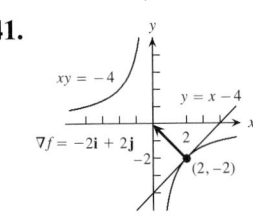

43. $x = 1,\ y = 1 + 2t,\ z = 1 - 2t$

45. $x = 1 - 2t,\ y = 1,\ z = \dfrac{1}{2} + 2t$

47. $x = 1 + 90t,\ y = 1 - 90t,\ z = 3$

49. $\mathbf{u} = \dfrac{7}{\sqrt{53}}\mathbf{i} - \dfrac{2}{\sqrt{53}}\mathbf{j},\ -\mathbf{u} = -\dfrac{7}{\sqrt{53}}\mathbf{i} + \dfrac{2}{\sqrt{53}}\mathbf{j}$

51. No

53. $-7/\sqrt{5}$

55. (a) $0.935°$ C/m   (b) $1.87°$ C/sec

57. $t = -\dfrac{\pi}{4}:\dfrac{-\pi}{2\sqrt{2}};\ t = 0:0;\ t = \dfrac{\pi}{4}:\dfrac{\pi}{2\sqrt{2}}$

63. $\nabla f \cdot \mathbf{u} = (D_{\mathbf{u}}f)_{P_0}$

Section 13.7 (pp. 736–745)

Quick Review

1. $(1, 1/2),\ (-1, -1/2)$

2. $-1/2$ at $x = -1$ **3.** 1 at $x = 1$

4. Absolute maximum of $1/2$ at $x = 1$, absolute minimum of $-1/2$ at $x = -1$

5. $(1, 4)$

6. -5 at $x = -2$, -12 at $x = 5$

7. 4 at $x = 1$

8. Absolute maximum of 4 at $x = 1$, absolute minimum of -12 at $x = 5$

9. $x = -2,\ y = -2$

10. $x = 9/11,\ y = -7/11$

Exercises

1. Local minimum of -5 at $(-3, 3)$

3. Saddle point at $(-2, 1)$

5. Saddle point at $(6/5, 69/25)$

7. Local minimum of -6 at $(2, -1)$

9. Saddle point at $(1, 2)$

11. Local minimum of 0 at $(0, 0)$, saddle point at $(1, -1)$

13. Local maximum of 2 at $(1, 1)$ and $(-1, -1)$, saddle point at $(0, 0)$

15. Local maximum of -1 at $(0, 0)$

17. Saddle points at $(n\pi, 0)$ for every integer n

19. Absolute maximum of 1 at $(0, 0)$, absolute minimum of -5 at $(1, 2)$

21. Absolute maximum of 11 at $(0, -3)$, absolute minimum of -10 at $(4, -2)$

23. Absolute maximum of 4 at $(2, 0)$, absolute minimum of $3\sqrt{2}/2$ at $(3, \pm\pi/4)$ and $(1, \pm\pi/4)$

25. Absolute maximum of 1 at $(1, 0)$, $(0, 1)$, and $(0, -1)$, absolute minimum of -1 at $(-1, 0)$

27. $a = -3$ and $b = 2$

29. Hottest: $2\tfrac{1}{4}°$ at $(-1/2, \pm\sqrt{3}/2)$, coldest: $-\tfrac{1}{4}°$ at $(1/2, 0)$

31. (a) Saddle point at $(0, 0)$
   (b) Local minimum at $(1, 2)$
   (c) Local minimum at $(1, -2)$, saddle point at $(-1, -2)$
35. (a) No   **(b)** Saddle point at (a, b)
37. $(1/6, 1/3, 355/36)$
39. No
41. (a) (i) Absolute minimum of -2 at $(-2, 0)$ when $t = \pi$; absolute maximum of $2\sqrt{2}$ at $(\sqrt{2}, \sqrt{2})$ when $t = \pi/4$
   (ii) Absolute minimum of 2 at $(2, 0)$ and $(0, 2)$ when $t = 0, \pi/2$, respectively; absolute maximum of $2\sqrt{2}$ at $(\sqrt{2}, \sqrt{2})$ when $t = \pi/4$
   (b) (i) Absolute minimum of -2 at $(-\sqrt{2}, \sqrt{2})$ when $t = 3\pi/4$; absolute maximum of 2 at $(\sqrt{2}, \sqrt{2})$ when $t = \pi/4$
   (ii) Absolute minimum of 0 at $(2, 0)$ and $(0, 2)$ when $t = 0, \pi/2$, respectively; absolute maximum of 2 at $(\sqrt{2}, \sqrt{2}$ when $t = \pi/4$
   (c) (i) Absolute minimum of 4 at $(0, 2)$ when $t = \pi/2$; absolute maximum of 8 at $(2, 0)$ and $(-2, 0)$ when $t = 0, \pi$, respectively
   (ii) Absolute minimum of 4 at $(0, 2)$ when $t = \pi/2$; absolute maximum of 8 at $(2, 0)$ when $t = 0$
43. (i) Absolute minimum of $-1/2$ at $(-1, 1/2)$ when $t = -1/2$; no absolute maximum
   (ii) Absolute minimum of $-1/2$ at $(-1, 1/2)$ when $t = -1/2$; absolute maximum of 0 at $(0, 1)$ and $(-2, 0)$ when $t = -1, 0$, respectively
   (iii) Absolute minimum of 0 at $(0, 1)$ when $t = 0$; absolute maximum of 4 at $(2, 2)$ when $t = 1$

Section 13.8 (pp. 746–756)

Quick Review

1. Local maximum of $-28/9$ at $(4/9, 2/9)$
2. Local maximum of 4 at $(0, 1)$
3. Local minimum of -63 at $(15, -8)$
4. Local maximum of 1 at $(-1/2, -1)$ [$(0, 0)$ is a saddle point]
5. $\nabla g = 4y\mathbf{j} - 6z\mathbf{k}$
6. $\nabla k = (9x^2 y - 2y^3 + y\cos xy)\mathbf{i} + (3x^3 - 6xy^2 + x\cos xy)\mathbf{j}$
7. $\nabla h = 3\mathbf{i} - 2\mathbf{j} + 4z\mathbf{k}$ **8.** Perpendicular
9. Perpendicular   **10.** Parallel

Exercises

1. $(\pm\sqrt{2}, \pm1/2)$
3. 39

5. $(3, \pm3\sqrt{2})$
7. (a) Minimum value of 8 at $(4, 4)$
   (b) Maximum value of 64 at $(8, 8)$
9. $r = 2$ cm, $h = 4$ cm, area $= 24\pi$ cm^2
11. $4\sqrt{2}$ by $3\sqrt{2}$
13. Maximum of 20 at $(2, 4)$, minimum of 0 at $(0, 0)$
15. Maximum of $125°$ at $(2\sqrt{5}, -\sqrt{5})$ and $(-2\sqrt{5}, \sqrt{5})$, minimum of $0°$ at $(\sqrt{5}, 2\sqrt{5})$ and $(-\sqrt{5}, -2\sqrt{5})$
17. $(3/2, 2, 5/2)$
19. Minimum distance is 1 at $(\pm1, 0, 0)$
21. $(0, 0, 2)$, $(0, 0, -2)$
23. Maximum value of 30 at $(1, -2, 5)$, minimum value of -30 at $(-1, 2, -5)$
25. 3, 3, and 3
27. $2/\sqrt{3}$ by $2/\sqrt{3}$ by $2/\sqrt{3}$
29. $(\pm4/3, -4/3, -4/3)$, temperature $= 642\frac{2}{3}°$
31. \$128 at $(8, 14)$
33. Maximum value of $4/3$ at $(2/3, 4/3, -4/3)$
35. $(2, 4, 4)$
37. Maximum of $1 + 6\sqrt{3}$ at $(\pm\sqrt{6}, \sqrt{3}, 1)$ minimum of $1 - 6\sqrt{3}$ at $(\pm\sqrt{6}, -\sqrt{3}, 1)$
39. Maximum value of 4 at $(0, 0, \pm2)$, minimum value of 2 at $(\pm\sqrt{2}, \pm\sqrt{2}, 0)$
42. $A = -1/2$, $B = 3/2$, $C = -1/4$

Chapter 13 Review Exercises
(pp. 756–759)

1. Domain: all (x, y); range: $z \geq 0$; level curves are ellipses with major axis along the y-axis and minor axis along the x-axis

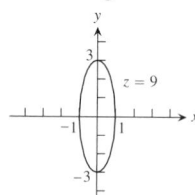

2. Domain: all (x, y); range: $0 < z < \infty$; level curves are the straight lines $x + y = \ln z$ with slope -1 and $z > 0$

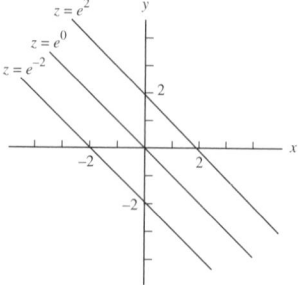

3. Domain: all (x, y) such that $x \neq 0$ and $y \neq 0$; range: $z \neq 0$; level curves are hyperbolas with the x- and y-axis as asymptotes

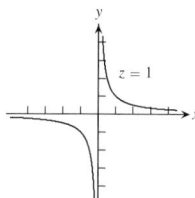

4. Domain: all (x, y) such that $x^2 - y \geq 0$; range: $z \geq 0$; level curves are the parabolas $y = x^2 - c$, $c \geq 0$

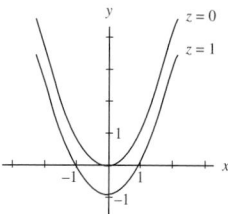

5. Domain: all (x, y, z); range: all real numbers; level curves are paraboloids of revolutions with the z-axis as axis

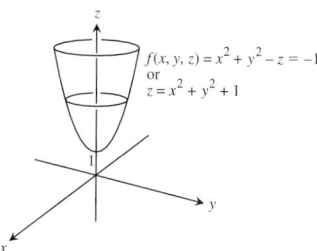

6. Domain: all (x, y, z); range: all nonnegative real numbers; level curves are ellipsoids with center $(0, 0, 0)$

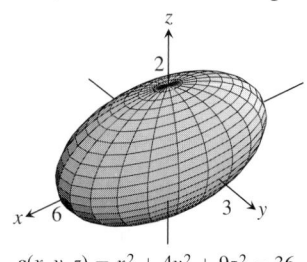

$g(x, y, z) = x^2 + 4y^2 + 9z^2 = 36$

7. Domain: all $(x, y, z) \neq (0, 0, 0)$; range: all positive real numbers; level curves are spheres with center $(0, 0, 0)$ and radius $r > 0$

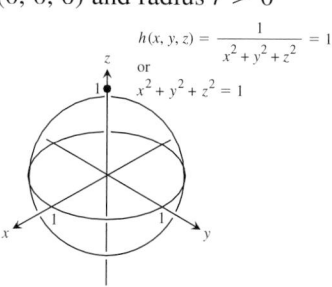

$h(x, y, z) = \dfrac{1}{x^2 + y^2 + z^2} = 1$
or
$x^2 + y^2 + z^2 = 1$

8. Domain: all (x, y, z); range: $(0, 1)$; level curves are spheres with center $(0, 0, 0)$ and radius $r > 0$

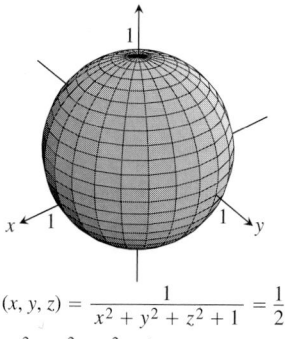

$k(x, y, z) = \dfrac{1}{x^2 + y^2 + z^2 + 1} = \dfrac{1}{2}$
or $x^2 + y^2 + z^2 = 1$

9. -2 **10.** 2

11. $1/2$ **12.** 3

13. 1 **14.** $-\pi/4$

17. (a) No   **(b)** No

18. (a) 1 **(b)** 0 **(c)** 0

19. $\dfrac{\partial g}{\partial r} = \cos\theta + \sin\theta, \dfrac{\partial g}{\partial \theta} = -r\sin\theta + r\cos\theta$

20. $\dfrac{\partial f}{\partial x} = \dfrac{x - y}{x^2 + y^2}, \dfrac{\partial f}{\partial y} = \dfrac{x + y}{x^2 + y^2}$

21. $\dfrac{\partial f}{\partial R_1} = -\dfrac{1}{R_1{}^2}, \dfrac{\partial f}{\partial R_2} = -\dfrac{1}{R_2{}^2}, \dfrac{\partial f}{\partial R_3} = -\dfrac{1}{R_3{}^2}$

22. $\dfrac{\partial h}{\partial x} = 2\pi\cos(2\pi x + y - 3z)$,

$\dfrac{\partial h}{\partial y} = \cos(2\pi x + y - 3z)$,

$\dfrac{\partial h}{\partial z} = -3\cos(2\pi x + y - 3z)$

23. $\dfrac{\partial P}{\partial n} = \dfrac{RT}{V}, \dfrac{\partial P}{\partial R} = \dfrac{nT}{V}, \dfrac{\partial P}{\partial T} = \dfrac{nR}{V}, \dfrac{\partial P}{\partial V} = -\dfrac{nRT}{V^2}$

24. $\dfrac{\partial f}{\partial r} = -\dfrac{1}{2r^2 l}\sqrt{\dfrac{T}{\pi w}}, \dfrac{\partial f}{\partial l} = -\dfrac{1}{2rl^2}\sqrt{\dfrac{T}{\pi w}}$,

$\dfrac{\partial f}{\partial T} = \dfrac{1}{4rlT}\sqrt{\dfrac{T}{\pi w}}, \dfrac{\partial f}{\partial w} = -\dfrac{1}{4rlw}\sqrt{\dfrac{T}{\pi w}}$

25. $\dfrac{\partial^2 g}{\partial x^2} = 0, \dfrac{\partial^2 g}{\partial y^2} = \dfrac{2x}{y^3}, \dfrac{\partial^2 g}{\partial y \partial x} = \dfrac{\partial^2 g}{\partial x \partial y} = -\dfrac{1}{y^2}$

26. $\dfrac{\partial^2 g}{\partial x^2} = e^x - y\sin x, \dfrac{\partial^2 g}{\partial y^2} = 0$,

$\dfrac{\partial^2 g}{\partial y \partial x} = \dfrac{\partial^2 g}{\partial x \partial y} = \cos x$

27. $\dfrac{\partial^2 f}{\partial x^2} = -30x + \dfrac{2 - 2x^2}{(x^2 + 1)^2}, \dfrac{\partial^2 f}{\partial y^2} = 0$,

$\dfrac{\partial^2 f}{\partial y \partial x} = \dfrac{\partial^2 f}{\partial x \partial y} = 1$

28. $\dfrac{\partial^2 f}{\partial x^2} = 0, \dfrac{\partial^2 f}{\partial y^2} = 2 - \cos y + 7e^y,$

$\dfrac{\partial^2 f}{\partial y \partial x} = \dfrac{\partial^2 f}{\partial x \partial y} = -3$

29. $L(x, y) = \dfrac{1}{2} + \dfrac{1}{2}x - \dfrac{1}{2}y.$ Upper bound for E depends on the bound M used for $|f_{xx}|, |f_{xy}|,$ and $|f_{yy}|.$ With $M = \sqrt{2}/2, |E| \le 0.0142$; with $M = 1,$ $|E| \le 0.02.$

30. $L(x, y) = x - 5y + 4.$ Maximum of $|f_{xx}|, |f_{xy}|,$ and $|f_{yy}|$ is 6, so $|E| \le 0.27.$

31. $L(x, y, z) = y - 3z$ at $(1, 0, 0)$; $L(x, y, z) = x + y - z - 1$ at $(1, 1, 0)$

32. $L(x, y, z) = 1 + y + z - \pi/4$ at $(0, 0, \pi/4)$;

$L(x, y, z) = \dfrac{\sqrt{2}}{2} - \dfrac{\sqrt{2}}{2}x + \dfrac{\sqrt{2}}{2}y + \dfrac{\sqrt{2}}{2}z$ at

$(\pi/4, \pi/4, 0)$

33. Diameter   **34.** y

35. I increases by 0.038 amp; % change in V about -4.17%; % change in R is -20%; % change in I about 15.83%

36. $\approx 0.8125\%$ **37. (a)** 5%

38. A 1-cm error in height **39.** -1

40. $\dfrac{\partial w}{\partial u} = \dfrac{2}{5}, \dfrac{\partial w}{\partial v} = 0$ at $u = v = 0$

41. $-(\sin 1 + \cos 2)(\sin 1) + (\cos 1 + \cos 2)$ $(\cos 1) - 2(\sin 1 + \cos 1)(\sin 2) \approx -2.80$

43. -1 **44.** $-(\ln 2 + 1)$

45. Increases most rapidly: $\mathbf{u} = -\dfrac{\sqrt{2}}{2}\mathbf{i} - \dfrac{\sqrt{2}}{2}\mathbf{j}$;

decreases most rapidly: $-\mathbf{u} = \dfrac{\sqrt{2}}{2}\mathbf{i} + \dfrac{\sqrt{2}}{2}\mathbf{j}$;

$(D_{\mathbf{u}}f)_{P_0} = \sqrt{2}/2; (D_{-\mathbf{u}}f)_{P_0} = -\sqrt{2}/2$; with

$\mathbf{u}_1 = \dfrac{\mathbf{A}}{|\mathbf{A}|}, (D_{\mathbf{u}_1}f)_{P_0} = -7/10$

46. Increases most rapidly: $\mathbf{u} = \dfrac{1}{\sqrt{2}}\mathbf{i} - \dfrac{1}{\sqrt{2}}\mathbf{j}$;

decreases most rapidly: $-\mathbf{u} = -\dfrac{1}{\sqrt{2}}\mathbf{i} + \dfrac{1}{\sqrt{2}}\mathbf{j}$;

$(D_{\mathbf{u}}f)_{P_0} = 2\sqrt{2}; (D_{-\mathbf{u}}f)_{P_0} = -2\sqrt{2}$; with

$\mathbf{u}_1 = \dfrac{\mathbf{A}}{|\mathbf{A}|}, (D_{\mathbf{u}_1}f)_{P_0} = 0$

47. Increases most rapidly: $\mathbf{u} = \dfrac{2}{7}\mathbf{i} + \dfrac{3}{7}\mathbf{j} + \dfrac{6}{7}\mathbf{k}$;

decreases most rapidly: $-\mathbf{u} = -\dfrac{2}{7}\mathbf{i} - \dfrac{3}{7}\mathbf{j} - \dfrac{6}{7}\mathbf{k}$;

$(D_{\mathbf{u}}f)_{P_0} = 7; (D_{-\mathbf{u}}f)_{P_0} = -7$; with

$\mathbf{u}_1 = \dfrac{\mathbf{A}}{|\mathbf{A}|}, (D_{\mathbf{u}_1}f)_{P_0} = 7$

48. Increases most rapidly: $\mathbf{u} = \dfrac{2}{\sqrt{5}}\mathbf{j} + \dfrac{1}{\sqrt{5}}\mathbf{k}$;

decreases most rapidly: $-\mathbf{u} = -\dfrac{2}{\sqrt{5}}\mathbf{j} - \dfrac{1}{\sqrt{5}}\mathbf{k}$;

$(D_{\mathbf{u}}f)_{P_0} = \sqrt{5}; (D_{-\mathbf{u}}f)_{P_0} = -\sqrt{5}$; with

$\mathbf{u}_1 = \dfrac{\mathbf{A}}{|\mathbf{A}|}, (D_{\mathbf{u}_1}f)_{P_0} = \sqrt{3}$

49. $\pi/\sqrt{2}$ **50.** $\sqrt{3}$

51. (a) $f_x(1, 2) = f_y(1, 2) = 2$

(b) $14/5$

52. (a) True   **(b)** False

(c) True   **(d)** True

53.

$x^2 + y^2 + z^2 = 0$

$\nabla f|_{(0, -1, 1)} = \mathbf{j} + 2\mathbf{k}$

$\nabla f|_{(0, 0, 0)} = \mathbf{j}$

$\nabla f|_{(0, -1, -1)} = \mathbf{j} - 2\mathbf{k}$

54.

$\nabla f|_{(2, 0, 2)} = 4\mathbf{k}$

$\nabla f|_{(-2, -2, 0)} = -4\mathbf{j}$

$y^2 + z^2 = 4$

$\nabla f|_{(2, 2, 0)} = 4\mathbf{j}$

$\nabla f|_{(2, 0, -2)} = -4\mathbf{k}$

55. $4x - y - 5z = 4; x = 2 + 4t, y = -1 - t,$ $z = 1 - 5t$

56. $2x + 2y + z = 6; x = 1 + 2t, y = 1 + 2t,$ $z = 2 + t$

57. $2y - z = 2$ **58.** $x + y + 2z = 3$

59. Tangent line: $x + y = \pi + 1$; normal line: $y = x - \pi + 1$

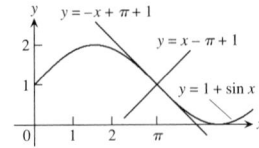

60. Tangent line: $y = \dfrac{1}{2}x + \dfrac{3}{2}$; normal line:

$y = -2x + 4$

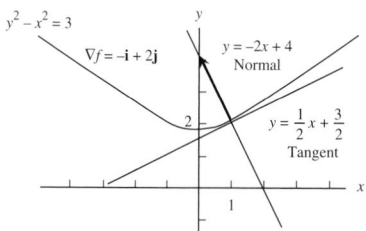

61. $x = 1 - 2t$, $y = 1$, $z = \dfrac{1}{2} + 2t$

62. $x = \dfrac{1}{2} - t$, $y = 1$, $z = \dfrac{1}{2} + t$

63. Local minimum of -8 at $(-2, -2)$

64. Saddle point at $(0, -1)$ with $f(0, -1) = 2$

65. Saddle point at $(0, 0)$ with $f(0, 0) = 0$; local maximum of $1/4$ at $(-1/2, -1/2)$

66. Saddle point at $(0, 0)$ with $f(0, 0) = 15$; local minimum of 14 at $(1, 1)$

67. Saddle points at $(0, 0)$ and $(-2, 2)$ with $f(0, 0) = f(-2, 2) = 0$; local minimum of -4 at $(0, 2)$; local maximum value of 4 at $(-2, 0)$

68. Saddle point at $(0, 1)$ with $f(0, 1) = -3$; local minimum of -19 at $(2, 1)$ and $(-2, 1)$

69. Absolute maximum of 28 at $(0, 4)$; absolute minimum of $-9/4$ at $(3/2, 0)$

70. Absolute maximum of 13 at $(4, 2)$; absolute minimum of 0 at $(1, 0)$

71. Absolute maximum of 18 at $(2, -2)$; absolute minimum of $-17/4$ at $(-2, 1/2)$

72. Absolute maximum of 2 at $(1, 1)$; absolute minimum of 0 at $(0, 0)$, $(0, 2)$, $(2, 2)$ and $(2, 0)$

73. Absolute maximum of 8 at $(-2, 0)$; absolute minimum of -1 at $(1, 0)$

74. Absolute maximum of 18 at $(1, 1)$ and $(-1, -1)$; absolute minimum of -32 at $(2, -2)$

75. Absolute maximum of 4 at $(1, 0)$; absolute minimum of -4 at $(0, -1)$

76. Absolute maximum of 6 at $(1, 1)$; absolute minimum of -2 at $(1, -1)$ and $(-1, 1)$

77. Absolute maximum of 1 at $(0, \pm 1)$ and $(1, 0)$; absolute minimum of -1 at $(-1, 0)$

78. Absolute maximum of $1/2$ at $\left(\dfrac{1}{\sqrt{2}}, \dfrac{1}{\sqrt{2}}\right)$ and $\left(-\dfrac{1}{\sqrt{2}}, -\dfrac{1}{\sqrt{2}}\right)$; absolute minimum of $-1/2$ at $\left(-\dfrac{1}{\sqrt{2}}, \dfrac{1}{\sqrt{2}}\right)$ and $\left(\dfrac{1}{\sqrt{2}}, -\dfrac{1}{\sqrt{2}}\right)$

79. Absolute maximum of 5 at $(0, 1)$; absolute minimum of $-1/3$ at $(0, -1/3)$

80. Absolute maximum of $9 + \dfrac{27\sqrt{3}}{4} \approx 20.69$ at $\left(-\dfrac{3\sqrt{3}}{2}, \dfrac{3}{2}\right)$; absolute minimum of -3 at $(2, 1)$

81. Absolute maximum of $\sqrt{3}$ at $\left(\dfrac{1}{\sqrt{3}}, -\dfrac{1}{\sqrt{3}}, \dfrac{1}{\sqrt{3}}\right)$; absolute minimum of $-\sqrt{3}$ at $\left(-\dfrac{1}{\sqrt{3}}, \dfrac{1}{\sqrt{3}}, -\dfrac{1}{\sqrt{3}}\right)$

82. $(0, 0, -2)$, $(0, 0, 2)$

83. Width $= \left(\dfrac{c^2 V}{ab}\right)^{1/3}$, depth $= \left(\dfrac{b^2 V}{ac}\right)^{1/3}$, height $= \left(\dfrac{a^2 V}{bc}\right)^{1/3}$

84. $x + 2y + z = 6$

85. Absolute maximum of $3/2$ at $\left(\dfrac{1}{\sqrt{2}}, \dfrac{1}{\sqrt{2}}, \sqrt{2}\right)$ and $\left(-\dfrac{1}{\sqrt{2}}, -\dfrac{1}{\sqrt{2}}, -\sqrt{2}\right)$; absolute minimum of $1/2$ at $\left(-\dfrac{1}{\sqrt{2}}, \dfrac{1}{\sqrt{2}}, -\sqrt{2}\right)$ and $\left(\dfrac{1}{\sqrt{2}}, -\dfrac{1}{\sqrt{2}}, \sqrt{2}\right)$

86. $(1/4, 1/4, 1/2)$

87. $\dfrac{\partial w}{\partial x} = (\cos \theta)\dfrac{\partial w}{\partial r} - \left(\dfrac{\sin \theta}{r}\right)\dfrac{\partial w}{\partial \theta}$, $\dfrac{\partial w}{\partial y} = (\sin \theta)\dfrac{\partial w}{\partial r} + \left(\dfrac{\cos \theta}{r}\right)\dfrac{\partial w}{\partial \theta}$

88. $z_x = af_u + af_v$, $z_y = bf_u - bf_v$

89. $w_r = w_s = \dfrac{2}{r + s}$ **90.** Angle is $\pi/2$

91. $(t, -t \pm 4, t)$, t a real number

92. $(-1/2, 1/2, 1/2)$, $(0, 1, 0)$

96. (a)

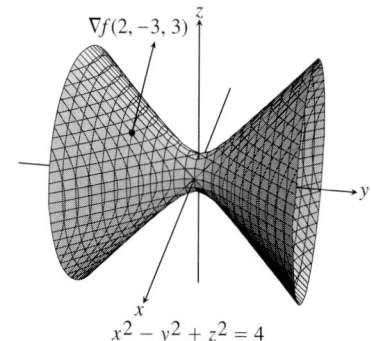

(b) $\nabla f = 4\mathbf{i} + 6\mathbf{j} + 6\mathbf{k}$
(c) $2x + 3y + 3z = 4$; $x = 2 + 4t$, $y = -3 + 6t$, $z = 3 + 6t$

Chapter 14

Section 14.1 (pp. 761–773)

Quick Review

1. $\int_0^3 x^3\,dx$

2. $\int_{-1}^1 \sqrt[3]{2-x^2}\,dx$

3. $e^{\sin x} + C$

4. $-2\ln\left(|x-1|\right) + C$

5. $e^x + \dfrac{5}{x} + C$

6. $\sqrt{2} - 1$

7. 11/2

8. ≈ 1.89

9. 4

10. $512\pi/15 \approx 107.23$

Exercises

1. 16

3. 1

5. $\dfrac{\pi^2}{2} + 2$

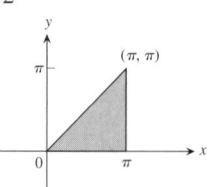

7. $8\ln 8 - 16 + e \approx 3.35$

9. $e - 2$

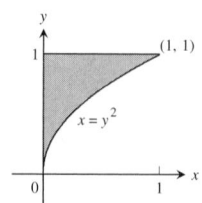

11. $\dfrac{3}{2}\ln 2$

13. 1/6

15. $-1/10$

17. 8

19. 2π

21. $\int_2^4 \int_0^{(4-y)/2} dx\,dy$

23. $\int_0^1 \int_{x^2}^x dy\,dx$

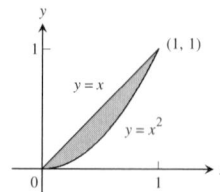

25. $\int_1^e \int_{\ln y}^1 dx\, dy$

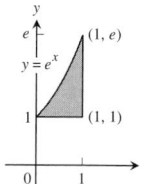

27. $\int_0^9 \int_0^{\frac{1}{2}\sqrt{9-y}} 16x\, dx\, dy$

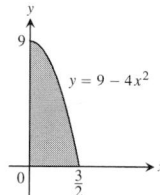

29. $\int_{-1}^1 \int_0^{\sqrt{1-x^2}} 3y\, dy\, dx$

31. 2

33. $\dfrac{e-2}{2}$

35. 2

37. $\dfrac{1}{80\pi}$

39. $-2/3$

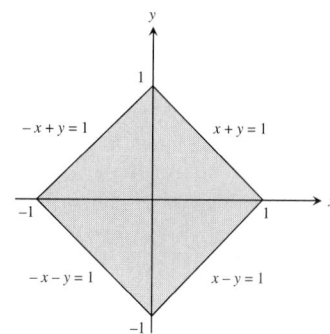

41. $4/3$

43. $625/12$

45. 16

47. 20

49. $2(1 + \ln 2) \approx 3.39$

51. 1

53. π^2

55. $1/16$

57. $20\sqrt{3}/9 \approx 3.85$

59. $\int_0^1 \int_x^{2-x} (x^2 + y^2)\, dy\, dx = 4/3$

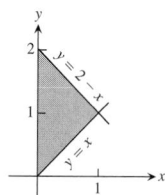

61. $x^2 + 2y^2 \le 4$

63. No

67. ≈ 0.60

69. ≈ 0.23

Section 14.2 (pp. 773–783)

Quick Review

1. $e - 1$ **2.** 1

3. $\int_{-2}^2 (2 - x^2 + 2)\, dx = 32/3$

4. $\int_{-2}^{2} 2\sqrt{2-y}\, dy = 32/3$

5. $2/\pi \approx 0.64$ **6.** 32/3
9. $\bar{x} = 0, \bar{y} = 12/5$ **10.** $\bar{x} = 4/3, \bar{y} = 0$

Exercises

1. $\int_{0}^{2}\int_{0}^{2-x} dy\, dx = \int_{0}^{2}\int_{0}^{2-y} dx\, dy = 2$

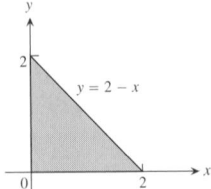

3. $\int_{-2}^{1}\int_{y-2}^{-y^2} dx\, dy = 9/2$

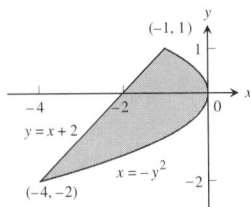

5. $\int_{0}^{\ln 2}\int_{0}^{e^x} dy\, dx = 1$

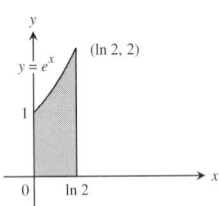

7. $\int_{0}^{1}\int_{y^2}^{2y-y^2} dx\, dy = 1/3$

9. 12

11. $\sqrt{2} - 1$

13. 3/2

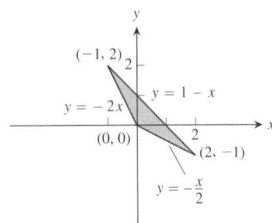

15. (a) 0   **(b)** $4/\pi^2$
17. 8/3
19. $\bar{x} = 5/14, \bar{y} = 38/35$
21. $\bar{x} = 64/35, \bar{y} = 5/7$
23. $\bar{x} = 0, \bar{y} = \dfrac{4}{3\pi}$

25. $\bar{x} = \bar{y} = \dfrac{4a}{3\pi}$

27. $\bar{x} = \pi/2, \bar{y} = \pi/8$
29. $\bar{x} = -1, \bar{y} = 1/4$

31. $I_x = 64/105, R_x = 2\sqrt{\dfrac{2}{7}}$

33. $\bar{x} = 3/8, \bar{y} = 17/16$
35. $\bar{x} = 11/3, \bar{y} = 14/27, I_y = 432, R_y = 4$
37. $\bar{x} = 0, \bar{y} = 13/31, I_y = 7/5, R_y = \sqrt{21/31}$
39. $\bar{x} = 0, \bar{y} = 7/10, I_x = 9/10, R_x = 3\sqrt{6}/10,$
     $I_y = 3/10, R_y = 3\sqrt{2}/10, I_0 = 6/5, R_0 = 3\sqrt{2}/5$
41. $\approx 43{,}329$
43. $0 < a \le 5/2$
45. $\bar{x} = 2/\pi, \bar{y} = 0$
47. (a) 3/2   **(b)** Values are the same
53. (a) $\bar{x} = 7/5, \bar{y} = 31/10$
     (b) $\bar{x} = 19/7, \bar{y} = 18/7$
     (c) $\bar{x} = 9/2, \bar{y} = 19/8$
     (d) $\bar{x} = 11/4, \bar{y} = 43/16$
55. On common boundary: $h = a\sqrt{2}$;
     inside T: $h > a\sqrt{2}$

Section 14.3 (pp. 784–790)

Quick Review

1. $r^2 = 9$ **2.** $r\cos\theta = 3$
3. $r = -4\sin\theta$ **4.** $(x-2)^2 + y^2 = 4$
5. $x = 1$ **6.** $0 < r \le 2, \dfrac{\pi}{2} < \theta < \pi$

7. $0 < r \le 2$, $\pi < \theta < 2\pi$, $\theta \ne \dfrac{3\pi}{2}$

8. 6π

9. 2

10. $5\pi - 8$

Exercises

1. $\displaystyle\int_0^\pi \int_0^1 r\, dr\, d\theta = \dfrac{\pi}{2}$

3. $\displaystyle\int_0^{\pi/2} \int_0^1 r^3\, dr\, d\theta = \dfrac{\pi}{8}$

5. $\displaystyle\int_0^{2\pi} \int_0^a r\, dr\, d\theta = \pi a^2$

7. $\displaystyle\int_{\pi/4}^{\pi/2} \int_0^{6\csc\theta} r^2 \cos\theta\, dr\, d\theta = 36$

9. $\displaystyle\int_\pi^{3\pi/2} \int_0^1 \dfrac{2r}{1+r}\, dr\, d\theta = (1 - \ln 2)\,\pi \approx 0.96$

11. $\displaystyle\int_0^{\pi/2} \int_0^{\ln 2} r e^r\, dr\, d\theta = \dfrac{\pi}{2}(2\ln 2 - 1) \approx 0.61$

13. $\displaystyle\int_0^{\pi/2} \int_0^{2\cos\theta} (\cos\theta + \sin\theta)\, dr\, d\theta = \dfrac{\pi}{2} + 1$

15. $4\displaystyle\int_0^{\pi/2} \int_0^1 \ln(r^2 + 1)\, r\, dr\, d\theta = \pi(\ln 4 - 1) \approx 1.21$

17. $2(\pi - 1)$

19. 12π

21. $\dfrac{3\pi}{8} + 1 \approx 2.18$

23. 4

25. $6\sqrt{3} - 2\pi \approx 4.11$

27. $\bar{x} = 5/6$, $\bar{y} = 0$

29. $2a/3$

31. $2a/3$

33. $2\pi(2 - \sqrt{e}) \approx 2.21$

35. $\dfrac{4}{3} + \dfrac{5\pi}{8} \approx 3.30$

37. (a) $I = \dfrac{\sqrt{\pi}}{2} \approx 0.89$ **(b)** 1

39. $\pi \ln 4$, no

41. $\dfrac{1}{2}(a^2 + 2h^2)$

Section 14.4 (pp. 790–799)

Quick Review

1. 4

2. $13/2$

5. 0

6. 4

7. 10

8. $2r + 2s$

9. $2/\pi$

10. $-4/\pi^2$

Exercises

1. 1

3. $\displaystyle\int_0^1 \int_0^{2-2x} \int_0^{3-3x-3y/2} dz\, dy\, dx = 1,$

$\displaystyle\int_0^2 \int_0^{1-y/2} \int_0^{3-3x-3y/2} dz\, dx\, dy,$

$\displaystyle\int_0^1 \int_0^{3-3x} \int_0^{2-2x-2z/3} dy\, dz\, dx,$

$\displaystyle\int_0^3 \int_0^{1-z/3} \int_0^{2-2x-2z/3} dy\, dx\, dz,$

$\displaystyle\int_0^2 \int_0^{3-3y/2} \int_0^{1-y/2-z/3} dx\, dz\, dy,$

$\displaystyle\int_0^3 \int_0^{2-2z/3} \int_0^{1-y/2-z/3} dx\, dy\, dz$

5. $\displaystyle\int_{-2}^2 \int_{-\sqrt{4-x^2}}^{\sqrt{4-x^2}} \int_{x^2+y^2}^{8-x^2-y^2} dz\, dy\, dx = 16\pi,$

$\displaystyle\int_{-2}^2 \int_{-\sqrt{4-y^2}}^{\sqrt{4-y^2}} \int_{x^2+y^2}^{8-x^2-y^2} dz\, dx\, dy,$

$\displaystyle\int_{-2}^2 \int_{y^2}^4 \int_{-\sqrt{z-y^2}}^{\sqrt{z-y^2}} dx\, dz\, dy + \int_{-2}^2 \int_4^{8-y^2} \int_{-\sqrt{8-z-y^2}}^{\sqrt{8-z-y^2}} dx\, dz\, dy,$

$\displaystyle\int_0^4 \int_{-\sqrt{z}}^{\sqrt{z}} \int_{-\sqrt{z-y^2}}^{\sqrt{z-y^2}} dx\, dy\, dz + \int_4^8 \int_{-\sqrt{8-z}}^{\sqrt{8-z}} \int_{-\sqrt{8-z-y^2}}^{\sqrt{8-z-y^2}} dx\, dy\, dz,$

$\displaystyle\int_{-2}^2 \int_{x^2}^4 \int_{-\sqrt{z-x^2}}^{\sqrt{z-x^2}} dy\, dz\, dx + \int_{-2}^2 \int_4^{8-x^2} \int_{-\sqrt{8-z-x^2}}^{\sqrt{8-z-x^2}} dy\, dz\, dx,$

$\displaystyle\int_0^4 \int_{-\sqrt{z}}^{\sqrt{z}} \int_{-\sqrt{z-x^2}}^{\sqrt{z-x^2}} dy\, dx\, dz + \int_4^8 \int_{-\sqrt{8-z}}^{\sqrt{8-z}} \int_{-\sqrt{8-z-x^2}}^{\sqrt{8-z-x^2}} dy\, dx\, dz$

7. 1

9. 1

11. $\dfrac{\pi^3}{2}(1 - \cos 1) \approx 7.13$

13. 18

15. $7/6$

17. 0

19. $1/2 - \pi/8$

21. (a) $\displaystyle\int_{-1}^1 \int_0^{1-x^2} \int_{x^2}^{1-z} dy\, dz\, dx$

(b) $\displaystyle\int_0^1 \int_{-\sqrt{1-z}}^{\sqrt{1-z}} \int_{x^2}^{1-z} dy\, dx\, dz$

(c) $\displaystyle\int_0^1 \int_0^{1-z} \int_{-\sqrt{y}}^{\sqrt{y}} dx\, dy\, dz$

(d) $\displaystyle\int_0^1 \int_0^{1-y} \int_{-\sqrt{y}}^{\sqrt{y}} dx\, dz\, dy$

(e) $\displaystyle\int_0^1 \int_{-\sqrt{y}}^{\sqrt{y}} \int_0^{1-y} dz\, dx\, dy$

23. 2/3
25. 20/3
27. 1
29. 16/3
31. $8\pi - \dfrac{32}{2} \approx 14.47$
33. 2
35. 4π
37. 31/3
39. 1
41. $2 \sin 4 \approx -1.51$
43. 4
45. 13/3 or 3
47. $4x^2 + 4y^2 + z^2 \leq 4$
49. $\pi/48 \approx 0.07$
51. ≈ 1.84

Section 14.5 (pp. 799–804)

Quick Review

1. 13/3

2. 158/15

3. 3/2

4. $\bar{x} = \dfrac{9}{26}, \bar{y} = \dfrac{158}{65}$

5. 6091/210
7. 1042/35
9. $\sqrt{23/130}$

6. 23/30
8. $\sqrt{6091/910}$
10. $\sqrt{3126/455}$

Exercises

1. $R_x = \sqrt{\dfrac{b^2 + c^2}{12}}, R_y = \sqrt{\dfrac{a^2 + c^2}{12}},$
$R_z = \sqrt{\dfrac{a^2 + b^2}{12}}$

3. $I_x = \dfrac{abc(b^2 + c^2)}{3}, I_y = \dfrac{abc(a^2 + c^2)}{3},$
$I_z = \dfrac{abc(a^2 + b^2)}{3}$

5. $\bar{x} = \bar{y} = 0, \bar{z} = 12/5, I_x = 7904/105,$
$I_y = 4832/63, I_z = 256/45$

7. (a) $\bar{x} = \bar{y} = 0, \bar{z} = 8/3$ **(b)** $2\sqrt{2}$

9. $I_L = 1386, R_L = \sqrt{77}/2$

11. $I_L = 40/3, R_L = \sqrt{5/3}$

13. (a) 4/3
(b) $\bar{x} = 4/5, \bar{y} = \bar{z} = 2/5$

15. (a) 5/2
(b) $\bar{x} = \bar{y} = \bar{z} = 8/15$
(c) $I_x = I_y = I_z = 11/6$
(d) $R_x = R_y = R_z = \sqrt{11/15}$

17. 3

19. (a) $\dfrac{4}{3} g$

(b) $\dfrac{4}{3} g$

23. (a) $I_{\text{c.m.}} = \dfrac{abc(a^2 + b^2)}{12}, R_{\text{c.m.}} = \sqrt{\dfrac{a^2 + b^2}{12}}$

(b) $I_L = \dfrac{abc(a^2 + 7b^2)}{3}, R_L = \sqrt{\dfrac{a^2 + 7b^2}{3}}$

27. (a) $h = a\sqrt{3}$ **(b)** $h = a\sqrt{2}$

Section 14.6 (pp. 805–816)

Quick Review

1. $z = r, 1 \leq r \leq 2$ **2.** $z^2 - x^2 - y^2 = 1$
3. Right circular cylinder parallel to the z-axis generated by the circle $r = 2 \sin \theta$ in the $r\theta$-plane
4. Cylinder of lines parallel to the z-axis generated by the cardioid $r = 1 + \cos \theta$
5. $z = x^2$ **6.** $z = y^2$
7. The origin (0, 0) **8.** $z = |x|$
9. $z = |y|$
10. Circle with radius 2 and center (0, 0, 2) in the plane $z = 2$

Exercises

1. (0, 0, 0)
3. (1, 0, 0)
5. $(0, -2\sqrt{2}, 0)$
7. Rectangular: $x^2 + y^2 = 0$; spherical: $\phi = 0$ or $\phi = \pi$; the z-axis
9. Cylindrical or rectangular: $z = 0$; spherical: $\phi = \pi/2$; the xy-plane
11. Cylindrical: $z = r, 0 \leq r \leq 1$; spherical: $\phi = \pi/4$, $0 \leq \rho \leq \sqrt{2}$; a (finite) cone
13. Rectangular: $x = 0$; cylindrical: $\theta = \pi/2$; the yz-plane
15. Rectangular: $x^2 + y^2 + (z - 5/2)^2 = 25/4$; cylindrical: $r^2 + z^2 = 5z$; a sphere of radius 5/2 centered at (0, 0, 5/2) (rectangular)
17. Cylindrical: $r^2 + z^2 = 2z, z \leq 1$; spherical: $\rho = 2 \cos \phi, \pi/4 \leq \phi \leq \pi/2$; the lower half (hemisphere) of the sphere of radius 1 centered at (0, 0, 1) (rectangular)
19. Rectangular: $z = -\sqrt{x^2 + y^2}, -1 \leq z \leq 0$; cylindrical: $r = -z$; a cone with vertex at the origin and base the circle $x^2 + y^2 = 1$ in the plane $z = -1$
21. Cardioid of revolution symmetric about the z-axis, cusp at the origin pointing down

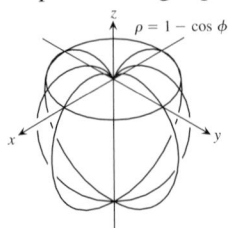

23. $\rho \sin \theta = a$

25. $\dfrac{4\pi(\sqrt{2} - 1)}{3} \approx 1.74$

27. $\pi(6\sqrt{2} - 8) \approx 1.52$

29. $3\pi/10$

31. $\pi/3$

33. (a) $\displaystyle\int_{0}^{2\pi}\int_{0}^{1}\int_{0}^{\sqrt{4-r^2}} dz\, r\, dr\, d\theta$

(b) $\displaystyle\int_{0}^{2\pi}\int_{0}^{\sqrt{3}}\int_{0}^{1} r\, dr\, dz\, d\theta$

$+ \displaystyle\int_{0}^{2\pi}\int_{\sqrt{3}}^{2}\int_{0}^{\sqrt{4-z^2}} r\, dr\, dz\, d\theta$

(c) $\displaystyle\int_{0}^{1}\int_{0}^{\sqrt{4-r^2}}\int_{0}^{2\pi} r\, d\theta\, dz\, dr$

35. $\displaystyle\int_{-\pi/2}^{\pi/2}\int_{0}^{\cos\theta}\int_{0}^{3r^2} f(r,\theta,z)\, dz\, r\, dr\, d\theta$

37. $\displaystyle\int_{0}^{\pi}\int_{0}^{2\sin\theta}\int_{0}^{4-r\sin\theta} f(r,\theta,z)\, dz\, r\, dr\, d\theta$

39. $\displaystyle\int_{-\pi/2}^{\pi/2}\int_{1}^{1+\cos\theta}\int_{0}^{4} f(r,\theta,z)\, dz\, r\, dr\, d\theta$

41. $\displaystyle\int_{0}^{\pi/4}\int_{0}^{\sec\theta}\int_{0}^{2-r\sin\theta} f(r,\theta,z)\, dz\, r\, dr\, d\theta$

43. π^2

45. 5π

47. 2π

49. $\dfrac{(4\sqrt{2} - 5)\pi}{\sqrt{2}} \approx 1.46$

51. (a) $\displaystyle\int_{0}^{2\pi}\int_{0}^{\pi/6}\int_{0}^{2} \rho^2 \sin\phi\, d\rho\, d\phi\, d\theta$

$+ \displaystyle\int_{0}^{2\pi}\int_{\pi/6}^{\pi/2}\int_{0}^{\csc\phi} \rho^2 \sin\phi\, d\rho\, d\phi\, d\theta$

(b) $\displaystyle\int_{0}^{2\pi}\int_{1}^{2}\int_{\pi/6}^{\sin^{-1}(1/\rho)} \rho^2 \sin\phi\, d\phi\, d\rho\, d\theta$

$+ \displaystyle\int_{0}^{2\pi}\int_{0}^{2}\int_{0}^{\pi/6} \rho^2 \sin\phi\, d\phi\, d\rho\, d\theta$

53. (a) $\displaystyle\int_{0}^{2\pi}\int_{0}^{\pi/2}\int_{\cos\phi}^{2} \rho^2 \sin\phi\, d\rho\, d\phi\, d\theta$

(b) $31\pi/6 \approx 16.23$

55. (a) $\displaystyle\int_{0}^{2\pi}\int_{0}^{\pi}\int_{0}^{1-\cos\phi} \rho^2 \sin\phi\, d\rho\, d\phi\, d\theta$

(b) $8\pi/3 \approx 8.38$

57. (a) $\displaystyle\int_{0}^{2\pi}\int_{\pi/4}^{\pi/2}\int_{0}^{2\cos\phi} \rho^2 \sin\phi\, d\rho\, d\phi\, d\theta$

(b) $\pi/3$

59. (a) $8\displaystyle\int_{0}^{\pi/2}\int_{0}^{\pi/2}\int_{0}^{2} \rho^2 \sin\phi\, d\rho\, d\phi\, d\theta$

(b) $8\displaystyle\int_{0}^{\pi/2}\int_{0}^{2}\int_{0}^{\sqrt{4-r^2}} dz\, r\, dr\, d\theta$

(c) $8\displaystyle\int_{0}^{2}\int_{0}^{\sqrt{4-x^2}}\int_{0}^{\sqrt{4-x^2-y^2}} dz\, dy\, dx$

61. (a) $\displaystyle\int_{0}^{\pi/2}\int_{0}^{\pi/3}\int_{\sec\phi}^{2} \rho^2 \sin\phi\, d\rho\, d\phi\, d\theta$

(b) $\displaystyle\int_{0}^{2\pi}\int_{0}^{\sqrt{3}}\int_{1}^{\sqrt{4-r^2}} dz\, r\, dr\, d\theta$

(c) $\displaystyle\int_{-\sqrt{3}}^{\sqrt{3}}\int_{-\sqrt{3-x^2}}^{\sqrt{3-x^2}}\int_{1}^{\sqrt{4-x^2-y^2}} dz\, dy\, dx$

(d) $5\pi/3$

63. $8\pi/3$

65. $9/4$

67. $\dfrac{3\pi - 4}{18} \approx 0.30$

69. $2\pi a^3/3$

71. $7\pi/3$

73. $\dfrac{4\pi(2\sqrt{2} - 1)}{3} \approx 7.66$

75. 16π

77. $2/3$

79. $\bar{x} = \bar{y} = 0, \bar{z} = 3/8$

81. $\bar{x} = \bar{y} = 0, \bar{z} = 3/8$

83. $\bar{x} = \bar{y} = 0, \bar{z} = 5/6$

85. $I_z = 8\pi a^5/15$

87. $4\pi\mu_0\left(\dfrac{R^2}{c} + \dfrac{2R}{c^2} + \dfrac{2}{c^3}\right)$

Section 14.7 (pp. 816–825)

Quick Review

1. 17 $\quad\quad$ **2.** 9

3. $x = 2, y = -1$ $\quad$ **4.** $x = -2, y = 1, z = 3$

5. $x = \dfrac{u + 2v}{5}, y = \dfrac{u - 3v}{5}$

6. $x = u + v, y = 2v$

7. $x = \dfrac{u - v}{3}, y = \dfrac{2u + v}{3}$

8. $x = \dfrac{u + 4v - w - 3}{10}, y = \dfrac{-u + 2v - w - 3}{2},$

$z = \dfrac{-3u + 8v - 7w - 11}{10}$

9. $x = u + v, y = 2v, z = 3w$

10. $\dfrac{\partial x}{\partial u} = \cos(vw), \dfrac{\partial x}{\partial v} = -uw\sin(vw),$

$\dfrac{\partial x}{\partial w} = -uv\sin(vw)$

Exercises

1. (a) $x = \frac{1}{3}(u + v)$, $y = \frac{1}{3}(-2u + v)$, $\frac{\partial(x, y)}{\partial(u, v)} = \frac{1}{3}$

(b)

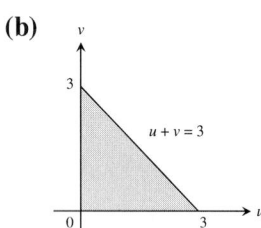

3. (a) $x = \frac{1}{5}(2u - v)$, $y = \frac{1}{10}(-u + 3v)$,

$\frac{\partial(x, y)}{\partial(u, v)} = \frac{1}{10}$

(b)

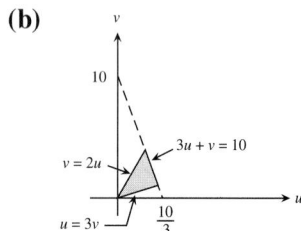

5. (a) u **(b)** $-u$

9. 64/5

11. $\int_1^3 \int_1^2 (v + u)\left(\frac{2u}{v}\right) dv\, du = 8 + \frac{52}{3} \ln 2 \approx 20.01$

13. $\frac{ab\pi(a^2 + b^2)}{4}$

15. $\int_0^2 \int_0^u ue^{-v}\left|-\frac{1}{3}\right| dv\, du = \frac{1}{3}(3e^{-2} + 1) \approx 0.47$

19. $4\pi abc/3$

21. $2 + \ln 8 \approx 4.08$

Chapter 14 Review Exercises
(pp. 825–827)

1. $9e - 9 \approx 15.46$

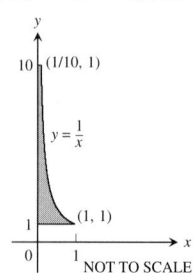

2. $\frac{e - 2}{2} \approx 0.36$

3. 9/2

4. 1/5

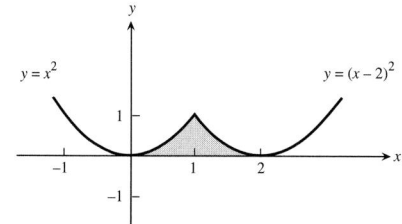

5. $\int_{-2}^0 \int_{2x+4}^{4-x^2} dy\, dx = 4/3$

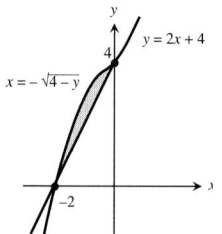

6. $\int_0^1 \int_y^{\sqrt{y}} dx\, dy = 4/35$

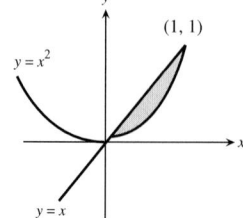

7. $\displaystyle\int_{-3}^{3}\int_{0}^{(1/2)\sqrt{9-x^2}} y \, dy \, dx = 9/2$

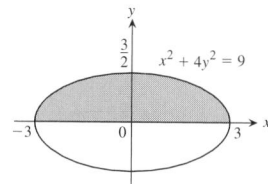

$x^2 + 4y^2 = 9$

8. $\displaystyle\int_{0}^{4}\int_{0}^{\sqrt{4-y}} 2x \, dx \, dy = 8$

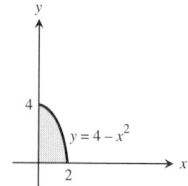

$y = 4 - x^2$

9. $\sin 4$ **10.** $e - 1$

11. $\dfrac{\ln 17}{4} \approx 0.71$ **12.** 2

13. $4/3$ **14.** $37/6$

15. $4/3$ **16.** $125/4$

17. $1/4$ **18.** $\dfrac{1}{2\pi}$

19. $\bar{x} = \bar{y} = \dfrac{1}{2 - \ln 4} \approx 1.63$

20. $\bar{x} = -12/5, \bar{y} = 2$ **21.** 104

22. (a) $40/3$ **(b)** $\dfrac{4ab(b^2 + a^2)}{3}$

23. $I_x = 2\delta, R_x = \sqrt{2/3}$

24. $\bar{x} = 8/15, \bar{y} = 13/30; I_x = 17/280; R_x = \sqrt{17/70};$
$I_y = 1/12; R_y = \sqrt{1/3}$

25. $M = 4; M_x = 0; M_y = 0$

26. $I_x = \delta bh^3/12; R_x = h/\sqrt{6}$

27. π **28.** $(\ln 4 - 1)\pi \approx 1.21$

29. $\bar{x} = 3\sqrt{3}/\pi, \bar{y} = 0$ **30.** $\bar{x} = \bar{y} = \dfrac{13}{3\pi}$

31. (a) $\bar{x} = \dfrac{15\pi + 32}{6\pi + 48} \approx 1.18, \bar{y} = 0$

(b)

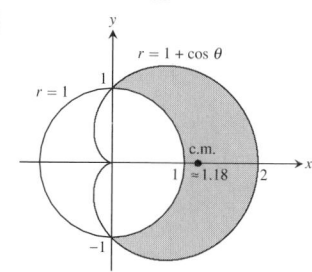

$r = 1 + \cos\theta$
$r = 1$
c.m.
≈ 1.18

32. (a) $\bar{x} = \dfrac{2a\sin\alpha}{3\alpha}, \bar{y} = 0; \lim_{\alpha \to \pi^-} \bar{x} = 0$

(b) $\bar{x} = \dfrac{2a}{5\pi}, \bar{y} = 0$

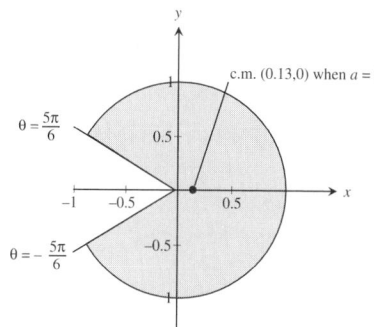

c.m. $(0.13,0)$ when $a = 1$
$\theta = \dfrac{5\pi}{6}$
$\theta = -\dfrac{5\pi}{6}$

33. $\dfrac{\pi - 2}{4}$

34. (a) $\dfrac{\sqrt{2}}{4} \tan^{-1}\sqrt{\dfrac{3}{2}} \approx 0.31$

(b) $\pi/4$

35. 0 **36.** 1

37. $8/35$ **38.** 1

39. $\pi/2$ **40.** 12π

41. $\dfrac{1}{3}\left[2(31 - 3^{5/2})\right] \approx 10.27$

42. $3a/4$

43. Cylindrical: $r^2 + z^2 = -2z$; spherical: $\rho = -2\cos\phi$; a sphere of radius 1 centered at $(0, 0, -1)$ (rectangular)

44. Cylindrical: $z = 1 - r^2$; rectangular: $z = 1 - x^2 - y^2$; a circular paraboloid opening downward from the point $(0, 0, 1)$ (rectangular), axis along the z-axis

45. Rectangular: $(x - 2)^2 + y^2 = 4$; spherical: $\rho \sin\phi = 4\cos\theta$; a circular cylinder parallel to the z-axis generated by the circle $(x - 2)^2 + y^2 = 4$ in the xy-plane

46. Cylindrical: $z = -r, r \geq 0$; rectangular: $z = -\sqrt{x^2 + y^2}$; the lower nappe of a cone making an angle of $3\pi/4$ with the positive z-axis and having vertex at the origin

47. (a) $\displaystyle\int_{-\sqrt{2}}^{\sqrt{2}}\int_{-\sqrt{2-y^2}}^{\sqrt{2-y^2}}\int_{\sqrt{x^2+y^2}}^{\sqrt{4-x^2-y^2}} 3 \, dz \, dx \, dy$

(b) $\displaystyle\int_{0}^{2\pi}\int_{0}^{\pi/4}\int_{0}^{2} 3\rho^2 \sin\phi \, d\rho \, d\phi \, d\theta$

(c) $2\pi(8 - 4\sqrt{2}) \approx 14.72$

48. (a) $\displaystyle\int_{-\pi/2}^{\pi/2}\int_{0}^{1}\int_{-r^2}^{r^2} 21r^3 \cos\theta \sin^2\theta \, dz \, r \, dr \, d\theta$

(b) 4

49. (a) $\displaystyle\int_{0}^{2\pi}\int_{0}^{\pi/4}\int_{0}^{\sec\phi} \rho^2 \sin\phi \, d\rho \, d\phi \, d\theta$

(b) $\pi/3$

50. (a) $\int_0^1 \int_0^{\sqrt{1-x^2}} \int_0^{\sqrt{x^2+y^2}} (6 + 4y) \, dz \, dy \, dx$

(b) $\int_0^{\pi/2} \int_0^1 \int_0^r (6 + 4r \sin \theta) \, dz \, r \, dr \, d\theta$

(c) $\int_0^{\pi/2} \int_{\pi/4}^{\pi/2} \int_0^{\csc \phi} (6 + 4 \rho \sin \phi \sin \theta) \cdot$
$(\rho^2 \sin \phi) \, d\rho \, d\phi \, d\theta$

(d) $\pi + 1$

51. $\int_0^1 \int_{\sqrt{1-x^2}}^{\sqrt{3-x^2}} \int_1^{\sqrt{4-x^2-y^2}} z^2 yx \, dz \, dy \, dx$
$+ \int_1^{\sqrt{3}} \int_0^{\sqrt{3-x^2}} \int_1^{\sqrt{4-x^2-y^2}} z^2 yx \, dz \, dy \, dx$

52. (a) Bounded on the top and bottom by the sphere $x^2 + y^2 + z^2 = 4$, on the right by the right circular cylinder $(x - 1)^2 + y^2 = 1$, on the left by the plane $y = 0$

(b) $\int_0^{\pi/2} \int_0^{2 \cos \theta} \int_{-\sqrt{4-r^2}}^{\sqrt{4-r^2}} dz \, r \, dr \, d\theta$

53. (a) $\int_{-\sqrt{3}}^{\sqrt{3}} \int_{-\sqrt{3-x^2}}^{\sqrt{3-x^2}} \int_1^{\sqrt{4-x^2-y^2}} dz \, dy \, dx$

(b) $\int_0^{2\pi} \int_0^{\sqrt{3}} \int_1^{\sqrt{4-r^2}} dz \, r \, dr \, d\theta$

(c) $\int_0^{2\pi} \int_0^{\pi/3} \int_{\sec \phi}^2 \rho^2 \sin \phi \, d\rho d\phi d \theta$

54. (a) $\int_{-1}^1 \int_{-\sqrt{1-x^2}}^{\sqrt{1-x^2}} \int_0^{\sqrt{1-x^2-y^2}} (x^2 + y^2) \, dz \, dy \, dx$

(b) $\int_0^{2\pi} \int_0^1 \int_0^{\sqrt{1-r^2}} r^3 \, dz \, dr \, d\theta$

(c) $\int_0^{2\pi} \int_0^{\pi/2} \int_0^1 \rho^4 \sin^3 \phi \, d\rho d\phi d \theta$

55. (a) and (b) $\dfrac{8\pi(4\sqrt{2} - 5)}{3} \approx 5.50$

56. $8\pi/3$

57. $\dfrac{8\pi\delta(b^5 - a^5)}{15}$

58. $64\pi/35$

60. $ac - b^2 = \pi^2$

Chapter 15

Section 15.1 (pp. 829–835)

Quick Review

1. $\dfrac{1}{3} \sin 3x + C$ **2.** $-\dfrac{1}{2} \cos 2x + C$

3. $-\dfrac{x^4}{4} + \dfrac{x^3}{3} + 2x + C$ **4.** 21/4

5. 0

6. $x = 2t, y = 3t, 0 \le t \le 1$

7. $x = at, y = bt, 0 \le t \le 1$

8. $x = -2 + 5t, y = 1 - 3t, 0 \le t \le 1$

9. $x = 2t, y = 3t, z = 4t, 0 \le t \le 1$

10. $x = -1 + 4t, y = 1 + 2t, z = 2 - 4t, 0 \le t \le 1$

Exercises

1. c

3. g

5. d

7. f

9. $\sqrt{2}$

11. 13/2

13. $3\sqrt{14}$

15. $\dfrac{5}{3}\sqrt{5} + \dfrac{4}{3} \approx 5.06$

17. $\sqrt{3} \ln\left(\dfrac{b}{a}\right)$

19. $\dfrac{10\sqrt{5} - 2}{3} \approx 6.79$

21. 8

23. $2\sqrt{2} - 1 \approx 1.83$

25. (a) $4\sqrt{2} - 2 \approx 3.66$
(b) $\sqrt{2} + \ln (1 + \sqrt{2}) \approx 2.30$

27. $I_z = 2\pi\delta a^3, R_z = a$

29. (a) $I_z = 2\pi\delta\sqrt{2}, R_z = 1$
(b) $I_z = 4\pi\delta\sqrt{2}, R_z = 1$

31. $I_x = 2\pi - 2, R_x = 1$

33. 326/3

35. $-\dfrac{25\pi^3\sqrt{29}}{8} \approx -521.79$

Section 15.2 (pp. 835–845)

Quick Review

1. $\dfrac{\partial f}{\partial x} = \sin y - y \sin x, \dfrac{\partial f}{\partial y} = x \cos y + \cos x$

2. $\dfrac{\partial f}{\partial x} = \sin (yz) + yz, \dfrac{\partial f}{\partial y} = xz \cos (yz) + xz,$
$\dfrac{\partial f}{\partial z} = xy \cos (yz) + xy$

3. $(3x^2 + 2 \sin x) \, dx$ **4.** $(\sin x + x \cos x) \, dx$

5. $dy = 3 \cos t, dt, dx = -2 \sin t \, dt$

6. $t^2 \sin t$

7. Counterclockwise

8. Clockwise

9. Counterclockwise

10. Clockwise

Exercises

1. $\nabla f = \dfrac{-x\mathbf{i} - y\mathbf{j} - z\mathbf{k}}{(x^2 + y^2 + z^2)^{3/2}}$

3. $\nabla g = -\left(\dfrac{2x}{x^2 + y^2}\right)\mathbf{i} - \left(\dfrac{2y}{x^2 + y^2}\right)\mathbf{j} + e^z\mathbf{k}$

5. $\mathbf{F} = -\dfrac{kx}{(x^2 + y^2)^{3/2}}\mathbf{i} - \dfrac{ky}{(x^2 + y^2)^{3/2}}\mathbf{j},\ k > 0$

7. (a) 9/2   **(b)** 13/3
  (c) 9/2

9. (a) 1/3   **(b)** −1/5
  (c) 0

11. (a) 2   **(b)** 3/2
  (c) 1/2

13. 1/2

15. $-\pi$

17. 69/4

19. $-39/2$

21. 25/6

23. (a) $\mathrm{Circ}_1 = 0,\ \mathrm{Circ}_2 = 2\pi,\ \mathrm{Flux}_1 = 2\pi,\ \mathrm{Flux}_2 = 0$
  (b) $\mathrm{Circ}_1 = 0,\ \mathrm{Circ}_2 = 8\pi,\ \mathrm{Flux}_1 = 8\pi,\ \mathrm{Flux}_2 = 0$

25. $\mathrm{Circ} = 0,\ \mathrm{Flux} = a^2\pi$

27. $\mathrm{Circ} = a^2\pi,\ \mathrm{Flux} = 0$

29. (a) $-\pi/2$ **(b)** 0
  (c) 1

31.

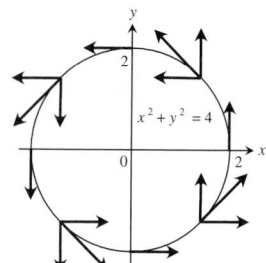

33. (a) $\mathbf{G} = -y\mathbf{i} + x\mathbf{j}$ **(b)** $\mathbf{G} = \left(\sqrt{x^2 + y^2}\right)\mathbf{F}$

35. $\mathbf{F} = -\dfrac{x\mathbf{i} + y\mathbf{j}}{\sqrt{x^2 + y^2}}$

37. 48

39. π

41. 0   **43.** 1/2

45. Yes, work and area are equal.

47. 12π

49. -6π

51. $\pi/2$

Section 15.3 (pp. 845–855)

Quick Review

1. 1   **2.** 2

3. $-\dfrac{yz}{\sqrt{1 - x^2y^2z^2}}$ **4.** $-\dfrac{xz}{\sqrt{1 - x^2y^2z^2}}$

5. $-\dfrac{xy}{\sqrt{1 - x^2y^2z^2}}$

6. $-\dfrac{z}{\sqrt{1 - x^2y^2z^2}} - \dfrac{x^2y^2z^3}{(1 - x^2y^2z^2)^{3/2}}$

7. $-\dfrac{y}{\sqrt{1 - x^2y^2z^2}} - \dfrac{x^2y^3z^2}{(1 - x^2y^2z^2)^{3/2}}$

8. $g(x, y, z) = h(y, z),\ h(y, z)$ and function of y and z

9. $g(x, y, z) = y^2 + h(x, z),\ h(x, z)$ any function of x and z

10. $g(x, y, z) = \dfrac{z^2}{2} + h(x, y),\ h(x, y)$ any function of x and y

Exercises

1. Conservative

3. Not conservative

5. Not conservative

7. $x^2 + \dfrac{3}{2}y^2 + 2z^2 + C$

9. $xe^{y+2z} + C$

11. $\dfrac{1}{2}\ln(y^2 + z^2) + x\ln x - x + \tan(x + y) + C$

13. 49

15. -16

17. 1

19. $9\ln 2$

21. 0

27. $\mathbf{F} = \nabla\left(\dfrac{x^2 - 1}{y}\right)$

29. (a) 1   **(b)** 1
  (c) 1

31. (a) 2

33. $\dfrac{GmM}{(x^2 + y^2 + z^2)^{1/2}}$

35. (a) $b = c = 2a$ **(b)** $a = 1,\ b = c = 2$

37. The path will not matter.

39. (b) $\dfrac{16}{3}g$

Section 15.4 (pp. 855–867)

Quick Review

1. $x = t,\ y = 0,\ 0 \le t \le 1$

2. $x = 1,\ y = t,\ 0 \le t \le 1$

3. $x = 1 - t,\ y = 1,\ 0 \le t \le 1$

4. $x = 0,\ y = 1 - t,\ 0 \le t \le 1$

5. 0   **6.** 1/2

7. 1   **8.** 0

9. $\dfrac{2xy}{(x^2 + y^2)^2}$ **10.** $\dfrac{y^2 - x^2}{(x^2 + y^2)^2}$

Exercises

5. Circ $= 0$, Flux $= 2$

7. Circ $= 9$, Flux $= -9$

9. Circ $= \dfrac{1}{2}$, Flux $= \dfrac{1}{2}$

11. Circ $= -\dfrac{1}{12}$, Flux $= \dfrac{1}{5}$

13. 0

15. 2/33

17. 0

19. -16π

21. (b) πa^2

23. $3\pi/8$

25. (a) 0

(b) $(h - k)$ (area of the region)

27. Its zero for any simple closed curve C.

31. (b) $-\dfrac{1}{2}\oint_C y^2\, dx = \oint_C xy\, dy$

$\qquad = \dfrac{1}{3}\oint xy\, dy - y^2\, dx = A\bar{y}$

35. (a) 0

(b) 0 when $(0, 0)$ is in the region; 4π when $(0, 0)$ is not in the region

39. Its zero.

Section 15.5 (pp. 867–877)

Quick Review

1. $-\dfrac{\sqrt{2}}{2}$ **2.** $-\dfrac{\sqrt{2}}{2}$

3. $\dfrac{\sqrt{2}}{2}\mathbf{i} + \dfrac{\sqrt{2}}{2}\mathbf{j} + \mathbf{k}$ **4.** $x + y + \sqrt{2}z = 4$

5. $x = 1 + \dfrac{\sqrt{2}}{2}t$, $y = 1 + \dfrac{\sqrt{2}}{2}t$, $z = \sqrt{2} + t$

6. $\dfrac{(17\sqrt{17} - 1)\pi}{6} \approx 36.18$

7. $2(\sqrt{2} - 1)\pi \approx 2.60$ **8.** π

9. $(e^2 - 1)^2 \approx 40.82$ **10.** ≈ 0.66

Exercises

1. $13\pi/3$

3. 4

5. $6\sqrt{6} - 2\sqrt{2} \approx 11.87$

7. $\pi\sqrt{c^2 + 1}$

9. $\dfrac{\pi}{6}(17\sqrt{17} - 5\sqrt{5}) \approx 30.85$

11. $3 + 2\ln 2 \approx 4.39$

13. $9a^3$

15. $\dfrac{abc(ab + ac + bc)}{4}$

17. 2

19. 18

21. $\pi a^3/6$

23. $\pi a^2/4$

25. $\pi a^3/2$

27. -32

29. -4

31. $3a^4$

33. $\bar{x} = \bar{y} = \bar{z} = a/2$

35. $\bar{x} = \bar{y} = 0$, $\bar{z} = 14/9$, $I_z = 15\pi\delta\sqrt{2}/2$, $R_z = \sqrt{10}/2$

37. (a) $8\pi a^4\delta/3$ **(b)** $20\pi a^4\delta/3$

39. $\dfrac{\pi}{6}(13\sqrt{13} - 1) \approx 24.02$

41. $5\pi\sqrt{2}$

43. $\dfrac{2}{3}(5\sqrt{5} - 1) \approx 6.79$

Section 15.6 (pp. 878–886)

Quick Review

1. $(\sqrt{2}, \pi/4, 1)$

2. $\left(\sqrt{3}, \dfrac{\pi}{2} - \tan^{-1}\left(\dfrac{1}{\sqrt{2}}\right), \dfrac{\pi}{4}\right)$ and $\dfrac{\pi}{2} - \tan^{-1}\left(\dfrac{1}{\sqrt{2}}\right) \approx 54.74°$

3. Right circular cylinder parallel to the z-axis generated by the circle $x^2 + (y + 2)^2 = 4$ in the xy-plane

4. (a) $r = -4\sin\theta$ **(b)** $\rho\sin\phi + 4\sin\theta = 0$

5. The sphere with center at the origin and radius 3

6. (a) $r^2 + z^2 = 9$ **(b)** $\rho = 3$

7. The plane $y = 2$

8. (a) $y = 2$ **(b)** $\rho\sin\phi\sin\theta = 2$

9. $2\mathbf{i} + 2\mathbf{j} + 2\mathbf{k}$ **10.** $-4\mathbf{i} - 4\mathbf{j} - 4\mathbf{k}$

Exercises

1. $\mathbf{r}(r, \theta) = (r\cos\theta)\mathbf{i} + (r\sin\theta)\mathbf{j} + r^2\,\mathbf{k}$, $0 \le r \le 2$, $0 \le \theta \le 2\pi$

3. $\mathbf{r}(r, \theta) = (r\cos\theta)\mathbf{i} + (r\sin\theta)\mathbf{j} + \left(\dfrac{r}{2}\right)\mathbf{k}$, $0 \le r \le 6$, $0 \le \theta \le \pi/2$

5. $\mathbf{r}(r, \theta) = (r\cos\theta)\mathbf{i} + (r\sin\theta)\mathbf{j} + \sqrt{9 - r^2}\mathbf{k}$, $0 \le r \le 3/\sqrt{2}$, $0 \le \theta \le 2\pi$

7. $\mathbf{r}(\phi, \theta) = (\sqrt{3}\sin\phi\cos\theta)\mathbf{i} + (\sqrt{3}\sin\phi\sin\theta)\mathbf{j} + (\sqrt{3}\cos\phi)\mathbf{k}$, $\dfrac{\pi}{3} \le \phi \le \dfrac{2\pi}{3}$, $0 \le \theta \le 2\pi$

9. $\mathbf{r}(x, y) = x\mathbf{i} + y\mathbf{j} + (4 - y^2)\mathbf{k}$, $-2 \le y \le 2$, $0 \le x \le 2$

11. $\mathbf{r}(u, v) = u\mathbf{i} + (3\cos v)\mathbf{j} + (3\sin v)\mathbf{k}$, $0 \le u \le 3$, $0 \le v \le 2\pi$

13. (a) $\mathbf{r}(r, \theta) = (r\cos\theta)\mathbf{i} + (r\sin\theta)\mathbf{j} + (1 - r\cos\theta - r\sin\theta)\mathbf{k}$, $0 \le \theta \le 2\pi$, $0 \le r \le 3$

(b) $\mathbf{r}(u, v) = (1 - u \cos v - u \sin v)\mathbf{i} +$
$(u \cos v)\mathbf{j} + (u \sin v)\mathbf{k}, 0 \leq u \leq 3,$
$0 \leq v \leq 2\pi$

15. $\mathbf{r}(u, v) = (4 \cos^2 v)\mathbf{i} + u\mathbf{j} + (4 \cos v \sin v)\mathbf{k},$
$-\dfrac{\pi}{2} \leq v \leq \dfrac{\pi}{2}, 0 \leq u \leq 3$

17. $\displaystyle\int_0^{2\pi} \int_0^1 \dfrac{\sqrt{5}}{2} r \, dr \, d\theta = \pi \dfrac{\sqrt{5}}{2} \approx 3.51$

19. $\displaystyle\int_0^{2\pi} \int_1^3 r\sqrt{5} \, dr \, d\theta = 8\pi\sqrt{5} \approx 56.20$

21. $\displaystyle\int_0^{2\pi} \int_1^4 dr \, d\theta = 6\pi$

23. $\displaystyle\int_0^{2\pi} \int_0^1 r\sqrt{4r^2 + 1} \, dr \, d\theta = \dfrac{\pi}{6}(5\sqrt{5} - 1) \approx 5.33$

25. $\displaystyle\int_0^{2\pi} \int_{\pi/4}^{\pi} 2 \sin \phi \, d\phi \, d\theta = (4 + 2\sqrt{2})\pi \approx 21.45$

27. $\dfrac{17\sqrt{17} - 1}{4} \approx 17.27$

29. $4\pi/3$
31. $3\sqrt{3}$
33. $11\pi/12 \approx 2.88$
35. -32
37. $\pi a^3/6$
39. $13a^4/6$
41. $2\pi/3$
43. $-73\pi/6 \approx -38.22$
45. $\bar{x} = \bar{y} = \bar{z} = a/2$
47. $I_z = 8\delta\pi a^4/3$
49. Tangent plane: $x + y - \sqrt{2}z = 0$; surface:
$z = \sqrt{x^2 + y^2}$

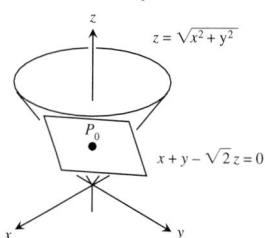

51. Tangent plane: $\sqrt{3}x + y = 9$; surface:
$x^2 + (y - 3)^2 = 9$

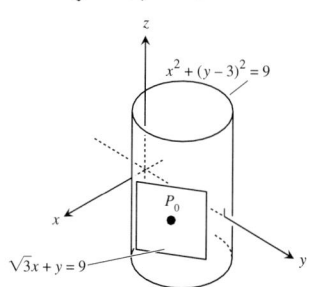

Section 15.7 (pp. 886–896)

Quick Review

1. $\dfrac{e^{-xyz}}{x + y + z} - yze^{-xyz} \ln(x + y + z)$

2. $\dfrac{e^{-xyz}}{x + y + z} - xye^{-xyz} \ln(x + y + z)$

3. $(xyz^2 - z)e^{-xyz} \ln(x + y + z)$
$- \dfrac{(y^2z + 2xyz + yz^2 + x^2z + xz^2 + 1)e^{-xyz}}{(x + y + z)^2}$

4. $(xyz^2 - z)e^{-xyz} \ln(x + y + z)$
$- \dfrac{(y^2z + 2xyz + yz^2 + x^2z + xz^2 + 1)e^{-xyz}}{(x + y + z)^2}$

5. $(x^2yz - x)e^{-xyz} \ln(x + y + z)$
$- \dfrac{(xz^2 + x^2z + 2xyz + x^2y + xy^2 + 1)e^{-xyz}}{(x + y + z)^2}$

6. $(x^2yz - x)e^{-xyz} \ln(x + y + z)$
$- \dfrac{(xz^2 + x^2z + 2xyz + x^2y + xy^2 + 1)e^{-xyz}}{(x + y + z)^2}$

7. $(cy + bz)\mathbf{i} - (cx + az)\mathbf{j} + (bx - ay)\mathbf{k}$
8. $-(cy + bz)\mathbf{i} + (cx + az)\mathbf{j} - (bx - ay)\mathbf{k}$
9. $\nabla f = (\sin(y + z))\mathbf{i} + (x \cos(y + z))\mathbf{j} +$
$(x \cos(y + z))\mathbf{k}$

10. $\dfrac{-1 \pm \sqrt{5}}{2}$

Exercises

1. 4π
3. $-5/6$
5. 0
7. -6π
9. $2\pi a^2$
13. 12π
15. $-\pi/4$
17. -15π
21. Its value is -2 times the area of the region.
23. No such field exists.
25. $16I_y + 16I_x$

Section 15.8 (pp. 896–907)

Quick Review

1. $4/3$ **2.** $2/3$
3. $2/3$ **4.** $2/3$
5. $\displaystyle\int_0^1 \int_0^1 \int_{\sqrt{z}}^1 dy \, dz \, dx$ **6.** $\displaystyle\int_0^1 \int_0^1 \int_{\sqrt{z}}^1 dy \, dx \, dz$

7. $\int_0^1 \int_{\sqrt{z}}^1 \int_0^1 dx\, dy\, dz$ **8.** $\int_0^1 \int_0^{y^2} \int_0^1 dx\, dz\, dy$

9. $\cos^{-1}\left(\dfrac{1}{\sqrt{3}}\right) \approx 54.74°$ **10.** $\cos^{-1}\left(\dfrac{1}{\sqrt{3}}\right) \approx 54.74°$

Exercises

1. 0
3. 0
5. -16
7. -8π
9. 3π
11. $-40/3$
13. 12π
15. $(48\sqrt{2} - 12)\pi \approx 175.56$
17. (b) Its value is zero.
21. Its value never exceeds the surface area of S.

31. (a) $\displaystyle\iiint_D p(t, x, y, z)\, dV$ represents the mass of the

fluid at any time t, and the instantaneous rate of change of mass is flux of the fluid through the surface S.

Chapter 15 Review Exercises
(pp. 908–911)

1. Path 1: $2\sqrt{3}$; path 2: $3\sqrt{2} + 1$

2. Path 1: 10/3; path 2: $\dfrac{5\sqrt{2} + 9}{6} \approx 2.68$;

path 3: $-2/3$

3. $4a^2$ **4.** 7/3
5. 0 **6.** 5
7. 0 **8.** 0
9. 0 **10.** 0
11. $\pi\sqrt{3}$

12. $\dfrac{\pi}{6}(7\sqrt{21} - 9) \approx 12.08$

13. $2\pi\left(1 - \dfrac{1}{\sqrt{2}}\right) \approx 1.84$

14. (a) $4\pi - 8$ **(b)** 8

15. $\dfrac{1}{2} abc \sqrt{\dfrac{1}{a^2} + \dfrac{1}{b^2} + \dfrac{1}{c^2}}$

16. (a) 0 **(b)** -4
17. 50 **18.** $\approx 97{,}751$ square miles
19. $\mathbf{r}(\phi, \theta) = (6 \sin \phi \cos \theta)\mathbf{i} + (6 \sin \phi \sin \theta)\mathbf{j} +$

$(6 \cos \phi)\mathbf{k}, \dfrac{\pi}{6} \le \phi \le \dfrac{2\pi}{3}, 0 \le \theta \le 2\pi$

20. $\mathbf{r}(r, \theta) = (r \cos \theta)\mathbf{i} + (r \sin \theta)\mathbf{j} - \left(\dfrac{r^2}{2}\right)\mathbf{k},$

$0 \le r \le 2, 0 \le \theta \le 2\pi$

21. $\mathbf{r}(r, \theta) = (r \cos \theta)\mathbf{i} + (r \sin \theta)\mathbf{j} + (1 + r)\mathbf{k},$

$0 \le r \le 2, 0 \le \theta \le 2\pi$

22. $\mathbf{r}(x, y) = x\mathbf{i} + y\mathbf{j} + \left(3 - x - \dfrac{y}{2}\right)\mathbf{k}, 0 \le x \le 2,$

$0 \le y \le 2$

23. $\mathbf{r}(u, v) = (u \cos v)\mathbf{i} + 2u^2\mathbf{j} + (u \sin v)\mathbf{k},$

$0 \le u \le 1, 0 \le v \le \pi$

24. $\mathbf{r}(\phi, \theta) = (\sqrt{10} \sin \phi \cos \theta)\mathbf{i} +$

$(\sqrt{10} \sin \phi \sin \theta)\mathbf{j} + (\sqrt{10} \cos \phi)\mathbf{k},$

$0 \le \phi \le \dfrac{\pi}{2}, 0 \le \theta \le \dfrac{\pi}{2}$

25. $\sqrt{6}$ **26.** $-\sqrt{2/3}$
27. $\pi(\sqrt{2} + \ln(1 + \sqrt{2})) \approx 7.21$
28. $8\pi/3$ **29.** Conservative
30. Conservative **31.** Not conservative
32. Conservative **33.** $2x + y^2 + zy + z + C$
34. $\sin(xz) + e^y + C$ **35.** Path 1: 2; Path 2: 8/3
36. Path 1: 2; Path 2: 2
37. (a) $1 - e^{-2\pi} \approx 1.00$ **(b)** $1 - e^{-2\pi} \approx 1.00$
38. (a) 0 **(b)** 2π
39. (a) $-\pi/2$ **(b)** 0
(c) 1
40. 0 **41.** 0
42. 0
43. (a) $4\sqrt{2} - 2 \approx 3.66$
(b) $\sqrt{2} + \ln(1 + \sqrt{2}) \approx 2.30$
44. $\bar{x} = 19/18, \bar{y} = 19/9, \bar{z} = 4\sqrt{2}/7$
45. $\bar{x} = 1, \bar{y} = 16/15, \bar{z} = 2/3, I_x + 232/45,$

$I_y = 64/15, I_z = 56/9, R_x = \sqrt{116/45},$

$R_y = \sqrt{32/15}, R_z = \sqrt{28/9}$

46. $\bar{x} = 0, \bar{y} = \dfrac{8a - a\pi}{4\pi - 4a}, \bar{z} = 0$

47. $\bar{z} = 3/2, I_z = 7\sqrt{3}/3, R_z = \sqrt{7/3}$
48. $M = 2\pi\delta\sqrt{13}, \bar{x} = \bar{y} = 0, \bar{z} = 3\pi$
49. $\bar{x} = \bar{y} = 0, \bar{z} = 49/12, I_z = 640\pi, R_z = 2\sqrt{2}$
50. 14/3
51. Circ $= -1/2$, Flux $= 3/2$
52. Circ $= 0$, Flux $= -11/3$
55. 3 **56.** 128π

57. $\dfrac{2}{3}\pi(7 - 8\sqrt{2}) \approx -9.03$

58. $\pi + 1$ **59.** 0
60. 8/3 **61.** π
62. (a) $2\pi a^3$ **(b)** $2\pi a^3$

Cumulative Review Exercises

Cumulative Review (pp. 573–576)

1. 0

2. $\dfrac{3}{4}$

3. -1

4. -1

5. 3

6. 1

7. e^2

8. 3

9. (a) 1 **(b)** 1
(c) 1 **(d)** Yes
(e) No

10. All $x \le -2$ and $x \ge 2$

11. Horizontal: $y = 0$
Vertical: $x = 0,\ x = \dfrac{1}{2}$

12.

$[-10, 10]$ by $[-4, 4]$

13. $\dfrac{1}{5}$

14. $-\dfrac{3}{(x-2)^2}$

15. $\dfrac{3 \sin (\sqrt{1-3x})}{2\sqrt{1-3x}}$

16. $\sin x \sec^2 x + \cos x \tan x = \dfrac{(\sin x)(1 + \cos^2 x)}{\cos^2 x}$

17. $\dfrac{2x}{x^2 + 1}$

18. $(2x - 1)e^{x^2 - x}$

19. $2x \tan^{-1} x + \dfrac{x^2}{1 + x^2}$

20. $(x^{-3} - 3x^{-4})e^x$

21. $\dfrac{3 \csc^2 x}{(1 + \cos x)^4}(1 - \csc x \cot x - \cos x \csc x \cot x)$
$= \dfrac{3(1 - 2\cos x)}{(\sin^4 x)(1 + \cos x)^3}$

22. $-\dfrac{1}{\sqrt{1 - x^2}} + \dfrac{1}{1 + x^2}$

23. $\dfrac{1 + xy \sin (xy)}{2xy - x^2 \sin (xy)}$

24. $\dfrac{|x|}{2x\sqrt{|x|}}$

25. $\cot t = \dfrac{x - 1}{1 - y}$

26. $(\cos x)^{x-1}(\cos x \ln (\cos x) - x \sin x)$

27. $\sqrt{1 + x^3}$

28. $2x \sin (x^2) - 2 \sin (2x)$

29. $\dfrac{(2y + 2)^2(\sec^3 x + \sec x \tan^2 x) - 2 \sec^2 x \tan^2 x}{(2y + 2)^3}$

30. -1

31. (a) $v = 3t^2 - 12t + 9,\ a = 6t - 12$
(b) $t = 1$ or $t = 3$
(c) Right: $0 \le t < 1,\ 3 < t \le 5$
Left: $1 < t < 3$
(d) -3 m/sec

32. (a) $y = -2x + 1$
(b) $y = \dfrac{1}{2}x - \dfrac{3}{2}$

33. (a) $y = \left(\dfrac{3 - \pi\sqrt{3}}{6}\right)\left(x - \dfrac{\pi}{3}\right) + \dfrac{\pi}{6}$
$\approx -0.407x + 0.950$
(b) $y = \left(\dfrac{6}{\pi\sqrt{3} - 3}\right)\left(x - \dfrac{\pi}{3}\right) + \dfrac{\pi}{6}$
$\approx 2.458x - 2.050$

34. (a) $y = -\dfrac{\sqrt{3}}{2}x + 2\sqrt{3} \approx -0.866x + 3.464$
(b) $y = \dfrac{2}{\sqrt{3}}x + \dfrac{5}{2\sqrt{3}} \approx 1.155x + 1.443$

35. (a) $y = -\dfrac{\sqrt{3}}{2}x + 2\sqrt{3} \approx -0.866x + 3.464$
(b) $y = \dfrac{2}{\sqrt{3}}x + \dfrac{5}{2\sqrt{3}} \approx 1.155x + 1.443$

36. (a) $y = \sqrt{2}x - 1 \approx 1.414x - 1$
(b) $y = -\dfrac{1}{\sqrt{2}}x + 2 \approx -0.707x + 2$

37. $f(x) = \begin{cases} -x + 4, & x \le 3 \\ 2x - 5 & x > 3 \end{cases}$

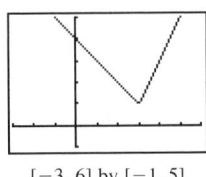

$[-3, 6]$ by $[-1, 5]$

38. (a) $x \ne 0, 2$ **(b)** $x = 0$
(c) $x = 2$
(d) Absolute maximum of 2 at $x = 0$
Absolute minimum of 0 at $x = -2, 2, 3$

39. According to the Mean Value Theorem the
driver's speed at some time was $\dfrac{111}{1.5} = 74$ mph.

40. (a) Increasing in $[-0.7, 2]$, decreasing in $[-2, -0.7]$, and has a local minimum at $x = -0.7$.

(b) $y \approx -2x^2 + 3x + 3$

$[-3, 3]$ by $[-15, 10]$

(c) $f(x) = -\dfrac{2}{3}x^3 + \dfrac{3}{2}x^2 + 3x + 1$

41. $f(x) = x^2 - 3x - \cos x - 1$

42. (a) $\left[-2, -\dfrac{2\sqrt{6}}{3}\right], \left[0, \dfrac{2\sqrt{6}}{3}\right]$

(b) $\left[-\dfrac{2\sqrt{6}}{3}, 0\right], \left[\dfrac{2\sqrt{6}}{3}, 2\right]$

(c) $\approx(-1.042, 1.042)$

(d) $\approx(-2, -1.042), \approx(1.042, 2)$

(e) Local max of approximately 3.079 at $x = -\dfrac{2\sqrt{6}}{3}$ and $x = \dfrac{2\sqrt{6}}{3}$ local min of 0 at $x = 0$

(f) Points of inflection at about $x = \pm 1.042$

43. (a) f has an absolute maximum at $x = 1$ and an absolute minimum at $x = 3$.

(b) f has a point of inflection at $x = 2$.

(c)

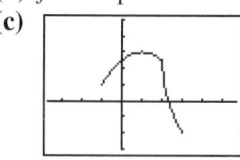

$[-3.7, 5.7]$ by $[-3,5]$

44. Dimensions: $4\sqrt{2}$ by $\sqrt{2}$, area: 8

45. $y = \sqrt{2}\left(x - \dfrac{\pi}{4}\right) + \sqrt{2} \approx 1.414x + 0.303$

46. 3%

47. (a) About 1.9 ft/sec

(b) About 0.15 rad/sec

48. (a) $\dfrac{9}{16\pi} \approx 0.179$ in./min

(b) $\dfrac{36}{25\pi} \approx 0.458$ in./min

49. (a) 165 in. **(b)** 165 in.

50. 2.5 **51.** 2π

52. $\ln 3 + \dfrac{26}{3} \approx 9.765$

53. 1 **54.** 7

55. $\dfrac{\ln 2}{1 + \ln 2} \approx 0.409$

56. $-2\mathbf{i} + \ln (3)\mathbf{j}$

57. $-\cot (e^x + 1) - e^x + C$

58. $\dfrac{1}{2} \tan^{-1}\left(\dfrac{s}{2}\right) + C$

59. $\dfrac{1}{2 \cos^2 (x - 3)} + C$

60. $\dfrac{e^{-x}}{5}(2 \sin 2x - \cos 2x) + C$

61. $\dfrac{8}{7} \ln |x - 6| - \dfrac{1}{7} \ln |x + 1| + C$

$= \dfrac{1}{7} \ln \dfrac{(x - 6)^8}{|x + 1|} + C$

62. 8975 ft³

63. $y = -\dfrac{1}{t + 1} - \dfrac{1}{2}e^{-2t} + \dfrac{7}{2}$

64. $y = -\dfrac{1}{4} \sin 2\theta + \cos \theta + \dfrac{1}{2}\theta - \dfrac{\pi}{4}$

65. $\int x^2 \sin x \, dx = (2 - x^2) \cos x + 2x \sin x + C$

The graph of the slope field of the differential equation $\dfrac{dy}{dx} = x^2 \sin x$ and the antiderivative $y = (2 - x^2) \cos x + 2x \sin x$ is shown below.

$[-5, 5]$ by $[-10, 10]$

66. $\int xe^x \, dx = e^x(x - 1) + C$

$\dfrac{d}{dx}(e^x(x - 1) + C) = e^x$

67. (a) $y = 4268e^{kt}, k = \dfrac{\ln (5/3)}{3} \approx 0.170$

(b) About 4268

68. (a) About 6.13°C

(b) About 2 hours and 27 minutes after it was 65°C above room temperature, or about 2 hours and 12 minutes after it was 50°C above room temperature.

69. $y = \dfrac{500}{1 + Ce^{-0.08x}}$

70. $y = Ce^{x^2/2 + 3x} + 4$

71.

x	y
0	0
0.1	0.1
0.2	0.2095
0.3	0.3285
0.4	0.4568
0.5	0.5946
0.6	0.7418
0.7	0.8986
0.8	1.0649
0.9	1.2411
1.0	1.4273

72. 4

73. $\dfrac{64}{3}$

74. ≈ 16.039

75. ≈ 42.412

76. $\dfrac{\pi}{14} \approx 0.224$

77. $\dfrac{128\pi}{3} \approx 134.041$

78. ≈ 0.763

79. ≈ 2.556

80. 4

81. ≈ 6.110

82. ≈ 8.423

83. ≈ 3.470

84. ≈ 32.683

85. ≈ 0.101

86. ≈ 5.394

87. **(a)** 1.2 m **(b)** 180 J

88. **(a)** 70,686 ft-lb **(b)** 4 min, 17 sec

89. 12.166 ft^3

90. $f(x) = \ln x$ grows slower than $g(x) = \sqrt{x}$.

91. Converges **92.** Diverges

93. Converges **94.** Converges

95. Diverges **96.** Converges

97. $1 - 2x + 4x^2 - 8x^3 + \cdots + (-1)^n 2^n x^n + \cdots$,
$-\dfrac{1}{2} < x < \dfrac{1}{2}$

98. **(a)** $x - \dfrac{x^5}{5 \cdot 2!} + \dfrac{x^9}{9 \cdot 4!} - \dfrac{x^{13}}{13 \cdot 6!} + \cdots$
$+ (-1)^n \dfrac{x^{4n+1}}{(4n+1) \cdot (2n)!} + \cdots$

(b) $-\infty < x < \infty$; Since the cosine series converges for all real numbers, so does the integrated series, by Theorem 2, the term-by-term integration theorem.

99. $\ln(2 + 2x) = \ln 2 + \ln(1 + x) = \ln 2 + x - \dfrac{x^2}{2} + \dfrac{x^3}{3} - \dfrac{x^4}{4} + \cdots + (-1)^{n-1}\dfrac{x^n}{n} + \cdots$, which converges for $-1 < x \le 1$.

100. $(x - 2\pi) - \dfrac{(x - 2\pi)^3}{3!} + \dfrac{(x - 2\pi)^5}{5!} - \cdots$
$+ (-1)^n \dfrac{(x - 2\pi)^{2n+1}}{(2n + 1)!} + \cdots$

101. $P_6(x) = 1 - x + \dfrac{x^2}{2!} - \dfrac{x^3}{3!} + \dfrac{x^4}{4!} - \dfrac{x^5}{5!} + \dfrac{x^6}{6!}$

By the Alternating Series Estimation Theorem,
$|\text{error}| \le \left|\dfrac{x^7}{7!}\right| \le \dfrac{1}{7!} < 0.001$.

102. $1 + \dfrac{1}{3}x - \dfrac{2}{2! \cdot 3^2}x^2 + \dfrac{2 \cdot 5}{3! \cdot 3^3}x^3 - \dfrac{2 \cdot 5 \cdot 8}{4! \cdot 3^4}x^4 + \cdots$
$+ (-1)^{n-1}\dfrac{2 \cdot 5 \cdot \cdots \cdot (3n - 4)}{n! \cdot 3^n}x^n + \cdots$

$R = 1$

103. Converges **104.** Diverges

105. Converges **106.** Converges

107. **(a)** $R = 1$; $-3 < x \le -1$
(b) $-3 < x < -1$
(c) At $x = -1$

108. **(a)** $R = 1$; $-1 \le x \le 1$
(b) $-1 \le x \le 1$
(c) Nowhere

109. $\left\langle \dfrac{2}{\sqrt{13}}, -\dfrac{3}{\sqrt{13}} \right\rangle$

110. $\left\langle \dfrac{1}{2}, \dfrac{\sqrt{3}}{2} \right\rangle$

111. Tangent: $\left\langle -\dfrac{4}{5}, -\dfrac{3}{5} \right\rangle, \left\langle \dfrac{4}{5}, \dfrac{3}{5} \right\rangle$

normal: $\left\langle \dfrac{3}{5}, -\dfrac{4}{5} \right\rangle, \left\langle -\dfrac{3}{5}, \dfrac{4}{5} \right\rangle$

112. **(a)** $\mathbf{v}(t) = (-\cos t)\mathbf{i} + (1 + \sin t)\mathbf{j}$
$\mathbf{a}(t) = (\sin t)\mathbf{i} + (\cos t)\mathbf{j}$
(b) 4

113. Yes. When $x = 130$ ft,
$t = \dfrac{130}{100} \cos 45° \approx 1.838$ sec and $y \approx 75.9$ ft, high enough to easily clear the 35-ft tree.

114. $x - y = 2$; A line with slope 1 and y-intercept -2.

115.

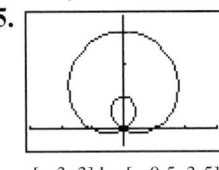

[-3, 3] by [-0.5, 3.5]

2π

116. Horizontal tangent line at:

$(0, 0)\ [y = 0]$, $\left(\dfrac{3}{2}, \dfrac{2\pi}{3}\right)\left[y = \dfrac{3\sqrt{3}}{4}\right]$,

$\left(\dfrac{3}{2}, \dfrac{4\pi}{3}\right)\left[y = -\dfrac{3\sqrt{3}}{4}\right]$, $(0, 2\pi)\ [y = 0]$

Vertical tangent lines at:

$(2, \pi)[x = -2]$, $\left(\dfrac{1}{2}, \dfrac{\pi}{3}\right)\left[x = \dfrac{1}{4}\right]$, $\left(\dfrac{1}{2}, \dfrac{5\pi}{3}\right)\left[x = \dfrac{1}{4}\right]$

Chapters 11–15

1. (a) 3   **(b)** $\sqrt{3}$
   (c) -1 **(d)** -1
   (e) $\mathbf{i} + 3\mathbf{j} + 4\mathbf{k}$ **(f)** $-\mathbf{i} - 3\mathbf{j} - 4\mathbf{k}$

   (g) $\cos^{-1}\left(\dfrac{-1}{3\sqrt{3}}\right) \approx 1.76$ rad

   (h) $-\dfrac{2}{9}\mathbf{i} + \dfrac{2}{9}\mathbf{j} - \dfrac{1}{9}\mathbf{k}$

2. ± 2
3. (a) 5   **(b)** 5

4. $\mathbf{B} = \left(\dfrac{4}{3}\mathbf{i} + \dfrac{4}{3}\mathbf{j} + \dfrac{4}{3}\mathbf{k}\right) + \left(\dfrac{2}{3}\mathbf{i} + \dfrac{5}{3}\mathbf{j} - \dfrac{7}{3}\mathbf{k}\right)$

5. $x = -2 + t$, $y = 1 - 2t$, $z = -1 + 3t$
6. $x = 2 - t$, $y = 1 + t$, $z = -1 + 5t$, $0 \le t \le 1$
7. $\sqrt{22/3}$ **8.** $\sqrt{61/14}$
9. $3x - y + 2z = 5$ **10.** $3x - 2y + 2z = -4$
11. $x - y + z = 0$
12. $x = t$, $y = -1 - t$, $z = -3t$
13. $\theta = \pi - \cos^{-1}(-2/\sqrt{42}) \approx 1.26$ rad
14. The line $x - 2y = 1$ in the plane, the plane $x - 2y = 1$ in three-dimensional space.
15. $x^2 + (y - 2)^2 = 4$; a circular cylinder parallel to the z-axis

16.

17.

18.

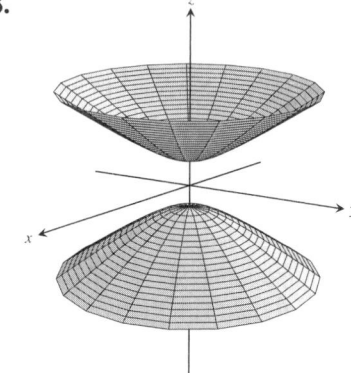

19. $\mathbf{r}(t) = (e^t + 1)\mathbf{i} + (\sin t + 3)\mathbf{j} - t^3\mathbf{k}$
20. ≈ 10.63

21. $\mathbf{T}(0) = \dfrac{1}{\sqrt{3}}\mathbf{i} + \dfrac{1}{\sqrt{3}}\mathbf{j} + \dfrac{1}{\sqrt{3}}\mathbf{k}$,

   $\mathbf{N}(0) = \dfrac{1}{\sqrt{2}}\mathbf{i} - \dfrac{1}{\sqrt{2}}\mathbf{j}$, $\mathbf{B}(0) = -\dfrac{1}{\sqrt{6}}\mathbf{i} + \dfrac{1}{\sqrt{6}}\mathbf{j}$,

   $\kappa = 1/3$, $\tau(0) = -2/3$

22. $\mathbf{a}(0) = \dfrac{2}{\sqrt{5}}\mathbf{T} + \dfrac{4}{\sqrt{5}}\mathbf{N}$

23. 0, $\pi/2$, π, $3\pi2$, 2π
24. Normal plane: $x - 2y + 6z = 15$;
   rectifying plane: $38x - 35y - 18z = 37$;
   osculating plane: $6x + 6y + z = 2$

25. (a) $\mathbf{r}(t) = \left(-\dfrac{8}{15}t^3 + 4t^2\right)\mathbf{i} + (100 - 20t)\mathbf{j}$

   (b) 100/3 m
26. (a) $\mathbf{v} = 4\mathbf{i} + 2\mathbf{j}$

   (b) $\mathbf{r}(t) = 4t\mathbf{i} + \left(2t + \dfrac{4t^3}{25} - \dfrac{6}{5}t^2\right)\mathbf{j}$, $0 \le t \le 5$

(c)

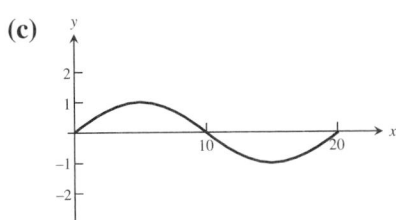

27. $\mathbf{v} = \dfrac{\sqrt{3}}{2}\mathbf{i} + \dfrac{\sqrt{3}}{2}\mathbf{j} - \sqrt{3}\mathbf{k}$

28. Ellipsoid; $\dfrac{x^2}{4} + \dfrac{y^2}{9} + z^2 = 1$

29. (a) $2\sqrt{\dfrac{\pi g b}{a^2 + b^2}}$

 (b) $\theta = \dfrac{gbt^2}{2(a^2 + b^2)}, z = \dfrac{gb^2t^2}{2(a^2 + b^2)}$

 (c) $\dfrac{d\mathbf{r}}{dt} = \dfrac{gbt}{\sqrt{a^2 + b^2}}\mathbf{T},$

 $\dfrac{d^2\mathbf{r}}{dt^2} = \dfrac{gb}{\sqrt{a^2 + b^2}}\mathbf{T} + \left(\dfrac{gbt}{a^2 + b^2}\right)\mathbf{N}$

 No component in the direction of **B**.

30. $d\mathbf{L}/dt = \mathbf{0}$ **31.** 1/4

32. Does not exist **33.** Yes, $f(0, 0) = 2$

34. $f_x = e^{-(x^2+y^2+z^2)} - 2x^2 e^{-(x^2+y^2+z^2)};$
 $f_y = -2xy e^{-(x^2+y^2+z^2)}; f_z = -2xz e^{-(x^2+y^2+z^2)};$
 $f_{xy} = 2y(2x^2 - 1)e^{-(x^2+y^2+z^2)}$

35. $L(x, y) = \dfrac{1}{2}x + \dfrac{\sqrt{3}}{2}y + \dfrac{6 - \sqrt{3}\pi}{12}$

36. $L(x, y, z) = x + \dfrac{\sqrt{2}}{2}y + 2z + \dfrac{\pi - 12}{4}$

37. $L(x, y) = 4x + 5y - 8; |E| \leq 0.086$

38. 5

39. $\partial w/\partial r = 2, \partial w/\partial s = 2 - \pi$

40. $w = x + e^x \cos y + y^2 - 2$

41. $f(x, y) = \dfrac{1}{2}y + 4, g(x, y) = \dfrac{1}{2}x + \dfrac{9}{2}$

42. $\pi/\sqrt{2}$

43. (a) $\mathbf{u} = \dfrac{1}{\sqrt{53}}(2\mathbf{i} + 7\mathbf{j})$

 (b) $\mathbf{u} = -\dfrac{1}{\sqrt{29,097}}(98\mathbf{i} - 127\mathbf{j} + 58\mathbf{k})$

44. $x + y - z = 0$

45. Local maximum of 20 at $(0, -2)$; saddle points at $(0, 1)$ and $(3, -2), f(0, 1) = -7, f(3, -2) = -7$; local minimum of -34 at $(3, 1)$

46. Absolute maximum of 17 at $(0, 4)$ and $(4, 4)$; absolute minimum of 1 at $(0, 0)$

47. Absolute maximum of 1 at $(\pm 1, 0), (0, 1)$; absolute minimum of -1 at $(0, -1)$

48. Absolute maximum of 3/2 at $\left(\dfrac{1}{\sqrt{2}}, \dfrac{1}{\sqrt{2}}, \sqrt{2}\right)$

 and $\left(-\dfrac{1}{\sqrt{2}}, -\dfrac{1}{\sqrt{2}}, -\sqrt{2}\right)$; absolute minimum of

 1/2 at $\left(\dfrac{1}{\sqrt{2}}, -\dfrac{1}{\sqrt{2}}, -\sqrt{2}\right)$ and $\left(-\dfrac{1}{\sqrt{2}}, \dfrac{1}{\sqrt{2}}, \sqrt{2}\right)$

49. $1/e^2$ at $(1/2, 1/3)$ **50.** $\dfrac{\sqrt{3}}{2}abc$

51. $\displaystyle\int_0^1 \int_y^{\sqrt{y}} f(x, y)\, dx\, dy$

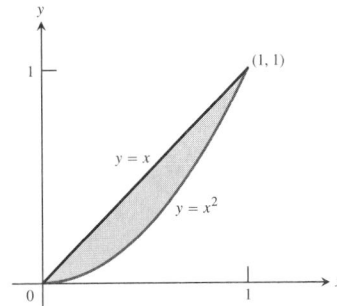

52. 243/8 **53.** 2416/105

54. $\dfrac{9\pi(8\sqrt{2} - 7)}{2} \approx 60.98$

55. $\pi/3$ **56.** 9/2

57. (a) $\displaystyle\int_{-3}^2 \int_x^{6-x^2} x^2\, dy\, dx$ **(b)** $\displaystyle\int_{-3}^2 \int_x^{6-x^2} \int_0^{x^2} dz\, dy\, dx$

 (c) 125/4

58. $3\pi/2$ **59.** $2\pi^2$

60. (a) The radius of the hole is 1, and the radius of the sphere is 2.

 (b) $4\sqrt{3}\,\pi$

61. $1/\sqrt[4]{3}$ **62.** 104

63. $\bar{x} = \dfrac{15\pi + 32}{6\pi + 48} \approx 1.18, \bar{y} = 0$

64. $R_x = \dfrac{a\sqrt{3\pi - 8}}{2\sqrt{6}} \approx 0.24a,$

 $R_y = \dfrac{a\sqrt{3\pi + 8}}{2\sqrt{6}} \approx 0.85a$

65. (a) $\bar{z} = 4/5, \bar{x} = \bar{y} = 0; I_z = \pi/12; R_z = 1/\sqrt{3}$
   (b) $\bar{z} = 5/6, \bar{x} = \bar{y} = 0; I_z = \pi/14; R_z = \sqrt{5/14}$

66. (a) $\frac{1}{3}(u_1 + u_2)$ **(b)** Average $= u_1\bar{x} + u_2\bar{y}$

67. $2\pi k R^3/3$

68. $h = \sqrt{20}$ for 1 in., $h = \sqrt{60}$ for 3 in.

69. $2\pi\left(\frac{1}{3} - \frac{\sqrt{2}}{6}\right) \approx 0.61$ **70.** $\sqrt{6}/6$

71. $6\pi^2\sqrt{13}$ **72.** π

73. Circulation $= -18$, flux $= 64$

74. -4π

75. $f(x, y, z) = \ln(y^2 + z^2) + \sin(x + z) + x^2yz + C$

76. 49

77. Circulation $= 0$, flux $= 16$

78. $\dfrac{80 - 9\pi^2}{18} \approx -0.49$ **79.** 8π

80. 465/2

81. (a) 18π **(b)** 135π

82. (b)

$$\int_0^{2\pi}\int_0^{\pi} [a^2b^2 \sin^2\phi \cos^2\phi + b^2c^2 \cos^4\phi \cos^2\theta + a^2c^2 \cos^4\phi \sin^2\theta]^{1/2}\, d\phi\, d\theta$$

83. $\left(\dfrac{R}{2}, \dfrac{R}{2}, \pm\dfrac{\sqrt{2}R}{2}\right), \left(-\dfrac{R}{2}, -\dfrac{R}{2}, \pm\dfrac{\sqrt{2}R}{2}\right),$
$\left(\dfrac{R}{2}, -\dfrac{R}{2}, \pm\dfrac{\sqrt{2}R}{2}\right), \left(-\dfrac{R}{2}, \dfrac{R}{2}, \pm\dfrac{\sqrt{2}R}{2}\right)$

84. $16\pi R^3/3$ **85.** $\dfrac{4\pi}{3}(2\sqrt{2} - 1) \approx 7.66$

86. $a = 2, b = 1$, least flux is -4

87. $x + y + z = 0$

Appendix A5.4

1. $r\cos\left(\theta - \dfrac{\pi}{6}\right) = 5; y = -\sqrt{3}x + 10$

3. $r\cos\left(\theta - \dfrac{4\pi}{3}\right) = 3; y = -\dfrac{\sqrt{3}}{3}x - 2\sqrt{3}$

5. $y = 2 - x$

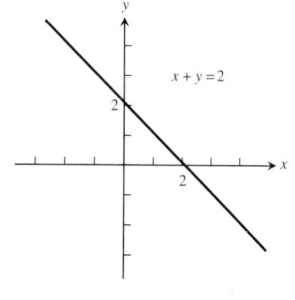

7. $y = \dfrac{\sqrt{3}}{3}x + 2\sqrt{3}$

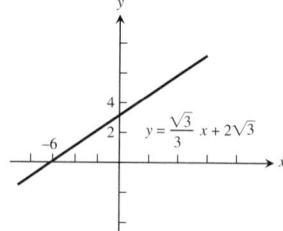

9. $r\cos\left(\theta - \dfrac{\pi}{4}\right) = 3$

11. $r\cos\left(\theta + \dfrac{\pi}{2}\right) = 5$

13. $r = 8\cos\theta$

15. $r = 2\sqrt{2}\sin\theta$

17.

19.

21.

23.

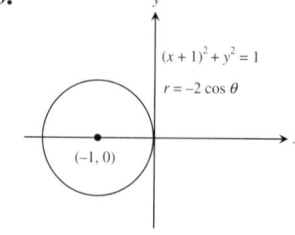

25. $r = \dfrac{2}{1 + \cos \theta}$

27. $r = \dfrac{30}{1 - 5 \sin \theta}$

29. $r = \dfrac{1}{2 + \cos \theta}$

31. $r = \dfrac{10}{5 - \sin \theta}$

33.

35.

37.

39.

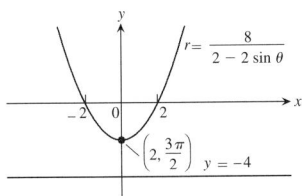

41. (a) $x^2 + y^2 = 4y; x = \sqrt{3}$

(b)

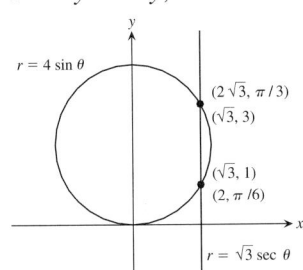

43. $r = \dfrac{4}{1 + \cos \theta}$

49.

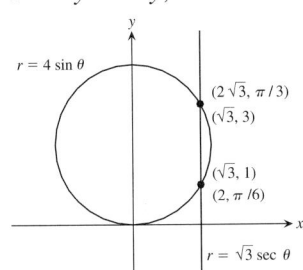

51. (b)

Planet	Perihelion	Aphelion
Mercury	0.3075 AU	0.4667 AU
Venus	0.7184 AU	0.7282 AU
Earth	0.9833 AU	1.0167 AU
Mars	1.3817 AU	1.6663 AU
Jupiter	4.9512 AU	5.4548 AU
Saturn	9.0210 AU	10.0570 AU
Uranus	18.2977 AU	20.0623 AU
Neptune	29.8135 AU	30.3065 AU
Pluto	29.6549 AU	49.2251 AU

53. (b) 2 in. apart

Appendix A8

1. -5

3. 1

5. -7

7. 38

9. $x = -4, y = 1$

11. $x = 3, y = 2$

13. $x = 3, y = -2, z = 2$

15. $x = 2, y = 0, z = -1$

17. (a) $h = 6, k = 4$ **(b)** $h = 6, k \neq 4$

APPLICATIONS INDEX

Note: Numbers in parentheses refer to exercises on the pages indicated.

Business, Economics, Industry, Labor

Airport parking, 81(44)

Average daily inventory and holding costs, 285, 288(59), 299(44)

Break-even, 211, 219(51)

Bus fares, 174(74)

Construction workers' average annual compensation, 8(40)

Container designs (best)

boxes, 206, 214(7), 215(18), 244(51, 57)

cheapest rectangular box, 759(83)

1-liter can, 207, 215(16, 17)

rainwater tank, 215(12)

soda cans, 714(37)

storage bin, 244(47)

suitcase, 215(19)

tank for storing LPG, 754(16)

vat, 244(48)

watering trough, 218(42)

Continuous price discounting, 348(22)

Cost from marginal cost

manufacturing, 284, 314(51)

printing posters, 288(57)

Dayton Power and Light transmission costs, 19(66)

Debit card transactions, 243(41)

Economic Injury Level, 54

Future costs of intranets, 93(53)

Inflation, 373(31, 32)

Inflection points and marginal cost, 199

Manufacturing tires, 244(55)

Marginal cost, 128, 130(10), 199

Marginal profit from marginal revenue and marginal cost, 133(35)

Marginal revenue, 129, 130(11)

bus fares, 174(74)

Maximizing utility, 755(31)

Minting coins to size, 231(46)

Monthly profit from software, 130(15)

Newsletter printing costs, 299(43)

NFL average salaries, 200

Oil depletion, 339(26)

Oil production

Canada, 40(50)

Indonesia, 38

Saudi Arabia, 40(49)

Orchard farming, 121(38)

Piping from an offshore drilling rig, 244(53)

Production levels

least costly, 212, 219(52)

most profitable, 211

Revenue from marginal revenue, 314(50)

eggbeater factory, 288(58)

Sales of prepaid phone cards, 205(49)

Sheetmetal fabrication, 400(23)

Stringing telephone cable, 245(63)

Taxi fares, 93(50)

Total production from labor, 241(41)

Tour service rates, 219(49)

Trans-Alaska Pipeline, 44

U.S. Postal Service package requirements, 216(25)

Variable inflation rates, 349(32)

Welder salary negotiation, 81(43)

Wilson lot size formula, 219(50), 701(30)

Wilson lot size formula (economic order quantity), 714(40)

CBL™ (Computer Based Laboratory) Data

Analyzing motion detector data, 193(52)

Coasting on in-line skates, 337, 340(33), 361(61)

Temperature probe, 334, 340(34), 361(60)

Tone of a tuning fork, 45, 46, 49(33)

Computer Science, Geometry, Probability, Simulation, and Other Mathematics

Angles between

diagonals of a cube, 592(15)

differentiable curves, 593(41–44)

intersecting lines, 593(35–40)

intersecting planes, 606(37)

vectors, 586

Approximation of π, 456

Archimedes' area formula for parabolas, 287(52)

Area

of a parallelogram, 595

of a triangle, 597

Arithmetic mean, 149(72)

Changing

cube, 245(60)

rectangle, 237(9)

rectangular box, 238(10)

Circumscribing an ellipse, 220(55)

Closest point of approach, 217(36, 37)

Computer graphics

hidden lines, 607(62)

perspective, 607(61)

Determinant equation for a plane, 628(65)

Direction angles, 592(13)

Distance

between lines in space, 628(57)

point to line, 605(27), 628(66)

point to plane, 605(31)

Distance from a point to a line, 9(52)

Effects of changing the dimensions of a box, 722(40)

Effect of punching out infinitely many triangles, 511(68)

Evaluating nonelementary integrals with infinite series, 510(63)

Exploring cross sections of spheres, 585(62)

Finding a least-squares plane, 755(42)

Function that cannot be a derivative, 112(30)

Geometric mean, 149(72)

Geometric mean of three nonnegative numbers does not exceed their arithmetic mean, 755(43)

Geometric models for $\tan^{-1} 1$, $\tan^{-1} 2$, and $\tan^{-1} 3$, 163(34, 35)

Geometry of curves in space. *See Chapter 12*

Improving approximations to π, 487(31)

Inscribing geometric figures

box of largest volume in a sphere, 754(27)

INDEX